instructors ... get resources to help you teach

Visit www.mhhe.com/raven6

Online Learning Center

A multitude of tools awaits you on *Biology's* Online Learning Center. You'll want to take advantage of our electronic illustrations and photographs from the text; classroom activities; lecture outlines; and access to PageOut: Course Website Development Center—all available anytime you want them.

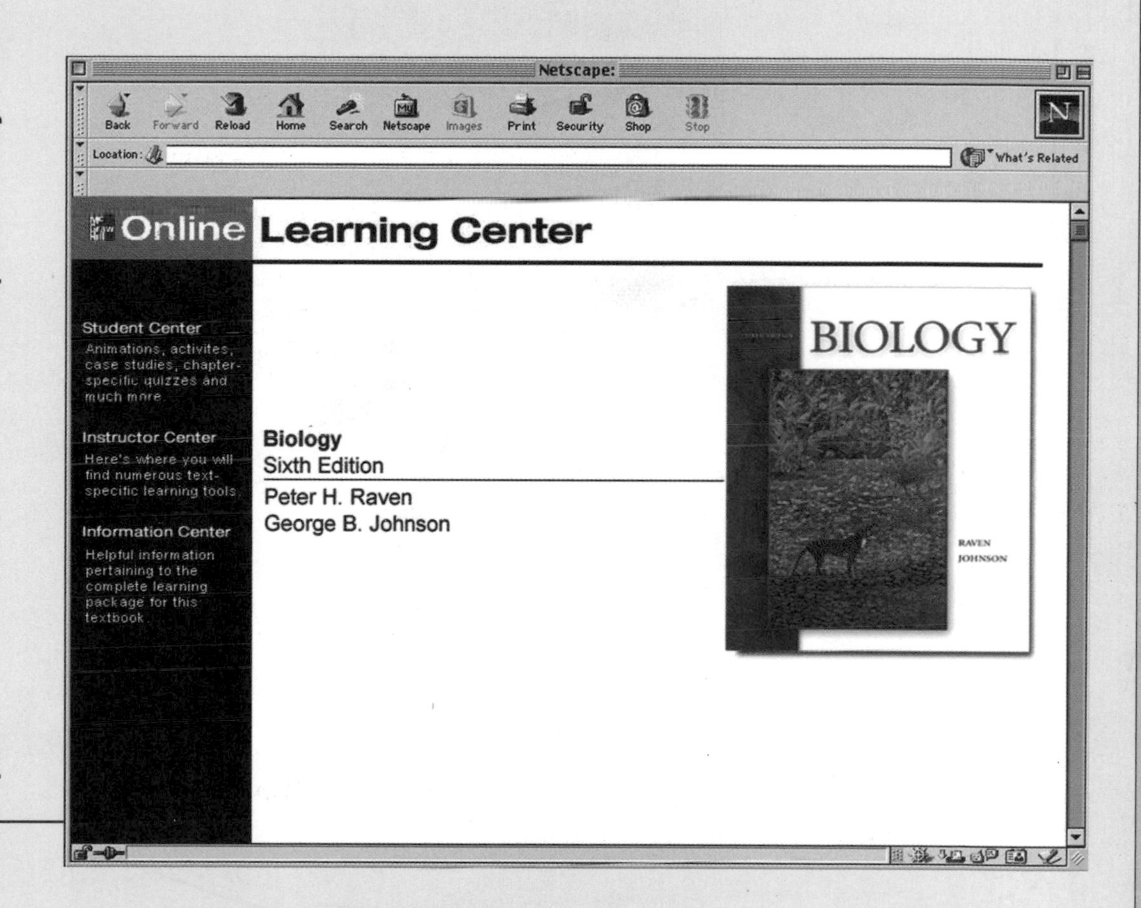

Go to the Online Learning Center to find:

- Instructor's Manual with links to Lab Supplies and Classroom Media Sources.
- Course Integration Guide.
- PowerPoint Lecture Outlines.
- NEW! Visual Resource Library including all art and all photos from the text.
- PageOut.
- Bioethics Case Studies.
- NEW! Genomics Chapter.
- NEW! Short Courses.
- Links to Professional Resources.
- Link to BioCourse.com.

For more information, visit *www.mhhe.com*.

Great news!

With the purchase of a new *Biology,* Sixth Edition, by Raven/Johnson, you receive a free 12-month subscription to the book-specific ONLINE LEARNING CENTER. The Online Learning Center offers quizzing, learning activities, animations and study aids developed to assist you in mastering biology.

Instructions to register:

1. Use a web browser to go to: http://register.mcgraw-hill.com

2. Enter your special code in the space provided exactly as it appears in the box below.

Special Code:

orchestral-928286

3. After you have entered your special code, click on the "REGISTER" button to continue the registration process.

Your ID and password can be used only once to establish your subscription, which is not transferable.

If you did not purchase this book new, this ID and password may not be valid. However, if your instructor is recommending or requiring use of the Online Learning Center, you can find more information on purchasing a subscription directly at www.mhhe.com/raven6 or through your college bookstore.

BIOLOGY

Sixth Edition

Peter H. Raven
Director, Missouri Botanical Gardens;
Engelmann Professor of Botany
Washington University

George B. Johnson
Professor of Biology
Washington University

Significant contributions by
Susan R. Singer
Carleton College

Jonathan B. Losos
Washington University

Original artwork by
William C. Ober, M.D.
and
Claire W. Garrison, R.N.

Boston Burr Ridge, IL Dubuque, IA Madison, WI New York San Francisco St. Louis
Bangkok Bogotá Caracas Kuala Lumpur Lisbon London Madrid Mexico City
Milan Montreal New Delhi Santiago Seoul Singapore Sydney Taipei Toronto

McGraw-Hill Higher Education

A Division of The **McGraw-Hill** *Companies*

BIOLOGY, SIXTH EDITION

Published by McGraw-Hill, a business unit of The McGraw-Hill Companies, Inc., 1221 Avenue of the Americas, New York, NY 10020.

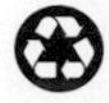

This book is printed on recycled, acid-free paper containing 10% postconsumer waste.

1 2 3 4 5 6 7 8 9 0 VNH/VNH 0 9 8 7 6 5 4 3 2 1

ISBN 0-07-303120-8
ISBN 0-07-112261-3 (ISE)

Sponsoring editor: *Patrick E. Reidy*
Developmental editor: *Lu Ann Weiss*
Off-site development editors: *Megan Jackman/Elizabeth Sievers*
Senior marketing manager: *Lisa L. Gottschalk*
Senior project manager: *Peggy J. Selle*
Production supervisor: *Kara Kudronowicz*
Design manager: *Stuart D. Paterson*
Cover designer: *Christopher Reese*
Interior designer: *Kathleen Theis*
Cover image: *Lobos Tasmanos by Alfredo Arreguin*
Senior photo research coordinator: *Lori Hancock*
Photo researcher: *Meyers Photo-Art*
Senior supplement producer: *Audrey A. Reiter*
Executive producer: *Linda Meehan Avenarius*
Compositor: *Carlisle Communications Ltd.*
Typeface: *10/12 Janson*
Printer: *Von Hoffmann Press, Inc.*

The credits section for this book begins on page C-1 and is considered an extension of the copyright page.

Library of Congress Cataloging-in-Publication Data

Raven, Peter H.
Biology / Peter H. Raven, George B. Johnson.—6th ed.
p. cm.
Includes bibliographical references and index.
ISBN 0-07-303120-8
1. Biology. I. Johnson, George B. (George Brooks), 1942- . II. Title.

QH308.2 .R38 2002
570—dc21 2001030052
CIP

INTERNATIONAL EDITION ISBN 0-07-112261-3

www.mhhe.com

Brief Contents

Section A. General Principles

Section B. Organismal Biology

Contents

Chemical Biology

Part I The Origin of Living Things

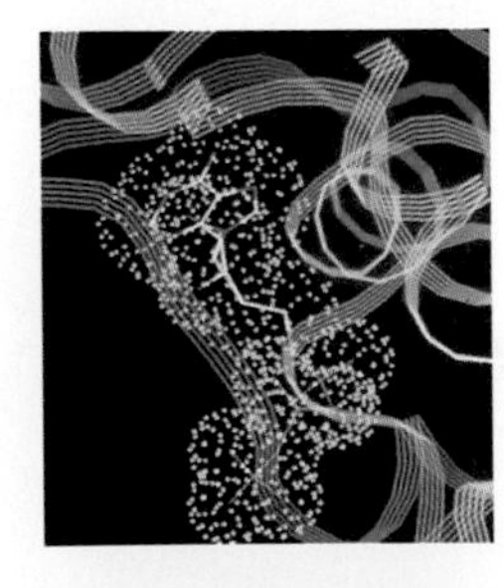

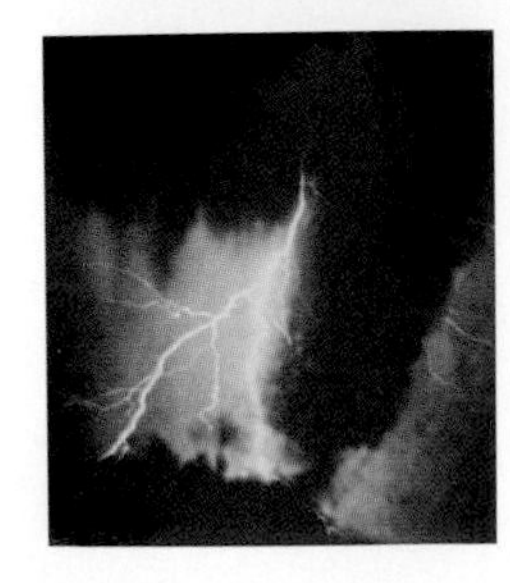

Cell Biology

Part II Biology of the Cell

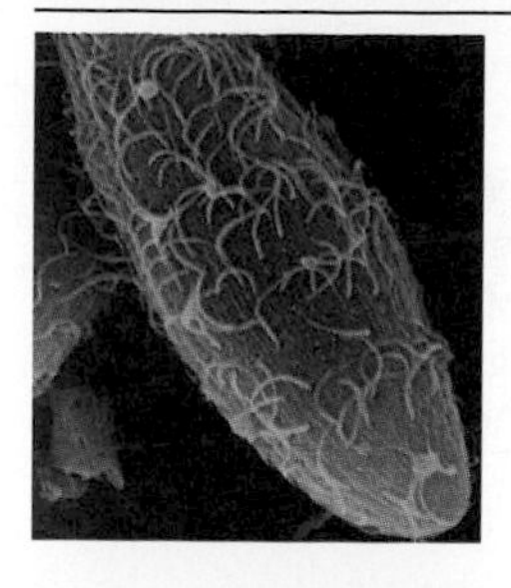

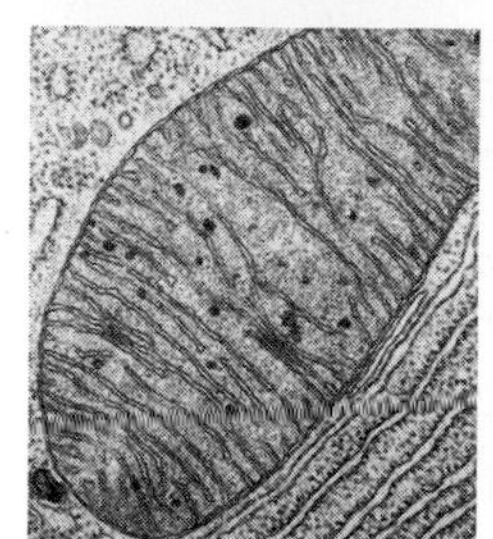

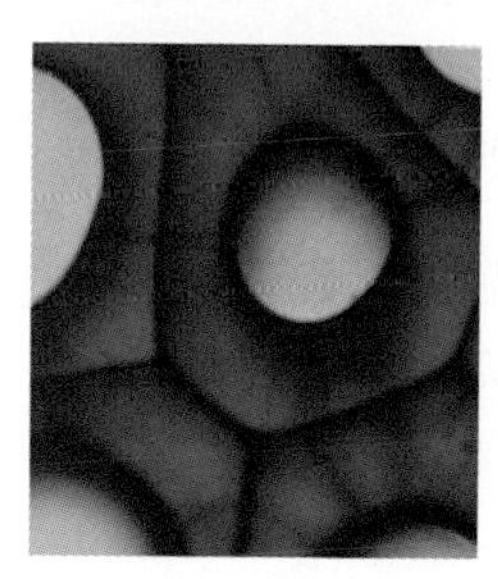

Cell Biology

Part III Energetics

Genetics

Part IV Reproduction and Heredity

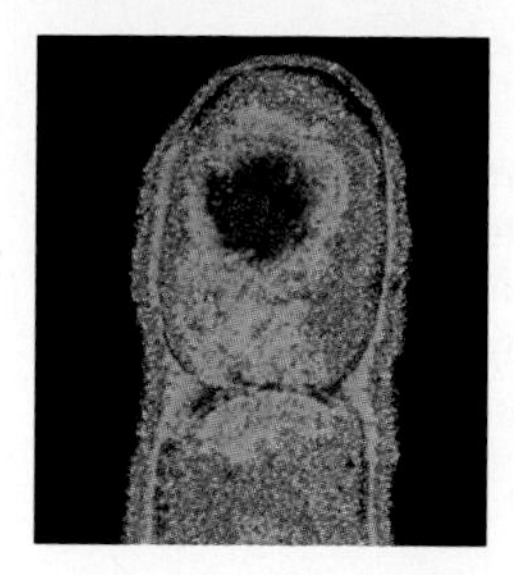

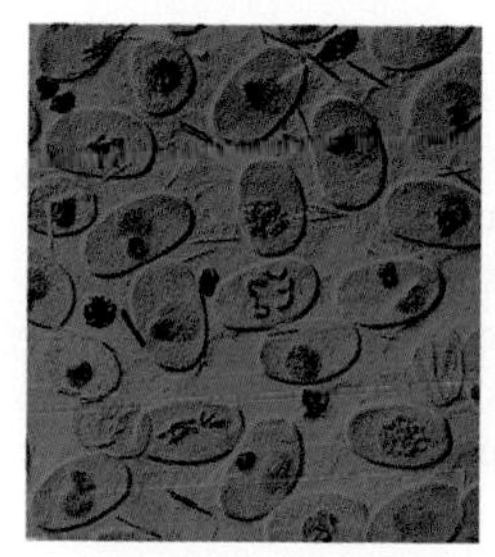

Genetics

Part V Molecular Genetics

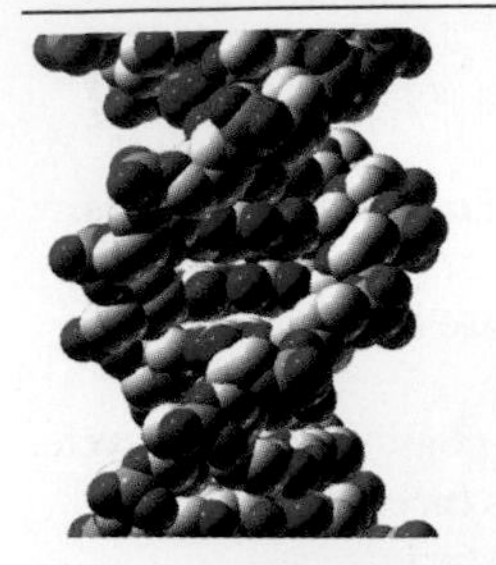

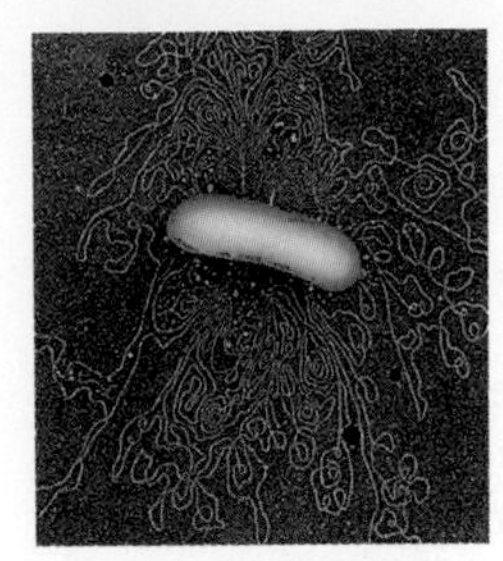

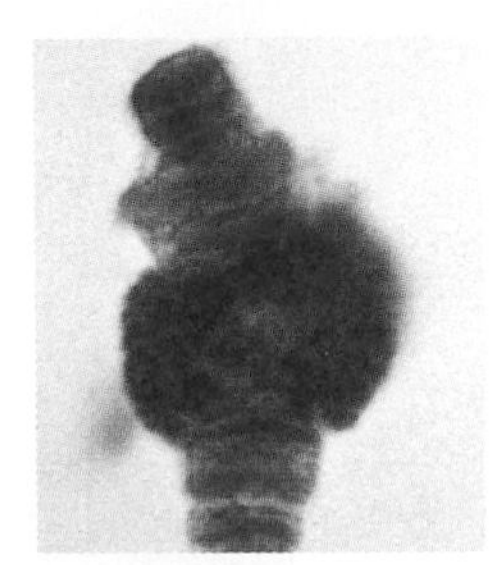

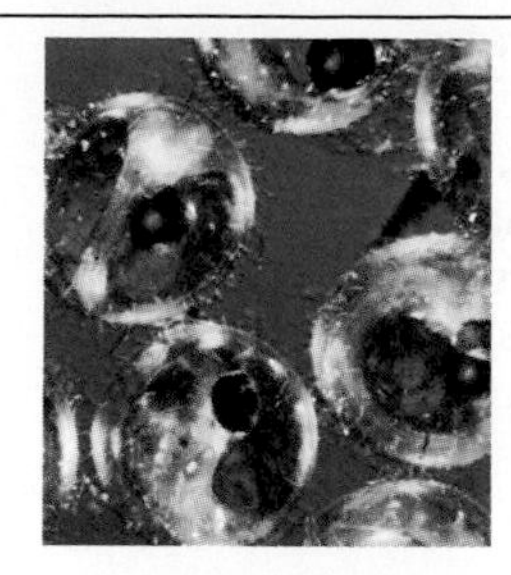

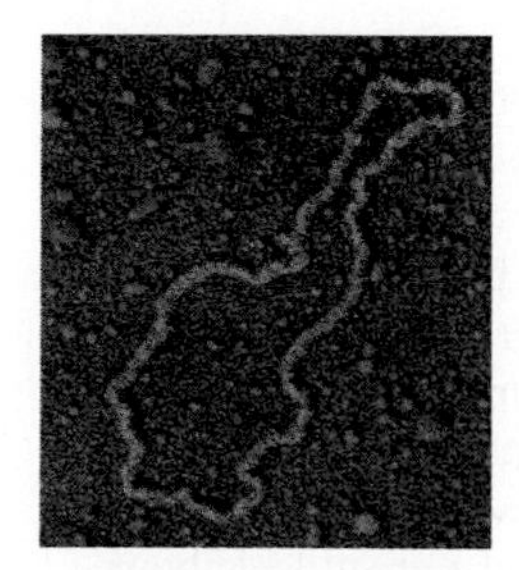

Evolution and Ecology

Part VI Evolution

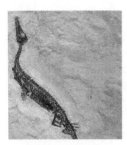

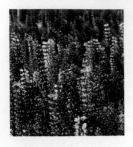

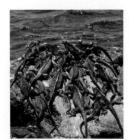

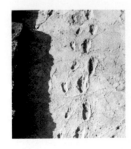

Evolution and Ecology

Part VIII The Global Environment

28

Dynamics of Ecosystems 571

Ecological systems are dynamic "machines" in which chemicals cycle between organisms and the environment and energy is used as it passes through the system. Much energy is lost to heat at each step of its journey through living things. In general, communities with more species interacting with one another are more stable.

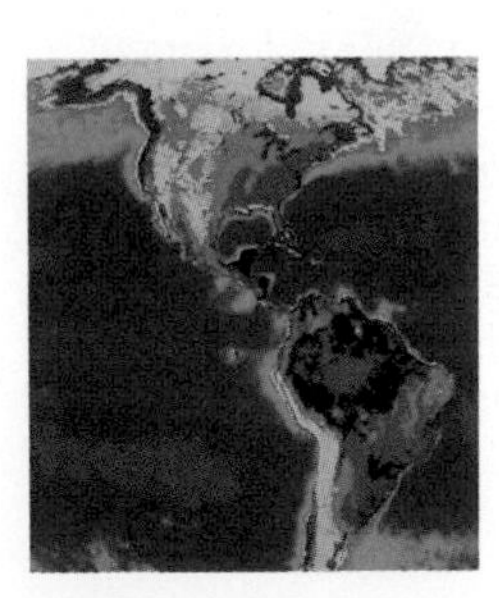

29

The Biosphere 591

The sun powers major movements of the atmosphere that circulate heat and moisture over the surface of the globe, creating different climates in different locales. Each characteristic climate favors a particular community of organisms adapted to living there.

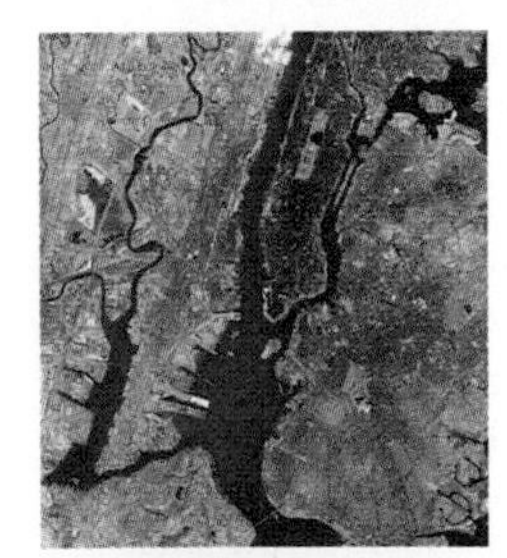

30

The Future of the Biosphere 611

The world's human population is growing at an explosive rate, placing a severe strain on the world's ecosystems. No one knows if the world can sustain the 6 billion people it now contains—and the population will soon grow much larger.

31

Conservation Biology 625

The successive efforts to halt the world's alarming loss of biodiversity will depend critically on our gaining a better understanding of what forces threaten species survival, and how they can be counteracted.

Part IX Viruses and Simple Organisms

32 How We Classify Organisms 649

The living world is a rich tapestry of diversity, teeming with different kinds of organisms. One of the great challenges of biology is to find sensible ways to name and classify kinds of organisms, ways that tell us about them and how they relate to each other.

32.1 Biologists name organisms in a systematic way.
32.2 Scientists construct phylogenies to understand the evolutionary relationships among organisms.
32.3 All living organisms are grouped into one of a few major categories.

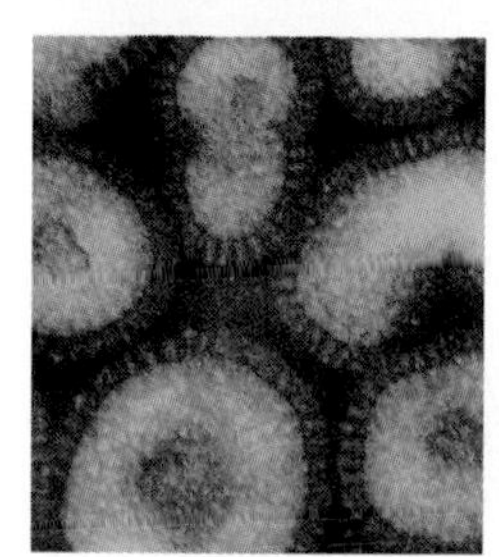

33 Viruses 665

Viruses are not considered organisms because they lack cellular structure. However, because they can replicate themselves within the cells of organisms, viruses are able to evolve, and are responsible for a broad array of serious diseases.

33.1 Viruses are strands of nucleic acid encased within a protein coat.
33.2 Bacterial viruses exhibit two sorts of reproductive cycles.
33.3 HIV is a complex animal virus.
33.4 Non-living infectious agents are responsible for many human diseases.

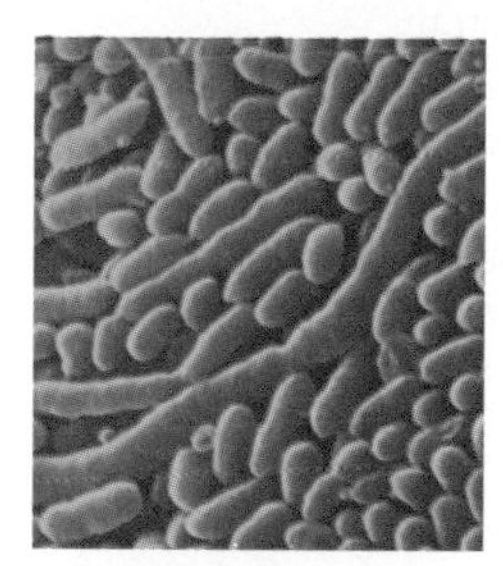

34 Bacteria 679

Bacteria are the simplest organisms, composed of single cells that lack the complex internal organization seen in all other cells. Biologists believe bacteria were the first organisms to evolve on earth, and they are easily the most numerous and successful.

34.1 Bacteria are the smallest and most numerous organisms.
34.2 Bacterial cell structure is more complex than commonly supposed.
34.3 Bacteria exhibit considerable diversity in both structure and metabolism.
34.4 Bacteria are responsible for many diseases but also make important contributions to ecosystems.

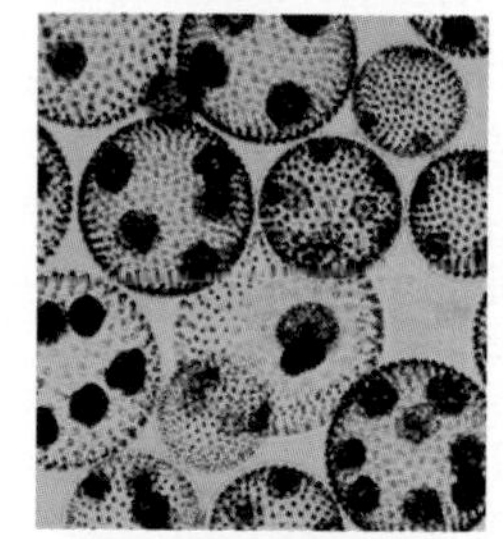

35 Protists 693

All eukaryotic organisms that are not fungi, plants, or animals are lumped together in a catch-all category called protists. Except for the marine algae, all protists are unicellular, but the diversity among protists is so great that it is difficult to compare them.

35.1 Eukaryotes probably arose by endosymbiosis.
35.1 The kingdom Protista is by far the most diverse of any kingdom.
35.2 Protists can be categorized into five groups.

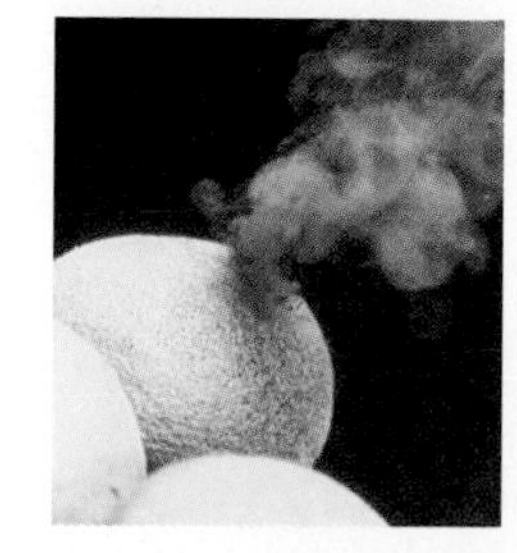

36 Fungi 719

Fungi are extraordinarily strange multicellular organisms whose cells share cytoplasm and nuclei. Fungi absorb their food from their surroundings, first excreting digestive enzymes, then soaking up the resulting soup of molecular fragments.

36.1 Fungi are unlike any other kind of organism.
36.2 Fungi are classified by their reproductive structures.
36.3 Fungi form two key mutualistic symbiotic associations.

Plants

Part X Plant Form and Function

Part XI Plant Growth and Reproduction

40 *Early Plant Development* 795

The plant body develops in modules of leaf, shoot, and root. The course of the plant body's development is strongly influenced by the environment. Many plant structures, including seeds and fruits, have evolved to aid dispersal of offspring to new locations.

40.1 Plant embryo development establishes a basic body plan.
40.2 The seed protects the dormant embryo from water loss.
40.3 Fruit formation enhances the dispersal of seeds.
40.4 Germination initiates post-seed development.

41 *How Plants Grow in Response to Their Environment* 807

Every plant cell contains a full set of hereditary information. Which genes are expressed is controlled by a set of plant hormones, including auxin, ethylene, and many others. These hormones interact with each other and with the environment to determine the pattern of growth.

41.1 Plant growth is often guided by environmental cues.
41.2 The hormones that guide growth are keyed to the environment.
41.3 The environment influences flowering.
41.4 Many short-term responses to the environment do not require growth.

42 *Plant Reproduction* 837

Many plants can clone themselves by asexual reproduction. Some plants use the wind to transport pollen from male to female. A much more efficient delivery system is employed by flowering plants, which use insects to carry pollen from flower to flower.

42.1 Angiosperms have been incredibly successful, in part, because of their reproductive strategies.
42.2 Flowering plants use animals or wind to transfer pollen between flowers.
42.3 Many plants can clone themselves by asexual reproduction.
42.4 How long do plants and plant organs live?

43 *Plant Genomics* 853

Plants organize their hereditary material in a more complex way than animals, with extensive regions of repetitive DNA and often many copies of chromosomes. Cloning plants in tissue culture and altering their genes have led to major agricultural advances.

43.1 Genomic organization is much more varied in plants than in animals.
43.2 Advances in plant tissue culture are revolutionizing agriculture.
43.3 Plant biotechnology now affects every aspect of agriculture.

Animals

Part XII Animal Diversity

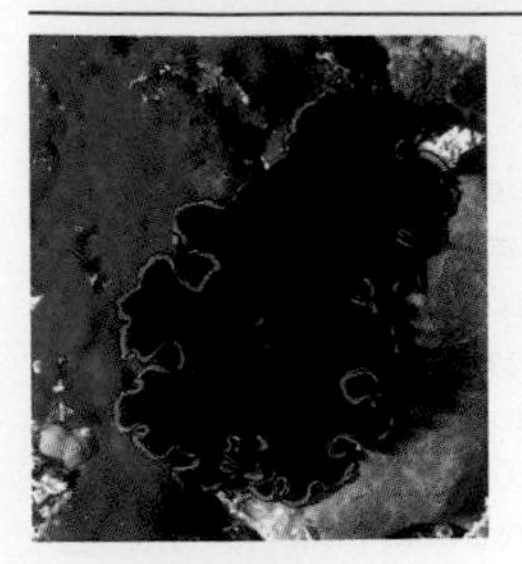

44 ***The Noncoelomate Animals*** **875**

The simplest animals lack a coelomic body cavity. The most ancient are the sponges, which lack tissues. Many of these evolutionarily old noncoelomate animals are called "worms," but this simple designation hides a great diversity of form and function.

44.1 Animals are multicellular heterotrophs without cell walls.
44.2 The simplest animals are not bilaterally symmetrical.
44.3 Acoelomates are solid worms that lack a body cavity.
44.4 Pseudocoelomates have a simple body cavity.
44.5 The coming revolution in animal taxonomy will likely alter traditional phylogenies.

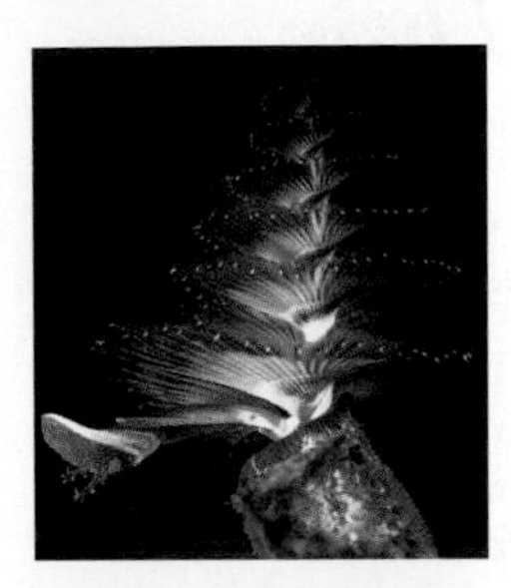

45 ***Mollusks and Annelids*** **899**

Mollusks and annelid worms, both of which originated in the sea and have similar larvae, are also very successful on land. Indeed, there are more terrestrial species of mollusks than there are terrestrial vertebrate species!

45.1 Mollusks were among the first coelomates.
45.2 Annelids were the first segmented animals.
45.3 Lophophorates appear to be a transitional group.

46 ***Arthropods*** **913**

Arthropods are easily the most successful animal group, particularly the insects. Along with the jointed appendages that are the hallmark of arthropods, insects evolved wings. There are more species of beetles than there are of any nonarthropod animal phylum.

46.1 The evolution of jointed appendages has made arthropods very successful.
46.2 The chelicerates all have fangs or pincers.
46.3 Crustaceans have branched appendages.
46.4 Insects are the most diverse of all animal groups.

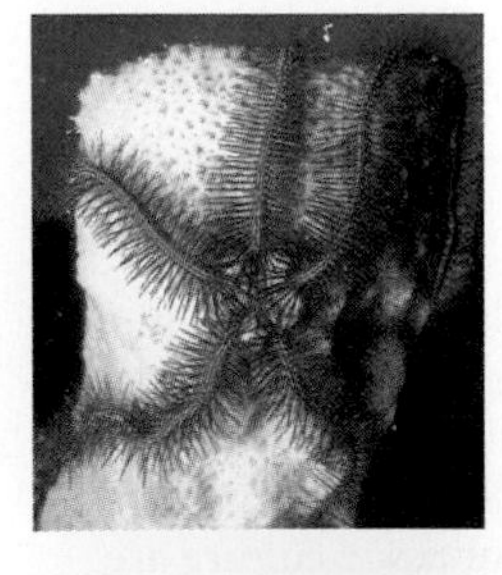

47 ***Echinoderms*** **933**

The vertebrates closest relatives among the animals are echinoderms, what most people think of as "starfish." Like vertebrates, they have deuterostome development and an endoskeleton. Unlike vertebrates, however, echinoderms are radially symmetrical as adults.

47.1 The embryos of deuterostomes develop quite differently from those of protostomes.
47.2 Echinoderms are deuterostomes with an endoskeleton.
47.3 The six classes of echinoderms are all radially symmetrical as adults.

48 ***Vertebrates*** **945**

Vertebrates are members of the phylum Chordata, all of whose members have a notochord at some stage in their development. The hallmark of vertebrates is an interior skeleton of bone which provides a superb framework for muscle attachment.

48.1 Attaching muscles to an internal framework greatly improves movement.
48.2 Nonvertebrate chordates have a notochord but no backbone.
48.3 The vertebrates have an interior framework of bone.
48.4 The evolution of vertebrates involves invasions of sea, land, and air.

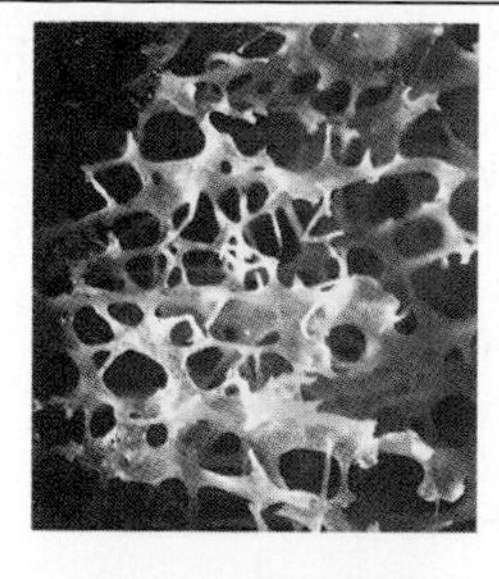

49 *Organization of the Animal Body* 983

Although they at first glance seem quite different from one another, all vertebrates are basically similar in body design, using the same array of tissues to form a similar array of organs. Vertebrates differ principally in the degree of terrestrial adaptation.

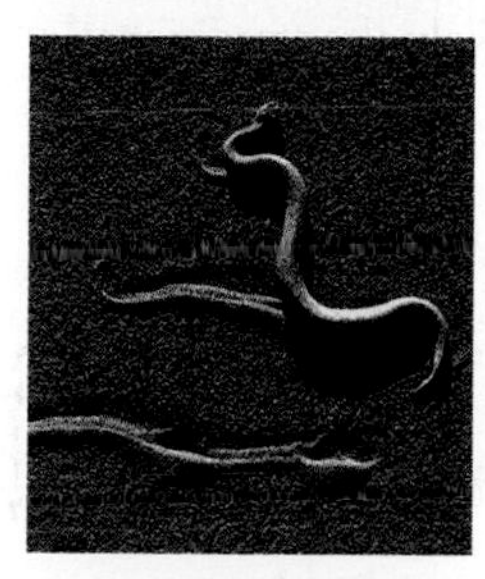

50 *Locomotion* 999

Muscle tissue, because it can contract, allows animals to move about. Some animals, like fish and snakes, move their entire bodies, while others move limbs such as arms, legs, or wings. The contraction of muscle is powered by ATP and controlled by the nervous system.

51 *Fueling Body Activities: Digestion* 1017

Before the cells of an animal can assimilate fatty acids, sugars, and amino acids to use in manufacturing ATP, these "food" molecules must be cleaved out of far larger fats, carbohydrates, and proteins. This process of digestion takes place in the stomach and small intestine.

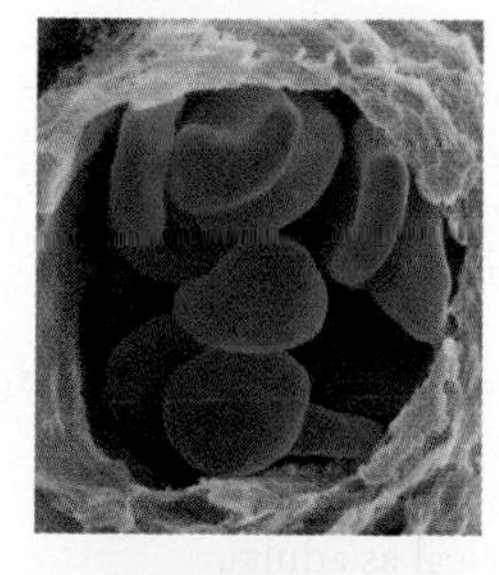

52 *Circulation* 1037

The highway that carries materials from one place to another in the vertebrate body is a network of flexible pipes called arteries and veins. A muscular pump called the heart pushes a rich fluid called blood through these pipes to every organ in the body.

53 *Respiration* 1053

One of the most important cargos carried by blood are two gases, oxygen and carbon dioxide. In the lungs, blood picks up oxygen from the air and dumps carbon dioxide acquired from the body's tissues as a by-product of their respiratory metabolism.

Animals

Part XIV Regulating the Animal Body

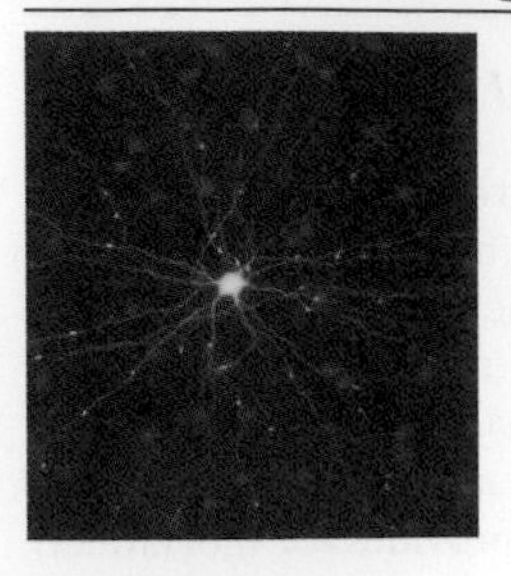

54 *The Nervous System* *1073*

The activities of the many organs of the vertebrate body are coordinated by the nervous system, an extensive network of neuron cells connecting all organs of the body. At a central coordinating center, the brain, information is processed, associations made, and commands given.

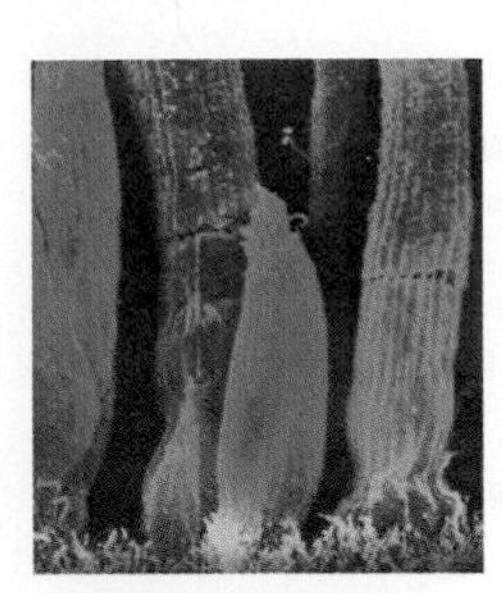

55 *Sensory Systems* *1103*

Information used by the nervous system to coordinate the body's activities comes from a system of sensors that collects information about the body's condition, its position in space, and what is going on around it. Vertebrates differ substantially in which sensors they employ.

56 *The Endocrine System* *1125*

The vertebrate nervous system uses long-lasting chemical signals called hormones to coordinate many physiological and developmental processes. The cells that secrete these hormones are often regulated by a set of command hormones released by the brain.

57 *The Immune System* *1147*

Vertebrates defend themselves from infection by viruses, bacteria, and other microbes with an army of circulating cells that seek out and check the identity of all cells in the body. When a stranger is identified, other "killer" cells are called in to deal with the invader.

58 *Maintaining the Internal Environment* 1173

Vertebrates maintain internal conditions at constant values, often quite different from those of their environment. This is particularly true of temperature and salt levels. Vertebrates go to great pains to maintain body temperature and avoid water loss or gain.

58.1 The regulatory systems of the body maintain homeostasis.
58.2 The extracellular fluid concentration is constant in most vertebrates.
58.3 The functions of the vertebrate kidney are performed by nephrons.
58.4 The kidney is regulated by hormones.

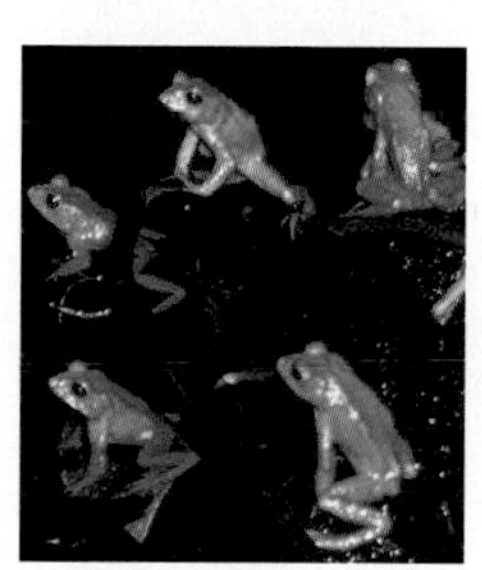

59 *Sex and Reproduction* 1195

Vertebrates usually reproduce sexually, although asexual reproduction is not impossible for them. In the sea, females release eggs into the water for external fertilization by the males' sperm. On land, evolution has favored internal fertilization.

59.1 Animals employ both sexual and asexual reproductive strategies.
59.2 The evolution of reproduction among the vertebrates has led to internalization of fertilization and development.
59.3 Male and female reproductive systems are specialized for different functions.
59.4 The physiology of human sexual intercourse is becoming better known.

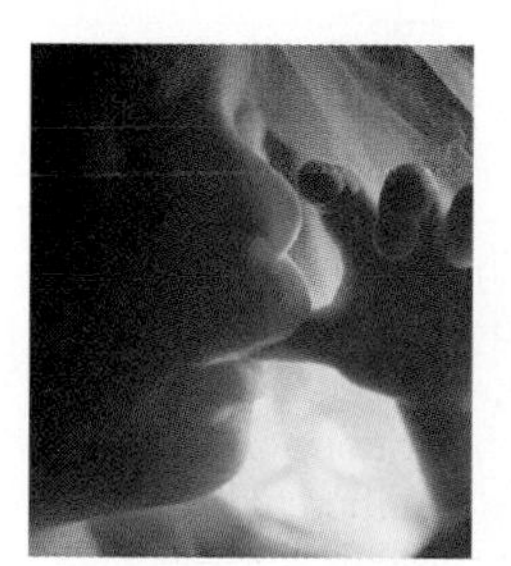

60 *Vertebrate Development* 1215

In vertebrates, the path of development from fertilization to birth is well known. The development of the embryo undergoes its key stages early on, determining the basic architecture of the body and its tissues. Thereafter, much of what occurs is growth.

60.1 Fertilization is the initial event in development.
60.2 Cell cleavage and the formation of a blastula set the stage for later development.
60.3 Gastrulation forms the three germ layers of the embryo.
60.4 Body architecture is determined during the next stages of embryonic development.
60.5 Human development is divided into trimesters.

Preface

We enter a new century with this sixth edition of *Biology*, one that is exploding with excitement in biology. This first year of the new millennium has seen the completion of the Human Genome Project, with the full sequence of the human genome now available for research and exploration. Embryonic stem cells were cloned for the first time in the year 2000, and offer the potential for curing a wide range of ills, from spinal cord injuries to diabetes. Golden rice, a genetically modified crop to which has been added a battery of genes that overcome deficiencies in vitamin A and iron, was planted for the first time in Asian fields. Neurobiologists for the first time caught a glimpse of the molecular basis of learning. Even taxonomy, that bastion of conservative judgments, seems to be undergoing a sea change, with molecular phylogenies forcing the redrawing of many family trees, from angiosperm plants to insects and other arthropods.

There probably has never been a more exciting time to learn biology. Adding together the years, Dr. Raven and I have been teaching biology for more than 70 years, neither of us can remember any time as fraught with promise as today. We started teaching in the sixties, also exciting times. In those revolutionary years the black box surrounding the gene machine was stripped away, revealing for the first time how DNA achieves the constancy and diversity that are the hallmarks of life. For 40 years researchers have been amplifying that picture, learning in ever-greater detail how life works.

In the last few decades, the pace of biological research has accelerated, as we have learned for the first time how to manipulate genes. In agriculture this has led to waves of controversy, in medicine to advances universally applauded. But no matter how one views genetic engineering, no one questions that it is changing the science of biology in profound ways.

What is important about these changes in biology, what excites us like no past year, is the potential to influence our health, and that of our world. Biology as a science can—indeed, must—be more than simply a trip to the zoo, an investigation of what living things are like and how they work. These things are important parts of biology, of course, the knowledge that provides the core of the science. But it can't stop there. The knowledge of biology that has been gained, especially in the last decade, provides us with a tool of unprecedented power to improve the human condition and lessen human impact on the world we share with life's other creatures.

It is with this sense of a science alive with promise that we set out in the first year of this new century to produce the sixth edition of *Biology*.

Significant Enhancements to the Sixth Edition

Every revision of a successful text starts with a plan to update areas where advances have occurred. Thus the initial plans for this sixth edition of *Biology* were to correct any errors detected by its many users, and to incorporate new findings in rapidly advancing areas of research. In publishing terms, this was to have been a "light" revision. However, that is not what happened. Inspired by the suggestions of reviewers, we found ourselves adding chapters, overhauling the way in which key chapters were organized, adding material and then more material—soon we were knee-deep in a significant revision.

Much of the focus of this sixth edition revision was on evolution, ecology, and botany, areas where there was an opportunity for exciting improvement. To revise these chapters, we recruited two young energetic biologists to provide fresh perspective. They brought with them new approaches, fresh ideas, and up-to-date knowledge of their areas of expertise. Indeed, it has been so much fun to work with them that in future editions they will join us as full coauthors of the text.

Ecology and Evolution

Professor Jonathan Losos, our colleague at Washington University, has revised the evolution and ecology sections of the text, bringing more experimental science into our discussions. Presentation of the experimental data used to derive key conclusions and concepts is key to this revision. Our goal is to better aid students to understand how the concepts arose from the research. For this reason, you will see that graphs and charts are more plentiful in these chapters.

Botany

Professor Susan Singer of Carleton College has revised the botany chapters. The botany sections have benefited from a new approach where plant development takes center stage. A plant developmental biologist, she has placed the traditional discussions of evolutionary influences on plant form and function into a developmental context. Thus while evolution is still presented as the underlying explanation for the character of vascular tissue, seeds, flowers, and fruits, the developmental processes that produce these organs are now given more prominence. This does not lessen the evolutionary character of the treatment, but rather serves to amplify it. Throughout all the botany chapters, there is an enhanced emphasis on the molecular aspects of plant life. Understanding the molecular underpinnings of plant form and function allows students to more clearly understand the evolutionary changes that have shaped them.

New Chapter: Conservation Biology (Chapter 31)

In the fifth edition, we presented a discussion of conservation biology on the *Biology* web site, as an "enhancement chapter." The response to this material was so overwhelming that we have included such a chapter in this edition of our text. In our own classroom teaching we find students to be keenly aware of the problems of dwindling natural resources, and the need to tackle the issue concretely. We feel a chapter focusing on conservation biology will be appreciated by students and useful to professors.

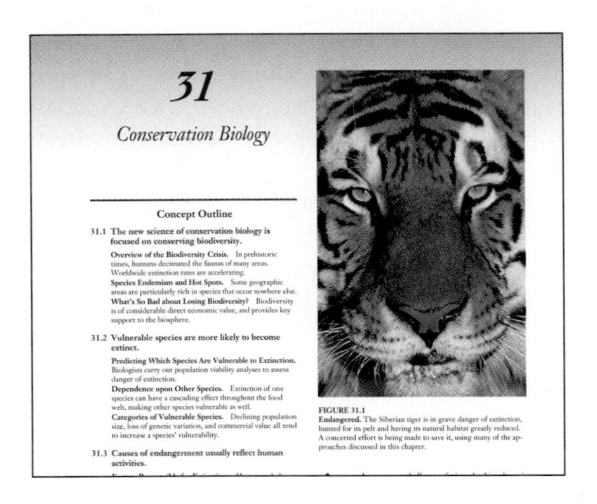
31

Conservation Biology

Concept Outline

31.1 The new science of conservation biology is focused on conserving biodiversity.

Overview of the Biodiversity Crisis. In prehistoric times, humans decimated the faunas of many areas. Worldwide extinction rates are accelerating.

Species Endemism and Hot Spots. Some geographic areas are particularly rich in species that occur nowhere else.

What's So Bad about Losing Biodiversity? Biodiversity is of considerable direct economic value, and provides key support to the biosphere.

31.2 Vulnerable species are more likely to become extinct.

Predicting Which Species Are Vulnerable to Extinction. Biologists carry out population viability analyses to assess danger of extinction.

Dependence upon Other Species. Extinction of one species can have a cascading effect throughout the food web, making other species vulnerable as well.

Categories of Vulnerable Species. Declining population size, loss of genetic variation, and commercial value all tend to increase a species' vulnerability.

31.3 Causes of endangerment usually reflect human activities.

FIGURE 31.1
Endangered. The Siberian tiger is in grave danger of extinction, hunted for its pelt and having its natural habitat greatly reduced. A concerted effort is being made to save it, using many of the approaches discussed in this chapter.

Genomics "Enhancement Chapter"

The rapidly advancing field of genomics is so key to the future of biology that we felt it necessary to discuss it in some way in this sixth edition. Including a chapter in the text seemed rather pointless—so much of what we would cover will have changed after the first year. So we turn again to an "enhancement chapter." We used enhancement chapters to expand information for the fifth edition of *Biology*, and as you see from above, after fine-tuning the conservation biology chapter, we now include it in this edition. The enhancement chapter on genomics can be found at *http://www/mhhe.com/raven6* This new chapter expands upon the discussion of gene technology to present and explain the advances now being made with genomics. While the chapter discusses the technology involved and the genomes that have been uncoded, it focuses on the significance of this information to biology as a science, and on what it could mean to the future of medicine, agriculture, and many other fields.

Real People Doing Real Science

We have added an inquiry-based learning experience at the beginning of every Part that walks a student through the process of scientific inquiry by examining a particular experiment. We have titled this feature "Real People Doing Real Science." After briefly reviewing the significance of the experimental question being addressed, we take the student through the actual experiment, discussing experimental design in depth, and then briefly describe the results and conclusion. This is but the first part of the learning experience. The student is then directed to the *Biology* sixth edition web site for an in-depth examination of the experiment. There a student can read the actual published research paper, allowing students to become more familiar with the primary literature. Then the student can carry out a "Virtual Experiment" where he or she is able to manipulate the parameters of the experiment and obtain data for analysis. We provide on-line questions and discussions to help the student better understand the thought process behind the experiment.

To explore this experiment further,
go to the Virtual Lab at
www.mhhe.com/raven6/vlab8.mhtml

A Thorough Revision

In addition to the extensive revisions of the ecology, evolution, and botany sections of the text, and the new chapter on conservation biology, we have thoroughly revised the rest of the text as well. Many chapters now sport radically different organizations, benefiting from extensive reviewer input. Pedagogy has been improved as well. We have included phylogenetic guideposts throughout the discussions of diversity to clarify for the student where each group fits in the tree of life. (You will find these guideposts in chapters 35, 36, 37, and 44–48.)

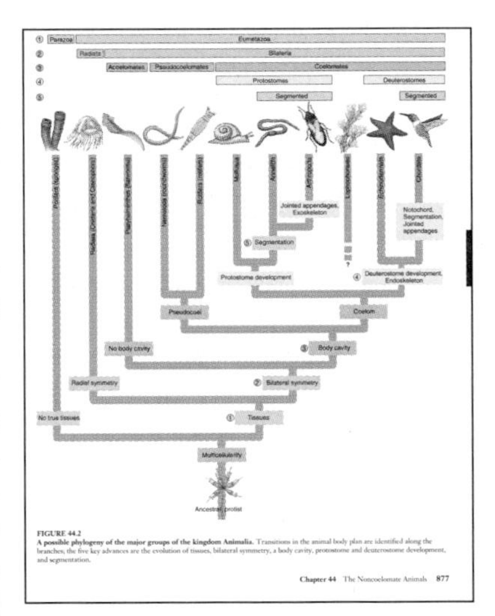

FIGURE 44.2
A possible phylogeny of the major groups of the kingdom Animalia. Transitions in the animal body plan are identified along the branches; the five key advances are the evolution of tissues, bilateral symmetry, a body cavity, protostome and deuterostome development, and segmentation.

Chapter 44 The Noncoelomate Animals 877

The Chemical Building Blocks of Life (Chapter 3)

The organization of this chapter has been turned on its head, presenting lipids before carbohydrates. This gives a greatly improved sense of the relative biological importance of these macromolecules, and actually makes the material easier to learn.

The Origin and Early History of Life (Chapter 4)

The discussion of ideas about the origin of life is now much more open-ended, stressing competing hypotheses and the key role of assumptions for which there is little data.

Photosynthesis (Chapter 10)

The internal organization of this chapter has been reworked to make it easier for students to understand how the many concepts covered in this chapter relate to one another.

Patterns of Inheritance (Chapter 13)

This chapter has been reorganized to incorporate the discussion of human genetics earlier in the chapter and then to use human examples as a means of explaining Mendelian principles.

Cellular Mechanisms of Development (Chapter 17)

We have moved the discussion of cellular development up earlier in the text, immediately following the discussion of gene expression, to reinforce key molecular concepts.

Altering the Genetic Message (Chapter 18)

Many recent advances in cancer research are highlighted, with greater emphasis on genes governing metastasis and angiogenesis.

Gene Technology (Chapter 19)

New topics such as biochips and transgenic rice have been included and rapidly advancing areas such as stem cells and ethics and regulations have been updated.

The Evidence for Evolution (Chapter 21)

We have expanded this chapter to include a complete discussion of the evolution of the horse, and have expanded the discussion of artificial selection as a means of showing the power of selection on the evolution of species.

Population Ecology (Chapter 24)

We have added and expanded the discussions of population distributions, ranges, dispersal mechanisms and human effects in examples replete with actual data.

Animal Behavior (Chapter 26) and Behavioral Ecology (Chapter 27)

We have amplified these two chapters, moving them to the ecology section, a more logical place to teach these topics.

Dynamics of Ecosystems (Chapter 28)

We have greatly expanded discussions of interactions among trophic levels and the controversial matter of how species richness influences community stability.

The Biosphere (Chapter 29)

We have expanded the discussion of evolutionary responses to environmental variation.

Evolutionary History of Plants (Chapter 37)

We now include a discussion of the green algal origin of all plants.

The Plant Body (Chapter 38)

We include a discussion of the genes involved in development of stomata, trichomes, root tissues and leaves.

How Plants Grow in Response to Their Environment (Chapter 41)

This chapter was extensively reworked and many new topics were added and expanded such as acid growth hypothesis of auxin actions, plant defense responses, cytokinin involvement in organ regeneration and crown gall tumors, brassinosteroids and oligosaccharins, transgenic tomatoes, initiating flowering, and circadian clocks.

The Noncoelomate Animals (Chapter 44)

This chapter now includes a molecular reevaluation of the evolution of the metazoan body plan.

Arthropods (Chapter 46)

New molecular data calls into question traditional classification of arthropods based on external characteristics.

Locomotion (Chapter 50)

We have added a discussion of modes of locomotion that ties together the concepts presented in the chapter.

Circulation (Chapter 52)

We have added a section on heart disease, explaining that heart disease is preventable and begins with establishing a heart-healthy lifestyle early.

Sensory Systems (Chapter 55)

We have broadened the coverage in this chapter to include more examples of nonmammalian sensory systems.

The Immune System (Chapter 57)

This chapter has been completely reorganized to improve clarity and understanding. The presentation of topics now more logically follows the process of the immune response in the body.

Real People Doing Real Science

Each of the fourteen parts of this text is introduced with a detailed look at an experiment—not a famous one, but rather the kind of experiment that real scientists do each day. There is no better way for a student to appreciate how scientific progress occurs than to get down in the trenches with the researchers doing the work.

Part I Kellar Autumn (Lewis & Clark College) and **Robert Full** (University of California, Berkeley)—*Unraveling the Mystery of How Geckos Defy Gravity.*

Part II Richard Cyr (Pennsylvania State University)—*How Do the Cells of a Growing Plant Know in Which Direction to Elongate?*

Part III Andrew Webber (Arizona State University)—*How Do Proteins Help Chlorophyll Carry Out Photosynthesis?*

Part IV Julian Adams (University of Michigan)—*Why Do Some Genes Maintain More Than One Common Allele in a Population?*

Part V Randall Johnson (University of California, San Diego)—*Can Cancer Tumors Be Starved to Death?*

Part VI John Endler (University of California, Santa Barbara) and **David Reznick** (University of California, Riverside)—*Catching Evolution in Action.*

Part VII Mark Boyce (University of Alberta, Edmonton)—*Why Do Tropical Songbirds Lay Fewer Eggs?*

Part VIII Andrew Blaustein (Oregon State University)—*Identifying the Environmental Culprit Harming Amphibians.*

Part IX Michael Houghton (Chiron)—*Discovering the Virus Responsible for Hepatitis C.*

Part X Robert Boyd (Auburn University) and **Scott Martens** (University of California, Davis)—*Why Do Some Plants Accumulate Toxic Levels of Metals?*

Part XI John Schiefelbein (University of Michigan)—*The Control of Patterning in Plant Root Development.*

Part XII Jon Harrison (Arizona State University)—*How Honeybees Keep Their Cool.*

Part XIII Elizabeth Brainerd (University of Massachusetts, Amherst)—*Why Some Lizards Take a Deep Breath.*

Part XIV Louis Guillette (University of Florida)—*Are Pollutants Affecting the Sexual Development of Florida's Alligators?*

Virtual Lab

To allow students to explore further, each of these fourteen experiments is linked to a far richer presentation on the internet. As an example, consider Part VIII, an experiment attempting to gain a better understanding of why many amphibian populations today are exhibiting decreasing numbers and numerous individuals with severe developmental deformities. By going to the *BIOLOGY* 6/e virtual lab devoted to this experiment (www.mhhe.com/raven6/vlab8.mhtml), a student can:

READ THE ORIGINAL RESEARCH PAPER Blaustein, Andrew R. et al., "Ambient UV-B radiation causes deformities in amphibian embryos," *Proc. Natl. Acad. Sci. USA* 1997 vol. 94: 13735–13737.

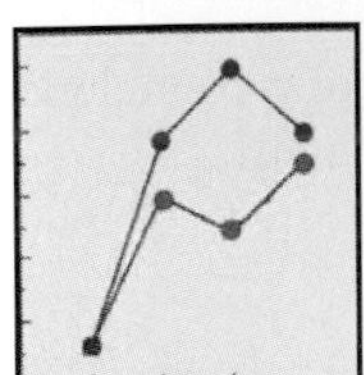

RUN A VIRTUAL EXPERIMENT EXPLORING THE ORIGINAL PAPER The student runs a virtual experiment, collects and plots data, and answers questions about the significance of the results.

MEET THE INVESTIGATOR An interview with the principle investigator, Andrew Blaustein, with a short bio and links to his home page and publication list.

READ A RELATED PAPER Blaustein, Andrew et al., "UV repair and resistance to solar UV-B in amphibian eggs: A link to population declines?" *Proc. Natl. Acad. Sci. USA* 1994 vol. 91:1791–1795.

RUN A VIRTUAL EXPERIMENT EXPLORING THE RELATED PAPER The student is presented with a second hypothesis, to be tested with another virtual experiment.

READINGS AND ADDITIONAL RESOURCES Links to other related papers, to web sites of interest, and to relevant "*ON SCIENCE*" articles written by the author George Johnson.

Acknowledgments

William C. Ober and Claire Garrison have again enhanced the art program for this text with many new and revised full-color illustrations. Bill's artistic skills, knowledge of biology, and experience gained from an earlier career as a practicing physician have enriched this text through six of its editions. Claire practiced pediatric and obstetric nursing before turning to scientific illustration as a full-time career. Texts illustrated by Bill and Claire have received national recognition and won awards from the Association of Medical Illustrators, American Institute of Graphic Arts, Chicago Book Clinic, Printing Industries of America, and Bookbuilders West. They are also recipients of the Art Directors Award.

Our goal for *Biology* has always been to present the science in an interesting and engaging manner while maintaining a comprehensive and authoritative text. This is a lofty goal considering the mountains of information and research authors must go through just to update the text from one edition to the next. This sixth edition would not have been possible without the contributions of many. As you will see on the title page and the "Meet the Authors" section of this Preface, two new contributors joined us for the revision of this new edition of *Biology*. Jonathan Losos brought major contributions to the evolution and ecology sections, increasing the authoritativeness of the text by adding more original research to the discussions. Susan Singer had the formidable responsibility of reevaluating the botany sections to give the chapters a new and more current approach. Without Jonathan and Susan, this sixth edition would not have been possible. Eric Strauss also provided extensive reviews of the diversity chapters with recommendations for revision and modifications. His comments were greatly appreciated. The visuals are so important in a biology textbook and the superb illustrations were conceived and rendered by Bill Ober and Claire Garrison. We also thank Don and Joan Murie of Meyers Photo-Art for their excellent research of new photographs for this and past editions. Of course we are also indebted to our colleagues from across the country and around the globe that provided numerous suggestions on how to improve the sixth edition. Every one of you has our heartfelt thanks.

A major feature of *Biology* continues to be the presentation of the information into conceptual modules. It is no small feat to take the information written by four individuals along with their suggestions for figures and tables and present it in a conceptual module. This formidable task would not have been possible without the efforts of Megan Jackman, our off-site developmental editor. Her intelligence and perseverance played a major role in the high quality of this book. Liz Sievers joined our off-site development team during the revision process, and her help and support was greatly appreciated. As any author knows, a textbook is made not by a writer but by a publishing team, a group of people that guide the raw book written by the authors through a year-long process of reviewing, editing, fine-tuning, and production. This edition was particularly fortunate in its book team, led by Patrick Reidy, sponsoring editor, Lu Ann Weiss, developmental editor, Peggy Selle, project manager, Stuart Paterson, design manager, Lori Hancock, photo research coordinator, and many, many more people behind the scenes.

As always, we have had the support of wives and family who have seen less of us than they might like because of the pressures of getting this revision done. They have become accustomed to the many hours this book draws us away from them, a hidden price of textbook writing of which they are fully aware.

Acknowledgments would not be complete without thanking the generations of students who have used the many editions of this text. They have taught us at least as much as we have taught them.

Finally, we need to thank our reviewers. Every text owes a great deal to those faculty across the country who review it. Serving as sensitive antennae for errors and sounding boards for new approaches, reviewers are among the most valuable tools at an author's disposal. Many improvements in this edition are the direct result of their suggestions. Every one of them has our sincere thanks.

Reviewers of Sixth Edition

Michael Adams *Pasco-Hernando Community College*
Sylvester Allred *Northern Arizona University*
Lon Alterman *Clarke College*
Elena Amesbury *University of Florida*
William Anyonge *University of California–Los Angeles*
Amir Assadirad *Delta College*
Gary I. Baird *Brigham Young University*
Ellen Baker *Santa Monica College*
Stephen W. Banks *Louisiana State University–Shreveport*
Ruth Beattie *University of Kentucky*
Samuel N. Beshers *University of Illinois*
Christine Konicki Bieszczad *Saint Joseph College*
John Birdsell *University of Arizona*
Brenda C. Blackwelder *Central Piedmont Community College*
Sandra Bobrick *Community College of Allegheny County Allegheny Campus*
Randall Breitwisch *University of Dayton*
Mark Browning *Purdue University*
Roger Buckanan *Arkansas State University*
Theodore Burk *Creighton University*
John S. Campbell *Northwest College*
John R. Capeheart *University of Houston–Downtown*
Michael S. Capp *Carlow College*
Jeff Carmichael *University of North Dakota*
George P. Chamuris *Bloomsburg University*
Susan Cockayne *Brigham Young University*
William Cohen *University of Kentucky*
W. Wade Cooper *Shelton State Community College*
Lisa M. Coussens *University of California–San Francisco, Cancer Research Institute*
Wilson Crone *Hudson Valley Community College*
Paul V. Cupp Jr. *Eastern Kentucky University*
Richard Cyr *The Pennsylvania State University*
Grayson Davis *Trinity University*
Mark A. DeCrosta *University of Tampa*

David L. Denlinger *Ohio State University*
C. Lynn Dorn *Valencia Community College*
Charles D. Drewes *Iowa State University*
Sondra Dubowsky *Allen County Community College*
Peter I. Ekechukwu *Horry-Georgetown Technical College*
Dennis Emery *Iowa State University*
Frederick B. Essig *University of South Florida*
Bruce Evans *Huntington College*
Deborah Fahey *Wheaton College*
Linda E. Fisher *University of Michigan–Dearborn*
Rob Fitch *Wenatchee Valley College*
Robert Fogel *University of Michigan*
James Franzen *University of Pittsburgh–Pittsburgh Campus*
William Friedman *University of Colorado*
Lawrence Fritz *Northern Arizona University*
Bernard Frye *University of Texas at Arlington*
Robert J. Full *University of California–Berkeley*
Warren Gallin *University of Alberta*
Darrell Galloway *The Ohio State University*
Ted Gish *St. Mary's College*
Donald Glassman *Des Moines Area Community College*
Jim Glenn *Red Deer College*
Jim R. Goetze *Laredo Community College*
Jack M. Goldberg *University of California–Davis*
Elizabeth Godrick *Boston University*
Dalton Gossett *Louisiana State University–Shreveport*
John Griffis *Joliet Junior College*
Kathryn Gronlund *New Mexico State University–Carlsbad*
Elizabeth L. Gross *The Ohio State University*
Patricia A. Grove *College of Mount St. Vincent*
Randolph Hampton *University of California–San Diego*
Sehoya E. Harris *The Pennsylvania State University*
Carla Ann Hass *The Pennsylvania State University*
Chris Haynes *Shelton State Community College*
Albert A. Herrera *University of Southern California*
Pamela Higgins *Allentown College of St. Francis DeSales*
Richard Hill *Michigan State University*
Phyllis Hirsch *East Los Angeles College*
Victoria Hittinger *Rhode Island College*
Nan Ho *Las Positas College*
Leland N. Holland, Jr. *Pasco-Hernando Community College–West Campus*
Elisabeth A. Hooper *Truman State University*
Terry L. Hufford *The George Washington University*
Allen Hunt *Elizabethtown Community College*
Sobrasua E. M. Ibin *Morris Brown College*
Louis Irwin *University of Texas at El Paso*
Laurie E. Iten *Purdue University*
Jeffrey Jack *College of Arts & Sciences*
James B. Jensen *Brigham Young University*
Judy Jernstedt *University of California - Davis*
George P. Johnson *Arkansas Tech University*
Kenneth V. Kardong *Washington State University*
Cheryl Kerfeld *University of California–Los Angeles*
Joanne M. Kilpatrick *Auburn University at Montgomery*
Peter King *Francis Marion University*
Edward C. Kisailus *Canisius College*
Robert M. Kitchin *University of Wyoming*
Will Kleinelp *Middlesex County College*
Kenton Ko *Queen's University*
Ross E. Koning *Eastern Connecticut State University*
Karen L. Koster *University of South Dakota*
V.A. Langman *Louisiana State University–Shreveport*
Simon Lawrance *Otterbein College*
Jeffrey N. Lee *Essex County College*
Laura G. Leff *Kent State University*
Mary E. Lehman *Longwood College*
Niles Lehman *University at Albany SUNY*
Michael Lema *Midlands Technical College*
Charles Kingsley Levy *Boston University*
Leslie Lichtenstein *Massasoit Community College*
Harvey Liftin *Broward Community College*
Richard Londraville *University of Akron*
Sonja L. Maki *Clemson University*
Bradford D. Martin *La Sierra University*
Barbara Maynard *Colorado State University*
Deanna McCullough *University of Houston Downtown*
L. R. McEdward *University of Florida*
Michael Ray Meighan *University of California–Berkeley*
John Merrill *Michigan State University*
Harry A. Meyer *McNeese State University*
Dennis J. Minchella *Purdue University*
Jonathan D. Monroe *James Madison University*
David L. Moore *Utica College of Syracuse University*
Tony E. Morris *Fairmont State College*
Roger N. Morrissette *Framingham State College*
Richard Mortensen *Albion College*
William H. Nelson *Morgan State University*
Peter H. Niewiarowski *University of Akron*
Colleen J. Nolan *St. Mary's University*
John C. Osterman *University of Nebraska–Lincoln*
Thomas G. Owens *Cornell University*
Bruce Parker Utah *Valley State University*
Dustin Penn *University of Utah*
Stacia Pieffer-Schneider *Marquette University*
Carl S. Pike *Franklin and Marshall College*
Nancy A. Perigo *Willamette University*
Greg Phillips *Blinn College–Brenham Campus*
Jon Pigage *University of Colorado at Colorado Springs*
Barbara Pleasants *Iowa State University*
John Pleasants *Iowa State University*
Peggy Pollack *Northern Arizona University*
Mitch Price *The Pennsylvania State University*
Margene Ranieri *Bob Jones University*
Arthur Raske *Northland Baptist Bible College*
Keith Redetzke *University of Texas at El Paso*
Peter J. Rizzo *Texas A&M University*
Ellison Robinson *Midlands Technical College*
Lyndell P. Robinson *Lincoln Land Community College*
Angel M. Rodriguez *Broward Community College*
June R. P. Ross *Western Washington University*
Patricia Rugaber *Coastal Georgia Community College*
Connie Rye *Bevill State Community College*
Nancy K. Sanders *Truman State University*
Robert B. Sanders *University of Kansas–Main Campus*
Lisa M. Sardinia *Pacific University*
Brian W. Schwartz *Columbus State University*
Bruce S. Serlin *DePauw University*
Mark A. Sheridan *North Dakota State University*
Janet Anne Sherman *Penn College of Technology*
Louis Sherman *Purdue University*
Jim Shinkle *Trinity University*
Richard Shippee *Vincennes University*
Brian Shmaefsky *Kingwood College*

Michele Shuster *University of Pittsburgh*
Robert C. Sizemore *Alcorn State University*
Mark Smith *Victor Valley College*
Nancy Solomon *Miami University*
Norm Stacey *University of Alberta*
Ruth Stutts-Moseley *Bishop State Community College*
Kathy Sympson *Florida Keys Community College*
Stan Szarek *Arizona State University*
Robert H. Tamarin *University of Massachusetts Lowell*
Michael Tenneson *Evangel University*
Sharon Thoma *Edgewood College*
Joanne Kivela Tillotson *Purchase College State University of New York*
Maurice Thomas *Palm Beach Atlantic College*
Thomas Tomasi *Southwest Missouri State University*
Leslie Towill *Arizona State University*
Akif Uzman *University of Houston–Downtown*
Thomas J. Volk *University of Wisconsin–La Crosse*
Keith D. Waddington *University of Miami*
D. Alexander Wait *Southwest Missouri State University*
Timothy S. Wakefield *Auburn University*
Charles Walcott *Cornell University*
Eileen Walsh *Westchester Community College*
Frederick Wasserman *Boston University*
Steven A. Wasserman *University of California–San Diego*
Robert F. Weaver *University of Kansas*
Andrew N. Webber *Arizona State University*
Harold J. Webster *Penn State DuBois*
Mark Wheelis *University of California–Davis*
Lynn D. Wike *University of South Carolina at Aiken*
William Williams *Saint Mary's College of Maryland*
Mary L. Wilson *Gordon College*
Kevin Winterling *Emory & Henry College*
E. William Wischusen *Louisiana State University and Agricultural and Mechanical College*
Kenneth Wunch *Tulane University*
Mark L. Wygoda *McNeese State University*
Roger Young *Drury College*

In June 1999, at the McGraw-Hill General Biology Symposium in St. Louis, Missouri, a talented group of instructors helped us map out a plan for the revision:

Ruth Beattie *University of Kentucky*
Douglas Gaffin *University of Oklahoma*
Jon C. Glase *Cornell University*
Randy Hampton *University of California-San Diego*
Marielle Hoefnagels *University of Oklahoma*
Laurie Iten *Purdue University*
Randall S. Johnson *University of California-San Diego*
Jonathan Losos *Washington University*
Michael Meighan *University of CA-Berkeley*
Craig Peebles *University of Pittsburgh*
Susan Singer *Carleton College*
Eric Strauss *Boston College*

We wish to thank our supplement authors who worked relentlessly to prepare new materials and who so kindly provided feedback on the page proofs of *Biology*, Sixth Edition:

Margaret Gould Burke *California Academy of Sciences*
Ron M. Taylor *Professor Emeritus, Lansing Community College*
Linda Van Thiel *Wayne State University*
Sylvester Allred *Northern Arizona University*
William Anyonge *UCLA*

A Guide to the Learning System

Summary, Questions, and Media Resources

Located at the end of each chapter, the Summary Page links to an abundance of chapter-related learning tools.

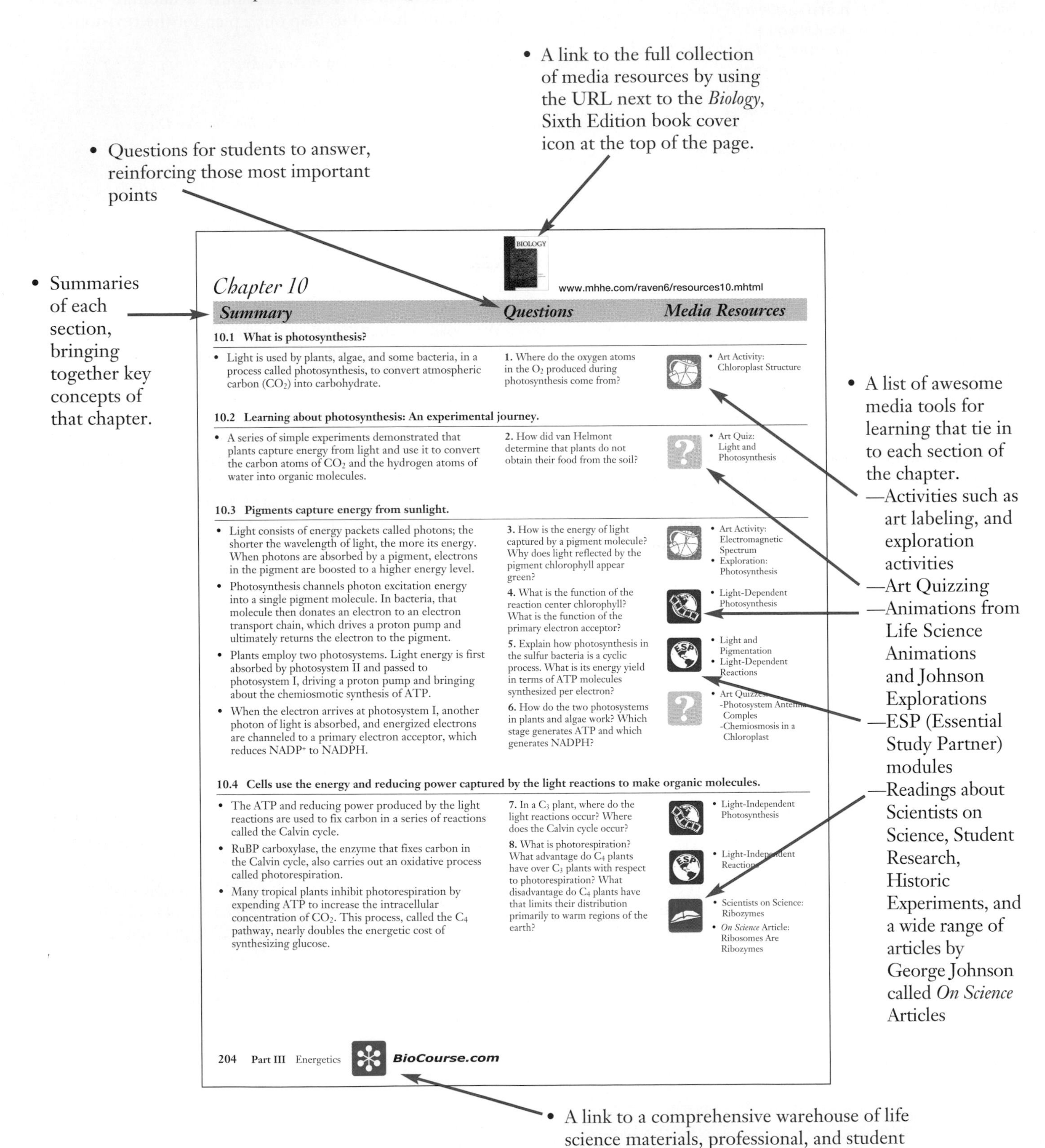

• A link to the full collection of media resources by using the URL next to the *Biology*, Sixth Edition book cover icon at the top of the page.
• Questions for students to answer, reinforcing those most important points
• Summaries of each section, bringing together key concepts of that chapter.
• A list of awesome media tools for learning that tie in to each section of the chapter.
—Activities such as art labeling, and exploration activities
—Art Quizzing
—Animations from Life Science Animations and Johnson Explorations
—ESP (Essential Study Partner) modules
—Readings about Scientists on Science, Student Research, Historic Experiments, and a wide range of articles by George Johnson called *On Science* Articles
• A link to a comprehensive warehouse of life science materials, professional, and student resources using the BioCourse icon at the bottom of the page.

BIOLOGY

Chapter 10

www.mhhe.com/raven6/resources10.mhtml

Summary	*Questions*	*Media Resources*
10.1 What is photosynthesis?		
• Light is used by plants, algae, and some bacteria, in a process called photosynthesis, to convert atmospheric carbon (CO_2) into carbohydrate.	**1.** Where do the oxygen atoms in the O_2 produced during photosynthesis come from?	• Art Activity: Chloroplast Structure
10.2 Learning about photosynthesis: An experimental journey.		
• A series of simple experiments demonstrated that plants capture energy from light and use it to convert the carbon atoms of CO_2 and the hydrogen atoms of water into organic molecules.	**2.** How did van Helmont determine that plants do not obtain their food from the soil?	• Art Quiz: Light and Photosynthesis
10.3 Pigments capture energy from sunlight.		
• Light consists of energy packets called photons; the shorter the wavelength of light, the more its energy. When photons are absorbed by a pigment, electrons in the pigment are boosted to a higher energy level.	**3.** How is the energy of light captured by a pigment molecule? Why does light reflected by the pigment chlorophyll appear green?	• Art Activity: Electromagnetic Spectrum • Exploration: Photosynthesis
• Photosynthesis channels photon excitation energy into a single pigment molecule. In bacteria, that molecule then donates an electron to an electron transport chain, which drives a proton pump and ultimately returns the electron to the pigment.	**4.** What is the function of the reaction center chlorophyll? What is the function of the primary electron acceptor?	• Light-Dependent Photosynthesis
• Plants employ two photosystems. Light energy is first absorbed by photosystem II and passed to photosystem I, driving a proton pump and bringing about the chemiosmotic synthesis of ATP.	**5.** Explain how photosynthesis in the sulfur bacteria is a cyclic process. What is its energy yield in terms of ATP molecules synthesized per electron?	• Light and Pigmentation • Light-Dependent Reactions
• When the electron arrives at photosystem I, another photon of light is absorbed, and energized electrons are channeled to a primary electron acceptor, which reduces $NADP^+$ to NADPH.	**6.** How do the two photosystems in plants and algae work? Which stage generates ATP and which generates NADPH?	• Art Quizzes: -Photosystem Antenna Complex -Chemiosmosis in a Chloroplast
10.4 Cells use the energy and reducing power captured by the light reactions to make organic molecules.		
• The ATP and reducing power produced by the light reactions are used to fix carbon in a series of reactions called the Calvin cycle.	**7.** In a C_3 plant, where do the light reactions occur? Where does the Calvin cycle occur?	• Light-Independent Photosynthesis
• RuBP carboxylase, the enzyme that fixes carbon in the Calvin cycle, also carries out an oxidative process called photorespiration.	**8.** What is photorespiration? What advantage do C_4 plants have over C_3 plants with respect to photorespiration? What disadvantage do C_4 plants have that limits their distribution primarily to warm regions of the earth?	• Light-Independent Reactions
• Many tropical plants inhibit photorespiration by expending ATP to increase the intracellular concentration of CO_2. This process, called the C_4 pathway, nearly doubles the energetic cost of synthesizing glucose.		• Scientists on Science: Ribozymes • *On Science* Article: Ribosomes Are Ribozymes

204 **Part III** Energetics **BioCourse.com**

Modular Learning System

To help students focus on concepts, this text is organized so that each chapter is a series of discrete learning modules occupying one or two pages and ending with a summary. The concept outline at the beginning of each chapter represents the conceptual skeleton of the chapter, allowing students to readily grasp how concepts relate to one another and to the overall theme of the chapter. This sort of modular organization has proven to be a very effective way to focus students on key ideas.

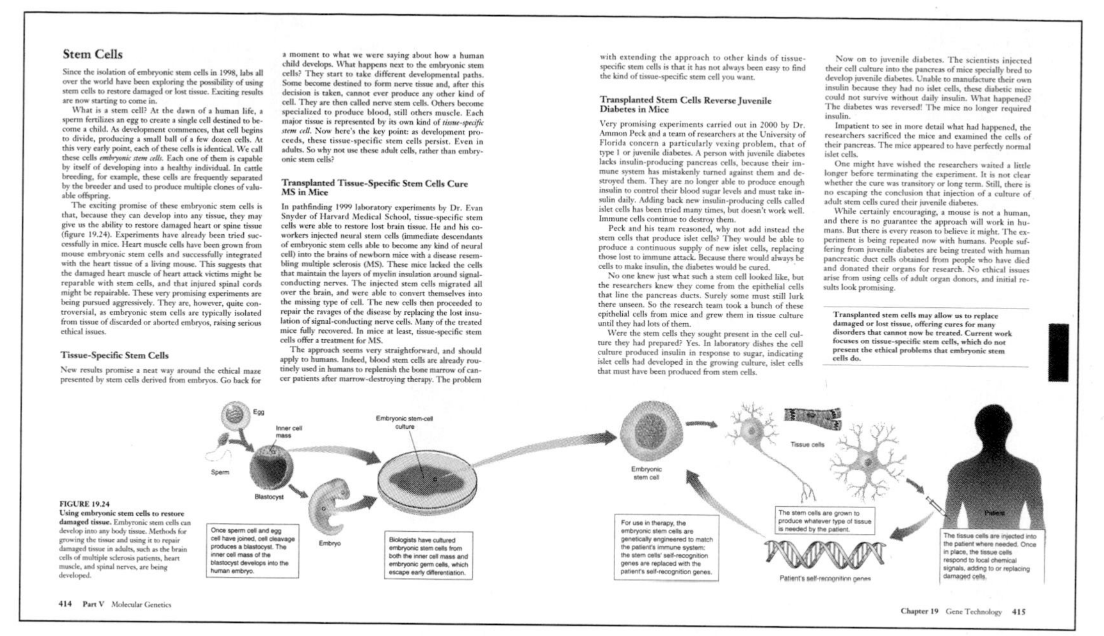

Stem Cells

Since the isolation of embryonic stem cells in 1998, labs all over the world have been exploring the possibility of using stem cells to restore damaged or lost tissue. Exciting results are now starting to come in.

What is a stem cell? At the dawn of a human life, a sperm fertilizes an egg to create a single cell destined to become a child. As development commences, that cell begins to divide, producing a small ball of a few dozen cells. At this very early point, each of these cells is identical. We call these cells *embryonic stem cells*. Each one of them is capable by itself of developing into a healthy individual. In cattle breeding, for example, these cells are frequently separated by the breeder and used to produce multiple clones of valuable offspring.

The exciting promise of these embryonic stem cells is that, because they can develop into any tissue, they may give us the ability to restore damaged heart or spine tissue (figure 19.24). Experiments have already been tried successfully in mice. Heart muscle cells have been grown from mouse embryonic stem cells and successfully integrated with the heart tissue of a living mouse. This suggests that the damaged heart muscle of heart attack victims might be reparable with stem cells, and that injured spinal cords might be repairable. These very promising experiments are being pursued aggressively. They are, however, quite controversial, as embryonic stem cells are typically isolated from tissue of discarded or aborted embryos, raising serious ethical issues.

Tissue-Specific Stem Cells

New results promise a neat way around the ethical maze presented by stem cells derived from embryos. Go back for a moment to what we were saying about how a human child develops. What happens next to the embryonic stem cells? They start to take different developmental paths. Some become destined to form nerve tissue and, after this decision is taken, cannot ever produce any other kind of cell. They are then called nerve stem cells. Others become specialized to produce blood, still others muscle. Each major tissue is represented by its own kind of *tissue-specific stem cell*. Now here's the key point: as development proceeds, these tissue-specific stem cells persist. Even in adults. So why not use these adult cells, rather than embryonic stem cells?

Transplanted Tissue-Specific Stem Cells Cure MS in Mice

In pathfinding 1999 laboratory experiments by Dr. Evan Snyder of Harvard Medical School, tissue-specific stem cells were able to restore lost brain tissue. He and his co-workers injected neural stem cells (immediate descendants of embryonic stem cells able to become any kind of neural cell) into the brains of newborn mice with a disease resembling multiple sclerosis (MS). These mice lacked the cells that maintain the layers of myelin insulation around signal-conducting nerves. The injected stem cells migrated all over the brain, and were able to convert themselves into the missing type of cell. The new cells then proceeded to repair the ravages of the disease by replacing the lost insulation of signal-conducting nerve cells. Many of the treated mice fully recovered. In mice at least, tissue-specific stem cells offer a treatment for MS.

The approach seems very straightforward, and should apply to humans. Indeed, blood stem cells are already routinely used in humans to replenish the bone marrow of cancer patients after marrow-destroying therapy. The problem with extending the approach to other kinds of tissue-specific stem cells is that it has not always been easy to find the kind of tissue-specific stem cell you want.

Transplanted Stem Cells Reverse Juvenile Diabetes in Mice

Very promising experiments carried out in 2000 by Dr. Ammon Peck and a team of researchers at the University of Florida concern a particularly vexing problem, that of type 1 or juvenile diabetes. A person with juvenile diabetes lacks insulin-producing pancreas cells, because their immune system has mistakenly turned against them and destroyed them. They are no longer able to produce enough insulin to control their blood sugar levels and must take insulin daily. Adding back new insulin-producing cells called islet cells has been tried many times, but doesn't work well. Immune cells continue to destroy them.

Peck and his team reasoned, why not add instead the stem cells that produce islet cells? They would be able to produce a continuous supply of new islet cells, replacing those lost to immune attack. Because there would always be cells to make insulin, the diabetes would be cured.

No one knew just what such a stem cell looked like, but the researchers knew they come from the epithelial cells that line the pancreas ducts. Surely some must still lurk there unseen. So the research team took a bunch of these epithelial cells from mice and grew them in tissue culture until they had lots of them.

Were the stem cells they sought present in the cell culture they had prepared? Yes. In laboratory dishes the cell culture produced insulin in response to sugar, indicating islet cells had developed in the growing culture, islet cells that must have been produced from stem cells.

Now on to juvenile diabetes. The scientists injected their cell culture into the pancreas of mice specially bred to develop juvenile diabetes. Unable to manufacture their own insulin because they had no islet cells, these diabetic mice could not survive without daily insulin. What happened? The diabetes was reversed! The mice no longer required insulin.

Impatient to see in more detail what had happened, the researchers sacrificed the mice and examined the cells of their pancreas. The mice appeared to have perfectly normal islet cells.

One might have wished the researchers waited a little longer before terminating the experiment. It is not clear whether the cure was transitory or long term. Still, there is no escaping the conclusion that injection of a culture of adult stem cells cured their juvenile diabetes.

While certainly encouraging, a mouse is not a human, and there is no guarantee the approach will work in humans. But there is every reason to believe it might. The experiment is being repeated now with humans. People suffering from juvenile diabetes are being treated with human pancreatic duct cells obtained from people who have died and donated their organs for research. No ethical issues arise from using cells of adult organ donors, and initial results look promising.

Transplanted stem cells may allow us to replace damaged or lost tissue, offering cures for many disorders that cannot now be treated. Current work focuses on tissue-specific stem cells, which do not present the ethical problems that embryonic stem cells do.

Egg
Inner cell mass
Sperm
Blastocyst
Embryo
Embryonic stem-cell culture
Once sperm cell and egg cell have joined, cell cleavage produces a blastocyst. The inner cell mass of the blastocyst develops into the human embryo.
Biologists have cultured embryonic stem cells from both the inner cell mass and embryonic germ cells, which escape early differentiation.
Embryonic stem cell
Tissue cells
Patient
For use in therapy, the embryonic stem cells are genetically engineered to match the patient's immune system: the stem cells' self-recognition genes are replaced with the patient's self-recognition genes.
The stem cells are grown to produce whatever type of tissue is needed by the patient.
Patient's self-recognition genes
The tissue cells are injected into the patient where needed. Once in place, the tissue cells respond to local chemical signals, adding to or replacing damaged cells.

FIGURE 19.24
Using embryonic stem cells to restore damaged tissue. Embryonic stem cells can develop into any body tissue. Methods for growing the tissue and using it to repair damaged tissue in adults, such as the brain cells of multiple sclerosis patients, heart muscle, and spinal nerves, are being developed.

414 Part V Molecular Genetics

Chapter 19 Gene Technology 415

Ancillaries

- The ***Instructor's Manual*** contains chapter synopses, chapter objectives, key terms, chapter outline, instructional strategy, and a list of related visual resources. The manual also provides links to biological supply companies and media resources as well as answers to review questions in the text. An extended chapter outline can be found on the Instructor's portion of the Online Learning Center.
- The ***Test Item File*** has been updated for a better balance between rote and concept-type questions, testing for both recall and understanding of the chapters' concepts. It is available in softcover or hybrid CD-ROM for both Macintosh and Windows platforms.
- The ***Student Study Guide*** has become more focused on student understanding and success. It contains: Tips for Mastering the Concepts of the Chapter, Concept Maps, Key Terms, Learning by Experience, Exercising Your Knowledge, and Assessing Your Knowledge.
- The ***Course Integration Guide*** helps professors correlate all of the ancillary materials to the chapters in the book. The guide will also be available on-line.
- The ***PowerPoint Outlines*** for instructors follow the structure of the text and can be manipulated to add your own topics, information, and images from the Visual Resource Library or other resources.
- ***BioCourse.com*** is an electronic meeting place for students and instructors. It provides a comprehensive set of resources in one easy place that is up-to-date and easy to navigate.
- ***Transparencies*** for instructors include all line art that is found in the textbook with better contrast and visibility than ever before. Labels are large and bold for clear projection.
- A multitude of tools awaits you on *Biology*'s password-protected ***Online Learning Center***. Simply visit *www.mhhe.com/raven6*.
- The ***Interactive e-Text*** CD-ROM is an electronic version of *Biology*, sixth edition, that is full of animations, activities, quizzes, and links to reference readings.
- The ***Life Science Animations*** CD-ROM includes over 200 animations for classroom presentations.
- The ***Visual Resource Library*** CD-ROM includes all of the photographs and illustrations from the textbook. The Visual Resource Library will also be available on-line.
- ***Vectorized Art*** files allow instructors to manipulate art and adapt figures to meet the particular needs of the lecture environment.
- The ***Biology Laboratory Manual*** (Vodopich Moore, Sixth Edition) is a full-color manual that contains approximately 50 laboratory exercises.
- The laboratory manual ***Biological Investigations: Form, Function, Diversity, and Process*** emphasizes investigative, quantitative, and comparative approaches to studying the life sciences.
- With the ***Life Science Animations Videotapes*** you can review 42 animations of the most difficult concepts to learn.

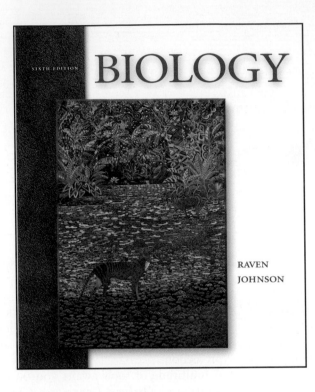

BioCourse.com

The number one source for your biology course.

BioCourse.com is an electronic meeting place for students and instructors. It provides a comprehensive set of resources in one place that is up-to-date and easy to navigate. You can access **BioCourse.com** from *Biology's* Online Learning Center.

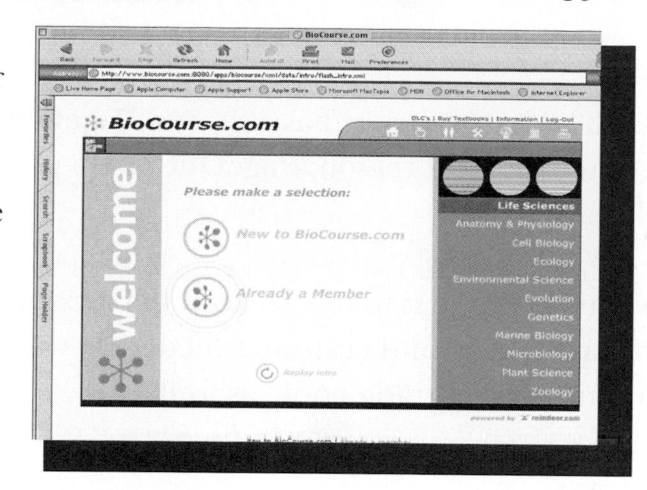

Here is what you will find at BioCourse.com:

Faculty Club is an array of information and links to related sites for instructors. Resources that you will find include:
- Teaching tips and basic information on pedagogy, assessment, etc.
- Suggestions for classroom and lecture activities.
- Reference searches and literature for faculty.
- Presentation tools.
- Test bank.
- Help for new instructors and teaching assistants.
- Information on available jobs, grant writing, and available funding.
- Case studies.

Student Center contains a wide range of materials to help biology students improve their study skills and achieve success in college and beyond. Examples of materials that will be available:
- Study aids.
- Résumé writing and information on jobs and internships.
- Graduate school options.
- Information for MCAT and other tests.
- Links to content websites by topic.

Briefing Room offers instructors and students up-to-date news articles, a selection of background readings and links to journal search tools and biology magazines. Users can e-mail articles to others, link to search engines, and read primary sources online.

BioLabs features materials for lab students and instructors. Some tools you will find include:

For students:
- Dissection techniques.
- Equipment tutorials.
- Safety and setup procedures.

For instructors:
- Lab preparations.
- Lab support.
- Simulations.

The Quad is a powerful indexing tool and hierarchical outline of content resources for searching by students and faculty. Users can search by topic through a "content warehouse" featuring text material, activities, visuals, and animations to learn more about a selected topic.

R & D Center features our newest simulations, animations, and other teaching and learning tools. This portion of our site will allow faculty members and students to try out our materials as they are being developed.

Contact your McGraw-Hill sales representative for more information or visit *www.mhhe.com*.

Visual Resource Library CD-ROMs

These two CD-ROM products contain textual images and life science animations that instructors can use to enhance their lectures. View, sort, search, and print catalog images, play chapter-specific slideshows using PowerPoint, or create customized presentations when you:

- Find and sort thumbnail image records by name, type, location, and user-defined keywords.
- Search using keywords or terms.
- View multiple images at the same time with the Small Gallery View.
- Select and view images at full size.
- Display all the important file information for easy file identification.
- Drag and place or copy and paste into virtually any graphics, desktop publishing, presentation, or multimedia application.

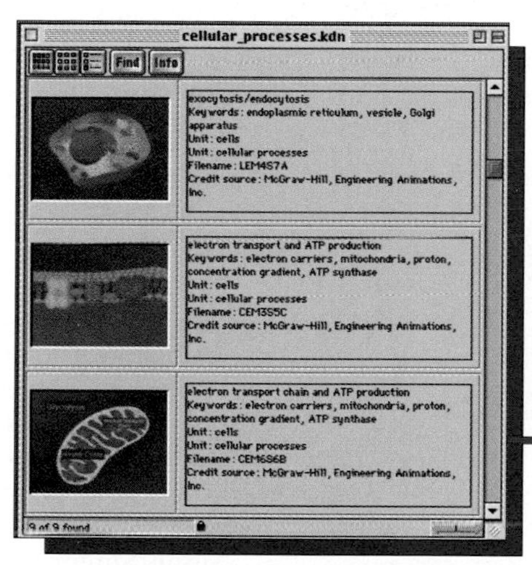

Life Science Animations Visual Resource Library CD-ROM, 2.0

(also available on the Online Learning Center)

This instructor's tool, containing more than 200 animations of important biological concepts and processes—found in the *Essential Study Partner and Dynamic Human CD-ROMs*—is perfect to support your lecture. The animations contained in this library are not limited to subjects covered in the text, but include a variety of general life science topics.

Visual Resource Library CD-ROM

This helpful CD-ROM contains more than 1,500 photographs and illustrations from *Biology*. You'll be able to create interesting multimedia presentations with the use of these images, and students will have the ability to easily access the same images in their texts to later review the content covered in class.

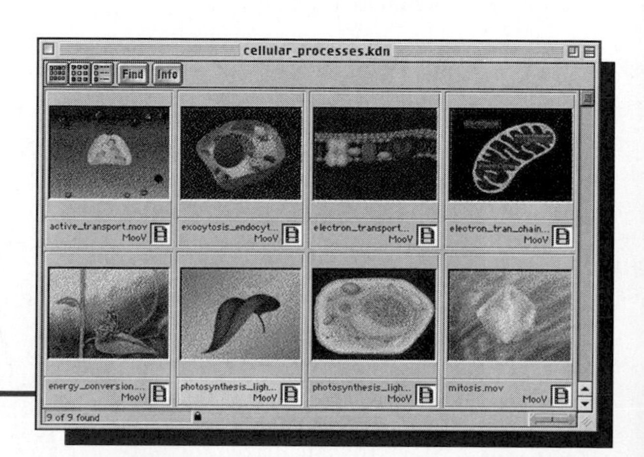

Contact your McGraw-Hill sales representative for more information or visit ***www.mhhe.com***.

About the Authors

Meet the Authors

Dr. Peter Raven is director of the Missouri Botanical Garden and Engelmann Professor of Botany at Washington University. A distinguished scientist, Dr. Raven is a member of the National Academy of Sciences, the National Research Council, and is a MacArthur and a Guggenheim fellow. He has received numerous honors and awards for his botanical research and work in tropical conservation, including the National Medal of Science. In addition to coauthoring this text with Dr. George Johnson, Dr. Raven has authored twenty other books and several hundred scientific articles.

George Johnson is professor of biology at Washington University in St. Louis, where he has taught genetics and general biology to undergraduates for 29 years. Also professor of genetics at Washington University School of Medicine, he is a student of population genetics and evolution. He has authored more than fifty scientific publications, and several high school and college texts, including *The Living World*, a very successful non-majors college biology text. He has pioneered the development of interactive CD-ROM and web-based investigations for biology teaching.

Jonathan Losos is an associate professor in the Department of Biology at Washington University and is also director of the university's Tyson Research Center. An evolutionary biologist, Losos's research has focused on studying patterns of adaptive radiation and evolutionary diversification in lizards. The recipient of several awards including the prestigious Theodosius Dobzhanksy and David Starr Jordan Prizes for outstanding young evolutionary biologists, Losos has published more than seventy scientific articles.

Susan Singer is professor of biology at Carleton College in Northfield, Minnesota, where she has taught introductory biology, plant biology, plant development, and developmental genetics for 15 years. Her research interests are focused on the development and evolution of flowering plants. Susan has authored numerous scientific publications on plant development, contributed chapters to developmental biology texts, and been actively involved with the education efforts of several professional societies. She is also the coordinator for the Perlman Center for Learning and Teaching at Carleton.

Part

I

The Origin of Living Things

Defying gravity. This gecko lizard is able to climb walls and walk upside down across ceilings. Learning how geckos do this is a fascinating bit of experimental science.

Real People Doing Real Science

Unraveling the Mystery of How Geckos Defy Gravity

Science is most fun when it tickles your imagination. This is particularly true when you see something your common sense tells you just *can't* be true. Imagine, for example, you are lying on a bed in a tropical hotel room. A little lizard, a blue gecko about the size of a toothbrush, walks up the wall beside you and upside down across the ceiling, stopping for a few moments over your head to look down at you, and then trots over to the far wall and down.

There is nothing at all unusual in what you have just imagined. Geckos are famous for strolling up walls in this fashion. How do geckos perform this gripping feat? Investigators have puzzled over the adhesive properties of geckos for decades. What force prevents gravity from dropping the gecko on your nose?

The most reasonable hypothesis seemed suction—salamanders' feet form suction cups that let them climb walls, so maybe geckos' do too. The way to test this is to see if the feet adhere in a vacuum, with no air to create suction. Salamander feet don't, but gecko feet do. It's not suction.

How about friction? Cockroaches climb using tiny hooks that grapple onto irregularities in the surface, much as rock-climbers use crampons. Geckos, however, happily run up walls of smooth polished glass that no cockroach can climb. It's not friction.

Electrostatic attraction? Clothes in a dryer stick together because of electrical charges created by their rubbing together. You can stop this by adding a "static remover" like a Cling-free sheet that is heavily ionized. But a gecko's feet still adhere in ionized air. It's not electrostatic attraction.

Could it be glue? Many insects use adhesive secretions from glands in their feet to aid climbing. But there are no glands cells in the feet of a gecko, no secreted chemicals, no footprints left behind. It's not glue.

There is one tantalizing clue, however, the kind that experimenters love. Gecko feet seem to get stickier on some surfaces than others. They are less sticky on low-energy surfaces like Teflon, and more sticky on surfaces made of polar molecules. This suggests that geckos are tapping directly into the molecular structure of the surfaces they walk on!

Tracking down this clue, Kellar Autumn of Lewis & Clark College in Portland, Oregon, and Robert Full of the University of California, Berkeley, took a closer look at gecko feet. Geckos have rows of tiny hairs called setae on the bottoms of their feet, like the bristles of some trendy toothbrush. When you look at these hairs under the microscope, the end of each seta is divided into 400 to 1000 fine projections called spatulae. There are about half a million of these setae on each foot, each only one-tenth the diameter of a human hair.

Autumn and Full put together an interdisciplinary team of scientists and set out to measure the force produced by a single seta. To do this, they had to overcome two significant experimental challenges:

Isolating a single seta. No one had ever isolated a single seta before. They succeeded in doing this by surgically plucking a hair from a gecko foot under a microscope and bonding the hair onto a microprobe. The microprobe was fitted into a specially designed micromanipulator that can move the mounted hair in various ways.

Measuring a very small force. Previous research had shown that if you pull on a whole gecko, the adhesive force sticking each of the gecko's feet to the wall is about 10 Newtons (N), which is like supporting 1 kg. Because each foot has half a million setae, this predicts that a single seta would produce about 20 microNewtons of force. That's a very tiny amount to measure. To attempt the measurement, Autumn and Full recruited a mechanical engineer from Stanford, Thomas Kenny. Kenny is an expert at building instruments that can measure forces at the atomic level.

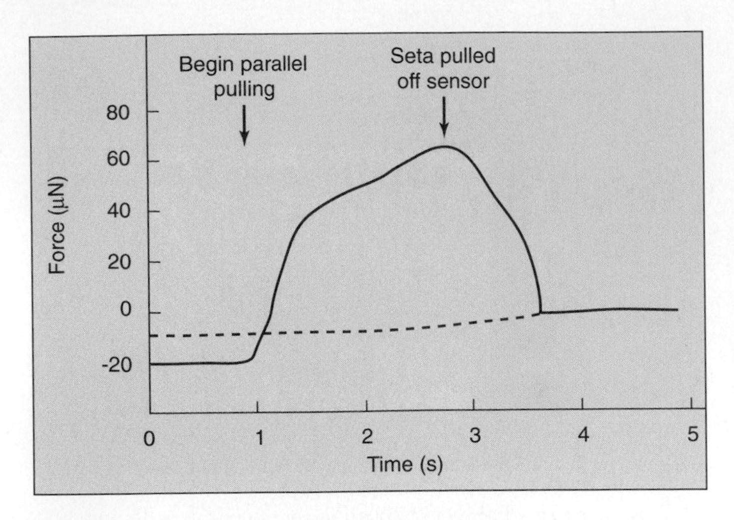

The sliding step experiment. The adhesive force of a single seta was measured. An initial push perpendicularly put the seta in contact with the sensor. Then, with parallel pulling, the force continued to increase over time to a value of 60 microNewtons (after this, the seta began to slide and pulled off the sensor). In a large number of similar experiments, adhesion forces typically approach 200 microNewtons.

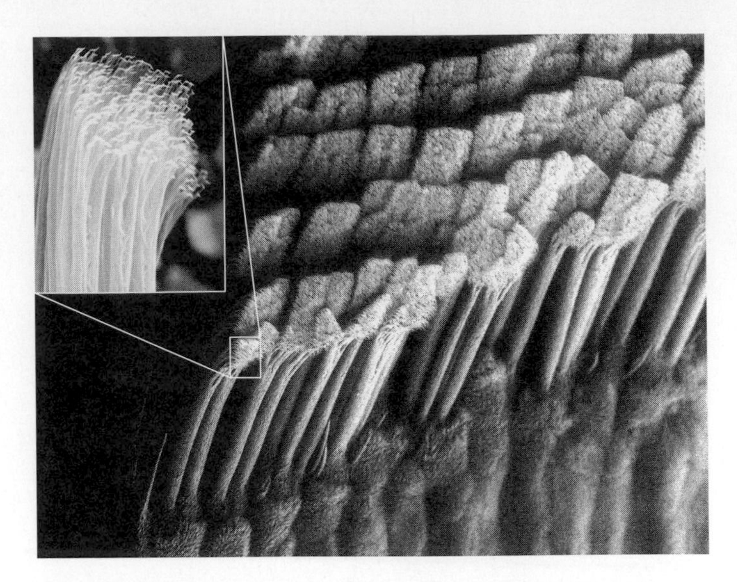

Closeup look at a gecko's foot. The setae on a gecko's foot are arranged in rows, and point backwards, away from the toenail. Each seta branches into several hundred spatulae (inset photo).

The Experiment

Once this team had isolated a seta and placed it in Kenny's device, "We had a real nasty surprise," says Autumn. For two months, pushing individual seta against a surface, they couldn't get the isolated hair to stick at all!

This forced the research team to stand back and think a bit. Finally it hit them. Geckos don't walk by pushing their feet down, like we do. Instead, when a gecko takes a step, it pushes the palm of the foot into the surface, then uncurls its toes, sliding them backwards onto the surface. This shoves the forest of tips *sideways* against the surface.

Going back to their instruments, they repeated their experiment, but this time they oriented the seta to approach the surface from the side rather than head-on. This had the effect of bringing the many spatulae on the tip of the seta into direct contact with the surface.

To measure these forces on the seta from the side, as well as the perpendicular forces they had already been measuring, the researchers constructed a micro-electromechanical cantilever. The apparatus consisted of two piezoresistive layers deposited on a silicon cantilever to detect force in both parallel and perpendicular angles.

The Results

With the seta oriented properly, the experiment yielded results. Fantastic results. The attachment force measured by the machine went up 600-fold from what the team had been measuring before. A single seta produced not the 20 microNewtons of force predicted by the whole-foot measurements, but up to an astonishing 200 microNewtons (see graph above)! Measuring many individual seta, adhesive forces averaged 194±25 microNewtons.

Two hundred microNewtons is a tiny force, but stupendous for a single hair only 100 microns long. Enough to hold up an ant. A million hairs could support a small child. A little gecko, ceiling walking with 2 million of them (see photos above), could theoretically carry a 90-pound backpack—talk about being over-engineered.

If a gecko's feet stick *that* good, how do geckos ever become unstuck? The research team experimented with unattaching individual seta; they used yet another micro-instrument, this one designed by engineer Ronald Fearing also from U.C. Berkeley, to twist the hair in various ways. They found that tipped past a critical angle, 30 degrees, the attractive forces between hair and surface atoms weaken to nothing. The trick is to tip a foot hair until its projections let go. Geckos release their feet by curling up each toe and peeling it off, just the way we remove tape.

What is the source of the powerful adhesion of gecko feet? The experiments do not reveal exactly what the attractive force is, but it seems almost certain to involve interactions at the atomic level. For a gecko's foot to stick, the hundreds of spatulae at the tip of each seta must butt up squarely against the surface, so the individual atoms of each spatula can come into play. When two atoms approach each other very closely—closer than the diameter of an atom—a subtle nuclear attraction called Van der Waals forces comes into play. These forces are individually very weak, but when lots of them add their little bits, the sum can add up to quite a lot.

Might robots be devised with feet tipped with artificial setae, able to walk up walls? Autumn and Full are working with a robotics company to find out. Sometimes science is not only fun, but can lead to surprising advances.

To explore this experiment further, go to the Virtual Lab at www.mhhe.com/raven6/vlab1.mhtml

1

The Science of Biology

Concept Outline

1.1 Biology is the science of life.

Properties of Life. Biology is the science that studies living organisms and how they interact with one another and their environment.

1.2 Scientists form generalizations from observations.

The Nature of Science. Science employs both deductive reasoning and inductive reasoning.

How Science Is Done. Scientists construct hypotheses from systematically collected objective data. They then perform experiments designed to disprove the hypotheses.

1.3 Darwin's theory of evolution illustrates how science works.

Darwin's Theory of Evolution. On a round-the-world voyage Darwin made observations that eventually led him to formulate the hypothesis of evolution by natural selection.

Darwin's Evidence. The fossil and geographic patterns of life he observed convinced Darwin that a process of evolution had occurred.

Inventing the Theory of Natural Selection. The Malthus idea that populations cannot grow unchecked led Darwin, and another naturalist named Wallace, to propose the hypothesis of natural selection.

Evolution After Darwin: More Evidence. In the century since Darwin, a mass of experimental evidence has supported his theory of evolution, now accepted by practically all practicing biologists.

1.4 This book is organized to help you learn biology.

Core Principles of Biology. The first half of this text is devoted to general principles that apply to all organisms, the second half to an examination of particular organisms.

FIGURE 1.1
A replica of the *Beagle*, off the southern coast of South America. The famous English naturalist, Charles Darwin, set forth on H.M.S. *Beagle* in 1831, at the age of 22.

You are about to embark on a journey—a journey of discovery about the nature of life. Nearly 180 years ago, a young English naturalist named Charles Darwin set sail on a similar journey on board H.M.S. *Beagle* (figure 1.1 shows a replica of the *Beagle*). What Darwin learned on his five-year voyage led directly to his development of the theory of evolution by natural selection, a theory that has become the core of the science of biology. Darwin's voyage seems a fitting place to begin our exploration of biology, the scientific study of living organisms and how they have evolved. Before we begin, however, let's take a moment to think about what biology is and why it's important.

1.1 Biology is the science of life.

Properties of Life

In its broadest sense, *biology is the study of living things—the science of life*. Living things come in an astounding variety of shapes and forms, and biologists study life in many different ways. They live with gorillas, collect fossils, and listen to whales. They isolate viruses, grow mushrooms, and examine the structure of fruit flies. They read the messages encoded in the long molecules of heredity and count how many times a hummingbird's wings beat each second.

What makes something "alive"? Anyone could deduce that a galloping horse is alive and a car is not, but *why?* We cannot say, "If it moves, it's alive," because a car can move, and gelatin can wiggle in a bowl. They certainly are not alive. What characteristics *do* define life? All living organisms share five basic characteristics:

1. **Order.** All organisms consist of one or more cells with highly ordered structures: atoms make up molecules, which construct cellular organelles, which are contained within cells. This hierarchical organization continues at higher levels in multicellular organisms and among organisms (figure 1.2).
2. **Sensitivity.** All organisms respond to stimuli. Plants grow toward a source of light, and your pupils dilate when you walk into a dark room.
3. **Growth, development, and reproduction.** All organisms are capable of growing and reproducing, and they all possess hereditary molecules that are passed to their offspring, ensuring that the offspring are of the same species. Although crystals also "grow," their growth does not involve hereditary molecules.
4. **Regulation.** All organisms have regulatory mechanisms that coordinate the organism's internal functions. These functions include supplying cells with nutrients, transporting substances through the organism, and many others.
5. **Homeostasis.** All organisms maintain relatively constant internal conditions, different from their environment, a process called homeostasis.

All living things share certain key characteristics: order, sensitivity, growth, development and reproduction, regulation, and homeostasis.

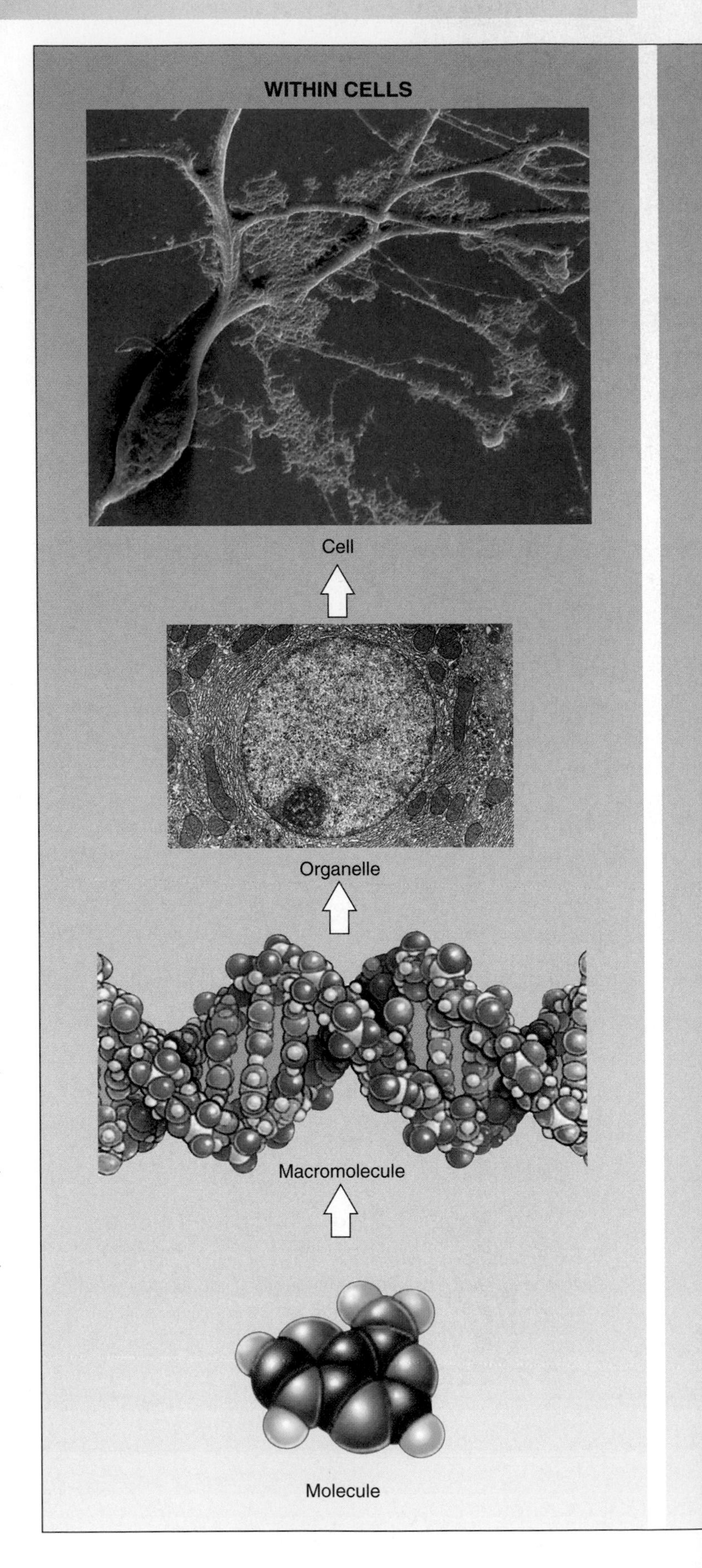

FIGURE 1.2
Hierarchical organization of living things. Life is highly organized—from small and simple to large and complex, within cells, within multicellular organisms, and among organisms.

WITHIN MULTICELLULAR ORGANISMS
Organism
Organ system
Organ
Tissue
AMONG ORGANISMS
Ecosystem
Community
Species
Population

1.2 Scientists form generalizations from observations.

The Nature of Science

Biology is a fascinating and important subject, because it dramatically affects our daily lives and our futures. Many biologists are working on problems that critically affect our lives, such as the world's rapidly expanding population and diseases like cancer and AIDS. The knowledge these biologists gain will be fundamental to our ability to manage the world's resources in a suitable manner, to prevent or cure diseases, and to improve the quality of our lives and those of our children and grandchildren.

Biology is one of the most successful of the "natural sciences," explaining what our world is like. To understand biology, you must first understand the nature of science. The basic tool a scientist uses is thought. To understand the nature of science, it is useful to focus for a moment on how scientists think. They reason in two ways: deductively and inductively.

Deductive Reasoning

Deductive reasoning applies general principles to predict specific results. Over 2200 years ago, the Greek Eratosthenes used deductive reasoning to accurately estimate the circumference of the earth. At high noon on the longest day of the year, when the sun's rays hit the bottom of a deep well in the city of Syene, Egypt, Eratosthenes measured the length of the shadow cast by a tall obelisk in Alexandria, about 800 kilometers to the north. Because he knew the distance between the two cities and the height of the obelisk, he was able to employ the principles of Euclidean geometry to correctly deduce the circumference of the earth (figure 1.3). This sort of analysis of specific cases using general principles is an example of deductive reasoning. It is the reasoning of mathematics and philosophy and is used to test the validity of general ideas in all branches of knowledge. General principles are constructed and then used as the basis for examining specific cases.

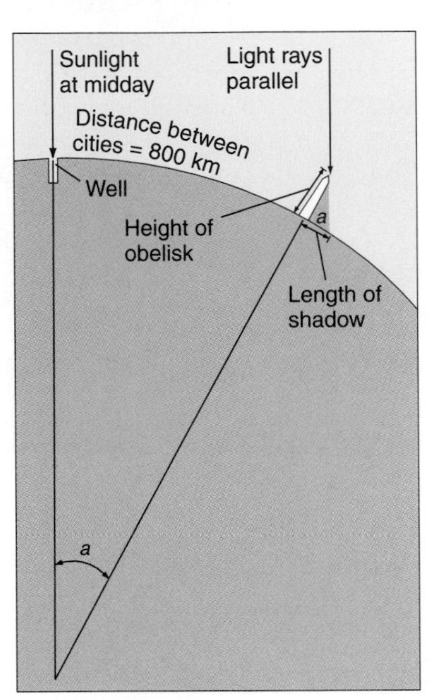

FIGURE 1.3
Deductive reasoning: How Eratosthenes estimated the circumference of the earth using deductive reasoning. 1. On a day when sunlight shone straight down a deep well at Syene in Egypt, Eratosthenes measured the length of the shadow cast by a tall obelisk in the city of Alexandria, about 800 kilometers away. **2.** The shadow's length and the obelisk's height formed two sides of a triangle. Using the recently developed principles of Euclidean geometry, he calculated the angle, *a*, to be 7° and 12′, exactly $\frac{1}{50}$ of a circle (360°). **3.** If angle $a = \frac{1}{50}$ of a circle, then the distance between the obelisk (in Alexandria) and the well (in Syene) must equal $\frac{1}{50}$ of the circumference of the earth. **4.** Eratosthenes had heard that it was a 50-day camel trip from Alexandria to Syene. Assuming that a camel travels about 18.5 kilometers per day, he estimated the distance between obelisk and well as 925 kilometers (using different units of measure, of course). **5.** Eratosthenes thus deduced the circumference of the earth to be 50 × 925 = 46,250 kilometers. Modern measurements put the distance from the well to the obelisk at just over 800 kilometers. Employing a distance of 800 kilometers, Eratosthenes's value would have been 50 × 800 = 40,000 kilometers. The actual circumference is 40,075 kilometers.

Inductive Reasoning

Inductive reasoning uses specific observations to construct general scientific principles. *Webster's Dictionary* defines science as systematized knowledge derived from observation and experiment carried on to determine the principles underlying what is being studied. In other words, a scientist determines principles from observations, discovering general principles by careful examination of specific cases. Inductive reasoning first became important to science in the 1600s in Europe, when Francis Bacon, Isaac Newton, and others began to use the results of particular experiments to infer general principles about how the world operates. If you release an apple from your hand, what happens? The apple falls to the ground. From a host of simple, specific observations like this, Newton inferred a general principle: all objects fall toward the center of the earth. What Newton did was construct a mental model of how the world works, a family of general principles consistent with what he could see and learn. Scientists do the same today. They use specific observations to build general models, and then test the models to see how well they work.

Science is a way of viewing the world that focuses on objective information, putting that information to work to build understanding.

How Science Is Done

How do scientists establish which general principles are true from among the many that might be true? They do this by systematically testing alternative proposals. If these proposals prove inconsistent with experimental observations, they are rejected as untrue. After making careful observations concerning a particular area of science, scientists construct a **hypothesis,** which is a suggested explanation that accounts for those observations. A hypothesis is a proposition that might be true. Those hypotheses that have not yet been disproved are retained. They are useful because they fit the known facts, but they are always subject to future rejection if, in the light of new information, they are found to be incorrect.

Testing Hypotheses

We call the test of a hypothesis an **experiment** (figure 1.4). Suppose that a room appears dark to you. To understand why it appears dark, you propose several hypotheses. The first might be, "There is no light in the room because the light switch is turned off." An alternative hypothesis might be, "There is no light in the room because the lightbulb is burned out." And yet another alternative hypothesis might be, "I am going blind." To evaluate these hypotheses, you would conduct an experiment designed to eliminate one or more of the hypotheses. For example, you might test your hypotheses by reversing the position of the light switch. If you do so and the light does not come on, you have disproved the first hypothesis. Something other than the setting of the light switch must be the reason for the darkness. Note that a test such as this does not prove that any of the other hypotheses are true; it merely demonstrates that one of them is not. A successful experiment is one in which one or more of the alternative hypotheses is demonstrated to be inconsistent with the results and is thus rejected.

As you proceed through this text, you will encounter many hypotheses that have withstood the test of experiment. Many will continue to do so; others will be revised as new observations are made by biologists. Biology, like all science, is in a constant state of change, with new ideas appearing and replacing old ones.

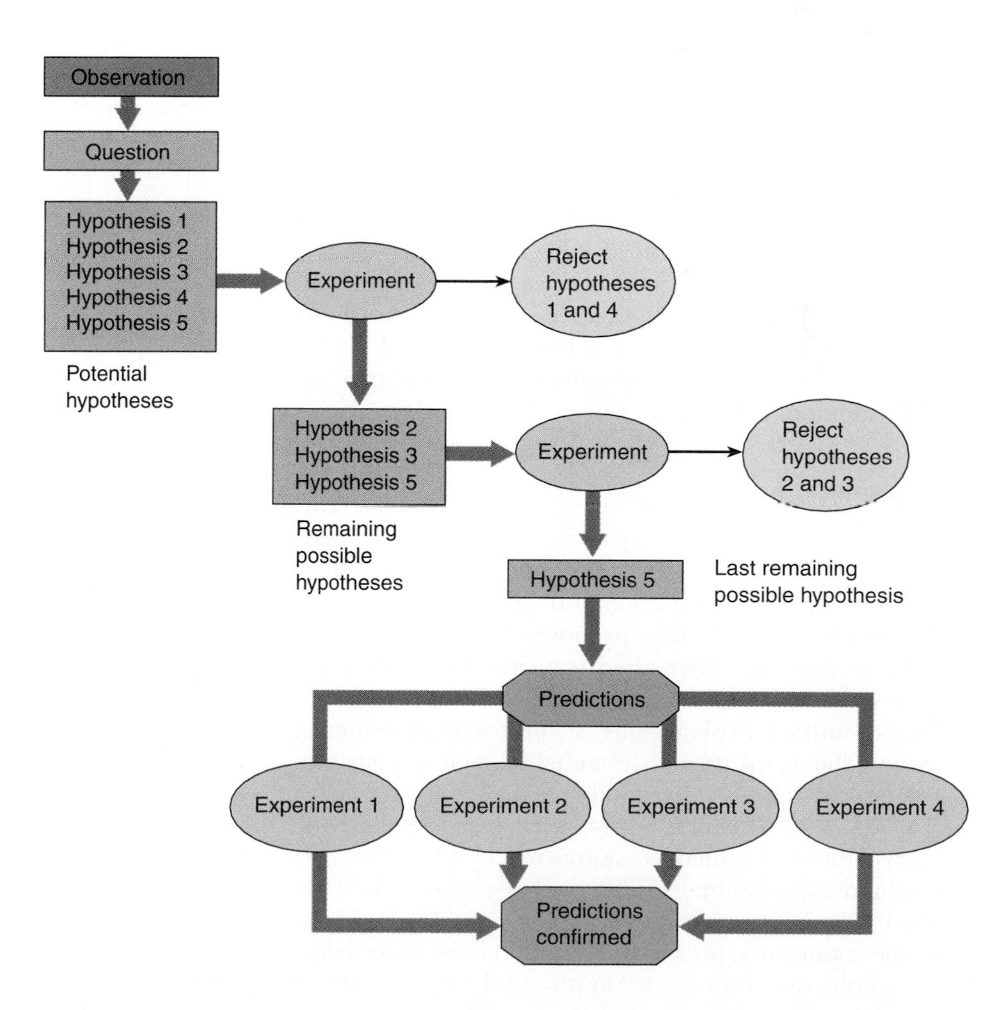

FIGURE 1.4
How science is done. This diagram illustrates the way in which scientific investigations proceed. First, scientists make observations that raise a particular question. They develop a number of potential explanations (hypotheses) to answer the question. Next, they carry out experiments in an attempt to eliminate one or more of these hypotheses. Then, predictions are made based on the remaining hypotheses, and further experiments are carried out to test these predictions. As a result of this process, the least unlikely hypothesis is selected.

Establishing Controls

Often we are interested in learning about processes that are influenced by many factors, or **variables.** To evaluate alternative hypotheses about one variable, all other variables must be kept constant. This is done by carrying out two experiments in parallel: in the first experiment, one variable is altered in a specific way to test a particular hypothesis; in the second experiment, called the **control experiment,** that variable is left unaltered. In all other respects the two experiments are identical, so any difference in the outcomes of the two experiments must result from the influence of the variable that was changed. Much of the challenge of experimental science lies in designing control experiments that isolate a particular variable from other factors that might influence a process.

Using Predictions

A successful scientific hypothesis needs to be not only valid but useful—it needs to tell you something you want to know. A hypothesis is most useful when it makes predictions, because those predictions provide a way to test the validity of the hypothesis. If an experiment produces results inconsistent with the predictions, the hypothesis must be rejected. On the other hand, if the predictions are supported by experimental testing, the hypothesis is supported. The more experimentally supported predictions a hypothesis makes, the more valid the hypothesis is. For example, Einstein's hypothesis of relativity was at first provisionally accepted because no one could devise an experiment that invalidated it. The hypothesis made a clear prediction: that the sun would bend the path of light passing by it. When this prediction was tested in a total eclipse, the light from background stars was indeed bent. Because this result was unknown when the hypothesis was being formulated, it provided strong support for the hypothesis, which was then accepted with more confidence.

Developing Theories

Scientists use the word **theory** in two main ways. A "theory" is a proposed explanation for some natural phenomenon, often based on some general principle. Thus one speaks of the principle first proposed by Newton as the "theory of gravity." Such theories often bring together concepts that were previously thought to be unrelated, and offer unified explanations of different phenomena. Newton's theory of gravity provided a single explanation for objects falling to the ground and the orbits of planets around the sun. "Theory" is also used to mean the body of interconnected concepts, supported by scientific reasoning and experimental evidence, that explains the facts in some area of study. Such a theory provides an indispensable framework for organizing a body of knowledge. For example, quantum theory in physics brings together a set of ideas about the nature of the universe, explains experimental facts, and serves as a guide to further questions and experiments.

To a scientist, such theories are the solid ground of science, that of which we are most certain. In contrast, to the general public, *theory* implies just the opposite—a *lack* of knowledge, or a guess. Not surprisingly, this difference often results in confusion. In this text, theory will always be used in its scientific sense, in reference to an accepted general principle or body of knowledge.

To suggest, as many critics outside of science do, that evolution is "just a theory" is misleading. The hypothesis that evolution has occurred is an accepted scientific fact; it is supported by overwhelming evidence. Modern evolutionary theory is a complex body of ideas whose importance spreads far beyond explaining evolution; its ramifications permeate all areas of biology, and it provides the conceptual framework that unifies biology as a science.

Research and the Scientific Method

It used to be fashionable to speak of the "scientific method" as consisting of an orderly sequence of logical "either/or" steps. Each step would reject one of two mutually incompatible alternatives, as if trial-and-error testing would inevitably lead one through the maze of uncertainty that always impedes scientific progress. If this were indeed so, a computer would make a good scientist. But science is not done this way. As British philosopher Karl Popper has pointed out, successful scientists without exception design their experiments with a pretty fair idea of how the results are going to come out. They have what Popper calls an "imaginative preconception" of what the truth might be. A hypothesis that a successful scientist tests is not just any hypothesis; rather, it is an educated guess or a hunch, in which the scientist integrates all that he or she knows and allows his or her imagination full play, in an attempt to get a sense of what *might* be true (see Box: How Biologists Do Their Work). It is because insight and imagination play such a large role in scientific progress that some scientists are so much better at science than others, just as Beethoven and Mozart stand out among most other composers.

Some scientists perform what is called basic research, which is intended to extend the boundaries of what we know. These individuals typically work at universities, and their research is usually financially supported by their institutions and by external sources, such as the government, industry, and private foundations. Basic research is as diverse as its name implies. Some basic scientists attempt to find out how certain cells take up specific chemicals, while others count the number of dents in tiger teeth. The information generated by basic research contributes to the growing body of scientific knowledge, and it provides the scientific foundation utilized by applied research. Scientists who conduct applied research are often employed in some kind of industry. Their work may involve the manu-

How Biologists Do Their Work

The Consent

Late in November, on a single night
Not even near to freezing, the ginkgo trees
That stand along the walk drop all their leaves
In one consent, and neither to rain nor to wind
But as though to time alone: the golden and green
Leaves litter the lawn today, that yesterday
Had spread aloft their fluttering fans of light.
What signal from the stars? What senses took it in?
What in those wooden motives so decided
To strike their leaves, to down their leaves,
Rebellion or surrender? And if this
Can happen thus, what race shall be exempt?
What use to learn the lessons taught by time,
If a star at any time may tell us: Now.

Howard Nemerov

FIGURE 1.A
A ginkgo tree.

What is bothering the poet Howard Nemerov is that life is influenced by forces he cannot control or even identify. It is the job of biologists to solve puzzles such as the one he poses, to identify and try to understand those things that influence life.

Nemerov asks why ginkgo trees (figure 1.A) drop all their leaves at once. To find an answer to questions such as this, biologists and other scientists pose *possible* answers and then try to determine which answers are false. Tests of alternative possibilities are called experiments. To learn why the ginkgo trees drop all their leaves simultaneously, a scientist would first formulate several possible answers, called hypotheses:

Hypothesis 1: Ginkgo trees possess an internal clock that times the release of leaves to match the season. On the day Nemerov describes, this clock sends a "drop" signal (perhaps a chemical) to all the leaves at the same time.

Hypothesis 2: The individual leaves of ginkgo trees are each able to sense day length, and when the days get short enough in the fall, each leaf responds independently by falling.

Hypothesis 3: A strong wind arose the night before Nemerov made his observation, blowing all the leaves off the ginkgo trees.

Next, the scientist attempts to eliminate one or more of the hypotheses by conducting an experiment. In this case, one might cover some of the leaves so that they cannot use light to sense day length. If hypothesis 2 is true, then the covered leaves should not fall when the others do, because they are not receiving the same information. Suppose, however, that despite the covering of some of the leaves, all the leaves still fall together. This result would eliminate hypothesis 2 as a possibility. Either of the other hypotheses, and many others, remain possibilities.

This simple experiment with ginkgoes points out the essence of scientific progress: science does not prove that certain explanations are true; rather, it proves that others are not. Hypotheses that are inconsistent with experimental results are rejected, while hypotheses that are not proven false by an experiment are provisionally accepted. However, hypotheses may be rejected in the future when more information becomes available, if they are inconsistent with the new information. Just as finding the correct path through a maze by trying and eliminating false paths, scientists work to find the correct explanations of natural phenomena by eliminating false possibilities.

facturing of food additives, creating of new drugs, or testing the quality of the environment.

After developing a hypothesis and performing a series of experiments, a scientist writes a paper carefully describing the experiment and its results. He or she then submits the paper for publication in a scientific journal, but before it is published, it must be reviewed and accepted by other scientists who are familiar with that particular field of research. This process of careful evaluation, called peer review, lies at the heart of modern science, fostering careful work, precise description, and thoughtful analysis. When an important discovery is announced in a paper, other scientists attempt to reproduce the result, providing a check on accuracy and honesty. Nonreproducible results are not taken seriously for long.

The explosive growth in scientific research during the second half of the twentieth century is reflected in the enormous number of scientific journals now in existence. Although some, such as *Science* and *Nature*, are devoted to a wide range of scientific disciplines, most are extremely specialized: *Cell Motility and the Cytoskeleton*, *Glycoconjugate Journal*, *Mutation Research*, and *Synapse* are just a few examples.

The scientific process involves the rejection of hypotheses that are inconsistent with experimental results or observations. Hypotheses that are consistent with available data are conditionally accepted. The formulation of the hypothesis often involves creative insight.

1.3 Darwin's theory of evolution illustrates how science works.

Darwin's Theory of Evolution

Darwin's theory of evolution explains and describes how organisms on earth have changed over time and acquired a diversity of new forms. This famous theory provides a good example of how a scientist develops a hypothesis and how a scientific theory grows and wins acceptance.

Charles Robert Darwin (1809–1882; figure 1.5) was an English naturalist who, after 30 years of study and observation, wrote one of the most famous and influential books of all time. This book, *On the Origin of Species by Means of Natural Selection, or The Preservation of Favoured Races in the Struggle for Life,* created a sensation when it was published, and the ideas Darwin expressed in it have played a central role in the development of human thought ever since.

FIGURE 1.5
Charles Darwin. This newly rediscovered photograph taken in 1881, the year before Darwin died, appears to be the last ever taken of the great biologist.

In Darwin's time, most people believed that the various kinds of organisms and their individual structures resulted from direct actions of the Creator (and to this day many people still believe this to be true). Species were thought to be specially created and unchangeable, or immutable, over the course of time. In contrast to these views, a number of earlier philosophers had presented the view that living things must have changed during the history of life on earth. Darwin proposed a concept he called natural selection as a coherent, logical explanation for this process, and he brought his ideas to wide public attention. His book, as its title indicates, presented a conclusion that differed sharply from conventional wisdom. Although his theory did not challenge the existence of a Divine Creator, Darwin argued that this Creator did not simply create things and then leave them forever unchanged. Instead, Darwin's God expressed Himself through the operation of natural laws that produced change over time, or **evolution.** These views put Darwin at odds with most people of his time, who believed in a literal interpretation of the Bible and accepted the idea of a fixed and constant world. His revolutionary theory deeply troubled not only many of his contemporaries but Darwin himself.

The story of Darwin and his theory begins in 1831, when he was 22 years old. On the recommendation of one of his professors at Cambridge University, he was selected to serve

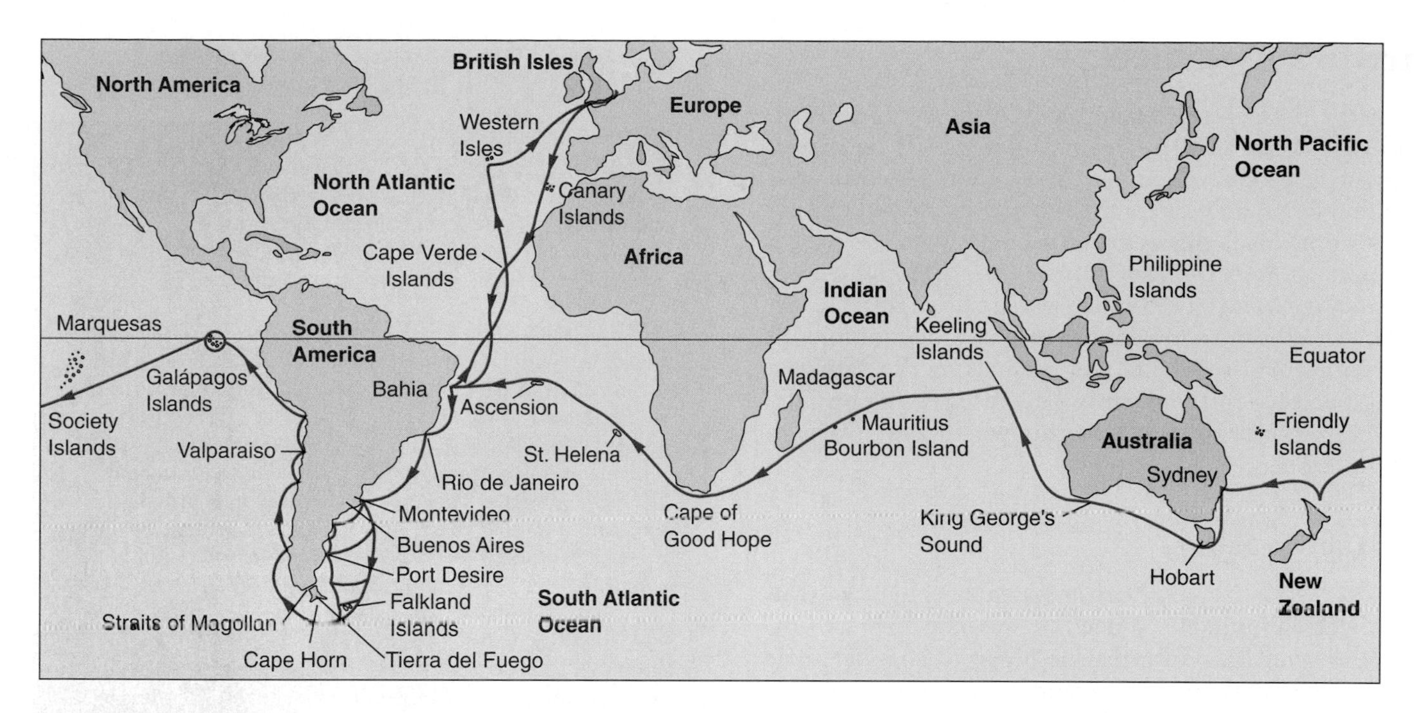

FIGURE 1.6
The five-year voyage of H.M.S. *Beagle*. Most of the time was spent exploring the coasts and coastal islands of South America, such as the Galápagos Islands. Darwin's studies of the animals of the Galápagos Islands played a key role in his eventual development of the theory of evolution by means of natural selection.

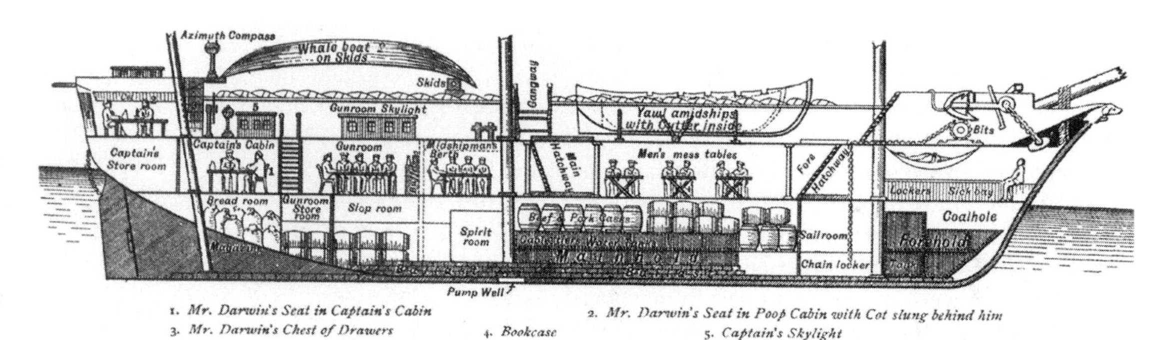

FIGURE 1.7
Cross section of the *Beagle*. A 10-gun brig of 242 tons, only 90 feet in length, the *Beagle* had a crew of 74 people! After he first saw the ship, Darwin wrote to his college professor Henslow: "The absolute want of room is an evil that nothing can surmount."

as naturalist on a five-year navigational mapping expedition around the coasts of South America (figure 1.6), aboard H.M.S. *Beagle* (figure 1.7). During this long voyage, Darwin had the chance to study a wide variety of plants and animals on continents and islands and in distant seas. He was able to explore the biological richness of the tropical forests, examine the extraordinary fossils of huge extinct mammals in Patagonia at the southern tip of South America, and observe the remarkable series of related but distinct forms of life on the Galápagos Islands, off the west coast of South America. Such an opportunity clearly played an important role in the development of his thoughts about the nature of life on earth.

When Darwin returned from the voyage at the age of 27, he began a long period of study and contemplation. During the next 10 years, he published important books on several different subjects, including the formation of oceanic islands from coral reefs and the geology of South America. He also devoted eight years of study to barnacles, a group of small marine animals with shells that inhabit rocks and pilings, eventually writing a four-volume work on their classification and natural history. In 1842, Darwin and his family moved out of London to a country home at Down, in the county of Kent. In these pleasant surroundings, Darwin lived, studied, and wrote for the next 40 years.

Darwin was the first to propose natural selection as an explanation for the mechanism of evolution that produced the diversity of life on earth. His hypothesis grew from his observations on a five-year voyage around the world.

Darwin's Evidence

One of the obstacles that had blocked the acceptance of any theory of evolution in Darwin's day was the incorrect notion, widely believed at that time, that the earth was only a few thousand years old. Evidence discovered during Darwin's time made this assertion seem less and less likely. The great geologist Charles Lyell (1797–1875), whose *Principles of Geology* (1830) Darwin read eagerly as he sailed on the *Beagle*, outlined for the first time the story of an ancient world of plants and animals in flux. In this world, species were constantly becoming extinct while others were emerging. It was this world that Darwin sought to explain.

What Darwin Saw

When the *Beagle* set sail, Darwin was fully convinced that species were immutable. Indeed, it was not until two or three years after his return that he began to consider seriously the possibility that they could change. Nevertheless, during his five years on the ship, Darwin observed a number of phenomena that were of central importance to him in reaching his ultimate conclusion (table 1.1). For example, in the rich fossil beds of southern South America, he observed fossils of extinct armadillos similar to the armadillos that still lived in the same area (figure 1.8). Why would similar living and fossil organisms be in the same area unless the earlier form had given rise to the other?

Repeatedly, Darwin saw that the characteristics of similar species varied somewhat from place to place. These geographical patterns suggested to him that organismal lineages change gradually as species migrate from one area to another. On the Galápagos Islands, off the coast of Ecuador, Darwin encountered giant land tortoises. Surprisingly, these tortoises were not all identical. In fact, local residents and the sailors who captured the tortoises for food could tell which island a particular tortoise had come from just by looking at its shell. This distribution of physical variation suggested that all of the tortoises were related, but that they had changed slightly in appearance after becoming isolated on different islands.

In a more general sense, Darwin was struck by the fact that the plants and animals on these relatively young volcanic islands resembled those on the nearby coast of South America. If each one of these plants and animals had been created independently and simply placed on the Galápagos Islands, why didn't they resemble the plants and animals of islands with similar climates, such as those off the coast of Africa, for example? Why did they resemble those of the adjacent South American coast instead?

The fossils and patterns of life that Darwin observed on the voyage of the *Beagle* eventually convinced him that evolution had taken place.

Table 1.1 Darwin's Evidence that Evolution Occurs

FOSSILS

1. Extinct species, such as the fossil armadillo in figure 1.8, most closely resemble living ones in the same area, suggesting that one had given rise to the other.
2. In rock strata (layers), progressive changes in characteristics can be seen in fossils from earlier and earlier layers.

GEOGRAPHICAL DISTRIBUTION

3. Lands with similar climates, such as Australia, South Africa, California, and Chile, have unrelated plants and animals, indicating that diversity is not entirely influenced by climate and environment.
4. The plants and animals of each continent are distinctive; all South American rodents belong to a single group, structurally similar to the guinea pigs, for example, while most of the rodents found elsewhere belong to other groups.

OCEANIC ISLANDS

5. Although oceanic islands have few species, those they do have are often unique (endemic) and show relatedness to one another, such as the Galápagos tortoises. This suggests that the tortoises and other groups of endemic species developed after their mainland ancestors reached the islands and are, therefore, more closely related to one another.
6. Species on oceanic islands show strong affinities to those on the nearest mainland. Thus, the finches of the Galápagos Islands closely resemble a finch seen on the western coast of South America. The Galápagos finches do *not* resemble the birds on the Cape Verde Islands, islands in the Atlantic Ocean off the coast of Africa that are similar to the Galápagos. Darwin visited the Cape Verde Islands and many other island groups personally and was able to make such comparisons on the basis of his own observations.

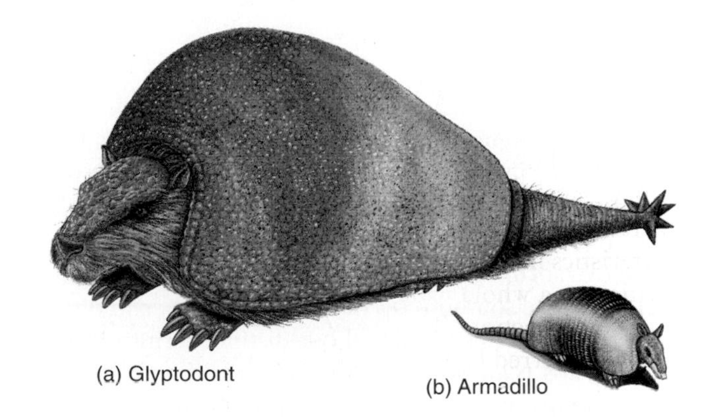

FIGURE 1.8
Fossil evidence of evolution. The now-extinct glyptodont (*a*) was a 2000-kilogram South American armadillo, much larger than the modern armadillo (*b*), which weighs an average of about 4.5 kilograms. (Drawings are not to scale.)

Inventing the Theory of Natural Selection

It is one thing to observe the results of evolution, but quite another to understand how it happens. Darwin's great achievement lies in his formulation of the hypothesis that evolution occurs because of natural selection.

Darwin and Malthus

Of key importance to the development of Darwin's insight was his study of Thomas Malthus's *Essay on the Principle of Population* (1798). In his book, Malthus pointed out that populations of plants and animals (including human beings) tend to increase geometrically, while the ability of humans to increase their food supply increases only arithmetically. A *geometric progression* is one in which the elements increase by a constant *factor;* for example, in the progression 2, 6, 18, 54, . . . , each number is three times the preceding one. An *arithmetic progression,* in contrast, is one in which the elements increase by a constant *difference;* in the progression 2, 4, 6, 8, . . . , each number is two greater than the preceding one (figure 1.9).

Because populations increase geometrically, virtually any kind of animal or plant, if it could reproduce unchecked, would cover the entire surface of the world within a surprisingly short time. Instead, populations of species remain fairly constant year after year, because death limits population numbers. Malthus's conclusion provided the key ingredient that was necessary for Darwin to develop the hypothesis that evolution occurs by natural selection.

Sparked by Malthus's ideas, Darwin saw that although every organism has the potential to produce more offspring than can survive, only a limited number actually do survive and produce further offspring. Combining this observation with what he had seen on the voyage of the *Beagle,* as well as with his own experiences in breeding domestic animals, Darwin made an important association (figure 1.10): Those individuals that possess superior physical, behavioral, or other attributes are more likely to survive than those that are not so well endowed. By surviving, they gain the opportunity to pass on their favorable characteristics to their offspring. As the frequency of these characteristics increases in the population, the nature of the population as a whole will gradually change. Darwin called this process selection. The driving force he identified has often been referred to as survival of the fittest.

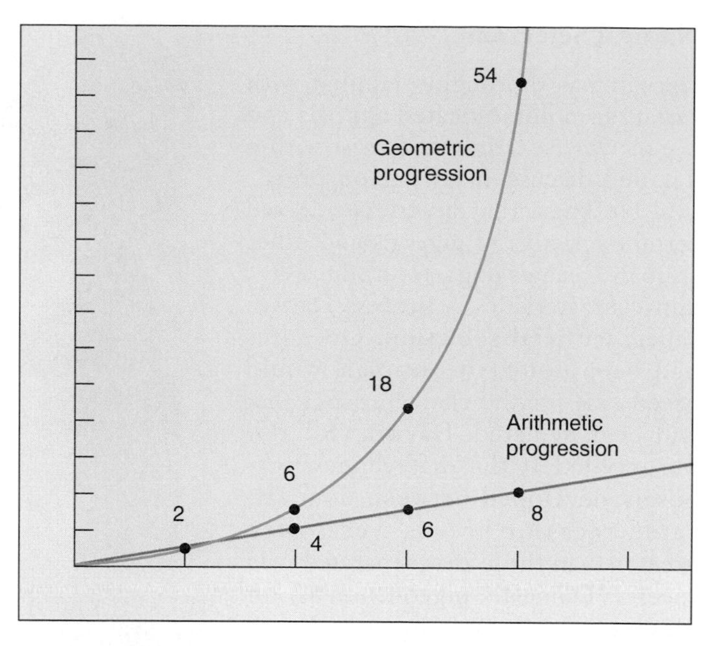

FIGURE 1.9
Geometric and arithmetic progressions. A geometric progression increases by a constant factor (for example, × 2 or × 3 or × 4), while an arithmetic progression increases by a constant difference (for example, units of 1 or 2 or 3). Malthus contended that the human growth curve was geometric, but the human food production curve was only arithmetic. Can you see the problems this difference would cause?

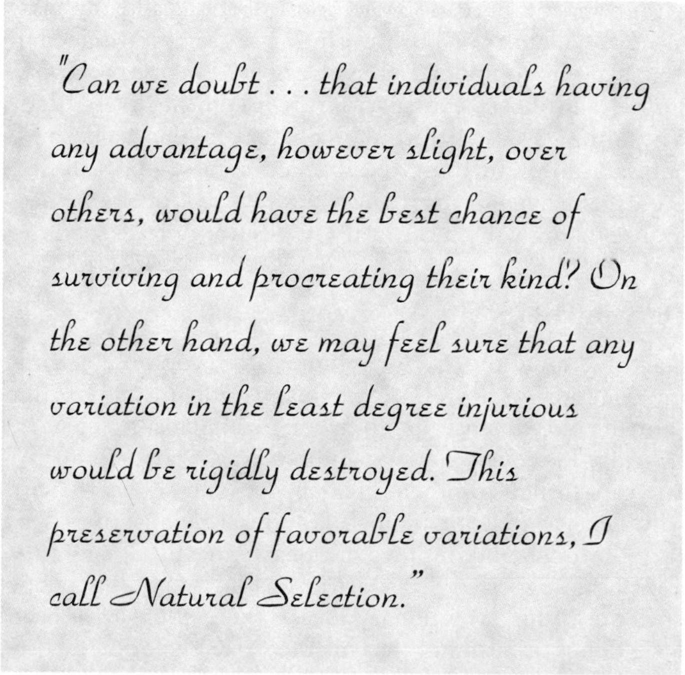

"Can we doubt . . . that individuals having any advantage, however slight, over others, would have the best chance of surviving and procreating their kind? On the other hand, we may feel sure that any variation in the least degree injurious would be rigidly destroyed. This preservation of favorable variations, I call Natural Selection."

FIGURE 1.10
An excerpt from Charles Darwin's *On the Origin of Species.*

Natural Selection

Darwin was thoroughly familiar with variation in domesticated animals and began *On the Origin of Species* with a detailed discussion of pigeon breeding. He knew that breeders selected certain varieties of pigeons and other animals, such as dogs, to produce certain characteristics, a process Darwin called **artificial selection.** Once this had been done, the animals would breed true for the characteristics that had been selected. Darwin had also observed that the differences purposely developed between domesticated races or breeds were often greater than those that separated wild species. Domestic pigeon breeds, for example, show much greater variety than all of the hundreds of wild species of pigeons found throughout the world. Such relationships suggested to Darwin that evolutionary change could occur in nature too. Surely if pigeon breeders could foster such variation by "artificial selection," nature could do the same, playing the breeder's role in selecting the next generation—a process Darwin called **natural selection.**

FIGURE 1.11
Darwin greets his monkey ancestor. In his time, Darwin was often portrayed unsympathetically, as in this drawing from an 1874 publication.

Darwin's theory thus incorporates the hypothesis of evolution, the process of natural selection, and the mass of new evidence for both evolution and natural selection that Darwin compiled. Thus, Darwin's theory provides a simple and direct explanation of biological diversity, or why animals are different in different places: because habitats differ in their requirements and opportunities, the organisms with characteristics favored locally by natural selection will tend to vary in different places.

Darwin Drafts His Argument

Darwin drafted the overall argument for evolution by natural selection in a preliminary manuscript in 1842. After showing the manuscript to a few of his closest scientific friends, however, Darwin put it in a drawer, and for 16 years turned to other research. No one knows for sure why Darwin did not publish his initial manuscript—it is very thorough and outlines his ideas in detail. Some historians have suggested that Darwin was shy of igniting public criticism of his evolutionary ideas—there could have been little doubt in his mind that his theory of evolution by natural selection would spark controversy. Others have proposed that Darwin was simply refining his theory all those years, although there is little evidence he altered his initial manuscript in all that time.

Wallace Has the Same Idea

The stimulus that finally brought Darwin's theory into print was an essay he received in 1858. A young English naturalist named Alfred Russel Wallace (1823–1913) sent the essay to Darwin from Malaysia; it concisely set forth the theory of evolution by means of natural selection, a theory Wallace had developed independently of Darwin. Like Darwin, Wallace had been greatly influenced by Malthus's 1798 essay. Colleagues of Wallace, knowing of Darwin's work, encouraged him to communicate with Darwin. After receiving Wallace's essay, Darwin arranged for a joint presentation of their ideas at a seminar in London. Darwin then completed his own book, expanding the 1842 manuscript which he had written so long ago, and submitted it for publication.

Publication of Darwin's Theory

Darwin's book appeared in November 1859 and caused an immediate sensation. Many people were deeply disturbed by the suggestion that human beings were descended from the same ancestor as apes (figure 1.11). Darwin did not actually discuss this idea in his book, but it followed directly from the principles he outlined. In a subsequent book, *The Descent of Man*, Darwin presented the argument directly, building a powerful case that humans and living apes have common ancestors. Although people had long accepted that humans closely resembled apes in many characteristics, the possibility that there might be a direct evolutionary relationship was unacceptable to many. Darwin's arguments for the theory of evolution by natural selection were so compelling, however, that his views were almost completely accepted within the intellectual community of Great Britain after the 1860s.

The fact that populations do not really expand geometrically implies that nature acts to limit population numbers. The traits of organisms that survive to produce more offspring will be more common in future generations—a process Darwin called natural selection.

Evolution After Darwin: More Evidence

More than a century has elapsed since Darwin's death in 1882. During this period, the evidence supporting his theory has grown progressively stronger. There have also been many significant advances in our understanding of how evolution works. Although these advances have not altered the basic structure of Darwin's theory, they have taught us a great deal more about the mechanisms by which evolution occurs. We will briefly explore some of this evidence here; in chapter 21 we will return to the theory of evolution and examine the evidence in more detail.

The Fossil Record

Darwin predicted that the fossil record would yield intermediate links between the great groups of organisms, for example, between fishes and the amphibians thought to have arisen from them, and between reptiles and birds. We now know the fossil record to a degree that was unthinkable in the nineteenth century. Recent discoveries of microscopic fossils have extended the known history of life on earth back to about 3.5 billion years ago. The discovery of other fossils has supported Darwin's predictions and has shed light on how organisms have, over this enormous time span, evolved from the simple to the complex. For vertebrate animals especially, the fossil record is rich and exhibits a graded series of changes in form, with the evolutionary parade visible for all to see (see Box: Why Study Fossils?).

The Age of the Earth

In Darwin's day, some physicists argued that the earth was only a few thousand years old. This bothered Darwin, because the evolution of all living things from some single original ancestor would have required a great deal more time. Using evidence obtained by studying the rates of radioactive decay, we now know that the physicists of Darwin's time were wrong, very wrong: the earth was formed about 4.5 billion years ago.

The Mechanism of Heredity

Darwin received some of his sharpest criticism in the area of heredity. At that time, no one had any concept of genes or of how heredity works, so it was not possible for Darwin to explain completely how evolution occurs. Theories of heredity in Darwin's day seemed to rule out the possibility of genetic variation in nature, a critical requirement of Darwin's theory. Genetics was established as a science only at the start of the twentieth century, 40 years after the publication of Darwin's *On the Origin of Species.* When scientists began to understand the laws of inheritance (discussed in chapter 13), the heredity problem with Darwin's theory vanished. Genetics accounts in a neat and orderly way for the production of new variations in organisms.

Comparative Anatomy

Comparative studies of animals have provided strong evidence for Darwin's theory. In many different types of vertebrates, for example, the same bones are present, indicating their evolutionary past. Thus, the forelimbs shown in figure 1.12 are all constructed from the same basic array of bones, modified in one way in the wing of a bat, in another way in the fin of a porpoise, and in yet another way in the leg of a horse. The bones are said to be **homologous** in the different vertebrates; that is, they have the same evolutionary origin, but they now differ in structure and function. This contrasts with **analogous** structures, such as the wings of birds and butterflies, which have similar structure and function but different evolutionary origins.

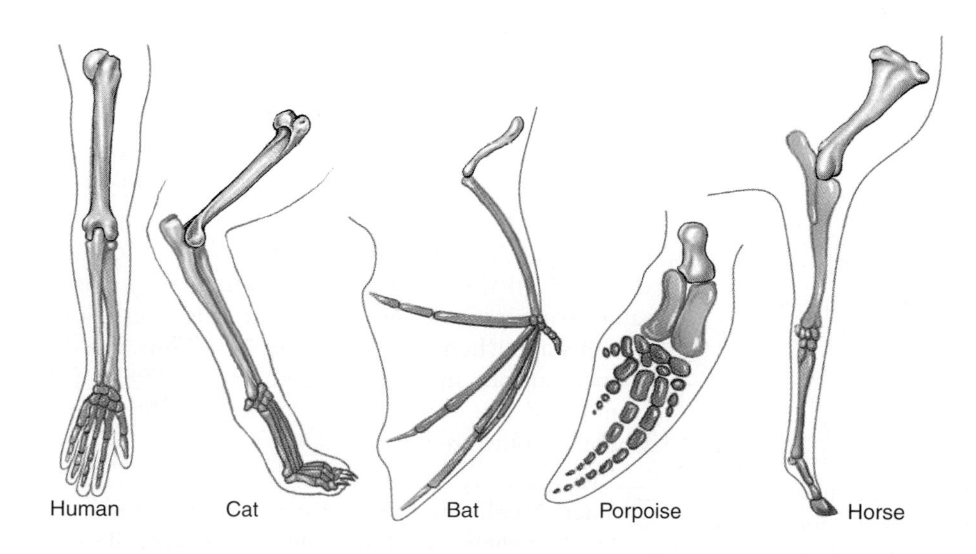

FIGURE 1.12
Homology among vertebrate limbs. The forelimbs of these five vertebrates show the ways in which the relative proportions of the forelimb bones have changed in relation to the particular way of life of each organism.

Why Study Fossils?

Flight has evolved three separate times among vertebrates. Birds and bats are still with us, but pterosaurs, such as the one pictured, became extinct with the dinosaurs about 65 million years ago.

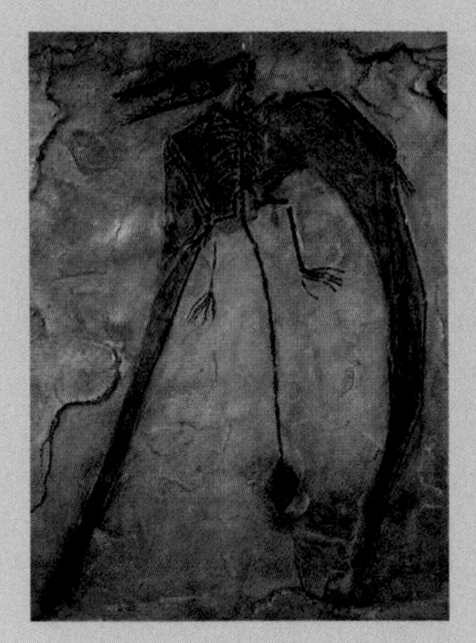

Stephen Jay Gould
Harvard University

I grew up on the streets of New York City, in a family of modest means and little formal education, but with a deep love of learning. Like many urban kids who become naturalists, my inspiration came from a great museum—in particular, from the magnificent dinosaurs on display at the American Museum of Natural History. As we all know from *Jurassic Park* and other sources, dinomania in young children (I was five when I saw my first dinosaur) is not rare—but nearly all children lose the passion, and the desire to become a paleontologist becomes a transient moment between policeman and fireman in a chronology of intended professions. But I persisted and became a professional paleontologist, a student of life's history as revealed by the evidence of fossils (though I ended up working on snails rather than dinosaurs!). Why?

I remained committed to paleontology because I discovered, still as a child, the wonder of one of the greatest transforming ideas ever discovered by science: evolution. I learned that those dinosaurs, and all creatures that have ever lived, are bound together in a grand family tree of physical relationships, and that the rich and fascinating changes of life, through billions of years in geological time, occur by a natural process of evolutionary transformation—"descent with modification," in Darwin's words. I was thrilled to learn that humans had arisen from apelike ancestors, who had themselves evolved from the tiny mouselike mammals that had lived in the time of dinosaurs and seemed then so inconspicuous, so unsuccessful, and so unpromising.

Now, at mid-career (I was born in 1941) I remain convinced that I made the right choice, and committed to learn and convey, as much as I can as long as I can, about evolution and the history of life. We can learn a great deal about the *process* of evolution by studying modern organisms. But history is complex and unpredictable—and principles of evolution (like natural selection) cannot specify the *pathway* that life's history has actually followed. Paleontology holds the archives of the pathway—the fossil record of past life, with its fascinating history of mass extinctions, periods of rapid change, long episodes of stability, and constantly changing patterns of dominance and diversity. Humans represent just one tiny, largely fortuitous, and late-arising twig on the enormously arborescent bush of life. Paleontology is the study of this grandest of all bushes.

Molecular Biology

Biochemical tools are now of major importance in efforts to reach a better understanding of how evolution occurs. Within the last few years, for example, evolutionary biologists have begun to "read" genes, much as you are reading this page. They have learned to recognize the order of the "letters" of the long DNA molecules, which are present in every living cell and which provide the genetic information for that organism. By comparing the sequences of "letters" in the DNA of different groups of animals or plants, we can specify the degree of relationship among the groups more precisely than by any other means. In many cases, detailed family trees can then be constructed. The consistent pattern emerging from a growing mountain of data is one of progressive change over time, with more distantly related species showing more differences in their DNA than closely related ones, just as Darwin's theory predicts. By measuring the degree of difference in the genetic coding, and by interpreting the information available from the fossil record, we can even estimate the *rates* at which evolution is occurring in different groups of organisms.

Development

Twentieth-century knowledge about growth and development further supports Darwin's theory of evolution. Striking similarities are seen in the developmental stages of many organisms of different species. Human embryos, for example, go through a stage in which they possess the same structures that give rise to the gills in fish, a tail, and even a stage when the embryo has fur! Thus, the development of an organism (its ontogeny) often yields information about the evolutionary history of the species as a whole (its phylogeny).

Since Darwin's time, new discoveries of the fossil record, genetics, anatomy, and development all support Darwin's theory.

1.4 This book is organized to help you learn biology.

Core Principles of Biology

From centuries of biological observation and inquiry, one organizing principle has emerged: biological diversity reflects history, a record of success, failure, and change extending back to a period soon after the formation of the earth. The explanation for this diversity, the theory of evolution by natural selection, will form the backbone of your study of biological science, just as the theory of the covalent bond is the backbone of chemistry, or the theory of quantum mechanics is that of physics. Evolution by natural selection is a thread that runs through everything you will learn in this book.

Basic Principles

The first half of this book is devoted to a description of the basic principles of biology, introduced through a levels-of-organization framework (see figure 1.2). At the molecular, organellar, and cellular levels of organization, you will be introduced to *cell biology*. You will learn how cells are constructed and how they grow, divide, and communicate. At the organismal level, you will learn the principles of *genetics*, which deal with the way that individual traits are transmitted from one generation to the next. At the population level, you will examine *evolution*, the gradual change in populations from one generation to the next, which has led through natural selection to the biological diversity we see around us. Finally, at the community and ecosystem levels, you will study *ecology*, which deals with how organisms interact with their environments and with one another to produce the complex communities characteristic of life on earth.

Organisms

The second half of the book is devoted to an examination of organisms, the products of evolution. It is estimated that at least 5 million different kinds of plants, animals, and microorganisms exist, and their diversity is *incredible* (figure 1.13). Later in the book, we will take a particularly detailed look at the vertebrates, the group of animals of which we are members. We will consider the vertebrate body and how it functions, as this information is of greatest interest and importance to most students.

As you proceed through this book, what you learn at one stage will give you the tools to understand the next. The core principle of biology is that biological diversity is the result of a long evolutionary journey.

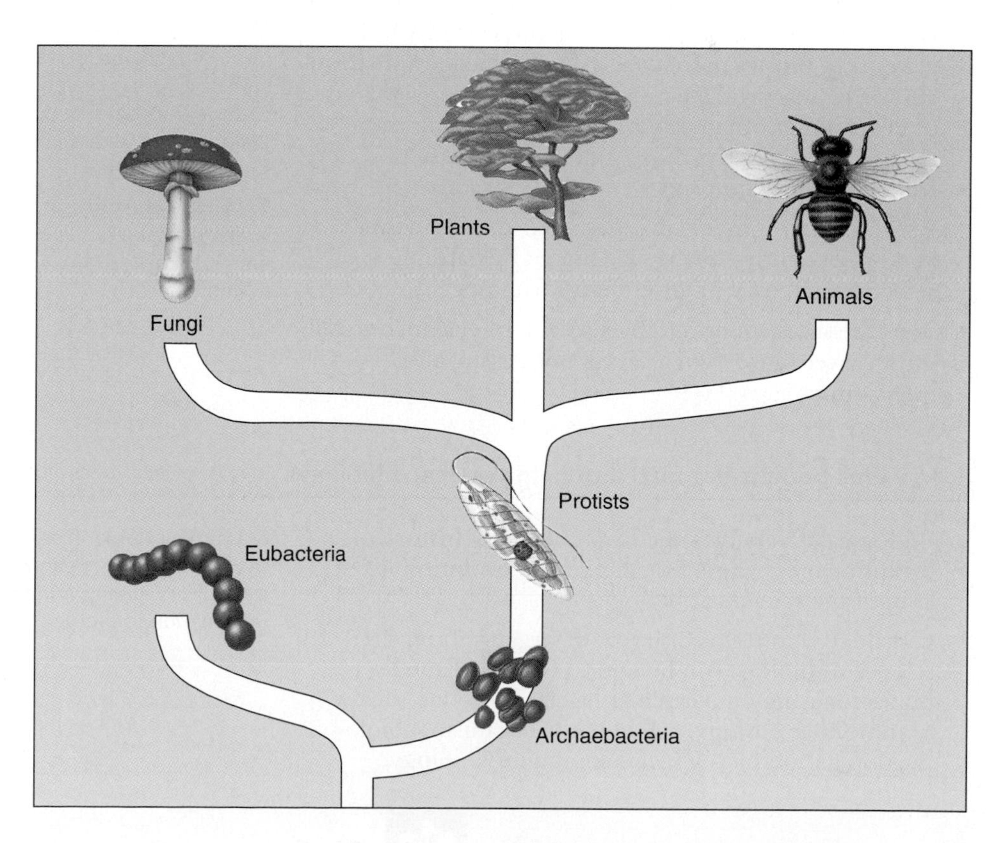

FIGURE 1.13
The diversity of life. Biologists categorize all living things into six major groups called kingdoms: archaebacteria, eubacteria, protists, fungi, plants, and animals.

Chapter 1

www.mhhe.com/raven6/resources1.mhtml

Summary	Questions	Media Resources

1.1 Biology is the science of life.

- Living things are highly organized, whether as single cells or as multicellular organisms, with several hierarchical levels.

1. What are the characteristics of living things?

- Art Activity: Biological Organization

1.2 Scientists form generalizations from observations.

- Science is the determination of general principles from observation and experimentation.
- Scientists select the best hypotheses by using controlled experiments to eliminate alternative hypotheses that are inconsistent with observations.
- A group of related hypotheses supported by a large body of evidence is called a theory. In science, a theory represents what we are most sure about. However, there are no absolute truths in science, and even theories are accepted only conditionally.
- Scientists conduct basic research, designed to gain information about natural phenomena in order to contribute to our overall body of knowledge, and applied research, devoted to solving specific problems with practical applications.

2. What is the difference between deductive and inductive reasoning? What is a hypothesis?

3. What are variables? How are control experiments used in testing hypotheses?

4. How does a hypothesis become a theory? At what point does a theory become accepted as an absolute truth, no longer subject to any uncertainty?

5. What is the difference between basic and applied research?

- Scientists on Science: Why Paleontology?
- How Scientists Think: Probability and Hypothesis Testing in Biology (Appendix)

- Art Quiz: Scientific Method

1.3 Darwin's theory of evolution illustrates how science works.

- One of the central theories of biology is Darwin's theory that evolution occurs by natural selection. It states that certain individuals have heritable traits that allow them to produce more offspring in a given kind of environment than other individuals lacking those traits. Consequently, those traits will increase in frequency through time.
- Because environments differ in their requirements and opportunities, the traits favored by natural selection will vary in different environments.
- This theory is supported by a wealth of evidence acquired over more than a century of testing and questioning.

6. Describe the evidence that led Darwin to propose that evolution occurs by means of natural selection. What evidence gathered since the publication of Darwin's theory has lent further support to the theory?

7. What is the difference between homologous and analogous structures? Give an example of each.

- Introduction to Evolution
- Before Darwin
- Voyage of the *Beagle*
- Natural Selection

- Student Research: The Search for Medicinal Plants
- *On Science* Articles:
  - –140 Years Without Darwin Are Enough
  - –Bird-Killing Cats: Nature's Way of Making Better Birds

- Art Quiz: Geometric and Arithmetic Progressions

1.4 This book is organized to help you learn biology.

- Biological diversity is the result of a long history of evolutionary change. For this reason evolution is the core of the science of biology.
- Considered in terms of levels-of-organization, the science of biology can be said to consist of subdisciplines focusing on particular levels. Thus one speaks of molecular biology, cell biology, organismal biology, population biology, and community biology.

8. Can you think of any alternatives to levels-of-organization as ways of organizing the mass of information in biology?

2

The Nature of Molecules

Concept Outline

2.1 Atoms are nature's building material.

Atoms. All substances are composed of tiny particles called atoms, each a positively charged nucleus around which orbit negative electrons.

Electrons Determine the Chemical Behavior of Atoms. Electrons orbit the nucleus of an atom; the closer an electron's orbit to the nucleus, the lower its energy level.

2.2 The atoms of living things are among the smallest.

Kinds of Atoms. Of the 92 naturally occurring elements, only 11 occur in organisms in significant amounts.

2.3 Chemical bonds hold molecules together.

Ionic Bonds Form Crystals. Atoms are linked together into molecules, joined by chemical bonds that result from forces like the attraction of opposite charges or the sharing of electrons.

Covalent Bonds Build Stable Molecules. Chemical bonds formed by the sharing of electrons can be very strong, and require much energy to break.

2.4 Water is the cradle of life.

Chemistry of Water. Water forms weak chemical associations that are responsible for much of the organization of living chemistry.

Water Atoms Act Like Tiny Magnets. Because electrons are shared unequally by the hydrogen and oxygen atoms of water, a partial charge separation occurs. Each water atom acquires a positive and negative pole and is said to be "polar."

Water Clings to Polar Molecules. Because the opposite partial charges of polar molecules attract one another, water tends to cling to itself and other polar molecules and to exclude nonpolar molecules.

Water Ionizes. Because its covalent bonds occasionally break, water contains a low concentration of hydrogen (H^+) and hydroxide (OH^-) ions, the fragments of broken water molecules.

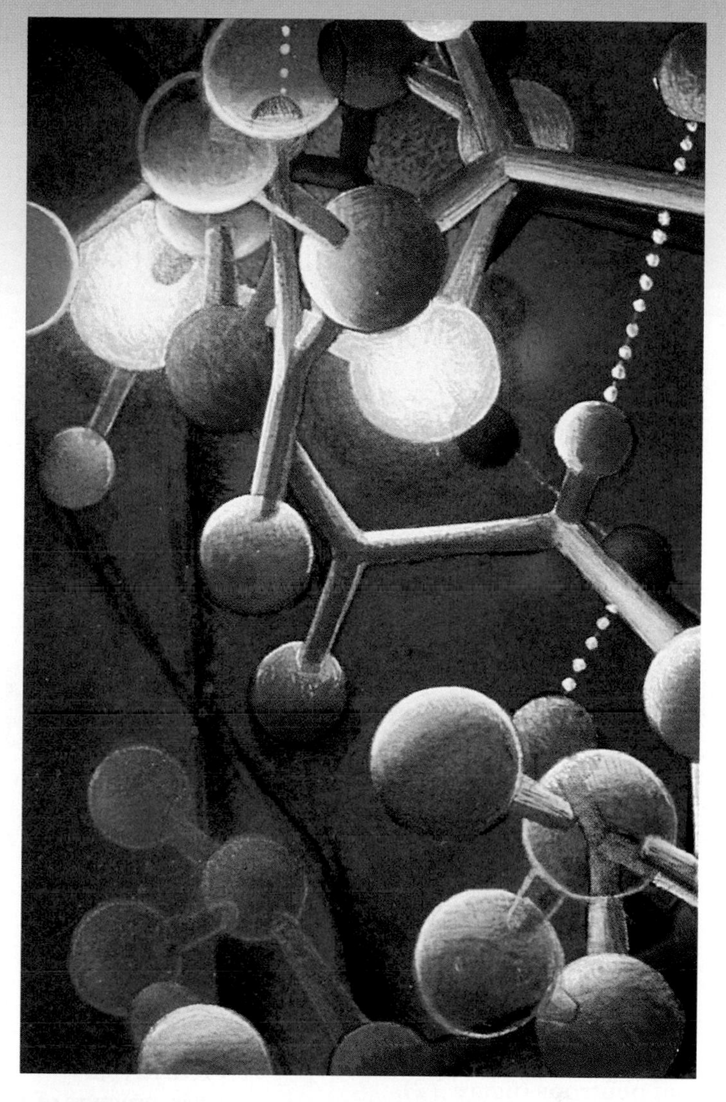

FIGURE 2.1
Cells are made of molecules. Specific, often simple, combinations of atoms yield an astonishing diversity of molecules within the cell, each with unique functional characteristics.

About 10 to 20 billion years ago, an enormous explosion likely marked the beginning of the universe. With this explosion began the process of evolution, which eventually led to the origin and diversification of life on earth. When viewed from the perspective of 20 billion years, life within our solar system is a recent development, but to understand the origin of life, we need to consider events that took place much earlier. The same processes that led to the evolution of life were responsible for the evolution of molecules (figure 2.1). Thus, our study of life on earth begins with physics and chemistry. As chemical machines ourselves, we must understand chemistry to begin to understand our origins.

2.1 Atoms are nature's building material.

Atoms

Any substance in the universe that has mass (see below) and occupies space is defined as **matter.** All matter is composed of extremely small particles called **atoms.** Because of their size, atoms are difficult to study. Not until early in this century did scientists carry out the first experiments suggesting what an atom is like.

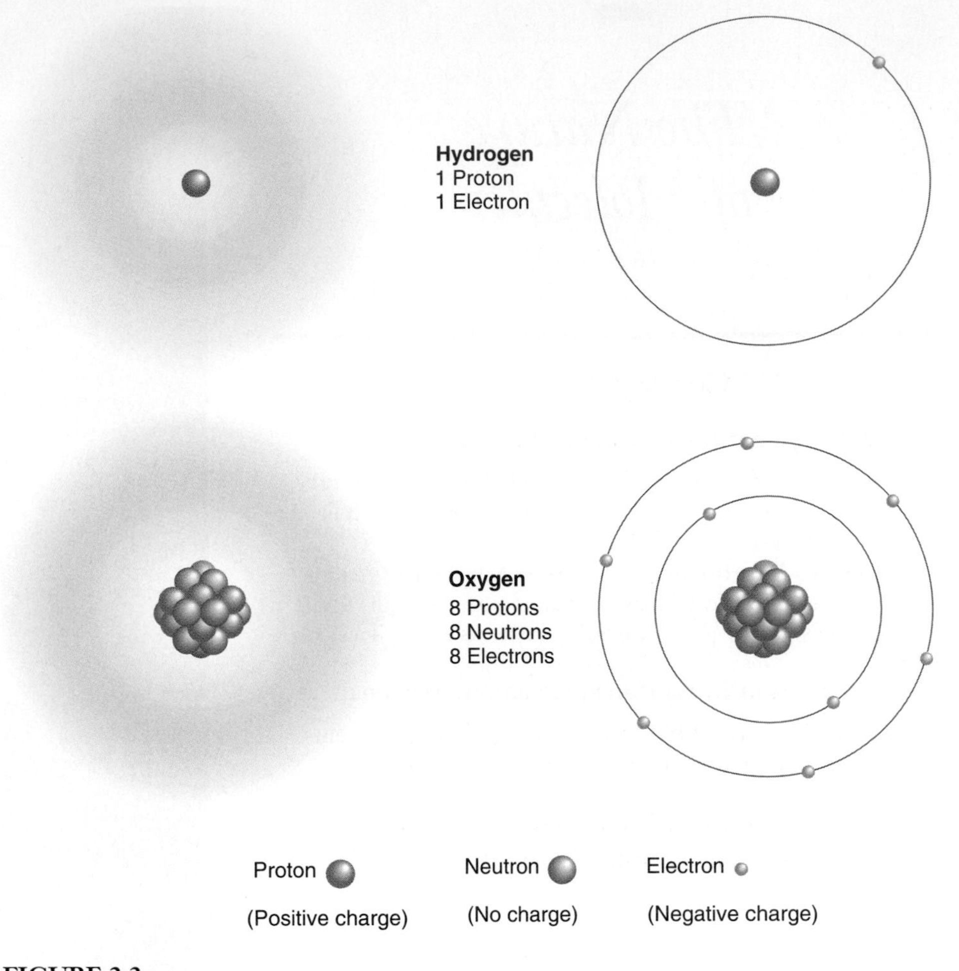

FIGURE 2.2
Basic structure of atoms. All atoms have a nucleus consisting of protons and neutrons, except hydrogen, the smallest atom, which has only one proton and no neutrons in its nucleus. Oxygen, for example, has eight protons and eight neutrons in its nucleus. Electrons spin around the nucleus a far distance away from the nucleus.

The Structure of Atoms

Objects as small as atoms can be "seen" only indirectly, by using very complex technology such as tunneling microscopy. We now know a great deal about the complexities of atomic structure, but the simple view put forth in 1913 by the Danish physicist Niels Bohr provides a good starting point. Bohr proposed that every atom possesses an orbiting cloud of tiny subatomic particles called **electrons** whizzing around a core like the planets of a miniature solar system. At the center of each atom is a small, very dense nucleus formed of two other kinds of subatomic particles, **protons** and **neutrons** (figure 2.2).

Within the nucleus, the cluster of protons and neutrons is held together by a force that works only over short subatomic distances. Each proton carries a positive (+) charge, and each electron carries a negative (–) charge. Typically an atom has one electron for each proton. The number of protons (the atom's **atomic number**) determines the chemical character of the atom, because it dictates the number of electrons orbiting the nucleus which are available for chemical activity. Neutrons, as their name implies, possess no charge.

Atomic Mass

The terms *mass* and *weight* are often used interchangeably, but they have slightly different meanings. *Mass* refers to the amount of a substance, while *weight* refers to the force gravity exerts on a substance. Hence, an object has the same *mass* whether it is on the earth or the moon, but its *weight* will be greater on the earth because the earth's gravitational force is greater than the moon's. The **atomic mass** of an atom is equal to the sum of the masses of its protons and neutrons. Atoms that occur naturally on earth contain from 1 to 92 protons and up to 146 neutrons.

The mass of atoms and subatomic particles is measured in units called *daltons.* To give you an idea of just how small these units are, note that it takes 602 million million billion (6.02×10^{23}) daltons to make 1 gram! A proton weighs approximately 1 dalton (actually 1.009 daltons), as does a neutron (1.007 daltons). In contrast, electrons weigh only $\frac{1}{1840}$ of a dalton, so their contribution to the overall mass of an atom is negligible.

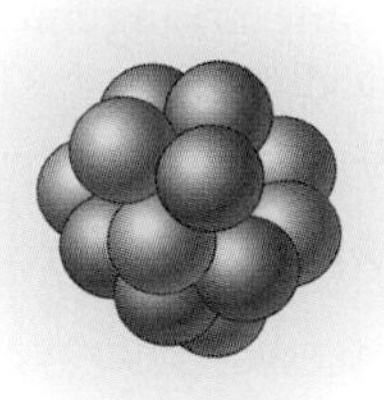

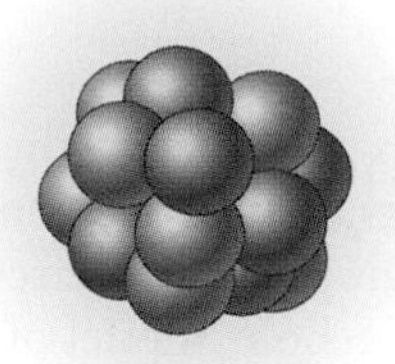

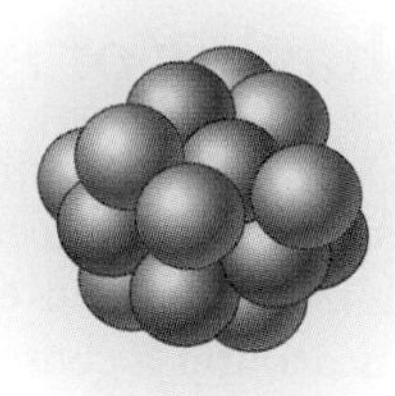

FIGURE 2.3
The three most abundant isotopes of carbon. Isotopes of a particular atom have different numbers of neutrons.

Isotopes

Atoms with the same atomic number (that is, the same number of protons) have the same chemical properties and are said to belong to the same **element.** Formally speaking, an element is any substance that cannot be broken down to any other substance by ordinary chemical means. However, while all atoms of an element have the same number of protons, they may not all have the same number of neutrons. Atoms of an element that possess different numbers of neutrons are called **isotopes** of that element. Most elements in nature exist as mixtures of different isotopes. Carbon (C), for example, has three isotopes, all containing six protons (figure 2.3). Over 99% of the carbon found in nature exists as an isotope with six neutrons. Because its total mass is 12 daltons (6 from protons plus 6 from neutrons), this isotope is referred to as carbon-12, and symbolized ^{12}C. Most of the rest of the naturally occurring carbon is carbon-13, an isotope with seven neutrons. The rarest carbon isotope is carbon-14, with eight neutrons. Unlike the other two isotopes, carbon-14 is unstable: its nucleus tends to break up into elements with lower atomic numbers. This nuclear breakup, which emits a significant amount of energy, is called radioactive decay, and isotopes that decay in this fashion are **radioactive isotopes.**

Some radioactive isotopes are more unstable than others and therefore decay more readily. For any given isotope, however, the rate of decay is constant. This rate is usually expressed as the **half-life,** the time it takes for one half of the atoms in a sample to decay. Carbon-14, for example, has a half-life of about 5600 years. A sample of carbon containing 1 gram of carbon-14 today would contain 0.5 gram of carbon-14 after 5600 years, 0.25 gram 11,200 years from now, 0.125 gram 16,800 years from now, and so on. By determining the ratios of the different isotopes of carbon and other elements in biological samples and in rocks, scientists are able to accurately determine when these materials formed.

While there are many useful applications of radioactivity, there are also harmful side effects that must be considered in any planned use of radioactive substances. Radioactive substances emit energetic subatomic particles that have the potential to severely damage living cells, producing mutations in their genes, and, at high doses, cell death. Consequently, exposure to radiation is now very carefully controlled and regulated. Scientists who work with radioactivity (basic researchers as well as applied scientists such as X-ray technologists) wear radiation-sensitive badges to monitor the total amount of radioactivity to which they are exposed. Each month the badges are collected and scrutinized. Thus, employees whose work places them in danger of excessive radioactive exposure are equipped with an "early warning system."

Electrons

The positive charges in the nucleus of an atom are counterbalanced by negatively charged electrons orbiting at varying distances around the nucleus. Thus, atoms with the same number of protons and electrons are electrically neutral, having no net charge.

Electrons are maintained in their orbits by their attraction to the positively charged nucleus. Sometimes other forces overcome this attraction and an atom loses one or more electrons. In other cases, atoms may gain additional electrons. Atoms in which the number of electrons does not equal the number of protons are known as **ions,** and they carry a net electrical charge. An atom that has more protons than electrons has a net positive charge and is called a **cation.** For example, an atom of sodium (Na) that has lost one electron becomes a sodium ion (Na^+), with a charge of +1. An atom that has fewer protons than electrons carries a net negative charge and is called an **anion.** A chlorine atom (Cl) that has gained one electron becomes a chloride ion (Cl^-), with a charge of –1.

An atom consists of a nucleus of protons and neutrons surrounded by a cloud of electrons. The number of its electrons largely determines the chemical properties of an atom. Atoms that have the same number of protons but different numbers of neutrons are called isotopes. Isotopes of an atom differ in atomic mass but have similar chemical properties.

Electrons Determine the Chemical Behavior of Atoms

The key to the chemical behavior of an atom lies in the arrangement of its electrons in their orbits. It is convenient to visualize individual electrons as following discrete circular orbits around a central nucleus, as in the Bohr model of the atom. However, such a simple picture is not realistic. It is not possible to precisely locate the position of any individual electron precisely at any given time. In fact, a particular electron can be anywhere at a given instant, from close to the nucleus to infinitely far away from it.

However, a particular electron is more likely to be located in some positions than in others. The area around a nucleus where an electron is most likely to be found is called the **orbital** of that electron (figure 2.4). Some electron orbitals near the nucleus are spherical (*s* orbitals), while others are dumbbell-shaped (*p* orbitals). Still other orbitals, more distant from the nucleus, may have different shapes. Regardless of its shape, no orbital may contain more than two electrons.

Almost all of the volume of an atom is empty space, because the electrons are quite far from the nucleus relative to its size. If the nucleus of an atom were the size of an apple, the orbit of the nearest electron would be more than 1600 meters away. Consequently, the nuclei of two atoms never come close enough in nature to interact with each other. It is for this reason that an atom's electrons, not its protons or neutrons, determine its chemical behavior. This also explains why the isotopes of an element, all of which have the same arrangement of electrons, behave the same way chemically.

Energy within the Atom

All atoms possess energy, defined as the ability to do work. Because electrons are attracted to the positively charged nucleus, it takes work to keep them in orbit, just as it takes work to hold a grapefruit in your hand against the pull of gravity. The grapefruit is said to possess *potential energy*, the ability to do work, because of its position; if you were to release it, the grapefruit would fall and its energy would be reduced. Conversely, if you were to move the grapefruit to the top of a building, you would increase its potential energy. Similarly, electrons have potential energy of position. To oppose the attraction of the nucleus and move the electron to a more distant orbital requires an input of energy and results in an electron with greater potential energy. This is how chlorophyll captures energy from light during photosynthesis (chapter 10)—the light excites electrons in the chlorophyll. Moving an electron closer to the nucleus has the opposite effect: energy is released, usually as heat, and the electron ends up with less potential energy (figure 2.5).

A given atom can possess only certain discrete amounts of energy. Like the potential energy of a grapefruit on a step of a staircase, the potential energy contributed by the position of an electron in an atom can have only certain values.

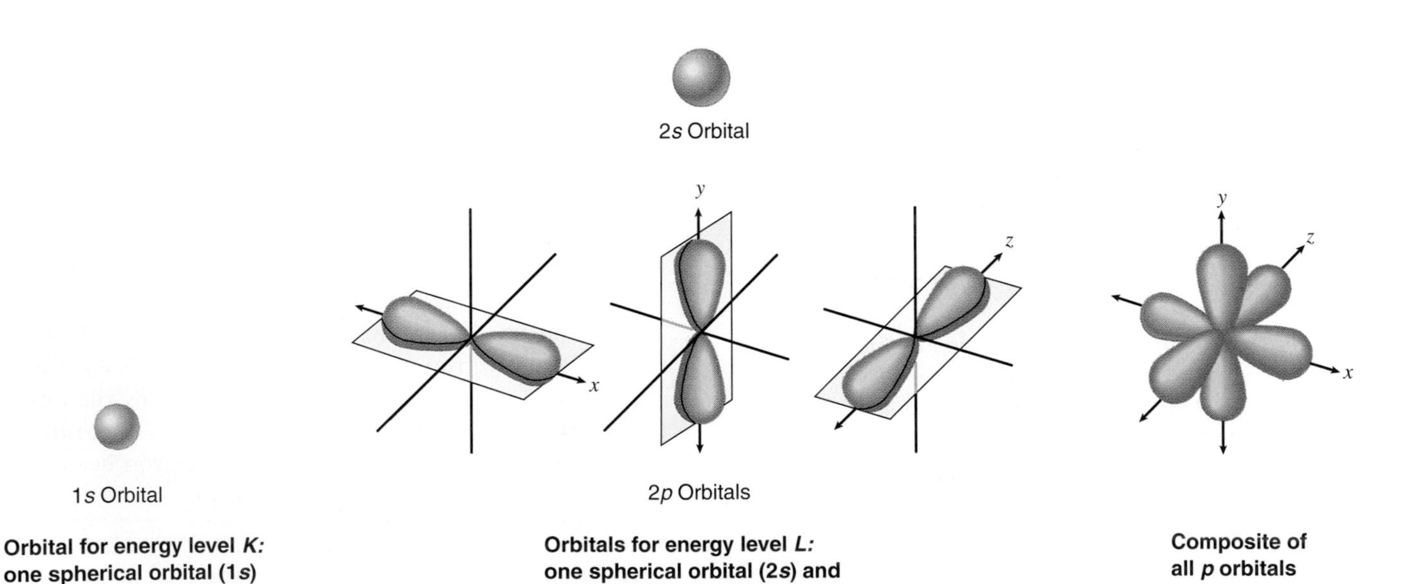

FIGURE 2.4
Electron orbitals. The lowest energy level or electron shell, which is nearest the nucleus, is level *K*. It is occupied by a single *s* orbital, referred to as 1*s*. The next highest energy level, *L*, is occupied by four orbitals: one *s* orbital (referred to as the 2*s* orbital) and three *p* orbitals (each referred to as a 2*p* orbital). The four *L*-level orbitals compactly fill the space around the nucleus, like two pyramids set base-to-base.

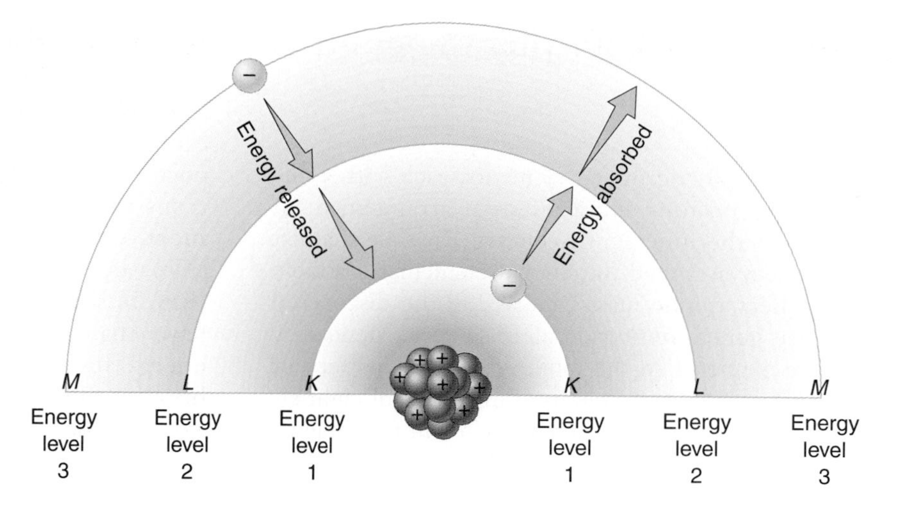

FIGURE 2.5
Atomic energy levels. When an electron absorbs energy, it moves to higher energy levels farther from the nucleus. When an electron releases energy, it falls to lower energy levels closer to the nucleus.

Every atom exhibits a ladder of potential energy values, rather than a continuous spectrum of possibilities, a discrete set of orbits at particular distances from the nucleus.

During some chemical reactions, electrons are transferred from one atom to another. In such reactions, the loss of an electron is called **oxidation,** and the gain of an electron is called **reduction** (figure 2.6). It is important to realize that when an electron is transferred in this way, it keeps its energy of position. In organisms, chemical energy is stored in high-energy electrons that are transferred from one atom to another in reactions involving oxidation and reduction.

Because the amount of energy an electron possesses is related to its distance from the nucleus, electrons that are the same distance from the nucleus have the same energy, even if they occupy different orbitals. Such electrons are said to occupy the same **energy level.** In a schematic diagram of an atom (figure 2.7), the nucleus is represented as a small circle and the electron energy levels are drawn as concentric rings, with the energy level increasing with distance from the nucleus. Be careful not to confuse energy levels, which are drawn as rings to indicate an electron's *energy,* with orbitals, which have a variety of three-dimensional shapes and indicate an electron's most likely *location.*

Electrons orbit a nucleus in paths called orbitals. No orbital can contain more than two electrons, but many orbitals may be the same distance from the nucleus and, thus, contain electrons of the same energy.

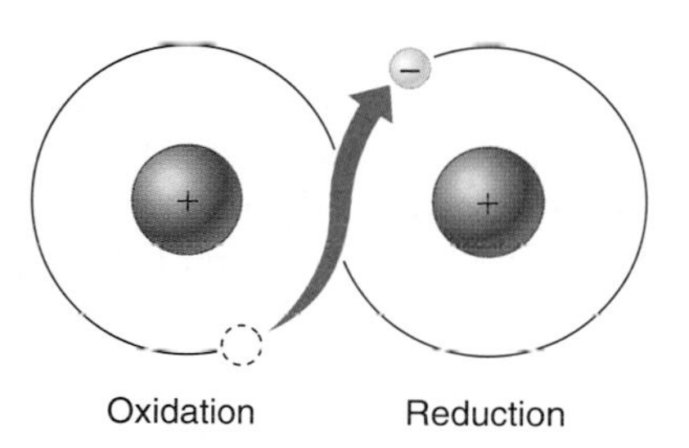

FIGURE 2.6
Oxidation and reduction. Oxidation is the loss of an electron; reduction is the gain of an electron.

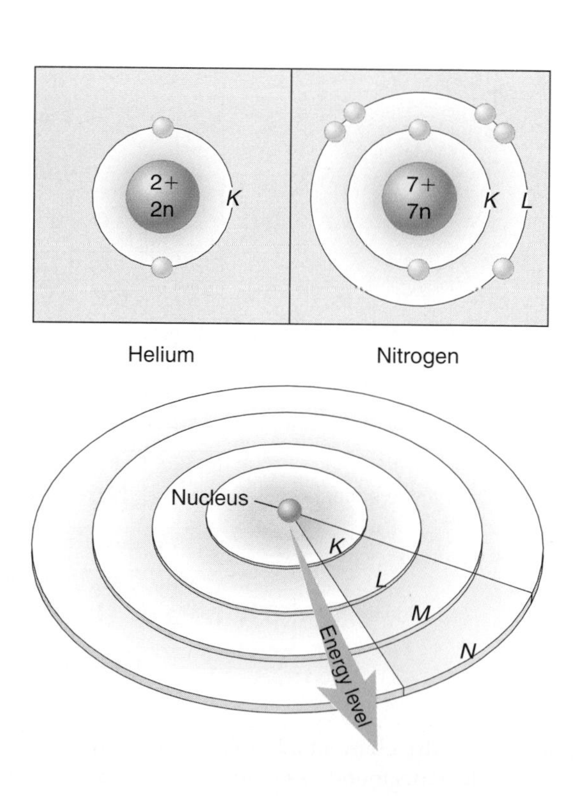

FIGURE 2.7
Electron energy levels for helium and nitrogen. Gold balls represent the electrons. Each concentric circle represents a different distance from the nucleus and, thus, a different electron energy level.

2.2 The atoms of living things are among the smallest.

Kinds of Atoms

There are 92 naturally occurring elements, each with a different number of protons and a different arrangement of electrons. When the nineteenth-century Russian chemist Dmitri Mendeleev arranged the known elements in a table according to their atomic mass (figure 2.8), he discovered one of the great generalizations in all of science. Mendeleev found that the elements in the table exhibited a pattern of chemical properties that repeated itself in groups of eight elements. This periodically repeating pattern lent the table its name: the periodic table of elements.

The Periodic Table

The eight-element periodicity that Mendeleev found is based on the interactions of the electrons in the outer energy levels of the different elements. These electrons are called **valence electrons** and their interactions are the basis for the differing chemical properties of the elements. For most of the atoms important to life, an outer energy level can contain no more than eight electrons; the chemical behavior of an element reflects how many of the eight positions are filled. Elements possessing all eight electrons in their outer energy level (two for helium) are **inert,** or nonreactive; they include helium (He), neon (Ne), argon (Ar), krypton (Kr), xenon (Xe), and radon (Rn). In sharp contrast, elements with seven electrons (one fewer than the maximum number of eight) in their outer energy level, such as fluorine (F), chlorine (Cl), and bromine (Br), are highly reactive. They tend to gain the extra electron needed to fill the energy level. Elements with only one electron in their outer energy level, such as lithium (Li), sodium (Na), and potassium (K), are also very reactive; they tend to lose the single electron in their outer level.

Mendeleev's periodic table thus leads to a useful generalization, the **octet rule** (Latin *octo,* "eight") or **rule of eight:** atoms tend to establish completely full outer energy levels. Most chemical behavior can be predicted quite accurately from this simple rule, combined with the tendency of atoms to balance positive and negative charges.

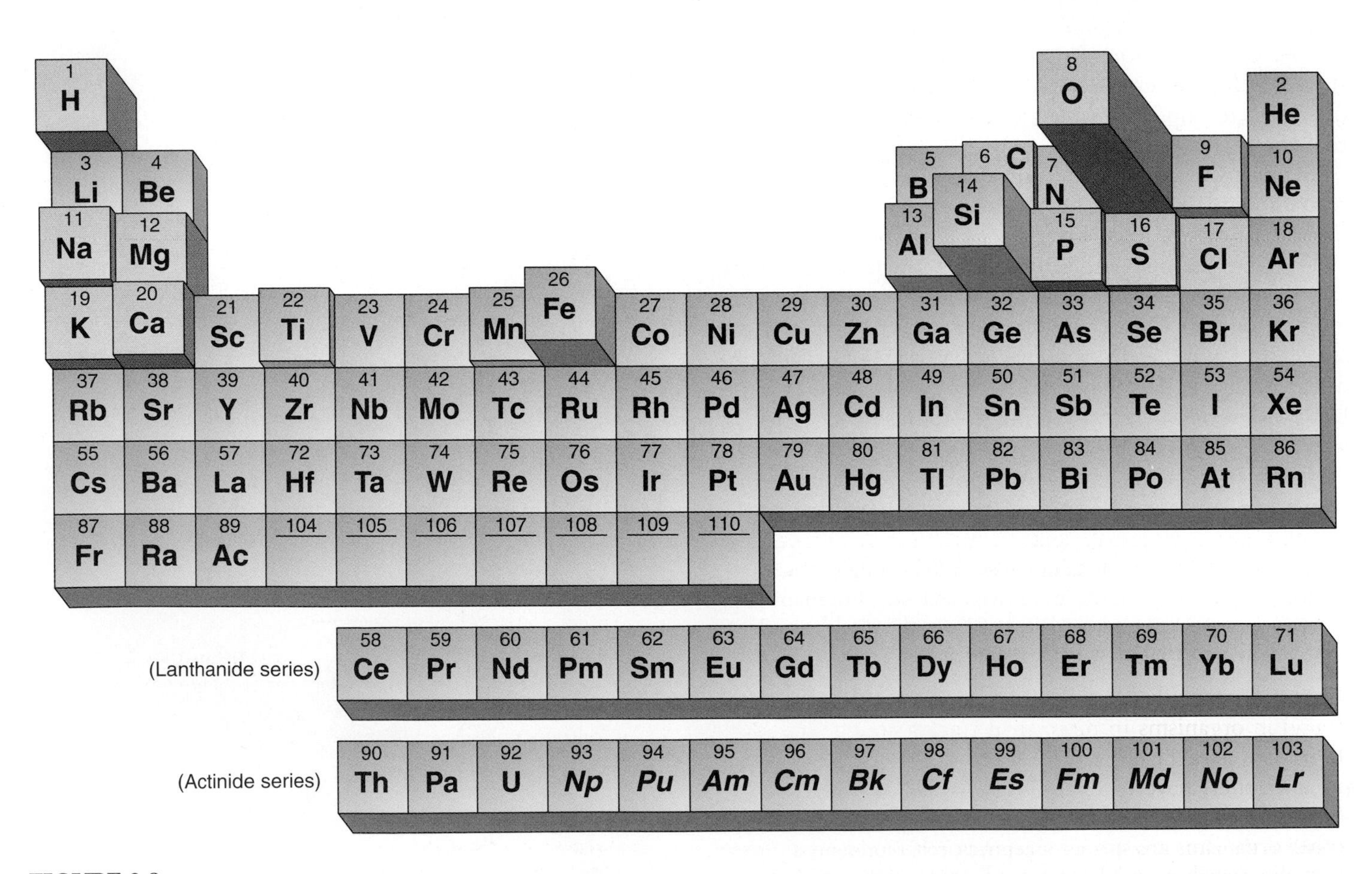

FIGURE 2.8
Periodic table of the elements. In this representation, the frequency of elements that occur in the earth's crust is indicated by the height of the block. Elements found in significant amounts in living organisms are shaded in *blue.*

Element	Symbol	Atomic Number	Approximate Percent of Earth's Crust by Weight	Percent of Human Body by Weight	Importance or Function
Oxygen	O	8	46.6	65.0	Required for cellular respiration; component of water
Silicon	Si	14	27.7	Trace	
Aluminum	Al	13	6.5	Trace	
Iron	Fe	26	5.0	Trace	Critical component of hemoglobin in the blood
Calcium	Ca	20	3.6	1.5	Component of bones and teeth; triggers muscle contraction
Sodium	Na	11	2.8	0.2	Principal positive ion outside cells; important in nerve function
Potassium	K	19	2.6	0.4	Principal positive ion inside cells; important in nerve function
Magnesium	Mg	12	2.1	0.1	Critical component of many energy-transferring enzymes
Hydrogen	H	1	0.14	9.5	Electron carrier; component of water and most organic molecules
Manganese	Mn	25	0.1	Trace	
Fluorine	F	9	0.07	Trace	
Phosphorus	P	15	0.07	1.0	Backbone of nucleic acids; important in energy transfer
Carbon	C	6	0.03	18.5	Backbone of organic molecules
Sulfur	S	16	0.03	0.3	Component of most proteins
Chlorine	Cl	17	0.01	0.2	Principal negative ion outside cells
Vanadium	V	23	0.01	Trace	
Chromium	Cr	24	0.01	Trace	
Copper	Cu	29	0.01	Trace	Key component of many enzymes
Nitrogen	N	7	Trace	3.3	Component of all proteins and nucleic acids
Boron	B	5	Trace	Trace	
Cobalt	Co	27	Trace	Trace	
Zinc	Zn	30	Trace	Trace	Key component of some enzymes
Selenium	Se	34	Trace	Trace	
Molybdenum	Mo	42	Trace	Trace	Key component of many enzymes
Tin	Sn	50	Trace	Trace	
Iodine	I	53	Trace	Trace	Component of thyroid hormone

Table 2.1 The Most Common Elements on Earth and Their Distribution in the Human Body

Distribution of the Elements

Of the 92 naturally occurring elements on earth, only 11 are found in organisms in more than trace amounts (0.01% or higher). These 11 elements have atomic numbers less than 21 and, thus, have low atomic masses. Table 2.1 lists the levels of various elements in the human body; their levels in other organisms are similar. Inspection of this table suggests that the distribution of elements in living systems is by no means accidental. The most common elements inside organisms are not the elements that are most abundant in the earth's crust. For example, silicon, aluminum, and iron constitute 39.2% of the earth's crust, but they exist in trace amounts in the human body. On the other hand, carbon atoms make up 18.5% of the human body but only 0.03% of the earth's crust.

Ninety-two elements occur naturally on earth; only eleven of them are found in significant amounts in living organisms. Four of them—oxygen, hydrogen, carbon, nitrogen—constitute 96.3% of the weight of your body.

Ionic Bonds Form Crystals

A group of atoms held together by energy in a stable association is called a **molecule.** When a molecule contains atoms of more than one element, it is called a **compound.** The atoms in a molecule are joined by **chemical bonds;** these bonds can result when atoms with opposite charges attract (ionic bonds), when two atoms share one or more pairs of electrons (covalent bonds), or when atoms interact in other ways. We will start by examining **ionic bonds,** which form when atoms with opposite electrical charges (ions) attract.

A Closer Look at Table Salt

Common table salt, sodium chloride (NaCl), is a lattice of ions in which the atoms are held together by ionic bonds (figure 2.9). Sodium has 11 electrons: 2 in the inner energy level, 8 in the next level, and 1 in the outer (valence) level. The valence electron is unpaired (free) and has a strong tendency to join with another electron. A stable configuration can be achieved if the valence electron is lost to another atom that also has an unpaired electron. The loss of this electron results in the formation of a positively charged sodium ion, Na^+.

The chlorine atom has 17 electrons: 2 in the inner energy level, 8 in the next level, and 7 in the outer level. Hence, one of the orbitals in the outer energy level has an unpaired electron. The addition of another electron to the outer level fills that level and causes a negatively charged chloride ion, Cl^-, to form.

When placed together, metallic sodium and gaseous chlorine react swiftly and explosively, as the sodium atoms donate electrons to chlorine, forming Na^+ and Cl^- ions. Because opposite charges attract, the Na^+ and Cl^- remain associated in an **ionic compound,** NaCl, which is electrically neutral. However, the electrical attractive force holding NaCl together is not directed specifically between particular Na^+ and Cl^- ions, and no discrete sodium chloride molecules form. Instead, the force exists between any one ion and all neighboring ions of the opposite charge, and the ions aggregate in a crystal matrix with a precise geometry. Such aggregations are what we know as salt crystals. If a salt such as NaCl is placed in water, the electrical attraction of the water molecules, for reasons we will point out later in this chapter, disrupts the forces holding the ions in their crystal matrix, causing the salt to dissolve into a roughly equal mixture of free Na^+ and Cl^- ions.

An ionic bond is an attraction between ions of opposite charge in an ionic compound. Such bonds are not formed between particular ions in the compound; rather, they exist between an ion and all of the oppositely charged ions in its immediate vicinity.

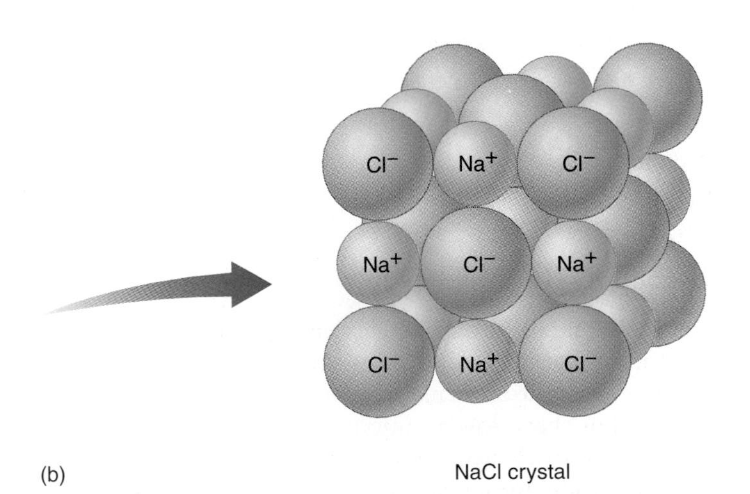

FIGURE 2.9
The formation of ionic bonds by sodium chloride. (*a*) When a sodium atom donates an electron to a chlorine atom, the sodium atom becomes a positively charged sodium ion, and the chlorine atom becomes a negatively charged chloride ion. (*b*) Sodium chloride forms a highly regular lattice of alternating sodium ions and chloride ions.

Covalent Bonds Build Stable Molecules

Covalent bonds form when two atoms share one or more pairs of valence electrons. Consider hydrogen (H) as an example. Each hydrogen atom has an unpaired electron and an unfilled outer energy level; for these reasons the hydrogen atom is unstable. When two hydrogen atoms are close to each other, however, each atom's electron can orbit both nuclei. In effect, the nuclei are able to share their electrons. The result is a diatomic (two-atom) molecule of hydrogen gas (figure 2.10).

The molecule formed by the two hydrogen atoms is stable for three reasons:

1. **It has no net charge.** The diatomic molecule formed as a result of this sharing of electrons is not charged, because it still contains two protons and two electrons.
2. **The octet rule is satisfied.** Each of the two hydrogen atoms can be considered to have two orbiting electrons in its outer energy level. This satisfies the octet rule, because each shared electron orbits both nuclei and is included in the outer energy level of *both* atoms.
3. **It has no free electrons.** The bonds between the two atoms also pair the two free electrons.

Unlike ionic bonds, covalent bonds are formed between two specific atoms, giving rise to true, discrete molecules. While ionic bonds can form regular crystals, the more specific associations made possible by covalent bonds allow the formation of complex molecular structures.

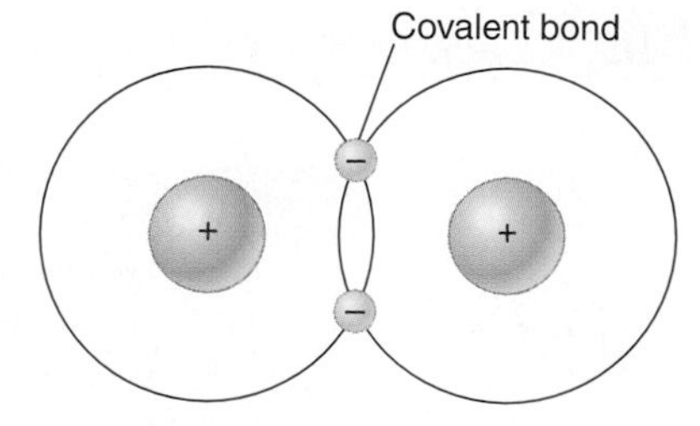

FIGURE 2.10
Hydrogen gas. (*a*) Hydrogen gas is a diatomic molecule composed of two hydrogen atoms, each sharing its electron with the other. (*b*) The flash of fire that consumed the *Hindenburg* occurred when the hydrogen gas that was used to inflate the dirigible combined explosively with oxygen gas in the air to form water.

Covalent Bonds Can Be Very Strong

The strength of a covalent bond depends on the number of shared electrons. Thus **double bonds,** which satisfy the octet rule by allowing two atoms to share *two* pairs of electrons, are stronger than **single bonds,** in which only one electron pair is shared. This means more chemical energy is required to break a double bond than a single bond. The strongest covalent bonds are **triple bonds,** such as those that link the two nitrogen atoms of nitrogen gas molecules. Covalent bonds are represented in chemical formulations as lines connecting atomic symbols, where each line between two bonded atoms represents the sharing of one pair of electrons. The **structural formulas** of hydrogen gas and oxygen gas are H—H and O=O, respectively, while their **molecular formulas** are H_2 and O_2.

Molecules with Several Covalent Bonds

Molecules often consist of more than two atoms. One reason that larger molecules may be formed is that a given atom is able to share electrons with more than one other atom. An atom that requires two, three, or four additional electrons to fill its outer energy level completely may acquire them by sharing its electrons with two or more other atoms.

For example, the carbon atom (C) contains six electrons, four of which are in its outer energy level. To satisfy the octet rule, a carbon atom must gain access to four additional electrons; that is, it must form four covalent bonds. Because four covalent bonds may form in many ways, carbon atoms are found in many different kinds of molecules.

Chemical Reactions

The formation and breaking of chemical bonds, the essence of chemistry, is called a **chemical reaction.** All chemical reactions involve the shifting of atoms from one molecule or ionic compound to another, without any change in the number or identity of the atoms. For convenience, we refer to the original molecules before the reaction starts as **reactants,** and the molecules resulting from the chemical reaction as **products.** For example:

$$\underbrace{A{-}B + C{-}D}_{\text{reactants}} \rightarrow \underbrace{A{-}C + B + D}_{\text{products}}$$

The extent to which chemical reactions occur is influenced by several important factors:

1. **Temperature.** Heating up the reactants increases the rate of a reaction (as long as the temperature isn't so high as to destroy the molecules).
2. **Concentration of reactants and products.** Reactions proceed more quickly when more reactants are available. An accumulation of products typically speeds reactions in the reverse direction.
3. **Catalysts.** A catalyst is a substance that increases the rate of a reaction. It doesn't alter the reaction's equilibrium between reactants and products, but it does shorten the time needed to reach equilibrium, often dramatically. In organisms, proteins called enzymes catalyze almost every chemical reaction.

A covalent bond is a stable chemical bond formed when two atoms share one or more pairs of electrons.

2.4 Water is the cradle of life.

(a)

(b)

(c)

FIGURE 2.11
Water takes many forms. As a liquid, water fills our rivers and runs down over the land to the sea. (*a*) The iceberg on which the penguins are holding their meeting was formed in Antarctica from a huge block of ice that broke away into the ocean water. (*b*) When water cools below 0°C, it forms beautiful crystals, familiar to us as snow and ice. However, water is not always plentiful. (*c*) At Badwater, in Death Valley, California, there is no hint of water except for the broken patterns of dried mud.

Chemistry of Water

Of all the molecules that are common on earth, only **water** exists as a liquid at the relatively low temperatures that prevail on the earth's surface, three-fourths of which is covered by liquid water (figure 2.11). When life was originating, water provided a medium in which other molecules could move around and interact without being held in place by strong covalent or ionic bonds. Life evolved as a result of these interactions, and it is still inextricably tied to water. Life began in water and evolved there for 3 billion years before spreading to land. About two-thirds of any organism's body is composed of water, and no organism can grow or reproduce in any but a water-rich environment. It is no accident that tropical rain forests are bursting with life, while dry deserts appear almost lifeless except when water becomes temporarily plentiful, such as after a rainstorm.

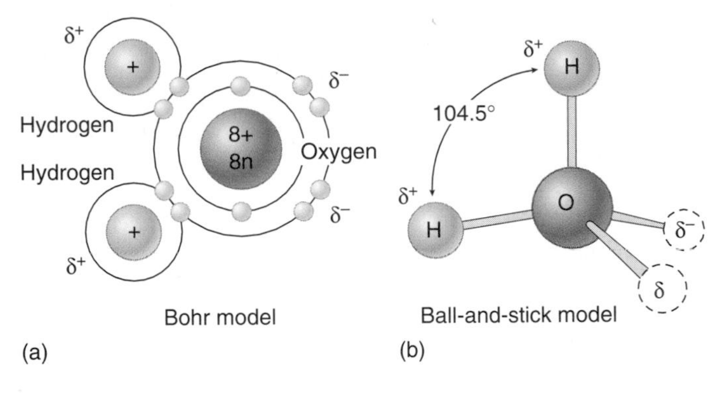

FIGURE 2.12
Water has a simple molecular structure. (*a*) Each molecule is composed of one oxygen atom and two hydrogen atoms. The oxygen atom shares one electron with each hydrogen atom. (*b*) The greater electronegativity of the oxygen atom makes the water molecule polar: water carries two partial negative charges (δ^-) near the oxygen atom and two partial positive charges (δ^+), one on each hydrogen atom.

The Atomic Structure of Water

Water has a simple atomic structure. It consists of an oxygen atom bound to two hydrogen atoms by two single covalent bonds (figure 2.12*a*). The resulting molecule is stable: it satisfies the octet rule, has no unpaired electrons, and carries no net electrical charge.

The single most outstanding chemical property of water is its ability to form weak chemical associations with only 5 to 10% of the strength of covalent bonds. This property, which derives directly from the structure of water, is responsible for much of the organization of living chemistry.

The chemistry of life is water chemistry. The way in which life first evolved was determined in large part by the chemical properties of the liquid water in which that evolution occurred.

Water Atoms Act Like Tiny Magnets

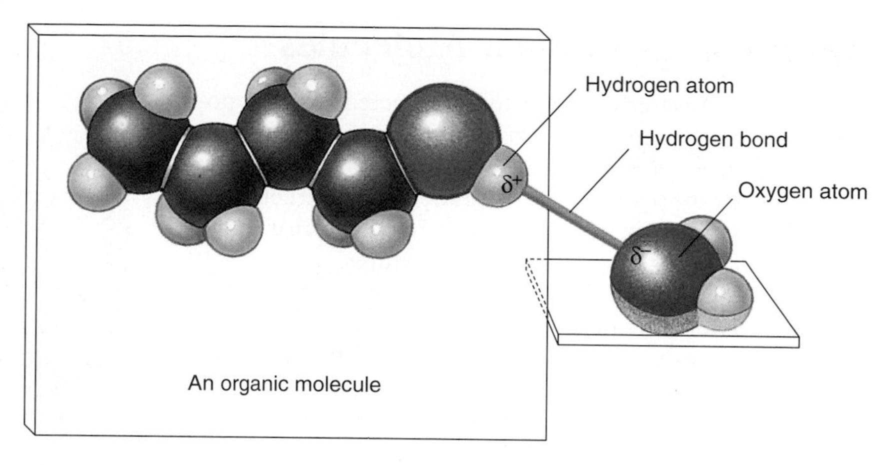

FIGURE 2.13
Structure of a hydrogen bond.

Both the oxygen and the hydrogen atoms attract the electrons they share in the covalent bonds of a water molecule; this attraction is called **electronegativity.** However, the oxygen atom is more electronegative than the hydrogen atoms, so it attracts the electrons more strongly than do the hydrogen atoms. As a result, the shared electrons in a water molecule are far more likely to be found near the oxygen nucleus than near the hydrogen nuclei. This stronger attraction for electrons gives the oxygen atom two partial negative charges (δ^-), as though the electron cloud were denser near the oxygen atom than around the hydrogen atoms. Because the water molecule as a whole is electrically neutral, each hydrogen atom carries a partial positive charge (δ^+). The Greek letter delta (δ) signifies a partial charge, much weaker than the full unit charge of an ion.

What would you expect the shape of a water molecule to be? Each of water's two covalent bonds has a partial charge at each end, δ^- at the oxygen end and δ^+ at the hydrogen end. The most stable arrangement of these charges is a *tetrahedron,* in which the two negative and two positive charges are approximately equidistant from one another (figure 2.12*b*). The oxygen atom lies at the center of the tetrahedron, the hydrogen atoms occupy two of the apexes, and the partial negative charges occupy the other two apexes. This results in a bond angle of 104.5° between the two covalent oxygen-hydrogen bonds. (In a regular tetrahedron, the bond angles would be 109.5°; in water, the partial negative charges occupy more space than the hydrogen atoms, and, therefore, they compress the oxygen-hydrogen bond angle slightly.)

The water molecule, thus, has distinct "ends," each with a partial charge, like the two poles of a magnet. (These partial charges are much less than the unit charges of ions, however.) Molecules that exhibit charge separation are called **polar molecules** because of their magnet-like poles, and water is one of the most polar molecules known. *The polarity of water underlies its chemistry and the chemistry of life.*

Polar molecules interact with one another, as the δ^- of one molecule is attracted to the δ^+ of another. Because many of these interactions involve hydrogen atoms, they are called **hydrogen bonds** (figure 2.13). Each hydrogen bond is individually very weak and transient, lasting on average only $\frac{1}{100{,}000{,}000{,}000}$ second (10^{-11} sec). However, the cumulative effects of large numbers of these bonds can be enormous. Water forms an abundance of hydrogen bonds, which are responsible for many of its important physical properties (table 2.2).

The water molecule is very polar, with ends that exhibit partial positive and negative charges. Opposite charges attract, forming weak linkages called hydrogen bonds.

Table 2.2 The Properties of Water

Property	Explanation	Example of Benefit to Life
Cohesion	Hydrogen bonds hold water molecules together	Leaves pull water upward from the roots; seeds swell and germinate
High specific heat	Hydrogen bonds absorb heat when they break, and release heat when they form, minimizing temperature changes	Water stabilizes the temperature of organisms and the environment
High heat of vaporization	Many hydrogen bonds must be broken for water to evaporate	Evaporation of water cools body surfaces
Lower density of ice	Water molecules in an ice crystal are spaced relatively far apart because of hydrogen bonding	Because ice is less dense than water, lakes do not freeze solid
High polarity	Polar water molecules are attracted to ions and polar compounds, making them soluble	Many kinds of molecules can move freely in cells, permitting a diverse array of chemical reactions

Water Clings to Polar Molecules

The polarity of water causes it to be attracted to other polar molecules. When the other molecules are also water, the attraction is referred to as **cohesion.** When the other molecules are of a different substance, the attraction is called **adhesion.** It is because water is cohesive that it is a liquid, and not a gas, at moderate temperatures.

The cohesion of liquid water is also responsible for its **surface tension.** Small insects can walk on water (figure 2.14) because at the air-water interface all of the hydrogen bonds in water face downward, causing the molecules of the water surface to cling together. Water is adhesive to any substance with which it can form hydrogen bonds. That is why substances containing polar molecules get "wet" when they are immersed in water, while those that are composed of nonpolar molecules (such as oils) do not.

The attraction of water to substances like glass with surface electrical charges is responsible for capillary action: if a glass tube with a narrow diameter is lowered into a beaker of water, water will rise in the tube above the level of the water in the beaker, because the adhesion of water to the glass surface, drawing it upward, is stronger than the force of gravity, drawing it down. The narrower the tube, the greater the electrostatic forces between the water and the glass, and the higher the water rises (figure 2.15).

FIGURE 2.14
Cohesion. Some insects, such as this water strider, literally walk on water. In this photograph you can see the dimpling the insect's feet make on the water as its weight bears down on the surface. Because the surface tension of the water is greater than the force that one foot brings to bear, the strider glides atop the surface of the water rather than sinking.

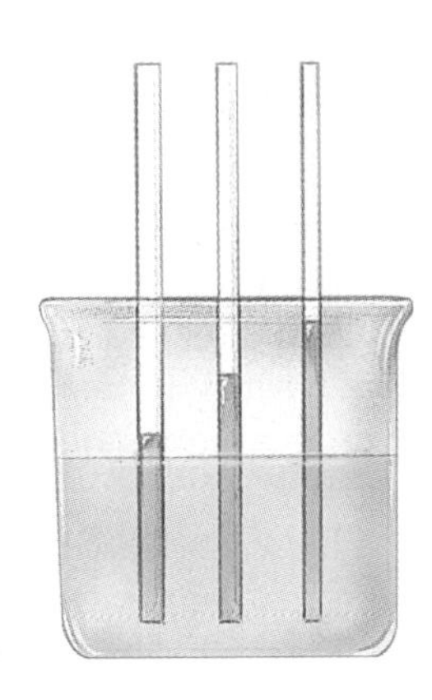

FIGURE 2.15
Capillary action. Capillary action causes the water within a narrow tube to rise above the surrounding water; the adhesion of the water to the glass surface, which draws water upward, is stronger than the force of gravity, which tends to draw it down. The narrower the tube, the greater the surface area available for adhesion for a given volume of water, and the higher the water rises in the tube.

Water Stores Heat

Water moderates temperature through two properties: its high specific heat and its high heat of vaporization. The temperature of any substance is a measure of how rapidly its individual molecules are moving. Because of the many hydrogen bonds that water molecules form with one another, a large input of thermal energy is required to break these bonds before the individual water molecules can begin moving about more freely and so have a higher temperature. Therefore, water is said to have a high **specific heat,** which is defined as the amount of heat that must be absorbed or lost by 1 gram of a substance to change its temperature by 1 degree Celsius (°C). Specific heat measures the extent to which a substance resists changing its temperature when it absorbs or loses heat. Because polar substances tend to form hydrogen bonds, and energy is needed to break these bonds, the more polar a substance is, the higher is its specific heat. The specific heat of water (1 calorie/gram/°C) is twice that of most carbon compounds and nine times that of iron. Only ammonia, which is more polar than water and forms very strong hydrogen bonds, has a higher specific heat than water (1.23 calories/gram/°C). Still, only 20% of the hydrogen bonds are broken as water heats from 0° to 100°C.

Because of its high specific heat, water heats up more slowly than almost any other compound and holds its temperature longer when heat is no longer applied. This characteristic enables organisms, which have a high water content, to maintain a relatively constant internal temperature. The heat generated by the chemical reactions inside cells would destroy the cells, if it were not for the high specific heat of the water within them.

A considerable amount of heat energy (586 calories) is required to change 1 gram of liquid water into a gas. Hence, water also has a high **heat of vaporization.** Because the transition of water from a liquid to a gas requires the input of energy to break its many hydrogen bonds, the evaporation of water from a surface causes cooling of that surface. Many organisms dispose of excess body heat by evaporative cooling; for example, humans and many other vertebrates sweat.

At low temperatures, water molecules are locked into a crystal-like lattice of hydrogen bonds, forming the solid we call ice (figure 2.16). Interestingly, ice is less dense than liquid water because the hydrogen bonds in ice space the water molecules relatively far apart. This unusual feature enables icebergs to float. Were it otherwise, ice would cover nearly all bodies of water, with only shallow surface melting annually.

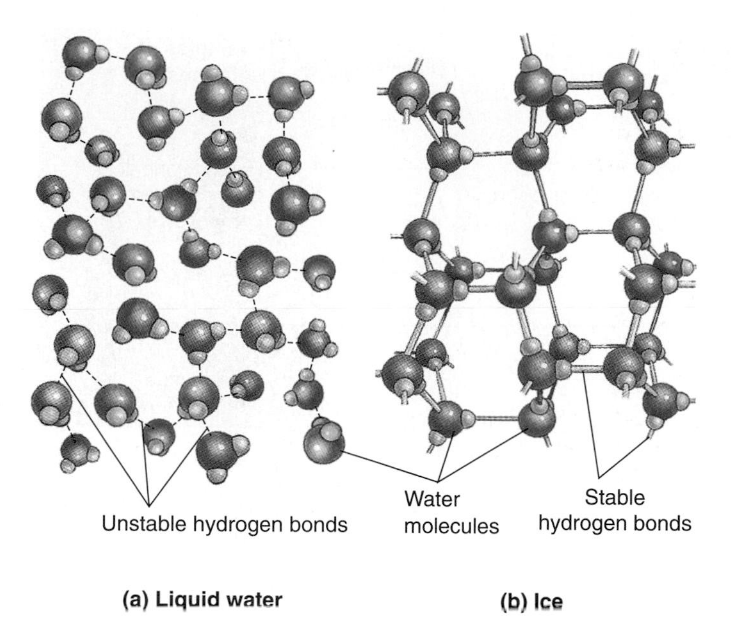

FIGURE 2.16
The role of hydrogen bonds in an ice crystal. (*a*) In liquid water, hydrogen bonds are not stable and constantly break and reform. (*b*) When water cools below 0°C, the hydrogen bonds are more stable, and a regular crystalline structure forms in which the four partial charges of one water molecule interact with the opposite charges of other water molecules. Because water forms a crystal latticework, ice is less dense than liquid water and floats. If it did not, inland bodies of water far from the earth's equator might never fully thaw.

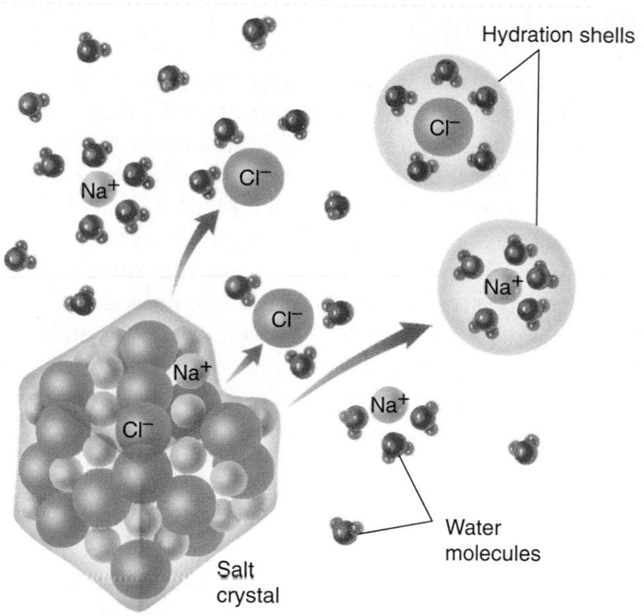

Hydration shells
Cl^-
Na^+
Salt crystal
Water molecules

FIGURE 2.17
Why salt dissolves in water. When a crystal of table salt dissolves in water, individual Na^+ and Cl^- ions break away from the salt lattice and become surrounded by water molecules. Water molecules orient around Cl^- ions so that their partial positive poles face toward the negative Cl^- ion; water molecules surrounding Na^+ ions orient in the opposite way, with their partial negative poles facing the positive Na^+ ion. Surrounded by hydration shells, Na^+ and Cl^- ions never reenter the salt lattice.

Water Is a Powerful Solvent

Water is an effective solvent because of its ability to form hydrogen bonds. Water molecules gather closely around any substance that bears an electrical charge, whether that substance carries a full charge (ion) or a charge separation (polar molecule). For example, sucrose (table sugar) is composed of molecules that contain slightly polar hydroxyl (OH) groups. A sugar crystal dissolves rapidly in water because water molecules can form hydrogen bonds with individual hydroxyl groups of the sucrose molecules. Therefore, sucrose is said to be *soluble* in water. Every time a sucrose molecule dissociates or breaks away from the crystal, water molecules surround it in a cloud, forming a **hydration shell** and preventing it from associating with other sucrose molecules. Hydration shells also form around ions such as Na^+ and Cl^- (figure 2.17).

Water Organizes Nonpolar Molecules

Water molecules always tend to form the maximum possible number of hydrogen bonds. When nonpolar molecules such as oils, which do not form hydrogen bonds, are placed in water, the water molecules act to exclude them. The nonpolar molecules are forced into association with one another, thus minimizing their disruption of the hydrogen bonding of water. In effect, they shrink from contact with water and for this reason they are referred to as **hydrophobic** (Greek *hydros*, "water" and *phobos*, "fearing"). In contrast, polar molecules, which readily form hydrogen bonds with water, are said to be **hydrophilic** ("water-loving").

The tendency of nonpolar molecules to aggregate in water is known as **hydrophobic exclusion.** By forcing the hydrophobic portions of molecules together, water causes these molecules to assume particular shapes. Different molecular shapes have evolved by alteration of the location and strength of nonpolar regions. As you will see, much of the evolution of life reflects changes in molecular shape that can be induced in just this way.

Water molecules, which are very polar, cling to one another, so that it takes considerable energy to separate them. Water also clings to other polar molecules, causing them to be soluble in water solution, but water tends to exclude nonpolar molecules.

Water Ionizes

The covalent bonds within a water molecule sometimes break spontaneously. In pure water at 25°C, only 1 out of every 550 million water molecules undergoes this process. When it happens, one of the protons (hydrogen atom nuclei) dissociates from the molecule. Because the dissociated proton lacks the negatively charged electron it was sharing in the covalent bond with oxygen, its own positive charge is no longer counterbalanced, and it becomes a positively charged ion, H^+. The rest of the dissociated water molecule, which has retained the shared electron from the covalent bond, is negatively charged and forms a **hydroxide ion** (OH^-). This process of spontaneous ion formation is called **ionization:**

$$\underset{\text{water}}{H_2O} \rightarrow \underset{\text{hydroxide ion}}{OH^-} + \underset{\text{hydrogen ion (proton)}}{H^+}$$

At 25°C, a liter of water contains $\frac{1}{10,000,000}$ (or 10^{-7}) mole of H^+ ions. (A **mole** is defined as the weight in grams that corresponds to the summed atomic masses of all of the atoms in a molecule. In the case of H^+, the atomic mass is 1, and a mole of H^+ ions would weigh 1 gram. One mole of any substance always contains 6.02×10^{23} molecules of the substance.) Therefore, the **molar concentration** of hydrogen ions (represented as $[H^+]$) in pure water is 10^{-7} mole/liter. Actually, the hydrogen ion usually associates with another water molecule to form a hydronium (H_3O^+) ion.

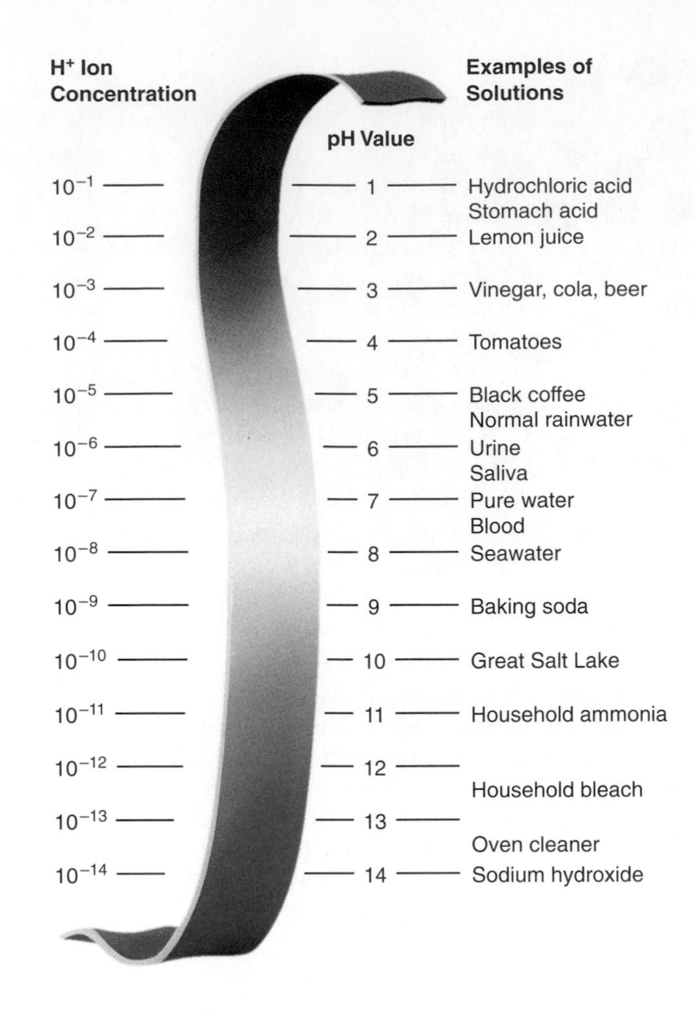

FIGURE 2.18
The pH scale. The pH value of a solution indicates its concentration of hydrogen ions. Solutions with a pH less than 7 are acidic, while those with a pH greater than 7 are basic. The scale is logarithmic, so that a pH change of 1 means a tenfold change in the concentration of hydrogen ions. Thus, lemon juice is 100 times more acidic than tomato juice, and seawater is 10 times more basic than pure water, which has a pH of 7.

pH

A more convenient way to express the hydrogen ion concentration of a solution is to use the **pH scale** (figure 2.18). This scale defines pH as the negative logarithm of the hydrogen ion concentration in the solution:

$$pH = -\log [H^+]$$

Because the logarithm of the hydrogen ion concentration is simply the exponent of the molar concentration of H^+, the pH equals the exponent times –1. Thus, pure water, with an $[H^+]$ of 10^{-7} mole/liter, has a pH of 7. Recall that for every H^+ ion formed when water dissociates, an OH^- ion is also formed, meaning that the dissociation of water produces H^+ and OH^- in equal amounts. Therefore, a pH value of 7 indicates neutrality—a balance between H^+ and OH^-—on the pH scale.

Note that the pH scale is *logarithmic,* which means that a difference of 1 on the scale represents a tenfold change in hydrogen ion concentration. This means that a solution with a pH of 4 has *10 times* the concentration of H^+ than is present in one with a pH of 5.

Acids. Any substance that dissociates in water to increase the concentration of H^+ ions is called an acid. Acidic solutions have pH values below 7. The stronger an acid is, the more H^+ ions it produces and the lower its pH. For example, hydrochloric acid (HCl), which is abundant in your stomach, ionizes completely in water. This means that 10^{-1} mole per liter of HCl will dissociate to form 10^{-1} mole per liter of H^+ ions, giving the solution a pH of 1. The pH of champagne, which bubbles because of the carbonic acid dissolved in it, is about 2.

Bases. A substance that combines with H^+ ions when dissolved in water is called a base. By combining with H^+ ions, a base lowers the H^+ ion concentration in the solution. Basic (or alkaline) solutions, therefore, have pH values above 7. Very strong bases, such as sodium hydroxide (NaOH), have pH values of 12 or more.

Buffers

The pH inside almost all living cells, and in the fluid surrounding cells in multicellular organisms, is fairly close to 7. Most of the biological catalysts (enzymes) in living systems are extremely sensitive to pH; often even a small change in pH will alter their shape, thereby disrupting their activities and rendering them useless. For this reason it is important that a cell maintain a constant pH level.

Yet the chemical reactions of life constantly produce acids and bases within cells. Furthermore, many animals eat substances that are acidic or basic; cola, for example, is a strong (although dilute) acidic solution. Despite such variations in the concentrations of H^+ and OH^-, the pH of an organism is kept at a relatively constant level by buffers (figure 2.19).

A **buffer** is a substance that acts as a reservoir for hydrogen ions, donating them to the solution when their concentration falls and taking them from the solution when their concentration rises. What sort of substance will act in this way? Within organisms, most buffers consist of pairs of substances, one an acid and the other a base. The key buffer in human blood is an acid-base pair consisting of carbonic acid (acid) and bicarbonate (base). These two substances interact in a pair of reversible reactions. First, carbon dioxide (CO_2) and H_2O join to form carbonic acid (H_2CO_3), which in a second reaction dissociates to yield bicarbonate ion (HCO_3^-) and H^+ (figure 2.20). If some acid or other substance adds H^+ ions to the blood, the HCO_3^- ions act as a base and remove the excess H^+ ions by forming H_2CO_3. Similarly, if a basic substance removes H^+ ions from the blood, H_2CO_3 dissociates, releasing more H^+ ions into the blood. The forward and reverse reactions that interconvert H_2CO_3 and HCO_3^- thus stabilize the blood's pH.

The reaction of carbon dioxide and water to form carbonic acid is important because it permits carbon, essential to life, to enter water from the air. As we will discuss in chapter 4, biologists believe that life first evolved in the early oceans. These oceans were rich in carbon because of the reaction of carbon dioxide with water.

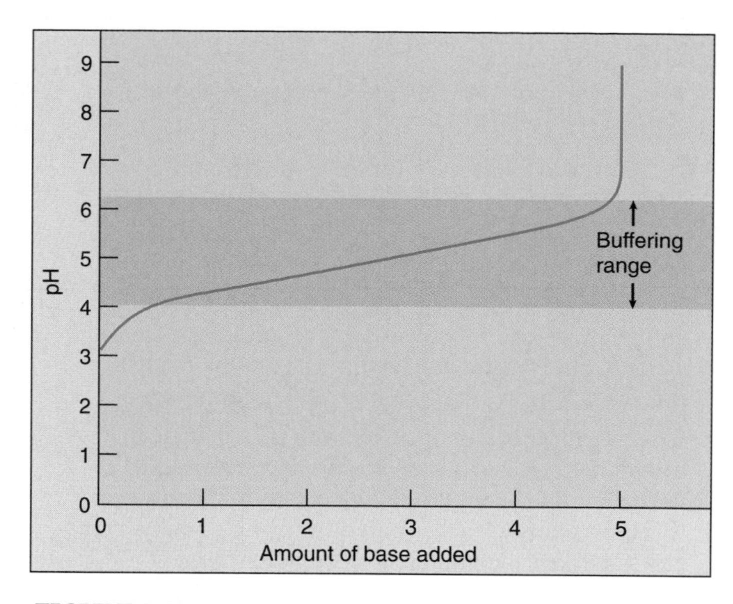

FIGURE 2.19
Buffers minimize changes in pH. Adding a base to a solution neutralizes some of the acid present, and so raises the pH. Thus, as the curve moves to the right, reflecting more and more base, it also rises to higher pH values. What a buffer does is to make the curve rise or fall very slowly over a portion of the pH scale, called the "buffering range" of that buffer.

In a condition called blood acidosis, human blood, which normally has a pH of about 7.4, drops 0.2 to 0.4 points on the pH scale. This condition is fatal if not treated immediately. The reverse condition, blood alkalosis, involves an increase in blood pH of a similar magnitude and is just as serious.

The pH of a solution is the negative logarithm of the H^+ ion concentration in the solution. Thus, low pH values indicate high H^+ concentrations (acidic solutions), and high pH values indicate low H^+ concentrations (basic solutions). Even small changes in pH can be harmful to life.

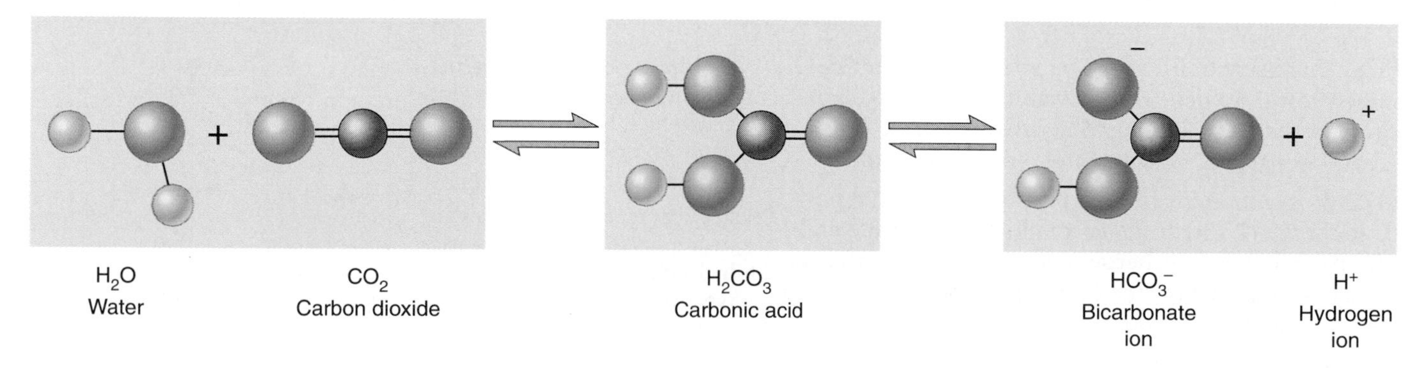

FIGURE 2.20
Buffer formation. Carbon dioxide and water combine chemically to form carbonic acid (H_2CO_3). The acid then dissociates in water, freeing H^+ ions. This reaction makes carbonated beverages acidic, and produced the carbon-rich early oceans that cradled life.

Chapter 2

www.mhhe.com/raven6/resources2.mhtml

Summary *Questions* *Media Resources*

2.1 Atoms are nature's building material.

- The smallest stable particles of matter are protons, neutrons, and electrons, which associate to form atoms.
- The core, or nucleus, of an atom consists of protons and neutrons; the electrons orbit around the nucleus in a cloud. The farther an electron is from the nucleus, the faster it moves and the more energy it possesses.
- The chemical behavior of an atom is largely determined by the distribution of its electrons and in particular by the number of electrons in its outermost (highest) energy level. There is a strong tendency for atoms to have a completely filled outer level; electrons are lost, gained, or shared until this condition is reached.

1. An atom of nitrogen has 7 protons and 7 neutrons. What is its atomic number? What is its atomic mass? How many electrons does it have?

2. How do the isotopes of a single element differ from each other?

3. The half-life of radium-226 is 1620 years. If a sample of material contains 16 milligrams of radium-226, how much will it contain in 1620 years? How much will it contain in 3240 years? How long will it take for the sample to contain 1 milligram of radium-226?

- Atomic Structure

- Basic Chemistry
- Atoms

- Art Quiz: Electron Energy Levels

2.2 The atoms of living things are among the smallest.

- More than 95% of the weight of an organism consists of oxygen, hydrogen, carbon, and nitrogen, all of which form strong covalent bonds with one another.

4. What is the octet rule, and how does it affect the chemical behavior of atoms?

- Art Quiz: Periodic Table

2.3 Chemical bonds hold molecules together.

- Ionic bonds form when electrons transfer from one atom to another, and the resulting oppositely charged ions attract one another.
- Covalent bonds form when two atoms share electrons. They are responsible for the formation of most biologically important molecules.

5. What is the difference between an ionic bond and a covalent bond? Give an example of each.

- Ionic Bonds
- Covalent Bonds

- Bonds
- Activity: Covalent Bonds Build Stable Molecules

2.4 Water is the cradle of life.

- The chemistry of life is the chemistry of water (H_2O). The central oxygen atom in water attracts the electrons it shares with the two hydrogen atoms. This charge separation makes water a polar molecule.
- A hydrogen bond is formed between the partial positive charge of a hydrogen atom in one molecule and the partial negative charge of another atom, either in another molecule or in a different portion of the same molecule.
- Water is cohesive and adhesive, has a great capacity for storing heat, is a good solvent for other polar molecules, and tends to exclude nonpolar molecules.
- The H^+ concentration in a solution is expressed by the pH scale, in which pH equals the negative logarithm of the H^+ concentration.

6. What types of atoms participate in the formation of hydrogen bonds? How do hydrogen bonds contribute to water's high specific heat?

7. What types of molecules are hydrophobic? What types are hydrophilic? Why do these two types of molecules behave differently in water?

8. What is the pH of a solution that has a hydrogen ion concentration of 10^{-3} mole/liter? Would such a solution be acidic or basic?

- Water
- pH Scale
- Activity: Water Clings to Polar Molecules

- Art Quizzes:
  - –Molecular Structure of Water
  - –Water as a Solvent
  - –pH Scale

3

The Chemical Building Blocks of Life

Concept Outline

3.1 Molecules are the building blocks of life.

The Chemistry of Carbon. Because individual carbon atoms can form multiple covalent bonds, organic molecules can be quite complex.

3.2 Proteins perform the chemistry of the cell.

The Many Functions of Proteins. Proteins can be catalysts, transporters, supporters, and regulators.
Amino Acids Are the Building Blocks of Proteins. Proteins are long chains of various combinations of amino acids.
A Protein's Function Depends on the Shape of the Molecule. A protein's shape is determined by its amino acid sequence.
How Proteins Fold Into Their Functional Shape. The distribution of nonpolar amino acids along a protein chain largely determines how the protein folds.
How Proteins Unfold. When conditions such as pH or temperature fluctuate, proteins may denature or unfold.

3.3 Nucleic acids store and transfer genetic information.

Information Molecules. Nucleic acids store information in cells. RNA is a single-chain polymer of nucleotides, while DNA possesses two chains twisted around each other.

3.4 Lipids make membranes and store energy.

Phospholipids Form Membranes. The spontaneous aggregation of phospholipids in water is responsible for the formation of biological membranes.
Fats and Other Kinds of Lipids. Organisms utilize a wide variety of water-insoluble molecules.
Fats as Food. Fats are very efficient energy storage molecules because of their high proportion of C—H bonds.

3.5 Carbohydrates store energy and provide building materials.

Simple Carbohydrates. Sugars are simple carbohydrates, often consisting of six-carbon rings.
Linking Sugars Together. Sugars can be linked together to form long polymers, or polysaccharides.
Structural Carbohydrates. Structural carbohydrates like cellulose are chains of sugars linked in a way that enzymes cannot easily attack.

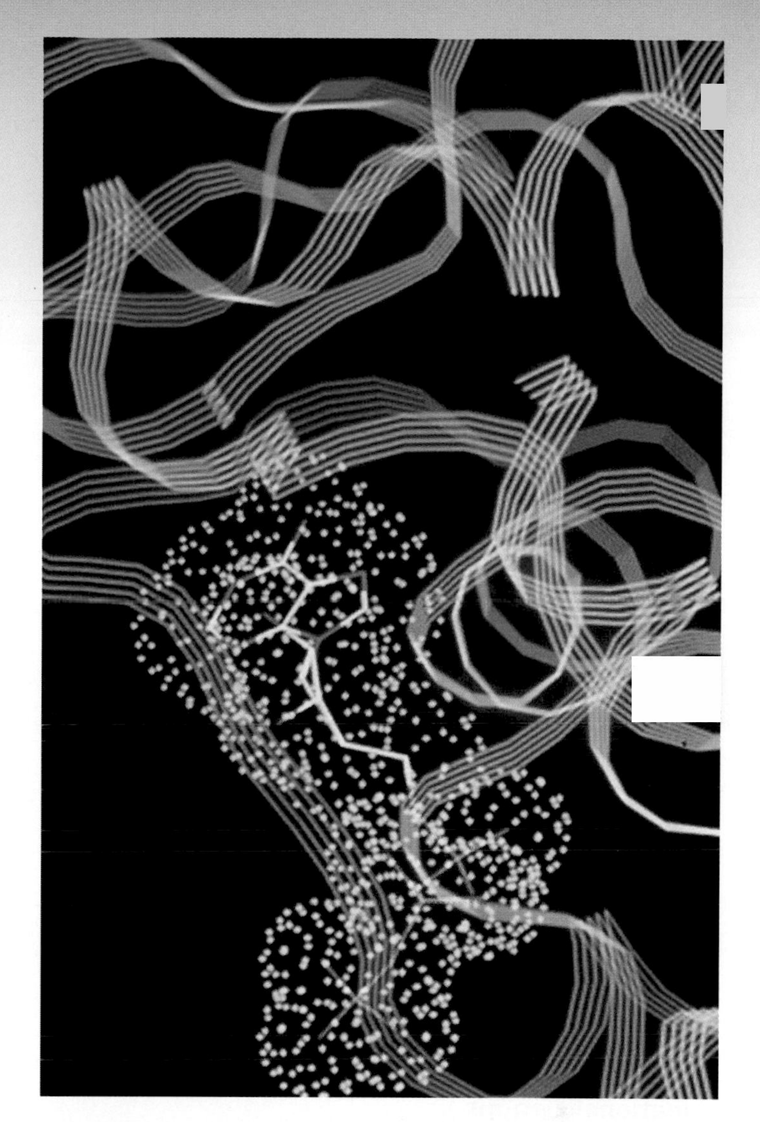

FIGURE 3.1
Computer-generated model of a macromolecule. Pictured is an enzyme responsible for releasing energy from sugar. This complex molecule consists of hundreds of different amino acids linked into chains that form the characteristic coils and folds seen here.

Molecules are extremely small compared with the familiar world we see about us. Imagine: there are more water molecules in a cup than there are stars in the sky. Many other molecules are gigantic, compared with water, consisting of thousands of atoms. These atoms are organized into hundreds of smaller molecules that are linked together into long chains (figure 3.1). These enormous molecules, almost always synthesized by living things, are called macromolecules. As we shall see, there are four general types of macromolecules, the basic chemical building blocks from which all organisms are assembled.

3.1 Molecules are the building blocks of life.

The Chemistry of Carbon

In chapter 2 we discussed how atoms combine to form molecules. In this chapter, we will focus on **organic molecules,** those chemical compounds that contain carbon. The frameworks of biological molecules consist predominantly of carbon atoms bonded to other carbon atoms or to atoms of oxygen, nitrogen, sulfur, or hydrogen. Because carbon atoms possess four valence electrons and so can form four covalent bonds, molecules containing carbon can form straight chains, branches, or even rings. As you can imagine, all of these possibilities generate an immense range of molecular structures and shapes.

Organic molecules consisting only of carbon and hydrogen are called **hydrocarbons.** Covalent bonds between carbon and hydrogen are energy-rich. We use hydrocarbons from fossil fuels as a primary source of energy today. Propane gas, for example, is a hydrocarbon consisting of a chain of three carbon atoms, with eight hydrogen atoms bound to it:

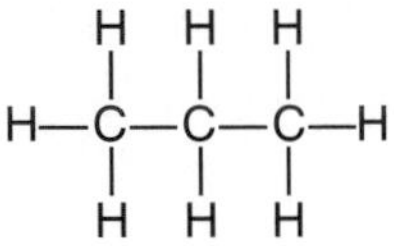

```
    H  H  H
    |  |  |
H—C—C—C—H
    |  |  |
    H  H  H
```

Because carbon-hydrogen covalent bonds store considerable energy, hydrocarbons make good fuels. Gasoline, for example, is rich in hydrocarbons.

Functional Groups

Carbon and hydrogen atoms both have very similar electronegativities, so electrons in C—C and C—H bonds are evenly distributed, and there are no significant differences in charge over the molecular surface. For this reason, hydrocarbons are nonpolar. Most organic molecules that are produced by cells, however, also contain other atoms. Because these other atoms often have different electronegativities, molecules containing them exhibit regions of positive or negative charge, and so are polar. These molecules can be thought of as a C—H core to which specific groups of atoms called **functional groups** are attached. For example, a hydrogen atom bonded to an oxygen atom (—OH) is a functional group called a *hydroxyl group.*

Functional groups have definite chemical properties that they retain no matter where they occur. The hydroxyl group, for example, is polar, because its oxygen atom, being very electronegative, draws electrons toward itself (as we saw in chapter 2). Figure 3.2 illustrates the hydroxyl group and other biologically important functional groups. Most chemical reactions that occur within organisms involve the transfer of a functional group as an intact unit from one molecule to another.

Biological Macromolecules

Some organic molecules in organisms are small and simple, containing only one or a few functional groups. Others are large complex assemblies called **macromolecules.** In many cases, these macromolecules are polymers, molecules built by linking together a large number of small, similar chemical subunits, like railroad cars coupled to form a train. For example, complex carbohydrates like starch are polymers of simple ring-shaped sugars, proteins are polymers of amino acids, and nucleic acids (DNA and RNA) are polymers of nucleotides. Biological macromolecules are traditionally grouped into four major categories: proteins, nucleic acids, lipids, and carbohydrates (table 3.1).

Group	Structural Formula	Ball-and-Stick Model	Found In:
Hydroxyl	—OH	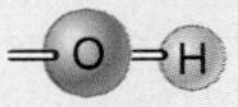 O H	Carbohydrates, alcohols
Carbonyl	C=O	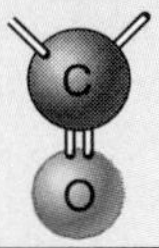 C O	Formaldehyde
Carboxyl	—C(=O)OH	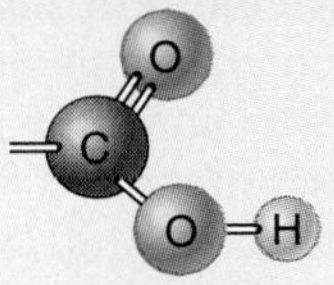 O C O H	Amino acids, vinegar
Amino	—N(H)H	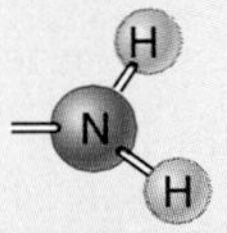 H N H	Ammonia
Sulfhydryl	—S—H	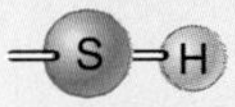 S H	Proteins, rubber
Phosphate	—O—P(O⁻)(=O)—O⁻	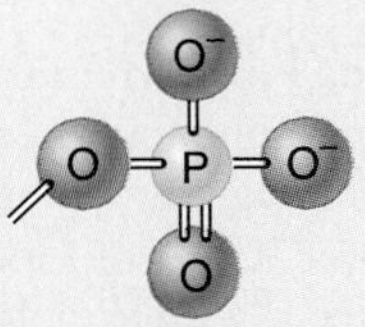 O⁻ O P O⁻ O	Phospholipids, nucleic acids, ATP
Methyl	—C(H)(H)—H	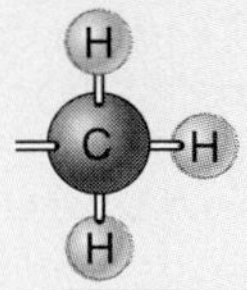 H C H H	Methane gas

FIGURE 3.2
The primary functional chemical groups. These groups tend to act as units during chemical reactions and confer specific chemical properties on the molecules that possess them. Amino groups, for example, make a molecule more basic, while carboxyl groups make a molecule more acidic.

Table 3.1 Macromolecules			
Macromolecule	**Subunit**	**Function**	**Example**
PROTEINS			
Globular	Amino acids	Catalysis; transport	Hemoglobin
Structural	Amino acids	Support	Hair; silk
NUCLEIC ACIDS			
DNA	Nucleotides	Encodes genes	Chromosomes
RNA	Nucleotides	Needed for gene expression	Messenger RNA
LIPIDS			
Fats	Glycerol and three fatty acids	Energy storage	Butter; corn oil; soap
Phospholipids	Glycerol, two fatty acids, phosphate, and polar R groups	Cell membranes	Lecithin
Prostaglandins	Five-carbon rings with two nonpolar tails	Chemical messengers	Prostaglandin E (PGE)
Steroids	Four fused carbon rings	Membranes; hormones	Cholesterol; estrogen
Terpenes	Long carbon chains	Pigments; structural	Carotene; rubber
CARBOHYDRATES			
Starch, glycogen	Glucose	Energy storage	Potatoes
Cellulose	Glucose	Cell walls	Paper; strings of celery
Chitin	Modified glucose	Structural support	Crab shells

Building Macromolecules

Although the four categories of macromolecules contain different kinds of subunits, they are all assembled in the same fundamental way: to form a covalent bond between two subunit molecules, an —OH group is removed from one subunit and a hydrogen atom (H) is removed from the other (figure 3.3*a*). This condensation reaction is called a **dehydration synthesis,** because the removal of the —OH group and H during the synthesis of a new molecule in effect constitutes the removal of a molecule of water (H_2O). For every subunit that is added to a macromolecule, one water molecule is removed. Energy is required to break the chemical bonds when water is extracted from the subunits, so cells must supply energy to assemble macromolecules. These and other biochemical reactions require that the reacting substances be held close together and that the correct chemical bonds be stressed and broken. This process of positioning and stressing, termed catalysis, is carried out in cells by a special class of proteins known as enzymes.

Cells disassemble macromolecules into their constituent subunits by performing reactions that are essentially the reverse of dehydration—a molecule of water is added instead of removed (figure 3.3*b*). In this process, which is called **hydrolysis** (Greek *hydro,* "water" + *lyse,* "break"), a hydrogen atom is attached to one subunit and a hydroxyl group to the other, breaking a specific covalent bond in the macromolecule. Hydrolytic reactions release the energy that was stored in the bonds that were broken.

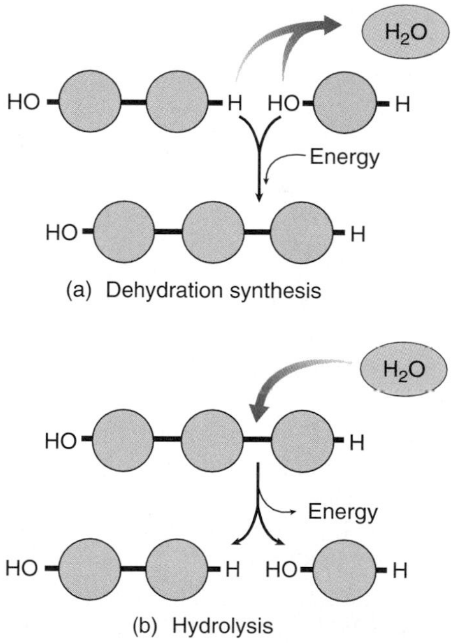

FIGURE 3.3
Making and breaking macromolecules. (*a*) Biological macromolecules are polymers formed by linking subunits together. The covalent bond between the subunits is formed by dehydration synthesis, an energy-requiring process that creates a water molecule for every bond formed. (*b*) Breaking the bond between subunits requires the returning of a water molecule with a subsequent release of energy, a process called hydrolysis.

Polymers are large molecules consisting of long chains of similar subunits joined by dehydration reactions. In a dehydration reaction, a hydroxyl (—OH) group is removed from one subunit and a hydrogen atom (H) is removed from the other.

3.2 Proteins perform the chemistry of the cell.

The Many Functions of Proteins

We will begin our discussion of macromolecules that make up the bodies of organisms with proteins (see table 3.1). The proteins within living organisms are immensely diverse in structure and function (table 3.2 and figure 3.4).

1. **Enzyme catalysis.** We have already encountered one class of proteins, enzymes, which are biological catalysts that facilitate specific chemical reactions. Because of this property, the appearance of enzymes was one of the most important events in the evolution of life. Enzymes are globular proteins, with a three-dimensional shape that fits snugly around the chemicals they work on, facilitating chemical reactions by stressing particular chemical bonds.
2. **Defense.** Other globular proteins use their shapes to "recognize" foreign microbes and cancer cells. These cell surface receptors form the core of the body's hormone and immune systems.
3. **Transport.** A variety of globular proteins transport specific small molecules and ions. The transport protein hemoglobin, for example, transports oxygen in the blood, and myoglobin, a similar protein, transports oxygen in muscle. Iron is transported in blood by the protein transferrin.

Table 3.2 The Many Functions of Proteins

Function	Class of Protein	Examples	Use
Metabolism (Catalysis)	Enzymes	Hydrolytic enzymes	Cleave polysaccharides
		Proteases	Break down proteins
		Polymerases	Produce nucleic acids
		Kinases	Phosphorylate sugars and proteins
Defense	Immunoglobulins	Antibodies	Mark foreign proteins for elimination
	Toxins	Snake venom	Block nerve function
Cell recognition	Cell surface antigens	MHC proteins	"Self" recognition
Transport throughout body	Globins	Hemoglobin	Carries O_2 and CO_2 in blood
		Myoglobin	Carries O_2 and CO_2 in muscle
		Cytochromes	Electron transport
Membrane transport	Transporters	Sodium-potassium pump	Excitable membranes
		Proton pump	Chemiosmosis
		Anion channels	Transport Cl^- ions
Structure/Support	Fibers	Collagen	Cartilage
		Keratin	Hair, nails
		Fibrin	Blood clot
Motion	Muscle	Actin	Contraction of muscle fibers
		Myosin	Contraction of muscle fibers
Osmotic regulation	Albumin	Serum albumin	Maintains osmotic concentration of blood
Regulation of gene action	Repressors	*lac* repressor	Regulates transcription
Regulation of body functions	Hormones	Insulin	Controls blood glucose levels
		Vasopressin	Increases water retention by kidneys
		Oxytocin	Regulates uterine contractions and milk production
Storage	Ion binding	Ferritin	Stores iron, especially in spleen
		Casein	Stores ions in milk
		Calmodulin	Binds calcium ions

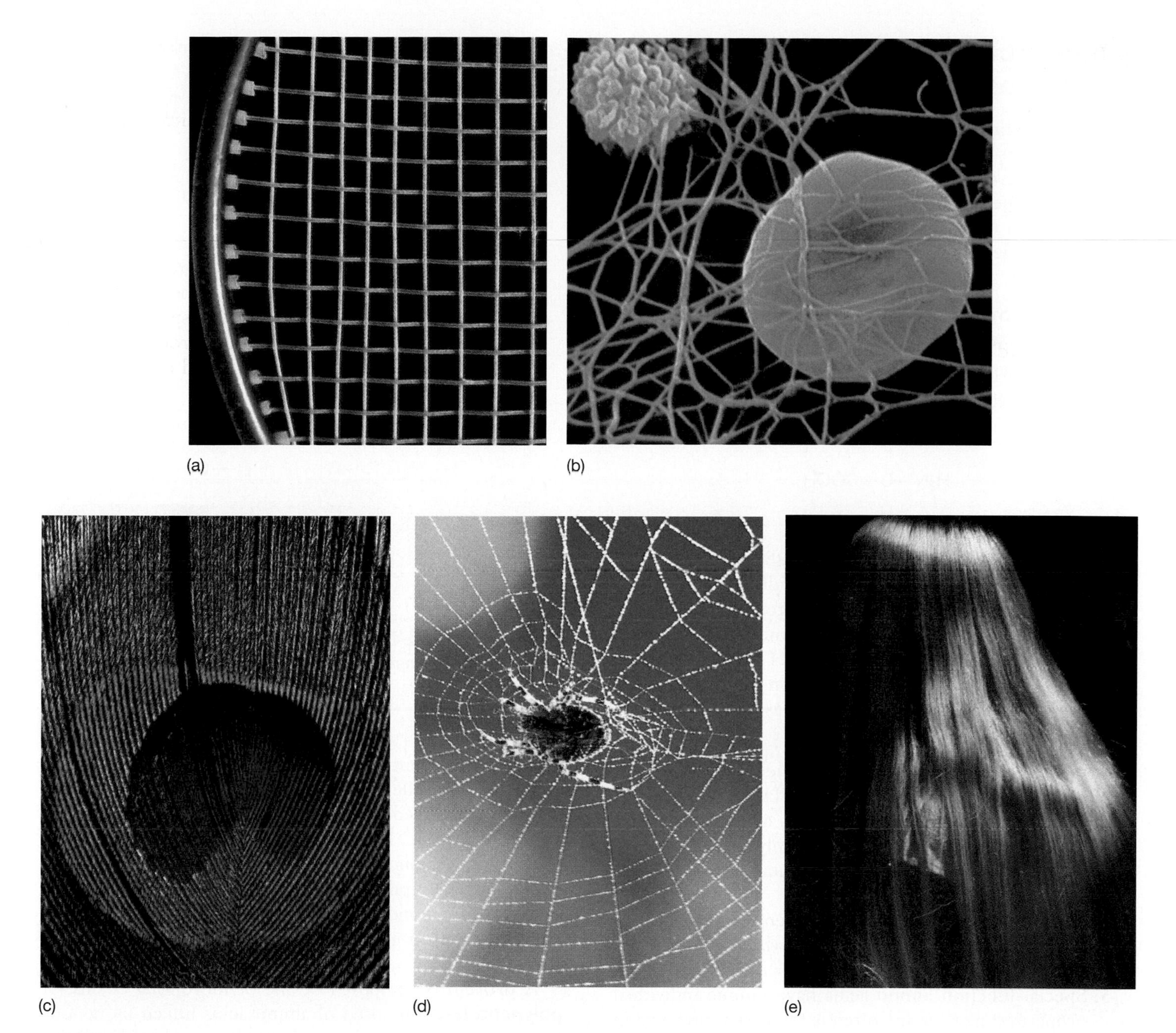

FIGURE 3.4
Some of the more common structural proteins. (*a*) Collagen: strings of a tennis racket from gut tissue; (*b*) fibrin: scanning electron micrograph of a blood clot (3000×); (*c*) keratin: a peacock feather; (*d*) silk: a spider's web; (*e*) keratin: human hair.

4. **Support.** Fibrous, or threadlike, proteins play structural roles; these structural proteins (see figure 3.4) include keratin in hair, fibrin in blood clots, and collagen, which forms the matrix of skin, ligaments, tendons, and bones and is the most abundant protein in a vertebrate body.
5. **Motion.** Muscles contract through the sliding motion of two kinds of protein filament: actin and myosin. **Contractile proteins** also play key roles in the cell's cytoskeleton and in moving materials within cells.
6. **Regulation.** Small proteins called hormones serve as **intercellular messengers** in animals. Proteins also play many regulatory roles within the cell, turning on and shutting off genes during development, for example. In addition, proteins also receive information, acting as cell surface receptors.

Proteins carry out a diverse array of functions, including catalysis, defense, transport of substances, support, motion, and regulation of cell and body functions.

Amino Acids Are the Building Blocks of Proteins

Although proteins are complex and versatile molecules, they are all polymers of only 20 amino acids, in a specific order. Many scientists believe amino acids were among the first molecules formed in the early earth. It seems highly likely that the oceans that existed early in the history of the earth contained a wide variety of amino acids.

Amino Acid Structure

An **amino acid** is a molecule containing an amino group (—NH_2), a carboxyl group (—COOH), and a hydrogen atom, all bonded to a central carbon atom:

$$H_2N\text{—}\underset{\displaystyle H}{\overset{\displaystyle R}{C}}\text{—}COOH$$

Each amino acid has unique chemical properties determined by the nature of the side group (indicated by R) covalently bonded to the central carbon atom. For example, when the side group is —CH_2OH, the amino acid (serine) is polar, but when the side group is —CH_3, the amino acid (alanine) is nonpolar. The 20 common amino acids are grouped into five chemical classes, based on their side groups:

1. Nonpolar amino acids, such as leucine, often have R groups that contain —CH_2 or —CH_3.
2. Polar uncharged amino acids, such as threonine, have R groups that contain oxygen (or only —H).
3. Ionizable amino acids, such as glutamic acid, have R groups that contain acids or bases.
4. Aromatic amino acids, such as phenylalanine, have R groups that contain an organic (carbon) ring with alternating single and double bonds.
5. Special-function amino acids have unique individual properties; methionine often is the first amino acid in a chain of amino acids, proline causes kinks in chains, and cysteine links chains together.

Each amino acid affects the shape of a protein differently depending on the chemical nature of its side group. Portions of a protein chain with numerous nonpolar amino acids, for example, tend to fold into the interior of the protein by hydrophobic exclusion.

Proteins Are Polymers of Amino Acids

In addition to its R group, each amino acid, when ionized, has a positive amino (NH_3^+) group at one end and a negative carboxyl (COO^-) group at the other end. The amino and carboxyl groups on a pair of amino acids can undergo a condensation reaction, losing a molecule of water and forming a covalent bond. A covalent bond that links two amino acids is called a **peptide bond** (figure 3.5). The two amino acids linked by such a bond are not free to rotate around the N—C linkage because the peptide bond has a partial double-bond character, unlike the N—C and C—C bonds to the central carbon of the amino acid. The stiffness of the peptide bond is one factor that makes it possible for chains of amino acids to form coils and other regular shapes.

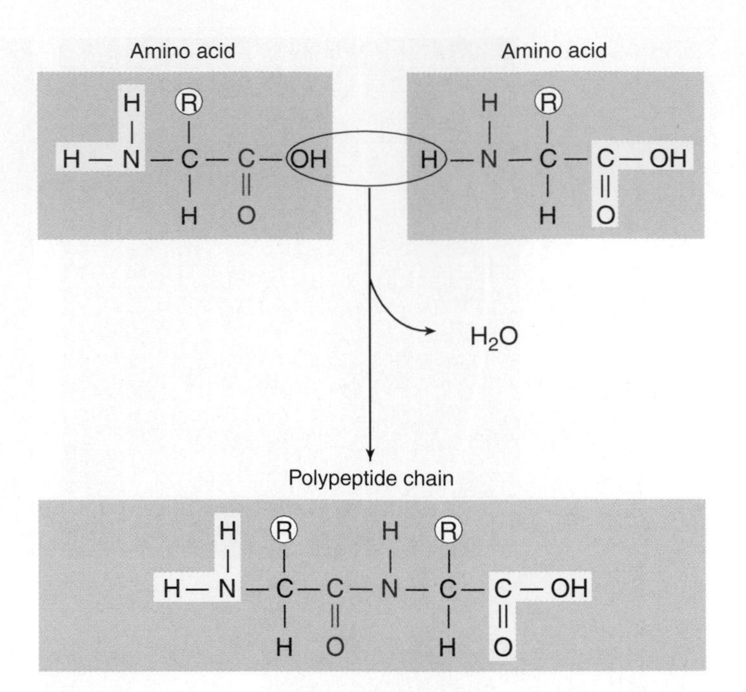

FIGURE 3.5
The peptide bond. A peptide bond forms when the —NH_2 end of one amino acid joins to the —COOH end of another. Because of the partial double-bond nature of peptide bonds, the resulting peptide chain cannot rotate freely around these bonds.

A **protein** is composed of one or more long chains, or **polypeptides,** composed of amino acids linked by peptide bonds. It was not until the pioneering work of Frederick Sanger in the early 1950s that it became clear that each kind of protein had a specific amino acid sequence. Sanger succeeded in determining the amino acid sequence of insulin and in so doing demonstrated clearly that this protein had a defined sequence, the same for all insulin molecules in the solution. Although many different amino acids occur in nature, only 20 commonly occur in proteins. Figure 3.6 illustrates these 20 "common" amino acids and their side groups.

A protein is a polymer containing a combination of up to 20 different kinds of amino acids. The amino acids fall into five chemical classes, each with different properties. These properties determine the nature of the resulting protein.

FIGURE 3.6
The 20 common amino acids. Each amino acid has the same chemical backbone, but differs in the side, or R, group it possesses. Six of the amino acids are nonpolar because they have —CH_2 or —CH_3 in their R groups. Two of the six are bulkier because they contain ring structures, which classifies them also as aromatic. Another six are polar because they have oxygen or just hydrogen in their R groups; these amino acids, which are uncharged, differ from one another in how polar they are. Five other amino acids are polar and, because they have a terminal acid or base in their R group, are capable of ionizing to a charged form. The remaining three have special chemical properties that allow them to help form links between protein chains or kinks in proteins.

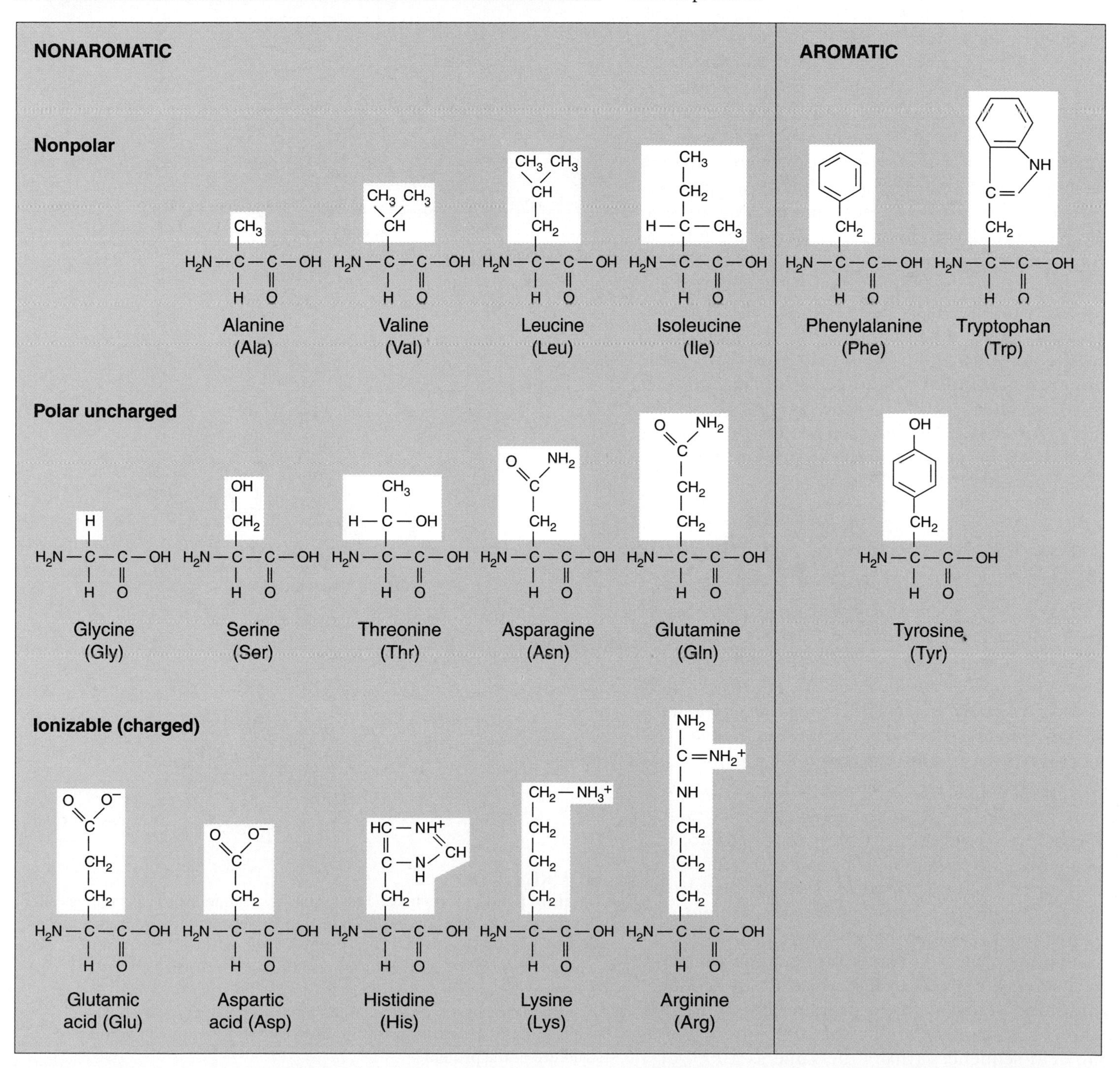

A Protein's Function Depends on the Shape of the Molecule

The shape of a protein is very important because it determines the protein's function. If we picture a polypeptide as a long strand similar to a reed, a protein might be the basket woven from it.

Overview of Protein Structure

Proteins consist of long amino acid chains folded into complex shapes. What do we know about the shape of these proteins? One way to study the shape of something as small as a protein is to look at it with very short wavelength energy—with X rays. X-ray diffraction is a painstaking procedure that allows the investigator to build up a three-dimensional picture of the position of each atom. The first protein to be analyzed in this way was myoglobin, soon followed by the related protein hemoglobin. As more and more proteins were added to the list, a general principle became evident: in every protein studied, essentially all the internal amino acids are nonpolar ones, amino acids such as leucine, valine, and phenylalanine. Water's tendency to hydrophobically exclude nonpolar molecules literally shoves the nonpolar portions of the amino acid chain into the protein's interior. This positions the nonpolar amino acids in close contact with one another, leaving little empty space inside. Polar and charged amino acids are restricted to the surface of the protein except for the few that play key functional roles.

Levels of Protein Structure

The structure of proteins is traditionally discussed in terms of four levels of structure, as *primary*, *secondary*, *tertiary*, and *quaternary* (figure 3.7). Because of progress in our knowledge of protein structure, two additional levels of structure are increasingly distinguished by molecular biologists: *motifs* and *domains.* Because these latter two elements play important roles in coming chapters, we introduce them here.

Primary Structure. The specific amino acid sequence of a protein is its primary structure. This sequence is determined by the nucleotide sequence of the gene that encodes the protein. Because the R groups that distinguish the various amino acids play no role in the peptide backbone of proteins, a protein can consist of any sequence of amino acids. Thus, a protein containing 100 amino acids could form any of 20^{100} different amino acid sequences (that's the same as 10^{130}, or 1

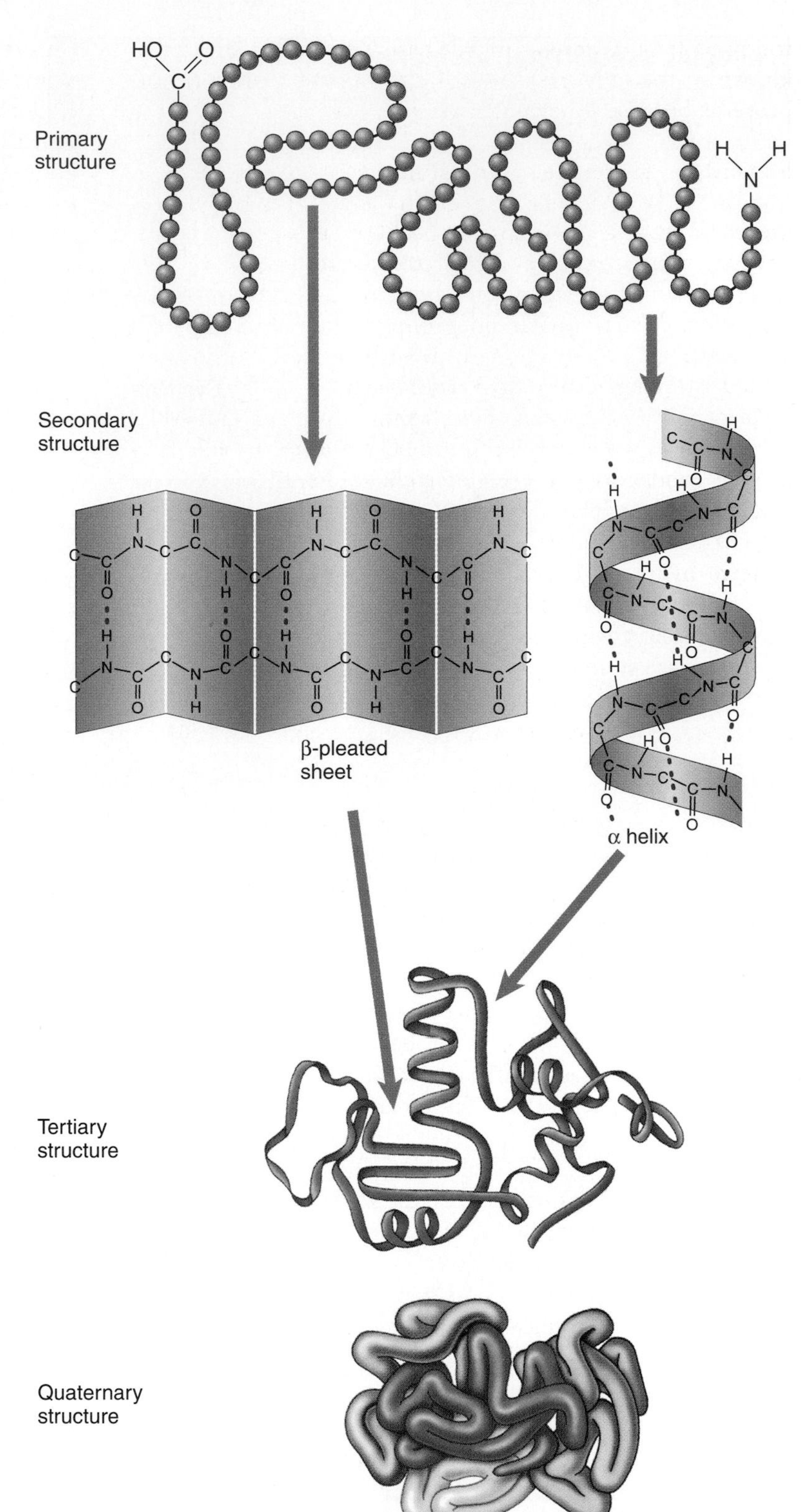

FIGURE 3.7
Levels of protein structure. The amino acid sequence of a protein is called its primary structure. Hydrogen bonds form between nearby amino acids, producing fold-backs called beta-pleated sheets and coils called alpha helices. These fold-backs and coils constitute the protein's secondary structure. A globular protein folds up on itself further to assume a three-dimensional tertiary structure. Many proteins aggregate with other polypeptide chains in clusters; this clustering is called the quaternary structure of the protein.

followed by 130 zeros—more than the number of atoms known in the universe). This is an important property of proteins because it permits such great diversity.

Secondary Structure. The amino acid side groups are not the only portions of proteins that form hydrogen bonds. The —COOH and —NH_2 groups of the main chain also form quite good hydrogen bonds—so good that their interactions with water might be expected to offset the tendency of nonpolar sidegroups to be forced into the protein interior. Inspection of the protein structures determined by X-ray diffraction reveals why they don't—the polar groups of the main chain form hydrogen bonds with each other! Two patterns of H bonding occur. In one, hydrogen bonds form along a single chain, linking one amino acid to another farther down the chain. This tends to pull the chain into a coil called an alpha (α) helix. In the other pattern, hydrogen bonds occur across two chains, linking the amino acids in one chain to those in the other. Often, many parallel chains are linked, forming a pleated, sheet-like structure called a β-pleated sheet. The folding of the amino acid chain by hydrogen bonding into these characteristic coils and pleats is called a protein's secondary structure.

Motifs. The elements of secondary structure can combine in proteins in characteristic ways called motifs, or sometimes "supersecondary structure." One very common motif is the $\beta\ \alpha\ \beta$ motif, which creates a fold or crease; the so-called "Rossmann fold" at the core of nucleotide binding sites in a wide variety of proteins is a $\beta\ \alpha\ \beta\ \alpha\ \beta$ motif. A second motif that occurs in many proteins is the β barrel, a β sheet folded around to form a tube. A third type of motif, the α turn α motif, is important because many proteins use it to bind the DNA double helix.

Tertiary Structure. The final folded shape of a globular protein, which positions the various motifs and folds nonpolar side groups into the interior, is called a protein's tertiary structure. A protein is driven into its tertiary structure by hydrophobic interactions with water. The final folding of a protein is determined by its primary structure—by the chemical nature of its side groups. Many proteins can be fully unfolded ("denatured") and will spontaneously refold back into their characteristic shape.

The stability of a protein, once it has folded into its 3-D shape, is strongly influenced by how well its interior fits together. When two nonpolar chains in the interior are in very close proximity, they experience a form of molecular attraction called van der Waal's forces. Individually quite weak, these forces can add up to a strong attraction when many of them come into play, like the combined strength of hundreds of hooks and loops on a strip of Velcro. They are effective forces only over short distances, however; there are no "holes" or cavities in the interior of proteins. That is why there are so many different nonpolar amino acids (alanine, valine, leucine, isoleucine). Each has a different-sized R group, allowing very precise fitting of nonpolar chains within the protein interior. Now you can understand why a mutation that converts one nonpolar amino acid within the protein interior (alanine) into another (leucine) very often disrupts the protein's stability; leucine is a lot bigger than alanine and disrupts the precise way the chains fit together within the protein interior. A change in even a single amino acid can have profound effects on protein shape and can result in loss or altered function of the protein.

Domains. Many proteins in your body are encoded within your genes in functional sections called exons (exons will be discussed in detail in chapter 15). Each exon-encoded section of a protein, typically 100 to 200 amino acids long, folds into a structurally independent functional unit called a **domain.** As the polypeptide chain folds, the domains fold into their proper shape, each more-or-less independent of the others. This can be demonstrated experimentally by artificially producing the fragment of polypeptide that forms the domain in the intact protein, and showing that the fragment folds to form the same structure as it does in the intact protein.

A single polypeptide chain connects the domains of a protein, like a rope tied into several adjacent knots. Often the domains of a protein have quite separate functions—one domain of an enzyme might bind a cofactor, for example, and another the enzyme's substrate.

Quaternary Structure. When two or more polypeptide chains associate to form a functional protein, the individual chains are referred to as subunits of the protein. The subunits need not be the same. Hemoglobin, for example, is a protein composed of two α-chain subunits and two β-chain subunits. A protein's subunit arrangement is called its quaternary structure. In proteins composed of subunits, the interfaces where the subunits contact one another are often nonpolar, and play a key role in transmitting information between the subunits about individual subunit activities.

A change in the identity of one of these amino acids can have profound effects. Sickle cell hemoglobin is a mutation that alters the identity of a single amino acid at the corner of the β subunit from polar glutamate to nonpolar valine. Putting a nonpolar amino acid on the surface creates a "sticky patch" that causes one hemoglobin molecule to stick to another, forming long nonfunctional chains and leading to the cell sickling characteristic of this hereditary disorder.

Protein structure can be viewed at six levels: 1. the amino acid sequence, or primary structure; 2. coils and sheets, called secondary structure; 3. folds or creases, called motifs; 4. the three-dimensional shape, called tertiary structure; 5. functional units, called domains; and 6. individual polypeptide subunits associated in a quaternary structure.

How Proteins Fold Into Their Functional Shape

How does a protein fold into a specific shape? Nonpolar amino acids play a key role. Until recently, investigators thought that newly made proteins fold spontaneously as hydrophobic interactions with water shove nonpolar amino acids into the protein interior. We now know this is too simple a view. Protein chains can fold in so many different ways that trial and error would simply take too long. In addition, as the open chain folds its way toward its final form, nonpolar "sticky" interior portions are exposed during intermediate stages. If these intermediate forms are placed in a test tube in the same protein environment that occurs in a cell, they stick to other unwanted protein partners, forming a gluey mess.

Chaperonins

How do cells avoid this? A vital clue came in studies of unusual mutations that prevented viruses from replicating in *E. coli* bacterial cells—it turned out the virus proteins could not fold properly! Further study revealed that normal cells contain special proteins called **chaperonins** that help new proteins fold correctly (figure 3.8). When the *E. coli* gene encoding its chaperone protein is disabled by mutation, the bacteria die, clogged with lumps of incorrectly folded proteins. Fully 30% of the bacteria's proteins fail to fold to the right shape.

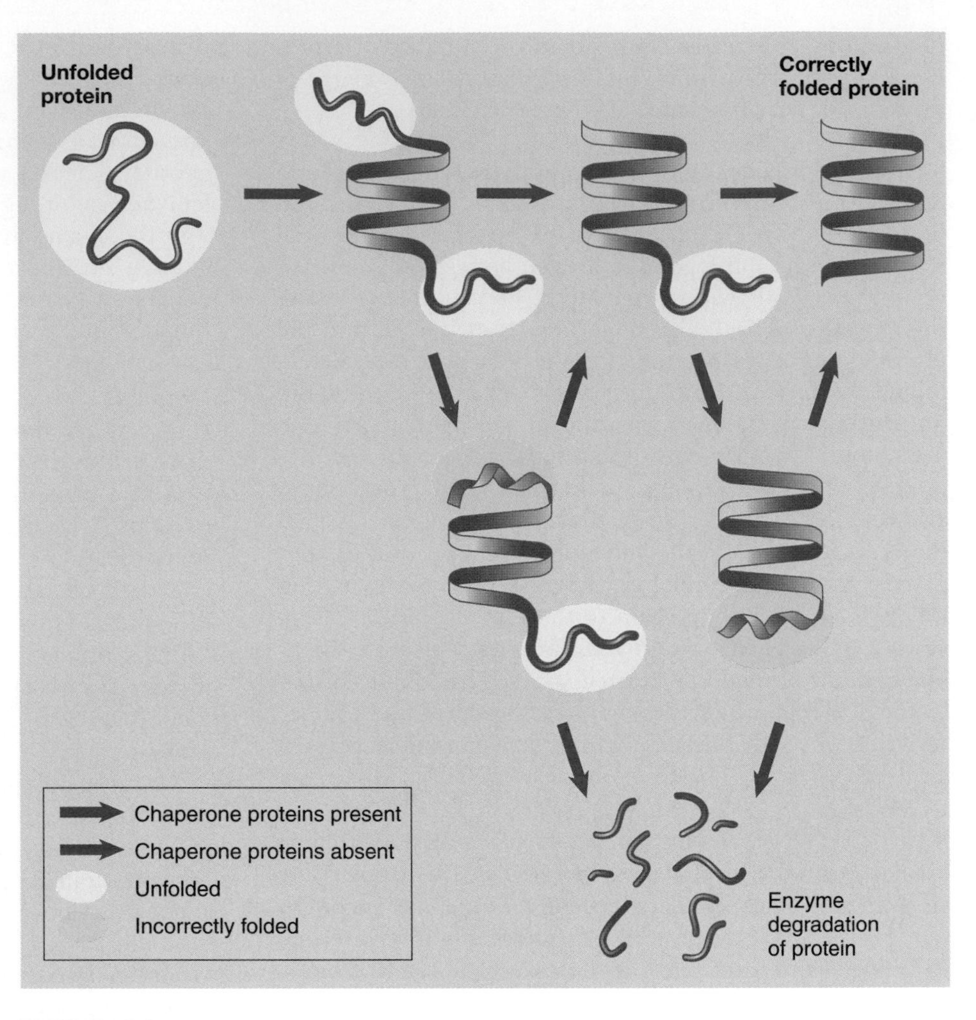

FIGURE 3.8
A current model of protein folding. A newly synthesized protein rapidly folds into characteristic motifs composed of α helices and β sheets, but these elements of structure are only roughly aligned in an open conformation. Subsequent folding occurs more slowly, by trial and error. This process is aided by chaperone proteins, which appear to recognize improperly folded proteins and unfold them, giving them another chance to fold properly. Eventually, if proper folding is not achieved, the misfolded protein is degraded by proteolytic enzymes.

Molecular biologists have now identified more than 17 kinds of proteins that act as molecular chaperones. Many are heat shock proteins, produced in greatly elevated amounts if a cell is exposed to elevated temperature; high temperatures cause proteins to unfold, and heat shock chaperonins help the cell's proteins refold.

There is considerable controversy about how chaperonins work. It was first thought that they provided a protected environment within which folding could take place unhindered by other proteins, but it now seems more likely that chaperonins rescue proteins that are caught in a wrongly folded state, giving them another chance to fold correctly. When investigators "fed" a deliberately misfolded protein called malate dehydrogenase to chaperonins, the protein was rescued, refolding to its active shape.

Protein Folding and Disease

There are tantalizing suggestions that chaperone protein deficiencies may play a role in certain diseases by failing to facilitate the intricate folding of key proteins. Cystic fibrosis is a hereditary disorder in which a mutation disables a protein that plays a vital part in moving ions across cell membranes. In at least some cases, the vital membrane protein appears to have the correct amino acid sequence, but fails to fold to its final form. It has also been speculated that chaperone deficiency may be a cause of the protein clumping in brain cells that produces the amyloid plaques characteristic of Alzheimer's disease.

Proteins called chaperones aid newly produced proteins to fold properly.

How Proteins Unfold

If a protein's environment is altered, the protein may change its shape or even unfold. This process is called **denaturation.** Proteins can be denatured when the pH, temperature, or ionic concentration of the surrounding solution is changed. When proteins are denatured, they are usually rendered biologically inactive. This is particularly significant in the case of enzymes. Because practically every chemical reaction in a living organism is catalyzed by a specific enzyme, it is vital that a cell's enzymes remain functional. That is the rationale behind traditional methods of salt-curing and pickling: prior to the ready availability of refrigerators and freezers, the only practical way to keep microorganisms from growing in food was to keep the food in a solution containing high salt or vinegar concentrations, which denatured the enzymes of microorganisms and kept them from growing on the food.

Most enzymes function within a very narrow range of physical parameters. Blood-borne enzymes that course through a human body at a pH of about 7.4 would rapidly become denatured in the highly acidic environment of the stomach. On the other hand, the protein-degrading enzymes that function at a pH of 2 or less in the stomach would be denatured in the basic pH of the blood. Similarly, organisms that live near oceanic hydrothermal vents have enzymes that work well at the temperature of this extreme environment (over 100°C). They cannot survive in cooler waters, because their enzymes would denature at lower temperatures. Any given organism usually has a "tolerance range" of pH, temperature, and salt concentration. Within that range, its enzymes maintain the proper shape to carry out their biological functions.

When a protein's normal environment is reestablished after denaturation, a small protein may spontaneously refold into its natural shape, driven by the interactions between its nonpolar amino acids and water (figure 3.9). Larger proteins can rarely refold spontaneously because of the complex nature of their final shape. It is important to distinguish denaturation from **dissociation.** The four subunits of hemoglobin (figure 3.10) may dissociate into four individual molecules (two α-globin and two β-globin) without denaturation of the folded globin proteins, and will readily reassume their four-subunit quaternary structure.

Every globular protein has a narrow range of conditions in which it folds properly; outside that range, proteins tend to unfold.

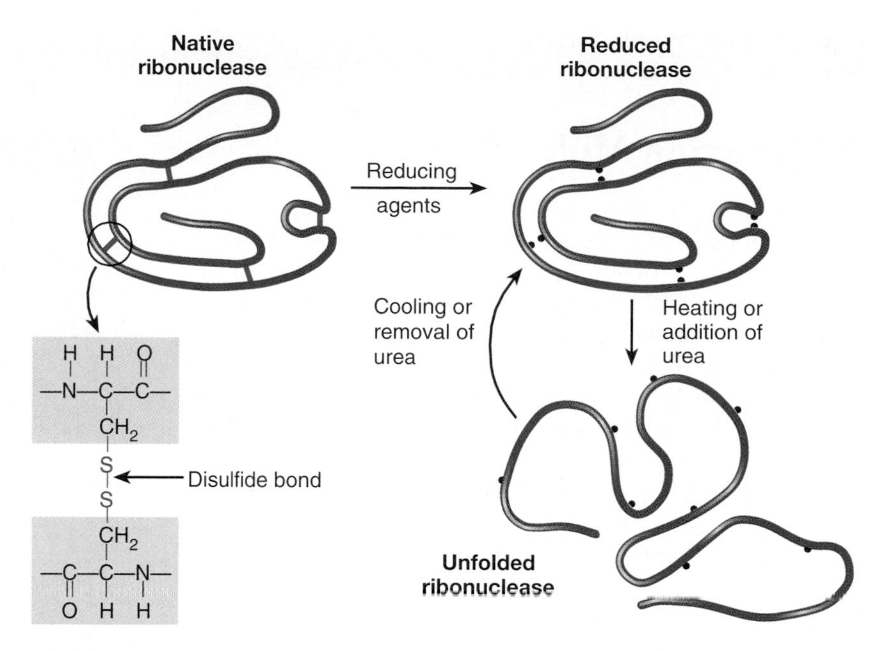

FIGURE 3.9
Primary structure determines tertiary structure. When the protein ribonuclease is treated with reducing agents to break the covalent disulfide bonds that cross-link its chains, and then placed in urea or heated, the protein denatures (unfolds) and loses its enzymatic activity. Upon cooling or removal of urea, it refolds and regains its enzymatic activity. This demonstrates that no information but the amino acid sequence of the protein is required for proper folding: the primary structure of the protein determines its tertiary structure.

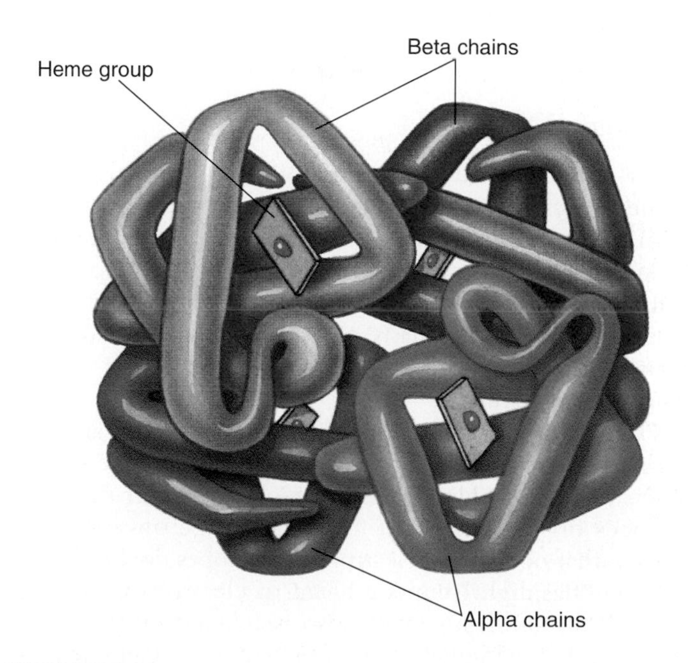

FIGURE 3.10
The four subunits of hemoglobin. The hemoglobin molecule is made of four globin protein subunits, informally referred to as polypeptide chains. The two lower α chains, identical α-globin proteins, are shaded *pink;* the two upper β chains, identical β-globin proteins, are shaded *blue.*

3.3 Nucleic acids store and transfer genetic information.

Information Molecules

The biochemical activity of a cell depends on production of a large number of proteins, each with a specific sequence. The ability to produce the correct proteins is passed between generations of organisms, even though the protein molecules themselves are not.

Nucleic acids are the information storage devices of cells, just as disks or tapes store the information that computers use, blueprints store the information that builders use, and road maps store the information that tourists use. There are two varieties of nucleic acids: *deoxyribonucleic acid* (**DNA;** figure 3.11) and *ribonucleic acid* (**RNA**). The way in which DNA encodes the information used to assemble proteins is similar to the way in which the letters on a page encode information (see chapter 14). Unique among macromolecules, nucleic acids are able to serve as templates to produce precise copies of themselves, so that the information that specifies what an organism is can be copied and passed down to its descendants. For this reason, DNA is often referred to as the hereditary material. Cells use the alternative form of nucleic acid, RNA, to read the cell's DNA-encoded information and direct the synthesis of proteins. RNA is similar to DNA in structure and is made as a transcripted copy of portions of the DNA. This transcript passes out into the rest of the cell, where it serves as a blueprint specifying a protein's amino acid sequence. This process will be described in detail in chapter 15.

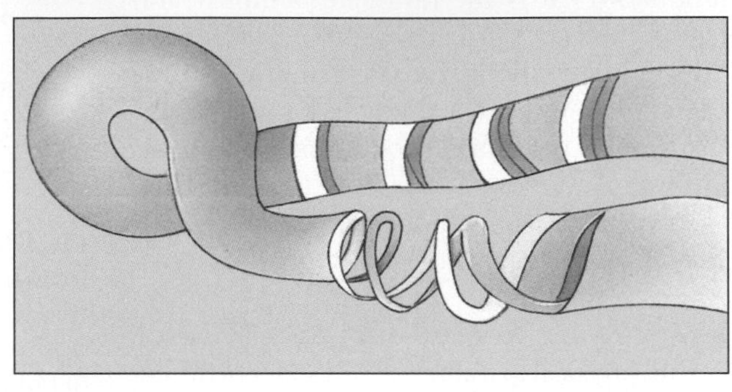

FIGURE 3.11
The first photograph of a DNA molecule. This micrograph, with sketch below, shows a section of DNA magnified a million times! The molecule is so slender that it would take 50,000 of them to equal the diameter of a human hair.

"Seeing" DNA

DNA molecules cannot be seen with an optical microscope, which is incapable of resolving anything smaller than 1000 atoms across. An electron microscope can image structures as small as a few dozen atoms across, but still cannot resolve the individual atoms of a DNA strand. This limitation was finally overcome in the last decade with the introduction of the scanning-tunneling microscope (figure 3.12).

How do these microscopes work? Imagine you are in a dark room with a chair. To determine the shape of the chair, you could shine a flashlight on it, so that the light bounces off the chair and forms an image on your eye. That's what optical and electron microscopes do; in the latter, the "flashlight" emits a beam of electrons instead of light. You could, however, also reach out and feel the chair's surface with your hand. In effect, you would be putting a probe (your hand) near the chair and measuring how far away the surface is. In a scanning-tunneling microscope, computers advance a probe over the surface of a molecule in steps smaller than the diameter of an atom.

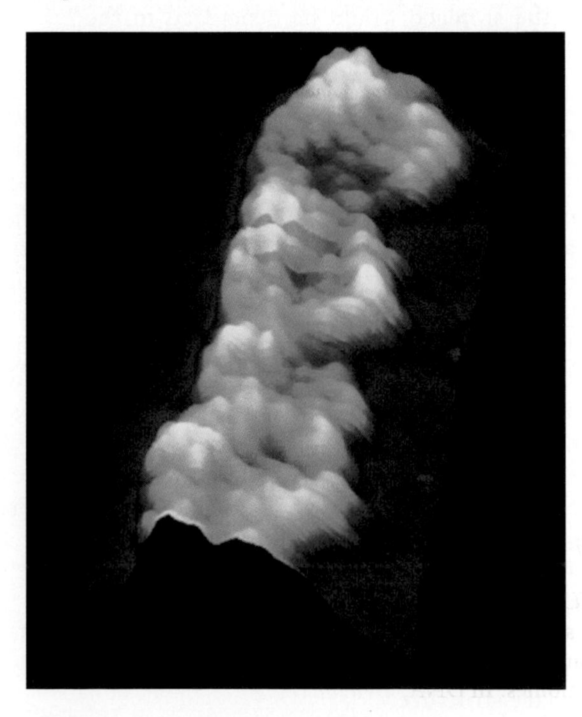

FIGURE 3.12
A scanning tunneling micrograph of DNA (false color; 2,000,000×). The micrograph shows approximately three turns of the DNA double helix (see figure 3.15).

The Structure of Nucleic Acids

Nucleic acids are long polymers of repeating subunits called **nucleotides.** Each nucleotide consists of three components: a five-carbon sugar (ribose in RNA and deoxyribose in DNA); a phosphate (—PO_4) group; and an organic nitrogen-containing base (figure 3.13). When a nucleic acid polymer forms, the phosphate group of one nucleotide binds to the hydroxyl group of another, releasing water and forming a phosphodiester bond. A **nucleic acid,** then, is simply a chain of five-carbon sugars linked together by phosphodiester bonds with an organic base protruding from each sugar (figure 3.14).

Two types of organic bases occur in nucleotides. The first type, *purines*, are large, double-ring molecules found in both DNA and RNA; they are adenine (A) and guanine (G). The second type, *pyrimidines*, are smaller, single-ring molecules; they include cytosine (C, in both DNA and RNA), thymine (T, in DNA only), and uracil (U, in RNA only).

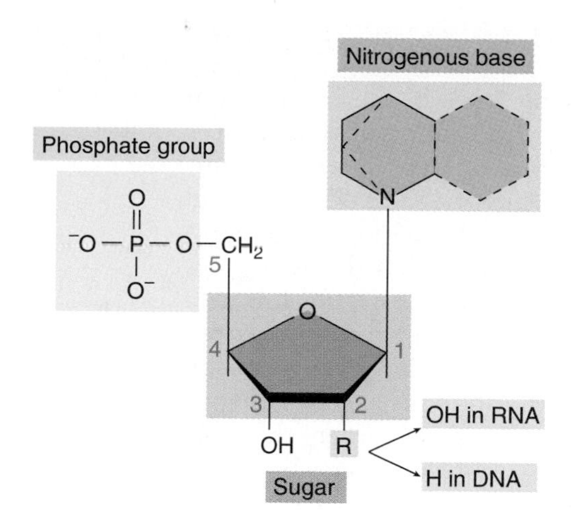

FIGURE 3.13
Structure of a nucleotide. The nucleotide subunits of DNA and RNA are made up of three elements: a five-carbon sugar, an organic nitrogenous base, and a phosphate group.

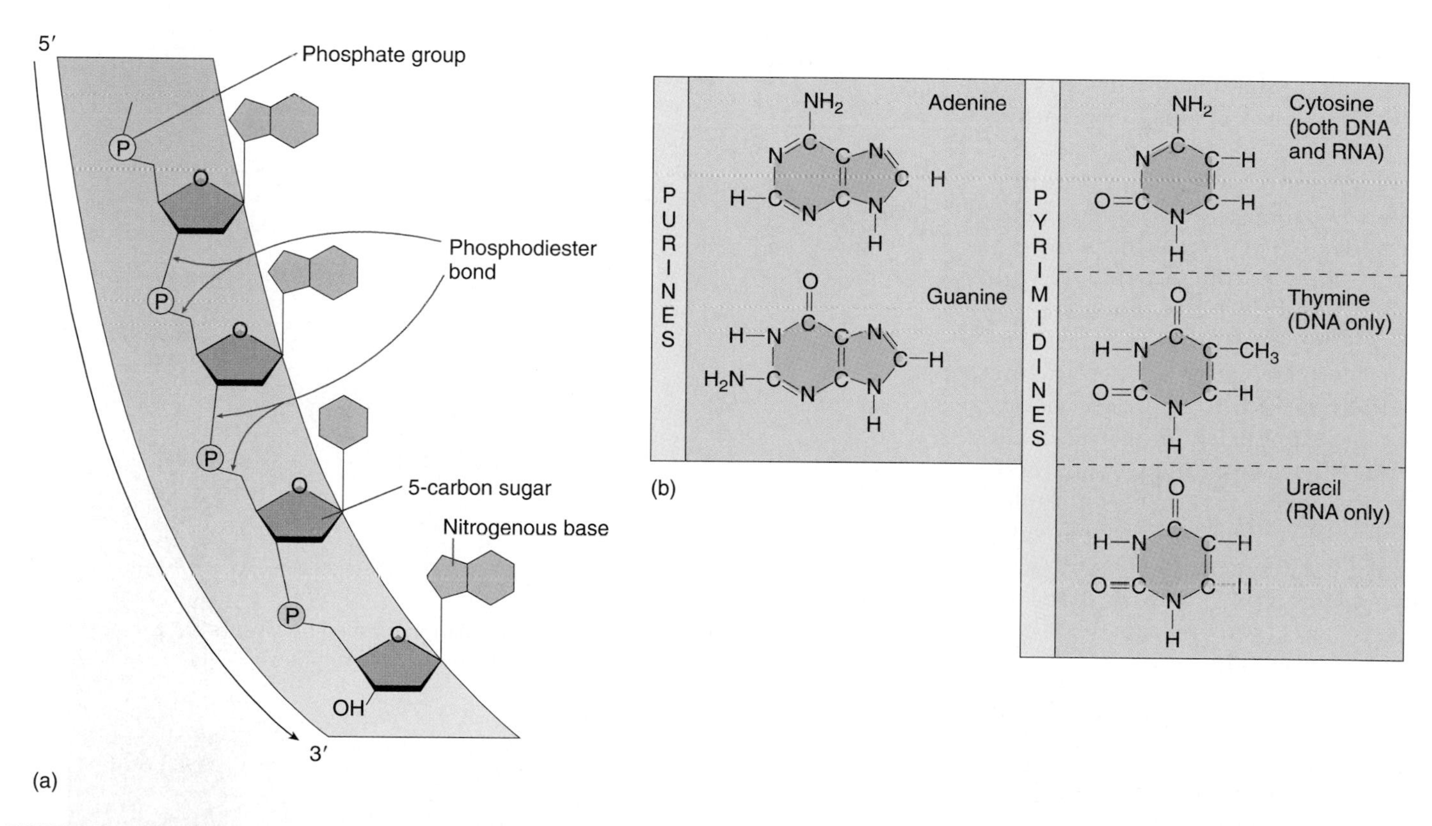

FIGURE 3.14
The structure of a nucleic acid and the organic nitrogen-containing bases. (*a*) In a nucleic acid, nucleotides are linked to one another via phosphodiester bonds, with organic bases protruding from the chain. (*b*) The organic nitrogenous bases can be either purines or pyrimidines. In DNA, thymine replaces the uracil found in RNA.

DNA

Organisms encode the information specifying the amino acid sequences of their proteins as sequences of nucleotides in the DNA. This method of encoding information is very similar to that by which the sequences of letters encode information in a sentence. While a sentence written in English consists of a combination of the 26 different letters of the alphabet in a specific order, the code of a DNA molecule consists of different combinations of the four types of nucleotides in specific sequences such as CGCTTACG. The information encoded in DNA is used in the everyday metabolism of the organism and is passed on to the organism's descendants.

DNA molecules in organisms exist not as single chains folded into complex shapes, like proteins, but rather as double chains. Two DNA polymers wind around each other like the outside and inside rails of a circular staircase. Such a winding shape is called a helix, and a helix composed of two chains winding about one another, as in DNA, is called a **double helix.** Each step of DNA's helical staircase is a base-pair, consisting of a base in one chain attracted by hydrogen bonds to a base opposite it on the other chain. These hydrogen bonds hold the two chains together as a duplex (figure 3.15). The base-pairing rules are rigid: adenine can pair only with thymine (in DNA) or with uracil (in RNA), and cytosine can pair only with guanine. The bases that participate in base-pairing are said to be **complementary** to each other. Additional details of the structure of DNA and how it interacts with RNA in the production of proteins are presented in chapters 14 and 15.

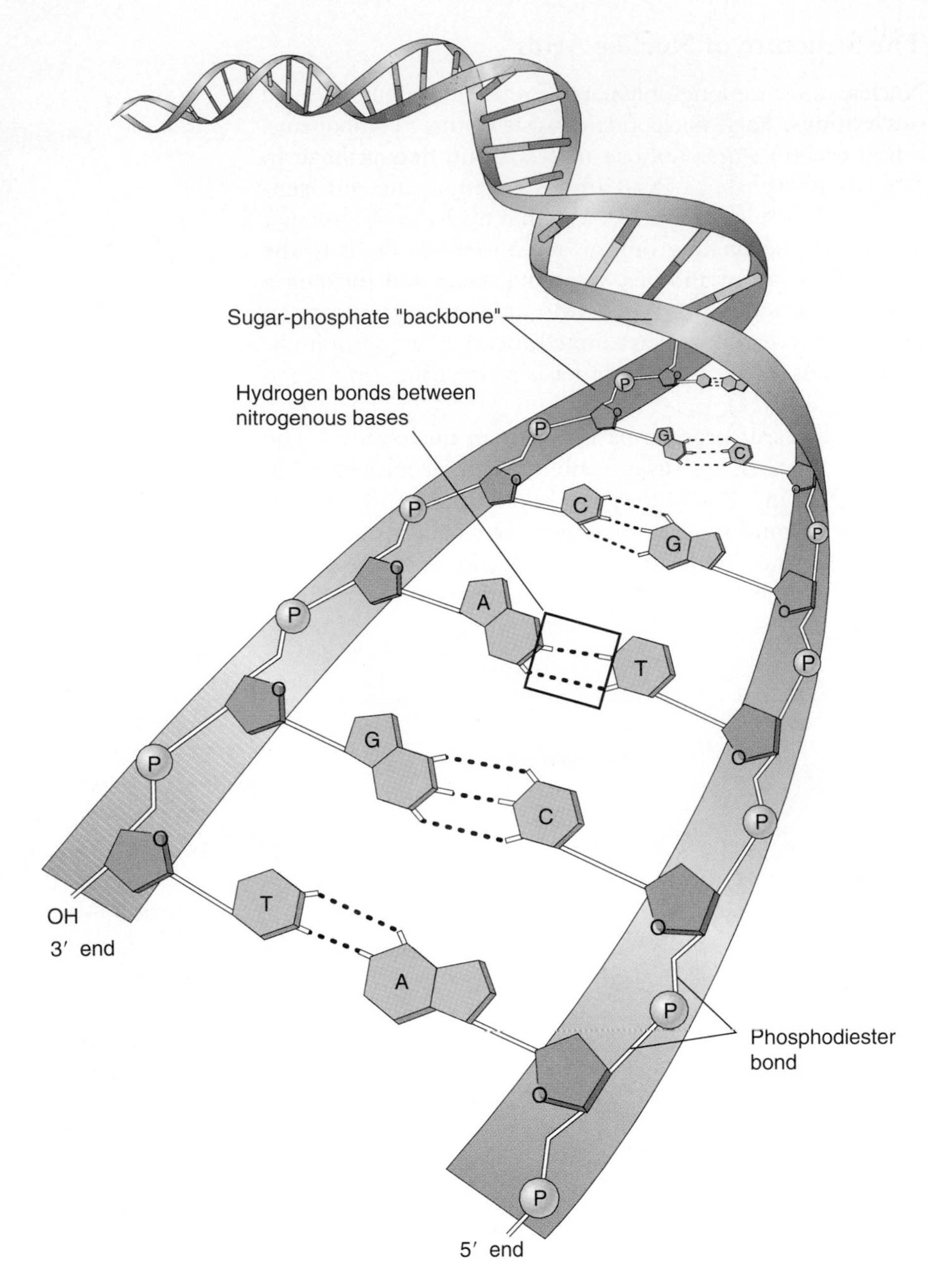

FIGURE 3.15
The structure of DNA. Hydrogen bond formation (dashed lines) between the organic bases, called base-pairing, causes the two chains of a DNA duplex to bind to each other and form a double helix.

RNA

RNA is similar to DNA, but with two major chemical differences. First, RNA molecules contain ribose sugars in which the number 2 carbon is bonded to a hydroxyl group. In DNA, this hydroxyl group is replaced by a hydrogen atom. Second, RNA molecules utilize uracil in place of thymine. Uracil has the same structure as thymine, except that one of its carbons lacks a methyl (—CH_3) group.

Transcribing the DNA message into a chemically different molecule such as RNA allows the cell to tell which is the original information storage molecule and which is the

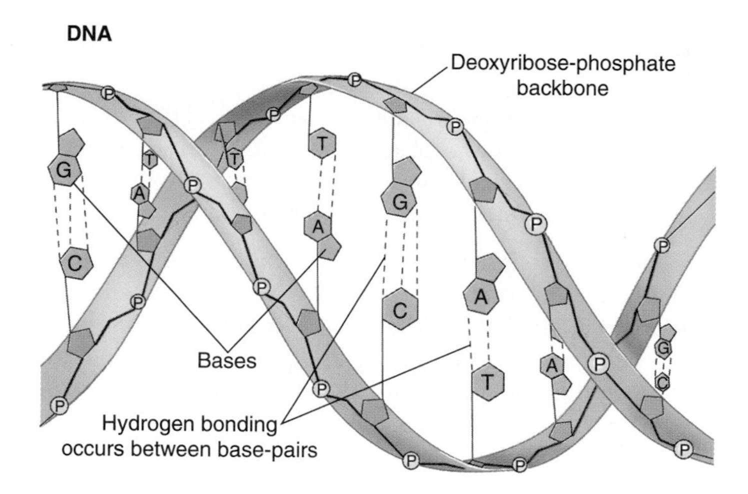

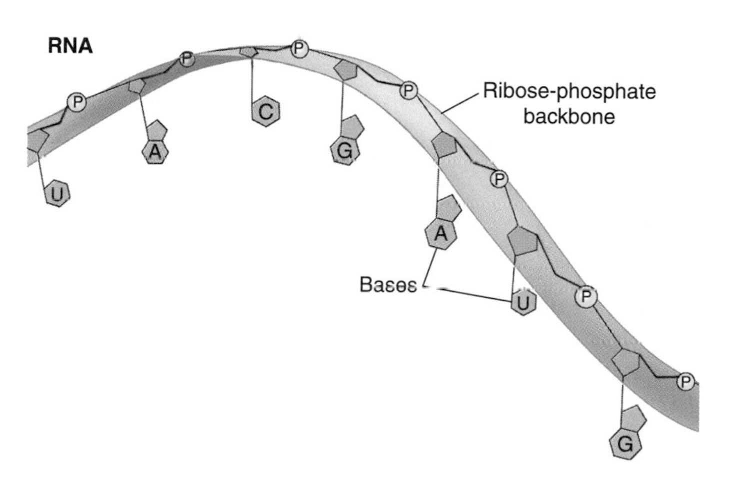

FIGURE 3.16
DNA versus RNA. DNA forms a double helix, uses deoxyribose as the sugar in its sugar-phosphate backbone, and utilizes thymine among its nitrogenous bases. RNA, on the other hand, is usually single-stranded, uses ribose as the sugar in its sugar-phosphate backbone, and utilizes uracil in place of thymine.

transcript. DNA molecules are always double-stranded (except for a few single-stranded DNA viruses that will be discussed in chapter 33), while the RNA molecules transcribed from DNA are typically single-stranded (figure 3.16). Although there is no chemical reason why RNA cannot form double helices as DNA does, cells do not possess the enzymes necessary to assemble double strands of RNA, as they do for DNA. Using two different molecules, one single-stranded and the other double-stranded, separates the role of DNA in storing hereditary information from the role of RNA in using this information to specify protein structure.

Which Came First, DNA or RNA?

The information necessary for the synthesis of proteins is stored in the cell's double-stranded DNA base sequences.

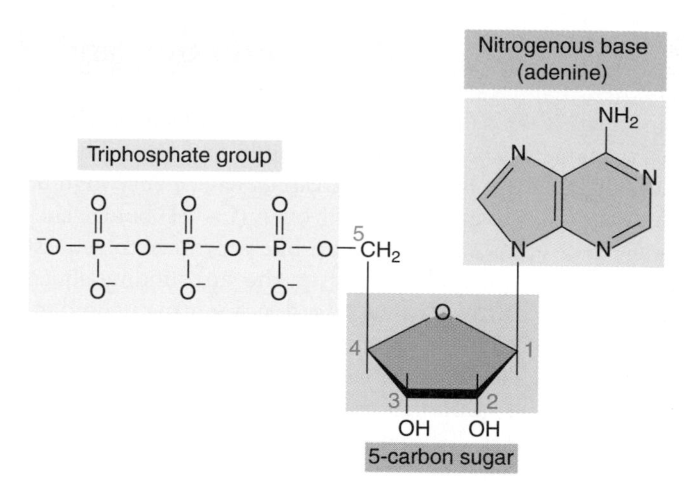

FIGURE 3.17
ATP. Adenosine triphosphate (ATP) contains adenine, a five-carbon sugar, and three phosphate groups. This molecule serves to transfer energy rather than store genetic information.

The cell uses this information by first making an RNA transcript of it: RNA nucleotides pair with complementary DNA nucleotides. By storing the information in DNA while using a complementary RNA sequence to actually direct protein synthesis, the cell does not expose the information-encoding DNA chain to the dangers of single-strand cleavage every time the information is used. Therefore, DNA is thought to have evolved from RNA as a means of preserving the genetic information, protecting it from the ongoing wear and tear associated with cellular activity. This genetic system has come down to us from the very beginnings of life.

The cell uses the single-stranded, short-lived RNA transcript to direct the synthesis of a protein with a specific sequence of amino acids. Thus, the information flows from DNA to RNA to protein, a process that has been termed the "central dogma" of molecular biology.

ATP

In addition to serving as subunits of DNA and RNA, nucleotide bases play other critical roles in the life of a cell. For example, adenine is a key component of the molecule *adenosine triphosphate* (ATP; figure 3.17), the energy currency of the cell. It also occurs in the molecules *nicotinamide adenine dinucleotide* (NAD^+) and *flavin adenine dinucleotide* (FAD^+), which carry electrons whose energy is used to make ATP.

A nucleic acid is a long chain of five-carbon sugars with an organic base protruding from each sugar. DNA is a double-stranded helix that stores hereditary information as a specific sequence of nucleotide bases. RNA is a single-stranded molecule that transcribes this information to direct protein synthesis.

3.4 Lipids make membranes and store energy.

Lipids are a loosely defined group of molecules with one main characteristic: they are insoluble in water. The most familiar lipids are fats and oils. Lipids have a very high proportion of nonpolar carbon-hydrogen (C—H) bonds, and so long-chain lipids cannot fold up like a protein to sequester their nonpolar portions away from the surrounding aqueous environment. Instead, when placed in water many lipid molecules will spontaneously cluster together and expose what polar groups they have to the surrounding water while sequestering the nonpolar parts of the molecules together within the cluster. This spontaneous assembly of lipids is of paramount importance to cells, as it underlies the structure of cellular membranes.

Phospholipids Form Membranes

Phospholipids are among the most important molecules of the cell, as they form the core of all biological membranes. An individual phospholipid is a composite molecule, made up of three kinds of subunits:

1. *Glycerol,* a three-carbon alcohol, with each carbon bearing a hydroxyl group. Glycerol forms the backbone of the phospholipid molecule.
2. *Fatty acids,* long chains of C—H bonds (hydrocarbon chains) ending in a carboxyl (—COOH) group. Two fatty acids are attached to the glycerol backbone in a phospholipid molecule.
3. *Phosphate group,* attached to one end of the glycerol. The charged phosphate group usually has a charged organic molecule linked to it, such as choline, ethanolamine, or the amino acid serine.

The phospholipid molecule can be thought of as having a polar "head" at one end (the phosphate group) and two long, very nonpolar "tails" at the other. In water, the nonpolar tails of nearby phospholipids aggregate away from the water, forming two layers of tails pointed toward each other—a lipid bilayer (figure 3.18). Lipid bilayers are the basic framework of biological membranes, discussed in detail in chapter 6.

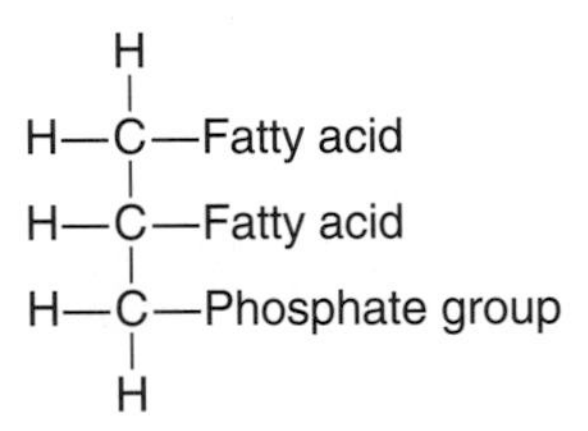

Because the C—H bonds in lipids are very nonpolar, they are not water-soluble, and aggregate together in water. This kind of aggregation by phospholipids forms biological membranes.

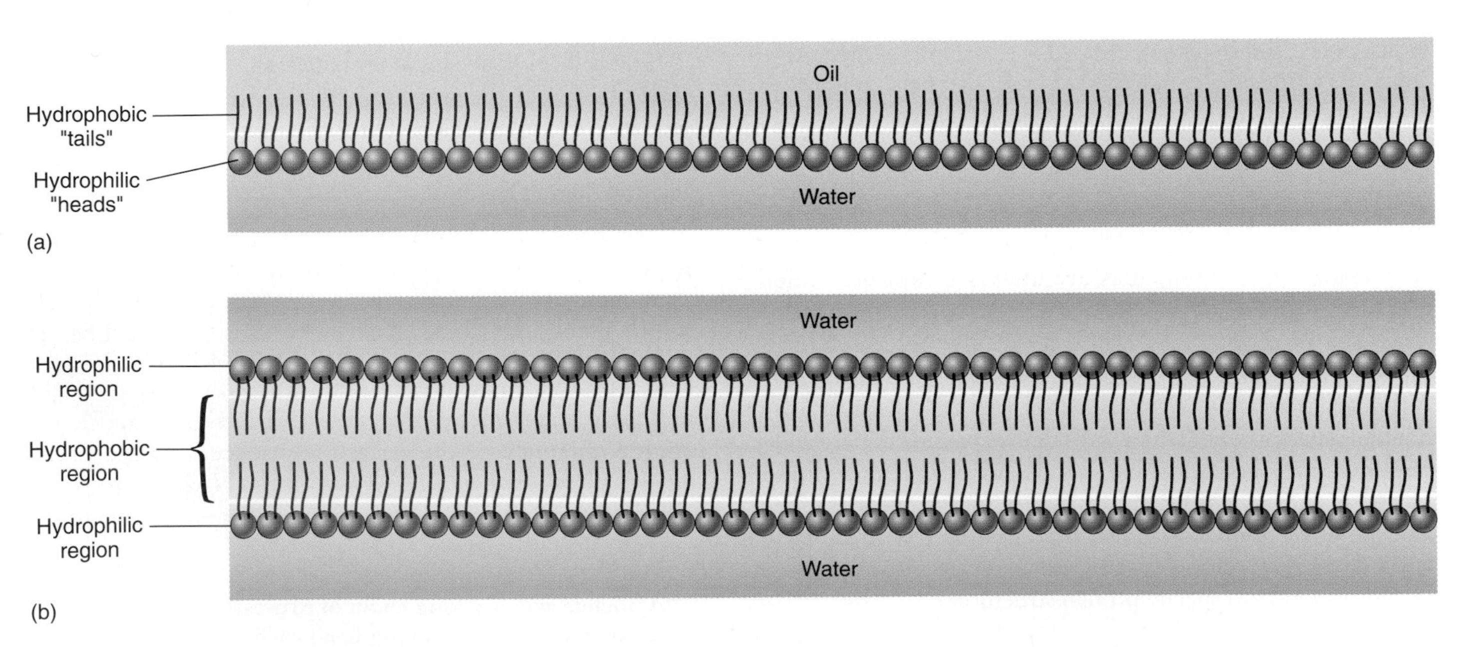

FIGURE 3.18
Phospholipids. (*a*) At an oil-water interface, phospholipid molecules will orient so that their polar (hydrophilic) heads are in the polar medium, water, and their nonpolar (hydrophobic) tails are in the nonpolar medium, oil. (*b*) When surrounded by water, phospholipid molecules arrange themselves into two layers with their heads extending outward and their tails inward.

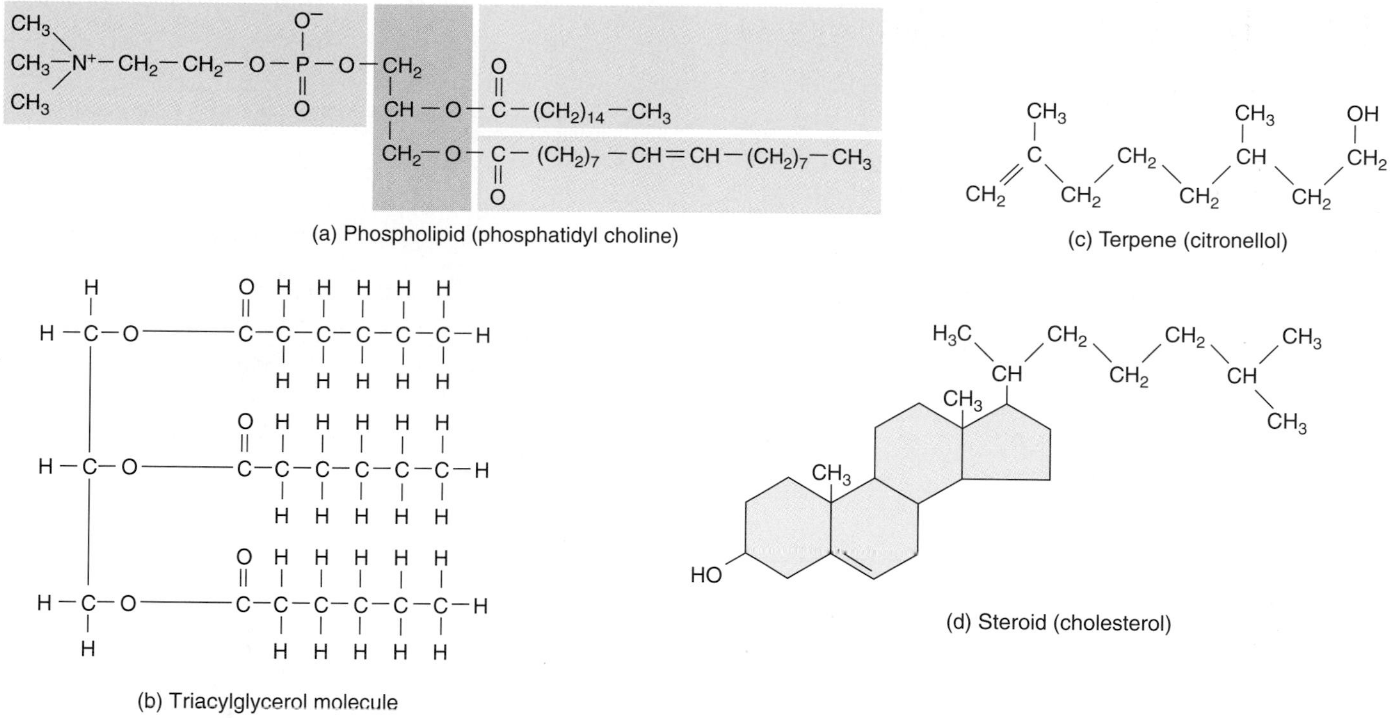

(a) Phospholipid (phosphatidyl choline)

(b) Triacylglycerol molecule

(c) Terpene (citronellol)

(d) Steroid (cholesterol)

FIGURE 3.19
Lipids. These structures represent four major classes of biologically important lipids: (*a*) phospholipids, (*b*) triacylglycerols (triglycerides), (*c*) terpenes, and (*d*) steroids.

Fats and Other Kinds of Lipids

Fats are another kind of lipid, but unlike phospholipids, fat molecules do not have a polar end. **Fats** consist of a glycerol molecule to which is attached three fatty acids, one to each carbon of the glycerol backbone. Because it contains three fatty acids, a fat molecule is called a **triglyceride**, or, more properly, a **triacylglycerol** (figure 3.19). The three fatty acids of a triglyceride need not be identical, and often they differ markedly from one another. Organisms store the energy of certain molecules for long periods in the many C—H bonds of fats.

Because triglyceride molecules lack a polar end, they are not soluble in water. Placed in water, they spontaneously clump together, forming fat globules that are very large relative to the size of the individual molecules. Because fats are insoluble, they can be deposited at specific locations within an organism.

Storage fats are one kind of lipid. Oils such as olive oil, corn oil, and coconut oil are also lipids, as are waxes such as beeswax and earwax. The hydrocarbon chains of fatty acids vary in length; the most common are even-numbered chains of 14 to 20 carbons. If all of the internal carbon atoms in the fatty acid chains are bonded to at least two hydrogen atoms, the fatty acid is said to be **saturated,** because it contains the maximum possible number of hydrogen atoms (figure 3.20). If a fatty acid has double bonds between one or more pairs of successive carbon atoms, the fatty acid is said to be **unsaturated.** If a given fatty acid has more than one double bond, it is said to be **polyunsaturated.** Fats made from polyunsaturated fatty acids have low melting points because their fatty acid chains bend at the double bonds, preventing the fat molecules from aligning closely with one another. Consequently, a polyunsaturated fat such as corn oil is usually liquid at room temperature and is called an oil. In contrast, most saturated fats such as those in butter are solid at room temperature.

Organisms contain many other kinds of lipids besides fats (see figure 3.19). *Terpenes* are long-chain lipids that are components of many biologically important pigments, such as chlorophyll and the visual pigment retinal. Rubber is also a terpene. *Steroids*, another type of lipid found in membranes, are composed of four carbon rings. Most animal cell membranes contain the steroid cholesterol. Other steroids, such as testosterone and estrogen, function in multicellular organisms as hormones. *Prostaglandins* are a group of about 20 lipids that are modified fatty acids, with two nonpolar "tails" attached to a five-carbon ring. Prostaglandins act as local chemical messengers in many vertebrate tissues.

Cells contain many kinds of molecules in addition to membrane phospholipids that are not soluble in water.

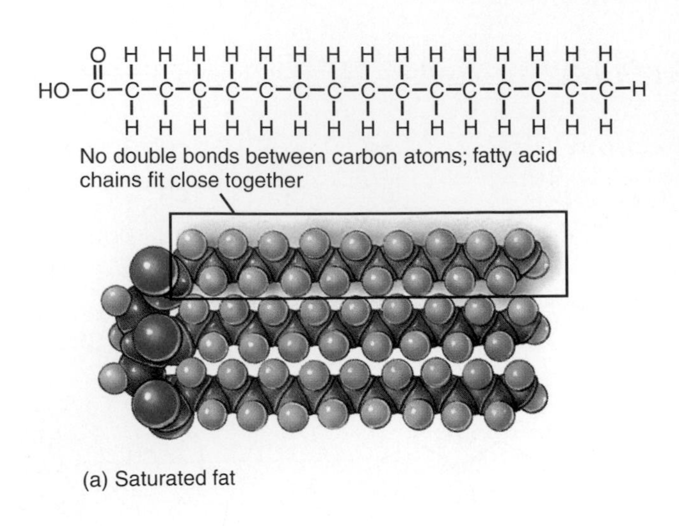

No double bonds between carbon atoms; fatty acid chains fit close together
(a) Saturated fat

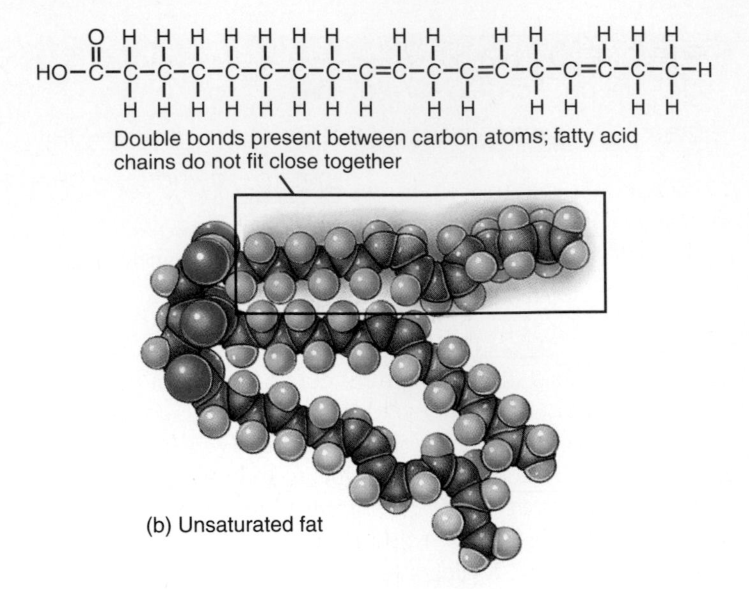

Double bonds present between carbon atoms; fatty acid chains do not fit close together
(b) Unsaturated fat

FIGURE 3.20
Saturated and unsaturated fats. (*a*) Palmitic acid, with no double bonds and, thus, a maximum number of hydrogen atoms bonded to the carbon chain, is a saturated fatty acid. Many animal triacylglycerols (fats) are saturated. Because their fatty acid chains can fit closely together, these triacylglycerols form immobile arrays called hard fat. (*b*) Linoleic acid, with three double bonds and, thus, fewer than the maximum number of hydrogen atoms bonded to the carbon chain, is an unsaturated fatty acid. Plant fats are typically unsaturated. The many kinks the double bonds introduce into the fatty acid chains prevent the triacylglycerols from closely aligning and produce oils, which are liquid at room temperature.

Fats as Food

Most fats contain over 40 carbon atoms. The ratio of energy-storing C—H bonds to carbon atoms in fats is more than twice that of carbohydrates (see next section), making fats much more efficient molecules for storing chemical energy. On the average, fats yield about 9 kilocalories (kcal) of chemical energy per gram, as compared with somewhat less than 4 kcal per gram for carbohydrates.

All fats produced by animals are saturated (except some fish oils), while most plant fats are unsaturated. The exceptions are the tropical oils (palm oil and coconut oil), which are saturated despite their fluidity at room temperature. It is possible to convert an oil into a solid fat by adding hydrogen. Peanut butter sold in stores is usually artificially hydrogenated to make the peanut fats solidify, preventing them from separating out as oils while the jar sits on the store shelf. However, artificially hydrogenating unsaturated fats seems to eliminate the health advantage they have over saturated fats, as it makes both equally rich in C—H bonds. Therefore, it now appears that margarine made from hydrogenated corn oil is no better for your health than butter.

When an organism consumes excess carbohydrate, it is converted into starch, glycogen, or fats and reserved for future use. The reason that many humans gain weight as they grow older is that the amount of energy they need decreases with age, while their intake of food does not. Thus, an increasing proportion of the carbohydrate they ingest is available to be converted into fat.

A diet rich in fats is one of several factors that are thought to contribute to heart disease, particularly to atherosclerosis, a condition in which deposits of fatty tissue called plaque adhere to the lining of blood vessels, blocking the flow of blood. Fragments of plaque, breaking off from a deposit, are a major cause of strokes.

Fats are efficient energy-storage molecules because of their high concentration of C—H bonds.

Simple Carbohydrates

Carbohydrates function as energy-storage molecules as well as structural elements. Some are small, simple molecules, while others form long polymers.

Sugars Are Simple Carbohydrates

The **carbohydrates** are a loosely defined group of molecules that contain carbon, hydrogen, and oxygen in the molar ratio 1:2:1. Their empirical formula (which lists the atoms in the molecule with subscripts to indicate how many there are of each) is $(CH_2O)_n$, where n is the number of carbon atoms. Because they contain many carbon-hydrogen (C—H) bonds, which release energy when they are broken, carbohydrates are well suited for energy storage.

Monosaccharides. The simplest of the carbohydrates are the simple sugars, or **monosaccharides** (Greek *mono*, "single" + Latin *saccharum*, "sugar"). Simple sugars may contain as few as three carbon atoms, but those that play the central role in energy storage have six (figure 3.21). The empirical formula of six-carbon sugars is:

$$C_6H_{12}O_6, \text{ or } (CH_2O)_6$$

Six-carbon sugars can exist in a straight-chain form, but in an aqueous environment they almost always form rings. The most important of these for energy storage is *glucose* (figure 3.22), a six-carbon sugar which has seven energy-storing C—H bonds.

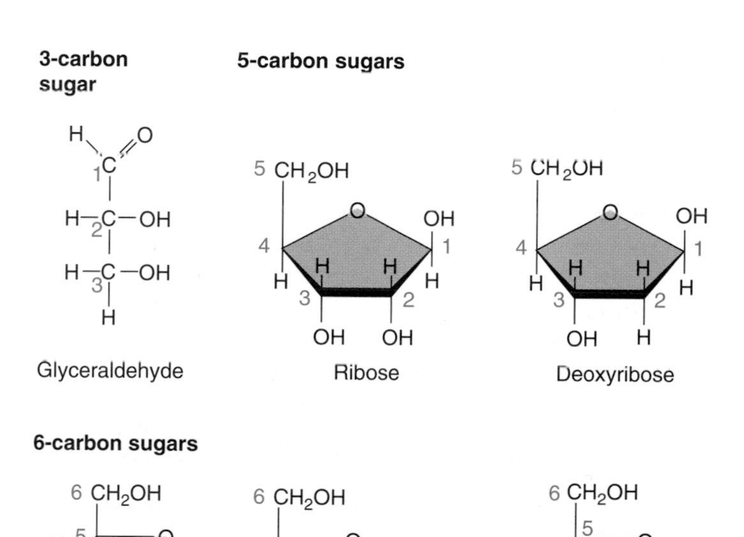

FIGURE 3.21
Monosaccharides. Monosaccharides, or simple sugars, can contain as few as three carbon atoms and are often used as building blocks to form larger molecules. The five-carbon sugars ribose and deoxyribose are components of nucleic acids (see figure 3.16). The six-carbon sugar glucose is a component of large energy-storage molecules.

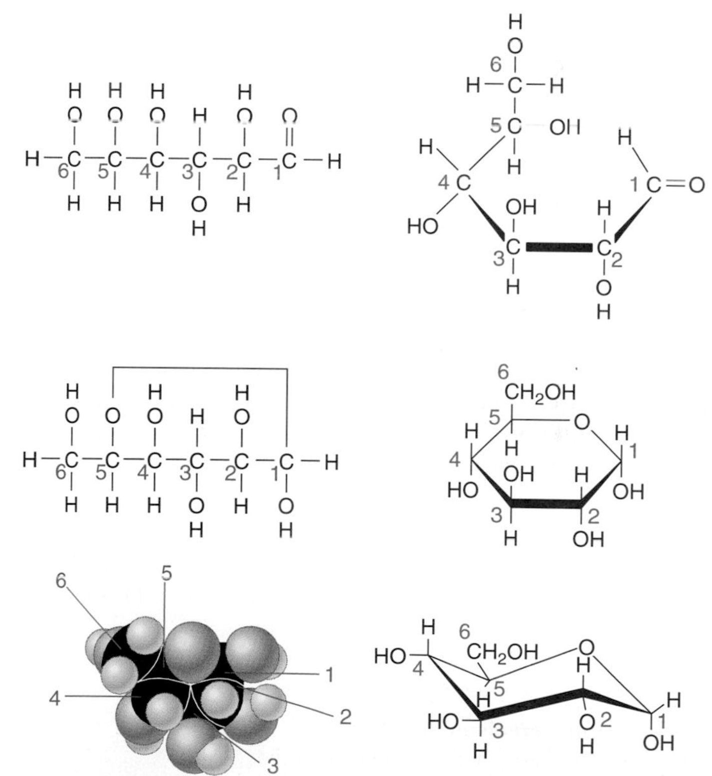

FIGURE 3.22
Structure of the glucose molecule. Glucose is a linear six-carbon molecule that forms a ring shape in solution. The structure of the ring can be represented in many ways; the ones shown here are the most common, with the carbons conventionally numbered (in *green*) so that the forms can be compared easily. The bold, darker lines represent portions of the molecule that are projecting out of the page toward you—remember, these are three-dimensional molecules!

Disaccharides. Many familiar sugars like sucrose are "double sugars," two monosaccharides joined by a covalent bond (figure 3.23). Called **disaccharides,** they often play a role in the transport of sugars, as we will discuss shortly.

Polysaccharides. Polysaccharides are macromolecules made up of monosaccharide subunits. Starch is a polysaccharide used by plants to store energy. It consists entirely of glucose molecules, linked one after another in long chains. Cellulose is a polysaccharide that serves as a structural building material in plants. It too consists entirely of glucose molecules linked together into chains, and special enzymes are required to break the links.

Sugar Isomers

Glucose is not the only sugar with the formula $C_6H_{12}O_6$. Other common six-carbon sugars such as fructose and galactose also have this same empirical formula (figure 3.24). These sugars are **isomers,** or alternative forms, of glucose. Even though isomers have the same empirical formula, their atoms are arranged in different ways; that is, their three-dimensional structures are different. These structural differences often account for substantial functional differences between the isomers. Glucose and fructose, for example, are *structural isomers.* In fructose, the double-bonded oxygen is attached to an internal carbon rather than to a terminal one. Your taste buds can tell the difference, as fructose tastes much sweeter than glucose, despite the fact that both sugars have the same chemical composition. This structural difference also has an important chemical consequence: the two sugars form different polymers.

Unlike fructose, galactose has the same bond structure as glucose; the only difference between galactose and glucose is the orientation of one hydroxyl group. Because the hydroxyl group positions are mirror images of each other, galactose and glucose are called *stereoisomers.* Again, this seemingly slight difference has important consequences, as this hydroxyl group is often involved in creating polymers with distinct functions, such as starch (energy storage) and cellulose (structural support).

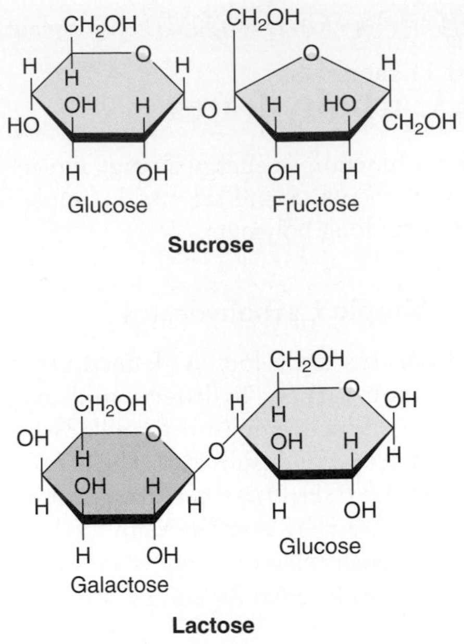

FIGURE 3.23
Disaccharides. Sugars like sucrose and lactose are disaccharides, composed of two monosaccharides linked by a covalent bond.

Sugars are among the most important energy-storage molecules in organisms, containing many energy-storing C—H bonds. The structural differences among sugar isomers can confer substantial functional differences upon the molecules.

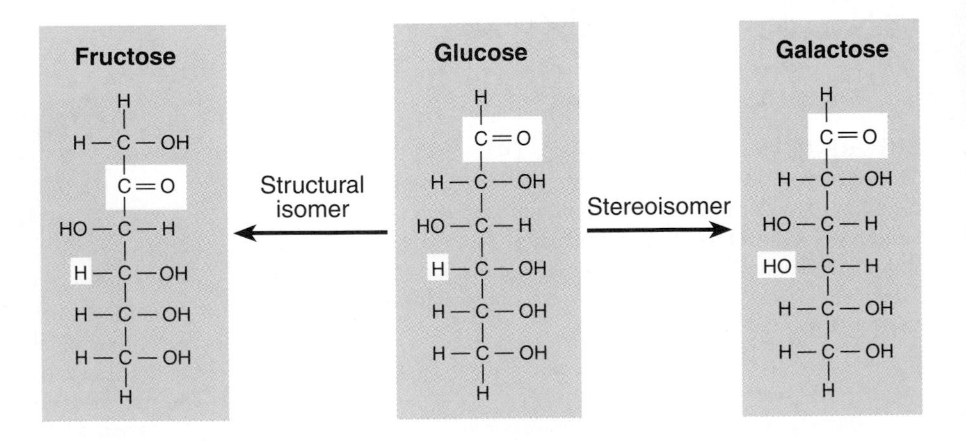

FIGURE 3.24
Isomers and stereoisomers. Glucose, fructose, and galactose are isomers with the empirical formula $C_6H_{12}O_6$. A structural isomer of glucose, such as fructose, has identical chemical groups bonded to different carbon atoms, while a stereoisomer of glucose, such as galactose, has identical chemical groups bonded to the same carbon atoms but in different orientations.

Linking Sugars Together

Transport Disaccharides

Most organisms transport sugars within their bodies. In humans, the glucose that circulates in the blood does so as a simple monosaccharide. In plants and many other organisms, however, glucose is converted into a transport form before it is moved from place to place within the organism. In such a form it is less readily metabolized (used for energy) during transport. Transport forms of sugars are commonly made by linking two monosaccharides together to form a disaccharide (Greek *di*, "two"). Disaccharides serve as effective reservoirs of glucose because the normal glucose-utilizing enzymes of the organism cannot break the bond linking the two monosaccharide subunits. Enzymes that can do so are typically present only in the tissue where the glucose is to be used.

Transport forms differ depending on which monosaccharides link to form the disaccharide. Glucose forms transport disaccharides with itself and many other monosaccharides, including fructose and galactose. When glucose forms a disaccharide with its structural isomer, fructose, the resulting disaccharide is *sucrose*, or table sugar (figure 3.25*a*). Sucrose is the form in which most plants transport glucose and the sugar that most humans (and other animals) eat. Sugarcane is rich in sucrose, and so are sugar beets.

When glucose is linked to its stereoisomer, galactose, the resulting disaccharide is *lactose*, or milk sugar. Many mammals supply energy to their young in the form of lactose. Adults have greatly reduced levels of lactase, the enzyme required to cleave lactose into its two monosaccharide components, and thus cannot metabolize lactose as efficiently. Most of the energy that is channeled into lactose production is therefore reserved for their offspring.

Storage Polysaccharides

Organisms store the metabolic energy contained in monosaccharides by converting them into disaccharides, such as *maltose* (figure 3.25*b*), which are then linked together into insoluble forms that are deposited in specific storage areas in their bodies. These insoluble polysaccharides are long polymers of monosaccharides formed by dehydration synthesis. Plant polysaccharides formed from glucose are called **starches.** Plants store starch as granules within chloroplasts and other organelles. Because glucose is a key metabolic fuel, the stored starch provides a reservoir of energy available for future needs. Energy for cellular work can be retrieved by hydrolyzing the links that bind the glucose subunits together.

The starch with the simplest structure is *amylose*, which is composed of many hundreds of glucose molecules linked together in long, unbranched chains. Each linkage occurs between the number 1 carbon of one glucose molecule and the number 4 carbon of another, so that amylose is, in effect, a longer form of maltose. The long chains of amylose tend to coil up in water (figure 3.26*a*), a property that renders amylose insoluble. Potato starch is about 20% amylose. When amylose is digested by a sprouting potato plant (or by an animal that eats a potato), enzymes first break it into fragments of random length, which are more soluble because they are shorter. Baking or boiling potatoes has the same effect, breaking the chains into fragments. Another enzyme then cuts these fragments into molecules of maltose. Finally, the maltose is cleaved into two glucose molecules, which cells are able to metabolize.

Most plant starch, including the remaining 80% of potato starch, is a somewhat more complicated variant of amylose called *amylopectin* (figure 3.26*b*). Pectins are branched polysaccharides. Amylopectin has short, linear amylose branches consisting of 20 to 30 glucose subunits.

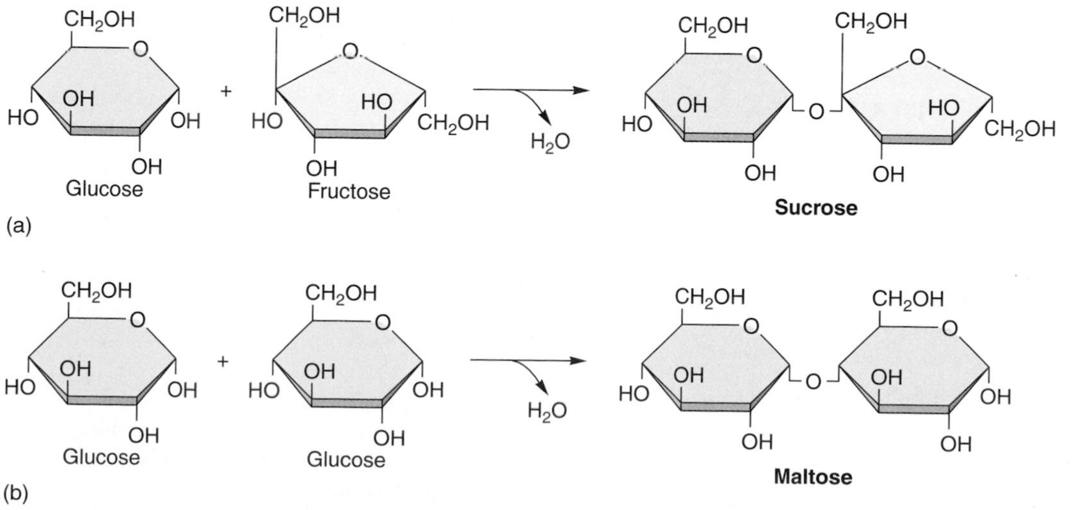

(a) Glucose + Fructose → Sucrose + H_2O

(b) Glucose + Glucose → Maltose + H_2O

FIGURE 3.25
How disaccharides form. Some disaccharides are used to transport glucose from one part of an organism's body to another; one example is sucrose (*a*), which is found in sugarcane. Other disaccharides, such as maltose in grain (*b*), are used for storage.

In some plants these chains are cross-linked. The cross-links create an insoluble mesh of glucose, which can be degraded only by another kind of enzyme. The size of the mesh differs from plant to plant; in rice about 100 amylose chains, each with one or two cross-links, forms the mesh.

The animal version of starch is glycogen. Like amylopectin, *glycogen* is an insoluble polysaccharide containing branched amylose chains. In glycogen, the average chain length is much greater and there are more branches than in plant starch (figure 3.26*c*). Humans and other vertebrates store excess food energy as glycogen in the liver and in muscle cells; when the demand for energy in a tissue increases, glycogen is hydrolyzed to release glucose.

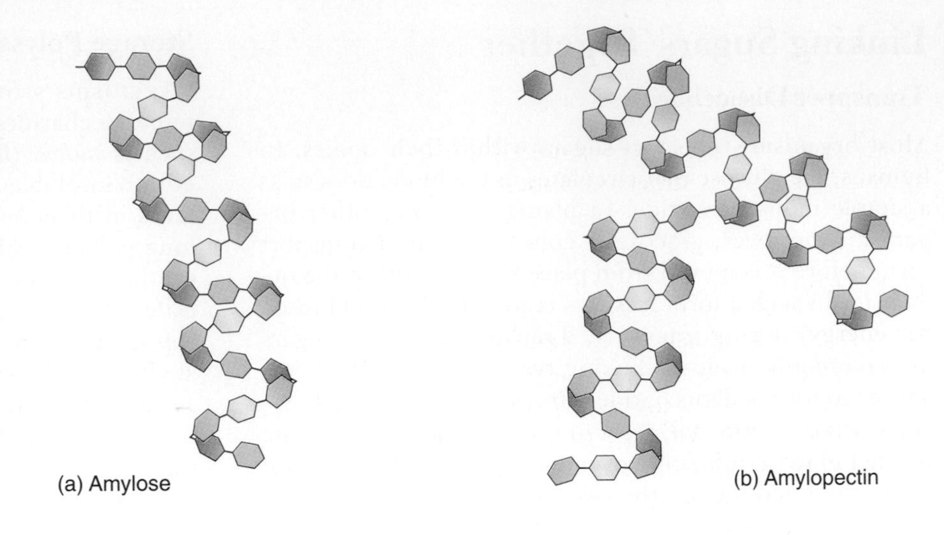

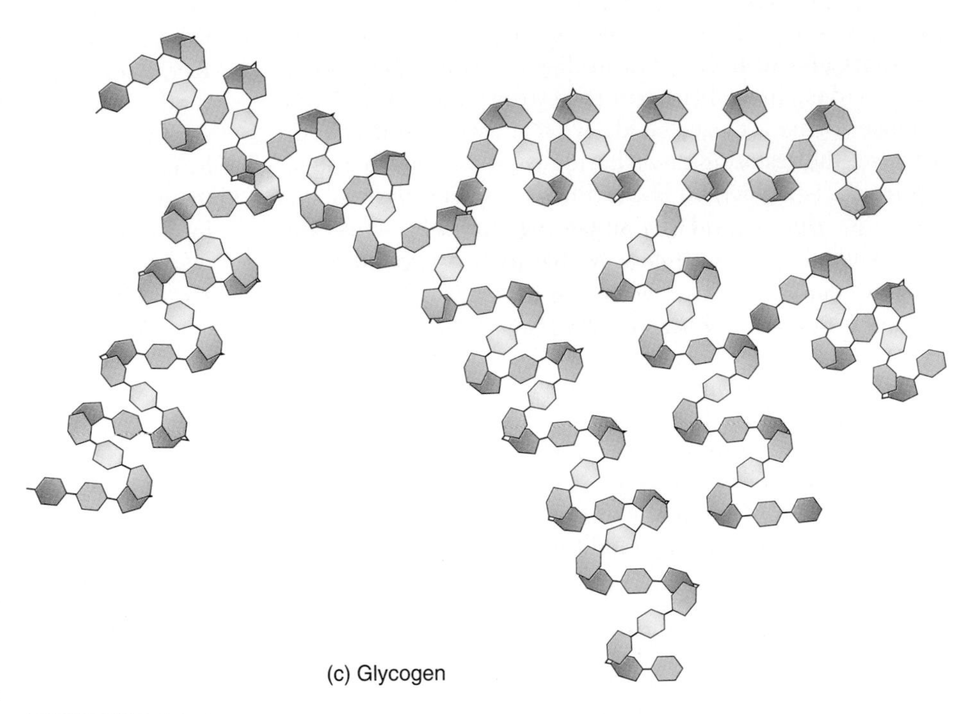

FIGURE 3.26
Storage polysaccharides. Starches are long glucose polymers that store energy in plants. (*a*) The simplest starches are long chains of maltose called amylose, which tend to coil up in water. (*b*) Most plants contain more complex starches called amylopectins, which are branched. (*c*) Animals store glucose in glycogen, which is more extensively branched than amylopectin and contains longer chains of amylose.

Nonfattening Sweets

Imagine a kind of table sugar that looks, tastes, and cooks like the real thing, but has no calories or harmful side effects. You could eat mountains of candy made from such sweeteners without gaining weight. As Louis Pasteur discovered in the late 1800s, most sugars are "right-handed" molecules, in that the hydroxyl group that binds a critical carbon atom is on the right side. However, "left-handed" sugars, in which the hydroxyl group is on the left side, can be made readily in the laboratory. These synthetic sugars are mirror-image chemical twins of the natural form, but the enzymes that break down sugars in the human digestive system can tell the difference. To digest a sugar molecule, an enzyme must first grasp it, much like a shoe fitting onto a foot, and all of the body's enzymes are right-handed! A left-handed sugar doesn't fit, any more than a shoe for the right foot fits onto a left foot.

The Latin word for "left" is *levo*, and left-handed sugars are called **levo-**, or **l-sugars.** They do not occur in nature except for trace amounts in red algae, snail eggs, and seaweed. Because they pass through the body without being used, they can let diet-conscious sweet-lovers have their cake and eat it, too. Nor will they contribute to tooth decay because bacteria cannot metabolize them, either.

Starches are glucose polymers. Most starches are branched and some are cross-linked. The branching and cross-linking render the polymer insoluble and protect it from degradation.

Structural Carbohydrates

While some chains of sugars store energy, others serve as structural material for cells.

Cellulose

For two glucose molecules to link together, the glucose subunits must be the same form. Glucose can form a ring in two ways, with the hydroxyl group attached to the carbon where the ring closes being locked into place either below or above the plane of the ring. If below, it is called the **alpha form,** and if above, the **beta form.** All of the glucose subunits of the starch chain are alpha-glucose. When a chain of glucose molecules consists of all beta-glucose subunits, a polysaccharide with very different properties results. This **structural polysaccharide** is *cellulose*, the chief component of plant cell walls (figure 3.27). Cellulose is chemically similar to amylose, with one important difference: the starch-degrading enzymes that occur in most organisms cannot break the bond between two beta-glucose sugars. This is not because the bond is stronger, but rather because its cleavage requires an enzyme most organisms lack. Because cellulose cannot be broken down readily, it works well as a biological structural material and occurs widely in this role in plants. Those few animals able to break down cellulose find it a rich source of energy. Certain vertebrates, such as cows, can digest cellulose by means of bacteria and protists they harbor in their intestines which provide the necessary enzymes.

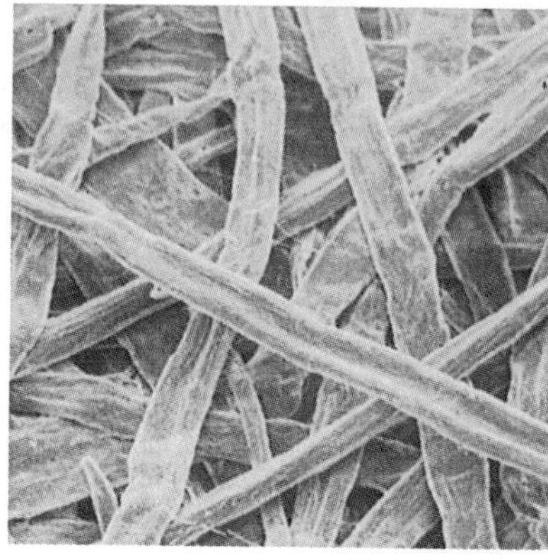

(a)

CH_2OH H H OH H 4 1 HO OH H OH

α form of glucose

(b)

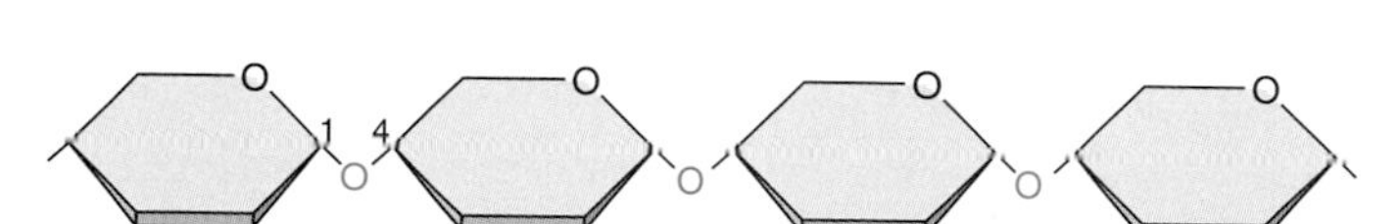

Starch: chain of α-glucose subunits

CH_2OH H OH H OH H 4 1 HO H H OH

β form of glucose

(c)

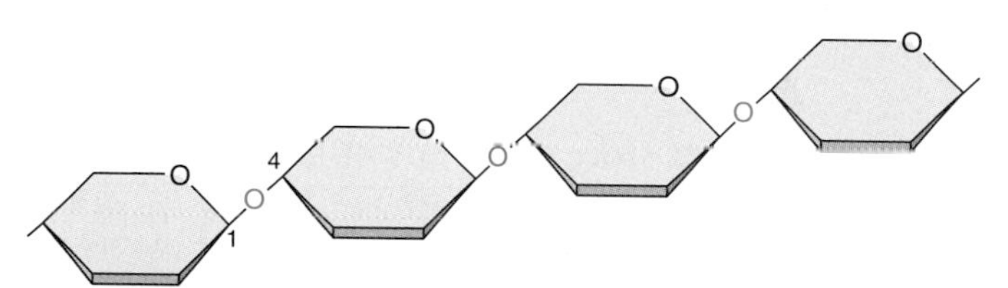

Cellulose: chain of β-glucose subunits

FIGURE 3.27
A journey into wood. This jumble of cellulose fibers (*a*) is from a yellow pine (*Pinus ponderosa*) (20×). (*b*) While starch chains consist of alpha-glucose subunits, (*c*) cellulose chains consist of beta-glucose subunits. Cellulose fibers can be very strong and are quite resistant to metabolic breakdown, which is one reason why wood is such a good building material.

Chitin

The structural building material in insects, many fungi, and certain other organisms is called chitin (figure 3.28). *Chitin* is a modified form of cellulose with a nitrogen group added to the glucose units. When cross-linked by proteins, it forms a tough, resistant surface material that serves as the hard exoskeleton of arthropods such as insects and crustaceans (see chapter 46). Few organisms are able to digest chitin.

Structural carbohydrates are chains of sugars that are not easily digested. They include cellulose in plants and chitin in arthropods and fungi.

FIGURE 3.28
Chitin. Chitin, which might be considered to be a modified form of cellulose, is the principal structural element in the external skeletons of many invertebrates, such as this lobster.

Chapter 3

www.mhhe.com/raven6/resources3.mhtml

Summary | ***Questions*** | ***Media Resources***

3.1 Molecules are the building blocks of life.

- The chemistry of living systems is the chemistry of carbon-containing compounds.
- Carbon's unique chemical properties allow it to polymerize into chains by dehydration synthesis, forming the four key biological macromolecules: proteins, nucleic acids, lipids, and carbohydrates.

1. What types of molecules are formed by dehydration reactions? What types of molecules are formed by hydrolysis?

- Organic Chemistry

3.2 Proteins perform the chemistry of the cell.

- Proteins are polymers of amino acids.
- Because the 20 amino acids that occur in proteins have side groups with different chemical properties, the function and shape of a protein are critically affected by its particular sequence of amino acids.

2. How are amino acids linked to form proteins?

3. Explain what is meant by the primary, secondary, tertiary, and quaternary structure of a protein.

- Explorations: How Proteins Function

- Proteins

3.3 Nucleic acids store and transfer genetic information.

- Hereditary information is stored as a sequence of nucleotides in a linear polymer called DNA, which exists in cells as a double helix.
- Cells use the information in DNA by producing a complementary single strand of RNA which directs the synthesis of a protein whose amino acid sequence corresponds to the nucleotide sequence of the DNA from which the RNA was transcribed.

4. What are the three components of a nucleotide? How are nucleotides linked to form nucleic acids?

5. Which of the purines and pyrimidines are capable of forming base-pairs with each other?

- Nucleic Acids

- Art Quiz: DNA Base-Pairing and Hydrogen Bonds

3.4 Lipids make membranes and store energy.

- Fats are one type of water-insoluble molecules called lipids.
- Fats are molecules that contain many energy-rich C—H bonds and, thus, provide an efficient form of long-term energy storage.
- Types of lipids include phospholipids, fats, terpenes, steroids, and prostaglandins.

6. What are the two kinds of subunits that make up a fat molecule, and how are they arranged in the molecule?

7. Describe the differences between a saturated and an unsaturated fat.

- Lipids

- How Scientists Think: Amino Acid Sequence Determines Protein Shape (Anfinsen)

3.5 Carbohydrates store energy and provide building materials.

- Carbohydrates store considerable energy in their carbon-hydrogen (C—H) bonds.
- The most metabolically important carbohydrate is glucose, a six-carbon sugar.
- Excess energy resources may be stored in complex sugar polymers called starches (in plants) and glycogen (in animals and fungi).

8. What does it mean to say that glucose, fructose, and galactose are isomers? Which two are structural isomers, and how do they differ from each other? Which two are stereoisomers, and how do they differ from each other?

- Carbohydrates

- Art Quiz: Disaccharides

4

The Origin and Early History of Life

Concept Outline

4.1 All living things share key characteristics.

What Is Life? All known organisms share certain general properties, and to a large degree these properties define what we mean by life.

4.2 There are many ideas about the origin of life.

Theories about the Origin of Life. There are both religious and scientific views about the origin of life. This text treats only the latter—only the scientifically testable.

Scientists Disagree about Where Life Started. The atmosphere of the early earth was rich in hydrogen, providing a ready supply of energetic electrons with which to build organic molecules.

The Miller-Urey Experiment. Experiments attempting to duplicate the conditions of early earth produce many of the key molecules of living organisms.

4.3 The first cells had little internal structure.

Theories about the Origin of Cells. The first cells are thought to have arisen spontaneously, but there is little agreement as to the mechanism.

The Earliest Cells. The earliest fossils are of bacteria too small to see with the unaided eye.

4.4 The first eukaryotic cells were larger and more complex than bacteria.

The First Eukaryotic Cells. Fossils of the first eukaryotic cells do not appear in rocks until 1.5 billion years ago, over 2 billion years after bacteria. Multicellular life is restricted to the four eukaryotic kingdoms of life.

Has Life Evolved Elsewhere? It seems probable that life has evolved on other worlds besides our own. The possible presence of life in the warm waters beneath the surface of Europa, a moon of Jupiter, is a source of current speculation.

FIGURE 4.1
The origin of life. The fortuitous mix of physical events and chemical elements at the right place and time created the first living cells on earth.

There are a great many scientists with intriguing ideas that explain how life may have originated on earth, but there is very little that we know for sure. New hypotheses are being proposed constantly, and old ones reevaluated. By the time this text is published, some of the ideas presented here about the origin of life will surely be obsolete. Thus, the contesting ideas are presented in this chapter in an open-ended format, attempting to make clear that there is as yet no one answer to the question of how life originated on earth. Although recent photographs taken by the Hubble Space Telescope have revived controversy about the age of the universe, it seems clear the earth itself was formed about 4.6 billion years ago. The oldest clear evidence of life—microfossils in ancient rock—are 3.5 billion years old. The origin of life seems to have taken just the right combination of physical events and chemical processes (figure 4.1).

4.1 All living things share key characteristics.

The earth formed as a hot mass of molten rock about 4.6 billion years ago. As the earth cooled, much of the water vapor present in its atmosphere condensed into liquid water, which accumulated on the surface in chemically rich oceans. One scenario for the origin of life is that it originated in this dilute, hot smelly soup of ammonia, formaldehyde, formic acid, cyanide, methane, hydrogen sulfide, and organic hydrocarbons. Whether at the oceans' edge, in hydrothermal deep-sea vents, or elsewhere, the consensus among researchers is that life arose spontaneously from these early waters less than 4 billion years ago. While the way in which this happened remains a puzzle, one cannot escape a certain curiosity about the earliest steps that eventually led to the origin of all living things on earth, including ourselves. How did organisms evolve from the complex molecules that swirled in the early oceans?

What Is Life?

Before we can address this question, we must first consider what qualifies something as "living." What *is* life? This is a difficult question to answer, largely because life itself is not a simple concept. If you try to write a definition of "life," you will find that it is not an easy task, because of the loose manner in which the term is used.

Imagine a situation in which two astronauts encounter a large, amorphous blob on the surface of a planet. How would they determine whether it is alive?

Movement. One of the first things the astronauts might do is observe the blob to see if it moves. Most animals move about (figure 4.2), but movement from one place to another in itself is not diagnostic of life. Most plants and even some animals do not move about, while numerous nonliving objects, such as clouds, do move. The criterion of movement is thus neither *necessary* (possessed by all life) nor *sufficient* (possessed only by life).

Sensitivity. The astronauts might prod the blob to see if it responds. Almost all living things respond to stimuli (figure 4.3). Plants grow toward light, and animals retreat from fire. Not all stimuli produce responses, however. Imagine kicking a redwood tree or singing to a hibernating bear. This criterion, although superior to the first, is still inadequate to define life.

Death. The astronauts might attempt to kill the blob. All living things die, while inanimate objects do not. Death is not easily distinguished from disorder, however; a car that breaks down has not died because it was never alive. Death is simply the loss of life, so this is a circular definition at best. Unless one can detect life, death is a meaningless concept, and hence a very inadequate criterion for defining life.

Complexity. Finally, the astronauts might cut up the blob, to see if it is complexly organized. All living things are complex. Even the simplest bacteria

FIGURE 4.2
Movement. Animals have evolved mechanisms that allow them to move about in their environment. While some animals, like this giraffe, move on land, others move through water or air.

FIGURE 4.3
Sensitivity. This father lion is responding to a stimulus: he has just been bitten on the rump by his cub. As far as we know, all organisms respond to stimuli, although not always to the same ones or in the same way. Had the cub bitten a tree instead of its father, the response would not have been as dramatic.

contain a bewildering array of molecules, organized into many complex structures. However a computer is also complex, but not alive. Complexity is a *necessary* criterion of life, but it is not *sufficient* in itself to identify living things because many complex things are not alive.

To determine whether the blob is alive, the astronauts would have to learn more about it. Probably the best thing they could do would be to examine it more carefully and determine whether it resembles the organisms we are familiar with, and if so, how.

Fundamental Properties of Life

As we discussed in chapter 1, all known organisms share certain general properties. To a large degree, these properties define what we mean by life. The following fundamental properties are shared by all organisms on earth.

Cellular organization. All organisms consist of one or more *cells*—complex, organized assemblages of molecules enclosed within membranes (figure 4.4).

Sensitivity. All organisms respond to stimuli—though not always to the same stimuli in the same ways.

Growth. All living things assimilate energy and use it to grow, a process called *metabolism*. Plants, algae, and some bacteria use sunlight to create covalent carbon-carbon bonds from CO_2 and H_2O through photosynthesis. This transfer of the energy in covalent bonds is essential to all life on earth.

Development. Multicellular organisms undergo systematic gene-directed changes as they grow and mature.

Reproduction. All living things reproduce, passing on traits from one generation to the next. Although some organisms live for a very long time, no organism lives forever, as far as we know. Because all organisms die, ongoing life is impossible without *reproduction*.

Regulation. All organisms have regulatory mechanisms that coordinate internal processes.

Homeostasis. All living things maintain relatively constant internal conditions, different from their environment.

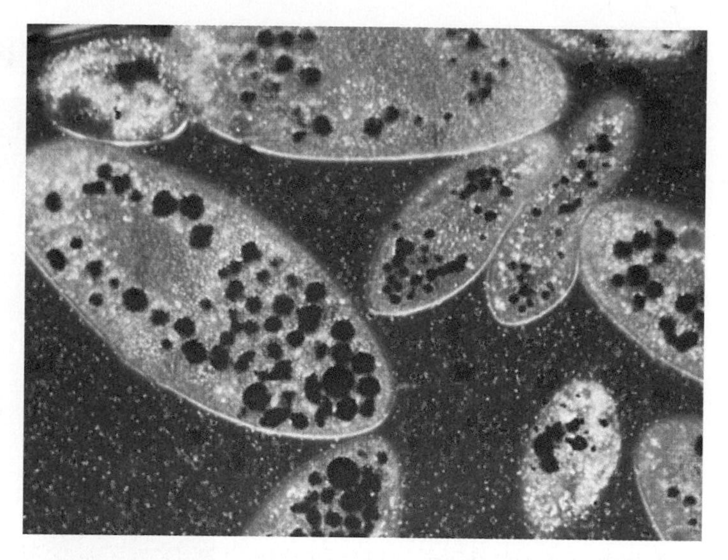

FIGURE 4.4
Cellular organization (150×). These *Paramecia*, which are complex, single-celled organisms called protists, have just ingested several yeast cells. The yeasts, stained *red* in this photograph, are enclosed within membrane-bounded sacs called digestive vacuoles. A variety of other organelles are also visible.

The Key Role of Heredity

Are these properties adequate to define life? Is a membrane-enclosed entity that grows and reproduces alive? Not necessarily. Soap bubbles and proteinoid microspheres spontaneously form hollow bubbles that enclose a small volume of water. These spheres can enclose energy-processing molecules, and they may also grow and subdivide. Despite these features, they are certainly not alive. Therefore, the criteria just listed, although necessary for life, are not sufficient to define life. One ingredient is missing—a mechanism for the preservation of improvement.

Heredity. All organisms on earth possess a *genetic system* that is based on the replication of a long, complex molecule called DNA. This mechanism allows for adaptation and evolution over time, also distinguishing characteristics of living things.

To understand the role of heredity in our definition of life, let us return for a moment to proteinoid microspheres. When we examine an individual microsphere, we see it at that precise moment in time but learn nothing of its predecessors. It is likewise impossible to guess what future droplets will be like. The droplets are the passive prisoners of a changing environment, and it is in this sense that they are not alive. The essence of being alive is the ability to encompass change and to reproduce the results of change permanently. Heredity, therefore, provides the basis for the great division between the living and the nonliving. Change does not become evolution unless it is passed on to a new generation. A genetic system is the sufficient condition of life. Some changes are preserved because they increase the chances of survival in a hostile world, while others are lost. Not only did life evolve—evolution is the very essence of life.

All living things on earth are characterized by cellular organization, heredity, and a handful of other characteristics that serve to define the term life.

4.2 There are many ideas about the origin of life.

Theories about the Origin of Life

The question of how life originated is not easy to answer because it is impossible to go back in time and observe life's beginnings; nor are there any witnesses. There is testimony in the rocks of the earth, but it is not easily read, and often it is silent on issues crying out for answers. There are, in principle, at least three possibilities:

1. **Special creation.** Life-forms may have been put on earth by supernatural or divine forces.
2. **Extraterrestrial origin.** Life may not have originated on earth at all; instead, life may have infected earth from some other planet.
3. **Spontaneous origin.** Life may have evolved from inanimate matter, as associations among molecules became more and more complex.

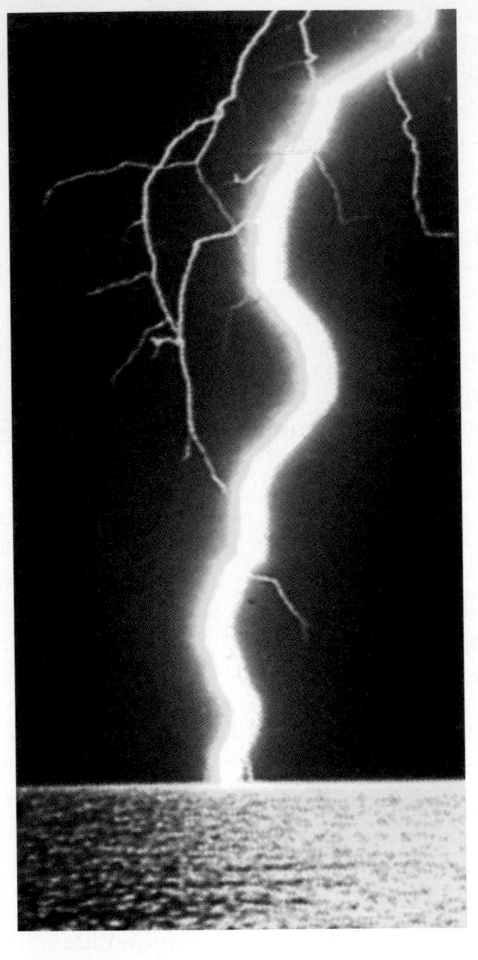

FIGURE 4.5
Lightning. Before life evolved, the simple molecules in the earth's atmosphere combined to form more complex molecules. The energy that drove these chemical reactions may have come from lightning and forms of geothermal energy.

Special Creation. The theory of special creation, that a divine God created life is at the core of most major religions. The oldest hypothesis about life's origins, it is also the most widely accepted. Far more Americans, for example, believe that God created life on earth than believe in the other two hypotheses. Many take a more extreme position, accepting the biblical account of life's creation as factually correct. This viewpoint forms the basis for the very unscientific "scientific creationism" viewpoint discussed in chapter 21.

Extraterrestrial Origin. The theory of **panspermia** proposes that meteors or cosmic dust may have carried significant amounts of complex organic molecules to earth, kicking off the evolution of life. Hundreds of thousands of meteorites and comets are known to have slammed into the early earth, and recent findings suggest that at least some may have carried organic materials. Nor is life on other planets ruled out. For example, the discovery of liquid water under the surface of Jupiter's ice-shrouded moon Europa and suggestions of fossils in rocks from Mars lend some credence to this idea. The hypothesis that an early source of carbonaceous material is extraterrestrial is testable, although it has not yet been proven. Indeed, NASA is planning to land on Europa, drill through the surface, and send a probe down to see if there is life.

Spontaneous Origin. Most scientists tentatively accept the theory of spontaneous origin, that life evolved from inanimate matter. In this view, the force leading to life was selection. As changes in molecules increased their stability and caused them to persist longer, these molecules could initiate more and more complex associations, culminating in the evolution of cells.

Taking a Scientific Viewpoint

In this book we will focus on the second and third possibilities, attempting to understand whether natural forces could have led to the origin of life and, if so, how the process might have occurred. This is not to say that the first possibility is definitely not the correct one. Any one of the three possibilities might be true. Nor do the second and third possibilities preclude religion (a divine agency might have acted via evolution, for example). However, we are limiting the scope of our inquiry to scientific matters, and only the second and third possibilities permit testable hypotheses to be constructed—that is, explanations that can be tested and potentially disproved.

In our search for understanding, we must look back to the early times. There are fossils of simple living things, bacteria, in rocks 3.5 billion years old. They tell us that life originated during the first billion years of the history of our planet. As we attempt to determine how this process took place, we will first focus on how organic molecules may have originated (figure 4.5), and then we will consider how those molecules might have become organized into living cells.

Panspermia and spontaneous origin are the only testable hypotheses of life's origin currently available.

Scientists Disagree about Where Life Started

While most researchers agree that life first appeared as the primitive earth cooled and its rocky crust formed, there is little agreement as to just where this occurred.

Did Life Originate at the Ocean's Edge?

The more we learn about earth's early history, the more likely it seems that earth's first organisms emerged and lived at very high temperatures. Rubble from the forming solar system slammed into early earth from 4.6 to 3.8 billion years ago, keeping the surface molten hot. As the bombardment slowed down, temperatures dropped. By about 3.8 billion years ago, ocean temperatures are thought to have dropped to a hot 49° to 88°C (120° to 190°F). Between 3.8 and 3.5 billion years ago, life first appeared, promptly after the earth was inhabitable. Thus, as intolerable as early earth's infernal temperatures seem to us today, they gave birth to life.

Very few geochemists agree on the exact composition of the early atmosphere. One popular view is that it contained principally carbon dioxide (CO_2) and nitrogen gas (N_2), along with significant amounts of water vapor (H_2O). It is possible that the early atmosphere also contained hydrogen gas (H_2) and compounds in which hydrogen atoms were bonded to the other light elements (sulfur, nitrogen, and carbon), producing hydrogen sulfide (H_2S), ammonia (NH_3), and methane (CH_4).

We refer to such an atmosphere as a reducing atmosphere because of the ample availability of hydrogen atoms and their electrons. In such a reducing atmosphere it would not take as much energy as it would today to form the carbon-rich molecules from which life evolved.

The key to this reducing atmosphere hypothesis is the assumption that there was very little oxygen around. In an atmosphere with oxygen, amino acids and sugars react spontaneously with the oxygen to form carbon dioxide and water. Therefore, the building blocks of life, the amino acids, would not last long and the spontaneous formation of complex carbon molecules could not occur. Our atmosphere changed once organisms began to carry out photosynthesis, harnessing the energy in sunlight to split water molecules and form complex carbon molecules, giving off gaseous oxygen molecules in the process. The earth's atmosphere is now approximately 21% oxygen.

Critics of the reducing atmosphere hypothesis point out that no carbonates have been found in rocks dating back to the early earth. This suggests that at that time carbon dioxide was locked up in the atmosphere, and if that was the case, then the prebiotic atmosphere would not have been reducing.

Another problem for the reducing atmosphere hypothesis is that because a prebiotic reducing atmosphere would have been oxygen free, there would have been no ozone. Without the protective ozone layer, any organic compounds that might have formed would have been broken down quickly by ultraviolet radiation.

Other Suggestions

If life did not originate at the ocean's edge under the blanket of a reducing atmosphere, where did it originate?

Under frozen oceans. One hypothesis proposes that life originated under a frozen ocean, not unlike the one that covers Jupiter's moon Europa today. All evidence suggests, however, that the early earth was quite warm and frozen oceans quite unlikely.

Deep in the earth's crust. Another hypothesis is that life originated deep in the earth's crust. In 1988 Gunter Wachtershauser proposed that life might have formed as a by-product of volcanic activity, with iron and nickel sulfide minerals acting as chemical catalysts to recombine gases spewing from eruptions into the building blocks of life. In later work he and coworkers were able to use this unusual chemistry to build precursors for amino acids (although they did not actually succeed in making amino acids), and to link amino acids together to form peptides. Critics of this hypothesis point out that the concentration of chemicals used in their experiments greatly exceed what is found in nature.

Within clay. Other researchers have proposed the unusual hypothesis that life is the result of silicate surface chemistry. The surface of clays have positive charges to attract organic molecules, and exclude water, providing a potential catalytic surface on which life's early chemistry might have occurred. While interesting conceptually, there is little evidence that this sort of process could actually occur.

At deep-sea vents. Becoming more popular is the hypothesis that life originated at deep-sea hydrothermal vents, with the necessary prebiotic molecules being synthesized on metal sulfides in the vents. The positive charge of the sulfides would have acted as a magnet for negatively charged organic molecules. In part, the current popularity of this hypothesis comes from the new science of genomics, which suggests that the ancestors of today's prokaryotes are most closely related to the bacteria that live on the deep-sea vents.

No one is sure whether life originated at the ocean's edge, under frozen ocean, deep in the earth's crust, within clay, or at deep-sea vents. Perhaps one of these hypotheses will be proven correct. Perhaps the correct theory has not yet been proposed.

When life first appeared on earth, the environment was very hot. All of the spontaneous origin hypotheses assume that the organic chemicals that were the building blocks of life arose spontaneously at that time. How is a matter of considerable disagreement.

The Miller-Urey Experiment

An early attempt to see what kinds of organic molecules might have been produced on the early earth was carried out in 1953 by Stanley L. Miller and Harold C. Urey. In what has become a classic experiment, they attempted to reproduce the conditions at ocean's edge under a reducing atmosphere. Even if this assumption proves incorrect—the jury is still out on this—their experiment is critically important, as it ushered in the whole new field of prebiotic chemistry.

To carry out their experiment, they (1) assembled a reducing atmosphere rich in hydrogen and excluding gaseous oxygen; (2) placed this atmosphere over liquid water, which would have been present at ocean's edge; (3) maintained this mixture at a temperature somewhat below 100°C; and (4) simulated lightning by bombarding it with energy in the form of sparks (figure 4.6).

They found that within a week, 15% of the carbon originally present as methane gas (CH_4) had converted into other simple carbon compounds. Among these compounds were formaldehyde (CH_2O) and hydrogen cyanide (HCN; figure 4.7). These compounds then combined to form simple molecules, such as formic acid (HCOOH) and urea (NH_2CONH_2), and more complex molecules containing carbon-carbon bonds, including the amino acids glycine and alanine.

As we saw in chapter 3, amino acids are the basic building blocks of proteins, and proteins are one of the major kinds of molecules of which organisms are composed. In similar experiments performed later by other scientists, more than 30 different carbon compounds were identified, including the amino acids glycine, alanine, glutamic acid, valine, proline, and aspartic acid. Other biologically important molecules were also formed in these experiments. For example, hydrogen cyanide contributed to the production of a complex ring-shaped molecule called adenine—one of the bases found in DNA and RNA. Thus, the key molecules of life could have formed in the atmosphere of the early earth.

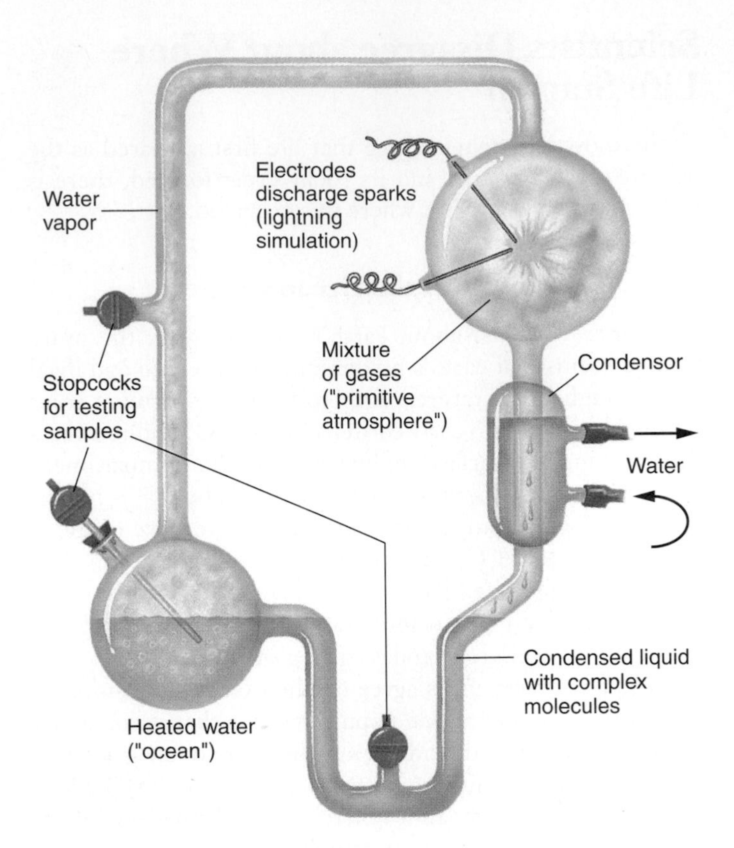

FIGURE 4.6
The Miller-Urey experiment. The apparatus consisted of a closed tube connecting two chambers. The upper chamber contained a mixture of gases thought to resemble the primitive earth's atmosphere. Electrodes discharged sparks through this mixture, simulating lightning. Condensers then cooled the gases, causing water droplets to form, which passed into the second heated chamber, the "ocean." Any complex molecules formed in the atmosphere chamber would be dissolved in these droplets and carried to the ocean chamber, from which samples were withdrawn for analysis.

The Path of Chemical Evolution

A raging debate among biologists who study the origin of life concerns which organic molecules came first, RNA or proteins. Scientists are divided into three camps, those that focus on RNA, protein, or a combination of the two. All three arguments have their strong points. Like the hypotheses that try to account for where life originated, these competing hypotheses are diverse and speculative.

An RNA World. The "RNA world" group feels that without a hereditary molecule, other molecules could not have formed consistently. The "RNA world" argument earned support when Thomas Cech at the University of Colorado discovered ribozymes, RNA molecules that can behave as enzymes, catalyzing their own assembly. Recent work has shown that the RNA contained in ribosomes (discussed in chapter 5) catalyzes the chemical reaction that links amino acids to form proteins. Therefore, the RNA in ribosomes also functions as an enzyme. If RNA has the ability to pass on inherited information and the capacity to act like an enzyme, were proteins really needed?

A Protein World. The "protein-first" group argues that without enzymes (which are proteins), nothing could replicate at all, heritable or not. The "protein-first" proponents argue that nucleotides, the individual units of nucleic acids such as RNA, are too complex to have formed spontaneously, and certainly too complex to form spontaneously again and again. While there is no doubt that simple proteins are easier to synthesize from abiotic components than nucleotides, both can form in the laboratory under the right conditions. Deciding which came first is a chicken-and-egg paradox. In an effort to shed light on this problem, Julius Rebek and a number of

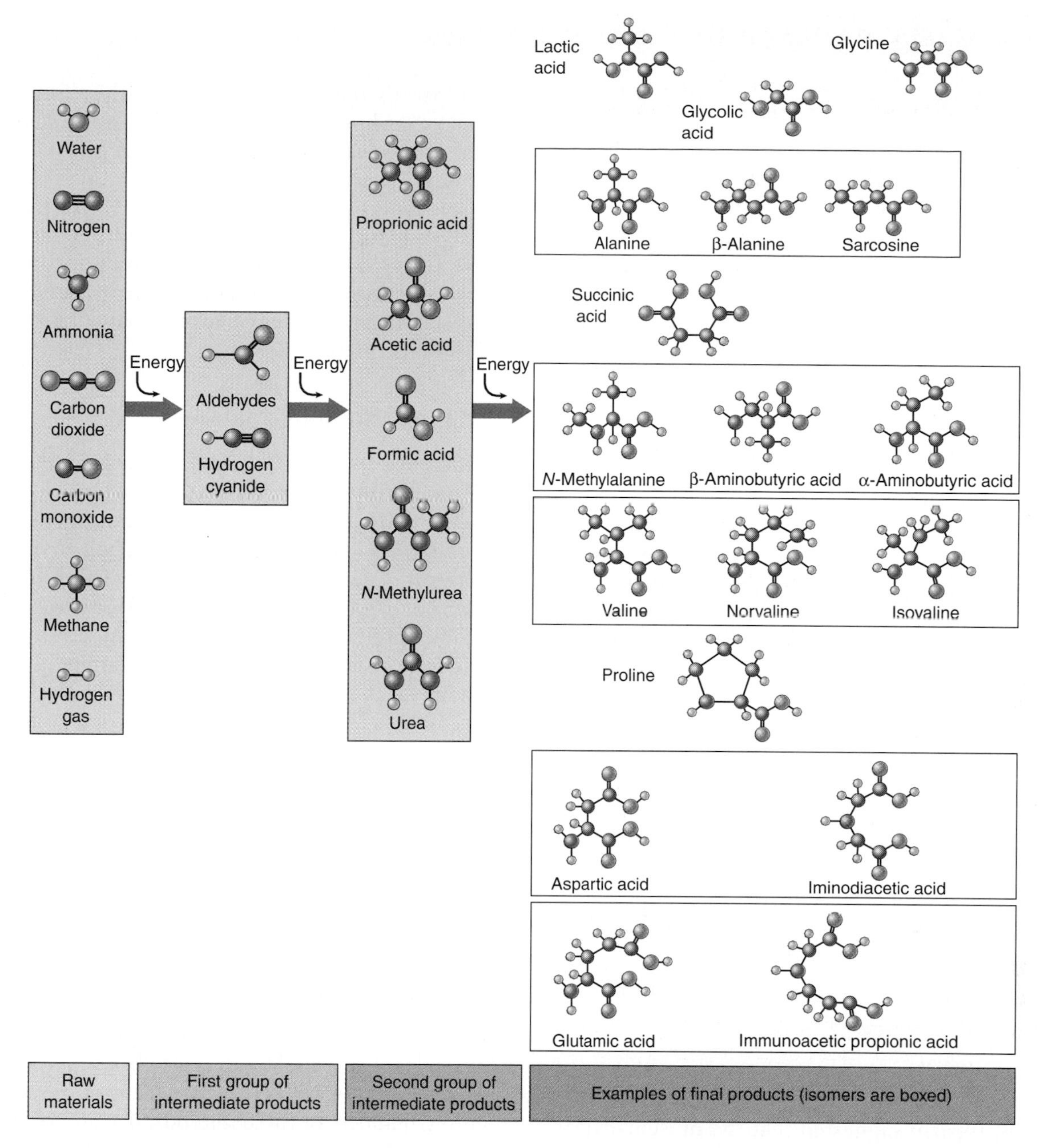

FIGURE 4.7
Results of the Miller-Urey experiment. Seven simple molecules, all gases, were included in the original mixture. Note that oxygen was not among them; instead, the atmosphere was rich in hydrogen. At each stage of the experiment, more complex molecules were formed: first aldehydes, then simple acids, then more complex acids. Among the final products, the molecules that are structural isomers of one another are grouped together in boxes. In most cases only one isomer of a compound is found in living systems today, although many may have been produced in the Miller-Urey experiment.

other chemists have created synthetic nucleotide-like molecules in the laboratory that are able to replicate. Moving even further, Rebek and his colleagues have created synthetic molecules that could replicate and "make mistakes." This simulates mutation, a necessary ingredient for the process of evolution.

A Peptide-Nucleic Acid World. Another important and popular theory about the first organic molecules assumes key roles for both peptides and nucleic acids. Because RNA is so complex and unstable, this theory assumes there must have been a pre-RNA world where the peptide-nucleic acid (PNA) was the basis for life. PNA is stable and simple enough to have formed spontaneously, and is also a self-replicator.

Molecules that are the building blocks of living organisms form spontaneously under conditions designed to simulate those of the primitive earth.

4.3 The first cells had little internal structure.

Theories about the Origin of Cells

The evolution of cells required early organic molecules to assemble into a functional, interdependent unit. Cells, discussed in the next chapter, are essentially little bags of fluid. What the fluid contains depends on the individual cell, but every cell's contents differ from the environment outside the cell. Therefore, an early cell may have floated along in a dilute "primordial soup," but its interior would have had a higher concentration of specific organic molecules.

Cell Origins: The Importance of Bubbles

How did these "bags of fluid" evolve from simple organic molecules? As you can imagine, the answer to this question is a matter for debate. Scientists favoring an "ocean's edge" scenario for the origin of life have proposed that bubbles may have played a key role in this evolutionary step. A bubble, such as those produced by soap solutions, is a hollow spherical structure. Certain molecules, particularly those with hydrophobic regions, will spontaneously form bubbles in water. The structure of the bubble shields the hydrophobic regions of the molecules from contact with water. If you have ever watched the ocean surge upon the shore, you may have noticed the foamy froth created by the agitated water. The edges of the primitive oceans were more than likely very frothy places bombarded by ultraviolet and other ionizing radiation, and exposed to an atmosphere that may have contained methane and other simple organic molecules.

Oparin's Bubble Theory

The first bubble theory is attributed to Alexander Oparin, a Russian chemist with extraordinary insight. In the mid-1930s, Oparin suggested that the present-day atmosphere was incompatible with the creation of life. He proposed that life must have arisen from nonliving matter under a set of very different environmental circumstances some time in the distant history of the earth. His was the theory of **primary abiogenesis** (primary because all living cells are now known to come from previously living cells, except in that first case). At the same time, J. B. S. Haldane, a British geneticist, was also independently espousing the same views. Oparin decided that in order for cells to evolve, they must have had some means of developing chemical complexity, separating their contents from their environment by means of a cell membrane, and concentrating materials within themselves. He termed these early, chemical-concentrating bubblelike structures **protobionts.**

Oparin's theories were published in English in 1938, and for awhile most scientists ignored them. However, Harold Urey, an astronomer at the University of Chicago, was quite taken with Oparin's ideas. He convinced one of his graduate students, Stanley Miller, to follow Oparin's rationale and see if he could "create" life. The Urey-Miller experiment has proven to be one of the most significant experiments in the history of science. As a result Oparin's ideas became better known and more widely accepted.

A Host of Bubble Theories

Different versions of "bubble theories" have been championed by numerous scientists since Oparin. The bubbles they propose go by a variety of names; they may be called *microspheres*, *protocells*, *protobionts*, *micelles*, *liposomes*, or *coacervates*, depending on the composition of the bubbles (lipid or protein) and how they form. In all cases, the bubbles are hollow spheres, and they exhibit a variety of cell-like properties. For example, the lipid bubbles called **coacervates** form an outer boundary with two layers that resembles a biological membrane. They grow by accumulating more subunit lipid molecules from the surrounding medium, and they can form budlike projections and divide by pinching in two, like bacteria. They also can contain amino acids and use them to facilitate various acid-base reactions, including the decomposition of glucose. Although they are not alive, they obviously have many of the characteristics of cells.

A Bubble Scenario

It is not difficult to imagine that a process of chemical evolution involving bubbles or microdrops preceded the origin of life (figure 4.8). The early oceans must have contained untold numbers of these microdrops, billions in a spoonful, each one forming spontaneously, persisting for a while, and then dispersing. Some would, by chance, have contained amino acids with side groups able to catalyze growth-promoting reactions. Those microdrops would have survived longer than ones that lacked those amino acids, because the persistence of both proteinoid microspheres and lipid coacervates is greatly increased when they carry out metabolic reactions such as glucose degradation and when they are actively growing.

Over millions of years, then, the complex bubbles that were better able to incorporate molecules and energy from the lifeless oceans of the early earth would have tended to persist longer than the others. Also favored would have been the microdrops that could use these molecules to expand in size, growing large enough to divide into "daughter"

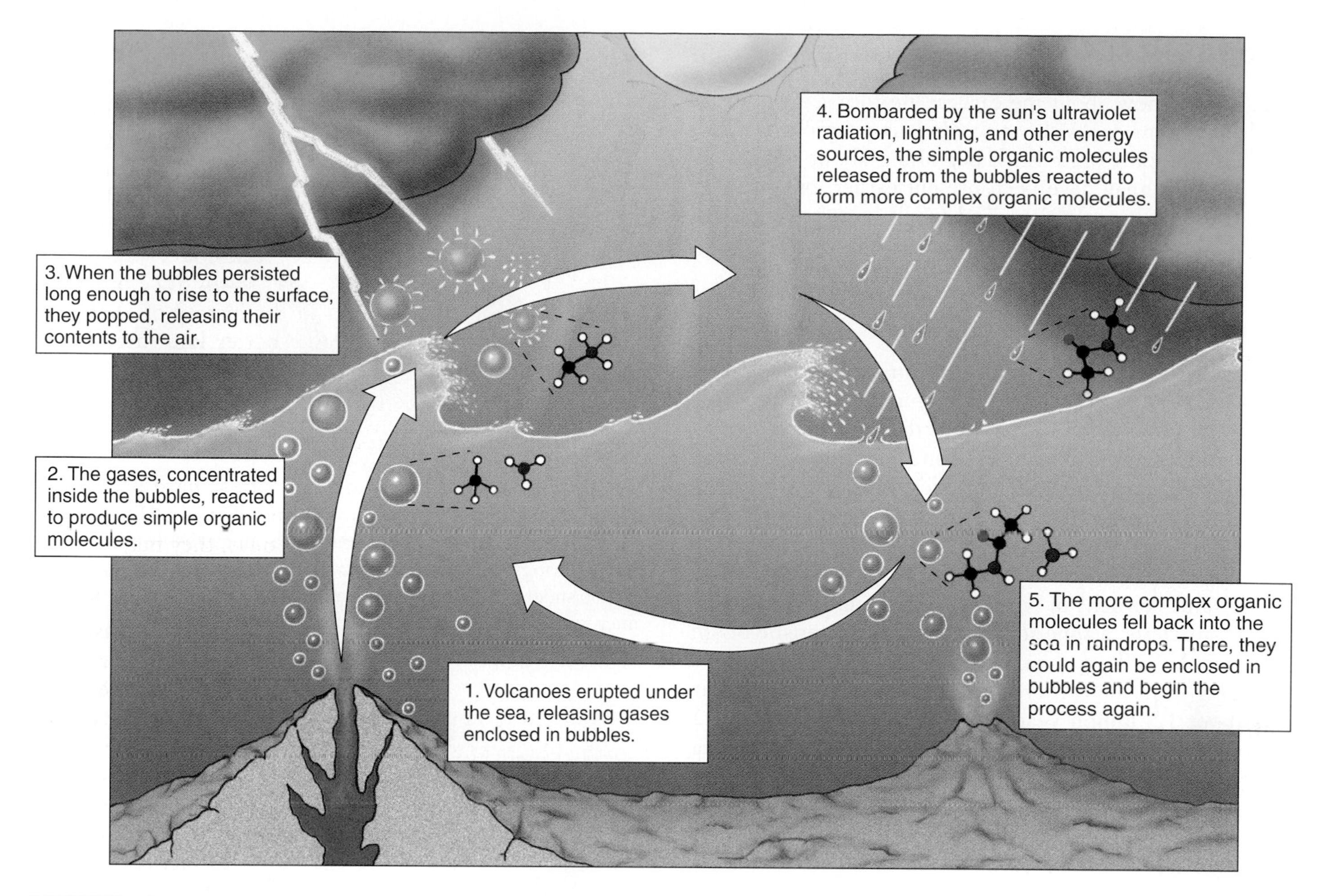

FIGURE 4.8
A current bubble hypothesis. In 1986 geophysicist Louis Lerman proposed that the chemical processes leading to the evolution of life took place within bubbles on the ocean's surface.

microdrops with features similar to those of their "parent" microdrop. The daughter microdrops have the same favorable combination of characteristics as their parent, and would have grown and divided, too. When a way to facilitate the reliable transfer of new ability from parent to offspring developed, heredity—and life—began.

Current Thinking

Whether the early bubbles that gave rise to cells were lipid or protein remains an unresolved argument. While it is true that lipid microspheres (coacervates) will form readily in water, there appears to be no mechanism for their heritable replication. On the other hand, one *can* imagine a heritable mechanism for protein microspheres. Although protein microspheres do not form readily in water, Sidney Fox and his colleagues at the University of Miami have shown that they can form under dry conditions.

The discovery that RNA can act as an enzyme to assemble new RNA molecules on an RNA template has raised the interesting possibility that neither coacervates nor protein microspheres were the first step in the evolution of life. Perhaps the first components were RNA molecules, and the initial steps on the evolutionary journey led to increasingly complex and stable RNA molecules. Later, stability might have improved further when a lipid (or possibly protein) microsphere surrounded the RNA. At present, those studying this problem have not arrived at a consensus about whether RNA evolved before or after a bubblelike structure that likely preceded cells.

Eventually, DNA took the place of RNA as the replicator in the cell and the storage molecule for genetic information. DNA, because it is a double helix, stores information in a more stable fashion than RNA, which is single-stranded.

Little is known about how the first cells originated. Current hypotheses involve chemical evolution within bubbles, but there is no general agreement about their composition, or about how the process occurred.

The Earliest Cells

What do we know about the earliest life-forms? The fossils found in ancient rocks show an obvious progression from simple to complex organisms, beginning about 3.5 billion years ago. Life may have been present earlier, but rocks of such great antiquity are rare, and fossils have not yet been found in them.

Microfossils

The earliest evidence of life appears in **microfossils,** fossilized forms of microscopic life (figure 4.9). Microfossils were small (1 to 2 micrometers in diameter) and single-celled, lacked external appendages, and had little evidence of internal structure. Thus, they physically resemble present-day **bacteria** (figure 4.10), although some ancient forms cannot be matched exactly. We call organisms with this simple body plan **prokaryotes,** from the Greek words meaning "before" and "kernel," or "nucleus." The name reflects their lack of a **nucleus,** a spherical organelle characteristic of the more complex cells of **eukaryotes.**

Judging from the fossil record, eukaryotes did not appear until about 1.5 billion years ago. Therefore, for at least 2 billion years—nearly a half of the age of the earth—bacteria were the only organisms that existed.

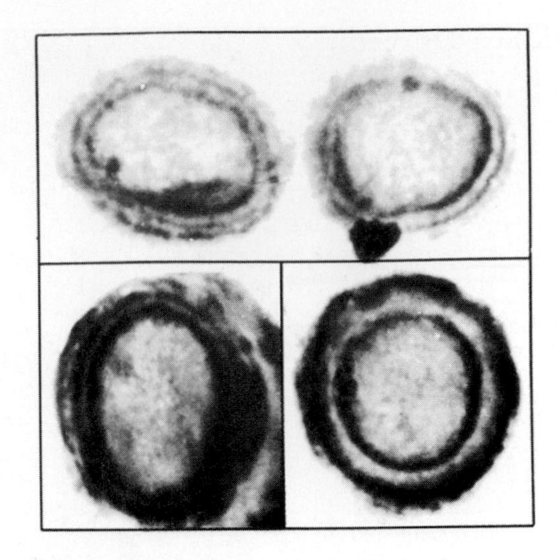

FIGURE 4.9
Cross-sections of fossil bacteria. These microfossils from the Bitter Springs formation of Australia are of ancient cyanobacteria, far too small to be seen with the unaided eye. In this electron micrograph, the cell walls are clearly evident.

Ancient Bacteria: Archaebacteria

Most organisms living today are adapted to the relatively mild conditions of present-day earth. However, if we look in unusual environments, we encounter organisms that are quite remarkable, differing in form and metabolism from other living things. Sheltered from evolutionary alteration in unchanging habitats that resemble earth's early environment, these living relics are the surviving representatives of the first ages of life on earth. In places such as the oxygen-free depths of the Black Sea or the boiling waters of hot springs and deep-sea vents, we can find bacteria living at very high temperatures without oxygen.

These unusual bacteria are called **archaebacteria,** from the Greek word for "ancient ones." Among the first to be studied in detail have been the **methanogens,** or methane-producing bacteria, among the most primitive bacteria that exist today. These organisms are typically simple in form and are able to grow only in an oxygen-free environment; in fact, oxygen poisons them. For this reason they are said to grow "without air," or **anaerobically** (Greek *an*, "without" + *aer*, "air" + *bios*, "life"). The methane-producing bacteria convert CO_2 and H_2 into methane gas (CH_4). Although primitive, they resemble all other bacteria in having DNA, a lipid cell membrane, an exterior cell wall, and a metabolism based on an energy-carrying molecule called ATP.

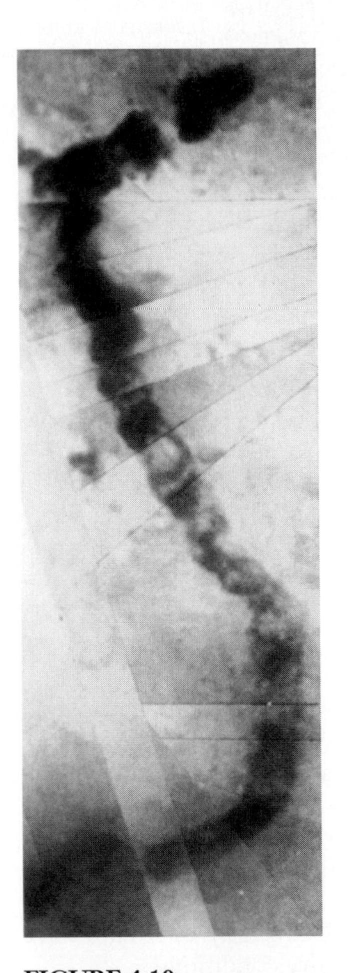

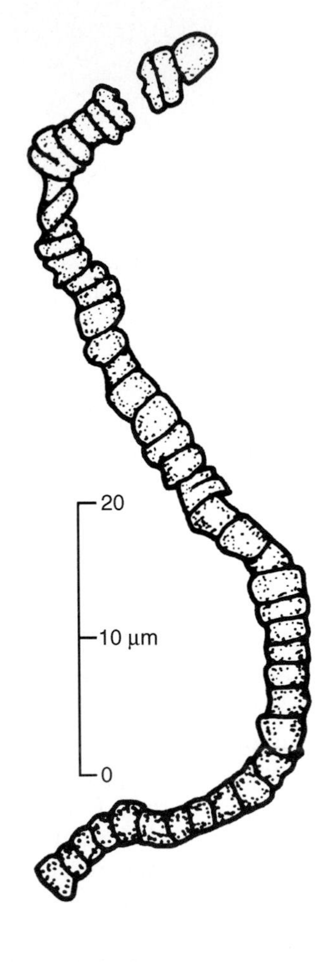

FIGURE 4.10
The oldest microfossil. This ancient bacterial fossil, discovered by J. William Schopf of UCLA in 3.5-billion-year-old rocks in western Australia, is similar to present-day cyanobacteria, as you can see by comparing it to figure 4.11.

Unusual Cell Structures

When the details of cell wall and membrane structure of the methane-producing bacteria were examined, they proved to be different from those of all other bacteria. Archaebacteria are characterized by a conspicuous lack of a protein cross-linked carbohydrate material called **peptidoglycan** in their cell walls, a key compound in the cell walls of most modern bacteria. Archaebacteria also have unusual lipids in their cell membranes that are not found in any other group of organisms. There are also major differences in some of the fundamental biochemical processes of metabolism, different from those of all other bacteria. The methane-producing bacteria are survivors from an earlier time when oxygen gas was absent.

Earth's First Organisms?

Other archaebacteria that fall into this classification are some of those that live in very salty environments like the Dead Sea (**extreme halophiles**—"salt lovers") or very hot environments like hydrothermal volcanic vents under the ocean (**extreme thermophiles**—"heat lovers"). Thermophiles have been found living comfortably in boiling water. Indeed, many kinds of thermophilic archaebacteria thrive at temperatures of 110°C (230°F). Because these thermophiles live at high temperatures similar to those that may have existed when life first evolved, microbiologists speculate that thermophilic archaebacteria may be relics of earth's first organisms.

Just how different are extreme thermophiles from other organisms? A methane-producing archaebacteria called *Methanococcus* isolated from deep-sea vents provides a startling picture. These bacteria thrive at temperatures of 88°C (185°F) and crushing pressures 245 times greater than at sea level. In 1996 molecular biologists announced that they had succeeded in determining the full nucleotide sequence of *Methanococcus*. This was possible because archaebacterial DNA is relatively small—it has only 1700 genes, coded in a DNA molecule only 1,739,933 nucleotides long (a human cell has 2000 times more!). The thermophile nucleotide sequence proved to be astonishingly different from the DNA sequence of any other organism ever studied; fully two-thirds of its genes are unlike any ever known to science before! Clearly these archaebacteria separated from other life on earth a long time ago. Preliminary comparisons to the gene sequences of other bacteria suggest that archaebacteria split from other types of bacteria over 3 billion years ago, soon after life began.

Eubacteria

The second major group of bacteria, the **eubacteria,** have very strong cell walls and a simpler gene architecture. Most bacteria living today are eubacteria. Included in this group are bacteria that have evolved the ability to capture the energy of light and transform it into the energy of chemical bonds within cells. These organisms are *photosynthetic*, as are plants and algae.

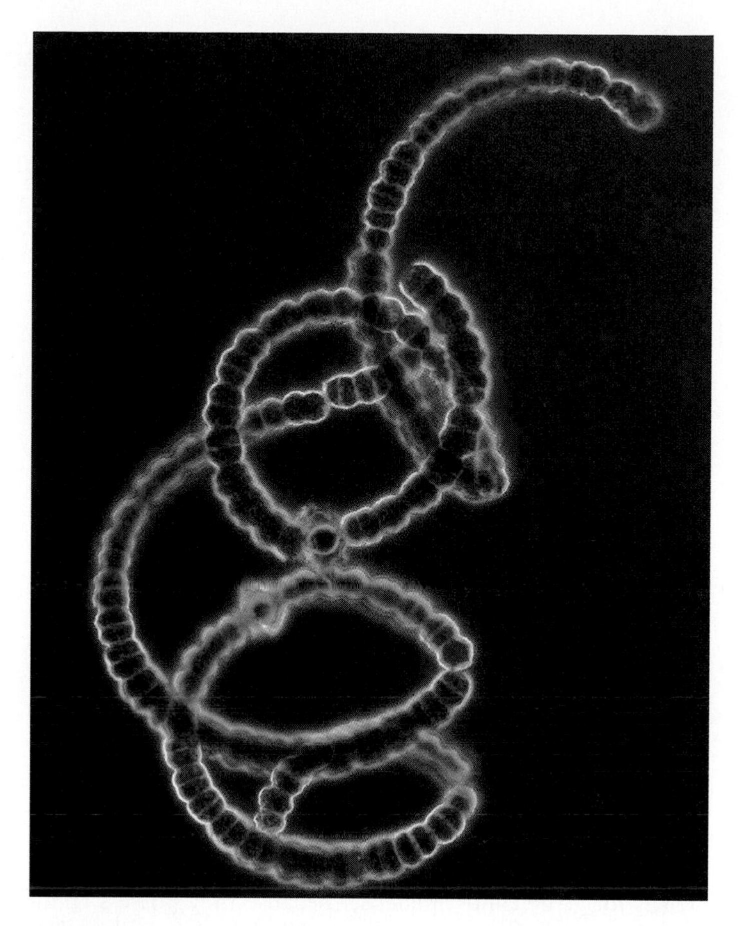

FIGURE 4.11
Living cyanobacteria. Although not multicellular, these bacteria often aggregate into chains such as those seen here.

One type of photosynthetic eubacteria that has been important in the history of life on earth is the cyanobacteria, sometimes called "blue-green algae" (figure 4.11). They have the same kind of chlorophyll pigment that is most abundant in plants and algae, as well as other pigments that are blue or red. Cyanobacteria produce oxygen as a result of their photosynthetic activities, and when they appeared at least 3 billion years ago, they played a decisive role in increasing the concentration of free oxygen in the earth's atmosphere from below 1% to the current level of 21%. As the concentration of oxygen increased, so did the amount of ozone in the upper layers of the atmosphere. The thickening ozone layer afforded protection from most of the ultraviolet radiation from the sun, radiation that is highly destructive to proteins and nucleic acids. Certain cyanobacteria are also responsible for the accumulation of massive limestone deposits.

All bacteria now living are members of either Archaebacteria or Eubacteria.

4.4 The first eukaryotic cells were larger and more complex than bacteria.

All fossils more than 1.5 billion years old are generally similar to one another structurally. They are small, simple cells; most measure 0.5 to 2 micrometers in diameter, and none are more than about 6 micrometers in diameter. These simple cells eventually evolved into larger, more complex forms—the first eukaryotic cells.

The First Eukaryotic Cells

In rocks about 1.5 billion years old, we begin to see the first microfossils that are noticeably different in appearance from the earlier, simpler forms (figure 4.12). These cells are much larger than bacteria and have internal membranes and thicker walls. Cells more than 10 micrometers in diameter rapidly increased in abundance. Some fossilized cells 1.4 billion years old are as much as 60 micrometers in diameter; others, 1.5 billion years old, contain what appear to be small, membrane-bound structures. Indirect chemical traces hint that eukaryotes may go as far back as 2.7 billion years, although no fossils as yet support such an early appearance of eukaryotes.

These early fossils mark a major event in the evolution of life: a new kind of organism had appeared (figure 4.13). These new cells are called **eukaryotes,** from the Greek words for "true" and "nucleus," because they possess an internal structure called a nucleus. All organisms other than the bacteria are eukaryotes.

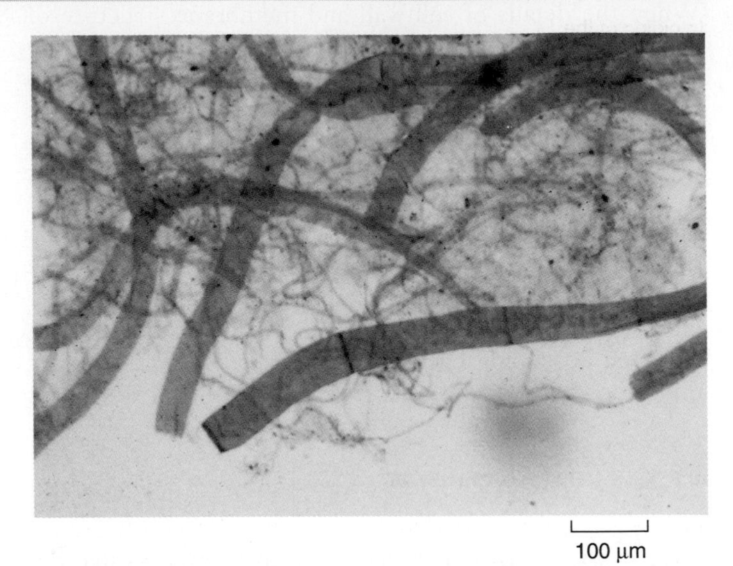

FIGURE 4.12
Microfossil of a primitive eukaryote. This multicellular alga is between 900 million and 1 billion years old.

Origin of the Nucleus and ER

Many bacteria have infoldings of their outer membranes extending into the cytoplasm and serving as passageways to the surface. The network of internal membranes in

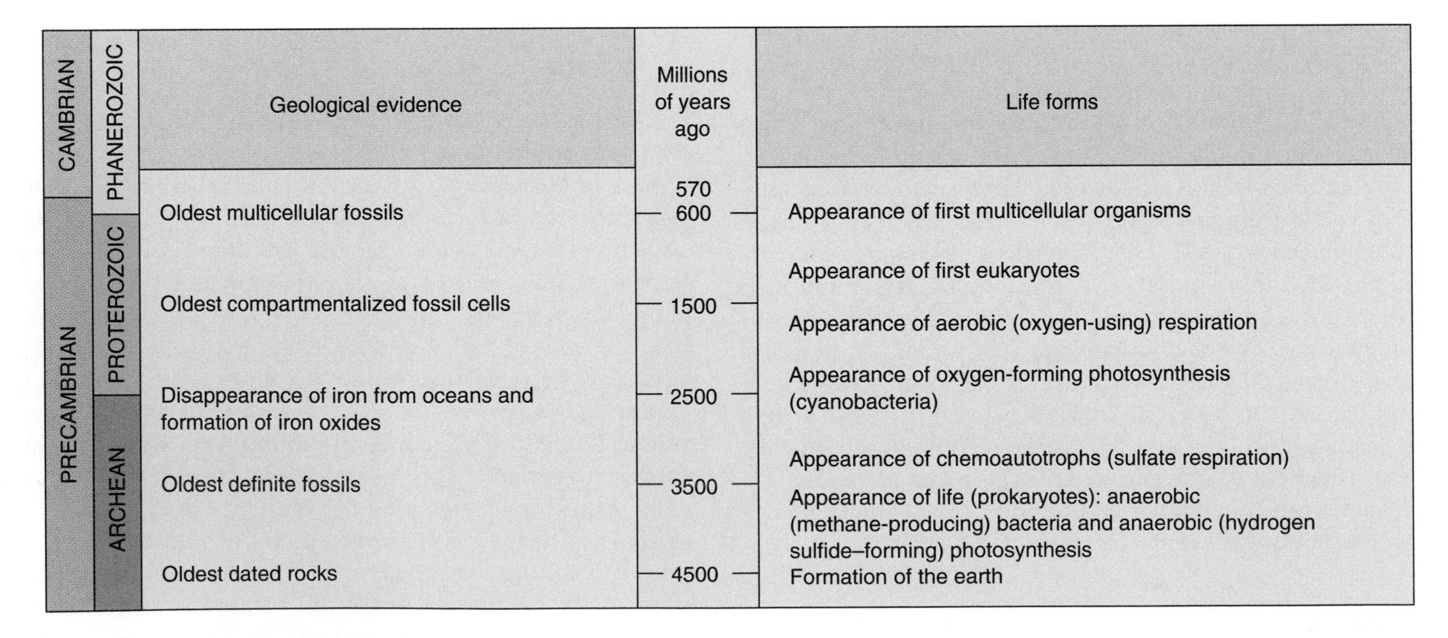

CAMBRIAN
PHANEROZOIC
PRECAMBRIAN
PROTEROZOIC
ARCHEAN
Geological evidence
Millions of years ago
Life forms
570
Oldest multicellular fossils
600
Appearance of first multicellular organisms
Appearance of first eukaryotes
Oldest compartmentalized fossil cells
1500
Appearance of aerobic (oxygen-using) respiration
Appearance of oxygen-forming photosynthesis (cyanobacteria)
Disappearance of iron from oceans and formation of iron oxides
2500
Appearance of chemoautotrophs (sulfate respiration)
Oldest definite fossils
3500
Appearance of life (prokaryotes): anaerobic (methane-producing) bacteria and anaerobic (hydrogen sulfide–forming) photosynthesis
Oldest dated rocks
4500
Formation of the earth

FIGURE 4.13
The geological timescale. The periods refer to different stages in the evolution of life on earth. The timescale is calibrated by examining rocks containing particular kinds of fossils; the fossils are dated by determining the degree of spontaneous decay of radioactive isotopes locked within rock when it was formed.

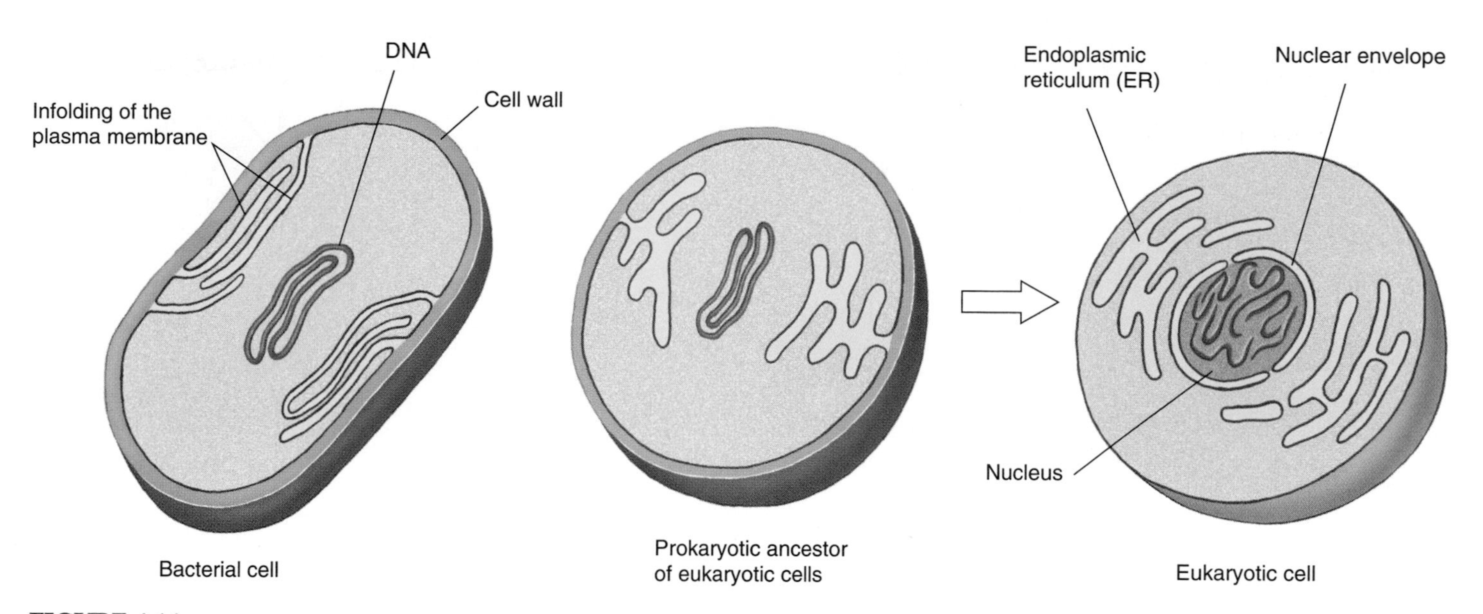

FIGURE 4.14
Origin of the nucleus and endoplasmic reticulum. Many bacteria today have infoldings of the plasma membrane (see also figure 34.7). The eukaryotic internal membrane system called the endoplasmic reticulum (ER) and the nuclear envelope may have evolved from such infoldings of the plasma membrane encasing prokaryotic cells that gave rise to eukaryotic cells.

eukaryotes called endoplasmic reticulum (ER) is thought to have evolved from such infoldings, as is the nuclear envelope, an extension of the ER network that isolates and protects the nucleus (figure 4.14).

Origin of Mitochondria and Chloroplasts

Bacteria that live within other cells and perform specific functions for their host cells are called *endosymbiotic bacteria*. Their widespread presence in nature led Lynn Margulis to champion the **endosymbiotic theory** in the early 1970s. This theory, now widely accepted, suggests that a critical stage in the evolution of eukaryotic cells involved endosymbiotic relationships with prokaryotic organisms. According to this theory, energy-producing bacteria may have come to reside within larger bacteria, eventually evolving into what we now know as mitochondria. Similarly, photosynthetic bacteria may have come to live within other larger bacteria, leading to the evolution of chloroplasts, the photosynthetic organelles of plants and algae. Bacteria with flagella, long whiplike cellular appendages used for propulsion, may have become symbiotically involved with nonflagellated bacteria to produce larger, motile cells. The fact that we now witness so many symbiotic relationships lends general support to this theory. Even stronger support comes from the observation that present-day organelles such as mitochondria, chloroplasts, and centrioles contain their own DNA, which is remarkably similar to the DNA of bacteria in size and character.

Sexual Reproduction

Eukaryotic cells also possess the ability to reproduce sexually, something prokaryotes cannot do effectively. **Sexual reproduction** is the process of producing offspring, with two copies of each chromosome, by fertilization, the union of two cells that each have one copy of each chromosome. The great advantage of sexual reproduction is that it allows for frequent genetic recombination, which generates the variation that is the raw material for evolution. Not all eukaryotes reproduce sexually, but most have the capacity to do so. The evolution of meiosis and sexual reproduction (discussed in chapter 12) led to the tremendous explosion of diversity among the eukaryotes.

Multicellularity

Diversity was also promoted by the development of **multicellularity.** Some single eukaryotic cells began living in association with others, in colonies. Eventually individual members of the colony began to assume different duties, and the colony began to take on the characteristics of a single individual. Multicellularity has arisen many times among the eukaryotes. Practically every organism big enough to see with the unaided eye is multicellular, including all animals and plants. The great advantage of multicellularity is that it fosters specialization; some cells devote all of their energies to one task, other cells to another. Few innovations have had as great an impact on the history of life as the specialization made possible by multicellularity.

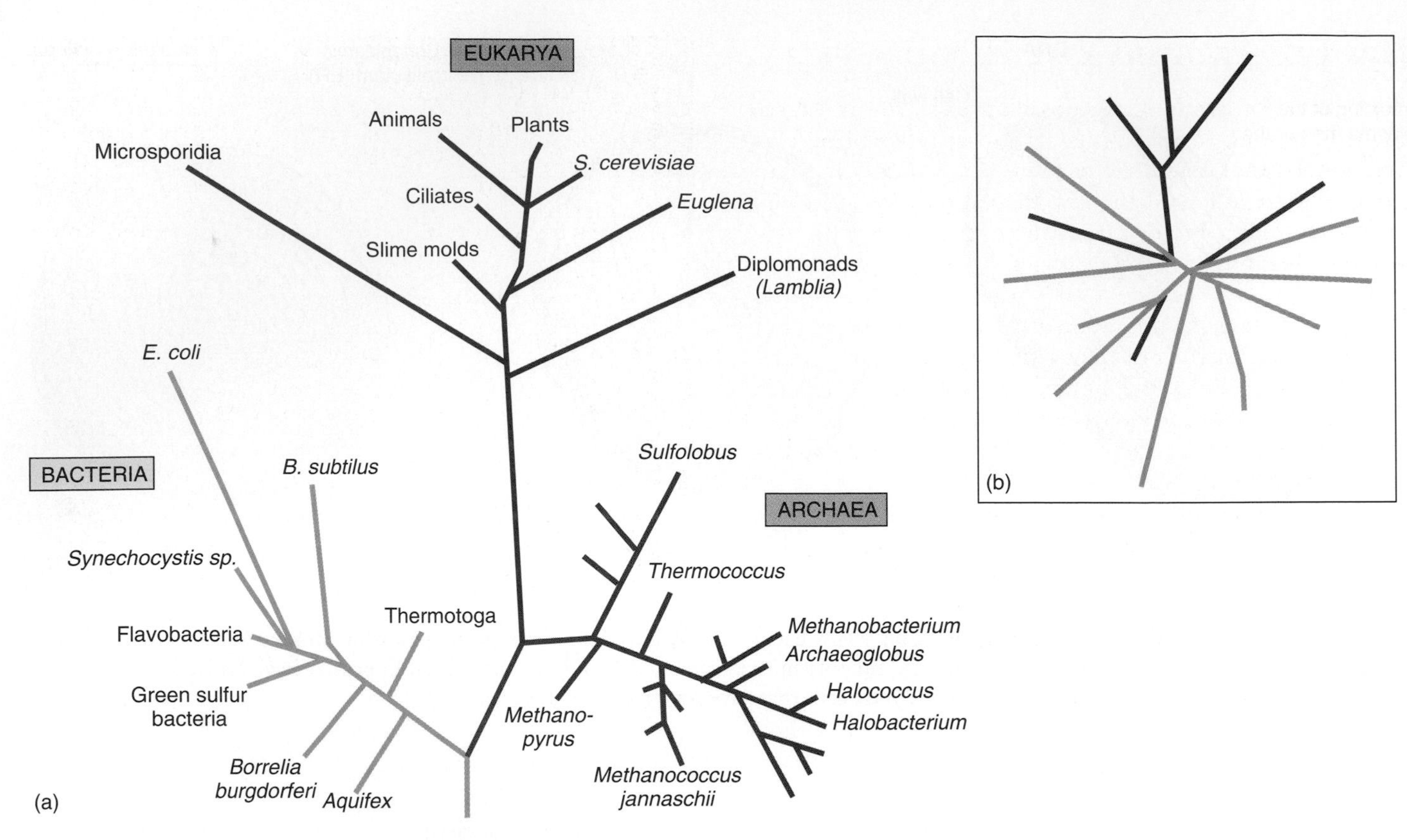

FIGURE 4.15
The three domains of life. The kingdoms Archaebacteria and Eubacteria are as different from each other as from eukaryotes, so biologists have assigned them a higher category, a "domain." (*a*) A three-domain tree of life based on ribosomal RNA consists of the Eukarya, Bacteria, and Archaea. (*b*) New analyses of complete genome sequences contradict the rRNA tree, and suggest other arrangements such as this one, which splits the Archaea. Apparently genes hopped from branch to branch as early organisms either stole genes from their food or swapped DNA with their neighbors, even distantly related ones.

The Kingdoms of Life

Confronted with the great diversity of life on earth today, biologists have attempted to categorize similar organisms in order to better understand them, giving rise to the science of taxonomy. In later chapters, we will discuss taxonomy and classification in detail, but for now we can generalize that all living things fall into one of three domains which include six kingdoms (figure 4.15):

Kingdom Archaebacteria: Prokaryotes that lack a peptidoglycan cell wall, including the methanogens and extreme halophiles and thermophiles.

Kingdom Eubacteria: Prokaryotic organisms with a peptidoglycan cell wall, including cyanobacteria, soil bacteria, nitrogen-fixing bacteria, and pathogenic (disease-causing) bacteria.

Kingdom Protista: Eukaryotic, primarily unicellular (although algae are multicellular), photosynthetic or heterotrophic organisms, such as amoebas and paramecia.

Kingdom Fungi: Eukaryotic, mostly multicellular (although yeasts are unicellular), heterotrophic, usually nonmotile organisms, with cell walls of chitin, such as mushrooms.

Kingdom Plantae: Eukaryotic, multicellular, nonmotile, usually terrestrial, photosynthetic organisms, such as trees, grasses, and mosses.

Kingdom Animalia: Eukaryotic, multicellular, motile, heterotrophic organisms, such as sponges, spiders, newts, penguins, and humans.

As more is learned about living things, particularly from the newer evidence that DNA studies provide, scientists will continue to reevaluate the relationships among the kingdoms of life.

For at least the first 1 billion years of life on earth, all organisms were bacteria. About 1.5 billion years ago, the first eukaryotes appeared. Biologists place living organisms into six general categories called kingdoms.

Has Life Evolved Elsewhere?

We should not overlook the possibility that life processes might have evolved in different ways on other planets. A functional genetic system, capable of accumulating and replicating changes and thus of adaptation and evolution, could theoretically evolve from molecules other than carbon, hydrogen, nitrogen, and oxygen in a different environment. Silicon, like carbon, needs four electrons to fill its outer energy level, and ammonia is even more polar than water. Perhaps under radically different temperatures and pressures, these elements might form molecules as diverse and flexible as those carbon has formed on earth.

The universe has 10^{20} (100,000,000,000,000,000,000) stars similar to our sun. We don't know how many of these stars have planets, but it seems increasingly likely that many do. Since 1996, astronomers have been detecting planets orbiting distant stars. At least 10% of stars are thought to have planetary systems. If only 1 in 10,000 of these planets is the right size and at the right distance from its star to duplicate the conditions in which life originated on earth, the "life experiment" will have been repeated 10^{15} times (that is, a million billion times). It does not seem likely that we are alone.

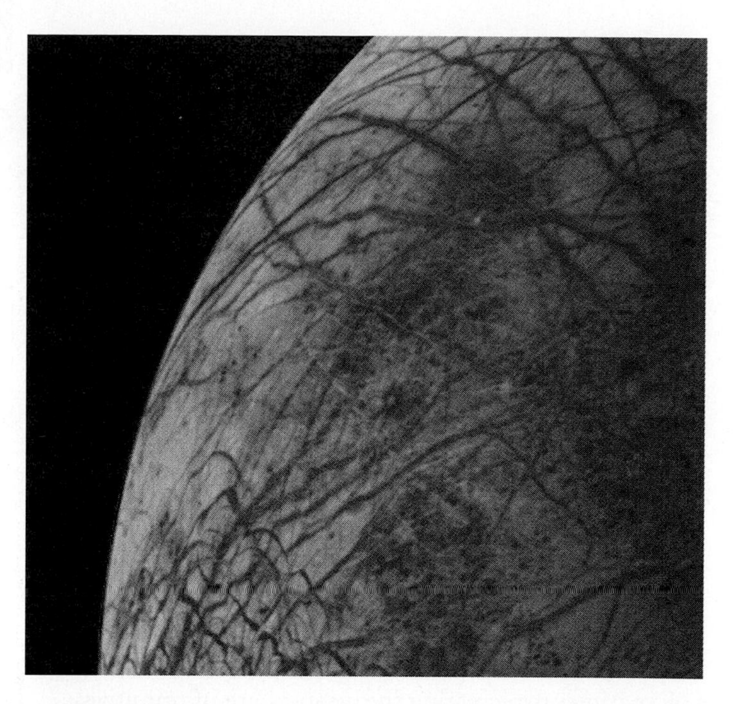

FIGURE 4.16
Is there life elsewhere? Currently the most likely candidate for life elsewhere within the solar system is Europa, one of the many moons of the large planet Jupiter.

Ancient Bacteria on Mars?

A dull gray chunk of rock collected in 1984 in Antarctica ignited an uproar about ancient life on Mars with the report that the rock contains evidence of possible life. Analysis of gases trapped within small pockets of the rock indicate it is a meteorite from Mars. It is, in fact, the oldest rock known to science—fully 4.5 billion years old. Back then, when this rock formed on Mars, that cold, arid planet was much warmer, flowed with water, and had a carbon dioxide atmosphere—conditions not too different from those that spawned life on earth.

When examined with powerful electron microscopes, carbonate patches within the meteorite exhibit what look like microfossils, some 20 to 100 nanometers in length. One hundred times smaller than any known bacteria, it is not clear they actually are fossils, but the resemblance to bacteria is striking.

Viewed as a whole, the evidence of bacterial life associated with the Mars meteorite is not compelling. Clearly, more painstaking research remains to be done before the discovery can claim a scientific consensus. However, while there is no conclusive evidence of bacterial life associated with this meteorite, it seems very possible that life has evolved on other worlds in addition to our own.

Deep-Sea Vents

The possibility that life on earth actually originated in the vicinity of deep-sea hydrothermal vents is gaining popularity. At the bottom of the ocean, where these vents spewed out a rich froth of molecules, the geological turbulence and radioactive energy battering the land was absent, and things were comparatively calm. The thermophilic archaebacteria found near these vents today are the most ancient group of organisms living on earth. Perhaps the gentler environment of the ocean depths was the actual cradle of life.

Other Planets

Has life evolved on other worlds within our solar system? There are planets other than ancient Mars with conditions not unlike those on earth. Europa, a large moon of Jupiter, is a promising candidate (figure 4.16). Europa is covered with ice, and photos taken in close orbit in the winter of 1998 reveal seas of liquid water beneath a thin skin of ice. Additional satellite photos taken in 1999 suggest that a few miles under the ice lies a liquid ocean of water larger than earth's, warmed by the push and pull of the gravitational attraction of Jupiter's many large satellite moons. The conditions on Europa now are far less hostile to life than the conditions that existed in the oceans of the primitive earth. In coming decades satellite missions are scheduled to explore this ocean for life.

There are so many stars that life may have evolved many times. Although evidence for life on Mars is not compelling, the seas of Europa offer a promising candidate which scientists are eager to investigate.

Chapter 4

www.mhhe.com/raven6/resources4.mhtml

Summary | **Questions** | **Media Resources**

4.1 All living things share key characteristics.

- All living things are characterized by cellular organization, growth, reproduction, and heredity.
- Other properties commonly exhibited by living things include movement and sensitivity to stimuli.

1. What characteristics of living things are *necessary* characteristics (possessed by all living things), and which are *sufficient* characteristics (possessed only by living things)?

4.2 There are many ideas about the origin of life.

- Of the many explanations of how life might have originated, only the theories of spontaneous and extraterrestrial origins provide scientifically testable explanations.
- Experiments recreating the atmosphere of primitive earth, with the energy sources and temperatures thought to be prevalent at that time, have led to the spontaneous formation of amino acids and other biologically significant molecules.

2. What molecules are thought to have been present in the atmosphere of the early earth? Which molecule that was notably absent then is now a major component of the atmosphere?

- Origin of Life

- Art Quizzes:
  - -Miller-Urey Experiment
  - -Miller-Urey Experiment Results

4.3 The first cells had little internal structure.

- The first cells are thought to have arisen from aggregations of molecules that were more stable and, therefore, persisted longer.
- It has been suggested that RNA may have arisen before cells did, and subsequently became packaged within a membrane.
- Bacteria were the only life-forms on earth for about 1 billion years. At least three kinds of bacteria were present in ancient times: methane utilizers, anaerobic photosynthesizers, and eventually O_2-forming photosynthesizers.

3. What evidence supports the argument that RNA evolved first on the early earth? What evidence supports the argument that proteins evolved first?

4. What are coacervates, and what characteristics do they have in common with organisms? Are they alive? Why or why not?

5. What were the earliest known organisms like, and when did they appear? What present-day organisms do they resemble?

- Art Quiz: Current Bubble Hypothesis

4.4 The first eukaryotic cells were larger and more complex than bacteria.

- The first eukaryotes can be seen in the fossil record about 1.5 billion years ago. All organisms other than bacteria are their descendants.
- Biologists group all living organisms into six "kingdoms," each profoundly different from the others.
- The two most ancient kingdoms contain prokaryotes (bacteria); the other four contain eukaryotes.
- There are approximately 10^{20} stars in the universe similar to our sun. It is almost certain that life has evolved on planets circling some of them.

6. When did the first eukaryotes appear? By what mechanism are they thought to have evolved from the earlier prokaryotes?

7. What sorts of organisms are contained in each of the six kingdoms of life recognized by biologists?

- Key Events in Earth's History

Part **II**

Biology of the Cell

Seeing cortical microtubules. Cortical microtubules in epidermal cells of a fava bean are tagged with a flourescent protein so that their ordered array can be seen.

Real People Doing Real Science

How Do the Cells of a Growing Plant Know in Which Direction to Elongate?

Sometimes questions that seem simple can be devilishly difficult to answer. Imagine, for example, that you are holding a green blade of grass in your hand. The grass blade has been actively growing, its cells dividing and then stretching and elongating as the blade lengthens. Did you ever wonder how the individual cells within the blade of grass know in what direction to grow?

To answer this deceptively simple question, we will first need to provide answers to several others. Like Sherlock Holmes following a trail of clues, we must approach the answer we seek in stages.

Question One. First, we need to ask how a blade of grass is able to grow at all. Plant cells are very different from animal cells in one key respect: every plant cell is encased within a tough cell wall made of cellulose and other tough building materials. This wall provides structural strength and protection to the plant cell, just as armor plate does for a battle tank. But battle tanks can't stretch into longer shapes! How is a plant cell able to elongate?

It works like this. A growing cell first performs a little chemistry to make its wall slightly acidic. The acidity activates enzymes that attack the cell wall from the inside, rearranging cellulose cross-links until the wall loses its rigidity. The cell wall is now able to stretch. The cell then sucks in water, creating pressure. Like blowing up a long balloon, the now-stretchable cell elongates.

Question Two. In a growing plant organ, like the blade of grass, each growing cell balloons out lengthwise. Stating this more formally, a botanist would say the cell elongates parallel to the axis along which the blade of grass is extending. This observation leads to the second question we must answer: How does an individual plant cell control the direction in which it elongates?

It works like this. Before the stretchable cell balloons out, tiny microfibrils of cellulose are laid down along its inside surface. On a per weight basis, these tiny fibrils have the tensile strength of steel! Arrays of these cellulose microfibrils are organized in bands perpendicular to the axis of elongation, like steel belts. These tough bands reinforce the plant cell wall laterally, so that when the cell sucks in water, there is only one way for the cell to expand—lengthwise, along the axis.

Question Three. Now we're getting somewhere. How are the newly made cellulose microfibrils laid down so that they are oriented correctly, perpendicular to the axis of elongation?

It works like this. The complicated enzymic machine that makes the cellulose microfibrils is guided by special guiderails that run like railroad tracks along the interior surface. The enzyme complex travels along these guiderails, laying down microfibrils as it goes. The guiderails are constructed of chainlike protein molecules called microtubules, assembled into overlapping arrays. Botanists call these arrays of microtubules associated with the interior of the cell surface "cortical microtubules."

Question Four. But we have only traded one puzzle for another. How are the cortical microtubules positioned correctly, perpendicular to the axis of elongation?

It works like this. In newly made cells, the microtubule assemblies are already present, but are not organized. They simply lie about in random disarray. As the cell prepares to elongate by lessening the rigidity of its cell wall, the microtubule assemblies become organized into the orderly transverse arrays we call cortical microtubules.

Question Five. Finally, we arrive at the question we had initially set out to answer. How are microtubule assemblies aligned properly? What sort of signal directs them to orient perpendicular to the axis of elongation? THAT is the question we need to answer.

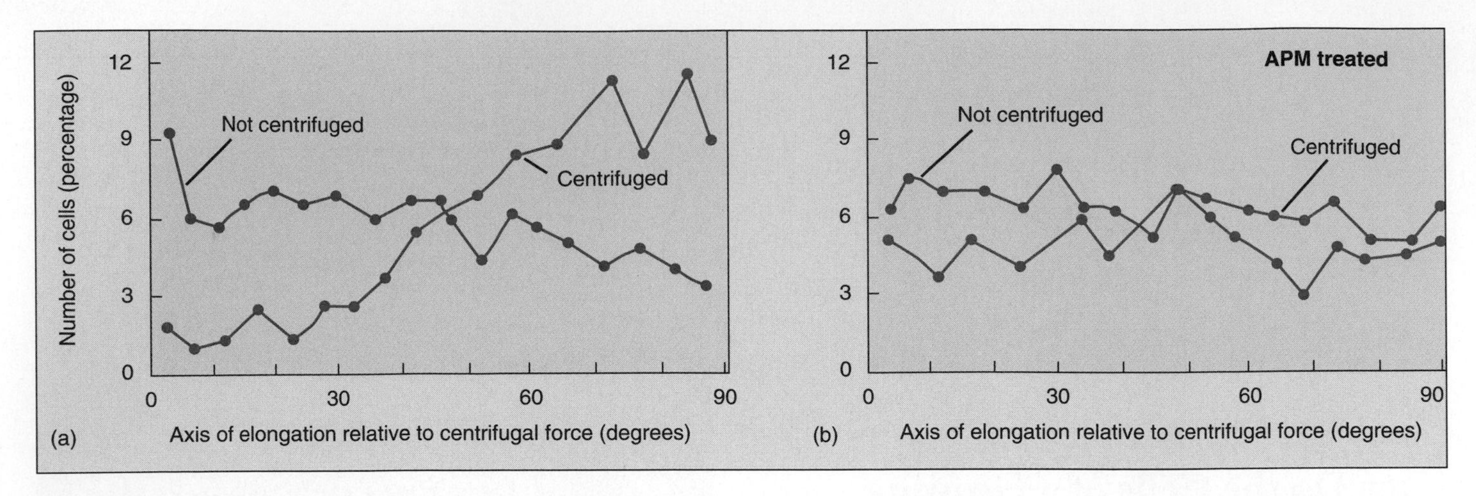

Effects of centrifugation on cell elongation. (*a*) Protoplasts (plant cells without cell walls) that were centrifuged showed preferential elongation in a direction approximately perpendicular to the direction of the force. (*b*) Protoplasts exposed to APM, a microtubule disrupting chemical, exhibited random cell elongation with or without centrifugation.

The Experiment

This question has been addressed experimentally in a simple and direct way in the laboratory of Richard Cyr at Pennsylvania State University. Rigid plant cells conduct mechanical force well from one cell to another, and Carol Wymer (then a graduate student in the Cyr lab) suspected some sort of mechanical force is the signal guiding cortical microtubule alignment

Wymer set out to test this hypothesis using centrifugation. If cortical microtubules are obtaining their positional information from an applied force, then their alignment should be affected by centrifugal force, and should be impossible if the integrity of the cell wall (which is supposedly transmitting the mechanical force) is perturbed with chemicals that prevent cell wall formation.

Wymer, along with others in the Cyr lab, started out with cells that were not elongated. She isolated protoplasts (cells without walls) from the tobacco plant, *Nicotiana tabacum*, by exposing the plant cells to enzymes that break down the cell wall, creating a spherical plant cell. If allowed to grow in culture, these protoplasts will eventually re-form their cell walls.

In order to examine the effects of directional force on the elongation patterns of plant cells, Wymer and co-workers exposed the tobacco protoplasts to a directional force generated by a centrifuge. Prior experiments had determined that centrifugation at the low speeds used in these experiments does not disrupt the integrity or shape of the protoplasts. The protoplasts were immobilized for centrifugation by embedding them in an agar medium supported in a mold. The embedded protoplasts were spun in a centrifuge at 450 rpm for 15 minutes. Following centrifugation, the embedded cells were cultured for 72 hours, allowing for cell elongation to occur.

Following centrifugation, fluorescently tagged microtubule antibody was applied to the protoplasts, which were then examined with immunofluorescence microscopy for microtubule orientation.

To confirm the involvement of microtubules as sensors of directional force in cell elongation, some protoplasts were incubated prior to centrifugation with a chemical herbicide, APM, which disrupts microtubules.

The Results

The biophysical force of centrifugation had significant effects on the pattern of elongation in the protoplasts following the 72-hour culturing period. The microtubules were randomly arranged in protoplasts that were not centrifuged but were more ordered in protoplasts that had been centrifuged. The microtubules in these cells were oriented parallel to the direction of the force, in a direction approximately perpendicular to the axis of elongation (graph *a* above). These results support the hypothesis that plant cell growth responds to an external biophysical force.

It is true that plant cells are not usually exposed to the type of mechanical force generated by centrifugation but this manipulation demonstrates how a physical force can affect cell growth, assumably by influencing the orientation of cortical microtubules. These could be small, transient biophysical forces acting at the subcellular level.

In preparations exposed to the microtubule disrupting chemical amiprophos-methyl (APM), directed elongation was blocked (graph *b* above). This suggests that reorientation of microtubules is indeed necessary to direct the elongation axis of the plant

Taken together, these experiments support the hypothesis that the microtubule reorientation that directs cell elongation may be oriented by a mechanical force. Just what the natural force might be is an open question, providing an opportunity for lots of interesting future experiments that are being pursued in the Cyr lab.

To explore this experiment further, go to the Virtual Lab at www.mhhe.com/raven6/vlab2.mhtml

5

Cell Structure

Concept Outline

5.1 All organisms are composed of cells.

Cells. A cell is a membrane-bounded unit that contains DNA and cytoplasm. All organisms are cells or aggregates of cells, descendants of the first cells.

Cells Are Small. The greater relative surface area of small cells enables more rapid communication between the cell interior and the environment.

5.2 Eukaryotic cells are far more complex than bacterial cells.

Bacteria Are Simple Cells. Bacterial cells are small and lack membrane-bounded organelles.

Eukaryotic Cells Have Complex Interiors. Eukaryotic cells are compartmentalized by membranes.

5.3 Take a tour of a eukaryotic cell.

The Nucleus: Information Center for the Cell. The nucleus of a eukaryotic cell isolates the cell's DNA.

The Endoplasmic Reticulum: Compartmentalizing the Cell. An extensive system of membranes subdivides the cell interior.

The Golgi Apparatus: Delivery System of the Cell. A system of membrane channels collects, modifies, packages, and distributes molecules within the cell.

Vesicles: Enzyme Storehouses. Sacs that contain enzymes digest or modify particles in the cell, while other vesicles transport substances in and out of cells.

Ribosomes: Sites of Protein Synthesis. An RNA-protein complex directs the production of proteins.

Organelles That Contain DNA. Some organelles with very different functions contain their own DNA.

The Cytoskeleton: Interior Framework of the Cell. A network of protein fibers supports the shape of the cell and anchors organelles.

Cell Movement. Eukaryotic cell movement utilizes cytoskeletal elements.

Special Things about Plant Cells. Plant cells have a large central vacuole and strong, multilayered cell walls.

5.4 Symbiosis played a key role in the origin of some eukaryotic organelles.

Endosymbiosis. Mitochondria and chloroplasts may have arisen from prokaryotes engulfed by other prokaryotes.

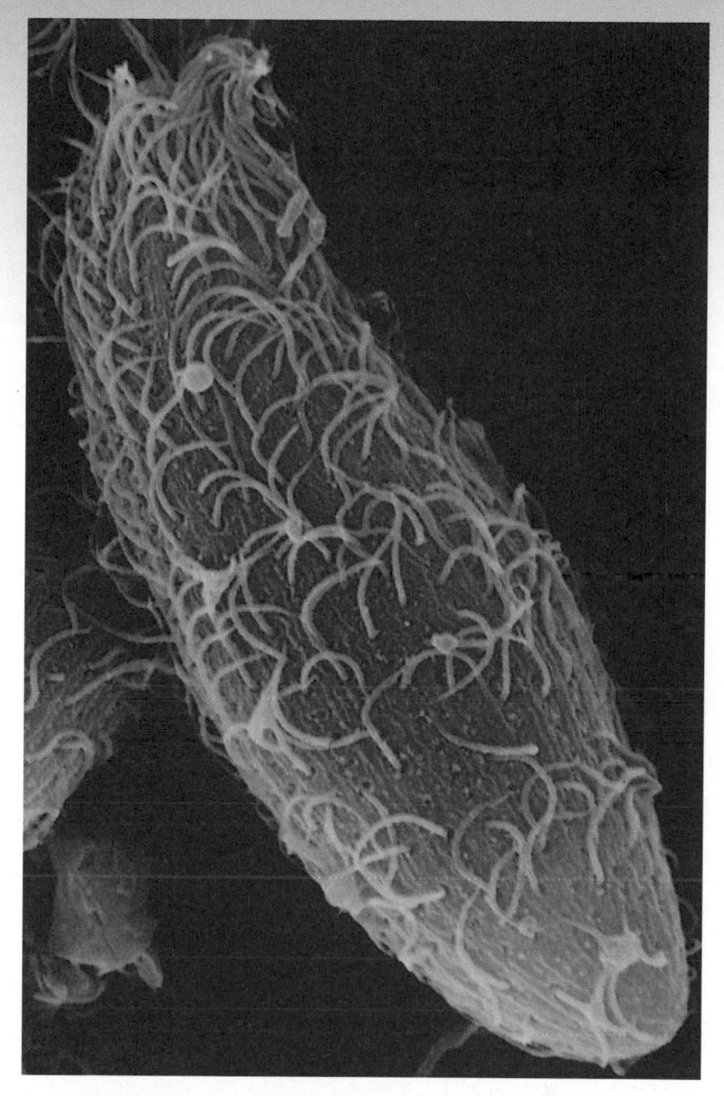

FIGURE. 5.1
The single-celled organism *Dileptus*. The hairlike projections that cover its surface are cilia, which it undulates to propel itself through the water (1000×).

All organisms are composed of cells. The gossamer wing of a butterfly is a thin sheet of cells, and so is the glistening outer layer of your eyes. The hamburger or tomato you eat is composed of cells, and its contents soon become part of your cells. Some organisms consist of a single cell too small to see with the unaided eye (figure 5.1), while others, like us, are composed of many cells. Cells are so much a part of life as we know it that we cannot imagine an organism that is not cellular in nature. In this chapter we will take a close look at the internal structure of cells. In the following chapters, we will focus on cells in action—on how they communicate with their environment, grow, and reproduce.

5.1 All organisms are composed of cells.

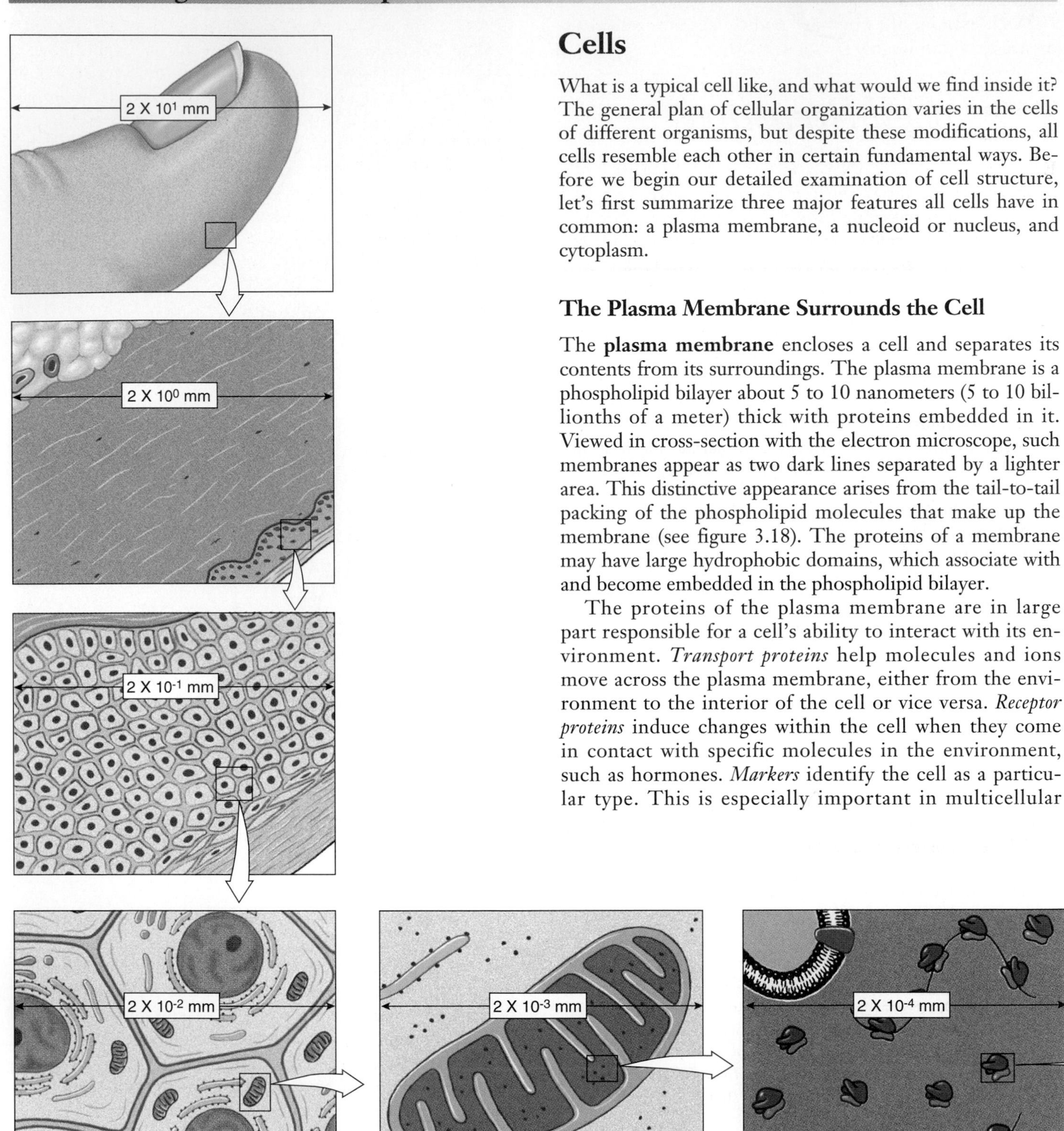

Cells

What is a typical cell like, and what would we find inside it? The general plan of cellular organization varies in the cells of different organisms, but despite these modifications, all cells resemble each other in certain fundamental ways. Before we begin our detailed examination of cell structure, let's first summarize three major features all cells have in common: a plasma membrane, a nucleoid or nucleus, and cytoplasm.

The Plasma Membrane Surrounds the Cell

The **plasma membrane** encloses a cell and separates its contents from its surroundings. The plasma membrane is a phospholipid bilayer about 5 to 10 nanometers (5 to 10 billionths of a meter) thick with proteins embedded in it. Viewed in cross-section with the electron microscope, such membranes appear as two dark lines separated by a lighter area. This distinctive appearance arises from the tail-to-tail packing of the phospholipid molecules that make up the membrane (see figure 3.18). The proteins of a membrane may have large hydrophobic domains, which associate with and become embedded in the phospholipid bilayer.

The proteins of the plasma membrane are in large part responsible for a cell's ability to interact with its environment. *Transport proteins* help molecules and ions move across the plasma membrane, either from the environment to the interior of the cell or vice versa. *Receptor proteins* induce changes within the cell when they come in contact with specific molecules in the environment, such as hormones. *Markers* identify the cell as a particular type. This is especially important in multicellular

FIGURE 5.2
The size of cells and their contents. This diagram shows the size of human skin cells, organelles, and molecules. In general, the diameter of a human skin cell is 20 micrometers (μm) or 2×10^{-2} mm, of a mitochondrion is 2 μm or 2×10^{-3} mm, of a ribosome is 20 nanometers (nm) or 2×10^{-5} mm, of a protein molecule is 2 nm or 2×10^{-6} mm, and of an atom is 0.2 nm or 2×10^{-7} mm.

organisms, whose cells must be able to recognize each other as they form tissues.

We'll examine the structure and function of cell membranes more thoroughly in chapter 6.

The Central Portion of the Cell Contains the Genetic Material

Every cell contains DNA, the hereditary molecule. In **prokaryotes** (bacteria), most of the genetic material lies in a single circular molecule of DNA. It typically resides near the center of the cell in an area called the **nucleoid,** but this area is not segregated from the rest of the cell's interior by membranes. By contrast, the DNA of **eukaryotes** is contained in the **nucleus,** which is surrounded by two membranes. In both types of organisms, the DNA contains the genes that code for the proteins synthesized by the cell.

The Cytoplasm Comprises the Rest of the Cell's Interior

A semifluid matrix called the **cytoplasm** fills the interior of the cell, exclusive of the nucleus (nucleoid in prokaryotes) lying within it. The cytoplasm contains the chemical wealth of the cell: the sugars, amino acids, and proteins the cell uses to carry out its everyday activities. In eukaryotic cells, the cytoplasm also contains specialized membrane-bounded compartments called **organelles.**

The Cell Theory

A general characteristic of cells is their microscopic size. While there are a few exceptions—the marine alga *Acetabularia* can be up to 5 centimeters long—a typical eukaryotic cell is 10 to 100 micrometers (10 to 100 millionths of a meter) in diameter (figure 5.2); most bacterial cells are only 1 to 10 micrometers in diameter.

Because cells are so small, no one observed them until microscopes were invented in the mid-seventeenth century. Robert Hooke first described cells in 1665, when he used a microscope he had built to examine a thin slice of cork, a nonliving tissue found in the bark of certain trees. Hooke observed a honeycomb of tiny, empty (because the cells were dead) compartments. He called the compartments in the cork *cellulae* (Latin, "small rooms"), and the term has come down to us as *cells.* The first living cells were observed a few years later by the Dutch naturalist Antonie van Leeuwenhoek, who called the tiny organisms that he observed "animalcules," meaning little animals. For another century and a half, however, biologists failed to recognize the importance of cells. In 1838, botanist Matthias Schleiden made a careful study of plant tissues and developed the first statement of the cell theory. He stated that all plants "are aggregates of fully individualized, independent, separate beings, namely the cells themselves." In 1839, Theodor Schwann reported that all animal tissues also consist of individual cells.

The **cell theory,** in its modern form, includes the following three principles:

1. All organisms are composed of one or more cells, and the life processes of metabolism and heredity occur within these cells.
2. Cells are the smallest living things, the basic units of organization of all organisms.
3. Cells arise only by division of a previously existing cell. Although life likely evolved spontaneously in the environment of the early earth, biologists have concluded that no additional cells are originating spontaneously at present. Rather, life on earth represents a continuous line of descent from those early cells.

A cell is a membrane-bounded unit that contains the DNA hereditary machinery and cytoplasm. All organisms are cells or aggregates of cells.

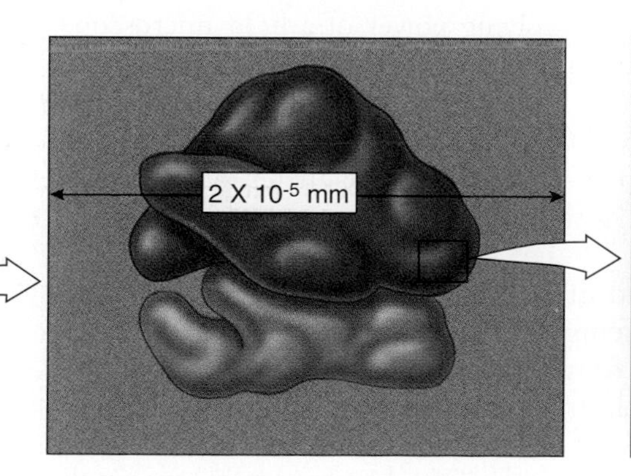

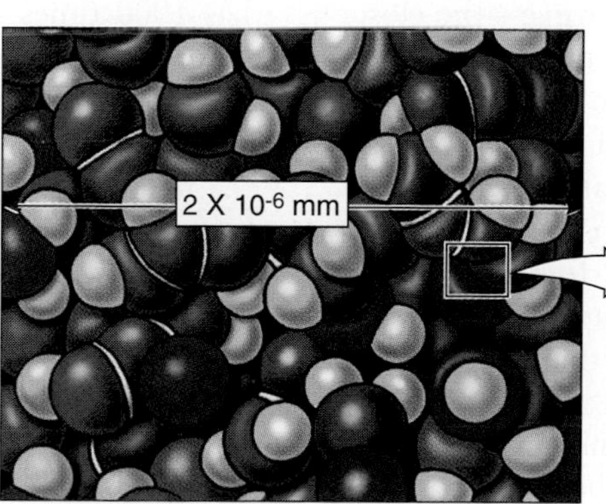

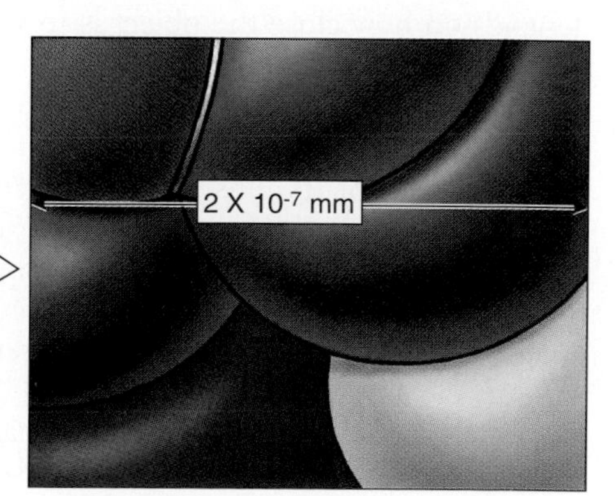

Cells Are Small

How many cells are big enough to see with the unaided eye? Other than egg cells, not many. Most are less than 50 micrometers in diameter, far smaller than the period at the end of this sentence.

The Resolution Problem

How do we study cells if they are too small to see? The key is to understand *why* we can't see them. The reason we can't see such small objects is the limited resolution of the human eye. **Resolution** is defined as the minimum distance two points can be apart and still be distinguished as two separated points. When two objects are closer together than about 100 micrometers, the light reflected from each strikes the same "detector" cell at the rear of the eye. Only when the objects are farther than 100 micrometers apart will the light from each strike different cells, allowing your eye to resolve them as two objects rather than one.

(a)

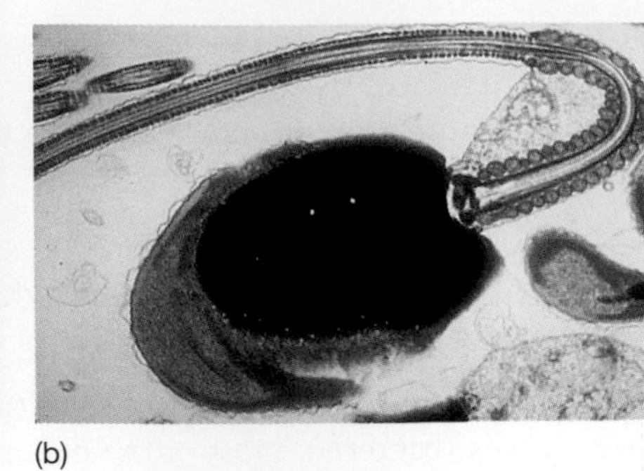

(b)

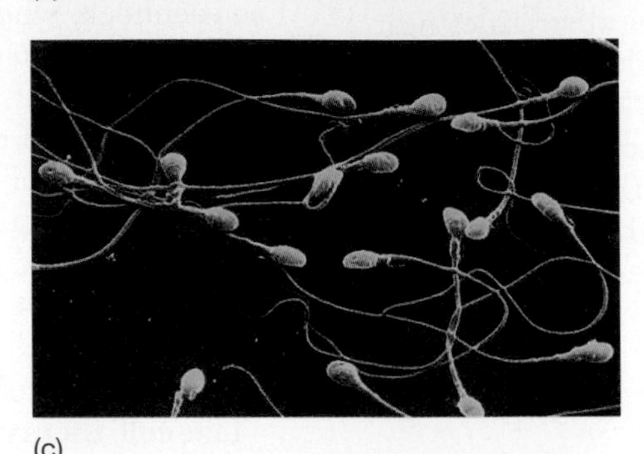

(c)

FIGURE 5.3
Human sperm cells viewed with three different microscopes. (*a*) Image of sperm taken with a light microscope. (*b*) Transmission electron micrograph of a sperm cell. (*c*) Scanning electron micrograph of sperm cells.

Microscopes

One way to increase resolution is to increase magnification, so that small objects appear larger. Robert Hooke and Antonie van Leeuwenhoek were able to see small cells by magnifying their size, so that the cells appeared larger than the 100-micrometer limit imposed by the human eye. Hooke and van Leeuwenhoek accomplished this feat with **microscopes** that magnified images of cells by bending light through a glass lens. The size of the image that falls on the sheet of detector cells lining the back of your eye depends on how close the object is to your eye—the closer the object, the bigger the image. Your eye, however, is incapable of focusing comfortably on an object closer than about 25 centimeters, because the eye is limited by the size and thickness of its lens. Hooke and van Leeuwenhoek assisted the eye by interposing a glass lens between object and eye. The glass lens adds additional focusing power. Because the glass lens makes the object appear closer, the image on the back of the eye is bigger than it would be without the lens.

Modern light microscopes use two magnifying lenses (and a variety of correcting lenses) that act like back-to-back eyes. The first lens focuses the image of the object on the second lens, which magnifies it again and focuses it on the back of the eye. Microscopes that magnify in stages using several lenses are called **compound microscopes.** They can resolve structures that are separated by more than 200 nm. An image from a compound microscope is shown in figure 5.3*a*.

Increasing Resolution

Light microscopes, even compound ones, are not powerful enough to resolve many structures within cells. For example, a membrane is only 5 nanometers thick. Why not just add another magnifying stage to the microscope and so increase its resolving power? Because when two objects are closer than a few hundred nanometers, the light beams reflecting from the two images start to overlap. The only way two light beams can get closer together and still be resolved is if their "wavelengths" are shorter.

One way to avoid overlap is by using a beam of electrons rather than a beam of light. Electrons have a much shorter wavelength, and a microscope employing electron beams has 1000 times the resolving power of a light microscope. **Transmission electron microscopes,** so called because the electrons used to visualize the specimens are transmitted through the material, are capable of resolving objects only 0.2 nanometer apart—just twice the diameter of a hydrogen atom! Figure 5.3*b* shows a transmission electron micrograph.

A second kind of electron microscope, the **scanning electron microscope,** beams the electrons onto the surface of the specimen from a fine probe that passes rapidly back and forth. The electrons reflected back from the surface of the specimen, together with other electrons that the specimen itself emits as a result of the bombardment, are amplified and transmitted to a television screen, where the image can be viewed and photographed. Scanning electron microscopy yields striking three-dimensional images and has improved our understanding of many biological and physical phenomena (figure 5.3*c*).

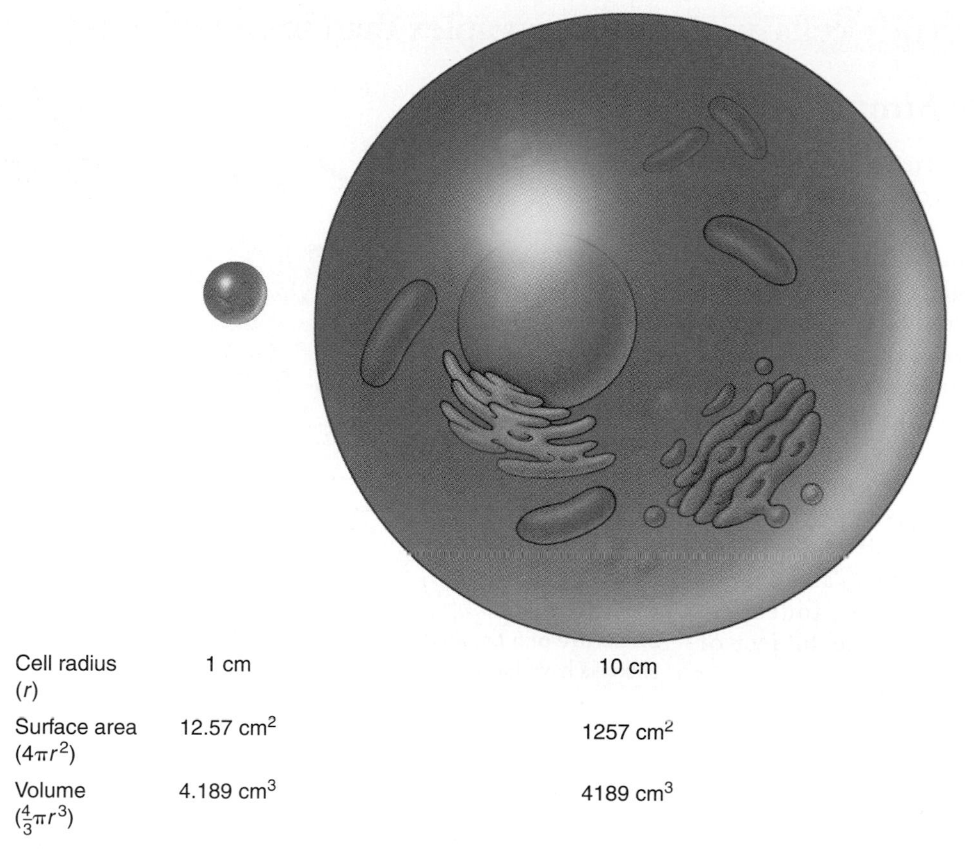

FIGURE 5.4
Surface area-to-volume ratio. As a cell gets larger, its volume increases at a faster rate than its surface area. If the cell radius increases by 10 times, the surface area increases by 100 times, but the volume increases by 1000 times. A cell's surface area must be large enough to meet the needs of its volume.

Why Aren't Cells Larger?

Most cells are not large for practical reasons. The most important of these is communication. The different regions of a cell need to communicate with one another in order for the cell as a whole to function effectively. Proteins and organelles are being synthesized, and materials are continually entering and leaving the cell. All of these processes involve the diffusion of substances at some point, and the larger a cell is, the longer it takes for substances to diffuse from the plasma membrane to the center of the cell. For this reason, an organism made up of many relatively small cells has an advantage over one composed of fewer, larger cells.

The advantage of small cell size is readily visualized in terms of the **surface area-to-volume ratio.** As a cell's size increases, its volume increases much more rapidly than its surface area. For a spherical cell, the increase in surface area is equal to the square of the increase in diameter, while the increase in volume is equal to the cube of the increase in diameter. Thus, if two cells differ by a factor of 10 cm in diameter, the larger cell will have 10^2, or 100 times, the surface area, but 10^3, or 1000 times, the volume, of the smaller cell (figure 5.4). A cell's surface provides its only opportunity for interaction with the environment, as all substances enter and exit a cell via the plasma membrane. This membrane plays a key role in controlling cell function, and because small cells have more surface area per unit of volume than large ones, the control is more effective when cells are relatively small.

Although most cells are small, some cells are nonetheless quite large and have apparently overcome the surface area-to-volume problem by one or more adaptive mechanisms. For example, some cells have more than one nucleus, allowing genetic information to be spread around a large cell. Also, some large cells actively move material around their cytoplasm so that diffusion is not a limiting factor. Lastly, some large cells, like your own neurons, are long and skinny so that any given point in the cytoplasm is close to the plasma membrane, and thus diffusion between the inside and outside of the cell can still be rapid.

Multicellular organisms usually consist of many small cells rather than a few large ones because small cells function more efficiently. They have a greater relative surface area, enabling more rapid communication between the center of the cell and the environment.

5.2 Eukaryotic cells are far more complex than bacterial cells.

Bacteria Are Simple Cells

Prokaryotes, the bacteria, are the simplest organisms. Prokaryotic cells are small, consisting of cytoplasm surrounded by a plasma membrane and encased within a rigid cell wall, with no distinct interior compartments (figure 5.5). A prokaryotic cell is like a one-room cabin in which eating, sleeping, and watching TV all occur in the same room. Bacteria are very important in the economy of living organisms. They harvest light in photosynthesis, break down dead organisms and recycle their components, cause disease, and are involved in many important industrial processes. Bacteria are the subject of chapter 34.

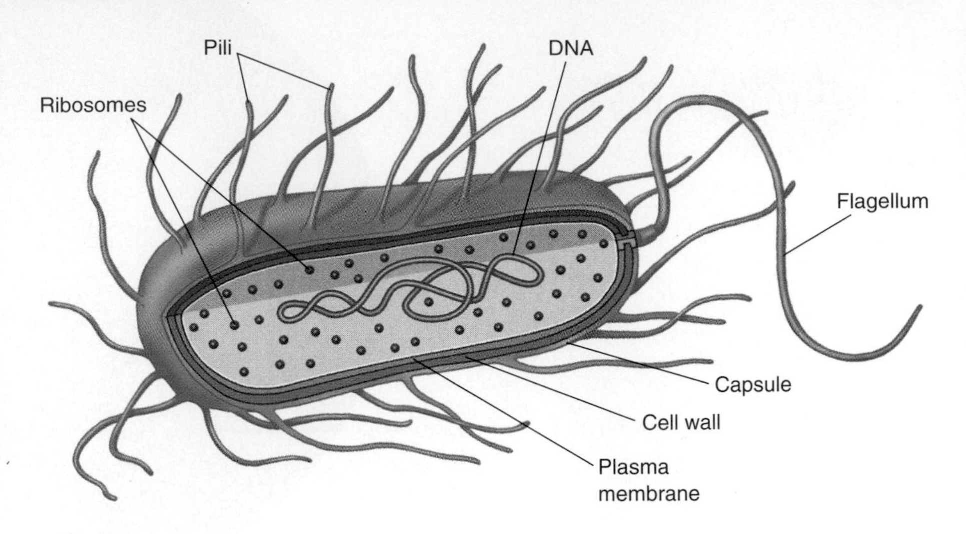

FIGURE 5.5
Structure of a bacterial cell. Generalized cell organization of a bacterium. Some bacteria have hairlike growths on the outside of the cell called pili.

Strong Cell Walls

Most bacteria are encased by a strong **cell wall** composed of *peptidoglycan*, which consists of a carbohydrate matrix (polymers of sugars) that is cross-linked by short polypeptide units. No eukaryotes possess cell walls with this type of chemical composition. With a few exceptions like TB and leprosy-causing bacteria, all bacteria may be classified into two types based on differences in their cell walls detected by the Gram staining procedure. The name refers to the Danish microbiologist Hans Christian Gram, who developed the procedure to detect the presence of certain disease-causing bacteria. **Gram-positive** bacteria have a thick, single-layered cell wall that retains a violet dye from the Gram stain procedure, causing the stained cells to appear purple under a microscope. More complex cell walls have evolved in other groups of bacteria. In them, the wall is multilayered and does not retain the purple dye after Gram staining; such bacteria exhibit the background red dye and are characterized as **gram-negative.**

The susceptibility of bacteria to antibiotics often depends on the structure of their cell walls. Penicillin and vancomycin, for example, interfere with the ability of bacteria to cross-link the peptide units that hold the carbohydrate chains of the wall together. Like removing all the nails from a wooden house, this destroys the integrity of the matrix, which can no longer prevent water from rushing in, swelling the cell to bursting.

Cell walls protect the cell, maintain its shape, and prevent excessive uptake of water. Plants, fungi, and most protists also have cell walls of a different chemical structure, which we will discuss in later chapters.

Long chains of sugars called polysaccharides cover the cell walls of many bacteria. They enable a bacterium to adhere to teeth, skin, food—practically any surface that will support their growth. Many disease-causing bacteria secrete a jellylike protective capsule of polysaccharide around the cell.

Rotating Flagella

Some bacteria use a flagellum (plural, flagella) to move. **Flagella** are long, threadlike structures protruding from the surface of a cell that are used in locomotion and feeding. Bacterial flagella are protein fibers that extend out from a bacterial cell. There may be one or more per cell, or none, depending on the species. Bacteria can swim at speeds up to 20 cell diameters per second by rotating their flagella like screws (figure 5.6). A "motor" unique to bacteria that is embedded within their cell walls and membranes powers the rotation. Only a few eukaryotic cells have structures that truly rotate.

Simple Interior Organization

If you were to look at an electron micrograph of a bacterial cell, you would be struck by the cell's simple organization. There are few, if any, internal compartments, and while they contain simple structures like ribosomes, most have no membrane-bounded organelles, the kinds so characteristic of eukaryotic cells. Nor do bacteria have a true nucleus. The entire cytoplasm of a bacterial cell is one unit with no internal support structure. Consequently,

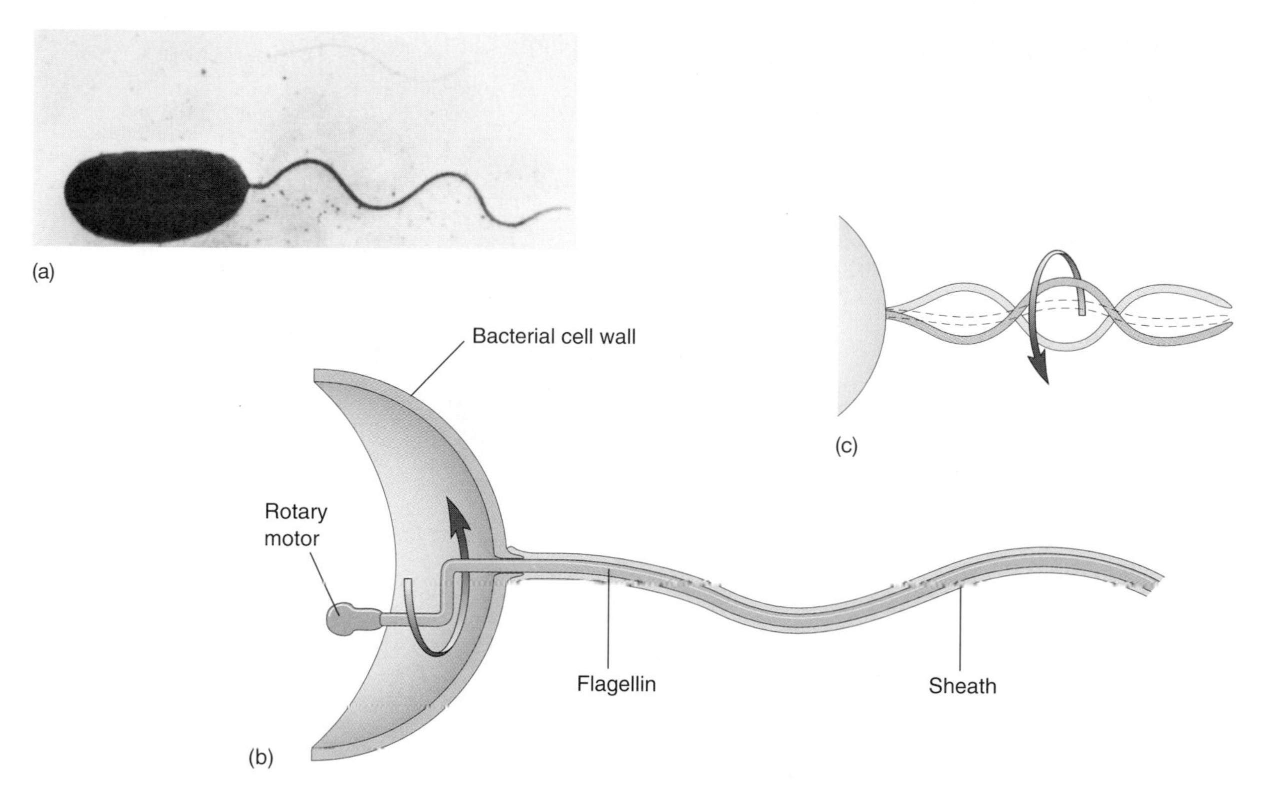

FIGURE 5.6
Bacteria swim by rotating their flagella. (*a*) The photograph is of *Vibrio cholerae*, the microbe that causes the serious disease cholera. The unsheathed core visible at the top of the photograph is composed of a single crystal of the protein flagellin. (*b*) In intact flagella, the core is surrounded by a flexible sheath. Imagine that you are standing inside the *Vibrio* cell, turning the flagellum like a crank. (*c*) You would create a spiral wave that travels down the flagellum, just as if you were turning a wire within a flexible tube. The bacterium creates this kind of rotary motion when it swims.

the strength of the cell comes primarily from its rigid wall (see figure 5.5).

The plasma membrane of a bacterial cell carries out some of the functions organelles perform in eukaryotic cells. When a bacterial cell divides, for example, the bacterial chromosome, a simple circle of DNA, replicates before the cell divides. The two DNA molecules that result from the replication attach to the plasma membrane at different points, ensuring that each daughter cell will contain one of the identical units of DNA. Moreover, some photosynthetic bacteria, such as cyanobacteria and *Prochloron* (figure 5.7), have an extensively folded plasma membrane, with the folds extending into the cell's interior. These membrane folds contain the bacterial pigments connected with photosynthesis.

Because a bacterial cell contains no membrane-bounded organelles, the DNA, enzymes, and other cytoplasmic constituents have access to all parts of the cell. Reactions are not compartmentalized as they are in eukaryotic cells, and the whole bacterium operates as a single unit.

Bacteria are small cells that lack interior organization. They are encased by an exterior wall composed of carbohydrates cross-linked by short polypeptides, and some are propelled by rotating flagella.

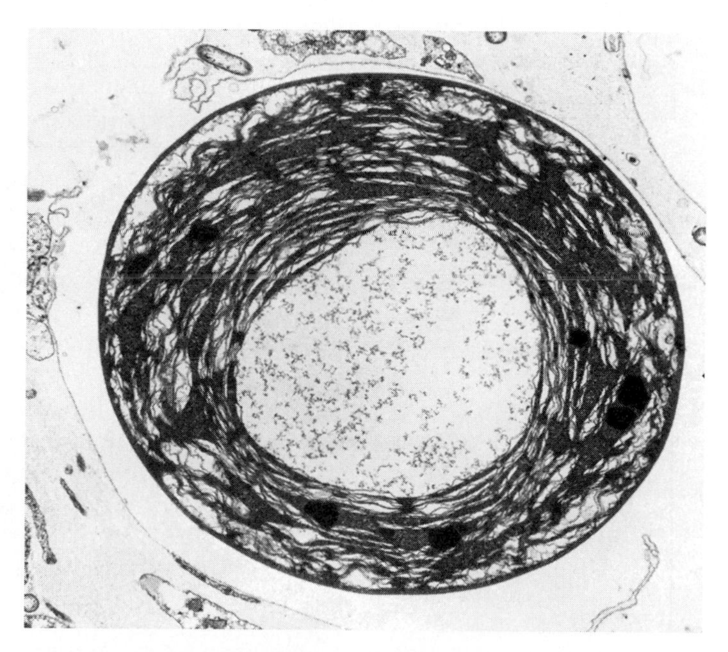

FIGURE 5.7
Electron micrograph of a photosynthetic bacterial cell. Extensive folded photosynthetic membranes are visible in this *Prochloron* cell (14,500×). The single, circular DNA molecule is located in the clear area in the central region of the cell.

Eukaryotic Cells Have Complex Interiors

Eukaryotic cells (figures 5.8 and 5.9) are far more complex than prokaryotic cells. The hallmark of the eukaryotic cell is compartmentalization. The interiors of eukaryotic cells contain numerous **organelles,** membrane-bounded structures that close off compartments within which multiple biochemical processes can proceed simultaneously and independently. Plant cells often have a large membrane-bounded sac called a **central vacuole,** which stores proteins, pigments, and waste materials. Both plant and animal cells contain **vesicles,** smaller sacs that store and transport a variety of materials. Inside the nucleus, the DNA is

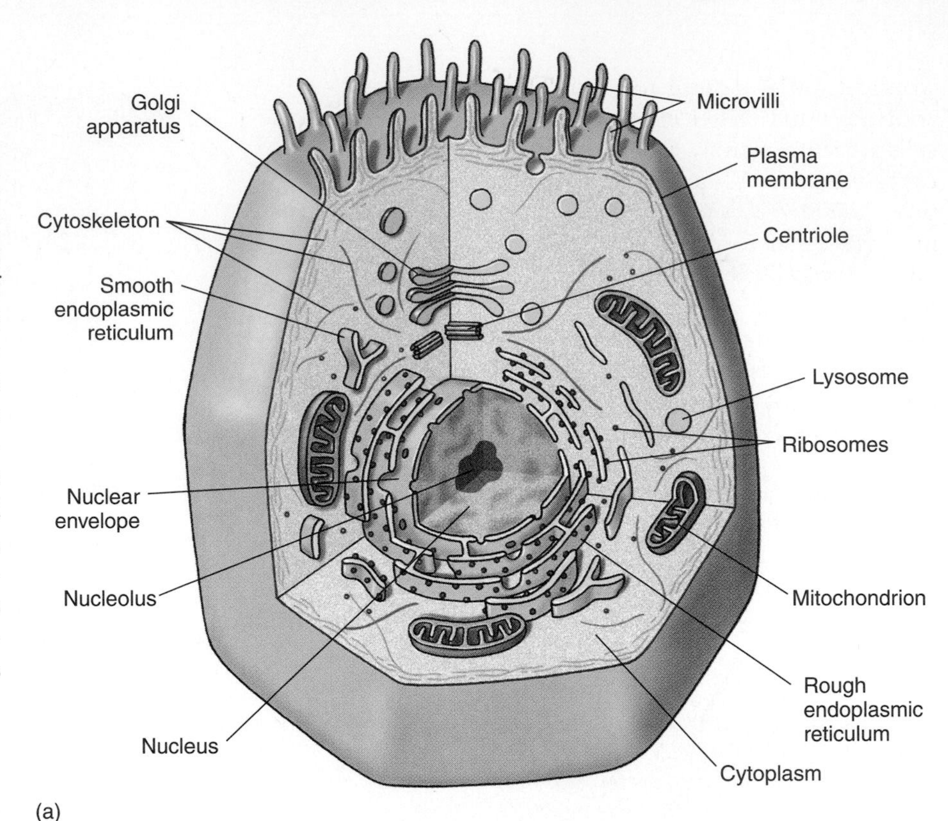

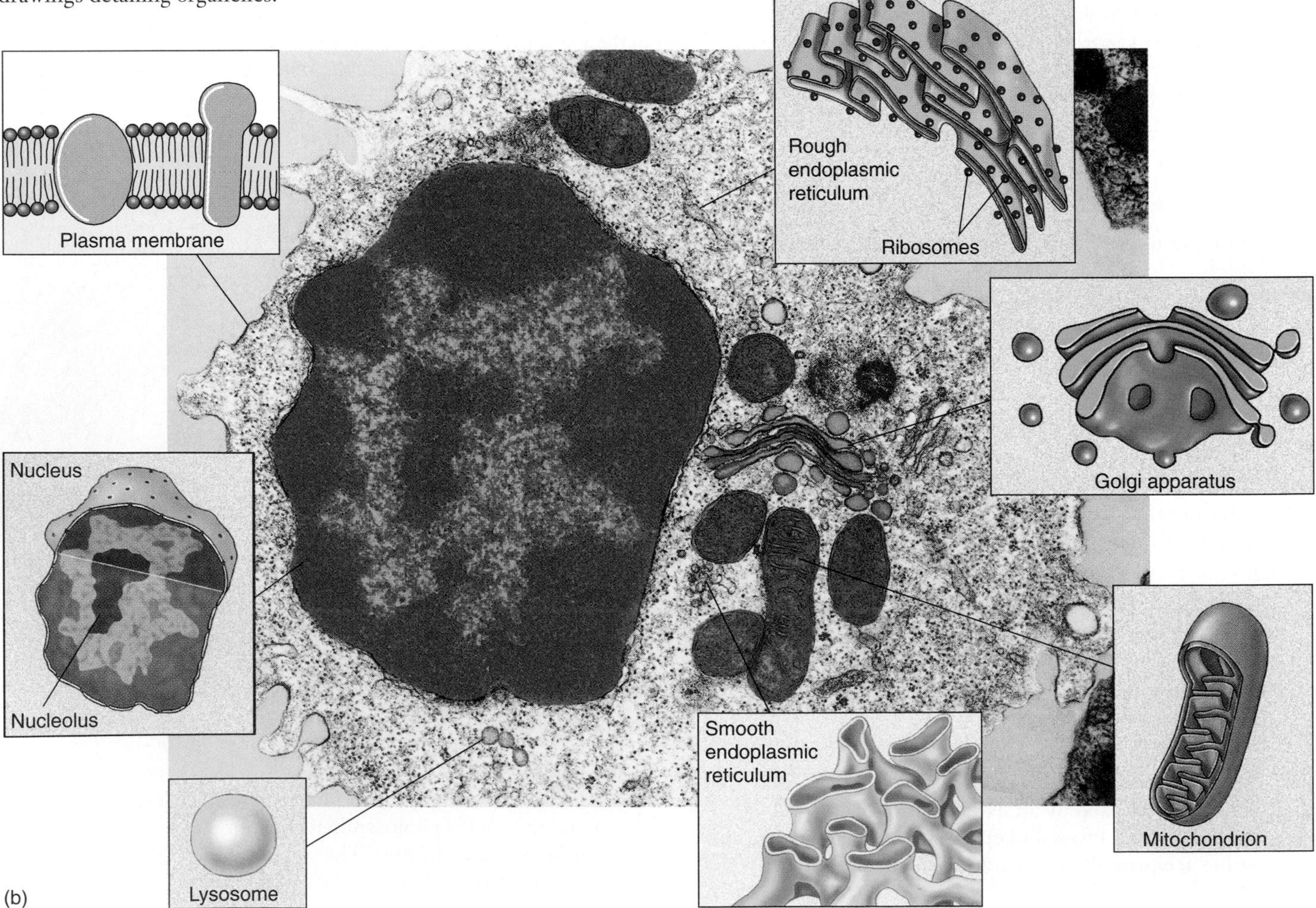

FIGURE 5.8
Structure of an animal cell. (*a*) A generalized diagram of an animal cell. (*b*) Micrograph of a human white blood cell (40,500×) with drawings detailing organelles.

wound tightly around proteins and packaged into compact units called **chromosomes.** All eukaryotic cells are supported by an internal protein scaffold, the **cytoskeleton.** While the cells of animals and some protists lack cell walls, the cells of fungi, plants, and many protists have strong **cell walls** composed of cellulose or chitin fibers embedded in a matrix of other polysaccharides and proteins. This composition is very different from the peptidoglycan that makes up bacterial cell walls. Let's now examine the structure and function of the internal components of eukaryotic cells in more detail.

Eukaryotic cells contain membrane-bounded organelles that carry out specialized functions.

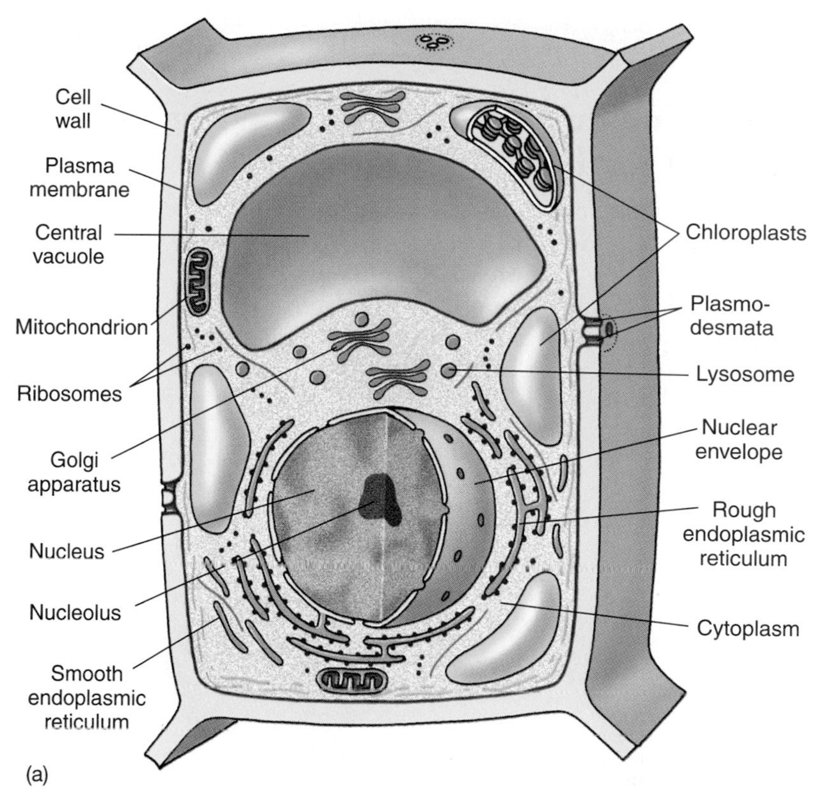

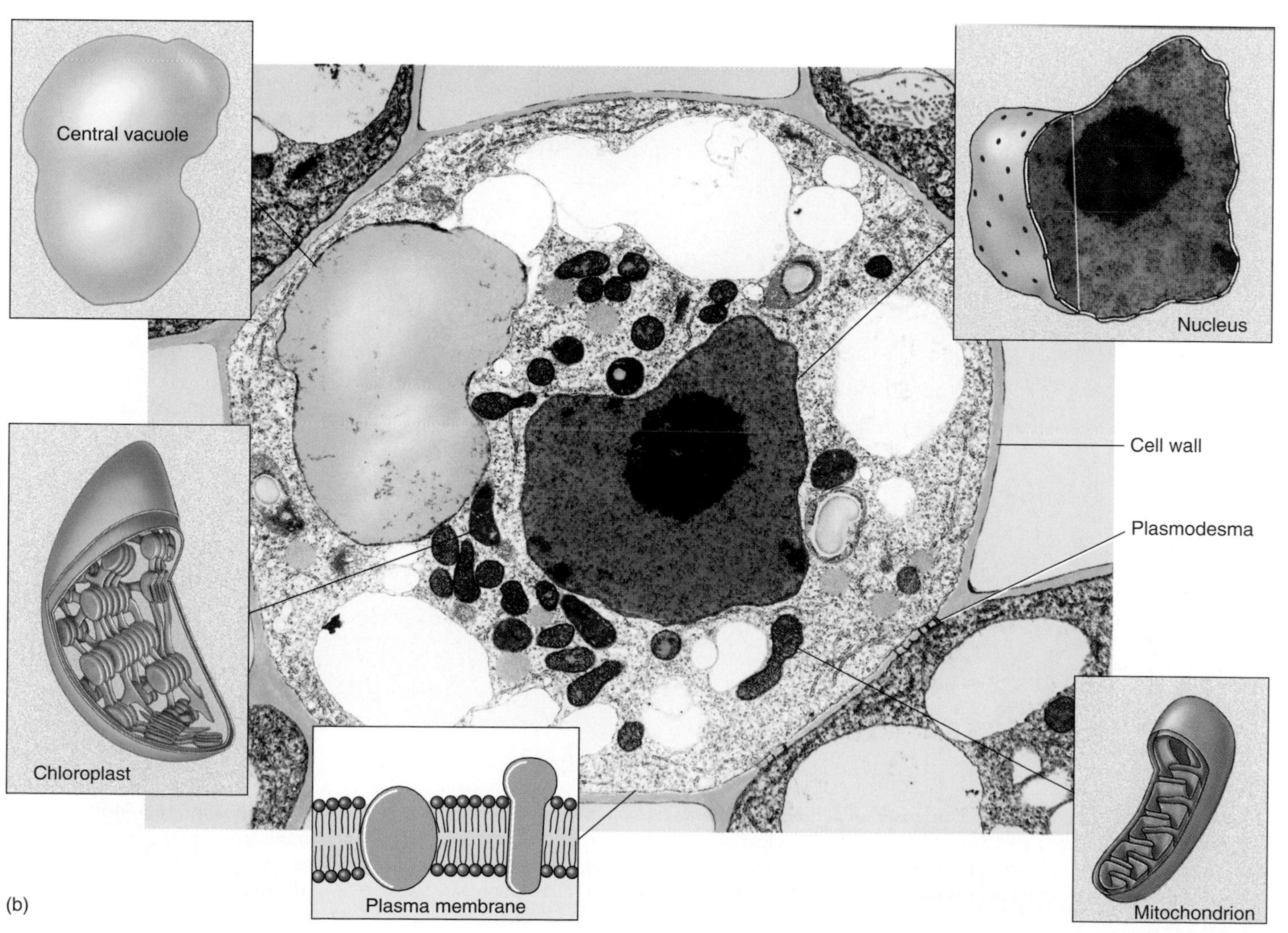

FIGURE 5.9
Structure of a plant cell. A generalized illustration (*a*) and micrograph (*b*) of a plant cell. Most mature plant cells contain large central vacuoles which occupy a major portion of the internal volume of the cell (14,000×).

5.3 Take a tour of a eukaryotic cell.

The Nucleus: Information Center for the Cell

The largest and most easily seen organelle within a eukaryotic cell is the **nucleus** (Latin, for kernel or nut), first described by the English botanist Robert Brown in 1831. Nuclei are roughly spherical in shape and, in animal cells, they are typically located in the central region of the cell (figure 5.10). In some cells, a network of fine cytoplasmic filaments seems to cradle the nucleus in this position. The nucleus is the repository of the genetic information that directs all of the activities of a living eukaryotic cell. Most eukaryotic cells possess a single nucleus, although the cells of fungi and some other groups may have several to many nuclei. Mammalian erythrocytes (red blood cells) lose their nuclei when they mature. Many nuclei exhibit a dark-staining zone called the nucleolus, which is a region where intensive synthesis of ribosomal RNA is taking place.

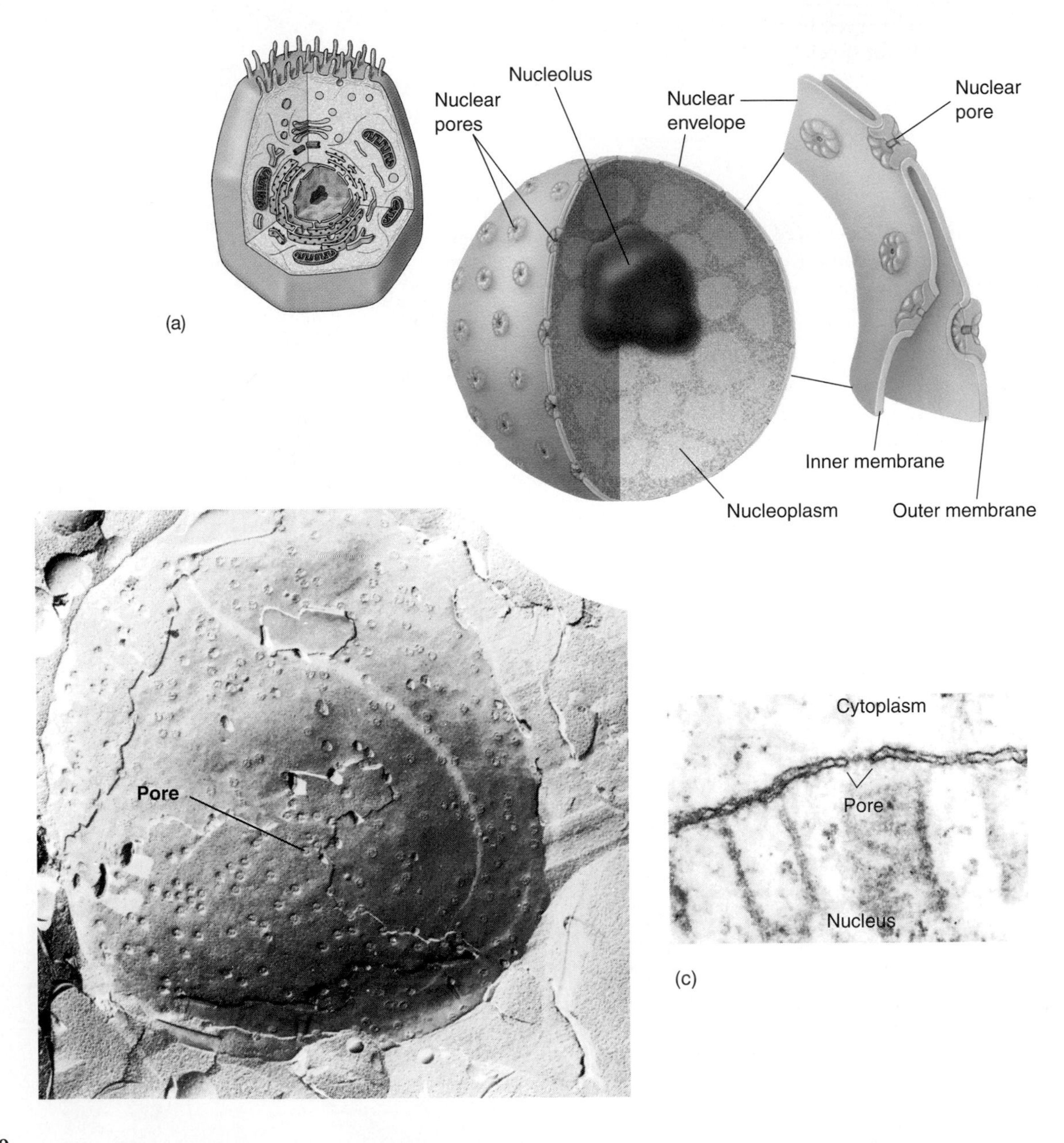

FIGURE 5.10
The nucleus. *(a)* The nucleus is composed of a double membrane, called a nuclear envelope, enclosing a fluid-filled interior containing the chromosomes. In cross-section, the individual nuclear pores are seen to extend through the two membrane layers of the envelope; the dark material within the pore is protein, which acts to control access through the pore. *(b)* A freeze-fracture scanning electron micrograph of a cell nucleus showing nuclear pores (9500×). *(c)* A transmission electron micrograph (see figure 6.6) of the nuclear membrane showing a nuclear pore.

The Nuclear Envelope: Getting In and Out

The surface of the nucleus is bounded by *two* phospholipid bilayer membranes, which together make up the **nuclear envelope** (see figure 5.10). The outer membrane of the nuclear envelope is continuous with the cytoplasm's interior membrane system, called the endoplasmic reticulum. Scattered over the surface of the nuclear envelope, like craters on the moon, are shallow depressions called **nuclear pores.** These pores form 50 to 80 nanometers apart at locations where the two membrane layers of the nuclear envelope pinch together. Rather than being empty, nuclear pores are filled with proteins that act as molecular channels, permitting certain molecules to pass into and out of the nucleus. Passage is restricted primarily to two kinds of molecules: (1) proteins moving into the nucleus to be incorporated into nuclear structures or to catalyze nuclear activities; and (2) RNA and protein-RNA complexes formed in the nucleus and exported to the cytoplasm.

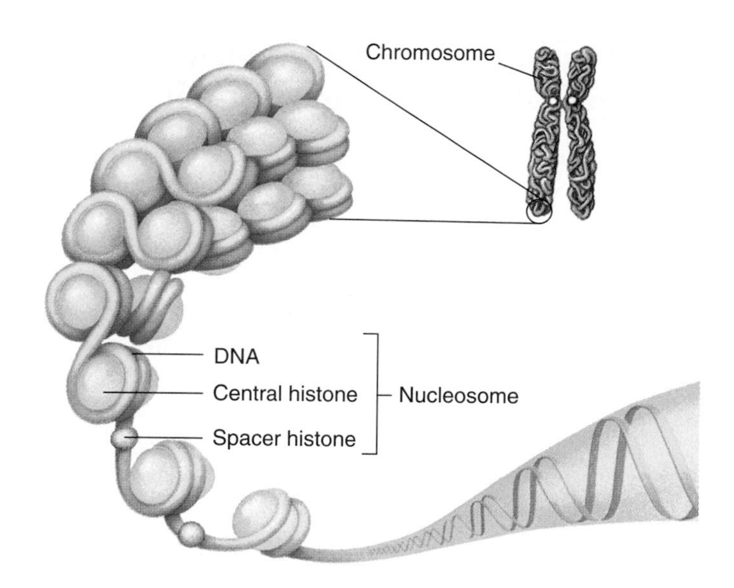

FIGURE 5.11
Nucleosomes. Each nucleosome is a region in which the DNA is wrapped tightly around a cluster of histone proteins.

The Chromosomes: Packaging the DNA

In both bacteria and eukaryotes, DNA contains the hereditary information specifying cell structure and function. However, unlike the circular DNA of bacteria, the DNA of eukaryotes is divided into several linear **chromosomes.** Except when a cell is dividing, its chromosomes are fully extended into threadlike strands, called **chromatin,** of DNA complexed with protein. This open arrangement allows proteins to attach to specific nucleotide sequences along the DNA. Without this access, DNA could not direct the day-to-day activities of the cell. The chromosomes are associated with packaging proteins called **histones.** When a cell prepares to divide, the DNA coils up around the histones into a highly condensed form. In the initial stages of this condensation, units of histone can be seen with DNA wrapped around like a sash. Called **nucleosomes,** these initial aggregations resemble beads on a string (figure 5.11). Coiling continues until the DNA is in a compact mass. Under a light microscope, these fully condensed chromosomes are readily seen in dividing cells as densely staining rods (figure 5.12). After cell division, eukaryotic chromosomes uncoil and can no longer be individually distinguished with a light microscope. Uncoiling the chromosomes into a more extended form permits enzymes to makes RNA copies of DNA. Only by means of these RNA copies can the information in the DNA be used to direct the synthesis of proteins.

The nucleus of a eukaryotic cell contains the cell's hereditary apparatus and isolates it from the rest of the cell. A distinctive feature of eukaryotes is the organization of their DNA into complex chromosomes.

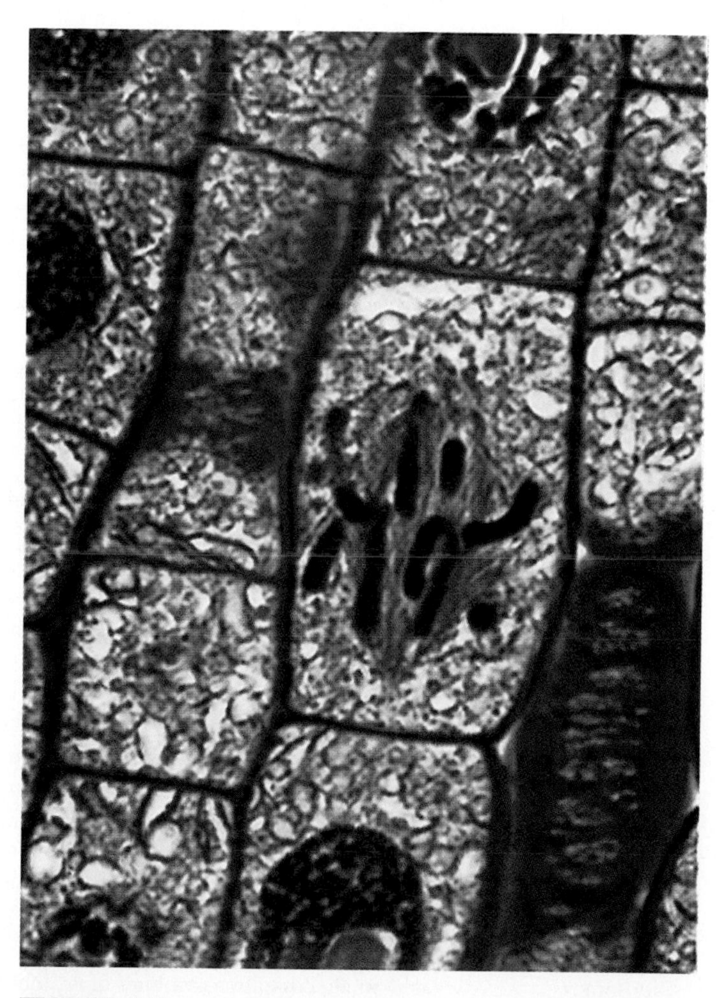

FIGURE 5.12
Eukaryotic chromosomes. These condensed chromosomes within an onion root tip are visible under the light microscope (500×).

The Endoplasmic Reticulum: Compartmentalizing the Cell

The interior of a eukaryotic cell is packed with membranes (table 5.1). So thin that they are invisible under the low resolving power of light microscopes, this **endomembrane system** fills the cell, dividing it into compartments, channeling the passage of molecules through the interior of the cell, and providing surfaces for the synthesis of lipids and some proteins. The presence of these membranes in eukaryotic cells constitutes one of the most fundamental distinctions between eukaryotes and prokaryotes.

The largest of the internal membranes is called the **endoplasmic reticulum (ER).** The term *endoplasmic* means "within the cytoplasm," and the term *reticulum* is Latin for "a little net." Like the plasma membrane, the ER is composed of a lipid bilayer embedded with proteins. It weaves in sheets through the interior of the cell, creating a series of channels between its folds (figure 5.13). Of the many compartments in eukaryotic cells, the two largest are the inner region of the ER, called the **cisternal space**, and the region exterior to it, the **cytosol**.

Rough ER: Manufacturing Proteins for Export

The ER surface regions that are devoted to protein synthesis are heavily studded with **ribosomes,** large molecular aggregates of protein and ribonucleic acid (RNA) that translate RNA copies of genes into protein (we will examine ribosomes in detail later in this chapter). Through the electron microscope, these ribosome-rich regions of the ER appear pebbly, like the surface of sandpaper, and they are therefore called **rough ER** (see figure 5.13).

The proteins synthesized on the surface of the rough ER are destined to be exported from the cell. Proteins to be exported contain special amino acid sequences called **signal sequences.** As a new protein is made by a free ribosome (one not attached to a membrane), the signal sequence of the growing polypeptide attaches to a recognition factor that carries the ribosome and its partially completed protein to a "docking site" on the surface of the ER. As the protein is assembled it passes through the ER membrane into the interior ER compartment, the cisternal space, from which it is transported by vesicles to the Golgi apparatus (figure 5.14). The protein then travels within vesicles to the inner surface of the plasma membrane, where it is released to the outside of the cell.

Table 5.1 Eukaryotic Cell Structures and Their Functions

Structure	Description	Function
Cell wall	Outer layer of cellulose or chitin; or absent	Protection; support
Cytoskeleton	Network of protein filaments	Structural support; cell movement
Flagella (cilia)	Cellular extensions with 9 + 2 arrangement of pairs of microtubules	Motility or moving fluids over surfaces
Plasma membrane	Lipid bilayer with embedded proteins	Regulates what passes into and out of cell; cell-to-cell recognition
Endoplasmic reticulum	Network of internal membranes	Forms compartments and vesicles; participates in protein and lipid synthesis
Nucleus	Structure (usually spherical) surrounded by double membrane that contains chromosomes	Control center of cell; directs protein synthesis and cell reproduction
Golgi apparatus	Stacks of flattened vesicles	Packages proteins for export from cell; forms secretory vesicles
Lysosomes	Vesicles derived from Golgi apparatus that contain hydrolytic digestive enzymes	Digest worn-out organelles and cell debris; play role in cell death
Microbodies	Vesicles formed from incorporation of lipids and proteins containing oxidative and other enzymes	Isolate particular chemical activities from rest of cell
Mitochondria	Bacteria-like elements with double membrane	"Power plants" of the cell; sites of oxidative metabolism
Chloroplasts	Bacteria-like elements with membranes containing chlorophyll, a photosynthetic pigment	Sites of photosynthesis
Chromosomes	Long threads of DNA that form a complex with protein	Contain hereditary information
Nucleolus	Site of genes for rRNA synthesis	Assembles ribosomes
Ribosomes	Small, complex assemblies of protein and RNA, often bound to endoplasmic reticulum	Sites of protein synthesis

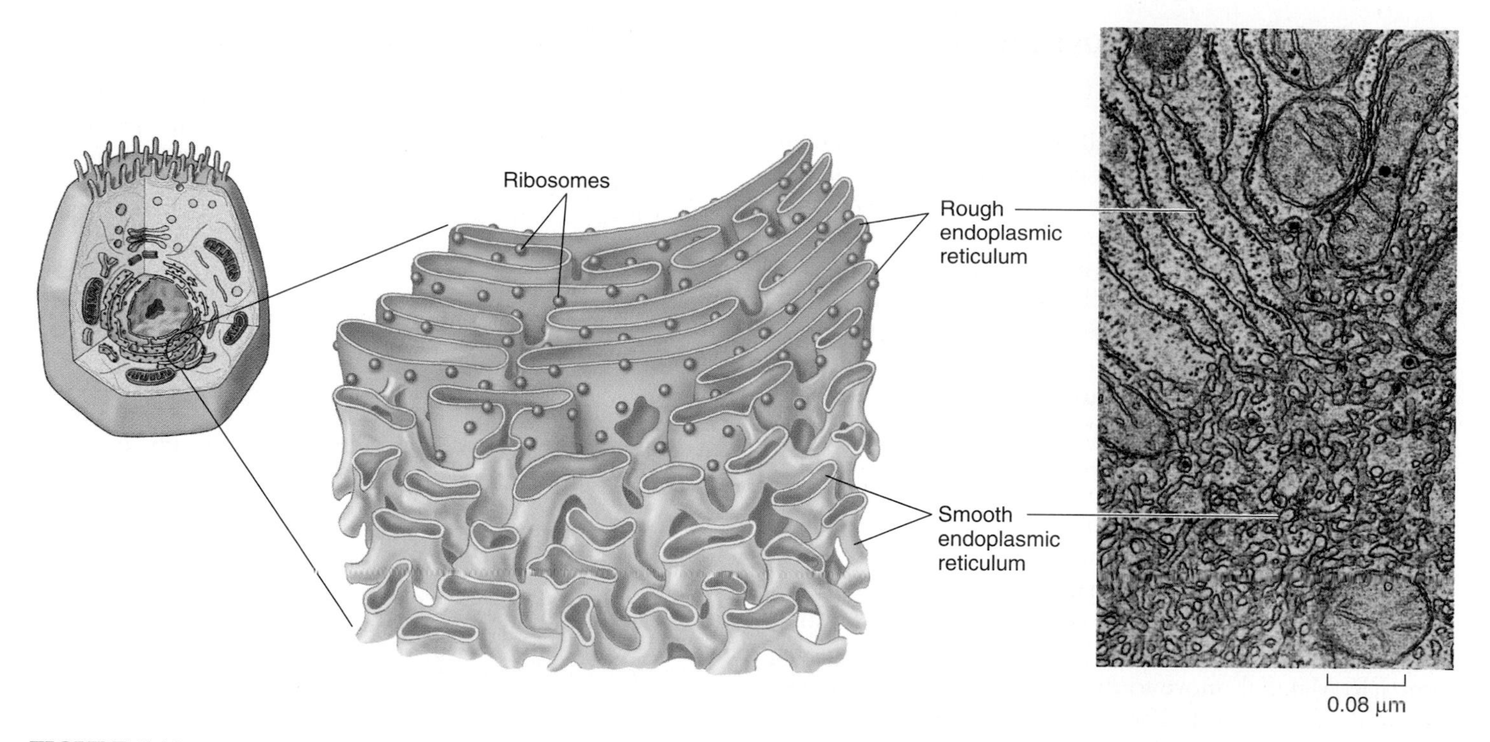

FIGURE 5.13
The endoplasmic reticulum. Ribosomes are associated with only one side of the rough ER; the other side is the boundary of a separate compartment within the cell into which the ribosomes extrude newly made proteins destined for secretion. Smooth endoplasmic reticulum has few to no bound ribosomes.

Smooth ER: Organizing Internal Activities

Regions of the ER with relatively few bound ribosomes are referred to as **smooth ER.** The membranes of the smooth ER contain many embedded enzymes, most of them active only when associated with a membrane. Enzymes anchored within the ER, for example, catalyze the synthesis of a variety of carbohydrates and lipids. In cells that carry out extensive lipid synthesis, such as those in the testes, intestine, and brain, smooth ER is particularly abundant. In the liver, the enzymes of the smooth ER are involved in the detoxification of drugs including amphetamines, morphine, codeine, and phenobarbital.

Some vesicles form at the plasma membrane by budding inward, a process called endocytosis. Some then move into the cytoplasm and fuse with the smooth endoplasmic reticulum. Others form secondary lysosomes or other interior vesicles.

The endoplasmic reticulum (ER) is an extensive system of folded membranes that divides the interior of eukaryotic cells into compartments and channels. Rough ER synthesizes proteins, while smooth ER organizes the synthesis of lipids and other biosynthetic activities.

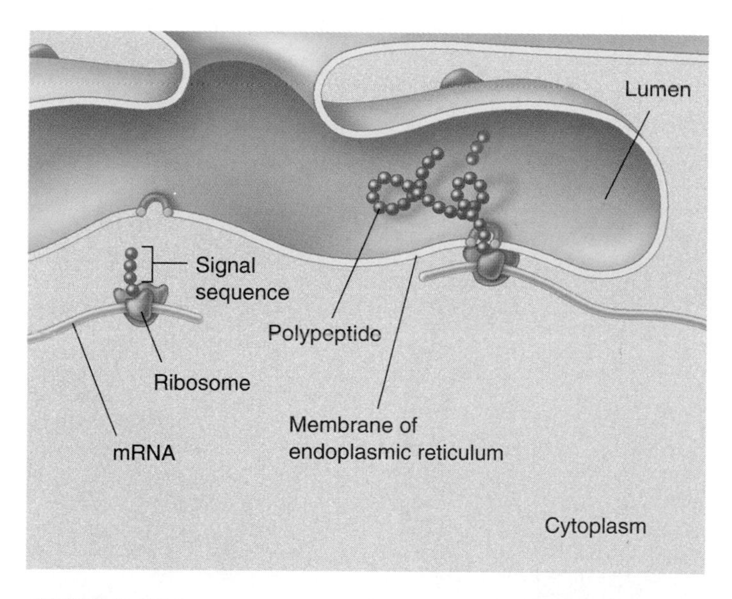

FIGURE 5.14
Signal sequences direct proteins to their destinations in the cell. In this example, a sequence of hydrophobic amino acids (the signal sequence) on a secretory protein attaches them (and the ribosomes making them) to the membrane of the ER. As the protein is synthesized, it passes into the lumen (internal chamber) of the ER. The signal sequence is clipped off after the leading edge of the protein enters the lumen.

The Golgi Apparatus: Delivery System of the Cell

At various locations within the endomembrane system, flattened stacks of membranes called **Golgi bodies** occur, often interconnected with one another. These structures are named for Camillo Golgi, the nineteenth-century Italian physician who first called attention to them. The numbers of Golgi bodies a cell contains ranges from 1 or a few in protists, to 20 or more in animal cells and several hundred in plant cells. They are especially abundant in glandular cells, which manufacture and secrete substances. Collectively the Golgi bodies are referred to as the **Golgi apparatus** (figure 5.15).

The Golgi apparatus functions in the collection, packaging, and distribution of molecules synthesized at one place in the cell and utilized at another location in the cell. A Golgi body has a front and a back, with distinctly different membrane compositions at the opposite ends. The front, or receiving end, is called the *cis* face, and is usually located near ER. Materials move to the *cis* face in transport vesicles that bud off of the ER. These vesicles fuse with the *cis* face, emptying their contents into the interior, or lumen, of the Golgi apparatus. These ER-synthesized molecules then pass through the channels of the Golgi apparatus until they reach the back, or discharging end, called the *trans* face, where they are discharged in secretory vesicles (figure 5.16).

Proteins and lipids manufactured on the rough and smooth ER membranes are transported into the Golgi apparatus and modified as they pass through it. The most common alteration is the addition or modification of short sugar chains, forming a *glycoprotein* when sugars are complexed to a protein and a *glycolipid* when sugars are bound to a lipid. In many instances, enzymes in the Golgi apparatus modify existing glycoproteins and glycolipids made in the ER by cleaving a sugar from their sugar chain or modifying one or more of the sugars.

The newly formed or altered glycoproteins and glycolipids collect at the ends of the Golgi bodies, in flattened stacked membrane folds called **cisternae** (Latin, "collecting vessels"). Periodically, the membranes of the cisternae push together, pinching off small, membrane-bounded secretory vesicles containing the glycoprotein and glycolipid molecules. These vesicles then move to other locations in the cell, distributing the newly synthesized molecules to their appropriate destinations. **Liposomes** are synthetically manufactured vesicles that contain any variety of desirable substances (such as drugs), and can be injected into the body. Because the membrane of liposomes is similar to plasma and organellar membranes, these liposomes serve as an effective and natural delivery system to cells and may prove to be of great therapeutic value.

The Golgi apparatus is the delivery system of the eukaryotic cell. It collects, packages, modifies, and distributes molecules that are synthesized at one location within the cell and used at another.

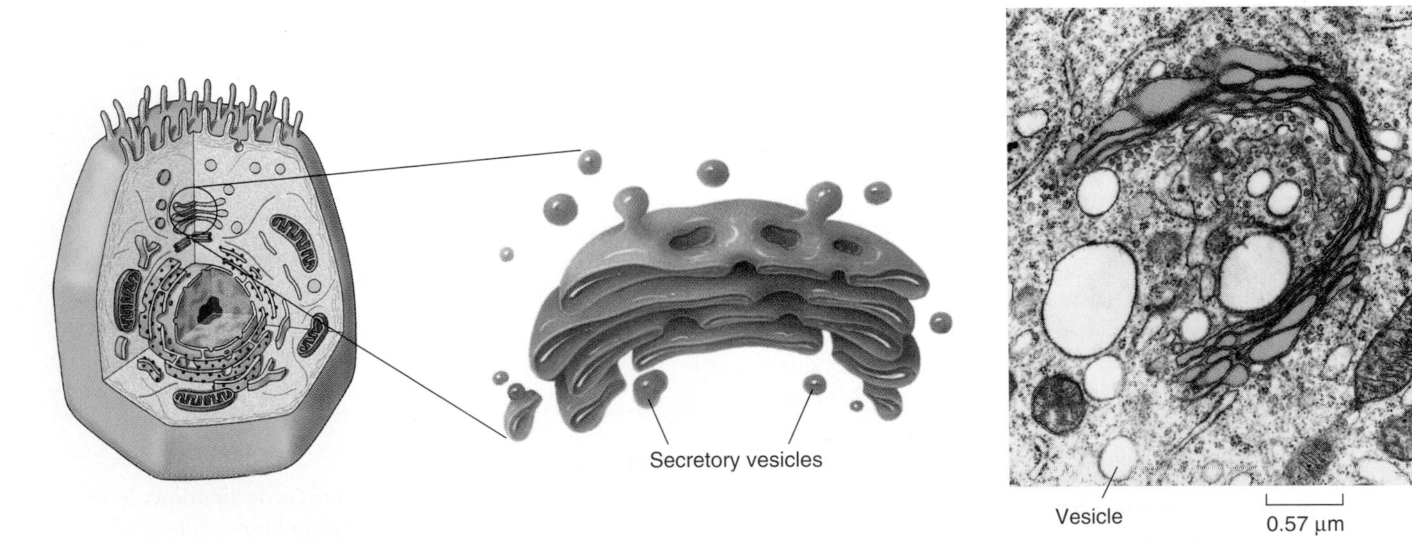

FIGURE 5.15
The Golgi apparatus. The Golgi apparatus is a smooth, concave membranous structure located near the middle of the cell. It receives material for processing on one surface and sends the material packaged in vesicles off the other. The substance in a vesicle could be for export out of the cell or for distribution to another region within the same cell.

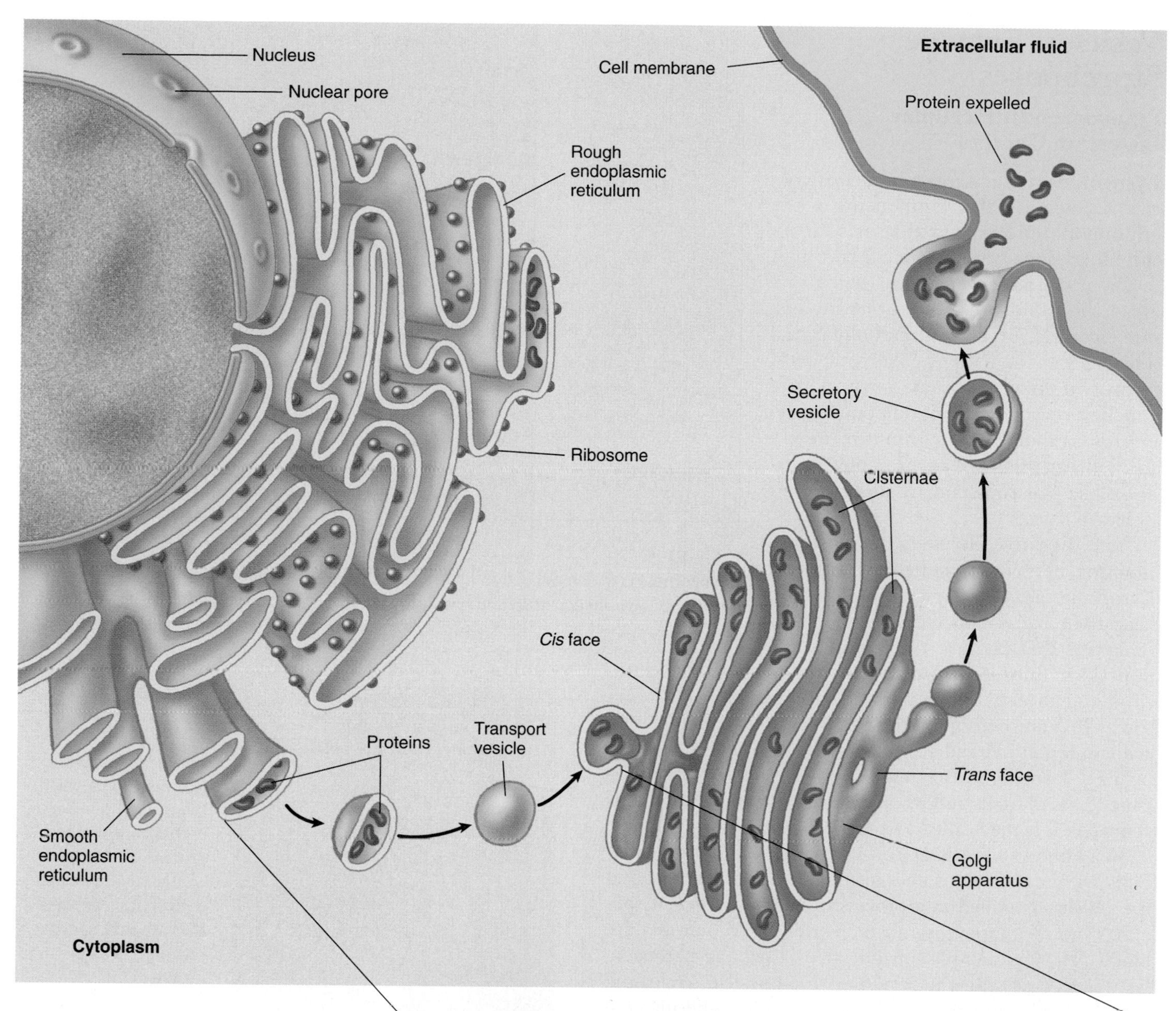

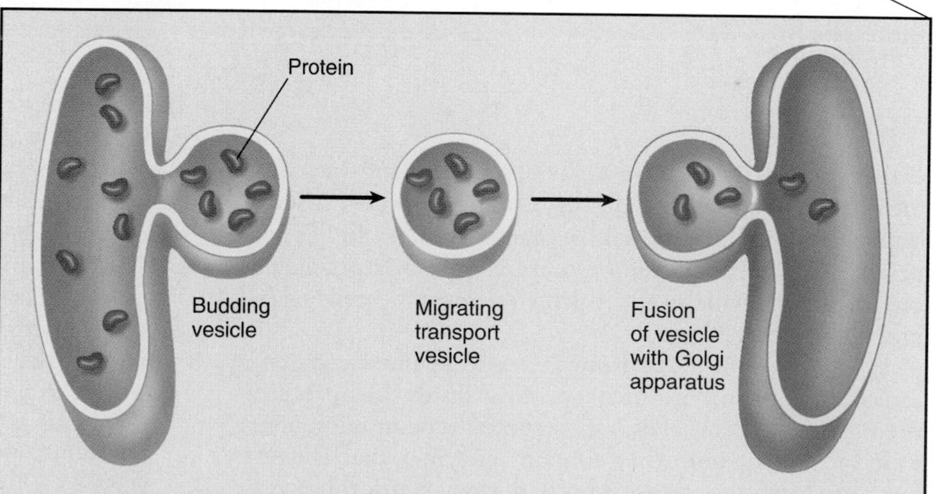

FIGURE 5.16
How proteins are transported within the cell. Proteins are manufactured at the ribosome and then released into the internal compartments of the rough ER. If the newly synthesized proteins are to be used at a distant location in or outside of the cell, they are transported within vesicles that bud off the rough ER and travel to the *cis* face, or receiving end, of the Golgi apparatus. There they are modified and packaged into secretory vesicles. The secretory vesicles then migrate from the *trans* face, or discharging end, of the Golgi apparatus to other locations in the cell, or they fuse with the cell membrane, releasing their contents to the external cellular environment.

Vesicles: Enzyme Storehouses

Lysosomes: Intracellular Digestion Centers

Lysosomes, membrane-bounded digestive vesicles, are also components of the endomembrane system that arise from the Golgi apparatus. They contain high levels of degrading enzymes, which catalyze the rapid breakdown of proteins, nucleic acids, lipids, and carbohydrates. Throughout the lives of eukaryotic cells, lysosomal enzymes break down old organelles, recycling their component molecules and making room for newly formed organelles. For example, mitochondria are replaced in some tissues every 10 days.

The digestive enzymes in lysosomes function best in an acidic environment. Lysosomes actively engaged in digestion keep their battery of hydrolytic enzymes (enzymes that catalyze the hydrolysis of molecules) fully active by pumping protons into their interiors and thereby maintaining a low internal pH. Lysosomes that are not functioning actively do not maintain an acidic internal pH and are called *primary lysosomes.* When a primary lysosome fuses with a food vesicle or other organelle, its pH falls and its arsenal of hydrolytic enzymes is activated; it is then called a *secondary lysosome.*

In addition to breaking down organelles and other structures within cells, lysosomes also eliminate other cells that the cell has engulfed in a process called *phagocytosis,* a specific type of endocytosis (see chapter 6). When a white blood cell, for example, phagocytizes a passing pathogen, lysosomes fuse with the resulting "food vesicle," releasing their enzymes into the vesicle and degrading the material within (figure 5.17).

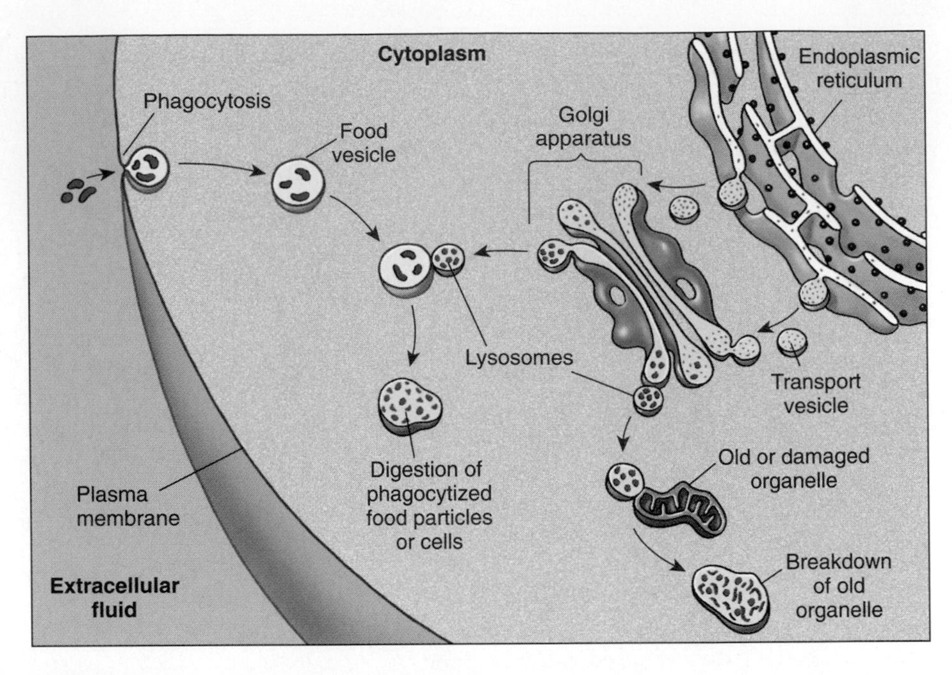

FIGURE 5.17
Lysosomes. Lysosomes contain hydrolytic enzymes that digest particles or cells taken into the cell by phagocytosis and break down old organelles.

Microbodies

Eukaryotic cells contain a variety of enzyme-bearing, membrane-enclosed vesicles called **microbodies.** Microbodies are found in the cells of plants, animals, fungi, and protists. The distribution of enzymes into microbodies is one of the principal ways in which eukaryotic cells organize their metabolism.

While lysosomes bud from the endomembrane system, microbodies grow by incorporating lipids and protein, then dividing. Plant cells have a special type of microbody called a **glyoxysome** that contains enzymes that convert fats into carbohydrates. Another type of microbody, a **peroxisome,** contains enzymes that catalyze the removal of electrons and associated hydrogen atoms (figure 5.18).

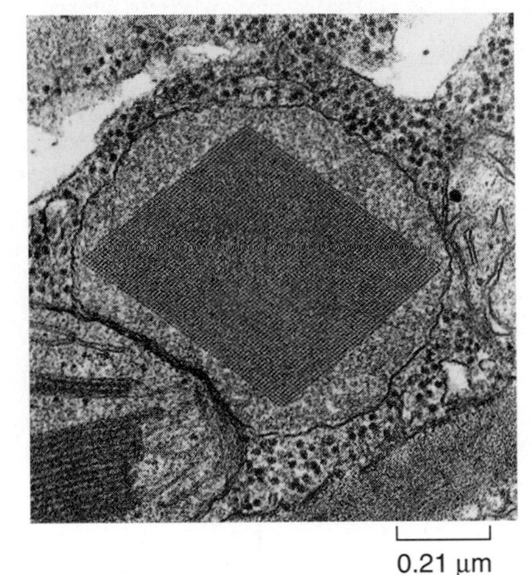

FIGURE 5.18
A peroxisome. Peroxisomes are spherical organelles that may contain a large diamond-shaped crystal composed of protein. Peroxisomes contain digestive and detoxifying enzymes that produce hydrogen peroxide as a by-product.

If these oxidative enzymes were not isolated within microbodies, they would tend to short-circuit the metabolism of the cytoplasm, which often involves adding hydrogen atoms to oxygen. The name *peroxisome* refers to the hydrogen peroxide produced as a by-product of the activities of the oxidative enzymes in the microbody. Hydrogen peroxide is dangerous to cells because of its violent chemical reactivity. However, peroxisomes also contain the enzyme catalase, which breaks down hydrogen peroxide into harmless water and oxygen.

Lysosomes and peroxisomes are vesicles that contain digestive and detoxifying enzymes. The isolation of these enzymes in vesicles protects the rest of the cell from inappropriate digestive activity.

Ribosomes: Sites of Protein Synthesis

Although the DNA in a cell's nucleus encodes the amino acid sequence of each protein in the cell, the proteins are not assembled there. A simple experiment demonstrates this: if a brief pulse of radioactive amino acid is administered to a cell, the radioactivity shows up associated with newly made protein, not in the nucleus, but in the cytoplasm. When investigators first carried out these experiments, they found that protein synthesis was associated with large RNA-protein complexes they called **ribosomes.**

Ribosomes are made up of several molecules of a special form of RNA called ribosomal RNA, or rRNA, bound within a complex of several dozen different proteins. Ribosomes are among the most complex molecular assemblies found in cells. Each ribosome is composed of two subunits (figure 5.19). The subunits join to form a functional ribosome only when they attach to another kind of RNA, called messenger RNA (mRNA) in the cytoplasm. To make proteins, the ribosome attaches to the mRNA, which is a transcribed copy of a portion of DNA, and uses the information to direct the synthesis of a protein.

Bacterial ribosomes are smaller than eukaryotic ribosomes. Also, a bacterial cell typically has only a few thousand ribosomes, while a metabolically active eukaryotic cell, such as a human liver cell, contains several million. Proteins that function in the cytoplasm are made by free ribosomes suspended there, while proteins bound within membranes or destined for export from the cell are assembled by ribosomes bound to rough ER.

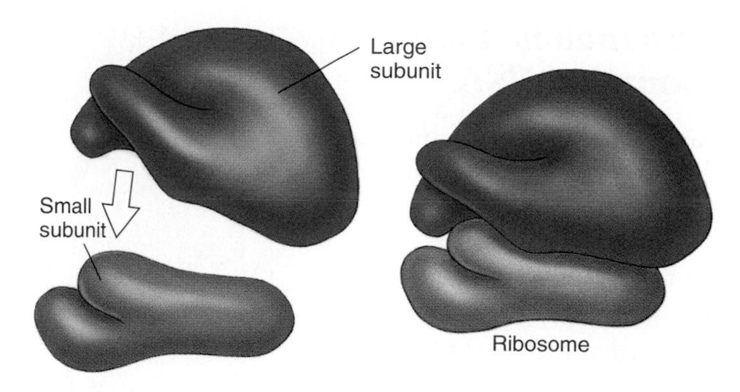

FIGURE 5.19
A ribosome. Ribosomes consist of a large and a small subunit composed of rRNA and protein. The individual subunits are synthesized in the nucleolus and then move through the nuclear pores to the cytoplasm, where they assemble. Ribosomes serve as sites of protein synthesis.

The Nucleolus Manufactures Ribosomal Subunits

When cells are synthesizing a large number of proteins, they must first make a large number of ribosomes. To facilitate this, many hundreds of copies of the portion of the DNA encoding the rRNA are clustered together on the chromosome. By transcribing RNA molecules from this cluster, the cell rapidly generates large numbers of the molecules needed to produce ribosomes.

At any given moment, many rRNA molecules dangle from the chromosome at the sites of these clusters of genes that encode rRNA. Proteins associate with the dangling rRNA molecules. These areas where ribosomes are being assembled are easily visible within the nucleus as one or more dark-staining regions, called **nucleoli** (singular, nucleolus; figure 5.20). Nucleoli can be seen under the light microscope even when the chromosomes are extended, unlike the rest of the chromosomes, which are visible only when condensed.

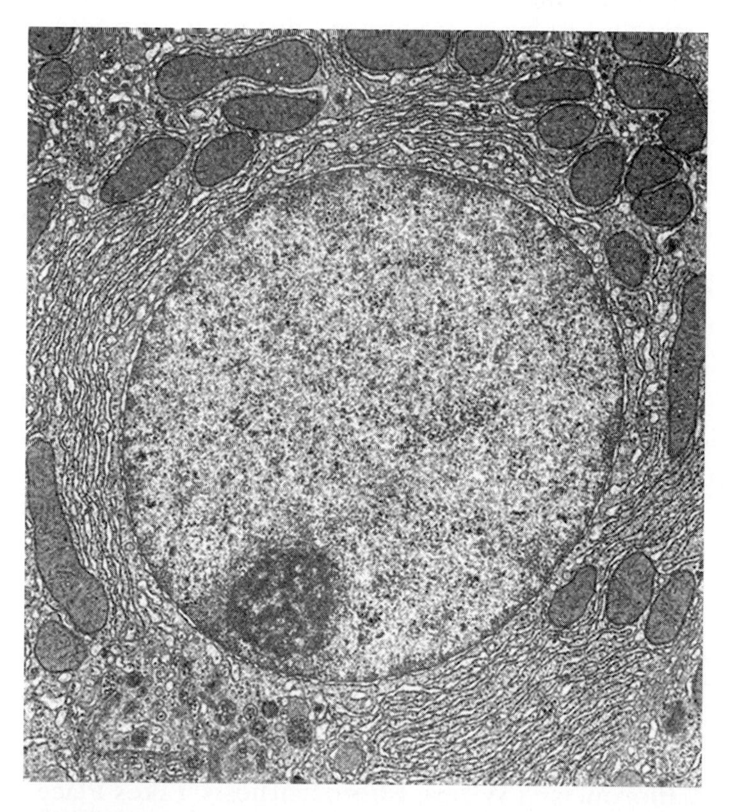

FIGURE 5.20
The nucleolus. This is the interior of a rat liver cell, magnified about 6000 times. A single large nucleus occupies the center of the micrograph. The electron-dense area in the lower left of the nucleus is the nucleolus, the area where the major components of the ribosomes are produced. Partly formed ribosomes can be seen around the nucleolus.

Ribosomes are the sites of protein synthesis in the cytoplasm.

Organelles That Contain DNA

Among the most interesting cell organelles are those in addition to the nucleus that contain DNA.

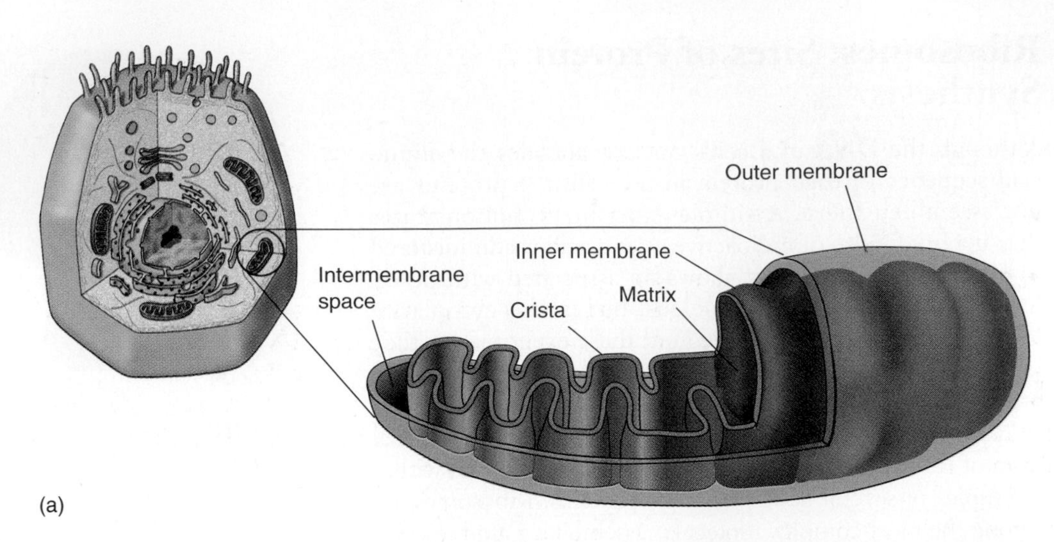

Mitochondria: The Cell's Chemical Furnaces

Mitochondria (singular, *mitochondrion*) are typically tubular or sausage-shaped organelles about the size of bacteria and found in all types of eukaryotic cells (figure 5.21). Mitochondria are bounded by two membranes: a smooth outer membrane and an inner one folded into numerous contiguous layers called **cristae** (singular, *crista*). The cristae partition the mitochondrion into two compartments: a **matrix,** lying inside the inner membrane; and an outer compartment, or **intermembrane space,** lying between the two mitochondrial membranes. On the surface of the inner membrane, and also embedded within it, are proteins that carry out oxidative metabolism, the oxygen-requiring process by which energy in macromolecules is stored in ATP.

Mitochondria have their own DNA; this DNA contains several genes that produce proteins essential to the mitochondrion's role in oxidative metabolism. All of these genes are copied into RNA and used to make proteins within the mitochondrion. In this process, the mitochondria employ small RNA molecules and ribosomal components that the mitochondrial DNA also encodes. However, most of the genes that produce the enzymes used in oxidative metabolism are located in the nucleus.

A eukaryotic cell does not produce brand new mitochondria each time the cell divides. Instead, the mitochondria themselves divide in two, doubling in number, and these are partitioned between the new cells. Most of the components required for mitochondrial division are encoded by genes in the nucleus and translated into proteins by cytoplasmic ribosomes. Mitochondrial replication is, therefore, impossible without nuclear participation, and mitochondria thus cannot be grown in a cell-free culture.

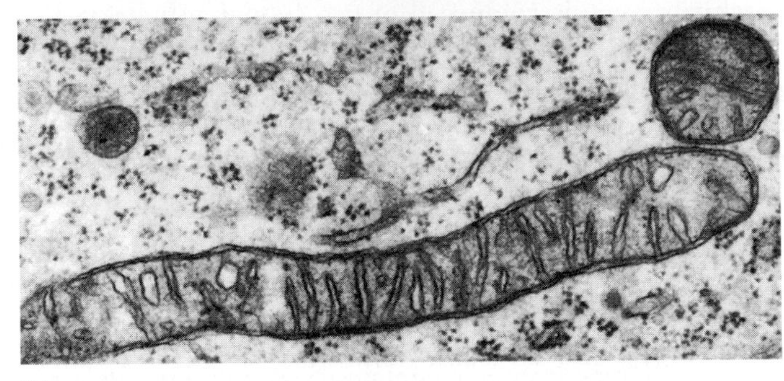

FIGURE 5.21
Mitochondria. (*a*) The inner membrane of a mitochondrion is shaped into folds called cristae, which greatly increase the surface area for oxidative metabolism. (*b*) Mitochondria in cross-section and cut lengthwise (70,000×).

Chloroplasts: Where Photosynthesis Takes Place

Plants and other eukaryotic organisms that carry out photosynthesis typically contain from one to several hundred **chloroplasts.** Chloroplasts bestow an obvious advantage on the organisms that possess them: they can manufacture their own food. Chloroplasts contain the photosynthetic pigment chlorophyll that gives most plants their green color.

The chloroplast body is enclosed, like the mitochondrion, within two membranes that resemble those of mitochondria (figure 5.22). However, chloroplasts are larger and more complex than mitochondria. In addition to the outer and inner membranes, which lie in close association with each other, chloroplasts have a closed compartment of stacked membranes called **grana** (singular, *granum*), which lie internal to the inner membrane. A chloroplast may contain a hundred or more grana, and each granum may contain from a few to several dozen disk-shaped structures called **thylakoids.** On the surface of the thylakoids are the light-capturing photosynthetic pigments, to be discussed in depth in chapter 10. Surrounding the thylakoid is a fluid matrix called the *stroma*.

Like mitochondria, chloroplasts contain DNA, but many of the genes that specify chloroplast components are also located in the nucleus. Some of the elements used in the photosynthetic process, including the specific protein components necessary to accomplish the reaction, are synthesized entirely within the chloroplast.

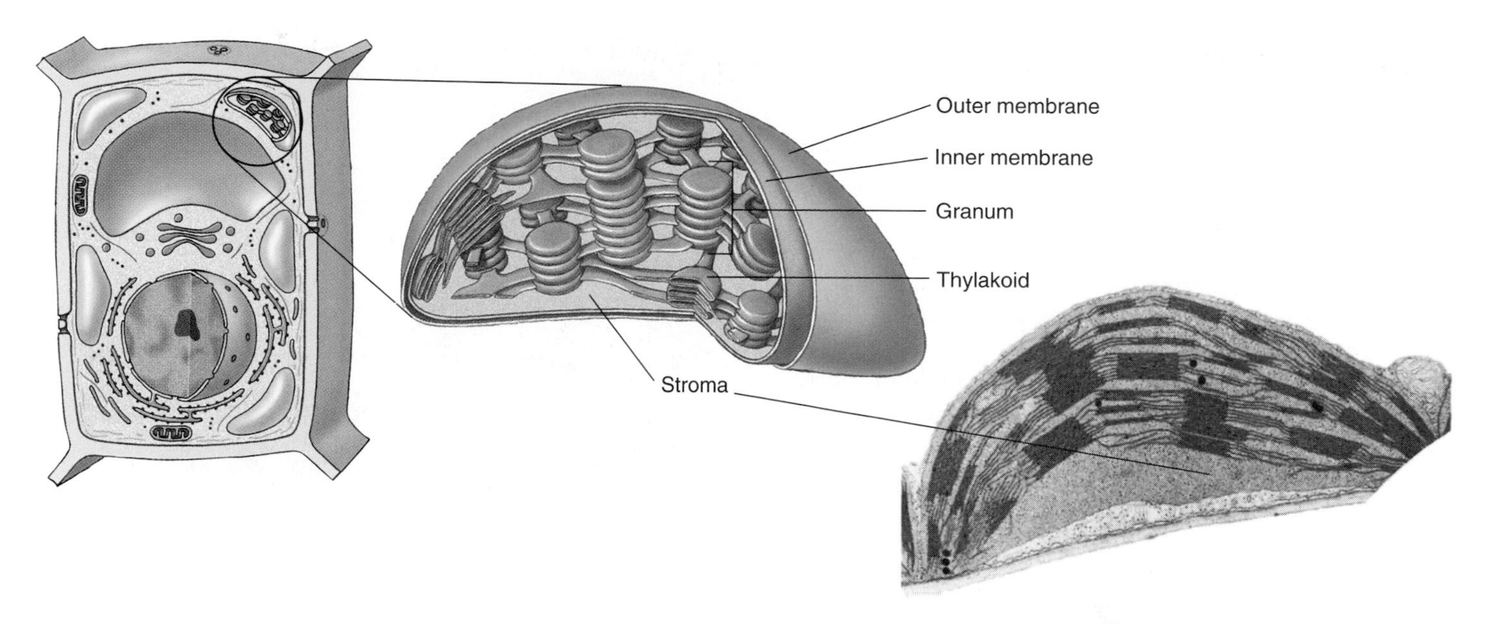

FIGURE 5.22
Chloroplast structure. The inner membrane of a chloroplast is fused to form stacks of closed vesicles called thylakoids. Within these thylakoids, photosynthesis takes place. Thylakoids are typically stacked one on top of the other in columns called grana.

Other DNA-containing organelles in plants are called leucoplasts, which lack pigment and a complex internal structure. In root cells and some other plant cells, leucoplasts may serve as starch storage sites. A leucoplast that stores starch (amylose) is sometimes termed an **amyloplast.** These organelles—chloroplasts, leucoplasts, and amyloplasts—are collectively called **plastids.** All plastids come from the division of existing plastids.

Centrioles: Microtubule Assembly Centers

Centrioles are barrel-shaped organelles found in the cells of animals and most protists. They occur in pairs, usually located at right angles to each other near the nuclear membranes (figure 5.23); the region surrounding the pair in almost all animal cells is referred to as a **centrosome.** Although the matter is in some dispute, at least some centrioles seem to contain DNA, which apparently is involved in producing their structural proteins. Centrioles help to assemble **microtubules,** long, hollow cylinders of the protein tubulin. Microtubules influence cell shape, move the chromosomes in cell division, and provide the functional internal structure of flagella and cilia, as we will discuss later. Centrioles may be contained in areas called **microtubule-organizing centers (MTOCs).** The cells of plants and fungi lack centrioles, and cell biologists are still in the process of characterizing their MTOCs.

Both mitochondria and chloroplasts contain specific genes related to some of their functions, but both depend on nuclear genes for other functions. Some centrioles also contain DNA, which apparently helps control the synthesis of their structural proteins.

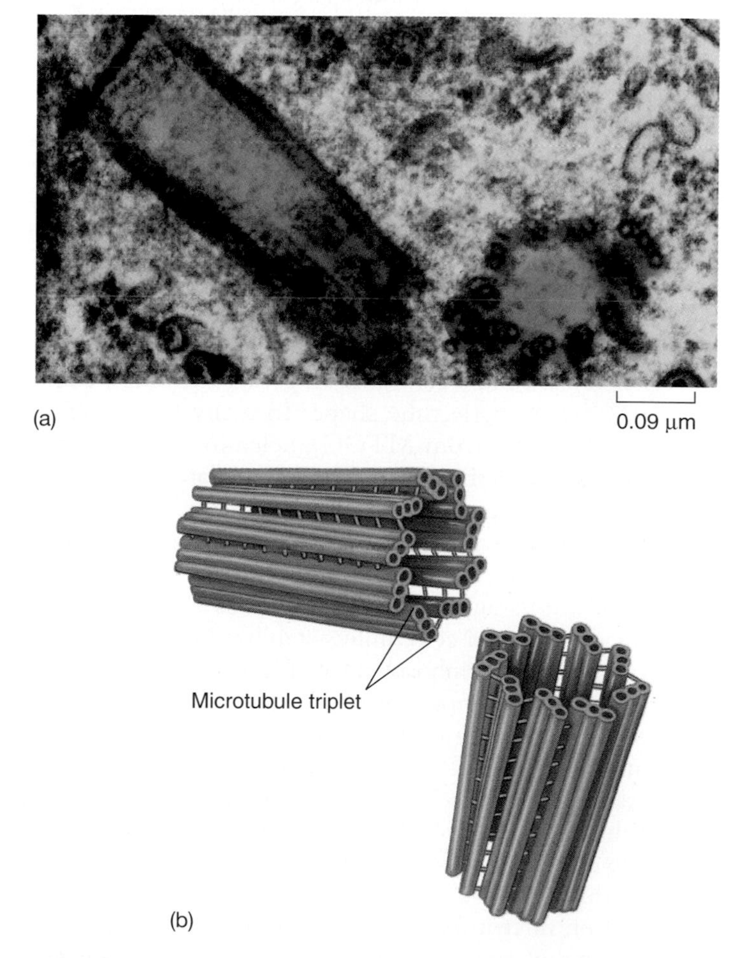

FIGURE 5.23
Centrioles. (*a*) This electron micrograph shows a pair of centrioles (75,000×). The round shape is a centriole in cross-section; the rectangular shape is a centriole in longitudinal section. (*b*) Each centriole is composed of nine triplets of microtubules.

The Cytoskeleton: Interior Framework of the Cell

The cytoplasm of all eukaryotic cells is crisscrossed by a network of protein fibers that supports the shape of the cell and anchors organelles to fixed locations. This network, called the **cytoskeleton** (figure 5.24), is a dynamic system, constantly forming and disassembling. Individual fibers form by **polymerization,** as identical protein subunits attract one another chemically and spontaneously assemble into long chains. Fibers disassemble in the same way, as one subunit after another breaks away from one end of the chain.

Eukaryotic cells may contain three types of cytoskeletal fibers, each formed from a different kind of subunit:

1. **Actin filaments.** Actin filaments are long fibers about 7 nanometers in diameter. Each filament is composed of two protein chains loosely twined together like two strands of pearls (figure 5.25*a*). Each "pearl," or subunit, on the chains is the globular protein **actin.** Actin molecules spontaneously form these filaments, even in a test tube; a cell regulates the rate of their formation through other proteins that act as switches, turning on polymerization when appropriate. Actin filaments are responsible for cellular movements such as contraction, crawling, "pinching" during division, and formation of cellular extensions.
2. **Microtubules.** Microtubules are hollow tubes about 25 nanometers in diameter, each composed of a ring of 13 protein protofilaments (figure 5.25*b*). Globular proteins consisting of dimers of alpha and beta *tubulin* subunits polymerize to form the 13 protofilaments. The protofilaments are arrayed side by side around a central core, giving the microtubule its characteristic tube shape. In many cells, microtubules form from MTOC nucleation centers near the center of the cell and radiate toward the periphery. They are in a constant state of flux, continually polymerizing and depolymerizing (the average half-life of a microtubule ranges from 10 minutes in a nondividing animal cell to as short as 20 seconds in a dividing animal cell), unless stabilized by the binding of guanosine triphosphate (GTP) to the ends, which inhibits depolymerization. The ends of the microtubule are designated as "+" (away from the nucleation center) or "–" (toward the nucleation center). Along with allowing for cellular movement, microtubules are responsible for moving materials within the cell itself. Special motor proteins, discussed later in this chapter, move cellular organelles around the cell on microtubular "tracks." *Kinesin* proteins move organelles toward the "+" end (toward the cell periphery), and *dyneins* move them toward the "–" end.

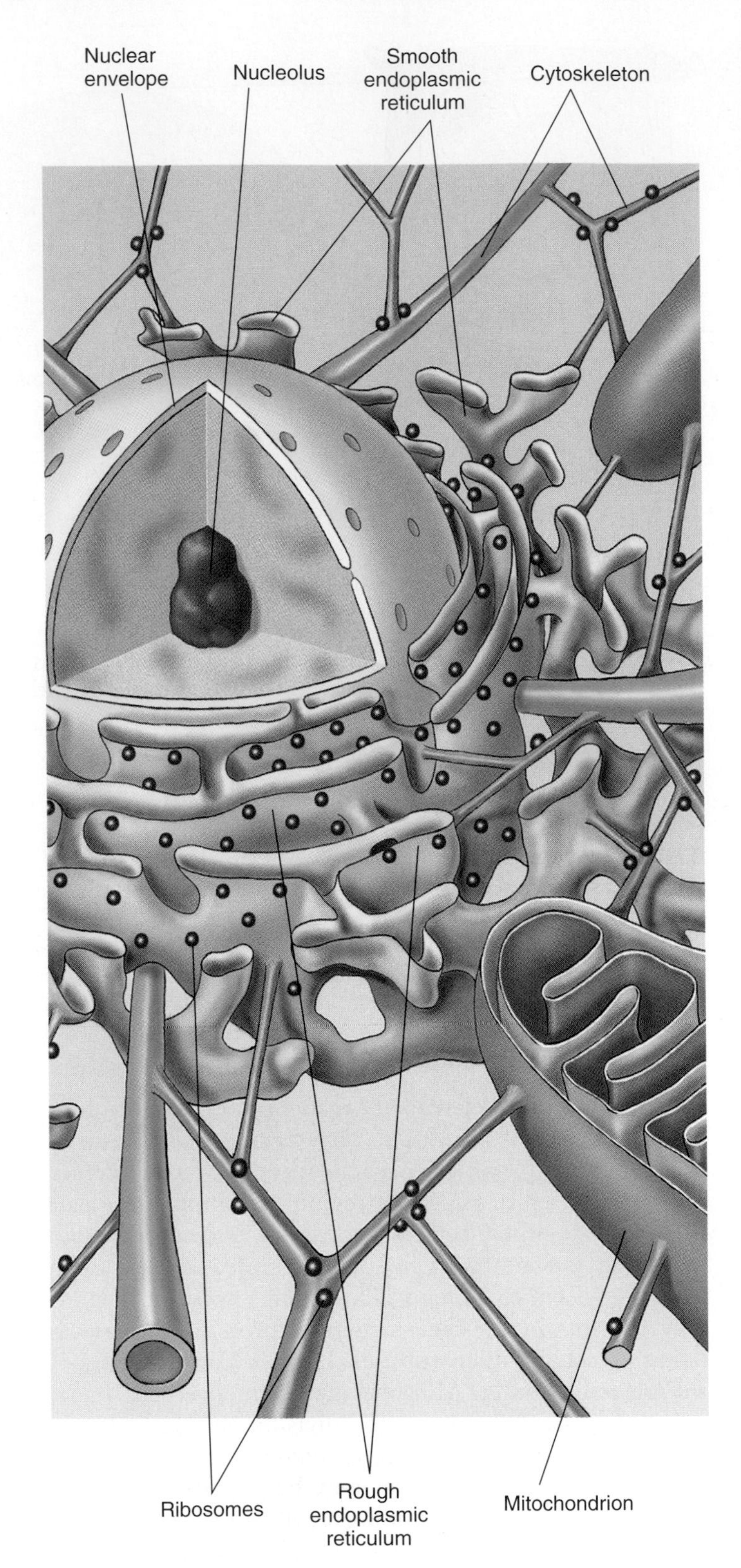

FIGURE 5.24
The cytoskeleton. In this diagrammatic cross-section of a eukaryotic cell, the cytoskeleton, a network of fibers, supports organelles such as mitochondria.

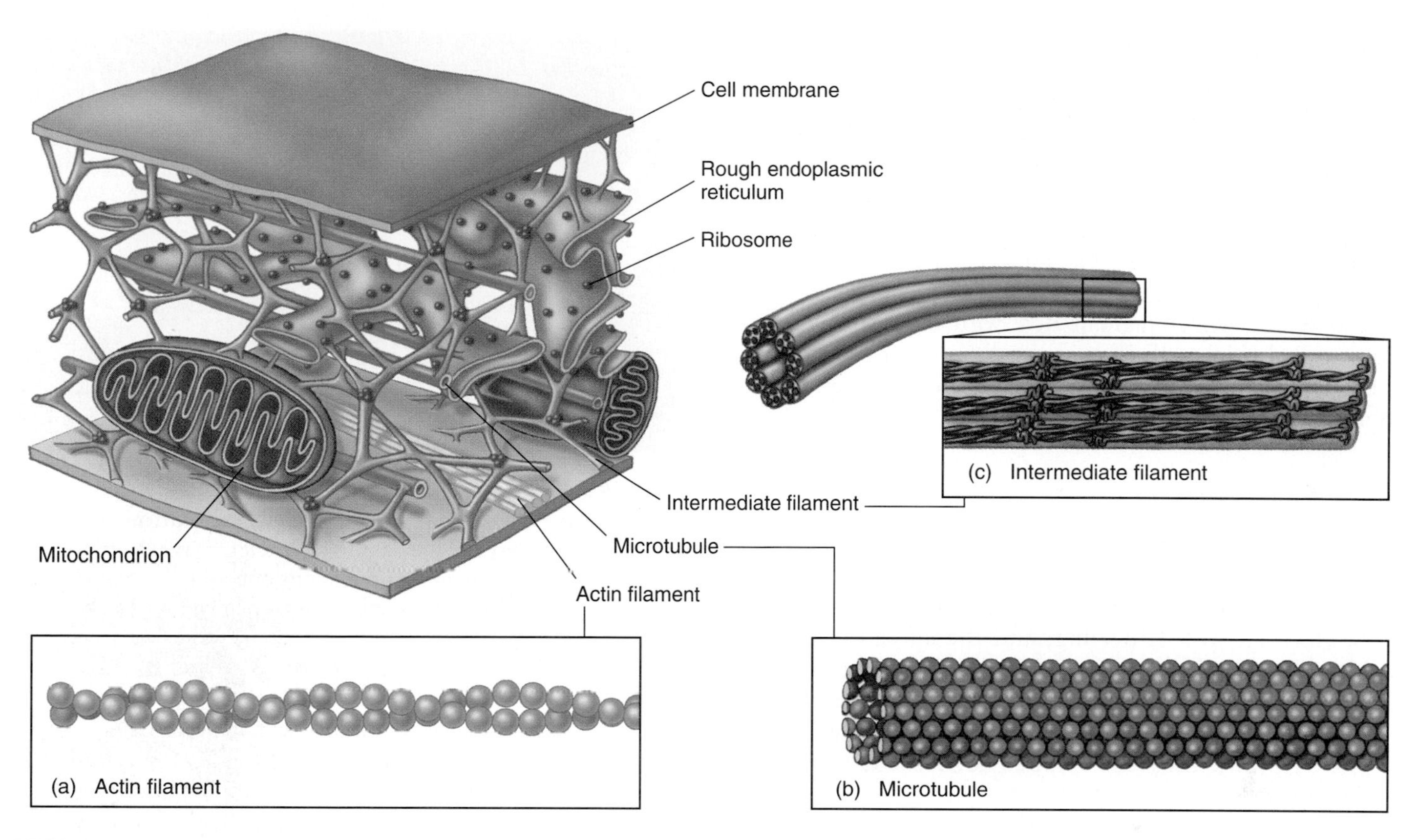

FIGURE 5.25

Molecules that make up the cytoskeleton. (*a*) Actin filaments. Actin filaments are made of two strands of the fibrous protein actin twisted together and usually occur in bundles. Actin filaments are ubiquitous, although they are concentrated below the plasma membrane in bundles known as stress fibers, which may have a contractile function. (*b*) Microtubules. Microtubules are composed of 13 stacks of tubulin protein subunits arranged side by side to form a tube. Microtubules are comparatively stiff cytoskeletal elements that serve to organize metabolism and intracellular transport in the nondividing cell. (*c*) Intermediate filaments. Intermediate filaments are composed of overlapping staggered tetramers of protein. This molecular arrangement allows for a ropelike structure that imparts tremendous mechanical strength to the cell.

3. **Intermediate filaments.** The most durable element of the cytoskeleton in animal cells is a system of tough, fibrous protein molecules twined together in an overlapping arrangement (figure 5.25*c*). These fibers are characteristically 8 to 10 nanometers in diameter, intermediate in size between actin filaments and microtubules (which is why they are called intermediate filaments). Once formed, intermediate filaments are stable and usually do not break down. Intermediate filaments constitute a heterogeneous group of cytoskeletal fibers. The most common type, composed of protein subunits called *vimentin*, provides structural stability for many kinds of cells. *Keratin*, another class of intermediate filament, is found in epithelial cells (cells that line organs and body cavities) and associated structures such as hair and fingernails. The intermediate filaments of nerve cells are called *neurofilaments*.

As we will discuss in the next section, the cytoskeleton provides an interior framework that supports the shape of the cell, stretching the plasma membrane much as the poles of a circus tent. Changing the relative length of cytoskeleton filaments allows cells to rapidly alter their shape, extending projections out or folding inward. Within the cell, the framework of filaments provides a molecular highway along which molecules can be transported.

Elements of the cytoskeleton crisscross the cytoplasm, supporting the cell shape and anchoring organelles in place. There are three principal types of fibers: actin filaments, microtubules, and intermediate filaments.

Cell Movement

Essentially all cell motion is tied to the movement of actin filaments, microtubules, or both. Intermediate filaments act as intracellular tendons, preventing excessive stretching of cells, and actin filaments play a major role in determining the shape of cells. Because actin filaments can form and dissolve so readily, they enable some cells to change shape quickly. If you look at the surfaces of such cells under a microscope, you will find them alive with motion, as projections, called **microvilli** in animal cells, shoot outward from the surface and then retract, only to shoot out elsewhere moments later (figure 5.26).

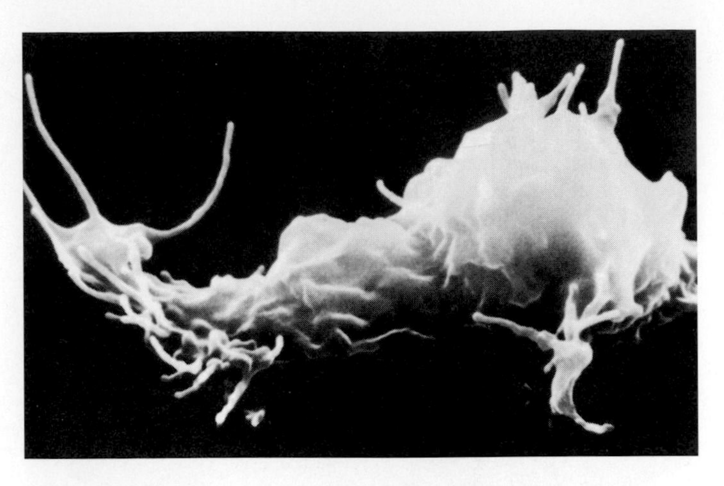

FIGURE 5.26
The surfaces of some cells are in constant motion. This amoeba, a single-celled protist, is advancing toward you, its advancing edges extending projections outward. The moving edges have been said to resemble the ruffled edges of a skirt.

Some Cells Crawl

It is the arrangement of actin filaments within the cell cytoplasm that allows cells to "crawl," *literally!* Crawling is a significant cellular phenomenon, essential to inflammation, clotting, wound healing, and the spread of cancer. White blood cells in particular exhibit this ability. Produced in the bone marrow, these cells are released into the circulatory system and then eventually crawl out of capillaries and into the tissues to destroy potential pathogens.

Cells exist in a *gel-sol* state; that is, at any given time, some regions of the cell are rigid (*gel*) and some are more fluid (*sol*). The cell is typically more sol-like in its interior, and more gel-like at its perimeter. To crawl, the cell creates a weak area in the gel perimeter, and then forces the fluid (sol) interior through the weak area, forming a **pseudopod** ("false foot"). As a result a large section of cytoplasm oozes off in a different direction, but still remains within the plasma membrane. Once extended, the pseudopod stabilizes into a gel state, assembling actin filaments. Specific membrane proteins in the pseudopod stick to the surface the cell is crawling on, and the rest of the cell is dragged in that direction. The pressure required to force out a developing pseudopod is created when actin filaments in the trailing end of the cell contract, just as squeezing a water balloon at one end forces the balloon to bulge out at the other end.

Moving Material within the Cell

Actin filaments and microtubules often orchestrate their activities to affect cellular processes. For example, during cell reproduction (see chapter 11), newly replicated chromosomes move to opposite sides of a dividing cell because they are attached to shortening microtubules. Then, in animal cells, a belt of actin pinches the cell in two by contracting like a purse string. Muscle cells also use actin filaments to contract their cytoskeletons. The fluttering of an eyelash, the flight of an eagle, and the awkward crawling of a baby all depend on these cytoskeletal movements within muscle cells.

Not only is the cytoskeleton responsible for the cell's shape and movement, but it also provides a scaffold that holds certain enzymes and other macromolecules in defined areas of the cytoplasm. Many of the enzymes involved in cell metabolism, for example, bind to actin filaments; so do ribosomes. By moving and anchoring particular enzymes near one another, the cytoskeleton, like the endoplasmic reticulum, organizes the cell's activities.

Intracellular Molecular Motors

Certain eukaryotic cells must move materials from one place to another in the cytoplasm. Most cells use the endomembrane system as an intracellular highway; the Golgi apparatus packages materials into vesicles that move through the channels of the endoplasmic reticulum to the far reaches of the cell. However, this highway is only effective over short distances. When a cell has to transport materials through long extensions like the axon of a nerve cell, the ER highways are too slow. For these situations, eukaryotic cells have developed high-speed locomotives that run along microtubular tracks.

Four components are required: (1) a **vesicle** or organelle that is to be transported, (2) a **motor molecule** that provides the energy-driven motion, (3) a **connector molecule** that connects the vesicle to the motor molecule, and (4) **microtubules** on which the vesicle will ride like a train on a rail. For example, embedded within the membranes of endoplasmic reticulum is a protein called *kinectin* that bind the ER vesicles to the motor protein *kinesin.* As nature's tiniest motors, these motor proteins literally pull the transport vesicles along the microtubular tracks. Kinesin uses ATP to power its movement toward

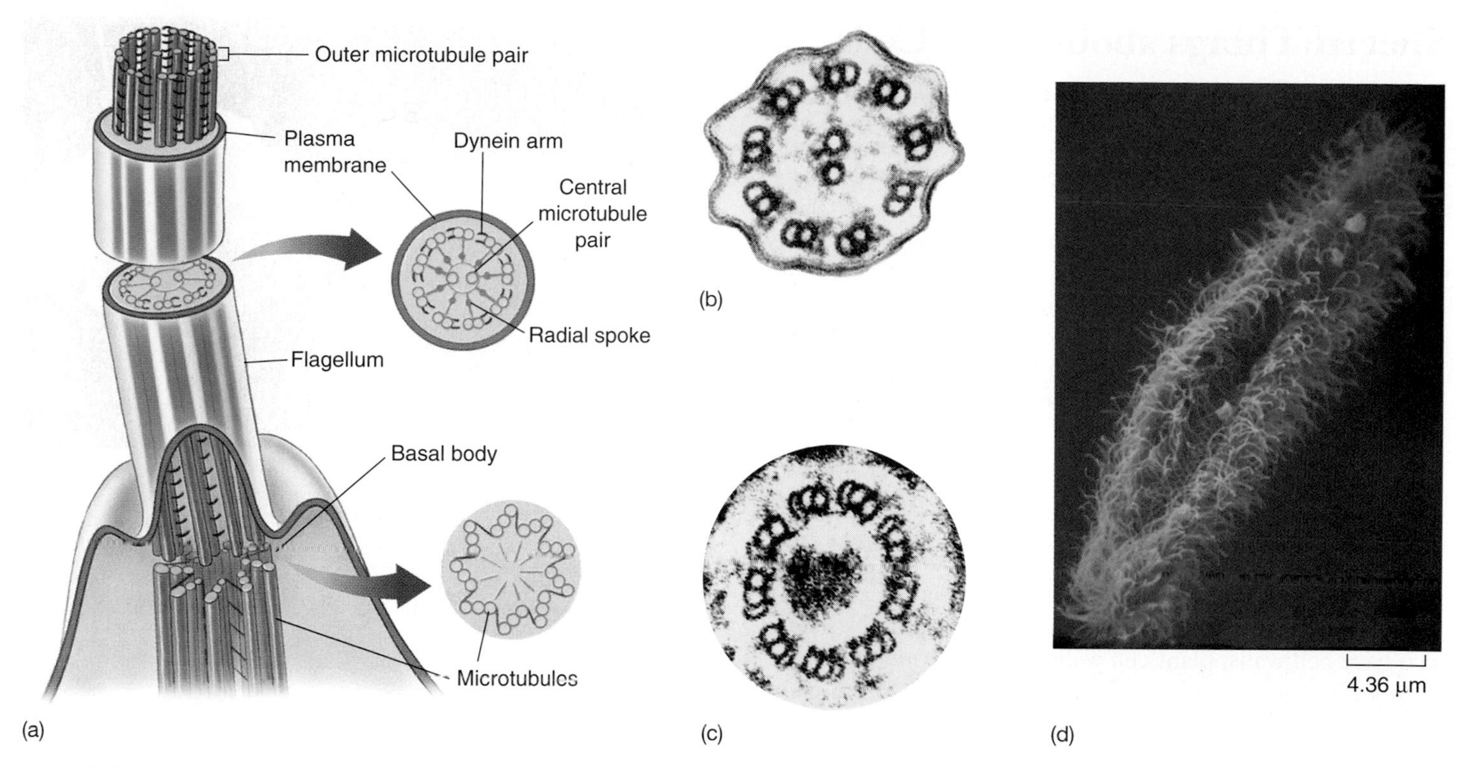

FIGURE 5.27
Flagella and cilia. (*a*) A eukaryotic flagellum originates directly from a basal body. (*b*) The flagellum has two microtubules in its core connected by radial spokes to an outer ring of nine paired microtubules with dynein arms. (*c*) The basal body consists of nine microtubule triplets connected by short protein segments. The structure of cilia is similar to that of flagella, but cilia are usually shorter. (*d*) The surface of this *Paramecium* is covered with a dense forest of cilia.

the cell periphery, dragging the vesicle with it as it travels along the microtubule. Another vesicle protein (or perhaps a slight modification of kinesin—further research is needed to determine which) binds vesicles to the motor protein dynein, which directs movement in the opposite direction, inward toward the cell's center. (Dynein is also involved in the movement of eukaryotic flagella, as discussed below.) The destination of a particular transport vesicle and its contents is thus determined by the nature of the linking protein embedded within the vesicle's membrane.

Swimming with Flagella and Cilia

Earlier in this chapter, we described the structure of bacterial flagella. Eukaryotic cells have a *completely different* kind of flagellum, consisting of a circle of nine microtubule pairs surrounding two central microtubules; this arrangement is referred to as the **9 + 2 structure** (figure 5.27). As pairs of microtubules move past one another using arms composed of the motor protein dynein, the eukaryotic flagellum undulates rather than rotates. When examined carefully, each flagellum proves to be an outward projection of the cell's interior, containing cytoplasm and enclosed by the plasma membrane. The microtubules of the flagellum are derived from a **basal body,** situated just below the point where the flagellum protrudes from the surface of the cell.

The flagellum's complex microtubular apparatus evolved early in the history of eukaryotes. Although the cells of many multicellular and some unicellular eukaryotes today no longer possess flagella and are nonmotile, an organization similar to the 9 + 2 arrangement of microtubules can still be found within them, in structures called **cilia** (singular, *cilium*). Cilia are short cellular projections that are often organized in rows (see figure 5.1). They are more numerous than flagella on the cell surface, but have the same internal structure. In many multicellular organisms, cilia carry out tasks far removed from their original function of propelling cells through water. In several kinds of vertebrate tissues, for example, the beating of rows of cilia moves water over the tissue surface. The sensory cells of the vertebrate ear also contain cilia; sound waves bend these cilia, the initial sensory input of hearing. Thus, the 9 + 2 structure of flagella and cilia appears to be a fundamental component of eukaryotic cells.

Some eukaryotic cells use pseudopodia to crawl about within multicellular organisms, while many protists swim using flagella and cilia. Materials are transported within cells by special motor proteins.

Special Things about Plant Cells

Vacuoles: A Central Storage Compartment

The center of a plant cell usually contains a large, apparently empty space, called the **central vacuole** (figure 5.28). This vacuole is not really empty; it contains large amounts of water and other materials, such as sugars, ions, and pigments. The central vacuole functions as a storage center for these important substances and also helps to increase the surface-to-volume ratio of the plant cell by applying pressure to the cell membrane. The cell membrane expands outward under this pressure, thereby increasing its surface area.

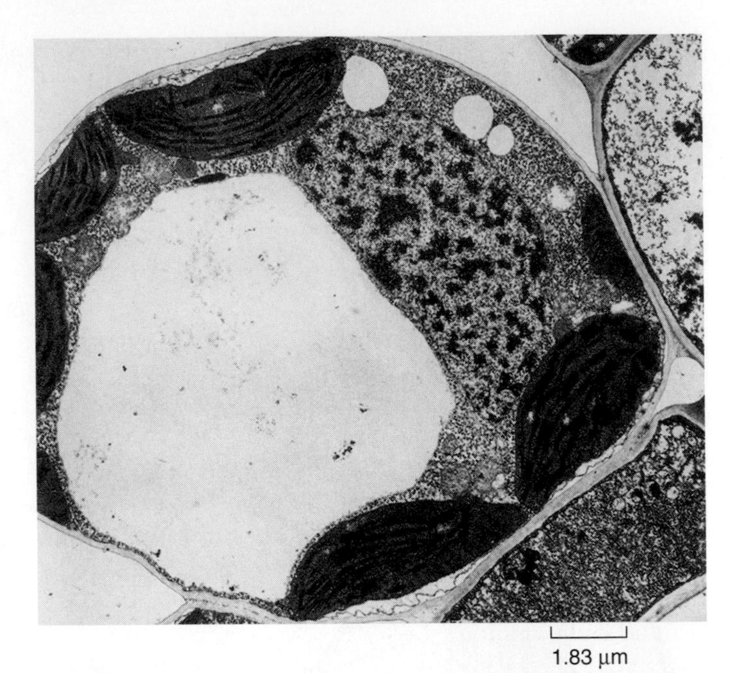

FIGURE 5.28
The central vacuole. A plant's central vacuole stores dissolved substances and can increase in size to increase the surface area of a plant cell.

Cell Walls: Protection and Support

Plant cells share a characteristic with bacteria that is not shared with animal cells—that is, plants have **cell walls,** which protect and support the plant cell. Although bacteria also have cell walls, plant cell walls are chemically and structurally different from bacterial cell walls. Cell walls are also present in fungi and some protists. In plants, cell walls are composed of fibers of the polysaccharide cellulose. **Primary walls** are laid down when the cell is still growing, and between the walls of adjacent cells is a sticky substance called the **middle lamella,** which glues the cells together (figure 5.29). Some plant cells produce strong **secondary walls,** which are deposited inside the primary walls of fully expanded cells.

Plant cells store substances in a large central vacuole, and encase themselves within a strong cellulose cell wall.

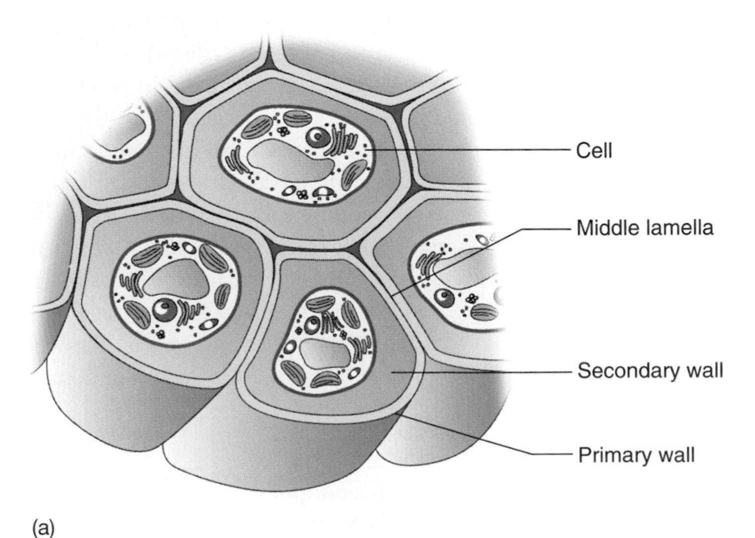

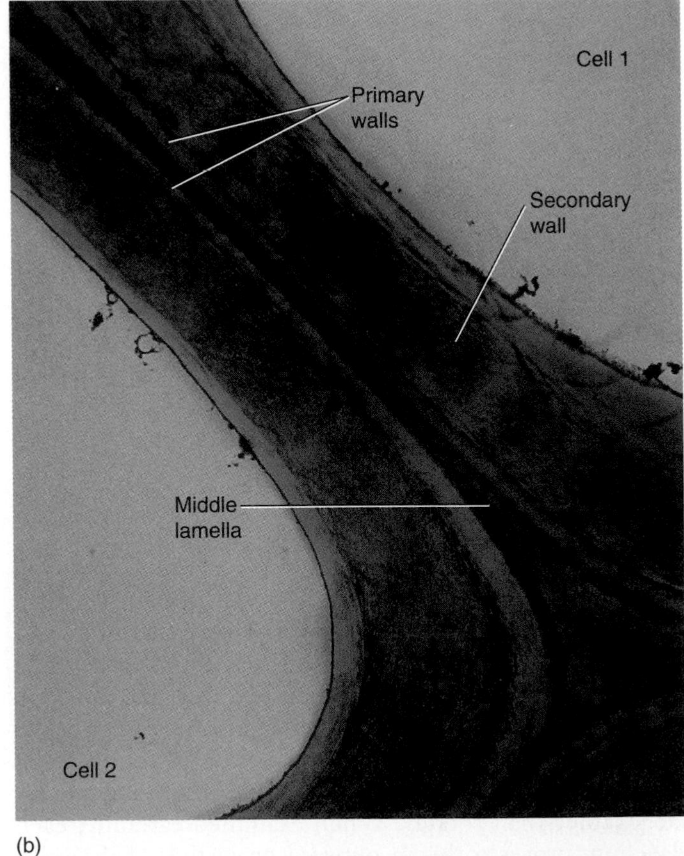

FIGURE 5.29
Cell walls in plants. As shown in this drawing (*a*) and transmission electron micrograph (*b*), plant cell walls are thicker, stronger, and more rigid than those of bacteria. Primary cell walls are laid down when the cell is young. Thicker secondary cell walls may be added later when the cell is fully grown.

5.4 Symbiosis played a key role in the origin of some eukaryotic organelles.

Endosymbiosis

Symbiosis is a close relationship between organisms of different species that live together. The theory of **endosymbiosis** proposes that some of today's eukaryotic organelles evolved by a symbiosis in which one species of prokaryote was engulfed by and lived inside another species of prokaryote that was a precursor to eukaryotes (figure 5.30). According to the endosymbiont theory, the engulfed prokaryotes provided their hosts with certain advantages associated with their special metabolic abilities. Two key eukaryotic organelles are believed to be the descendants of these endosymbiotic prokaryotes: mitochondria, which are thought to have originated as bacteria capable of carrying out oxidative metabolism; and chloroplasts, which apparently arose from photosynthetic bacteria.

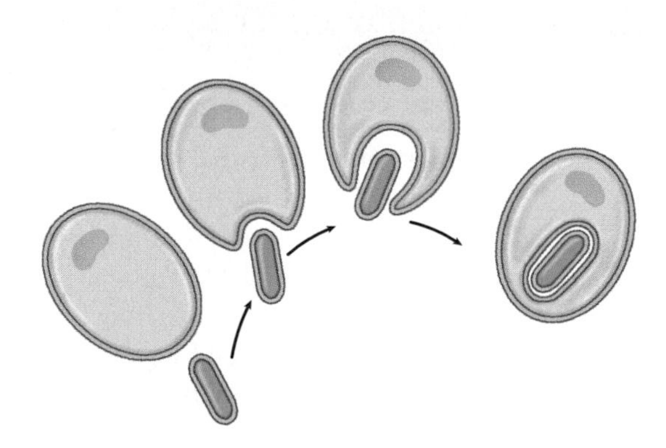

FIGURE 5.30
Endosymbiosis. This figure shows how a double membrane may have been created during the symbiotic origin of mitochondria or chloroplasts.

The endosymbiont theory is supported by a wealth of evidence. Both mitochondria and chloroplasts are surrounded by two membranes; the inner membrane probably evolved from the plasma membrane of the engulfed bacterium, while the outer membrane is probably derived from the plasma membrane or endoplasmic reticulum of the host cell. Mitochondria are about the same size as most bacteria, and the cristae formed by their inner membranes resemble the folded membranes in various groups of bacteria. Mitochondrial ribosomes are also similar to bacterial ribosomes in size and structure. Both mitochondria and chloroplasts contain circular molecules of DNA similar to those in bacteria. Finally, mitochondria divide by simple fission, splitting in two just as bacterial cells do, and they apparently replicate and partition their DNA in much the same way as bacteria. Table 5.2 compares and reviews the features of three types of cells.

Some eukaryotic organelles are thought to have arisen by endosymbiosis.

Table 5.2 A Comparison of Bacterial, Animal, and Plant Cells

	Bacterium	Animal	Plant
EXTERIOR STRUCTURES			
Cell wall	Present (protein-polysaccharide)	Absent	Present (cellulose)
Cell membrane	Present	Present	Present
Flagella	May be present (single strand)	May be present	Absent except in sperm of a few species
INTERIOR STRUCTURES			
ER	Absent	Usually present	Usually present
Ribosomes	Present	Present	Present
Microtubules	Absent	Present	Present
Centrioles	Absent	Present	Absent
Golgi apparatus	Absent	Present	Present
Nucleus	Absent	Present	Present
Mitochondria	Absent	Present	Present
Chloroplasts	Absent	Absent	Present
Chromosomes	A single circle of DNA	Multiple; DNA-protein complex	Multiple; DNA-protein complex
Lysosomes	Absent	Usually present	Present
Vacuoles	Absent	Absent or small	Usually a large single vacuole

Chapter 5

www.mhhe.com/raven6/resources5.mhtml

Summary | Questions | Media Resources

5.1 All organisms are composed of cells.

- The cell is the smallest unit of life. All living things are made of cells.
- The cell is composed of a nuclear region, which holds the hereditary apparatus, enclosed within the cytoplasm.
- In all cells, the cytoplasm is bounded by a membrane composed of phospholipid and protein.

1. What are the three principles of the cell theory?

2. How does the surface area-to-volume ratio of cells limit the size that cells can attain?

- Exploration: Cell Size

- Surface to Volume

5.2 Eukaryotic cells are far more complex than bacterial cells.

- Bacteria, which have prokaryotic cell structure, do not have membrane-bounded organelles within their cells. Their DNA molecule is circular.
- The eukaryotic cell is larger and more complex, with many internal compartments.

3. How are prokaryotes different from eukaryotes in terms of their cell walls, interior organization, and flagella?

- Art Activities:
  - -Animal Cell Structure
  - -Plant Cell Structure
  - -Nonphotosynthetic Bacterium
  - -Cyanobacterium

5.3 Take a tour of a eukaryotic cell.

- A eukaryotic cell is organized into three principal zones: the nucleus, the cytoplasm, and the plasma membrane. Located in the cytoplasm are numerous organelles, which perform specific functions for the cell.
- Many of these organelles, such as the endoplasmic reticulum, Golgi apparatus (which gives rise to lysosomes), and nucleus, are part of a complex endomembrane system.
- Mitochondria and chloroplasts are part of the energy-processing system of the cell.
- The cytoskeleton encompasses a variety of fibrous proteins that provide structural support and perform other functions for the cell.
- Many eukaryotic cells possess flagella or cilia having a 9 + 2 arrangement of microtubules; sliding of the microtubules past one another bends these cellular appendages.
- Cells transport materials long distances within the cytoplasm by packaging them into vesicles that are pulled by motor proteins along microtubule tracks.

4. What is the endoplasmic reticulum? What is its function? How does rough ER differ from smooth ER?

5. What is the function of the Golgi apparatus? How do the substances released by the Golgi apparatus make their way to other locations in the cell?

6. What types of eukaryotic cells contain mitochondria? What function do mitochondria perform?

7. What unique metabolic activity occurs in chloroplasts?

8. What cellular functions do centrioles participate in?

9. What kinds of cytoskeleton fibers are stable and which are changeable?

10. How do cilia compare with eukaryotic flagella?

- Art Activities:
  - -Anatomy of the Nucleus
  - -Golgi Apparatus Structure
  - -Mitochondrion Structure
  - -Organization of Cristae
  - -Chloroplast Structure
  - -The Cytoskeleton
  - -Plant Cell

- Endomembrane
- Energy Organelles
- Cytoskeleton

- Art Quizzes:
  - -Nucleosomes
  - -Rough ER and Protein Synthesis
  - -Protein Transport

5.4 Symbiosis played a key role in the origin of some eukaryotic organelles.

- Present-day mitochondria and chloroplasts probably evolved as a consequence of early endosymbiosis: the ancestor of the eukaryotic cell engulfed a bacterium, and the bacterium continued to function within the host cell.

11. What is the endosymbiont theory? What is the evidence supporting this theory?

- Scientists on Science: The Joy of Discovery

6

Membranes

Concept Outline

6.1 Biological membranes are fluid layers of lipid.

The Phospholipid Bilayer. Cells are encased by membranes composed of a bilayer of phospholipid.

The Lipid Bilayer Is Fluid. Because individual phospholipid molecules do not bind to one another, the lipid bilayer of membranes is a fluid.

6.2 Proteins embedded within the plasma membrane determine its character.

The Fluid Mosaic Model. A varied collection of proteins float within the lipid bilayer.

Examining Cell Membranes. Visualizing a plasma membrane requires a powerful electron microscope.

Kinds of Membrane Proteins. The proteins in a membrane function in support, transport, recognition, and reactions.

Structure of Membrane Proteins. Membrane proteins are anchored into the lipid bilayer by their nonpolar regions.

6.3 Passive transport across membranes moves down the concentration gradient.

Diffusion. Random molecular motion results in a net movement of molecules to regions of lower concentration.

Facilitated Diffusion. Passive movement across a membrane is often through specific carrier proteins.

Osmosis. Polar solutes interact with water and can affect the movement of water across semipermeable membranes.

6.4 Bulk transport utilizes endocytosis.

Bulk Passage Into and Out of the Cell. To transport large particles, membranes form vesicles.

6.5 Active transport across membranes is powered by energy from ATP.

Active Transport. Cells transport molecules up a concentration gradient using ATP-powered carrier proteins.

Coupled Transport. Active transport of ions drives coupled uptake of other molecules up their concentration gradients.

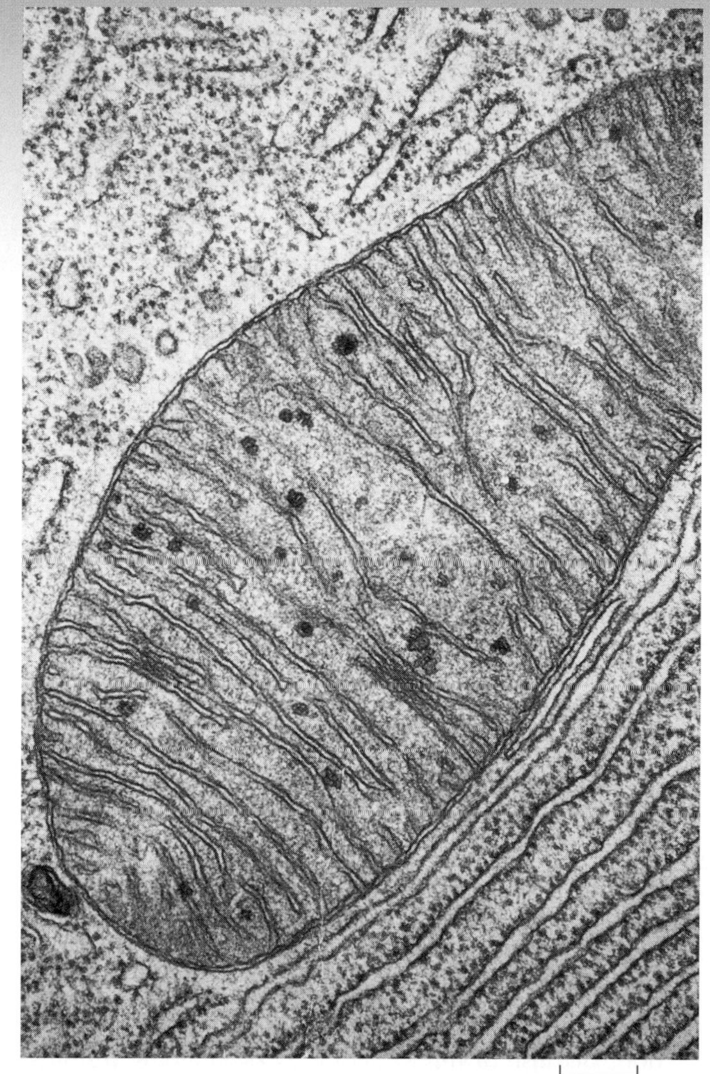

FIGURE 6.1
Membranes within a human cell. Sheets of endoplasmic reticulum weave through the cell interior. The large oval is a mitochondrion, itself filled with extensive internal membranes.

Among a cell's most important activities are its interactions with the environment, a give and take that never ceases. Without it, life could not persist. While living cells and eukaryotic organelles (figure 6.1) are encased within a lipid membrane through which few water-soluble substances can pass, the membrane contains protein passageways that permit specific substances to move in and out of the cell and allow the cell to exchange information with its environment. We call this delicate skin of protein molecules embedded in a thin sheet of lipid a **plasma membrane**. This chapter will examine the structure and function of this remarkable membrane.

6.1 Biological membranes are fluid layers of lipid.

The Phospholipid Bilayer

The membranes that encase all living cells are sheets of lipid only two molecules thick; more than 10,000 of these sheets piled on one another would just equal the thickness of this sheet of paper. The lipid layer that forms the foundation of a cell membrane is composed of molecules called **phospholipids** (figure 6.2).

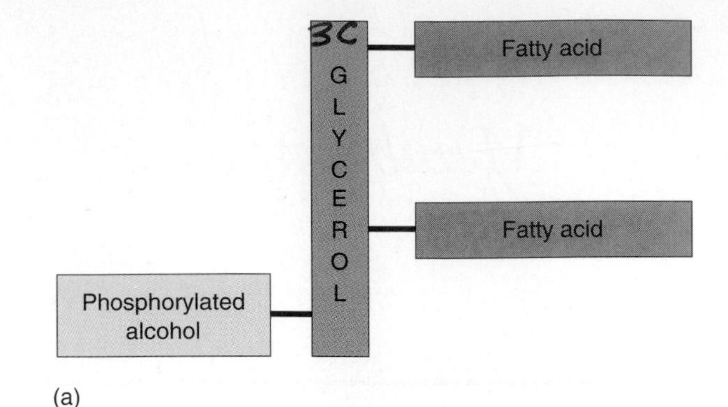

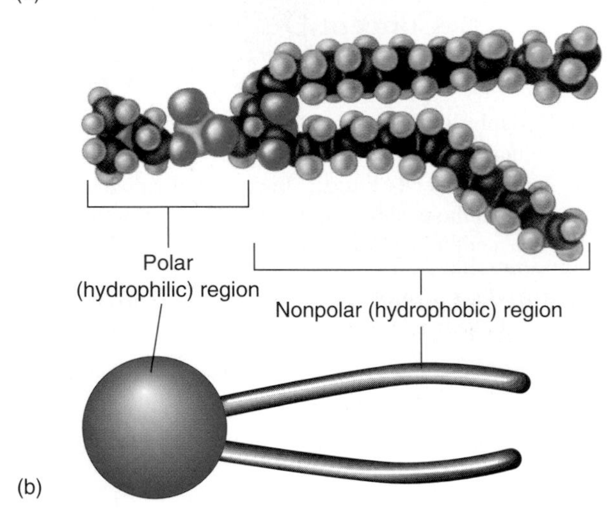

FIGURE 6.2
Phospholipid structure. (*a*) A phospholipid is a composite molecule similar to a triacylglycerol, except that only two fatty acids are bound to the glycerol backbone; a phosphorylated alcohol occupies the third position on the backbone. (*b*) Because the phosphorylated alcohol usually extends from one end of the molecule and the two fatty acid chains extend from the other, phospholipids are often diagrammed as a polar head with two nonpolar hydrophobic tails.

Phospholipids

Like the fat molecules you studied in chapter 3, a phospholipid has a backbone derived from a three-carbon molecule called glycerol. Attached to this backbone are fatty acids, long chains of carbon atoms ending in a carboxyl (—COOH) group. A fat molecule has three such chains, one attached to each carbon in the backbone; because these chains are nonpolar, they do not form hydrogen bonds with water, and the fat molecule is not water-soluble. A phospholipid, by contrast, has only two fatty acid chains attached to its backbone. The third carbon on the backbone is attached instead to a highly polar organic alcohol that readily forms hydrogen bonds with water. Because this alcohol is attached by a phosphate group, the molecule is called a *phospho*lipid.

One end of a phospholipid molecule is, therefore, strongly nonpolar (water-insoluble), while the other end is strongly polar (water-soluble). The two nonpolar fatty acids extend in one direction, roughly parallel to each other, and the polar alcohol group points in the other direction. Because of this structure, phospholipids are often diagrammed as a polar head with two dangling nonpolar tails (as in figure 6.2*b*).

Phospholipids Form Bilayer Sheets

What happens when a collection of phospholipid molecules is placed in water? The polar water molecules repel the long nonpolar tails of the phospholipids as the water molecules seek partners for hydrogen bonding. Due to the polar nature of the water molecules, the nonpolar tails of the phospholipids end up packed closely together, sequestered as far as possible from water. Every phospholipid molecule orients to face its polar head toward water and its nonpolar tails away. When *two* layers form with the tails facing each other, no tails ever come in contact with water. The resulting structure is called a **lipid bilayer** (figure 6.3). Lipid bilayers form spontaneously, driven by the tendency of water molecules to form the maximum number of hydrogen bonds.

The nonpolar interior of a lipid bilayer impedes the passage of any water-soluble substances through the bilayer, just as a layer of oil impedes the passage of a drop of water ("oil and water do not mix"). This barrier to the passage of water-soluble substances is the key biological property of the lipid bilayer. In addition to the phospholipid molecules that make up the lipid bilayer, the membranes of every cell also contain proteins that extend through the lipid bilayer, providing passageways across the membrane.

The basic foundation of biological membranes is a lipid bilayer, which forms spontaneously. In such a layer, the nonpolar hydrophobic tails of phospholipid molecules point inward, forming a nonpolar barrier to water-soluble molecules.

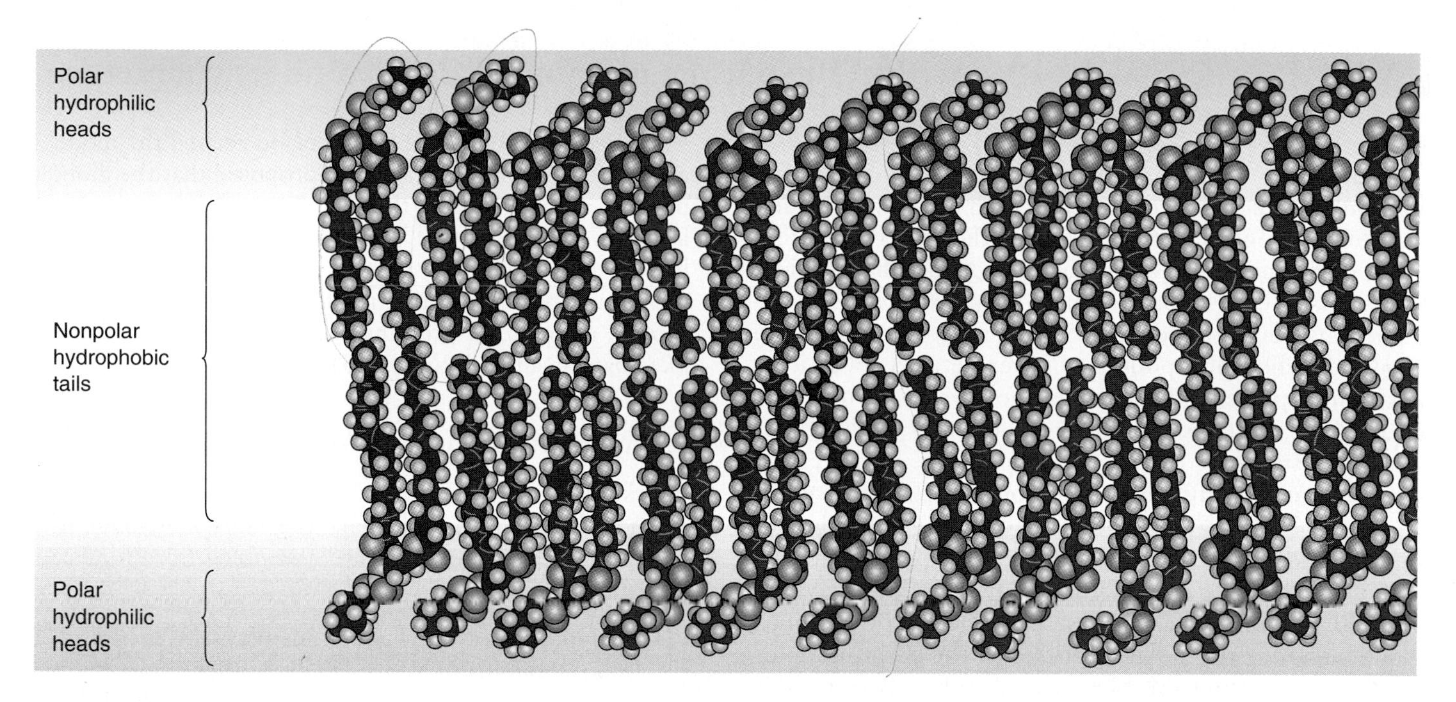

FIGURE 6.3
A phospholipid bilayer. The basic structure of every plasma membrane is a double layer of lipid, in which phospholipids aggregate to form a bilayer with a nonpolar interior. The phospholipid tails do not align perfectly and the membrane is "fluid." Individual phospholipid molecules can move from one place to another in the membrane.

The Lipid Bilayer Is Fluid

A lipid bilayer is stable because water's affinity for hydrogen bonding never stops. Just as surface tension holds a soap bubble together, even though it is made of a liquid, so the hydrogen bonding of water holds a membrane together. But while water continually drives phospholipid molecules into this configuration, it does not locate specific phospholipid molecules relative to their neighbors in the bilayer. As a result, individual phospholipids and unanchored proteins are free to move about within the membrane. This can be demonstrated vividly by fusing cells and watching their proteins reassort (figure 6.4).

Phospholipid bilayers are fluid, with the viscosity of olive oil (and like oil, their viscosity increases as the temperature decreases). Some membranes are more fluid than others, however. The tails of individual phospholipid molecules are attracted to one another when they line up close together. This causes the membrane to become less fluid, because aligned molecules must pull apart from one another before they can move about in the membrane. The greater the degree of alignment, the less fluid the membrane. Some phospholipid tails do not align well because they contain one or more double bonds between carbon atoms, introducing kinks in the tail. Membranes containing such phospholipids are more fluid than membranes that lack them. Most membranes also contain steroid lipids like cholesterol, which can either increase or decrease membrane fluidity, depending on temperature.

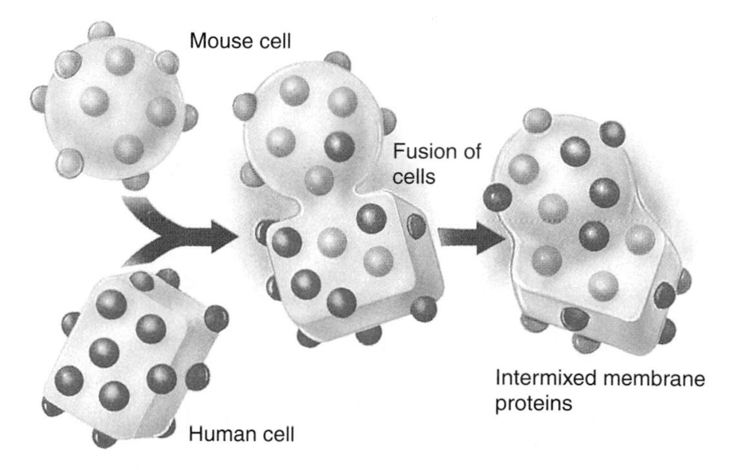

FIGURE 6.4
Proteins move about in membranes. Protein movement within membranes can be demonstrated easily by labeling the plasma membrane proteins of a mouse cell with fluorescent antibodies and then fusing that cell with a human cell. At first, all of the mouse proteins are located on the mouse side of the fused cell and all of the human proteins are located on the human side of the fused cell. However, within an hour, the labeled and unlabeled proteins are intermixed throughout the hybrid cell's plasma membrane.

The lipid bilayer is liquid like a soap bubble, rather than solid like a rubber balloon.

6.2 Proteins embedded within the plasma membrane determine its character.

The Fluid Mosaic Model

A plasma membrane is composed of both lipids and globular proteins. For many years, biologists thought the protein covered the inner and outer surfaces of the phospholipid bilayer like a coat of paint. The widely accepted Davson-Danielli model, proposed in 1935, portrayed the membrane as a sandwich: a phospholipid bilayer between two layers of globular protein. This model, however, was not consistent with what researchers were learning in the 1960s about the structure of membrane proteins. Unlike most proteins found within cells, membrane proteins are not very soluble in water—they possess long stretches of nonpolar hydrophobic amino acids. If such proteins indeed coated the surface of the lipid bilayer, as the Davson-Danielli model suggests, then their nonpolar portions would separate the polar portions of the phospholipids from water, causing the bilayer to dissolve! Because this doesn't happen, there is clearly something wrong with the model.

In 1972, S. Singer and G. Nicolson revised the model in a simple but profound way: they proposed that the globular proteins are *inserted* into the lipid bilayer, with their nonpolar segments in contact with the nonpolar interior of the bilayer and their polar portions protruding out from the membrane surface. In this model, called the **fluid mosaic model,** a mosaic of proteins float in the fluid lipid bilayer like boats on a pond (figure 6.5).

Components of the Cell Membrane

A eukaryotic cell contains many membranes. While they are not all identical, they share the same fundamental architecture. Cell membranes are assembled from four components (table 6.1):

1. **Lipid bilayer.** Every cell membrane is composed of phospholipids in a bilayer. The other components of the membrane are enmeshed within the bilayer, which provides a flexible matrix and, at the same time, imposes a barrier to permeability.

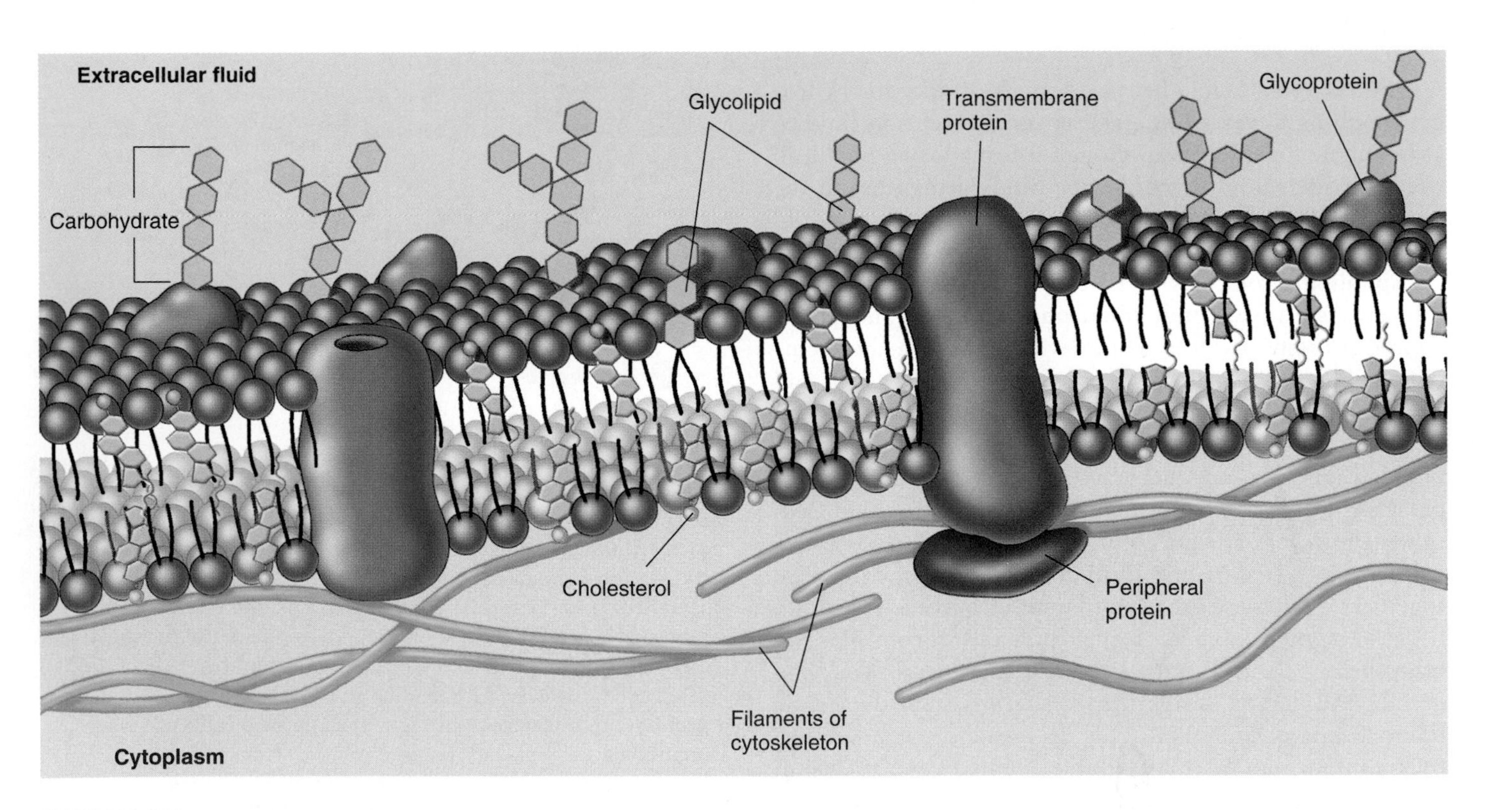

FIGURE 6.5
The fluid mosaic model of the plasma membrane. A variety of proteins protrude through the plasma membrane of animal cells, and nonpolar regions of the proteins tether them to the membrane's nonpolar interior. The three principal classes of membrane proteins are transport proteins, receptors, and cell surface markers. Carbohydrate chains are often bound to the extracellular portion of these proteins, as well as to the membrane phospholipids. These chains serve as distinctive identification tags, unique to particular cells.

Table 6.1 Components of the Cell Membrane

Component	Composition	Function	How It Works	Example
Phospholipid bilayer	Phospholipid molecules	Provides permeability barrier, matrix for proteins	Excludes water-soluble molecules from nonpolar interior of bilayer	Bilayer of cell is impermeable to water-soluble molecules, like glucose
Transmembrane proteins	Carriers	Active and passive transport of molecules across membrane	"Escort" molecules through the membrane in a series of conformational changes	Glycophorin carrier for sugar transport; sodium-potassium pump
	Channels	Passively transport molecules across membrane	Create a tunnel that acts as a passage through membrane	Sodium and potassium channels in nerve cells
	Receptors	Transmit information into cell	Signal molecules bind to cell-surface portion of the receptor protein; this alters the portion of the receptor protein within the cell, inducing activity	Specific receptors bind peptide hormones and neurotransmitters
Interior protein network	Spectrins	Determine shape of cell	Form supporting scaffold beneath membrane, anchored to both membrane and cytoskeleton	Red blood cell
	Clathrins	Anchor certain proteins to specific sites, especially on the exterior cell membrane in receptor-mediated endocytosis	Proteins line coated pits and facilitate binding to specific molecules	Localization of low-density lipoprotein receptor within coated pits
Cell surface markers	Glycoproteins	"Self"-recognition	Create a protein/carbohydrate chain shape characteristic of individual	Major histocompatibility complex protein recognized by immune system
	Glycolipid	Tissue recognition	Create a lipid/carbohydrate chain shape characteristic of tissue	A, B, O blood group markers

2. **Transmembrane proteins.** A major component of every membrane is a collection of proteins that float on or in the lipid bilayer. These proteins provide passageways that allow substances and information to cross the membrane. Many membrane proteins are not fixed in position; they can move about, as the phospholipid molecules do. Some membranes are crowded with proteins, while in others, the proteins are more sparsely distributed.
3. **Network of supporting fibers.** Membranes are structurally supported by intracellular proteins that reinforce the membrane's shape. For example, a red blood cell has a characteristic biconcave shape because a scaffold of proteins called spectrin links proteins in the plasma membrane with actin filaments in the cell's cytoskeleton. Membranes use networks of other proteins to control the lateral movements of some key membrane proteins, anchoring them to specific sites.
4. **Exterior proteins and glycolipids.** Membrane sections assemble in the endoplasmic reticulum, transfer to the Golgi complex, and then are transported to the plasma membrane. The endoplasmic reticulum adds chains of sugar molecules to membrane proteins and lipids, creating a "sugar coating" called the glycocalyx that extends from the membrane on the outside of the cell only. Different cell types exhibit different varieties of these glycoproteins and glycolipids on their surfaces, which act as cell identity markers.

The fluid mosaic model proposes that membrane proteins are embedded within the lipid bilayer. Membranes are composed of a lipid bilayer within which proteins are anchored. Plasma membranes are supported by a network of fibers and coated on the exterior with cell identity markers.

Examining Cell Membranes

Biologists examine the delicate, filmy structure of a cell membrane using electron microscopes that provide clear magnification to several thousand times. We discussed two types of electron microscopes in chapter 5: the transmission electron microscope (TEM) and the scanning electron microscope (SEM). When examining cell membranes with electron microscopy, specimens must be prepared for viewing.

In one method of preparing a specimen, the tissue of choice is embedded in a hard matrix, usually some sort of epoxy (figure 6.6). The epoxy block is then cut with a microtome, a machine with a very sharp blade that makes incredibly thin slices. The knife moves up and down as the specimen advances toward it, causing transparent "epoxy shavings" less than 1 micrometer thick to peel away from the block of tissue. These shavings are placed on a grid and a beam of electrons is directed through the grid with the TEM. At the high magnification an electron microscope provides, resolution is good enough to reveal the double layers of a membrane.

Freeze-fracturing a specimen is another way to visualize the inside of the membrane. The tissue is embedded in a medium and quick-frozen with liquid nitrogen. The frozen tissue is then "tapped" with a knife, causing a crack between the phospholipid layers of membranes. Proteins, carbohydrates, pits, pores, channels, or any other structure affiliated with the membrane will pull apart (whole, usually) and stick with one side of the split membrane. A very thin coating of platinum is then evaporated onto the fractured surface forming a replica or "cast" of the surface. Once the topography of the membrane has been preserved in the "cast," the actual tissue is dissolved away, and the "cast" is examined with electron microscopy, creating a strikingly different view of the membrane (see figure 5.10*b*).

Visualizing a plasma membrane requires a very powerful electron microscope. Electrons can either be passed through a sample or bounced off it.

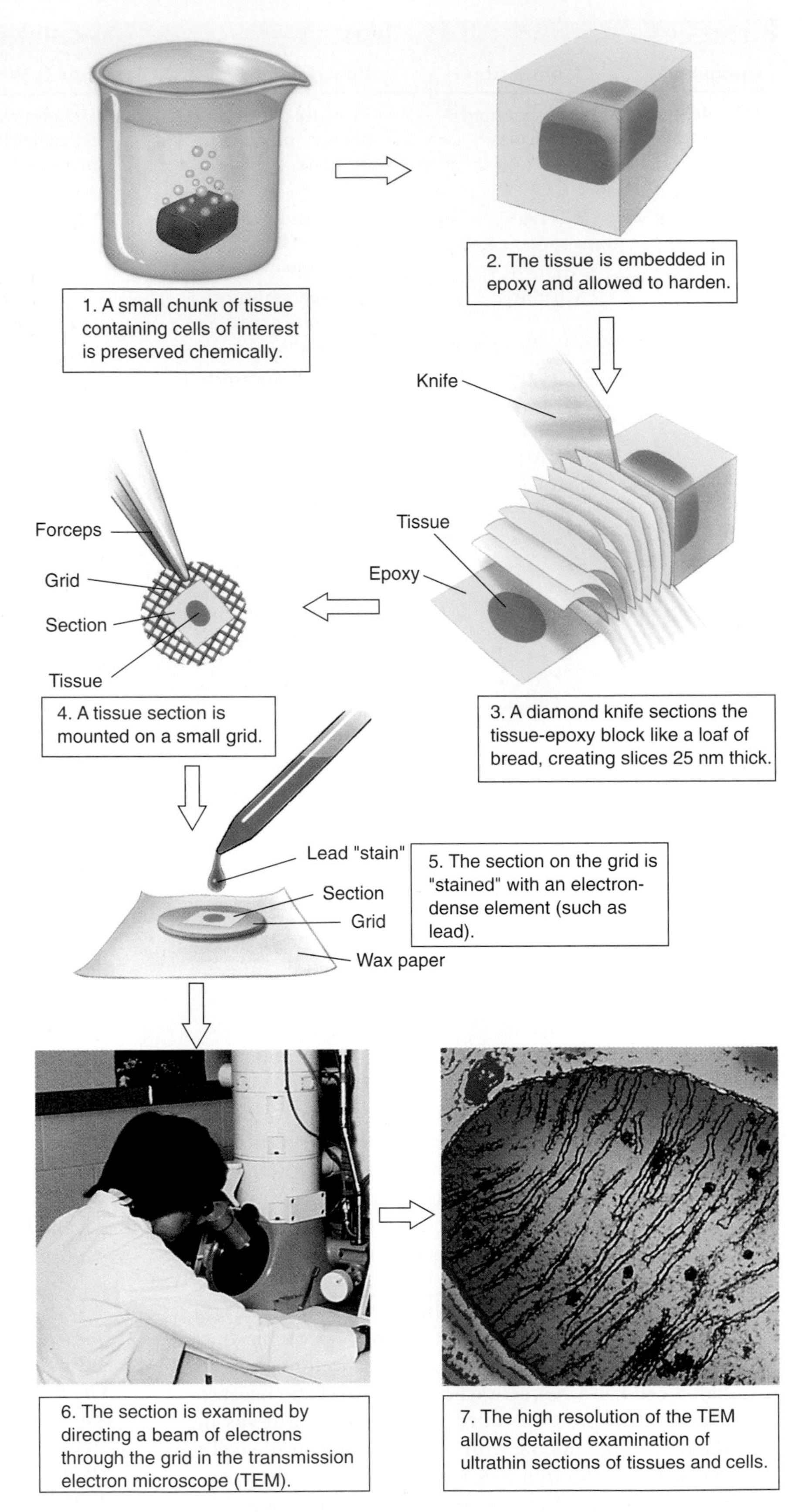

1. A small chunk of tissue containing cells of interest is preserved chemically.
2. The tissue is embedded in epoxy and allowed to harden.
Knife
Tissue
Epoxy
3. A diamond knife sections the tissue-epoxy block like a loaf of bread, creating slices 25 nm thick.
Forceps
Grid
Section
Tissue
4. A tissue section is mounted on a small grid.
Lead "stain"
Section
Grid
Wax paper
5. The section on the grid is "stained" with an electron-dense element (such as lead).
6. The section is examined by directing a beam of electrons through the grid in the transmission electron microscope (TEM).
7. The high resolution of the TEM allows detailed examination of ultrathin sections of tissues and cells.

FIGURE 6.6
Thin section preparation for viewing membranes with electron microscopy.

Kinds of Membrane Proteins

As we've seen, the plasma membrane is a complex assembly of proteins enmeshed in a fluid array of phospholipid molecules. This enormously flexible design permits a broad range of interactions with the environment, some directly involving membrane proteins (figure 6.7). Though cells interact with their environment through their plasma membranes in many ways, we will focus on six key classes of membrane protein in this and the following chapter (chapter 7).

1. **Transporters.** Membranes are very selective, allowing only certain substances to enter or leave the cell, either through channels or carriers. In some instances, they take up molecules already present in the cell in high concentration.
2. **Enzymes.** Cells carry out many chemical reactions on the interior surface of the plasma membrane, using enzymes attached to the membrane.
3. **Cell surface receptors.** Membranes are exquisitely sensitive to chemical messages, detecting them with receptor proteins on their surfaces that act as antennae.
4. **Cell surface identity markers.** Membranes carry cell surface markers that identify them to other cells. Most cell types carry their own ID tags, specific combinations of cell surface proteins characteristic of that cell type.
5. **Cell adhesion proteins.** Cells use specific proteins to glue themselves to one another. Some act like Velcro, while others form a more permanent bond.
6. **Attachments to the cytoskeleton.** Surface proteins that interact with other cells are often anchored to the cytoskeleton by linking proteins.

The many proteins embedded within a membrane carry out a host of functions, many of which are associated with transport of materials or information across the membrane.

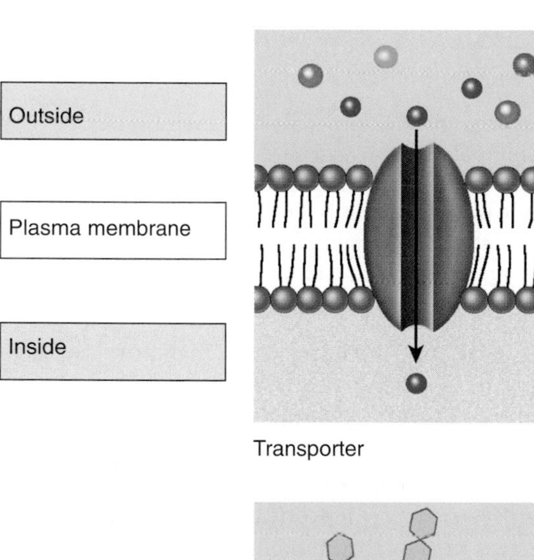

Transporter

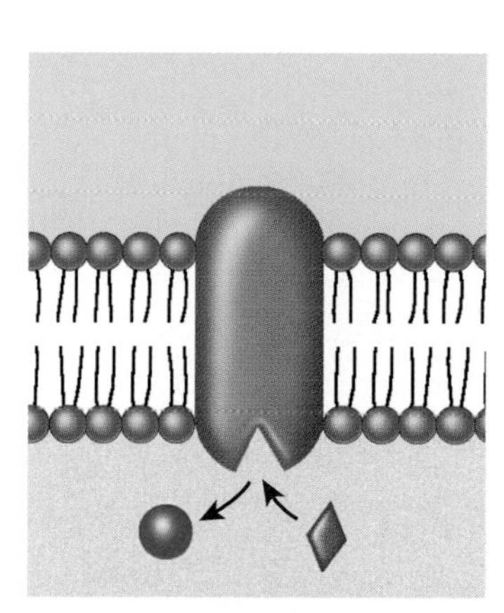

Enzyme

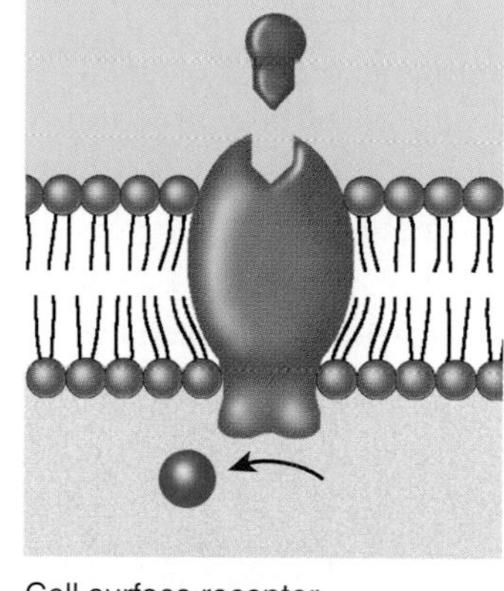

Cell surface receptor

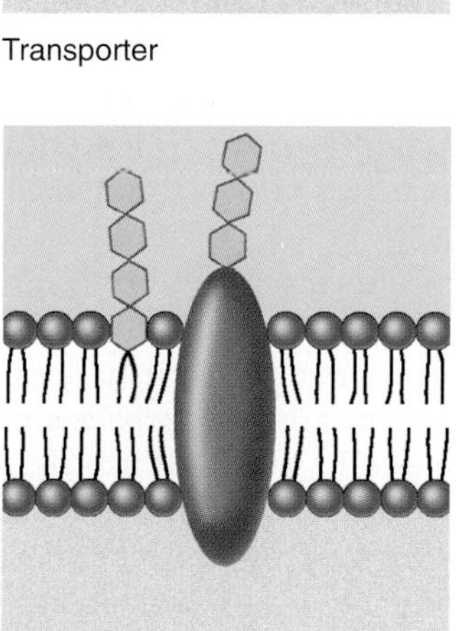

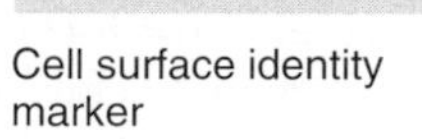

Cell surface identity marker

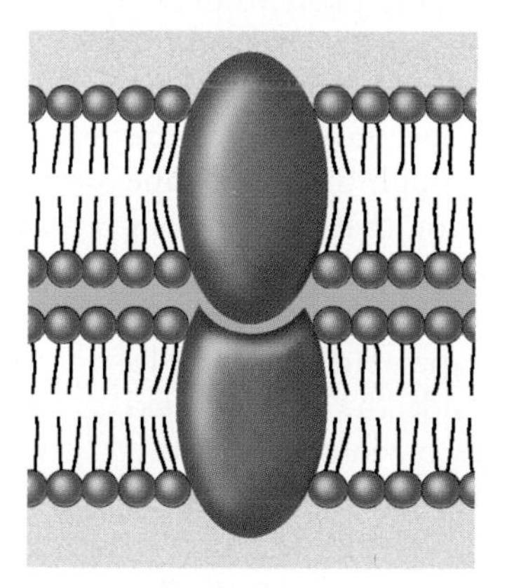

Cell adhesion

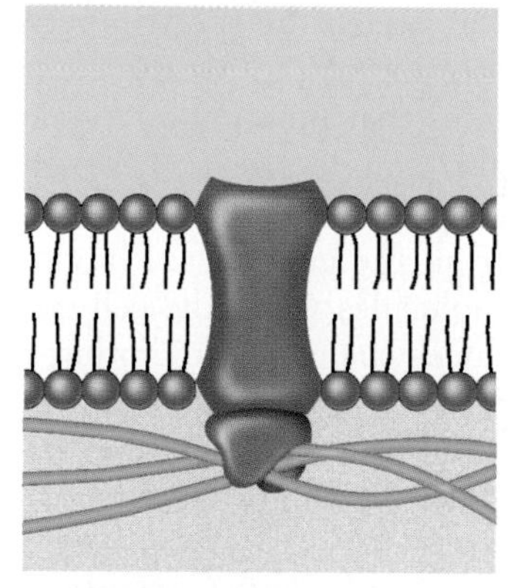

Attachment to the cytoskeleton

FIGURE 6.7
Functions of plasma membrane proteins. Membrane proteins act as transporters, enzymes, cell surface receptors, and cell surface markers, as well as aiding in cell-to-cell adhesion and securing the cytoskeleton.

Structure of Membrane Proteins

If proteins float on lipid bilayers like ships on the sea, how do they manage to extend through the membrane to create channels, and how can certain proteins be anchored into particular positions on the cell membrane?

Anchoring Proteins in the Bilayer

Many membrane proteins are attached to the surface of the membrane by special molecules that associate with phospholipids and thereby anchor the protein to the membrane. Like a ship tied up to a floating dock, these proteins are free to move about on the surface of the membrane tethered to a phospholipid.

In contrast, other proteins actually traverse the lipid bilayer. The part of the protein that extends through the lipid bilayer, in contact with the nonpolar interior, consists of one or more nonpolar helices or several β-pleated sheets of nonpolar amino acids (figure 6.8). Because water avoids nonpolar amino acids much as it does nonpolar lipid chains, the nonpolar portions of the protein are held within the interior of the lipid bilayer. Although the polar ends of the protein protrude from both sides of the membrane, the protein itself is locked into the membrane by its nonpolar segments. Any movement of the protein out of the membrane, in either direction, brings the nonpolar regions of the protein into contact with water, which "shoves" the protein back into the interior.

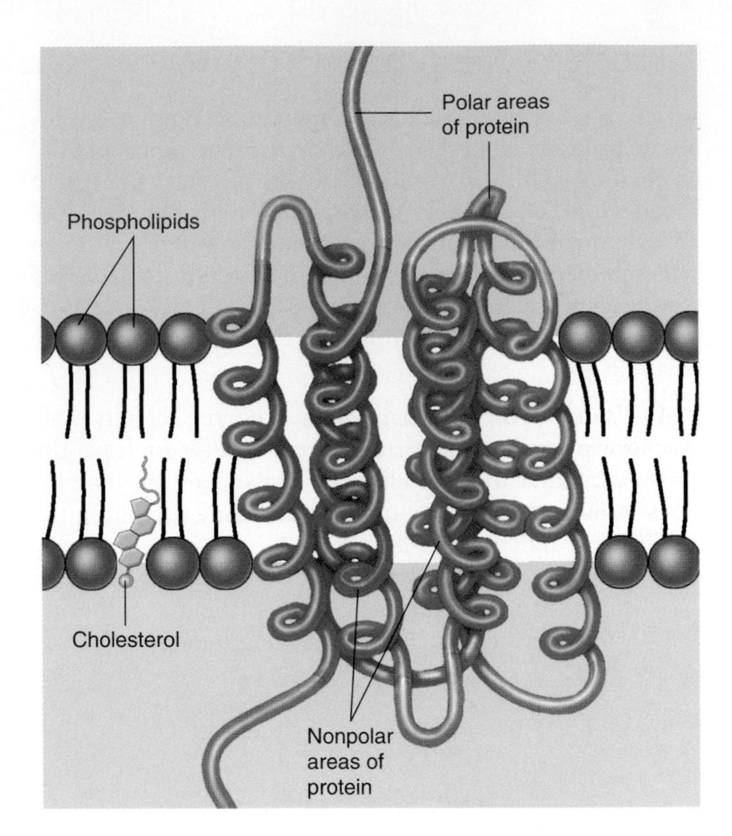

FIGURE 6.8
How nonpolar regions lock proteins into membranes. A spiral helix of nonpolar amino acids (*red*) extends across the nonpolar lipid interior, while polar (*purple*) portions of the protein protrude out from the bilayer. The protein cannot move in or out because such a movement would drag nonpolar segments of the protein into contact with water.

Extending Proteins across the Bilayer

Cells contain a variety of different **transmembrane proteins,** which differ in the way they traverse the bilayer, depending on their functions.

Anchors. A single nonpolar segment is adequate to anchor a protein in the membrane. Anchoring proteins of this sort attach the spectrin network of the cytoskeleton to the interior of the plasma membrane (figure 6.9). Many proteins that function as receptors for extracellular signals are also "single-pass" anchors that pass through the membrane only once. The portion of the receptor that extends out from the cell surface binds to specific hormones or other molecules when the cell encounters them; the binding induces changes at the other end of the protein, in the cell's interior. In this way, information outside the cell is translated into action within the cell. The mechanisms of cell signaling will be addressed in detail in chapter 7.

Channels and Carriers. Other proteins have several helical segments that thread their way back and forth through the membrane, forming a channel like the hole in a doughnut. For example, bacteriorhodopsin is one of the key transmembrane proteins that carries out photosynthesis in bacteria. It contains seven nonpolar helical segments that traverse the membrane, forming a circular pore through which protons pass during the light-driven pumping of protons (figure 6.10). Other transmembrane proteins do not create channels but rather act as carriers to transport molecules across the membrane. All water-soluble molecules or ions that enter or leave the cell are either transported by carriers or pass through channels.

Pores. Some transmembrane proteins have extensive nonpolar regions with secondary configurations of β-pleated sheets instead of α helices. The β sheets form a characteristic motif, folding back and forth in a circle so the sheets come to be arranged like the staves of a barrel. This so-called β barrel, open on both ends, is a common feature of the porin class of proteins that are found within the outer membrane of some bacteria (figure 6.11).

Transmembrane proteins are anchored into the bilayer by their nonpolar segments. While anchor proteins may pass through the bilayer only once, many channels and pores are created by proteins that pass back and forth through the bilayer repeatedly, creating a circular hole in the bilayer.

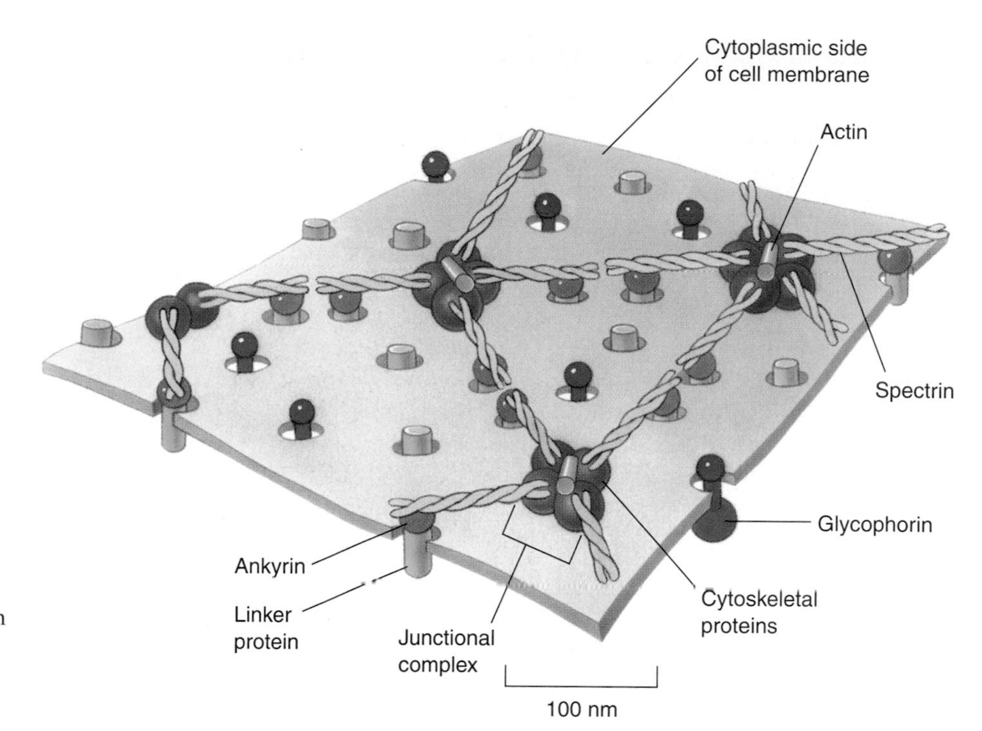

FIGURE 6.9
Anchoring proteins. Spectrin extends as a mesh anchored to the cytoplasmic side of a red blood cell plasma membrane. The spectrin protein is represented as a twisted dimer, attached to the membrane by special proteins such as junctional complexes and ankyrin; glycophorins can also be involved in attachments. This cytoskeletal protein network confers resiliency to cells like the red blood cell.

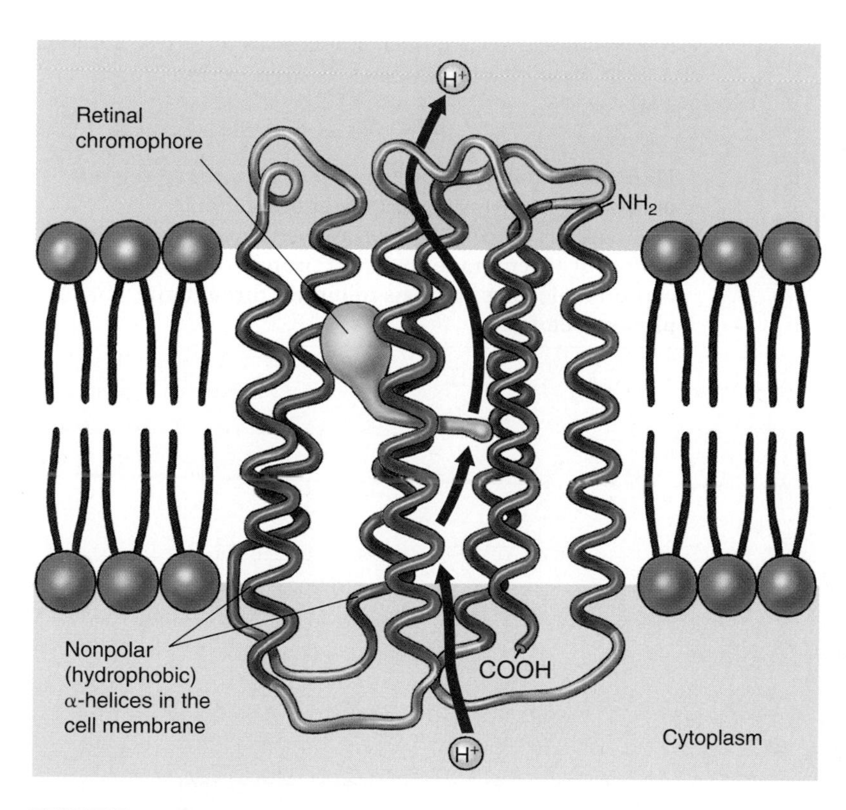

FIGURE 6.10
A channel protein. This transmembrane protein mediates photosynthesis in the bacterium *Halobacterium halobium*. The protein traverses the membrane seven times with hydrophobic helical strands that are within the hydrophobic center of the lipid bilayer. The helical regions form a channel across the bilayer through which protons are pumped by the retinal chromophore (*green*).

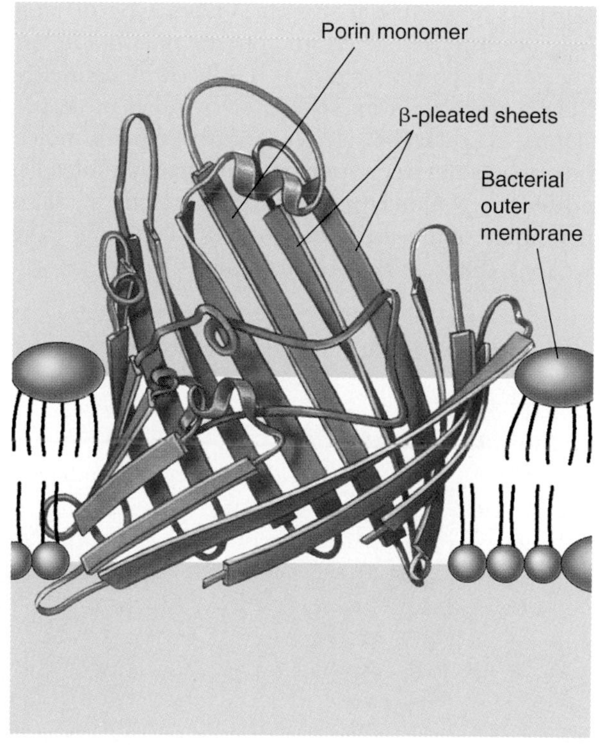

FIGURE 6.11
A pore protein. The bacterial transmembrane protein porin creates large open tunnels called pores in the outer membrane of a bacterium. Sixteen strands of β-pleated sheets run antiparallel to each other, creating a β barrel in the bacterial outer cell membrane. The tunnel allows water and other materials to pass through the membrane.

6.3 Passive transport across membranes moves down the concentration gradient.

Diffusion

Molecules and ions dissolved in water are in constant motion, moving about randomly. This random motion causes a net movement of these substances from regions where their concentration is high to regions where their concentration is lower, a process called **diffusion** (figure 6.12). Net movement driven by diffusion will continue until the concentrations in all regions are the same. You can demonstrate diffusion by filling a jar to the brim with ink, capping it, placing it at the bottom of a bucket of water, and then carefully removing the cap. The ink molecules will slowly diffuse out from the jar until there is a uniform concentration in the bucket and the jar. This uniformity in the concentration of molecules is a type of equilibrium.

Facilitated Transport Is Selective

Many molecules that cells require are polar and cannot pass through the nonpolar interior of the phospholipid bilayer. These molecules enter the cell through specific channels in the plasma membrane. The inside of the channel is polar and thus "friendly" to the polar molecules, facilitating their transport across the membrane. Each type of biomolecule that is transported across the plasma membrane has its own type of transporter (that is, it has its own channel which fits it like a glove and cannot be used by other molecules). Each channel is said to be selective for that type of molecule, and thus to be **selectively permeable**, as only molecules admitted by the channels it possesses can enter it. The plasma membrane of a cell has many types of channels, each selective for a different type of molecule.

Diffusion of Ions through Channels

One of the simplest ways for a substance to diffuse across a cell membrane is through a channel, as ions do. Ions are solutes (substances dissolved in water) with an unequal number of protons and electrons. Those with an excess of protons are positively charged and called *cations*. Ions with more electrons are negatively charged and called *anions*. Because they are charged, ions interact well with polar molecules like water but are repelled by the nonpolar interior of a phospholipid bilayer. Therefore, ions cannot move between the cytoplasm of a cell and the extracellular fluid without the assistance of membrane transport proteins. **Ion channels** possess a hydrated interior that spans the membrane. Ions can diffuse through the channel in either direction without coming into contact with the hydrophobic tails of the phospholipids in the membrane, and the transported ions do not bind to or otherwise interact with the channel proteins. Two conditions determine the direction of net movement of the ions: their relative concentrations on either side of the membrane, and the voltage across the membrane (a topic we'll explore in chapter 54). Each type of channel is specific for a particular ion, such as calcium (Ca^{++}) or chloride (Cl^{-}), or in some cases for a few kinds of ions. Ion channels play an essential role in signaling by the nervous system.

Diffusion is the net movement of substances to regions of lower concentration as a result of random spontaneous motion. It tends to distribute substances uniformly. Membrane transport proteins allow only certain molecules and ions to diffuse through the plasma membrane.

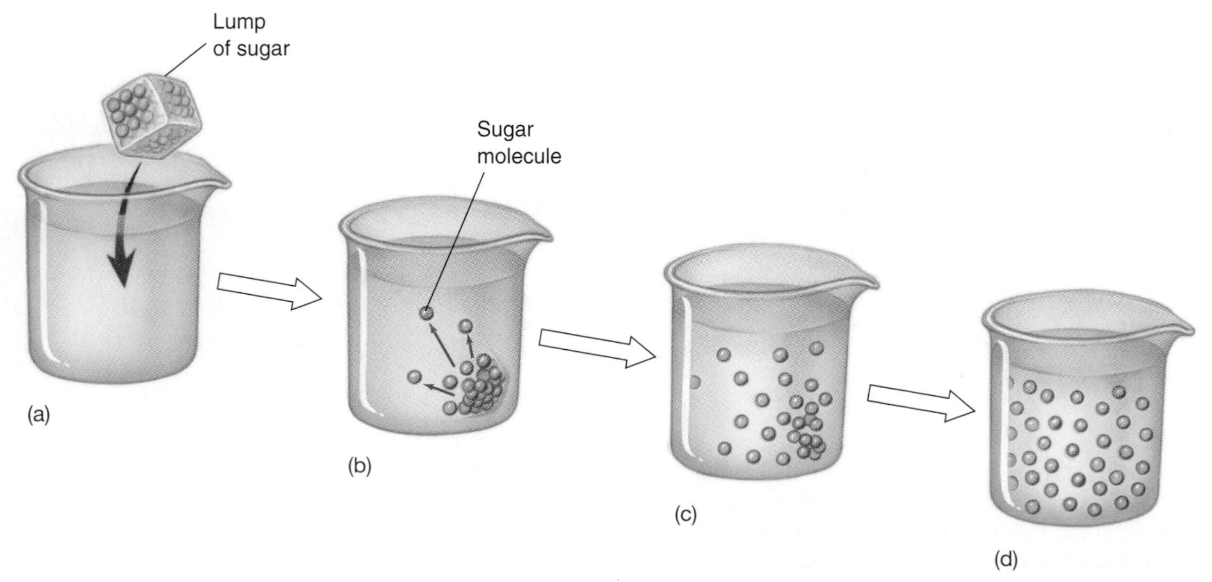

FIGURE 6.12
Diffusion. If a lump of sugar is dropped into a beaker of water (*a*), its molecules dissolve (*b*) and diffuse (*c*). Eventually, diffusion results in an even distribution of sugar molecules throughout the water (*d*).

Facilitated Diffusion

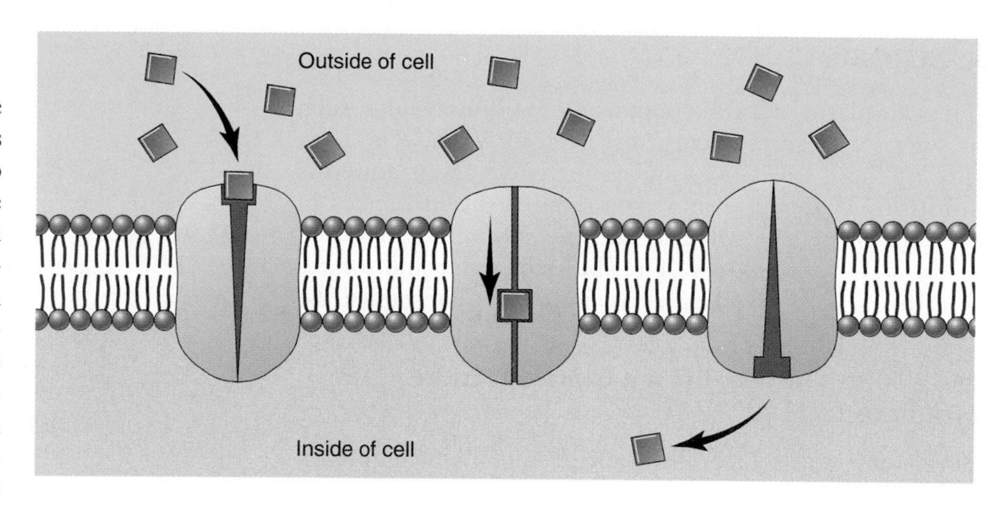

FIGURE 6.13
Facilitated diffusion is a carrier-mediated transport process. Molecules bind to a receptor on the extracellular side of the cell and are conducted through the plasma membrane by a membrane protein.

Carriers, another class of membrane proteins, transport ions as well as other solutes like sugars and amino acids across the membrane. Like channels, carriers are specific for a certain type of solute and can transport substances in either direction across the membrane. Unlike channels, however, they facilitate the movement of solutes across the membrane by physically binding to them on one side of the membrane and releasing them on the other. Again, the direction of the solute's net movement simply depends on its *concentration gradient* across the membrane. If the concentration is greater in the cytoplasm, the solute is more likely to bind to the carrier on the cytoplasmic side of the membrane and be released on the extracellular side. This will cause a net movement from inside to outside. If the concentration is greater in the extracellular fluid, the net movement will be from outside to inside. Thus, the net movement always occurs from areas of high concentration to low, just as it does in simple diffusion, but carriers facilitate the process. For this reason, this mechanism of transport is sometimes called **facilitated diffusion** (figure 6.13).

Facilitated Diffusion in Red Blood Cells

Several examples of facilitated diffusion by carrier proteins can be found in the membranes of vertebrate red blood cells (RBCs). One RBC carrier protein, for example, transports a different molecule in each direction: Cl^- in one direction and bicarbonate ion (HCO_3^-) in the opposite direction. As you will learn in chapter 52, this carrier is important in transporting carbon dioxide in the blood.

A second important facilitated diffusion carrier in RBCs is the glucose transporter. Red blood cells keep their internal concentration of glucose low through a chemical trick: they immediately add a phosphate group to any entering glucose molecule, converting it to a highly charged glucose phosphate that cannot pass back across the membrane. This maintains a steep concentration gradient for glucose, favoring its entry into the cell. The glucose transporter that carries glucose into the cell does not appear to form a channel in the membrane for the glucose to pass through. Instead, the transmembrane protein appears to bind the glucose and then flip its shape, dragging the glucose through the bilayer and releasing it on the inside of the plasma membrane. Once it releases the glucose, the glucose transporter reverts to its original shape. It is then available to bind the next glucose molecule that approaches the outside of the cell.

Transport through Carriers Saturates

A characteristic feature of transport through carriers is that its rate is saturable. In other words, if the concentration gradient of a substance is progressively increased, its rate of transport will also increase to a certain point and then level off. Further increases in the gradient will produce no additional increase in rate. The explanation for this observation is that there are a limited number of carriers in the membrane. When the concentration of the transported substance rises high enough, all of the carriers will be in use and the capacity of the transport system will be saturated. In contrast, substances that move across the membrane by simple diffusion (diffusion through channels in the bilayer without the assistance of carriers) do not show saturation.

Facilitated diffusion provides the cell with a ready way to prevent the buildup of unwanted molecules within the cell or to take up needed molecules, such as sugars, that may be present outside the cell in high concentrations. Facilitated diffusion has three essential characteristics:

1. **It is specific.** Any given carrier transports only certain molecules or ions.
2. **It is passive.** The direction of net movement is determined by the relative concentrations of the transported substance inside and outside the cell.
3. **It saturates.** If all relevant protein carriers are in use, increases in the concentration gradient do not increase the transport rate.

Facilitated diffusion is the transport of molecules and ions across a membrane by specific carriers in the direction of lower concentration of those molecules or ions.

Osmosis

The cytoplasm of a cell contains ions and molecules, such as sugars and amino acids, dissolved in water. The mixture of these substances and water is called an **aqueous solution.** Water, the most common of the molecules in the mixture, is the **solvent,** and the substances dissolved in the water are **solutes.** The ability of water and solutes to diffuse across membranes has important consequences.

Molecules Diffuse down a Concentration Gradient

Both water and solutes diffuse from regions of high concentration to regions of low concentration; that is, they diffuse down their concentration gradients. When two regions are separated by a membrane, what happens depends on whether or not the solutes can pass freely through that membrane. Most solutes, including ions and sugars, are not lipid-soluble and, therefore, are unable to cross the lipid bilayer of the membrane.

Even water molecules, which are very polar, cannot cross a lipid bilayer. Water flows through **aquaporins,** which are specialized channels for water. A simple experiment demonstrates this. If you place an amphibian egg in hypotonic spring water, it does not swell. If you then inject aquaporin mRNA into the egg, the channel proteins are expressed and the egg then swells.

Dissolved solutes interact with water molecules, which form hydration shells about the charged solute. When there is a concentration gradient of solutes, the solutes will move from a high to a low concentration, dragging with them their hydration shells of water molecules. When a membrane separates two solutions, hydration shell water molecules move with the diffusing ions, creating a net movement of water towards the low solute. This net water movement across a membrane by diffusion is called **osmosis** (figure 6.14).

The concentration of *all* solutes in a solution determines the **osmotic concentration** of the solution. If two solutions have unequal osmotic concentrations, the solution with the higher concentration is **hyperosmotic** (Greek *hyper,* "more than"), and the solution with the lower concentration is **hypoosmotic** (Greek *hypo,* "less than"). If the osmotic concentrations of two solutions are equal, the solutions are **isosmotic** (Greek *iso,* "the same").

In cells, a plasma membrane separates two aqueous solutions, one inside the cell (the cytoplasm) and one outside

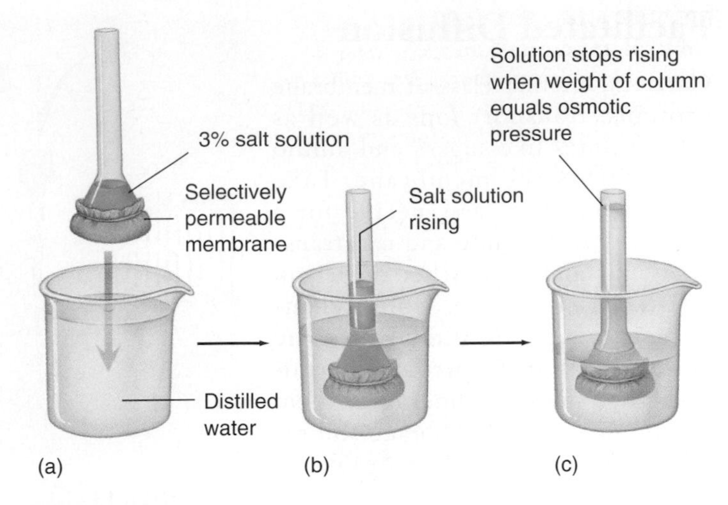

FIGURE 6.14
An experiment demonstrating osmosis. (*a*) The end of a tube containing a salt solution is closed by stretching a selectively permeable membrane across its face; the membrane allows the passage of water molecules but not salt ions. (*b*) When this tube is immersed in a beaker of distilled water, the salt cannot cross the membrane, but water can. The water entering the tube causes the salt solution to rise in the tube. (*c*) Water will continue to enter the tube from the beaker until the weight of the column of water in the tube exerts a downward force equal to the force drawing water molecules upward into the tube. This force is referred to as osmotic pressure.

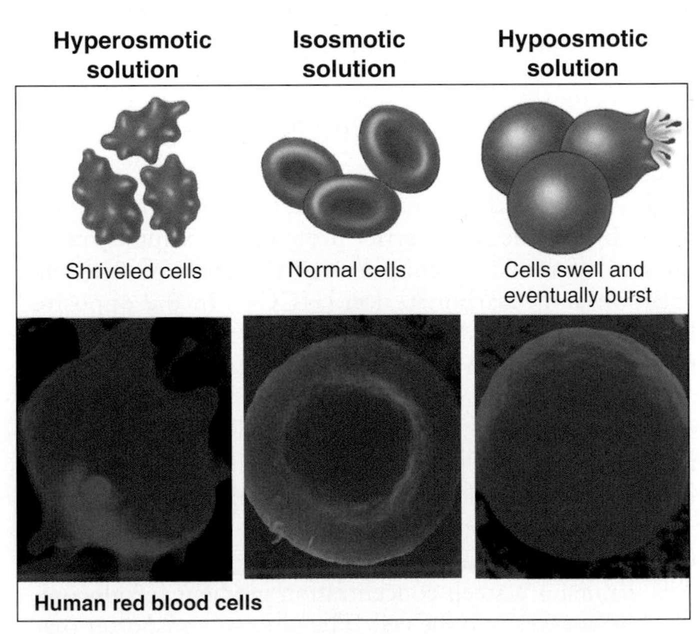

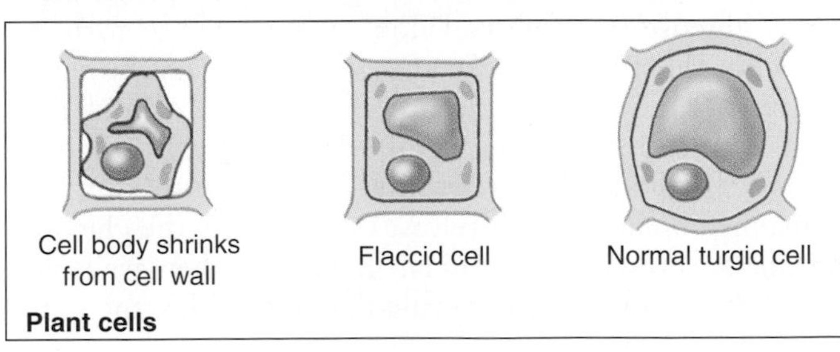

FIGURE 6.15
Osmosis. In a hyperosmotic solution water moves out of the cell toward the higher concentration of solutes, causing the cell to shrivel. In an isosmotic solution, the concentration of solutes on either side of the membrane is the same. Osmosis still occurs, but water diffuses into and out of the cell at the same rate, and the cell doesn't change size. In a hypoosmotic solution the concentration of solutes is higher within the cell than outside, so the net movement of water is into the cell.

FIGURE 6.16
How solutes create osmotic pressure. Charged or polar substances are soluble in water because they form hydrogen bonds with water molecules clustered around them. When a polar solute (illustrated here with urea) is added to the solution on one side of a membrane, the water molecules that gather around each urea molecule are no longer free to diffuse across the membrane; in effect, the polar solute has reduced the number of free water molecules on that side of the membrane increasing the osmotic pressure. Because the hypoosmotic side of the membrane (on the right, with less solute) has more unbound water molecules than the hyperosmotic side (on the left, with more solute), water moves by diffusion from the right to the left.

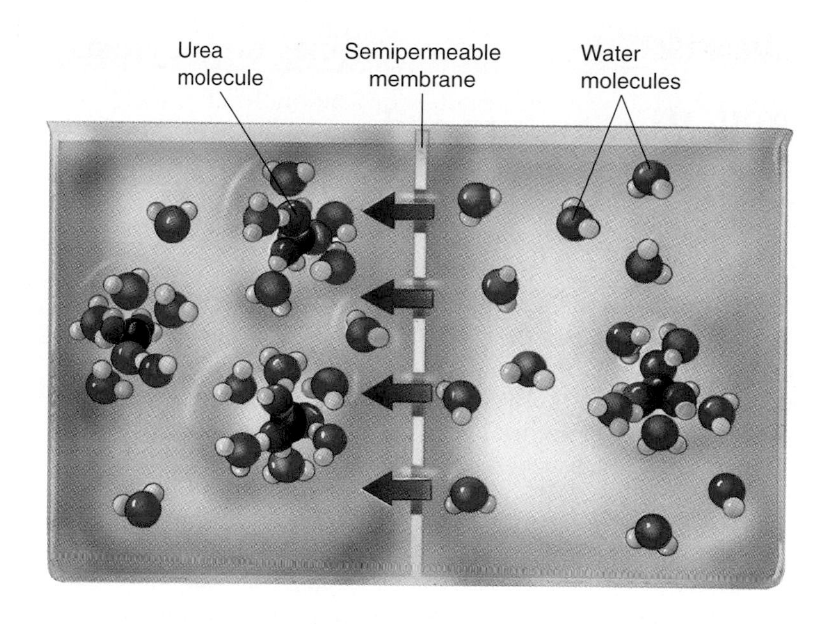

(the extracellular fluid). The direction of the net diffusion of water across this membrane is determined by the osmotic concentrations of the solutions on either side (figure 6.15). For example, if the cytoplasm of a cell were *hypo*osmotic to the extracellular fluid, water would diffuse out of the cell, toward the solution with the higher concentration of solutes (and, therefore, the lower concentration of unbound water molecules). This loss of water from the cytoplasm would cause the cell to shrink until the osmotic concentrations of the cytoplasm and the extracellular fluid become equal.

Osmotic Pressure

What would happen if the cell's cytoplasm were *hyper*osmotic to the extracellular fluid? In this situation, water would diffuse into the cell from the extracellular fluid, causing the cell to swell. The pressure of the cytoplasm pushing out against the cell membrane, or **hydrostatic pressure,** would increase. On the other hand, the **osmotic pressure** (figure 6.16), defined as the pressure that must be applied to stop the osmotic movement of water across a membrane, would also be at work. If the membrane were strong enough, the cell would reach an equilibrium, at which the osmotic pressure, which tends to drive water into the cell, is exactly counterbalanced by the hydrostatic pressure, which tends to drive water back out of the cell. However, a plasma membrane by itself cannot withstand large internal pressures, and an isolated cell under such conditions would burst like an overinflated balloon. Accordingly, it is important for animal cells to maintain isosmotic conditions. The cells of bacteria, fungi, plants, and many protists, in contrast, are surrounded by strong cell walls. The cells of these organisms can withstand high internal pressures without bursting.

Maintaining Osmotic Balance

Organisms have developed many solutions to the osmotic dilemma posed by being hyperosmotic to their environment.

Extrusion. Some single-celled eukaryotes like the protist *Paramecium* use organelles called contractile vacuoles to remove water. Each vacuole collects water from various parts of the cytoplasm and transports it to the central part of the vacuole, near the cell surface. The vacuole possesses a small pore that opens to the outside of the cell. By contracting rhythmically, the vacuole pumps the water out of the cell through the pore.

Isosmotic Solutions. Some organisms that live in the ocean adjust their internal concentration of solutes to match that of the surrounding seawater. Isosmotic with respect to their environment, there is no net flow of water into or out of these cells. Many terrestrial animals solve the problem in a similar way, by circulating a fluid through their bodies that bathes cells in an isosmotic solution. The blood in your body, for example, contains a high concentration of the protein albumin, which elevates the solute concentration of the blood to match your cells.

Turgor. Most plant cells are hyperosmotic to their immediate environment, containing a high concentration of solutes in their central vacuoles. The resulting internal hydrostatic pressure, known as **turgor pressure,** presses the plasma membrane firmly against the interior of the cell wall, making the cell rigid. The newer, softer portions of trees and shrubs depend on turgor pressure to maintain their shape, and wilt when they lack sufficient water.

Osmosis is the diffusion of water, but not solutes, across a membrane.

6.4 Bulk transport utilizes endocytosis.

Bulk Passage Into and Out of the Cell

Endocytosis

The lipid nature of their biological membranes raises a second problem for cells. The substances cells use as fuel are for the most part large, polar molecules that cannot cross the hydrophobic barrier a lipid bilayer creates. How do organisms get these substances into their cells? One process many single-celled eukaryotes employ is **endocytosis** (figure 6.17). In this process the plasma membrane extends outward and envelops food particles. Cells use three major types of endocytosis: phagocytosis, pinocytosis, and receptor-mediated endocytosis.

Phagocytosis and Pinocytosis. If the material the cell takes in is particulate (made up of discrete particles), such as an organism or some other fragment of organic matter (figure 6.17*a*), the process is called phagocytosis (Greek *phagein*, "to eat" + *cytos*, "cell"). If the material the cell takes in is liquid (figure 6.17*b*), it is called pinocytosis (Greek *pinein*, "to drink"). Pinocytosis is common among animal cells. Mammalian egg cells, for example, "nurse" from surrounding cells; the nearby cells secrete nutrients that the maturing egg cell takes up by pinocytosis. Virtually all eukaryotic cells constantly carry out these kinds of endocytosis, trapping particles and extracellular fluid in vesicles and ingesting them. Endocytosis rates vary from one cell type to another. They can be surprisingly high: some types of white blood cells ingest 25% of their cell volume each hour!

Receptor-Mediated Endocytosis. Specific molecules are often transported into eukaryotic cells through receptor-mediated endocytosis. Molecules to be transported first bind to specific receptors on the plasma membrane. The transport process is specific because only that molecule has a shape that fits snugly into the receptor. The plasma membrane of a particular kind of cell contains a characteristic battery of receptor types, each for a different kind of molecule.

The interior portion of the receptor molecule resembles a hook that is trapped in an indented pit coated with the protein clathrin. The pits act like molecular mousetraps, closing over to form an internal vesicle when the right molecule enters the pit (figure 6.18). The trigger that releases the trap is a receptor protein embedded in the membrane of the pit, which detects the presence of a particular target molecule and reacts by initiating endocytosis. The process is highly specific and very fast.

One type of molecule that is taken up by receptor-mediated endocytosis is called a low density lipoprotein (LDL). The LDL molecules bring cholesterol into the cell where it can be incorporated into membranes. Cholesterol plays a key role in determining the stiffness of the body's membranes. In the human genetic disease called hypercholesteremia, the receptors lack tails and so are never caught in the clathrin-coated pits and, thus, are never taken up by the cells. The cholesterol stays in the bloodstream of affected individuals, coating their arteries and leading to heart attacks.

Fluid-phase endocytosis is the receptor-mediated pinocytosis of fluids. It is important to understand that endocytosis in itself does not bring substances directly into the cytoplasm of a cell. The material taken in is still separated from the cytoplasm by the membrane of the vesicle.

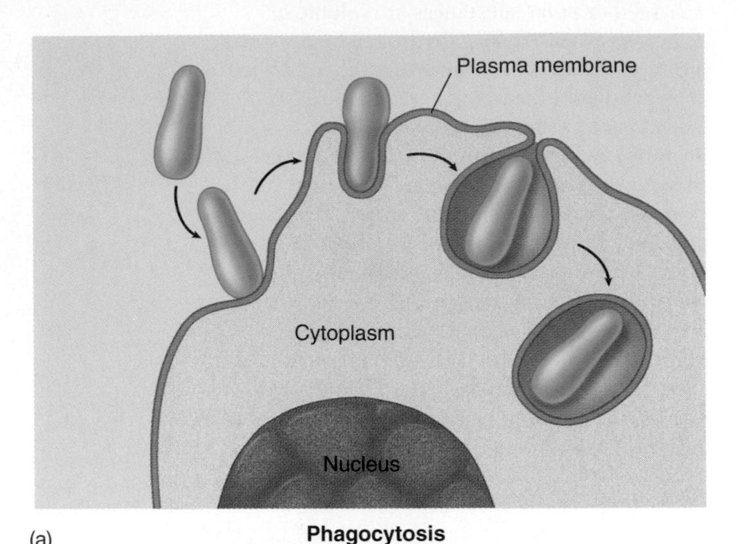

(a) Phagocytosis

(b) Pinocytosis

FIGURE 6.17
Endocytosis. Both phagocytosis (*a*) and pinocytosis (*b*) are forms of endocytosis.

FIGURE 6.18
Receptor-mediated endocytosis. (*a*) Cells that undergo receptor-mediated endocytosis have pits coated with the protein clathrin that initiate endocytosis when target molecules bind to receptor proteins in the plasma membrane. (*b*) A coated pit appears in the plasma membrane of a developing egg cell, covered with a layer of proteins (80,000×). When an appropriate collection of molecules gathers in the coated pit, the pit deepens (*c*) and seals off (*d*) to form a coated vesicle, which carries the molecules into the cell.

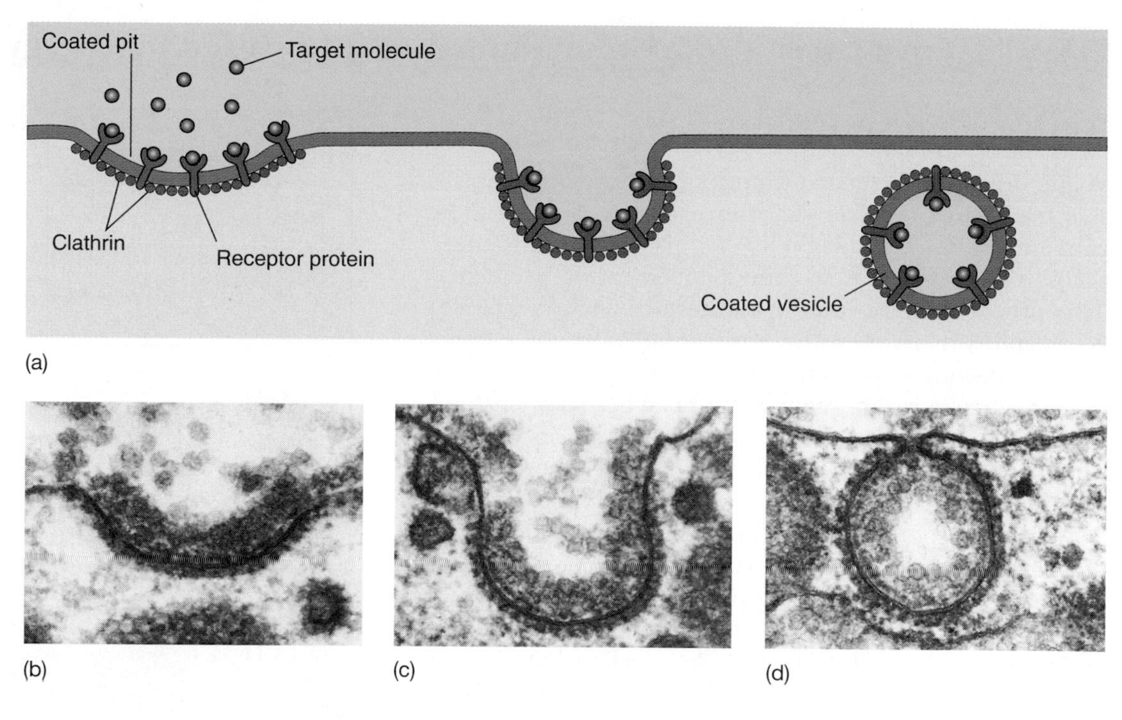

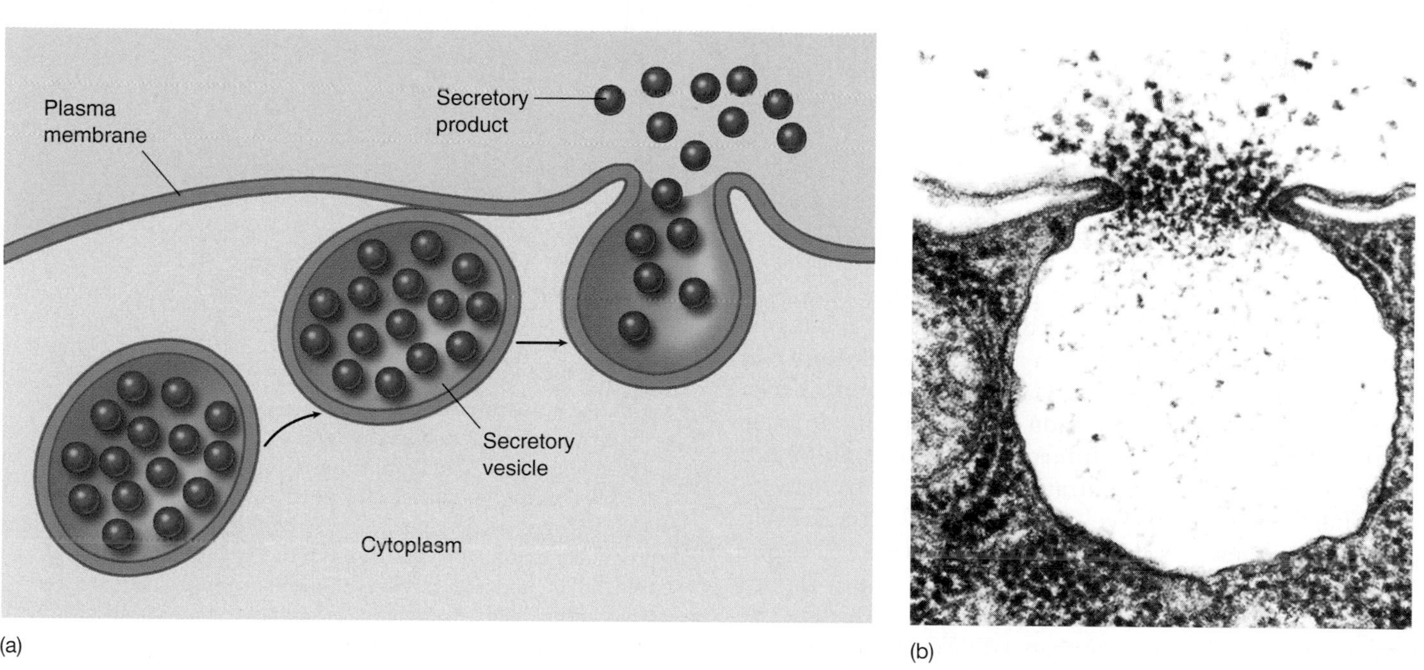

FIGURE 6.19
Exocytosis. (*a*) Proteins and other molecules are secreted from cells in small packets called vesicles, whose membranes fuse with the plasma membrane, releasing their contents to the cell surface. (*b*) A transmission electron micrograph showing exocytosis.

Exocytosis

The reverse of endocytosis is **exocytosis,** the discharge of material from vesicles at the cell surface (figure 6.19). In plant cells, exocytosis is an important means of exporting the materials needed to construct the cell wall through the plasma membrane. Among protists, contractile vacuole discharge is a form of exocytosis. In animal cells, exocytosis provides a mechanism for secreting many hormones, neurotransmitters, digestive enzymes, and other substances.

Cells import bulk materials by engulfing them with their plasma membranes in a process called endocytosis; similarly, they extrude or secrete material through exocytosis.

6.5 Active transport across membranes is powered by energy from ATP.

Active Transport

While diffusion, facilitated diffusion, and osmosis are passive transport processes that move materials down their concentration gradients, cells can also move substances across the membrane *up* their concentration gradients. This process requires the expenditure of energy, typically ATP, and is therefore called **active transport.** Like facilitated diffusion, active transport involves highly selective protein carriers within the membrane. These carriers bind to the transported substance, which could be an ion or a simple molecule like a sugar (figure 6.20), an amino acid, or a nucleotide to be used in the synthesis of DNA.

Active transport is one of the most important functions of any cell. It enables a cell to take up additional molecules of a substance that is already present in its cytoplasm in concentrations higher than in the extracellular fluid. Without active transport, for example, liver cells would be unable to accumulate glucose molecules from the blood plasma, as the glucose concentration is often higher inside the liver cells than it is in the plasma. Active transport also enables a cell to move substances from its cytoplasm to the extracellular fluid despite higher external concentrations.

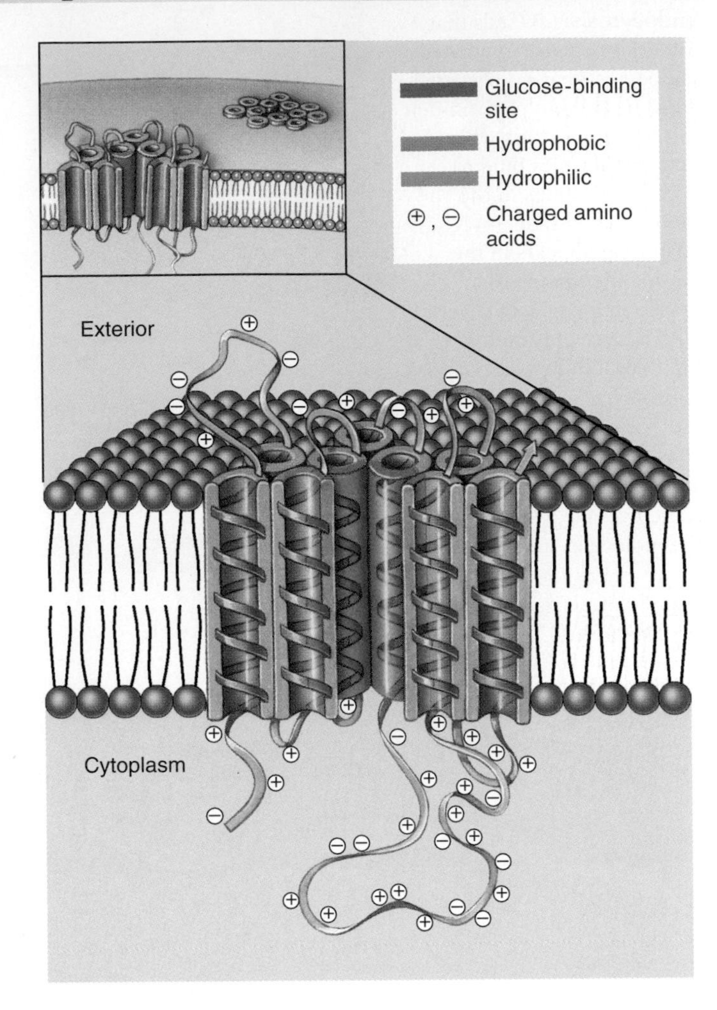

FIGURE 6.20
A glucose transport carrier. The molecular structure of this particular glucose transport carrier is known in considerable detail. The protein's 492 amino acids form a folded chain that traverses the lipid membrane 12 times. Amino acids with charged groups are less stable in the hydrophobic region of the lipid bilayer and are thus exposed to the cytoplasm or the extracellular fluid. Researchers think the center of the protein consists of five helical segments with glucose-binding sites (in *red*) facing inward. A conformational change in the protein transports glucose across the membrane by shifting the position of the glucose-binding sites.

The Sodium-Potassium Pump

The use of ATP in active transport may be direct or indirect. Let's first consider how ATP is used directly to move ions against their concentration gradient. More than one-third of all of the energy expended by an animal cell that is not actively dividing is used in the active transport of sodium (Na^+) and potassium (K^+) ions. Most animal cells have a low internal concentration of Na^+, relative to their surroundings, and a high internal concentration of K^+. They maintain these concentration differences by actively pumping Na^+ out of the cell and K^+ in. The remarkable protein that transports these two ions across the cell membrane is known as the **sodium-potassium pump** (figure 6.21). The cell obtains the energy it needs to operate the pump from adenosine triphosphate (ATP), a molecule we'll learn more about in chapter 8.

The important characteristic of the sodium-potassium pump is that it is an *active* transport process, transporting Na^+ and K^+ from areas of low concentration to areas of high concentration. This transport up their concentration gradients is the opposite of the passive transport in diffusion; it is achieved only by the constant expenditure of metabolic energy. The sodium-potassium pump works through a series of conformational changes in the transmembrane protein:

Step 1. Three sodium ions bind to the cytoplasmic side of the protein, causing the protein to change its conformation.

Step 2. In its new conformation, the protein binds a molecule of ATP and cleaves it into adenosine diphosphate and phosphate (ADP + P_i). ADP is released, but the phosphate group remains bound to the protein. The protein is now phosphorylated.

Step 3. The phosphorylation of the protein induces a second conformational change in the protein. This change translocates the three Na^+ across the membrane,

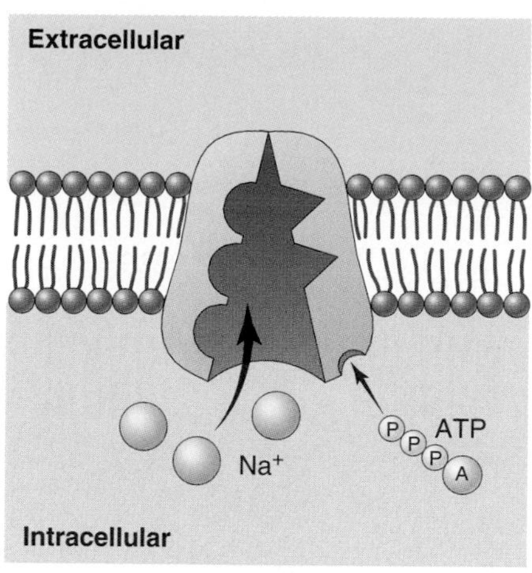

1. Protein in membrane binds intracellular sodium.

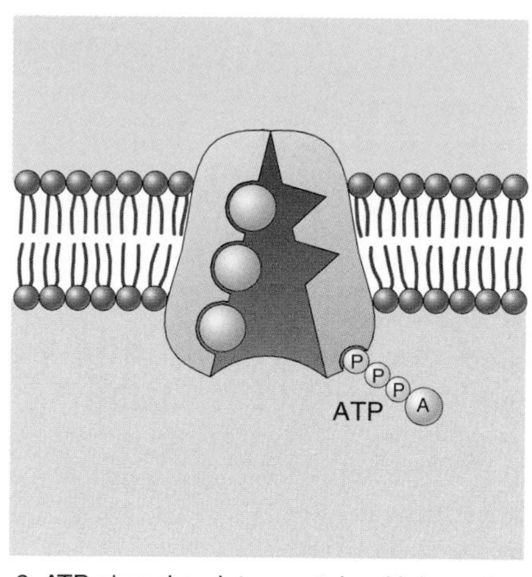

2. ATP phosphorylates protein with bound sodium.

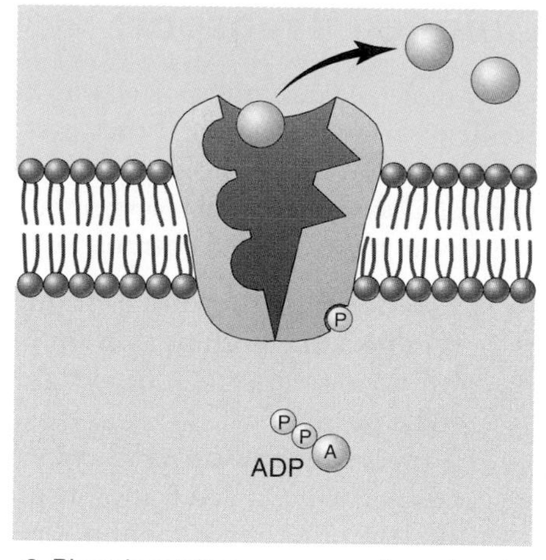

3. Phosphorylation causes conformational change in protein, allowing sodium to leave.

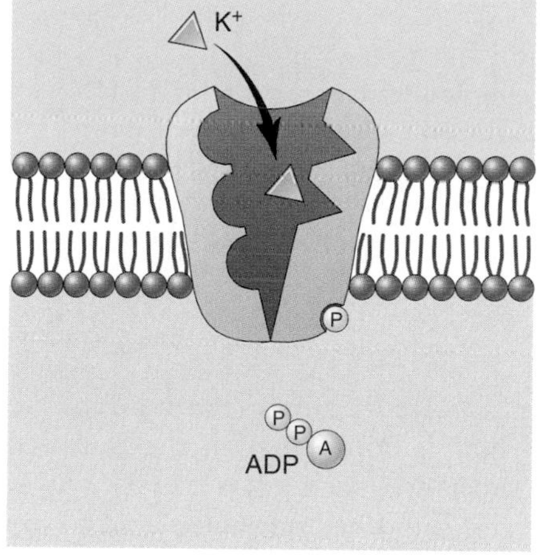

4. Extracellular potassium binds to exposed sites.

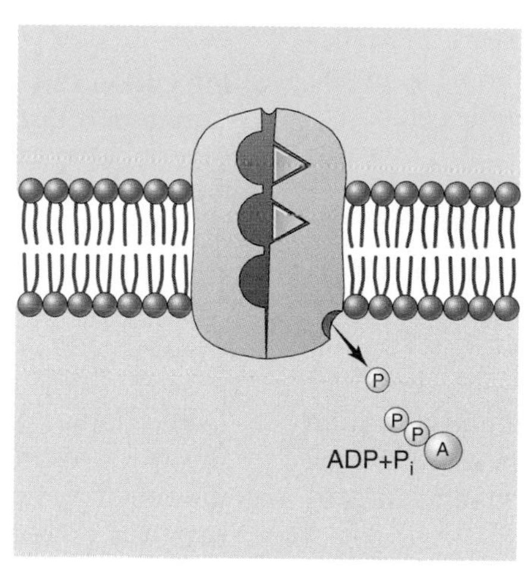

5. Binding of potassium causes dephosphorylation of protein.

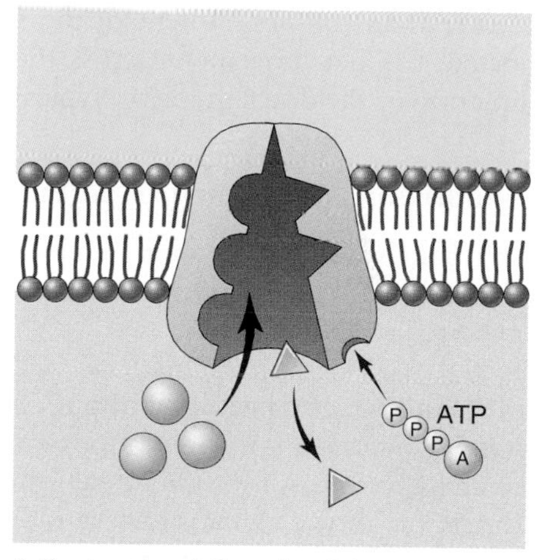

6. Dephosphorylation of protein triggers change back to original conformation, potassium moves into cell, and the cycle repeats.

FIGURE 6.21
The sodium-potassium pump. The protein carrier known as the sodium-potassium pump transports sodium (Na^+) and potassium (K^+) ions across the cell membrane. For every three Na^+ that are transported out of the cell, two K^+ are transported into the cell. The sodium-potassium pump is fueled by ATP.

so they now face the exterior. In this new conformation, the protein has a low affinity for Na^+, and the three bound Na^+ dissociate from the protein and diffuse into the extracellular fluid.

Step 4. The new conformation has a high affinity for K^+, two of which bind to the extracellular side of the protein as soon as it is free of the Na^+.

Step 5. The binding of the K^+ causes another conformational change in the protein, this time resulting in the dissociation of the bound phosphate group.

Step 6. Freed of the phosphate group, the protein reverts to its original conformation, exposing the two K^+ to the cytoplasm. This conformation has a low affinity for K^+, so the two bound K^+ dissociate from the protein and diffuse into the interior of the cell. The original conformation has a high affinity for Na^+; when these ions bind, they initiate another cycle.

Three Na^+ leave the cell and two K^+ enter in every cycle. The changes in protein conformation that occur during the cycle are rapid, enabling each carrier to transport as many as 300 Na^+ per second. The sodium-potassium pump appears to be ubiquitous in animal cells, although cells vary widely in the number of pump proteins they contain.

Active transport moves a solute across a membrane up its concentration gradient, using protein carriers driven by the expenditure of chemical energy.

Coupled Transport

Many molecules are transported into cells up a concentration gradient through a process that uses ATP indirectly. The molecules move hand-in-hand with sodium ions or protons that are moving *down* their concentration gradients. This type of active transport, called **cotransport,** has two components:

1. **Establishing the down gradient.** ATP is used to establish the sodium ion or proton *down* gradient, which is greater than the *up* gradient of the molecule to be transported.
2. **Traversing the up gradient.** Cotransport proteins (also called coupled transport proteins) carry the molecule and either a sodium ion or a proton together across the membrane.

Because the *down* gradient of the sodium ion or proton is greater than the *up* gradient of the molecule to be transported, the net movement across the membrane is in the direction of the down gradient, typically into the cell.

Establishing the *Down* Gradient

Either the sodium-potassium pump or the proton pump establishes the down gradient that powers most active transport processes of the cell.

The Sodium-Potassium Pump. The sodium-potassium pump actively pumps sodium ions out of the cell, powered by energy from ATP. This establishes a sodium ion concentration gradient that is lower inside the cell.

The Proton Pump. The **proton pump** pumps protons (H^+ ions) across a membrane using energy derived from energy-rich molecules or from photosynthesis. This creates a proton gradient, in which the concentration of protons is higher on one side of the membrane than the other. Membranes are impermeable to protons, so the only way protons can diffuse back down their concentration gradient is through a second cotransport protein.

Traversing the *Up* Gradient

Animal cells accumulate many amino acids and sugars against a concentration gradient: the molecules are transported into the cell from the extracellular fluid, even though their concentrations are higher inside the cell. These molecules couple with sodium ions to enter the cell down the Na^+ concentration gradient established by the sodium-potassium pump. In this **cotransport process,** Na^+ and a specific sugar or amino acid simultaneously bind to the same transmembrane protein on the outside of the cell, called a **symport** (figure 6.22). Both are then translocated to the inside of the cell, but in the process Na^+ moves *down* its concentration gradient while the sugar or amino acid moves *up* its concentration gradient. In effect, the cell uses some of the energy stored in the Na^+ concentration gradient to accumulate sugars and amino acids.

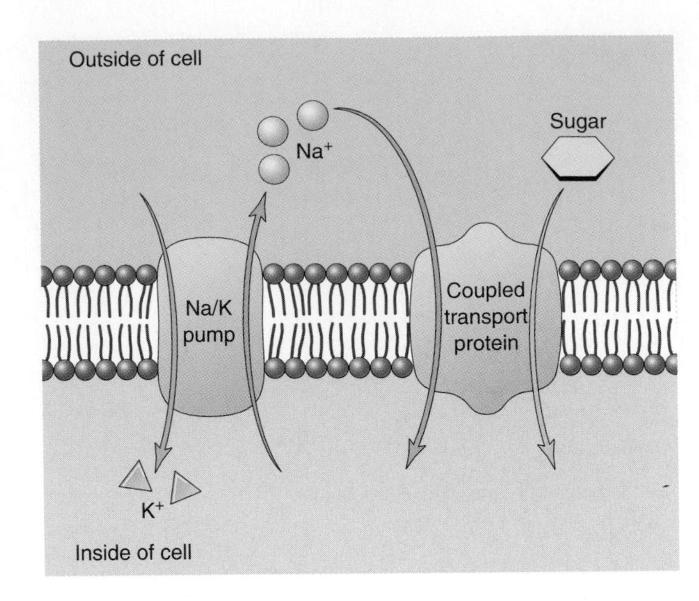

FIGURE 6.22
Cotransport through a coupled transport protein. A membrane protein transports sodium ions into the cell, down their concentration gradient, at the same time it transports a sugar molecule into the cell. The gradient driving the Na^+ entry is so great that sugar molecules can be brought in *against* their concentration gradient.

In a related process, called **countertransport,** the inward movement of Na^+ is coupled with the outward movement of another substance, such as Ca^{++} or H^+. As in cotransport, both Na^+ and the other substance bind to the same transport protein, in this case called an **antiport**, but in this case they bind on opposite sides of the membrane and are moved in opposite directions. In countertransport, the cell uses the energy released as Na^+ moves down its concentration gradient into the cell to extrude a substance up its concentration gradient.

The cell uses the proton *down* gradient established by the proton pump (figure 6.23) in ATP production. The movement of protons through their cotransport protein is coupled to the production of ATP, the energy-storing molecule we mentioned earlier. Thus, the cell expends energy to produce ATP, which provides it with a convenient energy storage form that it can employ in its many activities. The coupling of the proton pump to ATP synthesis, called **chemiosmosis,** is responsible for almost all of the ATP produced from food (see chapter 9) and all of the ATP produced by photosynthesis (see chapter 10). We know that proton pump proteins are ancient because they are present in bacteria as well as in eukaryotes. The mechanisms for transport across plasma membranes are summarized in table 6.2.

Many molecules are cotransported into cells up their concentration gradients by coupling their movement to that of sodium ions or protons moving down their concentration gradients.

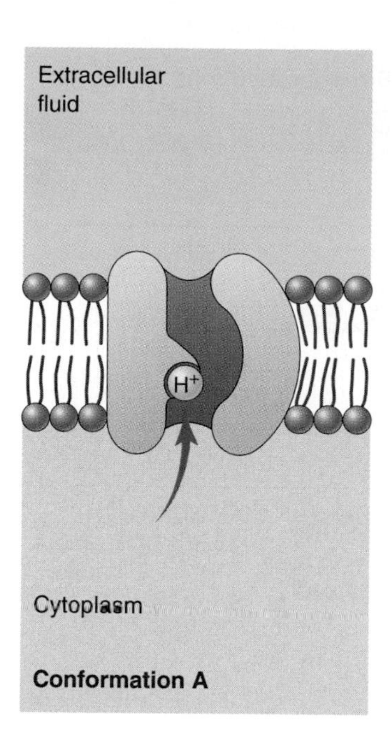

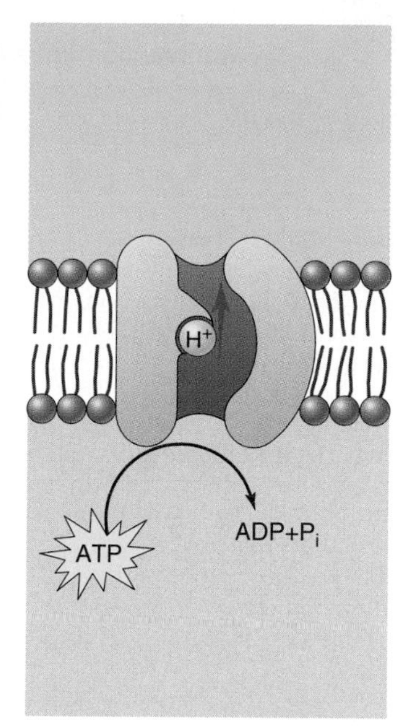

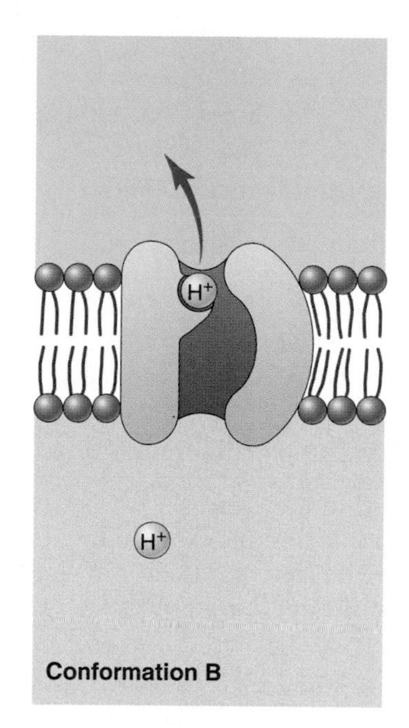

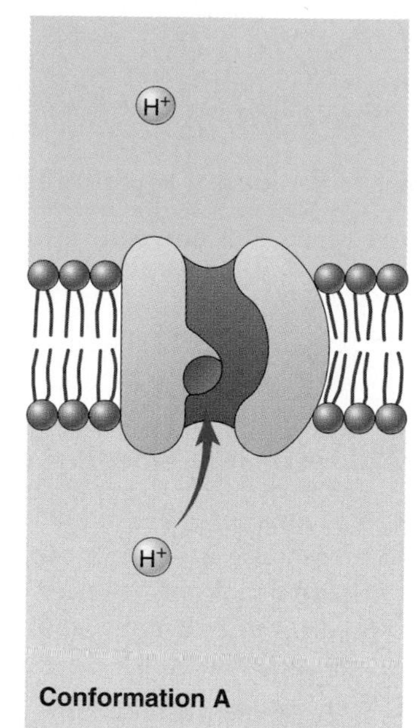

FIGURE 6.23
The proton pump. In this general model of energy-driven proton pumping, the transmembrane protein that acts as a proton pump is driven through a cycle of two conformations: A and B. The cycle A→B→A goes only one way, causing protons to be pumped from the inside to the outside of the membrane. ATP powers the pump.

Table 6.2 Mechanisms for Transport across Cell Membranes

Process	Passage through Membrane	How It Works	Example
PASSIVE PROCESSES			
Diffusion	Direct	Random molecular motion produces net migration of molecules toward region of lower concentration	Movement of oxygen into cells
	Protein channel	Polar molecules pass through a protein channel	Movement of ions in or out of cell
Facilitated diffusion	Protein carrier	Molecule binds to carrier protein in membrane and is transported across; net movement is toward region of lower concentration	Movement of glucose into cells
Osmosis	Direct	Diffusion of water across differentially permeable membrane	Movement of water into cells placed in a hypotonic solution
ACTIVE PROCESSES			
Endocytosis			
Phagocytosis	Membrane vesicle	Particle is engulfed by membrane, which folds around it and forms a vesicle	Ingestion of bacteria by white blood cells
Pinocytosis	Membrane vesicle	Fluid droplets are engulfed by membrane, which forms vesicles around them	"Nursing" of human egg cells
Carrier-mediated endocytosis	Membrane vesicle	Endocytosis triggered by a specific receptor	Cholesterol uptake
Exocytosis	Membrane vesicle	Vesicles fuse with plasma membrane and eject contents	Secretion of mucus
Active transport			
Na^+/K^+ pump	Protein carrier	Carrier expends energy to export Na^+ against a concentration gradient	Coupled uptake of glucose into cells against its concentration gradient
Proton pump	Protein carrier	Carrier expends energy to export protons against a concentration gradient	Chemiosmotic generation of ATP

Chapter 6

www.mhhe.com/raven6/resources6.mhtml

Summary | **Questions** | **Media Resources**

6.1 Biological membranes are fluid layers of lipid.

- Every cell is encased within a fluid bilayer sheet of phospholipid molecules called the plasma membrane.

1. How would increasing the number of phospholipids with double bonds between carbon atoms in their tails affect the fluidity of a membrane?

- Membrane Structure

6.2 Proteins embedded within the plasma membrane determine its character.

- Proteins that are embedded within the plasma membrane have their hydrophobic regions exposed to the hydrophobic interior of the bilayer, and their hydrophilic regions exposed to the cytoplasm or the extracellular fluid.
- Membrane proteins can transport materials into or out of the cell, they can mark the identity of the cell, or they can receive extracellular information.

2. Describe the two basic types of structures that are characteristic of proteins that span membranes.

- Art Activities:
  - -Fluid Mosaic Model
  - -Membrane Protein Diversity

6.3 Passive transport across membranes moves down the concentration gradient.

- Diffusion is the kinetic movement of molecules or ions from an area of high concentration to an area of low concentration.
- Facilitated diffusion involves a protein carrier that transports molcules down their concentration gradients.
- Osmosis is the diffusion of water. Because all organisms are composed of mostly water, maintaining osmotic balance is essential to life.

3. If a cell's cytoplasm were hyperosmotic to the extracellular fluid, how would the concentration of solutes in the cytoplasm compare with that in the extracellular fluid?

- Diffusion
- Osmosis

- Art Quiz: Osmotic Pressure

6.4 Bulk transport utilizes endocytosis.

- Materials or volumes of fluid that are too large to pass directly through the cell membrane can move into or out of cells through endocytosis or exocytosis, respectively.
- In these processes, the cell expends energy to change the shape of its plasma membrane, allowing the cell to engulf materials into a temporary vesicle (endocytosis), or eject materials by fusing a filled vesicle with the plasma membrane (exocytosis).

4. How do phagocytosis and pinocytosis differ?

5. Describe the mechanism of receptor-mediated endocytosis.

- Exocystosis/ endocytosis

- Exocystosis/ endocytosis
- Activity: Bulk Passage Into and Out of the Cell

6.5 Active transport across membranes is powered by energy from ATP.

- Cells use active transport to move substances across the plasma membrane against their concentration gradients, either accumulating them within the cell or extruding them from the cell. Active transport requires energy from ATP, either directly or indirectly.

6. In what two ways does facilitated diffusion differ from simple diffusion across a membrane?

7. How does active transport differ from facilitated diffusion? How is it similar to facilitated diffusion?

- Exploration: Active Transport

- Active Transport

7

Cell-Cell Interactions

Concept Outline

7.1 Cells signal one another with chemicals.

Receptor Proteins and Signaling between Cells. Receptor proteins embedded in the plasma membrane change shape when they bind specific signal molecules, triggering a chain of events within the cell.

Types of Cell Signaling. Cell signaling can occur between adjacent cells, although chemical signals called hormones act over long distances.

7.2 Proteins in the cell and on its surface receive signals from other cells.

Intracellular Receptors. Some receptors are located within the cell cytoplasm. These receptors respond to lipid-soluble signals, such as steroid hormones.

Cell Surface Receptors. Many cell-to-cell signals are water-soluble and cannot penetrate membranes. Instead, the signals are received by transmembrane proteins protruding out from the cell surface.

7.3 Follow the journey of information into the cell.

Initiating the Intracellular Signal. Cell surface receptors often use "second messengers" to transmit a signal to the cytoplasm.

Amplifying the Signal: Protein Kinase Cascades. Surface receptors and second messengers amplify signals as they travel into the cell, often toward the cell nucleus.

7.4 Cell surface proteins mediate cell-cell interactions.

The Expression of Cell Identity. Cells possess on their surfaces a variety of tissue-specific identity markers that identify both the tissue and the individual.

Intercellular Adhesion. Cells attach themselves to one another with protein links. Some of the links are very strong, others more transient.

Tight Junctions. Adjacent cells form a sheet when connected by tight junctions, and molecules are encouraged to flow through the cells, not between them.

Anchoring Junctions. The cytoskeleton of a cell is connected by an anchoring junction to the cytoskeleton of another cell or to the extracellular matrix.

Communicating Junctions. Many adjacent cells have direct passages that link their cytoplasms, permitting the passage of ions and small molecules.

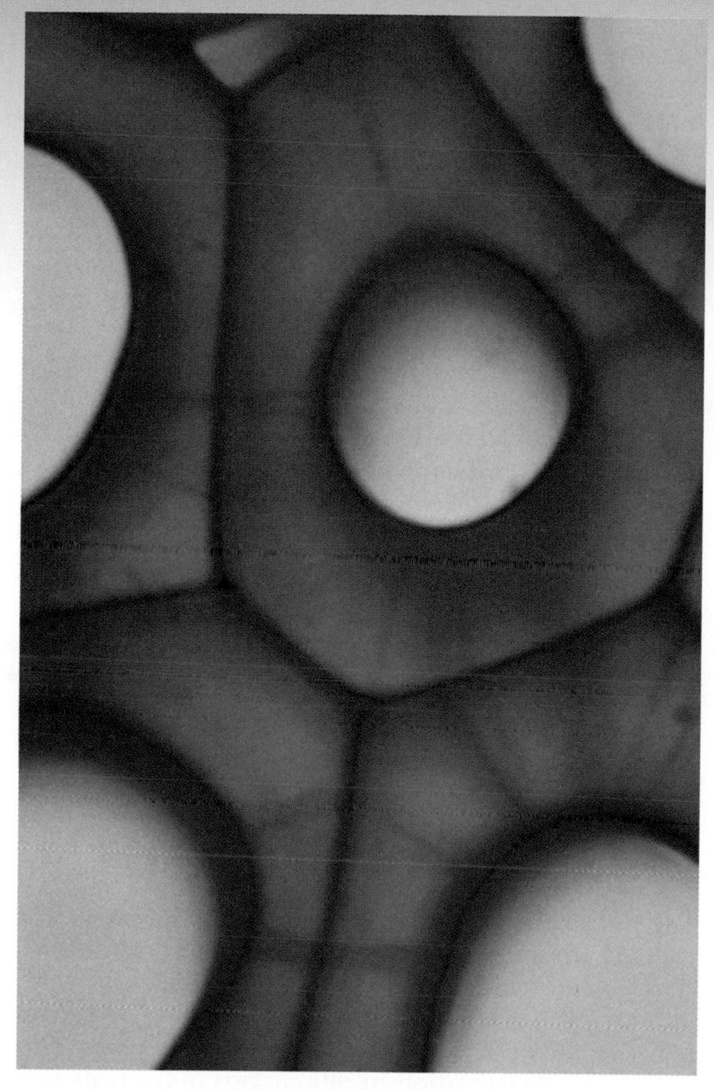

FIGURE 7.1
Persimmon cells in close contact with one another. These plant cells and all cells, no matter what their function, interact with their environment, including the cells around them.

Did you know that each of the 100 trillion cells of your body shares one key feature with the cells of tigers, bumblebees, and persimmons (figure 7.1)—a feature that most bacteria and protists lack? Your cells touch and communicate with one another. Sending and receiving a variety of chemical signals, they coordinate their behavior so that your body functions as an integrated whole, rather than as a massive collection of individual cells acting independently. The ability of cells to communicate with one another is the hallmark of multicellular organisms. In this chapter we will look in detail at how the cells of multicellular organisms interact with one another, first exploring how they signal one another with chemicals and then examining the ways in which their cell surfaces interact to organize tissues and body structures.

7.1 Cells signal one another with chemicals.

Receptor Proteins and Signaling between Cells

Communication between cells is common in nature. Cell signaling occurs in all multicellular organisms, providing an indispensable mechanism for cells to influence one another. The cells of multicellular organisms use a variety of molecules as signals, including not only peptides, but also large proteins, individual amino acids, nucleotides, steroids and other lipids.

Even dissolved gases are used as signals. Nitric oxide (NO) plays a role in mediating male erections (Viagra functions by stimulating NO release).

Some of these molecules are attached to the surface of the signaling cell; others are secreted through the plasma membrane or released by exocytosis.

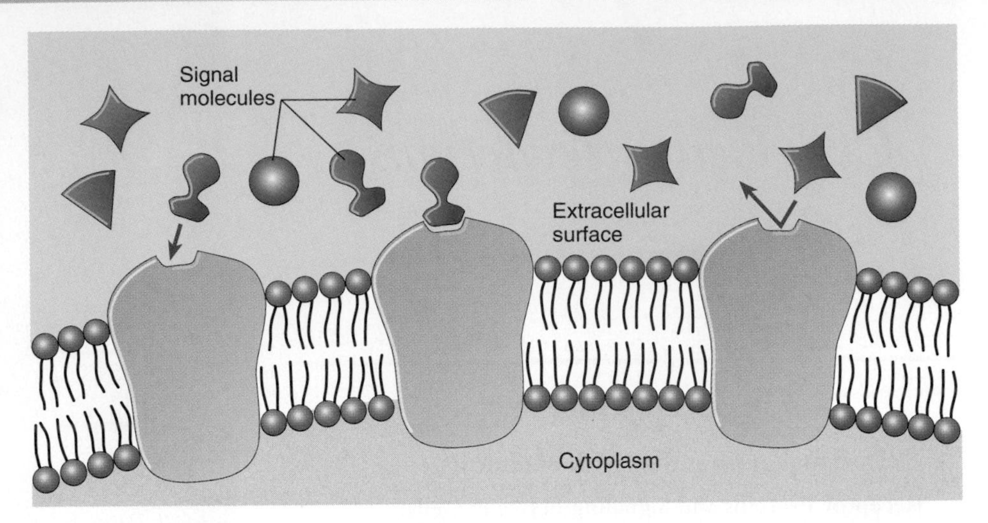

FIGURE 7.2
Cell surface receptors recognize only specific molecules. Signal molecules will bind only to those cells displaying receptor proteins with a shape into which they can fit snugly.

Cell Surface Receptors

Any given cell of a multicellular organism is exposed to a constant stream of signals. At any time, hundreds of different chemical signals may be in the environment surrounding the cell. However, each cell responds only to certain signals and ignores the rest (figure 7.2), like a person following the conversation of one or two individuals in a noisy, crowded room. How does a cell "choose" which signals to respond to? Located on or within the cell are **receptor proteins,** each with a three-dimensional shape that fits the shape of a specific signal molecule. When a signal molecule approaches a receptor protein of the right shape, the two can bind. This binding induces a change in the receptor protein's shape, ultimately producing a response in the cell. Hence, a given cell responds to the signal molecules that fit the particular set of receptor proteins it possesses, and ignores those for which it lacks receptors.

The Hunt for Receptor Proteins

The characterization of receptor proteins has presented a very difficult technical problem, because of their relative scarcity in the cell. Because these proteins may constitute less than 0.01% of the total mass of protein in a cell, purifying them is analogous to searching for a particular grain of sand in a sand dune! However, two recent techniques have enabled cell biologists to make rapid progress in this area.

Monoclonal antibodies. The first method uses *monoclonal antibodies.* An antibody is an immune system protein that, like a receptor, binds specifically to another molecule. Each individual immune system cell can make only one particular type of antibody, which can bind to only one specific target molecule. Thus, a cell-line derived from a single immune system cell (a clone) makes one specific antibody (a *monoclonal* antibody). Monoclonal antibodies that bind to particular receptor proteins can be used to isolate those proteins from the thousands of others in the cell.

Gene analysis. The study of mutants and isolation of gene sequences has had a tremendous impact on the field of receptor analysis. In chapter 19 we will present a detailed account of how this is done. These advances make it feasible to identify and isolate the many genes that encode for various receptor proteins.

Remarkably, these techniques have revealed that the enormous number of receptor proteins can be grouped into just a handful of "families" containing many related receptors. Later in this chapter we will meet some of the members of these receptor families.

Cells in a multicellular organism communicate with others by releasing signal molecules that bind to receptor proteins on the surface of the other cells. Recent advances in protein isolation have yielded a wealth of information about the structure and function of these proteins.

Types of Cell Signaling

Cells communicate through any of four basic mechanisms, depending primarily on the distance between the signaling and responding cells (figure 7.3). In addition to using these four basic mechanisms, some cells actually send signals to themselves, secreting signals that bind to specific receptors on their own plasma membranes. This process, called **autocrine signaling,** is thought to play an important role in reinforcing developmental changes.

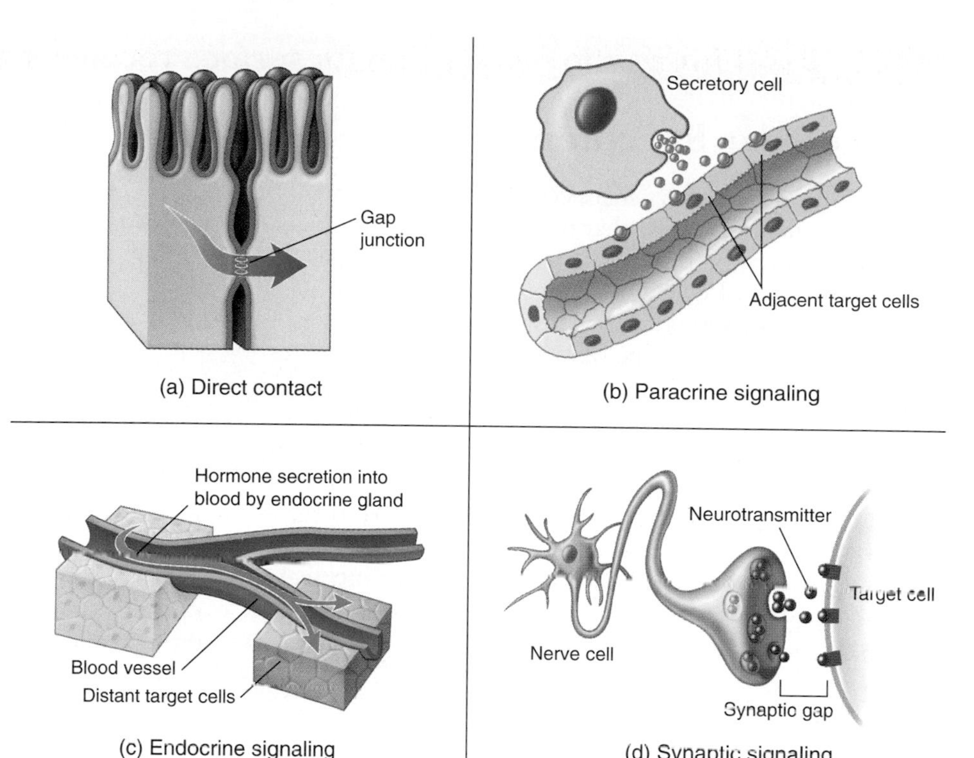

FIGURE 7.3
Four kinds of cell signaling. Cells communicate in several ways. (*a*) Two cells in *direct contact* with each other may send signals across gap junctions. (*b*) In *paracrine signaling*, secretions from one cell have an effect only on cells in the immediate area. (*c*) In *endocrine signaling*, hormones are released into the circulatory system, which carries them to the target cells. (*d*) Chemical *synaptic signaling* involves transmission of signal molecules, called neurotransmitters, from a neuron over a small synaptic gap to the target cell.

Direct Contact

As we saw in chapter 6, the surface of a eukaryotic cell is a thicket of proteins, carbohydrates, and lipids attached to and extending outward from the plasma membrane. When cells are very close to one another, some of the molecules on the cells' plasma membranes may bind together in specific ways. Many of the important interactions between cells in early development occur by means of direct contact between cell surfaces (figure 7.3*a*). We'll examine contact-dependent interactions more closely later in this chapter.

Paracrine Signaling

Signal molecules released by cells can diffuse through the extracellular fluid to other cells. If those molecules are taken up by neighboring cells, destroyed by extracellular enzymes, or quickly removed from the extracellular fluid in some other way, their influence is restricted to cells in the immediate vicinity of the releasing cell. Signals with such short-lived, local effects are called **paracrine** signals (figure 7.3*b*). Like direct contact, paracrine signaling plays an important role in early development, coordinating the activities of clusters of neighboring cells.

Endocrine Signaling

If a released signal molecule remains in the extracellular fluid, it may enter the organism's circulatory system and travel widely throughout the body. These longer lived signal molecules, which may affect cells very distant from the releasing cell, are called **hormones,** and this type of intercellular communication is known as **endocrine** signaling (figure 7.3*c*). Chapter 58 discusses endocrine signaling in detail. Both animals and plants use this signaling mechanism extensively.

Synaptic Signaling

In animals, the cells of the nervous system provide rapid communication with distant cells. Their signal molecules, **neurotransmitters,** do not travel to the distant cells through the circulatory system like hormones do. Rather, the long, fiberlike extensions of nerve cells release neurotransmitters from their tips very close to the target cells (figure 7.3*d*). The narrow gap between the two cells is called a **chemical synapse.** While paracrine signals move through the fluid between cells, neurotransmitters cross the synapse and persist only briefly. We will examine synaptic signaling more fully in chapter 54.

Adjacent cells can signal others by direct contact, while nearby cells that are not touching can communicate through paracrine signals. Two other systems mediate communication over longer distances: in endocrine signaling the blood carries hormones to distant cells, and in synaptic signaling nerve cells secrete neurotransmitters from long cellular extensions close to the responding cells.

7.2 Proteins in the cell and on its surface receive signals from other cells.

Intracellular Receptors

All cell signaling pathways share certain common elements, including a chemical signal that passes from one cell to another and a receptor that receives the signal in or on the target cell. We've looked at the sorts of signals that pass from one cell to another. Now let's consider the nature of the receptors that receive the signals. Table 7.1 summarizes the types of receptors we will discuss in this chapter.

Many cell signals are lipid-soluble or very small molecules that can readily pass across the plasma membrane of the target cell and into the cell, where they interact with a receptor. Some bind to protein receptors located in the cytoplasm; others pass across the nuclear membrane as well and bind to receptors within the nucleus. These **intracellular receptors** (figure 7.4) may trigger a variety of responses in the cell, depending on the receptor.

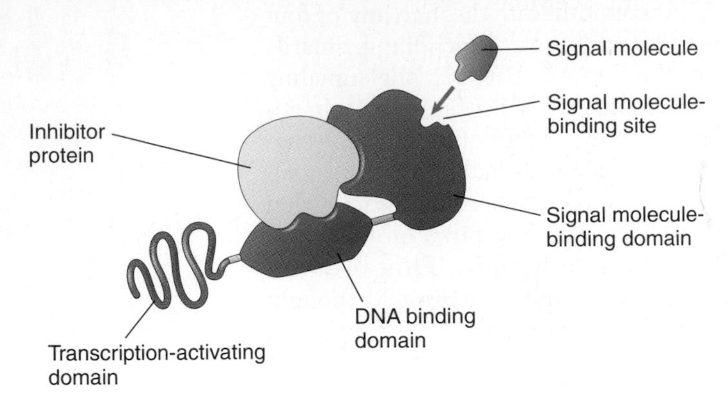

FIGURE 7.4
Basic structure of a gene-regulating intracellular receptor. These receptors are located within the cell and function in the reception of signals such as steroid hormones, vitamin D, and thyroid hormone.

Table 7.1 Cell Communicating Mechanisms

Mechanism	Structure	Function	Example
INTRACELLULAR RECEPTORS	No extracellular signal-binding site	Receives signals from lipid-soluble or noncharged, nonpolar small molecules	Receptors for NO, steroid hormone, vitamin D, and thyroid hormone
CELL SURFACE RECEPTORS			
Chemically gated ion channels	Multipass transmembrane protein forming a central pore	Molecular "gates" triggered chemically to open or close	Neurons
Enzymic receptors	Single-pass transmembrane protein	Binds signal extracellularly, catalyzes response intracellularly	Phosphorylation of protein kinases
G-protein-linked receptors	Seven-pass transmembrane protein with cytoplasmic binding site for G protein	Binding of signal to receptor causes GTP to bind a G protein; G protein, with attached GTP, detaches to deliver the signal inside the cell	Peptide hormones, rod cells in the eyes
PHYSICAL CONTACT WITH OTHER CELLS			
Surface markers	Variable; integral proteins or glycolipids in cell membrane	Identify the cell	MHC complexes, blood groups, antibodies
Tight junctions	Tightly bound, leakproof, fibrous protein "belt" that surrounds cell	Organizing junction: holds cells together such that material passes *through* but not *between* the cells	Junctions between epithelial cells in the gut
Desmosomes	Intermediate filaments of cytoskeleton linked to adjoining cells through cadherins	Anchoring junction: "buttons" cells together	Epithelium
Adherens junctions	Transmembrane fibrous proteins	Anchoring junction: "roots" extracellular matrix to cytoskeleton	Tissues with high mechanical stress, such as the skin
Gap junctions	Six transmembrane connexon proteins creating a "pipe" that connects cells	Communicating junction: allows passage of small molecules from cell to cell in a tissue	Excitable tissue such as heart muscle
Plasmodesmata	Cytoplasmic connections between gaps in adjoining plant cell walls	Communicating junction between plant cells	Plant tissues

Receptors That Act as Gene Regulators

Some intracellular receptors act as regulators of gene transcription. Among them are the receptors for steroid hormones, such as cortisol, estrogen, and progesterone, as well as the receptors for a number of other small, lipid-soluble signal molecules, such as vitamin D and thyroid hormone. All of these receptors have similar structures; the genes that code for them may well be the evolutionary descendants of a single ancestral gene. Because of their structural similarities, they are all part of the *intracellular receptor superfamily*.

Each of these receptors has a binding site for DNA. In its inactive state, the receptor typically cannot bind DNA because an inhibitor protein occupies the binding site. When the signal molecule binds to another site on the receptor, the inhibitor is released and the DNA binding site is exposed (figure 7.5). The receptor then binds to a specific nucleotide sequence on the DNA, which activates (or, in a few instances, suppresses) a particular gene, usually located adjacent to the regulatory site.

The lipid-soluble signal molecules that intracellular receptors recognize tend to persist in the blood far longer than water-soluble signals. Most water-soluble hormones break down within minutes, and neurotransmitters within seconds or even milliseconds. A steroid hormone like cortisol or estrogen, on the other hand, persists for hours.

The target cell's response to a lipid-soluble cell signal can vary enormously, depending on the nature of the cell. This is true even when different target cells have the same intracellular receptor, for two reasons: First, the binding site for the receptor on the target DNA differs from one cell type to another, so that different genes are affected when the signal-receptor complex binds to the DNA, and second, most eukaryotic genes have complex controls. We will discuss them in detail in chapter 16, but for now it is sufficient to note that several different regulatory proteins are usually involved in reading a eukaryotic gene. Thus the intracellular receptor interacts with different signals in different tissues. Depending on the cell-specific controls operating in different tissues, the effect the intracellular receptor produces when it binds with DNA will vary.

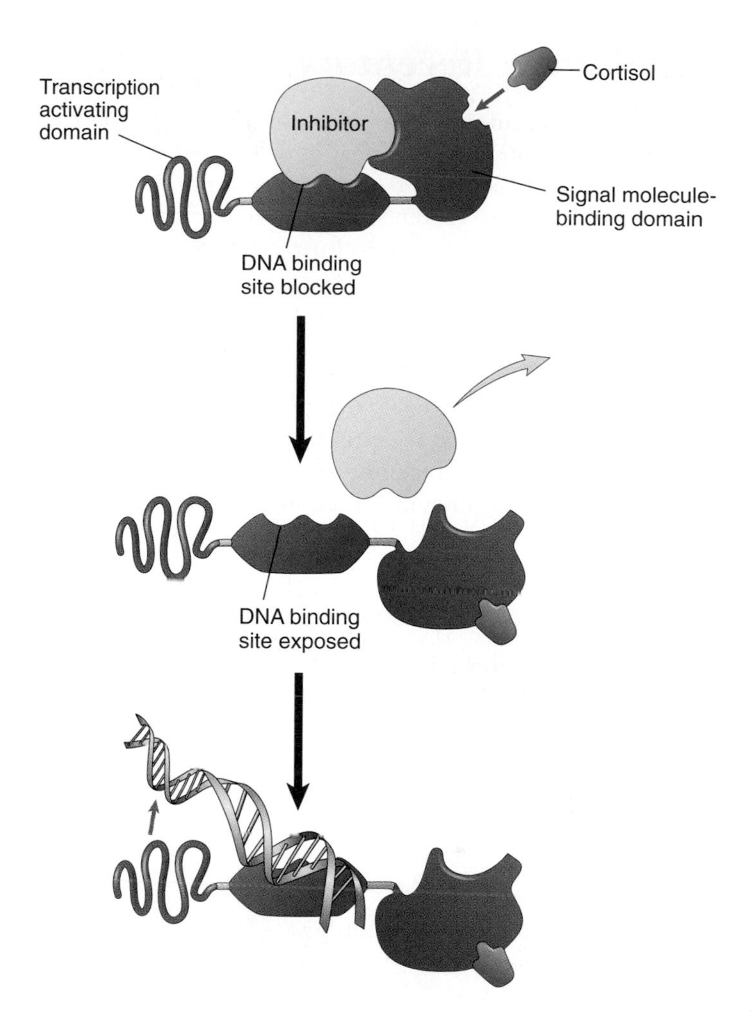

FIGURE 7.5
How intracellular receptors regulate gene transcription. In this model, the binding of the steroid hormone cortisol to a DNA regulatory protein causes it to alter its shape. The inhibitor is released, exposing the DNA binding site of the regulatory protein. The DNA binds to the site, positioning a specific nucleotide sequence over the transcription activating domain of the receptor and initiating transcription.

Receptors That Act as Enzymes

Other intracellular receptors act as enzymes. A very interesting example is the receptor for the signal molecule, **nitric oxide (NO).** A small gas molecule, NO diffuses readily out of the cells where it is produced and passes directly into neighboring cells, where it binds to the enzyme guanylyl cyclase. Binding of NO activates the enzyme, enabling it to catalyze the synthesis of cyclic guanosine monophosphate (GMP), an intracellular messenger molecule that produces cell-specific responses such as the relaxation of smooth muscle cells.

NO has only recently been recognized as a signal molecule in vertebrates. Already, however, a wide variety of roles have been documented. For example, when the brain sends a nerve signal relaxing the smooth muscle cells lining the walls of vertebrate blood vessels, the signal molecule acetylcholine released by the nerve near the muscle does not interact with the muscle cell directly. Instead, it causes nearby epithelial cells to produce NO, which then causes the smooth muscle to relax, allowing the vessel to expand and thereby increase blood flow.

Many target cells possess intracellular receptors, which are activated by substances that pass through the plasma membrane.

Cell Surface Receptors

Most signal molecules are water-soluble, including neurotransmitters, peptide hormones, and the many proteins that multicellular organisms employ as "growth factors" during development. Water-soluble signals cannot diffuse through cell membranes. Therefore, to trigger responses in cells, they must bind to receptor proteins on the surface of the cell. These **cell surface receptors** (figure 7.6) convert the extracellular signal to an intracellular one, responding to the binding of the signal molecule by producing a change within the cell's cytoplasm. Most of a cell's receptors are cell surface receptors, and almost all of them belong to one of three receptor superfamilies: chemically gated ion channels, enzymic receptors, and G-protein-linked receptors.

Chemically Gated Ion Channels

Chemically gated ion channels are receptor proteins that ions pass through. The receptor proteins that bind many neurotransmitters have the same basic structure (figure 7.6*a*). Each is a "multipass" transmembrane protein, meaning that the chain of amino acids threads back and forth across the plasma membrane several times. In the center of the protein is a pore that connects the extracellular fluid with the cytoplasm. The pore is big enough for ions to pass through, so the protein functions as an **ion channel.** The channel is said to be chemically gated because it opens when a chemical (the neurotransmitter) binds to it. The type of ion (sodium, potassium, calcium, chloride, for example) that flows across the membrane when a chemically gated ion channel opens depends on the specific three-dimensional structure of the channel.

Enzymic Receptors

Many cell surface receptors either act as enzymes or are directly linked to enzymes (figure 7.6*b*). When a signal molecule binds to the receptor, it activates the enzyme. In almost all cases, these enzymes are **protein kinases,** enzymes that add phosphate groups to proteins. Most enzymic receptors have the same general structure. Each is a single-pass transmembrane protein (the amino acid chain passes through the plasma membrane only once); the portion that binds the signal molecule lies outside the cell, and the portion that carries out the enzyme activity is exposed to the cytoplasm.

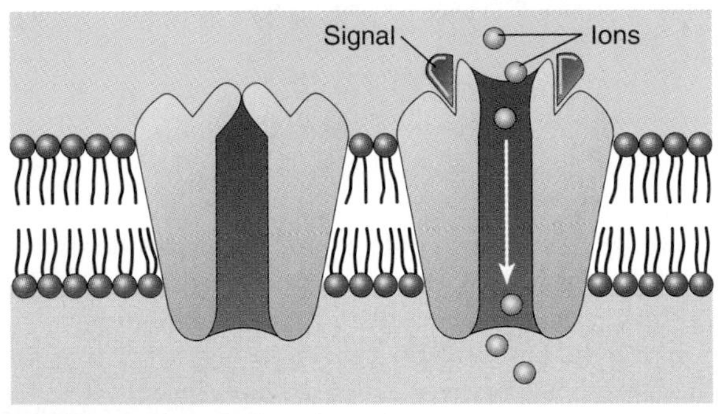

(a) Chemically gated ion channel

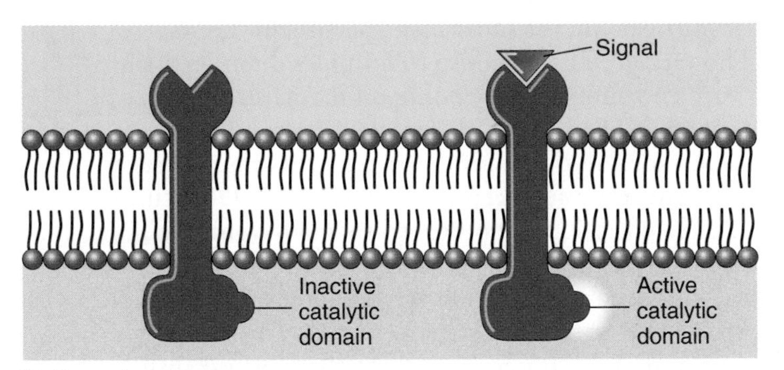

(b) Enzymic receptor

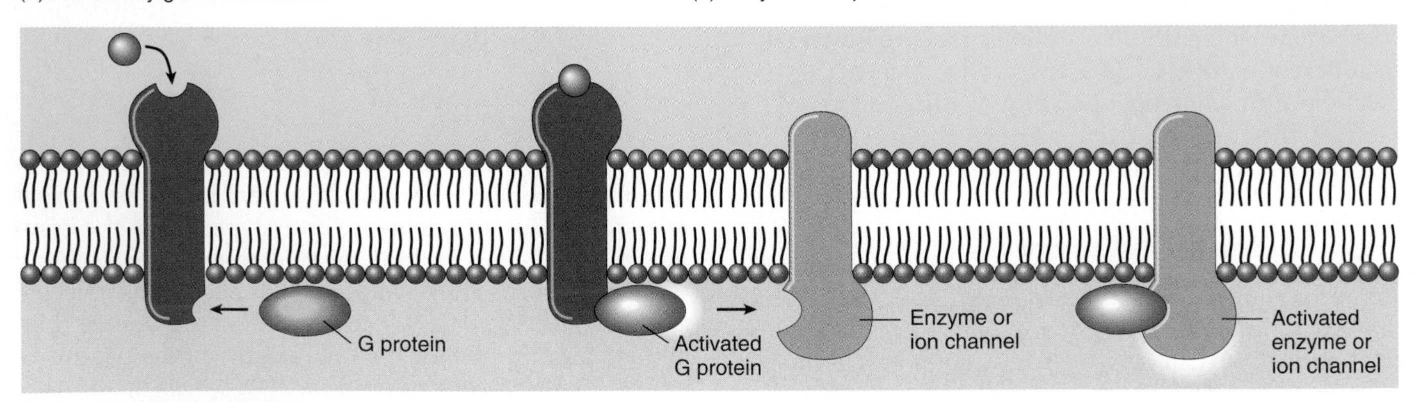

(c) G-protein-linked receptor

FIGURE 7.6
Cell surface receptors. (*a*) Chemically gated ion channels are multipass transmembrane proteins that form a pore in the cell membrane. This pore is opened or closed by chemical signals. (*b*) Enzymic receptors are single-pass transmembrane proteins that bind the signal on the extracellular surface. A catalytic region on their cytoplasmic portion then initiates enzymatic activity inside the cell. (*c*) G-protein-linked receptors bind to the signal outside the cell and to G proteins inside the cell. The G protein then activates an enzyme or ion channel, mediating the passage of a signal from the cell's surface to its interior.

G-Protein-Linked Receptors

A third class of cell surface receptors acts indirectly on enzymes or ion channels in the plasma membrane with the aid of an assisting protein, called a *guanosine triphosphate* (GTP)-*binding protein*, or **G protein** (figure 7.6*c*). Receptors in this category use G proteins to mediate passage of the signal from the membrane surface into the cell interior.

How G-Protein-Linked Receptors Work. G proteins are mediators that initiate a diffusible signal in the cytoplasm. They form a transient link between the receptor on the cell surface and the signal pathway within the cytoplasm. Importantly, this signal has a relatively short life span whose active age is determined by GTP. When a signal arrives, it finds the G protein nestled into the G-protein-linked receptor on the cytoplasmic side of the plasma membrane. Once the signal molecule binds to the receptor, the G-protein-linked receptor changes shape. This change in receptor shape twists the G protein, causing it to bind GTP. The G protein can now diffuse away from the receptor. The "activated" complex of a G protein with attached GTP is then free to initiate a number of events. However, this activation is short-lived, because GTP has a relatively short life span (seconds to minutes). This elegant arrangement allows the G proteins to activate numerous pathways, but only in a transient manner. In order for a pathway to "stay on," there must be a continuous source of incoming extracellular signals. When the rate of external signal drops off, the pathway shuts down.

The Largest Family of Cell Surface Receptors. Scientists have identified more than 100 different G-protein-linked receptors, more than any other kind of cell surface receptor. They mediate an incredible range of cell signals, including peptide hormones, neurotransmitters, fatty acids, and amino acids. Despite this great variation in specificity, however, all G-protein-linked receptors whose amino acid sequences are known have a similar structure. They are almost certainly closely related in an evolutionary sense, arising from a single ancestral sequence. Each of these G-protein-linked receptors is a seven-pass transmembrane protein (figure 7.7)—a single polypeptide chain that threads back and forth across the lipid bilayer seven times, creating a channel through the membrane.

Evolutionary Origin of G-Protein-Linked Receptors. As research revealed the structure of G-protein-linked receptors, an interesting pattern emerged: the same seven-pass structural motif is seen again and again, in sensory receptors such as the light-activated rhodopsin protein in the vertebrate eye, in the light-activated bacteriorhodopsin proton pump that plays a key role in bacterial photosynthesis, in the receptor that recognizes the yeast mating factor

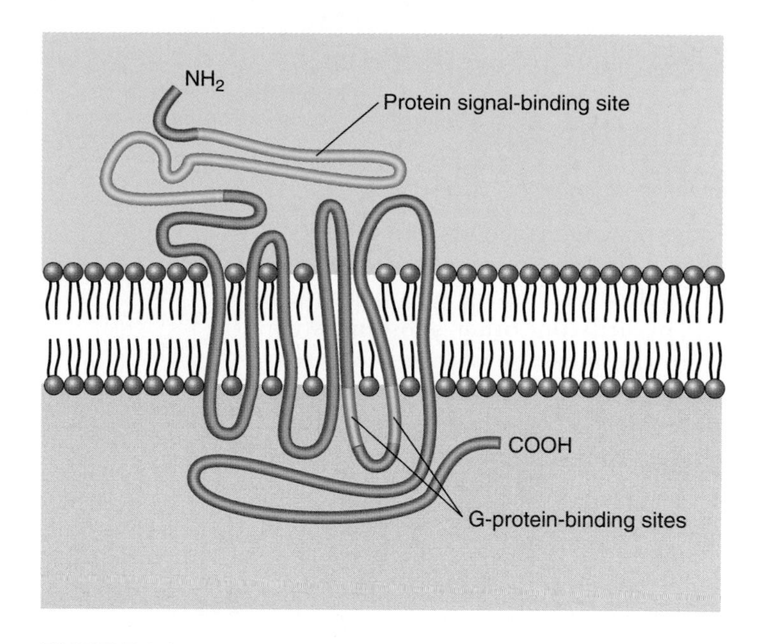

FIGURE 7.7
The G-protein-linked receptor is a seven-pass transmembrane protein.

protein, and in many other sensory receptors. Vertebrate rhodopsin is in fact a G-protein-linked receptor and utilizes a G protein. Bacteriorhodopsin is not. The occurrence of the seven-pass structural motif in both, and in so many other G-protein-linked receptors, suggests that this motif is a very ancient one, and that G-protein-linked receptors may have evolved from sensory receptors of single-celled ancestors.

Discovery of G Proteins. Martin Rodbell of the National Institute of Environmental Health Sciences and Alfred Gilman of the University of Texas Southwestern Medical Center received the 1994 Nobel Prize for Medicine or Physiology for their work on G proteins. Rodbell and Gilman's work has proven to have significant ramifications. G proteins are involved in the mechanism employed by over half of all medicines in use today. Studying G proteins will vastly expand our understanding of how these medicines work. Furthermore, the investigation of G proteins should help elucidate how cells communicate in general and how they contribute to the overall physiology of organism. As Gilman says, G proteins are "involved in everything from sex in yeast to cognition in humans."

Most receptors are located on the surface of the plasma membrane. Chemically gated ion channels open or close when signal molecules bind to the channel, allowing specific ions to diffuse through. Enzyme receptors typically activate intracellular proteins by phosphorylation. G-protein-linked receptors activate an intermediary protein, which then effects the intracellular change.

7.3 Follow the journey of information into the cell.

Initiating the Intracellular Signal

Some enzymic receptors and most G-protein-linked receptors carry the signal molecule's message into the target cell by utilizing other substances to relay the message within the cytoplasm. These other substances, small molecules or ions commonly called **second messengers** or intracellular mediators, alter the behavior of particular proteins by binding to them and changing their shape. The two most widely used second messengers are cyclic adenosine monophosphate (cAMP) and calcium.

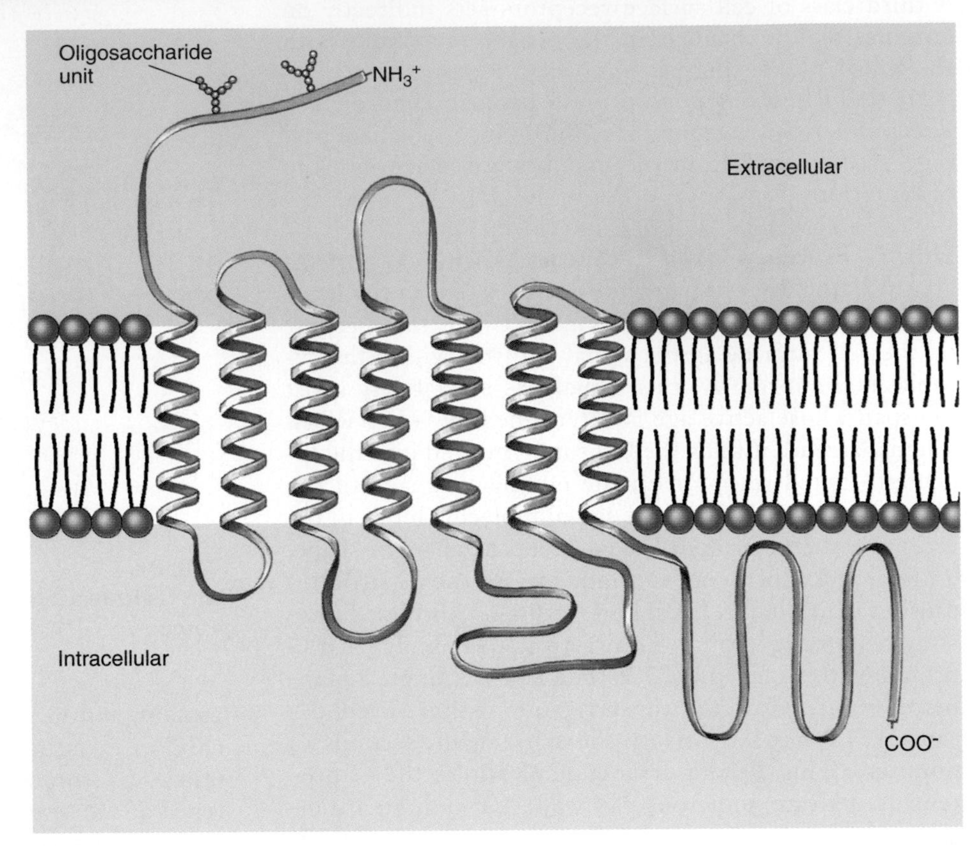

FIGURE 7.8
Structure of the β-adrenergic receptor. The receptor is a G-protein-linked molecule which, when it binds to an extracellular signal molecule, stimulates voluminous production of cAMP inside the cell, which then effects the cellular change.

cAMP

All animal cells studied thus far use **cAMP** as a second messenger (chapter 56 discusses cAMP in detail). To see how cAMP typically works as a messenger, let's examine what happens when the hormone epinephrine binds to a particular type of G-protein-linked receptor called the β-adrenergic receptor (figure 7.8). When epinephrine binds with this receptor, it activates a G protein, which then stimulates the enzyme **adenylyl cyclase** to produce large amounts of cAMP within the cell (figure 7.9*a*). The cAMP then binds to and activates the enzyme α-kinase, which adds phosphates to specific proteins in the cell. The effect this phosphorylation has on cell function depends on the identity of the cell and the proteins that are phosphorylated. In muscle cells, for example, the α-kinase phosphorylates and thereby activates enzymes that stimulate the breakdown of glycogen into glucose and inhibit the synthesis of glycogen from glucose. Glucose is then more available to the muscle cells for metabolism.

Calcium

Calcium (Ca^{++}) ions serve even more widely than cAMP as second messengers. Ca^{++} levels inside the cytoplasm of a cell are normally very low (less than 10^{-7} M), while outside the cell and in the endoplasmic reticulum Ca^{++} levels are quite high (about 10^{-3} M). Chemically gated calcium channels in the endoplasmic reticulum membrane act as switches; when they open, Ca^{++} rushes into the cytoplasm and triggers proteins sensitive to Ca^{++} to initiate a variety of activities. For example, the efflux of Ca^{++} from the endoplasmic reticulum causes skeletal muscle cells to contract and some endocrine cells to secrete hormones.

The gated Ca^{++} channels are opened by a G-protein-linked receptor. In response to signals from other cells, the receptor activates its G protein, which in turn activates the enzyme, *phospholipase C*. This enzyme catalyzes the production of *inositol trisphosphate* (*IP_3*) from phospholipids in the plasma membrane. The IP_3 molecules diffuse through the cytoplasm to the endoplasmic reticulum and bind to the Ca^{++} channels. This opens the channels and allows Ca^{++} to flow from the endoplasmic reticulum into the cytoplasm (figure 7.9*b*).

Ca^{++} initiates some cellular responses by binding to *calmodulin*, a 148-amino acid cytoplasmic protein that contains four binding sites for Ca^{++} (figure 7.10). When four Ca^{++} ions are bound to calmodulin, the calmodulin/Ca^{++} complex binds to other proteins, and activates them.

Cyclic AMP and Ca^{++} often behave as second messengers, intracellular substances that relay messages from receptors to target proteins.

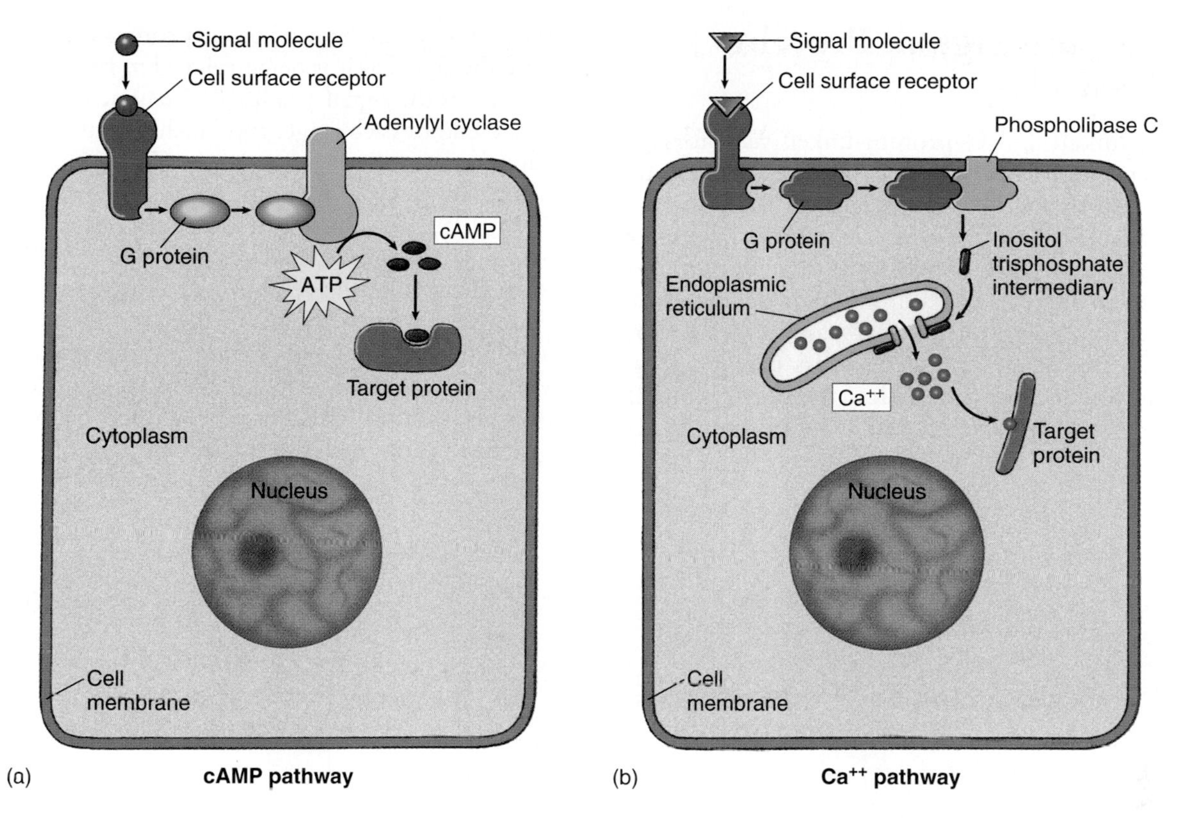

FIGURE 7.9
How second messengers work. (*a*) The cyclic AMP (cAMP) pathway. An extracellular receptor binds to a signal molecule and, through a G protein, activates the membrane-bound enzyme, adenylyl cyclase. This enzyme catalyzes the synthesis of cAMP, which binds to the target protein to initiate the cellular change. (*b*) The calcium (Ca^{++}) pathway. An extracellular receptor binds to another signal molecule and, through another G protein, activates the enzyme phospholipase C. This enzyme stimulates the production of inositol trisphosphate, which binds to and opens calcium channels in the membrane of the endoplasmic reticulum. Ca^{++} is released into the cytoplasm, effecting a change in the cell.

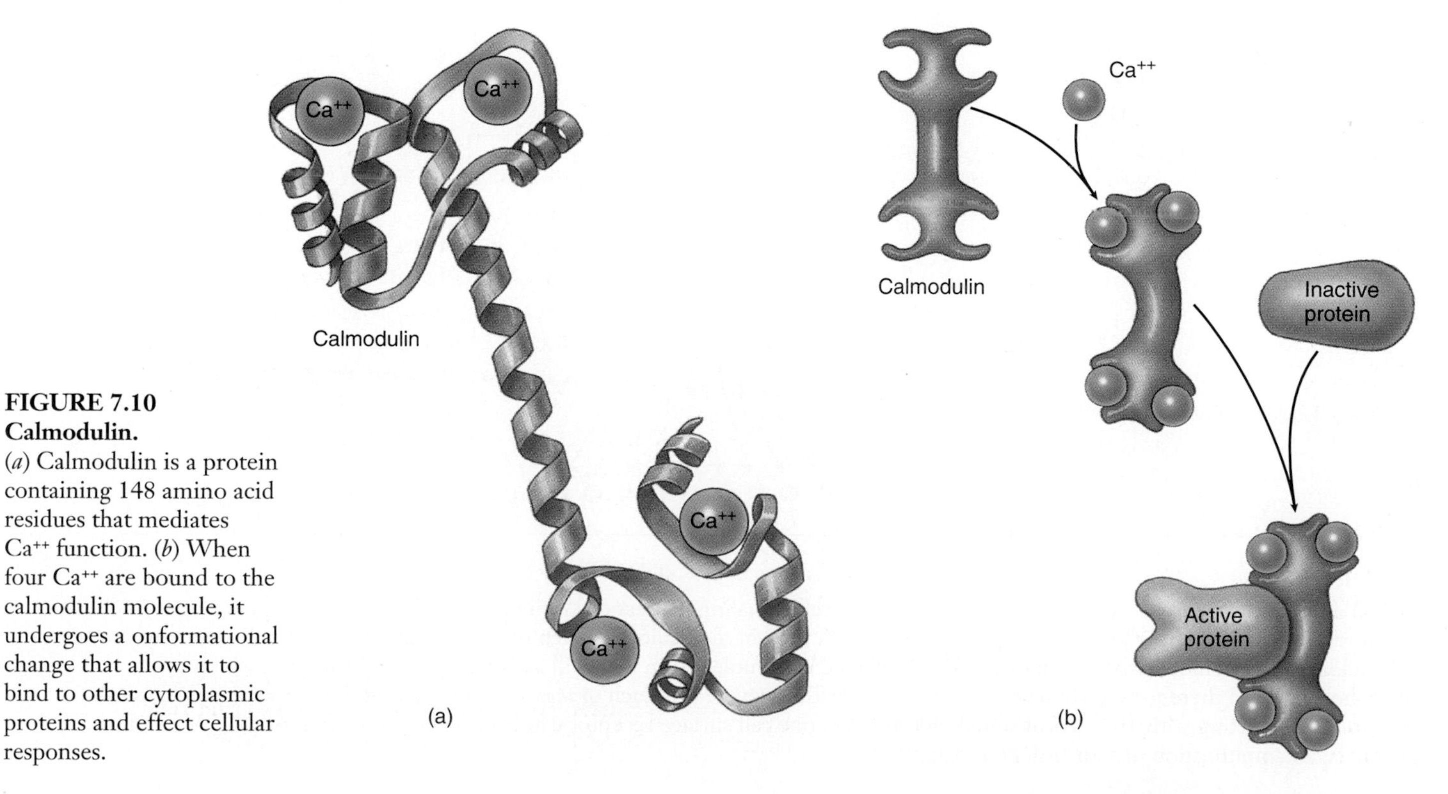

FIGURE 7.10
Calmodulin.
(*a*) Calmodulin is a protein containing 148 amino acid residues that mediates Ca^{++} function. (*b*) When four Ca^{++} are bound to the calmodulin molecule, it undergoes a onformational change that allows it to bind to other cytoplasmic proteins and effect cellular responses.

Amplifying the Signal: Protein Kinase Cascades

Both enzyme-linked and G-protein-linked receptors receive signals at the surface of the cell, but as we've seen, the target cell's response rarely takes place there. In most cases the signals are relayed to the cytoplasm or the nucleus by second messengers, which influence the activity of one or more enzymes or genes and so alter the behavior of the cell. But most signaling molecules are found in such low concentrations that their diffusion across the cytoplasm would take a great deal of time unless the signal is amplified. Therefore, most enzyme-linked and G-protein-linked receptors use a chain of other protein messengers to amplify the signal as it is being relayed to the nucleus.

How is the signal amplified? Imagine a relay race where, at the end of each stage, the finishing runner tags five new runners to start the next stage. The number of runners would increase dramatically as the race progresses: 1, then 5, 25, 125, and so on. The same sort of process takes place as a signal is passed from the cell surface to the cytoplasm or nucleus. First the receptor activates a stage-one protein, almost always by phosphorylating it. The receptor either adds a phosphate group directly, or, it activates a G protein that goes on to activate a second protein that does the phosphorylation. Once activated, each of these stage-one proteins in turn activates a large number of stage-two pro-

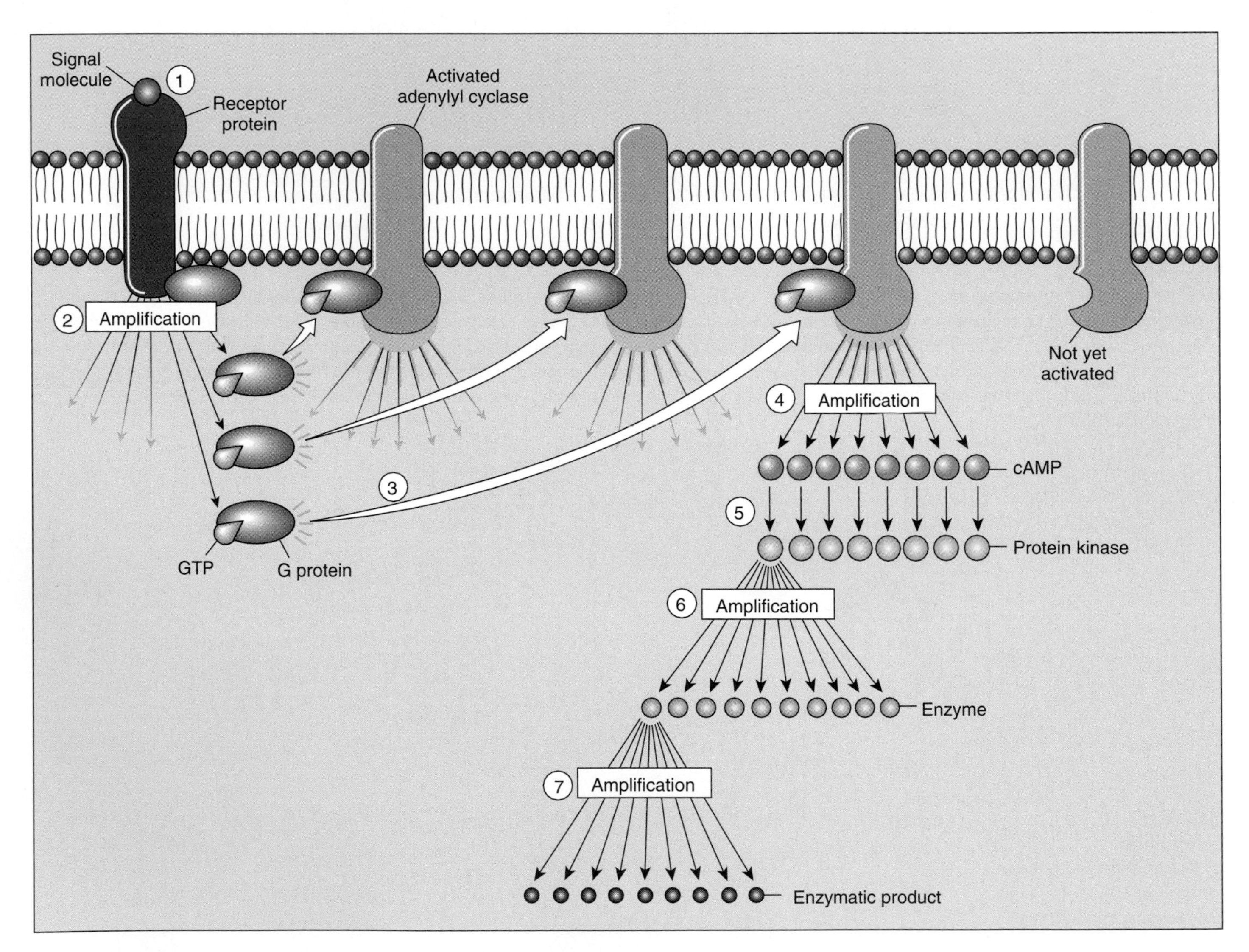

FIGURE 7.11

Signal amplification. Amplification at many steps of the cell-signaling process can ultimately produce a large response by the cell. One cell surface receptor (*1*), for example, may activate many G protein molecules (*2*), each of which activates a molecule of adenylyl cyclase (*3*), yielding an enormous production of cAMP (*4*). Each cAMP molecule in turn will activate a protein kinase (*5*), which can phosphorylate and thereby activate several copies of a specific enzyme (*6*). Each of *those* enzymes can then catalyze many chemical reactions (*7*). Starting with 10^{-10} M of signaling molecule, one cell surface receptor can trigger the production of 10^{-6} M of one of the products, an amplification of four orders of magnitude.

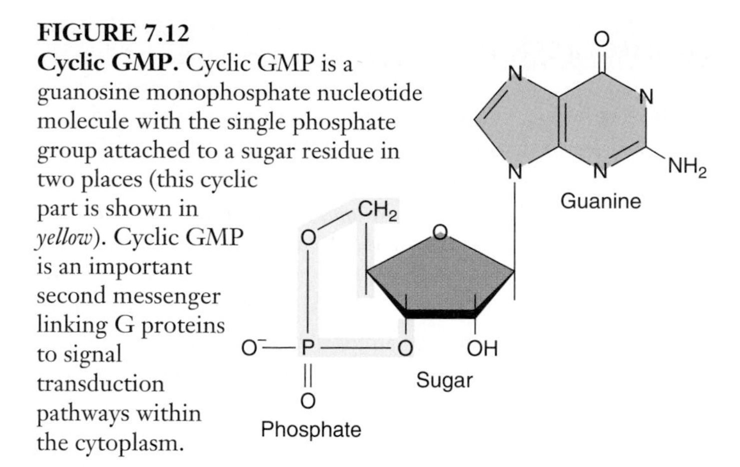
O
N
N
N
N
NH2
Guanine
CH2
O
O
O⁻— P — O
OH
O
Sugar
Phosphate

FIGURE 7.12
Cyclic GMP. Cyclic GMP is a guanosine monophosphate nucleotide molecule with the single phosphate group attached to a sugar residue in two places (this cyclic part is shown in *yellow*). Cyclic GMP is an important second messenger linking G proteins to signal transduction pathways within the cytoplasm.

teins; then each of them activates a large number of stage-three proteins, and so on (figure 7.11). A single cell surface receptor can thus stimulate a cascade of protein kinases to amplify the signal.

The Vision Amplification Cascade

Let's trace a protein amplification cascade to see exactly how one works. In vision, a single light-activated rhodopsin (a G-protein-linked receptor) activates hundreds of molecules of the G protein transducin in the first stage of the relay. In the second stage, each transducin causes an enzyme to modify thousands of molecules of a special inside-the-cell messenger called cyclic GMP (figure 7.12). (We will discuss cyclic GMP in more detail later.) In about 1 second, a single rhodopsin signal passing through this two-step cascade splits more than 10^5 (100,000) cyclic GMP molecules (figure 7.13)! The rod cells of humans are sufficiently sensitive to detect brief flashes of 5 photons.

The Cell Division Amplification Cascade

The amplification of signals traveling from the plasma membrane to the nucleus can be even more complex than the process we've just described. Cell division, for example, is controlled by a receptor that acts as a protein kinase. The receptor responds to growth-promoting signals by phosphorylating an intracellular protein called ras, which then activates a series of interacting phosphorylation cascades, some with five or more stages. If the ras protein becomes hyperactive for any reason, the cell acts as if it is being constantly stimulated to divide. Ras proteins were first discovered in cancer cells. A mutation of the gene that encodes ras had caused it to become hyperactive, resulting in unrestrained cell proliferation. Almost one-third of human cancers have such a mutation in a *ras* gene.

A small number of surface receptors can generate a vast intracellular response, as each stage of the pathway amplifies the next.

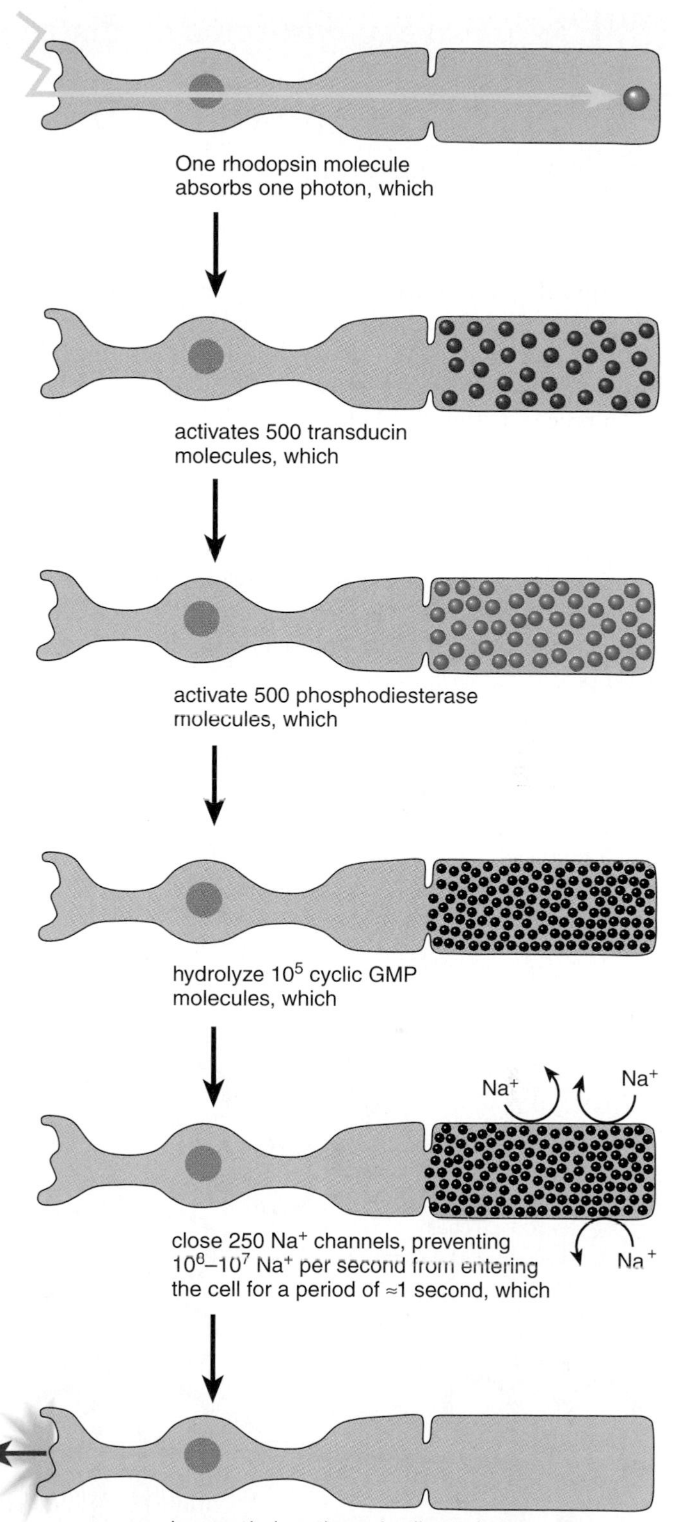
One rhodopsin molecule
absorbs one photon, which
activates 500 transducin
molecules, which
activate 500 phosphodiesterase
molecules, which
hydrolyze 10⁵ cyclic GMP
molecules, which
Na⁺
Na⁺
close 250 Na⁺ channels, preventing
10⁶–10⁷ Na⁺ per second from entering
the cell for a period of ≈1 second, which
Na⁺
hyperpolarizes the rod cell membrane
by 1 mV, sending a visual signal to the brain.

FIGURE 7.13
The role of signal amplification in vision. In this vertebrate rod cell (the cells of the eye responsible for interpreting light and dark), *one* single rhodopsin pigment molecule, when excited by a photon, ultimately yields *100,000* split cGMP molecules, which will then effect a change in the membrane of the rod cell, which will be interpreted by the organism as a visual event.

7.4 Cell surface proteins mediate cell-cell interactions.

The Expression of Cell Identity

With the exception of a few primitive types of organisms, the hallmark of multicellular life is the development of highly specialized groups of cells called **tissues,** such as blood and muscle. Remarkably, each cell within a tissue performs the functions of that tissue and no other, even though all cells of the body are derived from a single fertilized cell and contain the same genetic information. How do cells sense where they are, and how do they "know" which type of tissue they belong to?

Tissue-Specific Identity Markers

As it develops, each animal cell type acquires a unique set of cell surface molecules. These molecules serve as markers proclaiming the cells' tissue-specific identity. Other cells that make direct physical contact with them "read" the markers.

Glycolipids. Most tissue-specific cell surface markers are glycolipids, lipids with carbohydrate heads. The glycolipids on the surface of red blood cells are also responsible for the differences among A, B, and O blood types. As the cells in a tissue divide and differentiate, the population of cell surface glycolipids changes dramatically.

MHC Proteins. The immune system uses other cell surface markers to distinguish between "self" and "nonself" cells. All of the cells of a given individual, for example, have the same "self" markers, called *major histocompatibility complex* (MHC) *proteins.* Because practically every individual makes a different set of MHC proteins, they serve as distinctive identity tags for each individual. The MHC proteins and other self-identifying markers are single-pass proteins anchored in the plasma membrane, and many of them are members of a large superfamily of receptors, the immunoglobulins (figure 7.14). Cells of the immune system continually inspect the other cells they encounter in the body, triggering the destruction of cells that display foreign or "nonself" identity markers.

The immune systems of vertebrates, described in detail in chapter 57, shows an exceptional ability to distinguish self from nonself. However, other vertebrates and even some simple animals like sponges are able to make this distinction to some degree, even though they lack a complex immune system.

Every cell contains a specific array of marker proteins on its surface. These markers identify each type of cell in a very precise way.

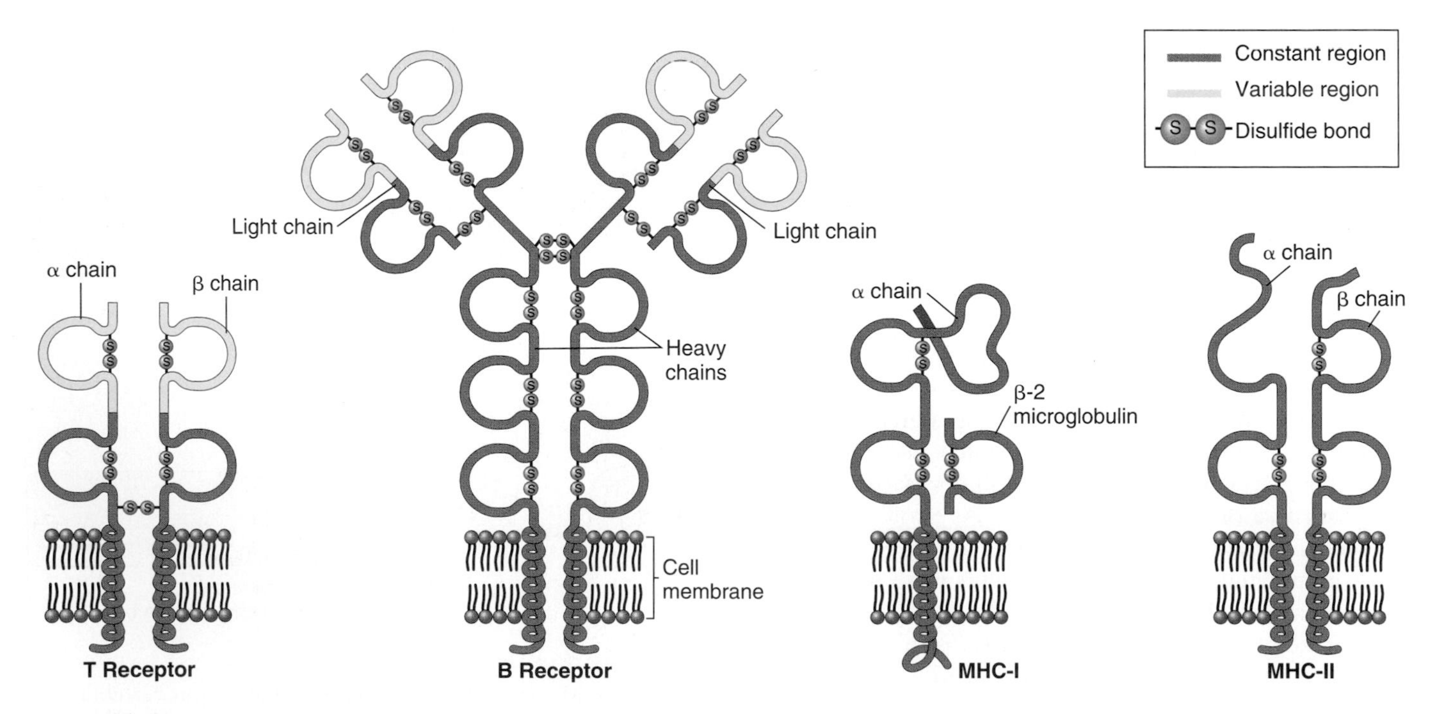

FIGURE 7.14
Structure of the immunoglobulin family of cell surface marker proteins. T and B cell receptors help mediate the immune response in organisms by recognizing and binding to foreign cell markers. MHC antigens label cells as "self," so that the immune system attacks only invading entities, such as bacteria, viruses, and usually even the cells of transplanted organs!

Intercellular Adhesion

Not all physical contacts between cells in a multicellular organism are fleeting touches. In fact, most cells are in physical contact with other cells at all times, usually as members of organized tissues such as those in the lungs, heart, or gut. These cells and the mass of other cells clustered around them form long-lasting or permanent connections with each other called **cell junctions** (figure 7.15). The nature of the physical connections between the cells of a tissue in large measure determines what the tissue is like. Indeed, a tissue's proper functioning often depends critically upon how the individual cells are arranged within it. Just as a house cannot maintain its structure without nails and cement, so a tissue cannot maintain its characteristic architecture without the appropriate cell junctions.

Cells attach themselves to one another with long-lasting bonds called cell junctions.

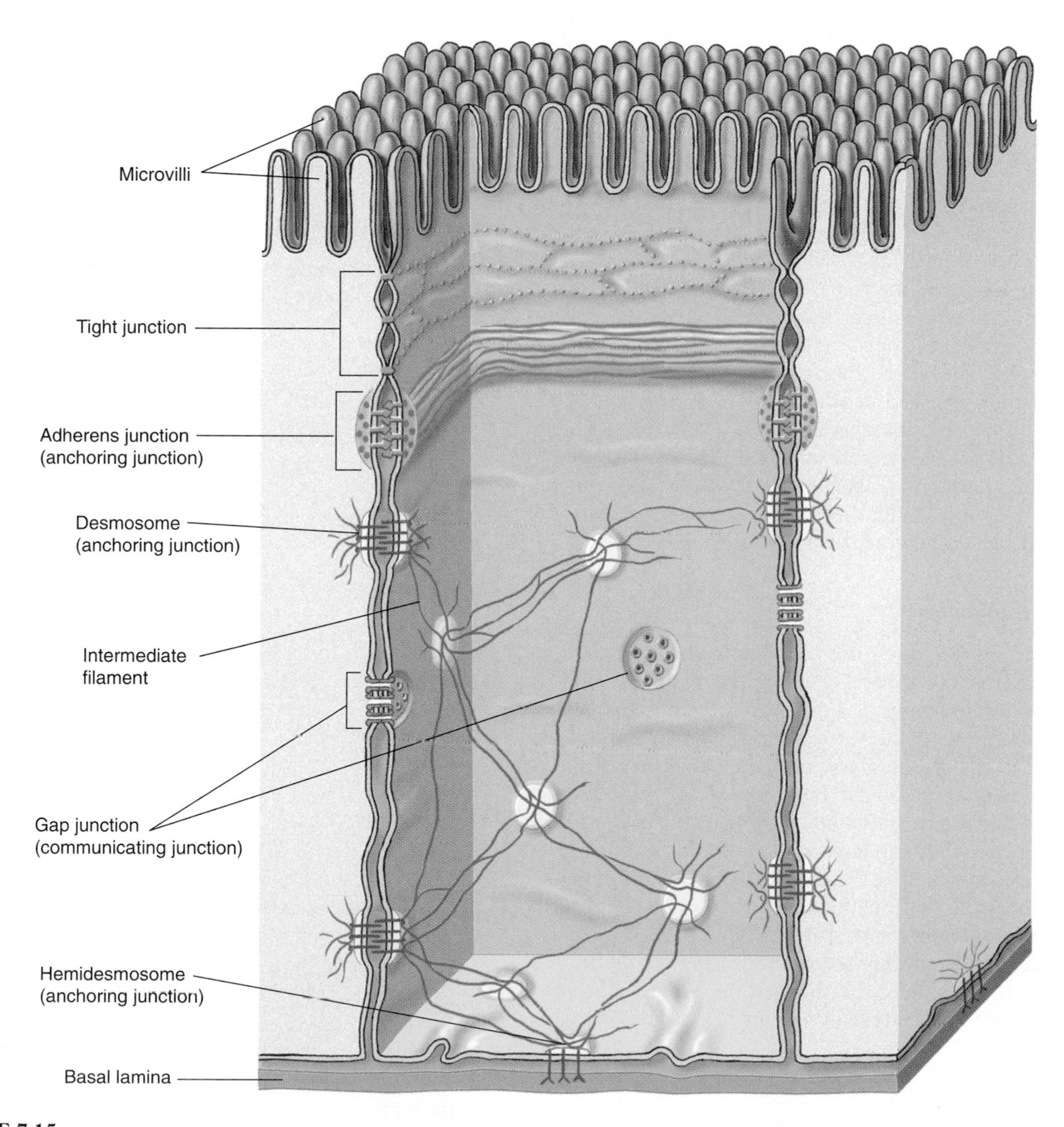

FIGURE 7.15
A summary of cell junction types. Gut epithelial cells are used here to illustrate the comparative structures and locations of common cell junctions.

Tight Junctions

Cell junctions are divided into three categories, based upon the functions they serve (figure 7.16): tight junctions, anchoring junctions, and communicating junctions.

Sometimes called occluding junctions, tight junctions connect the plasma membranes of adjacent cells in a sheet, preventing small molecules from leaking between the cells and through the sheet (figure 7.17). This allows the sheet of cells to act as a wall within the organ, keeping molecules on one side or the other.

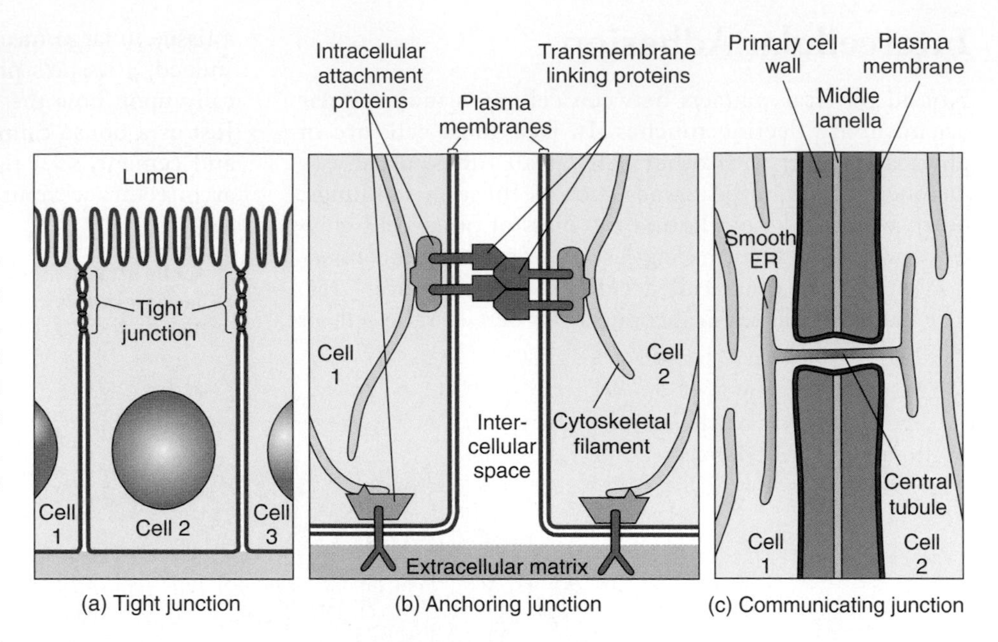

FIGURE 7.16
The three types of cell junctions. These three models represent current thinking on how the structures of the three major types of cell junctions facilitate their function: (*a*) tight junction; (*b*) anchoring junction; (*c*) communicating junction.

Creating Sheets of Cells

The cells that line an animal's digestive tract are organized in a sheet only one cell thick. One surface of the sheet faces the inside of the tract and the other faces the extracellular space where blood vessels are located. Tight junctions encircle each cell in the sheet, like a belt cinched around a pair of pants. The junctions between neighboring cells are so securely attached that there is no space between them for leakage. Hence, nutrients absorbed from the food in the digestive tract must pass directly through the cells in the sheet to enter the blood.

Partitioning the Sheet

The tight junctions between the cells lining the digestive tract also partition the plasma membranes of these cells into separate compartments. Transport proteins in the membrane facing the inside of the tract carry nutrients from that side to the cytoplasm of the cells. Other proteins, located in the membrane on the opposite side of the cells, transport those nutrients from the cytoplasm to the extracellular fluid, where they can enter the blood. For the sheet to absorb nutrients properly, these proteins must remain in the correct locations within the fluid membrane. Tight junctions effectively segregate the proteins on opposite sides of the sheet, preventing them from drifting within the membrane from one side of the sheet to the other. When tight junctions are experimentally disrupted, just this sort of migration occurs.

Tight junctions connect the plasma membranes of adjacent cells into sheets.

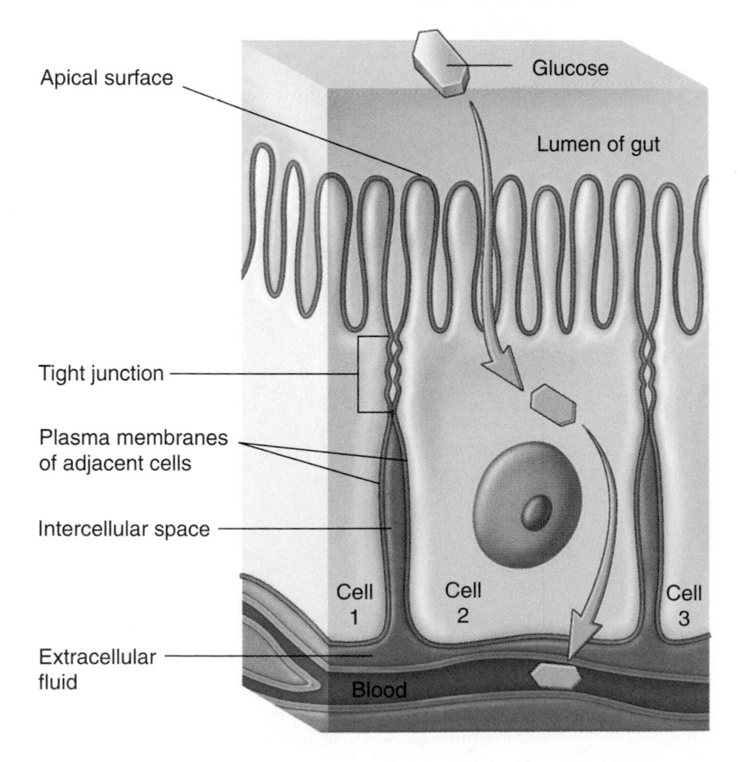

FIGURE 7.17
Tight junctions. Encircling the cell like a tight belt, these intercellular contacts ensure that materials move through the cells rather than between them.

Anchoring Junctions

Anchoring junctions mechanically attach the cytoskeleton of a cell to the cytoskeletons of other cells or to the extracellular matrix. They are commonest in tissues subject to mechanical stress, such as muscle and skin epithelium.

Cadherin and Intermediate Filaments: Desmosomes

Anchoring junctions called **desmosomes** connect the cytoskeletons of adjacent cells (figure 7.18), while hemidesmosomes anchor epithelial cells to a basement membrane. Proteins called cadherins, most of which are single-pass transmembrane glycoproteins, create the critical link. A variety of attachment proteins link the short cytoplasmic end of a cadherin to the intermediate filaments in the cytoskeleton. The other end of the cadherin molecule projects outward from the plasma membrane, joining directly with a cadherin protruding from an adjacent cell in a firm handshake binding the cells together.

Connections between proteins tethered to the intermediate filaments are much more secure than connections between free-floating membrane proteins. Proteins are suspended within the membrane by relatively weak interactions between the nonpolar portions of the protein and the membrane lipids. It would not take much force to pull an untethered protein completely out of the membrane, as if pulling an unanchored raft out of the water.

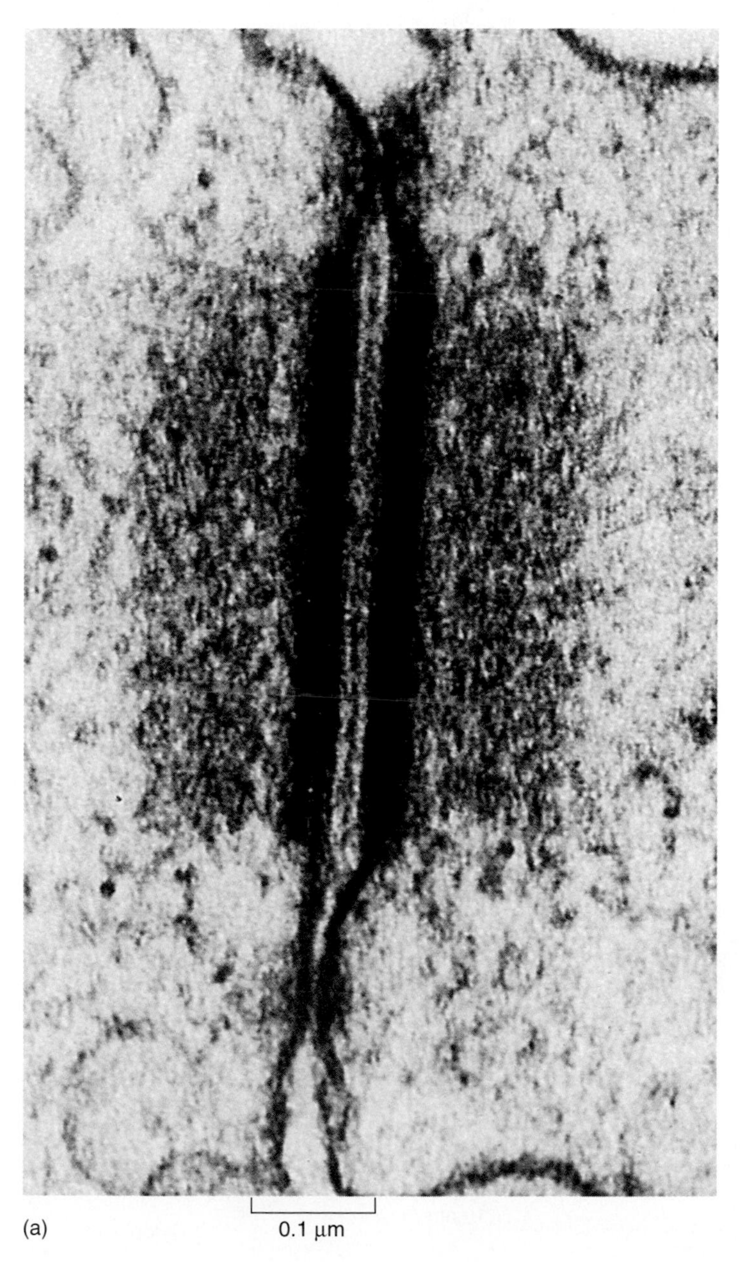

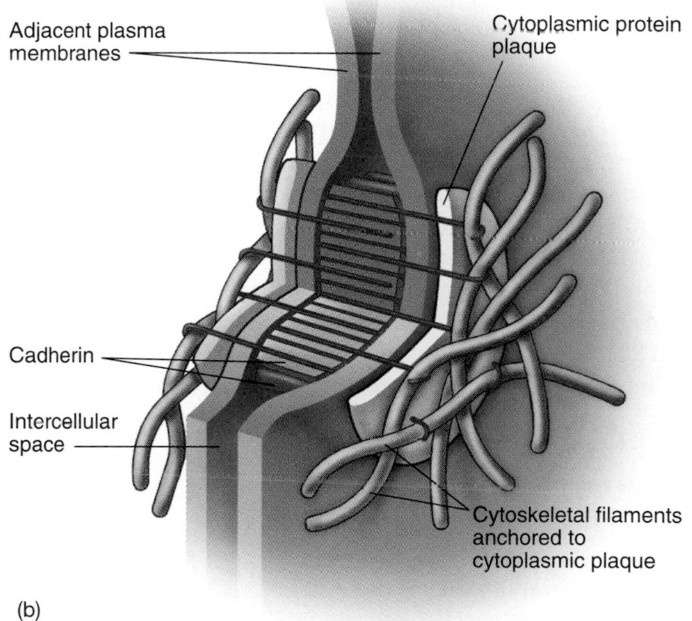

FIGURE 7.18
Desmosomes. (*a*) Desmosomes anchor adjacent cells to each other. (*b*) Cadherin proteins create the adhering link between adjoining cells.

Cadherin and Actin Filaments

Cadherins can also connect the actin frameworks of cells in cadherin-mediated junctions (figure 7.19). When they do, they form less stable links between cells than when they connect intermediate filaments. Many kinds of actin-linking cadherins occur in different tissues, as well as in the same tissue at different times. During vertebrate development, the migration of neurons in the embryo is associated with changes in the type of cadherin expressed on their plasma membranes. This suggests that gene-controlled changes in cadherin expression may provide the migrating cells with a "roadmap" to their destination.

Integrin-Mediated Links

Anchoring junctions called **adherens junctions** are another type of junction that connects the actin filaments of one cell with those of neighboring cells or with the extracellular matrix (figure 7.20). The linking proteins in these junctions are members of a large superfamily of cell surface receptors called integrins. Each integrin is a transmembrane protein composed of two different glycoprotein subunits that extend outward from the plasma membrane. Together, these subunits bind a protein component of the extracellular matrix, like two hands clasping a pole. There appear to be many different kinds of integrin (cell biologists have identified 20), each with a slightly different shaped "hand." The exact component of the matrix that a given cell binds to depends on which combination of integrins that cell has in its plasma membrane.

Anchoring junctions attach the cytoskeleton of a cell to the matrix surrounding the cell, or to the cytoskeleton of another cell.

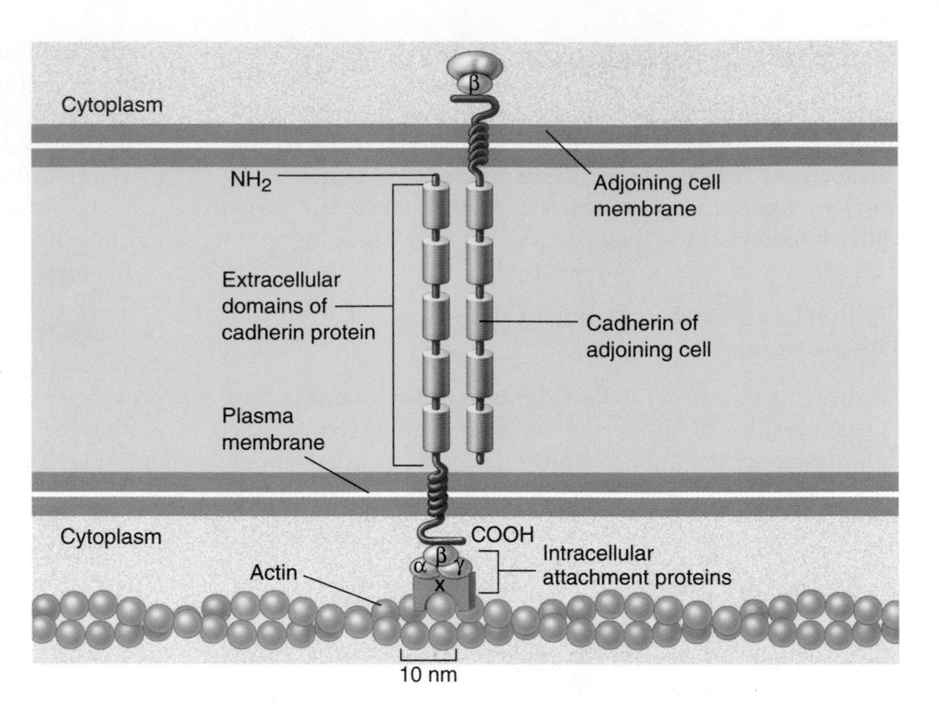

FIGURE 7.19
A cadherin-mediated junction. The cadherin molecule is anchored to actin in the cytoskeleton and passes through the membrane to interact with the cadherin of an adjoining cell.

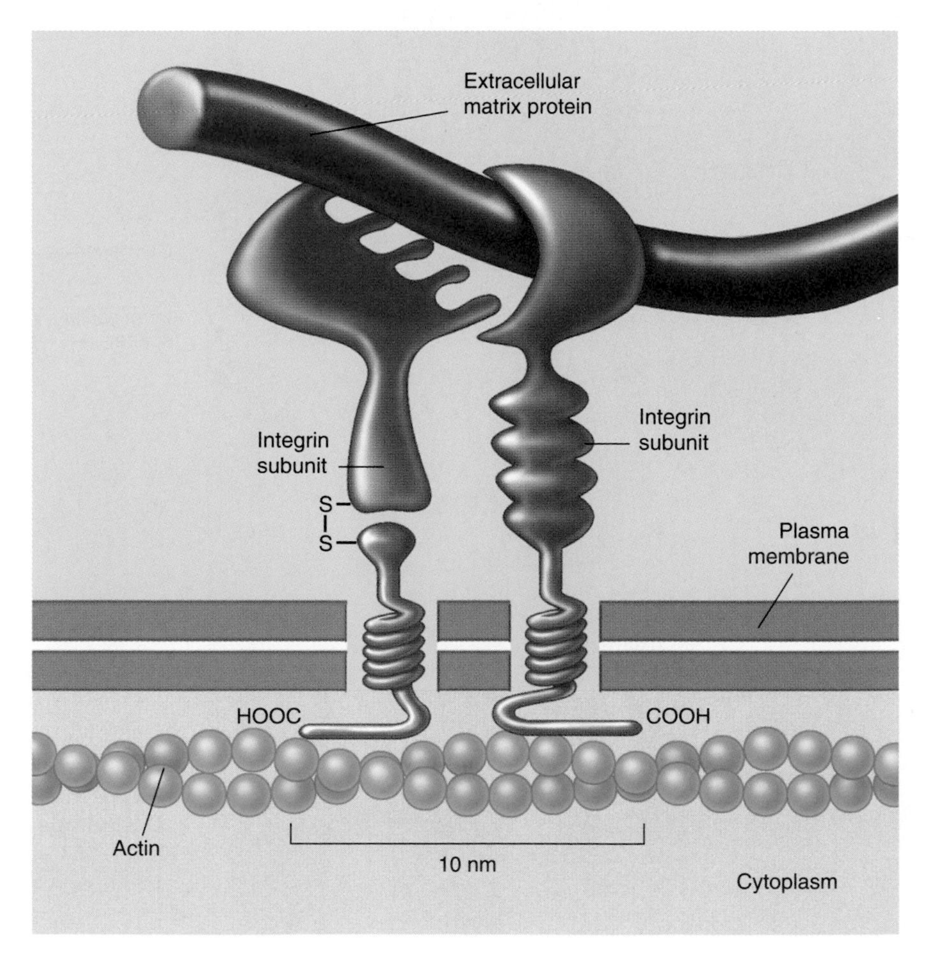

FIGURE 7.20
An integrin-mediated junction. These adherens junctions link the actin filaments inside cells to their neighbors and to the extracellular matrix.

Communicating Junctions

Many cells communicate with adjacent cells through direct connections, called **communicating junctions.** In these junctions, a chemical signal passes directly from one cell to an adjacent one. Communicating junctions establish direct physical connections that link the cytoplasms of two cells together, permitting small molecules or ions to pass from one to the other. In animals, these direct communication channels between cells are called gap junctions. In plants, they are called plasmodesmata.

Gap Junctions in Animals

Communicating junctions called gap junctions are composed of structures called connexons, complexes of six identical transmembrane proteins (figure 7.21). The proteins in a connexon are arranged in a circle to create a channel through the plasma membrane that protrudes several nanometers from the cell surface. A gap junction forms when the connexons of two cells align perfectly, creating an open channel spanning the plasma membranes of both cells. Gap junctions provide passageways large enough to permit small substances, such as simple sugars and amino acids, to pass from the cytoplasm of one cell to that of the next, yet small enough to prevent the passage of larger molecules such as proteins. The connexons hold the plasma membranes of the paired cells about 4 nanometers apart, in marked contrast to the more-or-less direct contact between the lipid bilayers in a tight junction.

Gap junction channels are dynamic structures that can open or close in response to a variety of factors, including Ca^{++} and H^{+} ions. This gating serves at least one important function. When a cell is damaged, its plasma membrane often becomes leaky. Ions in high concentrations outside the cell, such as Ca^{++}, flow into the damaged cell and shut its gap junction channels. This isolates the cell and so prevents the damage from spreading to other cells.

Plasmodesmata in Plants

In plants, cell walls separate every cell from all others. Cell-cell junctions occur only at holes or gaps in the walls, where the plasma membranes of adjacent cells can come into contact with each other. Cytoplasmic connections that form across the touching plasma membranes are called plasmodesmata (figure 7.22). The majority of living cells within a higher plant are connected with their neighbors by these junctions. Plasmodesmata function much like gap junctions in animal cells, although their structure is more complex. Unlike gap junctions, plasmodesmata are lined with plasma membrane and contain a central tubule that connects the endoplasmic reticulum of the two cells.

Communicating junctions permit the controlled passage of small molecules or ions between cells.

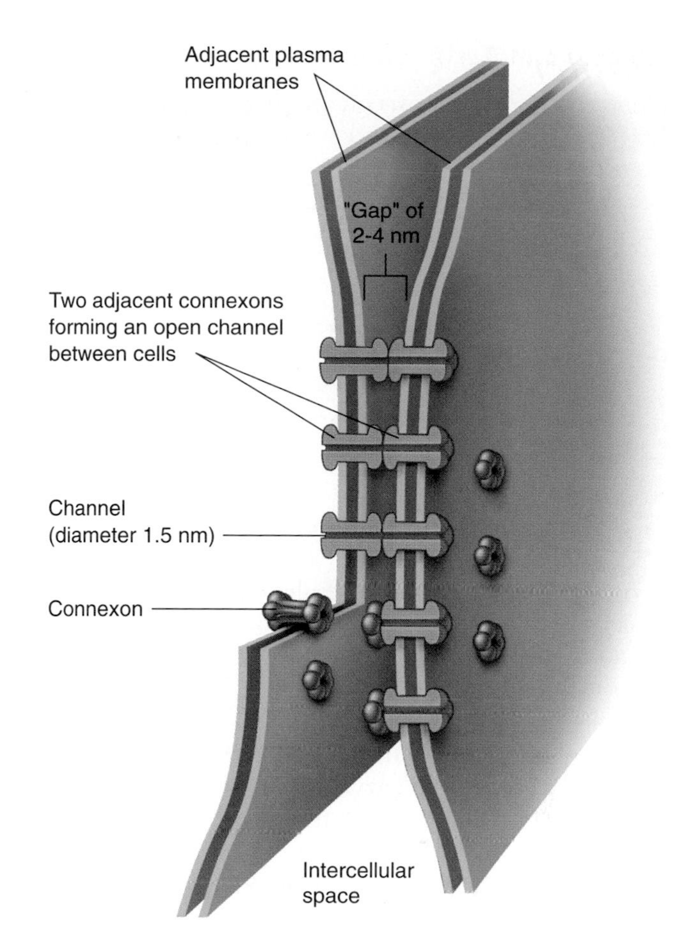

FIGURE 7.21
Gap junctions. Connexons in gap junctions create passageways that connect the cytoplasms of adjoining cells. Gap junctions readily allow the passage of small molecules and ions required for rapid communication (such as in heart tissue), but do not allow the passage of larger molecules like proteins.

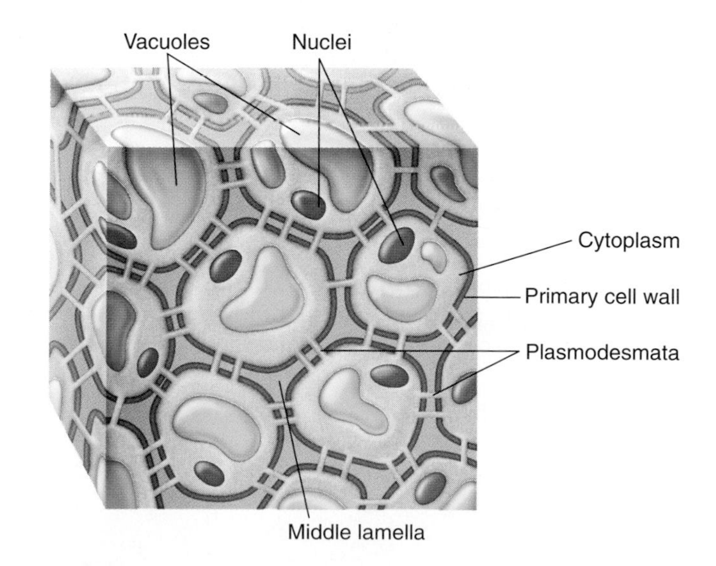

FIGURE 7.22
Plasmodesmata. Plant cells can communicate through specialized openings in their cell walls, called plasmodesmata, where the cytoplasms of adjoining cells are connected (see figure 7.16*c*).

Chapter 7

www.mhhe.com/raven6/resources7.mhtml

Summary	*Questions*	*Media Resources*

7.1 Cells signal one another with chemicals.

- Cell signaling is accomplished through the recognition of signal molecules by target cells.

1. What determines which signal molecules in the extracellular environment a cell will respond to?

2. How do paracrine, endocrine, and synaptic signaling differ?

- Cell Interactions

- Student Research: Retrograde Messengers between Nerve Cells

7.2 Proteins in the cell and on its surface receive signals from other cells.

- The binding of a signal molecule to an intracellular receptor usually initiates transcription of specific regions of DNA, ultimately resulting in the production of specific proteins.
- Cell surface receptors bind to specific molecules in the extracellular fluid. In some cases, this binding causes the receptor to enzymatically alter other (usually internal) proteins, typically through phosphorylation.
- G proteins behave as intracellular shuttles, moving from an activated receptor to other areas in the cell.

3. Describe two of the ways in which intracellular receptors control cell activities.

4. What structural features are characteristic of chemically gated ion channels, and how are these features related to the function of the channels?

5. What are G proteins? How do they participate in cellular responses mediated by G-protein-linked receptors?

- Scientists on Science: G Proteins

7.3 Follow the journey of information into the cell.

- There are usually several amplifying steps between the binding of a signal molecule to a cell surface receptor and the response of the cell. These steps often involve phosphorylation by protein kinases.

6. How does the binding of a single signal molecule to a cell surface receptor result in an amplified response within the target cell?

- Exploration: Cell-Cell Interactions

7.4 Cell surface proteins mediate cell-cell interactions.

- Cell surface markers, ususally glycolipids, identify a cell as belonging to a specific tissue and MHC proteins on the cell surface identify a cell as "self" or "nonself."
- Tight junctions and desmosomes enable cells to adhere in tight, leakproof sheets, holding the cells together such that materials cannot pass between them.
- Gap junctions (in animals) and plasmodesmata (in plants) permit small substances to pass directly from cell to cell through special passageways.

7. What are the functions of tight junctions? What are the functions of desmosomes and adherens junctions, and what proteins are involved in these junctions?

8. What are the molecular components that make up gap junctions? What sorts of substances can pass through gap junctions?

9. Where are plasmodesmata found? What cellular constituents are found in plasmodesmata?

- Art Quizzes:
  -Animal Cell Junctions
  -Plasmodesmata

Part **III**

Energetics

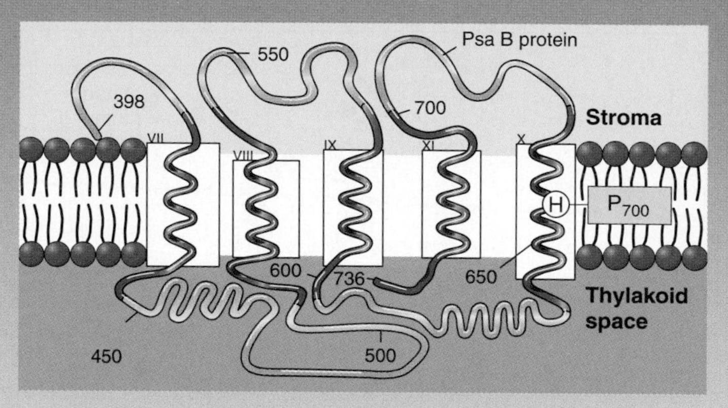

The proposed antenna complex of the PsaB protein. Position 656 is a histidine (H) in the tenth pass (helix X) of the PsaB protein across the thylakoid membrane within chloroplasts. This histidine is where the PsaB protein makes contact with a P_{700} chlorophyll molecule.

Real People Doing Real Science

How Do Proteins Help Chlorophyll Carry Out Photosynthesis?

Much public attention in recent years has been focused on high-profile science—headline-creating advances in the Human Genome Project, genetic engineering, and the battle against AIDS and cancer. Meanwhile, great advances have been made more quietly in other areas of biology. Among the greatest of these achievements has been the unmasking in the last decade of the underlying mechanism of photosynthesis.

In photosynthesis, photons of light are absorbed by chlorophyll molecules, causing them to donate a high-energy electron that is put to work making NADPH and pumping protons to produce ATP.

When researchers looked at the light-absorbing chlorophylls that carry out photosynthesis more closely, they found the chlorophylls to be arranged in clusters called photosystems, supported by proteins and accessory pigments. Within a photosystem, hundreds of chlorophyll molecules act like antennae, absorbing light and passing the energy they capture inward to a single chlorophyll molecule that acts as the reaction center. This chlorophyll acts as the primary electron donor of photosynthesis. Once it releases a light-energized electron, the complex series of chemical events we call photosynthesis begins, and, like a falling row of dominos, is difficult to stop.

Plants possess two kinds of photosystems that work together to harvest light energy. One of them, called photosystem I, is similar to a simpler photosystem found in photosynthetic bacteria, and is thought to have evolved from it.

Photosystem I has been the subject of intense research. In its reaction center, a pair of chlorophyll molecules act as the trap for photon energy, passing an excited electron on to an acceptor molecule outside the reaction center. This moves the photon energy away from the chlorophylls, and is the key conversion of light energy to chemical energy, the very heart of photosynthesis.

Because the pair of chlorophyll molecules in the reaction center of photosystem I absorb light at a wavelength of 700 nm, they are together given the name P_{700}. The P_{700} dimer is positioned within the photosystem by two related proteins that act as scaffolds. These proteins, discovered less than 10 years ago, turn out to play a pivotal role in the photosynthetic process. Passing back and forth across the internal chloroplast membranes 11 times, they form a molecular frame that positions P_{700} to accept energy from other chlorophyll molecules of the photosystem, and to donate a photo-excited electron to an acceptor molecule outside the photosystem.

Recent research suggests that the role of these scaffold proteins, called PsaA and PsaB, is far more active than the passive support provided by a scaffold. Analysis of highly purified photosystems carried out in 1995 revealed that the distribution of electric charge over the two halves of the P_{700} dimer is highly asymmetric—one chlorophyll molecule exhibits a far greater charge density than the other. Because the two chlorophyll molecules of P_{700} are themselves identical, this suggests that the PsaA and PsaB proteins are actively modulating the physicochemical properties of the chlorophyll.

How can a protein pull off this physical-chemical sleight-of-hand? Just what are these proteins *doing* to the chlorophyll molecules? To look more closely at what is going on, you have to first figure out what part of the protein to look at. One way to get a handle on this problem is to compare the amino acid sequences of PsaA and PsaB with that of the bacterial photosystem from which they are thought to have evolved. It is likely that such an important part of the sequence would have been conserved and will be found in all three.

Several sequences are indeed conserved, but most of them prove not to interact directly with chlorophyll. One, however, is a promising candidate. A single amino acid in the helix X domain (that is, the tenth pass of the PsaB protein across the membrane), dubbed His-656, is conserved in all sequences, and is positioned right where the PsaB protein touches the P_{700} chlorophyll (see above). This amino acid, a histidine, has become the focus of recent efforts to clarify how proteins help chlorophyll carry out photosynthesis.

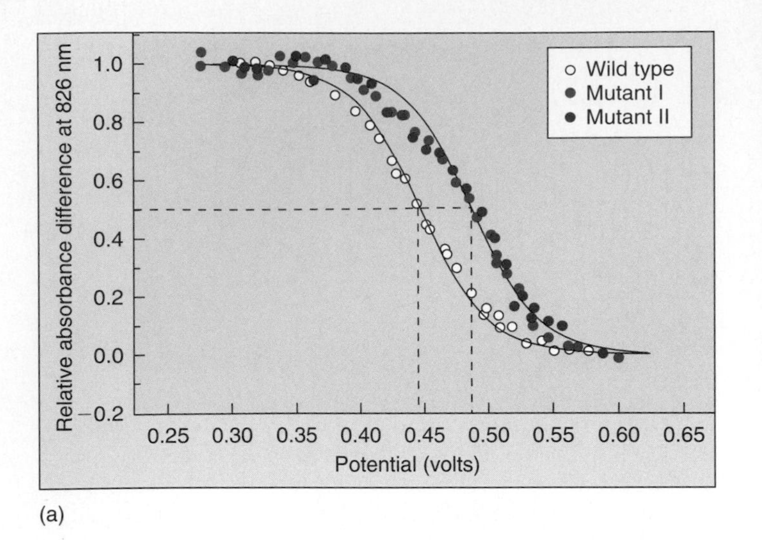

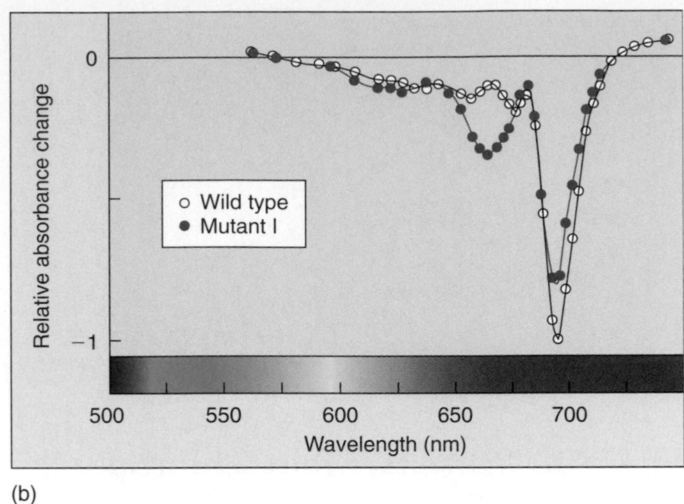

Effect of altering position 656. (*a*) When P_{700} interacts with normal and mutant forms of PsaB, the midpoint potentials are 447±6 mV in the wild type and 487±6 in the mutants, the mutant value being about 40 mV higher. (*b*) The bleaching band (dip in the absorbance) of P_{700} is shifted to the blue (left) and exhibits a new bleaching band at 667 nm when interacting with mutant forms of PsaB.

The Experiment

To determine the importance of His-656, and more generally of the helix X domain of the PsaB protein, Professor Andrew Webber of Arizona State University, working with his research team and the group of Professor Wolfgang Lubitz at Technische Universitat Berlin, has created site-directed mutations of His-656 in the photosynthetic protist *Chlamydomonas reinhardtii*. *C. reinhardtii* is widely used to study photosynthesis because of the ease with which lab experiments can be done.

Webber and his collaborators set out to change the amino acid located at position 656 of PsaB, and then to look and see what effect the change had on photosynthesis. If His-656 indeed plays a critical role in modifying the P_{700} chlorophylls, then a change at that position to a different amino acid should have profound effects.

Creating PsaB Proteins Mutant at Position 656. The first and key step in Webber's experimental approach was to genetically alter the chloroplast of *C. reinhardtii*, introducing a mutation of the PsaB gene at the His-656 position. To do this, the team employed site-specific mutagenesis to construct mutant plasmids pHN(B656) and pHS(B656), inserting a gene carrying either the Ser or Asn amino acids in place of His. The two mutation-carrying plasmids were then cloned into *C. reinhardtii*, cells carrying the mutant plasmids isolated, and presence of the mutated gene directly confirmed by sequencing the DNA.

Characterizing the Effects of 656 Mutations. Once researchers confirmed that the *C. reinhardtii* chloroplast DNA now contained the mutant forms of the PsaB gene, they proceeded to test the function of the mutated PsaB protein in coordinating P_{700}, examining interior thylakoid membranes isolated from the chloroplasts. To do this, the researchers measured the oxidation midpoint potentials of the P_{700} complexes, an indication of how tightly the chlorophyll molecules are holding onto their electrons.

The researchers further characterized the P_{700} complexes by measuring the changes in absorbance of the mutants versus the wild type to see if the mutations altered the spectral properties of the P_{700} chlorophylls.

The Results

The results of the examination of the oxidation midpoint potentials revealed that the influence of the PsaB protein on P_{700} had been profoundly altered by the mutations. The midpoint potential of P_{700} in the wild type was determined to be 447±6 mV, while the midpoint potential had increased to 487±6 mV in both the PsaB mutant I, HN(B656), and the PsaB mutant II, HS(B656) (see graph *a*). This increase in the oxidation midpoint potential by approximately 40 mV indicates that the mutations to the His residue significantly altered the redox property of P_{700} and, therefore, that His-656 is closely interacting with one of the chlorophyll molecules of the P_{700} dimer.

These results and this conclusion are further supported by changes observed in the spectral properties of the mutants and wild type (see graph *b*). There is a reduction and a slight shift in the 696 nm bleaching band (dip in absorbance) in PsaB mutant I toward the blue end of the spectrum and a new bleaching band appearing at 667 nm, both changes in the spectral properties of chlorophyll induced by the mutational changes in the PsaB protein.

Ultimately, the researchers conclude that the His-656 of PsaB directly coordinates the central magnesium atom of one of the two chlorophyll molecules of P_{700}. Their results are consistent with a model of photosystem I in which the first six spans of PsaB constitute an antenna domain for receiving energy from other chlorophylls and the last five membrane spans interact with the P_{700} reaction complex.

To explore this experiment further, go to the Virtual Lab at www.mhhe.com/raven6/vlab3.mhtml

Energy and Metabolism

Concept Outline

8.1 The laws of thermodynamics describe how energy changes.

The Flow of Energy in Living Things. Potential energy is present in the electrons of atoms, and so can be transferred from one molecule to another.

The Laws of Thermodynamics. Energy is never lost but as it is transferred, more and more of it dissipates as heat, a disordered form of energy.

Free Energy. In a chemical reaction, the energy released or supplied is the difference in bond energies between reactants and products, corrected for disorder.

Activation Energy. To start a chemical reaction, an input of energy is required to destabilize existing chemical bonds.

8.2 Enzymes are biological catalysts.

Enzymes. Globular proteins called enzymes catalyze chemical reactions within cells.

How Enzymes Work. Enzymes have sites on their surface shaped to fit their substrates snugly, forcing chemically reactive groups close enough to facilitate a reaction.

Enzymes Take Many Forms. Some enzymes are associated in complex groups; others are not even proteins.

Factors Affecting Enzyme Activity. Each enzyme works most efficiently at its optimal temperature and pH. Metal ions or other substances often help enzymes carry out their catalysis.

8.3 ATP is the energy currency of life.

What Is ATP? Cells store and release energy from the phosphate bonds of ATP, the energy currency of the cell.

8.4 Metabolism is the chemical life of a cell.

Biochemical Pathways: The Organizational Units of Metabolism. Biochemical pathways, where the product of one reaction becomes the substrate for the next, are the organizational units of metabolism.

The Evolution of Metabolism. The major metabolic processes evolved over a long period, building on what came before.

FIGURE 8.1
Lion at lunch. Energy that this lion extracts from its meal of giraffe will be used to power its roar, fuel its running, and build a bigger lion.

Life can be viewed as a constant flow of energy, channeled by organisms to do the work of living. Each of the significant properties by which we define life—order, growth, reproduction, responsiveness, and internal regulation—requires a constant supply of energy (figure 8.1). Deprived of a source of energy, life stops. Therefore, a comprehensive study of life would be impossible without discussing bioenergetics, the analysis of how energy powers the activities of living systems. In this chapter, we will focus on energy—on what it is and how organisms capture, store, and use it.

The Flow of Energy in Living Things

Energy is defined as the capacity to do work. It can be considered to exist in two states. **Kinetic energy** is the energy of motion (figure 8.2). Moving objects perform work by causing other matter to move. **Potential energy** is stored energy. Objects that are not actively moving but have the capacity to do so possess potential energy. A boulder perched on a hilltop has potential energy; as it begins to roll downhill, some of its potential energy is converted into kinetic energy. Much of the work that living organisms carry out involves transforming potential energy to kinetic energy.

Energy can take many forms: mechanical energy, heat, sound, electric current, light, or radioactive radiation. Because it can exist in so many forms, there are many ways to measure energy. The most convenient is in terms of heat, because all other forms of energy can be converted into heat. In fact, the study of energy is called **thermodynamics,** meaning heat changes. The unit of heat most commonly employed in biology is the **kilocalorie** (kcal). One kilocalorie is equal to 1000 calories (cal), and one calorie is the heat required to raise the temperature of one gram of water one degree Celsius (°C). (It is important not to confuse calories with a term related to diets and nutrition, the Calorie with a capital C, which is actually another term for kilocalorie.) Another energy unit, often used in physics, is the **joule;** one joule equals 0.239 cal.

Oxidation-Reduction

Energy flows into the biological world from the sun, which shines a constant beam of light on the earth. It is estimated that the sun provides the earth with more than 13×10^{23} calories per year, or 40 million billion calories per second! Plants, algae, and certain kinds of bacteria capture a fraction of this energy through photosynthesis. In photosynthesis, energy garnered from sunlight is used to combine small molecules (water and carbon dioxide) into more complex molecules (sugars). The energy is stored as potential energy in the covalent bonds between atoms in the sugar molecules. Recall from chapter 2 that an atom consists of a central nucleus surrounded by one or more orbiting electrons, and a covalent bond forms when two atomic nuclei share valence electrons. Breaking such a bond requires energy to pull the nuclei apart. Indeed, the strength of a covalent bond is measured by the amount of energy required to break it. For example, it takes 98.8 kcal to break one mole (6.023×10^{23}) of carbon-hydrogen (C—H) bonds.

During a chemical reaction, the energy stored in chemical bonds may transfer to new bonds. In some of these reactions, electrons actually pass from one atom or molecule to another. When an atom or molecule loses an electron, it is said to be oxidized, and the process by which this occurs is called **oxidation.** The name reflects the fact that in biological systems oxygen, which attracts electrons strongly, is the most common electron acceptor.

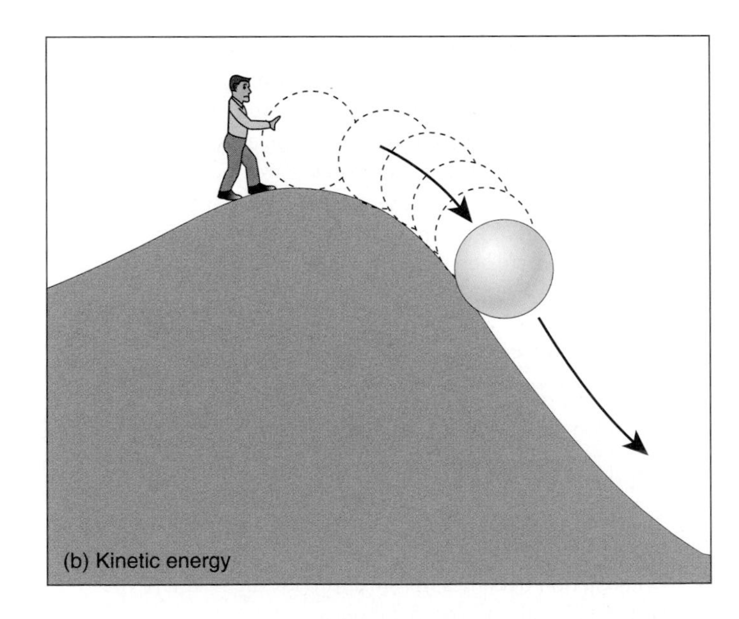

FIGURE 8.2
Potential and kinetic energy. (*a*) Objects that have the capacity to move but are not moving have potential energy. The energy required to move the ball up the hill is stored as potential energy. (*b*) Objects that are in motion have kinetic energy. The stored energy is released as kinetic energy as the ball rolls down the hill.

Energy-rich molecule

1. Enzymes that harvest hydrogen atoms have a binding site for NAD+ located near another binding site. NAD+ and an energy-rich molecule bind to the enzyme.

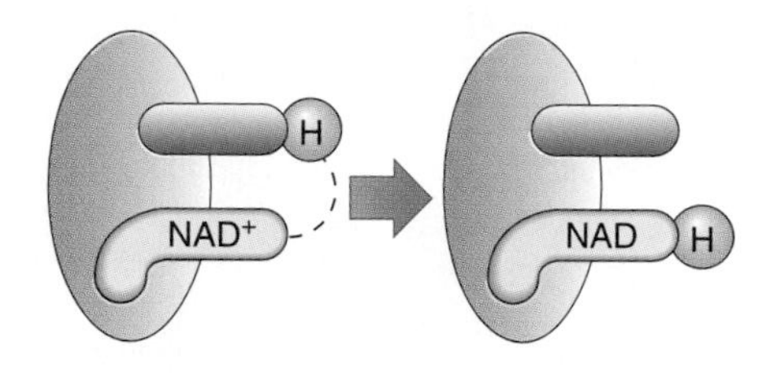

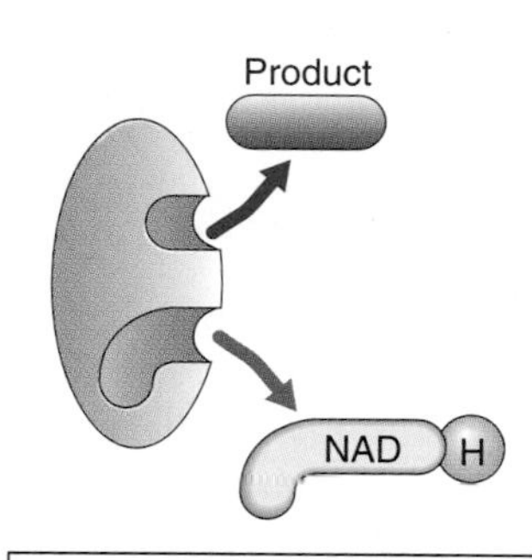

FIGURE 8.3
An oxidation-reduction reaction. Cells use a chemical called NAD^+ to carry out oxidation-reduction reactions. Energetic electrons are often paired with a proton as a hydrogen atom. Molecules that gain energetic electrons are said to be reduced, while ones that lose energetic electrons are said to be oxidized. NAD^+ oxidizes energy-rich molecules by acquiring their hydrogens (in the figure, this proceeds $1 \rightarrow 2 \rightarrow 3$) and then reduces other molecules by giving the hydrogens to them (in the figure, this proceeds $3 \rightarrow 2 \rightarrow 1$).

Conversely, when an atom or molecule gains an electron, it is said to be reduced, and the process is called **reduction.** Oxidation and reduction always take place together, because every electron that is lost by an atom through oxidation is gained by some other atom through reduction. Therefore, chemical reactions of this sort are called **oxidation-reduction (redox) reactions** (figure 8.3). Energy is transferred from one molecule to another via redox reactions. The reduced form of a molecule thus has a higher level of energy than the oxidized form (figure 8.4).

Oxidation-reduction reactions play a key role in the flow of energy through biological systems because the electrons that pass from one atom to another carry energy with them. The amount of energy an electron possesses depends on how far it is from the nucleus and how strongly the nucleus attracts it. Light (and other forms of energy) can add energy to an electron and boost it to a higher energy level. When this electron departs from one atom (oxidation) and moves to another (reduction), the electron's added energy is transferred with it, and the electron orbits the second atom's nucleus at the higher energy level. The added energy is stored as potential chemical energy that the atom can later release when the electron returns to its original energy level.

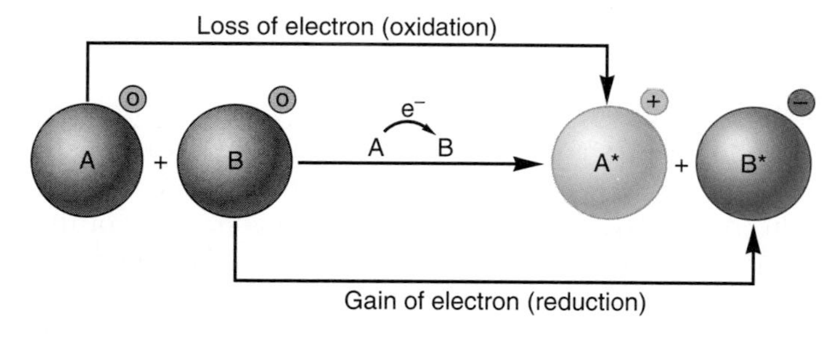

FIGURE 8.4
Redox reactions. Oxidation is the loss of an electron; reduction is the gain of an electron. In this example, the charges of molecules A and B are shown in small circles to the upper right of each molecule. Molecule A loses energy as it loses an electron, while molecule B gains energy as it gains an electron.

Energy is the capacity to do work, either actively (kinetic energy) or stored for later use (potential energy). Energy is transferred with electrons. Oxidation is the loss of an electron; reduction is the gain of one.

The Laws of Thermodynamics

Running, thinking, singing, reading these words—all activities of living organisms involve changes in energy. A set of universal laws we call the Laws of Thermodynamics govern all energy changes in the universe, from nuclear reactions to the buzzing of a bee.

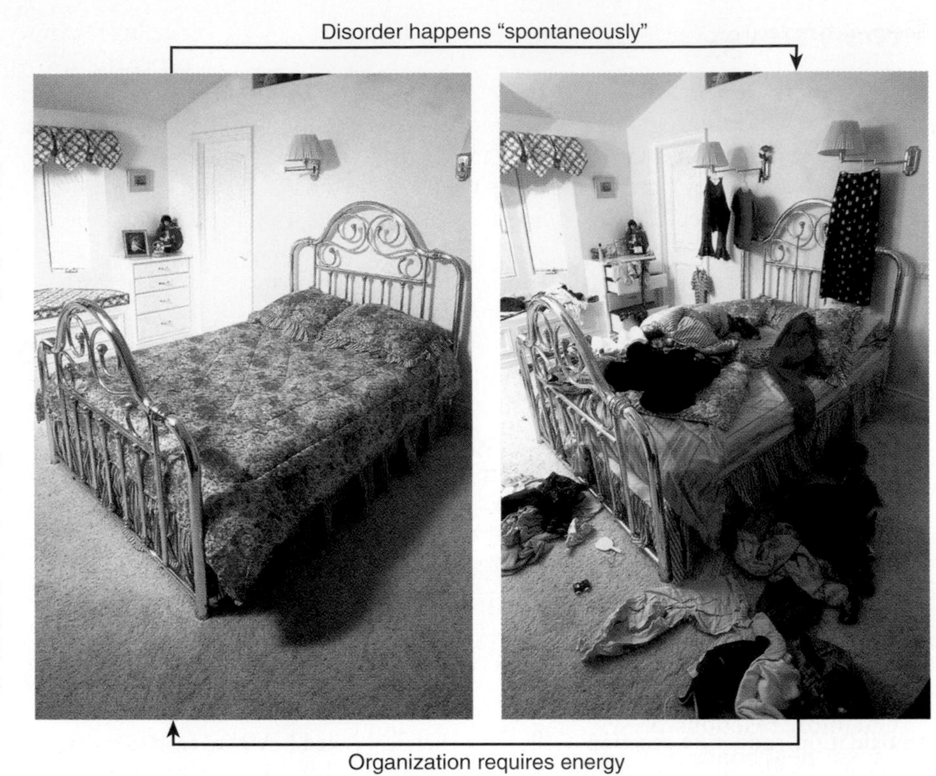
Disorder happens "spontaneously"
Organization requires energy

FIGURE 8.5
Entropy in action. As time elapses, a child's room becomes more disorganized. It takes effort to clean it up.

The First Law of Thermodynamics

The first of these universal laws, the **First Law of Thermodynamics,** concerns the amount of energy in the universe. It states that energy cannot be created or destroyed; it can only change from one form to another (from potential to kinetic, for example). The total amount of energy in the universe remains constant.

The lion eating a giraffe in figure 8.1 is in the process of acquiring energy. Rather than creating new energy or capturing the energy in sunlight, the lion is merely transferring some of the potential energy stored in the giraffe's tissues to its own body (just as the giraffe obtained the potential energy stored in the plants it ate while it was alive). Within any living organism, this chemical potential energy can be shifted to other molecules and stored in different chemical bonds, or it can convert into other forms, such as kinetic energy, light, or electricity. During each conversion, some of the energy dissipates into the environment as **heat,** a measure of the random motions of molecules (and, hence, a measure of one form of kinetic energy). Energy continuously flows through the biological world in one direction, with new energy from the sun constantly entering the system to replace the energy dissipated as heat.

Heat can be harnessed to do work only when there is a heat gradient, that is, a temperature difference between two areas (this is how a steam engine functions). Cells are too small to maintain significant internal temperature differences, so heat energy is incapable of doing the work of cells. Thus, although the total amount of energy in the universe remains constant, the energy available to do work decreases, as progressively more of it dissipates as heat.

The Second Law of Thermodynamics

The **Second Law of Thermodynamics** concerns this transformation of potential energy into heat, or random molecular motion. It states that the disorder (more formally called *entropy*) in the universe is continuously increasing. Put simply, disorder is more likely than order. For example, it is much more likely that a column of bricks will tumble over than that a pile of bricks will arrange themselves spontaneously to form a column. In general, energy transformations proceed spontaneously to convert matter from a more ordered, less stable form, to a less ordered, more stable form (figure 8.5).

Entropy

Entropy is a measure of the disorder of a system, so the Second Law of Thermodynamics can also be stated simply as "entropy increases." When the universe formed, it held all the potential energy it will ever have. It has become progressively more disordered ever since, with every energy exchange increasing the amount of entropy.

The First Law of Thermodynamics states that energy cannot be created or destroyed; it can only undergo conversion from one form to another. The Second Law of Thermodynamics states that disorder (entropy) in the universe is increasing. As energy is used, more and more of it is converted to heat, the energy of random molecular motion.

Free Energy

It takes energy to break the chemical bonds that hold the atoms in a molecule together. Heat energy, because it increases atomic motion, makes it easier for the atoms to pull apart. Both chemical bonding and heat have a significant influence on a molecule, the former reducing disorder and the latter increasing it. The net effect, the amount of energy actually available to break and subsequently form other chemical bonds, is called the **free energy** of that molecule. In a more general sense, free energy is defined as the energy available to do work in any system. In a molecule within a cell, where pressure and volume usually do not change, the free energy is denoted by the symbol G (for "Gibbs' free energy," which limits the system being considered to the cell). G is equal to the energy contained in a molecule's chemical bonds (called *enthalpy* and designated **H**) minus the energy unavailable because of disorder (called *entropy* and given the symbol **S**) times the absolute temperature, **T**, in degrees Kelvin (K = °C + 273):

$$G = H - TS$$

Chemical reactions break some bonds in the reactants and form new bonds in the products. Consequently, reactions can produce changes in free energy. When a chemical reaction occurs under conditions of constant temperature, pressure, and volume—as do most biological reactions—the change in free energy (ΔG) is simply:

$$\Delta G = \Delta H - T\,\Delta S$$

The change in free energy, or ΔG, is a fundamental property of chemical reactions.

In some reactions, the ΔG is positive. This means that the products of the reaction contain *more* free energy than the reactants; the bond energy (H) is higher or the disorder (S) in the system is lower. Such reactions do not proceed spontaneously because they require an input of energy. Any reaction that requires an input of energy is said to be **endergonic** ("inward energy").

In other reactions, the ΔG is negative. The products of the reaction contain less free energy than the reactants; either the bond energy is lower or the disorder is higher, or both. Such reactions tend to proceed spontaneously. Any chemical reaction will tend to proceed spontaneously if the difference in disorder ($T\,\Delta S$) is *greater* than the difference in bond energies between reactants and products (ΔH). Note that spontaneous does not mean the same thing as instantaneous. A spontaneous reaction may proceed very slowly. These reactions release the excess free energy as heat and are thus said to be **exergonic** ("outward energy"). Figure 8.6 sums up these reactions.

Free energy is the energy available to do work. Within cells, the change in free energy (ΔG) is the difference in bond energies between reactants and products (ΔH), minus any change in the degree of disorder of the system ($T\,\Delta S$). Any reaction whose products contain less free energy than the reactants (ΔG is negative) will tend to proceed spontaneously.

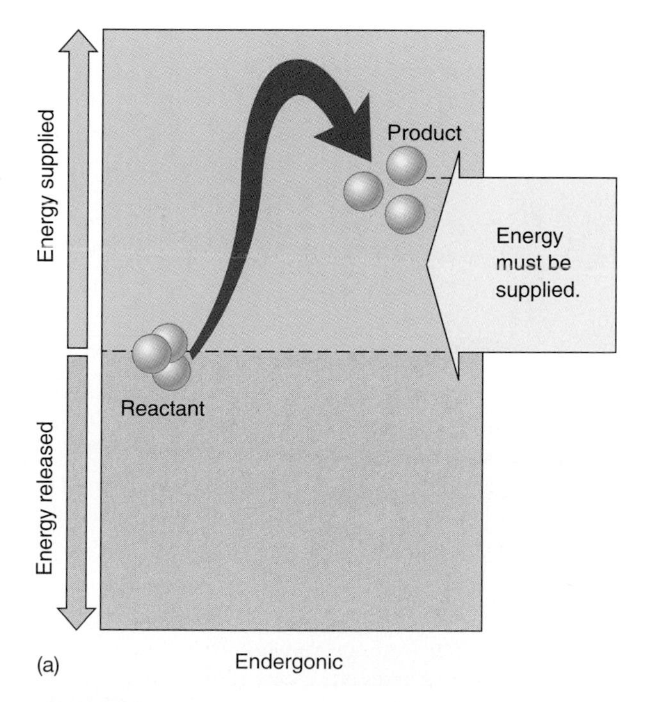

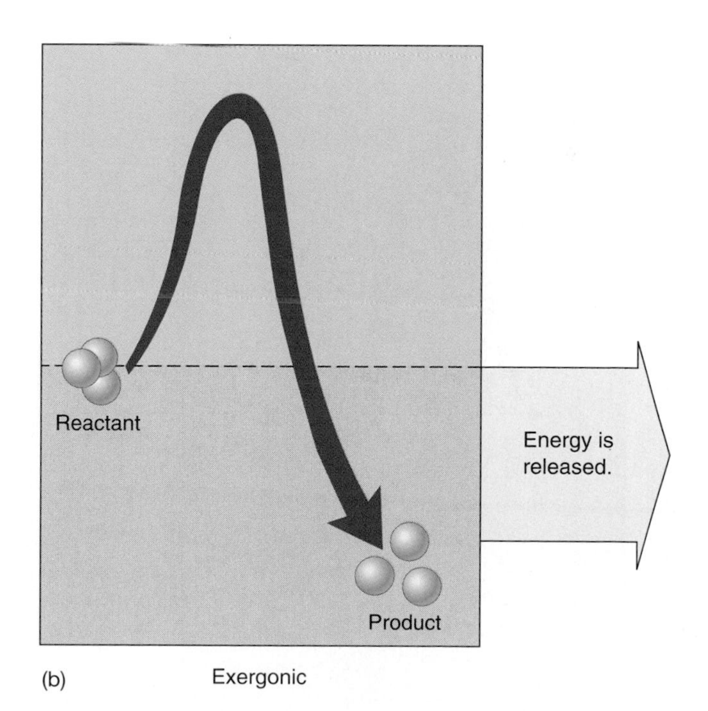

FIGURE 8.6
Energy in chemical reactions. (*a*) In an endergonic reaction, the products of the reaction contain more energy than the reactants, and the extra energy must be supplied for the reaction to proceed. (*b*) In an exergonic reaction, the products contain less energy than the reactants, and the excess energy is released.

Activation Energy

If all chemical reactions that release free energy tend to occur spontaneously, why haven't all such reactions already occurred? One reason they haven't is that most reactions require an input of energy to get started. Before it is possible to form new chemical bonds, even bonds that contain less energy, it is first necessary to break the existing bonds, and that takes energy. The extra energy required to destabilize existing chemical bonds and initiate a chemical reaction is called **activation energy** (figure 8.7*a*).

The rate of an exergonic reaction depends on the activation energy required for the reaction to begin. Reactions with larger activation energies tend to proceed more slowly because fewer molecules succeed in overcoming the initial energy hurdle. Activation energies are not constant, however. Stressing particular chemical bonds can make them easier to break. The process of influencing chemical bonds in a way that lowers the activation energy needed to initiate a reaction is called **catalysis,** and substances that accomplish this are known as catalysts (figure 8.7*b*).

Catalysts cannot violate the basic laws of thermodynamics; they cannot, for example, make an endergonic reaction proceed spontaneously. By reducing the activation energy, a catalyst accelerates both the forward and the reverse reactions by exactly the same amount. Hence, it does not alter the proportion of reactant ultimately converted into product.

To grasp this, imagine a bowling ball resting in a shallow depression on the side of a hill. Only a narrow rim of dirt below the ball prevents it from rolling down the hill. Now imagine digging away that rim of dirt. If you remove enough dirt from below the ball, it will start to roll down the hill—but removing dirt from below the ball will *never* cause the ball to roll UP the hill! Removing the lip of dirt simply allows the ball to move freely; gravity determines the direction it then travels. Lowering the resistance to the ball's movement will promote the movement dictated by its position on the hill.

Similarly, the direction in which a chemical reaction proceeds is determined solely by the difference in free energy. Like digging away the soil below the bowling ball on the hill, catalysts reduce the energy barrier preventing the reaction from proceeding. Catalysts don't favor endergonic reactions any more than digging makes the hypothetical bowling ball roll uphill. Only exergonic reactions can proceed spontaneously, and catalysts cannot change that. What catalysts *can* do is make a reaction proceed much faster.

The rate of a reaction depends on the activation energy necessary to initiate it. Catalysts reduce the activation energy and so increase the rates of reactions, although they do not change the final proportions of reactants and products.

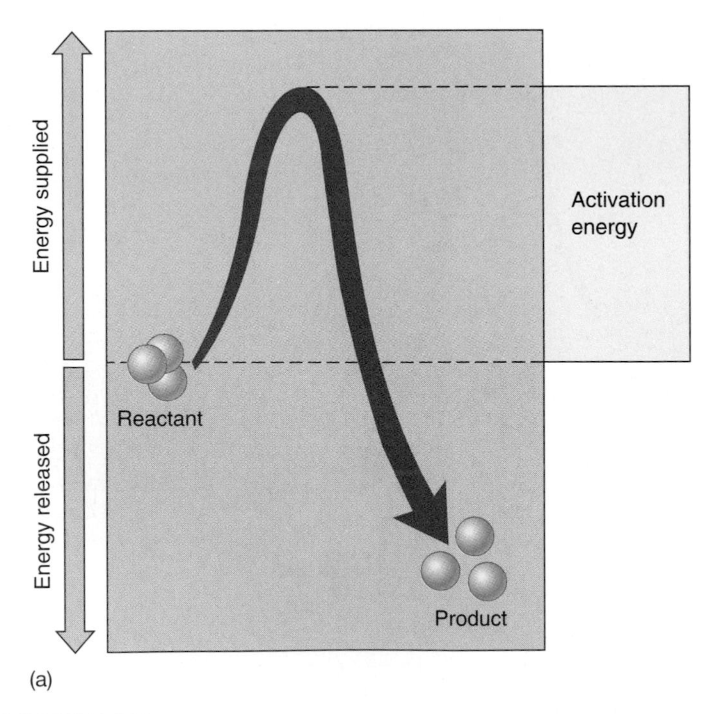

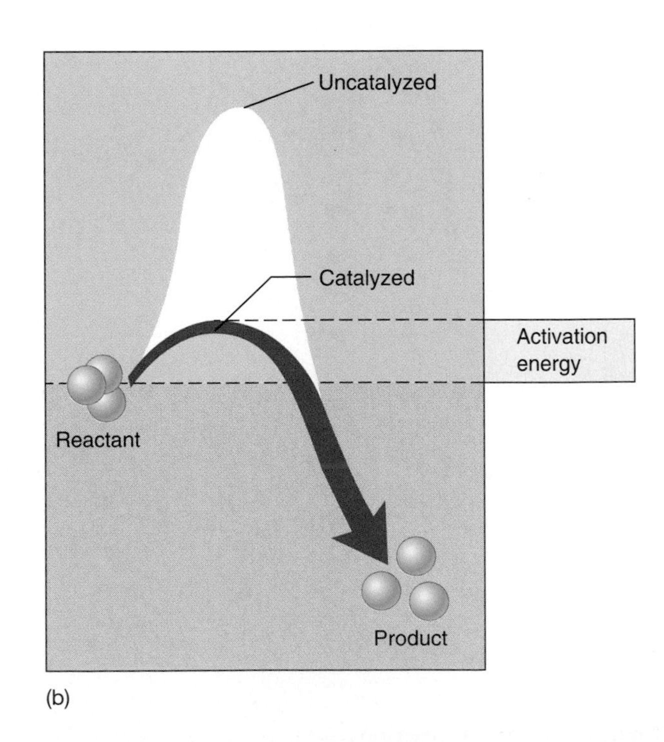

FIGURE 8.7
Activation energy and catalysis. (*a*) Exergonic reactions do not necessarily proceed rapidly because energy must be supplied to destabilize existing chemical bonds. This extra energy is the activation energy for the reaction. (*b*) Catalysts accelerate particular reactions by lowering the amount of activation energy required to initiate the reaction.

8.2 Enzymes are biological catalysts.

Enzymes

The chemical reactions within living organisms are regulated by controlling the points at which catalysis takes place. Life itself is, therefore, regulated by catalysts. The agents that carry out most of the catalysis in living organisms are proteins called **enzymes.** (There is increasing evidence that some types of biological catalysis are carried out by RNA molecules.) The unique three-dimensional shape of an enzyme enables it to stabilize a temporary association between **substrates,** the molecules that will undergo the reaction. By bringing two substrates together in the correct orientation, or by stressing particular chemical bonds of a substrate, an enzyme lowers the activation energy required for new bonds to form. The reaction thus proceeds much more quickly than it would without the enzyme. Because the enzyme itself is not changed or consumed in the reaction, only a small amount of an enzyme is needed, and it can be used over and over.

As an example of how an enzyme works, let's consider the reaction of carbon dioxide and water to form carbonic acid. This important enzyme-catalyzed reaction occurs in vertebrate red blood cells:

$$\underset{\text{carbon dioxide}}{CO_2} + \underset{\text{water}}{H_2O} \longrightarrow \underset{\text{carbonic acid}}{H_2CO_3}$$

This reaction may proceed in either direction, but because it has a large activation energy, the reaction is very slow in the absence of an enzyme: perhaps 200 molecules of carbonic acid form in an hour in a cell. Reactions that proceed this slowly are of little use to a cell. Cells overcome this problem by employing an enzyme within their cytoplasm called *carbonic anhydrase* (enzyme names usually end in "–ase"). Under the same conditions, but in the presence of carbonic anhydrase, an estimated 600,000 molecules of carbonic acid form every *second!* Thus, the enzyme increases the reaction rate more than 10 million times.

Thousands of different kinds of enzymes are known, each catalyzing one or a few specific chemical reactions. By facilitating particular chemical reactions, the enzymes in a cell determine the course of metabolism—the collection of all chemical reactions—in that cell. Different types of cells contain different sets of enzymes, and this difference contributes to structural and functional variations among cell types. The chemical reactions taking place within a red blood cell differ from those that occur within a nerve cell, in part because the cytoplasm and membranes of red blood cells and nerve cells contain different arrays of enzymes.

Cells use proteins called enzymes as catalysts to lower activation energies.

Catalysis: A Closer Look at Carbonic Anhydrase

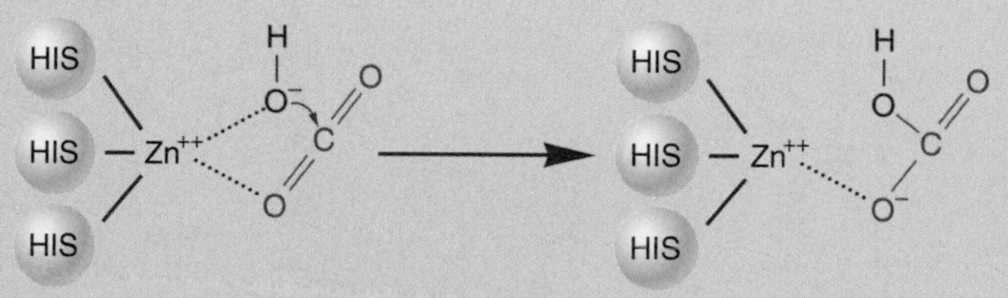

One of the most rapidly acting enzymes in the human body is carbonic anhydrase, which plays a key role in blood by converting dissolved CO_2 into carbonic acid, which dissociates into bicarbonate and hydrogen ions:

$$CO_2 + H_2O \rightarrow H_2CO_3 \rightarrow HCO_3^- + H^+$$

Fully 70% of the CO_2 transported by the blood is transported as bicarbonate ion. This reaction is exergonic, but its energy of activation is significant, so that little conversion to bicarbonate occurs spontaneously. In the presence of the enzyme carbonic anhydrase, however, the rate of the reaction accelerates by a factor of more than 10 million!

How does carbonic anhydrase catalyze this reaction so effectively? The active site of the enzyme is a deep cleft traversing the enzyme, as if it had been cut with the blade of an ax. Deep within the cleft, some 1.5 nm from the surface, are located three histidines, their imidazole (nitrogen ring) groups all pointed at the same place in the center of the cleft. Together they hold a zinc ion firmly in position. This zinc ion will be the cutting blade of the catalytic process.

Here is how the zinc catalyzes the reaction. Immediately adjacent to the position of the zinc atom in the cleft are a group of amino acids that recognize and bind carbon dioxide. The zinc atom interacts with this carbon dioxide molecule, orienting it in the plane of the cleft. Meanwhile, water bound to the zinc is rapidly converted to hydroxide ion. This hydroxide ion is now precisely positioned to attack the carbon dioxide. When it does so, HCO_3^- is formed—and the enzyme is unchanged (figure 8.A).

Carbonic anhydrase is an effective catalyst because it brings its two substrates into close proximity and optimizes their orientation for reaction. Other enzymes use other mechanisms. Many, for example, use charged amino acids to polarize substrates or electronegative amino acids to stress particular bonds. Whatever the details of the reaction, however, the precise positioning of substrates achieved by the particular shape of the enzyme always plays a key role.

FIGURE 8.A

How Enzymes Work

Most enzymes are globular proteins with one or more pockets or clefts on their surface called **active sites** (figure 8.8). Substrates bind to the enzyme at these active sites, forming an **enzyme-substrate complex.** For catalysis to occur within the complex, a substrate molecule must fit precisely into an active site. When that happens, amino acid side groups of the enzyme end up in close proximity to certain bonds of the substrate. These side groups interact chemically with the substrate, usually stressing or distorting a particular bond and consequently lowering the activation energy needed to break the bond. The substrate, now a product, then dissociates from the enzyme.

Proteins are not rigid. The binding of a substrate induces the enzyme to adjust its shape slightly, leading to a better *induced fit* between enzyme and substrate (figure 8.9). This interaction may also facilitate the binding of other substrates; in such cases, the substrate itself "activates" the enzyme to receive other substrates.

Enzymes typically catalyze only one or a few similar chemical reactions because they are specific in their choice of substrates. This specificity is due to the active site of the enzyme, which is shaped so that only a certain substrate molecule will fit into it.

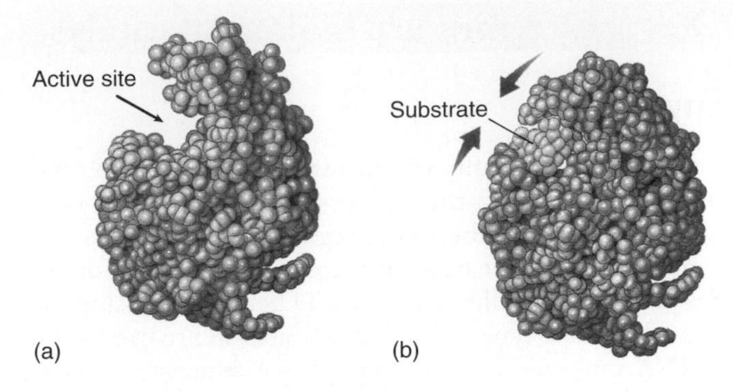

FIGURE 8.8
How the enzyme lysozyme works. (*a*) A groove runs through lysozyme that fits the shape of the polysaccharide (a chain of sugars) that makes up bacterial cell walls. (*b*) When such a chain of sugars, indicated in *yellow*, slides into the groove, its entry induces the protein to alter its shape slightly and embrace the substrate more intimately. This induced fit positions a glutamic acid residue in the protein next to the bond between two adjacent sugars, and the glutamic acid "steals" an electron from the bond, causing it to break.

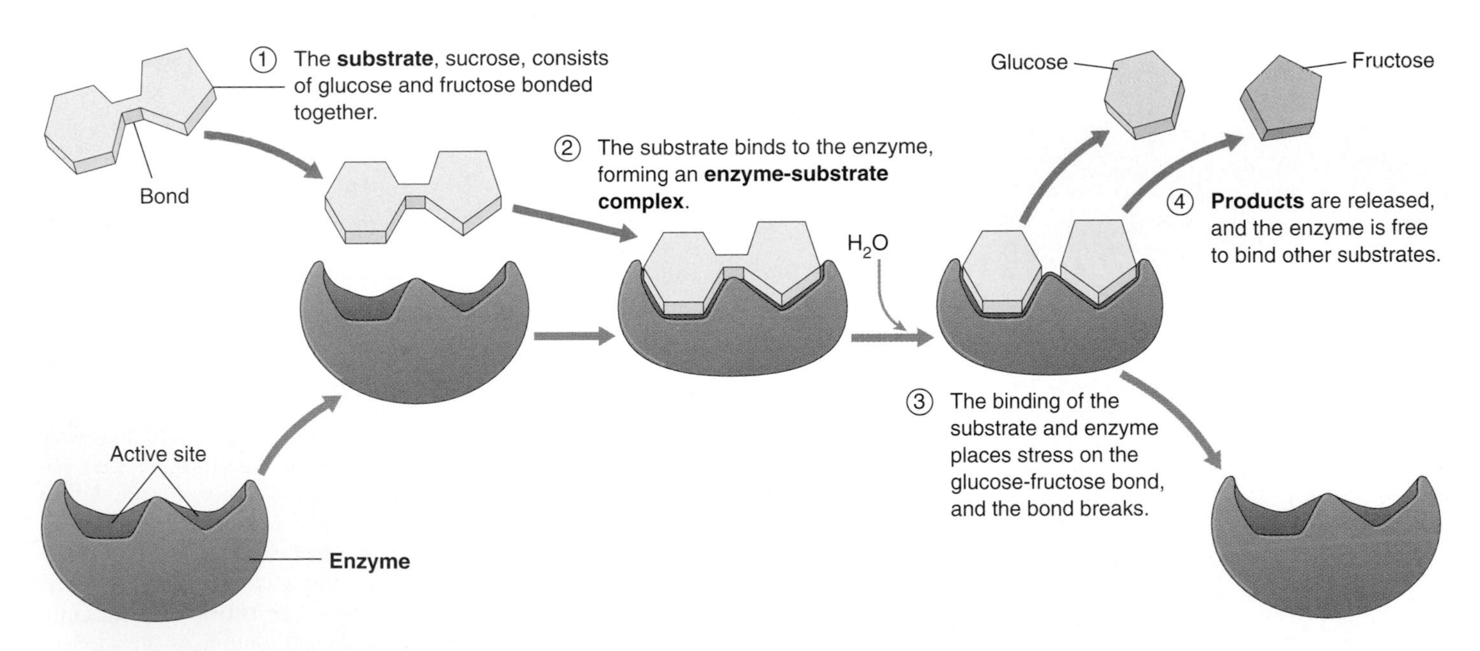

① The **substrate**, sucrose, consists of glucose and fructose bonded together.
Bond
② The substrate binds to the enzyme, forming an **enzyme-substrate complex**.
H_2O
Glucose
Fructose
④ **Products** are released, and the enzyme is free to bind other substrates.
③ The binding of the substrate and enzyme places stress on the glucose-fructose bond, and the bond breaks.
Active site
Enzyme

FIGURE 8.9
The catalytic cycle of an enzyme. Enzymes increase the speed with which chemical reactions occur, but they are not altered themselves as they do so. In the reaction illustrated here, the enzyme sucrase is splitting the sugar sucrose (present in most candy) into two simpler sugars: glucose and fructose. (*1*) First, the sucrose substrate binds to the active site of the enzyme, fitting into a depression in the enzyme surface. (*2*) The binding of sucrose to the active site forms an enzyme-substrate complex and induces the sucrase molecule to alter its shape, fitting more tightly around the sucrose. (*3*) Amino acid residues in the active site, now in close proximity to the bond between the glucose and fructose components of sucrose, break the bond. (*4*) The enzyme releases the resulting glucose and fructose fragments, the products of the reaction, and is then ready to bind another molecule of sucrose and run through the catalytic cycle once again. This cycle is often summarized by the equation: $E + S \leftrightarrow [ES] \leftrightarrow E + P$, where E = enzyme, S = substrate, ES = enzyme-substrate complex, and P = products.

Enzymes Take Many Forms

While many enzymes are suspended in the cytoplasm of cells, free to move about and not attached to any structure, other enzymes function as integral parts of cell structures and organelles.

Multienzyme Complexes

Often in cells the several enzymes catalyzing the different steps of a sequence of reactions are loosely associated with one another in noncovalently bonded assemblies called *multienzyme complexes*. The bacterial pyruvate dehydrogenase multienzyme complex seen in figure 8.10 contains enzymes that carry out three sequential reactions in oxidative metabolism. Each complex has multiple copies of each of the three enzymes—60 protein subunits in all. The many subunits work in concert, like a tiny factory.

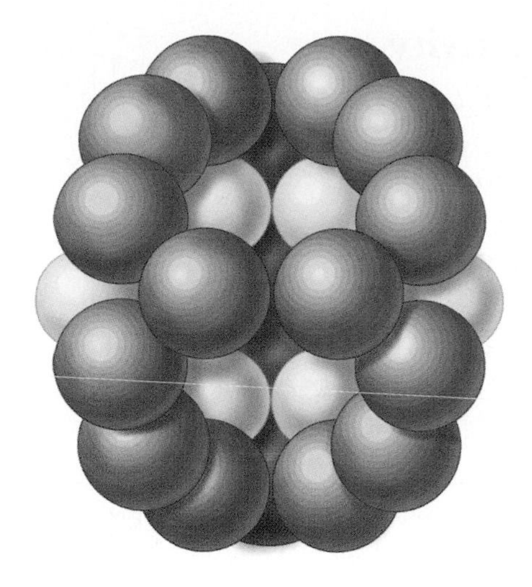
(a)

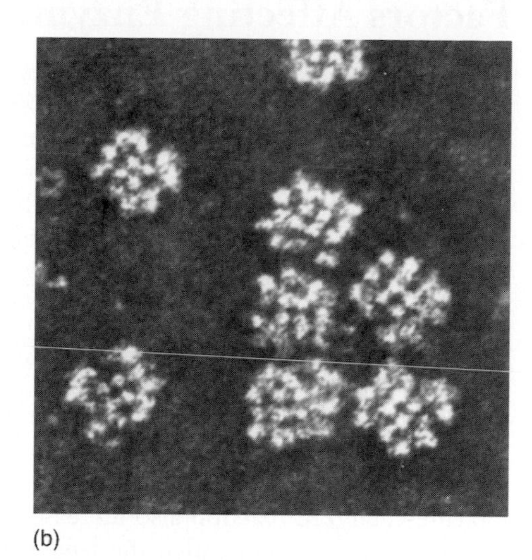
(b)

FIGURE 8.10
The enzyme pyruvate dehydrogenase. The enzyme (model, *a*) that carries out the oxidation of pyruvate is one of the most complex enzymes known—it has 60 subunits, many of which can be seen in the electron micrograph (*b*) (200,000×).

Multienzyme complexes offer significant advantages in catalytic efficiency:

1. The rate of any enzyme reaction is limited by the frequency with which the enzyme collides with its substrate. If a series of sequential reactions occurs within a multienzyme complex, the product of one reaction can be delivered to the next enzyme without releasing it to diffuse away.
2. Because the reacting substrate never leaves the complex during its passage through the series of reactions, the possibility of unwanted side reactions is eliminated.
3. All of the reactions that take place within the multienzyme complex can be controlled as a unit.

In addition to pyruvate dehydrogenase, which controls entry to the Krebs cycle, several other key processes in the cell are catalyzed by multienzyme complexes. One well-studied system is the fatty acid synthetase complex that catalyzes the synthesis of fatty acids from two-carbon precursors. There are seven different enzymes in this multienzyme complex, and the reaction intermediates remain associated with the complex for the entire series of reactions.

Not All Biological Catalysts Are Proteins

Until a few years ago, most biology textbooks contained statements such as "Enzymes are the catalysts of biological systems." We can no longer make that statement without qualification. As discussed in chapter 4, Tom Cech and his colleagues at the University of Colorado reported in 1981 that certain reactions involving RNA molecules appear to be catalyzed in cells by RNA itself, rather than by enzymes. This initial observation has been corroborated by additional examples of RNA catalysis in the last few years. Like enzymes, these RNA catalysts, which are loosely called "ribozymes," greatly accelerate the rate of particular biochemical reactions and show extraordinary specificity with respect to the substrates on which they act.

There appear to be at least two sorts of ribozymes. Those that carry out *intra*molecular catalysis have folded structures and catalyze reactions on themselves. Those that carry out *inter*molecular catalysis act on other molecules without themselves being changed in the process. Many important cellular reactions involve small RNA molecules, including reactions that chip out unnecessary sections from RNA copies of genes, that prepare ribosomes for protein synthesis, and that facilitate the replication of DNA within mitochondria. In all of these cases, the possibility of RNA catalysis is being actively investigated. It seems likely, particularly in the complex process of photosynthesis, that both enzymes and RNA play important catalytic roles.

The ability of RNA, an informational molecule, to act as a catalyst has stirred great excitement among biologists, as it appears to provide a potential answer to the "chicken-and-egg" riddle posed by the spontaneous origin of life hypothesis discussed in chapter 4. Which came first, the protein or the nucleic acid? It now seems at least possible that RNA may have evolved first and catalyzed the formation of the first proteins.

Not all biological catalysts float free in the cytoplasm. Some are part of other structures, and some are not even proteins.

Factors Affecting Enzyme Activity

The rate of an enzyme-catalyzed reaction is affected by the concentration of substrate, and of the enzyme that works on it. In addition, any chemical or physical factor that alters the enzyme's three-dimensional shape—such as temperature, pH, salt concentration, and the binding of specific regulatory molecules—can affect the enzyme's ability to catalyze the reaction.

Temperature

Increasing the temperature of an uncatalyzed reaction will increase its rate because the additional heat represents an increase in random molecular movement. The rate of an enzyme-catalyzed reaction also increases with temperature, but only up to a point called the *temperature optimum* (figure 8.11*a*). Below this temperature, the hydrogen bonds and hydrophobic interactions that determine the enzyme's shape are not flexible enough to permit the induced fit that is optimum for catalysis. Above the temperature optimum, these forces are too weak to maintain the enzyme's shape against the increased random movement of the atoms in the enzyme. At these higher temperatures, the enzyme denatures, as we described in chapter 3. Most human enzymes have temperature optima between 35°C and 40°C, a range that includes normal body temperature. Bacteria that live in hot springs have more stable enzymes (that is, enzymes held together more strongly), so the temperature optima for those enzymes can be 70°C or higher.

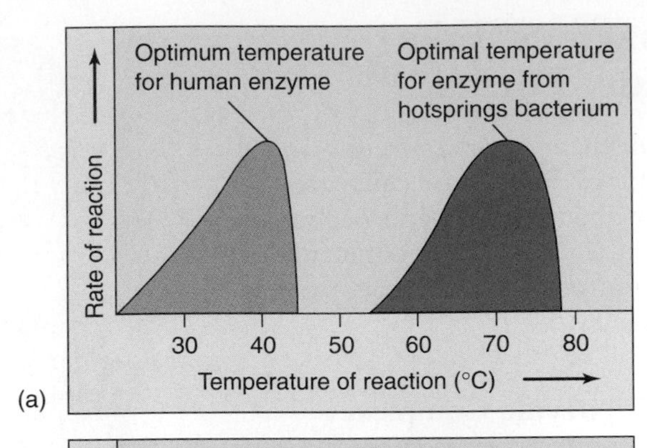

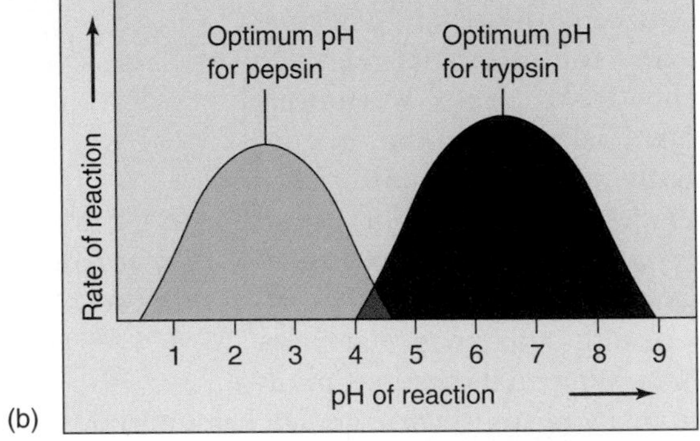

FIGURE 8.11
Enzymes are sensitive to their environment. The activity of an enzyme is influenced by both (*a*) temperature and (*b*) pH. Most human enzymes, such as the protein-degrading enzyme trypsin, work best at temperatures of about 40°C and within a pH range of 6 to 8.

pH

Ionic interactions between oppositely charged amino acid residues, such as glutamic acid (–) and lysine (+), also hold enzymes together. These interactions are sensitive to the hydrogen ion concentration of the fluid the enzyme is dissolved in, because changing that concentration shifts the balance between positively and negatively charged amino acid residues. For this reason, most enzymes have a **pH optimum** that usually ranges from pH 6 to 8. Those enzymes able to function in very acid environments are proteins that maintain their three-dimensional shape even in the presence of high levels of hydrogen ion. The enzyme pepsin, for example, digests proteins in the stomach at pH 2, a very acidic level (figure 8.11*b*).

Inhibitors and Activators

Enzyme activity is sensitive to the presence of specific substances that bind to the enzyme and cause changes in its shape. Through these substances, a cell is able to regulate which enzymes are active and which are inactive at a particular time. This allows the cell to increase its efficiency and to control changes in its characteristics during development. A substance that binds to an enzyme and *decreases* its activity is called an **inhibitor.** Very often, the end product of a biochemical pathway acts as an inhibitor of an early reaction in the pathway, a process called *feedback inhibition* (to be discussed later).

Enzyme inhibition occurs in two ways: **competitive inhibitors** compete with the substrate for the same binding site, displacing a percentage of substrate molecules from the enzymes; **noncompetitive inhibitors** bind to the enzyme in a location other than the active site, changing the shape of the enzyme and making it unable to bind to the substrate (figure 8.12). Most noncompetitive inhibitors bind to a specific portion of the enzyme called an **allosteric site** (Greek *allos*, "other" + *steros*, "form"). These sites serve as chemical on/off switches; the binding of a substance to the site can switch the enzyme between its active and inactive configurations. A substance that binds to an allosteric site and reduces enzyme activity is called an **allosteric inhibitor** (figure 8.12*b*). Alternatively, **activators** bind to allosteric sites and keep the enzymes in their active configurations, thereby *increasing* enzyme activity.

Enzyme Cofactors

Enzyme function is often assisted by additional chemical components known as **cofactors.** For example, the active sites of many enzymes contain metal ions that help draw electrons away from substrate molecules. The enzyme carboxypeptidase digests proteins by employing a zinc ion (Zn^{++}) in its active site to remove electrons from the bonds joining amino acids. Other elements, such as molybdenum and manganese, are also used as cofactors. Like zinc, these substances are required in the diet in small amounts. When the cofactor is a nonprotein organic molecule, it is called a **coenzyme.** Many vitamins are parts of coenzymes.

In numerous oxidation-reduction reactions that are catalyzed by enzymes, the electrons pass in pairs from the active site of the enzyme to a coenzyme that serves as the electron acceptor. The coenzyme then transfers the electrons to a different enzyme, which releases them (and the energy they bear) to the substrates in another reaction. Often, the electrons pair with protons (H^+) as hydrogen atoms. In this way, coenzymes shuttle energy in the form of hydrogen atoms from one enzyme to another in a cell.

One of the most important coenzymes is the hydrogen acceptor **nicotinamide adenine dinucleotide (NAD^+)** (figure 8.13). The NAD^+ molecule is composed of two nucleotides bound together. As you may recall from chapter 3, a nucleotide is a five-carbon sugar with one or more phosphate groups attached to one end and an organic base attached to the other end. The two nucleotides that make up NAD^+, nicotinamide monophosphate (NMP) and adenine monophosphate (AMP), are joined head-to-head by their phosphate groups. The two nucleotides serve different functions in the NAD^+ molecule: AMP acts as the core, providing a shape recognized by many enzymes; NMP is the active part of the molecule, contributing a site that is readily reduced (that is, easily accepts electrons).

When NAD^+ acquires an electron and a hydrogen atom (actually, two electrons and a proton) from the active site of an enzyme, it is reduced to NADH. The NADH molecule now carries the two energetic electrons and the proton. The oxidation of energy-containing molecules, which provides energy to cells, involves stripping electrons from those molecules and donating them to NAD^+. As we'll see, much of the energy of NADH is transferred to another molecule.

Enzymes have an optimum temperature and pH, at which the enzyme functions most effectively. Inhibitors decrease enzyme activity, while activators increase it. The activity of enzymes is often facilitated by cofactors, which can be metal ions or other substances. Cofactors that are nonprotein organic molecules are called coenzymes.

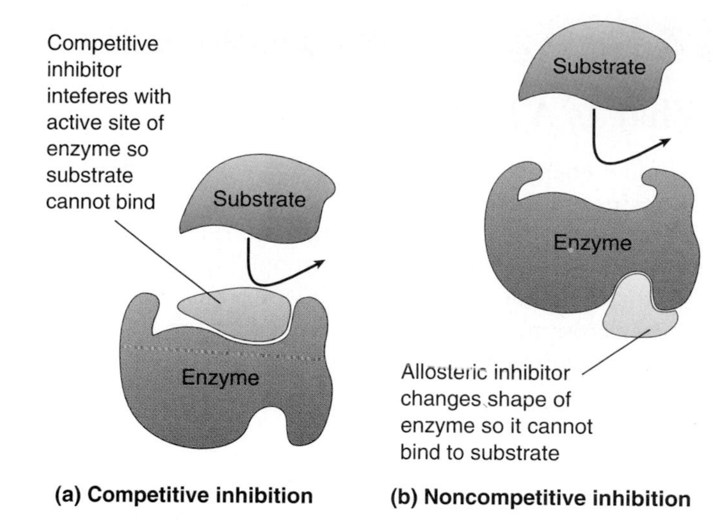

FIGURE 8.12
How enzymes can be inhibited. (*a*) In competitive inhibition, the inhibitor interferes with the active site of the enzyme. (*b*) In noncompetitive inhibition, the inhibitor binds to the enzyme at a place away from the active site, effecting a conformational change in the enzyme so that it can no longer bind to its substrate.

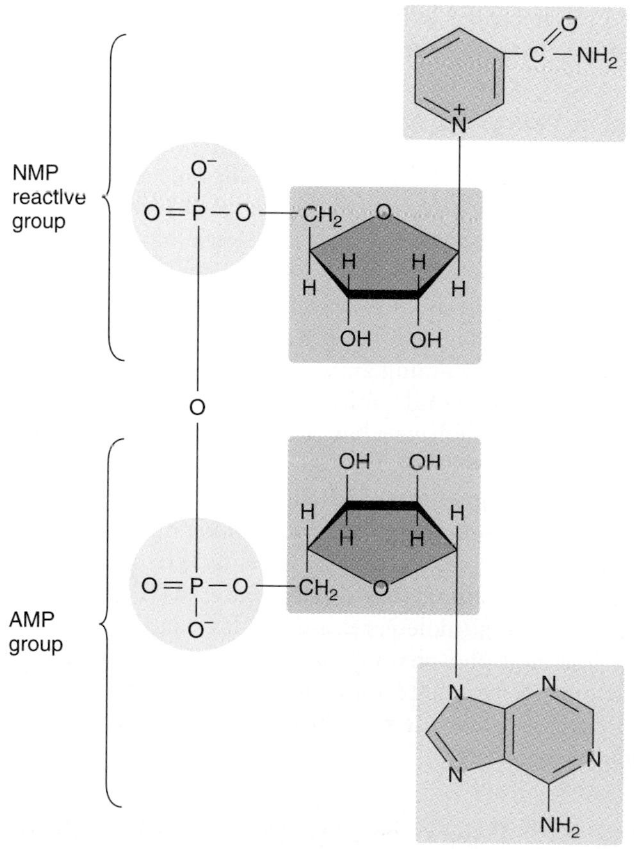

NMP reactive group

AMP group

FIGURE 8.13
The chemical structure of nicotinamide adenine dinucleotide (NAD^+). This key cofactor is composed of two nucleotides, NMP and AMP, attached head-to-head.

8.3 ATP is the energy currency of life.

What Is ATP?

The chief energy currency all cells use is a molecule called **adenosine triphosphate (ATP).** Cells use their supply of ATP to power almost every energy-requiring process they carry out, from making sugars, to supplying activation energy for chemical reactions, to actively transporting substances across membranes, to moving through their environment and growing.

Structure of the ATP Molecule

Each ATP molecule is a nucleotide composed of three smaller components (figure 8.14). The first component is a five-carbon sugar, ribose, which serves as the backbone to which the other two subunits are attached. The second component is adenine, an organic molecule composed of two carbon-nitrogen rings. Each of the nitrogen atoms in the ring has an unshared pair of electrons and weakly attracts hydrogen ions. Adenine, therefore, acts chemically as a base and is usually referred to as a nitrogenous base (it is one of the four nitrogenous bases found in DNA and RNA). The third component of ATP is a triphosphate group (a chain of three phosphates).

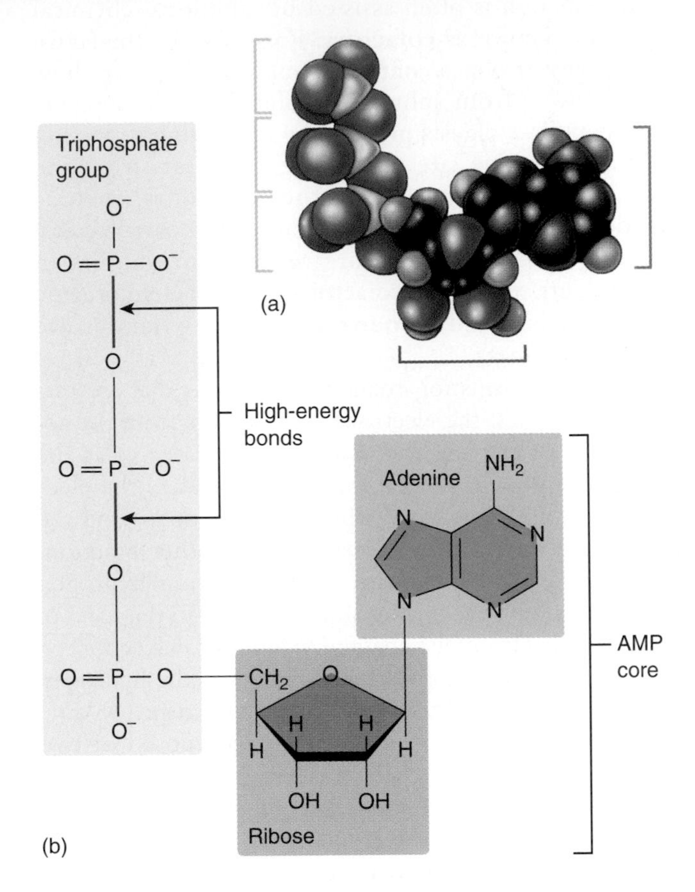

FIGURE 8.14
The ATP molecule. (*a*) The model and (*b*) structural diagram both show that like NAD^+, ATP has a core of AMP. In ATP the reactive group added to the end of the AMP phosphate group is not another nucleotide but rather a chain of two additional phosphate groups. The bonds connecting these two phosphate groups to each other and to AMP are energy-storing bonds.

How ATP Stores Energy

The key to how ATP stores energy lies in its triphosphate group. Phosphate groups are highly negatively charged, so they repel one another strongly. Because of the electrostatic repulsion between the charged phosphate groups, the two covalent bonds joining the phosphates are unstable. The ATP molecule is often referred to as a "coiled spring," the phosphates straining away from one another.

The unstable bonds holding the phosphates together in the ATP molecule have a low activation energy and are easily broken. When they break, they can transfer a considerable amount of energy. In most reactions involving ATP, only the outermost high-energy phosphate bond is hydrolyzed, cleaving off the phosphate group on the end. When this happens, ATP becomes **adenosine diphosphate (ADP),** and energy equal to 7.3 kcal/mole is released under standard conditions. The liberated phosphate group usually attaches temporarily to some intermediate molecule. When that molecule is dephosphorylated, the phosphate group is released as inorganic phosphate (P_i).

How ATP Powers Energy-Requiring Reactions

Cells use ATP to drive endergonic reactions. Such reactions do not proceed spontaneously, because their products possess more free energy than their reactants. However, if the cleavage of ATP's terminal high-energy bond releases more energy than the other reaction consumes, the overall energy change of the two coupled reactions will be exergonic (energy releasing) and they will both proceed. Because almost all endergonic reactions require less energy than is released by the cleavage of ATP, ATP can provide most of the energy a cell needs.

The same feature that makes ATP an effective energy donor—the instability of its phosphate bonds—precludes it from being a good long-term energy storage molecule. Fats and carbohydrates serve that function better. Most cells do not maintain large stockpiles of ATP. Instead, they typically have only a few seconds' supply of ATP at any given time, and they continually produce more from ADP and P_i.

The instability of its phosphate bonds makes ATP an excellent energy donor.

8.4 Metabolism is the chemical life of a cell.

Biochemical Pathways: The Organizational Units of Metabolism

This living chemistry, the total of all chemical reactions carried out by an organism, is called **metabolism** (Greek *metabole*, "change"). Those reactions that expend energy to make or transform chemical bonds are called *anabolic* reactions, or **anabolism.** Reactions that harvest energy when chemical bonds are broken are called *catabolic* reactions, or **catabolism.**

Organisms contain thousands of different kinds of enzymes that catalyze a bewildering variety of reactions. Many of these reactions in a cell occur in sequences called **biochemical pathways.** In such pathways, the product of one reaction becomes the substrate for the next (figure 8.15). Biochemical pathways are the organizational units of metabolism, the elements an organism controls to achieve coherent metabolic activity. Most sequential enzyme steps in biochemical pathways take place in specific compartments of the cell; the steps of the citric acid cycle (chapter 9), for example, occur inside mitochondria. By determining where many of the enzymes that catalyze these steps are located, we can "map out" a model of metabolic processes in the cell.

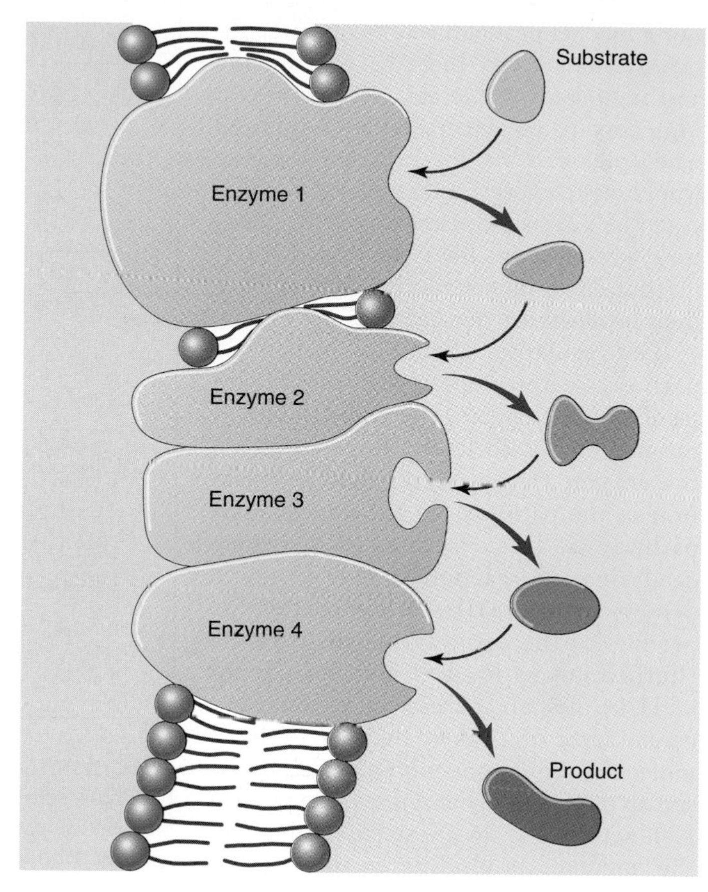

FIGURE 8.15
A biochemical pathway. The original substrate is acted on by enzyme 1, changing the substrate to a new form recognized by enzyme 2. Each enzyme in the pathway acts on the product of the previous stage.

How Biochemical Pathways Evolved

In the earliest cells, the first biochemical processes probably involved energy-rich molecules scavenged from the environment. Most of the molecules necessary for these processes are thought to have existed in the "organic soup" of the early oceans. The first catalyzed reactions are thought to have been simple, one-step reactions that brought these molecules together in various combinations. Eventually, the energy-rich molecules became depleted in the external environment, and only organisms that had evolved some means of making those molecules from other substances in the environment could survive. Thus, a hypothetical reaction,

$$\begin{matrix} \text{F} & \\ + & \longrightarrow \text{H} \\ \text{G} & \end{matrix}$$

where two energy-rich molecules (F and G) react to produce compound H and release energy, became more complex when the supply of F in the environment ran out. A new reaction was added in which the depleted molecule, F, is made from another molecule, E, which was also present in the environment:

$$\begin{matrix} \text{E} \longrightarrow & \text{F} & \\ & + & \longrightarrow \text{H} \\ & \text{G} & \end{matrix}$$

When the supply of E in turn became depleted, organisms that were able to make it from some other available precursor, D, survived. When D became depleted, those organisms in turn were replaced by ones able to synthesize D from another molecule, C:

$$\begin{matrix} \text{C} \longrightarrow \text{D} \longrightarrow \text{E} \longrightarrow & \text{F} & \\ & + & \longrightarrow \text{H} \\ & \text{G} & \end{matrix}$$

This hypothetical biochemical pathway would have evolved slowly through time, with the final reactions in the pathway evolving first and earlier reactions evolving later. Looking at the pathway now, we would say that the organism, starting with compound C, is able to synthesize H by means of a series of steps. This is how the biochemical pathways within organisms are thought to have evolved—not all at once, but one step at a time, backward.

How Biochemical Pathways Are Regulated

For a biochemical pathway to operate efficiently, its activity must be coordinated and regulated by the cell. Not only is it unnecessary to synthesize a compound when plenty is already present, doing so would waste energy and raw materials that could be put to use elsewhere. It is, therefore, advantageous for a cell to temporarily shut down biochemical pathways when their products are not needed.

The regulation of simple biochemical pathways often depends on an elegant feedback mechanism: the end product of the pathway binds to an allosteric site on the enzyme that catalyzes the first reaction in the pathway. In the hypothetical pathway we just described, the enzyme catalyzing the reaction C ⟶ D would possess an allosteric site for H, the end product of the pathway. As the pathway churned out its product and the amount of H in the cell increased, it would become increasingly likely that one of the H molecules would encounter the allosteric site on the C ⟶ D enzyme. If the product H functioned as an allosteric inhibitor of the enzyme, its binding to the enzyme would essentially shut down the reaction C ⟶ D. Shutting down this reaction, the first reaction in the pathway, effectively shuts down the whole pathway. Hence, as the cell produces increasing quantities of the product H, it automatically inhibits its ability to produce more. This mode of regulation is called **feedback inhibition** (figure 8.16).

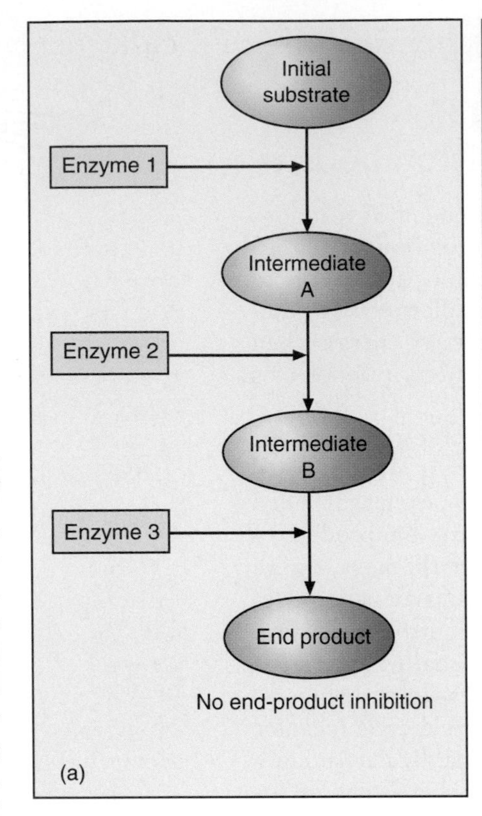

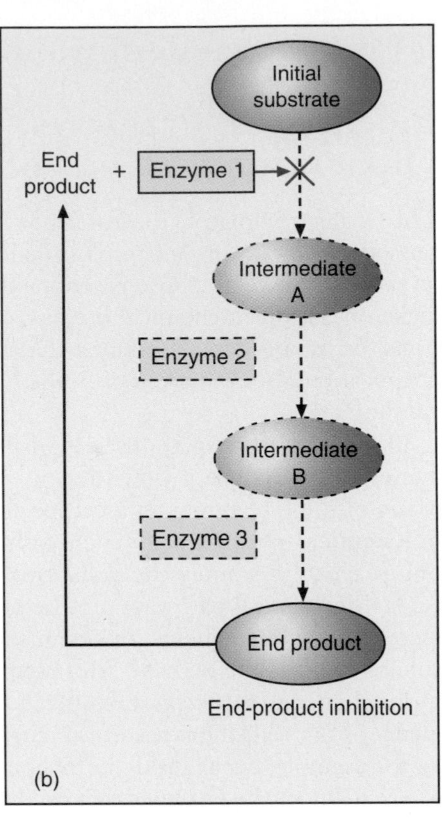

FIGURE 8.16
Feedback inhibition. (*a*) A biochemical pathway with no feedback inhibition. (*b*) A biochemical pathway in which the final end product becomes the allosteric effector for the first enzyme in the pathway. In other words, the formation of the pathway's final end product stops the pathway.

A biochemical pathway is an organized series of reactions, often regulated as a unit.

A Vocabulary of Metabolism

activation energy The energy required to destabilize chemical bonds and to initiate a chemical reaction.

catalysis Acceleration of the rate of a chemical reaction by lowering the activation energy.

coenzyme A nonprotein organic molecule that plays an accessory role in enzyme-catalyzed reactions, often by acting as a donor or acceptor of electrons. NAD^+ is a coenzyme.

endergonic reaction A chemical reaction to which energy from an outside source must be added before the reaction proceeds; the opposite of an exergonic reaction.

entropy A measure of the randomness or disorder of a system. In cells, it is a measure of how much energy has become so dispersed (usually as evenly distributed heat) that it is no longer available to do work.

exergonic reaction. An energy-yielding chemical reaction. Exergonic reactions tend to proceed spontaneously, although activation energy is required to initiate them.

free energy Energy available to do work.

kilocalorie 1000 calories. A calorie is the heat required to raise the temperature of 1 gram of water by 1°C.

metabolism The sum of all chemical processes occurring within a living cell or organism.

oxidation The loss of an electron by an atom or molecule. It occurs simultaneously with reduction of some other atom or molecule because an electron that is lost by one is gained by another.

reduction The gain of an electron by an atom or molecule. Oxidation-reduction reactions are an important means of energy transfer within living systems.

substrate A molecule on which an enzyme acts; the initial reactant in an enzyme-catalyzed reaction.

The Evolution of Metabolism

Metabolism has changed a great deal as life on earth has evolved. This has been particularly true of the reactions organisms use to capture energy from the sun to build organic molecules (anabolism), and then break down organic molecules to obtain energy (catabolism). These processes, the subject of the next two chapters, evolved in concert with each other.

Degradation

The most primitive forms of life are thought to have obtained chemical energy by degrading, or breaking down, organic molecules that were abiotically produced.

The first major event in the evolution of metabolism was the origin of the ability to harness chemical bond energy. At an early stage, organisms began to store this energy in the bonds of ATP, an energy carrier used by all organisms today.

Glycolysis

The second major event in the evolution of metabolism was glycolysis, the initial breakdown of glucose. As proteins evolved diverse catalytic functions, it became possible to capture a larger fraction of the chemical bond energy in organic molecules by breaking chemical bonds in a series of steps. For example, the progressive breakdown of the six-carbon sugar glucose into three-carbon molecules is performed in a series of 10 steps that results in the net production of two ATP molecules. The energy for the synthesis of ATP is obtained by breaking chemical bonds and forming new ones with less bond energy, the energy difference being channeled into ATP production. This biochemical pathway is called glycolysis.

Glycolysis undoubtedly evolved early in the history of life on earth, since this biochemical pathway has been retained by all living organisms. It is a chemical process that does not appear to have changed for well over 3 billion years.

Anaerobic Photosynthesis

The third major event in the evolution of metabolism was anaerobic photosynthesis. Early in the history of life, some organisms evolved a different way of generating ATP, called photosynthesis. Instead of obtaining energy for ATP synthesis by reshuffling chemical bonds, as in glycolysis, these organisms developed the ability to use light to pump protons out of their cells, and to use the resulting proton gradient to power the production of ATP, a process called chemiosmosis.

Photosynthesis evolved in the absence of oxygen and works well without it. Dissolved H_2S, present in the oceans beneath an atmosphere free of oxygen gas, served as a ready source of hydrogen atoms for building organic molecules. Free sulfur was produced as a by-product of this reaction.

Nitrogen Fixation

Nitrogen fixation was the fourth major step in the evolution of metabolism. Proteins and nucleic acids cannot be synthesized from the products of photosynthesis because both of these biologically critical molecules contain nitrogen. Obtaining nitrogen atoms from N_2 gas, a process called *nitrogen fixation*, requires the breaking of an $N{\equiv}N$ triple bond. This important reaction evolved in the hydrogen-rich atmosphere of the early earth, an atmosphere in which no oxygen was present. Oxygen acts as a poison to nitrogen fixation, which today occurs only in oxygen-free environments, or in oxygen-free compartments within certain bacteria.

Oxygen-Forming Photosynthesis

The substitution of H_2O for H_2S in photosynthesis was the fifth major event in the history of metabolism. Oxygen-forming photosynthesis employs H_2O rather than H_2S as a source of hydrogen atoms and their associated electrons. Because it garners its hydrogen atoms from reduced oxygen rather than from reduced sulfur, it generates oxygen gas rather than free sulfur.

More than 2 billion years ago, small cells capable of carrying out this oxygen-forming photosynthesis, such as cyanobacteria, became the dominant forms of life on earth. Oxygen gas began to accumulate in the atmosphere. This was the beginning of a great transition that changed conditions on earth permanently. Our atmosphere is now 20.9% oxygen, every molecule of which is derived from an oxygen-forming photosynthetic reaction.

Aerobic Respiration

Aerobic respiration is the sixth and final event in the history of metabolism. This cellular process harvests energy by stripping energetic electrons from organic molecules. Aerobic respiration employs the same kind of proton pumps as photosynthesis, and is thought to have evolved as a modification of the basic photosynthetic machinery. However, the hydrogens and their associated electrons are not obtained from H_2S or H_2O, as in photosynthesis, but rather from the breakdown of organic molecules.

Biologists think that the ability to carry out photosynthesis without H_2S first evolved among purple nonsulfur bacteria, which obtain their hydrogens from organic compounds instead. It was perhaps inevitable that among the descendants of these respiring photosynthetic bacteria, some would eventually do without photosynthesis entirely, subsisting only on the energy and hydrogens derived from the breakdown of organic molecules. The mitochondria within all eukaryotic cells are thought to be their descendants.

Six major innovations highlight the evolution of metabolism as we know it today.

Chapter 8

www.mhhe.com/raven6/resources8.mhtml

Summary	Questions	Media Resources

8.1 The laws of thermodynamics describe how energy changes.

- Energy is the capacity to bring about change, to provide motion against a force, or to do work.
- Kinetic energy is actively engaged in doing work, while potential energy has the capacity to do so.
- An oxidation-reduction (redox) reaction is one in which an electron is taken from one atom or molecule (oxidation) and donated to another (reduction).
- The First Law of Thermodynamics states that the amount of energy in the universe is constant; energy is neither lost nor created.
- The Second Law of Thermodynamics states that disorder in the universe (entropy) tends to increase.
- Any chemical reaction whose products contain less free energy than the original reactants can proceed spontaneously. However, the difference in free energy does not determine the rate of the reaction.
- The rate of a reaction depends on the amount of activation energy required to break existing bonds.
- Catalysis is the process of lowering activation energies by stressing chemical bonds.

1. Define oxidation and reduction. Why must these two reactions always occur in concert?

2. State the First and Second Laws of Thermodynamics.

3. What is heat? What is entropy? What is free energy?

4. What is the difference between an exergonic and an endergonic reaction? Which type of reaction tends to proceed spontaneously?

5. Define activation energy. How does a catalyst affect the final proportion of reactant converted into product?

- Energy Conversion
- Catalysis

- Thermodynamics
- Coupled Reactions

- Art Quizzes:
  - -Potential and Kinetic Energy
  - -Endergonic and Exergonic Reactions

8.2 Enzymes are biological catalysts.

- Enzymes are the major catalysts of cells; they affect the rate of a reaction but not the ultimate balance between reactants and products.
- Cells contain many different enzymes, each of which catalyzes a specific reaction.
- The specificity of an enzyme is due to its active site, which fits only one or a few types of substrate molecules.

6. How are the rates of enzyme-catalyzed reactions affected by temperature? What is the molecular basis for the effect on reaction rate?

7. What is the difference between the active site and an allosteric site on an enzyme?

- Explorations:
  - -Thermodynamics
  - -Kinetics

- Scientists on Science: Ribozymes
- *On Science* Article: Ribosomes Are Ribozymes

8.3 ATP is the energy currency of life.

- Cells obtain energy from photosynthesis and the oxidation of organic molecules and use it to manufacture ATP from ADP and phosphate.
- The energy stored in ATP is then used to drive endergonic reactions.

8. What part of the ATP molecule contains the bond that is employed to provide energy for most of the endergonic reactions in cells?

- ATP

8.4 Metabolism is the chemical life of a cell.

- Generally, the final reactions of a biochemical pathway evolved first; preceding reactions in the pathway were added later, one step at a time.

9. What is a biochemical pathway? How does feedback inhibition regulate the activity of a biochemical pathway?

- Feedback Inhibition

9

How Cells Harvest Energy

FIGURE 9.1
Harvesting chemical energy. Organisms such as these harvest mice depend on the energy stored in the chemical bonds of the food they eat to power their life processes.

Concept Outline

9.1 Cells harvest the energy in chemical bonds.

Using Chemical Energy to Drive Metabolism. The energy in C—H, C—O, and other chemical bonds can be captured and used to fuel the synthesis of ATP.

9.2 Cellular respiration oxidizes food molecules.

An Overview of Glucose Catabolism. The chemical energy in sugar is harvested by both substrate-level phosphorylation and by aerobic respiration.
Stage One: Glycolysis. The 10 reactions of glycolysis capture energy from glucose by reshuffling the bonds.
Stage Two: The Oxidation of Pyruvate. Pyruvate, the product of glycolysis, is oxidized to acetyl-CoA.
Stage Three: The Krebs Cycle. In a series of reactions, electrons are stripped from acetyl-CoA.
Harvesting Energy by Extracting Electrons. The respiration of glucose is a series of oxidation-reduction reactions which involve stripping electrons from glucose and using the energy of these electrons to power the synthesis of ATP.
Stage Four: The Electron Transport Chain. The electrons harvested from glucose pass through a chain of membrane proteins that use the energy to pump protons, driving the synthesis of ATP.
Summarizing Aerobic Respiration. The oxidation of glucose by aerobic respiration in eukaryotes produces up to three dozen ATP molecules, one third the energy in the chemical bonds of glucose.
Regulating Aerobic Respiration. High levels of ATP tend to shut down cellular respiration by feedback-inhibiting key reactions.

9.3 Catabolism of proteins and fats can yield considerable energy.

Glucose Is Not the Only Source of Energy. Proteins and fats are dismantled and the products fed into cellular respiration.

9.4 Cells can metabolize food without oxygen.

Fermentation. Fermentation allows continued metabolism in the absence of oxygen by donating the electrons harvested in glycolysis to organic molecules.

Life is driven by energy. All the activities organisms carry out—the swimming of bacteria, the purring of a cat, your reading of these words—use energy. In this chapter, we will discuss the processes all cells use to derive chemical energy from organic molecules and to convert that energy to ATP. We will consider photosynthesis, which uses light energy rather than chemical energy, in detail in chapter 10. We examine the conversion of chemical energy to ATP first because all organisms, both photosynthesizers and the organisms that feed on them (like the field mice in figure 9.1), are capable of harvesting energy from chemical bonds. As you will see, though, this process and photosynthesis have much in common.

9.1 Cells harvest the energy in chemical bonds.

Using Chemical Energy to Drive Metabolism

Plants, algae, and some bacteria harvest the energy of sunlight through photosynthesis, converting radiant energy into chemical energy. These organisms, along with a few others that use chemical energy in a similar way, are called **autotrophs** ("self-feeders"). All other organisms live on the energy autotrophs produce and are called **heterotrophs** ("fed by others"). At least 95% of the kinds of organisms on earth—all animals and fungi, and most protists and bacteria—are heterotrophs.

Where is the chemical energy in food, and how do heterotrophs harvest it to carry out the many tasks of living (figure 9.2)? Most foods contain a variety of carbohydrates, proteins, and fats, all rich in energy-laden chemical bonds. Carbohydrates and fats, for example, possess many carbon-hydrogen (C—H), as well as carbon-oxygen (C—O) bonds. The job of extracting energy from this complex organic mixture is tackled in stages. First, enzymes break the large molecules down into smaller ones, a process called **digestion.** Then, other enzymes dismantle these fragments a little at a time, harvesting energy from C—H and other chemical bonds at each stage. This process is called **catabolism.**

FIGURE 9.2
Start every day with a good breakfast. The carbohydrates, proteins, and fats in this fish contain energy that the bear's cells can use to power their daily activities.

While you obtain energy from many of the constituents of food, it is traditional to focus first on the catabolism of carbohydrates. We will follow the six-carbon sugar, glucose ($C_6H_{12}O_6$), as its chemical bonds are progressively harvested for energy. Later, we will come back and examine the catabolism of proteins and fats.

Cellular Respiration

The energy in a chemical bond can be visualized as potential energy borne by the electrons that make up the covalent bond. Cells harvest this energy by putting the electrons to work, often to produce ATP, the energy currency of the cell. Afterward, the energy-depleted electron (associated with a proton as a hydrogen atom) is donated to some other molecule. When oxygen gas (O_2) accepts the hydrogen atom, water forms, and the process is called **aerobic respiration.** When an inorganic molecule other than oxygen accepts the hydrogen, the process is called **anaerobic respiration.** When an organic molecule accepts the hydrogen atom, the process is called **fermentation.**

Chemically, there is little difference between the catabolism of carbohydrates in a cell and the burning of wood in a fireplace. In both instances, the reactants are carbohydrates and oxygen, and the products are carbon dioxide, water, and energy:

$$C_6H_{12}O_6 + 6\ O_2 \longrightarrow 6\ CO_2 + 6\ H_2O + \text{energy (heat or ATP)}$$

The change in free energy in this reaction is –720 kilocalories (–3012 kilojoules) per mole of glucose under the conditions found within a cell (the traditional value of –686 kilocalories, or –2870 kJ, per mole refers to standard conditions—room temperature, one atmosphere of pressure, etc.). This change in free energy results largely from the breaking of the six C—H bonds in the glucose molecule. The negative sign indicates that the products possess *less* free energy than the reactants. The same amount of energy is released whether glucose is catabolized or burned, but when it is burned most of the energy is released as heat. This heat cannot be used to perform work in cells. The key to a cell's ability to harvest useful energy from the catabolism of food molecules such as glucose is its conversion of a portion of the energy into a more useful form. Cells do this by using some of the energy to drive the production of ATP, a molecule that can power cellular activities.

The ATP Molecule

Adenosine triphosphate (ATP) is the energy currency of the cell, the molecule that transfers the energy captured during respiration to the many sites that use energy in the cell. How is ATP able to transfer energy so readily? Recall from chapter 8 that ATP is composed of a sugar (ribose) bound to an organic base (adenine) and a chain of three phosphate groups. As shown in figure 9.3, each phosphate group is negatively charged. Because like charges repel each other, the linked phosphate groups push against the bond that holds them together. Like a cocked mousetrap, the linked phosphates store the energy of their electrostatic repulsion. Transferring a phosphate group to another molecule relaxes the electrostatic spring of ATP, at the same time cocking the spring of the molecule that is phosphorylated. This molecule can then use the energy to undergo some change that requires work.

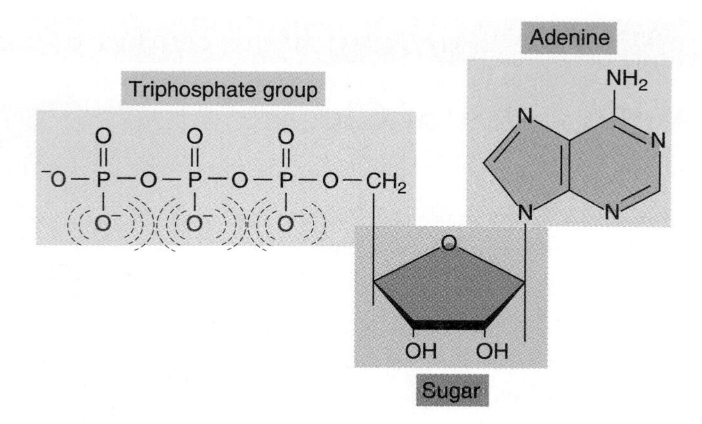

FIGURE 9.3
Structure of the ATP molecule. ATP is composed of an organic base and a chain of phosphates attached to opposite ends of a five-carbon sugar. Notice that the charged regions of the phosphate chain are close to one another. These like charges tend to repel one another, giving the bonds that hold them together a particularly high energy transfer potential.

How Cells Use ATP

Cells use ATP to do most of those activities that require work. One of the most obvious is movement. Some bacteria swim about, propelling themselves through the water by rapidly spinning a long, tail-like flagellum, much as a ship moves by spinning a propeller. During your development as an embryo, many of your cells moved about, crawling over one another to reach new positions. Movement also occurs within cells. Tiny fibers within muscle cells pull against one another when muscles contract. Mitochondria pass a meter or more along the narrow nerve cells that connect your feet with your spine. Chromosomes are pulled by microtubules during cell division. All of these movements by cells require the expenditure of ATP energy.

A second major way cells use ATP is to *drive endergonic reactions.* Many of the synthetic activities of the cell are endergonic, because building molecules takes energy. The chemical bonds of the products of these reactions contain more energy, or are more organized, than the reactants. The reaction can't proceed until that extra energy is supplied to the reaction. It is ATP that provides this needed energy.

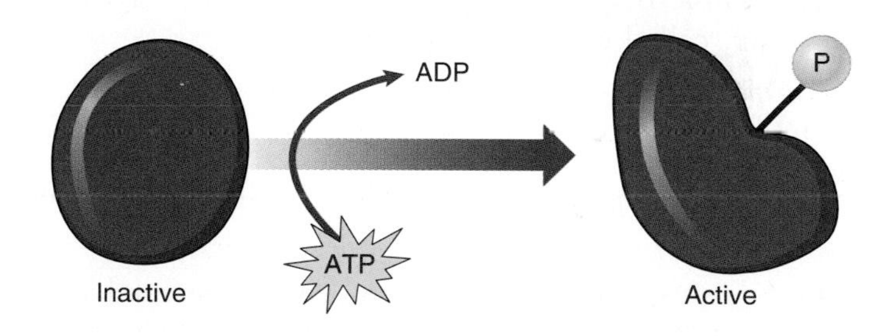

FIGURE 9.4
How ATP drives an endergonic reaction. In many cases, a phosphate group split from ATP activates a protein, catalyzing an endergonic process.

How ATP Drives Endergonic Reactions

How does ATP drive an endergonic reaction? The enzyme that catalyzes the endergonic reaction has *two* binding sites on its surface, one for the reactant and another for ATP. The ATP site splits the ATP molecule, liberating over 7 kcal (30 kJ) of chemical energy. This energy pushes the reactant at the second site "uphill," driving the endergonic reaction. (In a similar way, you can make water in a swimming pool leap straight up in the air, despite the fact that gravity prevents water from rising spontaneously—just jump in the pool! The energy you add going in more than compensates for the force of gravity holding the water back.)

When the splitting of ATP molecules drives an energy-requiring reaction in a cell, the two parts of the reaction—ATP-splitting and endergonic—take place in concert. In some cases, the two parts both occur on the surface of the same enzyme; they are physically linked, or "coupled," like two legs walking. In other cases, a high-energy phosphate from ATP attaches to the protein catalyzing the endergonic process, activating it (figure 9.4). Coupling energy-requiring reactions to the splitting of ATP in this way is one of the key tools cells use to manage energy.

The catabolism of glucose into carbon dioxide and water in living organisms releases about 720 kcal (3012 kJ) of energy per mole of glucose. This energy is captured in ATP, which stores the energy by linking charged phosphate groups near one another. When the phosphate bonds in ATP are hydrolyzed, energy is released and available to do work.

9.2 Cellular respiration oxidizes food molecules.

An Overview of Glucose Catabolism

Cells are able to make ATP from the catabolism of organic molecules in two different ways.

1. **Substrate-level phosphorylation.** In the first, called **substrate-level phosphorylation,** ATP is formed by transferring a phosphate group directly to ADP from a phosphate-bearing intermediate (figure 9.5). During glycolysis, discussed below, the chemical bonds of glucose are shifted around in reactions that provide the energy required to form ATP.
2. **Aerobic respiration.** In the second, called **aerobic respiration,** ATP forms as electrons are harvested, transferred along the electron transport chain, and eventually donated to oxygen gas. Eukaryotes produce the majority of their ATP from glucose in this way.

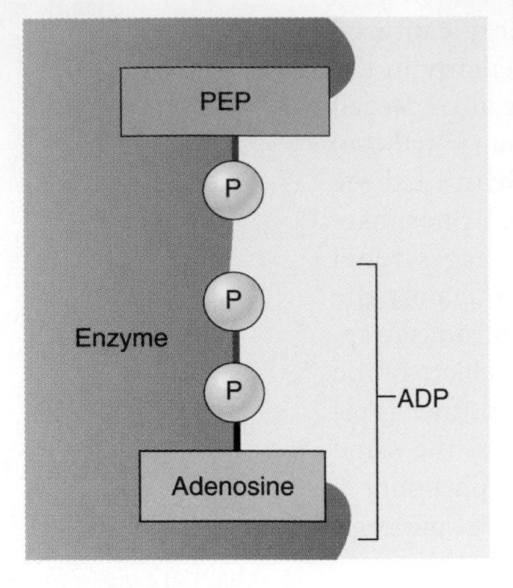

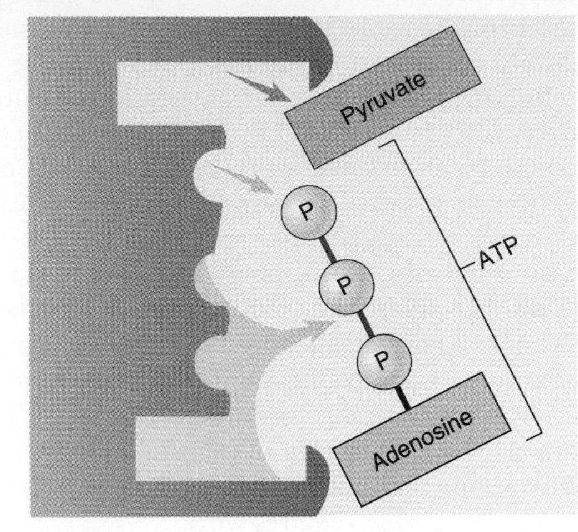

FIGURE 9.5
Substrate-level phosphorylation. Some molecules, such as phosphoenolpyruvate (PEP), possess a high-energy phosphate bond similar to the bonds in ATP. When PEP's phosphate group is transferred enzymatically to ADP, the energy in the bond is conserved and ATP is created.

In most organisms, these two processes are combined. To harvest energy to make ATP from the sugar glucose in the presence of oxygen, the cell carries out a complex series of enzyme-catalyzed reactions that occur in four stages: the first stage captures energy by substrate-level phosphorylation through glycolysis, the following three stages carry out aerobic respiration by oxidizing the end product of glycolysis.

Glycolysis

Stage One: Glycolysis. The first stage of extracting energy from glucose is a 10-reaction biochemical pathway called glycolysis that produces ATP by substrate-level phosphorylation. The enzymes that catalyze the glycolytic reactions are in the cytoplasm of the cell, not bound to any membrane or organelle. Two ATP molecules are used up early in the pathway, and four ATP molecules are formed by substrate-level phosphorylation. This yields a net of two ATP molecules for each molecule of glucose catabolized. In addition, four electrons are harvested as NADH that can be used to form ATP by aerobic respiration. Still, the total yield of ATP is small. When the glycolytic process is completed, the two molecules of pyruvate that are formed still contain most of the energy the original glucose molecule held.

Aerobic Respiration

Stage Two: Pyruvate Oxidation. In the second stage, pyruvate, the end product from glycolysis, is converted into carbon dioxide and a two-carbon molecule called acetyl-CoA. For each molecule of pyruvate converted, one molecule of NAD^+ is reduced to NADH.

Stage Three: The Krebs Cycle. The third stage introduces this acetyl-CoA into a cycle of nine reactions called the Krebs cycle, named after the British biochemist, Sir Hans Krebs, who discovered it. (The Krebs cycle is also called the citric acid cycle, for the citric acid, or citrate, formed in its first step, and less commonly, the tricarboxylic acid cycle, because citrate has three carboxyl groups.) In the Krebs cycle, two more ATP molecules are extracted by substrate-level phosphorylation, and a large number of electrons are removed by the reduction of NAD^+ to NADH.

Stage Four: Electron Transport Chain. In the fourth stage, the energetic electrons carried by NADH are employed to drive the synthesis of a large amount of ATP by the electron transport chain.

Pyruvate oxidation, the reactions of the Krebs cycle, and ATP production by electron transport chains occur within many forms of bacteria and inside the mitochondria of all eukaryotes. Recall from chapter 5 that mitochondria are thought to have evolved from bacteria. Although plants and algae can produce ATP by photosynthesis, they also produce ATP by aerobic respiration, just as animals and other nonphotosynthetic eukaryotes do. Figure 9.6 provides an overview of aerobic respiration.

Anaerobic Respiration

In the presence of oxygen, cells can respire aerobically, using oxygen to accept the electrons harvested from food molecules. In the absence of oxygen to accept the electrons, some organisms can still respire anaerobically, using inorganic molecules to accept the electrons. For example, many bacteria use sulfur, nitrate, or other inorganic compounds as the electron acceptor in place of oxygen.

Methanogens. Among the heterotrophs that practice anaerobic respiration are primitive archaebacteria such as the thermophiles discussed in chapter 4. Some of these, called methanogens, use CO_2 as the electron acceptor, reducing CO_2 to CH_4 (methane) with the hydrogens derived from organic molecules produced by other organisms.

Sulfur Bacteria. Evidence of a second anaerobic respiratory process among primitive bacteria is seen in a group of rocks about 2.7 billion years old, known as the Woman River iron formation. Organic material in these rocks is enriched for the light isotope of sulfur, ^{32}S, relative to the heavier isotope ^{34}S. No known geochemical process produces such enrichment, but biological sulfur reduction does, in a process still carried out today by certain primitive bacteria. In this sulfate respiration, the bacteria derive energy from the reduction of inorganic sulfates (SO_4) to H_2S. The hydrogen atoms are obtained from organic molecules other organisms produce. These bacteria thus do the same thing methanogens do, but they use SO_4 as the oxidizing (that is, electron-accepting) agent in place of CO_2.

The sulfate reducers set the stage for the evolution of photosynthesis, creating an environment rich in H_2S. As discussed in chapter 8, the first form of photosynthesis obtained hydrogens from H_2S using the energy of sunlight.

In aerobic respiration, the cell harvests energy from glucose molecules in a sequence of four major pathways: glycolysis, pyruvate oxidation, the Krebs cycle, and the electron transport chain. Oxygen is the final electron acceptor. Anaerobic respiration donates the harvested electrons to other inorganic compounds.

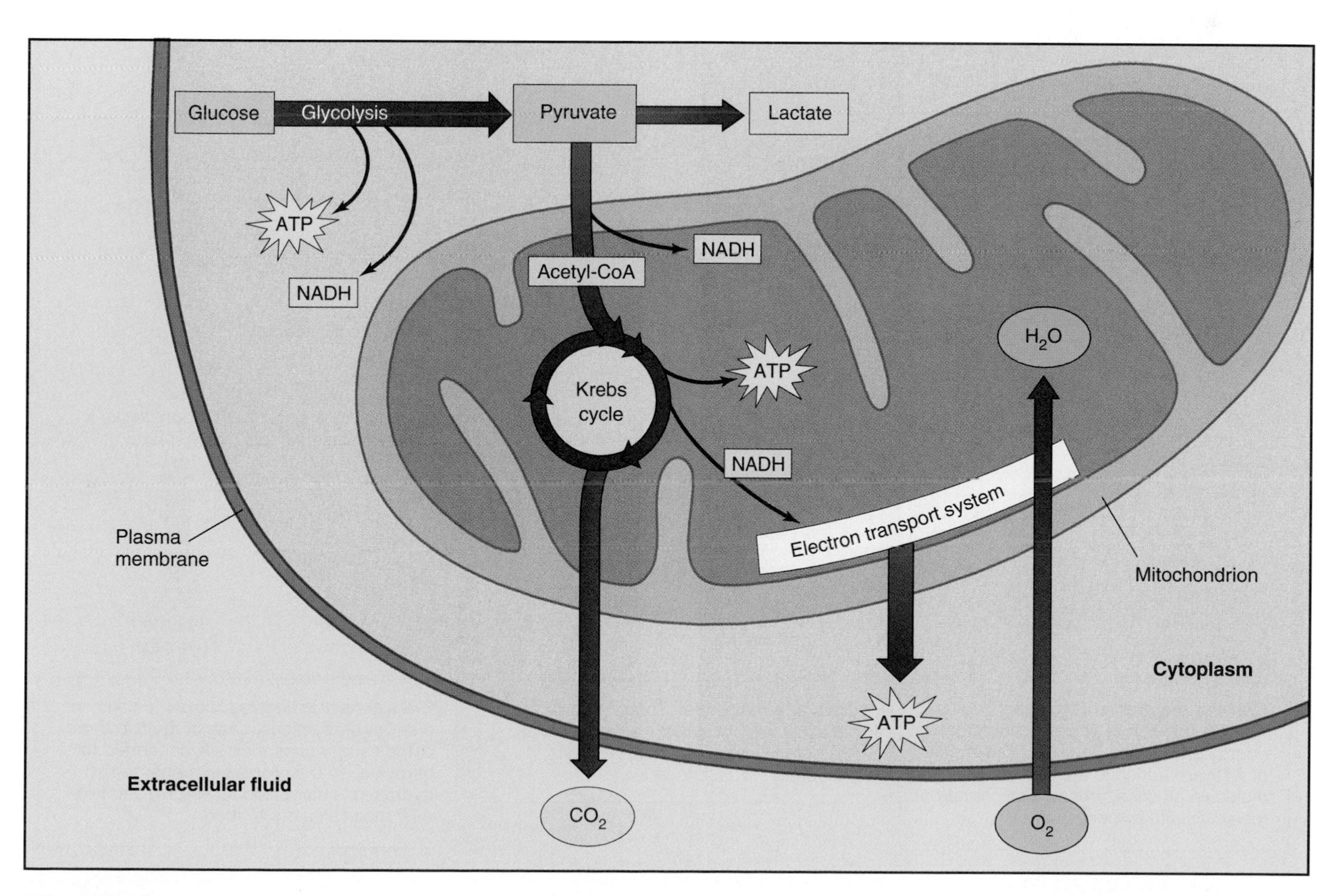

FIGURE 9.6
An overview of aerobic respiration.

Stage One: Glycolysis

The metabolism of primitive organisms focused on glucose. Glucose molecules can be dismantled in many ways, but primitive organisms evolved a glucose-catabolizing process that releases enough free energy to drive the synthesis of ATP in coupled reactions. This process, called glycolysis, occurs in the cytoplasm and involves a sequence of 10 reactions that convert glucose into 2 three-carbon molecules of pyruvate (figure 9.7). For each molecule of glucose that passes through this transformation, the cell nets two ATP molecules by substrate-level phosphorylation.

Priming

The first half of glycolysis consists of five sequential reactions that convert one molecule of glucose into two molecules of the three-carbon compound, glyceraldehyde 3-phosphate (G3P). These reactions demand the expenditure of ATP, so they are an energy-requiring process.

Step A: Glucose priming. Three reactions "prime" glucose by changing it into a compound that can be cleaved readily into 2 three-carbon phosphorylated molecules. Two of these reactions require the cleavage of ATP, so this step requires the cell to use two ATP molecules.

Step B: Cleavage and rearrangement. In the first of the remaining pair of reactions, the six-carbon product of step A is split into 2 three-carbon molecules. One is G3P, and the other is then converted to G3P by the second reaction (figure 9.8).

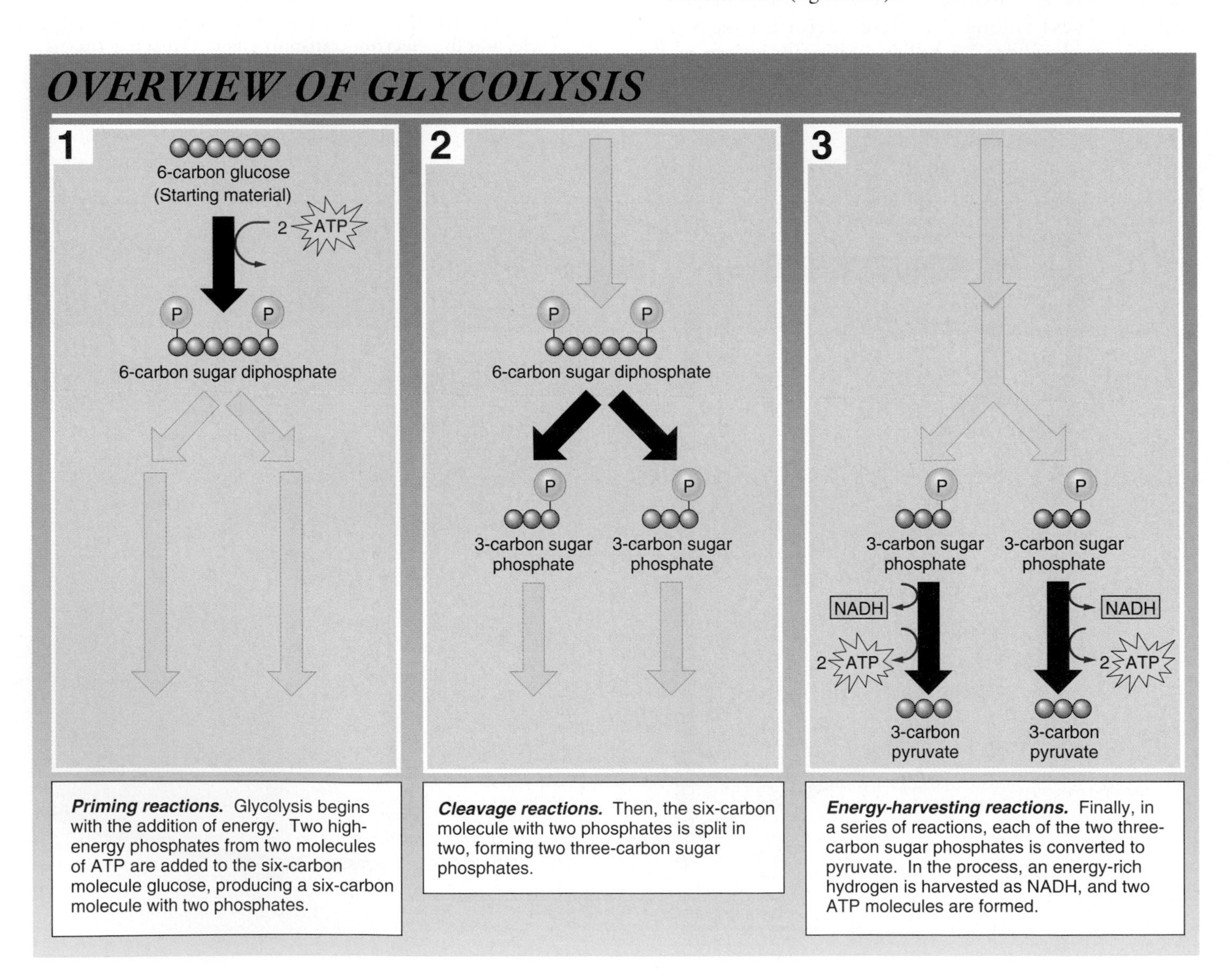

OVERVIEW OF GLYCOLYSIS
1
6-carbon glucose
(Starting material)
2 ATP
P
P
6-carbon sugar diphosphate
2
P
P
6-carbon sugar diphosphate
P
P
3-carbon sugar phosphate
3-carbon sugar phosphate
3
P
P
3-carbon sugar phosphate
3-carbon sugar phosphate
NADH
NADH
2 ATP
2 ATP
3-carbon pyruvate
3-carbon pyruvate
Priming reactions. Glycolysis begins with the addition of energy. Two high-energy phosphates from two molecules of ATP are added to the six-carbon molecule glucose, producing a six-carbon molecule with two phosphates.
Cleavage reactions. Then, the six-carbon molecule with two phosphates is split in two, forming two three-carbon sugar phosphates.
Energy-harvesting reactions. Finally, in a series of reactions, each of the two three-carbon sugar phosphates is converted to pyruvate. In the process, an energy-rich hydrogen is harvested as NADH, and two ATP molecules are formed.

FIGURE 9.7
How glycolysis works.

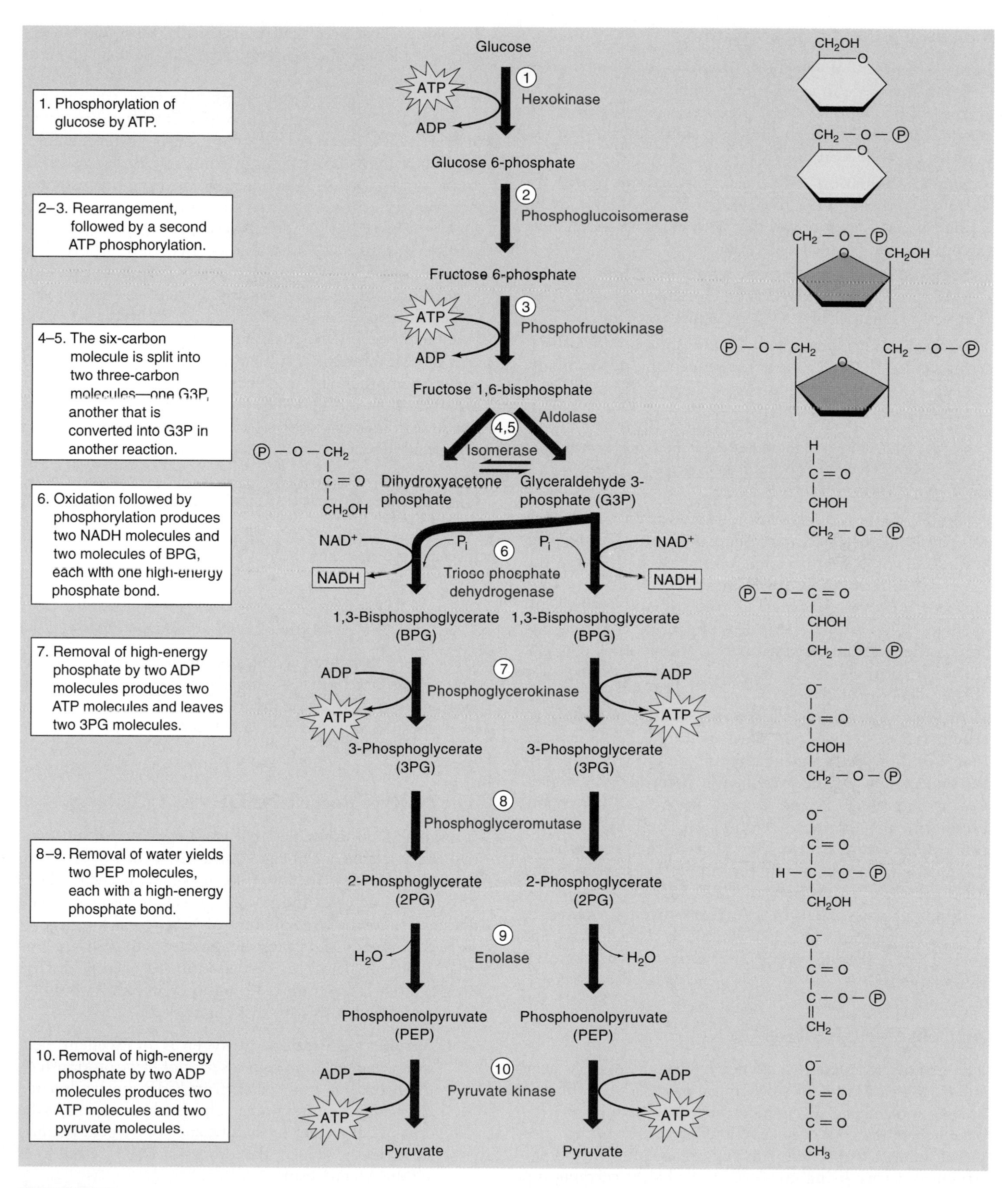

FIGURE 9.8
The glycolytic pathway. The first five reactions convert a molecule of glucose into two molecules of G3P. The second five reactions convert G3P into pyruvate.

Substrate-Level Phosphorylation

In the second half of glycolysis, five more reactions convert G3P into pyruvate in an energy-yielding process that generates ATP. Overall, then, glycolysis is a series of 10 enzyme-catalyzed reactions in which some ATP is invested in order to produce more.

Step C: Oxidation. Two electrons and one proton are transferred from G3P to NAD^+, forming NADH. Note that NAD^+ is an ion, and that both electrons in the new covalent bond come from G3P.

Step D: ATP generation. Four reactions convert G3P into another three-carbon molecule, pyruvate. This process generates two ATP molecules (see figure 9.5).

Because each glucose molecule is split into two G3P molecules, the overall reaction sequence yields two molecules of ATP, as well as two molecules of NADH and two of pyruvate:

4 ATP (2 ATP for each of the 2 G3P molecules in step D)
−2 ATP (used in the two reactions in step A)
2 ATP

Under the nonstandard conditions within a cell, each ATP molecule produced represents the capture of about 12 kcal (50 kJ) of energy per mole of glucose, rather than the 7.3 traditionally quoted for standard conditions. This means glycolysis harvests about 24 kcal/mole (100 kJ/mole). This is not a great deal of energy. The total energy content of the chemical bonds of glucose is 686 kcal (2870 kJ) per mole, so glycolysis harvests only 3.5% of the chemical energy of glucose.

Although far from ideal in terms of the amount of energy it releases, glycolysis does generate ATP. For more than a billion years during the anaerobic first stages of life on earth, it was the primary way heterotrophic organisms generated ATP from organic molecules. Like many biochemical pathways, glycolysis is believed to have evolved backward, with the last steps in the process being the most ancient. Thus, the second half of glycolysis, the ATP-yielding breakdown of G3P, may have been the original process early heterotrophs used to generate ATP. The synthesis of G3P from glucose would have appeared later, perhaps when alternative sources of G3P were depleted.

All Cells Use Glycolysis

The glycolytic reaction sequence is thought to have been among the earliest of all biochemical processes to evolve. It uses no molecular oxygen and occurs readily in an anaerobic environment. All of its reactions occur free in the cytoplasm; none is associated with any organelle or membrane structure. Every living creature is capable of carrying out glycolysis. Most present-day organisms, however, can extract considerably more energy from glucose through aerobic respiration.

Why does glycolysis take place even now, since its energy yield in the absence of oxygen is comparatively so paltry? The answer is that evolution is an incremental process: change occurs by improving on past successes. In catabolic metabolism, glycolysis satisfied the one essential evolutionary criterion: it was an improvement. Cells that could not carry out glycolysis were at a competitive disadvantage, and only cells capable of glycolysis survived the early competition of life. Later improvements in catabolic metabolism built on this success. Glycolysis was not discarded during the course of evolution; rather, it served as the starting point for the further extraction of chemical energy. Metabolism evolved as one layer of reactions added to another, just as successive layers of paint cover the walls of an old building. Nearly every present-day organism carries out glycolysis as a metabolic memory of its evolutionary past.

Closing the Metabolic Circle: The Regeneration of NAD^+

Inspect for a moment the net reaction of the glycolytic sequence:

$$\text{Glucose} + 2\ \text{ADP} + 2\ P_i + 2\ NAD^+ \longrightarrow 2\ \text{Pyruvate} + 2\ \text{ATP} + 2\ \text{NADH} + 2\ H^+ + 2\ H_2O$$

You can see that three changes occur in glycolysis: (1) glucose is converted into two molecules of pyruvate; (2) two molecules of ADP are converted into ATP via substrate level phosphorylation; and (3) two molecules of NAD^+ are reduced to NADH.

The Need to Recycle NADH

As long as food molecules that can be converted into glucose are available, a cell can continually churn out ATP to drive its activities. In doing so, however, it accumulates NADH and depletes the pool of NAD^+ molecules. A cell does not contain a large amount of NAD^+, and for glycolysis to continue, NADH must be recycled into NAD^+. Some other molecule than NAD^+ must ultimately accept the hydrogen atom taken from G3P and be reduced. Two molecules can carry out this key task (figure 9.9):

1. **Aerobic respiration.** Oxygen is an excellent electron acceptor. Through a series of electron transfers, the hydrogen atom taken from G3P can be donated to oxygen, forming water. This is what happens in the cells of eukaryotes in the presence of oxygen. Because air is rich in oxygen, this process is also referred to as aerobic metabolism.

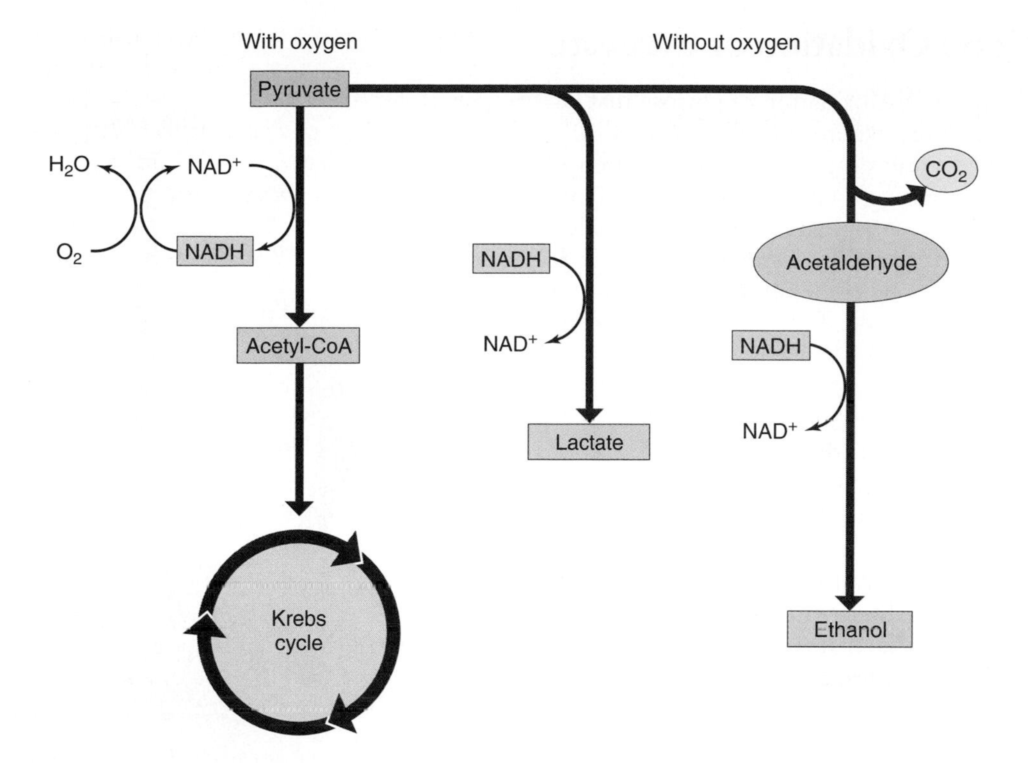

FIGURE 9.9
What happens to pyruvate, the product of glycolysis? In the presence of oxygen, pyruvate is oxidized to acetyl-CoA, which enters the Krebs cycle. In the absence of oxygen, pyruvate is instead reduced, accepting the electrons extracted during glycolysis and carried by NADH. When pyruvate is reduced directly, as in muscle cells, the product is lactate. When CO_2 is first removed from pyruvate and the product, acetaldehyde, is then reduced, as in yeast cells, the product is ethanol.

2. **Fermentation.** When oxygen is unavailable, an organic molecule can accept the hydrogen atom instead (figure 9.10). Such fermentation plays an important role in the metabolism of most organisms, even those capable of aerobic respiration.

The fate of the pyruvate that is produced by glycolysis depends upon which of these two processes takes place. The aerobic respiration path starts with the oxidation of pyruvate to a molecule called acetyl-CoA, which is then further oxidized in a series of reactions called the Krebs cycle. The fermentation path, by contrast, involves the reduction of all or part of pyruvate. We will start by examining aerobic respiration, then look briefly at fermentation.

Glycolysis generates a small amount of ATP by reshuffling the bonds of glucose molecules. In glycolysis, two molecules of NAD^+ are reduced to NADH. NAD^+ must be regenerated for glycolysis to continue unabated.

FIGURE 9.10
How wine is made. The conversion of pyruvate to ethanol takes place naturally in grapes left to ferment on vines, as well as in fermentation vats of crushed grapes. Yeasts carry out the process, but when their conversion increases the ethanol concentration to about 12%, the toxic effects of the alcohol kill the yeast cells. What is left is wine.

Stage Two: The Oxidation of Pyruvate

In the presence of oxygen, the oxidation of glucose that begins in glycolysis continues where glycolysis leaves off—with pyruvate. In eukaryotic organisms, the extraction of additional energy from pyruvate takes place exclusively inside mitochondria. The cell harvests pyruvate's considerable energy in two steps: first, by oxidizing pyruvate to form acetyl-CoA, and then by oxidizing acetyl-CoA in the Krebs cycle.

Producing Acetyl-CoA

Pyruvate is oxidized in a single "decarboxylation" reaction that cleaves off one of pyruvate's three carbons. This carbon then departs as CO_2 (figure 9.11, *top*). This reaction produces a two-carbon fragment called an acetyl group, as well as a pair of electrons and their associated hydrogen, which reduce NAD^+ to NADH. The reaction is complex, involving three intermediate stages, and is catalyzed within mitochondria by a *multienzyme complex*. As chapter 8 noted, such a complex organizes a series of enzymatic steps so that the chemical intermediates do not diffuse away or undergo other reactions. Within the complex, component polypeptides pass the substrates from one enzyme to the next, without releasing them. *Pyruvate dehydrogenase*, the complex of enzymes that removes CO_2 from pyruvate, is one of the largest enzymes known: it contains 60 subunits! In the course of the reaction, the acetyl group removed from pyruvate combines with a cofactor called coenzyme A (CoA), forming a compound known as **acetyl-CoA:**

$$\text{Pyruvate} + NAD^+ + \text{CoA} \longrightarrow \text{Acetyl-CoA} + \text{NADH} + CO_2$$

This reaction produces a molecule of NADH, which is later used to produce ATP. Of far greater significance than the reduction of NAD^+ to NADH, however, is the production of acetyl-CoA (figure 9.11, *bottom*). Acetyl-CoA is important because so many different metabolic processes generate it. Not only does the oxidation of pyruvate, an intermediate in carbohydrate catabolism, produce it, but the metabolic breakdown of proteins, fats, and other lipids also generate acetyl-CoA. Indeed, almost all molecules catabolized for energy are converted into acetyl-CoA. Acetyl-CoA is then channeled into fat synthesis or into ATP production, depending on the organism's energy requirements. Acetyl-CoA is a key point of focus for the many catabolic processes of the eukaryotic cell.

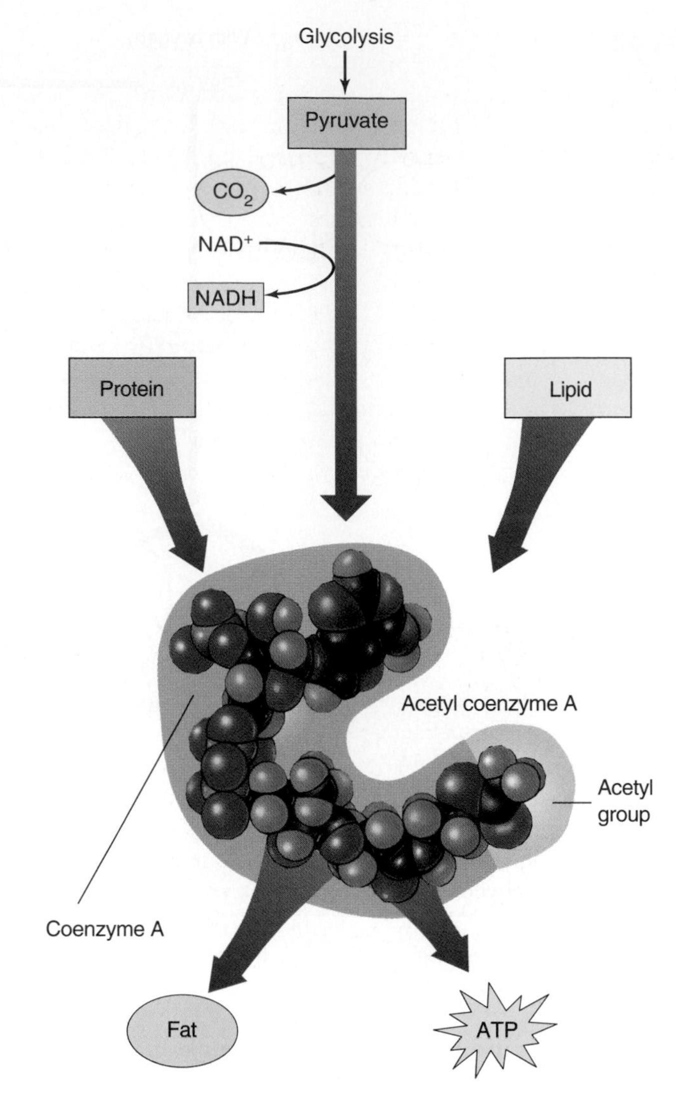

FIGURE 9.11
The oxidation of pyruvate. This complex reaction involves the reduction of NAD^+ to NADH and is thus a significant source of metabolic energy. Its product, acetyl-CoA, is the starting material for the Krebs cycle. Almost all molecules that are catabolized for energy are converted into acetyl-CoA, which is then channeled into fat synthesis or into ATP production.

Using Acetyl-CoA

Although the cell forms acetyl-CoA in many ways, only a limited number of processes use acetyl-CoA. Most of it is either directed toward energy storage (lipid synthesis, for example) or oxidized in the Krebs cycle to produce ATP. Which of these two options is taken depends on the level of ATP in the cell. When ATP levels are high, the oxidative pathway is inhibited, and acetyl-CoA is channeled into fatty acid synthesis. This explains why many animals (humans included) develop fat reserves when they consume more food than their bodies require. Alternatively, when ATP levels are low, the oxidative pathway is stimulated, and acetyl-CoA flows into energy-producing oxidative metabolism.

In the second energy-harvesting stage of glucose catabolism, pyruvate is decarboxylated, yielding acetyl-CoA, NADH, and CO_2. This process occurs within the mitochondrion.

Stage Three: The Krebs Cycle

After glycolysis catabolizes glucose to produce pyruvate, and pyruvate is oxidized to form acetyl-CoA, the third stage of extracting energy from glucose begins. In this third stage, acetyl-CoA is oxidized in a series of nine reactions called the Krebs cycle. These reactions occur in the matrix of mitochondria. In this cycle, the two-carbon acetyl group of acetyl-CoA combines with a four-carbon molecule called oxaloacetate (figure 9.12). The resulting six-carbon molecule then goes through a sequence of electron-yielding oxidation reactions, during which two CO_2 molecules split off, restoring oxaloacetate. The oxaloacetate is then recycled to bind to another acetyl group. In each turn of the cycle, a new acetyl group replaces the two CO_2 molecules lost, and more electrons are extracted to drive proton pumps that generate ATP.

Overview of the Krebs Cycle

The nine reactions of the Krebs cycle occur in two steps:

Step A: Priming. Three reactions prepare the six-carbon molecule for energy extraction. First, acetyl-CoA joins the cycle, and then chemical groups are rearranged.

Step B: Energy extraction. Four of the six reactions in this step are oxidations in which electrons are removed, and one generates an ATP equivalent directly by substrate-level phosphorylation.

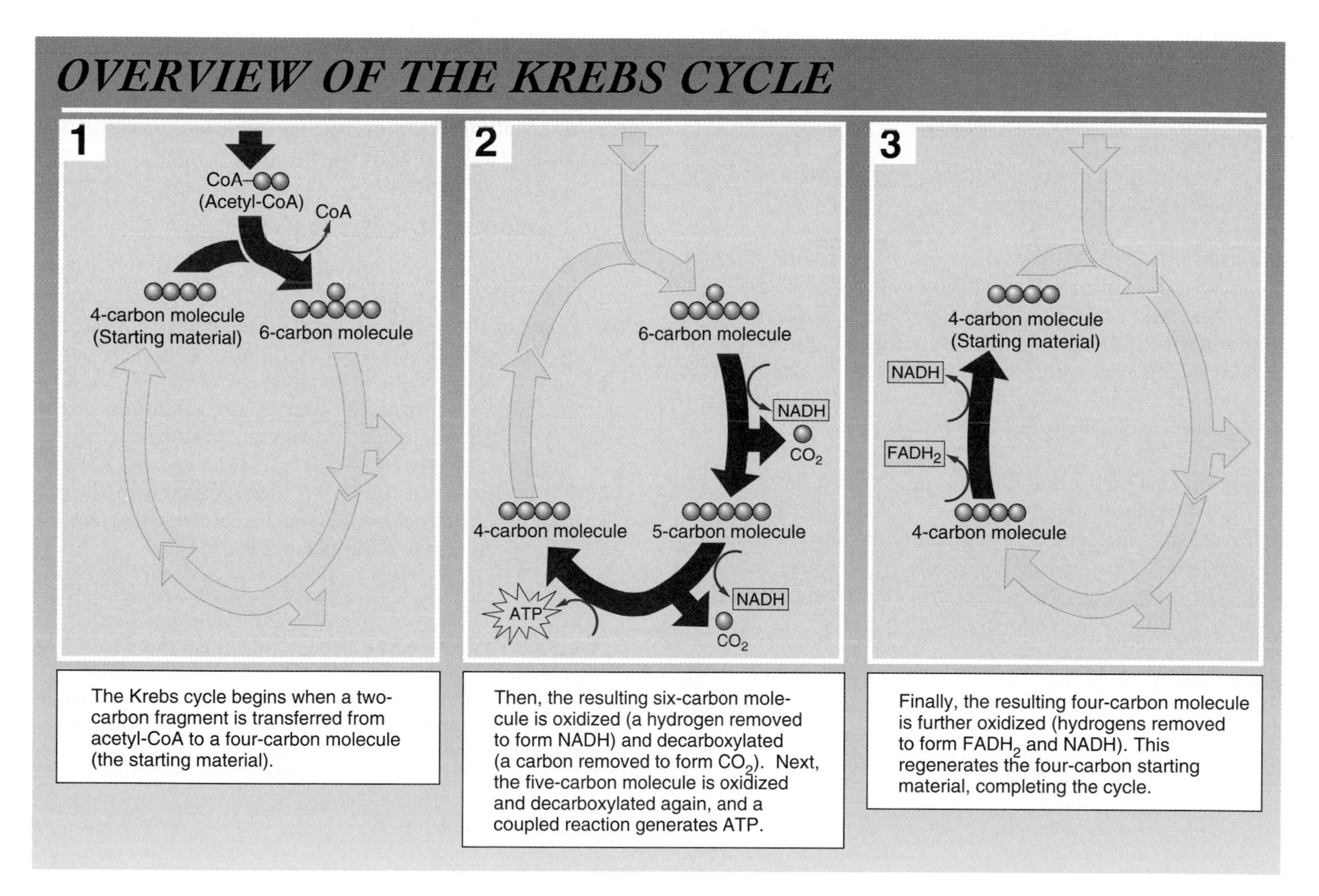

FIGURE 9.12
How the Krebs cycle works.

The Reactions of the Krebs Cycle

The **Krebs cycle** consists of nine sequential reactions that cells use to extract energetic electrons and drive the synthesis of ATP (figure 9.13). A two-carbon group from acetyl-CoA enters the cycle at the beginning, and two CO_2 molecules and several electrons are given off during the cycle.

Reaction 1: Condensation. The two-carbon group from acetyl-CoA joins with a four-carbon molecule, oxaloacetate, to form a six-carbon molecule, citrate. This condensation reaction is irreversible, committing the two-carbon acetyl group to the Krebs cycle. The reaction is inhibited when the cell's ATP concentration is high and stimulated when it is low. Hence, when the cell possesses ample amounts of ATP, the Krebs cycle shuts down and acetyl-CoA is channeled into fat synthesis.

Reactions 2 and 3: Isomerization. Before the oxidation reactions can begin, the hydroxyl (—OH) group of citrate must be repositioned. This is done in two steps: first, a water molecule is removed from one carbon; then, water is added to a different carbon. As a result, an —H group and an —OH group change positions. The product is an isomer of citrate called isocitrate.

Reaction 4: The First Oxidation. In the first energy-yielding step of the cycle, isocitrate undergoes an oxidative decarboxylation reaction. First, isocitrate is oxidized, yielding a pair of electrons that reduce a molecule of NAD^+ to NADH. Then the oxidized intermediate is decarboxylated; the central carbon atom splits off to form CO_2, yielding a five-carbon molecule called α-ketoglutarate.

Reaction 5: The Second Oxidation. Next, α-ketoglutarate is decarboxylated by a multienzyme complex similar to pyruvate dehydrogenase. The succinyl group left after the removal of CO_2 joins to coenzyme A, forming succinyl-CoA. In the process, two electrons are extracted, and they reduce another molecule of NAD^+ to NADH.

Reaction 6: Substrate-Level Phosphorylation. The linkage between the four-carbon succinyl group and CoA is a high-energy bond. In a coupled reaction similar to those that take place in glycolysis, this bond is cleaved, and the energy released drives the phosphorylation of guanosine diphosphate (GDP), forming guanosine triphosphate (GTP). GTP is readily converted into ATP, and the four-carbon fragment that remains is called succinate.

Reaction 7: The Third Oxidation. Next, succinate is oxidized to fumarate. The free energy change in this reaction is not large enough to reduce NAD^+. Instead, flavin adenine dinucleotide (FAD) is the electron acceptor. Unlike NAD^+, FAD is not free to diffuse within the mitochondrion; it is an integral part of the inner mitochondrial membrane. Its reduced form, $FADH_2$, contributes electrons to the electron transport chain in the membrane.

Reactions 8 and 9: Regeneration of Oxaloacetate. In the final two reactions of the cycle, a water molecule is added to fumarate, forming malate. Malate is then oxidized, yielding a four-carbon molecule of oxaloacetate and two electrons that reduce a molecule of NAD^+ to NADH. Oxaloacetate, the molecule that began the cycle, is now free to combine with another two-carbon acetyl group from acetyl-CoA and reinitiate the cycle.

The Products of the Krebs Cycle

In the process of aerobic respiration, glucose is entirely consumed. The six-carbon glucose molecule is first cleaved into a pair of three-carbon pyruvate molecules during glycolysis. One of the carbons of each pyruvate is then lost as CO_2 in the conversion of pyruvate to acetyl-CoA; two other carbons are lost as CO_2 during the oxidations of the Krebs cycle. All that is left to mark the passing of the glucose molecule into six CO_2 molecules is its energy, some of which is preserved in four ATP molecules and in the reduced state of 12 electron carriers. Ten of these carriers are NADH molecules; the other two are $FADH_2$.

The Krebs cycle generates two ATP molecules per molecule of glucose, the same number generated by glycolysis. More importantly, the Krebs cycle and the oxidation of pyruvate harvest many energized electrons, which can be directed to the electron transport chain to drive the synthesis of much more ATP.

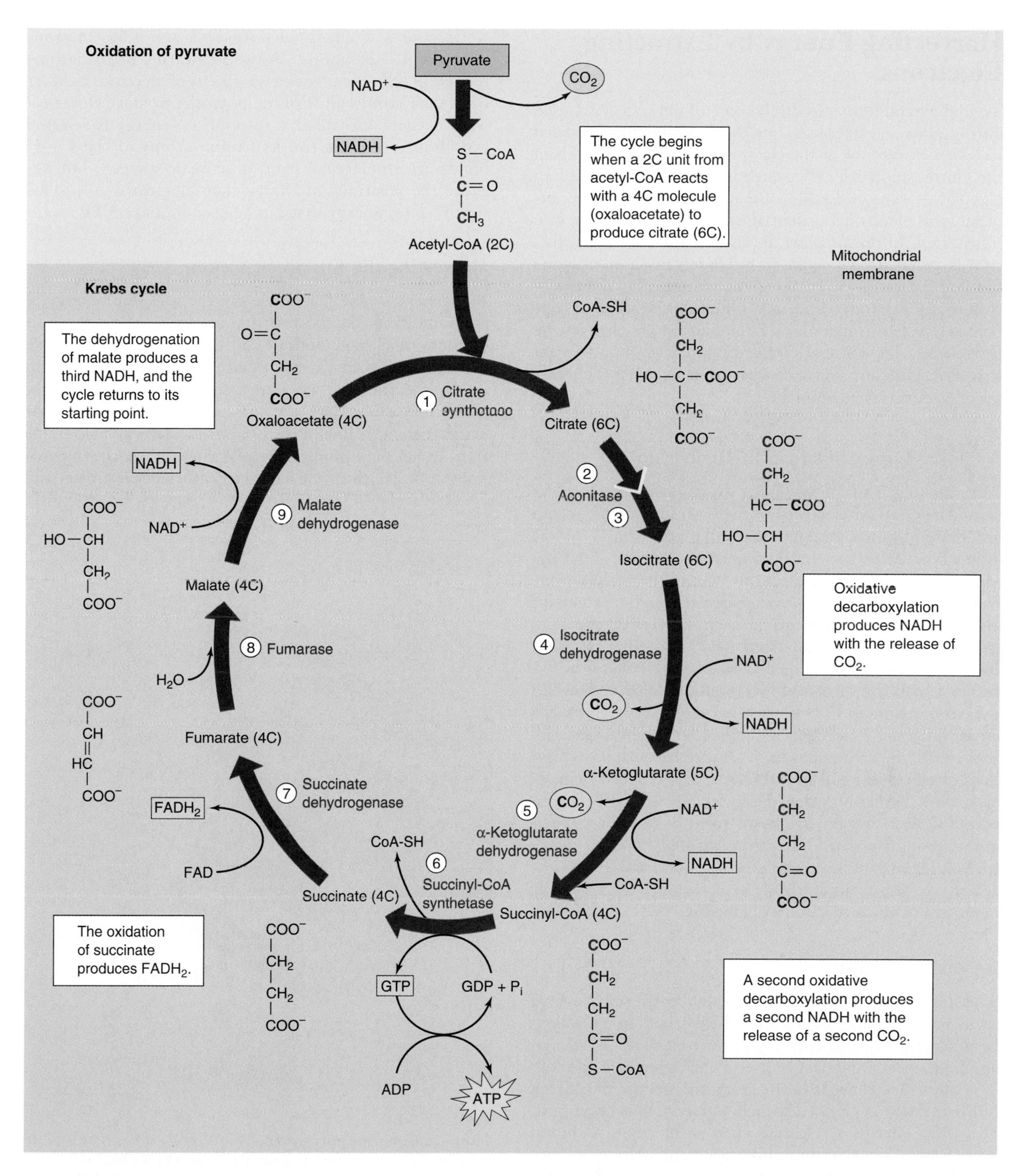

FIGURE 9.13

The Krebs cycle. This series of reactions takes place within the matrix of the mitochondrion. For the complete breakdown of a molecule of glucose, the two molecules of acetyl-CoA produced by glycolysis and pyruvate oxidation will each have to make a trip around the Krebs cycle. Follow the different carbons through the cycle, and notice the changes that occur in the carbon skeletons of the molecules as they proceed through the cycle.

Harvesting Energy by Extracting Electrons

To understand how cells direct some of the energy released during glucose catabolism into ATP production, we need to take a closer look at the electrons in the C—H bonds of the glucose molecule. We stated in chapter 8 that when an electron is removed from one atom and donated to another, the electron's potential energy of position is also transferred. In this process, the atom that receives the electron is reduced. We spoke of reduction in an all-or-none fashion, as if it involved the complete transfer of an electron from one atom to another. Often this is just what happens. However, sometimes a reduction simply changes the *degree of sharing* within a covalent bond. Let us now revisit that discussion and consider what happens when the transfer of electrons is incomplete.

A Closer Look at Oxidation-Reduction

The catabolism of glucose is an oxidation-reduction reaction. The covalent electrons in the C—H bonds of glucose are shared approximately equally between the C and H atoms because carbon and hydrogen nuclei have about the same affinity for valence electrons (that is, they exhibit similar *electronegativity*). However, when the carbon atoms of glucose react with oxygen to form carbon dioxide, the electrons in the new covalent bonds take a different position. Instead of being shared equally, the electrons that were associated with the carbon atoms in glucose shift far toward the oxygen atom in CO_2 because oxygen is very electronegative. Since these electrons are pulled farther from the carbon atoms, the carbon atoms of glucose have been oxidized (loss of electrons) and the oxygen atoms reduced (gain of electrons). Similarly, when the hydrogen atoms of glucose combine with oxygen atoms to form water, the oxygen atoms draw the shared electrons strongly toward them; again, oxygen is reduced and glucose is oxidized. In this reaction, oxygen is an oxidizing (electron-attracting) agent because it oxidizes the atoms of glucose.

Releasing Energy

The key to understanding the oxidation of glucose is to focus on the energy of the shared electrons. In a covalent bond, energy must be added to remove an electron from an atom, just as energy must be used to roll a boulder up a hill. The more electronegative the atom, the steeper the energy hill that must be climbed to pull an electron away from it. However, energy is released when an electron is shifted away from a less electronegative atom and *closer* to a more electronegative atom, just as energy is released when a boulder is allowed to roll down a hill. In the catabolism of glucose, energy is released when glucose is oxidized, as electrons relocate closer to oxygen (figure 9.14).

Glucose is an energy-rich food because it has an abundance of C—H bonds. Viewed in terms of oxidation-reduction, glucose possesses a wealth of electrons held far from their atoms, all with the potential to move closer toward oxygen. In oxidative respiration, energy is released not simply because the hydrogen atoms of the C—H bonds are transferred from glucose to oxygen, but because the positions of the valence electrons shift. This shift releases energy that can be used to make ATP.

Harvesting the Energy in Stages

It is generally true that the larger the release of energy in any single step, the more of that energy is released as heat (random molecular motion) and the less there is available to be channeled into more useful paths. In the combustion of gasoline, the same amount of energy is released whether all of the gasoline in a car's gas tank explodes at once, or whether the gasoline burns in a series of very small explosions inside the cylinders. By releasing the energy in gasoline a little at a time, the harvesting efficiency is greater and

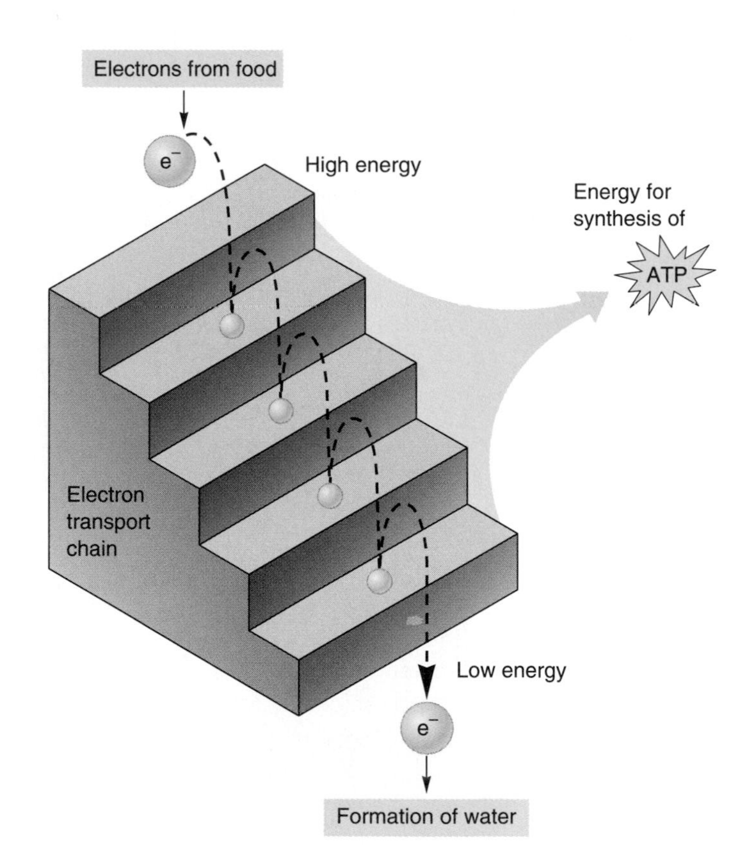

FIGURE 9.14
How electron transport works. This diagram shows how ATP is generated when electrons transfer from one energy level to another. Rather than releasing a single explosive burst of energy, electrons "fall" to lower and lower energy levels in steps, releasing stored energy with each fall as they tumble to the lowest (most electronegative) electron acceptor.

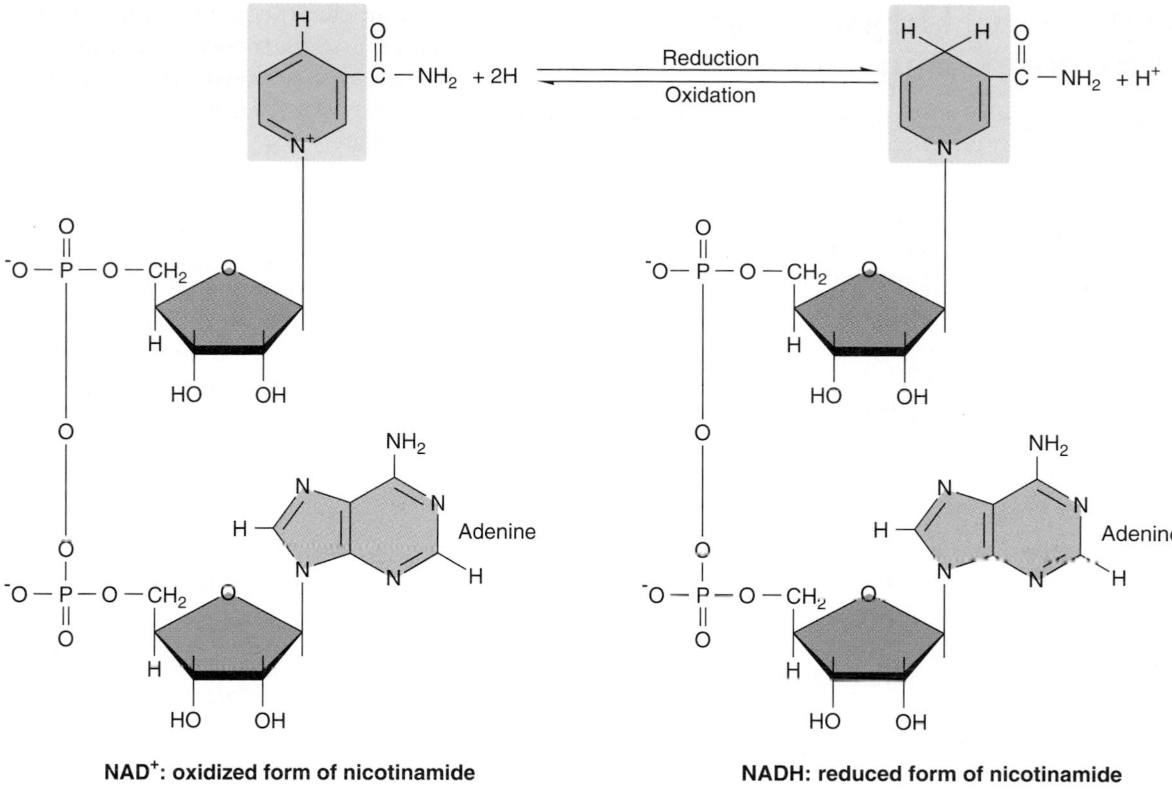

FIGURE 9.15
NAD^+ and NADH. This dinucleotide serves as an "electron shuttle" during cellular respiration. NAD^+ accepts electrons from catabolized macromolecules and is reduced to NADH.

NAD^+: oxidized form of nicotinamide

NADH: reduced form of nicotinamide

more of the energy can be used to push the pistons and move the car.

The same principle applies to the oxidation of glucose inside a cell. If all of the hydrogens were transferred to oxygen in one explosive step, releasing all of the free energy at once, the cell would recover very little of that energy in a useful form. Instead, cells burn their fuel much as a car does, a little at a time. The six hydrogens in the C—H bonds of glucose are stripped off in stages in the series of enzyme-catalyzed reactions collectively referred to as glycolysis and the Krebs cycle. We have had a great deal to say about these reactions already in this chapter. Recall that the hydrogens are removed by transferring them to a coenzyme carrier, NAD^+ (figure 9.15). Discussed in chapter 8, NAD^+ is a very versatile electron acceptor, shuttling energy-bearing electrons throughout the cell. In harvesting the energy of glucose, NAD^+ acts as the primary electron acceptor.

Following the Electrons

As you examine these reactions, try not to become confused by the changes in electrical charge. Always *follow the electrons.* Enzymes extract two hydrogens—that is, two electrons and two protons—from glucose and transfer both electrons and one of the protons to NAD^+. The other proton is released as a hydrogen ion, H^+, into the surrounding solution. This transfer converts NAD^+ into NADH; that is, two negative electrons and one positive proton are added to one positively charged NAD^+ to form NADH, which is electrically neutral.

Energy captured by NADH is *not* harvested all at once. Instead of being transferred directly to oxygen, the two electrons carried by NADH are passed along the **electron transport chain** if oxygen is present. This chain consists of a series of molecules, mostly proteins, embedded within the inner membranes of mitochondria. NADH delivers electrons to the top of the electron transport chain and oxygen captures them at the bottom. The oxygen then joins with hydrogen ions to form water. At each step in the chain, the electrons move to a slightly more electronegative carrier, and their positions shift slightly. Thus, the electrons move *down* an energy gradient. The entire process releases a total of 53 kcal/mole (222 kJ/mole) under standard conditions. The transfer of electrons along this chain allows the energy to be extracted gradually. In the next section, we will discuss how this energy is put to work to drive the production of ATP.

The catabolism of glucose involves a series of oxidation-reduction reactions that release energy by repositioning electrons closer to oxygen atoms. Energy is thus harvested from glucose molecules in gradual steps, using NAD^+ as an electron carrier.

Stage Four: The Electron Transport Chain

The NADH and $FADH_2$ molecules formed during the first three stages of aerobic respiration each contain a pair of electrons that were gained when NAD^+ and FAD were reduced. The NADH molecules carry their electrons to the inner mitochondrial membrane, where they transfer the electrons to a series of membrane-associated proteins collectively called the electron transport chain.

Moving Electrons through the Electron Transport Chain

The first of the proteins to receive the electrons is a complex, membrane-embedded enzyme called **NADH dehydrogenase.** A carrier called ubiquinone then passes the electrons to a protein-cytochrome complex called the *bc_1 complex.* This complex, along with others in the chain, operates as a proton pump, driving a proton out across the membrane. Cytochromes are respiratory proteins that contain heme groups, complex carbon rings with many alternating single and double bonds and an iron atom in the center.

The electron is then carried by another carrier, *cytochrome c,* to the cytochrome oxidase complex. This complex uses four such electrons to reduce a molecule of oxygen, each oxygen then combines with two hydrogen ions to form water:

$$O_2 + 4\,H^+ + 4\,e^- \longrightarrow 2\,H_2O$$

This series of membrane-associated electron carriers is collectively called the electron transport chain (figure 9.16).

NADH contributes its electrons to the first protein of the electron transport chain, NADH dehydrogenase. $FADH_2$, which is always attached to the inner mitochondrial membrane, feeds its electrons into the electron transport chain later, to ubiquinone.

It is the availability of a plentiful electron acceptor (often oxygen) that makes oxidative respiration possible. As we'll see in chapter 10, the electron transport chain used in aerobic respiration is similar to, and may well have evolved from, the chain employed in aerobic photosynthesis.

The electron transport chain is a series of five membrane-associated proteins. Electrons delivered by NADH and $FADH_2$ are passed from protein to protein along the chain, like a baton in a relay race.

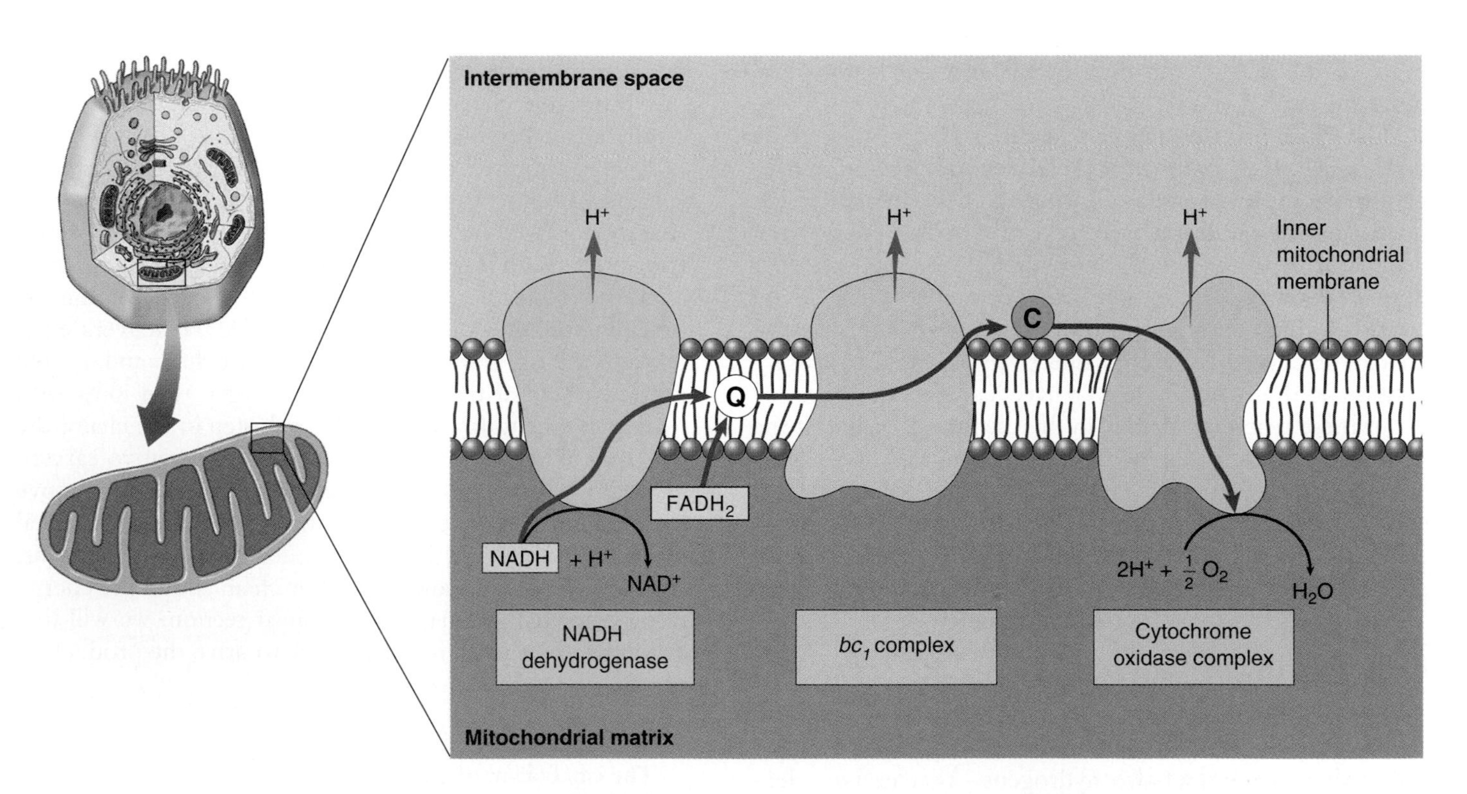

FIGURE 9.16
The electron transport chain. High-energy electrons harvested from catabolized molecules are transported (*red* arrows) by mobile electron carriers (ubiquinone, marked Q, and cytochrome *c,* marked C) along a chain of membrane proteins. Three proteins use portions of the electrons' energy to pump protons (*blue* arrows) out of the matrix and into the intermembrane space. The electrons are finally donated to oxygen to form water.

Building an Electrochemical Gradient

In eukaryotes, aerobic metabolism takes place within the mitochondria present in virtually all cells. The internal compartment, or matrix, of a mitochondrion contains the enzymes that carry out the reactions of the Krebs cycle. As the electrons harvested by oxidative respiration are passed along the electron transport chain, the energy they release transports protons out of the matrix and into the outer compartment, sometimes called the intermembrane space. Three transmembrane proteins in the inner mitochondrial membrane (see figure 9.16) actually accomplish the transport. The flow of excited electrons induces a change in the shape of these pump proteins, which causes them to transport protons across the membrane. The electrons contributed by NADH activate all three of these proton pumps, while those contributed by $FADH_2$ activate only two.

Producing ATP: Chemiosmosis

As the proton concentration in the intermembrane space rises above that in the matrix, the matrix becomes slightly negative in charge. This internal negativity attracts the positively charged protons and induces them to reenter the matrix. The higher outer concentration tends to drive protons back in by diffusion; since membranes are relatively impermeable to ions, most of the protons that reenter the matrix pass through special proton channels in the inner mitochondrial membrane. When the protons pass through, these channels synthesize ATP from ADP + P_i within the matrix. The ATP is then transported by facilitated diffusion out of the mitochondrion and into the cell's cytoplasm. Because the chemical formation of ATP is driven by a diffusion force similar to osmosis, this process is referred to as **chemiosmosis** (figure 9.17).

Thus, the electron transport chain uses electrons harvested in aerobic respiration to pump a large number of protons across the inner mitochondrial membrane. Their subsequent reentry into the mitochondrial matrix drives the synthesis of ATP by chemiosmosis. Figure 9.18 summarizes the overall process.

The electrons harvested from glucose are pumped out of the mitochondrial matrix by the electron transport chain. The return of the protons into the matrix generates ATP.

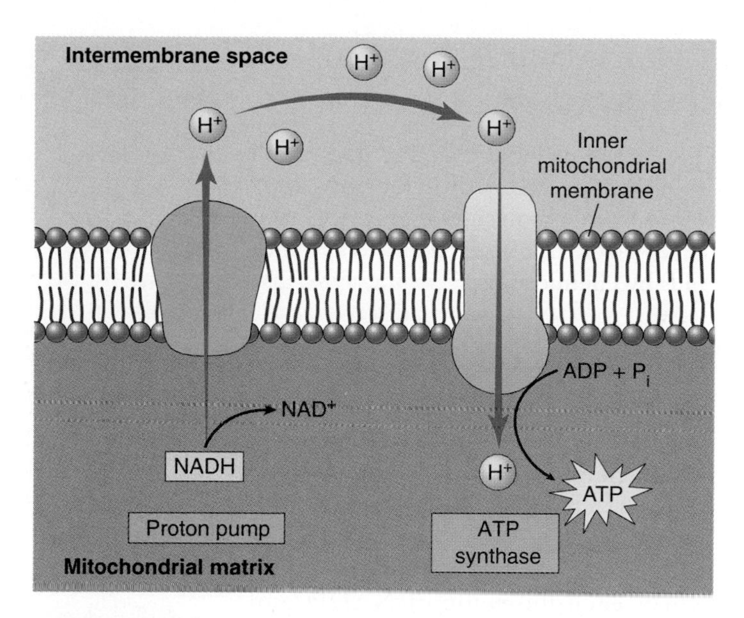

FIGURE 9.17
Chemiosmosis. NADH transports high-energy electrons harvested from the catabolism of macromolecules to "proton pumps" that use the energy to pump protons out of the mitochondrial matrix. As a result, the concentration of protons in the intermembrane space rises, inducing protons to diffuse back into the matrix. Many of the protons pass through special channels that couple the reentry of protons to the production of ATP.

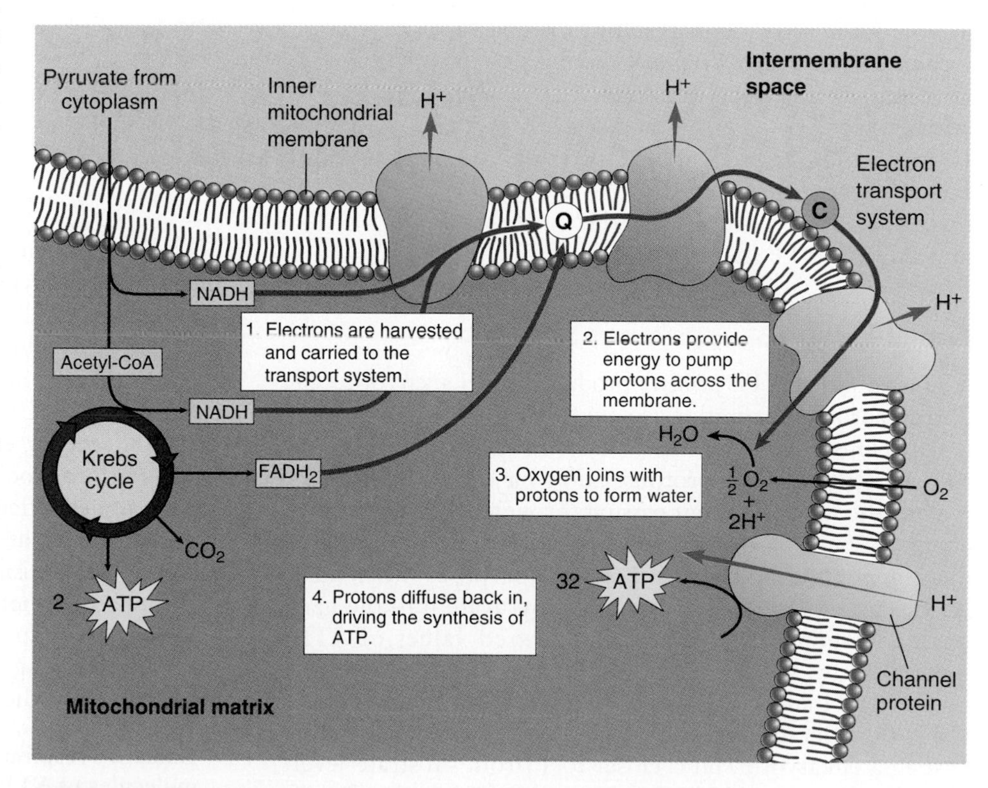

FIGURE 9.18
ATP generation during the Krebs cycle and electron transport chain. This process begins with pyruvate, the product of glycolysis, and ends with the synthesis of ATP.

Summarizing Aerobic Respiration

How much metabolic energy does a cell actually gain from the electrons harvested from a molecule of glucose, using the electron transport chain to produce ATP by chemiosmosis?

Theoretical Yield

The chemiosmotic model suggests that one ATP molecule is generated for each proton pump activated by the electron transport chain. As the electrons from NADH activate three pumps and those from $FADH_2$ activate two, we would expect each molecule of NADH and $FADH_2$ to generate three and two ATP molecules, respectively. However, because eukaryotic cells carry out glycolysis in their cytoplasm and the Krebs cycle within their mitochondria, they must transport the two molecules of NADH produced during glycolysis across the mitochondrial membranes, which requires one ATP per molecule of NADH. Thus, the net ATP production is decreased by two. Therefore, the overall ATP production resulting from aerobic respiration *theoretically* should be 4 (from substrate-level phosphorylation during glycolysis) + 30 (3 from each of 10 molecules of NADH) + 4 (2 from each of 2 molecules of $FADH_2$) – 2 (for transport of glycolytic NADH) = 36 molecules of ATP (figure 9.19).

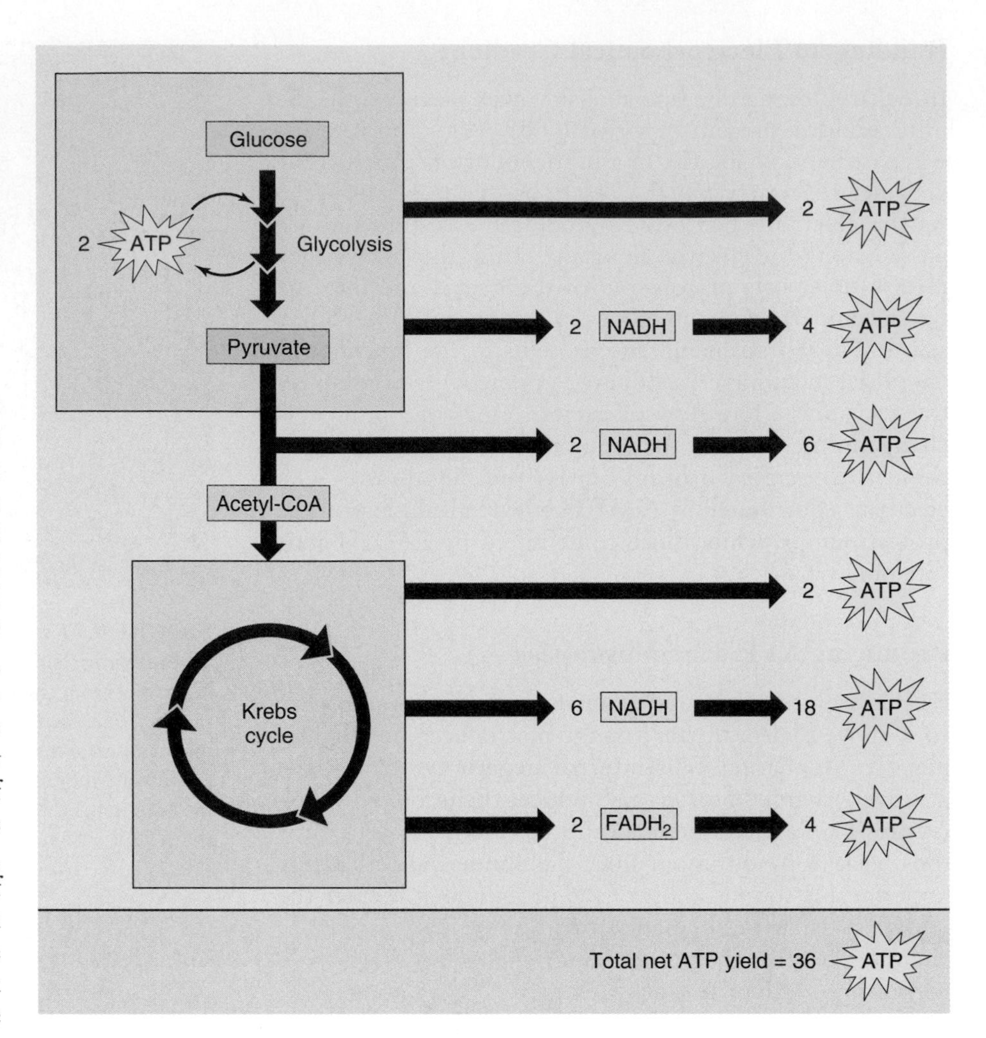

FIGURE 9.19
Theoretical ATP yield. The theoretical yield of ATP harvested from glucose by aerobic respiration totals 36 molecules.

Actual Yield

The amount of ATP *actually* produced in a eukaryotic cell during aerobic respiration is somewhat lower than 36, for two reasons. First, the inner mitochondrial membrane is somewhat "leaky" to protons, allowing some of them to reenter the matrix without passing through ATP-generating channels. Second, mitochondria often use the proton gradient generated by chemiosmosis for purposes other than ATP synthesis (such as transporting pyruvate into the matrix). Consequently, the actual measured values of ATP generated by NADH and $FADH_2$ are closer to 2.5 for each NADH and 1.5 for each $FADH_2$. With these corrections, the overall harvest of ATP from a molecule of glucose in a eukaryotic cell is closer to 4 (from substrate-level phosphorylation) + 25 (2.5 from each of 10 molecules of NADH) + 3 (1.5 from each of 2 molecules of $FADH_2$) – 2 (transport of glycolytic NADH) = 30 molecules of ATP.

The catabolism of glucose by aerobic respiration, in contrast to glycolysis, is quite efficient. Aerobic respiration in a eukaryotic cell harvests about $7.3 \times 30 \div 686 = 32\%$ of the energy available in glucose. (By comparison, a typical car converts only about 25% of the energy in gasoline into useful energy.) The efficiency of oxidative respiration at harvesting energy establishes a natural limit on the maximum length of food chains.

The higher efficiency of aerobic respiration was one of the key factors that fostered the evolution of heterotrophs. With this mechanism for producing ATP, it became feasible for nonphotosynthetic organisms to derive metabolic energy exclusively from the oxidative breakdown of other organisms. As long as some organisms captured energy by photosynthesis, others could exist solely by feeding on them.

Oxidative respiration produces approximately 30 molecules of ATP from each molecule of glucose in eukaryotic cells. This represents about one-third of the energy in the chemical bonds of glucose.

Regulating Aerobic Respiration

When cells possess plentiful amounts of ATP, the key reactions of glycolysis, the Krebs cycle, and fatty acid breakdown are inhibited, slowing ATP production. The regulation of these biochemical pathways by the level of ATP is an example of feedback inhibition. Conversely, when ATP levels in the cell are low, ADP levels are high; and ADP activates enzymes in the pathways of carbohydrate catabolism to stimulate the production of more ATP.

Control of glucose catabolism occurs at two key points of the catabolic pathway (figure 9.20). The control point in glycolysis is the enzyme phosphofructokinase, which catalyzes reaction 3, the conversion of fructose phosphate to fructose bisphosphate. This is the first reaction of glycolysis that is not readily reversible, committing the substrate to the glycolytic sequence. High levels of ADP relative to ATP (implying a need to convert more ADP to ATP) stimulate phosphofructokinase, committing more sugar to the catabolic pathway; so do low levels of citrate (implying the Krebs cycle is not running at full tilt and needs more input). The main control point in the oxidation of pyruvate occurs at the committing step in the Krebs cycle with the enzyme pyruvate decarboxylase. It is inhibited by high levels of NADH (implying no more is needed).

Another control point in the Krebs cycle is the enzyme citrate synthetase, which catalyzes the first reaction, the conversion of oxaloacetate and acetyl-CoA into citrate. High levels of ATP inhibit citrate synthetase (as well as pyruvate decarboxylase and two other Krebs cycle enzymes), shutting down the catabolic pathway.

Relative levels of ADP and ATP regulate the catabolism of glucose at key committing reactions.

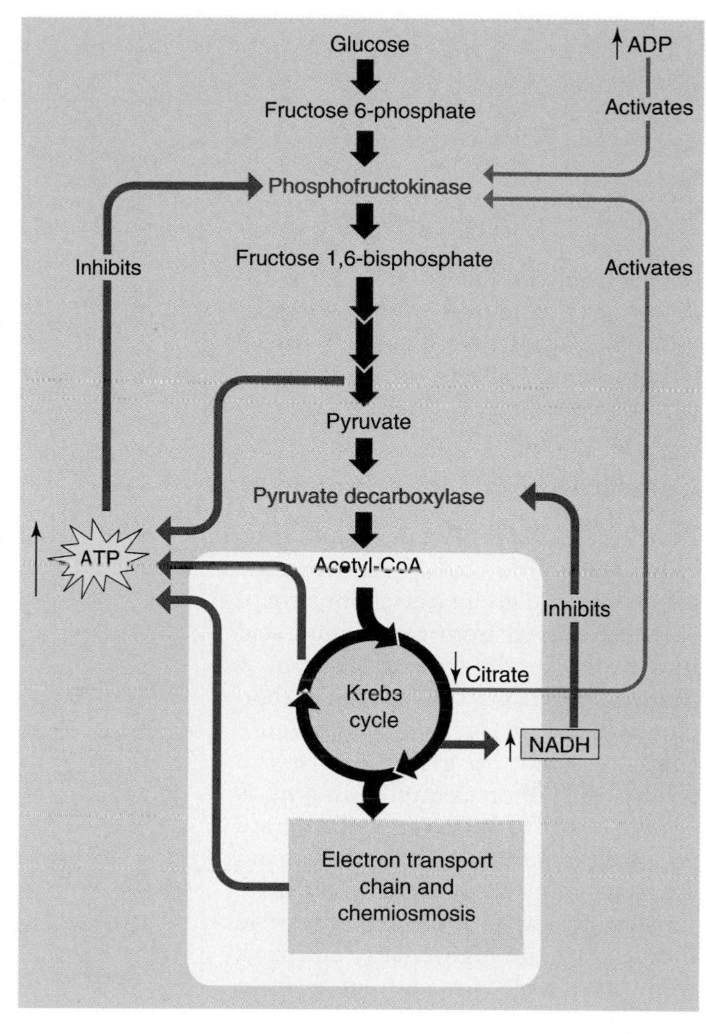

FIGURE 9.20
Control of glucose catabolism. The relative levels of ADP and ATP control the catabolic pathway at two key points: the committing reactions of glycolysis and the Krebs cycle.

A Vocabulary of ATP Generation

aerobic respiration The portion of cellular respiration that requires oxygen as an electron acceptor; it includes pyruvate oxidation, the Krebs cycle, and the electron transport chain.

anaerobic respiration Cellular respiration in which inorganic electron acceptors other than oxygen are used; it includes glycolysis.

cellular respiration The oxidation of organic molecules to produce ATP in which the final electron acceptor is oxygen, an inorganic molecule, or an organic molecule; it includes aerobic and anaerobic respiration.

chemiosmosis The passage of high-energy electrons along the electron transport chain, which is coupled to the pumping of protons across a membrane and the return of protons to the original side of the membrane through ATP-generating channels.

fermentation Alternative ATP-producing pathway performed by some cells in the absence of oxygen, in which the final electron acceptor is an organic molecule.

maximum efficiency The maximum number of ATP molecules generated by oxidizing a substance, relative to the free energy of that substance; in organisms, the actual efficiency is usually less than the maximum.

oxidation The loss of an electron. In cellular respiration, high-energy electrons are stripped from food molecules, oxidizing them.

photosynthesis The chemiosmotic generation of ATP and complex organic molecules powered by the energy derived from light.

substrate-level phosphorylation The generation of ATP by the direct transfer of a phosphate group to ADP from another phosphorylated molecule.

Glucose Is Not the Only Source of Energy

Thus far we have discussed oxidative respiration of glucose, which organisms obtain from the digestion of carbohydrates or from photosynthesis. Organic molecules other than glucose, particularly proteins and fats, are also important sources of energy (figure 9.21).

Cellular Respiration of Protein

Proteins are first broken down into their individual amino acids. The nitrogen-containing side group (the amino group) is then removed from each amino acid in a process called **deamination.** A series of reactions convert the carbon chain that remains into a molecule that takes part in glycolysis or the Krebs cycle. For example, alanine is converted into pyruvate, glutamate into α-ketoglutarate (figure 9.22), and aspartate into oxaloacetate. The reactions of glycolysis and the Krebs cycle then extract the high-energy electrons from these molecules and put them to work making ATP.

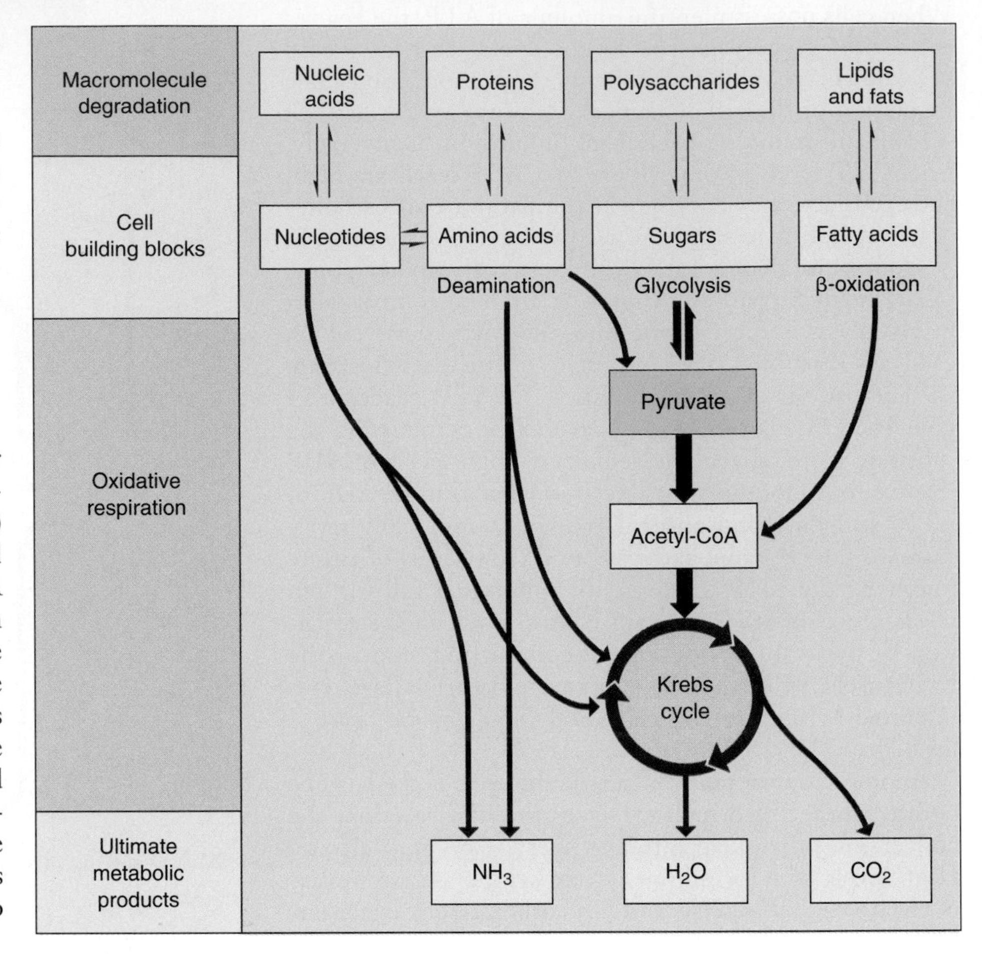

FIGURE 9.21
How cells extract chemical energy. All eukaryotes and many prokaryotes extract energy from organic molecules by oxidizing them. The first stage of this process, breaking down macromolecules into their constituent parts, yields little energy. The second stage, oxidative or aerobic respiration, extracts energy, primarily in the form of high-energy electrons, and produces water and carbon dioxide.

FIGURE 9.22
Deamination. After proteins are broken down into their amino acid constituents, the amino groups are removed from the amino acids to form molecules that participate in glycolysis and the Krebs cycle. For example, the amino acid glutamate becomes α-ketoglutarate, a Krebs cycle molecule, when it loses its amino group.

Cellular Respiration of Fat

Fats are broken down into fatty acids plus glycerol. The tails of fatty acids typically have 16 or more —CH_2 links, and the many hydrogen atoms in these long tails provide a rich harvest of energy. Fatty acids are oxidized in the matrix of the mitochondrion. Enzymes there remove the two-carbon acetyl groups from the end of each fatty acid tail until the entire fatty acid is converted into acetyl groups (figure 9.23). Each acetyl group then combines with coenzyme A to form acetyl-CoA. This process is known as **β-oxidation.**

How much ATP does the catabolism of fatty acids produce? Let's compare a hypothetical six-carbon fatty acid with the six-carbon glucose molecule, which we've said yields about 30 molecules of ATP in a eukaryotic cell. Two rounds of β-oxidation would convert the fatty acid into three molecules of acetyl-CoA. Each round requires one molecule of ATP to prime the process, but it also produces one molecule of NADH and one of $FADH_2$. These molecules together yield four molecules of ATP (assuming 2.5 ATPs per NADH and 1.5 ATPs per $FADH_2$). The oxidation of each acetyl-CoA in the Krebs cycle ultimately produces an additional 10 molecules of ATP. Overall, then, the ATP yield of a six-carbon fatty acid would be approximately 8 (from two rounds of β-oxidation) – 2 (for priming those two rounds) + 30 (from oxidizing the three acetyl-CoAs) = 36 molecules of ATP. Therefore, the respiration of a six-carbon fatty acid yields 20% more ATP than the respiration of glucose. Moreover, a fatty acid of that size would weigh less than two-thirds as much as glucose, so a gram of fatty acid contains more than twice as many kilocalories as a gram of glucose. That is why fat is a storage molecule for excess energy in many types of animals. If excess energy were stored instead as carbohydrate, as it is in plants, animal bodies would be much bulkier.

Proteins, fats, and other organic molecules are also metabolized for energy. The amino acids of proteins are first deaminated, while fats undergo a process called β-oxidation.

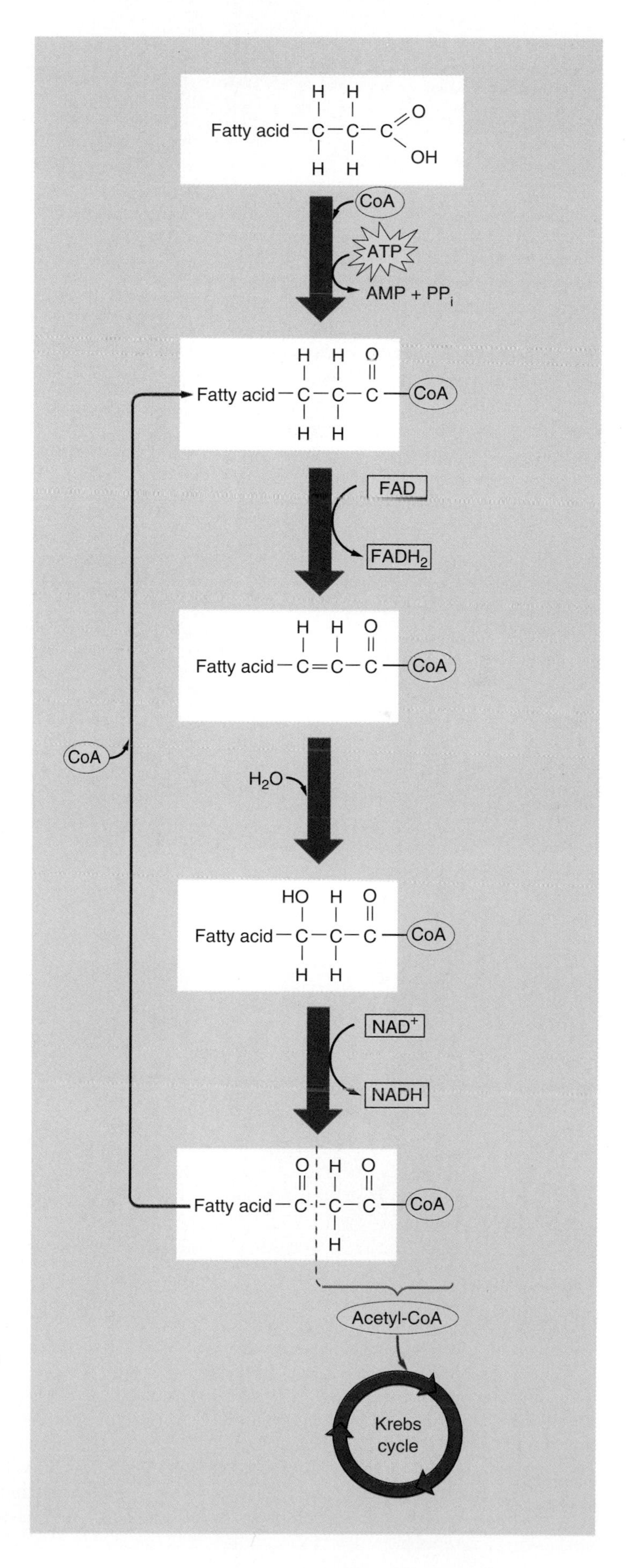

FIGURE 9.23
β-oxidation. Through a series of reactions known as β-oxidation, the last two carbons in a fatty acid tail combine with coenzyme A to form acetyl-CoA, which enters the Krebs cycle. The fatty acid, now two carbons shorter, enters the pathway again and keeps reentering until all its carbons have been used to form acetyl-CoA molecules. Each round of β-oxidation uses one molecule of ATP and generates one molecule each of $FADH_2$ and NADH, not including the molecules generated from the Krebs cycle.

Metabolic Efficiency and the Length of Food Chains

It has been estimated that a heterotroph limited to glycolysis captures only 3.5% of the energy in the food it consumes. Hence, if such a heterotroph preserves 3.5% of the energy in the autotrophs it consumes, then any other heterotrophs that consume the first hetertroph will capture through glycolysis 3.5% of the energy in it, or 0.12% of the energy available in the original autotrophs. A very large base of autotrophs would thus be needed to support a small number of heterotrophs.

When organisms became able to extract energy from organic molecules by oxidative metabolism, this constraint became far less severe, because the efficiency of oxidative respiration is estimated to be about 52 to 63%. This increased efficiency results in the transmission of much more energy from one trophic level to another than does glycolysis. (A trophic level is a step in the movement of energy through an ecosystem.) The efficiency of oxidative metabolism has made possible the evolution of food chains, in which autotrophs are consumed by heterotrophs, which are consumed by other heterotrophs, and so on. You will read more about food chains in chapter 28.

Even with oxidative metabolism, approximately two-thirds of the available energy is lost at each trophic level, and that puts a limit on how long a food chain can be. Most food chains, like the one illustrated in figure 9.A, involve only three or rarely four trophic levels. Too much energy is lost at each transfer to allow chains to be much longer than that. For example, it would be impossible for a large human population to subsist by eating lions captured from the Serengeti Plain of Africa; the amount of grass available there would not support enough zebras and other herbivores to maintain the number of lions needed to feed the human population. Thus, the ecological complexity of our world is fixed in a fundamental way by the chemistry of oxidative respiration.

Stage 1: Photosynthesizers

Stage 2: Herbivores

Stage 3: Carnivore

Stage 4: Scavengers

Stage 5: Refuse utilizers

FIGURE 9.A

A food chain in the savannas, or open grasslands, of East Africa. Stage 1: *Photosynthesizer.* The grass under these palm trees grows actively during the hot, rainy season, capturing the energy of the sun and storing it in molecules of glucose, which are then converted into starch and stored in the grass. Stage 2: *Herbivore.* These large antelopes, known as wildebeests, consume the grass and transfer some of its stored energy into their own bodies. Stage 3: *Carnivore.* The lion feeds on wildebeests and other animals, capturing part of their stored energy and storing it in its own body. Stage 4: *Scavenger.* This hyena and the vultures occupy the same stage in the food chain as the lion. They are also consuming the body of the dead wildebeest, which has been abandoned by the lion. Stage 5: *Refuse utilizer.* These butterflies, mostly *Precis octavia*, are feeding on the material left in the hyena's dung after the food the hyena consumed had passed through its digestive tract. At each of these four levels, only about a third or less of the energy present is used by the recipient.

9.4 Cells can metabolize food without oxygen.

Fermentation

In the absence of oxygen, aerobic metabolism cannot occur, and cells must rely exclusively on glycolysis to produce ATP. Under these conditions, the hydrogen atoms generated by glycolysis are donated to organic molecules in a process called **fermentation.**

Bacteria carry out more than a dozen kinds of fermentations, all using some form of organic molecule to accept the hydrogen atom from NADH and thus recycle NAD^+:

$$\text{Organic molecule} + \text{NADH} \longrightarrow \text{Reduced organic molecule} + NAD^+$$

Often the reduced organic compound is an organic acid—such as acetic acid, butyric acid, propionic acid, or lactic acid—or an alcohol.

Ethanol Fermentation

Eukaryotic cells are capable of only a few types of fermentation. In one type, which occurs in single-celled fungi called yeast, the molecule that accepts hydrogen from NADH is pyruvate, the end product of glycolysis itself. Yeast enzymes remove a terminal CO_2 group from pyruvate through decarboxylation, producing a two-carbon molecule called acetaldehyde. The CO_2 released causes bread made with yeast to rise, while bread made without yeast (unleavened bread) does not. The acetaldehyde accepts a hydrogen atom from NADH, producing NAD^+ and **ethanol** (ethyl alcohol) (figure 9.24). This particular type of fermentation is of great interest to humans, since it is the source of the ethanol in wine and beer. Ethanol is a by-product of fermentation that is actually toxic to yeast; as it approaches a concentration of about 12%, it begins to kill the yeast. That explains why naturally fermented wine contains only about 12% ethanol.

Lactic Acid Fermentation

Most animal cells regenerate NAD^+ without decarboxylation. Muscle cells, for example, use an enzyme called lactate dehydrogenase to transfer a hydrogen atom from NADH back to the pyruvate that is produced by glycolysis. This reaction converts pyruvate into lactic acid and regenerates NAD^+ from NADH. It therefore closes the metabolic circle, allowing glycolysis to continue as long as glucose is available. Circulating blood removes excess lactate (the ionized form of lactic acid) from muscles, but when removal cannot keep pace with production, the accumulating lactic acid interferes with muscle function and contributes to muscle fatigue.

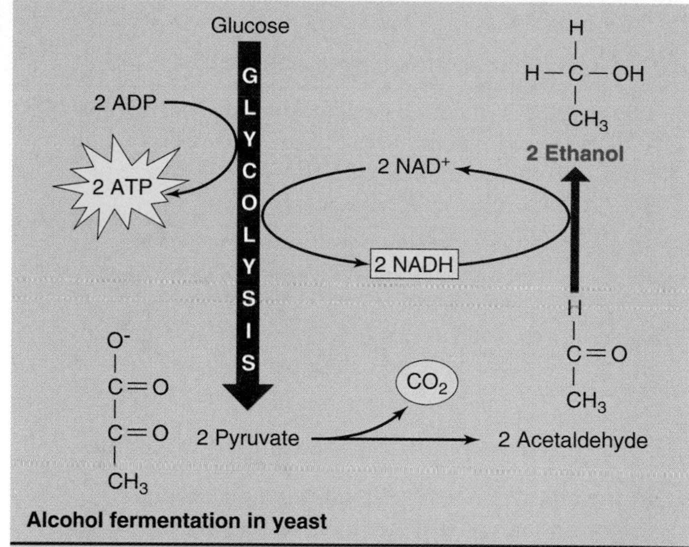

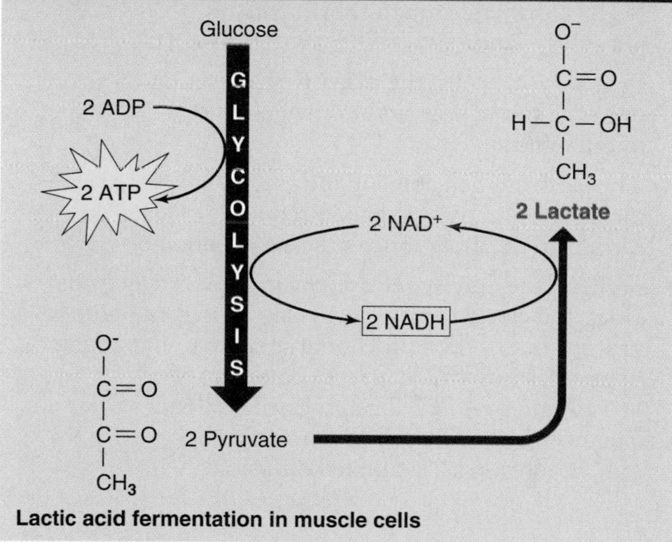

FIGURE 9.24
Fermentation. Yeasts carry out the conversion of pyruvate to ethanol (see also figure 9.10). Muscle cells convert pyruvate into lactate, which is less toxic than ethanol. However, lactate is still toxic enough to produce a painful sensation in muscles during heavy exercise, when oxygen in the muscles is depleted.

In fermentation, which occurs in the absence of oxygen, the electrons that result from the glycolytic breakdown of glucose are donated to an organic molecule, regenerating NAD^+ from NADH.

Chapter 9

www.mhhe.com/raven6/resources9.mhtml

Summary	*Questions*	*Media Resources*

9.1 Cells harvest the energy in chemical bonds.

- The reactions of cellular respiration are oxidation-reduction (redox) reactions. Those that require a net input of free energy are coupled to the cleavage of ATP, which releases free energy.
- The mechanics of cellular respiration are often dictated by electron behavior, which is in turn influenced by the presence of electron acceptors. Some atoms, such as oxygen, are very electronegative and thus behave as good oxidizing agents.

1. What is the difference between an autotroph and a heterotroph? How does each obtain energy?

2. What is the difference between digestion and catabolism? Which provides more energy?

9.2 Cellular respiration oxidizes food molecules.

- In the absence of oxygen, inorganic or organic molecules are used as the final electron acceptor and ATP is formed through substrate-level phosphorylation.
- In eukaryotic cells, the oxidative respiration of pyruvate takes place within the matrix of mitochondria.
- The electrons generated in the process are passed along the electron transport chain, a sequence of electron carriers in the inner mitochondrial membrane.
- Some of the energy released by passage of electrons along the electron transport chain is used to pump protons out of the mitochondrial matrix. The reentry of protons into the matrix is coupled to the production of ATP. This process is called chemiosmosis. The ATP then leaves the mitochondrion by facilitated diffusion.

3. Where in a eukaryotic cell does glycolysis occur? What is the net production of ATP during glycolysis, and why is it different from the number of ATP molecules synthesized during glycolysis?

4. By what two mechanisms can the NADH that results from glycolysis be converted back into NAD^+?

5. What is the theoretical maximum number of ATP molecules produced during the oxidation of a glucose molecule by these processes? Why is the actual number of ATP molecules produced usually lower than the theoretical maximum?

- Exploration: Oxidative Respiration
- Art Activity: Aerobic Cellular Respiration

- Electron Transport and ATP

- Glycolysis
- Krebs Cycle
- Electron Transport

- Art Quizzes:
  - -Substrate-Level Phosphorylation
  - -Pyruvate Pathways
  - -Electron Transport System
  - -Overview of ATP Synthesis

9.3 Catabolism of proteins and fats can yield considerable energy.

- The catabolism of proteins begins with the removal of the amino group from the amino acids through a process called deamination. The remaining carbon chain funnels into glycolysis or the Krebscycle.
- The catabolism of fatty acids begins with β-oxidation and provides more energy than the catabolism of carbohydrates.

6. How is acetyl-CoA produced during the aerobic oxidation of carbohydrates, and what happens to it? How is it produced during the aerobic oxidation of fatty acids, and what happens to it then?

- Art Quiz:
  - -Catabolism of Proteins and Fats

9.4 Cells can metabolize food without oxygen.

- Fermentation is an anaerobic process that uses an organic molecule instead of oxygen as a final electron acceptor.
- Fermentation occurs in bacteria as well as eukaryotic cells, including yeast and the muscle cells of animals.

7. What is the efficiency of aerobic oxidation of glucose compared to that of anaerobic fermentation?

- Fermentation

10

Photosynthesis

FIGURE 10.1
Capturing energy. These sunflower plants, growing vigorously in the August sun, are capturing light energy for conversion into chemical energy through photosynthesis.

Life on earth would be impossible without photosynthesis. Every oxygen atom in the air we breathe was once part of a water molecule, liberated by photosynthesis. The energy released by the burning of coal, firewood, gasoline, and natural gas, and by our bodies' burning of all the food we eat—all, directly or indirectly, has been captured from sunlight by photosynthesis. It is vitally important that we understand photosynthesis. Research may enable us to improve crop yields and land use, important goals in an increasingly crowded world. In the previous chapter we described how cells extract chemical energy from food molecules and use that energy to power their activities. In this chapter, we will examine photosynthesis, the process by which organisms capture energy from sunlight and use it to build food molecules rich in chemical energy (figure 10.1).

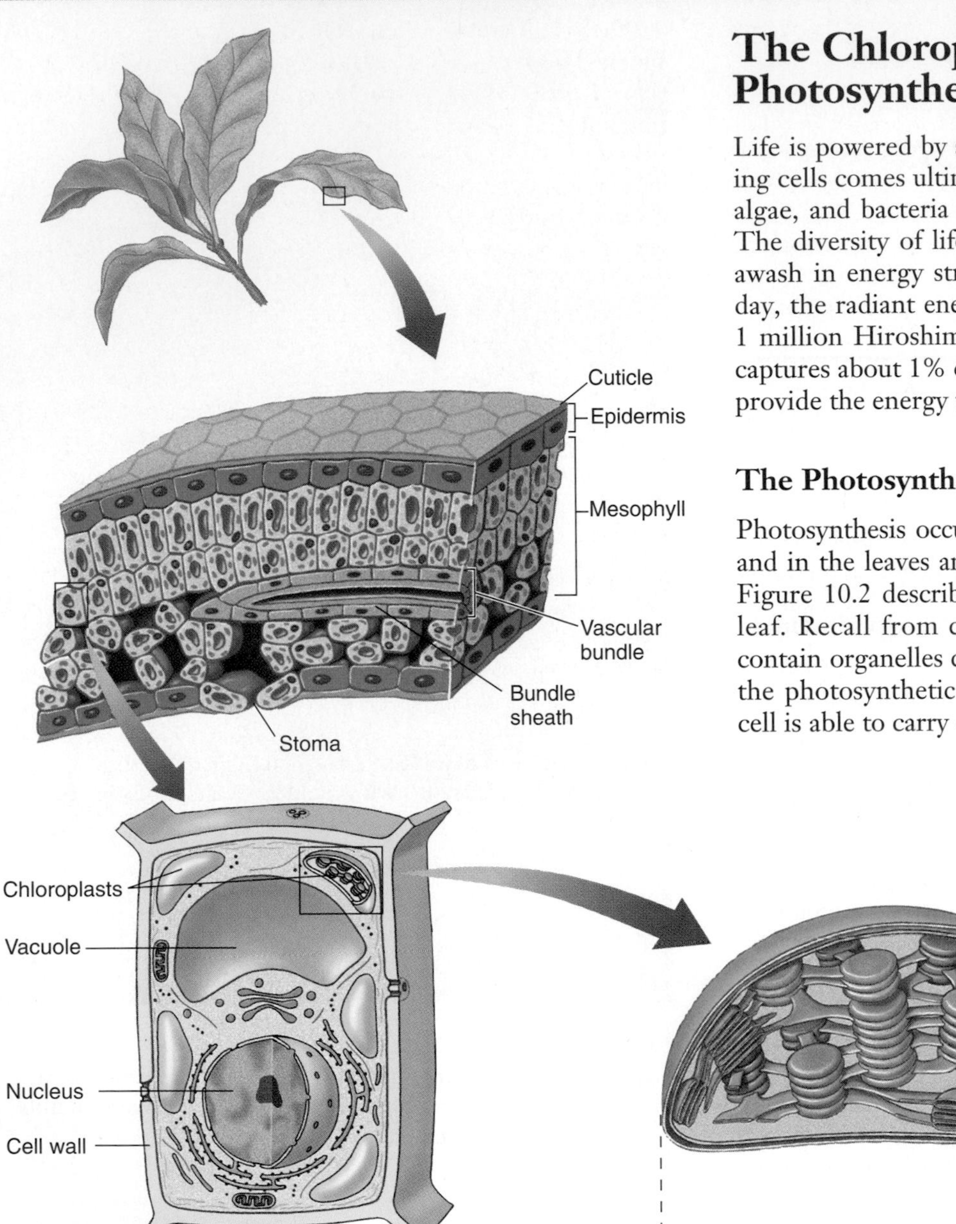

The Chloroplast as a Photosynthetic Machine

Life is powered by sunshine. The energy used by most living cells comes ultimately from the sun, captured by plants, algae, and bacteria through the process of photosynthesis. The diversity of life is only possible because our planet is awash in energy streaming earthward from the sun. Each day, the radiant energy that reaches the earth equals about 1 million Hiroshima-sized atomic bombs. Photosynthesis captures about 1% of this huge supply of energy, using it to provide the energy that drives all life.

The Photosynthetic Process: A Summary

Photosynthesis occurs in many kinds of bacteria and algae, and in the leaves and sometimes the stems of green plants. Figure 10.2 describes the levels of organization in a plant leaf. Recall from chapter 5 that the cells of plant leaves contain organelles called chloroplasts that actually carry out the photosynthetic process. No other structure in a plant cell is able to carry out photosynthesis. Photosynthesis takes

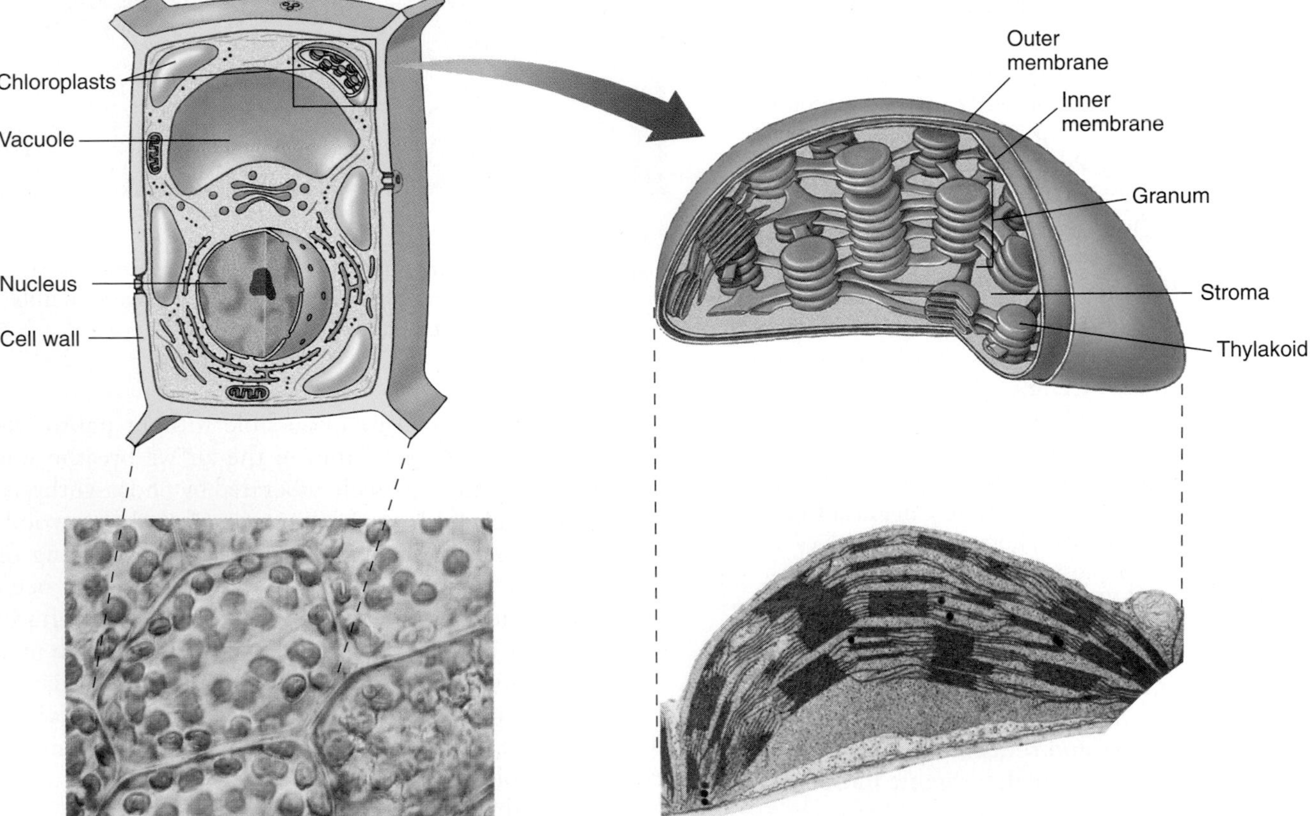

FIGURE 10.2
Journey into a leaf. A plant leaf possesses a thick layer of cells (the mesophyll) rich in chloroplasts. The flattened thylakoids in the chloroplast are stacked into columns called grana (singular, granum). The light reactions take place on the thylakoid

place in three stages: (1) capturing energy from sunlight; (2) using the energy to make ATP and reducing power in the form of a compound called NADPH; and (3) using the ATP and NADPH to power the synthesis of organic molecules from CO_2 in the air (carbon fixation).

The first two stages take place in the presence of light and are commonly called the **light reactions.** The third stage, the formation of organic molecules from atmospheric CO_2, is called the **Calvin cycle.** As long as ATP and NADPH are available, the Calvin cycle may occur in the absence of light.

The following simple equation summarizes the overall process of photosynthesis:

$$\underset{\text{carbon dioxide}}{6\,CO_2} + \underset{\text{water}}{12\,H_2O} + \text{light} \longrightarrow \underset{\text{glucose}}{C_6H_{12}O_6} + \underset{\text{water}}{6\,H_2O} + \underset{\text{oxygen}}{6\,O_2}$$

Inside the Chloroplast

The internal membranes of chloroplasts are organized into sacs called *thylakoids*, and often numerous thylakoids are stacked on one another in columns called *grana*. The thylakoid membranes house the photosynthetic pigments for capturing light energy and the machinery to make ATP. Surrounding the thylakoid membrane system is a semiliquid substance called *stroma*. The stroma houses the enzymes needed to assemble carbon molecules. In the membranes of thylakoids, photosynthetic pigments are clustered together to form a **photosystem.**

Each pigment molecule within the photosystem is capable of capturing *photons*, which are packets of energy. A lattice of proteins holds the pigments in close contact with one another. When light of a proper wavelength strikes a pigment molecule in the photosystem, the resulting excitation passes from one chlorophyll molecule to another. The excited electron is not transferred physically—it is the *energy* that passes from one molecule to another. A crude analogy to this form of energy transfer is the initial "break" in a game of pool. If the cue ball squarely hits the point of the triangular array of 15 pool balls, the two balls at the far corners of the triangle fly off, but none of the central balls move. The energy passes through the central balls to the most distant ones.

Eventually the energy arrives at a key chlorophyll molecule that is touching a membrane-bound protein. The energy is transferred as an excited electron to that protein, which passes it on to a series of other membrane proteins that put the energy to work making ATP and NADPH and building organic molecules. The photosystem thus acts as a large antenna, gathering the light harvested by many individual pigment molecules.

The reactions of photosynthesis take place within thylakoid membranes within chloroplasts in leaf cells.

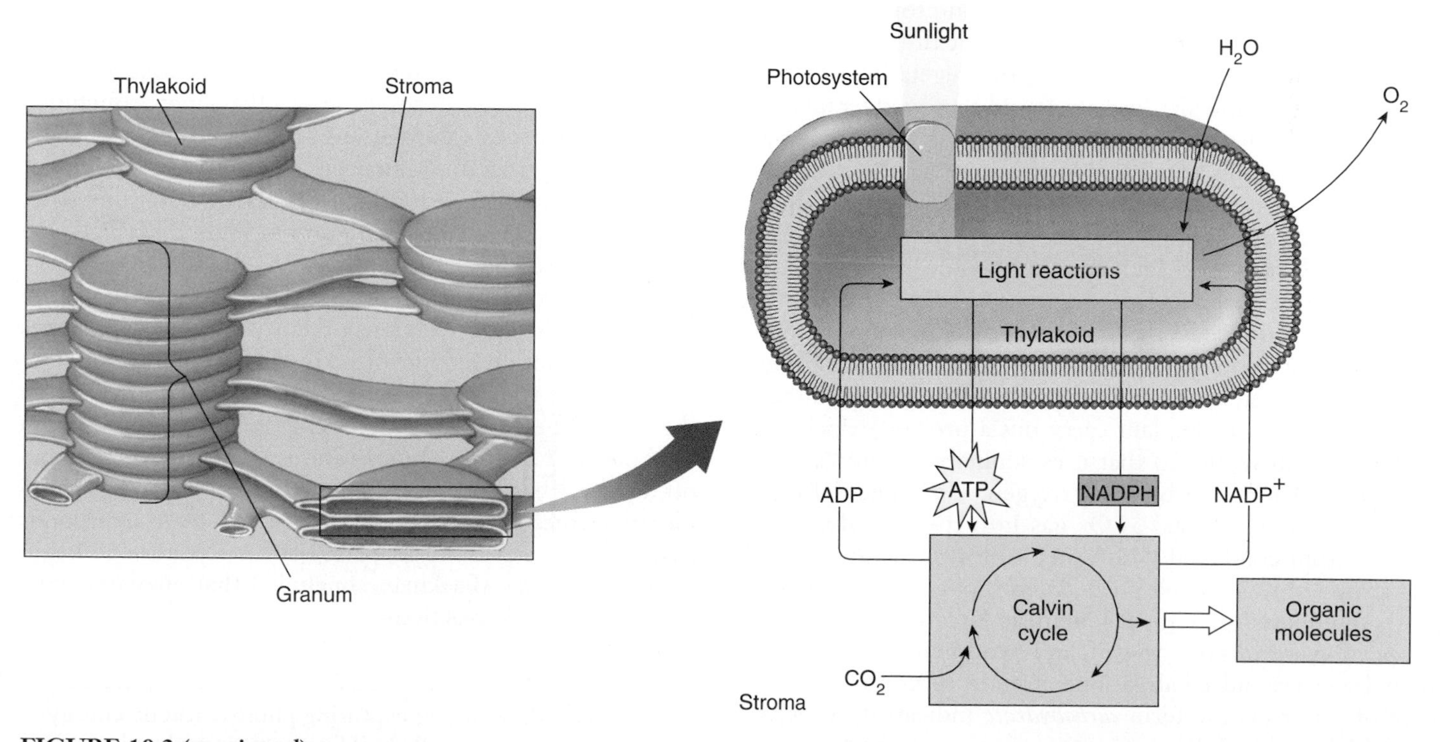

FIGURE 10.2 (continued)
membrane and generate the ATP and NADPH that fuel the Calvin cycle. The fluid interior matrix of a chloroplast, the stroma, contains the enzymes that carry out the Calvin cycle.

10.2 Learning about photosynthesis: An experimental journey.

The Role of Soil and Water

The story of how we learned about photosynthesis is one of the most interesting in science and serves as a good introduction to this complex process. The story starts over 300 years ago, with a simple but carefully designed experiment by a Belgian doctor, Jan Baptista van Helmont (1577–1644). From the time of the Greeks, plants were thought to obtain their food from the soil, literally sucking it up with their roots; van Helmont thought of a simple way to test the idea. He planted a small willow tree in a pot of soil after weighing the tree and the soil. The tree grew in the pot for several years, during which time van Helmont added only water. At the end of five years, the tree was much larger: its weight had increased by 74.4 kilograms. However, *all of this added mass could not have come from the soil*, because the soil in the pot weighed only 57 grams less than it had five years earlier! With this experiment, van Helmont demonstrated that the substance of the plant was not produced only from the soil. He incorrectly concluded that mainly the water he had been adding accounted for the plant's increased mass.

A hundred years passed before the story became clearer. The key clue was provided by the English scientist Joseph Priestly, in his pioneering studies of the properties of air. On the 17th of August, 1771, Priestly "accidentally hit upon a method of restoring air that had been injured by the burning of candles." He "put a [living] sprig of mint into air in which a wax candle had burnt out and found that, on the 27th of the same month, another candle could be burned in this same air." Somehow, the vegetation seemed to have restored the air! Priestly found that while a mouse could not breathe candle-exhausted air, air "restored" by vegetation was not "at all inconvenient to a mouse." The key clue was that living vegetation *adds something to the air*.

How does vegetation "restore" air? Twenty-five years later, Dutch physician Jan Ingenhousz solved the puzzle. Working over several years, Ingenhousz reproduced and significantly extended Priestly's results, demonstrating that air was restored only in the presence of sunlight, and only by a plant's green leaves, not by its roots. He proposed that the green parts of the plant carry out a process (which we now call photosynthesis) that uses sunlight to split carbon dioxide (CO_2) into carbon and oxygen. He suggested that the oxygen was released as O_2 gas into the air, while the carbon atom combined with water to form carbohydrates. His proposal was a good guess, even though the later step was subsequently modified. Chemists later found that the proportions of carbon, oxygen, and hydrogen atoms in carbohydrates are indeed about one atom of carbon per molecule of water (as the term *carbohydrate* indicates). A Swiss botanist found in 1804 that water was a necessary reactant. By the end of that century the overall reaction for photosynthesis could be written as:

$$CO_2 + H_2O + \text{light energy} \longrightarrow (CH_2O) + O_2$$

It turns out, however, that there's more to it than that. When researchers began to examine the process in more detail in the last century, the role of light proved to be unexpectedly complex.

Van Helmont showed that soil did not add mass to a growing plant. Priestly and Ingenhousz and others then worked out the basic chemical reaction.

Discovery of the Light-Independent Reactions

Ingenhousz's early equation for photosynthesis includes one factor we have not discussed: light energy. What role does light play in photosynthesis? At the beginning of the previous century, the English plant physiologist F. F. Blackman began to address the question of the role of light in photosynthesis. In 1905, he came to the startling conclusion that photosynthesis is in fact a two-stage process, only one of which uses light directly.

Blackman measured the effects of different light intensities, CO_2 concentrations, and temperatures on photosynthesis. As long as light intensity was relatively low, he found photosynthesis could be accelerated by increasing the amount of light, but not by increasing the temperature or CO_2 concentration (figure 10.3). At high light intensities, however, an increase in temperature or CO_2 concentration greatly accelerated photosynthesis. Blackman concluded that photosynthesis consists of an initial set of what he called "light" reactions, that are largely independent of temperature, and a second set of "dark" reactions, that seemed to be independent of light but limited by CO_2. Do not be confused by Blackman's labels—the so-called "dark" reactions occur in the light (in fact, they require the products of the light reactions); their name simply indicates that light is not directly involved in those reactions.

Blackman found that increased temperature increases the rate of the dark carbon-reducing reactions, but only up to about 35°C. Higher temperatures caused the rate to fall off rapidly. Because 35°C is the temperature at which many plant enzymes begin to be denatured (the hydrogen bonds that hold an enzyme in its particular catalytic shape begin to be disrupted), Blackman concluded that enzymes must carry out the dark reactions.

Blackman showed that capturing photosynthetic energy requires sunlight, while building organic molecules does not.

The Role of Light

The role of light in the so-called light and dark reactions was worked out in the 1930s by C. B. van Niel, then a graduate student at Stanford University studying photosynthesis in bacteria. One of the types of bacteria he was studying, the purple sulfur bacteria, does not release oxygen during photosynthesis; instead, they convert hydrogen sulfide (H_2S) into globules of pure elemental sulfur that accumulate inside themselves. The process that van Niel observed was

$$CO_2 + 2\ H_2S + \text{light energy} \rightarrow (CH_2O) + H_2O + 2\ S$$

The striking parallel between this equation and Ingenhousz's equation led van Niel to propose that the generalized process of photosynthesis is in fact

$$CO_2 + 2\ H_2A + \text{light energy} \rightarrow (CH_2O) + H_2O + 2\ A$$

In this equation, the substance H_2A serves as an electron donor. In photosynthesis performed by green plants, H_2A is water, while among purple sulfur bacteria, H_2A is hydrogen sulfide. The product, A, comes from the splitting of H_2A. Therefore, the O_2 produced during green plant photosynthesis results from splitting water, not carbon dioxide.

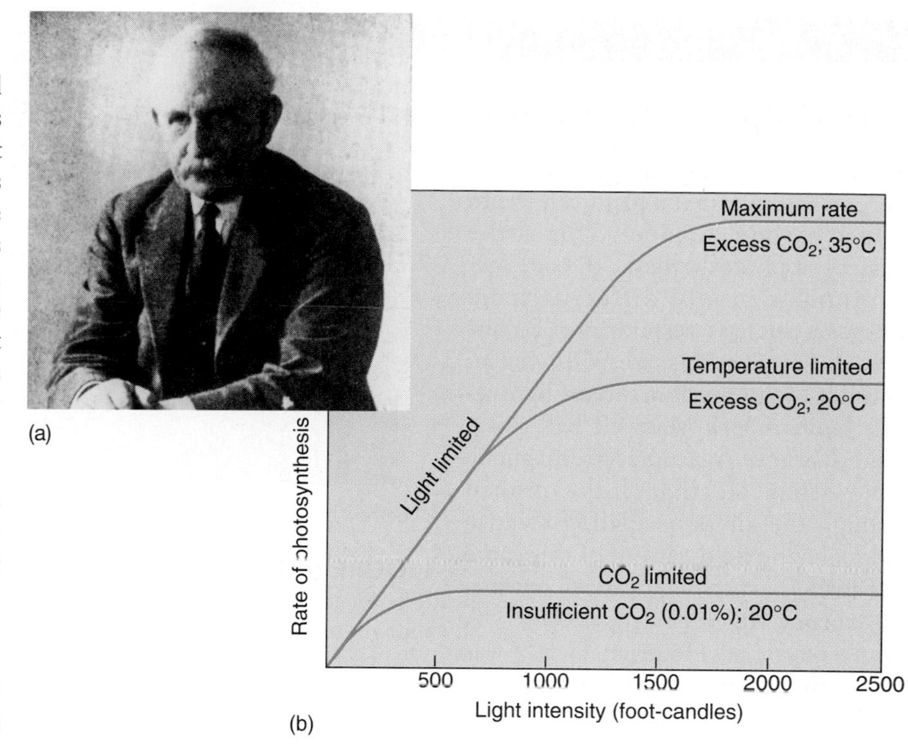

FIGURE 10.3
Discovery of the dark reactions. (*a*) Blackman measured photosynthesis rates under differing light intensities, CO_2 concentrations, and temperatures. (*b*) As this graph shows, light is the limiting factor at low light intensities, while temperature and CO_2 concentration are the limiting factors at higher light intensities.

When isotopes came into common use in biology in the early 1950s, it became possible to test van Niel's revolutionary proposal. Investigators examined photosynthesis in green plants supplied with ^{18}O water; they found that the ^{18}O label ended up in oxygen gas rather than in carbohydrate, just as van Niel had predicted:

$$CO_2 + 2\ H_2{}^{18}O + \text{light energy} \longrightarrow (CH_2O) + H_2O + {}^{18}O_2$$

In algae and green plants, the carbohydrate typically produced by photosynthesis is the sugar glucose, which has six carbons. The complete balanced equation for photosynthesis in these organisms thus becomes

$$6\ CO_2 + 12\ H_2O + \text{light energy} \longrightarrow C_6H_{12}O_6 + 6\ O_2 + 6\ H_2O.$$

We now know that the first stage of photosynthesis, the light reactions, uses the energy of light to reduce NADP (an electron carrier molecule) to NADPH and to manufacture ATP. The NADPH and ATP from the first stage of photosynthesis are then used in the second stage, the Calvin cycle, to reduce the carbon in carbon dioxide and form a simple sugar whose carbon skeleton can be used to synthesize other organic molecules.

Van Niel discovered that photosynthesis splits water molecules, incorporating the carbon atoms of carbon dioxide gas and the hydrogen atoms of water into organic molecules and leaving oxygen gas.

The Role of Reducing Power

In his pioneering work on the light reactions, van Niel had further proposed that the reducing power (H^+) generated by the splitting of water was used to convert CO_2 into organic matter in a process he called **carbon fixation**. Was he right?

In the 1950s Robin Hill demonstrated that van Niel was indeed right, and that light energy could be used to generate reducing power. Chloroplasts isolated from leaf cells were able to reduce a dye and release oxygen in response to light. Later experiments showed that the electrons released from water were transferred to $NADP^+$. Arnon and coworkers showed that illuminated chloroplasts deprived of CO_2 accumulate ATP. If CO_2 is then introduced, neither ATP nor NADPH accumulate, and the CO_2 is assimilated into organic molecules. These experiments are important for three reasons. First, they firmly demonstrate that photosynthesis occurs only within chloroplasts. Second, they show that the light-dependent reactions use light energy to reduce $NADP^+$ and to manufacture ATP. Thirdly, they confirm that the ATP and NADPH from this early stage of photosynthesis are then used in the later light-independent reactions to reduce carbon dioxide, forming simple sugars.

Hill showed that plants can use light energy to generate reducing power. The incorporation of carbon dioxide into organic molecules in the light-independent reactions is called carbon fixation.

10.3 Pigments capture energy from sunlight.

The Biophysics of Light

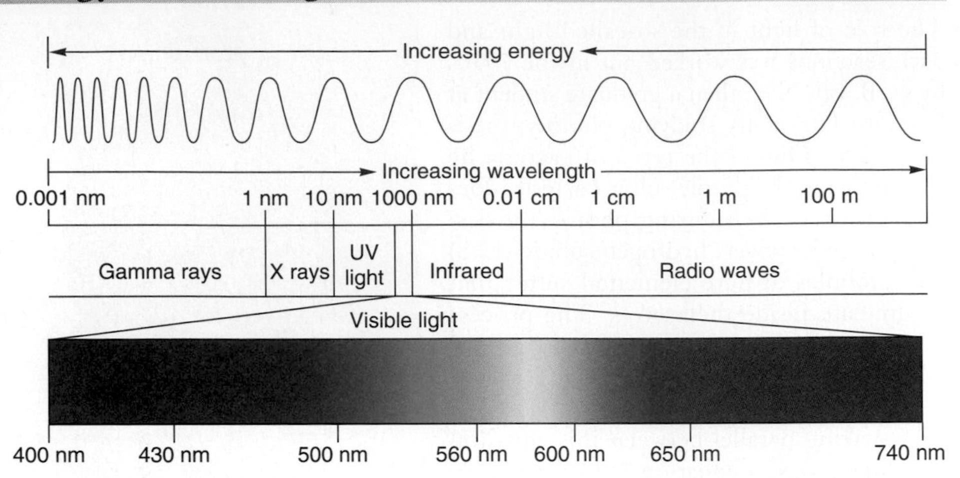

FIGURE 10.4
The electromagnetic spectrum. Light is a form of electromagnetic energy conveniently thought of as a wave. The shorter the wavelength of light, the greater its energy. Visible light represents only a small part of the electromagnetic spectrum between 400 and 740 nanometers.

Where is the energy in light? What is there in sunlight that a plant can use to reduce carbon dioxide? This is the mystery of photosynthesis, the one factor fundamentally different from processes such as respiration. To answer these questions, we will need to consider the physical nature of light itself. James Clerk Maxwell had theorized that light was an electromagnetic wave—that is, that light moved through the air as oscillating electric and magnetic fields. Proof of this came in a curious experiment carried out in a laboratory in Germany in 1887. A young physicist, Heinrich Hertz, was attempting to verify a highly mathematical theory that predicted the existence of electromagnetic waves. To see whether such waves existed, Hertz designed a clever experiment. On one side of a room he constructed a powerful spark generator that consisted of two large, shiny metal spheres standing near each other on tall, slender rods. When a very high static electrical charge was built up on one sphere, sparks would jump across to the other sphere.

After constructing this device, Hertz set out to investigate whether the sparking would create invisible electromagnetic waves, so-called radio waves, as predicted by the mathematical theory. On the other side of the room, he placed the world's first radio receiver, a thin metal hoop on an insulating stand. There was a small gap at the bottom of the hoop, so that the hoop did not quite form a complete circle. When Hertz turned on the spark generator across the room, he saw tiny sparks passing across the gap in the hoop! This was the first demonstration of radio waves. But Hertz noted another curious phenomenon. When UV light was shining across the gap on the hoop, the sparks were produced more readily. This unexpected facilitation, called the photoelectric effect, puzzled investigators for many years.

The photoelectric effect was finally explained using a concept proposed by Max Planck in 1901. Planck developed an equation that predicted the blackbody radiation curve based upon the assumption that light and other forms of radiation behaved as units of energy called photons. In 1905 Albert Einstein explained the photoelectric effect utilizing the photon concept. Ultraviolet light has photons of sufficient energy that when they fell on the loop, electrons were ejected from the metal surface. The photons had transferred their energy to the electrons, literally blasting them from the ends of the hoop and thus facilitating the passage of the electric spark induced by the radio waves. Visible wavelengths of light were unable to remove the electrons because their photons did not have enough energy to free the electrons from the metal surface at the ends of the hoop.

The Energy in Photons

Photons do not all possess the same amount of energy (figure 10.4). Instead, the energy content of a photon is inversely proportional to the wavelength of the light: short-wavelength light contains photons of higher energy than long-wavelength light. X rays, which contain a great deal of energy, have very short wavelengths—much shorter than visible light, making them ideal for high-resolution microscopes.

Hertz had noted that the strength of the photoelectric effect depends on the wavelength of light; short wavelengths are much more effective than long ones in producing the photoelectric effect. Einstein's theory of the photoelectric effect provides an explanation: sunlight contains photons of many different energy levels, only some of which our eyes perceive as visible light. The highest energy photons, at the short-wavelength end of the electromagnetic spectrum (see figure 10.4), are gamma rays, with wavelengths of less than 1 nanometer; the lowest energy photons, with wavelengths of up to thousands of meters, are radio waves. Within the visible portion of the spectrum, violet light has the shortest wavelength and the most energetic photons, and red light has the longest wavelength and the least energetic photons.

Ultraviolet Light

The sunlight that reaches the earth's surface contains a significant amount of ultraviolet (UV) light, which, because of its shorter wavelength, possesses considerably more energy than visible light. UV light is thought to have been an important source of energy on the primitive earth when life originated. Today's atmosphere contains ozone (derived from oxygen gas), which absorbs most of the UV photons in sunlight, but a considerable amount of UV light still manages to penetrate the atmosphere. This UV light is a potent force in disrupting the bonds of DNA, causing mutations that can lead to skin cancer. As we will describe in a later chapter, loss of atmospheric ozone due to human activities threatens to cause an enormous jump in the incidence of human skin cancers throughout the world.

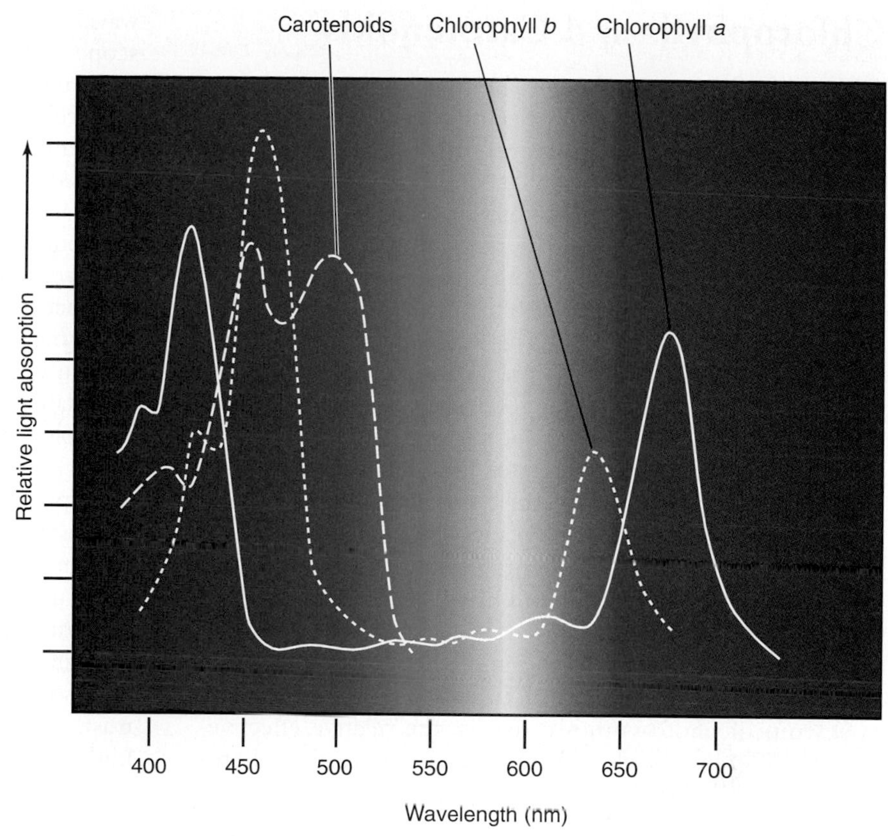

FIGURE 10.5
Absorption spectra of chlorophylls and carotenoids. The peaks represent wavelengths of sunlight absorbed by the two common forms of photosynthetic pigment, chlorophyll *a* and chlorophyll *b*, and carotenoids. Chlorophylls absorb predominately violet-blue and red light in two narrow bands of the spectrum and reflect the green light in the middle of the spectrum. Carotenoids absorb mostly blue and green light and reflect orange and yellow light.

Absorption Spectra and Pigments

How does a molecule "capture" the energy of light? A photon can be envisioned as a very fast-moving packet of energy. When it strikes a molecule, its energy is either lost as heat or absorbed by the electrons of the molecule, boosting those electrons into higher energy levels. Whether or not the photon's energy is absorbed depends on how much energy it carries (defined by its wavelength) and on the chemical nature of the molecule it hits. As we saw in chapter 2, electrons occupy discrete energy levels in their orbits around atomic nuclei. To boost an electron into a different energy level requires just the right amount of energy, just as reaching the next rung on a ladder requires you to raise your foot just the right distance. A specific atom can, therefore, absorb only certain photons of light—namely, those that correspond to the atom's available electron energy levels. As a result, each molecule has a characteristic **absorption spectrum,** the range and efficiency of photons it is capable of absorbing.

Molecules that are good absorbers of light in the visible range are called **pigments.** Organisms have evolved a variety of different pigments, but there are only two general types used in green plant photosynthesis: carotenoids and chlorophylls. Chlorophylls absorb photons within narrow energy ranges. Two kinds of chlorophyll in plants, chlorophylls *a* and *b*, preferentially absorb violet-blue and red light (figure 10.5). Neither of these pigments absorbs photons with wavelengths between about 500 and 600 nanometers, and light of these wavelengths is, therefore, reflected by plants. When these photons are subsequently absorbed by the pigment in our eyes, we perceive them as green.

Chlorophyll *a* is the main photosynthetic pigment and is the only pigment that can act directly to convert light energy to chemical energy. However, chlorophyll *b*, acting as an **accessory** or secondary light-absorbing pigment, complements and adds to the light absorption of chlorophyll *a*. Chlorophyll *b* has an absorption spectrum shifted toward the green wavelengths. Therefore, chlorophyll *b* can absorb photons chlorophyll *a* cannot. Chlorophyll *b* therefore greatly increases the proportion of the photons in sunlight that plants can harvest. An important group of accessory pigments, the carotenoids, assist in photosynthesis by capturing energy from light of wavelengths that are not efficiently absorbed by either chlorophyll.

In photosynthesis, photons of light are absorbed by pigments; the wavelength of light absorbed depends upon the specific pigment.

Chlorophylls and Carotenoids

Chlorophylls absorb photons by means of an excitation process analogous to the photoelectric effect. These pigments contain a complex ring structure, called a porphyrin ring, with alternating single and double bonds. At the center of the ring is a magnesium atom. Photons absorbed by the pigment molecule excite electrons in the ring, which are then channeled away through the alternating carbon-bond system. Several small side groups attached to the outside of the ring alter the absorption properties of the molecule in different kinds of chlorophyll (figure 10.6). The precise absorption spectrum is also influenced by the local microenvironment created by the association of chlorophyll with specific proteins.

Once Ingenhousz demonstrated that only the green parts of plants can "restore" air, researchers suspected chlorophyll was the primary pigment that plants employ to absorb light in photosynthesis. Experiments conducted in the 1800s clearly verified this suspicion. One such experiment, performed by T. W. Englemann in 1882 (figure 10.7), serves as a particularly elegant example, simple in design and clear in outcome. Englemann set out to characterize the **action spectrum** of photosynthesis, that is, the relative effectiveness of different wavelengths of light in promoting photosynthesis. He carried out the entire experiment utilizing a single slide mounted on a microscope. To obtain different wavelengths of light, he placed a prism under his microscope, splitting the light that illuminated the slide into a spectrum of colors. He then arranged a filament of green algal cells across the spectrum, so that different parts of the filament were illuminated with different wavelengths, and allowed the algae to carry out photosynthesis. To assess how fast photosynthesis was proceeding, Englemann chose to monitor the rate of oxygen production. Lacking a mass spectrometer and other modern instruments, he added aerotactic (oxygen-seeking) bacteria to the slide; he knew they would gather along the filament at locations where oxygen was being produced. He found that the bacteria accumulated in areas illuminated by red and violet light, the two colors most strongly absorbed by chlorophyll.

All plants, algae, and cyanobacteria use chlorophyll *a* as their primary pigments. It is reasonable to ask why these photosynthetic organisms do not use a pigment like retinal (the pigment in our eyes), which has a broad absorption spectrum that covers the range of 500 to 600 nanometers. The most likely hypothesis involves *photoefficiency*. Although retinal absorbs a broad range of wavelengths, it does so with relatively low efficiency. Chlorophyll, in contrast, absorbs in only two narrow bands, but does so with high efficiency. Therefore, plants and most other photosynthetic organisms achieve far higher overall photon capture rates with chlorophyll than with other pigments.

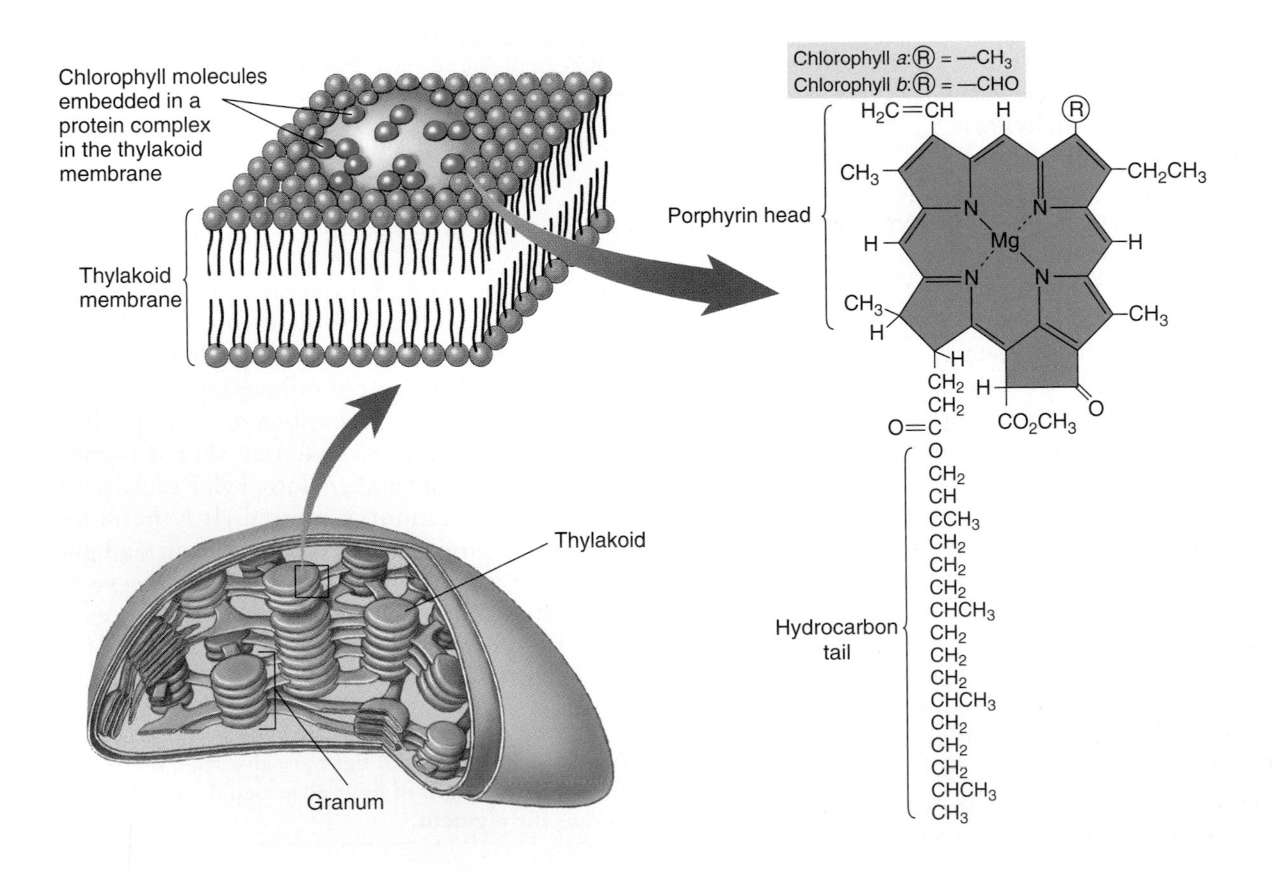

FIGURE 10.6 Chlorophyll. Chlorophyll molecules consist of a porphyrin head and a hydrocarbon tail that anchors the pigment molecule to hydrophobic regions of proteins embedded within the membranes of thylakoids. The only difference between the two chlorophyll molecules is the substitution of a —CHO (aldehyde) group in chlorophyll *b* for a $—CH_3$ (methyl) group in chlorophyll *a*.

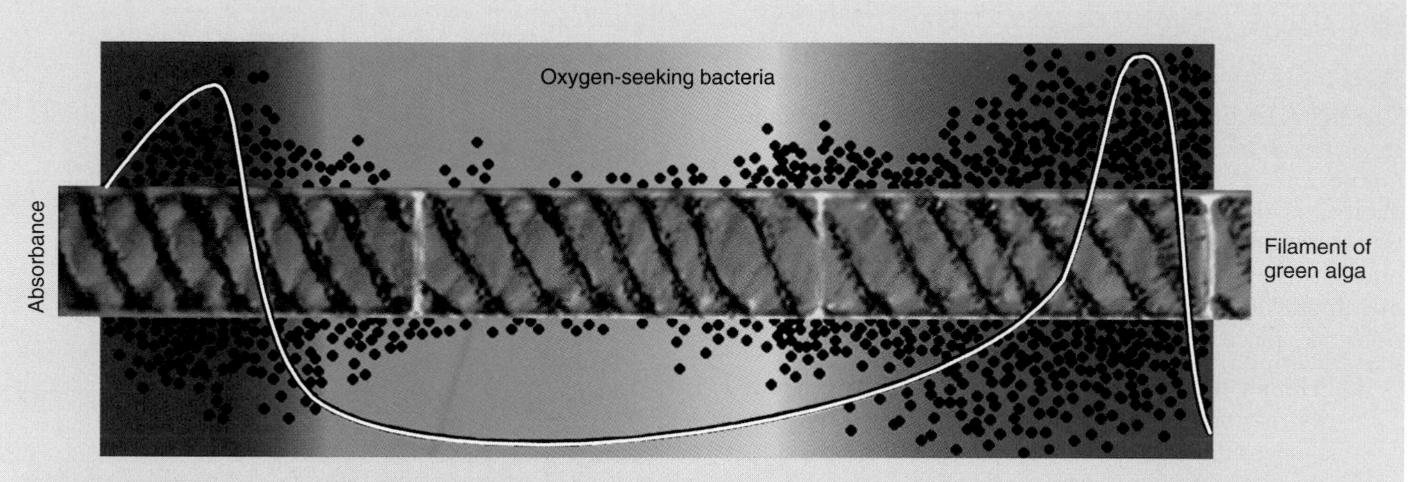

T.W. Englemann revealed the action spectrum of photosynthesis in the filamentous alga *Spirogyra* in 1882. Englemann used the rate of oxygen production to measure the rate of photosynthesis. As his oxygen indicator, he chose bacteria that are attracted by oxygen. In place of the mirror and diaphragm usually used to illuminate objects under view in his microscope, he substituted a "microspectral apparatus," which, as its name implies, produced a tiny spectrum of colors that it projected upon the slide under the microscope. Then he arranged a filament of algal cells parallel to the spread of the spectrum. The oxygen-seeking bacteria congregated mostly in the areas where the violet and red wavelengths fell upon the algal filament.

FIGURE 10.7
Constructing an action spectrum for photosynthesis. As you can see, the action spectrum for photosynthesis that Englemann revealed in his experiment parallels the absorption spectrum of chlorophyll (see figure 10.5).

FIGURE 10.8
Fall colors are produced by carotenoids and other accessory pigments. During the spring and summer, chlorophyll in leaves masks the presence of carotenoids and other accessory pigments. When cool fall temperatures cause leaves to cease manufacturing chlorophyll, the chlorophyll is no longer present to reflect green light, and the leaves reflect the orange and yellow light that carotenoids and other pigments do not absorb.

Carotenoids consist of carbon rings linked to chains with alternating single and double bonds. They can absorb photons with a wide range of energies, although they are not always highly efficient in transferring this energy. Carotenoids assist in photosynthesis by capturing energy from light of wavelengths that are not efficiently absorbed by chlorophylls (figure 10.8; see figure 10.5).

A typical carotenoid is β-carotene, whose two carbon rings are connected by a chain of 18 carbon atoms with alternating single and double bonds. Splitting a molecule of β-carotene into equal halves produces two molecules of vitamin A. Oxidation of vitamin A produces retinal, the pigment used in vertebrate vision. This explains why carrots, which are rich in β-carotene, enhance vision.

A pigment is a molecule that absorbs light. The wavelengths absorbed by a particular pigment depend on the available energy levels to which light-excited electrons can be boosted in the pigment.

Organizing Pigments into Photosystems

The light reactions of photosynthesis occur in membranes. In bacteria like those studied by van Niel, the plasma membrane itself is the photosynthetic membrane. In plants and algae, by contrast, photosynthesis is carried out by organelles that are the evolutionary descendants of photosynthetic bacteria, chloroplasts—the photosynthetic membranes exist *within* the chloroplasts. The light reactions take place in four stages:

1. **Primary photoevent.** A photon of light is captured by a pigment. The result of this primary photoevent is the excitation of an electron within the pigment.
2. **Charge separation.** This excitation energy is transferred to a specialized chlorophyll pigment termed a reaction center, which reacts by transferring an energetic electron to an acceptor molecule, thus initiating electron transport.
3. **Electron transport.** The excited electron is shuttled along a series of electron-carrier molecules embedded within the photosynthetic membrane. Several of them react by transporting protons across the membrane, generating a gradient of proton concentration. Its arrival at the pump induces the transport of a proton across the membrane. The electron is then passed to an acceptor.
4. **Chemiosmosis.** The protons that accumulate on one side of the membrane now flow back across the membrane through specific protein complexes where chemiosmotic synthesis of ATP takes place, just as it does in aerobic respiration.

Discovery of Photosystems

One way to study how pigments absorb light is to measure the dependence of the output of photosynthesis on the intensity of illumination—that is, how much photosynthesis is produced by how much light. When experiments of this sort are done on plants, they show that the output of photosynthesis increases linearly at low intensities but lessens at higher intensities, finally saturating at high-intensity light. Saturation occurs because all of the light-absorbing capacity of the plant is in use; additional light doesn't increase the output because there is nothing to absorb the added photons.

It is tempting to think that at saturation, all of a plant's pigment molecules are in use. In 1932 plant physiologists Robert Emerson and William Arnold set out to test this hypothesis in an organism where they could measure both the number of chlorophyll molecules and the output of photosynthesis. In their experiment, they measured the oxygen yield of photosynthesis when *Chlorella* (unicellular green algae) were exposed to very brief light flashes lasting only a few microseconds. Assuming the hypothesis of pigment saturation to be correct, they expected to find that as they increased the intensity of the flashes, the yield per flash would increase, until each chlorophyll molecule absorbed a photon, which would then be used in the light reactions, producing a molecule of O_2.

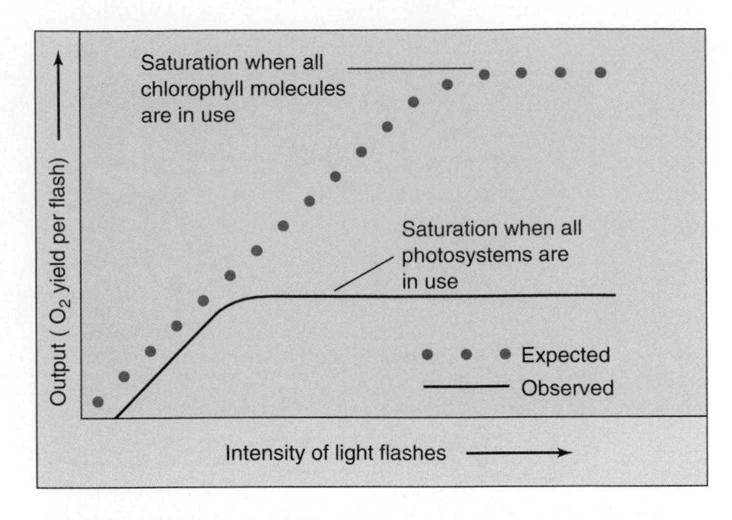

FIGURE 10.9
Emerson and Arnold's experiment. When photosynthetic saturation is achieved, further increases in intensity cause no increase in output.

Unexpectedly, this is not what happened. Instead, saturation was achieved much earlier, with only one molecule of O_2 per 2500 chlorophyll molecules (figure 10.9)! This led Emerson and Arnold to conclude that light is absorbed not by independent pigment molecules, but rather by clusters of chlorophyll and accessory pigment molecules which have come to be called *photosystems.* Light is absorbed by any one of the hundreds of pigment molecules in a photosystem, which transfer their excitation energy to one with a lower energy level than the others. This **reaction center** of the photosystem acts as an energy sink, trapping the excitation energy. It was the saturation of these reaction centers, not individual molecules, that was observed by Emerson and Arnold.

Architecture of a Photosystem

In chloroplasts and all but the most primitive bacteria, light is captured by such photosystems. Each photosystem is a network of chlorophyll *a* molecules, accessory pigments, and associated proteins held within a protein matrix on the surface of the photosynthetic membrane. Like a magnifying glass focusing light on a precise point, a photosystem channels the excitation energy gathered by any one of its pigment molecules to a specific molecule, the reaction center chlorophyll. This molecule then passes the energy out of the photosystem so it can be put to work driving the synthesis of ATP and organic molecules.

A photosystem thus consists of two closely linked components: (1) an *antenna complex* of hundreds of pigment molecules that gather photons and feed the captured light energy to the reaction center; and (2) a *reaction center*, consisting of one or more chlorophyll *a* molecules in a matrix of protein, that passes the energy out of the photosystem.

The Antenna Complex. The antenna complex captures photons from sunlight (figure 10.10). In chloroplasts, the antenna complex is a web of chlorophyll molecules linked together and held tightly on the thylakoid membrane by a matrix of proteins. Varying amounts of carotenoid accessory pigments may also be present. The protein matrix serves as a sort of scaffold, holding individual pigment molecules in orientations that are optimal for energy transfer. The excitation energy resulting from the absorption of a photon passes from one pigment molecule to an adjacent molecule on its way to the reaction center. After the transfer, the excited electron in each molecule returns to the low-energy level it had before the photon was absorbed. Consequently, it is energy, not the excited electrons themselves, that passes from one pigment molecule to the next. The antenna complex funnels the energy from many electrons to the reaction center.

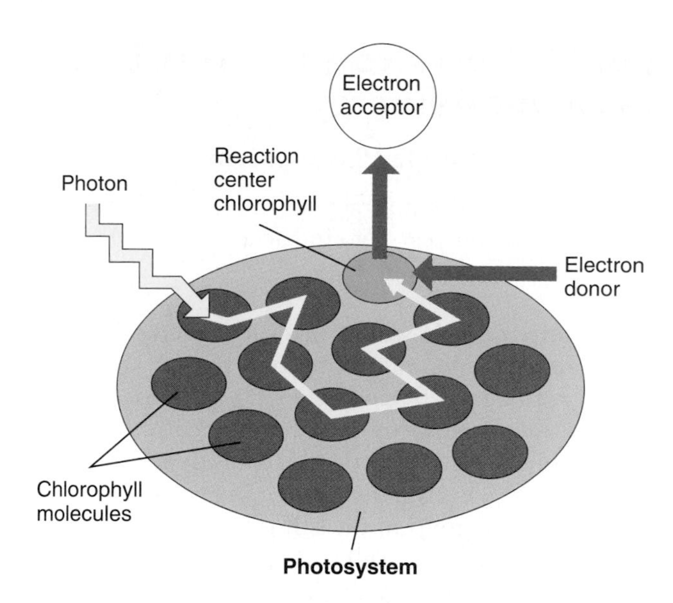

FIGURE 10.10
How the antenna complex works. When light of the proper wavelength strikes any pigment molecule within a photosystem, the light is absorbed by that pigment molecule. The excitation energy is then transferred from one molecule to another within the cluster of pigment molecules until it encounters the reaction center chlorophyll *a*. When excitation energy reaches the reaction center chlorophyll, electron transfer is initiated.

The Reaction Center. The reaction center is a transmembrane protein-pigment complex. In the reaction center of purple photosynthetic bacteria, which is simpler than in chloroplasts but better understood, a pair of chlorophyll *a* molecules acts as a trap for photon energy, passing an excited electron to an acceptor precisely positioned as its neighbor. Note that here the excited electron itself is transferred, not just the energy as we saw in pigment-pigment transfers. This allows the photon excitation to move away from the chlorophylls and is the key conversion of light to chemical energy.

Figure 10.11 shows the transfer of energy from the reaction center to the primary electron acceptor. By energizing an electron of the reaction center chlorophyll, light creates a strong electron donor where none existed before. The chlorophyll transfers the energized electron to the primary acceptor, a molecule of quinone, reducing the quinone and converting it to a strong electron donor. A weak electron donor then donates a low-energy electron to the chlorophyll, restoring it to its original condition. In plant chloroplasts, water serves as the electron donor.

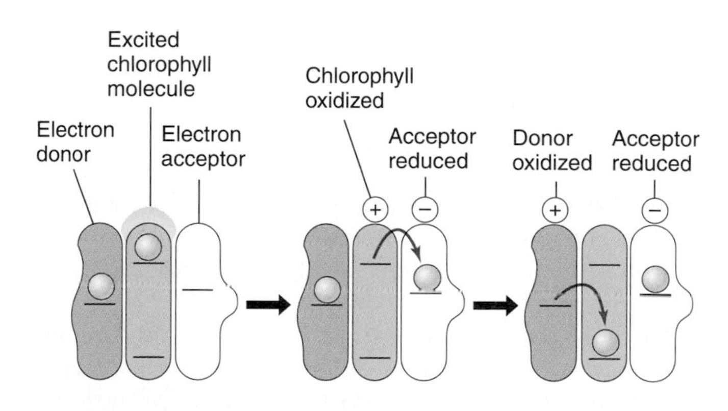

FIGURE 10.11
Converting light to chemical energy. The reaction center chlorophyll donates a light-energized electron to the primary electron acceptor, reducing it. The oxidized chlorophyll then fills its electron "hole" by oxidizing a donor molecule.

Photosystems contain pigments that capture photon energy from light. The pigments transfer the energy to reaction centers. There, the energy excites electrons, which are channeled away to do chemical work.

How Photosystems Convert Light to Chemical Energy

Bacteria Use a Single Photosystem

Photosynthetic pigment arrays are thought to have evolved more than 3 billion years ago in bacteria similar to the sulfur bacteria studied by van Niel.

1. Electron is joined with a proton to make hydrogen. In these bacteria, the absorption of a photon of light at a peak absorption of 870 nanometers (near infrared, not visible to the human eye) by the photosystem results in the transmission of an energetic electron along an electron transport chain, eventually combining with a proton to form a hydrogen atom. In the sulfur bacteria, the proton is extracted from hydrogen sulfide, leaving elemental sulfur as a by-product. In bacteria that evolved later, as well as in plants and algae, the proton comes from water, producing oxygen as a by-product.

2. Electron is recycled to chlorophyll. The ejection of an electron from the bacterial reaction center leaves it short one electron. Before the photosystem of the sulfur bacteria can function again, an electron must be returned. These bacteria channel the electron back to the pigment through an electron transport system similar to the one described in chapter 9; the electron's passage drives a proton pump that promotes the chemiosmotic synthesis of ATP. One molecule of ATP is produced for every three electrons that follow this path. Viewed overall (figure 10.12), the path of the electron is thus a circle. Chemists therefore call the electron transfer process leading to ATP formation **cyclic photophosphorylation.** Note, however, that the electron that left the P_{870} reaction center was a high-energy electron, boosted by the absorption of a photon of light, while the electron that returns has only as much energy as it had before the photon was absorbed. The difference in the energy of that electron is the photosynthetic payoff, the energy that drives the proton pump.

For more than a billion years, cyclic photophosphorylation was the only form of photosynthetic light reaction that organisms used. However, its major limitation is that it is geared only toward energy production, not toward biosynthesis. Most photosynthetic organisms incorporate atmospheric carbon dioxide into carbohydrates. Because the carbohydrate molecules are more reduced (have more hydrogen atoms) than carbon dioxide, a source of reducing power (that is, hydrogens) must be provided. Cyclic photophosphorylation does not do this. The hydrogen atoms extracted from H_2S are used as a source of protons, and are not available to join to carbon. Thus bacteria that are restricted to this process must scavenge hydrogens from other sources, an inefficient undertaking.

Why Plants Use Two Photosystems

After the sulfur bacteria appeared, other kinds of bacteria evolved an improved version of the photosystem that overcame the limitation of cyclic photophosphorylation in a neat and simple way: a second, more powerful photosystem using another arrangement of chlorophyll *a* was combined with the original.

In this second photosystem, called **photosystem II,** molecules of chlorophyll *a* are arranged with a different geometry, so that more shorter wavelength, higher energy photons are absorbed than in the ancestral photosystem, which is called **photosystem I.** As in the ancestral photosystem, energy is transmitted from one pigment molecule to another within the antenna complex of these photosystems until it reaches the reaction center, a particular pigment molecule positioned near a strong membrane-bound electron acceptor. In photosystem II, the absorption peak (that is, the wavelength of light most strongly absorbed) of the pigments is approximately 680 nanometers; therefore, the reaction center pigment is called P_{680}. The absorption peak of photosystem I pigments in plants is 700 nanometers, so its reaction center pigment is called P_{700}. Working together, the two photosystems carry out a noncyclic electron transfer.

When the rate of photosynthesis is measured using two light beams of different wavelengths (one red and the

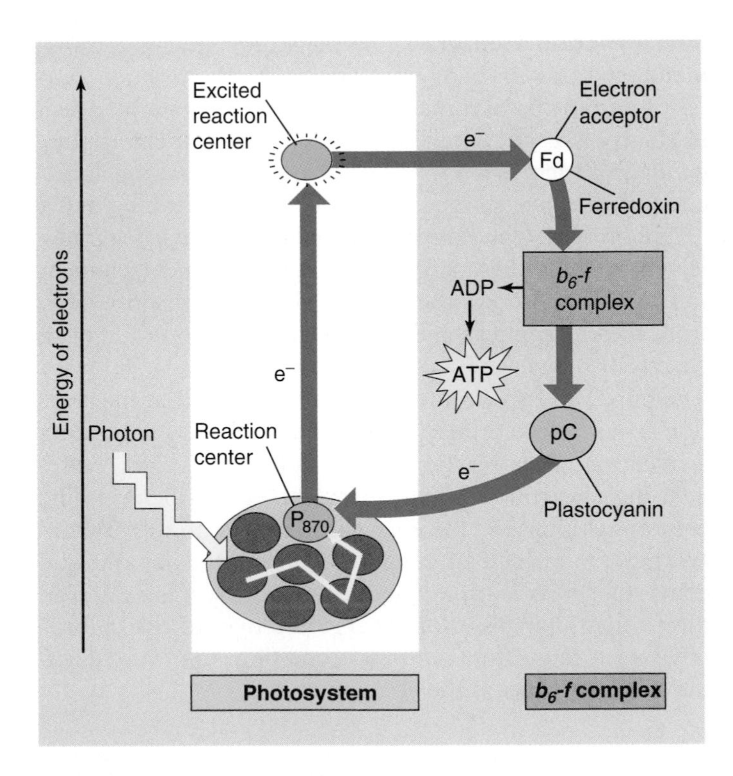

FIGURE 10.12
The path of an electron in purple sulfur bacteria. When a light-energized electron is ejected from the photosystem reaction center (P_{870}), it passes in a circle, eventually returning to the photosystem from which it was ejected.

other far-red), the rate was greater than the sum of the rates using individual beams of red and far-red light (figure 10.13). This surprising result, called the **enhancement effect,** can be explained by a mechanism involving two photosystems acting in series (that is, one after the other), one of which absorbs preferentially in the red, the other in the far-red.

The use of two photosystems solves the problem of obtaining reducing power in a simple and direct way, by harnessing the energy of two photosystems. The scheme shown in figure 10.14, called a *Z diagram,* illustrates the two electron-energizing steps, one catalyzed by each photosystem. The electrons originate from water, which holds onto its electrons very tightly (redox potential = +820 mV), and end up in NADPH, which holds its electrons much more loosely (redox potential = –320 mV).

In sulfur bacteria, excited electrons ejected from the reaction center travel a circular path, driving a proton pump and then returning to their original photosystem. Plants employ two photosystems in series, which generates power to reduce $NADP^+$ to NADPH with enough left over to make ATP.

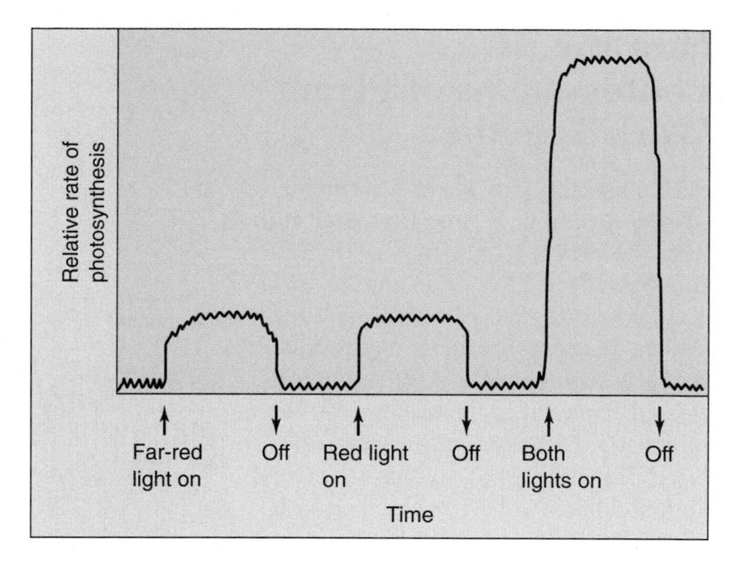

FIGURE 10.13
The "enhancement effect." The rate of photosynthesis when red and far-red light are provided together is greater than the sum of the rates when each wavelength is provided individually. This result baffled researchers in the 1950s. Today it provides the key evidence that photosynthesis is carried out by two photochemical systems with slightly different wavelength optima.

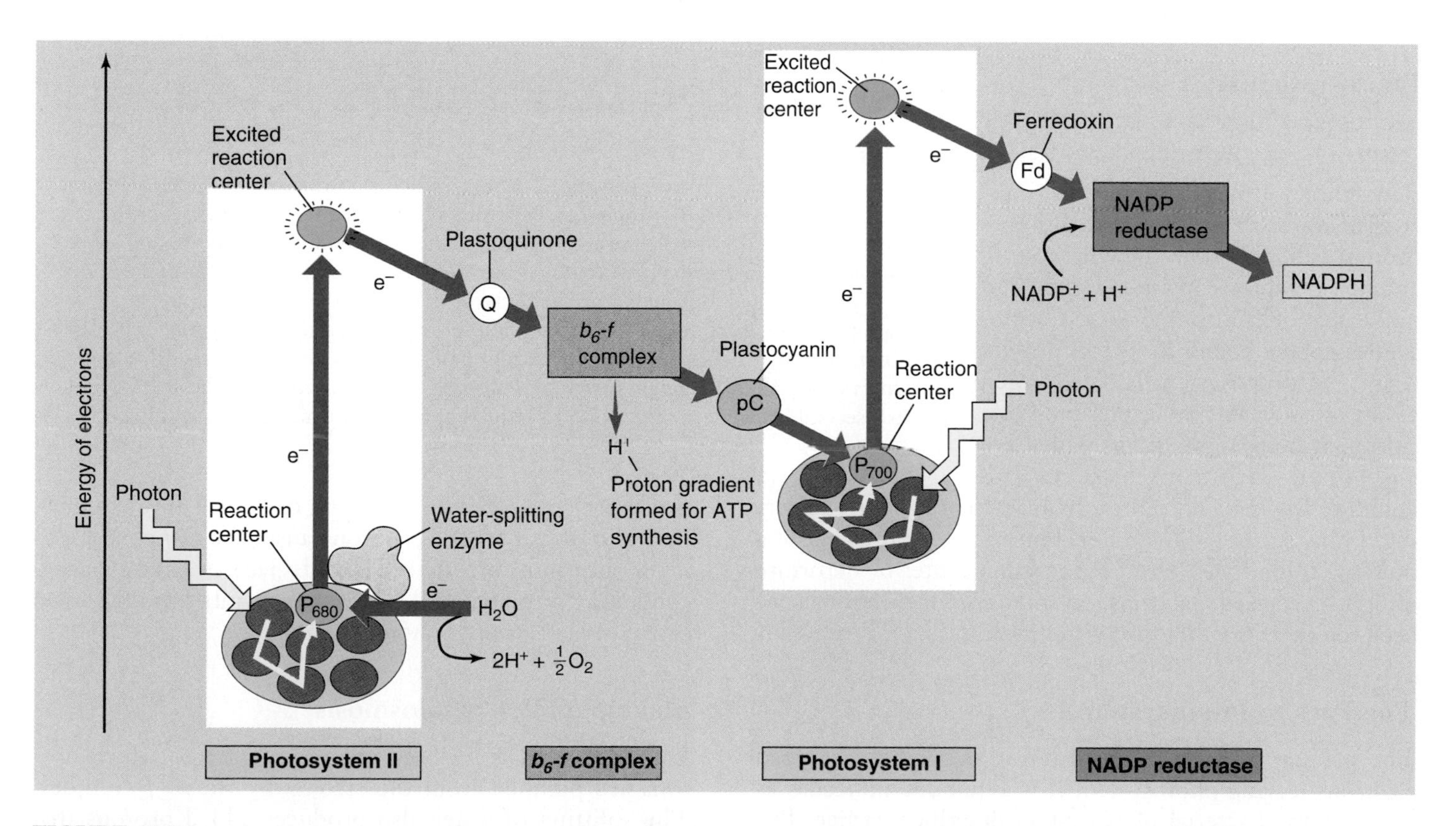

FIGURE 10.14
A Z diagram of photosystems I and II. Two photosystems work sequentially. First, a photon of light ejects a high-energy electron from photosystem II; that electron is used to pump a proton across the membrane, contributing chemiosmotically to the production of a molecule of ATP. The ejected electron then passes along a chain of cytochromes to photosystem I. When photosystem I absorbs a photon of light, it ejects a high-energy electron used to drive the formation of NADPH.

How the Two Photosystems of Plants Work Together

Plants use the two photosystems discussed earlier in series, first one and then the other, to produce both ATP and NADPH. This two-stage process is called **noncyclic photophosphorylation,** because the path of the electrons is not a circle—the electrons ejected from the photosystems do not return to them, but rather end up in NADPH. The photosystems are replenished instead with electrons obtained by splitting water. Photosystem II acts first. High-energy electrons generated by photosystem II are used to synthesize ATP and then passed to photosystem I to drive the production of NADPH. For every pair of electrons obtained from water, one molecule of NADPH and slightly more than one molecule of ATP are produced.

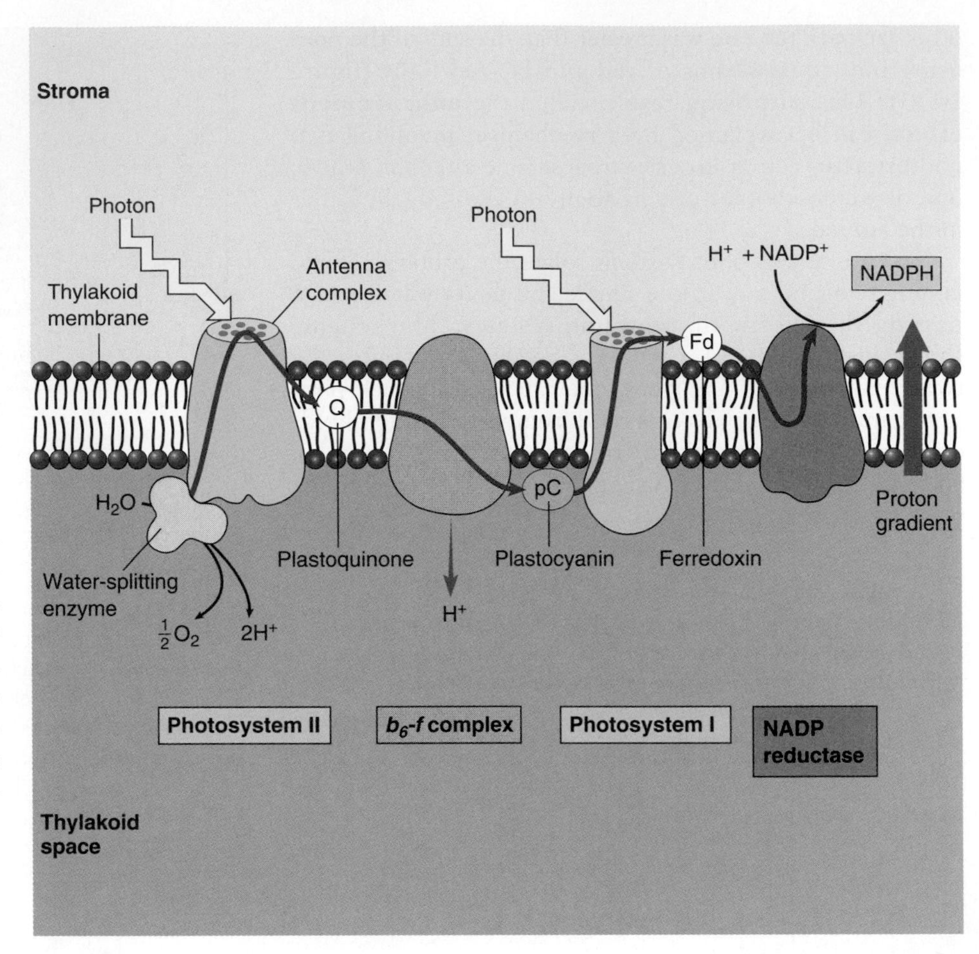

FIGURE 10.15
The photosynthetic electron transport system. When a photon of light strikes a pigment molecule in photosystem II, it excites an electron. This electron is coupled to a proton stripped from water by an enzyme and is passed along a chain of membrane-bound cytochrome electron carriers (*red arrow*). When water is split, oxygen is released from the cell, and the hydrogen ions remain in the thylakoid space. At the proton pump (b_6-f complex), the energy supplied by the photon is used to transport a proton across the membrane into the thylakoid. The concentration of hydrogen ions within the thylakoid thus increases further. When photosystem I absorbs another photon of light, its pigment passes a second high-energy electron to a reduction complex, which generates NADPH.

Photosystem II

The reaction center of photosystem II, called P_{680}, closely resembles the reaction center of purple bacteria. It consists of more than 10 transmembrane protein subunits. The light-harvesting antenna complex consists of some 250 molecules of chlorophyll *a* and accessory pigments bound to several protein chains. In photosystem II, the oxygen atoms of two water molecules bind to a cluster of manganese atoms which are embedded within an enzyme and bound to the reaction center. In a way that is poorly understood, this enzyme splits water, removing electrons one at a time to fill the holes left in the reaction center by departure of light-energized electrons. As soon as four electrons have been removed from the two water molecules, O_2 is released.

The Path to Photosystem I

The primary electron acceptor for the light-energized electrons leaving photosystem II is a quinone molecule, as it was in the bacterial photosystem described earlier. The reduced quinone which results (*plastoquinone*, symbolized Q) is a strong electron donor; it passes the excited electron to a proton pump called the *b_6-f complex* embedded within the thylakoid membrane (figure 10.15). This complex closely resembles the bc_1 complex in the respiratory electron transport chain of mitochondria discussed in chapter 9. Arrival of the energetic electron causes the b_6-f complex to pump a proton into the thylakoid space. A small copper-containing protein called *plastocyanin* (symbolized pC) then carries the electron to photosystem I.

Making ATP: Chemiosmosis

Each thylakoid is a closed compartment into which protons are pumped from the stroma by the b_6-f complex. The splitting of water also produces added protons that contribute to the gradient. The thylakoid membrane is impermeable to protons, so protons cross back out almost exclusively via the channels provided by *ATP synthases.* These channels protrude like knobs on the external surface of the thylakoid membrane. As protons pass out of the thylakoid through the ATP synthase channel, ADP is phosphorylated to ATP and released into the stroma, the fluid matrix inside the chloroplast (figure 10.16). The stroma contains the enzymes that catalyze the reactions of carbon fixation.

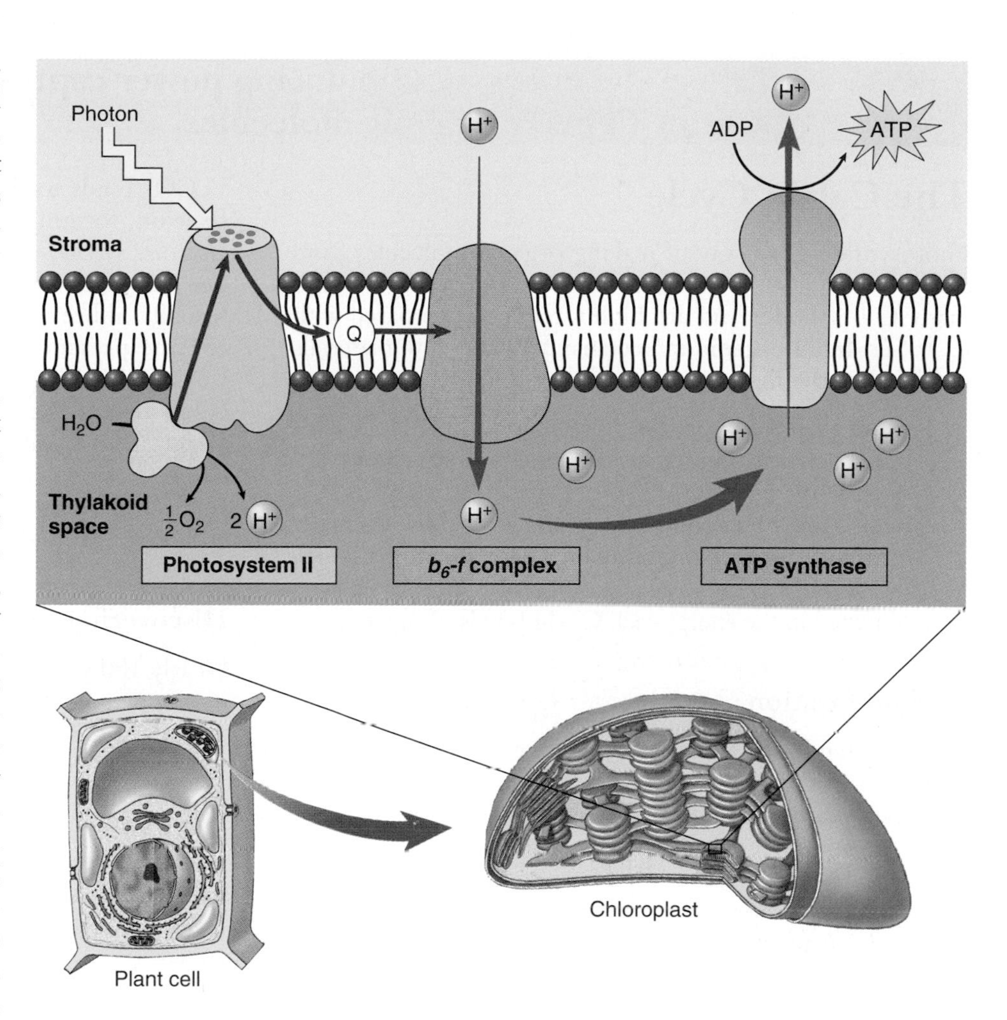

FIGURE 10.16
Chemiosmosis in a chloroplast. The b_6-f complex embedded in the thylakoid membrane pumps protons into the interior of the thylakoid. ATP is produced on the outside surface of the membrane (stroma side), as protons diffuse back out of the thylakoid through ATP synthase channels.

Photosystem I

The reaction center of photosystem I, called P_{700}, is a transmembrane complex consisting of at least 13 protein subunits. Energy is fed to it by an antenna complex consisting of 130 chlorophyll *a* and accessory pigment molecules. Photosystem I accepts an electron from plastocyanin into the hole created by the exit of a light-energized electron. This arriving electron has by no means lost all of its light-excited energy; almost half remains. Thus, the absorption of a photon of light energy by photosystem I boosts the electron leaving the reaction center to a very high energy level. Unlike photosystem II and the bacterial photosystem, photosystem I does not rely on quinones as electron acceptors. Instead, it passes electrons to an iron-sulfur protein called *ferredoxin* (Fd).

Making NADPH

Photosystem I passes electrons to ferredoxin on the stromal side of the membrane (outside the thylakoid). The reduced ferredoxin carries a very-high-potential electron. Two of them, from two molecules of reduced ferredoxin, are then donated to a molecule of $NADP^+$ to form NADPH. The reaction is catalyzed by the membrane-bound enzyme *NADP reductase.* Because the reaction occurs on the stromal side of the membrane and involves the uptake of a proton in forming NADPH, it contributes further to the proton gradient established during photosynthetic electron transport.

Making More ATP

The passage of an electron from water to NADPH in the noncyclic photophosphorylation described previously generates one molecule of NADPH and slightly more than one molecule of ATP. However, as you will learn later in this chapter, building organic molecules takes more energy than that—it takes one-and-a-half ATP molecules per NADPH molecule to fix carbon. To produce the extra ATP, many plant species are capable of short-circuiting photosystem I, switching photosynthesis into a *cyclic photophosphorylation* mode, so that the light-excited electron leaving photosystem I is used to make ATP instead of NADPH. The energetic electron is simply passed back to the b_6-f complex rather than passing on to $NADP^+$. The b_6-f complex pumps out a proton, adding to the proton gradient driving the chemiosmotic synthesis of ATP. The relative proportions of cyclic and noncyclic photophosphorylation in these plants determines the relative amounts of ATP and NADPH available for building organic molecules.

The electrons that photosynthesis strips from water molecules provide the energy to form ATP and NADPH. The residual oxygen atoms of the water molecules combine to form oxygen gas.

10.4 Cells use the energy and reducing power captured by the light reactions to make organic molecules.

The Calvin Cycle

Photosynthesis is a way of making organic molecules from carbon dioxide (CO_2). These organic molecules contain many C—H bonds and are highly reduced compared with CO_2. To build organic molecules, cells use raw materials provided by the light reactions:

1. **Energy.** ATP (provided by cyclic and noncyclic photophosphorylation) drives the endergonic reactions.
2. **Reducing power.** NADPH (provided by photosystem I) provides a source of hydrogens and the energetic electrons needed to bind them to carbon atoms. Much of the light energy captured in photosynthesis ends up invested in the energy-rich C—H bonds of sugars.

Carbon Fixation

The key step in the Calvin cycle—the event that makes the reduction of CO_2 possible—is the attachment of CO_2 to a very special organic molecule. Photosynthetic cells produce this molecule by reassembling the bonds of two intermediates in glycolysis, fructose 6-phosphate and glyceraldehyde 3-phosphate, to form the energy-rich five-carbon sugar, *ribulose 1,5-bisphosphate (RuBP)*, and a four-carbon sugar.

CO_2 binds to RuBP in the key process called **carbon fixation,** forming two three-carbon molecules of *phosphoglycerate (PGA)* (figure 10.17). The enzyme that carries out this reaction, *ribulose bisphosphate carboxylase/oxygenase* (usually abbreviated *rubisco*) is a very large four-subunit enzyme present in the chloroplast stroma. This enzyme works very sluggishly, processing only about three molecules of RuBP per second (a typical enzyme processes about 1000 substrate molecules per second). Because it works so slowly, many molecules of rubisco are needed. In a typical leaf, over 50% of all the protein is rubisco. It is thought to be the most abundant protein on earth.

Discovering the Calvin Cycle

Nearly 100 years ago, Blackman concluded that, because of its temperature dependence, photosynthesis might involve enzyme-catalyzed reactions. These reactions form a cycle of enzyme-catalyzed steps similar to the Krebs cycle. This cycle of reactions is called the **Calvin cycle**, after its discoverer, Melvin Calvin of the University of California, Berkeley. Because the cycle begins when CO_2 binds RuBP to form PGA, and PGA contains three carbon atoms, this process is also called **C_3 photosynthesis.**

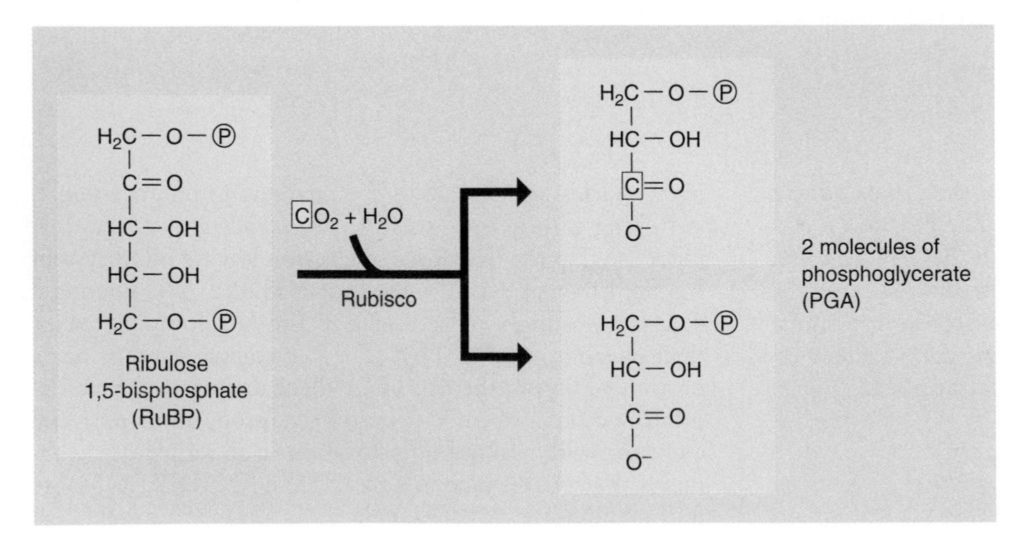

FIGURE 10.17
The key step in the Calvin cycle. Melvin Calvin and his coworkers at the University of California worked out the first step of what later became known as the Calvin cycle. They exposed photosynthesizing algae to radioactive carbon dioxide ($^{14}CO_2$). By following the fate of a radioactive carbon atom, they found that it first binds to a molecule of ribulose 1,5-bisphosphate (RuBP), then immediately splits, forming two molecules of phosphoglycerate (PGA). One of these PGAs contains the radioactive carbon atom. In 1948, workers isolated the enzyme responsible for this remarkable carbon-fixing reaction: rubisco.

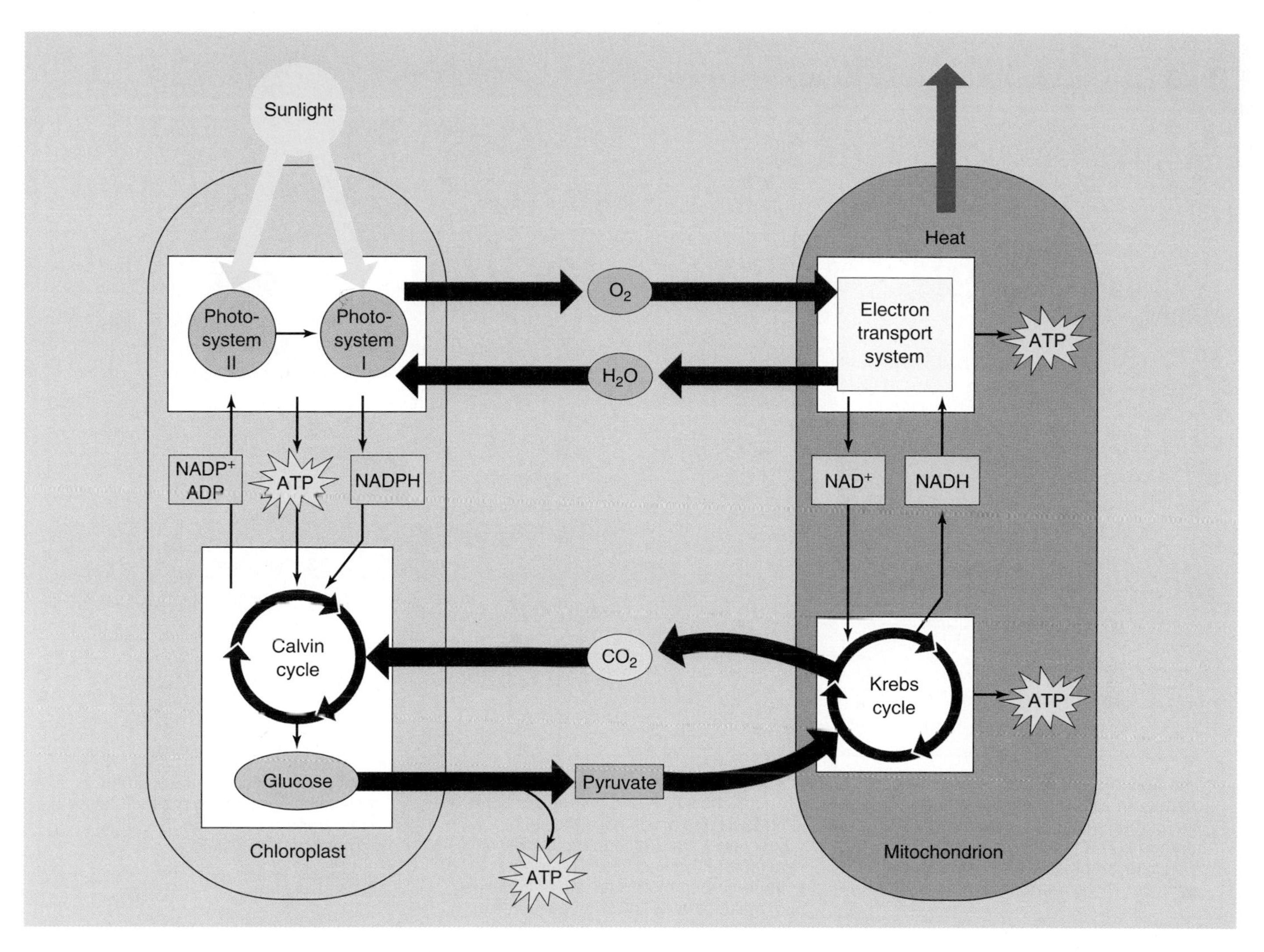

FIGURE 10.18
Chloroplasts and mitochondria: Completing an energy cycle. Water and oxygen gas cycle between chloroplasts and mitochondria within a plant cell, as do glucose and CO_2. Cells with chloroplasts require an outside source of CO_2 and water and generate glucose and oxygen. Cells without chloroplasts, such as animal cells, require an outside source of glucose and oxygen and generate CO_2 and water.

The Energy Cycle

The energy-capturing metabolisms of the chloroplasts studied in this chapter and the mitochondria studied in the previous chapter are intimately related. Photosynthesis uses the products of respiration as starting substrates, and respiration uses the products of photosynthesis as its starting substrates (figure 10.18). The Calvin cycle even uses part of the ancient glycolytic pathway, run in reverse, to produce glucose. And, the principal proteins involved in electron transport in plants are related to those in mitochondria, and in many cases are actually the same.

Photosynthesis is but one aspect of plant biology, although it is an important one. In chapters 37 through 43, we will examine plants in more detail. We have treated photosynthesis here, in a section devoted to cell biology, because photosynthesis arose long before plants did, and all organisms depend directly or indirectly on photosynthesis for the energy that powers their lives.

Chloroplasts put ATP and NADPH to work building carbon-based molecules, a process that essentially reverses the breakdown of such molecules that occurs in mitochondria. Taken together, chloroplasts and mitochondria carry out a cycle in which energy enters from the sun and leaves as heat and work.

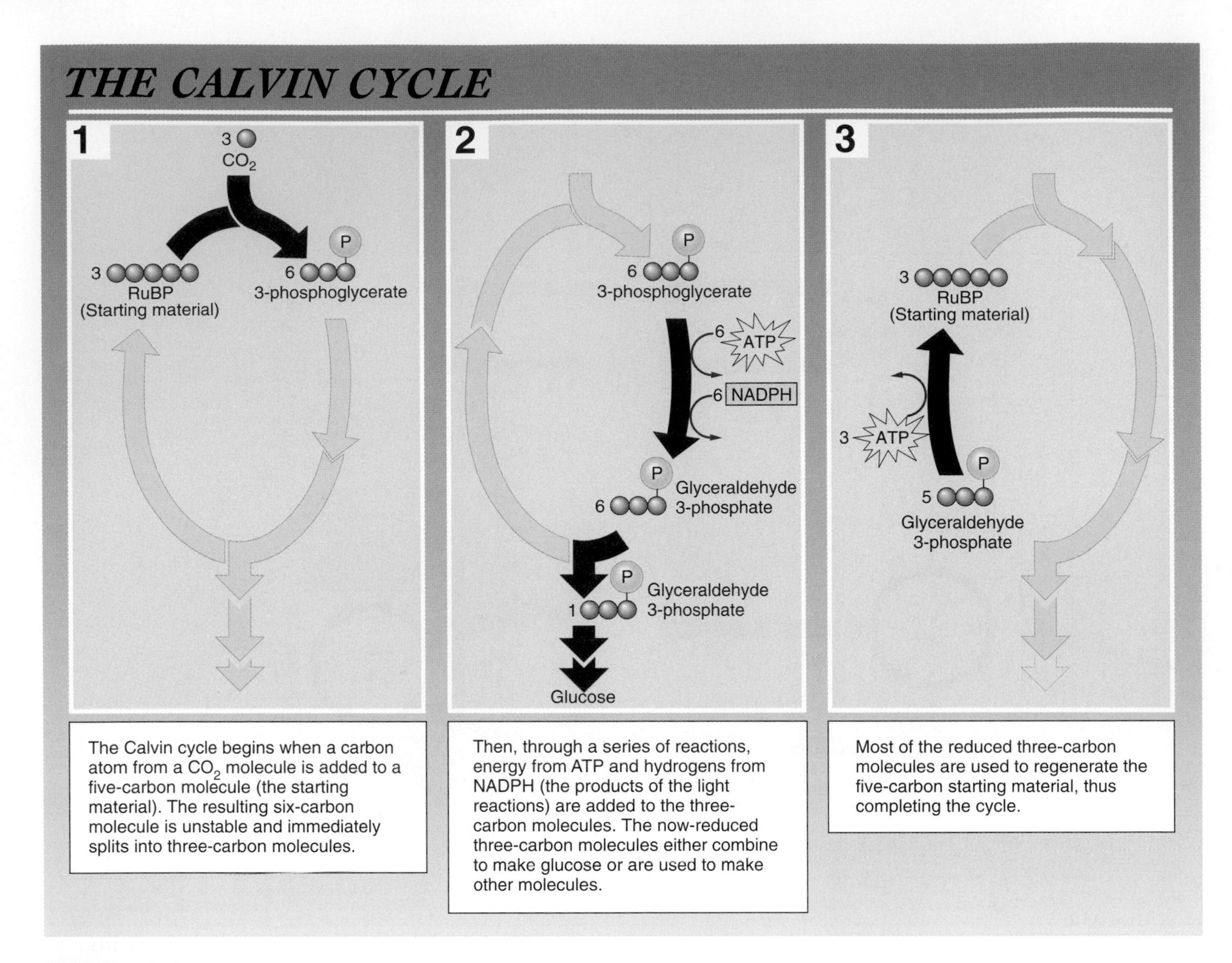
THE CALVIN CYCLE
1
3 CO_2
3 RuBP
(Starting material)
6 3-phosphoglycerate
P
The Calvin cycle begins when a carbon atom from a CO_2 molecule is added to a five-carbon molecule (the starting material). The resulting six-carbon molecule is unstable and immediately splits into three-carbon molecules.
2
6 3-phosphoglycerate
P
6 ATP
6 NADPH
6 Glyceraldehyde 3-phosphate
1 Glyceraldehyde 3-phosphate
Glucose
Then, through a series of reactions, energy from ATP and hydrogens from NADPH (the products of the light reactions) are added to the three-carbon molecules. The now-reduced three-carbon molecules either combine to make glucose or are used to make other molecules.
3
3 RuBP
(Starting material)
3 ATP
5 Glyceraldehyde 3-phosphate
P
Most of the reduced three-carbon molecules are used to regenerate the five-carbon starting material, thus completing the cycle.

FIGURE 10.19
How the Calvin cycle works.

Reactions of the Calvin Cycle

In a series of reactions (figure 10.19), three molecules of CO_2 are fixed by rubisco to produce six molecules of PGA (containing $6 \times 3 = 18$ carbon atoms in all, three from CO_2 and 15 from RuBP). The 18 carbon atoms then undergo a cycle of reactions that regenerates the three molecules of RuBP used in the initial step (containing $3 \times 5 = 15$ carbon atoms). This leaves one molecule of glyceraldehyde 3-phosphate (three carbon atoms) as the net gain.

The net equation of the Calvin cycle is:

$$3\ CO_2 + 9\ ATP + 6\ NADPH + \text{water} \longrightarrow \text{glyceraldehyde 3-phosphate} + 8\ P_i + 9\ ADP + 6\ NADP^+$$

With three full turns of the cycle, three molecules of carbon dioxide enter, a molecule of glyceraldehyde 3-phosphate (G3P) is produced, and three molecules of RuBP are regenerated (figure 10.20).

We now know that light is required *indirectly* for different segments of the CO_2 reduction reactions. Five of the Calvin cycle enzymes—including rubisco—are light activated; that is, they become functional or operate more efficiently in the presence of light. Light also promotes transport of three-carbon intermediates across chloroplast membranes that are required for Calvin cycle reactions. And finally, light promotes the influx of Mg^{++} into the chloroplast stroma, which further activates the enzyme rubisco.

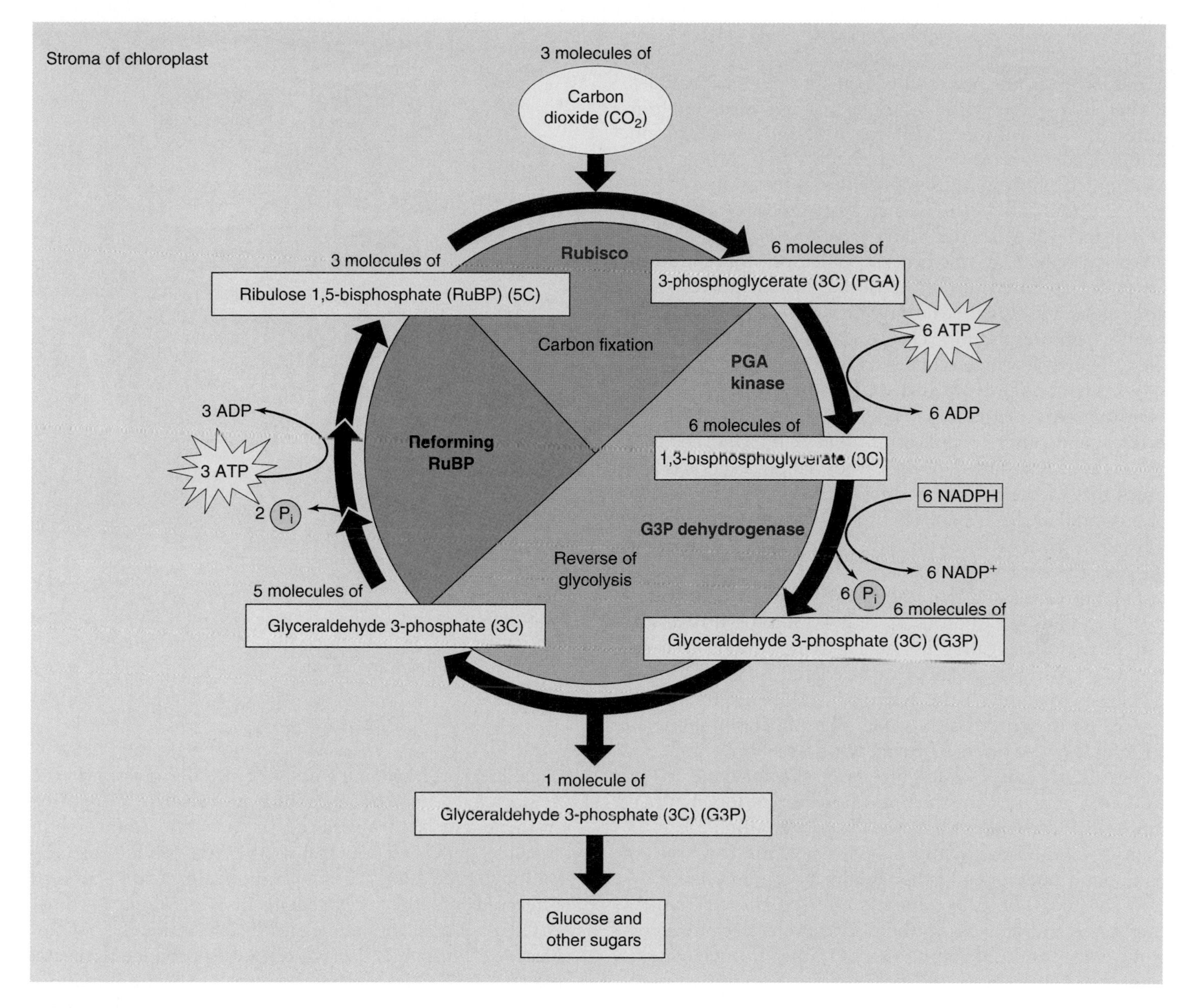

FIGURE 10.20

The Calvin cycle. For every three molecules of CO_2 that enter the cycle, one molecule of the three-carbon compound, glyceraldehyde 3-phosphate (G3P), is produced. Notice that the process requires energy stored in ATP and NADPH, which are generated by the light reactions. This process occurs in the stroma of the chloroplast.

Output of the Calvin Cycle

The glyceraldehyde 3-phosphate that is the product of the Calvin cycle is a three-carbon sugar that is a key intermediate in glycolysis. Much of it is exported from the chloroplast to the cytoplasm of the cell, where the reversal of several reactions in glycolysis allows it to be converted to fructose 6-phosphate and glucose 1-phosphate, and from that to sucrose, a major transport sugar in plants (sucrose, common table sugar, is a disaccharide made of fructose and glucose).

In times of intensive photosynthesis, glyceraldehyde 3-phosphate levels in the stroma of the chloroplast rise. As a consequence, some glyceraldehyde 3-phosphate in the chloroplast is converted to glucose 1-phosphate, in an analogous set of reactions to those done in the cytoplasm, by reversing several reactions similar to those of glycolysis. The glucose 1-phosphate is then combined into an insoluble polymer, forming long chains of starch stored as bulky starch grains in chloroplasts.

Plants incorporate carbon dioxide into sugars by means of a cycle of reactions called the Calvin cycle, which is driven by the ATP and NADPH produced in the light reactions which are consumed as CO_2 is reduced to G3P.

Photorespiration

Evolution does not necessarily result in optimum solutions. Rather, it favors workable solutions that can be derived from others that already exist. Photosynthesis is no exception. Rubisco, the enzyme that catalyzes the key carbon-fixing reaction of photosynthesis, provides a decidedly suboptimal solution. This enzyme has a second enzyme activity that interferes with the Calvin cycle, *oxidizing* ribulose 1,5-bisphosphate. In this process, called **photorespiration,** O_2 is incorporated into ribulose 1,5-bisphosphate, which undergoes additional reactions that actually release CO_2. Hence, photorespiration releases CO_2—essentially undoing the Calvin cycle which reduces CO_2 to carbohydrate.

The carboxylation and oxidation of ribulose 1,5-bisphosphate are catalyzed at the same active site on rubisco, and compete with each other. Under normal conditions at 25°C, the rate of the carboxylation reaction is four times that of the oxidation reaction, meaning that 20% of photosynthetically fixed carbon is lost to photorespiration. This loss rises substantially as temperature increases, because the rate of the oxidation reaction increases with temperature far faster than the carboxylation reaction rate.

Plants that fix carbon using only C_3 photosynthesis (the Calvin cycle) are called C_3 plants. In C_3 photosynthesis, ribulose 1,5-bisphosphate is carboxylated to form a three-carbon compound via the activity of rubisco. Other plants use **C_4 photosynthesis,** in which phosphoenolpyruvate, or PEP, is carboxylated to form a four-carbon compound using the enzyme PEP carboxylase. This enzyme has no oxidation activity, and thus no photorespiration. Furthermore, PEP carboxylase has a much greater affinity for CO_2 than does rubisco. In the C_4 pathway, the four-carbon compound undergoes further modification, only to be decarboxylated. The CO_2 which is released is then captured by rubisco and drawn into the Calvin cycle. Because an organic compound is donating the CO_2, the effective concentration of CO_2 relative to O_2 is increased, and photorespiration is minimized.

The loss of fixed carbon as a result of photorespiration is not trivial. C_3 plants lose between 25 and 50% of their photosynthetically fixed carbon in this way. The rate depends largely upon the temperature. In tropical climates, especially those in which the temperature is often above 28°C, the problem is severe, and it has a major impact on tropical agriculture.

The C_4 Pathway

Plants that adapted to these warmer environments have evolved two principal ways that use the C_4 pathway to deal with this problem. In one approach, plants conduct C_4 photosynthesis in the mesophyll cells and the Calvin cycle in the bundle sheath cells. This creates high local levels of CO_2 to favor the carboxylation reaction of rubisco. These plants are called C_4 plants and include corn, sugarcane, sorghum, and a number of other grasses. In the C_4 pathway, the three-carbon metabolite phosphoenolpyruvate is carboxylated to form the four-carbon molecule oxaloacetate, which is the first product of CO_2 fixation (figure 10.21). In C_4 plants, oxaloacetate is in turn converted into the intermediate malate, which is transported to an adjacent bundle-sheath cell. Inside the bundle-sheath cell, malate is decarboxylated to produce pyruvate, releasing CO_2. Because bundle-sheath cells are impermeable to CO_2, the CO_2 is retained within them in high concentrations. Pyruvate returns to the mesophyll cell, where two of the high-energy bonds in an ATP molecule are split to convert the pyruvate back into phosphoenolpyruvate, thus completing the cycle.

The enzymes that carry out the Calvin cycle in a C_4 plant are located within the bundle-sheath cells, where the increased CO_2 concentration decreases photorespiration. Because each CO_2 molecule is transported into the bundle-sheath cells at a cost of two high-energy ATP bonds, and because six carbons must be fixed to form a molecule of glucose, 12 additional molecules of ATP are required to form a molecule of glucose. In C_4 photosynthesis, the energetic cost of forming glucose is almost twice that of C_3 photosynthesis: 30 molecules of ATP versus 18. Nevertheless, C_4 photosynthesis is advantageous in a hot climate: photorespiration would otherwise remove more than half of the carbon fixed.

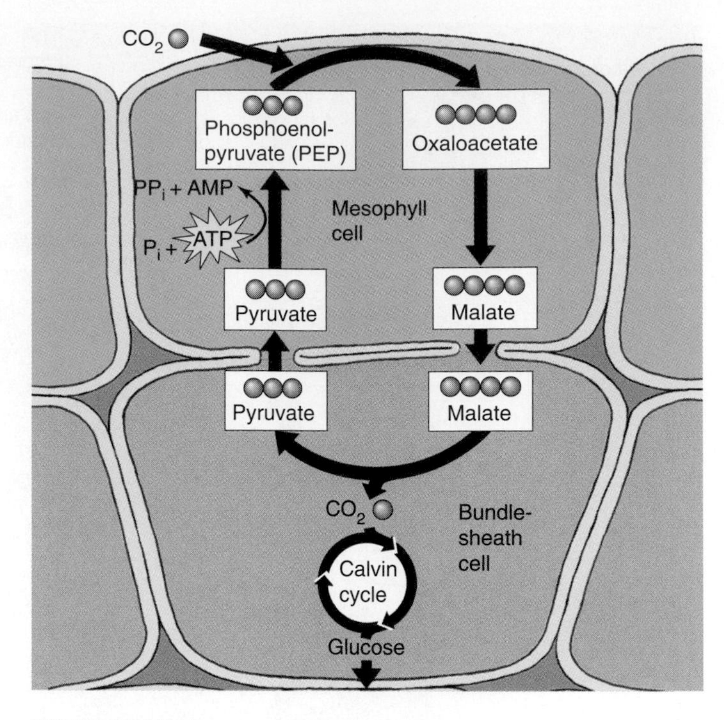

FIGURE 10.21
Carbon fixation in C_4 plants. This process is called the C_4 pathway because the first molecle formed, oxaloacetate, is a molecule containing four carbons.

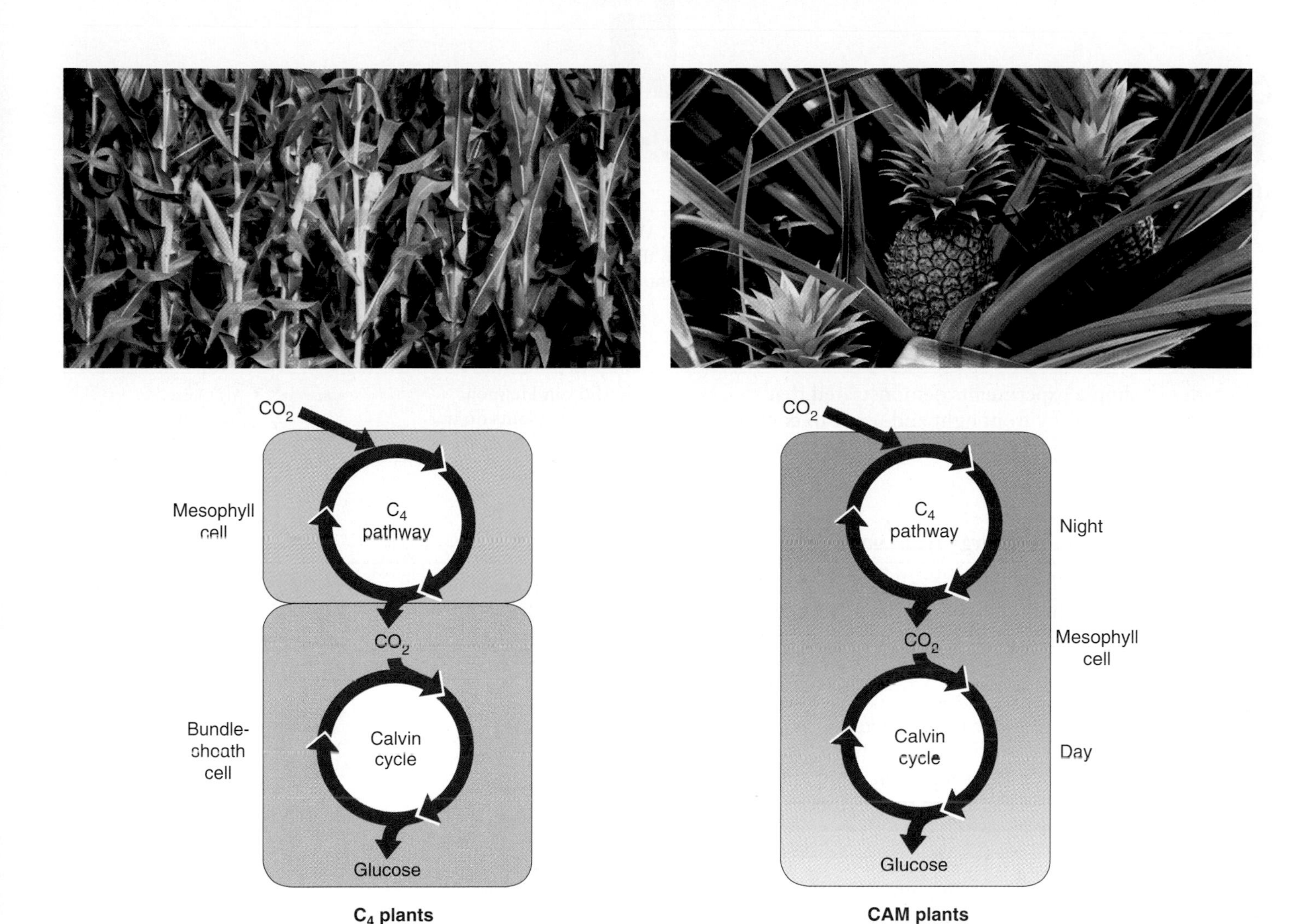

FIGURE 10.22
A comparison of C_4 and CAM plants. Both C_4 and CAM plants utilize the C_4 and the C_3 pathways. In C_4 plants, the pathways are separated spatially: the C_4 pathway takes place in the mesophyll cells and the C_3 pathway in the bundle-sheath cells. In CAM plants, the two pathways are separated temporally: the C_4 pathway is utilized at night and the C_3 pathway during the day.

The Crassulacean Acid Pathway

A second strategy to decrease photorespiration in hot regions has been adopted by many succulent (water-storing) plants such as cacti, pineapples, and some members of about two dozen other plant groups. This mode of initial carbon fixation is called **crassulacean acid metabolism (CAM),** after the plant family Crassulaceae (the stonecrops or hens-and-chicks), in which it was first discovered. In these plants, the stomata (singular, stoma), specialized openings in the leaves of all plants through which CO_2 enters and water vapor is lost, open during the night and close during the day. This pattern of stomatal opening and closing is the reverse of that in most plants. CAM plants open stomata at night and initially fix CO_2 into organic compounds using the C_4 pathway. These organic compounds accumulate throughout the night and are decarboxylated during the day to yield high levels of CO_2. In the day, these high levels of CO_2 drive the Calvin cycle and minimize photorespiration. Like C_4 plants, CAM plants use both C_4 and C_3 pathways. They differ from C_4 plants in that they use the C_4 pathway at night and the C_3 pathway during the day *within the same cells.* In C_4 plants, the two pathways take place in different cells (figure 10.22).

Photorespiration results in decreased yields of photosynthesis. C_4 and CAM plants circumvent this problem through modifications of leaf architecture and photosynthetic chemistry that locally increase CO_2 concentrations. C_4 plants isolate CO_2 production spatially, CAM plants temporally.

Chapter 10

www.mhhe.com/raven6/resources10.mhtml

Summary | Questions | Media Resources

10.1 What is photosynthesis?

- Light is used by plants, algae, and some bacteria, in a process called photosynthesis, to convert atmospheric carbon (CO_2) into carbohydrate.

1. Where do the oxygen atoms in the O_2 produced during photosynthesis come from?

- Art Activity: Chloroplast Structure

10.2 Learning about photosynthesis: An experimental journey.

- A series of simple experiments demonstrated that plants capture energy from light and use it to convert the carbon atoms of CO_2 and the hydrogen atoms of water into organic molecules.

2. How did van Helmont determine that plants do not obtain their food from the soil?

- Art Quiz: Light and Photosynthesis

10.3 Pigments capture energy from sunlight.

- Light consists of energy packets called photons; the shorter the wavelength of light, the more its energy. When photons are absorbed by a pigment, electrons in the pigment are boosted to a higher energy level.
- Photosynthesis channels photon excitation energy into a single pigment molecule. In bacteria, that molecule then donates an electron to an electron transport chain, which drives a proton pump and ultimately returns the electron to the pigment.
- Plants employ two photosystems. Light energy is first absorbed by photosystem II and passed to photosystem I, driving a proton pump and bringing about the chemiosmotic synthesis of ATP.
- When the electron arrives at photosystem I, another photon of light is absorbed, and energized electrons are channeled to a primary electron acceptor, which reduces $NADP^+$ to NADPH.

3. How is the energy of light captured by a pigment molecule? Why does light reflected by the pigment chlorophyll appear green?

4. What is the function of the reaction center chlorophyll? What is the function of the primary electron acceptor?

5. Explain how photosynthesis in the sulfur bacteria is a cyclic process. What is its energy yield in terms of ATP molecules synthesized per electron?

6. How do the two photosystems in plants and algae work? Which stage generates ATP and which generates NADPH?

- Art Activity: Electromagnetic Spectrum
- Exploration: Photosynthesis

- Light-Dependent Photosynthesis

- Light and Pigmentation
- Light-Dependent Reactions

- Art Quizzes:
  -Photosystem Antenna Complex
  -Chemiosmosis in a Chloroplast

10.4 Cells use the energy and reducing power captured by the light reactions to make organic molecules.

- The ATP and reducing power produced by the light reactions are used to fix carbon in a series of reactions called the Calvin cycle.
- RuBP carboxylase, the enzyme that fixes carbon in the Calvin cycle, also carries out an oxidative process called photorespiration.
- Many tropical plants inhibit photorespiration by expending ATP to increase the intracellular concentration of CO_2. This process, called the C_4 pathway, nearly doubles the energetic cost of synthesizing glucose.

7. In a C_3 plant, where do the light reactions occur? Where does the Calvin cycle occur?

8. What is photorespiration? What advantage do C_4 plants have over C_3 plants with respect to photorespiration? What disadvantage do C_4 plants have that limits their distribution primarily to warm regions of the earth?

- Light-Independent Photosynthesis

- Light-Independent Reactions

- Art Quiz: Comparison of C_3 and C_4 Plant Leaves

Part **IV**

Reproduction and Heredity

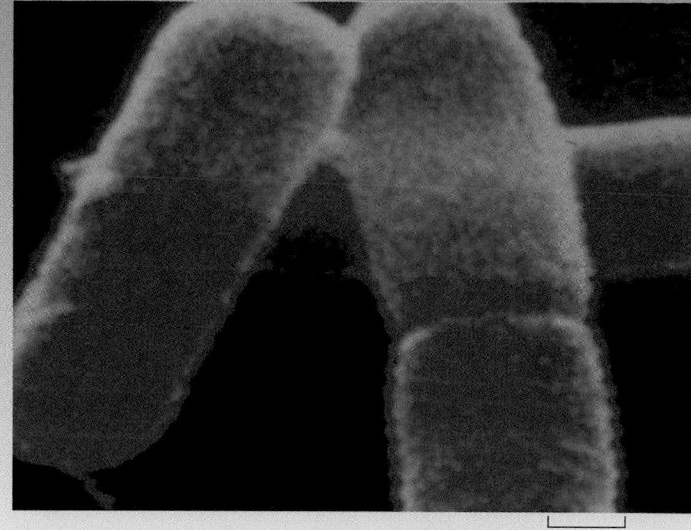

These bacterial cells are dividing. As the population grows, gene variants arise by mutation. Do the new variants persist, or are they eliminated by natural selection?

Why Do Some Genes Maintain More Than One Common Allele in a Population?

When Mendel did his crosses of pea plants, he knew what a pea plant was supposed to look like: a small plant with green leaves, purple flowers, and smooth seeds. But if all pea plants were like that, he would never have been able to sort out the rules of heredity—in a cross of green peas with green peas, there would have been no visible differences to reveal the 3:1 pattern of gene segregation. The variant alleles that Mendel employed in his studies—yellow leaves, white flowers, wrinkled seeds—were rare "accidents" maintained in seed collections for their novelty. In nature, such unusual kinds of peas had never been encountered by Mendel.

By the time Mendel's work was rediscovered in 1900, Darwin had provided a ready explanation of why alternative alleles seemed to be rare in natural populations. Natural selection was simply scouring the population, cleansing it in each generation of less fit alternatives. While recombination can complicate the process in interesting ways among sexual organisms like peas, asexual organisms like bacteria were predicted to be very sensitive to the effects of selection. Left to do its work, natural selection should crown as winner in bacterial population the best allele of each gene, producing a uniform population.

Why do populations contain variants at all? In 1932 the famous geneticist Herman Muller formulated what has come to be called the "classical model," explaining gene variation in natural populations of asexual organisms as a temporary, transient condition, new variations arising by random mutation only to be established or eliminated by selection. Except for the brief periods when populations are undergoing this periodic cleansing, they should remain genetically uniform.

The removal of variants was proposed to be a very straightforward process. During the periodic cleansing periods envisioned by Muller, his classical model operates under a "competitive exclusion" principle first proposed by Gause: whenever a new variant appears, it is weighed in the balance by natural selection, and either wins or loses. There are no ties. One version of the gene becomes universal in the population, and the other is eliminated.

Muller's classical model thus makes a very straightforward prediction: in nature, most populations of asexual organisms should be genetically uniform most of the time. However, this is not at all what is observed. Natural populations of most species, including asexual ones like bacteria, appear to have lots of common variants—they are said to be "polymorphic."

So where are all of these variants coming from? Variation in the environment, either spatial or temporal, can be used to explain how some polymorphisms arise. Selection favors one form at a particular place and time, a different form at a different place or time. In a nutshell, varying selection can encourage polymorphism.

Is that all there is to it? Is it really impossible for more than one variant to become common in a population, if the population lives in a constant uniform environment, an environment that does not vary from one place to another or from one time to another? Theory says so.

Biologists that study microbial communities have begun to report that bacteria are not aware of Muller's theory. Bacterial cultures started from a single cell living in simple unstructured environments rapidly become polymorphic.

There is a way to reconcile theory and experiment. Perhaps the variant individuals in the population are interacting with one another. Muller's theory assumes that every individual undergoes an independent trial by selection. But what if that's not so? What if different kinds of individuals help each other out? Stable coexistence of variants in a population might be possible if interactions between them contribute to the welfare of both (what a biologist calls mutualism) or favors one (what a biologist calls commensalism). In essence, cooperation would be counterbalancing the effects of competition.

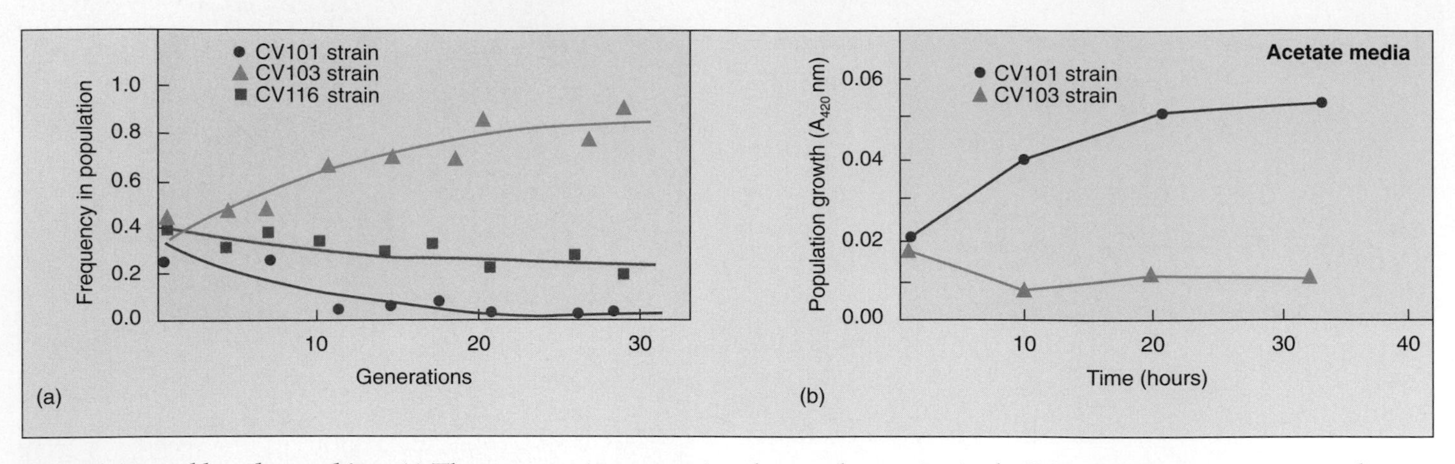

Maintaining stable polymorphism. (*a*) Three new strains emerge in culture and are maintained. (*b*) Two strains are grown on media containing acetate. The strain CV103 was found to excrete acetate, while the strain CV101 was found to thrive in media with acetate as the sole source of carbon. Population growth is measured by an increase in the turbidity of the liquid medium; turbidity is measured as an increase in light absorbance at a wavelength of 420 nm (A_{420} nm).

The Experiment

To investigate this intriguing possibility, Julian Adams and co-workers at the University of Michigan set out to see if polymorphism for metabolic abilities would develop spontaneously in bacteria growing in a uniform environment.

For a bacterial subject they chose *Escherichia coli* (*E. coli*), a widely studied bacterium whose growth under laboratory conditions is well understood. Cultures of *Escherichia coli* can be maintained in chemostat culture for many hundreds of generations. A chemostat is a large container holding liquid culture medium. A little bit of the liquid is continuously removed, and an equal amount of fresh culture medium added to replace what leaves. The growth of the *E. coli* culture is limited by the amount of glucose remaining in the culture medium to feed the growing cells.

Researchers inoculated a glucose-limited chemostat culture media with the *E. coli* strain JA122, and maintained the continuous culture for 773 generations. A sample was taken from the chemostat after 773 generations and analyzed for the presence of new strains of *E. coli*. Any variation among the cells in the sample would indicate that polymorphism had arisen.

To detect metabolic variation within the sample of growing cells, Adams's team analyzed the rate of glucose uptake and the concentration of acetate, among other variables. By examining such biochemical parameters, the researchers could determine if the different strains were filling different metabolic "niches"—that is, using the metabolic environment in different ways. Metabolic niches were characterized by looking at the normal products of aerobic fermentation, acetate and glycerol, which appear in the growth medium as a by-product of *E. coli* metabolism.

To further classify the strains, batch cultures containing two strains were established to analyze interactions between the two groups.

The Results

Three distinct variants were detected in the 773-generation *E. coli*, each being maintained at stable levels in the continuously growing culture. Clearly polymorphism *can* appear within an initially uniform bacterial population growing in a simple homogeneous environment.

When mixed together and allowed to compete, one strain does not drive the other two to extinction, as theory had predicted. Instead, the three new strains, CV101, CV103, and CV116, all persist (see graph *a* above).

The three strains were then analyzed to see how they differed. CV103 exhibited the highest rate of glucose uptake and produced the most acetate (an end product of glucose aerobic fermentation). Is this difference important? To see, the CV103 strain was co-cultured with CV101. They maintained stable growth levels, which indicated that the contribution of the third strain, CV116, was not required to maintain their growth.

What is the difference between CV101 and CV103? CV101 could grow in culture filtrate of CV103 but in the reverse situation, CV103 could not grow. This indicates that CV103 secretes a substance upon which CV101 can grow. Is CV101 utilizing the acetate produced by CV103 as its carbon source?

To test this possibility, CV101 and CV103 were grown together in media with acetate as the only carbon source. The results from this experiment are shown in graph *b* above and indicate that CV101 thrives on an acetate carbon source, while CV103 does not and requires an additional carbon source such as glucose.

These results indicate that two of the strains are maintained in polymorphism at stable levels because they have evolved different adaptations that allow them to coexist by filling different niches. One strain (CV101) is maintained in the population because it is able to use a metabolic by-product released by another strain (CV103).

To explore this experiment further, go to the Virtual Lab at www.mhhe.com/raven6/vlab4.mhtml

11

How Cells Divide

Concept Outline

11.1 Bacteria divide far more simply than do eukaryotes.

Cell Division in Prokaryotes. Bacterial cells divide by splitting in two.

11.2 Chromosomes are highly ordered structures.

Discovery of Chromosomes. All eukaryotic cells contain chromosomes, but different organisms possess differing numbers of chromosomes.

The Structure of Eukaryotic Chromosomes. Proteins play an important role in packaging DNA in chromosomes.

11.3 Mitosis is a key phase of the cell cycle.

Phases of the Cell Cycle. The cell cycle consists of three growth phases, a nuclear division phase, and a cytoplasmic division stage.

Interphase: Preparing for Mitosis. In interphase, the cell grows, replicates its DNA, and prepares for cell division.

Mitosis. In prophase, the chromosomes condense and microtubules attach sister chromosomes to opposite poles of the cell. In metaphase, chromosomes align along the center of the cell. In anaphase, the chromosomes separate; in telophase the spindle dissipates and the nuclear envelope reforms.

Cytokinesis. In cytokinesis, the cytoplasm separates into two roughly equal halves.

11.4 The cell cycle is carefully controlled.

General Strategy of Cell Cycle Control. At three points in the cell cycle, feedback from the cell determines whether the cycle will continue.

Molecular Mechanisms of Cell Cycle Control. Special proteins regulate the "checkpoints" of the cell cycle.

Cancer and the Control of Cell Proliferation. Cancer results from damage to genes encoding proteins that regulate the cell division cycle.

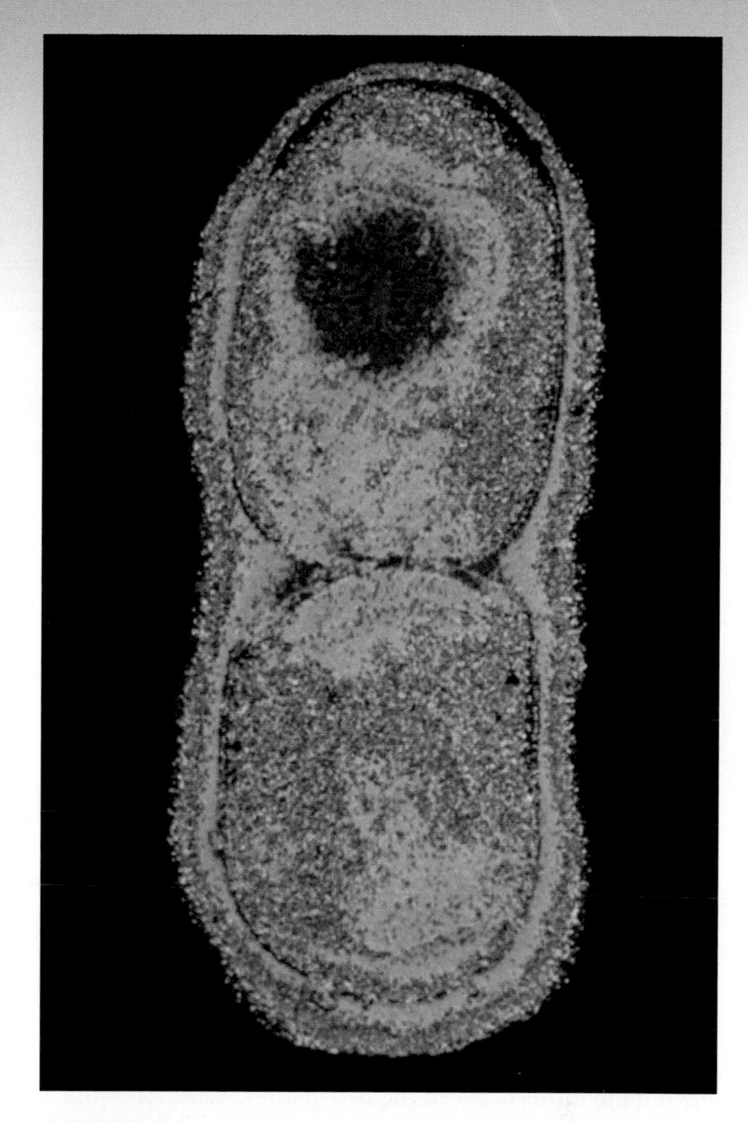

FIGURE 11.1
Cell division in bacteria. It's hard to imagine fecal coliform bacteria as beautiful, but here is *Escherichia coli,* inhabitant of the large intestine and the biotechnology lab, spectacularly caught in the act of fission.

All species of organisms—bacteria, alligators, the weeds in a lawn—grow and reproduce. From the smallest of creatures to the largest, all species produce offspring like themselves and pass on the hereditary information that makes them what they are. In this chapter, we begin our consideration of heredity with an examination of how cells reproduce (figure 11.1). The mechanism of cell reproduction and its biological consequences have changed significantly during the evolution of life on earth.

11.1 Bacteria divide far more simply than do eukaryotes.

Cell Division in Prokaryotes

In bacteria, which are prokaryotes and lack a nucleus, cell division consists of a simple procedure called **binary fission** (literally, "splitting in half"), in which the cell divides into two equal or nearly equal halves (figure 11.2). The genetic information, or *genome*, replicates early in the life of the cell. It exists as a single, circular, double-stranded DNA molecule. Fitting this DNA circle into the bacterial cell is a remarkable feat of packaging—fully stretched out, the DNA of a bacterium like *Escherichia coli* is about 500 times longer than the cell itself.

The DNA circle is attached at one point to the cytoplasmic surface of the bacterial cell's plasma membrane. At a specific site on the DNA molecule called the *replication origin,* a battery of more than 22 different proteins begins the process of copying the DNA (figure 11.3). When these enzymes have proceeded all the way around the circle of DNA, the cell possesses two copies of the genome. These "daughter" genomes are attached side-by-side to the plasma membrane.

The growth of a bacterial cell to about twice its initial size induces the onset of cell division. A wealth of recent evidence suggests that the two daughter chromosomes are actively partitioned during this process. As this process proceeds, the cell lays down new plasma membrane and cell wall materials in the zone between the attachment sites of the two daughter genomes. A new plasma membrane grows between the genomes; eventually, it reaches all the way into the center of the cell, dividing it in two. Because the membrane forms between the two genomes, each new cell is assured of retaining one of the genomes. Finally, a new cell wall forms around the new membrane.

The evolution of the eukaryotes introduced several additional factors into the process of cell division. Eukaryotic cells are much larger than bacteria, and their genomes contain much more DNA. Eukaryotic DNA is contained in a number of linear chromosomes, whose organization is much more complex than that of the single, circular DNA molecules in bacteria. In chromosomes, DNA forms a complex with packaging proteins called histones and is wound into tightly condensed coils.

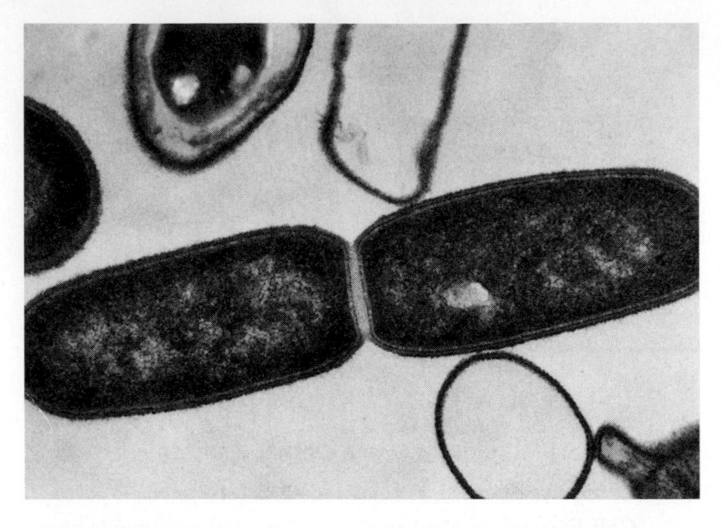

FIGURE 11.2
Fission (40,000×). Bacteria divide by a process of simple cell fission. Note the newly formed plasma membrane between the two daughter cells.

Bacteria divide by binary fission. Fission begins in the middle of the cell. An active partitioning process ensures that one genome will end up in each daughter cell.

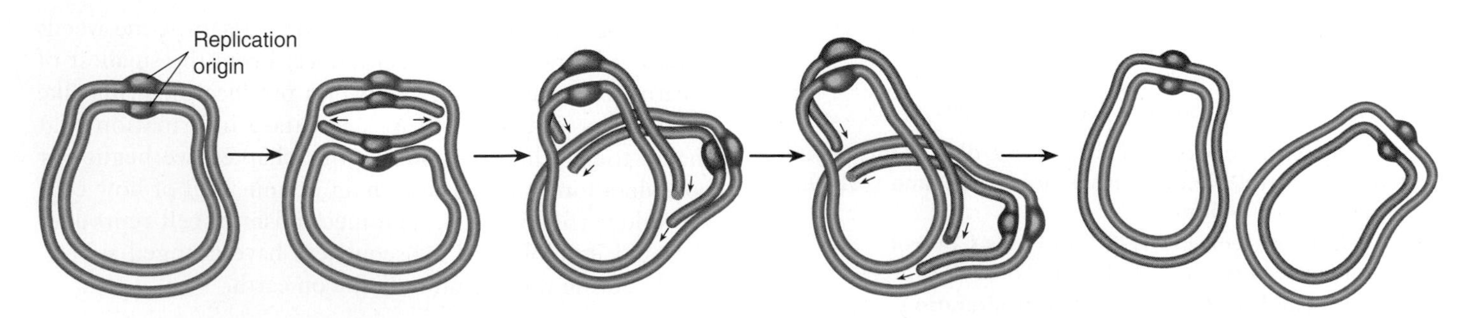

FIGURE 11.3
How bacterial DNA replicates. The replication of the circular DNA molecule (*blue*) that constitutes the genome of a bacterium begins at a single site, called the replication origin. The replication enzymes move out in both directions from that site and make copies (*red*) of each strand in the DNA duplex. When the enzymes meet on the far side of the molecule, replication is complete.

11.2 Chromosomes are highly ordered structures.

Discovery of Chromosomes

Chromosomes were first observed by the German embryologist Walther Fleming in 1882, while he was examining the rapidly dividing cells of salamander larvae. When Fleming looked at the cells through what would now be a rather primitive light microscope, he saw minute threads within their nuclei that appeared to be dividing lengthwise. Fleming called their division **mitosis,** based on the Greek word *mitos,* meaning "thread."

Chromosome Number

Since their initial discovery, chromosomes have been found in the cells of all eukaryotes examined. Their number may vary enormously from one species to another. A few kinds of organisms—such as the Australian ant *Myrmecia,* the plant *Haplopappus gracilis,* a relative of the sunflower that grows in North American deserts; and the fungus *Penicillium*—have only 1 pair of chromosomes, while some ferns have more than 500 pairs (table 11.1). Most eukaryotes have between 10 and 50 chromosomes in their body cells.

Human cells each have 46 chromosomes, consisting of 23 nearly identical pairs (figure 11.4). Each of these 46 chromosomes contains hundreds or thousands of genes that play important roles in determining how a person's body develops and functions. For this reason, possession of all the chromosomes is essential to survival. Humans missing even one chromosome, a condition called monosomy, do not survive embryonic development in most cases. Nor does the human embryo develop properly with an extra copy of any one chromosome, a condition called trisomy. For all but a few of the smallest chromosomes, trisomy is fatal, and even in those few cases, serious problems result. Individuals with an extra copy of the very small chromosome 21, for example, develop more slowly than normal and are mentally retarded, a condition called Down syndrome.

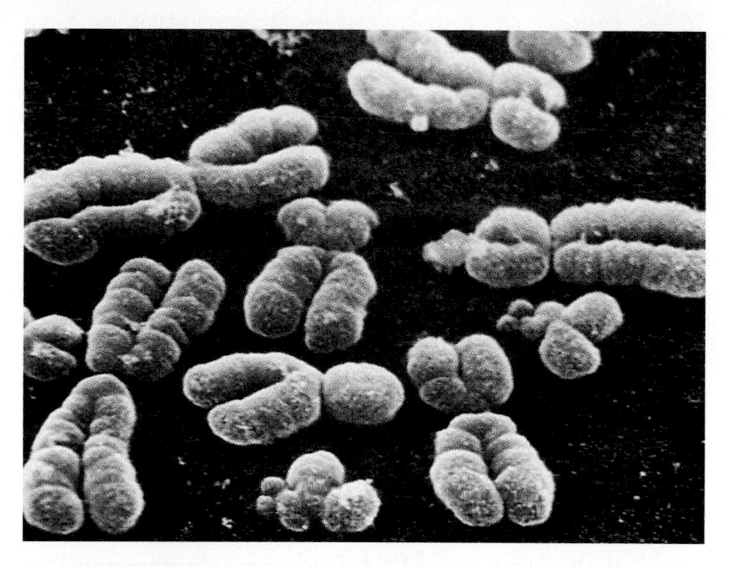

FIGURE 11.4
Human chromosomes. This photograph (950×) shows human chromosomes as they appear immediately before nuclear division. Each DNA molecule has already replicated, forming identical copies held together by a constriction called the centromere.

All eukaryotic cells store their hereditary information in chromosomes, but different kinds of organisms utilize very different numbers of chromosomes to store this information.

Table 11.1 Chromosome Number in Selected Eukaryotes

Group	Total Number of Chromosomes	Group	Total Number of Chromosomes	Group	Total Number of Chromosomes
FUNGI		**PLANTS**		**VERTEBRATES**	
Neurospora (haploid)	7	*Haplopappus gracilis*	2	Opossum	22
Saccharomyces (a yeast)	16	Garden pea	14	Frog	26
INSECTS		Corn	20	Mouse	40
Mosquito	6	Bread wheat	42	Human	46
Drosophila	8	Sugarcane	80	Chimpanzee	48
Honeybee	32	Horsetail	216	Horse	64
Silkworm	56	Adder's tongue fern	1262	Chicken	78
				Dog	78

The Structure of Chromosomes

In the century since discovery of chromosomes, we have learned a great deal about their structure and composition.

Composition of Chromatin

Chromosomes are composed of **chromatin,** a complex of DNA and protein; most are about 40% DNA and 60% protein. A significant amount of RNA is also associated with chromosomes because chromosomes are the sites of RNA synthesis. The DNA of a chromosome is one very long, double-stranded fiber that extends unbroken through the entire length of the chromosome. A typical human chromosome contains about 140 million (1.4×10^8) nucleotides in its DNA. The amount of information one chromosome contains would fill about 280 printed books of 1000 pages each, if each nucleotide corresponded to a "word" and each page had about 500 words on it. Furthermore, if the strand of DNA from a single chromosome were laid out in a straight line, it would be about 5 centimeters (2 inches) long. Fitting such a strand into a nucleus is like cramming a string the length of a football field into a baseball—and that's only 1 of 46 chromosomes! In the cell, however, the DNA is coiled, allowing it to fit into a much smaller space than would otherwise be possible.

Chromosome Coiling

How can this long DNA fiber coil so tightly? If we gently disrupt a eukaryotic nucleus and examine the DNA with an electron microscope, we find that it resembles a string of beads (figure 11.5). Every 200 nucleotides, the DNA duplex is coiled around a core of eight histone proteins, forming a complex known as a **nucleosome.** Unlike most proteins, which have an overall negative charge, histones are positively charged, due to an abundance of the basic amino acids arginine and lysine. They are thus strongly attracted to the negatively charged phosphate groups of the

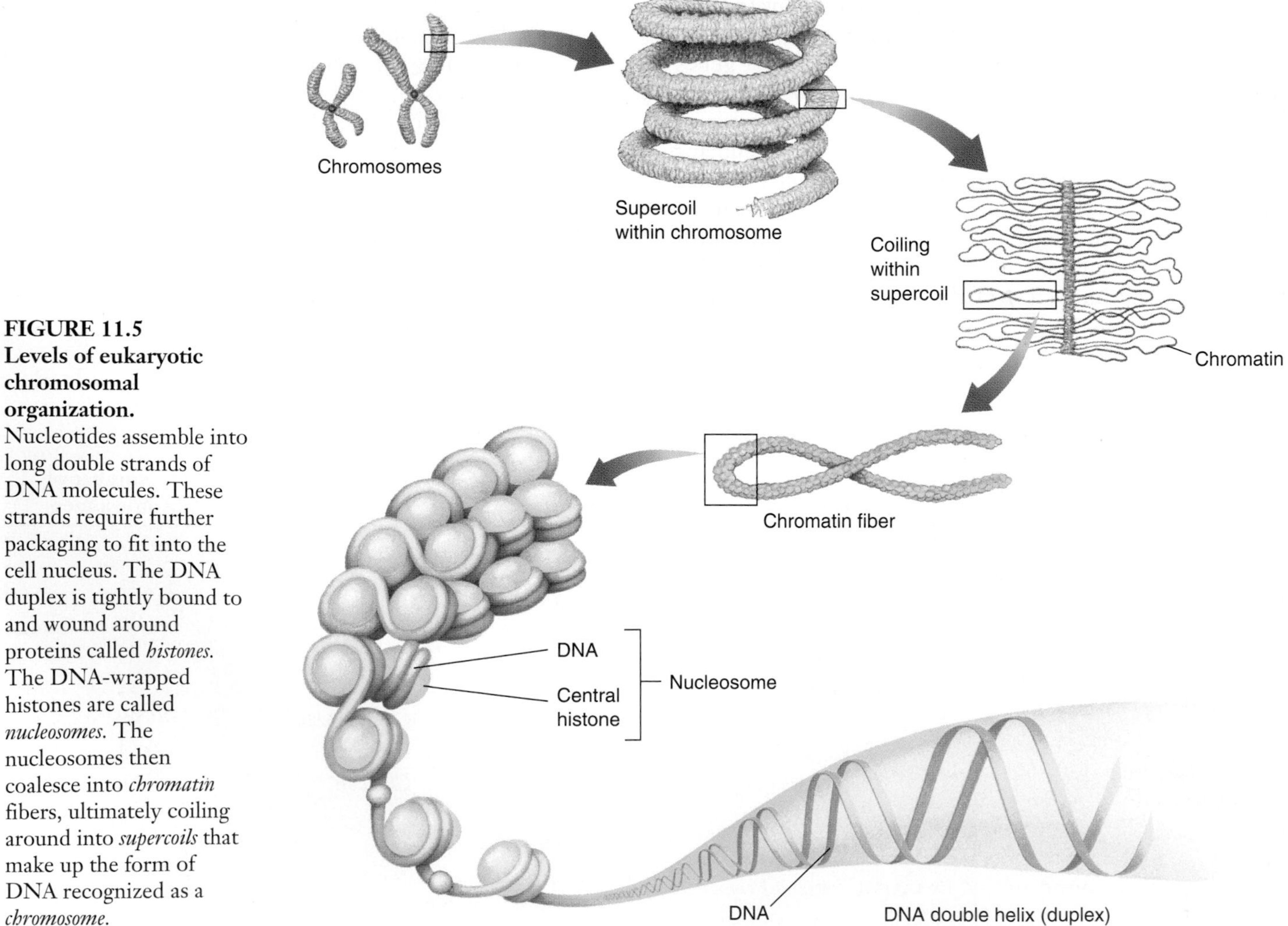

FIGURE 11.5
Levels of eukaryotic chromosomal organization.
Nucleotides assemble into long double strands of DNA molecules. These strands require further packaging to fit into the cell nucleus. The DNA duplex is tightly bound to and wound around proteins called *histones.* The DNA-wrapped histones are called *nucleosomes.* The nucleosomes then coalesce into *chromatin* fibers, ultimately coiling around into *supercoils* that make up the form of DNA recognized as a *chromosome.*

DNA. The histone cores thus act as "magnetic forms" that promote and guide the coiling of the DNA. Further coiling occurs when the string of nucleosomes wraps up into higher order coils called supercoils.

Highly condensed portions of the chromatin are called **heterochromatin.** Some of these portions remain permanently condensed, so that their DNA is never expressed. The remainder of the chromosome, called **euchromatin,** is condensed only during cell division, when compact packaging facilitates the movement of the chromosomes. At all other times, euchromatin is present in an open configuration, and its genes can be expressed. The way chromatin is packaged when the cell is not dividing is not well understood beyond the level of nucleosomes and is a topic of intensive research.

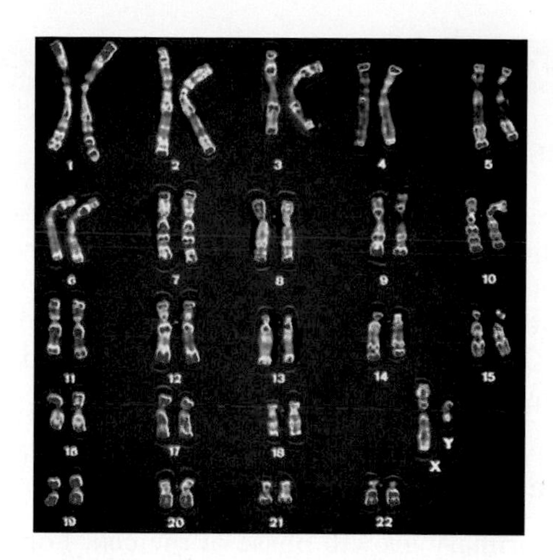

FIGURE 11.6
A human karyotype. The individual chromosomes that make up the 23 pairs differ widely in size and in centromere position. In this preparation, the chromosomes have been specifically stained to indicate further differences in their composition and to distinguish them clearly from one another.

Chromosome Karyotypes

Chromosomes may differ widely in appearance. They vary in size, staining properties, the location of the *centromere* (a constriction found on all chromosomes), the relative length of the two arms on either side of the centromere, and the positions of constricted regions along the arms. The particular array of chromosomes that an individual possesses is called its **karyotype** (figure 11.6). Karyotypes show marked differences among species and sometimes even among individuals of the same species.

To examine a human karyotype, investigators collect a cell sample from blood, amniotic fluid, or other tissue and add chemicals that induce the cells in the sample to divide. Later, they add other chemicals to stop cell division at a stage when the chromosomes are most condensed and thus most easily distinguished from one another. The cells are then broken open and their contents, including the chromosomes, spread out and stained. To facilitate the examination of the karyotype, the chromosomes are usually photographed, and the outlines of the chromosomes are cut out of the photograph and arranged in order (see figure 11.6).

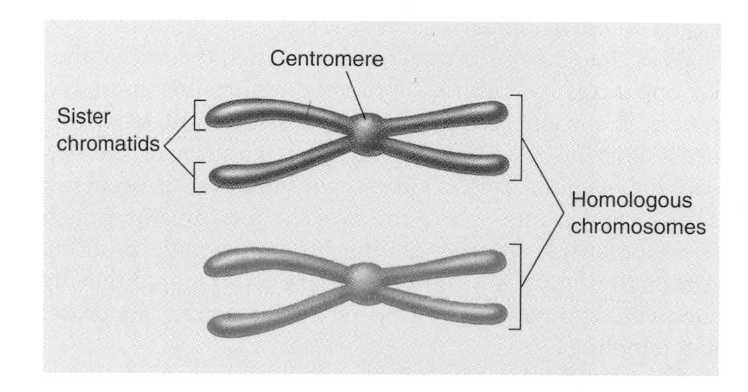

FIGURE 11.7
The difference between homologous chromosomes and sister chromatids. Homologous chromosomes are a pair of the same chromosome—say, chromosome number 16. Sister chromatids are the two replicas of a single chromosome held together by the centromeres after DNA replication.

How Many Chromosomes Are in a Cell?

With the exception of the **gametes** (eggs or sperm) and a few specialized tissues, every cell in a human body is **diploid (2*n*).** This means that the cell contains two nearly identical copies of each of the 23 types of chromosomes, for a total of 46 chromosomes. The **haploid (1*n*)** gametes contain only one copy of each of the 23 chromosome types, while certain tissues have unusual numbers of chromosomes—many liver cells, for example, have two nuclei, while mature red blood cells have no nuclei at all. The two copies of each chromosome in body cells are called **homologous chromosomes,** or **homologues** (Greek *homologia,* "agreement"). Before cell division, each homologue replicates, producing two identical **sister chromatids** joined at the **centromere,** a condensed area found on all eukaryotic chromosomes (figure 11.7). Hence, as cell division begins, a human body cell contains a total of 46 replicated chromosomes, each composed of two sister chromatids joined by one centromere. The cell thus contains 46 centromeres and 92 chromatids (2 sister chromatids for each of 2 homologues for each of 23 chromosomes). The cell is said to contain 46 chromosomes rather than 92 because, by convention, the number of chromosomes is obtained by counting centromeres.

Eukaryotic genomes are larger and more complex than those of bacteria. Eukaryotic DNA is packaged tightly into chromosomes, enabling it to fit inside cells. Haploid cells contain one set of chromosomes, while diploid cells contain two sets.

11.3 Mitosis is a key phase of the cell cycle.

Phases of the Cell Cycle

The increased size and more complex organization of eukaryotic genomes over those of bacteria required radical changes in the process by which the two replicas of the genome are partitioned into the daughter cells during cell division. This division process is diagrammed as a **cell cycle,** consisting of five phases (figure 11.8).

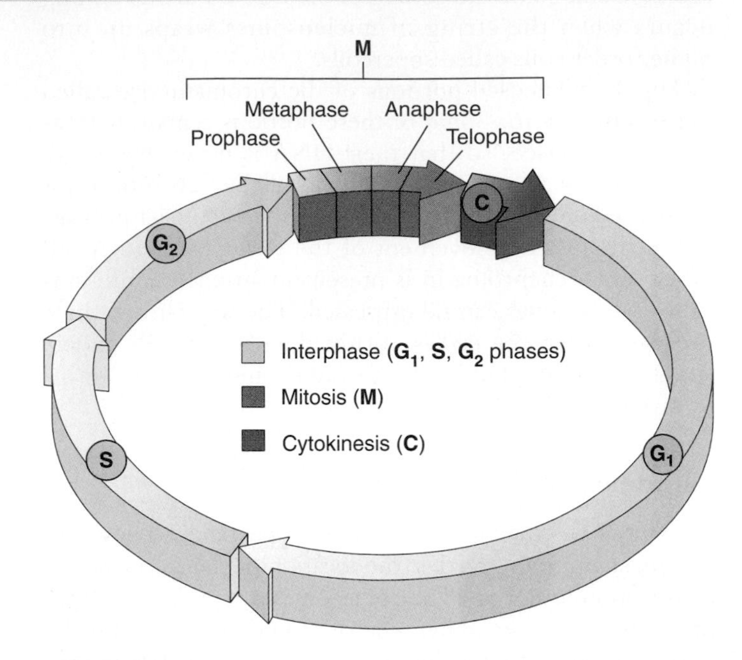

FIGURE 11.8
The cell cycle. This circle represents the 22-hour cell cycle in human cells growing in culture. G_1 represents the primary growth phase of the cell cycle, S the phase during which a replica of the genome is synthesized, and G_2 the second growth phase.

The Five Phases

G_1 is the primary growth phase of the cell. For many organisms, this encompasses the major portion of the cell's life span. **S** is the phase in which the cell synthesizes a replica of the genome. **G_2** is the second growth phase, in which preparations are made for genomic separation. During this phase, mitochondria and other organelles replicate, chromosomes condense, and microtubules begin to assemble at a spindle. G_1, S, and G_2 together constitute **interphase,** the portion of the cell cycle between cell divisions.

M is the phase of the cell cycle in which the microtubular apparatus assembles, binds to the chromosomes, and moves the sister chromatids apart. Called **mitosis,** this process is the essential step in the separation of the two daughter genomes. We will discuss mitosis as it occurs in animals and plants, where the process does not vary much (it is somewhat different among fungi and some protists). Although mitosis is a continuous process, it is traditionally subdivided into four stages: prophase, metaphase, anaphase, and telophase.

C is the phase of the cell cycle when the cytoplasm divides, creating two daughter cells. This phase is called **cytokinesis.** In animal cells, the microtubule spindle helps position a contracting ring of actin that constricts like a drawstring to pinch the cell in two. In cells with a cell wall, such as plant cells, a plate forms between the dividing cells.

Duration of the Cell Cycle

The time it takes to complete a cell cycle varies greatly among organisms. Cells in growing embryos can complete their cell cycle in under 20 minutes; the shortest known animal nuclear division cycles occur in fruit fly embryos (8 minutes). Cells such as these simply divide their nuclei as quickly as they can replicate their DNA, without cell growth. Half of the cycle is taken up by S, half by M, and essentially none by G_1 or G_2. Because mature cells require time to grow, most of their cycles are much longer than those of embryonic tissue. Typically, a dividing mammalian cell completes its cell cycle in about 24 hours, but some cells, like certain cells in the human liver, have cell cycles lasting more than a year. During the cycle, growth occurs throughout the G_1 and G_2 phases (referred to as "gap" phases, as they separate S from M), as well as during the S phase. The M phase takes only about an hour, a small fraction of the entire cycle.

Most of the variation in the length of the cell cycle from one organism or tissue to the next occurs in the G_1 phase. Cells often pause in G_1 before DNA replication and enter a resting state called **G_0 phase;** they may remain in this phase for days to years before resuming cell division. At any given time, most of the cells in an animal's body are in G_0 phase. Some, such as muscle and nerve cells, remain there permanently; others, such as liver cells, can resume G_1 phase in response to factors released during injury.

Most eukaryotic cells repeat a process of growth and division referred to as the cell cycle. The cycle can vary in length from a few minutes to several years.

Interphase: Preparing for Mitosis

The events that occur during interphase, made up of the G_1, S, and G_2 phases, are very important for the successful completion of mitosis. During G_1, cells undergo the major portion of their growth. During the S phase, each chromosome replicates to produce two sister chromatids, which remain attached to each other at the **centromere.** The centromere is a point of constriction on the chromosome, containing a specific DNA sequence to which is bound a disk of protein called a **kinetochore.** This disk functions as an attachment site for fibers that assist in cell division (figure 11.9). Each chromosome's centromere is located at a characteristic site.

The cell grows throughout interphase. The G_1 and G_2 segments of interphase are periods of active growth, when proteins are synthesized and cell organelles produced. The cell's DNA replicates only during the S phase of the cell cycle.

After the chromosomes have replicated in S phase, they remain fully extended and uncoiled. This makes them invisible under the light microscope. In G_2 phase, they begin the long process of **condensation,** coiling ever more tightly. Special *motor proteins* are involved in the rapid final condensation of the chromosomes that occurs early in mitosis. Also during G_2 phase, the cells begin to assemble the machinery they will later use to move the chromosomes to opposite poles of the cell. In animal cells, a pair of microtubule-organizing centers called **centrioles** replicate. All eukaryotic cells undertake an extensive synthesis of *tubulin,* the protein of which microtubules are formed.

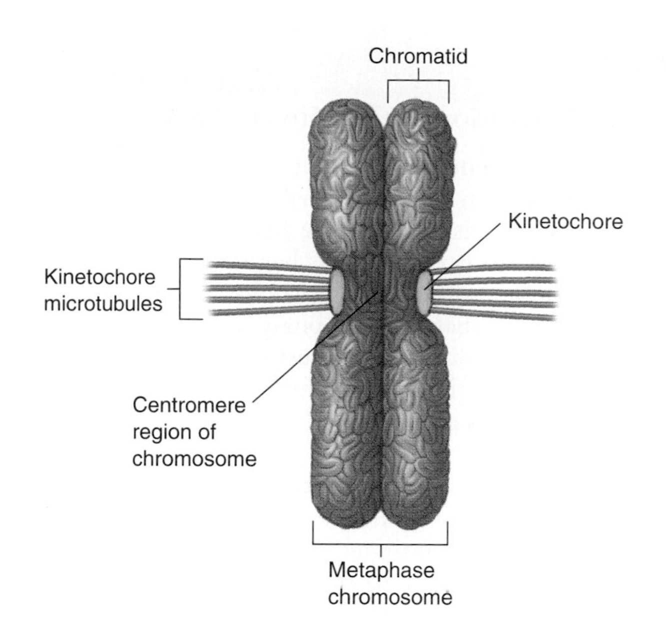

FIGURE 11.9
Kinetochores. In a metaphase chromosome, kinetochore microtubules are anchored to proteins at the centromere.

Interphase is that portion of the cell cycle in which the chromosomes are invisible under the light microscope because they are not yet condensed. It includes the G_1, S, and G_2 phases. In the G_2 phase, the cell mobilizes its resources for cell division.

A Vocabulary of Cell Division

binary fission Asexual reproduction of a cell by division into two equal or nearly equal parts. Bacteria divide by binary fission.

centromere A constricted region of a chromosome about 220 nucleotides in length, composed of highly repeated DNA sequences (satellite DNA). During mitosis, the centromere joins the two sister chromatids and is the site to which the kinetochores are attached.

chromatid One of the two copies of a replicated chromosome, joined by a single centromere to the other strand.

chromatin The complex of DNA and proteins of which eukaryotic chromosomes are composed.

chromosome The structure within cells that contains the genes. In eukaryotes, it consists of a single linear DNA molecule associated with proteins. The DNA is replicated during S phase, and the replicas separated during M phase.

cytokinesis Division of the cytoplasm of a cell after nuclear division.

euchromatin The portion of a chromosome that is extended except during cell division, and from which RNA is transcribed.

heterochromatin The portion of a chromosome that remains permanently condensed and, therefore, is not transcribed into RNA. Most centromere regions are heterochromatic.

homologues Homologous chromosomes; in diploid cells, one of a pair of chromosomes that carry equivalent genes.

kinetochore A disk of protein bound to the centromere and attached to microtubules during mitosis, linking each chromatid to the spindle apparatus.

microtubule A hollow cylinder, about 25 nanometers in diameter, composed of subunits of the protein tubulin. Microtubules lengthen by the addition of tubulin subunits to their end(s) and shorten by the removal of subunits.

mitosis Nuclear division in which replicated chromosomes separate to form two genetically identical daughter nuclei. When accompanied by cytokinesis, it produces two identical daughter cells.

nucleosome The basic packaging unit of eukaryotic chromosomes, in which the DNA molecule is wound around a cluster of histone proteins. Chromatin is composed of long strings of nucleosomes that resemble beads on a string.

Mitosis

Prophase: Formation of the Mitotic Apparatus

When the chromosome condensation initiated in G_2 phase reaches the point at which individual condensed chromosomes first become visible with the light microscope, the first stage of mitosis, **prophase,** has begun. The condensation process continues throughout prophase; consequently, some chromosomes that start prophase as minute threads appear quite bulky before its conclusion. Ribosomal RNA synthesis ceases when the portion of the chromosome bearing the rRNA genes is condensed.

Assembling the Spindle Apparatus. The assembly of the microtubular apparatus that will later separate the sister chromatids also continues during prophase. In animal cells, the two centriole pairs formed during G_2 phase begin to move apart early in prophase, forming between them an axis of microtubules referred to as spindle fibers. By the time the centrioles reach the opposite poles of the cell, they have established a bridge of microtubules called the spindle apparatus between them. In plant cells, a similar bridge of microtubular fibers forms between opposite poles of the cell, although centrioles are absent in plant cells.

During the formation of the spindle apparatus, the nuclear envelope breaks down and the endoplasmic reticulum reabsorbs its components. At this point, then, the microtubular spindle fibers extend completely across the cell, from one pole to the other. Their orientation determines the plane in which the cell will subsequently divide, through the center of the cell at right angles to the spindle apparatus.

In animal cell mitosis, the centrioles extend a radial array of microtubules toward the plasma membrane when they reach the poles of the cell. This arrangement of microtubules is called an **aster.** Although the aster's function is not fully understood, it probably braces the centrioles against the membrane and stiffens the point of microtubular attachment during the retraction of the spindle. Plant cells, which have rigid cell walls, do not form asters.

Linking Sister Chromatids to Opposite Poles. Each chromosome possesses two kinetochores, one attached to the centromere region of each sister chromatid (see figure 11.9). As prophase continues, a second group of microtubules appears to grow from the poles of the cell toward the centromeres. These microtubules connect the kinetochores on each pair of sister chromatids to the two poles of the spindle. Because microtubules extending from the two poles attach to opposite sides of the centromere, they attach one sister chromatid to one pole and the other sister chromatid to the other pole. This arrangement is absolutely critical to the process of mitosis; any mistakes in microtubule positioning can be disastrous. The attachment of the two sides of a centromere to the same pole, for example, leads to a failure of the sister chromatids to separate, so that they end up in the same daughter cell.

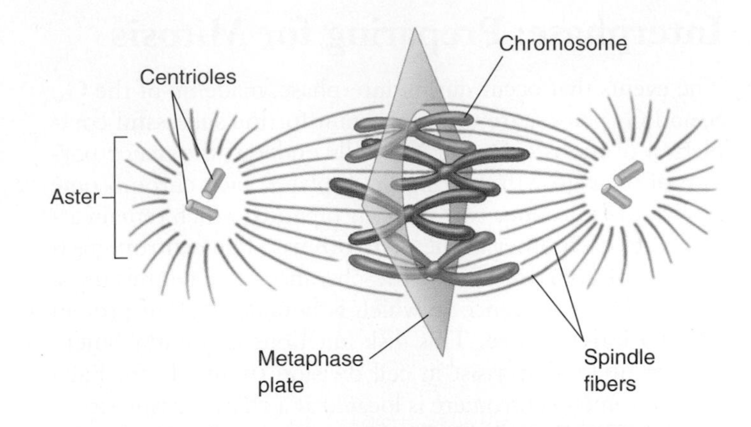

FIGURE 11.10
Metaphase. In metaphase, the chromosomes array themselves in a circle around the spindle midpoint.

Metaphase: Alignment of the Centromeres

The second stage of mitosis, **metaphase,** is the phase where the chromosomes align in the center of the cell. When viewed with a light microscope, the chromosomes appear to array themselves in a circle along the inner circumference of the cell, as the equator girdles the earth (figure 11.10). An imaginary plane perpendicular to the axis of the spindle that passes through this circle is called the *metaphase plate.* The metaphase plate is not an actual structure, but rather an indication of the future axis of cell division. Positioned by the microtubules attached to the kinetochores of their centromeres, all of the chromosomes line up on the metaphase plate (figure 11.11). At this point, which marks the end of metaphase, their centromeres are neatly arrayed in a circle, equidistant from the two poles of the cell, with microtubules extending back towards the opposite poles of the cell in an arrangement called a spindle because of its shape.

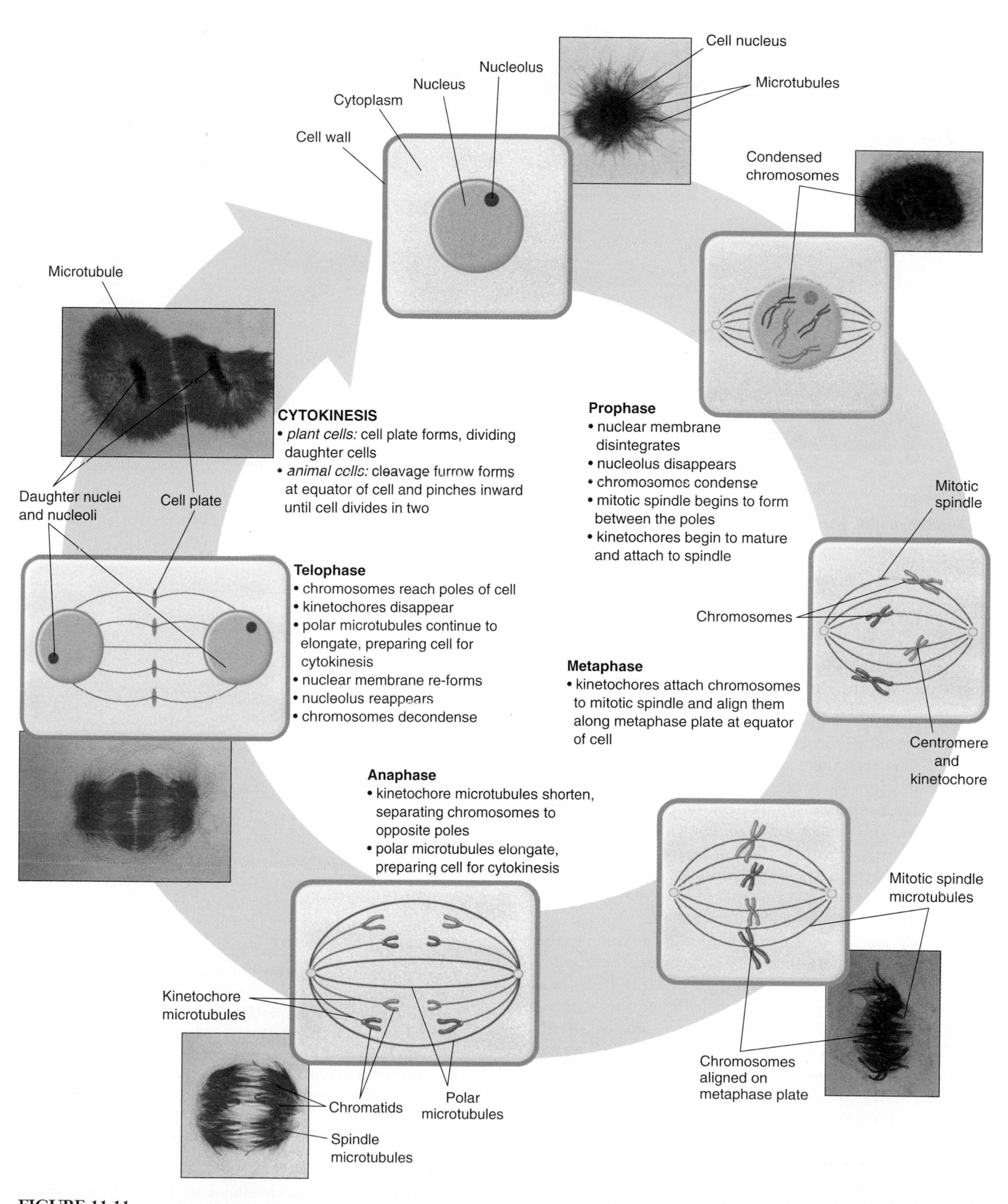
Cell nucleus
Microtubules
Nucleolus
Nucleus
Cytoplasm
Cell wall
Condensed chromosomes
Microtubule
CYTOKINESIS
• *plant cells:* cell plate forms, dividing daughter cells
• *animal cells:* cleavage furrow forms at equator of cell and pinches inward until cell divides in two
Prophase
• nuclear membrane disintegrates
• nucleolus disappears
• chromosomes condense
• mitotic spindle begins to form between the poles
• kinetochores begin to mature and attach to spindle
Daughter nuclei and nucleoli
Cell plate
Mitotic spindle
Telophase
• chromosomes reach poles of cell
• kinetochores disappear
• polar microtubules continue to elongate, preparing cell for cytokinesis
• nuclear membrane re-forms
• nucleolus reappears
• chromosomes decondense
Chromosomes
Metaphase
• kinetochores attach chromosomes to mitotic spindle and align them along metaphase plate at equator of cell
Centromere and kinetochore
Anaphase
• kinetochore microtubules shorten, separating chromosomes to opposite poles
• polar microtubules elongate, preparing cell for cytokinesis
Mitotic spindle microtubules
Kinetochore microtubules
Chromatids
Polar microtubules
Spindle microtubules
Chromosomes aligned on metaphase plate

FIGURE 11.11

Mitosis and cytokinesis. Mitosis (separation of the two genomes) occurs in four stages—prophase, metaphase, anaphase, and telophase—and is followed by cytokinesis (division into two separate cells). In this depiction, the chromosomes of the African blood lily, *Haemanthus katharinae*, are stained *blue*, and microtubules are stained *red*.

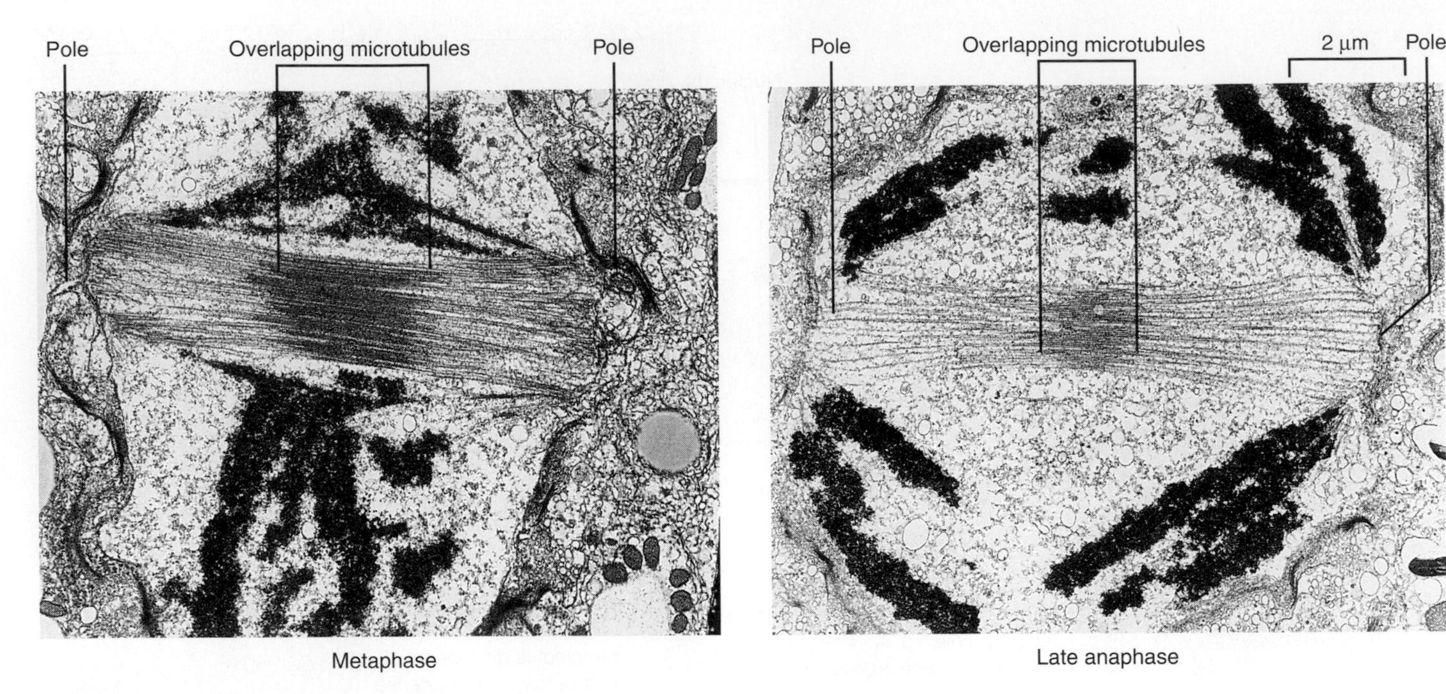

FIGURE 11.12
Microtubules slide past each other as the chromosomes separate. In these electron micrographs of dividing diatoms, the overlap of the microtubules lessens markedly during spindle elongation as the cell passes from metaphase to anaphase.

Anaphase and Telophase: Separation of the Chromatids and Reformation of the Nuclei

Of all the stages of mitosis, **anaphase** is the shortest and the most beautiful to watch. It starts when the centromeres divide. Each centromere splits in two, freeing the two sister chromatids from each other. The centromeres of all the chromosomes separate simultaneously, but the mechanism that achieves this synchrony is not known.

Freed from each other, the sister chromatids are pulled rapidly toward the poles to which their kinetochores are attached. In the process, two forms of movement take place simultaneously, each driven by microtubules.

First, *the poles move apart* as microtubular spindle fibers physically anchored to opposite poles slide past each other, away from the center of the cell (figure 11.12). Because another group of microtubules attach the chromosomes to the poles, the chromosomes move apart, too. If a flexible membrane surrounds the cell, it becomes visibly elongated.

Second, *the centromeres move toward the poles* as the microtubules that connect them to the poles shorten. This shortening process is not a contraction; the microtubules do not get any thicker. Instead, tubulin subunits are removed from the kinetochore ends of the microtubules by the organizing center. As more subunits are removed, the chromatid-bearing microtubules are progressively disassembled, and the chromatids are pulled ever closer to the poles of the cell.

When the sister chromatids separate in anaphase, the accurate partitioning of the replicated genome—the essential element of mitosis—is complete. In **telophase,** the spindle apparatus disassembles, as the microtubules are broken down into tubulin monomers that can be used to construct the cytoskeletons of the daughter cells. A nuclear envelope forms around each set of sister chromatids, which can now be called chromosomes because each has its own centromere. The chromosomes soon begin to uncoil into the more extended form that permits gene expression. One of the early group of genes expressed are the rRNA genes, resulting in the reappearance of the nucleolus.

During prophase, microtubules attach the centromeres joining pairs of sister chromatids to opposite poles of the spindle apparatus. During metaphase, each chromosome is drawn to a ring along the inner circumference of the cell by the microtubules extending from the centromere to the two poles of the spindle apparatus. During anaphase, the poles of the cell are pushed apart by microtubular sliding, and the sister chromatids are drawn to opposite poles by the shortening of the microtubules attached to them. During telophase, the spindle is disassembled, nuclear envelopes are reestablished, and the normal expression of genes present in the chromosomes is reinitiated.

Cytokinesis

Mitosis is complete at the end of telophase. The eukaryotic cell has partitioned its replicated genome into two nuclei positioned at opposite ends of the cell. While mitosis was going on, the cytoplasmic organelles, including mitochondria and chloroplasts (if present), were reassorted to areas that will separate and become the daughter cells. The replication of organelles takes place before cytokinesis, often in the S or G_2 phase. Cell division is still not complete at the end of mitosis, however, because the division of the cell proper has not yet begun. The phase of the cell cycle when the cell actually divides is called **cytokinesis.** It generally involves the cleavage of the cell into roughly equal halves.

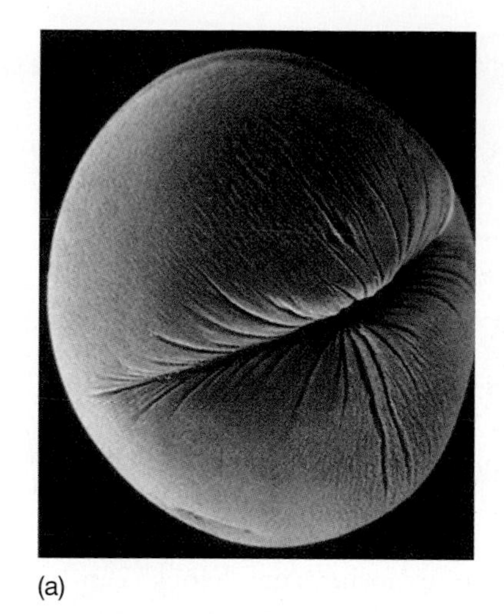

(a) (b)

FIGURE 11.13
Cytokinesis in animal cells.
(*a*) A cleavage furrow forms around a dividing sea urchin egg (30×). (*b*) The completion of cytokinesis in an animal cell. The two daughter cells are still joined by a thin band of cytoplasm occupied largely by microtubules.

Cytokinesis in Animal Cells

In animal cells and the cells of all other eukaryotes that lack cell walls, cytokinesis is achieved by means of a constricting belt of actin filaments. As these filaments slide past one another, the diameter of the belt decreases, pinching the cell and creating a *cleavage furrow* around the cell's circumference (figure 11.13*a*). As constriction proceeds, the furrow deepens until it eventually slices all the way into the center of the cell. At this point, the cell is divided in two (figure 11.13*b*).

Cytokinesis in Plant Cells

Plant cells possess a cell wall far too rigid to be squeezed in two by actin filaments. Instead, these cells assemble membrane components in their interior, at right angles to the spindle apparatus (figure 11.14). This expanding membrane partition, called a **cell plate,** continues to grow outward until it reaches the interior surface of the plasma membrane and fuses with it, effectively dividing the cell in two. Cellulose is then laid down on the new membranes, creating two new cell walls. The space between the daughter cells becomes impregnated with pectins and is called a **middle lamella.**

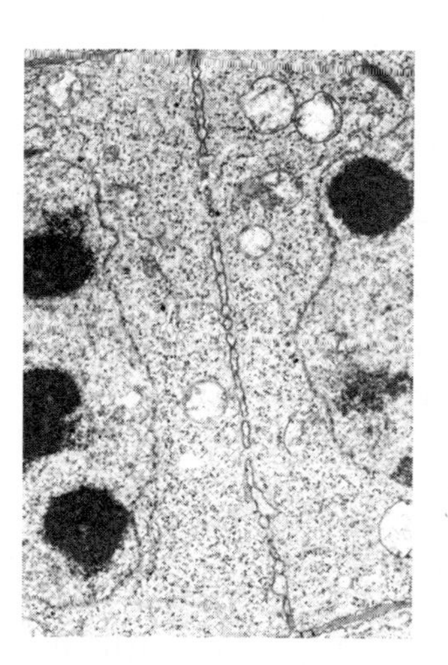

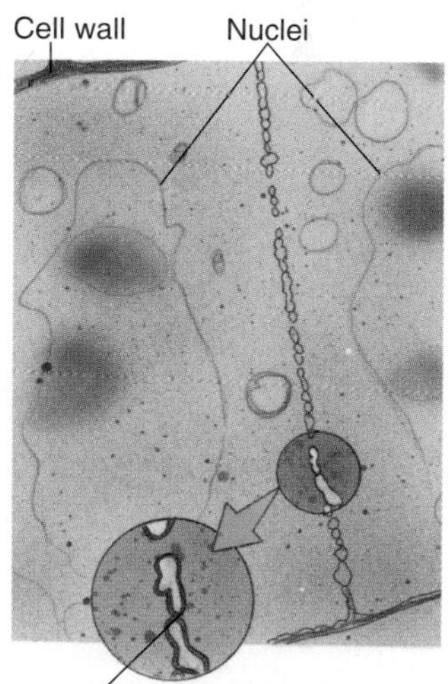

FIGURE 11.14
Cytokinesis in plant cells. In this photograph and companion drawing, a cell plate is forming between daughter nuclei. Once the plate is complete, there will be two cells.

Cytokinesis in Fungi and Protists

In fungi and some groups of protists, the nuclear membrane does not dissolve and, as a result, all the events of mitosis occurs entirely *within* the nucleus. Only after mitosis is complete in these organisms does the nucleus then divide into two daughter nuclei, and one nucleus goes to each daughter cell during cytokinesis. This separate nuclear division phase of the cell cycle does not occur in plants, animals, or most protists.

After cytokinesis in any eukaryotic cell, the two daughter cells contain all of the components of a complete cell. While mitosis ensures that both daughter cells contain a full complement of chromosomes, no similar mechanism ensures that organelles such as mitochondria and chloroplasts are distributed equally between the daughter cells. However, as long as some of each organelle are present in each cell, the organelles can replicate to reach the number appropriate for that cell.

Cytokinesis is the physical division of the cytoplasm of a eukaryotic cell into two daughter cells.

General Strategy of Cell Cycle Control

The events of the cell cycle are coordinated in much the same way in all eukaryotes. The control system human cells utilize first evolved among the protists over a billion years ago; today, it operates in essentially the same way in fungi as it does in humans.

The goal of controlling any cyclic process is to adjust the duration of the cycle to allow sufficient time for all events to occur. In principle, a variety of methods can achieve this goal. For example, an internal "clock" can be employed to allow adequate time for each phase of the cycle to be completed. This is how many organisms control their daily activity cycles. The disadvantage of using such a clock to control the cell cycle is that it is not very flexible. One way to achieve a more flexible and sensitive regulation of a cycle is simply to let the completion of each phase of the cycle trigger the beginning of the next phase, as a runner passing a baton starts the next leg in a relay race. Until recently, biologists thought this type of mechanism controlled the cell division cycle. However, we now know that eukaryotic cells employ a separate, centralized controller to regulate the process: at critical points in the cell cycle, further progress depends upon a central set of "go/no-go" switches that are regulated by feedback from the cell.

This mechanism is the same one engineers use to control many processes. For example, the furnace that heats a home in the winter typically goes through a daily heating cycle. When the daily cycle reaches the morning "turn on" checkpoint, sensors report whether the house temperature is below the set point (for example, 70°F). If it is, the thermostat triggers the furnace, which warms the house. If the house is already at least that warm, the thermostat does not start up the furnace. Similarly, the cell cycle has key checkpoints where feedback signals from the cell about its size and the condition of its chromosomes can either trigger subsequent phases of the cycle, or delay them to allow more time for the current phase to be completed.

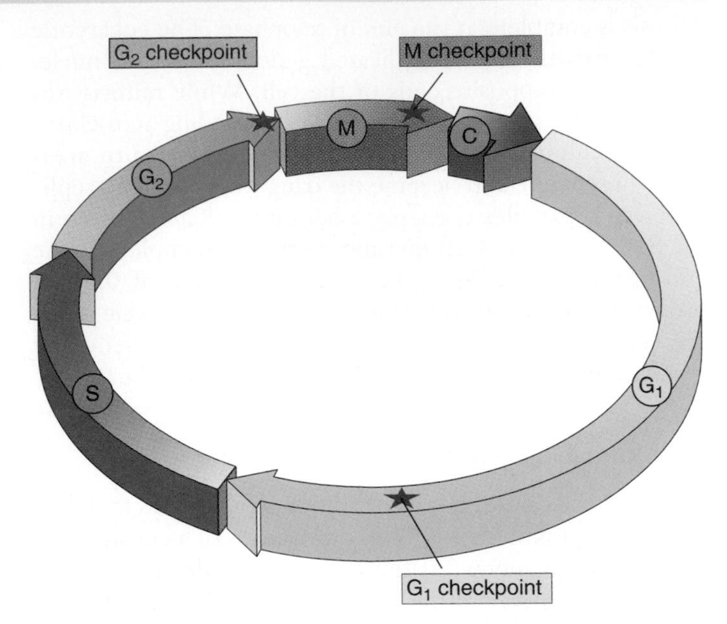

FIGURE 11.15
Control of the cell cycle. Cells use a centralized control system to check whether proper conditions have been achieved before passing three key "checkpoints" in the cell cycle.

FIGURE 11.16
The G_1 checkpoint. Feedback from the cell determines whether the cell cycle will proceed to the S phase, pause, or withdraw into G_0 for an extended rest period.

Architecture of the Control System

Three principal checkpoints control the cell cycle in eukaryotes (figure 11.15):

Cell growth is assessed at the G_1 checkpoint. Located near the end of G_1, just before entry into S phase, this checkpoint makes the key decision of whether the cell should divide, delay division, or enter a resting stage (figure 11.16). In yeasts, where researchers first studied this checkpoint, it is called START. If conditions are favorable for division, the cell begins to copy its DNA, initiating S phase. The G_1 checkpoint is where the more complex eukaryotes typically arrest the cell cycle if environmental conditions make cell division impossible, or if the cell passes into G_0 for an extended period.

The success of DNA replication is assessed at the G_2 checkpoint. The second checkpoint, which occurs at the end of G_2, triggers the start of M phase. If this checkpoint is passed, the cell initiates the many molecular processes that signal the beginning of mitosis.

Mitosis is assessed at the M checkpoint. Occurring at metaphase, the third checkpoint triggers the exit from mitosis and cytokinesis and the beginning of G_1.

The cell cycle is controlled at three checkpoints.

Molecular Mechanisms of Cell Cycle Control

Exactly how does a cell achieve central control of the division cycle? The basic mechanism is quite simple. A set of proteins sensitive to the condition of the cell interact at the checkpoints to trigger the next events in the cycle. Two key types of proteins participate in this interaction: cyclin-dependent protein kinases and cyclins (figure 11.17).

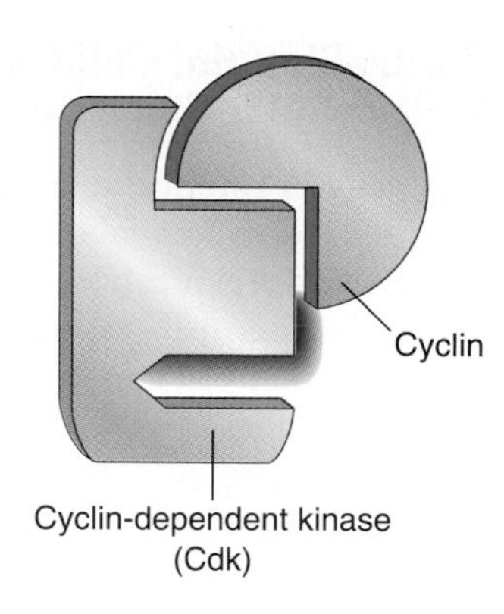

FIGURE 11.17
A complex of two proteins triggers passage through cell cycle checkpoints. Cdk is a protein kinase that activates numerous cell proteins by phosphorylating them. Cyclin is a regulatory protein required to activate Cdk; in other words, Cdk does not function unless cyclin is bound to it.

The Cyclin Control System

Cyclin-dependent protein kinases (Cdks) are enzymes that phosphorylate (add phosphate groups to) the serine and threonine amino acids of key cellular enzymes and other proteins. At the G_2 checkpoint, for example, Cdks phosphorylate histones, nuclear membrane filaments, and the microtubule-associated proteins that form the mitotic spindle. Phosphorylation of these components of the cell division machinery initiates activities that carry the cycle past the checkpoint into mitosis.

Cyclins are proteins that bind to Cdks, enabling the Cdks to function as enzymes. Cyclins are so named because they are destroyed and resynthesized during each turn of the cell cycle (figure 11.18). Different cyclins regulate the G_1 and G_2 cell cycle checkpoints.

The G_2 Checkpoint. During G_2, the cell gradually accumulates G_2 cyclin (also called mitotic cyclin). This cyclin binds to Cdk to form a complex called MPF (mitosis-promoting factor). At first, MPF is not active in carrying the cycle past the G_2 checkpoint. But eventually, other cellular enzymes phosphorylate and so activate a few molecules of MPF. These activated MPFs in turn increase the activity of the enzymes that phosphorylate MPF, setting up a positive feedback that leads to a very rapid increase in the cellular concentration of activated MPF. When the level of activated MPF exceeds the threshold necessary to trigger mitosis, G_2 phase ends.

MPF sows the seeds of its own destruction. The length of time the cell spends in M phase is determined by the activity of MPF, for one of its many functions is to activate proteins that destroy cyclin. As mitosis proceeds to the end of metaphase, Cdk levels stay relatively constant, but increasing amounts of G_2 cyclin are degraded, causing progressively less MPF to be available and so initiating the events that end mitosis. After mitosis, the gradual accumulation of new cyclin starts the next turn of the cell cycle.

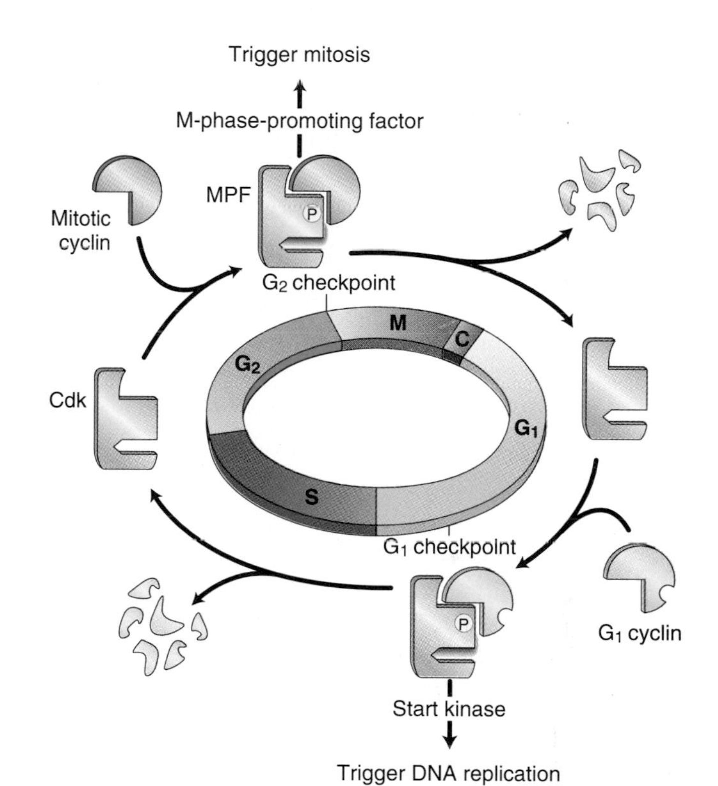

FIGURE 11.18
How cell cycle control works. As the cell cycle passes through the G_1 and G_2 checkpoints, Cdk becomes associated with different cyclins and, as a result, activates different cellular processes. At the completion of each phase, the cyclins are degraded, bringing Cdk activity to a halt until the next set of cyclins appears.

The G_1 Checkpoint. The G_1 checkpoint is thought to be regulated in a similar fashion. In unicellular eukaryotes such as yeasts, the main factor triggering DNA replication is cell size. Yeast cells grow and divide as rapidly as possible, and they make the START decision by comparing the volume of cytoplasm to the size of the genome. As a cell grows, its cytoplasm increases in size, while the amount of DNA remains constant. Eventually a threshold ratio is reached that promotes the production of cyclins and thus triggers the next round of DNA replication and cell division.

Controlling the Cell Cycle in Multicellular Eukaryotes

The cells of multicellular eukaryotes are not free to make individual decisions about cell division, as yeast cells are. The body's organization cannot be maintained without severely limiting cell proliferation, so that only certain cells divide, and only at appropriate times. The way that cells inhibit individual growth of other cells is apparent in mammalian cells growing in tissue culture: a single layer of cells expands over a culture plate until the growing border of cells comes into contact with neighboring cells, and then the cells stop dividing. If a sector of cells is cleared away, neighboring cells rapidly refill that sector and then stop dividing again. How are cells able to sense the density of the cell culture around them? Each growing cell apparently binds minute amounts of positive regulatory signals called **growth factors,** proteins that stimulate cell division (such as MPF). When neighboring cells have used up what little growth factor is present, not enough is left to trigger cell division in any one cell.

Growth Factors and the Cell Cycle

As you may recall from chapter 7 (cell-cell interactions), growth factors work by triggering intracellular signaling systems. Fibroblasts, for example, possess numerous receptors on their plasma membranes for one of the first growth factors to be identified: platelet-derived growth factor (PDGF). When PDGF binds to a membrane receptor, it initiates an amplifying chain of internal cell signals that stimulates cell division. PDGF was discovered when investigators found that fibroblasts would grow and divide in tissue culture only if the growth medium contained blood serum (the liquid that remains after blood clots); blood plasma (blood from which the cells have been removed without clotting) would not work. The researchers hypothesized that platelets in the blood clots were releasing into the serum one or more factors required for fibroblast growth. Eventually, they isolated such a factor and named it PDGF. Growth factors such as PDGF override cellular controls that otherwise inhibit cell division. When a tissue is injured, a blood clot forms and the release of PDGF triggers neighboring cells to divide, helping to heal the wound. Only a tiny amount of PDGF (approximately 10^{-10} M) is required to stimulate cell division.

Characteristics of Growth Factors. Over 50 different proteins that function as growth factors have been isolated (table 11.2 lists a few), and more undoubtedly exist. A specific cell surface receptor "recognizes" each growth factor, its shape fitting that growth factor precisely. When the growth factor binds with its receptor, the receptor reacts by triggering events within the cell (figure 11.19). The cellular selectivity of a particular growth factor depends upon which target cells bear its unique receptor. Some growth

Table 11.2 Growth Factors of Mammalian Cells

Growth Factor	Range of Specificity	Effects
Epidermal growth factor (EGF)	Broad	Stimulates cell proliferation in many tissues; plays a key role in regulating embryonic development
Erythropoietin	Narrow	Required for proliferation of red blood cell precursors and their maturation into erythrocytes (red blood cells)
Fibroblast growth factor (FGF)	Broad	Initiates the proliferation of many cell types; inhibits maturation of many types of stem cells; acts as a signal in embryonic development
Insulin-like growth factor	Broad	Stimulates metabolism of many cell types; potentiates the effects of other growth factors in promoting cell proliferation
Interleukin-2	Narrow	Triggers the division of activated T lymphocytes during the immune response
Mitosis-promoting factor (MPF)	Broad	Regulates entrance of the cell cycle into the M phase
Nerve growth factor (NGF)	Narrow	Stimulates the growth of neuron processes during neural development
Platelet-derived growth factor (PDGF)	Broad	Promotes the proliferation of many connective tissues and some neuroglial cells
Transforming growth factor β (TGF-β)	Broad	Accentuates or inhibits the responses of many cell types to other growth factors; often plays an important role in cell differentiation

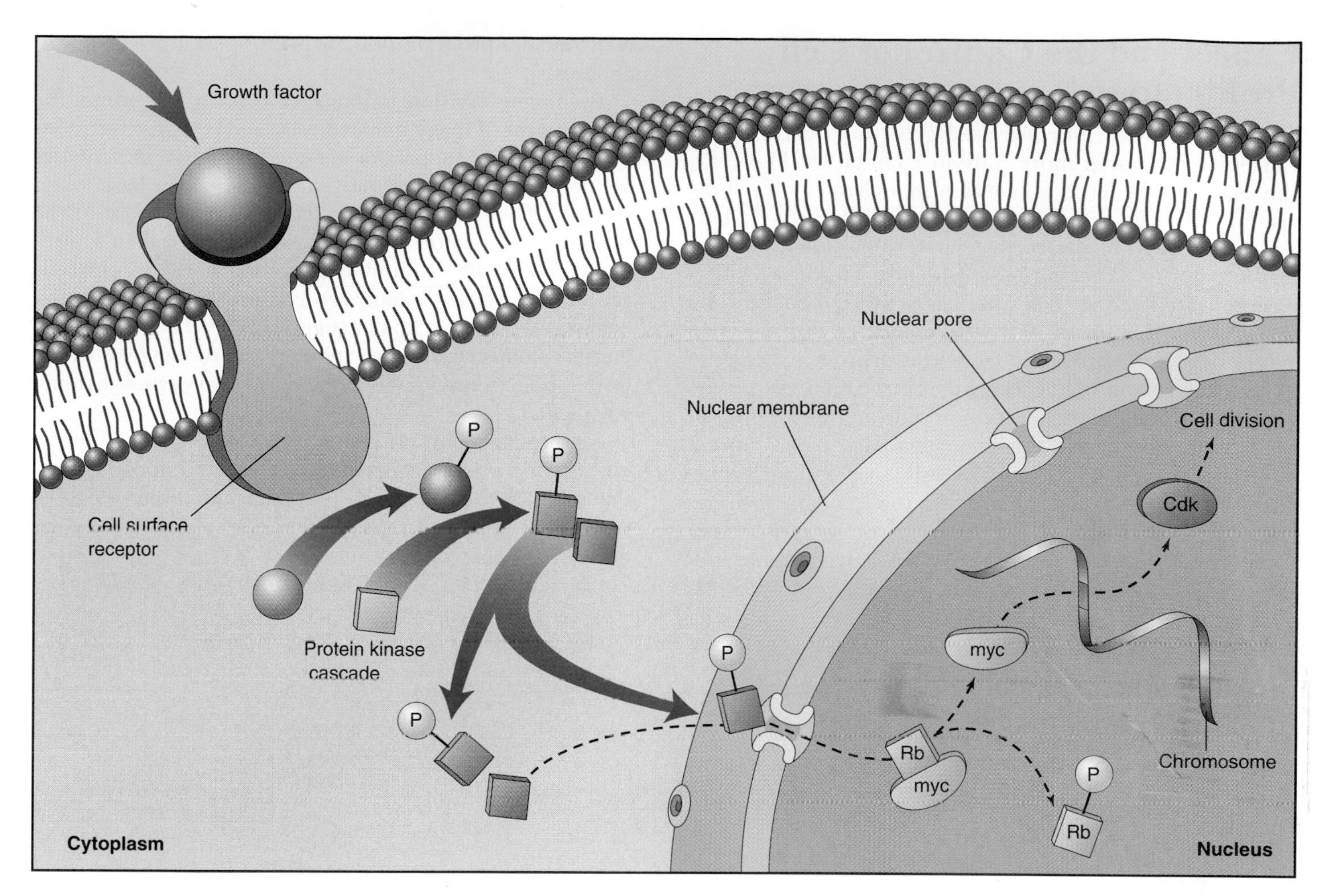

FIGURE 11.19
The cell proliferation-signaling pathway. Binding of a growth factor sets in motion a cascading intracellular signaling pathway (described in chapter 7), which activates nuclear regulatory proteins that trigger cell division. In this example, when the nuclear protein Rb is phosphorylated, another nuclear protein (myc) is released and is then able to stimulate the production of Cdk proteins.

factors, like PDGF and epidermal growth factor (EGF), affect a broad range of cell types, while others affect only specific types. For example, nerve growth factor (NGF) promotes the growth of certain classes of neurons, and erythropoietin triggers cell division in red blood cell precursors. Most animal cells need a combination of several different growth factors to overcome the various controls that inhibit cell division.

The G_0 Phase. If cells are deprived of appropriate growth factors, they stop at the G_1 checkpoint of the cell cycle. With their growth and division arrested, they remain in the G_0 phase, as we discussed earlier. This nongrowing state is distinct from the interphase stages of the cell cycle, G_1, S, and G_2.

It is the ability to enter G_0 that accounts for the incredible diversity seen in the length of the cell cycle among different tissues. Epithelial cells lining the gut divide more than twice a day, constantly renewing the lining of the digestive tract. By contrast, liver cells divide only once every year or two, spending most of their time in G_0 phase. Mature neurons and muscle cells usually never leave G_0.

Two groups of proteins, cyclins and Cdks, interact to regulate the cell cycle. Cells also receive protein signals called growth factors that affect cell division.

Cancer and the Control of Cell Proliferation

The unrestrained, uncontrolled growth of cells, called cancer, is addressed more fully in chapter 18. However, cancer certainly deserves mention in a chapter on cell division, as it is essentially a disease of cell division—a failure of cell division *control.* Recent work has identified one of the culprits. Working independently, cancer scientists have repeatedly identified what has proven to be the same gene! Officially dubbed *p53* (researchers italicize the gene symbol to differentiate it from the protein), this gene plays a key role in the G_1 checkpoint of cell division. The gene's product, the p53 protein, monitors the integrity of DNA, checking that it is undamaged. If the p53 protein detects damaged DNA, it halts cell division and stimulates the activity of special enzymes to repair the damage. Once the DNA has been repaired, *p53* allows cell division to continue. In cases where the DNA is irreparable, *p53* then directs the cell to kill itself, activating an apoptosis (cell suicide) program (see chapter 17 for a discussion of apoptosis).

By halting division in damaged cells, *p53* prevents the development of many mutated cells, and it is therefore considered a tumor-suppressor gene (even though its activities are not limited to cancer prevention). Scientists have found that *p53* is entirely absent or damaged beyond use in the majority of cancerous cells they have examined! It is precisely because *p53* is nonfunctional that these cancer cells are able to repeatedly undergo cell division without being halted at the G_1 checkpoint (figure 11.20). To test this, scientists administered healthy p53 protein to rapidly dividing cancer cells in a petri dish: the cells soon ceased dividing and died.

Scientists at Johns Hopkins University School of Medicine have further reported that cigarette smoke causes mutations in the *p53* gene. This study, published in 1995, reinforced the strong link between smoking and cancer described in chapter 18.

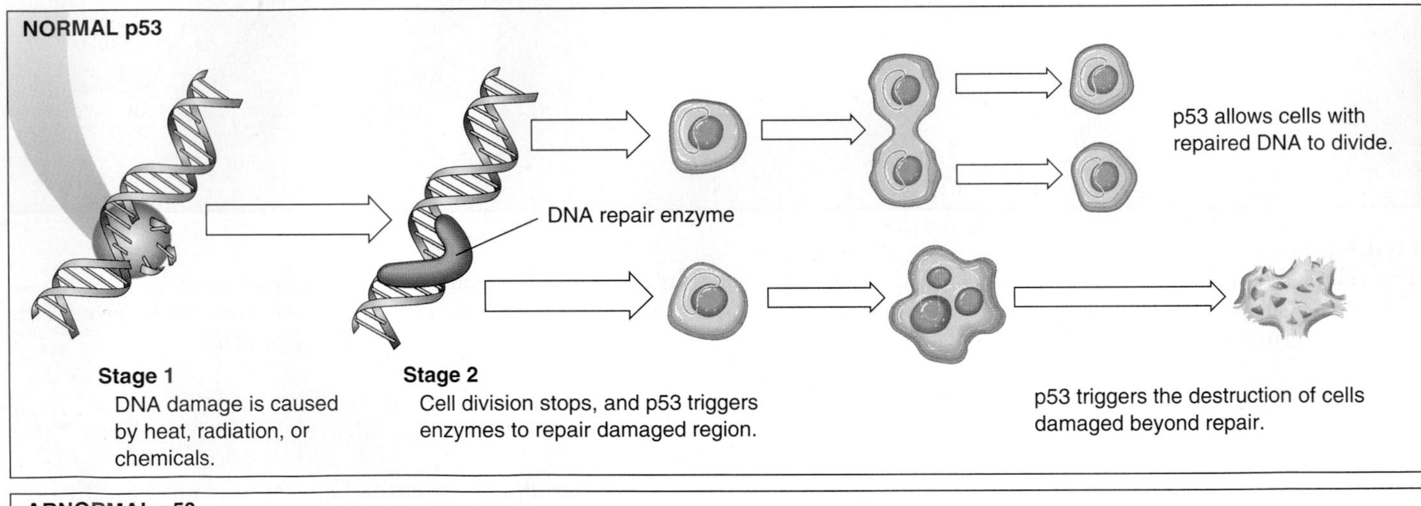

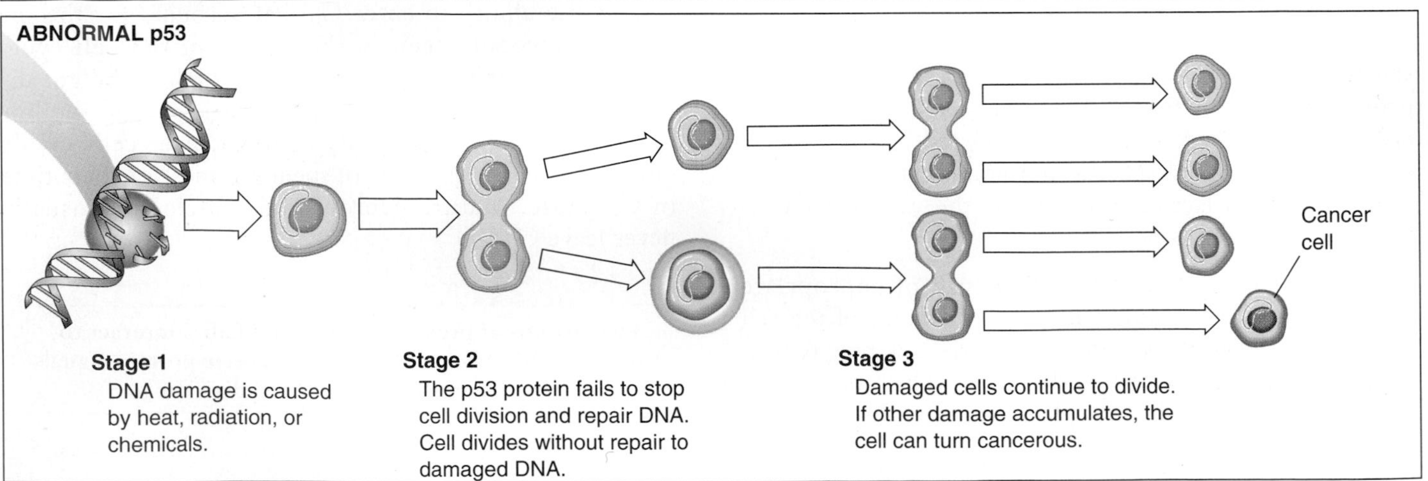

ABNORMAL p53
Cancer cell
Stage 1
DNA damage is caused by heat, radiation, or chemicals.
Stage 2
The p53 protein fails to stop cell division and repair DNA. Cell divides without repair to damaged DNA.
Stage 3
Damaged cells continue to divide. If other damage accumulates, the cell can turn cancerous.

FIGURE 11.20
Cell division and p53 protein. Normal p53 protein monitors DNA, destroying cells with irreparable damage to their DNA. Abnormal p53 protein fails to stop cell division and repair DNA. As damaged cells proliferate, cancer develops.

Growth Factors and Cancer

How do growth factors influence the cell cycle? As you have seen, there are two different approaches, one positive and the other negative.

Proto-oncogenes. PDGF and many other growth factors utilize the positive approach, stimulating cell division. They trigger passage through the G_1 checkpoint by aiding the formation of cyclins and so activating genes that promote cell division. Genes that normally stimulate cell division are sometimes called *proto-oncogenes* because mutations that cause them to be overexpressed or hyperactive convert them into oncogenes (Greek *onco*, "cancer"), leading to the excessive cell proliferation that is characteristic of cancer. Even a single mutation (creating a heterozygote) can lead to cancer if the other cancer-preventing genes are nonfunctional. Geneticists, using Mendel's terms, call such mutations of proto-oncogenes *dominant*.

Some 30 different proto-oncogenes are known. Some act very quickly after stimulation by growth factors. Among the most intensively studied of these are *myc*, *fos*, and *jun*, all of which cause unrestrained cell growth and division when they are overexpressed. In a normal cell, the *myc* proto-oncogene appears to be important in regulating the G_1 checkpoint. Cells in which *myc* expression is prevented will not divide, even in the presence of growth factors. A critical activity of *myc* and other genes in this group of immediately responding proto-oncogenes is to stimulate a second group of "delayed response" genes, including those that produce cyclins and Cdk proteins (figure 11.21).

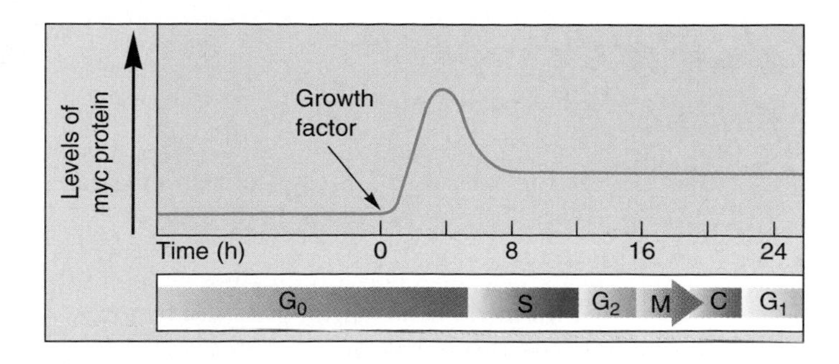

FIGURE 11.21
The role of *myc* in triggering cell division. The addition of a growth factor leads to transcription of the *myc* gene and rapidly increasing levels of the myc protein. This causes G_0 cells to enter the S phase and begin proliferating.

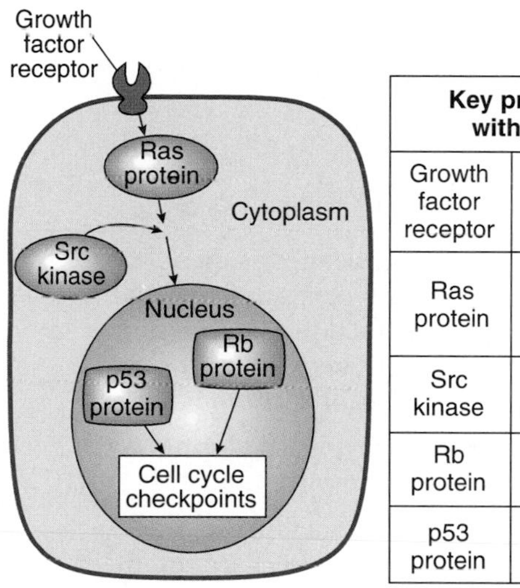

Key proteins associated with human cancers	
Growth factor receptor	More per cell in many breast cancers
Ras protein	Activated by mutations of *ras* in 20–30% of all cancers
Src kinase	Activated by mutations in 2–5% of all cancers
Rb protein	Mutated in 40% of all cancers
p53 protein	Mutated in 50% of all cancers

FIGURE 11.22
Mutations cause cancer. Mutations in genes encoding key components of the cell division-signaling pathway are responsible for many cancers. Among them are proto-oncogenes encoding growth factor receptors, such as ras protein, and kinase enzymes, such as src, that aid ras function. Mutations that disrupt tumor-suppressor proteins, such as Rb and p53, also foster cancer development.

Tumor-suppressor Genes. Other growth factors utilize a negative approach to cell cycle control. They block passage through the G_1 checkpoint by preventing cyclins from binding to Cdk, thus inhibiting cell division. Genes that normally inhibit cell division are called tumor-suppressor genes. When mutated, they can also lead to unrestrained cell division, but only if both copies of the gene are mutant. Hence, these cancer-causing mutations are *recessive*.

The most thoroughly understood of the tumor-suppressor genes is the retinoblastoma (*Rb*) gene. This gene was originally cloned from children with a rare form of eye cancer inherited as a recessive trait, implying that the normal gene product was a cancer suppressor that helped keep cell division in check. The *Rb* gene encodes a protein present in ample amounts within the nucleus. This protein interacts with many key regulatory proteins of the cell cycle, but how it does so depends upon its state of phosphorylation. In G_0 phase, the Rb protein is dephosphorylated. In this state, it binds to and ties up a set of regulatory proteins, like myc and fos, needed for cell proliferation, blocking their action and so inhibiting cell division (see figure 11.19). When phosphorylated, the Rb protein releases its captive regulatory proteins, freeing them to act and so promoting cell division. Growth factors lessen the inhibition the Rb protein imposes by activating kinases that phosphorylate it. Free of Rb protein inhibition, cells begin to produce cyclins and Cdk, pass the G_1 checkpoint, and proceed through the cell cycle. Figure 11.22 summarizes the types of genes that can cause cancer when mutated.

Some growth factors, proto-oncogenes, accelerate the cell cycle by promoting cyclins and Cdks, others, tumor-suppressor genes, suppress it by inhibiting their action. Cancer can result if mutations occur in either type of growth factor gene.

Chapter 11

www.mhhe.com/raven6/resources11.mhtml

Summary	Questions	Media Resources

11.1 Bacteria divide far more simply than do eukaryotes.

- Bacterial cells divide by simple binary fission.
- The two replicated circular DNA molecules attach to the plasma membrane at different points, and fission is initiated between those points.

1. How is the genome replicated prior to binary fission in a bacterial cell?

- Prokaryotes

11.2 Chromosomes are highly ordered structures.

- All eukaryotic cells contain chromosomes but they vary in the number of chromosomes.
- Eukaryotic DNA forms a complex with histones and other proteins and is packaged into chromosomes.

2. What are nucleosomes composed of, and how do they participate in the coiling of DNA?

3. What is a karyotype? How are chromosomes distinguished from one another in a karyotype?

- Cell Division Introduction
- Chromosomes

11.3 Mitosis is a key phase of the cell cycle.

- In eukaryotic cells, DNA replication is completed during the S phase of the cell cycle, and during the G_2 phase the cell makes its final preparation for mitosis. Along with G_1, these two phases constitute the portion of the cell cycle called interphase.
- The first stage of mitosis is prophase, during which the mitotic spindle apparatus forms.
- In the second stage of mitosis, metaphase, the chromosomes are arranged in a circle around the periphery of the cell.
- At the beginning of the third stage of mitosis, anaphase, the centromeres joining each pair of sister chromatids separate, freeing the sister chromatids from each other.
- After the chromatids physically separate, they are pulled to opposite poles of the cell by the microtubules attached to their centromeres.
- In the fourth and final stage of mitosis, telophase, the mitotic apparatus is disassembled, the nuclear envelope re-forms, and the chromosomes uncoil.
- When mitosis is complete, the cell divides in two, so that the two sets of chromosomes separated by mitosis end up in different daughter cells.

4. Which phases of the cell cycle is generally the longest in the cells of a mature eukaryote?

5. What happens to the chromosomes during S phase?

6. What changes with respect to ribosomal RNA occur during prophase?

7. What event signals the initiation of metaphase?

8. What molecular mechanism seems to be responsible for the movement of the poles during anaphase?

9. Describe three events that occur during telophase.

10. How is cytokinesis in animal cells different from that in plant cells?

- Art Activities:
  -Mitosis Overview
  -Plant Cell Mitosis

- Mitosis

- Mitosis
- Activity: Mitosis

- Student Research: Nuclear Division in *Drosophila*

- Art Quiz: Cell Cycle

11.4 The cell cycle is carefully controlled.

- The cell cycle is regulated by two types of proteins, cyclins and cyclin-dependent protein kinases, which permit progress past key "checkpoints" in the cell cycle only if the cell is ready to proceed further.
- Failures of cell cycle regulation can lead to uncontrolled cell growth and lie at the root of cancer.

11. What aspects of the cell cycle are controlled by the G_1, G_2, and M checkpoints? How are cyclins and cyclin-dependent protein kinases involved in cell cycle regulation at checkpoints?

- Exploration: Regulating the Cell Cycle

- Art Quiz: Cell Cycle Control

12

Sexual Reproduction and Meiosis

Concept Outline

12.1 Meiosis produces haploid cells from diploid cells.

Discovery of Reduction Division. Sexual reproduction does not increase chromosome number because gamete production by meiosis involves a decrease in chromosome number. Individuals produced from sexual reproduction inherit chromosomes from two parents.

12.2 Meiosis has three unique features.

Unique Features of Meiosis. Three unique features of meiosis are synapsis, homologous recombination, and reduction division.

12.3 The sequence of events during meiosis involves two nuclear divisions.

Prophase I. Homologous chromosomes pair intimately, and undergo crossing over that locks them together.
Metaphase I. Spindle microtubules align the chromosomes in the central plane of the cell in such a way that homologues separate at anaphase I, not sister chromatids.
Completing Meiosis. The second meiotic division is like a mitotic division, but has a very different outcome.

12.4 The evolutionary origin of sex is a puzzle.

Why Sex? Sex may have evolved as a mechanism to repair DNA, or perhaps as a means for contagious elements to spread. Sexual reproduction increases genetic variability by shuffling combinations of genes.

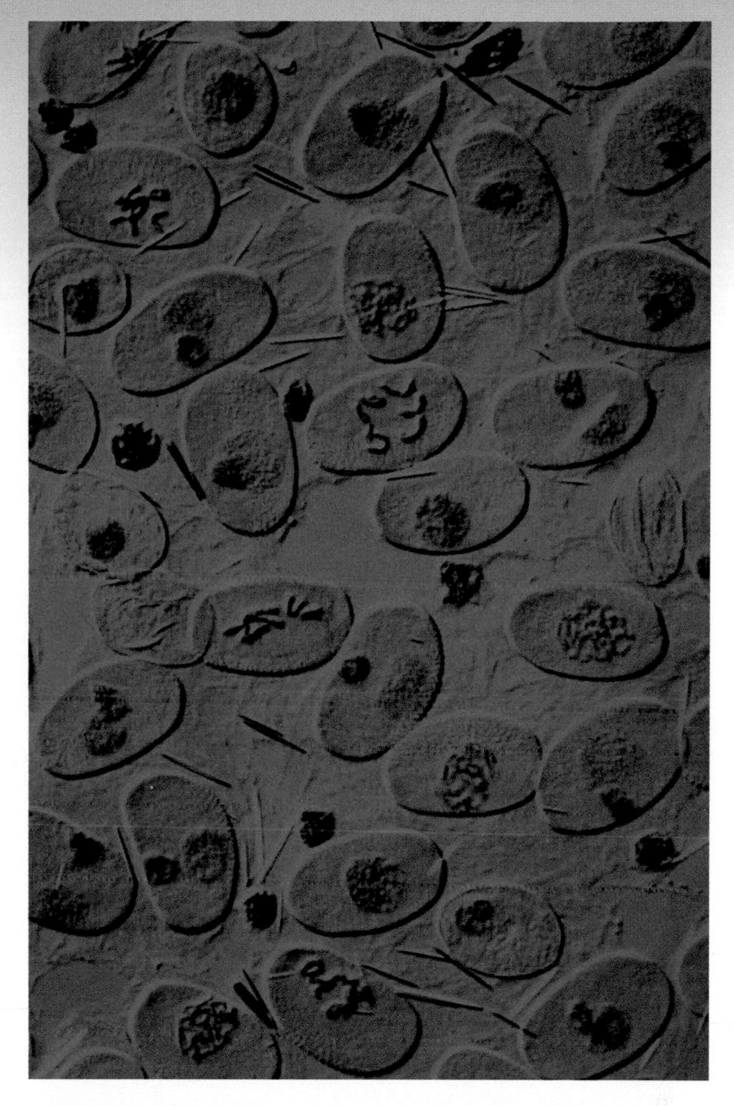

FIGURE 12.1
Plant cells undergoing meiosis (600×). This preparation of pollen cells of a spiderwort, *Tradescantia,* was made by freezing the cells and then fracturing them. It shows several stages of meiosis.

Most animals and plants reproduce sexually. Gametes of opposite sex unite to form a cell that, dividing repeatedly by mitosis, eventually gives rise to an adult body with some 100 trillion cells. The gametes that give rise to the initial cell are the products of a special form of cell division called meiosis (figure 12.1), the subject of this chapter. Far more intricate than mitosis, the details of meiosis are not as well understood. The basic process, however, is clear. Also clear are the profound consequences of sexual reproduction: it plays a key role in generating the tremendous genetic diversity that is the raw material of evolution.

12.1 Meiosis produces haploid cells from diploid cells.

Discovery of Reduction Division

Only a few years after Walther Fleming's discovery of chromosomes in 1882, Belgian cytologist Pierre-Joseph van Beneden was surprised to find different numbers of chromosomes in different types of cells in the roundworm *Ascaris*. Specifically, he observed that the **gametes** (eggs and sperm) each contained two chromosomes, while the **somatic** (nonreproductive) cells of embryos and mature individuals each contained four.

Fertilization

From his observations, van Beneden proposed in 1887 that an egg and a sperm, each containing half the complement of chromosomes found in other cells, fuse to produce a single cell called a **zygote.** The zygote, like all of the somatic cells ultimately derived from it, contains two copies of each chromosome. The fusion of gametes to form a new cell is called **fertilization,** or **syngamy.**

Reduction Division

It was clear even to early investigators that gamete formation must involve some mechanism that reduces the number of chromosomes to half the number found in other cells. If it did not, the chromosome number would double with each fertilization, and after only a few generations, the number of chromosomes in each cell would become impossibly large. For example, in just 10 generations, the 46 chromosomes present in human cells would increase to over 47,000 (46×2^{10}).

The number of chromosomes does not explode in this way because of a special reduction division that occurs during gamete formation, producing cells with half the normal number of chromosomes. The subsequent fusion of two of these cells ensures a consistent chromosome number from one generation to the next. This reduction division process, known as **meiosis,** is the subject of this chapter.

The Sexual Life Cycle

Meiosis and fertilization together constitute a cycle of reproduction. Two sets of chromosomes are present in the somatic cells of adult individuals, making them **diploid** cells (Greek *diploos,* "double" + *eidos,* "form"), but only one set is present in the gametes, which are thus **haploid** (Greek *haploos,* "single" + *ploion,* "vessel"). Reproduction that involves this alternation of meiosis and fertilization is called **sexual reproduction.** Its outstanding characteristic is that offspring inherit chromosomes from *two* parents (figure 12.2). You, for example, inherited 23 chromosomes from your mother, contributed by the egg fertilized at your conception, and 23 from your father, contributed by the sperm that fertilized that egg.

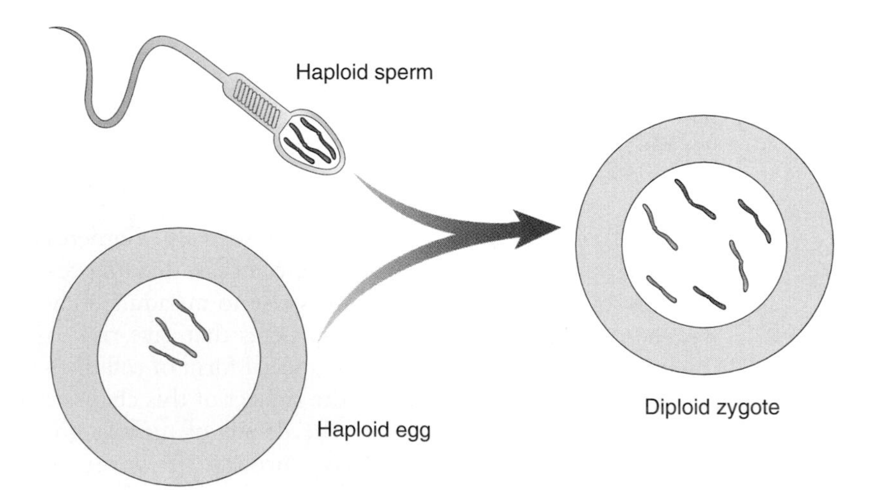

FIGURE 12.2
Diploid cells carry chromosomes from two parents. A diploid cell contains two versions of each chromosome, one contributed by the haploid egg of the mother, the other by the haploid sperm of the father.

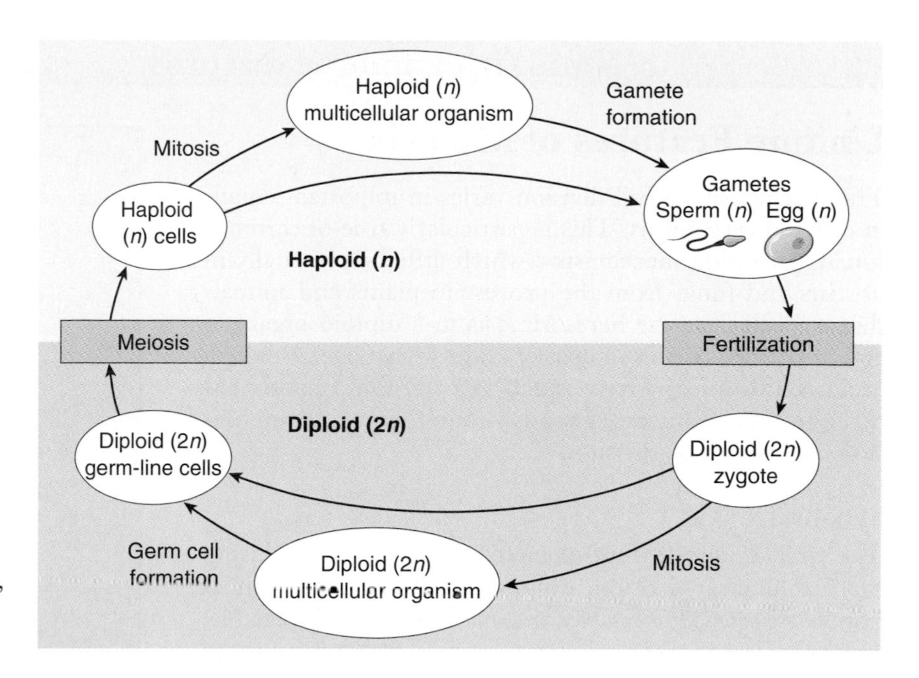

FIGURE 12.3
Alternation of generations. In sexual reproduction, haploid cells or organisms alternate with diploid cells or organisms.

Somatic Tissues. The life cycles of all sexually reproducing organisms follow the same basic pattern of alternation between the diploid and haploid chromosome numbers (figures 12.3 and 12.4). After fertilization, the resulting zygote begins to divide by mitosis. This single diploid cell eventually gives rise to all of the cells in the adult. These cells are called somatic cells, from the Latin word for "body." Except when rare accidents occur, or in special variation-creating situations such as occur in the immune system, every one of the adult's somatic cells is genetically identical to the zygote.

In unicellular eukaryotic organisms, including most protists, individual cells function as gametes, fusing with other gamete cells. The zygote may undergo mitosis, or it may divide immediately by meiosis to give rise to haploid individuals. In plants, the haploid cells that meiosis produces divide by mitosis, forming a multicellular haploid phase. Certain cells of this haploid phase eventually differentiate into eggs or sperm.

Germ-Line Tissues. In animals, the cells that will eventually undergo meiosis to produce gametes are set aside from somatic cells early in the course of development. These cells are often referred to as germ-line cells. Both the somatic cells and the gamete-producing germ-line cells are diploid, but while somatic cells undergo mitosis to form genetically identical, diploid daughter cells, gamete-producing germ-line cells undergo meiosis, producing haploid gametes.

Meiosis is a process of cell division in which the number of chromosomes in certain cells is halved during gamete formation. In the sexual life cycle, there is an alternation of diploid and haploid generations.

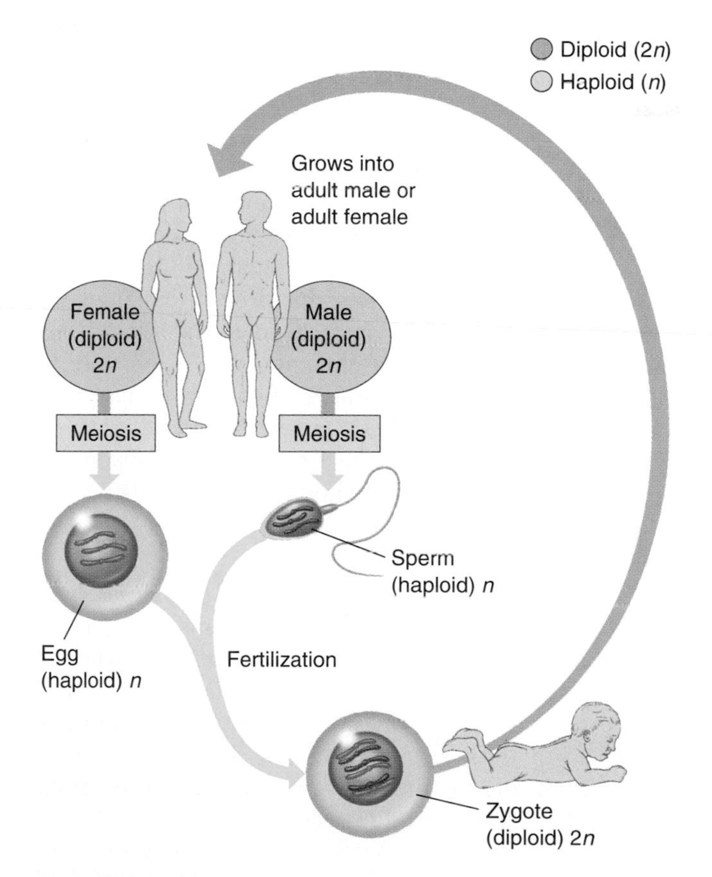

FIGURE 12.4
The sexual life cycle. In animals, the completion of meiosis is followed soon by fertilization. Thus, the vast majority of the life cycle is spent in the diploid stage.

12.2 Meiosis has three unique features.

Unique Features of Meiosis

The mechanism of cell division varies in important details in different organisms. This is particularly true of chromosomal separation mechanisms, which differ substantially in protists and fungi from the process in plants and animals that we will describe here. Meiosis in a diploid organism consists of two rounds of division, mitosis just one. Although meiosis and mitosis have much in common, meiosis has three unique features: synapsis, homologous recombination, and reduction division.

Synapsis

The first unique feature of meiosis happens early during the first nuclear division. Following chromosome replication, *homologous chromosomes*, or *homologues* (see chapter 11), *pair all along their length*. The process of forming these complexes of homologous chromosomes is called **synapsis**

Homologous Recombination

The second unique feature of meiosis is that *genetic exchange occurs between the homologous chromosomes* while they are thus physically joined (figure 12.5*a*). The exchange process that occurs between paired chromosomes is called **crossing over.** Chromosomes are then drawn together along the equatorial plane of the dividing cell; subsequently, homologues are pulled by microtubules toward opposite poles of the cell. When this process is complete, the cluster of chromosomes at each pole contains one of the two homologues of each chromosome. Each pole is haploid, containing half the number of chromosomes present in the original diploid cell. Sister chromatids do not separate from each other in the first nuclear division, so each homologue is still composed of two chromatids.

Reduction Division

The third unique feature of meiosis is that *the chromosomes do not replicate between the two nuclear divisions*, so that at the end of meiosis, each cell contains only half the original complement of chromosomes (figure 12.5*b*). In most respects, the second meiotic division is identical to a normal mitotic division. However, because of the crossing over that occurred during the first division, the sister chromatids in meiosis II are not identical to each other.

Meiosis is a continuous process, but it is most easily studied when we divide it into arbitrary stages. The stages of meiosis are traditionally called meiosis I and meiosis II. Like mitosis, each stage is subdivided further into prophase, metaphase, anaphase, and telophase (figure 12.6). In meiosis, however, prophase I is more complex than in mitosis.

In meiosis, homologous chromosomes become intimately associated and do not replicate between the two nuclear divisions.

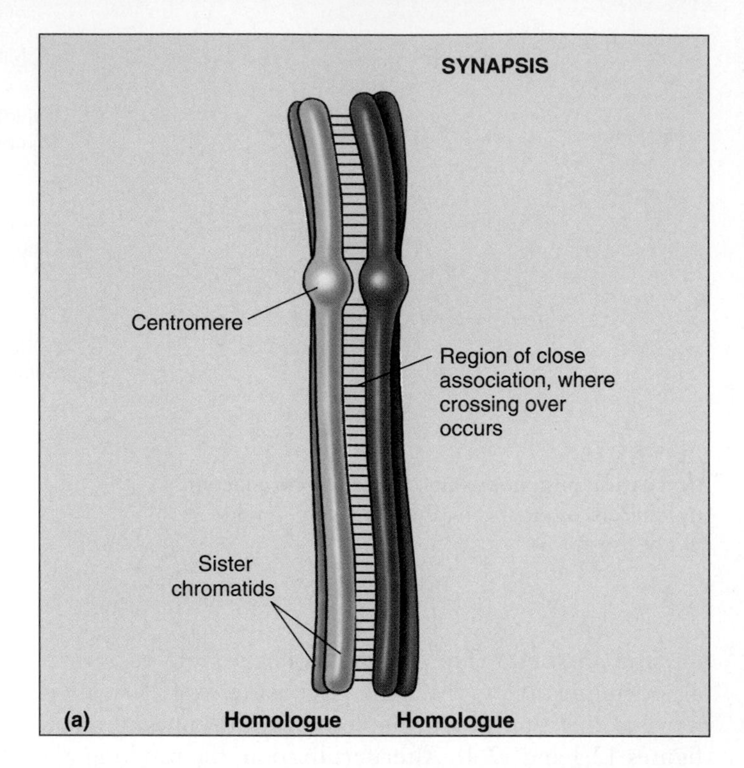

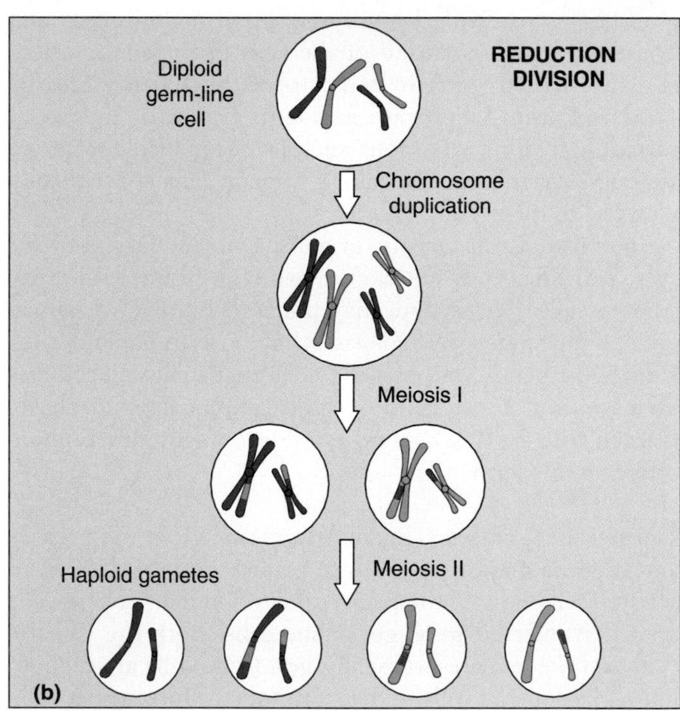

FIGURE 12.5
Unique features of meiosis. (*a*) Synapsis draws homologous chromosomes together, creating a situation where the two chromosomes can physically exchange parts, a process called crossing over. (*b*) Reduction division, by omitting a chromosome duplication before meiosis II, produces haploid gametes, thus ensuring that chromosome number remains stable during the reproduction cycle.

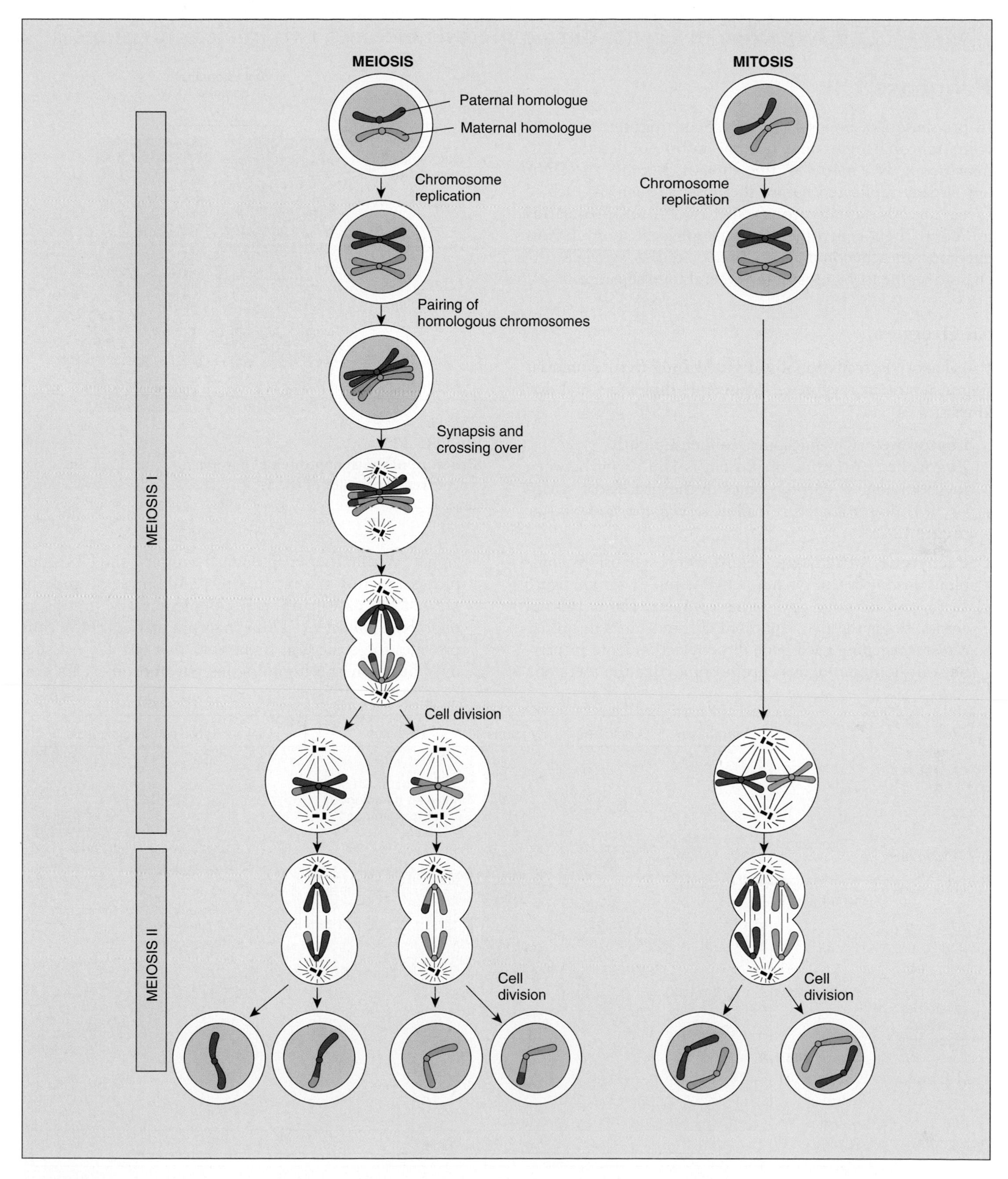

FIGURE 12.6
A comparison of meiosis and mitosis. Meiosis involves two nuclear divisions with no DNA replication between them. It thus produces four daughter cells, each with half the original number of chromosomes. Crossing over occurs in prophase I of meiosis. Mitosis involves a single nuclear division after DNA replication. It thus produces two daughter cells, each containing the original number of chromosomes.

12.3 The sequence of events during meiosis involves two nuclear divisions.

Prophase I

In prophase I of meiosis, the DNA coils tighter, and individual chromosomes first become visible under the light microscope as a matrix of fine threads. Because the DNA has already replicated before the onset of meiosis, each of these threads actually consists of two sister chromatids joined at their centromeres. In prophase I, homologous chromosomes become closely associated in synapsis, exchange segments by crossing over, and then separate.

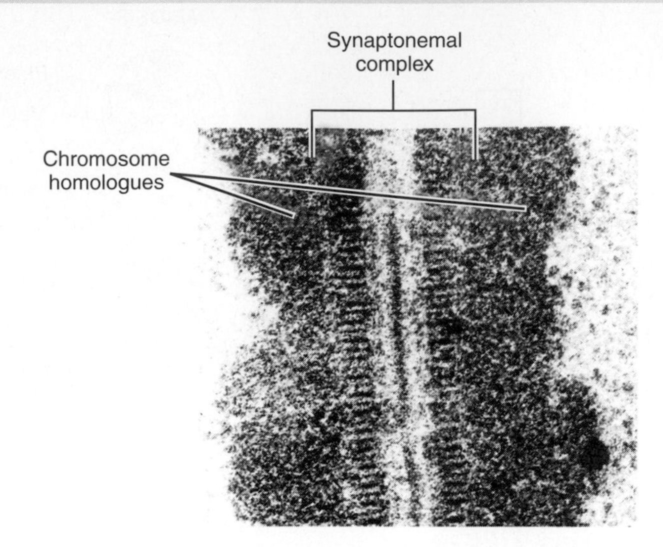

FIGURE 12.7
Structure of the synaptonemal complex. A portion of the synaptonemal complex of the ascomycete *Neotiella rutilans,* a cup fungus.

An Overview

Prophase I is traditionally divided into five sequential stages: leptotene, zygotene, pachytene, diplotene, and diakinesis.

Leptotene. Chromosomes condense tightly.

Zygotene. A lattice of protein is laid down between the homologous chromosomes in the process of synapsis, forming a structure called a *synaptonemal complex* (figure 12.7).

Pachytene. Pachytene begins when synapsis is complete (just after the synaptonemal complex forms; figure 12.8), and lasts for days. This complex, about 100 nm across, holds the two replicated chromosomes in precise register, keeping each gene directly across from its partner on the homologous chromosome, like the teeth of a zipper. Within the synaptonemal complex, the DNA duplexes unwind at certain sites, and single strands of DNA form base-pairs with complementary strands *on the other homologue.* The synaptonemal complex thus provides the structural framework that enables crossing over between the homologous chromosomes. As you

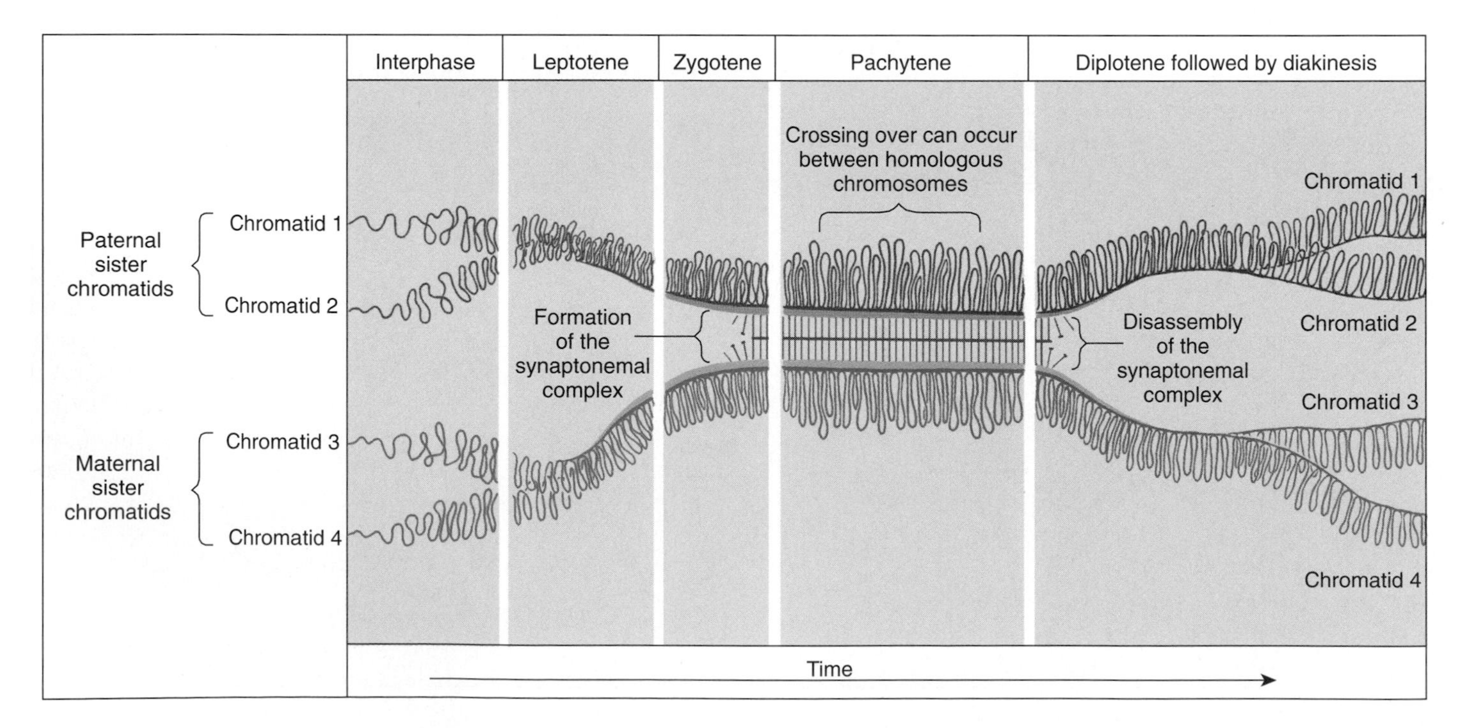

FIGURE 12.8
Time course of prophase I. The five stages of prophase I represent stages in the formation and subsequent disassembly of the synaptonemal complex, the protein lattice that holds homologous chromosomes together during synapsis.

will see, this has a key impact on how the homologues separate later in meiosis.

Diplotene. At the beginning of diplotene, the protein lattice of the synaptonemal complex disassembles. Diplotene is a period of intense cell growth. During this period the chromosomes decondense and become very active in transcription.

Diakinesis. At the beginning of diakinesis, the transition into metaphase, transcription ceases and the chromosomes recondense.

FIGURE 12.9
Chiasmata. This micrograph shows two distinct crossovers, or chiasmata.

Synapsis

During prophase, the ends of the chromatids attach to the nuclear envelope at specific sites. The sites the homologues attach to are adjacent, so that the members of each homologous pair of chromosomes are brought close together. They then line up side by side, apparently guided by heterochromatin sequences, in the process called synapsis.

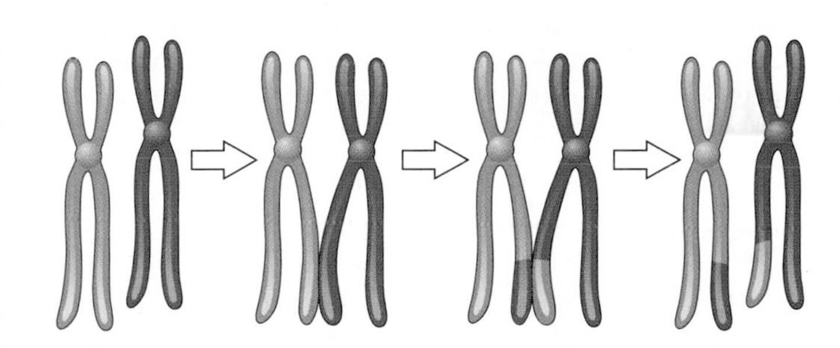

FIGURE 12.10
The results of crossing over. During crossing over, nonsister (shown above) or sister chromatids may exchange segments.

Crossing Over

Within the synaptonemal complex, recombination is thought to be carried out during pachytene by very large protein assemblies called **recombination nodules.** A nodule's diameter is about 90 nm, spanning the central element of the synaptonemal complex. Spaced along the synaptonemal complex, these recombination nodules act as large multienzyme "recombination machines," each nodule bringing about a recombination event. The details of the crossing over process are not well understood, but involve a complex series of events in which DNA segments are exchanged between nonsister or sister chromatids. In humans, an average of two or three such crossover events occur per chromosome pair.

When crossing over is complete, the synaptonemal complex breaks down, and the homologous chromosomes are released from the nuclear envelope and begin to move away from each other. At this point, there are four chromatids for each type of chromosome (two homologous chromosomes, each of which consists of two sister chromatids). The four chromatids do not separate completely, however, because they are held together in two ways: (1) the two sister chromatids of each homologue, recently created by DNA replication, are held near by their common centromeres; and (2) the paired homologues are held together at the points where crossing over occurred within the synaptonemal complex.

Chiasma Formation

Evidence of crossing over can often be seen under the light microscope as an X-shaped structure known as a **chiasma** (Greek, "cross"; plural, **chiasmata;** figure 12.9). The presence of a chiasma indicates that two chromatids (one from each homologue) have exchanged parts (figure 12.10). Like small rings moving down two strands of rope, the chiasmata move to the end of the chromosome arm as the homologous chromosomes separate.

Synapsis is the close pairing of homologous chromosomes that takes place early in prophase I of meiosis. Crossing over occurs between the paired DNA strands, creating the chromosomal configurations known as chiasmata. The two homologues are locked together by these exchanges and they do not disengage readily.

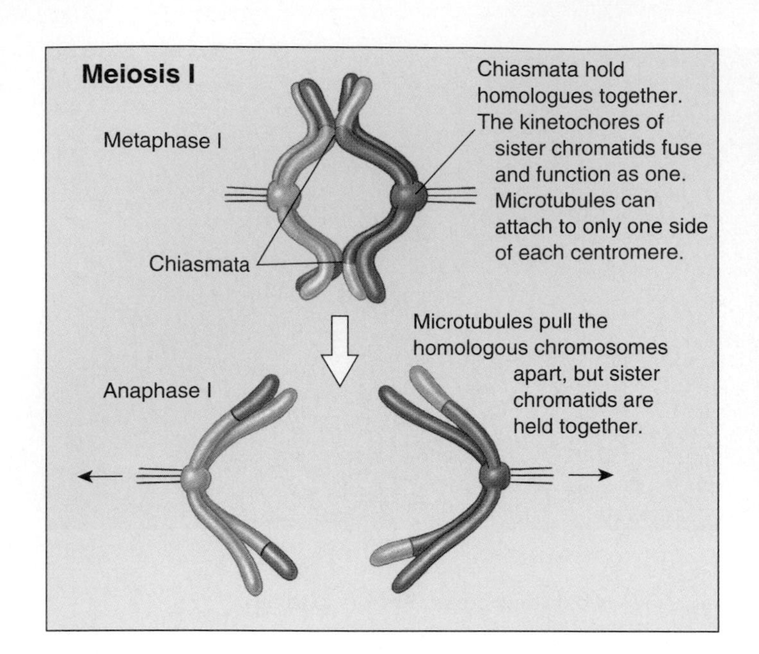
Meiosis I
Metaphase I
Chiasmata hold homologues together. The kinetochores of sister chromatids fuse and function as one. Microtubules can attach to only one side of each centromere.
Chiasmata
Microtubules pull the homologous chromosomes apart, but sister chromatids are held together.
Anaphase I

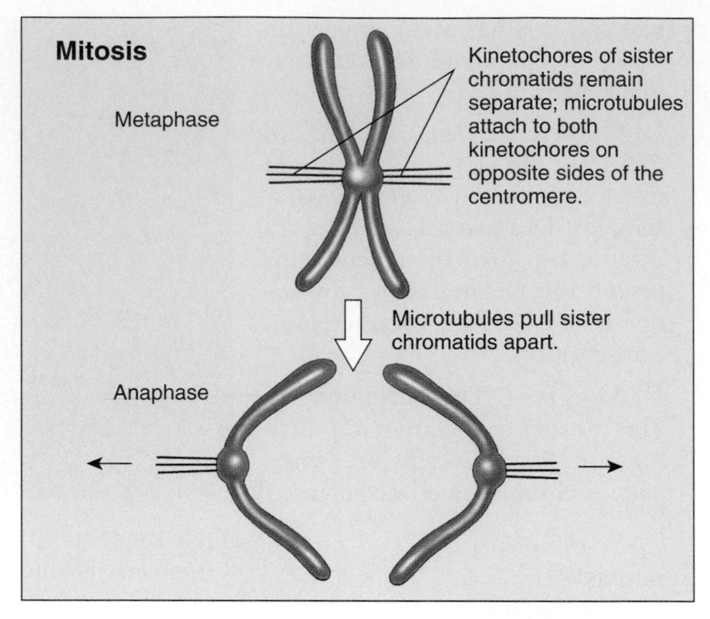
Mitosis
Metaphase
Kinetochores of sister chromatids remain separate; microtubules attach to both kinetochores on opposite sides of the centromere.
Microtubules pull sister chromatids apart.
Anaphase

FIGURE 12.11
Chiasmata created by crossing over have a key impact on how chromosomes align in metaphase I. In the first meiotic division, the chiasmata hold one sister chromatid to the other sister chromatid; consequently, the spindle microtubules can bind to only one side of each centromere, and the homologous chromosomes are drawn to opposite poles. In mitosis, microtubules attach to *both* sides of each centromere; when the microtubules shorten, the sister chromatids are split and drawn to opposite poles.

Metaphase I

By metaphase I, the second stage of meiosis I, the nuclear envelope has dispersed and the microtubules form a spindle, just as in mitosis. During diakinesis of prophase I, the chiasmata move down the paired chromosomes from their original points of crossing over, eventually reaching the ends of the chromosomes. At this point, they are called terminal chiasmata. Terminal chiasmata hold the homologous chromosomes together in metaphase I, so that only one side of each centromere faces outward from the complex; the other side is turned inward toward the other homologue (figure 12.11). Consequently, spindle microtubules are able to attach to kinetochore proteins only on the outside of each centromere, and the centromeres of the two homologues attach to microtubules originating from opposite poles. This one-sided attachment is in marked contrast to the attachment in mitosis, when kinetochores on *both* sides of a centromere bind to microtubules.

Each joined pair of homologues then lines up on the metaphase plate. The orientation of each pair on the spindle axis is random: either the maternal or the paternal homologue may orient toward a given pole (figure 12.12). Figure 12.13 illustrates the alignment of chromosomes during metaphase I.

Chiasmata play an important role in aligning the chromosomes on the metaphase plate.

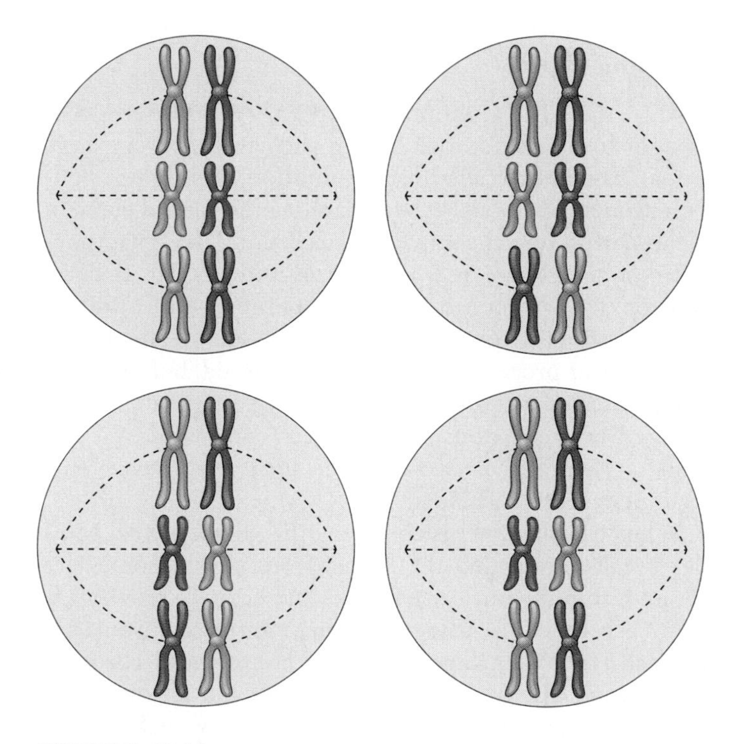

FIGURE 12.12
Random orientation of chromosomes on the metaphase plate. The number of possible chromosome orientations equals 2 raised to the power of the number of chromosome pairs. In this hypothetical cell with three chromosome pairs, eight (2^3) possible orientations exist, four of them illustrated here. Each orientation produces gametes with different combinations of parental chromosomes.

FIGURE 12.13
The stages of meiosis in a lily. Note the arrangement of chromosomes in metaphase I.

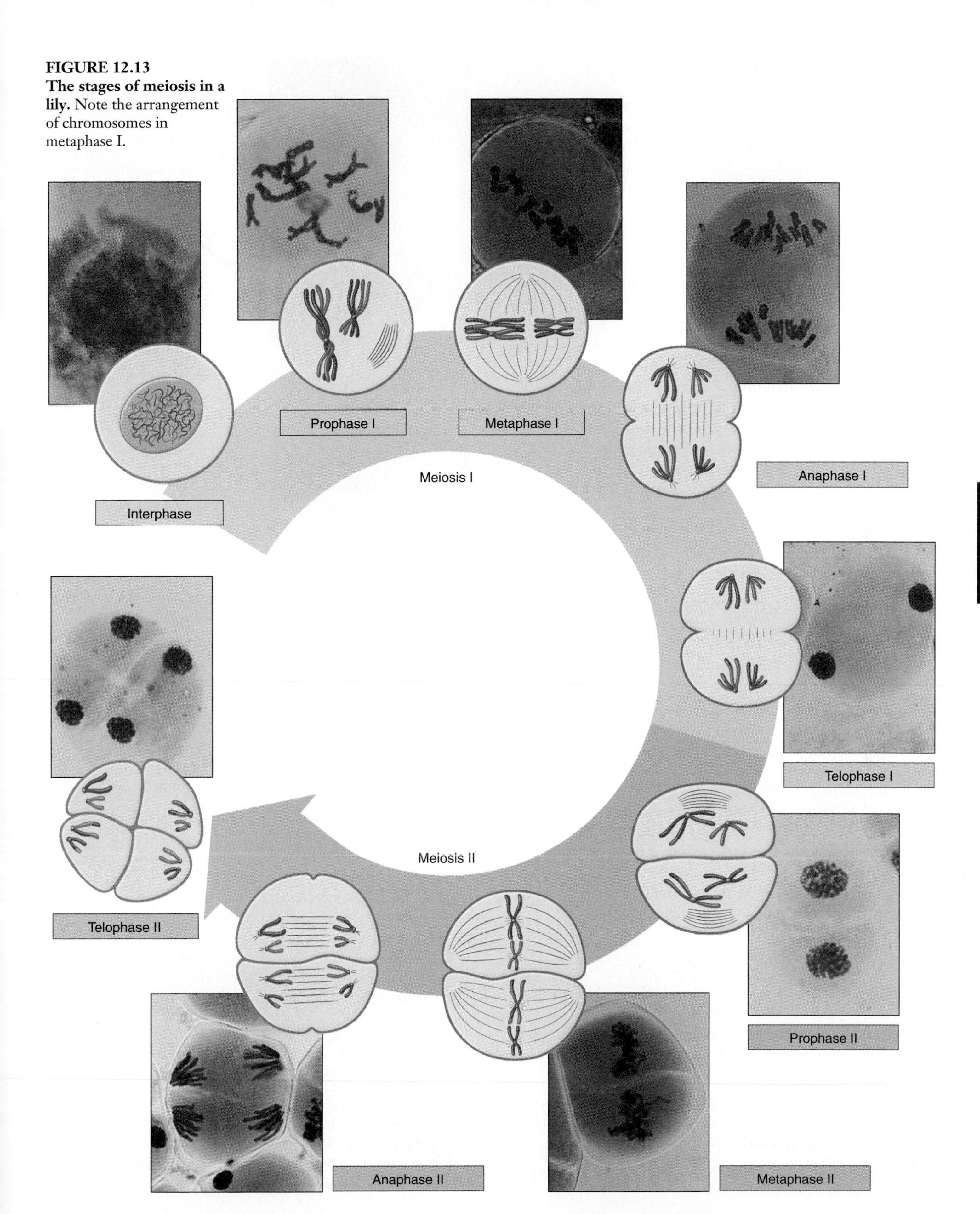

Completing Meiosis

After the long duration of prophase and metaphase, which together make up 90% or more of the time meiosis I takes, meiosis I rapidly concludes. Anaphase I and telophase I proceed quickly, followed—without an intervening period of DNA synthesis—by the second meiotic division.

Anaphase I

In anaphase I, the microtubules of the spindle fibers begin to shorten. As they shorten, they break the chiasmata and pull the centromeres toward the poles, dragging the chromosomes along with them. Because the microtubules are attached to kinetochores on only one side of each centromere, the individual centromeres are not pulled apart to form two daughter centromeres, as they are in mitosis. Instead, the entire centromere moves to one pole, taking both sister chromatids with it. When the spindle fibers have fully contracted, each pole has a complete haploid set of chromosomes consisting of one member of each homologous pair. Because of the random orientation of homologous chromosomes on the metaphase plate, a pole may receive either the maternal or the paternal homologue from each chromosome pair. As a result, the genes on different chromosomes assort independently; that is, meiosis I results in the **independent assortment** of maternal and paternal chromosomes into the gametes.

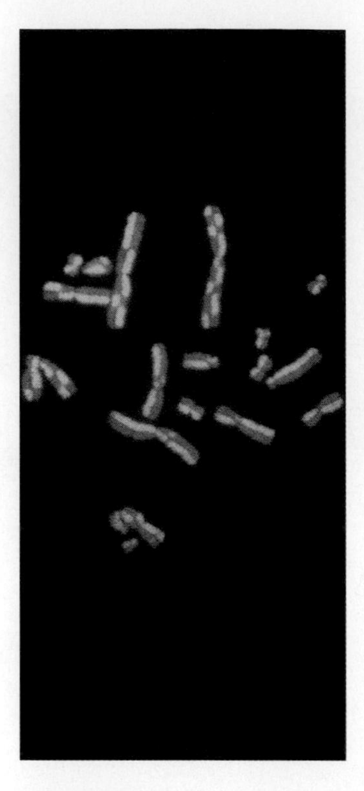
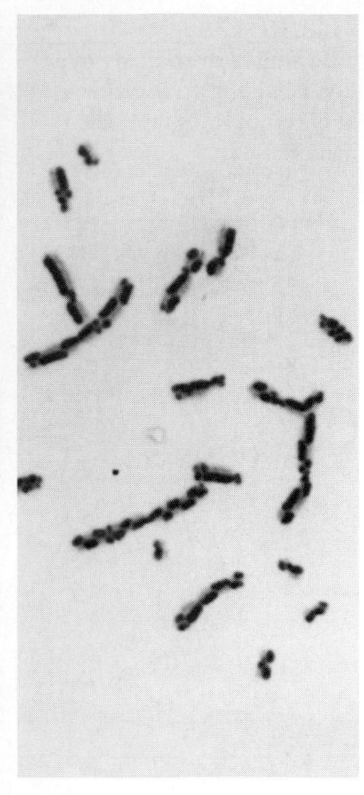

FIGURE 12.14
After meiosis I, sister chromatids are not identical. So-called "harlequin" chromosomes, each containing one fluorescent DNA strand, illustrate the reciprocal exchange of genetic material during meiosis I between sister chromatids.

Telophase I

By the beginning of telophase I, the chromosomes have segregated into two clusters, one at each pole of the cell. Now the nuclear membrane re-forms around each daughter nucleus. Because each chromosome within a daughter nucleus replicated before meiosis I began, each now contains two sister chromatids attached by a common centromere. Importantly, *the sister chromatids are no longer identical,* because of the crossing over that occurred in prophase I (figure 12.14). Cytokinesis may or may not occur after telophase I. The second meiotic division, meiosis II, occurs after an interval of variable length.

The Second Meiotic Division

After a typically brief interphase, in which no DNA synthesis occurs, the second meiotic division begins.

Meiosis II resembles a normal mitotic division. Prophase II, metaphase II, anaphase II, and telophase II follow in quick succession.

Prophase II. At the two poles of the cell the clusters of chromosomes enter a brief prophase II, each nuclear envelope breaking down as a new spindle forms.

Metaphase II. In metaphase II, spindle fibers bind to both sides of the centromeres.

Anaphase II. The spindle fibers contract, splitting the centromeres and moving the sister chromatids to opposite poles.

Telophase II. Finally, the nuclear envelope re-forms around the four sets of daughter chromosomes.

The final result of this division is four cells containing haploid sets of chromosomes (figure 12.15). No two are alike, because of the crossing over in prophase I. Nuclear envelopes then form around each haploid set of chromosomes. The cells that contain these haploid nuclei may develop directly into gametes, as they do in animals. Alternatively, they may themselves divide mitotically, as they do in plants, fungi, and many protists, eventually producing greater numbers of gametes or, as in the case of some plants and insects, adult individuals of varying ploidy.

During meiosis I, homologous chromosomes move toward opposite poles in anaphase I, and individual chromosomes cluster at the two poles in telophase I. At the end of meiosis II, each of the four haploid cells contains one copy of every chromosome in the set, rather than two. Because of crossing over, no two cells are the same. These haploid cells may develop directly into gametes, as in animals, or they may divide by mitosis, as in plants, fungi, and many protists.

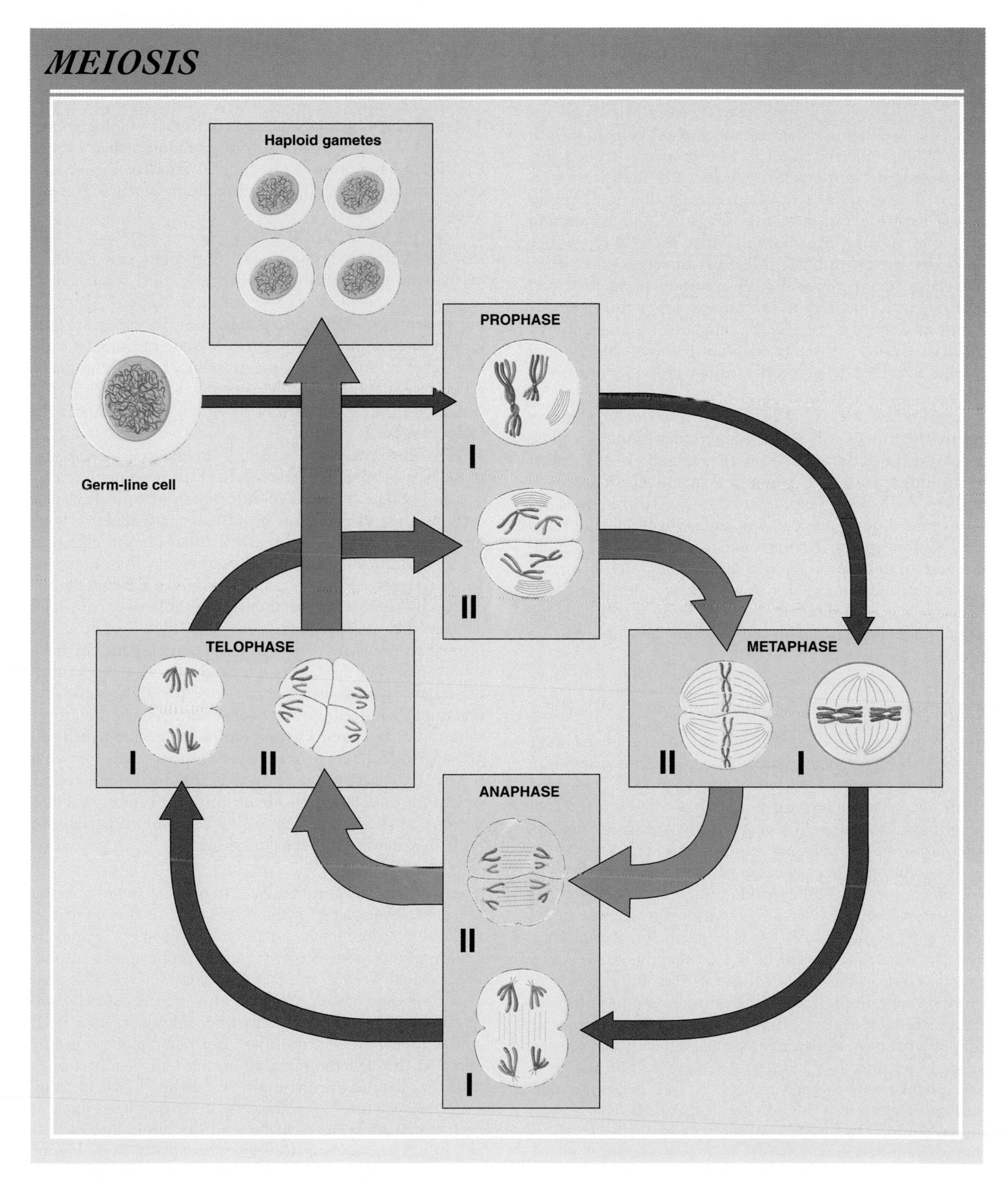

FIGURE 12.15
How meiosis works. Meiosis consists of two rounds of cell division and produces four haploid cells.

12.4 The evolutionary origin of sex is a puzzle.

Why Sex?

Not all reproduction is sexual. In **asexual reproduction,** an individual inherits all of its chromosomes from a single parent and is, therefore, genetically identical to its parent. Bacterial cells reproduce asexually, undergoing binary fission to produce two daughter cells containing the same genetic information. Most protists reproduce asexually except under conditions of stress; then they switch to sexual reproduction. Among plants, asexual reproduction is common, and many other multicellular organisms are also capable of reproducing asexually. In animals, asexual reproduction often involves the budding off of a localized mass of cells, which grows by mitosis to form a new individual.

Even when meiosis and the production of gametes occur, there may still be reproduction without sex. The development of an adult from an unfertilized egg, called **parthenogenesis,** is a common form of reproduction in arthropods. Among bees, for example, fertilized eggs develop into diploid females, but unfertilized eggs develop into haploid males. Parthenogenesis even occurs among the vertebrates. Some lizards, fishes, and amphibians are capable of reproducing in this way; their unfertilized eggs undergo a mitotic nuclear division without cell cleavage to produce a diploid cell, which then develops into an adult.

Recombination Can Be Destructive

If reproduction can occur without sex, why does sex occur at all? This question has generated considerable discussion, particularly among evolutionary biologists. Sex is of great evolutionary advantage for populations or species, which benefit from the variability generated in meiosis by random orientation of chromosomes and by crossing over. However, evolution occurs because of changes at the level of *individual* survival and reproduction, rather than at the population level, and no obvious advantage accrues to the progeny of an individual that engages in sexual reproduction. In fact, recombination is a destructive as well as a constructive process in evolution. The segregation of chromosomes during meiosis tends to disrupt advantageous combinations of genes more often than it creates new, better adapted combinations; as a result, some of the diverse progeny produced by sexual reproduction will not be as well adapted as their parents were. In fact, the more complex the adaptation of an individual organism, the less likely that recombination will improve it, and the more likely that recombination will disrupt it. It is, therefore, a puzzle to know what a well-adapted individual gains from participating in sexual reproduction, as *all* of its progeny could maintain its successful gene combinations if that individual simply reproduced asexually.

The Origin and Maintenance of Sex

There is no consensus among evolutionary biologists regarding the evolutionary origin or maintenance of sex. Conflicting hypotheses abound. Alternative hypotheses seem to be correct to varying degrees in different organisms.

The DNA Repair Hypothesis. If recombination is often detrimental to an individual's progeny, then what benefit promoted the evolution of sexual reproduction? Although the answer to this question is unknown, we can gain some insight by examining the protists. Meiotic recombination is often absent among the protists, which typically undergo sexual reproduction only occasionally. Often the fusion of two haploid cells occurs only under stress, creating a diploid zygote.

Why do some protists form a diploid cell in response to stress? Several geneticists have suggested that this occurs because only a diploid cell can effectively repair certain kinds of chromosome damage, particularly double-strand breaks in DNA. Both radiation and chemical events within cells can induce such breaks. As organisms became larger and longer-lived, it must have become increasingly important for them to be able to repair such damage. The synaptonemal complex, which in early stages of meiosis precisely aligns pairs of homologous chromosomes, may well have evolved originally as a mechanism for repairing double-strand damage to DNA, using the undamaged homologous chromosome as a template to repair the damaged chromosome. A transient diploid phase would have provided an opportunity for such repair. In yeast, mutations that inactivate the repair system for double-strand breaks of the chromosomes also prevent crossing over, suggesting a common mechanism for both synapsis and repair processes.

The Contagion Hypothesis. An unusual and interesting alternative hypothesis for the origin of sex is that it arose as a secondary consequence of the infection of eukaryotes by mobile genetic elements. Suppose a replicating transposable element were to infect a eukaryotic lineage. If it possessed genes promoting fusion with uninfected cells and synapsis, the transposable element could readily copy itself onto homologous chromosomes. It would rapidly spread by infection through the population, until all members contained it. The bizarre mating type "alleles" found in many fungi are very nicely explained by this hypothesis. Each of several mating types is in fact not an allele but an "idiomorph." Idiomorphs are genes occupying homologous positions on the chromosome but having such dissimilar sequences that they cannot be of homologous origin. These idiomorph genes may simply be the relics of several ancient infections by transposable elements.

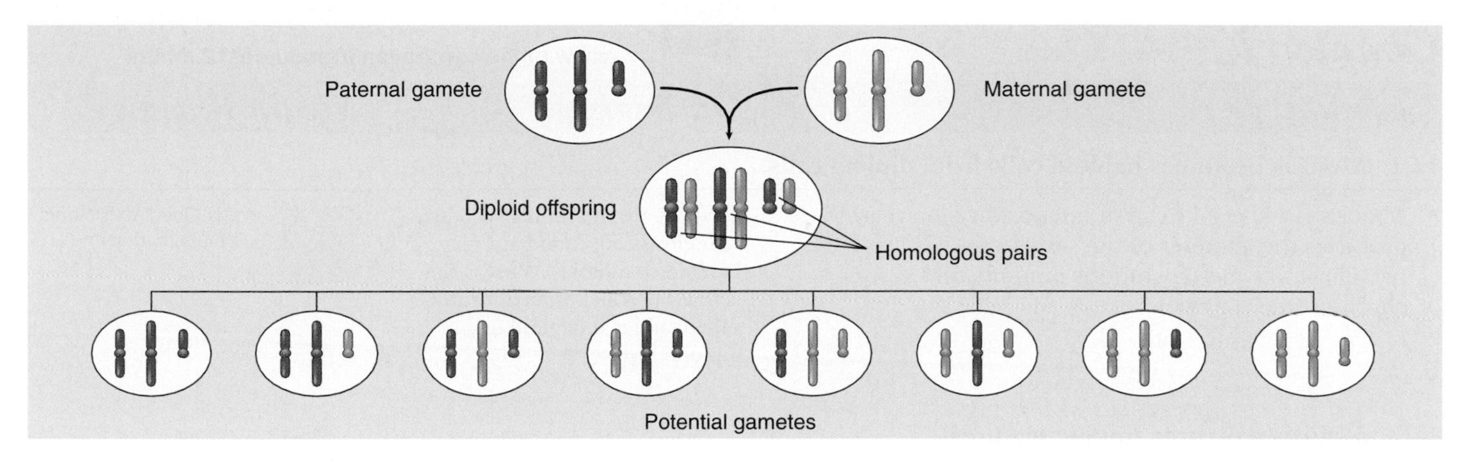

FIGURE 12.16
Independent assortment increases genetic variability. Independent assortment contributes new gene combinations to the next generation because the orientation of chromosomes on the metaphase plate is random. In the cells shown above with three chromosome pairs, eight different gametes can result, each with different combinations of parental chromosomes.

The Red Queen Hypothesis. One evolutionary advantage of sex may be that it allows populations to "store" recessive alleles that are currently bad but have promise for reuse at some time in the future. Because populations are constrained by a changing physical and biological environment, selection is constantly acting against such alleles, but in sexual species can never get rid of those sheltered in heterozygotes. The evolution of most sexual species, most of the time, thus manages to keep pace with ever-changing physical and biological constraints. This "treadmill evolution" is sometimes called the "Red Queen hypothesis," after the Queen of Hearts in Lewis Carroll's *Through the Looking Glass*, who tells Alice, "Now, here, you see, it takes all the running you can do, to keep in the same place."

Miller's Ratchet. The geneticist Herman Miller pointed out in 1965 that asexual populations incorporate a kind of mutational ratchet mechanism—once harmful mutations arise, asexual populations have no way of eliminating them, and they accumulate over time, like turning a ratchet. Sexual populations, on the other hand, can employ recombination to generate individuals carrying fewer mutations, which selection can then favor. Sex may just be a way to keep the mutational load down.

The Evolutionary Consequences of Sex

While our knowledge of how sex evolved is sketchy, it is abundantly clear that sexual reproduction has an enormous impact on how species evolve today, because of its ability to rapidly generate new genetic combinations. Independent assortment (figure 12.16), crossing over, and random fertilization each help generate genetic diversity.

Whatever the forces that led to sexual reproduction, its evolutionary consequences have been profound. No genetic process generates diversity more quickly; and, as you will see in later chapters, genetic diversity is the raw material of evolution, the fuel that drives it and determines its potential directions. In many cases, the pace of evolution appears to increase as the level of genetic diversity increases. Programs for selecting larger stature in domesticated animals such as cattle and sheep, for example, proceed rapidly at first, but then slow as the existing genetic combinations are exhausted; further progress must then await the generation of new gene combinations. Racehorse breeding provides a graphic example: thoroughbred racehorses are all descendants of a small initial number of individuals, and selection for speed has accomplished all it can with this limited amount of genetic variability—the winning times in major races ceased to improve decades ago.

Paradoxically, the evolutionary process is thus both revolutionary and conservative. It is revolutionary in that the pace of evolutionary change is quickened by genetic recombination, much of which results from sexual reproduction. It is conservative in that evolutionary change is not always favored by selection, which may instead preserve existing combinations of genes. These conservative pressures appear to be greatest in some asexually reproducing organisms that do not move around freely and that live in especially demanding habitats. In vertebrates, on the other hand, the evolutionary premium appears to have been on versatility, and sexual reproduction is the predominant mode of reproduction by an overwhelming margin.

The close association between homologous chromosomes that occurs during meiosis may have evolved as mechanisms to repair chromosomal damage, although several alternative mechanisms have also been proposed.

Chapter 12

www.mhhe.com/raven6/resources12.mhtml

Summary | **Questions** | **Media Resources**

12.1 Meiosis produces haploid cells from diploid cells.

- Meiosis is a special form of nuclear division that produces the gametes of the sexual cycle. It involves two chromosome separations but only one chromosome replication.

1. What are the cellular products of meiosis called, and are they haploid or diploid? What is the cellular product of fertilization called, and is it haploid or diploid?

- Art Quiz: Alternation of Generations

12.2 Meiosis has three unique features.

- The three unique features of meiosis are synapsis, homologous recombination, and reduction division.

2. What three unique features distinguish meiosis from mitosis?

- Review of Cell Division
- Meiosis/Mitosis

12.3 The sequence of events during meiosis involves two nuclear divisions.

- The crossing over that occurs between homologues during synapsis is an essential element of meiosis.
- Because crossing over binds the homologues together, only one side of each homologue is accessible to the spindle fibers. Hence, the spindle fibers separate the paired homologues rather than the sister chromatids.
- At the end of meiosis I, one homologue of each chromosome type is present at each of the two poles of the dividing nucleus. The homologues still consist of two chromatids, which may differ from each other as a result of crossing over that occurred during synapsis.
- No further DNA replication occurs before the second nuclear division, which is essentially a mitotic division occurring at each of the two poles.
- The sister chromatids of each chromosome are separated, resulting in the formation of four daughter nuclei, each with half the number of chromosomes that were present before meiosis.
- Cytokinesis typically but not always occurs at this point. When it does, each daughter nucleus has one copy of every chromosome.

3. What are synaptonemal complexes? How do they participate in crossing over? At what stage during meiosis are they formed?

4. How many chromatids are present for each type of chromosome at the completion of crossing over? What two structures hold the chromatids together at this stage?

5. How is the attachment of spindle microtubules to centromeres in metaphase I of meiosis different from that which occurs in metaphase of mitosis? What effect does this difference have on the movement of chromosomes during anaphase I?

6. What mechanism is responsible for the independent assortment of chromosomes?

- Art Activity: Meiosis I

- Meiosis

- Meiosis
- Recombination

12.4 The evolutionary origin of sex is a puzzle.

- In asexual reproduction, mitosis produces offspring genetically identical to the parent.
- Meiosis is thought to have evolved initially as a mechanism to repair double-strand breaks in DNA, in which the broken chromosome is paired with its homologue while it is being repaired.
- The evolutionary significance of meiosis is that it generates large amounts of recombination, rapidly reshuffling gene combinations, producing variability upon which evolutionary processes can act.

7. What is one of the current scientific explanations for the evolution of synapsis?

8. By what three mechanisms does sexual reproduction increase genetic variability? How does this increase in genetic variability affect the evolution of species?

- Evolution of Sex

13

Patterns of Inheritance

FIGURE 13.1
Human beings are extremely diverse in appearance. The differences between us are partly inherited and partly the result of environmental factors we encounter in our lives.

Every living creature is a product of the long evolutionary history of life on earth. While all organisms share this history, only humans wonder about the processes that led to their origin. We are still far from understanding everything about our origins, but we have learned a great deal. Like a partially completed jigsaw puzzle, the boundaries have fallen into place, and much of the internal structure is becoming apparent. In this chapter, we will discuss one piece of the puzzle—the enigma of heredity. Why do groups of people from different parts of the world often differ in appearance (figure 13.1)? Why do the members of a family tend to resemble one another more than they resemble members of other families?

13.1 Mendel solved the mystery of heredity.

Early Ideas about Heredity: The Road to Mendel

As far back as written records go, patterns of resemblance among the members of particular families have been noted and commented on (figure 13.2). Some familial features are unusual, such as the protruding lower lip of the European royal family Hapsburg, evident in pictures and descriptions of family members from the thirteenth century onward. Other characteristics, like the occurrence of redheaded children within families of redheaded parents, are more common (figure 13.3). Inherited features, the building blocks of evolution, will be our concern in this chapter.

FIGURE 13.2
Heredity is responsible for family resemblance. Family resemblances are often strong—a visual manifestation of the mechanism of heredity. This is the Johnson family, the wife and daughters of one of the authors. While each daughter is different, all clearly resemble their mother.

Classical Assumption 1: Constancy of Species

Two concepts provided the basis for most of the thinking about heredity before the twentieth century. The first is that *heredity occurs within species.* For a very long time people believed that it was possible to obtain bizarre composite animals by breeding (crossing) widely different species. The minotaur of Cretan mythology, a creature with the body of a bull and the torso and head of a man, is one example. The giraffe was thought to be another; its scientific name, *Giraffa camelopardalis,* suggests the belief that it was the result of a cross between a camel and a leopard. From the Middle Ages onward, however, people discovered that such extreme crosses were not possible and that variation and heredity occur mainly within the boundaries of a particular species. Species were thought to have been maintained without significant change from the time of their creation.

FIGURE 13.3
Red hair is inherited. Many different traits are inherited in human families. This redhead is exhibiting one of these traits.

Classical Assumption 2: Direct Transmission of Traits

The second early concept related to heredity is that *traits are transmitted directly.* When variation is inherited by offspring from their parents, *what* is transmitted? The ancient Greeks suggested that the parents' body parts were transmitted directly to their offspring. Hippocrates called this type of reproductive material *gonos,* meaning "seed." Hence, a characteristic such as a misshapen limb was the result of material that came from the misshapen limb of a parent. Information from each part of the body was supposedly passed along independently of the information from the other parts, and the child was formed after the hereditary material from all parts of the parents' bodies had come together.

This idea was predominant until fairly recently. For example, in 1868, Charles Darwin proposed that all cells and tissues excrete microscopic granules, or "gemmules," that are passed to offspring, guiding the growth of the corresponding part in the developing embryo. Most similar theories of the direct transmission of hereditary material assumed that the male and female contributions *blend* in the offspring. Thus, parents with red and brown hair would produce children with reddish brown hair, and tall and short parents would produce children of intermediate height.

Koelreuter Demonstrates Hybridization between Species

Taken together, however, these two concepts lead to a paradox. If no variation enters a species from outside, and if the variation within each species blends in every generation, then all members of a species should soon have the same appearance. Obviously, this does not happen. Individuals within most species differ widely from each other, and they differ in characteristics that are transmitted from generation to generation.

How could this paradox be resolved? Actually, the resolution had been provided long before Darwin, in the work of the German botanist Josef Koelreuter. In 1760, Koelreuter carried out successful **hybridizations** of plant species, crossing different strains of tobacco and obtaining fertile offspring. The hybrids differed in appearance from both parent strains. When individuals within the hybrid generation were crossed, their offspring were highly variable. Some of these offspring resembled plants of the hybrid generation (their parents), but a few resembled the original strains (their grandparents).

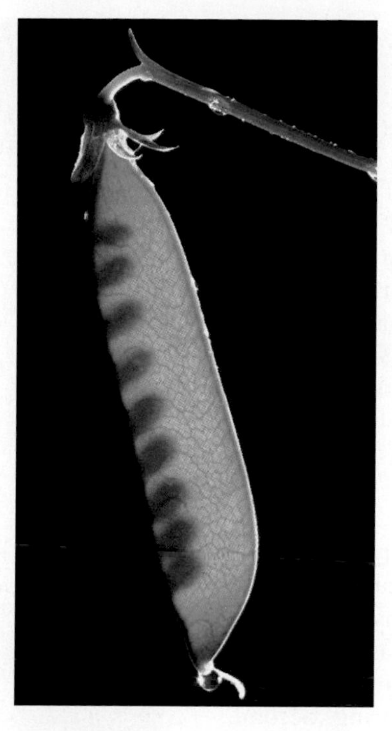

FIGURE 13.4
The garden pea, *Pisum sativum*. Easy to cultivate and able to produce many distinctive varieties, the garden pea was a popular experimental subject in investigations of heredity as long as a century before Gregor Mendel's experiments.

The Classical Assumptions Fail

Koelreuter's work represents the beginning of modern genetics, the first clues pointing to the modern theory of heredity. Koelreuter's experiments provided an important clue about how heredity works: the traits he was studying could be masked in one generation, only to reappear in the next. This pattern contradicts the theory of direct transmission. How could a trait that is transmitted directly disappear and then reappear? Nor were the traits of Koelreuter's plants blended. A contemporary account stated that the traits reappeared in the third generation "fully restored to all their original powers and properties."

It is worth repeating that the offspring in Koelreuter's crosses were not identical to one another. Some resembled the hybrid generation, while others did not. The alternative forms of the characters Koelreuter was studying were distributed among the offspring. Referring to a heritable feature as a **character,** a modern geneticist would say the alternative forms of each character were **segregating** among the progeny of a mating, meaning that some offspring exhibited one alternative form of a character (for example, hairy leaves), while other offspring from the same mating exhibited a different alternative (smooth leaves). This segregation of alternative forms of a character, or **traits,** provided the clue that led Gregor Mendel to his understanding of the nature of heredity.

Knight Studies Heredity in Peas

Over the next hundred years, other investigators elaborated on Koelreuter's work. Prominent among them were English gentleman farmers trying to improve varieties of agricultural plants. In one such series of experiments, carried out in the 1790s, T. A. Knight crossed two true-breeding varieties (varieties that remain uniform from one generation to the next) of the garden pea, *Pisum sativum* (figure 13.4). One of these varieties had purple flowers, and the other had white flowers. All of the progeny of the cross had purple flowers. Among the offspring of these hybrids, however, were some plants with purple flowers and others, less common, with white flowers. Just as in Koelreuter's earlier studies, a trait from one of the parents disappeared in one generation only to reappear in the next.

In these deceptively simple results were the makings of a scientific revolution. Nevertheless, another century passed before the process of gene segregation was fully appreciated. Why did it take so long? One reason was that early workers did not quantify their results. A numerical record of results proved to be crucial to understanding the process. Knight and later experimenters who carried out other crosses with pea plants noted that some traits had a "stronger tendency" to appear than others, but they did not record the numbers of the different classes of progeny. Science was young then, and it was not obvious that the numbers were important.

Early geneticists demonstrated that some forms of an inherited character (1) can disappear in one generation only to reappear unchanged in future generations; (2) segregate among the offspring of a cross; and (3) are more likely to be represented than their alternatives.

Mendel and the Garden Pea

The first quantitative studies of inheritance were carried out by Gregor Mendel, an Austrian monk (figure 13.5). Born in 1822 to peasant parents, Mendel was educated in a monastery and went on to study science and mathematics at the University of Vienna, where he failed his examinations for a teaching certificate. He returned to the monastery and spent the rest of his life there, eventually becoming abbot. In the garden of the monastery (figure 13.6), Mendel initiated a series of experiments on plant hybridization. The results of these experiments would ultimately change our views of heredity irrevocably.

FIGURE 13.5
Gregor Johann Mendel. Cultivating his plants in the garden of a monastery in Brunn, Austria (now Brno, Czech Republic), Mendel studied how differences among varieties of peas were inherited when the varieties were crossed. Similar experiments had been done before, but Mendel was the first to quantify the results and appreciate their significance.

Why Mendel Chose the Garden Pea

For his experiments, Mendel chose the garden pea, the same plant Knight and many others had studied earlier. The choice was a good one for several reasons. First, many earlier investigators had produced hybrid peas by crossing different varieties. Mendel knew that he could expect to observe segregation of traits among the offspring. Second, a large number of true-breeding varieties of peas were available. Mendel initially examined 32. Then, for further study, he selected lines that differed with respect to seven easily distinguishable traits, such as round versus wrinkled seeds and purple versus white flowers, a character that Knight had studied. Third, pea plants are small and easy to grow, and they have a relatively short generation time. Thus, one can conduct experiments involving numerous plants, grow several generations in a single year, and obtain results relatively quickly.

A fourth advantage of studying peas is that the sexual organs of the pea are enclosed within the flower (figure 13.7). The flowers of peas, like those of many flowering plants, contain both male and female sex organs. Furthermore, the gametes produced by the male and female parts of the same flower, unlike those of many flowering plants, can fuse to form viable offspring. Fertilization takes place automatically within an individual flower if it is not disturbed, resulting in offspring that are the progeny from a single individual. Therefore, one can either let individual flowers engage in **self-fertilization,** or remove the flower's male parts before fertilization and introduce pollen from a strain with a different trait, thus performing *cross-pollination* which results in **cross-fertilization.**

FIGURE 13.6
The garden where Mendel carried out his plant-breeding experiments. Gregor Mendel did his key scientific experiments in this small garden in a monastery.

Mendel's Experimental Design

Mendel was careful to focus on only a few specific differences between the plants he was using and to ignore the countless other differences he must have seen. He also had the insight to realize that the differences he selected to analyze must be comparable. For example, he appreciated that trying to study the inheritance of round seeds versus tall height would be useless.

Mendel usually conducted his experiments in three stages:

1. First, he allowed pea plants of a given variety to produce progeny by self-fertilization for several generations. Mendel thus was able to assure himself that the traits he was studying were indeed constant, transmitted unchanged from generation to generation. Pea plants with white flowers, for example, when crossed with each other, produced only offspring with white flowers, regardless of the number of generations.
2. Mendel then performed crosses between varieties exhibiting alternative forms of characters. For example, he removed the male parts from the flower of a plant that produced white flowers and fertilized it with pollen from a purple-flowered plant. He also carried out the reciprocal cross, using pollen from a white-flowered individual to fertilize a flower on a pea plant that produced purple flowers (figure 13.8).
3. Finally, Mendel permitted the hybrid offspring produced by these crosses to self-fertilize for several generations. By doing so, he allowed the alternative forms of a character to segregate among the progeny. This was the same experimental design that Knight and others had used much earlier. But Mendel went an important step farther: he counted the numbers of offspring exhibiting each trait in each succeeding generation. No one had ever done that before. The quantitative results Mendel obtained proved to be of supreme importance in revealing the process of heredity.

Mendel's experiments with the garden pea involved crosses between true-breeding varieties, followed by a generation or more of inbreeding.

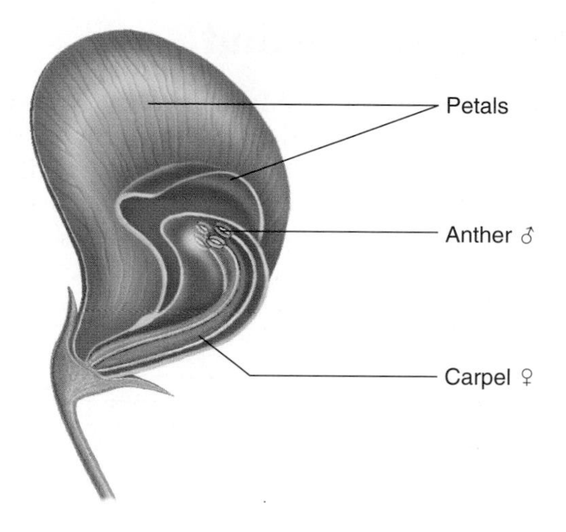

FIGURE 13.7
Structure of the pea flower (longitudinal section). In a pea plant flower, the petals enclose the male anther (containing pollen grains, which give rise to haploid sperm) and the female carpel (containing ovules, which give rise to haploid eggs). This ensures that self-fertilization will take place unless the flower is disturbed.

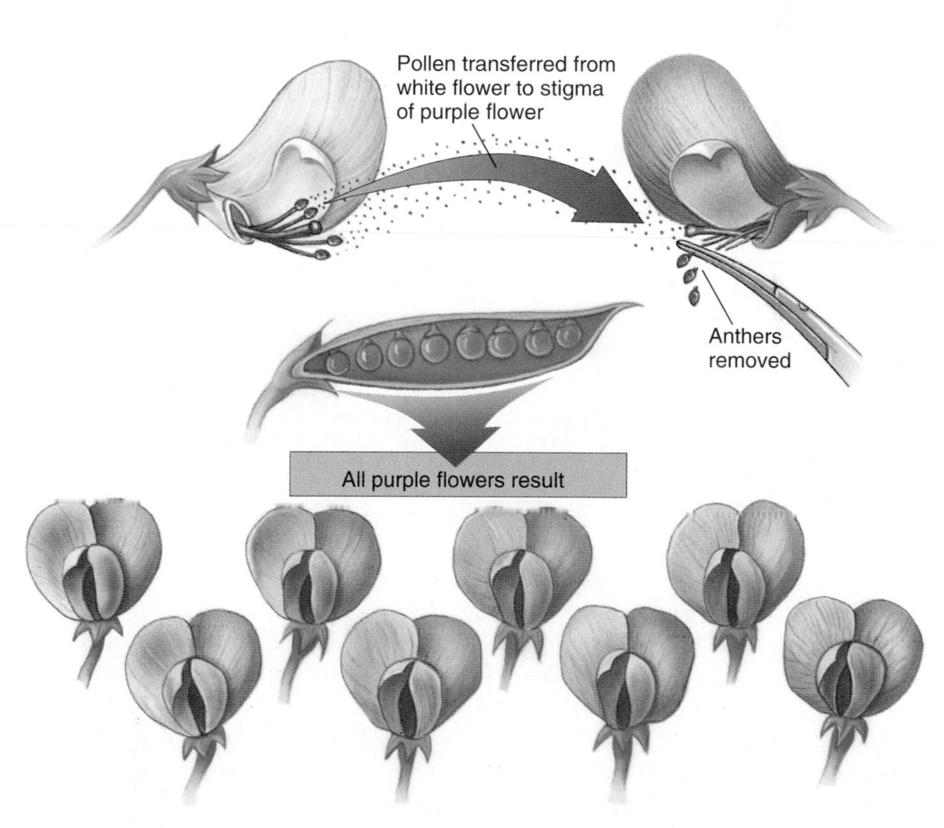

FIGURE 13.8
How Mendel conducted his experiments. Mendel pushed aside the petals of a white flower and collected pollen from the anthers. He then placed that pollen onto the stigma (part of the carpel) of a purple flower whose anthers had been removed, causing cross-fertilization to take place. All the seeds in the pod that resulted from this pollination were hybrids of the white-flowered male parent and the purple-flowered female parent. After planting these seeds, Mendel observed the pea plants they produced. All of the progeny of this cross had purple flowers.

What Mendel Found

The seven characters Mendel studied in his experiments possessed several variants that differed from one another in ways that were easy to recognize and score (figure 13.9). We will examine in detail Mendel's crosses with flower color. His experiments with other characters were similar, and they produced similar results.

The F_1 Generation

When Mendel crossed two contrasting varieties of peas, such as white-flowered and purple-flowered plants, the hybrid offspring he obtained did not have flowers of intermediate color, as the theory of blending inheritance would predict. Instead, in every case the flower color of the offspring resembled one of their parents. It is customary to refer to these offspring as the **first filial** (*filius* is

Character	Dominant vs. recessive trait	F_2 generation Dominant form	Recessive form	Ratio
Flower color	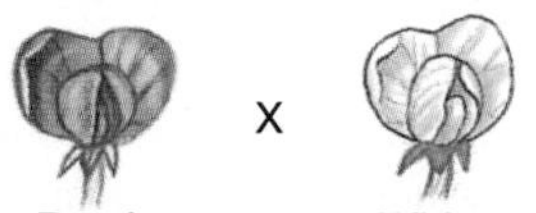 Purple x White	705	224	3.15:1
Seed color	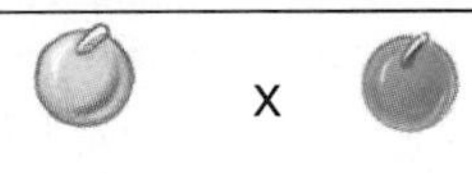 Yellow x Green	6022	2001	3.01:1
Seed shape	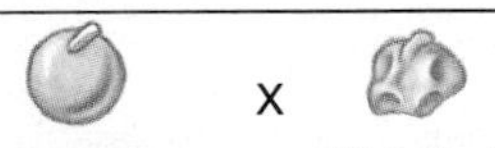 Round x Wrinkled	5474	1850	2.96:1
Pod color	 Green x Yellow	428	152	2.82:1
Pod shape	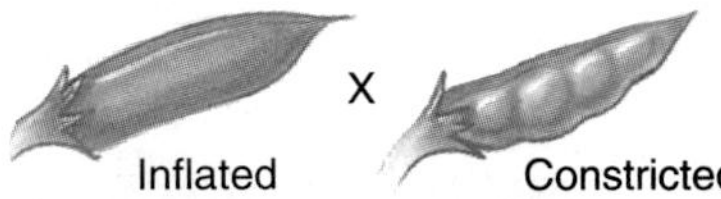 Inflated x Constricted	882	299	2.95:1
Flower position	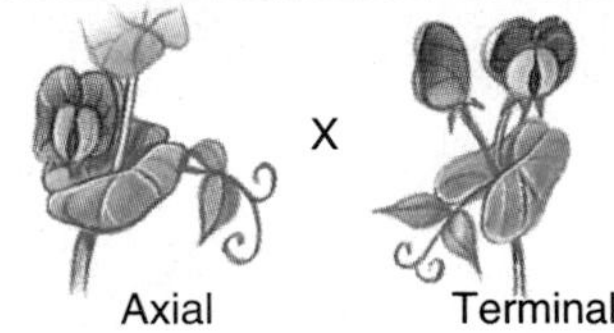 Axial x Terminal	651	207	3.14:1
Plant height	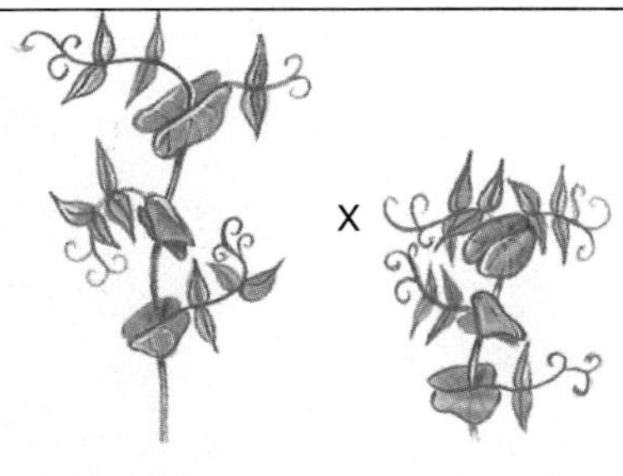 Tall x Dwarf	787	277	2.84:1

FIGURE 13.9
Mendel's experimental results. This table illustrates the seven characters Mendel studied in his crosses of the garden pea and presents the data he obtained from these crosses. Each pair of traits appeared in the F_2 generation in very close to a 3:1 ratio.

Latin for "son"), or **F_1**, generation. Thus, in a cross of white-flowered with purple-flowered plants, the F_1 offspring all had purple flowers, just as Knight and others had reported earlier.

Mendel referred to the trait expressed in the F_1 plants as **dominant** and to the alternative form that was not expressed in the F_1 plants as **recessive.** For each of the seven pairs of contrasting traits that Mendel examined, one of the pair proved to be dominant and the other recessive.

The F_2 Generation

After allowing individual F_1 plants to mature and self-fertilize, Mendel collected and planted the seeds from each plant to see what the offspring in the **second filial,** or **F_2,** generation would look like. He found, just as Knight had earlier, that some F_2 plants exhibited white flowers, the recessive trait. Hidden in the F_1 generation, the recessive form reappeared among some F_2 individuals.

Believing the proportions of the F_2 types would provide some clue about the mechanism of heredity, Mendel counted the numbers of each type among the F_2 progeny (figure 13.10). In the cross between the purple-flowered F_1 plants, he counted a total of 929 F_2 individuals (see figure 13.9). Of these, 705 (75.9%) had purple flowers and 224 (24.1%) had white flowers. Approximately ¼ of the F_2 individuals exhibited the recessive form of the character. Mendel obtained the same numerical result with the other six characters he examined: ¾ of the F_2 individuals exhibited the dominant trait, and ¼ displayed the recessive trait. In other words, the dominant:recessive ratio among the F_2 plants was always close to 3:1. Mendel carried out similar experiments with other traits, such as wrinkled versus round seeds (figure 13.11), and obtained the same result.

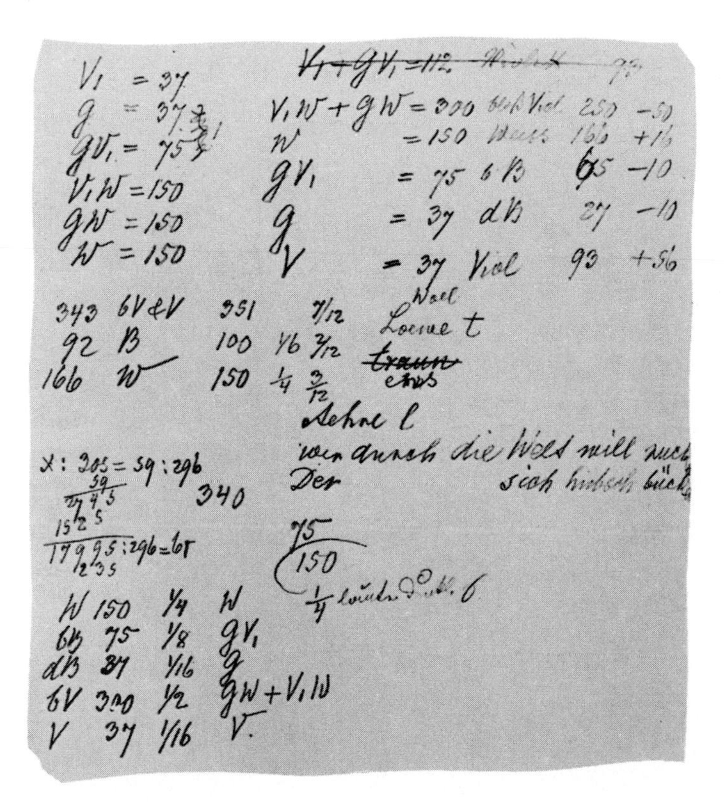

V1 = 37
g = 37
gV1 = 75
V1W = 150
gW = 150
W = 150
V1W + gW = 300
343 bV&V 351
92 B 100
166 W 150
W 150 1/4 W
bB 75 1/8 gV1
dB 37 1/16 g
bV 300 1/2 gW + V1W
V 37 1/16 V.

FIGURE 13.10
A page from Mendel's notebook.

FIGURE 13.11
Seed shape: a Mendelian character. One of the differences Mendel studied affected the shape of pea plant seeds. In some varieties, the seeds were round, while in others, they were wrinkled.

A Disguised 1:2:1 Ratio

Mendel went on to examine how the F_2 plants passed traits on to subsequent generations. He found that the recessive ¼ were always true-breeding. In the cross of white-flowered with purple-flowered plants, for example, the white-flowered F_2 individuals reliably produced white-flowered offspring when they were allowed to self-fertilize. By contrast, only ⅓ of the dominant purple-flowered F_2 individuals (¼ of all F_2 offspring) proved true-breeding, while ⅔ were not. This last class of plants produced dominant and recessive individuals in the third filial (F_3) generation in a 3:1 ratio. This result suggested that, for the entire sample, the 3:1 ratio that Mendel observed in the F_2 generation was really a disguised 1:2:1 ratio: ¼ pure-breeding dominant individuals, ½ not-pure-breeding dominant individuals, and ¼ pure-breeding recessive individuals (figure 13.12).

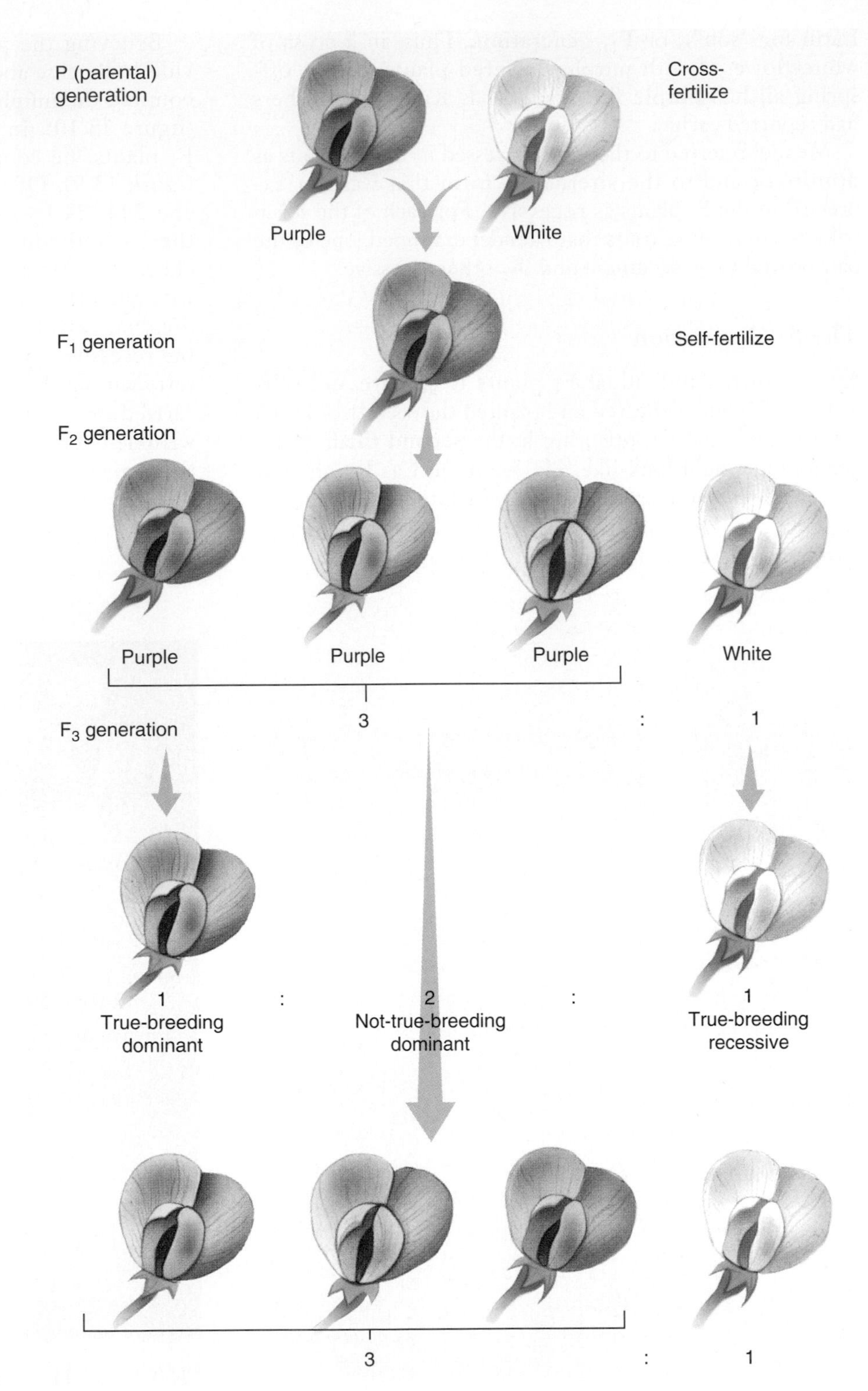

FIGURE 13.12
The F_2 generation is a disguised 1:2:1 ratio. By allowing the F_2 generation to self-fertilize, Mendel found from the offspring (F_3) that the ratio of F_2 plants was one true-breeding dominant, two not-true-breeding dominant, and one true-breeding recessive.

Mendel's Model of Heredity

From his experiments, Mendel was able to understand four things about the nature of heredity. *First*, the plants he crossed did not produce progeny of intermediate appearance, as a theory of blending inheritance would have predicted. Instead, different plants inherited each alternative intact, as a discrete characteristic that either was or was not visible in a particular generation. *Second*, Mendel learned that for each pair of alternative forms of a character, one alternative was not expressed in the F_1 hybrids, although it reappeared in some F_2 individuals. ***The trait that "disappeared" must therefore be latent (present but not expressed) in the F_1 individuals.*** *Third*, the pairs of alternative traits examined segregated among the progeny of a particular cross, some individuals exhibiting one trait, some the other. *Fourth*, these alternative traits were expressed in the F_2 generation in the ratio of ¾ dominant to ¼ recessive. This characteristic 3:1 segregation is often referred to as the **Mendelian ratio.**

Table 13.1 Some Dominant and Recessive Traits in Humans

Recessive Traits	Phenotypes	Dominant Traits	Phenotypes
Albinism	Lack of melanin pigmentation	Middigital hair	Presence of hair on middle segment of fingers
Alkaptonuria	Inability to metabolize homogenistic acid	Brachydactyly	Short fingers
Red-green color blindness	Inability to distinguish red or green wavelengths of light	Huntington's disease	Degeneration of nervous system, starting in middle age
Cystic fibrosis	Abnormal gland secretion, leading to liver degeneration and lung failure	Phenylthiocarbamide (PTC) sensitivity	Ability to taste PTC as bitter
Duchenne muscular dystrophy	Wasting away of muscles during childhood	Camptodactyly	Inability to straighten the little finger
Hemophilia	Inability to form blood clots	Hypercholesterolemia (the most common human Mendelian disorder—1 in 500)	Elevated levels of blood cholesterol and risk of heart attack
Sickle cell anemia	Defective hemoglobin that causes red blood cells to curve and stick together	Polydactyly	Extra fingers and toes

To explain these results, Mendel proposed a simple model. It has become one of the most famous models in the history of science, containing simple assumptions and making clear predictions. The model has five elements:

1. Parents do not transmit physiological traits directly to their offspring. Rather, they transmit discrete information about the traits, what Mendel called "factors." These factors later act in the offspring to produce the trait. In modern terms, we would say that information about the alternative forms of characters that an individual expresses is *encoded* by the factors that it receives from its parents.
2. Each individual receives two factors that may code for the same trait or for two alternative traits for a character. We now know that there are two factors for each character present in each individual because these factors are carried on chromosomes, and each adult individual is *diploid.* When the individual forms gametes (eggs or sperm), the gametes contain only one of each kind of chromosome (see chapter 12); the gametes are *haploid.* Therefore, only one factor for each character of the adult organism is contained in the gamete. Which of the two factors ends up in a particular gamete is randomly determined.
3. Not all copies of a factor are identical. In modern terms, the alternative forms of a factor, leading to alternative forms of a character, are called **alleles.** When two haploid gametes containing exactly the same allele of a factor fuse during fertilization to form a zygote, the offspring that develops from that zygote is said to be **homozygous;** when the two haploid gametes contain different alleles, the individual offspring is **heterozygous.**

   In modern terminology, Mendel's factors are called **genes.** We now know that each gene is composed of a particular DNA nucleotide sequence (see chapter 3). The particular location of a gene on a chromosome is referred to as the gene's **locus** (plural, loci).
4. The two alleles, one contributed by the male gamete and one by the female, do not influence each other in any way. In the cells that develop within the new individual, these alleles remain discrete. They neither blend with nor alter each other. (Mendel referred to them as "uncontaminated.") Thus, when the individual matures and produces its own gametes, the alleles for each gene segregate randomly into these gametes, as described in element 2.
5. The presence of a particular allele does not ensure that the trait encoded by it will be expressed in an individual carrying that allele. In heterozygous individuals, only one allele (the dominant one) is expressed, while the other (recessive) allele is present but unexpressed. To distinguish between the presence of an allele and its expression, modern geneticists refer to the totality of alleles that an individual contains as the individual's **genotype** and to the physical appearance of that individual as its **phenotype.** The phenotype of an individual is the observable outward manifestation of its genotype, the result of the functioning of the enzymes and proteins encoded by the genes it carries. In other words, the genotype is the blueprint, and the phenotype is the visible outcome.

These five elements, taken together, constitute Mendel's model of the hereditary process. Many traits in humans also exhibit dominant or recessive inheritance, similar to the traits Mendel studied in peas (table 13.1).

When Mendel crossed two contrasting varieties, he found all of the offspring in the first generation exhibited one (dominant) trait, and none exhibited the other (recessive) trait. In the following generation, 25% were pure-breeding for the dominant trait, 50% were hybrid for the two traits and exhibited the dominant trait, and 25% were pure-breeding for the recessive trait.

How Mendel Interpreted His Results

Does Mendel's model predict the results he actually obtained? To test his model, Mendel first expressed it in terms of a simple set of symbols, and then used the symbols to interpret his results. It is very instructive to do the same. Consider again Mendel's cross of purple-flowered with white-flowered plants. We will assign the symbol *P* to the dominant allele, associated with the production of purple flowers, and the symbol *p* to the recessive allele, associated with the production of white flowers. By convention, genetic traits are usually assigned a letter symbol referring to their more common forms, in this case "P" for purple flower color. The dominant allele is written in upper case, as *P*; the recessive allele (white flower color) is assigned the same symbol in lower case, *p*.

In this system, the genotype of an individual that is true-breeding for the recessive white-flowered trait would be designated *pp*. In such an individual, both copies of the allele specify the white-flowered phenotype. Similarly, the genotype of a true-breeding purple-flowered individual would be designated *PP*, and a heterozygote would be designated *Pp* (dominant allele first). Using these conventions, and denoting a cross between two strains with ×, we can symbolize Mendel's original cross as *pp* × *PP*.

The F_1 Generation

Using these simple symbols, we can now go back and reexamine the crosses Mendel carried out. Because a white-flowered parent (*pp*) can produce only *p* gametes, and a pure purple-flowered (homozygous dominant) parent (*PP*) can produce only *P* gametes, the union of an egg and a sperm from these parents can produce only heterozygous *Pp* offspring in the F_1 generation. Because the *P* allele is dominant, all of these F_1 individuals are expected to have purple flowers. The *p* allele is present in these heterozygous individuals, but it is not phenotypically expressed. This is the basis for the latency Mendel saw in recessive traits.

The F_2 Generation

When F_1 individuals are allowed to self-fertilize, the *P* and *p* alleles segregate randomly during gamete formation. Their subsequent union at fertilization to form F_2 individuals is also random, not being influenced by which alternative alleles the individual gametes carry. What will the F_2 individuals look like? The possibilities may be visualized in a simple diagram called a **Punnett square,** named after its originator, the English geneticist Reginald Crundall Punnett (figure 13.13). Mendel's model,

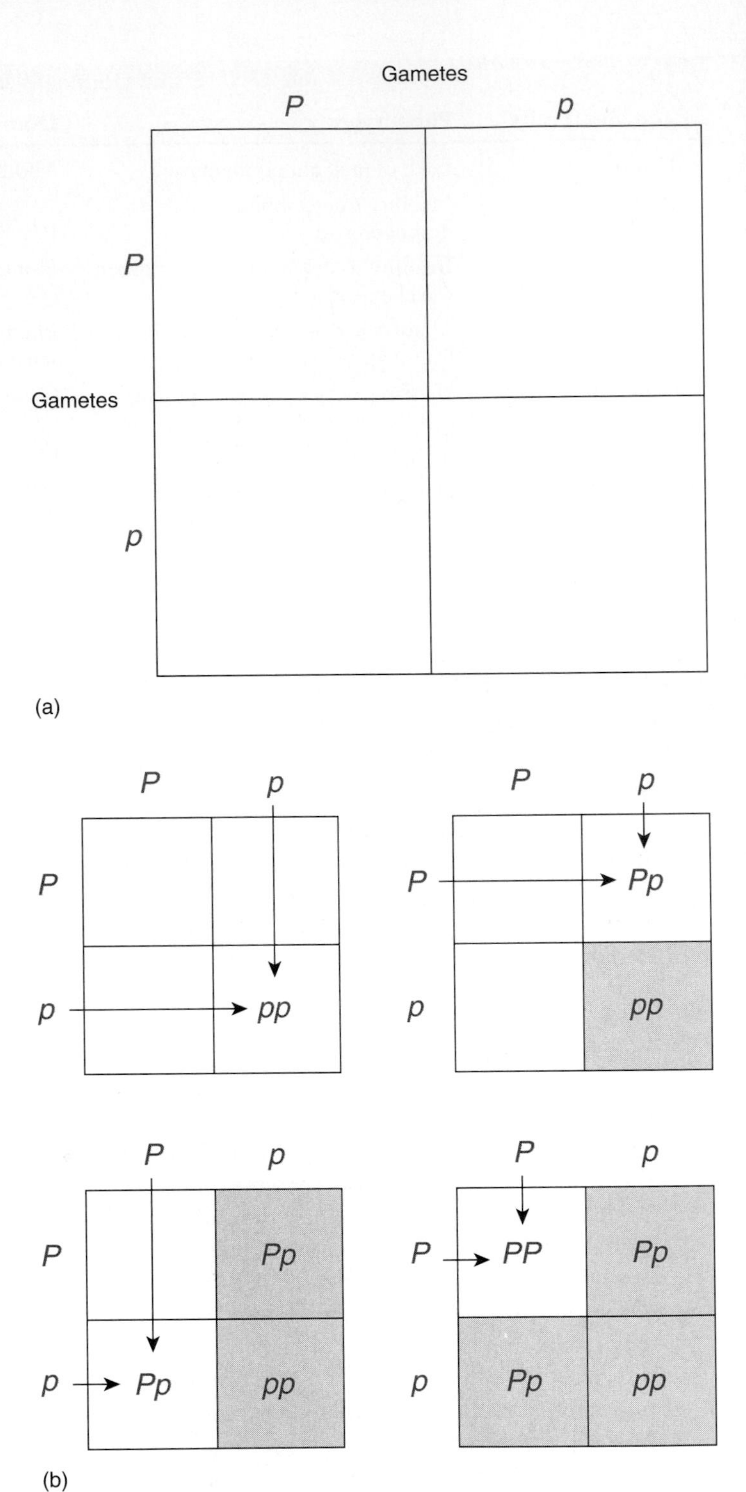

FIGURE 13.13
A Punnett square. (*a*) To make a Punnett square, place the different possible types of female gametes along one side of a square and the different possible types of male gametes along the other. (*b*) Each potential zygote can then be represented as the intersection of a vertical line and a horizontal line.

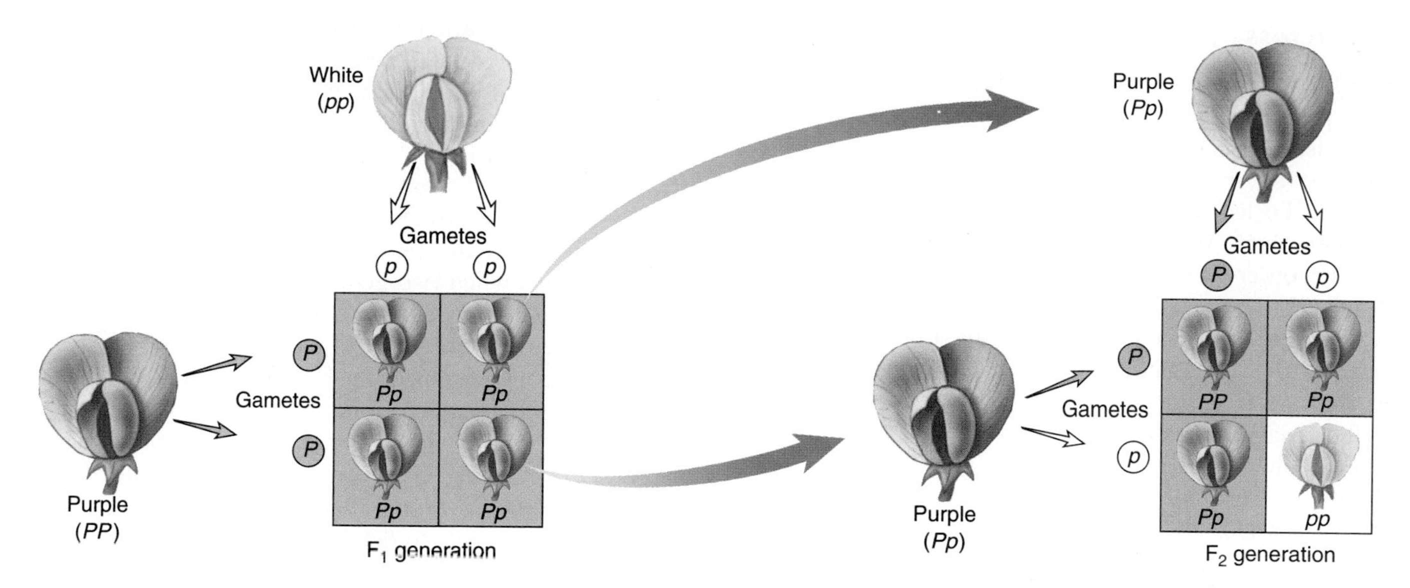

FIGURE 13.14
Mendel's cross of pea plants differing in flower color. All of the offspring of the first cross (the F_1 generation) are *Pp* heterozygotes with purple flowers. When two heterozygous F_1 individuals are crossed, three kinds of F_2 offspring are possible: *PP* homozygotes (purple flowers); *Pp* heterozygotes (also purple flowers); and *pp* homozygotes (white flowers). Therefore, in the F_2 generation, the ratio of dominant to recessive phenotypes is 3:1. However, the ratio of genotypes is 1:2:1 (1 *PP*: 2 *Pp*: 1 *pp*).

analyzed in terms of a Punnett square, clearly predicts that the F_2 generation should consist of ¾ purple-flowered plants and ¼ white-flowered plants, a phenotypic ratio of 3:1 (figure 13.14).

The Laws of Probability Can Predict Mendel's Results

A different way to express Mendel's result is to say that there are three chances in four (¾) that any particular F_2 individual will exhibit the dominant trait, and one chance in four (¼) that an F_2 individual will express the recessive trait. Stating the results in terms of probabilities allows simple predictions to be made about the outcomes of crosses. If both F_1 parents are *Pp* (heterozygotes), the probability that a particular F_2 individual will be *pp* (homozygous recessive) is the probability of receiving a *p* gamete from the male (½) times the probability of receiving a *p* gamete from the female (½), or ¼. This is the same operation we perform in the Punnett square illustrated in figure 13.13. The ways probability theory can be used to analyze Mendel's results is discussed in detail on page 251.

Further Generations

As you can see in figure 13.14, there are really three kinds of F_2 individuals: ¼ are pure-breeding, white-flowered individuals (*pp*); ½ are heterozygous, purple-flowered individuals (*Pp*); and ¼ are pure-breeding, purple-flowered individuals (*PP*). The 3:1 phenotypic ratio is really a disguised 1:2:1 genotypic ratio.

Mendel's First Law of Heredity: Segregation

Mendel's model thus accounts in a neat and satisfying way for the segregation ratios he observed. Its central assumption—that alternative alleles of a character segregate from each other in heterozygous individuals and remain distinct—has since been verified in many other organisms. It is commonly referred to as **Mendel's First Law of Heredity,** or the **Law of Segregation.** As you saw in chapter 12, the segregational behavior of alternative alleles has a simple physical basis, the alignment of chromosomes at random on the metaphase plate during meiosis I. It is a tribute to the intellect of Mendel's analysis that he arrived at the correct scheme with no knowledge of the cellular mechanisms of inheritance; neither chromosomes nor meiosis had yet been described.

The Testcross

To test his model further, Mendel devised a simple and powerful procedure called the **testcross.** Consider a purple-flowered plant. It is impossible to tell whether such a plant is homozygous or heterozygous simply by looking at its phenotype. To learn its genotype, you must cross it with some other plant. What kind of cross would provide the answer? If you cross it with a homozygous dominant individual, all of the progeny will show the dominant phenotype whether the test plant is homozygous or heterozygous. It is also difficult (but not impossible) to distinguish between the two possible test plant genotypes by crossing with a heterozygous individual. However, if you cross the test plant with a homozygous recessive individual, the two possible test plant genotypes will give totally different results (figure 13.15):

Alternative 1: unknown individual homozygous dominant (*PP*). *PP* × *pp:* all offspring have purple flowers (*Pp*)

Alternative 2: unknown individual heterozygous (*Pp*). *Pp* × *pp:* ½ of offspring have white flowers (*pp*) and ½ have purple flowers (*Pp*)

To perform his testcross, Mendel crossed heterozygous F_1 individuals back to the parent homozygous for the recessive trait. He predicted that the dominant and recessive traits would appear in a 1:1 ratio, and that is what he observed. For each pair of alleles he investigated, Mendel observed phenotypic F_2 ratios of 3:1 (see figure 13.14) and testcross ratios very close to 1:1, just as his model predicted.

Testcrosses can also be used to determine the genotype of an individual when two genes are involved. Mendel carried out many two-gene crosses, some of which we will discuss. He often used testcrosses to verify the genotypes of particular dominant-appearing F_2 individuals. Thus, an F_2 individual showing both dominant traits (*A_ B_*) might have any of the following genotypes: *AABB*, *AaBB*, *AABb*, or *AaBb*. By crossing dominant-appearing F_2 individuals with homozygous recessive individuals (that is, *A_ B_* × *aabb*), Mendel was able to determine if either or both of the traits bred true among the progeny, and so to determine the genotype of the F_2 parent:

AABB	trait A breeds true	trait B breeds true
AaBB	______________	trait B breeds true
AABb	trait A breeds true	______________
AaBb	______________	______________

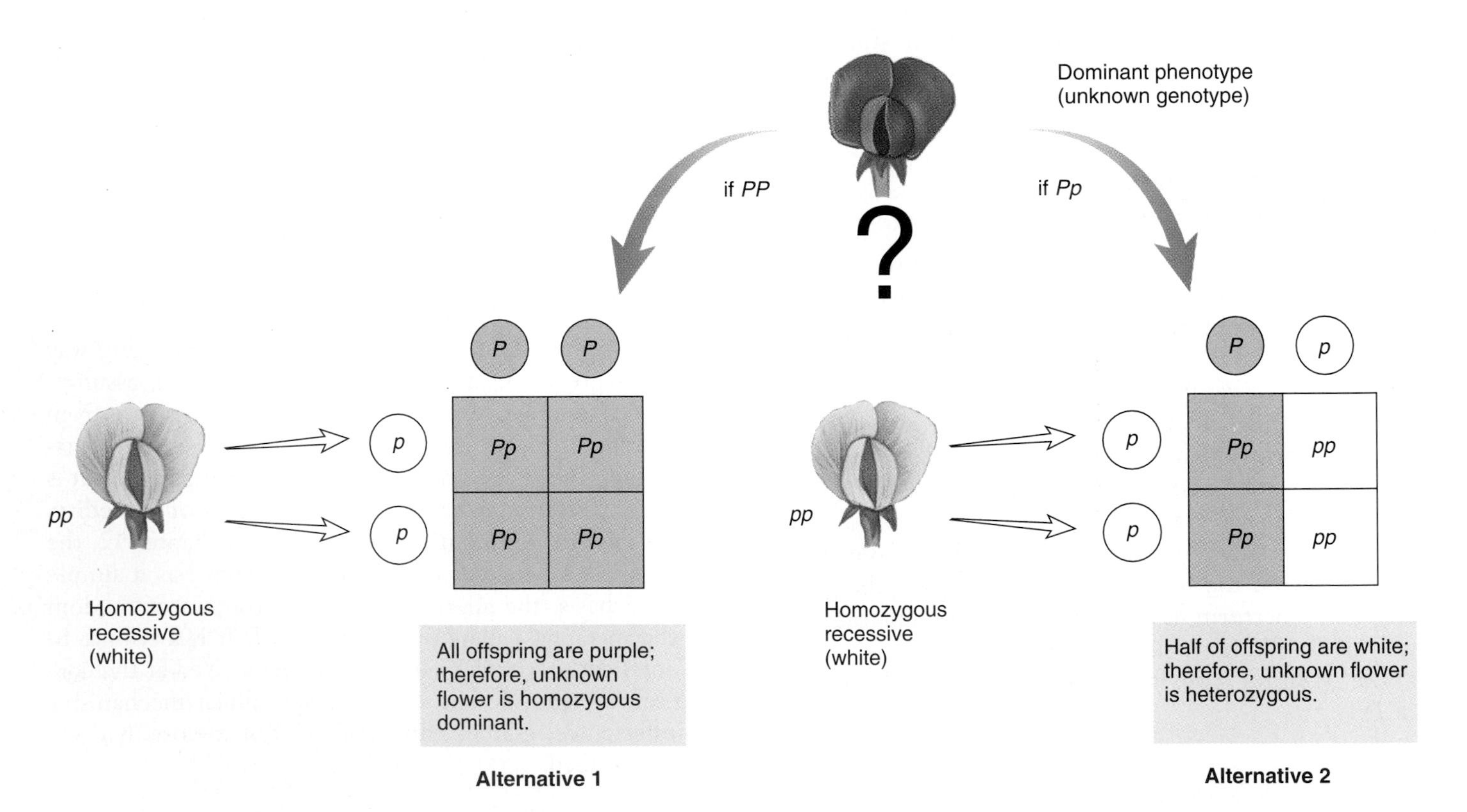

FIGURE 13.15
A testcross. To determine whether an individual exhibiting a dominant phenotype, such as purple flowers, is homozygous or heterozygous for the dominant allele, Mendel crossed the individual in question with a plant that he knew to be homozygous recessive, in this case a plant with white flowers.

Probability and Allele Distribution

Many, although not all, alternative alleles produce discretely different phenotypes. Mendel's pea plants were tall or dwarf, had purple or white flowers, and produced round or wrinkled seeds. The eye color of a fruit fly may be red or white, and the skin color of a human may be pigmented or albino. When only two alternative alleles exist for a given character, the distribution of phenotypes among the offspring of a cross is referred to as a **binomial distribution.**

As an example, consider the distribution of sexes in humans. Imagine that a couple has chosen to have three children. How likely is it that two of the children will be boys and one will be a girl? The frequency of any particular possibility is referred to as its **probability** of occurrence. Let p symbolize the probability of having a boy at any given birth and q symbolize the probability of having a girl. Since any birth is equally likely to produce a girl or boy:

$$p = q = \frac{1}{2}$$

Table 13.A shows eight possible gender combinations among the three children. The sum of the probabilities of the eight possible combinations must equal one. Thus:

$$p^3 + 3p^2q + 3pq^2 + q^3 = 1$$

The probability that the three children will be two boys and one girl is:

$$3p^2q = 3 \times (\tfrac{1}{2})^2 \times (\tfrac{1}{2}) = \tfrac{3}{8}$$

To test your understanding, try to estimate the probability that two parents heterozygous for the recessive allele producing albinism (a) will have one albino child in a family of three. First, set up a Punnett square:

		Father's Gametes *A*	*a*
Mother's Gametes	*A*	*AA*	*Aa*
	a	*Aa*	*aa*

You can see that one-fourth of the children are expected to be albino (*aa*). Thus, for any given birth the probability of an albino child is ¼. This probability can be symbolized by q. The probability of a nonalbino child is ¾, symbolized by p. Therefore, the probability that there will be one albino child among the three children is:

$$3p^2q = 3 \times (\tfrac{3}{4})^2 \times (\tfrac{1}{4}) = \tfrac{27}{64}, \text{ or } 42\%$$

This means that the chance of having one albino child in the three is 42%.

Table 13.A Binomial Distribution of the Sexes of Children in Human Families

Composition of Family	Order of Birth	Calculation	Probability	
3 boys	bbb	$p \times p \times p$	p^3	
2 boys and 1 girl	bbg	$p \times p \times q$	p^2q	$3p^2q$
	bgb	$p \times q \times p$	p^2q	
	gbb	$q \times p \times p$	p^2q	
1 boy and 2 girls	ggb	$q \times q \times p$	pq^2	$3pq^2$
	gbg	$q \times p \times q$	pq^2	
	bgg	$p \times q \times q$	pq^2	
3 girls	ggg	$q \times q \times q$	q^3	

Vocabulary of Genetics

allele One of two or more alternative forms of a gene.

diploid Having two sets of chromosomes, which are referred to as *homologues.* Animals and plants are diploid in the dominant phase of their life cycles as are some protists.

dominant allele An allele that dictates the appearance of heterozygotes. One allele is said to be dominant over another if a heterozygous individual with one copy of that allele has the same appearance as a homozygous individual with two copies of it.

gene The basic unit of heredity; a sequence of DNA nucleotides on a chromosome that encodes a polypeptide or RNA molecule and so determines the nature of an individual's inherited traits.

genotype The total set of genes present in the cells of an organism. This term is often also used to refer to the set of alleles at a single gene.

haploid Having only one set of chromosomes. Gametes, certain animals, protists and fungi, and certain stages in the life cycle of plants are haploid.

heterozygote A diploid individual carrying two different alleles of a gene on two homologous chromosomes. Most human beings are heterozygous for many genes.

homozygote A diploid individual carrying identical alleles of a gene on both homologous chromosomes.

locus The location of a gene on a chromosome.

phenotype The realized expression of the genotype; the observable manifestation of a trait (affecting an individual's structure, physiology, or behavior) that results from the biological activity of the DNA molecules.

recessive allele An allele whose phenotypic effect is masked in heterozygotes by the presence of a dominant allele.

Mendel's Second Law of Heredity: Independent Assortment

After Mendel had demonstrated that different traits of a given character (alleles of a given gene) segregate independently of each other in crosses, he asked whether different genes also segregate independently. Mendel set out to answer this question in a straightforward way. He first established a series of pure-breeding lines of peas that differed in just two of the seven characters he had studied. He then crossed contrasting pairs of the pure-breeding lines to create heterozygotes. In a cross involving different seed shape alleles (round, *R*, and wrinkled, *r*) and different seed color alleles (yellow, *Y*, and green, *y*), all the F_1 individuals were identical, each one heterozygous for both seed shape (*Rr*) and seed color (*Yy*). The F_1 individuals of such a cross are **dihybrids,** individuals heterozygous for both genes.

The third step in Mendel's analysis was to allow the dihybrids to self-fertilize. If the alleles affecting seed shape and seed color were segregating independently, then the probability that a particular pair of seed shape alleles would occur together with a particular pair of seed color alleles would be simply the product of the individual probabilities that each pair would occur separately. Thus, the probability that an individual with wrinkled green seeds (*rryy*) would appear in the F_2 generation would be equal to the probability of observing an individual with wrinkled seeds (¼) times the probability of observing one with green seeds (¼), or 1/16.

Because the gene controlling seed shape and the gene controlling seed color are each represented by a pair of alternative alleles in the dihybrid individuals, four types of gametes are expected: *RY*, *Ry*, *rY*, and *ry*. Therefore, in the F_2 generation there are 16 possible combinations of alleles, each of them equally probable (figure 13.16). Of these, 9 possess at least one dominant allele for each gene (signified *R*_*Y*_, where the dash indicates the presence of either allele) and, thus, should have round, yellow seeds. Of the rest, 3 possess at least one dominant *R* allele but are homozygous recessive for color (*R*_*yy*); 3 others possess at least one dominant *Y* allele but are homozygous recessive for shape (*rrY*_); and 1 combination among the 16 is homozygous recessive for both genes (*rryy*). The hypothesis that color and shape genes assort independently thus predicts that the F_2 generation will display a 9:3:3:1 phenotypic ratio: nine individuals with round, yellow seeds, three with round, green seeds, three with wrinkled, yellow seeds, and one with wrinkled, green seeds (see figure 13.16).

What did Mendel actually observe? From a total of 556 seeds from dihybrid plants he had allowed to self-fertilize, he observed: 315 round yellow (*R*_*Y*_), 108 round green (*R*_*yy*), 101 wrinkled yellow (*rrY*_), and 32 wrinkled green (*rryy*). These results are very close to a 9:3:3:1 ratio (which would be 313:104:104:35). Consequently, the two genes appeared to assort completely independently of each other.

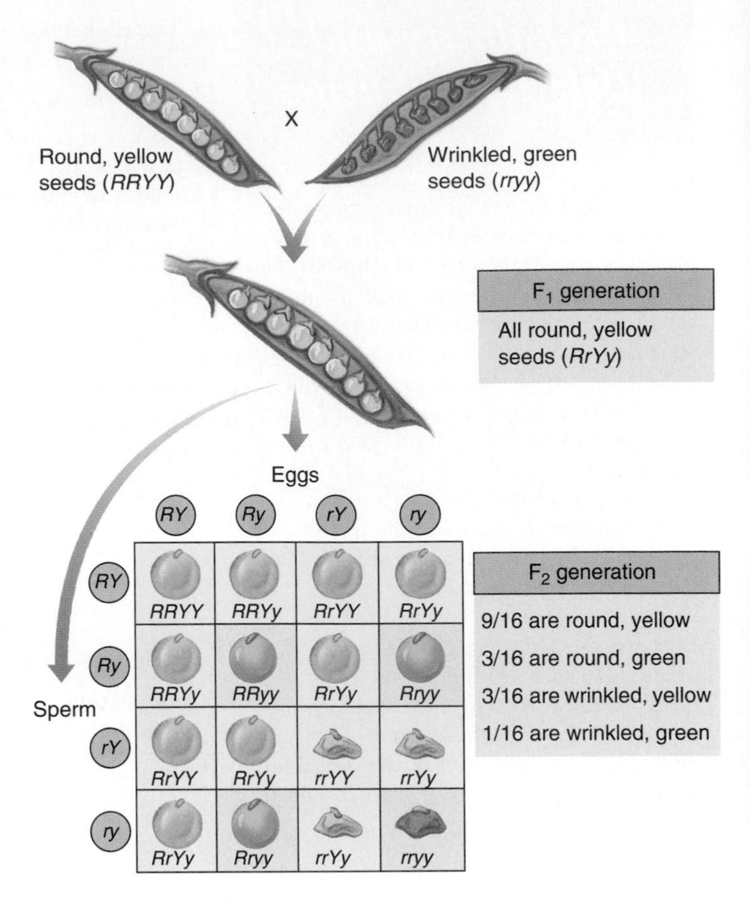

FIGURE 13.16
Analyzing a dihybrid cross. This Punnett square shows the results of Mendel's dihybrid cross between plants with round yellow seeds and plants with wrinkled green seeds. The ratio of the four possible combinations of phenotypes is predicted to be 9:3:3:1, the ratio that Mendel found.

Note that this independent assortment of different genes in no way alters the independent segregation of individual pairs of alleles. Round versus wrinkled seeds occur in a ratio of approximately 3:1 (423:133); so do yellow versus green seeds (416:140). Mendel obtained similar results for other pairs of traits.

Mendel's discovery is often referred to as **Mendel's Second Law of Heredity,** or the **Law of Independent Assortment.** Genes that assort independently of one another, like the seven genes Mendel studied, usually do so because they are located on different chromosomes, which segregate independently during the meiotic process of gamete formation. A modern restatement of Mendel's Second Law would be that *genes that are located on different chromosomes assort independently during meiosis.*

Mendel summed up his discoveries about heredity in two laws. Mendel's First Law of Heredity states that alternative alleles of a trait segregate independently; his Second Law of Heredity states that genes located on different chromosomes assort independently.

Mendelian Inheritance Is Not Always Easy to Analyze

Although Mendel's results did not receive much notice during his lifetime, three different investigators independently rediscovered his pioneering paper in 1900, 16 years after his death. They came across it while searching the literature in preparation for publishing their own findings, which closely resembled those Mendel had presented more than three decades earlier. In the decades following the rediscovery of Mendel, many investigators set out to test Mendel's ideas. However, scientists attempting to confirm Mendel's theory often had trouble obtaining the same simple ratios he had reported. Often, the expression of the genotype is not straightforward. Most phenotypes reflect the action of many genes that act sequentially or jointly, and the phenotype can be affected by alleles that lack complete dominance and the environment.

Continuous Variation

Few phenotypes are the result of the action of only one gene. Instead, most characters reflect the action of **polygenes,** many genes that act sequentially or jointly. When multiple genes act jointly to influence a character such as height or weight, the character often shows a range of small differences. Because all of the genes that play a role in determining phenotypes such as height or weight segregate independently of one another, one sees a gradation in the degree of difference when many individuals are examined (figure 13.17). We call this gradation **continuous variation.** The greater the number of genes that influence a character, the more continuous the expected distribution of the versions of that character.

How can one describe the variation in a character such as the height of the individuals in figure 13.17? Individuals range from quite short to very tall, with average heights more common than either extreme. What one often does is to group the variation into categories—in this case, by measuring the heights of the individuals in inches, rounding fractions of an inch to the nearest whole number. Each height, in inches, is a separate phenotypic category. Plotting the numbers in each height category produces a histogram, such as that in figure 13.17. The histogram approximates an idealized bell-shaped curve, and the variation can be characterized by the mean and spread of that curve.

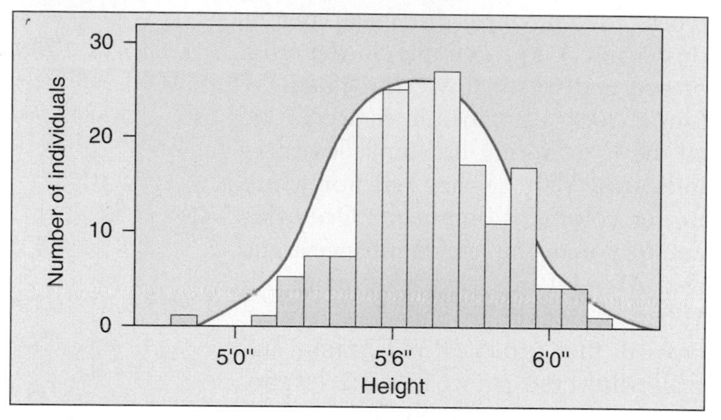

FIGURE 13.17
Height is a continuously varying trait. The photo shows variation in height among students of the 1914 class of the Connecticut Agricultural College. Because many genes contribute to height and tend to segregate independently of one another, the cumulative contribution of different combinations of alleles to height forms a *continuous* distribution of possible height, in which the extremes are much rarer than the intermediate values.

Pleiotropic Effects

Often, an individual allele will have more than one effect on the phenotype. Such an allele is said to be **pleiotropic.** When the pioneering French geneticist Lucien Cuenot studied yellow fur in mice, a dominant trait, he was unable to obtain a true-breeding yellow strain by crossing individual yellow mice with each other. Individuals homozygous for the yellow allele died, because the yellow allele was pleiotropic: one effect was yellow coat color, but another was a lethal developmental defect. A pleiotropic allele may be dominant with respect to one phenotypic consequence (yellow fur) and recessive with respect to another (lethal developmental defect). In pleiotropy, one gene affects many traits, in marked contrast to polygeny, where many genes affect one trait. Pleiotropic effects are difficult to predict, because the genes that affect a trait often perform other functions we may know nothing about.

Pleiotropic effects are characteristic of many inherited disorders, such as cystic fibrosis and sickle cell anemia, both discussed later in this chapter. In these disorders, multiple symptoms can be traced back to a single gene defect. In cystic fibrosis, patients exhibit clogged blood vessels, overly sticky mucus, salty sweat, liver and pancreas failure, and a battery of other symptoms. All are pleiotropic effects of a single defect, a mutation in a gene that encodes a chloride ion transmembrane channel. In sickle cell anemia, a defect in the oxygen-carrying hemoglobin molecule causes anemia, heart failure, increased susceptibility to pneumonia, kidney failure, enlargement of the spleen, and many other symptoms. It is usually difficult to deduce the nature of the primary defect from the range of a gene's pleiotropic effects.

Lack of Complete Dominance

Not all alternative alleles are fully dominant or fully recessive in heterozygotes. Some pairs of alleles instead produce a heterozygous phenotype that is either intermediate between those of the parents (incomplete dominance), or representative of both parental phenotypes (codominance, discussed later in this chapter). For example, in the cross of red and white flowering Japanese four o'clocks described in figure 13.18, all the F_1 offspring had pink flowers—indicating that neither red nor white flower color was dominant. Does this example of incomplete dominance argue that Mendel was wrong? Not at all. When two of the F_1 pink flowers were crossed, they produced red-, pink-, and white-flowered plants in a 1:2:1 ratio. Heterozygotes are simply intermediate in color.

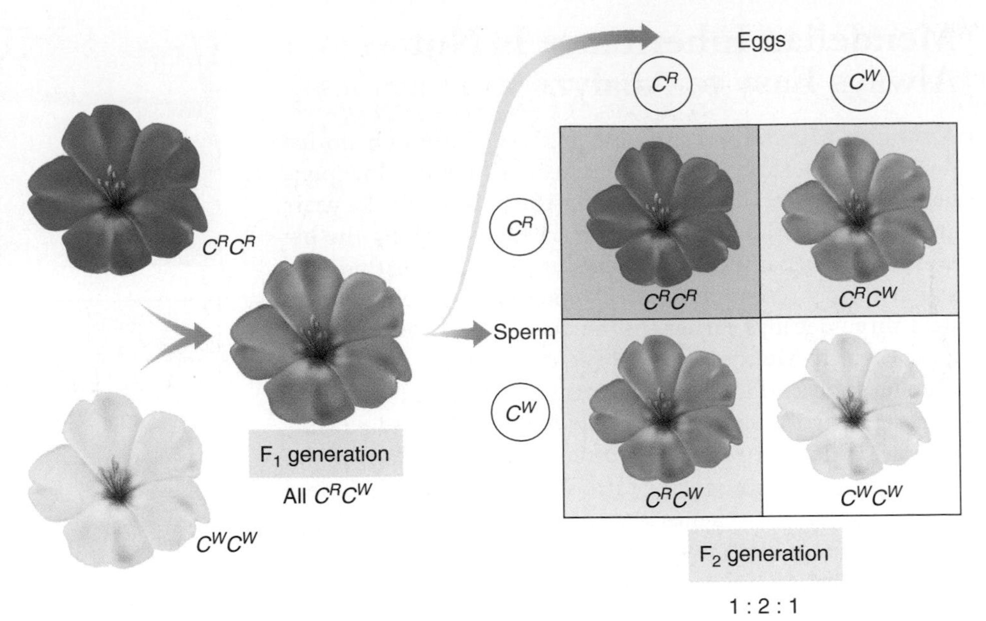

FIGURE 13.18
Incomplete dominance. In a cross between a red-flowered Japanese four o'clock, genotype C^RC^R, and a white-flowered one (C^WC^W), neither allele is dominant. The heterozygous progeny have pink flowers and the genotype C^RC^W. If two of these heterozygotes are crossed, the phenotypes of their progeny occur in a ratio of 1:2:1 (red:pink:white).

Environmental Effects

The degree to which an allele is expressed may depend on the environment. Some alleles are heat-sensitive, for example. Traits influenced by such alleles are more sensitive to temperature or light than are the products of other alleles. The arctic foxes in figure 13.19, for example, make fur pigment only when the weather is warm. Similarly, the *ch* allele in Himalayan rabbits and Siamese cats encodes a heat-sensitive version of tyrosinase, one of the enzymes mediating the production of melanin, a dark pigment. The ch version of the enzyme is inactivated at temperatures above about 33°C. At the surface of the body and head, the temperature is above 33°C and the tyrosinase enzyme is inactive, while it is more active at body extremities such as the tips of the ears and tail, where the temperature is below 33°C. The dark melanin pigment this enzyme produces causes the ears, snout, feet, and tail of Himalayan rabbits and Siamese cats to be black.

(a)

(b)

FIGURE 13.19
Environmental effects on an allele. (*a*) An arctic fox in winter has a coat that is almost white, so it is difficult to see the fox against a snowy background. (*b*) In summer, the same fox's fur darkens to a reddish brown, so that it resembles the color of the surrounding tundra. Heat-sensitive alleles control this color change.

Epistasis

In the tests of Mendel's ideas that followed the rediscovery of his work, scientists had trouble obtaining Mendel's simple ratios particularly with dihybrid crosses. Recall that when individuals heterozygous for two different genes mate (a dihybrid cross), four different phenotypes are possible among the progeny: offspring may display the dominant phenotype for both genes, either one of the genes, or for neither gene. Sometimes, however, it is not possible for an investigator to identify successfully each of the four phenotypic classes, because two or more of the classes look alike. Such situations proved confusing to investigators following Mendel.

One example of such difficulty in identification is seen in the analysis of particular varieties of corn, *Zea mays*. Some commercial varieties exhibit a purple pigment called anthocyanin in their seed coats, while others do not. In 1918, geneticist R. A. Emerson crossed two pure-breeding corn varieties, neither exhibiting anthocyanin pigment. Surprisingly, all of the F_1 plants produced purple seeds.

When two of these pigment-producing F_1 plants were crossed to produce an F_2 generation, 56% were pigment producers and 44% were not. What was happening? Emerson correctly deduced that two genes were involved in producing pigment, and that the second cross had thus been a dihybrid cross. Mendel had predicted 16 equally possible ways gametes could combine. How many of these were in each of the two types Emerson obtained? He multiplied the fraction that were pigment producers (0.56) by 16 to obtain 9, and multiplied the fraction that were not (0.44) by 16 to obtain 7. Thus, Emerson had a **modified ratio** of 9:7 instead of the usual 9:3:3:1 ratio.

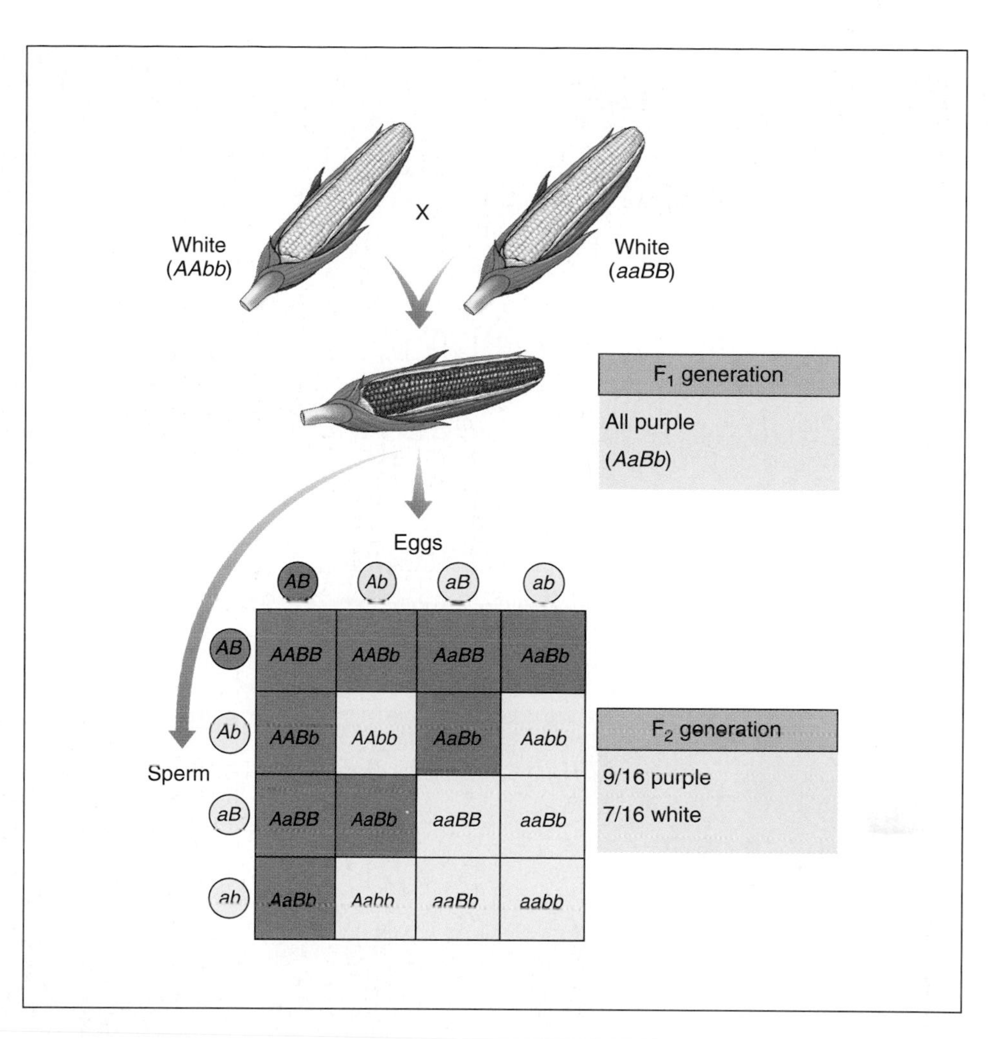

FIGURE 13.20
How epistasis affects grain color. The purple pigment found in some varieties of corn is the product of a two-step biochemical pathway. Unless both enzymes are active (the plant has a dominant allele for each of the two genes, *A* and *B*), no pigment is expressed.

Why Was Emerson's Ratio Modified? When genes act sequentially, as in a biochemical pathway, an allele expressed as a defective enzyme early in the pathway blocks the flow of material through the rest of the pathway. This makes it impossible to judge whether the later steps of the pathway are functioning properly. Such gene interaction, where one gene can interfere with the expression of another gene, is the basis of the phenomenon called **epistasis.**

The pigment anthocyanin is the product of a two-step biochemical pathway:

$$\textbf{Starting molecule} \xrightarrow{\text{Enzyme 1}} \textbf{Intermediate} \xrightarrow{\text{Enzyme 2}} \textbf{Anthocyanin}$$
(Colorless) (Colorless) (Purple)

To produce pigment, a plant must possess at least one functional copy of each enzyme gene (figure 13.20). The dominant alleles encode functional enzymes, but the recessive alleles encode nonfunctional enzymes. Of the 16 genotypes predicted by random assortment, 9 contain at least one dominant allele of both genes; they produce purple progeny. The remaining 7 genotypes lack dominant alleles at either or both loci (3 + 3 + 1 = 7) and so are phenotypically the same (nonpigmented), giving the phenotypic ratio of 9:7 that Emerson observed. The inability to see the effect of enzyme 2 when enzyme 1 is nonfunctional is an example of epistasis.

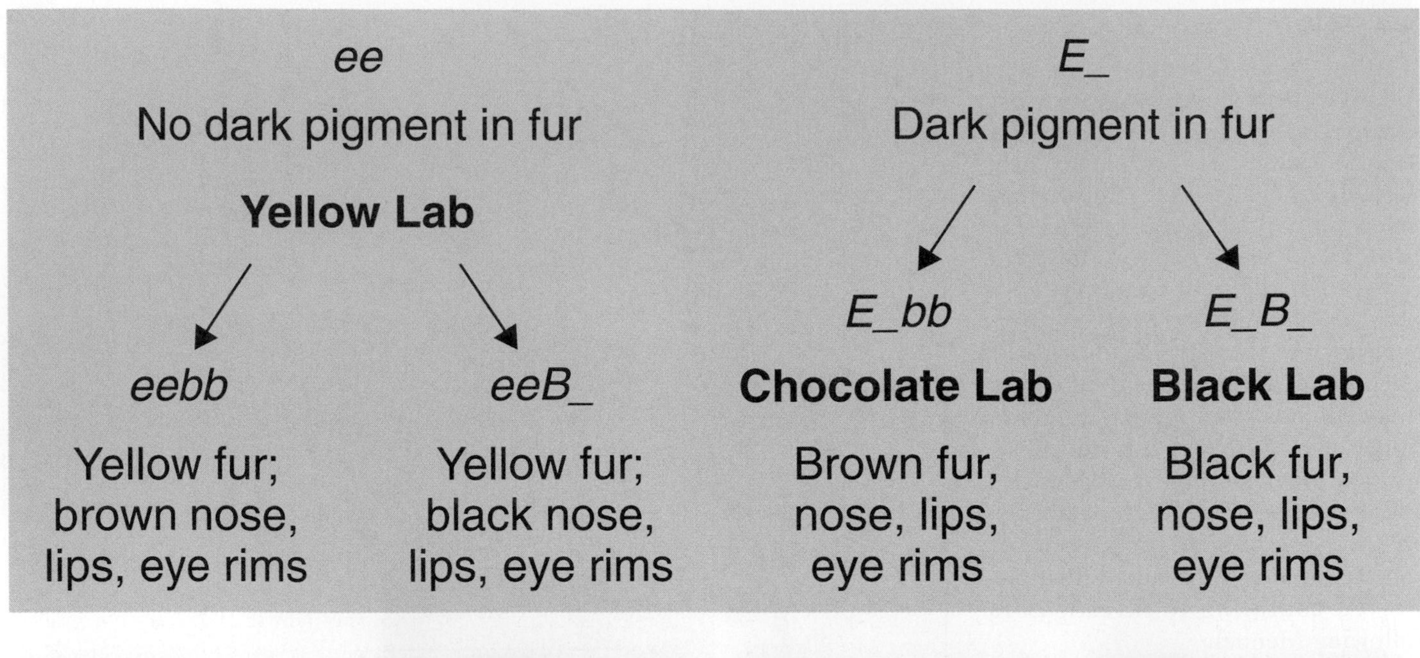

FIGURE 13.21
The effect of epistatic interactions on coat color in dogs. The coat color seen in Labrador retrievers is an example of the interaction of two genes, each with two alleles. The *E* gene determines if the pigment will be deposited in the fur, and the *B* gene determines how dark the pigment will be.

Other Examples of Epistasis

In many animals, coat color is the result of epistatic interactions among genes. Coat color in Labrador retrievers, a breed of dog, is due primarily to the interaction of two genes. The *E* gene determines if dark pigment (eumelanin) will be deposited in the fur or not. If a dog has the genotype *ee*, no pigment will be deposited in the fur, and it will be yellow. If a dog has the genotype *EE* or *Ee* (*E_*), pigment will be deposited in the fur.

A second gene, the *B* gene, determines how dark the pigment will be. This gene controls the distribution of melanosomes in a hair. Dogs with the genotype *E_bb* will have brown fur and are called chocolate labs. Dogs with the genotype *E_B_* will have black fur. But, even in yellow dogs, the *B* gene does have some effect. Yellow dogs with the genotype *eebb* will have brown pigment on their nose, lips, and eye rims, while yellow dogs with the genotype *eeB_* will have black pigment in these areas. The interaction among these alleles is illustrated in figure 13.21. The genes for coat color in this breed have been found, and a genetic test is available to determine the coat colors in a litter of puppies.

A variety of factors can disguise the Mendelian segregation of alleles. Among them are the continuous variation that results when many genes contribute to a trait, incomplete dominance that produce heterozygotes unlike either parent, environmental influences on the expression of phenotypes, and gene interactions that produce epistasis.

13.2 Human genetics follows Mendelian principles.

Random changes in genes, called mutations, occur in any population. These changes rarely improve the functioning of the proteins the genes encode, just as randomly changing a wire in a computer rarely improves the computer's functioning. Mutant alleles are usually recessive to other alleles. When two seemingly normal individuals who are heterozygous for such an allele produce offspring homozygous for the allele, the offspring suffer the detrimental effects of the mutant allele. When a detrimental allele occurs at a significant frequency in a population, the harmful effect it produces is called a **gene disorder.**

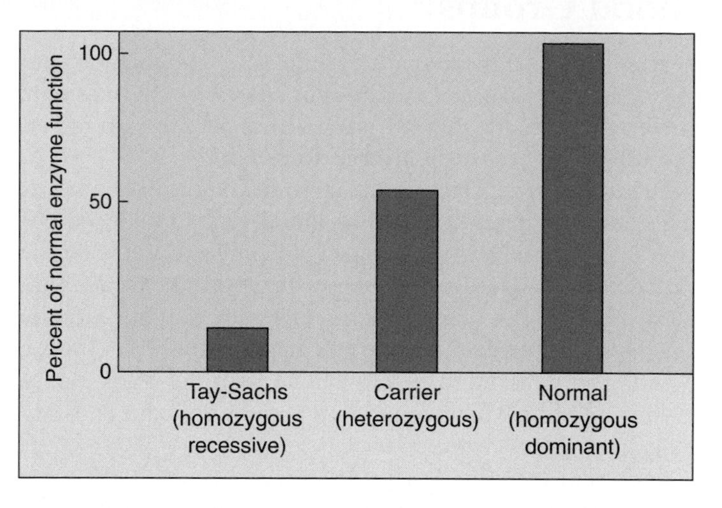

FIGURE 13.22
Tay-Sachs disease. Homozygous individuals (*left bar*) typically have less than 10% of the normal level of hexosaminidase A (*right bar*), while heterozygous individuals (*middle bar*) have about 50% of the normal level—enough to prevent deterioration of the central nervous system.

Most Gene Disorders Are Rare

Tay-Sachs disease is an incurable hereditary disorder in which the nervous system deteriorates. Affected children appear normal at birth and usually do not develop symptoms until about the eighth month, when signs of mental deterioration appear. The children are blind within a year after birth, and they rarely live past five years of age.

Tay-Sachs disease is rare in most human populations, occurring in only 1 of 300,000 births in the United States. However, the disease has a high incidence among Jews of Eastern and Central Europe (Ashkenazi), and among American Jews, 90% of whom trace their ancestry to Eastern and Central Europe. In these populations, it is estimated that 1 in 28 individuals is a heterozygous carrier of the disease, and approximately 1 in 3500 infants has the disease. Because the disease is caused by a recessive allele, most of the people who carry the defective allele do not themselves develop symptoms of the disease.

The Tay-Sachs allele produces the disease by encoding a nonfunctional form of the enzyme hexosaminidase A. This enzyme breaks down *gangliosides,* a class of lipids occurring within the lysosomes of brain cells (figure 13.22). As a result, the lysosomes fill with gangliosides, swell, and eventually burst, releasing oxidative enzymes that kill the cells. There is no known cure for this disorder.

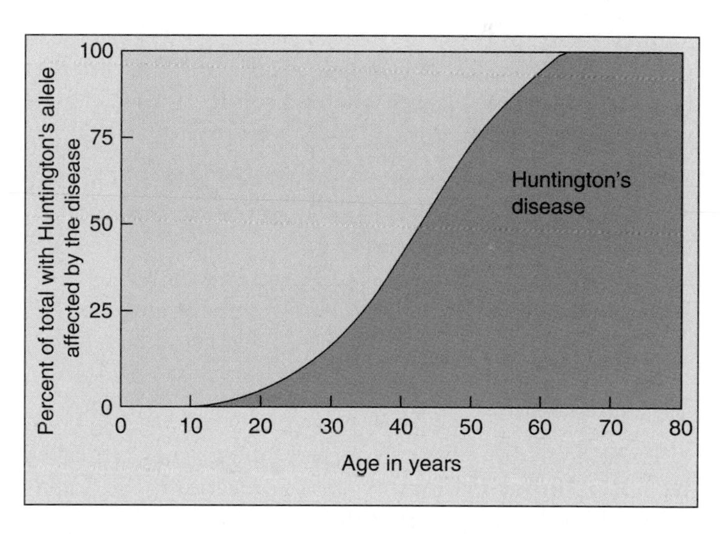

FIGURE 13.23
Huntington's disease is a dominant genetic disorder. It is because of the late age of onset of this disease that it persists despite the fact that it is dominant and fatal.

Not All Gene Defects Are Recessive

Not all hereditary disorders are recessive. **Huntington's disease** is a hereditary condition caused by a dominant allele that leads to the progressive deterioration of brain cells (figure 13.23). Perhaps 1 in 24,000 individuals develops the disorder. Because the allele is dominant, every individual that carries the allele expresses the disorder. Nevertheless, the disorder persists in human populations because its symptoms usually do not develop until the affected individuals are more than 30 years old, and by that time most of those individuals have already had children. Consequently, the allele is often transmitted before the lethal condition develops. A person who is heterozygous for Huntington's disease has a 50% chance of passing the *disease* to his or her children (even though the other parent does not have the disorder). In contrast, the carrier of a recessive disorder such as cystic fibrosis has a 50% chance of passing the *allele* to offspring and must mate with another carrier to risk bearing a child with the disease.

Most gene defects are rare recessives, although some are dominant.

Multiple Alleles: The ABO Blood Groups

A gene may have more than two alleles in a population, and most genes possess several different alleles. Often, no single allele is dominant; instead, each allele has its own effect, and the alleles are considered **codominant.**

A human gene with more than one codominant allele is the gene that determines ABO blood type. This gene encodes an enzyme that adds sugar molecules to lipids on the surface of red blood cells. These sugars act as recognition markers for the immune system. The gene that encodes the enzyme, designated I, has three common alleles: I^B, whose product adds galactose; I^A, whose product adds galactosamine; and i, which codes for a protein that does not add a sugar.

Different combinations of the three I gene alleles occur in different individuals because each person possesses two copies of the chromosome bearing the I gene and may be homozygous for any allele or heterozygous for any two. An individual heterozygous for the I^A and I^B alleles produces both forms of the enzyme and adds both galactose and galactosamine to the surfaces of red blood cells. Because both alleles are expressed simultaneously in heterozygotes, the I^A and I^B alleles are codominant. Both I^A and I^B are dominant over the i allele because both I^A or I^B alleles lead to sugar addition and the i allele does not. The different combinations of the three alleles produce four different phenotypes (figure 13.24):

1. Type A individuals add only galactosamine. They are either I^AI^A homozygotes or I^Ai heterozygotes.
2. Type B individuals add only galactose. They are either I^BI^B homozygotes or I^Bi heterozygotes.
3. Type AB individuals add both sugars and are I^AI^B heterozygotes.
4. Type O individuals add neither sugar and are ii homozygotes.

These four different cell surface phenotypes are called the **ABO blood groups** or, less commonly, the Landsteiner blood groups, after the man who first described them. As Landsteiner noted, a person's immune system can distinguish between these four phenotypes. If a type A individual receives a transfusion of type B blood, the recipient's immune system recognizes that the type B blood cells possess a "foreign" antigen (galactose) and attacks the donated blood cells, causing the cells to clump, or agglutinate. This also happens if the donated blood is type AB. However, if the donated blood is type O, no immune attack will occur, as there are no galactose antigens on the surfaces of blood cells produced by the type O donor. In general, any individual's immune system will tolerate a transfusion of type O blood. Because neither galactose nor galactosamine is foreign to type AB individuals (whose red blood cells have both sugars), those individuals may receive any type of blood.

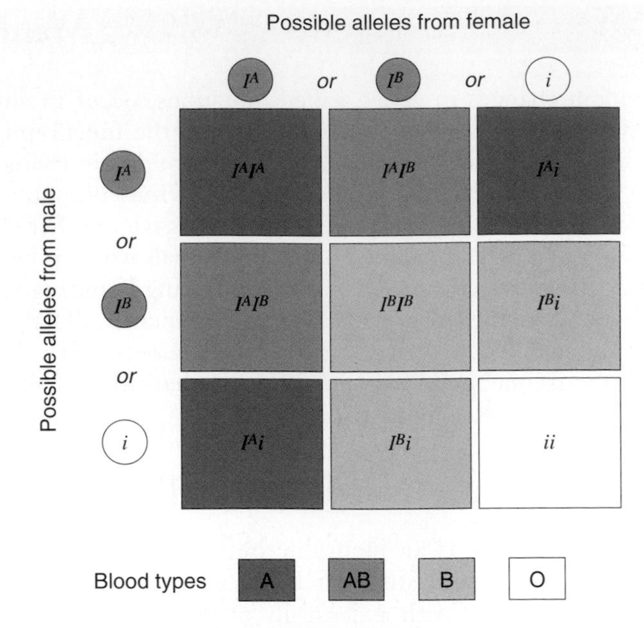

FIGURE 13.24
Multiple alleles control the ABO blood groups. Different combinations of the three I gene alleles result in four different blood type phenotypes: type A (either I^AI^A homozygotes or I^Ai heterozygotes), type B (either I^BI^B homozygotes or I^Bi heterozygotes), type AB (I^AI^B heterozygotes), and type O (ii homozygotes).

The Rh Blood Group

Another set of cell surface markers on human red blood cells are the **Rh blood group** antigens, named for the rhesus monkey in which they were first described. About 85% of adult humans have the Rh cell surface marker on their red blood cells, and are called Rh-positive. Rh-negative persons lack this cell surface marker because they are homozygous for the recessive gene encoding it.

If an Rh-negative person is exposed to Rh-positive blood, the Rh surface antigens of that blood are treated like foreign invaders by the Rh-negative person's immune system, which proceeds to make antibodies directed against the Rh antigens. This most commonly happens when an Rh-negative woman gives birth to an Rh-positive child (whose father is Rh-positive). At birth, some fetal red blood cells cross the placental barrier and enter the mother's bloodstream, where they induce the production of "anti-Rh" antibodies. In subsequent pregnancies, the mother's antibodies can cross back to the new fetus and cause its red blood cells to clump, leading to a potentially fatal condition called erythroblastosis fetalis.

Many blood group genes possess multiple alleles that are codominant, several of which may be common.

Patterns of Inheritance Can Be Deduced from Pedigrees

When a blood vessel ruptures, the blood in the immediate area of the rupture forms a solid gel called a clot. The clot forms as a result of the polymerization of protein fibers circulating in the blood. A dozen proteins are involved in this process, and all must function properly for a blood clot to form. A mutation causing any of these proteins to lose their activity leads to a form of **hemophilia,** a hereditary condition in which the blood is slow to clot or does not clot at all.

Hemophilias are recessive disorders, expressed only when an individual does not possess any copy of the normal allele and so cannot produce one of the proteins necessary for clotting. Most of the genes that encode the blood-clotting proteins are on autosomes, but two (designated VIII and IX) are on the X chromosome. These two genes are sex-linked: any male who inherits a mutant allele of either of the two genes will develop hemophilia because his other sex chromosome is a Y chromosome that lacks any alleles of those genes.

The most famous instance of hemophilia, often called the Royal hemophilia, is a sex-linked form that arose in one of the parents of Queen Victoria of England (1819–1901; figure 13.25). In the five generations since Queen Victoria, 10 of her male descendants have had hemophilia. The present British royal family has escaped the disorder because Queen Victoria's son, King Edward VII, did not inherit the defective allele, and all the subsequent rulers of England are his descendants. Three of Victoria's nine children did receive the defective allele, however, and they carried it by marriage into many of the other royal families of Europe (figure 13.26), where it is still being passed to future generations—except in Russia, where all of the five children of Victoria's granddaughter Alexandra were killed soon after the Russian revolution in 1917. (Speculation that one daughter, Anastasia, survived ended in 1999 when DNA analysis confirmed the identity of her remains.)

FIGURE 13.25
Queen Victoria of England, surrounded by some of her descendants in 1894. Of Victoria's four daughters who lived to bear children, two, Alice and Beatrice, were carriers of Royal hemophilia. Two of Alice's daughters are standing behind Victoria (wearing feathered boas): Princess Irene of Prussia (right), and Alexandra (left), who would soon become Czarina of Russia. Both Irene and Alexandra were also carriers of hemophilia.

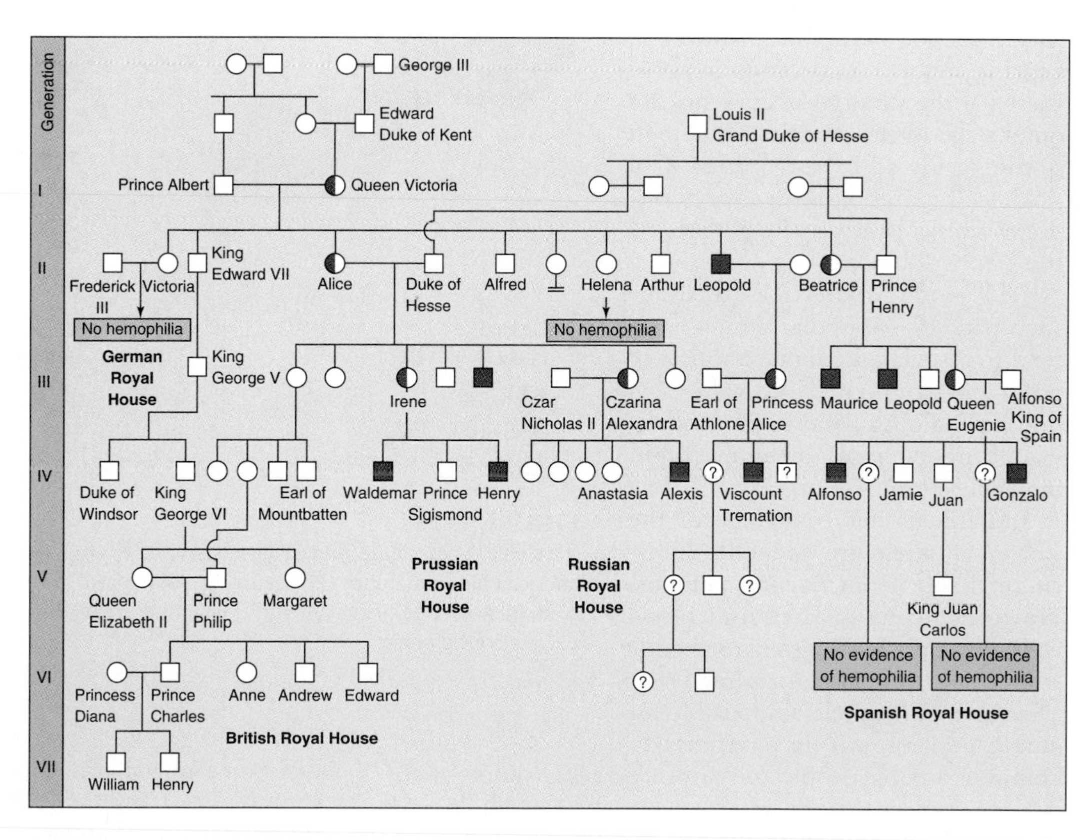

FIGURE 13.26
The Royal hemophilia pedigree. Queen Victoria's daughter Alice introduced hemophilia into the Russian and Austrian royal houses, and Victoria's daughter Beatrice introduced it into the Spanish royal house. Victoria's son Leopold, himself a victim, also transmitted the disorder in a third line of descent. Half-shaded symbols represent carriers with one normal allele and one defective allele; fully shaded symbols represent affected individuals.

Family pedigrees can reveal the mode of inheritance of a hereditary trait.

Gene Disorders Can Be Due to Simple Alterations of Proteins

Sickle cell anemia is a heritable disorder first noted in Chicago in 1904. Afflicted individuals have defective molecules of hemoglobin, the protein within red blood cells that carries oxygen. Consequently, these individuals are unable to properly transport oxygen to their tissues. The defective hemoglobin molecules stick to one another, forming stiff, rod-like structures and resulting in the formation of sickle-shaped red blood cells (figure 13.27). As a result of their stiffness and irregular shape, these cells have difficulty moving through the smallest blood vessels; they tend to accumulate in those vessels and form clots. People who have large proportions of sickle-shaped red blood cells tend to have intermittent illness and a shortened life span.

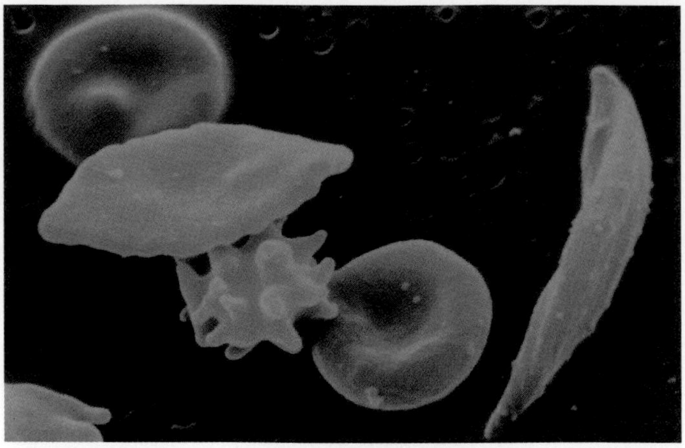

FIGURE 13.27
Sickle cell anemia. In individuals homozygous for the sickle cell trait, many of the red blood cells have sickle or irregular shapes, such as the cell on the far right.

The hemoglobin in the defective red blood cells differs from that in normal red blood cells in only one of hemoglobin's 574 amino acid subunits. In the defective hemoglobin, the amino acid valine replaces a glutamic acid at a single position in the protein. Interestingly, the position of the change is far from the active site of hemoglobin where the iron-bearing heme group binds oxygen. Instead, the change occurs on the outer edge of the protein. Why then is the result so catastrophic? The sickle cell mutation puts a very nonpolar amino acid on the surface of the hemoglobin protein, creating a "sticky patch" that sticks to other such patches—nonpolar amino acids tend to associate with one another in polar environments like water. As one hemoglobin adheres to another, ever-longer chains of hemoglobin molecules form.

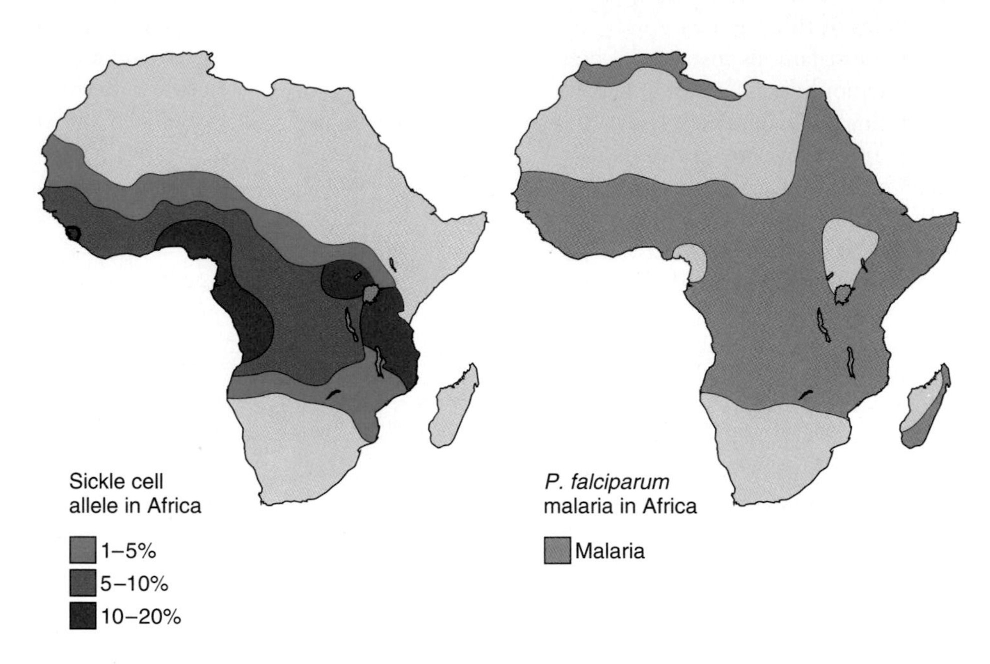

FIGURE 13.28
The sickle cell allele increases resistance to malaria. The distribution of sickle cell anemia closely matches the occurrence of malaria in central Africa. This is not a coincidence. The sickle cell allele, when heterozygous, increases resistance to malaria, a very serious disease.

Individuals heterozygous for the sickle cell allele are generally indistinguishable from normal persons. However, some of their red blood cells show the sickling characteristic when they are exposed to low levels of oxygen. The allele responsible for sickle cell anemia is particularly common among people of African descent; about 9% of African Americans are heterozygous for this allele, and about 0.2% are homozygous and therefore have the disorder. In some groups of people in Africa, up to 45% of all individuals are heterozygous for this allele, and 6% are homozygous. What factors determine the high frequency of sickle cell anemia in Africa? It turns out that heterozygosity for the sickle cell anemia allele increases resistance to malaria, a common and serious disease in central Africa (figure 13.28). We will discuss this situation in detail in chapter 21.

Sickle cell anemia is caused by a single-nucleotide change in the gene for hemoglobin, producing a protein with a nonpolar amino acid on its surface that tends to make the molecules clump together.

Table 13.2 Some Important Genetic Disorders

Disorder	Symptom	Defect	Dominant/ Recessive	Frequency among Human Births
Cystic fibrosis	Mucus clogs lungs, liver, and pancreas	Failure of chloride ion transport mechanism	Recessive	1/2500 (Caucasians)
Sickle cell anemia	Poor blood circulation	Abnormal hemoglobin molecules	Recessive	1/625 (African Americans)
Tay-Sachs disease	Deterioration of central nervous system in infancy	Defective enzyme (hexosaminidase A)	Recessive	1/3500 (Ashkenazi Jews)
Phenylketonuria	Brain fails to develop in infancy	Defective enzyme (phenylalanine hydroxylase)	Recessive	1/12,000
Hemophilia	Blood fails to clot	Defective blood clotting factor VIII	Sex-linked recessive	1/10,000 (Caucasian males)
Huntington's disease	Brain tissue gradually deteriorates in middle age	Production of an inhibitor of brain cell metabolism	Dominant	1/24,000
Muscular dystrophy (Duchenne)	Muscles waste away	Degradation of myelin coating of nerves stimulating muscles	Sex-linked recessive	1/3700 (males)
Hypercholesterolemia	Excessive cholesterol levels in blood, leading to heart disease	Abnormal form of cholesterol cell surface receptor	Dominant	1/500

Some Defects May Soon Be Curable

Some of the most common and serious gene defects result from single recessive mutations, including many of the defects listed in table 13.2. Recent developments in gene technology have raised the hope that this class of disorders may be curable. Perhaps the best example is **cystic fibrosis (CF),** the most common fatal genetic disorder among Caucasians.

Cystic fibrosis is a fatal disease in which the body cells of affected individuals secrete a thick mucus that clogs the airways of the lungs. These same secretions block the ducts of the pancreas and liver so that the few patients who do not die of lung disease die of liver failure. There is no known cure.

Cystic fibrosis results from a defect in a single gene, called *cf*, that is passed down from parent to child. One in 20 individuals possesses at least one copy of the defective gene. Most carriers are not afflicted with the disease; only those children who inherit a copy of the defective gene from each parent succumb to cystic fibrosis—about 1 in 2500 infants.

The function of the *cf* gene has proven difficult to study. In 1985 the first clear clue was obtained. An investigator, Paul Quinton, seized on a commonly observed characteristic of cystic fibrosis patients, that their sweat is abnormally salty, and performed the following experiment. He isolated a sweat duct from a small piece of skin and placed it in a solution of salt (NaCl) that was three times as concentrated as the NaCl inside the duct. He then monitored the movement of ions. Diffusion tends to drive both the sodium (Na^+) and the chloride (Cl^-) ions into the duct because of the higher outer ion concentrations. In skin isolated from normal individuals, Na^+ and Cl^- ions both entered the duct, as expected. In skin isolated from cystic fibrosis individuals, however, only Na^+ ions entered the duct—no Cl^- ions entered. For the first time, the molecular nature of cystic fibrosis became clear. Cystic fibrosis is a defect in a plasma membrane protein called CFTR (cystic fibrosis transmembrane conductance regulator) that normally regulates passage of Cl^- ions into and out of the body's cells. CFTR does not function properly in cystic fibrosis patients (see figure 4.8).

The defective *cf* gene was isolated in 1987, and its position on a particular human chromosome (chromosome 7) was pinpointed in 1989. In 1990 a working *cf* gene was successfully transferred via adenovirus into human lung cells growing in tissue culture. The defective cells were "cured," becoming able to transport chloride ions across their plasma membranes. Then in 1991, a team of researchers successfully transferred a normal human *cf* gene into the lung cells of a living animal—a rat. The *cf* gene was first inserted into a cold virus that easily infects lung cells, and the virus was inhaled by the rat. Carried piggyback, the *cf* gene entered the rat lung cells and began producing the normal human CFTR protein within these cells! Tests of gene transfer into CF patients were begun in 1993, and while a great deal of work remains to be done (the initial experiments were not successful), the future for cystic fibrosis patients for the first time seems bright.

Cystic fibrosis, and other genetic disorders, are potentially curable if ways can be found to successfully introduce normal alleles of the genes into affected individuals.

13.3 Genes are on chromosomes.

Chromosomes: The Vehicles of Mendelian Inheritance

Chromosomes are not the only kinds of structures that segregate regularly when eukaryotic cells divide. Centrioles also divide and segregate in a regular fashion, as do the mitochondria and chloroplasts (when present) in the cytoplasm. Therefore, in the early twentieth century it was by no means obvious that chromosomes were the vehicles of hereditary information.

The Chromosomal Theory of Inheritance

A central role for chromosomes in heredity was first suggested in 1900 by the German geneticist Karl Correns, in one of the papers announcing the rediscovery of Mendel's work. Soon after, observations that similar chromosomes paired with one another during meiosis led directly to the **chromosomal theory of inheritance,** first formulated by the American Walter Sutton in 1902.

Several pieces of evidence supported Sutton's theory. One was that reproduction involves the initial union of only two cells, egg and sperm. If Mendel's model were correct, then these two gametes must make equal hereditary contributions. Sperm, however, contain little cytoplasm, suggesting that the hereditary material must reside within the nuclei of the gametes. Furthermore, while diploid individuals have two copies of each pair of homologous chromosomes, gametes have only one. This observation was consistent with Mendel's model, in which diploid individuals have two copies of each heritable gene and gametes have one. Finally, chromosomes segregate during meiosis, and each pair of homologues orients on the metaphase plate independently of every other pair. Segregation and independent assortment were two characteristics of the genes in Mendel's model.

A Problem with the Chromosomal Theory

However, investigators soon pointed out one problem with this theory. If Mendelian characters are determined by genes located on the chromosomes, and if the independent assortment of Mendelian traits reflects the independent assortment of chromosomes in meiosis, why does the number of characters that assort independently in a given kind of organism often greatly exceed the number of chromosome pairs the organism possesses? This seemed a fatal objection, and it led many early researchers to have serious reservations about Sutton's theory.

Morgan's White-Eyed Fly

The essential correctness of the chromosomal theory of heredity was demonstrated long before this paradox was resolved. A single small fly provided the proof. In 1910 Thomas Hunt Morgan, studying the fruit fly *Drosophila melanogaster*, detected a **mutant** male fly, one that differed strikingly from normal flies of the same species: its eyes were white instead of red (figure 13.29).

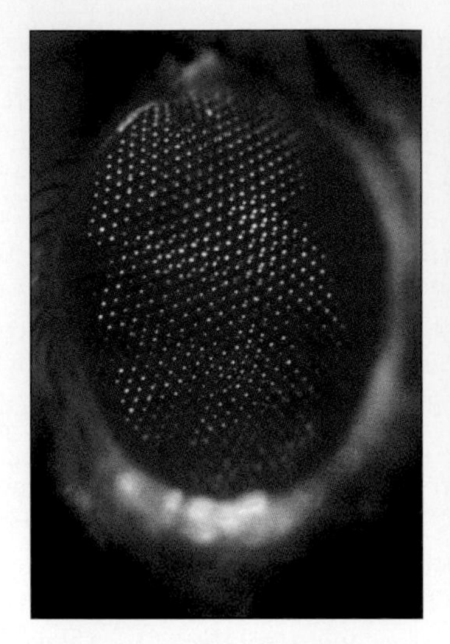
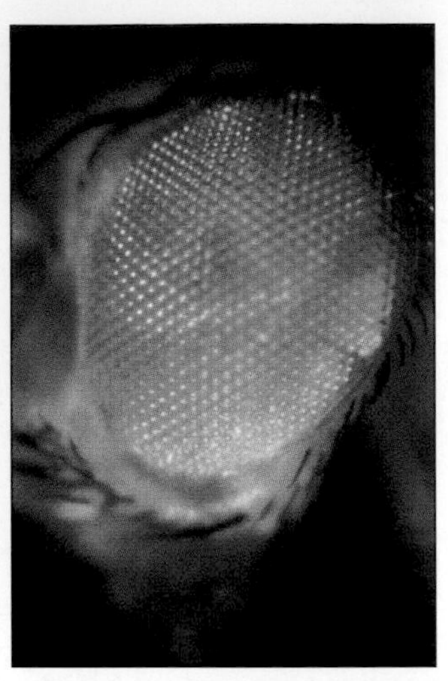

FIGURE 13.29
Red-eyed (normal) and white-eyed (mutant) *Drosophila*. The white-eyed defect is hereditary, the result of a mutation in a gene located on the X chromosome. By studying this mutation, Morgan first demonstrated that genes are on chromosomes.

Morgan immediately set out to determine if this new trait would be inherited in a Mendelian fashion. He first crossed the mutant male to a normal female to see if red or white eyes were dominant. All of the F_1 progeny had red eyes, so Morgan concluded that red eye color was dominant over white. Following the experimental procedure that Mendel had established long ago, Morgan then crossed the red-eyed flies from the F_1 generation with each other. Of the 4252 F_2 progeny Morgan examined, 782 (18%) had white eyes. Although the ratio of red eyes to white eyes in the F_2 progeny was greater than 3:1, the results of the cross nevertheless provided clear evidence that eye color segregates. However, there was something about the outcome that was strange and totally unpredicted by Mendel's theory—*all of the white-eyed F_2 flies were males!*

How could this result be explained? Perhaps it was impossible for a white-eyed female fly to exist; such individuals might not be viable for some unknown reason. To test this idea, Morgan testcrossed the female F_1 progeny with the original white-eyed male. He obtained both white-eyed and red-eyed males and females in a 1:1:1:1 ratio, just as Mendelian theory predicted. Hence, a female could have white eyes. Why, then, were there no white-eyed females among the progeny of the original cross?

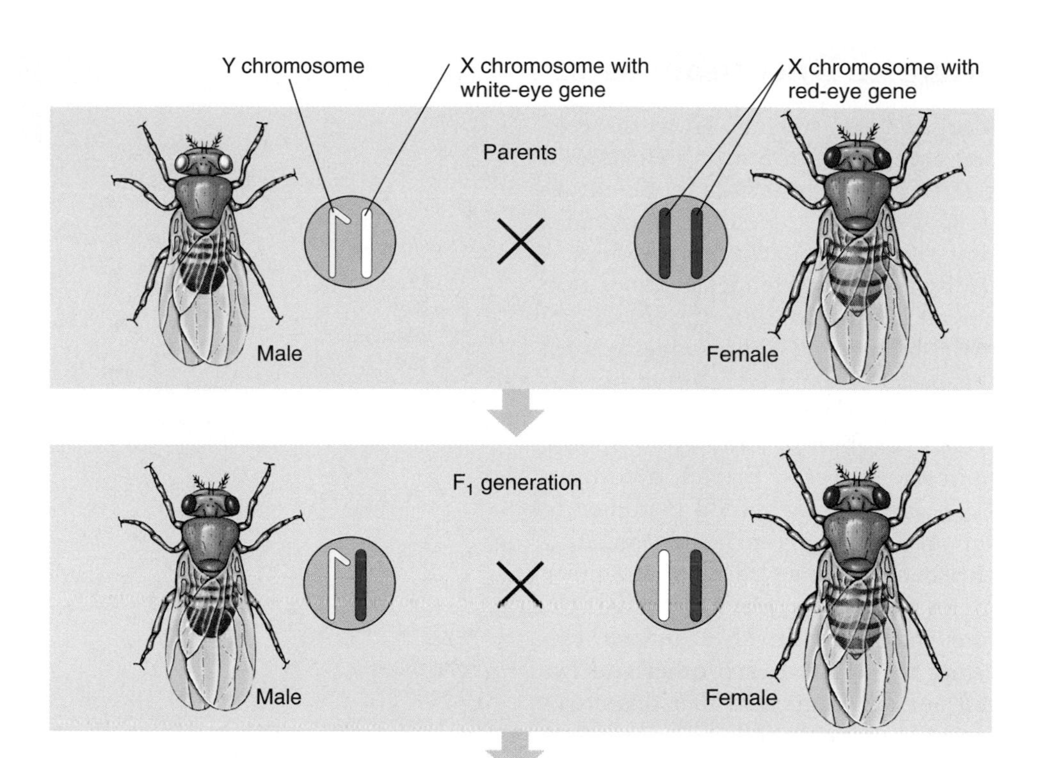

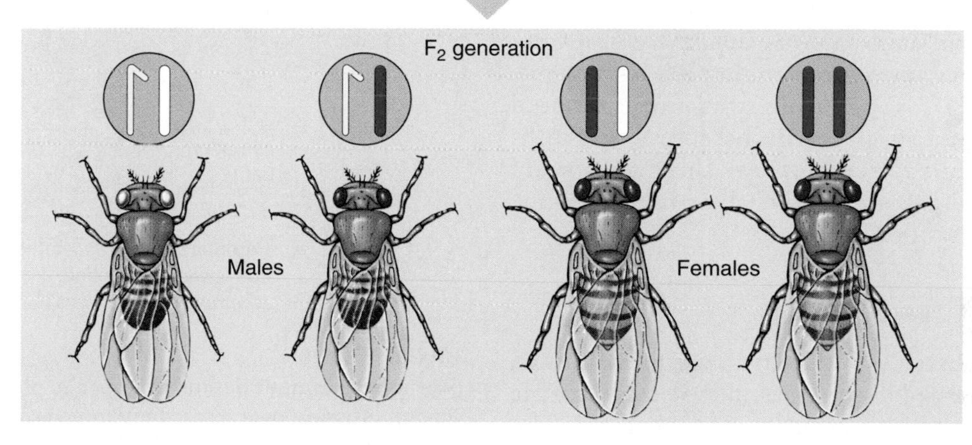

FIGURE 13.30
Morgan's experiment demonstrating the chromosomal basis of sex linkage in *Drosophila*. The white-eyed mutant male fly was crossed with a normal female. The F_1 generation flies all exhibited red eyes, as expected for flies heterozygous for a recessive white-eye allele. In the F_2 generation, all of the white-eyed flies were male.

Sex Linkage

The solution to this puzzle involved sex. In *Drosophila*, the sex of an individual is determined by the number of copies of a particular chromosome, the **X chromosome,** that an individual possesses. A fly with two X chromosomes is a female, and a fly with only one X chromosome is a male. In males, the single X chromosome pairs in meiosis with a dissimilar partner called the **Y chromosome.** The female thus produces only X gametes, while the male produces both X and Y gametes. When fertilization involves an X sperm, the result is an XX zygote, which develops into a female; when fertilization involves a Y sperm, the result is an XY zygote, which develops into a male.

The solution to Morgan's puzzle is that the gene causing the white-eye trait in *Drosophila* resides only on the X chromosome—it is absent from the Y chromosome. (We now know that the Y chromosome in flies carries almost no functional genes.) A trait determined by a gene on the X chromosome is said to be **sex-linked.** Knowing the white-eye trait is recessive to the red-eye trait, we can now see that Morgan's result was a natural consequence of the Mendelian assortment of chromosomes (figure 13.30).

Morgan's experiment was one of the most important in the history of genetics because it presented the first clear evidence that the genes determining Mendelian traits do indeed reside on the chromosomes, as Sutton had proposed. The segregation of the white-eye trait has a one-to-one correspondence with the segregation of the X chromosome. In other words, Mendelian traits such as eye color in *Drosophila* assort independently because chromosomes do. When Mendel observed the segregation of alternative traits in pea plants, he was observing a reflection of the meiotic segregation of chromosomes.

Mendelian traits assort independently because they are determined by genes located on chromosomes that assort independently in meiosis.

Genetic Recombination

Morgan's experiments led to the general acceptance of Sutton's chromosomal theory of inheritance. Scientists then attempted to resolve the paradox that there are many more independently assorting Mendelian genes than chromosomes. In 1903 the Dutch geneticist Hugo de Vries suggested that this paradox could be resolved only by assuming that homologous chromosomes exchange elements during meiosis. In 1909, French cytologist F. A. Janssens provided evidence to support this suggestion. Investigating chiasmata produced during amphibian meiosis, Janssens noticed that of the four chromatids involved in each chiasma, two crossed each other and two did not. He suggested that this crossing of chromatids reflected a switch in chromosomal arms between the paternal and maternal homologues, involving one chromatid in each homologue. His suggestion was not accepted widely, primarily because it was difficult to see how two chromatids could break and rejoin at exactly the same position.

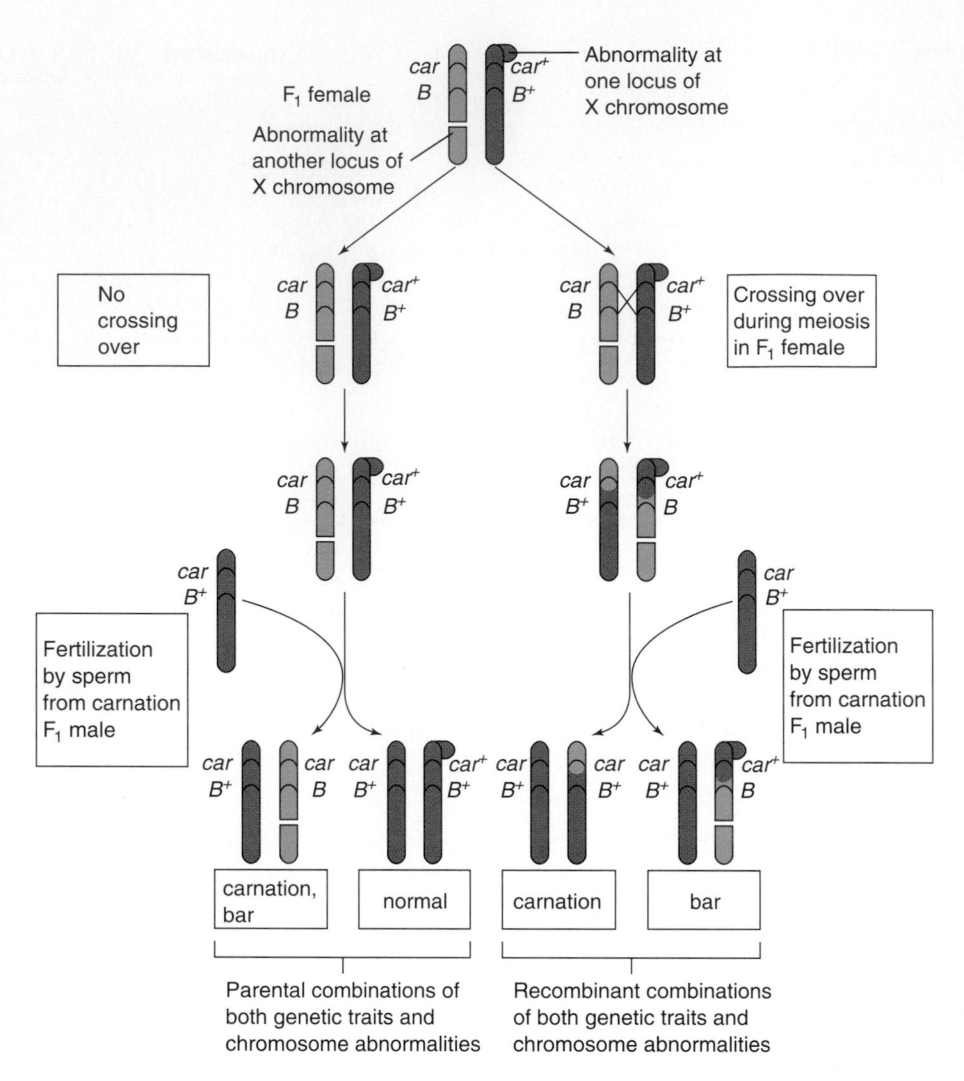

FIGURE 13.31
Stern's experiment demonstrating the physical exchange of chromosomal arms during crossing over. Stern monitored crossing over between two genes, the recessive carnation eye color (*car*) and the dominant bar-shaped eye (*B*), on chromosomes with physical peculiarities visible under a microscope. Whenever these genes recombined through crossing over, the chromosomes recombined as well. Therefore, the recombination of genes reflects a physical exchange of chromosome arms. The "+" notation on the alleles refers to the wild-type allele, the most common allele at a particular gene.

Crossing Over

Later experiments clearly established that Janssens was indeed correct. One of these experiments, performed in 1931 by American geneticist Curt Stern, is described in figure 13.31. Stern studied two sex-linked eye characters in *Drosophila* strains whose X chromosomes were visibly abnormal at both ends. He first examined many flies and identified those in which an exchange had occurred with respect to the two eye characters. He then studied the chromosomes of those flies to see if their X chromosomes had exchanged arms. Stern found that all of the individuals that had exchanged eye traits also possessed chromosomes that had exchanged abnormal ends. The conclusion was inescapable: genetic exchanges of characters such as eye color involve the physical exchange of chromosome arms, a phenomenon called **crossing over.** Crossing over creates new combinations of genes, and is thus a form of **genetic recombination.**

The chromosomal exchanges Stern demonstrated provide the solution to the paradox, because crossing over can occur between homologues anywhere along the length of the chromosome, in locations that seem to be randomly determined. Thus, if two different genes are located relatively far apart on a chromosome, crossing over is more likely to occur somewhere between them than if they are located close together. Two genes can be on the same chromosome and still show independent assortment if they are located so far apart on the chromosome that crossing over occurs regularly between them (figure 13.32).

Using Recombination to Make Genetic Maps

Because crossing over is more frequent between two genes that are relatively far apart than between two that are close together, the frequency of crossing over can be used to map the relative positions of genes on chromosomes. In a cross, the proportion of progeny exhibiting an exchange between two genes is a measure of the frequency of crossover events between them, and thus indicates the relative distance separating them. The results of such crosses can be used to construct a **genetic map** that measures distance between genes in terms of the frequency of recombination. One "map unit" is defined as the distance within which a crossover event is expected to occur in an average of 1% of gametes. A map unit is now called a **centimorgan,** after Thomas Hunt Morgan.

In recent times new technologies have allowed geneticists to create gene maps based on the relative positions of specific gene sequences called *restriction sites* because they are recognized by DNA-cleaving enzymes called restriction endonucleases. Restriction maps, discussed in chapter 19, have largely supplanted genetic recombination maps for detailed gene analysis because they are far easier to produce. Recombination maps remain the method of choice for genes widely separated on a chromosome.

The Three-Point Cross. In constructing a genetic map, one simultaneously monitors recombination among three or more genes located on the same chromosome, referred to as **syntenic** genes. When genes are close enough together on a chromosome that they do not assort independently, they are said to be **linked** to one another. A cross involving three linked genes is called a **three-point cross.** Data obtained by Morgan on traits encoded by genes on the X chromosome of *Drosophila* were used by his student A. H. Sturtevant, to draw the first genetic map (figure 13.33). By convention, the most common allele of a gene is often denoted with the symbol "+" and is designated as **wild type.** All other alleles are denoted with just the specific letters.

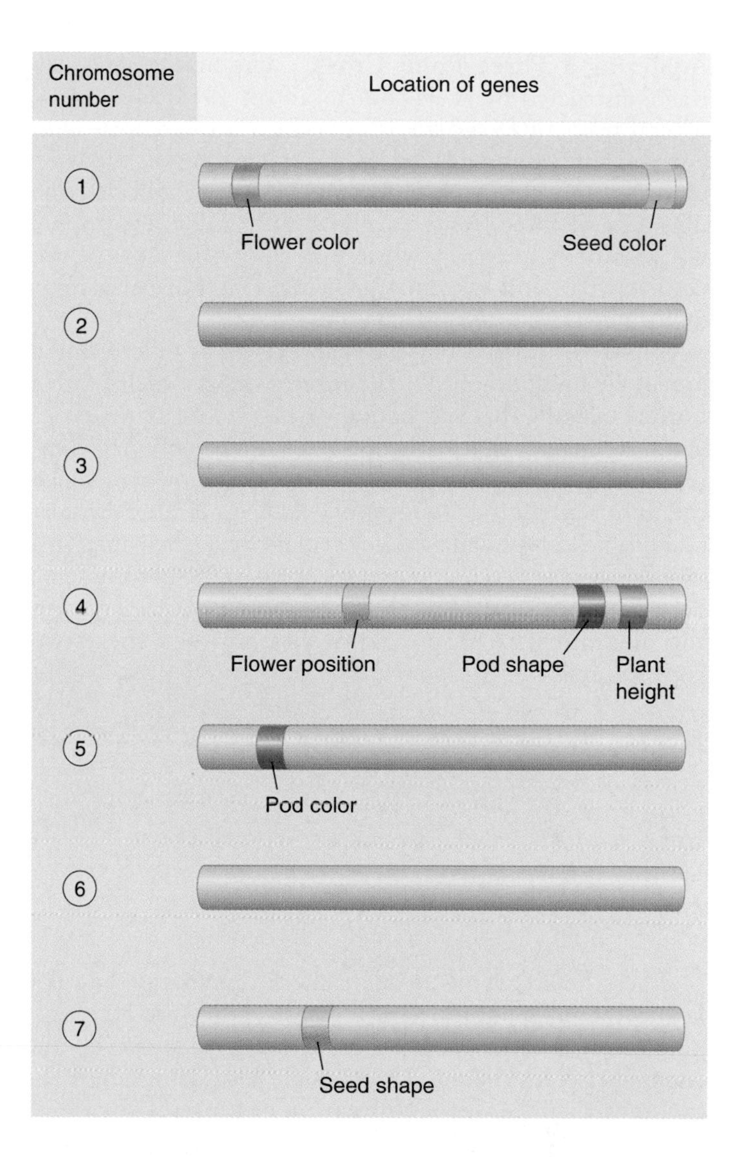

FIGURE 13.32
The chromosomal locations of the seven genes studied by Mendel in the garden pea. The genes for plant height and pod shape are very close to each other and rarely recombine. Plant height and pod shape were not among the characters Mendel examined in dihybrid crosses. One wonders what he would have made of the linkage he surely would have detected had he tested these characters.

Five traits

y	Yellow body color
w	White eye color
v	Vermilion eye color
m	Miniature wing
r	Rudimentary wing

Recombination frequencies

y and *w*	0.010
v and *m*	0.030
v and *r*	0.269
v and *w*	0.300
v and *y*	0.322
w and *m*	0.327
y and *m*	0.355
w and *r*	0.450

Genetic map

.58	*r*
.34	*m*
.31	*v*
.01	*w*
0	*y*

FIGURE 13.33
The first genetic map. This map of the X chromosome of *Drosophila* was prepared in 1913 by A. H. Sturtevant, a student of Morgan. On it he located the relative positions of five recessive traits that exhibited sex linkage by estimating their relative recombination frequencies in genetic crosses. Sturtevant arbitrarily chose the position of the *yellow* gene as zero on his map to provide a frame of reference. The higher the recombination frequency, the farther apart the two genes.

Analyzing a Three-Point Cross. The first genetic map was constructed by A. H. Sturtevant in 1913. He studied several traits of *Drosophila*, all of which exhibited sex linkage and thus were encoded by genes residing on the same chromosome (the X chromosome). Here we will describe his study of three traits: *y*, yellow body color (the normal body color is gray), *w*, white eye color (the normal eye color is red), and *m*, miniature wing (the normal wing is 50% longer).

Sturtevant carried out the mapping cross by crossing a female fly homozygous for the three recessive alleles with a normal male fly that carried none of them. All of the progeny were heterozygotes. Such a cross is conventionally represented by a diagram like the one that follows, in which the lines represent gene locations and + indicates the normal, or "wild-type" allele. Each female fly participating in a cross possesses two homologous copies of the chromosome being mapped, and both chromosomes are represented in the diagram. Crossing over occurs between these two copies in meiosis.

P generation $\dfrac{y\ w\ m}{y\ w\ m}$ × $\dfrac{y^+\ w^+\ m^+}{\longrightarrow}$ (Y chromosome)

↓

F₁ generation females $\dfrac{y\ w\ m}{y^+\ w^+\ m^+}$

These heterozygous females, the F_1 generation, are the key to the mapping procedure. Because they are heterozygous, any crossing over that occurs during meiosis will, if it occurs between where these genes are located, produce gametes with different combinations of alleles for these genes—in other words, recombinant chromosomes. Thus, a crossover between the homologous X chromosomes of such a female in the interval between the *y* and *w* genes will yield recombinant [$y\ w^+$] and [$y^+\ w$] chromosomes, which are different combinations than we started with. In the diagram below, the crossed lines between the chromosomes indicate where the crossover occurs. (In the parental chromosomes of this cross, *w* is always linked with *y* and y^+ linked with w^+.)

$\dfrac{y\ w\ m}{y^+\ w^+\ m^+}$ (crossover between *y* and *w*) → $\dfrac{y\ w^+\ m^+}{y^+\ w\ m}$

In order to see all the recombinant types that might be present among the gametes of these heterozygous flies, Sturtevant conducted a testcross. He crossed female heterozygous flies to males recessive for all three traits and examined the progeny. Because males contribute either a Y chromosome with no genes on it or an X chromosome with recessive alleles at all three loci, the male contribution does not disguise the potentially recombinant female chromosomes.

Table 13.3 summarizes the results Sturtevant obtained. The parentals are represented by the highest number of progeny and the double crossovers (progeny in which two crossovers occurred) by the lowest number. To analyze his data, Sturtevant considered the traits in pairs and determined which involved a crossover event.

1. For the body trait (*y*) and the eye trait (*w*), the first two classes, [$y^+\ w^+$] and [$y\ w$], involve no crossovers (they are parental combinations). In table 13.3, no progeny numbers are tabulated for these two classes on the "body-eye" column (a dash appears instead).
2. The next two classes have the same body-eye combination as the parents, [$y^+\ w^+$] and [$y\ w$], so again no numbers are entered as recombinants under body-eye crossover type.
3. The next two classes, [$y^+\ w$] and [$y\ w^+$], do *not* have the same body-eye combinations as the parent chromosomes, so the observed numbers of progeny are recorded, 16 and 12, respectively.
4. The last two classes also differ from parental chromosomes in body-eye combination, so again the observed numbers of each class are recorded, 1 and 0.
5. The sum of the numbers of observed progeny that are recombinant for body (*y*) and eye (*w*) is 16 + 12 + 1, or 29. Because the total number of progeny is 2205, this represents 29/2205, or 0.01315. The percentage of recombination between *y* and *w* is thus 1.315%, or 1.3 centimorgans.

To estimate the percentage of recombination between eye (*w*) and wing (*m*), one proceeds in the same manner, obtaining a value of 32.608%, or 32.6 centimorgans. Similarly, body (*y*) and wing (*m*) are separated by a recombination distance of 33.832%, or 33.8 centimorgans.

From this, then, we can construct our genetic map. The biggest distance, 33.8 centimorgans, separates the two outside genes, which are evidently *y* and *m*. The gene *w* is between them, near *y*.

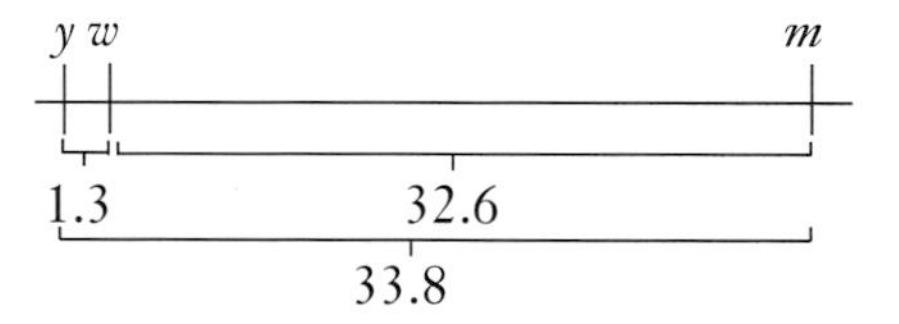

The two distances 1.3 and 32.6 do not add up to 33.8 but rather to 33.9. The difference, 0.1, represents chromosomes in which two crossovers occurred, one between *y* and *w* and another between *w* and *m*. These chromosomes do not exhibit recombination between *y* and *m*.

Genetic maps such as this are key tools in genetic analysis, permitting an investigator reliably to predict how a newly discovered trait, once it has been located on the chromosome map, will recombine with many others.

	Table 13.3 Sturtevant's Results						
	Phenotypes				Crossover Types		
	Body	Eye	Wing	Number of Progeny	Body-Eye	Eye-Wing	Body-Wing
Parental	y^+	w^+	m^+	758	—	—	—
	y	w	m	700	—	—	—
Single crossover	y^+	w^+	m	401	—	401	401
	y	w	m^+	317	—	317	317
	y^+	w	m	16	16	—	16
	y	w^+	m^+	12	12	—	12
Double crossover	y^+	w	m^+	1	1	1	—
	y	w^+	m	0	0	0	—
TOTAL				2205	29	719	746
Recombination frequency (%)					1.315	32.608	33.832

The Human Genetic Map

Genetic maps of human chromosomes (figure 13.34) are of great importance. Knowing where particular genes are located on human chromosomes can often be used to tell whether a fetus at risk of inheriting a genetic disorder actually has the disorder. The genetic-engineering techniques described in chapter 19 have begun to permit investigators to isolate specific genes and determine their nucleotide sequences. It is hoped that knowledge of differences at the gene level may suggest successful therapies for particular genetic disorders and that knowledge of a gene's location on a chromosome will soon permit the substitution of normal alleles for dysfunctional ones. Because of the great potential of this approach, investigators have sequenced the entire human genome, the **Human Genome Project,** described in chapter 18. Initially, this sequence allows researchers to rapidly screen a "library" of some 30,000 genes. Investigators wishing to study a particular gene can use techniques described in chapter 19 to screen a library of gene fragments and rapidly determine which fragment carries the gene of interest. They will then be able to analyze that fragment in detail. In parallel with this mammoth undertaking, the other smaller genomes have already been sequenced, including those of namatodes, *Drosophila*, mice, the plant *Arabidopsis*, yeasts and many bacteria.

Gene maps locate the relative positions of different genes on the chromosomes of an organism. Traditionally produced by analyzing the relative amounts of recombination in genetic crosses, gene maps are increasingly being made by analyzing the sizes of fragments made by restriction enzymes.

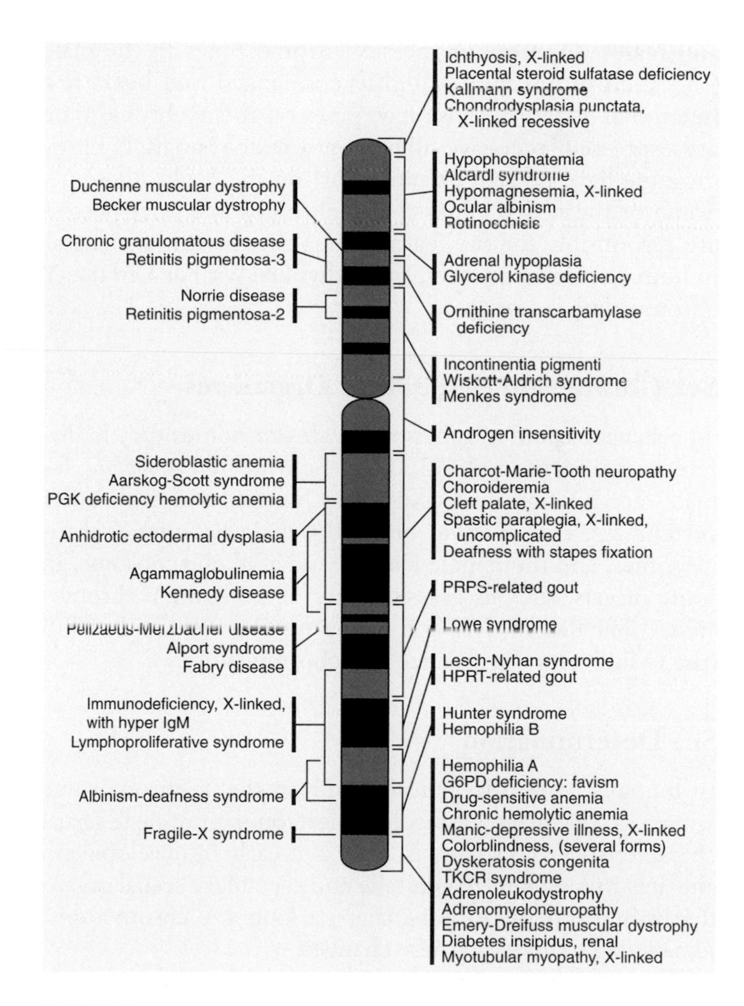

FIGURE 13.34
The human X chromosome gene map. Over 59 diseases have been traced to specific segments of the X chromosome. Many of these disorders are also influenced by genes on other chromosomes.

Human Chromosomes

Each human somatic cell normally has 46 chromosomes, which in meiosis form 23 pairs. By convention, the chromosomes are divided into seven groups (designated A through G), each characterized by a different size, shape, and appearance. The differences among the chromosomes are most clearly visible when the chromosomes are arranged in order in a karyotype (figure 13.35). Techniques that stain individual segments of chromosomes with different-colored dyes make the identification of chromosomes unambiguous. Like a fingerprint, each chromosome always exhibits the same pattern of colored bands.

Human Sex Chromosomes

Of the 23 pairs of human chromosomes, 22 are perfectly matched in both males and females and are called **autosomes.** The remaining pair, the **sex chromosomes,** consist of two similar chromosomes in females and two dissimilar chromosomes in males. In humans, females are designated XX and males XY. One of the sex chromosomes in the male (the Y chromosome) is highly condensed and bears few functional genes. Because few genes on the Y chromosome are expressed, recessive alleles on a male's single X chromosome have no *active* counterpart on the Y chromosome. Some of the active genes the Y chromosome does possess are responsible for the features associated with "maleness" in humans. Consequently, any individual with *at least* one Y chromosome is a male.

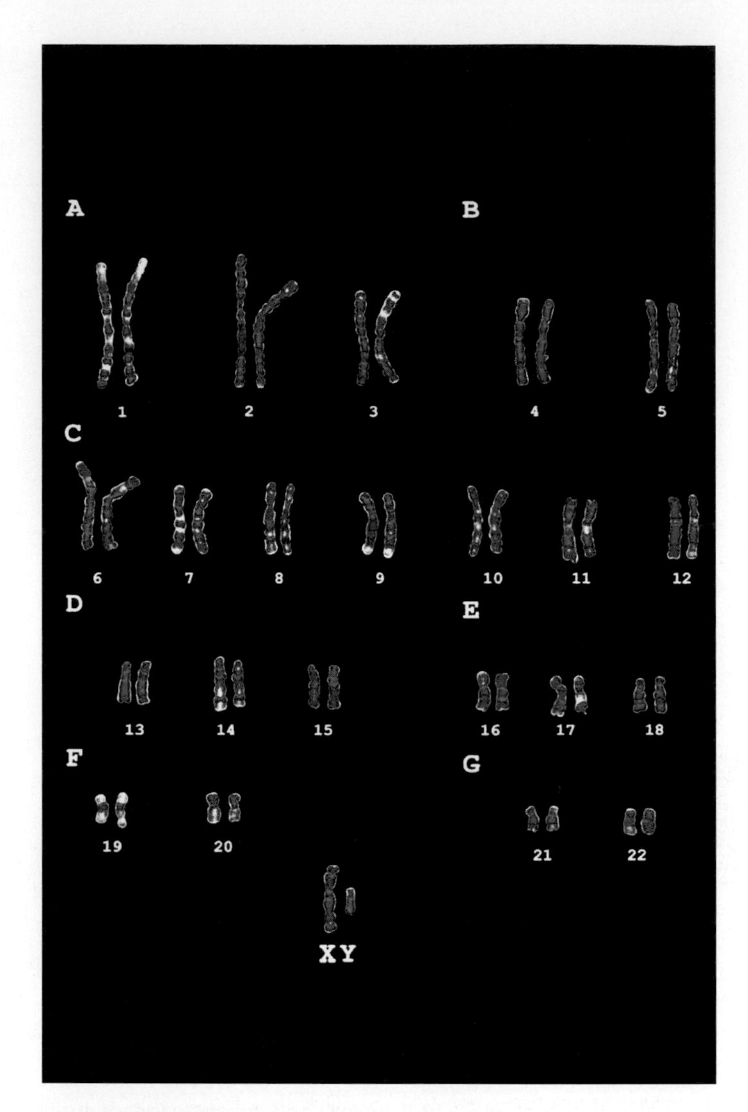

FIGURE 13.35
A human karyotype. This karyotype shows the colored banding patterns, arranged by class A–G.

Sex Chromosomes in Other Organisms

The structure and number of sex chromosomes vary in different organisms (table 13.4). In the fruit fly *Drosophila*, females are XX and males XY, as in humans and most other vertebrates. However, in birds, the male has two Z chromosomes, and the female has a Z and a W chromosome. In some insects, such as grasshoppers, there is no Y chromosome—females are XX and males are characterized as XO (the O indicates the absence of a chromosome).

Table 13.4 Sex Determination in Some Organisms

	Female	Male
Humans, *Drosophila*	XX	XY
Birds	ZW	ZZ
Grasshoppers	XX	XO
Honeybees	Diploid	Haploid

Sex Determination

In humans a specific gene located on the Y chromosome known as *SRY* plays a key role in development of male sexual characteristics. This gene is expressed early in development, and acts to masculinize genitalia and secondary sexual organs that would otherwise be female. Lacking a Y chromosome, females fail to undergo these changes.

Among fishes and in some species of reptiles, environmental changes can cause changes in the expression of this sex-determining gene, and thus of the sex of the adult individual.

Barr Bodies

Although males have only one copy of the X chromosome and females have two, female cells do not produce twice as much of the proteins encoded by genes on the X chromosome. Instead, one of the X chromosomes in females is inactivated early in embryonic development, shortly after the embryo's sex is determined. Which X chromosome is inactivated varies randomly from cell to cell. If a woman is heterozygous for a sex-linked trait, some of her cells will express one allele and some the other. The inactivated and highly condensed X chromosome is visible as a darkly staining **Barr body** attached to the nuclear membrane (figure 13.36).

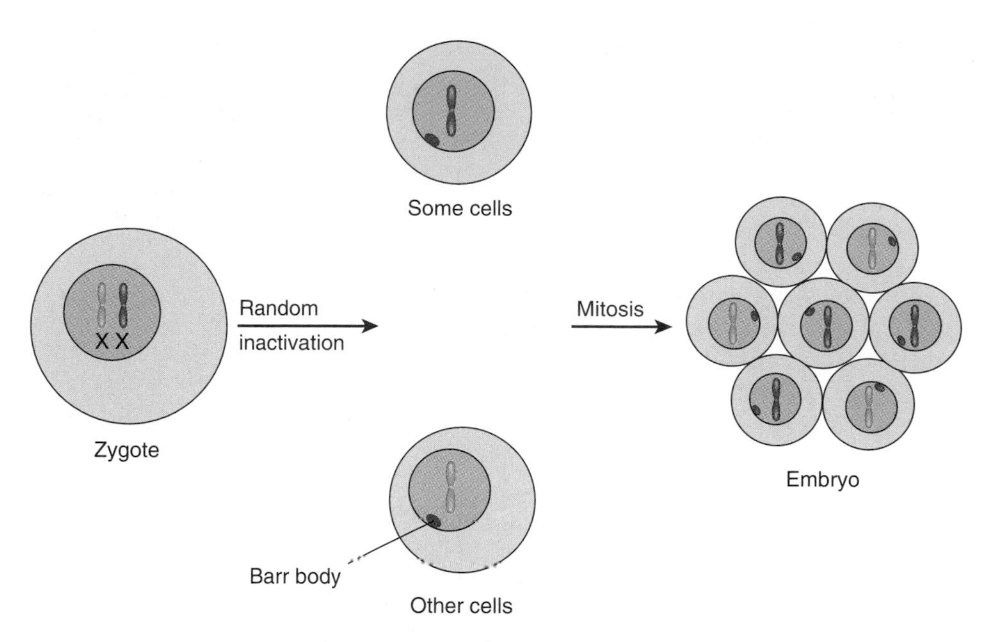

FIGURE 13.36
Barr bodies. In the developing female embryo, one of the X chromosomes (determined randomly) condenses and becomes inactivated. These condensed X chromosomes, called Barr bodies, then attach to the nuclear membrane.

X-inactivation is not restricted to humans. The patches of color on tortoiseshell and calico cats are a familiar result of this process. The gene for orange coat color is located on the X chromosome. The *O* allele specifies orange fur, and the *o* allele specifies black fur. Early in development, one X chromosome is inactivated in the cells that will become skin cells. If the remaining active X carries the *O* allele, then the patch of skin that results from that cell will have orange fur. If it carries the *o* allele, then the fur will be black. Because X-inactivation is a random process, the orange and black patches appear randomly in the cat's coat. Because only females have two copies of the X chromosome, only they can be heterozygous at the *O* gene, so almost all calico cats are females (figure 13.37). The exception is male cats that have the genotype XXY; the XXY genotype is discussed in the next section. The white on a calico cat is due to the action of an allele at another gene, the white spotting gene.

One of the 23 pairs of human chromosomes carries the genes that determine sex. The gene determining maleness is located on a version of the sex chromosome called Y, which has few other transcribed genes.

FIGURE 13.37
A calico cat. The coat coloration of this cat is due to the random inactivation of her X chromosome during early development. The female is heterozygous for orange coat color, but because only one coat color allele is expressed, she exhibits patches of orange and black fur.

Human Abnormalities Due to Alterations in Chromosome Number

Occasionally, homologues or sister chromatids fail to separate properly in meiosis, leading to the acquisition or loss of a chromosome in a gamete. This condition, called **primary nondisjunction,** can result in individuals with severe abnormalities if the affected gamete forms a zygote.

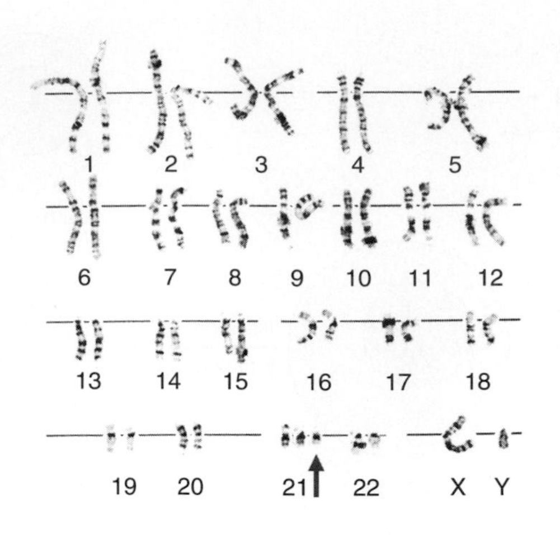

FIGURE 13.38
Down syndrome. As shown in this male karyotype, Down syndrome is associated with trisomy of chromosome 21. A child with Down syndrome sitting on his father's knee.

Nondisjunction Involving Autosomes

Almost all humans of the same sex have the same karyotype, for the same reason that all automobiles have engines, transmissions, and wheels: other arrangements don't work well. Humans who have lost even one copy of an autosome (called **monosomics**) do not survive development. In all but a few cases, humans who have gained an extra autosome (called **trisomics**) also do not survive. However, five of the smallest autosomes—those numbered 13, 15, 18, 21, and 22—can be present in humans as three copies and still allow the individual to survive for a time. The presence of an extra chromosome 13, 15, or 18 causes severe developmental defects, and infants with such a genetic makeup die within a few months. In contrast, individuals who have an extra copy of chromosome 21 or, more rarely, chromosome 22, usually survive to adulthood. In such individuals, the maturation of the skeletal system is delayed, so they generally are short and have poor muscle tone. Their mental development is also affected, and children with trisomy 21 or trisomy 22 are always mentally retarded.

Down Syndrome. The developmental defect produced by trisomy 21 (figure 13.38) was first described in 1866 by J. Langdon Down; for this reason, it is called **Down syndrome** (formerly "Down's syndrome"). About 1 in every 750 children exhibits Down syndrome, and the frequency is similar in all racial groups. Similar conditions also occur in chimpanzees and other related primates. In humans, the defect is associated with a particular small portion of chromosome 21. When this chromosomal segment is present in three copies instead of two, Down syndrome results. In 97% of the human cases examined, all of chromosome 21 is present in three copies. In the other 3%, a small portion of chromosome 21 containing the critical segment has been added to another chromosome by a process called *translocation* (see chapter 18); it exists along with the normal two copies of chromosome 21. This condition is known as *translocation Down syndrome.*

Not much is known about the developmental role of the genes whose extra copies produces Down syndrome, although clues are beginning to emerge from current research. Some researchers suspect that the gene or genes that produce Down syndrome are similar or identical to some of the genes associated with cancer and with Alzheimer's disease. The reason for this suspicion is that one of the human cancer-causing genes (to be described in chapter 18) and the gene causing Alzheimer's disease are located on the segment of chromosome 21 associated with Down syndrome. Moreover, cancer is more common in children with Down syndrome. The incidence of leukemia, for example, is 11 times higher in children with Down syndrome than in unaffected children of the same age.

How does Down syndrome arise? In humans, it comes about almost exclusively as a result of primary nondisjunction of chromosome 21 during egg formation. The cause of these primary nondisjunctions is not known, but their incidence, like that of cancer, increases with age (figure 13.39). In mothers younger than 20 years of age, the risk of giving birth to a child with Down syndrome is about 1 in 1700; in mothers 20 to 30 years old, the risk is only about 1 in 1400. In mothers 30 to 35 years old, however, the risk rises to 1 in 750, and by age 45, the risk is as high as 1 in 16!

Primary nondisjunctions are far more common in women than in men because all of the eggs a woman will ever produce have developed to the point of prophase in meiosis I by the time she is born. By the time she has children, her eggs are as old as she is. In contrast, men produce new sperm daily. Therefore, there is a much greater chance for problems of various kinds, including those that cause primary nondisjunction, to accumulate over time in the gametes of women than in those of men. For this reason, the age of the mother is more critical than that of the father in couples contemplating childbearing.

Nondisjunction Involving the Sex Chromosomes

Individuals that gain or lose a sex chromosome do not generally experience the severe developmental abnormalities caused by similar changes in autosomes. Such individuals may reach maturity, but they have somewhat abnormal features.

The X Chromosome. When X chromosomes fail to separate during meiosis, some of the gametes that are produced possess both X chromosomes and so are XX gametes; the other gametes that result from such an event have no sex chromosome and are designated "O" (figure 13.40).

If an XX gamete combines with an X gamete, the resulting XXX zygote develops into a female with one functional X chromosome and two Barr bodies. She is sterile but usually normal in other respects. If an XX gamete instead combines with a Y gamete, the effects are more serious. The resulting XXY zygote develops into a sterile male who has many female body characteristics and, in some cases, diminished mental capacity. This condition, called *Klinefelter syndrome*, occurs in about 1 out of every 500 male births.

If an O gamete fuses with a Y gamete, the resulting OY zygote is nonviable and fails to develop further because humans cannot survive when they lack the genes on the X chromosome. If, on the other hand, an O gamete fuses with an X gamete, the XO zygote develops into a sterile female of short stature, with a webbed neck and immature sex organs that do not undergo changes during puberty. The mental abilities of an XO individual are in the low-normal range. This condition, called *Turner syndrome*, occurs roughly once in every 5000 female births.

The Y Chromosome. The Y chromosome can also fail to separate in meiosis, leading to the formation of YY gametes. When these gametes combine with X gametes, the XYY zygotes develop into fertile males of normal appearance. The frequency of the XYY genotype (Jacob's syndrome) is about 1 per 1000 newborn males, but it is approximately 20 times higher among males in penal and mental institutions. This observation has led to the highly controversial suggestion that XYY males are inherently antisocial, a suggestion supported by some studies but not by others. In any case, most XYY males do not develop patterns of antisocial behavior.

Gene dosage plays a crucial role in development, so humans do not tolerate the loss or addition of chromosomes well. Autosome loss is always lethal, and an extra autosome is with few exceptions lethal too. Additional sex chromosomes have less serious consequences, although they can lead to sterility.

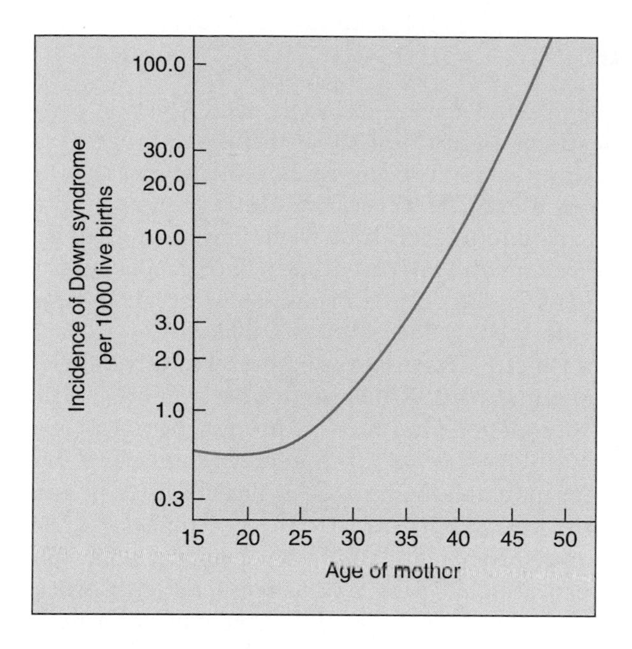

FIGURE 13.39
Correlation between maternal age and the incidence of Down syndrome. As women age, the chances they will bear a child with Down syndrome increase. After a woman reaches 35, the frequency of Down syndrome increases rapidly.

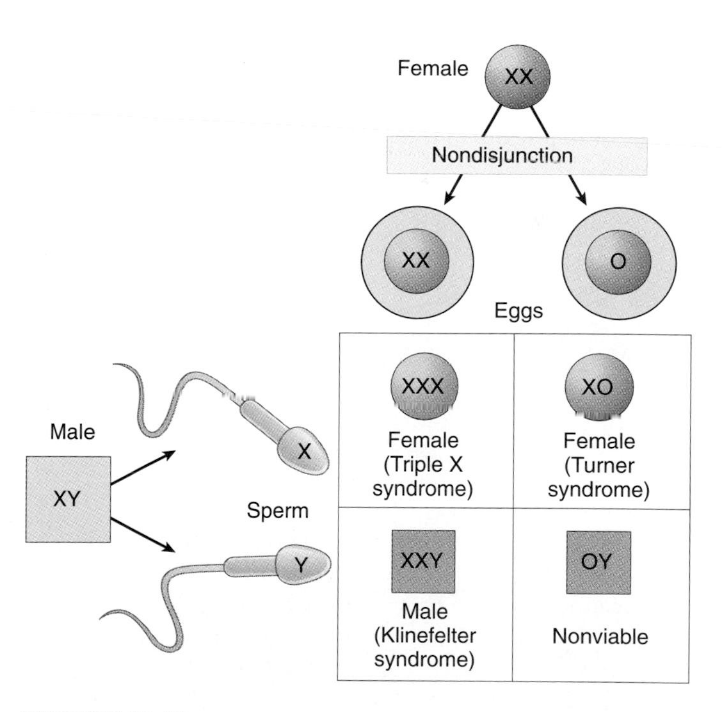

FIGURE 13.40
How nondisjunction can produce abnormalities in the number of sex chromosomes. When nondisjunction occurs in the production of female gametes, the gamete with two X chromosomes (XX) produces Klinefelter males (XXY) and XXX females. The gamete with no X chromosome (O) produces Turner females (XO) and nonviable OY males lacking any X chromosome.

Genetic Counseling

Although most genetic disorders cannot yet be cured, we are learning a great deal about them, and progress toward successful therapy is being made in many cases. In the absence of a cure, however, the only recourse is to try to avoid producing children with these conditions. The process of identifying parents at risk of producing children with genetic defects and of assessing the genetic state of early embryos is called *genetic counseling.*

If a genetic defect is caused by a recessive allele, how can potential parents determine the likelihood that they carry the allele? One way is through pedigree analysis, often employed as an aid in genetic counseling. By analyzing a person's pedigree, it is sometimes possible to estimate the likelihood that the person is a carrier for certain disorders. For example, if one of your relatives has been afflicted with a recessive genetic disorder such as cystic fibrosis, it is possible that you are a heterozygous carrier of the recessive allele for that disorder. When a couple is expecting a child, and pedigree analysis indicates that both of them have a significant probability of being heterozygous carriers of a recessive allele responsible for a serious genetic disorder, the pregnancy is said to be a **high-risk pregnancy.** In such cases, there is a significant probability that the child will exhibit the clinical disorder.

Another class of high-risk pregnancies is that in which the mothers are more than 35 years old. As we have seen, the frequency of birth of infants with Down syndrome increases dramatically in the pregnancies of older women (see figure 13.39).

When a pregnancy is diagnosed as being high-risk, many women elect to undergo *amniocentesis,* a procedure that permits the prenatal diagnosis of many genetic disorders. In the fourth month of pregnancy, a sterile hypodermic needle is inserted into the expanded uterus of the mother, removing a small sample of the amniotic fluid bathing the fetus (figure 13.41). Within the fluid are free-floating cells derived from the fetus; once removed, these cells can be grown in cultures in the laboratory. During amniocentesis, the position of the needle and that of the fetus are usually observed by means of *ultrasound.* The sound waves used in ultrasound are not harmful to mother or fetus, and they permit the person withdrawing the amniotic fluid to do so without damaging the fetus. In addition, ultrasound can be used to examine the fetus for signs of major abnormalities.

In recent years, physicians have increasingly turned to a new, less invasive procedure for genetic screening called **chorionic villi sampling.** In this procedure, the physician removes cells from the chorion, a membranous part of the placenta that nourishes the fetus. This procedure can be used earlier in pregnancy (by the eighth week) and yields results much more rapidly than does amniocentesis.

To test for certain genetic disorders, genetic counselors can look for three things in the cultures of cells obtained from amniocentesis or chorionic villi sampling. First, analysis of the karyotype can reveal aneuploidy (extra or missing chromosomes) and gross chromosomal alterations. Second, in many cases it is possible to test directly for the *proper functioning of enzymes* involved in genetic disorders. The lack of normal enzymatic activity signals the presence of the disorder. Thus, the lack of the enzyme responsible for breaking down phenylalanine signals PKU (phenylke-

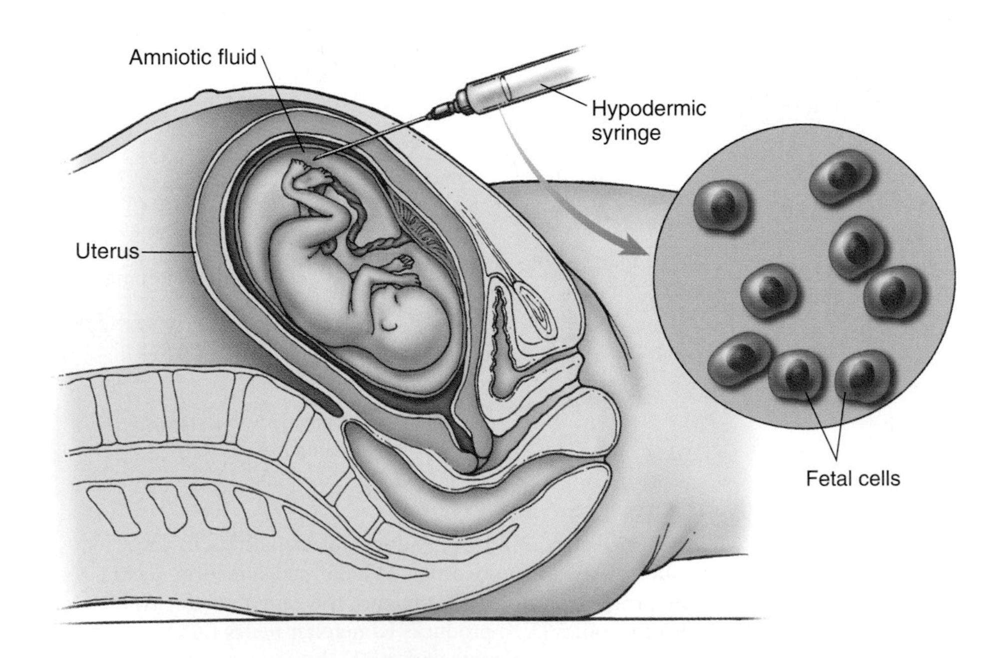

FIGURE 13.41
Amniocentesis. A needle is inserted into the amniotic cavity, and a sample of amniotic fluid, containing some free cells derived from the fetus, is withdrawn into a syringe. The fetal cells are then grown in culture and their karyotype and many of their metabolic functions are examined.

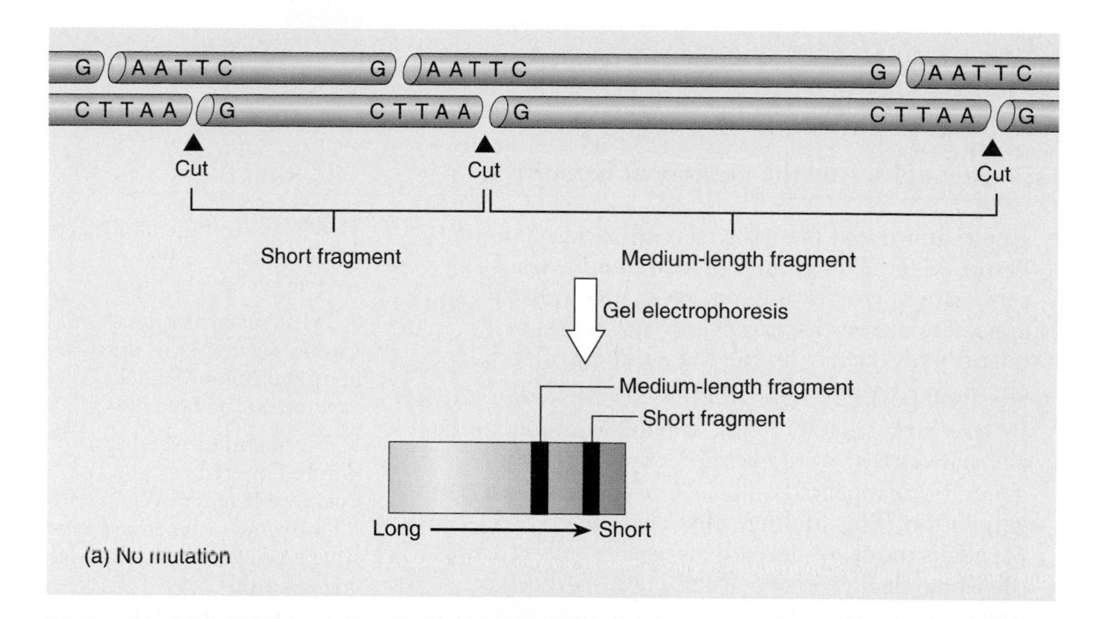

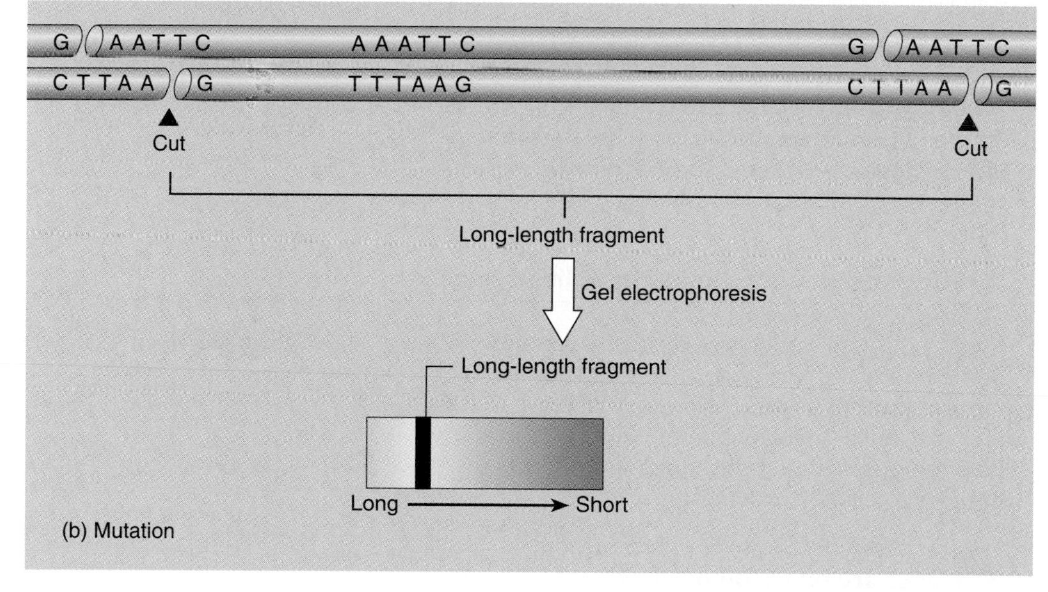

FIGURE 13.42
RFLPs. Restriction fragment length polymorphisms (RFLPs) are playing an increasingly important role in genetic identification. In (*a*), the restriction endonuclease cuts the DNA molecule in three places, producing two fragments. In (*b*), the mutation of a single nucleotide from G to A (see top fragment) alters a restriction endonuclease cutting site. Now the enzyme no longer cuts the DNA molecule at that site. As a result, a single long fragment is obtained, rather than two shorter ones. Such a change is easy to detect when the fragments are subjected to a technique called gel electrophoresis.

tonuria), the absence of the enzyme responsible for the breakdown of gangliosides indicates Tay-Sachs disease, and so forth.

Third, genetic counselors can look for an *association with known genetic markers.* For sickle cell anemia, Huntington's disease, and one form of muscular dystrophy (a genetic disorder characterized by weakened muscles), investigators have found other mutations on the same chromosomes that, by chance, occur at about the same place as the mutations that cause those disorders. By testing for the presence of these other mutations, a genetic counselor can identify individuals with a high probability of possessing the disorder-causing mutations. Finding such mutations in the first place is a little like searching for a needle in a haystack, but persistent efforts have proved successful in these three disorders. The associated mutations are detectable because they alter the length of the DNA segments that restriction enzymes produce when they cut strands of DNA at particular places (see chapter 19). Therefore, these mutations produce what are called **restriction fragment length polymorphisms,** or **RFLPs** (figure 13.42).

Many gene defects can be detected early in pregnancy, allowing for appropriate planning by the prospective parents.

Chapter 13

www.mhhe.com/raven6/resources13.mhtml

Summary	Questions	Media Resources

13.1 Mendel solved the mystery of heredity.

- Koelreuter noted the basic facts of heredity a century before Mendel. He found that alternative traits segregate in crosses and may mask each other's appearance. Mendel, however, was the first to quantify his data, counting the numbers of each alternative type among the progeny of crosses.
- By counting progeny types, Mendel learned that the alternatives that were masked in hybrids (the F_1 generation) appeared only 25% of the time in the F_2 generation. This finding, which led directly to Mendel's model of heredity, is usually referred to as the Mendelian ratio of 3:1 dominant-to-recessive traits.
- When two genes are located on different chromosomes, the alleles assort independently.
- Because phenotypes are often influenced by more than one gene, the ratios of alternative phenotypes observed in crosses sometimes deviate from the simple ratios predicted by Mendel.

1. Why weren't the implications of Koelreuter's results recognized for a century?

2. What characteristics of the garden pea made this organism a good choice for Mendel's experiments on heredity?

3. To determine whether a purple-flowered pea plant of unknown genotype is homozygous or heterozygous, what type of plant should it be crossed with?

4. In a dihybrid cross between two heterozygotes, what fraction of the offspring should be homozygous recessive for both traits?

- Introduction to Classic Genetics
- Monohybrid Cross
- Dihybrid Cross

- How Scientists Think: Probability and Hypothesis Testing in Biology (Appendix)

- Art Quizzes:
  - -Mendel's Expeimental Results
  - -Testcross
  - -Dihybrid Cross
  - -Continuous Variation
  - -Incomplete Dominance
  - -Epistasis

13.2 Human genetics follows Mendelian principles.

- Some genetic disorders are relatively common in human populations; others are rare. Many of the most important genetic disorders are associated with recessive alleles, which are not eliminated from the human population, even though their effects in homozygotes may be lethal.

5. Why is Huntington's disease maintained at its current frequency in human populations?

- *On Science* Article: Advances in Gene Therapy

13.3 Genes are on chromosomes.

- The first clear evidence that genes reside on chromosomes was provided by Thomas Hunt Morgan, who demonstrated that the segregation of the white-eye trait in *Drosophila* is associated with the segregation of the X chromosome, which is involved in sex determination.
- The first genetic evidence that crossing over occurs between chromosomes was provided by Curt Stern, who showed that when two Mendelian traits exchange during a cross, so do visible abnormalities on the ends of the chromosomes bearing those traits.
- The frequency of crossing over between genes can be used to construct genetic maps.
- Primary nondisjunction results when chromosomes do not separate during meiosis, leading to gametes with missing or extra chromosomes. In humans, the loss of an autosome is invariably fatal.

6. When Morgan crossed a white-eyed male fly with a normal red-eyed female, and then crossed two of the red-eyed progeny, about ¼ of the offspring were white-eyed—but they were ALL male! Why?

7. What is primary nondisjunction? How is it related to Down syndrome?

8. Is an individual with Klinefelter syndrome genetically male or female? Why?

- Explorations:
  - -Constructing a Genetic Map
  - -Down Syndrome
  - - Making a Restriction Map

- Introduction to Chromosomes
- Sex Chromosomes
- Recombination
- Abnormal Chromosomes

- Art Quizzes:
  - -Barr Bodies
  - -Nondisjunction and Sex Chromosomes

Mendelian Genetics Problems

1. The illustration below describes Mendel's cross of *wrinkled* and *round* seed characters. (Hint: Do you expect all the seeds in a pod to be the same?) What is wrong with this diagram?

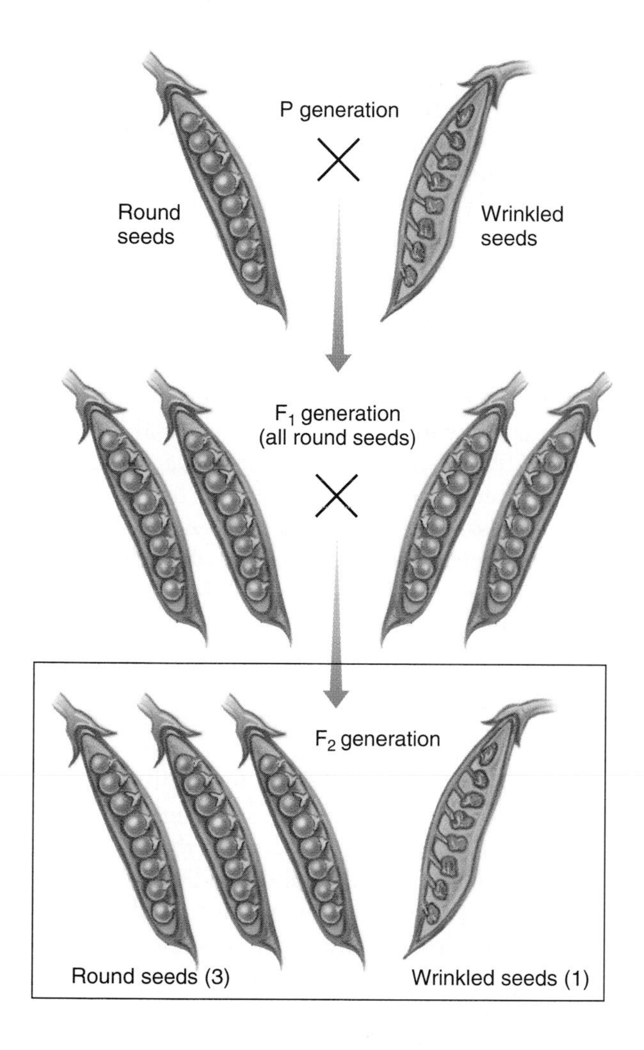

2. The annual plant *Haplopappus gracilis* has two pairs of chromosomes (1 and 2). In this species, the probability that two characters *a* and *b* selected at random will be on the same chromosome is equal to the probability that they will both be on chromosome 1 ($\frac{1}{2} \times \frac{1}{2} = \frac{1}{4}$, or 0.25), plus the probability that they will both be on chromosome 2 (also $\frac{1}{2} \times \frac{1}{2} = \frac{1}{4}$, or 0.25), for an overall probability of $\frac{1}{2}$, or 0.5. In general, the probability that two randomly selected characters will be on the same chromosome is equal to $\frac{1}{n}$ where n is the number of chromosome pairs. Humans have 23 pairs of chromosomes. What is the probability that any two human characters selected at random will be on the same chromosome?
3. Among Hereford cattle there is a dominant allele called *polled;* the individuals that have this allele lack horns. Suppose you acquire a herd consisting entirely of polled cattle, and you carefully determine that no cow in the herd has horns. Some of the calves born that year, however, grow horns. You remove them from the herd and make certain that no horned adult has gotten into your pasture. Despite your efforts, more horned calves are born the next year. What is the reason for the appearance of the horned calves? If your goal is to maintain a herd consisting entirely of polled cattle, what should you do?
4. An inherited trait among humans in Norway causes affected individuals to have very wavy hair, not unlike that of a sheep. The trait, called *woolly,* is very evident when it occurs in families; no child possesses woolly hair unless at least one parent does. Imagine you are a Norwegian judge, and you have before you a woolly-haired man suing his normal haired wife for divorce because their first child has woolly hair but their second child has normal hair. The husband claims this constitutes evidence of his wife's infidelity. Do you accept his claim? Justify your decision.
5. In human beings, Down syndrome, a serious developmental abnormality, results from the presence of three copies of chromosome 21 rather than the usual two copies. If a female exhibiting Down syndrome mates with a normal male, what proportion of her offspring would you expect to be affected?
6. Many animals and plants bear recessive alleles for *albinism,* a condition in which homozygous individuals lack certain pigments. An albino plant, for example, lacks chlorophyll and is white, and an albino human lacks melanin. If two normally pigmented persons heterozygous for the same albinism allele marry, what proportion of their children would you expect to be albino?
7. You inherit a racehorse and decide to put him out to stud. In looking over the stud book, however, you discover that the horse's grandfather exhibited a rare disorder that causes brittle bones. The disorder is hereditary and results from homozygosity for a recessive allele. If your horse is heterozygous for the allele, it will not be possible to use him for stud because the genetic defect may be passed on. How would you determine whether your horse carries this allele?
8. In the fly *Drosophila,* the allele for dumpy wings (d) is recessive to the normal long-wing allele (d^+), and the allele for white eye (w) is recessive to the normal red-eye allele (w^+). In a cross of $d^+d^+w^+w \times d^+dww$, what proportion of the offspring are expected to be "normal" (long wings, red eyes)? What proportion are expected to have dumpy wings and white eyes?
9. Your instructor presents you with a *Drosophila* with red eyes, as well as a stock of white-eyed flies and another stock of flies homozygous for the red-eye allele. You know that the presence of white eyes in *Drosophila* is caused by homozygosity for a recessive allele. How would you determine whether the single red-eyed fly was heterozygous for the white-eye allele?

Mendelian Genetics Problems (continued)

10. Some children are born with recessive traits (and, therefore, must be homozygous for the recessive allele specifying the trait), even though neither of the parents exhibits the trait. What can account for this?
11. You collect two individuals of *Drosophila*, one a young male and the other a young, unmated female. Both are normal in appearance, with the red eyes typical of *Drosophila*. You keep the two flies in the same bottle, where they mate. Two weeks later, the offspring they have produced all have red eyes. From among the offspring, you select 100 individuals, some male and some female. You cross each individually with a fly you know to be homozygous for the recessive allele *sepia*, which produces black eyes when homozygous. Examining the results of your 100 crosses, you observe that in about half of the crosses, only red-eyed flies were produced. In the other half, however, the progeny of each cross consists of about 50% red-eyed flies and 50% black-eyed flies. What were the genotypes of your original two flies?
12. Hemophilia is a recessive sex-linked human blood disease that leads to failure of blood to clot normally. One form of hemophilia has been traced to the royal family of England, from which it spread throughout the royal families of Europe. For the purposes of this problem, assume that it originated as a mutation either in Prince Albert or in his wife, Queen Victoria.
    a. Prince Albert did not have hemophilia. If the disease is a sex-linked recessive abnormality, how could it have originated in Prince Albert, a male, who would have been expected to exhibit sex-linked recessive traits?
    b. Alexis, the son of Czar Nicholas II of Russia and Empress Alexandra (a granddaughter of Victoria), had hemophilia, but their daughter Anastasia did not. Anastasia died, a victim of the Russian revolution, before she had any children. Can we assume that Anastasia would have been a carrier of the disease? Would your answer be different if the disease had been present in Nicholas II or in Alexandra?
13. In 1986, *National Geographic* magazine conducted a survey of its readers' abilities to detect odors. About 7% of Caucasians in the United States could not smell the odor of musk. If neither parent could smell musk, none of their children were able to smell it. On the other hand, if the two parents could smell musk, their children generally could smell it, too, but a few of the children in those families were unable to smell it. Assuming that a single pair of alleles governs this trait, is the ability to smell musk best explained as an example of dominant or recessive inheritance?
14. A couple with a newborn baby is troubled that the child does not resemble either of them. Suspecting that a mix-up occurred at the hospital, they check the blood type of the infant. It is type O. As the father is type A and the mother type B, they conclude a mix-up must have occurred. Are they correct?
15. Mabel's sister died of cystic fibrosis as a child. Mabel does not have the disease, and neither do her parents. Mabel is pregnant with her first child. If you were a genetic counselor, what would you tell her about the probability that her child will have cystic fibrosis?
16. How many chromosomes would you expect to find in the karyotype of a person with Turner syndrome?
17. A woman is married for the second time. Her first husband has blood type A and her child by that marriage has type O. Her new husband has type B blood, and when they have a child its blood type is AB. What is the woman's blood genotype and blood type?
18. Two intensely freckled parents have five children. Three eventually become intensely freckled and two do not. Assuming this trait is governed by a single pair of alleles, is the expression of intense freckles best explained as an example of dominant or recessive inheritance?
19. Total color blindness is a rare hereditary disorder among humans. Affected individuals can see no colors, only shades of gray. It occurs in individuals homozygous for a recessive allele, and it is not sex-linked. A man whose father is totally color blind intends to marry a woman whose mother is totally color blind. What are the chances they will produce offspring who are totally color blind?
20. A normally pigmented man marries an albino woman. They have three children, one of whom is an albino. What is the genotype of the father?
21. Four babies are born in a hospital, and each has a different blood type: A, B, AB, and O. The parents of these babies have the following pairs of blood groups: A and B, O and O, AB and O, and B and B. Which baby belongs to which parents?
22. A couple both work in an atomic energy plant, and both are exposed daily to low-level background radiation. After several years, they have a child who has Duchenne muscular dystrophy, a recessive genetic defect caused by a mutation on the X chromosome. Neither the parents nor the grandparents have the disease. The couple sue the plant, claiming that the abnormality in their child is the direct result of radiation-induced mutation of their gametes, and that the company should have protected them from this radiation. Before reaching a decision, the judge hearing the case insists on knowing the sex of the child. Which sex would be more likely to result in an award of damages, and why?

Part **V**

Molecular Genetics

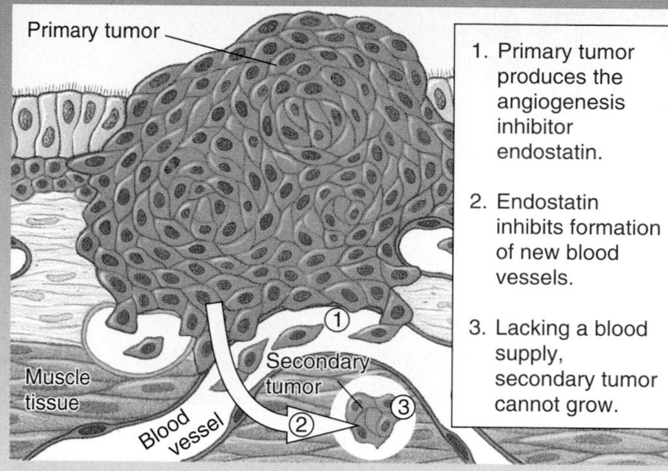
Primary tumor
1. Primary tumor produces the angiogenesis inhibitor endostatin.
2. Endostatin inhibits formation of new blood vessels.
3. Lacking a blood supply, secondary tumor cannot grow.
Muscle tissue
Blood vessel
Secondary tumor
1
2
3

How primary tumors kill off the competition. Tumors require an ample blood supply to fuel their growth. The growth of new blood vessels is called angiogenesis. Inhibiting angiogenesis offers a possible way to block tumor growth.

Can Cancer Tumors Be Starved to Death?

One of the most exciting recent developments in the war against cancer is the report that it might be possible to starve cancer tumors to death. Many laboratories have begun to look into this possibility, although it's not yet clear that the approach will actually work to cure cancer. One of the most exciting and frustrating things about watching a developing science story like this one is that you can't flip ahead and read the ending—in the real world of research, you never know how things are going to turn out.

This story starts when a Harvard University researcher, Dr. Judah Folkman, followed up on a familiar observation made by many oncologists (cancer specialists), that removal of a primary tumor often leads to more rapid growth of secondary tumors. "Perhaps," Folkman reasoned, "the primary tumor is producing some substance that inhibits the growth of the other tumors." Such a substance could be a powerful weapon against cancer.

Folkman set out to see if he could isolate a chemical from primary tumors that inhibited the growth of secondary ones. Three years ago he announced he had found two. He called them angiostatin and endostatin.

To understand how these two proteins work, put yourself in the place of a tumor. To grow, a tumor must obtain from the body's blood supply all the food and nutrients it needs to make more cancer cells. To facilitate this necessary grocery shopping, tumors leak out substances into the surrounding tissues that encourage angiogenesis, the formation of small blood vessels. This call for more blood vessels insures an ever-greater flow of blood to the tumor as it grows larger.

When examined, Folkman's two cancer inhibitors turned out to be angiogenesis inhibitors. Angiostatin and endostatin kill a tumor by cutting off its blood supply. This may sound like an unlikely approach to curing cancer, but think about it—the cells of a growing tumor require a plentiful supply of food and nutrients to fuel their production of new cancer cells. Cut this off, and the tumor cells die, literally starving to death.

By producing factors like angiostatin and endostatin, the primary tumor holds back the growth of any competing tumors, allowing the primary tumor to hog the available resources for its own use (see above).

In laboratory tests the angiogenesis inhibitors caused tumors in mice to regress to microscopic size, a result that electrified researchers all over the world. Other scientists were soon trying to replicate this exciting result. Some have succeeded, others not. Five major laboratories have isolated their own angiogenesis inhibitors and published findings of antitumor activity. The National Cancer Institute is proceeding with tests of angiostatin and other angiogenesis inhibitors in humans. Preliminary results are encouraging. While not a cure-all for all cancers, angiogenesis inhibitors seem very effective against some, particularly solid-tumor cancers.

Gaining a better understanding of how tumors induce angiogenesis has become a high priority of cancer research. One promising line of research concerns hypoxia. As a solid tumor grows and outstrips its blood supply, its interior becomes hypoxic (oxygen depleted). In response to hypoxia, it appears that genes are turned on that promote survival under low oxygen pressure, including ones that increase blood flow to the tumor by promoting angiogenesis. Understanding this process may give important clues as to how angiogenesis inhibitors work to inhibit tumor growth.

So how does a lowering of oxygen pressure within a tumor promote blood vessel formation? Dr. Randall Johnson of the University of California, San Diego, is studying one important response by a tumor to hypoxia—the induction of a gene-specific transcription factor (that is, a protein that activates the transcription of a particular gene) that promotes angiogenesis. Called HIF-1, for *hypoxia inducible factor-1*, this transcription factor appears to induce the transcription of genes necessary for blood vessel formation.

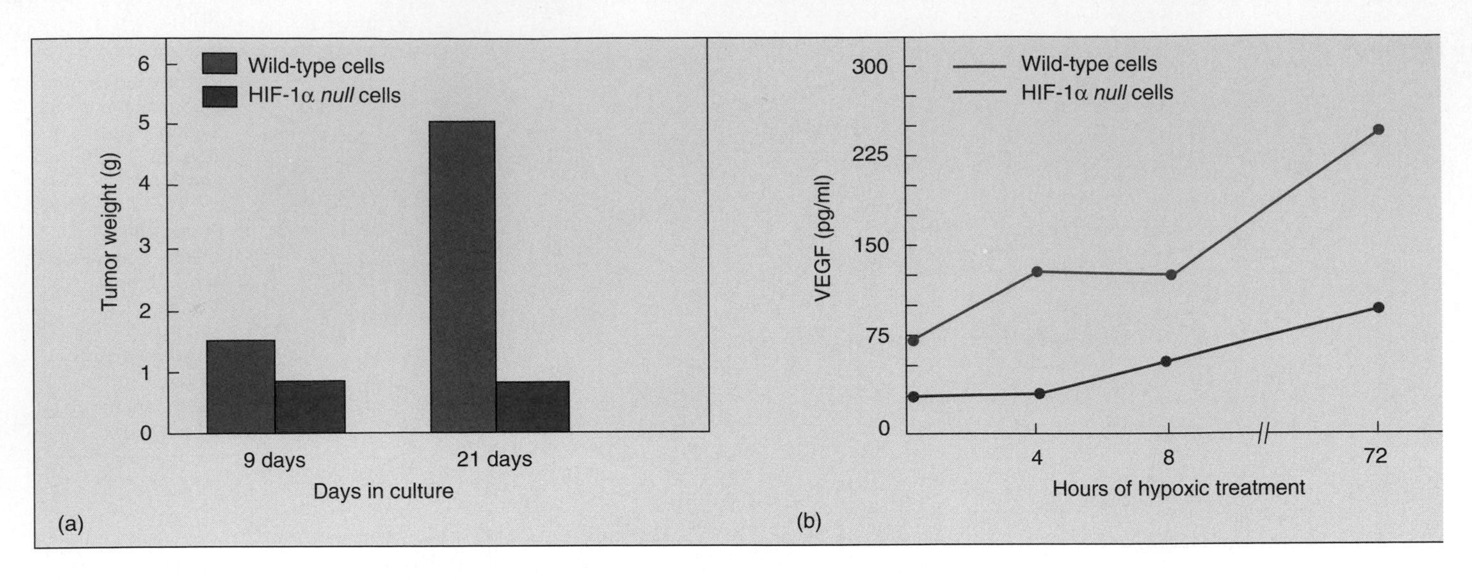

Tumor growth in HIF-1α *null* cells and wild-type cells. (*a*) The size of tumors formed by the HIF-1α *null* cells were significantly smaller compared to those formed by wild-type cells. (*b*) HIF-1α *null* cells had significantly lower levels of VEGF protein production under hypoxic conditions compared to wild-type cells. VEGF promotes the formation of capillaries.

The Experiment

In order to examine the involvement of the hypoxia-inducible transcriptional factor (HIF-1) in angiogenesis, Johnson and his co-workers were faced with the problem that HIF-1 has many other effects on cell growth. To get a clear look at its role in angiogenesis, the researchers turned to embryonic stem cells. Embryonic stem cells are cells harvested from early embryos, before they have differentiated, while they are still capable of unlimited division. Because such stem cells have the capacity to form tumors (teratocarcinomas) when injected into certain kinds of mice, they offer a good natural laboratory in which to study how HIF-1 might influence cancer growth. The research team first prepared a mutant HIF-1 embryonic stem cell line in which the function of the transcription factor encoded by HIF-1 was completely destroyed or *null*.

The researchers then grew these HIF-1 *null* stem cells under hypoxic conditions. If HIF-1 genes indeed foster tumor growth in normal cells by promoting angiogenesis, then it would be expected that these *null* cells would be unable to promote tumor growth in this way.

The researchers tested the ability of *null* cells to promote tumor growth by injecting HIF-1α *null* cells into laboratory mice, and in control experiments injecting wild-type stem cells. The injected cells were allowed to grow and form tumors in both *null* and control host animals. The tumors that formed were then examined and measured for differences.

To get a closer look at what was really going on, the *null* and wild-type cells were compared in their ability to actually form new blood vessels. This was done by examining levels of mRNA of a growth factor that plays a key role in the formation and growth of blood vessels. This factor is a protein called vascular endothelial growth factor (VEGF). The levels of VEGF mRNA in the cells were determined by hybridizing cDNA VEGF probes to mRNA isolated from tumors, and measuring in each instance how much tumor mRNA bound to the cDNA probe. In parallel studies, antibodies were used to determine levels of VEGF protein.

The Results

The researchers found that the *null* cells were greatly compromised in their ability to form tumors compared to the wild-type cells with the effects becoming more significant over time (see graph *a* above). Tumors were five times larger in wild-type cells than in the HIF-1 *null* cells after 21 days. Clearly knocking out HIF-1 retards tumor growth significantly.

This decrease in the size of tumors produced by *null* cells is further supported by the results of the VEGF protein analysis (see graph *b* above). Levels of the protein VEGF rise in wild-type cells under conditions of hypoxia, increasing the immediate availability of oxygen to the tumor by promoting capillary formation. The researchers found levels of VEGF protein were lower in *null* cell tumors, and responded to hypoxia at a lower rate.

Both the decrease in tumor size and the lower level of VEGF in the HIF-1 *null* cells supports the hypothesis that HIF-1 plays an essential role in promoting angiogenesis in a tumor, responding to a hypoxic condition by increasing the levels of VEGF.

Do the angiogenesis inhibitors like angiostatin, being tested as cancer cures, in fact act by inhibiting VEGF? The sorts of experiments being carried out in Johnson's laboratory, and in many other cancer centers, should soon cast light on this still-murky question.

To explore this experiment further, go to the Virtual Lab at www.mhhe.com/raven6/vlab5.mhtml

14

DNA: The Genetic Material

Concept Outline

14.1 What is the genetic material?

The Hammerling Experiment: Cells Store Hereditary Information in the Nucleus

Transplantation Experiments: Each Cell Contains a Full Set of Genetic Instructions

The Griffith Experiment: Hereditary Information Can Pass between Organisms

The Avery and Hershey-Chase Experiments: The Active Principle Is DNA

14.2 What is the structure of DNA?

The Chemical Nature of Nucleic Acids. Nucleic acids are polymers containing four nucleotides.

The Three-Dimensional Structure of DNA. The DNA molecule is a double helix, with two strands held together by base-pairing.

14.3 How does DNA replicate?

The Meselson–Stahl Experiment: DNA Replication Is Semiconservative

The Replication Process. DNA is replicated by the enzyme DNA polymerase III, working in concert with many other proteins. DNA replicates by assembling a complementary copy of each strand semidiscontinuously.

Eukaryotic DNA Replication. Eukaryotic chromosomes consist of many origins of replication.

14.4 What is a gene?

The One-Gene/One-Polypeptide Hypothesis. A gene encodes all the information needed to express a functional protein.

How DNA Encodes Protein Structure. The nucleotide sequence of a gene dictates the amino acid sequence of the protein it encodes.

FIGURE 14.1
DNA. The hereditary blueprint in each cell of all living organisms is a very long, slender molecule called deoxyribonucleic acid (DNA).

The realization that patterns of heredity can be explained by the segregation of chromosomes in meiosis raised a question that occupied biologists for over 50 years: What is the exact nature of the connection between hereditary traits and chromosomes? This chapter describes the chain of experiments that have led to our current understanding of the molecular mechanisms of heredity (figure 14.1). The experiments are among the most elegant in science. Just as in a good detective story, each conclusion has led to new questions. The intellectual path taken has not always been a straight one, the best questions not always obvious. But however erratic and lurching the course of the experimental journey, our picture of heredity has become progressively clearer, the image more sharply defined.

The Hammerling Experiment: Cells Store Hereditary Information in the Nucleus

Perhaps the most basic question one can ask about hereditary information is where it is stored in the cell. To answer this question, Danish biologist Joachim Hammerling, working at the Max Plank Institute for Marine Biology in Berlin in the 1930s, cut cells into pieces and observed the pieces to see which were able to express hereditary information. For this experiment, Hammerling needed cells large enough to operate on conveniently and differentiated enough to distinguish the pieces. He chose the unicellular green alga *Acetabularia,* which grows up to 5 cm, as a **model organism** for his investigations. Just as Mendel used pea plants and Sturtevant used fruit flies as model organisms, Hammerling picked an organism that was suited to the specific experimental question he wanted to answer, assuming that what he learned could then be applied to other organisms.

Individuals of the genus *Acetabularia* have distinct foot, stalk, and cap regions; all are differentiated parts of a single cell. The nucleus is located in the foot. As a preliminary experiment, Hammerling amputated the caps of some cells and the feet of others. He found that when he amputated the cap, a new cap regenerated from the remaining portions of the cell (foot and stalk). When he amputated the foot, however, no new foot regenerated from the cap and stalk. Hammerling, therefore, hypothesized that the hereditary information resided within the foot of *Acetabularia.*

Surgery on Single Cells

To test his hypothesis, Hammerling selected individuals from two species of the genus *Acetabularia* in which the caps look very different from one another: *A. mediterranea* has a disk-shaped cap, and *A. crenulata* has a branched, flower-like cap. Hammerling grafted a stalk from *A. crenulata* to a foot from *A. mediterranea* (figure 14.2). The cap that regenerated looked somewhat like the cap of *A. crenulata,* though not exactly the same.

Hammerling then cut off this regenerated cap and found that a disk-shaped cap exactly like that of *A. mediterranea* formed in the second regeneration and in every regeneration thereafter. This experiment supported Hammerling's hypothesis that the instructions specifying the kind of cap are stored in the foot of the cell, and that these instructions must pass from the foot through the stalk to the cap.

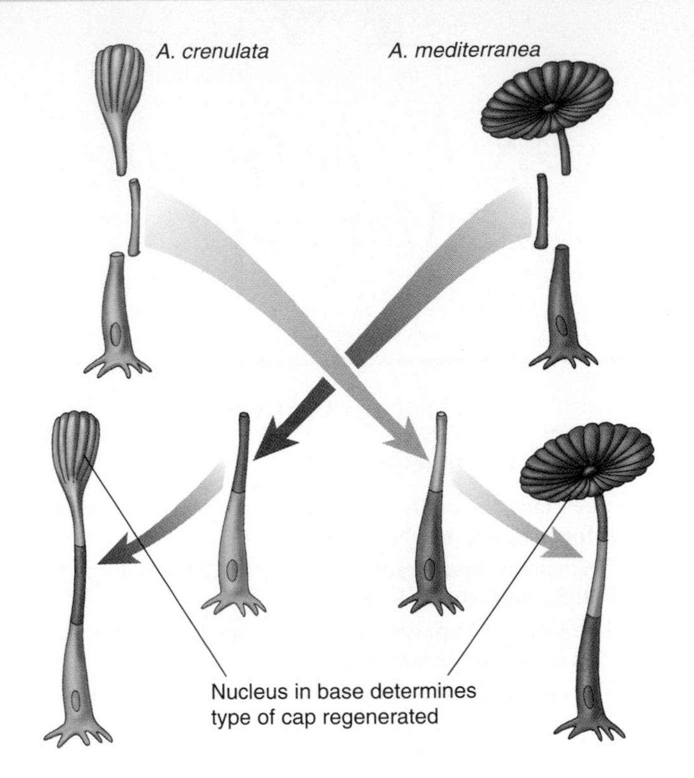

FIGURE 14.2
Hammerling's *Acetabularia* reciprocal graft experiment. Hammerling grafted a stalk of each species of *Acetabularia* onto the foot of the other species. In each case, the cap that eventually developed was dictated by the nucleus-containing foot rather than by the stalk.

In this experiment, the initial flower-shaped cap was somewhat intermediate in shape, unlike the disk-shaped caps of subsequent generations. Hammerling speculated that this initial cap, which resembled that of *A. crenulata,* was formed from instructions already present in the transplanted stalk when it was excised from the original *A. crenulata* cell. In contrast, all of the caps that regenerated subsequently used new information derived from the foot of the *A. mediterranea* cell the stalk had been grafted onto. In some unknown way, the original instructions that had been present in the stalk were eventually "used up." We now understand that genetic instructions (in the form of messenger RNA, discussed in chapter 15) pass from the nucleus in the foot upward *through the stalk* to the developing cap.

Hereditary information in *Acetabularia* is stored in the foot of the cell, where the nucleus resides.

Transplantation Experiments: Each Cell Contains a Full Set of Genetic Instructions

Because the nucleus is contained in the foot of *Acetabularia*, Hammerling's experiments suggested that the nucleus is the repository of hereditary information in a cell. A direct test of this hypothesis was carried out in 1952 by American embryologists Robert Briggs and Thomas King. Using a glass pipette drawn to a fine tip and working with a microscope, Briggs and King removed the nucleus from a frog egg. Without the nucleus, the egg did not develop. However, when they replaced the nucleus with one removed from a more advanced frog embryo cell, the egg developed into an adult frog. Clearly, the nucleus was directing the egg's development (figure 14.3).

Successfully Transplanting Nuclei

Can every nucleus in an organism direct the development of an entire adult individual? The experiment of Briggs and King did not answer this question definitively, because the nuclei they transplanted from frog embryos into eggs often caused the eggs to develop abnormally. Two experiments performed soon afterward gave a clearer answer to the question. In the first, John Gurdon, working with another species of frog at Oxford and Yale, transplanted nuclei from tadpole cells into eggs from which the nuclei had been removed. The experiments were difficult—it was necessary to synchronize the division cycles of donor and host. However, in many experiments, the eggs went on to develop normally, indicating that the nuclei of cells in later stages of development retain the genetic information necessary to direct the development of all other cells in an individual.

Totipotency in Plants

In the second experiment, F. C. Steward at Cornell University in 1958 placed small fragments of fully developed carrot tissue (isolated from a part of the vascular system called the phloem) in a flask containing liquid growth medium. Steward observed that when individual cells broke away from the fragments, they often divided and developed into multicellular roots. When he immobilized the roots by placing them in a solid growth medium, they went on to develop normally into entire, mature plants. Steward's experiment makes it clear that, even in adult tissues, the nuclei of individual plant cells are "totipotent"—each contains a full set of hereditary instructions and can generate an entire adult individual. As you will learn in chapter 19, animal cells, like plant cells, can be totipotent, and a single adult animal cell can generate an entire adult animal.

Hereditary information is stored in the nucleus of eukaryotic cells. Each nucleus in any eukaryotic cell contains a full set of genetic instructions.

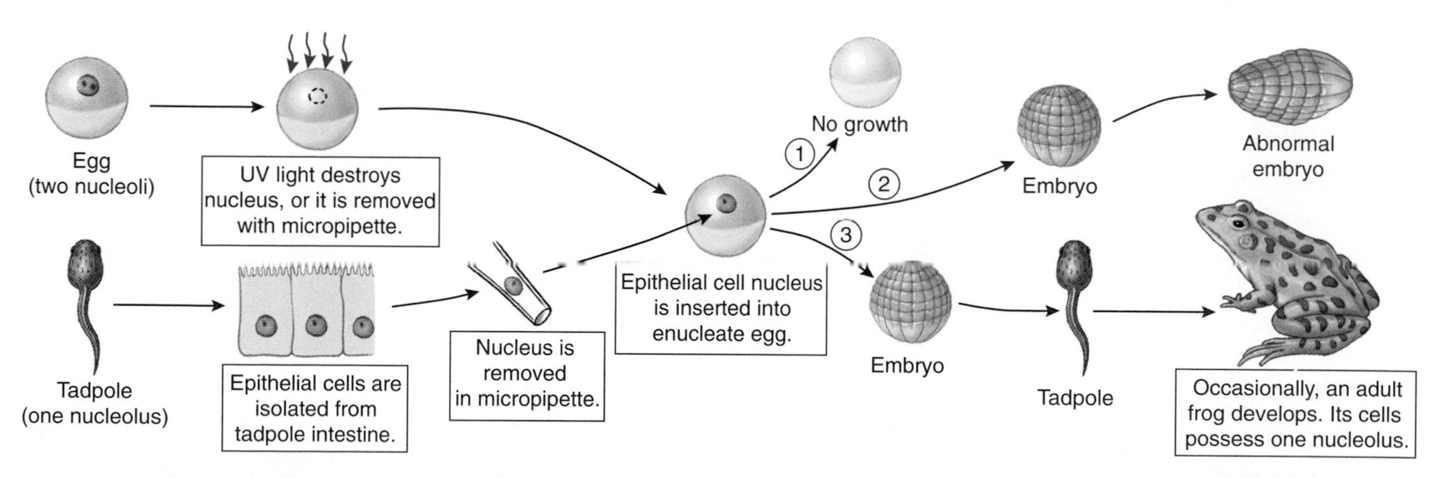

FIGURE 14.3
Briggs and King's nuclear transplant experiment. Two strains of frogs were used that differed from each other in the number of nucleoli their cells possessed. The nucleus was removed from an egg of one strain, either by sucking the egg nucleus into a micropipette or, more simply, by destroying it with ultraviolet light. A nucleus obtained from a differentiated cell of the other strain was then injected into this enucleate egg. The hybrid egg was allowed to develop. One of three results was obtained in individual experiments: (*1*) no growth occurred, perhaps reflecting damage to the egg cell during the nuclear transplant operation; (*2*) normal growth and development occurred up to an early embryo stage, but subsequent development was not normal and the embryo did not survive; and (*3*) normal growth and development occurred, eventually leading to the development of an adult frog. That frog was of the strain that contributed the nucleus and not of the strain that contributed the egg. Only a few experiments gave this third result, but they serve to clearly establish that the nucleus directs frog development.

The Griffith Experiment: Hereditary Information Can Pass between Organisms

The identification of the nucleus as the repository of hereditary information focused attention on the chromosomes, which were already suspected to be the vehicles of Mendelian inheritance. Specifically, biologists wondered how the **genes,** the units of hereditary information studied by Mendel, were actually arranged in the chromosomes. They knew that chromosomes contained both protein and deoxyribonucleic acid (DNA). Which of these held the genes? Starting in the late 1920s and continuing for about 30 years, a series of investigations addressed this question.

In 1928, British microbiologist Frederick Griffith made a series of unexpected observations while experimenting with pathogenic (disease-causing) bacteria. When he infected mice with a virulent strain of *Streptococcus pneumoniae* bacteria (then known as *Pneumococcus*), the mice died of blood poisoning. However, when he infected similar mice with a mutant strain of *S. pneumoniae* that lacked the virulent strain's polysaccharide coat, the mice showed no ill effects. The coat was apparently necessary for virulence. The normal pathogenic form of this bacterium is referred to as the S form because it forms smooth colonies on a culture dish. The mutant form, which lacks an enzyme needed to manufacture the polysaccharide coat, is called the R form because it forms rough colonies.

To determine whether the polysaccharide coat itself had a toxic effect, Griffith injected dead bacteria of the virulent S strain into mice; the mice remained perfectly healthy. As a control, he injected mice with a mixture containing dead S bacteria of the virulent strain and live coatless R bacteria, each of which by itself did not harm the mice (figure 14.4). Unexpectedly, the mice developed disease symptoms and many of them died. The blood of the dead mice was found to contain high levels of live, virulent *Streptococcus* type S bacteria, which had surface proteins characteristic of the live (previously R) strain. Somehow, the information specifying the polysaccharide coat had passed from the dead, virulent S bacteria to the live, coatless R bacteria in the mixture, permanently transforming the coatless R bacteria into the virulent S variety. **Transformation** is the transfer of genetic material from one cell to another and can alter the genetic makeup of the recipient cell.

Hereditary information can pass from dead cells to living ones, transforming them.

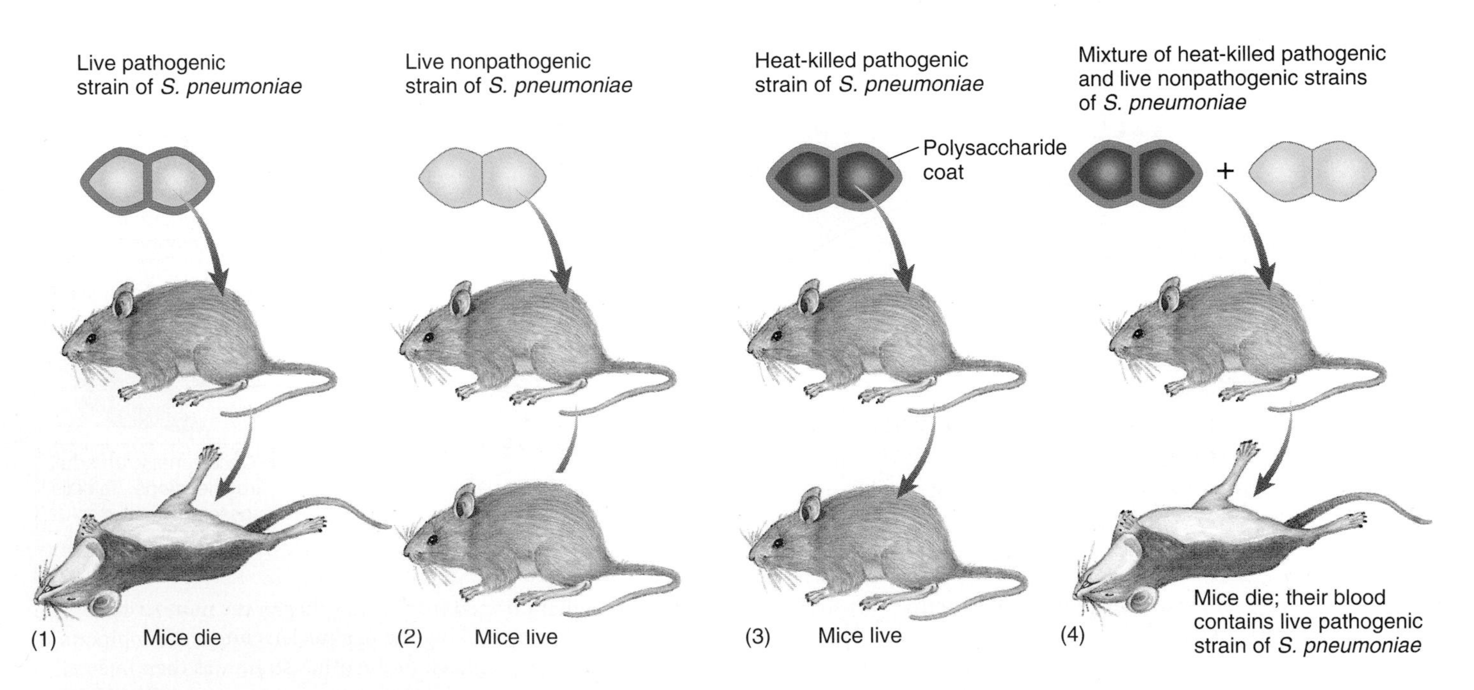

FIGURE 14.4
Griffith's discovery of transformation. (*1*) The pathogenic strain of the bacterium *Streptococcus pneumoniae* kills many of the mice it is injected into. The bacterial cells are covered with a polysaccharide coat, which the bacteria themselves synthesize. (*2*) Interestingly, an injection of live, coatless bacteria produced no ill effects. However, the coat itself is not the agent of disease. (*3*) When Griffith injected mice with dead bacteria that possessed polysaccharide coats, the mice were unharmed. (*4*) But when Griffith injected a mixture of dead bacteria with polysaccharide coats and live bacteria without such coats, many of the mice died, and virulent bacteria with coats were recovered. Griffith concluded that the live cells had been "transformed" by the dead ones; that is, genetic information specifying the polysaccharide coat had passed from the dead cells to the living ones.

The Avery and Hershey-Chase Experiments: The Active Principle Is DNA

The Avery Experiments

The agent responsible for transforming *Streptococcus* went undiscovered until 1944. In a classic series of experiments, Oswald Avery and his coworkers Colin MacLeod and Maclyn McCarty characterized what they referred to as the "transforming principle." They first prepared the mixture of dead S *Streptococcus* and live R *Streptococcus* that Griffith had used. Then Avery and his colleagues removed as much of the protein as they could from their preparation, eventually achieving 99.98% purity. Despite the removal of nearly all protein, the transforming activity was not reduced. Moreover, the properties of the transforming principle resembled those of DNA in several ways:

1. When the purified principle was analyzed chemically, the array of elements agreed closely with DNA.
2. When spun at high speeds in an ultracentrifuge, the transforming principle migrated to the same level (density) as DNA.
3. Extracting the lipid and protein from the purified transforming principle did not reduce its activity.
4. Protein-digesting enzymes did not affect the principle's activity; nor did RNA-digesting enzymes.
5. The DNA-digesting enzyme DNase destroyed all transforming activity.

The evidence was overwhelming. They concluded that "a nucleic acid of the deoxyribose type is the fundamental unit of the transforming principle of *Pneumococcus* Type III"—in essence, that DNA is the hereditary material.

The Hershey–Chase Experiment

Avery's results were not widely accepted at first, as many biologists preferred to believe that proteins were the repository of hereditary information. Additional evidence supporting Avery's conclusion was provided in 1952 by Alfred Hershey and Martha Chase, who experimented with **bacteriophages,** viruses that attack bacteria. Viruses, described in more detail in chapter 33, consist of either DNA or RNA (ribonucleic acid) surrounded by a protein coat. When a *lytic* (potentially cell-rupturing) bacteriophage infects a bacterial cell, it first binds to the cell's outer surface and then injects its hereditary information into the cell. There, the hereditary information directs the production of thousands of new viruses within the bacterium. The bacterial cell eventually ruptures, or lyses, releasing the newly made viruses.

To identify the hereditary material injected into bacterial cells at the start of an infection, Hershey and Chase used the bacteriophage T2, which contains DNA rather than RNA. They labeled the two parts of the viruses, the DNA and the protein coat, with different radioactive isotopes that

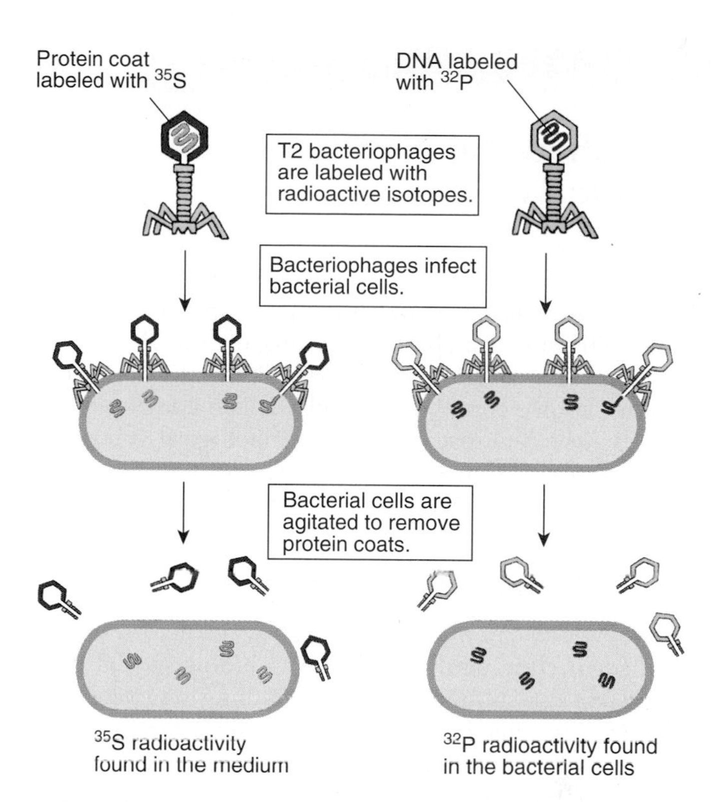

FIGURE 14.5
The Hershey and Chase experiment. Hershey and Chase found that ^{35}S radioactivity did not enter infected bacterial cells and ^{32}P radioactivity did. They concluded that viral DNA, not protein, was responsible for directing the production of new viruses.

would serve as tracers. In some experiments, the viruses were grown on a medium containing an isotope of phosphorus, ^{32}P, and the isotope was incorporated into the phosphate groups of newly synthesized DNA molecules. In other experiments, the viruses were grown on a medium containing ^{35}S, an isotope of sulfur, which is incorporated into the amino acids of newly synthesized protein coats. The ^{32}P and ^{35}S isotopes are easily distinguished from each other because they emit particles with different energies when they decay.

After the labeled viruses were permitted to infect bacteria, the bacterial cells were agitated violently to remove the protein coats of the infecting viruses from the surfaces of the bacteria. This procedure removed nearly all of the ^{35}S label (and thus nearly all of the viral protein) from the bacteria. However, the ^{32}P label (and thus the viral DNA) had transferred to the interior of the bacteria (figure 14.5) and was found in viruses subsequently released from the infected bacteria. Hence, the hereditary information injected into the bacteria that specified the new generation of viruses was DNA and not protein.

Avery's experiments demonstrate conclusively that DNA is Griffith's transforming material. The hereditary material of bacteriophages is DNA and not protein.

The Chemical Nature of Nucleic Acids

A German chemist, Friedrich Miescher, discovered DNA in 1869, only four years after Mendel's work was published. Miescher extracted a white substance from the nuclei of human cells and fish sperm. The proportion of nitrogen and phosphorus in the substance was different from that in any other known constituent of cells, which convinced Miescher that he had discovered a new biological substance. He called this substance "nuclein," because it seemed to be specifically associated with the nucleus.

Levene's Analysis: DNA Is a Polymer

Because Miescher's nuclein was slightly acidic, it came to be called **nucleic acid.** For 50 years biologists did little research on the substance, because nothing was known of its function in cells. In the 1920s, the basic structure of nucleic acids was determined by the biochemist P. A. Levene, who found that DNA contains three main components (figure 14.6): (1) phosphate (PO_4) groups; (2) five-carbon sugars; and (3) nitrogen-containing bases called **purines** (adenine, A, and guanine, G) and **pyrimidines** (thymine, T, and cytosine, C; RNA contains uracil, U, instead of T). From the roughly equal proportions of these components, Levene concluded correctly that DNA and RNA molecules are made of repeating units of the three components. Each unit, consisting of a sugar attached to a phosphate group and a base, is called a **nucleotide.** The identity of the base distinguishes one nucleotide from another.

To identify the various chemical groups in DNA and RNA, it is customary to number the carbon atoms of the base and the sugar and then refer to any chemical group attached to a carbon atom by that number. In the sugar, four of the carbon atoms together with an oxygen atom form a five-membered ring. As illustrated in figure 14.7, the carbon atoms are numbered 1′ to 5′, proceeding clockwise from the oxygen atom; the prime symbol (′) indicates that the number refers to a carbon in a sugar rather than a base. Under this numbering scheme, the phosphate group is attached to the 5′ carbon atom of the sugar, and the base is attached to the 1′ carbon atom. In addition, a free hydroxyl (—OH) group is attached to the 3′ carbon atom.

The 5′ phosphate and 3′ hydroxyl groups allow DNA and RNA to form long chains of nucleotides, because these two groups can react chemically with each other. The reaction between the phosphate group of one nucleotide and the hydroxyl group of another is a dehydration synthesis, eliminating a water molecule and forming a covalent bond that links the two groups (figure 14.8). The linkage is called a **phosphodiester bond** because the

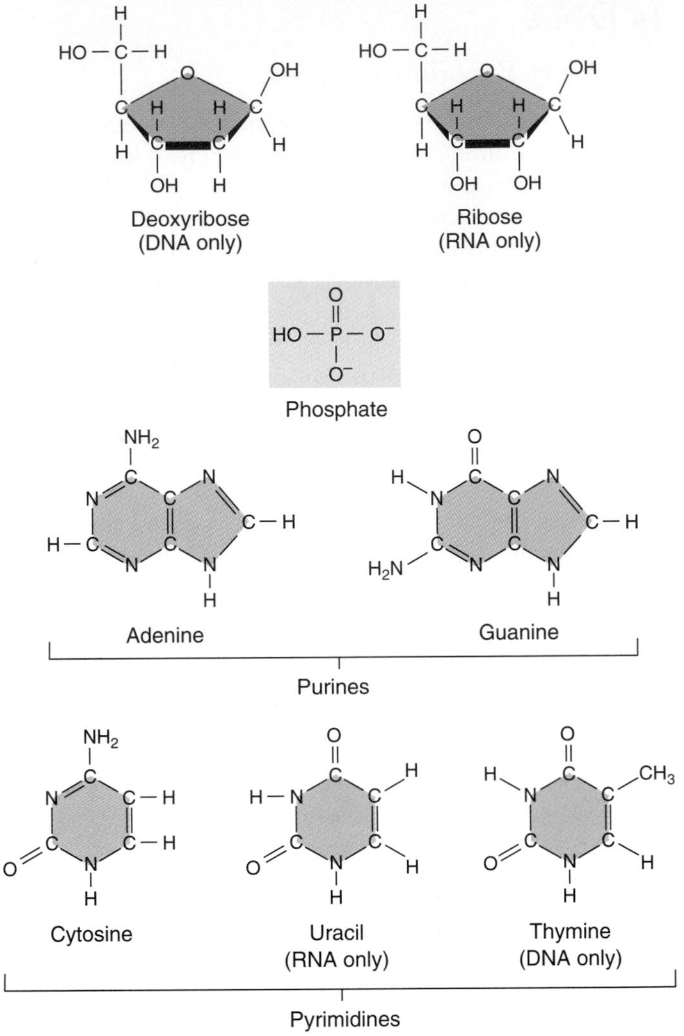

FIGURE 14.6
Nucleotide subunits of DNA and RNA. The nucleotide subunits of DNA and RNA are composed of three elements: a five-carbon sugar (deoxyribose in DNA and ribose in RNA), a phosphate group, and a nitrogenous base (either a purine or a pyrimidine).

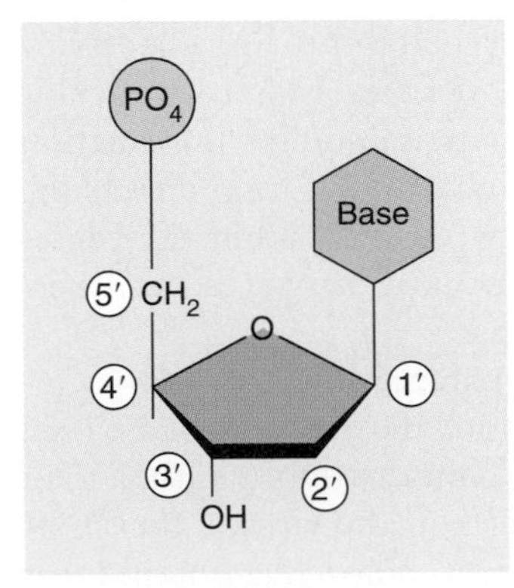

FIGURE 14.7
Numbering the carbon atoms in a nucleotide. The carbon atoms in the sugar of the nucleotide are numbered 1′ to 5′, proceeding clockwise from the oxygen atom. The "prime" symbol (′) indicates that the carbon belongs to the sugar rather than the base.

Table 14.1 Chargaff's Analysis of DNA Nucleotide Base Compositions

Organism	Base Composition (Mole Percent)			
	A	T	G	C
Escherichia coli (K12)	26.0	23.9	24.9	25.2
Mycobacterium tuberculosis	15.1	14.6	34.9	35.4
Yeast	31.3	32.9	18.7	17.1
Herring	27.8	27.5	22.2	22.6
Rat	28.6	28.4	21.4	21.5
Human	30.9	29.4	19.9	19.8

Source: Data from E. Chargaff and J. Davidson (editors), *The Nucleic Acids*, 1955, Academic Press, New York, NY.

phosphate group is now linked to the two sugars by means of a pair of ester (P—O—C) bonds. The two-unit polymer resulting from this reaction still has a free 5′ phosphate group at one end and a free 3′ hydroxyl group at the other, so it can link to other nucleotides. In this way, many thousands of nucleotides can join together in long chains.

Linear strands of DNA or RNA, no matter how long, will almost always have a free 5′ phosphate group at one end and a free 3′ hydroxyl group at the other. Therefore, every DNA and RNA molecule has an intrinsic directionality, and we can refer unambiguously to each end of the molecule. By convention, the sequence of bases is usually expressed in the 5′-to-3′ direction. Thus, the base sequence "GTCCAT" refers to the sequence,

5′ pGpTpCpCpApT—OH 3′

where the phosphates are indicated by "p." Note that this is not the same molecule as that represented by the reverse sequence:

5′ pTpApCpCpTpG—OH 3′

Levene's early studies indicated that all four types of DNA nucleotides were present in roughly equal amounts. This result, which later proved to be erroneous, led to the mistaken idea that DNA was a simple polymer in which the four nucleotides merely repeated (for instance, GCAT . . . GCAT . . . GCAT . . . GCAT . . .). If the sequence never varied, it was difficult to see how DNA might contain the hereditary information; this was why Avery's conclusion that DNA is the transforming principle was not readily accepted at first. It seemed more plausible that DNA was simply a structural element of the chromosomes, with proteins playing the central genetic role.

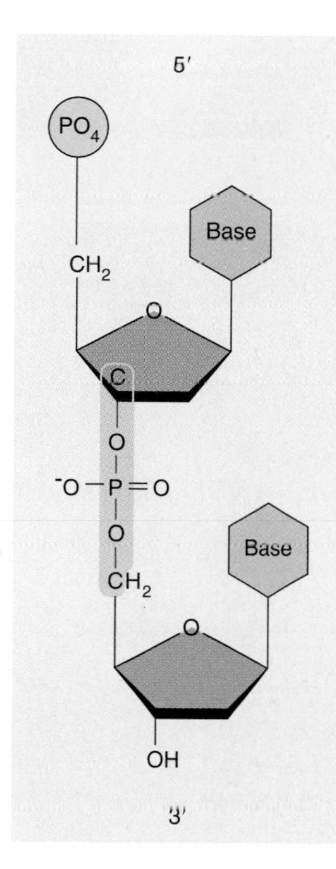

FIGURE 14.8
A phosphodiester bond.

Chargaff's Analysis: DNA Is Not a Simple Repeating Polymer

When Levene's chemical analysis of DNA was repeated using more sensitive techniques that became available after World War II, quite a different result was obtained. The four nucleotides were *not* present in equal proportions in DNA molecules after all. A careful study carried out by Erwin Chargaff showed that the nucleotide composition of DNA molecules varied in complex ways, depending on the source of the DNA (table 14.1). This strongly suggested that DNA was not a simple repeating polymer and might have the information-encoding properties genetic material must have. Despite DNA's complexity, however, Chargaff observed an important underlying regularity in double-stranded DNA: *the amount of adenine present in DNA always equals the amount of thymine, and the amount of guanine always equals the amount of cytosine.* These findings are commonly referred to as **Chargaff's rules:**

1. The proportion of A always equals that of T, and the proportion of G always equals that of C:

   A = T, and G = C.

2. It follows that there is always an equal proportion of purines (A and G) and pyrimidines (C and T).

A single strand of DNA or RNA consists of a series of nucleotides joined together in a long chain. In all natural double-stranded DNA molecules, the proportion of A equals that of T, and the proportion of G equals that of C.

The Three-Dimensional Structure of DNA

As it became clear that DNA was the molecule that stored the hereditary information, investigators began to puzzle over how such a seemingly simple molecule could carry out such a complex function.

Franklin: X-ray Diffraction Patterns of DNA

The significance of the regularities pointed out by Chargaff were not immediately obvious, but they became clear when a British chemist, Rosalind Franklin (figure 14.9*a*), carried out an X-ray diffraction analysis of DNA. In X-ray diffraction, a molecule is bombarded with a beam of X rays. When individual rays encounter atoms, their path is bent or diffracted, and the diffraction pattern is recorded on photographic film. The patterns resemble the ripples created by tossing a rock into a smooth lake (figure 14.9*b*). When carefully analyzed, they yield information about the three-dimensional structure of a molecule.

X-ray diffraction works best on substances that can be prepared as perfectly regular crystalline arrays. However, it was impossible to obtain true crystals of natural DNA at the time Franklin conducted her analysis, so she had to use DNA in the form of fibers. Franklin worked in the laboratory of British biochemist Maurice Wilkins, who was able to prepare more uniformly oriented DNA fibers than anyone had previously. Using these fibers, Franklin succeeded in obtaining crude diffraction information on natural DNA. The diffraction patterns she obtained suggested that the DNA molecule had the shape of a helix, or corkscrew, with a diameter of about 2 nanometers and a complete helical turn every 3.4 nanometers (figure 14.9*c*).

(a)

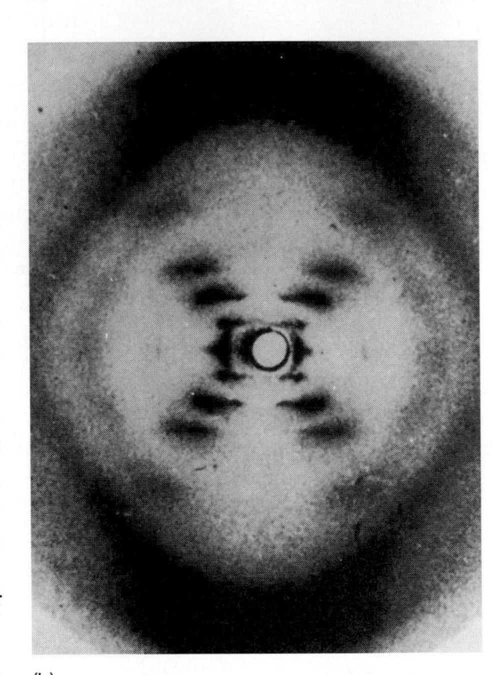

(b)

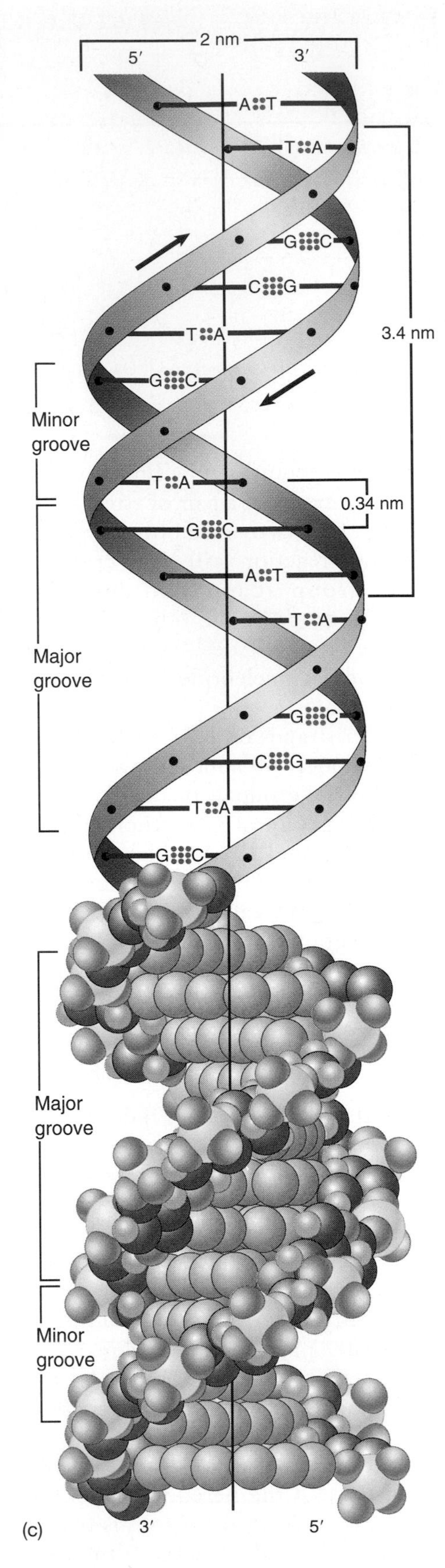

(c)

FIGURE 14.9
Rosalind Franklin's X-ray diffraction patterns suggested the shape of DNA. (*a*) Rosalind Franklin developed techniques for taking X-ray diffraction pictures of fibers of DNA. (*b*) This is the telltale X-ray diffraction photograph of DNA fibers made in 1953 by Rosalind Franklin in the laboratory of Maurice Wilkins. (*c*) These X-ray diffraction studies suggested the dimensions of the double helix.

Watson and Crick: A Model of the Double Helix

Learning informally of Franklin's results before they were published in 1953, James Watson and Francis Crick, two young investigators at Cambridge University, quickly worked out a likely structure for the DNA molecule (figure 14.10), which we now know was substantially correct. They analyzed the problem deductively, first building models of the nucleotides, and then trying to assemble the nucleotides into a molecule that matched what was known about the structure of DNA. They tried various possibilities before they finally hit on the idea that the molecule might be a simple **double helix,** with the bases of two strands pointed inward toward each other, forming **base-pairs.** In their model, base-pairs always consist of purines, which are large, pointing toward pyrimidines, which are small, keeping the diameter of the molecule a constant 2 nanometers. Because hydrogen bonds can form between the bases in a base-pair, the double helix is stabilized as a duplex DNA molecule composed of two **antiparallel strands,** one chain running 3′ to 5′ and the other 5′ to 3′. The base-pairs are planar (flat) and stack 0.34 nm apart as a result of hydrophobic interactions, contributing to the overall stability of the molecule.

The Watson–Crick model explained why Chargaff had obtained the results he had: in a double helix, adenine forms two hydrogen bonds with thymine, but it will not form hydrogen bonds properly with cytosine. Similarly, guanine forms three hydrogen bonds with cytosine, but it will not form hydrogen bonds properly with thymine. Consequently, adenine and thymine will always occur in the same proportions in any DNA molecule, as will guanine and cytosine, because of this base-pairing.

The DNA molecule is a double helix, the strands held together by base-pairing.

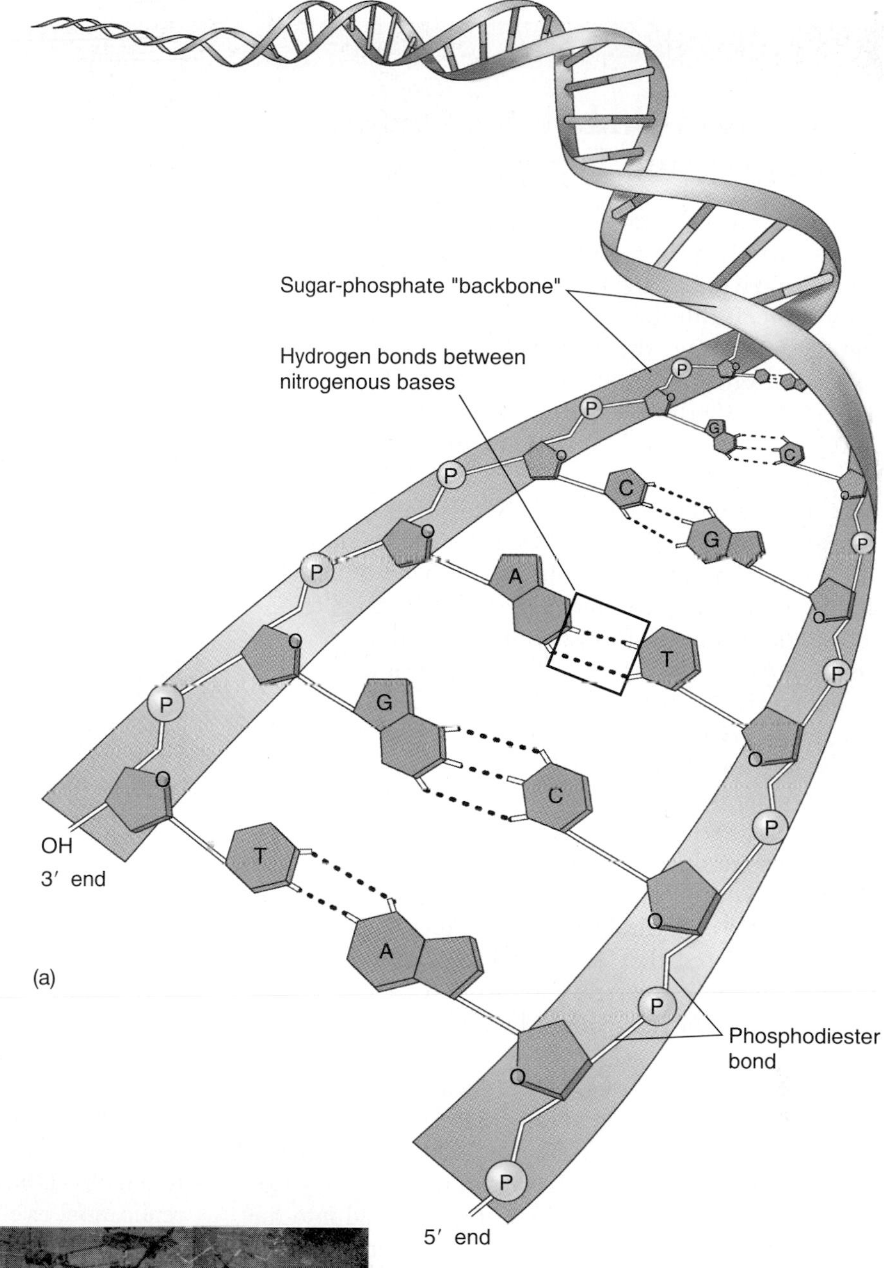

FIGURE 14.10
DNA is a double helix. (*a*) In a DNA duplex molecule, only two base-pairs are possible: adenine (A) can pair with thymine (T), and guanine (G) can pair with cytosine (C). An A-T base-pair has two hydrogen bonds, while a G-C base-pair has three. (*b*) James Watson (*far left*), and Francis Crick (*left*) deduced the structure of DNA in 1953 from Chargaff's rules and Franklin's diffraction studies.

The Meselson–Stahl Experiment: DNA Replication Is Semiconservative

The Watson–Crick model immediately suggested that the basis for copying the genetic information is **complementarity.** One chain of the DNA molecule may have any conceivable base sequence, but this sequence completely determines the sequence of its partner in the duplex. For example, if the sequence of one chain is 5′-ATTGCAT-3′, the sequence of its partner *must* be 3′-TAACGTA-5′. Thus, each chain in the duplex is a complement of the other.

The complementarity of the DNA duplex provides a ready means of accurately duplicating the molecule. If one were to "unzip" the molecule, one would need only to assemble the appropriate complementary nucleotides on the exposed single strands to form two daughter duplexes with the same sequence. This form of DNA replication is called **semiconservative,** because while the sequence of the original duplex is conserved after one round of replication, the duplex itself is not. Instead, each strand of the duplex becomes part of another duplex.

Two other hypotheses of gene replication were also proposed. The conservative model stated that the parental double helix would remain intact and generate DNA copies consisting of entirely new molecules. The dispersive model predicted that parental DNA would become dispersed throughout the new copy so that each strand of all the daughter molecules would be a mixture of old and new DNA.

The three hypotheses of DNA replication were evaluated in 1958 by Matthew Meselson and Franklin Stahl of the California Institute of Technology. These two scientists grew bacteria in a medium containing the heavy isotope of nitrogen, ^{15}N, which became incorporated into the bases of the bacterial DNA. After several generations, the DNA of these bacteria was denser than that of bacteria grown in a medium containing the lighter isotope of nitrogen, ^{14}N. Meselson and Stahl then transferred the bacteria from the ^{15}N medium to the ^{14}N medium and collected the DNA at various intervals.

By dissolving the DNA they had collected in a heavy salt called cesium chloride and then spinning the solution at very high speeds in an ultracentrifuge, Meselson and Stahl were able to separate DNA strands of different densities. The enormous centrifugal forces generated by the ultracentrifuge caused the cesium ions to migrate toward the bottom of the centrifuge tube, creating a gradient of cesium concentration, and thus of density. Each DNA strand floats or sinks in the gradient until it reaches the position where its density exactly matches the density of the cesium there. Because ^{15}N strands are denser than ^{14}N strands, they migrate farther down the tube to a denser region of the cesium gradient.

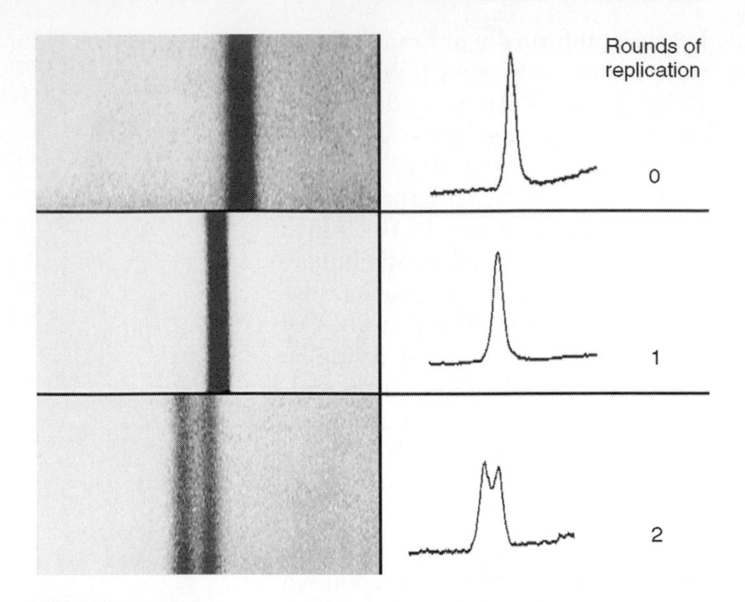

FIGURE 14.11
The key result of the Meselson and Stahl experiment. These bands of DNA, photographed on the left and scanned on the right, are from the density-gradient centrifugation experiment of Meselson and Stahl. At 0 generation, all DNA is heavy; after one replication all DNA has a hybrid density; after two replications, all DNA is hybrid or light.

The DNA collected immediately after the transfer was all dense. However, after the bacteria completed their first round of DNA replication in the ^{14}N medium, the density of their DNA had decreased to a value intermediate between ^{14}N-DNA and ^{15}N-DNA. After the second round of replication, two density classes of DNA were observed, one intermediate and one equal to that of ^{14}N-DNA (figure 14.11).

Meselson and Stahl interpreted their results as follows: after the first round of replication, each daughter DNA duplex was a hybrid possessing one of the heavy strands of the parent molecule and one light strand; when this hybrid duplex replicated, it contributed one heavy strand to form another hybrid duplex and one light strand to form a light duplex (figure 14.12). Thus, this experiment clearly confirmed the prediction of the Watson-Crick model that DNA replicates in a semiconservative manner.

The basis for the great accuracy of DNA replication is complementarity. A DNA molecule is a duplex, containing two strands that are complementary mirror images of each other, so either one can be used as a template to reconstruct the other.

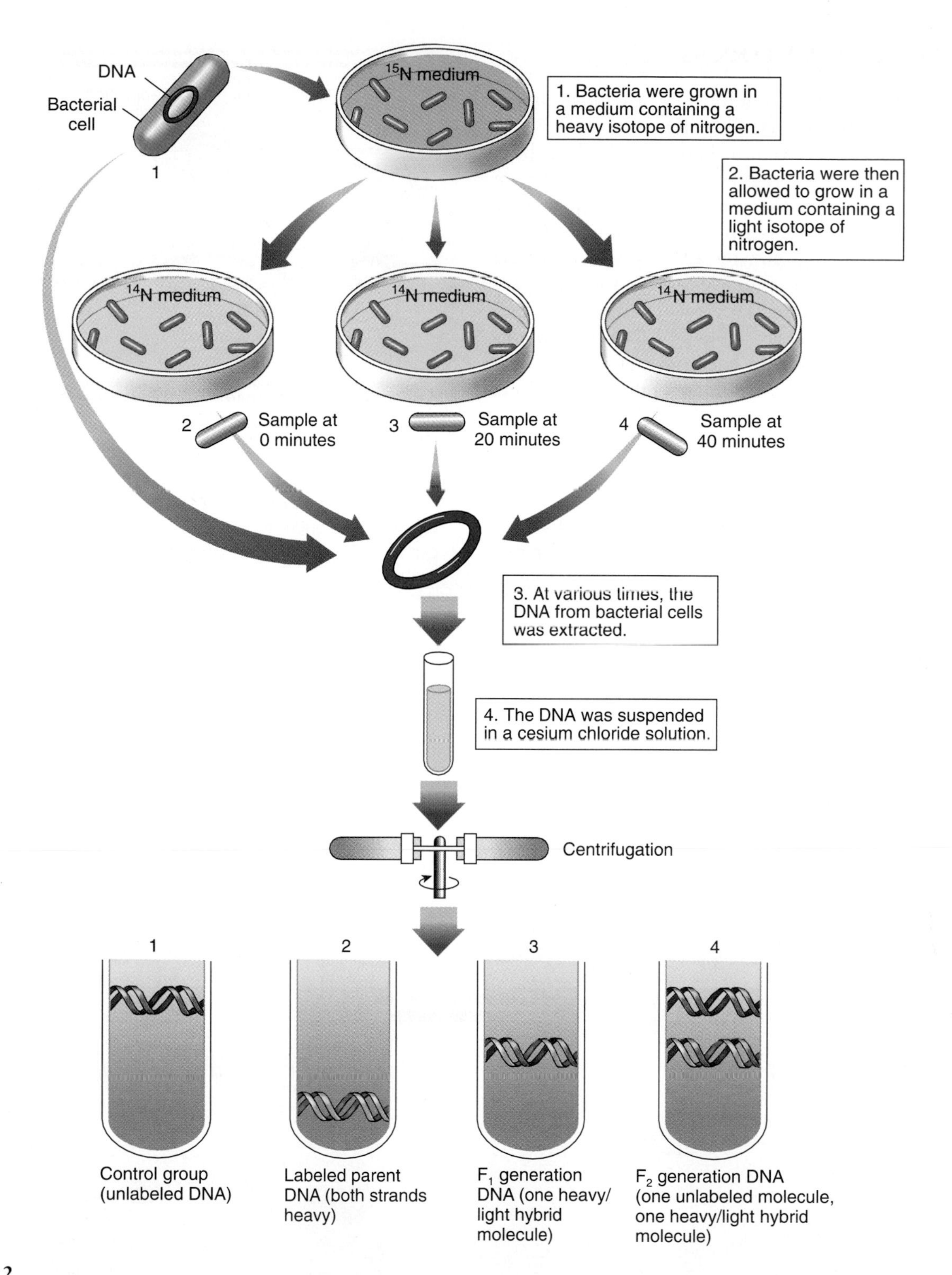

FIGURE 14.12

The Meselson and Stahl experiment: evidence demonstrating semiconservative replication. Bacterial cells were grown for several generations in a medium containing a heavy isotope of nitrogen (^{15}N) and then were transferred to a new medium containing the normal lighter isotope (^{14}N). At various times thereafter, samples of the bacteria were collected, and their DNA was dissolved in a solution of cesium chloride, which was spun rapidly in a centrifuge. Because the cesium ion is so massive, it tends to settle toward the bottom of the spinning tube, establishing a gradient of cesium density. DNA molecules sink in the gradient until they reach a place where their density equals that of the cesium; they then "float" at that position. DNA containing ^{15}N is denser than that containing ^{14}N, so it sinks to a lower position in the cesium gradient. After one generation in ^{14}N medium, the bacteria yielded a single band of DNA with a density between that of ^{14}N-DNA and ^{15}N-DNA, indicating that only one strand of each duplex contained ^{15}N. After two generations in ^{14}N medium, two bands were obtained: one of intermediate density (in which one of the strands contained ^{15}N) and one of low density (in which neither strand contained ^{15}N). Meselson and Stahl concluded that replication of the DNA duplex involves building new molecules by separating strands and assembling new partners on each of these templates.

The Replication Process

To be effective, DNA replication must be fast and accurate. The machinery responsible has been the subject of intensive study for 40 years, and we now know a great deal about it. The replication of DNA begins at one or more sites on the DNA molecule where there is a specific sequence of nucleotides called a **replication origin** (figure 14.13). There the DNA replicating enzyme **DNA polymerase III** and other enzymes begin a complex process that catalyzes the addition of nucleotides to the growing complementary strands of DNA (figure 14.14). Table 14.2 lists the proteins involved in DNA replication in bacteria. Before considering the replication process in detail, let's take a closer look at DNA polymerase III.

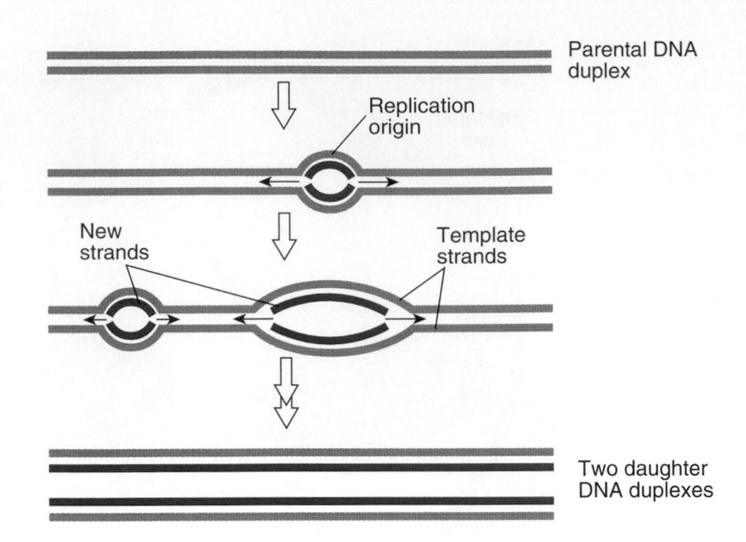

FIGURE 14.13
Origins of replication. At a site called the replication origin, the DNA duplex opens to create two separate strands, each of which can be used as a template for a new strand. Eukaryotic DNA has multiple origins of replication.

DNA Polymerase III

The first DNA polymerase enzyme to be characterized, DNA polymerase I of the bacterium *Escherichia coli*, is a relatively small enzyme that plays a key supporting role in

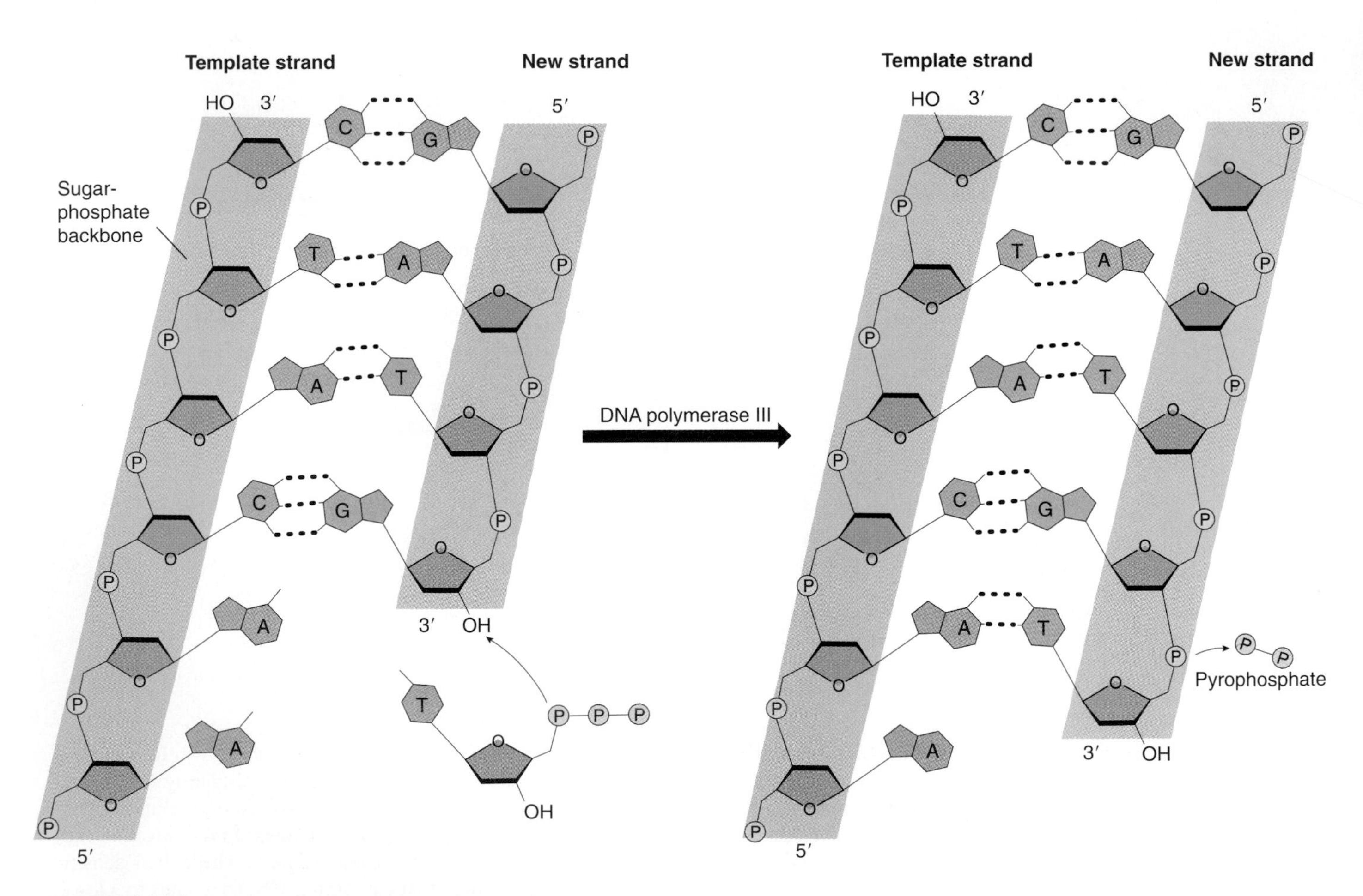

FIGURE 14.14
How nucleotides are added in DNA replication. DNA polymerase III, along with other enzymes, catalyzes the addition of nucleotides to the growing complementary strand of DNA. When a nucleotide is added, two of its phosphates are lost as pyrophosphate.

DNA replication. The true *E. coli* replicating enzyme, dubbed DNA polymerase III, is some 10 times larger and far more complex in structure. We know more about DNA polymerase III than any other organism's DNA polymerase, and so will describe it in detail here. Other DNA polymerases are thought to be broadly similar.

DNA polymerase III contains 10 different kinds of polypeptide chains, as illustrated in figure 14.15. The enzyme is a dimer, with two similar multisubunit complexes. Each complex catalyzes the replication of one DNA strand.

A variety of different proteins play key roles within each complex. The subunits include a single large catalytic α subunit that catalyzes 5′ to 3′ addition of nucleotides to a growing chain, a smaller ε subunit that proofreads 3′ to 5′ for mistakes, and a ring-shaped β_2 dimer subunit that clamps the polymerase III complex around the DNA double helix. Polymerase III progressively threads the DNA through the enzyme complex, moving it at a rapid rate, some 1000 nucleotides per second (100 full turns of the helix, 0.34 micrometers).

Table 14.2 DNA Replication Proteins of *E. coli*

Protein	Role	Size (kd)	Molecules per Cell
Helicase	Unwinds the double helix	300	20
Primase	Synthesizes RNA primers	60	50
Single-strand binding protein	Stabilizes single-stranded regions	74	300
DNA gyrase	Relieves torque	400	250
DNA polymerase III	Synthesizes DNA	≈900	20
DNA polymerase I	Erases primer and fills gaps	103	300
DNA ligase	Joins the ends of DNA segments	74	300

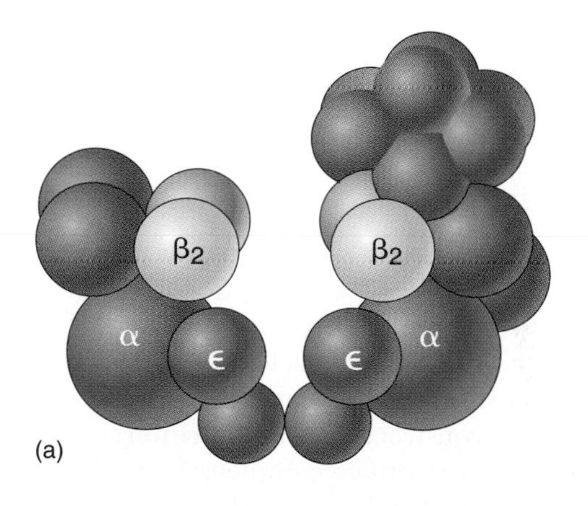

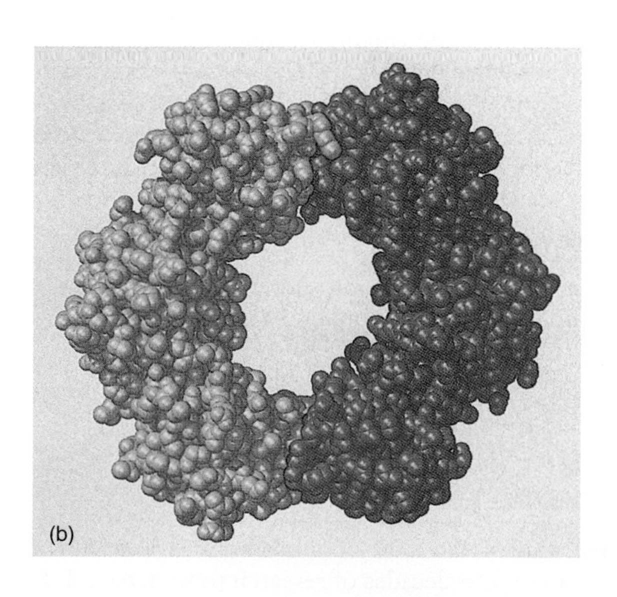

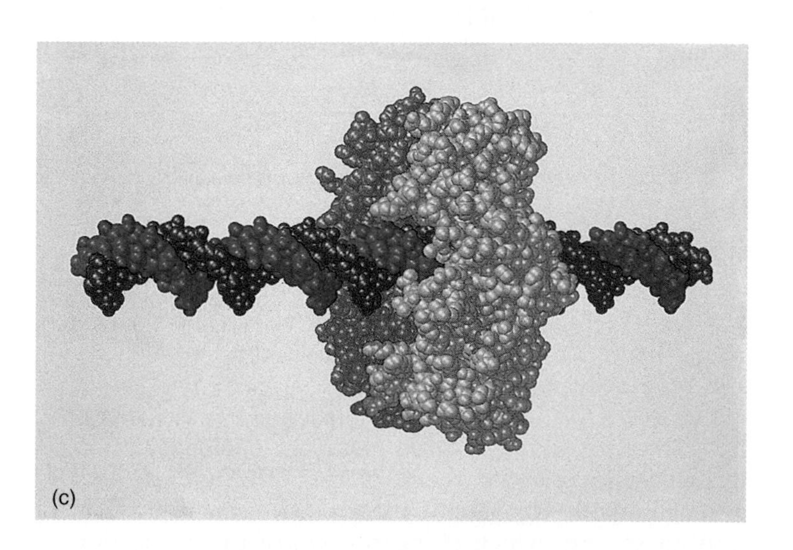

FIGURE 14.15
The DNA polymerase III complex. *(a)* The complex contains 10 kinds of protein chains. The protein is a dimer because both strands of the DNA duplex must be replicated simultaneously. The catalytic (α) subunits, the proofreading (ε) subunits, and the "sliding clamp" (β_2) subunits (*yellow* and *blue*) are labeled. *(b)* The "sliding clamp" units encircle the DNA template and *(c)* move it through the catalytic subunit like a rope drawn through a ring.

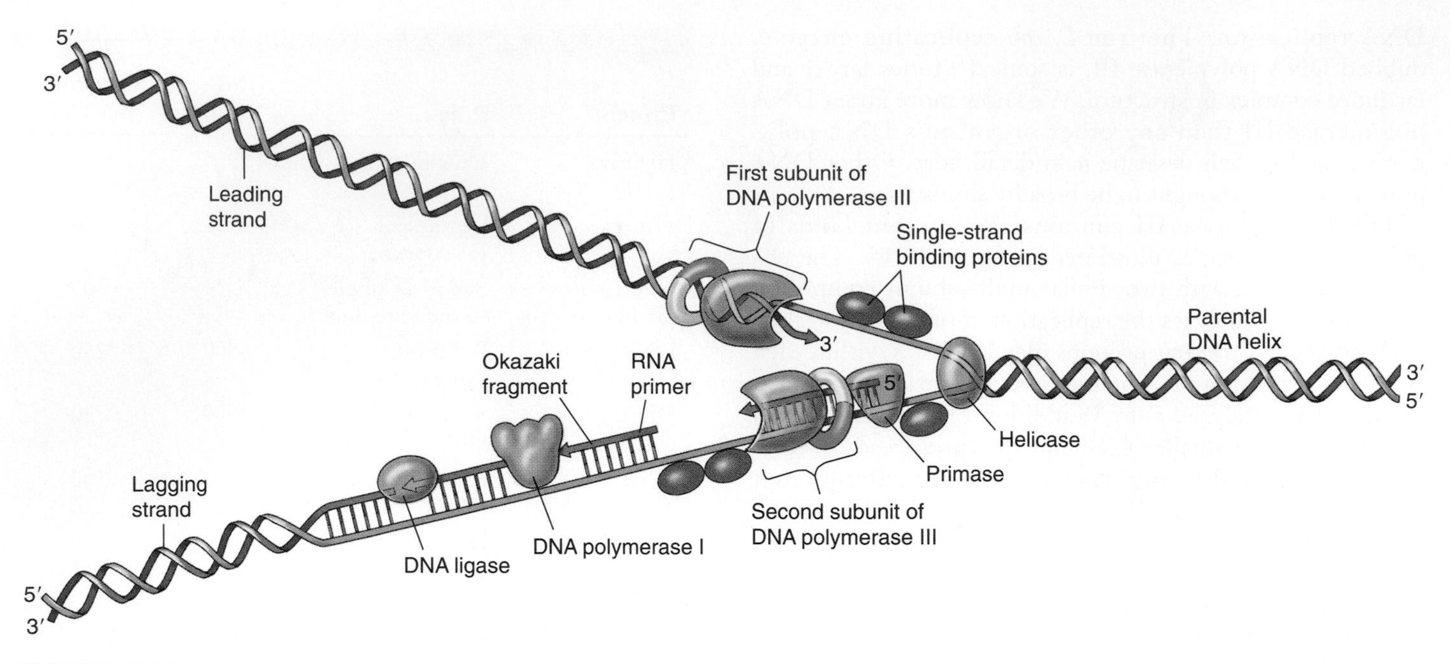

FIGURE 14.16
A DNA replication fork. Helicase enzymes separate the strands of the double helix, and single-strand binding proteins stabilize the single-stranded regions. Replication occurs by two mechanisms. (*1*) *Continuous synthesis:* After primase adds a short RNA primer, DNA polymerase III adds nucleotides to the 3′ end of the leading strand. DNA polymerase I then replaces the RNA primer with DNA nucleotides. (*2*) *Discontinuous synthesis:* Primase adds a short RNA primer (*green*) ahead of the 5′ end of the lagging strand. DNA polymerase III then adds nucleotides to the primer until the gap is filled in. DNA polymerase I replaces the primer with DNA nucleotides, and DNA ligase attaches the short segment of nucleotides to the lagging strand.

The Need for a Primer

One of the features of DNA polymerase III is that it can add nucleotides only to a chain of nucleotides that is already paired with the parent strands. Hence, DNA polymerase cannot link the first nucleotides in a newly synthesized strand. Instead, another enzyme, an RNA polymerase called **primase,** constructs an **RNA primer,** a sequence of about 10 RNA nucleotides complementary to the parent DNA template. DNA polymerase III recognizes the primer and adds DNA nucleotides to it to construct the new DNA strands. The RNA nucleotides in the primers are then replaced by DNA nucleotides.

The Two Strands of DNA Are Assembled in Different Ways

Another feature of DNA polymerase III is that it can add nucleotides only to the 3′ end of a DNA strand (the end with an —OH group attached to a 3′ carbon atom). This means that replication always proceeds in the $5' \rightarrow 3'$ direction on a growing DNA strand. Because the two parent strands of a DNA molecule are antiparallel, *the new strands are oriented in opposite directions* along the parent templates at each replication fork (figure 14.16). Therefore, the new strands must be elongated by different mechanisms! The **leading strand,** which elongates *toward* the replication fork, is built up simply by adding nucleotides continuously to its growing 3′ end. In contrast, the **lagging strand,** which elongates *away from* the replication fork, is synthesized discontinuously as a series of short segments that are later connected. These segments, called **Okazaki fragments,** are about 100 to 200 nucleotides long in eukaryotes and 1000 to 2000 nucleotides long in prokaryotes. Each Okazaki fragment is synthesized by DNA polymerase III in the $5' \rightarrow 3'$ direction, beginning at the replication fork and moving away from it. When the polymerase reaches the 5′ end of the lagging strand, another enzyme, **DNA ligase,** attaches the fragment to the lagging strand. The DNA is further unwound, new RNA primers are constructed, and DNA polymerase III then jumps ahead 1000 to 2000 nucleotides (toward the replication fork) to begin constructing another Okazaki fragment. If one looks carefully at electron micrographs showing DNA replication in progress, one can sometimes see that one of the parent strands near the replication fork appears single-stranded over a distance of about 1000 nucleotides. Because the synthesis of the leading strand is continuous, while that of the lagging strand is discontinuous, the overall replication of DNA is said to be **semidiscontinuous.**

The Replication Process

The replication of the DNA double helix is a complex process that has taken decades of research to understand. It takes place in five interlocking steps:

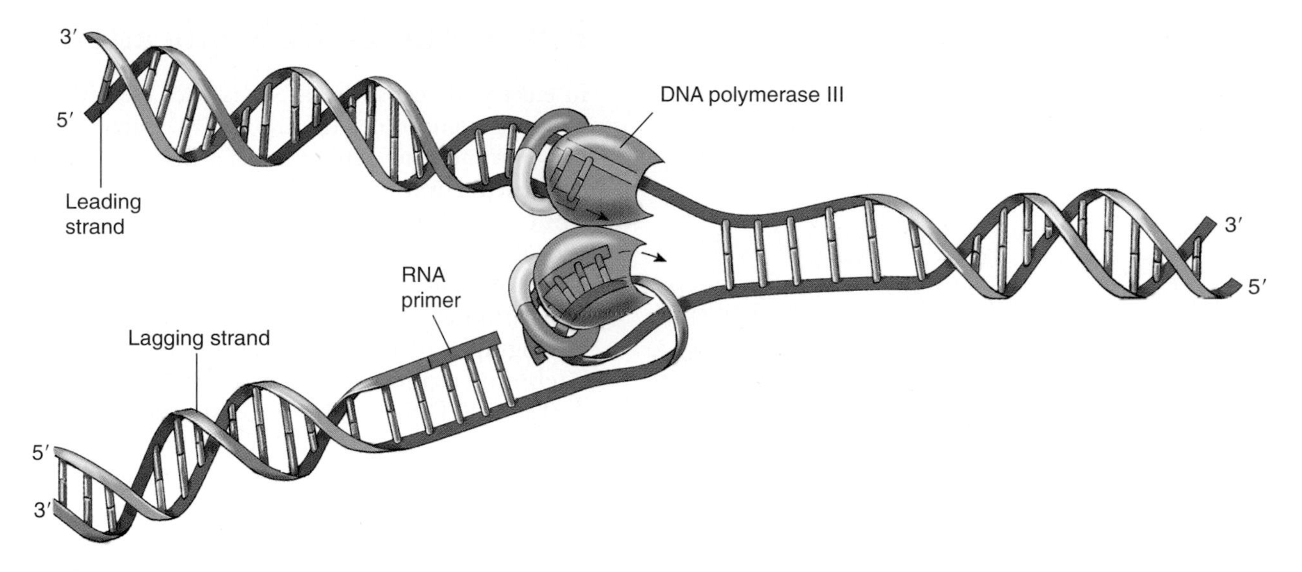

FIGURE 14.17
How DNA polymerase III works. This diagram presents a current view of how DNA polymerase III works. Note that the DNA on the lagging strand is folded to allow the dimeric DNA polymerase III molecule to replicate both strands of the parental DNA duplex simultaneously. This brings the 3′ end of each completed Okazaki fragment close to the start site for the next fragment.

1. **Opening up the DNA double helix.** The very stable DNA double helix must be opened up and its strands separated from each other for semiconservative replication to occur.

   Stage one: Initiating replication. The binding of **initiator proteins** to the replication origin starts an intricate series of interactions that opens the helix.

   Stage two: Unwinding the duplex. After initiation, "unwinding" enzymes called **helicases** bind to and move along one strand, shouldering aside the other strand as they go.

   Stage three: Stabilizing the single strands. The unwound portion of the DNA double helix is stabilized by **single-strand binding protein,** which binds to the exposed single strands, protecting them from cleavage and preventing them from rewinding.

   Stage four: Relieving the torque generated by unwinding. For replication to proceed at 1000 nucleotides per second, the parental helix ahead of the replication fork must rotate 100 revolutions per second! To relieve the resulting twisting, called torque, enzymes known as topisomerases—or, more informally, **gyrases**—cleave a strand of the helix, allow it to swivel around the intact strand, and then reseal the broken strand.

2. **Building a primer.** New DNA cannot be synthesized on the exposed templates until a primer is constructed, as DNA polymerases require 3′ primers to initiate replication. The necessary primer is a short stretch of RNA, added by a specialized RNA polymerase called *primase* in a multisubunit complex informally called a *primosome.* Why an RNA primer, rather than DNA? Starting chains on exposed templates introduces many errors; RNA marks this initial stretch as "temporary," making this error-prone stretch easy to excise later.
3. **Assembling complementary strands.** Next, the dimeric DNA polymerase III then binds to the replication fork. While the leading strand complexes with one half of the polymerase dimer, the lagging strand is thought to loop around and complex with the other half of the polymerase dimer (figure 14.17). Moving in concert down the parental double helix, DNA polymerase III catalyzes the formation of complementary sequences on each of the two single strands at the same time.
4. **Removing the primer.** The enzyme DNA polymerase I now removes the RNA primer and fills in the gap, as well as any gaps between Okazaki fragments.
5. **Joining the Okazaki fragments.** After any gaps between Okazaki fragments are filled in, the enzyme DNA ligase joins the fragments to the lagging strand.

DNA replication involves many different proteins that open and unwind the DNA double helix, stabilize the single strands, synthesize RNA primers, assemble new complementary strands on each exposed parental strand—one of them discontinuously—remove the RNA primer, and join new discontinuous segments on the lagging strand.

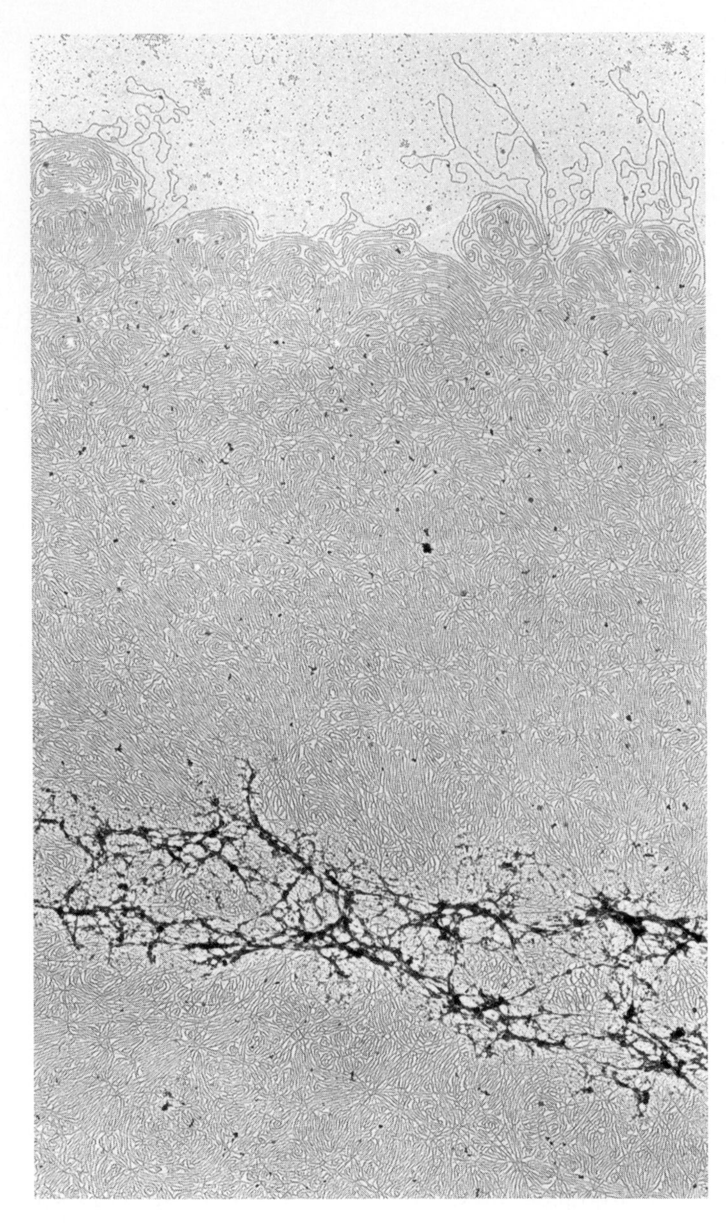

FIGURE 14.18
DNA of a single human chromosome. This chromosome has been "exploded," or relieved, of most of its packaging proteins. The residual protein scaffolding appears as the dark material in the lower part of the micrograph.

Eukaryotic DNA Replication

In eukaryotic cells, the DNA is packaged in nucleosomes within chromosomes (figure 14.18). Each individual zone of a chromosome replicates as a discrete section called a **replication unit,** or **replicon,** which can vary in length from 10,000 to 1 million base-pairs; most are about 100,000 base-pairs long. Each replication unit has its own origin of replication, and multiple units may be undergoing replication at any given time, as can be seen in electron micrographs of replicating chromosomes (figure 14.19). Each unit replicates in a way fundamentally similar to prokaryotic DNA replication, using similar enzymes. The advantage of having multiple origins of replication in eukaryotes is speed: replication takes approximately eight hours in humans cells, but if there were only one origin, it would take 100 times longer. Regulation of the replication process ensures that only one copy of the DNA is ultimately produced. How a cell achieves this regulation is not yet completely clear. It may involve periodic inhibitor or initiator proteins on the DNA molecule itself.

Eukaryotic chromosomes have multiple origins of replication.

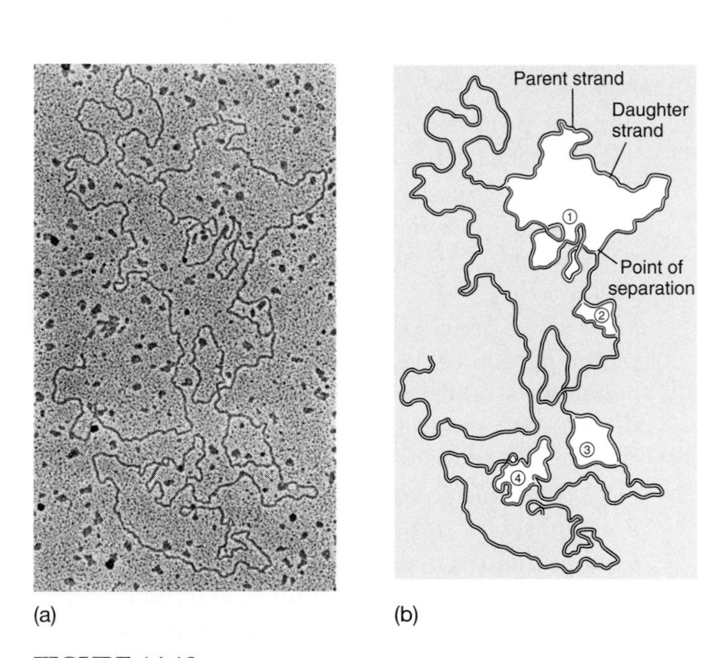

FIGURE 14.19
Eukaryotic chromosomes possess numerous replication forks spaced along their length. Four replication units (each with two replication forks) are producing daughter strands (*a*) in this electron micrograph, as indicated in *red* in the (*b*) corresponding drawing.

The One-Gene/One-Polypeptide Hypothesis

As the structure of DNA was being solved, other biologists continued to puzzle over how the genes of Mendel were related to DNA.

Garrod: Inherited Disorders Can Involve Specific Enzymes

In 1902, a British physician, Archibald Garrod, was working with one of the early Mendelian geneticists, his countryman William Bateson, when he noted that certain diseases he encountered among his patients seemed to be more prevalent in particular families. By examining several generations of these families, he found that some of the diseases behaved as if they were the product of simple recessive alleles. Garrod concluded that these disorders were Mendelian traits and that they had resulted from changes in the hereditary information in an ancestor of the affected families.

Garrod investigated several of these disorders in detail. In alkaptonuria the patients produced urine that contained homogentisic acid (alkapton). This substance oxidized rapidly when exposed to air, turning the urine black. In normal individuals, homogentisic acid is broken down into simpler substances. With considerable insight, Garrod concluded that patients suffering from alkaptonuria lacked the enzyme necessary to catalyze this breakdown. He speculated that many other inherited diseases might also reflect enzyme deficiencies.

Beadle and Tatum: Genes Specify Enzymes

From Garrod's finding, it took but a short leap of intuition to surmise that the information encoded within the DNA of chromosomes acts to specify particular enzymes. This point was not actually established, however, until 1941, when a series of experiments by Stanford University geneticists George Beadle and Edward Tatum provided definitive evidence on this point. Beadle and Tatum deliberately set out to create Mendelian mutations in chromosomes and then studied the effects of these mutations on the organism (figure 14.20).

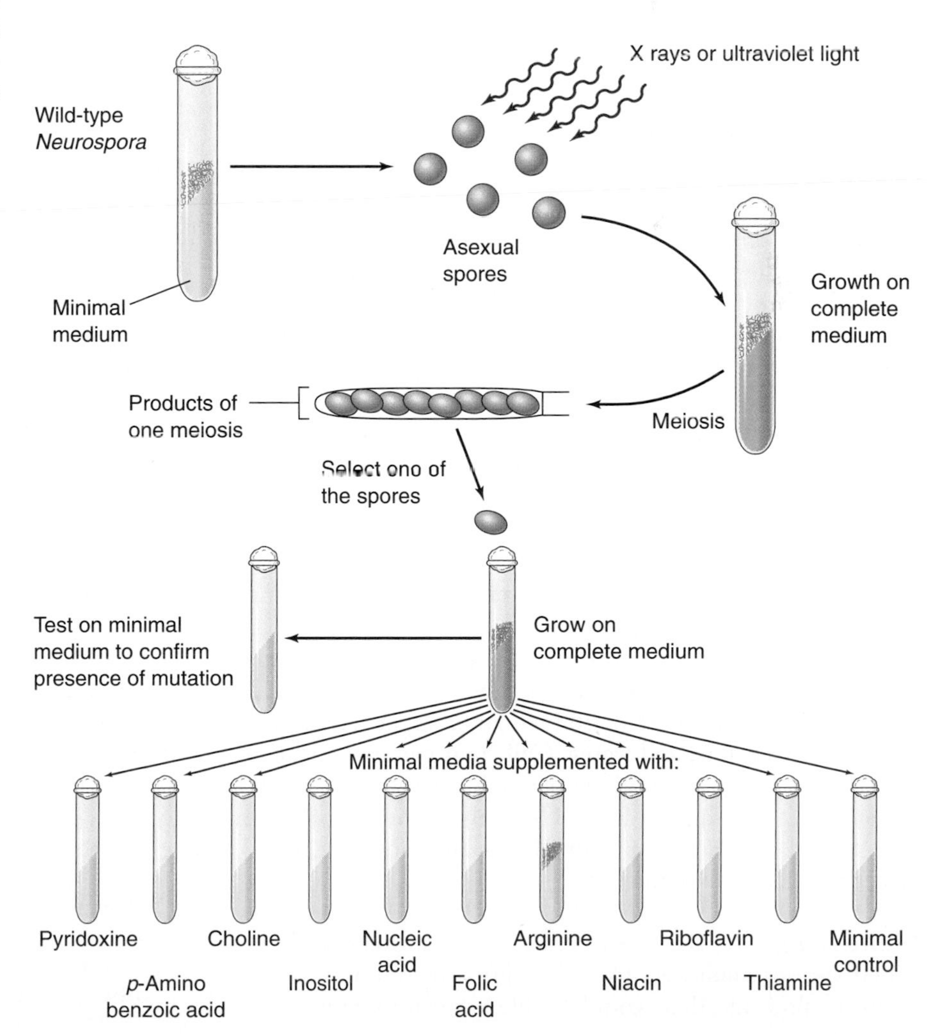

FIGURE 14.20
Beadle and Tatum's procedure for isolating nutritional mutants in *Neurospora*. This fungus grows easily on an artificial medium in test tubes. In this experiment, spores were irradiated to increase the frequency of mutation; they were then placed on a "complete" medium that contained all of the nutrients necessary for growth. Once the fungal colonies were established on the complete medium, individual spores were transferred to a "minimal" medium that lacked various substances the fungus could normally manufacture. Any spore that would not grow on the minimal medium but would grow on the complete medium contained one or more mutations in genes needed to produce the missing nutrients. To determine which gene had mutated, the minimal medium was supplemented with particular substances. The mutation illustrated here produced an arginine mutant, a collection of cells that lost the ability to manufacture arginine. These cells will not grow on minimal medium but will grow on minimal medium with only arginine added.

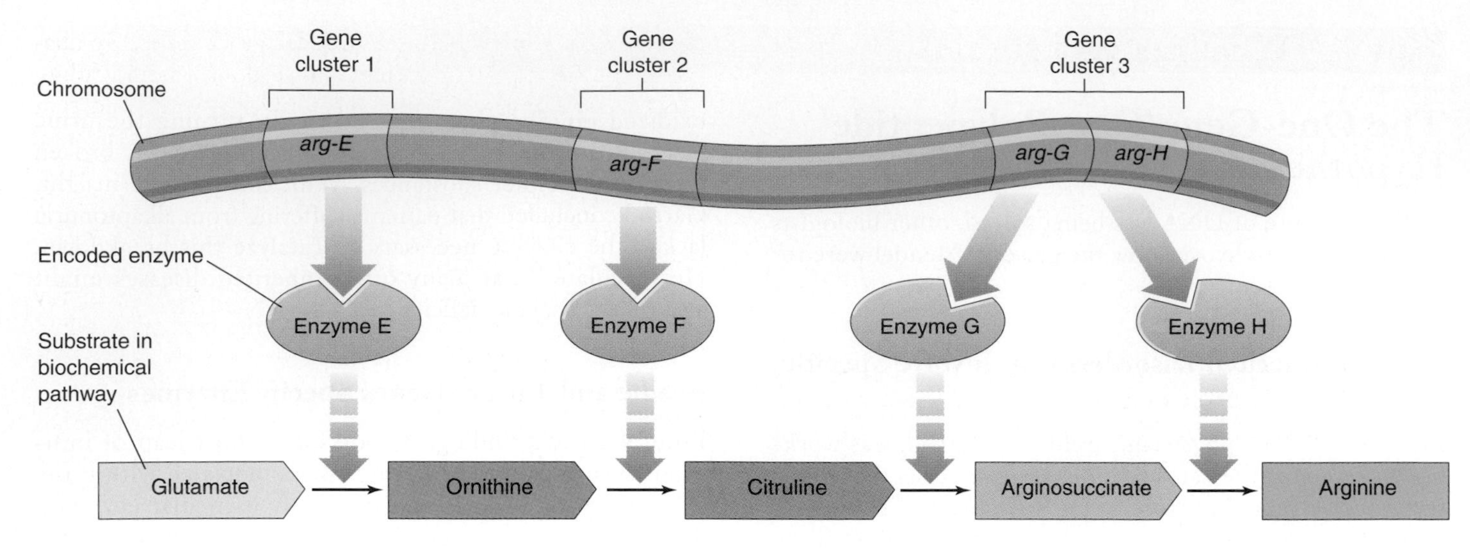

FIGURE 14.21
Evidence for the "one-gene/one-polypeptide" hypothesis. The chromosomal locations of the many arginine mutants isolated by Beadle and Tatum cluster around three locations. These locations correspond to the locations of the genes encoding the enzymes that carry out arginine biosynthesis.

A Defined System. One of the reasons Beadle and Tatum's experiments produced clear-cut results is that the researchers made an excellent choice of experimental organism. They chose the bread mold *Neurospora*, a fungus that can be grown readily in the laboratory on a defined medium (a medium that contains only known substances such as glucose and sodium chloride, rather than some uncharacterized mixture of substances such as ground-up yeasts). Beadle and Tatum exposed *Neurospora* spores to X rays, expecting that the DNA in some of the spores would experience damage in regions encoding the ability to make compounds needed for normal growth (see figure 14.20). DNA changes of this kind are called mutations, and organisms that have undergone such changes (in this case losing the ability to synthesize one or more compounds) are called mutants. Initially, they allowed the progeny of the irradiated spores to grow on a defined medium containing all of the nutrients necessary for growth, so that any growth-deficient mutants resulting from the irradiation would be kept alive.

Isolating Growth-Deficient Mutants. To determine whether any of the progeny of the irradiated spores had mutations causing metabolic deficiencies, Beadle and Tatum placed subcultures of individual fungal cells on a "minimal" medium that contained only sugar, ammonia, salts, a few vitamins, and water. Cells that had lost the ability to make other compounds necessary for growth would not survive on such a medium. Using this approach, Beadle and Tatum succeeded in identifying and isolating many growth-deficient mutants.

Identifying the Deficiencies. Next the researchers added various chemicals to the minimal medium in an attempt to find one that would enable a given mutant strain to grow. This procedure allowed them to pinpoint the nature of the biochemical deficiency that strain had. The addition of arginine, for example, permitted several mutant strains, dubbed *arg* mutants, to grow. When their chromosomal positions were located, the *arg* mutations were found to cluster in three areas.

One-Gene/One-Polypeptide

For each enzyme in the arginine biosynthetic pathway, Beadle and Tatum were able to isolate a mutant strain with a defective form of that enzyme, and the mutation was always located at one of a few specific chromosomal sites. Most importantly, they found there was a different site for each enzyme. Thus, each of the mutants they examined had a defect in a single enzyme, caused by a mutation at a single site on one chromosome. Beadle and Tatum concluded that genes produce their effects by specifying the structure of enzymes and that each gene encodes the structure of one enzyme (figure 14.21). They called this relationship the **one-gene/one-enzyme hypothesis.** Because many enzymes contain multiple protein or polypeptide subunits, each encoded by a separate gene, the relationship is today more commonly referred to as **"one-gene/one-polypeptide."**

Enzymes are responsible for catalyzing the synthesis of all the parts of an organism. They mediate the assembly of nucleic acids, proteins, carbohydrates, and lipids. Therefore, by encoding the structure of enzymes and other proteins, DNA specifies the structure of the organism itself.

Genetic traits are expressed largely as a result of the activities of enzymes. Organisms store hereditary information by encoding the structures of enzymes and other proteins in their DNA.

How DNA Encodes Protein Structure

What kind of information must a gene encode to specify a protein? For some time, the answer to that question was not clear, as protein structure seemed impossibly complex.

Sanger: Proteins Consist of Defined Sequences of Amino Acids

The picture changed in 1953, the same year in which Watson and Crick unraveled the structure of DNA. That year, the English biochemist Frederick Sanger, after many years of work, announced the complete sequence of amino acids in the protein insulin. Insulin, a small protein hormone, was the first protein for which the amino acid sequence was determined. Sanger's achievement was extremely significant because it demonstrated for the first time that proteins consisted of definable sequences of amino acids—for any given form of insulin, every molecule has the same amino acid sequence. Sanger's work soon led to the sequencing of many other proteins, and it became clear that all enzymes and other proteins are strings of amino acids arranged in a certain definite order. The information needed to specify a protein such as an enzyme, therefore, is an ordered list of amino acids.

Ingram: Single Amino Acid Changes in a Protein Can Have Profound Effects

Following Sanger's pioneering work, Vernon Ingram in 1956 discovered the molecular basis of sickle cell anemia, a protein defect inherited as a Mendelian disorder. By analyzing the structures of normal and sickle cell hemoglobin, Ingram, working at Cambridge University, showed that sickle cell anemia is caused by a change from glutamic acid to valine at a single position in the protein (figure 14.22). The alleles of the gene encoding hemoglobin differed only in their specification of this one amino acid in the hemoglobin amino acid chain.

These experiments and other related ones have finally brought us to a clear understanding of the unit of heredity. The characteristics of sickle cell anemia and most other hereditary traits are defined by changes in protein structure brought about by an alteration in the sequence of amino acids that make up the protein. This sequence in turn is dictated by the order of nucleotides in a particular region of the chromosome. For example, the critical change leading to sickle cell disease is a mutation that replaces a single thymine with an adenine at the position that codes for glutamic acid, converting the position to valine. The sequence of nucleotides that determines the amino acid sequence of a protein is called a **gene**. Although most genes encode proteins or subunits of proteins, some genes are devoted to the production of special forms of RNA, many of which play important roles in protein synthesis themselves.

A half-century of experimentation has made clear that DNA is the molecule responsible for the inheritance of traits, and that this molecule is divided into functional units called genes.

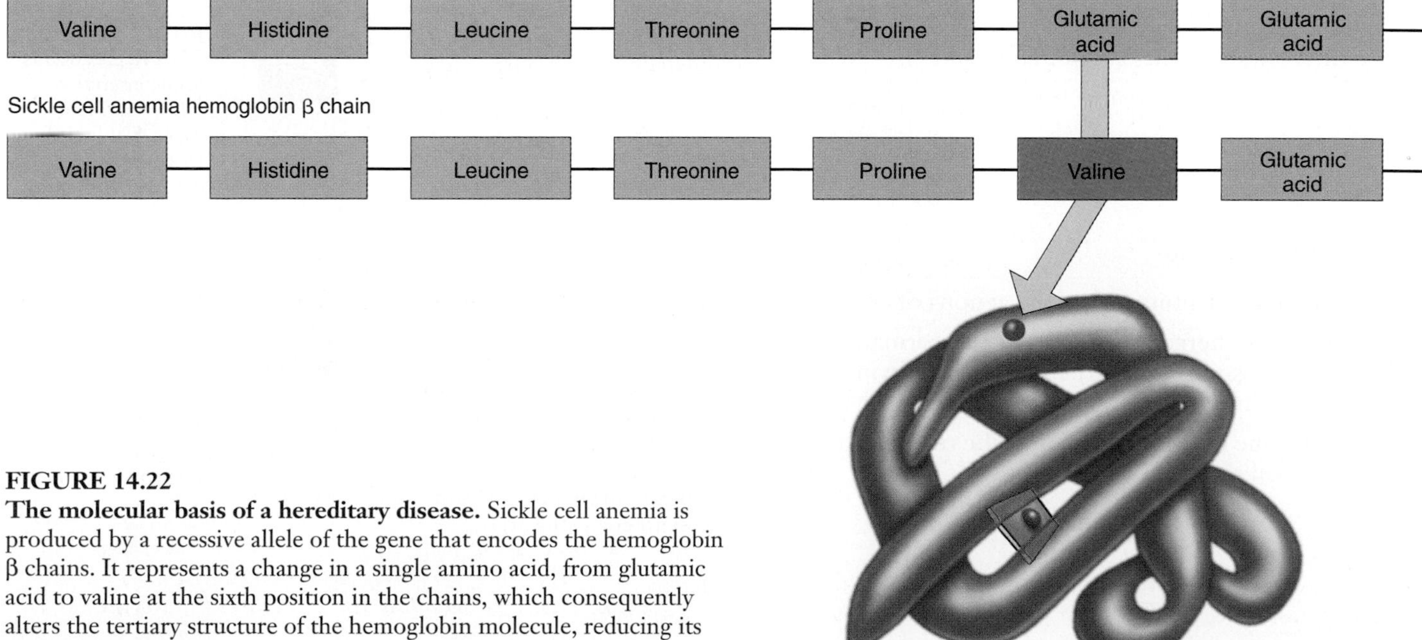

FIGURE 14.22
The molecular basis of a hereditary disease. Sickle cell anemia is produced by a recessive allele of the gene that encodes the hemoglobin β chains. It represents a change in a single amino acid, from glutamic acid to valine at the sixth position in the chains, which consequently alters the tertiary structure of the hemoglobin molecule, reducing its ability to carry oxygen.

Chapter 14

www.mhhe.com/raven6/resources14.mhtml

Summary	Questions		Media Resources
14.1 What is the genetic material?			
• Eukaryotic cells store hereditary information within the nucleus. • In viruses, bacteria, and eukaryotes, the hereditary information resides in nucleic acids. The transfer of nucleic acids can lead to the transfer of hereditary traits. • When radioactively labeled DNA viruses infect bacteria, the DNA but not the protein coat of the viruses enters the bacterial cells, indicating that the hereditary material is DNA rather than protein.	**1.** In Hammerling's experiments on *Acetabularia*, what happened when a stalk from *A. crenulata* was grafted to a foot from *A. mediterranea?* **2.** How did Hershey and Chase determine which component of bacterial viruses contains the viruses' hereditary information?	 	• How Scientists Think: DNA Is the Genetic Material (Griffith/ Hershey/Chase) • Art Quizzes: -Griffith-Experiment of Transformation -Hershey/Chase Experiment
14.2 What is the structure of DNA?			
• Chargaff showed that the proportion of adenine in DNA always equals that of thymine, and the proportion of guanine always equals that of cytosine. • DNA has the structure of a double helix, consisting of two chains of nucleotides held together by hydrogen bonds between adenines and thymines, and between guanines and cytosines.	**3.** What is the three-dimensional shape of DNA, and how does this shape fit with Chargaff's observations on the proportions of purines and pyrimidines in DNA?	 	• DNA Structure • Nucleic Acid • DNA Structure
14.3 How does DNA replicate?			
• The hereditary message in DNA is replicated with great accuracy through semiconservative replication. • During replication, the DNA duplex is unwound, and two new strands are assembled in opposite directions along the original strands. One strand elongates by the continuous addition of nucleotides to its growing end; the other is constructed by the addition of segments containing 100 to 2000 nucleotides, which are then joined to the end of that strand.	**4.** How did Meselson and Stahl show that DNA replication is semiconservative? **5.** How is the leading strand of a DNA duplex replicated? How is the lagging strand replicated? What is the basis for the requirement that the leading and lagging strands be replicated by different mechanisms?	  	• DNA Replication • DNA Replication • Activity: The Replication Process • How Scientists Think: -DNA Replication Is Semiconservative (Meselson/Stahl) - DNA Synthesis Is Discontinous (Okazaki)
14.4 What is a gene?			
• Most hereditary traits reflect the actions of enzymes. • The traits are hereditary because the information necessary to specify the structure of the enzymes is stored within the DNA. • Each enzyme is encoded by a specific region of the DNA called a gene.	**6.** What hypothesis did Beadle and Tatum test in their experiments on *Neurospora?* What did they do to change the DNA in individuals of this organism? How did they determine whether any of these changes affected enzymes in biosynthetic pathways?	 	• How Scientists Think: Genes Encode Enzymes (Ephrussi/ Beadle/Tatum) • Art Quiz: One-gene/One-Polypeptide Evidence

15

Genes and How They Work

Concept Outline

15.1 The Central Dogma traces the flow of gene-encoded information.

Cells Use RNA to Make Protein. The information in genes is expressed in two steps, first being transcribed into RNA, and the RNA then being translated into protein.

15.2 Genes encode information in three-nucleotide code words.

The Genetic Code. The sequence of amino acids in a protein is encoded in the sequence of nucleotides in DNA, three nucleotides encoding an amino acid.

15.3 Genes are first transcribed, then translated.

Transcription. The enzyme RNA polymerase unwinds the DNA helix and synthesizes an RNA copy of one strand.

Translation. mRNA is translated by activating enzymes that select tRNAs to match amino acids. Proteins are synthesized on ribosomes, which provide a framework for the interaction of tRNA and mRNA.

15.4 Eukaryotic gene transcripts are spliced.

The Discovery of Introns. Eukaryotic genes contain extensive material that is not translated.

Differences between Bacterial and Eukaryotic Gene Expression. Gene expression is broadly similar in bacteria and eukaryotes, although it differs in some respects.

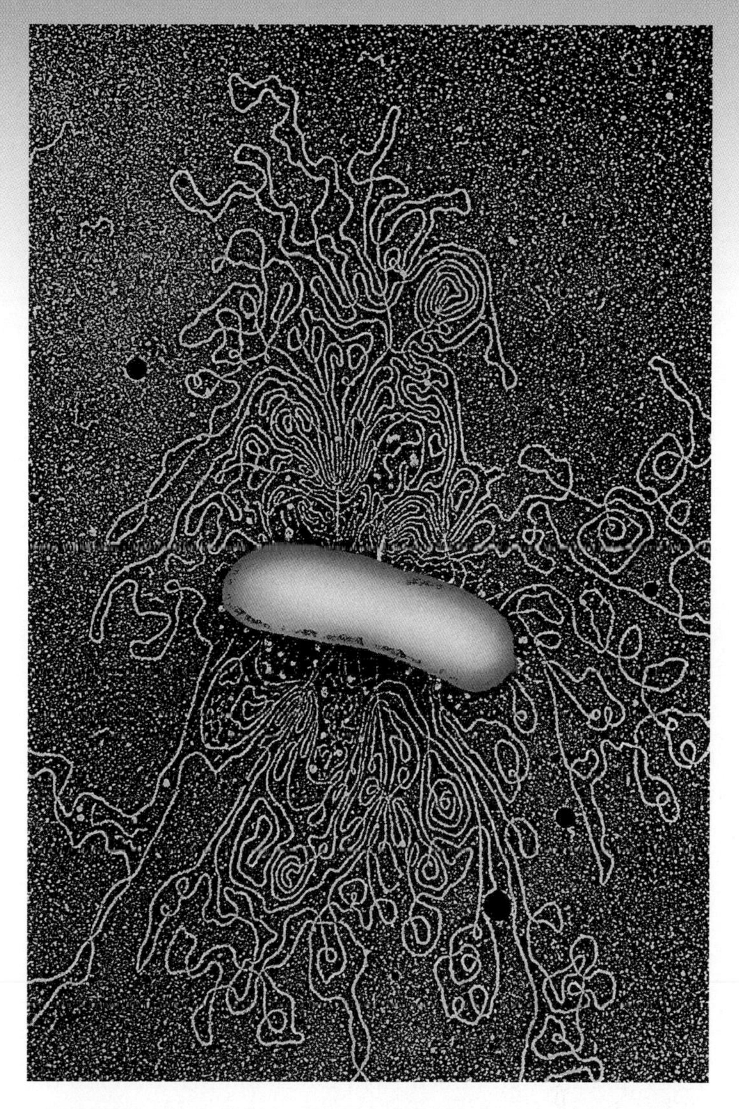

FIGURE 15.1
The unraveled chromosome of an *E. coli* bacterium. This complex tangle of DNA represents the full set of assembly instructions for the living organism *E. coli.*

Every cell in your body contains the hereditary instructions specifying that you will have arms rather than fins, hair rather than feathers, and two eyes rather than one. The color of your eyes, the texture of your fingernails, and all of the other traits you receive from your parents are recorded in the cells of your body. As we have seen, this information is contained in long molecules of DNA (figure 15.1). The essence of heredity is the ability of cells to use the information in their DNA to produce particular proteins, thereby affecting what the cells will be like. In that sense, proteins are the tools of heredity. In this chapter, we will examine how proteins are synthesized from the information in DNA, using both prokaryotes and eukaryotes as models.

Cells Use RNA to Make Protein

To find out how a eukaryotic cell uses its DNA to direct the production of particular proteins, you must first ask where in the cell the proteins are made. We can answer this question by placing cells in a medium containing radioactively labeled amino acids for a short time. The cells will take up the labeled amino acids and incorporate them into proteins. If we then look to see where in the cells radioactive proteins first appear, we will find that it is not in the nucleus, where the DNA is, but rather in the cytoplasm, on large RNA-protein aggregates called **ribosomes** (figure 15.2). These polypeptide-making factories are very complex, composed of several RNA molecules and over 50 different proteins (figure 15.3). Protein synthesis involves three different sites on the ribosome surface, called the P, A, and E sites, discussed later in this chapter.

Kinds of RNA

The class of RNA found in ribosomes is called **ribosomal RNA (rRNA).** During polypeptide synthesis, rRNA provides the site where polypeptides are assembled. In addition to rRNA, there are two other major classes of RNA in cells. **Transfer RNA (tRNA)** molecules both transport the amino acids to the ribosome for use in building the polypeptides and position each amino acid at the correct place on the elongating polypeptide chain (figure 15.4). Human cells contain about 45 different kinds of tRNA molecules. **Messenger RNA (mRNA)** molecules are long strands of RNA that are transcribed from DNA and that travel to the ribosomes to direct precisely *which* amino acids are assembled into polypeptides.

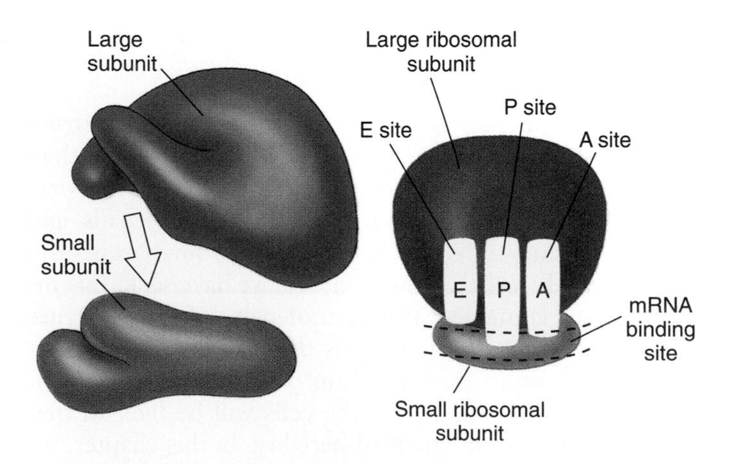

FIGURE 15.2
A ribosome is composed of two subunits. The smaller subunit fits into a depression on the surface of the larger one. The A, P, and E sites on the ribosome, discussed later in this chapter, play key roles in protein synthesis.

These RNA molecules, together with ribosomal proteins and certain enzymes, constitute a system that reads the genetic messages encoded by nucleotide sequences in the DNA and produces the polypeptides that those sequences specify. As we will see, biologists have also learned to read these messages. In so doing, they have learned a great deal about what genes are and how they are able to dictate what a protein will be like and when it will be made.

The Central Dogma

All organisms, from the simplest bacteria to ourselves, use the same basic mechanism of reading and expressing genes, so fundamental to life as we know it that it is often referred to as the "Central Dogma": Information passes from the genes (DNA) to an RNA copy of the gene, and the RNA copy directs the sequential assembly of a chain of amino acids (figure 15.5). Said briefly,

$$\text{DNA} \rightarrow \text{RNA} \rightarrow \text{protein}$$

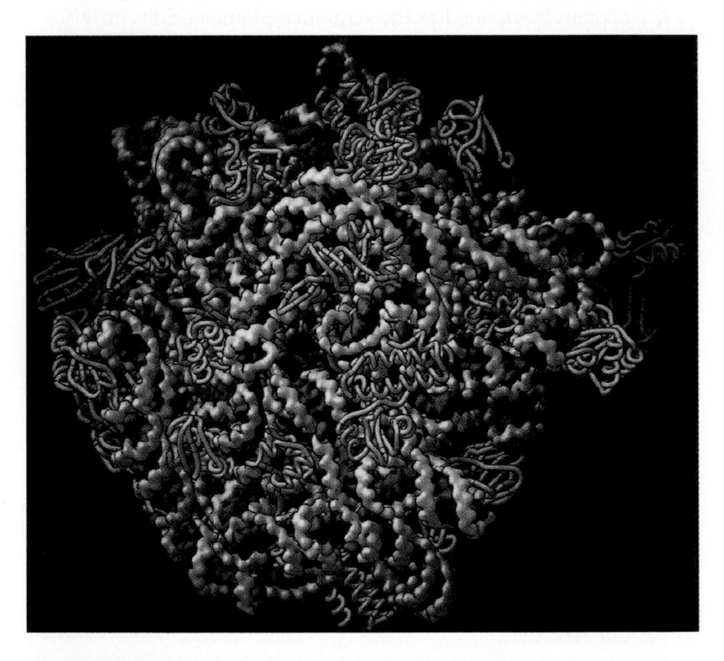

FIGURE 15.3
Ribosomes are very complex machines. The complete atomic structure of a bacterial large ribosomal subunit has recently been determined at 2.4 Å resolution. The RNA of the subunit is shown in *gray* and the proteins in *gold*. The subunit's RNA is twisted into irregular shapes that fit together like a three-dimensional jigsaw puzzle. The chemical reactions which form the peptide bond in protein synthesis are carried out deep in the interior by ribosomal RNA. The ribosome is thus a ribozyme. Proteins are absent from the active site but abundant everywhere on the surface. The proteins stabilize the structure by interacting with adjacent RNA strands.

Transcription: An Overview

The first step of the Central Dogma is the transfer of information from DNA to RNA, which occurs when an mRNA copy of the gene is produced. Like all classes of RNA, mRNA is formed on a DNA template. Because the DNA sequence in the gene is transcribed into an RNA sequence, this stage is called **transcription.** Transcription is initiated when the enzyme **RNA polymerase** binds to a particular binding site called a **promoter** located at the beginning of a gene. Starting there, the RNA polymerase moves along the strand into the gene. As it encounters each DNA nucleotide, it adds the corresponding complementary RNA nucleotide to a growing mRNA strand. Thus, guanine (G), cytosine (C), thymine (T), and adenine (A) in the DNA would signal the addition of C, G, A, and uracil (U), respectively, to the mRNA.

When the RNA polymerase arrives at a transcriptional "stop" signal at the opposite end of the gene, it disengages from the DNA and releases the newly assembled RNA chain. This chain is a complementary transcript of the gene from which it was copied.

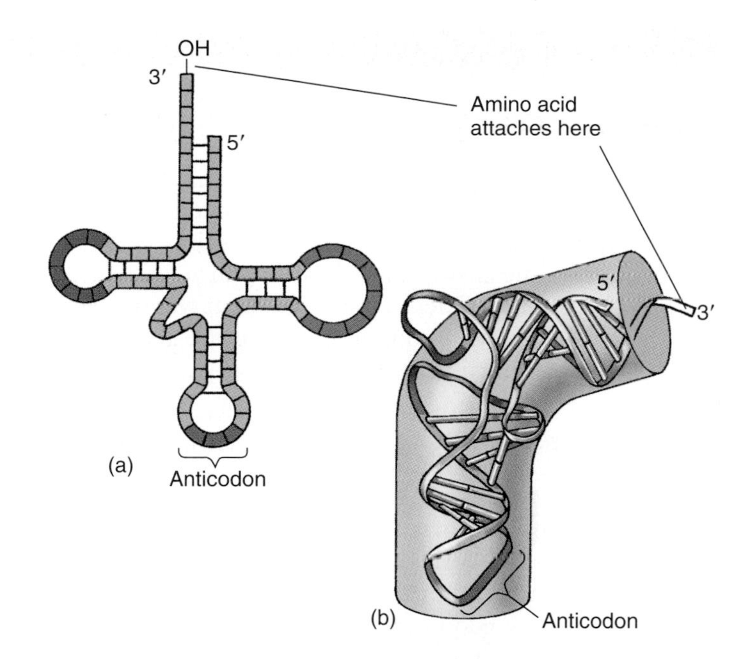

FIGURE 15.4
The structure of tRNA. (*a*) In the two-dimensional schematic, the three loops of tRNA are unfolded. Two of the loops bind to the ribosome during polypeptide synthesis, and the third loop contains the anticodon sequence, which is complementary to a three-base sequence on messenger RNA. Amino acids attach to the free, single-stranded —OH end. (*b*) In the three-dimensional structure, the loops of tRNA are folded.

Translation: An Overview

The second step of the Central Dogma is the transfer of information from RNA to protein, which occurs when the information contained in the mRNA transcript is used to direct the sequence of amino acids during the synthesis of polypeptides by ribosomes. This process is called **translation** because the nucleotide sequence of the mRNA transcript is translated into an amino acid sequence in the polypeptide. Translation begins when an rRNA molecule within the ribosome recognizes and binds to a "start" sequence on the mRNA. The ribosome then moves along the mRNA molecule, three nucleotides at a time. Each group of three nucleotides is a codeword that specifies which amino acid will be added to the growing polypeptide chain. The ribosome continues in this fashion until it encounters a translational "stop" signal; then it disengages from the mRNA and releases the completed polypeptide.

The two steps of the Central Dogma, taken together, are a concise summary of the events involved in the expression of an active gene. Biologists refer to this process as **gene expression.**

The information encoded in genes is expressed in two phases: transcription, in which an RNA polymerase enzyme assembles an mRNA molecule whose nucleotide sequence is complementary to the DNA nucleotide sequence of the gene; and translation, in which a ribosome assembles a polypeptide, whose amino acid sequence is specified by the nucleotide sequence in the mRNA.

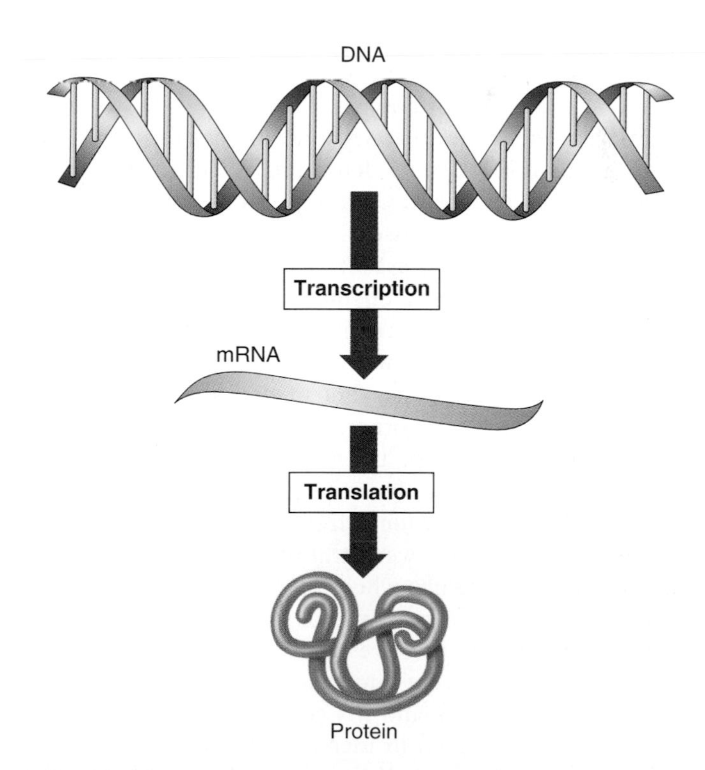

FIGURE 15.5
The Central Dogma of gene expression. DNA is transcribed to make mRNA, which is translated to make a protein.

15.2 Genes encode information in three-nucleotide code words.

The Genetic Code

The essential question of gene expression is, "How does the *order* of nucleotides in a DNA molecule encode the information that specifies the order of amino acids in a polypeptide?" The answer came in 1961, through an experiment led by Francis Crick. That experiment was so elegant and the result so critical to understanding the genetic code that we will describe it in detail.

Proving Code Words Have Only Three Letters

Crick and his colleagues reasoned that the genetic code most likely consisted of a series of blocks of information called **codons,** each corresponding to an amino acid in the encoded protein. They further hypothesized that the information within one codon was probably a sequence of three nucleotides specifying a particular amino acid. They arrived at the number three, because a two-nucleotide codon would not yield enough combinations to code for the 20 different amino acids that commonly occur in proteins. With four DNA nucleotides (G, C, T, and A), only 4^2, or 16, different pairs of nucleotides could be formed. However, these same nucleotides can be arranged in 4^3, or 64, different combinations of three, more than enough to code for the 20 amino acids.

In theory, the codons in a gene could lie immediately adjacent to each other, forming a continuous sequence of transcribed nucleotides. Alternatively, the sequence could be punctuated with untranscribed nucleotides between the codons, like the spaces that separate the words in this sentence. It was important to determine which method cells employ because these two ways of transcribing DNA imply different translating processes.

To choose between these alternative mechanisms, Crick and his colleagues used a chemical to delete one, two, or three nucleotides from a viral DNA molecule and then asked whether a gene downstream of the deletions was transcribed correctly. When they made a single deletion or two deletions near each other, the **reading frame** of the genetic message shifted, and the downstream gene was transcribed as nonsense. However, when they made three deletions, the correct reading frame was restored, and the sequences downstream were transcribed correctly. They obtained the same results when they made additions to the DNA consisting of one, two, or three nucleotides. As shown in figure 15.6, these results could not have been obtained if the codons were punctuated by untranscribed nucleotides. Thus, Crick and his colleagues concluded that the genetic code is read in increments consisting of three nucleotides (in other words, it is a **triplet code**) and that reading occurs continuously without punctuation between the three-nucleotide units.

Delete 1 base

WHYDIDTHEREDBATEATTHEFATRAT?
Hypothesis A : unpunctuated Delete T
WHY DID HER EDB ATE ATT HEF ATR AT?
(Nonsense)

WHYODIDOTHEOREDOBATOEATOTHEOFATORAT?
Hypothesis B : punctuated Delete T
O O R B E T F R
WHY DID HEO EDO ATO ATO HEO ATO AT?
(Nonsense)

Delete 3 bases

Hypothesis A : unpunctuated WHYDIDTHEREDBATEATTHEFATRAT?
Delete T,R,and A
WHY DID HEE DBT EAT THE FAT RAT?
(Nonsense) (Sense)

WHYODIDOTHEOREDOBATOEATOTHEOFATORAT?
Hypothesis B : punctuated Delete T,R,and A
O O E T T E T T
WHY DID HEO DOB OEA OTH OFA ORA?
(Nonsense)

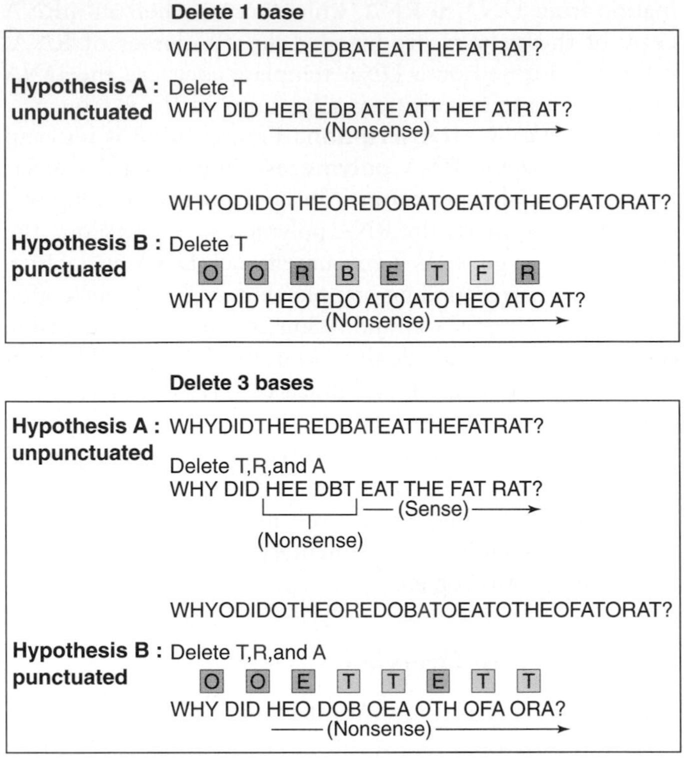

FIGURE 15.6
Using frame-shift alterations of DNA to determine if the genetic code is punctuated. The hypothetical genetic message presented here is "Why did the red bat eat the fat rat?" Under hypothesis B, which proposes that the message is punctuated, the three-letter words are separated by nucleotides that are not read (indicated by the letter "O").

Breaking the Genetic Code

Within a year of Crick's experiment, other researchers succeeded in determining the amino acids specified by particular three-nucleotide units. Marshall Nirenberg discovered in 1961 that adding the synthetic mRNA molecule polyU (an RNA molecule consisting of a string of uracil nucleotides) to cell-free systems resulted in the production of the polypeptide polyphenylalanine (a string of phenylalanine amino acids). Therefore, one of the three-nucleotide sequences specifying phenylalanine is UUU. In 1964, Nirenberg and Philip Leder developed a powerful **triplet binding assay** in which a specific triplet was tested to see which radioactive amino acid (complexed to tRNA) it would bind. Some 47 of the 64 possible triplets gave unambiguous results. Har Gobind Khorana decoded the remaining 17 triplets by constructing artificial mRNA molecules of defined sequence and examining what polypeptides they directed. In these ways, all 64 possible three-nucleotide sequences were tested, and the full genetic code was determined (table 15.1).

Table 15.1 The Genetic Code

First Letter	Second Letter: U		Second Letter: C		Second Letter: A		Second Letter: G		Third Letter
U	UUU	Phenylalanine	UCU	Serine	UAU	Tyrosine	UGU	Cysteine	U
	UUC		UCC		UAC		UGC		C
	UUA	Leucine	UCA		UAA	Stop	UGA	Stop	A
	UUG		UCG		UAG	Stop	UGG	Tryptophan	G
C	CUU	Leucine	CCU	Proline	CAU	Histidine	CGU	Arginine	U
	CUC		CCC		CAC		CGC		C
	CUA		CCA		CAA	Glutamine	CGA		A
	CUG		CCG		CAG		CGG		G
A	AUU	Isoleucine	ACU	Threonine	AAU	Asparagine	AGU	Serine	U
	AUC		ACC		AAC		AGC		C
	AUA		ACA		AAA	Lysine	AGA	Arginine	A
	AUG	Methionine; Start	ACG		AAG		AGG		G
G	GUU	Valine	GCU	Alanine	GAU	Aspartate	GGU	Glycine	U
	GUC		GCC		GAC		GGC		C
	GUA		GCA		GAA	Glutamate	GGA		A
	GUG		GCG		GAG		GGG		G

A codon consists of three nucleotides read in the sequence shown. For example, ACU codes for threonine. The first letter, A, is in the First Letter column; the second letter, C, is in the Second Letter column; and the third letter, U, is in the Third Letter column. Each of the mRNA codons is recognized by a corresponding anticodon sequence on a tRNA molecule. Some tRNA molecules recognize more than one codon in mRNA, but they always code for the same amino acid. In fact, most amino acids are specified by more than one codon. For example, threonine is specified by four codons, which differ only in the third nucleotide (ACU, ACC, ACA, and ACG).

The Code Is Practically Universal

The genetic code is the same in almost all organisms. For example, the codon AGA specifies the amino acid arginine in bacteria, in humans, and in all other organisms whose genetic code has been studied. The universality of the genetic code is among the strongest evidence that all living things share a common evolutionary heritage. Because the code is universal, genes transcribed from one organism can be translated in another; the mRNA is fully able to dictate a functionally active protein. Similarly, genes can be transferred from one organism to another and be successfully transcribed and translated in their new host. This universality of gene expression is central to many of the advances of genetic engineering. Many commercial products such as the insulin used to treat diabetes are now manufactured by placing human genes into bacteria, which then serve as tiny factories to turn out prodigious quantities of insulin.

But Not Quite

In 1979, investigators began to determine the complete nucleotide sequences of the mitochondrial genomes in humans, cattle, and mice. It came as something of a shock when these investigators learned that the genetic code used by these mammalian mitochondria was not quite the same as the "universal code" that has become so familiar to biologists. In the mitochondrial genomes, what should have been a "stop" codon, UGA, was instead read as the amino acid tryptophan, AUA was read as methionine rather than isoleucine; and AGA and AGG were read as "stop" rather than arginine. Furthermore, minor differences from the universal code have also been found in the genomes of chloroplasts and ciliates (certain types of protists).

Thus, it appears that the genetic code is not quite universal. Some time ago, presumably after they began their endosymbiotic existence, mitochondria and chloroplasts began to read the code differently, particularly the portion of the code associated with "stop" signals.

Within genes that encode proteins, the nucleotide sequence of DNA is read in blocks of three consecutive nucleotides, without punctuation between the blocks. Each block, or codon, codes for one amino acid.

15.3 Genes are first transcribed, then translated.

Transcription

The first step in gene expression is the production of an RNA copy of the DNA sequence encoding the gene, a process called **transcription.** To understand the mechanism behind the transcription process, it is useful to focus first on RNA polymerase, the remarkable enzyme responsible for carrying it out (figure 15.7).

RNA Polymerase

RNA polymerase is best understood in bacteria. Bacterial RNA polymerase is very large and complex, consisting of five subunits: two α subunits bind regulatory proteins, a β' subunit binds the DNA template, a β subunit binds RNA nucleoside subunits, and a σ subunit recognizes the promoter and initiates synthesis. Only one of the two strands of DNA, called the **template strand,** is transcribed. The RNA transcript's sequence is complementary to the template strand. The strand of DNA that is not transcribed is called the **coding strand.** It has the same sequence as the RNA transcript, except T takes the place of U. The coding strand is also known as the sense (+) strand, and the template strand as the antisense (–) strand.

In both bacteria and eukaryotes, the polymerase adds ribonucleotides to the growing 3′ end of an RNA chain. No primer is needed, and synthesis proceeds in the $5' \rightarrow 3'$ direction. Bacteria contain only one RNA polymerase enzyme, while eukaryotes have three different RNA polymerases: RNA polymerase I synthesizes rRNA in the nucleolus; RNA polymerase II synthesizes mRNA; and RNA polymerase III synthesizes tRNA.

Promoter

Transcription starts at RNA polymerase binding sites called **promoters** on the DNA template strand. A promoter is a short sequence that is not itself transcribed by the polymerase that binds to it. Striking similarities are evident in the sequences of different promoters. For example, two six-base sequences are common to many bacterial promoters, a TTGACA sequence called the **–35 sequence,** located 35 nucleotides upstream of the position where transcription actually starts, and a TATAAT sequence called the **–10 sequence,** located 10 nucleotides upstream of the start site. In eukaryotic DNA, the sequence TATAAA, called the **TATA box,** is located at –25 and is very similar to the prokaryotic –10 sequence but is farther from the start site.

Promoters differ widely in efficiency. Strong promoters cause frequent initiations of transcription, as often as every 2 seconds in some bacteria. Weak promoters may transcribe only once every 10 minutes. Most strong promoters have unaltered –35 and –10 sequences, while weak promoters often have substitutions within these sites.

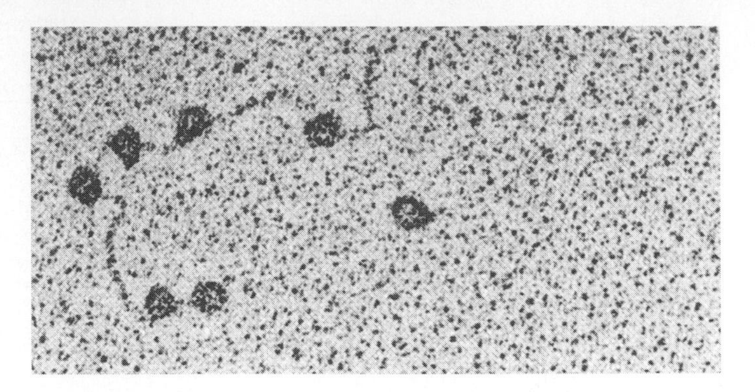

FIGURE 15.7
RNA polymerase. In this electron micrograph, the dark circles are RNA polymerase molecules bound to several promoter sites on bacterial virus DNA.

Initiation

The binding of RNA polymerase to the promoter is the first step in gene transcription. In bacteria, a subunit of RNA polymerase called **σ (sigma)** recognizes the –10 sequence in the promoter and binds RNA polymerase there. Importantly, this subunit can detect the –10 sequence without unwinding the DNA double helix. In eukaryotes, the –25 sequence plays a similar role in initiating transcription, as it is the binding site for a key protein factor. Other eukaryotic factors then bind one after another, assembling a large and complicated **transcription complex.** The eukaryotic transcription complex is described in detail in the following chapter.

Once bound to the promoter, the RNA polymerase begins to unwind the DNA helix. Measurements indicate that bacterial RNA polymerase unwinds a segment approximately 17 base-pairs long, nearly two turns of the DNA double helix. This sets the stage for the assembly of the RNA chain.

Elongation

The transcription of the RNA chain usually starts with ATP or GTP. One of these forms the 5′ end of the chain, which grows in the $5' \rightarrow 3'$ direction as ribonucleotides are added. Unlike DNA synthesis, a primer is not required. The region containing the RNA polymerase, DNA, and growing RNA transcript is called the **transcription bubble** because it contains a locally unwound "bubble" of DNA (figure 15.8). Within the bubble, the first 12 bases of the

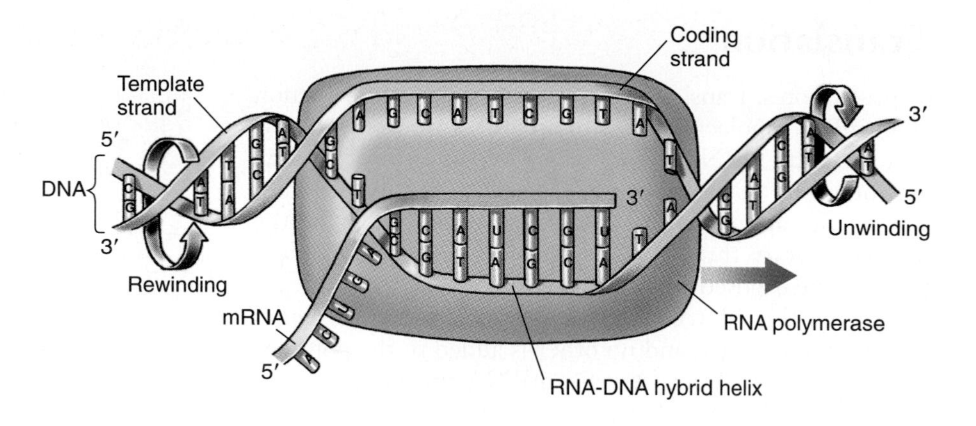

FIGURE 15.8
Model of a transcription bubble. The DNA duplex unwinds as it enters the RNA polymerase complex and rewinds as it leaves. One of the strands of DNA functions as a template, and nucleotide building blocks are assembled into RNA from this template.

newly synthesized RNA strand temporarily form a helix with the template DNA strand. Corresponding to not quite one turn of the helix, this stabilizes the positioning of the 3′ end of the RNA so it can interact with an incoming ribonucleotide. The RNA-DNA hybrid helix rotates each time a nucleotide is added so that the 3′ end of the RNA stays at the catalytic site.

The transcription bubble moves down the DNA at a constant rate, about 50 nucleotides per second, leaving the growing RNA strand protruding from the bubble. After the transcription bubble passes, the now transcribed DNA is rewound as it leaves the bubble.

Unlike DNA polymerase, RNA polymerase has no proofreading capability. Transcription thus produces many more copying errors than replication. These mistakes, however, are not transmitted to progeny. Most genes are transcribed many times, so a few faulty copies are not harmful.

Termination

At the end of a gene are "stop" sequences that cause the formation of phosphodiester bonds to cease, the RNA-DNA hybrid within the transcription bubble to dissociate, the RNA polymerase to release the DNA, and the DNA within the transcription bubble to rewind. The simplest stop signal is a series of GC base-pairs followed by a series of AT base-pairs. The RNA transcript of this stop region forms a GC hairpin (figure 15.9), followed by four or more U ribonucleotides. How does this structure terminate transcription? The hairpin causes the RNA polymerase to pause immediately after the polymerase has synthesized it, placing the polymerase directly over the run of four uracils. The pairing of U with DNA's A is the weakest of the four hybrid base-pairs and is not strong enough to hold the hybrid strands together during the long pause. Instead, the RNA strand dissociates from the DNA within the transcription bubble, and transcription stops. A variety of protein factors aid hairpin loops in terminating transcription of particular genes.

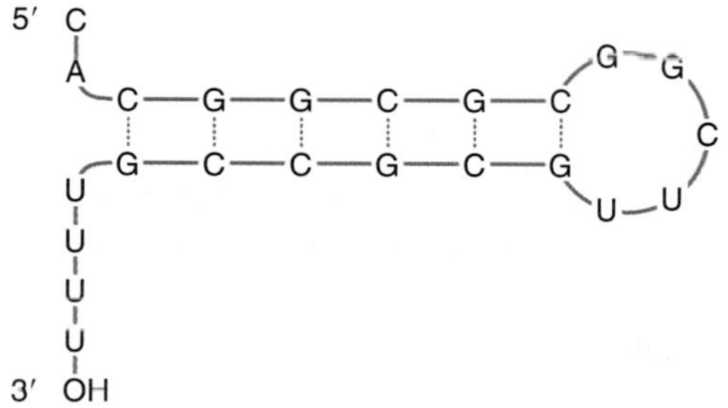

```
5′ C
   A
    C—G—G—C—G—C   G G
                      C
    G—C—C—G—C—G   U U
   U
   U
   U
   U
3′ OH
```

FIGURE 15.9
A GC hairpin. This structure stops gene transcription.

Posttranscriptional Modifications

In eukaryotes, every mRNA transcript must travel a long journey out from the nucleus into the cytoplasm before it can be translated. Eukaryotic mRNA transcripts are modified in several ways to aid this journey:

5′ caps. Transcripts usually begin with A or G, and, in eukaryotes, the terminal phosphate of the 5′ A or G is removed, and then a very unusual 5′-5′ linkage forms with GTP. Called a **5′ cap,** this structure protects the 5′ end of the RNA template from nucleases and phosphatases during its long journey through the cytoplasm. Without these caps, RNA transcripts are rapidly degraded.

3′ poly-A tails. The 3′ end of eukaryotic transcript is cleaved off at a specific site, often containing the sequence AAUAAA. A special poly-A polymerase enzyme then adds about 250 A ribonucleotides to the 3′ end of the transcript. Called a **3′ poly-A tail,** this long string of A's protects the transcript from degradation by nucleases. It also appears to make the transcript a better template for protein synthesis.

Transcription is carried out by the enzyme RNA polymerase, aided in eukaryotes by many other proteins.

Translation

In prokaryotes, translation begins when the initial portion of an mRNA molecule binds to an rRNA molecule in a ribosome. The mRNA lies on the ribosome in such a way that only one of its codons is exposed at the polypeptide-making site at any time. A tRNA molecule possessing the complementary three-nucleotide sequence, or anticodon, binds to the exposed codon on the mRNA.

Because this tRNA molecule carries a particular amino acid, that amino acid and no other is added to the polypeptide in that position. As the mRNA molecule moves through the ribosome, successive codons on the mRNA are exposed, and a series of tRNA molecules bind one after another to the exposed codons. Each of these tRNA molecules carries an attached amino acid, which it adds to the end of the growing polypeptide chain (figure 15.10).

There are about 45 different kinds of tRNA molecules. Why are there 45 and not 64 tRNAs (one for each codon)? Because the third base-pair of a tRNA anticodon allows some "wobble," some tRNAs recognize more than one codon.

How do particular amino acids become associated with particular tRNA molecules? The key translation step, which pairs the three-nucleotide sequences with appropriate amino acids, is carried out by a remarkable set of enzymes called activating enzymes.

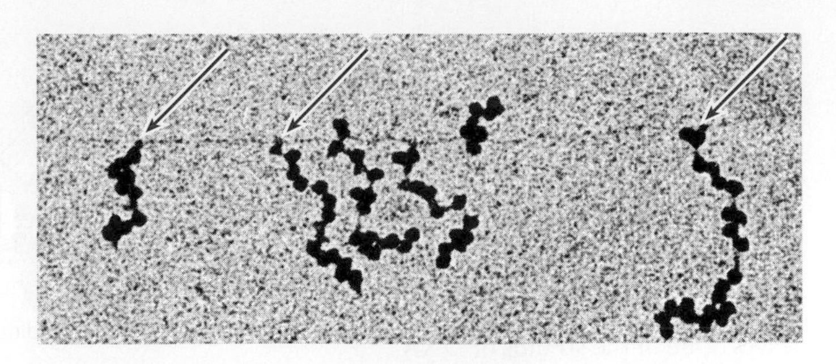

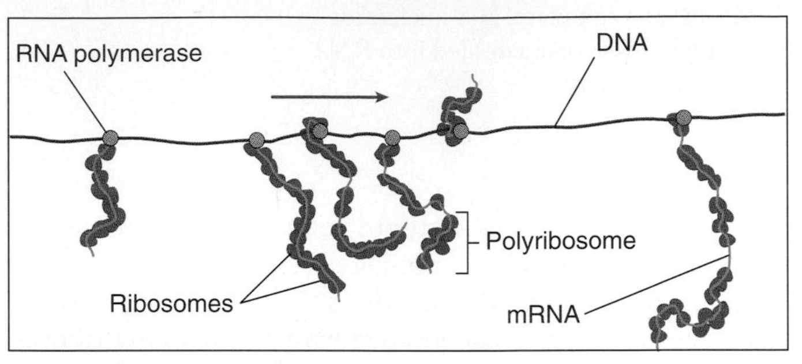

FIGURE 15.10
Translation in action. Bacteria have no nucleus and hence no membrane barrier between the DNA and the cytoplasm. In this electron micrograph of genes being transcribed in the bacterium *Escherichia coli*, you can see every stage of the process. The arrows point to RNA polymerase enzymes. From each mRNA molecule dangling from the DNA, a series of ribosomes is assembling polypeptides. These clumps of ribosomes are sometimes called "polyribosomes."

Activating Enzymes

Particular tRNA molecules become attached to specific amino acids through the action of activating enzymes called **aminoacyl-tRNA synthetases,** one of which exists for each of the 20 common amino acids (figure 15.11). Therefore, these enzymes must correspond to specific anticodon sequences on a tRNA molecule as well as particular amino acids. Some activating enzymes correspond to only one anticodon and thus only one tRNA molecule. Others recognize two, three, four, or six different tRNA molecules, each with a different anticodon but coding for the same amino acid (see table 15.1). If one considers the nucleotide sequence of mRNA a coded message, then the 20 activating enzymes are responsible for decoding that message.

"Start" and "Stop" Signals

There is no tRNA with an anticodon complementary to three of the 64 codons: UAA, UAG, and UGA. These codons, called **nonsense codons,** serve as "stop" signals in the mRNA message, marking the end of a polypeptide. The "start" signal that marks the beginning of a polypeptide within an mRNA message is the codon AUG, which also encodes the amino acid methionine. The ribosome will usually use the first AUG that it encounters in the mRNA to signal the start of translation.

Initiation

In prokaryotes, polypeptide synthesis begins with the formation of an **initiation complex.** First, a tRNA molecule carrying a chemically modified methionine called *N*-formylmethionine ($tRNA^{fMet}$) binds to the small ribosomal subunit. Proteins called **initiation factors** position the $tRNA^{fMet}$ on the ribosomal surface at the *P site* (for peptidyl), where peptide bonds will form. Nearby, two other sites will form: the *A site* (for aminoacyl), where successive amino acid-bearing tRNAs will bind, and the *E site* (for exit), where empty tRNAs will exit the ribosome (figure 15.12). This initiation complex, guided by another initiation factor, then binds to the anticodon AUG on the mRNA. Proper positioning of the mRNA is critical because it determines the reading frame—that is, which groups of three nucleotides will be read as codons. Moreover, the complex must bind to the beginning of the mRNA molecule, so that all of the transcribed gene will be translated. In bacteria, the beginning of each mRNA molecule is marked by a *leader sequence* complementary to one of the rRNA molecules on the ribosome. This complementarity ensures that the mRNA is read from the beginning. Bacteria often include several genes within a single mRNA transcript (polycistronic mRNA), while each eukaryotic gene is transcribed on a separate mRNA (monocistronic mRNA).

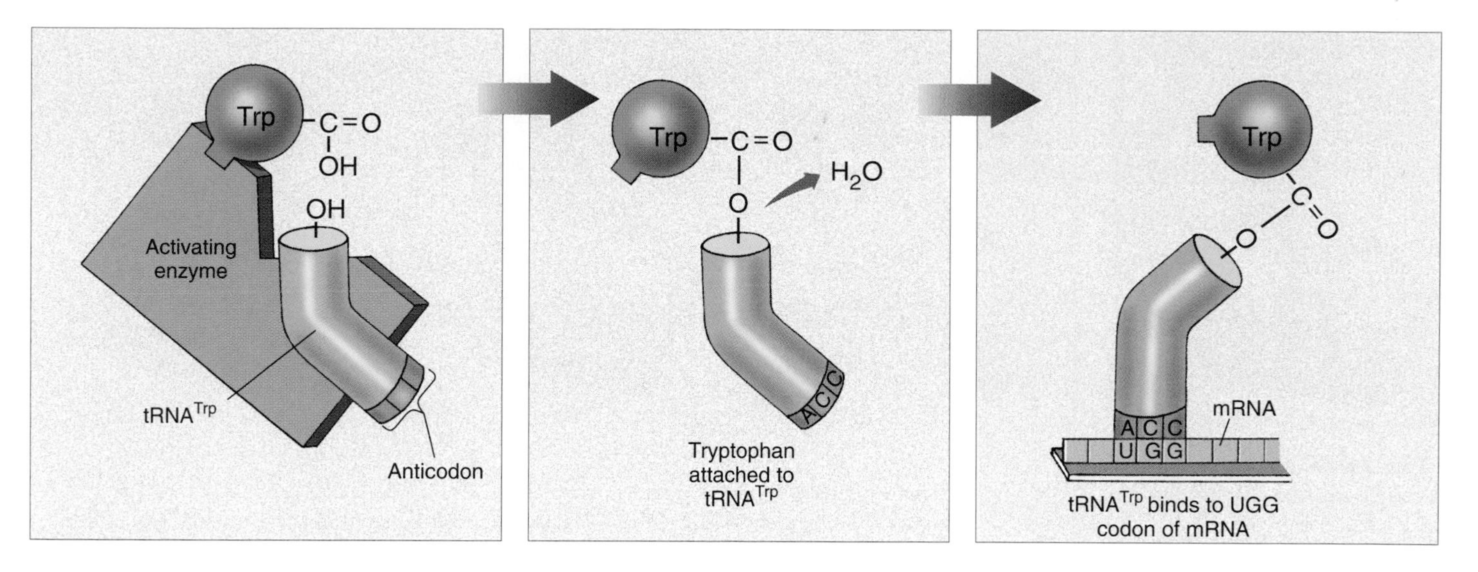

FIGURE 15.11
Activating enzymes "read" the genetic code. Each kind of activating enzyme recognizes and binds to a specific amino acid, such as tryptophan; it also recognizes and binds to the tRNA molecules with anticodons specifying that amino acid, such as ACC for tryptophan. In this way, activating enzymes link the tRNA molecules to specific amino acids.

Initiation in eukaryotes is similar, although it differs in two important ways. First, in eukaryotes, the initiating amino acid is methionine rather than *N*-formylmethionine. Second, the initiation complex is far more complicated than in bacteria, containing nine or more protein factors, many consisting of several subunits. Eukaryotic initiation complexes are discussed in detail in the following chapter.

Elongation

After the initiation complex has formed, the large ribosome subunit binds, exposing the mRNA codon adjacent to the initiating AUG codon, and so positioning it for interaction with another amino acid-bearing tRNA molecule. When a tRNA molecule with the appropriate anticodon appears, proteins called elongation factors assist in binding it to the exposed mRNA codon at the A site. When the second tRNA binds to the ribosome, it places its amino acid directly adjacent to the initial methionine, which is still attached to its tRNA molecule, which in turn is still bound to the ribosome. The two amino acids undergo a chemical reaction, catalyzed by the large ribosomal subunit, which releases the initial methionine from its tRNA and attaches it instead by a peptide bond to the second amino acid.

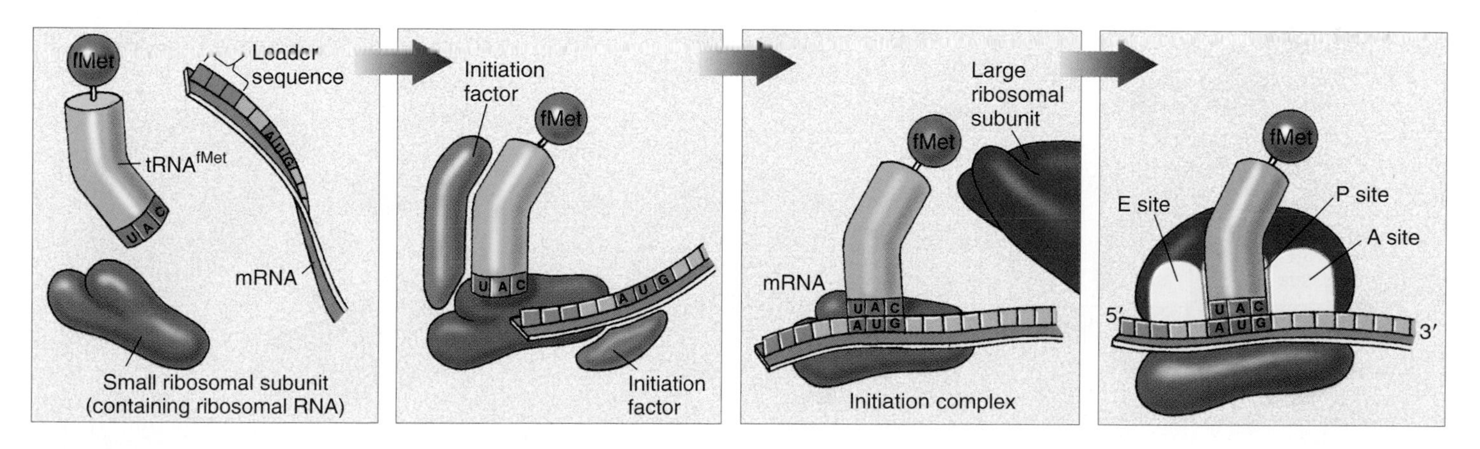

FIGURE 15.12
Formation of the initiation complex. In prokaryotes, proteins called initiation factors play key roles in positioning the small ribosomal subunit and the *N*-formylmethionine tRNA, or $tRNA^{fMet}$, molecule at the beginning of the mRNA. When the $tRNA^{fMet}$ is positioned over the first AUG codon of the mRNA, the large ribosomal subunit binds, forming the P, A, and E sites where successive tRNA molecules bind to the ribosomes, and polypeptide synthesis begins.

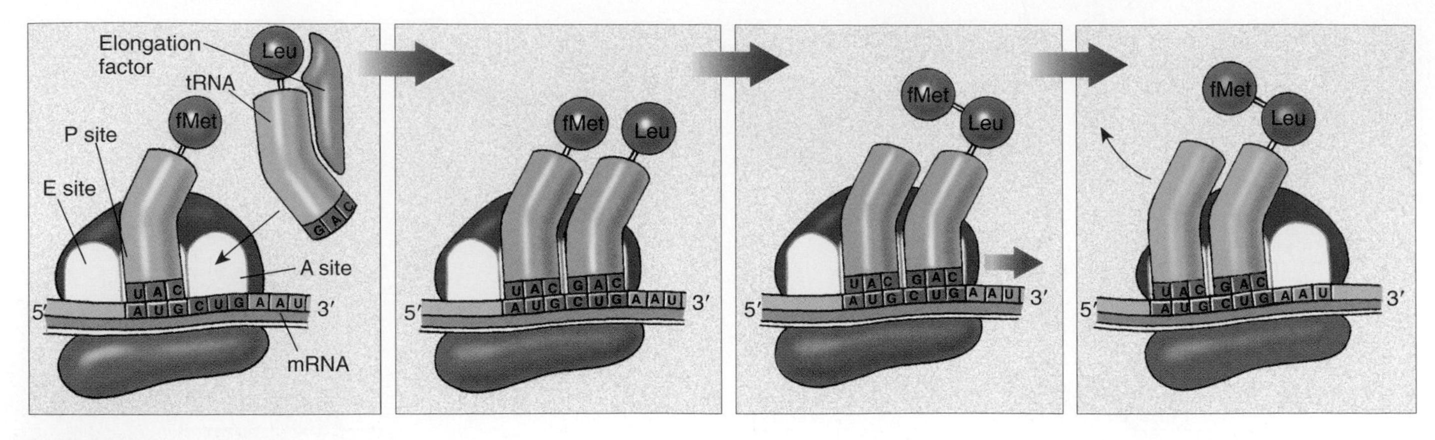

FIGURE 15.13
Translocation. The initiating tRNA^{fMet} in prokaryotes (tRNA^{Met} in eukaryotes) occupies the P site, and a tRNA molecule with an anticodon complementary to the exposed mRNA codon binds at the A site. fMet is transferred to the incoming amino acid (Leu), as the ribosome moves three nucleotides to the right along the mRNA. The empty tRNA^{fMet} moves to the E site to exit the ribosome, the growing polypeptide chain moves to the P site, and the A site is again exposed and ready to bind the next amino acid–laden tRNA.

Translocation

In a process called **translocation** (figure 15.13), the ribosome now moves (translocates) three more nucleotides along the mRNA molecule in the $5' \rightarrow 3'$ direction, guided by other elongation factors. This movement relocates the initial tRNA to the E site and ejects it from the ribosome, repositions the growing polypeptide chain (at this point containing two amino acids) to the P site, and exposes the next codon on the mRNA at the A site. When a tRNA molecule recognizing that codon appears, it binds to the codon at the A site, placing its amino acid adjacent to the growing chain. The chain then transfers to the new amino acid, and the entire process is repeated.

Termination

Elongation continues in this fashion until a chain-terminating nonsense codon is exposed (for example, UAA in figure 15.14). Nonsense codons do not bind to tRNA, but they are recognized by **release factors,** proteins that release the newly made polypeptide from the ribosome.

The first step in protein synthesis is the formation of an initiation complex. Each step of the ribosome's progress exposes a codon, to which a tRNA molecule with the complementary anticodon binds. The amino acid carried by each tRNA molecule is added to the end of the growing polypeptide chain.

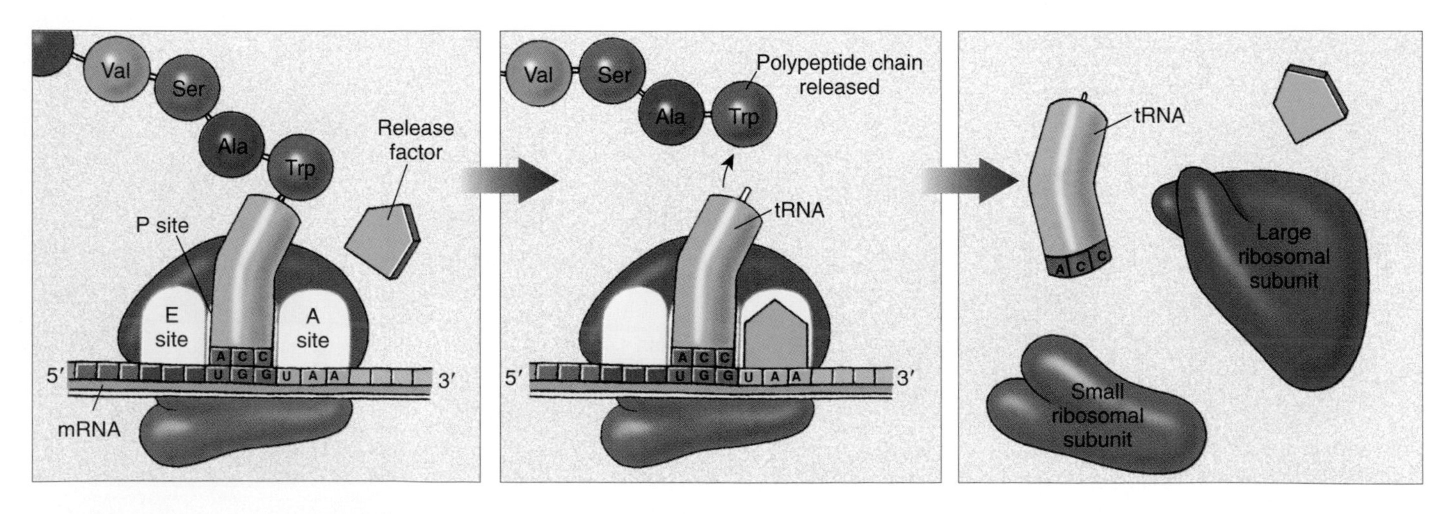

FIGURE 15.14
Termination of protein synthesis. There is no tRNA with an anticodon complementary to any of the three termination signal codons, such as the UAA nonsense codon illustrated here. When a ribosome encounters a termination codon, it therefore stops translocating. A specific release factor facilitates the release of the polypeptide chain by breaking the covalent bond that links the polypeptide to the P-site tRNA.

15.4 Eukaryotic gene transcripts are spliced.

The Discovery of Introns

While the mechanisms of protein synthesis are similar in bacteria and eukaryotes, they are not identical. One difference is of particular importance. Unlike bacterial genes, most eukaryotic genes are far larger than they need to be to produce the polypeptides they code for.

A typical eukaryotic gene is not simply a straight sequence of DNA, the order of its units corresponding to the sequence of amino acids in a protein. Instead, a eukaryotic gene is fragmented. The sequence of DNA units that specifies a protein is broken into many bits called **exons** that are scattered about within a gene among much longer segments of noncoding DNA called **introns**. Imagine looking at an interstate highway from a satellite. Scattered randomly along the thread of concrete would be cars, some moving in clusters, others individually; most of the road would be bare. That is what a eukaryotic gene is like, scattered exons embedded within much longer sequences of introns. In humans, only 1% to 1.5% of the genome is devoted to the exons that encode proteins, while 24% is devoted to the noncoding introns within which these exons are embedded.

When a eukaryotic cell transcribes a gene, it first produces a **primary RNA transcript** of the entire gene. The primary transcript is then processed. First, enzyme-RNA complexes called small nuclear ribonucleoproteins or **snRNPs** (pronounced "snurps) recognize short nucleotide sequences at the ends of introns. Several different snRNPs then associate with proteins to form a large assembly called a **spliceosome**, almost as big as a ribosome. Within the spliceosome, the introns of the primary RNA transcript become folded into loops, shoving the exons close to one another. Cutting the DNA at these sites, the spliceosome excises out the introns and joins together the exons to form the much shorter mature RNA transcript that is actually translated into protein (figure 15.15).

Because introns are excised from the primary RNA transcript before the transcript is translated into protein, introns do not affect the structure of the protein encoded by the gene in which they occur—despite comprising over 90% of a typical human gene.

Why introns? It appears that many human genes can be spliced together by spliceosomes in more than one way. In many instances, exons are not just a random fragments, but rather functional modules. One exon may encode a straight stretch of protein, another a curve, yet another a flat place. Like mixing tinker toy parts, you can construct quite different assemblies by employing the same exons in different combinations and orders.

With this sort of **alternative splicing**, the 30,000 genes of the human genome seem to encode as many as 120,000 different expressed messenger RNAs. It seems we humans achieved added complexity not by gaining more gene parts (we have only twice as many genes as a fruit fly), but rather by learning new ways to put them together.

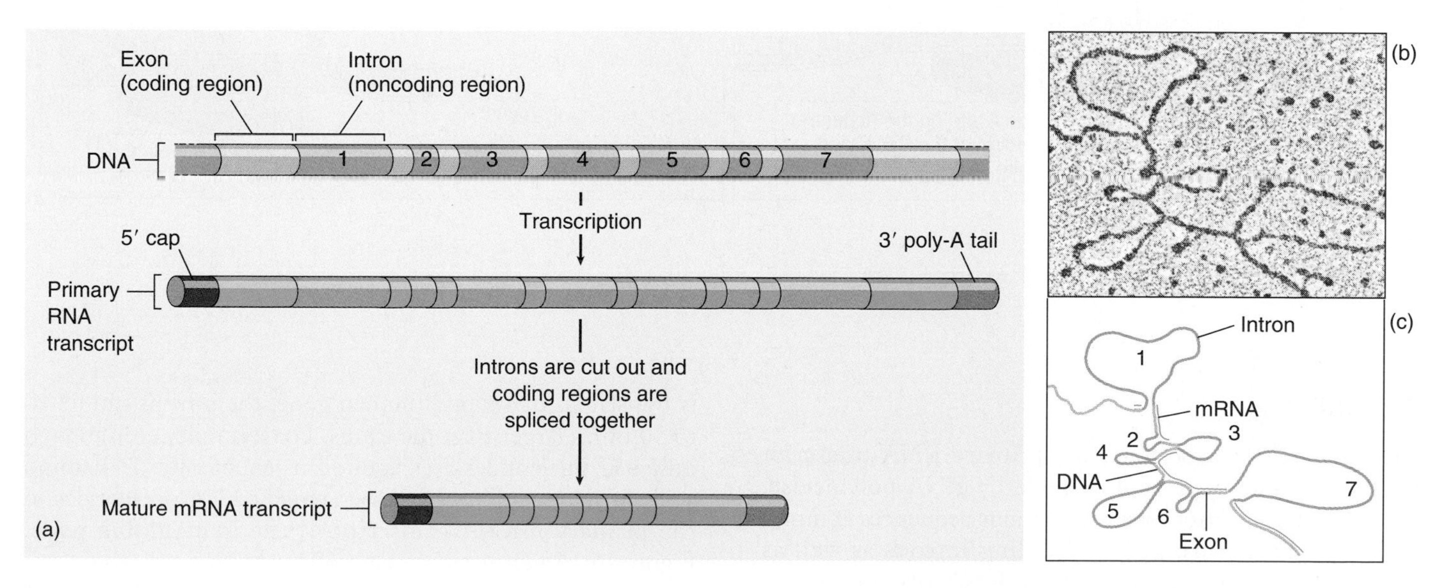

FIGURE 15.15
The eukaryotic ovalbumin gene is fragmented. (*a*) The ovalbumin gene and its primary RNA transcript contain seven segments not present in the mRNA the ribosomes use to direct protein synthesis. Spliceosomes cut these segments (introns) out and splice together the remaining segments (exons). (*b*) By hybridizing the processed transcript to the DNA, introns within the DNA sequence can be visualized directly. The seven loops in the electron micrograph are the seven introns represented in the schematic drawing (*c*).

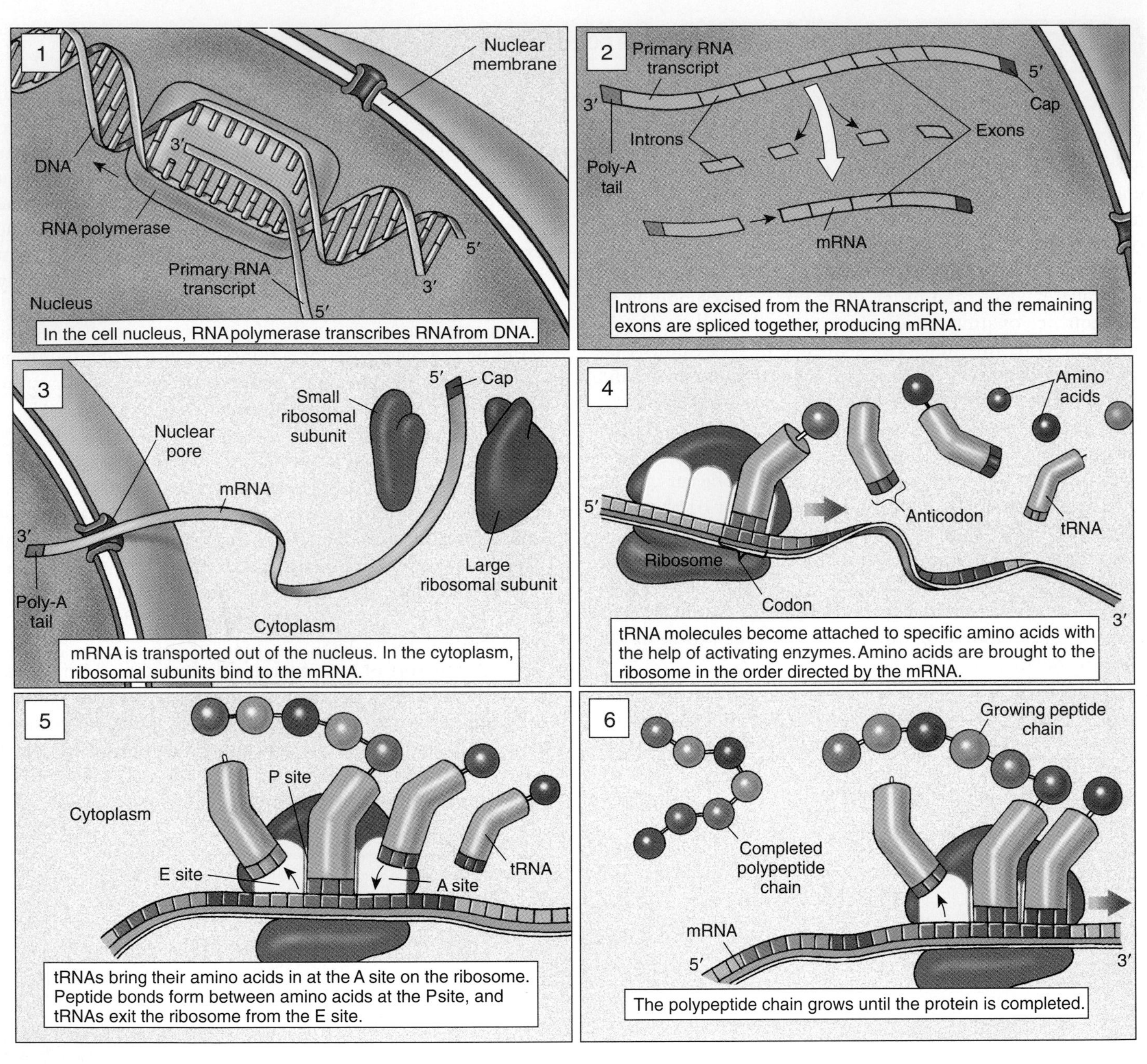

FIGURE 15.16
An overview of gene expression in eukaryotes.

RNA Splicing

When a gene is transcribed, the primary RNA transcript (that is, the gene copy as it is made by RNA polymerase, before any modification occurs) contains sequences complementary to the entire gene, including introns as well as exons. However, in a process called **RNA processing,** or **splicing,** the intron sequences are cut out of the primary transcript before it is used in polypeptide synthesis; therefore, those sequences are not translated. The remaining sequences, which correspond to the exons, are spliced together to form the final, "processed" mRNA molecule that is translated. In a typical human gene, the introns can be 10 to 30 times larger than the exons. For example, even though only 432 nucleotides are required to encode the 144 amino acids of hemoglobin, there are actually 1356 nucleotides in the primary mRNA transcript of the hemoglobin gene. Figure 15.16 summarizes eukaryotic protein synthesis.

Much of a eukaryotic gene is not translated. Noncoding segments scattered throughout the gene are removed from the primary transcript before the mRNA is translated.

Differences between Bacterial and Eukaryotic Gene Expression

1. Most eukaryotic genes possess introns. With the exception of a few genes in the Archaebacteria, prokaryotic genes lack introns (figure 15.17).
2. Individual bacterial mRNA molecules often contain transcripts of several genes. By placing genes with related functions on the same mRNA, bacteria coordinate the regulation of those functions. Eukaryotic mRNA molecules rarely contain transcripts of more than one gene. Regulation of eukaryotic gene expression is achieved in other ways.
3. Because eukaryotes possess a nucleus, their mRNA molecules must be completely formed and must pass across the nuclear membrane before they are translated. Bacteria, which lack nuclei, often begin translation of an mRNA molecule before its transcription is completed.
4. In bacteria, translation begins at an AUG codon preceded by a special nucleotide sequence. In eukaryotic cells, mRNA molecules are modified at the 5′ leading end after transcription, adding a 5′ cap, a methylated guanosine triphosphate. The cap initiates translation by binding the mRNA, usually at the first AUG, to the small ribosomal subunit.
5. Eukaryotic mRNA molecules are modified before they are translated: introns are cut out, and the remaining exons are spliced together; a 5′ cap is added; and a 3′ poly-A tail consisting of some 200 adenine (A) nucleotides is added. These modifications can delay the destruction of the mRNA by cellular enzymes.
6. The ribosomes of eukaryotes are a little larger than those of bacteria.

Gene expression is broadly similar in bacteria and eukaryotes, although it differs in some details.

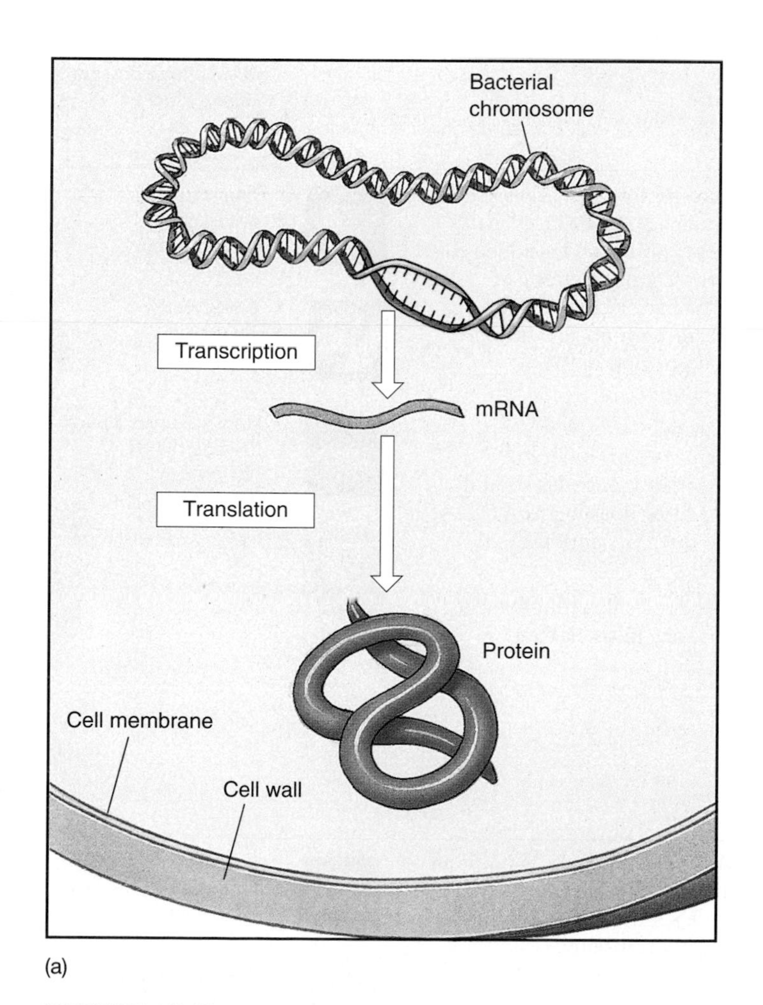

(a)

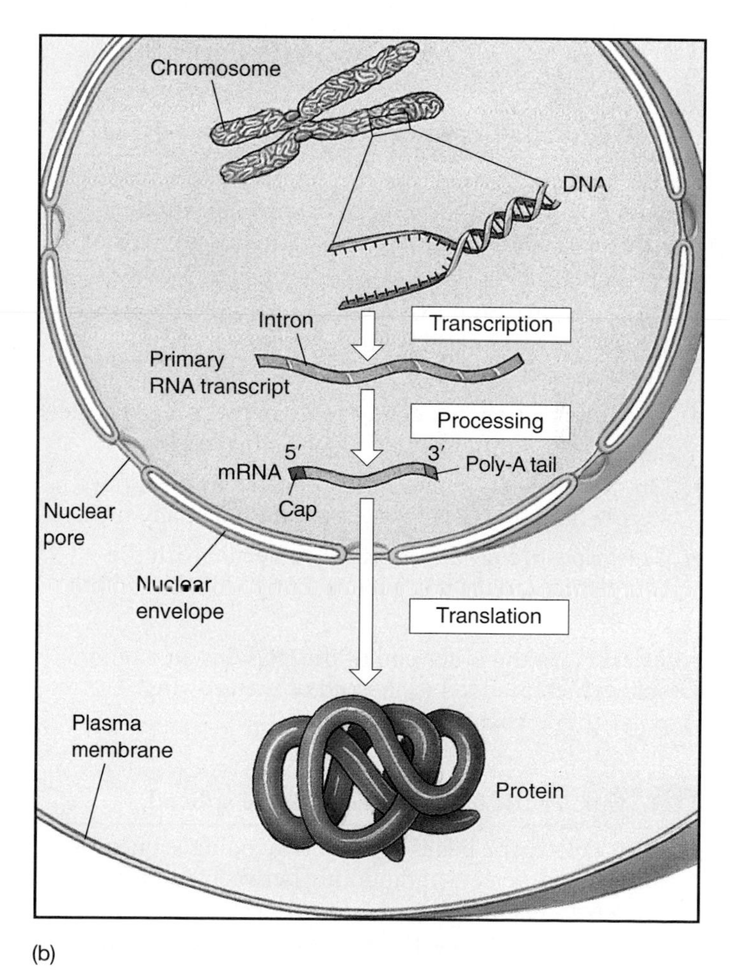

(b)

FIGURE 15.17
Gene information is processed differently in prokaryotes and eukaryotes. (*a*) Bacterial genes are transcribed into mRNA, which is translated immediately. Hence, the sequence of DNA nucleotides corresponds exactly to the sequence of amino acids in the encoded polypeptide. (*b*) Eukaryotic genes are typically different, containing long stretches of nucleotides called introns that do not correspond to amino acids within the encoded polypeptide. Introns are removed from the primary RNA transcript of the gene and a 5' cap and 3' poly-A tail are added before the mRNA directs the synthesis of the polypeptide.

Chapter 15

www.mhhe.com/raven6/resources15.mhtml

Summary	Questions	Media Resources

15.1 The Central Dogma traces the flow of gene-encoded information.

- The Central Dogma, DNA → RNA → protein requires three principal kinds of RNA: messenger RNA (mRNA), transcripts of genes used to direct the assembly of amino acids into proteins; ribosomal RNA (rRNA), which combines with proteins to make up the ribosomes that carry out the assembly process; and transfer RNA (tRNA), molecules that transport the amino acids to the ribosome for assembly into proteins.

1. What are the three major classes of RNA? What is the function of each type?

2. What is the function of RNA polymerase in transcription? What determines where RNA polymerase begins and ends its function?

- Gene Activity

- How Scientists Think: Discovery of Messenger RNA (mRNA) (Jacob/Meselson/Brenner)

15.2 Genes encode information in three-nucleotide code words.

- The sequence of nucleotides in DNA encodes the sequence of amino acids in proteins. The mRNA transcribed from the DNA is read by ribosomes in increments of three nucleotides called codons.

3. How did Crick and his colleagues determine how many nucleotides are used to specify each amino acid? What is an anticodon?

- How Scientists Think:
  - -Breaking the Genetic Code (Nirenberg/Khorana)
  - -The Genetic Code Is Read Three Bases at a Time (Crick)

15.3 Genes are first transcribed, then translated.

- During transcription, the enzyme RNA polymerase manufactures mRNA molecules with nucleotide sequences complementary to particular segments of the DNA.
- During translation, the mRNA sequences direct the assembly of amino acids into proteins on cytoplasmic ribosomes.
- The information in a gene and in an mRNA molecule is read in three-nucleotide blocks called codons.
- On the ribosome, the mRNA molecule is positioned so that only one of its codons is exposed at any time.
- This exposure permits a tRNA molecule with the complementary base sequence (anticodon) to bind to it.
- Attached to the other end of the tRNA is an amino acid, which is added to the end of the growing polypeptide chain.

4. During protein synthesis, what mechanism ensures that only one amino acid is added to the growing polypeptide at a time? What mechanism ensures the correct amino acid is added at each position in the polypeptide?

5. How does an mRNA molecule specify where the polypeptide it encodes should begin? How does it specify where the polypeptide should end?

6. What roles do elongation factors play in translation?

- Transcription
- Translation
- Polyribosomes

- Transcription
- Translation

- How Scientists Think: Proving the tRNA Hypothesis (Chapeville)

- Art Quiz: Translation

15.4 Eukaryotic gene transcripts are spliced.

- Most eukaryotic genes contain noncoding sequences (introns) interspersed randomly between coding sequences (exons).
- The portions of an mRNA molecule corresponding to the introns are removed from the primary RNA transcript before the remainder is translated.

7. What is an intron? What is an exon? How is each involved in the mRNA molecule that is ultimately translated?

- How Scientists Think: Discovery of Introns (Chambon)

- Art Quiz: Introns and Exons

16

Control of Gene Expression

Concept Outline

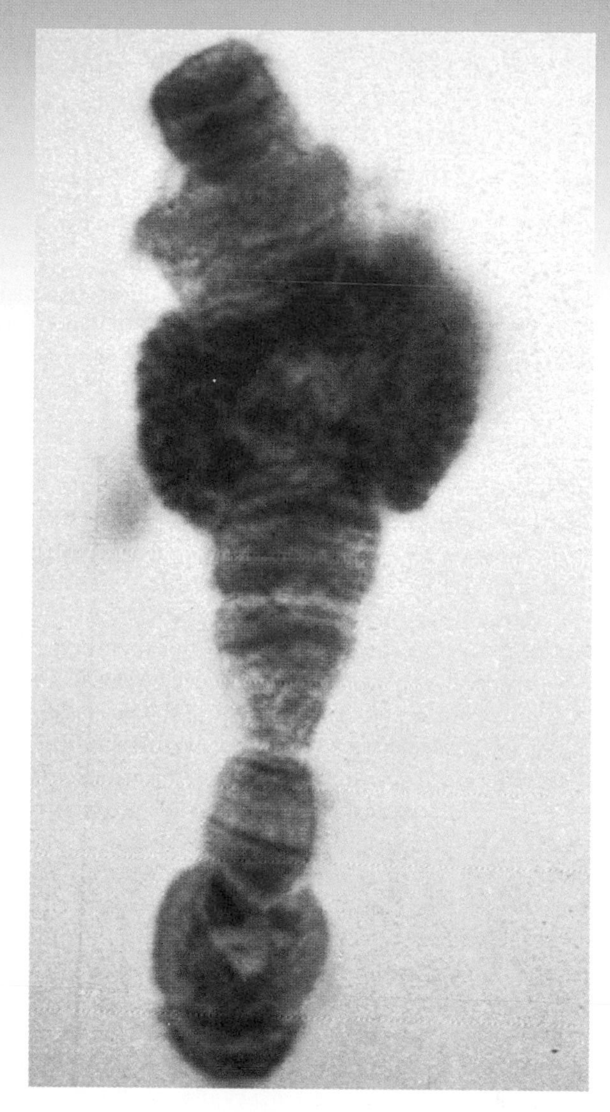

FIGURE 16.1
Chromosome puffs. In this chromosome of the fly *Drosophila melanogaster*, individual active genes can be visualized as "puffs" on the chromosomes. The RNA being transcribed from the DNA template has been radioactively labeled, and the dark specks indicate its position on the chromosome.

In an orchestra, all of the instruments do not play all the time; if they did, all that would be produced is noise. Instead, a musical score determines which instruments in the orchestra play when. Similarly, all of the genes in an organism are not expressed at the same time, each gene producing the protein it encodes full tilt. Instead, different genes are expressed at different times, with a genetic score written in regulatory regions of the DNA determining which genes are active when (figure 16.1).

16.1 Gene expression is controlled by regulating transcription.

An Overview of Transcriptional Control

Control of gene expression is essential to all organisms. In bacteria, it allows the cell to take advantage of changing environmental conditions. In multicellular organisms, it is critical for directing development and maintaining homeostasis.

Regulating Promoter Access

One way to control transcription is to regulate the initiation of transcription. In order for a gene to be transcribed, RNA polymerase must have access to the DNA helix and must be capable of binding to the gene's **promoter,** a specific sequence of nucleotides at one end of the gene that tells the polymerase where to begin transcribing. How is the initiation of transcription regulated? Protein-binding nucleotide sequences on the DNA regulate the initiation of transcription by modulating the ability of RNA polymerase to bind to the promoter. These protein-binding sites are usually only 10 to 15 nucleotides in length (even a large regulatory protein has a "footprint," or binding area, of only about 20 nucleotides). Hundreds of these regulatory sequences have been characterized, and each provides a binding site for a specific protein able to recognize the sequence. Binding the protein to the regulatory sequence either *blocks* transcription by getting in the way of RNA polymerase, or *stimulates* transcription by facilitating the binding of RNA polymerase to the promoter.

Transcriptional Control in Prokaryotes

Control of gene expression is accomplished very differently in bacteria than in the cells of complex multicellular organisms. Bacterial cells have been shaped by evolution to grow and divide as rapidly as possible, enabling them to exploit transient resources. In bacteria, the primary function of gene control is to adjust the cell's activities to its immediate environment. Changes in gene expression alter which enzymes are present in the cell in response to the quantity and type of available nutrients and the amount of oxygen present. Almost all of these changes are fully reversible, allowing the cell to adjust its enzyme levels up or down as the environment changes.

Transcriptional Control in Eukaryotes

The cells of multicellular organisms, on the other hand, have been shaped by evolution to be protected from transient changes in their immediate environment. Most of them experience fairly constant conditions. Indeed, **homeostasis**—the maintenance of a constant internal environment—is considered by many to be the hallmark of multicellular organisms. Although cells in such organisms still respond to signals in their immediate environment (such as growth factors and hormones) by altering gene expression, in doing so they participate in regulating the body as a whole. In multicellular organisms with relatively constant internal environments, the primary function of gene control in a cell is not to respond to that cell's immediate environment, but rather to participate in regulating the body as a whole.

Some of these changes in gene expression compensate for changes in the physiological condition of the body. Others mediate the decisions that *produce* the body, ensuring that the right genes are expressed in the right cells at the right time during development. The growth and development of multicellular organisms entail a long series of biochemical reactions, each catalyzed by a specific enzyme. Once a particular developmental change has occurred, these enzymes cease to be active, lest they disrupt the events that must follow. To produce these enzymes, genes are transcribed in a carefully prescribed order, each for a specified period of time. In fact, many genes are activated only once, producing irreversible effects. In many animals, for example, **stem cells** develop into differentiated tissues like skin cells or red blood cells, following a fixed genetic program that often leads to programmed cell death. The one-time expression of the genes that guide this program is fundamentally different from the reversible metabolic adjustments bacterial cells make to the environment. In all multicellular organisms, changes in gene expression within particular cells serve the needs of the whole organism, rather than the survival of individual cells.

Posttranscriptional Control

Gene expression can be regulated at many levels. By far the most common form of regulation in both bacteria and eukaryotes is **transcriptional control,** that is, control of the transcription of particular genes by RNA polymerase. Other less common forms of control occur after transcription, influencing the mRNA that is produced from the genes or the activity of the proteins encoded by the mRNA. These controls, collectively referred to as **posttranscriptional controls,** will be discussed briefly later in this chapter.

Gene expression is controlled at the transcriptional and posttranscriptional levels. Transcriptional control, more common, is effected by the binding of proteins to regulatory sequences within the DNA.

How to Read a Helix without Unwinding It

It is the ability of certain proteins to bind to *specific* DNA regulatory sequences that provides the basic tool of gene regulation, the key ability that makes transcriptional control possible. To understand how cells control gene expression, it is first necessary to gain a clear picture of this molecular recognition process.

Looking into the Major Groove

Molecular biologists used to think that the DNA helix had to unwind before proteins could distinguish one DNA sequence from another; only in this way, they reasoned, could regulatory proteins gain access to the hydrogen bonds between base-pairs. We now know it is unnecessary for the helix to unwind because proteins can bind to its outside surface, where the edges of the base-pairs are exposed. Careful inspection of a DNA molecule reveals two helical grooves winding round the molecule, one deeper than the other. Within the deeper groove, called the **major groove,** the nucleotides' hydrophobic methyl groups, hydrogen atoms, and hydrogen bond donors and acceptors protrude. The pattern created by these chemical groups is unique for each of the four possible base-pair arrangements, providing a ready way for a protein nestled in the groove to read the sequence of bases (figure 16.2).

DNA-Binding Motifs

Protein-DNA recognition is an area of active research; so far, the structures of over 30 regulatory proteins have been analyzed. Although each protein is unique in its fine details, the part of the protein that actually binds to the DNA is much less variable. Almost all of these proteins employ one of a small set of **structural,** or **DNA-binding, motifs,** particular bends of the protein chain that permit it to interlock with the major groove of the DNA helix.

Regulatory proteins identify specific sequences on the DNA double helix, without unwinding it, by inserting DNA-binding motifs into the major groove of the double helix where the edges of the bases protrude.

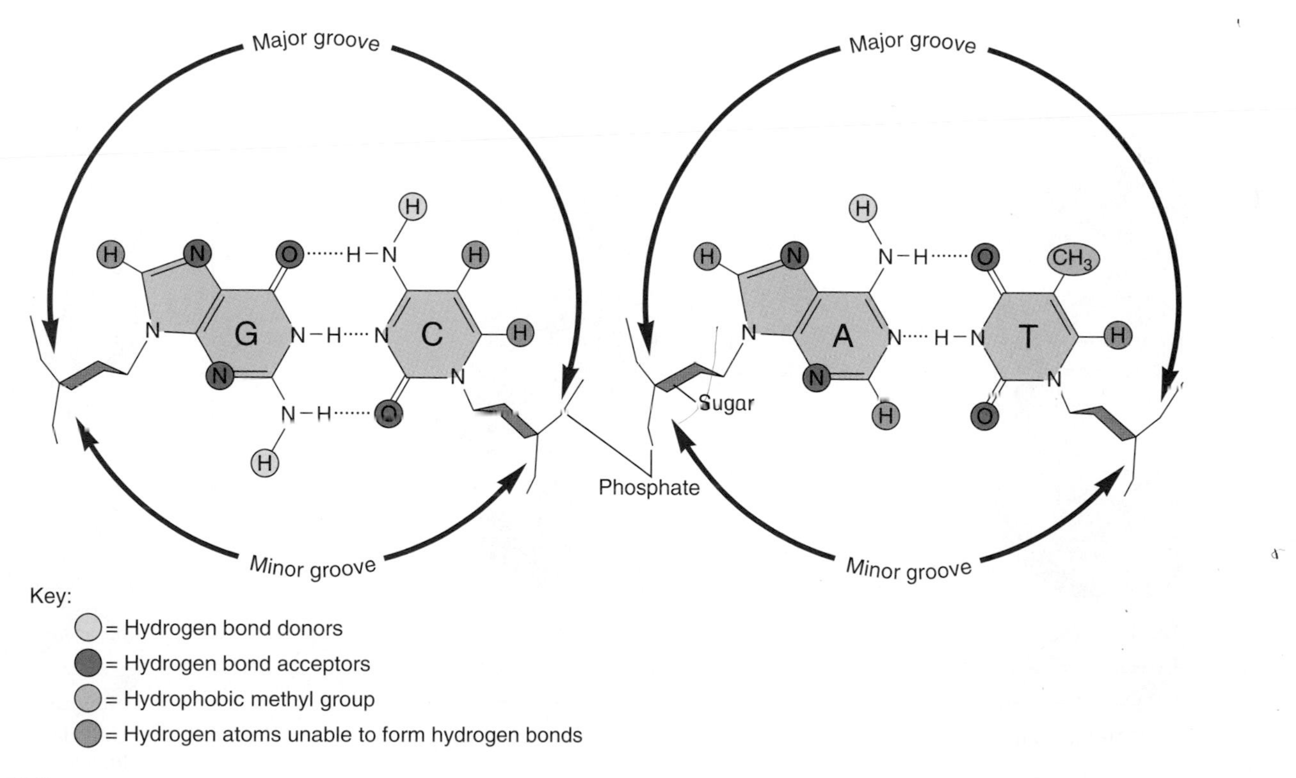

FIGURE 16.2
Reading the major groove of DNA. Looking down into the major groove of a DNA helix, we can see the edges of the bases protruding into the groove. Each of the four possible base-pair arrangements (two are shown here) extends a unique set of chemical groups into the groove, indicated in this diagram by differently colored balls. A regulatory protein can identify the base-pair arrangement by this characteristic signature.

Four Important DNA-Binding Motifs

The Helix-Turn-Helix Motif

The most common DNA-binding motif is the **helix-turn-helix,** constructed from two α-helical segments of the protein linked by a short nonhelical segment, the "turn" (figure 16.3). The first DNA-binding motif recognized, the helix-turn-helix motif has since been identified in hundreds of DNA-binding proteins.

A close look at the structure of a helix-turn-helix motif reveals how proteins containing such motifs are able to interact with the major groove of DNA. Interactions between the helical segments of the motif hold them at roughly right angles to each other. When this motif is pressed against DNA, one of the helical segments (called the recognition helix) fits snugly in the major groove of the DNA molecule, while the other butts up against the outside of the DNA molecule, helping to ensure the proper positioning of the recognition helix. Most DNA regulatory sequences recognized by helix-turn-helix motifs occur in symmetrical pairs. Such sequences are bound by proteins containing two helix-turn-helix motifs separated by 3.4 nm, the distance required for one turn of the DNA helix (figure 16.4). Having *two* protein/DNA-binding sites doubles the zone of contact between protein and DNA and so greatly strengthens the bond that forms between them.

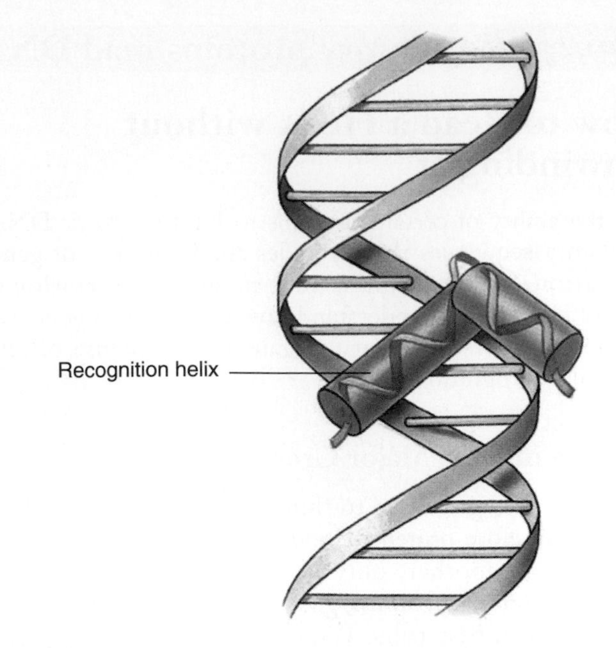

FIGURE 16.3
The helix-turn-helix motif. The recognition helix, one helical region of the motif, actually fits into the major groove of DNA. There it contacts the edges of base-pairs, enabling it to recognize specific sequences of DNA bases.

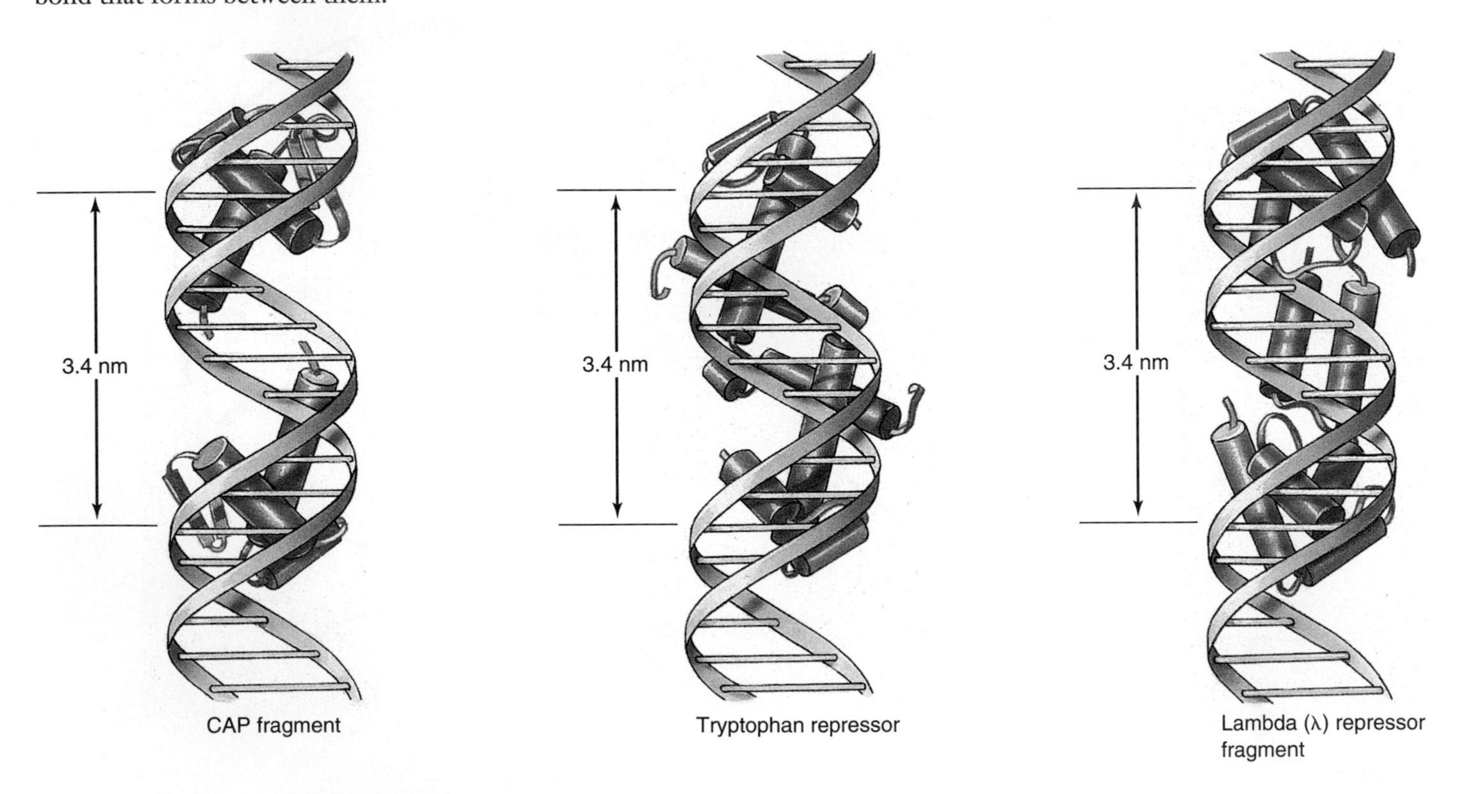

FIGURE 16.4
How the helix-turn-helix binding motif works. The three regulatory proteins illustrated here (*purple*) all bind to DNA using a pair of helix-turn-helix binding motifs. In each case, the two copies of the motif (*red*) are separated by 3.4 nm, precisely the spacing of one turn of the DNA helix. This allows the regulatory proteins to slip into two adjacent portions of the major groove in DNA, providing a strong attachment.

The Homeodomain Motif

A special class of helix-turn-helix motifs plays a critical role in development in a wide variety of eukaryotic organisms, including humans. These motifs were discovered when researchers began to characterize a set of homeotic mutations in *Drosophila* (mutations that alter how the parts of the body are assembled). They found that the mutant genes encoded regulatory proteins whose normal function was to initiate key stages of development by binding to developmental switch-point genes. More than 50 of these regulatory proteins have been analyzed, and they all contain a nearly identical sequence of 60 amino acids, the **homeodomain** (figure 16.5*b*). The center of the homeodomain is occupied by a helix-turn-helix motif that binds to the DNA. Surrounding this motif within the homeodomain is a region that always presents the motif to the DNA in the same way.

The Zinc Finger Motif

A different kind of DNA-binding motif uses one or more zinc atoms to coordinate its binding to DNA. Called **zinc fingers** (figure 16.5*c*), these motifs exist in several forms. In one form, a zinc atom links an α-helical segment to a β sheet segment so that the helical segment fits into the major groove of DNA. This sort of motif often occurs in clusters, the β sheets spacing the helical segments so that each helix contacts the major groove. The more zinc fingers in the cluster, the stronger the protein binds to the DNA. In other forms of the zinc finger motif, the β sheet's place is taken by another helical segment.

The Leucine Zipper Motif

In yet another DNA-binding motif, two different protein subunits cooperate to create a single DNA-binding site. This motif is created where a region on one of the subunits containing several hydrophobic amino acids (usually leucines) interacts with a similar region on the other subunit. This interaction holds the two subunits together at those regions, while the rest of the subunits are separated. Called a **leucine zipper,** this structure has the shape of a "Y," with the two arms of the Y being helical regions that fit into the major groove of DNA (figure 16.5*d*). Because the two subunits can contribute quite different helical regions to the motif, leucine zippers allow for great flexibility in controlling gene expression.

Regulatory proteins bind to the edges of base-pairs exposed in the major groove of DNA. Most contain structural motifs such as the helix-turn-helix, homeodomain, zinc finger, or leucine zipper.

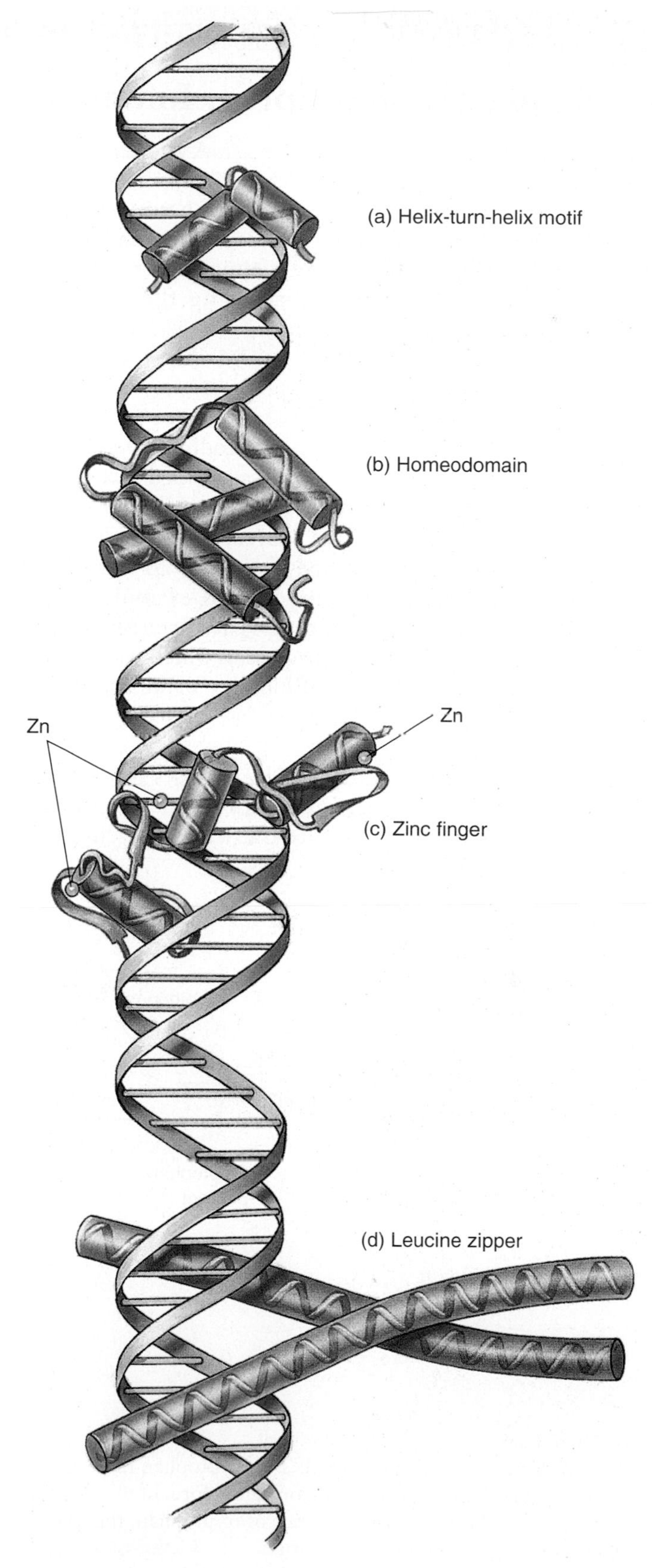

FIGURE 16.5
Major DNA-binding motifs.

16.3 Bacteria limit transcription by blocking RNA polymerase.

Controlling Transcription Initiation

How do organisms use regulatory DNA sequences and the proteins that bind them to control when genes are transcribed? The same basic controls are used in bacteria and eukaryotes, but eukaryotes employ several additional elements that reflect their more elaborate chromosomal structure. We will begin by discussing the relatively simple controls found in bacteria.

Repressors Are *OFF* Switches

A typical bacterium possesses genes encoding several thousand proteins, but only some are transcribed at any one time; the others are held in reserve until needed. When the cell encounters a potential food source, for example, it begins to manufacture the enzymes necessary to metabolize that food. Perhaps the best-understood example of this type of transcriptional control is the regulation of tryptophan-producing genes (*trp* genes), which was investigated in the pioneering work of Charles Yanofsky and his students at Stanford University.

Operons. The bacterium *Escherichia coli* uses proteins encoded by a cluster of five genes to manufacture the amino acid tryptophan. All five genes are transcribed together as a unit called an **operon,** producing a single, long piece of mRNA. RNA polymerase binds to a promoter located at the beginning of the first gene, and then proceeds down the DNA, transcribing the genes one after another. Regulatory proteins shut off transcription by binding to an operator site immediately in front of the promoter and often overlapping it.

When tryptophan is present in the medium surrounding the bacterium, the cell shuts off transcription of the *trp* genes by means of a tryptophan **repressor,** a helix-turn-helix regulatory protein that binds to the operator site located within the *trp* promoter (figure 16.6). Binding of the repressor to the operator prevents RNA polymerase from binding to the promoter. The key to the functioning of this control mechanism is that the tryptophan repressor cannot bind to DNA unless it has first bound to two molecules of tryptophan. The binding of tryptophan to the repressor alters the orientation of a pair of helix-turn-helix motifs in the repressor, causing their recognition helices to fit into adjacent major grooves of the DNA (figure 16.7).

Thus, the bacterial cell's synthesis of tryptophan depends upon the absence of tryptophan in the environment. When the environment lacks tryptophan, there is nothing to activate the repressor, so the repressor cannot prevent

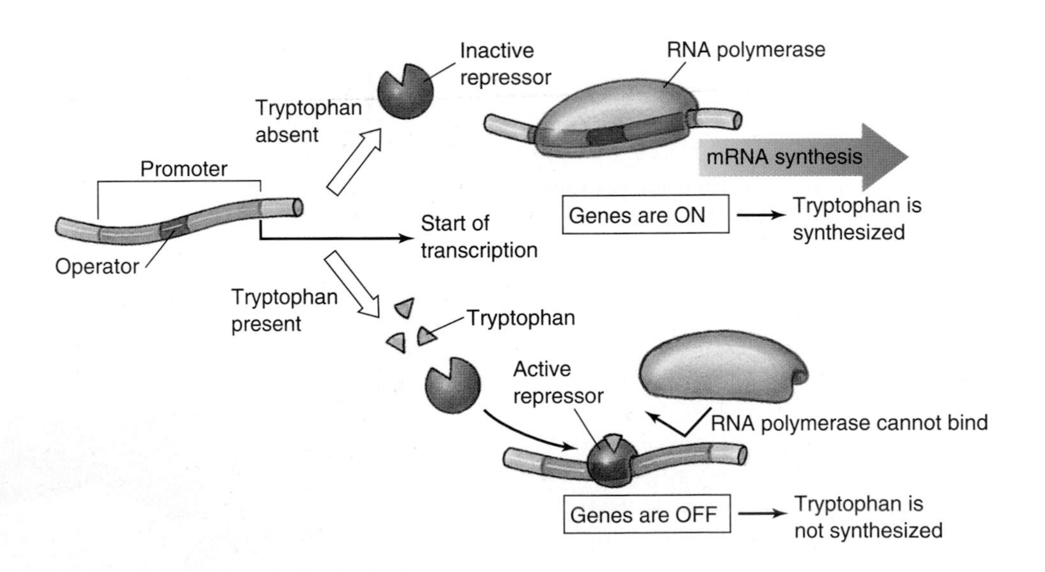

FIGURE 16.6
How the *trp* operon is controlled. The tryptophan repressor cannot bind the operator (which is located *within* the promoter) unless tryptophan first binds to the repressor. Therefore, in the absence of tryptophan, the promoter is free to function, and RNA polymerase transcribes the operon. In the presence of tryptophan, the tryptophan-repressor complex binds tightly to the operator, preventing RNA polymerase from initiating transcription.

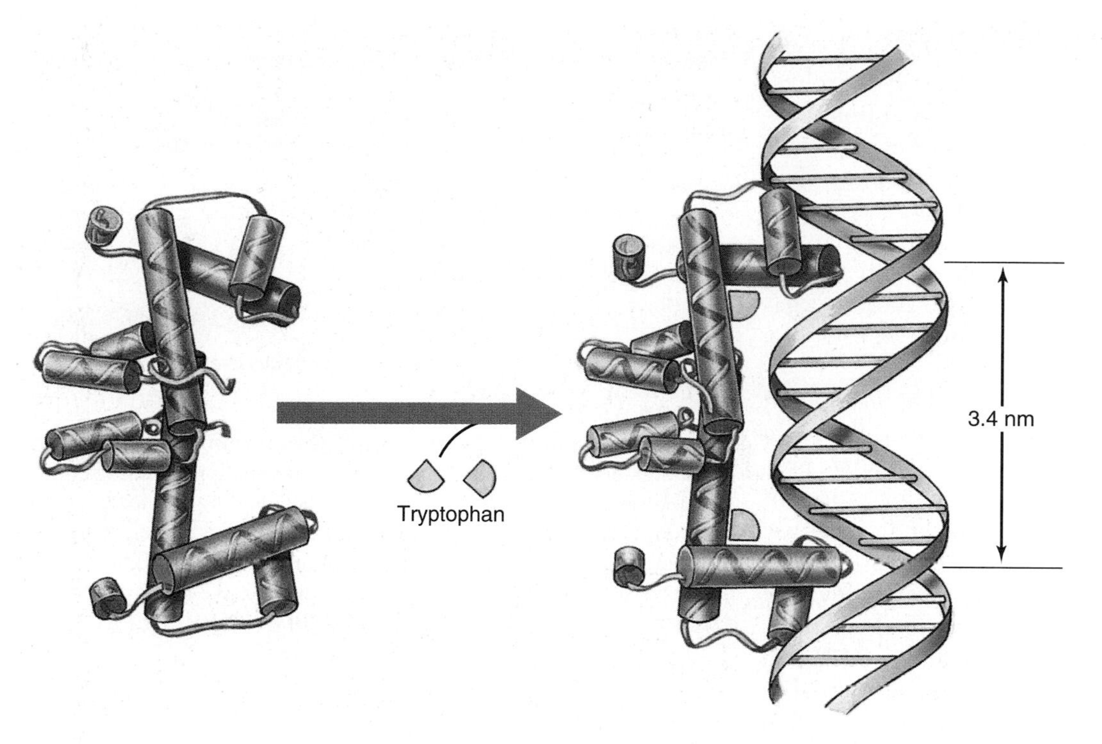

FIGURE 16.7
How the tryptophan repressor works. The binding of tryptophan to the repressor increases the distance between the two recognition helices in the repressor, allowing the repressor to fit snugly into two adjacent portions of the major groove in DNA.

RNA polymerase from binding to the *trp* promoter. The *trp* genes are transcribed, and the cell proceeds to manufacture tryptophan from other molecules. On the other hand, when tryptophan is present in the environment, it binds to the repressor, which is then able to bind to the *trp* promoter. This blocks transcription of the *trp* genes, and the cell's synthesis of tryptophan halts.

Activators Are *ON* Switches

Not all regulatory switches shut genes off—some turn them on. In these instances, bacterial promoters are deliberately constructed to be poor binding sites for RNA polymerase, and the genes these promoters govern are thus rarely transcribed—unless something happens to improve the promoter's ability to bind RNA polymerase. This can happen if a regulatory protein called a **transcriptional activator** binds to the DNA nearby. By contacting the polymerase protein itself, the activator protein helps hold the polymerase against the DNA promoter site so that transcription can begin.

A well-understood transcriptional activator is the catabolite activator protein (CAP) of *E. coli*, which initiates the transcription of genes that allow *E. coli* to use other molecules as food when glucose is not present. Falling levels of glucose lead to higher intracellular levels of the signaling molecule, cyclic AMP (cAMP), which binds to the CAP protein. When cAMP binds to it, the CAP protein changes shape, enabling its helix-turn-helix motif to bind to the DNA near any of several promoters. Consequently, those promoters are activated and their genes can be transcribed (figure 16.8).

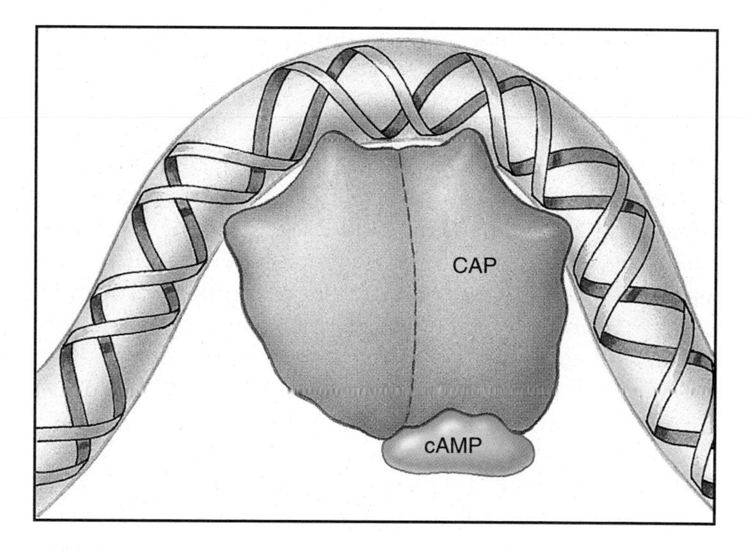

FIGURE 16.8
How CAP works. Binding of the catabolite activator protein (CAP) to DNA causes the DNA to bend around it. This increases the activity of RNA polymerase.

Combinations of Switches

By combining ON and OFF switches, bacteria can create sophisticated transcriptional control systems. A particularly well-studied example is the ***lac* operon** of *E. coli* (figure 16.9). This operon is responsible for producing three proteins that import the disaccharide lactose into the cell and break it down into two monosaccharides: glucose and galactose.

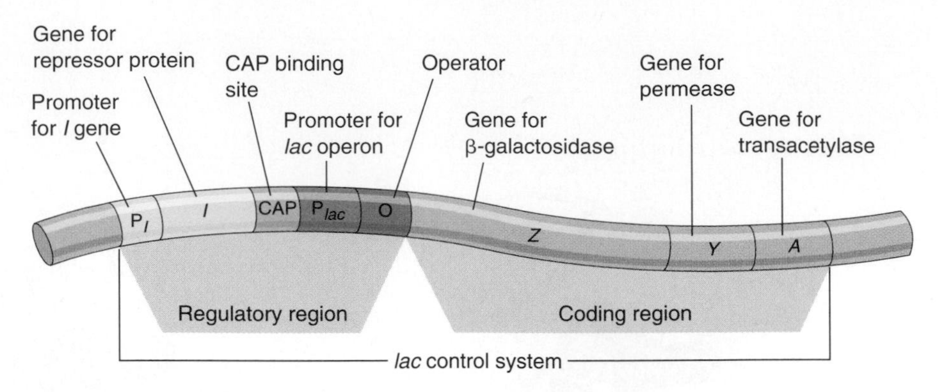

FIGURE 16.9
The *lac* region of the *Escherichia coli* chromosome. The *lac* operon consists of a promoter, an operator, and three genes that code for proteins required for the metabolism of lactose. In addition, there is a binding site for the catabolite activator protein (CAP), which affects whether or not RNA polymerase will bind to the promoter. Gene *I* codes for a repressor protein, which will bind to the operator and block transcription of the *lac* genes. The genes *Z*, *Y*, and *A* encode the two enzymes and the permease involved in the metabolism of lactose.

The Activator Switch. The *lac* operon possesses two regulatory sites. One is a CAP site located adjacent to the *lac* promoter. It ensures that the *lac* genes are not transcribed effectively when ample amounts of glucose are already present. In the absence of glucose, a high level of cAMP builds up in the cell. Consequently, cAMP is available to bind to CAP and allow it to change shape, bind to the DNA, and activate the *lac* promoter (figure 16.10). In the presence of glucose, cAMP levels are low, CAP is unable to bind to the DNA, and the *lac* promoter is not activated.

The Repressor Switch. Whether the *lac* genes are actually transcribed in the absence of glucose is determined by the second regulatory site, the **operator,** which is located adjacent to the promoter. A protein called the *lac* repressor is capable of binding to the operator, but only when lactose is absent. Because the operator and the promoter are close together, the repressor covers part of the promoter when it binds to the operator, preventing RNA polymerase from proceeding and so blocking transcription of the *lac* genes. These genes are then said to be "repressed" (figure 16.11). As a result, the cell does not transcribe genes whose products it has no use for. However, when lactose is present, a lactose isomer binds to the repressor, twisting its binding motif away from the major groove of the DNA. This prevents the repressor from binding to the operator and so allows RNA polymerase to bind to the promoter and transcribe the *lac* genes. Transcription of the *lac* operon is said to have been "induced" by lactose.

This two-switch control mechanism thus causes the cell to produce lactose-utilizing proteins whenever lactose is present but glucose is not, enabling it to make a metabolic decision to produce only what the cell needs, conserving its resources (figure 16.12).

Bacteria regulate gene expression transcriptionally through the use of repressor and activator "switches," such as the *trp* repressor and the CAP activator. The transcription of some clusters of genes, such as the *lac* operon, is regulated by both repressors and activators.

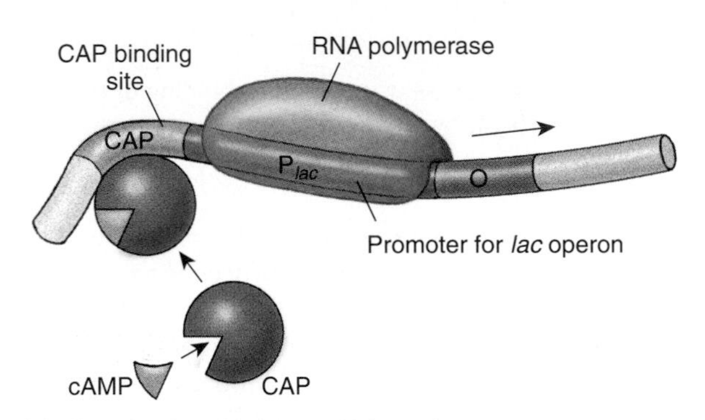

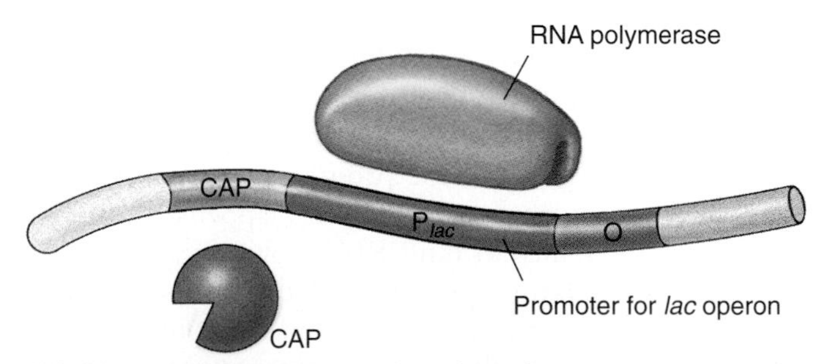

FIGURE 16.10
How the CAP site works. The CAP molecule can attach to the CAP binding site only when the molecule is bound to cAMP. (*a*) When glucose levels are low, cAMP is abundant and binds to CAP. The cAMP-CAP complex binds to the CAP site, bends in the DNA, and gives RNA polymerase access to the promoter. (*b*) When glucose levels are high, cAMP is scarce, and CAP is unable to activate the promoter.

RNA polymerase cannot transcribe *lac* genes

RNA polymerase

Repressor

P_{lac}

O

Promoter for *lac* operon

DNA helix

(a)

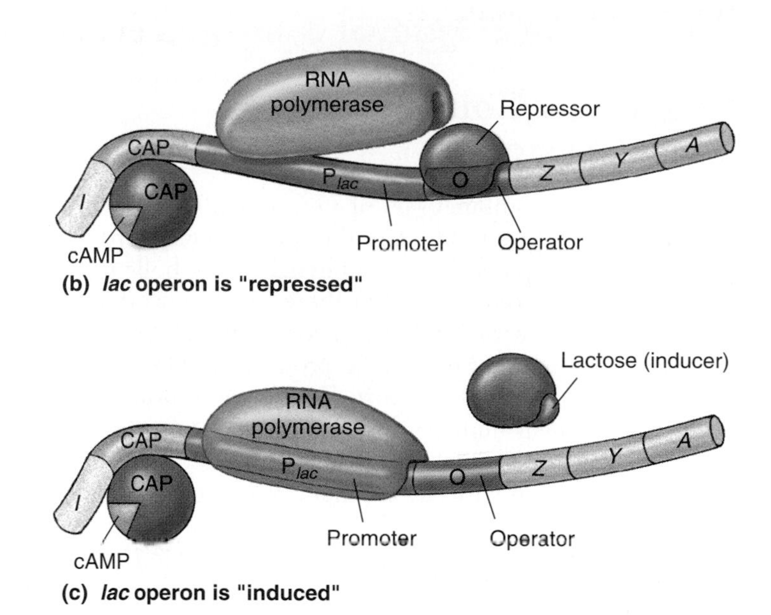

FIGURE 16.11
How the *lac* repressor works. (*a*) The *lac* repressor. Because the repressor fills the major groove of the DNA helix, RNA polymerase cannot fully attach to the promoter, and transcription is blocked. (*b*) The *lac* operon is shut down ("repressed") when the repressor protein is bound to the operator site. Because promoter and operator sites overlap, RNA polymerase and the repressor cannot functionally bind at the same time, any more than two people can sit in the same chair at once. (*c*) The *lac* operon is transcribed ("induced") when CAP is bound and when lactose binding to the repressor changes its shape so that it can no longer sit on the operator site and block RNA polymerase activity.

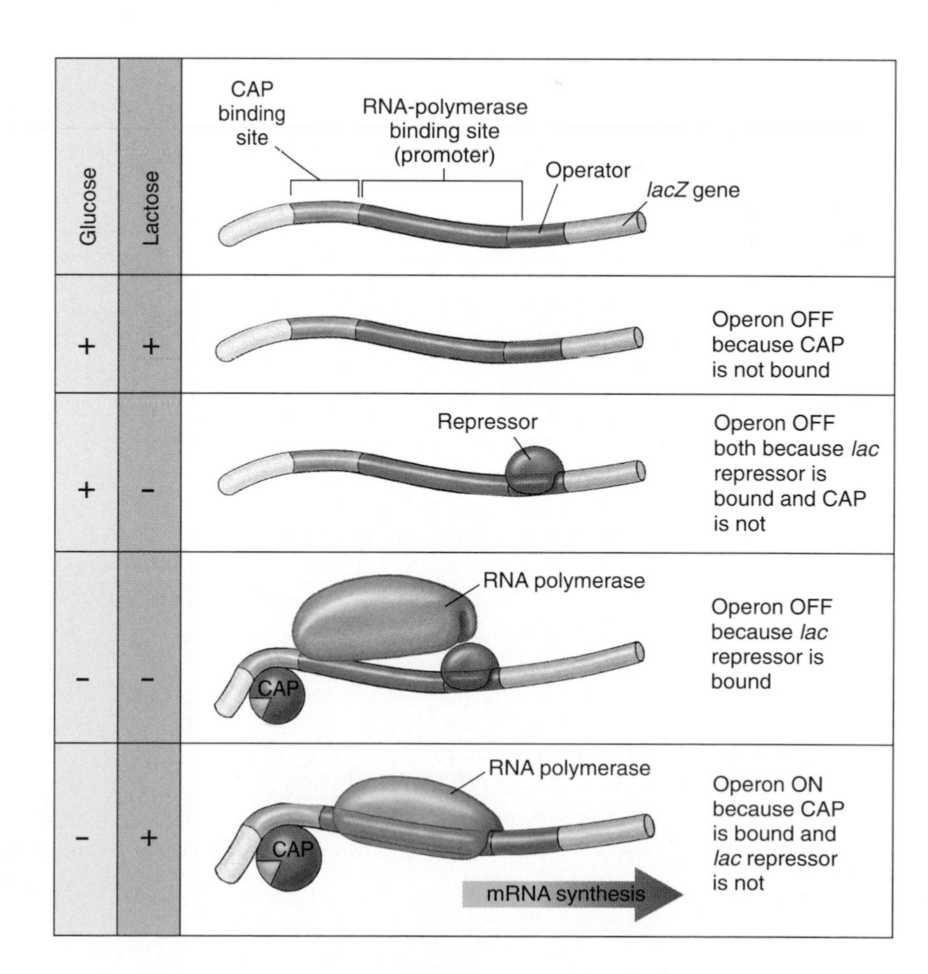

FIGURE 16.12
Two regulatory proteins control the *lac* operon. Together, the *lac* repressor and CAP provide a very sensitive response to the cell's need to utilize lactose-metabolizing enzymes.

16.4 Transcriptional control in eukaryotes operates at a distance.

Designing a Complex Gene Control System

As we have seen, combinations of ON and OFF control switches allow bacteria to regulate the transcription of particular genes in response to the immediate metabolic demands of their environment. All of these switches work by interacting directly with RNA polymerase, either blocking or enhancing its binding to specific promoters. There is a limit to the complexity of this sort of regulation, however, because only a small number of switches can be squeezed into and around one promoter. In a eukaryotic organism that undergoes a complex development, many genes must interact with one another, requiring many more interacting elements than can fit around a single promoter (table 16.1).

In eukaryotes, this physical limitation is overcome by having distant sites on the chromosome exert control over the transcription of a gene (figure 16.13). In this way, many regulatory sequences scattered around the chromosomes can influence a particular gene's transcription. This "control-at-a-distance" mechanism includes two features: a set of proteins that help bind RNA polymerase to the promoter, and modular regulatory proteins that bind to distant sites. These two features produce a truly flexible control system.

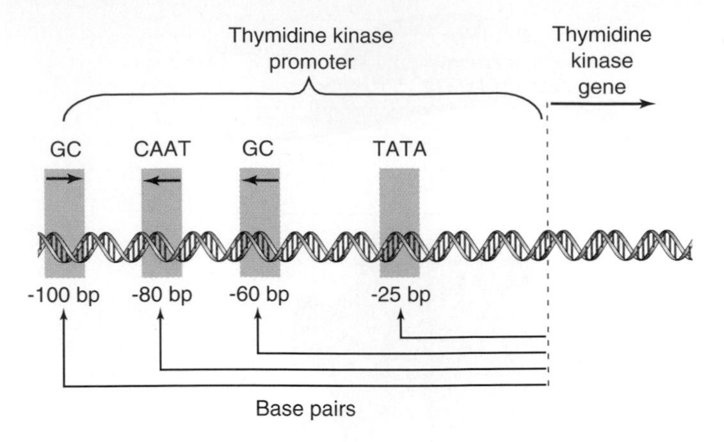

FIGURE 16.13
A eukaryotic promoter. This promoter for the gene encoding the enzyme thymidine kinase contains the TATA box that the initiation factor binds to, as well as three other DNA sequences that direct the binding of other elements of the transcription complex.

Table 16.1 Some Gene Regulatory Proteins and the DNA Sequences They Recognize

Regulatory Proteins of Species	DNA Sequence Recognized*	Regulatory Proteins of Species	DNA Sequence Recognized*
ESCHERICHIA COLI		***DROSOPHILA MELANOGASTER***	
lac repressor	AATTGTGAGCGGATAACAATT TTAACACTCGCCTATTGTTAA	Krüppel	AACGGGTTAA TTGCCCAATT
CAP	TGTGAGTTAGCTCACT ACACTCAATCGAGTGA	bicoid	GGGATTAGA CCCTAATCT
λ repressor	TATCACCGCCAGAGGTA ATAGTGGCGGTCTCCAT		
YEAST		**HUMAN**	
GAL4	CGGAGGACTGTCCTCCG GCCTCCTGACAGGAGGC	Spl	GGGCGG CCCGCC
MAT α2	CATGTAATT GTACATTAA	Oct-1	ATGCAAAT TACGTTTA
GCN4	ATGACTCAT TACTGAGTA	GATA-1	TGATAG ACTATC

*Each regulatory protein is able to recognize a family of closely related DNA sequences; only one member of each family is listed here.

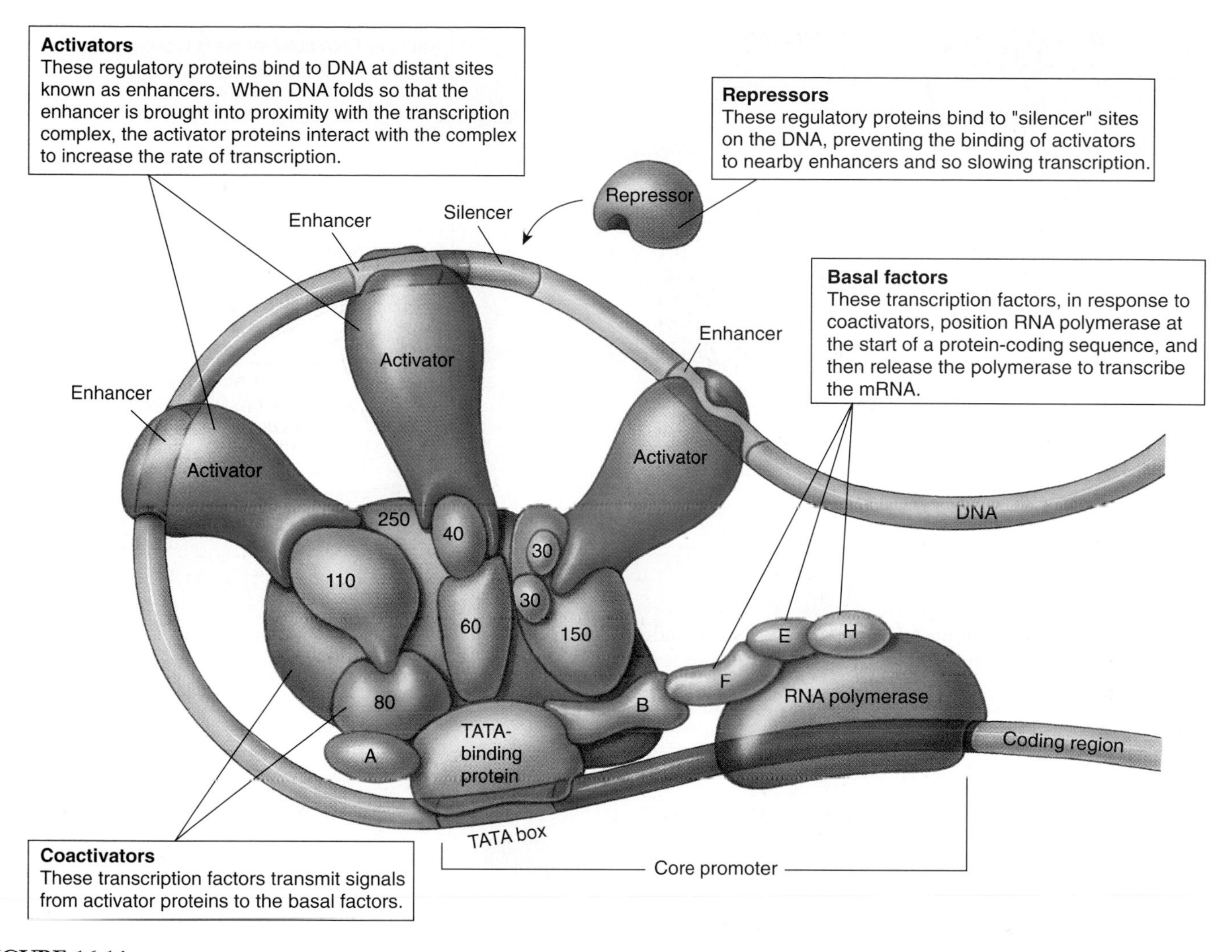

FIGURE 16.14

The structure of a human transcription complex. The transcription complex that positions RNA polymerase at the beginning of a human gene consists of four kinds of proteins. Basal factors (the *green* shapes at bottom of complex with letter names) are transcription factors that are essential for transcription but cannot by themselves increase or decrease its rate. They include the TATA-binding protein, the first of the basal factors to bind to the core promoter sequence. Coactivators (the *tan* shapes that form the bulk of the transcription complex, named according to their molecular weights) are transcription factors that link the basal factors with regulatory proteins called activators (the *red* shapes). The activators bind to enhancer sequences at other locations on the DNA. The interaction of individual basal factors with particular activator proteins is necessary for proper positioning of the polymerase, and the rate of transcription is regulated by the availability of these activators. When a second kind of regulatory protein called a repressor (the *purple* shape) binds to a so-called "silencer" sequence located adjacent to or overlapping an enhancer sequence, the corresponding activator that would normally have bound that enhancer is no longer able to do so. The activator is thus unavailable to interact with the transcription complex and initiate transcription.

Eukaryotic Transcription Factors

For RNA polymerase to successfully bind to a eukaryotic promoter and initiate transcription, a set of proteins called **transcription factors** must first assemble on the promoter, forming a complex that guides and stabilizes the binding of the polymerase (figure 16.14). The assembly process begins some 25 nucleotides upstream from the transcription start site, where a transcription factor composed of many subunits binds to a short TATA sequence (discussed in chapter 15). Other transcription factors then bind, eventually forming a full transcription factor complex able to capture RNA polymerase. In many instances, the transcription factor complex then phosphorylates the bound polymerase, disengaging it from the complex so that it is free to begin transcription.

The binding of several different transcription factors provides numerous points where control over transcription may be exerted. Anything that reduces the availability of a particular factor (for example, by regulating the promoter that governs the expression and synthesis of that factor) or limits its ease of assembly into the transcription factor complex will inhibit transcription.

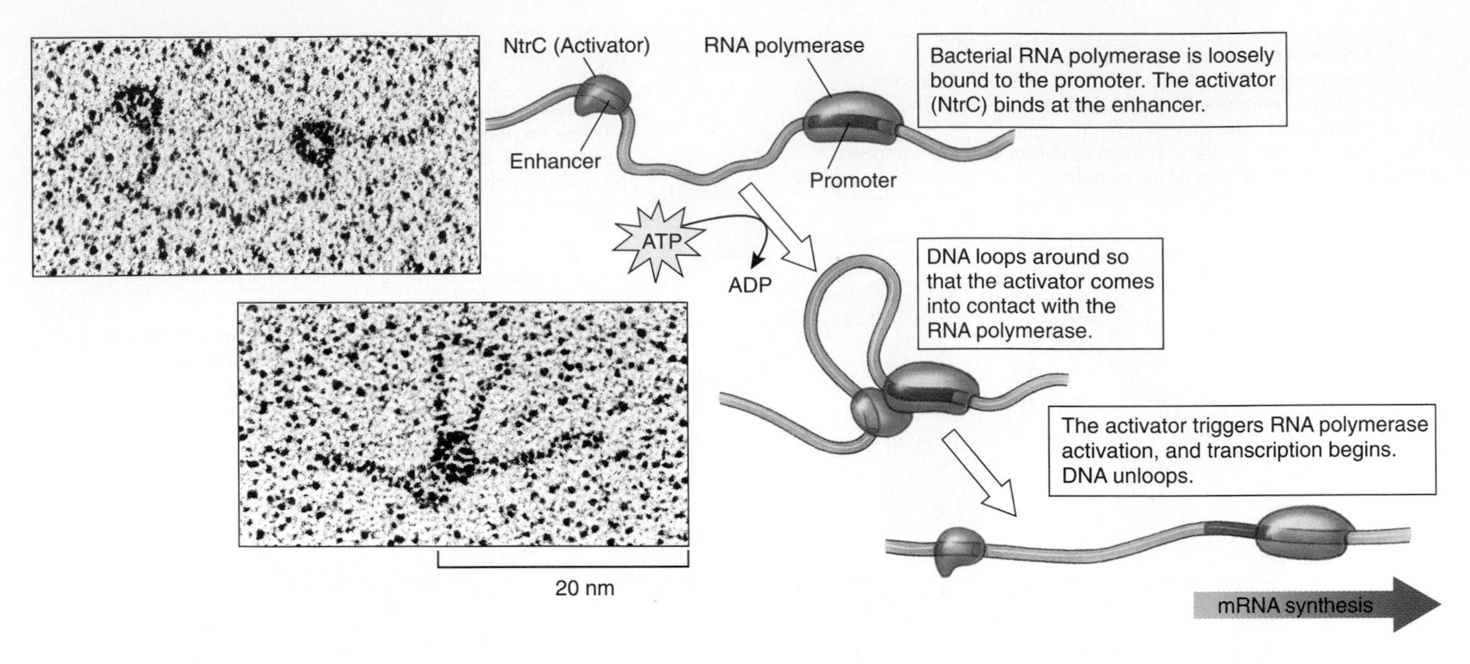

FIGURE 16.15
An enhancer in action. When the bacterial activator NtrC binds to an enhancer, it causes the DNA to loop over to a distant site where RNA polymerase is bound, activating transcription. While such enhancers are rare in bacteria, they are common in eukaryotes.

Enhancers

A key advance in the evolution of eukaryotic gene transcription was the advent of regulatory proteins composed of two distinct modules, or domains. The **DNA-binding domain** physically attaches the protein to the DNA at a specific site, using one of the structural motifs discussed earlier, while the **regulatory domain** interacts with other regulatory proteins.

The great advantage of this modular design is that it uncouples regulation from DNA binding, allowing a regulatory protein to bind to a specific DNA sequence at one site on a chromosome and exert its regulation over a promoter at another site, which may be thousands of nucleotides away. The distant sites where these regulatory proteins bind are called **enhancers.** Although enhancers also occur in exceptional instances in bacteria (figure 16.15), they are the rule rather than the exception in eukaryotes.

How can regulatory proteins affect a promoter when they bind to the DNA at enhancer sites located far from the promoter? Apparently the DNA loops around so that the enhancer is positioned near the promoter. This brings the regulatory domain of the protein attached to the enhancer into direct contact with the transcription factor complex attached to the promoter (figure 16.16).

The enhancer mode of transcriptional control that has evolved in eukaryotes adds a great deal of flexibility to the control process. The positioning of regulatory sites at a distance permits a large number of different regulatory sequences scattered about the DNA to influence a particular gene.

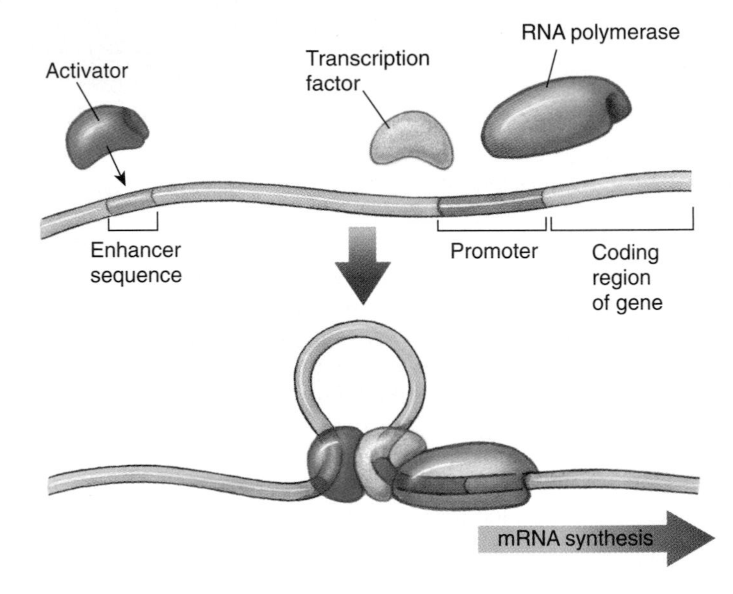

FIGURE 16.16
How enhancers work. The enhancer site is located far away from the gene being regulated. Binding of an activator (*red*) to the enhancer allows the activator to interact with the transcription factors (*green*) associated with RNA polymerase, activating transcription.

Transcription factors and enhancers confer great flexibility on the control of gene expression in eukaryotes.

The Effect of Chromosome Structure on Gene Regulation

The way DNA is packaged into chromosomes can have a profound effect on gene expression. As we saw in chapter 11, the DNA of eukaryotes is packaged in a highly compact form that enables it to fit into the cell nucleus. DNA is wrapped tightly around histone proteins to form nucleosomes (figure 16.17) and then the strand of nucleosomes is twisted into 30-nm filaments.

Promoter Blocking by Nucleosomes

Intensive study of eukaryotic chromosomes has shown that histones positioned over promoters block the assembly of transcription factor complexes. Therefore, transcription factors appear unable to bind to a promoter packaged in a nucleosome. In this way, nucleosomes may prevent continuous transcription initiation. On the other hand, nucleosomes do *not* inhibit activators and RNA polymerase. The regulatory domains of activators attached to enhancers apparently are able to displace the histones that block a promoter. In fact, this displacement of histones and the binding of activator to promoter are required for the assembly of the transcription factor complex. Once transcription has begun, RNA polymerase seems to push the histones aside as it traverses the nucleosome.

DNA Methylation

Chemical **methylation** of the DNA was once thought to play a major role in gene regulation in vertebrate cells. The addition of a methyl group to cytosine creates 5-methylcytosine but has no effect on base-pairing with guanine (figure 16.18), just as the addition of a methyl group to uracil produces thymine without affecting base-pairing with adenine. Many inactive mammalian genes are methylated, and it was tempting to conclude that methylation caused the inactivation. However, methylation is now viewed as having a less direct role, blocking accidental transcription of "turned-off" genes. Vertebrate cells apparently possess a protein that binds to clusters of 5-methylcytosine, preventing transcriptional activators from gaining access to the DNA. DNA methylation in vertebrates thus ensures that once a gene is turned off, it stays off.

Transcriptional control of gene expression occurs in eukaryotes despite the tight packaging of DNA into nucleosomes.

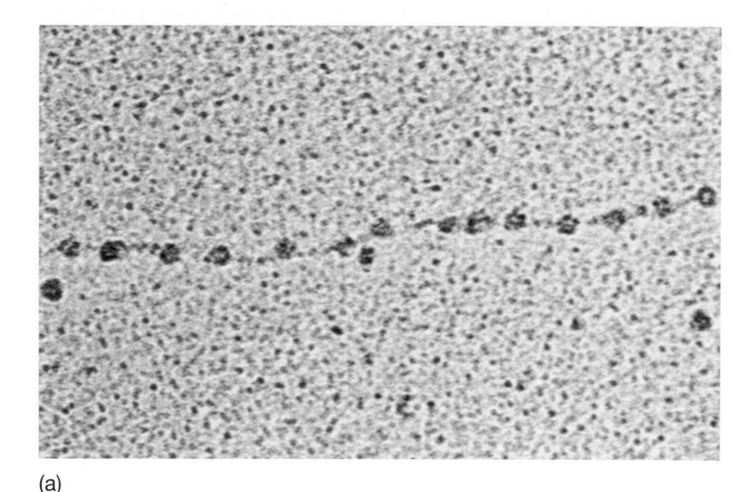

FIGURE 16.17
Nucleosomes. (*a*) In the electron micrograph, the individual nucleosomes have diameters of about 10 nm. (*b*) In the diagram of a nucleosome, the DNA double helix is wound around a core complex of eight histones; one additional histone binds to the outside of the nucleosome, exterior to the DNA.

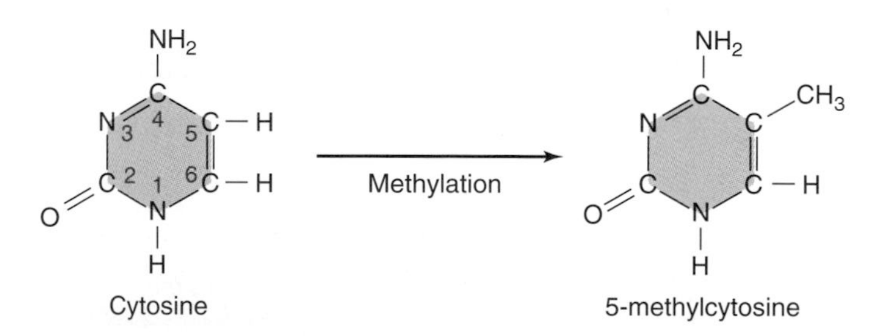

FIGURE 16.18
DNA methylation. Cytosine is methylated, creating 5-methylcytosine. Because the methyl group is positioned to the side, it does not interfere with the hydrogen bonds of a GC base-pair.

Posttranscriptional Control in Eukaryotes

Thus far we have discussed gene regulation entirely in terms of transcription initiation, that is, when and how often RNA polymerase starts "reading" a particular gene. Most gene regulation appears to occur at this point. However, there are many other points after transcription where gene expression could be regulated in principle, and all of them serve as control points for at least some eukaryotic genes. In general, these posttranscriptional control processes involve the recognition of specific sequences on the primary RNA transcript by regulatory proteins or other RNA molecules.

Processing of the Primary Transcript

As we learned in chapter 15, most eukaryotic genes have a patchwork structure, being composed of numerous short coding sequences (exons) embedded within long stretches of noncoding sequences (introns). The initial mRNA molecule copied from a gene by RNA polymerase, the **primary transcript,** is a faithful copy of the entire gene, including introns as well as exons. Before the primary transcript is translated, the introns, which comprise on average 90% of the transcript, are removed in a process called **RNA processing,** or **RNA splicing.** Particles called *small nuclear ribonucleoproteins,* or *snRNPs* (more informally, **snurps**), are thought to play a role in RNA splicing. These particles reside in the nucleus of a cell and are composed of proteins and a special type of RNA called *small nuclear RNA,* or *snRNA.* One kind of snRNP contains snRNA that can bind to the 5′ end of an intron by forming base-pairs with complementary sequences on the intron. When multiple snRNPs combine to form a larger complex called a **spliceosome,** the intron loops out and is excised (figure 16.19).

RNA splicing provides a potential point where the expression of a gene can be controlled, because exons can be

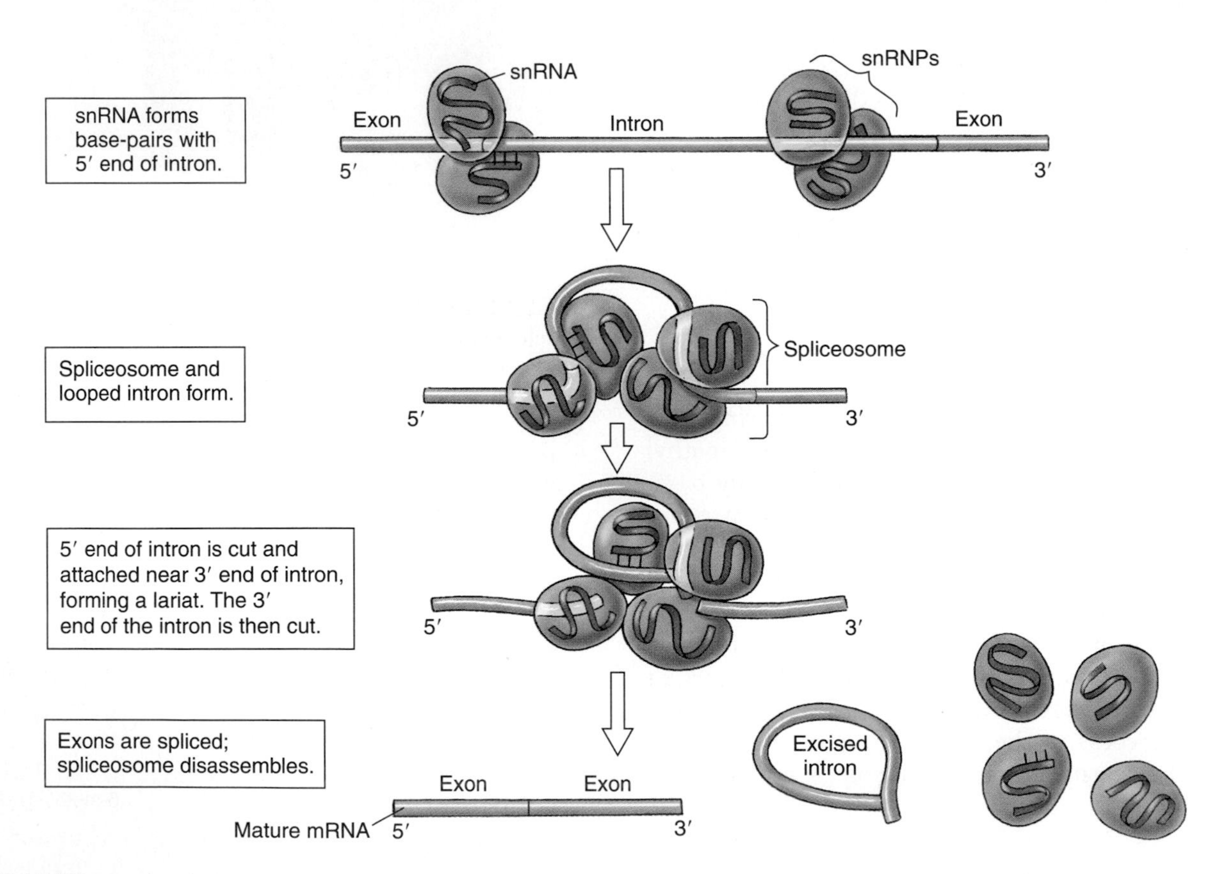

FIGURE 16.19
How spliceosomes process RNA. Particles called snRNPs contain snRNA that interacts with the 5' end of an intron. Several snRNPs come together and form a spliceosome. As the intron forms a loop, the 5' end is cut and linked to a site near the 3' end of the intron. The intron forms a lariat that is excised, and the exons are spliced together. The spliceosome then disassembles and releases the mature mRNA.

spliced together in different ways, allowing a variety of different polypeptides to be assembled from the same gene! Alternative splicing is common in insects and vertebrates, with two or three different proteins produced from one gene. In many cases, gene expression is regulated by changing which splicing event occurs during different stages of development or in different tissues.

An excellent example of alternative splicing in action is found in two different human organs, the thyroid and the hypothalamus. The thyroid gland (see chapter 56) is responsible for producing hormones that control processes such as metabolic rate. The hypothalamus, located in the brain, collects information from the body (for example, salt balance) and releases hormones that in turn regulate the release of hormones from other glands, such as the pituitary gland (see chapter 56). The two organs produce two distinct hormones, calcitonin and CGRP (calcitonin gene-related peptide) as part of their function. Calcitonin is responsible for controlling the amount of calcium we take up from our food and the balance of calcium in tissues like bone and teeth. CGRP is involved in a number of neural and endocrine functions. Although these two hormones are used for very different physiological purposes, the hormones are made using the same transcript (figure 16.20). The appearance of one product versus another is determined by tissue-specific factors that regulate the processing of the primary transcript. This ability offers another powerful way to control the expression of gene products, ranging from proteins with subtle differences to totally unrelated proteins.

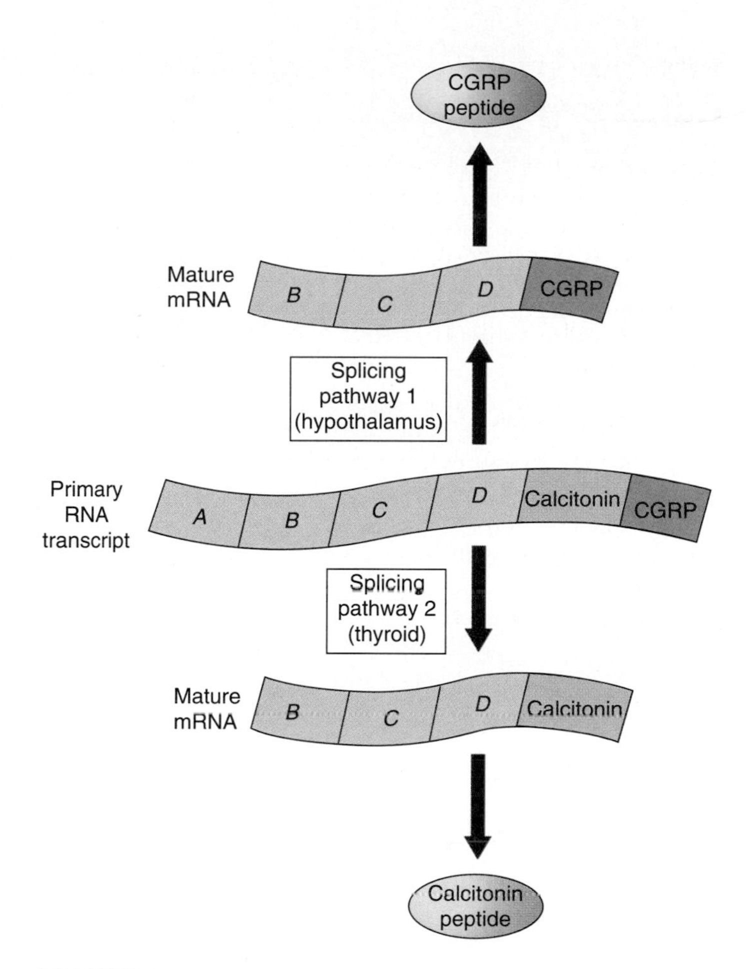

FIGURE 16.20
Alternative splicing products. The same transcript made from one gene can be spliced differently to give rise to two very distinct protein products, calcitonin and CGRP.

Transport of the Processed Transcript Out of the Nucleus

Processed mRNA transcripts exit the nucleus through the nuclear pores described in chapter 5. The passage of a transcript across the nuclear membrane is an active process that requires that the transcript be recognized by receptors lining the interior of the pores. Specific portions of the transcript, such as the poly-A tail, appear to play a role in this recognition. The transcript cannot move through a pore as long as any of the splicing enzymes remain associated with the transcript, ensuring that partially processed transcripts are not exported into the cytoplasm.

There is little hard evidence that gene expression is regulated at this point, although it could be. On average, about 10% of transcribed genes are exon sequences, but only about 5% of the total mRNA produced as primary transcript ever reaches the cytoplasm. This suggests that about half of the exon primary transcripts never leave the nucleus, but it is not clear whether the disappearance of this mRNA is selective.

Selecting Which mRNAs Are Translated

The translation of a processed mRNA transcript by the ribosomes in the cytoplasm involves a complex of proteins called translation factors. In at least some cases, gene expression is regulated by modification of one or more of these factors. In other instances, **translation repressor proteins** shut down translation by binding to the beginning of the transcript, so that it cannot attach to the ribosome. In humans, the production of ferritin (an iron-storing protein) is normally shut off by a translation repressor protein called aconitase. Aconitase binds to a 30-nucleotide sequence at the beginning of the ferritin mRNA, forming a stable loop to which ribosomes cannot bind. When the cell encounters iron, the binding of iron to aconitase causes the aconitase to dissociate from the ferritin mRNA, freeing the mRNA to be translated and increasing ferritin production 100-fold.

A Vocabulary of Gene Expression

activator A regulatory protein that promotes gene transcription by binding to DNA sequences upstream of a promoter. Activator binding stimulates RNA polymerase activity.

anticodon The three-nucleotide sequence on one end of a tRNA molecule that is complementary to and base-pairs with an amino acid–specifying codon in mRNA.

codon The basic unit of the genetic code; a sequence of three adjacent nucleotides in DNA or mRNA that codes for one amino acid or for polypeptide termination.

exon A segment of eukaryotic DNA that is both transcribed into mRNA and translated into protein. Exons are typically scattered within much longer stretches of nontranslated intron sequences.

intron A segment of eukaryotic DNA that is transcribed into mRNA but removed before translation.

nonsense codon A codon (UAA, UAG, or UGA) for which there is no tRNA with a complementary andicodon; a chain-terminating codon often called a "stop" codon.

operator A site of negative gene regulation; a sequence of nucleotides near or within the promoter that is recognized by a repressor. Binding of the repressor to the operator prevents the functional binding of RNA polymerase to the promoter and so blocks transcription.

operon A cluster of functionally related genes transcribed into a single mRNA molecule. A common mode of gene regulation in prokaryotes, it is rare in eukaryotes other than fungi.

promoter A site upstream from a gene to which RNA polymerase attaches to initiate transcription.

repressor A protein that regulates transcription by binding to the operator and so preventing RNA polymerase from initiating transcription from the promoter.

RNA polymerase The enzyme that transcribes DNA into RNA.

transcription The RNA polymerase-catalyzed assembly of an RNA molecule complementary to a strand of DNA.

translation The assembly of a polypeptide on the ribosomes, using mRNA to direct the sequence of amino acids.

Selectively Degrading mRNA Transcripts

Another aspect that affects gene expression is the stability of mRNA transcripts in the cell cytoplasm (figure 16.21). Unlike bacterial mRNA transcripts, which typically have a half-life of about 3 minutes, eukaryotic mRNA transcripts are very stable. For example, β-globin gene transcripts have a half-life of over 10 hours, an eternity in the fast-moving metabolic life of a cell. The transcripts encoding regulatory proteins and growth factors, however, are usually much less stable, with half-lives of less than 1 hour. What makes these particular transcripts so unstable? In many cases, they contain specific sequences near their 3′ ends that make them attractive targets for enzymes that degrade mRNA. A sequence of A and U nucleotides near the 3′ poly-A tail of a transcript promotes removal of the tail, which destabilizes the mRNA. Histone transcripts, for example, have a half-life of about 1 hour in cells that are actively synthesizing DNA; at other times during the cell cycle, the poly-A tail is lost and the transcripts are degraded within minutes. Other mRNA transcripts contain sequences near their 3′ ends that are recognition sites for endonucleases, which causes these transcripts to be digested quickly. The short half-lives of the mRNA transcripts of many regulatory genes are critical to the function of those genes, as they enable the levels of regulatory proteins in the cell to be altered rapidly.

An Example of a Complex Gene Control System

Sunlight is an important gene-controlling signal for plants, from germination to seed formation. Plants must regulate their genes according to the presence of sunlight, the quality of the light source, the time of day, and many other environmental signals. The combination of these responses culminate in the way the genes are regulated, such as the genes *cab* (a chlorophyll-binding photosynthetic protein) and *rbcS* (a subunit of a carbon-fixing enzyme). For instance, photosynthesis-related genes tend to express early in the day, to carry out photosynthesis, and begin to shut down later in the day. Expression levels may also be regulated according to lighting conditions, such as cloudy days versus sunny days. When darkness arrives, the transcripts must be degraded in preparation for the next day. This is an example of how complex a gene control system can be, and scientists are just beginning to understand parts of such a complicated system.

Although less common than transcriptional control, posttranscriptional control of gene expression occurs in eukaryotes via RNA splicing, translation repression, and selective degradation of mRNA transcripts.

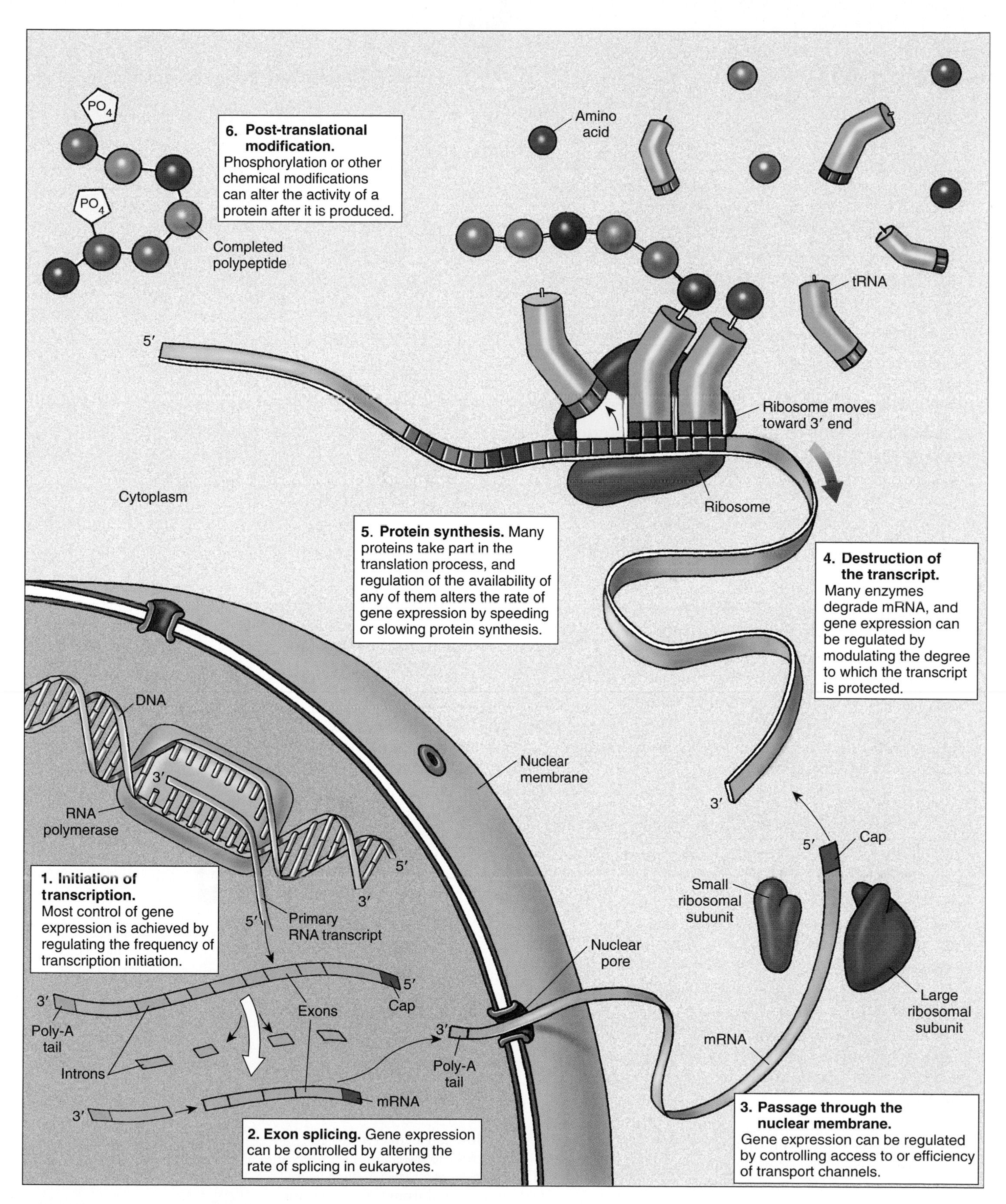

6. Post-translational modification.
Phosphorylation or other chemical modifications can alter the activity of a protein after it is produced.
PO4
PO4
Completed polypeptide
Amino acid
tRNA
5′
Ribosome moves toward 3′ end
Ribosome
Cytoplasm
5. Protein synthesis. Many proteins take part in the translation process, and regulation of the availability of any of them alters the rate of gene expression by speeding or slowing protein synthesis.
4. Destruction of the transcript.
Many enzymes degrade mRNA, and gene expression can be regulated by modulating the degree to which the transcript is protected.
DNA
3′
RNA polymerase
Nuclear membrane
3′
5′
Cap
5′
Small ribosomal subunit
1. Initiation of transcription.
Most control of gene expression is achieved by regulating the frequency of transcription initiation.
3′
5′
Primary RNA transcript
Nuclear pore
5′
3′
Exons
Cap
Poly-A tail
3′
Large ribosomal subunit
mRNA
Introns
Poly-A tail
3′
mRNA
2. Exon splicing. Gene expression can be controlled by altering the rate of splicing in eukaryotes.
3. Passage through the nuclear membrane.
Gene expression can be regulated by controlling access to or efficiency of transport channels.

FIGURE 16.21
Six levels where gene expression can be controlled in eukaryotes.

Chapter 16

www.mhhe.com/raven6/resources16.mhtml

Summary	Questions	Media Resources

16.1 Gene expression is controlled by regulating transcription.

- Regulatory sequences are short stretches of DNA that function in transcriptional control but are not transcribed themselves.
- Regulatory proteins recognize and bind to specific regulatory sequences on the DNA.

- Exploration: Gene Regulation

16.2 Regulatory proteins read DNA without unwinding it.

- Regulatory proteins possess structural motifs that allow them to fit snugly into the major groove of DNA, where the sides of the base-pairs are exposed.
- Common structural motifs include the helix-turn-helix, homeodomain, zinc finger, and leucine zipper.

1. How do regulatory proteins identify specific nucleotide sequences without unwinding the DNA?

2. What is a helix-turn-helix motif? What sort of developmental events are homeodomain motifs involved in?

- Exploration: Reading DNA

16.3 Bacteria limit transcription by blocking RNA polymerase.

- Many genes are transcriptionally regulated through repressors, proteins that bind to the DNA at or near the promoter and thereby inhibit transcription of the gene.
- Genes may also be transcriptionally regulated through activators, proteins that bind to the DNA and thereby stimulate the binding of RNA polymerase to the promoter.
- Transcription is often controlled by a *combination* of repressors and activators.

3. Describe the mechanism by which the transcription of *trp* genes is regulated in *Escherichia coli* when tryptophan is present in the environment.

4. Describe the mechanism by which the transcription of *lac* genes is regulated in *E. coli* when glucose is absent but lactose is present in the environment.

- Exploration: Gene Regulation
- Art Activity: The *lac* operon

- Regulation of *E.coli lac* operon
- Regulation of *E.coli trp* operon

16.4 Transcriptional control in eukaryotes operates at a distance.

- In eukaryotes, RNA polymerase cannot bind to the promoter unless aided by a family of transcription factors.
- Anything that interferes with the activity of the transcription factors can block or alter gene expression.
- Eukaryotic DNA is packaged tightly in nucleosomes within chromosomes. This packaging appears to provide some inhibition of transcription, although regulatory proteins and RNA polymerase can still activate specific genes even when they are so packaged.
- Gene expression can also be regulated at the posttranscriptional level, through RNA splicing, translation repressor proteins, and the selective degradation of mRNA transcripts.

5. How do transcription factors promote transcription in eukaryotic cells? How do the enhancers of eukaryotic cells differ from most regulatory sites on bacterial DNA?

6. What role does the methylation of DNA likely play in transcriptional control?

7. How does the *primary* RNA transcript of a eukaryotic gene differ from the mRNA transcript of that gene as it is translated in the cytoplasm?

8. How can a eukaryotic cell control the translation of mRNA transcripts after they have been transported from the nucleus to the cytoplasm?

- Gene Regulation

- Art Quiz: RNA Processing by Spliceosome

17

Cellular Mechanisms of Development

Concept Outline

17.1 Development is a regulated process.

Overview of Development. Studies of cellular mechanisms have focused on mice, fruit flies, nematodes, and flowering plants.

Vertebrate Development. Vertebrates develop in a highly orchestrated fashion.

Insect Development. Insect development is highly specialized, many key events occurring in a fused mass of cells.

Plant Development. Unlike animal development, which is buffered from the environment, plant development is sensitive to environmental influences.

17.2 Multicellular organisms employ the same basic mechanisms of development.

Cell Movement and Induction. Animal cells move by extending protein cables that they use to pull themselves past surrounding cells. Transcription within cells is influenced by signal molecules from other cells.

Determination. Cells become reversibly committed to particular developmental paths.

Pattern Formation. Diffusion of chemical inducers governs pattern formation in fly embryos.

Expression of Homeotic Genes. Master genes determine the form body segments will take.

Programmed Cell Death. Some genes, when activated, kill their cells.

17.3 Four model developmental systems have been extensively researched.

The Mouse. *Mus musculus.*

The Fruit Fly. *Drosophila melanogaster.*

The Nematode. *Caenorhabditis elegans.*

The Flowering Plant. *Arabidopsis thaliana.*

17.4 Aging can be considered a developmental process.

Theories of Aging. While there are many ideas about why cells age, no one theory of aging is widely accepted.

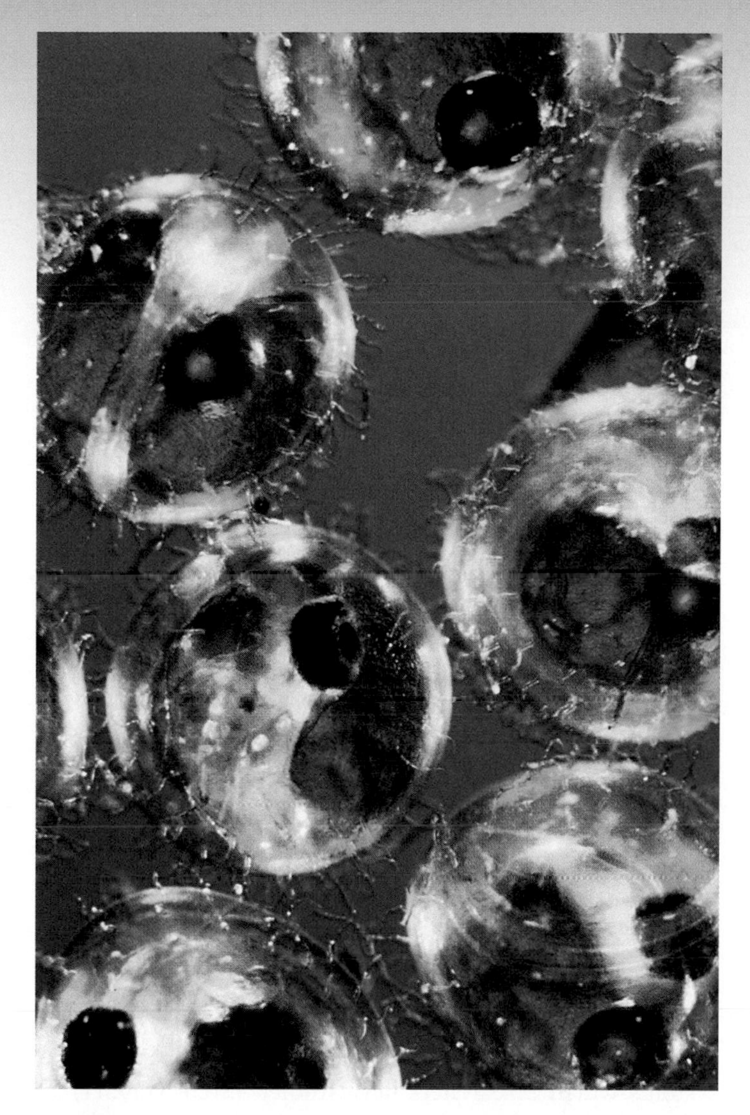

FIGURE 17.1
A collection of future fish undergo embryonic development. Inside a transparent fish egg, a single cell becomes millions of cells that form eyes, fins, gills, and other body parts.

In the previous chapter, we explored gene expression from the perspective of an individual cell, examining the diverse mechanisms that may be employed by a cell to control the transcription of particular genes. Now we will broaden our perspective and look at the unique challenge posed by the development of a cell into a multicellular organism (figure 17.1). In the course of this developmental journey, a pattern of decisions about transcription are made that cause particular lines of cells to proceed along different paths, spinning an incredibly complex web of cause and effect. Yet, for all its complexity, this developmental program works with impressive precision. In this chapter, we will explore the mechanisms used by multicellular organisms to control their development and achieve this precision.

17.1 Development is a regulated process.

Overview of Development

Organisms in all three multicellular kingdoms—fungi, plants, and animals—realize cell specialization by orchestrating gene expression. That is, different cells express different genes at different times. To understand development, we need to focus on how cells determine which genes to activate, and when.

Among the fungi, the specialized cells are largely limited to reproductive cells. In basidiomycetes and ascomycetes (the so-called higher fungi), certain cells produce hormones that influence other cells, but the basic design of all fungi is quite simple. For most of its life, a fungus has a two-dimensional body, consisting of long filaments of cells that are only imperfectly separated from each other. Fungal maturation is primarily a process of growth rather than specialization.

Development is far more complex in plants, where the adult individuals contain a variety of specialized cells organized into tissues and organs. A hallmark of plant development is flexibility; as a plant develops, the precise array of tissues it achieves is greatly influenced by its environment.

In animals, development is complex and rigidly controlled, producing a bewildering array of specialized cell types through mechanisms that are much less sensitive to the environment. The subject of intensive study, animal development has in the last decades become relatively well understood.

Here we will focus our attention on four developmental systems which researchers have studied intensively: (1) an animal with a very complexly arranged body, a mammal; (2) a less complex animal with an intricate developmental cycle, an insect; (3) a very simple animal, a nematode; and (4) a flowering plant (figure 17.2).

To begin our investigation of development, we will first examine the overall process of development in three quite different organisms, so we can sort through differences in the gross process to uncover basic similarities in underlying mechanisms. We will start by describing the overall process in vertebrates, because it is the best understood among the animals. Then we will examine the very different developmental process carried out by insects, in which genetics has allowed us to gain detailed knowledge of many aspects of the process. Finally we will look at development in a third very different organism, a flowering plant.

Almost all multicellular organisms undergo development. The process has been well studied in animals, especially in mammals, insects, nematodes, and in flowering plants.

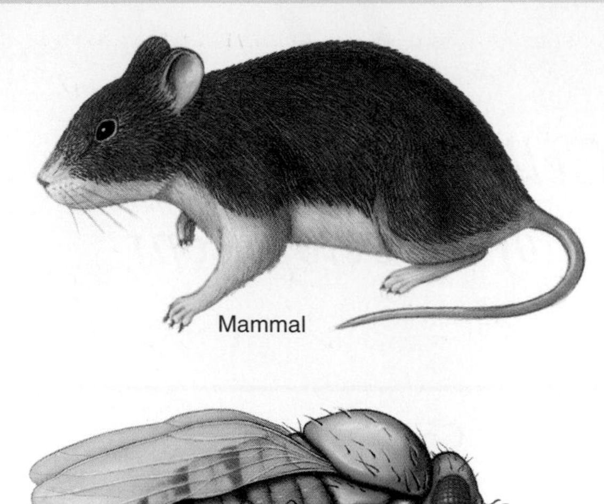

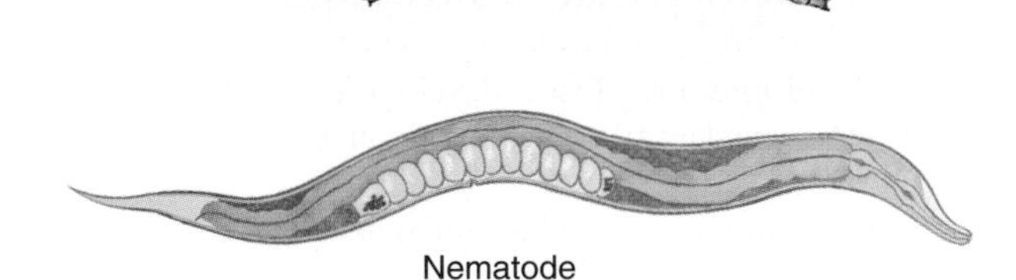

FIGURE 17.2
Four developmental systems. Researchers studying the cellular mechanisms of development have focused on these four organisms.

Vertebrate Development

Vertebrate development is a dynamic process in which cells divide rapidly and move over each other as they first establish the basic geometry of the body (figure 17.3). At different sites, particular cells then proceed to form the body's organs, and then the body grows to a size and shape that will allow it to survive after birth. The entire process, described more fully in chapter 60, is traditionally divided into phases. As in mitosis, however, the boundaries between phases are somewhat artificial, and the phases, in fact, grade into one another.

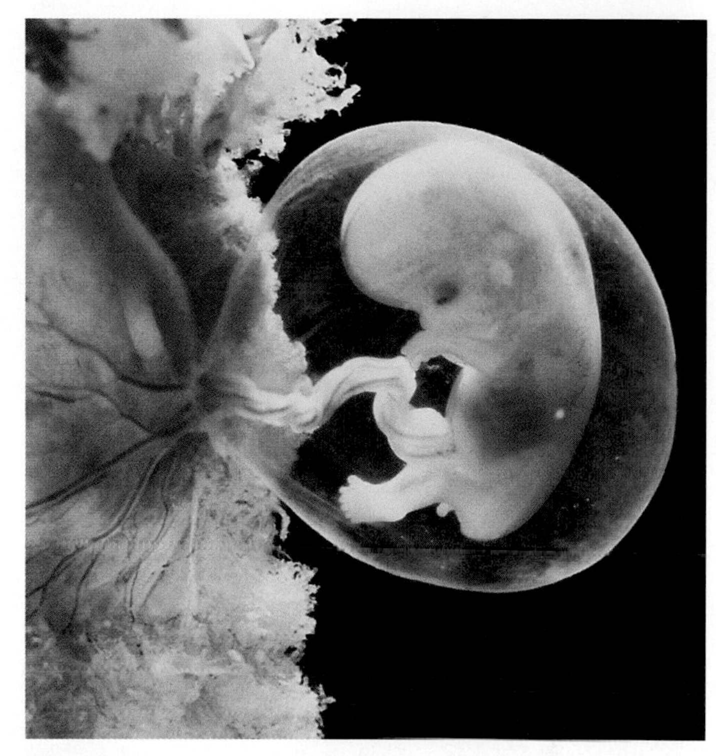

FIGURE 17.3
The miracle of development. This nine-week-old human fetus started out as a single cell: a fertilized egg, or zygote. The zygote's daughter cells have been repeatedly dividing and specializing to produce the distinguishable features of a fetus.

Cleavage

Vertebrates begin development as a single fertilized egg, the zygote. Within an hour after fertilization, the zygote begins to divide rapidly into a larger and larger number of smaller and smaller cells called **blastomeres,** until a solid ball of cells is produced (figure 17.4). This initial period of cell division, termed **cleavage,** is not accompanied by any increase in the overall size of the embryo; rather, the contents of the zygote are simply partitioned into the daughter cells. The two ends of the zygote are traditionally referred to as the **animal** and **vegetal poles.** In general, the blastomeres of the animal pole will go on to form the external tissues of the body, while those of the vegetal pole will form the internal tissues. The initial top-bottom (dorsal-ventral) orientation of the embryo is determined at fertilization by the location where the sperm nucleus enters the egg, a point that corresponds roughly to the future belly. After about 12 divisions, the burst of cleavage divisions slows, and transcription of key genes begins within the embryo cells.

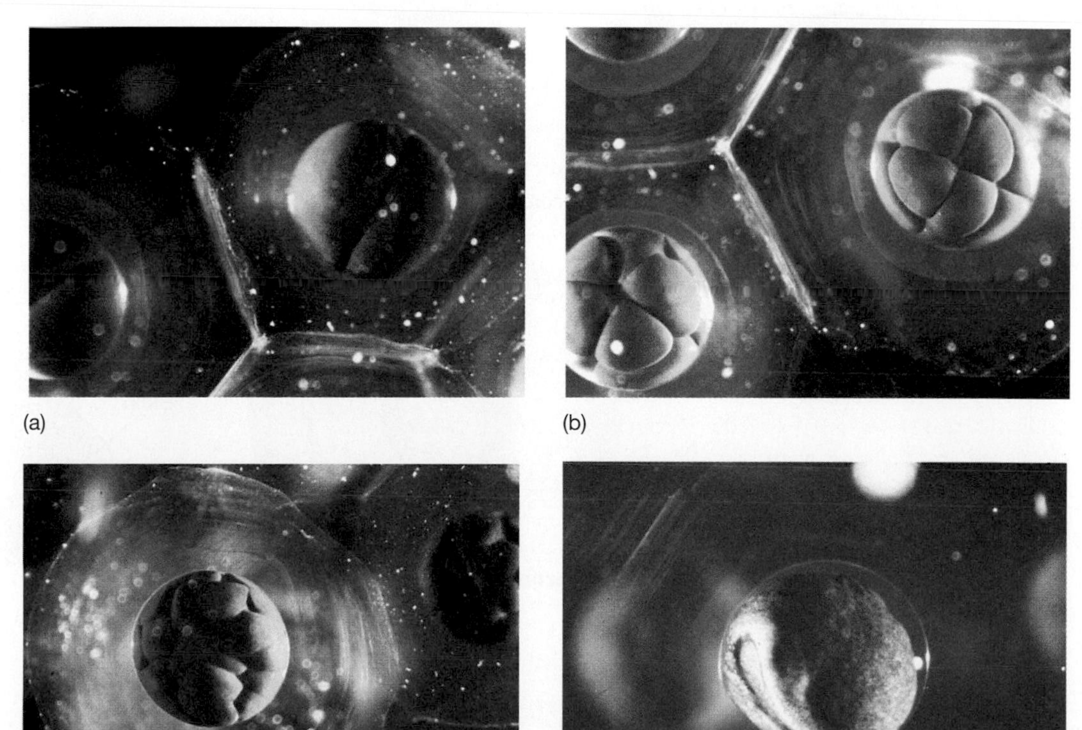

FIGURE 17.4
Cleavage divisions producing a frog embryo. (*a*) The initial divisions are, in this case, on the side of the embryo facing you, producing (*b*) a cluster of cells on this side of the embryo, which soon expands to become a (*c*) compact mass of cells. (*d*) This mass eventually invaginates into the interior of the embryo, forming a gastrula, then a neurula.

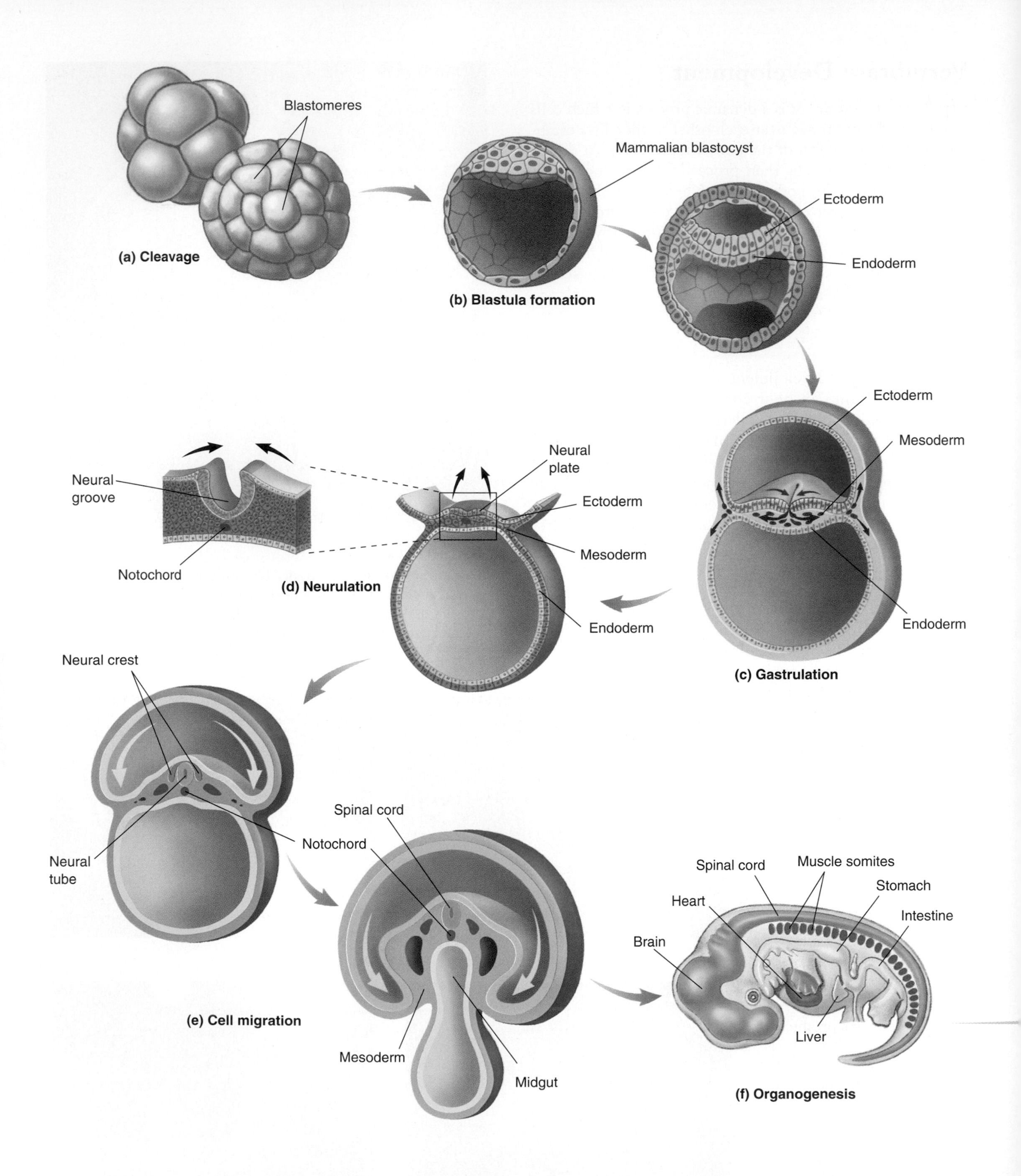

FIGURE 17.5
The path of vertebrate development. An illustration of the major events in the development of *Mus musculus*, the house mouse. (*a*) Cleavage. (*b*) Formation of blastula. (*c*) Gastrulation. (*d*) Neurulation. (*e*) Cell migration. (*f*) Organogenesis. (*g*) Growth.

Formation of the Blastula

The outermost blastomeres (figure 17.5*a*) in the ball of cells produced during cleavage are joined to one another by tight junctions, which, as you may recall from chapter 7, are belts of protein that encircle a cell and weld it firmly to its neighbors. These tight junctions create a seal that isolates the interior of the cell mass from the surrounding medium. At about the 16-cell stage, the cells in the interior of the mass begin to pump Na^+ from their cytoplasm into the spaces between cells. The resulting osmotic gradient causes water to be drawn into the center of the cell mass, enlarging the intercellular spaces. Eventually, the spaces coalesce to form a single large cavity within the cell mass. The resulting hollow ball of cells is called a **blastula,** or **blastocyst** in mammals (figure 17.5*b*).

Gastrulation

Some cells of the blastula then push inward, forming a **gastrula** that is invaginated. Cells move by using extensions called lamellipodia to crawl over neighboring cells, which respond by forming lamellipodia of their own. Soon a sheet of cells contracts on itself and shoves inward, starting the invagination. Called **gastrulation** (figure 17.5*c*), this process creates the main axis of the vertebrate body, converting the blastula into a bilaterally symmetrical embryo with a central gut. From this point on, the embryo has three **germ layers** whose organization foreshadows the future organization of the adult body. The cells that invaginate and form the tube of the primitive gut are endoderm; they give rise to the stomach, lungs, liver, and most of the other internal organs. The cells that remain on the exterior are ectoderm, and their derivatives include the skin on the outside of the body and the nervous system. The cells that break away from the invaginating cells and invade the space between the gut and the exterior wall are mesoderm; they eventually form the notochord, bones, blood vessels, connective tissues, and muscles.

Neurulation

Soon after gastrulation is complete, a broad zone of ectoderm begins to thicken on the dorsal surface of the embryo, an event triggered by the presence of the notochord beneath it. The thickening is produced by the elongation of certain ectodermal cells. Those cells then assume a wedge shape by contracting bundles of actin filaments at one end. This change in shape causes the neural tissue to roll up into a tube, which eventually pinches off from the rest of the ectoderm and gives rise to the brain and spinal cord. This tube is called the **neural tube,** and the process by which it forms is termed **neurulation** (figure 17.5*d*).

Cell Migration

During the next stage of vertebrate development, a variety of cells migrate to form distant tissues, following specific paths through the embryo to particular locations (figure 17.5*e*). These migrating cells include those of the **neural crest,** which pinch off from the neural tube and form a number of structures, including some of the body's sense organs; cells that migrate from central blocks of muscle tissue called **somites** and form the skeletal muscles of the body; and the precursors of blood cells and gametes. When a migrating cell reaches its destination, receptor proteins on its surface interact with proteins on the surfaces of cells in the destination tissue, triggering changes in the cytoskeleton of the migrating cell that cause it to cease moving.

Organogenesis and Growth

At the end of this wave of cell migration and colonization, the basic vertebrate body plan has been established, although the embryo is only a few millimeters long and has only about 10^5 cells. Over the course of subsequent development, tissues will develop into organs (figure 17.5*f*), and the embryo will grow to be a hundred times larger, with a million times as many cells (figure 17.5*g*).

(g) Growth

Vertebrates develop in a highly orchestrated fashion. The zygote divides rapidly, forming a hollow ball of cells that then pushes inward, forming the main axis of an embryo that goes on to develop tissues, and after a process of cell migration, organs.

Insect Development

Like all animals, insects develop through an orchestrated series of cell changes, but the path of development is quite different from that of a vertebrate. Many insects produce two different kinds of bodies during their development, the first a tubular eating machine called a **larva,** and the second a flying machine with legs and wings. The passage from one body form to the other is called **metamorphosis** and involves a radical shift in development. Here we will describe development in the fruit fly *Drosophila* (figure 17.6), which is the subject of much genetic research.

FIGURE 17.6
The fruit fly, *Drosophila melanogaster*. A dorsal view of *Drosophila*, one of the most intensively studied animals in development.

Maternal Genes

The development of an insect like *Drosophila* begins before fertilization, with the construction of the egg. Specialized nurse cells that help the egg to grow move some of their own mRNA into the end of the egg nearest them (figure 17.7*a*). As a result, mRNAs produced by maternal genes are positioned in particular locations in the egg, so that after repeated divisions subdivide the fertilized egg, different daughter cells will contain different maternal products. Thus, the action of maternal (rather than zygotic) genes determines the initial course of development.

Syncytial Blastoderm

After fertilization, 12 rounds of nuclear division without cytokinesis produce about 6000 nuclei, all within a single cytoplasm. All of the nuclei within this **syncytial blastoderm** (figure 17.7*b*) can freely communicate with one another, but nuclei located in different sectors of the egg experience different maternal products. The nuclei then space themselves evenly along the surface of the blastoderm, and membranes grow between them. Folding of the embryo and primary tissue development soon follow, in a process fundamentally similar to that seen in vertebrate development. The tubular body that results within a day of fertilization is a larva.

Larval Instars

The larva begins to feed immediately, and as it does so, it grows. Its chitinous exoskeleton cannot stretch much, however, and within a day it sheds the exoskeleton. Before the new exoskeleton has had a chance to harden, the larva expands in size. A total of three larval stages, or **instars,** are produced over a period of four days (figure 17.7*c*).

Imaginal Discs

During embryonic growth, about a dozen groups of cells called **imaginal discs** are set aside in the body of the larva (figure 17.7*d*). Imaginal discs play no role in the life of the larva, but are committed to form key parts of the adult fly's body.

Metamorphosis

After the last larval stage, a hard outer shell forms, and the larva is transformed into a **pupa** (figure 17.7*e*). Within the pupa, the larval cells break down and release their nutrients, which are used in the growth and development of the various imaginal discs (eye discs, wing discs, leg discs, and so on). The imaginal discs then associate with one another, assembling themselves into the body of the adult fly (figure 17.7*f*). The metamorphosis of a *Drosophila* larva into a pupa and then into adult fly takes about four days, after which the pupal shell splits and the fly emerges.

***Drosophila* development proceeds through two discrete phases, the first a larval phase that gathers food, then an adult phase that is capable of flight and reproduction.**

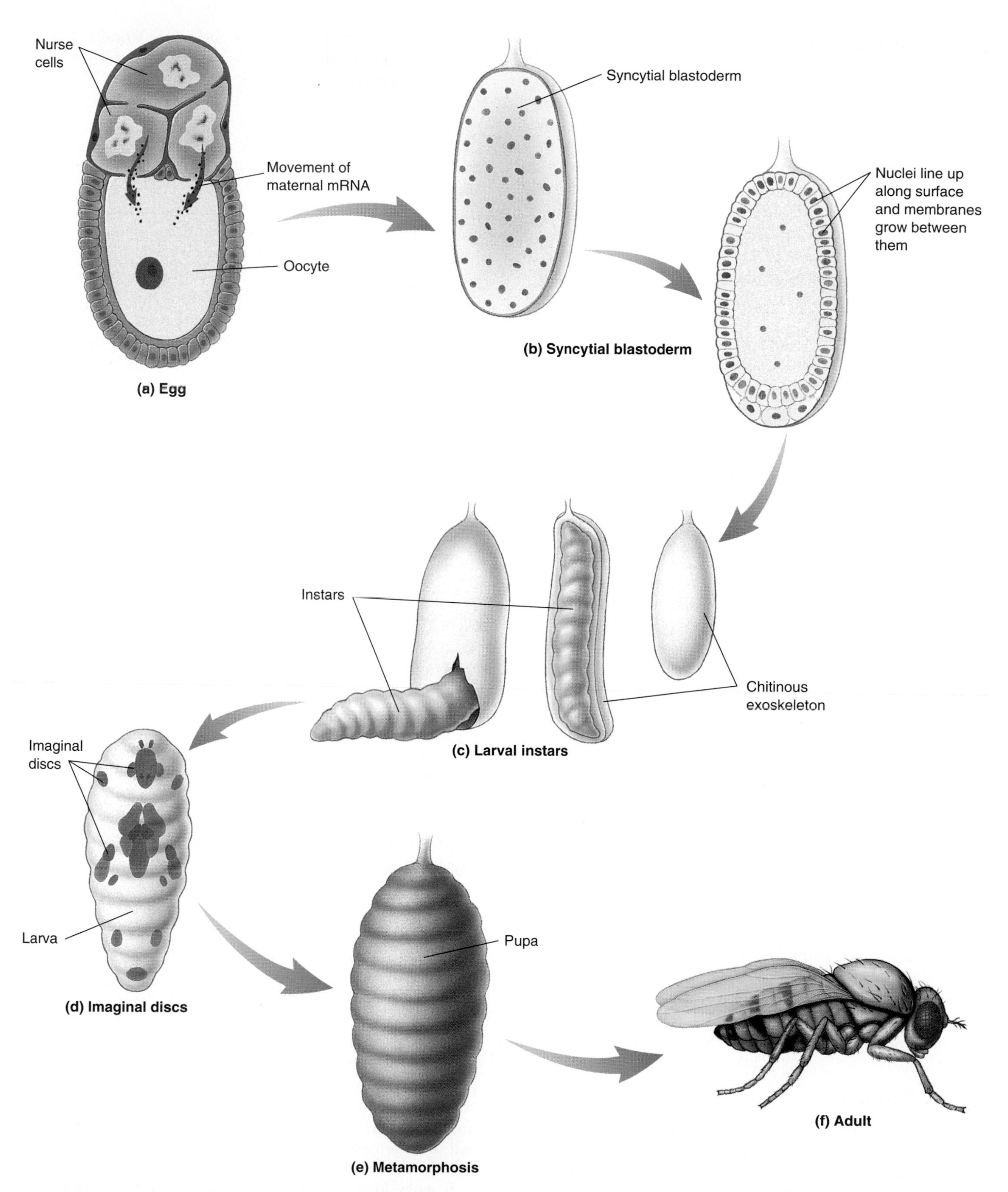

FIGURE 17.7
The path of insect development. An illustration of the major events in the development of *Drosophila melanogaster*. (*a*) Egg. (*b*) Syncytial blastoderm. (*c*) Larval instars. (*d*) Imaginal discs. (*e*) Metamorphosis. (*f*) Adult.

Plant Development

At the most basic level, the developmental paths of plants and animals share many key elements. However, the mechanisms used to achieve body form are quite different. While animal cells follow an orchestrated series of movements during development, plant cells are encased within stiff cellulose walls, and, therefore, cannot move. Each cell in a plant is fixed into position when it is created. Instead of using cell migration, plants develop by building their bodies outward, creating new parts from special groups of self-renewing cells called **meristems.** As meristem cells continually divide, they produce cells that can differentiate into the tissues of the plant.

Another major difference between animals and plants is that most animals are mobile and can move away from unfavorable circumstances, while plants are anchored in position and must simply endure whatever environment they experience. Plants compensate for this restriction by relaxing the rules of development to accommodate local circumstances. Instead of creating a body in which every part is specified to have a fixed size and location, a plant assembles its body from a few types of modules, such as leaves, roots, branch nodes, and flowers. Each module has a rigidly controlled structure and organization, but how the modules are utilized is quite flexible. As a plant develops, it simply adds more modules, with the environment having a major influence on the type, number, size, and location of what is added. In this way the plant is able to adjust the path of its development to local circumstances.

Early Cell Division

The first division of the fertilized egg in a flowering plant is off-center, so that one of the daughter cells is small, with dense cytoplasm (figure 17.8*a*). That cell, the future embryo, begins to divide repeatedly, forming a ball of cells. The other daughter cell also divides repeatedly, forming an elongated structure called a **suspensor,** which links the embryo to the nutrient tissue of the seed. The suspensor also provides a route for nutrients to reach the developing embryo. Just as the animal embryo acquires its initial axis as a cell mass formed during cleavage divisions, so the plant embryo forms its root-shoot axis at this time. Cells near the suspensor are destined to form a root, while those at the other end of the axis ultimately become a shoot.

Tissue Formation

Three basic tissues differentiate while the plant embryo is still a ball of cells (figure 17.8*b*), analogous to the formation of the three germ layers in animal embryos, although in plants, no cell movements are involved. The outermost cells in a plant embryo become **epidermal cells.** The bulk of the embryonic interior consists of **ground tissue** cells that eventually function in food and water storage. Lastly, cells at the core of the embryo are destined to form the future **vascular tissue.**

Seed Formation

Soon after the three basic tissues form, a flowering plant embryo develops one or two seed leaves called **cotyledons.** At this point, development is arrested, and the embryo is now either surrounded by nutritive tissue or has amassed stored food in its cotyledons (figure 17.8*c*). The resulting package, known as a *seed,* is resistant to drought and other unfavorable conditions; in its dormant state, it is a vehicle for dispersing the embryo to distant sites and allows a plant embryo to survive in environments that might kill a mature plant.

Germination

A seed germinates in response to changes in its environment brought about by water, temperature, or other factors. The embryo within the seed resumes development and grows rapidly, its roots extending downward and its leaf-bearing shoots extending upward (figure 17.8*d*).

Meristematic Development

Plant development exhibits its great flexibility during the assembly of the modules that make up a plant body. Apical meristems at the root and shoot tips generate the large numbers of cells needed to form leaves, flowers, and all other components of the mature plant (figure 17.8*e*). At the same time, meristems ensheathing the stems and roots produce the wood and other tissues that allow growth in circumference. A variety of hormones produced by plant tissues influence meristem activity and, thus, the development of the plant body. Plant hormones (see chapter 41) are the tools that allow plant development to adjust to the environment.

Morphogenesis

The form of a plant body is largely determined by controlled changes in cell shape as they expand osmotically after they form (see figure 17.8*e*). Plant growth-regulating hormones and other factors influence the orientation of bundles of microtubules on the interior of the plasma membrane. These microtubules seem to guide cellulose deposition as the cell wall forms around the outside of a new cell. The orientation of the cellulose fibers, in turn, determines how the cell will elongate as it increases in volume, and so determines the cell's final shape.

In a developing plant, leaves, flowers, and branches are added to the growing body in ways that are strongly influenced by the environment.

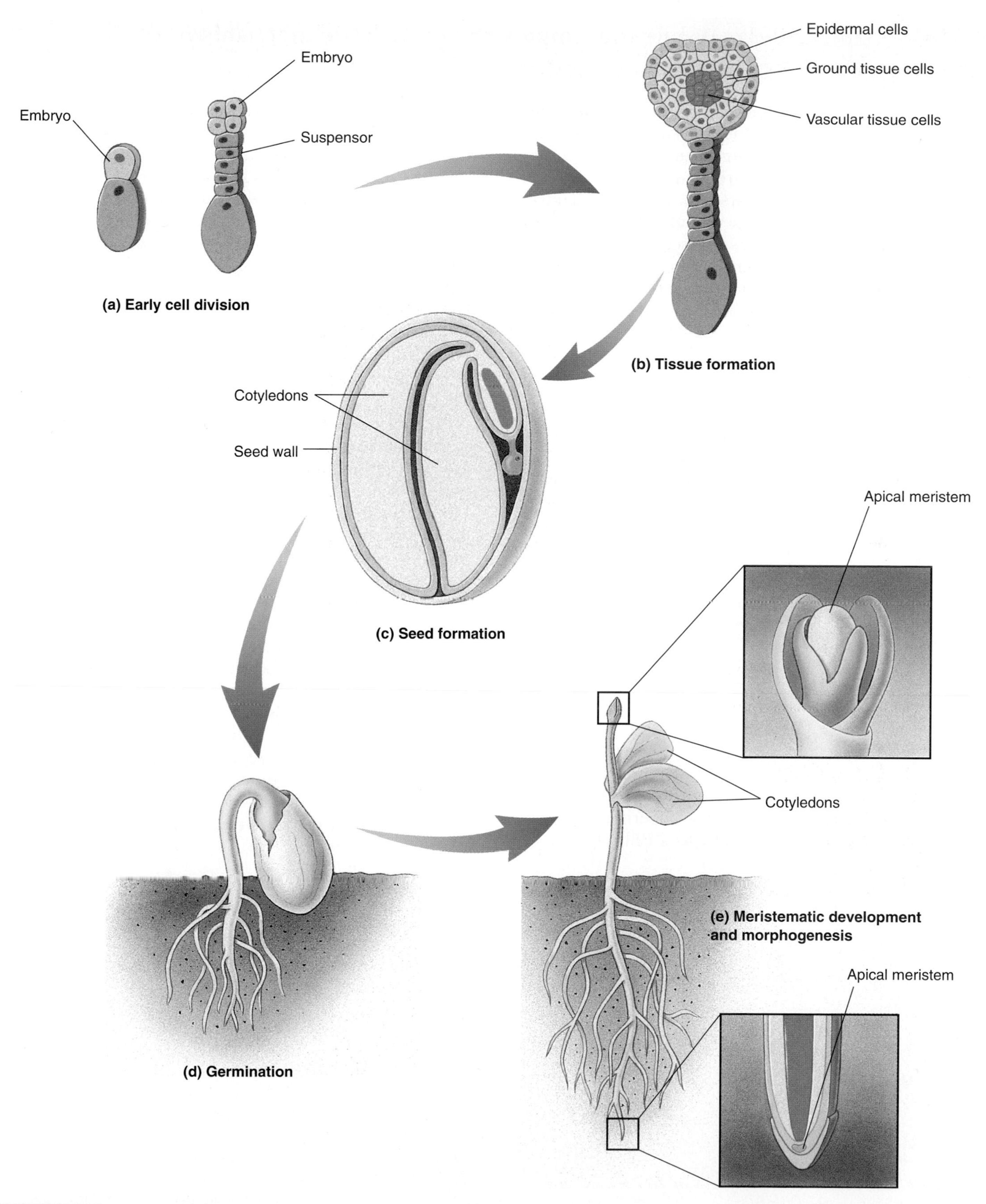

FIGURE 17.8
The path of plant development. An illustration of the developmental stages of *Arabidopsis thaliana*. (*a*) Early cell division. (*b*) Tissue formation. (*c*) Seed formation. (*d*) Germination. (*e*) Meristematic development and morphogenesis.

17.2 Multicellular organisms employ the same basic mechanisms of development.

Despite the many differences in the three developmental paths we have just discussed, it is becoming increasingly clear that most multicellular organisms develop according to molecular mechanisms that are fundamentally very similar. This observation suggests that these mechanisms evolved very early in the history of multicellular life. Here, we will focus on six mechanisms that seem to be of particular importance in the development of a wide variety of organisms. We will consider them in roughly the order in which they first become important during development.

Cell Movement and Induction

Cell Movement

Cells migrate during many stages in animal development, sometimes traveling great distances before reaching the site where they are destined to develop. By the time vertebrate development is complete, most tissues contain cells that originated from quite different parts of the early embryo. One way cells move is by pulling themselves along using cell adhesion molecules, such as the cadherin proteins you read about in chapter 7. Cadherins span the plasma membrane, protruding into the cytoplasm and extending out from the cell surface. The cytoplasmic portion of the molecule is attached to actin or intermediate filaments of the cytoskeleton, while the extracellular portion has five 100-amino acid segments linked end-to-end; three or more of these segments have Ca^{++} binding sites that play a critical role in the attachment of the cadherin to other cells. Over a dozen different cadherins have been discovered to date. Each type of cadherin attaches to others of its own type at its terminal segments, forming a two-cadherin link between the cytoskeletons of adjacent cells. As a cell migrates to a different tissue, the nature of the cadherin it expresses changes, and if cells expressing two different cadherins are mixed, they quickly sort themselves out, aggregating into two separate masses. This is how the different imaginal discs of a *Drosophila* larva assemble into an adult. Other calcium-independent cell adhesion molecules, such as the neural cell adhesion molecules (N CAMs) expressed by migrating nerve cells, reinforce the associations made by cadherins, but cadherins play the major role in holding aggregating cells together.

In some tissues, such as connective tissue, much of the volume of the tissue is taken up by the spaces *between* cells. These spaces are not vacant, however. Rather, they are filled with a network of molecules secreted by surrounding cells, principally, a matrix of long polysaccharide chains covalently linked to proteins (proteoglycans), within which are embedded strands of fibrous protein (collagen, elastin, and fibronectin). Migrating cells traverse this matrix by binding to it with cell surface proteins called integrins, which was also described in chapter 7. Integrins are attached to actin filaments of the cytoskeleton and protrude out from the cell surface in pairs, like two hands. The "hands" grasp a specific component of the matrix such as collagen or fibronectin, thus linking the cytoskeleton to the fibers of the matrix. In addition to providing an anchor, this binding can initiate changes within the cell, alter the growth of the cytoskeleton, and change the way in which the cell secretes materials into the matrix.

Thus, cell migration is largely a matter of changing patterns of cell adhesion. As a migrating cell travels, it continually extends projections that probe the nature of its environment. Tugged this way and that by different tentative attachments, the cell literally feels its way toward its ultimate target site.

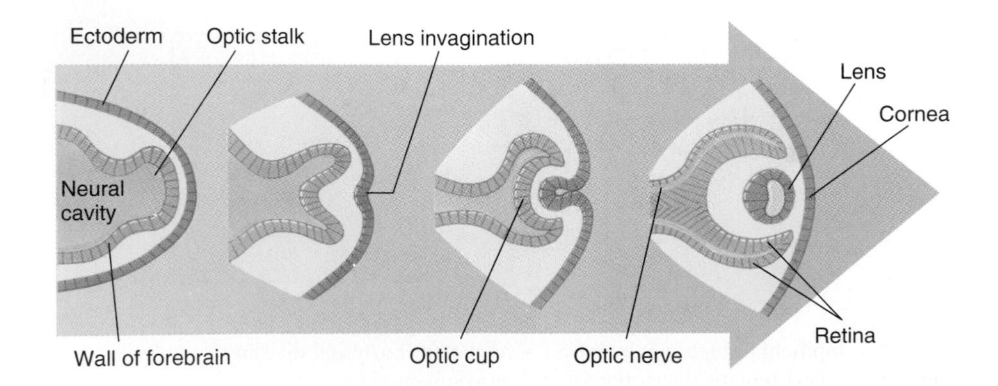

FIGURE 17.9
Development of the vertebrate eye proceeds by induction. The eye develops as an extension of the forebrain called the optic stalk that grows out until it contacts the ectoderm. This contact induces the formation of a lens from the ectoderm.

Induction

In *Drosophila* the initial cells created by cleavage divisions contain different developmental signals (called **determinants**) from the egg, setting individual cells off on different developmental paths. This pattern of development is called **mosaic development.** In mammals, by contrast, all of the blastomeres receive equivalent sets of determinants; body form is determined by cell-cell interactions, a pattern called **regulative development.**

We can demonstrate the importance of cell-cell interactions in development by separating the cells of an early blastula and allowing them to develop independently. Under these conditions, animal pole blastomeres develop features of ectoderm and vegetal pole blastomeres develop features of endoderm, but none of the cells ever develop features characteristic of mesoderm. However, if animal pole and vegetal pole cells are placed next to each other, some of the animal pole cells will develop as mesoderm. The interaction between the two cell types triggers a switch in the developmental path of the cells! When a cell switches from one path to another as a result of interaction with an adjacent cell, **induction** has taken place (figure 17.9).

How do cells induce developmental changes in neighboring cells? Apparently, the inducing cells secrete proteins that act as intercellular signals. Signal molecules, which we discussed in detail in chapter 7, are capable of producing abrupt changes in the patterns of gene transcription.

In some cases, particular groups of cells called **organizers** produce diffusible signal molecules that convey positional information to other cells. Organizers can have a profound influence on the development of surrounding tissues (see chapter 60). Working as signal beacons, they inform surrounding cells of their distance from the organizer. The closer a particular cell is to an organizer, the higher the concentration of the signal molecule, or **morphogen,** it experiences (figure 17.10). Although only a few morphogens have been isolated, they are thought to be part of a widespread mechanism for determining relative position during development.

A single morphogen can have different effects, depending upon how far away from the organizer the affected cell is located. Thus, low levels of the morphogen activin will cause cells of the animal pole of an early *Xenopus* embryo to develop into epidermis, while slightly higher levels will induce the cells to develop into muscles, and levels a little higher than that will induce them to form notochord (figure 17.11).

Cells migrate by extending probes to neighboring cells which they use to pull themselves along. Interactions between cells strongly influence the developmental paths they take. Signal molecules from an inducing cell alter patterns of transcription in cells which come in contact with it.

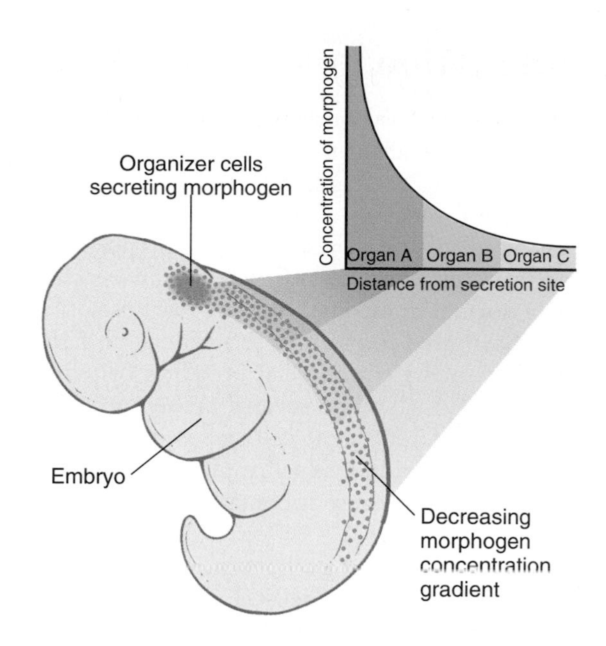

FIGURE 17.10
An organizer creates a morphogen gradient. As a morphogen diffuses from the organizer site, it becomes less concentrated. Different concentrations of the morphogen stimulate the development of different organs.

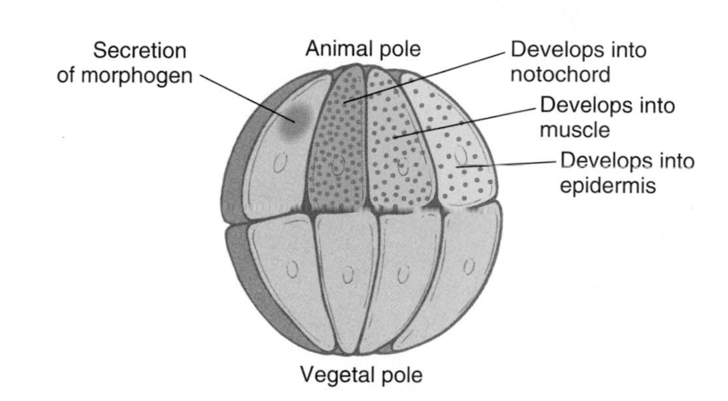

FIGURE 17.11
Fate of cells in an early *Xenopus* embryo. The fates of the individual cells are determined by the concentration of morphogen washing over them.

Determination

The mammalian egg is symmetrical in its contents as well as its shape, so that all of the cells of an early blastoderm are equivalent up to the eight-cell stage. The cells are said to be **totipotent,** meaning that they are potentially capable of expressing all of the genes of their genome. If they are separated from one another, any one of them can produce a completely normal individual. Indeed, just this sort of procedure has been used to produce sets of four or eight identical offspring in the commercial breeding of particularly valuable lines of cattle. The reverse process works, too; if cells from two different eight-cell-stage embryos are combined, a single normal individual results. Such an individual is called a **chimera,** because it contains cells from different genetic lines (figure 17.12).

Mammalian cells start to become different after the eight-cell stage as a result of cell-cell interactions like those we just discussed. At this point, the pathway that will influence the future developmental fate of the cells is determined. The commitment of a particular cell to a specialized developmental path is called **determination.** A cell in the prospective brain region of an amphibian embryo at the early gastrula stage has not yet been determined; if transplanted elsewhere in the embryo, it will develop like its new neighbors (see chapter 60). By the late gastrula stage, however, determination has taken place, and the cell will develop as neural tissue no matter where it is transplanted. Determination must be carefully distinguished from **differentiation,** which is the cell specialization that occurs at the end of the developmental path. Cells may become *determined* to give rise to particular tissues long before they actually *differentiate* into those tissues. The cells of a *Drosophila* eye imaginal disc, for example, are fully determined to produce an eye, but they remain totally undifferentiated during most of the course of larval development.

The Mechanism of Determination

What is the molecular mechanism of determination? The gene regulatory proteins discussed in detail in chapter 16 are the tools used by cells to initiate developmental changes. When genes encoding these proteins are activated, one of their effects is to reinforce their own activation. This makes the developmental switch deterministic, initiating a chain of events that leads down a particular developmental pathway. Cells in which a set of regulatory genes have been activated may not actually undergo differentiation until some time later, when other factors interact with the regulatory protein and cause it to activate still other genes. Nevertheless, once the initial "switch" is thrown, the cell is fully committed to its future developmental path.

Often, before a cell becomes fully committed to a particular developmental path, it first becomes partially committed, acquiring **positional labels** that reflect its location in the embryo. These labels can have a great influence on how the pattern of the body subsequently develops. In a chicken embryo, if tissue at the base of the leg bud (which would normally give rise to the thigh) is transplanted to the tip of the identical-looking wing bud (which would normally give rise to the wing tip), that tissue will develop into a toe rather than a thigh! The tissue has already been determined as leg but is not yet committed to being a particular part of the leg. Therefore, it can be influenced by the positional signaling at the tip of the wing bud to form a tip (in this case a tip of leg).

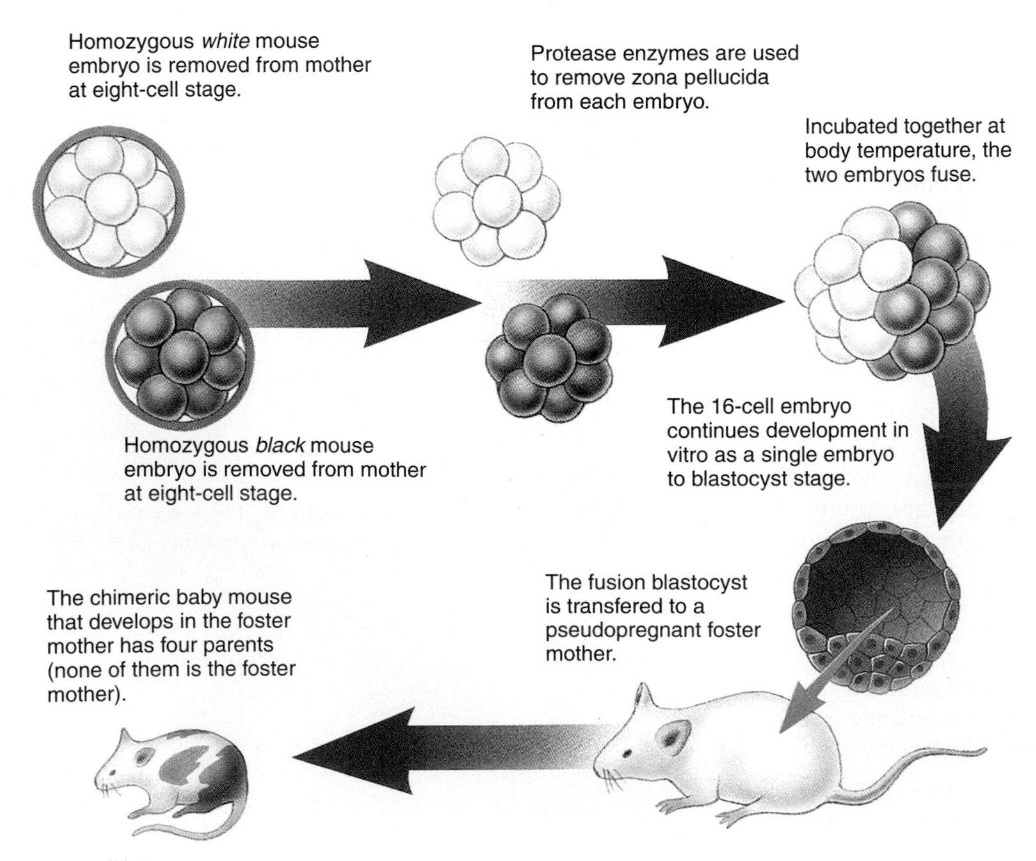

Homozygous white mouse embryo is removed from mother at eight-cell stage.
Protease enzymes are used to remove zona pellucida from each embryo.
Incubated together at body temperature, the two embryos fuse.
Homozygous black mouse embryo is removed from mother at eight-cell stage.
The 16-cell embryo continues development in vitro as a single embryo to blastocyst stage.
The chimeric baby mouse that develops in the foster mother has four parents (none of them is the foster mother).
The fusion blastocyst is transfered to a pseudopregnant foster mother.

FIGURE 17.12
Constructing a chimeric mouse. Cells from two eight-cell individuals fuse to form a single individual.

Is Determination Irreversible?

Until very recently, biologists thought determination was irreversible. Experiments carried out in the 1950s and 1960s by John Gurdon and others made what seemed a convincing case: using very fine pipettes (hollow glass tubes) to suck the nucleus out of a frog or toad egg, these researchers replaced the egg nucleus with a nucleus sucked out of a body cell taken from another individual (see figure 14.3). If the transplanted nucleus was obtained from an advanced embryo, the egg went on to develop into a tadpole, but most died before becoming an adult.

Nuclear transplant experiments were attempted without success by many investigators, until finally, in 1984, Steen Willadsen, a Danish embryologist working in Texas, succeeded in cloning a sheep using the nucleus from a cell of an early embryo. The key to his success was in picking a cell very early in development. This exciting result was soon replicated by others in a host of other organisms, including pigs and monkeys.

Only early embryo cells seemed to work, however. Researchers became convinced, after many attempts to transfer older nuclei, that animal cells become irreversibly committed after the first few cell divisions of the developing embryo.

We now know this conclusion to have been unwarranted. The key advance unraveling this puzzle was made in Scotland by geneticists Keith Campbell and Ian Wilmut, who reasoned that perhaps the egg and the donated nucleus needed to be at the same stage in the cell cycle. They removed mammary cells from the udder of a six-year-old sheep. The origin of these cells gave the clone its name, "Dolly," after the country singer Dolly Parton. The cells were grown in tissue culture; then, in preparation for cloning, the researchers substantially reduced for five days the concentration of serum nutrients on which the sheep mammary cells were subsisting. Starving the cells caused them to pause at the beginning of the cell cycle. In parallel preparation, eggs obtained from a ewe were enucleated (figure 17.13).

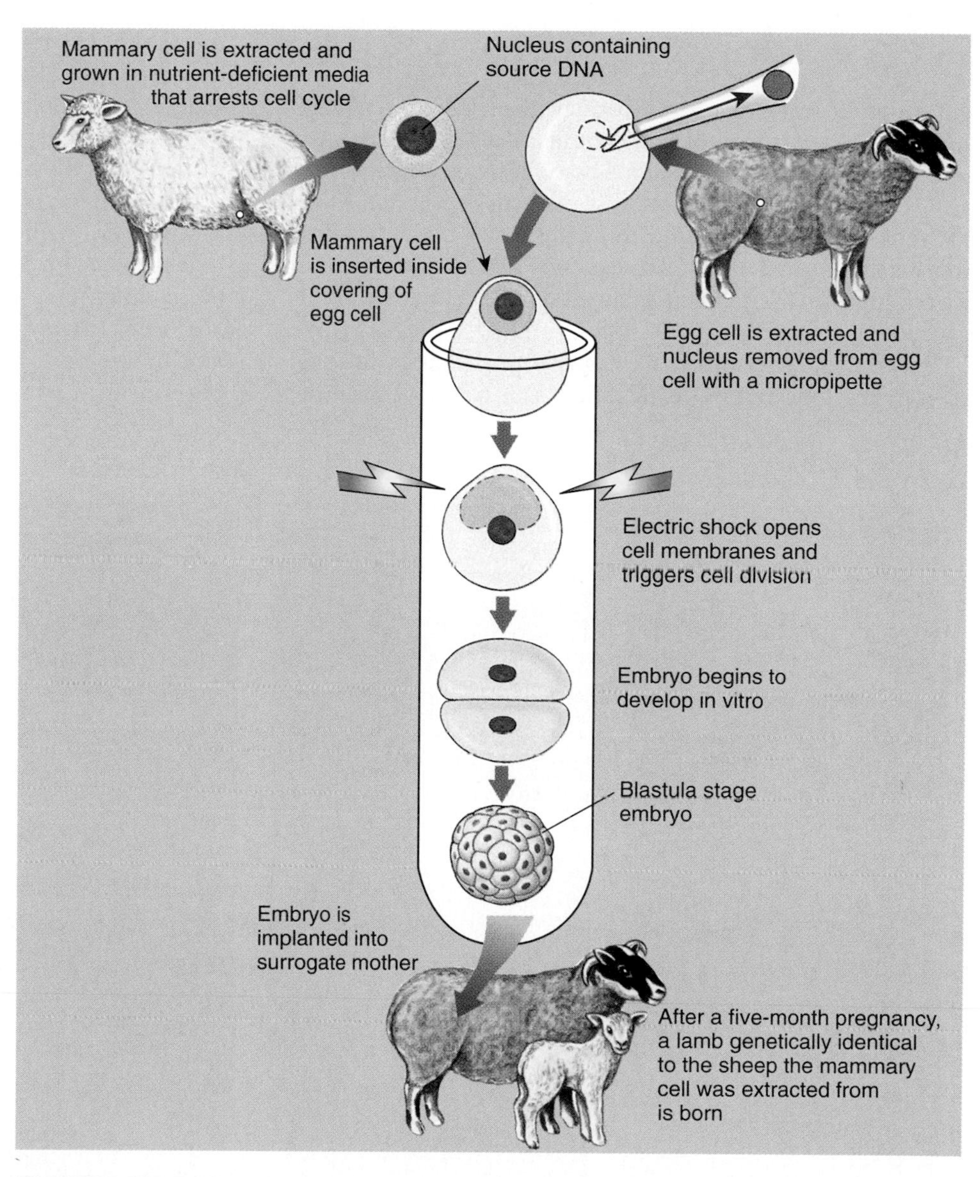

FIGURE 17.13
Proof that determination is reversible. This experiment by Campbell and Wilmut was the first successful cloning of an adult animal.

Mammary cells and egg cells were then surgically combined in January of 1996, inserting the mammary cells inside the covering around the egg cell. The researchers then applied a brief electrical shock. This caused the plasma membranes surrounding the two cells to become leaky, so that the nucleus of the mammary cell passed into the egg cell—a neat trick. The shock also kick-started the cell cycle, causing the cell to begin to divide.

After six days, in 30 of 277 tries, the dividing embryo reached the hollow-ball "blastula" stage, and 29 of these were transplanted into surrogate mother sheep. A little over five months later, on July 5, 1996, one sheep gave birth to a lamb, Dolly, the first clone generated from a fully differentiated animal cell. Dolly established beyond all dispute that determination is reversible, that with the right techniques the fate of a fully differentiated cell *can* be altered.

The commitment of particular cells to certain developmental fates is fully reversible.

Pattern Formation

All animals seem to use positional information to determine the basic pattern of body compartments and, thus, the overall architecture of the adult body. How is positional information encoded in labels and read by cells? To answer this question, let us consider how positional labels are used in pattern formation in *Drosophila*. The Nobel Prize in Physiology or Medicine was awarded in 1995 for the unraveling of this puzzle.

As we noted previously, a *Drosophila* egg acquires an initial asymmetry long before fertilization as a result of maternal mRNA molecules that are deposited in one end of the egg by nurse cells. Part of this maternal mRNA, from a gene called *bicoid*, remains near its point of entry, marking what will become the embryo's front end. Fertilization causes this mRNA to be translated into bicoid protein, which diffuses throughout the syncytial blastoderm, forming a morphogen gradient. Mothers unable to make bicoid protein produce embryos without a head or thorax (in effect, these embryos are two-tailed, or bicaudal—hence the name "bicoid"). Bicoid protein establishes the anterior (front) end of the embryo. If bicoid protein is injected into the anterior end of mutant embryos unable to make it, the

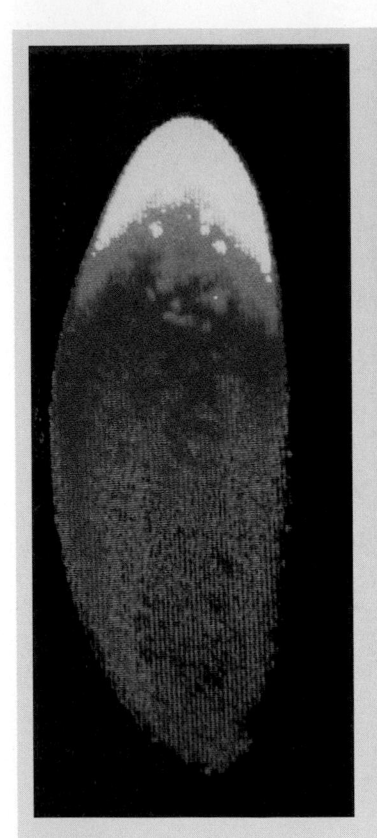

Establishing polarity of the embryo: Fertilization of the egg triggers the production of bicoid protein from maternal RNA in the egg. The bicoid protein diffuses through the egg, forming a gradient. This gradient determines the polarity of the embryo, with the head and thorax developing in the zone of high concentration (*yellow* through *red*).

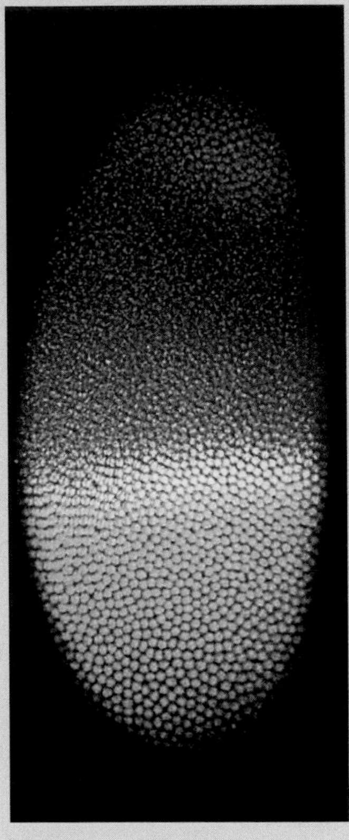

Setting the stage for segmentation: About 2½ hours after fertilization, bicoid protein turns on a series of brief signals from so-called *gap* genes. The gap proteins act to divide the embryo into large blocks. In this photo, fluorescent dyes in antibodies that bind to the gap proteins Krüppel *(red)* and hunchback *(green)* make the blocks visible; the region of overlap is yellow.

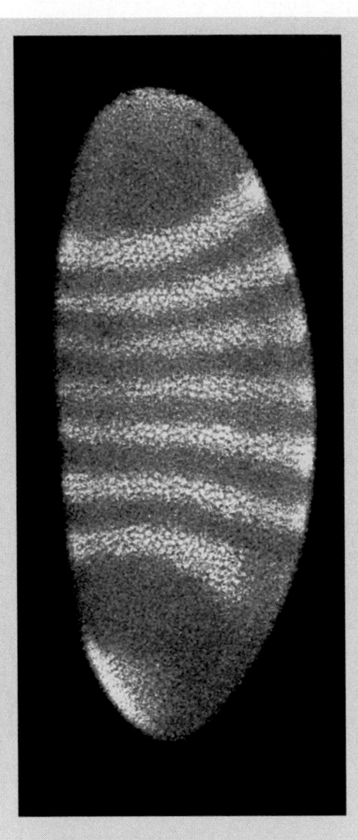

Laying down the fundamental regions: About ½ hour later, the *gap* genes switch on a so-called "pair-rule" gene called *hairy. Hairy* produces a series of boundaries within each block, dividing the embryo into seven fundamental regions.

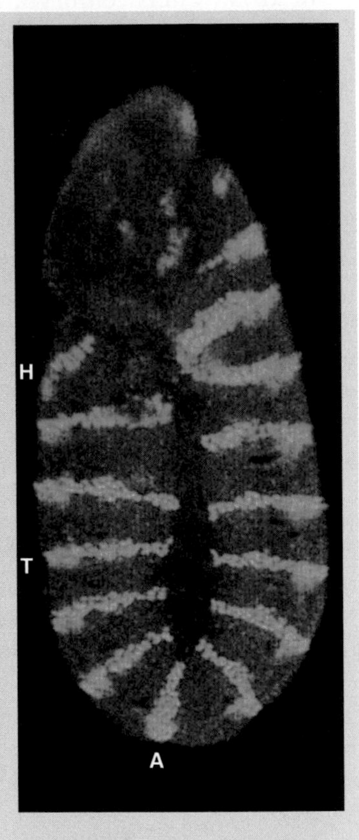

Forming the segments: The final stage of segmentation occurs when a "segment-polarity" gene called *engrailed* divides each of the seven regions into halves, producing 14 narrow compartments. Each compartment corresponds to one segment of the future body. There are three head segments (H, top left), three thoracic segments (T, lower left), and eight abdominal segments (A, from bottom left to upper right).

FIGURE 17.14
Body organization in an early *Drosophila* embryo. In these images by 1995 Nobel laureate, Christiane Nüsslein-Volhard, and Sean Carroll, we watch a *Drosophila* egg pass through the early stages of development, in which the basic segmentation pattern of the embryo is established.

embryos will develop normally. If it is injected into the opposite (posterior) end of normal embryos, a head and thorax will develop at that end.

Bicoid protein exerts this profound effect on the organization of the embryo by activating genes that encode the first mRNAs to be transcribed after fertilization. Within the first two hours, before cellularization of the syncytial blastoderm, a group of six genes called the **gap genes** begins to be transcribed. These genes map out the coarsest subdivision of the embryo (figure 17.14). One of them is a gene called *hunchback* (because an embryo without *hunchback* lacks a thorax and so, takes on a hunched shape). Although *hunchback* mRNA is distributed throughout the embryo, its translation is controlled by the protein product of another maternal mRNA called *nanos* (named after the Greek word for "dwarf," as mutants without *nanos* genes lack abdominal segments and hence, are small). The **nanos** protein binds to *hunchback* mRNA, preventing it from being translated. The only place in the embryo where there is too little nanos protein to block translation of *hunchback* mRNA is the far anterior end. Consequently, hunchback protein is made primarily at the anterior end of the embryo. As it diffuses back toward the posterior end, it sets up a second morphogen gradient responsible for establishing the thoracic and abdominal segments.

Other gap genes act in more posterior regions of the embryo. They, in turn, activate 11 or more **pair-rule genes.** (When mutated, each of these genes alters every other body segment.) One of the pair-rule genes, named *hairy*, produces seven bands of protein, which look like stripes when visualized with fluorescent markers. These bands establish boundaries that divide the embryo into seven zones. Finally, a group of 16 or more **segment polarity genes** subdivide these zones. The *engrailed* gene, for example, divides each of the seven zones established by *hairy* into anterior and posterior compartments. The 14 compartments that result correspond to the three head segments, three thoracic segments, and eight abdominal segments of the embryo.

Thus, within three hours after fertilization, a highly orchestrated cascade of segmentation gene activity produces the fly embryo's basic body plan. The activation of these and other developmentally important genes (figure 17.15) depends upon the free diffusion of morphogens that is possible within a syncytial blastoderm. In mammalian embryos with cell partitions, other mechanisms must operate.

In *Drosophila* diffusion of chemical inducers produces the embryo's basic body plan, a cascade of genes dividing it into 14 compartments.

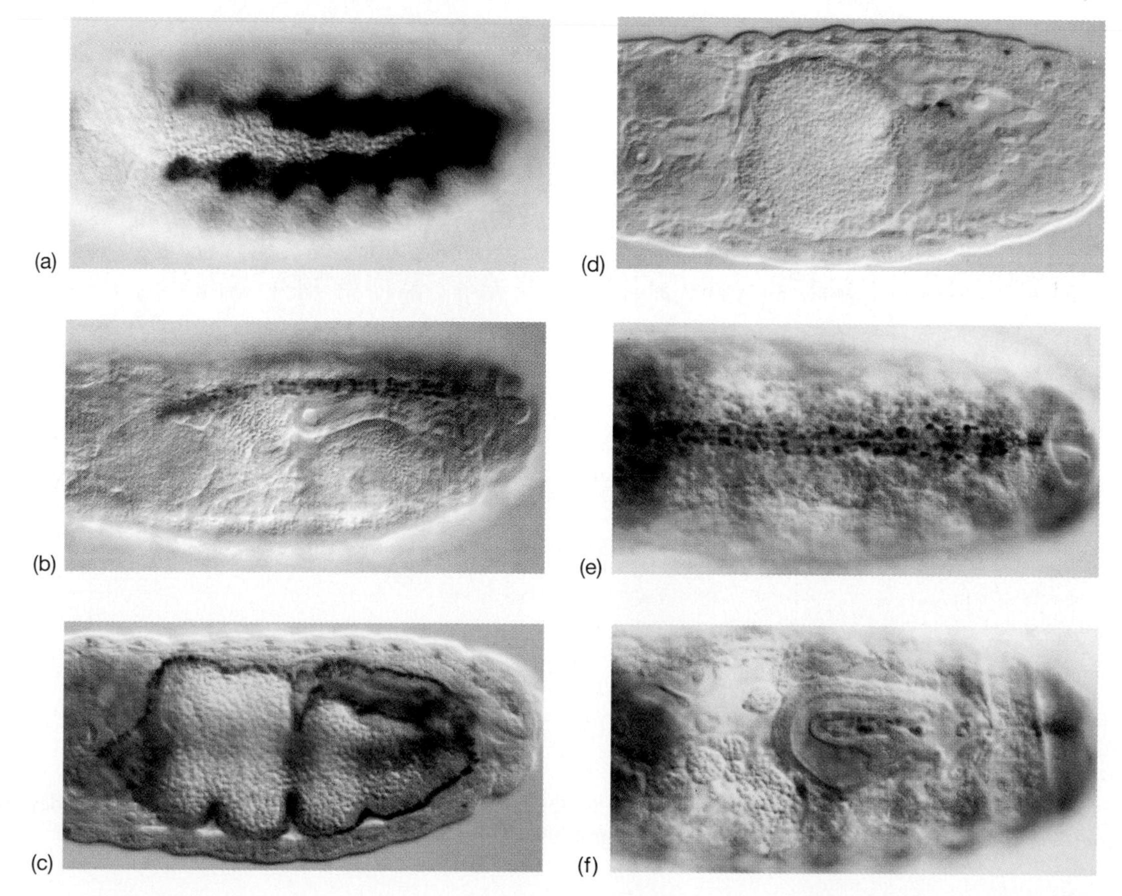

FIGURE 17.15
A gene controlling organ formation in *Drosophila*. Called *tinman*, this gene is responsible for the formation of gut musculature and the heart. The dye shows expression of *tinman* in five-hour (*a*) and seventeen-hour (*b*) *Drosophila* embryos. The gut musculature then appears along the edges of normal embryos (*c*) but is not present in embryos in which the gene has been mutated (*d*). The heart tissue develops along the center of normal embryos (*e*) but is missing in *tinman* mutant embryos (*f*).

Expression of Homeotic Genes

After pattern formation has successfully established the number of body segments in *Drosophila*, a series of **homeotic genes** act as master switches to determine the forms these segments will assume. Homeotic genes code for proteins that function as transcription factors. Each homeotic gene activates a particular module of the genetic program, initiating the production of specific body parts within each of the 14 compartments.

Homeotic Mutations

Mutations in homeotic genes lead to the appearance of perfectly normal body parts in unusual places. Mutations in *bithorax* (figure 17.16), for example, cause a fly to grow an extra pair of wings, as if it had a double thoracic segment, and mutations in *Antennapedia* cause legs to grow out of the head in place of antennae! In the early 1950s, geneticist Edward Lewis discovered that several homeotic genes, including *bithorax*, map together on the third chromosome of *Drosophila*, in a tight cluster called the **bithorax complex.**

FIGURE 17.16
Mutations in homeotic genes. Three separate mutations in the *bithorax* gene caused this fruit fly to develop an extra thoracic segment, with accompanying wings. Compare this photograph with that of the normal fruit fly in figure 17.6.

FIGURE 17.17
***Drosophila* homeotic genes.** Called the homeotic gene complex, or HOM complex, the genes are grouped into two clusters, the Antennapedia complex (anterior) and the bithorax complex (posterior).

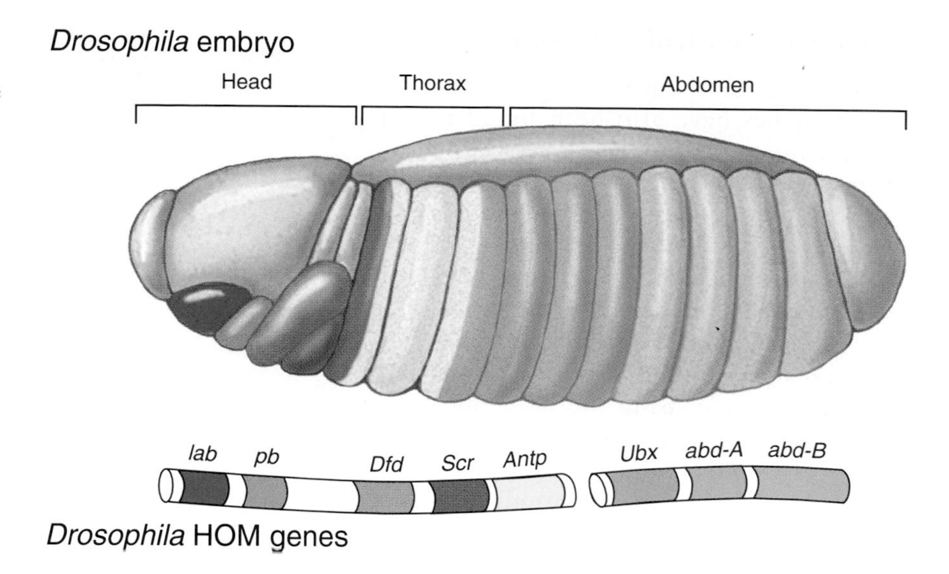

Mutations in these genes all affect body parts of the thoracic and abdominal segments, and Lewis concluded that the genes of the bithorax complex control the development of body parts in the rear half of the thorax and all of the abdomen. Most interestingly, the order of the genes in the bithorax complex mirrors the order of the body parts they control, as if the genes are activated serially! Genes at the beginning of the cluster switch on development of the thorax, those in the middle control the anterior part of the abdomen, and those at the end affect the tip of the abdomen. A second cluster of homeotic genes, the **Antennapedia complex,** was discovered in 1980 by Thomas Kaufmann. The Antennapedia complex governs the anterior end of the fly, and the order of genes in it also corresponds to the order of segments they control (figure 17.17).

The Homeobox

Drosophila homeotic genes typically contain the **homeobox,** a sequence of 180 nucleotides that codes for a 60-amino acid DNA-binding peptide domain called the homeodomain (figure 17.18). As we saw in chapter 16, proteins that contain the homeodomain function as transcription factors, ensuring that developmentally related genes are transcribed at the appropriate time. Segmentation genes such as *bicoid* and *engrailed* also contain the homeobox sequence. Clearly, the homeobox distinguishes those portions of the genome devoted to pattern formation.

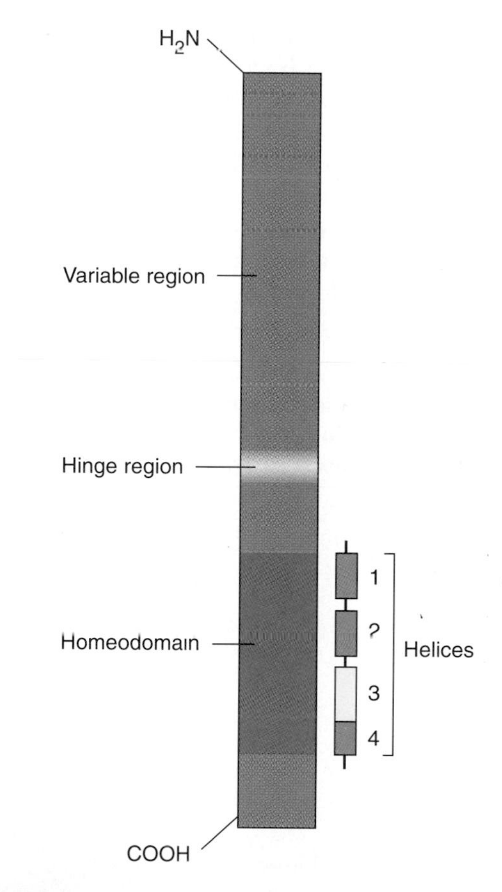

FIGURE 17.18
Homeodomain protein. This protein plays an important regulatory role when it binds to DNA and regulates expression of specific genes. The variable region of the protein determines the specific activity of the protein. Also included in this protein is a small hinge region and the homeodomain, a 60-amino-acid sequence common to all proteins of this type. The homeodomain region of the protein is coded for by the homeobox region of genes and is composed of four α helices. One of the helices recognizes and binds to a specific DNA sequence in target genes.

Evolution of Homeobox Genes

Since their initial discovery in *Drosophila*, homeotic genes have also been found in mice and humans, which are separated from insects by over 600 million years of evolution. Their presence in mammals and insects indicates that homeotic genes governing the positioning of body parts must have arisen very early in the evolutionary history of animals. Similar genes also appear to operate in flowering plants. Gene probes made using the homeobox sequence of *Drosophila* have been used to identify very similar sequences in a wide variety of other organisms, including frogs, mice, humans, cows, chickens, beetles, and even earthworms. Mice and humans have four clusters of homeobox-containing genes, called *Hox* genes in mice. Just as in flies, the homeotic genes of mammals appear to be lined up in the same order as the segments they control (figure 17.19). Thus, the ordered nature of homeotic gene clusters is highly conserved in evolution (figure 17.20). There is a total of 38 *Hox* genes in the four homeotic clusters of a mouse, and we are only beginning to understand how they interact.

Homeotic genes encode transcription factors that activate blocks of genes specifying particular body parts.

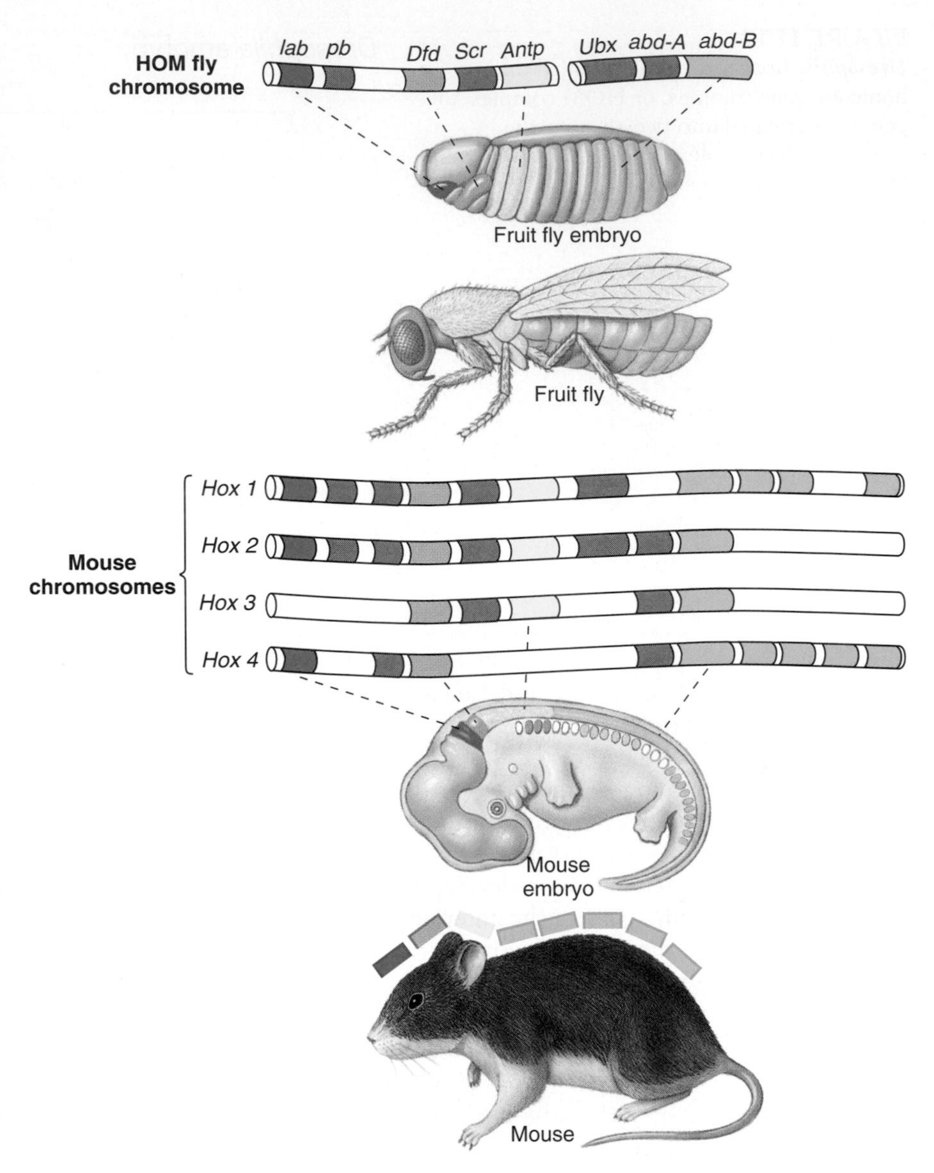

FIGURE 17.19
A comparison of homeotic gene clusters in the fruit fly *Drosophila melanogaster* and the mouse *Mus musculus*. Similar genes, the *Drosophila* HOM genes and the mouse *Hox* genes, control the development of front and back parts of the body. These genes are located on a single chromosome in the fly, and on four separate chromosomes in mammals. The genes are color-coded to match the parts of the body in which they are expressed.

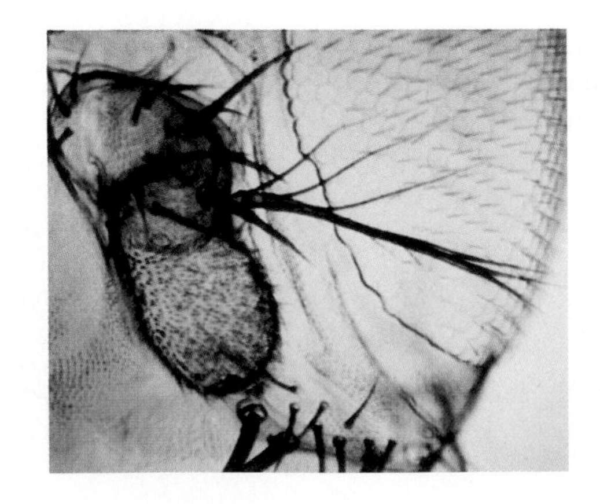

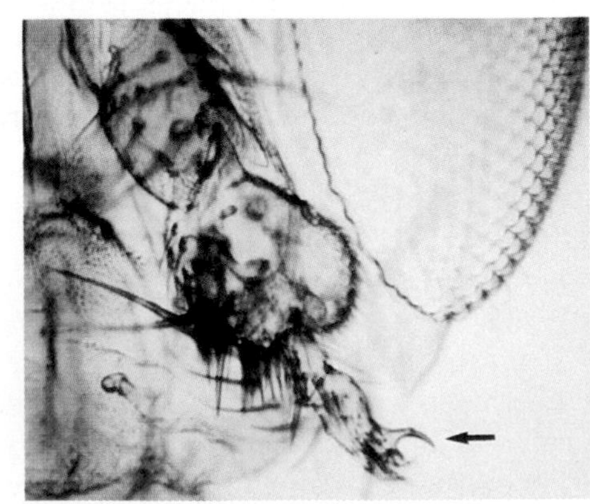

FIGURE 17.20
The remarkably conserved homeobox series. By inserting a mouse homeobox-containing gene into a fruit fly, a mutant fly (*right*) can be manufactured with a leg (*arrow*) growing from where its antenna would be in a normal fly (*left*).

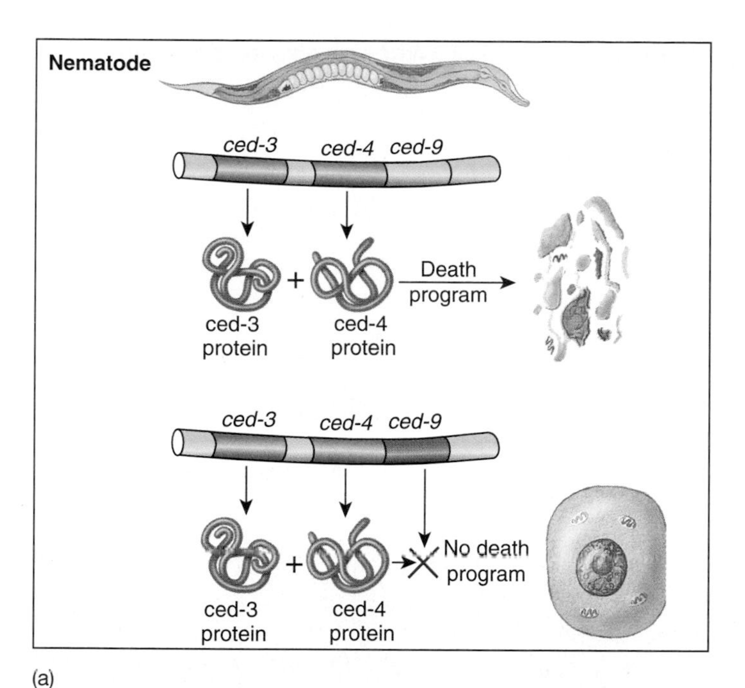

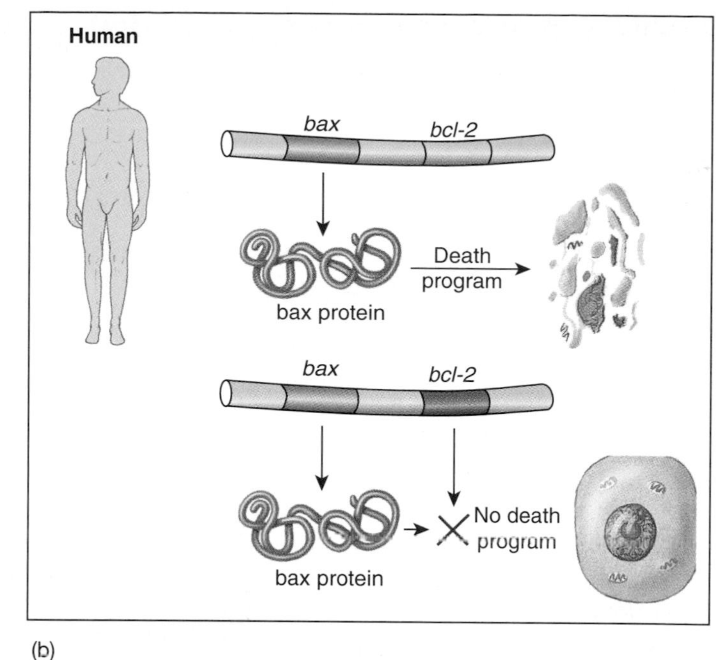

FIGURE 17.21
Programmed cell death. Apoptosis, or programmed cell death, is necessary for normal development of all animals. (*a*) In the developing nematode, for example, two genes, *ced-3* and *ced-4*, code for proteins that cause the programmed cell death of 131 specific cells. In the other cells of the developing nematode, the product of a third gene, *ced-9*, represses the death program encoded by *ced-3* and *ced-4*. (*b*) In developing humans, the product of a gene called *bax* causes a cell death program in some cells and is blocked by the *bcl-2* gene in other cells.

Programmed Cell Death

Not every cell that is produced during development is destined to survive. For example, the cells between your fingers and toes die; if they did not, you would have paddles rather than digits. Vertebrate embryos produce a very large number of neurons, ensuring that there are enough neurons to make all of the necessary synaptic connections, but over half of these neurons never make connections and die in an orderly way as the nervous system develops. Unlike accidental cell deaths due to injury, these cell deaths are planned for and indeed required for proper development. Cells that die due to injury typically swell and burst, releasing their contents into the extracellular fluid. This form of cell death is called **necrosis.** In contrast, cells programmed to die shrivel and shrink in a process called **apoptosis** (from the Greek word meaning shedding of leaves in autumn), and their remains are taken up by surrounding cells.

Gene Control of Apoptosis

This sort of developmentally regulated cell suicide occurs when a "death program" is activated. All animal cells appear to possess such programs. In the nematode worm, for example, the same 131 cells always die during development in a predictable and reproducible pattern of apoptosis. Three genes govern this process. Two (*ced-3* and *ced-4*) constitute the death program itself; if either is mutant, those 131 cells do not die, and go on instead to form nervous and other tissue. The third gene (*ced-9*) represses the death program encoded by the other two (figure 17.21*a*). The same sorts of apoptosis programs occur in human cells: the *bax* gene encodes the cell death program, and another, an oncogene called *bcl-2*, represses it (figure 17.21*b*). The mechanism of apoptosis appears to have been highly conserved during the course of animal evolution. The protein made by *bcl-2* is 25% identical in amino acid sequence to that made by *ced-9*. If a copy of the human *bcl-2* gene is transferred into a nematode with a defective *ced-9* gene, *bcl-2* suppresses the cell death program of *ced-3* and *ced-4!*

How does *bax* kill a cell? The bax protein seems to induce apoptosis by binding to the permeability pore of the cell's mitochondria, increasing its permeability and in doing so triggering cell death. How does *bcl-2* prevent cell death? One suggestion is that it prevents damage from free radicals, highly reactive fragments of atoms that can damage cells severely. Proteins or other molecules that destroy free radicals are called **antioxidants.** Antioxidants are almost as effective as *bcl-2* in blocking apoptosis.

Animal development involves programmed cell death (apoptosis), in which particular genes, when activated, kill their cells.

The Mouse

Some of the most elegant investigations of the cellular mechanisms of development are being done with mammals, particularly the mouse *Mus musculus.* Mice have a battery of homeotic genes, the *Hox* genes (figure 17.22), which seem to be closely related to the homeotic genes of *Drosophila*. Very interestingly, not only do the same genes occur, but they also seem to operate in the same order! Clearly, the homeotic gene system has been highly conserved during the course of animal evolution.

What lends great power to this developmental model system is the ability to create chimeric mice containing cells from two different genetic lines. Mammalian embryos are unusual among vertebrates in that they arise from symmetrical eggs; there are no chemical gradients, and during the initial cleavage divisions, all of the daughter cells are identical. Up to the eight-cell stage, any one of the cells, if isolated, will form a normal adult. Moreover, two different eight-cell-stage embryos can be fused to form a single embryo that will go on to form a normal adult. The resulting adult is a chimera, containing cells from both embryos. In a very real sense, these chimeric mice each have four parents!

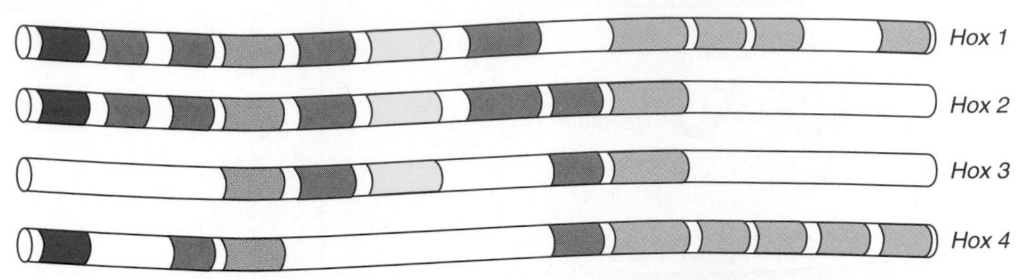

FIGURE 17.22
Studying development in the mouse.

The *Hox* genes control body part development in mice.

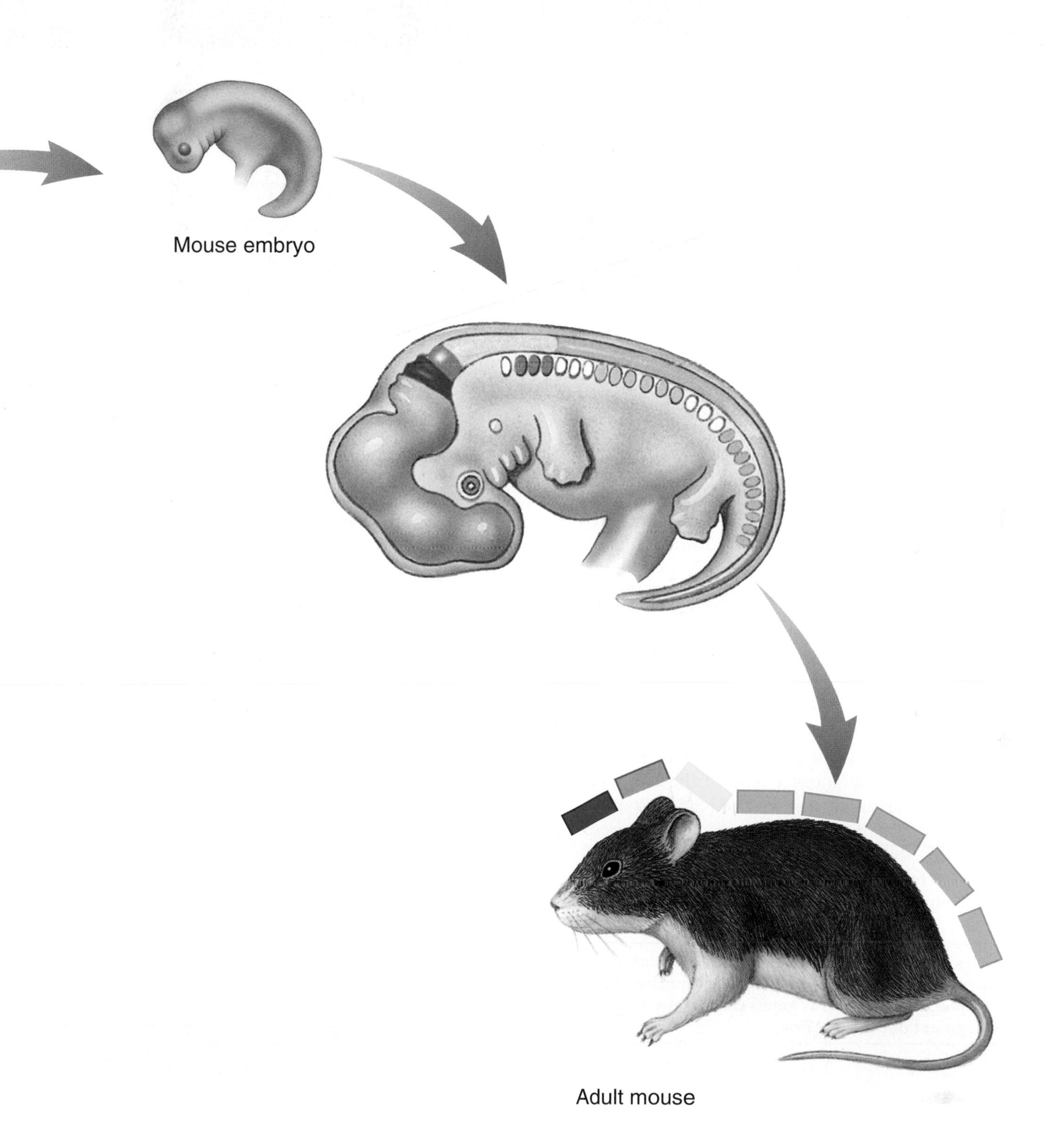
Mouse embryo
Adult mouse

The Fruit Fly

The tiny fruit fly *Drosophila melanogaster* has been a favorite of geneticists for over 90 years and is now playing a key role in our growing understanding of the cellular mechanisms of development. Over the last 10 years, researchers have pieced together a fairly complete picture of how genes expressed early in fruit fly development determine the pattern of the adult body (figure 17.23). The major parts of the adult body are determined as patches of tissue called imaginal discs that float within the body of the larva; during the pupal stage, these discs grow, develop, and associate to form the adult body.

The adult *Drosophila* body is divided into 17 segments, some bearing jointed appendages such as wings or legs. These segments are established during very early development, before the many nuclei of the blastoderm are fully separated from one another. Chemical gradients, established within the egg by material from the mother, create a polarity that directs embryonic development. Reacting to this gradient, a series of segmentation genes progressively subdivide the embryo, first into four broad stripes, and then into 7, 14, and finally 17 segments.

Within each segment, the development of key body parts is under the control of homeotic genes that determine where the body part will form. As we have seen, there are two clusters of homeotic genes, one called Antennapedia that governs the front (anterior) end of the body, and another called bithorax that governs the rear (posterior) end. The organization of genes within each cluster corresponds nicely with the order of the segments they affect. A very similar set of homeotic genes governs body architecture in mice and humans.

A series of segmentation genes divides a *Drosophila* embryo into parts; Antennapedia genes control anterior development, and bithorax genes control the development of the posterior.

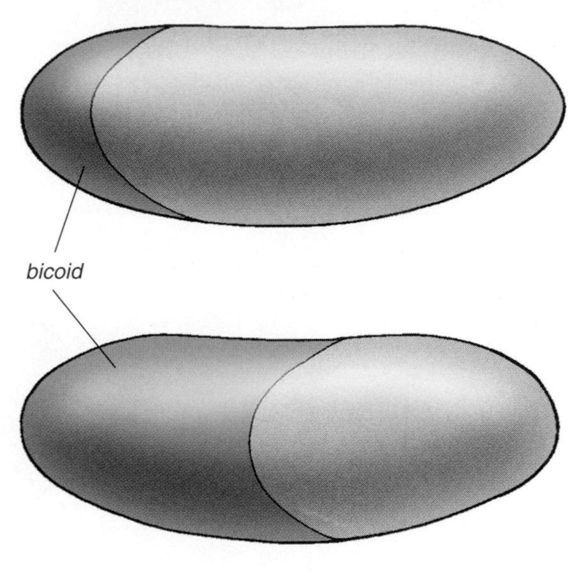

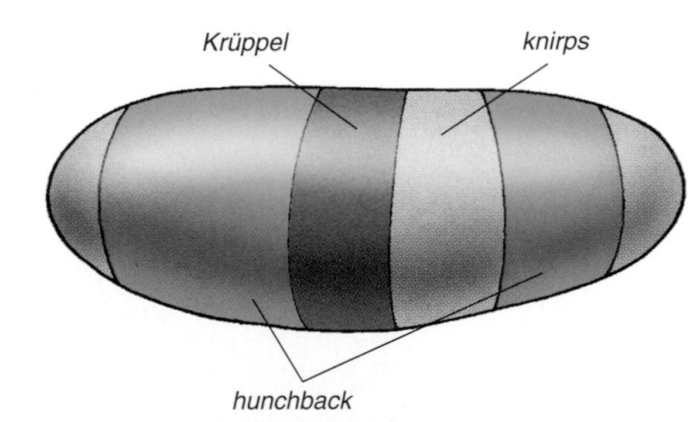

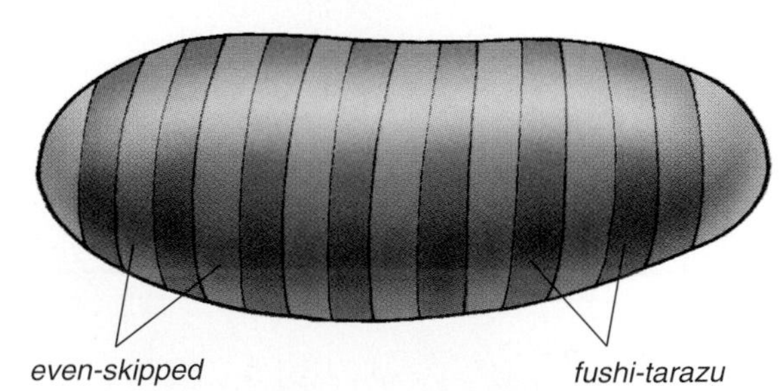

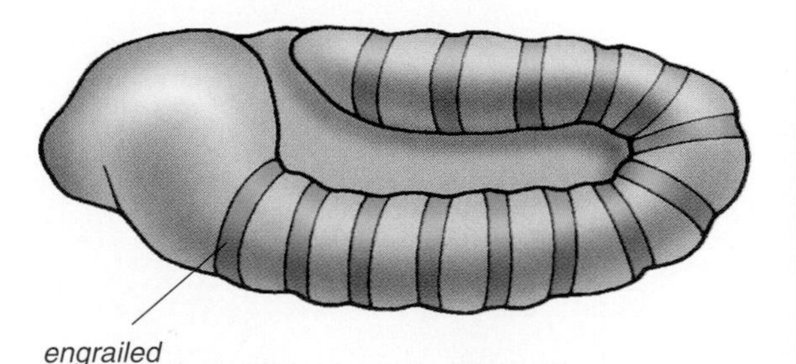

FIGURE 17.23
Studying development in the fruit fly.

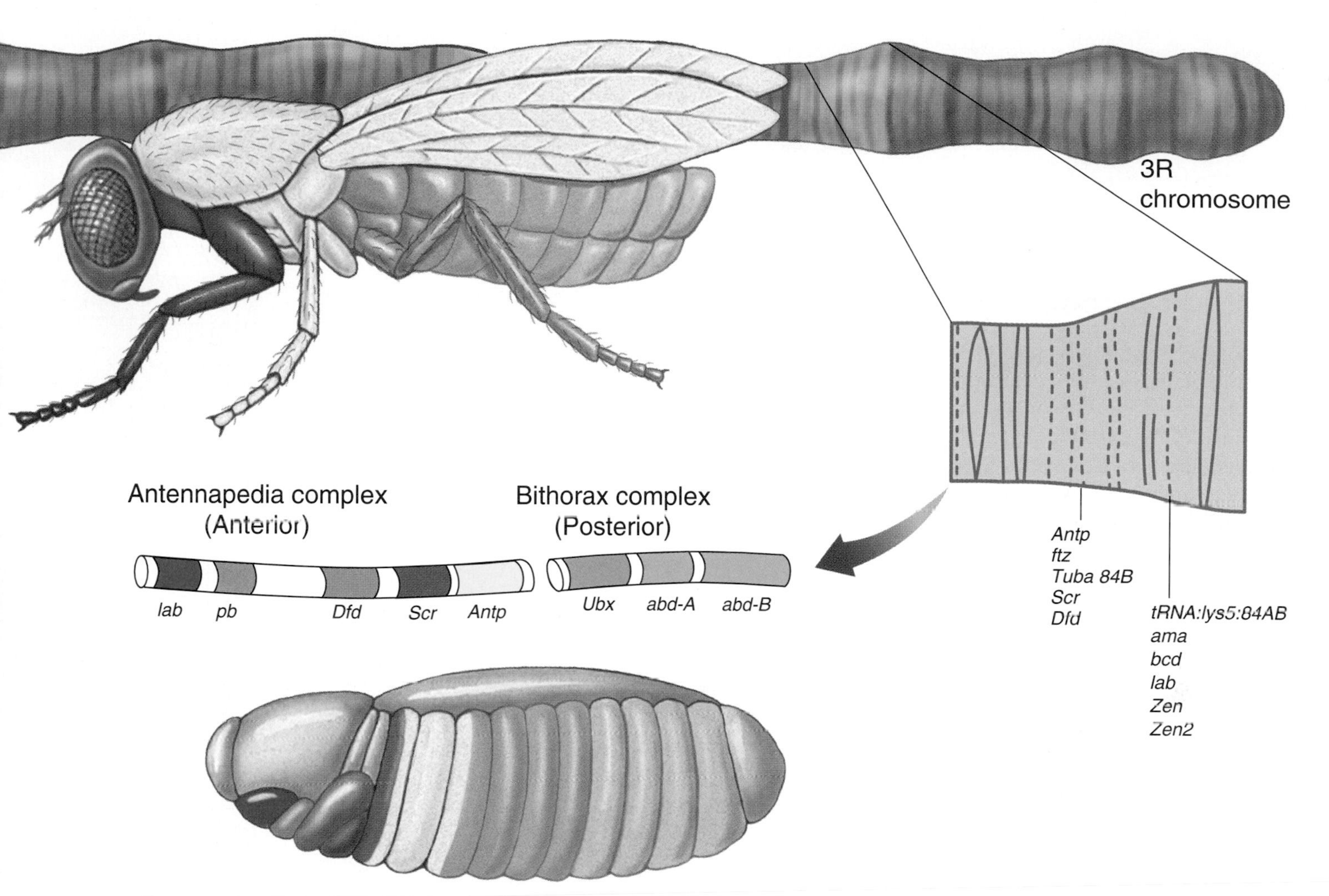
3R chromosome
Antp
ftz
Tuba 84B
Scr
Dfd
tRNA:lys5:84AB
ama
bcd
lab
Zen
Zen2
Antennapedia complex (Anterior)
Bithorax complex (Posterior)
lab
pb
Dfd
Scr
Antp
Ubx
abd-A
abd-B
Drosophila embryo

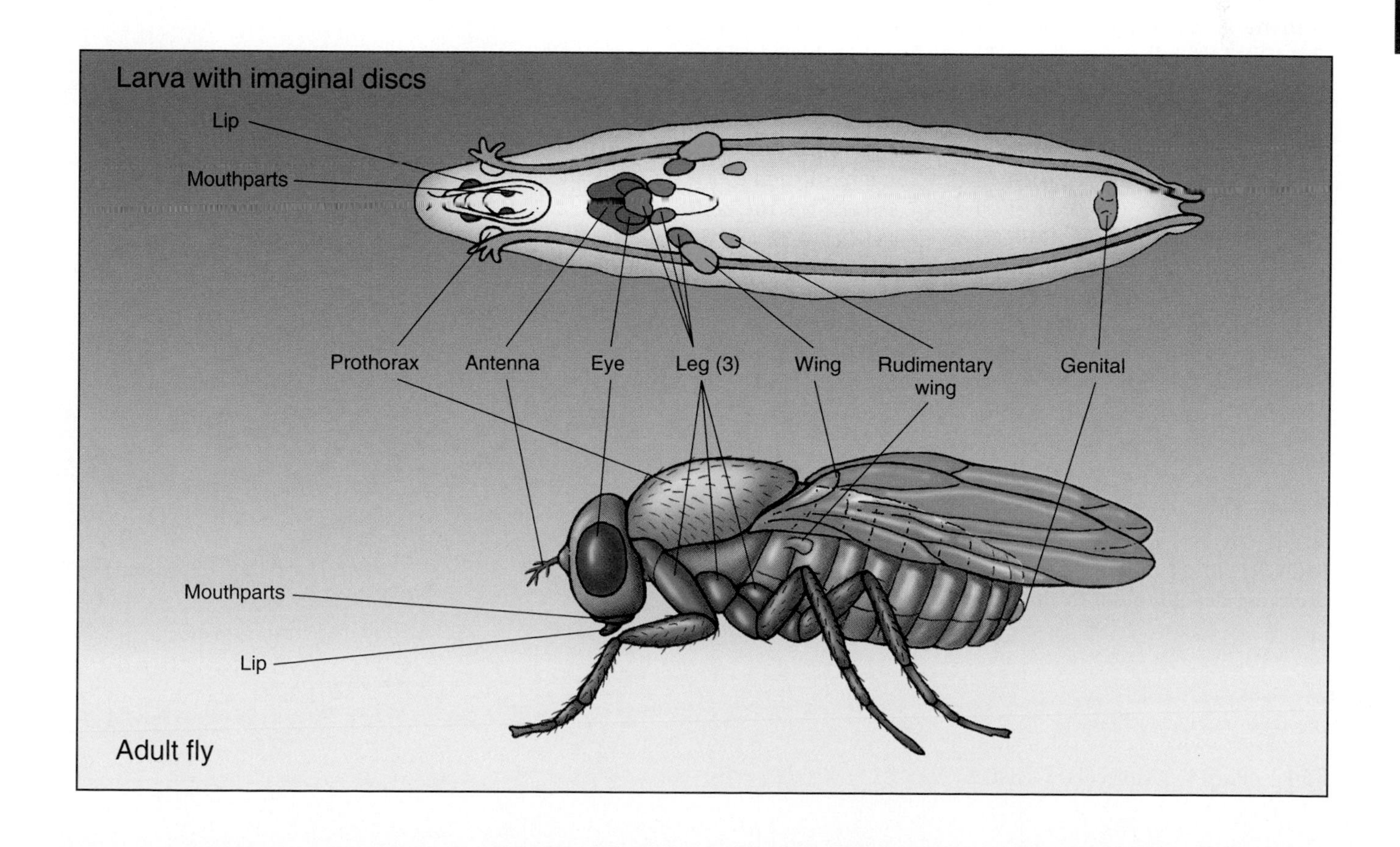
Larva with imaginal discs
Lip
Mouthparts
Prothorax
Antenna
Eye
Leg (3)
Wing
Rudimentary wing
Genital
Mouthparts
Lip
Adult fly

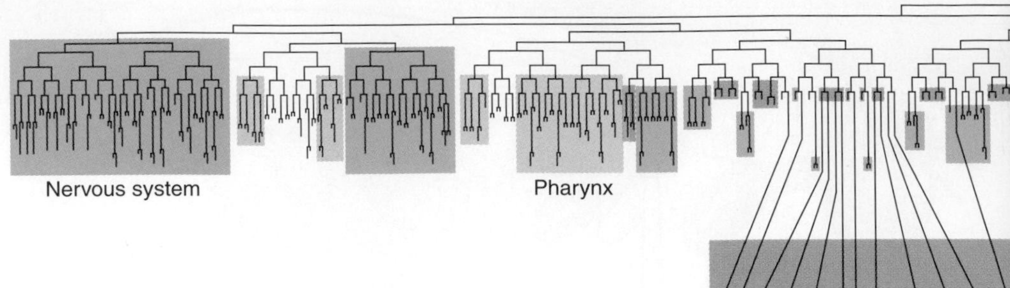

The Nematode

One of the most powerful models of animal development is the tiny nematode *Caenorhabditis elegans.* Only about 1 mm long, it consists of 959 somatic cells and has about the same amount of DNA as *Drosophila.* The entire genome has been mapped as a series of overlapping fragments, and a serious effort is underway to determine the complete DNA sequence of the genome.

Because *C. elegans* is transparent, individual cells can be followed as they divide. By observing them, researchers have learned how each of the cells that make up the adult worm is derived from the fertilized egg. As shown on this lineage map (figure 17.24), the egg divides into two, and then its daughter cells continue to divide. Each horizontal line on the map represents one round of cell division. The length of each vertical line represents the time between cell divisions, and the end of each vertical line represents one fully differentiated cell. In figure 17.24, the major organs of the worm are color-coded to match the colors of the corresponding groups of cells on the lineage map.

Some of these differentiated cells, such as some of the cells that generate the worm's external cuticle, are "born" after only 8 rounds of cell division; other cuticle cells require as many as 14 rounds. The cells that make up the worm's pharynx, or feeding organ, are born after 9 to 11 rounds of division, while cells in the gonads require up to 17 divisions.

Exactly 302 nerve cells are destined for the worm's nervous system. Exactly 131 cells are programmed to die, mostly within minutes of their birth. The fate of each cell is the same in every *C. elegans* individual, except for the cells that will become eggs and sperm.

The nematode develops 959 somatic cells from a single fertilized egg in a carefully orchestrated series of cell divisions which have been carefully mapped by researchers.

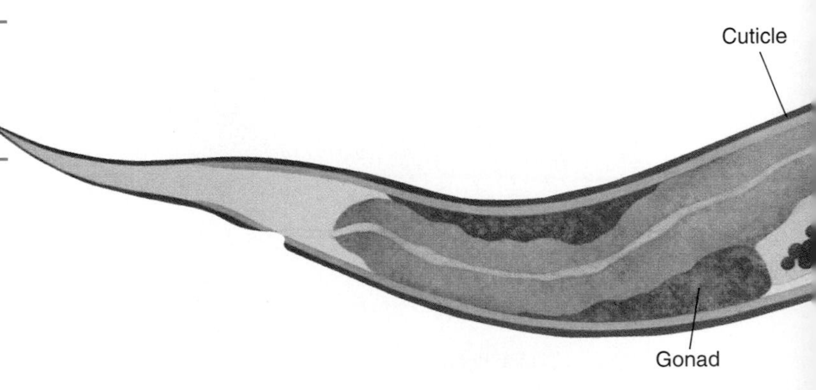

FIGURE 17.24
Studying development in the nematode.

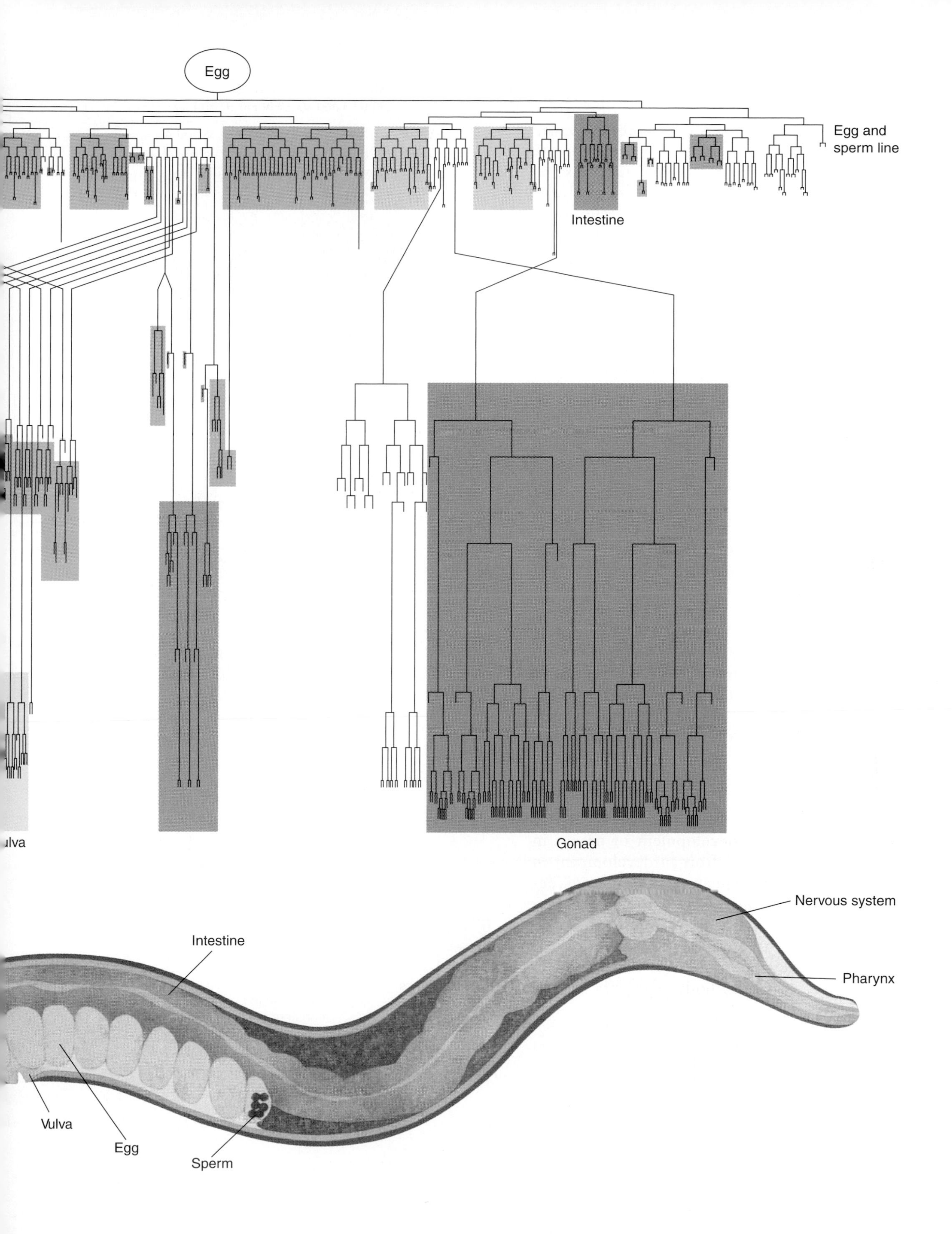

Egg
Egg and sperm line
Intestine
ulva
Gonad
Nervous system
Intestine
Pharynx
Vulva
Egg
Sperm

The Flowering Plant

Scientists are only beginning to unravel the molecular biology of plant development, largely through intensive recent study of a small weedy relative of the mustard plant, the wall cress *Arabidopsis thaliana.* Easy to grow and cross, and with a short generation time, *Arabidopsis* makes an ideal model for investigating plant development. It is able to self-fertilize, like Mendel's pea plants, making genetic analysis convenient. *Arabidopsis* can be grown indoors in test tubes, a single plant producing thousands of offspring after only two months. Its genome is approximately the same size as those of the nematode *Caenorhabditis elegans* and the fruit fly *Drosophila melanogaster.* An extensive library of *Arabidopsis* gene clones is now available to researchers, keyed to the complete genome sequence.

Pattern Formation

Much of the current work investigating *Arabidopsis* development has centered on obtaining and studying mutations that alter the plant's development. Many different sorts of mutations have been identified. Some of the most interesting of them alter the basic architecture of the embryo, the pattern of tissues laid down as the embryo first forms. Mutations in over 50 different genes that alter pattern formation in *Arabidopsis* embryos are now known, affecting every stage of development. While work in this area is still very preliminary, it appears that the mechanisms that establish patterns in the early *Arabidopsis* embryo are broadly similar to those known to function in animal development.

Organ Formation

Importantly, the subsequent development of organs in *Arabidopsis* also seems to parallel organ development in animals, and a similar set of regulatory genes control development in *Arabidopsis, Drosophila,* and mice. *Arabidopsis* flowers, for example, are modified leaves formed as four whorls in a specific order, and homeotic mutations have been identified that convert one part of the pattern to another, just as they do in the body segments of a fly (figure 17.25).

Scientists are only beginning to understand the molecular biology of plant development. In broad outline, it appears quite similar to the development in animals. The genes that determine pattern formation and organ development, for example, operate in the same way in plants and animals.

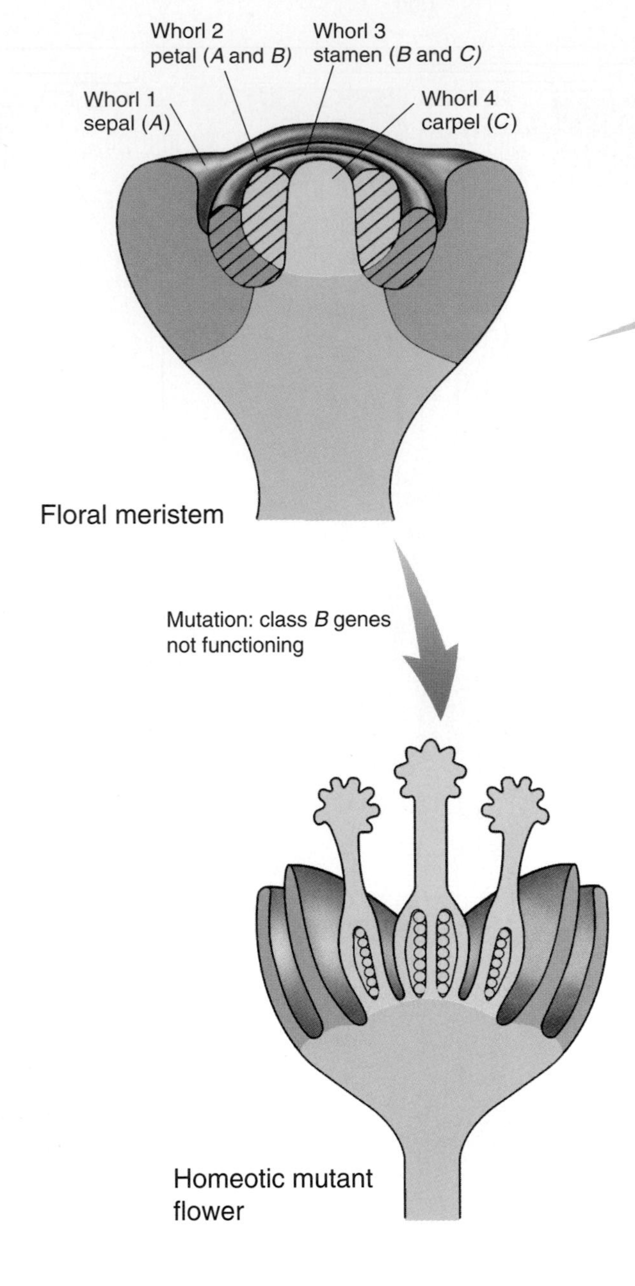

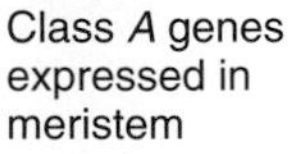
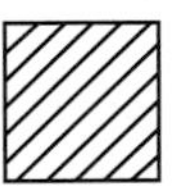
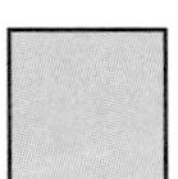

FIGURE 17.25
Studying development in a flowering plant.

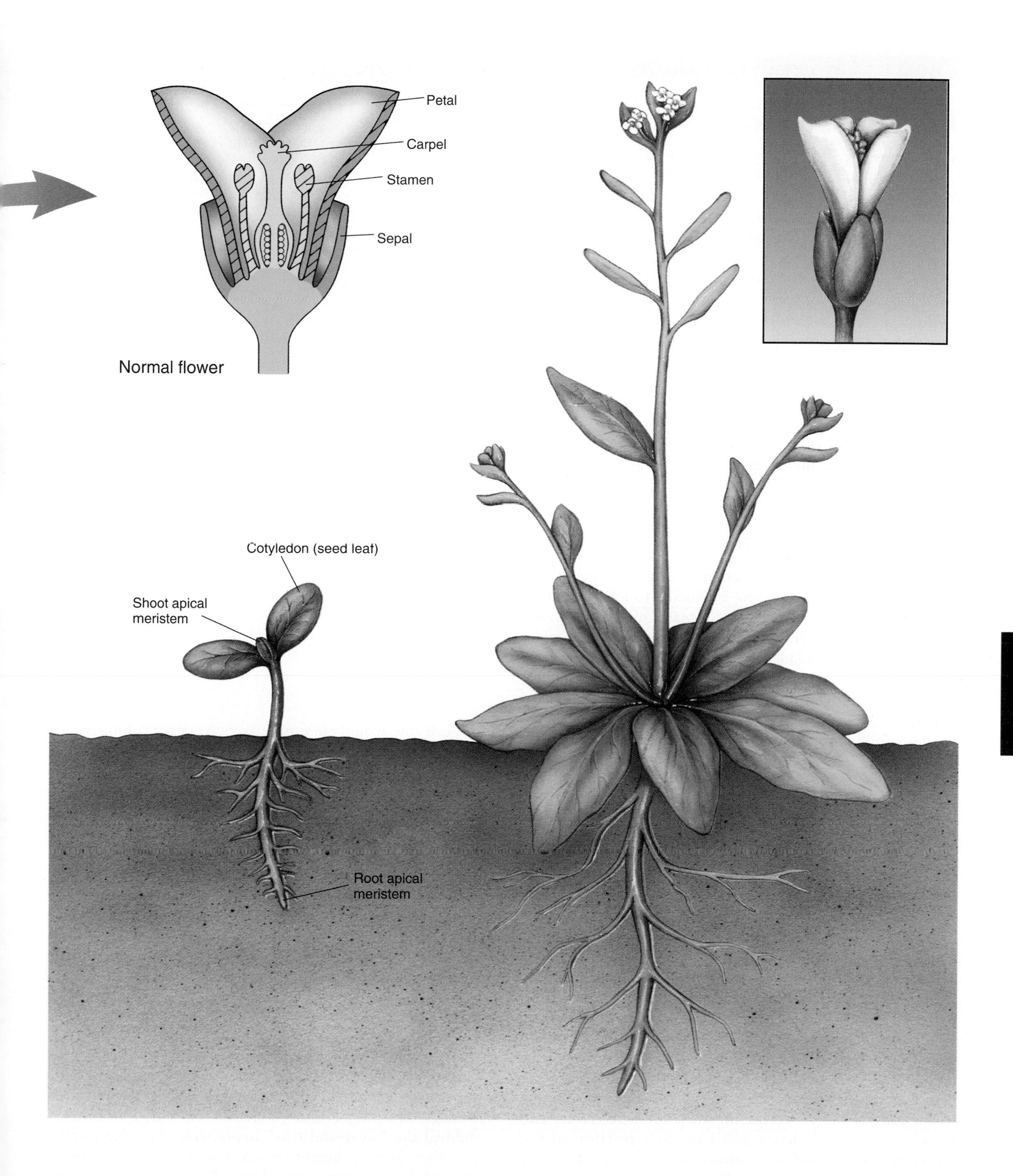
Petal
Carpel
Stamen
Sepal
Normal flower
Cotyledon (seed leaf)
Shoot apical
meristem
Root apical
meristem

17.4 Aging can be considered a developmental process.

Theories of Aging

All humans die. The oldest documented person, Jeanne Louise Calment of Arles, France, reached the age of 122 years before her death in 1997. The "safest" age is around puberty. As you can see in figure 17.26, 10- to 15-year-olds have the lowest risk of dying. The death rate begins to increase rapidly after puberty; the rate of mortality then begins to increase as an exponential function of increasing age. Plotted on a log scale as in figure 17.26 (in a so-called Gompertz plot), the mortality rate increases as a straight line from about 15 to 90 years, doubling about every eight years (the "Gompertz number"). By the time we reach 100, age has taken such a toll that the risk of dying reaches 50% per year.

A wide variety of theories have been advanced to explain why humans and other animals age. No one theory has gained general acceptance, but the following four are being intensively investigated:

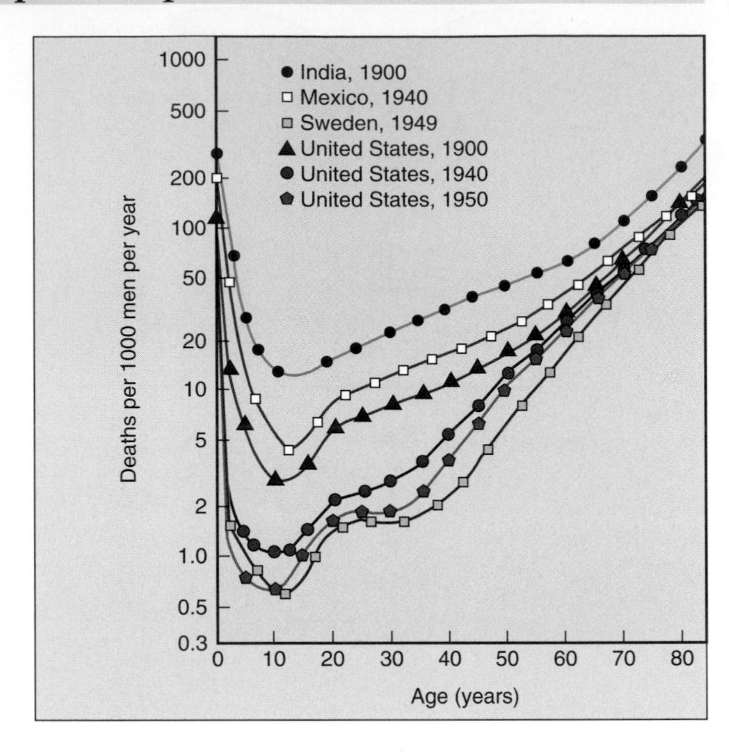

FIGURE 17.26
Gompertz curves. While human populations may differ 25-fold in their mortality rates before puberty, the slopes of their Gompertz curves are about the same in later years.

Accumulated Mutation Hypothesis

The oldest general theory of aging is that cells accumulate mutations as they age, leading eventually to lethal damage. Careful studies have shown that somatic mutations do indeed accumulate during aging. As cells age, for example, they tend to accumulate the modified base 8-hydroxyguanine, in which an —OH group is added to the base guanine. There is little direct evidence, however, that these mutations *cause* aging. No acceleration in aging occurred among survivors of Hiroshima and Nagasaki despite their enormous added mutation load, arguing against any general relationship between mutation and aging.

Telomere Depletion Hypothesis

In a seminal experiment carried out in 1961, Leonard Hayflick demonstrated that fibroblast cells growing in tissue culture will divide only a certain number of times (figure 17.27). After about 50 population doublings, cell division stops, the cell cycle blocked just before DNA replication. If a cell sample is taken after 20 doublings and frozen, when thawed it resumes growth for 30 more doublings, then stops.

An explanation of the "Hayflick limit" was suggested in 1986 when Howard Cooke first glimpsed an extra length of DNA at the ends of chromosomes. These **telomeric regions,** repeats of the sequence TTAGGG, were found to be substantially shorter in older somatic tissue, and Cooke speculated that a 100 base-pair portion of the telomere cap was lost by a chromosome during each cycle of DNA replication. Eventually, after some 50 replication cycles, the protective telomeric cap would be used up, and the cell line would then enter senescence, no longer able to proliferate. Cancer cells appear to avoid telomeric shortening.

Research reported in 1998 has confirmed Cooke's hypothesis, providing direct evidence for a causal relation between telomeric shortening and cell senescence. Using genetic engineering, researchers transferred into human primary cell cultures a gene that leads to expression of telomerase, an enzyme that builds TTAGGG telomeric caps. The result was unequivocal. New telomeric caps were added to the chromosomes of the cells, and the cells with the artificially elongated telomeres did not senesce at the Hayflick limit, continuing to divide in a healthy and vigorous manner for more than 20 additional generations.

Wear-and-Tear Hypothesis

Numerous theories of aging focus in one way or another on the general idea that cells wear out over time, accumulating damage until they are no longer able to function. Loosely dubbed the "wear-and-tear" hypothesis, this idea implies that there is no inherent designed-in limit to aging, just a statistical one—over time, disruption, wear, and damage eventually erode a cell's ability to function properly.

There is considerable evidence that aging cells do accumulate damage. Some of the most interesting evidence concerns free radicals, fragments of molecules or atoms that contain an unpaired electron. Free radicals are very reactive chemically and can be quite destructive in a cell. Free radicals are produced as natural by-products of oxidative metabolism, but most are mopped up by special enzymes that function to sweep the cell interior free of their destructive effects.

One of the most damaging free radical reactions that occurs in cells causes glucose to become linked to proteins, a nonenzymatic process called glycation. Two of the most commonly glycated proteins are collagen and elastin, key components of the connective tissues in our joints. Glycated collagen and elastin are not replaced, and individual molecules may be as old as the individual.

Glycation of collagen, elastin, and a diverse collection of other proteins within the cell produces a complex mixture of glucose-linked proteins called advanced glycosylation end products (AGEs). AGEs can cross-link to one another, reducing the flexibility of connective tissues in the joints and producing many of the other characteristic symptoms of aging.

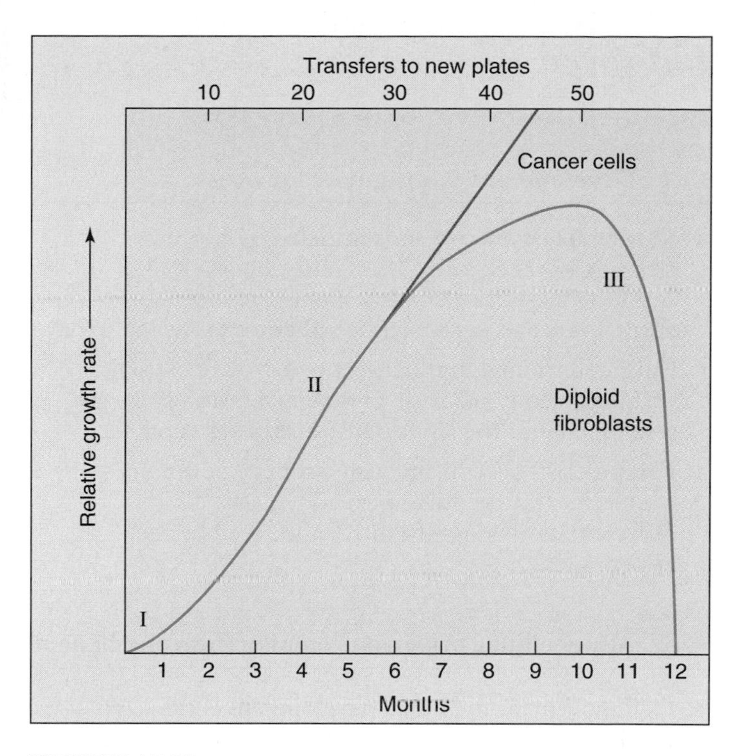

FIGURE 17.27
Hayflick's experiment. Fibroblast cells stop growing after about 50 doublings. Growth is rapid in phases I and II, but slows in phase III, as the culture becomes senescent, until the final doubling. Cancer cells, by contrast, do not "age."

Gene Clock Hypothesis

There is very little doubt that at least some aspects of aging are under the direct control of genes. Just as genes regulate the body's development, so they appear to regulate its rate of aging. Mutations in these genes can produce premature aging in the young. In the very rare recessive Hutchinson-Gilford syndrome, growth, sexual maturation, and skeletal development are retarded; atherosclerosis and strokes usually lead to death by age 12 years. Only some 20 cases have ever been described.

The similar Werner's syndrome is not as rare, affecting some 10 people per million worldwide. The syndrome is named after Otto Werner, who in Germany in 1904 reported a family affected by premature aging and said a genetic component was at work. Werner's syndrome makes its appearance in adolescence, usually producing death before age 50 of heart attack or one of a variety of rare connective tissue cancers. The gene responsible for Werner's syndrome was identified in 1996. Located on the short arm of chromosome 8, it seems to affect a helicase enzyme involved in the repair of DNA. The gene, which codes for a 1432-amino-acid protein, has been fully sequenced, and four mutant alleles identified. Helicase enzymes are needed to unwind the DNA double helix whenever DNA has to be replicated, repaired, or transcribed. The high incidence of certain cancers among Werner's syndrome patients leads investigators to speculate that the mutant helicase may fail to activate critical tumor suppressor genes. The potential role of helicases in aging is the subject of heated research.

Research on aging in other animals strongly supports the hypothesis that genes regulate the rate of aging. A *Drosophila melanogaster* gene mutation called *Indy* (I'm not dead yet") doubles the fruitfly life span from the usual 37 days to an average of 70 days. When researchers isolated the DNA of the *Indy* gene, and compared its DNA sequence with the human genome project sequences, they found that the *Indy* gene is 50% similar to a human gene called *dicarboxylate cotransporter*. In humans, dicarboxylate cotransporter proteins move preliminary products of food metabolism (dicarboxylic acids of the Krebs Cycle) across membranes to where the food's processing takes place. In mutant *Indy* flies, poor dicarboxylic acid pumping means that less metabolic energy can be gleaned from the fly's food. In essence, the *Indy* mutation is the genetic equivalent of caloric restriction. Starving is known to prolong life in the nematode *Caenorhabditis elegans*, but *Indy's* caloric restriction does not involve the unpleasantness of starving. The *Indy* mutation in effect puts flies on a severe diet, while the flies eat as much as normal and lead a normal vigorous life—for far longer.

Among the many theories advanced to explain aging, many involve the progressive accumulation of damage to DNA. When genes affecting aging have been isolated, they affect DNA repair processes.

Chapter 17

www.mhhe.com/raven6/resources17.mhtml

Summary	Questions	Media Resources

17.1 Development is a regulated process.

- Vertebrate development is initiated by a rapid cleavage of the fertilized egg into a hollow ball of cells, the blastula. Cell movements then form primary germ layers and organize the structure of the embryo.
- Cells determined in the insect embryo are carried within the body of larvae as imaginal discs, which are assembled into the adult body during pupation.
- Plant meristems continuously produce new tissues, which then differentiate into body parts. This differentiation is significantly influenced by the environment.

1. What is cleavage? How does the type of cleavage influence subsequent embryonic development?

2. What is a blastula? How does it form and what does it turn into?

3. What is a gastrula? Where are the germ layers in a gastrula?

4. What is neurulation? How and when does it occur?

- Introduction to Development

17.2 Multicellular organisms employ the same basic mechanisms of development.

- Cell movement in animal development is carried out by altering a cell's complement of surface adhesion molecules, which it uses to pull itself over other cells.
- A key to animal development is the ability of cells to alter the developmental paths of adjacent cells, a process called induction. Induction is achieved by diffusible chemicals called morphogens.
- Determination of a cell's ultimate developmental fate often involves the addition to it of positional labels that reflect its location in the embryo.
- The location of structures within body segments is dictated by a spatially organized assembly of homeotic genes, first discovered in *Drosophila* but now known to occur in all animals.
- Many cells are genetically programmed to die, usually soon after they are formed during development, in a process called apoptosis.

5. What role do cadherins and integrins play in cell movement?

6. What is the difference between mosaic development and regulative development?

7. How do organizers and morphogens participate in induction?

8. How is determination distinguished from differentiation?

9. What role does maternal mRNA play in the development of a *Drosophila* embryo?

10. What are homeotic genes and what do they do?

- Induction
- Pattern Formation

- Student Research:
  - -Vertebrate Limb Formation
  - -Homeobox Genes in the Medicinal Leech

- Art Quiz: Cloning Experiment

17.3 Four model developmental systems have been extensively researched.

- The four most intensively studied model systems of development are the mouse *Mus musculus*, the fruit fly *Drosophila melanogaster*, the nematode *Caenorhabditis elegans*, and the flowering plant *Arabidopsis thaliana*.

11. What are the major differences between vertebrate, insect, and plant developmental pathways? What are the similarities?

17.4 Aging can be considered a developmental process.

- Aging is not well understood, although not for want of theories, most of which involve progressive damage to DNA.

12. Cancer cell cultures never seem to grow old, dividing ceaselessly. What can you deduce about the state of their telomerase gene?

- *On Science* Article:
  - -Unraveling the Mystery of Aging
  - -I'm Not Dead Yet

18

Altering the Genetic Message

Concept Outline

18.1 Mutations are changes in the genetic message.

Mutations Are Rare But Important. Changes in genes provide the raw material for evolution.
Kinds of Mutation. Some mutations alter genes themselves, others alter the positions of genes.
Point Mutations. Radiation damage or chemical modification can change one or a few nucleotides.
Changes in Gene Position. Chromosomal rearrangement and insertional inactivation reflect changes in gene position.

18.2 Cancer results from mutation of growth-regulating genes.

What Is Cancer? Cancer is a growth disorder of cells.
Kinds of Cancer. Cancer occurs in almost all tissues, but more in some than others.
Some Tumors Are Caused by Chemicals. Chemicals that mutate DNA cause cancer.
Other Tumors Result from Viral Infection. Viruses carrying growth-regulating genes can cause cancer.
Cancer and the Cell Cycle. Cancer results from mutations of genes regulating cell proliferation.
Smoking and Cancer. Smoking causes lung cancer.
Curing Cancer. New approaches offer promise of a cure.

18.3 Recombination alters gene location.

An Overview of Recombination. Recombination is produced by gene transfer and by reciprocal recombination.
Gene Transfer. Many genes move within small circles of DNA called plasmids. Plasmids can move between bacterial cells and carry bacterial genes. Some gene sequences move from one location to another on a chromosome.
Reciprocal Recombination. Reciprocal recombination can alter genes in several ways.
Trinucleotide Repeats. Increases in the number of repeated triplets can produce gene disorders.

18.4 Genomes are continually evolving.

Classes of Eukaryotic DNA. Unequal crossing over expands eukaryotic genomes.

FIGURE 18.1
Cancer. A scanning electron micrograph of deadly cancer cells (8000×).

In general, the genetic message can be altered in two broad ways: mutation and recombination. A change in the content of the genetic message—the base sequence of one or more genes—is referred to as a mutation. Some mutations alter the identity of a particular nucleotide, while others remove or add nucleotides to a gene. A change in the position of a portion of the genetic message is referred to as recombination. Some recombination events move a gene to a different chromosome; others alter the location of only part of a gene. In this chapter, we will first consider gene mutation, using cancer as a focus for our inquiry (figure 18.1). Then we will turn to recombination, focusing on how it has affected the organization of the eukaryotic genome.

18.1 Mutations are changes in the genetic message.

Mutations Are Rare But Important

The cells of eukaryotes contain an enormous amount of DNA. If the DNA in all of the cells of an adult human were lined up end-to-end, it would stretch nearly 100 billion kilometers—60 times the distance from Earth to Jupiter! The DNA in any multicellular organism is the final result of a long series of replications, starting with the DNA of a single cell, the fertilized egg. Organisms have evolved many different mechanisms to avoid errors during DNA replication and to preserve the DNA from damage. Some of these mechanisms "proofread" the replicated DNA strands for accuracy and correct any mistakes. The proofreading is not perfect, however. If it were, no variation in the nucleotide sequences of genes would be generated.

FIGURE 18.2
Mutation. Normal fruit flies have one pair of wings extending from the thorax. This fly is a mutant because of changes in *bithorax*, a gene region regulating a critical stage of development; it possesses two thoracic segments and thus two sets of wings.

Mistakes Happen

In fact, cells do make mistakes during replication, and damage to the genetic message also occurs, causing mutation (figure 18.2). However, change is rare. Typically, a particular gene is altered in only one of a million gametes. If changes were common, the genetic instructions encoded in DNA would soon degrade into meaningless gibberish. Limited as it might seem, the steady trickle of change that does occur is the very stuff of evolution. Every difference in the genetic messages that specify different organisms arose as the result of genetic change.

The Importance of Genetic Change

All evolution begins with alterations in the genetic message: mutation creates new alleles, gene transfer and transposition alter gene location, reciprocal recombination shuffles and sorts these changes, and chromosomal rearrangement alters the organization of entire chromosomes. Some changes in germ-line tissue produce alterations that enable an organism to leave more offspring, and those changes tend to be preserved as the genetic endowment of future generations. Other changes reduce the ability of an organism to leave offspring. Those changes tend to be lost, as the organisms that carry them contribute fewer members to future generations.

Evolution can be viewed as the selection of particular combinations of alleles from a pool of alternatives. The rate of evolution is ultimately limited by the rate at which these alternatives are generated. Genetic change through mutation and recombination provides the raw material for evolution.

Genetic changes in somatic cells do not pass on to offspring, and so have less evolutionary consequence than germ-line change. However, changes in the genes of somatic cells can have an important immediate impact, particularly if the gene affects development or is involved with regulation of cell proliferation.

Rare changes in genes, called mutations, can have significant effects on the individual when they occur in somatic tissue, but are only inherited if they occur in germ-line tissue. Inherited changes provide the raw material for evolution.

Kinds of Mutation

Because mutations can occur randomly anywhere in a cell's DNA, mutations can be detrimental, just as making a random change in a computer program or a musical score usually worsens performance. The consequences of a detrimental mutation may be minor or catastrophic, depending on the function of the altered gene.

Mutations in Germ-Line Tissues

The effect of a mutation depends critically on the identity of the cell in which the mutation occurs. During the embryonic development of all multicellular organisms, there comes a point when cells destined to form gametes **(germ-line cells)** are segregated from those that will form the other cells of the body **(somatic cells).** Only when a mutation occurs within a germ-line cell is it passed to subsequent generations as part of the hereditary endowment of the gametes derived from that cell.

Mutations in Somatic Tissues

Mutations in germ-line tissue are of enormous biological importance because they provide the raw material from which natural selection produces evolutionary change. Change can occur only if there are new, different allele combinations available to replace the old. Mutation produces new alleles, and recombination puts the alleles together in different combinations. In animals, it is the occurrence of these two processes in germ-line tissue that is important to evolution, as mutations in somatic cells **(somatic mutations)** are not passed from one generation to the next. However, a somatic mutation may have drastic effects on the individual organism in which it occurs, as it *is* passed on to all of the cells that are descended from the original mutant cell. Thus, if a mutant lung cell divides, all cells derived from it will carry the mutation. Somatic mutations of lung cells are, as we shall see, the principal cause of lung cancer in humans.

Point Mutations

One category of mutational changes affects the message itself, producing alterations in the sequence of DNA nucleotides (table 18.1 summarizes the sources and types of mutations). If alterations involve only one or a few base-pairs in the coding sequence, they are called **point mutations.** While some point mutations arise due to spontaneous pairing errors that occur during DNA replication, others result from damage to the DNA caused by **mutagens,** usually radiation or chemicals. The latter class of mutations is of particular practical importance because modern industrial societies often release many chemical mutagens into the environment.

Table 18.1 Types of Mutation

Mutation	Example result
NO MUTATION A B C	Normal B protein is produced by the *B* gene.
POINT MUTATION	
Base substitution Substitution of one or a few bases A B C	B protein is inactive because changed amino acid disrupts function.
Insertion Addition of one or a few bases A C	B protein is inactive because inserted material disrupts proper shape.
Deletion Loss of one or a few bases A C	B protein is inactive because portion of protein is missing.
CHANGES IN GENE POSITION	
Transposition A C B	*B* gene or B protein may be regulated differently because of change in gene position.
Chromosomal rearrangement A C B	*B* gene may be inactivated or regulated differently in its new location on chromosome.

Changes in Gene Position

Another category of mutations affects the way the genetic message is organized. In both bacteria and eukaryotes, individual genes may move from one place in the genome to another by **transposition.** When a particular gene moves to a different location, its expression or the expression of neighboring genes may be altered. In addition, large segments of chromosomes in eukaryotes may change their relative locations or undergo duplication. Such **chromosomal rearrangements** often have drastic effects on the expression of the genetic message.

Point mutations are changes in the hereditary message of an organism. They may result from spontaneous errors during DNA replication or from DNA damage due to mutagens. Changes in gene position may affect expression of genes.

Point Mutations

Physical Damage to DNA

Ionizing Radiation. High-energy forms of radiation, such as X rays and gamma rays, are highly mutagenic. When such radiation reaches a cell, it is absorbed by the atoms it encounters, imparting energy to the electrons in their outer shells. These energized electrons are ejected from the atoms, leaving behind free radicals, ionized atoms with unpaired electrons. Free radicals react violently with other molecules, including DNA.

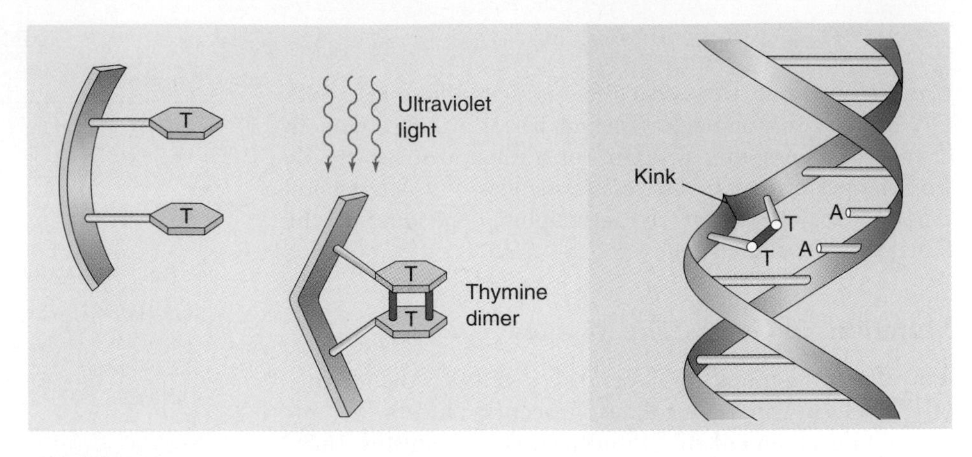

FIGURE 18.3
Making a pyrimidine dimer. When two pyrimidines, such as two thymines, are adjacent to each other in a DNA strand, the absorption of UV radiation can cause covalent bonds to form between them—creating a pyrimidine dimer. The dimer introduces a "kink" into the double helix that prevents replication of the duplex by DNA polymerase.

When a free radical breaks *both* phosphodiester bonds of a DNA helix, causing a **double-strand break,** the cell's usual mutational repair enzymes cannot fix the damage. The two fragments created by the break must be aligned while the phosphodiester bonds between them form again. Bacteria have no mechanism to achieve this alignment, and double-strand breaks are lethal to their descendants. In eukaryotes, which almost all possess multiple copies of their chromosomes, the synaptonemal complex assembled in meiosis is used to pair the fragmented chromosome with its homologue. In fact, it is speculated that meiosis may have evolved initially as a mechanism to repair double-strand breaks in DNA (see chapter 12).

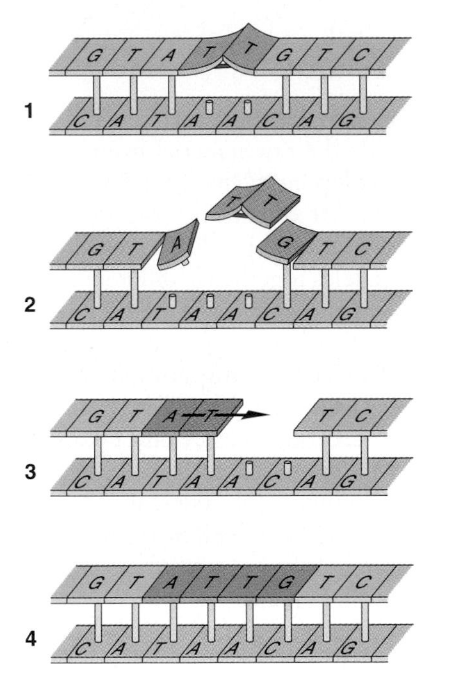

FIGURE 18.4
Repair of a pyrimidine dimer. Some pyrimidine dimers are repaired by excising the dimer, as well as a short run of nucleotides on either side of it, and then filling in the gap using the other strand as a template.

Ultraviolet Radiation. Ultraviolet (UV) radiation, the component of sunlight that tans (and burns), contains much less energy than ionizing radiation. It does not induce atoms to eject electrons, and thus it does not produce free radicals. The only molecules capable of absorbing UV radiation are certain organic ring compounds, whose outer-shell electrons become reactive when they absorb UV energy.

DNA strongly absorbs UV radiation in the pyrimidine bases, thymine and cytosine. If one of the nucleotides on either side of the absorbing pyrimidine is also a pyrimidine, a double covalent bond forms between them. The resulting cross-link between adjacent pyrimidines is called a **pyrimidine dimer** (figure 18.3). In most cases, cellular UV repair systems either cleave the bonds that link the adjacent pyrimidines or excise the entire pyrimidine dimer from the strand and fill in the gap, using the other strand as a template (figure 18.4). In those rare instances in which a pyrimidine dimer goes unrepaired, DNA polymerase may fail to replicate the portion of the strand that includes the dimer, skipping ahead and leaving the problem area to be filled in later. This filling-in process is often error-prone, however, and it may create mutational changes in the base sequence of the gap region. Some unrepaired pyrimidine dimers block DNA replication altogether, which is lethal to the cell.

Sunlight can wreak havoc on the cells of the skin because its UV light causes mutations. Indeed, a strong and direct association exists between exposure to bright sunlight, UV-induced DNA damage, and skin cancer. A deep tan is *not* healthy! A rare hereditary disorder among humans called **xeroderma pigmentosum** causes these problems after a lesser exposure to UV. Individuals with this disorder develop extensive skin tumors after exposure to sunlight because they lack a mechanism for repairing the DNA damage UV radiation causes. Because of the many different proteins involved in excision and repair of pyrimidine dimers, mutations in as many as eight different genes cause the disease.

Chemical Modification of DNA

Many mutations result from direct chemical modification of the DNA. The chemicals that act on DNA fall into three classes: (1) chemicals that resemble DNA nucleotides but pair incorrectly when they are incorporated into DNA (figure 18.5). Some of the new AIDS chemotherapeutic drugs are analogues of nitrogenous bases that are inserted into the viral or infected cell DNA. This DNA cannot be properly transcribed, so viral growth slows; (2) chemicals that remove the amino group from adenine or cytosine, causing them to mispair; and (3) chemicals that add hydrocarbon groups to nucleotide bases, also causing them to mispair. This last group includes many particularly potent mutagens commonly used in laboratories, as well as compounds sometimes released into the environment, such as mustard gas.

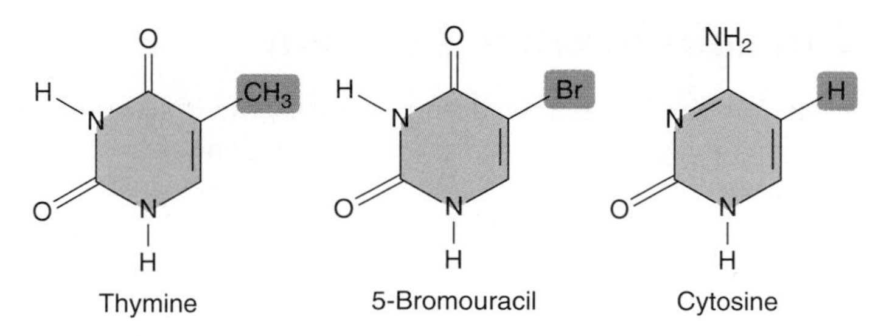

FIGURE 18.5
Chemicals that resemble DNA bases can cause mutations. For example, DNA polymerase cannot distinguish between thymine and 5-bromouracil, which are similar in shape. Once incorporated into a DNA molecule, however, 5-bromouracil tends to rearrange to a form that resembles cytosine and pairs with guanine. When this happens, what was originally an A-T base-pair becomes a G-C base-pair.

Spontaneous Mutations

Many point mutations occur spontaneously, without exposure to radiation or mutagenic chemicals. Sometimes nucleotide bases spontaneously shift to alternative conformations, or isomers, which form different kinds of hydrogen bonds than the normal conformations. During replication, DNA polymerase pairs a different nucleotide with the isomer than it would have otherwise selected. Unrepaired spontaneous errors occur in fewer than one in a billion nucleotides per generation, but they are still an important source of mutation.

Sequences sometimes misalign when homologous chromosomes pair, causing a portion of one strand to loop out. These misalignments, called **slipped mispairing,** are usually only transitory, and the chromosomes quickly revert to the normal arrangement (figure 18.6). If the error-correcting system of the cell encounters a slipped mispairing before it reverts, however, the system will attempt to "correct" it, usually by excising the loop. This may result in a **deletion** of several hundred nucleotides from one of the chromosomes. Many of these deletions start or end in the middle of a codon, thereby shifting the reading frame by one or two bases. These so-called **frame-shift mutations** cause the gene to be read in the wrong three-base groupings, distorting the genetic message, just as the deletion of the letter F from the sentence, THE FAT CAT ATE THE RAT shifts the reading frame of the sentence, producing the meaningless message, THE ATC ATA TET HER AT. Some chemicals specifically promote deletions and frame-shift mutations by stabilizing the loops produced during slipped mispairing, thus increasing the time the loops are vulnerable to excision.

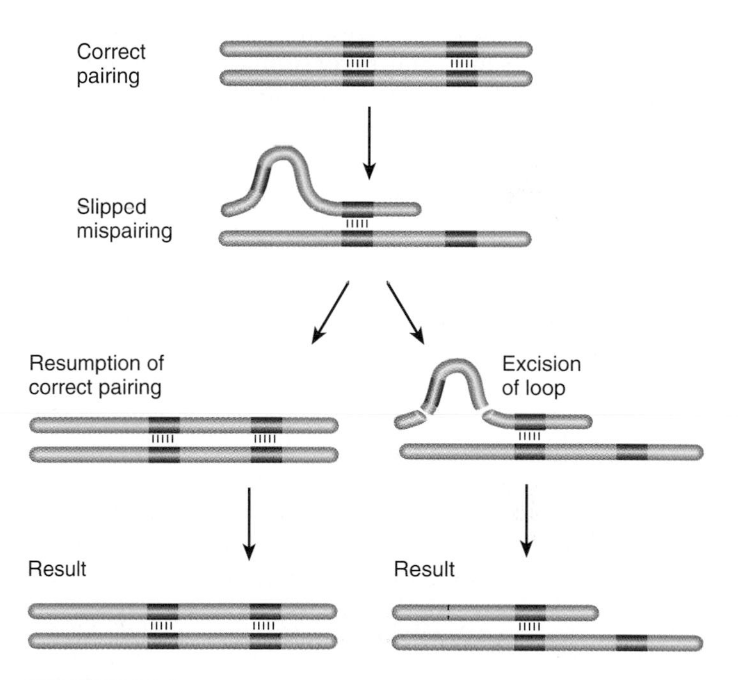

FIGURE 18.6
Slipped mispairing. Slipped mispairing occurs when a sequence is present in more than one copy on a chromosome and the copies on homologous chromosomes pair out of register, like a shirt buttoned wrong. The loop this mistake produces is sometimes excised by the cell's repair enzymes, producing a short deletion and often altering the reading frame. Any chemical that stabilizes the loop increases the chance it will be excised.

The major sources of physical damage to DNA are ionizing radiation, which breaks the DNA strands; ultraviolet radiation, which creates nucleotide cross-links whose removal often leads to errors in base selection; and chemicals that modify DNA bases and alter their base-pairing behavior. Unrepaired spontaneous errors in DNA replication occur rarely.

Changes in Gene Position

Chromosome location is an important factor in determining whether genes are transcribed. Some genes cannot be transcribed if they are adjacent to a tightly coiled region of the chromosome, even though the same gene can be transcribed normally in any other location. Transcription of many chromosomal regions appears to be regulated in this manner; the binding of specific proteins regulates the degree of coiling in local regions of the chromosome, determining the accessibility RNA polymerase has to genes located within those regions.

Chromosomal Rearrangements

Chromosomes undergo several different kinds of gross physical alterations that have significant effects on the locations of their genes. The two most important are **translocations,** in which a segment of one chromosome becomes part of another chromosome, and **inversions,** in which the orientation of a portion of a chromosome is reversed. Translocations often have significant effects on gene expression. Inversions, on the other hand, usually do not alter gene expression, but they are nonetheless important. Recombination within a region that is inverted on one homologue but not the other (figure 18.7) leads to serious problems: none of the gametes that contain chromatids produced following such a crossover event will have a complete set of genes.

Other chromosomal alterations change the number of gene copies an individual possesses. Particular genes or segments of chromosomes may be deleted or duplicated, whole chromosomes may be lost or gained *(aneuploidy)*, and entire sets of chromosomes may be added *(polyploidy)*. Most deletions are harmful because they halve the number of gene copies within a diploid genome and thus seriously affect the level of transcription. Duplications cause gene imbalance and are also usually harmful.

Insertional Inactivation

Many small segments of DNA are capable of moving from one location to another in the genome, using an enzyme to cut and paste themselves into new genetic neighborhoods. We call these mobile bits of DNA transposable elements, or **transposons.** Transposons select their new locations at random, and are as likely to enter one segment of a chromosome as another. Inevitably, some transposons end up inserted into genes, and this almost always inactivates the gene. The encoded protein now has a large meaningless chunk inserted within it, disrupting its structure. This form of mutation, called **insertional inactivation,** is common in nature. Indeed, it seems to be one of the most significant causes of mutation. The original white-eye mutant of *Drosophila* discovered by Morgan (see chapter 13) is the result of a transposition event, a transposon nested within a gene encoding a pigment-producing enzyme.

As you might expect, a variety of human gene disorders are the result of transposition. The human transposon called *Alu*, for example, is responsible for an X-linked hemophilia, inserting into clotting factor IX and placing a premature stop codon there. It also causes inherited high levels of cholesterol (hypercholesterolemia), *Alu* elements inserting into the gene encoding the low density lipoprotein (LDL) receptor. In one very interesting case, a *Drosophila* transposon called *Mariner* proves responsible for a rare human neurological disorder called Charcot-Marie-Tooth disease, in which the muscles and nerves of the legs and feet gradually wither away. The Mariner transposon is inserted into a key gene called *CMT* on chromosome 17, creating a weak site where the chromosome can break. No one knows how the *Drosophila* transposon got into the human genome.

Many mutations result from changes in gene location or from insertional inactivation.

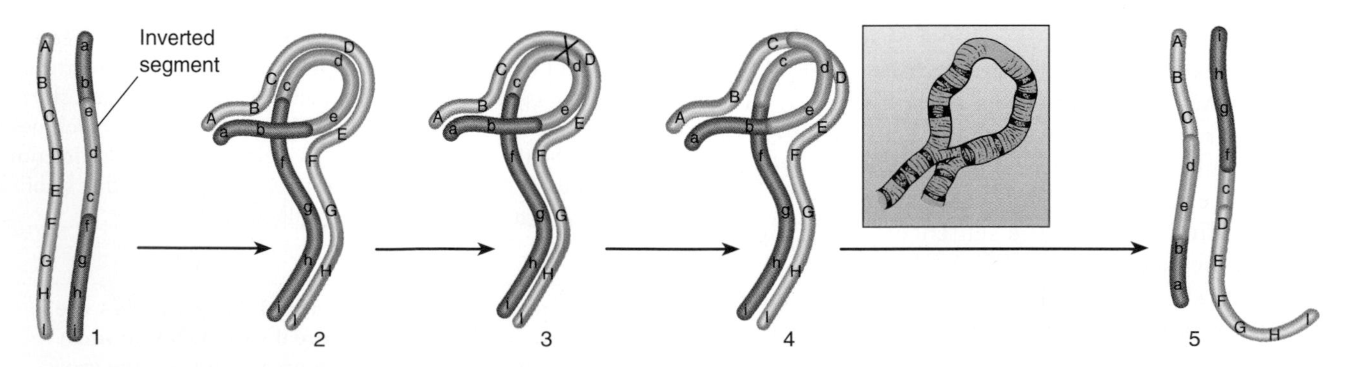

FIGURE 18.7
The consequence of inversion. (*1*) When a segment of a chromosome is inverted, (*2*) it can pair in meiosis only by forming an internal loop. (*3*) Any crossing over that occurs within the inverted segment during meiosis will result in nonviable gametes; some genes are lost from each chromosome, while others are duplicated (*4* and *5*). For clarity, only two strands are shown, although crossing over occurs in the four-strand stage. The pairing that occurs between inverted segments is sometimes visible under the microscope as a characteristic loop (*inset*).

18.2 Cancer results from mutation of growth-regulating genes.

What Is Cancer?

Cancer is a growth disorder of cells. It starts when an apparently normal cell begins to grow in an uncontrolled and invasive way (figure 18.8). The result is a cluster of cells, called a **tumor,** that constantly expands in size. Cells that leave the tumor and spread throughout the body, forming new tumors at distant sites, are called **metastases** (figure 18.9). Cancer is perhaps the most pernicious disease. Of the children born in 1999, one-third will contract cancer at some time during their lives; one-fourth of the male children and one-third of the female children will someday die of cancer. Most of us have had family or friends affected by the disease. In 2000, 552,200 Americans died of cancer.

Not surprisingly, researchers are expending a great deal of effort to learn the cause of this disease. Scientists have made a great deal of progress in the last 20 years using molecular biological techniques, and the rough outlines of understanding are now emerging. We now know that cancer is a gene disorder of somatic tissue, in which damaged genes fail to properly control cell proliferation. The cell division cycle is regulated by a sophisticated group of proteins described in chapter 11. Cancer results from the mutation of the genes encoding these proteins.

Cancer can be caused by chemicals that mutate DNA or in some instances by viruses that circumvent the cell's normal proliferation controls. Whatever the immediate cause, however, all cancers are characterized by unrestrained growth and division. Cell division never stops in a cancerous line of cells. Cancer cells are virtually immortal—until the body in which they reside dies.

Cancer is unrestrained cell proliferation caused by damage to genes regulating the cell division cycle.

FIGURE 18.8
Lung cancer cells (530×). These cells are from a tumor located in the alveolus (air sac) of a lung.

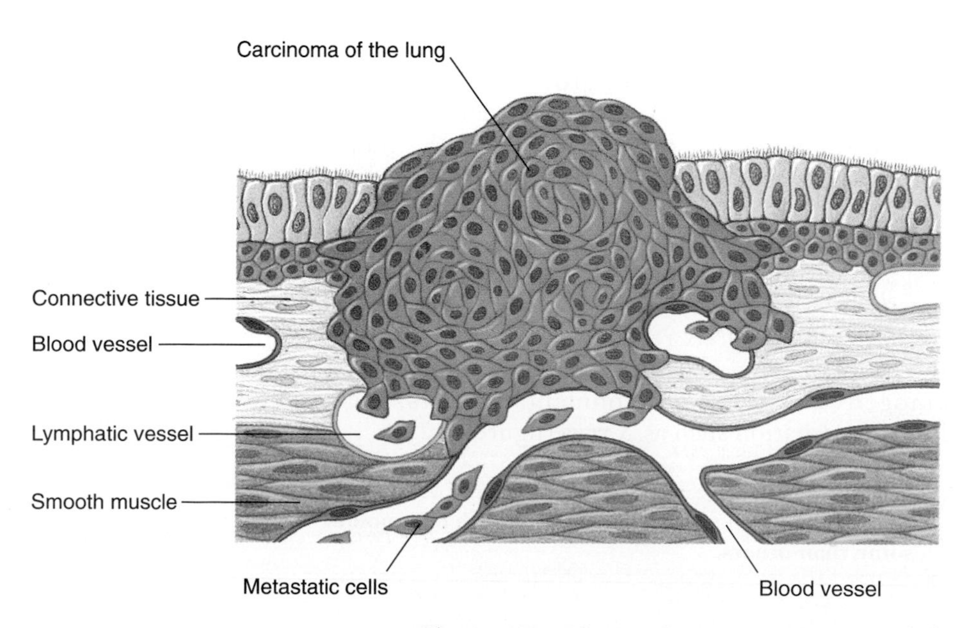

FIGURE 18.9
Portrait of a cancer. This ball of cells is a carcinoma (cancer tumor) developing from epithelial cells that line the interior surface of a human lung. As the mass of cells grows, it invades surrounding tissues, eventually penetrating lymphatic and blood vessels, both plentiful within the lung. These vessels carry metastatic cancer cells throughout the body, where they lodge and grow, forming new masses of cancerous tissue.

Kinds of Cancer

Cancer can occur in almost any tissue, so a bewildering number of different cancers occur. Tumors arising from cells in connective tissue, bone, or muscle are known as **sarcomas,** while those that originate in epithelial tissue such as skin are called **carcinomas.** In the United States, the three deadliest human cancers are lung cancer, cancer of the colon and rectum, and breast cancer (table 18.2). Lung cancer, responsible for the most cancer deaths, is largely preventable; most cases result from smoking cigarettes. Colorectal cancers appear to be fostered by the high-meat diets so favored in the United States. The cause of breast cancer is still a mystery, although in 1994 and 1995 researchers isolated two genes responsible for hereditary susceptibility to breast cancer, *BRCA1* and *BRCA2* (Breast Cancer genes #1 and #2 located on human chromosomes 17 and 13); their discovery offers hope that researchers will soon be able to unravel the fundamental mechanism leading to hereditary breast cancer, about one-third of all breast cancers.

Table 18.2 Incidence of Cancer in the United States in 2000

Type of Cancer	New Cases	Deaths	% of Cancer Deaths
Lung	164,100	156,900	28
Colon and rectum	130,200	56,300	10
Leukemia/lymphoma	93,100	49,200	9
Breast	184,200	41,200	8
Prostate	180,400	31,900	7
Pancreas	28,300	28,200	5
Ovary	23,100	14,000	3
Stomach	21,500	13,000	2
Liver	15,300	13,800	2
Nervous system/eye	18,700	13,200	2
Bladder	53,200	12,200	2
Oral cavity	30,200	7,800	2
Kidney	31,200	11,900	2
Cervix/uterus	48,900	11,100	2
Malignant melanoma	47,700	7,700	1
Sarcoma (connective tissue)	10,600	6,000	1
All other cancers	139,400	77,800	14

In the United States in 2000 there were 1,220,100 reported cases of new cancers and 552,200 cancer deaths, indicating that roughly half the people who develop cancer die from it.

Source: Data from the American Cancer Society, Inc., 2000.

The association of particular chemicals with cancer, particularly chemicals that are potent mutagens, led researchers early on to the suspicion that cancer might be caused, at least in part, by chemicals, the so-called **chemical carcinogenesis theory.** Agents thought to cause cancer are called **carcinogens.** A simple and effective way to test if a chemical is mutagenic is the Ames test (figure 18.10), named for its developer, Bruce Ames. The test uses a strain of *Salmonella* bacteria that has a defective histidine-synthesizing gene. Because these bacteria cannot make histidine, they cannot grow on media without it. Only a back-mutation that restores the ability to manufacture histidine will permit growth. Thus the number of colonies of these bacteria that grow on histidine-free medium is a measure of the frequency of back-mutation. A majority of chemicals that cause back-mutations in this test are carcinogenic, and vice versa. To increase the sensitivity of the test, the strains of bacteria are altered to disable their DNA repair machinery. The search for the cause of cancer has focused in part on chemical carcinogens and other environmental factors, including ionizing radiation such as X rays (figure 18.11).

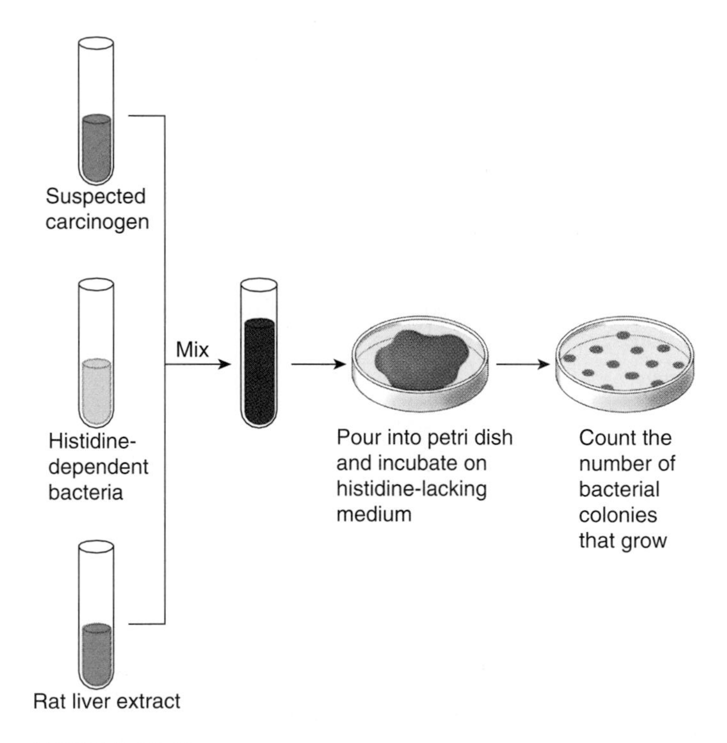

FIGURE 18.10
The Ames test. This test uses a strain of *Salmonella* bacteria that requires histidine in the growth medium due to a mutated gene. If a suspected carcinogen is mutagenic, it can reverse this mutation. Rat liver extract is added because it contains enzymes that can convert carcinogens into mutagens. The mutagenicity of the carcinogen can be quantified by counting the number of bacterial colonies that grow on a medium lacking histidine.

Cancers occur in all tissues, but are more common in some than others.

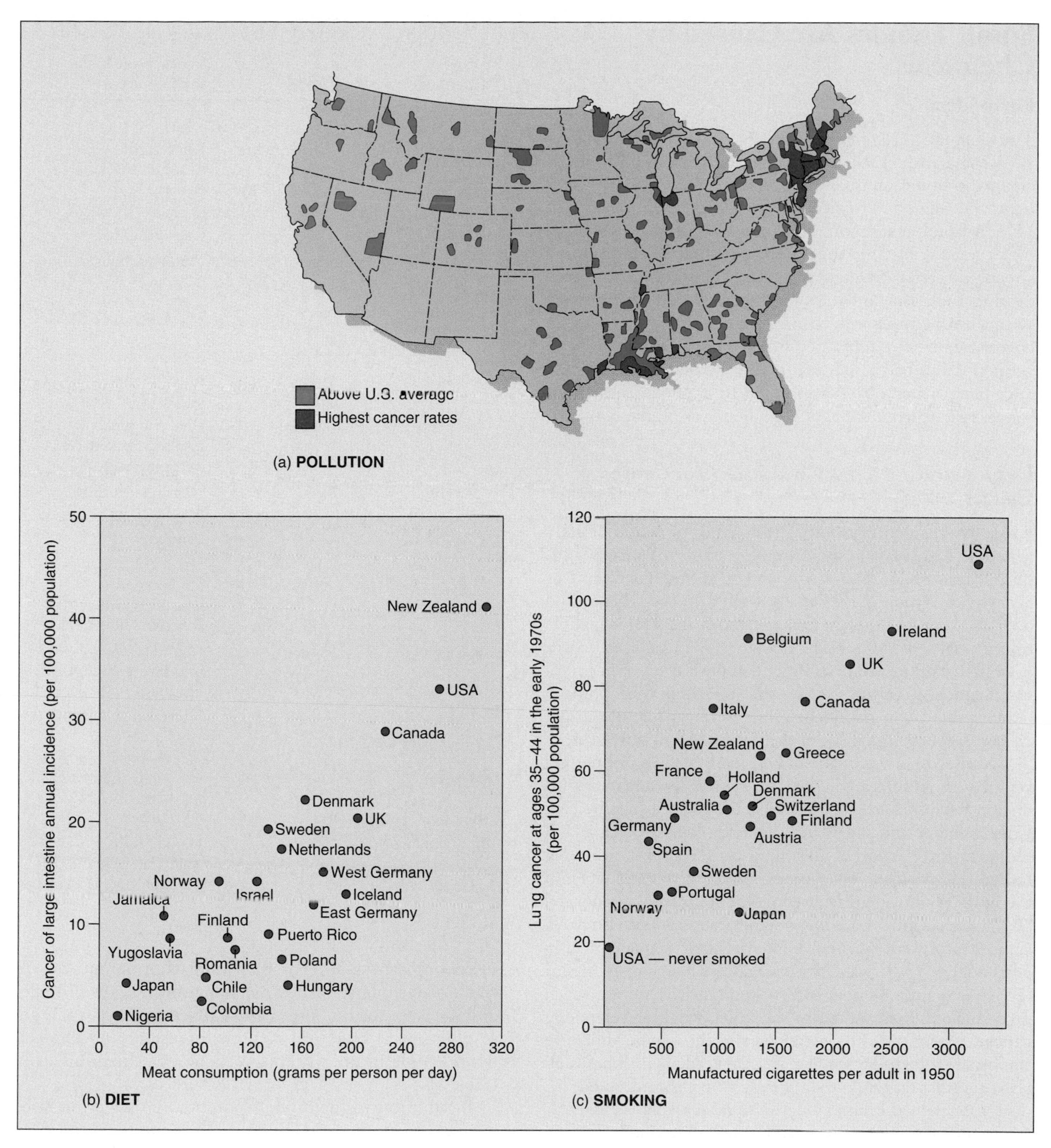

FIGURE 18.11

Potential cancer-causing agents. (*a*) The incidence of cancer per 1000 people is not uniform throughout the United States. The incidence is higher in cities and in the Mississippi Delta, suggesting that pollution and pesticide runoff may contribute to the development of cancer. (*b*) One of the deadliest cancers in the United States, cancer of the large intestine, is uncommon in many other countries. Its incidence appears to be related to the amount of meat a person consumes: a high-meat diet slows the passage of food through the intestine, prolonging exposure of the intestinal wall to digestive waste. (*c*) The biggest killer among cancers is lung cancer, and the most deadly environmental agent producing lung cancer is cigarette smoke. The incidence of lung cancer among men 35 to 44 years of age in various countries strongly correlates with the cigarette consumption in that country 20 years earlier.

Some Tumors Are Caused by Chemicals

Early Ideas

The chemical carcinogenesis theory was first advanced over 200 years ago in 1761 by Dr. John Hill, an English physician, who noted unusual tumors of the nose in heavy snuff users and suggested tobacco had produced these cancers. In 1775, a London surgeon, Sir Percivall Pott, made a similar observation, noting that men who had been chimney sweeps exhibited frequent cancer of the scrotum, and suggesting that soot and tars might be responsible. British sweeps washed themselves infrequently and always seemed covered with soot. Chimney sweeps on the continent, who washed daily, had much less of this scrotal cancer. These and many other observations led to the hypothesis that cancer results from the action of chemicals on the body.

Demonstrating That Chemicals Can Cause Cancer

It was over a century before this hypothesis was directly tested. In 1915, Japanese doctor Katsusaburo Yamagiwa applied extracts of coal tar to the skin of 137 rabbits every 2 or 3 days for 3 months. Then he waited to see what would happen. After a year, cancers appeared at the site of application in seven of the rabbits. Yamagiwa had induced cancer with the coal tar, the first direct demonstration of chemical carcinogenesis. In the decades that followed, this approach demonstrated that many chemicals were capable of causing cancer. Importantly, most of them were potent mutagens.

Because these were lab studies, many people did not accept that the results applied to real people. Do tars in fact induce cancer in humans? In 1949, the American physician Ernst Winder and the British epidemiologist Richard Doll independently reported that lung cancer showed a strong link to the smoking of cigarettes, which introduces tars into the lungs. Winder interviewed 684 lung cancer patients and 600 normal controls, asking whether each had ever smoked. Cancer rates were 40 times higher in heavy smokers than in nonsmokers. Doll's study was even more convincing. He interviewed a large number of British physicians, noting which ones smoked, then waited to see which would develop lung cancer. Many did. Overwhelmingly, those who did were smokers. From these studies, it seemed likely as long as 50 years ago that tars and other chemicals in cigarette smoke induce cancer in the lungs of persistent smokers. While this suggestion was (and is) resisted by the tobacco industry, the evidence that has accumulated since these pioneering studies makes a clear case, and there is no longer any real doubt. Chemicals in cigarette smoke cause cancer.

Carcinogens Are Common

In ongoing investigations over the last 50 years, many hundreds of synthetic chemicals have been shown capable of causing cancer in laboratory animals. Among them are trichloroethylene, asbestos, benzene, vinyl chloride, arsenic, arylamide, and a host of complex petroleum products with chemical structures resembling chicken wire. People in the workplace encounter chemicals daily (table 18.3).

In addition to identifying potentially dangerous substances, what have the studies of potential carcinogens told us about the nature of cancer? What do these cancer-causing chemicals have in common? *They are all mutagens, each capable of inducing changes in DNA.*

Table 18.3 Chemical Carcinogens in the Workplace

Chemical	Cancer	Workers at Risk for Exposure
COMMON EXPOSURE		
Benzene	Myelogenous leukemia	Painters; dye users; furniture finishers
Diesel exhaust	Lung	Railroad and bus-garage workers; truckers; miners
Mineral oils	Skin	Metal machinists
Pesticides	Lung	Sprayers
Cigarette tar	Lung	Smokers
UNCOMMON EXPOSURE		
Asbestos	Mesothelioma, lung	Brake-lining, insulation workers
Synthetic mineral fibers	Lung	Wall and pipe insulation and duct wrapping users
Hair dyes	Bladder	Hairdressers and barbers
Paint	Lung	Painters
Polychlorinated biphenyls	Liver, skin	Users of hydraulic fluids and lubricants, inks, adhesives, insecticides
Soot	Skin	Chimney sweeps; bricklayers; firefighters; heating-unit service workers
RARE EXPOSURE		
Arsenic	Lung, skin	Insecticide/herbicide sprayers; tanners; oil refiners
Formaldehyde	Nose	Wood product, paper, textiles, and metal product workers

Chemicals that produce mutations in DNA are often potent carcinogens. Tars in cigarette smoke, for example, are the direct cause of most lung cancers.

Other Tumors Result from Viral Infection

Chemical mutagens are not the only carcinogens, however. Some tumors seem almost certainly to result from viral infection. Viruses can be isolated from certain tumors, and these viruses cause virus-containing tumors to develop in other individuals. About 15% of human cancers are associated with viruses.

A Virus That Causes Cancer

In 1911, American medical researcher Peyton Rous reported that a virus, subsequently named **Rous avian sarcoma virus (RSV),** was associated with chicken sarcomas. He found that RSV could infect and initiate cancer in chicken fibroblast (connective tissue) cells growing in culture; from those cancerous cells, more viruses could be isolated. Rous was awarded the 1966 Nobel Prize in Physiology or Medicine for this discovery. RSV proved to be a kind of RNA virus called a **retrovirus.** When retroviruses infect a cell, they make a DNA copy of their RNA genome and insert that copy into the host cell's DNA.

How RSV Causes Cancer

How does RSV initiate cancer? When RSV was compared to a closely related virus, RAV-O, which is unable to transform normal chicken cells into cancerous cells, the two viruses proved to be identical except for one gene that was present in RSV but absent from RAV-O. That gene was called the *src* gene, short for sarcoma.

How do viral genes cause cancer? An essential clue came in 1970, when temperature-sensitive RSV mutants were isolated. These mutants would transform tissue culture cells into cancer cells at 35°C, but not at 41°C. Temperature sensitivity of this kind is almost always associated with proteins. It seemed likely, therefore, that the *src* gene was actively transcribed by the cell, rather than serving as a recognition site for some sort of regulatory protein. This was an exciting result, suggesting that the protein specified by this cancer-causing gene, or **oncogene,** could be isolated and its properties studied.

The *src* protein was first isolated in 1977 by J. Michael Bishop and Harold Varmus, who won the Nobel Prize for their efforts. It turned out to be an enzyme of moderate size that phosphorylates (adds a phosphate group to) the tyrosine amino acids of proteins. Such enzymes, called **tyrosine kinases,** are quite common in animal cells. One example is an enzyme that also serves as a plasma membrane receptor for **epidermal growth factor,** a protein that signals the initiation of cell division. This finding raised the exciting possibility that RSV may cause cancer by introducing into cells an active form of a normally quiescent growth-promoting enzyme. Later experiments showed this is indeed the case.

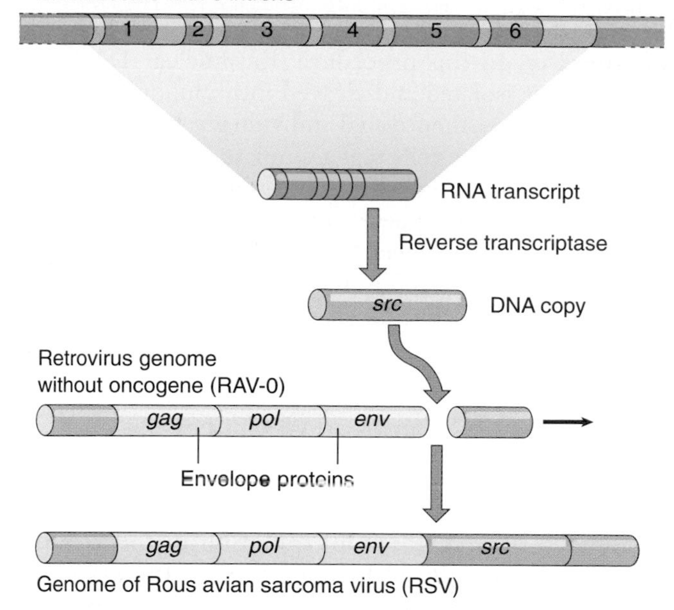

FIGURE 18.12
How a chicken gene got into the RSV genome. RSV contains only a few genes: *gag* and *env*, which encode the viral protein coat and envelope proteins, and *pol*, which encodes reverse transcriptase. It also contains the *src* gene that causes sarcomas, which the RAV-O virus lacks. RSV originally obtained its *src* gene from chickens, where a copy of the gene occurs normally and is controlled by the chicken's regulatory genes.

Origin of the *src* Gene

Does the *src* gene actually integrate into the host cell's chromosome along with the rest of the RSV genome? One way to answer this question is to prepare a radioactive version of the gene, allow it to bind to complementary sequences on the chicken chromosomes, and examine where the chromosomes become radioactive. The result of this experiment is that radioactive *src* DNA does in fact bind to the site where RSV DNA is inserted into the chicken genome—but it also binds to a second site where there is no part of the RSV genome!

The explanation for the second binding site is that the *src* gene is not exclusively a viral gene. It is also a growth-promoting gene that evolved in and occurs normally in chickens. This normal chicken gene is the second site where *src* binds to chicken DNA. Somehow, an ancestor of RSV picked up a copy of the normal chicken gene in some past infection. Now part of the virus, the gene is transcribed under the control of viral promoters rather than under the regulatory system of the chicken genome (figure 18.12).

Studies of RSV reveal that cancer results from the inappropriate activity of growth-promoting genes that are less active or completely inactive in normal cells.

Cancer and the Cell Cycle

An important technique used to study tumors is called **transfection.** In this procedure, the nuclear DNA from tumor cells is isolated and cleaved into random fragments. Each fragment is then tested individually for its ability to induce cancer in the cells that assimilate it.

Using transfection, researchers have discovered that most human tumors appear to result from the mutation of genes that regulate the cell cycle. Sometimes the mutation of only one or two gene is all that is needed to transform normally dividing cells into cancerous cells in tissue culture (table 18.4).

Point Mutations Can Lead to Cancer

The difference between a normal gene encoding a protein that regulates the cell cycle and a cancer-inducing version can be a single point mutation in the DNA. In one case of *ras*-induced bladder cancer, for example, a single DNA base change from guanine to thymine converts a glycine in the normal *ras* protein into a valine in the cancer-causing version. Several other *ras*-induced human carcinomas have been shown to also involve single nucleotide substitutions.

Telomerase and Cancer

Telomeres are short sequences of nucleotides repeated thousands of times on the ends of chromosomes. Because DNA polymerase is unable to copy chromosomes all the way to the tip (there is no place for the primer necessary to copy the last Okazaki fragment), telomeric segments are lost every time a cell divides.

In healthy cells a tumor suppressor inhibits production of a special enzyme called telomerase that adds the lost telomere material back to the tip. Without this enzyme, a cell's chromosomes lose material from their telomeres with each replication. Every time a chromosome is copied as the cell prepares to divide, more of the tip is lost. After some 30 divisions, so much is lost that copying is no longer possible. Cells in the tissues of an adult human have typically undergone 25 or more divisions. Cancer can't get very far with only the 5 remaining cell divisions. Were cancer to start, it would grind to a halt after only a few divisions for lack of telomere.

Thus, we see that the cell's inhibition of telomerase in somatic cells is a very effective natural brake on the cancer process. Any mutation that destroys the telomerase inhibitor releases that brake, making cancer possible. Thus, when researchers looked for telomerase in human ovarian tumor cells, they found it. These cells contained mutations that had inactivated the cell control that blocks the transcription of the telomerase gene. Telomerase produced in these cells reversed normal telomere shortening, allowing the cells to proliferate and gain the immortality of cancer cells.

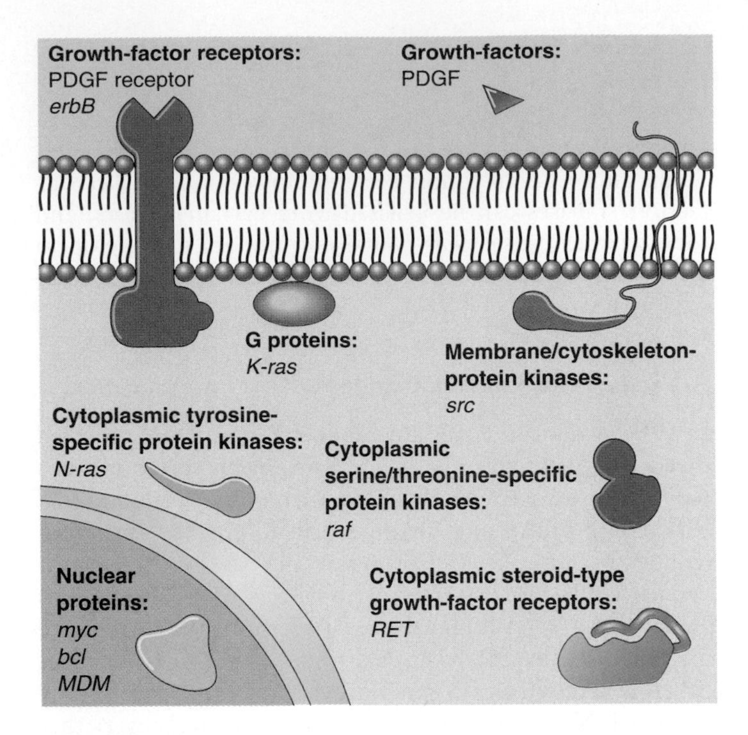

FIGURE 18.13
The main classes of oncogenes. Before they are altered by mutation to their cancer-causing condition, oncogenes are called proto-oncogenes (that is, genes able to become oncogenes). Illustrated here are the principal classes of proto-oncogenes, with some typical representatives indicated.

Mutations in Proto-Oncogenes: Accelerating the Cell Cycle

Most cancers are the direct result of mutations in growth-regulating genes. There are two general classes of cancer-inducing mutations: mutations of proto-oncogenes and mutations of tumor-suppressor genes.

Genes known as **proto-oncogenes** encode proteins that stimulate cell division. Mutations that overactivate these stimulatory proteins cause the cells that contain them to proliferate excessively. Mutated proto-oncogenes become cancer-causing genes called **oncogenes** (Greek *onco-*, "tumor") (figure 18.13). Often the induction of these cancers involves changes in the activity of intracellular signalling molecules associated with receptors on the surface of the plasma membrane. In a normal cell, the signalling pathways activated by these receptors trigger passage of the G_1 checkpoint of cell proliferation (see figure 11.17).

The mutated alleles of these oncogenes are genetically dominant. Among the most widely studied are *myc* and *ras*. Expression of *myc* stimulates the production of cyclins and cyclin-dependent protein kinases (Cdks), key elements in regulating the checkpoints of cell division.

Table 18.4 Some Genes Implicated in Human Cancers

Gene	Product	Cancer
ONCOGENES		
Genes Encoding Growth Factors or Their Receptors		
erb-B	Receptor for epidermal growth factor	Glioblastoma (a brain cancer); breast cancer
erb-B2	A growth factor receptor (gene also called *neu*)	Breast cancer; ovarian cancer; salivary gland cancer
PDGF	Platelet-derived growth factor	Glioma (a brain cancer)
RET	A growth factor receptor	Thyroid cancer
Genes Encoding Cytoplasmic Relays in Intracellular Signaling Pathways		
K-ras	Protein kinase	Lung cancer; colon cancer; ovarian cancer; pancreatic cancer
N-ras	Protein kinase	Leukemias
Genes Encoding Transcription Factors That Activate Transcription of Growth-Promoting Genes		
c-myc	Transcription factor	Lung cancer; breast cancer; stomach cancer; leukemias
L-myc	Transcription factor	Lung cancer
N-myc	Transcription factor	Neuroblastoma (a nerve cell cancer)
Genes Encoding Other Kinds of Proteins		
bcl-2	Protein that blocks cell suicide	Follicular B cell lymphoma
bcl-1	Cyclin D1, which stimulates the cell cycle clock (gene also called *PRAD1*)	Breast cancer; head and neck cancers
MDM2	Protein antagonist of p53 tumor-supressor protein	Wide variety of sarcomas (connective tissue cancers)
TUMOR-SUPRESSOR GENES		
Genes Encoding Cytoplasmic Proteins		
APC	Step in a signaling pathway	Colon cancer; stomach cancer
DPC4	A relay in signaling pathway that inhibits cell division	Pancreatic cancer
NF-1	Inhibitor of ras, a protein that stimulates cell division	Neurofibroma; myeloid leukemia
NF-2	Inhibitor of ras	Meningioma (brain cancer); schwannoma (cancer of cells supporting peripheral nerves)
Genes Encoding Nuclear Proteins		
MTS1	p16 protein, which slows the cell cycle clock	A wide range of cancers
p53	p53 protein, which halts cell division at the G_1 checkpoint	A wide range of cancers
Rb	Rb protein, which acts as a master brake of the cell cycle	Retinoblastoma; breast cancer; bone cancer; bladder cancer
Genes Encoding Proteins of Unknown Cellular Locations		
BRCA1	?	Breast cancer; ovarian cancer
BRCA2	?	Breast cancer
VHL	?	Renal cell cancer

The *ras* gene product is involved in the cellular response to a variety of growth factors such as EGF, an intercellular signal that normally initiates cell proliferation. When EGF binds to a specific receptor protein on the plasma membrane of epithelial cells, the portion of the receptor that protrudes into the cytoplasm stimulates the *ras* protein to bind to GTP. The *ras* protein/GTP complex in turn recruits and activates a protein called Raf to the inner surface of the plasma membrane, which in turn activates cytoplasmic kinases and so triggers an intracellular signaling system (see chapter 7). The final step in the pathway is the activation of transcription factors that trigger cell proliferation. Cancer-causing mutations in *ras* greatly reduce the amount of EGF necessary to initiate cell proliferation.

Mutations in Tumor-Suppressor Genes: Inactivating the Cell's Inhibitors of Proliferation

If the first class of cancer-inducing mutations "steps on the accelerator" of cell division, the second class of cancer-inducing mutations "removes the brakes." Cell division is normally turned off in healthy cells by proteins that prevent cyclins from binding to Cdks. The genes that encode these proteins are called **tumor-suppressor genes.** Their mutant alleles are genetically recessive.

Among the most widely studied tumor-suppressor genes are *Rb*, *p16*, *p21*, and *p53*. The unphosphorylated product of the *Rb* gene ties up transcription factor E2F, which transcribes several genes required for passage through the G_1 checkpoint into S phase of the cell cycle (figure 18.14). The proteins encoded by *p16* and *p21* reinforce the tumor-suppressing role of the Rb protein, preventing its phosphorylation by binding to the appropriate Cdk/cyclin complex and inhibiting its kinase activity. The p53 protein senses the integrity of the DNA and is activated if the DNA is damaged (figure 18.15). It appears to act by inducing the transcription of *p21*, which binds to cyclins and Cdk and prevents them from interacting. One of the reasons repeated smoking leads inexorably to lung cancer is that it induces *p53* mutations. Indeed, almost half of all cancers involve mutations of the *p53* gene.

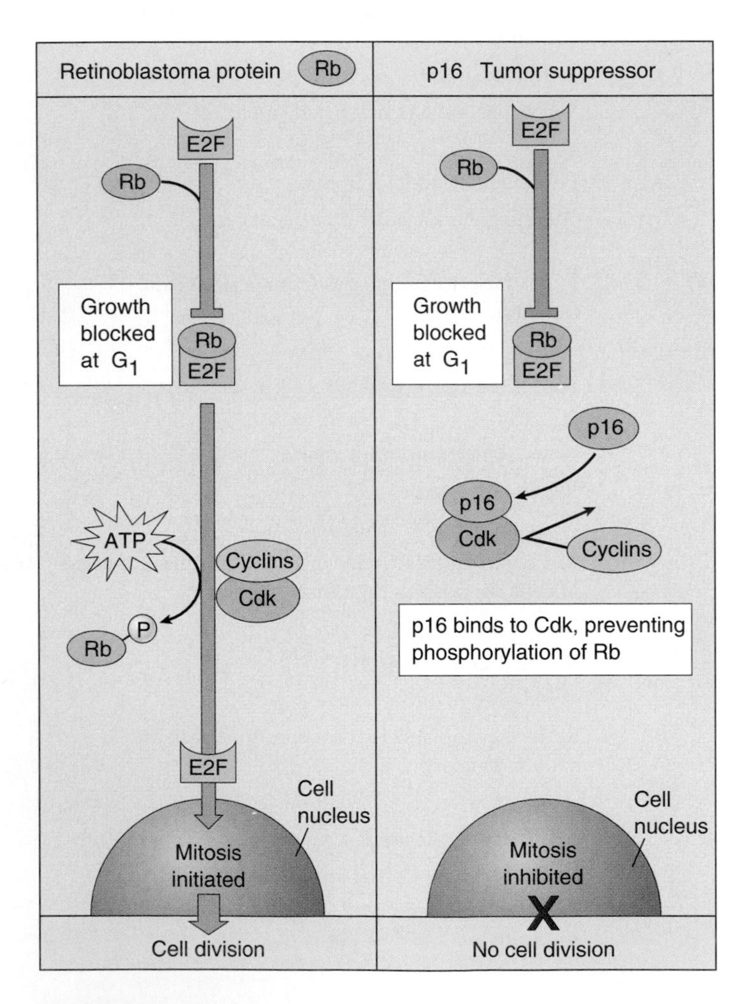

FIGURE 18.14
How the tumor-suppressor genes *Rb* and *p16* interact to block cell division. The retinoblastoma protein (Rb) binds to the transcription factor (E2F) that activates genes in the nucleus, preventing this factor from initiating mitosis. The G_1 checkpoint is passed when Cdk interacts with cyclins to phosphorylate Rb, releasing E2F. The p16 tumor-suppressor protein reinforces Rb's inhibitory action by binding to Cdk so that Cdk is not available to phosphorylate Rb.

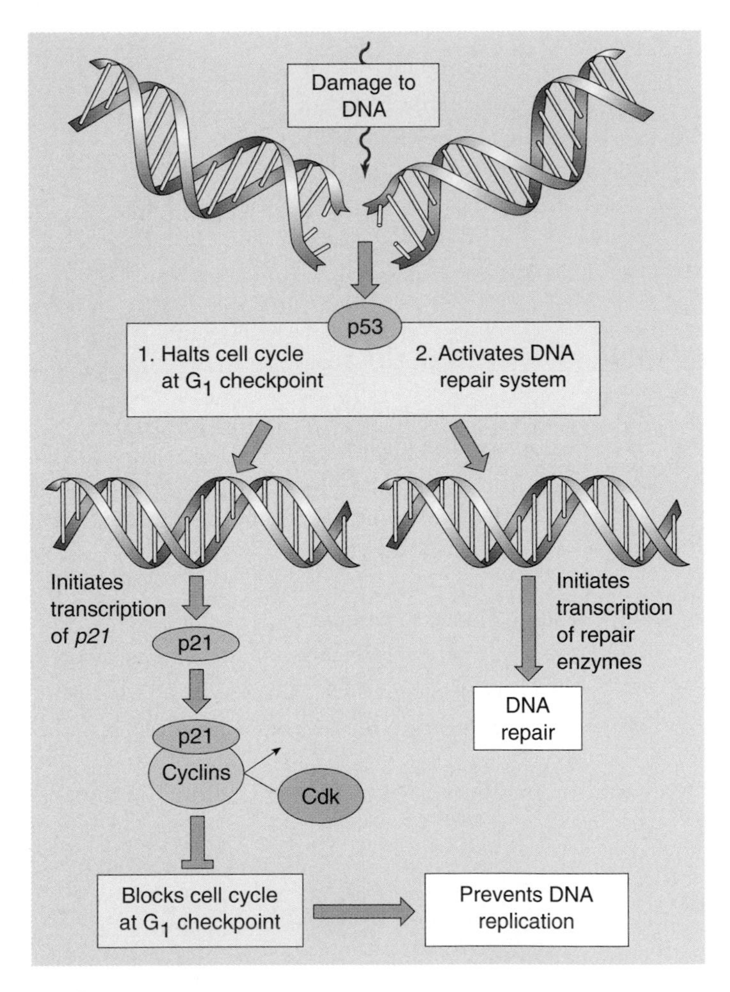

FIGURE 18.15
The role of tumor-suppressor *p53* in regulating the cell cycle. The p53 protein works at the G_1 checkpoint to check for DNA damage. If the DNA is damaged, p53 activates the DNA repair system and stops the cell cycle at the G_1 checkpoint (before DNA replication). This allows time for the damage to be repaired. p53 stops the cell cycle by inducing the transcription of *p21*. The p21 protein then binds to cyclins and prevents them from complexing with Cdk.

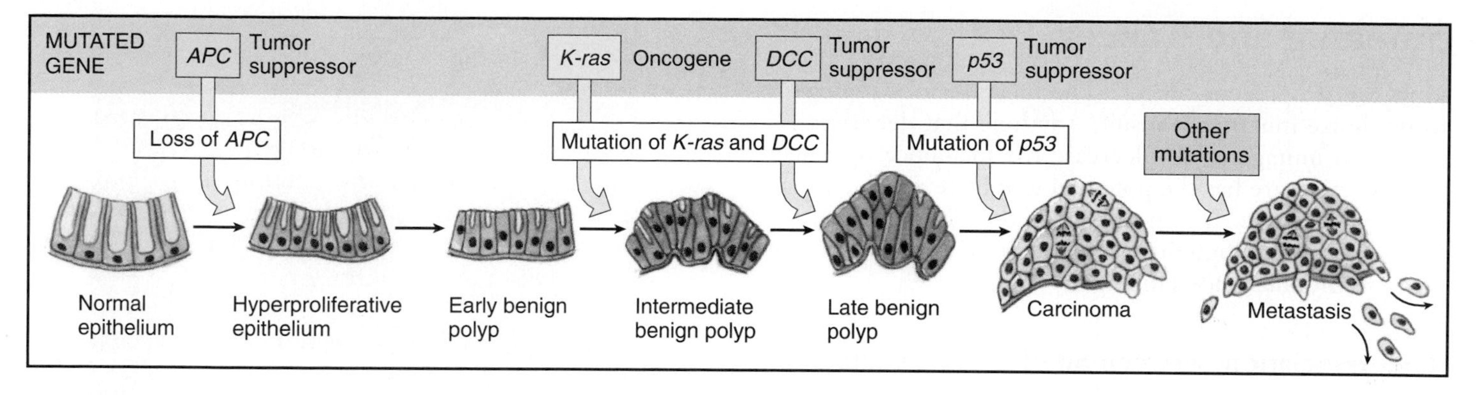

FIGURE 18.16
The progression of mutations that commonly lead to colorectal cancer. The fatal metastasis is the last of six serial changes that the epithelial cells lining the rectum undergo. One of these changes is brought about by mutation of a proto-oncogene, and three of them involve mutations that inactivate tumor-suppressor genes.

Cancer-Causing Mutations Accumulate over Time

Cells control proliferation at several checkpoints, and all of these controls must be inactivated for cancer to be initiated. Therefore, the induction of most cancers involves the mutation of multiple genes; four is a typical number (figure 18.16). In many of the tissue culture cell lines used to study cancer, most of the controls are already inactivated, so that mutations in only one or a few genes transform the line into cancerous growth. The need to inactivate several regulatory genes almost certainly explains why most cancers occur in people over 40 years old (figure 18.17); in older persons, there has been more time for individual cells to accumulate multiple mutations. It is now clear that mutations, including those in potentially cancer-causing genes, do accumulate over time. Using the polymerase chain reaction (PCR), researchers in 1994 searched for a certain cancer-associated gene mutation in the blood cells of 63 cancer-free people. They found that the mutation occurred 13 times more often in people over 60 years old than in people under 20.

Cancer is a disease in which the controls that normally restrict cell proliferation do not operate. In some cases, cancerous growth is initiated by the inappropriate activation of proteins that regulate the cell cycle; in other cases, it is initiated by the inactivation of proteins that normally suppress cell division.

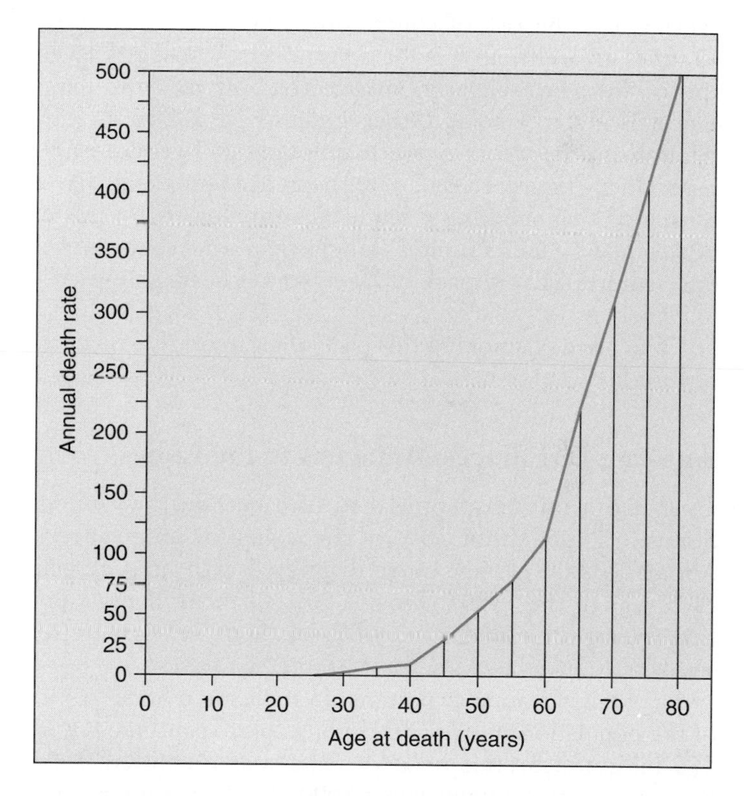

FIGURE 18.17
The annual death rate from cancer climbs with age. The rate of cancer deaths increases steeply after age 40 and even more steeply after age 60, suggesting that several independent mutations must accumulate to give rise to cancer.

Smoking and Cancer

How can we prevent cancer? The most obvious strategy is to minimize mutational insult. Anything that decreases exposure to mutagens can decrease the incidence of cancer because exposure has the potential to mutate a normal gene into an oncogene. It is no accident that the most reliable tests for the carcinogenicity of a substance are tests that measure the substance's mutagenicity.

The Association between Smoking and Cancer

About a third of all cases of cancer in the United States are directly attributable to cigarette smoking. The association between smoking and cancer is particularly striking for lung cancer (figure 18.18). Studies of male smokers show a highly positive correlation between the number of cigarettes smoked per day and the incidence of lung cancer (figure 18.19). For individuals who smoke two or more packs a day, the risk of contracting lung cancer is at least 40 times greater than it is for nonsmokers, whose risk level approaches zero. Clearly, an effective way to avoid lung cancer is not to smoke. Other studies have shown a clear relationship between cigarette smoking and reduced life expectancy (figure 18.20). Life insurance companies have calculated that smoking a single cigarette lowers one's life expectancy by 10.7 minutes (longer than it takes to smoke the cigarette)! Every pack of 20 cigarettes bears an unwritten label:

"The price of smoking this pack of cigarettes is 3½ hours of your life."

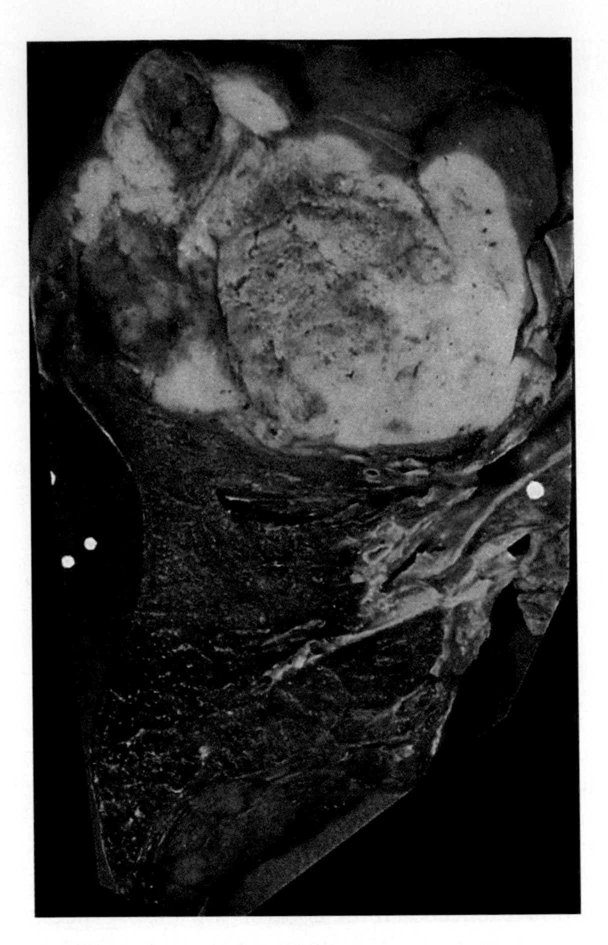

FIGURE 18.18
Photo of a cancerous human lung. The bottom half of the lung is normal, while a cancerous tumor has completely taken over the top half. The cancer cells will eventually break through into the lymph and blood vessels and spread through the body.

Smoking Introduces Mutagens to the Lungs

Over half a million people died of cancer in the United States in 2000; about 28% of them died of lung cancer. About 140,000 persons were diagnosed with lung cancer each year in the 1980s. Around 90% of them died within three years after diagnosis; 96% of them were cigarette smokers.

Smoking is a popular pastime. In the United States, 24% of the population smokes, and U.S. smokers consumed over 450 billion cigarettes in 1999. The smoke emitted from these cigarettes contains some 3000 chemical components, including vinyl chloride, benzo[*a*]pyrenes, and nitroso-*nor*-nicotine, all potent mutagens. Smoking places these mutagens into direct contact with the tissues of the lungs.

Mutagens in the Lung Cause Cancer

Introducing powerful mutagens to the lungs causes considerable damage to the genes of the epithelial cells that line the lungs and are directly exposed to the chemicals. Among the genes that are mutated as a result are some whose normal function is to regulate cell proliferation. When these genes are damaged, lung cancer results.

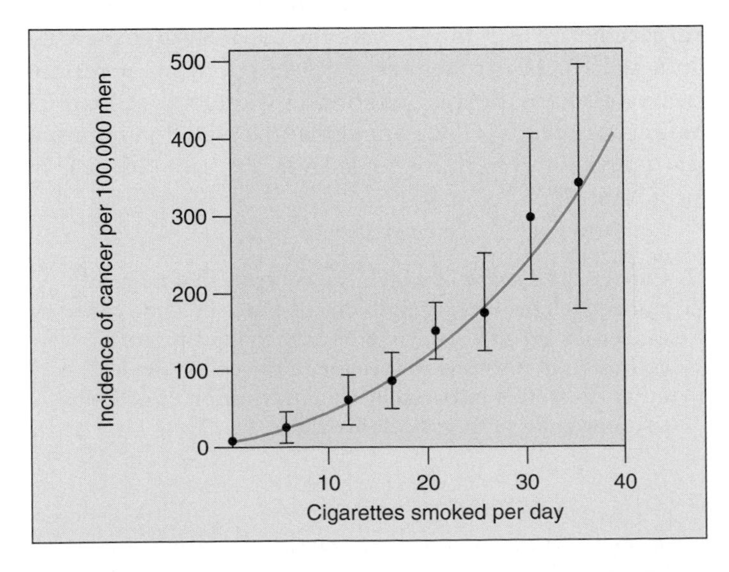

FIGURE 18.19
Smoking causes cancer. The annual incidence of lung cancer per 100,000 men clearly increases with the number of cigarettes smoked per day.

This process has been clearly demonstrated for benzo[*a*]pyrene (BP), one of the potent mutagens released into cigarette smoke from tars in the tobacco. The epithelial cells of the lung absorb BP from tobacco smoke and chemically alter it to a derivative form. This derivative form, benzo[*a*]pyrene-diolepoxide (BPDE), binds directly to the tumor-suppressor gene *p53* and mutates it to an inactive form. The protein encoded by *p53* oversees the G_1 cell cycle checkpoint described in chapter 11 and is one of the body's key mechanisms for preventing uncontrolled cell proliferation. The destruction of *p53* in lung epithelial cells greatly hastens the onset of lung cancer—*p53* is mutated to an inactive form in over 70% of lung cancers. When examined, the *p53* mutations in cancer cells almost all occur at one of three "hot spots." The key evidence linking smoking and cancer is that when the mutations of *p53* caused by BPDE from cigarettes are examined, they occur at the *same* three specific "hot spots!"

The Incidence of Cancer Reflects Smoking

Cigarette manufacturers argue that the causal connection between smoking and cancer has not been proved, and that somehow the relationship is coincidental. Look carefully at the data presented in figure 18.21 and see if you agree. The upper graph, compiled from data on American men, shows the incidence of smoking from 1900 to 1990 and the incidence of lung cancer over the same period. Note that as late as 1920, lung cancer was a rare disease. About 20 years after the incidence of smoking began to increase among men, lung cancer also started to become more common.

Now look at the lower graph, which presents data on American women. Because of social mores, significant numbers of American women did not smoke until after World War II, when many social conventions changed. As late as 1963, when lung cancer among males was near current levels, this disease was still rare in women. In the United States that year, only 6588 women died of lung cancer. But as more women smoked, more developed lung cancer, again with a lag of about 20 years. American women today have achieved equality with men in the numbers of cigarettes they smoke, and their lung cancer death rates are today approaching those for men. In 1990, more than 49,000 women died of lung cancer in the United States. The current annual rate of deaths from lung cancer in male and female smokers is 180 per 100,000, or about 2 out of every 1000 smokers *each year*.

The easiest way to avoid cancer is to avoid exposure to mutagens. The single greatest contribution one can make to a longer life is not to smoke.

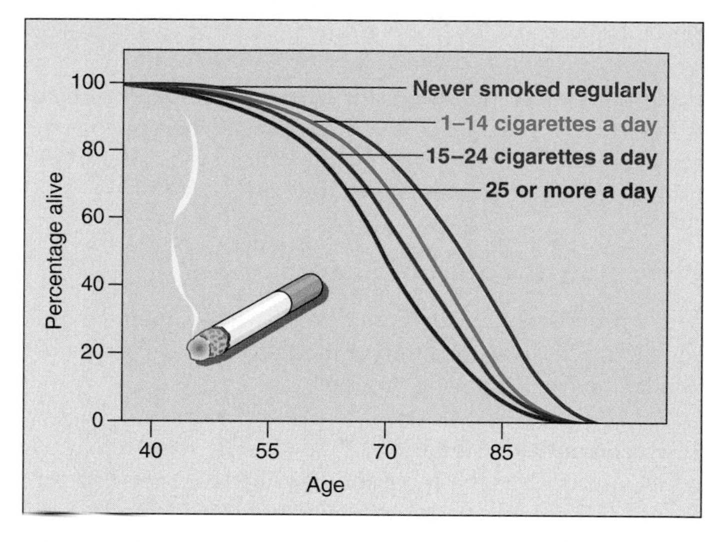

FIGURE 18.20
Tobacco reduces life expectancy. The world's longest-running survey of smoking, begun in 1951 in Britain, revealed that by 1994 the death rate for smokers had climbed to three times the rate for nonsmokers among men 35 to 69 years of age.
Source: Data from *New Scientist,* October 15, 1994.

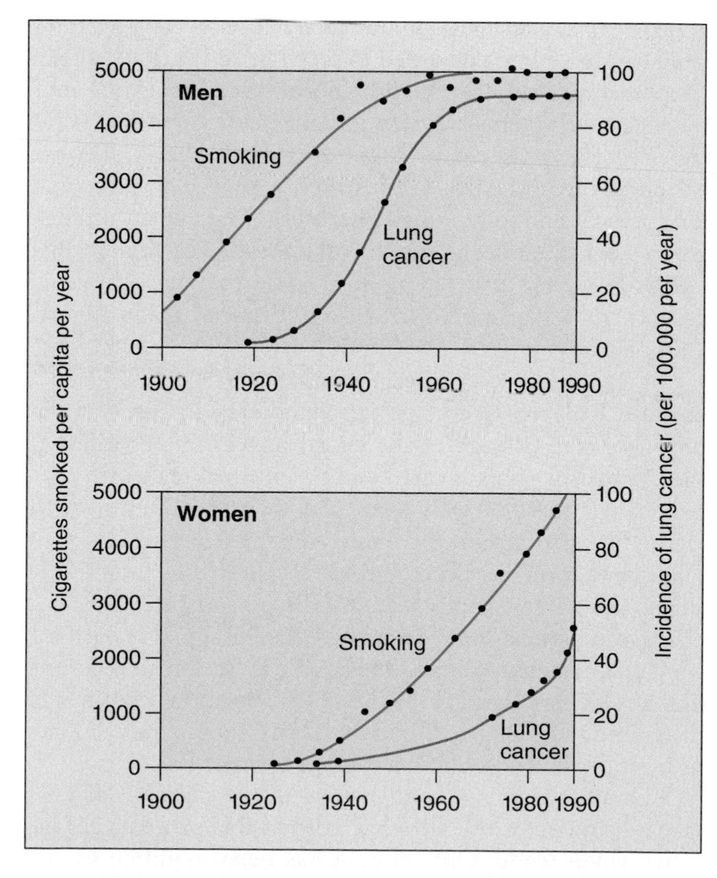

FIGURE 18.21
The incidence of lung cancer in men and women. What do these graphs indicate about the connection between smoking and lung cancer?

Curing Cancer

Potential cancer therapies are being developed on many fronts (figure 18.22). Some act to prevent the start of cancer within cells. Others act outside cancer cells, preventing tumors from growing and spreading.

Preventing the Start of Cancer

Many promising cancer therapies act within potential cancer cells, focusing on different stages of the cell's "Shall I divide?" decision-making process.

1. Receiving the Signal to Divide. The first step in the decision process is the reception of a "divide" signal, usually a small protein called a growth factor released from a neighboring cell. The growth factor is received by a protein receptor on the cell surface. Mutations that increase the number of receptors on the cell surface amplify the division signal and so lead to cancer. Over 20% of breast cancer tumors prove to overproduce a protein called HER2 associated with the receptor for epidermal growth factor.

Therapies directed at this stage of the decision process utilize the human immune system to attack cancer cells. Special protein molecules called "monoclonal antibodies," created by genetic engineering, are the therapeutic agents. These monoclonal antibodies are designed to seek out and stick to HER2. Like waving a red flag, the presence of the monoclonal antibody calls down attack by the immune system on the HER2 cell. Because breast cancer cells overproduce HER2, they are killed preferentially. Genentech's recently approved monoclonal antibody, called "herceptin," has given promising results in clinical tests. In other tests, the monoclonal antibody C225, directed against epidermal growth factor receptors, has succeeded in curing advanced colon cancer. Clinical trials of C225 have begun.

2. The Relay Switch. The second step in the decision process is the passage of the signal into the cell's interior, the cytoplasm. This is carried out in normal cells by a protein called Ras that acts as a relay switch. When growth factor binds to a receptor like EGF, the adjacent Ras protein acts like it has been "goosed," contorting into a new shape. This new shape is chemically active, and initiates a chain of reactions that passes the "divide" signal inward toward the nucleus. Mutated forms of the Ras protein behave like a relay switch stuck in the "ON" position, continually instructing the cell to divide when it should not. Thirty percent of all cancers have a mutant form of Ras.

Therapies directed at this stage of the decision process take advantage of the fact that normal Ras proteins are inactive when made. Only after it has been modified by the special enzyme *farnesyl transferase* does Ras protein become able to function as a relay switch. In tests on animals, farnesyl transferase inhibitors induce the regression of tumors and prevent the formation of new ones.

3. Amplifying the Signal. The third step in the decision process is the amplification of the signal within the cytoplasm. Just as a TV signal needs to be amplified in order to be received at a distance, so a "divide" signal must be amplified if it is to reach the nucleus at the interior of the cell, a very long journey at a molecular scale. Cells use an ingenious trick to amplify the signal. Ras, when "ON," activates an enzyme, a protein kinase. This protein kinase activates other protein kinases that in their turn activate still others. The trick is that once a protein kinase enzyme is activated, it goes to work like a demon, activating hoards of others every second! And each and every one it activates behaves the same way too, activating still more, in a cascade of ever-widening effect. At each stage of the relay, the signal is amplified a thousand-fold. Mutations stimulating any of the protein kinases can dangerously increase the already amplified signal and lead to cancer. Five percent of all cancers, for example, have a mutant hyperactive form of the protein kinase Src.

Therapies directed at this stage of the decision process employ so-called "anti-sense RNA" directed specifically against Src or other cancer-inducing kinase mutations. The idea is that the *src* gene uses a complementary copy of itself to manufacture the Src protein (the "sense" RNA or messenger RNA), and a mirror image complementary copy of the sense RNA ("anti-sense RNA") will stick to it, gumming it up so it can't be used to make Src protein. The approach appears promising. In tissue culture, anti-sense RNAs inhibit the growth of cancer cells, and some also appear to block the growth of human tumors implanted in laboratory animals. Human clinical trials are underway.

4. Releasing the Brake. The fourth step in the decision process is the removal of the "brake" the cell uses to restrain cell division. In healthy cells this brake, a tumor suppressor protein called Rb, blocks the activity of a transcription factor protein called E2F. When free, E2F enables the cell to copy its DNA. Normal cell division is triggered to begin when Rb is inhibited, unleashing E2F. Mutations which destroy Rb release E2F from its control completely, leading to ceaseless cell division. Forty percent of all cancers have a defective form of Rb.

Therapies directed at this stage of the decision process are only now being attempted. They focus on drugs able to inhibit E2F, which should halt the growth of tumors arising from inactive Rb. Experiments in mice in which the E2F genes have been destroyed provide a model system to study such drugs, which are being actively investigated.

5. Checking That Everything Is Ready. The fifth step in the decision process is the mechanism used by the cell to ensure that its DNA is undamaged and ready to divide. This job is carried out in healthy cells by the tumor-suppressor protein p53, which inspects the integrity of the DNA. When it detects damaged or foreign DNA, p53 stops cell division and activates the cell's DNA repair systems. If the damage doesn't get repaired in a reasonable time, p53 pulls the plug, triggering events that kill the cell. In this way, mutations such as those that cause cancer are either repaired or the cells containing them eliminated. If p53 is itself destroyed by mutation, future damage accumulates unrepaired. Among this damage are mutations that lead to cancer. Fifty percent of all cancers have a disabled p53. Fully 70 to 80% of lung cancers have a mutant inactive p53—the chemical benzo[*a*]pyrene in cigarette smoke is a potent mutagen of p53.

A promising new therapy using adenovirus (responsible for mild colds) is being targeted at cancers with a mutant p53. To grow in a host cell, adenovirus must use the product of its gene *E1B* to block the host cell's p53, thereby enabling replication of the adenovirus DNA. This means that while mutant adenovirus without *E1B* cannot grow in healthy cells, the mutants should be able to grow in, and destroy, cancer cells with defective p53. When human colon and lung cancer cells are introduced into mice lacking an immune system and allowed to produce substantial tumors, 60% of the tumors simply disappear when treated with E1B-deficient adenovirus, and do not reappear later. Initial clinical trials are less encouraging, as many people possess antibodies to adenovirus.

6. Stepping on the Gas. Cell division starts with replication of the DNA. In healthy cells, another tumor suppressor "keeps the gas tank nearly empty" for the DNA replication process by inhibiting production of an enzyme called telomerase. Without this enzyme, a cell's chromosomes lose material from their tips, called telomeres. Every time a chromosome is copied, more tip material is lost. After some thirty divisions, so much is lost that copying is no longer possible. Cells in the tissues of an adult human have typically undergone twenty five or more divisions. Cancer can't get very far with only the five remaining cell divisions, so inhibiting telomerase is a very effective natural break on the cancer process. It is thought that almost all cancers involve a mutation that destroys the telomerase inhibitor, releasing this break and making cancer possible. It should be possible to block cancer by reapplying this inhibition. Cancer therapies that inhibit telomerase are just beginning clinical trials.

Preventing the Spread of Cancer

7. Tumor Growth. Once a cell begins cancerous growth, it forms an expanding tumor. As the tumor grows ever-larger, it requires an increasing supply of food and nutrients, obtained from the body's blood supply. To facilitate this necessary grocery shopping, tumors leak out substances into the surrounding tissues that encourage angiogenesis, the formation of small blood vessels. Chemicals that inhibit this process are called angiogenesis inhibitors. In mice, two such angiogenesis inhibitors, angiostatin and endostatin, caused tumors to regress to microscopic size. This very exciting result has proven controversial, but initial human trials seem promising.

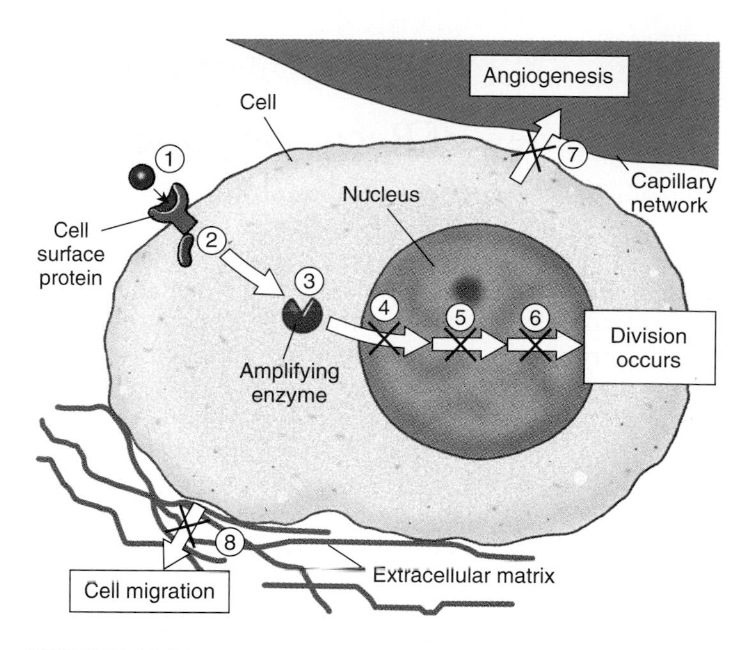

FIGURE 18.22
New molecular therapies for cancer target eight different stages in the cancer process. (*1*) On the cell surface, a growth factor signals the cell to divide. (*2*) Just inside the cell, a protein relay switch passes on the divide signal. (*3*) In the cytoplasm, enzymes amplify the signal. In the nucleus, (*4*) a "brake" preventing DNA replication is released, (*5*) proteins check that the replicated DNA is not damaged, and (*6*) other proteins rebuild chromosome tips so DNA can replicate. (*7*) The new tumor promotes angiogenesis, the formation of growth-promoting blood vessels. (*8*) Some cancer cells break away from the extracellular matrix and invade other parts of the body.

8. Metastasis. If cancerous tumors simply continued to grow where they form, many could be surgically removed, and far fewer would prove fatal. Unfortunately, many cancerous tumors eventually metastasize, individual cancer cells breaking their moorings to the extracellular matrix and spreading to other locations in the body where they initiate formation of secondary tumors. This process involves metal-requiring protease enzymes that cleave the cell-matrix linkage, components of the extracellular matrix such as fibronectin that also promote the migration of several noncancerous cell types, and RhoC, a GTP-hydrolyzing enzyme that promotes cell migration by providing needed GTP. All of these components offer promising targets for future anti-cancer therapy.

Therapies such as those described here are only part of a wave of potential treatments under development and clinical trial. The clinical trials will take years to complete, but in the coming decade we can expect cancer to become a curable disease.

Understanding of how mutations produce cancer has progressed to the point where promising potential therapies can be tested.

18.3 Recombination alters gene location.

An Overview of Recombination

Mutation is a change in the *content* of an organism's genetic message, but it is not the only source of genetic diversity. Diversity is also generated when existing elements of the genetic message move around within the genome. As an analogy, consider the pages of this book. A point mutation would correspond to a change in one or more of the letters on the pages. For example, " . . . in one or more of the letters *of* the pages" is a mutation of the previous sentence, in which an "n" is changed to an "f." A significant alteration is also achieved, however, when we move the position of words, as in " . . . in one or more of the pages on the letters." The change alters (and destroys) the meaning of the sentence by exchanging the position of the words "letters" and "pages." This second kind of change, which represents an alteration in the genomic *location* of a gene or a fragment of a gene, demonstrates **genetic recombination.**

FIGURE 18.23
A Nobel Prize for discovering gene transfer by transposition. Barbara McClintock receiving her Nobel Prize in 1983.

Gene Transfer

Viewed broadly, genetic recombination can occur by two mechanisms (table 18.5). In **gene transfer,** one chromosome or genome donates a segment to another chromosome or genome. The transfer of genes from the human immunodeficiency virus (HIV) to a human chromosome is an example of gene transfer. Because gene transfer occurs in both prokaryotes and eukaryotes, it is thought to be the more primitive of the two mechanisms.

Reciprocal Recombination

Reciprocal recombination is when two chromosomes trade segments. It is exemplified by the crossing over that occurs between homologous chromosomes during meiosis. Independent assortment during meiosis is another form of reciprocal recombination. Discussed in chapters 12 and 13, it is responsible for the 9:3:3:1 ratio of phenotypes in a dihybrid cross and occurs only in eukaryotes.

Genetic recombination is a change in the genomic association among genes. It often involves a change in the position of a gene or portion of a gene. Recombination of this sort may result from one-way gene transfer or reciprocal gene exchange.

Table 18.5 Classes of Genetic Recombination

Class	Occurrence
GENE TRANSFERS	
Conjugation	Occurs predominantly but not exclusively in bacteria and is targeted to specific locations in the genome
Transposition	Common in both bacteria and eukaryotes; genes move to new genomic locations, apparently at random
RECIPROCAL RECOMBINATIONS	
Crossing over	Requires the pairing of homologous chromosomes and may occur anywhere along their length
Unequal crossing over	The result of crossing over between mismatched segments; leads to gene duplication and deletion
Gene conversion	Occurs when homologous chromosomes pair and one is "corrected" to resemble the other
Independent assortment	Haploid cells produced by meiosis contain only one randomly selected member of each pair of homologous chromosomes

Gene Transfer

Genes are not fixed in their locations on chromosomes or the circular DNA molecules of bacteria; they can move around. Some genes move because they are part of small, circular, extrachromosomal DNA segments called **plasmids.** Plasmids enter and leave the main genome at specific places where a nucleotide sequence matches one present on the plasmid. Plasmids occur primarily in bacteria, in which the main genomic DNA can interact readily with other DNA fragments. About 5% of the DNA that occurs in a bacterium is plasmid DNA. Some plasmids are very small, containing only one or a few genes, while others are quite complex and contain many genes. Other genes move within **transposons,** which jump from one genomic position to another at random in both bacteria and eukaryotes.

Gene transfer by plasmid movement was discovered by Joshua Lederberg and Edward Tatum in 1947. Three years later, transposons were discovered by Barbara McClintock. However, her work implied that the position of genes in a genome need not be constant. Researchers accustomed to viewing genes as fixed entities, like beads on a string, did not readily accept the idea of transposons. Therefore, while Lederberg and Tatum were awarded a Nobel Prize for their discovery in 1958, McClintock did not receive the same recognition for hers until 1983 (figure 18.23).

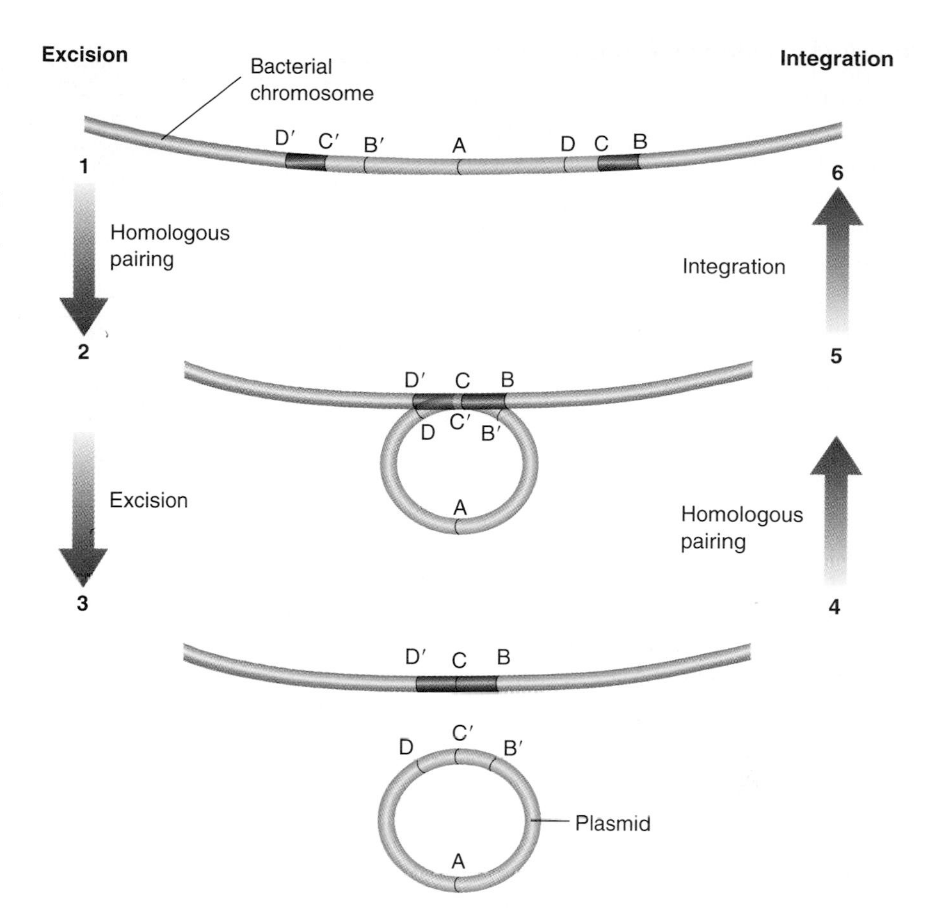

FIGURE 18.24
Integration and excision of a plasmid. Because the ends of the two sequences in the bacterial genome are the same (D′, C′, B′, and D, C, B), it is possible for the two ends to pair. Steps 1–3 show the sequence of events if the strands exchange during the pairing. The result is excision of the loop and a free circle of DNA—a plasmid. Steps 4–6 show the sequence when a plasmid integrates itself into a bacterial genome.

Plasmid Creation

To understand how plasmids arise, consider a hypothetical stretch of bacterial DNA that contains two copies of the same nucleotide sequence. It is possible for the two copies to base-pair with each other and create a transient "loop," or double duplex. All cells have recombination enzymes that can cause such double duplexes to undergo a **reciprocal exchange,** in which they exchange strands. As a result of the exchange, the loop is freed from the rest of the DNA molecule and becomes a plasmid (figure 18.24, steps 1–3). Any genes between the duplicated sequences (such as gene A in figure 18.24) are transferred to the plasmid.

Once a plasmid has been created by reciprocal exchange, DNA polymerase will replicate it if it contains a replication origin, often without the controls that restrict the main genome to one replication per cell division. Consequently, some plasmids may be present in multiple copies, others in just a few copies, in a given cell.

Integration

A plasmid created by recombination can reenter the main genome the same way it left. Sometimes the region of the plasmid DNA that was involved in the original exchange, called the **recognition site,** aligns with a matching sequence on the main genome. If a recombination event occurs anywhere in the region of alignment, the plasmid will integrate into the genome (figure 18.24, steps 4–6). Integration can occur wherever any shared sequences exist, so plasmids may be integrated into the main genome at positions other than the one from which they arose. If a plasmid is integrated at a new position, it transfers its genes to that new position.

Transposons and plasmids transfer genes to new locations on chromosomes. Plasmids can arise from and integrate back into a genome wherever DNA sequences in the genome and in the plasmid match.

Gene Transfer by Conjugation

One of the startling discoveries Lederberg and Tatum made was that plasmids can pass from one bacterium to another. The plasmid they studied was part of the genome of *Escherichia coli.* It was given the name F for fertility factor because only cells which had that plasmid integrated into their DNA could act as plasmid donors. These cells are called Hfr cells (for "high-frequency recombination"). The F plasmid contains a DNA replication origin and several genes that promote its transfer to other cells. These genes encode protein subunits that assemble on the surface of the bacterial cell, forming a hollow tube called a **pilus.**

When the pilus of one cell (F^+) contacts the surface of another cell that lacks a pilus, and therefore does not contain an F plasmid (F^-), the pilus draws the two cells close together so that DNA can be exchanged (figure 18.25). First, the F plasmid binds to a site on the interior of the F^+ cell just beneath the pilus. Then, by a process called **rolling-circle replication,** the F plasmid begins to copy its DNA at the binding point. As it is replicated, the single-stranded copy of the plasmid passes into the other cell. There a complementary strand is added, creating a new, stable F plasmid (figure 18.26). In this way, genes are passed from one bacterium to another. This transfer of genes between bacteria is called **conjugation.**

In an Hfr cell, with the F plasmid integrated into the main bacterial genome rather than free in the cytoplasm, the F plasmid can still organize the transfer of genes. In this case, the integrated F region binds beneath the pilus and initiates the *replication of the bacterial genome*, transferring the newly replicated portion to the recipient cell. Transfer proceeds as if the bacterial genome were simply a part of the F plasmid. By studying this phenomenon, researchers have been able to locate the positions of different genes in bacterial genomes (figure 18.27).

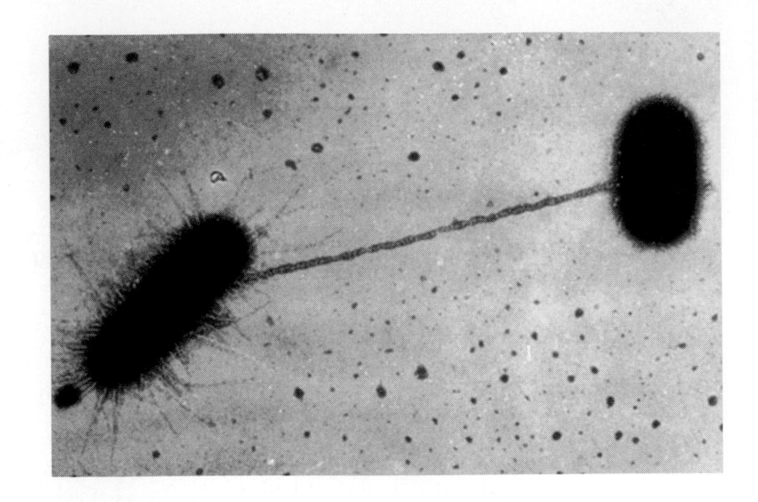

FIGURE 18.25
Contact by a pilus. The pilus of an F^+ cell connects to an F^- cell and draws the two cells close together so that DNA transfer can occur.

Gene Transfer by Transposition

Like plasmids, transposons (figure 18.28) move from one genomic location to another. After spending many generations in one position, a transposon may abruptly move to a new position in the genome, carrying various genes along with it. Transposons encode an enzyme called **transposase,** that inserts the transposon into the genome (figure 18.29). Because this enzyme usually does not recognize any particular sequence on the genome, transposons appear to move to random destinations.

The movement of any given transposon is relatively rare: it may occur perhaps once in 100,000 cell generations. Although low, this rate is still about 10 times as frequent as

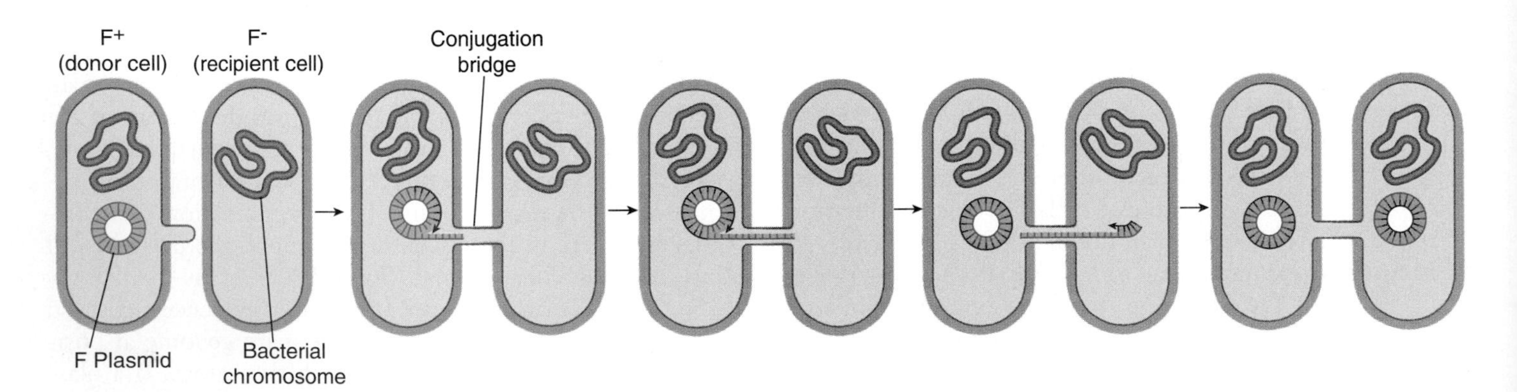

FIGURE 18.26
Gene transfer between bacteria. Donor cells (F^+) contain an F plasmid that recipient cells (F^-) lack. The F plasmid replicates itself and transfers the copy across a conjugation bridge. The remaining strand of the plasmid serves as a template to build a replacement. When the single strand enters the recipient cell, it serves as a template to assemble a double-stranded plasmid. When the process is complete, both cells contain a complete copy of the plasmid.

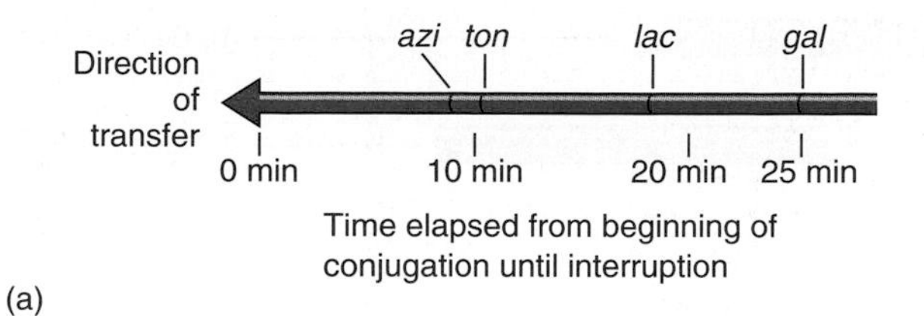

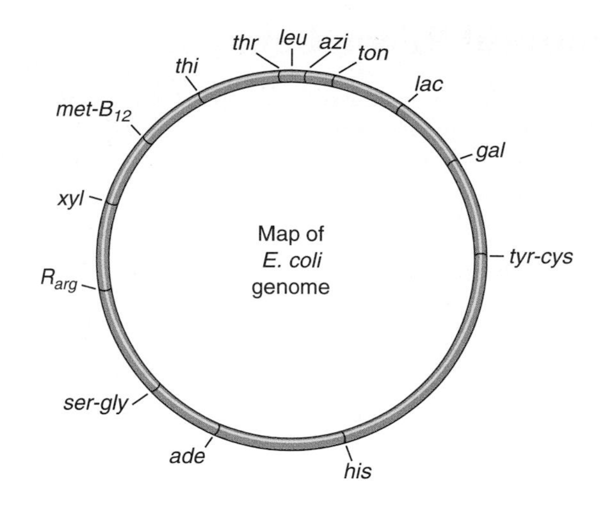

FIGURE 18.27
A conjugation map of the *E. coli* chromosome. Scientists have been able to break the *Escherichia coli* conjugation bridges by agitating the cell suspension rapidly in a blender. By agitating at different intervals after the start of conjugation, investigators can locate the positions of various genes along the bacterial genome. (*a*) The closer the genes are to the origin of replication, the sooner one has to turn on the blender to block their transfer. (*b*) Map of the *E. coli* genome developed using this method.

the rate at which random mutational changes occur. Furthermore, there are many transposons in most cells. Hence, over long periods of time, transposition can have an enormous evolutionary impact.

One way this impact can be felt is through mutation. The insertion of a transposon into a gene often destroys the gene's function, resulting in what is termed **insertional inactivation.** This phenomenon is thought to be the cause of a significant number of the spontaneous mutations observed in nature.

Transposition can also facilitate **gene mobilization,** the bringing together in one place of genes that are usually located at different positions in the genome. In bacteria, for example, a number of genes encode enzymes that make the bacteria resistant to antibiotics such as penicillin, and many of these genes are located on plasmids. The simultaneous exposure of bacteria to multiple antibiotics, a common medical practice some years ago, favors the persistence of plasmids that have managed to acquire several resistance genes. Transposition can rapidly generate such composite plasmids, called **resistance transfer factors** (RTFs), by moving antibiotic resistance genes from several plasmids to one. Bacteria possessing RTFs are thus able to survive treatment with a wide variety of antibiotics. RTFs are thought to be responsible for much of the recent difficulty in treating hospital-engendered *Staphylococcus aureus* infections and the new drug-resistant strains of tuberculosis.

Plasmids transfer copies of bacterial genes (and even entire genomes) from one bacterium to another. Transposition is the one-way transfer of genes to a randomly selected location in the genome. The genes move because they are associated with mobile genetic elements called transposons.

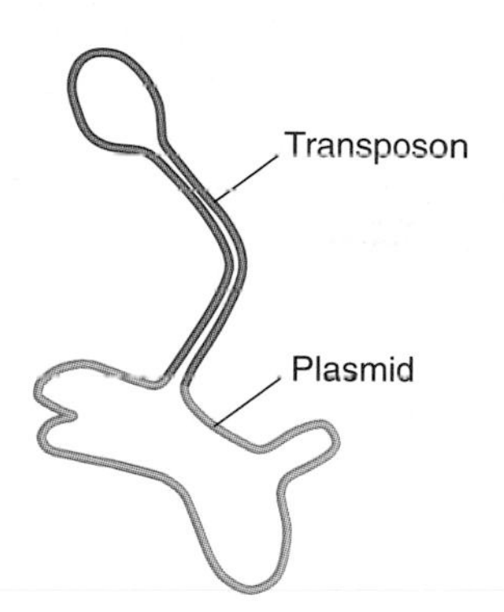

FIGURE 18.28
Transposon. Transposons form characteristic stem-and-loop structures called "lollipops" because their two ends have the same nucleotide sequence as inverted repeats. These ends pair together to form the stem of the lollipop.

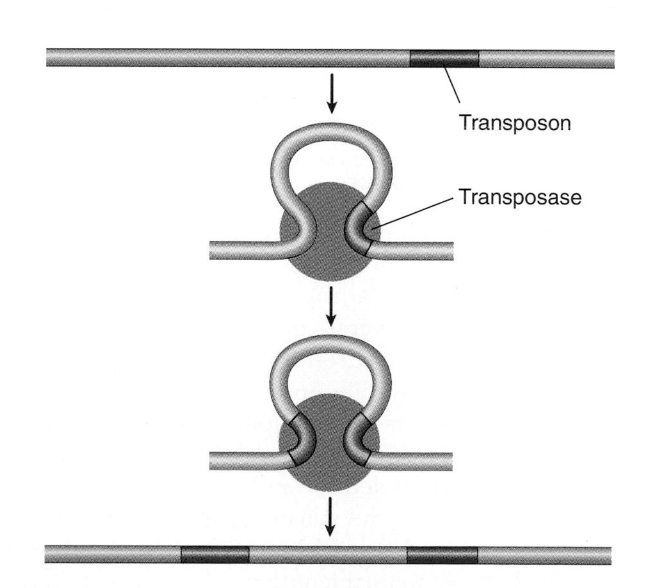

FIGURE 18.29
Transposition. Transposase does not recognize any particular DNA sequence; rather, it selects one at random, moving the transposon to a random location. Some transposons leave a copy of themselves behind when they move.

Reciprocal Recombination

In the second major mechanism for producing genetic recombination, reciprocal recombination, two homologous chromosomes exchange all or part of themselves during the process of meiosis.

Crossing Over

As we saw in chapter 12, crossing over occurs in the first prophase of meiosis, when two homologous chromosomes line up side by side within the synaptonemal complex. At this point, the homologues exchange DNA strands at one or more locations. This exchange of strands can produce chromosomes with new combinations of alleles.

Imagine, for example, that a giraffe has genes encoding neck length and leg length at two different loci on one of its chromosomes. Imagine further that a recessive mutation occurs at the neck length locus, leading after several rounds of independent assortment to some individuals that are homozygous for a variant "long-neck" allele. Similarly, a recessive mutation at the leg length locus leads to homozygous "long-leg" individuals.

It is very unlikely that these two mutations would arise at the same time in the same individual because the probability of two independent events occurring together is the product of their individual probabilities. If the spontaneous occurrence of both mutations in a single individual were the only way to produce a giraffe with both a long neck and long legs, it would be extremely unlikely that such an individual would ever occur. Because of recombination, however, a crossover in the interval between the two genes could in one meiosis produce a chromosome bearing both variant alleles. This ability to reshuffle gene combinations rapidly is what makes recombination so important to the production of natural variation.

Unequal Crossing Over

Reciprocal recombination can occur in any region along two homologous chromosomes with sequences similar enough to permit close pairing. Mistakes in pairing occasionally happen when several copies of a sequence exist in different locations on a chromosome. In such cases, one copy of a sequence may line up with one of the duplicate copies instead of with its homologous copy. Such misalignment causes slipped mispairing, which, as we discussed earlier, can lead to small deletions and frame-shift mutations. If a crossover occurs in the pairing region, it will result in unequal crossing over because the two homologues will exchange segments of unequal length.

In unequal crossing over, one chromosome gains extra copies of the multicopy sequences, while the other chromosome loses them (figure 18.30). This process can generate a chromosome with hundreds of copies of a particular gene, lined up side by side in tandem array.

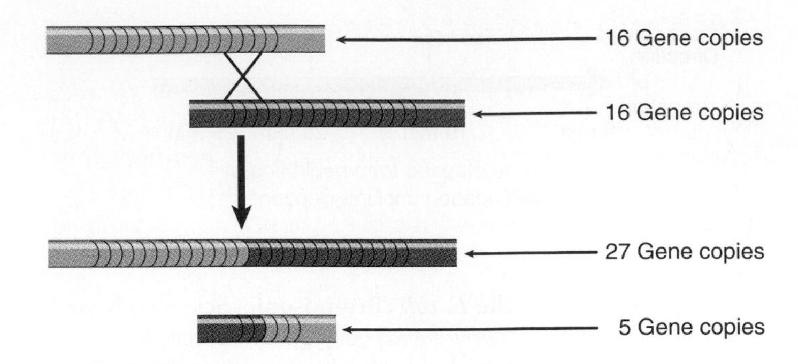

FIGURE 18.30
Unequal crossing over. When a repeated sequence pairs out of register, a crossover within the region will produce one chromosome with fewer gene copies and one with more. Much of the gene duplication that has occurred in eukaryotic evolution may well be the result of unequal crossing over.

Because the genomes of most eukaryotes possess multiple copies of transposons scattered throughout the chromosomes, unequal crossing over between copies of transposons located in different positions has had a profound influence on gene organization in eukaryotes. As we shall see later, most of the genes of eukaryotes appear to have been duplicated one or more times during their evolution.

Gene Conversion

Because the two homologues that pair within a synaptonemal complex are not identical, some nucleotides in one homologue are not complementary to their counterpart in the other homologue with which it is paired. These occasional nonmatching pairs of nucleotides are called **mismatch pairs.**

As you might expect, the cell's error-correcting machinery is able to detect mismatch pairs. If a mismatch is detected during meiosis, the enzymes that "proofread" new DNA strands during DNA replication correct it. The mismatched nucleotide in one of the homologues is excised and replaced with a nucleotide complementary to the one in the other homologue. Its base-pairing partner in the first homologue is then replaced, producing two chromosomes with the same sequence. This error correction causes one of the mismatched sequences to convert into the other, a process called **gene conversion.**

Unequal crossing over is a crossover between chromosomal regions that are similar in nucleotide sequence but are not homologous. Gene conversion is the alteration of one homologue by the cell's error-detection and repair system to make it resemble the other homologue.

Trinucleotide Repeats

In 1991, a new kind of change in the genetic material was reported, one that involved neither changes in the identity of nucleotides (mutation) nor changes in the position of nucleotide sequences (recombination), but rather an increase in the number of copies of repeated trinucleotide sequences. Called **trinucleotide repeats,** these changes appear to be the root cause of a surprisingly large number of inherited human disorders.

The first examples of disorders resulting from the expansion of trinucleotide repeat sequences were reported in individuals with *fragile X syndrome* (the most common form of developmental disorder) and *spinal muscular atrophy*. In both disorders, genes containing runs of repeated nucleotide triplets (CGG in fragile X syndrome and CAG in spinal muscular atrophy) exhibit large increases in copy number. In individuals with fragile X syndrome, for example, the CGG sequence is repeated hundreds of times (figure 18.31), whereas in normal individuals it repeats only about 30 times.

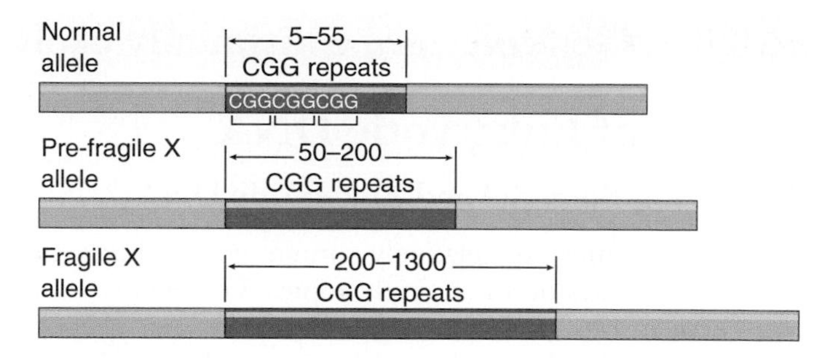

FIGURE 18.31
CGG repeats in fragile X alleles. The CGG triplet is repeated approximately 30 times in normal alleles. Individuals with pre-fragile X alleles show no detectable signs of the syndrome but do have increased numbers of CGG repeats. In fragile X alleles, the CGG triplet repeats hundreds of times.

Ten additional human genes are now known to have alleles with expanded trinucleotide repeats (figure 18.32). Many (but not all) of these alleles are GC-rich. A few of the alleles appear benign, but most are associated with heritable disorders, including Huntington's disease, myotonic dystrophy, and a variety of neurological ataxias. In each case, the expansion transmits as a dominant trait. Often the repeats are found within the exons of their genes, but sometimes, as in the case of fragile X syndrome where the repeats appear in an untranslated leader region of the first exon, they are located outside the coding segment. Furthermore, although the repeat number is stably transmitted in normal families, it shows marked instability once it has abnormally expanded. Siblings often exhibit unique repeat lengths.

As the repeat number increases, disease severity tends to increase in step. In fragile X syndrome, the CGG triplet number first increases from the normal stable range of 5 to 55 times (the most common allele has 29 repeats) to an unstable number of repeats ranging from 50 to 200, with no detectable effect. In offspring, the number increases markedly, with copy numbers ranging from 200 to 1300, with significant mental retardation (see figure 18.31). Similarly, the normal allele for myotonic dystrophy has 5 GTC repeats. Mildly affected individuals have about 50, and severely affected individuals have up to 1000.

Trinucleotide repeats appear common in human genes, but their function is unknown. Nor do we know the mechanism behind trinucleotide repeat expansion. It may involve unequal crossing over, which can readily produce copy-number expansion, or perhaps some sort of stutter in the DNA polymerase when it encounters a run of triplets. The fact that di- and tetranucleotide repeat expansions are not found seems an important clue. Undoubtedly, further examples of this remarkable class of genetic change will be reported in the future. Considerable research is currently focused on this extremely interesting area.

Many human genes contain runs of a trinucleotide sequence. Their function is unknown, but if the copy number expands, hereditary disorders often result.

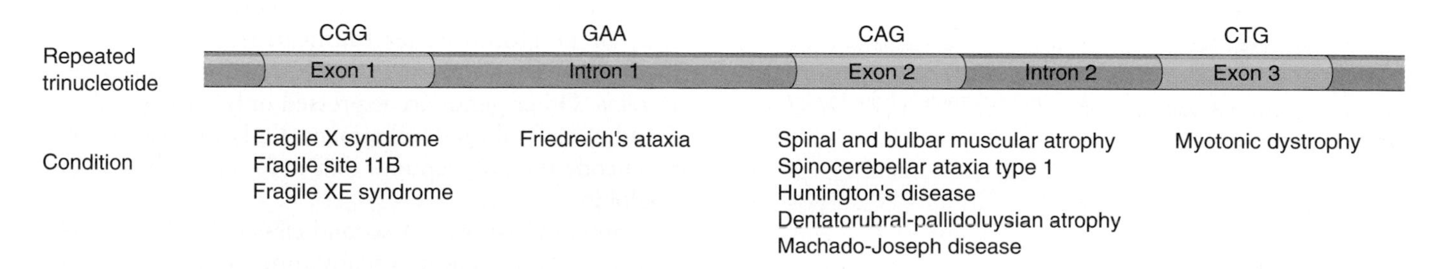

FIGURE 18.32
A hypothetical gene showing the locations and types of trinucleotide repeats associated with various human diseases. The CGG repeats of fragile X syndrome, fragile XE syndrome, and fragile site 11B occur in the first exon of their respective genes. GAA repeats characteristic of Friedreich's ataxia exist in the first intron of its gene. The genes for five different diseases, including Huntington's disease, have CAG repeats within their second exons. Lastly, the myotonic dystrophy gene contains CTG repeats within the third exon.

Classes of Eukaryotic DNA

Comparing Bacterial and Eukaryotic DNA

Bacterial genomes are relatively simple, containing genes that almost always occur as single copies. Unequal crossing over between repeated transposition elements in their circular DNA molecules tends to *delete* material, fostering the maintenance of a minimum genome size (figure 18.33*a*). For this reason, these genomes are very tightly packed, with few or no noncoding nucleotides. Recall the efficient use of space in the organization of the *lac* genes described in chapter 16.

In eukaryotes, by contrast, the introduction of *pairs* of homologous chromosomes has led to a radically different situation. Unequal crossing over between homologous chromosomes tends to promote the *duplication* of material rather than its reduction (figure 18.33*b*). Consequently, eukaryotic genomes have been in a constant state of flux during the course of their evolution. Multiple copies of genes have evolved, some of them subsequently diverging in sequence to become different genes, which in turn have duplicated and diverged.

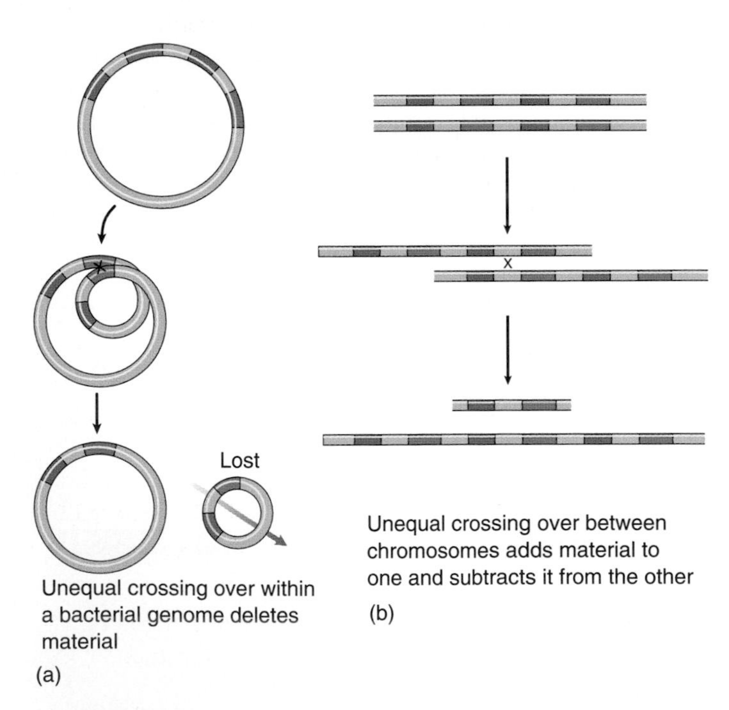
Lost
Unequal crossing over within a bacterial genome deletes material
(a)
x
Unequal crossing over between chromosomes adds material to one and subtracts it from the other
(b)

FIGURE 18.33
Unequal crossing over has different consequences in bacteria and eukaryotes. (*a*) Bacteria have a circular DNA molecule, and a crossover between duplicate regions within the molecule deletes the intervening material. (*b*) In eukaryotes, with two versions of each chromosome, crossing over adds material to one chromosome.

DNA That Codes for Proteins

Four different classes of eukaryotic genes are commonly recognized, differing largely in gene copy number.

Single-copy genes. Many eukaryotic genes exist as single copies at a particular location on a chromosome. Mutations in these genes produce recessive Mendelian inheritance. Silent copies, inactivated by mutation, are called *pseudogenes.*

Segmental duplications. Sequencing of the human genome has revealed that human chromosomes contain many *segmental duplications*, where whole blocks of genes have been copied over from one chromosome to another. Blocks of similar genes in the same order are found throughout the genome, indicating segmental duplication has been a major factor in determining the human genome's present architecture. Chromosome 19 seems to have been the biggest borrower, with blocks of genes shared with 16 other chromosomes.

Multigene families. As we have learned more about the nucleotide sequences of eukaryotic genomes, it has become apparent that many genes exist as parts of *multigene families*, groups of related but distinctly different genes that often occur together in a cluster. Multigene families differ from tandem clusters in that they contain far fewer genes (from three to several hundred), and those genes differ much more from one another than the genes in tandem clusters. Despite their differences, the genes in a multigene family are clearly related in their sequences, making it likely that they arose from a single ancestral sequence through a series of unequal crossing over events. For example, studies of the evolution of the hemoglobin multigene family indicate that the ancestral globin gene is at least 800 million years old. By the time modern fishes evolved, this ancestral gene had already duplicated, forming the α and β forms. Later, after the evolutionary divergence of amphibians and reptiles, these two globin gene forms moved apart on the chromosome; the mechanism of this movement is not known, but it may have involved transposition. In mammals, two more waves of duplication occurred to produce the array of 11 globin genes found in the human genome. Three of these genes are silent, encoding nonfunctional proteins. Other genes are expressed only during embryonic (ζ and ε) or fetal (γ) development. Only four (δ, β, α_1, and α_2) encode the polypeptides that make up adult human hemoglobin.

Tandem clusters. A second class consists of DNA sequences that are repeated many times, one copy following another in tandem array. By transcribing all of the copies in these *tandem clusters* simultaneously, a cell can rapidly obtain large amounts of the product they encode. For example, the genes encoding rRNA are typically present in clusters of several hundred copies.

Table 18.6 Classes of DNA Sequences Found in the Human Genome

Class	Frequency	Description
Protein-encoding genes	1%	Transcribed portions of the 30,000 genes scattered about the chromosomes
Introns	24%	Noncoding DNA that comprises the great majority of each human gene
Structural DNA	20%	Constitutive heterochromatin, localized near centromeres and telomeres
Repeated sequences	3%	Stuttering repeats of a few nucleotides like CGG, repeated thousands of times
Transposable elements	45%	20% - Long interspersed elements (LINEs), which are active transposons
		15% - Other less active transposable elements, like LTRs
		10% - The parasite sequence ALU, present in half a million copies

Noncoding DNA

Sequencing of several eukaryotic genomes has now been completed, and one of the most notable characteristics is the amount of noncoding DNA they possess. The recent sequencing of the human genome reveals a particularly startling picture. Each of your cells has about six feet of DNA stuffed into it, but of that, less than one inch is devoted to genes! Nearly 99% of the DNA in your cells has little or nothing to do with the instructions that make you you.

True genes are scattered about the human genome in clumps among the much larger amount of noncoding DNA, like isolated hamlets in a desert. There are four major sorts of noncoding human DNA (Table 18.6).

Noncoding DNA within genes. As discussed in chapter 15, a human gene is not simply a stretch of DNA, like the letters of a word. Instead, a human gene is made up of numerous fragments of protein-encoding information (exons) embedded within a much larger matrix of noncoding DNA (introns). Together, introns make up about 24% of the human genome, and exons about 1.5%.

Structural DNA. Some regions of the chromosomes remain highly condensed, tightly coiled, and untranscribed throughout the cell cycle. Called *constitutive heterochromatin*, these portions tend to be localized around the centromere, or located near the ends of the chromosome, at the telomeres.

Repeated sequences. Scattered about chromosomes are *simple sequence repeats* (SSRs). An SSR is a two- or three-nucleotide sequence like CA or CGG, repeated like a broken record thousands and thousands of times. SSRs make up about 3% of the human genome. An additional 7% is devoted to other sorts of duplicated sequences.

Transposable elements. Fully 45% of the human genome consists of mobile bits of DNA called transposable elements. Discovered by Barbara McClintock in 1950 (she won the Nobel Prize for her discovery in 1983), transposable elements are bits of DNA that are able to jump from one location on a chromosome to another, tiny molecular versions of Mexican jumping beans.

How do they accomplish this remarkable feat? While the details of the transposition mechanism are complex and vary for different kinds of element, the basic process is quite simple. The two ends of a transposable element have similar DNA sequences. Because the two sequences are similar, the strands of DNA from one end can pair up with the strands from the other. When its ends associate in this way, the element forms a loop. Mark a piece of string with black ink at two places, and let the black sections touch, and you will see how a loop forms. Now if a nick is made in the DNA of the loop, an enzyme called reverse transcriptase can make an RNA copy of the DNA in the loop, a copy which is free to move away and reenter the chromosome elsewhere.

Human chromosomes contain five sorts of transposable elements. Fully 20% of the genome consists of *long interspersed elements* (LINEs). An ancient and very successful element, LINEs are about 6KB (6 thousand DNA bases) long, and contain all the equipment needed for transposition, including genes for a DNA-loop-nicking enzyme and a reverse transcriptase.

Nested within the genome's LINEs are over half a million copies of a parasitic element called ALU, 10% of the human genome. ALU is only about 300 bases long, and has no transposition machinery of its own -- like a flea on a dog, ALU moves with the LINE it resides within. Just as a flea sometimes jumps to a different dog, so ALU sometimes uses the enzymes of its LINE to move to a new chromosome location. Often jumping right into genes, ALU transpositions cause many harmful mutations.

Three other sorts of transposable elements are also present in the human genome: 8% of the genome is devoted to *long terminal repeats* (LTRs), also called "retroposons." 3% is devoted to *DNA transposons*, which copy themselves as DNA rather than RNA. Some 4% is devoted to dead transposons, elements that have lost the signals for replication and so can no longer jump.

Gene sequences in eukaryotes vary greatly in copy number, some occurring many thousands of times, others only once. Only about 1% of the human genome is devoted to protein-encoding genes. Much of the rest is comprised of transposable elements.

Chapter 18

www.mhhe.com/raven6/resources18.mhtml

Summary | **Questions** | **Media Resources**

18.1 Mutations are changes in the genetic message.

- A mutation is any change in the hereditary message.
- Mutations that change one or a few nucleotides are called point mutations. They may arise as a result of damage from ionizing or ultraviolet radiation, chemical mutagens, or errors in pairing during DNA replication.

1. What are pyrimidine dimers? How do they form? How are they repaired? What may happen if they are not repaired?

2. Explain how slipped mispairing can cause deletions and frame-shift mutations.

- Mutations
- DNA Repair

- How Scientists Think: Mutations Occur at Random (Luria/ Delbrück)

18.2 Cancer results from mutation of growth-regulating genes.

- Cancer is a disease in which the regulatory controls that normally restrain cell division are disrupted.
- A variety of environmental factors, including ionizing radiation, chemical mutagens, and viruses, have been implicated in causing cancer.
- The best way to avoid getting cancer is to avoid exposure to mutagens, for example, those in cigarette smoke.

3. What is transfection? What has it revealed about the genetic basis of cancer?

4. About how many genes can be mutated to cause cancer? Why do most cancers require mutations in multiple genes?

- *On Science* Articles:
  - -Understanding Cancer
  - -Evidence Links Cigarette Smoking to Lung Cancer
  - -Deadly Cancer is Becoming More Common

18.3 Recombination alters gene location.

- Recombination is the creation of new gene combinations. It includes changes in the position of genes or fragments of genes as well as the exchange of entire chromosomes during meiosis.
- Genes may be transferred between bacteria when they are included within small circles of DNA called plasmids.
- Transposition is the random movement of genes within transposons to new locations in the genome. It is responsible for many naturally occurring mutations, as the insertion of a transposon into a gene often inactivates the gene.
- Crossing over involves a physical exchange of genetic material between homologous chromosomes during the close pairing that occurs in meiosis. It may produce chromosomes that have different combinations of alleles.

5. What is genetic recombination? What mechanisms produce it? Which of these mechanisms occurs in prokaryotes, and which occurs in eukaryotes?

6. What is a plasmid? What is a transposon? How are plasmids and transposons similar, and how are they different?

7. What are mismatched pairs? How are they corrected? What effect does this correction have on the genetic message?

- How Scientists Think: Genetic Recombination Involves Physical Exchange (McClintock/ Stern)

18.4 Genomes are continually evolving.

- There are four different classes of eukaryotic genes that differ primarily in gene copy number.
- Many eukaryotic genes exist as single-copy genes. Segmental duplications contain whole blocks of genes that have been copied. Multigene families are clusters of related but different genes. A gene copied many times in series is a tandem cluster.
- Noncoding sequences comprise a large portion of the eukaryotic genome.

8. What kinds of genes exist in multigene families? How are these families thought to have evolved?

- Student Research: DNA Repair in Fish

19

Gene Technology

Concept Outline

19.1 The ability to manipulate DNA has led to a new genetics.

Restriction Endonucleases. Enzymes that cleave DNA at specific sites allow DNA segments from different sources to be spliced together.

Using Restriction Endonucleases to Manipulate Genes. Fragments produced by cleaving DNA with restriction endonucleases can be spliced into plasmids, which can be used to insert the DNA into host cells.

19.2 Genetic engineering involves easily understood procedures.

The Four Stages of a Genetic Engineering Experiment. Gene engineers cut DNA into fragments that they splice into vectors that carry the fragments into cells.

Working with Gene Clones. Gene technology is used in a variety of procedures involving DNA manipulation.

19.3 Biotechnology is producing a scientific revolution.

DNA Sequence Technology. The complete nucleotide sequence of the genomes of many organisms are now known and can be compared.

Biochips. Biochips are squares of glass etched with DNA strands and can be used for genetic screening.

Medical Applications. Many drugs and vaccines are now produced with gene technology.

Agricultural Applications. Gene engineers have developed crops resistant to pesticides and pests, as well as commercially superior animals.

Cloning. Recent experiments show it is possible to clone agricultural animals, a result with many implications for both agriculture and society.

Stem Cells. Both embryonic stem cells and tissue-specific stem cells can potentially be used to repair or replace damaged tissue.

Ethics and Regulation. Genetic engineering raises important questions about danger and privacy.

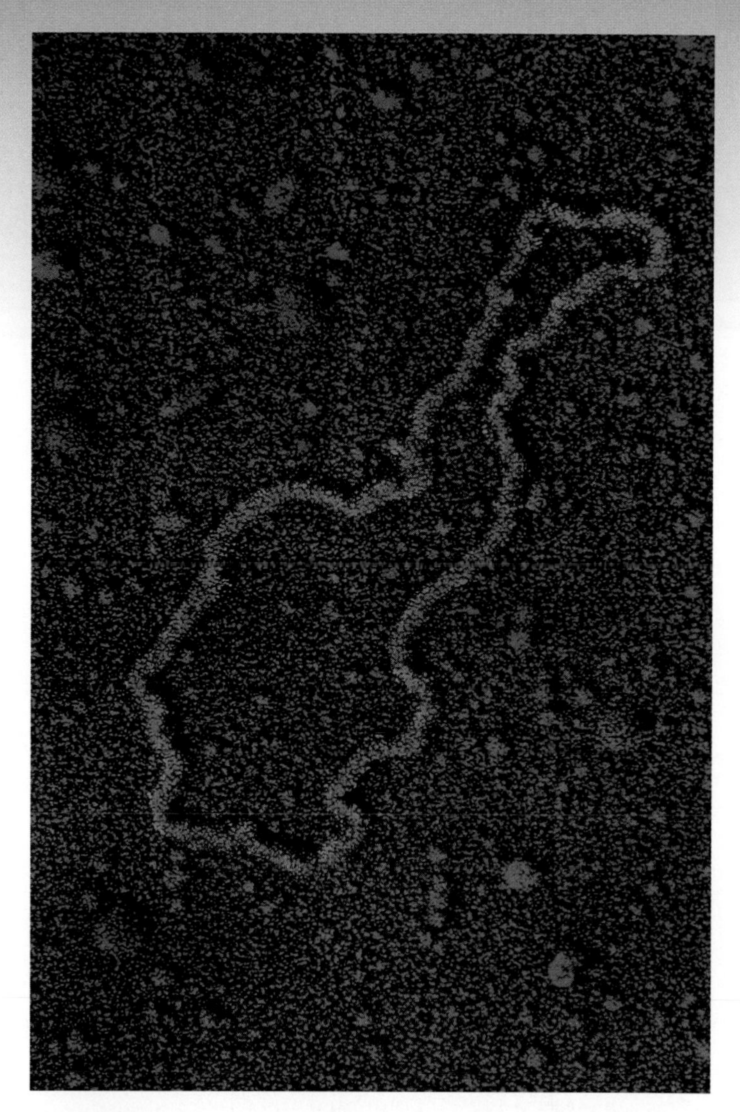

FIGURE 19.1
A famous plasmid. The circular molecule in this electron micrograph is pSC101, the first plasmid used successfully to clone a vertebrate gene. Its name comes from the fact that it was the one-hundred-and-first plasmid isolated by Stanley Cohen.

Over the past decades, the development of new and powerful techniques for studying and manipulating DNA has revolutionized genetics (figure 19.1). These techniques have allowed biologists to intervene directly in the genetic fate of organisms for the first time. In this chapter, we will explore these technologies and consider how they apply to specific problems of great practical importance. Few areas of biology will have as great an impact on our future lives.

19.1 The ability to manipulate DNA has led to a new genetics.

Restriction Endonucleases

In 1980, geneticists used the relatively new technique of gene splicing, which we will describe in this chapter, to introduce the human gene that encodes **interferon** into a bacterial cell's genome. Interferon is a rare blood protein that increases human resistance to viral infection, and medical scientists have been interested in its possible usefulness in cancer therapy. This possibility was difficult to investigate before 1980, however, because purification of the large amounts of interferon required for clinical testing would have been prohibitively expensive, given interferon's scarcity in the blood. An inexpensive way to produce interferon was needed, and introducing the gene responsible for its production into a bacterial cell made that possible. The cell that had acquired the human interferon gene proceeded to produce interferon at a rapid rate, and to grow and divide. Soon there were millions of interferon-producing bacteria in the culture, all of them descendants of the cell that had originally received the human interferon gene.

The Advent of Genetic Engineering

This procedure of producing a line of genetically identical cells from a single altered cell, called **cloning,** made every cell in the culture a miniature factory for producing interferon. The human insulin gene has also been cloned in bacteria, and now large amounts of insulin, a hormone essential for treating some forms of diabetes, can be manufactured at relatively little expense. Beyond these clinical applications, cloning and related molecular techniques are used to obtain basic information about how genes are put together and regulated. The interferon experiment and others like it marked the beginning of a new genetics, **genetic engineering.**

The essence of genetic engineering is the ability to cut DNA into recognizable pieces and rearrange those pieces in different ways. In the interferon experiment, a piece of DNA carrying the interferon gene was inserted into a plasmid, which then carried the gene into a bacterial cell. Most other genetic engineering approaches have used the same general strategy, bringing the gene of interest into the target cell by first incorporating it into a plasmid or an infective virus. To make these experiments work, one must be able to cut the source DNA (human DNA in the interferon experiment, for example) and the plasmid DNA in such a way that the desired fragment of source DNA can be spliced permanently into the plasmid. This cutting is performed by enzymes that recognize and cleave specific sequences of nucleotides in DNA. These enzymes are the basic tools of genetic engineering.

Discovery of Restriction Endonucleases

Scientific discoveries often have their origins in seemingly unimportant observations that receive little attention by researchers before their general significance is appreciated. In the case of genetic engineering, the original observation was that bacteria use enzymes to defend themselves against viruses.

Most organisms eventually evolve means of defending themselves from predators and parasites, and bacteria are no exception. Among the natural enemies of bacteria are bacteriophages, viruses that infect bacteria and multiply within them. At some point, they cause the bacterial cells to burst, releasing thousands more viruses. Through natural selection, some types of bacteria have acquired powerful weapons against these viruses: they contain enzymes called **restriction endonucleases** that fragment the viral DNA as soon as it enters the bacterial cell. Many restriction endonucleases recognize specific nucleotide sequences in a DNA strand, bind to the DNA at those sequences, and cleave the DNA at a particular place within the recognition sequence.

Why don't restriction endonucleases cleave the bacterial cells' own DNA as well as that of the viruses? The answer to this question is that bacteria modify their own DNA, using other enzymes known as **methylases** to add methyl (—CH_3) groups to some of the nucleotides in the bacterial DNA. When nucleotides within a restriction endonuclease's recognition sequence have been methylated, the endonuclease cannot bind to that sequence. Consequently, the bacterial DNA is protected from being degraded at that site. Viral DNA, on the other hand, has not been methylated and therefore is not protected from enzymatic cleavage.

How Restriction Endonucleases Cut DNA

The sequences recognized by restriction endonucleases are typically four to six nucleotides long, and they are often palindromes. This means the nucleotides at one end of the recognition sequence are complementary to those at the other end, so that the two strands of the DNA duplex have the same nucleotide sequence running in opposite directions for the length of the recognition sequence. Two important consequences arise from this arrangement of nucleotides.

First, because the same recognition sequence occurs on both strands of the DNA duplex, the restriction endonuclease can bind to and cleave both strands, effectively cutting the DNA in half. This ability to cut across both strands is almost certainly the reason that restriction endonucleases have evolved to recognize nucleotide sequences with twofold rotational symmetry.

Second, because the bond cleaved by a restriction endonuclease is typically not positioned in the center of the recognition sequence to which it binds, and because the DNA strands are antiparallel, the cut sites for the two strands of a duplex are offset from each other (figure 19.2). After cleavage, each DNA fragment has a single-stranded end a few nucleotides long. The single-stranded ends of the two fragments are complementary to each other.

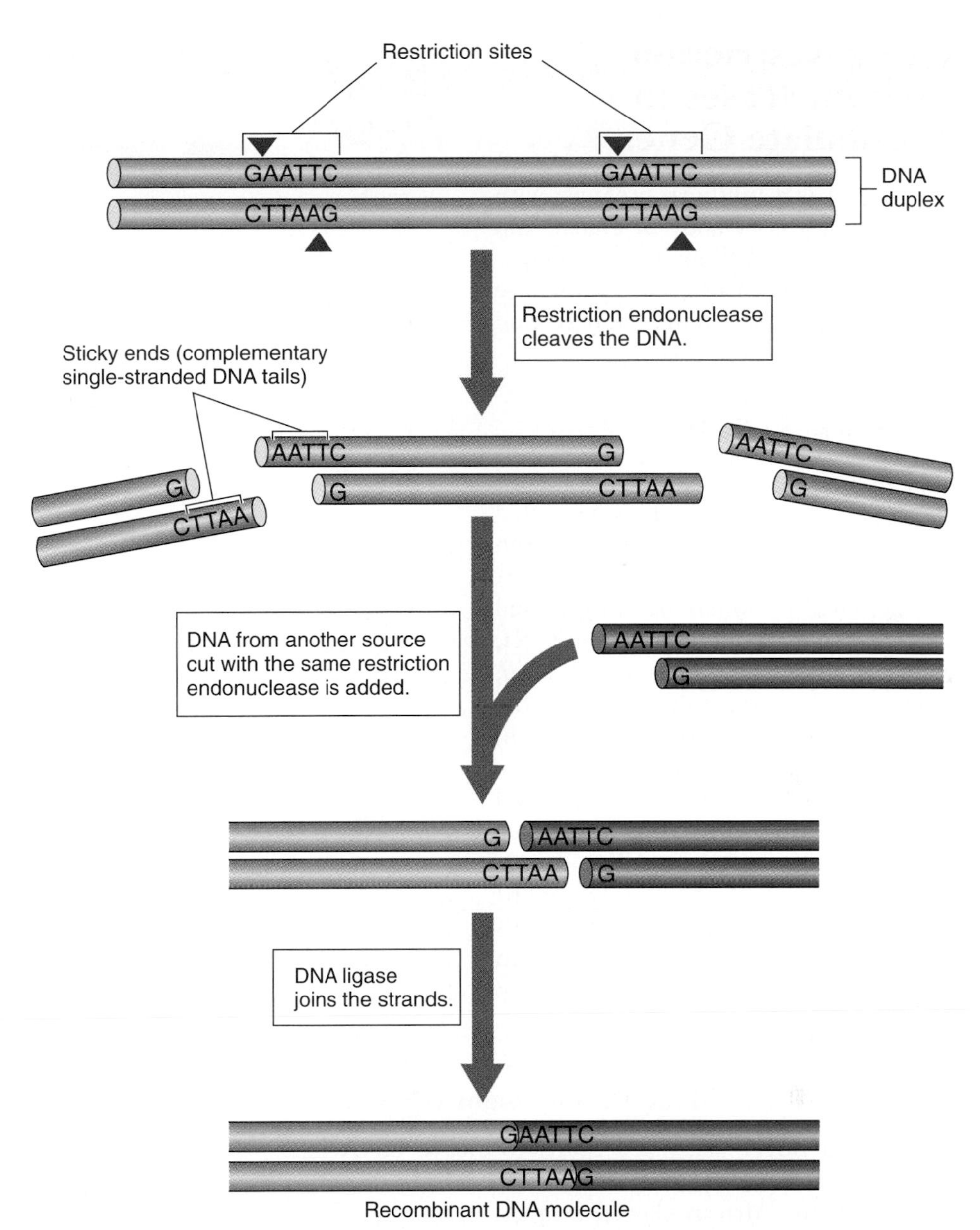

FIGURE 19.2
Many restriction endonucleases produce DNA fragments with "sticky ends." The restriction endonuclease *Eco*RI always cleaves the sequence GAATTC between G and A. Because the same sequence occurs on both strands, both are cut. However, the two sequences run in opposite directions on the two strands. As a result, single-stranded tails are produced that are complementary to each other, or "sticky."

Why Restriction Endonucleases Are So Useful

There are hundreds of bacterial restriction endonucleases, and each one has a specific recognition sequence. By chance, a particular endonuclease's recognition sequence is likely to occur somewhere in any given sample of DNA; the shorter the sequence, the more often it will arise by chance within a sample. Therefore, a given restriction endonuclease can probably cut DNA from any source into fragments. Each fragment will have complementary single-stranded ends characteristic of that endonuclease. Because of their complementarity, these single-stranded ends can pair with each other (consequently, they are sometimes called "sticky ends"). Once their ends have paired, two fragments can then be joined together with the aid of the enzyme **DNA ligase,** which re-forms the phosphodiester bonds of DNA. What makes restriction endonucleases so valuable for genetic engineering is the fact that *any* two fragments produced by the same restriction endonuclease can be joined together. Fragments of elephant and ostrich DNA cleaved by the same endonuclease can be joined to one another as readily as two bacterial DNA fragments.

Genetic engineering involves manipulating specific genes by cutting and rearranging DNA. A restriction endonuclease cleaves DNA at a specific site, generating in most cases two fragments with short single-stranded ends. Because these ends are complementary to each other, any pair of fragments produced by the same endonuclease, from any DNA source, can be joined together.

Using Restriction Endonucleases to Manipulate Genes

A **chimera** is a mythical creature with the head of a lion, body of a goat, and tail of a serpent. Although no such creatures existed in nature, biologists have made chimeras of a more modest kind through genetic engineering.

Constructing pSC101

One of the first chimeras was manufactured from a bacterial plasmid called a resistance transfer factor by American geneticists Stanley Cohen and Herbert Boyer in 1973. Cohen and Boyer used a restriction endonuclease called *Eco*RI, which is obtained from *Escherichia coli,* to cut the plasmid into fragments. One fragment, 9000 nucleotides in length, contained both the origin of replication necessary for replicating the plasmid and a gene that conferred resistance to the antibiotic tetracycline (tet^r). Because both ends of this fragment were cut by the same restriction endonuclease, they could be ligated to form a circle, a smaller plasmid Cohen dubbed pSC101.

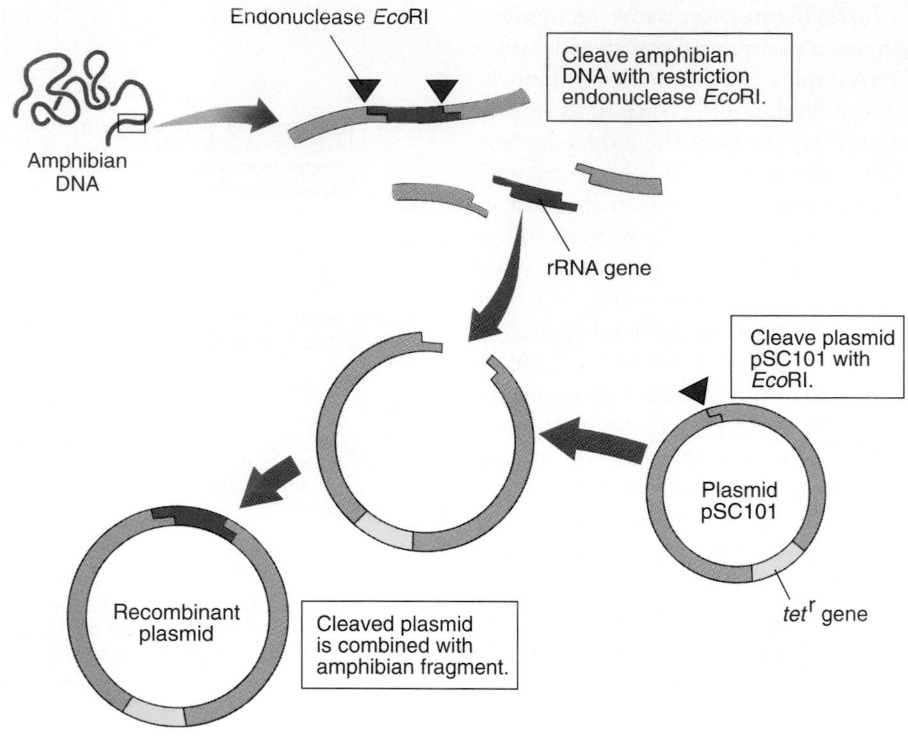

FIGURE 19.3
One of the first genetic engineering experiments. This diagram illustrates how Cohen and Boyer inserted an amphibian gene encoding rRNA into pSC101. The plasmid contains a single site cleaved by the restriction endonuclease *Eco*RI; it also contains tet^r, a gene which confers resistance to the antibiotic tetracycline. The rRNA-encoding gene was inserted into pSC101 by cleaving the amphibian DNA and the plasmid with *Eco*RI and allowing the complementary sequences to pair.

Using pSC101 to Make Recombinant DNA

Cohen and Boyer also used *Eco*RI to cleave DNA that coded for rRNA that they had isolated from an adult amphibian, the African clawed frog, *Xenopus laevis.* They then mixed the fragments of *Xenopus* DNA with pSC101 plasmids that had been "reopened" by *Eco*RI and allowed bacterial cells to take up DNA from the mixture (figure 19.3). Some of the bacterial cells immediately became resistant to tetracycline, indicating that they had incorporated the pSC101 plasmid with its antibiotic-resistance gene. Furthermore, some of these pSC101-containing bacteria also began to produce frog ribosomal RNA! Cohen and Boyer concluded that the frog rRNA gene must have been inserted into the pSC101 plasmids in those bacteria. In other words, the two ends of the pSC101 plasmid, produced by cleavage with *Eco*RI, had joined to the two ends of a frog DNA fragment that contained the rRNA gene, also cleaved with *Eco*RI.

The pSC101 plasmid containing the frog rRNA gene is a true chimera, an entirely new genome that never existed in nature and never would have evolved by natural means. It is a form of **recombinant DNA**—that is, DNA created in the laboratory by joining together pieces of different genomes to form a novel combination.

Other Vectors

The introduction of foreign DNA fragments into host cells has become common in molecular genetics. The genome that carries the foreign DNA into the host cell is called a **vector.** Plasmids, with names like pUC18 can be induced to make hundreds of copies of themselves and thus of the foreign genes they contain. Much larger pieces of DNA can be introduced using YAKs (yeast artificial chromosomes) as a vector instead of a plasmid. Not all vectors have bacterial targets. Animal viruses such as the human cold virus adenovirus, for example, are serving as vectors to carry genes into monkey and human cells, and animal genes have even been introduced into plant cells.

One of the first recombinant genomes produced by genetic engineering was a bacterial plasmid into which an amphibian ribosomal RNA gene was inserted. Viruses can also be used as vectors to insert foreign DNA into host cells and create recombinant genomes.

Examples of Gene Manipulation

HERMAN THE WONDER BULL

GenPharm, a California biotechnology company, engineered Herman, a bull that possesses the gene for human lactoferrin (HLF). HLF confers antibacterial and iron transport properties to humans. Many of Herman's female offspring now produce milk containing HLF, and GenPharm intends to build a herd of transgenic cows for the large-scale commercial production of HLF.

WILT-PROOF FLOWERS

Ethylene, the plant hormone that causes fruit to ripen, also causes flowers to wilt. Researchers at Purdue have found the gene that makes flower petals respond to ethylene by wilting and replaced it with a gene insensitive to ethylene. The transgenic carnations they produced lasted for 3 weeks after cutting, while normal carnations last only 3 days.

SUPER SALMON!

Canadian fisheries scientists have inserted recombinant growth hormone genes into developing salmon embryos, creating the first transgenic salmon. Not only do these transgenic fish have shortened production cycles, they are, on an average, 11 times heavier than nontransgenic salmon! The implications for the fisheries industry and for worldwide food production are obvious.

WEEVIL-PROOF PEAS

Not only has gene technology afforded agriculture viral and pest control in the field, it has also provided a pest control technique for the storage bin. A team of U.S. and Australian scientists have engineered a gene that is expressed only in the seed of the pea plant. The enzyme inhibitor encoded by this gene inhibits feeding by weevils, one of the most notorious pests affecting stored crops. The worldwide ramifications are significant as up to 40% of stored grains are lost to pests.

The Four Stages of a Genetic Engineering Experiment

Like the experiment of Cohen and Boyer, most genetic engineering experiments consist of four stages: DNA cleavage, production of recombinant DNA, cloning, and screening.

Stage 1: DNA Cleavage

A restriction endonuclease is used to cleave the source DNA into fragments. Because the endonuclease's recognition sequence is likely to occur many times within the source DNA, cleavage will produce a large number of different fragments. A different set of fragments will be obtained by employing endonucleases that recognize different sequences. The fragments can be separated from one another according to their size by electrophoresis (figure 19.4).

Stage 2: Production of Recombinant DNA

The fragments of DNA are inserted into plasmids or viral vectors, which have been cleaved with the same restriction endonuclease as the source DNA.

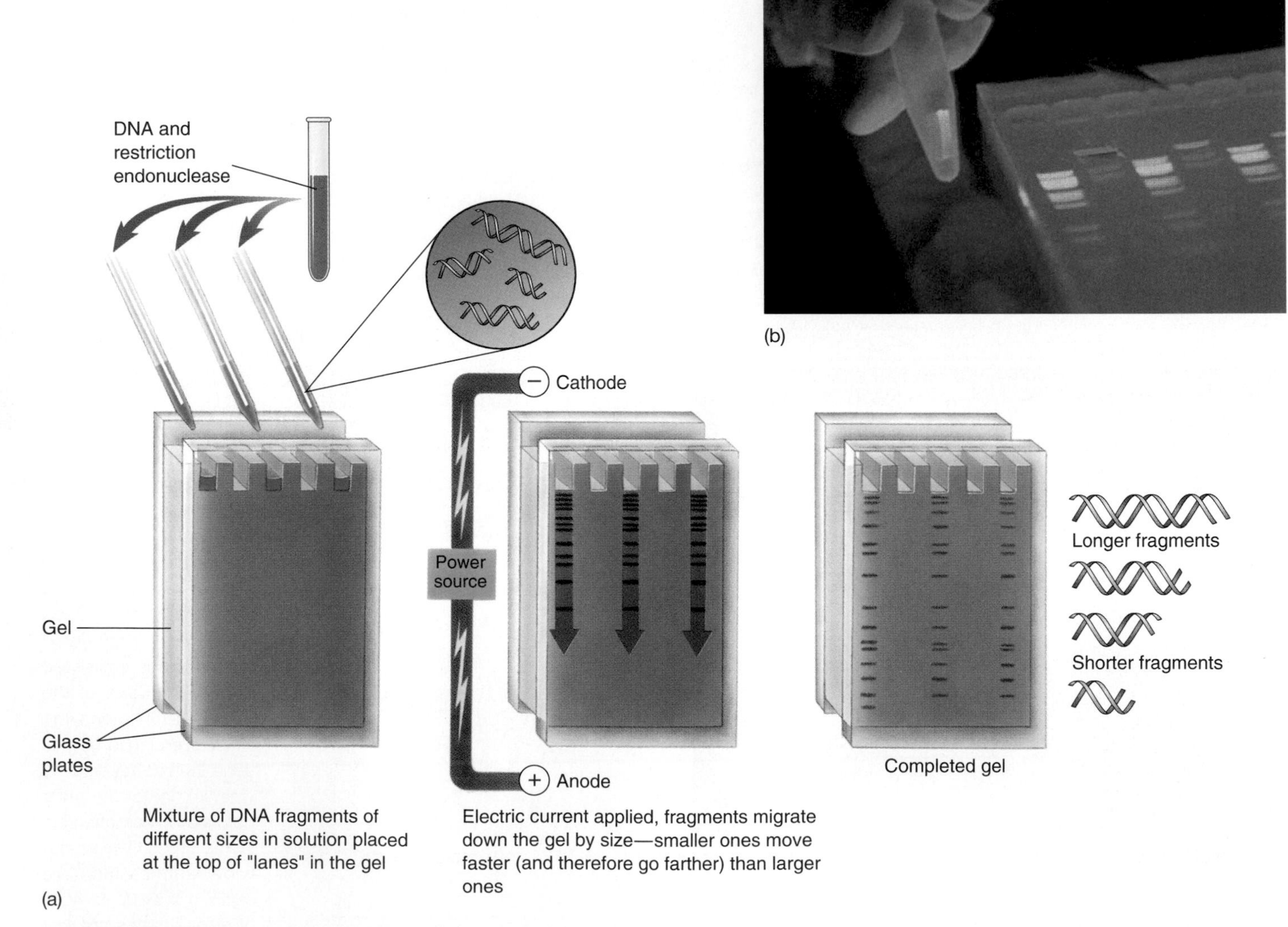

FIGURE 19.4

Gel electrophoresis. (*a*) After restriction endonucleases have cleaved the DNA, the fragments are loaded on a gel, and an electric current is applied. The DNA fragments migrate through the gel, with bigger ones moving more slowly. The fragments can be visualized easily, as the migrating bands fluoresce in UV light when stained with ethidium bromide. (*b*) In the photograph, one band of DNA has been excised from the gel for further analysis and can be seen glowing in the tube the technician holds.

Stage 3: Cloning

The plasmids or viruses serve as vectors that can introduce the DNA fragments into cells—usually, but not always, bacteria (figure 19.5). As each cell reproduces, it forms a clone of cells that all contain the fragment-bearing vector. Each clone is maintained separately, and all of them together constitute a clone library of the original source DNA.

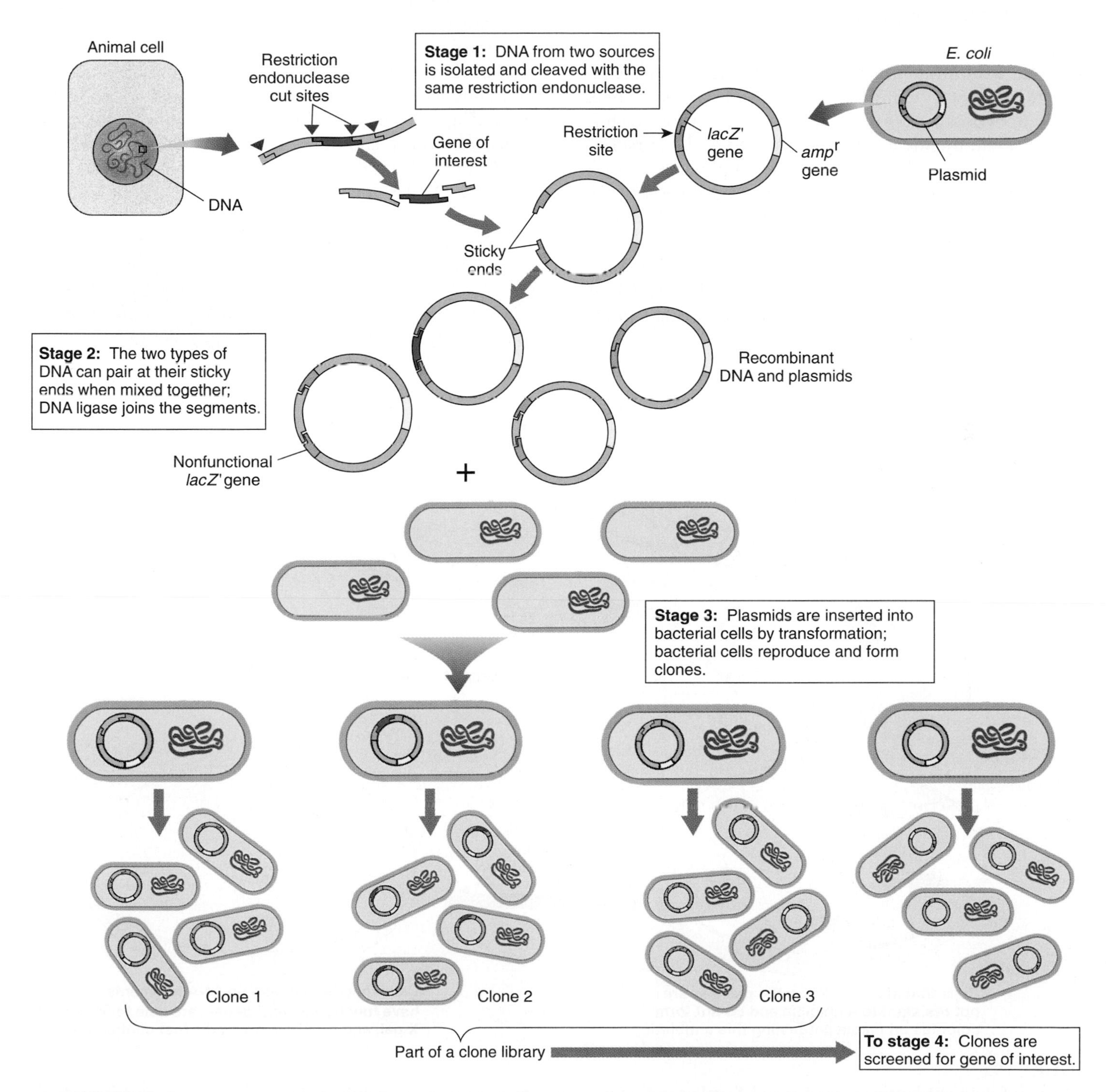

FIGURE 19.5

Stages in a genetic engineering experiment. In stage 1, DNA containing the gene of interest (in this case, from an animal cell) and DNA from a plasmid are cleaved with the same restriction endonuclease. The genes *amp*^{r} and *lacZ'* are contained within the plasmid and used for screening a clone (stage 4). In stage 2, the two cleaved sources of DNA are mixed together and pair at their sticky ends. In stage 3, the recombinant DNA is inserted into a bacterial cell, which reproduces and forms clones. In stage 4, the bacterial clones will be screened for the gene of interest.

Stage 4: Screening

The clones containing a specific DNA fragment of interest, often a fragment that includes a particular gene, are identified from the clone library. Let's examine this stage in more detail, as it is generally the most challenging in any genetic engineering experiment.

4–I: The Preliminary Screening of Clones. Investigators initially try to eliminate from the library any clones that do not contain vectors, as well as clones whose vectors do not contain fragments of the source DNA. The first category of clones can be eliminated by employing a vector with a gene that confers resistance to a specific antibiotic, such as tetracycline, penicillin, or ampicillin. In figure 19.6*a*, the gene *amp*^{r} is incorporated into the plasmid and confers resistance to the antibiotic ampicillin. When the clones are exposed to a medium containing that antibiotic, only clones that contain the vector will be resistant to the antibiotic and able to grow.

One way to eliminate clones with vectors that do not have an inserted DNA fragment is to use a vector that, in addition to containing antibiotic resistance genes, contains the *lacZ'* gene which is required to produce β-galactosidase, an enzyme that enables the cells to metabolize the sugar, X-gal. Metabolism of X-gal results in the formation of a blue reaction product, so any cells whose vectors contain a functional version of this gene will turn blue in the presence of X-gal (figure 19.6*b*). However, if one uses a restriction endonuclease whose recognition sequence lies within the *lacZ'* gene, the gene will be interrupted when recombinants are formed, and the cell will be unable to metabolize X-gal. Therefore, cells with vectors that contain a fragment of source DNA should remain colorless in the presence of X-gal.

Any cells that are able to grow in a medium containing the antibiotic but don't turn blue in the medium with X-gal must have incorporated a vector with a fragment of source DNA. Identifying cells that have a *specific* fragment of the source DNA is the next step in screening clones.

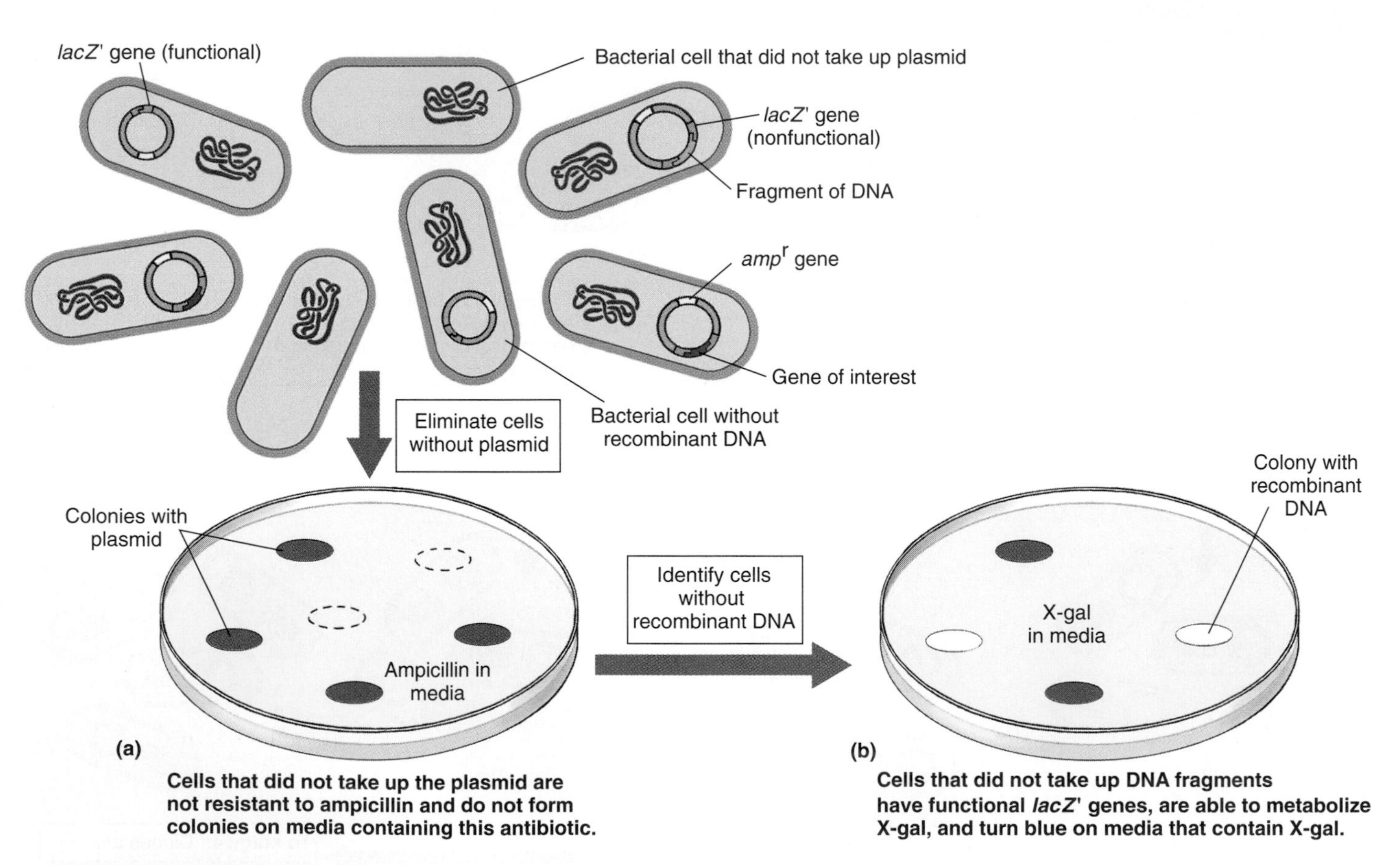

lacZ' gene (functional)
Bacterial cell that did not take up plasmid
lacZ' gene (nonfunctional)
Fragment of DNA
amp^r gene
Gene of interest
Eliminate cells without plasmid
Bacterial cell without recombinant DNA
Colonies with plasmid
Ampicillin in media
Identify cells without recombinant DNA
Colony with recombinant DNA
X-gal in media
(a)
Cells that did not take up the plasmid are not resistant to ampicillin and do not form colonies on media containing this antibiotic.
(b)
Cells that did not take up DNA fragments have functional lacZ' genes, are able to metabolize X-gal, and turn blue on media that contain X-gal.

FIGURE 19.6
Stage 4-I: Using antibiotic resistance and X-gal as preliminary screens of restriction fragment clones. Bacteria are transformed with recombinant plasmids that contain a gene (*amp*^{r}) that confers resistance to the antibiotic ampicillin and a gene (*lacZ'*) that is required to produce β-galactosidase, the enzyme which enables the cells to metabolize the sugar X-gal. (*a*) Only those bacteria that have incorporated a plasmid will be resistant to ampicillin and will grow on a medium that contains the antibiotic. (*b*) Ampicillin-resistant bacteria will be able to metabolize X-gal if their plasmid does *not* contain a DNA fragment inserted in the *lacZ'* gene; such bacteria will turn blue when grown on a medium containing X-gal. Bacteria with a plasmid that has a DNA fragment inserted within the *lacZ'* gene will not be able to metabolize X-gal and, therefore, will remain colorless in the presence of X-gal.

4–II: Finding the Gene of Interest. A clone library may contain anywhere from a few dozen to many thousand individual fragments of source DNA. Many of those fragments will be identical, so to assemble a complete library of the entire source genome, several hundred thousand clones could be required. A complete *Drosophila* (fruit fly) library, for example, contains more than 40,000 different clones; a complete human library consisting of fragments 20 kilobases long would require close to a million clones. To search such an immense library for a clone that contains a fragment corresponding to a particular gene requires ingenuity, but many different approaches have been successful.

The most general procedure for screening clone libraries to find a particular gene is **hybridization** (figure 19.7). In this method, the cloned genes form base-pairs with complementary sequences on another nucleic acid. The complementary nucleic acid is called a **probe** because it is used to probe for the presence of the gene of interest. At least part of the nucleotide sequence of the gene of interest must be known to be able to construct the probe.

In this method of screening, bacterial colonies containing an inserted gene are grown on agar. Some cells are transferred to a filter pressed onto the colonies, forming a replica of the plate. The filter is then treated with a solution that denatures the bacterial DNA and that contains a radioactively labeled probe. The probe hybridizes with complementary single-stranded sequences on the bacterial DNA.

When the filter is laid over photographic film, areas that contain radioactivity will expose the film (autoradiography). Only colonies which contain the gene of interest hybridize with the radioactive probe and emit radioactivity onto the film. The pattern on the film is then compared to the original master plate, and the gene-containing colonies may be identified.

Genetic engineering generally involves four stages: cleaving the source DNA; making recombinants; cloning copies of the recombinants; and screening the cloned copies for the desired gene. Screening can be achieved by making the desired clones resistant to certain antibiotics and giving them other properties that make them readily identifiable.

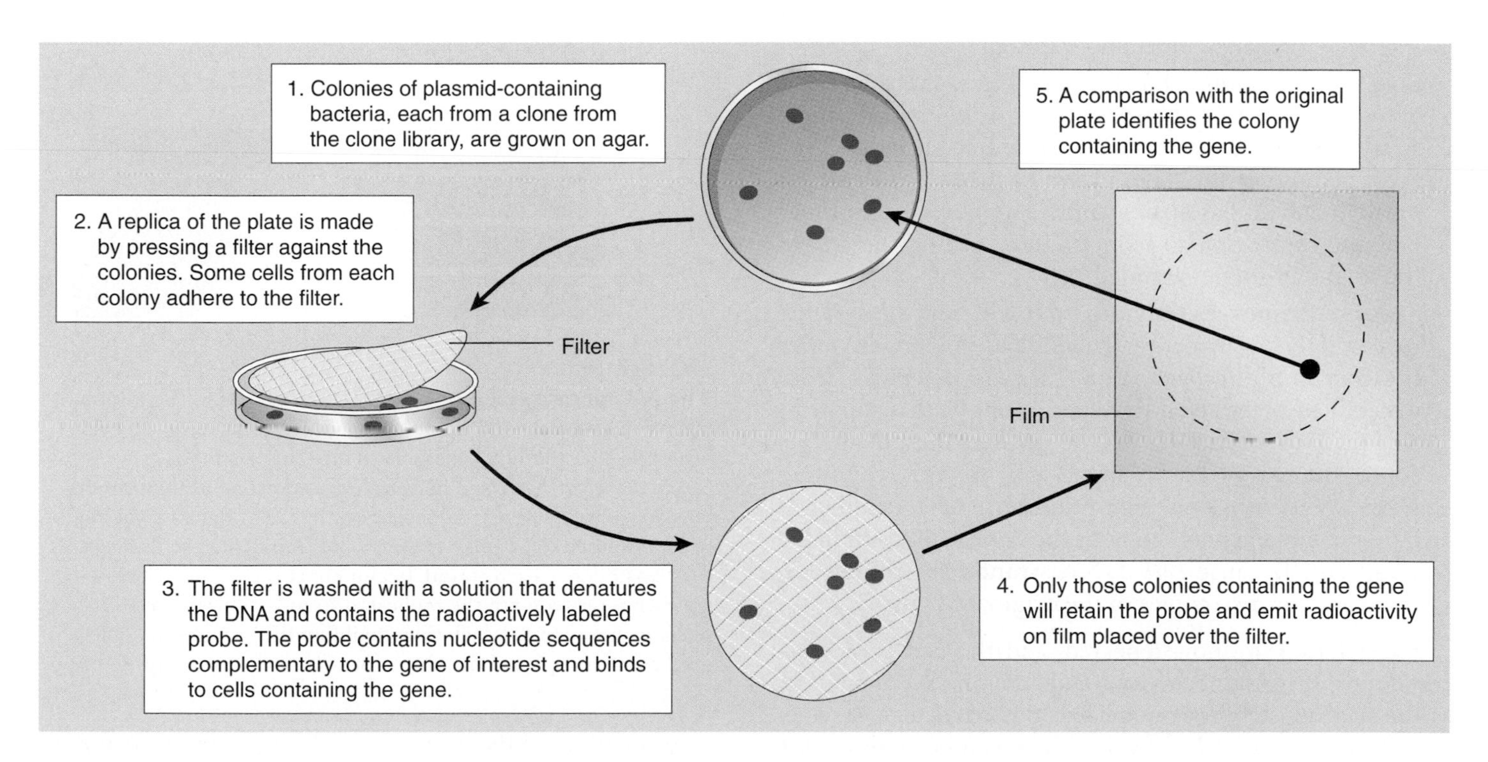

1. Colonies of plasmid-containing bacteria, each from a clone from the clone library, are grown on agar.
2. A replica of the plate is made by pressing a filter against the colonies. Some cells from each colony adhere to the filter.
Filter
3. The filter is washed with a solution that denatures the DNA and contains the radioactively labeled probe. The probe contains nucleotide sequences complementary to the gene of interest and binds to cells containing the gene.
4. Only those colonies containing the gene will retain the probe and emit radioactivity on film placed over the filter.
Film
5. A comparison with the original plate identifies the colony containing the gene.

FIGURE 19.7
Stage 4-II: Using hybridization to identify the gene of interest. (*1*) Each of the colonies on these bacterial culture plates represents millions of clones descended from a single cell. To test whether a certain gene is present in any particular clone, it is necessary to identify colonies whose cells contain DNA that hybridizes with a probe containing DNA sequences complementary to the gene. (*2*) Pressing a filter against the master plate causes some cells from each colony to adhere to the filter. (*3*) The filter is then washed with a solution that denatures the DNA and contains the radioactively labeled probe. (*4*) Only those colonies that contain DNA that hybridizes with the probe, and thus contain the gene of interest, will expose film in autoradiography. (*5*) The film is then compared to the master plate to identify the gene-containing colony.

Working with Gene Clones

Once a gene has been successfully cloned, a variety of procedures are available to characterize it.

Getting Enough DNA to Work with: The Polymerase Chain Reaction

Once a particular gene is identified within the library of DNA fragments, the final requirement is to make multiple copies of it. One way to do this is to insert the identified fragment into a bacterium; after repeated cell divisions, millions of cells will contain copies of the fragment. A far more direct approach, however, is to use DNA polymerase to copy the gene sequence of interest through the **polymerase chain reaction (PCR;** figure 19.8). Kary Mullis developed PCR in 1983 while he was a staff chemist at the Cetus Corporation; in 1993, it won him the Nobel Prize in Chemistry. PCR can amplify specific sequences or add sequences (such as endonuclease recognition sequences) as primers to cloned DNA. There are three steps in PCR:

Step 1: Denaturation. First, an excess of primer (typically a synthetic sequence of 20 to 30 nucleotides) is mixed with the DNA fragment to be amplified. This mixture of primer and fragment is heated to about 98° C. At this temperature, the double-stranded DNA fragment dissociates into single strands.

Step 2: Annealing of Primers. Next, the solution is allowed to cool to about 60°C. As it cools, the single strands of DNA reassociate into double strands. However, because of the large excess of primer, each strand of the fragment base-pairs with a complementary primer flanking the region to be amplified, leaving the rest of the fragment single-stranded.

Step 3: Primer Extension. Now a very heat-stable type of DNA polymerase, called Taq polymerase (after the thermophilic bacterium *Thermus aquaticus*, from which Taq is extracted) is added, along with a supply of all four nucleotides. Using the primer, the polymerase copies the rest of the fragment as if it were replicating DNA. When it is done, the primer has been lengthened into a complementary copy of the entire single-stranded fragment. Because *both* DNA strands are replicated, there are now two copies of the original fragment.

Steps 1 to 3 are now repeated, and the two copies become four. It is not necessary to add any more polymerase, as the heating step does not harm this particular enzyme. Each heating and cooling cycle, which can be as short as 1 or 2 minutes, doubles the number of DNA molecules. After 20 cycles, a single fragment produces more than one million (2^{20}) copies! In a few hours, 100 billion copies of the fragment can be manufactured.

PCR, now fully automated, has revolutionized many aspects of science and medicine because it allows the investigation of minute samples of DNA. In criminal investigations, "DNA fingerprints" are prepared from the cells in a tiny speck of dried blood or at the base of a single human hair. Physicians can detect genetic defects in very early embryos by collecting a few sloughed-off cells and amplifying their DNA. PCR could also be used to examine the DNA of historical figures such as Abraham Lincoln and of now-extinct species, as long as even a minuscule amount of their DNA remains intact.

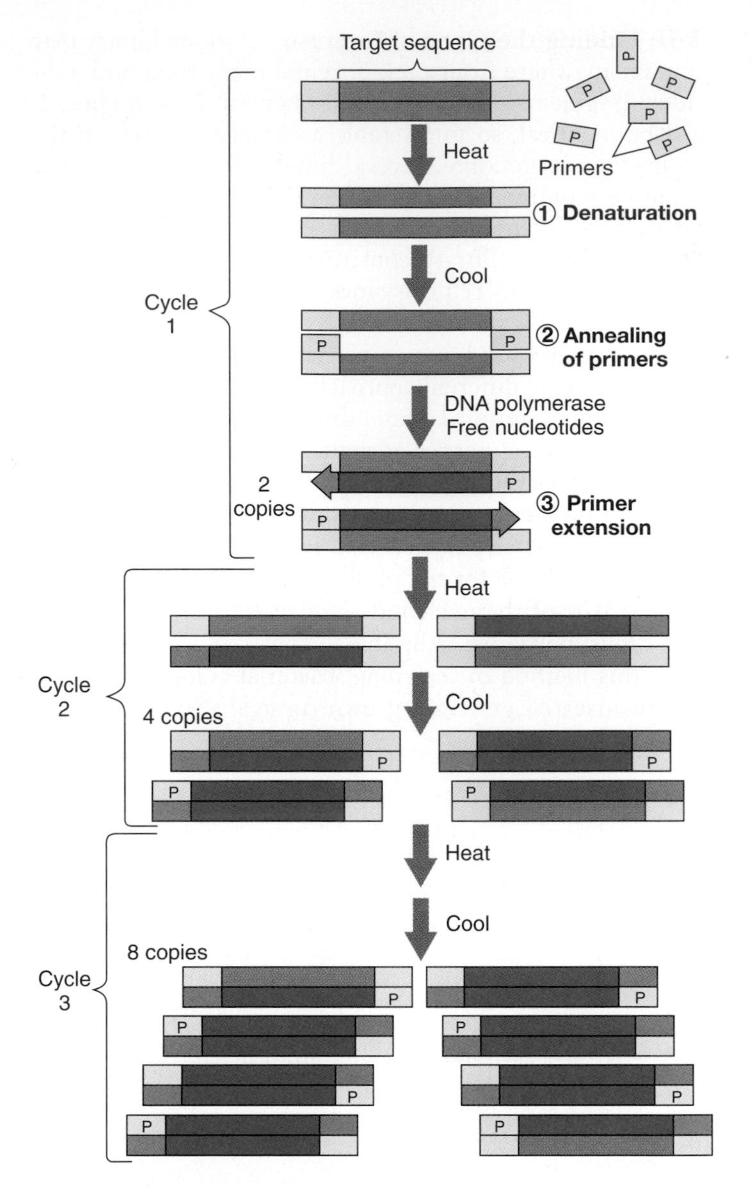

FIGURE 19.8
The polymerase chain reaction. *(1) Denaturation.* A solution containing primers and the DNA fragment to be amplified is heated so that the DNA dissociates into single strands. *(2) Annealing of primers.* The solution is cooled, and the primers bind to complementary sequences on the DNA flanking the region to be amplified. *(3) Primer extension.* DNA polymerase then copies the remainder of each strand, beginning at the primer. Steps 1–3 are then repeated with the replicated strands. This process is repeated many times, each time doubling the number of copies, until enough copies of the DNA fragment exist for analysis.

Identifying DNA: Southern Blotting

Once a gene has been cloned, it may be used as a probe to identify the same or a similar gene in another sample (figure 19.9). In this procedure, called a **Southern blot,** DNA from the sample is cleaved into restriction fragments with a restriction endonuclease, and the fragments are spread apart by gel electrophoresis. The double-stranded helix of each DNA fragment is then denatured into single strands by making the pH of the gel basic, and the gel is "blotted" with a sheet of nitrocellulose, transferring some of the DNA strands to the sheet. Next, a probe consisting of purified, single-stranded DNA corresponding to a specific gene (or mRNA transcribed from that gene) is poured over the sheet. Any fragment that has a nucleotide sequence complementary to the probe's sequence will hybridize (base-pair) with the probe. If the probe has been labeled with ^{32}P, it will be radioactive, and the sheet will show a band of radioactivity where the probe hybridized with the complementary fragment.

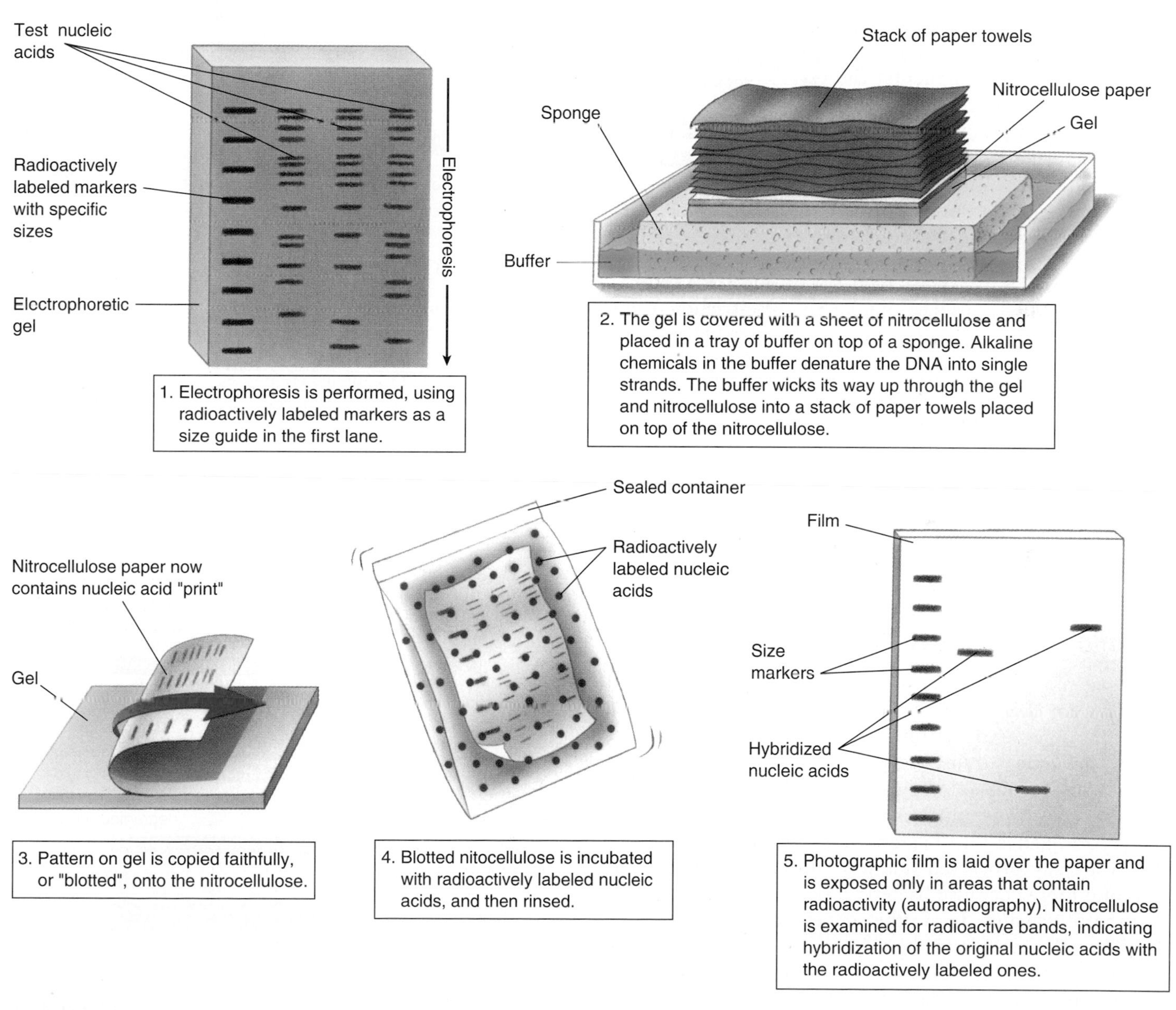

Test nucleic acids
Radioactively labeled markers with specific sizes
Elcctrophoretic gel
Electrophoresis
1. Electrophoresis is performed, using radioactively labeled markers as a size guide in the first lane.
Stack of paper towels
Nitrocellulose paper
Gel
Sponge
Buffer
2. The gel is covered with a sheet of nitrocellulose and placed in a tray of buffer on top of a sponge. Alkaline chemicals in the buffer denature the DNA into single strands. The buffer wicks its way up through the gel and nitrocellulose into a stack of paper towels placed on top of the nitrocellulose.
Nitrocellulose paper now contains nucleic acid "print"
Gel
3. Pattern on gel is copied faithfully, or "blotted", onto the nitrocellulose.
Sealed container
Radioactively labeled nucleic acids
4. Blotted nitocellulose is incubated with radioactively labeled nucleic acids, and then rinsed.
Film
Size markers
Hybridized nucleic acids
5. Photographic film is laid over the paper and is exposed only in areas that contain radioactivity (autoradiography). Nitrocellulose is examined for radioactive bands, indicating hybridization of the original nucleic acids with the radioactively labeled ones.

FIGURE 19.9
The Southern blot procedure. E. M. Southern developed this procedure in 1975 to enable DNA fragments of interest to be visualized in a complex sample containing many other fragments of similar size. The DNA is separated on a gel, then transferred ("blotted") onto a solid support medium such as nitrocellulose paper or a nylon membrane. It is then incubated with a radioactive single-strand copy of the gene of interest, which hybridizes to the blot at the location(s) where there is a fragment with a complementary sequence. The positions of radioactive bands on the blot identify the fragments of interest.

Distinguishing Differences in DNA: RFLP Analysis

Often a researcher wishes not to find a specific gene, but rather to identify a particular *individual* using a specific gene as a marker. One powerful way to do this is to analyze **restriction fragment length polymorphisms,** or **RFLPs** (figure 19.10). Point mutations, sequence repetitions, and transposons (see chapter 18) that occur within or between the restriction endonuclease recognition sites will alter the length of the DNA fragments (restriction fragments) the restriction endonucleases produce. DNA from different individuals rarely has exactly the same array of restriction sites and distances between sites, so the population is said to be polymorphic (having many forms) for their restriction fragment patterns. By cutting a DNA sample with a particular restriction endonuclease, separating the fragments according to length on an electrophoretic gel, and then using a radioactive probe to identify the fragments on the gel, one can obtain a pattern of bands often unique for each region of DNA analyzed. These "DNA fingerprints" are used in forensic analysis during criminal investigations. RFLPs are also useful as markers to identify particular groups of people at risk for some genetic disorders.

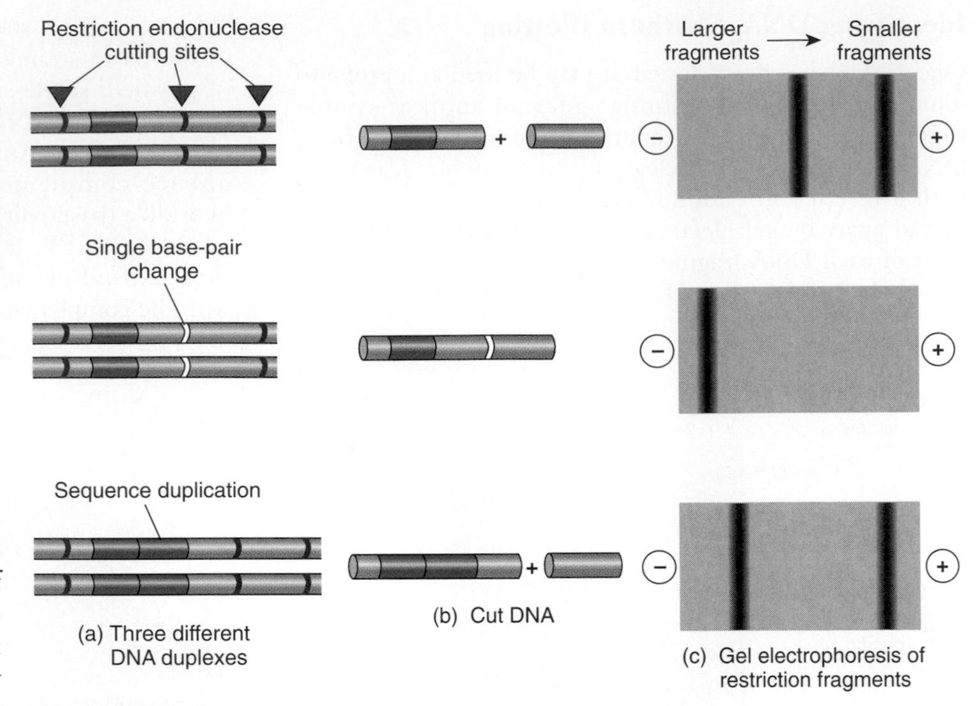

FIGURE 19.10
Restriction fragment length polymorphism (RFLP) analysis. (*a*) Three samples of DNA differ in their restriction sites due to a single base-pair substitution in one case and a sequence duplication in another case. (*b*) When the samples are cut with a restriction endonuclease, different numbers and sizes of fragments are produced. (*c*) Gel electrophoresis separates the fragments, and different banding patterns result.

Making an Intron-Free Copy of a Eukaryotic Gene

Recall from chapter 15 that eukaryotic genes are encoded in exons separated by numerous nontranslated introns. When the gene is transcribed to produce the primary transcript, the introns are cut out during RNA processing to produce the mature mRNA transcript. When transferring eukaryotic genes into bacteria, it is desirable to transfer DNA already processed this way, instead of the raw eukaryotic DNA, because bacteria lack the enzymes to carry out the processing. To do this, genetic engineers first isolate from the cytoplasm the mature mRNA corresponding to a particular gene. They then use an enzyme called reverse transcriptase to make a DNA version of the mature mRNA transcript (figure 19.11). The single strand of DNA can then serve as a template for the synthesis of a complementary strand. In this way, one can produce a double-stranded molecule of DNA that contains a gene lacking introns. This molecule is called **complementary DNA,** or **cDNA.**

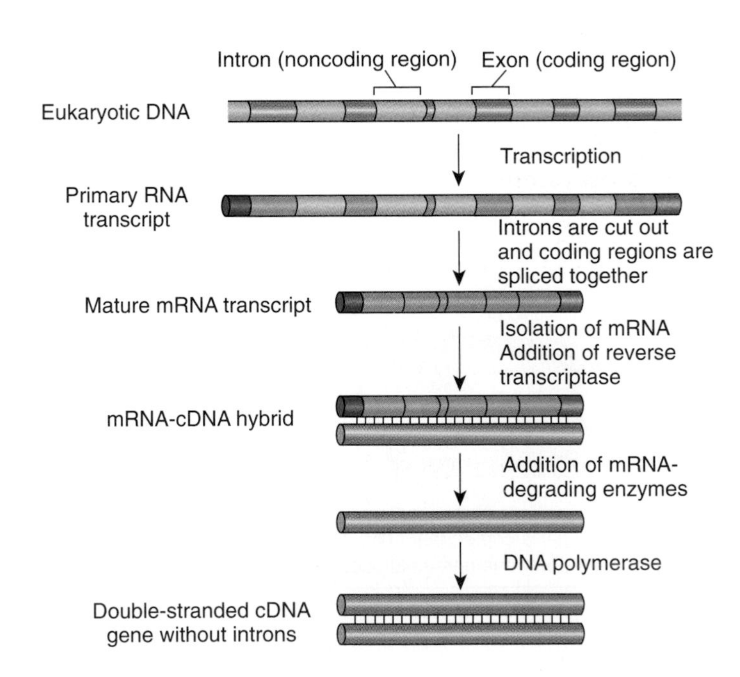

FIGURE 19.11
The formation of cDNA. A mature mRNA transcript is isolated from the cytoplasm of a cell. The enzyme reverse transcriptase is then used to make a DNA strand complementary to the processed mRNA. That newly made strand of DNA is the template for the enzyme DNA polymerase, which assembles a complementary DNA strand along it, producing cDNA, a double-stranded DNA version of the intron-free mRNA.

DNA Fingerprinting

As stated previously, two individuals rarely produce identical RFLP analyses, therefore these DNA fingerprints can be used in criminal investigations. Figure 19.12 shows the DNA fingerprints a prosecuting attorney presented in a rape trial in 1987. They consisted of autoradiographs, parallel bars on X-ray film resembling the line patterns of the universal price code found on groceries. Each bar represents the position of a DNA restriction endonuclease fragment produced by techniques similar to those described in figures 19.4 and 19.10. The lane with many bars represents a standardized control. Two different probes were used to identify the restriction fragments. A vaginal swab had been taken from the victim within hours of her attack; from it semen was collected and the semen DNA analyzed for its restriction endonuclease patterns.

Compare the restriction endonuclease patterns of the semen to that of the suspect Andrews. You can see that the suspect's two patterns match that of the rapist (and are not at all like those of the victim). Clearly the semen collected from the rape victim and the blood sample from the suspect came from the same person. The suspect was Tommie Lee Andrews, and on November 6, 1987, the jury returned a verdict of guilty. Andrews became the first person in the United States to be convicted of a crime based on DNA evidence.

Since the Andrews verdict, DNA fingerprinting has been admitted as evidence in more than 2000 court cases (figure 19.13). While some probes highlight profiles shared by many people, others are quite rare. Using several probes, identity can be clearly established or ruled out.

Just as fingerprinting revolutionized forensic evidence in the early 1900s, so DNA fingerprinting is revolutionizing it today. A hair, a minute speck of blood, a drop of semen can all serve as sources of DNA to damn or clear a suspect. As the man who analyzed Andrews' DNA says: "It's like leaving your name, address, and social security number at the scene of the crime. It's that precise." Of course, laboratory analyses of DNA samples must be carried out properly—sloppy procedures could lead to a wrongful conviction. After widely publicized instances of questionable lab procedures, national standards are being developed.

Techniques such as Southern blotting and PCR enable investigators to identify specific genes and produce them in large quantities, while RFLP analysis and DNA fingerprinting identify individuals and unknown gene sequences.

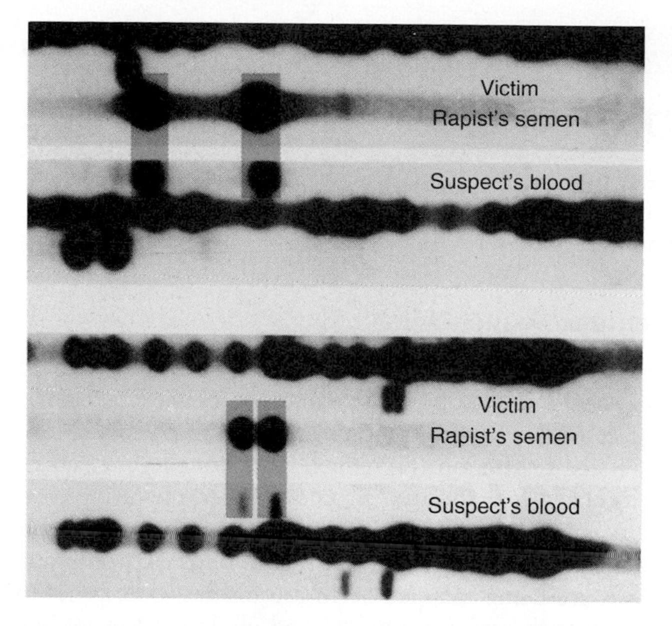

FIGURE 19.12
Two of the DNA profiles that led to the conviction of Tommie Lee Andrews for rape in 1987. The two DNA probes seen here were used to characterize DNA isolated from the victim, the semen left by the rapist, and the suspect. The dark channels are multiband controls. There is a clear match between the suspect's DNA and the DNA of the rapist's semen in these.

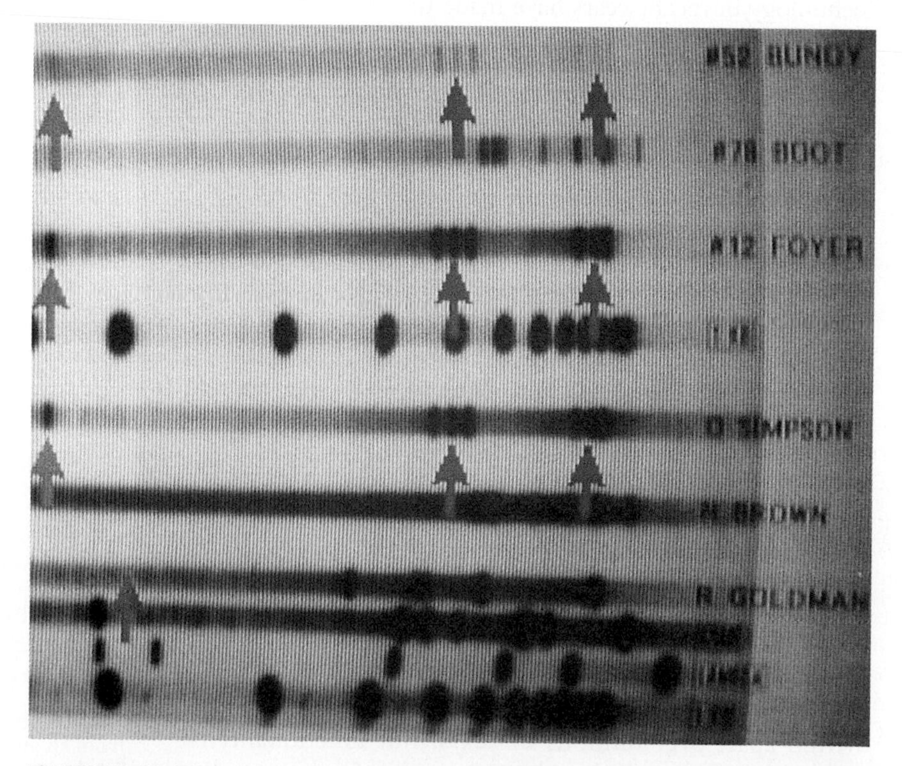

FIGURE 19.13
The DNA profiles of O. J. Simpson and blood samples from the murder scene of his former wife from his highly publicized and controversial murder trial in 1995.

19.3 Biotechnology is producing a scientific revolution.

DNA Sequence Technology

Recent years have seen an explosion of interest in **biotechnology**, the application of genetic engineering to practical human problems.

Genome Sequencing

Genetic engineering techniques are enabling us to learn a great deal more about the human genome. Several clonal libraries of the human genome have been assembled, using large-size restriction fragments. Any cloned gene can now be localized to a specific chromosomal site by using probes to detect in situ hybridization (that is, binding between the probe and a complementary sequence on the chromosome). Genes are being mapped at an astonishing rate: genes that contribute to dyslexia, obesity, and cholesterol-proof blood are but a few of the important ones. With an understanding of where specific genes are located in the human genome and how they work, it is not difficult to imagine a future in which virtually any genetic disease could be treated or perhaps even cured with gene therapy. As we mentioned in chapter 13, some success has already been reported in treating patients who have cystic fibrosis with a genetically corrected version of the *cystic fibrosis* gene.

Exciting scientific advances in DNA sequencing technology in recent years have made the DNA sequencing of entire genomes practical (table 19.1). Researchers first focused on microorganisms with relatively small genomes, on the order of a few million nucleotide base-pairs (Mb). In general, about half of the genes prove to have a known function. What the other half of the genes are doing is a complete mystery.

The first eukaryotic genomes to be sequenced in their entirety were the microbes brewer's yeast *Saccharomyces cerevisiae* (13 Mb) and the malarial *Plasmodium* parasite (30 Mb). The first animal to have its DNA completely read was the nematode *C. elegans* (100 Mb) in December 1998, followed by the fruit fly *Drosophila* (120 Mb), and the mouse (300 Mb). The plants *Arabidopsis* (100 Mb) (figure 19.14) and rice (430 Mb) were completed in the year 2001.

One of the most surprising results revealed by these genome sequences has been the discovery of just how similar living things are to one another at the genetic level. Forty-two percent of the genes discovered in *C. elegans* had some sort of match to genes in other organisms only distantly related. Fully 83% of *Drosophila* genes match those of other species. The matches are not perfect however—the DNA sequences of genes that do the same job have drifted apart over millions of years. But functionality is maintained. For example, when a gene involved in eye development in mice was substituted for its homologue in *Drosophila*, the flies were born with normal, functional eyes.

Table 19.1 Genome Sequencing Projects

Organism	Genome Size (Mb)	Description
ARCHAEBACTERIA		
Methanococcus jannaschii	1.7	Extreme thermophile
EUBACTERIA		
Mycoplasma genitalium	0.6	Smallest known organism
Helicobacter pylori	1.7	Ulcers
Vibrio cholerae	4.0	Often fatal disease
Mycobacterium tuberculosis	4.4	Tuberculosis
Escherichia coli	4.6	Laboratory standard
FUNGI		
Saccharomyces cerevisiae	13	Baker's yeast
PROTIST		
Plasmodium	30	Malarial parasite
PLANT		
Arabidopsis thaliana	100	Relative of mustard plant
Oryza sativa	430	Commercial rice
ANIMAL		
Caenorhabditis elegans	100	Nematode
Drosophila melanogaster	120	Fruit fly
Mus musculus	300	Mouse
Homo sapiens	3200	Human

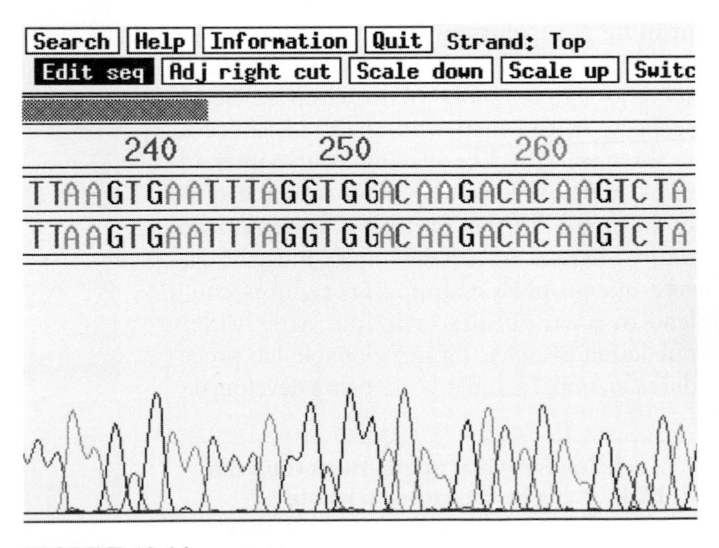

FIGURE 19.14
Part of the genome sequence of the plant *Arabidopsis*. Data from an automated DNA-sequencing run shows the nucleotide sequence for a small section of the *Arabidopsis* genome. Automated DNA sequencing has greatly increased the speed at which genomes can be sequenced.

The Human Genome Project

The human genome contains some 3.2 billion bases (3200 Mb), so its sequencing has presented no small challenge. Rapid progress was made possible by the use of so-called shotgun cloning techniques, in which the entire genome is first fragmented, then each of the fragments is sequenced by automated machines, and finally computers use overlaps to order the fragments. All but a small portion of the sequence was completed by the year 2000.

The number of genes in the human genome is only about 30,000 (figure 19.15). This is barely a third more than in nematodes, scarcely double the number of genes in *Drosophila*, and but a quarter of the number that had been anticipated. We intuitively think of ourselves as far more complex than a nematode or fruitfly—clearly it does not take a lot of genes to achieve complexity.

Compared to other eukaryotes, human genes are quite fragmented, split into numerous exons scattered within much larger regions of noncoding introns. It seems likely that humans achieve added diversity at the protein level by splicing exons together in different combinations.

Genes are not distributed evenly over the genome. The small chromosome number 19 is packed densely with genes, transcription factors, and other functional elements. The much larger chromosomes numbers 4 and 8, by contrast, have few genes. On most chromosomes, vast stretches of seemingly barren DNA fill the chromosomes between scattered clusters rich in genes.

Only 1 to 1.5 percent of the human genome is devoted to genes encoding proteins. Fully one third of the genome is composed of noncoding repeated sequences. Repetitive sequences with excess C and G tend to be found in the neighborhood of genes, while A and T rich repeats dominate the non-gene deserts. The light bands on chromosome karyotypes now have an explanation—they are regions rich in GC and genes. Dark bands signal neighborhoods rich in AT and thin on genes. Chromosome 19, dense with genes, has few dark bands. Roughly 25% of the human genome has no genes at all.

Perhaps the most striking lesson learned from the sequence of the human genome is how very like other organisms humans are. More than half of the genes of *Drosophila* have human counterparts. Among mammals the differences are even fewer. Humans have only 300 genes that have no counterpart in the mouse genome. This suggests that when an ape genome is sequenced, it will possess practically all of the genes that humans do.

The evolutionary history of eukaryotes can be read in the human genome. Some 230 bacterial genes are nested in our chromosomes, borrowed in the distant past from bacteria. The gene encoding monoamine oxidase, an important degradative enzyme for the central nervous system, was bequeathed by bacteria. Other genes were contributed by the symbiotic bacteria that became mitochondria.

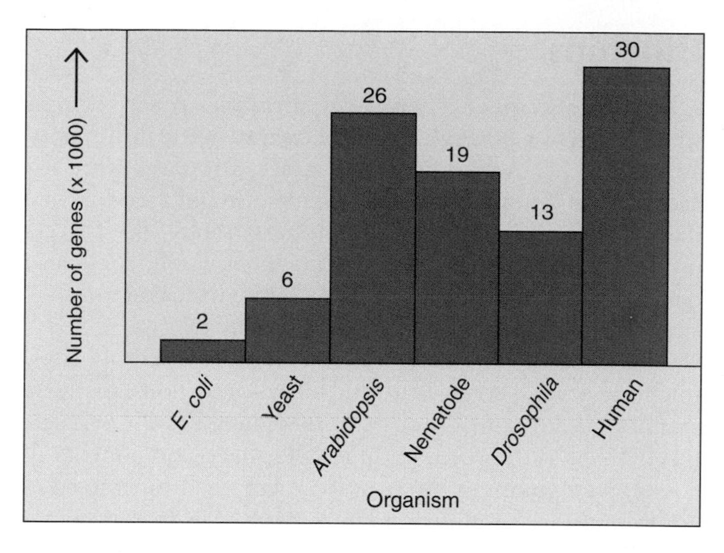

FIGURE 19.15
What the human genome is like. The human genome has an unexpectedly small number of genes, some 30,000. This is not many more than the plant *Arabidopsis*, and only a third more than nematode worms.

Proteomics: The Next Frontier

With the sequencing of the human genome now essentially complete, researchers have begun an even more challenging task: the cataloging and analysis of every protein in the human body, an endeavor called **proteomics.** Each gene's nucleotide sequence specifies an amino acid sequence that folds in a certain way, producing a protein whose shape gives it a particular function. Only by understanding the protein shapes that genes produce can we begin to make sense of the human genome.

Ideally, a researcher would like to be able to examine a nucleotide sequence and know what sort of functional protein the gene specifies. However, efforts to calculate the shape a protein based on knowledge of its amino acid sequence have proven difficult, even with the aid of large computers. Now, by looking at the proteins that are produced by the genes of the human genome, researchers are beginning to get a clearer picture of how gene sequence relates to protein shape and function.

Fortunately, while there may be as many as a million different proteins, most are just variations on a handful of themes. The same shared structural motifs—barrels, helices, molecular zippers—are found in the proteins of plants, insects, and humans. The maximum number of distinct motifs has been estimated as fewer than 5000. About 1000 of these motifs have already been cataloged. Both public and privately-financed efforts are now underway to detail the shapes of all the common motifs.

Genome sequencing has revealed that organisms have many genes in common. Surprisingly, humans have a small genome, only a third larger than a nematode worm.

Biochips

A biochip, also called a gene microarray, is a square of glass smaller than a postage stamp, covered with millions of strands of DNA like blades of grass. Biochips were invented nine years ago by gene scientist Stephen Fodor. In a flash of insight, he saw that photolithography, the process used to etch semiconductor circuits into silicon, could also be used to assemble particular DNA molecules on a chip—a biochip.

Think of the chip surface as a field of assembly sites, much as a TV screen is a field of colored dots. Just as a scanning beam moves over each individual TV dot instructing it to be red, green, or blue (the three components of color), so a scanning beam moves over each biochip spot, commanding the addition there of a base to a growing strand of DNA. A computer, by varying the wavelength of the scanning beam, determines which of four possible nucleotides is added to the growing DNA strand anchored to each spot. When the entire chip has been scanned, each DNA strand has been lengthened one nucleotide unit. The computer repeats the process, layer by layer, until each DNA strand is an entire gene or gene fragment. One biochip made in this way contains hundreds of thousands of specific gene sequences.

How could you use such a biochip to delve into a person's genes? All you would have to do is to obtain a little of the person's DNA, say from a blood sample or even a bit of hair. Flush fluid containing the DNA over the biochip surface. Every place that the DNA has a gene matching one of the biochip strands, it will stick to it in a way the computer can detect.

Now here is where it gets interesting. The mad rush to sequence the human genome is over. The gene research firm Celera has recently announced it has essentially completed the sequence, with over 90% of genes done. Already the researchers are busily comparing their consensus "reference sequence" to the DNA of individual people, and noting any differences they detect. Called single nucleotide polymorphisms, or SNPs (pronounced "snips"), these spot differences in the identity of particular nucleotides collectively record every way in which a particular individual differs from the reference sequence. Some SNPs cause diseases like cystic fibrosis or sickle cell anemia. Others may give you red hair or elevated cholesterol in your blood. As the human genome project charges toward completion, its researchers are excitedly assembling a huge database of SNPs. Research indicates that SNPs can be expected to occur at a frequency of about one per thousand nucleotides, scattered about randomly over the chromosomes. Each of us thus differs from the standard "type sequence" in several thousand nucleotide SNPs. Everything genetic about you that is different from a stranger you meet is caused by a few thousand SNPs; otherwise you and that stranger are identical.

How Biochips Can Be Used to Screen for Cancer

One of the biggest decisions facing an oncologist (cancer doctor) treating a tumor is to select the proper treatment. Most cancer cells look alike, although the tumors may in fact be caused by quite different forms of cancer. If the oncologist could clearly identify the cancer, very targeted therapies might be possible. Unable to tell the difference for sure, however, oncologists take no chances. Tumors are treated with therapy that attacks all cancers, usually with severe side effects.

This year Boston researchers Todd Golub and Eric Lander took a vital step towards treating cancer, using new DNA technology to sniff out the differences between different forms of a deadly cancer of the immune system. Golub and Lander worked with biochips.

The way to tell the difference between two kinds of cancer is to compare the mutations that led to the cancer in the first place. Biologists call such gene changes mutations. The mutations that cause many lung cancers are caused by a tobacco-induced alteration of a single DNA nucleotide in one gene. Such spot differences between the version of a gene one person has and another person has, or a cancer patient has, are examples of SNPs.

Golub and Lander obtained bone marrow cells from patients with two types of leukemia (cancer of white blood cells), and exposed DNA from each to biochips containing all known human genes, 6817 in all (figure 19.16). Using high-speed computer programs, Golub and Lander examined each of the 6817 positions on the chip. The two forms of leukemia each showed gene changes from normal, but, importantly, the changes were different in each case! Each had their own characteristic SNP.

Biochips thus may offer a quick and reliable way to identify any type of cancer. Just look and see what SNP is present.

The Use of Gene Chips Will Soon Be Widespread

Biochip technology is likely to dominate medicine in the coming millennium, a prospect both exciting and scary. Researchers have announced plans to compile a database of hundreds of thousands of SNPs over the next two years. Screening SNPs and comparing them to known SNP databases will soon allow doctors to screen each of us for copies of genes leading to genetic diseases. Many genetic diseases are associated with SNPs, including cystic fibrosis and muscular dystrophy.

Biochips Raise Critical Issues of Personal Privacy

The scary part is SNPs on chips. Researchers plan to have identified some 300,000 different SNPs by 2001, all of which could reside on a single biochip. When your DNA is flushed over a SNP biochip, the sequences that light up

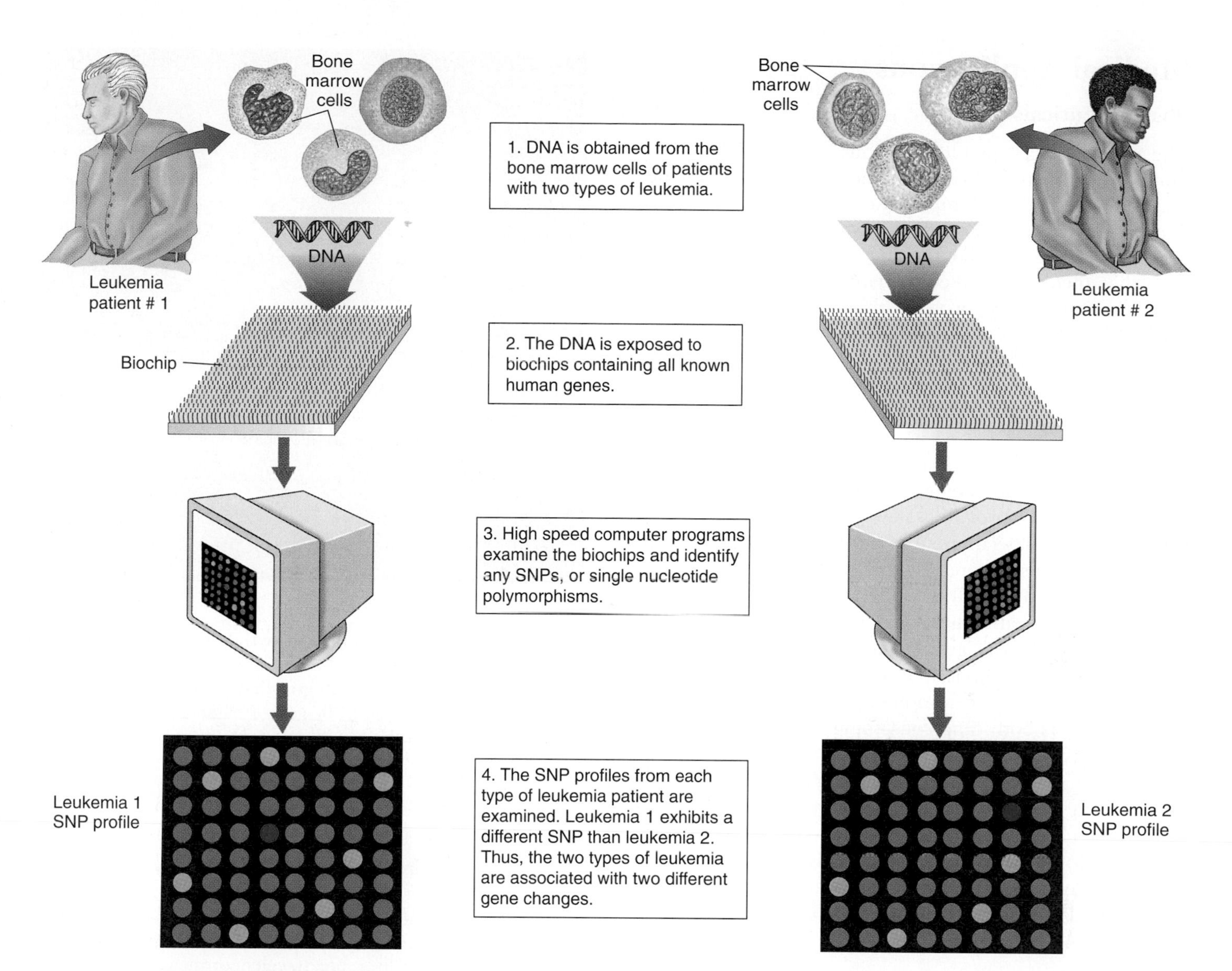
Bone
marrow
cells
Leukemia
patient # 1
DNA
Biochip
1. DNA is obtained from the bone marrow cells of patients with two types of leukemia.
2. The DNA is exposed to biochips containing all known human genes.
3. High speed computer programs examine the biochips and identify any SNPs, or single nucleotide polymorphisms.
4. The SNP profiles from each type of leukemia patient are examined. Leukemia 1 exhibits a different SNP than leukemia 2. Thus, the two types of leukemia are associated with two different gene changes.
Bone
marrow
cells
DNA
Leukemia
patient # 2
Leukemia 1
SNP profile
Leukemia 2
SNP profile

FIGURE 19.16
Biochips can help in identifying precise forms of cancer.

will instantly reveal your SNP profile. The genetic characteristics that make you you, genes that might affect your health, your behavior, your future potential—all are there to be read by anyone clever enough to interpret the profile.

To what extent are your genes you? Scientists fight about this question, and no one really knows the answer. It is clear that much of what each of us is like is strongly affected by our genetic makeup. Researchers have proven beyond any real dispute that intelligence and major personality traits like aggressiveness and inquisitiveness are about 80% heritable (that is, 80% of the variation in these traits reflects variation in genes).

Your SNP profile will reflect all of this variation, a table of contents of your chromosomes, a molecular window to who you are. When millions of such SNP profiles have been gathered over the coming years, computers will be able to identify other individuals with profiles like yours, and, by examining health records, standard personality tests, and the like, correlate parts of your profile with particular traits. Even behavioral characteristics involving many genes, which until now have been thought too complex to ever analyze, cannot resist a determined assault by a computer comparing SNP profiles.

A biochip is a discrete collection of gene fragments on a stamp-sized chip that can be used to screen for the presence of particular gene variants. Biochips allow rapid screening of gene profiles, a tool that promises to have a revolutionary impact on medicine and society.

Medical Applications

Pharmaceuticals

The first and perhaps most obvious commercial application of genetic engineering was the introduction of genes that encode clinically important proteins into bacteria. Because bacterial cells can be grown cheaply in bulk (fermented in giant vats, like the yeasts that make beer), bacteria that incorporate recombinant genes can synthesize large amounts of the proteins those genes specify. This method has been used to produce several forms of human insulin and interferon, as well as other commercially valuable proteins such as growth hormone (figure 19.17) and erythropoietin, which stimulates red blood cell production.

Among the medically important proteins now manufactured by these approaches are **atrial peptides,** small proteins that may provide a new way to treat high blood pressure and kidney failure. Another is **tissue plasminogen activator,** a human protein synthesized in minute amounts that causes blood clots to dissolve and may be effective in preventing and treating heart attacks and strokes.

A problem with this general approach has been the difficulty of separating the desired protein from the others the bacteria make. The purification of proteins from such complex mixtures is both time-consuming and expensive, but it is still easier than isolating the proteins from the tissues of animals (for example, insulin from hog pancreases), which is how such proteins used to be obtained. Recently, however, researchers have succeeded in producing RNA transcripts of cloned genes; they can then use the transcripts to produce only these proteins in a test tube containing the transcribed RNA, ribosomes, cofactors, amino acids, tRNA, and ATP.

Gene Therapy

In 1990, researchers first attempted to combat genetic defects by the transfer of human genes. When a hereditary disorder is the result of a single defective gene, an obvious way to cure the disorder is to add a working copy of the gene. This approach is being used in an attempt to combat cystic fibrosis, and it offers potential for treating muscular dystrophy and a variety of other disorders (table 19.2). One of the first successful attempts was the transfer of a gene encoding the enzyme adenosine deaminase into the bone marrow of two girls suffering from a rare blood disorder caused by the lack of this enzyme. However, while many clinical trials are underway, no others have yet proven successful. This extremely promising approach will require a lot of additional effort.

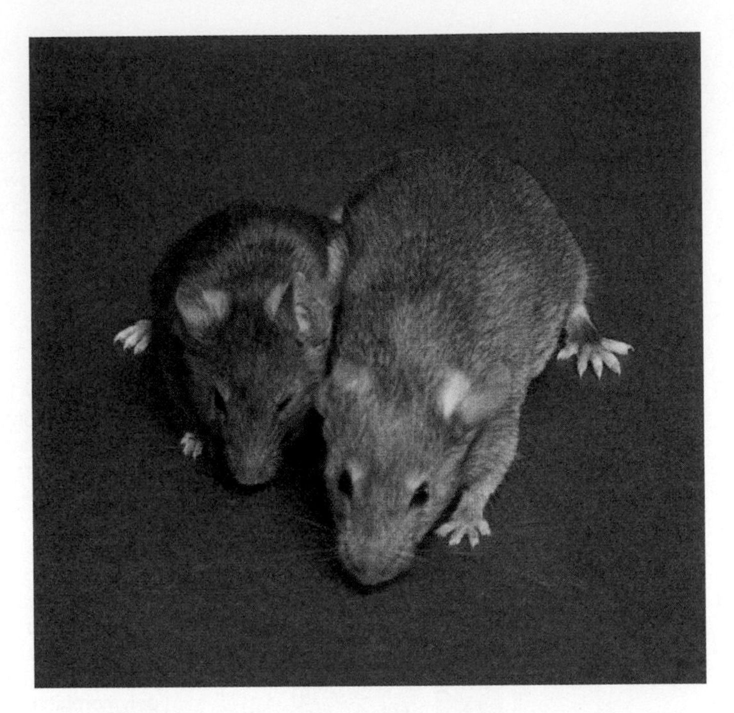

FIGURE 19.17
Genetically engineered human growth hormone. These two mice are genetically identical, but the large one has one extra gene: the gene encoding human growth hormone. The gene was added to the mouse's genome by genetic engineers and is now a stable part of the mouse's genetic endowment.

Table 19.2 Diseases Being Treated in Clinical Trials of Gene Therapy

Disease
Cancer (melanoma, renal cell, ovarian, neuroblastoma, brain, head and neck, lung, liver, breast, colon, prostate, mesothelioma, leukemia, lymphoma, multiple myeloma)
SCID (severe combined immunodeficiency)
Cystic fibrosis
Gaucher's disease
Familial hypercholesterolemia
Hemophilia
Purine nucleoside phosphorylase deficiency
Alpha-1 antitrypsin deficiency
Fanconi's anemia
Hunter's syndrome
Chronic granulomatous disease
Rheumatoid arthritis
Peripheral vascular disease
AIDS

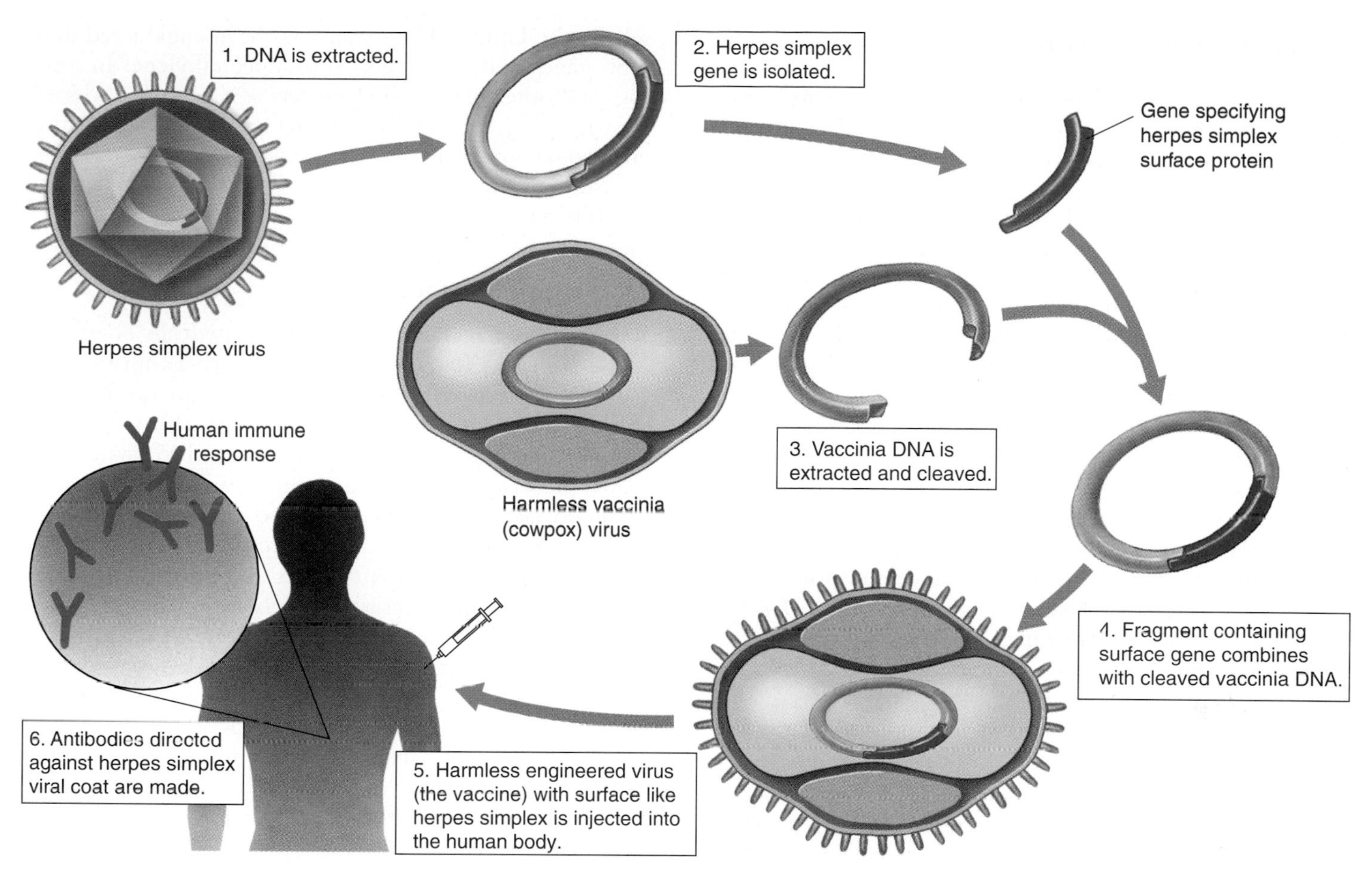
1. DNA is extracted.
2. Herpes simplex gene is isolated.
Gene specifying herpes simplex surface protein
Herpes simplex virus
3. Vaccinia DNA is extracted and cleaved.
Harmless vaccinia (cowpox) virus
Human immune response
4. Fragment containing surface gene combines with cleaved vaccinia DNA.
5. Harmless engineered virus (the vaccine) with surface like herpes simplex is injected into the human body.
6. Antibodies directed against herpes simplex viral coat are made.

FIGURE 19.18
Strategy for constructing a subunit vaccine for herpes simplex.

Piggyback Vaccines

Another area of potential significance involves the use of genetic engineering to produce **subunit vaccines** against viruses such as those that cause herpes and hepatitis. Genes encoding part of the protein-polysaccharide coat of the herpes simplex virus or hepatitis B virus are spliced into a fragment of the vaccinia (cowpox) virus genome (figure 19.18). The vaccinia virus, which British physician Edward Jenner used almost 200 years ago in his pioneering vaccinations against smallpox, is now used as a vector to carry the herpes or hepatitis viral coat gene into cultured mammalian cells. These cells produce many copies of the recombinant virus, which has the outside coat of a herpes or hepatitis virus. When this recombinant virus is injected into a mouse or rabbit, the immune system of the infected animal produces antibodies directed against the coat of the recombinant virus. It therefore develops an immunity to herpes or hepatitis virus. Vaccines produced in this way are harmless because the vaccinia virus is benign and only a small fragment of the DNA from the disease-causing virus is introduced via the recombinant virus.

The great attraction of this approach is that it does not depend upon the nature of the viral disease. In the future, similar recombinant viruses may be injected into humans to confer resistance to a wide variety of viral diseases.

In 1995, the first clinical trials began of a novel new kind of **DNA vaccine,** one that depends not on antibodies but rather on the second arm of the body's immune defense, the so-called cellular immune response, in which blood cells known as killer T cells attack infected cells. The infected cells are attacked and destroyed when they stick fragments of foreign proteins onto their outer surfaces that the T cells detect (the discovery by Peter Doherty and Rolf Zinkernagel that infected cells do so led to their receiving the Nobel Prize in Physiology or Medicine in 1996). The first DNA vaccines spliced an influenza virus gene encoding an internal nucleoprotein into a plasmid, which was then injected into mice. The mice developed strong cellular immune responses to influenza. New and controversial, the approach offers great promise.

Genetic engineering has produced commercially valuable proteins, gene therapies, and, possibly, new and powerful vaccines.

Agricultural Applications

Another major area of genetic engineering activity is manipulation of the genes of key crop plants. In plants the primary experimental difficulty has been identifying a suitable vector for introducing recombinant DNA. Plant cells do not possess the many plasmids that bacteria do, so the choice of potential vectors is limited. The most successful results thus far have been obtained with the **Ti** (tumor-inducing) **plasmid** of the plant bacterium *Agrobacterium tumefaciens,* which infects broadleaf plants such as tomato, tobacco, and soybean. Part of the Ti plasmid integrates into the plant DNA, and researchers have succeeded in attaching other genes to this portion of the plasmid (figure 19.19). The characteristics of a number of plants have been altered using this technique, which should be valuable in improving crops and forests. Among the features scientists would like to affect are resistance to disease, frost, and other forms of stress; nutritional balance and protein content; and herbicide resistance. Unfortunately, *Agrobacterium* generally does not infect cereals such as corn, rice, and wheat, but alternative methods can be used to introduce new genes into them.

A recent advance in genetically manipulated fruit is Calgene's "Flavr Savr" tomato, which has been approved for sale by the USDA. The tomato has been engineered to inhibit genes that cause cells to produce ethylene. In tomatoes and other plants, ethylene acts as a hormone to speed fruit ripening. In Flavr Savr tomatoes, inhibition of ethylene production delays ripening. The result is a tomato that can stay on the vine longer and that resists overripening and rotting during transport to market.

Herbicide Resistance

Recently, broadleaf plants have been genetically engineered to be resistant to **glyphosate,** the active ingredient in Roundup, a powerful, biodegradable herbicide that kills most actively growing plants (figure 19.20). Glyphosate works by inhibiting an enzyme called EPSP synthetase, which plants require to produce aromatic amino acids. Humans do not make aromatic amino acids; they get them from their diet, so they are unaffected by glyphosate. To make glyphosate-resistant plants, agricultural scientists used a Ti plasmid to insert extra copies of the EPSP synthetase genes into plants. These engineered plants produce 20 times the normal level of EPSP synthetase, enabling them to synthesize proteins and grow despite glyphosate's suppression of the enzyme. In later experiments, a bacterial form of the EPSP synthetase gene that differs from the

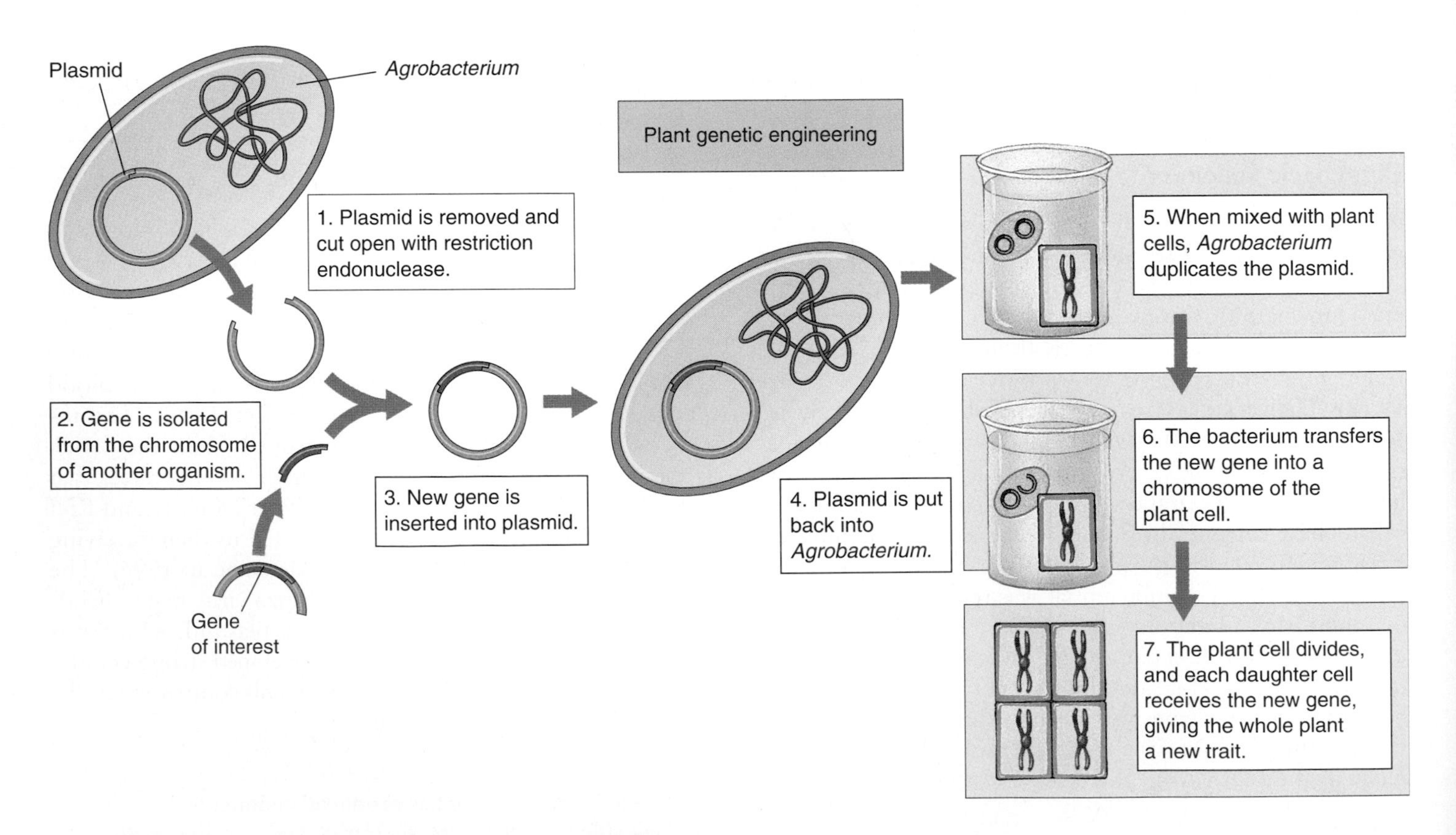

FIGURE 19.19
The Ti plasmid. This *Agrobacterium tumefaciens* plasmid is used in plant genetic engineering.

plant form by a single nucleotide was introduced into plants via Ti plasmids; the bacterial enzyme in these plants is not inhibited by glyphosate.

These advances are of great interest to farmers because a crop resistant to Roundup would never have to be weeded if the field were simply treated with the herbicide. Because Roundup is a broad-spectrum herbicide, farmers would no longer need to employ a variety of different herbicides, most of which kill only a few kinds of weeds. Furthermore, glyphosate breaks down readily in the environment, unlike many other herbicides commonly used in agriculture. A plasmid is actively being sought for the introduction of the EPSP synthetase gene into cereal plants, making them also glyphosate-resistant.

FIGURE 19.20
Genetically engineered herbicide resistance. All four of these petunia plants were exposed to equal doses of the herbicide Roundup. The two on top were genetically engineered to be resistant to glyphosate, the active ingredient of Roundup, while the two on the bottom were not.

Nitrogen Fixation

A long-range goal of agricultural genetic engineering is to introduce the genes that allow soybeans and other legume plants to "fix" nitrogen into key crop plants. These so-called ***nif* genes** are found in certain symbiotic root-colonizing bacteria. Living in the root nodules of legumes, these bacteria break the powerful triple bond of atmospheric nitrogen gas, converting N_2 into NH_3 (ammonia). The plants then use the ammonia to make amino acids and other nitrogen-containing molecules. Other plants lack these bacteria and cannot fix nitrogen, so they must obtain their nitrogen from the soil. Farmland where these crops are grown soon becomes depleted of nitrogen, unless nitrogenous fertilizers are applied. Worldwide, farmers applied over 60 million metric tons of such fertilizers in 1987, an expensive undertaking. Farming costs would be much lower if major crops like wheat and corn could be engineered to carry out biological nitrogen fixation. However, introducing the nitrogen-fixing genes from bacteria into plants has proved difficult because these genes do not seem to function properly in eukaryotic cells. Researchers are actively experimenting with other species of nitrogen-fixing bacteria whose genes might function better in plant cells.

Insect Resistance

Many commercially important plants are attacked by insects, and the traditional defense against such attacks is to apply insecticides. Over 40% of the chemical insecticides used today are targeted against boll weevils, bollworms, and other insects that eat cotton plants. Genetic engineers are now attempting to produce plants that are resistant to insect pests, removing the need to use many externally applied insecticides.

The approach is to insert into crop plants genes encoding proteins that are harmful to the insects that feed on the plants but harmless to other organisms. One such insecticidal protein has been identified in *Bacillus thuringiensis*, a soil bacterium. When the tomato hornworm caterpillar ingests this protein, enzymes in the caterpillar's stomach convert it into an insect-specific toxin, causing paralysis and death. Because these enzymes are not found in other animals, the protein is harmless to them. Using the Ti plasmid, scientists have transferred the gene encoding this protein into tomato and tobacco plants. They have found that these **transgenic** plants are indeed protected from attack by the insects that would normally feed on them. In 1995, the EPA approved altered forms of potato, cotton, and corn. The genetically altered potato can kill the Colorado potato beetle, a common pest. The altered cotton is resistant to cotton bollworm, budworm, and pink bollworm. The corn has been altered to resist the European corn borer and other mothlike insects.

Monsanto scientists screening natural compounds extracted from plant and soil samples have recently isolated a new insect-killing compound from a fungus, the enzyme cholesterol oxidase. Apparently, the enzyme disrupts membranes in the insect gut. The fungus gene, called the Bollgard gene after its discoverer, has been successfully inserted into a variety of crops. It kills a wide range of insects, including the cotton boll weevil and the Colorado potato beetle, both serious agricultural pests. Field tests began in 1996.

Some insect pests attack plant roots, and *B. thuringiensis* is being employed to counter that threat as well. This bacterium does not normally colonize plant roots, so biologists have introduced the *B. thuringiensis* insecticidal protein gene into root-colonizing bacteria, especially strains of *Pseudomonas*. Field testing of this promising procedure has been approved by the Environmental Protection Agency.

The Real Promise of Plant Genetic Engineering

In the last decade the cultivation of genetically modified crops of corn, cotton, and soybeans has become commonplace in the United States—in 1999, over half of the 72 million acres planted with soybeans in the United States were planted with seeds genetically modified to be herbicide resistant, with the result that less tillage has been needed, and as a consequence soil erosion has been greatly lessened. These benefits, while significant, have been largely confined to farmers, making their cultivation of crops cheaper and more efficient. The food that the public gets is the same, it just costs less to get it to the table.

Like the first act of a play, these developments have served mainly to set the stage for the real action, which is only now beginning to happen. The real promise of plant genetic engineering is to produce genetically modified plants with desirable traits that directly benefit the consumer.

One recent advance, nutritionally improved rice, gives us a hint of what is to come. In developing countries large numbers of people live on simple diets that are poor sources of vitamins and minerals (what botanists called "micronutrients"). Worldwide, the two major micronutrient deficiencies are iron, which affects 1.4 billion women, 24% of the world population, and vitamin A, affecting 40 million children, 7% of the world population. The deficiencies are especially severe in developing countries where the major staple food is rice. In recent research, Swiss bioengineer Ingo Potrykus and his team at the Institute of Plant Sciences, Zurich, have gone a long way towards solving this problem. Supported by the Rockefeller Foundation and with results to be made free to developing countries, the work is a model of what plant genetic engineering can achieve.

To solve the problem of dietary iron deficiency among rice eaters, Potrykus first asked why rice is such a poor source of dietary iron. The problem, and the answer, proved to have three parts:

1. *Too little iron.* The proteins of rice endosperm have unusually low amounts of iron. To solve this problem, a ferritin gene was transferred into rice from beans (figure 19.21). Ferritin is a protein with an extraordinarily high iron content, and so greatly increased the iron content of the rice.
2. *Inhibition of iron absorption by the intestine.* Rice contains an unusually high concentration of a chemical called phytate, which inhibits iron reabsorption in the intestine—it stops your body from taking up the iron in the rice. To solve this problem, a gene encoding an enzyme that destroys phytate was transferred into rice from a fungus.
3. *Too little sulfur for efficient iron absorption.* Sulfur is required for iron uptake, and rice has very little of it. To solve this problem, a gene encoding a particularly sulfur-rich metallothionin protein was transferred into rice from wild rice.

To solve the problem of vitamin A deficiency, the same approach was taken. First, the problem was identified. It turns out rice only goes part way toward making beta-carotene (provitamin A); there are no enzymes in rice to catalyze the last four steps. To solve the problem, genes encoding these four enzymes were added to rice from a familiar flower, the daffodil.

Potrykus's development of transgenic rice to combat dietary deficiencies involved no subtle tricks, just straightforward bioengineering and the will to get the job done. The transgenic rice he has developed will directly improve the lives of millions of people. His work is

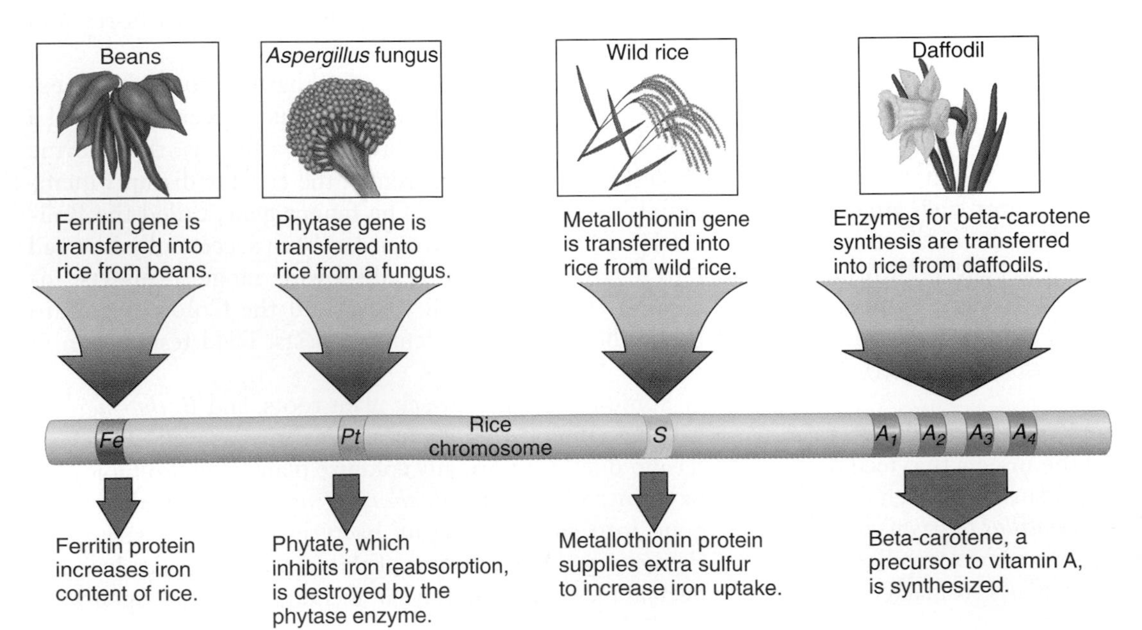

FIGURE 19.21
Transgenic rice. Developed by Swiss bioengineer Ingo Potrykus, transgenic rice offers the promise of improving the diets of people in rice-consuming developing countries, where iron and vitamin A deficiencies are a serious problem.

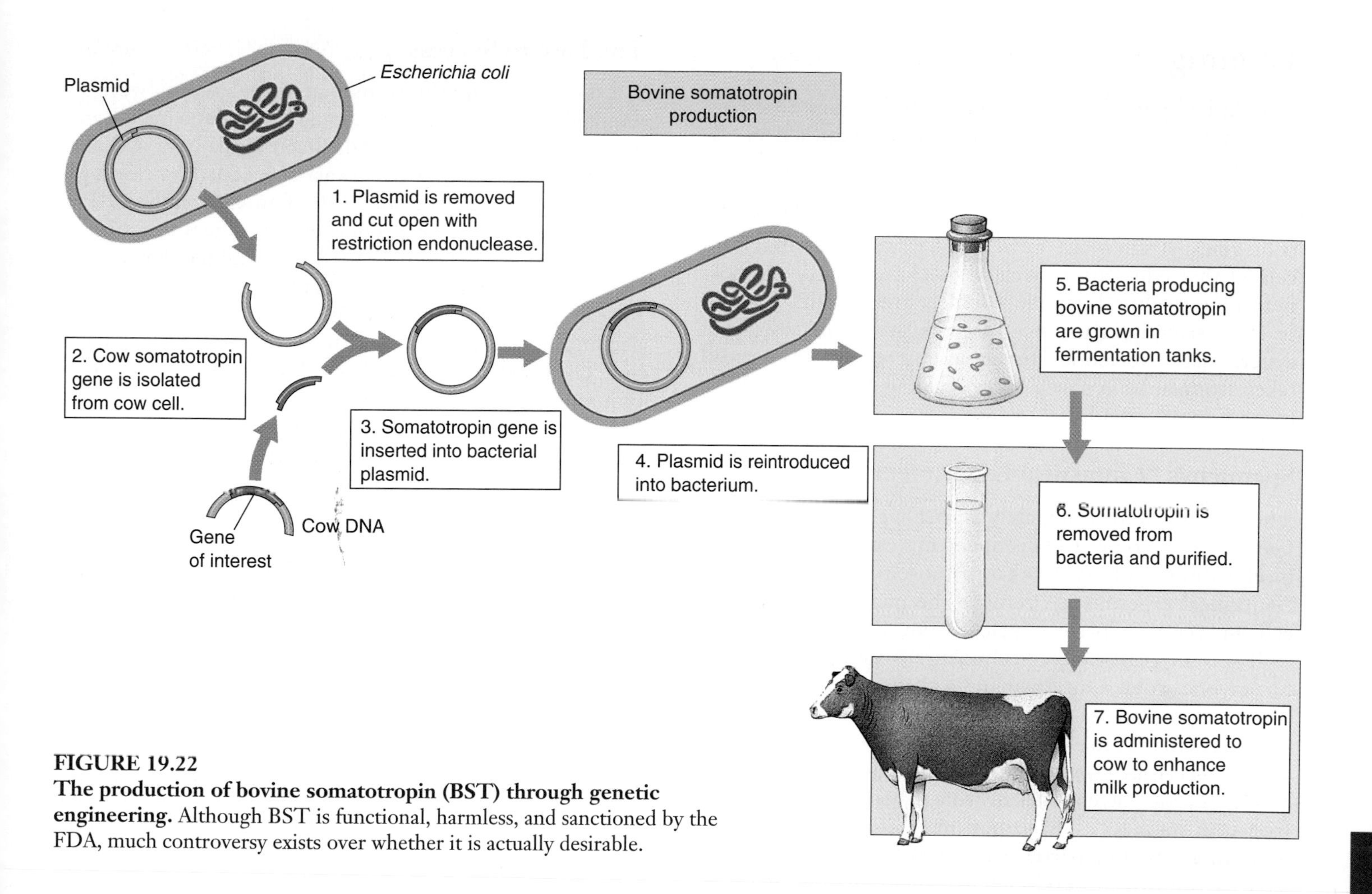
Plasmid
Escherichia coli
Bovine somatotropin production
1. Plasmid is removed and cut open with restriction endonuclease.
2. Cow somatotropin gene is isolated from cow cell.
3. Somatotropin gene is inserted into bacterial plasmid.
4. Plasmid is reintroduced into bacterium.
5. Bacteria producing bovine somatotropin are grown in fermentation tanks.
6. Somatotropin is removed from bacteria and purified.
7. Bovine somatotropin is administered to cow to enhance milk production.
Gene of interest
Cow DNA

FIGURE 19.22
The production of bovine somatotropin (BST) through genetic engineering. Although BST is functional, harmless, and sanctioned by the FDA, much controversy exists over whether it is actually desirable.

representative of the very real promise of genetic engineering to help meet the challenges of the coming new millennium.

The list of gene modifications that directly aid consumers will only grow. In Holland, Dutch bioengineers announced last month that they are genetically engineering plants to act as vaccine-producing factories! To petunias they have added a gene for a vaccine against dog parvovirus, hiding the gene within the petunia genes that direct nectar production. The drug is produced in the nectar, collected by bees, and extracted from the honey. It is hard to believe this isn't science fiction. Clearly, the real promise of plant genetic engineering lies ahead, and not very far.

Farm Animals

The gene encoding the growth hormone somatotropin was one of the first to be cloned successfully. In 1994, Monsanto received federal approval to make its recombinant bovine somatotropin (BST) commercially available, and dairy farmers worldwide began to add the hormone as a supplement to their cows' diets, increasing the animals' milk production (figure 19.22). Genetically engineered somatotropin is also being tested to see if it increases the muscle weight of cattle and pigs, and as a treatment for human disorders in which the pituitary gland fails to make adequate levels of somatotropin, producing dwarfism. BST ingested in milk or meat has no effect on humans, because it is a protein and is digested in the stomach. Nevertheless, BST has met with some public resistance, due primarily to generalized fears of gene technology. Some people mistrust milk produced through genetic engineering, even though the milk itself is identical to other milk. Problems concerning public perception are not uncommon as gene technology makes an even greater impact on our lives.

Transgenic animals engineered to have specific desirable genes are becoming increasingly available to breeders. Now, instead of selectively breeding for several generations to produce a racehorse or a stud bull with desirable qualities, the process can be shortened by simply engineering such an animal right at the start.

Gene technology is revolutionizing agriculture, increasing yields and resistance to pests, improving nutritional value, and producing animals with desirable traits.

Cloning

The difficulty in using transgenic animals to improve livestock is in getting enough of them. Breeding produces offspring only slowly, and recombination acts to undo the painstaking work of the genetic engineer. Ideally, one would like to "Xerox" many exact genetic copies of the transgenic strain—but until 1997 it was commonly accepted that adult animals can't be cloned. Now the holy grail of agricultural genetic engineers seems within reach. In 1997, scientists announced the first successful cloning of differentiated vertebrate tissue, a lamb grown from a cell taken from an adult sheep. This startling result promises to revolutionize agricultural science.

Spemann's "Fantastical Experiment"

The idea of cloning animals was first suggested in 1938 by German embryologist Hans Spemann (called the "father of modern embryology"), who proposed what he called a "fantastical experiment": remove the nucleus from an egg cell, and put in its place a nucleus from another cell.

It was 14 years before technology advanced far enough for anyone to take up Spemann's challenge. In 1952, two American scientists, Robert Briggs and T. J. King, used very fine pipettes to suck the nucleus from a frog egg (frog eggs are unusually large, making the experiment feasible) and transfer a nucleus sucked from a body cell of an adult frog into its place. The experiment did not work when done this way, but partial success was achieved 18 years later by the British developmental biologist John Gurdon, who in 1970 inserted nuclei from advanced frog embryos rather than adult tissue. The frog eggs developed into tadpoles, but died before becoming adults.

The Path to Success

For 14 years, nuclear transplant experiments were attempted without success. Technology continued to advance however, until finally in 1984, Steen Willadsen, a Danish embryologist working in Texas, succeeded in cloning a sheep using a nucleus from a cell of an early embryo. This exciting result was soon replicated by others in a host of other organisms, including cattle, pigs, and monkeys.

Only early embryo cells seemed to work, however. Researchers became convinced that animal embryo cells become irreversibly "committed" after the first few cell divisions. After that, nuclei from differentiated animal cells cannot be used to clone entire organisms.

We now know this conclusion to have been unwarranted. The key advance for unraveling this puzzle was made in Scotland by geneticist Keith Campbell, a specialist in studying the cell cycle of agricultural animals. By the early 1990s, knowledge of how the cell cycle is controlled, advanced by cancer research, had led to an understanding that cells don't divide until conditions are appropriate. Just as a washing machine checks that the water has completely emptied before initiating the spin cycle, so the cell checks that everything needed is on hand before initiating cell division. Campbell reasoned: "Maybe the egg and the donated nucleus need to be at the same stage in the cell cycle."

This proved to be a key insight. In 1994 researcher Neil First, and in 1995 Campbell himself working with reproductive biologist Ian Wilmut, succeeded in cloning farm animals from advanced embryos by first starving the cells, so that they paused at the beginning of the cell cycle at the G_1 checkpoint. Two starved cells are thus synchronized at the same point in the cell cycle.

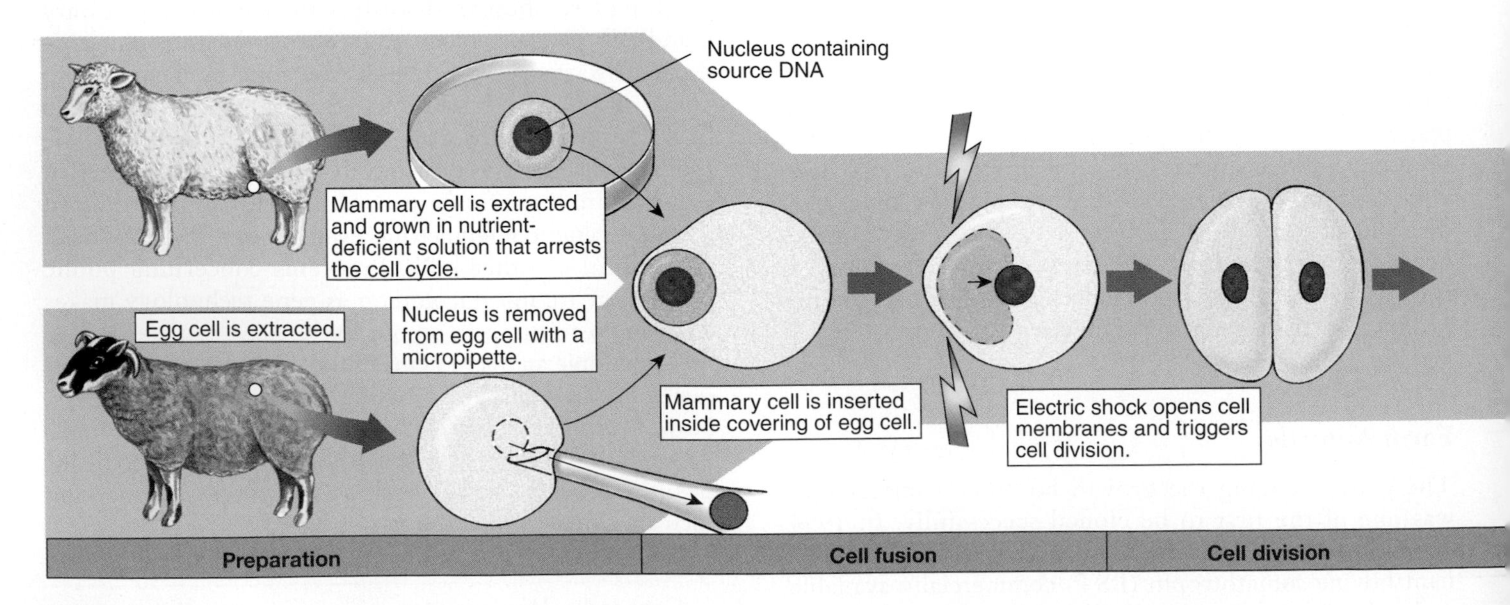

FIGURE 19.23
Wilmut's animal cloning experiment. Wilmut combined a nucleus from a mammary cell and an egg cell (with its nucleus removed) to successfully clone a sheep.

Wilmut's Lamb

Wilmut then set out to attempt the key breakthrough, the experiment that had eluded researchers since Spemann proposed it 59 years before: to transfer the nucleus from an adult differentiated cell into an enucleated egg, and allow the resulting embryo to grow and develop in a surrogate mother, hopefully producing a healthy animal.

Wilmut removed mammary cells from the udder of a six-year-old sheep (figure 19.23). The origin of these cells, gave the clone its name, "Dolly" after the country singer Dolly Parton. The cells were grown in tissue culture, and some frozen so that in the future it would be possible with genetic fingerprinting to prove that a clone was indeed genetically identical with the six-year-old sheep.

In preparation for cloning, Wilmut's team reduced for five days the concentration of serum on which the sheep mammary cells were subsisting. In parallel preparation, eggs obtained from a ewe were enucleated, the nucleus of each egg carefully removed with a micropipette.

Mammary cells and egg cells were then surgically combined in January of 1996, the mammary cells inserted inside the covering around the egg cell. Wilmut then applied a brief electrical shock. A neat trick, this causes the plasma membranes surrounding the two cells to become leaky, so that the contents of the mammary cell passes into the egg cell. The shock also kick-starts the cell cycle, causing the cell to begin to divide.

After six days, in 30 of 277 tries, the dividing embryo reached the hollow-ball "blastula" stage, and 29 of these were transplanted into surrogate mother sheep. A little over five months later, on July 5, 1997, one sheep gave birth to a lamb. This lamb, "Dolly," was the first successful clone generated from a differentiated animal cell.

The Future of Cloning

Wilmut's successful cloning of fully differentiated sheep cells is a milestone event in gene technology. Even though his procedure proved inefficient (only one of 277 trials succeeded), it established the point beyond all doubt that cloning of adult animal cells *can* be done. In the following four years researchers succeeded in greatly improving the efficiency of cloning. Seizing upon the key idea in Wilmut's experiment, to clone a resting-stage cell, they have returned to the nuclear transplant procedure pioneered by Briggs and King. It works well. Many different mammals have been successfully cloned including mice, pigs, and cattle.

Transgenic cloning can be expected to have a major impact on medicine as well as agriculture. Animals with human genes can be used to produce rare hormones. For example, sheep that have recently been genetically engineered to secrete a protein called alpha-1 antitrypsin (helpful in relieving the symptoms of cystic fibrosis) into their milk may be cloned, greatly cheapening the production of this expensive drug.

It is impossible not to speculate on the possibility of cloning a human. There is no reason to believe such an experiment would not work, but many reasons to question whether it should be done. Because much of Western thought is based on the concept of human individuality, we can expect the possibility of human cloning to engender considerable controversy.

Recent experiments have demonstrated the possibility of cloning differentiated mammalian tissue, opening the door for the first time to practical transgenic cloning of farm animals.

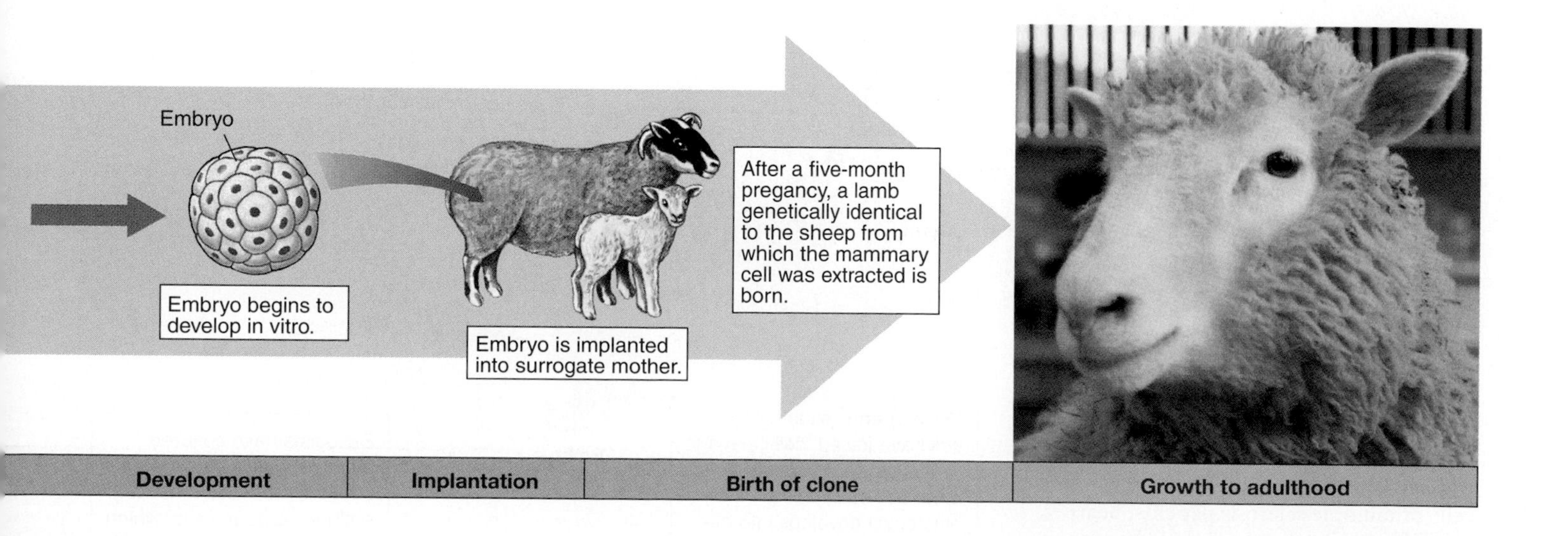
Embryo
Embryo begins to develop in vitro.
Embryo is implanted into surrogate mother.
After a five-month pregancy, a lamb genetically identical to the sheep from which the mammary cell was extracted is born.
Development
Implantation
Birth of clone
Growth to adulthood

Stem Cells

Since the isolation of embryonic stem cells in 1998, labs all over the world have been exploring the possibility of using stem cells to restore damaged or lost tissue. Exciting results are now starting to come in.

What is a stem cell? At the dawn of a human life, a sperm fertilizes an egg to create a single cell destined to become a child. As development commences, that cell begins to divide, producing a small ball of a few dozen cells. At this very early point, each of these cells is identical. We call these cells *embryonic stem cells*. Each one of them is capable by itself of developing into a healthy individual. In cattle breeding, for example, these cells are frequently separated by the breeder and used to produce multiple clones of valuable offspring.

The exciting promise of these embryonic stem cells is that, because they can develop into any tissue, they may give us the ability to restore damaged heart or spine tissue (figure 19.24). Experiments have already been tried successfully in mice. Heart muscle cells have been grown from mouse embryonic stem cells and successfully integrated with the heart tissue of a living mouse. This suggests that the damaged heart muscle of heart attack victims might be reparable with stem cells, and that injured spinal cords might be repairable. These very promising experiments are being pursued aggressively. They are, however, quite controversial, as embryonic stem cells are typically isolated from tissue of discarded or aborted embryos, raising serious ethical issues.

Tissue-Specific Stem Cells

New results promise a neat way around the ethical maze presented by stem cells derived from embryos. Go back for a moment to what we were saying about how a human child develops. What happens next to the embryonic stem cells? They start to take different developmental paths. Some become destined to form nerve tissue and, after this decision is taken, cannot ever produce any other kind of cell. They are then called nerve stem cells. Others become specialized to produce blood, still others muscle. Each major tissue is represented by its own kind of *tissue-specific stem cell*. Now here's the key point: as development proceeds, these tissue-specific stem cells persist. Even in adults. So why not use these adult cells, rather than embryonic stem cells?

Transplanted Tissue-Specific Stem Cells Cure MS in Mice

In pathfinding 1999 laboratory experiments by Dr. Evan Snyder of Harvard Medical School, tissue-specific stem cells were able to restore lost brain tissue. He and his coworkers injected neural stem cells (immediate descendants of embryonic stem cells able to become any kind of neural cell) into the brains of newborn mice with a disease resembling multiple sclerosis (MS). These mice lacked the cells that maintain the layers of myelin insulation around signal-conducting nerves. The injected stem cells migrated all over the brain, and were able to convert themselves into the missing type of cell. The new cells then proceeded to repair the ravages of the disease by replacing the lost insulation of signal-conducting nerve cells. Many of the treated mice fully recovered. In mice at least, tissue-specific stem cells offer a treatment for MS.

The approach seems very straightforward, and should apply to humans. Indeed, blood stem cells are already routinely used in humans to replenish the bone marrow of cancer patients after marrow-destroying therapy. The problem

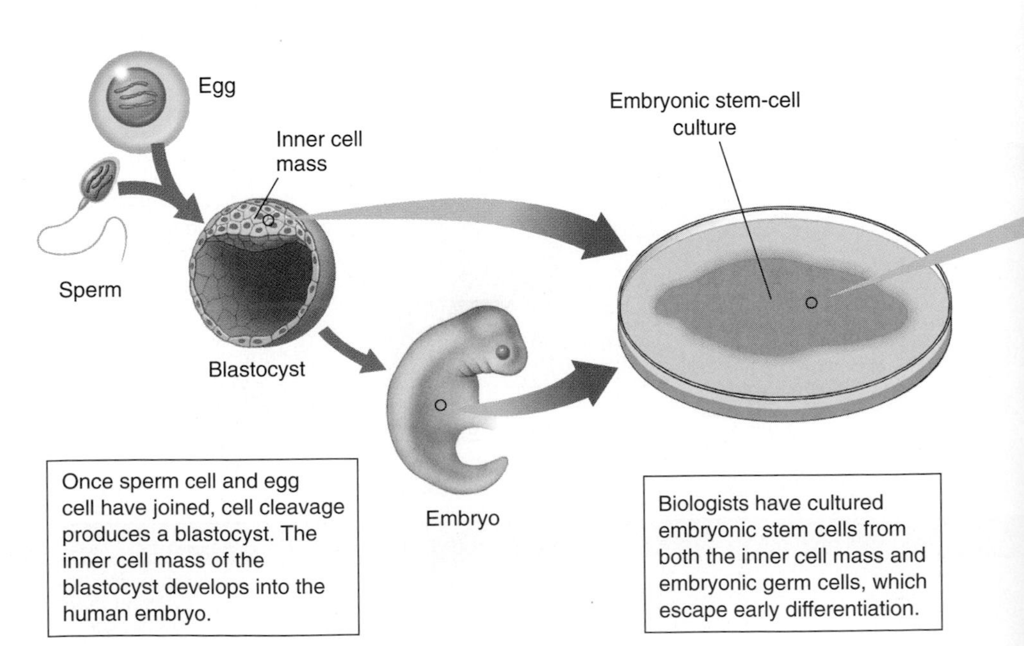
Egg
Inner cell mass
Sperm
Blastocyst
Embryo
Embryonic stem-cell culture
Once sperm cell and egg cell have joined, cell cleavage produces a blastocyst. The inner cell mass of the blastocyst develops into the human embryo.
Biologists have cultured embryonic stem cells from both the inner cell mass and embryonic germ cells, which escape early differentiation.

FIGURE 19.24
Using embryonic stem cells to restore damaged tissue. Embryronic stem cells can develop into any body tissue. Methods for growing the tissue and using it to repair damaged tissue in adults, such as the brain cells of multiple sclerosis patients, heart muscle, and spinal nerves, are being developed.

with extending the approach to other kinds of tissue-specific stem cells is that it has not always been easy to find the kind of tissue-specific stem cell you want.

Transplanted Stem Cells Reverse Juvenile Diabetes in Mice

Very promising experiments carried out in 2000 by Dr. Ammon Peck and a team of researchers at the University of Florida concern a particularly vexing problem, that of type 1 or juvenile diabetes. A person with juvenile diabetes lacks insulin-producing pancreas cells, because their immune system has mistakenly turned against them and destroyed them. They are no longer able to produce enough insulin to control their blood sugar levels and must take insulin daily. Adding back new insulin-producing cells called islet cells has been tried many times, but doesn't work well. Immune cells continue to destroy them.

Peck and his team reasoned, why not add instead the stem cells that produce islet cells? They would be able to produce a continuous supply of new islet cells, replacing those lost to immune attack. Because there would always be cells to make insulin, the diabetes would be cured.

No one knew just what such a stem cell looked like, but the researchers knew they come from the epithelial cells that line the pancreas ducts. Surely some must still lurk there unseen. So the research team took a bunch of these epithelial cells from mice and grew them in tissue culture until they had lots of them.

Were the stem cells they sought present in the cell culture they had prepared? Yes. In laboratory dishes the cell culture produced insulin in response to sugar, indicating islet cells had developed in the growing culture, islet cells that must have been produced from stem cells.

Now on to juvenile diabetes. The scientists injected their cell culture into the pancreas of mice specially bred to develop juvenile diabetes. Unable to manufacture their own insulin because they had no islet cells, these diabetic mice could not survive without daily insulin. What happened? The diabetes was reversed! The mice no longer required insulin.

Impatient to see in more detail what had happened, the researchers sacrificed the mice and examined the cells of their pancreas. The mice appeared to have perfectly normal islet cells.

One might have wished the researchers waited a little longer before terminating the experiment. It is not clear whether the cure was transitory or long term. Still, there is no escaping the conclusion that injection of a culture of adult stem cells cured their juvenile diabetes.

While certainly encouraging, a mouse is not a human, and there is no guarantee the approach will work in humans. But there is every reason to believe it might. The experiment is being repeated now with humans. People suffering from juvenile diabetes are being treated with human pancreatic duct cells obtained from people who have died and donated their organs for research. No ethical issues arise from using cells of adult organ donors, and initial results look promising.

Transplanted stem cells may allow us to replace damaged or lost tissue, offering cures for many disorders that cannot now be treated. Current work focuses on tissue-specific stem cells, which do not present the ethical problems that embryonic stem cells do.

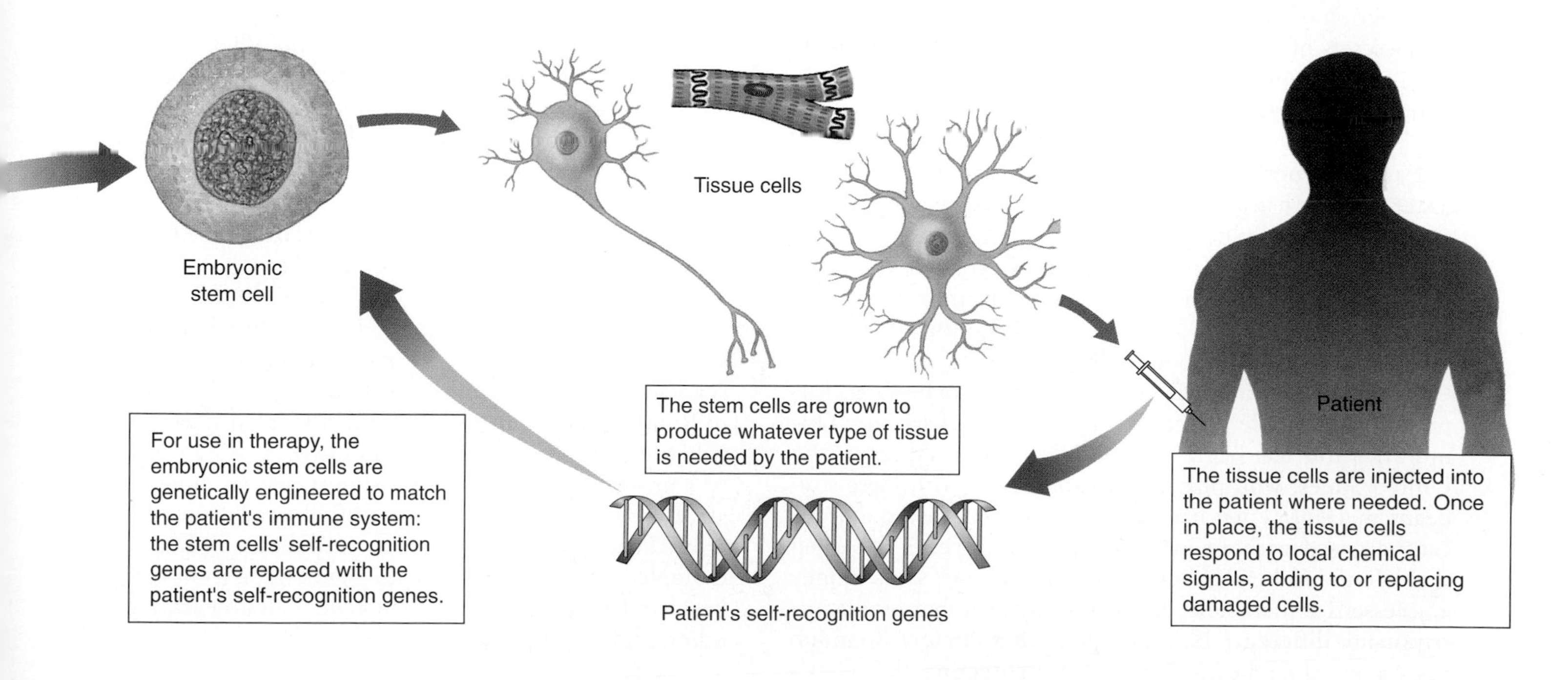
Embryonic stem cell
Tissue cells
Patient
The stem cells are grown to produce whatever type of tissue is needed by the patient.
For use in therapy, the embryonic stem cells are genetically engineered to match the patient's immune system: the stem cells' self-recognition genes are replaced with the patient's self-recognition genes.
The tissue cells are injected into the patient where needed. Once in place, the tissue cells respond to local chemical signals, adding to or replacing damaged cells.
Patient's self-recognition genes

Ethics and Regulation

The advantages afforded by genetic engineering are revolutionizing our lives. But what are the disadvantages, the potential costs and dangers of genetic engineering? Many people, including influential activists and members of the scientific community, have expressed concern that genetic engineers are "playing God" by tampering with genetic material. For instance, what would happen if one fragmented the DNA of a cancer cell, and then incorporated the fragments at random into vectors that were propagated within bacterial cells? Might there not be a danger that some of the resulting bacteria would transmit an infective form of cancer? Could genetically engineered products administered to plants or animals turn out to be dangerous for consumers after several generations? What kind of unforeseen impact on the ecosystem might "improved" crops have? Is it ethical to create "genetically superior" organisms, including humans?

How Do We Measure the Potential Risks of Genetically Modified Crops?

While the promise of genetic engineering is very much in evidence, this same genetic engineering has been the cause of outright war between researchers and protesters. In June 1999, British protesters attacked an experimental plot of genetically modified (GM) sugar beets; the following August they destroyed a test field of GM canola (used for cooking oil and animal feed). The contrast could not be more marked between American acceptance of genetically modified crops on the one hand, and European distrust of genetically modified foods, on the other. The intense feelings generated by this dispute point to the need to understand how we measure the risks associated with the genetic engineering of plants.

Two sets of risks need to be considered. The first stems from eating genetically modified foods, the other concerns potential ecological effects.

Is Eating Genetically Modified Food Dangerous? Protesters worry that genetically modified food may have been rendered somehow dangerous. To sort this out, it is useful to bear in mind that bioengineers modify crops in two quite different ways. One class of gene modification makes the crop easier to grow; a second class of modification is intended to improve the food itself.

The introduction of Roundup-resistant soybeans to Europe is an example of the first class of modification. This modification has been very popular with farmers in the United States, who planted half their crop with these soybeans in 1999. They like GM soybeans because the beans can be raised without intense cultivation (weeds are killed with Roundup herbicide instead), which both saves money and lessens soil erosion. But is the soybean that results nutritionally different? No. The gene that confers Roundup resistance in soybeans does so by protecting the plant's ability to manufacture so-called "aromatic" amino acids. In unprotected weeds, by contrast, Roundup blocks this manufacturing process, killing the weed. Because humans don't make any aromatic amino acids anyway (we get them in our diets), Roundup doesn't hurt us. The GM soybean we eat is nutritionally the same as an "organic" one, just cheaper to produce.

In the second class of modification, where a gene is added to improve the nutritional character of some food, the food will be nutritionally different. In each of these instances, it is necessary to examine the possibility that consumers may prove allergic to the product of the introduced gene. In one instance, for example, addition of a methionine-enhancing gene from Brazil nut into soybeans (which are deficient in this amino acid) was discontinued when six of eight individuals allergic to Brazil nuts produced antibodies to the GM soybeans, suggesting the possibility of a reverse reaction. Instead, methionine levels in GM crops are being increased with genes from sunflowers. Screening for allergy problems is now routine.

On both scores, then, the risk of bioengineering to the food supply seems to be very slight. GM foods to date seem completely safe.

Are GM Crops Harmful to the Environment? What are we to make of the much-publicized report that Monarch butterflies might be killed by eating pollen blowing out of fields planted with GM corn? First, it should come as no surprise. The GM corn (so-called Bt corn) was engineered to contain an insect-killing toxin (harmless to people) in order to combat corn borer pests. Of course it will kill any butterflies or other insects in the immediate vicinity of the field. However, focus on the fact that the GM corn fields do not need to be sprayed with pesticide to control the corn borer. An estimated $9 billion in damage is caused annually by the application of pesticides in the United States, and billions of insects and other animals, including an estimated 67 million birds, are killed each year. This pesticide-induced murder of wildlife is far more damaging ecologically than any possible effects of GM crops on butterflies.

Will pests become resistant to the GM toxin? Not nearly as fast as they now become resistant to the far higher levels of chemical pesticide we spray on crops.

How about the possibility that introduced genes will pass from GM crops to their wild or weedy relatives? This sort of gene flow happens naturally all the time, and so this is a legitimate question. But so what if genes for resistance to Roundup herbicide spread from cultivated sugar beets to wild populations of sugar beets in Europe? Why would that be a problem? Besides, there is almost never a potential relative around to receive the modified gene from the GM crop. There are no wild relatives of soybeans in Europe, for example. Thus, there can be no gene escape from GM soybeans in Europe, any more than genes can flow from you to other kinds of animals.

Calvin and Hobbes by Bill Watterson

On either score, then, the risk of bioengineering to the environment seems to be very slight. Indeed, in some cases it lessens the serious environmental damage produced by cultivation and agricultural pesticides.

Should We Label Genetically Modified Foods?

While there seems little tangible risk in the genetic modification of crops, public assurance that these risks are being carefully assessed is important. Few issues manage to raise the temperature of discussions about plant genetic engineering more than labeling of genetically modified (GM) crops. Agricultural producers have argued that there are no demonstrable risks, so that a GM label can only have the function of scaring off wary consumers. Consumer advocates respond that consumers have every right to make that decision, and to the information necessary to make it.

In considering this matter, it is important to separate two quite different issues, the *need* for a label, and the *right* of the public to have one. Every serious scientific investigation of the risks of GM foods has concluded that they are safe—indeed, in the case of soybeans and many other crops modified to improve cultivation, the foods themselves are not altered in any detectable way, and no nutritional test could distinguish them from "organic" varieties. So there seems to be little if any health need for a GM label for genetically engineered foods.

The right of the public to know what it is eating is a very different issue. There is widespread fear of genetic manipulation in Europe, because it is unfamiliar. People there don't trust their regulatory agencies as we do here, because their agencies have a poor track record of protecting them. When they look at genetically modified foods, they are haunted by past experiences of regulatory ineptitude. In England they remember British regulators' failure to protect consumers from meat infected with mad cow disease.

It does no good whatsoever to tell a fearful European that there is no evidence to warrant fear, no trace of data supporting danger from GM crops. A European consumer will simply respond that the harm is not yet evident, that we don't know enough to see the danger lurking around the corner. "Slow down," the European consumers say. "Give research a chance to look around all the corners. Lets be sure." No one can argue against caution, but it is difficult to imagine what else researchers can look into—safety has been explored very thoroughly. The fear remains, though, for the simple reason that no amount of information can remove it. Like a child scared of a monster under the bed, looking under the bed again doesn't help—the monster still might be there next time. And that means we are going to have to have GM labels, for people have every right to be informed about something they fear.

What should these labels be like? A label that only says "GM FOOD" simply acts as a brand—like a POISON label, it shouts a warning to the public of lurking danger. Why not instead have a GM label that provides information to the consumer, that tells the consumer what regulators know about that product?

(*For Bt corn*): The production of this food was made more efficient by the addition of genes that made plants resistant to pests so that less pesticides were required to grow the crop.

(*For Roundup-ready soybeans*): Genes have been added to this crop to render it resistant to herbicides—this reduces soil erosion by lessening the need for weed-removing cultivation.

(*For high beta-carotene rice*): Genes have been added to this food to enhance its beta-carotene content and so combat vitamin A deficiency.

GM food labels that in each instance actually tell consumers what has been done to the gene-modified crop would go a long way toward hastening public acceptance of gene technology in the kitchen.

Genetic engineering affords great opportunities for progress in medicine and food production, although many are concerned about possible risks. On balance, the risks appear slight, and the potential benefits substantial.

Chapter 19

www.mhhe.com/raven6/resources19.mhtml

Summary | **Questions** | **Media Resources**

19.1 The ability to manipulate DNA has led to a new genetics.

- Genetic engineering involves the isolation of specific genes and their transfer to new genomes.
- An important component of genetic engineering technology is a special class of enzymes called restriction endonucleases, which cleave DNA molecules into fragments.
- The first such recombinant DNA was made by Cohen and Boyer in 1973, when they inserted a frog ribosomal RNA gene into a bacterial plasmid.

1. Why do the ends of the DNA fragments created by restriction endonucleases enable fragments from different genomes to be spliced together?

- Recombinant DNA Technology

- How Scientists Think: The First Genetically Engineered Organism (Cohen/Boyer/Berg)

19.2 Genetic engineering involves easily understood procedures.

- Genetic engineering experiments consist of four stages: isolation of DNA, production of recombinant DNA, cloning, and screening for the gene(s) of interest.
- Preliminary screening can be accomplished by making the desired clones resistant to an antibiotic; hybridization can then be employed to identify the gene of interest.
- Gene technologies, including PCR, Southern blotting, and RFLP analysis, enable researchers to isolate genes and produce them in large quantities.

2. Describe the procedure used to eliminate clones that have not incorporated a vector in a genetic engineering experiment.

3. What is used as a probe in a Southern blot? With what does the probe hybridize? How are the regions of hybridization visualized?

- Polymerase Chain Reaction

- *On Science* Articles:
  - -How Genetic Engineering Is Done
  - -Altering ANDi: Genetic Engineering Gets a (Small) Step Closer to Humans

19.3 Biotechnology is producing a scientific revolution.

- Extensive research on the human genome has yielded important information about the location of genes and the comparisons of the human genome with other organisms.
- Biochip technology, also called gene microarray, will expand the analysis of DNA.
- Gene splicing holds great promise as a clinical tool, particularly in the prevention of disease with bioengineered vaccines.
- A major focus of genetic engineering activity has been agriculture, where genes conferring resistance to herbicides or insect pests have been incorporated into crop plants.
- Recent experiments open the way for cloning of genetically altered animals and suggest that human cloning is feasible.
- The impact of genetic engineering has skyrocketed over the past decade, providing many useful innovations for society; its moral and ethical aspects still provide a topic for heated debates.

4. What is the primary vector used to introduce genes into plant cells? What types of plants are generally infected by this vector? Describe three examples of how this vector has been used for genetic engineering, and explain the agricultural significance of each example.

5. How is the genetic engineering of bovine somatotropin (BST) used to increase milk production in the dairy industry? What effect would BST in milk have on persons who drink it?

- Exploration: DNA from Real Court Cases

- *On Science* Articles:
  - -Who Should Own the Secrets of Your Genes?
  - -The Real Promise of Plant Genetic Engineering
  - -The Search for the Natural Relatives of Cassava
  - - Should a Clone Have Rights?
  - -Renouncing the Terminator
  - -Biochips and Personal Privacy

- Art Quizzes:
  - -Bovine Somatotropin
  - -Cloning Experiment

Part **VI**

Evolution

Catching Evolution in Action

To study evolution, biologists have traditionally investigated what has happened in the past, sometimes many millions of years ago. To learn about dinosaurs, a paleontologist looks at dinosaur fossils. To study human evolution, an anthropologist looks at human fossils and, increasingly, examines the "family tree" of mutations that have accumulated in human DNA over millions of years. For the biologists taking this traditional approach, evolutionary biology is similar to astronomy and history, relying on observation and deduction rather than experiment and induction to examine ideas about past events.

However, evolutionary biology is not entirely an observational science. In recent years many case studies of natural populations have demonstrated that in some circumstances evolutionary change can occur rapidly. In these instances, it is possible to establish experimental studies to directly test evolutionary hypotheses. Although laboratory studies on fruit flies and other organisms have been common for more than 50 years, it has only been in recent years that scientists have started conducting experimental studies of evolution in nature.

To conduct experimental tests of evolution, it is first necessary to identify a population in nature upon which selection might be operating. By manipulating the strength of the selection, an investigator can predict what outcome selection might produce, then look and see the actual effect on the population.

Guppies offer an excellent experimental opportunity. The guppy, *Poecilia reticulata*, is found in small streams in Venezuela and the nearby island of Trinidad. In Trinidad, guppies are found in many mountain streams. One interesting feature of several streams is that they have waterfalls. Amazingly, guppies are capable of colonizing portions of the stream above the waterfall. During flood seasons, rivers sometimes swell, reducing the depth of waterfalls. During these occasions, guppies may be able to jump these barriers and invade pools above waterfalls. By contrast, not all species are capable of such dispersal and thus are only found in these streams below the first waterfall. One species whose distribution is restricted by waterfalls is the pike cichlid, *Crenicichla alta*, a voracious predator that feeds on guppies and other fish.

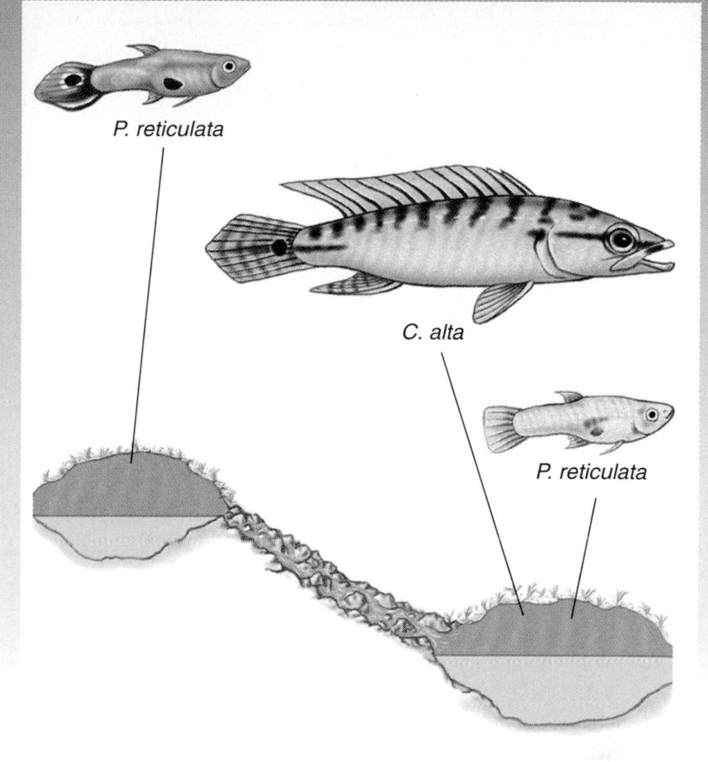

The evolution of protective coloration in guppies. In pools below waterfalls where predation is high, guppies (*Poecilia reticulata*) are drab colored. In the absence of the highly predatory pike cichlid (*Crenicichla alta*), guppies in pools above waterfalls are much more colorful and attractive to females. The evolution of these differences can be experimentally tested.

Because of these barriers to dispersal, guppies can be found in two very different environments. In pools just below the waterfalls, predation is a substantial risk and rates of survival are relatively low. By contrast, in similar pools just above the waterfall, few predators prey on guppies. As a result, guppy populations above and below waterfalls have evolved many differences. In the high-predation pools, guppies exhibit drab coloration. Moreover, they tend to reproduce at a younger age.

The differences suggest the action of natural selection. Perhaps as a result of shunting energy to reproduction rather than growth, the fish in high-predation pools attain relatively smaller adult sizes. By contrast, male fish above the waterfall display gaudy colors and spots that they use to court females (see figure above). Adults there mature later and grow to larger sizes.

Evolution does not offer the only explanation for these observations. Perhaps, for example, only very large fish are capable of jumping past the waterfall to colonize pools. If this were the case, then a founder effect would occur in which the new population was established solely by individuals with genes for large size.

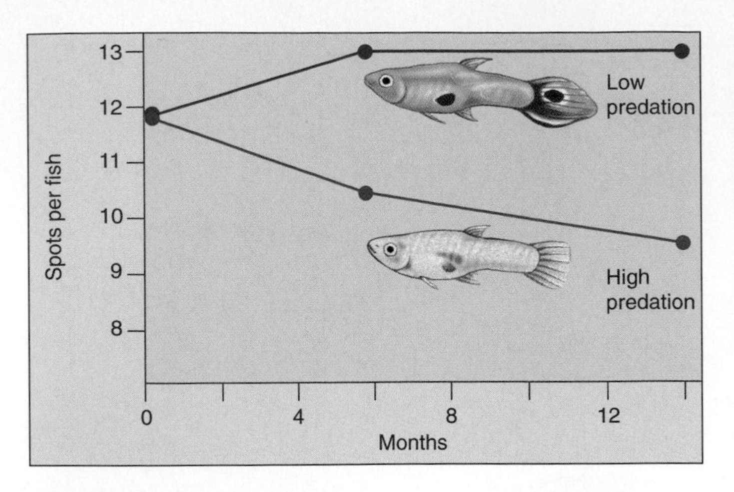

Evolutionary change in spot number. Guppies raised in the low-predation environment had a greater number of spots, whereas selection in more dangerous environments, like the predator-filled pools, led to less conspicuous fish (*above left*). The same results are seen in field experiments in pools above and below waterfalls (*above right*).

The Experiments

The only way to rule out such alternative possibilities is to conduct a controlled experiment. A classic set of laboratory and field experiments carried out by John Endler in the late 1970s (now at the University of California, Santa Barbara) first attempted to demonstrate that natural selection was acting on these Trinidad guppies.

Laboratory Experiment. Endler constructed a series of ten large artificial ponds in a laboratory greenhouse, with size and color of gravel designed to mimic the different background patterns found in the natural streams of Trinidad. In each pond he raised a diverse population of guppies, mixing the ten populations as they grew so that all the populations had a similar range of genetic diversity.

He then added cichlid predators (*C. alta*) to four of the pools and killifish (which rarely prey on guppies) to another four, with the remaining pools left as "no predator" controls. The populations were then allowed to evolve with or without predators.

Field Experiment. In a parallel field experiment, Endler captured drab guppies from a stream they shared with cichlid predators, and released them upstream above a waterfall where there were no cichlids or guppies.

As a key follow-up, David Reznick of the University of California, Riverside, reexamined the evolving population of Trinidad guppies 11 years after its initial transfer by Endler from a high-predation community to a low-predation one.

Results

Laboratory Experiment. Fourteen months after adding the predator fish to the ponds (which corresponds to 10 guppy generations), Endler compared the populations (see graph). The guppies in the killifish and control pools were indistinguishable, all brightly colored. In contrast, the surviving guppies in the pike cichlid pools were drab in coloration. Predation greatly reduced bright coloration in guppies.

Field Experiment. Two years after transferring the drab guppies to a predator-free environment—that is, after 15 generations of relaxed selection pressures—the drab color patterns of the guppy population had shifted toward the more complex and colorful pattern typical of guppy populations living where there are no predators.

Endler's initial field results offer an exciting but sketchy picture of evolution in action. Many questions immediately suggest themselves. Is color pattern the only trait under selection by predation? What about the number of offspring, and their growth rate? Can we actually measure how fast selection is acting?

These questions were addressed in an extensive follow-up analysis by David Reznick of the University of California, Riverside, Frank and Ruth Shaw of the University of Minnesota, and Helen Rodd of the University of California, Davis. They reexamined the guppy population 11 years after transfer, examining a wide array of characters designed to reveal not only the physical appearance of the guppies, but also their investment in reproduction.

Reznick's team found that the descendants of the transplanted guppies were not only more brightly colored, they matured at a later age and were larger in size than the control population living below the barrier waterfall with cichlid predators. They also produced fewer but larger offspring per litter, devoting a smaller proportion of their resources to reproduction. In a word, their life histories had evolved to resemble that of guppies living in low-predation or predator-free communities.

The speed of evolutionary change is measured in darwins, the proportional amount of change per unit time. Reznick's team estimates the guppies evolved at a rate of up to 45,000 darwins. By comparison, rates of change measured in the fossil record are only one-tenth to 1 darwin. Apparently evolution can sometimes act very fast.

To explore this experiment further, go to the Virtual Lab at www.mhhe.com/raven6/vlab6.mhtml

20

Genes within Populations

FIGURE 20.1
Genetic variation. The range of genetic material in a population is expressed in a variety of ways—including color.

Concept Outline

20.1 Genes vary in natural populations.

Gene Variation Is the Raw Material of Evolution. Selection acts on the genetic variation present in populations, favoring variants that increase the likelihood of survival and reproduction.

Gene Variation in Nature. Natural populations contain considerable amounts of variation, present at the DNA level and expressed in proteins.

20.2 Why do allele frequencies change in populations?

The Hardy–Weinberg Principle. The proportion of homozygotes and heterozygotes in a population is not altered by meiosis or sexual reproduction.

Five Agents of Evolutionary Change. The frequency of alleles in a population can be changed by evolutionary forces like mutation, gene flow, non random mating, genetic drift, and selection.

Identifying the Evolutionary Forces Maintaining Polymorphism. A number of processes can influence allele frequencies in natural populations, but it is difficult to ascertain their relative importance.

Heterozygote Advantage. In some cases, heterozygotes are superior to either type of homozygote. The gene for sickle cell anemia is one particularly well-understood example.

20.3 Selection can act on traits affected by many genes.

Forms of Selection. Selection can act on traits like height or weight to stabilize or change the level at which the trait is expressed.

Limits to What Selection Can Accomplish. Selection cannot act on traits with little or no genetic variation.

No other human being is exactly like you (unless you have an identical twin). Often the particular characteristics of an individual have an important bearing on its survival, on its chances to reproduce, and on the success of its offspring. Evolution is driven by such consequences. Genetic variation that influences these characteristics provides the raw material for natural selection, and natural populations contain a wealth of such variation. In plants (figure 20.1), insects, and vertebrates, practically every gene exhibits some level of variation. In this chapter, we will explore genetic variation in natural populations and consider the evolutionary forces that cause allele frequencies in natural populations to change. These deceptively simple matters lie at the core of evolutionary biology.

20.1 Genes vary in natural populations.

Gene Variation Is the Raw Material of Evolution

Evolution Is Descent with Modification

The word *"evolution"* is widely used in the natural and social sciences. It refers to how an entity—be it a social system, a gas, or a planet—changes through time. Although development of the modern concept of evolution in biology can be traced to Darwin's *On the Origin of Species*, the first five editions of this book never actually used the term! Rather, Darwin used the phrase "descent with modification." Although many more complicated definitions have been proposed, Darwin's phrase probably best captures the essence of biological evolution: all species arise from other, pre-existing species. However, through time, they accumulate differences such that ancestral and descendant species are not identical.

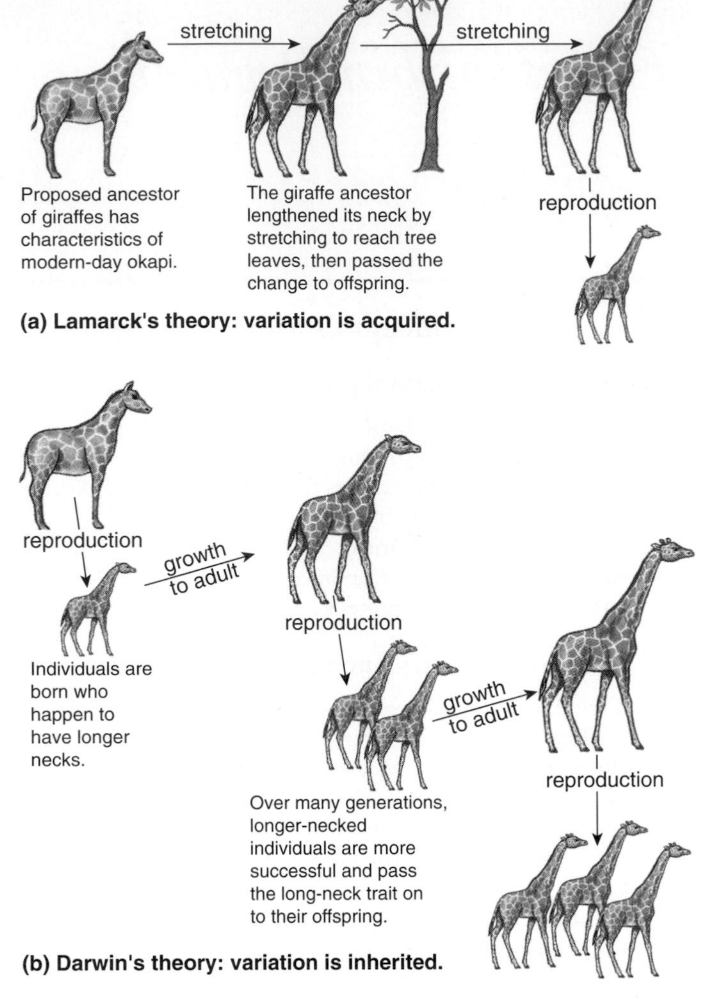

stretching
stretching
Proposed ancestor of giraffes has characteristics of modern-day okapi.
The giraffe ancestor lengthened its neck by stretching to reach tree leaves, then passed the change to offspring.
reproduction
(a) Lamarck's theory: variation is acquired.
reproduction
growth to adult
Individuals are born who happen to have longer necks.
reproduction
growth to adult
reproduction
Over many generations, longer-necked individuals are more successful and pass the long-neck trait on to their offspring.
(b) Darwin's theory: variation is inherited.

FIGURE 20.2
How did giraffes evolve a long neck?

Natural Selection Is an Important Mechanism of Evolutionary Change

Darwin was not the first to propose a theory of evolution. Rather, he followed a long line of earlier philosophers and naturalists who deduced that the many kinds of organisms around us were produced by a process of evolution. Unlike his predecessors, however, Darwin proposed **natural selection** as the mechanism of evolution. Natural selection produces evolutionary change when in a population some individuals, which possess certain inherited characteristics, produce more surviving offspring than individuals lacking these characteristics. As a result, the population will gradually come to include more and more individuals with the advantageous characteristics. In this way, the population evolves and becomes better adapted to its local circumstances.

Natural selection was by no means the only evolutionary mechanism proposed. A rival theory, championed by the prominent biologist Jean-Baptiste Lamarck, was that evolution occurred by the **inheritance of acquired characteristics.** According to Lamarck, individuals passed on to offspring body and behavior changes acquired during their lives. Thus, Lamarck proposed that ancestral giraffes with short necks tended to stretch their necks to feed on tree leaves, and this extension of the neck was passed on to subsequent generations, leading to the long-necked giraffe (figure 20.2*a*). In Darwin's theory, by contrast, the variation is not created by experience, but is the result of preexisting genetic differences among individuals (figure 20.2*b*).

Although the efficacy of natural selection is now widely accepted, it is not the only process that can lead to changes in the genetic makeup of populations. Allele frequencies can also change as the result of repeated mutations from one allele to another and from migrants bringing alleles into a population. In addition, when populations are small, the frequencies of alleles can change randomly as the result of chance events. Evolutionary biologists debate the relative strengths of these processes. Although no one denies that natural selection is a powerful force leading to adaptive change, the importance of other processes is less certain.

Darwin proposed that natural selection on variants within populations leads to the evolution of different species.

Gene Variation in Nature

Evolution within a species may result from any process that causes a change in the genetic composition of a population. In considering this theory of population genetics, it is best to start by looking at the genetic variation present among individuals within a species. This is the raw material available for the selective process.

Measuring Levels of Genetic Variation

As we saw in chapter 13, a natural population can contain a great deal of genetic variation. This is true not only of humans, but of all organisms. How much variation usually occurs? Biologists have looked at many different genes in an effort to answer this question:

1. **Blood groups.** Chemical analysis has revealed the existence of more than 30 blood group genes in humans, in addition to the ABO locus. At least a third of these genes are routinely found in several alternative allelic forms in human populations. In addition to these, there are more than 45 variable genes encoding other proteins in human blood cells and plasma which are not considered blood groups. Thus, there are more than 75 genetically variable genes in this one system alone.
2. **Enzymes.** Alternative alleles of genes specifying particular enzymes are easy to distinguish by measuring how fast the alternative proteins migrate in an electric field (a process called **electrophoresis**). A great deal of variation exists at enzyme-specifying loci. About 5% of the enzyme loci of a typical human are heterozygous: if you picked an individual at random, and in turn selected one of the enzyme-encoding genes of that individual at random, the chances are 1 in 20 (5%) that the gene you selected would be heterozygous in that individual.

Considering the entire human genome, it is fair to say that almost all people are different from one another. This is also true of other organisms, except for those that reproduce asexually. In nature, genetic variation is the rule.

Enzyme Polymorphism

Many loci in a given population have more than one allele at frequencies significantly greater than would occur from mutation alone. Researchers refer to a locus with more variation than can be explained by mutation as **polymorphic** (*poly*, "many," *morphic*, "forms") (figure 20.3). The extent of such variation within natural populations was not even suspected a few decades ago, until modern techniques such as gel electrophoresis made it possible to examine enzymes and other proteins directly. We now know that most populations of insects and plants are polymorphic (that is, have more than one allele occurring at a frequency greater than 5%) at more than half of their enzyme-encoding loci, although vertebrates are somewhat less polymorphic. **Heterozygosity** (that is, the probability that a randomly selected gene will be heterozygous for a randomly selected individual) is about 15% in *Drosophila* and other invertebrates, between 5% and 8% in vertebrates, and around 8% in outcrossing plants. These high levels of genetic variability provide ample supplies of raw material for evolution.

FIGURE 20.3
Polymorphic variation. These Australian snails, all of the species *Bankivia fasciata*, exhibit considerable variation in pattern and color. Individual variations are heritable and passed on to offspring.

DNA Sequence Polymorphism

With the advent of gene technology, it has become possible to assess genetic variation even more directly by sequencing the DNA itself. In a pioneering study in 1989, Martin Kreitman sequenced ADH genes isolated from 11 individuals of the fruit fly *Drosophila melanogaster*. He found 43 variable sites, only one of which had been detected by protein electrophoresis! In the following decade, numerous other studies of variation at the DNA level have confirmed these findings: abundant variation exists in both the coding regions of genes and in their nontranslated introns—considerably more variation than we can detect examining enzymes with electrophoresis.

Natural populations contain considerable amounts of genetic variation—more than can be accounted for by mutation alone.

20.2 Why do allele frequencies change in populations?

Population genetics is the study of the properties of genes in populations. Genetic variation within natural populations was a puzzle to Darwin and his contemporaries. The way in which meiosis produces genetic segregation among the progeny of a hybrid had not yet been discovered. Selection, scientists then thought, should always favor an optimal form, and so tend to eliminate variation. Moreover, the theory of **blending inheritance**—in which offspring were expected to be phenotypically intermediate relative to their parents—was widely accepted. If blending inheritance were correct, then the effect of any new genetic variant would quickly be diluted to the point of disappearance in subsequent generations.

The Hardy–Weinberg Principle

Following the rediscovery of Mendel's research, two people in 1908 independently solved the puzzle of why genetic variation persists—G. H. Hardy, an English mathematician, and G. Weinberg, a German physician. They pointed out that the original proportions of the genotypes in a population will remain constant from generation to generation, as long as the following assumptions are met:

1. The population size is very large.
2. Random mating is occurring.
3. No mutation takes place.
4. No genes are input from other sources (no immigration takes place).
5. No selection occurs.

Dominant alleles do not, in fact, replace recessive ones. Because their proportions do not change, the genotypes are said to be in **Hardy–Weinberg equilibrium.**

In algebraic terms, the Hardy–Weinberg principle is written as an equation. Consider a population of 100 cats, with 84 black and 16 white cats. In statistics, **frequency** is defined as the proportion of individuals falling within a certain category in relation to the total number of individuals under consideration. In this case, the respective frequencies would be 0.84 (or 84%) and 0.16 (or 16%). Based on these phenotypic frequencies, can we deduce the underlying frequency of genotypes? If we assume that the white cats are homozygous recessive for an allele we designate *b*, and the black cats are therefore either homozygous dominant *BB* or heterozygous *Bb*, we can calculate the **allele frequencies** of the two alleles in the population from the proportion of black and white individuals. Let the letter p designate the frequency of one allele and the letter q the frequency of the alternative allele. Because there are only two alleles, p plus q must always equal 1.

The Hardy-Weinberg equation can now be expressed in the form of what is known as a binomial expansion:

$$(p + q)^2 = p^2 + 2pq + q^2$$

p^2	$2pq$	q^2
(Individuals homozygous for allele *B*)	(Individuals heterozygous with alleles *B* + *b*)	(Individuals homozygous for allele *b*)

If $q^2 = 0.16$ (the frequency of white cats), then $q = 0.4$. Therefore, p, the frequency of allele *B*, would be 0.6 ($1.0 - 0.4 = 0.6$). We can now easily calculate the **genotype frequencies:** there are $p^2 = (0.6)^2 \times 100$ (the number of cats in the total population), or 36 homozygous dominant *BB* individuals. The heterozygous individuals have the *Bb* genotype, and there would be $2pq$, or $(2 \times 0.6 \times 0.4) \times 100$, or 48 heterozygous *Bb* individuals.

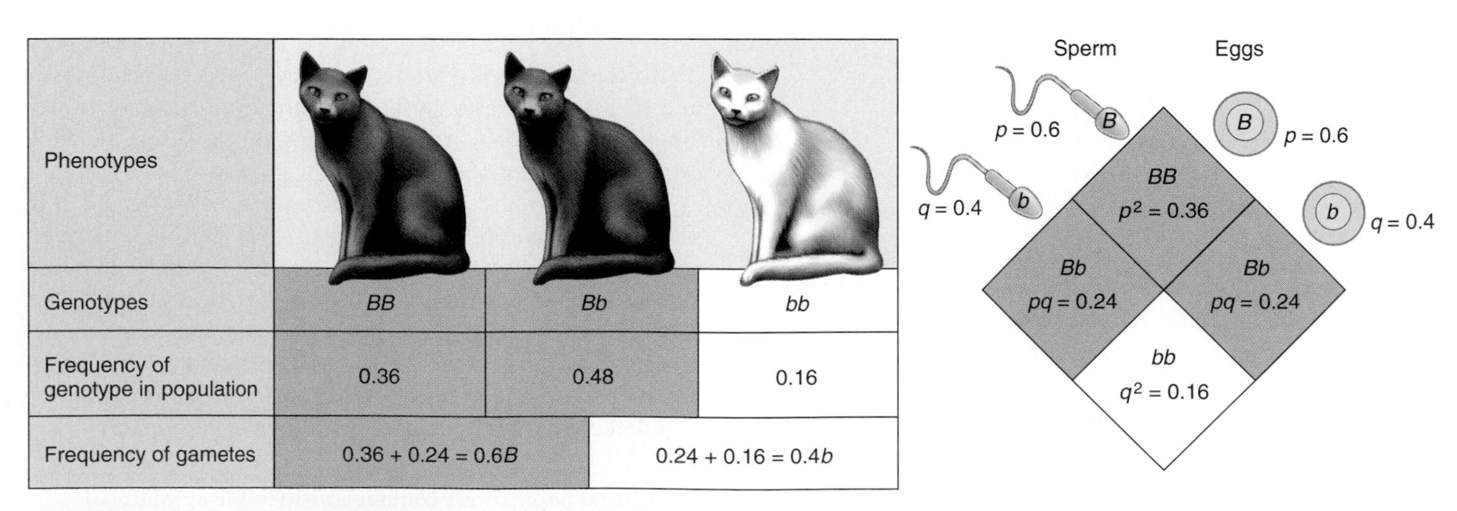

FIGURE 20.4
The Hardy–Weinberg equilibrium. In the absence of factors that alter them, the frequencies of gametes, genotypes, and phenotypes remain constant generation after generation.

Using the Hardy–Weinberg Equation

The Hardy–Weinberg equation is a simple extension of the Punnett square described in chapter 13, with two alleles assigned frequencies p and q. Figure 20.4 allows you to trace genetic reassortment during sexual reproduction and see how it affects the frequencies of the *B* and *b* alleles during the next generation. In constructing this diagram, we have assumed that the union of sperm and egg in these cats is random, so that all combinations of *b* and *B* alleles occur. For this reason, the alleles are mixed randomly and represented in the next generation in proportion to their original representation. Each individual egg or sperm in each generation has a 0.6 chance of receiving a *B* allele ($p = 0.6$) and a 0.4 chance of receiving a *b* allele ($q = 0.4$).

In the next generation, therefore, the chance of combining two *B* alleles is p^2, or 0.36 (that is, 0.6×0.6), and approximately 36% of the individuals in the population will continue to have the *BB* genotype. The frequency of *bb* individuals is q^2 (0.4×0.4) and so will continue to be about 16%, and the frequency of *Bb* individuals will be $2pq$ ($2 \times 0.6 \times 0.4$), or approximately 48%. Phenotypically, if the population size remains at 100 cats, we will still see approximately 84 black individuals (with either *BB* or *Bb* genotypes) and 16 white individuals (with the *bb* genotype) in the population. Allele, genotype, and phenotype frequencies have remained unchanged from one generation to the next.

This simple relationship has proved extraordinarily useful in assessing actual situations. Consider the recessive allele responsible for the serious human disease cystic fibrosis. This allele is present in North Americans of Caucasian descent at a frequency q of about 22 per 1000 individuals, or 0.022. What proportion of North American Caucasians, therefore, is expected to express this trait? The frequency of double recessive individuals (q^2) is expected to be 0.022×0.022, or 1 in every 2000 individuals. What proportion is expected to be heterozygous carriers? If the frequency of the recessive allele q is 0.022, then the frequency of the dominant allele p must be $1 - 0.022$, or 0.978. The frequency of heterozygous individuals ($2pq$) is thus expected to be $2 \times 0.978 \times 0.022$, or 43 in every 1000 individuals.

How valid are these calculated predictions? For many genes, they prove to be very accurate. As we will see, for some genes the calculated predictions do *not* match the actual values. The reasons they do not tell us a great deal about evolution.

Why Do Allele Frequencies Change?

According to the Hardy–Weinberg principle, both the allele and genotype frequencies in a large, random-mating population will remain constant from generation to generation if no mutation, no gene flow, and no selection occur. The stipulations tacked onto the end of the statement are important. In fact, they are the key to the importance of the Hardy–Weinberg principle, because individual allele frequencies often change in natural populations, with some alleles becoming more common and others decreasing in frequency. The Hardy–Weinberg principle establishes a convenient baseline against which to measure such changes. By looking at how various factors alter the proportions of homozygotes and heterozygotes, we can identify the forces affecting particular situations we observe.

Many factors can alter allele frequencies. Only five, however, alter the proportions of homozygotes and heterozygotes enough to produce significant deviations from the proportions predicted by the Hardy–Weinberg principle: mutation, gene flow (including both immigration into and emigration out of a given population), nonrandom mating, genetic drift (random change in allele frequencies, which is more likely in small populations), and selection (table 20.1). Of these, only selection produces adaptive evolutionary change because only in selection does the result depend on the nature of the environment. The other factors operate relatively independently of the environment, so the changes they produce are not shaped by environmental demands.

Table 20.1 Agents of Evolutionary Change

Factor	Description
Mutation	The ultimate source of variation. Individual mutations occur so rarely that mutation alone does not change allele frequency much.
Gene flow	A very potent agent of change. Populations exchange members.
Nonrandom mating	Inbreeding is the most common form. It does not alter allele frequency but decreases the proportion of heterozygotes.
Genetic drift	Statistical accidents. Usually occurs only in very small populations.
Selection	The only form that produces *adaptive* evolutionary changes.

The Hardy–Weinberg principle states that in a large population mating at random and in the absence of other forces that would change the proportions of the different alleles at a given locus, the process of sexual reproduction (meiosis and fertilization) alone will not change these proportions.

Five Agents of Evolutionary Change

1. Mutation

Mutation from one allele to another can obviously change the proportions of particular alleles in a population. Mutation rates are generally so low that they have little effect on the Hardy–Weinberg proportions of common alleles. A single gene may mutate about 1 to 10 times per 100,000 cell divisions (although *some* genes mutate much more frequently than that). Because most environments are constantly changing, it is rare for a population to be stable enough to accumulate changes in allele frequency produced by a process this slow. Nonetheless, mutation is the ultimate source of genetic variation and thus makes evolution possible. It is important to remember, however, that the likelihood of a particular mutation occurring is not affected by natural selection; that is, mutations do not occur more frequently in situations in which they would be favored by natural selection.

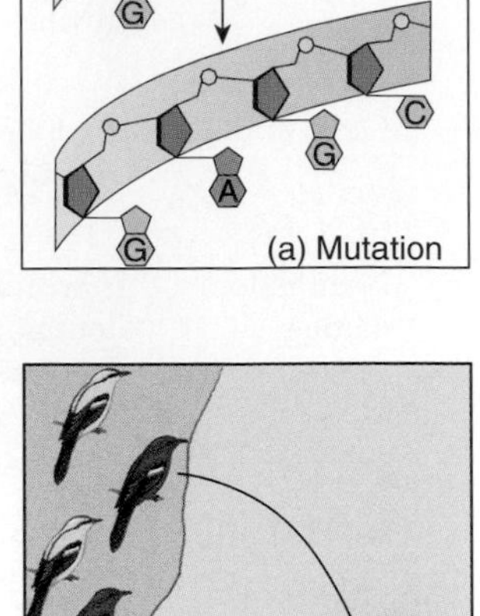

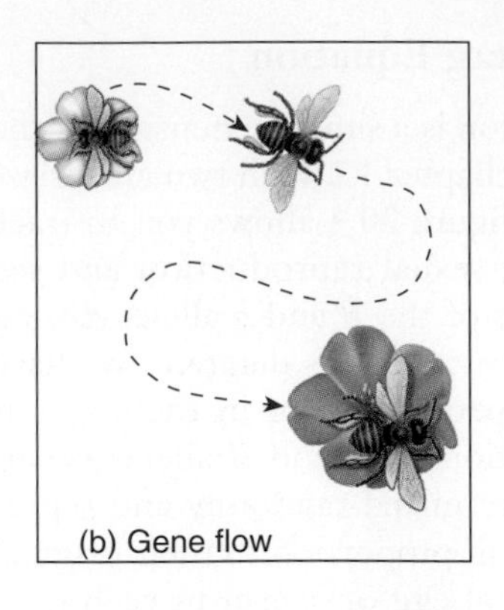

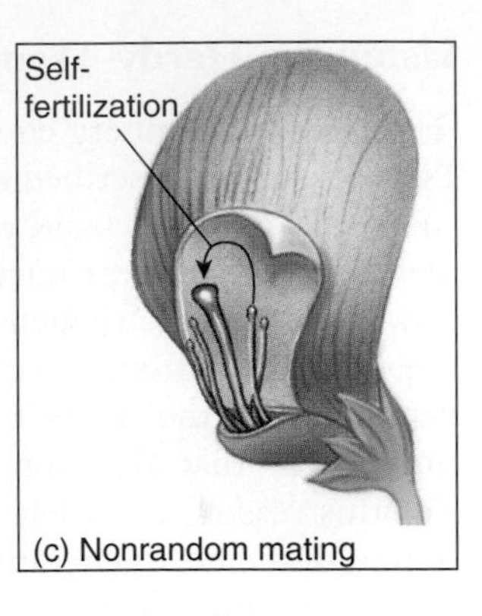

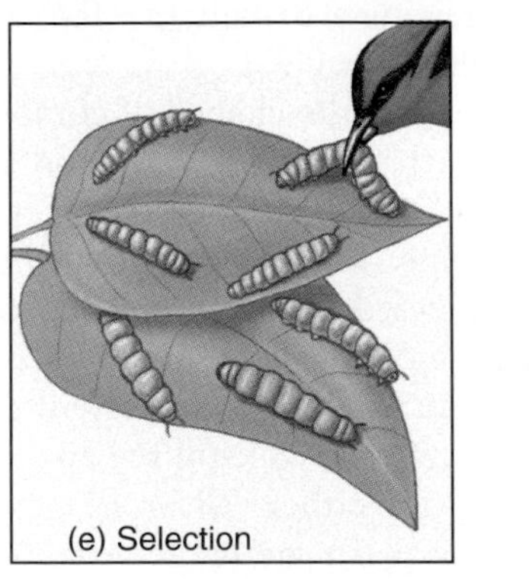

FIGURE 20.5
Five agents of evolutionary change.
(*a*) Mutation, (*b*) gene flow, (*c*) nonrandom mating, (*d*) genetic drift, and (*e*) selection.

2. Gene Flow

Gene flow is the movement of alleles from one population to another. It can be a powerful agent of change because members of two different populations may exchange genetic material. Sometimes gene flow is obvious, as when an animal moves from one place to another. If the characteristics of the newly arrived animal differ from those of the animals already there, and if the newcomer is adapted well enough to the new area to survive and mate successfully, the genetic composition of the receiving population may be altered. Other important kinds of gene flow are not as obvious. These subtler movements include the drifting of gametes or immature stages of plants or marine animals from one place to another (figure 20.5). Male gametes of flowering plants are often carried great distances by insects and other animals that visit their flowers. Seeds may also blow in the wind or be carried by animals or other agents to new populations far from their place of origin. In addition, gene flow may also result from the mating of individuals belonging to adjacent populations.

However it occurs, gene flow can alter the genetic characteristics of populations and prevent them from maintaining Hardy–Weinberg equilibrium. In addition, even low levels of gene flow tend to homogenize allele frequencies among populations and thus keep the populations from diverging genetically. In some situations, gene flow can counter the effect of natural selection by bringing an allele into a population at a rate greater than that at which the allele is removed by selection.

3. Nonrandom Mating

Individuals with certain genotypes sometimes mate with one another more commonly than would be expected on a random basis, a phenomenon known as nonrandom mating. **Inbreeding** (mating with relatives) is a type of nonrandom mating that causes the frequencies of particular genotypes to differ greatly from those predicted by the Hardy–Weinberg principle. Inbreeding does not change the frequency of the alleles, but rather increases the proportion of homozygous individuals because relatives are likely be genetically similar and thus produce offspring with two copies of the same allele. This is why populations of self-fertilizing plants consist primarily of homozygous individuals, whereas **outcrossing** plants, which interbreed with individuals different from themselves, have a higher proportion of heterozygous individuals.

By increasing homozygosity in a population, inbreeding increases the expression of recessive alleles. It is for this reason that marriage between close relatives is discouraged and to some degree outlawed—it increases the possibility of producing children homozygous for an allele associated with one or more of the recessive genetic disorders discussed in chapter 13.

4. Genetic Drift

In small populations, frequencies of particular alleles may change drastically by chance alone. Such changes in allele frequencies occur randomly, as if the frequencies were drifting, and are thus known as **genetic drift.** For this reason, a population must be large to be in Hardy–Weinberg equilibrium. If the gametes of only a few individuals form the next generation, the alleles they carry may by chance not be representative of the parent population from which they were drawn, as illustrated in figure 20.6, where a small number of individuals are removed from a bottle containing many. By chance, most of the individuals removed are blue, so the new population has a much higher population of blue individuals than the parent one had.

A set of small populations that are isolated from one another may come to differ strongly as a result of genetic drift even if the forces of natural selection do not differ between the populations. Indeed, because of genetic drift, harmful alleles may increase in frequency in small populations, despite selective disadvantage, and favorable alleles may be lost even though selectively advantageous. It is interesting to realize that humans have lived in small groups for much of the course of their evolution; consequently, genetic drift may have been a particularly important factor in the evolution of our species.

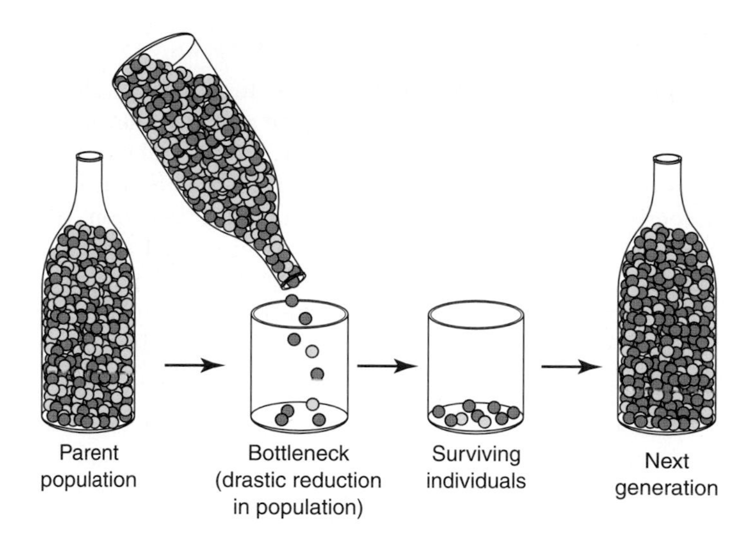

FIGURE 20.6
Genetic drift: The bottleneck effect. The parent population contains roughly equal numbers of blue and yellow individuals. By chance, the few remaining individuals that comprise the next generation are mostly blue. The bottleneck occurs because so few individuals form the next generation, as might happen after an epidemic or catastrophic storm.

Even large populations may feel the effect of genetic drift. Large populations may have been much smaller in the past, and genetic drift may have greatly altered allele frequencies at that time. Imagine a population containing only two alleles of a gene, *B* and *b*, in equal frequency (that is, $p = q = 0.5$). In a large Hardy–Weinberg population, the genotype frequencies are expected to be 0.25 *BB*, 0.50 *Bb*, and 0.25 *bb*. If only a small sample produces the next generation, large deviations in these genotype frequencies can occur by chance. Imagine, for example, that four individuals form the next generation, and that by chance they are two *Bb* heterozygotes and two *BB* homozygotes—the allele frequencies in the next generation are $p = 0.75$ and $q = 0.25$! If you were to replicate this experiment 1000 times, each time randomly drawing four individuals from the parental population, one of the two alleles would be missing entirely from about 8 of the 1000 populations. This leads to an important conclusion: genetic drift leads to the loss of alleles in isolated populations. Two related causes of decreases in a population's size are founder effects and bottlenecks.

Founder Effects. Sometimes one or a few individuals disperse and become the founders of a new, isolated population at some distance from their place of origin. These pioneers are not likely to have all the alleles present in the source population. Thus, some alleles may be lost from the new population and others may change drastically in frequency. In some cases, previously rare alleles in the source population may be a significant fraction of the new population's genetic endowment. This phenomenon is called the founder effect. Founder effects are not rare in nature. Many self-pollinating plants start new populations from a single seed.

Founder effects have been particularly important in the evolution of organisms on distant oceanic islands, such as the Hawaiian Islands and the Galápagos Islands visited by Darwin. Most of the organisms in such areas probably derive from one or a few initial "founders." In a similar way, isolated human populations are often dominated by genetic features characteristic of their particular founders.

The Bottleneck Effect. Even if organisms do not move from place to place, occasionally their populations may be drastically reduced in size. This may result from flooding, drought, epidemic disease, and other natural forces, or from progressive changes in the environment. The few surviving individuals may constitute a random genetic sample of the original population (unless some individuals survive specifically because of their genetic makeup). The resultant alterations and loss of genetic variability has been termed the **bottleneck effect.**

Some living species appear to be severely depleted genetically and have probably suffered from a bottleneck effect in the past. For example, the northern elephant seal, which breeds on the western coast of North America and nearby islands, was nearly hunted to extinction in the nineteenth century and was reduced to a single population containing perhaps no more than 20 individuals on the island of Guadalupe off the coast of Baja, California. As a result of this bottleneck, even though the seal populations have rebounded and now number in the tens of thousands, this species has lost almost all of its genetic variation.

5. Selection

As Darwin pointed out, some individuals leave behind more progeny than others, and the rate at which they do so is affected by phenotype and behavior. We describe the results of this process as **selection** and speak of both **artificial selection** and **natural selection.** In artificial selection, the breeder selects for the desired characteristics. In natural selection, environmental conditions determine which individuals in a population produce the most offspring. For natural selection to occur and result in evolutionary change, three conditions must be met:

1. **Variation must exist among individuals in a population.** Natural selection works by favoring individuals with some traits over individuals with alternative traits. If no variation exists, natural selection cannot operate.
2. **Variation among individuals results in differences in number of offspring surviving in the next generation.** This is the essence of natural selection. Because of their phenotype or behavior, some individuals are more successful than others in producing offspring and thus passing their genes on to the next generation.
3. **Variation must be genetically inherited.** For natural selection to result in evolutionary change, the selected differences must have a genetic basis. However, not all variation has a genetic basis—even genetically identical individuals may be phenotypically quite distinctive if they grow up in different environments. Such environmental effects are common in nature. In many turtles, for example, individuals that hatch from eggs laid in moist soil are heavier, with longer and wider shells, than individuals from nests in drier areas. As a result of these environmental effects, variation within a population does not always indicate the existence of underlying genetic variation. When phenotypically different individuals do not differ genetically, then differences in the number of their offspring will not alter the genetic composition of the population in the next generation and, thus, no evolutionary change will have occurred.

It is important to remember that natural selection and evolution are not the same—the two concepts often are incorrectly equated. Natural selection is a process, whereas evolution is the historical record of change through time. Evolution is an outcome, not a process. Natural selection (the process) can lead to evolution (the outcome), but natural selection is only one of several processes that can produce evolutionary change. Moreover, natural selection can occur without producing evolutionary change; only if variation is genetically based will natural selection lead to evolution.

Selection to Avoid Predators. Many of the most dramatic documented instances of adaptation involve genetic changes which decrease the probability of capture by a predator. The caterpillar larvae of the common sulphur butterfly *Colias eurytheme* usually exhibit a dull Kelly green color, providing excellent camouflage on the alfalfa plants on which they feed. An alternative bright blue color morph is kept at very low frequency because this color renders the larvae highly visible on the food plant, making it easier for bird predators to see them. In a similar fashion, the way the shell markings in the land snail *Cepaea nemoralis* match its background habitat reflects the same pattern of avoiding predation by camouflage.

One of the most dramatic examples of background matching involves ancient lava flows in the middle of deserts in the American southwest. In these areas, the black rock formations produced when the lava cooled contrasts starkly to the surrounding bright glare of the desert sand. Populations of many species of animals—including lizards, rodents, and a variety of insects—occurring on these rocks are dark in color, whereas sand-dwelling populations in surrounding areas are much lighter (figure 20.7). Predation is the likely cause selecting for these differences in color. Laboratory studies have confirmed that predatory birds are adept at picking out individuals occurring on backgrounds to which they are not adapted.

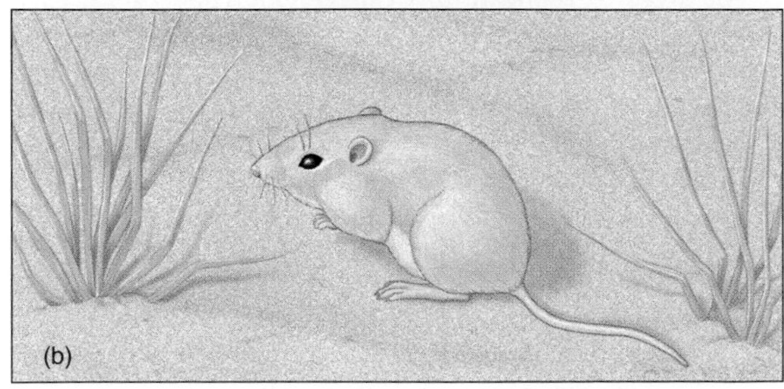

FIGURE 20.7
Pocket mice from the Tularosa Basin of New Mexico whose color matches their background. (*a*) The rock pocket mouse lives on lava, (*b*) while the Apache pocket mouse lives on white sand.

Selection to Match Climatic Conditions. Many studies of selection have focused on genes encoding enzymes because in such cases the investigator can directly assess the consequences to the organism of changes in the frequency of alternative enzyme alleles. Often investigators find that enzyme allele frequencies vary latitudinally, with one allele more common in northern populations but progressively less common at more southern locations. A superb example is seen in studies of a fish, the mummichog, *Fundulus heteroclitus*, which ranges along the eastern coast of North America. In this fish, allele frequencies of the gene that produces the enzyme lactase dehydrogenase, which catalyzes the conversion of pyruvate to lactate, vary geographically (figure 20.8). Biochemical studies show that the enzymes formed by these alleles function differently at different temperatures, thus explaining their geographic distributions. For example, the form of the enzyme that is more frequent in the north is a better catalyst at low temperatures than the enzyme from the south. Moreover, functional studies indicate that at low temperatures, individuals with the northern allele swim faster, and presumably survive better, than individuals with the alternative allele.

Selection for Pesticide Resistance. A particularly clear example of selection in action in natural populations is provided by studies of pesticide resistance in insects. The widespread use of insecticides has led to the rapid evolution of resistance in more than 400 pest species. For example, the resistance allele at the *pen* gene decreases the uptake of insecticide, whereas alleles at the *kdr* and *dld-r* genes decrease the number of target sites, thus decreasing the binding ability of the insecticide (figure 20.9). Other alleles enhance the ability of the insects' enzymes to identify and detoxify insecticide molecules.

Single genes are also responsible for resistance in other organisms. The pigweed, *Amaranthus hybridus*, is one of about 28 agricultural weeds that have evolved resistance to the herbicide Triazine. Triazine inhibits photosynthesis by binding to a protein in the chloroplast membrane. Single amino acid substitutions in the gene encoding the protein diminish the ability of Triazine to decrease the plant's photosynthetic capabilities. Similarly, Norway rats are normally susceptible to the pesticide Warfarin, which diminishes the clotting ability of the rat's blood and leads to fatal hemorrhaging. However, a resistance allele at a single gene alters a metabolic pathway and renders Warfarin ineffective.

Five factors can bring about a deviation from the proportions of homozygotes and heterozygotes predicted by the Hardy-Weinberg principle. Only selection regularly produces adaptive evolutionary change, but the genetic constitution of individual populations, and thus the course of evolution, can also be affected by mutation, gene flow, nonrandom mating, and genetic drift.

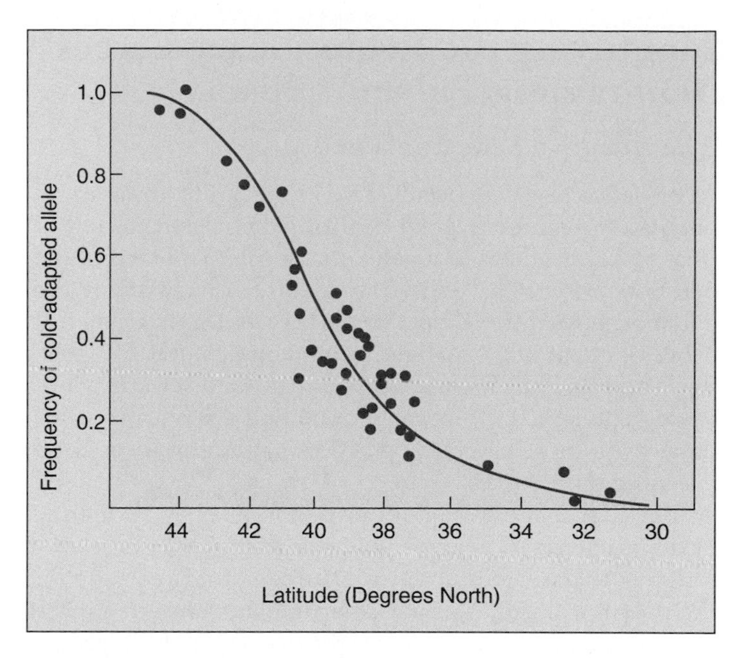

FIGURE 20.8
Selection to match climatic conditions. Frequency of the cold-adapted allele for lactase dehydrogenase in a type of fish (the mummichog *Fundulus heteroclitus*) decreases at lower latitudes, which are warmer.

Source: Data from P.A. Powers, et. al, "A Multidisciplinary Approach to the Selectionist/Neutralist Controversy". *Oxford Surveys in Evolutionary Biology.* Oxford University Press, 1993.

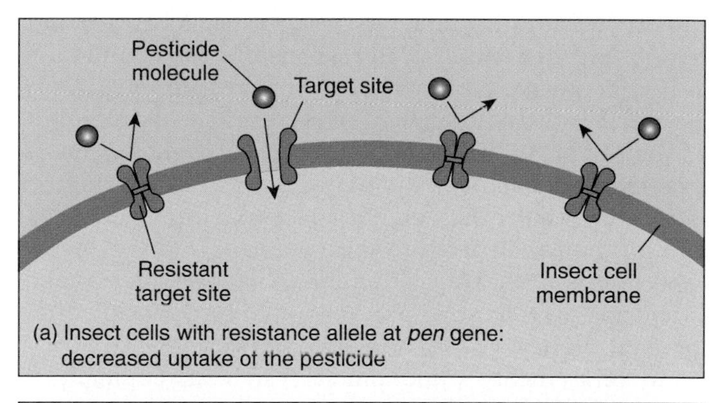

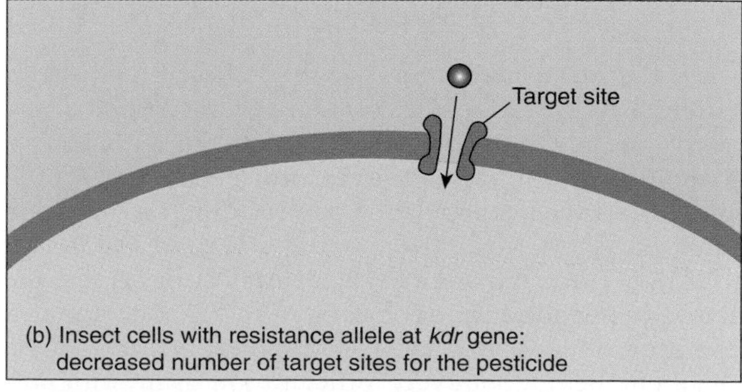

Target site
(b) Insect cells with resistance allele at *kdr* gene: decreased number of target sites for the pesticide

FIGURE 20.9
Selection for pesticide resistance. Resistance alleles at genes like *pen* and *kdr* allow insects to be more resistant to pesticides. Insects that possess these resistance alleles have become more common through selection.

Identifying the Evolutionary Forces Maintaining Polymorphism

The Adaptive Selection Theory

As evidence began to accumulate in the 1970s that natural populations exhibit a great deal of genetic polymorphism (that is, many alleles of a gene exist in the population), the question arose: What evolutionary force is maintaining the polymorphism? As we have seen, there are in principle five processes that act on allele frequencies: mutation, gene flow, nonrandom mating, genetic drift, and selection. Because gene flow and nonrandom mating are not major influences in most natural populations, attention focused on the other three forces.

The first suggestion, advanced by R. C. Lewontin (one of the discovers of enzyme polymorphism) and many others, was that selection was the force acting to maintain the polymorphism. Natural environments are often quite heterogeneous, so selection might reasonably be expected to pull gene frequencies in different directions within different microhabitats, generating a condition in which many alleles persist. This proposal is called the **adaptive selection theory.**

The Neutral Theory

A second possibility, championed by the great Japanese geneticist Moto Kimura, was that a balance between mutation and genetic drift is responsible for maintaining polymorphism. Kimura used elegant mathematics to demonstrate that, even in the absence of selection, natural populations could be expected to contain considerable polymorphism if mutation rates (generating the variation) were high enough and population sizes (promoting genetic drift) were small enough. In this proposal, selection is not acting, differences between alleles being "neutral to selection." The proposal is thus called the **neutral theory.**

Kimura's theory, while complex, can be stated simply:

$$\bar{H} = 1/(4N_e\mu + 1)$$

$\bar{H}$, the mean heterozygosity, is the likelihood that a randomly selected member of the population will be heterozygous at a randomly selected locus. In a population without selection, this value is influenced by two variables, the effective population size (N_e) and the mutation rate (μ).

The peculiar difficulty of the neutral theory is that the level of polymorphism, as measured by $\bar{H}$, is determined by the product of a very large number, N_e, and a very small number, μ, both very difficult to measure with precision. As a result, the theory can account for almost any value of $\bar{H}$, making it very difficult to prove or disprove. As you might expect, a great deal of controversy has resulted.

Testing the Neutral Theory

Choosing between the adaptive selection theory and the neutral theory is not simple, for they both appear to account for much of the data on gene polymorphism in natural populations. A few well-characterized instances where selection acts on enzyme alleles do not settle the more general issue. An attempt to test the neutral theory by examining large-scale patterns of polymorphism sheds light on the difficulty of choosing between the two theories:

Population size: According to the neutral theory, polymorphism as measured by $\bar{H}$ should be proportional to the effective population size N_e, assuming the mutation rate among neutral alleles μ is constant. Thus, $\bar{H}$ should be much greater for insects than humans, as there are far more individuals in an insect population than in a human one. When DNA sequence variation is examined, the fruit fly *Drosophila melanogaster* indeed exhibits sixfold higher levels of variation, as the theory predicts; but when enzyme polymorphisms are examined, levels of variation in fruit flies and humans are similar. If the level of DNA variation correctly mirrors the predictions of the neutral theory, then something (selection?) is increasing variation at the enzyme level in humans. These sorts of patterns argue for rejection of the neutral theory.

The nearly neutral model: One way to rescue the neutral theory from these sorts of difficulties is to retreat from the assumption of strict neutrality, modifying the theory to assume that many of the variants are slightly deleterious rather than strictly neutral to selection. With this adjustment, it is possible to explain many of the population-size-dependent large-scale patterns. However, little evidence exists that the wealth of enzyme polymorphism in natural populations is in fact slightly deleterious.

As increasing amounts of DNA sequence data become available, a detailed picture of variation at the DNA level is emerging. It seems clear that most nucleotide substitutions that change amino acids are disadvantageous and are eliminated by selection. But what about the many protein alleles that are seen in natural populations? Are they nearly neutral or advantageous? No simple answer is yet available, although the question is being actively investigated. Levels of polymorphism at enzyme-encoding genes may depend on both the action of selection on the gene (the adaptive selection theory) and on the population dynamics of the species (the nearly neutral theory), with the relative contribution varying from one gene to the next.

Adaptive selection clearly maintains some enzyme polymorphisms in natural populations. Genetic drift seems to play a major role in producing the variation we see at the DNA level. For most enzyme-level polymorphism, investigators cannot yet choose between the selection theory and the nearly neutral theory.

Interactions among Evolutionary Forces

When alleles are not selectively neutral, levels of variation retained in a population may be determined by the relative strength of different evolutionary processes. In theory, for example, if allele *B* mutates to allele *b* at a high enough rate, allele *b* could be maintained in the population even if natural selection strongly favored allele *B*. In nature, however, mutation rates are rarely high enough to counter the effects of natural selection.

The effect of natural selection also may be countered by genetic drift. Both processes may act to remove variation from a population. However, whereas selection is a deterministic process that operates to increase the representation of alleles that enhance survival and reproductive success, drift is a random process. Thus, in some cases, drift may lead to a decrease in the frequency of an allele that is favored by selection. In some extreme cases, drift may even lead to the loss of a favored allele from a population. Remember, however, that the magnitude of drift is negatively related to population size; consequently, natural selection is expected to overwhelm drift except when populations are very small.

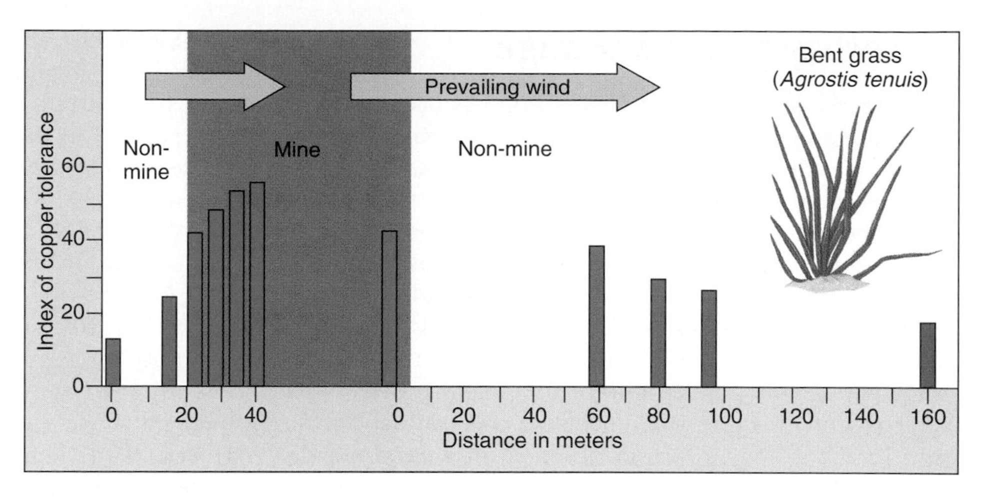

FIGURE 20.10
Degree of copper tolerance in grass plants on and near ancient mine sites. Individuals with tolerant alleles have decreased growth rates on unpolluted soil. Thus, we would expect copper tolerance to be 100% on mine sites and 0% on non-mine sites. However, prevailing winds blow pollen containing nontolerant alleles onto the mine site and tolerant alleles beyond the sites borders.

Source: Data from M.R. MacNair in J.M. Bishops & L.M. Cook, *Genetic Consequences of Man-Made Change. Academic Press*, 1981, pp. 177–207.

Gene Flow versus Natural Selection

Gene flow can be either a constructive or a constraining force. On one hand, gene flow can increase the adaptedness of a species by spreading a beneficial mutation that arises in one population to other populations within a species. On the other hand, gene flow can act to impede adaptation within a population by continually importing inferior alleles from other populations. Consider two populations of a species that live in different environments. In this situation, natural selection might favor different alleles—*B* and *b*—in the different populations. In the absence of gene flow and other evolutionary processes, the frequency of *B* would be expected to reach 100% in one population and 0% in the other. However, if gene flow were going on between the two populations, then the less favored allele would continually be reintroduced into each population. As a result, the frequency of the two alleles in each population would reflect a balance between the rate at which gene flow brings the inferior allele into a population versus the rate at which natural selection removes it.

A classic example of gene flow opposing natural selection occurs on abandoned mine sites in Great Britain. Although mining activities ceased hundreds of years ago, the concentration of metal ions in the soil is still much greater than in surrounding areas. Heavy metal concentrations are generally toxic to plants, but alleles at certain genes confer resistance. The ability to tolerate heavy metals comes at a price, however; individuals with the resistance allele exhibit lower growth rates on non-polluted soil. Consequently, we would expect the resistance allele to occur with a frequency of 100% on mine sites and 0% elsewhere. Heavy metal tolerance has been studied particularly intensively in the slender bent grass, *Agrostis tenuis*, in which researchers have found that the resistance allele occurs at intermediate levels in many areas (figure 20.10). The explanation relates to the reproductive system of this grass in which pollen, the male gamete (that is, the floral equivalent of sperm), is dispersed by the wind. As a result, pollen—and the alleles it carries—can be blown for great distances, leading to levels of gene flow between mine sites and unpolluted areas high enough to counteract the effects of natural selection.

In general, the extent to which gene flow can hinder the effects of natural selection should depend on the relative strengths of the two processes. In species in which gene flow is generally strong, such as birds and wind-pollinated plants, the frequency of the less favored allele may be relatively high, whereas in more sedentary species which exhibit low levels of gene flow, such as salamanders, the favored allele should occur at a frequency near 100%.

Evolutionary processes may act to either remove or maintain genetic variation within a population. Allele frequency sometimes may reflect a balance between opposed processes, such as gene flow and natural selection. In such cases, observed frequencies will depend on the relative strength of the processes.

Heterozygote Advantage

In the previous pages, natural selection has been discussed as a process that removes variation from a population by favoring one allele over others at a genetic locus. However, if heterozygotes are favored over homozygotes, then natural selection actually will tend to maintain variation in the population. The reason is simple. Instead of tending to remove less successful alleles from a population, such **heterozygote advantage** will favor individuals with copies of both alleles, and thus will work to maintain both alleles in the population. Some evolutionary biologists believe that heterozygote advantage is pervasive and can explain the high levels of polymorphism observed in natural populations. Others, however, believe that it is relatively rare.

Sickle Cell Anemia

The best documented example of heterozygote advantage is sickle cell anemia, a hereditary disease affecting hemoglobin in humans. Individuals with sickle cell anemia exhibit symptoms of severe anemia and contain abnormal red blood cells which are irregular in shape, with a great number of long and sickle-shaped cells. The disease is particularly common among African Americans. In chapter 13, we noted that this disorder, which affects roughly 3 African Americans out of every 1000, is associated with a particular recessive allele. Using the Hardy–Weinberg equation, you can calculate the frequency of the sickle cell allele in the African-American population; this frequency is the square root of 0.003, or approximately 0.054. In contrast, the frequency of the allele among white Americans is only about 0.001.

Sickle cell anemia is often fatal. Until therapies were developed to more effectively treat its symptoms, almost all affected individuals died as children. Even today, 31% of patients in the United States die by the age of 15. The disease occurs because of a single amino acid change, repeated in the two beta chains of the hemoglobin molecule. In this change, a valine replaces the usual glutamic acid at a location on the surface of the protein. Unlike glutamic acid, valine is nonpolar (hydrophobic). Its presence on the surface of the molecule creates a "sticky" patch that attempts to escape from the polar water environment by binding to another similar patch. As long as oxygen is bound to the hemoglobin molecule there is no problem, because the hemoglobin atoms shield the critical area of the surface. When oxygen levels fall, such as after exercise or when an individual is stressed, oxygen is not so readily bound to hemoglobin and the exposed sticky patch binds to similar patches on other hemoglobin molecules, eventually producing long, fibrous clumps (figure 20.11). The result is a deformed, "sickle-shaped" red blood cell.

Individuals who are heterozygous or homozygous for the valine-specifying allele (designated allele *S*) are said to possess the sickle cell trait. Heterozygotes produce some sickle-shaped red blood cells, but only 2% of the number seen in homozygous individuals. The reason is that in heterozygotes, one-half of the molecules do not contain valine at the critical location. Consequently, when a molecule produced by the non-sickle cell allele is added to the chain, there is no further "sticky" patch available to add additional molecules and chain elongation stops. Hence, most chains in heterozygotes are too short to produce sickling of the cell.

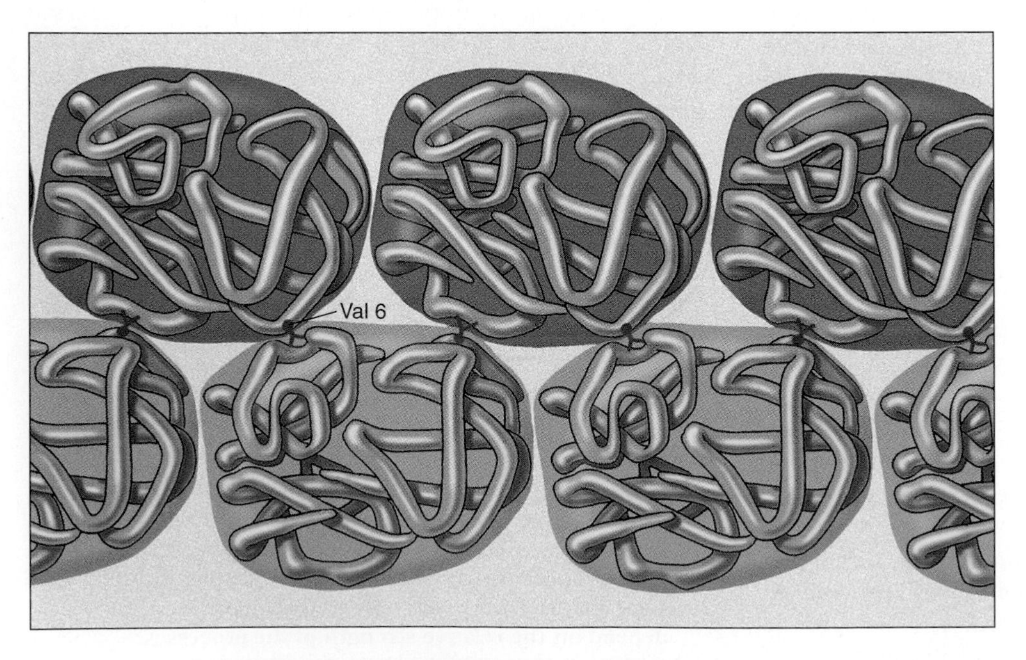

FIGURE 20.11
Why the sickle cell mutation causes hemoglobin to clump. The sickle cell mutation changes the sixth amino acid in the hemoglobin β chain (position B6) from glutamic acid (very polar) to valine (nonpolar). The unhappy result is that the nonpolar valine at position B6, protruding from a corner of the hemoglobin molecule, fits into a nonpolar pocket on the opposite side of another hemoglobin molecule, causing the two molecules to clump together. As each molecule has both a B6 valine and an opposite nonpolar pocket, long chains form. When polar glutamic acid (the normal allele) occurs at position B6, it is not attracted to the nonpolar pocket, and no clumping occurs. Copyright © Irving Geis.

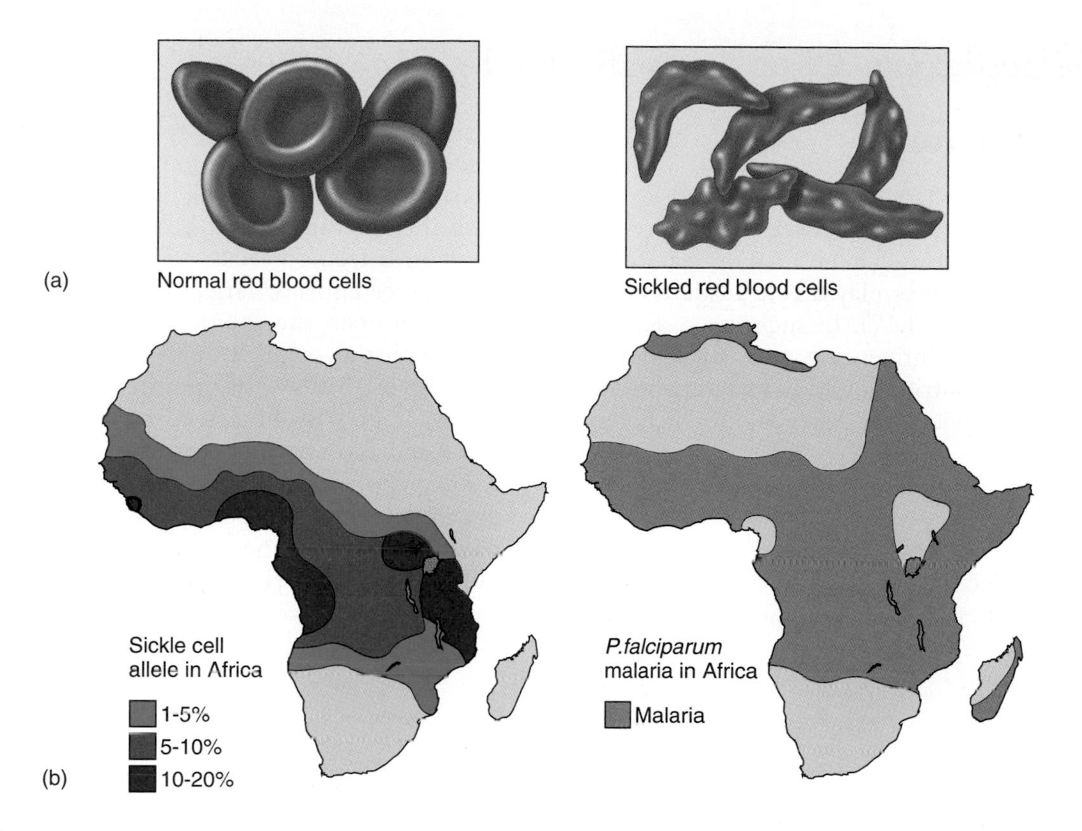

FIGURE 20.12
Frequency of sickle cell allele and distribution of *Plasmodium falciparum* malaria. (*a*)The red blood cells of people homozygous for the sickle cell allele collapse into sickled shapes when the oxygen level in the blood is low. (*b*) The distribution of the sickle cell allele in Africa coincides closely with that of *P. falciparum* malaria.

Malaria and Heterozygote Advantage

The average incidence of the *S* allele in the Central African population is about 0.12, far higher than that found among African Americans. From the Hardy–Weinberg principle, you can calculate that 1 in 5 Central African individuals are heterozygous at the *S* allele, and 1 in 100 develops the fatal form of the disorder. People who are homozygous for the sickle cell allele almost never reproduce because they usually die before they reach reproductive age. Why is the *S* allele not eliminated from the Central African population by selection, rather than being maintained at such high levels? People who are heterozygous for the sickle cell allele are much less susceptible to malaria—one of the leading causes of illness and death in Central Africa, especially among young children—in the areas where the allele is common. The reason is that when the parasite that causes malaria, *Plasmodium falciparum*, enters a red blood cell, it causes extremely low oxygen tension in the cell, which leads to cell sickling even in heterozygotes. Such cells are quickly filtered out of the bloodstream by the spleen, thus eliminating the parasite (the spleen's filtering effect is what leads to anemia in homozygotes as large numbers of red blood cells are removed).

Consequently, even though most homozygous recessive individuals die before they have children, the sickle cell allele is maintained at high levels in these populations (it is selected for) because of its association with resistance to malaria in heterozygotes and also, for reasons not yet fully understood, with increased fertility in female heterozygotes.

For people living in areas where malaria is common, having the sickle cell allele in the heterozygous condition has adaptive value (figure 20.12). Among African Americans, however, many of whose ancestors have lived for some 15 generations in a country where malaria has been relatively rare and is now essentially absent, the environment does not place a premium on resistance to malaria. Consequently, no adaptive value counterbalances the ill effects of the disease; in this nonmalarial environment, selection is acting to eliminate the *S* allele. Only 1 in 375 African Americans develop sickle cell anemia, far less than in Central Africa.

The hemoglobin allele *S*, responsible for sickle cell anemia in homozygotes, is maintained by heterozygote advantage in Central Africa, where heterozygotes for the *S* allele are resistant to malaria.

Forms of Selection

In nature many traits, perhaps most, are affected by more than one gene. The interactions between genes are typically complex, as you saw in chapter 13. For example, alleles of many different genes play a role in determining human height (see figure 13.17). In such cases, selection operates on all the genes, influencing most strongly those that make the greatest contribution to the phenotype. How selection changes the population depends on which genotypes are favored.

Disruptive Selection

In some situations, selection acts to eliminate rather than to favor intermediate types. A clear example is the different beak sizes of the African fire-bellied seedcracker finch *Pyronestes ostrinus.* Populations of these birds contain individuals with large and small beaks, but very few individuals with intermediate-sized beaks. As their name implies, these birds feed on seeds, and the available seeds fall into two size categories: large and small. Only large-beaked birds can open the tough shells of large seeds, whereas birds with the smallest beaks are most adept at handling small seeds. Birds with intermediate-sized beaks are at a disadvantage with both seed types: unable to open large seeds and too clumsy to efficiently process small seeds. Consequently, selection acts to eliminate the intermediate phenotypes, in effect partitioning the population into two phenotypically distinct groups. This form of selection is called **disruptive selection** (figure 20.13*a*).

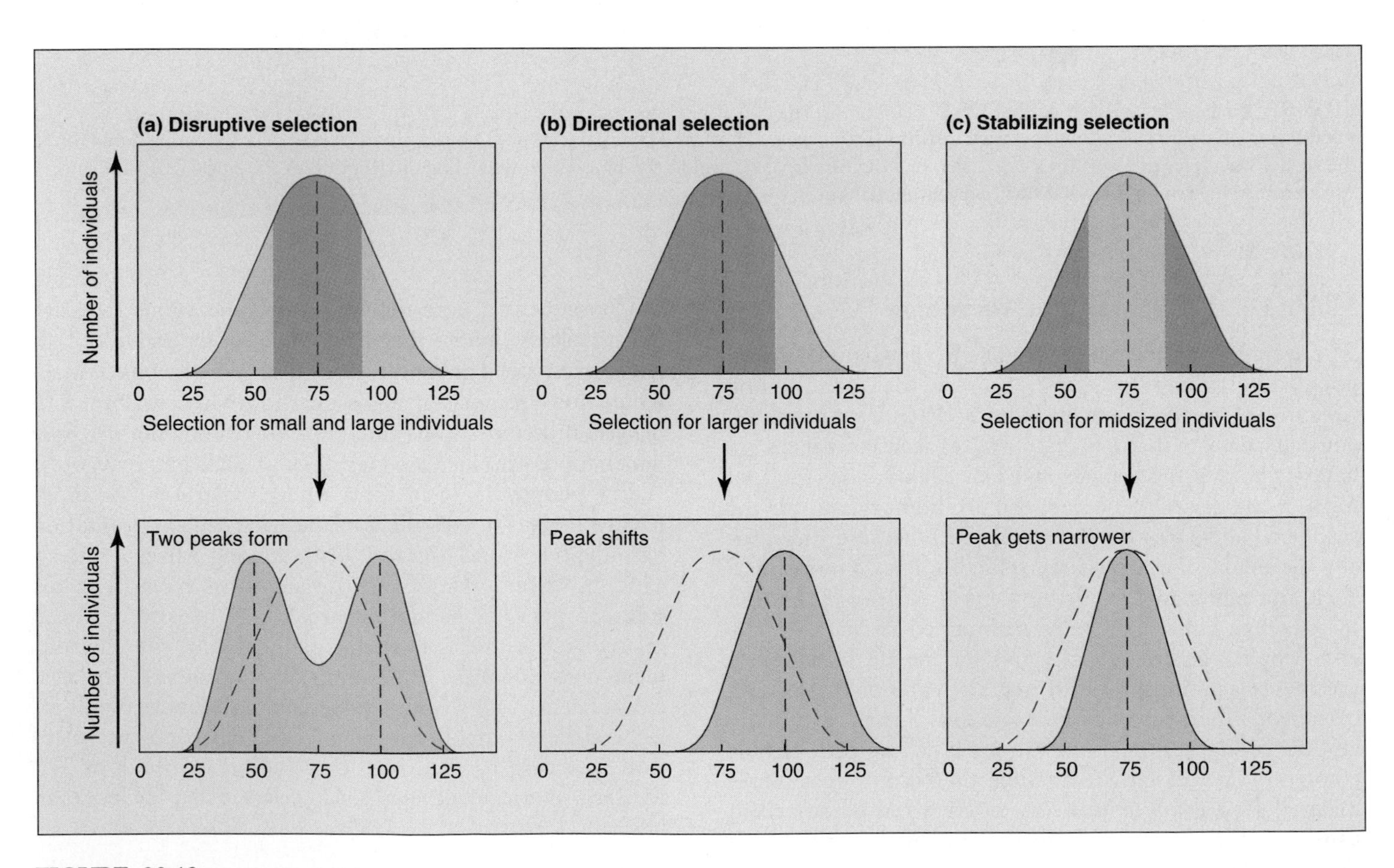

FIGURE 20.13
Three kinds of natural selection. The top panels show the populations before selection has occurred, with the forms that will be selected against shaded *red* and the forms that will be favored shaded *blue*. The bottom panels indicate what the populations will look like after selection has occurred. (*a*) In *disruptive selection*, individuals in the middle of the range of phenotypes of a certain trait are selected against (*red*), and the extreme forms of the trait are favored (*blue*). (*b*) In *directional selection*, individuals concentrated toward one extreme of the array of phenotypes are favored. (*c*) In *stabilizing selection*, individuals with midrange phenotypes are favored, with selection acting against both ends of the range of phenotypes.

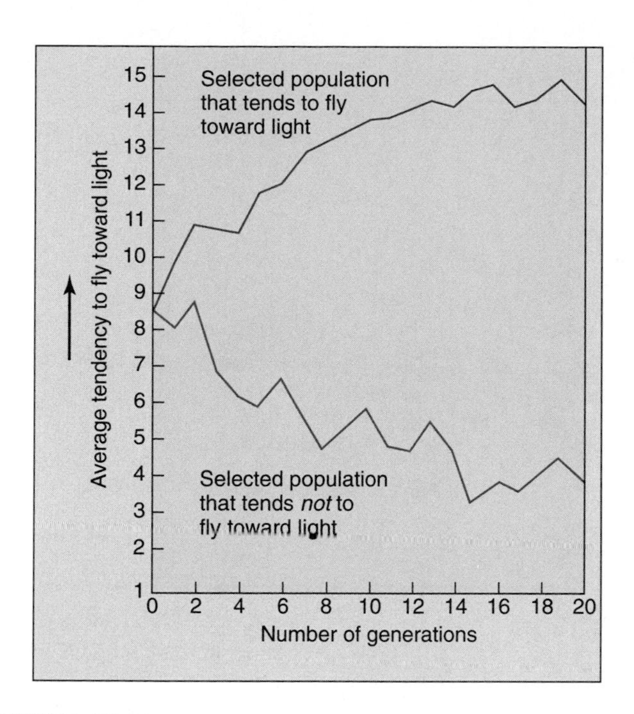

FIGURE 20.14
Directional selection for phototropism in *Drosophila.* In generation after generation, individuals of the fly *Drosophila* were selectively bred to obtain two populations. When flies with a strong tendency to fly toward light (positive phototropism) were used as parents for the next generation, their offspring had a greater tendency to fly toward light (top curve). When flies that tended *not* to fly toward light were used as parents for the next generation, their offspring had an even greater tendency not to fly toward light (bottom curve).

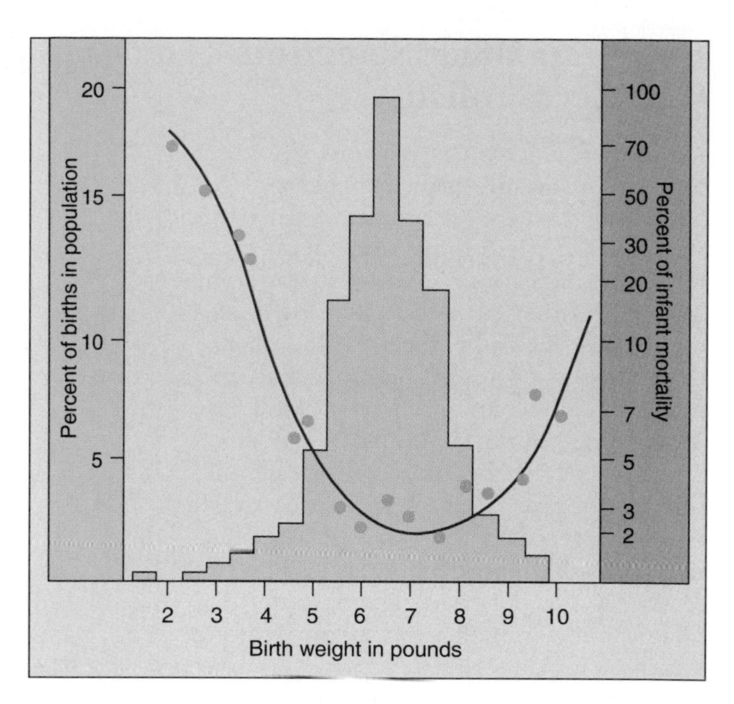

FIGURE 20.15
Stabilizing selection for birth weight in human beings. The death rate among babies (*red* curve; right *y*-axis) is lowest at an intermediate birth weight; both smaller and larger babies have a greater tendency to die than those around the optimum weight (*blue* area; left *y*-axis) of between 7 and 8 pounds.

Directional Selection

When selection acts to eliminate one extreme from an array of phenotypes (figure 20.13*b*), the genes promoting this extreme become less frequent in the population. Thus, in the *Drosophila* population illustrated in figure 20.14, the elimination of flies that move toward light causes the population to contain fewer individuals with alleles promoting such behavior. If you were to pick an individual at random from the new fly population, there is a smaller chance it would spontaneously move toward light than if you had selected a fly from the old population. Selection has changed the population in the direction of lower light attraction. This form of selection is called **directional selection.**

Stabilizing Selection

When selection acts to eliminate *both* extremes from an array of phenotypes (figure 20.13*c*), the result is to increase the frequency of the already common intermediate type. In effect, selection is operating to prevent change away from this middle range of values. Selection does not change the most common phenotype of the population, but rather makes it even more common by eliminating extremes. Many examples are known. In humans, infants with intermediate weight at birth have the highest survival rate (figure 20.15). In ducks and chickens, eggs of intermediate weight have the highest hatching success. This form of selection is called **stabilizing selection.**

Components of Fitness

Natural selection occurs when individuals with one phenotype leave more surviving offspring in the next generation than individuals with an alternative phenotype. Evolutionary biologists quantify reproductive success as **fitness,** the number of surviving offspring left in the next generation. Although selection is often characterized as "survival of the fittest," differences in survival are only one component of fitness. Even if no differences in survival occur, selection may operate if some individuals are more successful than others in attracting mates. In many territorial animal species, large males mate with many females and small males rarely get to mate. In addition, the number of offspring produced per mating is also important. Large female frogs and fish lay more eggs than smaller females and thus may leave more offspring in the next generation.

Selection on traits affected by many genes can favor both extremes of the trait, or intermediate values, or only one extreme.

Limits to What Selection Can Accomplish

Although selection is perhaps the most powerful of the five principal agents of genetic change, there are limits to what it can accomplish. These limits arise because alternative alleles may interact in different ways with other genes and because alleles often affect multiple aspects of the phenotype (the phenomena of epistasis and pleiotropy discussed in chapter 13). These interactions tend to set limits on how much a phenotype can be altered. For example, selecting for large clutch size in barnyard chickens eventually leads to eggs with thinner shells that break more easily. For this reason, we do not have gigantic cattle that yield twice as much meat as our leading strains, chickens that lay twice as many eggs as the best layers do now, or corn with an ear at the base of every leaf, instead of just at the base of a few leaves.

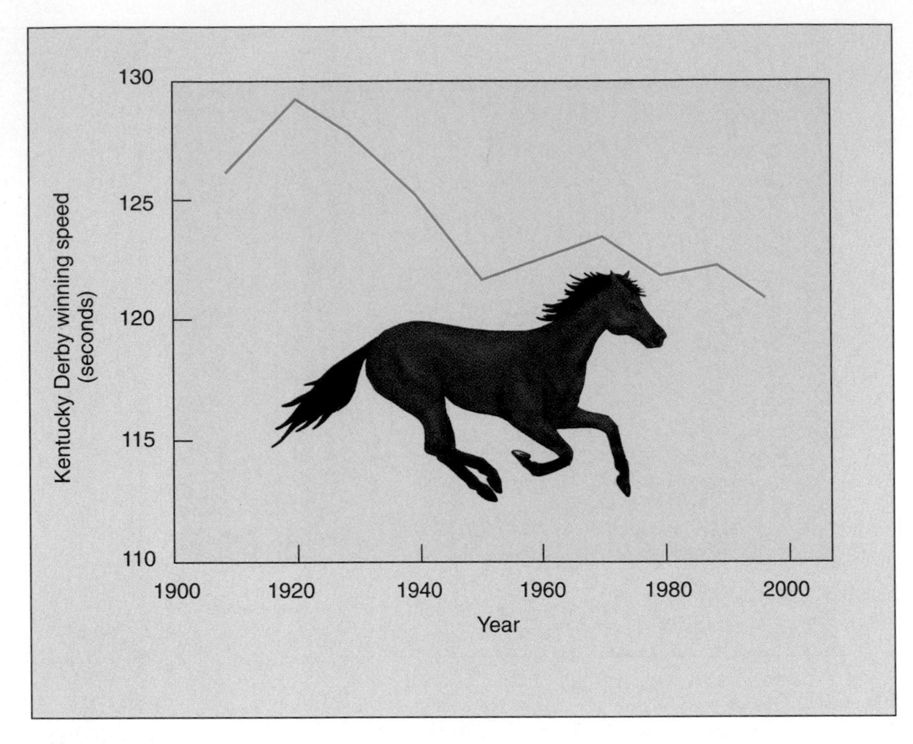

FIGURE 20.16
Selection for increased speed in racehorses is no longer effective. Kentucky Derby winning speeds have not improved significantly since 1950.

Evolution Requires Genetic Variation

Over 80% of the gene pool of the thoroughbred horses racing today goes back to 31 known ancestors from the late eighteenth century. Despite intense directional selection on thoroughbreds, their performance times have not improved for the last 50 years (figure 20.16). Years of intense selection presumably have removed variation from the population at a rate greater than it could be replenished by mutation such that now no genetic variation remains and evolutionary change is not possible.

In some cases, phenotypic variation for a trait may never have had a genetic basis. The compound eyes of insects are made up of hundreds of visual units, termed ommatidia. In some individuals, the left eye contains more ommatidia than the right eye. In other individuals, the right eye contains more than the left. However, despite intense selection in the laboratory, scientists have never been able to produce a line of fruit flies that consistently have more ommatidia in the left eye than in the right. The reason is that separate genes do not exist for the left and right eyes. Rather, the same genes affect both eyes, and differences in the number of ommatidia result from differences that occur as the eyes are formed in the development process (figure 20.17). Thus, despite the existence of phenotypic variation, no genetic variation is available for selection to favor.

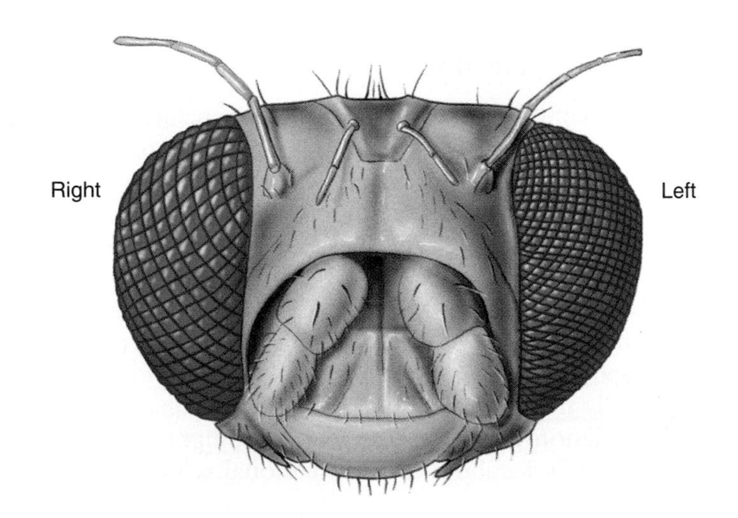

FIGURE 20.17
Phenotypic variation in insect ommatidia. In some individuals, the number of ommatidia in the left eye is greater than the number in the right eye. However, this difference is not genetically based; developmental processes cause the difference.

Selection against Rare Alleles

A second factor limits what selection can accomplish: selection acts only on phenotypes. For this reason, selection does not operate efficiently on rare recessive alleles, simply because there is no way to select against them unless they come together as homozygotes. For example, when a recessive allele *a* is present at a frequency *q* equal to 0.2, only four out of a hundred individuals (q^2) will be double recessive and display the phenotype associated with this allele (figure 20.18). For lower allele frequencies, the effect is even more dramatic: if the frequency in the population of the recessive allele $q = 0.01$, the frequency of recessive homozygotes in that population will be only 1 in 10,000.

The fact that selection acts on phenotypes rather than genotypes means that selection against undesirable genetic traits in humans or domesticated animals is difficult unless the heterozygotes can also be detected. For example, if a particular recessive allele *r* ($q = 0.01$) was considered undesirable, and none of the homozygotes for this allele were allowed to breed, it would take 1000 generations, or about 25,000 years in humans, to lower the allele frequency by half to 0.005. At this point, after 25,000 years of work, the frequency of homozygotes would still be 1 in 40,000, or 25% of what it was initially.

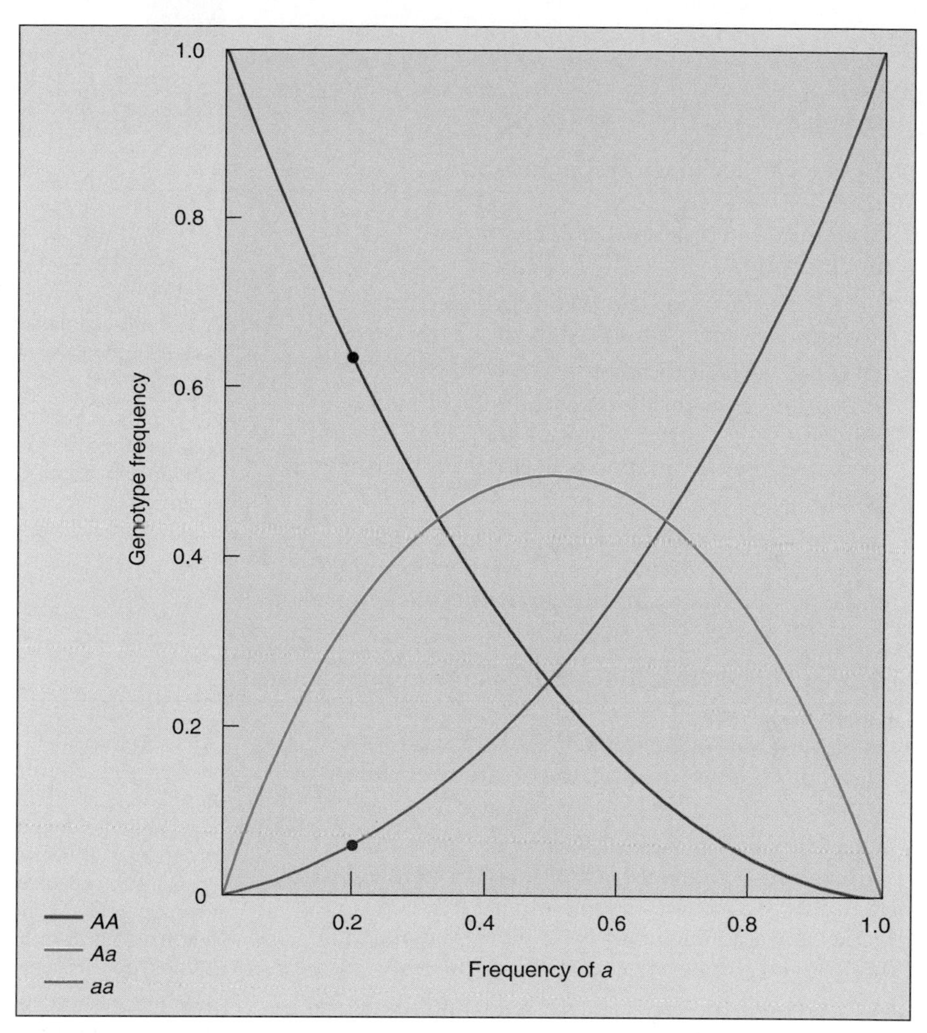

FIGURE 20.18
The relationship between allele frequency and genotype frequency. If allele *a* is present at a frequency of 0.2, the double recessive genotype *aa* is only present at a frequency of 0.04. In other words, only 4 in 100 individuals will have a homozygous recessive genotype, while 64 in 100 will have a homozygous dominant genotype.

Selection in Laboratory Environments

One way to assess the action of selection is to carry out artificial selection in the laboratory. Strains that are genetically identical except for the gene subject to selection can be crossed so that the possibility of linkage disequilibrium does not confound the analysis. Populations of bacteria provide a particularly powerful tool for studying selection in the laboratory because bacteria have a short generation time (less than an hour) and can be grown in huge numbers in growth vats called chemostats. In pioneering studies, Dan Hartl and coworkers backcrossed bacteria with different alleles of the enzyme 6-PGD into a homogeneous genetic background, and then compared the growth of the different strains when they were fed only gluconate, the enzyme's substrate. Hartl found that all of the alleles grew at the same rate! The different alleles were thus selectively neutral in a normal genetic background. However, when Hartl disabled an alternative biochemical pathway for the metabolism of gluconate, so that only 6-PGD mediated the utilization of this sole source of carbon, he obtained very different results: several alleles were markedly superior to others. Selection was clearly able to operate on these alleles, but only under certain conditions.

The ability of selection to produce evolutionary change is hindered by a variety of factors, including multiple effects of single genes, gene interactions, and lack of genetic variation. Moreover, selection can only eliminate rare recessive alleles very slowly.

Chapter 20

www.mhhe.com/raven6/resources20.mhtml

Summary | **Questions** | **Media Resources**

20.1 Genes vary in natural populations.

- Evolution is best defined as "descent with modification."
- Darwin's primary insight was to propose that evolutionary change resulted from the operation of natural selection.
- Selection is a powerful force leading to changes in populations but other processes may contribute. In any case, genetic variation is necessary for changes to occur.
- Invertebrates and outcrossing plants are often heterozygous at about 8 to 15% of their loci; the corresponding value for vertebrates is about 5 to 8%.

1. What is the difference between natural selection and evolution?

2. What is adaptation? How does it fit into Darwin's concept of evolution?

3. What is genetic polymorphism? What has polymorphism to do with evolution?

- Scientists on Science: From Butterflies to Global Preservation
- Student Research: Cotton Boll Weevil

20.2 Why do allele frequencies change in populations?

- Studies of how allele frequencies shift within populations allow investigators to study evolution in action.
- Meiosis does not alter allele frequencies within populations. The Hardy–Weinberg Principle states that unless selection or some other force acts on the genes, the frequencies of their alleles remain unchanged from one generation to the next.
- A variety of processes can lead to evolutionary change within a population, including mutation, gene flow, nonrandom mating, genetic drift, and selection.
- For evolution to occur by natural selection, three conditions must be met: 1. variation must exist in the population; 2. variation must be related to the number of offspring left in the next generation; 3. the variation must have a genetic basis.
- Natural selection can usually overpower the effects of genetic drift, except in very small populations.
- Natural selection can overwhelm the effects of gene flow in some cases, but not in others.

4. Given that allele A is present in a large random-mating population at a frequency of 54 per 100 individuals, what is the proportion of individuals in that population expected to be heterozygous for the allele? homozygous dominant? homozygous recessive?

5. Why does the founder effect have such a profound influence on a population's genetic makeup? How does the bottleneck effect differ from the founder effect?

6. What effect does inbreeding have on allele frequency? Why is marriage between close relatives discouraged?

- Hardy–Weinberg Equilibrium

- Natural Selection
- Other Processes of Evolution
- Evolutionary Variation
- Activities:
  - -Natural Selection
  - -Allele Frequencies
  - -Genetic Drift

- Art Quiz: Hardy–Weinberg Equilibrium

20.3 Selection can act on traits affected by many genes.

- Disruptive selection acts to eliminate rather than to favor the intermediate type; directional selection acts to eliminate one extreme from an array of phenotypes; and stabilizing selection acts to eliminate *both* extremes.
- Natural selection is not all powerful; genetic variation is required for natural selection to produce evolutionary change.

7. Define *selection*. How does it alter allele frequencies? What are the three types of selection? Give an example of each.

8. Why are there limitations to the success of selection?

- Types of Selection

- Art Quiz: Types of Natural Selection

21

The Evidence for Evolution

Concept Outline

21.1 Fossil evidence indicates that evolution has occurred.

The Fossil Record. When fossils are arranged in the order of their age, a continual series of change is seen, new changes being added at each stage.

The Evolution of Horses. The record of horse evolution is particularly well-documented and instructive.

21.2 Natural selection can produce evolutionary change.

The Beaks of Darwin's Finches. Natural selection favors stouter bills in dry years, when large tough-to-crush seeds are the only food available to finches.

Peppered Moths and Industrial Melanism. Natural selection favors dark-colored moths in areas of heavy pollution, while light-colored moths survive better in unpolluted areas.

Artificial Selection. Artificial selection practiced in laboratory studies, agriculture, and domestication demonstrate that selection can produce substantial evolutionary change.

21.3 Evidence for evolution can be found in other fields of biology.

The Anatomical Record. When anatomical features of living animals are examined, evidence of shared ancestry is often apparent.

The Molecular Record. When gene or protein sequences from organisms are arranged, species thought to be closely related based on fossil evidence are seen to be more similar than species thought to be distantly related.

Convergent and Divergent Evolution. Evolution favors similar forms under similar circumstances.

21.4 The theory of evolution has proven controversial.

Darwin's Critics. Critics have raised seven objections to Darwin's theory of evolution by natural selection.

FIGURE 21.1
A window into the past. The fossil remains of the now-extinct reptile Mesosaurus found in Permian sediments in Africa and South America provided one of the earliest clues to a former connection between the two continents. Mesosaurus was a freshwater species and so clearly incapable of a transatlantic swim. Therefore, it must have lived in the lakes and rivers of a formerly contiguous landmass that later became divided as Africa and South America drifted apart in the Cretaceous.

Of all the major ideas of biology, the theory that today's organisms evolved from now-extinct ancestors (figure 21.1) is perhaps the best known to the general public. This is not because the average person truly understands the basic facts of evolution, but rather because many people mistakenly believe that it represents a challenge to their religious beliefs. Similar highly publicized criticisms of evolution have occurred ever since Darwin's time. For this reason, it is important that, during the course of your study of biology, you address the issue squarely: Just what is the evidence for evolution?

21.1 Fossil evidence indicates that evolution has occurred.

At its core, the case for evolution is built upon two pillars: first, evidence that natural selection can produce evolutionary change and, second, evidence from the fossil record that evolution has occurred. In addition, information from many different areas of biology—including fields as different as embryology, anatomy, molecular biology, and biogeography (the study of the geographic distribution of species)—can only be interpreted sensibly as the outcome of evolution.

The Fossil Record

The most direct evidence that evolution has occurred is found in the fossil record. Today we have a far more complete understanding of this record than was available in Darwin's time. Fossils are the preserved remains of once-living organisms. Fossils are created when three events occur. First, the organism must become buried in sediment; then, the calcium in bone or other hard tissue must mineralize; and, finally, the surrounding sediment must eventually harden to form rock. The process of fossilization probably occurs rarely. Usually, animal or plant remains will decay or be scavenged before the process can begin. In addition, many fossils occur in rocks that are inaccessible to scientists. When they do become available, they are often destroyed by erosion and other natural processes before they can be collected. As a result, only a fraction of the species that have ever existed (estimated by some to be as many as 500 million) are known from fossils. Nonetheless, the fossils that have been discovered are sufficient to provide detailed information on the course of evolution through time.

Dating Fossils

By dating the rocks in which fossils occur, we can get an accurate idea of how old the fossils are. In Darwin's day, rocks were dated by their position with respect to one another (*relative dating*); rocks in deeper strata are generally older. Knowing the relative positions of sedimentary rocks and the rates of erosion of different kinds of sedimentary rocks in different environments, geologists of the nineteenth century derived a fairly accurate idea of the relative ages of rocks.

Today, rocks are dated by measuring the degree of decay of certain radioisotopes contained in the rock (*absolute dating*); the older the rock, the more its isotopes have decayed. Because radioactive isotopes decay at a constant rate unaltered by temperature or pressure, the isotopes in a rock act as an internal clock, measuring the time since the rock was formed. This is a more accurate way of dating rocks and provides dates stated in millions of years, rather than relative dates.

A History of Evolutionary Change

When fossils are arrayed according to their age, from oldest to youngest, they often provide evidence of successive evolutionary change. At the largest scale, the fossil record documents the progression of life through time, from the origin of eukaryotic organisms, through the evolution of fishes, the rise of land-living organisms, the reign of the dinosaurs, and on to the origin of humans (figure 21.2).

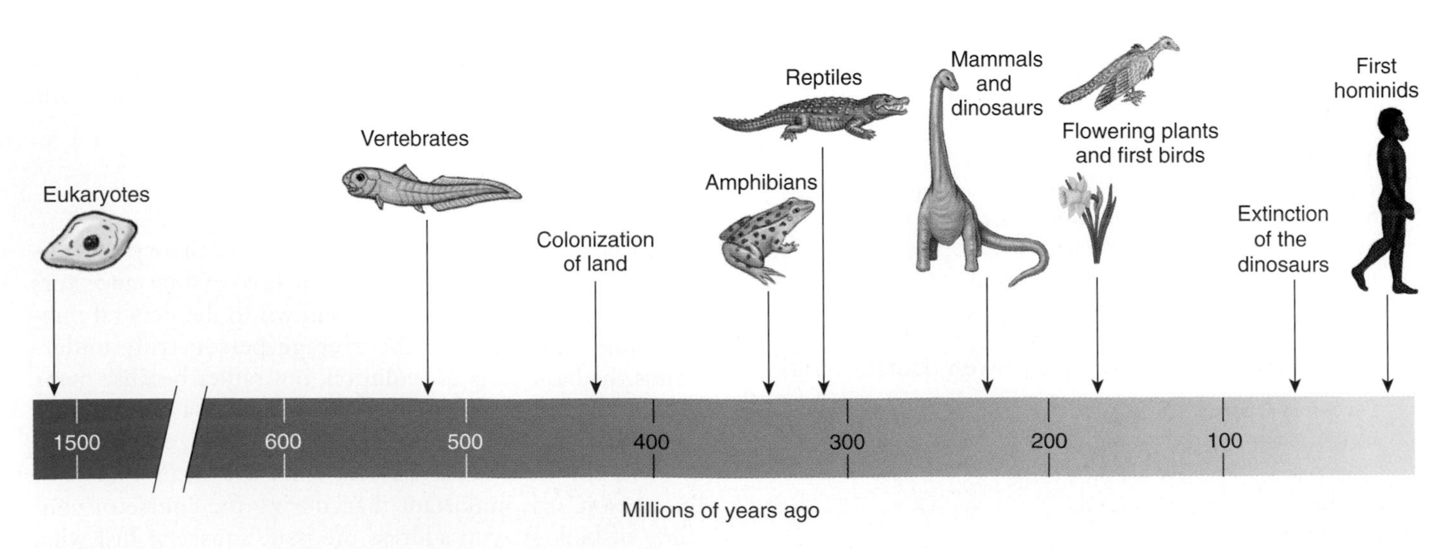

FIGURE 21.2
Timeline of the history of life as revealed by the fossil record.

Gaps in the Fossil Record

This is not to say that the fossil record is complete. Given the low likelihood of fossil preservation and recovery, it is not surprising that there are gaps in the fossil record. Nonetheless, paleontologists (the scientists who study fossils) continue to fill in the gaps in the fossil record. While many gaps interrupted the fossil record in Darwin's era, even then, scientists knew of the *Archaeopteryx* fossil transitional between dinosaurs and birds. Today, the fossil record is far more complete, particularly among the vertebrates; fossils have been found linking all the major groups. Recent years have seen spectacular discoveries closing some of the major remaining gaps in our understanding of vertebrate evolution. For example, recently a four-legged aquatic mammal was discovered that provides important insights concerning the evolution of whales and dolphins from land-living, hoofed ancestors (figure 21.3). Similarly, a fossil snake with legs has shed light on the evolution of snakes, which are descended from lizards that gradually became more and more elongated with simultaneous reduction and eventual disappearance of the limbs.

On a finer scale, evolutionary change within some types of animals is known in exceptional detail. For example, about 200 million years ago, oysters underwent a change from small curved shells to larger, flatter ones, with progressively flatter fossils being seen in the fossil record over a period of 12 million years (figure 21.4). A host of other examples all illustrate a record of successive change. The demonstration of this successive change is one of the strongest lines of evidence that evolution has occurred.

The fossil record provides a clear record of the major evolutionary transitions that have occurred through time.

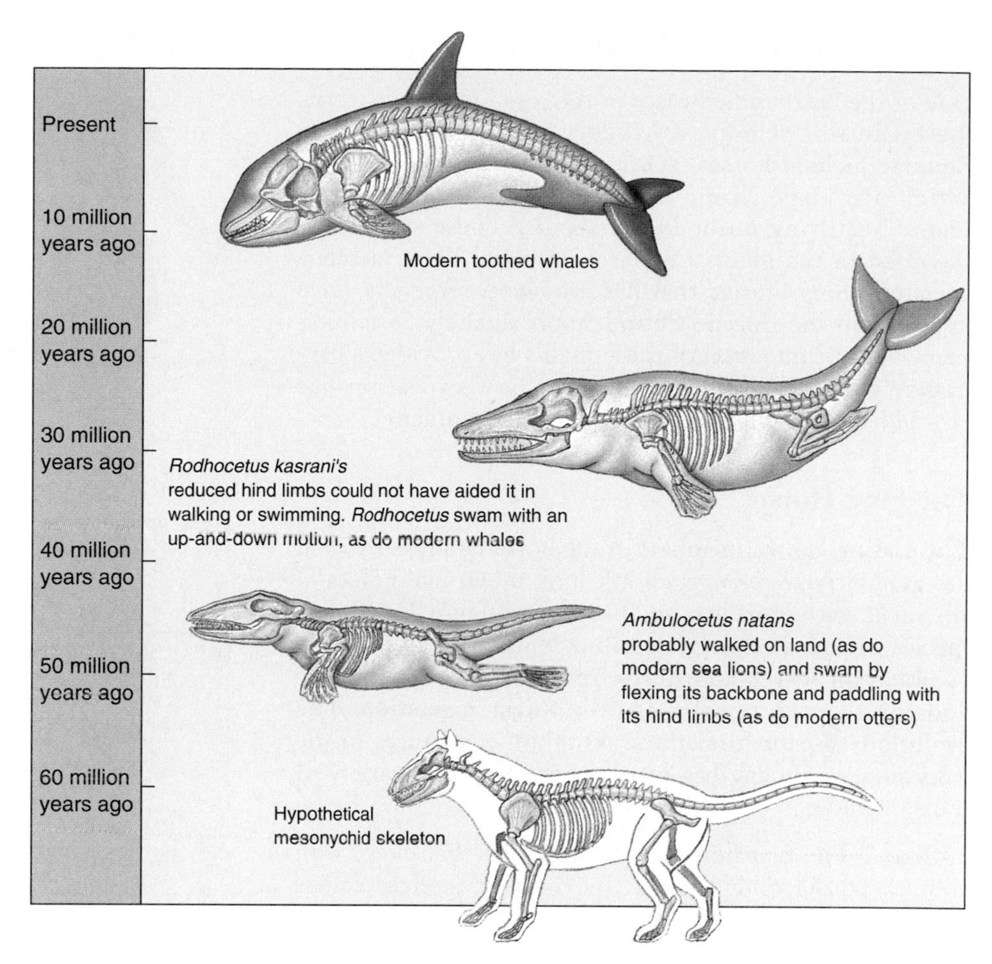

FIGURE 21.3
Whale "missing links." The recent discoveries of *Ambulocetus* and *Rodhocetus* have filled in the gaps between the mesonychids, the hypothetical ancestral link between the whales and the hoofed mammals, and present-day whales.

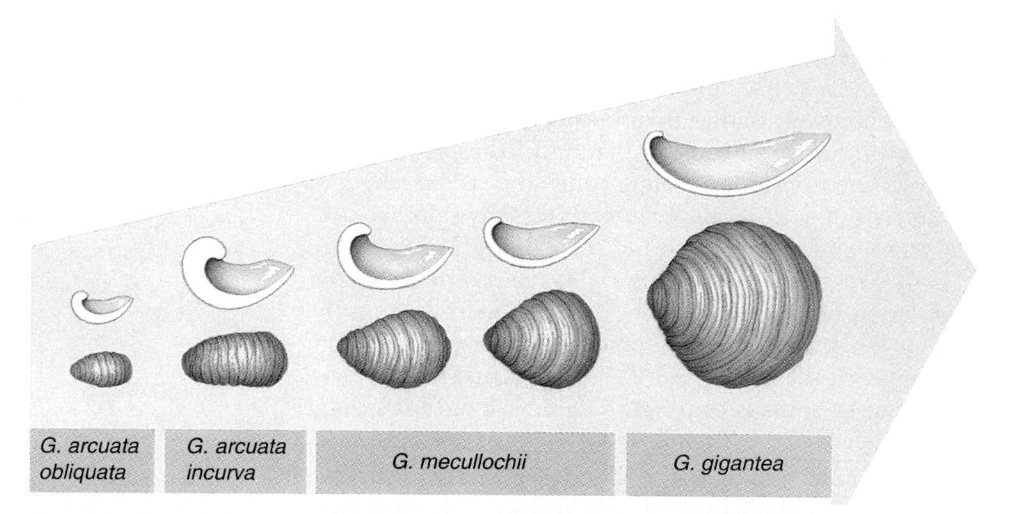

FIGURE 21.4
Evolution of shell shape in oysters. Over 12 million years of the Early Jurassic Period, the shells of this group of coiled oysters became larger, thinner, and flatter. These animals rested on the ocean floor in a special position called the "life position," and it may be that the larger, flatter shells were more stable in disruptive water movements.

The Evolution of Horses

One of the best-studied cases in the fossil record concerns the evolution of horses. Modern-day members of the Equidae include horses, zebras, donkeys and asses, all of which are large, long-legged, fast-running animals adapted to living on open grasslands. These species, all classified in the genus *Equus*, are the last living descendants of a long lineage that has produced 34 genera since its origin in the Eocene Period, approximately 55 million years ago. Examination of these fossils has provided a particularly well-documented case of how evolution has proceeded by adaptation to changing environments.

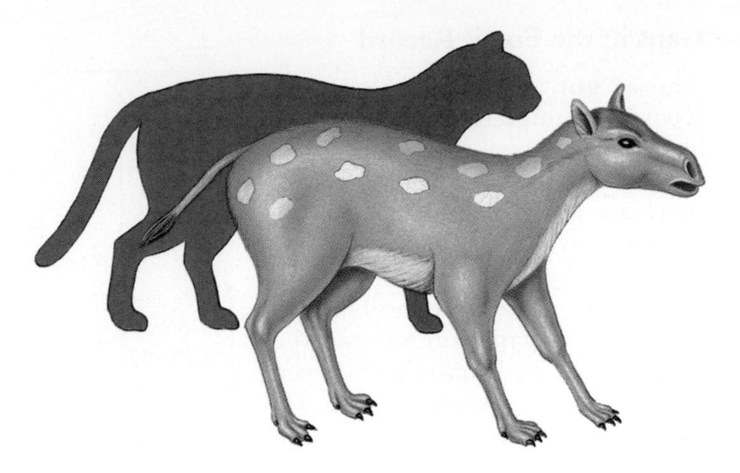

FIGURE 21.5
***Hyracotherium sandrae*, one of the earliest horses, was the size of a housecat.**

The First Horse

The earliest known members of the horse family, species in the genus *Hyracotherium*, didn't look much like horses at all. Small, with short legs and broad feet (figure 21.5), these species occurred in wooded habitats, where they probably browsed on leaves and herbs and escaped predators by dodging through openings in the forest vegetation. The evolutionary path from these diminutive creatures to the workhorses of today has involved changes in a variety of traits, including:

Size. The first horses were no bigger than dogs, with some considerably smaller. By contrast, modern equids can weigh more than a half ton. Examination of the fossil record reveals that horses changed little in size for their first 30 million years, but since then, a number of different lineages exhibited rapid and substantial increases. However, trends toward decreased size were also exhibited among some branches of the equid evolutionary tree (figure 21.6).

Toe reduction. The feet of modern horses have a single toe, enclosed in a tough, bony hoof. By contrast, *Hyracotherium* had four toes on its front feet and three on its hindfeet. Rather than hooves, these toes were encased in fleshy pads. Examination of the fossils clearly shows the transition through time: increase in length of the central toe, development of the bony hoof, and reduction and loss of the other toes (figure 21.7). As with body size, these trends occurred concurrently on several different branches of the horse evolutionary tree. At the same time as these developments, horses were evolving changes in the length and skeletal structure of the limbs, leading to animals capable of running long distances at high speeds.

Tooth size and shape. The teeth of *Hyracotherium* were small and relatively simple in shape. Through time, horse teeth have increased greatly in length and have developed a complex pattern of ridges on their molars and premolars (figure 21.7). The effect of these changes is to produce teeth better capable of chewing tough and gritty vegetation, such as grass, which tends to wear teeth down. Accompanying these changes have been alterations in the shape of the skull that strengthened the skull to withstand the stresses imposed by continual chewing. As with body size, evolutionary change has not been constant through time. Rather, much of the change in tooth shape has occurred within the past 20 million years.

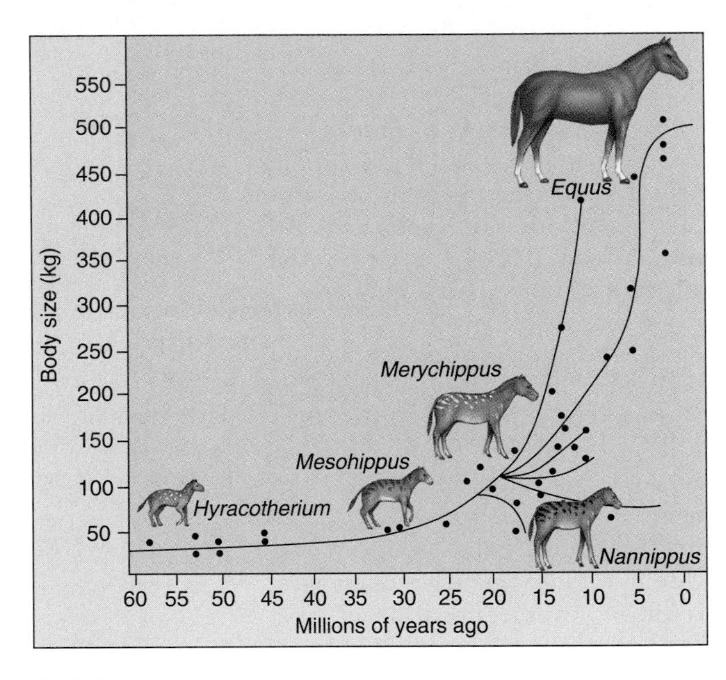

FIGURE 21.6
Evolutionary change in body size of horses. Lines show the broad outline of evolutionary relationships. Although most change involved increases in size, some decreases also occurred.

All of these changes may be understood as adaptations to changing global climates. In particular, during the late

Miocene and early Oligocene (approximately 20 to 25 million years ago), grasslands became widespread in North America, where much of horse evolution occurred. As horses adapted to these habitats, long-distance and high-speed locomotion probably became more important to escape predators and travel great distances. By contrast, the greater flexibility provided by multiple toes and shorter limbs, which was advantageous for ducking through complex forest vegetation, was no longer beneficial. At the same time, horses were eating grasses and other vegetation that contained more grit and other hard substances, thus favoring teeth and skulls better suited for withstanding such materials.

Evolutionary Trends

For many years, horse evolution was held up as an example of constant evolutionary change through time. Some even saw in the record of horse evolution evidence for a progressive, guiding force, consistently pushing evolution to move in a single direction. We now know that such views are misguided; evolutionary change over millions of years is rarely so simple.

Rather, the fossils demonstrate that, although there have been overall trends evident in a variety of characteristics, evolutionary change has been far from constant and uniform through time. Instead, rates of evolution have varied widely, with long periods of little change and some periods of great change. Moreover, when changes happen, they often occur simultaneously in different lineages of the horse evolutionary tree. Finally, even when a trend exists, exceptions, such as the evolutionary decrease in body size exhibited by some lineages, are not uncommon. These patterns, evident in our knowledge of horse evolution, are usually discovered for any group of plants and animals for which we have an extensive fossil record, as we shall see when we discuss human evolution in chapter 23.

Horse Diversity

One reason that horse evolution was originally conceived of as linear through time may be that modern horse diversity is relatively limited. Thus, it is easy to mentally picture a straight line from *Hyracotherium* to modern-day *Equus*. However, today's limited horse diversity—only one surviving genus—is unusual. Indeed, at the peak of horse diversity in the Miocene, as many as 13 genera of horses could be found in North America alone. These species differed in body size and in a wide variety of other characteristics. Presumably, they lived in different habitats and exhibited different dietary preferences. Had this diversity existed to modern times, early workers presumably would have had a different outlook on horse evolution, a situation that is again paralleled by the evolution of humans.

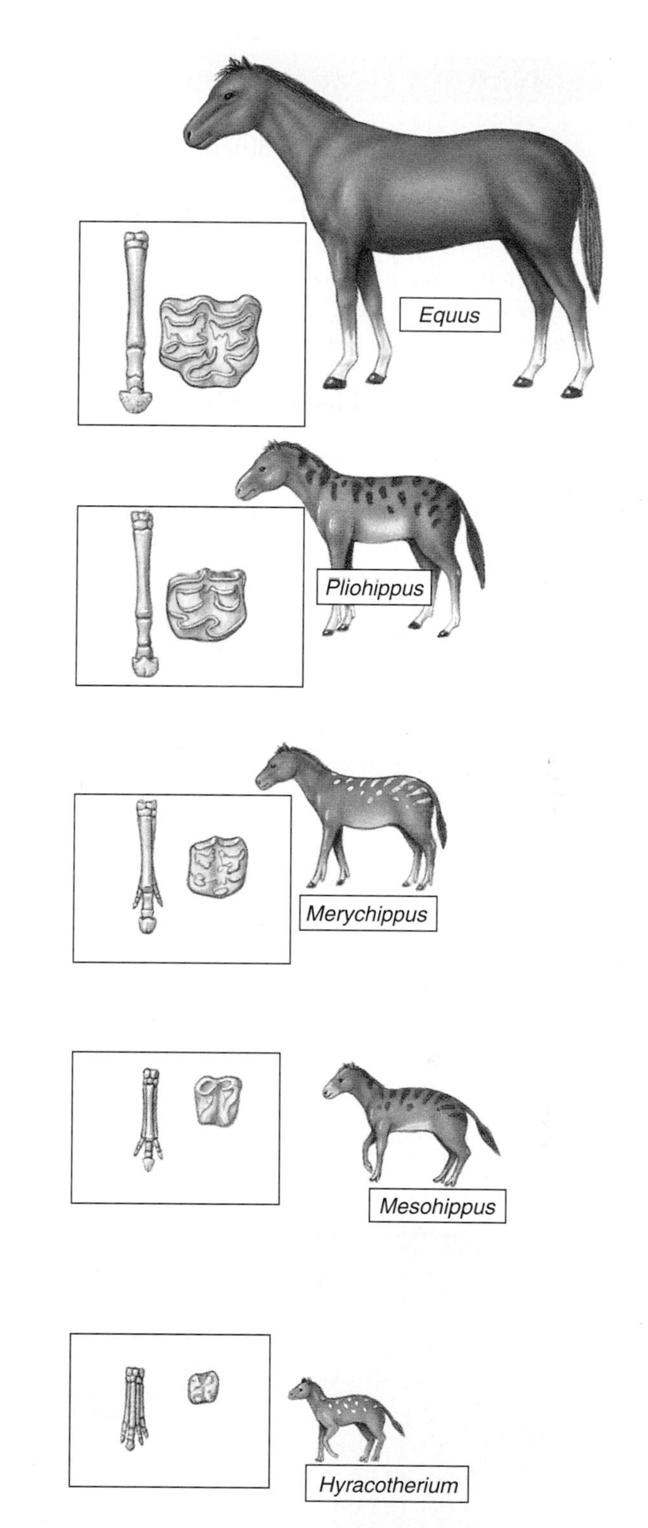

FIGURE 21.7
Evolutionary changes in horses through time.

The extensive fossil record for horses provides a detailed view of the evolutionary diversification of this group from small forest dwellers to the large and fast modern grassland species.

21.2 Natural selection can produce evolutionary change.

As we saw in chapter 20, a variety of different processes can result in evolutionary change. Nonetheless, in agreement with Darwin, most evolutionary biologists would agree that natural selection is the process responsible for most of the major evolutionary changes that have occurred through time. Although we cannot travel back through time, a variety of modern-day evidence confirms the power of natural selection as an agent of evolutionary change. These data come from both the field and the laboratory and from natural and human-altered situations.

The Beaks of Darwin's Finches

Darwin's finches are a classic example of evolution by natural selection. Darwin collected 31 specimens of finch from three islands when he visited the Galápagos Islands off the coast of Ecuador in 1835. Darwin, not an expert on birds, had trouble identifying the specimens, believing by examining their bills that his collection contained wrens, "grossbeaks," and blackbirds. You can see Darwin's sketches of four of these birds in figure 21.8.

The Importance of the Beak

Upon Darwin's return to England, ornithologist John Gould examined the finches. Gould recognized that Darwin's collection was in fact a closely related group of distinct species, all similar to one another except for their bills. In all, there were 13 species. The two ground finches with the larger bills in figure 21.8 feed on seeds that they crush in their beaks, whereas the two with narrower bills eat insects. One species is a fruit eater, another a cactus eater, yet another a "vampire" that creeps up on seabirds and uses its sharp beak to drink their blood. Perhaps most remarkable are the tool users, woodpecker finches that pick up a twig, cactus thorn, or leaf stalk, trim it into shape with their bills, and then poke it into dead branches to pry out grubs.

The correspondence between the beaks of the 13 finch species and their food source immediately suggested to Darwin that evolution had shaped them:

> "Seeing this gradation and diversity of structure in one small, intimately related group of birds, one might really fancy that from an original paucity of birds in this archipelago, one species has been taken and modified for different ends."

Was Darwin Wrong?

If Darwin's suggestion that the beak of an ancestral finch had been "modified for different ends" is correct, then it ought to be possible to see the different species of finches acting out their evolutionary roles, each using their bills to acquire their particular food specialty. The four species that crush seeds within their bills, for example, should feed on different seeds, those with stouter beaks specializing on harder-to-crush seeds.

FIGURE 21.8
Darwin's own sketches of Galápagos finches. From Darwin's *Journal of Researches:* (*1*) large ground finch *Geospiza magnirostris;* (*2*) medium ground finch *Geospiza fortis;* (*3*) small tree finch *Camarhynchus parvulus;* (*4*) warbler finch *Certhidea olivacea.*

Many biologists visited the Galápagos after Darwin, but it was 100 years before any tried this key test of his hypothesis. When the great naturalist David Lack finally set out to do this in 1938, observing the birds closely for a full five months, his observations seemed to contradict Darwin's proposal! Lack often observed many different species of finch feeding together on the same seeds. His data indicated that the stout-beaked species and the slender-beaked species were feeding on the very same array of seeds.

We now know that it was Lack's misfortune to study the birds during a wet year, when food was plentiful. The finch's beak is of little importance in such flush times; small seeds are so abundant that birds of all species are able to get enough to eat. Later work has revealed a very different picture during leaner, dry years, when few seeds are available and the difference between survival and starvation depends on being able to efficiently gather enough to eat. In such times, having beaks designed to be maximally effective for a particular type of food becomes critical and the species diverge in their diet, each focusing on the type of food to which it is specialized.

A Closer Look

The key to successfully testing Darwin's proposal that the beaks of Galápagos finches are adaptations to different food sources proved to be patience. Starting in 1973, Peter and Rosemary Grant of Princeton University and generations of their students have studied the medium ground finch *Geospiza fortis* on a tiny island in the center of the Galápagos called Daphne Major. These finches feed preferentially on small tender seeds, produced in abundance by plants in wet years. The birds resort to larger, drier seeds, which are harder to crush, only when small seeds become depleted during long periods of dry weather, when plants produce few seeds.

The Grants quantified beak shape among the medium ground finches of Daphne Major by carefully measuring beak depth (width of beak, from top to bottom, at its base) on individual birds. Measuring many birds every year, they were able to assemble for the first time a detailed portrait of evolution in action. The Grants found that beak depth changed from one year to the next in a predictable fashion. During droughts, plants produced few seeds and all available small seeds quickly were eaten, leaving large seeds as the major remaining source of food. As a result, birds with large beaks survived better, because they were better able to break open these large seeds. Consequently, the average beak depth of birds in the population increased the next year, only to decrease again when wet seasons returned (figure 21.9).

Could these changes in beak dimension reflect the action of natural selection? An alternative possibility might be that the changes in beak depth do not reflect changes in gene frequencies, but rather are simply a response to diet—perhaps during lean times the birds become malnourished and then grow stouter beaks, for example. To rule out this possibility, the Grants measured the relation of parent bill size to offspring bill size, examining many broods over several years. The depth of the bill was passed down faithfully from one generation to the next, regardless of environmental conditions, suggesting that the differences in bill size indeed reflected genetic differences.

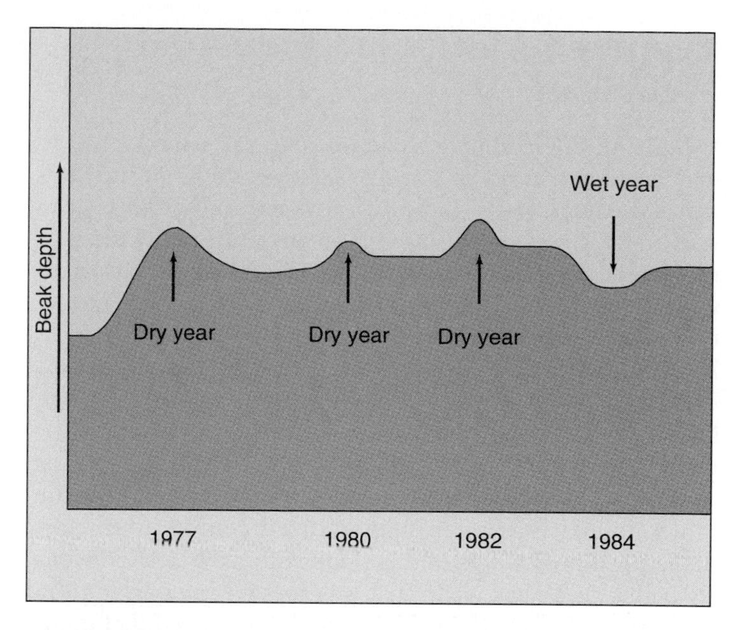

FIGURE 21.9
Evidence that natural selection alters beak size in *Geospiza fortis*. In dry years, when only large, tough seeds are available, the mean beak size increases. In wet years, when many small seeds are available, smaller beaks become more common.

Source: Data from Grant, "Nature Selection and Darwin's Finches" in *Scientific American*, October 1991.

Darwin Was Right After All

If the year-to-year changes in beak depth indeed reflect genetic changes, as now seems likely, and these changes can be predicted by the pattern of dry years, then Darwin was right after all—natural selection does seem to be operating to adjust the beak to its food supply. Birds with stout beaks have an advantage during dry periods, for they can break the large, dry seeds that are the only food available. When small seeds become plentiful once again with the return of wet weather, a smaller beak proves a more efficient tool for harvesting the more abundant smaller seeds.

Among Darwin's finches, natural selection adjusts the shape of the beak in response to the nature of the available food supply, adjustments that can be seen to be occurring even today.

Peppered Moths and Industrial Melanism

When the environment changes, natural selection often may favor new traits in a species. The example of the Darwin's finches clearly indicates how natural variation can lead to evolutionary change. Humans are greatly altering the environment in many ways; we should not be surprised to see organisms attempting to adapt to these new conditions. One classic example concerns the peppered moth, *Biston betularia*. Until the mid-nineteenth century, almost every individual of this species captured in Great Britain had light-colored wings with black specklings (hence the name "peppered" moth). From that time on, individuals with dark-colored wings increased in frequency in the moth populations near industrialized centers until they made up almost 100% of these populations. Black individuals had a dominant allele that was present but very rare in populations before 1850. Biologists soon noticed that in industrialized regions where the dark moths were common, the tree trunks were darkened almost black by the soot of pollution. Dark moths were much less conspicuous resting on them than were light moths. In addition, the air pollution that was spreading in the industrialized regions had killed many of the light-colored lichens on tree trunks, making the trunks darker.

Selection for Melanism

Can Darwin's theory explain the increase in the frequency of the dark allele? Why did dark moths gain a survival advantage around 1850? An amateur moth collector named J. W. Tutt proposed what became the most commonly accepted hypothesis explaining the decline of the light-colored moths. He suggested that peppered forms were more visible to predators on sooty trees that have lost their lichens. Consequently, birds ate the peppered moths resting on the trunks of trees during the day. The black forms, in contrast, were at an advantage because they were camouflaged (figure 21.10). Although Tutt initially had no evidence, British ecologist Bernard Kettlewell tested the hypothesis in the 1950s by rearing populations of peppered moths with equal numbers of dark and light individuals. Kettlewell then released these populations into two sets of woods: one, near heavily polluted Birmingham, the other, in unpolluted Dorset. Kettlewell set up rings of traps around the woods to see how many of both kinds of moths survived. To evaluate his results, he had marked the released moths with a dot of paint on the underside of their wings, where birds could not see it.

In the polluted area near Birmingham, Kettlewell trapped 19% of the light moths, but 40% of the dark ones. This indicated that dark moths had a far better chance of surviving in these polluted woods, where the tree trunks were dark. In the relatively unpolluted Dorset woods, Kettlewell recovered 12.5% of the light moths but only 6% of the dark ones. This indicated that where the tree trunks were still light-colored, light moths had a much better chance of survival. Kettlewell later solidified his argument by placing hidden blinds in the woods and actually filming birds eating the moths. Sometimes the birds Kettlewell observed actually passed right over a moth that was the same color as its background.

FIGURE 21.10
Tutt's hypothesis explaining industrial melanism. These photographs show color variants of the peppered moth, *Biston betularia.* Tutt proposed that the dark moth is more visible to predators on unpolluted trees (*top*), while the light moth is more visible to predators on bark blackened by industrial pollution (*bottom*).

Industrial Melanism

Industrial melanism is a term used to describe the evolutionary process in which darker individuals come to predominate over lighter individuals since the industrial revolution as a result of natural selection. The process is widely believed to have taken place because the dark organisms are better concealed from their predators in habitats that have been darkened by soot and other forms of industrial pollution, as suggested by Kettlewell's research.

Dozens of other species of moths have changed in the same way as the peppered moth in industrialized areas throughout Eurasia and North America, with dark forms becoming more common from the mid-nineteenth century onward as industrialization spread.

Selection against Melanism

In the second half of the twentieth century, with the widespread implementation of pollution controls, these trends are reversing, not only for the peppered moth in many areas in England, but also for many other species of moths throughout the northern continents. These examples provide some of the best documented instances of changes in allelic frequencies of natural populations as a result of natural selection due to specific factors in the environment.

In England, the pollution promoting industrial melanism began to reverse following enactment of Clean Air legislation in 1956. Beginning in 1959, the *Biston* population at Caldy Common outside Liverpool has been sampled each year. The frequency of the melanic (dark) form has dropped from a high of 94% in 1960 to a 1994 low of 19% (figure 21.11). Similar reversals have been documented at numerous other locations throughout England. The drop correlates well with a drop in air pollution, particularly with tree-darkening sulfur dioxide and suspended particulates.

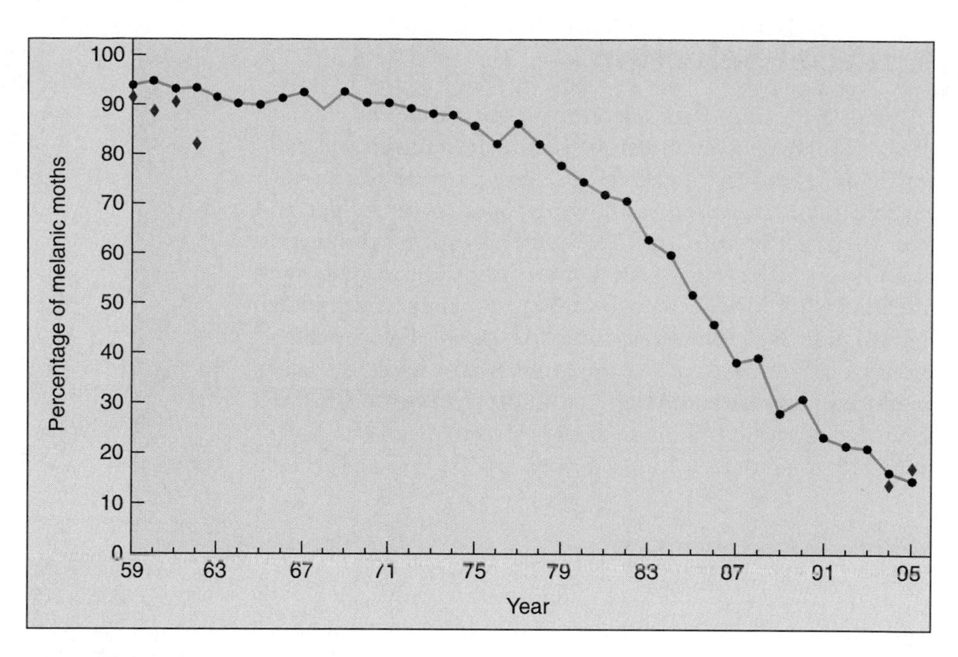

FIGURE 21.11
Selection against melanism. The circles indicate the frequency of melanic Biston moths at Caldy Common in England, sampled continuously from 1959 to 1995. Diamonds indicate frequencies in Michigan from 1959 to 1962 and from 1994 to 1995.
Source: Data from Grant, et al., "Parallel Rise and Fall of Melanic Peppered Moths" in *Journal of Heredity*, vol. 87, 1996, Oxford University Press.

Interestingly, the same reversal of industrial melanism appears to have occurred in America during the same time that it was happening in England. Industrial melanism in the American subspecies of the peppered moth was not as widespread as in England, but it has been well-documented at a rural field station near Detroit. Of 576 peppered moths collected there from 1959 to 1961, 515 were melanic, a frequency of 89%. The American Clean Air Act, passed in 1963, led to significant reductions in air pollution. Resampled in 1994, the Detroit field station peppered moth population had only 15% melanic moths (see figure 21.11)! The moths in Liverpool and Detroit, both part of the same natural experiment, exhibit strong evidence of natural selection.

Reconsidering the Target of Natural Selection

Tutt's hypothesis, widely accepted in the light of Kettlewell's studies, is currently being reevaluated. The problem is that the recent selection against melanism does not appear to correlate with changes in tree lichens. At Caldy Common, the light form of the peppered moth began its increase in frequency long before lichens began to reappear on the trees. At the Detroit field station, the lichens never changed significantly as the dark moths first became dominant and then declined over the last 30 years. In fact, investigators have not been able to find peppered moths on Detroit trees at all, whether covered with lichens or not. Wherever the moths rest during the day, it does not appear to be on tree bark. Some evidence suggests they rest on leaves on the treetops, but no one is sure.

The action of selection may depend less on the presence of lichens and more on other differences in the environment resulting from industrial pollution. Pollution tends to cover all objects in the environment with a fine layer of particulate dust, which tends to decrease how much light surfaces reflect. In addition, pollution has a particularly severe effect on birch trees, which are light in color. Both effects would tend to make the environment darker and thus would favor darker color in moths.

Natural selection has favored the dark form of the peppered moth in areas subject to severe air pollution, perhaps because on darkened trees they are less easily seen by moth-eating birds. Selection has in turn favored the light form as pollution has abated.

Artificial Selection

Humans have imposed selection upon plants and animals since the dawn of civilization. Just as in natural selection, artificial selection operates by favoring individuals with certain phenotypic traits, allowing them to reproduce and pass their genes into the next generation. Assuming that phenotypic differences are genetically determined, such selection should lead to evolutionary change and, indeed, it has. Artificial selection, imposed in laboratory experiments, agriculture, and the domestication process, has produced substantial change in almost every case in which it has been applied. This success is strong proof that selection is an effective evolutionary process.

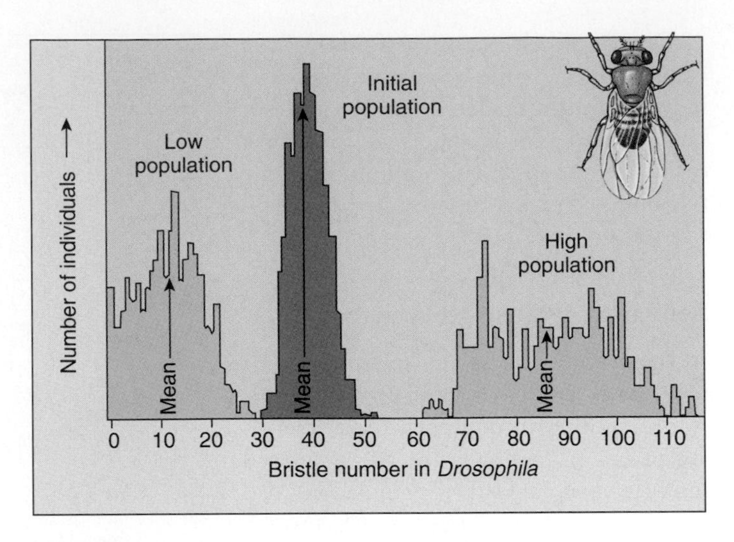

FIGURE 21.12
Artificial selection in the laboratory. In this experiment, one population of *Drosophila* was selected for low numbers of bristles and the other for high numbers. Note that not only did the means of the populations change greatly in 35 generations, but also that all individuals in both experimental populations lie outside the range of the initial population.

Source: Data from G. Dayton and A. Roberson, *Journal of Genetics*, Vol.55, p. 154, 1957.

Laboratory Experiments

With the rise of genetics as a field of science in the 1920s and 1930s, researchers began conducting experiments to test the hypothesis that selection can produce evolutionary change. A favorite subject was the now-famous laboratory fruit fly, *Drosophila melanogaster*. Geneticists have imposed selection on just about every conceivable aspect of the fruit fly—including body size, eye color, growth rate, life span, and exploratory behavior—with a consistent result: selection for a trait leads to strong and predictable evolutionary response.

In one classic experiment, scientists selected for fruit flies with many bristles (stiff, hairlike structures) on their abdomen. At the start of the experiment, the average number of bristles was 9.5. Each generation, scientists picked out the 20% of the population with the greatest number of bristles and allowed them to reproduce, thus establishing the next generation. After 86 generations of such selection, the average number of bristles had quadrupled, to nearly 40. In a similar experiment, fruit flies were selected for either the most or the fewest numbers of bristles. Within 35 generations, the populations did not overlap at all in range of variation (figure 21.12).

Similar experiments have been conducted on a wide variety of other laboratory organisms. For example, by selecting for rats that were resistant to tooth decay, scientists were able to increase in less than 20 generations the average time for onset of decay from barely over 100 days to greater than 500 days.

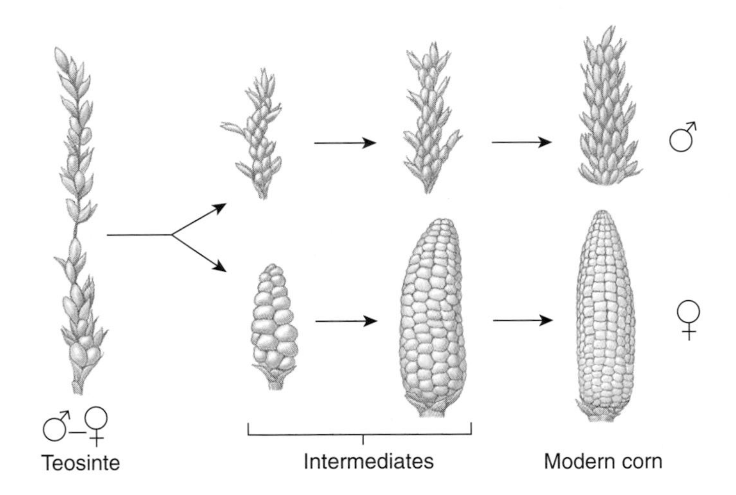

FIGURE 21.13
Corn looks very different from its ancestor. The tassels and seeds of a wild grass, such as teosinte, evolved into the male tassels and female ears of modern corn.

Agriculture

Similar methods have been practiced in agriculture for many centuries. Familiar livestock, such as cattle and pigs, and crops, like corn and strawberries, are greatly different from their wild ancestors (figure 21.13). These differences have resulted from generations of selection for desirable traits like milk production and corn stalk size.

An experimental study with corn demonstrates the ability of artificial selection to rapidly produce major change in crop plants. In 1896, agricultural scientists began selecting on oil content of corn kernels, which initially was 4.5%. As in the fruit fly experiments, the top 20% of all individuals were allowed to reproduce. In addition, a parallel experiment selected for the individuals with the lowest oil content. By 1986, at which time 90 generations had passed, average oil content had increased approximately 450% in the high-content experiment; by contrast, oil content in the low experiment had decreased to about 0.5%, a level at which it is difficult to get accurate readings.

Domestication

Artificial selection has also been responsible for the great variety of breeds of cats, dogs (figure 21.14), pigeons, cattle and other domestic animals. In some cases, breeds have been developed for particular purposes. Greyhound dogs, for example, were bred by selecting for maximal running abilities, with the end result being an animal with long legs and tail (the latter used as a rudder), an arched back (to increase the length of its stride), and great muscle mass. By contrast, the odd proportions of the ungainly basset hound resulted from selection for dogs that could enter narrow holes in pursuit of rabbits and other small game. In other cases, breeds have been developed primarily for their appearance, such as the many colorful and ornamented varieties of pigeons or the breeds of cats.

Domestication also has led to unintentional selection for some traits. In recent years, as part of an attempt to domesticate the silver fox, Russian scientists each generation have chosen the most docile animals and allowed them to reproduce. Within 40 years, the vast majority of foxes born were exceptionally docile, not only allowing themselves to be petted, but also whimpering to get attention and sniffing and licking their caretakers. In many respects, they had become no different than domestic dogs! However, it was not only behavior that changed. These foxes also began to exhibit different color patterns, floppy ears, curled tails, and shorter legs and tails. Presumably, the genes responsible for docile behavior have other effects as well (the phenomenon of pleiotropy discussed in the last chapter); as selection has favored docile animals, it has also led to the evolution of these other traits.

FIGURE 21.14
Breeds of dogs. The differences between these dogs are greater than the differences displayed between any wild species of canids.

Can Selection Produce Major Evolutionary Changes?

Given that we can observe the results of selection operating over relatively short periods of time, most scientists believe that natural selection is the process responsible for the evolutionary changes documented in the fossil record. Some critics of evolution accept that selection can lead to changes within a species, but contend that such changes are relatively minor in scope and not equivalent to the substantial changes documented in the fossil record. In other words, it is one thing to change the number of bristles on a fruit fly or the size of a corn stalk, and quite another to produce an entirely new species.

This argument does not fully appreciate the extent of change produced by artificial selection. Consider, for example, the breeds of dogs, all of which have been produced since wolves were first domesticated, perhaps 10,000 years ago. If the various dog breeds did not exist and a paleontologist found fossils of animals similar to dachshunds, greyhounds, mastiffs, chihuahuas, and pomeranians, there is no question that they would be considered different species. Indeed, these breeds are so different that they would probably be classified in different genera. In fact, the diversity exhibited by dog breeds far outstrips the differences observed among wild members of the family Canidae—such as coyotes, jackals, foxes, and wolves. Consequently, the claim that artificial selection produces only minor changes is clearly incorrect. Indeed, if selection operating over a period of only 10,000 years can produce such substantial differences, then it would seem powerful enough, over the course of many millions of years, to produce the diversity of life we see around us today.

Artificial selection often leads to rapid and substantial results over short periods of time, thus demonstrating the power of selection to produce major evolutionary change.

21.3 Evidence for evolution can be found in other fields of biology.

The Anatomical Record

Much of the power of the theory of evolution is its ability to provide a sensible framework for understanding the diversity of life. Many observations from a wide variety of fields of biology simply cannot be understood in any meaningful way except as a result of evolution.

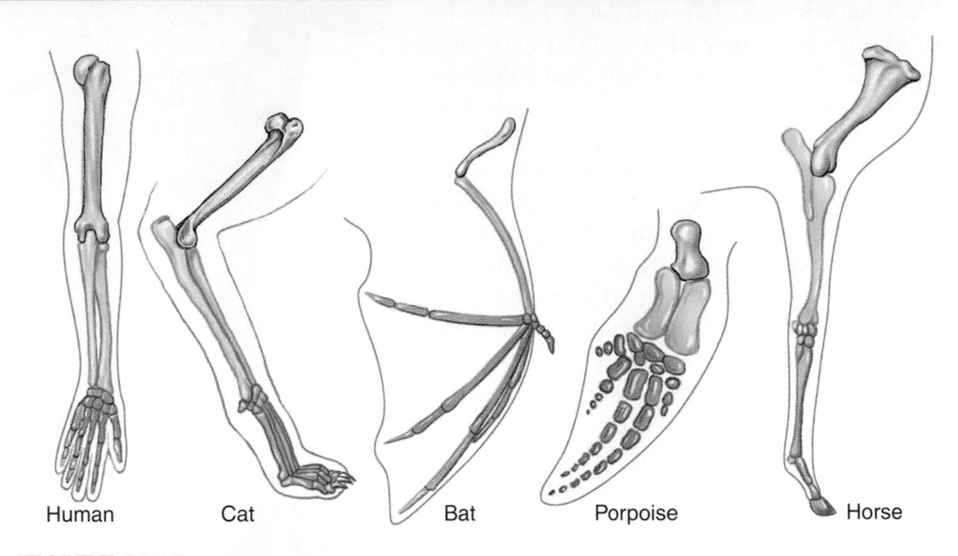

FIGURE 21.15
Homology among the bones of the forelimb. Although these structures show considerable differences in form and function, the same basic bones are present in the forelimbs of humans, cats, bats, porpoises, and horses.

Homology

As vertebrates evolved, the same bones were sometimes put to different uses. Yet the bones are still seen, their presence betraying their evolutionary past. For example, the forelimbs of vertebrates are all **homologous structures,** that is, structures with different appearances and functions that all derived from the same body part in a common ancestor. You can see in figure 21.15 how the bones of the forelimb have been modified in different ways for different vertebrates. Why should these very different structures be composed of the same bones? If evolution had not occurred, this would indeed be a riddle. But when we consider that all of these animals are descended from a common ancestor, it is easy to understand that natural selection has modified the same initial starting blocks to serve very different purposes.

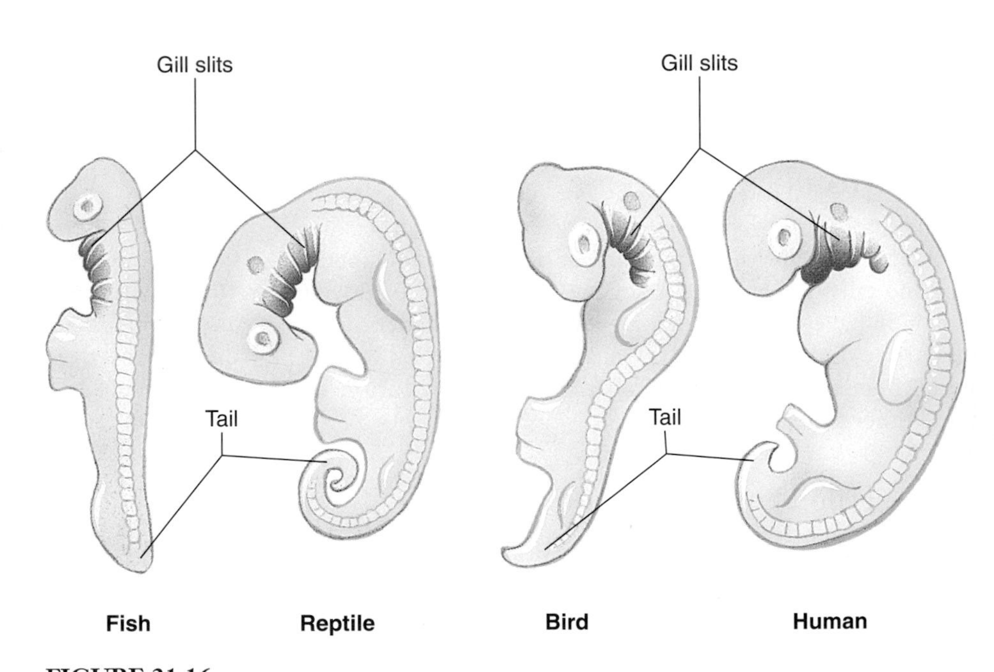

FIGURE 21.16
Our embryos show our evolutionary history. The embryos of various groups of vertebrate animals show the features they all share early in development, such as gill slits (*in purple*) and a tail.

Development

Some of the strongest anatomical evidence supporting evolution comes from comparisons of how organisms develop. In many cases, the evolutionary history of an organism can be seen to unfold during its development, with the embryo exhibiting characteristics of the embryos of its ancestors (figure 21.16). For example, early in their development, human embryos possess gill slits, like a fish; at a later stage, every human embryo has a long bony tail, the vestige of which we carry to adulthood as the coccyx at the end of our spine. Human fetuses even possess a fine fur (called *lanugo*) during the fifth month of development. These relict developmental forms suggest strongly that our development has evolved, with new instructions layered on top of old ones.

The observation that seemingly different organisms may exhibit similar embryological forms provides indirect but convincing evidence of a past evolutionary relationship. Slugs and giant ocean squids, for example, do not bear much superficial resemblance to each other, but the similarity of their embryological forms provides convincing evidence that they are both mollusks.

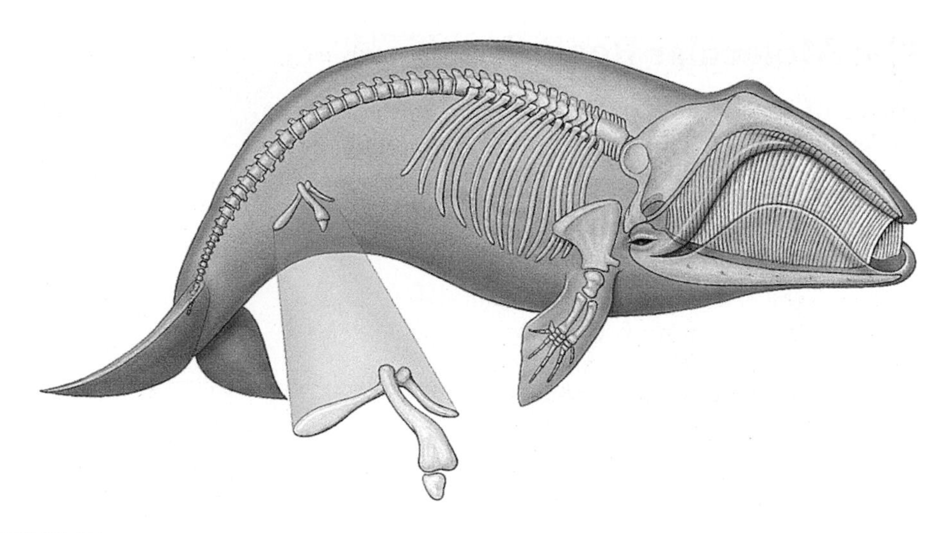

FIGURE 21.17

Vestigial features. The skeleton of a baleen whale, a representative of the group of mammals that contains the largest living species, contains pelvic bones. These bones resemble those of other mammals, but are only weakly developed in the whale and have no apparent function.

Vestigial Structures

Many organisms possess **vestigial structures** that have no apparent function, but that resemble structures their presumed ancestors had. Humans, for example, possess a complete set of muscles for wiggling their ears, just as a coyote does (table 21.1). Boa constrictors have hip bones and rudimentary hind legs. Manatees (a type of aquatic mammal often referred to as "sea cows") have fingernails on their fins (which evolved from legs). Figure 21.17 illustrates the skeleton of a baleen whale, which contains pelvic bones, as other mammal skeletons do, even though such bones serve no known function in the whale. The human vermiform appendix is apparently vestigial; it represents the degenerate terminal part of the cecum, the blind pouch or sac in which the large intestine begins. In other mammals such as mice, the cecum is the largest part of the large intestine and functions in storage—usually of bulk cellulose in herbivores. Although some suggestions have been made, it is difficult to assign any current function to the vermiform appendix. In many respects, it is a dangerous organ: quite often it becomes infected, leading to an inflammation called appendicitis; without surgical removal, the appendix may burst, allowing the contents of the gut to come in contact with the lining of the body cavity, a potentially fatal event. It is difficult to understand vestigial structures such as these as anything other than evolutionary relicts, holdovers from the evolutionary past. They argue strongly for the common ancestry of the members of the groups that share them, regardless of how different they have subsequently become.

Comparisons of the anatomy of different living animals often reveal evidence of shared ancestry. In some instances, the same organ has evolved to carry out different functions, in others, an organ loses its function altogether. Sometimes, different organs evolve in similar ways when exposed to the same selective pressures.

Table 21.1 Some Vestigial Traits in Humans

Trait	Description
Ear-wiggling muscles	Three small muscles around each ear that are large and important in some mammals, such as dogs, turning the ears toward a source of sound. Few people can wiggle their ears, and none can turn them toward sound.
Tail	Present in human and all vertebrate embryos. In humans, the tail is reduced; most adults only have three to five tiny tail bones and, occasionally, a trace of a tail-extending muscle.
Appendix	Structure which presumably had a digestive function in some of our ancestors, like the cecum of some herbivores. In humans, it varies in length from 5–15 cm, and some people are born without one.
Wisdom teeth	Molars that are often useless and sometimes even trapped in the jawbone. Some people never develop wisdom teeth.

Based on a suggestion by Dr. Leslie Dendy, Department of Science and Technology, University of New Mexico, Los Alamos.

The Molecular Record

Traces of our evolutionary past are also evident at the molecular level. If you think about it, the fact that organisms have evolved successively from relatively simple ancestors implies that a record of evolutionary change is present in the cells of each of us, in our DNA. When an ancestral species gives rise to two or more descendants, those descendants will initially exhibit fairly high overall similarity in their DNA. However, as the descendants evolve independently, they will accumulate more and more differences in their DNA. Consequently, organisms that are more distantly related would be expected to accumulate a greater number of evolutionary differences, whereas two species that are more closely related should share a greater portion of their DNA.

To examine this hypothesis, we need an estimate of evolutionary relationships that has been developed from data other than DNA (it would be a circular argument to use DNA to estimate relationships and then conclude that closely related species are more similar in their DNA than are distantly related species). Such an hypothesis of evolutionary relationships is provided by the fossil record, which indicates when particular types of organisms evolved. In addition, by examining the anatomical structures of fossils and of modern species, we can infer how closely species are related to each other.

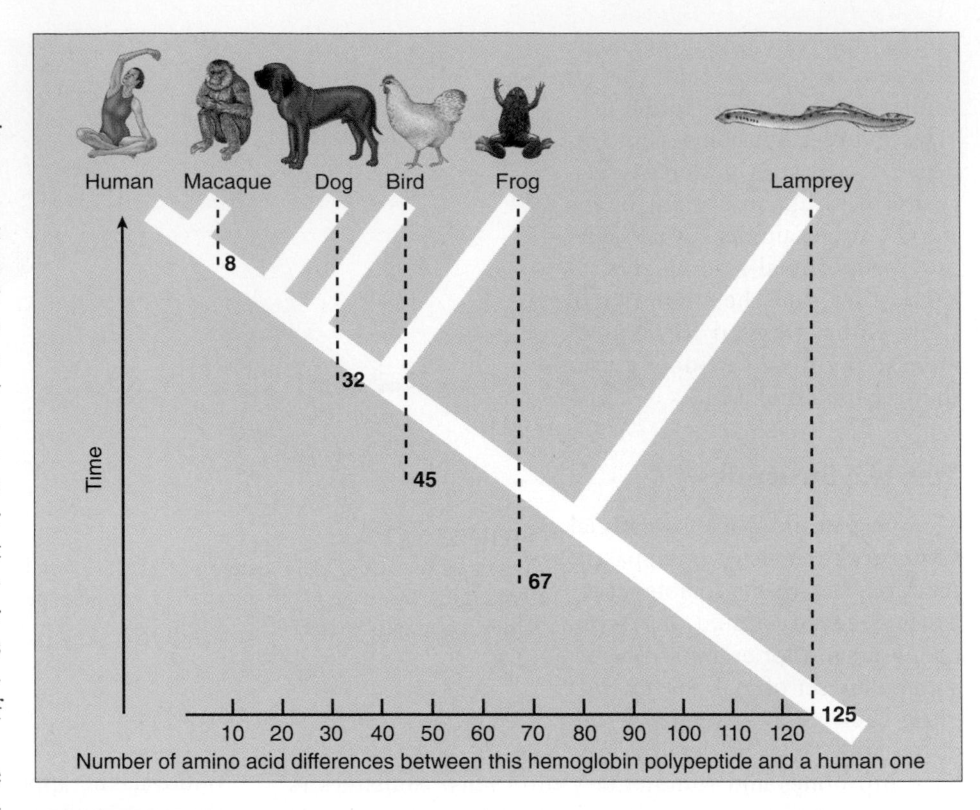

FIGURE 21.18
Molecules reflect evolutionary divergence. You can see that the greater the evolutionary distance from humans (white cladogram), the greater the number of amino acid differences in the vertebrate hemoglobin polypeptide.

When degree of genetic similarity is compared with our ideas of evolutionary relationships based on fossils, a close match is evident. For example, when the human hemoglobin polypeptide is compared to the corresponding molecule in other species, closely related species are found to be more similar. Chimpanzees, gorillas, orangutans, and macaques, vertebrates thought to be more closely related to humans, have fewer differences from humans in the 146-amino-acid hemoglobin β chain than do more distantly related mammals, like dogs. Nonmammalian vertebrates differ even more, and nonvertebrate hemoglobins are the most different of all (figure 21.18). Similar patterns are also evident when the DNA itself is compared. For example, chimps and humans, which are thought to have descended from a common ancestor that lived approximately 6 million years ago, exhibit few differences in their DNA.

Why should closely related species be similar in DNA? Because DNA is the genetic code that produces the structure of living organisms, one might expect species that are similar in overall appearance and structure, such as humans and chimpanzees, to be more similar in DNA than are more dissimilar species, such as humans and frogs. This expectation would hold true even if evolution had not occurred. However, there are some noncoding stretches of DNA (sometimes called "junk DNA") that have no function and appear to serve no purpose. If evolution had not occurred, there would be no reason to expect similar-appearing species to be similar in their junk DNA. However, comparisons of such stretches of DNA provide the same results as for other parts of the genome: more closely related species are more similar, an observation that only makes sense if evolution has occurred.

Comparison of the DNA of different species provides strong evidence for evolution. Species deduced from the fossil record to be closely related are more similar in their DNA than are species thought to be more distantly related.

Convergent and Divergent Evolution

Different geographical areas sometimes exhibit groups of plants and animals of strikingly similar appearance, even though the organisms may be only distantly related. It is difficult to explain so many similarities as the result of coincidence. Instead, natural selection appears to have favored parallel evolutionary adaptations in similar environments. Because selection in these instances has tended to favor changes that made the two groups more alike, their phenotypes have converged. This form of evolutionary change is referred to as **convergent evolution,** or sometimes, **parallel evolution.**

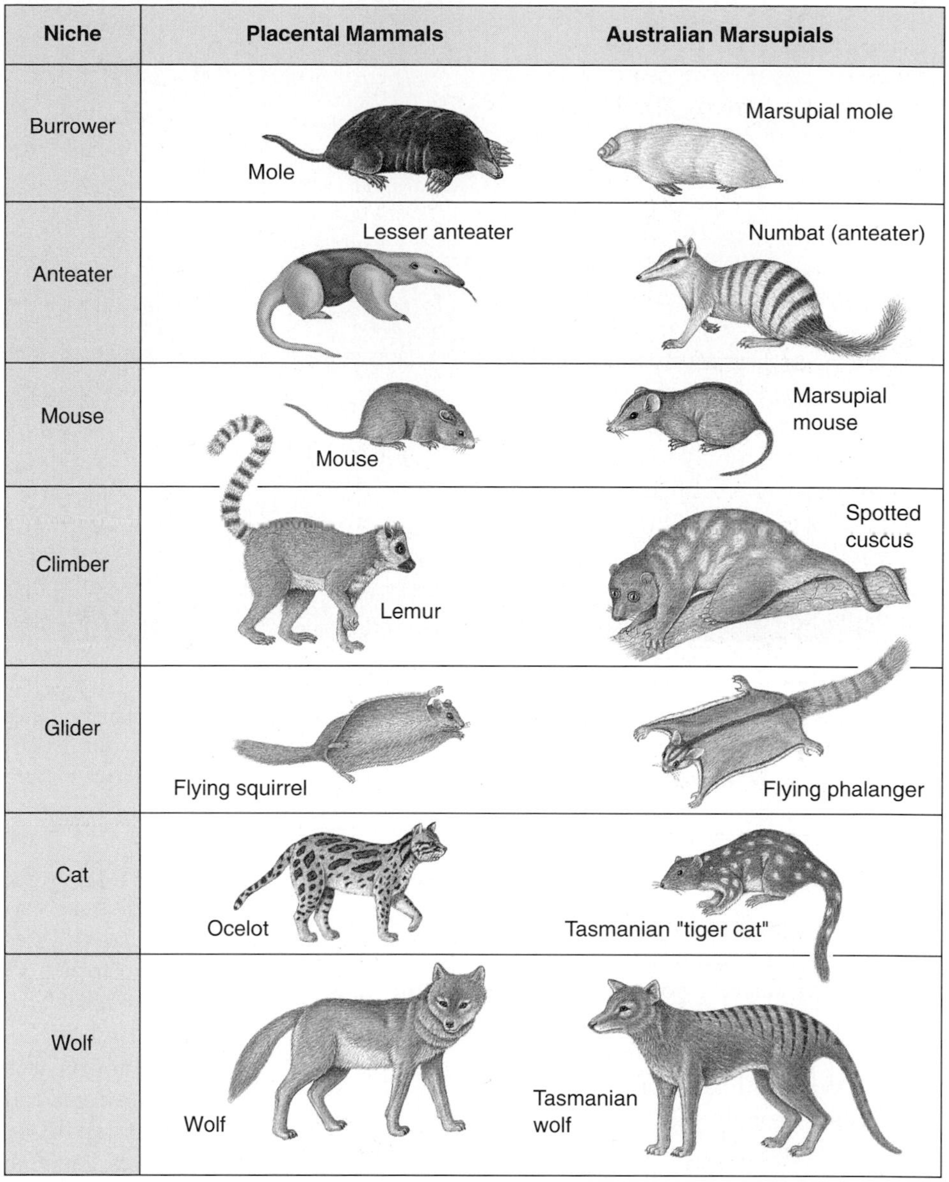

The Marsupial-Placental Convergence

In the best-known case of convergent evolution, two major groups of mammals, marsupials and placentals, have evolved in a very similar way, even though the two lineages have been living independently on separate continents. Australia separated from the other continents more than 50 million years ago, after marsupials had evolved but before the appearance of placental mammals. As a result, the only mammals in Australia (other than bats and a few colonizing rodents) have been marsupials, members of a group in which the young are born in a very immature condition and held in a pouch until they are ready to emerge into the outside world. Thus, even though placental mammals are the dominant mammalian group throughout most of the world, marsupials retained supremacy in Australia.

Niche	Placental Mammals	Australian Marsupials
Burrower	Mole	Marsupial mole
Anteater	Lesser anteater	Numbat (anteater)
Mouse	Mouse	Marsupial mouse
Climber	Lemur	Spotted cuscus
Glider	Flying squirrel	Flying phalanger
Cat	Ocelot	Tasmanian "tiger cat"
Wolf	Wolf	Tasmanian wolf

FIGURE 21.19
Convergent evolution. Marsupials in Australia resemble placental mammals in the rest of the world. They evolved in isolation after Australia separated from other continents.

What are the Australian marsupials like? To an astonishing degree, they resemble the placental mammals living today on the other continents (figure 21.19). The similarity between some individual members of these two sets of mammals argues strongly that they are the result of convergent evolution, similar forms having evolved in different, isolated areas because of similar selective pressures in similar environments.

Homology versus Analogy

How do we know when two similar characters are homologous and when they are analogous? As we have seen, adaptation favoring different functions can obscure homologies, while convergent evolution can create analogues that appear as similar as homologues. There is no hard-and-fast answer to this question; the determination of homologues is often a thorny issue in biological classification. As we have seen in comparing vertebrate embryos, and again in comparing slugs and squids, studies of embryonic development often reveal features not apparent when studying adult organisms. In general, the more complex two structures are, the less likely they evolved independently.

FIGURE 21.20
A Galápagos tortoise most closely resembles South American tortoises. Isolated on these remote islands, the Galápagos tortoise has evolved distinctive forms. This natural experiment is being terminated, however. Since Darwin's time, much of the natural habitat of the larger islands has been destroyed by human intrusion. Goats introduced by settlers, for example, have drastically altered the vegetation.

Darwin and Patterns of Recent Divergence

Darwin was the first to present evidence that animals and plants living on oceanic islands resemble most closely the forms on the nearest continent—a relationship that only makes sense as reflecting common ancestry. The Galápagos tortoise in figure 21.20 is more similar to South American tortoises than to those of any other continent. This kind of relationship strongly suggests that the island forms evolved from individuals that came from the adjacent mainland at some time in the past. Thus, the Galápagos finches of figure 21.8 have different beaks than their South American relatives. In the absence of evolution, there seems to be no logical explanation of why individual kinds of island plants and animals would be clearly related to others on the nearest mainland, but still have some divergent features. As Darwin pointed out, this relationship provides strong evidence that macroevolution, evolution among species, has occurred.

A similar resemblance to mainland birds can be seen in an island finch Darwin never saw—a solitary finch species living on Cocos Island, a tiny, remote volcanic island located 630 kilometers to the northeast of the Galápagos. This finch does not resemble the finches of Europe, Australia, Africa, or North America. Instead, it resembles the finches of Costa Rica, 500 kilometers to the east.

Of course, because of adaptation to localized habitats, island forms are not identical to those on the nearby continents. The tortoises have evolved different shell shapes, for example; those living in moist habitats have dome-shaped shells while others living in dry places have low, saddle-backed shells with the front of the shell bent up to expose the head and neck. Similarly, the Galápagos finches have evolved from a single presumptive ancestor into 13 species, each specialized in a different way. These Galápagos tortoises and finches have evolved in concert with the continental forms, from the same ancestors, but the two lineages have diverged rather than converged.

It is fair to ask how Darwin knew that the Galápagos tortoises and finches do not represent the convergence of unrelated island and continental forms (analogues) rather than the divergence of recently isolated groups (homologues). While either hypothesis would argue for natural selection, Darwin chose divergence of homologues as by far the simplest explanation, because the tortoises and finches differ by only a few traits, and are similar in many.

In sum total, the evidence for macroevolution is overwhelming. In the next chapter, we will consider Darwin's proposal that microevolutionary changes, evolution within a species, have led directly to macroevolutionary changes, the key argument in his theory that evolution occurs by natural selection.

Evolution favors similar forms under similar circumstances. Convergence is the evolution of similar forms in different lineages when exposed to the same selective pressures. Divergence is the evolution of different forms in the same lineage when exposed to different selective pressures.

21.4 The theory of evolution has proven controversial.

Darwin's Critics

In the century since he proposed it, Darwin's theory of evolution by natural selection has become nearly universally accepted by biologists, but has proven controversial among some of the general public. Darwin's critics raise seven principal objections to teaching evolution:

1. **Evolution is not solidly demonstrated.** *"Evolution is just a theory,"* Darwin's critics point out, as if theory meant lack of knowledge, some kind of guess. Scientists, however, use the word theory in a very different sense than the general public does. Theories are the solid ground of science, that of which we are most certain. Few of us doubt the theory of gravity because it is "just a theory."
2. **There are no fossil intermediates.** *"No one ever saw a fin on the way to becoming a leg,"* critics claim, pointing to the many gaps in the fossil record in Darwin's day. Since then, however, most fossil intermediates in vertebrate evolution have indeed been found. A clear line of fossils now traces the transition between whales and hoofed mammals, between reptiles and mammals, between dinosaurs and birds, between apes and humans. The fossil evidence of evolution between major forms is compelling.
3. **The intelligent design argument.** *"The organs of living creatures are too complex for a random process to have produced—the existence of a clock is evidence of the existence of a clockmaker."* Biologists do not agree. The intermediates in the evolution of the mammalian ear can be seen in fossils, and many intermediate "eyes" are known in various invertebrates. These intermediate forms arose because they have value—being able to detect light a little is better than not being able to detect it at all. Complex structures like eyes evolved as a progression of slight improvements.
4. **Evolution violates the Second Law of Thermodynamics.** *"A jumble of soda cans doesn't by itself jump neatly into a stack—things become more disorganized due to random events, not more organized."* Biologists point out that this argument ignores what the second law really says: disorder increases in a closed system, which the earth most certainly is not. Energy continually enters the biosphere from the sun, fueling life and all the processes that organize it.
5. **Proteins are too improbable.** *"Hemoglobin has 141 amino acids. The probability that the first one would be leucine is 1/20, and that all 141 would be the ones they are by chance is $(1/20)^{141}$, an impossibly rare event."* This is statistical foolishness—you cannot use probability to argue backwards. The probability that a student in a classroom has a particular birthday is 1/365; arguing this way, the probability that everyone in a class of 50 would have the birthdays they do is $(1/365)^{50}$, and yet there the class sits.
6. **Natural selection does not imply evolution.** *"No scientist has come up with an experiment where fish evolve into frogs and leap away from predators."* Is microevolution (evolution within a species) the mechanism that has produced macroevolution (evolution among species)? Most biologists that have studied the problem think so. Some kinds of animals produced by artificial selection are remarkably distinctive, such as Chihuahuas, dachshunds, and greyhounds. While all dogs are in fact the same species and can interbreed, laboratory selection experiments easily create forms that cannot interbreed and thus would in nature be considered different species. Thus, production of radically different forms has indeed been observed, repeatedly. To object that evolution still does not explain really major differences, like between fish and amphibians, simply takes us back to point 2—these changes take millions of years, and are seen clearly in the fossil record.
7. **The irreducible complexity argument.** *The intricate molecular machinery of the cell cannot be explained by evolution from simpler stages. Because each part of a complex cellular process like blood clotting is essential to the overall process, how can natural selection fashion any one part?* What's wrong with this argument is that each part of a complex molecular machine evolves as part of the system. Natural selection can act on a complex system because at every stage of its evolution, the system functions. Parts that improve function are added, and, because of later changes, become essential. The mammalian blood clotting system, for example, has evolved from much simpler systems. The core clotting system evolved at the dawn of the vertebrates 600 million years ago, and is found today in lampreys, the most primitive fish. One hundred million years later, as vertebrates evolved, proteins were added to the clotting system making it sensitive to substances released from damaged tissues. Fifty million years later, a third component was added, triggering clotting by contact with the jagged surfaces produced by injury. At each stage as the clotting system evolved to become more complex, its overall performance came to depend on the added elements. Thus, blood clotting has become "irreducibly complex"—as the result of Darwinian evolution.

Darwin's theory of evolution has proven controversial among some of the general public, although the commonly raised objections are without scientific merit.

Chapter 21

www.mhhe.com/raven6/resources21.mhtml

Summary	Questions	Media Resources

21.1 Fossil evidence indicates that evolution has occurred.

- Fossils of many extinct species have never been discovered. Nonetheless, the fossil record is complete enough to allow a detailed understanding of the evolution of life through time. The evolution of the major vertebrate groups is quite well known.
- Although evolution of groups like horses may appear to be a straight-line progression, in fact there have been many examples of parallel evolution, and even reversals from overall trends.

1. Why do gaps exist in the fossil record? What lessons can be learned from the fossil record of horse evolution?

2. How did scientists date fossils in Darwin's day? Why are scientists today able to date rocks more accurately?

- Book Reviews:
  - *In Search of Deep Time* by Gee
  - *Digging Dinosaurs* by Horner

21.2 Natural selection can produce evolutionary change.

- Natural populations provide clear evidence of evolutionary change.
- Darwin's finches have different-sized beaks, which are adaptations to eating different kinds of seeds. In particularly dry years, natural selection favors birds with stout beaks within one species, *Geospiza fortis*. As a result, the average bill size becomes larger in the next generation.
- The British populations of the peppered moth, *Biston betularia*, consisted mostly of light-colored individuals before the Industrial Revolution. Over the last two centuries, populations that occur in heavily polluted areas, where the tree trunks are darkened with soot, have come to consist mainly of dark-colored (melanic) individuals—a result of rapid natural selection.

3. Why did the average beak size of the medium ground finch increase after a particularly dry year?

4. Why did the frequency of light-colored moths decrease and that of dark-colored moths increase with the advent of industrialism? What is industrial melanism?

5. What can artificial selection tell us about evolution? Is artificial selection a good analogy for the selection that occurs in nature?

- The Process of Natural Selection
- Fossils
- Activity: Evolution of Fish

- Book Reviews:
  - *The Beak of the Finch* by Weiner
  - *Darwin's Ghost* by Jones

21.3 Evidence for evolution can be found in other fields of biology.

- Several indirect lines of evidence argue that macroevolution has occurred, including successive changes in homologous structures, developmental patterns, vestigial structures, parallel patterns of evolution, and patterns of distribution.
- When differences in genes or proteins are examined, species that are thought to be closely related based on the fossil record may be more similar than species thought to be distantly related.

6. What is homology? How does it support evolutionary theory?

7. What is convergent evolution? Give examples.

8. How did Darwin's studies of island populations provide evidence for evolution?

- Molecular Clock

- Evidence for Evolution
- Activity: Divergence

- Art Quiz: Embryos and Evolutionary History

21.4 The theory of evolution has proven controversial.

- The objections raised by Darwin's critics are easily answered.

9. Is "Darwinism" really science? Explain.

- *On Science* Article: Answering Evolution's Critics
- Bioethics Case Study: Creationism

22

The Origin of Species

Concept Outline

22.1 Species are the basic units of evolution.

The Nature of Species. Species are groups of actually or potentially interbreeding natural populations which are reproductively isolated from other such groups and that maintain connectedness over geographic distances.

22.2 Species maintain their genetic distinctiveness through barriers to reproduction.

Prezygotic Isolating Mechanisms. Some breeding barriers prevent the formation of zygotes.

Postzygotic Isolating Mechanisms. Other breeding barriers prevent the proper development or reproduction of the zygote after it forms.

22.3 We have learned a great deal about how species form.

Reproductive Isolation May Evolve as a By-Product of Evolutionary Change. Speciation can occur in the absence of natural selection, but reproductive isolation generally occurs more quickly when populations are adapting to different environments.

The Geography of Speciation. Speciation occurs most readily when populations are geographically isolated. Sympatric speciation can occur by polyploidy and, perhaps, by other means.

22.4 Clusters of species reflect rapid evolution.

Darwin's Finches. Thirteen species of finches, all descendants of one ancestral finch, occupy diverse niches.

Hawaiian *Drosophila*. More than a quarter of the world's fruit fly species are found on the Hawaiian Islands.

Lake Victoria Cichlid Fishes. Isolation has led to extensive species formation among these small fishes.

New Zealand Alpine Buttercups. Repeated glaciations have fostered waves of species formation in alpine plants.

Diversity of Life through Time. The number of species has increased through time, despite a number of mass extinction events.

The Pace of Evolution. The idea that evolution occurs in spurts is controversial.

Problems with the Biological Species Concept. This concept is not as universal as previously thought.

FIGURE 22.1
A group of Galápagos marine iguanas bask in the sun on their isolated island. How does geographic isolation contribute to the formation of new species?

Although Darwin titled his book *On the Origin of Species*, he never actually discussed what he referred to as that "mystery of mysteries" of how one species gives rise to another. Rather, his argument concerned evolution by natural selection; that is, how one species evolves through time to adapt to its changing environment. Although of fundamental importance to evolutionary biology, the process of adaptation does not explain how one species becomes another (figure 22.1); much less can it explain how one species can give rise to many descendant species. As we shall see, adaptation may be involved in this process of **speciation**, but it need not be.

22.1 Species are the basic units of evolution.

The Nature of Species

Before we can discuss how one species gives rise to another, we need to understand exactly what a species is. Even though definition of what constitutes a species is of fundamental importance to evolutionary biology, this issue has still not been completely settled and is currently the subject of considerable research and debate. However, any concept of a species must account for two phenomena: the distinctiveness of species that occur together at a single locality, and the connection that exists among populations of the same species that are geographically separated.

The Distinctiveness of Sympatric Species

Put out a birdfeeder on your balcony or back porch and you will attract a wide variety of different types of birds (especially if you put out a variety of different kinds of foods). In the midwestern United States, for example, you might routinely see cardinals, blue jays, downy woodpeckers, house finches—even hummingbirds in the summer (figure 22.2). Although it might take a few days of careful observation, you would soon be able to readily distinguish the many different species. The reason is that species that occur together (termed **sympatric** from the Greek *sym* for "same" and *patria* for "species") are distinctive entities that are phenotypically different, utilize different parts of the habitat, and behave separately. This observation is generally true not only for birds, but also for most other types of organisms in most places.

Occasionally, two species occur together that appear to be nearly identical, and are thus called **sibling species**. In most cases, however, our inability to distinguish the two reflects our own reliance on vision as our primary sense. When the mating calls or chemicals exuded by such species are examined, they usually reveal great differences. In other words, even though we have trouble separating them, the animals themselves have no such difficulties!

Geographic Variation within Species

Within the units classified as species, populations that occur in different areas may be more or less distinct from one another. Such groups of distinctive individuals may be classified taxonomically as **subspecies** or **varieties** (the vague term "race" has a similar connotation, but is no longer commonly used). In areas where these populations approach one another, individuals often exhibit combinations of features characteristic of both populations. In other words, even though geographically distant populations may appear distinct, they usually are connected by intervening populations that are intermediate in their characteristics (figure 22.3).

The Biological Species Concept

What can account both for the distinctiveness of sympatric species and the connectedness of geographic populations of the same species? One obvious possibility is that each species exchanges genetic material only with other members of its species. If sympatric species commonly

FIGURE 22.2
Common birds in the midwestern United States. No one would doubt that these birds are distinct species. Each can be distinguished from the others by many ecological, behavioral, and phenotypic traits.

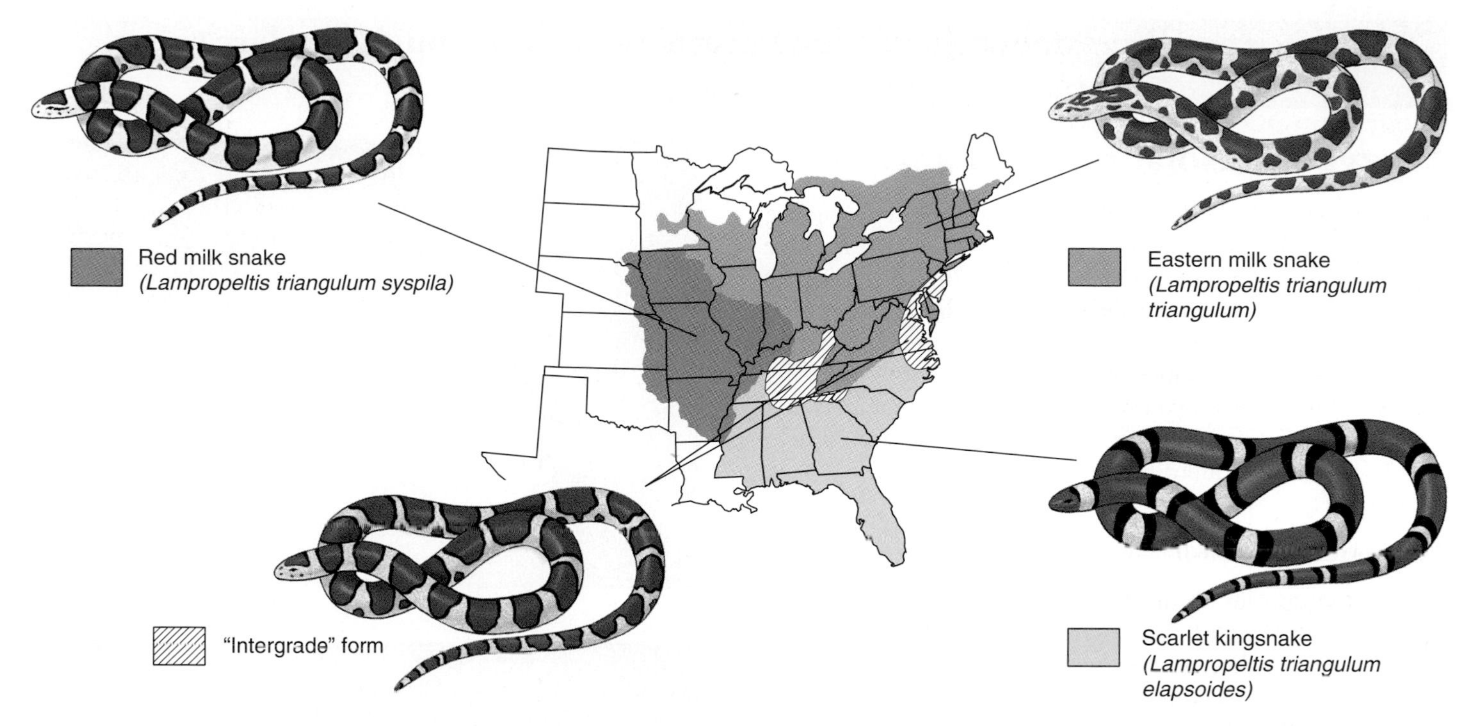

FIGURE 22.3
Geographic variation in the milk snake, *Lampropeltis triangulum*. Although each subspecies appears phenotypically quite distinctive from the others, they are connected by populations that are phenotypically intermediate.
Source: Date from R. Conant & J.T. Collins, *Reptiles & Amphibians of Eastern/Central North America*, 3rd edition, 1991, Houghton Mifflin Company.

exchanged genes, we might expect such species to rapidly lose their distinctions as the **gene pools** of the different species became homogenized. Conversely, the ability of geographically distant populations to share genes through the process of gene flow may keep these populations integrated as members of the same species. Based on these ideas, the evolutionary biologist Ernst Mayr coined the **biological species concept** which defines species as:

". . . groups of actually or potentially interbreeding natural populations which are reproductively isolated from other such groups."

In other words, the biological species concept says that a species is all individuals that are capable of interbreeding and producing fertile offspring. Conversely, individuals that cannot produce fertile offspring are said to be **reproductively isolated** and, thus, members of different species.

Occasionally, members of different species will interbreed, a process termed **hybridization**. If the species are reproductively isolated, either no offspring will result, or if offspring are produced, they will be either unhealthy or sterile. In this way, genes from one species generally will not be able to enter the gene pool of another species.

Problems with Applying the Biological Species Concept

The biological species concept has proven to be an effective way of understanding the existence of species in nature. Nonetheless, the concept has some practical difficulties. For example, it can be difficult to apply the concept to populations that do not occur together in nature (and are thus said to be **allopatric**). Because individuals of these populations do not encounter each other, it is not possible to observe whether they would interbreed naturally. Although experiments can determine whether fertile hybrids can be produced, this information is not enough. The reason is that many species that will coexist without interbreeding in nature will readily hybridize in the artificial settings of the laboratory or zoo. Consequently, evaluating whether allopatric populations constitute different species is ultimately a judgment call.

In addition, the concept is more limited than its name would imply. Many organisms are asexual and reproduce without mating; reproductive isolation has no meaning for such organisms.

Moreover, despite its name, the concept is really a zoological species concept and applies less readily to plants. Even among animals, the biological species concept appears to apply more successfully to some groups than to others. As we will see in section 22.4, biologists are currently reevaluating this and other approaches to the study of species.

Species are groups of organisms that are distinct from other co-occurring species and that are interconnected geographically. The ability to exchange genes appears to be a hallmark of such species.

22.2 Species maintain their genetic distinctiveness through barriers to reproduction.

Prezygotic Isolating Mechanisms

How do species keep their separate identities? Reproductive isolating mechanisms fall into two categories: **prezygotic isolating mechanisms,** which prevent the formation of zygotes; and **postzygotic isolating mechanisms,** which prevent the proper functioning of zygotes after they form. In the following sections we will discuss various isolating mechanisms in these two categories and offer examples that illustrate how the isolating mechanisms operate to help species retain their identities.

Ecological Isolation

Even if two species occur in the same area, they may utilize different portions of the environment and thus not hybridize because they do not encounter each other. For example, in India, the ranges of lions and tigers overlapped until about 150 years ago. Even when they did, however, there were no records of natural hybrids. Lions stayed mainly in the open grassland and hunted in groups called prides; tigers tended to be solitary creatures of the forest (figure 22.4). Because of their ecological and behavioral differences, lions and tigers rarely came into direct contact with each other, even though their ranges overlapped over thousands of square kilometers.

In another example, the ranges of two toads, *Bufo woodhousei* and *B. americanus,* overlap in some areas. Although these two species can produce viable hybrids, they usually do not interbreed because they utilize different portions of the habitat for breeding. Whereas *B. woodhousei* prefers to breed in streams, *B. americanus* breeds in rainwater puddles. Similarly, the ranges of two species of dragonflies overlap in Florida. However, the dragonfly *Progomphus obscurus* lives near rivers and streams, and *P. alachuenis* lives near lakes.

Similar situations occur among plants. Two species of oaks occur widely in California: the valley oak, *Quercus lobata,* and the scrub oak, *Q. dumosa.* The valley oak, a graceful deciduous tree that can be as tall as 35 meters, occurs in the fertile soils of open grassland on gentle slopes and valley floors. In contrast, the scrub oak is an evergreen shrub, usually only 1 to 3 meters tall, which often forms the kind of dense scrub known as chaparral. The scrub oak is found on steep slopes in less fertile soils. Hybrids between these different oaks do occur and are fully fertile, but they are rare. The sharply distinct habitats of their parents limit their occurrence together, and there is no intermediate habitat where the hybrids might flourish.

(a)

(b)

(c)

FIGURE 22.4
Lions and tigers are ecologically isolated. The ranges of lions and tigers used to overlap in India. However, lions and tigers do not hybridize in the wild because they utilize different portions of the habitat. (*a*) Lions live in open grassland. (*b*) Tigers are solitary animals that live in the forest. (*c*) Hybrids, such as this tiglon, have been successfully produced in captivity, but hybridization does not occur in the wild.

Behavioral Isolation

In chapter 27, we will consider the often elaborate courtship and mating rituals of some groups of animals. Related species of organisms such as birds often differ in their courtship rituals, which tends to keep these species distinct in nature even if they inhabit the same places (figure 22.5). For example, mallard and pintail ducks are perhaps the two most common freshwater ducks in North America. In captivity, they produce completely fertile offspring, but in nature they nest side-by-side and only rarely hybridize.

More than 500 species of flies of the genus *Drosophila* live in the Hawaiian Islands. This is one of the most remarkable concentrations of species in a single animal genus found anywhere. The genus occurs throughout the world, but nowhere are the flies more diverse in external appearance or behavior than in Hawaii. Many of these flies differ greatly from other species of *Drosophila*, exhibiting characteristics that can only be described as bizarre.

The Hawaiian species of *Drosophila* are long-lived and often very large compared with their relatives on the mainland. The females are more uniform than the males, which are often bizarrely distinctive. The males display complex territorial behavior and elaborate courtship rituals.

The mating behavior patterns among Hawaiian species of *Drosophila* are of great importance in maintaining the distinctiveness of the individual species. For example, despite the great differences between them, *D. heteroneura* and *D. silvestris* are very closely related. Hybrids between them are fully fertile. The two species occur together over a wide area on the island of Hawaii, yet hybridization has been observed at only one locality. The very different and complex behavioral characteristics of these flies obviously play a major role in maintaining their distinctiveness.

Other Prezygotic Isolating Mechanisms

Temporal Isolation. *Lactuca graminifolia* and *L. canadensis*, two species of wild lettuce, grow together along roadsides throughout the southeastern United States. Hybrids between these two species are easily made experimentally and are completely fertile. But such hybrids are rare in nature because *L. graminifolia* flowers in early spring and *L. canadensis* flowers in summer. When their blooming periods overlap, as they do occasionally, the two species do form hybrids, which may become locally abundant.

Many species of closely related amphibians have different breeding seasons that prevent hybridization between the species. For example, five species of frogs of the genus *Rana* occur together in most of the eastern United States, but hybrids are rare because the peak breeding time is different for each of them.

FIGURE 22.5
Differences in courtship rituals can isolate related bird species. These Galápagos blue-footed boobies select their mates only after an elaborate courtship display. This male is lifting his feet in a ritualized high-step that shows off his bright blue feet. The display behavior of other species of boobies, some of which also occur in the Galápagos, is much different.

Mechanical Isolation. Structural differences prevent mating between some related species of animals. Aside from such obvious features as size, the structure of the male and female copulatory organs may be incompatible. In many insect and other arthropod groups, the sexual organs, particularly those of the male, are so diverse that they are used as a primary basis for classification.

Similarly, flowers of related species of plants often differ significantly in their proportions and structures. Some of these differences limit the transfer of pollen from one plant species to another. For example, bees may pick up the pollen of one species on a certain place on their bodies; if this area does not come into contact with the receptive structures of the flowers of another plant species, the pollen is not transferred.

Prevention of Gamete Fusion. In animals that shed their gametes directly into water, eggs and sperm derived from different species may not attract one another. Many land animals may not hybridize successfully because the sperm of one species may function so poorly within the reproductive tract of another that fertilization never takes place. In plants, the growth of pollen tubes may be impeded in hybrids between different species. In both plants and animals the operation of such isolating mechanisms prevents the union of gametes even following successful mating.

Prezygotic isolating mechanisms lead to reproductive isolation by preventing the formation of hybrid zygotes.

Postzygotic Isolating Mechanisms

All of the factors we have discussed up to this point tend to prevent hybridization. If hybrid matings do occur and zygotes are produced, many factors may still prevent those zygotes from developing into normally functioning, fertile individuals. Development in any species is a complex process. In hybrids, the genetic complements of two species may be so different that they cannot function together normally in embryonic development. For example, hybridization between sheep and goats usually produces embryos that die in the earliest developmental stages.

Leopard frogs (*Rana pipiens* complex) of the eastern United States are a group of similar species, assumed for a long time to constitute a single species (figure 22.6). However, careful examination revealed that although the frogs appear similar, successful mating between them is rare because of problems that occur as the fertilized eggs develop. Many of the hybrid combinations cannot be produced even in the laboratory.

Examples of this kind, in which similar species have been recognized only as a result of hybridization experiments, are common in plants. Sometimes the hybrid embryos can be removed at an early stage and grown in an artificial medium. When these hybrids are supplied with extra nutrients or other supplements that compensate for their weakness or inviability, they may complete their development normally.

Even if hybrids survive the embryo stage, however, they may not develop normally. If the hybrids are weaker than their parents, they will almost certainly be eliminated in nature. Even if they are vigorous and strong, as in the case of the mule, a hybrid between a horse and a donkey, they may still be sterile and thus incapable of contributing to succeeding generations. Sterility may result in hybrids because the development of sex organs may be abnormal, the chromosomes derived from the respective parents may not pair properly, or from a variety of other causes.

Postzygotic isolating mechanisms are those in which hybrid zygotes fail to develop or develop abnormally, or in which hybrids cannot become established in nature.

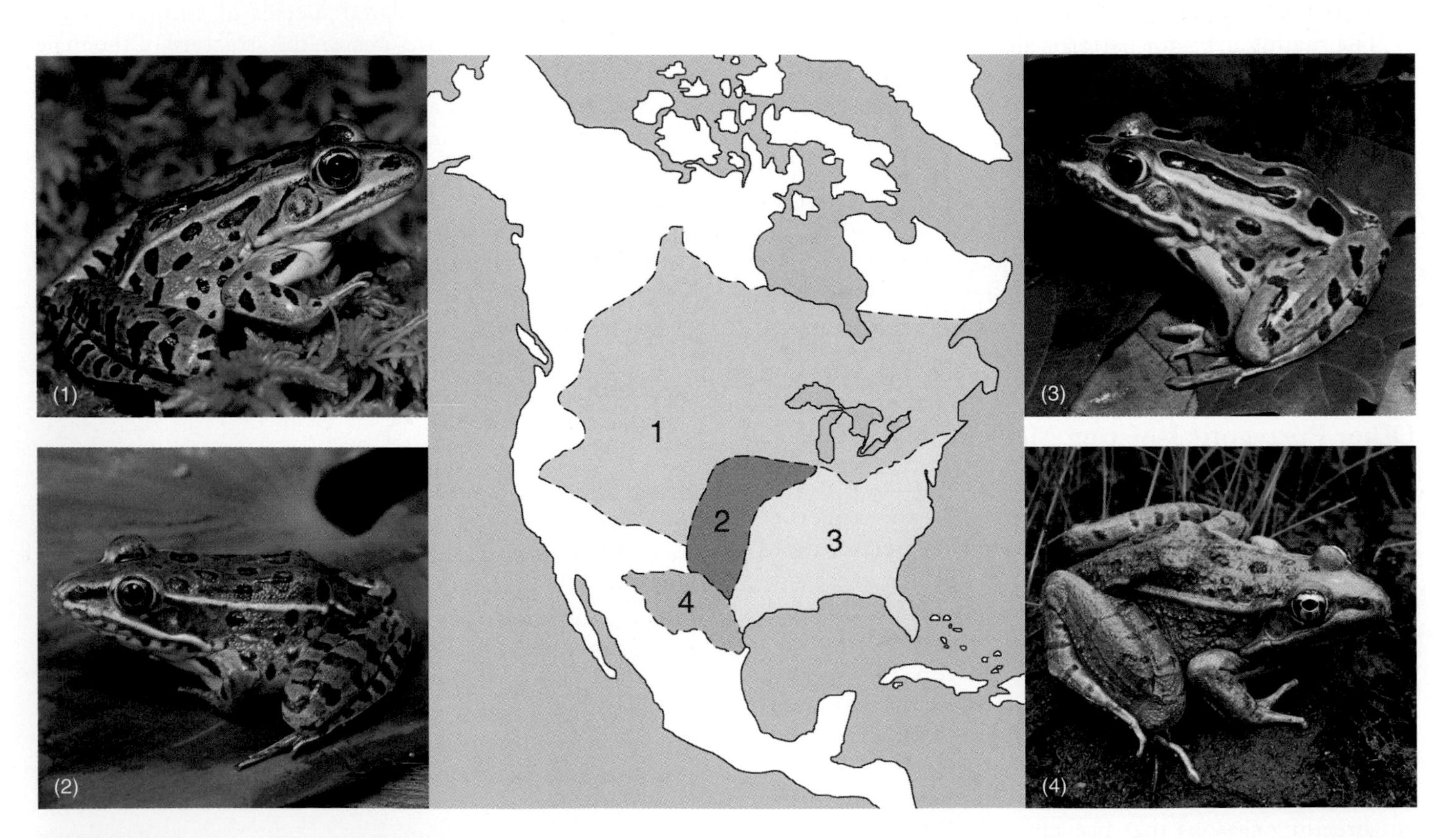

FIGURE 22.6
Postzygotic isolation in leopard frogs. Numbers indicate the following species in the geographical ranges shown: (*1*) *Rana pipiens;* (*2*) *Rana blairi;* (*3*) *Rana utricularia;* (*4*) *Rana berlandieri.* These four species resemble one another closely in their external features. Their status as separate species was first suspected when hybrids between them produced defective embryos in some combinations. Subsequent research revealed that the mating calls of the four species differ substantially, indicating that the species have both pre- and postzygotic isolating mechanisms.

22.3 We have learned a great deal about how species form.

One of the oldest questions in the field of evolution is: how does one ancestral species become divided into two descendant species? If species are defined by the existence of reproductive isolation, then the process of speciation equates with the evolution of reproductive isolating mechanisms. How do reproductive isolating mechanisms evolve?

Reproductive Isolation May Evolve as a By-Product of Evolutionary Change

Most reproductive isolating mechanisms initially arise for some reason other than to provide reproductive isolation. For example, a population that colonizes a new habitat may evolve adaptations for living in that habitat. As a result, individuals from that population might never encounter individuals from the ancestral population. Even if they do meet, the population in the new habitat may have evolved new phenotypes or behavior so that members of the two populations no longer recognize each other as potential mates (figure 22.7). For this reason, some biologists believe that the term "isolating mechanisms" is misguided, because it implies that the traits evolved specifically for the purpose of genetically isolating a species, which in most cases is probably incorrect.

Species 1
Species 2
Prezygotic isolating mechanisms
Geographic isolation
Species occur in different places
Ecological isolation
Species utilize different resources in the habitat
Behavioral isolation
Species have different mating rituals
Temporal isolation
Mating or flowering occur during different seasons or at different times of the day
Mating
Mechanical isolation
Structural differences prevent mating or pollen transfer
Prevention of gamete fusion
Gametes fail to attract each other or function poorly
Fertilization
Postzygotic isolating mechanisms
Hybrid embryos do not develop properly
Hybrid adults do not survive in nature
Hybrid adults are sterile or have reduced fertility
Fertile hybrid offspring

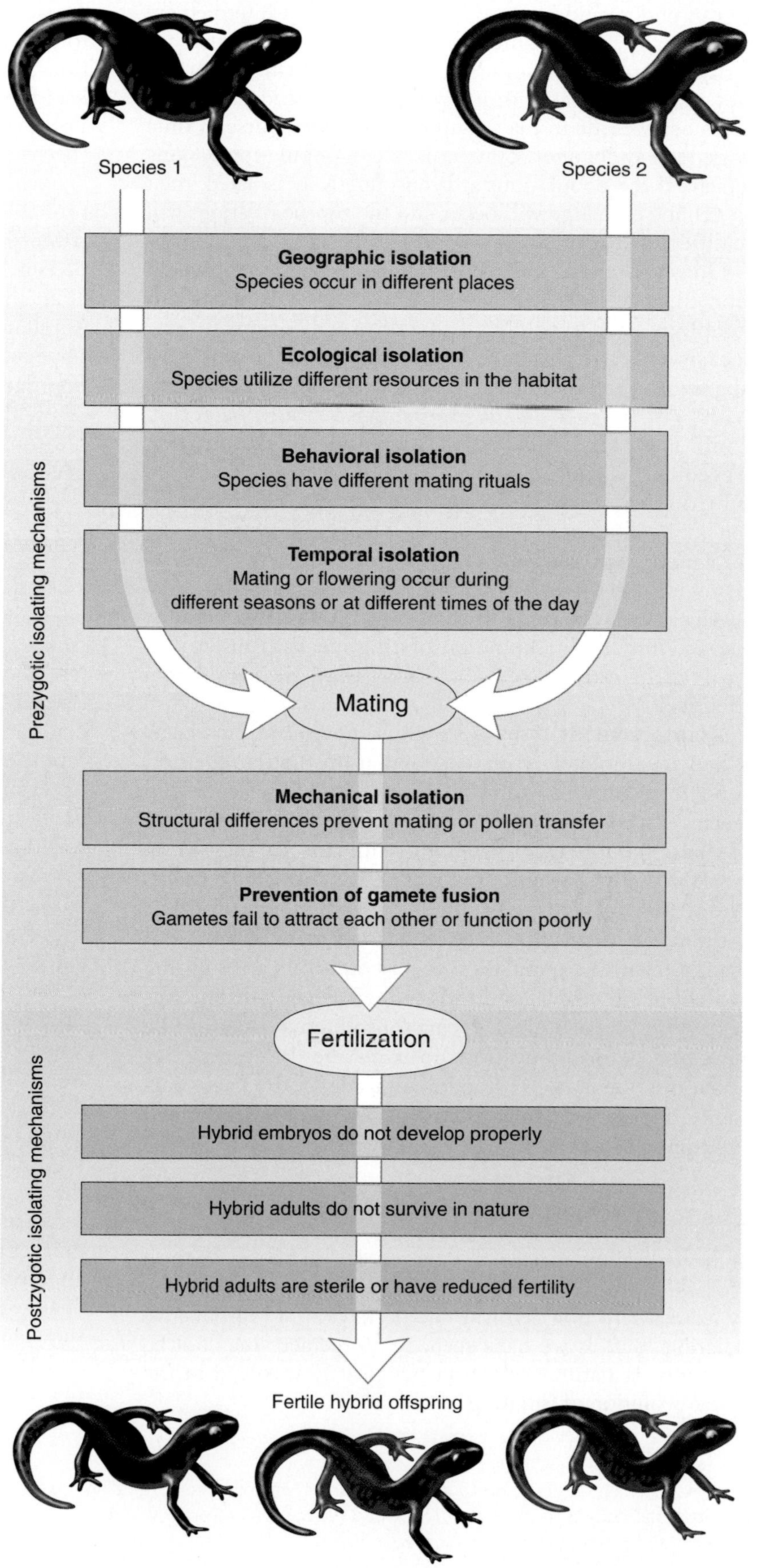

FIGURE 22.7
Reproductive isolating mechanisms. A variety of different mechanisms can prevent successful reproduction between individuals of different species.

Selection May Reinforce Isolating Mechanisms

The formation of species is a continuous process, one that we can understand because of the existence of intermediate stages at all levels of differentiation. If populations that are partly differentiated come into contact with one another, they may still be able to interbreed freely, and the differences between them may disappear over the course of time as genetic exchange homogenizes the populations. Conversely, if the populations are reproductively isolated, then no genetic exchange will occur and the two populations will be different species.

However, there is an intermediate situation in which reproductive isolation has partially evolved, but is not complete. As a result, hybridization will occur at least occasionally. If they are partly sterile, or not as well adapted to the existing habitats as their parents, these hybrids will be at a disadvantage. As a result, selection would favor any alleles in the parental populations that prevented hybridization because individuals that avoided hybridizing would be more successful in passing their genes on to the next generation. The result would be the continual improvement of prezygotic isolating mechanisms until the two populations were completely reproductively isolated. This process is termed **reinforcement** because initially incomplete isolating mechanisms are reinforced by natural selection until they are completely effective.

Reinforcement is by no means inevitable, however. When incompletely isolated populations come together, gene flow immediately begins to occur between the species. Although hybrids may be inferior, they are not, in this case, completely inviable or infertile (if they were, then the species would be reproductively isolated); hence, when these hybrids reproduce with members of either population, they will serve as a conduit of genetic exchange from one population to the other. As a result, the two populations will tend to lose their genetic distinctiveness. Thus, a race ensues: can reproductive isolation be perfected before gene flow destroys the differences between the populations? Experts disagree on the likely outcome, but many believe that reinforcement is the much less common outcome.

The Role of Natural Selection in Speciation

What role does natural selection play in the speciation process? Certainly, the process of reinforcement is driven by natural selection favoring the perfection of reproductive isolation. But, as we have seen, reinforcement may not be common. Is natural selection necessarily involved in the initial evolution of isolating mechanisms?

Random Changes May Cause Reproductive Isolation

As we discussed in chapter 20, populations may diverge for purely random reasons. Genetic drift in small populations, founder effects, and population bottlenecks all may lead to changes in traits that cause reproductive isolation. For example, in the Hawaiian Islands, closely related species of *Drosophila* often differ greatly in their courtship behavior. Colonization of new islands by these fruit flies probably involves a founder effect, in which one or a few fruit flies—perhaps only a single pregnant female—is blown by strong winds to a new island. Changes in courtship behavior between ancestor and descendant populations may be the result of such founder events. Given long enough periods of time, any two isolated populations will diverge due to genetic drift. In some cases, this random divergence may affect traits responsible for reproductive isolation, and speciation will have occurred.

Adaptation and Speciation

Nonetheless, adaptation and speciation are probably related in many cases. As species adapt to different circumstances, they will accumulate many differences that may lead to reproductive isolation. For example, if one population of flies adapts to wet conditions and another to dry conditions, then the populations will evolve a variety of differences in physiological and sensory traits; these differences may promote ecological and behavioral isolation and may cause any hybrids they produce to be poorly adapted to either habitat.

Selection might also act directly on mating behavior. Male *Anolis* lizards, for example, court females by extending a colorful flap of skin, called a "dewlap," that is located under their throat (figure 22.8). The ability of one lizard to see the dewlap of another lizard depends not only on the color of the dewlap, but the environment in which they occur. As a result, a light-colored dewlap is most effective in reflecting light in a dim forest, whereas dark colors are more apparent in the bright glare of open habitats. As a result, when these lizards occupy new habitats, natural selection will favor evolutionary change in dewlap color because males whose dewlaps cannot be seen will attract few mates. However, the lizards also distinguish members of their own species from those of other species by the color of the dewlap. Hence, adaptive change in mating behavior could have the incidental consequence of causing speciation.

Laboratory scientists have conducted experiments on fruit flies and other organisms in which they isolate populations

(a) *Anolis carolinensis.*

(b) *Anolis sagrei.*

(c) *Anolis grahami.*

(d) *Anolis mestrei.*

FIGURE 22.8

Dewlaps of several different species of Caribbean *Anolis* lizards. Males use their dewlaps in both territorial and courtship displays. Coexisting species almost always differ in their dewlaps, which are used in species recognition. Some dewlaps are easier to see in open habitats, whereas others are more visible in shaded environments.

in different laboratory chambers and measure how much reproductive isolation evolves. These experiments indicate that genetic drift by itself can lead to some degree of reproductive isolation, but, in general, reproductive isolation evolves more rapidly when the populations are forced to adapt to different laboratory environments (such as temperature or food type).

Reproductive isolating mechanisms can evolve either through random changes or as an incidental by-product of adaptive evolution. Under some circumstances, however, natural selection can directly select for traits that increase the reproductive isolation of a species.

The Geography of Speciation

Speciation is a two-part process. First, initially identical populations must diverge and, second, reproductive isolation must evolve to maintain these differences. The difficulty with this process, as we have seen, is that the homogenizing effect of gene flow between populations will constantly be acting to erase any differences that may arise, either by genetic drift or natural selection. Of course, gene flow only occurs between populations that are in contact. Consequently, evolutionary biologists have long recognized that speciation is much more likely in geographically isolated populations.

Allopatric Divergence Is the Primary Means of Speciation

Ernst Mayr was the first biologist to strongly make the case for allopatric speciation. Marshalling data from a wide variety of organisms and localities, Mayr was clearly able to demonstrate that geographically separated populations appear much more likely to have evolved substantial differences leading to speciation. For example, the Papuan kingfisher, *Tanysiptera hydrocharis*, varies little throughout its wide range in New Guinea despite the great variation in the island's topography and climate. By contrast, isolated populations on nearby islands are strikingly different from each other and from the mainland population (figure 22.9).

Many other examples indicate that speciation can occur in allopatry. Given that one would expect isolated populations to diverge over time by either drift or selection, this result is not surprising. Rather, the question becomes: Is geographic isolation required for speciation to occur?

Whether Speciation Can Occur in Sympatry Is Controversial

As we saw in chapter 20, disruptive selection can cause a population to contain individuals exhibiting two different phenotypes. One might think that if selection were strong enough, these two phenotypes would evolve into different species. However, before the two phenotypes could become different species, they would have to evolve reproductive isolating mechanisms. Because the two phenotypes would initially not be reproductively isolated at all, genetic exchange between individuals of the two phenotypes would tend to prevent genetic divergence in mating preferences or other isolating mechanisms. As a result, the two

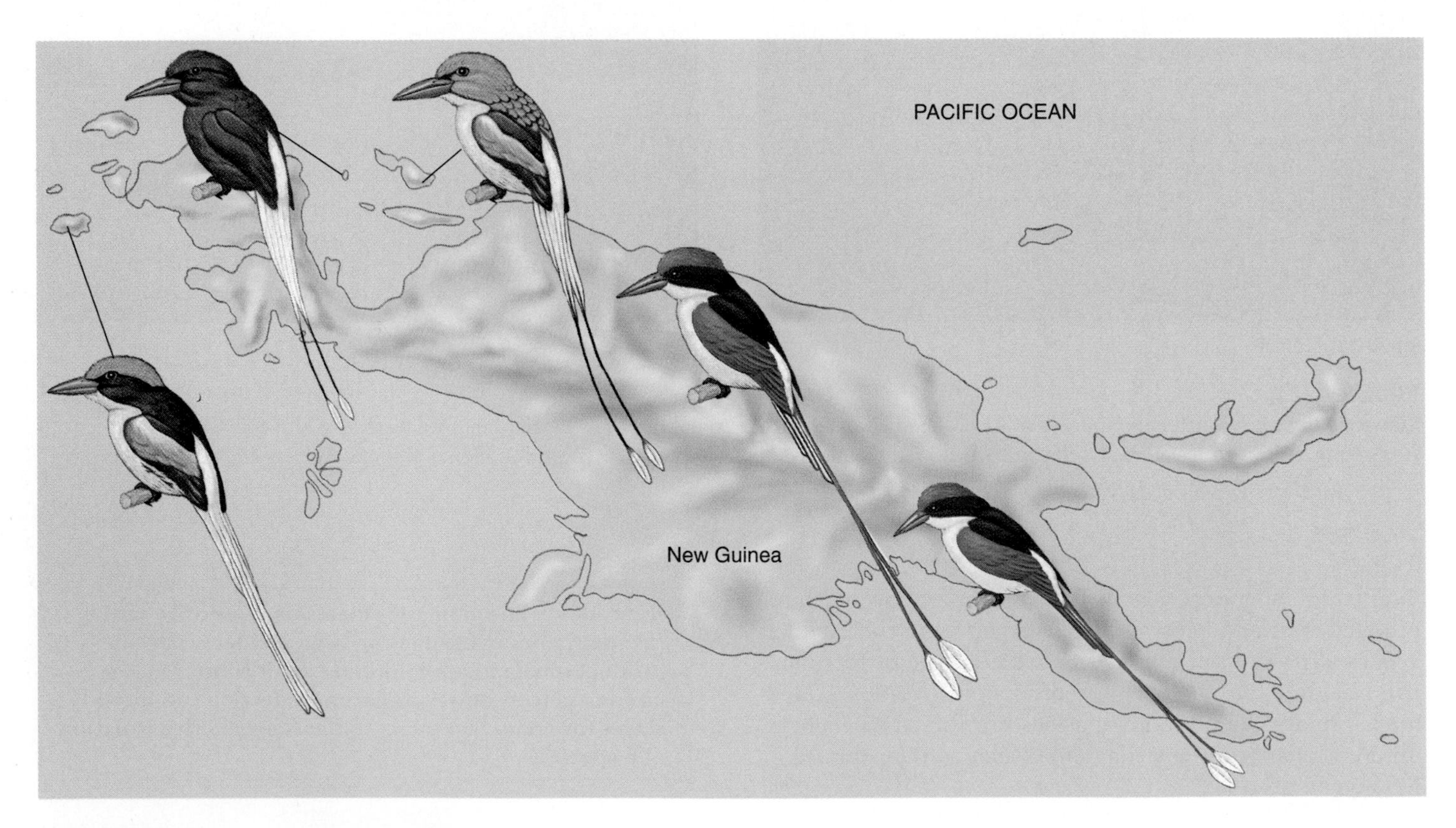

FIGURE 22.9
Phenotypic differentiation in the Papuan kingfisher in New Guinea. Isolated island populations (*above left*) are quite distinctive, showing variation in tail feather structure and length, plumage coloration, and bill size, whereas kingfishers on the mainland (*above right*) show little variation.
Source: Data from B.M. Bechler, et al., *Birds of New Guinea*, 1986, Princeton University Press.

FIGURE 22.10
Differences between species can result from a few genes that have major effects. (*a*) *Mimulus lewisii* has pink flowers and concentrated nectar, which are optimal for attracting bumblebees to serve as pollinators. (*b*) By contrast, *M. cardinalis* has the red flowers and copious dilute nectar typical of hummingbird-pollinated plants. Differences in flower shape and color are the result of a few genes of large effect.

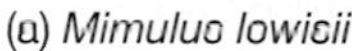

(a) *Mimulus lewisii*

(b) *Mimulus cardinalis*

phenotypes would be retained as polymorphisms within a single population. For this reason, most biologists consider sympatric speciation a rare event.

Nonetheless, in recent years, a number of cases have appeared that appear difficult to interpret in any way other than sympatric speciation. For example, the volcanic crater lake Barombi Mbo in Cameroon is extremely small and ecologically homogeneous, with no opportunity for within-lake isolation. Nonetheless, 11 species of closely related cichlid fish occur in the lake; all of the species are more closely related evolutionarily to each other than to any species outside of the crater. The most reasonable explanation is that an ancestral species colonized the crater and subsequently speciated in sympatry multiple times.

Genetic Changes Underlying Speciation

How much divergence does it take to create a new species? How many gene changes does it take? Since Darwin, the traditional view has been that new species arise by the accumulation of many small genetic differences. While there is little doubt that many species have formed in this gradual way, new techniques of molecular biology suggest that in at least some cases, the evolution of a new species may involve very few genes. Studying two species of monkeyflower found in the western United States, researchers found that only a few genes separate the two species, even though at first glance the two species appear to be very different (figure 22.10). Using gene technologies like those described in chapter 19, the researchers found that all of the major differences in the flowers, including not only flower shape and color, but also nectar production, were attributable to several genes, each of which had great phenotypic effects. Because individual genes have such powerful effects, species as different as these two can evolve in relatively few steps.

The Role of Polyploidy in Species Formation

Among plants, fertile individuals often arise from sterile ones through **polyploidy,** which doubles the chromosome number of the original sterile hybrid individual. A polyploid cell, tissue, or individual has more than two sets of chromosomes. Polyploid cells and tissues occur spontaneously and reasonably often in all organisms, although in many they are soon eliminated. A hybrid may be sterile simply because its sets of chromosomes, derived from male and female parents of different species, do not pair with one another. If the chromosome number of such a hybrid doubles, the hybrid, as a result of the doubling, will have a duplicate of each chromosome. In that case, the chromosomes will pair, and the fertility of the polyploid hybrid individual may be restored. It is estimated that about half of the approximately 260,000 species of plants have a polyploid episode in their history, including many of great commercial importance, such as bread wheat, cotton, tobacco, sugarcane, bananas, and potatoes. As you might imagine, the advantages a polyploid plant offers for natural selection can be substantial as a result of their great levels of genetic variation; hence, the significance of polyploidy in the evolution of plants.

Because polyploid plants cannot reproduce with their ancestors, reproductive isolation can evolve in one step. Consequently, speciation by polyploidy is one uncontroversial means of sympatric speciation. Although much rarer than in plants, speciation by polyploidy is also known from a variety of animals, including insects, fish, and salamanders.

Speciation occurs much more readily in the absence of gene flow among populations. However, speciation can occur in sympatry by means of polyploidy.

22.4 Clusters of species reflect rapid evolution.

Darwin's Finches

One of the most visible manifestations of evolution is the existence of groups of closely related species that have recently evolved from a common ancestor by occupying different habitats. This type of **adaptive radiation** occurred among the 13 species of Darwin's finches on the Galápagos Islands. Presumably, the ancestor of Darwin's finches reached these islands before other land birds, and all of the types of habitats where birds occur on the mainland were unoccupied. As the new arrivals moved into these vacant ecological niches and adopted new lifestyles, they were subjected to diverse sets of selective pressures. Under these circumstances, and aided by the geographic isolation afforded by the many islands of the Galápagos archipelago, the ancestral finches rapidly split into a series of diverse populations, some of which evolved into separate species. These species now occupy many different kinds of habitats on the Galápagos Islands (figure 22.11), habitats comparable to those several distinct groups of birds occupy on the mainland. The 13 species comprise four groups:

1. **Ground finches.** There are six species of *Geospiza* ground finches. Most of the ground finches feed on seeds. The size of their bills is related to the size of the seeds they eat. Some of the ground finches feed primarily on cactus flowers and fruits and have a longer, larger, more pointed bill than the others.
2. **Tree finches.** There are five species of insect-eating tree finches. Four species have bills that are suitable for feeding on insects. The woodpecker finch has a chisel-like beak. This unusual bird carries around a twig or a cactus spine, which it uses to probe for insects in deep crevices.
3. **Warbler finch.** This unusual bird plays the same ecological role in the Galápagos woods that warblers play on the mainland, searching continually over the leaves and branches for insects. It has a slender, warbler-like beak.
4. **Vegetarian finch.** The very heavy bill of this bud-eating bird is used to wrench buds from branches.

Darwin's finches, all derived from one similar mainland species, have radiated widely on the Galápagos Islands in the absence of competition.

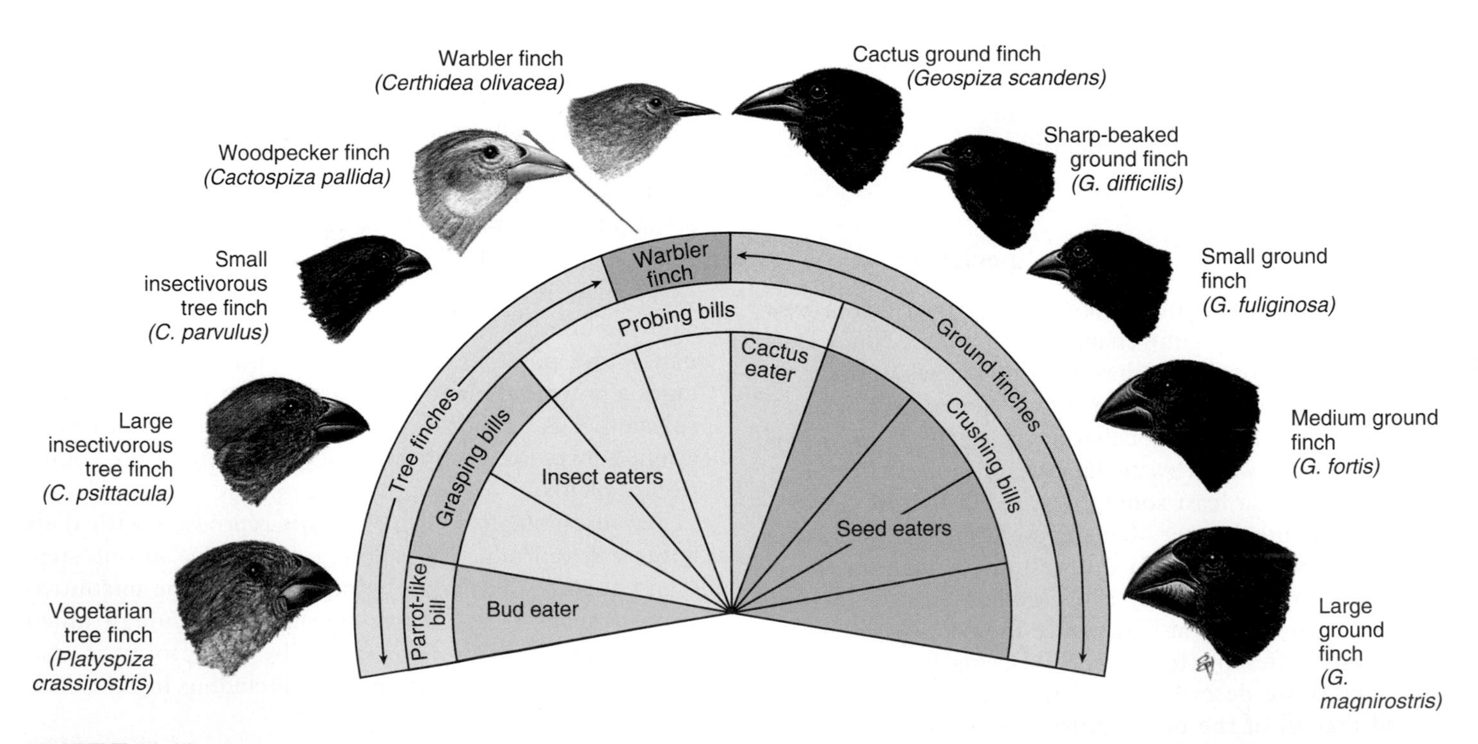

FIGURE 22.11
Darwin's finches. Ten of the 13 Galápagos species of Darwin's finches occur on Isla Santa Cruz, one of the Galápagos Islands. These species show differences in bills and feeding habits. The bills of several of these species resemble those of distinct families of birds on the mainland. This condition presumably arose when the finches evolved new species in habitats lacking small birds. The woodpecker finch uses cactus spines to probe in crevices of bark and rotten wood for food. Scientists believe all of these birds derived from a single common ancestor.

Hawaiian *Drosophila*

Our second example of a cluster of species is the fly genus *Drosophila* on the Hawaiian Islands, which we mentioned earlier as an example of behavioral isolation. There are at least 1250 species of this genus throughout the world, and more than a quarter are found only in the Hawaiian Islands (figure 22.12). New species of *Drosophila* are still being discovered in Hawaii, although the rapid destruction of the native vegetation is making the search more difficult. Aside from their sheer number, Hawaiian *Drosophila* species are unusual because of the morphological and behavioral traits mentioned earlier. No comparable species of *Drosophila* are found anywhere else in the world.

A second, closely related genus of flies, *Scaptomyza*, also forms a species cluster in Hawaii, where it is represented by as many as 300 species. A few species of *Scaptomyza* are found outside of Hawaii, but the genus is better represented there than elsewhere. In addition, species intermediate between *Scaptomyza* and *Drosophila* exist in Hawaii, but nowhere else. The genera are so closely related that scientists have suggested that all of the estimated 800 species of these two genera that occur in Hawaii may have derived from a single common ancestor.

The native Hawaiian flies are closely associated with the remarkable native plants of the islands and are often abundant in the native vegetation. Evidently, when their ancestors first reached these islands, they encountered many "empty" habitats that other kinds of insects and other animals occupied elsewhere. The evolutionary opportunities the ancestral *Drosophila* flies found were similar to those the ancestors of Darwin's finches in the Galápagos Islands encountered, and both groups evolved in a similar way. Many of the Hawaiian *Drosophila* species are highly selective in their choice of host plants for their larvae and in the part of the plant they use. The larvae of various species live in rotting stems, fruits, bark, leaves, or roots, or feed on sap.

New islands have continually arisen from the sea in the region of the Hawaiian Islands. As they have done so, they appear to have been invaded successively by the various *Drosophila* groups present on the older islands. New species have evolved as new islands have been colonized. The Hawaiian species of *Drosophila* have had even greater evolutionary opportunities than Darwin's finches because of their restricted ecological niches and the variable ages of the islands. They clearly tell one of the most unusual evolutionary stories found anywhere in the world.

(a) *Drosophila mulli*

(b) *Drosophila digressa*

FIGURE 22.12
Hawaiian *Drosophila*. The hundreds of species that have evolved on the Hawaiian Islands are extremely variable in appearance, although genetically almost identical.

The adaptive radiation of about 800 species of the flies *Drosophila* and *Scaptomyza* on the Hawaiian Islands, probably from a single common ancestor, is one of the most remarkable examples of intensive species formation found anywhere on earth.

Lake Victoria Cichlid Fishes

Lake Victoria is an immense shallow freshwater sea about the size of Switzerland in the heart of equatorial East Africa, until recently home to an incredibly diverse collection of over 300 species of cichlid fishes.

Recent Radiation

This cluster of species appears to have evolved recently and quite rapidly. By sequencing the cytochrome *b* gene in many of the lake's fish, scientists have been able to estimate that the first cichlids entered Lake Victoria only 200,000 years ago, colonizing from the Nile. Dramatic changes in water level encouraged species formation. As the lake rose, it flooded new areas and opened up new habitat. Many of the species may have originated after the lake dried down 14,000 years ago, isolating local populations in small lakes until the water level rose again.

Cichlid Diversity

These small, perchlike fishes range from 2 to 10 inches in length, and the males come in endless varieties of colors. The ecological and morphological diversity of these fish is remarkable, particularly given the short span over which it has evolved. We can gain some sense of the vast range of types by looking at how different species eat. There are mud biters, algae scrapers, leaf chewers, snail crushers, snail shellers (who pounce on slow-crawling snails and spear their soft parts with long curved teeth before the snail can retreat into its shell), zooplankton eaters, insect eaters, prawn eaters, and fish eaters. Scale-scraping cichlids rasp slices of scales off of other fish. There are even cichlid species that are "pedophages," eating the young of other cichlids.

Cichlid fish have a remarkable trait that may have been instrumental in this evolutionary radiation: a second set of functioning jaws occurs in the throats of cichlid fish (figure 22.13)! The ability of these jaws to manipulate and process food has freed the oral jaws to evolve for other purposes, and the result has been the incredible diversity of ecological roles filled by these fish.

Abrupt Extinction

Much of this diversity is gone. In the 1950s, the Nile perch, a commercial fish with a voracious appetite, was introduced on the Ugandan shore of Lake Victoria. Since then it has spread through the lake, eating its way through the cichlids. By 1990 all the open-water cichlid species were extinct, as well as many living in rocky shallow regions. Over 70% of all the named Lake Victoria cichlid species had disappeared, as well as untold numbers of species that had yet to be described.

Very rapid speciation occurred among cichlid fishes isolated in Lake Victoria, but widespread extinction followed with the introduction of a predator into the lake.

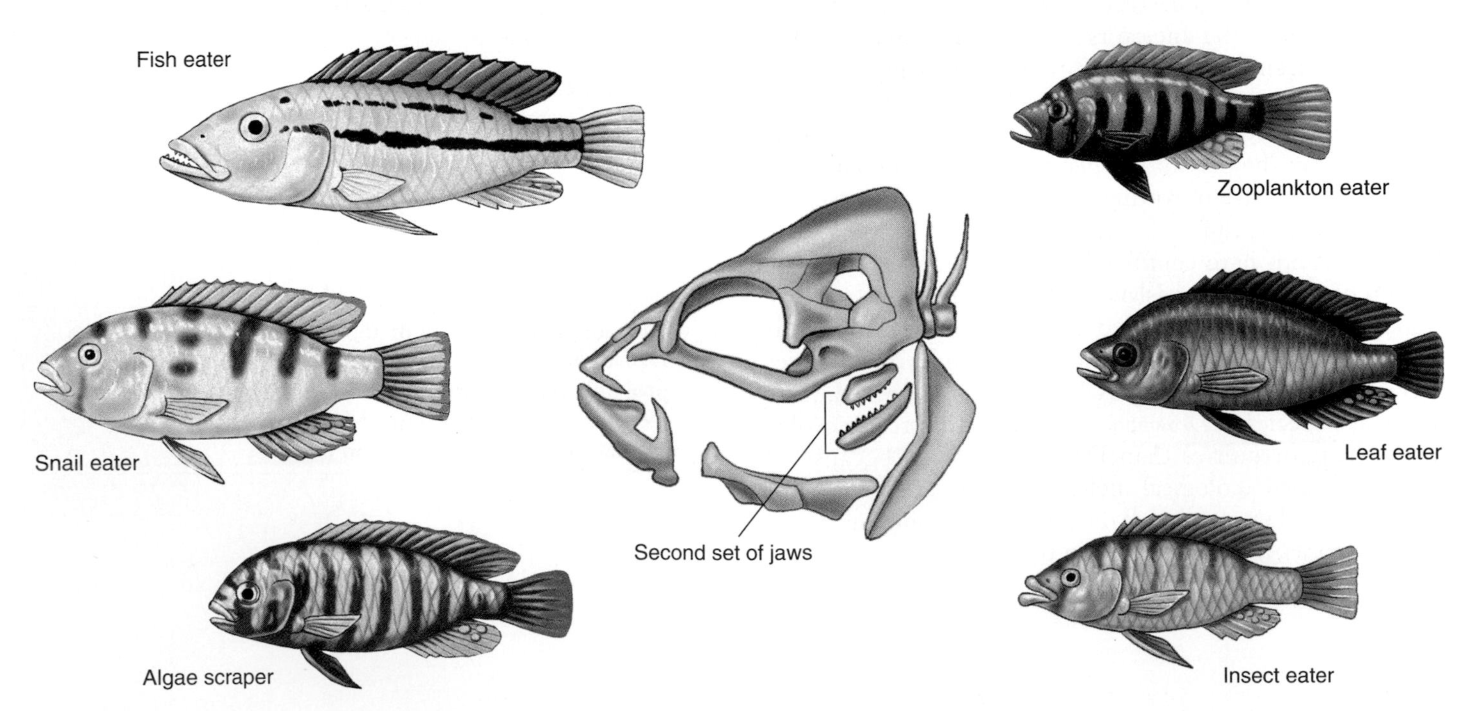

FIGURE 22.13
Cichlid fishes of Lake Victoria. These fishes have evolved adaptations to use a variety of different habitats. The second set of jaws located in the throat of these fish has provided evolutionary flexibility, allowing oral jaws to be modified in many ways.

New Zealand Alpine Buttercups

Adaptive radiations as we have described in Galápagos finches, Hawaiian *Drosophila*, and cichlid fishes seem to be favored by *periodic isolation*. Finches and *Drosophila* invade new islands, local species evolve, and they in turn reinvade the home island, in a cycle of expanding diversity. Similarly, cichlids become isolated by falling water levels, evolving separate species in isolated populations that later are merged when the lake's water level rises again.

A clear example of the role periodic isolation plays in species formation can be seen in the alpine buttercups (genus *Ranunculus*) which grow among the glaciers of New Zealand (figure 22.14). More species of alpine buttercup grow on the two islands of New Zealand than in all of North and South America combined. Detailed studies by the Canadian taxonomist Fulton Fisher revealed that the evolutionary mechanism responsible for inducing this diversity is recurrent isolation associated with the recession of glaciers. The 14 species of alpine *Ranunculus* occupy five distinctive habitats within glacial areas: *snowfields* (rocky crevices among outcrops in permanent snowfields at 7000 to 9000 feet elevation); *snowline fringe* (rocks at lower margin of snowfields between 4000 and 7000 ft); *stony debris* (slopes of exposed loose rocks at 2000 to 6000 ft); *sheltered situations* (shaded by rock or shrubs at 1000 to 6000 ft); and *boggy habitats* (sheltered slopes and hollows, poorly drained tussocks at elevations between 2500 and 5000 ft).

Ranunculus speciation and diversification has been promoted by repeated cycles of glacial advance and retreat. As the glaciers retreat, populations become isolated on mountain peaks, permitting speciation (figure 22.15). In the next advance, these new species can expand throughout the mountain range, coming into contact with their close relatives. In this way, one initial species could give rise to many descendants. Moreover, on isolated mountaintops during glacial retreats, species have convergently evolved to occupy similar habitats; these distantly related but ecologically similar species have then been brought back into contact in subsequent glacial advances.

FIGURE 22.14
A New Zealand alpine buttercup. Fourteen species of alpine *Ranunculus* grow among the glaciers and mountains of New Zealand, including this *R. lyallii*, the giant buttercup.

Recurrent isolation promotes species formation.

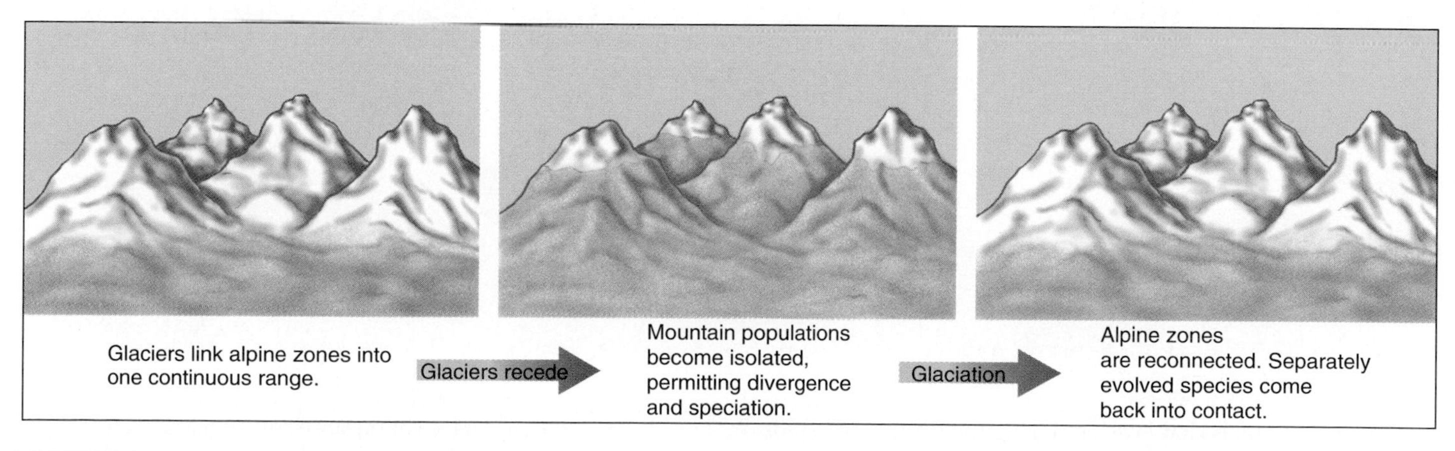

Glaciers link alpine zones into one continuous range.
Glaciers recede
Mountain populations become isolated, permitting divergence and speciation.
Glaciation
Alpine zones are reconnected. Separately evolved species come back into contact.

FIGURE 22.15
Periodic glaciation encouraged species formation among alpine buttercups in New Zealand. The formation of extensive glaciers during the Pleistocene linked the alpine zones of many mountains together. When the glaciers receded, these alpine zones were isolated from one another, only to become reconnected with the advent of the next glacial period. During periods of isolation, populations of alpine buttercups diverged in the isolated habitats.

Diversity of Life through Time

Although eukaryotes evolved nearly 3 billion years ago, the diversity of life didn't increase substantially until approximately 550 million years ago. Then, almost all of the extant types of animals evolved in a geologically short period termed the "Cambrian explosion." In addition to organisms whose descendants are recognizable today, a wide variety of other types of organisms also evolved (figure 22.16). The biology of these creatures, which quickly disappeared without leaving any descendants, is poorly understood. The Cambrian explosion seems to have been a time of evolutionary experimentation and innovation, in which many types of organisms appeared, but most were quickly weeded out. What prompted this explosion of diversity is still a subject of considerable controversy.

FIGURE 22.16
Diversity of animals that evolved during the Cambrian explosion. In addition to the appearance of the ancestors of many present-day groups, such as insects and vertebrates, a variety of bizarre creatures evolved that left no descendants, such as *Wiwaxia*, *Marrella*, *Opabinia*, and the aptly named *Hallucigenia*. The natural history of these species is open to speculation. Key: (*1*) *Amiskwia*, (*2*) *Odontogriphus*, (*3*) *Eldonia*, (*4*) *Halichondrites*, (*5*) *Anomalocaris canadensis*, (*6*) *Pikaia*, (*7*) *Canadia*, (*8*) *Marrella splendens*, (*9*) *Opabinia*, (*10*) *Ottoia*, (*11*) *Wiwaxia*, (*12*) *Yohoia*, (*13*) *Xianguangia*, (*14*) *Aysheaia*, (*15*) *Sidneyia*, (*16*) *Dinomischus*, (*17*) *Hallucigenia*.

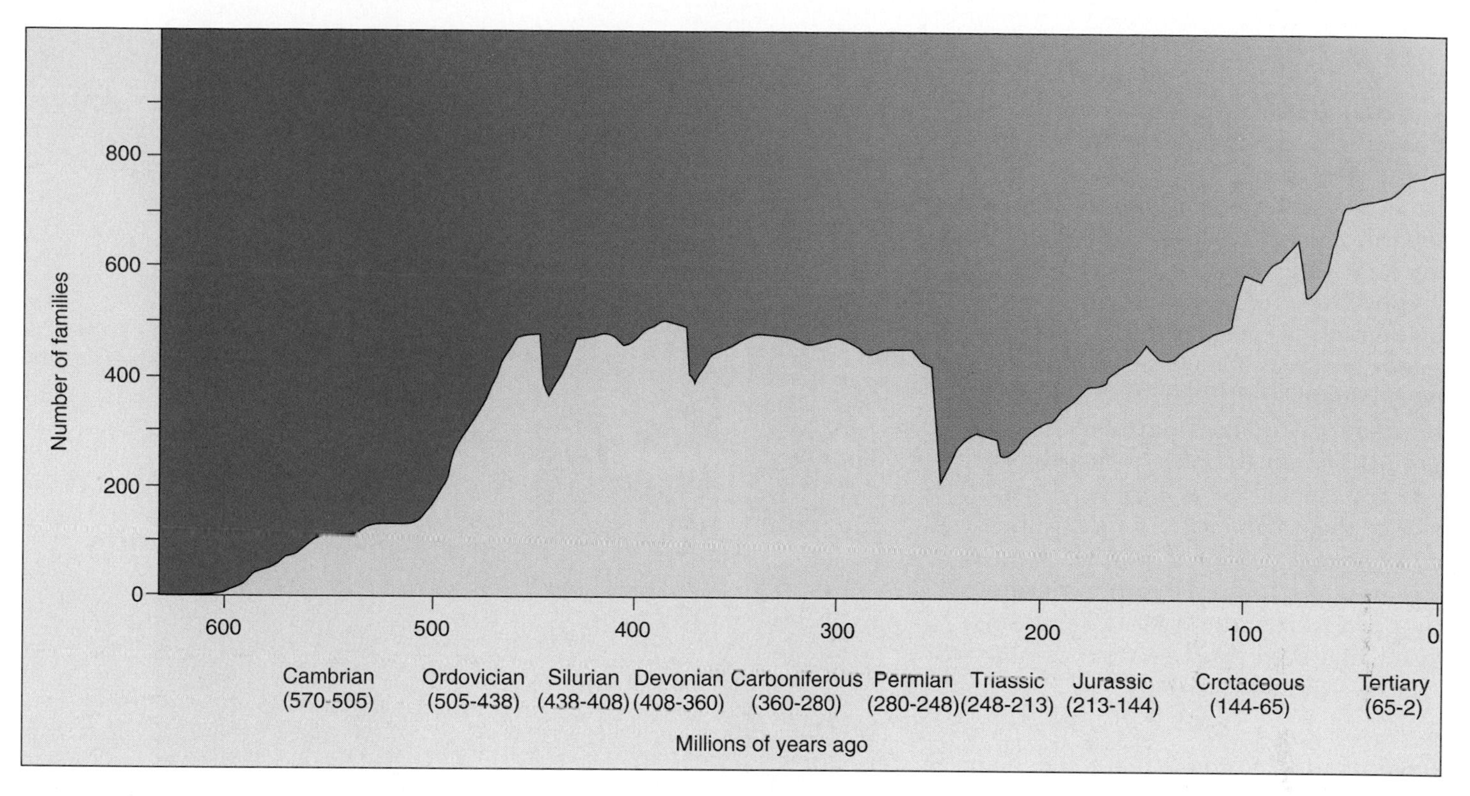

FIGURE 22.17
Diversity through time. Taxonomic diversity of families of marine animals since the Cambrian Period. The fossil record is most complete for marine organisms because they are more readily fossilized than terrestrial species. Families are shown, rather than species, because many species are known from only one specimen, thus introducing error into estimates of time of extinction.
Source: Data from D. Futuyma, *Evolutionary Biology*, 1998, Sinauer.

Trends in Species Diversity

The number of species in the world has increased vastly since the Cambrian. However, the trend has been far from consistent (figure 22.17). After a rapid rise, the number of species reached a plateau for about 200 million years; since then, the number has risen steadily.

Interspersed in these patterns, however, have been a number of major setbacks, termed **mass extinctions,** in which the number of species has greatly decreased. Five major mass extinctions have been identified, the most severe of which occurred at the end of the Permian Period, approximately 250 million years ago, at which time more than half of all families and as many as 96% of all species may have perished.

The most famous and well-studied extinction, though not as drastic, occurred at the end of the Cretaceous Period (65 million years ago), at which time the dinosaurs and a variety of other organisms went extinct. Recent studies have provided support for the hypothesis that this extinction event was triggered by a large asteroid which slammed into the earth, perhaps causing global forest fires and obscuring the sun for months by throwing particles into the air. This mass extinction did have one positive effect, though: with the disappearance of dinosaurs, mammals, which previously had been small and inconspicuous, quickly experienced a vast evolutionary radiation, which ultimately produced a wide variety of organisms, including elephants, tigers, whales, and humans. Indeed, a general observation is that biological diversity tends to rebound quickly after mass extinctions, reaching comparable levels of species richness within a few million years, even if the organisms making up that diversity are not the same.

A Sixth Extinction

The number of species in the world in recent times is greater than it has ever been. Unfortunately, that number is decreasing at an alarming rate due to human activities. Some estimate that as many as one-fourth of all species will become extinct in the next 50 years, a rate of extinction not seen on earth since the Cretaceous mass extinction.

The number of species has increased through time, although not at constant rates. Several major extinction events have substantially, though briefly, reduced the number of species.

The Pace of Evolution

Different kinds of organisms evolve at different rates. Mammals, for example, evolve relatively slowly. On the basis of a relatively complete fossil record, it has been estimated that an average value for the duration of a "typical" mammal species, from formation of the species to its extinction, might be about 200,000 years. American paleontologist George Gaylord Simpson has pointed out that certain groups of animals, such as lungfishes, are apparently evolving even more slowly than mammals. In fact, Simpson estimated that there has been little evolutionary change among lungfishes over the past 150 million years, and even slower rates of evolution occur in other groups.

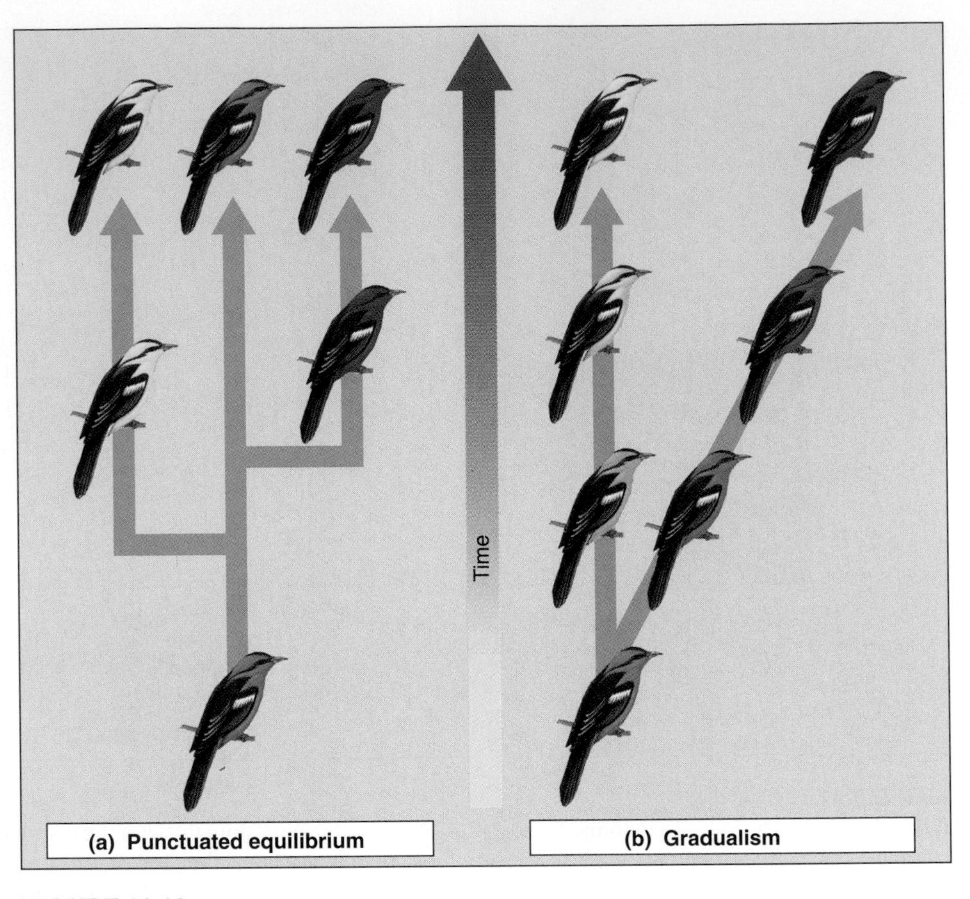

FIGURE 22.18
Two views of the pace of macroevolution. (*a*) Punctuated equilibrium surmises that species formation occurs in bursts, separated by long periods of quiet, while (*b*) gradualism surmises that species formation is constantly occurring.

Evolution in Spurts?

Not only does the rate of evolution differ greatly from group to group, but evolution within a group apparently proceeds rapidly during some periods and relatively slowly during others. The fossil record provides evidence for such variability in evolutionary rates, and evolutionists are very interested in understanding the factors that account for it. In 1972, paleontologists Niles Eldredge of the American Museum of Natural History in New York and Stephen Jay Gould of Harvard University proposed that evolution normally proceeds in spurts. They claimed that the evolutionary process is a series of **punctuated equilibria.** Evolutionary innovations would occur and give rise to new lines; then these lines might persist unchanged for a long time, in "equilibrium." Eventually there would be a new spurt of evolution, creating a "punctuation" in the fossil record. Eldredge and Gould contrast their theory of punctuated equilibrium with that of **gradualism,** or gradual evolutionary change, which they claimed was what Darwin and most earlier students of evolution had considered normal (figure 22.18).

Eldredge and Gould proposed that **stasis,** or lack of evolutionary change, would be expected in large populations experiencing stabilizing selection over long periods of time. In contrast, rapid evolution of new species might occur if populations colonized new areas. Such populations would be small, isolated, and possibly already differing from their parental population as a result of the founder effect. This, combined with selective pressures from a new environment, could bring about rapid change.

Unfortunately, the distinctions are not as clear-cut as implied by this discussion. Some well-documented groups such as African mammals clearly have evolved gradually, and not in spurts. Other groups, like marine bryozoa, seem to show the irregular pattern of evolutionary change the punctuated equilibrium model predicts. It appears, in fact, that gradualism and punctuated equilibrium are two ends of a continuum. Although some groups appear to have evolved solely in a gradual manner and others only in a punctuated mode, many other groups appear to show evidence of both gradual and punctuated episodes at different times in their evolutionary history.

The punctuated equilibrium model assumes that evolution occurs in spurts, between which there are long periods in which there is little evolutionary change. The gradualism model assumes that evolution proceeds gradually, with successive change in a given evolutionary line.

Problems with the Biological Species Concept

Since the biological species concept was first proposed by Ernst Mayr in the 1940s, it has been the predominant idea of how to recognize and define species. However, in recent years, workers from a variety of fields have begun to question how universally applicable the concept really is.

The Extent of Hybridization

The crux of the matter concerns hybridization. Biological species are reproductively isolated, so that hybridization should be rare. If hybridization is common, one would expect one of two quick outcomes: either reinforcement would occur, leading to the perfection of isolating mechanisms and an end to hybridization, or the two populations would merge together into a single homogeneous gene pool.

However, in recent years biologists have detected much greater amounts of hybridization than previously realized between populations that seem to neither be experiencing reinforcement nor losing their specific identities. Botanists have always been aware that species can often experience substantial amounts of hybridization. One study found that more than 50% of the plant species surveyed in California were not well defined by genetic isolation. For example, the fossil record indicates that balsam poplars and cottonwoods have been phenotypically distinct for 12 million years, but throughout this time, they have routinely produced hybrids. Consequently, for many years, many botanists have felt that the biological species concept only applies to animals.

What is becoming increasingly evident, however, is that hybridization is not all that uncommon in animals, either. One recent survey indicated that almost 10% of the world's 9500 bird species are known to have hybridized in nature. Recent years have seen the documentation of more and more cases in which substantial hybridization occurs between animal species. Again, the Galápagos finches provide a particularly well-studied example. Three species on the island of Daphne Major—the medium ground finch, the cactus finch, and the small ground finch—are clearly distinct morphologically and occupy different ecological niches. Careful studies over the past 20 years by Peter and Rosemary Grant found that, on average, 2% of the medium ground finches and 1% of the cactus finches mated with other species every year. Furthermore, hybrid offspring appeared to be at no disadvantage either in terms of survival or subsequent reproduction. This is not a trivial amount of genetic exchange, and one might expect to see the species coalescing into one variable population, but the species are nonetheless maintaining their distinctiveness.

Alternatives to the Biological Species Concept

This is not to say hybridization is rampant throughout the animal world. As the bird survey indicated, 90% of bird species are not known to hybridize, and even fewer probably experience significant amounts of hybridization. Still, it is a common enough occurrence to cast doubt about whether reproductive isolation is the only force maintaining the integrity of species.

An alternative hypothesis is that the distinctions among species are maintained by natural selection. The idea is that each species has adapted to its own specific part of the environment. Stabilizing selection then maintains the species' adaptations; hybridization has little effect because alleles introduced into the gene pool from other species quickly would be eliminated by natural selection.

We have already seen in chapter 20 that the interaction between gene flow and natural selection can have many outcomes. In some cases, strong selection can overwhelm any effects of gene flow, but in other situations, gene flow can prevent populations from eliminating less successful alleles from a population. As a general explanation, then, natural selection is not likely to have any fewer exceptions than the biological species concept, although it may prove more successful for certain types of organisms or habitats.

A variety of other ideas have been put forward to establish criteria for defining species. Many of these are specific to a particular type of organism and none has universal applicability. In truth, it may be that there is no single explanation for what maintains the identity of species. Given the incredible variation evident in plants, animals, and microorganisms in all aspects of their biology, it is perhaps not surprising that different processes are operating in different organisms. This is an area of active research that demonstrates the dynamic nature of the field of evolutionary biology.

Hybridization has always been recognized to be widespread among plants, but recent research reveals that it is surprisingly high in animals, too. Because of the diversity of living organisms, no single definition of what constitutes a species may be universally applicable.

Chapter 22

Summary | **Questions** | **Media Resources**

22.1 Species are the basic units of evolution.

- Species are groups of organisms that differ from one another in one or more characteristics and do not hybridize freely when they come into contact in their natural environment. Many species cannot hybridize with one another at all.
- Species exhibit geographic variation, yet phenotypically distinctive populations are connected by intermediate forms.

1. Define the term *sympatry.* Why is sympatric speciation thought by many to be unlikely?

2. What is the biological species concept?

22.2 Species maintain their genetic distinctiveness through barriers to reproduction.

- Among the factors that separate populations and species are geographical, ecological, behavioral, temporal, and mechanical isolation, as well as factors that inhibit the fusion of gametes or the normal development of the hybrid organisms.
- Some isolating mechanisms (prezygotic) prevent hybrid formation; others (postzygotic) prevent hybrids from surviving and reproducing.

3. What is the difference between prezygotic and postzygotic isolating mechanisms?

4. What barriers exist to hybrid formation and success? Which are prezygotic and which are postzygotic isolating mechanisms? Why do some people think the term "isolating mechanism" is misleading?

22.3 We have learned a great deal about how species form.

- Reproductive isolation can arise as populations differentiate by adaptation to different environments, as well as by random genetic drift, founder effects, or population bottlenecks.
- Natural selection may favor changes in the mating system when a species occupies a new habitat, so that the species becomes reproductively isolated from other species.
- When two species are not completely reproductively isolated, natural selection may favor the evolution of more effective isolating mechanisms to prevent hybridization, a process termed "reinforcement."

5. How does selection relate to population divergence?

6. How many genes are involved in the speciation process?

7. When are hybrids at a disadvantage? What can be the result of this disadvantage?

8. Define the term *polyploidy.*

- Introduction to Speciation
- Allopatric Speciation
- Sympatric Speciation

- *On Science* Article: Lonesome George
- Student Research: Hybrids and Evolution in Fragile Ferns

22.4 Clusters of species reflect rapid evolution.

- Clusters of species arise when populations differentiate to fill several niches. On islands, differentiation is often rapid because of numerous open habitats.
- The pace of evolution is not constant among all organisms. Some scientists believe it occurs in spurts, others argue that it proceeds gradually.
- Hybridization occurs commonly among plants and even among animals. The biological species concept may not apply to all organisms.

9. What is adaptive radiation? What types of habitats encourage it? Why?

10. What is the difference between gradualism and punctuated equilibrium?

11. Why is the biological species concept no longer considered to be universally applicable?

- Extinctions
- Evolutionary Trends

- Art Quiz: Evolutionary Pace

23

How Humans Evolved

FIGURE 23.1
The trail of our ancestors. These fossil footprints, made in Africa 3.7 million years ago, look as if they might have been left by a mother and child walking on the beach. But these tracks, preserved in volcanic ash, are not human. They record the passage of two individuals of the genus *Australopithecus*, the group from which our genus, *Homo*, evolved.

Concept Outline

23.1 The evolutionary path to humans starts with the advent of primates.

The Evolutionary Path to Apes. Primates first evolved 65 million years ago, giving rise first to prosimians and then to monkeys.

How the Apes Evolved. Apes, including our closest relatives, the chimpanzees, arose from an ancestor common to Old World monkeys.

23.2 The first hominids to evolve were australopithecines.

An Evolutionary Tree with Many Branches. The first hominids were australopithecines, of which there were several different kinds.

The Beginning of Hominid Evolution. The ability to walk upright on two legs marks the beginning of hominid evolution. One can draw the hominid family tree in two very different ways, either lumping variants together or splitting them into separate species.

23.3 The genus *Homo* evolved in Africa.

African Origin: Early *Homo*. There may have been several species of early *Homo*, with brains significantly larger than those of australopithecines.

Out of Africa: *Homo erectus*. The first hominid species to leave Africa was the relatively large-brained *H. erectus*, the longest lived species of *Homo*.

23.4 Modern humans evolved quite recently.

The Last Stage of Hominid Evolution. Modern humans evolved within the last 600,000 years, our own species within the last 150,000 years.

Our Own Species: *Homo sapiens*. Our species appears to have evolved in Africa, and then migrated to Europe and Asia.

Human Races. Our species is unique in evolving culturally. Differences in populations in skin color reflect adaptation to different environments, rather than genetic differentiation among populations.

In 1871 Charles Darwin published another groundbreaking book, *The Descent of Man*. In this book, he suggested that humans evolved from the same African ape ancestors that gave rise to the gorilla and the chimpanzee. Although little fossil evidence existed at that time to support Darwin's case, numerous fossil discoveries made since then strongly support his hypothesis (figure 23.1). Human evolution is the part of the evolution story that often interests people most, and it is also the part about which we know the most. In this chapter we follow the evolutionary journey that has led to humans, telling the story chronologically. It is an exciting story, replete with controversy.

The Evolutionary Path to Apes

The story of human evolution begins around 65 million years ago, with the explosive radiation of a group of small, arboreal mammals called the Archonta. These primarily insectivorous mammals had large eyes and were most likely nocturnal (active at night). Their radiation gave rise to different types of mammals, including bats, tree shrews, and primates, the order of mammals that contains humans.

The Earliest Primates

Primates are mammals with two distinct features that allowed them to succeed in the arboreal, insect-eating environment:

1. **Grasping fingers and toes.** Unlike the clawed feet of tree shrews and squirrels, primates have grasping hands and feet that let them grip limbs, hang from branches, seize food, and, in some primates, use tools. The first digit in many primates is opposable and at least some, if not all, of the digits have nails.
2. **Binocular vision.** Unlike the eyes of shrews and squirrels, which sit on each side of the head so that the two fields of vision do not overlap, the eyes of primates are shifted forward to the front of the face. This produces overlapping binocular vision that lets the brain judge distance precisely—important to an animal moving through the trees.

Other mammals have binocular vision, but only primates have both binocular vision and grasping hands, making them particularly well adapted to their environment. While early primates were mostly insectivorous, their dentition began to change from the shearing, triangular-shaped molars specialized for insect eating to the more flattened, square-shaped molars and rodentlike incisors specialized for plant eating. Primates that evolved later also show a continuous reduction in snout length and number of teeth.

The Evolution of Prosimians

About 40 million years ago, the earliest primates split into two groups: the prosimians and the anthropoids. The prosimians ("before monkeys") looked something like a cross between a squirrel and a cat and were common in North America, Europe, Asia, and Africa. Only a few prosimians survive today, lemurs, lorises and tarsiers (figure 23.2). In addition to having grasping digits and binocular vision, prosimians have large eyes with increased visual acuity. Most prosimians are nocturnal, feeding on fruits, leaves, and flowers, and many lemurs have long tails for balancing.

FIGURE 23.2
A prosimian. This tarsier, a prosimian native to tropical Asia, shows the characteristic features of primates: grasping fingers and toes and binocular vision.

Origin of the Anthropoids

The anthropoids, or higher primates, include monkeys, apes, and humans (figure 23.3). Anthropoids are almost all diurnal—that is, active during the day—feeding mainly on fruits and leaves. Evolution favored many changes in eye design, including color vision, that were adaptations to daytime foraging. An expanded brain governs the improved senses, with the braincase forming a larger portion of the head. Anthropoids, like the relatively few diurnal prosimians, live in groups with complex social interactions. In addition, the anthropoids tend to care for their young for prolonged periods, allowing for a long childhood of learning and brain development.

The early anthropoids, now extinct, are thought to have evolved in Africa. Their direct descendants are a very successful group of primates, the monkeys.

New World Monkeys. About 30 million years ago, some anthropoids migrated to South America, where they evolved in isolation. Their descendants, known as the **New World monkeys**, are easy to identify: all are arboreal, they have flat spreading noses, and many of them grasp objects with long prehensile tails (figure 23.4*a*).

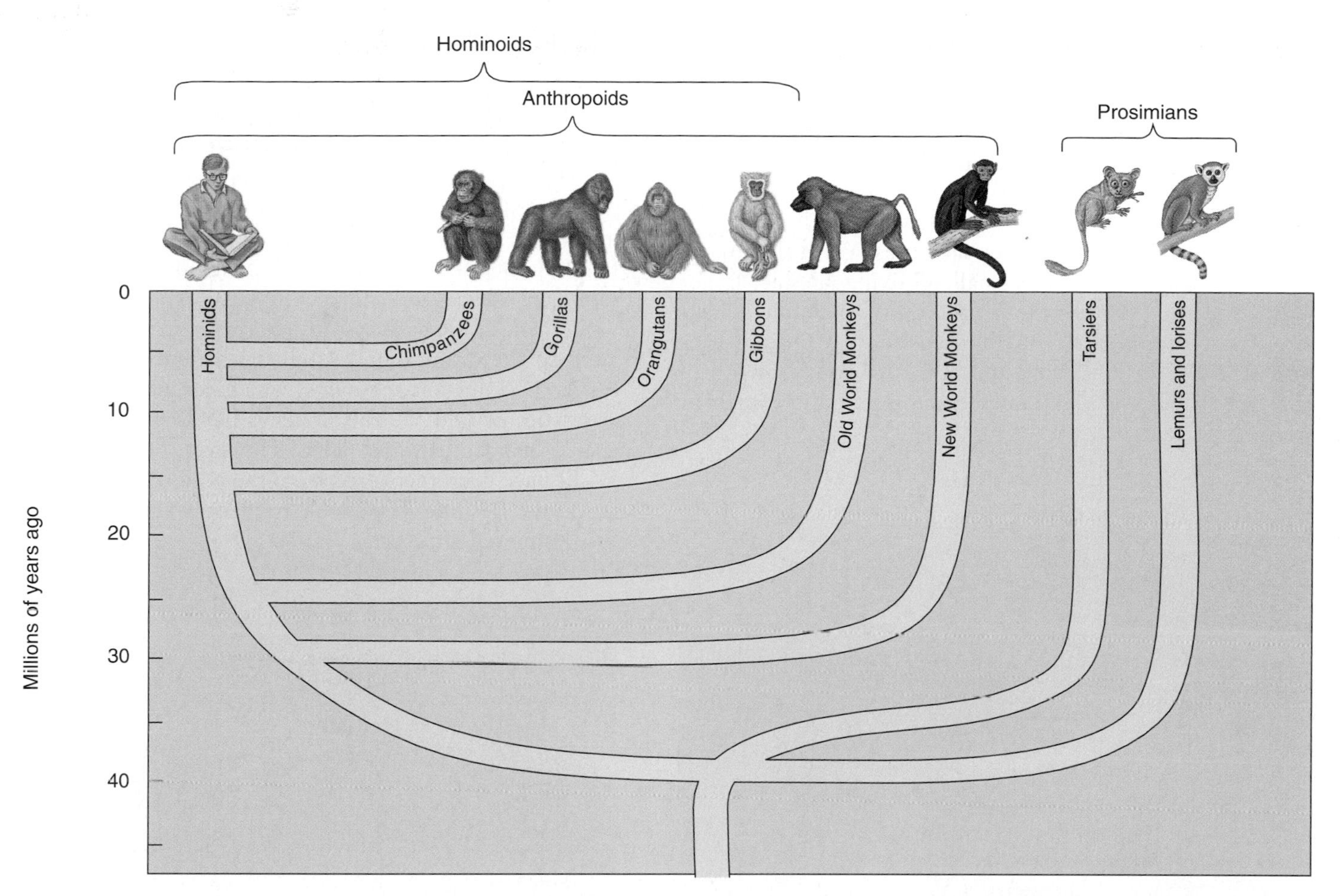

FIGURE 23.3
A primate evolutionary tree. The most ancient of the primates are the prosimians, while the hominids were the most recent to evolve.

Old World Monkeys. Around 25 million years ago, anthropoids that remained in Africa split into two lineages: one gave rise to the **Old World monkeys** and one gave rise to the hominoids (see page 480). Old World monkeys include ground-dwelling as well as arboreal species. None of the Old World monkeys have prehensile tails. Their nostrils are close together, their noses point downward, and some have toughened pads of skin for prolonged sitting (figure 23.4*b*).

The earliest primates arose from small, tree-dwelling, insect-eaters and gave rise to prosimians and then anthropoids. Early anthropoids gave rise to New World monkeys and Old World monkeys.

(a)

(b)

FIGURE 23.4
New and Old World monkeys. (*a*) New World monkeys, such as this golden lion tamarin, are arboreal, and many have prehensile tails. (*b*) Old World monkeys lack prehensile tails, and many are ground dwellers.

How the Apes Evolved

The other African anthropoid lineage is the **hominoids,** which includes the **apes** and the **hominids** (humans and their direct ancestors). The living apes consist of the gibbon (genus *Hylobates*), orangutan (*Pongo*), gorilla (*Gorilla*), and chimpanzee (*Pan*) (figure 23.5). Apes have larger brains than monkeys, and they lack tails. With the exception of the gibbon, which is small, all living apes are larger than any monkey. Apes exhibit the most adaptable behavior of any mammal except human beings. Once widespread in Africa and Asia, apes are rare today, living in relatively small areas. No apes ever occurred in North or South America.

The First Hominoid

Considerable controversy exists about the identity of the first hominoid. During the 1980s it was commonly believed that the common ancestor of apes and hominids was a late Miocene ape living 5 to 10 million years ago. In 1932, a candidate fossil, an 8-million-year-old jaw with teeth, was unearthed in India. It was called *Ramapithecus* (after the Hindi deity Rama). However, these fossils have never been found in Africa, and more complete fossils discovered in 1981 made it clear that *Ramapithecus* is in fact closely related to the orangutan. Attention has now shifted to an earlier Miocene ape, *Proconsul*, which has many of the characteristics of Old World monkeys but lacks a tail and has apelike hands, feet, and pelvis. However, because very few fossils have been recovered from the period 5 to 10 million years ago, it is not yet possible to identify with certainty the first hominoid ancestor.

FIGURE 23.5
The living apes. (*a*) Mueller gibbon, *Hylobates muelleri*. (*b*) Orangutan, *Pongo pygmaeus*. (*c*) Gorilla, *Gorilla gorilla*. (*d*) Chimpanzee, *Pan troglodytes*.

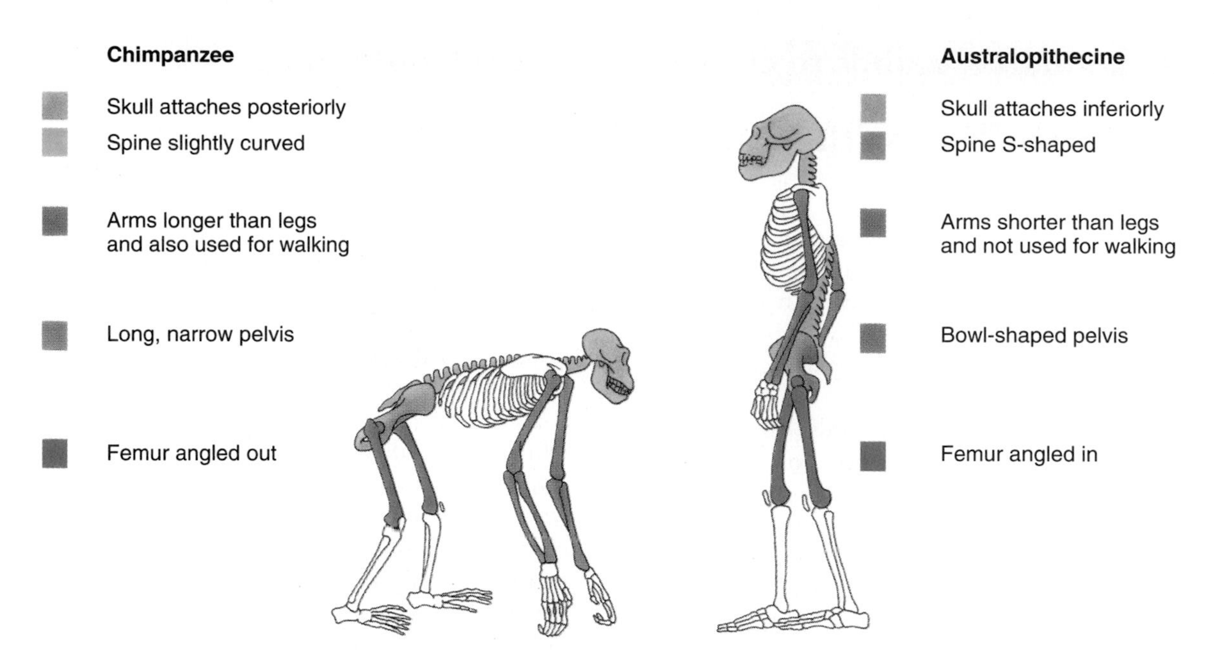

FIGURE 23.6
A comparison of ape and hominid skeletons. Early humans, such as australopithecines, were able to walk upright because their arms were shorter, their spinal cord exited from the bottom of the skull, their pelvis was bowl-shaped and centered the body weight over the legs, and their femurs angled inward, directly below the body, to carry its weight.

Which Ape Is Our Closest Relative?

Studies of ape DNA have explained a great deal about how the living apes evolved. The Asian apes evolved first. The line of apes leading to gibbons diverged from other apes about 15 million years ago, while orangutans split off about 10 million years ago (see figure 23.3). Neither are closely related to humans.

The African apes evolved more recently, between 6 and 10 million years ago. These apes are the closest living relatives to humans; some taxonomists have even advocated placing humans and the African apes in the same zoological family, the Hominidae. Fossils of the earliest hominids, described later in the chapter, suggest that the common ancestor of the hominids was more like a chimpanzee than a gorilla. Based on genetic differences, scientists estimate that gorillas diverged from the line leading to chimpanzees and humans some 8 million years ago.

Sometime after the gorilla lineage diverged, the common ancestor of all hominids split off from the chimpanzee line to begin the evolutionary journey leading to humans. Because this split was so recent, the genes of humans and chimpanzees have not had time to evolve many genetic differences. For example, a human hemoglobin molecule differs from its chimpanzee counterpart in only a single amino acid. In general, humans and chimpanzees exhibit a level of genetic similarity normally found between closely related sibling species of the same genus!

Comparing Apes to Hominids

The common ancestor of apes and hominids is thought to have been an arboreal climber. Much of the subsequent evolution of the hominoids reflected different approaches to locomotion. Hominids became **bipedal**, walking upright, while the apes evolved knuckle-walking, supporting their weight on the back sides of their fingers (monkeys, by contrast, use the palms of their hands).

Humans depart from apes in several areas of anatomy related to bipedal locomotion (figure 23.6). Because humans walk on two legs, their vertebral column is more curved than an ape's, and the human spinal cord exits from the bottom rather than the back of the skull. The human pelvis has become broader and more bowl-shaped, with the bones curving forward to center the weight of the body over the legs. The hip, knee, and foot (in which the human big toe no longer splays sideways) have all changed proportions. Being bipedal, humans carry much of the body's weight on the lower limbs, which comprise 32 to 38% of the body's weight and are longer than the upper limbs; human upper limbs do not bear the body's weight and make up only 7 to 9% of human body weight. African apes walk on all fours, with the upper and lower limbs both bearing the body's weight; in gorillas, the longer upper limbs account for 14 to 16% of body weight, the somewhat shorter lower limbs for about 18%.

Hominoids, the apes and hominids, arose from Old World monkeys. Among living apes, chimpanzees seem the most closely related to humans.

23.2 The first hominids to evolve were australopithecines.

An Evolutionary Tree with Many Branches

Five to 10 million years ago, the world's climate began to get cooler, and the great forests of Africa were largely replaced with savannas and open woodland. In response to these changes, a new kind of hominoid was evolving, one that was bipedal. These new hominoids are classified as hominids—that is, of the human line.

There are two major groups of hominids: three to seven species of the genus *Homo* (depending how you count them) and seven species of the older, smaller-brained genus *Australopithecus*. In every case where the fossils allow a determination to be made, the hominids are bipedal, walking upright. Bipedal locomotion is the hallmark of hominid evolution. We will first discuss *Australopithecus*, and then *Homo*.

Discovery of *Australopithecus*

The first hominid was discovered in 1924 by Raymond Dart, an anatomy professor at Johannesburg in South Africa. One day, a mine worker brought him an unusual chunk of rock—actually, a rock-hard mixture of sand and soil. Picking away at it, Professor Dart uncovered a skull unlike that of any ape he had ever seen. Beautifully preserved, the skull was of a five-year-old individual, still with its milk teeth. While the skull had many apelike features such as a projecting face and a small brain, it had distinctly human features as well—for example, a rounded jaw unlike the pointed jaw of apes. The ventral position of the foramen magnum (the hole at the base of the skull from which the spinal cord emerges) suggested that the creature had walked upright. Dart concluded it was a human ancestor.

What riveted Dart's attention was that the rock in which the skull was embedded had been collected near other fossils that suggested that the rocks and their fossils were several million years old! At that time, the oldest reported fossils of hominids were less than 500,000 years old, so the antiquity of this skull was unexpected and exciting. Scientists now estimate Dart's skull to be 2.8 million years old. Dart called his find *Australopithecus africanus* (from the Latin *australo*, meaning "southern" and the Greek *pithecus*, meaning "ape"), the ape from the south of Africa.

Today, fossils are dated by the relatively new process of single-crystal laser-fusion dating. A laser beam melts a single potassium feldspar crystal, releasing argon gas, which is measured in a gas mass spectrometer. Because the argon in the crystal has accumulated at a known rate, the amount released reveals the age of the rock and thus of nearby fossils. The margin of error is less than 1%.

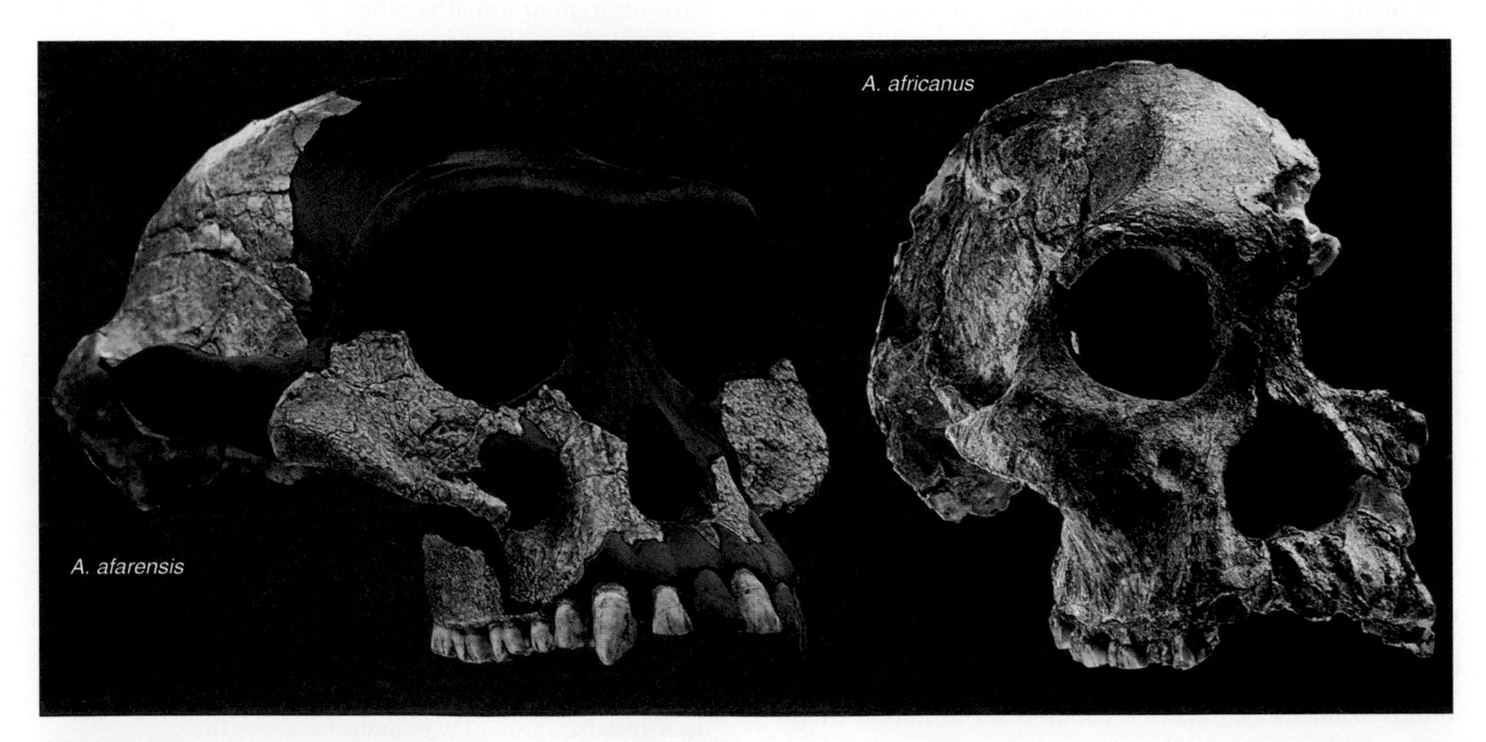

FIGURE 23.7
Nearly human. These four skulls, all photographed from the same angle, are among the best specimens available of the key *Australopithecus* species.

Other Kinds of *Australopithecus*

In 1938, a second, stockier kind of *Australopithecus* was unearthed in South Africa. Called *A. robustus*, it had massive teeth and jaws. In 1959, in East Africa, Mary Leakey discovered a third kind of *Australopithecus*—*A. boisei* (after Charles Boise, an American-born businessman who contributed to the Leakeys' projects)—who was even more stockily built. Like the other australopithecines, *A. boisei* was very old—almost 2 million years. Nicknamed "Nutcracker man," *A. boisei* had a great bony ridge—a Mohawk haircut of bone—on the crest of the head to anchor its immense jaw muscles (figure 23.7).

In 1974, anthropologist Don Johanson went to the remote Afar Desert of Ethiopia in search of early human fossils and hit the jackpot. He found the most complete, best preserved australopithecine skeleton known. Nicknamed "Lucy," the skeleton was 40% complete and over 3 million years old. The skeleton and other similar fossils have been assigned the scientific name *Australopithecus afarensis* (from the Afar Desert). The shape of the pelvis indicated that Lucy was a female, and her leg bones proved she walked upright. Her teeth were distinctly hominid, but her head was shaped like an ape's, and her brain was no larger than that of a chimpanzee, about 400 cubic centimeters, about the size of a large orange. More than 300 specimens of *A. afarensis* have since been discovered.

In the last 10 years, three additional kinds of australopithecines have been reported. These seven species provide ample evidence that australopithecines were a diverse group, and additional species will undoubtedly be described by future investigators. The evolution of hominids seems to have begun with an initial radiation of numerous species.

Early Australopithecines Were Bipedal

We now know australopithecines from hundreds of fossils. The structure of these fossils clearly indicates that australopithecines walked upright. These early hominids weighed about 18 kilograms and were about 1 meter tall. Their dentition was distinctly hominid, but their brains were not any larger than those of apes, generally 500 cc or less. *Homo* brains, by comparison, are usually larger than 600 cc; modern *H. sapiens* brains average 1350 cc. Australopithecine fossils have been found only in Africa. Although all the fossils to date come from sites in South and East Africa (except for one specimen from Chad), it is probable that they lived over a much broader area of Africa. Only in South and East Africa, however, are sediments of the proper age exposed to fossil hunters.

The australopithecines were hominids that walked upright and lived in Africa over 3 million years ago.

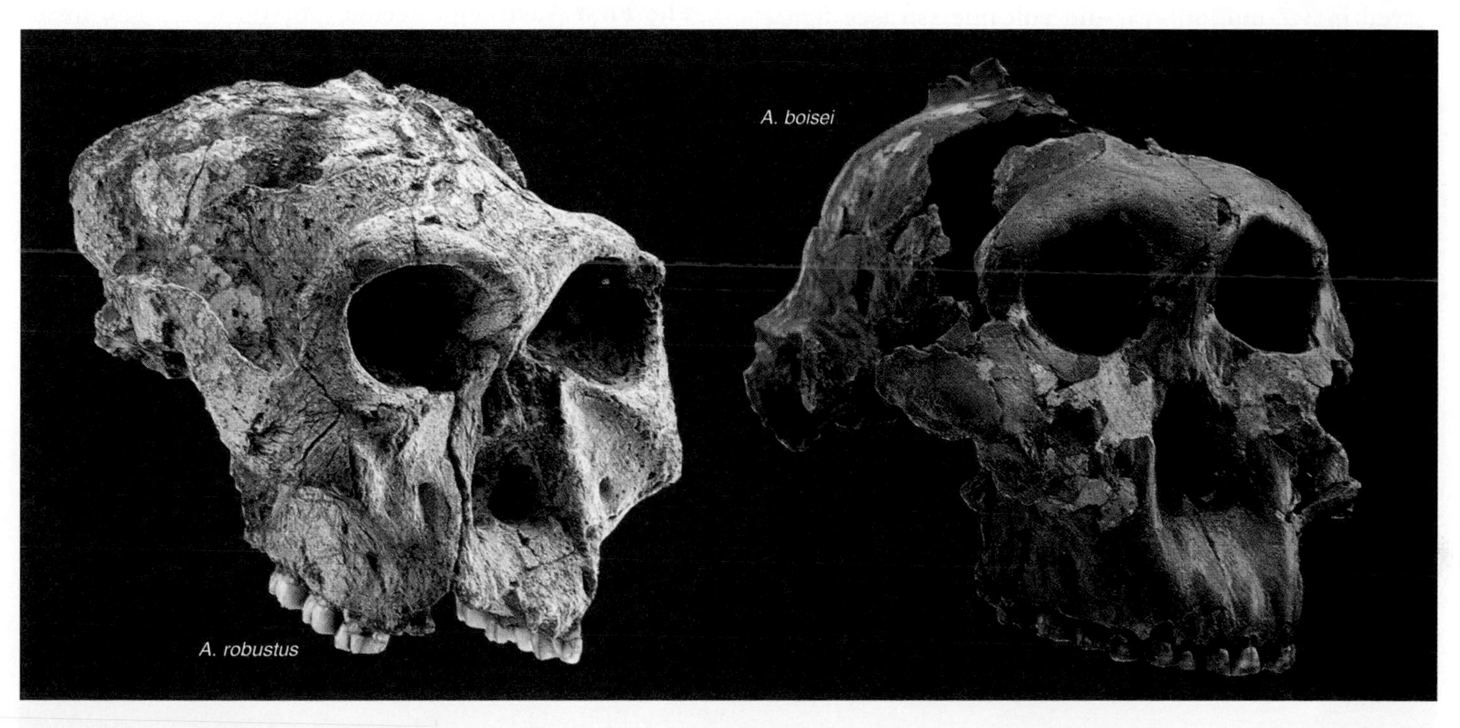

FIGURE 23.7 (continued).

The Beginning of Hominid Evolution

The Origins of Bipedalism

For much of this century, biologists have debated the sequence of events that led to the evolution of hominids. A key element may have been bipedalism. Bipedalism seems to have evolved as our ancestors left dense forests for grasslands and open woodland (figure 23.8). One school of thought proposes that hominid brains enlarged first, and then hominids became bipedal. Another school sees bipedalism as a precursor to bigger brains. Those who favor the brain-first hypothesis speculate that human intelligence was necessary to make the decision to walk upright and move out of the forests and onto the grassland. Those who favor the bipedalism-first hypothesis argue that bipedalism freed the forelimbs to manufacture and use tools, favoring the subsequent evolution of bigger brains.

A treasure trove of fossils unearthed in Africa has settled the debate. These fossils demonstrate that bipedalism extended back 4 million years ago; knee joints, pelvis, and leg bones all exhibit the hallmarks of an upright stance. Substantial brain expansion, on the other hand, did not appear until roughly 2 million years ago. In hominid evolution, upright walking clearly preceded large brains.

Remarkable evidence that early hominids were bipedal is a set of some 69 hominid footprints found at Laetoli, East Africa. Two individuals, one larger than the other, walked upright side-by-side for 27 meters, their footprints preserved in 3.7-million-year-old volcanic ash (see figure 23.1). Importantly, the big toe is not splayed out to the side as in a monkey or ape—the footprints were clearly made by a hominid.

The evolution of bipedalism marks the beginning of the hominids. The reason why bipedalism evolved in hominids remains a matter of controversy. No tools appeared until 2.5 million years ago, so toolmaking seems an unlikely cause. Alternative ideas suggest that walking upright is faster and uses less energy than walking on four legs; that an upright posture permits hominids to pick fruit from trees and see over tall grass; that being upright reduces the body surface exposed to the sun's rays; that an upright stance aided the wading of aquatic hominids, and that bipedalism frees the forelimbs of males to carry food back to females, encouraging pair-bonding. All of these suggestions have their proponents, and none are universally accepted. The origin of bipedalism, the key event in the evolution of hominids, remains a mystery.

The Root of the Hominid Tree

The Oldest Known Hominid. In 1994, a remarkable, nearly complete fossil skeleton was unearthed in Ethiopia. The skeleton is still being painstakingly assembled, but it seems almost certainly to have been bipedal; the foramen magnum, for example, is situated far forward, as in other bipedal hominids. Some 4.4 million years old, it is the most ancient hominid yet discovered. It is significantly more apelike than any australopithecine and so has been assigned to a new genus, *Ardipithecus* from the local Afar language *ardi* for "ground" and the Greek *pithecus* for "ape" (figure 23.9*a*).

The First Australopithecine. In 1995, hominid fossils of nearly the same age, 4.2 million years old, were found in the Rift Valley in Kenya. The fossils are fragmentary, but they include complete upper and lower jaws, a piece of the

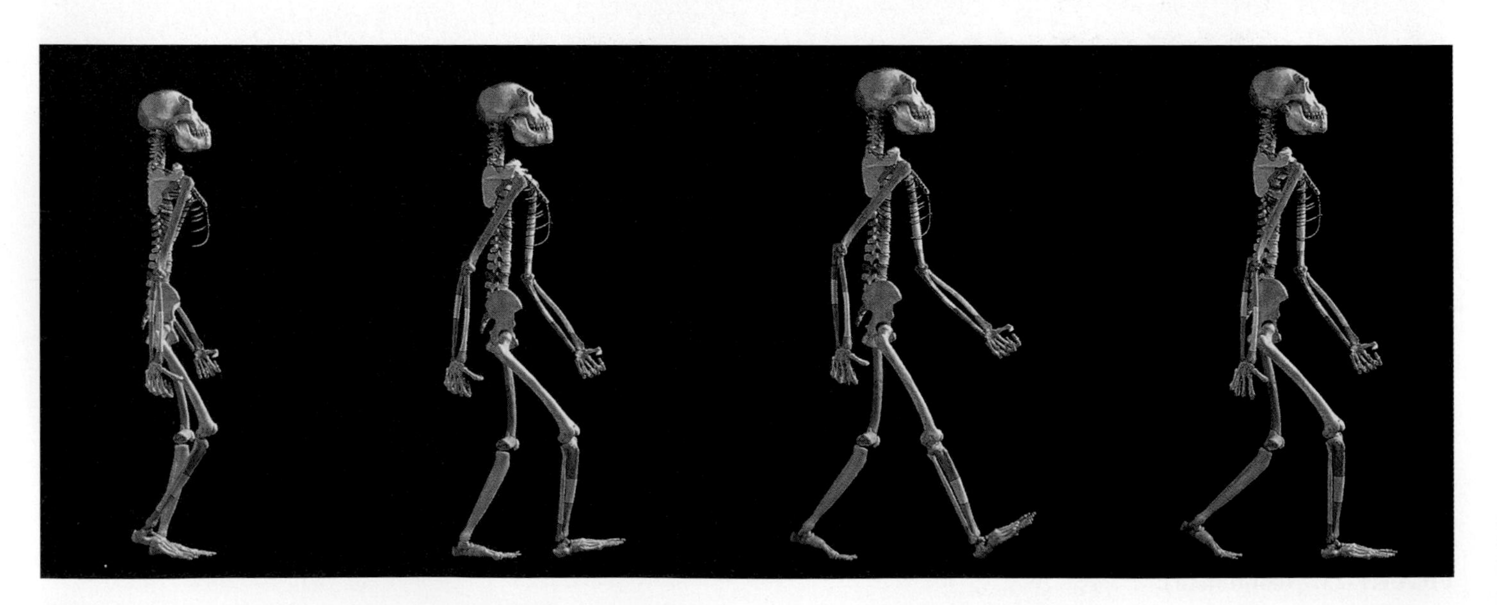

FIGURE 23.8
A reconstruction of an early hominid walking upright. These articulated plaster skeletons, made by Owen Lovejoy and his students at Kent State University, depict an early hominid (*Australopithecus afarensis*) walking upright.

(a)

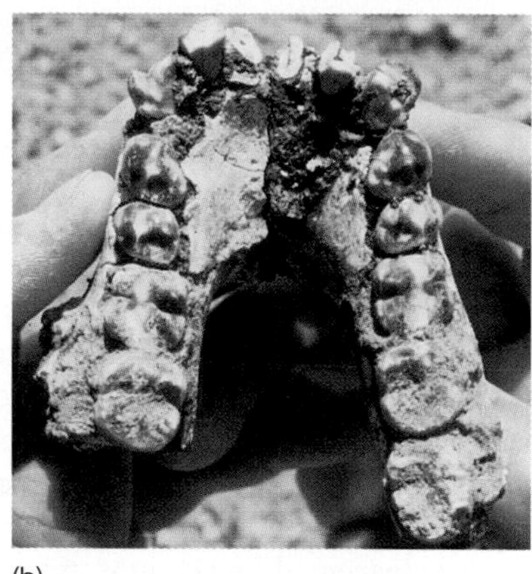

(b)

FIGURE 23.9
Hominid fossils. (*a*) Our earliest known ancestor. A tooth from *Ardipithecus ramidus*, discovered in 1994. The name *ramidus* is from the Latin word for "root," as this is thought to be the root of the hominid family tree. The earliest known hominid, at 4.4 million years old, *A. ramidus* was about the size of a chimpanzee and apparently could walk upright. (*b*) The earliest australopithecine. This fossil jaw of *Australopithecus* is about 4.2 million years old, making *A. anamensis* the oldest known australopithecine.

skull, arm bones, and a partial leg bone. The fossils were assigned to the species *Australopithecus anamensis* (figure 23.9*b*); *anam* is the Turkana word for lake. They were categorized in the genus *Australopithecus* rather than *Ardipithecus* because the fossils have bipedal characteristics and are much less apelike than *A. ramidus*. Although clearly australopithecine, the fossils are intermediate in many ways between apes and *A. afarensis*. Numerous fragmentary specimens of *A. anamensis* have since been found. Most researchers agree that these slightly built *A. anamensis* individuals represent the true base of our family tree, the first members of the genus *Australopithecus*, and thus ancestor to *A. afarensis* and all other australopithecines.

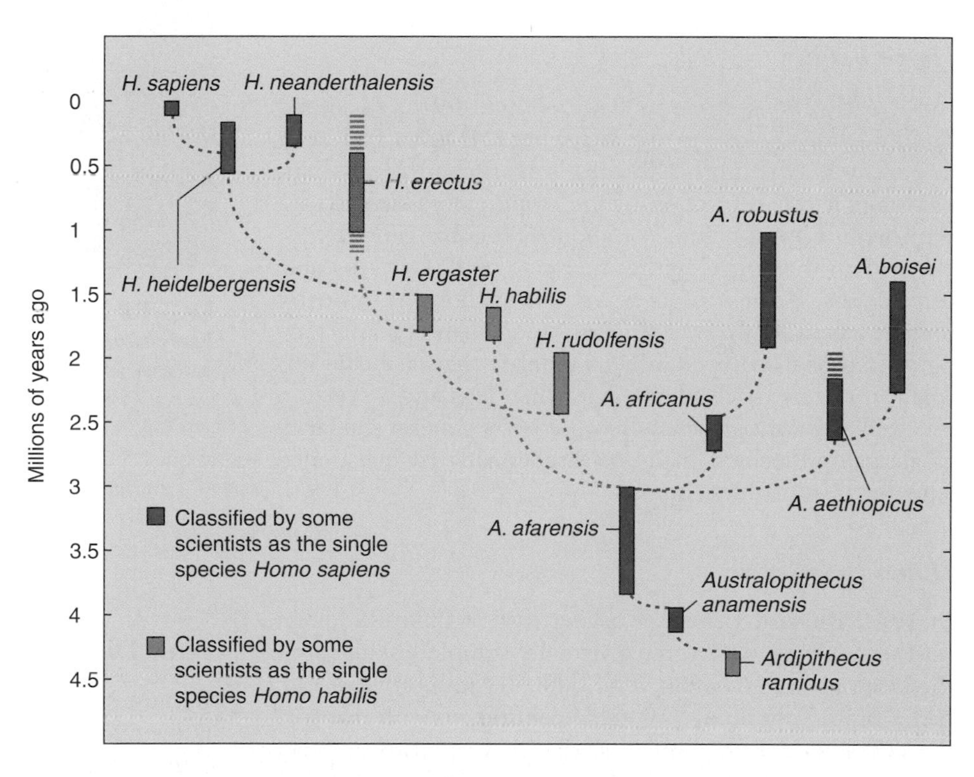

FIGURE 23.10
A hominid evolutionary tree. In this tree, the most widely accepted, the vertical bars show the known dates of first and last appearances of proposed species; bars are broken where dates are uncertain. Six species of *Australopithecus* and seven of *Homo* are included.

Differing Views of the Hominid Family Tree

Investigators take two different philosophical approaches to characterizing the diverse group of African hominid fossils. One group focuses on common elements in different fossils, and tends to lump together fossils that share key characters. Differences between the fossils are attributed to diversity within the group. Other investigators focus more pointedly on the differences between hominid fossils. They are more inclined to assign fossils that exhibit differences to different species. The hominid phylogenetic tree in figure 23.10 presents such a view. Where the "lumpers" tree presents three species of *Homo*, for example, the "splitters" tree presents no fewer than seven! At this point, it is not possible to decide which view is correct; more fossils are needed to determine how much of the differences between fossils represents within-species variation and how much characterizes between-species differences.

The evolution of bipedalism—walking upright—marks the beginning of hominid evolution, although no one is quite sure why bipedalism evolved. The root of the hominid evolutionary tree is only imperfectly known. The earliest australopithecine yet described is *A. anamensis*, over 4 million years old.

23.3 The genus *Homo* evolved in Africa.

African Origin: Early *Homo*

The first humans evolved from australopithecine ancestors about 2 million years ago. The exact ancestor has not been clearly defined, but is commonly thought to be *A. afarensis*. Only within the last 30 years have a significant number of fossils of early *Homo* been uncovered. An explosion of interest has fueled intensive field exploration in the last few years, and new finds are announced regularly; every year, our picture of the base of the human evolutionary tree grows clearer. The account given here will undoubtedly be outdated by future discoveries, but it provides a good example of science at work.

Homo habilis

In the early 1960s, stone tools were found scattered among hominid bones close to the site where *A. boisei* had been unearthed. Although the fossils were badly crushed, painstaking reconstruction of the many pieces suggested a skull with a brain volume of about 680 cubic centimeters, larger than the australopithecine range of 400 to 550 cubic centimeters. Because of its association with tools, this early human was called *Homo habilis*, meaning "handy man." Partial skeletons discovered in 1986 indicate that *H. habilis* was small in stature, with arms longer than legs and a skeleton much like *Australopithecus*. Because of its general similarity to australopithecines, many researchers at first questioned whether this fossil was human.

Homo rudolfensis

In 1972, Richard Leakey, working east of Lake Rudolf in northern Kenya, discovered a virtually complete skull about the same age as *H. habilis*. The skull, 1.9 million years old, had a brain volume of 750 cubic centimeters and many of the characteristics of human skulls—it was clearly human and not australopithecine. Some anthropologists assign this skull to *H. habilis*, arguing it is a large male. Other anthropologists assign it to a separate species, *H. rudolfensis*, because of its substantial brain expansion.

Homo ergaster

Some of the early *Homo* fossils being discovered do not easily fit into either of these species (figure 23.11). They tend to have even larger brains than *H. rudolfensis*, with skeletons less like an australopithecine and more like a modern human in both size and proportion. Interestingly, they also have small cheek teeth, as modern humans do. Some anthropologists have placed these specimens in a third species of early *Homo*, *H. ergaster* (*ergaster* is from the Greek for "workman").

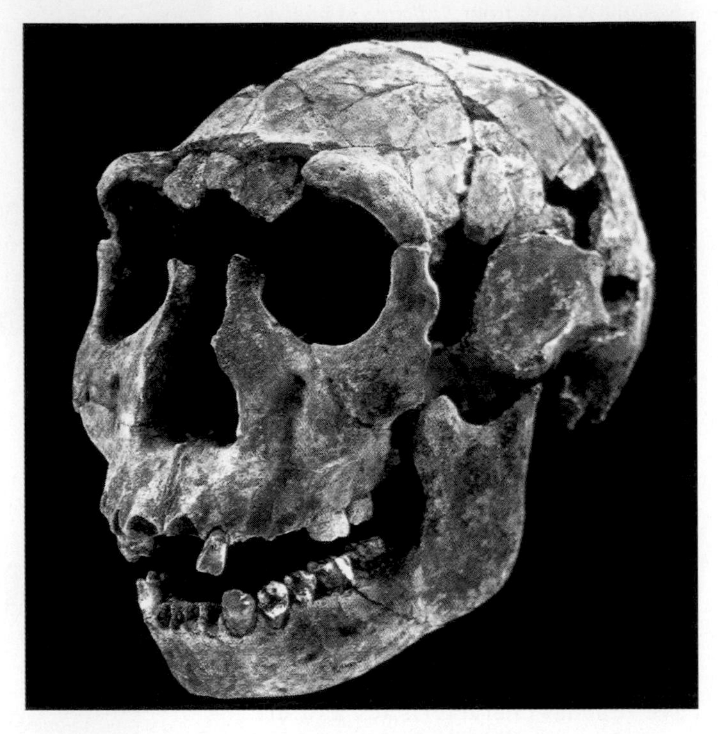

FIGURE 23.11
Early *Homo*. This skull of a boy, who apparently died in early adolescence, is 1.6 million years old and has been assigned to the species *Homo ergaster* (a form of *Homo habilis* recognized by some as a separate species). Much larger than earlier hominids, he was about 1.5 meters in height and weighed approximately 47 kilograms.

How Diverse Was Early *Homo*?

Because so few fossils have been found of early *Homo*, there is lively debate about whether they should all be lumped into *H. habilis* or split into the three species *H. rudolfensis*, *H. habilis*, and *H. ergaster*. If the three species designations are accepted, as increasing numbers of researchers are doing, then it would appear that *Homo* underwent an adaptive radiation (as described in chapter 22) with *H. rudolfensis* the most ancient species, followed by *H. habilis* and then *H. ergaster*. Because of its modern skeleton, *H. ergaster* is thought the most likely ancestor to later species of *Homo* (see figure 23.10).

Early species of *Homo*, the oldest members of our genus, had a distinctly larger brain than australopithecines and most likely used tools. There may have been several different species.

Out of Africa: *Homo erectus*

Our picture of what early *Homo* was like lacks detail, because it is based on only a few specimens. We have much more information about the species that replaced it, *Homo erectus*.

Java Man

After the publication of Darwin's book *On the Origin of Species* in 1859, there was much public discussion about "the missing link," the fossil ancestor common to both humans and apes. Puzzling over the question, a Dutch doctor and anatomist named Eugene Dubois decided to seek fossil evidence of the missing link in the home country of the orangutan, Java. Dubois set up practice in a river village in eastern Java. Digging into a hill that villagers claimed had "dragon bones," he unearthed a skull cap and a thighbone in 1891. He was very excited by his find, informally called Java man, for three reasons:

1. The structure of the thigh bone clearly indicated that the individual had long, straight legs and was an excellent walker.
2. The size of the skull cap suggested a very large brain, about 1000 cubic centimeters.
3. Most surprisingly, the bones seemed as much as 500,000 years old, judged by other fossils Dubois unearthed with them.

The fossil hominid that Dubois had found was far older than any discovered up to that time, and few scientists were willing to accept that it was an ancient species of human.

Peking Man

Another generation passed before scientists were forced to admit that Dubois had been right all along. In the 1920s a skull was discovered near Peking (now Beijing), China, that closely resembled Java man. Continued excavation at the site eventually revealed 14 skulls, many excellently preserved. Crude tools were also found, and most important of all, the ashes of campfires. Casts of these fossils were distributed for study to laboratories around the world. The originals were loaded onto a truck and evacuated from Peking at the beginning of World War II, only to disappear into the confusion of history. No one knows what happened to the truck or its priceless cargo. Fortunately, Chinese scientists have excavated numerous additional skulls of Peking man since 1949.

A Very Successful Species

Java man and Peking man are now recognized as belonging to the same species, *Homo erectus*. *Homo erectus* was a lot larger than *Homo habilis*—about 1.5 meters tall. It had a large brain, about 1000 cubic centimeters (figure 23.12), and walked erect. Its skull had prominent brow ridges and, like modern humans, a rounded jaw. Most interesting of all, the shape of the skull interior suggests that *H. erectus* was able to talk.

Where did *H. erectus* come from? It should come as no surprise to you that it came out of Africa. In 1976 a complete *H. erectus* skull was discovered in East Africa. It was 1.5 million years old, a million years older than the Java and Peking finds. Far more successful than *H. habilis*, *H. erectus* quickly became widespread and abundant in Africa, and within 1 million years had migrated into Asia and Europe. A social species, *H. erectus* lived in tribes of 20 to 50 people, often dwelling in caves. They successfully hunted large animals, butchered them using flint and bone tools, and cooked them over fires—the site in China contains the remains of horses, bears, elephants, deer, and rhinoceroses.

Homo erectus survived for over a million years, longer than any other species of human. These very adaptable humans only disappeared in Africa about 500,000 years ago, as modern humans were emerging. Interestingly, they survived even longer in Asia, until about 250,000 years ago.

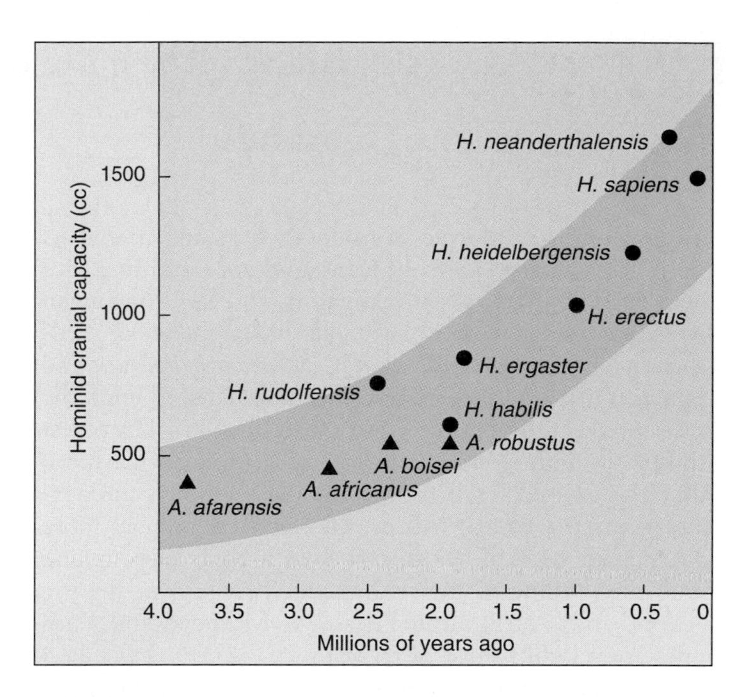

FIGURE 23.12
Brain size increased as hominids evolved. *Homo erectus* had a larger brain than early *Homo*, which in turn had larger brains than those of the australopithecines with which they shared East African grasslands. Maximum brain size (and apparently body size) was attained by *H. neanderthalensis*. Both brain and body size appear to have declined some 10% in recent millennia.

Homo erectus **evolved in Africa, and migrated from there to Europe and Asia.**

23.4 Modern humans evolved quite recently.

The Last Stage of Hominid Evolution

The evolutionary journey to modern humans entered its final phase when modern humans first appeared in Africa about 600,000 years ago. Investigators who focus on human diversity consider there to have been three species of modern humans: *Homo heidelbergensis*, *H. neanderthalensis*, and *H. sapiens* (see figure 23.10). Other investigators lump the three species into one, *H. sapiens* ("wise man"). The oldest modern human, *Homo heidelbergensis*, is known from a 600,000-year-old fossil from Ethiopia. Although it coexisted with *H. erectus* in Africa, *H. heidelbergensis* has more advanced anatomical features, such as a bony keel running along the midline of the skull, a thick ridge over the eye sockets, and a large brain. Also, its forehead and nasal bones are very like those of *H. sapiens*.

As *H. erectus* was becoming rarer, about 130,000 years ago, a new species of human arrived in Europe from Africa. *Homo neanderthalensis* likely branched off from the ancestral line leading to modern humans as long as 500,000 years ago. Compared with modern humans, Neanderthals were short, stocky, and powerfully built. Their skulls were massive, with protruding faces, heavy, bony ridges over the brows (figure 23.13), and larger braincases.

Out of Africa—Again?

The oldest fossil known of *Homo sapiens*, our own species, is from Ethiopia and is about 130,000 years old. Other fossils from Israel appear to be between 100,000 and 120,000 years old. Outside of Africa and the Middle East, there are no clearly dated *H. sapiens* fossils older than roughly 40,000 years of age. The implication is that *H. sapiens* evolved in Africa, then migrated to Europe and Asia, the Recently-Out-of-Africa model. An opposing view, the Multiregional hypothesis, argues that the human races independently evolved from *H. erectus* in different parts of the world.

Recently, scientists studying human mitochondrial DNA have added fuel to the fire of this controversy. Because DNA accumulates mutations over time, the oldest populations should show the greatest genetic diversity. Researchers sequencing the entire mDNA from 53 individuals of differing ethnic backgrounds found that modern humans shared a common ancestor 170,000 years ago, confirming that *H. sapiens* originated in Africa at about this time. The data also reveal a distinct branch on the human family tree 52,000 years ago, separating Africans from non-Africans. This is consistent with the hypothesis that humans originated in Africa, from there spreading to all parts of the world, retracing the path taken by *H. erectus* half a million years before (figure 23.14).

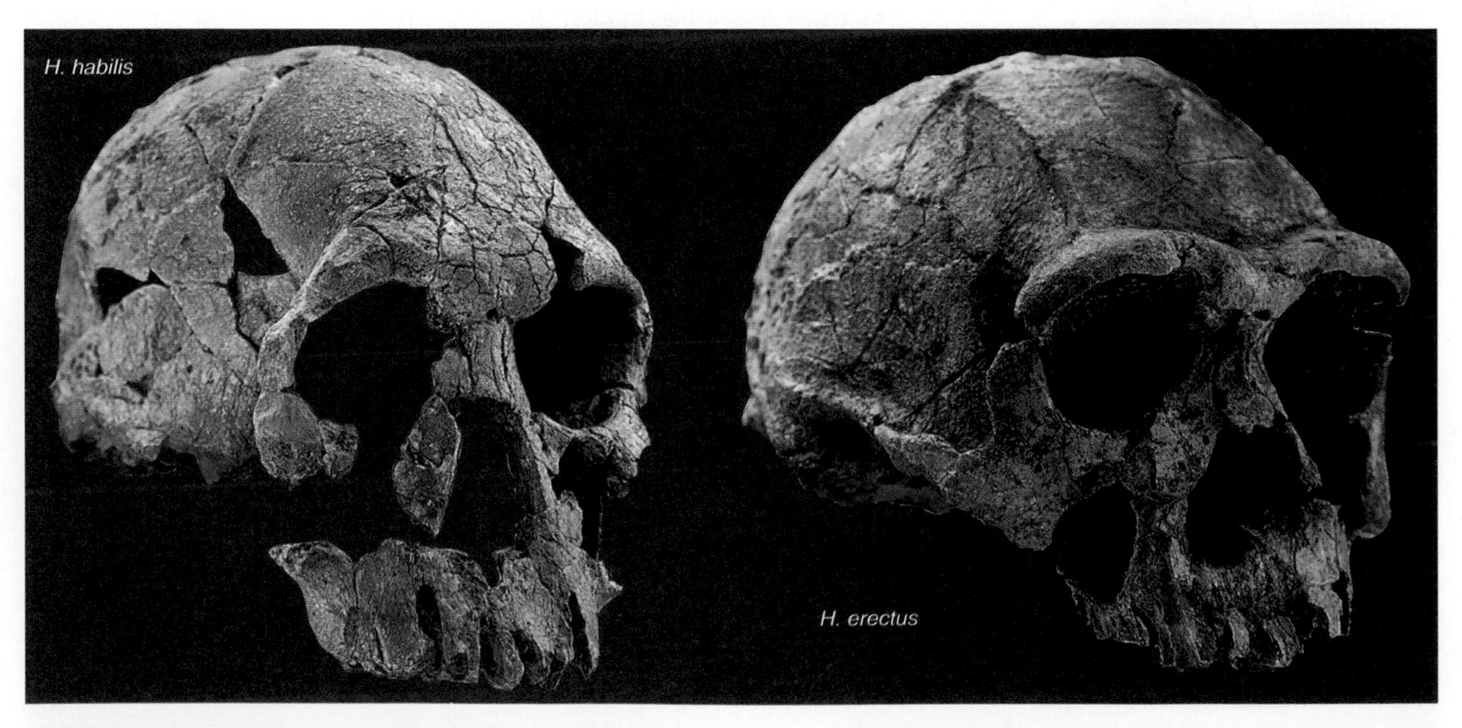

FIGURE 23.13
Our own genus. These four skulls illustrate the changes that have occurred during the evolution of the genus *Homo*. The *Homo sapiens* is essentially the same as human skulls today. The skulls were photographed from the same angle.

Another way to examine the human family tree is to look at genes on the Y chromosome, which does not undergo recombination. Y chromosomes pass down unchanged in males from one generation to the next. Any new changes that arise during evolution are easy to track on the family tree, as they too are passed down unchanged. The researchers in this large study looked at the pattern of gene variation among more than a thousand European males. While they identified many different patterns of variation, fully 80% of European males shared a single pattern, suggesting modern Europeans have a common ancestor. The data indicate the pattern arose some 40,000 to 50,000 years ago. In other words, our species came to Europe recently.

By both these sets of evidence, the multiregional hypothesis is wrong. Our family tree has a single stem.

Homo sapiens*, our species, seems to have evolved in Africa and then, like *H. erectus* before it, migrated to Europe and Asia.

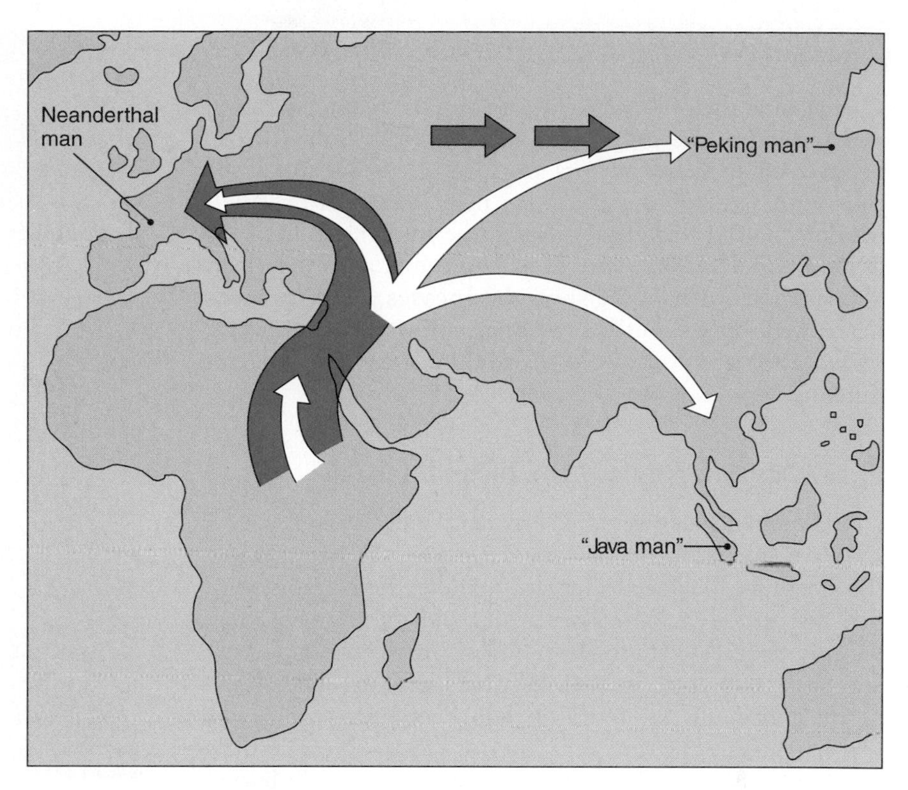

FIGURE 23.14
Out of Africa—many times. A still-controversial theory suggests that *Homo* spread from Africa to Europe and Asia repeatedly. First, *Homo erectus* (*white* arrow) spread as far as Java and China. Later, *H. erectus* was followed and replaced by *Homo neanderthalensis*, a pattern repeated again still later *by Homo sapiens* (*red* arrow).

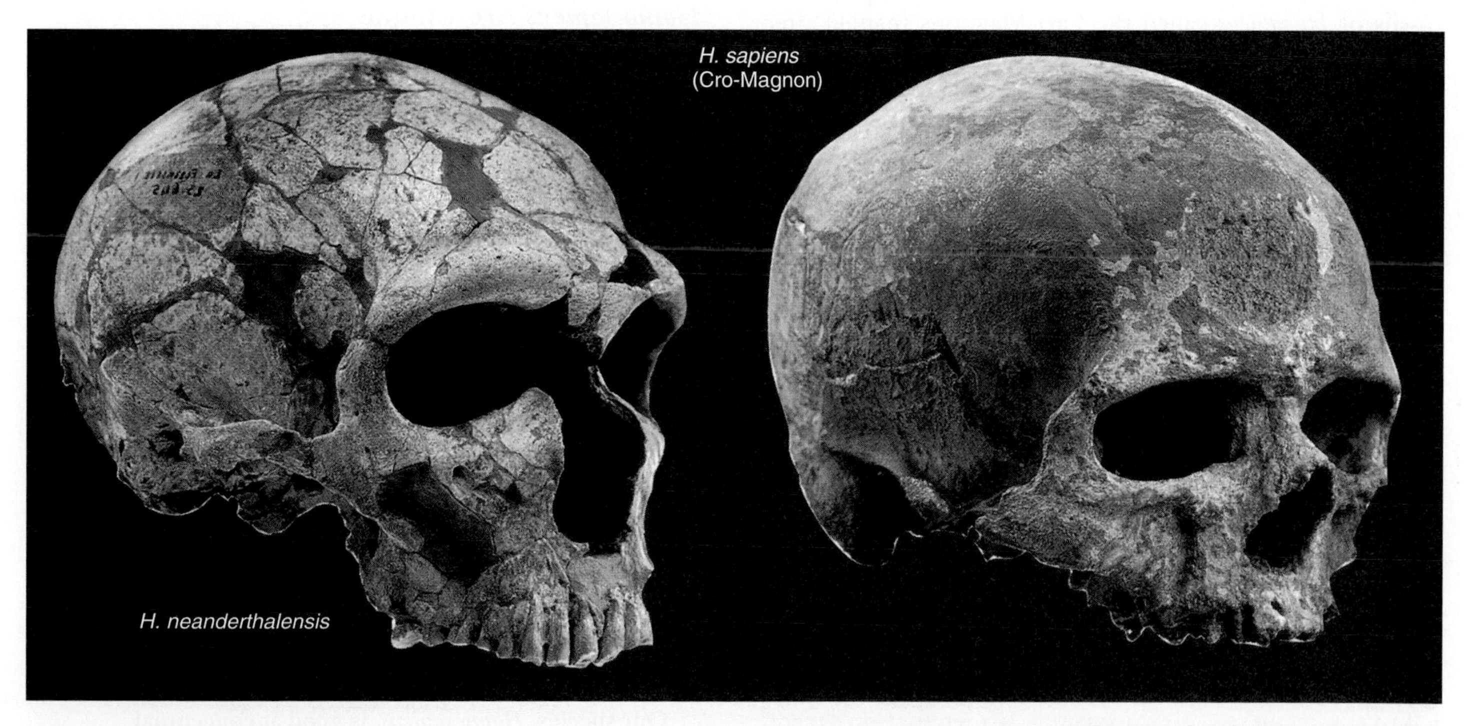

FIGURE 23.13 (continued)

Our Own Species: *Homo sapiens*

H. sapiens is the only surviving species of the genus *Homo*, and indeed is the only surviving hominid. Some of the best fossils of *Homo sapiens* are 20 well-preserved skeletons with skulls found in a cave near Nazareth in Israel. Modern dating techniques date these humans to between 90,000 and 100,000 years old. The skulls are modern in appearance, with high, short braincases, vertical foreheads with only slight brow ridges, and a cranial capacity of roughly 1550 cc, well within the range of modern humans.

Cro-Magnons Replace the Neanderthals

The Neanderthals (classified by many paleontologists as a separate species *Homo neanderthalensis*) were named after the Neander Valley of Germany where their fossils were first discovered in 1856. Rare at first outside of Africa, they became progressively more abundant in Europe and Asia, and by 70,000 years ago had become common. The Neanderthals made diverse tools, including scrapers, spearheads, and handaxes. They lived in huts or caves. Neanderthals took care of their injured and sick and commonly buried their dead, often placing food, weapons, and even flowers with the bodies. Such attention to the dead strongly suggests that they believed in a life after death. This is the first evidence of the symbolic thinking characteristic of modern humans.

Fossils of *H. neanderthalensis* abruptly disappear from the fossil record about 34,000 years ago and are replaced by fossils of *H. sapiens* called the Cro-Magnons (named after the valley in France where their fossils were first discovered). We can only speculate why this sudden replacement occurred, but it was complete all over Europe in a short period. A variety of evidence indicates that Cro-Magnons came from Africa—fossils of essentially modern aspect but as much as 100,000 years old have been found there. Cro-Magnons seem to have replaced the Neanderthals completely in the Middle East by 40,000 years ago, and then spread across Europe, coexisting with the Neanderthals for several thousand years. Recent analyses of Neanderthal DNA reveals it to be quite distinct from Cro-Magnon DNA, indicating the two species did not interbreed. Neanderthals are our cousins, not our ancestors. The Cro-Magnons that replaced the Neanderthals had a complex social organization and are thought to have had full language capabilities. Elaborate and often beautiful cave paintings made by Cro-Magnons can be seen throughout Europe (figure 23.15).

Humans of modern appearance eventually spread across Siberia to North America, which they reached at least 13,000 years ago, after the ice had begun to retreat and a land bridge still connected Siberia and Alaska. By 10,000 years ago, about 5 million people inhabited the entire world (compared with more than 6 billion today).

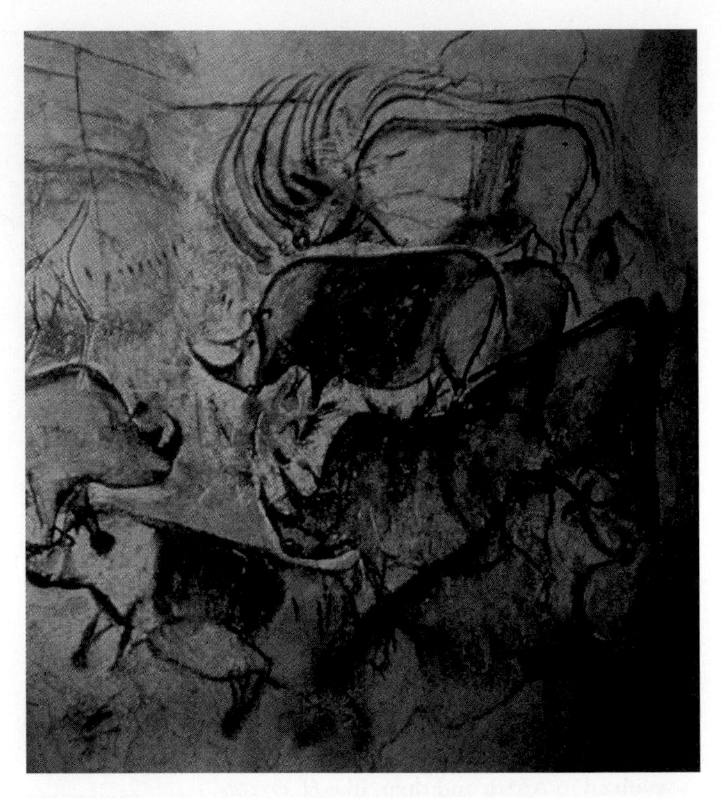

FIGURE 23.15
Cro-Magnon art. Rhinoceroses are among the animals depicted in this remarkable cave painting found in 1995 near Vallon-Pont d'Arc, France.

Homo sapiens Are Unique

We humans are animals and the product of evolution. Our evolution has been marked by a progressive increase in brain size, distinguishing us from other animals in several ways. First, humans are able to make and use tools effectively—a capability that, more than any other factor, has been responsible for our dominant position in the animal kingdom. Second, although not the only animal capable of conceptual thought, we have refined and extended this ability until it has become the hallmark of our species. Lastly, we use symbolic language and can with words shape concepts out of experience. Our language capability has allowed the accumulation of experience, which can be transmitted from one generation to another. Thus, we have what no other animal has ever had: extensive cultural evolution. Through culture, we have found ways to change and mold our environment, rather than changing evolutionarily in response to the demands of the environment. We control our biological future in a way never before possible—an exciting potential and frightening responsibility.

Our species, *Homo sapiens*, is good at conceptual thought and tool use, and is the only animal that uses symbolic language.

Human Races

Human beings, like all other species, have differentiated in their characteristics as they have spread throughout the world. Local populations in one area often appear significantly different from those that live elsewhere. For example, northern Europeans often have blond hair, fair skin, and blue eyes, whereas Africans often have black hair, dark skin, and brown eyes. These traits may play a role in adapting the particular populations to their environments. Blood groups may be associated with immunity to diseases more common in certain geographical areas, and dark skin shields the body from the damaging effects of ultraviolet radiation, which is much stronger in the tropics than in temperate regions.

All human beings are capable of mating with one another and producing fertile offspring. The reasons that they do or do not choose to associate with one another are purely psychological and behavioral (cultural). The number of groups into which the human species might logically be divided has long been a point of contention. Some contemporary anthropologists divide people into as many as 30 "races," others as few as three: Caucasoid, Negroid, and Oriental. American Indians, Bushmen, and Aborigines are examples of particularly distinctive subunits that are sometimes regarded as distinct groups.

The problem with classifying people or other organisms into races in this fashion is that the characteristics used to define the races are usually not well correlated with one another, and so the determination of race is always somewhat arbitrary. Humans are visually oriented; consequently, we have relied on visual cues—primarily skin color—to define races. However, when other types of characters, such as blood groups, are examined, patterns of variation correspond very poorly with visually determined racial classes. Indeed, if one were to break the human species into subunits based on overall genetic similarity, the groupings would be very different than those based on skin color or other visual features (figure 23.16).

In human beings, it is simply not possible to delimit clearly defined races that reflect biologically differentiated and well-defined groupings. The reason is simple: different groups of people have constantly intermingled and interbred with one another during the entire course of history. This constant gene flow has prevented the human species from fragmenting in highly differentiated subspecies. Those characteristics that are differentiated among populations, such as skin color, represent classic examples of the antagonism between gene flow and natural selection. As we saw in chapter 20, when selection is strong enough, as it is for dark coloration in tropical regions, populations can differentiate even in the presence of gene flow. However, even in cases such as this, gene flow will still ensure that populations are relatively homogeneous for genetic variation at other loci.

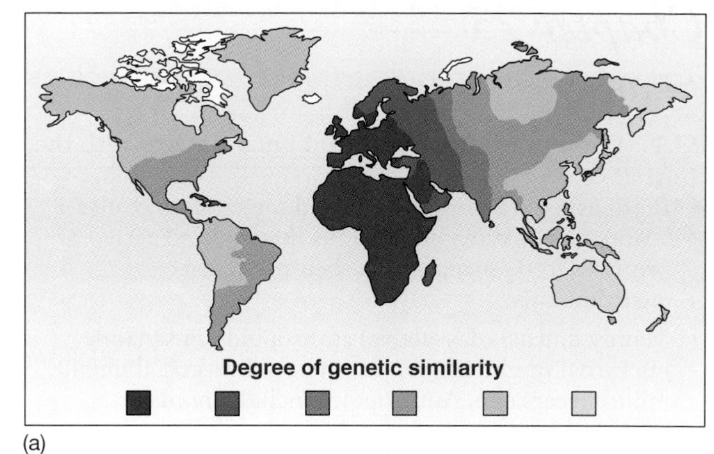

(a)

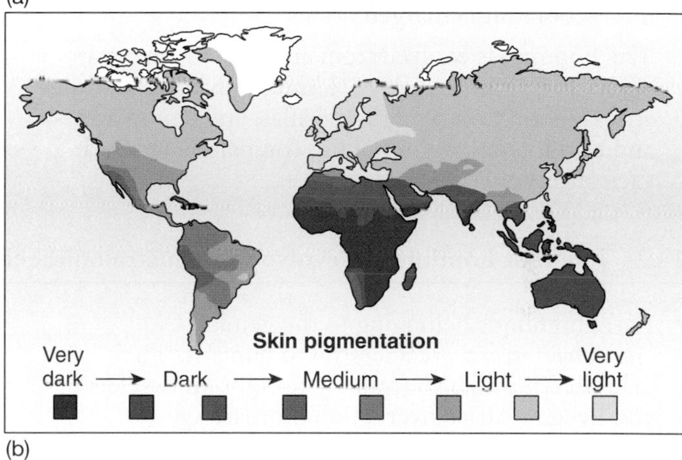

(b)

FIGURE 23.16
Patterns of genetic variation in human populations differ from patterns of skin color variation. (*a*) Genetic variation among *Homo sapiens*. Categories of humans were recognized based on overall similarity at many enzyme and blood group genetic loci. The code below the figure is arranged by degree of similarity. (*b*) Similarity among *Homo sapiens* based on skin color. The categories are arranged by amount of pigmentation in the skin.
Source: Data from Cavalli-Sforza, et al., *The History & Geography of Human Genes*, 1996, Princeton University Press.

For this reason, relatively little of the variation in the human species represents differences between the described races. Indeed, one study calculated that only 8% of all genetic variation among humans could be accounted for as differences that exist among racial groups; in other words, the human racial categories do a very poor job in describing the vast majority of genetic variation that exists in humans. For this reason, most modern biologists reject human racial classifications as reflecting patterns of biological differentiation in the human species. This is a sound biological basis for dealing with each human being on his or her own merits and not as a member of a particular "race."

Human races do not reflect significant patterns of underlying biological differentiation.

Chapter 23

www.mhhe.com/raven6/resources23.mhtml

Summary | **Questions** | **Media Resources**

23.1 The evolutionary path to humans starts with the advent of primates.

- Prehensile (grasping) fingers and toes and binocular vision were distinct adaptations that allowed early primates to be successful in their particular environments.
- Mainly diurnal (day-active) anthropoids and mainly nocturnal (night-active) prosimians diverged about 40 million years ago. Anthropoids include monkeys, apes, and humans, and all exhibit complex social interactions and enlarged brains.
- The hominoids evolved from anthropoid ancestors about 25 million years ago. Hominoids consist of the apes (gibbons, orangutans, gorillas, and chimpanzees) and upright-walking hominids (human beings and their direct ancestors).

1. Which characteristics were selected for in the earliest primates to allow them to become successful in their environment?

2. How do monkeys differ from prosimians?

3. How are apes distinguished from monkeys?

4. What is the best explanation for why humans and chimpanzees are so similar genetically?

- Evolution of Primates

- Art Quizzes:
  - -Primate Evolutionary Tree
  - -Ape and Hominid Skeletons

23.2 The first hominids to evolve were australopithecines.

- Early hominids belonging to the genus *Australopithecus* were ancestral to humans. They exhibited bipedalism (walking upright on two feet) and lived in Africa over 4 million years ago.

5. When did the first hominids appear? What were they called? What distinguished them from the apes?

23.3 The genus *Homo* evolved in Africa.

- Hominids with an enlarged brain and the ability to use tools belong to the genus *Homo.* Species of early *Homo* appeared in Africa about 2 million years ago and became extinct about 1.5 million years ago.
- *Homo erectus* appeared in Africa at least 1.5 million years ago and had a much larger brain than early species of *Homo. Homo erectus* also walked erect and presumably was able to talk. Within a million years, *Homo erectus* migrated from Africa to Europe and Asia.

6. Why is there some doubt in the scientific community that *Homo habilis* was a true human?

7. How did *Homo erectus* differ from *Homo habilis?*

- *On Science* Articles:
  - -Human Evolution
  - -Humans Evolved Recently

23.4 Modern humans evolved quite recently.

- The modern species of *Homo* appeared about 600,000 years ago in Africa and then migrated from there to Europe and Asia.
- The Neanderthals appeared in Europe about 130,000 years ago. They made diverse tools and showed evidence of symbolic thinking.
- Studies of mitochondrial DNA suggest (but do not yet prove) that all of today's human races originated from Africa.
- Categorization of humans into races does not adequately reflect patterns of genetic differentiation among people in different parts of the world.

8. The greatest number of different mitochondrial DNA sequences in humans occurs in Africa. What does this tell us about human evolution?

9. How did Cro-Magnons differ from Neanderthals? Is there any evidence that they coexisted with Neanderthals? If so, where and when?

10. Are the commonly recognized human races equivalent to subspecies of other plant and animal species?

- Hominid History

Part **VII**

Ecology and Behavior

This Kentucky warbler is tending her nest of eggs. A similar species in the tropics would lay fewer eggs. Why?

Why Do Tropical Songbirds Lay Fewer Eggs?

Sometimes odd generalizations in science lead to unexpected places. Take, for example, a long obscure monograph published in 1944 by British ornithologist (bird expert) Reginald Moreau in the journal *Ibis* on bird eggs. Moreau had worked in Africa for many years before moving to a professorship in England in the early 1940s. He was not in England long before noting that the British songbirds seemed to lay more eggs than he was accustomed to seeing in nests in Africa. He set out to gather information on songbird clutch size (that is, the number of eggs in a nest) all over the world.

Wading through a mountain of data (his *Ibis* paper is 51 pages long!), Moreau came to one of these odd generalizations: songbirds in the tropics lay fewer eggs than their counterparts at higher latitudes (see above *right*). Tropical songbirds typically lay a clutch of 2 or 3 eggs, on average, while songbirds in temperate and subarctic regions generally lay clutches of 4 to 6 eggs, and some species as many as 10. The trend is general, affecting all groups of songbirds in all regions of the world.

What is a biologist to make of such a generalization? At first glance, we would expect natural selection to maximize evolutionary fitness—that is, songbirds the world over should have evolved to produce as many eggs as possible. Clearly, the birds living in the tropics have not read Darwin, as they are producing only half as many eggs as they are capable of doing.

A way out of this quandary was proposed by ornithologist Alexander Skutch in 1949. He argued that birds produced just enough offspring to offset deaths in the population. Any extra offspring would be wasteful of individuals, and so minimized by natural selection. An interesting idea, but it didn't hold water. Bird populations are not smaller in the tropics, or related to the size of the populations there.

A second idea, put forward a few years earlier in 1947 by a colleague of Moreau's, David Lack, was more promising. Lack, one of the twentieth-century's great biologists, argued that few if any birds ever produce as many eggs as they might under ideal conditions, for the simple reason that conditions in nature are rarely ideal. Natural selection will indeed tend to maximize reproductive rate (that is, the number of eggs laid in clutches) as Darwin predicted, but only to the greatest level possible within the limits of resources. There is nothing here that would have surprised Darwin. Birds lay fewer eggs in the tropics simply because parents can gather fewer resources to provide their young there—competition is just too fierce, resources too scanty.

Lack went on to construct a general theory of clutch size in birds. He started with the sensible assumption that in a resource-limited environment birds can supply only so much food to their young. Thus, the more offspring they have, the less they can feed each nestling. As a result, Lack proposed that natural selection will favor a compromise between offspring number and investment in each offspring, which maximizes the number of offspring that are fed enough to survive to maturity.

The driving force behind Lack's theory of optimal clutch size is his idea that broods with too many offspring would be undernourished, reducing the probability that the chicks would survive. In Lack's own words:

"The average clutch-size is ultimately determined by the average maximum number of young which the parents can successfully raise in the region and at the season in question, i.e. ... natural selection eliminates a disproportionately large number of young in those clutches which are higher than the average, through the inability of the parents to get enough food for their young, so that some or all of the brood die before or soon after fledging (leaving the nest), with the result that few or no descendants are left with their parent's propensity to lay a larger clutch."

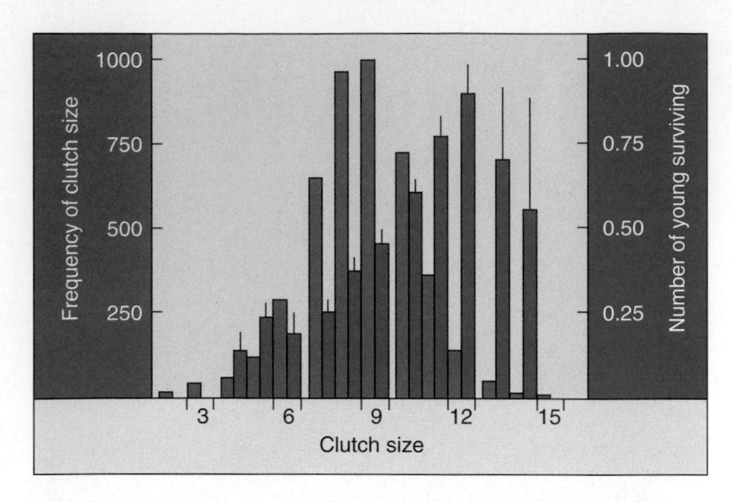

Testing Lack's theory of optimum clutch size. In this study from woods near Oxford, England, researchers found that the most common clutch size was 8, even though clutches of 12 produced the greatest number of surviving offspring. (After Boyce and Perrins, 1987.)

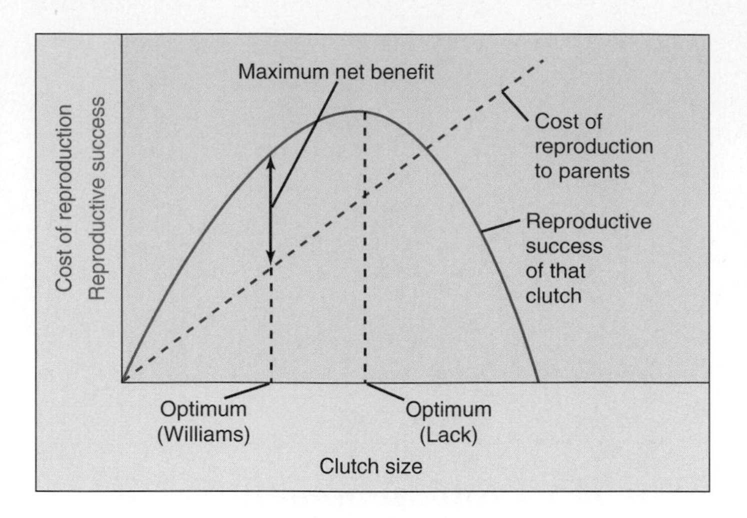

Two theories of optimum clutch size. David Lack's theory predicts that optimum clutch size will be where reproductive success of the clutch is greatest. George Williams' theory predicts that optimum clutch size will be where the *net* benefit is greatest—that is, where the difference between the cost of reproduction and the reproductive success of the clutch is greatest.

The Experiment

Lack's theory is attractive because of its simplicity and common sense—but is it right? Many studies have been conducted to examine this hypothesis. Typically, experimenters would remove eggs from nests, and look to see if this improved the survivorship of the remaining offspring. If Lack is right, then it should, as the remaining offspring will have access to a larger share of what the parents can provide. Usually, however, removal of eggs did not seem to make any difference. Parents just adjusted down the amount of food they provided. The situation was clearly more complicated than Lack's simple theory envisioned.

One can always argue with tests such as these, however, as they involve direct interference with the nests, potentially having a major influence on how the birds behave. It is hard to believe that a bird caring for a nest of six eggs would not notice when one turned up missing. A clear test of Lack's theory would require avoiding all intervention.

Just such a test was completed in 1987 in the woods near Oxford, England. Over many years, Oxford University researchers including Professor Mark Boyce (now at the University of Alberta, Edmonton) carefully monitored nests of a songbird, the great tit, very common in the English countryside. They counted the number of eggs laid in each nest (the clutch size) and then watched to see how many of the offspring survived after flying away from the nest. Over 22 years, they patiently examined 4489 nests and in 603 of these nests the brood size was manipulated. Some nests received additional young whereas young were removed from other nests.

The Results

The Oxford researchers found that the average clutch size was 8 eggs, but that nests with the greatest number of surviving offspring had not 8 but 12 eggs in them! Clearly, Lack's theory is wrong. These birds are not producing as many offspring as natural selection to maximize fitness (that is, number of surviving offspring) would predict (see above *left*).

Lack's proposal had seemed eminently sensible. What was wrong? In 1966, the evolutionary theorist George Williams suggested the problem was that Lack's theory ignores the cost of reproduction (see above *right*). If a bird spends too much energy feeding one brood, then it may not survive to raise another. Looking after a large clutch may extract too high a price in terms of future reproductive success of the parent. The clutch size actually favored by natural selection is adjusted for the wear-and-tear on the parents, so that it is almost always smaller than the number that would produce the most offspring in that nest—just what the Oxford researchers observed.

However, even Williams's "cost-of-reproduction" is not enough to completely explain Boyce's great tit data. There were marked fluctuations in the weather over the years that the Oxford researchers gathered their data, and they observed that harsh years decreased survival of the young in large nests more than in small ones. This "bad-year" effect reduces the fitness of individuals laying larger clutches, and Boyce argues that it probably contributes at least as much as cost-of-reproduction in making it more advantageous, in the long term, for birds to lay clutches smaller than the Lack optimum.

To explore this experiment further, go to the Virtual Lab at www.mhhe.com/raven6/vlab7.mhtml

24

Population Ecology

Concept Outline

24.1 Populations are individuals of the same species that live together.

Population Ecology. The borders of populations are determined by areas in which individuals cannot survive and reproduce. Population ranges expand and contract through time as conditions change.

Population Dispersion. The distribution of individuals in a population can be random, uniform, or clumped.

Metapopulations. Sometimes, populations are arranged in networks connected by the exchange of individuals.

24.2 Population dynamics depend critically upon age distribution.

Demography. The growth rate of a population is a sensitive function of its age structure; populations with many young individuals grow rapidly as these individuals enter reproductive age.

24.3 Life histories often reflect trade-offs between reproduction and survival.

The Cost of Reproduction. Evolutionary success is a trade-off between investment in current reproduction and in growth that promotes future reproduction.

24.4 Population growth is limited by the environment.

Biotic Potential. Populations grow if the birthrate exceeds the death rate until they reach the carrying capacity of their environment.

The Influence of Population Density. Some of the factors that regulate a population's growth depend upon the size of the population; others do not.

Population Growth Rates and Life History Models. Some species have adaptations for rapid, exponential population growth, whereas other species exhibit slower population growth and have intense competition for resources.

24.5 The human population has grown explosively in the last three centuries.

The Advent of Exponential Growth. Human populations have been growing exponentially since the 1700s and will continue to grow in developing countries because of the number of young people entering their reproductive years.

FIGURE 24.1
Life takes place in populations. This population of gannets is subject to the rigorous effects of reproductive strategy, competition, predation, and other limiting factors.

Ecology, the study of how organisms relate to one another and to their environments, is a complex and fascinating area of biology that has important implications for each of us. In our exploration of ecological principles, we will first consider the properties of populations, emphasizing population dynamics (figure 24.1). In chapter 25, we will discuss communities and the interactions that occur in them. Chapter 26 moves on to focus on animals and how and why they behave as they do. Chapter 27 then deals with behavior in an environmental context, the extent to which natural selection has molded behaviors adaptively.

24.1 Populations are individuals of the same species that live together.

Population Ecology

Organisms live as members of **populations,** groups of individuals of a species that live together. In this chapter, we will consider the properties of populations, focusing on elements that influence whether a population will grow or shrink, and at what rate. The explosive growth of the world's human population in the last few centuries provides a focus for our inquiry.

A population consists of the individuals of a given species that occur together at one place and time. This flexible definition allows us to speak in similar terms of the world's human population, the population of protozoa in the gut of an individual termite, or the population of deer that inhabit a forest. Sometimes the boundaries defining a population are sharp, such as the edge of an isolated mountain lake for trout, and sometimes they are more fuzzy, such as when individuals readily move back and forth between areas, like deer in two forests separated by a cornfield.

Three aspects of populations are particularly important: the range throughout which a population occurs, the dispersion of individuals within that range, and the size a population attains.

Population Distributions

No population, not even of humans, occurs in all habitats throughout the world. Most species, in fact, have relatively limited geographic ranges. The Devil's Hole pupfish, for example, lives in a single hot water spring in southern Nevada, and the Socorro isopod is known from a single spring system in Socorro, New Mexico (figure 24.2). At the other extreme, some species are widely distributed. Populations of some whales, for example, are found throughout all of the oceans of the northern or southern hemisphere.

In chapter 29 we will discuss the variety of environmental challenges facing organisms. Suffice it to say for now that no population contains individuals adapted to live in all of the different environments on the earth. Polar bears are exquisitely adapted to survive the cold of the Arctic, but you won't find them in the tropical rain forest. Certain bacteria can live in the near boiling waters of Yellowstone's geysers, but they do not occur in cooler streams that are nearby. Each population has its own requirements—temperature, humidity, certain types of food, and a host of other factors—that determine where it can live and reproduce and where it can't. In addition, in places that are otherwise suitable, the presence of predators, competitors, or parasites may prevent a population from occupying an area.

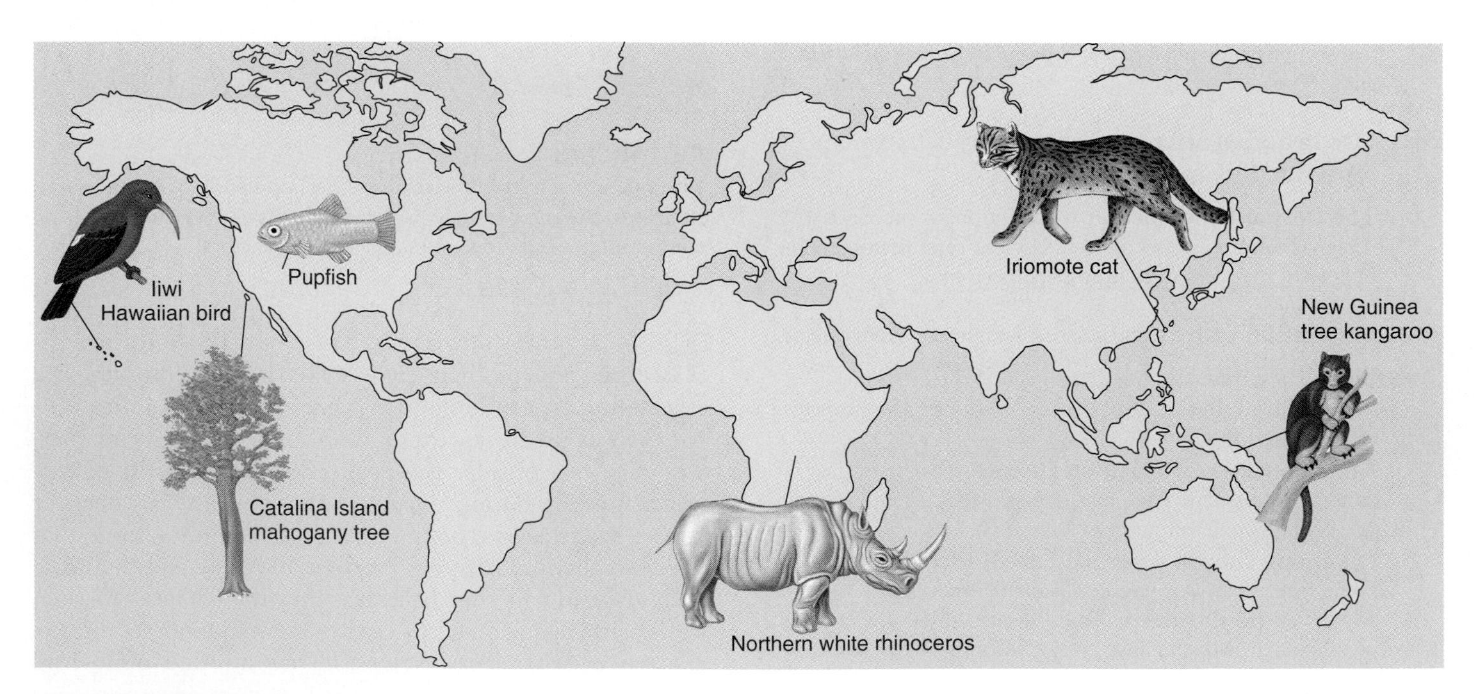

FIGURE 24.2
Species that occur in only one place. These species, and many others, are only found in a single population. All are endangered species, and should anything happen to their single habitat, the population—and the species—would go extinct.
Source: After E.R. Pianka, *Evolutionary Ecology*, 4th edition, New York, Harper & Row, 1987.

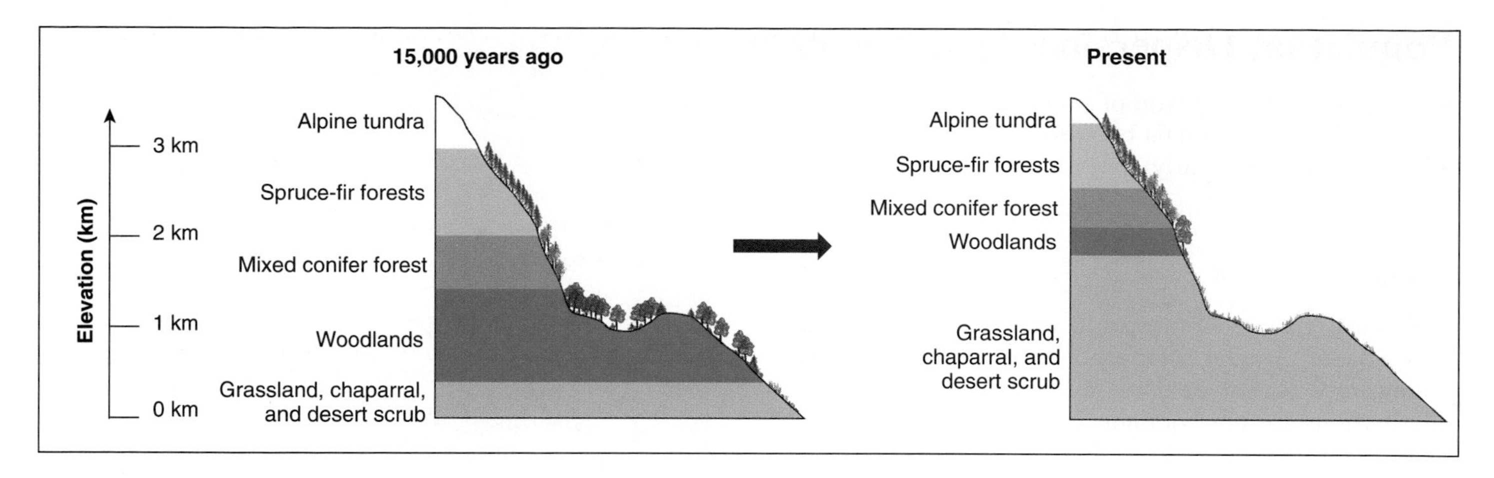

FIGURE 24.3
Altitudinal shifts in population ranges. During the glacial period 15,000 years ago, conditions were cooler than they are now. As the climate has warmed, tree species that require colder temperatures have shifted their distributional range upward in altitude so that they live in the climatic conditions to which they are adapted.
Source: Data from Brown & Lomolino, *Biogeography*, 3rd edition, 1998, Sinauer Associates, Inc.

Range Expansions and Contractions

Population ranges are not static, but, rather, change through time. These changes occur for two reasons. In some cases, the environment changes. For example, as the glaciers retreated at the end of the last ice age, approximately 10,000 years ago, many North American plant and animal populations expanded northward. At the same time, as climates have warmed, species have experienced shifts in the elevation at which they are found on mountains (figure 24.3).

In addition, populations can expand their ranges when they are able to circumvent inhospitable habitat to colonize suitable, previously unoccupied areas. For example, the cattle egret is native to Africa. Some time in the late 1800s, these birds appeared in northern South America, having made the nearly 2000-mile transatlantic crossing, perhaps aided by strong winds. Since then, they have steadily expanded their range such that they now can be found throughout most of the United States (figure 24.4).

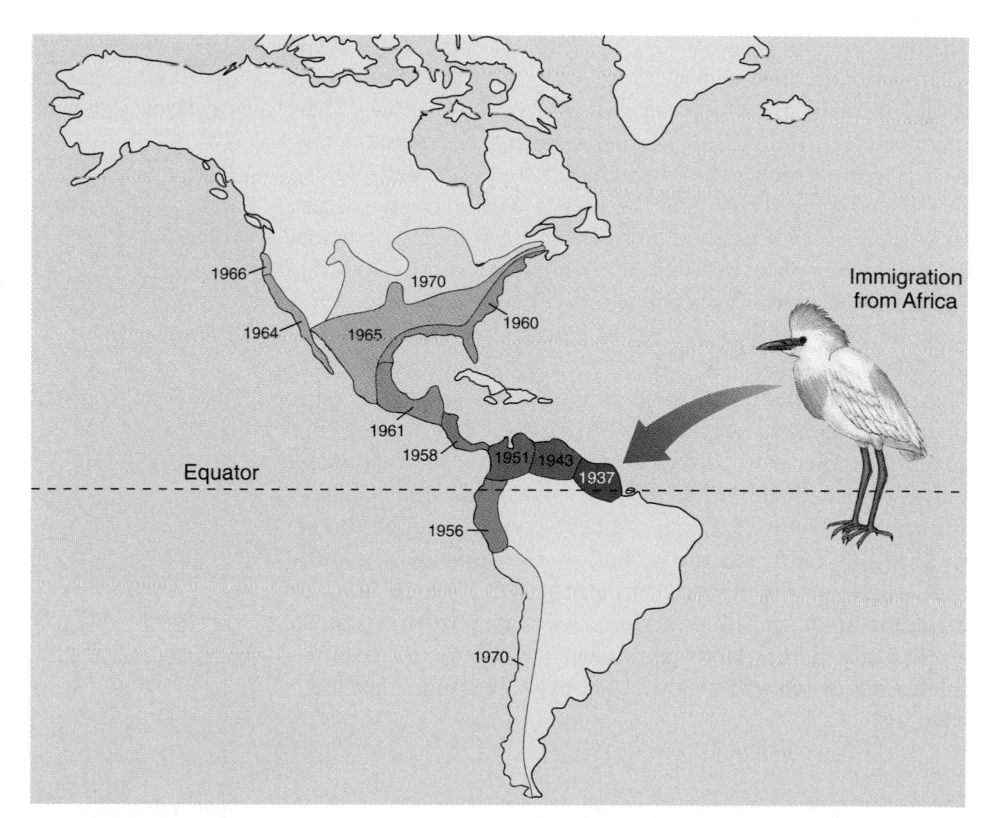

FIGURE 24.4
Range expansion of the cattle egret. Although the cattle egret—so-named because it follows cattle and other hoofed animals, catching any insects or small vertebrates that they disturb—first arrived in South America in the late 1800s, the oldest preserved specimen dates from the 1930s. Since then, the range expansion of this species has been well documented, as it has moved westward and up into much of North America, as well as down the western side of the Andes to near the southern tip of South America.
Source: Data from Brown & Lomolino, *Biogeography*, 3rd edition, 1998, Sinauer Associates; After A.T. Smith *Ecology*, 1974.

A population is a group of individuals of the same species existing together in an area. Its range, the area a population occupies, changes over time.

Population Dispersion

Another key characteristic of population structure is the way in which individuals of a population are arranged. They may be randomly spaced, uniformly spaced, or clumped (figure 24.5).

Randomly Spaced

Individuals are randomly spaced within populations when they do not interact strongly with one another or with nonuniform aspects of their microenvironment. Random distributions are not common in nature. Some species of trees, however, appear to exhibit random distributions in Amazonian rain forests (figure 24.5*b*).

Uniformly Spaced

Individuals often are uniformly spaced within a population. This spacing may often, but not always, result from competition for resources. The means by which it is accomplished, however, varies.

In animals, uniform spacing often results from behavioral interactions, which we will discuss in chapter 27. In many species, individuals of one or both sexes defend a territory from which other individuals are excluded. These territories serve to provide the owner with exclusive access to resources such as food, water, hiding refuges, or mates and tend to space individuals evenly across the habitat. Even in nonterritorial species, individuals often maintain a defended space into which other animals are not allowed to intrude.

Among plants, uniform spacing also is a common result of competition for resources. In this case, however, the spacing results from direct competition for the resources. Closely spaced individual plants will contest for available sunlight, nutrients, or water. These contests can be direct, such as one plant casting a shadow over another, or indirect, such as two plants competing to see which is more efficient at extracting nutrients or water from a shared area. Only plants that are spaced an adequate distance from each other will be able to coexist, leading to uniform spacing.

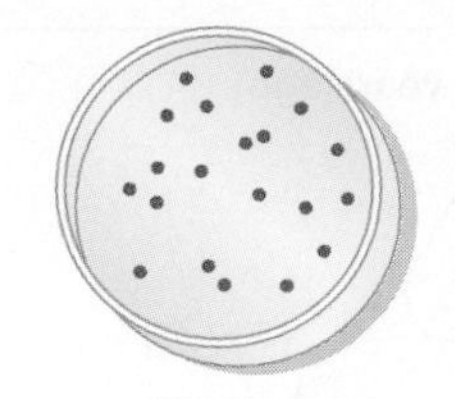
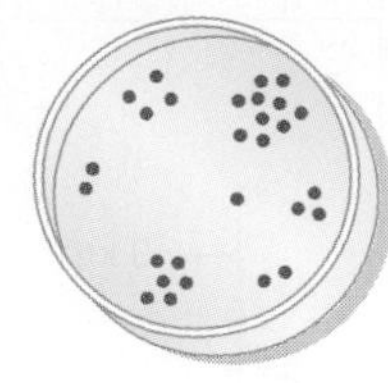

Random Uniform Clumped

(a) Bacterial colonies

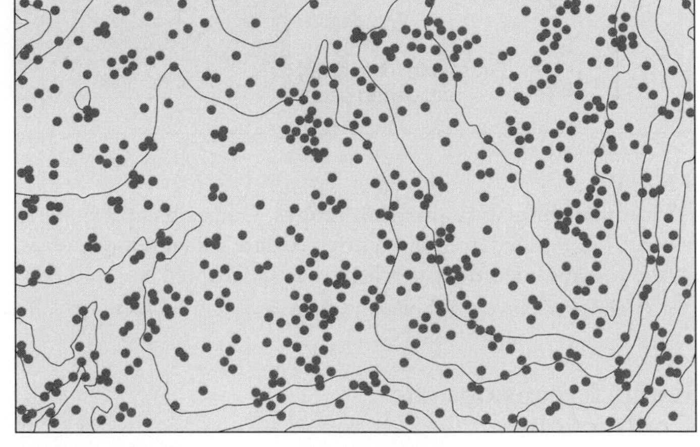

(b) Random distribution of *Brosimum alicastrum*

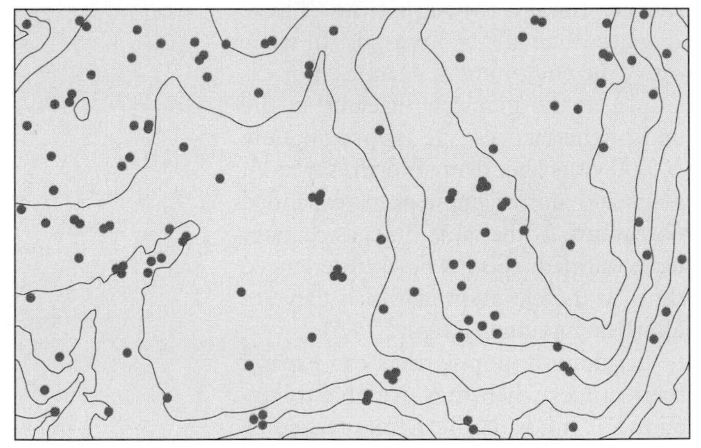

(c) Uniform distribution of *Coccoloba coronata*

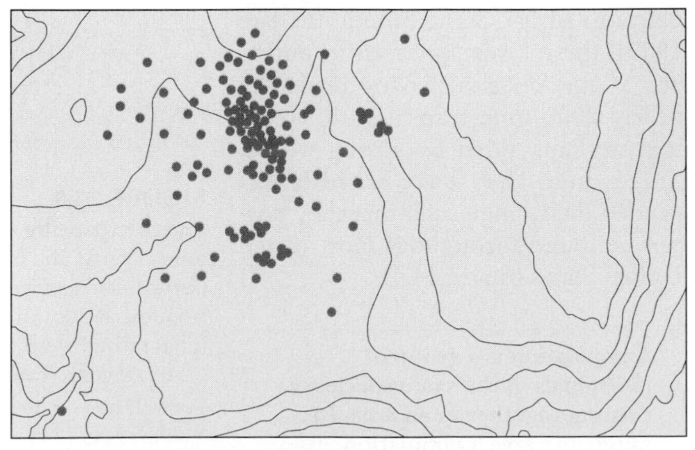

(d) Clumped distribution of *Chamguava schippii*

FIGURE 24.5
Population dispersion. The different patterns of dispersion are exhibited by (*a*) different arrangements of bacterial colonies and three different species of trees from the same locality in Panama. (*b*) *Brosimum alicastrum* is randomly dispersed, (*c*) *Coccoloba coronata* is uniformly dispersed, and (*d*) *Chamguava schippii* exhibits a clumped distribution.

Source: Data from Elizabeth Losos, Center for Tropical Forest Science, Smithsonian Tropical Research Institute.

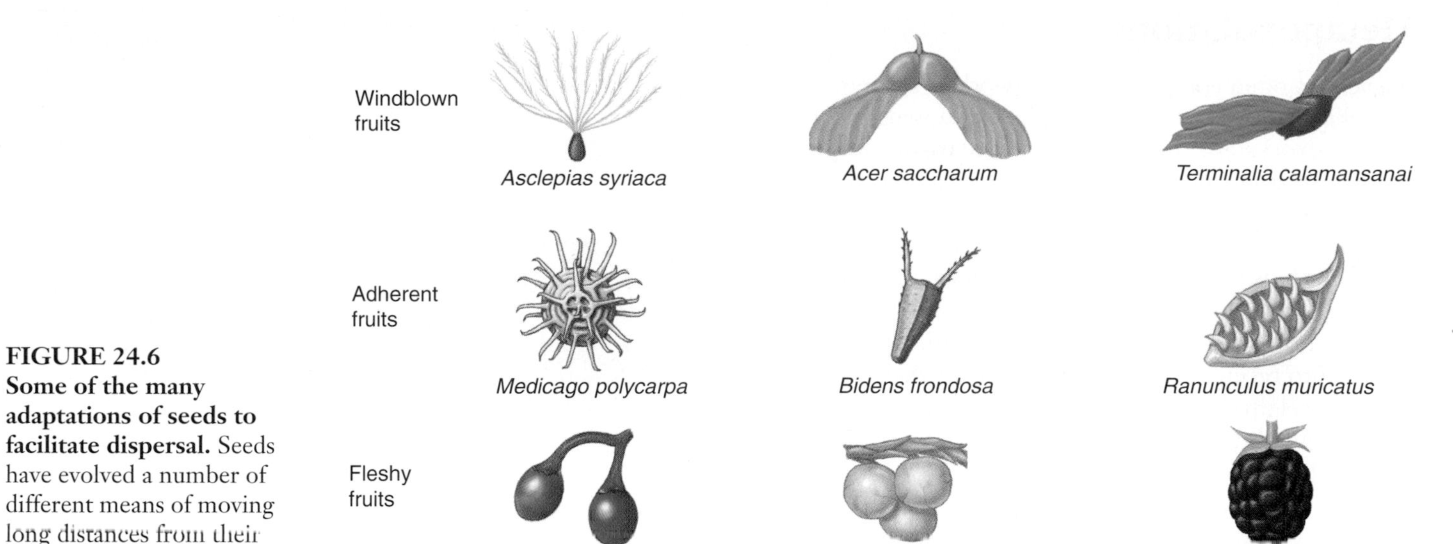

FIGURE 24.6
Some of the many adaptations of seeds to facilitate dispersal. Seeds have evolved a number of different means of moving long distances from their maternal plant.

Clumped Spacing

Individuals clump into groups or clusters in response to uneven distribution of resources in their immediate environments. Clumped distributions are common in nature because individual animals, plants, and microorganisms tend to prefer microhabitats defined by soil type, moisture, or certain kinds of host trees.

Social interactions also can lead to clumped distributions. Many species live and move around in large groups, which go by a variety of names (examples include herds of antelope, flocks of birds, gaggles of geese, packs of wolves, prides of lions). Such groupings can provide many advantages, including increased awareness of and defense against predators, decreased energetic cost of moving through air and water, and access to the knowledge of all group members.

At a broader scale, populations are often most densely populated in the interior of their range and less densely distributed toward the edges. Such patterns usually result from the manner in which the environment changes in different areas. Populations are often best adapted to the conditions in the interior of their distribution. As environmental conditions change, individuals are less well adapted and thus densities decrease. Ultimately, the point is reached at which individuals cannot persist at all; this marks the edge of a population's range.

The Human Effect

By altering the environment, we have allowed some species, such as coyotes, to expand their ranges, although, sadly, for most species the effect has been detrimental. Moreover, humans have served as an agent of dispersal for many species. Some of these transplants have been widely successful. For example, 100 starlings were introduced into New York City in 1896 in a misguided attempt to establish every species of bird mentioned by Shakespeare. Their population steadily spread such that by 1980, they occurred throughout the United States. Similar stories could be told for countless numbers of plants and animals, and the list increases every year. Unfortunately, the success of these invaders often comes at the expense of native species.

Dispersal Mechanisms

Dispersal to new areas can occur in many ways. Lizards, for example, have colonized many distant islands, probably by individuals or their eggs floating or drifting on vegetation. Seeds of many plants are designed to disperse in many ways (figure 24.6). Some seeds are aerodynamically designed to be blown long distances by the wind. Others have structures that stick to the fur or feathers of animals, so that they are carried long distances before falling to the ground. Still others are enclosed in fruits. These seeds can pass through the digestive systems of mammals or birds and then germinate at the spot upon which they are defecated. Finally, seeds of *Arceuthobium* are violently propelled from the base of the fruit in an explosive discharge. Although the probability of long-distance dispersal events occurring and leading to successful establishment of new populations is slim, over millions of years, many such dispersals have occurred.

The distribution of individuals within a population can be random, uniform or clumped and is determined in part, by the availability of resources.

Metapopulations

Species are often composed of a network of distinct populations that interact with each other by exchanging individuals. Such networks are termed **metapopulations** and usually occur in areas in which suitable habitat is patchily distributed and separated by intervening stretches of unsuitable habitat.

To what degree populations within a metapopulation interact depends on the amount of dispersal and is often not symmetrical: populations increasing in size may tend to send out many dispersers, whereas populations at low levels will tend to receive more immigrants than they send off. In addition, relatively isolated populations will tend to receive relatively few arrivals.

Not all suitable habitats within a metapopulation's area may be occupied at any one time. For various reasons, some individual populations may go extinct, perhaps as a result of an epidemic disease, a catastrophic fire, or inbreeding depression. However, because of dispersal from other populations, such areas may eventually be recolonized. In some cases, the number of habitats occupied in a metapopulation may represent an equilibrium in which the rate of extinction of existing populations is balanced by the rate of colonization of empty habitats.

A second type of metapopulation structure occurs in areas in which some habitats are suitable for long-term population maintenance, whereas others are not. In these situations, termed **source-sink metapopulations**, the populations in the better areas (the sources) continually send out dispersers that bolster the populations in the poorer habitats (the sinks). In the absence of such continual replenishment, sink populations would have a negative growth rate and would eventually become extinct.

Metapopulations of butterflies have been studied particularly intensively (figure 24.7). In one study, Ilkka Hanski and colleagues at the University of Helsinki sampled populations of the glanville fritillary butterfly at 1600 meadows in southwestern Finland. On average, every year, 200 populations became extinct, but 114 empty meadows were colonized. A variety of factors seemed to increase the likelihood of a population's extinction, including small population size, isolation from sources of immigrants, low resource availability (as indicated by the number of flowers on a meadow), and lack of genetic variation present within the population. The researchers attribute the greater number of extinctions than colonizations to a string of very dry summers. Because none of the populations is large enough to survive on its own, continued survival of the species in southwestern Finland would appear to require the continued existence of a metapopulation network in which new populations are continually created and existing populations are supplemented by emigrants. Continued bad weather thus may doom the species, at least in this part of its range.

Metapopulations, where they occur, can have two important implications for the range of a species. First, by continual colonization of empty patches, they prevent long-term extinction. If no such dispersal existed, then each population might eventually perish, leading to disappearance of the species from the entire area. Moreover, in source-sink metapopulations, the species as a whole occupies a larger area than it otherwise might occupy. For these reasons, the study of metapopulations has become very important in conservation biology as natural habitats become increasingly fragmented.

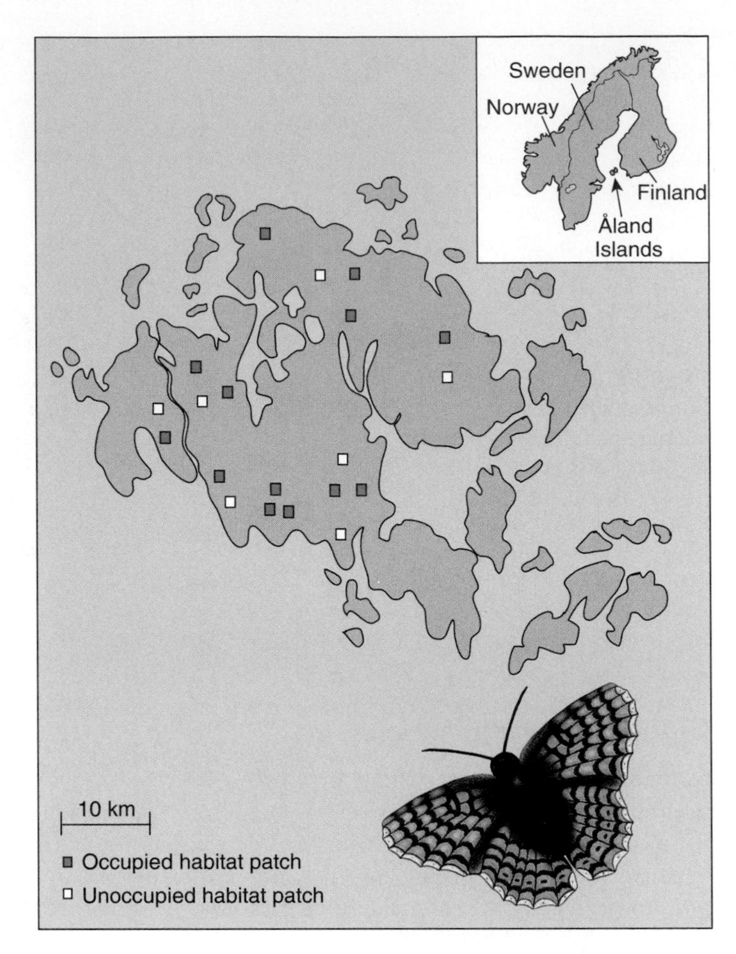

FIGURE 24.7
Metapopulations of butterflies. The glanville fritillary butterfly occurs in metapopulations in southwestern Finland on the Åland Islands. None of the populations is large enough to survive for long on its own, but continual emigration of individuals from other populations allows some populations to survive. In addition, continual establishment of new populations tends to offset extinction of established populations, although in recent years, extinctions have outnumbered colonizations.

Source: Data from *Patch Occupation and Population size of the Glanville Fritillary in the Aland Islands*, Metapopulation Research Group, Helsinki, Finland.

Across broader areas, individuals may occur in populations that are loosely interconnected, termed metapopulations.

24.2 Population dynamics depend critically upon age distribution.

One of the important features of any population is its size. Population size has a direct bearing on the ability of a given population to survive: for a variety of reasons that will be discussed in chapter 31, smaller populations are at a greater risk of disappearing than large populations. In addition, the interactions that occur between members of a population also depend critically on a population's size and density, the number of individuals per unit area.

Demography

Demography (from the Greek *demos*, "the people," + *graphos*, "measurement") is the statistical study of populations. How the size of a population changes through time can be studied at two levels, as a whole or broken down into parts. At the most inclusive level, we can study the population as a whole to determine whether it is increasing, decreasing, or remaining constant. Populations grow if births outnumber deaths and shrink if deaths outnumber births. Understanding these trends is often easier if we break a population down into its constituent parts and analyze each separately.

Factors Affecting Population Growth Rates

The proportion of males and females in a population is its **sex ratio.** The number of births in a population is usually directly related to the number of females, but may not be as closely related to the number of males in species in which a single male can mate with several females. In many species, males compete for the opportunity to mate with females (a situation we discuss in chapter 27); consequently, a few males get many matings, whereas many males do not mate at all. In such species, a female-biased sex ratio would not affect population growth rates; reduction in the number of males simply changes the identities of the reproductive males without reducing the number of births. Among monogamous species like many birds, by contrast, in which pairs form long-lasting reproductive relationships, a reduction in the number of males can directly reduce the number of births.

Generation time, defined as the average interval between the birth of an individual and the birth of its offspring, can also affect population growth rates. Species differ greatly in generation time. Differences in body size can explain much of this variation—mice go through approximately 100 generations during the course of one elephant generation—but not all of it (figure 24.8). Newts, for example, are smaller than mice, but have considerably longer generation times. Everything else equal, populations with shorter generations can increase in size more quickly than populations with long generations.

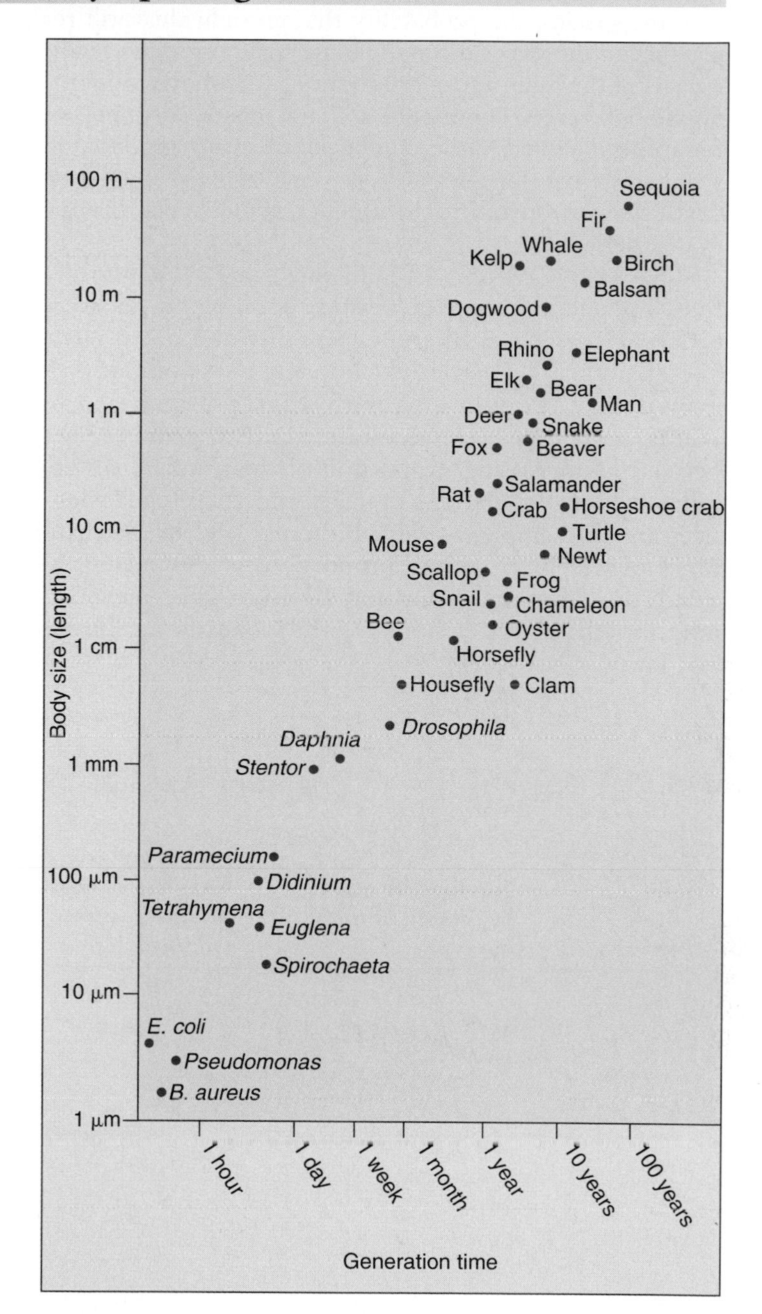

FIGURE 24.8
The relationship between body size and generation time. In general, larger animals have longer generation times, although there are exceptions.
Source: Data from Bonner, 1965

Conversely, because generation time and life span are usually closely correlated, populations with short generation times may also diminish in size more rapidly if birthrates suddenly decrease.

Age Structure

In most species, the probability that an individual will reproduce or die varies through its life span. A group of individuals of the same age is referred to as a **cohort.** Within a population, every cohort has a characteristic birthrate, or **fecundity,** defined as the number of offspring produced in a standard time (for example, per year), and a characteristic death rate, or **mortality,** the number of individuals that die in that period.

The relative number of individuals in each cohort defines a population's **age structure.** Because individuals of different ages have different fecundity and death rates, age structure has a critical impact on a population's growth rate. Populations with a large proportion of young individuals, for example, tend to grow rapidly because an increasing proportion of their individuals are reproductive. Populations in many underdeveloped countries are an example, as we will discuss later in the chapter. Conversely, if a large proportion of a population is relatively old, populations may decline. This phenomenon now characterizes some wealthy countries in Europe and Japan.

Life Tables and Population Change through Time

Ecologists use **life tables** to assess how populations in nature are changing. Life tables can be constructed by following the fate of a cohort from birth until death, noting the number of offspring produced and individuals that die each year. A very nice example of a life table analysis is exhibited in a study of the meadow grass *Poa annua*. This study follows the fate of 843 individuals through time, charting how many survive in each interval and how many offspring each survivor produces (table 24.1).

In table 24.1, the first column indicates the age of the cohort (that is, the number of 3-month intervals from the start of the study). The second and third columns indicate the number of survivors and the proportion of the original cohort still alive at the beginning of that interval. The fourth column presents the **mortality rate,** the proportion of individuals that started that interval alive but died by the end of it. The fifth column indicates the average number of seeds produced by each surviving individual in that interval, and the last column presents the number of seeds produced relative to the size of the original cohort.

Table 24.1 Life Table for a Cohort of the Grass *Poa annua*

Age (in 3-month intervals)	Number alive at beginning of time interval	Proportion of cohort surviving to beginning of time interval (survivorship)	Mortality rate during time interval	Seeds produced per surviving individual (fecundity)	Fecundity × survivorship
0	843	1.000	0.143	0.00	0.00
1	722	0.857	0.271	0.42	0.36
2	527	0.625	0.400	1.18	0.74
3	316	0.375	0.544	1.36	0.51
4	144	0.171	0.626	1.46	0.25
5	54	0.064	0.722	1.11	0.07
6	15	0.018	0.800	2.00	0.04
7	3	0.004	1.000	3.33	0.01
8	0	0.000			Total = 1.98

Modified from Ricklefs, 1997.

Much can be learned from examination of life tables. In this particular case, we see that the probability of dying increases steadily with age, whereas the number of offspring produced increases with age. By adding up the numbers in the last column, we get the total number of offspring produced per individual in the initial cohort. This number is almost 2, which means that for every original member of the cohort, on average two individuals have been produced. A figure of 1.0 would be the break-even number, the point at which the population was neither growing nor shrinking. In this case, the population appears to be growing rapidly.

In most cases, life table analysis is more complicated than this. First, except for organisms with short life spans, it is difficult to track the fate of a cohort from birth until death of the last individual. An alternative approach is to construct a cross-sectional study, examining the fate of all cohorts over a single year. In addition, many factors—such as offspring reproducing before all members of their parental generation's cohort have died—complicate the interpretation of whether populations are growing or shrinking.

Survivorship Curves

One way to express some aspects of the age distribution characteristics of populations is through a **survivorship curve.** Survivorship is defined as the percentage of an original population that survives to a given age. Examples of different kinds of survivorship curves are shown in figure 24.9. In hydra, animals related to jellyfish, individuals are equally likely to die at any age, as indicated by the straight survivorship curve (type II). Oysters, like plants, produce vast numbers of offspring, only a few of which live to reproduce. However, once they become established and grow into reproductive individuals, their mortality rate is extremely low (type III survivorship curve). Finally, even though human babies are susceptible to death at relatively high rates, mortality rates in humans, as in many animals and protists, rise steeply in the postreproductive years (type I survivorship curve). Examination of the data for *Poa annua* reveals that it approximates a type II survivorship curve (figure 24.10).

The growth rate of a population is a sensitive function of its age structure. The age structure of a population and the manner in which mortality and birthrates vary among different age cohorts determine whether a population will increase or decrease in size.

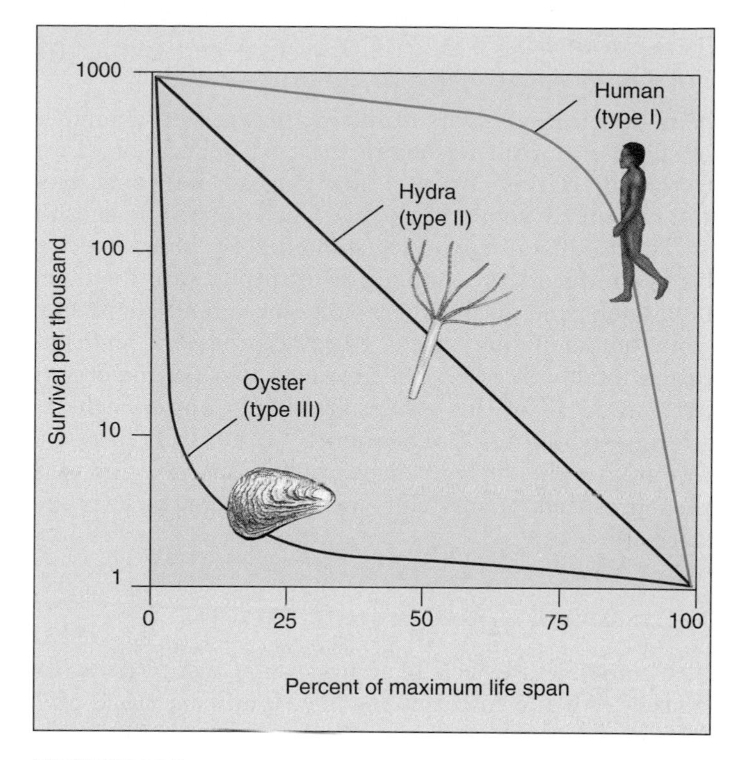

FIGURE 24.9
Survivorship curves. By convention, survival (the vertical axis) is plotted on a log scale. Humans have a type I life cycle, the hydra (an animal related to jellyfish) type II, and oysters type III.

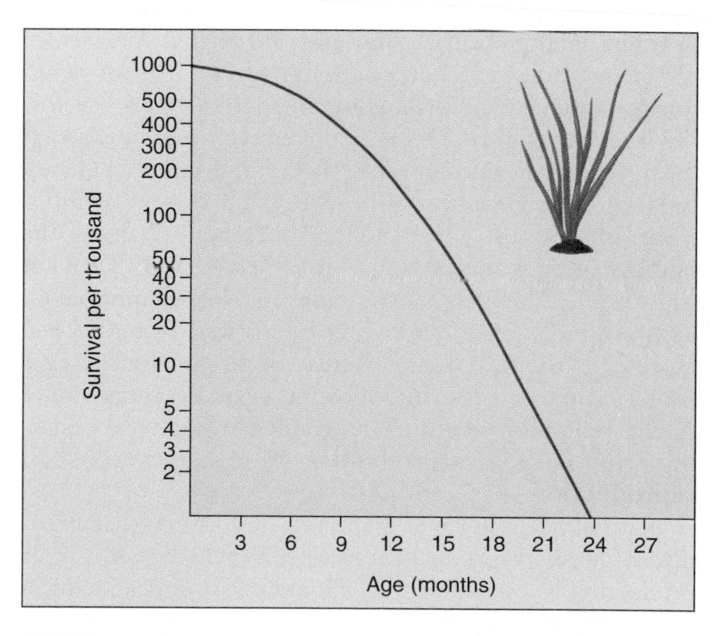

FIGURE 24.10
Survivorship curve for a cohort of the meadow grass, *Poa annua*. Mortality increases at a constant rate through time.
Source: Data from Ricklefs, 1997.

24.3 Life histories often reflect trade-offs between reproduction and survival.

Natural selection favors traits that maximize the number of surviving offspring left in the next generation. Two factors affect this quantity: how long an individual lives and how many young it produces each year. Why doesn't every organism reproduce immediately after its own birth, produce large families of offspring, care for them intensively, and do this repeatedly throughout a long life, while outcompeting others, escaping predators, and capturing food with ease? The answer is that no one organism can do all of this—there are simply not enough resources available. Consequently, organisms allocate resources either to current reproduction or to increase their prospects of surviving and reproducing at later life stages.

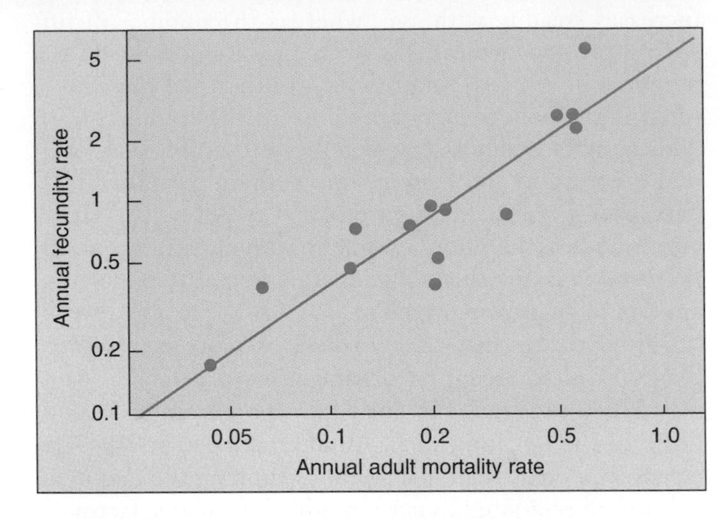

FIGURE 24.11
Reproduction has a price. Increased fecundity in birds correlates with higher mortality in several populations of birds ranging from albatross (low) to sparrow (high). Birds that raise more offspring per year have a higher probability of dying during that year.

The Cost of Reproduction

The complete life cycle of an organism constitutes its **life history**. All life histories involve significant trade-offs. Because resources are limited, a change that increases reproduction may decrease survival and reduce future reproduction. Thus, a Douglas fir tree that produces more cones increases its current reproductive success, but it also grows more slowly; because the number of cones produced is a function of how large a tree is, this diminished growth will decrease the number of cones it can produce in the future. Similarly, birds that have more offspring each year have a higher probability of dying during that year or producing smaller clutches the following year (figure 24.11). Conversely, individuals that delay reproduction may grow faster and larger, enhancing future reproduction.

In one elegant experiment, researchers changed the number of eggs in nests of a bird, the collared flycatcher (figure 24.12). Birds whose clutch size (the number of eggs produced in one breeding event) was decreased laid more eggs the next year, whereas those given more eggs produced fewer eggs the following year. Ecologists refer to the reduction in future reproductive potential resulting from current reproductive efforts as the **cost of reproduction.**

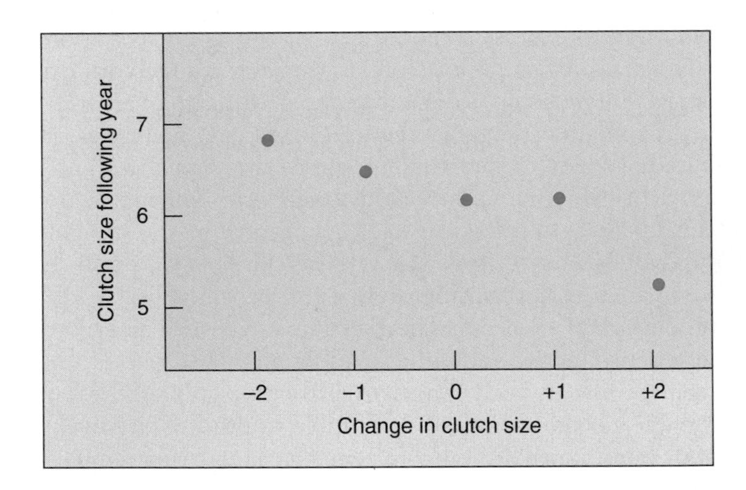

FIGURE 24.12
Reproductive events per lifetime. Adding eggs to nests of collared flycatchers (which increases the reproductive efforts of the female rearing the young) decreases clutch size the following year; removing eggs from the nest increases the next year's clutch size. This experiment demonstrates the tradeoff between current reproductive effort and future reproductive success.

Natural selection will favor the life history that maximizes lifetime reproductive success. When the cost of reproduction is low, individuals should invest in producing as many offspring as possible because there is little cost. Low costs of reproduction may occur when resources are abundant, such that producing offspring does not impair survival or the ability to produce many offspring in subsequent years. Costs of reproduction will also be low when overall mortality rates are high. In such cases, individuals may be unlikely to survive to the next breeding season anyway, so the incremental effect of increased reproductive efforts may not make a difference in future survival.

Alternatively, when costs of reproduction are high, lifetime reproductive success may be maximized by deferring or minimizing current reproduction to enhance growth and survival rates. This may occur when costs of reproduction significantly affect the ability of an individual to survive or decrease the number of offspring that can be produced in the future.

Investment per Offspring

In terms of natural selection, the number of offspring produced is not as important as how many of those offspring themselves survive to reproduce.

A key reproductive trade-off concerns how many resources to invest in producing any single offspring. Assuming that the amount of energy to be invested in offspring is limited, a trade-off must exist between the number of offspring produced and the size of each offspring (figure 24.13). This trade-off has been experimentally demonstrated in the side-blotched lizard, *Uta stansburiana*, which normally lays on average four and a half eggs at a time. When some of the eggs are removed surgically early in the reproductive cycle, the female lizard produces only 1 to 3 eggs, but supplies each of these eggs with greater amounts of yolk, producing eggs that are much larger than normal.

In many species, the size of offspring critically affects their survival prospects—larger offspring have a greater chance of survival. Producing many offspring with little chance of survival might not be the best strategy, but producing only a single, extraordinarily robust offspring also would not maximize the number of surviving offspring. Rather, an intermediate situation, in which several fairly large offspring are produced, should maximize the number of surviving offspring. This example is fundamentally the same as the trade-off between clutch size and parental investment discussed above; in this case, the parental investment is simply how many resources can be invested in each offspring before they are born.

Reproductive Events per Lifetime

The trade-off between age and fecundity plays a key role in many life histories. Annual plants and most insects focus all of their reproductive resources on a single large event and then die. This life history adaptation is called **semelparity** (from the Latin *semel*, "once," + *parito*, "to beget"). Organisms that produce offspring several times over many seasons exhibit a life history adaptation called **iteroparity** (from the Latin *itero*, "to repeat"). Species that reproduce yearly must avoid overtaxing themselves in any one reproductive episode so that they will be able to survive and reproduce in the future. Semelparity, or "big bang" reproduction, is usually found in short-lived species in which the probability of staying alive between broods is low, such as plants growing in harsh climates. Semelparity is also favored when fecundity entails large reproductive cost, as when Pacific salmon migrate upriver to their spawning grounds. In these species, rather than investing some resources in an unlikely bid to survive until the next breeding season, individuals place all their resources into reproduction.

FIGURE 24.13
The relationship between clutch size and offspring size. In great tits, the size of nestlings is inversely related to the number of eggs laid. The more mouths they have to feed, the less the parents can provide to any one nestling.
Source: Data from C.M. Perrins, *Animal Ecology*, 1965.

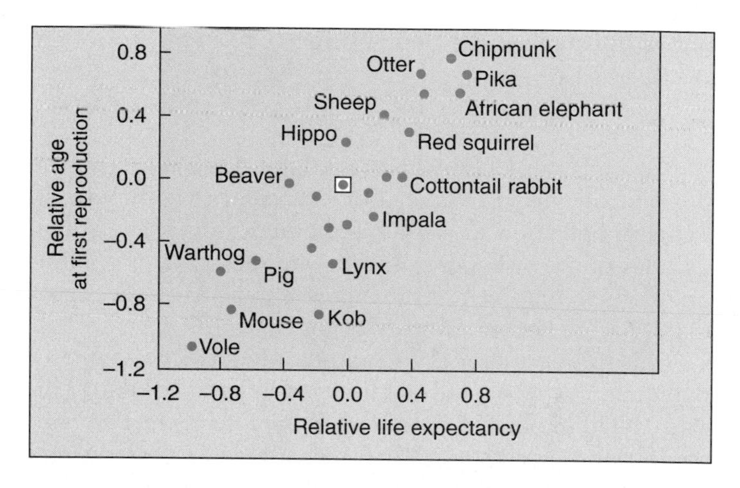

FIGURE 24.14
Age at first reproduction. Among mammals, compensating for the effects of size, age at first reproduction increases with life expectancy at birth. Each dot represents a species. Values are relative to each species' body size. (After Begon et al., 1996.)

Age at First Reproduction

Among mammals and many other animals, longer-lived species reproduce later (figure 24.14). Birds, for example, gain experience as juveniles before expending the high costs of reproduction. In long-lived animals, the relative advantage of juvenile experience outweighs the energy investment in survival and growth. In shorter-lived animals, on the other hand, quick reproduction is more critical than juvenile training, and reproduction tends to occur earlier.

Life history adaptations involve many trade-offs between reproductive cost and investment in survival. Different kinds of animals and plants employ quite different approaches.

24.4 Population growth is limited by the environment.

Biotic Potential

Populations often remain at a relatively constant size, regardless of how many offspring they produce. As you saw in chapter 1, Darwin based his theory of natural selection partly on this seeming contradiction. Natural selection occurs because of checks on reproduction, with some individuals reproducing less often than others. To understand populations, we must consider how they grow and what factors in nature limit population growth.

The Exponential Growth Model

The actual rate of population increase, r, is defined as the difference between the birthrate (b) and the death rate (d) corrected for any movement of individuals in or out of the population, whether net emigration (e, movement out of the area) or net immigration (i, movement into the area). Thus,

$$r = (b - d) + (i - e)$$

Movements of individuals can have a major impact on population growth rates. For example, the increase in human population in the United States during the closing decades of the twentieth century was mostly due to immigrants. Less than half of the increase came from the reproduction of the people already living there.

The simplest model of population growth assumes a population growing without limits at its maximal rate. This rate, called the **biotic potential,** is the rate at which a population of a given species will increase when no limits are placed on its rate of growth. In mathematical terms, this is defined by the following formula:

$$\frac{dN}{dt} = r_i N$$

where N is the number of individuals in the population, dN/dt is the rate of change in its numbers over time, and r_i is the intrinsic rate of natural increase for that population—its innate capacity for growth.

The innate capacity for growth of any population is exponential (red line in figure 24.15). Even when the *rate* of increase remains constant, the actual increase in the *number* of individuals accelerates rapidly as the size of the population grows. The result of unchecked exponential growth is a population explosion. A single pair of houseflies, laying 120 eggs per generation, could produce more than 5 trillion descendants in a year. In 10 years, their descendants would form a swarm more than 2 meters thick over the entire surface of the earth! In practice, such patterns of unrestrained growth prevail only for short periods, usually when an organism reaches a new habitat with abundant resources (figure 24.16). Natural examples include dandelions reaching the fields, lawns, and meadows

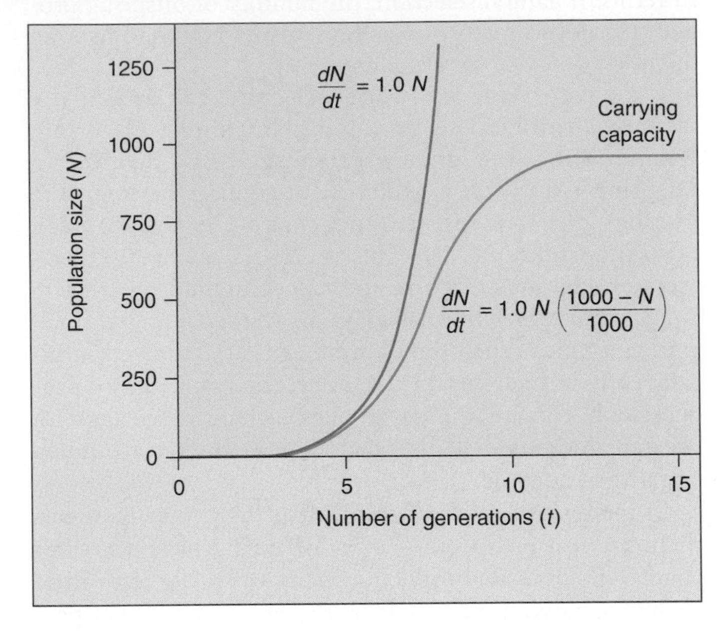

FIGURE 24.15
Two models of population growth. The red line illustrates the exponential growth model for a population with an r of 1.0. The blue line illustrates the logistic growth model in a population with $r = 1.0$ and $K = 1000$ individuals. At first, logistic growth accelerates exponentially, then, as resources become limiting, the death rate increases and growth slows. Growth ceases when the death rate equals the birthrate. The carrying capacity (K) ultimately depends on the resources available in the environment.

FIGURE 24.16
An example of a rapidly increasing population. European purple loosestrife, *Lythrum salicaria*, became naturalized over thousands of square miles of marshes and other wetlands in North America. It was introduced sometime before 1860 and has had a negative impact on many native plants and animals.

of North America from Europe for the first time; algae colonizing a newly formed pond; or the first terrestrial immigrants arriving on an island recently thrust up from the sea.

Carrying Capacity

No matter how rapidly populations grow, they eventually reach a limit imposed by shortages of important environmental factors, such as space, light, water, or nutrients. A population ultimately may stabilize at a certain size, called the **carrying capacity** of the particular place where it lives. The carrying capacity, symbolized by K, is the maximum number of individuals that the environment can support.

The Logistic Growth Model

As a population approaches its carrying capacity, its rate of growth slows greatly, because fewer resources remain for each new individual to use. The growth curve of such a population, which is always limited by one or more factors in the environment, can be approximated by the following logistic growth equation:

$$\frac{dN}{dt} = rN\left(\frac{K - N}{K}\right)$$

In this logistic model of population growth, the growth rate of the population (dN/dt) equals its rate of increase (r multiplied by N, the number of individuals present at any one time), adjusted for the amount of resources available. The adjustment is made by multiplying rN by the fraction of K still unused (K minus N, divided by K). As N increases (the population grows in size), the fraction by which r is multiplied (the remaining resources) becomes smaller and smaller, and the rate of increase of the population declines.

In mathematical terms, as N approaches K, the *rate* of population growth (dN/dt) begins to slow, reaching 0 when $N = K$ (blue line in figure 24.15). Graphically, if you plot N versus t (time) you obtain an S-shaped **sigmoid growth curve** characteristic of many biological populations. The curve is called "sigmoid" because its shape has a double curve like the letter S. As the size of a population stabilizes at the carrying capacity, its rate of growth slows down, eventually coming to a halt (figure 24.17*a*).

In many cases, real populations display trends corresponding to a logistic growth curve. This is true not only in the laboratory, but also in natural populations (figure 24.17*b*). In some cases, however, the fit is not perfect (figure 24.17*c*) and, as we shall see shortly, many populations exhibit other patterns.

The size at which a population stabilizes in a particular place is defined as the carrying capacity of that place for that species. Populations often grow to the carrying capacity of their environment.

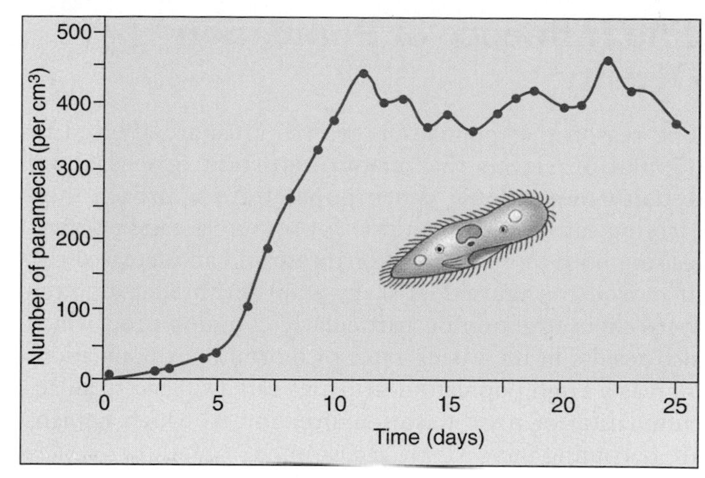

(a)

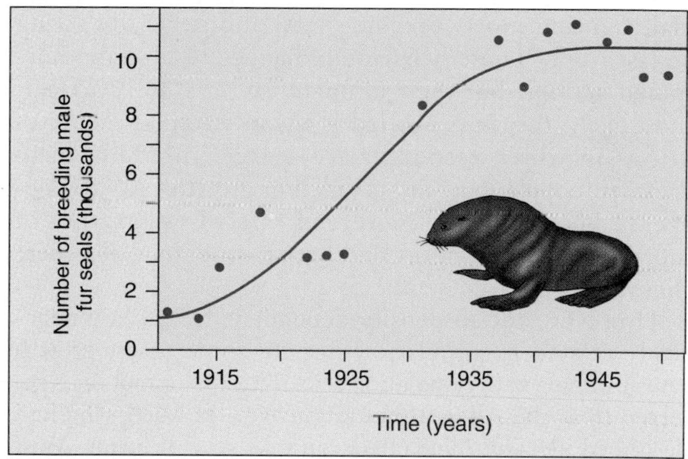

(b)

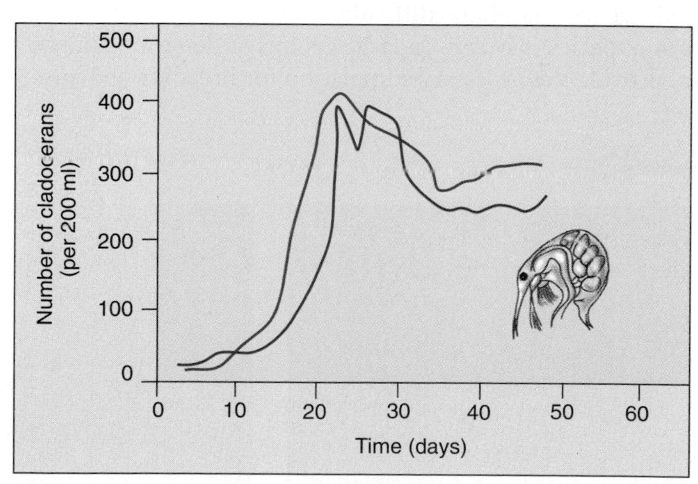

(c)

FIGURE 24.17

Most natural populations exhibit logistic growth. (*a*) *Paramecium* grown in a laboratory environment. (*b*) A fur seal (*Callorhinus ursinus*) population on St. Paul Island, Alaska. (*c*) Laboratory populations of two populations of the cladoceran *Bosmina longirsotris*. Note that the populations first exceeded the carrying capacity, before decreasing to a size which was then maintained.

Source: (*a*) Data from Gause 1934. (*c*) Data from C.E. Goulden, L.L. Henry, and A.J. Tessier, *Ecology*, 1982.

The Influence of Population Density

The reason that population growth rates are affected by population size is that many important processes are **density-dependent**. When populations approach their carrying capacity, competition for resources can be severe, leading both to a decreased birthrate and an increased risk of mortality (figure 24.18). In addition, predators often focus their attention on particularly common prey, which also results in increasing rates of mortality as populations increase. High population densities can also lead to an accumulation of toxic wastes, a situation to which humans are becoming increasingly accustomed.

Behavioral changes may also affect population growth rates. Some species of rodents, for example, become antisocial, fighting more, breeding less, and generally acting stressed-out. These behavioral changes result from hormonal actions, but their ultimate cause is not yet clear; most likely, they have evolved as adaptive responses to situations in which resources are scarce. In addition, in crowded populations, the population growth rate may decrease because of an increased rate of emigration of individuals attempting to find better conditions elsewhere (figure 24.19).

However, not all density-dependent factors are negatively related to population size. In some cases, growth rates increase with population size. This phenomenon is referred to as the **Allee effect** (after Warder Allee, who first described it). The Allee effect can take several forms. Most obviously, in populations that are too sparsely distributed, individuals may have difficulty finding mates. Moreover, some species may rely on large groups to deter predators or to provide the necessary stimulation for breeding activities.

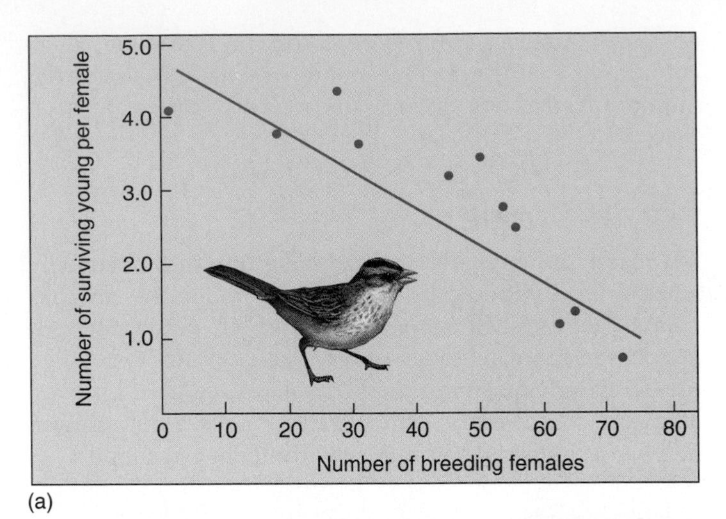

(a)

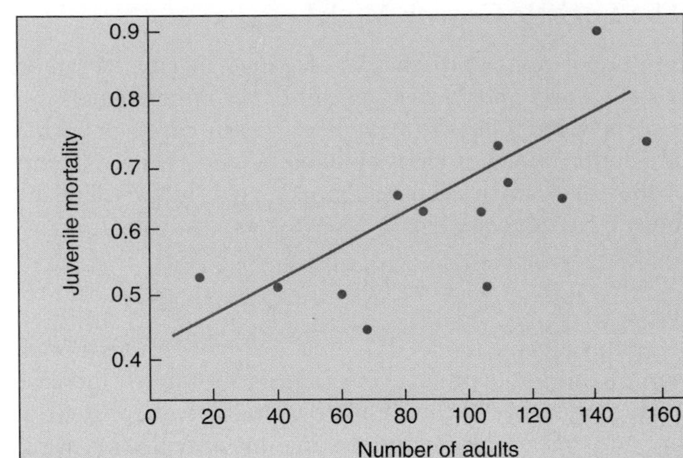

(b)

FIGURE 24.18
Density dependence in the song sparrow (*Melospiza melodia*) on Mandarte Island. Reproductive success decreases (*a*) and mortality rates increase (*b*) as population size increases.

Source: Data from Arcese & Smith J. *Animal Biology*, vol. 57, pp. 119-136, 1989 and Smith et al. 1991.

FIGURE 24.19
Density-dependent effects. Migratory locusts, *Locusta migratoria*, are a legendary plague of large areas of Africa and Eurasia. At high population densities, the locusts have different hormonal and physical characteristics and take off as a swarm. The most serious infestation of locusts in 30 years occurred in North Africa in 1988.

Density-Independent Effects

Growth rates in populations sometimes do not correspond to the logistic growth equation. In many cases, such patterns result because growth is under the control of **density-independent effects.** In other words, the rate of growth of a population at any instant is limited by something other than the size of the population.

A variety of factors may affect populations in a density-independent manner. Most of these are aspects of the external environment. Extremely cold winters, droughts, storms, volcanic eruptions—individuals often will be affected by these activities regardless of the size of the population. Populations that occur in areas in which such events occur relatively frequently will display erratic population growth patterns in which the populations increase rapidly when conditions are benign, but suffer extreme reductions whenever the environment turns hostile.

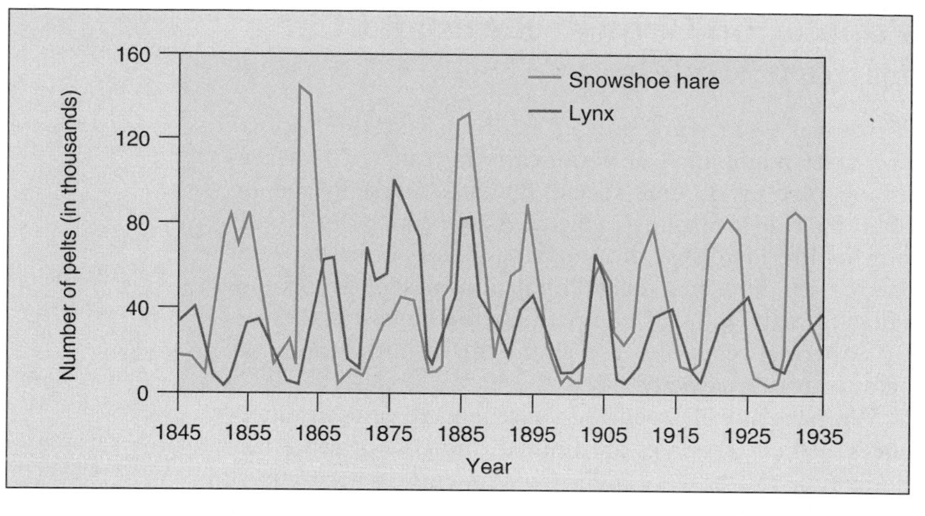

FIGURE 24.20
Linked population cycles of the snowshoe hare and the northern lynx. These data are based on records of fur returns from trappers in the Hudson Bay region of Canada. The lynx populations carefully track the snowshoe hares, but lag behind them slightly.

Population Cycles

Some populations exhibit another type of pattern inconsistent with simple logistic equations: they exhibit cyclic patterns of increase and decrease. Ecologists have studied cycles in hare populations since the 1820s. They have found that the North American snowshoe hare (*Lepus americanus*) follows a "10-year cycle" (in reality, it varies from 8 to 11 years). Its numbers fall tenfold to 30-fold in a typical cycle, and 100-fold changes can occur. Two factors appear to be generating the cycle: food plants and predators.

Food plants. The preferred foods of snowshoe hares are willow and birch twigs. As hare density increases, the quantity of these twigs decreases, forcing the hares to feed on high-fiber (low-quality) food. Lower birthrates, low juvenile survivorship, and low growth rates follow. The hares also spend more time searching for food, exposing them more to predation. The result is a precipitous decline in willow and birch twig abundance, and a corresponding fall in hare abundance. It takes two to three years for the quantity of mature twigs to recover.

Predators. A key predator of the snowshoe hare is the Canada lynx (*Lynx canadensis*). The Canada lynx shows a "10-year" cycle of abundance that seems remarkably entrained to the hare abundance cycle (figure 24.20). As hare numbers increase, lynx numbers do, too, rising in response to the increased availability of lynx food. When hare numbers fall, so do lynx numbers, their food supply depleted.

Which factor is responsible for the predator-prey oscillations? Do increasing numbers of hares lead to overharvesting of plants (a hare-plant cycle) or do increasing numbers of lynx lead to overharvesting of hares (a hare-lynx cycle)? Field experiments carried out by C. Krebs and coworkers in 1992 provide an answer. Krebs set up experimental plots in Canada's Yukon-containing hare populations. If food is added (no food effect) and predators excluded (no predator effect) from an experimental area, hare numbers increase tenfold and stay there—the cycle is lost. However, the cycle is retained if either of the factors is allowed to operate alone: exclude predators but don't add food (food effect alone), or add food in presence of predators (predator effect alone). Thus, both factors can affect the cycle, which, in practice, seems to be generated by the interaction between the two factors.

Population cycles traditionally have been considered to occur rarely. However, a recent review of nearly 700 long-term (25 years or more) studies of trends within populations found that cycles were not uncommon; nearly 30% of the studies—including birds, mammals, fish, and crustaceans—provided evidence of some cyclic pattern in population size through time, although most of these cycles are nowhere near as dramatic in amplitude as the snowshoe hare and lynx cycles.

Density-dependent effects are caused by factors that come into play particularly when the population size is larger; density-independent effects are controlled by factors that operate regardless of population size.

Population Growth Rates and Life History Models

As we have seen, some species usually have stable population sizes maintained near the carrying capacity, whereas the populations of other species fluctuate markedly and are often far below carrying capacity. As we saw in our discussion of life histories, the selective factors affecting such species will differ markedly. Populations near their carrying capacity may face stiff competition for limited resources. By contrast, resources are abundant in populations far below carrying capacity.

We have already seen the consequences of such differences. When resources are limited, the cost of reproduction often will be very high. Consequently, selection will favor individuals that can compete effectively and utilize resources efficiently. Such adaptations often come at the cost of lowered reproductive rates. Such populations are termed **K-selected** because they are adapted to thrive when the population is near its carrying capacity (K). Table 24.2 lists some of the typical features of K-selected populations. Examples of K-selected species include coconut palms, whooping cranes, whales, and humans.

By contrast, in populations far below the carrying capacity, resources may be abundant. Costs of reproduction will be low, and selection will favor those individuals that can produce the maximum number of offspring. Selection here favors individuals with the highest reproductive rates; such populations are termed **r-selected**. Examples of organisms displaying r-selected life history adaptations include dandelions, aphids, mice, and cockroaches (figure 24.21).

Most natural populations show life history adaptations that exist along a continuum ranging from completely r-selected traits to completely K-selected traits. Although these tendencies hold true as generalities, few populations are purely r- or K-selected and show all of the traits listed in table 24.2. These attributes should be treated as generalities, with the recognition that many exceptions do exist.

Some life history adaptations favor near-exponential growth, others the more competitive logistic growth. Most natural populations exhibit a combination of the two.

Table 24.2 r-Selected and K-Selected Life History Adaptations

Adaptation	r-Selected Populations	K-Selected Populations
Age at first reproduction	Early	Late
Life span	Short	Long
Maturation time	Short	Long
Mortality rate	Often high	Usually low
Number of offspring produced per reproductive episode	Many	Few
Number of reproductions per lifetime	Usually one	Often several
Parental care	None	Often extensive
Size of offspring or eggs	Small	Large

Source: After E. R. Pianka, *Evolutionary Ecology*, 4th edition, 1987, New York, Harper & Row, 1987.

FIGURE 24.21
The consequences of exponential growth. All organisms have the potential to produce populations larger than those that actually occur in nature. The German cockroach (*Blatella germanica*), a major household pest, produces 80 young every six months. If every cockroach that hatched survived for three generations, kitchens might look like this theoretical culinary nightmare concocted by the Smithsonian Museum of Natural History.

24.5 The human population has grown explosively in the last three centuries.

The Advent of Exponential Growth

Humans exhibit many *K*-selected life history traits, including small brood size, late reproduction, and a high degree of parental care. These life history traits evolved during the early history of hominids, when the limited resources available from the environment controlled population size. Throughout most of human history, our populations have been regulated by food availability, disease, and predators. Although unusual disturbances, including floods, plagues, and droughts no doubt affected the pattern of human population growth, the overall size of the human population grew only slowly during our early history. Two thousand years ago, perhaps 130 million people populated the earth. It took a thousand years for that number to double, and it was 1650 before it had doubled again, to about 500 million. For over 16 centuries, the human population was characterized by very slow growth. In this respect, human populations resembled many other species with predominantly *K*-selected life history adaptations.

Starting in the early 1700s, changes in technology have given humans more control over their food supply, enabled them to develop superior weapons to ward off predators, and led to the development of cures for many diseases. At the same time, improvements in shelter and storage capabilities have made humans less vulnerable to climatic uncertainties. These changes allowed humans to expand the carrying capacity of the habitats in which they lived, and thus to escape the confines of logistic growth and reenter the exponential phase of the sigmoidal growth curve.

Responding to the lack of environmental constraints, the human population has grown explosively over the last 300 years. While the birthrate has remained unchanged at about 30 per 1000 per year over this period, the death rate has fallen dramatically, from 20 per 1000 per year to its present level of 13 per 1000 per year. The difference between birth and death rates meant that the population grew as much as 2% per year, although the rate has now declined to 1.4% per year.

A 1.4% annual growth rate may not seem large, but it has produced a current human population of 6 billion people (figure 24.22)! At this growth rate, 77 million people are added to the world population annually, and the human population will double in 39 years. As we will discuss in chapter 30, both the current human population level and the projected growth rate have potential consequences for our future that are extremely grave.

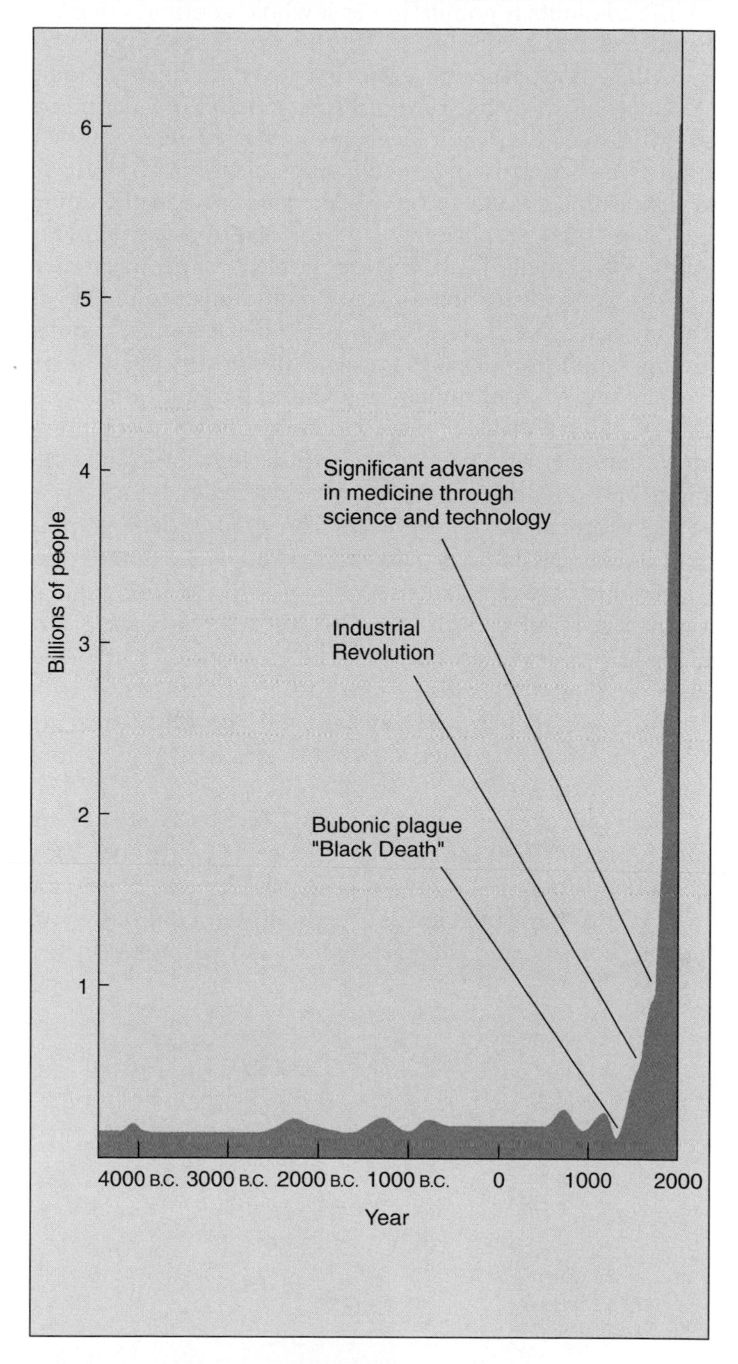

FIGURE 24.22
History of human population size. Temporary increases in death rate, even severe ones like the Black Death of the 1400s, have little lasting impact. Explosive growth began with the Industrial Revolution in the 1700s, which produced a significant long-term lowering of the death rate. The current population is 6 billion, and at the current rate will double in 39 years.

Population Pyramids

While the human population as a whole continues to grow rapidly at the beginning of the twenty-first century, this growth is not occurring uniformly over the planet. Some countries, like Mexico, are growing rapidly, their birthrate greatly exceeding their death rate (figure 24.23). Other countries are growing much more slowly. The rate at which a population can be expected to grow in the future can be assessed graphically by means of a **population pyramid**—a bar graph displaying the numbers of people in each age category. Males are conventionally shown to the left of the vertical age axis, females to the right. A human population pyramid thus displays the age composition of a population by sex. In most human population pyramids, the number of older females is disproportionately large compared to the number of older males, because females in most regions have a longer life expectancy than males.

Viewing such a pyramid, one can predict demographic trends in births and deaths. In general, rectangular "pyramids" are characteristic of countries whose populations are stable, their numbers neither growing nor shrinking. A triangular pyramid is characteristic of a country that will exhibit rapid future growth, as most of its population has not yet entered the child-bearing years. Inverted triangles are characteristic of populations that are shrinking.

Examples of population pyramids for the United States and Kenya in 1990 are shown in figure 24.24. In the nearly rectangular population pyramid for the United States, the cohort (group of individuals) 55 to 59 years old represents people born during the Depression and is smaller in size than the cohorts in the preceding and following years. The cohorts 25 to 44 years old represent the "baby boom." The rectangular shape of the population pyramid indicates that the population of the United States is not expanding rapidly. The very triangular pyramid of Kenya, by contrast, predicts explosive future growth. The population of Kenya is predicted to double in less than 20 years.

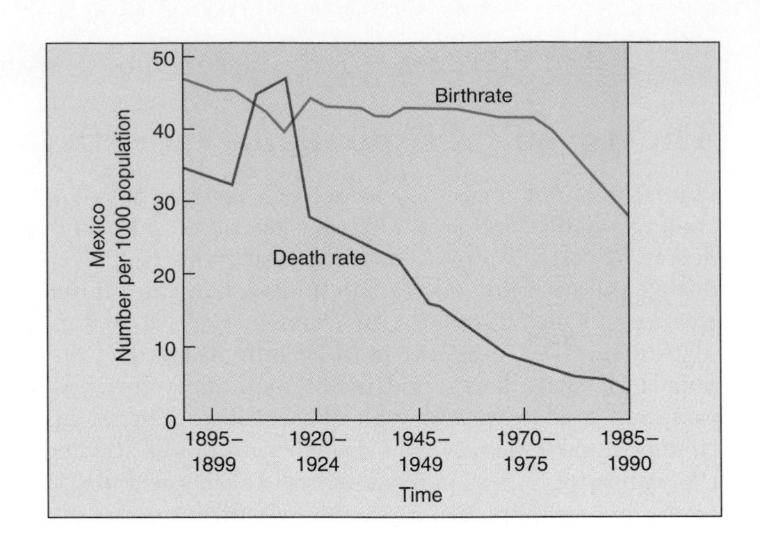

FIGURE 24.23
Why the population of Mexico is growing. The death rate (*red line*) in Mexico fell steadily throughout the last century, while the birthrate (*blue line*) remained fairly steady until 1970. The difference between birth and death rates has fueled a high growth rate. Efforts begun in 1970 to reduce the birthrate have been quite successful, although the growth rate remains rapid.

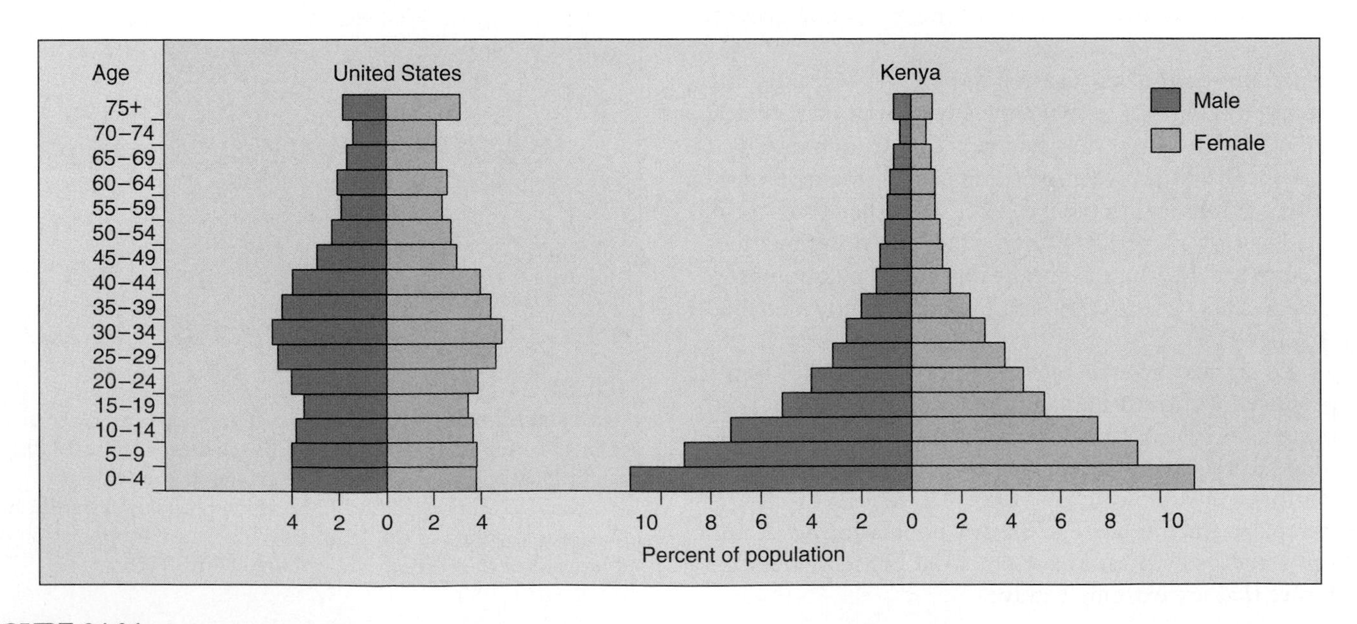

FIGURE 24.24
Population pyramids from 1990. Population pyramids are graphed according to a population's age distribution. Kenya's pyramid has a broad base because of the great number of individuals below child-bearing age. When all of the young people begin to bear children, the population will experience rapid growth. The U.S. pyramid demonstrates a larger number of individuals in the "baby boom" cohort—the pyramid bulges because of an increase in births between 1945 and 1964.

Table 24.3 A Comparison of 1996 Population Data in Developed and Developing Countries

	United States (*highly developed*)	Brazil (*moderately developed*)	Ethiopia (*poorly developed*)
Fertility rate	2.0	2.8	6.8
Doubling time at current rate (yr)	114	41	23
Infant mortality rate (per 1000 births)	7.5	58	120
Life expectancy at birth (yrs)	76	66	50
Per capita GNP (U.S. $; 1994)	$25,860	$3,370	$130

An Uncertain Future

The earth's rapidly growing human population constitutes perhaps the greatest challenge to the future of the biosphere, the world's interacting community of living things. Humanity is adding 77 million people a year to the earth's population—a million every five days, 150 every minute! In more rapidly growing countries, the resulting population increase is staggering (table 24.3). India, for example, had a population of 853 million in 1996; by 2020 its population will exceed 1.4 billion!

A key element in the world's population growth is its uneven distribution among countries. Of the billion people added to the world's population in the 1990s, 90% live in developing countries (figure 24.25). This is leading to a major reduction in the fraction of the world's population that lives in industrialized countries. In 1950, fully one-third of the world's population lived in industrialized countries; by 1996 that proportion had fallen to one-quarter; in 2020 the proportion will have fallen to one-sixth. Thus the world's population growth will be centered in the parts of the world least equipped to deal with the pressures of rapid growth.

Rapid population growth in developing countries has the harsh consequence of increasing the gap between rich and poor. Today 23% of the world's population lives in the industrialized world with a per capita income of $17,900, while 77% of the world's population lives in developing countries with a per capita income of only $810. The disproportionate wealth of the industrialized quarter of the world's population is evidenced by the fact that 85% of the world's capital wealth is in the industrial world, only 15% in developing countries. Eighty percent of all the energy used today is consumed by the industrial world, only 20% by developing countries. Perhaps most worrisome for the future, fully 94% of all scientists and engineers reside in the industrialized world, only 6% in developing countries. Thus the problems created by the future's explosive population growth will be faced by countries with little of the world's scientific or technological expertise.

No one knows whether the world can sustain today's population of 6 billion people, much less the far greater populations expected in the future. As chapter 30 outlines, the world ecosystem is already under considerable stress.

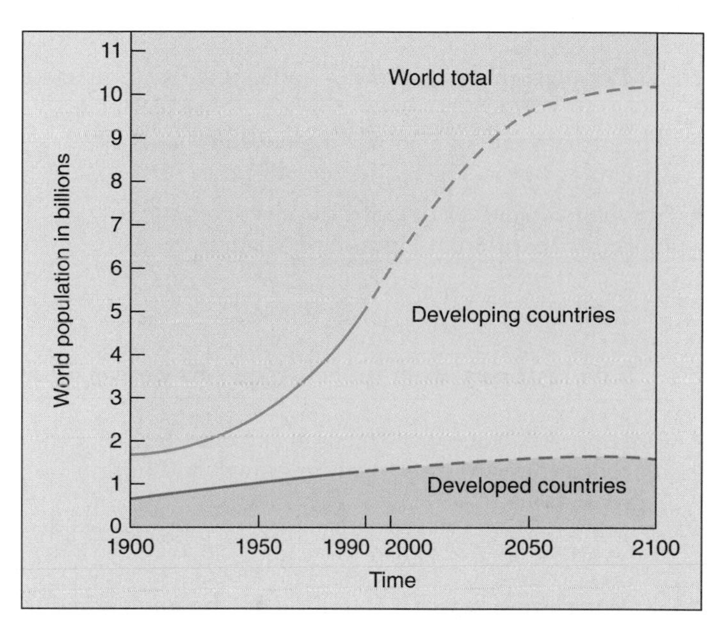

FIGURE 24.25
Most of the worldwide increase in population since 1950 has occurred in developing countries. The age structures of developing countries indicate that this trend will increase in the near future. The stabilizing of the world's population at about 10 billion (shown here) is an optimistic World Bank/United Nations prediction that assumes significant worldwide reductions in growth rate. If the world's population continues to increase at its 1996 rate, there will be over 30 billion humans by 2100!

We cannot reasonably expect to continue to expand its carrying capacity indefinitely, and indeed we already seem to be stretching the limits. It seems unavoidable that to restrain the world's future population growth, birth and death rates must be equalized. If we are to avoid catastrophic increases in the death rate, the birthrates must fall dramatically. Faced with this grim dichotomy, significant efforts are underway worldwide to lower birthrates.

The human population has been growing rapidly for 300 years, since technological innovations dramatically reduced the death rate.

Chapter 24

www.mhhe.com/raven6/resources24.mhtml

Summary	Questions	Media Resources

24.1 Populations are individuals of the same species that live together.

- Populations are individuals of the same species living together in one place. A population's range, the area in which the population exists, can change over time.
- Populations may be dispersed in a random, uniform, or clumped manner.

1. What are the three types of dispersion in a population? Which type is most frequently seen in nature? Why?

2. What are some causes of clumped distributions?

- Introduction to Populations
- Population Characteristics

- *On Science* Article: Snakes in Ireland

24.2 Population dynamics depend critically upon age distribution.

- The growth rate of a population depends on its age structure, and to a lesser degree, sex ratio.
- Survivorship curves describe the characteristics of mortality in different kinds of populations.

3. What is survivorship? Describe the three types of survivorship curves and give examples of each.

4. What is demography? How does a life table work?

- *On Science* Article: Deer Hunting

* Art Quiz: Survivorship Curves

24.3 Life histories often reflect trade-offs between reproduction and survival.

- Organisms balance investment in current reproduction with investment in growth and future reproduction.

5. Why do some birds lay fewer than the optimal number of eggs as predicted by David Lack?

24.4 Population growth is limited by the environment.

- Population size will change if birth and death rates differ, or if there is net migration into or out of the population. The intrinsic rate of increase of a population is defined as its biotic potential.
- Many populations exhibit a sigmoid growth curve, with a relatively slow start in growth, a rapid increase, and then a leveling off when the carrying capacity of the environment is reached.
- Density-dependent factors affect a population that is approaching its carrying capacity.
- Density-independent factors have the same impact on a population no matter what its density.
- Large broods and rapid rates of population growth characterize *r*-strategists. *K*-strategists are limited in population size by the carrying capacity of their environments; they tend to have fewer offspring and slower rates of population growth.

6. Define the biotic potential of a population. What is the definition for the actual rate of population increase? What other two factors affect it?

7. What is an exponential capacity for growth? When does this type of growth naturally occur? Give an example.

8. What is carrying capacity? Is this a static or dynamic measure? Why?

9. What is the difference between *r*- and *K*-selected populations?

- Stages of Population Growth

- Size Regulation

- Scientists on Science: Coral Reefs Threatened
- *On Science* Article: Was Malthus Mistaken?

- Art Quiz: Population Growth

24.5 The human population has grown explosively in the last three centuries.

- Exponential growth of the world's human population is placing severe strains on the global environment.

10. How do population pyramids predict whether a population is likely to grow or shrink?

- Population Growth
- Human Population

25

Community Ecology

Concept Outline

25.1 Interactions among competing species shape ecological niches.

The Realized Niche. Interspecific interactions often limit the portion of their niche that they can actually use.

Gause and the Principle of Competitive Exclusion. No two species can occupy the same niche indefinitely without competition driving one to extinction.

Resource Partitioning. Species that live together partition the available resources, reducing competition.

Detecting Interspecific Competition. Experiments are often the best way to detect competition, but they have their limitations.

25.2 Predators and their prey coevolve.

Predation and Prey Populations. Predators can limit the size of populations and sometimes even eliminate a species from a community.

Plant Defenses against Herbivores. Plants use chemicals to defend themselves against animals trying to eat them.

Animal Defenses against Predators. Animals defend themselves with camouflage, chemicals, and stings.

Mimicry. Sometimes a species copies the appearance of another protected one.

25.3 Evolution sometimes fosters cooperation.

Coevolution and Symbiosis. Organisms have evolved many adjustments and accommodations to living together.

Commensalism. Some organisms use others, neither hurting or helping their benefactors.

Mutualism. Often species interact in ways that benefit both.

Parasitism. Sometimes one organism serves as the food supply of another much smaller one.

Interactions among Ecological Processes. Multiple processes may occur simultaneously within a community.

25.4 Ecological succession may increase species richness.

Succession. Communities change through time.

The Role of Disturbance. Disturbances can disrupt successional change. In some cases, moderate amounts of disturbance increase species diversity.

FIGURE 25.1
Communities involve interactions between disparate groups. This clownfish is one of the few species that can nestle safely among the stinging tentacles of the sea anemone—a classic example of a symbiotic relationship.

All the organisms that live together in a place are called a community. The myriad of species that inhabit a tropical rain forest are a community. Indeed, every inhabited place on earth supports its own particular array of organisms. Over time, the different species have made many complex adjustments to community living (figure 25.1), evolving together and forging relationships that give the community its character and stability. Both competition and cooperation have played key roles; in this chapter, we will look at these and other factors in community ecology.

25.1 Interactions among competing species shape ecological niches.

The Realized Niche

Each organism in an ecosystem confronts the challenge of survival in a different way. The **niche** an organism occupies is the sum total of all the ways it utilizes the resources of its environment. A niche may be described in terms of space utilization, food consumption, temperature range, appropriate conditions for mating, requirements for moisture, and other factors. *Niche* is not synonymous with **habitat,** the place where an organism lives. *Habitat* is a place, *niche* a pattern of living.

Sometimes species are not able to occupy their entire niche because of the presence or absence of other species. Species can interact with each other in a number of ways, and these interactions can either have positive or negative effects. One type of interaction is **interspecific competition,** which occurs when two species attempt to utilize the same resource when there is not enough of the resource to satisfy both. Fighting over resources is referred to as **interference competition;** consuming shared resources is called **exploitative competition.**

The entire niche that a species is capable of using, based on its physiological requirements and resource needs, is called the **fundamental niche.** The actual niche the species occupies is called its **realized niche.** Because of interspecific interactions, the realized niche of a species may be considerably smaller than its fundamental niche.

In a classic study, J. H. Connell of the University of California, Santa Barbara investigated competitive interactions between two species of barnacles that grow together on rocks along the coast of Scotland. Of the two species Connell studied, *Chthamalus stellatus* lives in shallower water, where tidal action often exposed it to air, and *Semibalanus balanoides* (called *Balanus balanoides* prior to 1995) lives lower down, where it is rarely exposed to the atmosphere (figure 25.2). In the deeper zone, *Semibalanus* could always outcompete *Chthamalus* by crowding it off the rocks, undercutting it, and replacing it even where it had begun to grow, an example of interference competition. When Connell removed *Semibalanus* from the area, however, *Chthamalus* was easily able to occupy the deeper zone, indicating that no physiological or other general obstacles prevented it from becoming established there. In contrast, *Semibalanus* could not survive in the shallow-water habitats where *Chthamalus* normally occurs; it evidently does not have the special adaptations that allow *Chthamalus* to occupy this zone. Thus, the fundamental niche of the barnacle *Chthamalus* included both shallow and deeper zones, but its realized niche was much narrower because *Chthamalus* was outcompeted by *Semibalanus* in parts of its fundamental niche. By contrast, the realized and fundamental niches of *Semibalanus* appear to be identical.

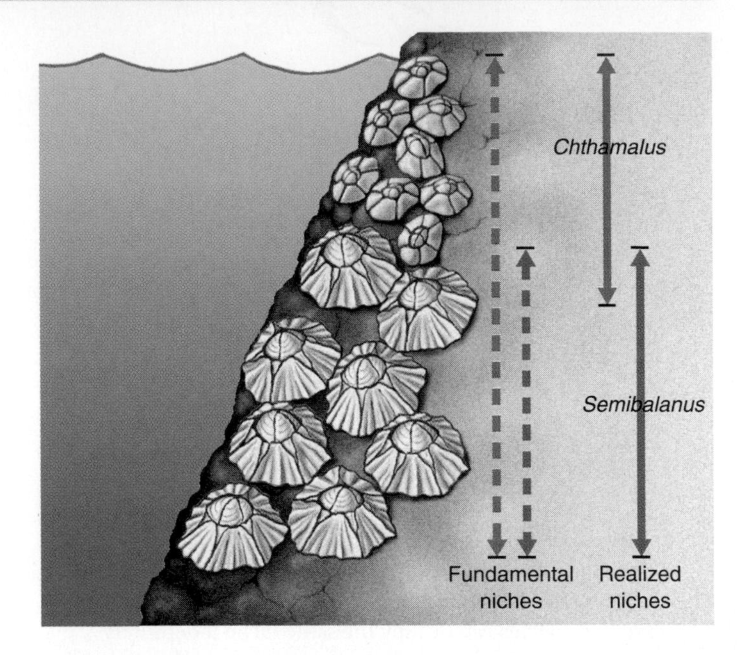

FIGURE 25.2
Competition among two species of barnacles limits niche use. *Chthamalus* can live in both deep and shallow zones (its fundamental niche), but *Semibalanus* forces *Chthamalus* out of the part of its fundamental niche that overlaps the realized niche of *Semibalanus.*

Processes other than competition can also restrict the realized niche of a species. For example, a plant, the St. John's-wort, was introduced and became widespread in open rangeland habitats in California until a specialized beetle was introduced to control it. Populations of the plant quickly decreased and it is now only found in shady sites where the beetle cannot thrive. In this case, the presence of a predator limits the realized niche of a plant.

In some cases, the absence of another species leads to a smaller realized niche. For example, many North American plants depend on the American honeybee for pollination. The honeybee's population is currently declining for a variety of reasons. Conservationists are concerned that if the honeybee disappears from some habitats, the niche of these plant species will decrease or even disappear entirely. In this case, then, the absence—rather than the presence—of another species will be the cause of a relatively small realized niche.

A niche may be defined as the way in which an organism utilizes its environment. Interspecific interactions may cause a species' realized niche to be smaller than its fundamental niche.

Gause and the Principle of Competitive Exclusion

In classic experiments carried out in 1934 and 1935, Russian ecologist G. F. Gause studied competition among three species of *Paramecium*, a tiny protist. All three species grew well alone in culture tubes, preying on bacteria and yeasts that fed on oatmeal suspended in the culture fluid (figure 25.3*a*). However, when Gause grew *P. aurelia* together with *P. caudatum* in the same culture tube, the numbers of *P. caudatum* always declined to extinction, leaving *P. aurelia* the only survivor (figure 25.3*b*). Why? Gause found *P. aurelia* was able to grow six times faster than its competitor *P. caudatum* because it was able to better utilize the limited available resources, an example of exploitative competition.

From experiments such as this, Gause formulated what is now called the **principle of competitive exclusion.** This principle states that if two species are competing for a limited resource, the species that uses the resource more efficiently will eventually eliminate the other locally—no two species with the same niche can coexist when resources are limiting.

Niche Overlap

In a revealing experiment, Gause challenged *Paramecium caudatum*—the defeated species in his earlier experiments—with a third species, *P. bursaria.* Because he expected these two species to also compete for the limited bacterial food supply, Gause thought one would win out, as had happened in his previous experiments. But that's not what happened. Instead, both species survived in the culture tubes; the paramecia found a way to divide the food resources. How did they do it? In the upper part of the culture tubes, where the oxygen concentration and bacterial density were high, *P. caudatum* dominated because it was better able to feed on bacteria. However, in the lower part of the tubes, the lower oxygen concentration favored the growth of a different potential food, yeast, and *P. bursaria* was better able to eat this food. The fundamental niche of each species was the whole culture tube, but the realized niche of each species was only a portion of the tube. Because the niches of the two species did not overlap too much, both species were able to survive. However, competition did have a negative effect on the participants (figure 25.3*c*). When grown without a competitor, both species reached densities three times greater than when they were grown with a competitor.

Competitive Exclusion

Gause's principle of competitive exclusion can be restated to say that *no two species can occupy the same niche indefinitely when resources are limiting.* Certainly species can and do coexist while competing for some of the same resources. Nevertheless, Gause's theory predicts that when two species coexist on a long-term basis, either resources must not be limited or their niches will always differ in one or more features; otherwise, one species will outcompete the other and the extinction of the second species will inevitably result, a process referred to as **competitive exclusion.**

If resources are limiting, no two species can occupy the same niche indefinitely without competition driving one to extinction.

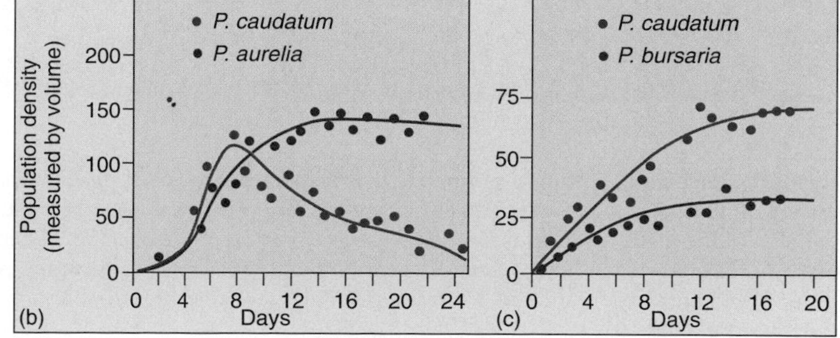

FIGURE 25.3
Competitive exclusion among three species of *Paramecium*. In the microscopic world, *Paramecium* is a ferocious predator. (*a*) In his experiments, Gause found that three species of *Paramecium* grew well alone in culture tubes. (*b*) However, *Paramecium caudatum* declined to extinction when grown with *P. aurelia* because they shared the same realized niche, and *P. aurelia* outcompeted *P. caudatum* for food resources. (*c*) *P. caudatum* and *P. bursaria* were able to coexist because the two have different realized niches and thus avoided competition.

Source: Data from Begon et al., *Ecology*, 1996. After: W.B. Clapham, *Natural Ecosystems*, Clover, Macmillan.

Resource Partitioning

Gause's exclusion principle has a very important consequence: persistent competition between two species is rare in natural communities. Either one species drives the other to extinction, or natural selection reduces the competition between them. When the late Princeton ecologist Robert MacArthur studied five species of warblers, small insect-eating forest songbirds, he found that they all appeared to be competing for the same resources. However, when he studied them more carefully, he found that each species actually fed in a different part of spruce trees and so ate different subsets of insects. One species fed on insects near the tips of branches, a second within the dense foliage, a third on the lower branches, a fourth high on the trees and a fifth at the very apex of the trees. Thus, each species of warbler had evolved so as to utilize a different portion of the spruce tree resource. They *subdivided the niche*, partitioning the available resource so as to avoid direct competition with one another.

Resource partitioning is often seen in similar species that occupy the same geographical area. Such **sympatric species** often avoid competition by living in different portions of the habitat or by utilizing different food or other resources (figure 25.4). This pattern of resource partitioning is thought to result from the process of natural selection causing initially similar species to diverge in resource use in order to reduce competitive pressures.

Evidence for the role of evolution comes from comparison of species whose ranges are only partially overlapping. Where the two species co-occur, they tend to exhibit greater differences in **morphology** (the form and structure of an organism) and resource use than do their allopatric populations. Called **character displacement,** the differences evident between sympatric species are thought to have been favored by natural selection as a mechanism to facilitate habitat partitioning and thus reduce competition. Thus, the two Darwin's finches in figure 25.5 have bills of similar size where the finches are allopatric, each living on an island where the other does not occur. On islands where they are sympatric, the two species have evolved beaks of different sizes, one adapted to larger seeds, the other to smaller ones.

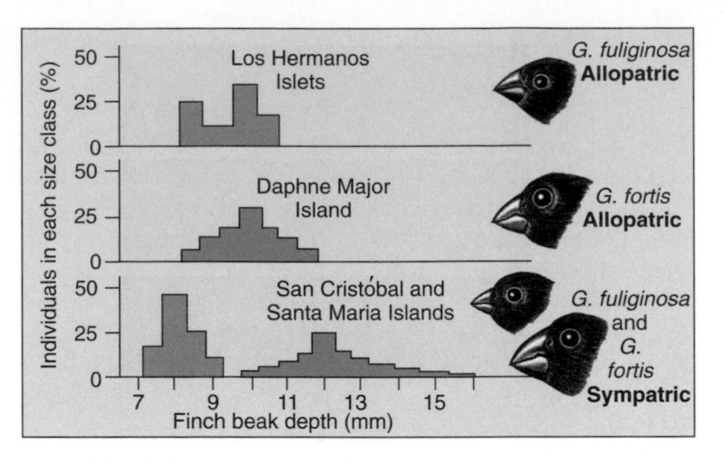

FIGURE 25.5
Character displacement in Darwin's finches. These two species of finches (genus *Geospiza*) have bills of similar size when allopatric, but different size when sympatric.
Source: Data from E.J. Heske et al, *Ecology*, 1994.

Sympatric species partition available resources, reducing competition between them.

FIGURE 25.4
Resource partitioning among sympatric lizard species. Species of *Anolis* lizards on Caribbean islands partition their tree habitats in a variety of ways. Some species of anoles occupy the canopy of trees (*a*), others use twigs on the periphery (*b*), and still others are found at the base of the trunk (*c*). In addition, some use grassy areas in the open (*d*). When two species occupy the same part of the tree, they either utilize different-sized insects as food or partition the thermal microhabitat; for example, one might only be found in the shade, whereas the other would only bask in the sun. Most interestingly, the same pattern of resource partitioning has evolved independently on different Caribbean islands.

Detecting Interspecific Competition

It is not simple to determine when two species are competing. The fact that two species use the same resources need not imply competition if that resource is not in limited supply. If the population sizes of two species are negatively correlated, such that where one species has a large population, the other species has a small population and vice versa, the two species need not be competing for the same limiting resource. Instead, the two species might be independently responding to the same feature of the environment—perhaps one species thrives best in warm conditions and the other in cool conditions.

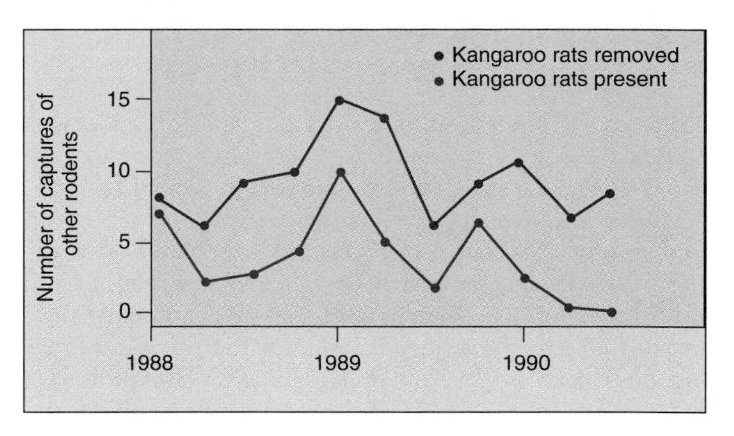

FIGURE 25.6
Detecting interspecific competition. This experiment tests the effect of removal of kangaroo rats on the population size of other rodents. Immediately after kangaroo rats were removed, the number of rodents increased relative to the enclosures that still had kangaroo rats. Notice that population sizes (as estimated by number of captures) increased and decreased in synchrony in the two treatments, probably reflecting changes in the weather.
Source: Data from E.J. Heske et al, *Ecology*, 1994.

Experimental Studies of Competition

Some of the best evidence for the existence of competition comes from experimental field studies. By setting up experiments in which two species either occur alone or together, scientists can determine whether the presence of one species has a negative effect on a population of a second species. For example, a variety of seed-eating rodents occur in the Chihuahuan Desert of the southwestern part of North America. In 1988, researchers set up a series of 50 meter × 50 meter enclosures to investigate the effect of kangaroo rats on other, smaller seed-eating rodents. Kangaroo rats were removed from half of the enclosures, but not from the other enclosures. The walls of all of the enclosures had holes in them that allowed rodents to come and go, but in the kangaroo rat removal plots, the holes were too small to allow the kangaroo rats to enter. Over the course of the next three years, the researchers monitored the number of the other, smaller seed-eating rodents present in the plots. As figure 25.6 illustrates, the number of other rodents was substantially higher in the absence of kangaroo rats, indicating that kangaroo rats compete with the other rodents and limit their population sizes.

A great number of similar experiments have indicated that interspecific competition occurs between many species of plants and animals. Effects of competition can be seen in aspects of population biology other than population size, such as behavior and individual growth rates. For example, two species of *Anolis* lizards occur on the island of St. Maarten. When one of the species, *A. gingivinus,* is placed in 12 m × 12 m enclosures without the other species, individual lizards grow faster and perch lower than lizards of the same species do when placed in enclosures in which *A. pogus* is also present.

Caution Is Necessary

Although experimental studies can be a powerful means of understanding the interactions that occur between coexisting species, they have their limitations.

First, care is necessary in interpreting the results of field experiments. Negative effects of one species on another do not automatically indicate the existence of competition. For example, many similar-sized fish have a negative effect on each other, but it results not from competition, but from the fact that adults of each species will prey on juveniles of the other species. In addition, the presence of one species may attract predators, which then also prey on the second species. In this case, the second species may have a lower population size in the presence of the first species due to the presence of predators, even if they are not competing at all. Thus, experimental studies are most effective when they are combined with detailed examination of the ecological mechanism causing the negative effect of one species on another species.

In addition, experimental studies are not always feasible. For example, the coyote has increased its population in the United States in recent years simultaneously with the decline of the grey wolf. Is this trend an indication that the species compete? Because of the size of the animals and the large geographic areas occupied by each individual, manipulative experiments involving fenced areas with only one or both species—with each experimental treatment replicated several times for statistical analysis—are not practical. Similarly, studies of slow-growing trees might require many centuries to detect competition between adult trees. In such cases, detailed studies of the ecological requirements of the species are our best bet to understanding interspecific interactions.

Experimental studies can provide strong tests of the hypothesis that interspecific competition occurs, but such studies have limitations. Detailed ecological studies are important regardless of whether experiments are conducted.

25.2 Predators and their prey coevolve.

Predation is the consuming of one organism by another. In this sense, predation includes everything from a leopard capturing and eating an antelope, to a deer grazing on spring grass. When experimental populations are set up under simple laboratory conditions, the predator often exterminates its prey and then becomes extinct itself, having nothing left to eat (figure 25.7). However, if refuges are provided for the prey, its population will drop to low levels but not to extinction. Low prey population levels will then provide inadequate food for the predators, causing the predator population to decrease. When this occurs, the prey population can recover.

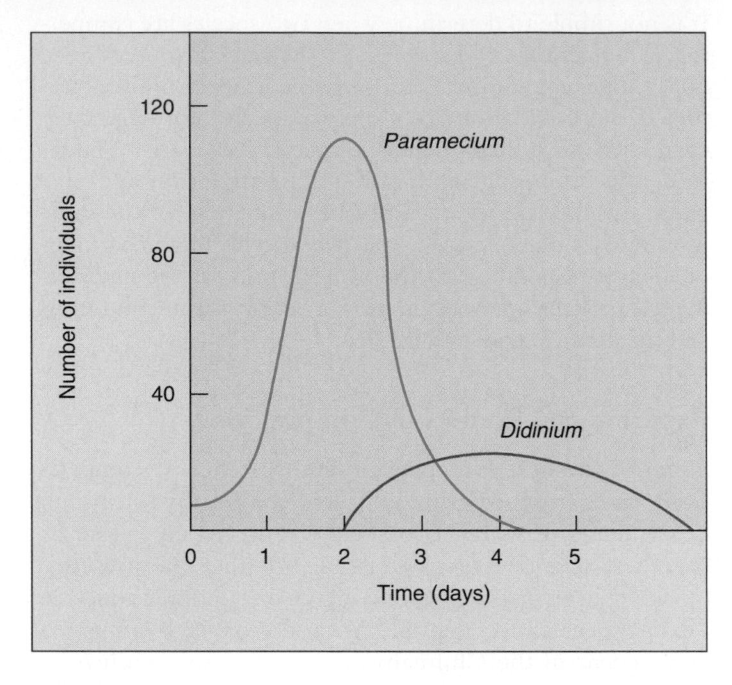

FIGURE 25.7
Predator-prey in the microscopic world. When the predatory *Didinium* is added to a *Paramecium* population, the numbers of *Didinium* initially rise, while the numbers of *Paramecium* steadily fall. When the *Paramecium* population is depleted, however, the *Didinium* individuals also die.

Predation and Prey Populations

In nature, predators can often have large effects on prey populations. Some of the most dramatic examples involve situations in which humans have either added or eliminated predators from an area. For example, the elimination of large carnivores from much of the eastern United States has led to population explosions of white-tailed deer, which strip the habitat of all edible plant life. Similarly, when sea otters were hunted to near extinction on the western coast of the United States, sea urchin populations exploded.

Conversely, the introduction of rats, dogs, and cats to many islands around the world has led to the decimation of native faunas. Populations of Galápagos tortoises on several islands are endangered, for example, by introduced rats, dogs, and cats, which eat eggs and young tortoises. Similarly, several species of birds and reptiles have been eradicated by rat predation from New Zealand and now only occur on a few offshore islands that the rats have not reached. In addition, on Stephens Island, near New Zealand, every individual of the now extinct Stephen Island wren was killed by a single lighthouse keeper's cat!

A classic example of the role predation can play in a community involves the introduction of prickly pear cactus to Australia in the nineteenth century. In the absence of predators, the cactus spread rapidly, by 1925 occupying 12 million hectares of rangeland in an impenetrable morass of spines that made cattle ranching difficult. To control the cactus, a predator from its natural habitat in Argentina, the moth *Cactoblastis cactorum*, was introduced beginning in 1926. By 1940, cactus populations had been decimated, and it now generally occurs in small populations.

Predation and Evolution

Predation provides strong selective pressures on prey populations. Any feature that would decrease the probability of capture should be strongly favored. In the next three pages, we discuss a number of defense mechanisms in plants and animals. In turn, the evolution of such features will cause natural selection to favor counteradaptations in predator populations. In this way, a coevolutionary arms race may ensue in which predators and prey are constantly evolving better defenses and better means of circumventing these defenses.

One example comes from the fossil record of mollusks and their predators. During the Mesozoic period (approximately 65 to 225 million years ago), new forms of predatory fish and crustaceans evolved that were able to crush or tear open shells. As a result, a variety of defensive measures evolved in mollusks, including thicker shells, spines, and shells too smooth for predators to be able to grasp. In turn, these adaptations may have pressured predators to evolve ever more effective predatory adaptations and tactics.

Predation can have substantial effects on prey populations. As a result prey species often evolve defensive adaptations.

Plant Defenses against Herbivores

Plants have evolved many mechanisms to defend themselves from herbivores. The most obvious are **morphological defenses:** thorns, spines, and prickles play an important role in discouraging browsers, and plant hairs, especially those that have a glandular, sticky tip, deter insect herbivores. Some plants, such as grasses, deposit silica in their leaves, both strengthening and protecting themselves. If enough silica is present in their cells, these plants are simply too tough to eat.

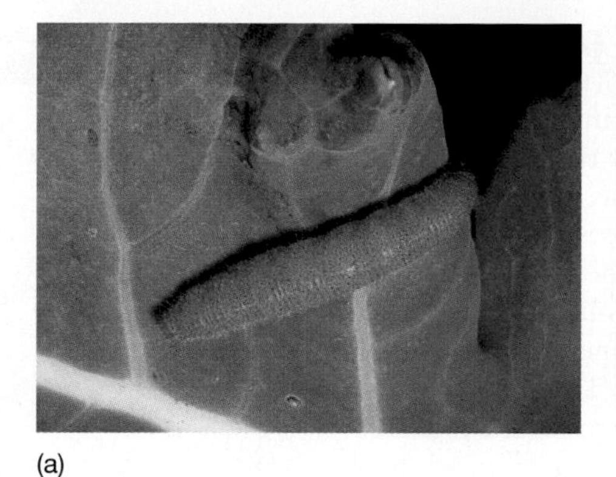

(a)

(b)

FIGURE 25.8
Insect herbivores are well suited to their hosts. (*a*) The green caterpillars of the cabbage butterfly, *Pieris rapae*, are camouflaged on the leaves of cabbage and other plants on which they feed. Although mustard oils protect these plants against most herbivores, the cabbage butterfly caterpillars are able to break down the mustard oil compounds. (*b*) An adult cabbage butterfly.

Chemical Defenses

Significant as these morphological adaptations are, the chemical defenses that occur so widely in plants are even more crucial. Best known and perhaps most important in the defenses of plants against herbivores are **secondary chemical compounds.** These are distinguished from primary compounds, which are regular components of the major metabolic pathways, such as respiration. Many plants, and apparently many algae as well, contain very structurally diverse secondary compounds that are either toxic to most herbivores or disturb their metabolism greatly, preventing, for example, the normal development of larval insects. Consequently, most herbivores tend to avoid the plants that possess these compounds.

The mustard family (Brassicaceae) is characterized by a group of chemicals known as mustard oils. These are the substances that give the pungent aromas and tastes to such plants as mustard, cabbage, watercress, radish, and horseradish. The same tastes we enjoy signal the presence of chemicals that are toxic to many groups of insects. Similarly, plants of the milkweed family (Asclepiadaceae) and the related dogbane family (Apocynaceae) produce a milky sap that deters herbivores from eating them. In addition, these plants usually contain cardiac glycosides, molecules named for their drastic effect on heart function in vertebrates.

The Evolutionary Response of Herbivores

Certain groups of herbivores are associated with each family or group of plants protected by a particular kind of secondary compound. These herbivores are able to feed on these plants without harm, often as their exclusive food source. For example, cabbage butterfly caterpillars (subfamily Pierinae) feed almost exclusively on plants of the mustard and caper families, as well as on a few other small families of plants that also contain mustard oils (figure 25.8). Similarly, caterpillars of monarch butterflies and their relatives (subfamily Danainae) feed on plants of the milkweed and dogbane families. How do these animals manage to avoid the chemical defenses of the plants, and what are the evolutionary precursors and ecological consequences of such patterns of specialization?

We can offer a potential explanation for the evolution of these particular patterns. Once the ability to manufacture mustard oils evolved in the ancestors of the caper and mustard families, the plants were protected for a time against most or all herbivores that were feeding on other plants in their area. At some point, certain groups of insects—for example, the cabbage butterflies—evolved the ability to break down mustard oils and thus feed on these plants without harming themselves. Having developed this ability, the butterflies were able to use a new resource without competing with other herbivores for it. Often, in groups of insects such as cabbage butterflies, sense organs have evolved that are able to detect the secondary compounds that their food plants produce. Clearly, the relationship that has formed between cabbage butterflies and the plants of the mustard and caper families is an example of **coevolution**.

The members of many groups of plants are protected from most herbivores by their secondary compounds. Once the members of a particular herbivore group evolve the ability to feed on them, these herbivores gain access to a new resource, which they can exploit without competition from other herbivores.

Animal Defenses against Predators

Some animals that feed on plants rich in secondary compounds receive an extra benefit. When the caterpillars of monarch butterflies feed on plants of the milkweed family, they do not break down the cardiac glycosides that protect these plants from herbivores. Instead, the caterpillars concentrate and store the cardiac glycosides in fat bodies; they then pass them through the chrysalis stage to the adult and even to the eggs of the next generation. The incorporation of cardiac glycosides thus protects all stages of the monarch life cycle from predators. A bird that eats a monarch butterfly quickly regurgitates it (figure 25.9) and in the future avoids the conspicuous orange-and-black pattern that characterizes the adult monarch. Some birds, however, appear to have acquired the ability to tolerate the protective chemicals. These birds eat the monarchs.

(a)

(b)

FIGURE 25.9
A blue jay learns that monarch butterflies taste bad. (*a*) This cage-reared jay had never seen a monarch butterfly before it tried eating one. (*b*) The same jay regurgitated the butterfly a few minutes later. This bird will probably avoid trying to capture all orange-and-black insects in the future.

Defensive Coloration

Many insects that feed on milkweed plants are brightly colored; they advertise their poisonous nature using an ecological strategy known as **warning coloration,** or **aposematic coloration.** Showy coloration is characteristic of animals that use poisons and stings to repel predators, while organisms that lack specific chemical defenses are seldom brightly colored. In fact, many have **cryptic coloration**—color that blends with the surroundings and thus hides the individual from predators (figure 25.10). Camouflaged animals usually do not live together in groups because a predator that discovers one individual gains a valuable clue to the presence of others.

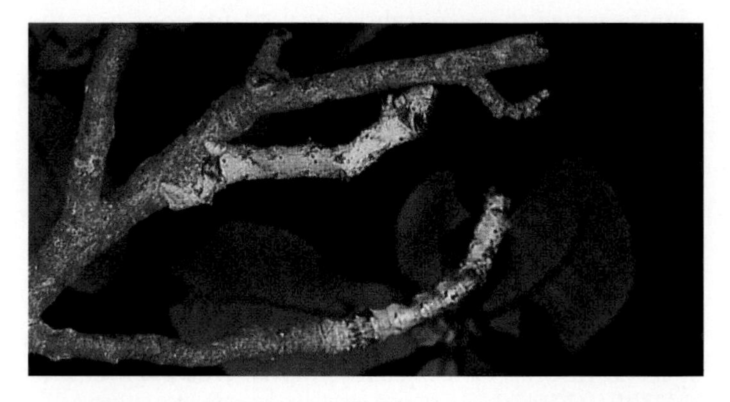

FIGURE 25.10
Cryptic coloration. An inchworm caterpillar (*Necophora quernaria*) (hanging from the upper twig) closely resembles a twig.

Chemical Defenses

Animals also manufacture and use a startling array of substances to perform a variety of defensive functions. Bees, wasps, predatory bugs, scorpions, spiders, and many other arthropods use chemicals to defend themselves and to kill their prey. In addition, various chemical defenses have evolved among marine animals and the vertebrates, including venomous snakes, lizards, fishes, and some birds. The poison-dart frogs of the family Dendrobatidae produce toxic alkaloids in the mucus that covers their brightly colored skin (figure 25.11). Some of these toxins are so powerful that a few micrograms will kill a person if injected into the bloodstream. More than 200 different alkaloids have been isolated from these frogs, and some are playing important roles in neuromuscular research. There is an intensive investigation of marine animals, algae, and flowering plants for new drugs to fight cancer and other diseases, or as sources of antibiotics.

FIGURE 25.11
Vertebrate chemical defenses. Frogs of the family Dendrobatidae, abundant in the forests of Latin America, are extremely poisonous to vertebrates. Dendrobatids advertise their toxicity with aposematic coloration, as shown here.

Animals defend themselves against predators with warning coloration, camouflage, and chemical defenses such as poisons and stings.

Mimicry

During the course of their evolution, many species have come to resemble distasteful ones that exhibit aposematic coloration. The mimic gains an advantage by looking like the distasteful model. Two types of mimicry have been identified: Batesian and Müllerian mimicry.

Batesian Mimicry

Batesian mimicry is named for Henry Bates, the British naturalist who first brought this type of mimicry to general attention in 1857. In his journeys to the Amazon region of South America, Bates discovered many instances of palatable insects that resembled brightly colored, distasteful species. He reasoned that the mimics would be avoided by predators, who would be fooled by the disguise into thinking the mimic actually is the distasteful model.

Many of the best-known examples of Batesian mimicry occur among butterflies and moths. Obviously, predators in systems of this kind must use visual cues to hunt for their prey; otherwise, similar color patterns would not matter to potential predators. There is also increasing evidence indicating that Batesian mimicry can also involve nonvisual cues, such as olfaction, although such examples are less obvious to humans.

The kinds of butterflies that provide the models in Batesian mimicry are, not surprisingly, members of groups whose caterpillars feed on only one or a few closely related plant families. The plant families on which they feed are strongly protected by toxic chemicals. The model butterflies incorporate the poisonous molecules from these plants into their bodies. The mimic butterflies, in contrast, belong to groups in which the feeding habits of the caterpillars are not so restricted. As caterpillars, these butterflies feed on a number of different plant families unprotected by toxic chemicals.

One often-studied mimic among North American butterflies is the viceroy, *Limenitis archippus* (figure 25.12*a*). This butterfly, which resembles the poisonous monarch, ranges from central Canada through much of the United States and into Mexico. The caterpillars feed on willows and cottonwoods, and neither caterpillars nor adults were thought to be distasteful to birds, although recent findings may dispute this. Interestingly, the Batesian mimicry seen in the adult viceroy butterfly does not extend to the caterpillars: viceroy caterpillars are camouflaged on leaves, resembling bird droppings, while the monarch's distasteful caterpillars are very conspicuous.

(a) Batesian mimicry: Monarch (*Danaus*) is poisonous; viceroy (*Limenitis*) is palatable mimic

(b) Müllerian mimicry: two pairs of mimics; all are distasteful

FIGURE 25.12
Mimicry. (*a*) Batesian mimicry. Monarch butterflies (*Danaus plexippus*) are protected from birds and other predators by the cardiac glycosides they incorporate from the milkweeds and dogbanes they feed on as larvae. Adult monarch butterflies advertise their poisonous nature with warning coloration. Viceroy butterflies (*Limenitis archippus*) are Batesian mimics of the poisonous monarch. (*b*) Pairs of Müllerian mimics. *Heliconius erato* and *H. melpomene* are sympatric, and *H. sapho* and *H. cydno* are sympatric. All of these butterflies are distasteful. They have evolved similar coloration patterns in sympatry to minimize predation; predators need only learn one pattern to avoid.

Müllerian Mimicry

Another kind of mimicry, **Müllerian mimicry,** was named for German biologist Fritz Müller, who first described it in 1878. In Müllerian mimicry, several unrelated but protected animal species come to resemble one another (figure 25.12*b*). If animals that resemble one another are all poisonous or dangerous, they gain an advantage because a predator will learn more quickly to avoid them. In some cases, predator populations even evolve an innate avoidance of species; such evolution may occur more quickly when multiple dangerous prey look alike.

In both Batesian and Müllerian mimicry, mimic and model must not only look alike but also act alike if predators are to be deceived. For example, the members of several families of insects that closely resemble wasps behave surprisingly like the wasps they mimic, flying often and actively from place to place.

In Batesian mimicry, unprotected species resemble others that are distasteful. Both species exhibit aposematic coloration. In Müllerian mimicry, two or more unrelated but protected species resemble one another, thus achieving a kind of group defense.

25.3 Evolution sometimes fosters cooperation.

Coevolution and Symbiosis

The plants, animals, protists, fungi, and bacteria that live together in communities have changed and adjusted to one another continually over a period of millions of years. For example, many features of flowering plants have evolved in relation to the dispersal of the plant's gametes by animals (figure 25.13). These animals, in turn, have evolved a number of special traits that enable them to obtain food or other resources efficiently from the plants they visit, often from their flowers. While doing so, the animals pick up pollen, which they may deposit on the next plant they visit, or seeds, which may be left elsewhere in the environment, sometimes a great distance from the parental plant.

Such interactions, which involve the long-term, mutual evolutionary adjustment of the characteristics of the members of biological communities, are examples of **coevolution,** a phenomenon we have already seen in predator-prey interactions.

FIGURE 25.13
Pollination by bat. Many flowers have coevolved with other species to facilitate pollen transfer. Insects are widely known as pollinators, but they're not the only ones. Notice the cargo of pollen on the bat's snout.

Symbiosis Is Widespread

Another type of coevolution involves **symbiotic relationships** in which two or more kinds of organisms live together in often elaborate and more-or-less permanent relationships. All symbiotic relationships carry the potential for coevolution between the organisms involved, and in many instances the results of this coevolution are fascinating. Examples of symbiosis include *lichens,* which are associations of certain fungi with green algae or cyanobacteria. Lichens are discussed in more detail in chapter 36. Another important example are *mycorrhizae,* the association between fungi and the roots of most kinds of plants. The fungi expedite the plant's absorption of certain nutrients, and the plants in turn provide the fungi with carbohydrates. Similarly, root nodules that occur in legumes and certain other kinds of plants contain bacteria that fix atmospheric nitrogen and make it available to their host plants.

In the tropics, leafcutter ants are often so abundant that they can remove a quarter or more of the total leaf surface of the plants in a given area. They do not eat these leaves directly; rather, they take them to underground nests, where they chew them up and inoculate them with the spores of particular fungi. These fungi are cultivated by the ants and brought from one specially prepared bed to another, where they grow and reproduce. In turn, the fungi constitute the primary food of the ants and their larvae. The relationship between leafcutter ants and these fungi is an excellent example of symbiosis.

Kinds of Symbiosis

The major kinds of symbiotic relationships include (1) **commensalism,** in which one species benefits while the other neither benefits nor is harmed; (2) **mutualism,** in which both participating species benefit; and (3) **parasitism,** in which one species benefits but the other is harmed. Parasitism can also be viewed as a form of predation, although the organism that is preyed upon does not necessarily die.

Coevolution is a term that describes the long-term evolutionary adjustments of species to one another. In symbiosis two or more species interact closely, with at least one species benefitting.

Commensalism

Commensalism is a symbiotic relationship that benefits one species and neither hurts nor helps the other. In nature, individuals of one species are often physically attached to members of another. For example, epiphytes are plants that grow on the branches of other plants. In general, the host plant is unharmed, while the epiphyte that grows on it benefits. Similarly, various marine animals, such as barnacles, grow on other, often actively moving sea animals like whales and thus are carried passively from place to place. These "passengers" presumably gain more protection from predation than they would if they were fixed in one place, and they also reach new sources of food. The increased water circulation that such animals receive as their host moves around may be of great importance, particularly if the passengers are filter feeders. The gametes of the passenger are also more widely dispersed than would be the case otherwise.

FIGURE 25.14
Commensalism in the sea. Clownfishes, such as this *Amphiprion perideraion* in Guam, often form symbiotic associations with sea anemones, gaining protection by remaining among their tentacles and gleaning scraps from their food. Different species of anemones secrete different chemical mediators; these attract particular species of fishes and may be toxic to the fish species that occur symbiotically with other species of anemones in the same habitat. There are 26 species of clownfishes, all found only in association with sea anemones; 10 species of anemones are involved in such associations, so that some of the anemone species are host to more than one species of clownfish.

Examples of Commensalism

The best-known examples of commensalism involve the relationships between certain small tropical fishes and sea anemones, marine animals that have stinging tentacles (see chapter 44). These fish have evolved the ability to live among the tentacles of sea anemones, even though these tentacles would quickly paralyze other fishes that touched them (figure 25.14). The anemone fishes feed on the detritus left from the meals of the host anemone, remaining uninjured under remarkable circumstances.

On land, an analogous relationship exists between birds called oxpeckers and grazing animals such as cattle or rhinoceros. The birds spend most of their time clinging to the animals, picking off parasites and other insects, carrying out their entire life cycles in close association with the host animals.

When Is Commensalism Commensalism?

In each of these instances, it is difficult to be certain whether the second partner receives a benefit or not; there is no clear-cut boundary between commensalism and mutualism. For instance, it may be advantageous to the sea anemone to have particles of food removed from its tentacles; it may then be better able to catch other prey. Similarly, while often thought of as commensalism, the association of grazing mammals and gleaning birds is actually an example of mutualism. The mammal benefits by having parasites and other insects removed from its body, but the birds also benefit by gaining a dependable source of food.

On the other hand, commensalism can easily transform itself into parasitism. For example, oxpeckers are also known to pick not only parasites, but also scabs off their grazing hosts. Once the scab is picked, the birds drink the blood that flows from the wound. Occasionally, the cumulative effect of persistent attacks can greatly weaken the herbivore, particularly when conditions are not favorable, such as during droughts.

Commensalism is the benign use of one organism by another.

Mutualism

Mutualism is a symbiotic relationship among organisms in which both species benefit. Examples of mutualism are of fundamental importance in determining the structure of biological communities. Some of the most spectacular examples of mutualism occur among flowering plants and their animal visitors, including insects, birds, and bats. As we will see in chapter 37, during the course of their evolution, the characteristics of flowers have evolved in large part in relation to the characteristics of the animals that visit them for food and, in doing so, spread their pollen from individual to individual. At the same time, characteristics of the animals have changed, increasing their specialization for obtaining food or other substances from particular kinds of flowers.

Another example of mutualism involves ants and aphids. Aphids, also called greenflies, are small insects that suck fluids from the phloem of living plants with their piercing mouthparts. They extract a certain amount of the sucrose and other nutrients from this fluid, but they excrete much of it in an altered form through their anus. Certain ants have taken advantage of this—in effect, domesticating the aphids. The ants carry the aphids to new plants, where they come into contact with new sources of food, and then consume as food the "honeydew" that the aphids excrete.

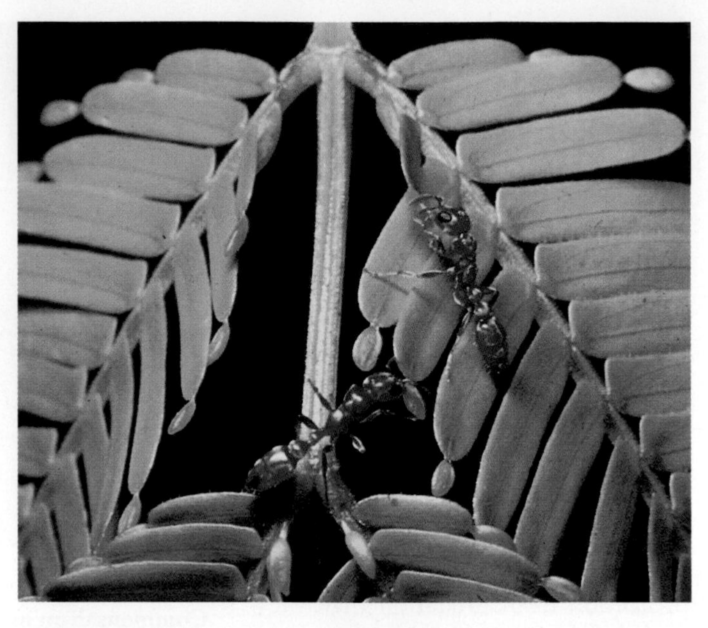

FIGURE 25.15
Mutualism: ants and acacias. Ants of the genus *Pseudomyrmex* live within the hollow thorns of certain species of acacia trees in Latin America. The nectaries at the bases of the leaves and the Beltian bodies at the ends of the leaflets provide food for the ants. The ants, in turn, supply the acacias with organic nutrients and protect the acacia from herbivores and shading from other plants.

Ants and Acacias

A particularly striking example of mutualism involves ants and certain Latin American species of the plant genus *Acacia.* In these species, certain leaf parts, called stipules, are modified as paired, hollow thorns. The thorns are inhabited by stinging ants of the genus *Pseudomyrmex*, which do not nest anywhere else (figure 25.15). Like all thorns that occur on plants, the acacia thorns serve to deter herbivores.

At the tip of the leaflets of these acacias are unique, protein-rich bodies called Beltian bodies, named after the nineteenth-century British naturalist Thomas Belt. Beltian bodies do not occur in species of *Acacia* that are not inhabited by ants, and their role is clear: they serve as a primary food for the ants. In addition, the plants secrete nectar from glands near the bases of their leaves. The ants consume this nectar as well, feeding it and the Beltian bodies to their larvae.

Obviously, this association is beneficial to the ants, and one can readily see why they inhabit acacias of this group. The ants and their larvae are protected within the swollen thorns, and the trees provide a balanced diet, including the sugar-rich nectar and the protein-rich Beltian bodies. What, if anything, do the ants do for the plants?

Whenever any herbivore lands on the branches or leaves of an acacia inhabited by ants, the ants, which continually patrol the acacia's branches, immediately attack and devour the herbivore. The ants that live in the acacias also help their hosts to compete with other plants. The ants cut away any branches of other plants that touch the acacia in which they are living. They create, in effect, a tunnel of light through which the acacia can grow, even in the lush deciduous forests of lowland Central America. In fact, when an ant colony is experimentally removed from a tree, the acacia is unable to compete successfully in this habitat. Finally, the ants bring organic material into their nests. The parts they do not consume, together with their excretions, provide the acacias with an abundant source of nitrogen.

As with commensalism, however, things are not always as they seem. Ant-acacia mutualisms also occur in Africa. In Kenya, several species of acacia ants occur, but only one species occurs on any tree. One species, *Crematogaster nigriceps*, is competitively inferior to two of the other species. To prevent invasion by other ant species, *C. nigriceps* prunes the branches of the acacia, preventing it from coming into contact with branches of other trees, which would serve as a bridge for invaders. Although this behavior is beneficial to the ant, it is detrimental to the tree, as it destroys the tissue from which flowers are produced, essentially sterilizing the tree. In this case, what has initially evolved as a mutualistic interaction has instead become a parasitic one.

Mutualism involves cooperation between species, to the mutual benefit of both.

Parasitism

Parasitism may be regarded as a special form of symbiosis in which the predator, or parasite, is much smaller than the prey and remains closely associated with it. Parasitism is harmful to the prey organism and beneficial to the parasite. The concept of parasitism seems obvious, but individual instances are often surprisingly difficult to distinguish from predation and from other kinds of symbiosis.

External Parasites

Parasites that feed on the exterior surface of an organism are external parasites, or **ectoparasites.** Many instances of external parasitism are known (figure 25.16). Lice, which live on the bodies of vertebrates—mainly birds and mammals—are normally considered parasites. Mosquitoes are not considered parasites, even though they draw food from birds and mammals in a similar manner to lice, because their interaction with their host is so brief.

Parasitoids are insects that lay eggs on living hosts. This behavior is common among wasps, whose larvae feed on the body of the unfortunate host, often killing it.

FIGURE 25.16
An external parasite. The flowering plant dodder (*Cuscuta*) is a parasite and has lost its chlorophyll and its leaves in the course of its evolution. Because it is heterotrophic, unable to manufacture its own food, dodder obtains its food from the host plants it grows on.

Internal Parasites

Vertebrates are parasitized internally by **endoparasites,** members of many different phyla of animals and protists. Invertebrates also have many kinds of parasites that live within their bodies. Bacteria and viruses are not usually considered parasites, even though they fit our definition precisely.

Internal parasitism is generally marked by much more extreme specialization than external parasitism, as shown by the many protist and invertebrate parasites that infect humans. The more closely the life of the parasite is linked with that of its host, the more its morphology and behavior are likely to have been modified during the course of its evolution. The same is true of symbiotic relationships of all sorts. Conditions within the body of an organism are different from those encountered outside and are apt to be much more constant. Consequently, the structure of an internal parasite is often simplified, and unnecessary armaments and structures are lost as it evolves.

FIGURE 25.17
Brood parasitism. This bird, a dunnock, is feeding a cuckoo chick in its nest. The cuckoo chick is larger than the adult bird, but the bird does not recognize that the cuckoo is not its own offspring. Cuckoo mothers sneak into the nests of other birds and lay an egg, entrusting the care of their offspring to an unwitting bird of another species.

Brood Parasitism

Not all parasites consume the body of their host. In brood parasitism, birds like cowbirds and European cuckoos lay their eggs in the nests of other species. The host parents raise the brood parasite as if it were one of their own clutch, in many cases investing more in feeding the imposter than in feeding their own offspring (figure 25.17). The brood parasite reduces the reproductive success of the foster parent hosts, so it is not surprising that in some cases natural selection has fostered the hosts' ability to detect parasite eggs and reject them. What is more surprising is that in many other species, the ability to detect parasite eggs has not evolved.

In parasitism, one organism serves as a host to another organism, usually to the host's disadvantage.

Interactions among Ecological Processes

We have seen the different ways in which species within a community can interact with each other. In nature, however, more than one type of interaction usually occurs at the same time. In many cases, the outcome of one type of interaction is modified or even reversed when another type of interaction is also occurring.

Predation Reduces Competition

When resources are limiting, a superior competitor can eliminate other species from a community. However, predators can prevent or greatly reduce competitive exclusion by reducing the numbers of individuals of competing species. A given predator may often feed on two, three, or more kinds of plants or animals in a given community. The predator's choice depends partly on the relative abundance of the prey options. In other words, a predator may feed on species *A* when it is abundant and then switch to species *B* when *A* is rare. Similarly, a given prey species may be a primary source of food for increasing numbers of species as it becomes more abundant. In this way, superior competitors may be prevented from outcompeting other species.

Such patterns are often characteristic of biological communities in marine intertidal habitats. For example, in preying selectively on bivalves, starfish prevent bivalves from monopolizing such habitats, opening up space for many other organisms (figure 25.18). When starfish are removed from a habitat, species diversity falls precipitously, the seafloor community coming to be dominated by a few species of bivalves. Because predation tends to reduce competition in natural communities, it is usually a mistake to attempt to eliminate a major predator such as wolves or mountain lions from a community. The result is to decrease rather than increase the biological diversity of the community, the opposite of what is intended.

(a)

(b)

FIGURE 25.18
Predation reduces competition. (*a*) In a controlled experiment in a coastal ecosystem, an investigator removed a key predator (*Pisaster*). (*b*) In response, fiercely competitive mussels exploded in growth, effectively crowding out seven other indigenous species.

Parasitism May Counter Competition

Parasites may effect sympatric species differently and thus influence the outcome of interspecific interactions. In a classic experiment, Thomas Park of the University of Chicago investigated interactions between two flour beetles, *Tribolium castaneum* and *T. confusum* with a parasite, *Adelina*. In the absence of the parasite, *T. castaneum* is dominant and *T. confusum* normally goes extinct. When the parasite is present, however, the outcome is reversed and *T. castaneum* perishes. Similar effects of parasites in natural systems have been observed in many species. For example, in the *Anolis* lizards of St. Maarten mentioned previously, the competitively inferior species is resistant to malaria, whereas the other species is highly susceptible. Only in areas in which the malaria parasite occurs are the two species capable of coexisting.

Indirect Effects

In some cases, species may not directly interact, yet the presence of one species may effect a second species by way of interactions with a third species. Such effects are termed **indirect effects.** For example, in the Chihuahuan Desert, rodents and ants both eat seeds. Thus, one might expect them to compete with each other. However, when all rodents were completely removed from large enclosures (unlike the experiment discussed above, there were no holes in

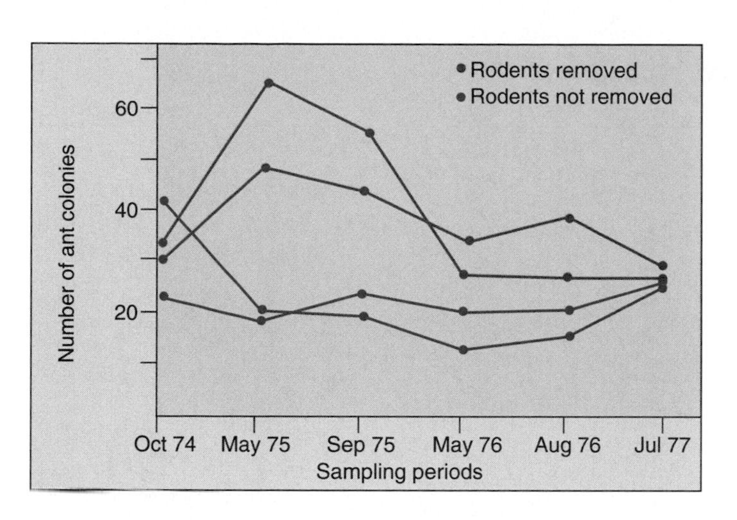

FIGURE 25.19
Change in ant population size after the removal of rodents. Ants initially increased in population size relative to ants in the enclosures from which rodents weren't removed, but then these ant populations declined.

Source: Data from D.W. Davidson et al., "Granivory in a desert Ecosystem", *Ecology*, 1984.

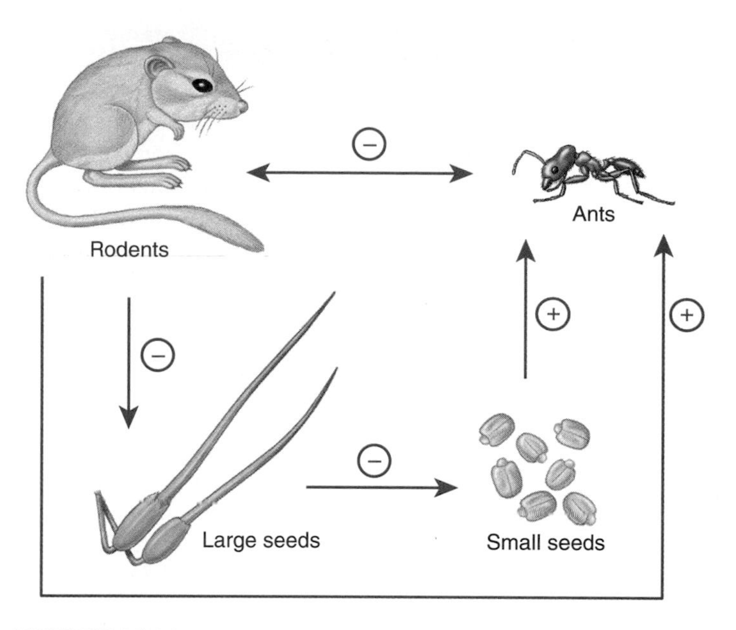

FIGURE 25.20
Rodent-ant interactions. Rodents and ants both eat seeds, so the presence of rodents has a negative effect on ants and vice versa. However, the presence of rodents has a negative effect on large seeds. In turn, the number of plants with large seeds has a negative effect on plants that produce small seeds. Hence, the presence of rodents should increase the number of small seeds. In turn, the number of small seeds has a positive effect on ant populations. Thus, indirectly, the presence of rodents has a positive effect on ant population size.

the enclosure walls, so once removed, rodents couldn't get back in), ant populations first increased, but then declined (figure 25.19). The initial increase was the expected result of removing a competitor; why did it reverse? The answer reveals the intricacies of natural ecosystems (figure 25.20). Rodents prefer large seeds, whereas ants prefer smaller seeds. Further, in this system plants with large seeds are competitively superior to plants with small seeds. Thus, the removal of rodents leads to an increase in the number of plants with large seeds, which reduces the number of small seeds available to ants, which thus leads to a decline in ant populations. Thus, the effect of rodents on ants is complicated: a direct negative effect of resource competition and an indirect, positive effect mediated by plant competition.

Keystone Species

Species that have particularly strong effects on the composition of communities are termed **keystone species.** Predators, such as the starfish, can often serve as keystone species by preventing one species from outcompeting others, thus maintaining high levels of species richness in a community.

There are, however, a wide variety of other types of keystone species. Some species manipulate the environment in ways that create new habitats for other species. Beavers, for example, change running streams into small impoundments, changing the flow of water and flooding areas (figure 25.21). Similarly, alligators excavate deep holes at the bottoms of lakes. In times of drought, these holes are the only areas in which water remains, thus allowing aquatic species that otherwise would perish to persist until the drought ends and the lake refills.

FIGURE 25.21
Example of a keystone species. Beavers, by constructing dams and transforming flowing streams into ponds, create new habitats for many plant and animal species.

Many different processes are likely to be occurring simultaneously within communities. Only by understanding how these processes interact will we be able to understand how communities function.

25.4 Ecological succession may increase species richness.

Even when the climate of an area remains stable year after year, ecosystems have a tendency to change from simple to complex in a process known as **succession.** This process is familiar to anyone who has seen a vacant lot or cleared woods slowly become occupied by an increasing number of plants, or a pond become dry land as it is filled with vegetation encroaching from the sides.

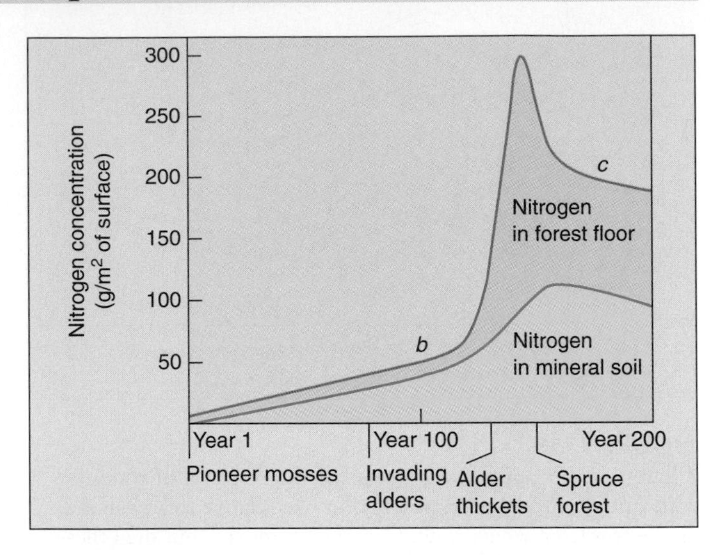

FIGURE 25.22
Plant succession produces progressive changes in the soil. Initially, the glacial moraine at Glacier Bay, Alaska, portrayed in figure 25.23, had little soil nitrogen, but nitrogen-fixing alders led to a buildup of nitrogen in the soil, encouraging the subsequent growth of the conifer forest. Letters in the graph correspond to photographs in parts *b* and *c* of figure 25.23.

Succession

If a wooded area is cleared and left alone, plants will slowly reclaim the area. Eventually, traces of the clearing will disappear and the area will again be woods. This kind of succession, which occurs in areas where an existing community has been disturbed, is called **secondary succession.**

In contrast, **primary succession** occurs on bare, lifeless substrate, such as rocks, or in open water, where organisms gradually move into an area and change its nature. Primary succession occurs in lakes left behind after the retreat of glaciers, on volcanic islands that rise above the sea, and on land exposed by retreating glaciers (figure 25.22). Primary succession on glacial moraines provides an example (figure 25.23). On bare, mineral-poor soil, lichens grow first, forming small pockets of soil. Acidic secretions from the lichens help to break down the substrate and add to the accumulation of soil. Mosses then colonize these pockets of soil, eventually building up enough nutrients in the soil for alder shrubs to take hold. Over a hundred years, the alders build up the soil nitrogen levels until spruce are able to thrive, eventually crowding out the alder and forming a dense spruce forest.

In a similar example, an **oligotrophic** lake—one poor in nutrients—may gradually, by the accumulation of organic matter, become **eutrophic**—rich in nutrients. As this occurs, the composition of communities will change, first increasing in species richness and then declining.

Primary succession in different habitats often eventually arrives at the same kinds of vegetation—vegetation characteristic of the region as a whole. This relationship led American ecologist F. E. Clements, at about the turn of the century, to propose the concept of a final **climax community.** With an increasing realization that (1) the climate keeps changing, (2) the process of succession is often very slow, and (3) the nature of a region's vegetation is being determined to an increasing extent by human activities, ecologists do not consider the concept of "climax community" to be as useful as they once did.

Why Succession Happens

Succession happens because species alter the habitat and the resources available in it in ways that favor other species. Three dynamic concepts are of critical importance in the process: tolerance, facilitation, and inhibition.

1. **Tolerance.** Early successional stages are characterized by weedy *r*-selected species that are tolerant of the harsh, abiotic conditions in barren areas.
2. **Facilitation.** The weedy early successional stages introduce local changes in the habitat that favor other, less weedy species. Thus, the mosses in the Glacier Bay succession convert nitrogen to a form that allows alders to invade. The alders in turn lower soil pH as their fallen leaves decompose, and spruce and hemlock, which require acidic soil, are able to invade.
3. **Inhibition.** Sometimes the changes in the habitat caused by one species, while favoring other species, inhibit the growth of the species that caused them. Alders, for example, do not grow as well in acidic soil as the spruce and hemlock that replace them.

Over the course of succession, the number of species typically increases as the environment becomes more hospitable. In some cases, however, as ecosystems mature, more *K*-selected species replace *r*-selected ones, and superior competitors force out other species, leading ultimately to a decline in species richness.

Communities evolve to have greater total biomass and species richness in a process called succession.

(a)

(b)

(c)

FIGURE 25.23
Primary succession at Alaska's Glacier Bay. (*a*) The sides of the glacier have been retreating at a rate of some 8 meters a year, leaving behind exposed soil from which nitrogen and other minerals have been leached out. The first invaders of these exposed sites are pioneer lichen and moss species with nitrogen-fixing mutualistic microbes. Within 20 years, young alder shrubs take hold. (*b*) Rapidly fixing nitrogen, they soon form dense thickets. As soil nitrogen levels rise, (*c*) spruce crowd out the mature alders, forming a forest.

The Role of Disturbance

Disturbances often interrupt the succession of plant communities. Depending on the magnitude of the disturbance, communities may revert to earlier stages of succession or even, in extreme cases, begin at the earliest stages of primary succession. Disturbances severe enough to disrupt succession include calamities such as forest fires, drought, and floods. Animals may also cause severe disruptions. Gypsy moths can devastate a forest by consuming its trees. Unregulated deer populations may grow explosively, the deer overgrazing and so destroying the forest they live in, in the same way too many cattle overgraze a pasture by eating all available grass down to the ground.

FIGURE 25.24
Intermediate disturbance. A single fallen tree creates a small light gap in the tropical rain forest of Panama. Such gaps play a key role in maintaining the high species diversity of the rain forest.

Intermediate Disturbance Hypothesis

In some cases, disturbance may act to increase the species richness of an area. According to the intermediate disturbance hypothesis, communities experiencing moderate amounts of disturbance will have higher levels of species richness than communities experiencing either little or great amounts of disturbance. Two factors could account for this pattern. First, in communities in which moderate amounts of disturbance occur, patches of habitat will exist at different successional stages. Thus, within the area as a whole, species diversity will be greatest because the full range of species—those characteristic of all stages of succession—will be present. For example, a pattern of intermittent episodic disturbance that produces gaps in the rain forest (like when a tree falls) allows invasion of the gap by other species (figure 25.24). Eventually, the species inhabiting the gap will go through a successional sequence, one tree replacing another, until a canopy tree species comes again to occupy the gap. But if there are lots of gaps of different ages in the forest, many different species will coexist, some in young gaps, others in older ones.

Second, moderate levels of disturbance may prevent communities from reaching the final stages of succession, in which a few dominant competitors eliminate most of the other species. On the other hand, too much disturbance might leave the community continually in the earliest stages of succession, when species richness is relatively low.

Ecologists are increasingly realizing that disturbance is the norm, rather than the exception, in many communities. As a result, the idea that communities inexorably move along a successional trajectory culminating in the development of a climax community is no longer widely accepted. Rather, predicting the state of a community in the future may be difficult because the unpredictable occurrence of disturbances will often counter successional changes. Understanding the role that disturbances play in structuring communities is currently an important area of investigation in ecology.

Succession is often disrupted by natural or human causes. In some cases, intermediate levels of disturbance may maximize the species richness of a community.

Chapter 25

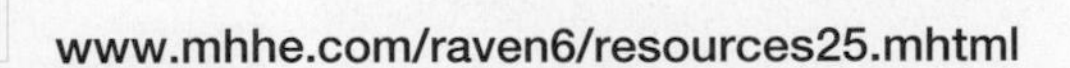

Summary | **Questions** | **Media Resources**

25.1 Interactions among competing species shape ecological niches.

- Each species plays a specific role in its ecosystem; this role is called its niche.
- An organism's fundamental niche is the total niche that the organism would occupy in the absence of competition. Its realized niche is the actual niche it occupies in nature.
- Two species cannot occupy the same niche for long if resources are limiting; one will outcompete the other, driving it to extinction.
- Species can coexist by partitioning resources to minimize competition.

1. What do you think would be the difference between interspecific competition and intraspecific competition? What is Gause's principle of competitive exclusion?

2. Is the term niche synonymous with the term *habitat?* Why or why not? How does an organism's fundamental niche differ from its realized niche?

- Introduction to Communities

- *On Science* Article: Killer Bees

- Art Quiz: Competition and Niches

25.2 Predators and their prey coevolve.

- Predator-prey relationships are of crucial importance in limiting population sizes in nature.
- Plants are often protected from herbivores by chemicals they manufacture.
- Warning, or aposematic, coloration is characteristic of organisms that are poisonous, sting, or are otherwise harmful. In contrast, cryptic coloration, or camouflage, is characteristic of nonpoisonous organisms.

3. What morphological defenses do plants use to defend themselves against herbivores?

4. Consider aposematic coloration, cryptic coloration, and Batesian mimicry. Which would be associated with an adult viceroy butterfly? Which would be associated with a larval monarch butterfly? Which would be associated with a larval viceroy butterfly?

- Art Quiz: Predator-Prey Cycle

25.3 Evolution sometimes fosters cooperation.

- Coevolution occurs when different kinds of organisms evolve adjustments to one another over long periods of time.
- Many organisms have coevolved to a point of dependence. In commensalism, only one organism benefits while the other is unharmed; in mutualism the relationship is mutually beneficial; and in parasitism one organism serves as a host to another to the host's disadvantage.

5. Why is eliminating predators a bad idea for species richness?

6. How can predation and competition interact in regulating species diversity of a community?

- Community Organization

- Student Research: Hermit Crab–Sea Anemone Associations

25.4 Ecological succession may increase species richness.

- Primary succession takes place in barren areas, like rocks or open water. Secondary succession takes place in areas where the original communities of organisms have been disturbed.
- Succession occurs because of tolerance, facilitation, and inhibition.
- Disturbance can disrupt successional changes. In some cases, disturbance can increase species richness of a community.

7. Why have scientists altered the concept of a final, climax vegetation in a given ecosystem? What types of organisms are often associated with early stages of succession? What is the role of disturbance in succession?

- Succession

26

Animal Behavior

FIGURE 26.1
Rearing offspring involves complex behaviors. Living in groups called prides makes lions better mothers. Females share the responsibilities of nursing and protecting the pride's young, increasing the probability that the youngsters will survive into adulthood.

Concept Outline

26.1 Ethology focuses on the natural history of behavior.

Approaches to the Study of Behavior. Field biologists focus on evolutionary aspects of behavior.
Behavioral Genetics. At least some behaviors are genetically determined.

26.2 Comparative psychology focuses on how learning influences behavior.

Learning. Association plays a major role in learning.
The Development of Behavior. Parent-offspring interactions play a key role in the development of behavior.
The Physiology of Behavior. Hormones influence many behaviors, particularly reproductive ones.
Behavioral Rhythms. Many behaviors are governed by innate biological clocks.

26.3 Communication is a key element of many animal behaviors.

Courtship. Animals use many kinds of signals to court one another.
Communication in Social Groups. Bees and other social animals communicate in complex ways.

26.4 Migratory behavior presents many puzzles.

Orientation and Migration. Animals use many cues from the environment to navigate during migrations.

26.5 To what degree animals "think" is a subject of lively dispute.

Animal Cognition. It is not clear to what degree animals "think."

Organisms interact with their environment in many ways. To understand these interactions, we need to appreciate both the internal factors that shape the way an animal behaves, as well as aspects of the external environment that affect individuals and organisms. In this chapter, we explore the mechanisms that determine an animal's behavior (figure 26.1), as well as the ways in which behavior develops in an individual. In the next chapter, we will consider the field of behavioral ecology, which investigates how natural selection has molded behavior through evolutionary time.

26.1 Ethology focuses on the natural history of behavior.

Approaches to the Study of Behavior

During the past two decades, the study of animal behavior has emerged as an important and diverse science that bridges several disciplines within biology. Evolution, ecology, physiology, genetics, and psychology all have natural and logical linkages with the study of behavior, each discipline adding a different perspective and addressing different questions.

Research in animal behavior has made major contributions to our understanding of nervous system organization, child development, and human communication, as well as the process of speciation, community organization, and the mechanism of natural selection itself. The study of the behavior of nonhuman animals has led to the identification of general principles of behavior, which have been applied, often controversially, to humans. This has changed the way we think about the origins of human behavior and the way we perceive ourselves.

Behavior can be defined as the way an organism responds to stimuli in its environment. These stimuli might be as simple as the odor of food. In this sense, a bacterial cell "behaves" by moving toward higher concentrations of sugar. This behavior is very simple and well-suited to the life of bacteria, allowing these organisms to live and reproduce. As animals evolved, they occupied different environments and faced diverse problems that affected their survival and reproduction. Their nervous systems and behavior concomitantly became more complex. Nervous systems perceive and process information concerning environmental stimuli and trigger adaptive motor responses, which we see as patterns of behavior.

When we observe animal behavior, we can explain it in two different ways. First, we might ask *how* it all works, that is, how the animal's senses, nerve networks, or internal state provide a physiological basis for the behavior. In this way, we would be asking a question of **proximate causation.** To analyze the proximate cause of behavior, we might measure hormone levels or record the impulse activity of neurons in the animal. We could also ask *why* the behavior evolved, that is, what is its adaptive value? This is a question concerning **ultimate causation.** To study the ultimate cause of a behavior, we would attempt to determine how it influenced the animal's survival or reproductive success. Thus, a male songbird may sing during the breeding season because it has a level of the steroid sex hormone, testosterone, which binds to hormone receptors in the brain and triggers the production of song; this would be the proximate cause of the male bird's song. But the male sings to defend a territory from other males and to attract a female to reproduce; this is the ultimate, or evolutionary, explanation for the male's vocalization.

The study of behavior has had a long history of controversy. One source of controversy has been the question of whether behavior is determined more by an individual's genes or its learning and experience. In other words, is behavior the result of nature (instinct) or nurture (experience)? In the past, this question has been considered an "either/or" proposition, but we now know that instinct and experience both play significant roles, often interacting in complex ways to produce the final behavior. The scientific study of instinct and learning, as well as their interrelationship, has led to the growth of several scientific disciplines, including ethology, behavioral genetics, behavioral neuroscience, and comparative psychology.

Ethology

Ethology is the study of the natural history of behavior. Early ethologists (figure 26.2) were trained in zoology and evolutionary biology, fields that emphasize the study of animal behavior under natural conditions. As a result of this training, they believed that behavior is largely instinctive, or innate—the product of natural selection. Because behavior is often **stereotyped** (appearing in the same way in different individuals of a species), they argued that it must be based on preset paths in the nervous system. In their view, these paths are structured from genetic blueprints and cause animals to show a relatively complete behavior the first time it is produced.

The early ethologists based their opinions on behaviors such as egg retrieval by geese. Geese incubate their eggs in a nest. If a goose notices that an egg has been knocked out of the nest, it will extend its neck toward the egg, get up, and roll the egg back into the nest with a side-to-side motion of its neck while the egg is tucked beneath its bill. Even if the egg is removed during retrieval, the goose completes the behavior, as if driven by a program released by the initial sight of the egg outside the nest. According to ethologists, egg retrieval behavior is triggered by a **sign stimulus** (also called a **key stimulus**), the appearance of an egg out of the nest; a component of the goose's nervous system, the **innate releasing mechanism,** provides the neural instructions for the motor program, or **fixed action pattern** (figure 26.3). More generally, the sign stimulus is a "signal" in the environment that triggers a behavior. The innate releasing mechanism is the sensory mechanism that detects the signal, and the fixed action pattern is the stereotyped act. Similarly, a

FIGURE 26.2
The founding fathers of ethology: Karl von Frisch, Konrad Lorenz, and Niko Tinbergen pioneered the study of behavioral science. In 1973, they received the Nobel Prize in Physiology or Medicine for their path-making contributions. Von Frisch led the study of honeybee communication and sensory biology. Lorenz focused on social development (imprinting) and the natural history of aggression. Tinbergen examined the functional significance of behavior and was the first behavioral ecologist.

frog unfolds its long, sticky tongue at the sight of a moving insect, and a male stickleback fish will attack another male showing a bright red underside. Such responses certainly appear to be programmed and instinctive, but what evidence supports the ethological view that behavior has an underlying neural basis?

Behavior as a Response to Stimuli in the Environment

In the example of egg retrieval behavior in geese, a goose must first perceive that an egg is outside of the nest. To respond to this stimulus, it must convert one form of energy which is an input to its visual system—the energy of photons of light—into a form of energy its nervous system can understand and use to respond—the electrical energy of a nerve impulse. Animals need to respond to other stimuli in the environment as well. For an animal to orient from a food source back to its nest, it might rely on the position of the sun. To find a mate, an animal might use a particular chemical scent. The electromagnetic energy of light and the chemical energy of an odor must be converted to the electrical energy of a nerve impulse. This is done through **transduction,** the conversion of energy in the environment to an action potential, and the first step in the processing of stimuli perceived by the senses. For example, rhodopsin is responsible for the transduction of visual stimuli. Rhodopsin is made of *cis*-retinal and the protein opsin. Light is absorbed by the visual pigment *cis*-retinal causing it to change its shape to *trans*-retinal (see chapter 55). This in turn changes the shape of the companion protein opsin, and induces the first step in a cascade of molecular events that finally triggers a nerve impulse. Sound, odor, and tastes are transduced to nerve impulses by similar processes.

FIGURE 26.3
Lizard prey capture. The complex series of movements of the tongue this chameleon uses to capture an insect represents a fixed action pattern.

Ethologists study behavior from an evolutionary perspective, focusing on the neural basis of behaviors.

Behavioral Genetics

In a famous experiment carried out in the 1940s, Robert Tryon studied the ability of rats to find their way through a maze with many blind alleys and only one exit, where a reward of food awaited. Some rats quickly learned to zip right through the maze to the food, making few incorrect turns, while other rats took much longer to learn the correct path (figure 26.4). Tryon bred the fast learners with one another to establish a "maze-bright" colony, and he similarly bred the slow learners with one another to establish a "maze-dull" colony. He then tested the offspring in each colony to see how quickly they learned the maze. The offspring of maze-bright rats learned even more quickly than their parents had, while the offspring of maze-dull parents were even poorer at maze learning. After repeating this procedure over several generations, Tryon was able to produce two behaviorally distinct types of rat with very different maze-learning abilities. Clearly the ability to learn the maze was to some degree hereditary, governed by genes passed from parent to offspring. Furthermore, those genes were specific to this behavior, as the two groups of rats did not differ in their ability to perform other behavioral tasks, such as running a completely different kind of maze. Tryon's research demonstrates how a study can reveal that behavior has a heritable component.

Further support for the genetic basis of behavior has come from studies of hybrids. William Dilger of Cornell University has examined two species of lovebird (genus *Agapornis*), which differ in the way they carry twigs, paper, and other materials used to build a nest. *A. personata* holds nest material in its beak, while *A. roseicollis* carries material tucked under its flank feathers (figure 26.5). When Dilger crossed the two species to produce hybrids, he found that the hybrids carry nest material in a way that seems intermediate between that of the parents: they repeatedly shift material between the bill and the flank feathers. Other studies conducted on courtship songs in crickets and tree frogs also demonstrate the intermediate nature of hybrid behavior.

The role of genetics can also be seen in humans by comparing the behavior of identical twins. Identical twins are, as their name implies, genetically identical. However, most sets of identical twins are raised in the same environment, so it is not possible to determine whether similarities in behavior result from their genetic similarity or from experiences shared as they grew up (the classic nature versus nurture debate). However, in some cases, twins have been separated at birth. A recent study of 50 such sets of twins revealed many similarities in personality, temperament, and even leisure-time activities, even though the twins were often raised in very different environments. These similarities indicate that genetics plays a role in determining behavior even in humans, although the relative importance of genetics versus environment is still hotly debated.

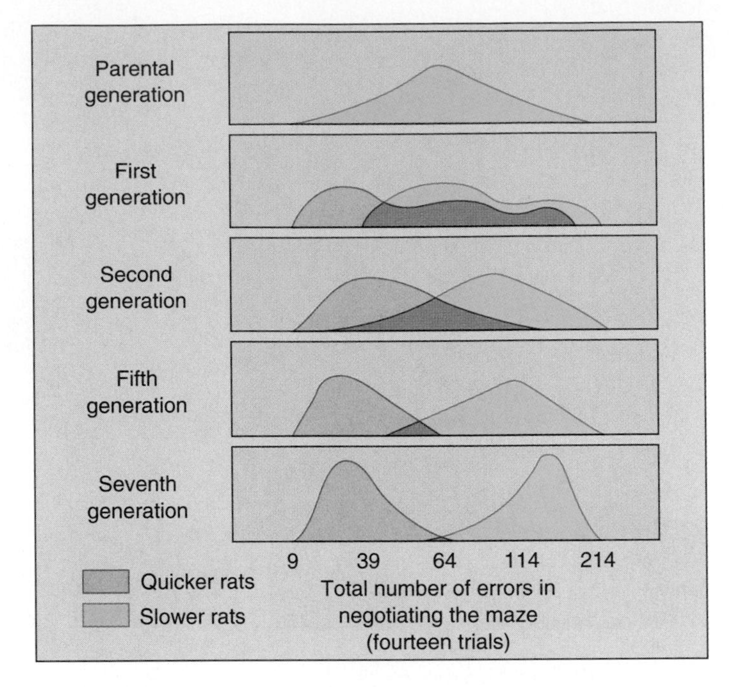

FIGURE 26.4
The genetics of learning. Tryon selected rats for their ability to learn to run a maze and demonstrated that this ability is influenced by genes. He tested a large group of rats, selected those that ran the maze in the shortest time, and let them breed with one another. He then tested their progeny and again selected those with the quickest maze-running times for breeding. After seven generations, he had succeeded in halving the average time an inexperienced rat required to negotiate the maze. Parallel "artificial selection" for slow running time more than doubled the average running time.

FIGURE 26.5
Genetics of lovebird behavior. Lovebirds inherit the tendency to carry nest material, such as these paper strips, under their flank feathers.

(a)

(b)

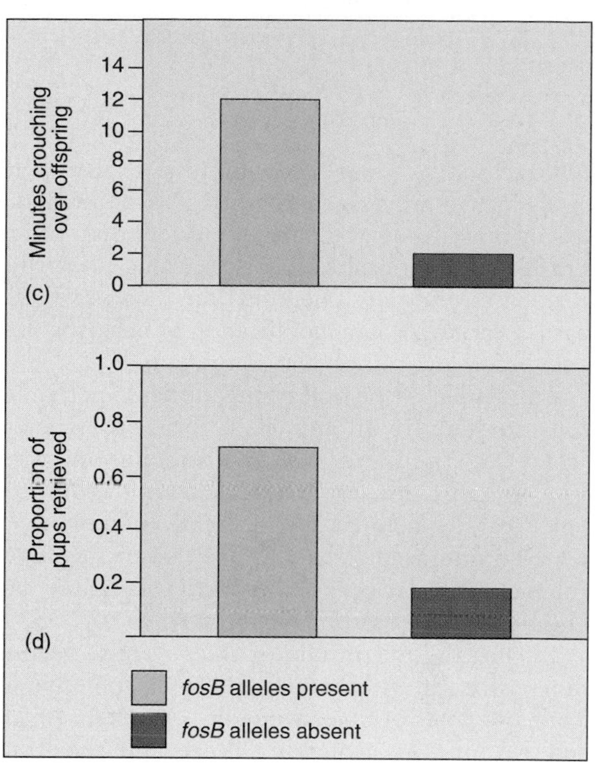

FIGURE 26.6
Genetically caused defect in maternal care. (*a*) In mice, normal mothers take very good care of their offspring, retrieving them if they move away and crouching over them. (*b*) Mothers with the mutant *fosB* allele perform neither of these behaviors, leaving their pups exposed. (*c*) Amount of time female mice were observed crouching in a nursing posture over offspring. (*d*) Proportion of pups retrieved when they were experimentally moved.

Source: (*c* & *d*) Data from J.R. Brown et al, "A Defect in Nurturing in Mice Lacking the Immediate Early Gene for *fosB*", *Cell*, 1996.

Single Gene Effects on Behavior

The maze-learning, hybrid, and identical twins studies just described suggest genes play a role in behavior, but recent research has provided much greater detail on the genetic basis of behavior. In the fruit fly *Drosophila*, and in mice, many mutations have been associated with particular behavioral abnormalities.

In fruit flies, for example, individuals that possess alternative alleles for a single gene differ greatly in their feeding behavior as larvae; larvae with one allele move around a great deal as they eat, whereas individuals with the alternative allele move hardly at all. A wide variety of mutations at other genes are now known in *Drosophila* which affect almost every aspect of courtship behavior.

The ways in which genetic differences affect behavior have been worked out for several mouse genes. For example, some mice with one mutation have trouble remembering information that they learned two days earlier about where objects are located. This difference appears to result because the mutant mice do not produce the enzyme α-calcium-calmodulin-dependent kinase II, which plays an important role in the functioning of a part of the brain, the hippocampus, that is important for spatial learning.

Modern molecular biology techniques allow the role of genetics in behavior to be investigated with ever greater precision. For example, male mice genetically engineered (as "knock-outs") to lack the ability to synthesize nitric oxide, a brain neurotransmitter, show increased aggressive behavior.

A particularly fascinating breakthrough occurred in 1996, when scientists using the knock-out technique discovered a new gene, *fosB*, that seems to determine whether or not female mice will nurture their young. Females with both *fosB* alleles knocked out will initially investigate their newborn babies, but then ignore them, in stark contrast to the caring and protective maternal behavior provided by normal females (figure 26.6).

The cause of this inattentiveness appears to result from a chain reaction. When mothers of new babies initially inspect them, information from their auditory, olfactory, and tactile senses are transmitted to the hypothalamus, where *fosB* alleles are activated, producing a particular protein, which in turn activates other enzymes and genes that affect the neural circuitry within the hypothalamus. These modifications within the brain cause the female to react maternally toward her offspring. In contrast, in mothers lacking the *fosB* alleles, this reaction is stopped midway. No protein is activated, the brain's neural circuitry is not rewired, and maternal behavior does not result.

As these genetic techniques are becoming used more widely, the next few years should see similar dramatic advances in our knowledge of how genes affect various human behaviors.

The genetic basis of behavior is supported by artificial selection experiments, hybridization studies, and studies on the behavior of mutants. Research has also identified specific genes that control behavior.

26.2 Comparative psychology focuses on how learning influences behavior.

Learning

While ethologists were attempting to explain behavior as an instinctive process, **comparative psychologists** focused heavily on learning as the major element that shapes behavior. These behavioral scientists, working primarily on rats in laboratory settings, identified the ways in which animals learn. Learning is any modification of behavior that results from experience rather than maturation.

The simplest type of learning, **nonassociative learning,** does not require an animal to form an association between two stimuli or between a stimulus and a response. One form of nonassociative learning is **habituation,** which can be defined as a decrease in response to a repeated stimulus that has no positive or negative consequences (that is, no *reinforcement*). In many cases, the stimulus evokes a strong response when it is first encountered, but the magnitude of the response gradually declines with repeated exposure. For example, young birds see many types of objects moving overhead. At first, they may respond by crouching down and remaining still. Some of the objects, like falling leaves or members of their own species flying by, are seen very frequently and have no positive or negative consequence to the nestlings. Over time, the young birds may habituate to such stimuli and stop responding. Thus, habituation can be thought of as learning *not* to respond to a stimulus. Being able to ignore unimportant stimuli is critical to an animal confronting a barrage of stimuli in a complex environment. Another form of nonassociative learning is **sensitization,** characterized by an increased responsiveness to a stimulus. Sensitization is essentially the opposite of habituation.

A change in behavior that involves an association between two stimuli or between a stimulus and a response is termed **associative learning** (figure 26.7). The behavior is modified, or **conditioned,** through the association. This form of learning is more complex than habituation or sensitization. The two major types of associative learning are called **classical conditioning** and **operant conditioning;** they differ in the way the associations are established.

Classical Conditioning

In classical conditioning, the paired presentation of two different kinds of stimuli causes the animal to form an association between the stimuli. Classical conditioning is also called **Pavlovian conditioning,** after Russian psychologist Ivan Pavlov, who first described it. Pavlov presented meat powder, an *unconditioned stimulus,* to a dog and noted that the dog responded by salivating, an *unconditioned response.* If an unrelated stimulus, such as the ringing of a bell, was

(a)

(b)

(c)

FIGURE 26.7
Learning what is edible. Associative learning is involved in predator-prey interactions. (*a*) A naive toad is offered a bumblebee as food. (*b*) The toad is stung, and (*c*) subsequently avoids feeding on bumblebees or any other insects having their black-and-yellow coloration. The toad has associated the appearance of the insect with pain, and modifies its behavior.

presented at the same time as the meat powder, over repeated trials the dog would salivate in response to the sound of the bell *alone*. The dog had learned to associate the unrelated sound stimulus with the meat powder stimulus. Its response to the sound stimulus was, therefore, conditioned, and the sound of the bell is referred to as a *conditioned stimulus*.

Operant Conditioning

In operant conditioning, an animal learns to associate its behavioral response with a reward or punishment. American psychologist B. F. Skinner studied operant conditioning in rats by placing them in an apparatus that came to be called a "Skinner box." As the rat explored the box, it would occasionally press a lever by accident, causing a pellet of food to appear. At first, the rat would ignore the lever, eat the food pellet, and continue to move about. Soon, however, it learned to associate pressing the lever (the behavioral response) with obtaining food (the reward). When it was hungry, it would spend all its time pressing the lever. This sort of trial-and-error learning is of major importance to most vertebrates.

Comparative psychologists used to believe that any two stimuli could be linked in classical conditioning and that animals could be conditioned to perform any learnable behavior in response to any stimulus by operant conditioning. As you will see below, this view has changed. Today, it is thought that instinct guides learning by determining what type of information can be learned through conditioning.

FIGURE 26.8
The Clark's nutcracker has an extraordinary memory. A Clark's nutcracker can remember the locations of up to 2000 seed caches months after hiding them. After conducting experiments, scientists have concluded that the birds use features of the landscape and other surrounding objects as spatial references to memorize the locations of the caches.

Instinct

It is now clear that some animals have innate predispositions toward forming certain associations. For example, if a rat is offered a food pellet at the same time it is exposed to X rays (which *later* produces nausea), the rat will remember the taste of the food pellet but not its size. Conversely, if a rat is given a food pellet at the same time an electric shock is delivered (which *immediately* causes pain), the rat will remember the size of the pellet but not its taste. Similarly, pigeons can learn to associate *food* with colors but not with sounds; on the other hand, they can associate *danger* with sounds but not with colors.

These examples of **learning preparedness** demonstrate that what an animal can learn is biologically influenced—that is, learning is possible only within the boundaries set by instinct. Innate programs have evolved because they underscore adaptive responses. Rats, which forage at night and have a highly developed sense of smell, are better able to identify dangerous food by its odor than by its size or color. The seed a pigeon eats may have a distinctive color that the pigeon can see, but it makes no sound the pigeon can hear. The study of learning has expanded to include its ecological significance, so that we are now able to consider the "evolution of learning." An animal's ecology, of course, is key to understanding what an animal is capable of learning. Some species of birds, like Clark's nutcracker, feed on seeds. Birds store seeds in caches they bury when seeds are abundant so they will have food during the winter. Thousands of seed caches may be buried and then later recovered. One would expect the birds to have an extraordinary spatial memory, and this is indeed what has been found (figure 26.8). Clark's nutcracker, and other seed-hoarding birds, have an unusually large hippocampus, the center for memory storage in the brain (see chapter 54).

Habituation and sensitization are simple forms of learning in which there is no association between stimuli and responses. In contrast, associative learning (classical and operant conditioning) involves the formation of an association between two stimuli or between a stimulus and a response.

The Development of Behavior

Behavioral biologists now recognize that behavior has both genetic and learned components, and the schools of ethology and psychology are less polarized than they once were. Thus far in this chapter we have discussed the influence of genes and learning separately. As we will see, these factors interact during development to shape behavior.

FIGURE 26.9
An unlikely parent. The eager goslings follow Konrad Lorenz as if he were their mother. He is the first object they saw when they hatched, and they have used him as a model for imprinting.

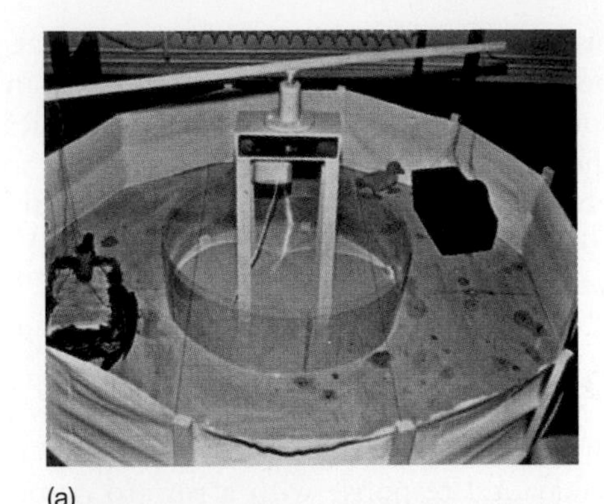

(a)

(b)

FIGURE 26.10
How imprinting is studied. Ducklings will imprint on the first object they see, even (*a*) a black box or (*b*) a white sphere.

Parent-Offspring Interactions

As an animal matures, it may form social attachments to other individuals or form preferences that will influence behavior later in life. This process, called **imprinting,** is sometimes considered a type of learning. In **filial imprinting,** social attachments form between parents and offspring. For example, young birds of some species begin to follow their mother within a few hours after hatching, and their following response results in a bond between mother and young. However, the young birds' initial experience determines how this imprint is established. The German ethologist Konrad Lorenz showed that birds will follow the first object they see after hatching and direct their social behavior toward that object. Lorenz raised geese from eggs, and when he offered himself as a model for imprinting, the goslings treated him as if he were their parent, following him dutifully (figure 26.9). Black boxes, flashing lights, and watering cans can also be effective imprinting objects (figure 26.10). Imprinting occurs during a **sensitive phase,** or a **critical period** (roughly 13 to 16 hours after hatching in geese), when the success of imprinting is highest.

Several studies demonstrate that the social interactions that occur between parents and offspring during the critical period are key to the normal development of behavior. The psychologist Harry Harlow gave orphaned rhesus monkey infants the opportunity to form social attachments with two surrogate "mothers," one made of soft cloth covering a wire frame and the other made only of wire. The infants chose to spend time with the cloth mother, even if only the wire mother provided food, indicating that texture and tactile contact, rather than providing food, may be among the key qualities in a mother that promote infant social attachment. If infants are deprived of normal social contact, their development is abnormal. Greater degrees of deprivation lead to greater abnormalities in social behavior during childhood and adulthood. Studies on orphaned human infants suggest that a constant "mother figure" is required for normal growth and psychological development.

Recent research has revealed a biological need for the stimulation that occurs during parent-offspring interactions early in life. Female rats lick their pups after birth, and this stimulation inhibits the release of an endorphin (see chapter 56) that can block normal growth. Pups that receive normal tactile stimulation also have more brain receptors for glucocorticoid hormones, longer-lived brain neurons, and a greater tolerance for stress. Premature human infants who are massaged gain weight rapidly. These studies indicate that the need for normal social interaction is based in the brain and that touch and other aspects of contact between parents and offspring are important for physical as well as behavioral development.

Sexual imprinting is a process in which an individual learns to direct its sexual behavior at members of its own species. **Cross-fostering** studies, in which individuals of one species are raised by parents of another species, reveal that this form of imprinting also occurs early in life. In most species of birds, these studies have shown that the fostered bird will attempt to mate with members of its foster species when it is sexually mature.

Interaction between Instinct and Learning

The work of Peter Marler and his colleagues on the acquisition of courtship song by white-crowned sparrows provides an excellent example of the interaction between instinct and learning in the development of behavior. Courtship songs are sung by mature males and are species-specific. By rearing male birds in soundproof incubators provided with speakers and microphones, Marler could control what a bird heard as it matured and record the song it produced as an adult. He found that white-crowned sparrows that heard no song at all during development, or that heard only the song of a different species, the song sparrow, sang a poorly developed song as adults (figure 26.11). But birds that heard the song of their own species, or that heard the songs of *both* the white-crowned sparrow and the song sparrow, sang a fully developed, white-crowned sparrow song as adults. These results suggest that these birds have a **genetic template,** or instinctive program, that guides them to learn the appropriate song. During a critical period in development, the template will accept the correct song as a model. Thus, song acquisition depends on learning, but only the song of the correct species can be learned. The genetic template for learning is *selective.* However, learning plays a prominent role as well. If a young white-crowned sparrow is surgically deafened *after* it hears its species' song during the critical period, it will also sing a poorly developed song as an adult. Therefore, the bird must "practice" listening to himself sing, matching what he hears to the model his template has accepted.

Although this explanation of song development stood unchallenged for many years, recent research has shown that white-crowned sparrow males *can* learn another species' song under certain conditions. If a live male strawberry finch is placed in a cage next to a young male sparrow, the young sparrow will learn to sing the strawberry finch's song! This finding indicates that social stimuli may be more effective than a tape-recorded song in overriding the innate program that guides song development. Furthermore, the males of some bird species have no opportunity to hear the song of their own species. In such cases, it appears that the males instinctively "know" their own species' song. For example, cuckoos are brood parasites; females lay their eggs in the nest of another species of bird, and the young that hatch are reared by the foster parents (figure 26.12). When the cuckoos become adults, they sing the song of their own species rather than that of their foster parents. Because male brood parasites would most likely hear the song of their host species during development, it is adaptive for them to ignore such "incorrect" stimuli. They hear no adult males of their own species singing, so no correct song models are available. In these species, natural selection has programmed the male with a genetically guided song.

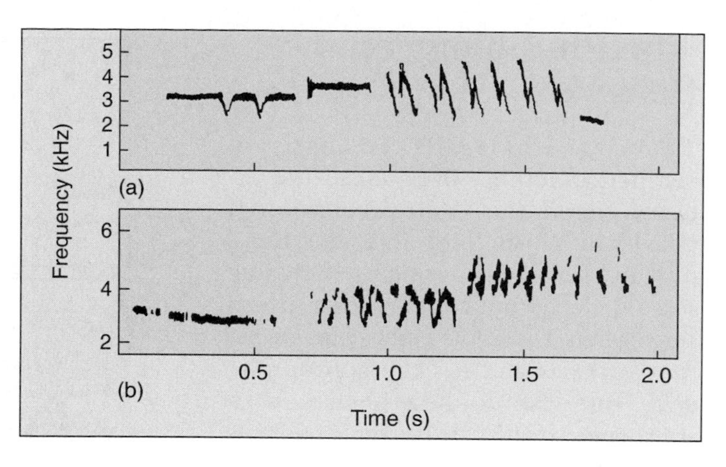

FIGURE 26.11
Song development in birds. (*a*) The sonograms of songs produced by male white-crowned sparrows that had been exposed to their own species' song during development are different from (*b*) those of male sparrows that heard no song during rearing. This difference indicates that the genetic program itself is insufficient to produce a normal song.

FIGURE 26.12
Brood parasite. Cuckoos lay their eggs in the nests of other species of birds. Because the young cuckoos (large bird to the right) are raised by a different species (like this meadow pipit, smaller bird to the left), they have no opportunity to *learn* the cuckoo song; the cuckoo song they later sing is innate.

Interactions that occur during sensitive phases of imprinting are critical to normal behavioral development. Physical contact plays an important role in the development of psychological well-being and growth.

The Physiology of Behavior

The early ethologists' emphasis on instinct sometimes overlooked the internal factors that control behavior. If asked why a male bird defends a territory and sings only during the breeding season, they would answer that a bird sings when it is in the right *motivational state* or *mood* and has the appropriate *drive.* But what do these phrases mean in terms of physiological control mechanisms?

Part of our understanding of the physiological control of behavior has come from the study of reproductive behavior. Animals show reproductive behaviors such as courtship only during the breeding season. Research on lizards, birds, rats, and other animals has revealed that hormones play an important role in these behaviors. Changes in day length trigger the secretion of gonadotropin-releasing hormone by the hypothalamus, which stimulates the release of the gonadotropins, follicle-stimulating hormone (FSH) and luteinizing hormone, by the anterior pituitary gland. These hormones cause the development of reproductive tissues to ready the animal for breeding. The gonadotropins, in turn, stimulate the secretion of the steroid sex hormones, estrogens and progesterone in females and testosterone in males. The sex hormones act on the brain to trigger behaviors associated with reproduction. For example, birdsong and territorial behavior depend upon the level of testosterone in the male, and the receptivity of females to male courtship depends upon estrogen levels.

Hormones have both organizational and activational effects. In the example of birdsong given above, estrogen in the male causes the development of the song system, which is composed of neural tissue in the forebrain and its connections to the spinal cord and the syrinx (a structure like our larynx that allows the bird to sing). Early in a male's development, the gonads produce estrogen, which stimulates neuron growth in the brain. In the mature male, the testes produce testosterone, which activates song. Thus, the development of the neural systems that are responsible for behavior is first organized, then activated by hormones.

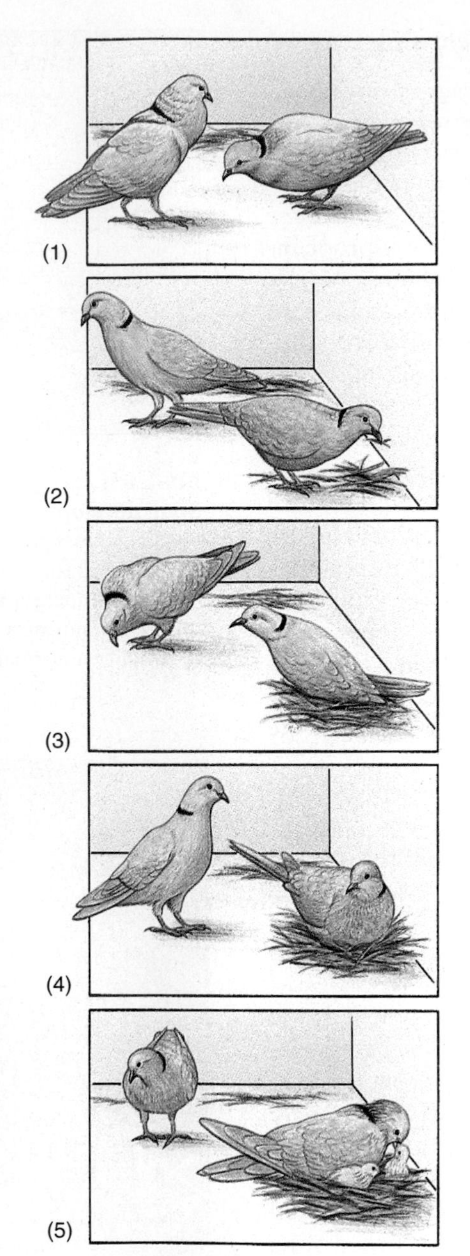

FIGURE 26.13
Hormonal control of reproductive behavior. Reproduction in the ring dove involves a sequence of behaviors regulated by hormones: (*1*) courtship and copulation; (*2*) nest building; (*3*) egg laying; (*4*) incubation; and (*5*) feeding crop milk to the young after they hatch.

Research on the physiology of reproductive behavior shows that there are important interactions among hormones, behavior, and stimuli in both the physical and social environments of an individual. Daniel Lehrman's work on reproduction in ring doves provides an excellent example of how these factors interact (figure 26.13). Male courtship behavior is controlled by testosterone and related steroid hormones. The male's behavior causes the release of FSH in the female, and FSH promotes the growth of the ovarian follicles (see chapter 59). The developing follicles release estrogens, which affect other reproductive tissues. Nest construction follows after one or two days. The presence of the nest then triggers the secretion of progesterone in the female, initiating incubation behavior after the egg is laid. Feeding occurs once the eggs hatch, and this behavior is also hormonally controlled.

The research of Lehrman and his colleagues paved the way for many additional investigations in **behavioral endocrinology,** the study of the hormonal regulation of behavior. For example, male *Anolis* lizards begin courtship after a seasonal rise in temperature, and the male's courtship is needed to stimulate the growth of ovarian follicles in the female. These and other studies demonstrate the interactive effects of the physical environment (for example, temperature and day length) and the social environment (such as the presence of a nest and the courtship display of a mate) on the hormonal condition of an animal. Hormones are, therefore, a proximate cause of behavior. To control reproductive behavior, they must be released when the conditions are most favorable for the growth of young. Other behaviors, such as territoriality and dominance behavior, also have hormonal correlates.

Hormones may interact with neurotransmitters to alter behavior. Estrogen affects the neurotransmitter serotonin in female mice, and may be in part responsible for the "mood swings" experienced by some human females during the menstrual cycle.

Hormones have important influences on reproductive and social behavior.

Behavioral Rhythms

Many animals exhibit behaviors that vary at regular intervals of time. Geese migrate south in the fall, birds sing in the early morning, bats fly at night rather than during the day, and most humans sleep at night and are active in the daytime. Some behaviors are timed to occur in concert with lunar or tidal cycles (figure 26.14). Why do regular repeating patterns of behavior occur, and what determines when they occur? The study of questions like these has revealed that rhythmic animal behaviors are based on both **endogenous** (internal) **rhythms** and **exogenous** (external) **timers.**

Most studies of behavioral rhythms have focused on behaviors that appear to be keyed to a daily cycle, such as sleeping. Rhythms with a period of about 24 hours are called **circadian** ("about a day") **rhythms.** Many of these behaviors have a strong endogenous component, as if they were driven by a **biological clock.** Such behaviors are said to be *free-running*, continuing on a regular cycle even in the absence of any cues from the environment. Almost all fruit fly pupae hatch in the early morning, for example, even if they are kept in total darkness throughout their week-long development. They keep track of time with an internal clock whose pattern is determined by a single gene, called the *per* (for *period*) gene. Different mutations of the *per* gene shorten or lengthen the daily rhythm. The *per* gene produces a protein in a regular 24-hour cycle in the brain, serving as the fly's pacemaker of activity. The protein probably affects the expression of other genes that ultimately regulate activity. As the per protein accumulates, it seems to turn off the gene. In mice, the *clock* gene is responsible for regulating the animal's daily rhythm.

Most biological clocks do not exactly match the rhythms of the environment. Therefore, the behavioral rhythm of an individual deprived of external cues gradually drifts out of phase with the environment. Exposure to an environmental cue resets the biological clock and keeps the behavior properly synchronized with the environment. Light is the most common cue for resetting circadian rhythms.

The most obvious circadian rhythm in humans is the sleep-activity cycle. In controlled experiments, humans have lived for months in underground apartments, where all light is artificial and there are no external cues whatsoever indicating day length. Left to set their own schedules, most of these people adopt daily activity patterns (one phase of activity plus one phase of sleep) of about 25 hours, although there is considerable variation. Some individuals exhibit 50-hour clocks, active for as long as 36 hours during each period! Under normal circumstances, the day-night cycle resets an individual's free-running clock every day to a cycle period of 24 hours.

What constitutes an animal's biological clock? In some insects, the clock is thought to be located in the optic lobes of the brain, and timekeeping appears to be based on hormones. In mammals, including humans, the biological clock lies in a specific region of the hypothalamus called the **suprachiasmatic nucleus (SCN).** The SCN is a self-sustaining oscillator, which means it undergoes spontaneous, cyclical changes in activity. This oscillatory activity helps the SCN to act as a pacemaker for circadian rhythms, but in order for the rhythms to be entrained to external light-dark cycles, the SCN must be influenced by light. In fact, there are both direct and indirect neural projections from the retina to the SCN.

FIGURE 26.14
Tidal rhythm. Oysters open their shells for feeding when the tide is in and close them when the tide is out.

The SCN controls circadian rhythms by regulating the secretion of the hormone **melatonin** by the **pineal gland.** During the daytime, the SCN suppresses melatonin secretion. Consequently, more melatonin is secreted over a 24-hour period during short days than during long days. Variations in melatonin secretion thus serve as an indicator of seasonal changes in day length, and these variations participate in timing the seasonal reproductive behavior of many mammals. Disturbances in melatonin secretion may be partially responsible for the "jet-lag" people experience when air travel suddenly throws their internal clocks out of register with the day-night cycle.

Many important behavioral rhythms have cycle periods longer than 24 hours. For example, **circannual behaviors** such as breeding, hibernation, and migration occur on a yearly cycle. These behaviors seem to be largely timed by hormonal and other physiological changes keyed to exogenous factors such as day length. The degree to which endogenous biological clocks underlie circannual rhythms is not known, as it is very difficult to perform constant-environment experiments of several years' duration. The mechanism of the biological clock remains one of the most tantalizing puzzles in biology today.

Endogenous circadian rhythms have free-running cycle periods of approximately 24 hours; they are entrained to a more exact 24-hour cycle period by environmental cues.

26.3 Communication is a key element of many animal behaviors.

Much of the research in animal behavior is devoted to analyzing the nature of communication signals, determining how they are perceived, and identifying the ecological roles they play and their evolutionary origins.

Courtship

During courtship, animals produce signals to communicate with potential mates and with other members of their own sex. A **stimulus-response chain** sometimes occurs, in which the behavior of one individual in turn releases a behavior by another individual (figure 26.15).

Courtship Signaling

A male stickleback fish will defend the nest it builds on the bottom of a pond or stream by attacking *conspecific* males (that is, males of the same species) that approach the nest. Niko Tinbergen studied the social releasers responsible for this behavior by making simple clay models. He found that a model's shape and degree of resemblance to a fish were unimportant; any model with a red underside (like the underside of a male stickleback) could release the attack behavior. Tinbergen also used a series of clay models to demonstrate that a male stickleback recognizes a female by her abdomen, swollen with eggs.

Courtship signals are often *species-specific,* limiting communication to members of the same species and thus playing a key role in reproductive isolation. The flashes of fireflies (which are actually beetles) are such species-specific signals. Females recognize conspecific males by their flash pattern (figure 26.16), and males recognize conspecific females by their flash response. This series of reciprocal responses provides a continuous "check" on the species identity of potential mates.

Visual courtship displays sometimes have more than one component. The male *Anolis* lizard extends and retracts his fleshy and often colorful dewlap while perched on a branch in his territory (figure 26.17). The display thus involves both color and movement (the extension of the dewlap as well as a series of lizard "push-ups"). To which component of the display does the female respond? Experiments in which the dewlap color is altered with ink show that color is unimportant for some species; that is, a female can be courted successfully by a male with an atypically colored dewlap.

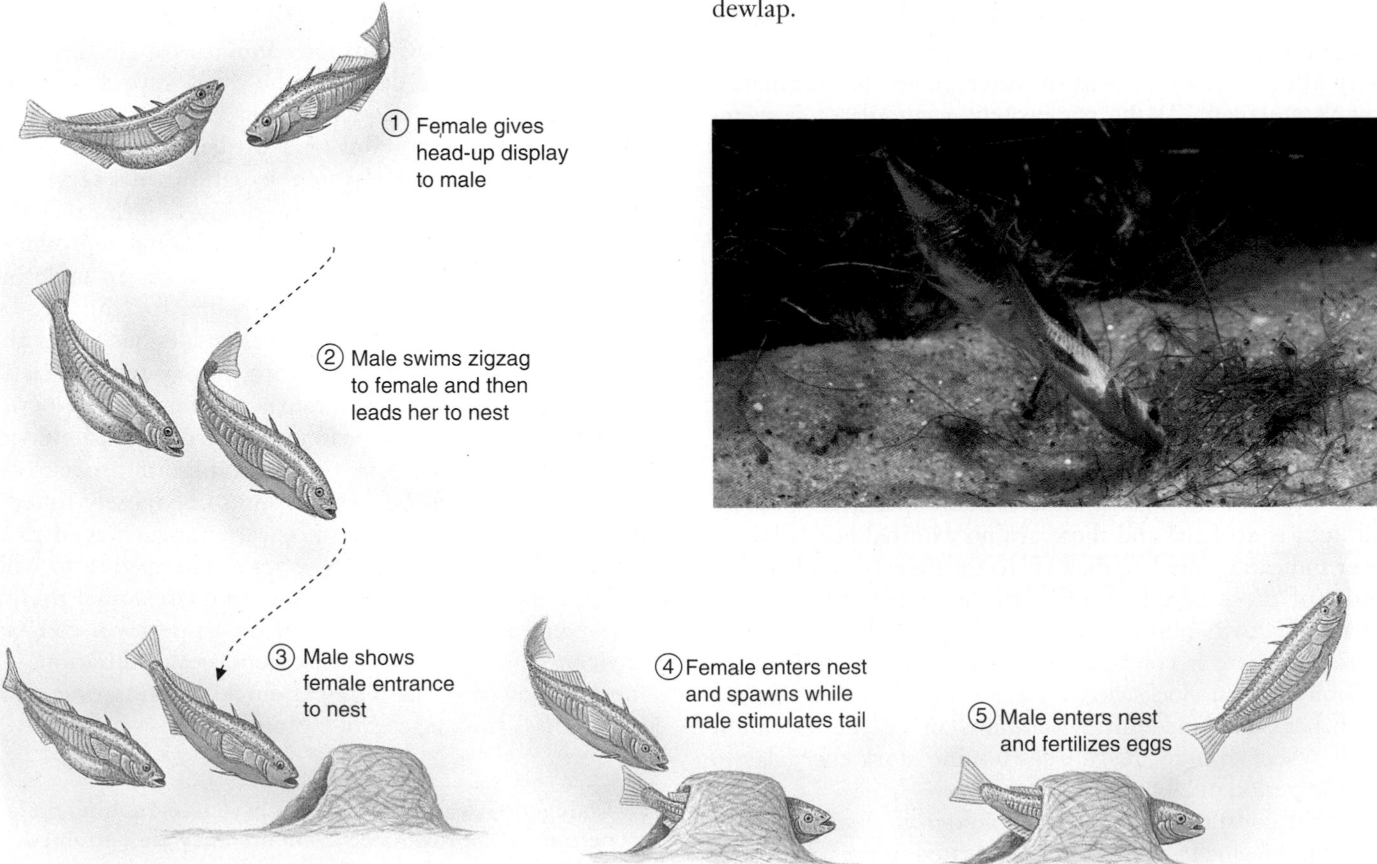

① Female gives head-up display to male
② Male swims zigzag to female and then leads her to nest
③ Male shows female entrance to nest
④ Female enters nest and spawns while male stimulates tail
⑤ Male enters nest and fertilizes eggs

FIGURE 26.15
A stimulus-response chain. Stickleback courtship involves a sequence of behaviors leading to the fertilization of eggs.

Pheromones and Acoustic Signals

Chemical signals also mediate interactions between males and females. **Pheromones,** chemical messengers used for communication between individuals of the same species, serve as sex attractants among other functions in many animals. Even the human egg produces a chemical attractant to communicate with sperm! Female silk moths *(Bombyx mori)* produce a sex pheromone called *bombykol* in a gland associated with the reproductive system. Neurophysiological studies show that the male's antennae contain numerous sensory receptors specific for bombykol. These receptors are extraordinarily sensitive, enabling the male to respond behaviorally to concentrations of bombykol as low as one molecule in 10^{17} molecules of oxygen in the air!

Many insects, amphibians, and birds produce species-specific acoustic signals to attract mates. Bullfrog males call to females by inflating and discharging air from their vocal sacs, located beneath the lower jaw. The female can distinguish a conspecific male's call from the call of other frogs that may be in the same habitat and mating at the same time. Male birds produce songs, complex sounds composed of notes and phrases, to advertise their presence and to attract females. In many bird species, variations in the males' songs identify *particular* males in a population. In these species, the song is individually specific as well as species-specific.

FIGURE 26.16
Firefly fireworks. The bioluminescent displays of these lampyrid beetles are species-specific and serve as behavioral mechanisms of reproductive isolation. Each number represents the flash pattern of a male of a different species.

Level of Specificity

Why should different signals have different levels of specificity? The **level of specificity** relates to the function of the signal. Many courtship signals are species-specific to help animals avoid making errors in mating that would produce inviable hybrids or otherwise waste reproductive effort. A male bird's song is individually specific because it allows his presence (as opposed to simply the presence of an unidentifiable member of the species) to be recognized by neighboring birds. When territories are being established, males may sing and aggressively confront neighboring conspecifics to defend their space. Aggression carries the risk of injury, and it is energetically costly to sing. After territorial borders have been established, intrusions by neighbors are few because the outcome of the contests have already been determined. Each male then "knows" his neighbor by the song he sings, and also "knows" that male does not constitute a threat because they have already settled their territorial contests. So, all birds in the population can lower their energy costs by identifying their neighbors through their individualistic songs. In a similar way, mammals mark their territories with pheromones that signal individual identity, which may be encoded as a blend of a number of chemicals. Other signals, such as the mobbing and alarm calls of birds, are anonymous, conveying no information about the identity of the sender. These signals may permit communication about the presence of a predator common to several bird species.

FIGURE 26.17
Dewlap display of a male *Anolis* lizard. Under hormonal stimulation, males extend their fleshy, colored dewlaps to court females. This behavior also stimulates hormone release and egg-laying in the female.

Courtship behaviors are keyed to species-specific visual, chemical, and acoustic signals.

Communication in Social Groups

Many insects, fish, birds, and mammals live in social groups in which information is communicated between group members. For example, some individuals in mammalian societies serve as "guards." When a predator appears, the guards give an **alarm call,** and group members respond by seeking shelter. Social insects, such as ants and honeybees, produce **alarm pheromones** that trigger attack behavior. Ants also deposit **trail pheromones** between the nest and a food source to induce cooperation during foraging (figure 26.18). Honeybees have an extremely complex **dance language** that directs nestmates to rich nectar sources.

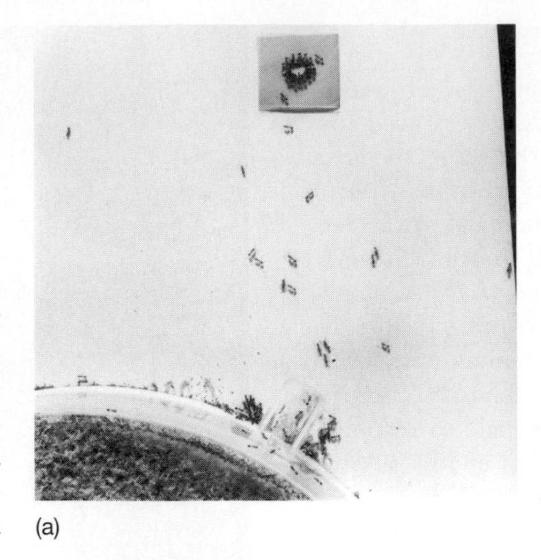
(a)

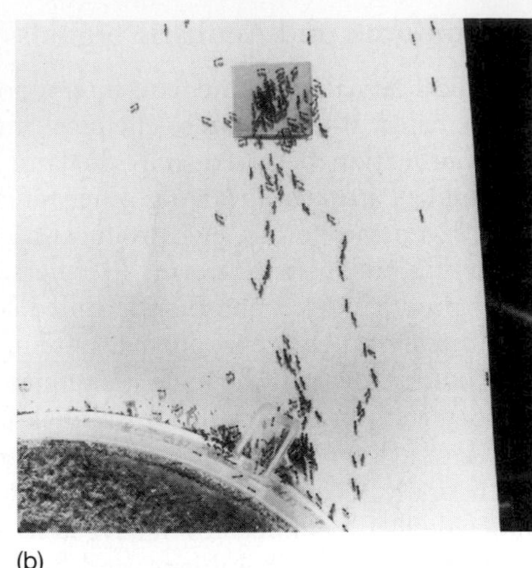
(b)

FIGURE 26.18
The chemical control of fire ant foraging. Trail pheromones, produced in an accessory gland near the fire ant's sting, organize cooperative foraging. The trails taken by the first ants to travel to a food source (*a*) are soon followed by most of the other ants (*b*).

The Dance Language of the Honeybee

The European honeybee, *Apis mellifera,* lives in hives consisting of 30,000 to 40,000 individuals whose behaviors are integrated into a complex colony. Worker bees may forage for miles from the hive, collecting nectar and pollen from a variety of plants and switching between plant species and populations on the basis of how energetically rewarding their food is. The food sources used by bees tend to occur in patches, and each patch offers much more food than a single bee can transport to the hive. A colony is able to exploit the resources of a patch because of the behavior of scout bees, which locate patches and communicate their location to hivemates through a *dance language.* Over many years, Nobel laureate Karl von Frisch was able to unravel the details of this communication system.

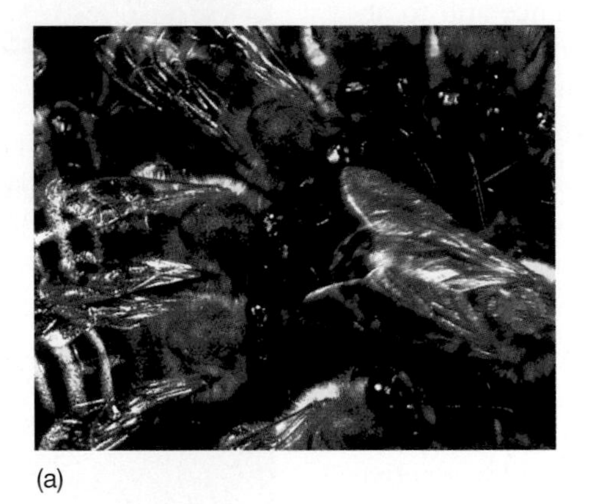
(a)

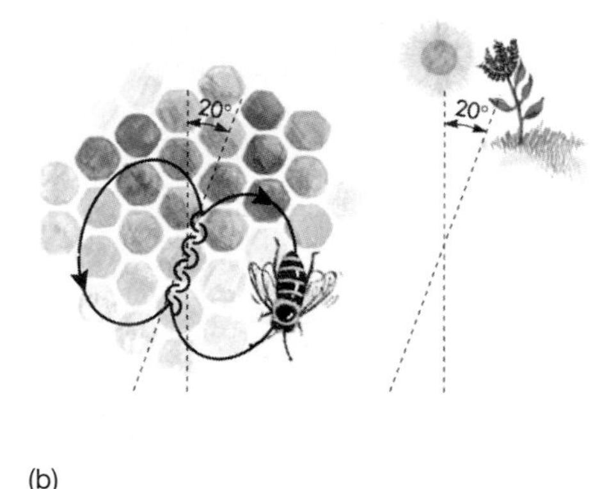

(b)

FIGURE 26.19
The waggle dance of honeybees. (*a*) A scout bee dances on a comb in the hive. (*b*) The angle between the food source and the nest is represented by a dancing bee as the angle between the straight part of the dance and vertical. The food is 20° to the right of the sun, and the straight part of the bee's dance on the hive is 20° to the right of vertical.

After a successful scout bee returns to the hive, she performs a remarkable behavior pattern called a *waggle dance* on a vertical comb (figure 26.19). The path of the bee during the dance resembles a figure-eight. On the straight part of the path, the bee vibrates or waggles her abdomen while producing bursts of sound. She may stop periodically to give her hivemates a sample of the nectar she has carried back to the hive in her crop. As she dances, she is followed closely by other bees, which soon appear as foragers at the new food source.

Von Frisch and his colleagues claimed that the other bees use information in the waggle dance to locate the food source. According to their explanation, the scout bee indicates the *direction* of the food source by representing the angle between the food source and the hive in reference to the sun as the angle between the straight part of the dance and vertical in the hive. The *distance* to the food source is indicated by the tempo, or degree of vigor, of the dance.

Adrian Wenner, a scientist at the University of California, did not believe that the dance language communicated anything about the location of food, and he challenged von Frisch's explanation. Wenner maintained that flower odor was the most important cue allowing recruited bees to arrive at a new food source. A heated controversy ensued as

(a)

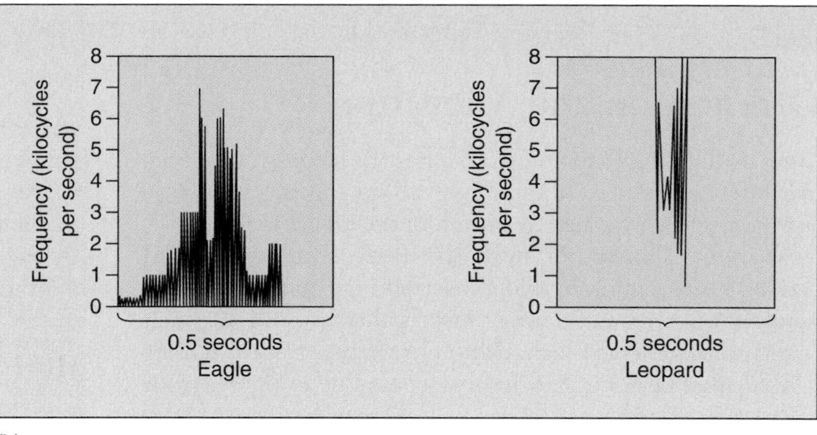

(b)

FIGURE 26.20
Primate semantics. (*a*) Predators, like this leopard, attack and feed on vervet monkeys. (*b*) The monkeys give different alarm calls when eagles, leopards, and snakes are sighted by troupe members. Each distinctive call elicits a different and adaptive escape behavior.

the two groups of researchers published articles supporting their positions.

The "dance language controversy" was resolved (in the minds of most scientists) in the mid-1970s by the creative research of James L. Gould. Gould devised an experiment in which hive members were tricked into misinterpreting the directions given by the scout bee's dance. As a result, Gould was able to manipulate where the hive members would go if they were using visual signals. If odor were the cue they were using, hive members would have appeared at the food source, but instead they appeared exactly where Gould predicted. This confirmed von Frisch's ideas.

Recently, researchers have extended the study of the honeybee dance language by building robot bees whose dances can be completely controlled. Their dances are programmed by a computer and perfectly reproduce the natural honeybee dance; the robots even stop to give food samples! The use of robot bees has allowed scientists to determine precisely which cues direct hivemates to food sources.

Primate Language

Some primates have a "vocabulary" that allows individuals to communicate the identity of specific predators. The vocalizations of African vervet monkeys, for example, distinguish eagles, leopards, and snakes (figure 26.20). Chimpanzees and gorillas can learn to recognize a large number of symbols and use them to communicate abstract concepts. The complexity of human language would at first appear to defy biological explanation, but closer examination suggests that the differences are in fact superficial—all languages share many basic structural similarities. All of the roughly 3000 languages draw from the same set of 40 consonant sounds (English uses two dozen of them), and any human can learn them. Researchers believe these similarities reflect the way our brains handle abstract information, a genetically determined characteristic of all humans.

Language develops at an early age in humans. Human infants are capable of recognizing the 40 consonant sounds characteristic of speech, including those not present in the particular language they will learn, while they ignore other sounds. In contrast, individuals who have not heard certain consonant sounds as infants can only rarely distinguish or produce them as adults. That is why English speakers have difficulty mastering the throaty French "r," French speakers typically replace the English "th" with "z," and native Japanese often substitute "r" for the unfamiliar English "l." Children go through a "babbling" phase, in which they learn by trial and error how to make the sounds of language. Even deaf children go through a babbling phase using sign language. Next, children quickly and easily learn a vocabulary of thousands of words. Like babbling, this phase of rapid learning seems to be genetically programmed. It is followed by a stage in which children form simple sentences which, though they may be grammatically incorrect, can convey information. Learning the rules of grammar constitutes the final step in language acquisition.

While language is the primary channel of human communication, odor and other nonverbal signals (such as "body language") may also convey information. However, it is difficult to determine the relative importance of these other communication channels in humans.

The study of animal communication involves analysis of the specificity of signals, their information content, and the methods used to produce and receive them.

26.4 Migratory behavior presents many puzzles.

Orientation and Migration

Animals may travel to and from a nest to feed or move regularly from one place to another. To do so, they must orient themselves by tracking stimuli in the environment.

Movement toward or away from some stimulus is called **taxis.** The attraction of flying insects to outdoor lights is an example of *positive phototaxis.* Insects that avoid light, such as the common cockroach, exhibit *negative phototaxis.* Other stimuli may be used as orienting cues. For example, trout orient themselves in a stream so as to face against the current. However, not all responses involve a specific orientation. Some animals just become more active when stimulus intensity increases, a responses called **kineses.**

Long-range, two-way movements are known as **migrations.** In many animals, migrations occur circannually. Ducks and geese migrate along flyways from Canada across the United States each fall and return each spring. Monarch butterflies migrate each fall from central and eastern North America to several small, geographically isolated areas of coniferous forest in the mountains of central Mexico (figure 26.21). Each August, the butterflies begin a flight southward to their overwintering sites. At the end of winter, the monarchs begin the return flight to their summer breeding ranges. What is amazing about the migration of the monarch, however, is that two to five generations may be produced as the butterflies fly north. The butterflies that migrate in the autumn to the precisely located overwintering grounds in Mexico have never been there before.

When colonies of bobolinks became established in the western United States, far from their normal range in the Midwest and East, they did not migrate directly to their winter range in South America. Instead, they migrated east to their ancestral range and then south along the original flyway (figure 26.22). Rather than changing the original migration pattern, they simply added a new pattern.

How Migrating Animals Navigate

Biologists have studied migration with great interest, and we now have a good understanding of how these feats of navigation are achieved. It is important to understand the distinction between **orientation** (the ability to follow a bearing) and **navigation** (the ability to set or adjust a bearing, and then follow it). The former is analogous to using a compass, while the latter is like using a compass in conjunction with a map. Experiments on starlings indicate that inexperienced birds migrate by orientation, while older birds that have migrated previously use true navigation (figure 26.23).

Birds and other animals navigate by looking at the sun and the stars. The indigo bunting, which flies during the day and uses the sun as a guide, compensates for the movement of the sun in the sky as the day progresses by reference to the north star, which does not move in the sky. Buntings also use the positions of the constellations and the position of the pole star in the night sky, cues they learn as young birds. Starlings and certain other birds compensate for the sun's apparent movement in the

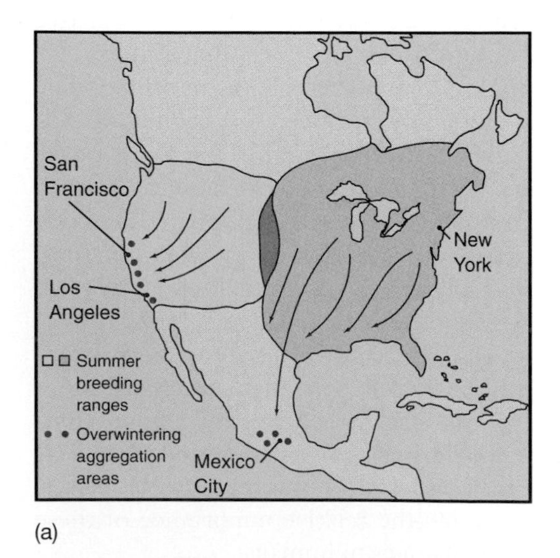

(a)

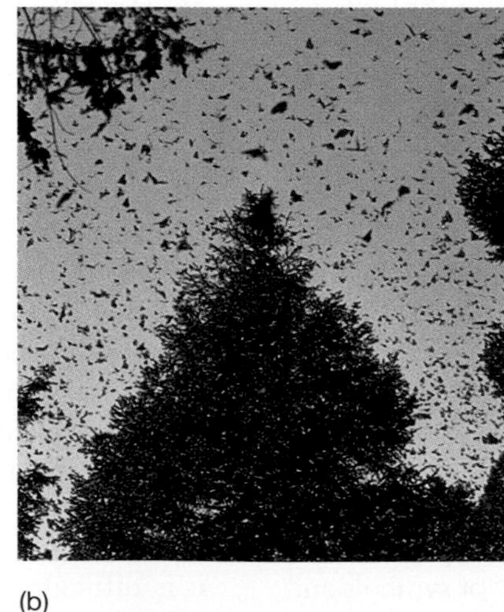

(b)

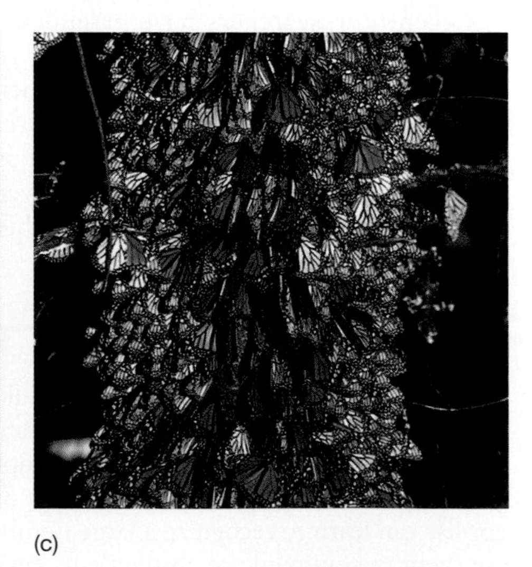

(c)

FIGURE 26.21
Migration of monarch butterflies. (*a*) Monarchs from western North America overwinter in areas of mild climate along the Pacific Coast. Those from the eastern United States and southeastern Canada migrate to Mexico, a journey of over 3000 kilometers that takes from two to five generations to complete. (*b*) Monarch butterflies arriving at the remote fir forests of the overwintering grounds and (*c*) forming aggregations on the tree trunks.

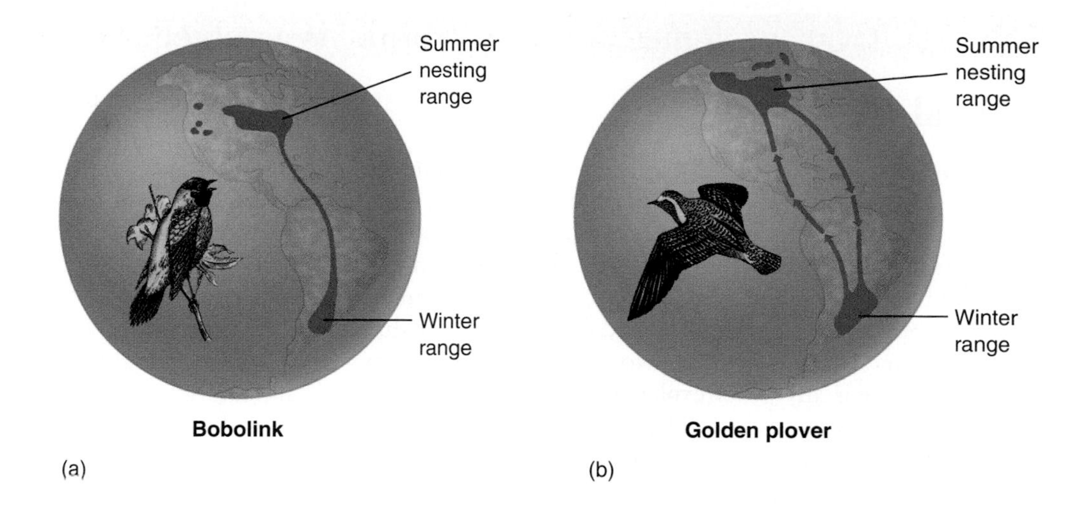

FIGURE 26.22
Birds on the move. (*a*) The summer range of bobolinks recently extended to the far West from their more established range in the Midwest. When they migrate to South America in the winter, bobolinks that nested in the West do not fly directly to the winter range; instead, they fly to the Midwest first and then use the ancestral flyway. (*b*) The golden plover has an even longer migration route that is circular. These birds fly from Arctic breeding grounds to wintering areas in southeastern South America, a distance of some 13,000 kilometers.

sky by using an internal clock. If such birds are shown an experimental sun in a fixed position while in captivity, they will change their orientation to it at a constant rate of about 15° per hour.

Many migrating birds also have the ability to detect the earth's magnetic field and to orient themselves with respect to it. In a closed indoor cage, they will attempt to move in the correct geographical direction, even though there are no visible external cues. However, the placement of a powerful magnet near the cage can alter the direction in which the birds attempt to move. Magnetite, a magnetized iron ore, has been found in the heads of some birds, but the sensory receptors birds employ to detect magnetic fields have not been identified.

It appears that the first migration of a bird is innately guided by both celestial cues (the birds fly mainly at night) and the earth's magnetic field. These cues give the same information about the general direction of the migration, but the information about direction provided by the stars seems to dominate over the magnetic information when the two cues are experimentally manipulated to give conflicting directions. Recent studies, however, indicate that celestial cues tell northern hemisphere birds to move south when they begin their migration, while magnetic cues give them the direction for the specific migratory path (perhaps a southeast turn the bird must make midroute). In short, these new data suggest that celestial and magnetic cues interact during development to fine-tune the bird's navigation.

We know relatively little about how other migrating animals navigate. For instance, green sea turtles migrate from Brazil halfway across the Atlantic Ocean to Ascension Island, where the females lay their eggs. How do they find this tiny island in the middle of the ocean, which they haven't seen for perhaps 30 years? How do the young that hatch on the island know how to find their way to Brazil? Recent studies suggest that wave action is an important cue.

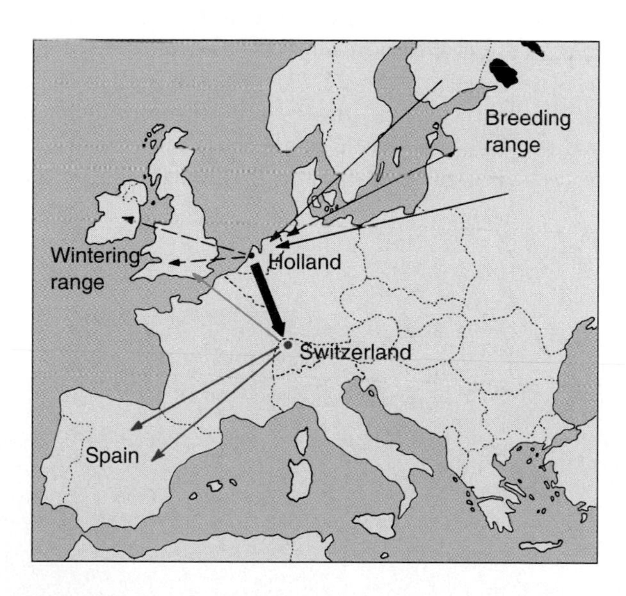

FIGURE 26.23
Migratory behavior of starlings. The navigational abilities of inexperienced birds differ from those of adults who have made the migratory journey before. Starlings were captured in Holland, halfway along their full migratory route from Baltic breeding grounds to wintering grounds in the British Isles; these birds were transported to Switzerland and released. Experienced older birds compensated for the displacement and flew toward the normal wintering grounds (*blue arrow*). Inexperienced young birds kept flying in the same direction, on a course that took them toward Spain (*red arrow*). These observations imply that inexperienced birds fly by orientation, while experienced birds learn true navigation.

Many animals migrate in predictable ways, navigating by looking at the sun and stars, and in some cases by detecting magnetic fields.

26.5 To what degree animals "think" is a subject of lively dispute.

Animal Cognition

It is likely each of us could tell an anecdotal story about the behavior of a pet cat or dog that would seem to suggest that the animal had a degree of reasoning ability or was capable of thinking. For many decades, however, students of animal behavior flatly rejected the notion that nonhuman animals can think. In fact, behaviorist Lloyd Morgan stated that one should never assume a behavior represents conscious thought if there is any other explanation that precludes the assumption of consciousness. The prevailing approach was to treat animals as though they responded to the environment through reflexlike behaviors.

In recent years, serious attention has been given to the topic of animal awareness. The central question is whether animals show **cognitive behavior**—that is, do they process information and respond in a manner that suggests thinking (figure 26.24)? What kinds of behavior would demonstrate cognition? Some birds in urban areas remove the foil caps from nonhomogenized milk bottles to get at the cream beneath, and this behavior is known to have spread within a population to other birds. Japanese macaques learned to wash potatoes and float grain to separate it from sand. A chimpanzee pulls the leaves off of a tree branch and uses the stick to probe the entrance to a termite nest and gather termites. As we saw earlier, vervet monkeys have a vocabulary that identifies specific predators.

Only a few experiments have tested the thinking ability of nonhuman animals. Some of these studies suggest that animals may give false information (that is, they "lie"). Currently, researchers are trying to determine if some primates deceive others to manipulate the behavior of the other members of their troop. There are many anecdotal accounts that appear to support the idea that deception occurs in some nonhuman primate species such as baboons and chimpanzees, but it has been difficult to devise field-based experiments to test this idea. Much of this type of research on animal cognition is in its infancy, but it is sure to grow and to raise controversy. In any case, there is nothing to be gained by a dogmatic denial of the *possibility* of animal consciousness.

(a)

(b)

FIGURE 26.24
Animal thinking? (*a*) This chimpanzee is stripping the leaves from a twig, which it will then use to probe a termite nest. This behavior strongly suggests that the chimpanzee is consciously planning ahead, with full knowledge of what it intends to do. (*b*) This sea otter is using a rock as an "anvil," against which it bashes a clam to break it open. A sea otter will often keep a favorite rock for a long time, as though it has a clear idea of what it is going to use the rock for. Behaviors such as these suggest that animals have cognitive abilities.

FIGURE 26.25
Problem solving by a chimpanzee. Unable to get the bananas by jumping, the chimpanzee devises a solution.

In any case, some examples, particularly those involving problem-solving, are hard to explain in any way other than as a result of some sort of mental process. For example, in a series of classic experiments conducted in the 1920s, a chimpanzee was left in a room with a banana hanging from the ceiling out of reach. Also in the room were several boxes, each lying on the floor. After unsuccessful attempts to jump up and grab the bananas, the chimpanzee suddenly looks at the boxes and immediately proceeds to move them underneath the banana, stack one on top of another, and climb up to claim its prize (figure 26.25).

Perhaps it is not so surprising to find obvious intelligence in animals as closely related to us as chimpanzees. But recent studies have found that other animals also show evidence of cognition. Ravens have always been considered among the most intelligent of birds. Bernd Heinrich of the University of Vermont recently conducted an experiment using a group of hand-reared ravens that lived in an outdoor aviary. Heinrich placed a piece of meat on the end of a string and hung it from a branch in the aviary. The birds liked to eat meat, but had never seen string before and were unable to get at the meat. After several hours, during which time the birds periodically looked at the meat but did nothing else, one bird flew to the branch, reached down, grabbed the string, pulled it up, and placed it under his foot. He then reached down and grabbed another piece of the string, repeating this action over and over, each time bringing the meat closer (figure 26.26). Eventually, the meat was within reach and was grasped. The raven, presented with a completely novel problem, had devised a solution. Eventually, three of the other five ravens also figured out how to get to the meat. Heinrich has conducted other similarly creative experiments that can leave little doubt that ravens have advanced cognitive abilities.

FIGURE 26.26
Problem solving by a raven. Confronted with a problem it had never previously confronted, the raven figures out how to get the meat at the end of the string by repeatedly pulling up a bit of string and stepping on it.

Research on the cognitive behavior of animals is in its infancy, but some examples are compelling.

Chapter 26

www.mhhe.com/raven6/resources26.mhtml

Summary | **Questions** | **Media Resources**

26.1 Ethology focuses on the natural history of behavior.

- Behavior is an adaptive response to stimuli in the environment. An animal's sensory systems detect and process information about these stimuli.

1. How does a hybrid lovebird's method of carrying nest materials compare with that of its parents? What does this comparison suggest about whether the behavior is instinctive or learned?

- Student Research: Behavior in Voles
- *On Science* Article: Polyandry in Hawks

26.2 Comparative psychology focuses on how learning influences behavior.

- Behavior is both instinctive (influenced by genes) and learned through experience. Genes are thought to limit the extent to which behavior can be modified and the types of associations that can be made.
- The simplest forms of learning involve habituation and sensitization. More complex associative learning, such as classical and operant conditioning, involves a connection being made between two stimuli or a stimulus and a response.
- An animal's internal state influences when and how a behavior will occur. Hormones can change an animal's behavior and perception of stimuli in a way that facilitates reproduction.

2. How does associative learning differ from nonassociative learning? How does classical conditioning differ from operant conditioning?

3. What is filial imprinting? What is sexual imprinting? Why do some young animals imprint on objects like a moving box?

4. How does Marler's work on song development in white-crowned sparrows indicate that behavior is shaped by learning? How does it indicate that behavior is shaped by instinct?

- *On Science* Article: Repetition and Learning

26.3 Communication is a key element of many animal behaviors.

- Animals communicate by producing visual, acoustic, and chemical signals. These signals are involved in mating, finding food, defense against predators, and other social situations.

5. How do communication signals participate in reproductive isolation? Give one example of a signal that is species-specific. Why are some signals individually specific?

26.4 Migratory behavior presents many puzzles.

- Animals use cues such as the position of the sun and stars to orient during daily activities and to navigate during long-range migrations.

6. What is the definition of taxis? What are kineses? What cues do migrating birds use to orient and navigate during their migrations?

26.5 To what degree animals "think" is a subject of lively dispute.

- Many anecdotal accounts point to animal cognition, but research is in its infancy.

7. What evidence would you accept that an animal is "thinking"?

- *On Science* Article: Can Dogs Think

27

Behavioral Ecology

Concept Outline

27.1 Evolutionary forces shape behavior.

Behavioral Ecology. Behavior is shaped by natural selection.

Foraging Behavior. Natural selection favors the most efficient foraging behavior.

Territorial Behavior. Animals defend territory to increase reproductive advantage and foraging efficiency.

27.2 Reproductive behavior involves many choices influenced by natural selection.

Parental Investment and Mate Choice. The degree of parental investment strongly influences other reproductive behaviors.

Reproductive Competition and Sexual Selection. Mate choice affects reproductive success, and so is a target of natural selection.

Mating Systems. Mating systems are reproductive solutions to particular ecological challenges.

27.3 There is considerable controversy about the evolution of social behavior.

Factors Favoring Altruism and Group Living. Many explanations have been put forward to explain the evolution of altruism.

Examples of Kin Selection. One explanation for altruism is that individuals can increase the extent to which their genes are passed on to the next generation by aiding their relatives.

Group Living and the Evolution of Social Systems. Insect societies exhibit extreme cooperation and altruism, perhaps as a result of close genetic relationship of society members.

27.4 Vertebrates exhibit a broad range of social behaviors.

Vertebrate Societies. Many vertebrate societies exhibit altruism.

Human Sociobiology. Human behavior, like that of other vertebrates, is influenced by natural selection.

FIGURE 27.1
A snake in the throes of death—or is it? When threatened, many organisms feign death, as this snake is doing—foaming at the mouth and going limp or looking paralyzed.

Animal behavior can be investigated in a variety of ways. An investigator can ask, how did the behavior develop? What is the physiology behind the behavior? Or what is the function of the behavior (figure 27.1), and does it confer an advantage to the animal? The field of behavioral ecology deals with the last two questions. Specifically, *behavioral ecologists* study the ways in which behavior may be adaptive by allowing an animal to increase or even maximize its reproductive success. This chapter examines both of these aspects of behavioral ecology.

27.1 Evolutionary forces shape behavior.

Behavioral Ecology

In an important essay, Nobel laureate Niko Tinbergen outlined the different types of questions biologists can ask about animal behavior. In essence, he divided the investigation of behavior into the study of its development, physiological basis, and function (evolutionary significance). One type of evolutionary analysis pioneered by Tinbergen himself was the study of the **survival value** of behavior. That is, how does an animal's behavior allow it to stay alive or keep its offspring alive? For example, Tinbergen observed that after gull nestlings hatch, the parents remove the eggshells from the nest. To understand *why* this behavior occurs, he camouflaged chicken eggs by painting them to resemble the natural background where they would lie and distributed them throughout the area in which the gulls were nesting (figure 27.2). He placed broken eggshells next to some of the eggs, and as a control, he left other camouflaged eggs alone without eggshells. He then noted which eggs were found more easily by crows. Because the crows could use the white interior of a broken eggshell as a cue, they ate more of the camouflaged eggs that were near eggshells. Thus, Tinbergen concluded that eggshell removal behavior is *adaptive:* it reduces predation and thus increases the offspring's chances of survival.

Tinbergen is credited with being one of the founders of the field of **behavioral ecology,** the study of how natural selection shapes behavior. This branch of ecology examines the **adaptive significance** of behavior, or how behavior may increase survival and reproduction. Current research in behavioral ecology focuses on the contribution behavior makes to an animal's reproductive success, or **fitness.** As we saw in chapter 26, differences in behavior among individuals often result from genetic differences. Thus, natural selection operating on behavior has the potential to produce evolutionary change. To study the relation between behavior and fitness, then, is to study the process of adaptation itself.

Consequently, the field of behavioral ecology is concerned with two questions. First, is behavior adaptive? Although it is tempting to assume that the behavior produced by individuals must in some way represent an adaptive response to the environment, this need not be the case. As we saw in chapter 20, traits can evolve for many reasons other than natural selection, such as genetic drift or gene flow. Moreover, traits may be present in a population because they evolved as adaptations in the past, but no longer are useful. These possibilities hold true for behavioral traits as much as they do for any other kind of trait.

If a trait is adaptive, the question then becomes: how is it adaptive? Although the ultimate criterion is reproductive success, behavioral ecologists are interested in how a trait can lead to greater reproductive success. By enhancing energy intake, thus increasing the number of offspring produced? By improving success in getting more matings? By decreasing the chance of predation? The job of a behavioral ecologist is to determine the effect of a behavioral trait on each of these activities and then to discover whether increases in, for example, foraging efficiency, translate into increased fitness.

FIGURE 27.2
The adaptive value of egg coloration. Niko Tinbergen painted chicken eggs to resemble the mottled brown camouflage of gull eggs. The eggs were used to test the hypothesis that camouflaged eggs are more difficult for predators to find and thus increase the young's chances of survival.

Behavioral ecology is the study of how natural selection shapes behavior.

Foraging Behavior

The best introduction to behavioral ecology is the examination of one well-defined behavior in detail. While many behaviors might be chosen, we will focus on foraging behavior. For many animals, food comes in a variety of sizes. Larger foods may contain more energy but may be harder to capture and less abundant. In addition, some types of food may be farther away than other types. Hence, foraging for these animals involves a trade-off between a food's energy content and the cost of obtaining it. The *net energy* (in calories or Joules) gained by feeding on each size prey is simply the energy content of the prey minus the energy costs of pursuing and handling it. According to **optimal foraging theory,** natural selection favors individuals whose foraging behavior is as energetically efficient as possible. In other words, animals tend to feed on prey that maximize their net energy intake per unit of foraging time.

A number of studies have demonstrated that foragers do preferentially utilize prey that maximize the energy return. Shore crabs, for example, tend to feed primarily on intermediate-sized mussels which provide the greatest energetic return; larger mussels provide more energy, but also take considerably more energy to crack open (figure 27.3).

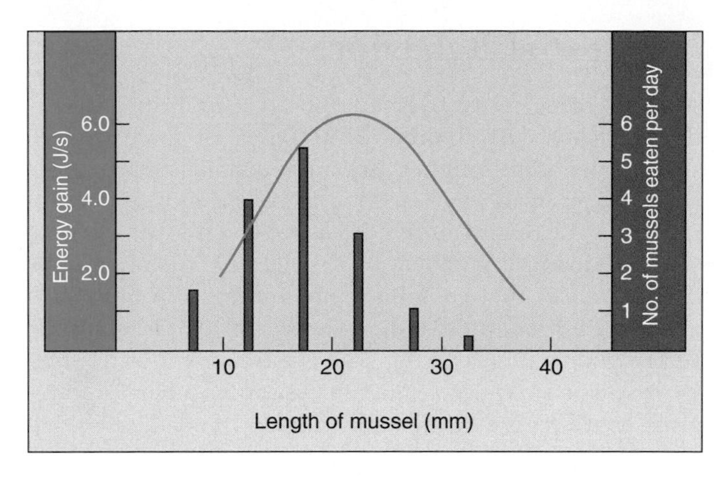

FIGURE 27.3
Optimal diet. The shore crab selects a diet of energetically profitable prey. The curve describes the net energy gain (equal to energy gained minus energy expended) derived from feeding on different sizes of mussels. The bar graph shows the numbers of mussels of each size in the diet. Shore crabs most often feed on those mussels that provide the most energy.

This optimal foraging approach makes two assumptions. First, natural selection will only favor behavior that maximizes energy acquisition if increased energy reserves lead to increases in reproductive success. In some cases, this is true. For example, in both Columbian ground squirrels and captive zebra finches, a direct relationship exists between net energy intake and the number of offspring raised; similarly, the reproductive success of orb-weaving spiders is related to how much food they can capture.

However, animals have other needs beside energy acquisition, and sometimes these needs come in conflict. One obvious alternative is avoiding predators: oftentimes the behavior that maximizes energy intake is not the one that minimizes predation risk. Thus, the behavior that maximizes fitness often may reflect a trade-off between obtaining the most energy at the least risk of being eaten. Not surprisingly, many studies have shown that a wide variety of animal species alter their foraging behavior when predators are present. Still another alternative is finding mates: males of many species, for example, will greatly reduce their feeding rate in order to enhance their ability to attract and defend females.

The second assumption of optimal foraging theory is that it has resulted from natural selection. As we have seen, natural selection can lead to evolutionary change only when differences among individuals have a genetic basis. Few studies have investigated whether differences among individuals in their ability to maximize energy intake is the result of genetic differences, but there are some exceptions. For example, one study found that female zebra finches that were particularly successful in maximizing net energy intake tended to have offspring that were similarly successful. Because birds were removed from their mothers before they left the nest, this similarity likely reflected a genetic basis for foraging behavior, rather than being a result of young birds learning to forage from their mothers.

Differences among individuals in foraging behavior may also be a function of age. Inexperienced yellow-eyed juncos (a small North American bird), for example, have not learned how to handle large prey items efficiently. As a result, the energetic costs of eating such prey are higher than the benefits, and as a result they tend to focus on smaller prey. Only when the birds are older and more experienced do they learn to easily dispatch these prey, which are then included in the diet.

Natural selection may favor the evolution of foraging behaviors that maximize the amount of energy gained per unit time spent foraging. Animals that acquire energy efficiently during foraging may increase their fitness by having more energy available for reproduction, but other considerations, such as avoiding predators, also are important in determining reproductive success.

Territorial Behavior

Animals often move over a large area, or **home range,** during their daily course of activity. In many animal species, the home ranges of several individuals may overlap in time or in space, but each individual defends a *portion* of its home range and uses it *exclusively*. This behavior, in which individual members of a species maintain exclusive use of an area that contains some limiting resource, such as foraging ground, food, or potential mates, is called **territoriality** (figure 27.4). The critical aspect of territorial behavior is *defense* against intrusion by other individuals. Territories are defended by displays that advertise that the territories are occupied and by overt aggression. A bird sings from its perch within a territory to prevent a takeover by a neighboring bird. If an intruder is not deterred by the song, it may be attacked. However, territorial defense has its costs. Singing is energetically expensive, and attacks can lead to injury. In addition, advertisement through song or visual display can reveal one's position to a predator.

Why does an animal bear the costs of territorial defense? Over the past two decades, it has become increasingly clear that an *economic* approach can be useful in answering this question. Although there are costs to defending a territory, there are also benefits; these benefits may take the form of increased food intake, exclusive access to mates, or access to refuges from predators. Studies of nectar-feeding birds like hummingbirds and sunbirds provide an example (figure 27.5). A bird benefits from having the exclusive use of a patch of flowers because it can efficiently harvest the nectar they produce. In order to maintain exclusive use, however, the bird must actively defend the flowers. The benefits of exclusive use outweigh the costs of defense only under certain conditions. Sunbirds, for example, expend 3000 calories per hour chasing intruders from a territory. Whether or not the benefit of defending a territory will exceed this cost depends upon the amount of nectar in the flowers and how efficiently the bird can collect it. If flowers are very scarce or nectar levels are very low, for example, a nectar-feeding bird may not gain enough energy to balance the energy used in defense. Under this circumstance, it is not advantageous to be territorial. Similarly, if flowers are very abundant, a bird can efficiently meet its daily energy requirements without behaving territorially and adding the costs of defense. From an energetic standpoint, defending abundant resources isn't worth the cost. Territoriality thus only occurs at intermediate levels of flower availability and higher levels of nectar production, where the benefits of defense outweigh the costs.

In many species, exclusive access to females is a more important determinant of territory size of males than is food availability. In some lizards, for example, males maintain enormous territories during the breeding season. These territories, which encompass the territories of several females, are much larger than what is required to supply enough food and are defended vigorously. In the nonbreeding season, by contrast, male territory size decreases dramatically, as does aggressive territorial behavior.

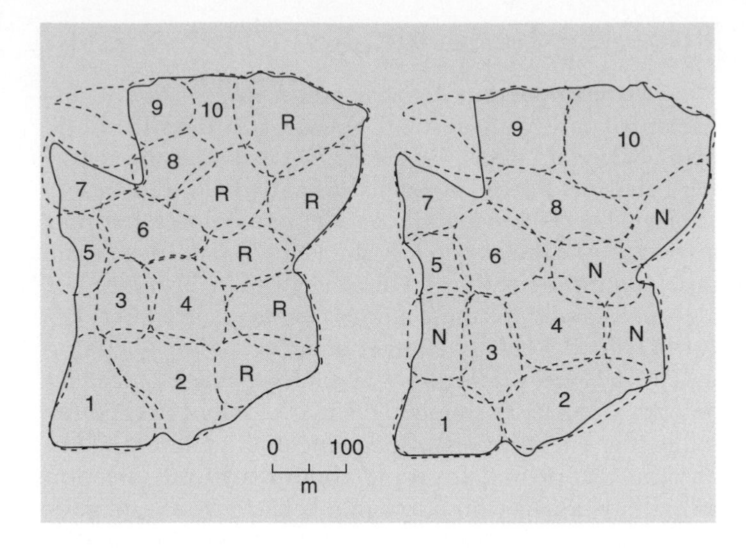

FIGURE 27.4
Competition for space. Territory size in birds is adjusted according to the number of competitors. When six pairs of great tits *(Parus major)* were removed from their territories (indicated by R in the left figure), their territories were taken over by other birds in the area and by four new pairs (indicated by N in the right figure). Numbers correspond to the birds present before and after.

FIGURE 27.5
The benefit of territoriality. Sunbirds increase nectar availability by defending flowers.

An economic approach can be used to explain the evolution and ecology of reproductive behaviors such as territoriality. This approach assumes that animals that gain more energy from a behavior than they expend will have an advantage in survival and reproduction over animals that behave in less efficient ways.

27.2 Reproductive behavior involves many choices influenced by natural selection.

Searching for a place to nest, finding a mate, and rearing young involve a collection of behaviors loosely referred to as reproductive behavior. These behaviors often involve seeking and defending a particular territory, making choices about mates and about the amount of energy to devote to the rearing of young. Mate selection, in particular, often involves intense natural selection. We will look briefly at each of these components of reproductive behavior.

During the breeding season, animals make several important "decisions" concerning their choice of mates, how many mates to have, and how much time and energy to devote to rearing offspring. These decisions are all aspects of an animal's **reproductive strategy,** a set of behaviors that presumably have evolved to maximize reproductive success. Reproductive strategies have evolved partly in response to the energetic costs of reproduction and the way food resources, nest sites, and members of the opposite sex are spatially distributed in the environment.

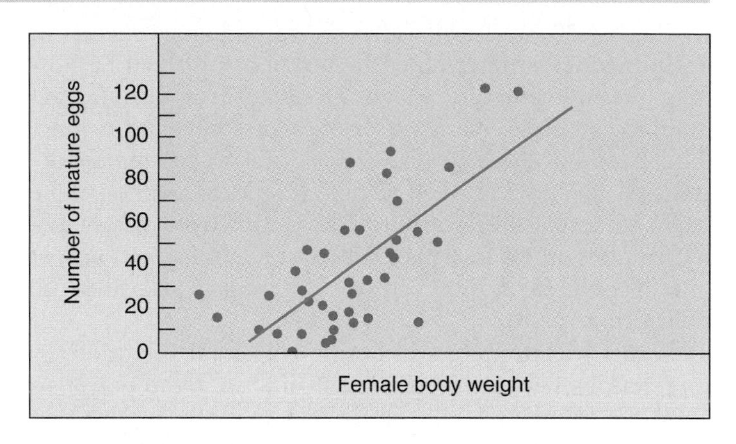

FIGURE 27.6
The advantage of male mate choice. Male mormon crickets choose heavier females as mates, and larger females have more eggs. Thus, male mate selection increases fitness.

Parental Investment and Mate Choice

Males and females usually differ in their reproductive strategies. Darwin was the first to observe that females often do not simply mate with the first male they encounter, but instead seem to evaluate a male's quality and then decide whether to mate. This behavior, called **mate choice**, has since been described in many invertebrate and vertebrate species.

By contrast, mate choice by males is much less common. Why should this be? Many of the differences in reproductive strategies between the sexes can be understood by comparing the parental investment made by males and females. **Parental investment** refers to the contributions each sex makes in producing and rearing offspring; it is, in effect, an estimate of the energy expended by males and females in each reproductive event.

Many studies have shown that parental investment is high in females. One reason is that eggs are much larger than sperm—195,000 times larger in humans! Eggs contain proteins and lipids in the yolk and other nutrients for the developing embryo, but sperm are little more than mobile DNA. Furthermore, in some groups of animals, females are responsible for gestation and lactation, costly reproductive functions only they can carry out.

The consequence of such great disparities in reproductive investment is that the sexes should face very different selective pressures. Because any single reproductive event is relatively cheap for males, they can best increase their fitness by mating with as many females as possible—male fitness is rarely limited by the amount of sperm they can produce. By contrast, each reproductive event for females is much more costly and the number of eggs that can be produced often does limit reproductive success. For this reason, females have an incentive to be choosy, trying to pick the male the can provide the greatest benefit to her offspring. As we shall see, this benefit can take a number of different forms.

These conclusions only hold when female reproductive investment is much greater than that of males. In species with parental care, males may contribute equally to the cost of raising young; in this case, the degree of mate choice should be equal between the sexes.

Furthermore, in some cases, male investment exceeds that of females. For example, male mormon crickets transfer a protein-containing spermatophore to females during mating. Almost 30% of a male's body weight is made up by the spermatophore, which provides nutrition for the female, and helps her develop her eggs. As one might expect, in this case it is the females that compete with each other for access to males, and the males that are the choosy sex. Indeed, males are quite selective, favoring heavier females. The selective advantage of this strategy results because heavier females have more eggs; thus, males that choose larger females leave more offspring (figure 27.6).

Reproductive investment by the sexes is strongly influenced by differences in the degree of parental investment.

Reproductive Competition and Sexual Selection

In chapter 20, we learned that the reproductive success of an individual is determined by a number of factors: how long the individual lives, how successful it is in obtaining matings, and how many offspring it produces per mating. The second of these factors, competition for mating opportunities, has been termed **sexual selection**. Some people consider sexual selection to be distinctive from natural selection, but others see it as a subset of natural selection, just one of a number of ways in which organisms can increase their fitness.

Sexual selection involves both **intrasexual selection,** or interactions between members of one sex ("the power to conquer other males in battle," as Darwin put it), and **intersexual selection,** essentially mate choice ("the power to charm"). Sexual selection thus leads to the evolution of structures used in combat with other males, such as a deer's antlers and a ram's horns, as well as ornamentation used to "persuade" members of the opposite sex to mate, such as long tail feathers and bright plumage (figure 27.7*a*). These traits are called **secondary sexual characteristics.**

Intrasexual Selection

In many species, individuals of one sex—usually males—compete with each other for the opportunity to mate with individuals of the other sex. These competitions may take place over ownership of a territory in which females reside or direct control of the females themselves. The latter case is exemplified by many species, such as impala, in which females travel in large groups with a single male that gets exclusive rights to mate with the females and thus strives vigorously to defend these rights against other males which would like to supplant him.

In mating systems such as these, a few males may get an inordinate number of matings and most males do not mate at all. In elephant seals, in which males control territories on the breeding beaches, a few dominant males do most of the breeding. On one beach, for example, eight males impregnated 348 females, while the remaining males got very little action (or, we could say, while the remaining males mated rarely, if at all).

For this reason, selection will strongly favor any trait that confers greater ability to outcompete other males. In many cases, size determines mating success: the larger male is able to dominate the smaller one. As a result, in many territorial species, males have evolved to be considerably larger than females, for the simple reason that the largest males are the ones that get to mate. Such differences between the sexes are referred to as **sexual dimorphism**. In other species, males have evolved structures used for fighting, such as horns, antlers, and large canine teeth. These traits are also often sexually dimorphic and may have evolved because of the advantage they give in intrasexual conflicts.

Intersexual Selection

Peahens prefer to mate with peacocks that have more spots in their long tail feathers (figure 27.7*b*,*c*). Similarly, female frogs prefer to mate with males with more complex calls. Why did such mating preferences evolve?

(a)

(b)

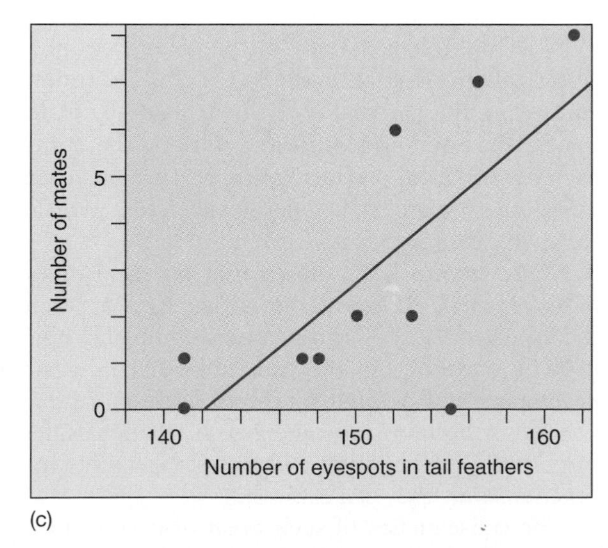

(c)

FIGURE 27.7
Products of sexual selection. Attracting mates with long feathers is common in bird species such as the African paradise whydah (*a*) and the peacock (*b*), which show pronounced sexual dimorphism. (*c*) Female peahens prefer to mate with males with greater numbers of eyespots in their tail feathers.
Source: Data from M. Petrie, et al "Peahens Prefer Peacocks with Elaborate Trains", *Animal Behavior*, 1991.

The Benefits of Mate Choice

In some cases, the benefits are obvious. In many species of birds and mammals, and some species of other types of animals, males help raise the offspring. In these cases, females would benefit by choosing the male that can provide the best care—the better the parent, the more offspring she is likely to rear.

In other species, males provide no care, but maintain territories that provide food, nesting sites, and predator refuges. In such species, females that choose males with the best territories will maximize their reproductive success.

Indirect Benefits

In other species, however, males provide no direct benefits of any kind to females. In such cases, it is not intuitively obvious what females have to gain by being choosy. Moreover, what could be the possible benefit of choosing a male with an extremely long tail or a complex song?

A number of theories have been proposed to explain the evolution of such preferences. One idea is that females choose the male that is the healthiest or oldest. Large males, for example, have probably been successful at living long, acquiring a lot of food and resisting parasites and disease. Similarly, in guppies and some birds, the brightness of a male's color is a reflection of the quality of his diet and overall health. Females may gain two benefits from mating with large or colorful males. First, to the extent that the males' success in living long and prospering is the result of a good genetic makeup, the female will be ensuring that her offspring receive good genes from their father. Indeed, several studies have demonstrated that males that are preferred by females produce offspring that are more vigorous and survive better than offspring of males that are not preferred. Second, healthy males are less likely to be carrying diseases, which might be transmitted to the female during mating.

A variant of this theory goes one step further. In some cases, females prefer mates with traits that are detrimental to survival (figure 27.8). The long tail of the peacock is a hindrance in flying and makes males more vulnerable to predators. Why should females prefer males with such traits? The **handicap hypothesis** states that only genetically superior mates can survive with such a handicap. By choosing a male with the largest handicap, the female is ensuring that her offspring will receive these quality genes. Of course, the male offspring will also inherit the genes for the handicap. For this reason, evolutionary biologists are still debating the merits of this hypothesis.

Other courtship displays appear to have evolved from a predisposition in the female's sensory system to be stimulated by a certain type of stimulus. For example, females may be better able to detect particular colors or sounds at a certain frequency. Sensory exploitation involves the evolution in males of an attractive signal that "exploits" these preexisting biases—if females are particularly adept at detecting red objects, for example, then males will evolve red coloration. Consider the vocalizations of the Túngara frog (*Physalaemus pustulosus*) (see figure 27.8). Unlike related species, males include a "chuck" in their calls. Recent research suggests that even females of related species are particularly attracted to calls of this sort, even though males of these species do not produce "chucks." Why this preference evolved is unknown, but males of the Túngara frog have evolved to take advantage of it.

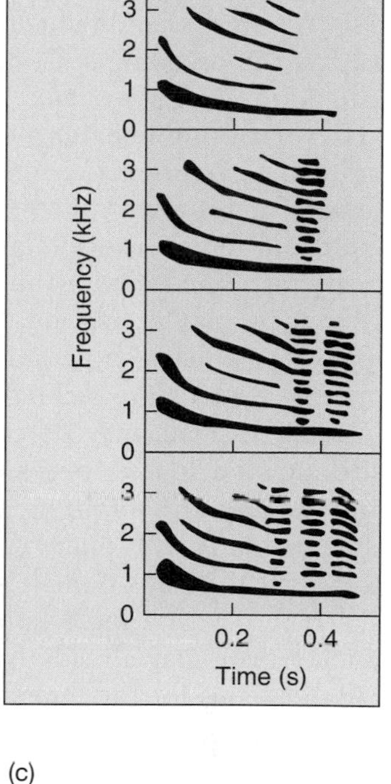

FIGURE 27.8
The benefits and costs of vocalizing. (*a*) The male Túngara frog, *Physalaemus pustulosus*. (*b*) The males' calls attract females as well as predatory bats. Calls of greater complexity are represented from top to bottom in (*c*). Females prefer more complex calls, but these calls are detected particularly well by bats. Consequently, males that females prefer are at the greatest risk of being captured.

A great variety of other theories have been proposed to explain the evolution of mating preferences. Many of these theories may be correct in some circumstances and none seems capable of explaining all of the variation in mating behavior in the animal world. This is an area of vibrant research with new discoveries appearing regularly.

Natural selection has favored the evolution of behaviors that maximize the reproductive success of males and females. By evaluating and selecting mates with superior qualities, an animal can increase its reproductive success.

Mating Systems

The number of individuals with which an animal mates during the breeding season varies throughout the animal kingdom. Mating systems such as monogamy (one male mates with one female); polygyny (one male mates with more than one female; figure 27.9), and polyandry (one female mates with more than one male) are aspects of male and female reproductive strategy that concern how many mates an individual has during the breeding season. Like mate choice, mating systems have evolved to maximize reproductive fitness. Much research has shown that mating systems are strongly influenced by ecology. For instance, a male may defend a territory that holds nest sites or food sources necessary for a female to reproduce, and the territory might have resources sufficient for more than one female. If males differ in the quality of the territories they hold, a female's fitness will be maximized if she mates with a male holding a high-quality territory. Such a male may already have a mate, but it is still more advantageous for the female to breed with that male than with an unmated male that defends a low-quality territory. In this way, natural selection would favor the evolution of polygyny.

Mating systems are also constrained by the needs of offspring. If the presence of both parents is necessary for young to be reared successfully, then monogamy may be favored. This is generally the case in birds, in which over 90% of all species are monogamous. A male may either remain with his mate and provide care for the offspring or desert that mate to search for others; both strategies may increase his fitness. The strategy that natural selection will favor depends upon the requirement for male assistance in feeding or defending the offspring. In some species, offspring are **altricial**—they require prolonged and extensive care. In these species, the need for care by two parents will reduce the tendency for the male to desert his mate and seek other matings. In species where the young are **precocial** (requiring little parental care), males may be more likely to be polygynous.

Although polygyny is much more common, polyandrous systems—in which one female mates with several males—are known in a variety of animals. For example, in spotted sandpipers, males take care of all incubation and parenting, and females mate and leave eggs with two or more males.

In recent years, researchers have uncovered many unexpected aspects of animal reproductive systems. Some of these discoveries have resulted from the application of new technologies, whereas others have come from detailed and intensive field studies.

FIGURE 27.9
Female defense polygyny in bats. The male at the lower right is guarding a group of females.

Extra-Pair Copulations

In chapter 19, we saw how DNA fingerprinting can be used to identify blood samples. Another common use of this technology is to establish paternity. DNA fingerprints are so variable that each individual's is unique. Thus, by comparing the DNA of a man and a child, experts can establish with a relatively high degree of confidence whether the man is the child's father.

This approach is now commonly used in paternity lawsuits, but it has also become a standard weapon in the arsenal of behavioral ecologists. By establishing paternity, researchers can precisely quantify the reproductive success of individual males and thus assess how successful their particular reproductive strategies have been (figure 27.10*a*). In one classic study of red-winged blackbirds (figure 27.10*b*), researchers established that half of all nests contained at least one bird fertilized by a male other than the territory owner; overall, 20% of the offspring were the result of such **extra-pair copulations** (EPCs).

Studies such as this have established that EPCs—"cheating"—are much more pervasive in the bird world than originally suspected. Even in some species that were believed to be monogamous on the basis of behavioral observations, the incidence of offspring being fathered by a male other than the female's mate is sometimes surprisingly high.

Why do individuals have extra-pair copulations? For males, the answer is obvious: increased reproductive success. For females, it is less clear, as in most cases, it does not result in an increased number of offspring. One possibility is that females mate with genetically superior individuals, thus enhancing the genes passed on to their offspring. Another possibility is that females can increase the amount of help they get in raising their offspring. If a female mates with more than one male, each male may help raise the offspring. This is exactly what happens in a common English bird, the dunnock. Females mate not only with the territory owner, but also with subordinate males that hang around the edge of the territory. If these subordinates mate enough with a female, they will help raise her young, presumably because some of these young may have been fathered by this male.

Alternative Mating Tactics

Natural selection has led to the evolution of a variety of other means of increasing reproductive success. For example, in many species of fish, there are two genetic classes of males. One group is large and defends territories to obtain matings. The other type of male is small and adopts a completely different strategy. They do not maintain territories, but hang around the edge of the territories of large males. Just at the end of a male's courtship, when the female is laying her eggs and the territorial male is depositing sperm, the smaller male will dart in and release its own sperm into the water, thus fertilizing some of the eggs. If this strategy is successful, natural selection will favor the evolution of these two different male reproductive strategies.

Similar patterns are seen in other organisms. In some dung beetles, territorial males have large horns that they use to guard the chambers in which females reside, whereas genetically small males don't have horns; instead, they dig side tunnels and attempt to intercept the female inside her chamber. In isopods, there are three genetic size classes. The medium-sized males pass for females and enter a large male's territory in this way; the smallest class are so tiny, they are able to sneak in completely undetected.

This is just a glimpse of the rich diversity in mating systems that have evolved. The bottom line is: if there is a way of increasing reproductive success, natural selection will favor its evolution.

Mating systems represent reproductive adaptations to ecological conditions. The need for parental care and the ability of both sexes to provide it are important influences on the evolution of monogamy, polygyny, and polyandry. Detailed study of animal mating systems, along with the use of modern molecular techniques, are revealing many surprises in animal mating systems. This diversity is a testament to the power of natural selection to favor any trait that increases an animal's fitness.

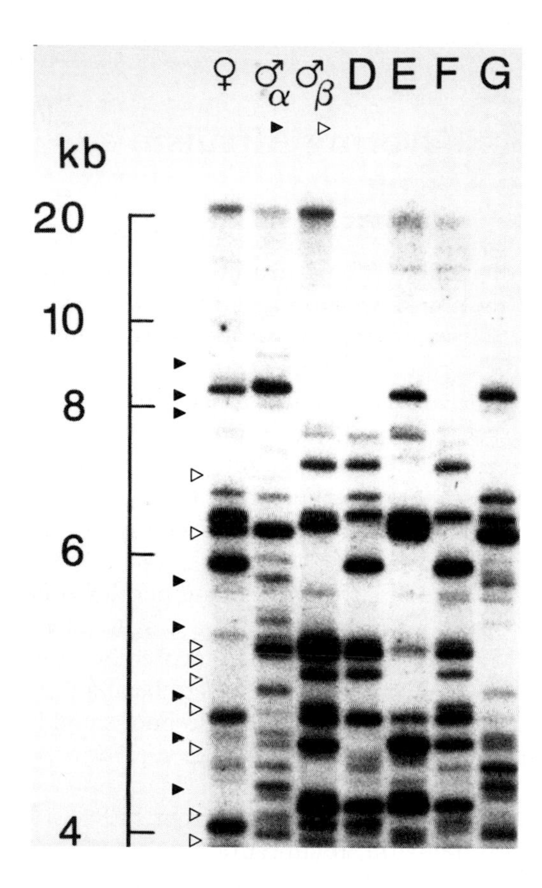

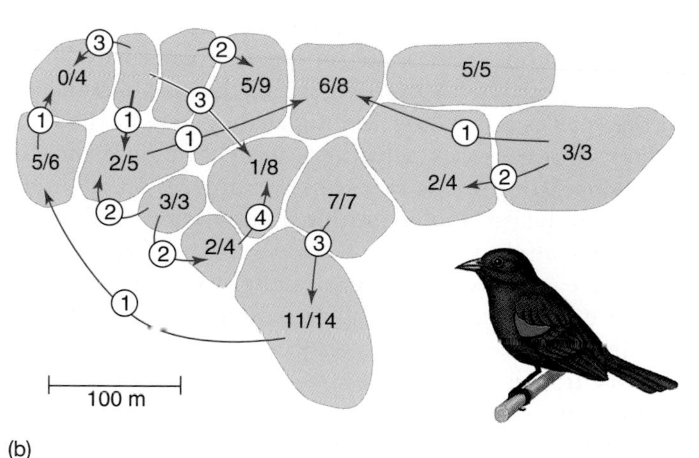

FIGURE 27.10
The study of paternity. (*a*) A DNA fingerprinting gel from the dunnock. The bands represent fragments of DNA of different lengths. The four nestlings (D-G) were in the nest of the female. By comparing the bands present in the two males, we can determine which male fathered which offspring. The triangles point to the bands which are diagnostic for one male and not the other. In this case, the beta male fathered three (D, E, F, but not G) of the four offspring. (*b*) Results of a DNA fingerprinting study in red-winged blackbirds. Fractions indicate the proportion of offspring fathered by the male in whose territory the nest occurred. Arrows indicate how many offspring were fathered by particular males outside of each territory. Nests on some territories were not sampled.

Factors Favoring Altruism and Group Living

Altruism—the performance of an action that benefits another individual at a cost to the actor—occurs in many guises in the animal world. In many bird species, for example, parents are assisted in raising their young by other birds, which are called **helpers at the nest**. In species of both mammals and birds, individuals that spy a predator will give an alarm call, alerting other members of their group, even though such an act would seem to call the predator's attention to the caller. Finally, lionesses with cubs will allow all cubs in the pride to nurse, including cubs of other females.

The existence of altruism has long perplexed evolutionary biologists. If altruism imposes a cost to an individual, how could an allele for altruism be favored by natural selection? One would expect such alleles to be at a disadvantage and thus their frequency in the gene pool should decrease through time.

A number of explanations have been put forward to explain the evolution of altruism. One suggestion often heard on television documentaries is that such traits evolve for the good of the species. The problem with such explanations is that natural selection operates on individuals within species, not on species themselves. Thus, it is even possible for traits to evolve that are detrimental to the species as a whole, as long as they benefit the individual. In some cases, selection can operate on groups of individuals, but this is rare. For example, if an allele for supercannibalism evolved within a population, individuals with that allele would be favored, as they would have more to eat; however, the group might eventually eat itself to extinction, and the allele would be removed from the species. In certain circumstances, such **group selection** can occur, but the conditions for it to occur are rarely met in nature. In most cases, consequently, the "good of the species" cannot explain the evolution of altruistic traits.

Another possibility is that seemingly altruistic acts aren't altruistic after all. For example, helpers at the nest are often young and gain valuable parenting experience by assisting established breeders. Moreover, by hanging around an area, such individuals may inherit the territory when the established breeders die. Similarly, alarm callers may actually benefit by causing other animals to panic. In the ensuing confusion, the caller may be able to slip off undetected. Detailed field studies in recent years have demonstrated that some acts truly are altruistic, but others are not as they seemed.

Reciprocity

Robert Trivers, now of Rutgers University, proposed that individuals may form "partnerships" in which mutual exchanges of altruistic acts occur, because it benefits both participants to do so. In the evolution of such **reciprocal altruism,** "cheaters" (nonreciprocators) are discriminated against and are cut off from receiving future aid. According to Trivers, if the altruistic act is relatively inexpensive, the small benefit a cheater receives by not reciprocating is far outweighed by the potential cost of not receiving future aid. Under these conditions, cheating should not occur.

Vampire bats roost in hollow trees in groups of 8 to 12 individuals. Because these bats have a high metabolic rate, individuals that have not fed recently may die. Bats that have found a host imbibe a great deal of blood; giving up a small amount presents no great energy cost to the donor, and it can keep a roostmate from starvation. Vampire bats tend to share blood with past reciprocators. If an individual fails to give blood to a bat from which it had received blood in the past, it will be excluded from future bloodsharing.

Kin Selection

The most influential explanation for the origin of altruism was presented by William D. Hamilton in 1964. It is perhaps best introduced by quoting a passing remark made in a pub in 1932 by the great population geneticist J. B. S. Haldane. Haldane said that he would willingly lay down his life for *two brothers* or *eight first cousins.* Evolutionarily speaking, Haldane's statement makes sense, because for each allele Haldane received from his parents, his brothers each had a 50% chance of receiving the same allele (figure 27.11). Consequently, it is statistically expected that two of his brothers would pass on as many of Haldane's particular combination of alleles to the next generation as Haldane himself would. Similarly, Haldane and a first cousin would share an eighth of their alleles (see figure 27.11). Their parents, which are siblings, would each share half their alleles, and each of their children would receive half of these, of which half on the average would be in common: one-half × one-half × one-half = one-eighth. Eight first cousins would therefore pass on as many of those alleles to the next generation as Haldane himself would. Hamilton saw Haldane's point clearly: natural selection will favor any strategy that increases the net flow of an individual's alleles to the next generation.

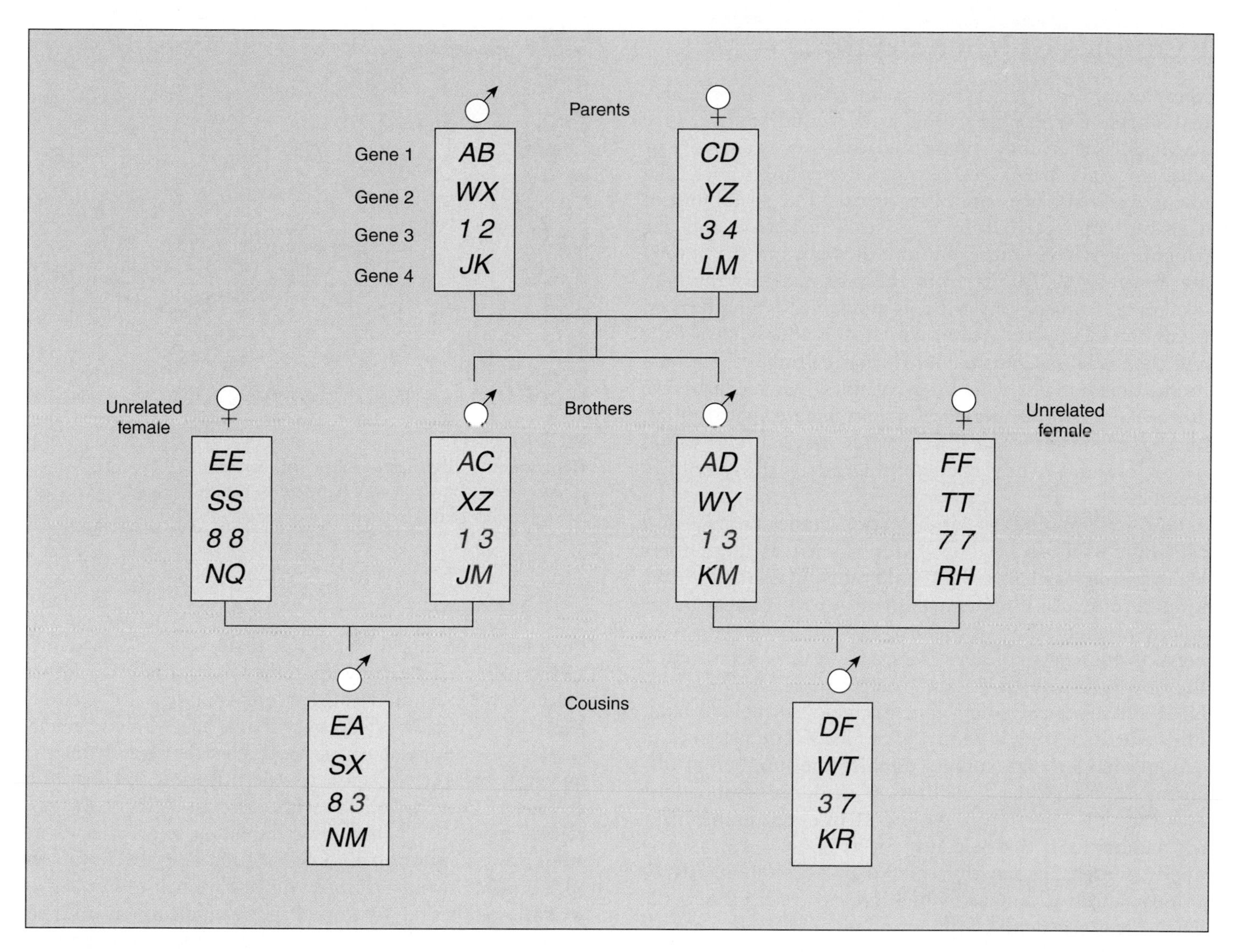

FIGURE 27.11
Hypothetical example of genetic relationship. On average, full siblings will share half of their alleles. By contrast, cousins will, on average, only share one-eighth of their alleles.

Hamilton showed that by directing aid toward kin, or close genetic relatives, an altruist may increase the reproductive success of its relatives enough to compensate for the reduction in its own fitness. Because the altruist's behavior increases the propagation of alleles in relatives, it will be favored by natural selection. Selection that favors altruism directed toward relatives is called **kin selection.** Although the behaviors being favored are cooperative, the genes are actually "behaving selfishly," because they encourage the organism to support copies of themselves in other individuals.

Hamilton's kin selection model predicts that altruism is likely to be directed toward close relatives. The more closely related two individuals are, the greater the potential genetic payoff. This relationship is described by **Hamilton's rule**, which states that altruistic acts are favored when $b/c > 1/r$. In this expression, b and c are the benefits and costs of the altruistic act, respectively, and r is the coefficient of relatedness, the proportion of alleles shared by two individuals through common descent. For example, an individual should be willing to have one less child if such actions allow a half-sibling, which shares one-quarter of its genes, to have more than four additional offspring.

Many factors could be responsible for the evolution of altruistic behaviors.

Examples of Kin Selection

Many examples of kin selection are known from the animal world. For example, Belding's ground squirrel give alarm calls when they spot a predator such as a coyote or a badger. Such predators may attack a calling squirrel, so giving a signal places the caller at risk. The social unit of a ground squirrel colony consists of a female and her daughters, sisters, aunts, and nieces. Males in the colony are not genetically related to these females. By marking all squirrels in a colony with an individual dye pattern on their fur and by recording which individuals gave calls and the social circumstances of their calling, researchers found that females who have relatives living nearby are more likely to give alarm calls than females with no kin nearby. Males tend to call much less frequently as would be expected as they are not related to most colony members.

Another example of kin selection comes from a bird called the white-fronted bee-eater which lives along rivers in Africa in colonies of 100 to 200 birds. In contrast to the ground squirrels, it is the males that usually remain in the colony in which they were born, and the females that disperse to join new colonies. Many bee-eaters do not raise their own offspring, but rather help others. Many of these birds are relatively young, but helpers also include older birds whose nesting attempts have failed. The presence of a single helper, on average, doubles the number of offspring that survive. Two lines of evidence support the idea that kin selection is important in determining helping behavior in this species. First, helpers are usually males, which are usually related to other birds in the colony, and not females, which are not related. Second, when birds have the choice of helping different parents, they almost invariably choose the parents to which they are most closely related.

Haplodiploidy and Hymenopteran Social Evolution

Probably the most famous application of kin selection theory has been to social insects. A hive of honeybees consists of a single queen, who is the sole egg-layer, and up to 50,000 of her offspring, nearly all of whom are female workers with nonfunctional ovaries (figure 27.12), a situation termed **eusociality**. The sterility of the workers is altruistic: these offspring gave up their personal reproduction to help their mother rear more of their sisters.

Hamilton explained the origin of altruism in hymenopterans (that is, bees, wasps, and ants) with his kin selection model. In these insects, males are haploid and females are diploid. This unusual system of sex determination, called **haplodiploidy,** leads to an unusual situation. If the queen is fertilized by a single male, then all female offspring will inherit exactly the same alleles from their father (because he is haploid and has only one copy of each allele). These female offspring will also share among themselves, on average, half of the alleles they get from the queen. Consequently, each female offspring will share on average 75% of her alleles with each sister (to verify this, rework figure 27.11, but allow the father to only have one allele for each gene). By contrast, should she have offspring of her own, she would only share half of her alleles with these offspring (the other half would come from their father). Thus, because of this close genetic relatedness, *workers propagate more alleles by giving up their own reproduction to assist their mother in rearing their sisters, some of whom will be new queens and start new colonies and reproduce*. Thus, this unusual haplodiploid system may have set the stage for the evolution of eusociality in hymenopterans and, indeed, such systems have evolved as many as 12 or more times in the Hymenoptera.

FIGURE 27.12
Reproductive division of labor in honeybees. The queen (shown here with a red spot painted on her thorax) is the sole egg-layer. Her daughters are sterile workers.

One wrinkle in this theory, however, is that eusocial systems have evolved in several other groups, including thrips, termites, and naked mole rats. Although thrips are also haplodiploid, both termites and naked mole rats are not. Thus, although haplodiploidy may have facilitated the evolution of eusociality, it is not a necessary prerequisite.

Kin selection is a potent force favoring, in some situations, the evolution of altruism and even complex social systems.

Group Living and the Evolution of Social Systems

Organisms as diverse as bacteria, cnidarians, insects, fish, birds, prairie dogs, lions, whales, and chimpanzees exist in social groups. To encompass the wide variety of social phenomena, we can broadly define a society as *a group of organisms of the same species that are organized in a cooperative manner.*

Why have individuals in some species given up a solitary existence to become members of a group? We have just seen that one explanation is kin selection: groups may be composed of close relatives. In other cases, individuals may benefit directly from social living. For example, a bird that joins a flock may receive greater protection from predators. As flock size increases, the risk of predation decreases because there are more individuals to scan the environment for predators (figure 27.13). A member of a flock may also increase its feeding success if it can acquire information from other flock members about the location of new, rich food sources. In some predators, hunting in groups can increase success and allow the group to tackle prey too large for any one individual.

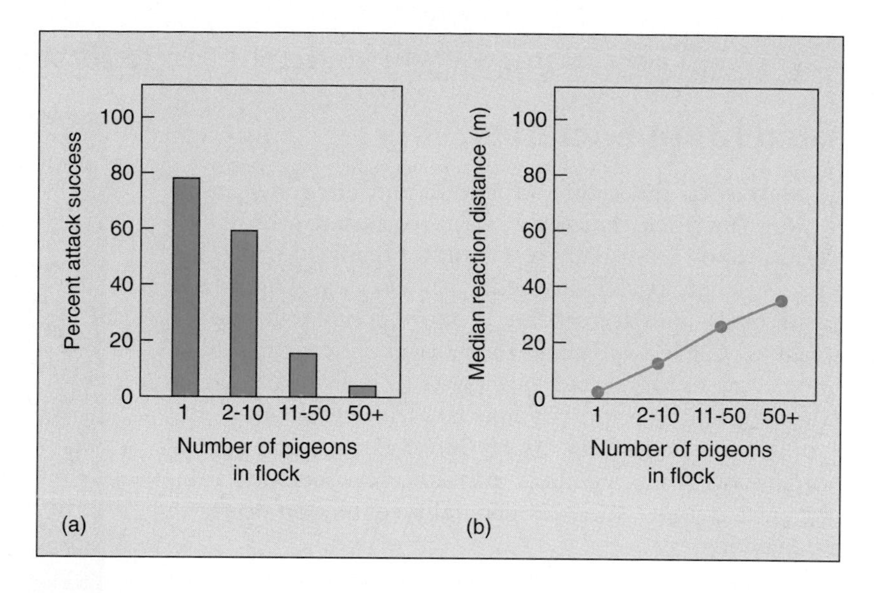

FIGURE 27.13
Flocking behavior decreases predation. (*a*) As the size of a pigeon flock increases, hawks are less successful at capturing pigeons. (*b*) When more pigeons are present in the flock, they can detect hawks at greater distances, thus allowing more time for the pigeons to escape.

Insect Societies

In insects, sociality has chiefly evolved in two orders, the Hymenoptera (ants, bees, and wasps) and the Isoptera (termites), although a few other insect groups include social species. All ants, some bees, some wasps, and all termites are **eusocial** (truly social): they have a division of labor in reproduction (a fertile queen and sterile workers), cooperative care of brood and an overlap of generations so that the queen lives alongside her offspring. Social insect colonies are composed of different **castes** of workers that differ in size and morphology and have different tasks they perform, such as workers and soldiers.

In honeybees, the queen maintains her dominance in the hive by secreting a pheromone, called "queen substance," that suppresses development of the ovaries in other females, turning them into sterile workers. Drones (male bees) are produced only for purposes of mating. When the colony grows larger in the spring, some members do not receive a sufficient quantity of queen substance, and the colony begins preparations for swarming. Workers make several new queen chambers, in which new queens begin to develop. Scout workers look for a new nest site and communicate its location to the colony. The old queen and a swarm of female workers then move to the new site. Left behind, a new queen emerges, kills the other potential queens, flies out to mate, and returns to assume "rule" of the hive.

The leafcutter ants provide another fascinating example of the remarkable lifestyles of social insects. Leafcutters live in colonies of up to several million individuals, growing crops of fungi beneath the ground. Their mound-like nests are underground "cities" covering more than 100 square meters, with hundreds of entrances and chambers as deep as 5 meters beneath the ground. The division of labor among the worker ants is related to their size. Every day, workers travel along trails from the nest to a tree or a bush, cut its leaves into small pieces, and carry the pieces back to the nest. Smaller workers chew the leaf fragments into a mulch, which they spread like a carpet in the underground fungus chambers. Even smaller workers implant fungal hyphae in the mulch. Soon a luxuriant garden of fungi is growing. While other workers weed out undesirable kinds of fungi, nurse ants carry the larvae of the nest to choice spots in the garden, where the larvae graze. This elaborate social system has evolved to produce reproductive queens that will disperse from the parent nest and start new colonies, repeating the cycle.

Eusocial insect workers exhibit an advanced social structure that includes division of labor in reproduction and workers with different tasks.

27.4 Vertebrates exhibit a broad range of social behaviors.

Vertebrate Societies

In contrast to the highly structured and integrated insect societies and their remarkable forms of altruism, vertebrate social groups are usually less rigidly organized and cohesive. It seems paradoxical that vertebrates, which have larger brains and are capable of more complex behaviors, are generally less altruistic than insects. Nevertheless, in some complex vertebrate social systems individuals may be exhibiting both reciprocity and kin-selected altruism. But vertebrate societies also display more conflict and aggression among group members than do insect societies. Conflict in vertebrate societies generally centers on access to food and mates.

Vertebrate societies, like insect societies, have particular types of organization. Each social group of vertebrates has a certain size, stability of members, number of breeding males and females, and type of mating system. Behavioral ecologists have learned that the way a group is organized is influenced most often by ecological factors such as food type and predation (figure 27.14).

African weaver birds, which construct nests from vegetation, provide an excellent example to illustrate the relationship between ecology and social organization. Their roughly 90 species can be divided according to the type of social group they form. One set of species lives in the forest and builds camouflaged, solitary nests. Males and females are monogamous; they forage for insects to feed their young. The second group of species nests in colonies in trees on the savanna. They are polygynous and feed in flocks on seeds. The feeding and nesting habits of these two sets of species are correlated with their mating systems. In the forest, insects are hard to find, and both parents must cooperate in feeding the young. The camouflaged nests do not call the attention of predators to their brood. On the open savanna, building a hidden nest is not an option. Rather, savanna-dwelling weaver birds protect their young from predators by nesting in trees which are not very abundant. This shortage of safe nest sites means that birds must nest together in colonies. Because seeds occur abundantly, a female can acquire all the food needed to rear young without a male's help. The male, free from the duties of parenting, spends his time courting many females—a polygynous mating system.

One exception to the general rule that vertebrate societies are not organized like those of insects is the naked mole rat, a small, hairless rodent that lives in and near East Africa. Unlike other kinds of mole rats, which live alone or in small family groups, naked mole rats form large underground colonies with a far-ranging system of tunnels and a central nesting area. It is not unusual for a colony to contain 80 individuals.

FIGURE 27.14
Foraging and predator avoidance. A meerkat sentinel on duty. Meerkats, *Suricata suricata*, are a species of highly social mongoose living in the semiarid sands of the Kalahari Desert. This meerkat is taking its turn to act as a lookout for predators. Under the security of its vigilance, the other group members can focus their attention on foraging.

Naked mole rats feed on bulbs, roots and tubers, which they locate by constant tunneling. As in insect societies, there is a division of labor among the colony members, with some mole rats working as tunnelers while others perform different tasks, depending upon the size of their body. Large mole rats defend the colony and dig tunnels.

Naked mole rat colonies have a *reproductive* division of labor similar to the one normally associated with the eusocial insects. All of the breeding is done by a single female or "queen," who has one or two male consorts. The workers, consisting of both sexes, keep the tunnels clear and forage for food.

Social behavior in vertebrates is often characterized by kin-selected altruism. Altruistic behavior is involved in cooperative breeding in birds and alarm-calling in mammals.

Human Sociobiology

As a social species, humans have an unparalleled complexity. Indeed, we are the only species with the intelligence to contemplate the social behavior of other animals. Intelligence is just one human trait. If an ethologist were to take an inventory of human behavior, he or she would list kin-selected altruism; reciprocity and other elaborate social contracts; extensive parental care; conflicts between parents and offspring; violence and warfare; infanticide; a variety of mating systems, including monogamy, polygyny, and polyandry; along with sexual behaviors such as extra-pair copulation ("adultery") and homosexuality; and behaviors like adoption that appear to defy evolutionary explanation. This incredible variety of behaviors occurs *in one species*, and any trait can change within *any individual*. Are these behaviors rooted in human biology?

Biological and Cultural Evolution

During the course of human evolution and the emergence of civilization, two processes have led to adaptive change. One is **biological evolution.** We have a primate heritage, reflected in the extensive amount of genetic material we share with our closest relatives, the chimpanzees. Our upright posture, bipedal locomotion, and powerful, precise hand grips are adaptations whose origins are traceable through our primate ancestors. Kin-selected and reciprocal altruism, as well as other shared traits like aggression and different types of mating systems, can also be seen in nonhuman primates, in whom we can demonstrate that these social traits are adaptive. We may speculate, based on various lines of evidence, that similar traits evolved in early humans. If individuals with certain social traits had an advantage in reproduction over other individuals that lacked the traits, and if these traits had a genetic basis, then the alleles for their expression would now be expected to be part of the human genome and to influence our behavior.

The second process that has underscored the emergence of civilization and led to adaptive change is **cultural evolution,** the transfer across generations of information necessary for survival. This is a nongenetic mode of adaptation. Many adaptations—the use of tools, the formation of cooperative hunting groups, the construction of shelters, and marriage practices—do not follow Mendelian rules of inheritance and are passed from generation to generation by *tradition*. Nonetheless, cultural inheritance is as valid a way to convey adaptations across generations as genetic inheritance. Human cultures are also extraordinarily diverse. The ways in which children are socialized among Trobriand Islanders, Pygmies, and Yanomamo Indians are very different. Again, we must remember that this fantastic variation occurs within one species, and that individual behavior is very flexible.

Identifying the Biological Components of Human Behavior

Given this great flexibility, how can the biological components of human behavior be identified? One way is to look for common patterns that appear in a wide variety of cultures, that is, to study behaviors that are cross-cultural. In spite of cultural variation, there are some traits that characterize all human societies. For example, all cultures have an incest taboo, forbidding marriages between close relatives. Incestuous matings lead to a greater chance of exposing disorders such as mental retardation and hemophilia. Natural selection may have acted to create a behavioral disposition against incest, and that disposition is now a cultural norm. Genes responsible for guiding this behavior may have become fixed in human populations because of their adaptive effects. Genes thus guide the direction of culture.

Although human mating systems vary, polygyny is found to be the most common among all cultures. Because most mammalian species are polygynous, the human pattern seems to reflect our mammalian evolutionary heritage and thus is a part of our biology. This conclusion is drawn from using the comparative approach, common in evolutionary science. Nonverbal communication patterns, like smiling and raising the hand in a greeting, also occur in many cultures. Perhaps these behaviors represent a common human heritage.

The explanations sociobiology offers to understand human behavior have been and continue to be controversial. For example, the new discipline of **evolutionary psychology** seeks to understand the origins of the human mind. Human behaviors are viewed as being extensions of our genes. The diversity of human cultures are thought to have a common core of characteristics that are generated by our psychology, which evolved as an adaptation to the lifestyle of our hunter-gatherer ancestors during the Pleistocene. Much of human behavior is seen as reflecting ancient, adaptive traits, now expressed in the context of modern civilization. In this controversial view, human behaviors such as jealousy and infidelity are viewed as adaptations; these behaviors increased the fitness of our ancestors, and thus are now part of the human psyche.

Sociobiology offers general explanations of human behavior that are controversial, but are becoming more generally accepted than in the past.

Chapter 27

www.mhhe.com/raven6/resources27.mhtml

Summary | **Questions** | **Media Resources**

27.1 Evolutionary forces shape behavior.

- Many behaviors are ecologically important and serve as adaptations.
- Foraging and territorial behaviors have evolved because they allow animals to use resources efficiently.

1. What does optimal foraging theory predict about an animal's foraging behavior? What factors unrelated to this theory may also influence an animal's foraging choices?

2. What are the benefits of territorial behavior, and what are its costs? Under what circumstances is territorial behavior disadvantageous?

27.2 Reproductive behavior involves many choices influenced by natural selection.

- Male and female animals maximize their fitness with different reproductive behaviors. The differences relate to the extent to which each sex provides care for offspring.
- Usually, males are competitive and females show mate choice because females have higher reproductive costs.
- A species' mating system is related to its ecology.

3. Why does natural selection favor mate choice? What factor is most important in determining which sex exhibits mate choice?

4. In birds, how does the amount of parental care required by the offspring affect the evolution of a species' mating system?

27.3 There is considerable controversy about the evolution of social behavior.

- Many animals show altruistic, or self-sacrificing, behavior. Altruism may evolve through reciprocity or be directed toward relatives. Cooperative behavior often increases an individual's fitness.
- Individuals form social groups because it is advantageous for them to do so.
- Animal societies are characterized by cooperation and conflict. The organization of a society is related to the ecology of a species.

5. What is reciprocal altruism? What is kin selection? How does kin selection increase an individual's success in passing its genes on to the next generation?

27.4 Vertebrates exhibit a broad range of social behaviors.

- Human behavior is extremely rich and varied and may result from both biology and culture.
- Evolutionary theory can give us important insight into human nature, but such an approach to the study of human behavior may be controversial.

6. In vertebrate societies, what are the costs to an individual who makes an alarm call? Based on research in ground squirrels, which individuals are most likely to make alarm calls, and what benefits do they receive by doing so?

- Bioethics Case Study: Behavior Disordered Students
- *On Science* Article: Flipper, A Senseless Killer?

Part **VIII**

The Global Environment

Disappearing amphibians. Populations of amphibians, like this Western toad (*Bufo boreas*) are declining in numbers in many regions.

Real People Doing Real Science

Identifying the Environmental Culprit Harming Amphibians

What started out as a relatively standard biology field trip in Minnesota in 1995 to collect frogs turned into a bizarre experience. Approximately half of the frogs students collected were deformed, with extra legs or missing legs or no eyes. Turning to the Internet, the class soon discovered that the problem was not isolated to Minnesota. Neighboring states were reporting the same phenomenon—an alarming number of deformed frogs, all across the United States and Canada.

Some environmental scientists suspected that chemical pollutants in the water might be causing the deformities and that the widespread occurrence of deformed frogs might well be an early warning of potential future problems in other species, including humans.

Other scientists cautioned that a different factor might be responsible. Although chemicals such as pesticides certainly *could* produce deformities in localized situations, say near a chemical spill, so too could other environmental factors affecting local habitats, such as parasitic infections. Demonstrating this point, researchers in 1999 showed that the multilimb and missing limb phenomenon in frogs can be caused by parasites that infect the developing tadpoles, disrupting development of their limbs.

Responding to this alternative suggestion, those scientists nominating pesticides as the principal culprit have cautioned that showing parasites can have a significant effect on local populations is not the same thing as demonstrating that they have in fact done so. And, they add, it certainly doesn't rule out a major contribution to the problem by environmental pollutants.

Although deformed frogs quickly received national attention, they are but the tip of the iceberg, a global problem of declining amphibian populations. During the past 50 years, there has been a worldwide catastrophic decline in amphibian populations. In some cases we can point to specific local human activities as the cause: habitat destruction, the introduction of competitive species, and industrial pollution.

However, because the problem appears to be global in nature, we must also consider the possibility that the decline of amphibians reflects some more global environmental change. Chemical contamination of water by acid rain, increasing ionizing radiation (UV-B) resulting from ozone layer depletion, changes in weather patterns caused by the warming of the atmosphere—all have been seriously proposed as potential global causes of declining amphibian populations.

Amphibians are not the only species experiencing declining populations. Songbirds in the eastern United States, for example, have declined precipitously in the last few decades, largely due to habitat destruction. So why focus on amphibians? While amphibians play a significant role in the ecological balance of many habitats, and have proven effective in the development of new drugs, their real importance to biologists is that amphibians are particularly sensitive indicators of the environment. Their semiaquatic mode of living, depending on a watery environment to reproduce and keep their skin moist, means that they are exposed to all types of environmental changes.

Amphibians are particularly vulnerable during early development, when their fertilized eggs lay in water, exposed to acid, chemical pollutants, and UV-B radiation. While numerous experiments performed under laboratory conditions have demonstrated the power of these factors to produce developmental deformities, and in so doing to reduce population survival rates, it is important to understand that "can" does not equal "does." To learn what is in fact going on, it is necessary to also examine the effects of these factors on amphibian development in the natural environment.

In a particularly clear example of the sort of investigation that will be needed to sort out this complex issue, Andrew Blaustein of Oregon State University headed a team of scientists that set out to examine the effects of UV-B radiation on amphibians in natural populations. In a series of experiments carried out in the field, they attempted to assess the degree to which UV-B radiation promoted amphibian developmental deformities under natural conditions.

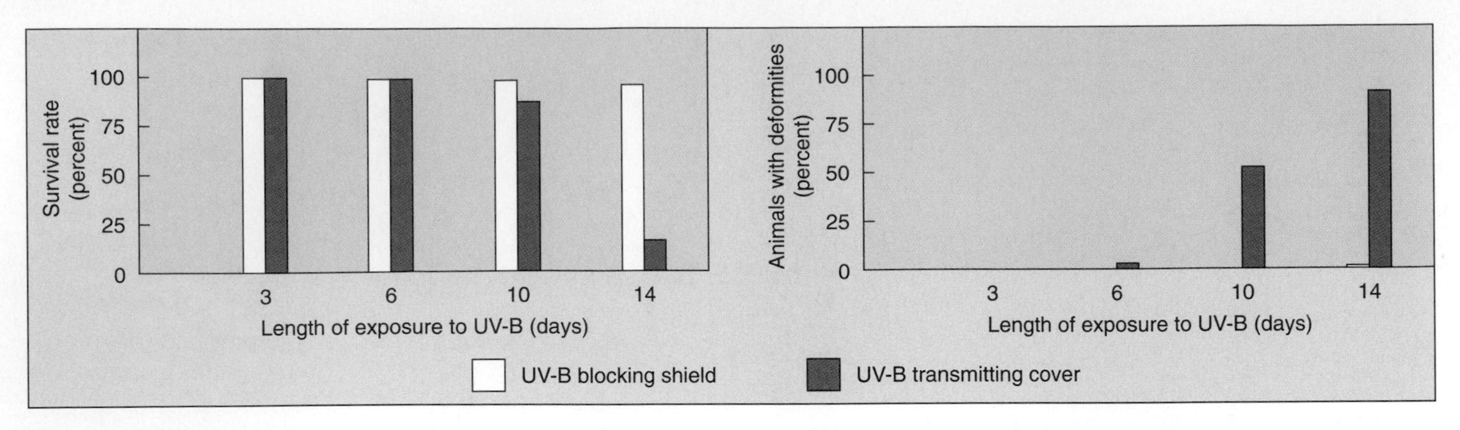

Blaustein's UV-B experiment. In the group of salamanders whose eggs were protected from UV-B radiation, hatching rates were higher and deformity rates were lower.

The Experiment

Laboratory experiments examining the affects of UV-B on amphibian development had already shown a significant increase in embryonic mortality in some amphibian species, and not in others. Why only in some?

Perhaps behaviors shared by many amphibian species might lead to an increased susceptibility to damage from UV-B radiation, behaviors such as laying eggs in open, shallow waters that offer significant exposure to UV-B radiation. Perhaps physiological traits of certain species make them particularly susceptible to damage from UV-B radiation, traits such as low levels of photolyase, an enzyme that removes harmful photoproducts induced by UV light.

Blaustein's group selected specimens for two different experiments that exhibit these factors, the long-toed salamander, *Ambystoma macrodactylum* and several species of frogs.

The goal of Blaustein's field experiment was to allow fertilized eggs to develop in their natural environment with and without a UV-B protective shield. Eggs in both groups were monitored for the appearance of deformities and for survival rates. Eggs were collected from natural shallow water sites (<20 cm deep) and randomly placed within enclosures containing either a UV-B blocking Mylar shield or a UV-B transmitting acetate cover (50 eggs per each enclosure replicated four times). The enclosures were placed in small, unperforated plastic pools containing pond water and the pools were placed back in the pond thereby exposing the eggs and developing embryos to ambient conditions. The UV-B blocking Mylar shield filtered out more than 94% of ambient UV-B radiation, while the UV-B transmitting acetate cover allowed about 90% of ambient UV-B radiation to pass through.

The Results

Embryos under the UV-B shields had significantly higher hatching rates and fewer deformities compared with those under the UV-B transmitting acetate covers (see graphs above). Of the 29 UV-B exposed individuals that hatched, 25 had deformities. This is significant compared to the 190 UV-B shielded individuals that hatched and only 1 showed deformities. These results support the hypothesis that naturally occurring UV-B radiation can adversely affect development in some amphibians, inducing deformities.

Blaustein's group speculates that the higher mortality rates and deformities in frogs and other amphibian species might in fact be due to lower than normal levels of photolyase activity in their developing eggs and embryos, low levels such as found naturally in salamanders.

Laboratory and field experiments seem to support this idea. For one thing, frog species that are not sensitive to UV-B have very high photolyase activity levels. Evaluating 10 different species, Blaustein's team found a strong correlation between species exhibiting little UV-B radiation effects and higher levels of photolyase activity in developing eggs and embryos.

In these experiments, the Pacific tree frog (*Hyla regilla*)—whose populations have not been declining—exhibited the highest photolyase activity and was not affected by UV-B radiation, showing no significant increases in mortality rates in UV-B exposed individuals.

In parallel experiments, the Cascades frog (*Rana cascadae*) and the Western toad (*Bufo boreas*)—both of whose populations have been experiencing markedly declining numbers—had less than one-third the photolyase activity seen in *Hyla*, and were strongly affected by UV-B radiation, showing significant increases in mortality rates when exposed to UV-B radiation.

These results suggest that increased level of UV-B radiation resulting from ozone depletion may indeed be a major contributor to amphibian decline—in populations with low photolyase activity. Could chemical pollutants be acting to lower activity levels of this key enzyme? The investigation continues. Undoubtedly, many factors are contributing to declining amphibian population, and there are not going to be many simple answers.

To explore this experiment further,
go to the Virtual Lab at
www.mhhe.com/raven6/vlab8.mhtml

28

Dynamics of Ecosystems

Concept Outline

28.1 Chemicals cycle within ecosystems.

The Water Cycle. Water cycles between the atmosphere and the oceans, although deforestation has broken the cycle in some ecosystems.

The Carbon Cycle. Photosynthesis captures carbon from the atmosphere; respiration returns it.

The Nitrogen Cycle. Nitrogen is captured from the atmosphere by the metabolic activities of bacteria; other bacteria degrade organic nitrogen, returning it to the atmosphere.

The Phosphorus Cycle. Of all nutrients that plants require, phosphorus tends to be the most limiting.

Biogeochemical Cycles Illustrated: Recycling in a Forest Ecosystem. In a classic experiment, the role of forests in retaining nutrients is assessed.

28.2 Ecosystems are structured by who eats whom.

Trophic Levels. Energy passes through ecosystems in a limited number of steps, typically three or four.

28.3 Energy flows through ecosystems.

Primary Productivity. Plants produce biomass by photosynthesis, while animals produce biomass by consuming plants or other animals.

The Energy in Food Chains. As energy passes through an ecosystem, a good deal is lost at each step.

Ecological Pyramids. The biomass of a trophic level is less, the farther it is from the primary production of photosynthesizers.

Interactions among Different Trophic Levels. Processes on one trophic level can have effects on higher or lower levels of the food chain.

28.4 Biodiversity promotes ecosystem stability.

Effects of Species Richness. Species-rich communities are more productive and resistant to disturbance.

Causes of Species Richness. Ecosystem productivity, spatial heterogeneity, and climate all affect the number of species in an ecosystem.

Biogeographic Patterns of Species Diversity. Many more species occur in the tropics than in temperate regions.

Island Biogeography. Species richness on islands may be a dynamic equilibrium between extinction and colonization.

FIGURE 28.1
Mushrooms serve a greater function than haute cuisine. Mushrooms and other organisms are crucial recyclers in ecosystems, breaking down dead and decaying material and releasing critical elements such as carbon and nitrogen back into nutrient cycles.

The earth is a closed system with respect to chemicals, but an open system in terms of energy. Collectively, the organisms in ecosystems regulate the capture and expenditure of energy and the cycling of chemicals (figure 28.1). As we will see in this chapter, all organisms, including humans, depend on the ability of other organisms—plants, algae, fungi, and some bacteria—to recycle the basic components of life. In chapter 29, we consider the many different types of ecosystems that constitute the biosphere. Chapters 30 and 31 then discuss the many threats to the biosphere and the species it contains.

28.1 Chemicals cycle within ecosystems.

All of the chemical elements that occur in organisms cycle through ecosystems in **biogeochemical cycles,** cyclical paths involving both biological and chemical processes. On a global scale, only a very small portion of these substances is contained within the bodies of organisms; almost all exists in nonliving reservoirs: the atmosphere, water, or rocks. Carbon (in the form of carbon dioxide), nitrogen, and oxygen enter the bodies of organisms primarily from the atmosphere, while phosphorus, potassium, sulfur, magnesium, calcium, sodium, iron, and cobalt come from rocks. All organisms require carbon, hydrogen, oxygen, nitrogen, phosphorus, and sulfur in relatively large quantities; they require other elements in smaller amounts.

The cycling of materials in ecosystems begins when these chemicals are incorporated into the bodies of organisms from nonliving reservoirs such as the atmosphere or the waters of oceans or rivers. Many minerals, for example, first enter water from weathered rock, then pass into organisms when they drink the water. Materials pass from the organisms that first acquire them into the bodies of other organisms that eat them, until ultimately, through decomposition, they complete the cycle and return to the nonliving world.

The Water Cycle

The water cycle (figure 28.2) is the most familiar of all biogeochemical cycles. All life depends directly on the presence of water; the bodies of most organisms consist mainly of this substance. Water is the source of hydrogen ions, whose movements generate ATP in organisms. For that reason alone, it is indispensable to their functioning.

The Path of Free Water

The oceans cover three-fourths of the earth's surface. From the oceans, water evaporates into the atmosphere, a process powered by energy from the sun. Over land approximately 90% of the water that reaches the atmosphere is moisture that evaporates from the surface of plants through a process called transpiration (see chapter 39). Most precipitation falls directly into the oceans, but some falls on land, where it passes into surface and subsurface bodies of fresh water. Only about 2% of all the water on earth is captured in any form—frozen, held in the soil, or incorporated into the bodies of organisms. All of the rest is free water, circulating between the atmosphere and the oceans.

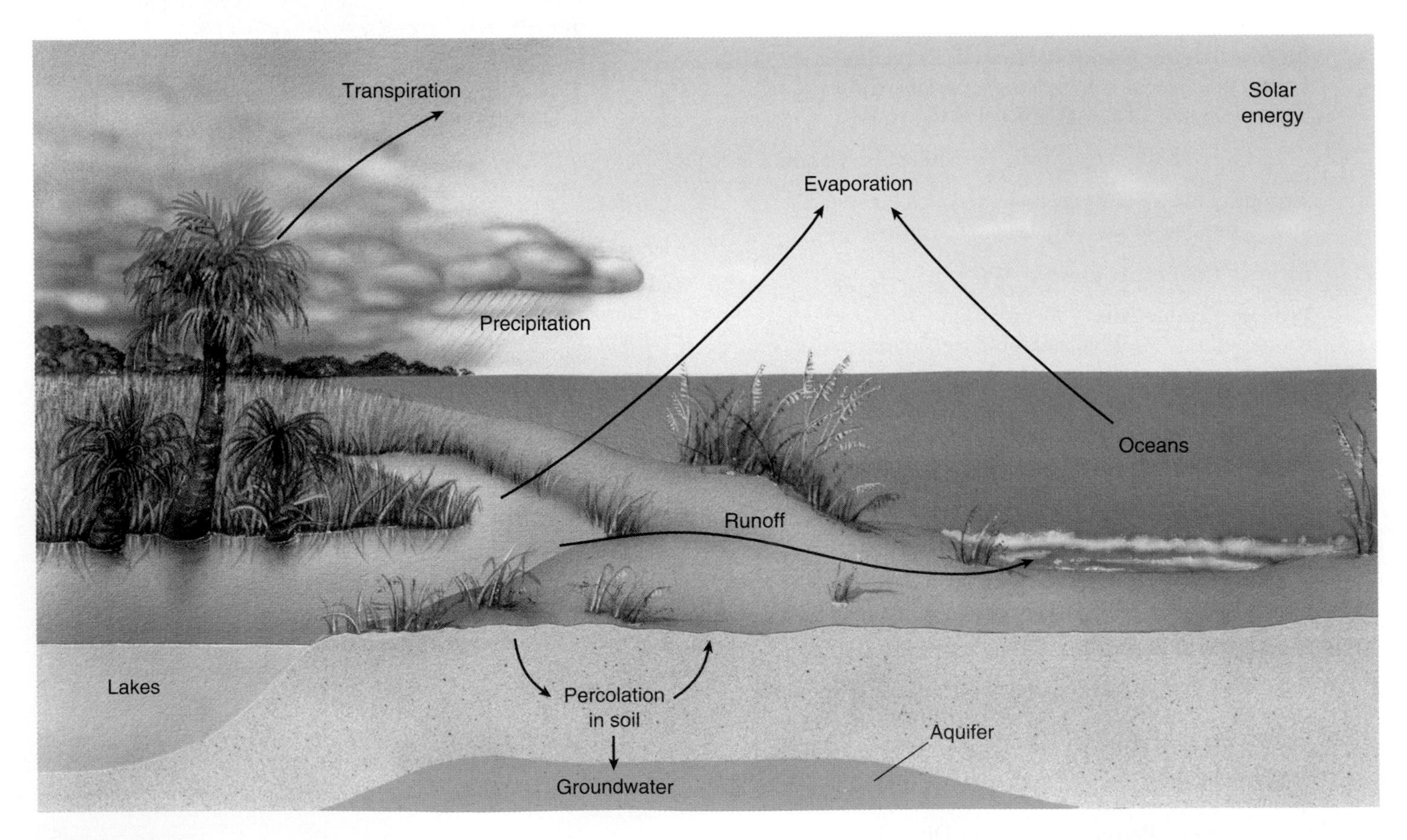

FIGURE 28.2
The water cycle. Water circulates from atmosphere to earth and back again.

The Importance of Water to Organisms

Organisms live or die on the basis of their ability to capture water and incorporate it into their bodies. Plants take up water from the earth in a continuous stream. Crop plants require about 1000 kilograms of water to produce one kilogram of food, and the ratio in natural communities is similar. Animals obtain water directly or from the plants or other animals they eat. The amount of free water available at a particular place often determines the nature and abundance of the living organisms present there.

Groundwater

Much less obvious than surface water, which we see in streams, lakes, and ponds, is groundwater, which occurs in aquifers—permeable, saturated, underground layers of rock, sand, and gravel. In many areas, groundwater is the most important reservoir of water. It amounts to more than 96% of all fresh water in the United States. The upper, unconfined portion of the groundwater constitutes the water table, which flows into streams and is partly accessible to plants; the lower confined layers are generally out of reach, although they can be "mined" by humans. The water table is recharged by water that percolates through the soil from precipitation as well as by water that seeps downward from ponds, lakes, and streams. The deep aquifers are recharged very slowly from the water table.

Groundwater flows much more slowly than surface water, anywhere from a few millimeters to a meter or so per day. In the United States, groundwater provides about 25% of the water used for all purposes and provides about 50% of the population with drinking water. Rural areas tend to depend almost exclusively on wells to access groundwater, and its use is growing at about twice the rate of surface water use. In the Great Plains of the central United States, the extensive use of the Ogallala Aquifer as a source of water for agricultural needs as well as for drinking water is depleting it faster than it can be naturally recharged. This seriously threatens the agricultural production of the area and similar problems are appearing throughout the drier portions of the globe.

Because of the greater rate of groundwater use, and because it flows so slowly, the increasing chemical pollution of groundwater is also a very serious problem. It is estimated that about 2% of the groundwater in the United States is already polluted, and the situation is worsening. Pesticides, herbicides, and fertilizers have become a serious threat to water purity. Another key source of groundwater pollution consists of the roughly 200,000 surface pits, ponds, and lagoons that are actively used for the disposal of chemical wastes in the United States alone. Because of the large volume of water, its slow rate of turnover, and its inaccessibility, removing pollutants from aquifers is virtually impossible.

FIGURE 28.3
Deforestation breaks the water cycle. As time goes by, the consequences of tropical deforestation may become even more severe, as the extensive erosion in this deforested area of Madagascar shows.

Breaking the Water Cycle

In dense forest ecosystems such as tropical rainforests, more than 90% of the moisture in the ecosystem is taken up by plants and then transpired back into the air. Because so many plants in a rainforest are doing this, the vegetation is the primary source of local rainfall. In a very real sense, these plants create their own rain: the moisture that travels up from the plants into the atmosphere falls back to earth as rain.

Where forests are cut down, the organismic water cycle is broken, and moisture is not returned to the atmosphere. Water drains away from the area to the sea instead of rising to the clouds and falling again on the forest. As early as the late 1700s, the great German explorer Alexander von Humbolt reported that stripping the trees from a tropical rainforest in Colombia prevented water from returning to the atmosphere and created a semiarid desert. It is a tragedy of our time that just such a transformation is occurring in many tropical areas, as tropical and temperate rainforests are being clear-cut or burned in the name of "development" (figure 28.3). Much of Madagascar, a California-sized island off the east coast of Africa, has been transformed in this century from lush tropical forest into semiarid desert by deforestation. Because the rain no longer falls, there is no practical way to reforest this land. The water cycle, once broken, cannot be easily reestablished.

Water cycles between oceans and atmosphere. Some 96% of the fresh water in the United States consists of groundwater, which provides 25% of all the water used in this country.

The Carbon Cycle

The **carbon cycle** is based on carbon dioxide, which makes up only about 0.03% of the atmosphere (figure 28.4). Worldwide, the synthesis of organic compounds from carbon dioxide and water through photosynthesis (see chapter 10) utilizes about 10% of the roughly 700 billion metric tons of carbon dioxide in the atmosphere each year. This enormous amount of biological activity takes place as a result of the combined activities of photosynthetic bacteria, protists, and plants. All terrestrial heterotrophic organisms obtain their carbon indirectly from photosynthetic organisms. When the bodies of dead organisms decompose, microorganisms release carbon dioxide back to the atmosphere. From there, it can be reincorporated into the bodies of other organisms.

Most of the organic compounds formed as a result of carbon dioxide fixation in the bodies of photosynthetic organisms are ultimately broken down and released back into the atmosphere or water. Certain carbon-containing compounds, such as cellulose, are more resistant to breakdown than others, but certain bacteria and fungi, as well as a few kinds of insects, are able to accomplish this feat. Some cellulose, however, accumulates as undecomposed organic matter such as peat. The carbon in this cellulose may eventually be incorporated into fossil fuels such as oil or coal.

In addition to the roughly 700 billion metric tons of carbon dioxide in the atmosphere, approximately 1 trillion metric tons are dissolved in the ocean. More than half of this quantity is in the upper layers, where photosynthesis takes place. The fossil fuels, primarily oil and coal, contain more than 5 trillion additional metric tons of carbon, and between 600 million and 1 trillion metric tons are locked up in living organisms at any one time. In global terms, respiration and photosynthesis (see chapters 9 and 10) are approximately balanced, but the balance has been shifted recently because of the consumption of fossil fuels. The combustion of coal, oil, and natural gas has released large stores of carbon into the atmosphere as carbon dioxide. The increase of carbon dioxide in the atmosphere appears to be changing global climates, and may do so even more rapidly in the future, as we will discuss in chapter 30.

About 10% of the estimated 700 billion metric tons of carbon dioxide in the atmosphere is fixed annually by the process of photosynthesis.

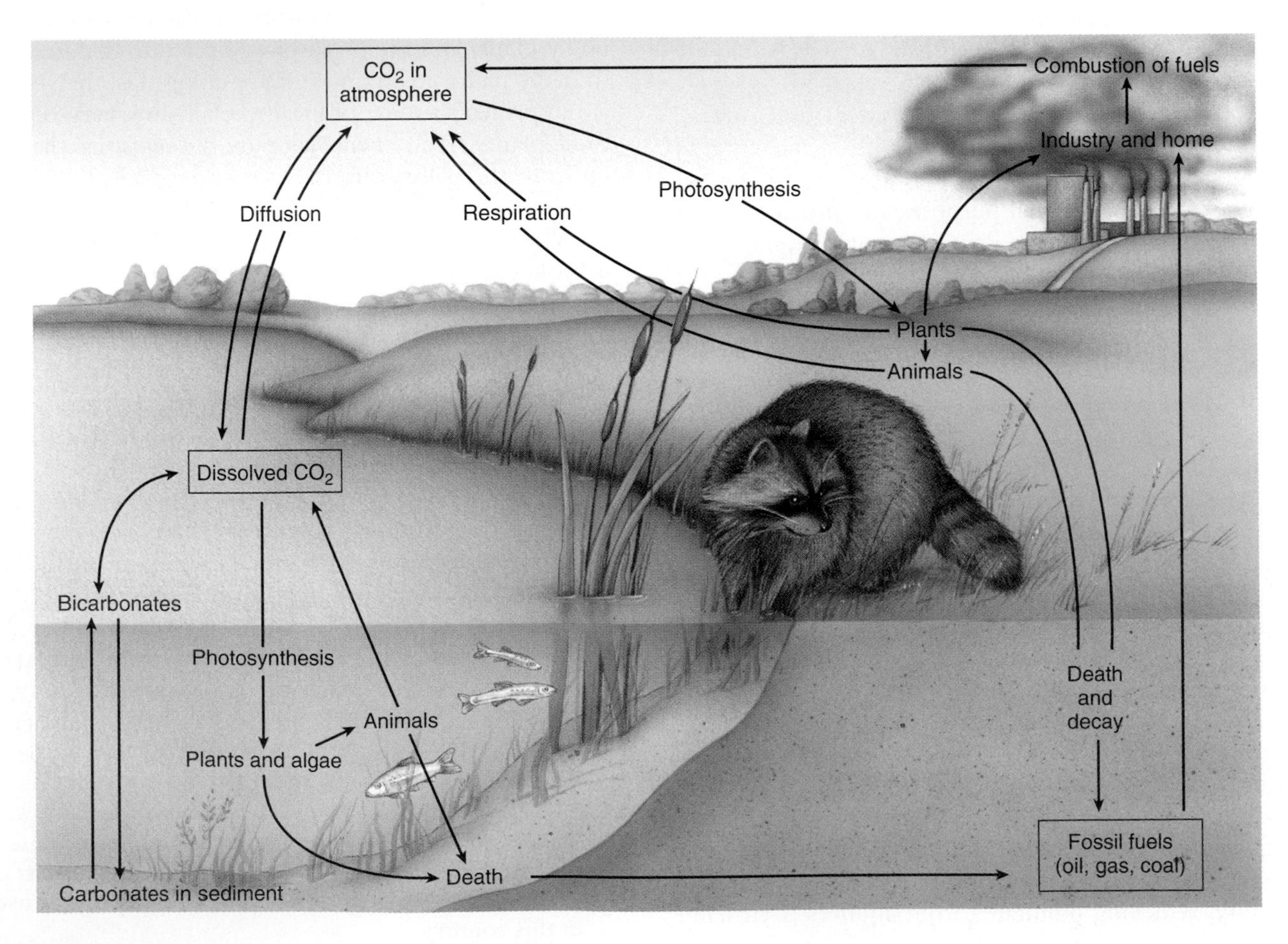

FIGURE 28.4
The carbon cycle. Photosynthesis captures carbon; respiration returns it to the atmosphere.

The Nitrogen Cycle

Relatively few kinds of organisms—all of them bacteria—can convert, or fix, atmospheric nitrogen (78% of the earth's atmosphere) into forms that can be used for biological processes via the **nitrogen cycle** (figure 28.5). The triple bond that links together the two atoms that make up diatomic atmospheric nitrogen (N_2) makes it a very stable molecule. In living systems the cleavage of atmospheric nitrogen is catalyzed by a complex of three proteins—ferredoxin, nitrogen reductase, and nitrogenase. This process uses ATP as a source of energy, electrons derived from photosynthesis or respiration, and a powerful reducing agent. The overall reaction of nitrogen fixation is written:

$$N_2 + 3H_2 \rightarrow 2NH_3$$

Some genera of bacteria have the ability to fix atmospheric nitrogen. Most are free-living, but some form symbiotic relationships with the roots of legumes (plants of the pea family, Fabaceae) and other plants. Only the symbiotic bacteria fix enough nitrogen to be of major significance in nitrogen production. Because of the activities of such organisms in the past, a large reservoir of ammonia and nitrates now exists in most ecosystems. This reservoir is the immediate source of much of the nitrogen used by organisms.

Nitrogen-containing compounds, such as proteins in plant and animal bodies, are decomposed rapidly by certain bacteria and fungi. These bacteria and fungi use the amino acids they obtain through decomposition to synthesize their own proteins and to release excess nitrogen in the form of ammonium ions (NH_4^+), a process known as **ammonification.** The ammonium ions can be converted to soil nitrites and nitrates by certain kinds of organisms and which then can be absorbed by plants.

A certain proportion of the fixed nitrogen in the soil is steadily lost. Under anaerobic conditions, nitrate is often converted to nitrogen gas (N_2) and nitrous oxide (N_2O), both of which return to the atmosphere. This process, which several genera of bacteria carry out, is called **denitrification.**

Nitrogen becomes available to organisms almost entirely through the metabolic activities of bacteria, some free-living and others which live symbiotically in the roots of legumes and other plants.

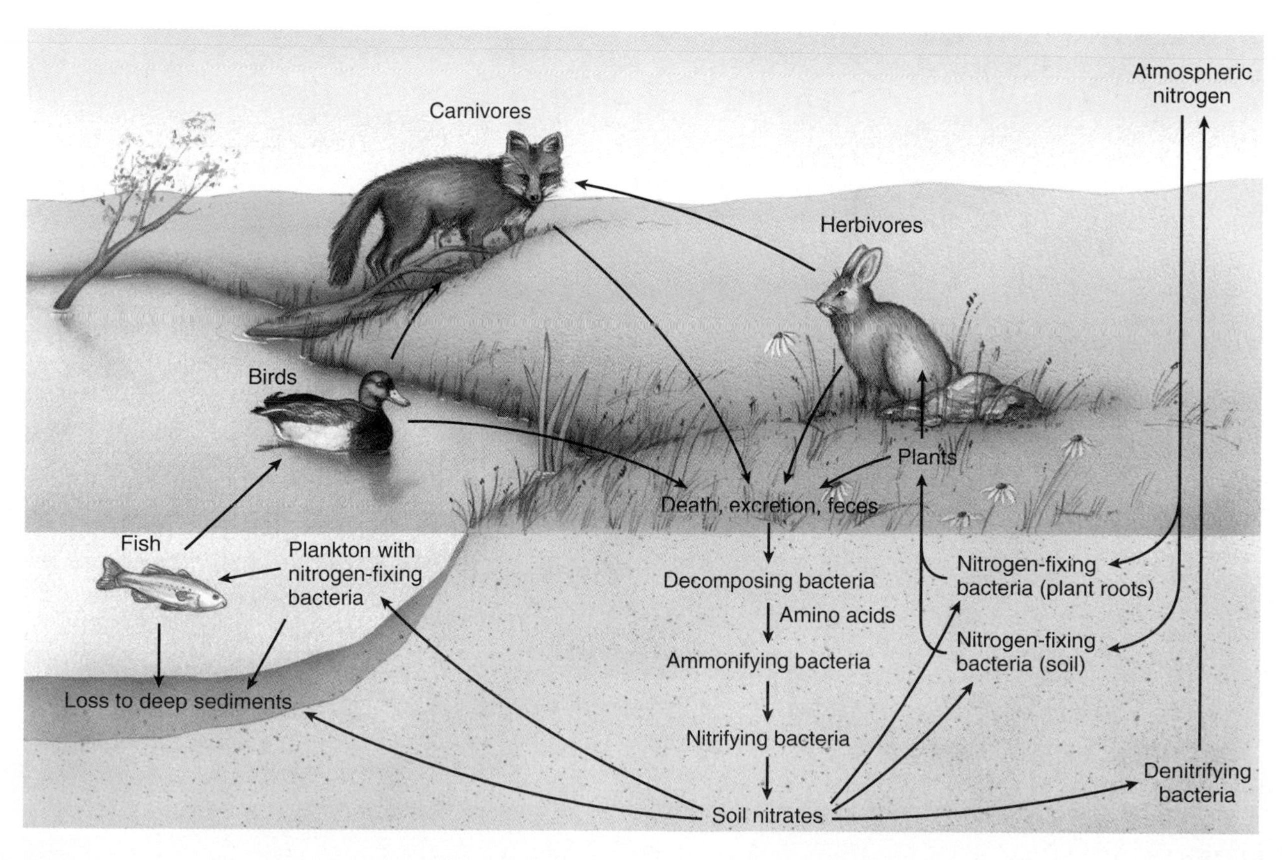

FIGURE 28.5
The nitrogen cycle. Certain bacteria fix atmospheric nitrogen, converting it to a form living organisms can use. Other bacteria decompose nitrogen-containing compounds from plant and animal materials, returning it to the atmosphere.

The Phosphorus Cycle

In all biogeochemical cycles other than those involving water, carbon, oxygen, and nitrogen, the reservoir of the nutrient exists in mineral form, rather than in the atmosphere. The **phosphorus cycle** (figure 28.6) is presented as a representative example of all other mineral cycles. Phosphorus, a component of ATP, phospholipids, and nucleic acid, plays a critical role in plant nutrition.

Of all the required nutrients other than nitrogen, phosphorus is the most likely to be scarce enough to limit plant growth. Phosphates, in the form of phosphorus anions, exist in soil only in small amounts. This is because they are relatively insoluble and are present only in certain kinds of rocks. As phosphates weather out of soils, they are transported by rivers and streams to the oceans, where they accumulate in sediments. They are naturally brought back up again only by the uplift of lands, such as occurs along the Pacific coast of North and South America, creating upwelling currents. Phosphates brought to the surface are assimilated by algae, and then by fish, which are in turn eaten by birds. Seabirds deposit enormous amounts of guano (feces) rich in phosphorus along certain coasts. Guano deposits have traditionally been used for fertilizer. Crushed phosphate-rich rocks, found in certain regions, are also used for fertilizer. The seas are the only inexhaustible source of phosphorus, making deep-seabed mining look increasingly commercially attractive.

Every year, millions of tons of phosphate are added to agricultural lands in the belief that it becomes fixed to and enriches the soil. In general, three times more phosphate than a crop requires is added each year. This is usually in the form of **superphosphate,** which is soluble calcium dihydrogen phosphate, $Ca(H_2PO_4)_2$, derived by treating bones or apatite, the mineral form of calcium phosphate, with sulfuric acid. But the enormous quantities of phosphates that are being added annually to the world's agricultural lands are not leading to proportionate gains in crops. Plants can apparently use only so much of the phosphorus that is added to the soil.

Phosphates are relatively insoluble and are present in most soils only in small amounts. They often are so scarce that their absence limits plant growth.

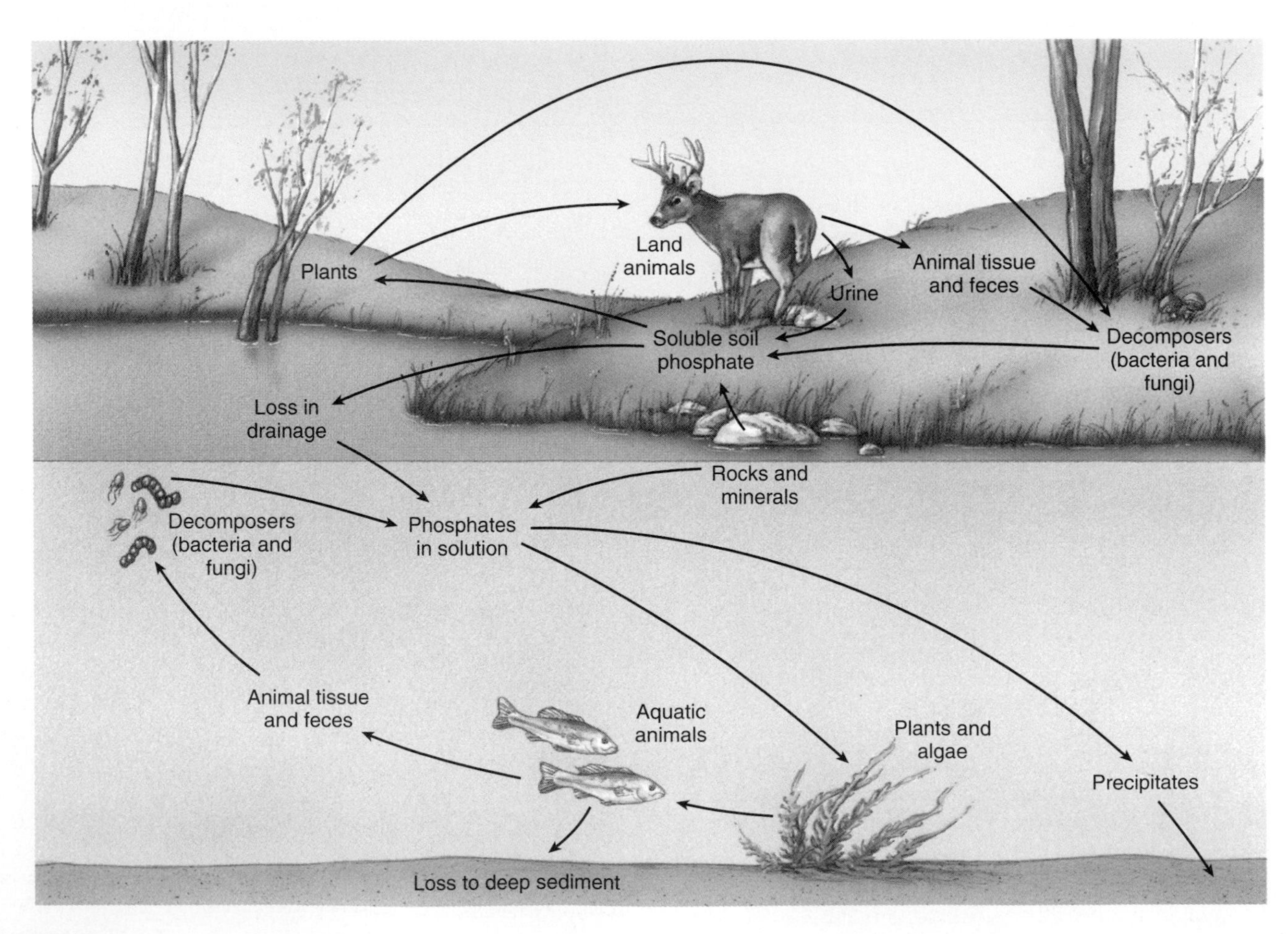

FIGURE 28.6
The phosphorus cycle. Phosphates weather from soils into water, enter plants and animals, and are redeposited in the soil when plants and animals decompose.

Biogeochemical Cycles Illustrated: Recycling in a Forest Ecosystem

An ongoing series of studies conducted at the Hubbard Brook Experimental Forest in New Hampshire has revealed in impressive detail the overall recycling pattern of nutrients in an ecosystem. The way this particular ecosystem functions, and especially the way nutrients cycle within it, has been studied since 1963 by Herbert Bormann of the Yale School of Forestry and Environmental Studies, Gene Likens of the Institute of Ecosystem Studies, and their colleagues. These studies have yielded much of the available information about the cycling of nutrients in forest ecosystems. They have also provided the basis for the development of much of the experimental methodology that is being applied successfully to the study of other ecosystems.

Hubbard Brook is the central stream of a large watershed that drains a region of temperate deciduous forest. To measure the flow of water and nutrients within the Hubbard Brook ecosystem, concrete weirs with V-shaped notches were built across six tributary streams. All of the water that flowed out of the valleys had to pass through the notches, as the weirs were anchored in bedrock. The researchers measured the precipitation that fell in the six valleys, and determined the amounts of nutrients that were present in the water flowing in the six streams. By these methods, they demonstrated that the undisturbed forests in this area were very efficient at retaining nutrients; the small amounts of nutrients that precipitated from the atmosphere with rain and snow were approximately equal to the amounts of nutrients that ran out of the valleys. These quantities were very low in relation to the total amount of nutrients in the system. There was a small net loss of calcium—about 0.3% of the total calcium in the system per year—and small net gains of nitrogen and potassium.

In 1965 and 1966, the investigators felled all the trees and shrubs in one of the six watersheds and then prevented regrowth by spraying the area with herbicides. The effects were dramatic. The amount of water running out of that valley increased by 40%. This indicated that water that previously would have been taken up by vegetation and ultimately evaporated into the atmosphere was now running off. For the four-month period from June to September 1966, the runoff was four times higher than it had been during comparable periods in the preceding years. The amounts of nutrients running out of the system also greatly increased; for example, the loss of calcium was 10 times higher than it had been previously. Phosphorus, on the other hand, did not increase in the stream water; it apparently was locked up in the soil.

The change in the status of nitrogen in the disturbed valley was especially striking (figure 28.7). The undisturbed ecosystem in this valley had been accumulating nitrogen at a rate of about 2 kilograms per hectare per year, but the deforested ecosystem *lost* nitrogen at a rate of about 120 kilograms per hectare per year. The nitrate level of the water rapidly increased to a level exceeding that judged safe for human consumption, and the stream that drained the area generated massive blooms of cyanobacteria and algae. In other words, the fertility of this logged-over valley decreased rapidly, while at the same time the danger of flooding greatly increased. This experiment is particularly instructive at the start of the twenty-first century, as large areas of tropical rain forest are being destroyed to make way for cropland, a topic that will be discussed further in chapter 30.

(a)

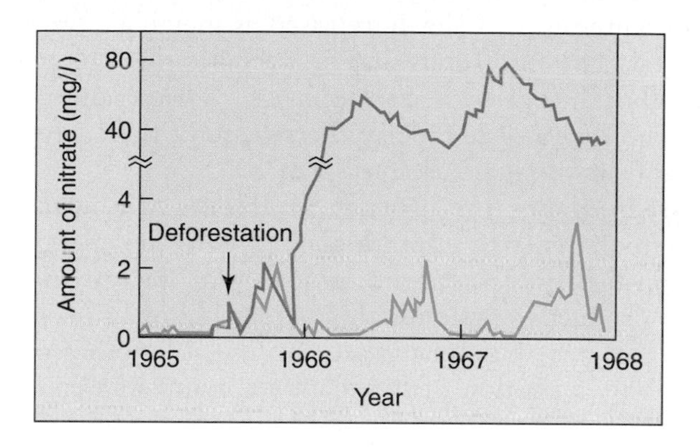

(b)

FIGURE 28.7
The Hubbard Brook experiment. (*a*) A 38-acre watershed was completely deforested, and the runoff monitored for several years. (*b*) Deforestation greatly increased the loss of minerals in runoff water from the ecosystem. The *red* curve represents nitrate in the runoff water from the deforested watershed; the *blue* curve, nitrate in runoff water from an undisturbed neighboring watershed.

When the trees and shrubs in one of the valleys in the Hubbard Brook watershed were cut down and the area was sprayed with herbicide, water runoff and the loss of nutrients from that valley increased. Nitrogen, which had been accumulating at a rate of about 2 kilograms per hectare per year, was lost at a rate of 120 kilograms per hectare per year.

28.2 Ecosystems are structured by who eats whom.

Trophic Levels

An ecosystem includes autotrophs and heterotrophs. **Autotrophs** are plants, algae, and some bacteria that are able to capture light energy and manufacture their own food. To support themselves, **heterotrophs,** which include animals, fungi, most protists and bacteria, and nongreen plants, must obtain organic molecules that have been synthesized by autotrophs. Autotrophs are also called **primary producers,** and heterotrophs are also called **consumers.**

Once energy enters an ecosystem, usually as the result of photosynthesis, it is slowly released as metabolic processes proceed. The autotrophs that first acquire this energy provide all of the energy heterotrophs use. The organisms that make up an ecosystem delay the release of the energy obtained from the sun back into space.

Green plants, the primary producers of a terrestrial ecosystem, generally capture about 1% of the energy that falls on their leaves, converting it to food energy. In especially productive systems, this percentage may be a little higher. When these plants are consumed by other organisms, only a portion of the plant's accumulated energy is actually converted into the bodies of the organisms that consume them.

Several different levels of consumers exist. The **primary consumers,** or herbivores, feed directly on the green plants. **Secondary consumers,** carnivores and the parasites of animals, feed in turn on the herbivores. **Decomposers** break down the organic matter accumulated in the bodies of other organisms. Another more general term that includes decomposers is **detritivores.** Detritivores live on the refuse of an ecosystem. They include large scavengers, such as crabs, vultures, and jackals, as well as decomposers.

All of these categories occur in any ecosystem. They represent different **trophic levels,** from the Greek word *trophos,* which means "feeder." Organisms from each trophic level, feeding on one another, make up a series called a **food chain** (figure 28.8). The length and complexity of food chains vary greatly. In real life, it is rather rare for a given kind of organism to feed only on one other type of organism. Usually, each organism feeds on two or more kinds and in turn is eaten by several other kinds of organisms. When diagrammed, the relationship appears as a series of branching lines, rather than a straight line; it is called a **food web** (figure 28.9).

A certain amount of the chemical-bond energy ingested by the organisms at a given trophic level goes toward staying alive (for example, carrying out mechanical motion). Using the chemical-bond energy converts it to heat, which organisms cannot use to do work. Another portion of the chemical-bond energy taken in is retained as chemical-bond energy within the organic molecules produced by growth. Usually 40% or less of the energy ingested is stored by growth. An invertebrate typically uses about a quarter of this 40% for growth; in other words, about 10% of the food an invertebrate eats is turned into its own body and thus into potential food for its predators. Although the comparable figure varies from approximately 5% in carnivores to nearly 20% for herbivores, 10% is a good average value for the amount of organic matter that reaches the next trophic level.

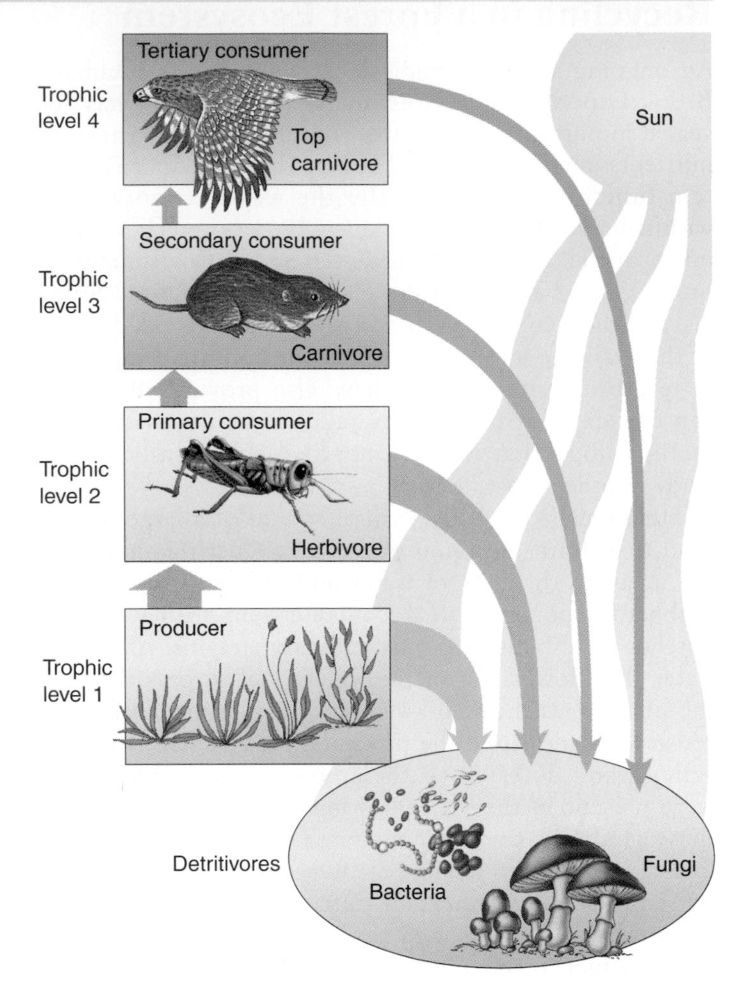

FIGURE 28.8
Trophic levels within a food chain. Plants obtain their energy directly from the sun, placing them at trophic level 1. Animals that eat plants, such as grasshoppers, are primary consumers or herbivores and are at trophic level 2. Animals that eat plant-eating animals, such as shrews, are carnivores and are at trophic level 3 (secondary consumers); animals that eat carnivorous animals, such as hawks, are tertiary consumers at trophic level 4. Detritivores use all trophic levels for food.

Energy passes through ecosystems, a good deal being lost at each step.

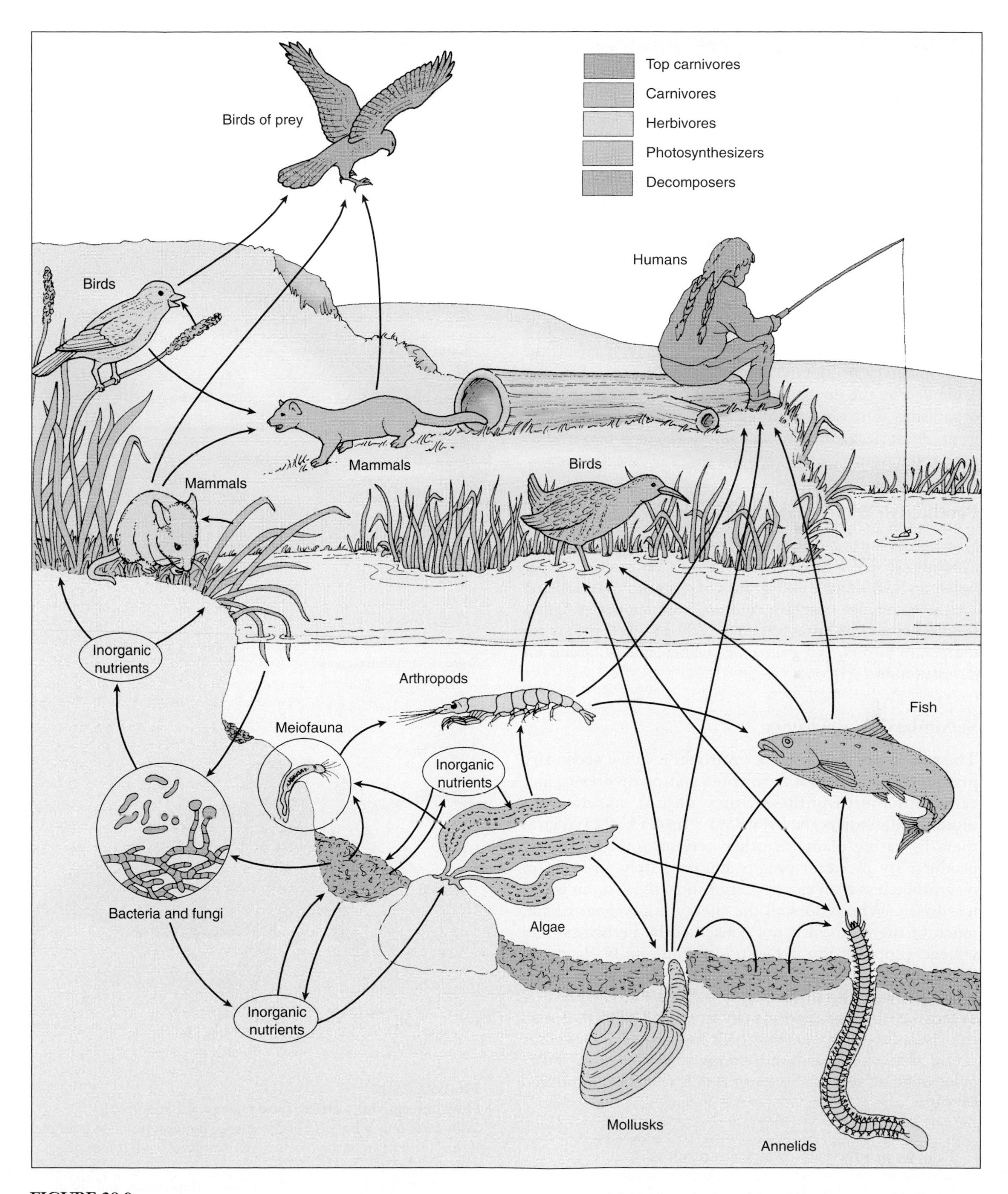

FIGURE 28.9
The food web in a salt marsh shows the complex interrelationships among organisms. The meiofauna are very small animals that live between the grains of sand.

28.3 Energy flows through ecosystems.

Primary Productivity

Approximately 1 to 5% of the solar energy that falls on a plant is converted to the chemical bonds of organic material. **Primary production** or **primary productivity** are terms used to describe the amount of organic matter produced from solar energy in a given area during a given period of time. **Gross primary productivity** is the total organic matter produced, including that used by the photosynthetic organism for respiration. **Net primary productivity (NPP),** therefore, is a measure of the amount of organic matter produced in a community in a given time that is available for heterotrophs. It equals the gross primary productivity minus the amount of energy expended by the metabolic activities of the photosynthetic organisms. The net weight of all of the organisms living in an ecosystem, its **biomass,** increases as a result of its net production.

Productive Biological Communities

Some ecosystems have a high net primary productivity. For example, tropical forests and wetlands normally produce between 1500 and 3000 grams of organic material per square meter per year. By contrast, corresponding figures for other communities include 1200 to 1300 grams for temperate forests, 900 grams for savanna, and 90 grams for deserts (table 28.1).

Table 28.1 Terrestrial Ecosystem Productivity Per Year

Ecosystem Type	Net Primary Productivity (NPP) NPP per Unit Area (g/m^2)	World NPP (10^9 tons)
Tropical rain forest	2200	37.4
Wetlands	2000	4.0
Tropical seasonal forest	1600	12.0
Temperate evergreen forest	1300	6.5
Temperate deciduous forest	1200	8.4
Savanna	900	13.5
Boreal forest	800	9.6
Woodland and shrubland	700	6.0
Cultivated land	650	9.1
Temperate grassland	600	5.4
Tundra and alpine	140	1.1
Desert and semidesert shrub	90	1.6
Extreme desert, rock, sand, and ice	3	0.07

Source: After Whittaker, 1975.

Secondary Productivity

The rate of production by heterotrophs is called **secondary productivity.** Because herbivores and carnivores cannot carry out photosynthesis, they do not manufacture biomolecules directly from CO_2. Instead, they obtain them by eating plants or other heterotrophs. Secondary productivity by herbivores is approximately an order of magnitude less than the primary productivity upon which it is based. Where does all the energy in plants go? First, much of the biomass is not consumed by herbivores and instead supports the decomposer community (bacteria, fungi and detritivorous animals). Second, some energy is not assimilated by the herbivore's body but is passed on as feces to the decomposers (figure 28.10). Third, not all the chemical-bond energy which herbivores assimilate is retained as chemical-bond energy in the organic molecules of their tissues. Some of it is lost as heat produced by work.

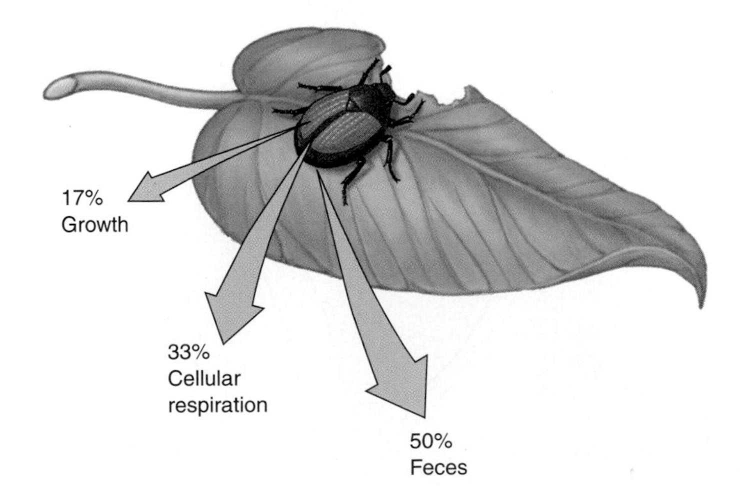

FIGURE 28.10
How heterotrophs utilize food energy. A heterotroph assimilates only a fraction of the energy it consumes. For example, if the "bite" of an herbivorous insect comprises 500 Joules of energy (1 Joule = 0.239 calories), about 50%, 250 J, is lost in feces, about 33%, 165 J, is used to fuel cellular respiration, and about 17%, 85 J, is converted into insect biomass. Only this 85 J is available to the next trophic level.

Primary productivity occurs as a result of photosynthesis, which is carried out by green plants, algae, and some bacteria. Secondary productivity is the production of new biomass by heterotrophs.

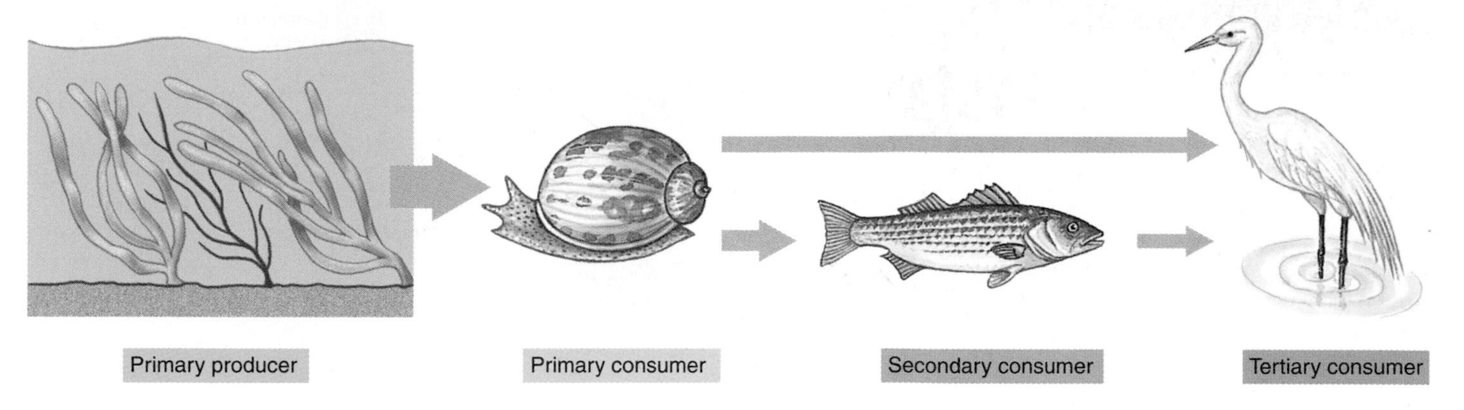

FIGURE 28.11
A food chain. Because so much energy is lost at each step, food chains usually consist of just three or four steps.

The Energy in Food Chains

Food chains generally consist of only three or four steps (figure 28.11). So much energy is lost at each step that very little usable energy remains in the system after it has been incorporated into the bodies of organisms at four successive trophic levels.

Community Energy Budgets

Lamont Cole of Cornell University studied the flow of energy in a freshwater ecosystem in Cayuga Lake in upstate New York. He calculated that about 150 of each 1000 calories of potential energy fixed by algae and cyanobacteria are transferred into the bodies of small heterotrophs (figure 28.12). Of these, about 30 calories are incorporated into the bodies of smelt, small fish that are the principal secondary consumers of the system. If humans eat the smelt, they gain about 6 of the 1000 calories that originally entered the system. If trout eat the smelt and humans eat the trout, humans gain only about 1.2 calories.

Factors Limiting Community Productivity

Communities with higher productivity can in theory support longer food chains. The limit on a community's productivity is determined ultimately by the amount of sunlight it receives, for this determines how much photosynthesis can occur. This is why in the deciduous forests of North America the net primary productivity increases as the growing season lengthens. NPP is higher in warm climates than cold ones not only because of the longer growing seasons, but also because more nitrogen tends to be available in warm climates, where nitrogen-fixing bacteria are more active.

Considerable energy is lost at each stage in food chains, which limits their length. In general, more productive communities can support longer food chains.

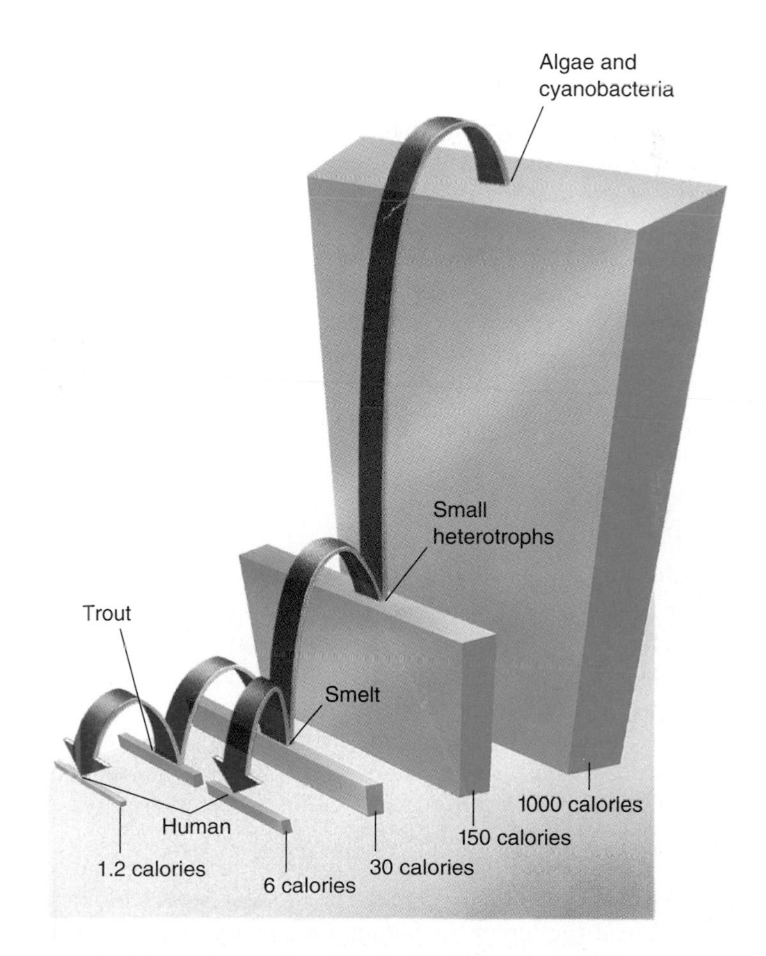

FIGURE 28.12
The food web in Cayuga Lake. Autotrophic plankton (algae and cyanobacteria) fix the energy of the sun, heterotrophic plankton feed on them, and both are consumed by smelt. The smelt are eaten by trout, with about a fivefold loss in fixed energy; for humans, the amount of smelt biomass is at least five times greater than that available in trout, although humans prefer to eat trout.

Ecological Pyramids

A plant fixes about 1% of the sun's energy that falls on its green parts. The successive members of a food chain, in turn, process into their own bodies about 10% of the energy available in the organisms on which they feed. For this reason, there are generally far more individuals at the lower trophic levels of any ecosystem than at the higher levels. Similarly, the biomass of the primary producers present in a given ecosystem is greater than the biomass of the primary consumers, with successive trophic levels having a lower and lower biomass and correspondingly less potential energy.

These relationships, if shown diagrammatically, appear as pyramids (figure 28.13). We can speak of "pyramids of biomass," "pyramids of energy," "pyramids of number," and so forth, as characteristic of ecosystems.

Inverted Pyramids

Some aquatic ecosystems have inverted biomass pyramids. For example, in a planktonic ecosystem—dominated by small organisms floating in water—the turnover of photosynthetic phytoplankton at the lowest level is very rapid, with zooplankton consuming phytoplankton so quickly that the phytoplankton (the producers at the base of the food chain) can never develop a large population size. Because the phytoplankton reproduce very rapidly, the community can support a population of heterotrophs that is larger in biomass and more numerous than the phytoplankton (see figure 28.13*b*).

Top Carnivores

The loss of energy that occurs at each trophic level places a limit on how many top-level carnivores a community can support. As we have seen, only about one-thousandth of the energy captured by photosynthesis passes all the way through a three-stage food chain to a tertiary consumer such as a snake or hawk. This explains why there are no predators that subsist on lions or eagles—the biomass of these animals is simply insufficient to support another trophic level.

In the pyramid of numbers, top-level predators tend to be fairly large animals. Thus, the small residual biomass available at the top of the pyramid is concentrated in a relatively small number of individuals.

Because energy is lost at every step of a food chain, the biomass of primary producers (photosynthesizers) tends to be greater than that of the herbivores that consume them, and herbivore biomass greater than the biomass of the predators that consume them.

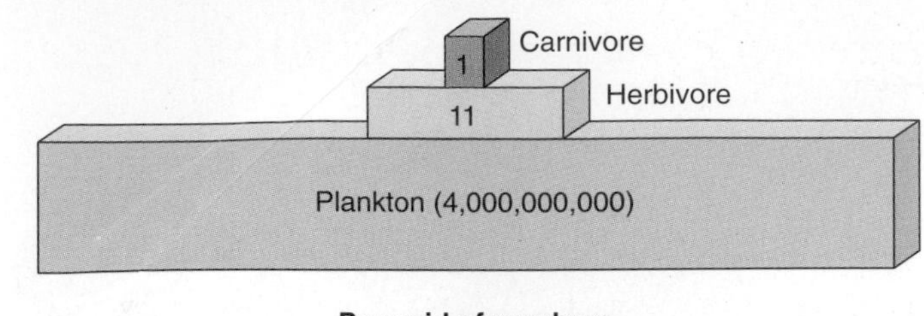

Pyramid of numbers

(a)

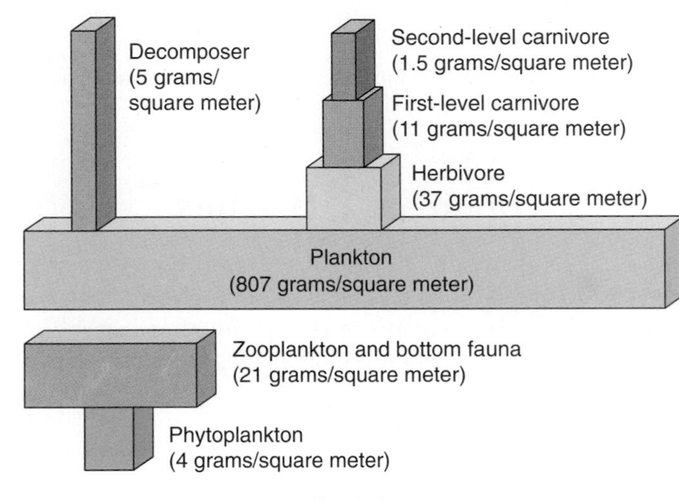

Pyramid of biomass

(b)

Decomposer
(3890 kilocalories/
square meter/year)

First-level carnivore
(48 kilocalories/
square meter/year)

Herbivore
(596 kilocalories/
square meter/year)

Plankton
(36,380 kilocalories/square meter/year)

Pyramid of energy

(c)

FIGURE 28.13
Ecological pyramids. Ecological pyramids measure different characteristics of each trophic level. (*a*) Pyramid of numbers. (*b*) Pyramids of biomass, both normal (*top*) and inverted (*bottom*). (*c*) Pyramid of energy.

Interactions among Different Trophic Levels

The existence of food webs creates the possibility of interactions among species at different trophic levels. Predators will not only have effects on the species upon which they prey, but also, indirectly, upon the plants eaten by these prey. Conversely, increases in primary productivity will not only provide more food for herbivores but, indirectly, lead also to more food for carnivores.

Trophic Cascades

When we look at the world around us, we see a profusion of plant life. Why is this? Why don't herbivore populations increase to the extent that all available vegetation is consumed? The answer, of course, is that predators keep the herbivore populations in check, thus allowing plant populations to thrive. This phenomenon, in which the effect of one trophic level flows down to lower levels, is called a **trophic cascade.**

Experimental studies have confirmed the existence of trophic cascades. For example, in one study in New Zealand, sections of a stream were isolated with a mesh that prevented fish from entering. In some of the enclosures, brown trout were added, whereas other enclosures were left without large fish. After 10 days, the number of invertebrates in the trout enclosures was one-half of that in the controls (figure 28.14). In turn, the biomass of algae, which invertebrates feed upon, was five times greater in the trout enclosures than in the controls.

The logic of trophic cascades leads to the prediction that a fourth trophic level, carnivores that preyed on other carnivores, would also lead to cascading effects. In this case, the top predators would keep lower-level predator populations in check, which should lead to a profusion of herbivores and a paucity of vegetation. In an experiment similar to the one just described, enclosures were created in free-flowing streams in northern California. In this case, large predatory fish were added to some enclosures and not others. In the large fish enclosures, the number of smaller predators, such as damselfly nymphs was greatly reduced, leading to an increase in their prey, including algae-eating insects, which lead, in turn, to decreases in the biomass of algae (figure 28.15).

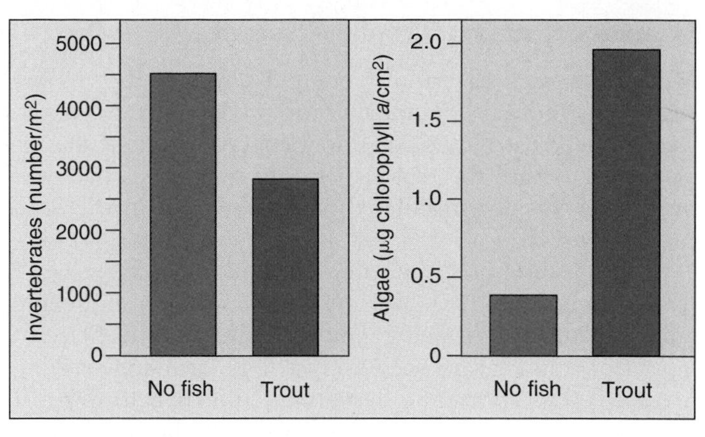

FIGURE 28.14
Trophic cascades. Streams with trout have fewer herbivorous invertebrates and more algae than streams without trout.
Source: Data from Flecker, A.S. and Townsend, C.R., "Community-wide Consequences of Trout Introduction in New Zealand Streams." In *Ecosystem Management: Selected Readings,* F.B. Samson and F.L. Knopf eds., Springer-Verlag, New York, 1996.

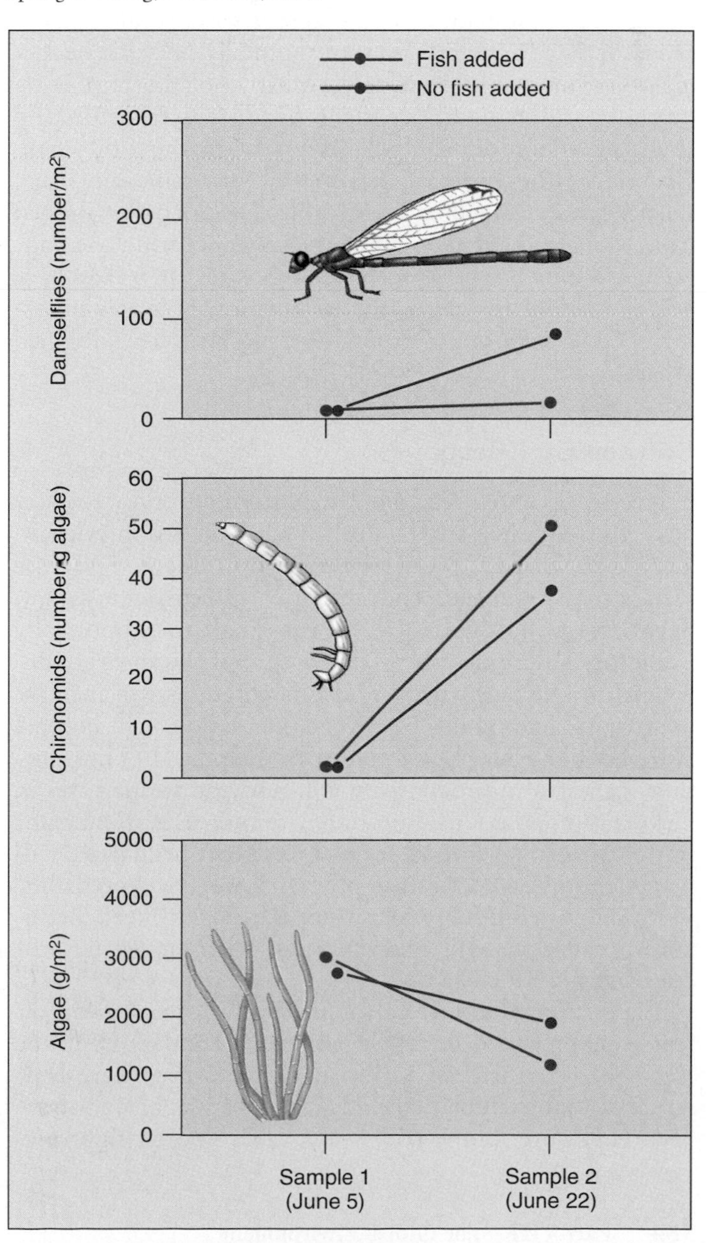

FIGURE 28.15
Four-level trophic cascades. Streams with fish have fewer lower-level predators, such as damselflies, more herbivorous insects (exemplified by the number of chironomids, a type of aquatic insect), and lower levels of algae.
Source: Data from M. Power "Habitat Heterogenicity and the Functional Significance of Fish," *Ecology,* 1997.

Human Effects on Trophic Cascades

Humans have inadvertently created a test of the trophic cascade hypothesis by removing top predators from ecosystems. The great naturalist Aldo Leopold captured the results long before the trophic cascade hypothesis had ever been scientifically articulated when he wrote in the *Sand County Almanac:*

"I have lived to see state after state extirpate its wolves. I have watched the face of many a new wolfless mountain, and seen the south-facing slopes wrinkle with a maze of new deer trails. I have seen every edible bush and seedling browsed, first to anemic desuetude, and then to death. I have seen every edible tree defoliated to the height of a saddle horn."

Many similar examples exist in nature in which the removal of predators has led to cascading effects on lower trophic levels. On Barro Colorado Island, a hilltop turned into an island by the construction of the Panama Canal at the beginning of the last century, large predators such as jaguars and mountain lions are absent. As a result, smaller predators whose populations are normally held in check—including monkeys, peccaries (a relative of the pig), coatimundis and armadillos—have become extraordinarily abundant. These animals will eat almost anything they find. Ground-nesting birds are particularly vulnerable, and many species have declined; at least 15 bird species have vanished from the island entirely. Similarly, in woodlots in the midwestern United States, raccoons, opossums, and foxes have become abundant due to the elimination of larger predators, and populations of ground-nesting birds have declined greatly.

Bottom-Up Effects

Conversely, factors acting at the bottom of food webs may have consequences that ramify to higher trophic levels, leading to what are termed **bottom-up effects**. The basic idea is that when the productivity of an ecosystem is low, herbivore populations will be too small to support any predators. Increases in productivity will be entirely devoured by the herbivores, whose populations will increase in size. At some point, herbivore populations will become large enough that predators can be supported. Thus, further increases in productivity will not lead to increases in herbivore populations, but, rather to increases in predator populations. Again, at some level, top predators will become established that can prey on lower-level predators. With the lower-level predator populations in check, herbivore populations will again increase with increasing productivity (figure 28.16).

Experimental evidence for the role of bottom-up effects was provided in an elegant study conducted on the Eel River in northern California. Enclosures were constructed that excluded large fish. A roof was placed above each enclosure. Some roofs were clear and let light pass

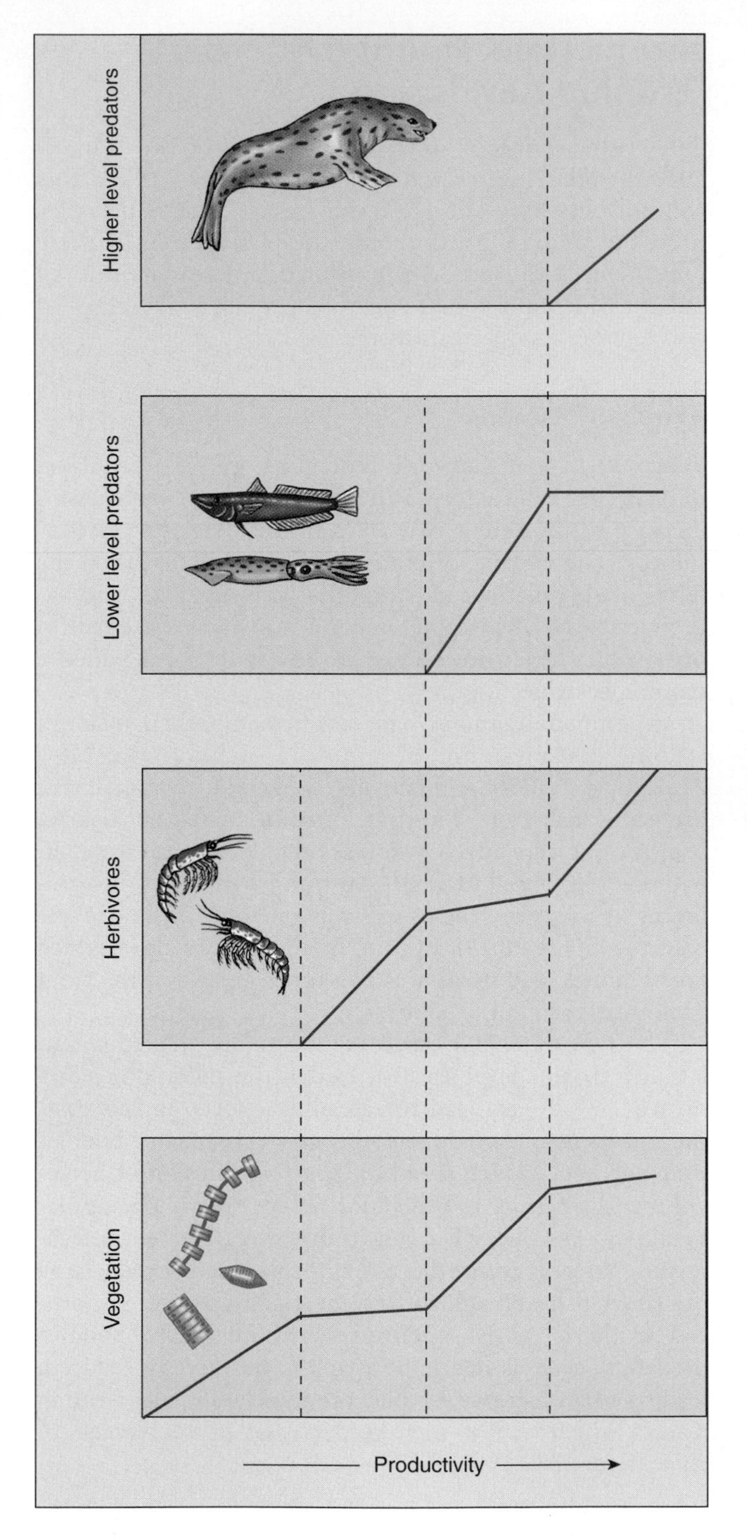

FIGURE 28.16
Bottom-up effects. At low levels of productivity, herbivore populations cannot be maintained. Above some threshold, increases in productivity lead to increases in herbivore biomass; vegetation biomass no longer increases with productivity because it is converted into herbivore biomass. Similarly, above another threshold, herbivore biomass gets converted to carnivore biomass. At this point, vegetation biomass is no longer constrained by herbivores, and so again increases with increasing productivity.

through, whereas others produced little light or deep shade. The result was that the enclosures differed in the amount of sunlight reaching them. As one might expect, the primary productivity differed and was greatest in the unshaded enclosures. This increased productivity led to both more vegetation and more predators, but the trophic level sandwiched in between, the herbivores, did not increase, precisely as the bottom-up hypothesis predicted (figure 28.17).

Relative Importance of Trophic Cascades and Bottom-Up Effects

Neither trophic cascades nor bottom-up effects are inevitable. For example, if two species of herbivores exist in an ecosystem and compete strongly, and if one species is much more vulnerable to predation than the other, then top-down effects will not propagate to the next lower trophic level. Rather, increased predation will simply decrease the population of the vulnerable species while increasing the population of its competitor, with potentially no net change on the vegetation in the next lower trophic level.

Similarly, productivity increases might not move up through all trophic levels. In some cases, for example, prey populations increase so quickly that their predators cannot control them. In such cases, increases in productivity would not move up the food chain.

In other cases, trophic cascades and bottom-up effects may reinforce each other. In one experiment, large fish were removed from one lake, leaving only minnows, which ate most of the algae-eating zooplankton. By contrast, in the other lake, there were few minnows and much zooplankton. The researchers then added nutrients to both lakes. In the minnow lake, there were few zooplankton, so the resulting increase in algal productivity did not propagate up the food chain and large mats of algae formed. By comparison, in the large fish lake, increased productivity moved up the food chain and algae populations were controlled. In this case, both top-down and bottom-up processes were operating.

Nature, of course, is not always so simple. In some cases, species may simultaneously operate on multiple trophic levels, such as the jaguar which eats both smaller carnivores and herbivores, or the bear which eats both fish and berries. Nature is often much more complicated than a simple, linear food chain, as figure 28.8 indicates. Ecologists are currently working to apply theories of food chain interactions to these more complicated situations.

Because of the linked nature of food webs, species on different trophic levels will effect each other, and these effects can promulgate both up and down the food web.

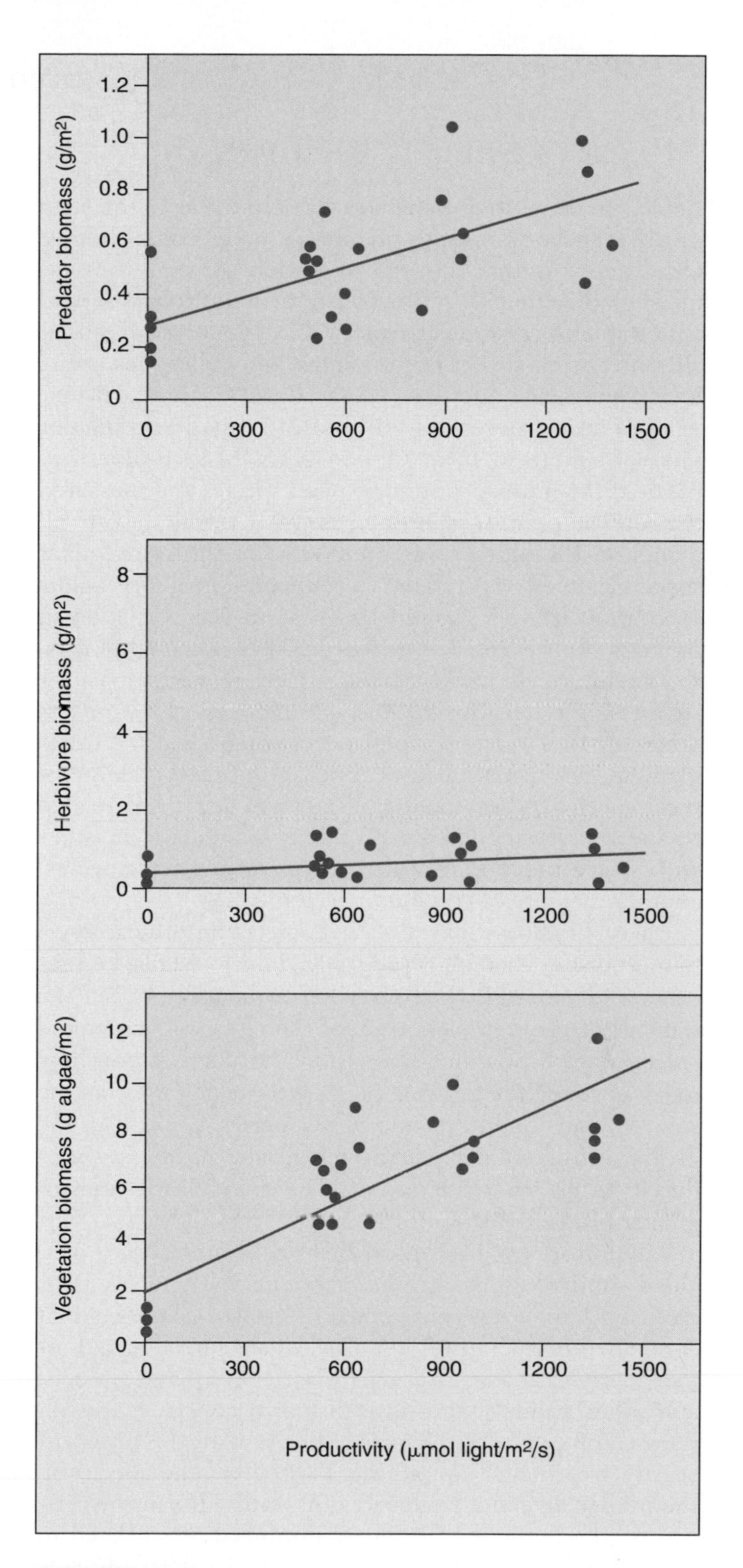

FIGURE 28.17
Bottom-up effects on a stream ecosystem. As predicted, increases in productivity—which are a function of the amount of light hitting the stream and leading to photosynthesis—lead to increases in the amount of vegetation. However, herbivore biomass does not increase with increased productivity because it is converted into predator biomass.

Source: Data from J.T. Wooten & M.E. Power, "Productivity, Consumers & the Structure of a River Food chain," *Proceedings National Academic Sciences*, 1993.

28.4 Biodiversity promotes ecosystem stability.

Effects of Species Richness

Ecologists have long debated what are the consequences of differences in species richness among communities. One theory is that more species-rich communities are more stable; that is, more constant in composition and better able to resist disturbance. This hypothesis has been elegantly studied by David Tilman and colleagues at the University of Minnesota's Cedar Creek Natural History Area. These workers monitored 207 small rectangular plots of land (8 to 16 m^2) for 11 years. In each plot, they counted the number of prairie plant species and measured the total amount of plant biomass (that is, the mass of all plants on the plot). Over the course of the study, plant species richness was related to community stability—plots with more species showed less year-to-year variation in biomass (figure 28.18). Moreover, in two drought years, the decline in biomass was negatively related to species richness; in other words, plots with more species were less affected. In a related experiment, when seeds of other plant species were added to different plots, the ability of these species to become established was negatively related to species richness. More diverse communities, in other words, are more resistant to invasion by new species, another measure of community stability.

Species richness may also have effects on other ecosystem processes. In a follow-up study, Tilman established another 147 plots in which they experimentally varied the number of plant species. Each of the plots was monitored to estimate how much growth was occurring and how much nitrogen the growing plants were taking up from the soil. Tilman found that the more species a plot had, the greater nitrogen uptake and total amount of biomass produced. In his study, increased biodiversity clearly leads to greater productivity (figure 28.19).

Laboratory studies on artificial ecosystems have provided similar results. In one elaborate study, ecosystems covering 1 m^2 were constructed in growth chambers that controlled temperature, light levels, air currents, and atmospheric gas concentrations. A variety of plants, insects, and other animals were introduced to construct ecosystems composed of 9, 15, or 31 species with the lower diversity treatments containing a subset of the species in the higher diversity enclosures. As with Tilman's experiments, the amount of biomass produced was related to species richness, as was the amount of carbon dioxide consumed, a measure of respiration occurring in the ecosystem.

Tilman's conclusion that healthy ecosystems depend on diversity is not accepted by all ecologists. Critics question the validity and relevance of these biodiversity studies, claiming their experimental design is critically flawed. Tilman's Cedar Creek result was a statistical artifact, they argue—the more species you add to a mix, the greater the probability that you will add a highly productive one. Adding taller or highly productive plants of course increases productivity, they explain. To show a real benefit from diversity, experimental plots would have to exhibit "overyielding"—plot productivity would have to be greater than that of the single most productive species grown in isolation. The long-simmering debate continues.

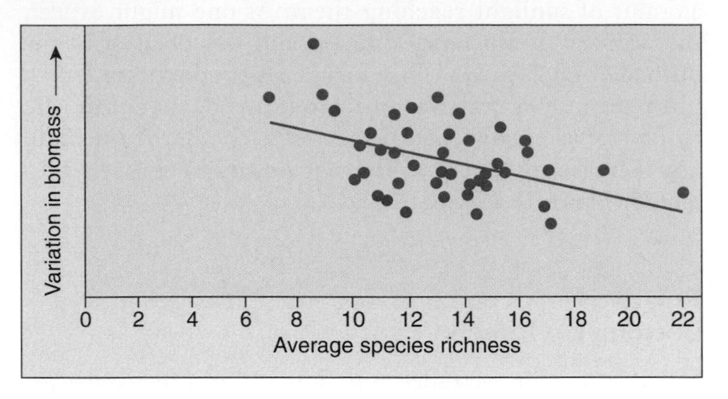

FIGURE 28.18
Effect of species richness on ecosystem stability. Each dot represents data from a 100-square-foot experimental plot in the Cedar Creek experimental fields. Experimental plots with more plant species seem to show less variation in the total amount of biomass produced, and thus more community stability.
Source: Data from D. Tilman, "Biodiversity & Populations Versus Ecosystem Stability," *Ecology*, 1996.

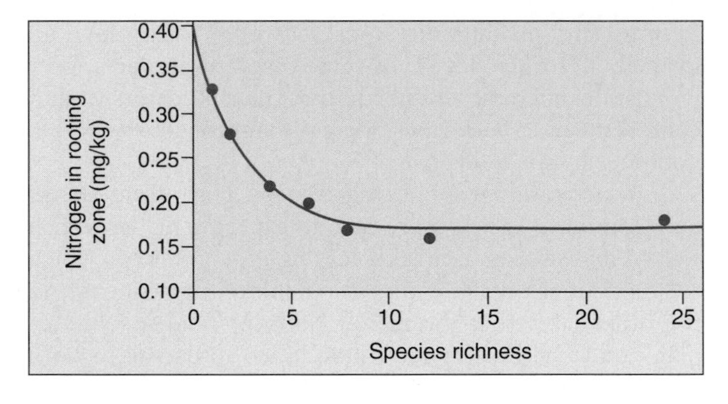

FIGURE 28.19
Effect of species richness on productivity. In Tilman's experimental studies, plots with more species took up more nitrogen from the soil, leaving less in the rooting zone. The increased amount of nitrogen absorption is an indicator of increased growth, increased biomass, and thus increased productivity.
Source: Data from D. Tilman, "Productivity & Sustainability Influenced by Biodiversity in Grassland Ecosystems," *Nature*, 1996.

Controversial experimental field studies support the hypothesis that species-rich communities are more stable.

Causes of Species Richness

While ecologists still argue about why some ecosystems are more stable than others—better able to avoid permanent change and return to normal after disturbances like land clearing, fire, invasion by plagues of insects, or severe storm damage—most ecologists now accept as a working hypothesis that biologically diverse ecosystems are generally more stable than simple ones. Ecosystems with many different kinds of organisms support a more complex web of interactions, and an alternative niche is thus more likely to exist to compensate for the effect of a disruption.

Factors Promoting Species Richness

How does the number of species in a community affect the functioning of the ecosystem? How does ecosystem functioning affect the number of species in a community? It is often extremely difficult to sort out the relative contributions of different factors. With regard to determinants of species richness in a community, of the many variables that may play a role, we will discuss three: ecosystem productivity, spatial heterogeneity, and climate. Two additional factors that may play an important role, the evolutionary age of the community and the degree to which the community has been disturbed, will be examined later in this chapter.

Ecosystem Productivity. Ecosystems differ in productivity, which is a measure of how much new growth they can produce. Surprisingly, the relationship between productivity and species richness is not linear. Rather, ecosystems with intermediate levels of productivity tend to have the most species (figure 28.20). Why this is so is a topic of considerable current debate. One possibility is that levels of productivity are linked to numbers of predators. At low productivity, there are few predators and superior competitors eliminate most species, whereas at high productivity, there are so many predators that only the most predation-resistant species survive. At intermediate levels, however, predators may act as keystone species, maintaining species richness.

Spatial Heterogeneity. Environments that are more spatially heterogeneous—that contain more soil types, topographies, and other habitat variations—can be expected to accommodate more species because they provide a greater variety of microhabitats, microclimates, places to hide from predators, and so on. In general, the species richness of animals tends to reflect the species richness of the plants in their community, while plant species richness reflects the spatial heterogeneity of the ecosystem. The plants provide a biologically derived spatial heterogeneity of microhabitats to the animals. Thus, the number of lizard species in the American Southwest mirrors the structural diversity of the plants (figure 28.21).

Climate. The role of climate is more difficult to assess. On the one hand, more species might be expected to coexist in a seasonal environment than in a constant one, because a changing climate may favor different species at different times of the year. On the other hand, stable environments are able to support specialized species that would be unable to survive where conditions fluctuate. Thus, the number of mammal species along the west coast of North America increases as the temperature range decreases (figure 28.22).

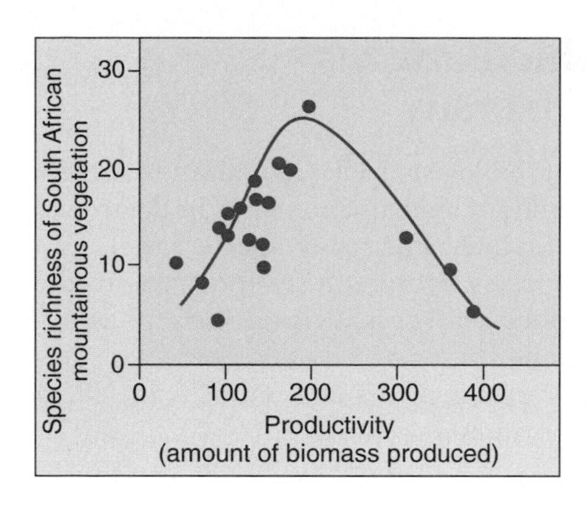

FIGURE 28.20 Productivity. In fynbos plant communities of mountainous areas of South Africa, species richness of plants peaks at intermediate levels of productivity (biomass).

Source: Data from P. Morrin, *Community Ecology*, Blackwell, 1999.

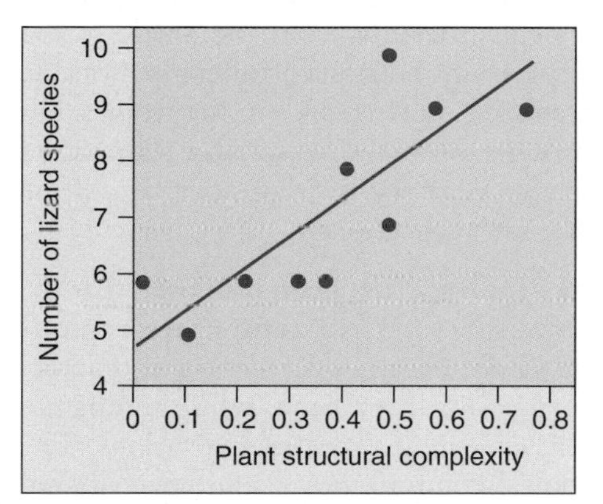

FIGURE 28.21 Spatial heterogeneity. The species richness of desert lizards is positively correlated with the structural complexity of the plant cover in desert sites in the American Southwest.

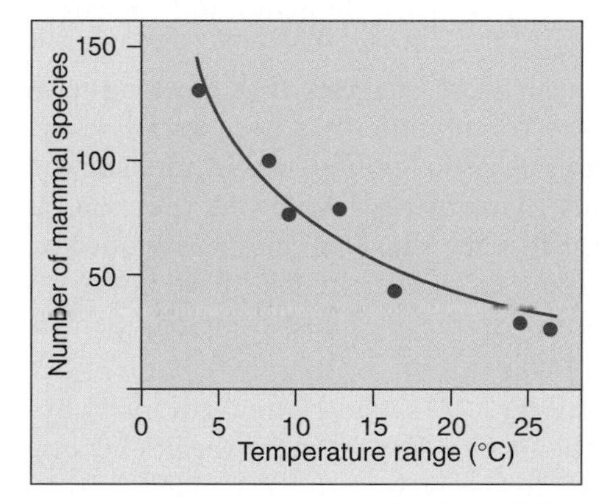

FIGURE 28.22 Climate. The species richness of mammals is inversely correlated with monthly mean temperature range along the west coast of North America.

Species richness promotes ecosystem productivity and is fostered by spatial heterogeneity and stable climate.

Biogeographic Patterns of Species Diversity

Since before Darwin, biologists have recognized that there are more different kinds of animals and plants in the tropics than in temperate regions. For many species, there is a steady increase in species richness from the arctic to the tropics. Called a **species diversity cline,** such a biogeographic gradient in numbers of species correlated with latitude has been reported for plants and animals, including birds (figure 28.23), mammals, reptiles.

Why Are There More Species in the Tropics?

For the better part of a century, ecologists have puzzled over the cline in species diversity from the arctic to the tropics. The difficulty has not been in forming a reasonable hypothesis of why there are more species in the tropics, but rather in sorting through the many reasonable hypotheses that suggest themselves. Here we will consider five of the most commonly discussed suggestions:

Evolutionary age. It has often been proposed that the tropics have more species than temperate regions because the tropics have existed over long and uninterrupted periods of evolutionary time, while temperate regions have been subject to repeated glaciations. The greater age of tropical communities would have allowed complex population interactions to coevolve within them, fostering a greater variety of plants and animals in the tropics.

However, recent work suggests that the long-term stability of tropical communities has been greatly exaggerated. An examination of pollen within undisturbed soil cores reveals that during glaciations the tropical forests contracted to a few small refuges surrounded by grassland. This suggests that the tropics have not had a continuous record of species richness over long periods of evolutionary time.

Higher productivity. A second often-advanced hypothesis is that the tropics contain more species because this part of the earth receives more solar radiation than temperate regions do. The argument is that more solar energy, coupled to a year-round growing season, greatly increases the overall photosynthetic activity of plants in the tropics. If we visualize the tropical forest as a pie (total resources) being cut into slices (species niches), we can see that a larger pie accommodates more slices. However, many field studies have indicated that species richness is highest at intermediate levels of productivity. Accordingly, increasing productivity would be expected to lead to lower, not higher, species richness. Perhaps the long column of vegetation down through which light passes in a tropical forest produces a wide range of frequencies and intensities, creating a greater variety of light environments and so promoting species diversity.

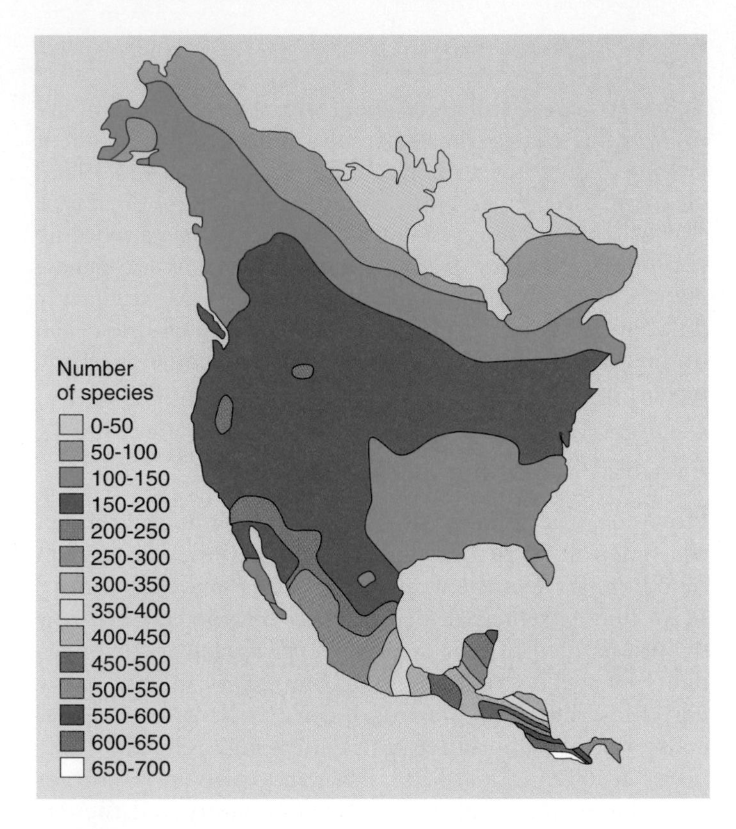

FIGURE 28.23
A latitudinal cline in species richness. Among North and Central American birds, a marked increase in the number of species occurs as one moves toward the tropics. Fewer than 100 species are found at arctic latitudes, while more than 600 species live in southern Central America.

Predictability. There are no winters in the tropics. Tropical temperatures are stable and predictable, one day much like the next. These unchanging environments might encourage specialization, with niches subdivided to partition resources and so avoid competition. The expected result would be a larger number of more specialized species in the tropics, which is what we see. Many field tests of this hypothesis have been carried out, and almost all support it, reporting larger numbers of narrower niches in tropical communities than in temperate areas.

Predation. Many reports indicate that predation may be more intense in the tropics. In theory, more intense predation could reduce the importance of competition, permitting greater niche overlap and thus promoting greater species richness.

Spatial heterogeneity. As noted earlier, spatial heterogeneity promotes species richness. Tropical forests, by virtue of their complexity, create a variety of microhabitats and so may foster larger numbers of species.

No one really knows why there are more species in the tropics, but there are plenty of suggestions.

Island Biogeography

One of the most reliable patterns in ecology is the observation that larger islands contain more species than smaller islands. In 1967, Robert MacArthur of Princeton University and Edward O. Wilson of Harvard University proposed that this **species-area relationship** was a result of the effect of area and isolation on the likelihood of species extinction and colonization.

The Equilibrium Model

MacArthur and Wilson reasoned that species are constantly being dispersed to islands, so islands have a tendency to accumulate more and more species. At the same time that new species are added, however, other species are lost by extinction. As the number of species on an initially empty island increases, the rate of colonization must decrease as the pool of potential colonizing species not already present on the island becomes depleted. At the same time, the rate of extinction should increase—the more species on an island, the greater the likelihood that any given species will perish. As a result, at some point, the number of extinctions and colonizations should be equal and the number of species should then remain constant. Every island of a given size, then, has a characteristic equilibrium number of species that tends to persist through time (the intersection point in figure 28.24*a*), although the individual species will change as species become extinct and new species colonize.

MacArthur and Wilson's equilibrium theory proposes that island species richness is a dynamic equilibrium between colonization and extinction. Both island size and distance from the mainland would play important roles. We would expect smaller islands to have higher rates of extinction because their population sizes would, on average, be smaller. Also, we would expect fewer colonizers to reach islands that lie farther from the mainland. Thus, small islands far from the mainland have the fewest species; large islands near the mainland have the most (figure 28.24*b*).

The predictions of this simple model bear out well in field data. Asian Pacific bird species (figure 28.24*c*) exhibit a positive correlation of species richness with island size, but a negative correlation of species richness with distance from the mainland.

Testing the Equilibrium Model

Field studies in which small islands have been censused, cleared, and allowed to recolonize tend to support the equilibrium model. However, long-term experimental field studies are suggesting that the situation is more complicated than MacArthur and Wilson envisioned. Their theory predicts a high level of **species turnover** as some species perish and others arrive. However, studies on island birds and spiders indicate that very little turnover occurs from year to year. Moreover, those species that do come and go are a subset of species that never attain high populations. A substantial proportion of the species appears to maintain high populations and rarely go extinct. These studies, of course, have only been going on for 20 years or less. It is possible that over periods of centuries, the equilibrium theory is a good description of what determines island species richness.

Species richness on islands is a dynamic equilibrium between colonization and extinction.

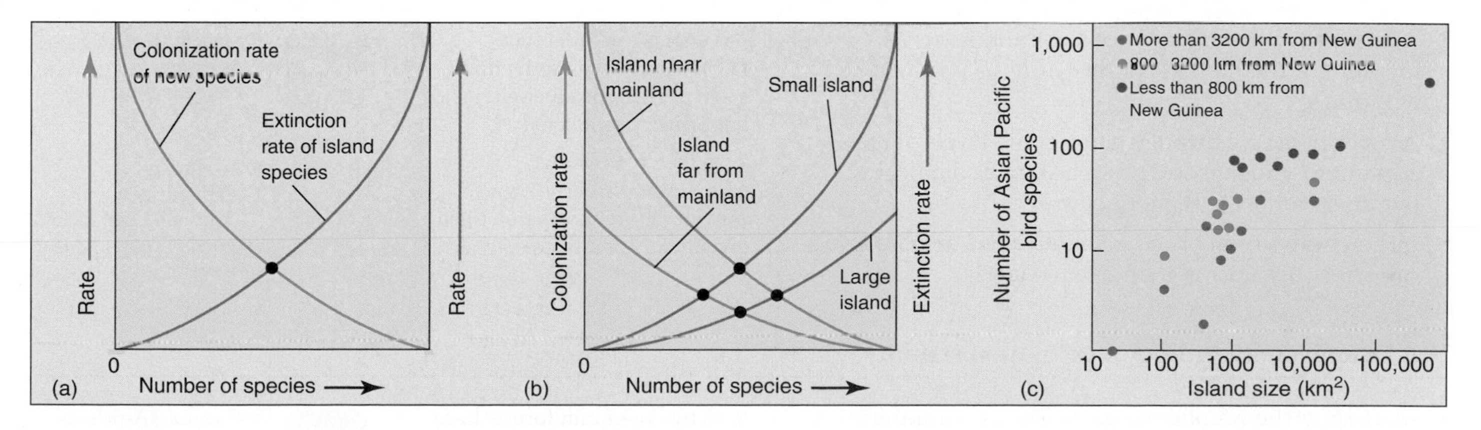

FIGURE 28.24

The equilibrium model of island biogeography. (*a*) Island species richness reaches an equilibrium (*black dot*) when the colonization rate of new species equals the extinction rate of species on the island. (*b*) The equilibrium shifts depending on the rate of colonization, the size of an island, and its distance to sources of colonists. Species richness is positively correlated with island size and inversely correlated with distance from the mainland. Smaller islands have higher extinction rates, shifting the equilibrium point to the left. Similarly, more distant islands have lower colonization rates, again shifting the equilibrium point leftward. (*c*) The effect of distance from a larger island—which can be the source of colonizing species—is readily apparent. More distant islands have fewer species compared to nearer islands of the same size.

Chapter 28

www.mhhe.com/raven6/resources28.mhtml

Summary | **Questions** | **Media Resources**

28.1 Chemicals cycle within ecosystems.

- Fully 98% of the water on earth is free water cycling through the atmosphere. In the United States, 96% of the fresh water is groundwater.
- About 10% of the roughly 700 billion metric tons of free carbon dioxide in the atmosphere is fixed each year through photosynthesis. About as much carbon exists in living organisms at any one time as is dissolved in the oceans.
- Nitrogen is made available to organisms by cycling through nitrogen-fixing bacteria located in the soil or associated with the roots of some plants.
- Phosphorus is a key component of many biological molecules; it weathers out of soils and is transported to the world's oceans.

1. What are the primary reservoirs for the chemicals in biogeochemical cycles? Are more of the life-sustaining chemicals found in these reservoirs or in the earth's living organisms?

2. What is denitrification? Which organisms carry it out?

3. How is the phosphorus cycle different from the water, carbon, nitrogen, and oxygen cycles? What are the natural sources for phosphorus?

4. What effect does deforestation have on the water cycle and overall fertility of the land?

- Carbon Cycle
- Nitrogen Cycle

- Nutrient Cycles
- Water Cycle
- Groundwater

- Art Quizzes:
  - -The Water Cycle
  - -Carbon Cycle
  - -The Nitrogen Cycle

28.2 Ecosystems are structured by who eats whom.

- Plants are autotrophs that convert the light energy that falls on their leaves to food energy. Producers, the herbivores that eat them, and the carnivores that eat the herbivores constitute three trophic levels.
- At each level, only about 10% of the energy available in the food is fixed in the body of the consumer. For this reason, food chains are always relatively short.

5. How might an increase in the number of predators affect lower levels of a food chain. How might an increase in nutrients affect upper levels?

- Ecosystem Introduction
- Energy Flow/Trophic Levels

- Art Quiz: Trophic Levels Within a Food Chain

28.3 Energy flows through ecosystems.

- The primary productivity of a community is a measure of the biomass photosynthesis produces within it.
- As energy passes through the trophic levels of an ecosystem, much is lost at each step. Ecological pyramids reflect this energy loss.
- Interactions among species on different trophic levels can affect organisms at all trophic levels.

6. What is the difference between primary productivity, gross primary productivity, and net primary productivity?

7. Which type of diet, carnivorous or herbivorous, provides more food value to any given living organism?

- Student Research: Assessing Paleoenvironments

- Art Quizzes:
  - -Path of Energy Through Food Webs
  - -Food Web in Cayuga Lake

28.4 Biodiversity promotes ecosystem stability.

- Increasing the number of species in a community seems to promote ecosystem productivity. Controversial experiments suggest that communities with increased species richness are more stable and less vulnerable to disturbance.

8. Why might rain forests have high levels of species diversity?

9. Why do distant islands tend to have fewer species than nearer islands of the same size? Why do different-sized islands tend to differ in species number?

- *On Science* Article: Is Biodiversity Good?
- Bioethics Case Study: Wolves in Yellowstone

- Art Quiz:
  - -Island Biogeography–Equilibrium Model

29

The Biosphere

Concept Outline

29.1 Organisms must cope with a varied environment.

The Environmental Challenge. Habitats vary in ways important to survival. Organisms cope with environmental variation with physiological, morphological, and behavioral adaptations.

29.2 Climate shapes the character of ecosystems.

The Sun and Atmospheric Circulation. The sun powers major movements in atmospheric circulation.
Atmospheric Circulation, Precipitation, and Climate. Latitude and elevation have important effects on climate, although other factors affect regional climate.

29.3 Biomes are widespread terrestrial ecosystems.

The Major Biomes. Characteristic communities called biomes occur in different climatic regions. Variations in temperature and precipitation are good predictors of what biomes will occur where. Major biomes include tropical rain forest, savanna, desert, temperate grassland, temperate deciduous forest, temperate evergreen forest, taiga, and tundra.

29.4 Aquatic ecosystems cover much of the earth.

Patterns of Circulation in the Oceans. The world's oceans circulate in huge circles deflected by the continents.
Life in the Oceans. Most of the major groups of organisms originated and are still represented in the sea.
Marine Ecosystems. The communities of the ocean are delineated primarily by depth.
Freshwater Habitats. Like miniature oceans, ponds and lakes support different communities at different depths.
Productivity of Freshwater Ecosystems. Freshwater ecosystems are often highly productive.

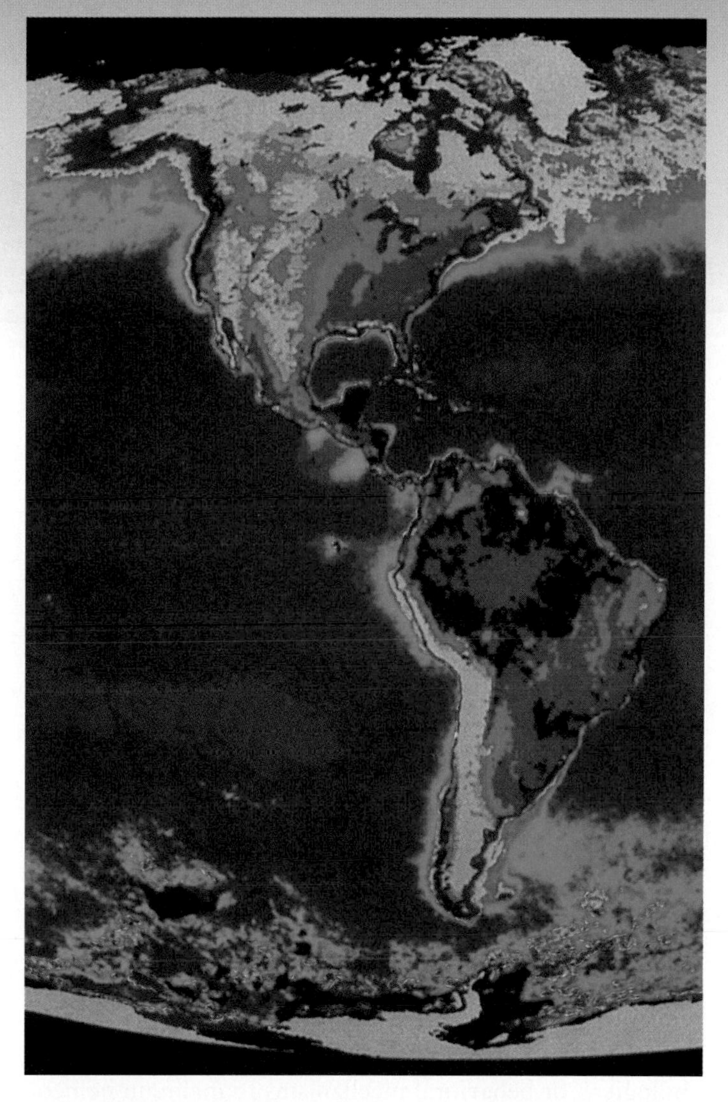

FIGURE 29.1
Life in the biosphere. In this satellite image, orange zones are largely arid. Almost every environment on earth can be described in terms of temperature and moisture. These physical parameters have great bearing on the forms of life that are able to inhabit a particular region.

The biosphere includes all living communities on earth, from the profusion of life in the tropical rain forests to the photosynthetic phytoplankton in the world's oceans. In a very general sense, the distribution of life on earth reflects variations in the world's environments, principally in temperature and the availability of water. Figure 29.1 is a satellite image of North and South America, collected over eight years, the colors keyed to the relative abundance of chlorophyll, a good indicator of rich biological communities. Phytoplankton and algae produce the dark red zones in the oceans and along the seacoasts. Green and dark green areas on land are dense forests, while orange areas like the deserts of western South America are largely barren of life.

29.1 Organisms must cope with a varied environment.

The Environmental Challenge

How Environments Vary

The nature of the physical environment in large measure determines what organisms live in a place. Key elements include:

Temperature. Most organisms are adapted to live within a relatively narrow range of temperatures and will not thrive if temperatures are colder or warmer. The growing season of plants, for example, is importantly influenced by temperature.

Water. Plants and all other organisms require water. On land, water is often scarce, so patterns of rainfall have a major influence on life.

Sunlight. Almost all ecosystems rely on energy captured by photosynthesis; the availability of sunlight influences the amount of life an ecosystem can support, particularly below the surface in marine communities.

Soil. The physical consistency, pH, and mineral composition of soil often severely limit plant growth, particularly the availability of nitrogen and phosphorus.

FIGURE 29.2
Meeting the challenge of obtaining moisture in a desert. On the dry sand dunes of the Namib Desert in southwestern Africa, the beetle *Onymacris unguicularis* collects moisture from the fog by holding its abdomen up at the crest of a dune to gather condensed water.

Table 29.1 Physiological changes at high altitude that increase the amount of oxygen delivered to body tissues

Increased rate of breathing
Increased erythrocyte production, increasing the amount of hemoglobin in the blood
Decreased binding capacity of hemoglobin, thus increasing the rate at which oxygen is unloaded in body tissues
Increased density of mitochondria, capillaries, and muscle myoglobin

Active and Passive Approaches to Coping with Environmental Variation

An individual encountering environmental variation may choose to maintain a "steady-state" internal environment, an approach known as maintaining **homeostasis.** Many animals and plants actively employ physiological, morphological, or behavioral mechanisms to maintain homeostasis. The beetle in figure 29.2 is using a behavioral mechanism to cope with drastic changes in water availability. Other animals and plants simply conform to the environment in which they find themselves, their bodies adopting the temperature, salinity, and other aspects of their surroundings.

Responses to environmental variation can be seen over both the short and the long term. In the short term, spanning periods of a few minutes to an individual's lifetime, organisms have a variety of ways of coping with environmental change. Over longer periods, natural selection can operate to make a population better adapted to the environment.

Individual Response to Environmental Change

Physiology. Many organisms are able to adapt to environmental change by making physiological adjustments. Thus, your body constricts the blood vessels on the surface of your face on a cold day, reducing heat loss (and also giving your face a "flush"). Similarly, humans who visit high altitudes, such as in the Andes, initially experience altitude sickness—the symptoms of which include heart palpitations, nausea, fatigue, headache, mental impairment and, in serious cases, pulmonary edema—because of the lower atmospheric pressure and consequent lower oxygen availability in the air. After several days, however, the same people will feel fine, because of a number of physiological changes that increase the delivery of oxygen to body (table 29.1).

Some insects avoid freezing in the winter by adding glycerol "antifreeze" to their blood; others tolerate freezing by converting much of their glycogen reserves into alcohols that protect their cell membranes from freeze damage.

Morphology. Animals that maintain a constant internal temperature (endotherms) in a cold environment have adaptations that tend to minimize energy expenditure. Many other mammals grower thicker coats during the winter, utilizing their fur as insulation to retain body heat during the winter. In general, the thicker the fur, the greater the insulation (figure 29.3). Thus, a wolf's fur is some three times as thick in winter as summer and insulates more than twice as well. Other mammals escape some of the costs of maintaining a constant body temperature during winter by hibernating during the coldest season, behaving, in effect, like conformers.

Behavior. Many animals deal with variation in the environment by moving from one patch of habitat to another, avoiding areas that are unsuitable. The tropical lizards in figure 29.4 manage to maintain a fairly uniform body temperature in an open habitat by basking in patches of sun, retreating to the shade when they become too hot. By contrast, in shaded forests, the same lizards do not have the opportunity to regulate their body temperature through behavioral means. Thus, they become conformers and adopt the temperature of their surroundings.

Behavioral adaptations can be extreme. The spadefoot toad *Scaphiophus*, which lives in the deserts of North America, can burrow nearly a meter below the surface and remain there for as long as nine months of each year, its metabolic rate greatly reduced, living on fat reserves. When moist cool conditions return, the toads emerge and breed. The young toads mature rapidly and burrow back underground.

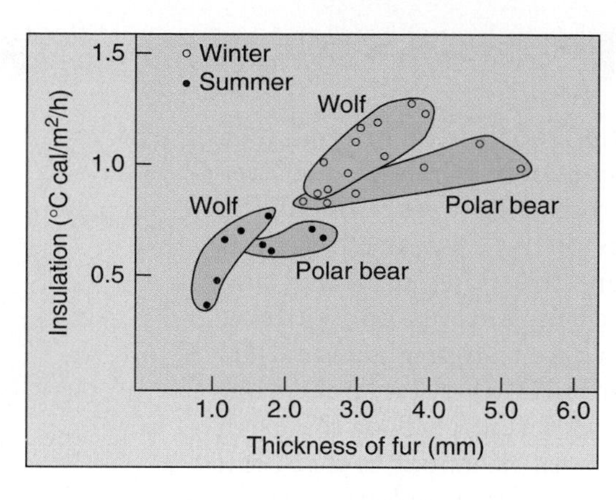

FIGURE 29.3
Morphological adaptation. Fur thickness in North American mammals has a major impact on the degree of insulation the fur provides.

Evolutionary Responses to Environmental Variation

These examples represent different ways in which organisms may adjust to changing environmental conditions. The ability of an individual to alter its physiology, morphology, or behavior is itself an evolutionary adaptation, the result of natural selection. The results of natural selection can also be detected by comparing closely related species that live in different environments. In such cases, species often exhibit striking adaptations to the particular environment in which they live.

For example, animals that live in different climates show many differences. Mammals from colder climates tend to have shorter ears and limbs (Allen's Rule) and larger bodies (Bergmann's Rule) to limit heat loss. Both mechanisms reduce the surface area across which animals lose heat. Lizards that live in different climates exhibit physiological adaptations for coping with life at different temperatures. Desert lizards are unaffected by high temperatures that would kill a lizard from northern Europe, but the northern lizards are capable of running, capturing prey, and digesting food at cooler temperatures at which desert lizards would be completely immobilized.

Many species also exhibit adaptations to living in areas where water is scarce. Everyone knows of the camel, and other desert animals, which can go extended periods without drinking water. Another example of desert adaptation is seen in frogs. Most frogs have moist skins through which water permeates readily. Such animals could not survive in arid climates because they would rapidly dehydrate and dry. However, some frogs have solved this problem by greatly reducing the rate of water loss through the skin. One species, for example, secretes a waxy substance from specialized glands that waterproofs its skin and reduces rates of water loss by 95%.

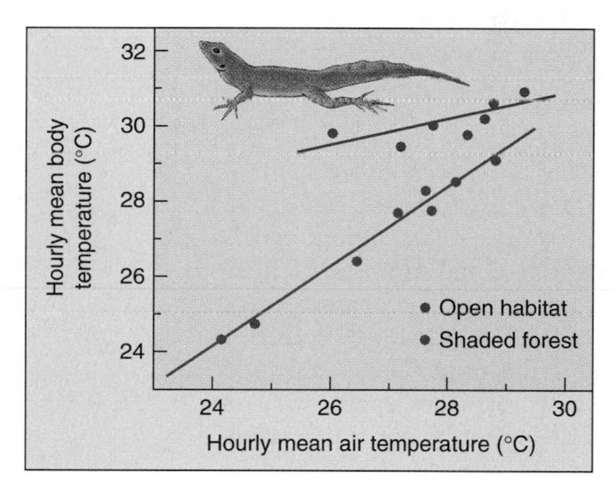

FIGURE 29.4
Behavioral adaptation. The Puerto Rican lizard *Anolis cristatellus* maintains a relatively constant temperature by seeking out and basking in patches of sunlight; in shaded forests, this behavior is not possible and body temperature conforms to the surroundings.

Adaptation to different environments can also be studied experimentally. For example, when strains of *E. coli* are grown at high temperatures (42°C), the speed at which resources are utilized improves through time. After 2000 generations, this ability increased 30% over what it had been when the experiment started. The mechanism by which efficiency of resource use was increased is still unknown and is the focus of current research.

Organisms use a variety of physiological, morphological, and behavioral mechanisms to adjust to environmental variation. Over time, species evolve adaptations to living in different environments.

29.2 Climate shapes the character of ecosystems.

The distribution of biomes (see discussion later in this chapter) results from the interaction of the features of the earth itself, such as different soil types or the occurrence of mountains and valleys, with two key physical factors: (1) the amount of solar heat that reaches different parts of the earth and seasonal variations in that heat; and (2) global atmospheric circulation and the resulting patterns of oceanic circulation. Together these factors dictate local climate, and so determine the amounts and distribution of precipitation.

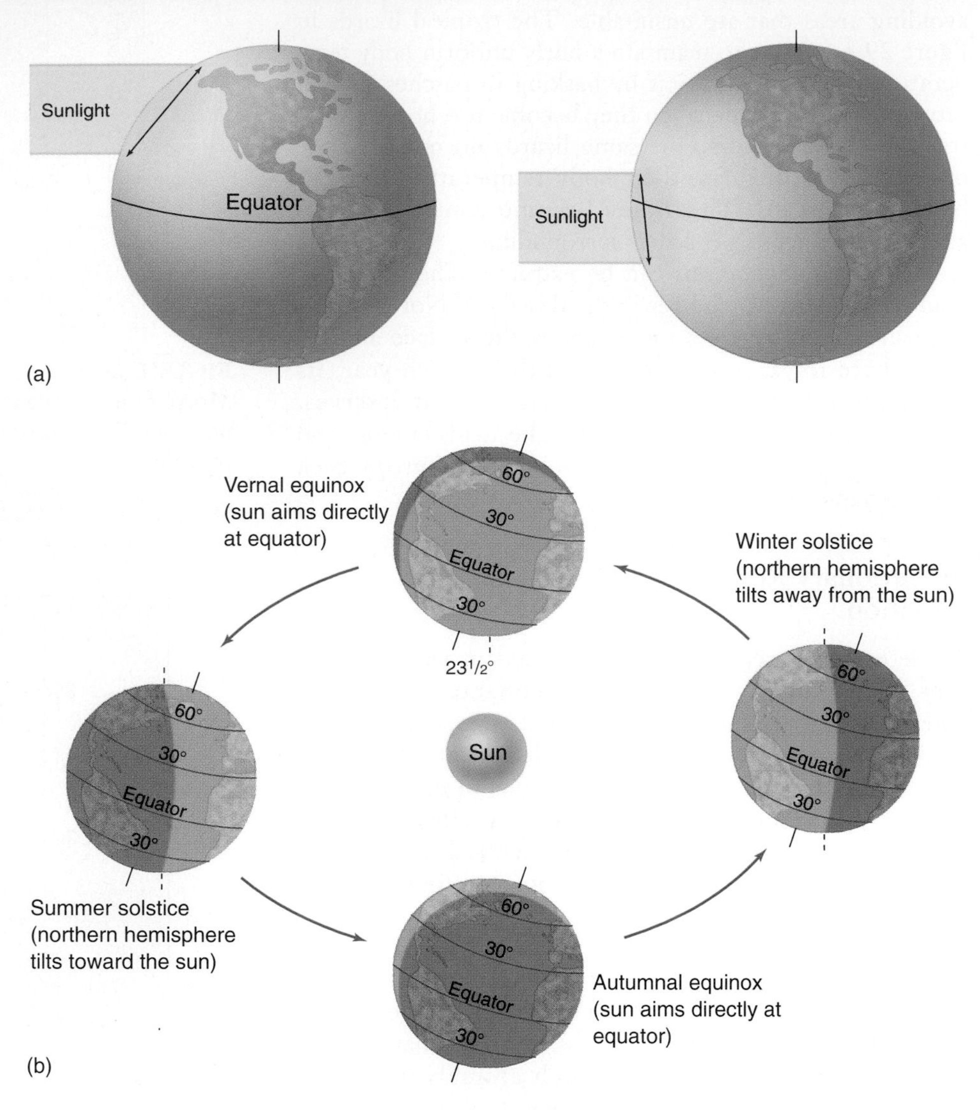

FIGURE 29.5
Relationships between the earth and the sun are critical in determining the nature and distribution of life on earth. (*a*) A beam of solar energy striking the earth in the middle latitudes spreads over a wider area of the earth's surface than a similar beam striking the earth near the equator. (*b*) The rotation of the earth around the sun has a profound effect on climate. In the northern and southern hemispheres, temperatures change in an annual cycle because the earth tilts slightly on its axis in relation to the path around the sun.

The Sun and Atmospheric Circulation

The earth receives an enormous quantity of heat from the sun in the form of shortwave radiation, and it radiates an equal amount of heat back to space in the form of longwave radiation. About 10^{24} calories arrive at the upper surface of the earth's atmosphere each year, or about 1.94 calories per square centimeter per minute. About half of this energy reaches the earth's surface. The wavelengths that reach the earth's surface are not identical to those that reach the outer atmosphere. Most of the ultraviolet radiation is absorbed by the oxygen (O_2) and ozone (O_3) in the atmosphere. As we will see in chapter 30, the depletion of the ozone layer, apparently as a result of human activities, poses serious ecological problems.

Why the Tropics Are Warmer

The world contains a great diversity of biomes because its climate varies so much from place to place. On a given day, Miami, Florida, and Bangor, Maine, often have very different weather. There is no mystery about this. Because the earth is a sphere, some parts of it receive more energy from the sun than others. This variation is responsible for many of the major climatic differences that occur over the earth's surface, and, indirectly, for much of the diversity of biomes. The tropics are warmer than temperate regions because the sun's rays arrive almost perpendicular to regions near the equator. Near the poles the angle of incidence of the sun's rays spreads them out over a much greater area, providing less energy per unit area (figure 29.5*a*).

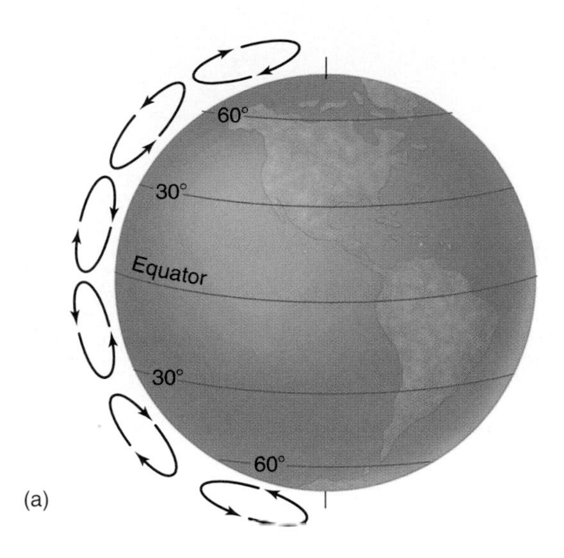

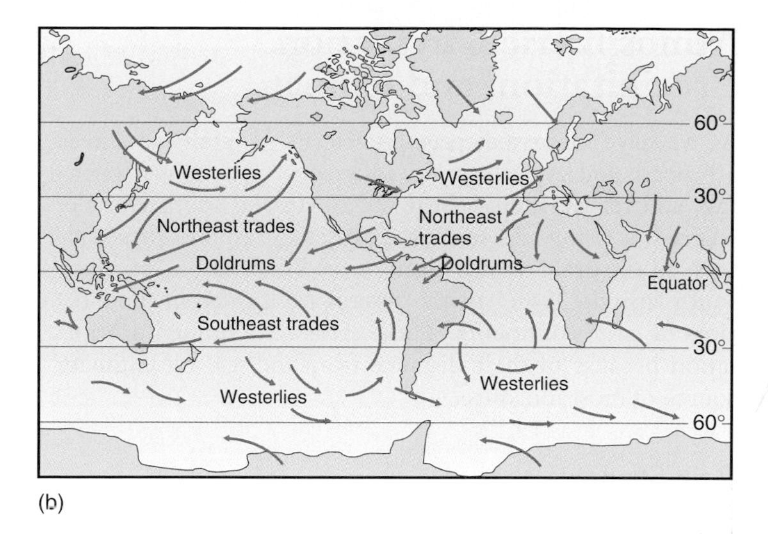

FIGURE 29.6
General patterns of atmospheric circulation. (*a*) The pattern of air movement toward and away from the earth's surface. (*b*) The major wind currents across the face of the earth.

The earth's annual orbit around the sun and its daily rotation on its own axis are both important in determining world climate (figure 29.5*b*). Because of the annual cycle, and the inclination of the earth's axis at approximately 23.5° from its plane of revolution around the sun, there is a progression of seasons in all parts of the earth away from the equator. One pole or the other is tilted closer to the sun at all times except during the spring and autumn equinoxes.

Major Atmospheric Circulation Patterns

The moisture-holding capacity of air increases when it warms and decreases when it cools. High temperatures near the equator encourage evaporation and create warm, moist air. As this air rises and flows toward the poles, it cools and loses most of its moisture (figure 29.6). Consequently, the greatest amounts of precipitation on earth fall near the equator. This equatorial region of rising air is one of low pressure, called the **doldrums,** which draws air from both north and south of the equator. When the air masses that have risen reach about 30° north and south latitude, the dry air, now cooler, sinks and becomes reheated. As the air reheats, its evaporative capacity increases, creating a zone of decreased precipitation. The air, still warmer than in the polar regions, continues to flow toward the poles. It rises again at about 60° north and south latitude, producing another zone of high precipitation. At this latitude there is another low-pressure area, the polar front. Some of this rising air flows back to the equator and some continues north and south, descending near the poles and producing another zone of low precipitation before it returns to the equator.

Air Currents Generated by the Earth's Rotation

Related to these bands of north-south circulation are three major air currents generated mainly by the interaction of the earth's rotation with patterns of worldwide heat gain. Between about 30° north latitude and 30° south latitude, the trade winds blow, from the east-southeast in the southern hemisphere and from the east-northeast in the northern hemisphere. The trade winds blow all year long and are the steadiest winds found anywhere on earth. They are stronger in winter and weaker in summer. Between 30° and 60° north and south latitude, strong prevailing westerlies blow from west to east and dominate climatic patterns in these latitudes, particularly along the western edges of the continents. Weaker winds, blowing from east to west, occur farther north and south in their respective hemispheres.

Warm air rises near the equator, descends and produces arid zones at about 30° north and south latitude, flows toward the poles, then rises again at about 60° north and south latitude, and moves back toward the equator. Part of this air, however, moves toward the poles, where it produces zones of low precipitation.

Atmospheric Circulation, Precipitation, and Climate

As we have discussed, precipitation is generally low near 30° north and south latitude, where air is falling and warming, and relatively high near 60° north and south latitude, where it is rising and cooling. Partly as a result of these factors, all the great deserts of the world lie near 30° north or south latitude. Other major deserts are formed in the interiors of large continents. These areas have limited precipitation because of their distance from the sea, the ultimate source of most moisture.

Rain Shadows

Other deserts occur because mountain ranges intercept moisture-laden winds from the sea. When this occurs, the air rises and the moisture-holding capacity of the air decreases, resulting in increased precipitation on the windward side of the mountains—the side from which the wind is blowing. As the air descends the other side of the mountains, the leeward side, it is warmed, and its moisture-holding capacity increases, tending to block precipitation. In California, for example, the eastern sides of the Sierra Nevada Mountains are much drier than the western sides, and the vegetation is often very different. This phenomenon is called the **rain shadow effect** (figure 29.7).

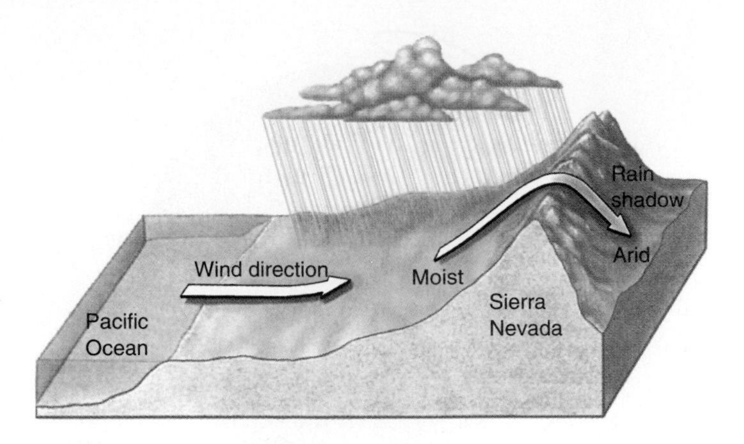

FIGURE 29.7
The rain shadow effect. Moisture-laden winds from the Pacific Ocean rise and are cooled when they encounter the Sierra Nevada Mountains. As their moisture-holding capacity decreases, precipitation occurs, making the middle elevation of the range one of the snowiest regions on earth; it supports tall forests, including those that contain the famous giant sequoias (*Sequoiadendron giganteum*). As the air descends on the east side of the range, its moisture-holding capacity increases again, and the air picks up rather than releases moisture from its surroundings. As a result, desert conditions prevail on the east side of the mountains.

Regional Climates

Four relatively small areas, each located on a different continent, share a climate that resembles that of the Mediterranean region. So-called Mediterranean climates are found in portions of Baja, California, and Oregon; in central Chile; in southwestern Australia; and in the Cape region of South Africa. In all of these areas, the prevailing westerlies blow during the summer from a cool ocean onto warm land. As a result, the air's moisture-holding capacity increases, the air absorbing moisture and creating hot rainless summers. Such climates are unusual on a world scale. In the five regions where they occur, many unique kinds of plants and animals, often local in distribution, have evolved. Because of the prevailing westerlies, the great deserts of the world (other than those in the interiors of continents) and the areas of Mediterranean climate lie on the western sides of the continents.

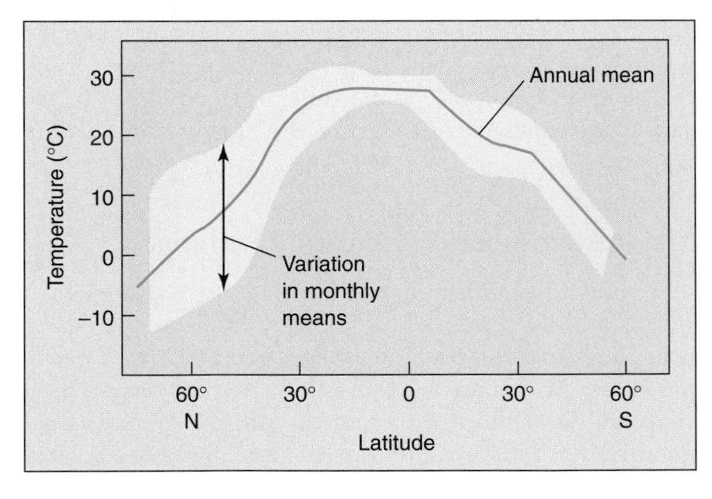

FIGURE 29.8
Temperature varies with latitude. The *blue* line represents the annual mean temperature at latitudes from the North Pole to Antarctica.

Another kind of regional climate occurs in southern Asia. The monsoon climatic conditions characteristic of India and southern Asia occur during the summer months. During the winter, the trade winds blow from the east-northeast off the cool land onto the warm sea. From June to October, though, when the land is heated, the direction of the air flow reverses, and the winds veer around to blow onto the Indian subcontinent and adjacent areas from the southwest bringing rain. The duration and strength of the monsoon winds spell the difference between food sufficiency and starvation for hundreds of millions of people in this region each year.

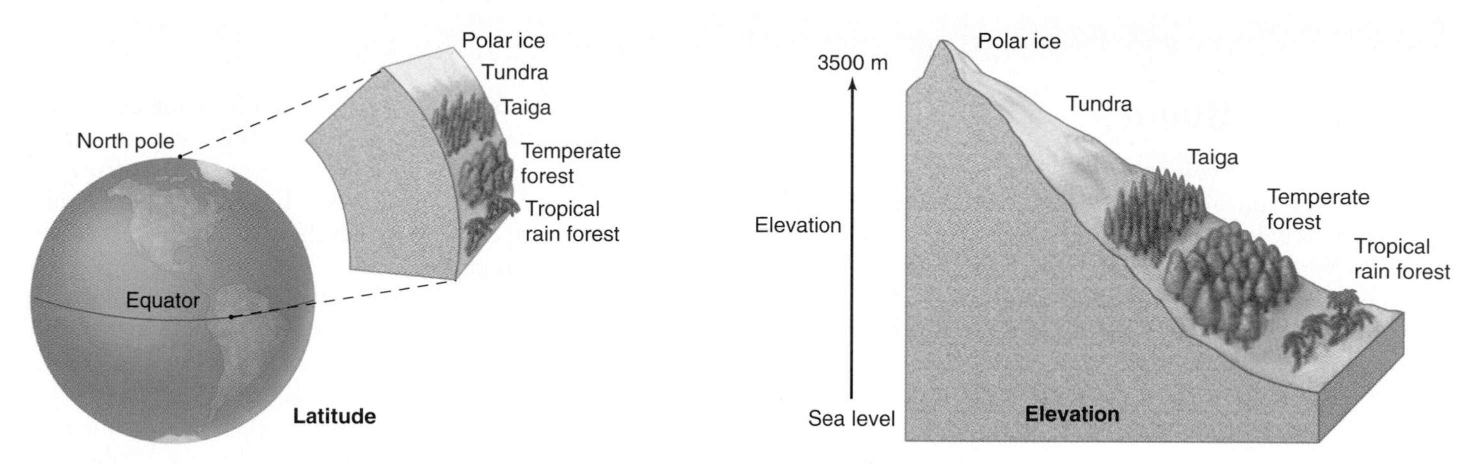

FIGURE 29.9
Elevation affects the distribution of biomes much as latitude does. Biomes that normally occur far north and far south of the equator at sea level also occur in the tropics at high mountain elevations. Thus, on a tall mountain in southern Mexico or Guatemala, one might see a sequence of biomes like the one illustrated here.

Latitude

Temperatures are higher in tropical ecosystems for a simple reason: more sunlight per unit area falls on tropical latitudes. Solar radiation is most intense when the sun is directly overhead, and this only occurs in the tropics, where sunlight strikes the equator perpendicularly. As figure 29.8 shows, the highest mean global temperatures occur near the equator (that is, 0 latitude). Because there are no seasons in the tropics, there is little variation in mean monthly temperature in tropical ecosystems. As you move from the equator into temperate latitudes, sunlight strikes the earth at a more oblique angle, so that less falls on a given area. As a result, mean temperatures are lower. At temperate latitudes, temperature variation increases because of the increasingly marked seasons.

Seasonal changes in wind circulation produce corresponding changes in ocean currents, sometimes causing nutrient-rich cold water to well up from ocean depths. This produces "blooms" among the plankton and other organisms living near the surface. Similar turnover occurs seasonally in freshwater lakes and ponds, bringing nutrients from the bottom to the surface in the fall and again in the spring.

Elevation

Temperature also varies with elevation, with higher altitudes becoming progressively colder. At any given latitude, air temperature falls about 6°C for every 1000-meter increase in elevation. The ecological consequences of temperature varying with elevation are the same as temperature varying with latitude (figure 29.9). Thus, in North America a 1000-meter increase in elevation results in a temperature drop equal to that of an 880-kilometer increase in latitude. This is one reason "timberline" (the elevation above which trees do not grow) occurs at progressively lower elevations as one moves farther from the equator.

Microclimate

Climate also varies on a very fine scale within ecosystems. Within the litter on a forest floor, there is considerable variation in shading, local temperatures, and rates of evaporation from the soil. Called **microclimate,** these very localized climatic conditions can be very different from those of the overhead atmosphere. Gardeners spread straw over newly seeded lawns to create such a moisture-retaining microclimate.

The great deserts and associated arid areas of the world mostly lie along the western sides of continents at about 30° north and south latitude. Mountain ranges tend to intercept rain, creating deserts in their shadow. In general, temperatures are warmer in the tropics and at lower elevations.

29.3 Biomes are widespread terrestrial ecosystems.

The Major Biomes

Biomes are major communities of organisms that have a characteristic appearance and that are distributed over a wide land area defined largely by regional variations in climate. As you might imagine from such a broad definition, there are many ways to classify biomes, and different ecologists may assign the same community to different biomes. There is little disagreement, however, about the reality of biomes as major biological communities—only about how to best describe them.

Distribution of the Major Biomes

Eight major biome categories are presented in this text: tropical rain forest, savanna, desert, temperate grassland, temperate deciduous forest, temperate evergreen forest, taiga, and tundra. These biomes occur worldwide, occupying large regions that can be defined by rainfall and temperature.

Six additional biomes are considered by some ecologists to be subsets of the eight major ones: polar ice, mountain zone, chaparral, warm moist evergreen forest, tropical monsoon forest, and semidesert. They vary remarkably from one another because they have evolved in regions with very different climates.

Distributions of the 14 biomes are mapped in figure 29.10. Although each is by convention named for the dominant vegetation (deciduous forest, evergreen forest, grassland, and so on) each biome is also characterized by particular animals, fungi, and microorganisms adapted to live as members of that community. Wolves, caribou or reindeer, polar bears, hares, lynx, snowy owls, deer flies, and mosquitoes inhabit the tundra all over the world and are as much a defining characteristic of the tundra biomes as the low, shrubby, matlike vegetation.

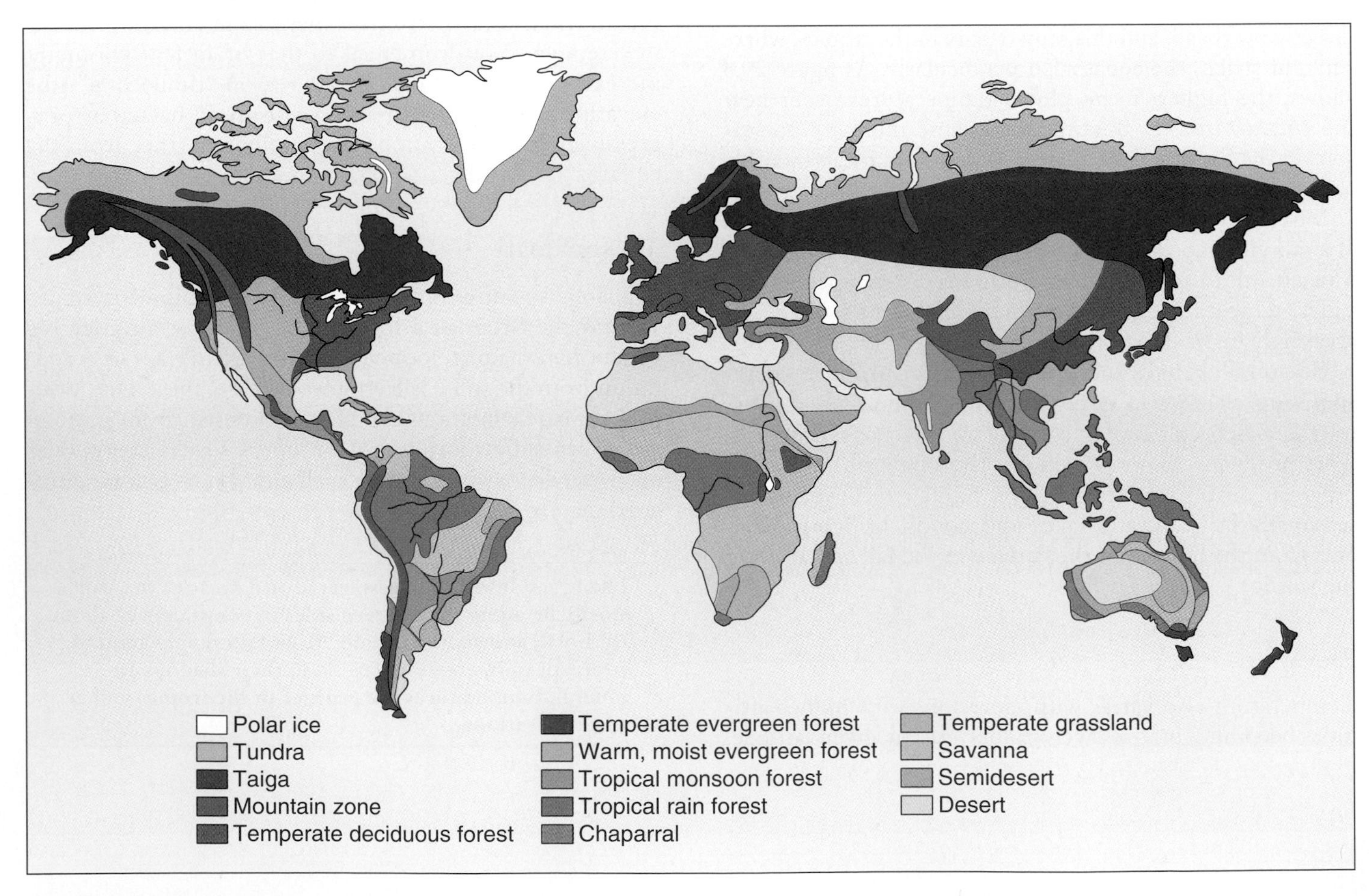

FIGURE 29.10
The distribution of biomes. Each biome is similar in structure and appearance wherever it occurs on earth.

Biomes and Climate

Many different environmental factors play a role in determining which biomes are found where. Two key parameters are available moisture and temperature. Figure 29.11 presents data on ecosystem productivity as a function of annual precipitation and of annual mean temperature: ecosystem productivity is strongly influenced by both. This is not to say that other factors such as soil structure and its mineral composition (discussed in detail in chapter 39), or seasonal versus constant climate, are not also important. Different places with the same annual precipitation and temperature sometimes support different biomes, so other factors must also be important. Nevertheless, these two variables do a fine job of predicting what biomes will occur in most places, as figure 29.12 illustrates.

If there were no mountains and no climatic effects caused by the irregular outlines of continents and by different sea temperatures, each biome would form an even belt around the globe, defined largely by latitude. In truth, these other factors also greatly affect the distribution of biomes. Distance from the ocean has a major impact on rainfall, and elevation affects temperature—the summits of the Rocky Mountains are covered with a vegetation type that resembles the tundra which normally occurs at a much higher latitude.

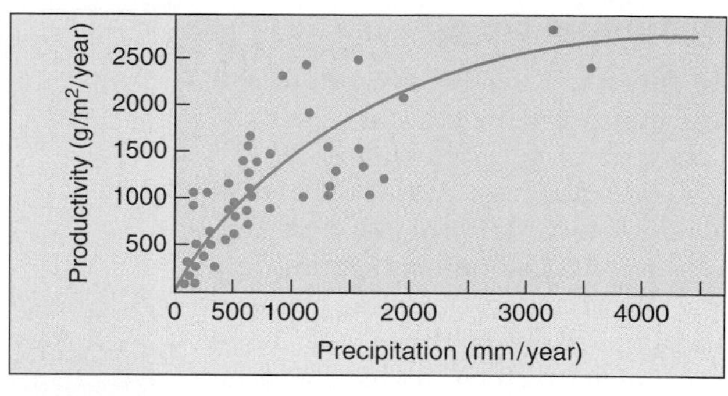

(a)

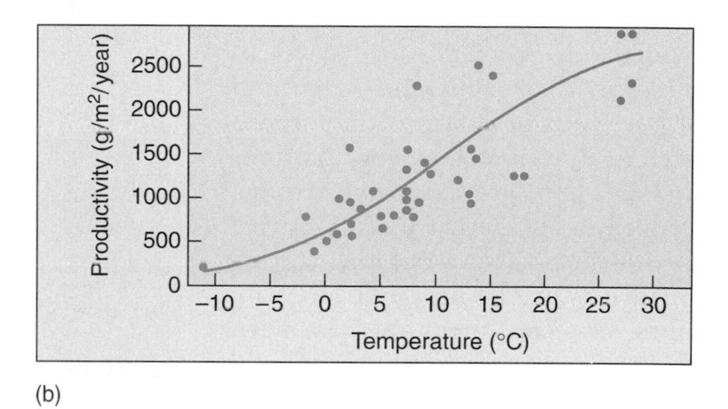

(b)

FIGURE 29.11
The effects of precipitation and temperature on primary productivity. The net primary productivity of ecosystems at 52 locations around the globe depends significantly upon (*a*) mean annual precipitation and (*b*) mean annual temperature.

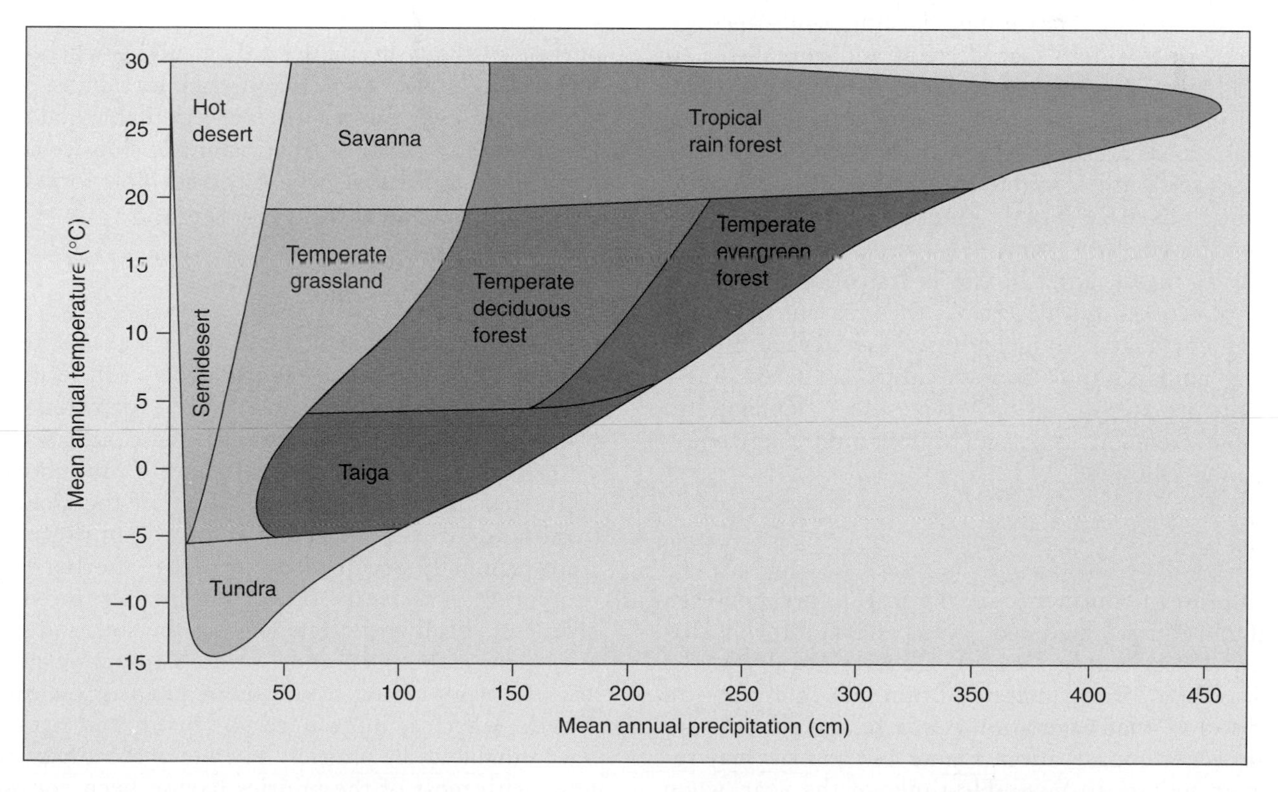

FIGURE 29.12
Temperature and precipitation are excellent predictors of biome distribution. At mean annual precipitations between 50 and 150 cm, other factors such as seasonal drought, fire, and grazing also have a major influence on biome distribution.

Tropical Rain Forests

Rain forests, which receive 140 to 450 centimeters of rain a year, are the richest ecosystems on earth (figure 29.13). They contain at least half of the earth's species of terrestrial plants and animals—more than 2 million species! In a single square mile of tropical forest in Rondonia, Brazil, there are 1200 species of butterflies—twice the total number found in the United States and Canada combined. The communities that make up tropical rain forests are diverse in that each kind of animal, plant, or microorganism is often represented in a given area by very few individuals. There are extensive tropical rain forests in South America, Africa, and Southeast Asia. But the world's rain forests are being destroyed, and countless species, many of them never seen by humans, are disappearing with them. A quarter of the world's species will disappear with the rain forests during the lifetime of many of us.

FIGURE 29.13
Tropical rain forest.

Savannas

In the dry climates that border the tropics are the world's great grasslands, called **savannas.** Savanna landscapes are open, often with widely spaced trees, and rainfall (75 to 125 centimeters annually) is seasonal. Many of the animals and plants are active only during the rainy season. The huge herds of grazing animals that inhabit the African savanna are familiar to all of us. Such animal communities lived in North America during the Pleistocene epoch but have persisted mainly in Africa. On a global scale, the savanna biome is transitional between tropical rain forest and desert. As these savannas are increasingly converted to agricultural use to feed rapidly expanding human populations in subtropical areas, their inhabitants are struggling to survive. The elephant and rhino are now endangered species; lion, giraffe, and cheetah will soon follow.

Deserts

In the interior of continents are the world's great deserts, especially in Africa (the Sahara), Asia (the Gobi) and Australia (the Great Sandy Desert). **Deserts** are dry places where less than 25 centimeters of rain falls in a year—an amount so low that vegetation is sparse and survival depends on water conservation. Plants and animals may restrict their activity to favorable times of the year, when water is present. To avoid high temperatures, most desert vertebrates live in deep, cool, and sometimes even somewhat moist burrows. Those that are active over a greater portion of the year emerge only at night, when temperatures are relatively cool. Some, such as camels, can drink large quantities of water when it is available and then survive long, dry periods. Many animals simply migrate to or through the desert, where they exploit food that may be abundant seasonally.

Temperate Grasslands

Halfway between the equator and the poles are temperate regions where rich **grasslands** grow. These grasslands once covered much of the interior of North America, and they were widespread in Eurasia and South America as well. Such grasslands are often highly productive when converted to agricultural use. Many of the rich agricultural lands in the United States and southern Canada were originally occupied by **prairies,** another name for temperate grasslands. The roots of perennial grasses characteristically penetrate far into the soil, and grassland soils tend to be deep and fertile. Temperate grasslands are often populated by herds of grazing mammals. In North America, huge herds of bison and pronghorns once inhabited the prairies. The herds are almost all gone now, with most of the prairies having been converted to the richest agricultural region on earth.

Temperate Deciduous Forests

Mild climates (warm summers and cool winters) and plentiful rains promote the growth of **deciduous** (hardwood) **forests** in Eurasia, the northeastern United States, and eastern Canada (figure 29.14). A deciduous tree is one that drops its leaves in the winter. Deer, bears, beavers, and raccoons are familiar animals of the temperate regions. Because the temperate deciduous forests represent the remnants of more extensive forests that stretched across North America and Eurasia several million years ago, the remaining areas in eastern Asia and eastern North America share animals and plants that were once more widespread. Alligators, for example, are found today only in China and in the southeastern United States. The deciduous forest in eastern Asia is rich in species because climatic conditions have historically remained constant. Many perennial herbs live in areas of temperate deciduous forest.

FIGURE 29.14
Temperate deciduous forest.

Temperate Evergreen Forests

Temperate evergreen forests occur in regions where winters are cold and there is a strong, seasonal dry period. The pine forests of the western United States and California oak woodlands are typical temperate evergreen forests. Temperate evergreen forests are characteristic of regions with nutrient-poor soils. Temperate-mixed evergreen forests represent a broad transitional zone between temperate deciduous forests to the south and taiga to the north. Many of these forests are endangered by overlogging, particularly in the western United States.

Taiga

A great ring of northern forests of coniferous trees (spruce, hemlock, and fir) extends across vast areas of Asia and North America. Coniferous trees are ones with leaves like needles that are kept all year long. This ecosystem, called **taiga,** is one of the largest on earth. Here, the winters are long and cold, and most of the limited precipitation falls in the summer. Because the taiga has too short a growing season for farming, few people live there. Many large mammals, including elk, moose, deer, and such carnivores as wolves, bears, lynx, and wolverines, live in the taiga. Traditionally, fur trapping has been extensive in this region, as has lumber production. Marshes, lakes, and ponds are common and are often fringed by willows or birches. Most of the trees occur in dense stands of one or a few species.

Tundra

In the far north, above the great coniferous forests and south of the polar ice, few trees grow. There the grassland, called **tundra,** is open, windswept, and often boggy. Enormous in extent, this ecosystem covers one-fifth of the earth's land surface. Very little rain or snow falls. When rain does fall during the brief arctic summer, it sits on frozen ground, creating a sea of boggy ground. **Permafrost,** or permanent ice, usually exists within a meter of the surface. Trees are small and are mostly confined to the margins of streams and lakes. As in taiga, herbs of the tundra are perennials that grow rapidly during the brief summers. Large grazing mammals, including musk-oxen, caribou, reindeer, and carnivores, such as wolves, foxes, and lynx, live in the tundra. Lemming populations rise and fall on a long-term cycle, with important effects on the animals that prey on them.

Major biological communities called biomes can be distinguished in different climatic regions. These communities, which occur in regions of similar climate, are much the same wherever they are found. Variation in annual mean temperature and precipitation are good predictors of what biome will occur where.

29.4 Aquatic ecosystems cover much of the earth.

Patterns of Circulation in the Oceans

Patterns of ocean circulation are determined by the patterns of atmospheric circulation, but they are modified by the locations of landmasses. Oceanic circulation is dominated by huge surface gyres (figure 29.15), which move around the subtropical zones of high pressure between approximately 30° north and 30° south latitude. These gyres move clockwise in the northern hemisphere and counterclockwise in the southern hemisphere. The ways they redistribute heat profoundly affects life not only in the oceans but also on coastal lands. For example, the Gulf Stream, in the North Atlantic, swings away from North America near Cape Hatteras, North Carolina, and reaches Europe near the southern British Isles. Because of the Gulf Stream, western Europe is much warmer and more temperate than eastern North America at similar latitudes. As a general principle, western sides of continents in temperate zones of the northern hemisphere are warmer than their eastern sides; the opposite is true of the southern hemisphere. In addition, winds passing over cold water onto warm land increase their moisture-holding capacity, limiting precipitation.

In South America, the Humboldt Current carries phosphorus-rich cold water northward up the west coast. Phosphorus is brought up from the ocean depths by the upwelling of cool water that occurs as offshore winds blow from the mountainous slopes that border the Pacific Ocean. This nutrient-rich current helps make possible the abundance of marine life that supports the fisheries of Peru and northern Chile. Marine birds, which feed on these organisms, are responsible for the commercially important, phosphorus-rich, guano deposits on the seacoasts of these countries.

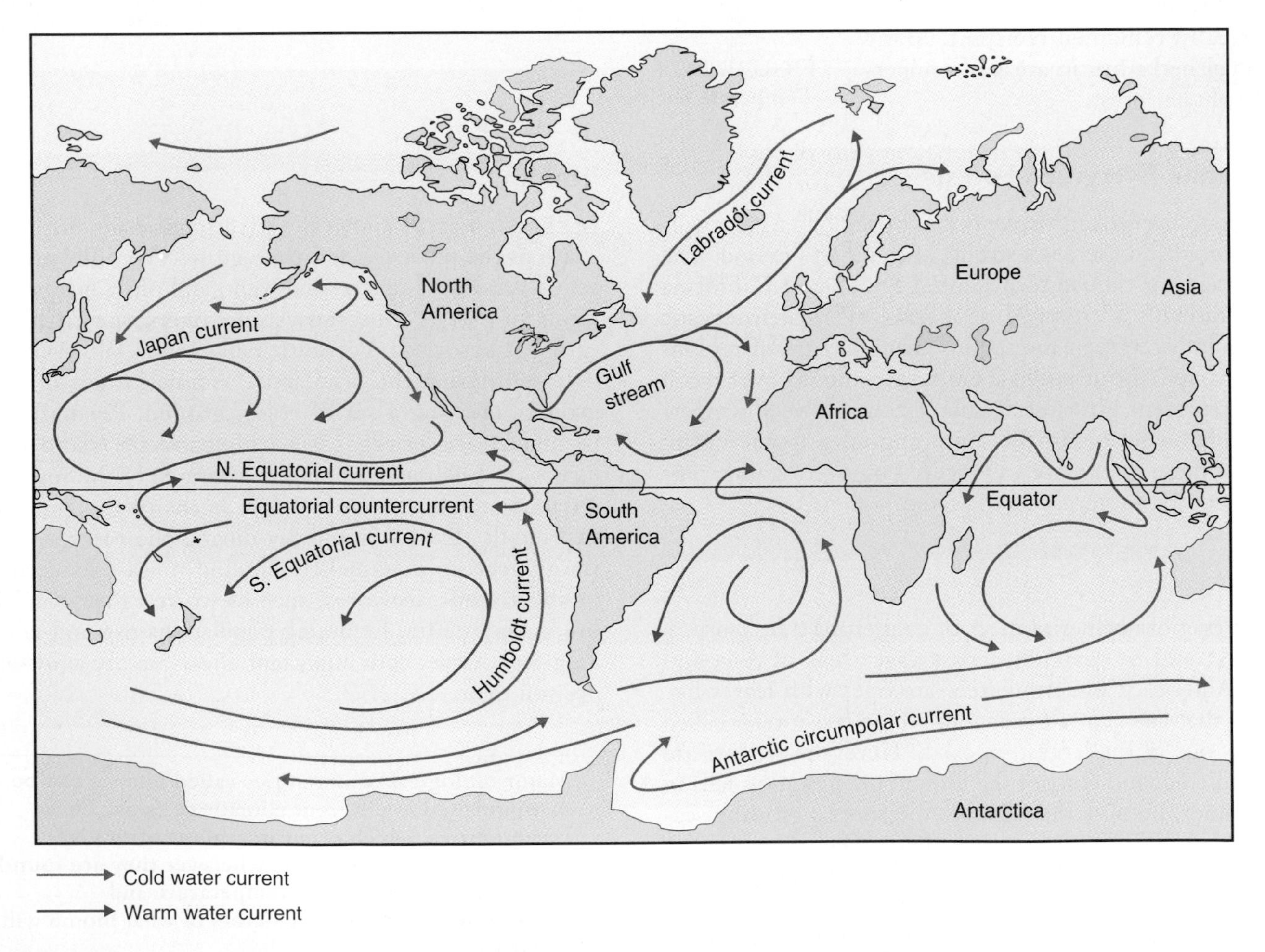

FIGURE 29.15
Ocean circulation. Water moves in the oceans in great surface spiral patterns called gyres; they profoundly affect the climate on adjacent lands.

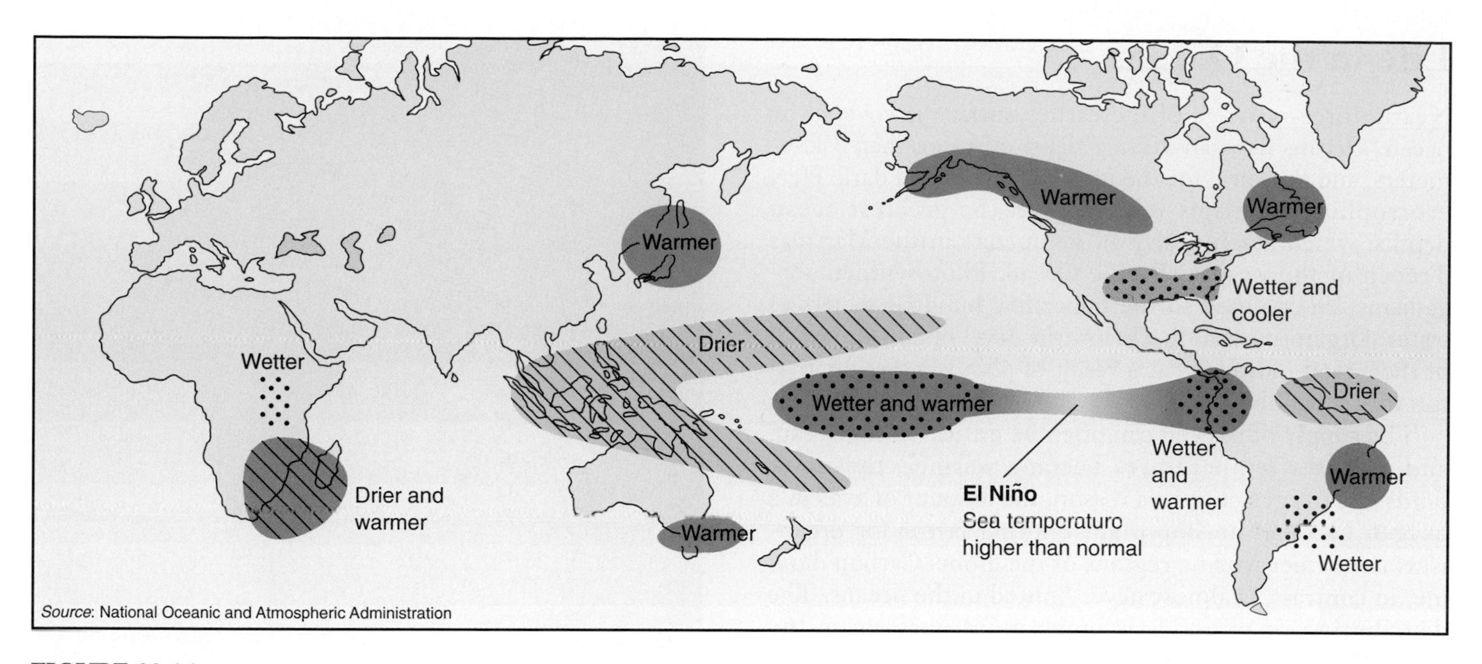

FIGURE 29.16
An El Niño winter. El Niño currents produce unusual weather patterns all over the world as warm waters from the western Pacific move eastward.

El Niño and Ocean Ecology

Every Christmas a tepid current sweeps down the coast of Peru and Ecuador from the tropics, reducing the fish population slightly and giving local fishermen some time off. The local fishermen named this Christmas current *El Niño* (literally, "the child," after "the Christ Child"). Now, though, the term is reserved for a catastrophic version of the same phenomenon, one that occurs every two to seven years and is felt not only locally but on a global scale.

Scientists now have a pretty good idea of what goes on in an El Niño. Normally the Pacific Ocean is fanned by constantly blowing east-to-west trade winds that push warm surface water away from the ocean's eastern side (Peru, Ecuador, and Chile) and allow cold water to well up from the depths in its place, carrying nutrients that feed plankton and hence fish. This surface water piles up in the west, around Australia and the Philippines, making it several degrees warmer and a meter or so higher than the eastern side of the ocean. But if the winds slacken briefly, warm water begins to slosh back across the ocean.

Once this happens, ocean and atmosphere conspire to ensure it keeps happening. The warmer the eastern ocean gets, the warmer and lighter the air above it becomes, and hence more similar to the air on the western side. This reduces the difference in pressure across the ocean. Because a pressure difference is what makes winds blow, the easterly trades weaken further, letting the warm water continue its eastward advance.

The end result is to shift the weather systems of the western Pacific Ocean 6000 km eastward. The tropical rainstorms that usually drench Indonesia and the Philippines occur when warm seawater abutting these islands causes the air above it to rise, cool, and condense its moisture into clouds. When the warm water moves east, so do the clouds, leaving the previously rainy areas in drought. Conversely, the western edge of South America, its coastal waters usually too cold to trigger much rain, gets a soaking, while the upwelling slows down. During an El Niño, commercial fish stocks virtually disappear from the waters of Peru and northern Chile, and plankton drop to a twentieth of their normal abundance.

That is just the beginning. El Niño's effects are propagated across the world's weather systems (figure 29.16). Violent winter storms lash the coast of California, accompanied by flooding, and El Niño produces colder and wetter winters than normal in Florida and along the Gulf Coast. The American midwest experiences heavier-than-normal rains, as do Israel and its neighbors.

Though the effects of El Niños are now fairly clear, what triggers them still remains a mystery. Models of these weather disturbances suggest that the climatic change that triggers El Niño is "chaotic." Wind and ocean currents return again and again to the same condition, but never in a regular pattern, and small nudges can send them off in many different directions—including an El Niño.

The world's oceans circulate in huge gyres deflected by continental landmasses. Circulation of ocean water redistributes heat, warming the western side of continents. Disturbances in ocean currents like El Niño can have profound influences on world climate.

Life in the Oceans

Nearly three-quarters of the earth's surface is covered by ocean. Oceans have an *average* depth of more than 3 kilometers, and they are, for the most part, cold and dark. Heterotrophic organisms inhabit even the greatest ocean depths, which reach nearly 11 kilometers in the Marianas Trench of the western Pacific Ocean. Photosynthetic organisms are confined to the upper few hundred meters of water. Organisms that live below this level obtain almost all of their food indirectly, as a result of photosynthetic activities that occur above.

The supply of oxygen can often be critical in the ocean, and as water temperatures become warmer, the water holds less oxygen. For this reason, the amount of available oxygen becomes an important limiting factor for organisms in warmer marine regions of the globe. Carbon dioxide, in contrast, is almost never limited in the oceans. The distribution of minerals is much more uniform in the ocean than it is on land, where individual soils reflect the composition of the parent rocks from which they have weathered.

Frigid and bare, the floors of the deep sea have long been considered a biological desert. Recent close-up looks taken by marine biologists, however, paint a different picture (figure 29.17). The ocean floor is teeming with life. Often miles deep, thriving in pitch darkness under enormous pressure, crowds of marine invertebrates have been found in hundreds of deep samples from the Atlantic and Pacific. Rough estimates of deep-sea diversity have soared to millions of species. Many appear endemic (local). The diversity of species is so high it may rival that of tropical rain forests! This profusion is unexpected. New species usually require some kind of barrier in order to diverge (see chapter 22), and the ocean floor seems boringly uniform. However, little migration occurs among deep populations, thus allowing populations to diverge and encouraging local specialization and species formation. A patchy environment may also contribute to species formation there; deep-sea ecologists find evidence that fine but nonetheless formidable resource barriers arise in the deep sea.

FIGURE 29.17
Food comes to the ocean floor from above. Looking for all the world like some undersea sunflower, the two sea anemones (actually animals) use a glass-sponge stalk to catch "marine snow," food particles raining down on the ocean floor from the ocean surface miles above.

Another conjecture is that the extra billion years or so that life has been evolving in the sea compared with land may be a factor in the unexpected biological richness of its deep recesses.

Despite the many new forms of small invertebrates now being discovered on the seafloor, and the huge biomass that occurs in the sea, more than 90% of all *described* species of organisms occur on land. Each of the largest groups of organisms, including insects, mites, nematodes, fungi, and plants has marine representatives, but they constitute only a very small fraction of the total number of described species. There are two reasons for this. First, barriers between habitats are sharper on land, and variations in elevation, parent rock, degree of exposure, and other factors have been crucial to the evolution of the millions of species of terrestrial organisms. Second, there are simply few taxonomists actively classifying the profusion of ocean floor life being brought to the surface.

In terms of higher level diversity, the pattern is quite different. Of the major groups of organisms—phyla—most originated in the sea, and almost every one has representatives in the sea. Only a few phyla have been successful on land or in freshwater habitats, but these have given rise to an extraordinarily large number of described species.

Although representatives of almost every phylum occur in the sea, an estimated 90% of living species of organisms are terrestrial. This is because of the enormous evolutionary success of a few phyla on land, where the boundaries between different habitats are sharper than they are in the sea.

Marine Ecosystems

The marine environment consists of three major habitats: (1) the **neritic zone,** the zone of shallow waters along the coasts of continents; (2) the **pelagic zone,** the area of water above the ocean floor; and (3) the **benthic zone,** the actual ocean floor (figure 29.18). The part of the ocean floor that drops to depths where light does not penetrate is called the **abyssal zone.**

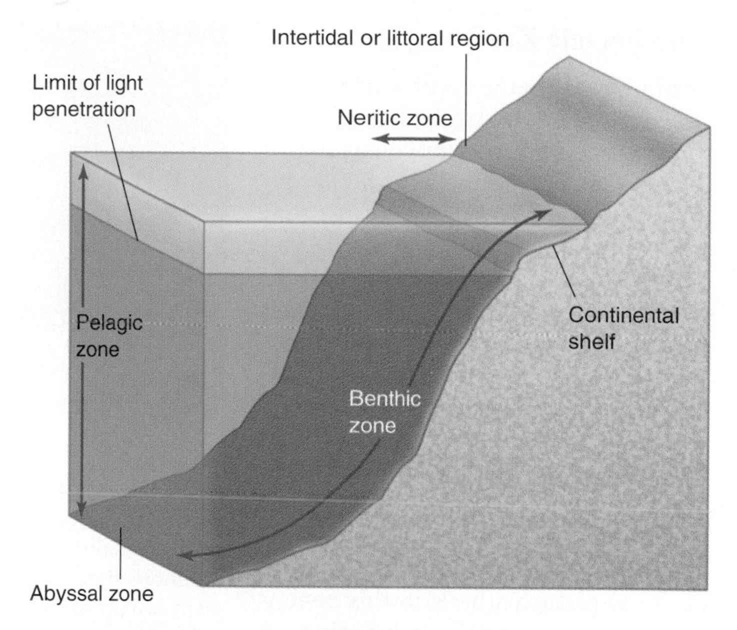

FIGURE 29.18
Marine ecosystems. Ecologists classify marine communities into neritic, pelagic, benthic, and abyssal zones, according to depth (which affects how much light penetrates) and distance from shore.

The Neritic Zone

The **neritic zone** of the ocean is the area less than 300 meters below the surface along the coasts of continents and islands. The zone is small in area, but it is inhabited by large numbers of species (figure 29.19). The intense and sometimes violent interaction between sea and land in this zone gives a selective advantage to well-secured organisms that can withstand being washed away by the continual beating of the waves. Part of this zone, the **intertidal region,** sometimes called the **littoral region,** is exposed to the air whenever the tides recede.

The world's great fisheries are in shallow waters over continental shelves, either near the continents themselves or in the open ocean, where huge banks come near the surface. Nutrients, derived from land, are much more abundant in coastal and other shallow regions, where upwelling from the depths occurs, than in the open ocean. This accounts for the great productivity of the continental shelf fisheries. The preservation of these fisheries, a source of high-quality protein exploited throughout the world, has become a growing concern. In Chesapeake Bay, where complex systems of rivers enter the ocean from heavily populated areas, environmental stresses have become so severe that they not only threaten the continued existence of formerly highly productive fisheries, but also diminish the quality of human life in these regions. Increased runoff from farms and sewage effluent in areas like Chesapeake Bay add large amounts of nutrients to the water. This increased nutrient supply allows an increase in the numbers of some marine organisms. The increased populations then use up more and more of the oxygen in the water and thus may disturb established populations of organisms such as oysters. Climatic shifts may magnify these effects, and large numbers of marine animals die suddenly as a result.

About three-fourths of the surface area of the world's oceans are located in the tropics. In these waters, where the water temperature remains about 21°C, coral reefs can grow. These highly productive ecosystems can successfully concentrate nutrients, even from the relatively nutrient-poor waters characteristic of the tropics.

FIGURE 29.19
Diversity is great in coastal regions. Fishes and many other kinds of animals find food and shelter among the kelp beds in the coastal waters of temperate regions.

The Pelagic Zone

Drifting freely in the upper waters of the pelagic zone, a diverse biological community exists, primarily consisting of microscopic organisms called **plankton.** Fish and other larger organisms that swim in these waters constitute the *nekton,* whose members feed on plankton and one another. Some members of the plankton, including protists and some bacteria, are photosynthetic. Collectively, these organisms account for about 40% of all photosynthesis that takes place on earth. Most plankton live in the top 100 meters of the sea, the zone into which light from the surface penetrates freely. Perhaps half of the total photosynthesis in this zone is carried out by organisms less than 10 micrometers in diameter—at the lower limits of size for organisms—including cyanobacteria and algae, organisms so small that their abundance and ecological importance have been unappreciated until relatively recently.

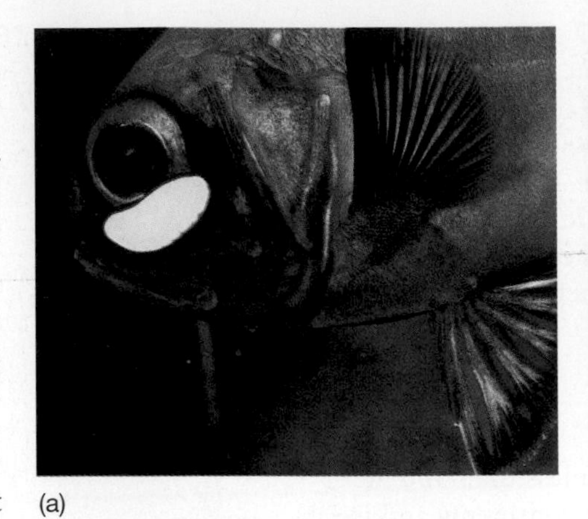
(a)

(b)

FIGURE 29.20
Life in the abyssal and benthic zones. (*a*) The luminous spot below the eye of this deep-sea fish results from the presence of a symbiotic colony of luminous bacteria. Similar luminous signals are a common feature of deep-sea animals that move about. (*b*) These giant beardworms live along vents where water jets from fissures at 350°C and then cools to the 2°C of the surrounding water.

Many heterotrophic protists and animals live in the plankton and feed directly on photosynthetic organisms and on one another. Gelatinous animals, especially jellyfish and ctenophores, are abundant in the plankton. The largest animals that have ever existed on earth, baleen whales, graze on plankton and nekton as do a number of other organisms, such as fishes and crustaceans.

Populations of organisms that make up plankton can increase rapidly, and the turnover of nutrients in the sea is great, although the productivity in these systems is quite low. Because nitrogen and phosphorus are often present in only small amounts and organisms may be relatively scarce, this productivity reflects rapid use and recycling rather than an abundance of these nutrients.

The Benthic Zone

The seafloor at depths below 1000 meters, the **abyssal zone,** has about twice the area of all the land on earth. The seafloor itself, sometimes called the **benthic zone,** is a thick blanket of mud, consisting of fine particles that have settled from the overlying water and accumulated over millions of years. Because of high pressures (an additional atmosphere of pressure for every 10 meters of depth), cold temperatures (2° to 3°C), darkness, and lack of food, biologists thought that nothing could live on the seafloor. In fact, recent work has shown that the number of species that live at great depth is quite high. Most of these animals are only a few millimeters in size, although larger ones also occur in these regions. Some of the larger ones are bioluminescent (figure 29.20*a*) and thus are able to communicate with one another or attract their prey.

Animals on the sea bottom depend on the meager leftovers from organisms living kilometers overhead. The low densities and small size of most deep-sea bottom animals is in part a consequence of this limited food supply. In 1977, oceanographers diving in a research submarine were surprised to find dense clusters of large animals living on geothermal energy at a depth of 2500 meters. These deep-sea oases occur where seawater circulates through porous rock at sites where molten material from beneath the earth's crust comes close to the rocky surface. A series of these areas occur on the Mid-Ocean Ridge, where basalt erupts through the ocean floor.

This water is heated to temperatures in excess of 350°C and, in the process, becomes rich in reduced compounds. These compounds, such as hydrogen sulfide, provide energy for bacterial primary production through chemosynthesis instead of photosynthesis. Mussels, clams, and large red-plumed worms in a phylum unrelated to any shallow-water invertebrates cluster around the vents (figure 29.20*b*). Bacteria live symbiotically within the tissues of these animals. The animal supplies a place for the bacteria to live and transfers CO_2, H_2S, and O_2 to them for their growth; the bacteria supply the animal with organic compounds to use as food. Polychaete worms (see chapter 46), anemones, and limpets live on free-living chemosynthetic bacteria. Crabs act as scavengers and predators, and some of the fish are also predators. This is one of the few ecosystems on earth that does not depend on the sun's energy.

About 40% of the world's photosynthetic productivity is estimated to occur in the oceans. The turnover of nutrients in the plankton is much more rapid than in most other ecosystems, and the total amounts of nutrients are very low.

Freshwater Habitats

Freshwater habitats are distinct from both marine and terrestrial ones, but they are limited in area. Inland lakes cover about 1.8% of the earth's surface, and running water (streams and rivers) covers about 0.3%. All freshwater habitats are strongly connected with terrestrial ones, with marshes and swamps constituting intermediate habitats. In addition, a large amount of organic and inorganic material continuously enters bodies of fresh water from communities growing on the land nearby (figure 29.21*a*). Many kinds of organisms are restricted to freshwater habitats (figure 29.21*b*,*c*). When organisms live in rivers and streams, they must be able to swim against the current or attach themselves in such a way as to resist the effects of current, or risk being swept away.

(a)

(b)

(c)

FIGURE 29.21
A nutrient-rich stream. (*a*) Much organic material falls or seeps into streams from communities along the edges. This input increases the stream's biological productivity. (*b*) This speckled darter and (*c*) this giant waterbug with eggs on its back can only live in freshwater habitats.

Ponds and Lakes

Small bodies of fresh water are called ponds, and larger ones lakes. Because water absorbs light passing through it at wavelengths critical to photosynthesis (every meter absorbs 40% of the red and about 2% of the blue), the distribution of photosynthetic organisms is limited to the upper **photic zone;** heterotrophic organisms occur in the lower **aphotic zone** where very little light penetrates.

Ponds and lakes, like the ocean, have three zones where organisms occur, distributed according to the depth of the water and its distance from shore (figure 29.22). The *littoral zone* is the shallow area along the shore. The *limnetic zone* is the well-illuminated surface water away from the shore, inhabited by plankton and other organisms that live in open water. The *profundal zone* is the area below the limits where light can effectively penetrate.

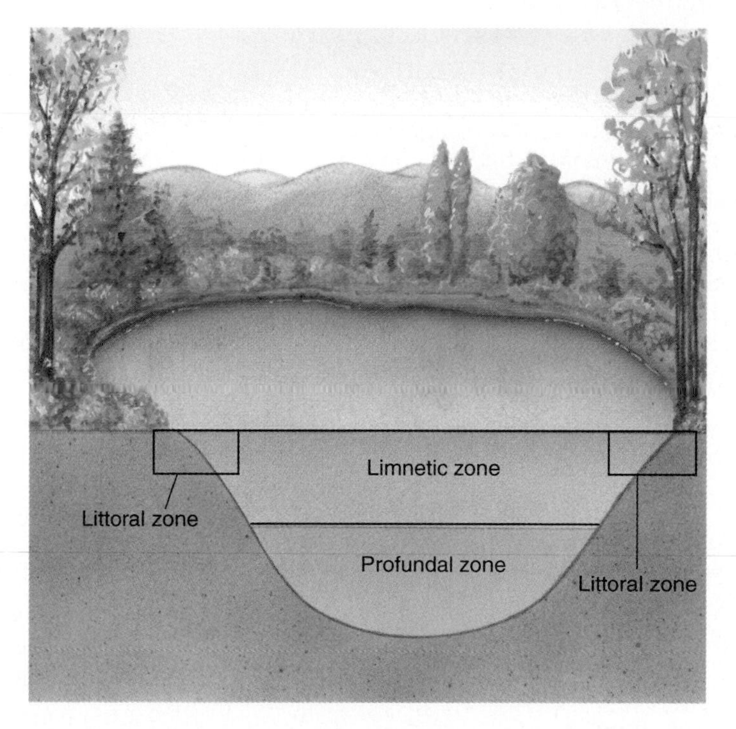

FIGURE 29.22
The three zones in ponds and lakes. A shallow "edge" (littoral) zone lines the periphery of the lake, where attached algae and their insect herbivores live. An open-water surface (limnetic) zone lies across the entire lake and is inhabited by floating algae, zooplankton, and fish. A dark, deep-water (profundal) zone overlies the sediments at the bottom of the lake and contains numerous bacteria and wormlike organisms that consume dead debris settling at the bottom of the lake.

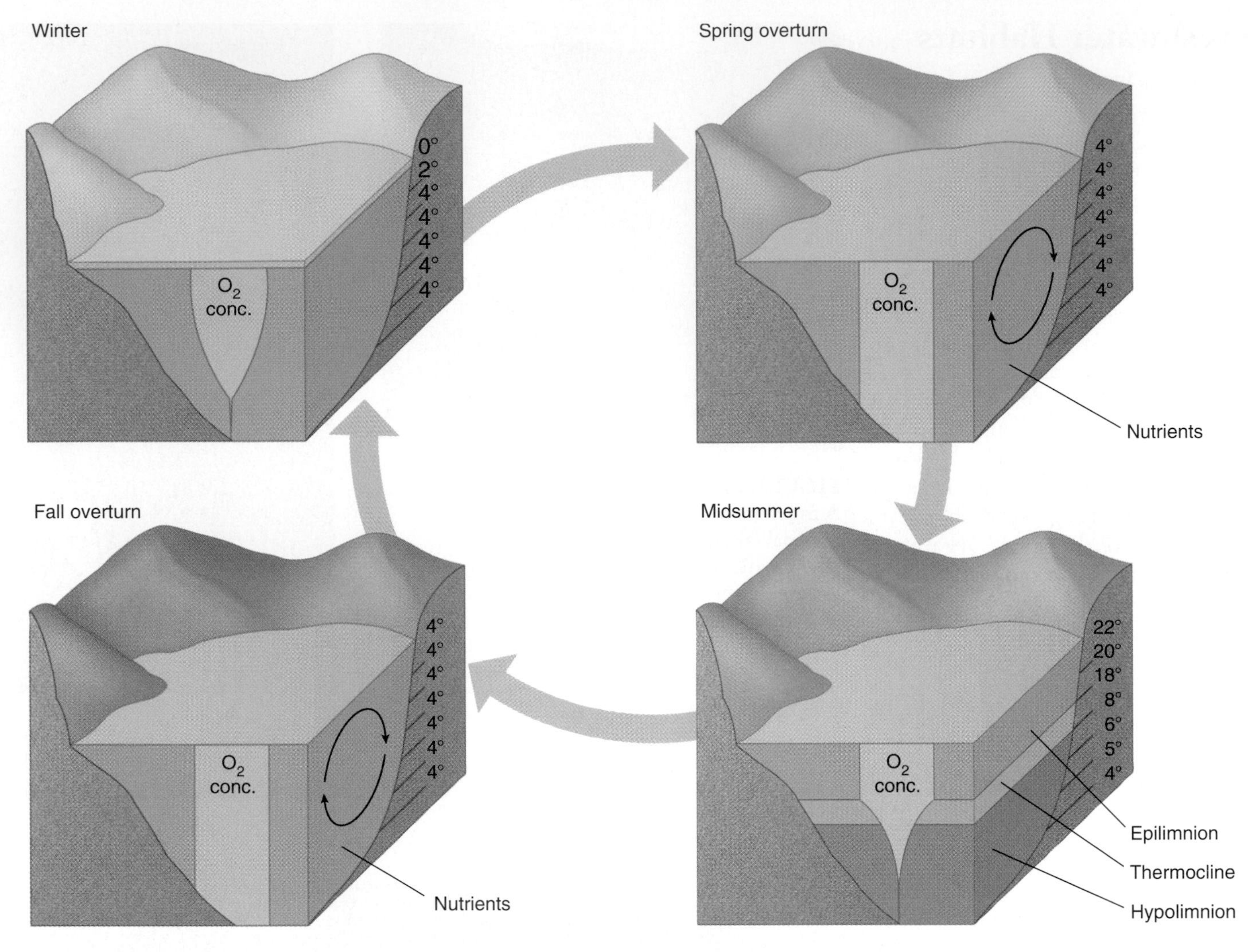

FIGURE 29.23
Stratification in fresh water. The pattern of stratification in a large pond or lake in temperate regions is upset in the spring and fall overturns. Of the three layers of water shown, the hypolimnion consists of the densest water, at 4°C; the epilimnion consists of warmer water that is less dense; and the thermocline is the zone of abrupt change in temperature that lies between them. If you have dived into a pond in temperate regions in the summer, you have experienced the existence of these layers.

Thermal Stratification

Thermal stratification is characteristic of larger lakes in temperate regions (figure 29.23). In summer, warmer water forms a layer at the surface known as the *epilimnion*. Cooler water, called the *hypolimnion* (about 4°C), lies below. An abrupt change in temperature, the thermocline, separates these two layers. Depending on the climate of the particular area, the epilimnion may become as much as 20 meters thick during the summer.

In autumn the temperature of the epilimnion drops until it is the same as that of the hypolimnion, 4°C. When this occurs, epilimnion and hypolimnion mix—a process called fall overturn. Because water is densest at about 4°C, further cooling of the water as winter progresses creates a layer of cooler, lighter water, which freezes to form a layer of ice at the surface. Below the ice, the water temperature remains between 0° and 4°C, and plants and animals can survive. In spring, the ice melts, and the surface water warms up. When it warms back to 4°C, it again mixes with the water below. This process is known as spring overturn. When lake waters mix in the spring and fall, nutrients formerly held in the depths of the lake are returned to the surface, and oxygen from surface waters is carried to the depths.

Freshwater habitats include several distinct life zones. These zones shift seasonally in temperate lakes and ponds. In the spring and fall, when their temperatures are equal, shallower and deeper waters of the lake mix, with oxygen being carried to the depths and nutrients being brought to the surface.

Productivity of Freshwater Ecosystems

Some aquatic communities, such as fast-moving streams, are not highly productive. Because the moving water washes away plankton, the photosynthesis that supports the community is limited to algae attached to the surface and to rooted plants.

The Productivity of Lakes

Lakes can be divided into two categories based on their production of organic matter. **Eutrophic lakes** contain an abundant supply of minerals and organic matter. As the plentiful organic material drifts below the thermocline from the well-illuminated surface waters of the lake, it provides a source of energy for other organisms. Most of these are oxygen-requiring organisms that can easily deplete the oxygen supply below the thermocline during the summer months. The oxygen supply of the deeper waters cannot be replenished until the layers mix in the fall. This lack of oxygen in the deeper waters of some lakes may have profound effects, such as allowing relatively harmless materials such as sulfates and nitrates to convert into toxic materials such as hydrogen sulfide and ammonia.

In **oligotrophic lakes,** organic matter and nutrients are relatively scarce. Such lakes are often deeper than eutrophic lakes and have very clear blue water. Their hypolimnetic water is always rich in oxygen.

Human activities can transform oligotrophic lakes into eutrophic ones. In many lakes, phosphorus is in short supply and is the nutrient that limits growth. When excess phosphorus from sources such as fertilizer runoff, sewage, and detergents enters lakes, it can quickly lead to harmful effects. In many cases, this leads to perfect conditions for the growth of blue-green algae, which proliferate immensely. Soon, larger plants are outcompeted and disappear, along with the animals that live on them. In addition, as these phytoplankton die and decompose, oxygen in the water is used up, killing the natural fish and invertebrate populations. This situation can be remedied if the continual input of phosphorus can be diminished. Given time, lakes can recover and return to prepollution states, as happened with Lake Washington pictured in figure 29.24.

The Productivity of Wetlands

Swamps, marshes, bogs, and other **wetlands** covered with water support a wide variety of water-tolerant plants, called hydrophytes ("water plants"), and a rich diversity of invertebrates, birds, and other animals. Wetlands are among the most productive ecosystems on earth (table 29.2). They also play a key ecological role by providing water storage basins that moderate flooding. Many wetlands are being disrupted by human "development" of what is sometimes perceived as otherwise useless land, but government efforts are now underway to protect the remaining wetlands.

Table 29.2 The Most Productive Ecosystems

Ecosystem	Net Primary Productivity per Unit Area (g/m^2)
Coral reefs	2500
Tropical rain forest	2200
Wetlands	2000
Tropical seasonal forest	1600
Estuaries	1500
Temperate evergreen forest	1300
Temperate deciduous forest	1200
Savanna	900
Boreal forest	800
Cultivated land	650
Continental shelf	360
Lake and stream	250
Open ocean	125
Extreme desert, rock, sand, and ice	3

Source: Whitaker, 1975.

FIGURE 29.24
Oligotrophic lakes are highly susceptible to pollution. Lake Washington is an oligotrophic lake near Seattle, Washington. The drainage from fertilizers applied to the plantings around residences, business concerns, and recreational facilities bordering the lake poses an ever-present threat to its deep blue water. By supplying phosphorus, the drainage promotes algal growth. Aerobic bacteria decomposing dead algae can deplete the lake's oxygen, killing much of the lake's life.

The most productive freshwater ecosystems are wetlands. Most lakes are far less productive, limited by lack of nutrients.

Chapter 29

www.mhhe.com/raven6/resources29.mhtml

Summary | **Questions** | **Media Resources**

29.1 Organisms must cope with a varied environment.

- Organisms employ physiological, morphological, and behavioral mechanisms to cope with variations in the environment.

1. What are several ways that individual organisms adjust to changes in temperature during the course of a year?

- Biosphere Introduction

29.2 Climate shapes the character of ecosystems.

- Warm air rises near the equator and flows toward the poles, descending at about 30° north and south latitude. Because the air falls in these regions, it is warmed, and its moisture-holding capacity increases. The great deserts of the world are formed in these drier latitudes.

2. Why are the majority of great deserts located near 30° north and south latitude? Is it more likely that a desert will form in the interior or at the edge of a continent? Explain why.

- Global Air Circulation
- Rainshadow Effect

- Art Quizzes:
  -Rainshadow Effect
  -Latitude, Elevation, and Biome Distribution

29.3 Biomes are widespread terrestrial ecosystems.

- The world's major biomes, or terrestrial communities, can be grouped into eight major categories. These are (1) tropical rain forest; (2) savanna; (3) desert; (4) temperate grassland; (5) temperate deciduous forest; (6) temperate evergreen forest; (7) taiga; and (8) tundra.

3. What is a biome? What are the two key physical factors that affect the distribution of biomes across the earth?

- Biomes/Climate
- Soils
- Land Biomes
- Tropical Forests
- Temperate Forests

- Art Quiz: Classification of Biomes

29.4 Aquatic ecosystems cover much of the earth.

- The ocean contains three major environments: the neritic zone, the pelagic zone, and the benthic zone.
- The neritic zone, which lies along the coasts, is small in area but very productive and rich in species.
- The surface layers of the pelagic zone are home to plankton (drifting organisms) and nekton (actively swimming ones). The productivity of this zone has been underestimated because of the very small size (less than 10 mm) of many of its key organisms and because of its rapid turnover of nutrients.
- The benthic zone is home to a surprising number of species.
- Freshwater habitats constitute only about 2.1% of the earth's surface; most are ponds and lakes. These possess a littoral zone, a limnetic zone, and a profundal zone. The waters in these zones mix seasonally, delivering oxygen to the bottom and nutrients to the surface.

4. What is the difference between plankton and nekton in the ocean's pelagic zone? How important are the photosynthetic plankton to the survival of the earth? Is the turnover of nutrients in the surface zone slow or fast?

5. What conditions of the abyssal zone led early deep-sea biologists to believe nothing lived there? What provides the energy for the deep-sea communities found around thermal vents? What kind of organisms live there?

6. Does much diversity occur in the abyssal zone? How are such ecosystems supported in the absence of light?

7. What is the difference between a eutrophic and an oligotrophic lake? Why have humans increased the frequency of lakes becoming eutrophied?

- El Niño Southern Oscillation

- Aquatic Systems

- Student Research: Exotic Species and Freshwater Ecology
- *On Science* Article: Our Winters May Get Colder

- Art Quizzes:
  -Marine Ecosystems
  -Thermal Stratification

30

The Future of the Biosphere

Concept Outline

30.1 The world's human population is growing explosively.

A Growing Population. The world's population of 6 billion people is growing rapidly and at current rates will double in 39 years.

30.2 Improvements in agriculture are needed to feed a hungry world.

The Future of Agriculture. Much of the effort in searching for new sources of food has focused on improving the productivity of existing crops.

30.3 Human activity is placing the environment under increasing stress.

Nuclear Power. Nuclear power, a plentiful source of energy, is neither cheap nor safe.

Carbon Dioxide and Global Warming. The world's industrialization has led to a marked increase in the atmosphere's level of CO_2, with resulting warming of climates.

Pollution. Human industrial and agricultural activity introduces significant levels of many harmful chemicals into ecosystems.

Acid Precipitation. Burning of cheap high-sulfur coal has introduced sulfur to the upper atmosphere, where it combines with water to form sulfuric acid that falls back to earth, harming ecosystems.

The Ozone Hole. Industrial chemicals called CFCs are destroying the atmosphere's ozone layer, removing an essential shield from the sun's UV radiation.

Destruction of the Tropical Forests. Much of the world's tropical forest is being destroyed by human activity.

30.4 Solving environmental problems requires individual involvement.

Environmental Science. The commitment of one person often makes a key difference in solving environmental problems.

Preserving Nonreplaceable Resources. Three key nonreplaceable resources are topsoil, groundwater, and biodiversity.

FIGURE 30.1
New York City by satellite.

The view of New York City in figure 30.1 was photographed from a satellite in the spring of 1985. At the moment this picture was taken, millions of people within its view were talking, hundreds of thousands of cars were struggling through traffic, hearts were being broken, babies born, and dead people buried. Our futures and those of everyone on the planet are linked to the unseen millions in this photograph, for we share the earth with them. A lot of people consume a lot of food and water, use a great deal of energy and raw materials, and produce a great deal of waste. They also have the potential to solve the problems that arise in an increasingly crowded world. In this chapter, we will study how human life affects the environment, and the efforts being mounted to lessen the adverse impact and increase the potential benefits of our burgeoning population.

30.1 The world's human population is growing explosively.

A Growing Population

The current world population of 6 billion people is placing severe strains on the biosphere. How did it grow so large? For the past 300 years, the human birthrate (as a global average) has remained nearly constant, at about 30 births per year per 1000 people. Today it is about 25 births per year per 1000 people. However, at the same time, better sanitation and improved medical techniques have caused the death rate to fall steadily, from about 29 deaths per 1000 people per year to 13 per 1000 per year. Thus, while the birthrate has remained fairly constant and may have even decreased slightly, the tremendous fall in the death rate has produced today's enormous population. The difference between the birth and death rates amounts to an annual worldwide increase of approximately 1.4%. This rate of increase may seem relatively small, but it would double the world's population in only 39 years!

The *annual* increase in world population today is nearly 77 million people, about equal to the current population of Germany. Two hundred ten thousand people are added to the world each day, or more than 140 every minute! The world population is expected to continue beyond its current level of 6 billion people, perhaps stabilizing at a figure anywhere between 8.5 billion and 20 billion during this century.

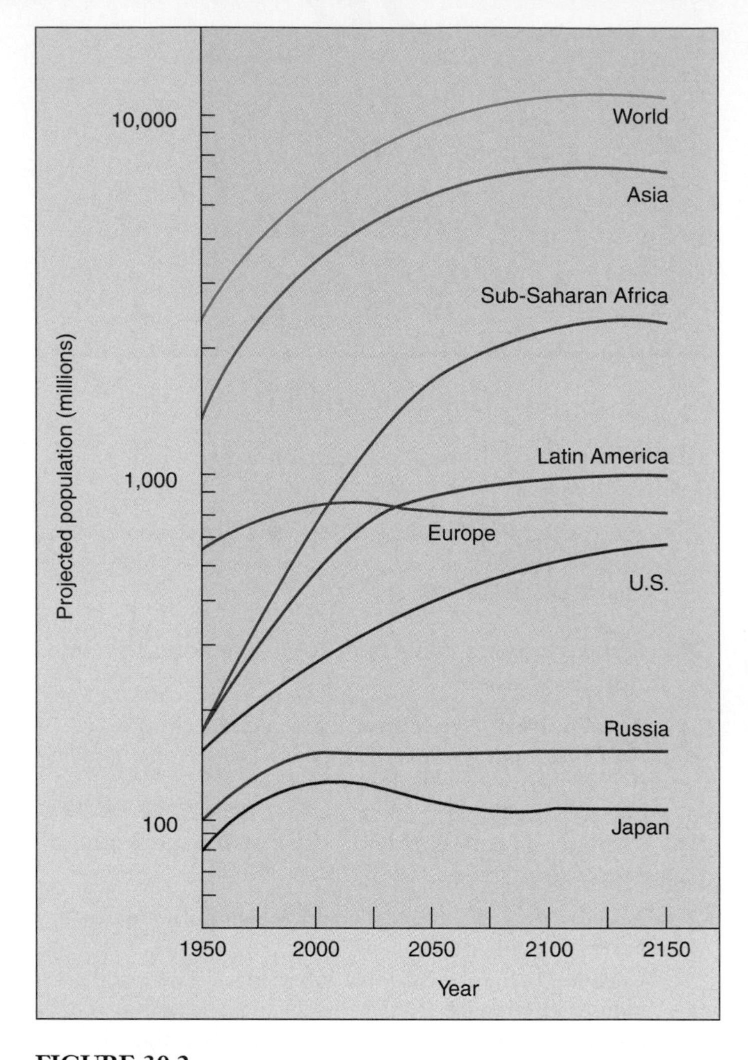

FIGURE 30.2
Anticipated growth of the global human population. Despite considerable progress in lowering birthrates, the human population will continue to grow this century (data are presented above on a log scale). Much of the growth will center in sub-Saharan Africa, the poorest region on the globe, where the population could reach over 2 billion. Fertility rates there currently range from 3 to more than 5 children per woman, compared to fewer than 2.1 in Europe and the United States.
Source: U.S. data based on Census Bureau projections through 2050. All other data from Edward Bros. et al., World Population Projects in 1994-1995 Edition. John Hopkins University Press for the World Bank, 1994.

The Future Situation

About 60% of the people in the world live in tropical or subtropical regions (figure 30.2). An additional 20% are living in China, and the remaining 20% in the developed or industrialized countries: Europe, the successor states of the Soviet Union, Japan, United States, Canada, Australia, and New Zealand. Although populations of industrialized countries are growing at an annual rate of about 0.3%, those of the developing, mostly tropical countries (excluding China) are growing at an annual rate estimated in 1995 to be about 2.2%. For every person living in an industrialized country like the United States in 1950, there were two people living elsewhere; in 2020, just 70 years later, there will be five.

As you learned in chapter 24, the age structure of a population determines how fast the population will grow. To predict the future growth patterns of a population, it is essential to know what proportion of its individuals have not yet reached childbearing age. In industrialized countries such as the United States, about a fifth of the population is under 15 years of age; in developing countries such as Mexico, the proportion is typically about twice as high. Even if most tropical and subtropical countries consistently carry out the policies they have established to limit population growth, their populations will continue to grow well into the twenty-first century (figure 30.3), and industrialized countries will constitute a smaller and smaller proportion of the world's population. If India, with a 1995 population level of about 930 million people (36% under 15 years old), managed to reach a simple replacement reproductive rate by the year 2000, its population would still not stop growing until the middle of this century. At present rates of growth, India will have a population of nearly 1.4 billion people by 2025 and will still be growing rapidly.

Population Growth Rate Starting to Decline

The United Nations has announced that the world population growth rate continues to decline, down from a high of 2.0% in the period 1965–1970 to 1.4% in 1998. Nonetheless, because of the larger population, this amounts to an increase of 77 million people per year to the world population, compared to 53 million per year in the 1960s.

The U.N. attributes the decline to increased family planning efforts and the increased economic power and social status of women. While the U.N. applauds the United States for leading the world in funding family planning programs abroad, some oppose spending money on international family planning. The opposition states that money is better spent on improving education and the economy in other countries, leading to an increased awareness and lowered fertility rates. The U.N. certainly supports the improvement of education programs in developing countries, but, interestingly, it has reported increased education levels *following* a decrease in family size as a result of family planning.

Most countries are devoting considerable attention to slowing the growth rate of their populations, and there are genuine signs of progress. If these efforts are maintained, the world population may stabilize sometime in this century. No one knows how many people the planet can support, but we clearly already have more people than can be sustainably supported with current technologies.

However, population size is not the only factor that determines resource use; per capita consumption is also important. In this respect, we in the developing world need to pay more attention to lessening the impact each of us makes, because, even though the vast majority of the world's population is in developing countries, the vast majority of resource consumption occurs in the developed world. Indeed, the wealthiest 20% of the world's population accounts for 86% of the world's consumption of resources and produces 53% of the world's carbon dioxide emissions, whereas the poorest 20% of the world is responsible for only 1.3% of consumption and 3% of CO_2 emissions. Looked at another way, in terms of resource use, a child born today in the developed world will consume as many resources over the course of his or her life as 30 to 50 children born in the developing world.

Building a sustainable world is the most important task facing humanity's future. The quality of life available to our children will depend to a large extent on our success both in limiting population growth and the amount of per capita resource consumption.

In 1998, the global human population of 6 billion people was growing at a rate of approximately 1.4% annually. At that rate, the population would double in 39 years. Growth rates, however, are declining, but consumption per capita in the developed world is also a significant drain on resources.

FIGURE 30.3
Population growth is highest in tropical and subtropical countries. Mexico City, the world's largest city, has well over 20 million inhabitants.

30.2 Improvements in agriculture are needed to feed a hungry world.

The Future of Agriculture

One of the greatest and most immediate challenges facing today's world is producing enough food to feed our expanding population. This problem is often not appreciated by economists, who estimate that world food production has expanded 2.6 times since 1950, more rapidly than the human population. However, virtually all land that can be cultivated is already in use, and much of the world is populated by large numbers of hungry people who are rapidly destroying the sustainable productivity of the lands they inhabit. Well over 20% of the world's topsoil has been lost from agricultural lands since 1950. In the face of these massive problems, we need to consider what the prospects are for increased agricultural productivity in the future.

Finding New Food Plants

How many food plants do we use at present? Just three species—rice, wheat, and corn—supply more than half of all human energy requirements. Just over 100 kinds of plants supply over 90% of the calories we consume. Only about 5000 have ever been used for food. There may be tens of thousands of additional kinds of plants, among the 250,000 known species, that could be used for human food if their properties were fully explored and they were brought into cultivation (figure 30.4).

Agricultural scientists are attempting to identify such new crops, especially ones that will grow well in the tropics and subtropics, where the world's population is expanding most rapidly. Nearly all major crops now grown in the world have been cultivated for hundreds or even thousands of years. Only a few, including rubber and oil palms, have entered widespread cultivation since 1800.

One key feature for which nearly all of our important crops were first selected was ease of growth by relatively simple methods. Today, however, techniques of cultivation are far more sophisticated and are able to improve soil fertility and combat pests. This enables us to consider many more plants as potential crops. Agricultural scientists are searching systematically for new crops that fit the multiple needs of modern society, in ways that would not have been considered earlier.

Improving the Productivity of Today's Crops

Searching for new crops is not a quick process. While the search proceeds, the most promising strategy to quickly expand the world food supply is to improve the productivity of crops that are already being grown. Much of the

(a)

(b)

FIGURE 30.4
New food plants. (*a*) Grain amaranths (*Amaranthus* spp.) were important crops in the Latin American highlands during the days of the Incas and Aztecs. Grain amaranths are fast-growing plants that produce abundant grain rich in lysine, an amino acid rare in plant proteins but essential for animal nutrition. (*b*) The winged bean (*Psophocarpus tetragonolobus*) is a nitrogen-fixing tropical vine that produces highly nutritious leaves and tubers whose seeds produce large quantities of edible oil. First cultivated in New Guinea and Southeast Asia, the winged bean has spread since the 1970s throughout the tropics.

improvement in food production must take place in the tropics and subtropics, where the rapidly growing majority of the world's population lives, including most of those enduring a life of extreme poverty. These people cannot be fed by exports from industrial nations, which contribute only about 8% of their total food at present and whose agricultural lands are already heavily exploited. During the 1950s and 1960s, the so-called Green Revolution introduced new, improved strains of wheat and rice. The production of wheat in Mexico increased nearly tenfold between 1950 and 1970, and Mexico temporarily became an exporter of wheat rather than an importer. During the same decades, food production in India was largely able to outstrip even a population growth of approximately 2.3% annually, and China became self-sufficient in food.

Despite the apparent success of the Green Revolution, improvements were limited. Raising the new agricultural strains of plants requires the expenditure of large amounts of energy and abundant supplies of fertilizers, pesticides, and herbicides, as well as adequate machinery. For example, in the United States it requires about 1000 times as much energy to produce the same amount of wheat produced from traditional farming methods in India.

Biologists are playing a crucial role in improving existing crops and in developing new ones by applying traditional methods of plant breeding and selection to many new, nontraditional crops in the tropics and subtropics (see figure 30.4).

Genetic Engineering to Improve Crops

Genetic engineering techniques (discussed in chapter 19) make it possible to produce plants resistant to specific herbicides. These herbicides can then control weeds much more effectively, without damaging crop plants. Genetic engineers are also developing new strains of plants that will grow successfully in areas where they previously could not grow. Desirable characteristics are being introduced into important crop plants. Genetically modified rice, for example, is no longer deficient in ascorbic acid and iron, providing a major improvement in human nutrition. Other modifications allow crops to tolerate irrigation with salt water, fix nitrogen, carry out C_4 photosynthesis, and produce substances that deter pests and diseases.

Genetically modified crops (so-called GM foods) have proven to be a highly controversial issue, one currently being debated in legislative bodies all over the globe. Critics fear loss of genetic diversity, escape of engineered varieties into the environment, harm to insects feeding near GM crops, undo influence of seed companies, and many other real or imagined potential problems. The issue of risk is assessed in chapter 19. While risks must be carefully considered, the ability to transfer genes between organisms, first accomplished in a laboratory in 1973, has tremendous potential for the improvement of crop plants as the twenty-first century opens.

New Approaches to Cultivation

Several new approaches may improve crop production. "No-till" agriculture, spreading widely in the United States and elsewhere in the 1990s, conserves topsoil and so is a desirable agricultural practice for many areas. On the other hand, **hydroponics,** the cultivation of plants in water containing an appropriate mixture of nutrients, holds less promise. It does not differ remarkably in its requirements and challenges from growing plants on land. It requires as much fertilizer and other chemicals, as well as the water itself.

The oceans were once regarded as an inexhaustible source of food, but overexploitation of their resources is limiting the world catch more each year, and these catches are costing more in terms of energy. Mismanagement of fisheries, mainly through overfishing, local pollution, and the destruction of fish breeding and feeding grounds, has lowered the catch of fish in the sea by about 20% from maximum levels. Many fishing areas that were until recently important sources of food have been depleted or closed. For example, the Grand Banks in the North Atlantic Ocean off Newfoundland, a major source of cod and other fish, are now nearly depleted. In 1994, the Canadian government prohibited all cod fishing there indefinitely, throwing 27,000 fishermen out of work, and the United States government banned all fishing on Georges Bank and other defined New England waters. Populations of Atlantic bluefin tuna have dropped 90% since 1975. The United Nations Food and Agriculture Organization estimated in 1993 that 13 of 17 major ocean fisheries are in trouble, with the annual marine fish catch dropping from 86 million metric tons in 1986 to 82.5 million tons by 1992 and continuing to fall each year as the intensity of the fishing increases.

The development of new kinds of food, such as microorganisms cultured in nutrient solutions, should definitely be pursued. For example, the photosynthetic, nitrogen-fixing cyanobacterium *Spirulina* is being investigated in several countries as a possible commercial food source. It is a traditional food in Africa, Mexico, and other regions. *Spirulina* thrives in very alkaline water, and it has a higher protein content than soybeans. Ponds in which it grows are 10 times more productive than wheat fields. Such protein-rich concentrates of microorganisms could provide important nutritional supplements. However, psychological barriers must be overcome to persuade people to eat such foods, and the processing required tends to be energy-expensive.

Just over 100 kinds of plants, out of the roughly 250,000 known, supply more than 90% of all the calories we consume. Many more could be developed by a careful search for new crops.

30.3 Human activity is placing the environment under increasing stress.

The simplest way to gain a feeling for the dimensions of the global environmental problem we face is simply to scan the front pages of any newspaper or news magazine or to watch television. Although they are only a sampling, features selected by these media teach us a great deal about the scale and complexity of the challenge we face. We will discuss a few of the most important issues here.

Nuclear Power

At 1:24 A.M. on April 26, 1986, one of the four reactors of the Chernobyl nuclear power plant blew up. Located in Ukraine 100 kilometers north of Kiev, Chernobyl was one of the largest nuclear power plants in Europe, producing 1000 megawatts of electricity, enough to light a medium-sized city. Before dawn on April 26, workers at the plant hurried to complete a series of tests of how Reactor Number 4 performed during a power reduction and took a foolish short-cut: they shut off all the safety systems. Reactors at Chernobyl were graphite reactors designed with a series of emergency systems that shut the reactors down at low power, because the core is unstable then—and these are the emergency systems the workers turned off. A power surge occurred during the test, and there was nothing to dampen it. Power zoomed to hundreds of times the maximum, and a white-hot blast with the force of a ton of dynamite partially melted the fuel rods and heated a vast head of steam that blew the reactor apart.

The explosion and heat sent up a plume 5 kilometers high, carrying several tons of uranium dioxide fuel and fission products. The blast released over 100 megacuries of radioactivity, making it the largest nuclear accident ever reported; by comparison, the Three Mile Island accident in Pennsylvania in 1979 released 17 curies, millions of times less. This cloud traveled first northwest, then southeast, spreading the radioactivity in a band across central Europe from Scandinavia to Greece. Within a 30-kilometer radius of the reactor, at least one-fifth of the population, some 24,000 people, received serious radiation doses (greater than 45 rem). Thirty-one individuals died as a direct result of radiation poisoning, most of them firefighters who succeeded in preventing the fire from spreading to nearby reactors.

The rest of Europe received a much lower but still significant radiation dose. Data indicate that, because of the large numbers of people exposed, radiation outside of the immediate Chernobyl area can be expected to cause from 5000 to 75,000 cancer deaths.

The Promise of Nuclear Power

Our industrial society has grown for over 200 years on a diet of cheap energy. Until recently, much of this energy has been derived from burning wood and fossil fuels: coal, gas, and oil. However, as these sources of fuel become increasingly scarce and the cost of locating and extracting new deposits becomes more expensive, modern society is being forced to look elsewhere for energy. The great promise of nuclear power is that it provides an alternative source of plentiful energy. Although nuclear power is not cheap—power plants are expensive to build and operate—its raw material, uranium ore, is so common in the earth's crust that it is unlikely we will ever run out of it.

Burning coal and oil to obtain energy produces two undesirable chemical by-products: sulfur and carbon dioxide. The sulfur emitted from burning coal is a principal cause of acid rain, while the CO_2 produced from burning all fossil fuels is a major greenhouse gas (see the discussion of global warming in the next section). For these reasons, we need to find replacements for fossil fuels.

For all of its promise of plentiful energy, nuclear power presents several new problems that must be addressed before its full potential can be realized. First, safe operation of the world's approximately 390 nuclear reactors must be ensured. A second challenge is the need to safely dispose of the radioactive wastes produced by the plants and to safely decommission plants that have reached the end of their useful lives (about 25 years). In 1997, over 35 plants were more than 25 years old, and not one has been safely decommissioned, its nuclear wastes disposed of. A third challenge is the need to guard against terrorism and sabotage, because the technology of nuclear power generation is closely linked to that of nuclear weapons.

For these reasons, it is important to continue to investigate and develop other promising alternatives to fossil fuels, such as solar energy and wind energy. The generation of electricity by burning fossil fuels accounts for up to 15% of global warming gas emissions in the United States. As much as 75% of the electricity produced in the United States and Canada currently is wasted through the use of inefficient appliances, according to scientists at Lawrence Berkeley Laboratory. Using highly efficient motors, lights, heaters, air conditioners, refrigerators, and other technologies already available could save huge amounts of energy and greatly reduce global warming gas emission. For example, a new, compact fluorescent light bulb uses only 20% of the amount of electricity a conventional light bulb uses, provides equal or better lighting, lasts up to 13 times longer, and provides substantial cost savings.

Nuclear power offers plentiful energy for the world's future, but its use involves significant problems and dangers.

Carbon Dioxide and Global Warming

By studying earth's history and making comparisons with other planets, scientists have determined that concentrations of gases in the atmosphere, particularly carbon dioxide, maintain the average temperature on earth about 25°C higher than it would be if these gases were absent. Carbon dioxide and other gases trap the longer wavelengths of infrared light, or heat, radiating from the surface of the earth, creating what is known as a **greenhouse effect** (figure 30.5). The atmosphere acts like the glass of a gigantic greenhouse surrounding the earth.

Roughly seven times as much carbon dioxide is locked up in fossil fuels as exists in the atmosphere today. Before industrialization, the concentration of carbon dioxide in the atmosphere was approximately 260 to 280 parts per million (ppm). Since the extensive use of fossil fuels, the amount of carbon dioxide in the atmosphere has been increasing rapidly. During the 25-year period starting in 1958, the concentration of carbon dioxide increased from 315 ppm to more than 340 ppm and continues to rise. Climatologists have calculated that the actual mean global temperature has increased about 1°C since 1900, a change known as **global warming**.

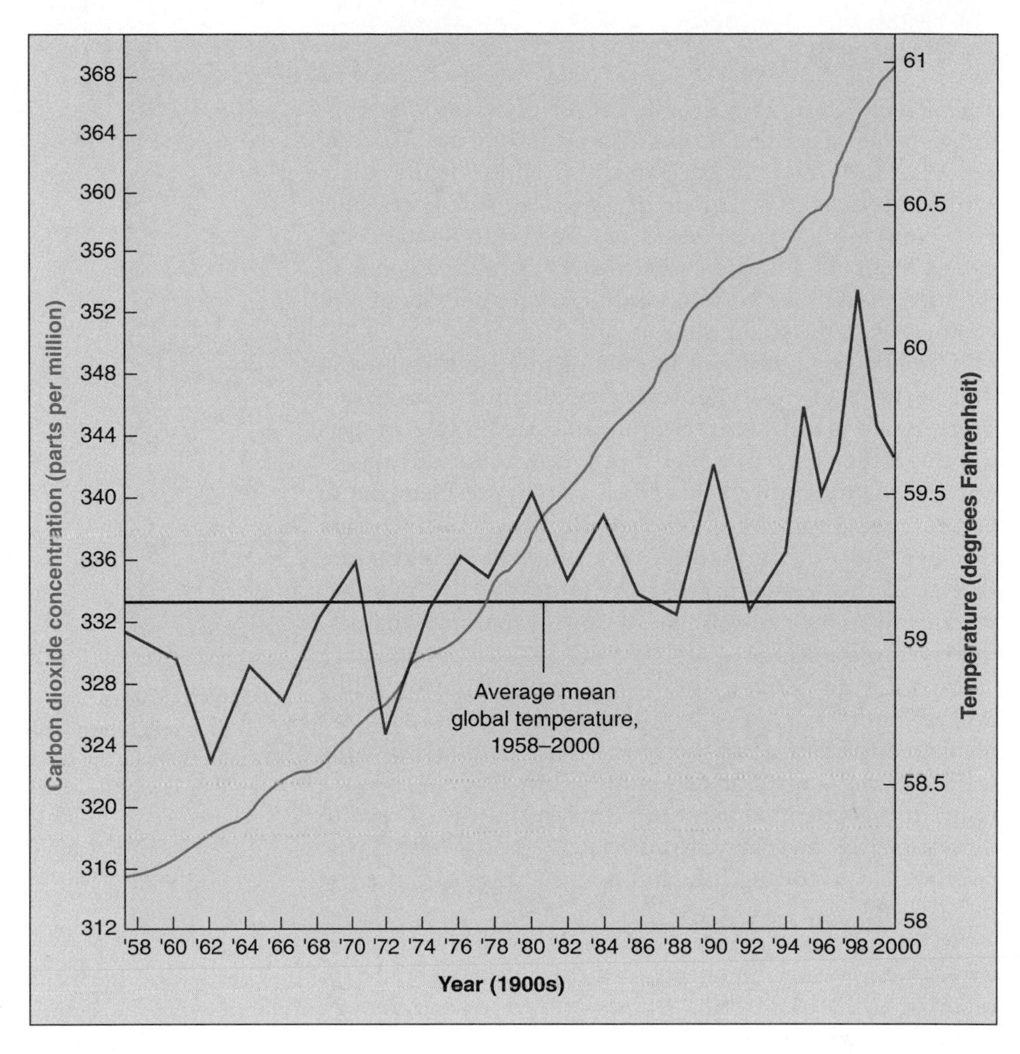

FIGURE 30.5

The greenhouse effect. The concentration of carbon dioxide in the atmosphere has steadily increased since the 1950s (*blue line*). The *red line* shows the general increase in average global temperature for the same period of time.

Source: Data from Geophysical Monograph, American Geophysical Union, National Academy of Sciences, and National Center for Atmospheric Research.

In a recent study, the U.S. National Research Council estimated that the concentration of carbon dioxide in the atmosphere would pass 600 ppm (roughly double the current level) by the third quarter of this century, and might exceed that level as soon as 2035. These concentrations of carbon dioxide, if actually reached, would warm global surface air by between 1.5° and 4.5°C. The actual increase might be considerably greater, however, because a number of trace gases, such as nitrous oxide, methane, ozone, and chlorofluorocarbons, are also increasing rapidly in the atmosphere as a result of human activities. These gases have warming, or "greenhouse," effects similar to those of carbon dioxide. One, methane, increased from 1.14 ppm in the atmosphere in 1951 to 1.68 ppm in 1986—nearly a 50% increase.

Major problems associated with climatic warming include rising sea levels. Sea levels may have already risen 2 to 5 centimeters from global warming. If the climate becomes so warm that the polar ice caps melt, sea levels would rise by more than 150 meters, flooding the entire Atlantic coast of North America for an average distance of several hundred kilometers inland.

Changes in the distribution of precipitation are difficult to model. Certainly, changing climatic patterns are likely to make some of the best farmlands much drier than they are at present. If the climate warms as rapidly as many scientists project, the next 50 years may see greatly altered weather patterns, a rising sea level, and major shifts of deserts and fertile regions.

As the global concentration of carbon dioxide increases, the world's temperature is rising, with great potential impact on the world's climate.

Pollution

The River Rhine is a broad ribbon of water that runs through the heart of Europe. From high in the Alps that separate Italy and Switzerland, the Rhine flows north across the industrial regions of Germany before reaching Holland and the sea. Where it crosses the mountains between Mainz and Coblenz, Germany, the Rhine is one of the most beautiful rivers on earth. On the first day of November 1986, the Rhine almost died.

The blow that struck at the life of the Rhine did not at first seem deadly. Firefighters were battling a blaze that morning in Basel, Switzerland. The fire was gutting a huge warehouse, into which firefighters shot streams of water to dampen the flames. The warehouse belonged to Sandoz, a giant chemical company. In the rush to contain the fire, no one thought to ask what chemicals were stored in the warehouse. By the time the fire was out, streams of water had washed 30 tons of mercury and pesticides into the Rhine.

Flowing downriver, the deadly wall of poison killed everything it passed. For hundreds of kilometers, dead fish blanketed the surface of the river. Many cities that use the water of the Rhine for drinking had little time to make other arrangements. Even the plants in the river began to die. All across Germany, from Switzerland to the sea, the river reeked of rotting fish, and not one drop of water was safe to drink.

Six months later, Swiss and German environmental scientists monitoring the effects of the accident were able to report that the blow to the Rhine was not mortal. Enough small aquatic invertebrates and plants had survived to provide a basis for the eventual return of fish and other water life, and the river was rapidly washing out the remaining residues from the spill. A lesson difficult to ignore, the spill on the Rhine has caused the governments of Germany and Switzerland to intensify efforts to protect the river from future industrial accidents and to regulate the growth of chemical and industrial plants on its shores.

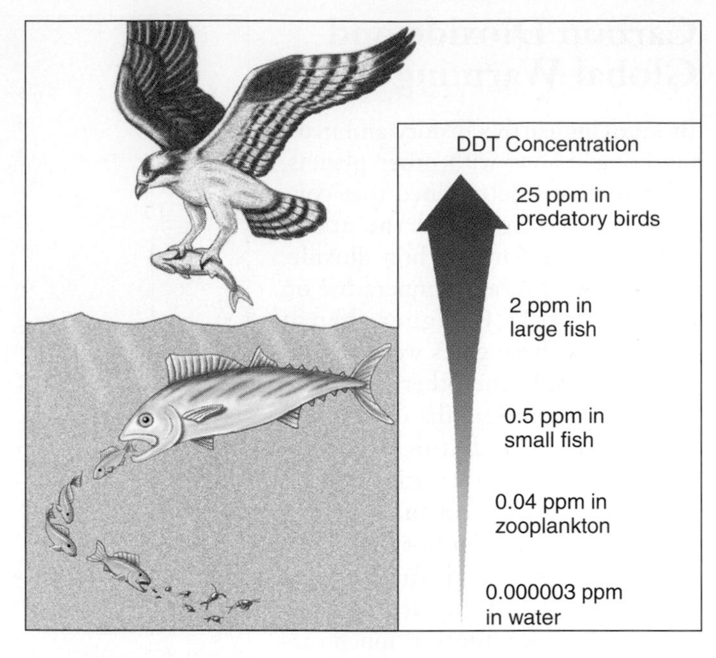

FIGURE 30.6
Biological magnification of DDT. Because DDT accumulates in animal fat, the compound becomes increasingly concentrated in higher levels of the food chain. Before DDT was banned in the United States, predatory bird populations drastically declined because DDT made their eggshells thin and fragile enough to break during incubation.

The Threat of Pollution

The pollution of the Rhine is a story that can be told countless times in different places in the industrial world, from Love Canal in New York to the James River in Virginia to the town of Times Beach in Missouri. Nor are all pollutants that threaten the sustainability of life immediately toxic. Many forms of pollution arise as by-products of industry. For example, the polymers known as plastics, which we produce in abundance, break down slowly in nature. Much is burned or otherwise degraded to smaller vinyl chloride units. Virtually all of the plastic that has ever been produced is still with us, in one form or another. Collectively, it constitutes a new form of pollution.

Widespread agriculture, carried out increasingly by modern methods, introduces large amounts of many new kinds of chemicals into the global ecosystem, including pesticides, herbicides, and fertilizers. Industrialized countries like the United States now attempt to carefully monitor side effects of these chemicals. Unfortunately, large quantities of many toxic chemicals no longer manufactured still circulate in the ecosystem.

For example, the chlorinated hydrocarbons, a class of compounds that includes DDT, chlordane, lindane, and dieldrin, have all been banned for normal use in the United States, where they were once widely used. They are still manufactured in the United States, however, and exported to other countries, where their use continues. Chlorinated hydrocarbon molecules break down slowly and accumulate in animal fat. Furthermore, as they pass through a food chain, they become increasingly concentrated in a process called **biological magnification** (figure 30.6). DDT caused serious problems by leading to the production of thin, fragile eggshells in many predatory bird species in the United States and elsewhere until the late 1960s, when it was banned in time to save the birds from extinction. Chlorinated compounds have other undesirable side effects and exhibit hormonelike activities in the bodies of animals, disrupting normal hormonal cycles with sometimes potentially serious consequences.

Chemical pollution is causing ecosystems to accumulate many harmful substances, as the result of spills and runoff from agricultural or urban use.

Acid Precipitation

The Four Corners power plant in New Mexico burns coal, sending smoke up high into the atmosphere through its smokestacks, each over 65 meters tall. The smoke the stacks belch out contains high concentrations of sulfur dioxide and other sulfates, which produce acid when they combine with water vapor in the air. The intent of those who designed the plant was to release the sulfur-rich smoke high in the atmosphere, where winds would disperse and dilute it, carrying the acids far away.

Environmental effects of this acidity are serious. Sulfur introduced into the upper atmosphere combines with water vapor to produce sulfuric acid, and when the water later falls as rain or snow, the precipitation is acid. Natural rainwater rarely has a pH lower than 5.6; in the northeastern United States, however, rain and snow now have a pH of about 3.8, roughly 100 times as acidic (figure 30.7).

Acid precipitation destroys life. Thousands of lakes in southern Sweden and Norway no longer support fish; these lakes are now eerily clear. In the northeastern United States and eastern Canada, tens of thousands of lakes are dying biologically as a result of acid precipitation. At pH levels below 5.0, many fish species and other aquatic animals die, unable to reproduce. In southern Sweden and elsewhere, groundwater now has a pH between 4.0 and 6.0, as acid precipitation slowly filters down into the underground reservoirs.

There has been enormous forest damage in the Black Forest in Germany and in the forests of the eastern United States and Canada. It has been estimated that at least 3.5 million hectares of forest in the northern hemisphere are being affected by acid precipitation (figure 30.8), and the problem is clearly growing.

Its solution at first seems obvious: capture and remove the emissions instead of releasing them into the atmosphere. However, there are serious difficulties in executing this solution. First, it is expensive. The costs of installing and maintaining the necessary "scrubbers" in the United States are estimated to be 4 to 5 billion dollars per year. An additional difficulty is that the polluter and the recipient of the pollution are far from each other, and neither wants to pay for what they view as someone else's problem. The Clean Air Act revisions of 1990 addressed this problem in the United States significantly for the first time, and substantial worldwide progress has been made in implementing a solution.

Industrial pollutants such as nitric and sulfuric acids, introduced into the upper atmosphere by factory smokestacks, are spread over wide areas by the prevailing winds and fall to earth with precipitation called "acid rain," lowering the pH of water on the ground and killing life.

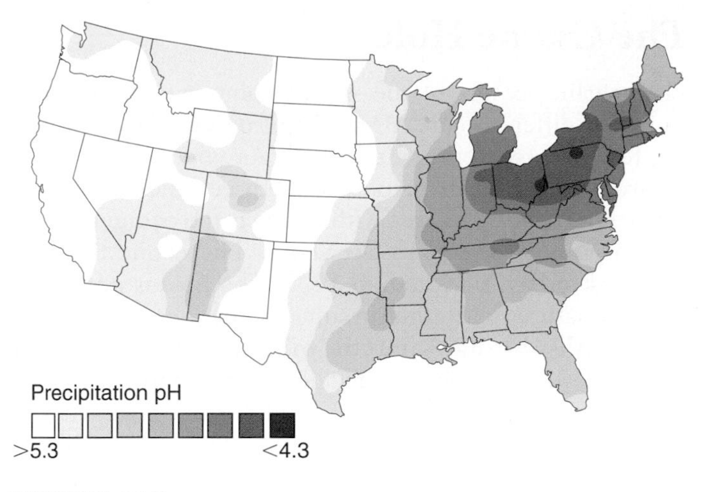

FIGURE 30.7
pH values of rainwater in the United States. Precipitation in the United States, especially in the Northeast, is more acidic than that of natural rainwater, which has a pH of about 5.6.

FIGURE 30.8
Damage to trees at Clingman's Dome, Tennessee. Acid precipitation weakens trees and makes them more susceptible to pests and predators.

The Ozone Hole

The swirling colors of the satellite photos in figure 30.9 represent different concentrations of **ozone** (O_3), a different form of oxygen gas than O_2. As you can see, over Antarctica there is an "ozone hole" three times the size of the United States, an area within which the ozone concentration is much less than elsewhere. The ozone thinning appeared for the first time in 1975. The hole is not a permanent feature, but rather becomes evident each year for a few months during Antarctic winter. Every September from 1975 onward, the ozone "hole" has reappeared. Each year the layer of ozone is thinner and the hole is larger.

The major cause of the ozone depletion had already been suggested in 1974 by Sherwood Roland and Mario Molina, who were awarded the Nobel Prize for their work in 1995. They proposed that chlorofluorocarbons (CFCs), relatively inert chemicals used in cooling systems, fire extinguishers, and Styrofoam containers, were percolating up through the atmosphere and reducing O_3 molecules to O_2. One chlorine atom from a CFC molecule could destroy 100,000 ozone molecules in the following mechanism:

UV radiation causes CFCs to release Cl atoms:

$$CCl_3F \xrightarrow{UV} Cl + CCl_2F$$

UV creates oxygen free radicals:

$$O_2 \longrightarrow 2O$$

Cl atoms and O free radicals interact with ozone:

$$2Cl + 2O_3 \longrightarrow 2ClO + 2O_2$$
$$2ClO + 2O \longrightarrow 2Cl + 2O_2$$

Net reaction: $2O_3 \longrightarrow 3O_2$

Although other factors have also been implicated in ozone depletion, the role of CFCs is so predominant that worldwide agreements have been signed to phase out their production. The United States banned the production of CFCs and other ozone-destroying chemicals after 1995. Nonetheless, the CFCs that were manufactured earlier are moving slowly upward through the atmosphere. The ozone layer will be further depleted before it begins to form again.

Thinning of the ozone layer in the stratosphere, 25 to 40 kilometers above the surface of the earth, is a matter of serious concern. This layer protects life from the harmful ultraviolet rays that bombard the earth continuously from the sun. Life appeared on land only after the oxygen layer was sufficiently thick to generate enough ozone to shield the surface of the earth from these destructive rays.

Ultraviolet radiation is a serious human health concern. Every 1% drop in atmospheric ozone is estimated to lead to a 6% increase in the incidence of skin cancers. At middle latitudes, the approximately 3% drop that has already occurred worldwide is estimated to have increased skin cancers by as much as 20%. A type of skin cancer (melanoma) is one of the more lethal human diseases.

Industrial CFCs released into the atmosphere react at very cold temperatures with ozone, converting it to oxygen gas. This has the effect of destroying the earth's ozone shield and exposing the earth's surface to increased levels of harmful UV radiation.

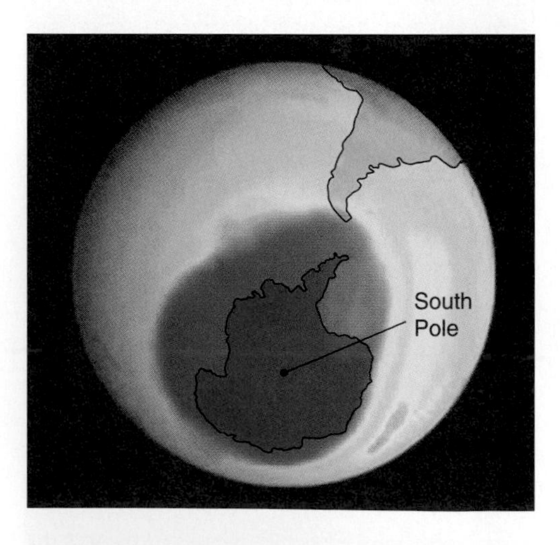

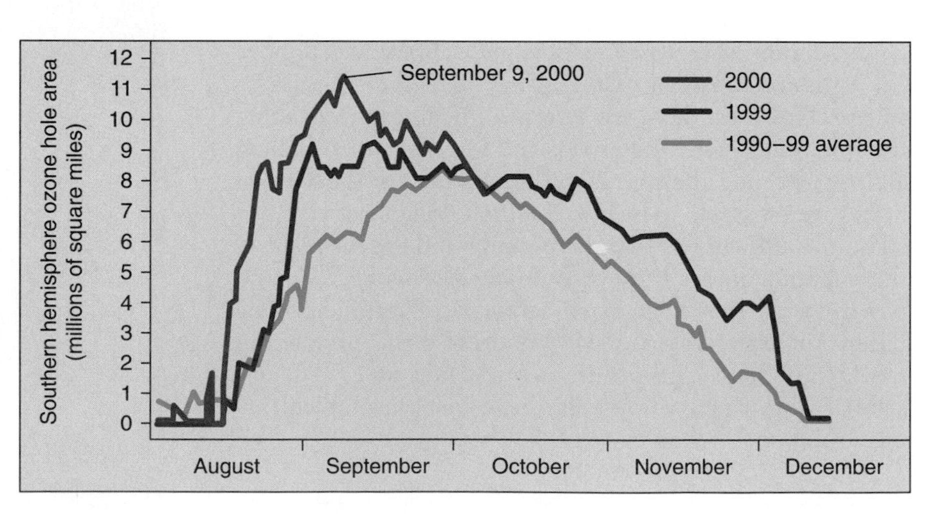

FIGURE 30.9
The ozone hole over Antarctica is still growing. For decades NASA satellites have tracked the extent of ozone depletion over Antarctica. Every year since 1975 an ozone "hole" has appeared in August when sunlight triggers chemical reactions in cold air trapped over the South Pole during Antarctic winter. The hole intensifies during September before tailing off as temperatures rise in November-December. In 2000, the 11.4 million square-mile hole (dark *purple* in the satellite image) covered an area larger than the United States, Canada, and Mexico combined, the largest hole ever recorded. In September 2000, the hole extended over Punta Arenas, a city of about 120,000 people in southern Chile, exposing residents to very high levels of UV radiation.

Destruction of the Tropical Forests

More than half of the world's human population lives in the tropics, and this percentage is increasing rapidly. For global stability, and for the sustainable management of the world ecosystem, it will be necessary to solve the problems of food production and regional stability in these areas. World trade, political and economic stability, and the future of most species of plants, animals, fungi, and microorganisms depend on our addressing these problems.

(a) (b)

FIGURE 30.10
Destroying the tropical forests. (*a*) When tropical forests are cleared, the ecological consequences can be disastrous. These fires are destroying rain forest in Brazil and clearing it for cattle pasture. (*b*) The consequences of deforestation can be seen on these middle-elevation slopes in Ecuador, which now support only low-grade pastures and permit topsoil to erode into the rivers (note the color of the water, stained brown by high levels of soil erosion). These areas used to support highly productive forest, which protected the watersheds of the area, in the 1970s.

Rain Forests Are Rapidly Disappearing

Tropical rain forests are biologically the richest of the world's biomes. Most other kinds of tropical forest, such as seasonally dry forests and savanna forests, have already been largely destroyed—because they tend to grow on more fertile soils, they were exploited by humans a long time ago. Now the rain forests, which grow on poor soils, are being destroyed. In the mid-1990s, it is estimated that only about 5.5 million square kilometers of tropical rain forest still exist in a relatively undisturbed form. This area, about two-thirds of the size of the United States (excluding Alaska), represents about half of the original extent of the rain forest. From it, about 160,000 square kilometers are being clear-cut every year, with perhaps an equivalent amount severely disturbed by shifting cultivation, firewood gathering, and the clearing of land for cattle ranching. The total area of tropical rain forest destroyed—and therefore permanently removed from the world total—amounts to an area greater than the size of Indiana each year. At this rate, all of the tropical rain forest in the world will be gone in about 30 years; but in many regions, the rate of destruction is much more rapid. As a result of this overexploitation, experts predict there will be little undisturbed tropical forest left anywhere in the world by early in this century. Many areas now occupied by dense, species-rich forests may still be tree-covered, but the stands will be sparse and species-poor.

A Serious Matter

Not only does the disappearance of tropical forests represent a tragic loss of largely unknown biodiversity, but the loss of the forests themselves is ecologically a serious matter. Tropical forests are complex, productive ecosystems that function well in the areas where they have evolved. When people cut a forest or open a prairie in the north temperate zone, they provide farmland that we know can be worked for generations. In most areas of the tropics, people are unable to engage in continuous agriculture. When they clear a tropical forest, they engage in a one-time consumption of natural resources that will never be available again (figure 30.10). The complex ecosystems built up over millions of years are now being dismantled, in almost complete ignorance, by humans.

What biologists must do is to learn more about the construction of sustainable agricultural ecosystems that will meet human needs in tropical and subtropical regions. The ecological concepts we have been reviewing in the last two chapters are universal principles. The undisturbed tropical rain forest has one of the highest rates of net primary productivity of any plant community on earth, and it is therefore imperative to develop ways that it can be harvested for human purposes in a sustainable, intelligent way.

More than half of the tropical rain forests have been destroyed by human activity, and the rate of loss is accelerating.

30.4 Solving environmental problems requires individual involvement.

Environmental Science

Environmental scientists attempt to find solutions to environmental problems, considering them in a broad context. Unlike biology or ecology, sciences that seek to learn general principles about how life functions, environmental science is an applied science dedicated to solving practical problems. Its basic tools are derived from ecology, geology, meteorology, social sciences, and many other areas of knowledge that bear on the functioning of the environment and our management of it. Environmental science addresses the problems created by rapid human population growth: an increasing need for energy, a depletion of resources, and a growing level of pollution.

Solving Environmental Problems

The problems our severely stressed planet faces are not insurmountable. A combination of scientific investigation and public action, when brought to bear effectively, can solve environmental problems that seem intractable. Viewed simply, there are five components to solving any environmental problem:

1. **Assessment.** The first stage in addressing any environmental problem is scientific analysis, the gathering of information. Data must be collected and experiments performed to construct a model that describes the situation. This model can be used to make predictions about the future course of events.
2. **Risk analysis.** Using the results of scientific analysis as a tool, it is possible to analyze what could be expected to happen if a particular course of action were followed. It is necessary to evaluate not only the potential for solving the environmental problem, but also any adverse effects a plan of action might create.
3. **Public education.** When a clear choice can be made among alternative courses of action, the public must be informed. This involves explaining the problem in terms the public can understand, presenting the alternatives available, and explaining the probable costs and results of the different choices.
4. **Political action.** The public, through its elected officials selects a course of action and implements it. Choices are particularly difficult to implement when environmental problems transcend national boundaries.
5. **Follow-through.** The results of any action should be carefully monitored to see whether the environmental problem is being solved as well as to evaluate and improve the initial modeling of the problem. Every environmental intervention is an experiment, and we need the knowledge gained from each one to better address future problems.

Individuals Can Make the Difference

The development of appropriate solutions to the world's environmental problems must rest partly on the shoulders of politicians, economists, bankers, engineers—many kinds of public and commercial activity will be required. However, it is important not to lose sight of the key role often played by informed individuals in solving environmental problems. Often one person has made the difference; two examples serve to illustrate the point.

The Nashua River. Running through the heart of New England, the Nashua River was severely polluted by mills established in Massachusetts in the early 1900s. By the 1960s, the river was clogged with pollution and declared ecologically dead. When Marion Stoddart moved to a town along the river in 1962, she was appalled. She approached the state about setting aside a "greenway" (trees running the length of the river on both sides), but the state wasn't interested in buying land along a filthy river. So Stoddart organized the Nashua River Cleanup Committee and began a campaign to ban the dumping of chemicals and wastes into the river. The committee presented bottles of dirty river water to politicians, spoke at town meetings, recruited businesspeople to help finance a waste treatment plant, and began to clean garbage from the Nashua's banks. This citizen's campaign, coordinated by Stoddart, greatly aided passage of the Massachusetts Clean Water Act of 1966. Industrial dumping into the river is now banned, and the river has largely recovered.

Lake Washington. A large, 86 km^2 freshwater lake east of Seattle, Lake Washington became surrounded by Seattle suburbs in the building boom following the Second World War. Between 1940 and 1953, a ring of 10 municipal sewage plants discharged their treated effluent into the lake. Safe enough to drink, the effluent was believed "harmless." By the mid-1950s a great deal of effluent had been dumped into the lake (try multiplying 80 million liters/day × 365 days/year × 10 years). In 1954, an ecology professor at the University of Washington in Seattle, W. T. Edmondson, noted that his research students were reporting filamentous blue-green algae growing in the lake. Such algae require plentiful nutrients, which deep freshwater lakes usually lack—the sewage had been fertilizing the lake! Edmondson, alarmed, began a campaign in 1956 to educate public officials to the danger: bacteria decomposing dead algae would soon so deplete the lake's oxygen that the lake would die. After five years, joint municipal taxes financed the building of a sewer to carry the effluent out to sea. The lake is now clean.

In solving environmental problems, the commitment of one person can make a critical difference.

Preserving Nonreplaceable Resources

Among the many ways ecosystems are suffering damage, one class of problem stands out as more serious than the rest: consuming or destroying resources that we cannot replace in the future. While a polluted stream can be cleaned up, no one can restore an extinct species. In the United States, we are consuming three nonreplaceable resources at alarming rates: topsoil, groundwater, and biodiversity. We will briefly discuss the first two of these in this chapter, with a more detailed discussion of biodiversity in the following chapter.

Topsoil

The United States is one of the most productive agricultural countries on earth, largely because much of it is covered with particularly fertile soils. Our midwestern farm belt sits astride what was once a great prairie. The **topsoil** of that ecosystem accumulated bit by bit from countless generations of animals and plants until, by the time humans began to plow it, the rich soil extended down several feet.

We can never replace this rich topsoil, the capital upon which our country's greatness is built, yet we are allowing it to be lost at a rate of centimeters every decade. By repeatedly tilling (turning the soil over) to eliminate weeds, we permit rain to wash more and more of the topsoil away, into rivers, and eventually out to sea. Our country has lost one-quarter of its topsoil since 1950! New approaches are desperately needed to lessen our reliance on intensive cultivation. Some possible solutions include using genetic engineering to make crops resistant to weed-killing herbicides and terracing to recapture lost topsoil.

Groundwater

A second resource we cannot replace is **groundwater,** water trapped beneath the soil within porous rock reservoirs called aquifers (figure 30.11). This water seeped into its underground reservoir very slowly during the last ice age over 12,000 years ago. We should not waste this treasure, for we cannot replace it.

In most areas of the United States, local governments exert relatively little control over the use of groundwater. As a result, a large portion is wasted watering lawns, washing cars, and running fountains. A great deal more is inadvertently polluted by poor disposal of chemical wastes—and once pollution enters the groundwater, there is no effective means of removing it.

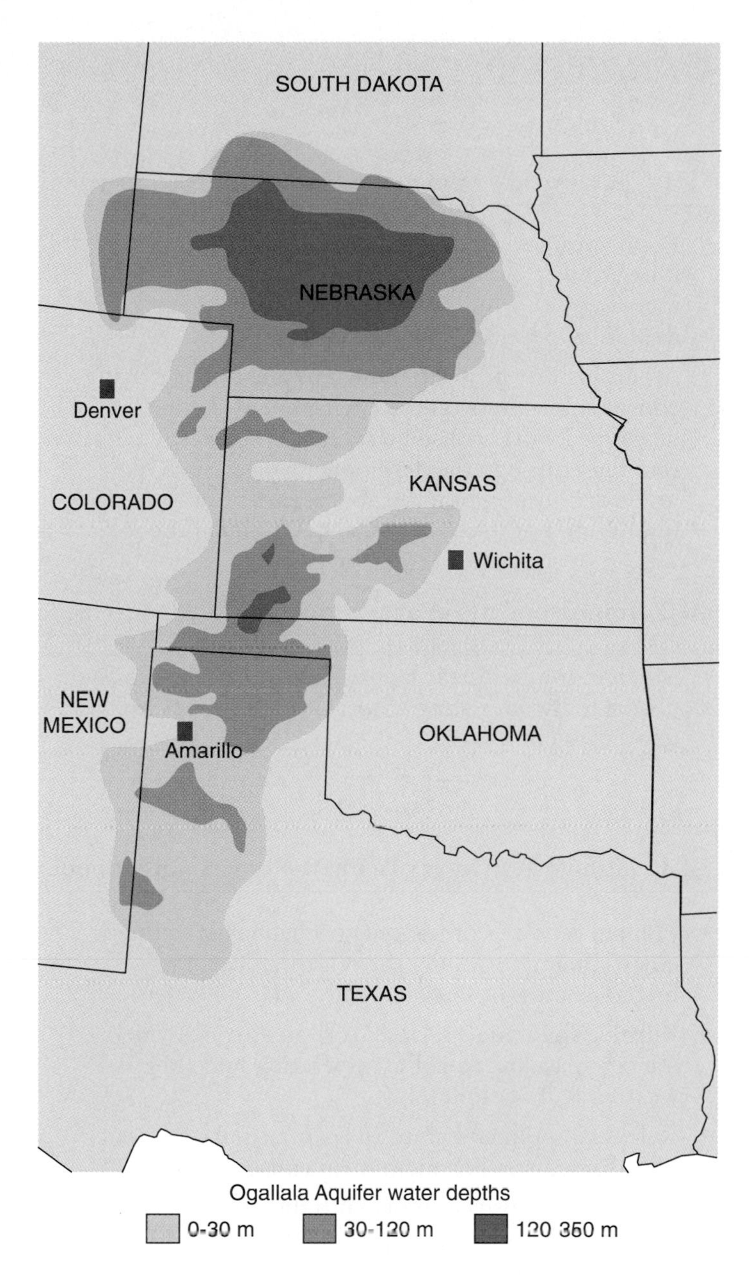

FIGURE 30.11
The Ogallala Aquifer. This massive deposit of groundwater lies under eight states, mainly Texas, Kansas, and Nebraska. Excessive pumping of this water for irrigation and other uses has caused the water level to drop 30 meters in some places. Continued excessive use of this kind endangers the survival of the Ogallala Aquifer, as it takes hundreds or even thousands of years for aquifers to recharge.

Source: Data from Raven, Berg & Johnson, *Environment*, 3rd edition, Saunders, 1998.

Topsoil and groundwater are essential for agriculture and other human activities. Replenishment of these resources occurs at a very slow rate. Current levels of consumption are not sustainable and will cause serious problems in the near future.

Chapter 30

www.mhhe.com/raven6/resources30.mhtml

Summary	Questions	Media Resources

30.1 The world's human population is growing explosively.

- Population growth rates are declining throughout much of the world, but still the human population increases by 77 million people per year. At this rate, the global population will double in 39 years.
- An explosively growing human population is placing considerable stress on the environment. People in the developed world consume resources at a vastly higher rate than those in the developing world. Such high levels of consumption are not sustainable and are as important a problem as global overpopulation.

1. Why, in some respects, is the population size of the developed world more of a consideration in discussing resource use than the population of the developing world?

30.2 Improvements in agriculture are needed to feed a hungry world.

- Much current effort is focused on improving the productivity of existing crops, although the search for new crops continues.

2. What three species supply more than half of the human energy requirements on earth? How many plants supply over 90%?

- Food Production

30.3 Human activity is placing the environment under increasing stress.

- Human activities present many challenges to the environment, including the release of harmful materials into the environment.
- Burning fossil fuels releases carbon dioxide, which may increase the world's temperature and alter weather and ocean levels.
- Release of pollutants into rivers may make the water unfit for aquatic life and human consumption.
- Release of industrial smoke into the upper atmosphere leads to acid precipitation that kills forests and lakes.
- Release of chemicals such as chlorofluorocarbons may destroy the atmosphere's ozone and expose the world to dangerous levels of ultraviolet radiation.
- Cutting and burning the tropical rain forests of the world to make pasture and cropland is producing a massive wave of extinction.

3. What problems must we master before nuclear power's full potential can be realized?

4. Why were chlorinated hydrocarbons banned in the United States? Why can you still find them as contaminants on fruits and vegetables?

5. How does acid precipitation form? Why has it been difficult to implement solutions to this problem?

6. What is the ozone layer? How is it formed? What are the harmful effects of decreasing the earth's ozone layer? What may be the primary cause of this damage?

- Global Warming
- Bioaccumulation
- Acid Rain
- Ozone Layer Depletion

- Human Impact
- Nuclear Power
- Fossil Fuels
- Water Quality
- Biomagnification
- Pesticides

- Art Quizzes:
  - -Formation of Acid Precipitation
  - -Greenhouse Effect

30.4 Solving environmental problems requires individual involvement.

- All of these challenges to our future can and must be addressed. Today, environmental scientists and concerned citizens are actively searching for constructive solutions to these problems.

7. What sort of action might you take that would make a significant contribution to solving the world's environmental problems?

- Scientists on Science: History of Life
- *On Science* Article: Do-It-Yourself Environmentalism

31

Conservation Biology

FIGURE 31.1
Endangered. The Siberian tiger is in grave danger of extinction, hunted for its pelt and having its natural habitat greatly reduced. A concerted effort is being made to save it, using many of the approaches discussed in this chapter.

Concept Outline

31.1 The new science of conservation biology is focused on conserving biodiversity.

Overview of the Biodiversity Crisis. In prehistoric times, humans decimated the faunas of many areas. Worldwide extinction rates are accelerating.

Species Endemism and Hot Spots. Some geographic areas are particularly rich in species that occur nowhere else.

What's So Bad about Losing Biodiversity? Biodiversity is of considerable direct economic value, and provides key support to the biosphere.

31.2 Vulnerable species are more likely to become extinct.

Predicting Which Species Are Vulnerable to Extinction. Biologists carry out population viability analyses to assess danger of extinction.

Dependence upon Other Species. Extinction of one species can have a cascading effect throughout the food web, making other species vulnerable as well.

Categories of Vulnerable Species. Declining population size, loss of genetic variation, and commercial value all tend to increase a species' vulnerability.

31.3 Causes of endangerment usually reflect human activities.

Factors Responsible for Extinction. Most recorded extinctions can be attributed to a few causes.

Habitat Loss. Without a place to live, species cannot survive.

Case Study: Overexploitation

Case Study: Introduced Species

Case Study: Disruption of Ecological Relationships

Case Study: Loss of Genetic Variation

Case Study: Habitat Loss and Fragmentation

31.4 Successful recovery plans will need to be multidimensional.

Many Approaches Exist for Preserving Endangered Species. Species recovery requires restoring degraded habitats, breeding in captivity, maintaining population diversity, and maintaining keystone species.

Conservation of Ecosystems. Maintaining large preserves and focusing on the health of the entire ecosystem may be the best means of preserving biodiversity.

Among the greatest challenges facing the biosphere is the accelerating pace of species extinctions—not since the Cretaceous have so many species become extinct in so short a period of time (figure 31.1). This challenge has led to the emergence in the last decade of the new discipline of conservation biology. Conservation biology is an applied discipline that seeks to learn how to preserve species, communities, and ecosystems. It both studies the causes of declines in species richness and attempts to develop methods to prevent such declines. In this chapter we will first examine the biodiversity crisis and its importance. Then, we will assess the sorts of species which seem vulnerable to extinction. Using case histories, we go on to identify and study five factors that have played key roles in many extinctions. We finish with a review of recovery efforts at the species and community level.

31.1 The new science of conservation biology is focused on conserving biodiversity.

Overview of the Biodiversity Crisis

Extinction is a fact of life, as normal and necessary as species formation is to a stable world ecosystem. Most species, probably all, go extinct eventually. More than 99% of species known to science (most from the fossil record) are now extinct. However, current rates are alarmingly high. Taking into account the rapid and accelerating loss of habitat that is occurring at present, especially in the tropics, it has been calculated that as much as 20% of the world's biodiversity may be lost during the next 30 years. In addition, many of these species may be lost before we are even aware of their extinction. Scientists estimate that no more than 15% of the world's eukaryotic organisms have been discovered and given scientific names, and this proportion probably is much lower for tropical species.

FIGURE 31.2
North America before human inhabitants. Animals found in North America prior to the migration of humans included large mammals and birds such as the ancient North American camel, saber-toothed cat, giant ground-sloth, and the teratorn vulture.

These losses will not just affect poorly known groups. As many as 50,000 species of the world's total of 250,000 species of plants, 4000 of the world's 20,000 species of butterflies, and nearly 2000 of the world's 9000 species of birds could be lost during this short period of time. Considering that our species has been in existence for only 500,000 years of the world's 4.5-billion-year history, and that our ancestors developed agriculture only about 10,000 years ago, this is an astonishing—and dubious—accomplishment.

Extinctions Due to Prehistoric Humans

A great deal can be learned about current rates of extinction by studying the past, and in particular the impact of human-caused extinctions. In prehistoric times, *Homo sapiens* wreaked havoc whenever they entered a new area. For example, at the end of the last ice age, approximately 12,000 years ago, the fauna of North America was composed of a diversity of large mammals similar to Africa today: mammoths and mastodons, horses, camels, giant ground-sloths, saber-toothed cats, and lions, among others (figure 31.2). Shortly after humans arrived, 74 to 86% of the megafauna (that is, animals weighing more than 100 pounds) became extinct. These extinctions are thought to have been caused by hunting, and indirectly by burning and clearing forests (some scientists attribute these extinctions to climate change, but that hypothesis doesn't explain why the end of earlier ice ages was not associated with mass extinctions, nor does it explain why extinctions occurred primarily among larger animals, with smaller species relatively unaffected).

Similar results have followed the arrival of humans around the globe. Forty thousand years ago, Australia was occupied by a wide variety of large animals, including marsupials similar in size and ecology to hippos and leopards, a kangaroo nine feet tall, and a 20-foot-long monitor lizard. These all disappeared, at approximately the same time as humans arrived. Smaller islands have also been devastated. On Madagscar, at least 15 species of lemurs, including one the size of a gorilla, a pygmy hippopotamus , and the flightless elephant bird, *Aepyornis*, the largest bird to ever live (more than 3 meters tall and weighing 450 kilograms) all perished. On New Zealand, 30 species of birds went extinct, including all 13 species of moas, another group of large, flightless birds. Interestingly, one continent that seems to have been spared these megafaunal extinctions is Africa. Scientists speculate that this lack of extinction in prehistoric Africa may have resulted because much of human evolution occurred in Africa. Consequently, other African species had been coevolving with humans for several million years and thus had evolved counteradaptations to human predation.

Extinctions in Historical Time

Historical extinction rates are best known for birds and mammals because these species are conspicuous—relatively large and well studied. Estimates of extinction rates for other species are much rougher. The data presented

Table 31.1 Recorded Extinctions since 1600 a.d.

Taxon	Recorded Extinctions: Mainland	Island	Ocean	Total	Approximate Number of Species	Percent of Taxon Extinct
Mammals	30	51	4	85	4,000	2.1
Birds	21	92	0	113	9,000	1.3
Reptiles	1	20	0	21	6,300	0.3
Amphibians*	2	0	0	2	4,200	0.05
Fish	22	1	0	23	19,100	0.1
Invertebrates	49	48	1	98	1,000,000+	0.01
Flowering plants	245	139	0	384	250,000	0.2

*There has been an alarming decline in amphibian populations recently, and many species may be on the verge of extinction.
Source: Data from Reid and Miller, 1989.

in table 31.1, based on the best available evidence, show recorded extinctions from 1600 to the present. These estimates indicate that about 85 species of mammals and 113 species of birds have become extinct since the year 1600. That is about 2.1% of known mammal species and 1.3% of known birds. The majority of extinctions have come in the last 150 years. The extinction rate for birds and mammals was about one species every decade from 1600 to 1700, but it rose to one species every year during the period from 1850 to 1950, and four species per year between 1986 and 1990 (figure 31.3). It is this increase in the rate of extinction that is the heart of the biodiversity crisis.

The majority of historic extinctions—though by no means all of them—have occurred on islands. For example, of the 90 species of mammals that have gone extinct in the last 500 years, 73% lived on islands (and another 19% on Australia). The particular vulnerability of island species probably results from a number of factors: such species have often evolved in the absence of predators and so have lost their ability to escape both humans and introduced predators such as rats and cats. In addition, humans have introduced competitors and diseases (avian malaria, for example has devastated the bird fauna of the Hawaiian Islands). Finally, island populations are often relatively small, and thus particularly vulnerable to extinction, as we shall see later in the chapter.

In recent years, however, the extinction crisis has moved from islands to continents. Most species now threatened with extinction occur on continents, and it is these areas which will bear the brunt of the extinction crisis in this century.

Some people have argued that we should not be concerned because extinctions are a natural event and mass extinctions have occurred in the past. Indeed, as we saw in chapter 21, mass extinctions have occurred several times over the past half billion years. However, the current mass extinction event is notable in several respects. First, it is the only such event triggered by a single species. Moreover, although species diversity usually recovers after a few million years, this is a long time to deny our descendants the benefits and joys of biodiversity. In addition, it is not clear that biodiversity will rebound this time. After previous mass extinction events, new species have evolved to utilize resources available due to species extinctions. Today, however, such resources are unlikely to be available, because humans are destroying the habitats and taking the resources for their own use.

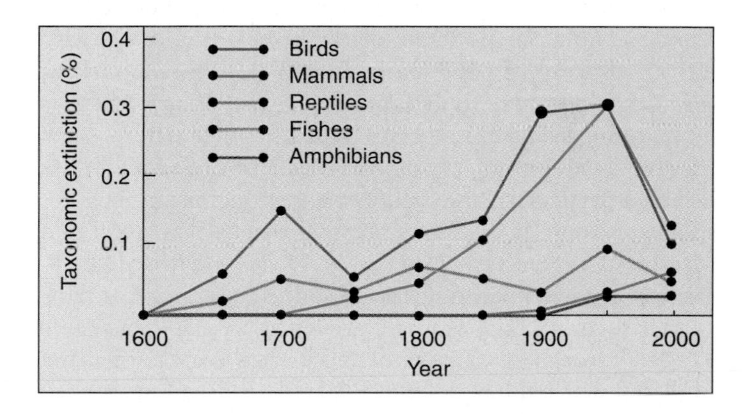

FIGURE 31.3
Trends in species loss. The graphs above present data on recorded animal extinctions since 1600. The majority of extinctions have occurred on islands, with birds and mammals particularly affected (although this may reflect to some degree our more limited knowledge of other groups).
Source: After Smith et al., 1993.

Biologists estimate rates of extinction both by studying recorded extinction events and by analyzing trends in habitat loss and disruption. Since prehistoric times, humans have had a devastating effect on biodiversity almost everywhere in the world.

Species Endemism and Hot Spots

A species found naturally in only one geographic area and no place else is said to be **endemic** to that area. The area over which an endemic species is found may be very large. The black cherry tree (*Prunus serotina*), for example, is endemic to all of temperate North America. More typically, however, endemic species occupy restricted ranges. The Komodo dragon (*Varanus komodoensis*) lives only on a few small islands in the Indonesian archipelago, while the Mauna Kea silversword (*Argyroxiphium sandwicense*) lives in a single volcano crater on the island of Hawaii.

Isolated geographical areas, such as oceanic islands, lakes, and mountain peaks, often have high percentages of endemic species, often in significant danger of extinction. The number of endemic plant species varies greatly in the United States from one state to another. Thus, 379 plant species are found in Texas and nowhere else, whereas New York has only one endemic plant species. California, with its varied array of habitats, including deserts, mountains, seacoast, old growth forests, grasslands, and many others, is home to more endemic species than any other state.

Worldwide, notable concentrations of endemic species occur in particular "hot spots" of high endemism. Such hot spots are found in Madagascar, in a variety of tropical rain forests, in the eastern Himalayas, in areas with Mediterranean climates like California, South Africa, and Australia, and in several other climatic areas (figure 31.4 and table 31.2). Unfortunately, many of these areas are experiencing high rates of habitat destruction with consequent species extinctions. In Madagascar, it is estimated that 90% of the original forest has already been lost, this in an island in which 85% of the species are found nowhere else in the world. In the forests of the Atlantic coast of Brazil, the extent of deforestation is even higher: 95% of the original forest is gone.

Table 31.2 Numbers of Endemic Vertebrate Species in Some "Hot Spot" Areas

Region	Mammals	Reptiles	Amphibians
Atlantic coastal Brazil	73	60	253
South American Chocó	60	63	210
Philippines	111	159	65
Northern Borneo	42	69	47
Southwestern Australia	7	50	24
Madagascar	84	301	187
Cape region (South Africa)	9	19	19
California Floristic Province	30	16	17
New Caledonia	6	56	0
South Central China	75	16	51

Source: Data from Myers et al 2000.

Some areas of the earth have particularly high levels of species endemism. Unfortunately, many of these areas are currently in great jeopardy due to habitat destruction with correspondingly high rates of species extinction.

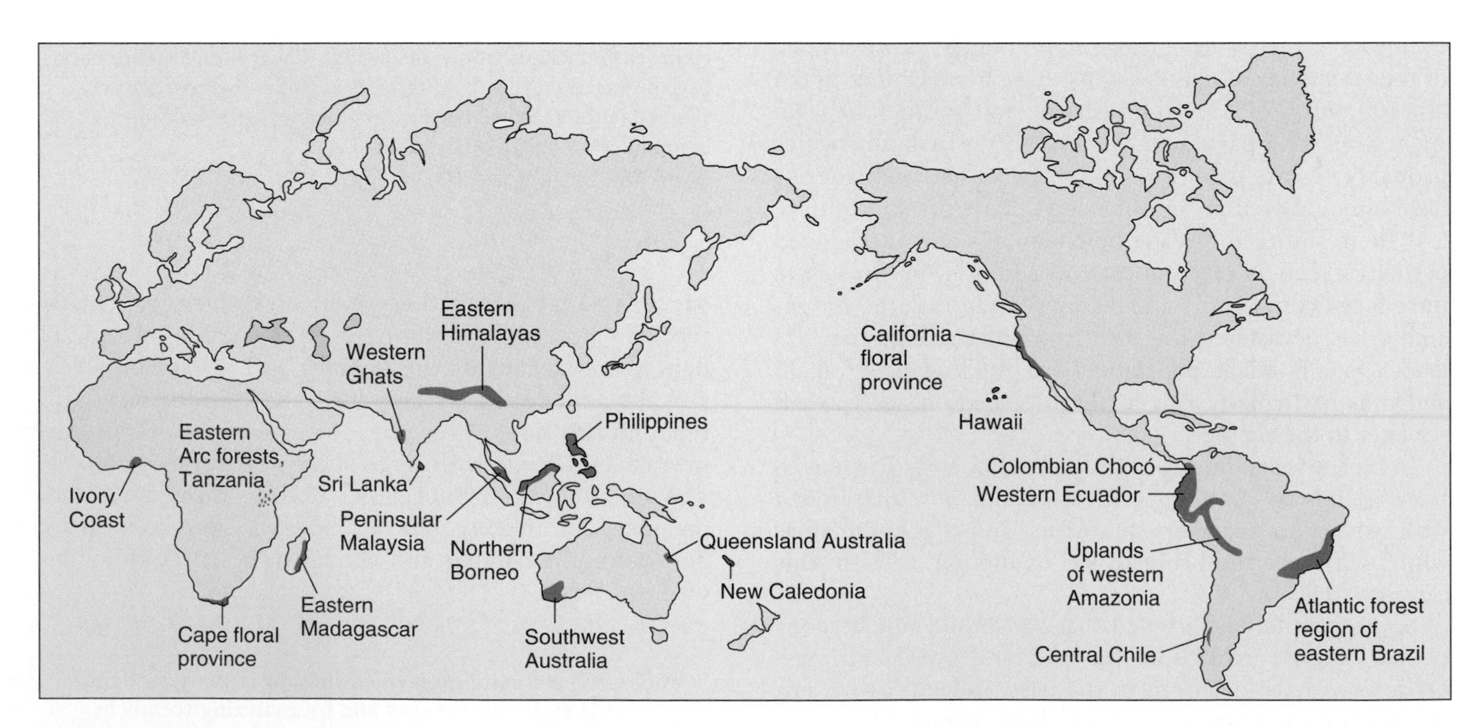

FIGURE 31.4
"Hot spots" of high endemism. These areas are rich in endemic species under significant threat of imminent extinction.
Source: After Myers, 1988, 1991.

What's So Bad about Losing Biodiversity?

What's so bad about losing species? What is the value of biodiversity? Its value can be divided into three principal components: (1) *direct economic value* of products we obtain from species of plants, animals, and other groups; (2) *indirect economic value* of benefits produced by species without our consuming them; and (3) *ethical* and *aesthetic* value.

Direct Economic Value

Many species have direct value, as sources of food, medicine, clothing, biomass (for energy and other purposes), and shelter. Most of the world's food, for example, is derived from a small number of plants that were originally domesticated from wild plants in tropical and semi-arid regions. In the future, wild strains of these species may be needed for their genetic diversity if we are to improve yields, or find a way to breed resistance to new pests.

About 40% of the prescription and nonprescription drugs used today have active ingredients extracted from plants or animals. Aspirin, the world's most widely used drug, was first extracted from the leaves of the tropical willow, *Salix alba*. The rosy periwinkle, *Catharanthus roseus*, from Madagascar has yielded potent drugs for combating leukemia (figure 31.5).

Only in the last few decades have biologists perfected the techniques that make possible the transfer of genes from one kind of organism to another. We are just beginning to be able to use genes obtained from other species to our advantage, as explored at length in chapter 19. So-called **gene prospecting** of the genomes of plants and animals for useful genes has only begun. We have been able to examine only a minute proportion of the world's organisms so far, to see whether any of their genes have useful properties. By conserving biodiversity we maintain the option of finding useful benefit in the future.

Indirect Economic Value

Diverse biological communities are of vital importance to healthy ecosystems, in maintaining the chemical quality of natural water, in buffering ecosystems against floods and drought, in preserving soils and preventing loss of minerals and nutrients, in moderating local and regional climate, in absorbing pollution, and in promoting the breakdown of organic wastes and the cycling of minerals. By destroying biodiversity, we are creating conditions of instability and lessened productivity and promoting desertification, waterlogging, mineralization, and many other undesirable outcomes throughout the world.

Given the major role played by many species in maintaining healthy ecosystems, it is alarming how little we know about the details of how ecosystems and communities function. It is impossible to predict all the consequences of removing a species, or to be sure that some of them will not be catastrophic. Imagine taking a parts list for an airliner, and randomly changing a digit in one of the part numbers: you might change a cushion to a roll of toilet paper—but you might as easily change a key bolt holding up the wing to a pencil. The point is, you shouldn't gamble if you cannot afford to lose, and in removing biodiversity we are gambling with the future of ecosystems upon which we depend, and upon whose functioning we understand very little.

FIGURE 31.5
The rosy periwinkle. Two drugs extracted from the Madagascar periwinkle *Catharanthus roseus*, vinblastine and vincristine, effectively treat common forms of childhood leukemia, increasing chances of survival from 20% to over 95%.

Ethical and Aesthetic Value

Many people believe that preserving biodiversity is an ethical issue, feeling that every species is of value in its own right, even if humans are not able to exploit or benefit from it. It is clear that humans have the power to exploit and destroy other species, but it is not as ethically clear that they have the *right* to do so. Many people believe that along with power comes responsibility: as the only organisms capable of eliminating species and entire ecosystems, and as the only organisms capable of reflecting upon what we are doing, we should act as guardians or stewards for the diversity of life around us.

Almost no one would deny the aesthetic value of biodiversity, of a beautiful flower or noble elephant, but how do we place a value on beauty? Perhaps the best we can do is to appreciate the deep sense of loss we feel at its permanent loss.

Biodiversity is of great value, for the products with which it provides us, for its contributions to the health of the ecosystems upon which we all depend, and for the beauty it provides us, as well as being valuable in its own right.

31.2 Vulnerable species are more likely to become extinct.

Predicting Which Species Are Vulnerable to Extinction

How can a biologist assess whether a particular species is vulnerable to extinction? To get some handle on this, conservation biologists look for changes in population size and habitat availability. Species whose populations are shrinking rapidly, whose habitats are being destroyed (figure 31.6), or which are endemic to small areas can be considered to be endangered.

FIGURE 31.6
Habitat removal. In this clear-cut lumbering of National Forest land in Washington State, few if any trees have been left standing, removing as well the home of the deer, birds, and other animal inhabitants of temperate forest. Until a replacement habitat is provided by replanting, this is a truly "lost" habitat.

Population Viability Analysis

Quantifying the risk faced by a particular species is not a simple or precise enterprise. Increasingly, conservation biologists make a rough estimate of a population's risk of local extinction in terms of a **minimum viable population** (MVP), the estimated number or density of individuals necessary for the population to maintain or increase its numbers.

Some small populations are at high risk of extinction, while other populations equally small are at little or no risk. Conservation biologists carry out a **population viability analysis** (PVA) to assess how the size of a population influences its risk of becoming extinct over a specific time period, often 100 years. Many factors must be taken into account in a PVA. Two components of particular importance are *demographic stochasticity* (the amount of random variation in birth and death rates) and *genetic stochasticity* (fluctuations in a population's level of genetic variation). Demographic stochasticity refers to random events that affect a population. The smaller the population, the more likely it is that a random event, such as a disease epidemic or an environmental disturbance (such as a flood or a fire) could decimate a population and lead to extinction. Similarly, small populations are most likely to lose genetic variation due to genetic drift (see chapter 20) and thus be vulnerable to both the short- and long-term consequences of genetic uniformity. For these reasons, small populations are at particularly great risk of extinction.

Many species are distributed as metapopulations, collections of small populations each occupying a suitable patch of habitat in an otherwise unsuitable landscape (see chapter 24). Each individual subpopulation may be quite small and in real threat of extinction, but the metapopulation may be quite safe from extinction so long as individuals from other populations repopulate the habitat patches vacated by extinct populations. The extent of this rescue effect is an important component of the PVA of such species; if rates of population extinction increase, there may not be enough surviving populations to found new populations, and the species as a whole may slide toward extinction.

Small populations are particularly in danger of extinction. To assess the risk of local extinction of a particular species, conservation biologists carry out a population viability analysis that takes into account demographic and genetic variation.

Dependence upon Other Species

Species often become vulnerable to extinction when their web of ecological interactions becomes seriously disrupted. A recent case in point are the sea otters that live in the cold waters off Alaska and the Aleutian Islands. A keystone species in the kelp forest ecosystem, the otter populations have declined sharply in recent years. In a 500-mile stretch of coastline, otter numbers dropped to an estimated 6000 from 53,000 in the 1970s, a plunge of nearly 90%. Investigating this catastrophic decline, marine ecologists uncovered a chain of interactions among the species of the ocean and kelp forest ecosystems, a falling domino chain of lethal effects.

The first in a series of events leading to the sea otter's decline seems to have been the heavy commercial harvesting of whales (see the case history later in this chapter). Without whales to keep their numbers in check, ocean zooplankton thrived, leading in turn to proliferation of a species of fish called pollock that feed on the now-abundant zooplankton. Given this ample food supply, the pollock proved to be very successful competitors of other northern Pacific fish like herring and ocean perch, so that levels of these other fish fell steeply in the 1970s.

Now the falling chain of dominoes begins to accelerate. The decline in the nutritious forage fish led to an ensuing crash in Alaskan populations of Steller's sea lions and harbor seals, for which pollock did not provide sufficient nourishment. Numbers of these pinniped species have fallen precipitously since the 1970s.

Pinnipeds are the major food of orcas, also called killer whales. Faced with a food shortage, some orcas seem to have turned to the next best thing: sea otters. In one bay where the entrance from the sea was too narrow and shallow for orcas to enter, only 12% of the sea otters have disappeared, while in a similar bay which orcas could enter easily, two-thirds of the otters disappeared in a year's time.

Without otters to eat them, the population of sea urchins in the ecosystem exploded, eating the kelp and so "deforesting" the kelp forests and denuding the ecosystem (figure 31.7). As a result, fish species that live in the kelp forest, like sculpins and greenlings (a cod relative), are declining. This chain reaction demonstrates why sea otters are considered to be a keystone species.

Commercial whaling appears to have initiated a series of changes that have led to orcas feeding on sea otters, with disastrous effects on their kelp forest ecosystem.

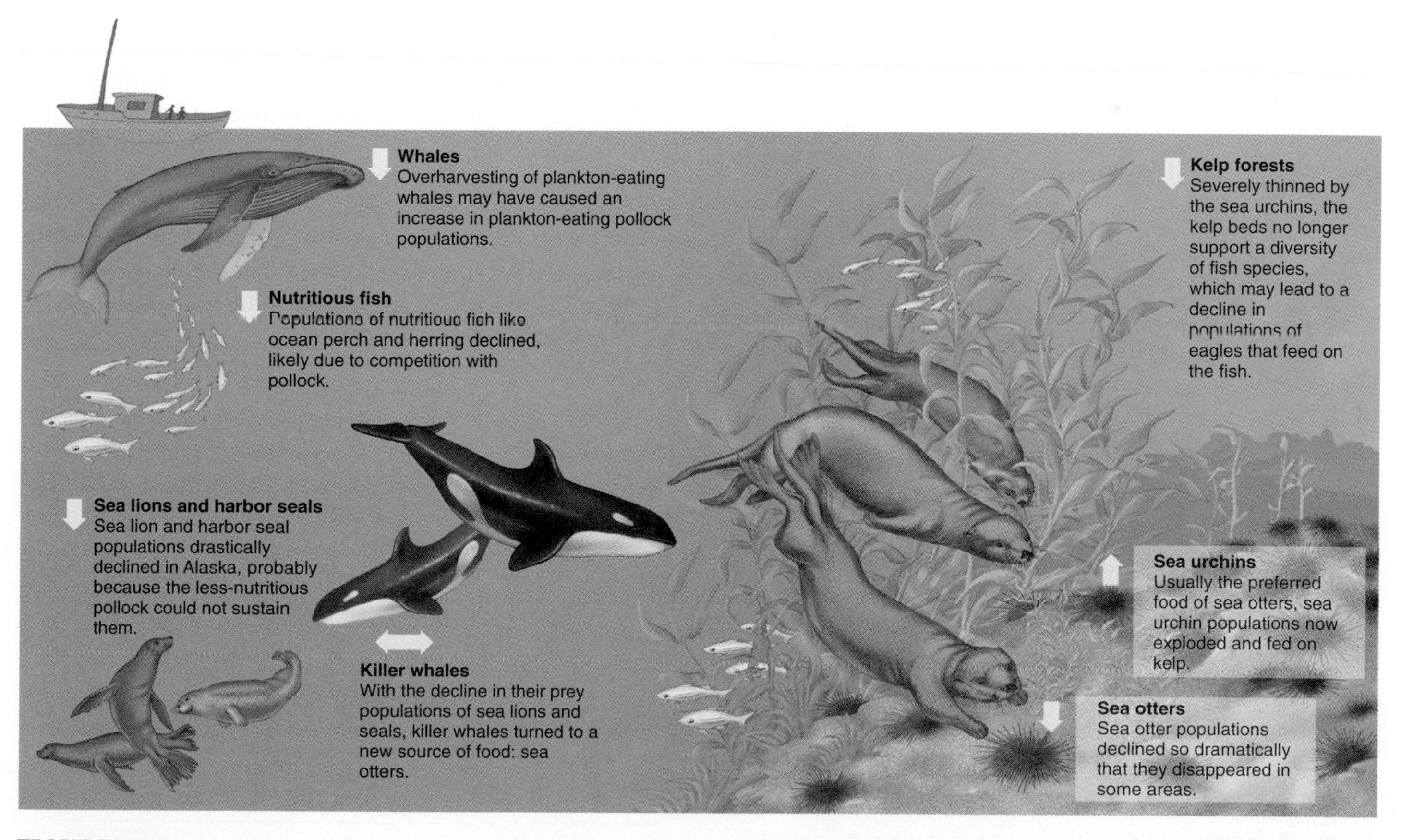

FIGURE 31.7
Disruption of the kelp forest ecosystem. Overharvesting by commercial whalers altered the balance of fish in the ocean ecosystem, inducing killer whales to feed on sea otters, a keystone species of the kelp forest ecosystem.

Categories of Vulnerable Species

Studying past extinctions of species and using population viability analyses of threatened ones, conservation biologists have observed that some categories of species are particularly vulnerable to extinction.

Local Endemic Distribution

Local endemic species typically occur at only one or a few sites in a restricted geographical range, which makes them particularly vulnerable to anything that harms the site, such as destruction of habitat by human activity. Bird species on oceanic islands have often become extinct as humans affect the island habitats. Many endemic fish species confined to a single lake undergo similar fates.

Local endemic species often have small population sizes, placing them at particular risk of extinction because of their greater vulnerability to demographic and genetic fluctuations. Indeed, population size by itself seems to be one of the best predictors of the extinction risk of populations.

Local endemic species often have quite specialized niche requirements. Once a habitat is altered, it may no longer be able to support a particular local endemic, while remaining satisfactory for species with less particular requirements. For example, wetland plants that require very specific and regular changes in water level may be rapidly eliminated when human activity affects the hydrology of an area.

Declining Population Size

Species in which population size is declining are often at grave risk of extinction, particularly if the decline in numbers of individuals is severe. Although there is no hard rule, population trends in nature tend to continue, so a population showing significant signs of decline should be considered at risk of extinction unless the cause of the decline is identified and corrected. Darwin makes this point very clearly in *On the Origin of Species*:

"To admit that species generally become rare before they become extinct, to feel no surprise at the rarity of the species, and yet to marvel greatly when the species ceases to exist, is much the same as to admit that sickness in the individual is the forerunner of death—to feel no surprise at sickness, but when the sick man dies, to wonder and to suspect that he dies of some deed of violence."

Although long-term trends toward smaller population numbers suggest that a species may be at risk in future years, abrupt recent declines in population numbers, particularly when the population is small or locally endemic, fairly scream of risk of extinction. It is for this reason that PVA is best carried out with data on population sizes gathered over a period of time.

Lack of Genetic Variability

Species with little genetic variability are generally at significantly greater risk of extinction than more variable species, simply because they have a more limited arsenal with which to respond to the vagaries of environmental change. Species with extremely low genetic variability are particularly vulnerable when faced with a new disease, predator, or other environmental challenge. For example, the African cheetah (*Acinonyx jubatus*) has almost no genetic variability. This lack of genetic variability is considered to be a significant contributing factor to a lack of disease resistance in the cheetah—diseases that are of little consequence to other cat species can wipe out a colony of cheetahs (although environmental factors also seem to have played a key role in the cheetah's decline).

Hunted or Harvested by People

Species that are hunted or harvested by people have historically been at grave risk of extinction. Overharvesting of natural populations can rapidly reduce the population size of a species, even when that species is initially very abundant. A century ago the skies of North America were darkened by huge flocks of passenger pigeons; hunted as free and tasty food, they were driven to extinction. The buffalo that used to migrate in enormous herds across the central plains of North America only narrowly escaped the same fate, a few individuals preserved from this catastrophic exercise in overhunting founded today's modest herds.

The existence of a commercial market often leads to overexploitation of a species. The international trade in furs, for example, has severely reduced the numbers of chinchilla, vicuna, otter, and many wild cat species. The harvesting of commercially valuable trees provides another telling example: almost all West Indies mahogany (*Swietenia mahogani*) have been logged from the Caribbean islands, and the extensive cedar forests of Lebanon, once widespread at high elevations in the Middle East, now survive in only a few isolated groves.

A particularly telling example of overharvesting of a so-called commercial species is the commercial harvesting of fish in the North Atlantic. Fishing fleets continued to harvest large amounts of cod off Newfoundland during the 1980s, even as the population numbers declined precipitously. By 1992 the cod population had dropped to less than 1% of their original numbers. The American and Canadian governments have closed the fishery, but no one can predict if the fish populations will recover. The Atlantic bluefin tuna has experienced a 90% population decline in the last 10 years. The swordfish has declined even further. In both cases, the drop has led to even more intense fishing of the remaining populations.

A variety of factors can make a species particularly vulnerable to extinction.

31.3 Causes of endangerment usually reflect human activities.

Factors Responsible For Extinction

Because a species is rare does not necessarily mean that it is in danger of extinction. The habitat it utilizes may simply be in short supply, preventing population numbers from growing. In a similar way, shortage of some other resource may be limiting the size of populations. Secondary carnivores, for example, are usually rare because so little energy is available to support their populations. Nor are vulnerable species such as those categories discussed in the previous section always threatened with extinction. Many local endemics are quite stable and not at all threatened.

If it's not just size or vulnerability, what factors are responsible for extinction? Studying a wide array of recorded extinctions and many species currently threatened with extinction, conservation biologists have identified a few factors that seem to play a key role in many extinctions: overexploitation, introduced species, disruption of ecological relationships, loss of genetic variability, and habitat loss and fragmentation (figure 31.8 and table 31.3).

Most recorded extinctions can be attributed to one of five causes: overexploitation, introduced species, ecodisruption, loss of genetic variability, and habitat loss and fragmentation.

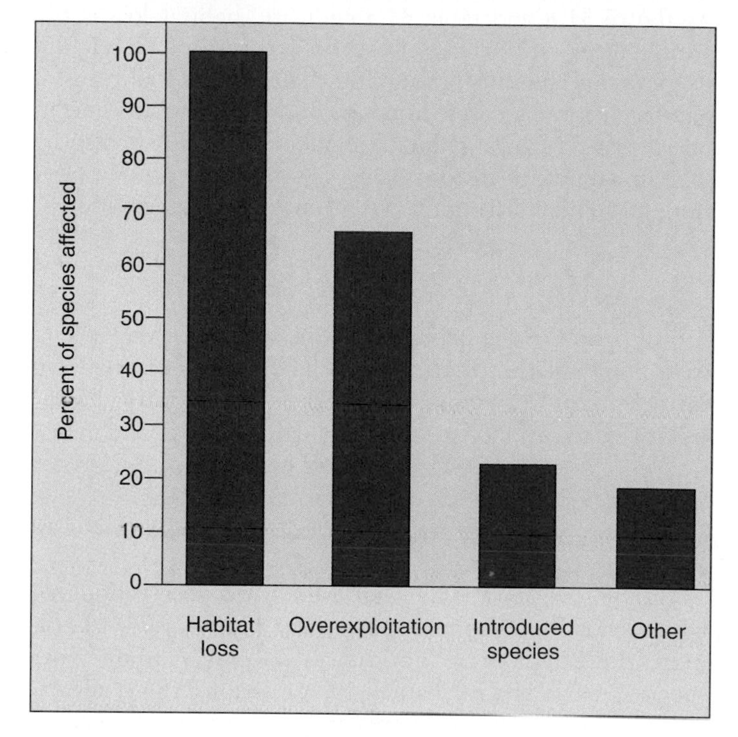

FIGURE 31.8
Factors responsible for animal extinction. These data represent known extinctions of mammals in Australasia and the Americas.
Source: Data from UNEP, 1994.

Table 31.3 Causes of Extinctions

Group	Habitat Loss	Overexploitation	Species Introduction	Predators	Other	Unknown
	Percentage of Species Influenced by the Given Factor*					
EXTINCTIONS						
Mammals	19	23	20	1	1	36
Birds	20	11	22	0	2	37
Reptiles	5	32	42	0	0	21
Fish	35	4	30	0	4	48
THREATENED EXTINCTIONS						
Mammals	68	54	6	8	12	—
Birds	58	30	28	1	1	—
Reptiles	53	63	17	3	6	—
Amphibians	77	29	14	—	3	—
Fish	78	12	28	—	2	—

*Some species may be influenced by more than one factor; thus, some rows may exceed 100%.
Source: Data from Reid and Miller, 1989.

Habitat Loss

As figure 31.8 and table 31.3 indicate, habitat loss is the single most important cause of extinction. Given the tremendous amounts of ongoing destruction of all types of habitat, from rain forest to ocean floor, this should come as no surprise. Natural habitats may be adversely affected by human influences in four ways: (1) destruction, (2) pollution, (3) human disruption, and (4) habitat fragmentation.

Destruction

A proportion of the habitat available to a particular species may simply be destroyed. This is a common occurrence in the "clear-cut" harvesting of timber, in the burning of tropical forest to produce grazing land, and in urban and industrial development. Forest clearance has been, and is, by far the most pervasive form of habitat disruption (figure 31.9). Many tropical forests are being cut or burned at a rate of 1% or more per year.

Biologists often use the well-established observation that larger areas support more species (see figure 28.24) to estimate the effect of reductions in habitat available to a species. As we saw in chapter 30, a relationship usually exists between the size of an area and the number of species it contains. Although this relationship varies according to geographic area, type of organism, and type of area (for example, oceanic islands, patches of habitat on the mainland), a general result is that a tenfold increase in area usually leads to approximately a doubling in number of species. This relationship suggests, conversely, that if the area of a habitat is reduced by 90%, so that only 10% remains, then half of all species will be lost. Evidence for this theory comes from a study of extinction rates of birds on habitat islands (that is, islands of a particular type of habitat surrounded by unsuitable habitat) in Finland where the extinction rate was found to be inversely proportional to island size (figure 31.10).

Pollution

Habitat may be degraded by pollution to the extent that some species can no longer survive there. Degradation occurs as a result of many forms of pollution, from acid rain to pesticides. Aquatic environments are particularly vulnerable; many northern lakes in both Europe and North America, for example, have been essentially sterilized by acid rain.

Human Disruption

Habitat may be so disturbed by human activities as to make it untenable for some species. For example, visitors to caves in Alabama and Tennessee produced significant population declines in bats over an eight-year period, some as great as 100%. When visits were fewer than one

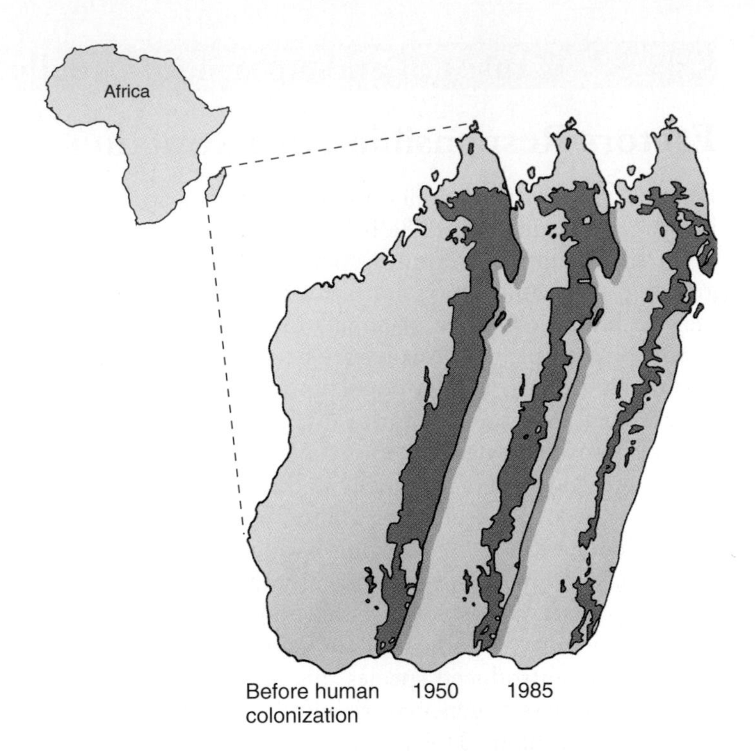

FIGURE 31.9
Extinction and habitat destruction. The rain forest covering the eastern coast of Madagascar, an island off the coast of East Africa, has been progressively destroyed as the island's human population has grown. Ninety percent of the original forest cover is now gone. Many species have become extinct, and many others are threatened, including 16 of Madagascar's 31 primate species.

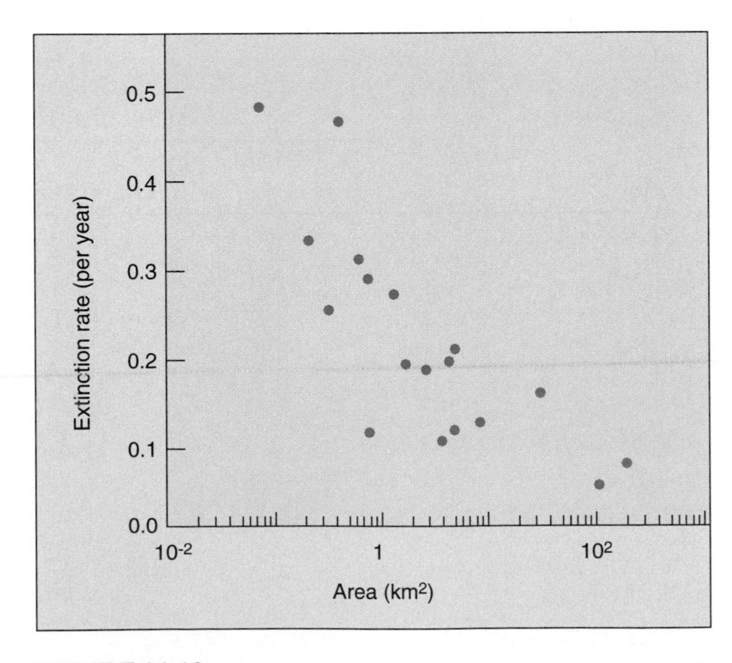

FIGURE 31.10
Extinction and the species-area relationship. The data present percent extinction rates as a function of habitat area for birds on a series of Finnish islands. Smaller islands experience far greater local extinction rates.

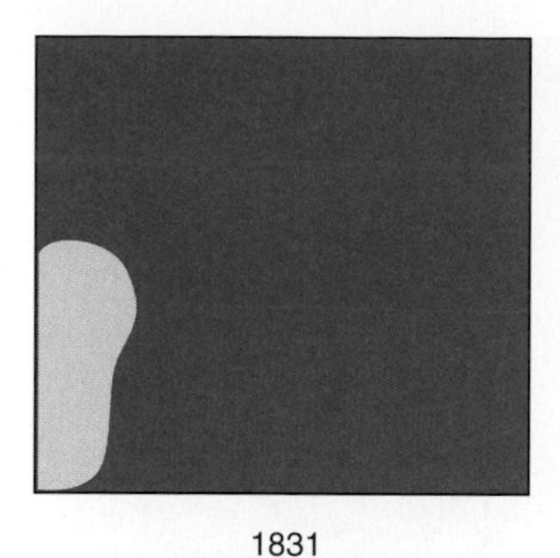

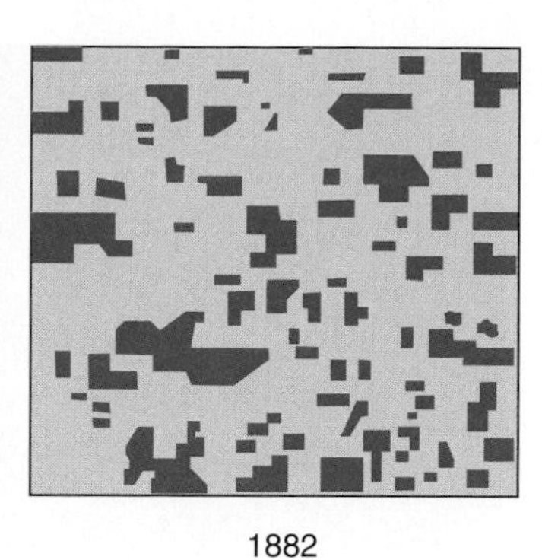

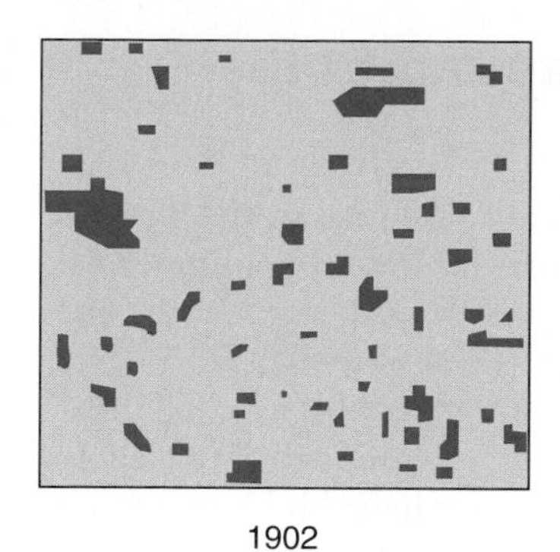

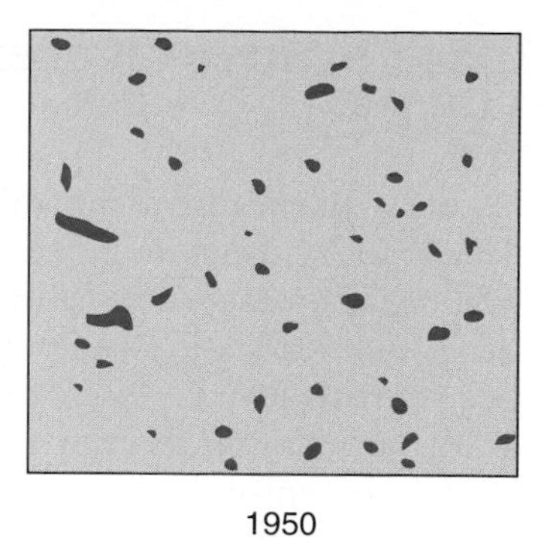

FIGURE 31.11
Fragmentation of woodland habitat. From the time of settlement of Cadiz Township, Wisconsin, the forest has been progressively reduced from a nearly continuous cover to isolated woodlots covering less than 1% of the original area.

per month, less than 20% of bats were lost, but caves with more than four visits per month suffered population declines of between 86 and 95%.

Habitat Fragmentation

Loss of habitat by a species frequently results not only in a lowering of population numbers, but also in fragmentation of the population into unconnected patches (figure 31.11).

A habitat may become fragmented in unobvious ways, such as when roads and habitation intrude into forest. The effect is to carve up the populations living in the habitat into a series of smaller populations, often with disastrous consequences. Although detailed data are not available, fragmentation of wildlife habitat in developed temperate areas is thought to be very substantial.

As habitats become fragmented and shrink in size, the relative proportion of the habitat that occurs on the boundary, or edge, increases. **Edge effects** can significantly degrade a population's chances of survival. Changes in microclimate (temperature, wind, humidity, etc.) near the edge may reduce appropriate habitat for many species more than the physical fragmentation suggests. In isolated fragments of rain forest, for example, trees on the edge are exposed to direct sunlight and, consequently, hotter and drier conditions than they are accustomed to in the cool, moist forest interior. As a result, within 100 meters of the forest edge, tree biomass decreased by 36% in the first 17 years after fragment isolation in one study.

Also, increasing habitat edges opens up opportunities for parasites and predators, both more effective at edges. As fragments decrease in size, the proportion of habitat that is distant from any edge decreases and, consequently, more and more of the habitat is within the range of these predators. Habitat fragmentation is thought to have been responsible for local extinctions in a wide range of species.

The impact of habitat fragmentation can be seen clearly in a major study done in Manaus, Brazil, as the rain forest was commercially logged. Landowners agreed to preserve patches of rain forest of various sizes, and censuses of these patches were taken before the logging started, while they were still part of a continuous forest. After logging, species began to disappear from the now-isolated patches (figure 31.12). First to go were the monkeys, which have large home ranges. Birds that prey on ant colonies followed, disappearing from patches too small to maintain enough ant colonies to support them.

FIGURE 31.12
A study of habitat fragmentation. Biodiversity was monitored in the isolated patches of rain forest in Manaus, Brazil, before and after logging. Fragmentation led to significant species loss within patches.

Because some species like monkeys require large patches, this means that large fragments are indispensable if we wish to preserve high levels of biodiversity. The take-home lesson is that preservation programs will need to provide suitably large habitat fragments to avoid this impact.

Habitat loss is probably the greatest cause of extinction. As habitats are destroyed, remaining habitat becomes fragmented, increasing the threat to many species.

Case Study: Overexploitation—Whales

Whales, the largest living animals, are rare in the world's oceans today, their numbers driven down by commercial whaling. Commercial whaling began in the sixteenth century, and reached its apex in the nineteenth and early twentieth centuries. Before the advent of cheap high-grade oils manufactured from petroleum in the early twentieth century, oil made from whale blubber was an important commercial product in the worldwide marketplace. In addition, the fine lattice-like structure used by baleen whales to filter-feed plankton from seawater (termed "baleen," but sometimes called "whalebone" even though it is actually made of keratin, like fingernails) was used in undergarments. Because a whale is such a large animal, each individual captured is of significant commercial value.

Right whales were the first to bear the brunt of commercial whaling. They were called right whales because they were slow, easy to capture, and provided up to 150 barrels of blubber oil and abundant whalebone, making them the "right" whale for a commercial whaler to hunt.

As the right whale declined in the eighteenth century, whalers turned to other species, the gray, humpback (figure 31.13), and bowhead. As their numbers declined, whalers turned to the blue, largest of all whales, and when they were decimated, to smaller whales: the fin, then the Sei, then the sperm whales. As each species of whale became the focus of commercial whaling, its numbers inevitably began a steep decline (figure 31.14).

Hunting of right whales was made illegal in 1935. By then, all three species had been driven to the brink of extinction, their numbers less than 5% of what they used to be. Protected since, their numbers have not recovered in either the North Atlantic or North Pacific. By 1946 several other species faced imminent extinction, and whaling nations formed the International Whaling Commission (IWC) to regulate commercial whale hunting. Like having the fox guard the hen house, the IWC for decades did little to limit whale harvests, and whale numbers continued a steep decline. Finally, in 1974, when numbers of all but the small minke whales had been driven down, the IWC banned hunting of blue, gray, and humpback whales, and instituted partial bans on other species. The rule was violated so often, however, that the IWC in 1986 instituted a worldwide moratorium on all commercial killing of whales. While some commercial whaling continues, often under the guise of harvesting for scientific studies, annual whale harvests have dropped dramatically in the last 15 years.

Some species appear to be recovering, while others do not. Humpback numbers have more than doubled since the early 1960s, increasing nearly 10% annually, and Pacific gray whales have fully recovered to their previous numbers of about 20,000 animals after being hunted to less than 1000. Right, sperm, fin, and blue whales have not recovered, and no one knows whether they will.

FIGURE 31.13
A humpback whale. Only 5000 to 10,000 humpback whales remain, out of a world population estimated to have been 100,000.

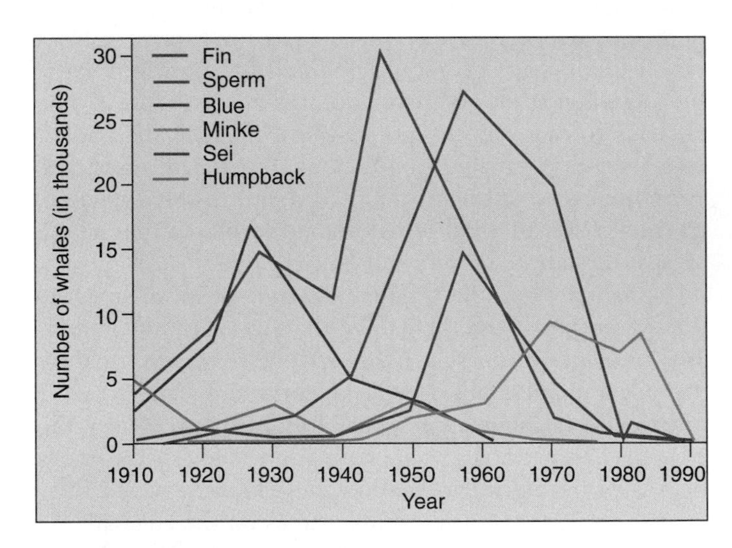

FIGURE 31.14
A history of commercial whaling. These data show the world catch of whales in the twentieth century. Each species is hunted in turn until its numbers fall so low that hunting it becomes commercially unprofitable.

Source: Data from UNEP, *Environmental Data Report*, 1993-1994.

Commercial whaling, by overharvesting, has driven most large whale species to the brink of extinction. Stopping the harvest has allowed recovery of some but not all species.

Case Study: Introduced Species—Lake Victoria Cichlids

Lake Victoria, an immense shallow freshwater sea about the size of Switzerland in the heart of equatorial East Africa, had until 1954 been home to an incredibly diverse collection of over 300 species of cichlid fishes (figure 31.15). These small, perchlike fishes range from 2 to 10 inches in length, with males coming in endless varieties of colors. Today, all of these cichlid species are threatened, endangered, or extinct.

What happened to bring about the abrupt loss of so many endemic cichlid species? In 1954, the Nile perch, *Lates niloticus*, a commercial fish with a voracious appetite, was introduced on the Ugandan shore of Lake Victoria. Nile perch, which grow to over 4 feet in length, were to form the basis of a new fishing industry (figure 31.16). For decades, these perch did not seem to have a significant impact—over 30 years later, in 1978, Nile perch still made up less than 2% of the fish harvested from the lake.

Then something happened to cause the Nile perch to explode and to spread rapidly through the lake, eating their way through the cichlids. By 1986, Nile perch constituted nearly 80% of the total catch of fish from the lake, and the endemic cichlid species were virtually gone. Over 70% of cichlid species disappeared, including all open-water species.

So what happened to kick-start the mass extinction of the cichlids? The trigger seems to have been eutrophication. Before 1978, Lake Victoria had high oxygen levels at all depths, down to the bottom layers exceeding 60 meters depth. However, by 1989 high inputs of nutrients from agricultural runoff and sewage from towns and villages had led to algal blooms that severely depleted oxygen levels in deeper parts of the lake. Cichlids feed on algae, and initially their population numbers are thought to have risen in response to this increase in their food supply, but unlike similar algal blooms of the past, the Nile perch was now present to take advantage of the situation. With a sudden increase in its food supply (cichlids), the numbers of Nile perch exploded, and the greater numbers of them simply ate all available cichlids.

Since 1990 the situation has been compounded by a second factor, the introduction into Lake Victoria of a floating water weed from South America, the water hyacinth *Eichornia crassipes*. Extremely fecund under eutrophic conditions, thick mats of water hyacinth soon covered entire bays and inlets, choking off the coastal habitats of non-open-water cichlids.

Lake Victoria's diverse collection of cichlid species is being driven to extinction by an introduced species, the Nile perch. A normal increase in cichlid numbers due to algal blooms led to an explosive increase in perch, which then ate their way through the cichlids.

FIGURE 31.15
Lake Victoria cichlids. Cichlid fishes are extremely diverse and occupy different niches. Some species feed on arthropods, others on dense stands of plants; there are fish-eaters, and still other species feed on fish eggs and larvae.

FIGURE 31.16
Victor and vanquished. The Nile perch (larger fishes in foreground), a commercial fish introduced into Lake Victoria as a potential food source, is responsible for the virtual extinction of hundreds of species of cichlid fishes (smaller fishes in tub).

Case Study: Disruption of Ecological Relationships—Black-Footed Ferrets

The black-footed ferret (*Mustela nigripes*) is one of the most attractive weasels of North America. A highly specialized predator, black-footed ferrets prey on prairie dogs, which live in large underground colonies connected by a maze of tunnels. These ferrets experienced a dramatic decline in their North American range during the past century, as agricultural development destroyed their prairie habitat, and particularly the prairie dogs on which they feed (figure 31.17). Prairie dogs once roamed freely over 100 million acres of the Great Plains states, but are now confined to under 700,000 acres (table 31.4). Their ecological niche devastated, populations of the black-footed ferret collapsed. Increasingly rare in the second half of the century, the black-footed ferret was thought to have gone extinct in the late 1970s, when the only known wild population—a small colony in South Dakota—died out.

In 1981, a colony of 128 animals was located in Meeteese, Wyoming. Left undisturbed for four years, the number of ferrets dropped by 50%, and the entire population seemed in immediate danger of extinction. A decision was made to capture some animals for a captive breeding program. The first six black-footed ferrets captured died of canine distemper, a disease present in the colony and probably responsible for its rapid decline.

At this point, drastic measures seemed called for. In the next year, a concerted effort was made to capture all the remaining ferrets in the Meeteese colony. A captive population of 18 individuals was established before the Meeteese colony died out. The breeding program proved a great success, the population jumping to 311 individuals by 1991.

In 1991, biologists began to attempt to reintroduce black-footed ferrets to the wild, releasing 49 animals in Wyoming. An additional 159 were released over the next two years. Six litters were born that year in the wild, and the reintroduction seemed a success. However, the released animals then underwent a drastic decline, and only ten individuals were still alive in the wild five years later in 1998. The reason for the decline is not completely understood, but predators such as coyotes appear to have played a large role. Current attempts at reintroduction involve killing the local coyotes. It is important that these attempts succeed, as numbers of offspring in the captive breeding colony are declining, probably as a result of intensive inbreeding. The black-footed ferret still teeters at the brink of extinction.

Loss of its natural prey has eliminated black-footed ferrets from the wild; attempts to reintroduce them have not yet proven successful.

FIGURE 31.17
Teetering on the brink. The black-footed ferret is a predator of prairie dogs. Loss of prairie dog habitat as agriculture came to dominate the plains states in the last century has led to a drastic decline in prairie dogs, and an even more drastic decline in the black-footed ferrets that feed on them. Attempts are now being made to reestablish natural populations of these ferrets, which became extinct in the wild in 1986.

Table 31.4 Acres of Prairie Dog Habitat

State	1899-1990	1998
Arizona	unknown	extinct
Colorado	7,000,000	44,000
Kansas	2,500,000	36,000
Montana	6,000,000	65,000
Nebraska	6,000,000	60,000
New Mexico	12,000,000	15,000
North Dakota	2,000,000	20,400
Oklahoma	950,000	9,500*
South Dakota	1,757,000	244,500
Texas	56,833,000	22,650
Wyoming	16,000,000	70,000–180,000
U.S. Total	**111,000,000**	**700,000**

Source: Data from National Wildlife Federation and U.S. Fish and Wildlife report, 1998. *1990.

Case Study: Loss of Genetic Variation—Prairie Chickens

The greater prairie chicken (*Tympanuchus cupido pinnatus*) is a showy 2-pound wild bird renowned for its flamboyant mating rituals (figure 31.18). Abundant in many midwestern states, the prairie chickens in Illinois have in the last six decades undergone a population collapse. Once, enormous numbers of birds covered the state, but with the introduction of the steel plow in 1837, the first that could slice through the deep dense root systems of prairie grasses, the Illinois prairie began to be replaced by farmland, and by the turn of the century the prairie had vanished. By 1931, the subspecies known as the heath hen (*Tympanuchus cupido cupido*) became extinct in Illinois. The greater prairie chicken fared little better in Illinois, its numbers falling to 25,000 statewide in 1933, then to 2000 in 1962. In surrounding states with less intensive agriculture, it continued to prosper.

In 1962, a sanctuary was established in an attempt to preserve the prairie chicken, and another in 1967. But privately owned grasslands kept disappearing, with their prairie chickens, and by the 1980s the birds were extinct in Illinois except for the two preserves. And there they were not doing well. Their numbers kept falling. By 1990, the egg hatching rate, which had averaged between 91 and 100%, had dropped to an extremely low 38%. By the mid-1990s, the count of males dropped to as low as six in each sanctuary.

What was wrong with the sanctuary populations? One reasonable suggestion was that because of very small population sizes and a mating ritual where one male may dominate a flock, the Illinois prairie chickens had lost so much genetic variability as to create serious inbreeding problems. To test this idea, biologists at the University of Illinois compared DNA from frozen tissue samples of birds that died in Illinois between 1974 and 1993 and found that in recent years, Illinois birds had indeed become genetically less diverse. Extracting DNA from tissue in the roots of feathers from stuffed birds collected in the 1930s from the same population, the researchers found little genetic difference between the Illinois birds of the 1930s and present-day prairie chickens of other states. However, present-day Illinois birds had lost fully one-third of the genetic diversity of birds living in the same place before the population collapse of the 1970s.

Now the stage was set to halt the Illinois prairie chicken's race toward extinction. Wildlife managers began to transplant birds from genetically diverse populations of Minnesota, Kansas, and Nebraska to Illinois. Between 1992 and 1996, a total of 518 out-of-state prairie chickens were brought in to interbreed with the Illinois birds, and hatching rates were back up to 94% by 1998. It looks like the Illinois prairie chickens have been saved from extinction.

The key lesson to be learned is the importance of not allowing things to go too far, not to drop down to a single isolated population (figure 31.19). Without the outlying genetically different populations, the prairie chickens in Illinois could not have been saved. The black-footed ferrets discussed on the previous page are particularly endangered because they exist as a single isolated population.

FIGURE 31.18
A male prairie chicken performing a mating ritual. He inflates bright orange air sacs, part of his esophagus, into balloons on each side of his head. As air is drawn into the sacs, it creates a three-syllable "boom-boom-boom" that can be heard for miles.

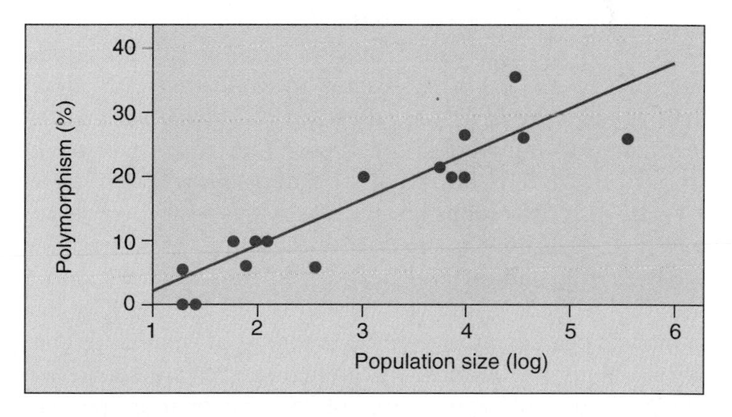

FIGURE 31.19
Small populations lose much of their genetic variability. The percentage of polymorphic genes in isolated populations of the tree *Halocarpus bidwilli* in the mountains of New Zealand is a sensitive function of population size.

Source: Data from H.L. Billington, "Effects of Population Size on a Genetic Variation in a Dioecious Conifer: in *Conservation Biology*, Blackwell Scientific Publication, Inc., 1991.

When their numbers fell, Illinois prairie chickens lost much of their genetic variability, resulting in reproductive failure and the threat of immediate extinction. Breeding with genetically more variable birds appears to have reversed the decline.

Case Study: Habitat Loss and Fragmentation—Songbirds

Every year since 1966, the U.S. Fish and Wildlife Service has organized thousands of amateur ornithologists and bird watchers in an annual bird count called the Breeding Bird Survey. In recent years, a shocking trend has emerged. While year-round residents that prosper around humans, like robins, starlings, and blackbirds, have increased their numbers and distribution over the last 30 years, forest songbirds have declined severely. The decline has been greatest among long-distance migrants such as thrushes, orioles, tanagers, catbirds, vireos, buntings, and warblers. These birds nest in northern forests in the summer, but spend their winters in South or Central America or the Caribbean Islands.

In many areas of the eastern United States, more than three-quarters of the neotropical migrant bird species have declined significantly. Rock Creek Park in Washington, D.C., for example, has lost 90 percent of its long distance migrants in the last 20 years. Nationwide, American redstarts declined about 50% in the single decade of the 1970s. Studies of radar images from National Weather Service stations in Texas and Louisiana indicate that only about half as many birds fly over the Gulf of Mexico each spring compared to numbers in the 1960s. This suggests a loss of about half a billion birds in total, a devastating loss.

The culprit responsible for this widespread decline appears to be habitat fragmentation and loss. Fragmentation of breeding habitat and nesting failures in the summer nesting grounds of the United States and Canada have had a major negative impact on the breeding of woodland songbirds. Many of the most threatened species are adapted to deep woods and need an area of 25 acres or more per pair to breed and raise their young. As woodlands are broken up by roads and developments, it is becoming increasingly difficult to find enough contiguous woods to nest successfully.

A second and perhaps even more important factor seems to be the availability of critical winter habitat in Central and South America. Living in densely crowded limited areas, the availability of high-quality food is critical. Studies of the American redstart clearly indicate that birds with better winter habitat have a superior chance of successfully migrating back to their breeding grounds in the spring. Peter Marra and Richard Holmes of Dartmouth College and Keith Hobson of the Canadian Wildlife Service captured birds, took blood samples, and measured the levels of the stable carbon isotope ^{13}C. Plants growing in the best overwintering habitats in Jamaica and Honduras (mangroves and wetland forests) have low levels of ^{13}C, and so do the redstarts that feed on them. Sixty-five percent of the wet forest birds maintained or gained weight over the winter. Plants growing in substandard dry scrub, by contrast, have high levels of ^{13}C, and so do the redstarts that feed on them. Scrub-dwelling birds lost up to 11% of their body mass over the winter. Now here's the key: birds that winter in the substandard scrub leave later in the spring on the long flight to northern breeding grounds, arrive at their summer homes, and have fewer young. You can see this clearly in the redstart study (figure 31.20): the proportion of ^{13}C carbon in birds arriving in New Hampshire breeding grounds increases as spring wears on and scrub-overwintering stragglers belatedly arrive. Thus, loss of mangrove habitat in the neotropics is having a real negative impact. As the best habitat disappears, overwintering birds fare poorly, and this leads to population declines. Unfortunately, the Caribbean lost about 10% of its mangroves in the 1980s, and continues to lose about 1% a year. This loss of key habitat appears to be a driving force in the looming extinction of songbirds.

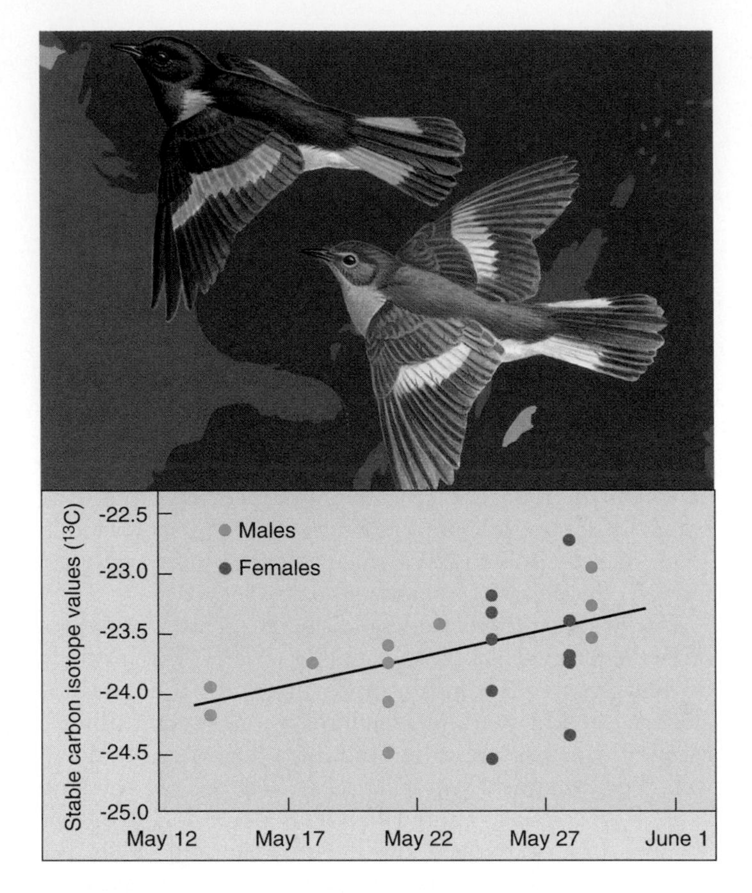

FIGURE 31.20
The American redstart, a migratory songbird whose numbers are in serious decline. The graph presents data on the level of ^{13}C in redstarts arriving at summer breeding grounds. Early arrivals, with the best shot at reproductive success, have lower levels of ^{13}C, indicating they wintered in more favorable mangrove-wetland forest habitats.
Source: Data from Marra, Hobson, Holmes, "Linking Winter & Summer Events", in *Science*, Dec. 1998.

Fragmentation of summer breeding grounds and loss of high-quality overwintering habitat seem both to be contributing to a marked decline in migratory songbird species.

31.4 Successful recovery plans will need to be multidimensional.

Many Approaches Exist for Preserving Endangered Species

Once you understand the reasons why a particular species is endangered, it becomes possible to think of designing a recovery plan. If the cause is commercial overharvesting, regulations can be designed to lessen the impact and protect the threatened species. If the cause is habitat loss, plans can be instituted to restore lost habitat. Loss of genetic variability in isolated subpopulations can be countered by transplanting individuals from genetically different populations. Populations in immediate danger of extinction can be captured, introduced into a captive breeding program, and later reintroduced to other suitable habitat.

Of course, all of these solutions are extremely expensive. As Bruce Babbitt, Interior Secretary in the Clinton administration, noted, it is much more economical to prevent such "environmental trainwrecks" from occurring than it is to clean them up afterwards. Preserving ecosystems and monitoring species before they are threatened is the most effective means of protecting the environment and preventing extinctions.

(a)

(b)

FIGURE 31.21
The University of Wisconsin-Madison Arboretum has pioneered restoration ecology. (*a*) The restoration of the prairie was at an early stage in November, 1935. (*b*) The prairie as it looks today. This picture was taken at approximately the same location as the 1935 photograph.

Habitat Restoration

Conservation biology typically concerns itself with preserving populations and species in danger of decline or extinction. Conservation, however, requires that there be something left to preserve, while in many situations, conservation is no longer an option. Species, and in some cases whole communities, have disappeared or have been irretrievably modified. The clear-cutting of the temperate forests of Washington State leaves little behind to conserve; nor does converting a piece of land into a wheat field or an asphalt parking lot. Redeeming these situations requires restoration rather than conservation.

Three quite different sorts of habitat restoration programs might be undertaken, depending very much on the cause of the habitat loss.

Pristine Restoration. In situations where all species have been effectively removed, one might attempt to restore the plants and animals that are believed to be the natural inhabitants of the area, when such information is available. When abandoned farmland is to be restored to prairie (figure 31.21), how do you know what to plant? Although it is in principle possible to reestablish each of the original species in their original proportions, rebuilding a community requires that you know the identity of all of the original inhabitants, and the ecologies of each of the species. We rarely ever have this much information, so no restoration is truly pristine.

Removing Introduced Species. Sometimes the habitat of a species has been destroyed by a single introduced species. In such a case, habitat restoration involves removal of the introduced species. Restoration of the once-diverse cichlid fishes to Lake Victoria will require more than breeding and restocking the endangered species. Eutrophication will have to be reversed, and the introduced water hyacinth and Nile perch populations brought under control or removed.

It is important to act quickly if an introduced species is to be removed. When aggressive African bees (the so-called "killer bees") were inadvertently released in Brazil, they remained in the local area only one season. Now they occupy much of the Western hemisphere.

Cleanup and Rehabilitation. Habitats seriously degraded by chemical pollution cannot be restored until the pollution is cleaned up. The successful restoration of the Nashua River in New England, discussed in chapter 30, is one example of how a concerted effort can succeed in restoring a heavily polluted habitat to a relatively pristine condition.

Captive Propagation

Recovery programs, particularly those focused on one or a few species, often must involve direct intervention in natural populations to avoid an immediate threat of extinction. Earlier we learned how introducing wild-caught individuals into captive breeding programs is being used in an attempt to save ferret and prairie chicken populations in immediate danger of disappearing. Several other such captive propagation programs have had significant success.

Case History: The Peregrine Falcon. American populations of birds of prey such as the peregrine falcon (*Falco peregrinus*) began an abrupt decline shortly after World War II. Of the approximately 350 breeding pairs east of the Mississippi River in 1942, all had disappeared by 1960. The culprit proved to be the chemical pesticide DDT (dichlorodiphenyltrichloroethane) and related organochlorine pesticides. Birds of prey are particularly vulnerable to DDT because they feed at the top of the food chain, where DDT becomes concentrated. DDT interferes with the deposition of calcium in the bird's eggshells, causing most of the eggs to break before they hatch.

The use of DDT was banned by federal law in 1972, causing levels in the eastern United States to fall quickly. There were no peregrine falcons left in the eastern United States to reestablish a natural population, however. Falcons from other parts of the country were used to establish a captive breeding program at Cornell University in 1970, with the intent of reestablishing the peregrine falcon in the eastern United States by releasing offspring of these birds By the end of 1986, over 850 birds had been released in 13 eastern states, producing an astonishingly strong recovery (figure 31.22).

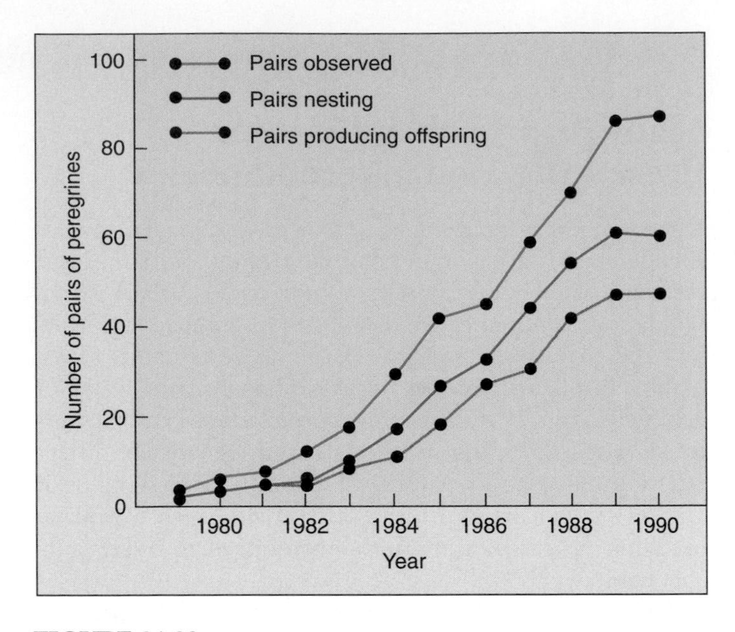

FIGURE 31.22
Captive propagation. The peregrine falcon has been reestablished in the eastern United States by releasing captive-bred birds over a period of 10 years.
Source: Data from The Peregrine Fund.

Case History: The California Condor. Numbers of the California condor (*Gymnogyps californianus*), a large vulturelike bird with a wingspan of nearly 3 meters, have been declining gradually for the last 200 years. By 1985 condor numbers had dropped so low the bird was on the verge of extinction. Six of the remaining 15 wild birds disappeared that year alone. The entire breeding population of the species consisted of the 6 birds remaining in the wild, and an additional 21 birds in captivity. In a last-ditch attempt to save the condor from extinction, the remaining birds were captured and placed in a captive breeding population. The breeding program was set up in zoos, with the goal of releasing offspring on a large 5300-ha ranch in prime condor habitat. Birds were isolated from human contact as much as possible, and closely related individuals were prevented from breeding. By the end of 1999 the captive population of California condors had reached over 110 individuals. Twenty-nine captive-reared condors have been released successfully in California at two sites in the mountains north of Los Angeles, after extensive prerelease training to avoid power poles and people. All of the released birds seem to be doing well. Twenty additional birds released into the Grand Canyon have adapted well. Biologists are waiting to see if the released condors will breed in the wild and successfully raise a new generation of wild condors.

Case History: Yellowstone Wolves. The ultimate goal of captive breeding programs is not simply to preserve interesting species, but rather to restore ecosystems to a balanced functional state. Yellowstone Park has been an ecosystem out of balance, due in large part to the systematic extermination of the gray wolf (*Canis lupus*) in the park early in this century. Without these predators to keep their numbers in check, herds of elk and deer expanded rapidly, damaging vegetation so that the elk themselves starve in times of scarcity. In an attempt to restore the park's natural balance, two complete wolf packs from Canada were released into the park in 1995 and 1996. The wolves adapted well, breeding so successfully that by 2001 the park contained 16 free-ranging packs, a total of 164 wolves.

While ranchers near the park have been unhappy about the return of the wolves, little damage to livestock has been noted, and the ecological equilibrium of Yellowstone Park seems well on the way to recovery. Elk are congregating in larger herds, and their populations are not growing as rapidly as in years past. Importantly, wolves are killing coyotes and their pups, driving them out of some areas. Coyotes, the top predators in the absence of wolves, are known to attack cattle on surrounding ranches, so reintroduction of wolves to the park may actually benefit the cattle ranchers that are opposed to it.

Sustaining Genetic Diversity

One of the chief obstacles to a successful species recovery program is that a species is generally in serious trouble by the time a recovery program is instituted. When populations become very small, much of their genetic diversity is lost (see figure 31.19), as we have seen clearly in our examination of the case histories of prairie chickens and black-footed ferrets. If a program is to have any chance of success, every effort must be made to sustain as much genetic diversity as possible.

Case History: The Black Rhino. All five species of rhinoceros are critically endangered. The three Asian species live in forest habitat that is rapidly being destroyed, while the two African species are illegally killed for their horns. Fewer than 11,000 individuals of all five species survive today. The problem is intensified by the fact that many of the remaining animals live in very small, isolated populations. The 2400 wild-living individuals of the black rhino, *Diceros bicornis*, live in approximately 75 small widely separated groups (figure 31.23) consisting of six subspecies adapted to local conditions throughout the species' range. All of these subspecies appear to have low genetic variability; in three of the subspecies, only a few dozen animals remain. Analysis of mitochondrial DNA suggests that in these populations most individuals are genetically very similar.

This lack of genetic variability represents the greatest challenge to the future of the species. Much of the range of the black rhino is still open and not yet subject to human encroachment. To have any significant chance of success, a species recovery program will have to find a way to sustain the genetic diversity that remains in this species. Heterozygosity could be best maintained by bringing all black rhinos together in a single breeding population, but this is not a practical possibility. A more feasible solution would be to move individuals between populations. Managing the black rhino populations for genetic diversity could fully restore the species to its original numbers and much of its range.

Placing black rhinos from a number of different locations together in a sanctuary to increase genetic diversity raises a potential problem: local subspecies may be adapted in different ways to their immediate habitats—what if these local adaptations are crucial to their survival? Homogenizing the black rhino populations by pooling their genes risks destroying such local adaptations, if they exist, perhaps at great cost to survival.

(a)

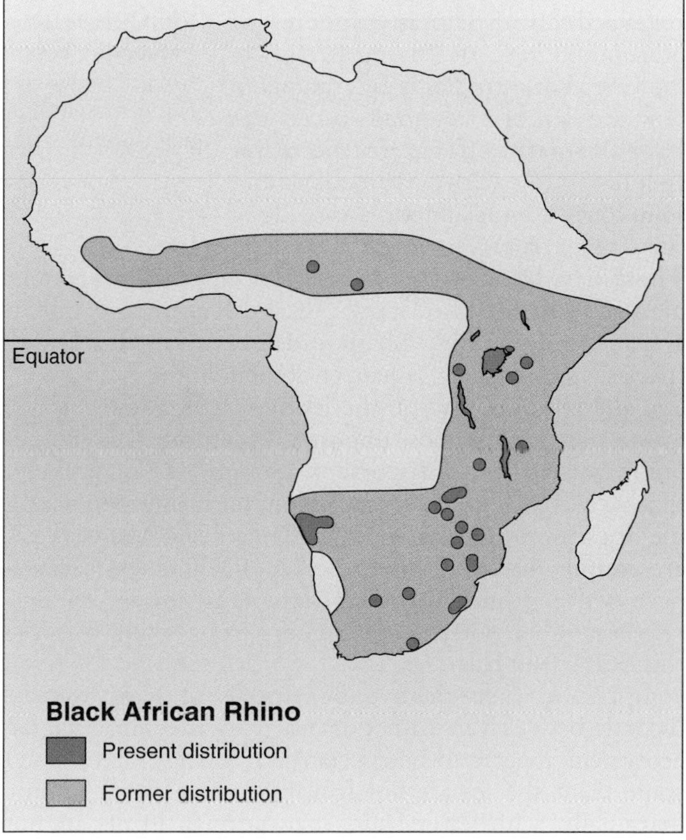

(b)

FIGURE 31.23
Sustaining genetic diversity. The black rhino (*a*) is highly endangered, living in 75 small, widely separated populations (*b*). Only about 2400 individuals survive in the wild.
Source: (*b*) Data from Caughley and Gum, 1996.

Preserving Keystone Species

Keystone species are species that exert a particularly strong influence on the structure and functioning of a particular ecosystem. The sea otters of figure 31.7 are a keystone species of the kelp forest ecosystem, and their removal can have disastrous consequences. There is no hard-and-fast line that allows us to clearly identify keystone species. It is rather a qualitative concept, a statement that a species plays a particularly important role in its community. Keystone species are usually characterized by measuring the strength of their impact on their community. **Community importance** measures the change in some quantitative aspect of the ecosystem (species richness, productivity, nutrient cycling) per unit of change in the abundance of a species.

FIGURE 31.24
Preserving keystone species. The flying fox is a keystone species in many Old World tropical islands. It pollinates many of the plants, and is a key disperser of seeds. Its elimination by hunting and habitat loss is having a devastating effect on the ecosystems of many South Pacific islands.

Case History: Flying Foxes. The severe decline of many species of pteropodid bats, or "flying foxes," in the Old World tropics is an example of how the loss of a keystone species can have dramatic effects on the other species living within an ecosystem, sometimes even leading to a cascade of further extinctions (figure 31.24). These bats have very close relationships with important plant species on the islands of the Pacific and Indian Oceans. The family Pteropodidae contains nearly 200 species, approximately a quarter of them in the genus *Pteropus*, and is widespread on the islands of the South Pacific, where they are the most important—and often the only—pollinators and seed dispersers. A study in Samoa found that 80 to 100% of the seeds landing on the ground during the dry season were deposited by flying foxes. Many species are entirely dependent on these bats for pollination. Some have evolved features like night-blooming flowers that prevent any other potential pollinators from taking over the role of the fruit bats.

In Guam, where the two local species of flying fox have recently been driven extinct or nearly so, the impact on the ecosystem appears to be substantial. Botanists have found some plant species are not fruiting, or are doing so only marginally, with fewer fruits than normal. Fruits are not being dispersed away from parent plants, so offspring shoots are being crowded out by the adults.

Flying foxes are being driven to extinction by human hunting. They are hunted for food, for sport, and by orchard farmers, who consider them pests. Flying foxes are particularly vulnerable because they live in large, easily seen groups of up to a million individuals. Because they move in regular and predictable patterns and can be easily tracked to their home roost, hunters can easily bag thousands at a time.

Species preservation programs aimed at preserving particular species of flying foxes are only just beginning. One particularly successful example is the program to save the Rodrigues fruit bat, *Pteropus rodricensis*, which occurs only on Rodrigues Island in the Indian Ocean near Madagascar. The population dropped from about 1000 individuals in 1955 to fewer than 100 by 1974, the drop reflecting largely the loss of the fruit bat's forest habitat to farming. Since 1974 the species has been legally protected, and the forest area of the island is being increased through a tree-planting program. Eleven captive breeding colonies have been established, and the bat population is now increasing rapidly. The combination of legal protection, habitat restoration, and captive breeding has in this instance produced a very effective preservation program.

Recovery programs at the species level must deal with habitat loss and fragmentation, and often with a marked reduction in genetic diversity. Captive breeding programs that stabilize genetic diversity and pay careful attention to habitat preservation and restoration are typically involved in successful recoveries.

Conservation of Ecosystems

Habitat fragmentation is one of the most pervasive enemies of biodiversity conservation efforts. As we have seen, some species simply require large patches of habitat to thrive, and conservation efforts that cannot provide suitable habitat of such a size are doomed to failure. As it has become clear that isolated patches of habitat lose species far more rapidly than large preserves do, conservation biologists have promoted the creation, particularly in the tropics, of so-called megareserves, large areas of land containing a core of one or more undisturbed habitats (figure 31.25). The key to devoting such large tracts of land to reserves successfully over a long period of time is to operate the reserve in a way compatible with local land use. Thus, while no economic activity is allowed in the core regions of the megareserve, the remainder of the reserve may be used for nondestructive harvesting of resources. Linking preserved areas to carefully managed land zones creates a much larger total "patch" of habitat than would otherwise be economically practical, and thus addresses the key problem created by habitat fragmentation. Pioneering these efforts, a series of eight such megareserves have been created in Costa Rica (figure 31.26) to jointly manage biodiversity and economic activity.

In addition to this focus on maintaining large enough reserves, in recent years, conservation biologists also have recognized that the best way to preserve biodiversity is to focus on preserving intact ecosystems, rather than focusing on particular species. For this reason, attention in many cases is turning to identifying those ecosystems most in need of preservation and devising the means to protect not only the species within the ecosystem, but the functioning of the ecosystem itself.

Efforts are being undertaken worldwide to preserve biodiversity in megareserves designed to counter the influences of habitat fragmentation. Focusing on the health of entire ecosystems, rather than of particular species, can often be a more effective means of preserving biodiversity.

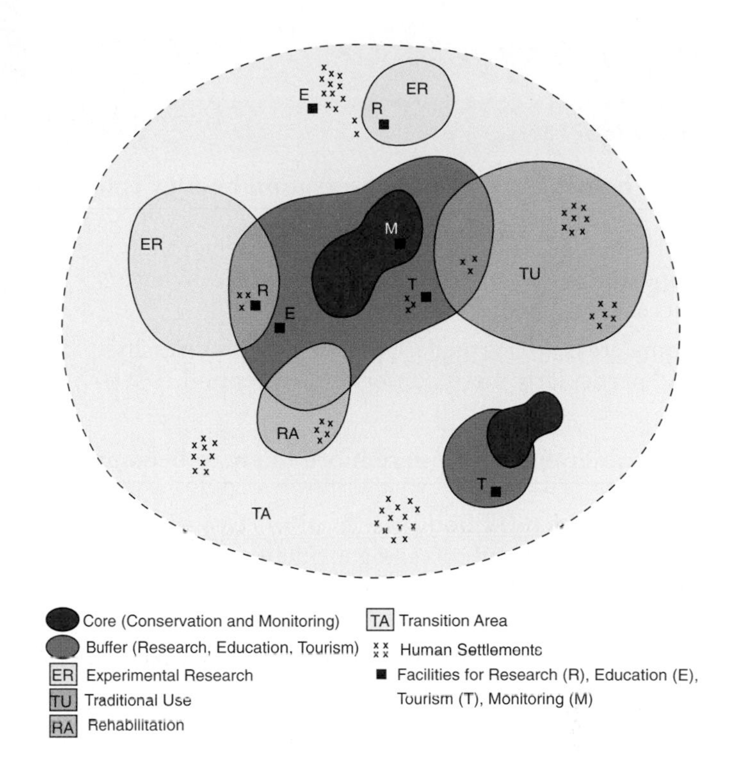

FIGURE 31.25
Design of a megareserve. A megareserve, or biosphere reserve, recognizes the need for people to have access to resources. Critical ecosystems are preserved in the core zone. Research and tourism is allowed in the buffer zone. Sustainable resource harvesting and permanent habitation is allowed in the multiple-use areas surrounding the buffer.

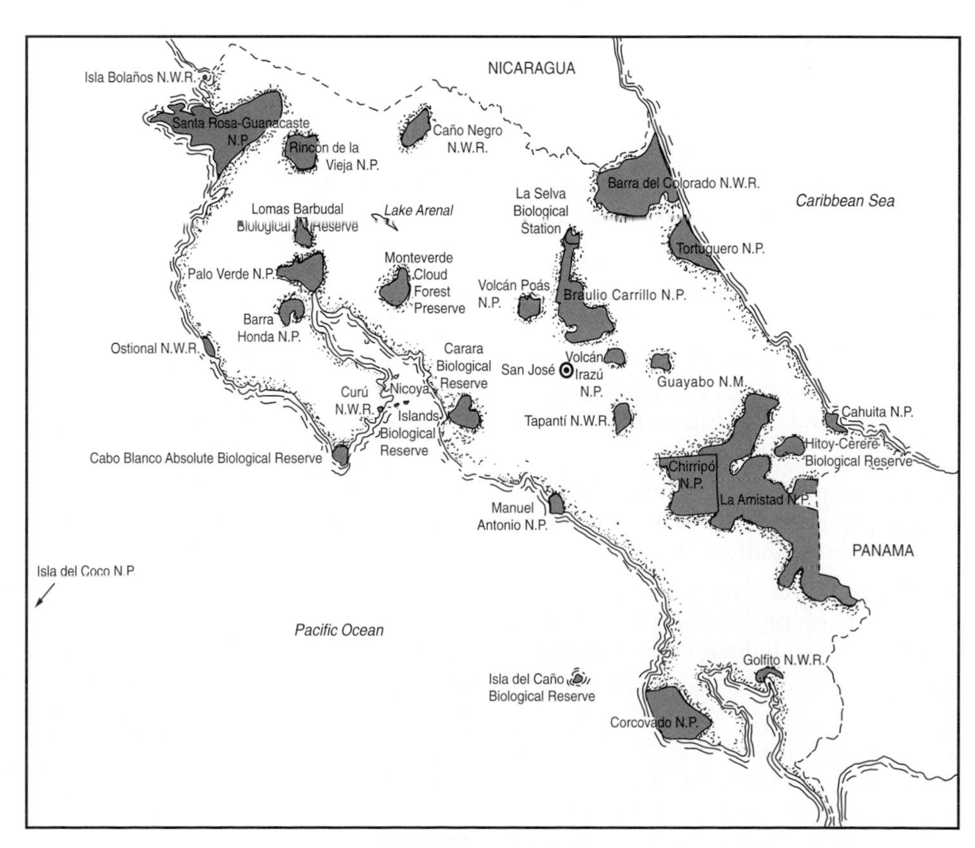

FIGURE 31.26
Biopreserves in Costa Rica. Costa Rica has placed about 12% of its land into national parks and eight megareserves.

Chapter 31

www.mhhe.com/raven6/resources31.mhtml

Summary | **Questions** | **Media Resources**

31.1 The new science of conservation biology is focusing on conserving biodiversity.

- Early humans caused many extinctions when they appeared in new areas, but rates of extinction have increased in modern times.
- Some areas are particularly rich in species diversity and particularly merit conservation attention.

1. What areas are particularly important for conserving biodiversity?

2. Describe some of the indirect economic values of biodiversity.

- Biodiversity

31.2 Vulnerable species are more likely to become extinct.

- Interdependence among species in an ecosystem leads to the possibility of cascading extinctions if removal of one species has major effects throughout the food web.
- Species are particularly vulnerable when they have localized distributions, are declining in population size, lack genetic variability, or are harvested or hunted by humans.

3. What factors contribute to the extinction rate on a particular piece of land?

4. How does a low genetic variability contribute to a species' greater risk of extinction?

- Wetlands

31.3 Causes of endangerment usually reflect human activities.

- Habitat loss is the single most important cause of species extinction.
- As suggested by the species-area relationship, a reduced habitat will support fewer numbers of species.
- This reduction in habitat can occur in four different ways: a habitat can be completely removed or destroyed, a habitat can become fragmented and disjunct, a habitat can be degraded or altered, or a habitat can become too frequently used by humans so as to disturb the species there.

5. How can problems resulting from lack of genetic diversity within a population be solved?

6. How can extinction of a keystone species be particularly disruptive to an ecosystem?

- *On Science* Article: What's Killing the Frogs?
- Global Environmental Issues in the News
- Regional Perspective in Environmental Science

31.4 Successful recovery plans will need to be multidimensional.

- Pristine restoration of a habitat may be attempted, but removing introduced species, rehabilitating the habitat, and cleaning up the habitat may be more feasible.
- Captive propagation, sustaining genetic variability, and preserving keystone species have been effective in preserving biodiversity.
- Megareserves have been successfully designed in many parts of the world to contain core areas of undisturbed habitat surrounded by managed land.

7. Why is maintaining large preserves particularly important?

8. Is captive propagation always an answer to species vulnerability?

9. Why is it important to attempt to eradicate introduced species soon after they appear?

- *On Science* Article: Biodiversity Behind Bars

Part **IX**

Viruses and Simple Organisms

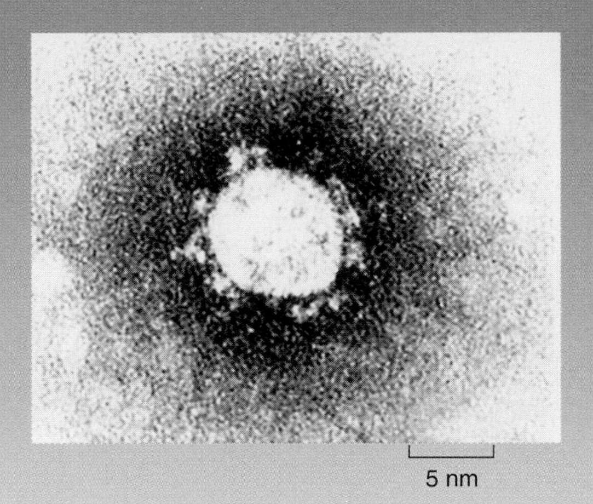

Electron micrograph of hepatitis C virus.

Real People Doing Real Science

Discovering the Virus Responsible for Hepatitis C

You may not be aware that our country is in the midst of an epidemic of a potentially fatal liver disease. Almost 4 million Americans are infected with the hepatitis C virus, most without knowing it. In the first years of this century, the number of annual United States deaths caused by hepatitis C is predicted to overtake deaths caused by AIDS.

Hepatitis is inflammation of the liver. There are three distinct forms. One, called infectious hepatitis or hepatitis A, is transmitted by contact with feces from infected individuals. A second form, called serum hepatitis or hepatitis B, is passed through blood and other body fluids. A third form, called hepatitis C virus (HCV), was only isolated in 1990 (see above photo). It too is passed through the blood.

HCV was difficult to isolate because it cannot be grown reliably in a laboratory culture of cells. Making the problem even more difficult, HCV is a strictly primate virus. It infects only humans and our close relatives, chimpanzees and tamarins. Because it is very expensive to maintain these animals in research laboratories, only small numbers of animals can be employed in any one study. Thus, the virus could not be isolated by the traditional means of purification from extracts of infected cells. What finally succeeded, after 15 years of failed attempts at isolation, was molecular technology. HCV was the first virus isolated entirely by cloning infectious nucleic acid.

The successful experiment was carried out by Michael Houghton and fellow researchers at Chiron, a California biotechnology company. What they did was shotgun clone the DNA of infected cells (that is, break the cell DNA into many pieces, and isolate each), and then screen each cloned piece of DNA for HCV.

The genetic material of HCV is RNA. So the first step was to convert HCV RNA to DNA, so that it could be cloned. There was no need to attempt to achieve entire faithful copies of the whole virus genome, a touchy and difficult task, because they did not wish to reproduce the HCV virus, only identify it. So the researchers took the far easier route of copying bits and pieces of the virus RNA, each piece carrying some part of the virus genome.

Next, they inserted these DNA copies of HCV genes into a bacteriophage, and allowed the bacteriophage to infect *Escherichia coli* bacteria. In a "shotgun" experiment like this, millions of bacterial cells are infected with bacteriophages. The researchers grew individual infected cells to form discrete colonies on plates of solid culture media. The colonies together constituted a "clone library." The problem then is to screen the library for colonies that had successfully received HCV.

To understand how they did this, focus on the quarry, a cell infected with an HCV gene. Once inside a bacterial cell, an HCV gene fragment becomes just so much more DNA, not particularly different from all the rest. The cellular machinery of the bacteria reads it just like bacterial genes, manufacturing the virus protein that the inserted HCV gene encodes. The secret is to look for cells with HCV proteins.

How to identify an HCV protein from among a background of thousands of bacterial proteins? Houghton and his colleagues tested each colony for its ability to cause a visible immune reaction with serum isolated from HCV-infected chimpanzees.

The test is a very simple and powerful one, because its success does not depend on knowing the identity of the genes you seek. The serum of HCV-infected animals should contain antibodies directed against a broad range of proteins, including HCV proteins encountered while combating the animal's HCV infection. Thus among the many proteins the serum can respond to in an antibody test will be some HCV proteins. The investigators can use the serum as a probe for the presence of HCV proteins in bacterial cells, which would not have any other animal proteins to confuse the meaning of a positive reaction.

Out of a million bacterial clones tested, just one was found that reacted with the chimp HCV serum, but not with serum from the same chimp before infection.

Using this clone as a toehold, the researchers were able to go back and fish out the rest of the virus genome from infected cells. From the virus genome, it was a straightforward matter to develop a diagnostic antibody test for the presence of the HCV virus.

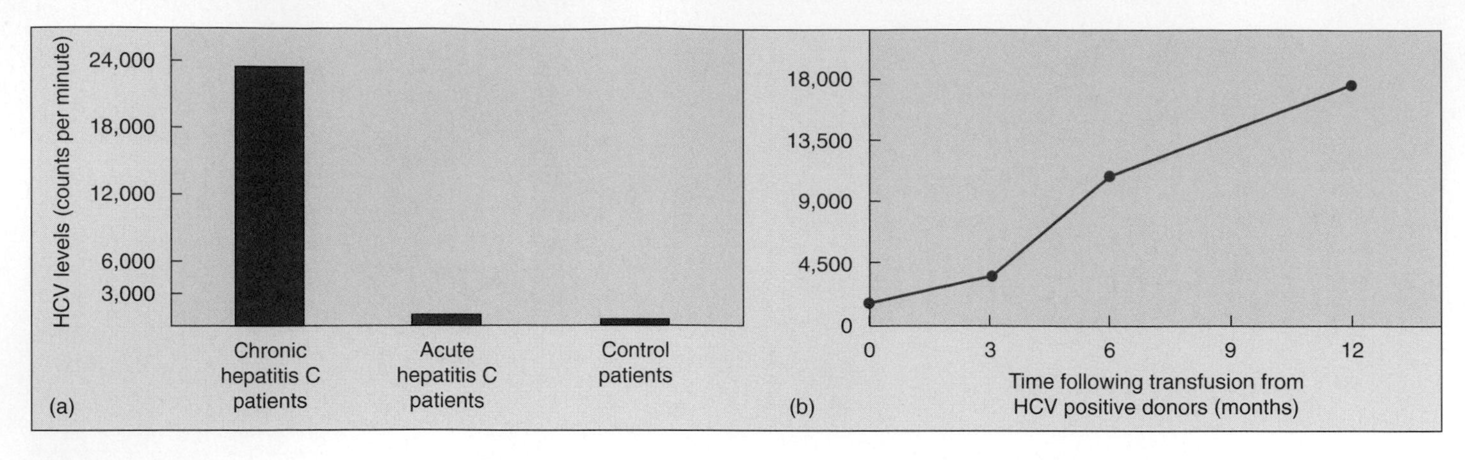

Detection of HCV antibodies. (*a*) Levels of HCV were measured using radioactive antibodies in three general groups of patients. The counts-per-minute measure indicates the sensitivity of the antibody (samples >3549 were considered positive). (*b*) Levels of HCV antibody were measured in the blood of individuals who had received blood transfusions from donors infected with the hepatitis C virus.

The Experiment

Using the diagnostic test, researchers found hepatitis C to be far more common than had been supposed. This is a problem of major proportions, because hepatitis C virus is unlike hepatitis A or B in a very important respect: it causes chronic disease. Most viruses cause a brief, intense infection and then are done. Hepatitis A, for example, typically lasts a few weeks. Ninety percent of people with hepatitis C have it for years, many of them for decades.

All during these long years of infection, damage is being done to the liver. Cells of the immune system called cytotoxic T cells recognize hepatitis C virus proteins on the surface of liver cells, and kill the infected cells. Over the years, many dead liver cells accumulate, and in response the cells around them begin to secrete collagen and other proteins to cover the mess. This eventually produces protein fibers interlacing the liver, fibers which disrupt the flow of materials through the liver's many internal passages. Imagine dropping bricks and rubble on a highway—it gets more and more difficult for traffic to move as the rubble accumulates.

If this fibrosis progresses far enough, it results in complete blockage, cirrhosis, a serious condition which may induce fatal liver failure, and which often induces primary liver cancer. About 20% of patients develop cirrhosis within 20 years of infection.

The development of a diagnostic antibody test was important for the screening of the blood supply for contamination. It was believed that contaminated blood through transfusions caused a large majority of new cases of hepatitis C. In order to develop a diagnostic antibody test an international research team headed by the Chiron group compared the DNA copies of several HCV clones and identified one common DNA sequence present in all clones. This section of DNA was reconstructed and then incorporated into the yeast genome along with another polypeptide gene, human superoxide dismutase (SOD), that assists in the incorporation of the HCV DNA copy into the yeast genome. Recombinant yeast cells express this SOD/HCV polypeptide, called C100-3, at high levels. C100-3 was used to coat the wells of test plates so that HCV antibodies in blood samples could be captured and measured.

The Results

To test the specificity and sensitivity of the antibody assay, HCV infected blood sera from patients with hepatitis C was tested (see graph *a* above). The researchers then assayed matched blood donors and recipients of both negative and positive donors and determined that recipients of negative donors did not develop hepatitis C. However, recipients of donors determined to be positive for HCV, based on reviewing stored samples, developed hepatitis C and showed positive HCV detection in the blood over a 12 month period (see graph *b* above).

These results indicated a specific association between the HCV antibody test and hepatitis C. Similar results obtained from patients in Italy and Japan confirmed the association of the HCV antibody and hepatitis C. These data supported the hypothesis that HCV is a primary cause of chronic hepatitis C and that use of this HCV antibody test would improve the safety of the world blood supply as well as become an important diagnostic tool.

Luckily, HCV is a very difficult virus to transmit. Unluckily, it is proving difficult to produce a vaccine directed against HCV. Antibodies directed against HCV are largely ineffective, and HCV mutates so frequently that, like AIDS, no vaccine seems likely to be effective. Attempts to combat hepatitis are focused on the virus itself. This virus carries just one gene, a very big one that is translated into a single immense "polyprotein" which enzymes cut into 10 functional pieces. Each of these 10 proteins is being investigated as a potential target for a drug to fight the virus, although no success is yet reported.

To explore this experiment further, go to the Virtual Lab at www.mhhe.com/raven6/vlab9.mhtml

32

How We Classify Organisms

Concept Outline

32.1 Biologists name organisms in a systematic way.

The Classification of Organisms. Biologists name organisms using a binomial system.

Species Names. Every kind of organism is assigned a unique name.

The Taxonomic Hierarchy. The higher groups into which an organism is placed reveal a great deal about the organism.

What Is a Species? Species are groups of similar organisms that tend not to interbreed with individuals of other groups.

32.2 Scientists construct phylogenies to understand the evolutionary relationships among organisms.

Evolutionary Classifications. Traditional and cladistic interpretations of evolution differ in the emphasis they place on particular traits.

32.3 All living organisms are grouped into one of a few major categories.

The Kingdoms of Life. Living organisms are grouped into three great groups called domains, and within domains into kingdoms.

Domain Archaea (Archaebacteria). The oldest domain consists of primitive bacteria that often live in extreme environments.

Domain Bacteria (Eubacteria). Too small to see with the unaided eye, eubacteria are more numerous than any other organism.

Domain Eukarya (Eukaryotes). There are four kingdoms of eukaryotes, three of them entirely or predominantly multicellular. Two of the most important characteristics to have evolved among the eukaryotes are multicellularity and sexuality.

Viruses: A Special Case. Viruses are not organisms, and thus do not belong to any kingdom.

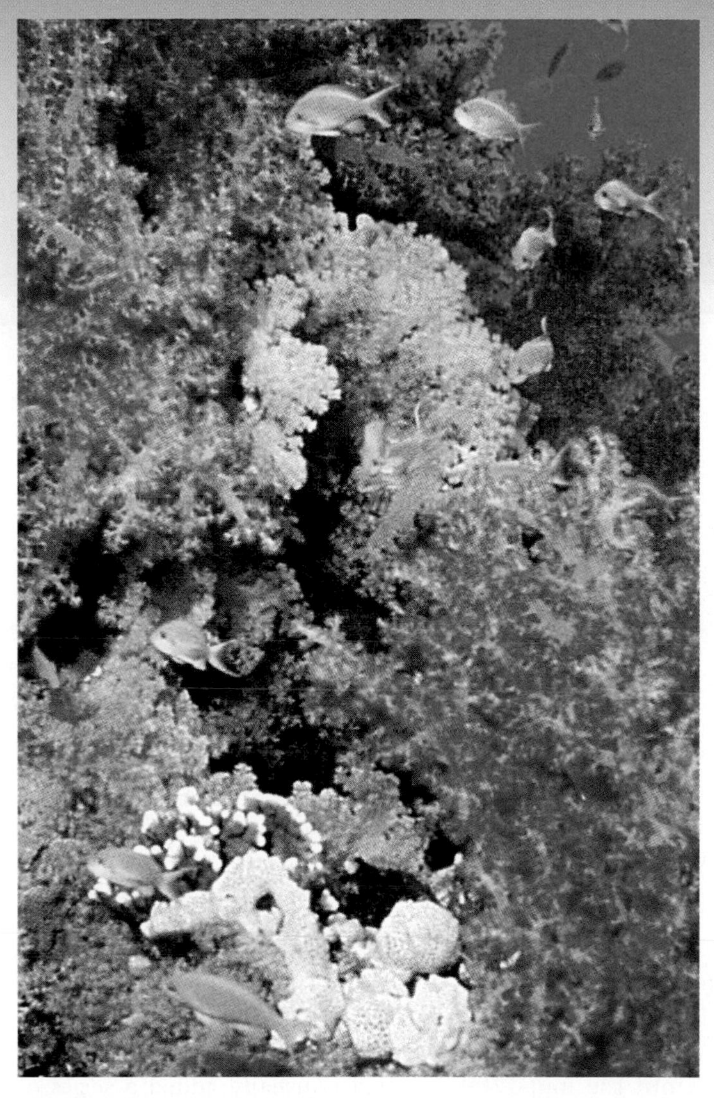

FIGURE 32.1
Biological diversity. All living things are assigned to particular classifications based on characteristics such as their anatomy, development, mode of nutrition, level of organization, and biochemical composition. Coral reefs, like the one seen here, are home to a variety of living things.

All organisms share many biological characteristics. They are composed of one or more cells, carry out metabolism and transfer energy with ATP, and encode hereditary information in DNA. All species have evolved from simpler forms and continue to evolve. Individuals live in populations. These populations make up communities and ecosystems, which provide the overall structure of life on earth. So far, we have stressed these common themes, considering the general principles that apply to all organisms. Now we will consider the diversity of the biological world and focus on the differences among groups of organisms (figure 32.1). For the rest of the text, we will examine the different kinds of life on earth, from bacteria and amoebas to blue whales and sequoia trees.

32.1 Biologists name organisms in a systematic way.

The Classification of Organisms

Organisms were first classified more than 2000 years ago by the Greek philosopher Aristotle, who categorized living things as either plants or animals. He classified animals as either land, water, or air dwellers, and he divided plants into three kinds based on stem differences. This simple classification system was expanded by the Greeks and Romans, who grouped animals and plants into basic units such as cats, horses, and oaks. Eventually, these units began to be called **genera** (singular, **genus**), the Latin word for "groups." Starting in the Middle Ages, these names began to be systematically written down, using Latin, the language used by scholars at that time. Thus, cats were assigned to the genus *Felis*, horses to *Equus*, and oaks to *Quercus*—names that the Romans had applied to these groups. For genera that were not known to the Romans, new names were invented.

FIGURE 32.2
Two species of oaks. (*a*) Willow oak, *Quercus phellos*. (*b*) Red oak, *Quercus rubra*. Although they are both oaks (*Quercus*), these two species differ sharply in leaf shape and size and in many other features, including geographical range.

The classification system of the Middle Ages, called the *polynomial system*, was used virtually unchanged for hundreds of years.

The Polynomial System

Until the mid-1700s, biologists usually added a series of descriptive terms to the name of the genus when they wanted to refer to a particular kind of organism, which they called a **species.** These phrases, starting with the name of the genus, came to be known as polynomials (*poly*, "many"; *nomial*, "name"), strings of Latin words and phrases consisting of up to 12 or more words. One name for the European honeybee, for example, was *Apis pubescens, thorace subgriseo, abdomine fusco, pedibus posticis glabris utrinque margine ciliatis.* As you can imagine, these polynomial names were cumbersome. Even more worrisome, the names were altered at will by later authors, so that a given organism really did not have a single name that was its alone.

The Binomial System

A much simpler system of naming animals, plants, and other organisms stems from the work of the Swedish biologist Carolus Linnaeus (1707–1778). Linnaeus devoted his life to a challenge that had defeated many biologists before him—cataloging all the different kinds of organisms. In the 1750s he produced several major works that, like his earlier books, employed the polynomial system. But as a kind of shorthand, Linnaeus also included in these books a two-part name for each species. For example, the honeybee became *Apis mellifera.* These two-part names, or **binomials** (*bi*, "two") have become our standard way of designating species.

A Closer Look at Linnaeus

To illustrate Linnaeus's work further, let's consider how he treated two species of oaks from North America, which by 1753 had been described by scientists. He grouped all oaks in the genus *Quercus*, as had been the practice since Roman times. Linnaeus named the willow oak of the southeastern United States (figure 32.2*a*) *Quercus foliis lanceolatis integerrimis glabris* ("oak with spear-shaped, smooth leaves with absolutely no teeth along the margins"). For the common red oak of eastern temperate North America (figure 32.2*b*), Linnaeus devised a new name, *Quercus foliis obtuse-sinuatis setaceo-mucronatis* ("oak with leaves with deep blunt lobes bearing hairlike bristles"). For each of these species, he also presented a shorthand designation, the binomial names *Quercus phellos* and *Quercus rubra.* These have remained the official names for these species since 1753, even though Linnaeus did not intend this when he first used them in his book. He considered the polynomials the true names of the species.

Two-part ("binomial") Latin names, first utilized by Linnaeus, are now universally employed by biologists to name particular organisms.

Species Names

Taxonomy is the science of classifying living things, and a group of organisms at a particular level in a classification system is called a **taxon** (plural, **taxa**). By agreement among taxonomists throughout the world, no two organisms can have the same name. So that no one country is favored, a language spoken by no country—Latin—is used for the names. Because the scientific name of an organism is the same anywhere in the world, this system provides a standard and precise way of communicating, whether the language of a particular biologist is Chinese, Arabic, Spanish, or English. This is a great improvement over the use of common names, which often vary from one place to the next. As you can see in figure 32.3, corn in Europe refers to the plant Americans call wheat; a bear is a large placental omnivore in the United States but a koala (a vegetarian marsupial) in Australia; and a robin is a very different bird in Europe and North America.

Also by agreement, the first word of the binomial name is the genus to which the organism belongs. This word is always capitalized. The second word refers to the particular species and is not capitalized. The two words together are called the **scientific name** and are written in italics or distinctive print: for example, *Homo sapiens*. Once a genus has been used in the body of a text, it is often abbreviated in later uses. For example, the dinosaur *Tyrannosaurus rex* becomes *T. rex*, and the potentially dangerous bacterium *Escherichia coli* is known as *E. coli*. The system of naming animals, plants, and other organisms established by Linnaeus has served the science of biology well for nearly 230 years.

By convention, the first part of a binomial species name identifies the genus to which the species belongs, and the second part distinguishes that particular species from other species in the genus.

FIGURE 32.3
Common names make poor labels. The common names corn (*a*), bear (*b*), and robin (*c*) bring clear images to our minds (photos on *left*), but the images are very different to someone living in Europe or Australia (photos on *right*). There, the same common names are used to label very different species.

The Taxonomic Hierarchy

In the decades following Linnaeus, taxonomists began to group organisms into larger, more inclusive categories. Genera with similar properties were grouped into a cluster called a **family,** and similar families were placed into the same **order** (figure 32.4). Orders with common properties were placed into the same **class,** and classes with similar characteristics into the same **phylum** (plural, **phyla**). For historical reasons, phyla may also be called *divisions* among plants, fungi, and algae. Finally, the phyla were assigned to one of several great groups, the **kingdoms.** Biologists currently recognize six kingdoms: two kinds of bacteria (Archaebacteria and Eubacteria), a largely unicellular group of eukaryotes (Protista), and three multicellular groups (Fungi, Plantae, and Animalia). In order to remember the seven categories of the taxonomic hierarchy in their proper order, it may prove useful to memorize a phrase such as "**k**indly **p**ay **c**ash **o**r **f**urnish **g**ood **s**ecurity" (**k**ingdom–**p**hylum–**c**lass–**o**rder–**f**amily–**g**enus–**s**pecies).

In addition, an eighth level of classification, called *domains*, is sometimes used. Biologists recognize three domains, which will be discussed later in this chapter. The scientific names of the taxonomic units higher than the genus level are capitalized but not printed distinctively, italicized, or underlined.

The categories at the different levels may include many, a few, or only one taxon. For example, there is only one living genus of the family Hominidae, but several living genera of Fagaceae. To someone familiar with classification or with access to the appropriate reference books, each taxon implies both a set of characteristics and a group of organisms belonging to the taxon. For example, a honeybee has the species (level 1) name *Apis mellifera.* Its genus name (level 2) *Apis* is a member of the family Apidae (level 3). All members of this family are bees, some solitary, others living in hives as *A. mellifera* does. Knowledge of its order (level 4), Hymenoptera, tells you that *A. mellifera* is likely able to sting and may live in colonies. Its class (level 5) Insecta indicates that *A. mellifera* has three major body segments, with wings and three pairs of legs attached to the middle segment. Its phylum (level 6), Arthropoda, tells us that the honeybee has a hard cuticle of chitin and jointed appendages. Its kingdom (level 7), Animalia, tells us that *A. mellifera* is a multicellular heterotroph whose cells lack cell walls.

Species are grouped into genera, genera into families, families into orders, orders into classes, and classes into phyla. Phyla are the basic units within kingdoms; such a system is hierarchical.

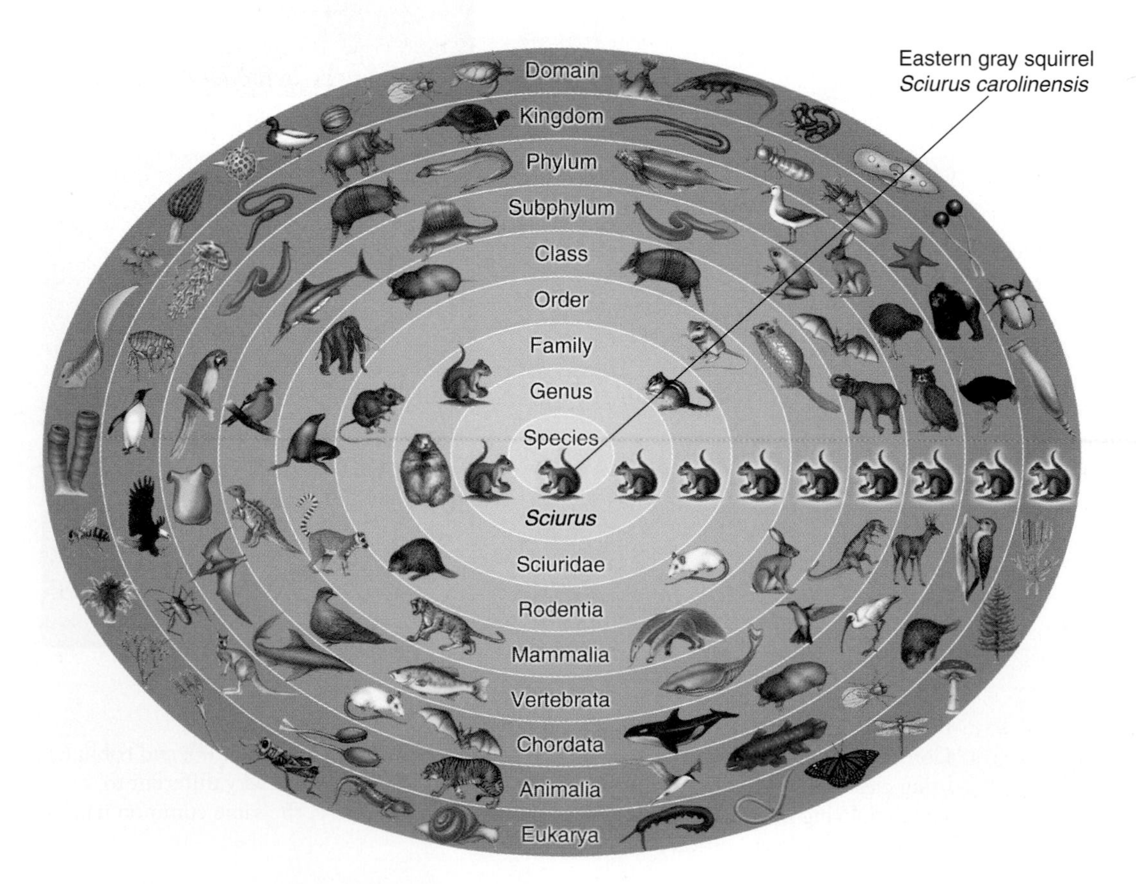

FIGURE 32.4
The hierarchical system used in classifying an organism. The organism is first recognized as a eukaryote (domain: Eukarya). Second, within this domain, it is an animal (kingdom: Animalia). Among the different phyla of animals, it is a vertebrate (phylum: Chordata, subphylum: Vertebrata). The organism's fur characterizes it as a mammal (class: Mammalia). Within this class, it is distinguished by its gnawing teeth (order: Rodentia). Next, because it has four front toes and five back toes, it is a squirrel (family: Sciuridae). Within this family, it is a tree squirrel (genus: *Sciurus*), with gray fur and white-tipped hairs on the tail (species: *Sciurus carolinensis*, the eastern gray squirrel).

What Is a Species?

In the previous section we discussed how species are named and grouped, but how do biologists decide when one organism is distinct enough from another to be called its own species? In chapter 22, we reviewed the nature of species and saw there are no absolute criteria for the definition of this category. Looking different, for example, is not a useful criterion: different individuals that belong to the same species (for example, dogs) may look very unlike one another, as different as a Chihuahua and a St. Bernard. These very different-appearing individuals are fully capable of hybridizing with one another.

The **biological species concept** (figure 32.5) essentially says that two organisms that cannot interbreed and produce fertile offspring are different species. This definition of a species can be useful in describing sexually reproducing species that regularly **outcross**—interbreed with individuals other than themselves. However, in many groups of organisms, including bacteria, fungi, and many plants and animals, **asexual reproduction**—reproduction without sex—predominates. Among them, hybridization cannot be used as a criterion for species recognition.

Defining Species

Despite such difficulties, biologists generally agree on the organisms they classify as species based on the similarity of morphological features and ecology. As a practical definition, we can say that species are groups of organisms that remain relatively constant in their characteristics, can be distinguished from other species, and do not normally interbreed with other species in nature.

Evolutionary Species Concept

This simple definition of species leaves many problems unsolved. How, for instance, are we to compare living species with seemingly similar ones now extinct? Much of the disagreement among alternative species concepts relates to solving this problem. When do we assign fossil specimens a unique species name, and when do we assign them to species living today? If we trace the lineage of two sister species backwards through time, how far must we go before the two species converge on their common ancestor? It is often very hard to know where to draw a sharp line between two closely related species.

To address this problem, biologists have added an evolutionary time dimension to the biological species concept. A current definition of an evolutionary species is *a single lineage of populations that maintains its distinctive identity from other such lineages.* Unlike the biological species concept, the evolutionary species concept applies to both asexual and sexually reproducing forms. Abrupt changes in diagnostic features mark the boundaries of different species in evolutionary time.

(a)

(b)

(c)

FIGURE 32.5
The biological species concept. Horses (*a*) and donkeys (*b*) are not the same species, because the offspring they produce when they interbreed, mules (*c*), are sterile.

How Many Species Are There?

Scientists have described and named a total of 1.5 million species, but doubtless many more actually exist. Some groups of organisms, such as flowering plants, vertebrate animals, and butterflies, are relatively well known with an estimated 90% of the total number of species that actually exist in these groups having already been described. Many other groups, however, are very poorly known. It is generally accepted that only about 5% of all species have been recognized for bacteria, nematodes (roundworms), fungi, and mites (a group of organisms related to spiders).

By taking representative samples of organisms from different environments, such as the upper branches of tropical trees or the deep ocean, scientists have estimated the total numbers of species that may actually exist to be about 10 million, about 15% of them marine organisms.

Most Species Live in the Tropics

Most species, perhaps 6 or 7 million, are tropical. Presently only 400,000 species have been named in tropical Asia, Africa, and Latin America combined, well under 10% of all species that occur in the tropics. This is an incredible gap in our knowledge concerning biological diversity in a world that depends on biodiversity for its sustainability.

These estimates apply to the number of eukaryotic organisms only. There is no functional way of estimating the numbers of species of prokaryotic organisms, although it is clear that only a very small fraction of all species have been discovered and characterized so far.

Species are groups of organisms that differ from one another in recognizable ways and generally do not interbreed with one another in nature.

32.2 Scientists construct phylogenies to understand the evolutionary relationships among organisms.

Evolutionary Classifications

After naming and classifying some 1.5 million organisms, what have biologists learned? One very important advantage of being able to classify particular species of plants, animals, and other organisms is that individuals of species that are useful to humans as sources of food and medicine can be identified. For example, if you cannot tell the fungus *Penicillium* from *Aspergillus*, you have little chance of producing the antibiotic penicillin. In a thousand ways, just having names for organisms is of immense importance in our modern world.

Taxonomy also enables us to glimpse the evolutionary history of life on earth. The more similar two taxa are, the more closely related they are likely to be. By looking at the differences and similarities between organisms, biologists can construct an evolutionary tree, or **phylogeny,** inferring which organisms evolved from which other ones, in what order, and when. The reconstruction and study of phylogenies is called **systematics.** Within a phylogeny, a grouping can be either monophyletic, paraphyletic, or polyphyletic. A **monophyletic** group includes the most recent common ancestor of the group and all of its descendants. A **paraphyletic** group includes the most recent common ancestor of the group but not all of its descendants. And, a **polyphyletic** group does not include the most recent common ancestor of all the members of the group. Monophyletic groups are commonly assigned names, but systematists will not assign a taxonomic classification to a polyphyletic group. Paraphyletic groups may be considered taxa by some scientists, although they do not accurately represent the evolutionary relationships among the members of the group (figure 32.6).

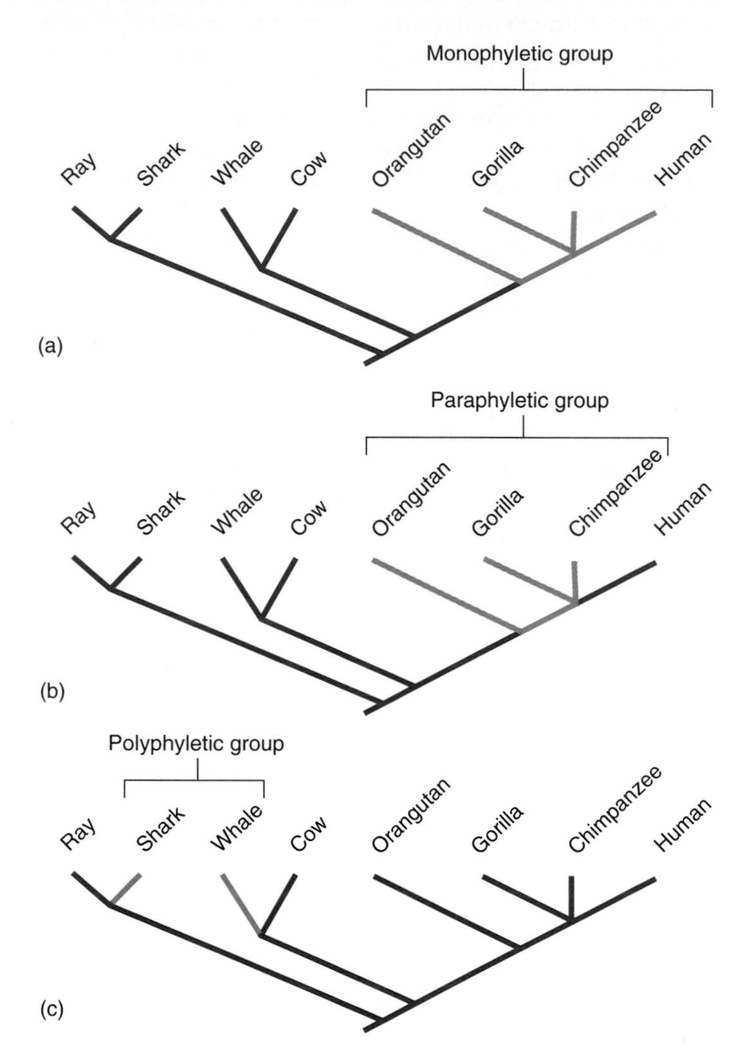

FIGURE 32.6
Monophyletic, paraphyletic, and polyphyletic groups. (*a*) A monophyletic group consists of the most recent common ancestor and all of its descendants. All taxonomists accept monophyletic groups in their classifications and in the above example would give the name "Apes" to the orangutans, gorillas, chimpanzees, and humans. (*b*) A paraphyletic group consists of the most recent common ancestor and some of its descendants. Taxonomists differ in their acceptance of paraphyletic groups. For example, some taxonomists arbitrarily group orangutans, gorillas, and chimpanzees into the paraphyletic family Pongidae, separate from humans. Other taxonomists do not use the family Pongidae in their classifications because gorillas and chimpanzees are more closely related to humans than to orangutans. (*c*) A polyphyletic group does not contain the most recent common ancestor of the group, and taxonomists do not assign taxa to polyphyletic groups. For example, sharks and whales could be classified in the same group because they have similar shapes, anatomical features, and habitats. However, their similarities reflect convergent evolution, not common ancestry.

Cladistics

A simple and objective way to construct a phylogenetic tree is to focus on key characters that a group of organisms share because they have inherited them from a common ancestor. A **clade** is a group of organisms related by descent, and this approach to constructing a phylogeny is called **cladistics.** Cladistics infers phylogeny (that is, builds family trees) according to similarities derived from a common ancestor, so-called derived characters. A derived character that is unique to a particular clade is sometimes called a **synapomorphy.** The key to the approach is being able to identify morphological, physiological, or behavioral traits that differ among the organisms being studied and can be attributed to a common ancestor. By examining the distribution of these traits among the organisms, it is possible to construct a **cladogram**

Traits: Organism	Jaws	Lungs	Amniotic membrane	Hair	No tail	Bipedal
Lamprey	0	0	0	0	0	0
Shark	1	0	0	0	0	0
Salamander	1	1	0	0	0	0
Lizard	1	1	1	0	0	0
Tiger	1	1	1	1	0	0
Gorilla	1	1	1	1	1	0
Human	1	1	1	1	1	1

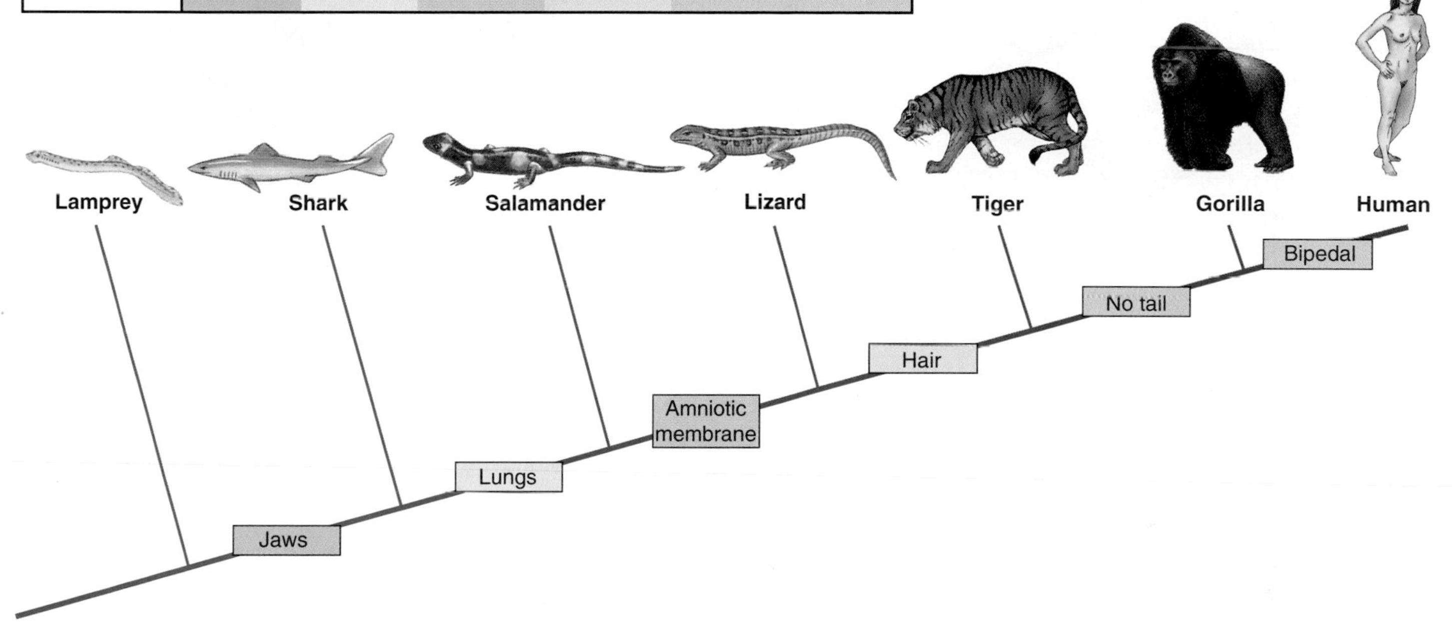

FIGURE 32.7
A cladogram. Morphological data for a group of seven vertebrates is tabulated. A "1" indicates the presence of a trait, or derived character, and a "0" indicates the absence of the trait. A tree, or cladogram, diagrams the proposed evolutionary relationships among the organisms based on the presence of derived characters. The derived characters between the cladogram branch points are shared by all organisms above the branch point and are not present in any below it. The outgroup, in this case the lamprey, does not possess any of the derived characters.

(figure 32.7), a branching diagram that represents the phylogeny.

In traditional phylogenies, proposed ancestors will often be indicated at the nodes between branches, and the lengths of branches correspond to evolutionary time, with extinct groups having shorter branches. In contrast, cladograms are not true family trees in that they do not identify ancestors, and the branch lengths do not reflect evolutionary time (see figure 32.6). Instead, they convey comparative information about relative relationships. Organisms that are closer together on a cladogram simply share a more recent common ancestor than those that are farther apart. Because the analysis is comparative, it is necessary to have something to anchor the comparison to, some solid ground against which the comparisons can be made. To achieve this, each cladogram must contain an **outgroup,** a rather different organism (but not too different) to serve as a baseline for comparisons among the other organisms being evaluated, the **ingroup.** For example, in figure 32.7, the lamprey is the outgroup to the clade of animals that have jaws.

Cladistics is a relatively new approach in biology and has become popular among students of evolution. This is because it does a very good job of portraying the order in which a series of evolutionary events have occurred. The great strength of a cladogram is that it can be completely objective. In fact, most cladistic analyses involve many characters, and computers are required to make the comparisons.

Sometime it is necessary to "weight" characters, or take into account the variation in the "strength" of a character, such as the size or location of a fin or the effectiveness of a lung. To reduce a systematist's bias even more, many analyses will be run through the computer with the traits weighted differently each time. Under this procedure, several different cladograms will be constructed, the goal being to choose the one that is the most **parsimonious,** or simplest and thus most likely. Reflecting the importance of evolutionary processes to all fields of biology, most taxonomy today includes at least some element of cladistic analysis.

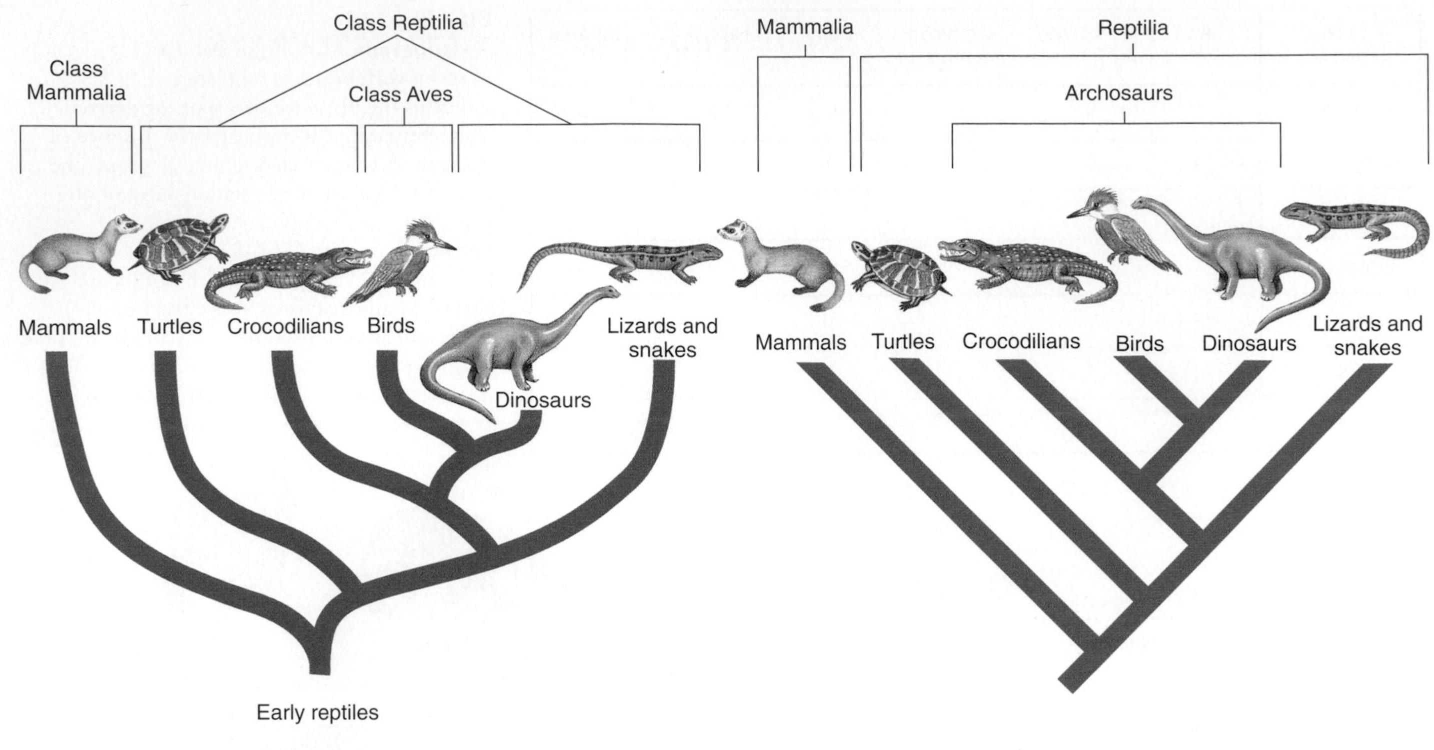

FIGURE 32.8
Traditional and cladistic interpretations of vertebrate classification. Traditional and cladistic taxonomic analyses of the same set of data often produce different results: in these two classifications of vertebrates, notice particularly the placement of the birds. (*a*) In the traditional analysis, key characteristics such as feathers and hollow bones are weighted more heavily than others, placing the birds in their own group and the reptiles in a paraphyletic group. (*b*) Cladistic analysis gives equal weight to these and many other characters and places birds in the same grouping with crocodiles and dinosaurs, reflecting the close evolutionary relationship between them. Also, in the traditional phylogeny, the branch leading to the dinosaurs is shorter because the length corresponds to evolutionary time. In cladograms, branch lengths do not correspond to evolutionary time.

Traditional Taxonomy

Weighting characters lies at the core of **traditional taxonomy.** In this approach, taxa are assigned based on a vast amount of information about the morphology and biology of the organism gathered over a long period of time. Traditional taxonomists consider both the common descent and amount of adaptive evolutionary change when grouping organisms. The large amount of information used by traditional taxonomists permits a knowledgeable weighting of characters according to their biological significance. In traditional taxonomy, the full observational power and judgment of the biologist is brought to bear—and also any biases he or she may have. For example, in classifying the terrestrial vertebrates, traditional taxonomists place birds in their own class (Aves), giving great weight to the characters that made powered flight possible, such as feathers. However, cladists (figure 32.8) lumps birds in among the reptiles with crocodiles. This accurately reflects their true ancestry but ignores the immense evolutionary impact of a derived character such as feathers.

Overall, classifications based on traditional taxonomy are information-rich, while classifications based on cladograms need not be. Traditional taxonomy is often used when a great deal of information is available to guide character weighting, while cladistics is a good approach when little information is available about how the character affects the life of the organism. DNA sequence comparisons, for example, lend themselves well to cladistics—you have a great many derived characters (DNA sequence differences) but little or no idea of what impact the sequence differences have on the organism.

A phylogeny may be represented as a cladogram based on the order in which groups evolved. Traditional taxonomists weight characters according to assumed importance.

32.3 All living organisms are grouped into one of a few major categories.

The Kingdoms of Life

The earliest classification systems recognized only two kingdoms of living things: animals and plants (figure 32.9*a*). But as biologists discovered microorganisms and learned more about other organisms, they added kingdoms in recognition of fundamental differences discovered among organisms (figure 32.9*b*). Most biologists now use a six-kingdom system first proposed by Carl Woese of the University of Illinois (figure 32.9*c*).

In this system, four kingdoms consist of eukaryotic organisms. The two most familiar kingdoms, **Animalia** and **Plantae,** contain only organisms that are multicellular during most of their life cycle. The kingdom **Fungi** contains multicellular forms and single-celled yeasts, which are thought to have multicellular ancestors. Fundamental differences divide these three kingdoms. Plants are mainly stationary, but some have motile sperm; fungi have no motile cells; animals are mainly motile. Animals ingest their food, plants manufacture it, and fungi digest it by means of secreted extracellular enzymes. Each of these kingdoms probably evolved from a different single-celled ancestor.

The large number of unicellular eukaryotes are arbitrarily grouped into a single kingdom called **Protista** (see chapter 35). This kingdom includes the algae, all of which are unicellular during parts of their life cycle.

The remaining two kingdoms, **Archaebacteria** and **Eubacteria,** consist of prokaryotic organisms, which are vastly different from all other living things (see chapter 34). Archaebacteria are a diverse group including the methanogens and extreme thermophiles, and differ from the other bacteria, members of the kingdom Eubacteria.

Domains

As biologists have learned more about the archaebacteria, it has become increasingly clear that this ancient group is very different from all other organisms. When the full genomic DNA sequences of an archaebacterium and a eubacterium were first compared in 1996, the differences proved striking. Archaebacteria are as different from eubacteria as eubacteria are from eukaryotes. Recognizing this, biologists are increasingly adopting a classification of living organisms that recognizes three **domains,** a taxonomic level higher than kingdom (figure 32.9*d*). Archaebacteria are in one domain, eubacteria in a second, and eukaryotes in the third.

Living organisms are grouped into three general categories called domains. One of the domains, the eukaryotes, is subdivided into four kingdoms: protists, fungi, plants, and animals.

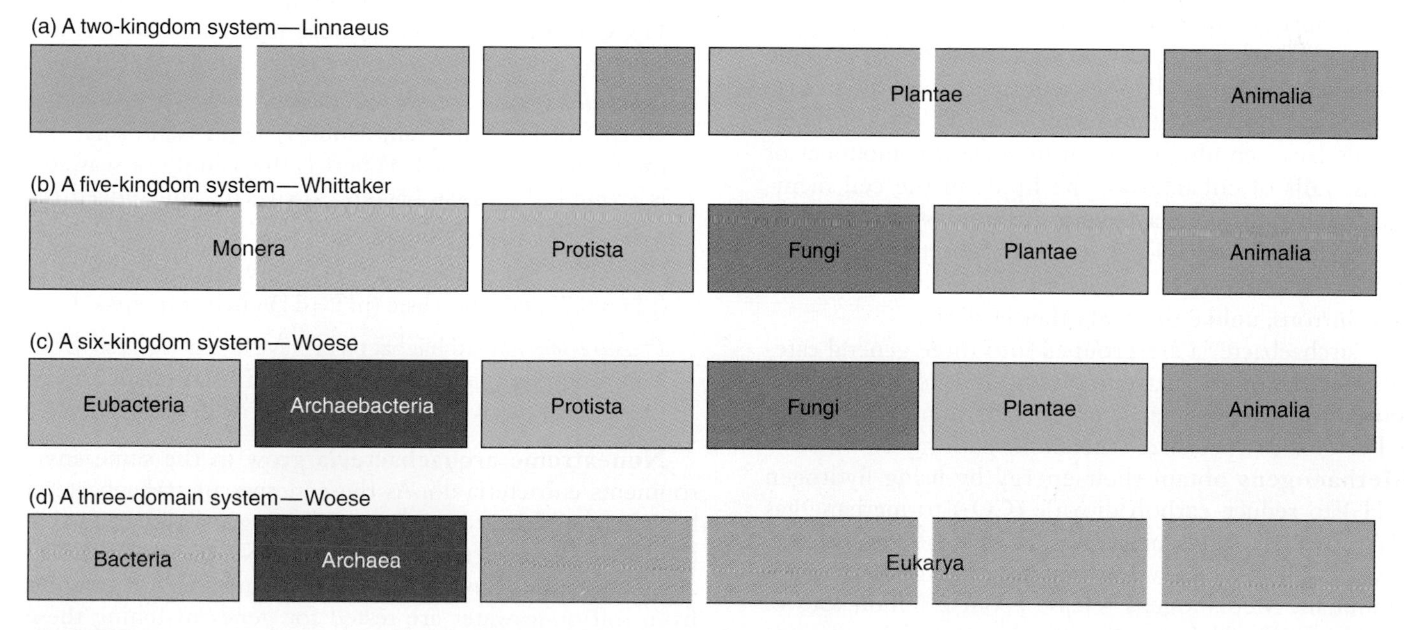

FIGURE 32.9
Different approaches to classifying living organisms. (*a*) Linnaeus popularized a two-kingdom approach, in which the fungi and the photosynthetic protists were classified as plants, and the nonphotosynthetic protists as animals; when bacteria were described, they too were considered plants. (*b*) Whittaker in 1969 proposed a five-kingdom system that soon became widely accepted. (*c*) Woese has championed splitting the bacteria into two kingdoms for a total of six kingdoms, or even assigning them separate domains (*d*).

Domain Archaea (Archaebacteria)

The term *archaebacteria* (Greek, *archaio*, ancient) refers to the ancient origin of this group of bacteria, which seem to have diverged very early from the eubacteria (figure 32.10). This conclusion comes largely from comparisons of genes that encode ribosomal RNAs. The last several years have seen an explosion of DNA sequence information from microorganisms, information which paints a more complex picture. It had been thought that by sequencing numerous microbes we could eventually come up with an accurate picture of the phylogeny of the earliest organisms on earth. The new whole-genome DNA sequence data described in chapter 19 tells us that it will not be that simple. Comparing whole-genome sequences leads evolutionary biologists to a variety of trees, some of which contradict each other. It appears that during their early evolution microorganisms have swapped genetic information, making constructing phylogenetic trees very difficult.

As an example of the problem, we can look at *Thermotoga*, a thermophile found on Volcano Island off Italy. The sequence of one of its RNAs places it squarely within the eubacteria near an ancient microbe called *Aquifex*. Recent DNA sequencing, however, fails to support any consistent relationship between the two microbes. There is disagreement as to the serious effect of gene swapping on the ability of evolutionary biologists to provide accurate phylogenies from molecular data. For now, we will provisionally accept the tree presented in figure 32.10. Over the next few years we can expect to see considerable change in accepted viewpoints as more and more data is brought to bear.

Today, archaebacteria inhabit some of the most extreme environments on earth. Though a diverse group, all archaebacteria share certain key characteristics (table 32.1). Their cell walls lack peptidoglycan (an important component of the cell walls of eubacteria), the lipids in the cell membranes of archaebacteria have a different structure than those in all other organisms, and archaebacteria have distinctive ribosomal RNA sequences. Some of their genes possess introns, unlike those of other bacteria.

The archaebacteria are grouped into three general categories, methanogens, extremophiles, and nonextreme archaebacteria, based primarily on the environments in which they live or their specialized metabolic pathways.

Methanogens obtain their energy by using hydrogen gas (H_2) to reduce carbon dioxide (CO_2) to methane gas (CH_4). They are strict anaerobes, poisoned by even traces of oxygen. They live in swamps, marshes, and the intestines of mammals. Methanogens release about 2 billion tons of methane gas into the atmosphere each year.

Extremophiles are able to grow under conditions that seem extreme to us.

Thermophiles ("heat lovers") live in very hot places, typically from 60° to 80°C. Many thermophiles are autotrophs and have metabolisms based on sulfur. Some thermophilic archaebacteria form the basis of food webs around deep-sea thermal vents where they must withstand extreme temperatures and pressures. Other types, like *Sulfolobus*, inhabit the hot sulfur springs of Yellowstone National Park at 70° to 75°C. The recently described *Pyrolobus fumarii* holds the current record for heat stability, with a 106°C temperature optimum and 113°C maximum—it is so heat tolerant that it is not killed by a one-hour treatment in an autoclave (121°C)!

Halophiles ("salt lovers") live in very salty places like the Great Salt Lake in Utah, Mono Lake in California, and the Dead Sea in Israel. Whereas the salinity of seawater is around 3%, these bacteria thrive in, and indeed require, water with a salinity of 15 to 20%.

pH-tolerant archaebacteria grow in highly acidic (pH = 0.7) and very basic (pH = 11) environments.

Pressure-tolerant archaebacteria have been isolated from ocean depths that require at least 300 atmospheres of pressure to survive, and tolerate up to 800 atmospheres!

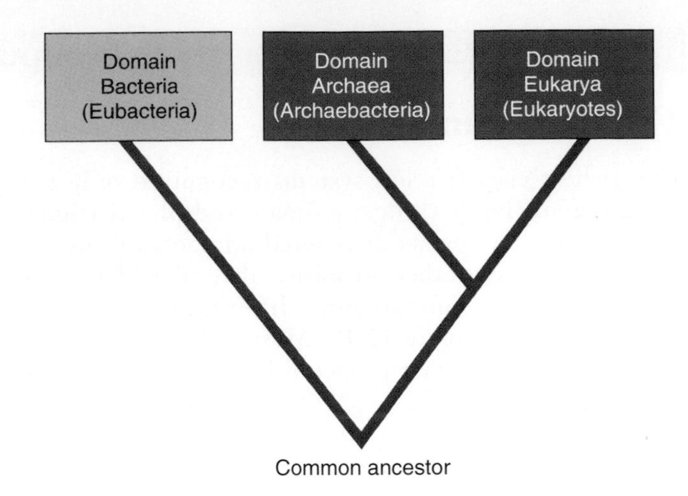

FIGURE 32.10
An evolutionary relationship among the three domains. Eubacteria are thought to have diverged early from the evolutionary line that gave rise to the archaebacteria and eukaryotes.

Nonextreme archaebacteria grow in the same environments eubacteria do. As the genomes of archaebacteria have become better known, microbiologists have been able to identify **signature sequences** of DNA present in all archaebacteria and in no other organisms. When samples from soil or seawater are tested for genes matching these signal sequences, many of the bacteria living there prove to be archaebacteria. Clearly, archaebacteria are not restricted to extreme habitats, as microbiologists used to think.

Archaebacteria are poorly understood bacteria that inhabit diverse environments, some of them extreme.

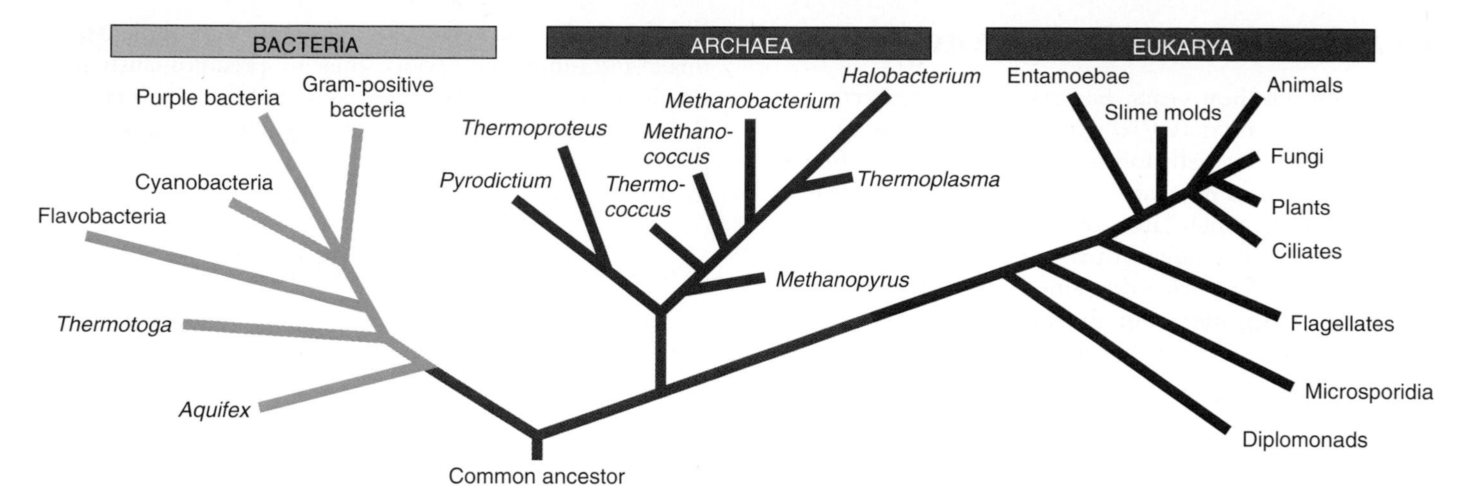

FIGURE 32.11
A tree of life. This phylogeny, prepared from rRNA analyses, shows the evolutionary relationships among the three domains. The base of the tree was determined by examining genes that are duplicated in all three domains, the duplication presumably having occurred in the common ancestor. When one of the duplicates is used to construct the tree, the other can be used to root it. This approach clearly indicates that the root of the tree is within the eubacterial domain. Archaebacteria and eukaryotes diverged later and are more closely related to each other than either is to eubacteria.

Domain Bacteria (Eubacteria)

The eubacteria are the most abundant organisms on earth. There are more living eubacteria in your mouth than there are mammals living on earth. Although too tiny to see with the unaided eye, eubacteria play critical roles throughout the biosphere. They extract from the air all the nitrogen used by organisms, and play key roles in cycling carbon and sulfur. Much of the world's photosynthesis is carried out by eubacteria. However, certain groups of eubacteria are also responsible for many forms of disease. Understanding their metabolism and genetics is a critical part of modern medicine.

There are many different kinds of eubacteria, and the evolutionary links between them are not well understood. While there is considerable disagreement among taxonomists about the details of bacterial classification, most recognize 12 to 15 major groups of eubacteria. Comparisons of the nucleotide sequences of ribosomal RNA (rRNA) molecules are beginning to reveal how these groups are related to one another and to the other two domains. One view of our current understanding of the "Tree of Life" is presented in figure 32.11. The oldest divergences represent the deepest rooted branches in the tree. The root of the tree is within the eubacterial domain. The archaebacteria and eukaryotes are more closely related to each other than to eubacteria and are on a separate evolutionary branch of the tree, even though archaebacteria and eubacteria are both prokaryotes.

Eubacteria are as different from archaebacteria as from eukaryotes.

Table 32.1 Features of the Domains of Life

Feature	Domain: Archaea	Domain: Bacteria	Domain: Eukarya
Amino acid that initiates protein synthesis	Methionine	Formyl-methionine	Methionine
Introns	Present in some genes	Absent	Present
Membrane-bounded organelles	Absent	Absent	Present
Membrane lipid structure	Branched	Unbranched	Unbranched
Nuclear envelope	Absent	Absent	Present
Number of different RNA polymerases	Several	One	Several
Peptidoglycan in cell wall	Absent	Present	Absent
Response to the antibiotics streptomycin and chloramphenicol	Growth not inhibited	Growth inhibited	Growth not inhibited

Domain Eukarya (Eukaryotes)

For at least 2 billion years, bacteria ruled the earth. No other organisms existed to eat them or compete with them, and their tiny cells formed the world's oldest fossils. The third great domain of life, the eukaryotes, appear in the fossil record much later, only about 1.5 billion years ago. Metabolically, eukaryotes are more uniform than bacteria. Each of the two domains of prokaryotic organisms has far more metabolic diversity than all eukaryotic organisms taken together. However, despite the metabolic similarity of eukaryotic cells, their structure and function allowed larger cell sizes and, eventually, multicellular life to evolve.

Four Kingdoms of Eukaryotes

The first eukaryotes were unicellular organisms. A wide variety of unicellular eukaryotes exist today, grouped together in the kingdom Protista on the basis that they do not fit into any of the other three kingdoms of eukaryotes. Protists are a fascinating group containing many organisms of intense interest and great biological significance. They vary from the relatively simple, single-celled amoeba to multicellular organisms like kelp that can be 20 meters long.

Fungi, plants, and animals are largely multicellular kingdoms, each a distinct evolutionary line from a single-celled ancestor that would be classified in the kingdom Protista. Because of the size and ecological dominance of plants, animals, and fungi, and because they are predominantly multicellular, we recognize them as kingdoms distinct from Protista, even though the amount of diversity among the protists is much greater than that within or between the fungi, plants, and animals.

Symbiosis and the Origin of Eukaryotes

The hallmark of eukaryotes is complex cellular organization, highlighted by an extensive endomembrane system that subdivides the eukaryotic cell into functional compartments. Not all of these compartments, however, are derived from the endomembrane system. With few exceptions, all modern eukaryotic cells possess energy-producing organelles, the mitochondria, and some eukaryotic cells possess chloroplasts, which are energy-harvesting organelles. Mitochondria and chloroplasts are both believed to have entered early eukaryotic cells by a process called **endosymbiosis** (*endo*, inside). We discussed the theory of the endosymbiotic origin of mitochondria and chloroplasts in chapter 5; also see figure 32.12. Both organelles contain their own ribosomes, which are more similar to bacterial ribosomes than to eukaryotic cytoplasmic ribosomes. They manufacture their own inner membranes. They divide independently of the cell and contain chromosomes similar to those in bacteria. Mitochondria are about the size of bacteria and contain DNA. Comparison of the nucleotide sequence of this DNA with that of a variety of organisms

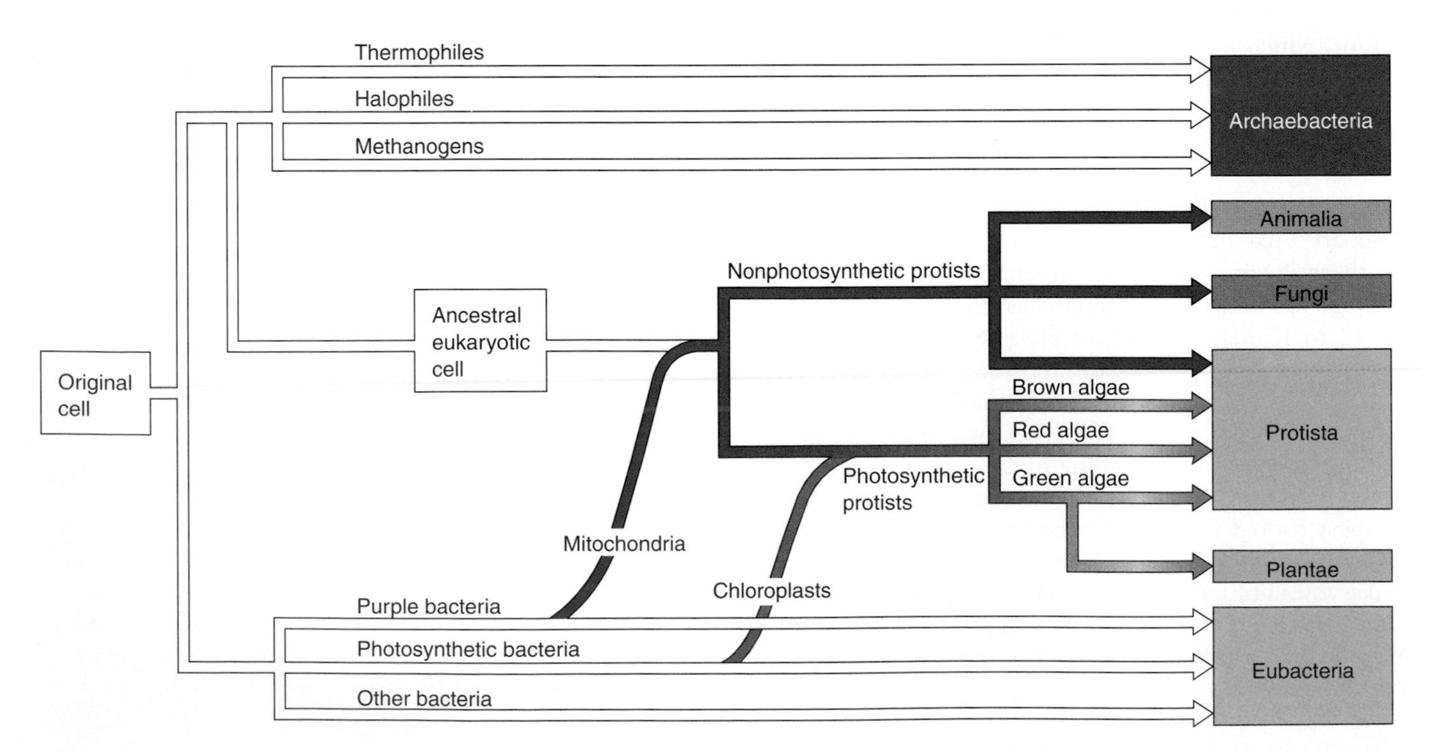

FIGURE 32.12
Diagram of the evolutionary relationships among the six kingdoms of organisms. The colored lines indicate symbiotic events.

indicates clearly that mitochondria are the descendants of purple bacteria that were incorporated into eukaryotic cells early in the history of the group. Chloroplasts are derived from cyanobacteria that became symbiotic in several groups of protists early in their history.

Some biologists suggest that basal bodies, centrioles, flagella, and cilia may have arisen from endosymbiotic spirochaete-like bacteria. Even today, so many bacteria and unicellular protists form symbiotic alliances that the incorporation of smaller organisms with desirable features into eukaryotic cells appears to be a relatively common process.

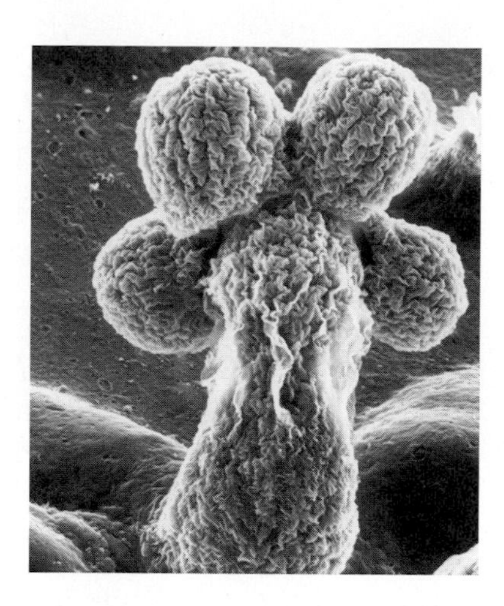

FIGURE 32.13
Colonial bacteria. No bacteria are truly multicellular. These gliding bacteria, *Stigmatella aurantiaca*, have aggregated into a structure called a fruiting body; within, some cells transform into spores.

Key Characteristics of Eukaryotes

Multicellularity. The unicellular body plan has been tremendously successful, with unicellular prokaryotes and eukaryotes constituting about half of the biomass on earth. Yet a single cell has limits. The evolution of multicellularity allowed organisms to deal with their environments in novel ways. Distinct types of cells, tissues, and organs can be differentiated within the complex bodies of multicellular organisms. With such a functional division within its body, a multicellular organism can do many things, like protect itself, resist drought efficiently, regulate its internal conditions, move about, seek mates and prey, and carry out other activities on a scale and with a complexity that would be impossible for its unicellular ancestors. With all these advantages, it is not surprising that multicellularity has arisen independently so many times.

True multicellularity, in which the activities of individual cells are coordinated and the cells themselves are in contact, occurs only in eukaryotes and is one of their major characteristics. The cell walls of bacteria occasionally adhere to one another, and bacterial cells may also be held together within a common sheath. Some bacteria form filaments, sheets, or three-dimensional aggregates (figure 32.13), but the individual cells remain independent of each other, reproducing and carrying on their metabolic functions and without coordinating with the other cells. Such bacteria are considered colonial, but none are truly multicellular. Many protists also form similar colonial aggregates of many cells with little differentiation or integration.

Other protists—the red, brown, and green algae, for example—have independently attained multicellularity. Certain forms of multicellular green algae were ancestors of the plants (see chapters 35 and 37), and, like the other photosynthetic protists, are considered plants in some classification schemes. In the system adopted here, the plant kingdom includes only multicellular land plants, a group that arose from a single ancestor in terrestrial habitats and that has a unique set of characteristics. Aquatic plants are recent derivatives.

Fungi and animals arose from unicellular protist ancestors with different characteristics. As we will see in subsequent chapters, the groups that seem to have given rise to each of these kingdoms are still in existence.

Sexuality. Another major characteristic of eukaryotic organisms as a group is sexuality. Although some interchange of genetic material occurs in bacteria (see chapter 34), it is certainly not a regular, predictable mechanism in the same sense that sex is in eukaryotes. The sexual cycle characteristic of eukaryotes alternates between **syngamy,** the union of male and female gametes producing a cell with two sets of chromosomes, and meiosis, cell division producing daughter cells with one set of chromosomes. This cycle differs sharply from any exchange of genetic material found in bacteria.

Except for gametes, the cells of most animals and plants are diploid, containing two sets of chromosomes, during some part of their life cycle. A few eukaryotes complete their life cycle in the haploid condition, with only one set of chromosomes in each cell. As we have seen, in diploid cells, one set of chromosomes comes from the male parent and one from the female parent. These chromosomes segregate during meiosis. Because crossing over frequently occurs during meiosis (see chapter 12), no two products of a single meiotic event are ever identical. As a result, the offspring of sexual, eukaryotic organisms vary widely, thus providing the raw material for evolution.

Sexual reproduction, with its regular alternation between syngamy and meiosis, produces genetic variation. Sexual organisms can adapt to the demands of their environments because they produce a variety of progeny.

In many of the unicellular phyla of protists, sexual reproduction occurs only occasionally. Meiosis may have originally evolved as a means of repairing damage to DNA, producing an organism better adapted to survive changing environmental conditions. The first eukaryotes were probably haploid. Diploids seem to have arisen on a number of separate occasions by the fusion of haploid cells, which then eventually divided by meiosis.

Key: Haploid Diploid

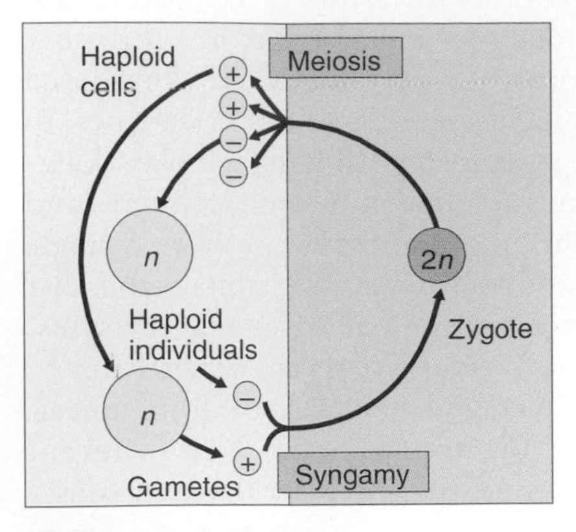

(a) Zygotic meiosis

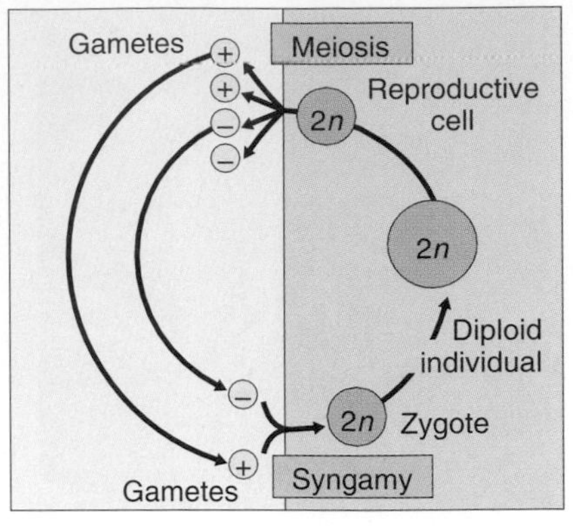

(b) Gametic meiosis

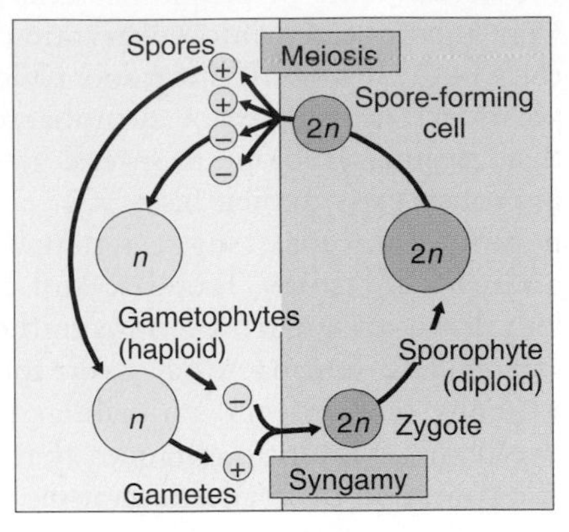

(c) Sporic meiosis

FIGURE 32.14
Diagrams of the three major kinds of life cycles in eukaryotes. (*a*) Zygotic meiosis, (*b*) gametic meiosis, and (*c*) sporic meiosis.

Eukaryotic Life Cycles

Eukaryotes are characterized by three major types of life cycles (figure 32.14):

1. In the simplest cycle, found in algae, the zygote is the only diploid cell. Such a life cycle is said to be characterized by **zygotic meiosis,** because the zygote immediately undergoes meiosis.
2. In most animals, the gametes are the only haploid cells. Animals exhibit **gametic meiosis,** meiosis producing gametes which fuse, giving rise to a zygote.
3. Plants show a regular *alternation of generations* between a multicellular haploid phase and a multicellular diploid phase. The diploid phase undergoes meiosis producing haploid spores that give rise to the haploid phase, and the haploid phase produces gametes that fuse to form the zygote. The zygote is the first cell of the multicellular diploid phase. This kind of life cycle is characterized by **alternation of generations** and has **sporic meiosis.**

The characteristics of the six kingdoms are outlined in table 32.2.

Eukaryotic cells acquired mitochondria and chloroplasts by endosymbiosis, mitochondria being derived from purple bacteria and chloroplasts from cyanobacteria. The complex differentiation that we associate with advanced life-forms depends on multicellularity and sexuality, which must have been highly advantageous to have evolved independently so often.

Table 32.2 Characteristics of the Six Kingdoms

Kingdom	Cell Type	Nuclear Envelope	Mitochondria	Chloroplasts	Cell Wall
Archaebacteria and Eubacteria	Prokaryotic	Absent	Absent	None (photosynthetic membranes in some types)	Noncellulose (polysaccharide plus amino acids)
Protista	Eukaryotic	Present	Present or absent	Present (some forms)	Present in some forms, various types
Fungi	Eukaryotic	Present	Present or absent	Absent	Chitin and other noncellulose polysaccharides
Plantae	Eukaryotic	Present	Present	Present	Cellulose and other polysaccharides
Animalia	Eukaryotic	Present	Present	Absent	Absent

Viruses: A Special Case

Viruses pose a challenge to biologists as they do not possess the fundamental characteristics of living organisms. Viruses appear to be fragments of nucleic acids originally derived from the genome of a living cell. Unlike all living organisms, viruses are acellular—that is, they are not cells and do not consist of cells. They do not have a metabolism; in other words, viruses do not carry out photosynthesis, cellular respiration, or fermentation. The one characteristic of life that they do display is reproduction, which they do by hijacking the metabolism of living cells.

Viruses thus present a special classification problem. Because they are not organisms, we cannot logically place them in any of the kingdoms. Viruses are really just complicated associations of molecules, bits of nucleic acids usually surrounded by a protein coat. But, despite their simplicity, viruses are able to invade cells and direct the genetic machinery of these cells to manufacture more of the molecules that make up the virus (figure 32.15). Viruses can infect organisms at all taxonomic levels.

Viruses are not organisms and are not classified in the kingdoms of life.

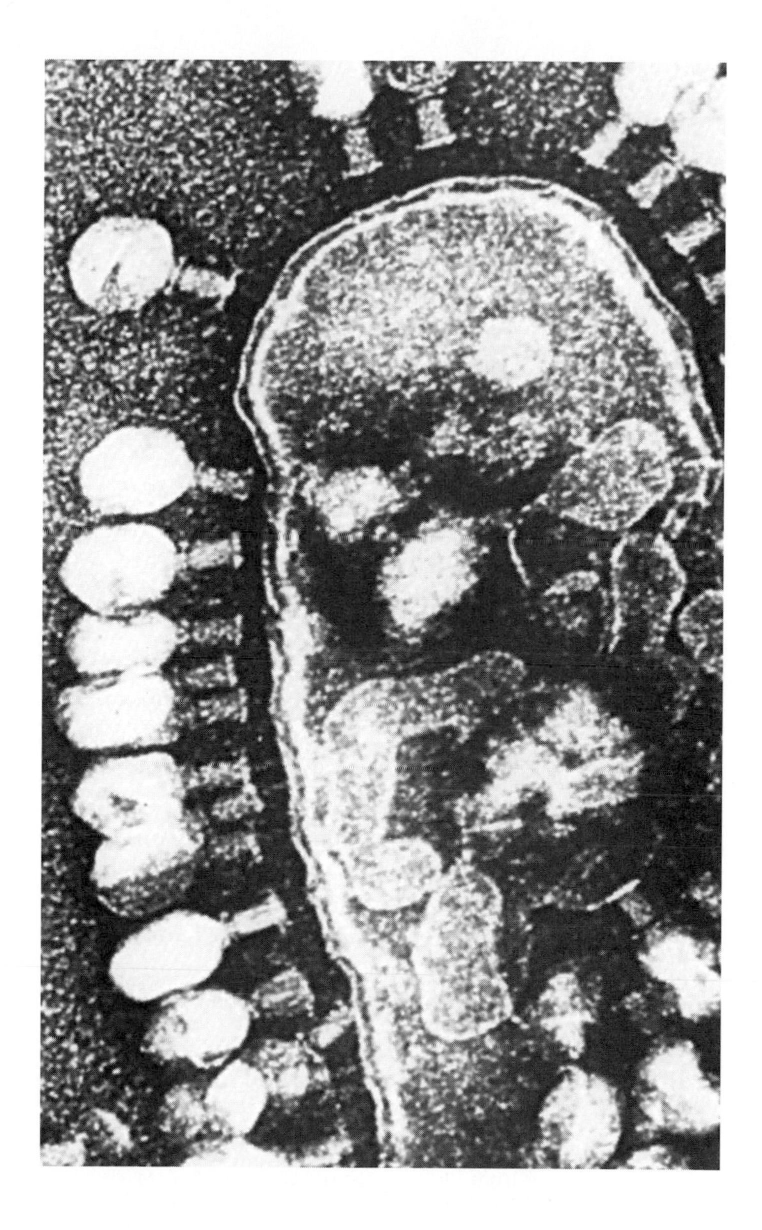

FIGURE 32.15
Viruses are cell parasites. In this micrograph, several T4 bacteriophages (viruses) are attacking an *Escherichia coli* bacterium. Some of the viruses have already entered the cell and are reproducing within it.

Means of Genetic Recombination, if Present	Mode of Nutrition	Motility	Multicellularity	Nervous System
Conjugation, transduction, transformation	Autotrophic (chemosynthetic, photosynthetic) or heterotrophic	Bacterial flagella, gliding or nonmotile	Absent	None
Fertilization and meiosis	Photosynthetic or heterotrophic, or combination of both	9 + 2 cilia and flagella; amoeboid, contractile fibrils	Absent in most forms	Primitive mechanisms for conducting stimuli in some forms
Fertilization and meiosis	Absorption	Nonmotile	Present in most forms	None
Fertilization and meiosis	Photosynthetic chlorophylls *a* and *b*	None in most forms, 9 + 2 cilia and flagella in gametes of some forms	Present in all forms	None
Fertilization and meiosis	Digestion	9 + 2 cilia and flagella, contractile fibrils	Present in all forms	Present, often complex

Chapter 32

www.mhhe.com/raven6/resources32.mhtml

Summary | Questions | Media Resources

32.1 Biologists name organisms in a systematic way.

- Biologists give every species a two-part (binomial) name that consists of the name of its genus plus a distinctive specific epithet.
- In the hierarchical system of classification used in biology, genera are grouped into families, families into orders, orders into classes, classes into phyla, and phyla into kingdoms.
- There are perhaps 10 million species of plants, animals, fungi, and eukaryotic microorganisms, but only about 1.5 million of them have been assigned names. About 15% of the total number of species are marine; the remainder are mostly terrestrial.

1. What was the polynomial system? Why didn't this system become the standard for naming particular species?

2. From the most specific to the most general, what are the names of the groups in the hierarchical taxonomic system? Which two are given special consideration in the way in which they are printed? What are these distinctions?

- Hierarchies

- Art Quiz: Taxonomic Hierarchy

32.2 Scientists construct phylogenies to understand the evolutionary relationships among organisms.

- Taxonomists may use different approaches to classify organisms.
- Cladistic systems of classification arrange organisms according to evolutionary relatedness based on the presence of shared, derived traits.
- Traditional taxonomy classifies organisms based on large amounts of information, giving due weight to the evolutionary significance of certain characters.

3. What types of features are emphasized in a cladistic classification system? What is the resulting relationship of organisms that are classified in this manner?

4. What does it mean when characters are weighted?

- Phylogeny

- Art Quiz: Traditional versus Cladistic Taxonomy

32.3 All living organisms are grouped into one of a few major categories.

- A fundamental division among organisms is between prokaryotes, which lack a true nucleus, and eukaryotes, which have a true nucleus and several membrane-bound organelles.
- Prokaryotes, or bacteria, are assigned to two quite different kingdoms, Archaebacteria and Eubacteria.
- The eukaryotic kingdoms are more closely related than are the two kingdoms of prokaryotes. Many distinctive evolutionary lines of unicellular eukaryotes exist, most are in the Protista kingdom.
- Three of the major evolutionary lines of eukaryotic organisms that consist principally or entirely of multicellular organisms are recognized as separate kingdoms: Plantae, Animalia, and Fungi.
- True multicellularity and sexuality are found only among eukaryotes. Multicellularity confers the advantages of functional specialization. Sexuality permits genetic variation among descendants.
- Viruses are not organisms and are not included in the classification of organisms. They are self-replicating portions of the genomes of organisms.

5. Is there a greater fundamental difference between plants and animals or between prokaryotes and eukaryotes? Explain.

6. From which of the four eukaryotic kingdoms have the other three evolved?

7. What is the apparent origin of the organelles found in almost all eukaryotes?

8. What defines if a collection of cells is truly multicellular? Did multicellularity arise once or many times in the evolutionary process? What advantages do multicellular organisms have over unicellular ones?

9. What are the three major types of life cycles in eukaryotes? Describe the major events of each.

- Art Activity: Organism Classification

- Kingdoms
- Three Domains

- Art Quizzes::
  - -Tree of Life
  - –Evolutionary Relationships Among Kingdoms

33

Viruses

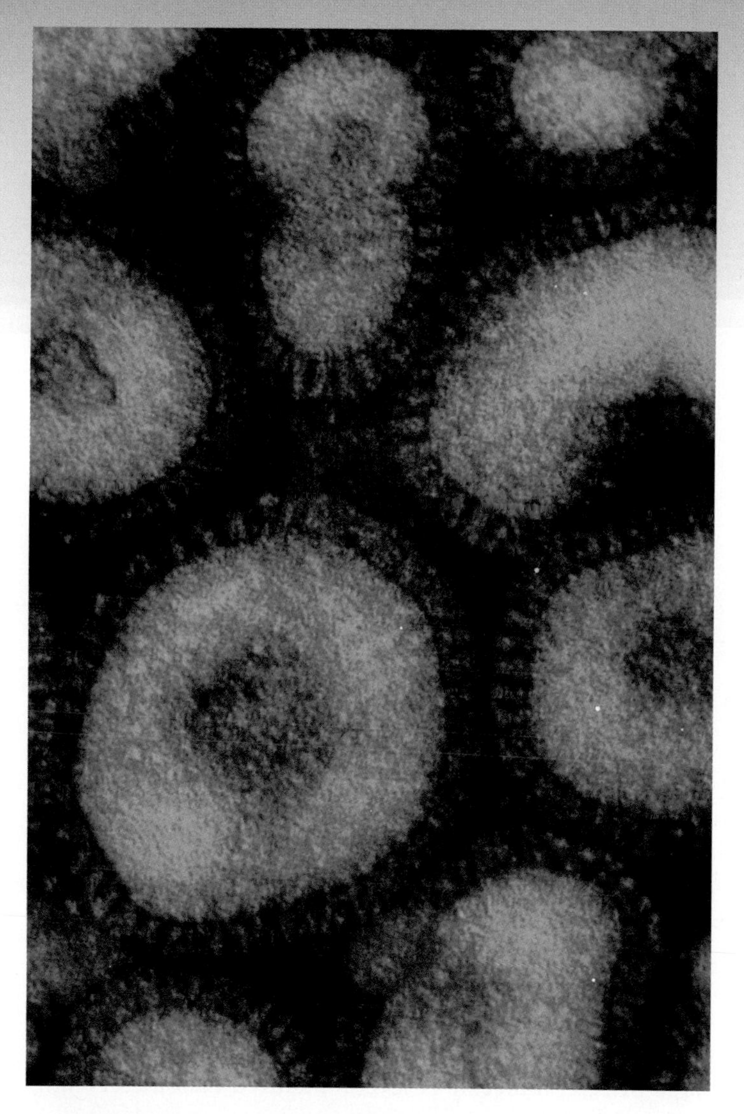

FIGURE 33.1
Influenza viruses (30,000×). A virus has been referred to as "a piece of bad news wrapped up in a protein." How can something as "simple" as a virus have such a profound effect on living organisms?

Concept Outline

33.1 Viruses are strands of nucleic acid encased within a protein coat.

The Discovery of Viruses. The first virus to be isolated proved to consist of two chemicals, one a protein and the other a nucleic acid.

The Nature of Viruses. Viruses occur in all organisms. Able to reproduce only within living cells, viruses are not themselves alive.

33.2 Bacterial viruses exhibit two sorts of reproductive cycles.

Bacteriophages. Some bacterial viruses, called bacteriophages, rupture the cells they infect, while others integrate themselves into the bacterial chromosome to become a stable part of the bacterial genome.

Cell Transformation and Phage Conversion. Integrated bacteriophages sometimes modify the host bacterium they infect.

33.3 HIV is a complex animal virus.

AIDS. The animal virus HIV infects certain key cells of the immune system, destroying the ability of the body to defend itself from cancer and disease. The HIV infection cycle is typically a lytic cycle, in which the HIV RNA first directs the production of a corresponding DNA, and this DNA then directs the production of progeny virus particles.

The Future of HIV Treatment. Combination therapies and chemokines offer promising avenues of AIDS therapy.

33.4 Nonliving infectious agents are responsible for many human diseases.

Disease Viruses. Some of the most serious viral diseases have only recently infected human populations, the result of transfer from other hosts.

Prions and Viroids. In some instances, proteins and "naked" RNA molecules can also transmit diseases.

We start our exploration of the diversity of life with viruses. Viruses are genetic elements enclosed in protein and are not considered to be organisms, as they cannot reproduce independently. Because of their disease-producing potential, viruses are important biological entities. The virus particles you see in figure 33.1 produce the important disease influenza. Other viruses cause AIDS, polio, flu, and some can lead to cancer. Many scientists have attempted to unravel the nature of viral genes and how they work. For more than four decades, viral studies have been thoroughly intertwined with those of genetics and molecular biology. In the future, it is expected that viruses will be one of the principal tools used to experimentally carry genes from one organism to another. Already, viruses are being employed in the treatment of human genetic diseases.

33.1 Viruses are strands of nucleic acid encased within a protein coat.

The Discovery of Viruses

The border between the living and the nonliving is very clear to a biologist. Living organisms are cellular and able to grow and reproduce independently, guided by information encoded within DNA. The simplest creatures living on earth today that satisfy these criteria are bacteria. Even simpler than bacteria are viruses. As you will learn in this section, viruses are so simple that they do not satisfy the criteria for "living."

Viruses possess only a portion of the properties of organisms. Viruses are literally "parasitic" chemicals, segments of DNA or RNA wrapped in a protein coat. They cannot reproduce on their own, and for this reason they are not considered alive by biologists. They can, however, reproduce within cells, often with disastrous results to the host organism. Earlier theories that viruses represent a kind of halfway point between life and nonlife have largely been abandoned. Instead, viruses are now viewed as detached fragments of the genomes of organisms due to the high degree of similarity found among some viral and eukaryotic genes.

Viruses vary greatly in appearance and size. The smallest are only about 17 nanometers in diameter, and the largest are up to 1000 nanometers (1 micrometer) in their greatest dimension (figure 33.2). The largest viruses are barely visible with a light microscope, but viral morphology is best revealed using the electron microscope. Viruses are so small that they are comparable to molecules in size; a hydrogen atom is about 0.1 nanometer in diameter, and a large protein molecule is several hundred nanometers in its greatest dimension.

Biologists first began to suspect the existence of viruses near the end of the nineteenth century. European scientists attempting to isolate the infectious agent responsible for hoof-and-mouth disease in cattle concluded that it was smaller than a bacterium. Investigating the agent further, the scientists found that it could not multiply in solution—it could only reproduce itself within living host cells that it infected. The infecting agents were called viruses.

The true nature of viruses was discovered in 1933, when the biologist Wendell Stanley prepared an extract of a plant virus called *tobacco mosaic virus* (TMV) and attempted to purify it. To his great surprise, the purified TMV preparation precipitated (that is, separated from solution) in the form of crystals. This was surprising because precipitation is something that only chemicals do—the TMV virus was acting like a chemical off the shelf rather than an organism. Stanley concluded that TMV is best regarded as just that—chemical matter rather than a living organism.

Within a few years, scientists disassembled the TMV virus and found that Stanley was right. TMV was not cellular but rather chemical. Each particle of TMV virus is in fact a mixture of two chemicals: RNA and protein. The TMV virus has the structure of a Twinkie, a tube made of an RNA core surrounded by a coat of protein. Later workers were able to separate the RNA from the protein and purify and store each chemical. Then, when they reassembled the two components, the reconstructed TMV particles were fully able to infect healthy tobacco plants and so clearly *were* the virus itself, not merely chemicals derived from it. Further experiments carried out on other viruses yielded similar results.

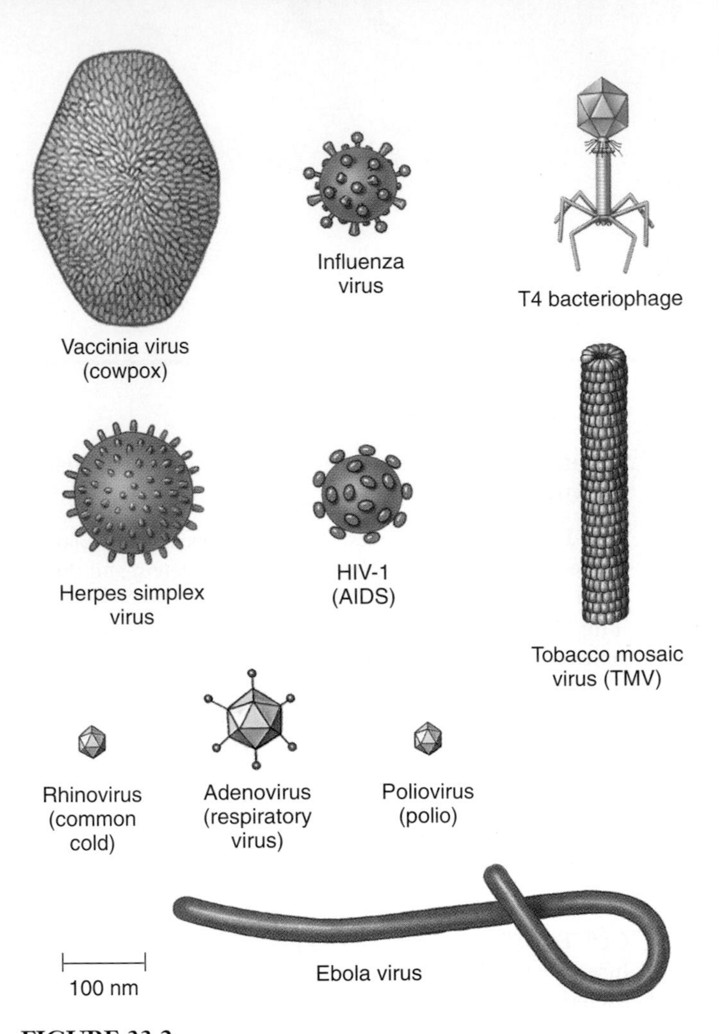

FIGURE 33.2
Viral diversity. A sample of the extensive diversity and small size of viruses is depicted. At the scale these viruses are shown, a human hair would be nearly 8 meters thick.

Viruses are chemical assemblies that can infect cells and replicate within them. They are not alive.

The Nature of Viruses

Viral Structure

All viruses have the same basic structure: a core of nucleic acid surrounded by protein. Individual viruses contain only a single type of nucleic acid, either DNA or RNA. The DNA or RNA genome may be linear or circular, and single-stranded or double-stranded. Viruses are frequently classified by the nature of their genomes. RNA-based viruses are known as **retroviruses.**

Nearly all viruses form a protein sheath, or **capsid,** around their nucleic acid core. The capsid is composed of one to a few different protein molecules repeated many times (figure 33.3) In some viruses, specialized enzymes are stored within the capsid. Many animal viruses form an **envelope** around the capsid rich in proteins, lipids, and glycoprotein molecules. While some of the material of the envelope is derived from the host cell's membrane, the envelope does contain proteins derived from viral genes as well.

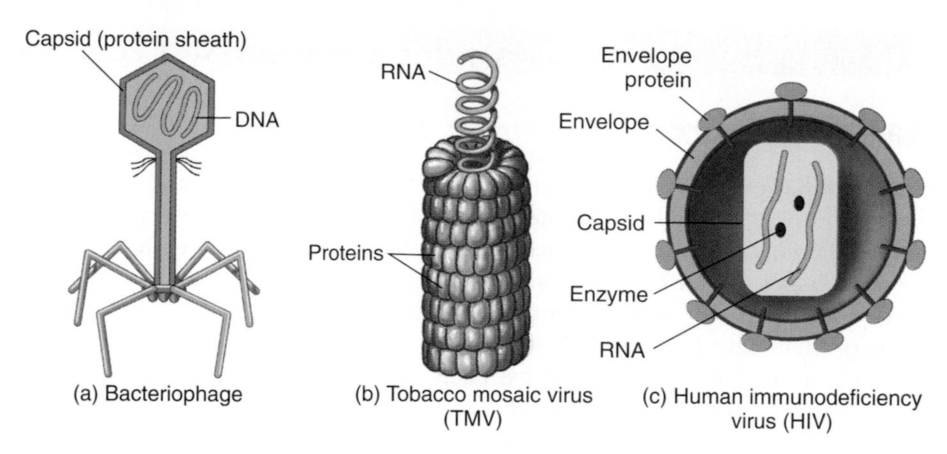

FIGURE 33.3
The structure of a bacterial, plant, and animal virus. (*a*) Bacterial viruses, called bacteriophages, often have a complex structure. (*b*) TMV infects plants and consists of 2130 identical protein molecules (*purple*) that form a cylindrical coat around the single strand of RNA (*green*). The RNA backbone determines the shape of the virus and is protected by the identical protein molecules packed tightly around it. (*c*) In the human immunodeficiency virus (HIV), the RNA core is held within a capsid that is encased by a protein envelope.

Viruses occur in virtually every kind of organism that has been investigated for their presence. However, each type of virus can replicate in only a very limited number of cell types. The suitable cells for a particular virus are collectively referred to as its **host range.** The size of the host range reflects the coevolved histories of the virus and its potential hosts. A recently discovered herpesvirus turned lethal when it expanded its host range from the African elephant to the Indian elephant, a situation made possible through cross species contacts between elephants in zoos. Some viruses wreak havoc on the cells they infect; many others produce no disease or other outward sign of their infection. Still other viruses remain dormant for years until a specific signal triggers their expression. A given organism often has more than one kind of virus. This suggests that there may be many more kinds of viruses than there are kinds of organisms—perhaps millions of them. Only a few thousand viruses have been described at this point.

Viral Replication

An infecting virus can be thought of as a set of instructions, not unlike a computer program. A computer's operation is directed by the instructions in its operating program, just as a cell is directed by DNA-encoded instructions. A new program can be introduced into the computer that will cause the computer to cease what it is doing and devote all of its energies to another activity, such as making copies of the introduced program. The new program is not itself a computer and cannot make copies of itself when it is outside the computer, lying on the desk. The introduced program, like a virus, is simply a set of instructions.

Viruses can reproduce only when they enter cells and utilize the cellular machinery of their hosts. Viruses code their genes on a single type of nucleic acid, either DNA or RNA, but viruses lack ribosomes and the enzymes necessary for protein synthesis. Viruses are able to reproduce because their genes are translated into proteins by the cell's genetic machinery. These proteins lead to the production of more viruses.

Viral Shape

Most viruses have an overall structure that is either **helical** or **isometric.** Helical viruses, such as the tobacco mosaic virus, have a rodlike or threadlike appearance. Isometric viruses have a roughly spherical shape whose geometry is revealed only under the highest magnification.

The only structural pattern found so far among isometric viruses is the **icosahedron,** a structure with 20 equilateral triangular facets, like the adenovirus shown in figure 33.2. Most viruses are icosahedral in basic structure. The icosahedron is the basic design of the geodesic dome. It is the most efficient symmetrical arrangement that linear subunits can take to form a shell with maximum internal capacity.

Viruses occur in all organisms and can only reproduce within living cells. Most are icosahedral in structure.

33.2 Bacterial viruses exhibit two sorts of reproductive cycles.

Bacteriophages

Bacteriophages are viruses that infect bacteria. They are diverse both structurally and functionally, and are united solely by their occurrence in bacterial hosts. Many of these bacteriophages, called *phages* for short, are large and complex, with relatively large amounts of DNA and proteins. Some of them have been named as members of a "T" series (T1, T2, and so forth); others have been given different kinds of names. To illustrate the diversity of these viruses, T3 and T7 phages are icosahedral and have short tails. In contrast, the so-called T-even phages (T2, T4, and T6) have an icosahedral head, a capsid that consists primarily of three proteins, a connecting neck with a collar and long "whiskers," a long tail, and a complex base plate (figure 33.4).

The Lytic Cycle

During the process of bacterial infection by T4 phage, at least one of the tail fibers of the phage—they are normally held near the phage head by the "whiskers"—contacts the lipoproteins of the host bacterial cell wall. The other tail fibers set the phage perpendicular to the surface of the bacterium and bring the base plate into contact with the cell surface. The tail contracts, and the tail tube passes through an opening that appears in the base plate, piercing the bacterial cell wall. The contents of the head, mostly DNA, are then injected into the host cytoplasm.

When a virus kills the infected host cell in which it is replicating, the reproductive cycle is referred to as a **lytic cycle** (figure 33.5). The T-series bacteriophages are all **virulent viruses,** multiplying within infected cells and eventually lysing (rupturing) them. However, they vary considerably as to when they become virulent within their host cells.

The Lysogenic Cycle

Many bacteriophages do not immediately kill the cells they infect, instead integrating their nucleic acid into the genome of the infected host cell. While residing there, it is called a **prophage.** Among the bacteriophages that do this is the lambda (λ) phage of *Escherichia coli.* We know as much about this bacteriophage as we do about virtually any other biological particle; the complete sequence of its 48,502 bases has been determined. At least 23 proteins are associated with the development and maturation of lambda phage, and many other enzymes are involved in the integration of these viruses into the host genome.

The integration of a virus into a cellular genome is called **lysogeny.** At a later time, the prophage may exit the genome and initiate virus replication. This sort of reproductive cycle, involving a period of genome integration, is called a **lysogenic cycle.** Viruses that become stably integrated within the genome of their host cells are called **lysogenic viruses** or **temperate viruses.**

Bacteriophages are a diverse group of viruses that attack bacteria. Some kill their host in a lytic cycle; others integrate into the host's genome, initiating a lysogenic cycle.

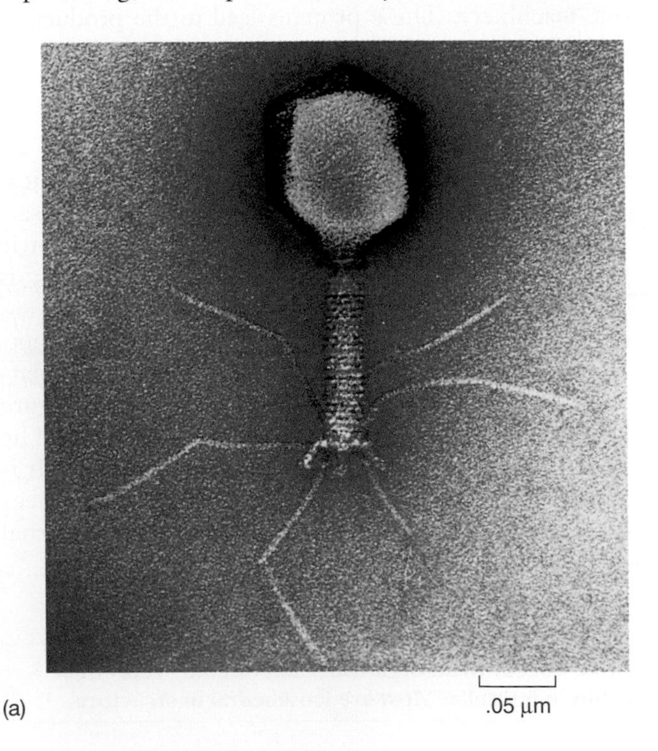

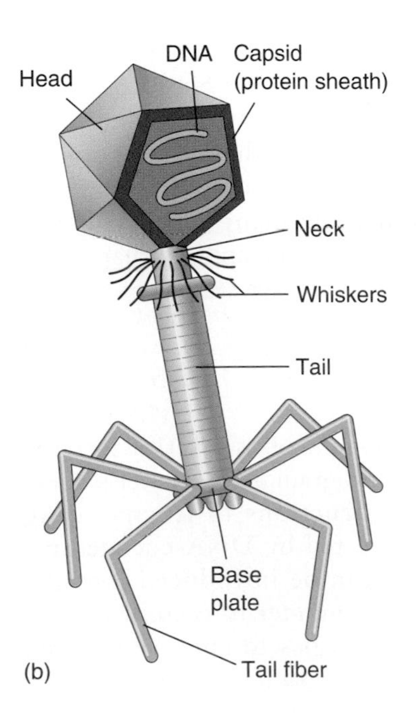

FIGURE 33.4
A bacterial virus. Bacteriophages exhibit a complex structure. (*a*) Electron micrograph and (*b*) diagram of the structure of a T4 bacteriophage.

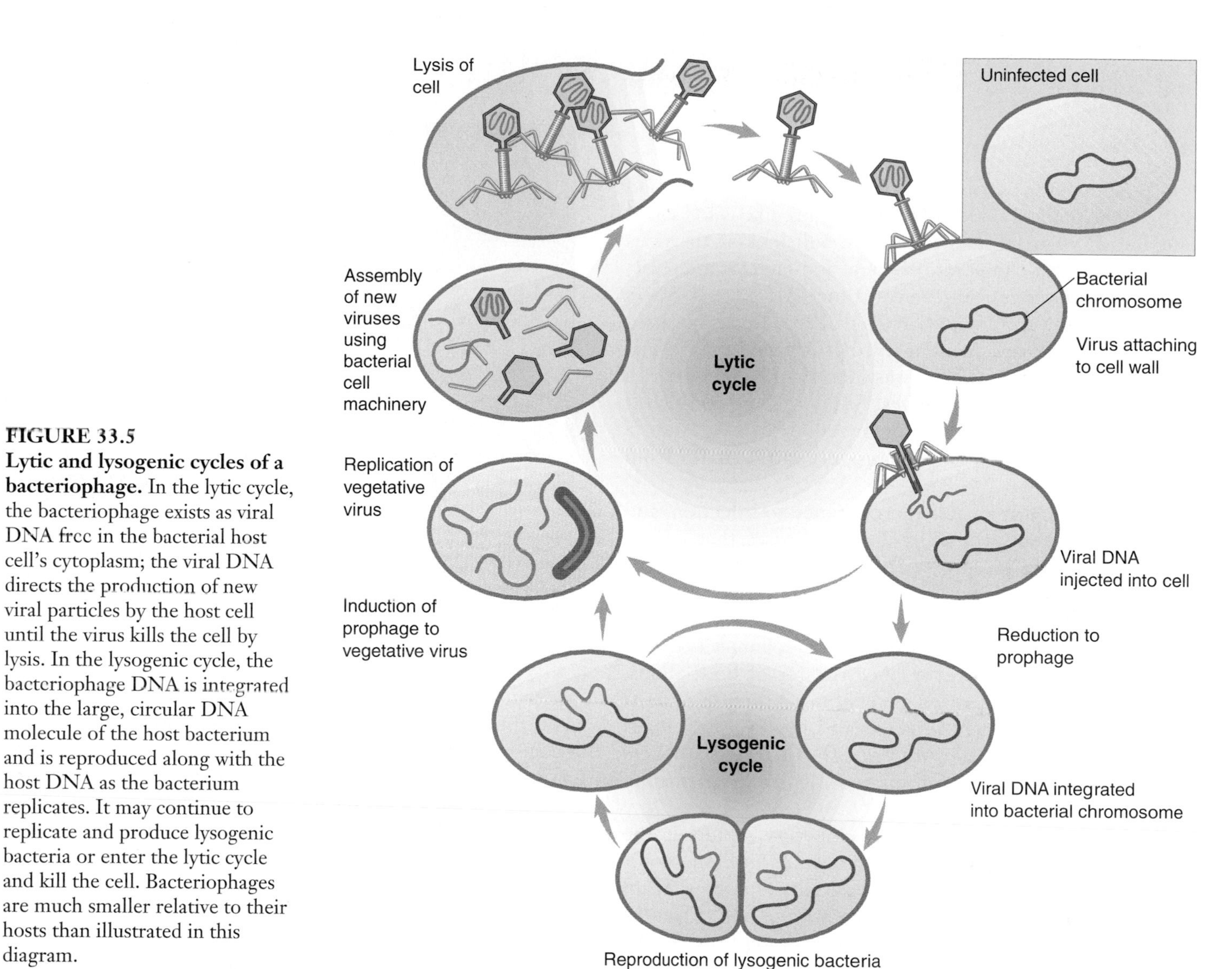

FIGURE 33.5
Lytic and lysogenic cycles of a bacteriophage. In the lytic cycle, the bacteriophage exists as viral DNA free in the bacterial host cell's cytoplasm; the viral DNA directs the production of new viral particles by the host cell until the virus kills the cell by lysis. In the lysogenic cycle, the bacteriophage DNA is integrated into the large, circular DNA molecule of the host bacterium and is reproduced along with the host DNA as the bacterium replicates. It may continue to replicate and produce lysogenic bacteria or enter the lytic cycle and kill the cell. Bacteriophages are much smaller relative to their hosts than illustrated in this diagram.

Cell Transformation and Phage Conversion

During the integrated portion of a lysogenic reproductive cycle, virus genes are often expressed. The RNA polymerase of the host cell reads the viral genes just as if they were host genes. Sometimes, expression of these genes has an important effect on the host cell, altering it in novel ways. The genetic alteration of a cell's genome by the introduction of foreign DNA is called **transformation.** When the foreign DNA is contributed by a bacterial virus, the alteration is called phage conversion.

Phage Conversion of the Cholera-Causing Bacterium

An important example of this sort of phage conversion directed by viral genes is provided by the bacterium responsible for an often-fatal human disease. The disease-causing bacteria *Vibrio cholerae* usually exists in a harmless form, but a second disease-causing, virulent form also occurs. In this latter form, the bacterium causes the deadly disease cholera, but how the bacteria changed from harmless to deadly was not known until recently. Research now shows that a bacteriophage that infects *V. cholerae* introduces into the host bacterial cell a gene that codes for the cholera toxin. This gene becomes incorporated into the bacterial chromosome, where it is translated along with the other host genes, thereby converting the benign bacterium to a disease-causing agent. The transfer occurs through bacterial pili (see chapter 34); in further experiments, mutant bacteria that did not have pili were resistant to infection by the bacteriophage. This discovery has important implications in efforts to develop vaccines against cholera, which have been unsuccessful up to this point.

Bacteriophages convert *Vibrio cholerae* bacteria from a harmless form into disease-causing agents.

AIDS

A diverse array of viruses occur among animals. A good way to gain a general idea of what they are like is to look at one animal virus in detail. Here we will look at the virus responsible for a comparatively new and fatal viral disease, acquired immunodeficiency syndrome (AIDS). AIDS was first reported in the United States in 1981. It was not long before the infectious agent, a retrovirus called human immunodeficiency virus (HIV), was identified by laboratories in France and the United States. Study of HIV revealed it to be closely related to a chimpanzee virus, suggesting a recent host expansion to humans in central Africa from chimpanzees.

Infected humans have little resistance to infection, and nearly all of them eventually die of diseases that noninfected individuals easily ward off. Few who contract AIDS survive more than a few years untreated. The risk of HIV transmission from an infected individual to a healthy one in the course of day-to-day contact is essentially nonexistent. However, the transfer of body fluids, such as blood, semen, or vaginal fluid, or the use of nonsterile needles, between infected and healthy individuals poses a severe risk. In addition, HIV-infected mothers can pass the virus on to their unborn children during fetal development.

The incidence of AIDS is growing very rapidly in the United States. It is estimated that over 33 million people worldwide are infected with HIV. Many—perhaps all of them—will eventually come down with AIDS. Over 16 million people have died already since the outbreak of the epidemic. AIDS incidence is already very high in many African countries and is growing at 20% worldwide. The AIDS epidemic is discussed further in chapter 57.

How HIV Compromises the Immune System

In normal individuals, an army of specialized cells (white blood cells) patrols the bloodstream, attacking and destroying any invading bacteria or viruses. In AIDS patients, this army of defenders is vanquished. One special kind of white blood cell, called a $CD4^+$ T cell (discussed further in chapter 57) is required to rouse the defending cells to action. In AIDS patients, the virus homes in on $CD4^+$ T cells, infecting and killing them until none are left (figure 33.6). Without these crucial immune system cells, the body cannot mount a defense against invading bacteria or viruses. AIDS patients die of infections that a healthy person could fight off.

Clinical symptoms typically do not begin to develop until after a long latency period, generally 8 to 10 years after the initial infection with HIV. During this long interval, carriers of HIV have no clinical symptoms but are apparently fully infectious, which makes the spread of HIV very difficult to control. The reason why HIV remains hidden for so long seems to be that its infection cycle continues throughout the 8- to 10-year latent period without doing serious harm to the infected person. Eventually, however, a random mutational event in the virus allows it to quickly overcome the immune defense, starting AIDS.

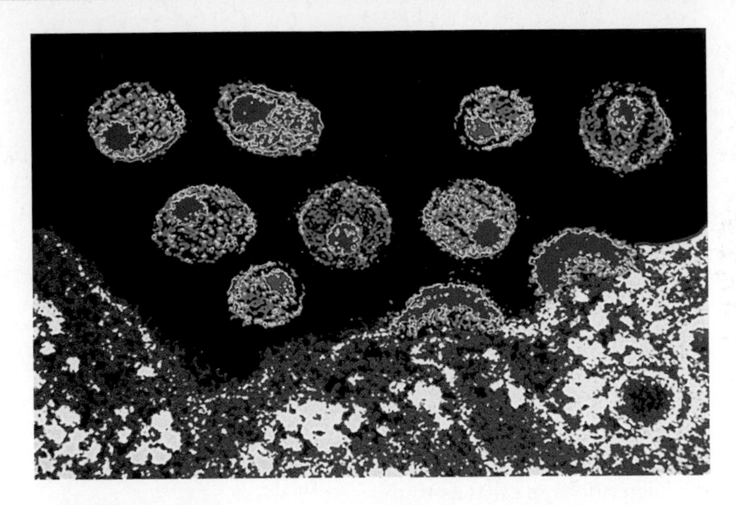

FIGURE 33.6
The AIDS virus. HIV particles exit an infected $CD4^+$ T cell (both shown in false color). The free virus particles are able to infect neighboring $CD4^+$ T cells.

The HIV Infection Cycle

The HIV virus infects and eliminates key cells of the immune system, destroying the body's ability to defend itself from cancer and infection. The way HIV infects humans (figure 33.7) provides a good example of how animal viruses replicate. Most other viral infections follow a similar course, although the details of entry and replication differ in individual cases.

Attachment. When HIV is introduced into the human bloodstream, the virus particle circulates throughout the entire body but will only infect $CD4^+$ cells. Most other animal viruses are similarly narrow in their requirements; hepatitis goes only to the liver, and rabies to the brain.

How does a virus such as HIV recognize a specific kind of target cell? Recall from chapter 7 that every kind of cell in the human body has a specific array of cell-surface glycoprotein markers that serve to identify them to other, similar cells. Each HIV particle possesses a glycoprotein (called **gp120**) on its surface that precisely fits a cell-surface marker protein called **CD4** on the surfaces of immune system cells called macrophages and T cells. Macrophages are infected first.

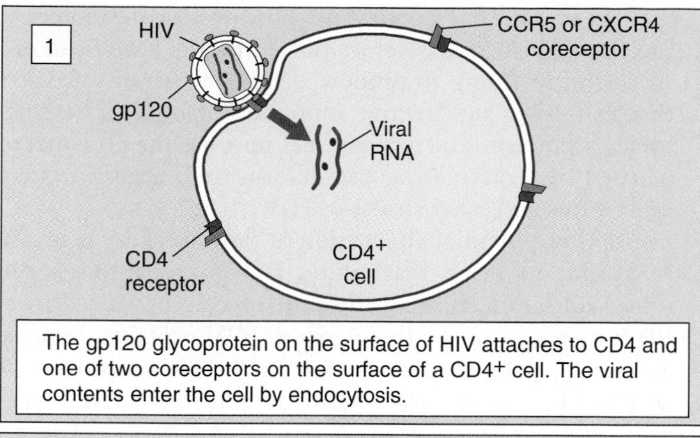

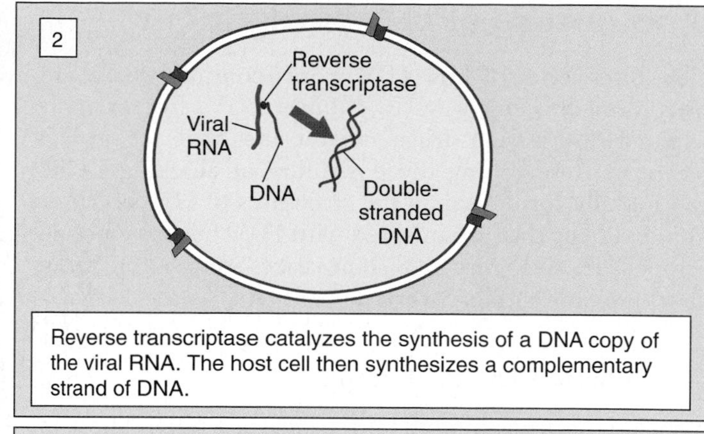

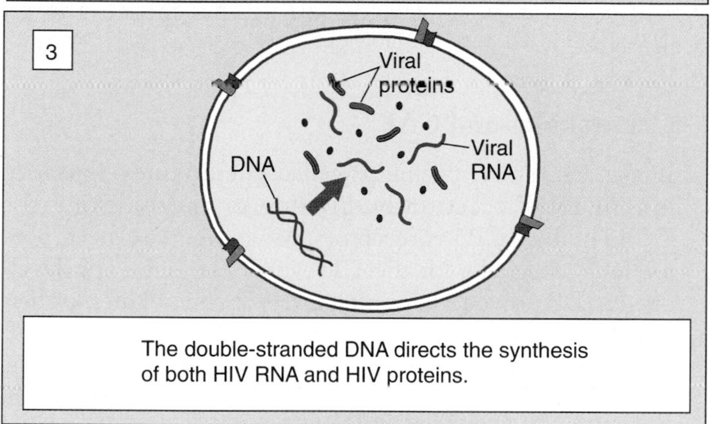

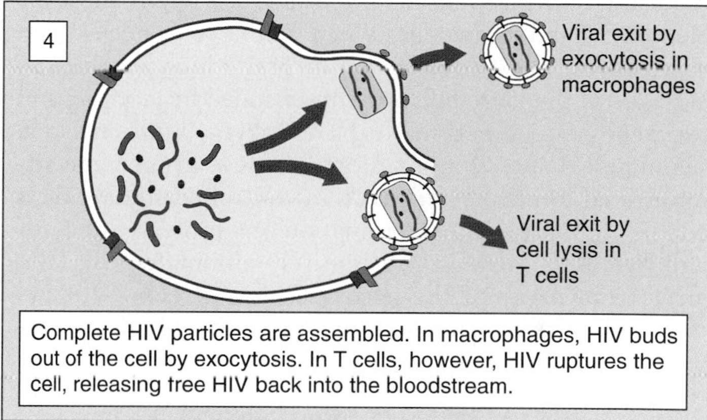

4
Viral exit by exocytosis in macrophages
Viral exit by cell lysis in T cells
Complete HIV particles are assembled. In macrophages, HIV buds out of the cell by exocytosis. In T cells, however, HIV ruptures the cell, releasing free HIV back into the bloodstream.

FIGURE 33.7
The HIV infection cycle. The cycle begins and ends with free HIV particles present in the bloodstream of its human host. These free viruses infect white blood cells called $CD4^+$ T cells.

Entry into Macrophages. After docking onto the CD4 receptor of a macrophage, HIV requires a second macrophage receptor, called CCR5, to pull itself across the cell membrane. After gp120 binds to CD4, it goes through a conformational change that allows it to bind to CCR5. The current model suggests that after the conformational change, the second receptor passes the gp120-CD4 complex through the cell membrane, triggering passage of the contents of the HIV virus into the cell by endocytosis, with the cell membrane folding inward to form a deep cavity around the virus.

Replication. Once inside the macrophage, the HIV particle sheds its protective coat. This leaves virus RNA floating in the cytoplasm, along with a virus enzyme that was also within the virus shell. This enzyme, called **reverse transcriptase,** synthesizes a double strand of DNA complementary to the virus RNA, often making mistakes and so introducing new mutations. This double-stranded DNA directs the host cell machinery to produce many copies of the virus. HIV does not rupture and kill the macrophage cells it infects. Instead, the new viruses are released from the cell by exocytosis. HIV synthesizes large numbers of viruses in this way, challenging the immune system over a period of years.

Entry into T Cells. During this time, HIV is constantly replicating and mutating. Eventually, by chance, HIV alters the gene for gp120 in a way that causes the gp120 protein to change its second-receptor allegiance. This new form of gp120 protein prefers to bind instead to a different second receptor, CXCR4, a receptor that occurs on the surface of T lymphocyte $CD4^+$ cells. Soon the body's T lymphocytes become infected with HIV. This has deadly consequences, as new viruses exit the cell by rupturing the cell membrane, effectively killing the infected T cell. Thus, the shift to the CXCR4 second receptor is followed swiftly by a steep drop in the number of T cells. This destruction of the body's T cells blocks the immune response and leads directly to the onset of AIDS, with cancers and opportunistic infections free to invade the defenseless body.

HIV, the virus that causes AIDS, is an RNA virus that replicates inside human cells by first making a DNA copy of itself. It is only able to gain entrance to those cells possessing a particular cell surface marker recognized by a glycoprotein on its own surface.

The Future of HIV Treatment

New discoveries of how HIV works continue to fuel research on devising ways to counter HIV. For example, scientists are testing drugs and vaccines that act on HIV receptors, researching the possibility of blocking CCR5, and looking for defects in the structures of HIV receptors in individuals that are infected with HIV but have not developed AIDS. Figure 33.8 summarizes some of the recent developments and discoveries.

Combination Drug Therapy

A variety of drugs inhibit HIV in the test tube. These include AZT and its analogs (which inhibit virus nucleic acid replication) and protease inhibitors (which inhibit the cleavage of the large polyproteins encoded by *gag*, *poll*, and *env* genes into functional capsid, enzyme, and envelope segments). When combinations of these drugs were administered to people with HIV in controlled studies, their condition improved. A combination of a protease inhibitor and two AZT analog drugs entirely eliminated the HIV virus from many of the patients' bloodstreams. Importantly, all of these patients began to receive the drug therapy within three months of contracting the virus, before their bodies had an opportunity to develop tolerance to any one of them. Widespread use of this **combination therapy** has cut the U.S. AIDS death rate by three-fourths since its introduction in the mid-1990s, from 49,000 AIDS deaths in 1995 to 36,000 in 1996, and just over 10,000 in 1999.

Unfortunately, this sort of combination therapy does not appear to actually succeed in eliminating HIV from the body. While the virus disappears from the bloodstream, traces of it can still be detected in lymph tissue of the patients. When combination therapy is discontinued, virus levels in the bloodstream once again rise. Because of demanding therapy schedules and many side effects, long-term combination therapy does not seem a promising approach.

Using a Defective HIV Gene to Combat AIDS

Recently, five people in Australia who are HIV-positive but have not developed AIDS in 14 years were found to have all received a blood transfusion from the same HIV-positive person, who also has not developed AIDS. This led scientists to believe that the strain of virus transmitted to these people has some sort of genetic defect that prevents it from effectively disabling the human immune system. In subsequent research, a defect was found in one of the nine genes present in this strain of HIV. This gene is called ***nef***, named for "negative factor," and the defective version of *nef* in the HIV strain that infected the six Australians seems to be missing some pieces. Viruses with the defective gene may have reduced reproductive capability, allowing the immune system to keep the virus in check.

This finding has exciting implications for developing a vaccine against AIDS. Before this, scientists have been unsuccessful in trying to produce a harmless strain of AIDS that can elicit an effective immune response. The Australian strain with the defective *nef* gene has the potential to be used in a vaccine that would arm the immune system against this and other strains of HIV.

Another potential application of this discovery is its use in developing drugs that inhibit HIV proteins that speed virus replication. It seems that the protein produced from the *nef* gene is one of these critical HIV proteins, because viruses with defective forms of *nef* do not reproduce, as seen in the cases of the six Australians. Research is currently underway to develop a drug that targets the *nef* protein.

Chemokines and CAF

In the laboratory, chemicals called **chemokines** appear to inhibit HIV infection by binding to and blocking the CCR5 and CXCR4 coreceptors. As you might expect, people long infected with the HIV virus who have not developed AIDS prove to have high levels of chemokines in their blood.

The search for HIV-inhibiting chemokines is intense. Not all results are promising. Researchers report that in their tests, the levels of chemokines were not different between patients in which the disease was not progressing and those in which it was rapidly progressing. More promising, levels of another factor called **CAF** ($CD8^+$ cell antiviral factor) *are* different between these two groups. Researchers have not yet succeeded in isolating CAF, which seems not to block receptors that HIV uses to gain entry to cells, but, instead, to prevent replication of the virus once it has infected the cells. Research continues on the use of chemokines in treatments for HIV infection, either increasing the amount of chemokines or disabling the CCR5 receptor. However, promising research on CAF suggests that it may be an even better target for treatment and prevention of AIDS.

One problem with using chemokines as drugs is that they are also involved in the inflammatory response of the immune system. The function of chemokines is to attract white blood cells to areas of infection. Chemokines work beautifully in small amounts and in local areas, but chemokines in mass numbers can cause an inflammatory response that is worse than the original infection. Injections of chemokines may hinder the immune system's ability to respond to local chemokines, or they may even trigger an out-of-control inflammatory response. Thus, scientists caution that injection of chemokines could make patients *more* susceptible to infections, and they continue to research other methods of using chemokines to treat AIDS.

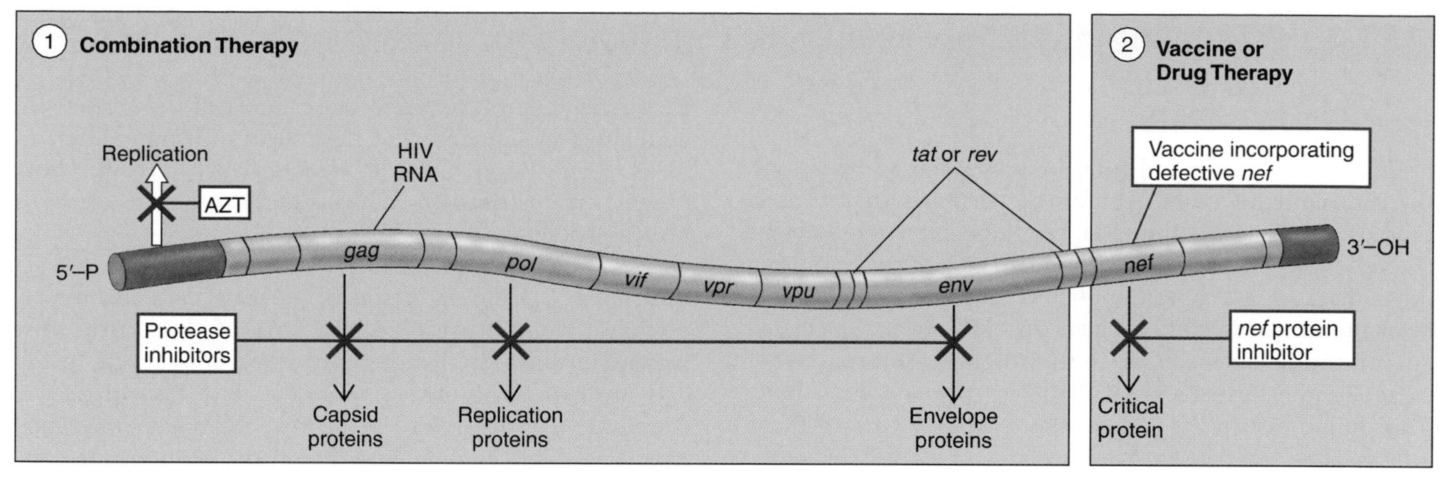

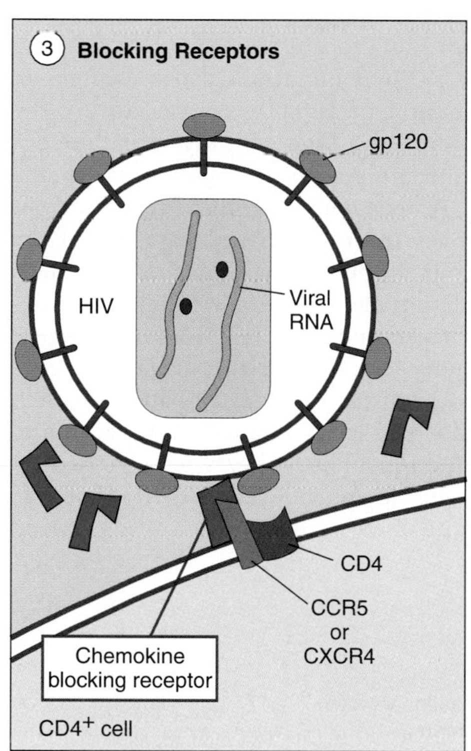

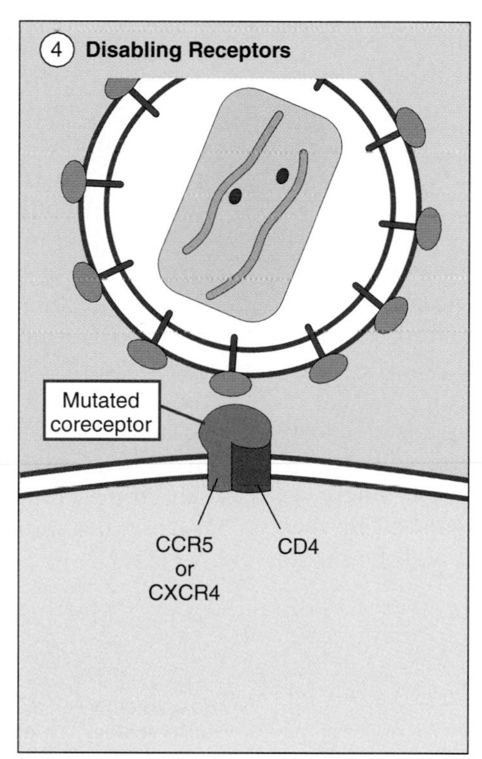

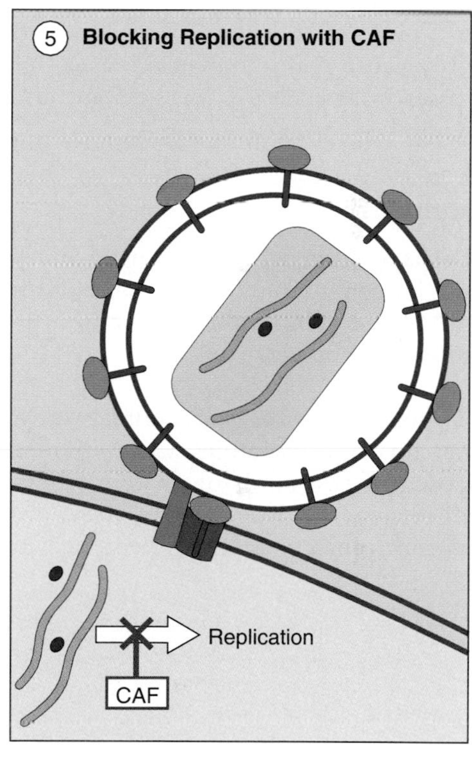

FIGURE 33.8
Research is currently underway to develop new treatments for HIV. Among them are these five: (*1*) *Combination therapy* involves using two drugs, AZT to block replication of the virus and protease inhibitors to block the production of critical viral proteins. (*2*) Using a defective form of the viral gene *nef*, scientists may be able to construct an HIV *vaccine*. Also, *drug therapy* that inhibits *nef's* protein product is being tested. (*3*) Other research focuses on the use of chemokine chemicals to *block receptors* (CXCR4 and CCR5), thereby disabling the mechanism HIV uses to enter CD4$^+$ T cells. (*4*) Producing mutations that will *disable receptors* may also be possible. (*5*) Lastly, CAF, an antiviral factor which acts inside the CD4$^+$ T cell, may be able to *block replication* of HIV.

Disabling Receptors

A 32-base-pair deletion in the gene that codes for the CCR5 receptor appears to block HIV infection. Individuals at high risk of HIV infection who are homozygous for this mutation do not seem to develop AIDS. In one study of 1955 people, scientists found no individuals who were infected and homozygous for the mutated allele. The allele seems to be more common in Caucasian populations (10 to 11%) than in African-American populations (2%), and absent in African and Asian populations. Treatment for AIDS involving disruption of CCR5 looks promising, as research indicates that people live perfectly well without CCR5. Attempts to block or disable CCR5 are being sought in numerous laboratories.

A cure for AIDS is not yet in hand, but many new approaches look promising.

Disease Viruses

Humans have known and feared diseases caused by viruses for thousands of years. Among the diseases that viruses cause (table 33.1) are influenza, smallpox, infectious hepatitis, yellow fever, polio, rabies, and AIDS, as well as many other diseases not as well known. In addition, viruses have been implicated in some cancers and leukemias. For many autoimmune diseases, such as multiple sclerosis and rheumatoid arthritis, and for diabetes, specific viruses have been found associated with certain cases. In view of their effects, it is easy to see why the late Sir Peter Medawar, Nobel laureate in Physiology or Medicine, wrote, "A virus is a piece of bad news wrapped in protein." Viruses not only cause many human diseases, but also cause major losses in agriculture, forestry, and in the productivity of natural ecosystems.

Influenza

Perhaps the most lethal virus in human history has been the influenza virus. Some 22 million Americans and Europeans died of flu within 18 months in 1918 and 1919, an astonishing number.

Types. Flu viruses are animal retroviruses. An individual flu virus resembles a rod studded with spikes composed of two kinds of protein (figure 33.9). There are three general "types" of flu virus, distinguished by their capsid (inner membrane) protein, which is different for each type: Type A flu virus causes most of the serious flu epidemics in humans, and also occurs in mammals and birds. Type B and Type C viruses, with narrower host ranges, are restricted to humans and rarely cause serious health problems.

Subtypes. Different strains of flu virus, called subtypes, differ in their protein spikes. One of these proteins, hemagglutinin (H) aids the virus in gaining access to the cell interior. The other, neuraminidase (N) helps the daughter virus break free of the host cell once virus replication has been completed. Parts of the H molecule contain "hot spots" that display an unusual tendency to change as a result of mutation of the virus RNA during imprecise replication. Point mutations cause changes in these spike proteins in 1 of 100,000 viruses during the course of each generation. These highly variable segments of the H molecule are targets against which the body's antibodies are directed. The high variability of these targets improves the reproductive capacity of the virus and hinders our ability to make perfect vaccines. Because of accumulating changes in the H and N molecules, different flu vaccines are required to protect against different subtypes. Type A flu viruses are currently classified into 13 distinct H subtypes and 9 distinct N subtypes, each of which requires a different vaccine to protect against infection. Thus, the type A virus that caused the Hong Kong flu epidemic of 1968 has type 3 H molecules and type 2 N molecules, and is called A(H3N2).

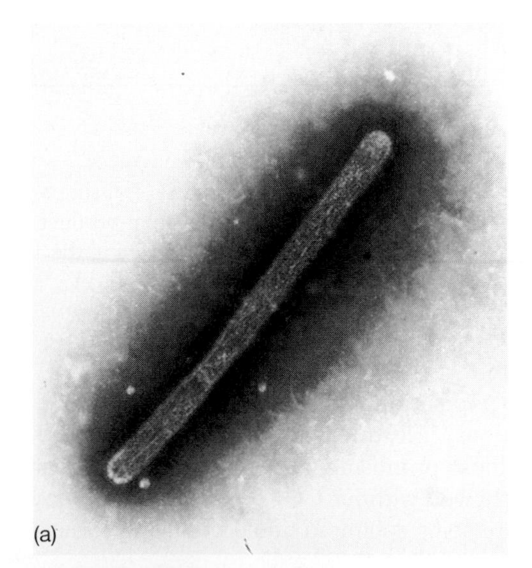

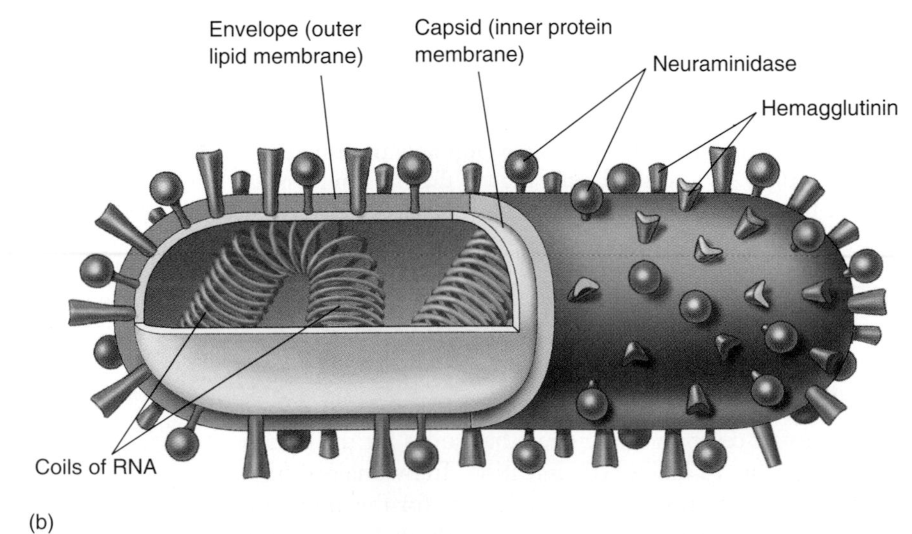

FIGURE 33.9
The influenza virus. (*a*) TEM of the so-called "bird flu" influenza virus, A(H5N1), which first infected humans in Hong Kong in 1997. (*b*) Diagram of an influenza virus. The coiled RNA has been revealed by cutting through the outer lipid-rich envelope, with its two kinds of projecting spikes, and the inner protein capsid.

Table 33.1 Important Human Viral Diseases			
Disease	**Pathogen**	**Reservoir**	**Vector/Epidemiology**
AIDS	HIV	STD	Destroys immune defenses, resulting in death by infection or cancer. Over 33 million cases worldwide by 1998.
Chicken pox	Human herpes-virus 3 (varicella-zoster)	Humans	Spread through contact with infected individuals. No cure. Rarely fatal. Vaccine approved in U.S. in early 1995.
Ebola	Filoviruses	Unknown	Acute hemorrhagic fever; virus attacks connective tissue, leading to massive hemorrhaging and death. Peak mortality is 50–90% if the disease goes untreated. Outbreaks confined to local regions of central Africa.
Hepatitus B (viral)	Hepatitis B virus (HBV)	Humans	Highly infectious through contact with infected body fluids. Approximately 1% of U.S. population infected. Vaccine available, no cure. Can be fatal.
Herpes	Herpes simplex virus (HSV)	Humans	Fever blisters; spread primarily through contact with infected saliva. Very prevalent worldwide. No cure. Exhibits latency—the disease can be dormant for several years.
Influenza	Influenza viruses	Humans, ducks	Historically a major killer (22 million died in 18 months in 1918–19); wild Asian ducks, chickens, and pigs are major reservoirs. The ducks are not affected by the flu virus, which shuffles its antigen genes while multiplying within them, leading to new flu strains.
Measles	Paramyxoviruses	Humans	Extremely contagious through contact with infected individuals. Vaccine available. Usually contracted in childhood, when it is not serious; more dangerous to adults.
Mononucleosis	Epstein-Barr virus (EBV)	Humans	Spread through contact with infected saliva. May last several weeks; common in young adults. No cure. Rarely fatal.
Mumps	Paramyxovirus	Humans	Spread through contact with infected saliva. Vaccine available; rarely fatal. No cure.
Pneumonia	Influenza virus	Humans	Acute infection of the lungs, often fatal without treatment.
Polio	Poliovirus	Humans	Acute viral infection of the CNS that can lead to paralysis and is often fatal. Prior to the development of Salk's vaccine in 1954, 60,000 people a year contracted the disease in the U.S. alone.
Rabies	Rhabdovirus	Wild and domestic Canidae (dogs, foxes, wolves, coyotes), bats, and raccoons	An acute viral encephalomyelitis transmitted by the bite of an infected animal. Fatal if untreated.
Smallpox	Variola virus	Formerly humans, now only exists in two research labs—may be eliminated	Historically a major killer; the last recorded case of smallpox was in 1977. A worldwide vaccination campaign wiped out the disease completely.
Yellow fever	Flavivirus	Humans, mosquitoes	Spread from individual to individual by mosquito bites; a notable cause of death during the construction of the Panama Canal. If untreated, this disease has a peak mortality rate of 60%.

Importance of Recombination. The greatest problem in combating flu viruses arises not through mutation, but through recombination. Viral genes are readily reassorted by genetic recombination, sometimes putting together novel combinations of H and N spikes unrecognizable by human antibodies specific for the old configuration. Viral recombination of this kind seems to have been responsible for the three major flu pandemics (that is, worldwide epidemics) that occurred in the twentieth century, by producing drastic shifts in H N combinations. The "killer flu" of 1918, A(H1N1), killed 40 million people. The Asian flu of 1957, A(H2N2), killed over 100,000 Americans. The Hong Kong flu of 1968, A(H3N2), infected 50 million people in the United States alone, of which 70,000 died.

It is no accident that new strains of flu usually originate in the far east. The most common hosts of influenza virus are ducks, chickens, and pigs, which in Asia often live in close proximity to each other and to humans. Pigs are subject to infection by both bird and human strains of the virus, and individual animals are often simultaneously infected with multiple strains. This creates conditions favoring genetic recombination between strains, producing new combinations of H and N subtypes. The Hong Kong flu, for example, arose from recombination between A(H3N8) [from ducks] and A(H2N2) [from humans]. The new strain of influenza, in this case A(H3N2), then passed back to humans, creating an epidemic because the human population has never experienced that H N combination before.

A potentially deadly new strain of flu virus emerged in Hong Kong in 1997, A(H5N1). Unlike all previous instances of new flu strains, A(H5N1) passed to humans directly from birds, in this case chickens. A(H5N1) was first identified in chickens in 1961, and in the spring of 1997 devastated flocks of chickens in Hong Kong. Fortunately, this strain of flu virus does not appear to spread easily from person to person, and the number of human infections by A(H5N1) remains small. Public health officials remain concerned that the genes of A(H5N1) could yet mix with those of a human strain to create a new strain that could spread widely in the human population, and to prevent this ordered the killing of all 1.2 million chickens in Hong Kong in 1997.

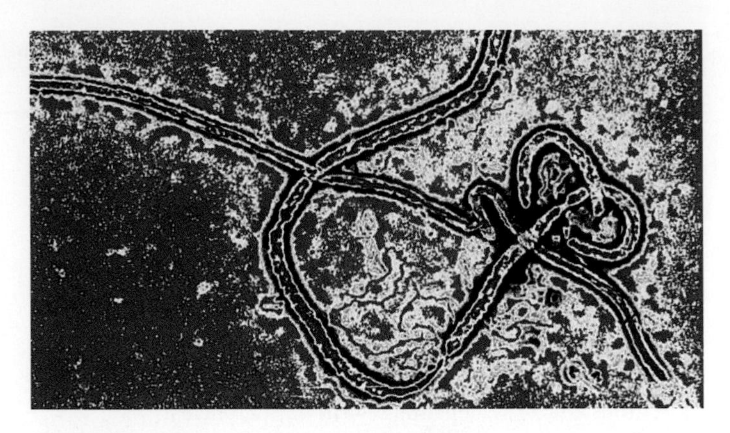

FIGURE 33.10
The Ebola virus. This virus, with a fatality rate that can exceed 90%, appears sporadically in West Africa. Health professionals are scrambling to identify the natural host of the virus, which is unknown, so they can devise strategies to combat transmission of the disease.

Emerging Viruses

Sometimes viruses that originate in one organism pass to another, thus expanding their host range. Often, this expansion is deadly to the new host. HIV, for example, arose in chimpanzees and relatively recently passed to humans. Influenza is fundamentally a bird virus. Viruses that originate in one organism and then pass to another and cause disease are called **emerging viruses** and represent a considerable threat in an age when airplane travel potentially allows infected individuals to move about the world quickly, spreading an infection.

Among the most lethal of emerging viruses are a collection of filamentous viruses arising in central Africa that cause severe hemorrhagic fever. With lethality rates in excess of 50%, these so-called filoviruses are among the most lethal infectious diseases known. One, Ebola virus (figure 33.10), has exhibited lethality rates in excess of 90% in isolated outbreaks in central Africa. The outbreak of Ebola virus in the summer of 1995 in Zaire killed 245 people out of 316 infected—a mortality rate of 78%. The latest outbreak occurred in Gabon, West Africa, in February 1996. The natural host of Ebola is unknown.

Another type of emerging virus caused a sudden outbreak of a hemorrhagic-type infection in the southwestern United States in 1993. This highly fatal disease was soon attributed to the hantavirus, a single-stranded RNA virus associated with rodents. The hantavirus is transmitted to humans through rodent fecal contamination in areas of human habitation. Although hantavirus has been known for some period of time, this particular outbreak was attributed to the presence of an unusually large rodent population in the area following a higher than normal amount of rainfall the previous winter.

Viruses and Cancer

Through epidemiological studies and research, scientists have established a link between some viral infections and the subsequent development of cancer. Examples include the association between chronic hepatitis B infections and the development of liver cancer and the development of cervical carcinoma following infections with certain strains of papillomaviruses. It has been suggested that viruses contribute to about 15% of all human cancer cases worldwide. Viruses are capable of altering the growth properties of human cells they infect by triggering the expression of oncogenes (cancer-causing genes). Certain viruses can either activate host proto-oncogenes (see chapter 18) or bring in viral oncogenes that become incorporated into the host genome. Virus-induced cancer is not simply a matter of infection. The disease involves complex interactions with cellular genes and requires a series of events in order to develop.

Viruses are responsible for some of the most lethal diseases of humans. Some of the most serious examples are viruses that have transferred to humans from some other host. Influenza, a bird virus, has been responsible for the most devastating epidemics in human history. Newly emerging viruses such as Ebola have received considerable public attention.

Prions and Viroids

For decades scientists have been fascinated by a peculiar group of fatal brain diseases. These diseases have the unusual property that it is years and often decades after infection before the disease is detected in infected individuals. The brains of infected individuals develop numerous small cavities as neurons die, producing a marked spongy appearance. Called **transmissible spongiform encephalopathies (TSEs)**, these diseases include scrapie in sheep, "mad cow" disease in cattle, and kuru and Creutzfeldt-Jakob disease in humans.

TSEs can be transmitted by injecting infected brain tissue into a recipient animal's brain. TSEs can also spread via tissue transplants and, apparently, food. Kuru was common in the Fore people of Papua New Guinea, when they practiced ritual cannibalism, literally eating the brains of infected individuals. Mad cow disease spread widely among the cattle herds of England in the 1990s because cows were fed bone meal prepared from cattle carcasses to increase the protein content of their diet. Like the Fore, the British cattle were literally eating the tissue of cattle that had died of the disease.

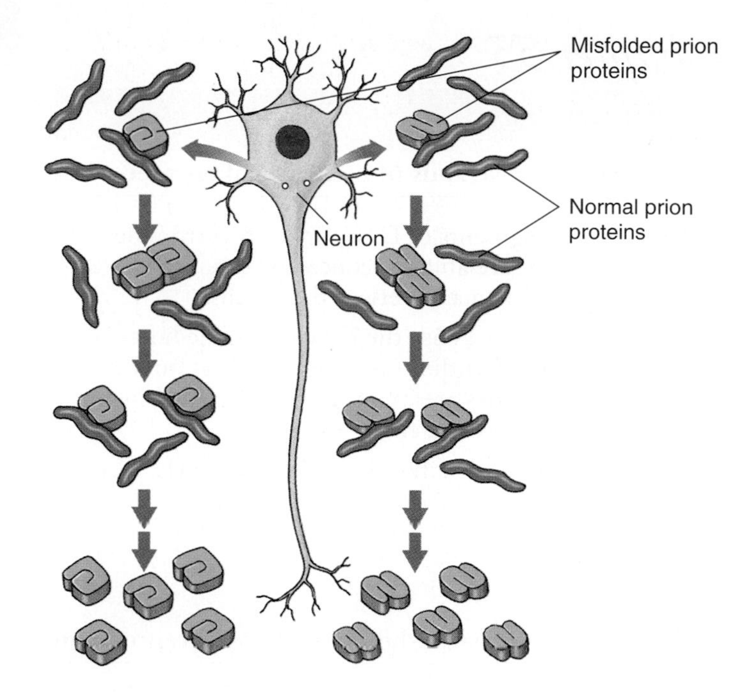

FIGURE 33.11
How prions arise. Misfolded prions seem to cause normal prion protein to misfold simply by contacting them. When prions misfolded in different ways (*blue*) contact normal prion protein (*purple*), the normal prion protein misfolds in the same way.

A Heretical Suggestion

In the 1960s, British researchers T. Alper and J. Griffith noted that infectious TSE preparations remained infectious even after exposed to radiation that would destroy DNA or RNA. They suggested that the infectious agent was a protein. Perhaps, they speculated, the protein usually preferred one folding pattern, but could sometimes misfold, and then catalyze other proteins to do the same, the misfolding spreading like a chain reaction. This heretical suggestion was not accepted by the scientific community, as it violates a key tenant of molecular biology: only DNA or RNA act as hereditary material, transmitting information from one generation to the next.

Prusiner's Prions

In the early 1970s, physician Stanley Prusiner, moved by the death of a patient from Creutzfeldt-Jakob disease, began to study TSEs. Prusiner became fascinated with Alper and Griffith's hypothesis. Try as he might, Prusiner could find no evidence of nucleic acids or viruses in the infectious TSE preparations, and concluded, as Alper and Griffith had, that the infectious agent was a *protein*, which in a 1982 paper he named a **prion**, for "proteinaceous infectious particle."

Prusiner went on to isolate a distinctive prion protein, and for two decades continued to amass evidence that prions play a key role in triggering TSEs. The scientific community resisted Prusiner's renegade conclusions, but eventually experiments done in Prusiner's and other laboratories began to convince many. For example, when Prusiner injected prions of a different abnormal conformation into several different hosts, these hosts developed prions with the same abnormal conformations as the parent prions (figure 33.11). In another important experiment, Charles Weissmann showed that mice genetically engineered to lack Prusiner's prion protein are immune to TSE infection. However, if brain tissue with the prion protein is grafted into the mice, the grafted tissue—but not the rest of the brain—can then be infected with TSE. In 1997, Prusiner was awarded the Nobel Prize in Physiology or Medicine for his work on prions.

Viroids

Viroids are tiny, naked molecules of RNA, only a few hundred nucleotides long, that are important infectious disease agents in plants. A recent viroid outbreak killed over ten million coconut palms in the Philippines. It is not clear how viroids cause disease. One clue is that viroid nucleotide sequences resemble the sequences of introns within ribosomal RNA genes. These sequences are capable of catalyzing excision from DNA—perhaps the viroids are catalyzing the destruction of chromosomal integrity.

Prions are infectious proteins that some scientists believe are responsible for serious brain diseases. In plants, naked RNA molecules called viroids can also transmit disease.

Chapter 33

www.mhhe.com/raven6/resources33.mhtml

Summary	Questions	Media Resources

33.1 Viruses are strands of nucleic acid encased within a protein coat.

- Viruses are fragments of DNA or RNA surrounded by protein that are able to replicate within cells by using the genetic machinery of those cells.
- The simplest viruses use the enzymes of the host cell for both protein synthesis and gene replication; the more complex ones are capable of synthesizing many structural proteins and enzymes.
- Viruses are basically either helical or isometric. Most isometric viruses are icosahedral in shape.

1. Why are viruses not considered to be living organisms?

2. How did early scientists come to the conclusion that the infectious agents associated with hoof-and-mouth disease in cattle were not bacteria?

3. What is the approximate size range of viruses and type of microscope is generally required to visualize viruses?

- Characteristics of Viruses

33.2 Bacterial viruses exhibit two sorts of reproductive cycles.

- Virulent bacteriophages infect bacterial cells by injecting their viral DNA or RNA into the cell, where it directs the production of new virus particles, ultimately lysing the cell.
- Temperate bacteriophages, upon entering a bacterial cell, insert their DNA into the cell genome, where they may remain integrated into the bacterial genome as a prophage for many generations.

4. What is a bacteriophage? How does a T4 phage infect a host cell?

33.3 HIV is a complex animal virus.

- AIDS, a viral infection that destroys the immune system, is caused by HIV (human immunodeficiency virus). After docking on a specific protein called CD4, HIV enters the cell and replicates, destroying the cell.
- Considerable progress has been made in the treatment of AIDS, particularly with drugs such as protease inhibitors that block cleavage of HIV polyproteins into functional segments.

5. What specific type of human cell does the AIDS virus infect? How does it recognize this specific kind of cell?

6. How do many animal viruses penetrate the host cell? How does a plant virus infect its host? How does a bacterial virus infect its host?

- Life Cycle of Viruses

- Bioethics Case Study: AIDS Vaccine
- *On Science* Article: Curing AIDS Just Got Harder

33.4 Nonliving infectious agents are responsible for many human diseases.

- Viruses are responsible for many serious human diseases. Influenza, the most lethal virus in human history, evades the human immune system through a recombination of surface proteins. Some of the most serious viral diseases, like AIDS and Ebola, have only recently transferred to humans from some other animal host.
- Proteins called prions may transmit serious brain diseases from one individual to another.

7. Why is it so much more difficult to treat a viral infection than a bacterial one? Is this different from treating bacterial infections?

8. What is a prion? How does it integrate into living systems?

- Scientists on Science: Prions
- Book Review: *The Coming Plague* by Garrett
- *On Science* Articles:
  - -Smallpox: Tomorrow's Nightmare?
  - -Mad Cow Epidemic in Europe

- Art Quiz: Maintaining Prions

34

Bacteria

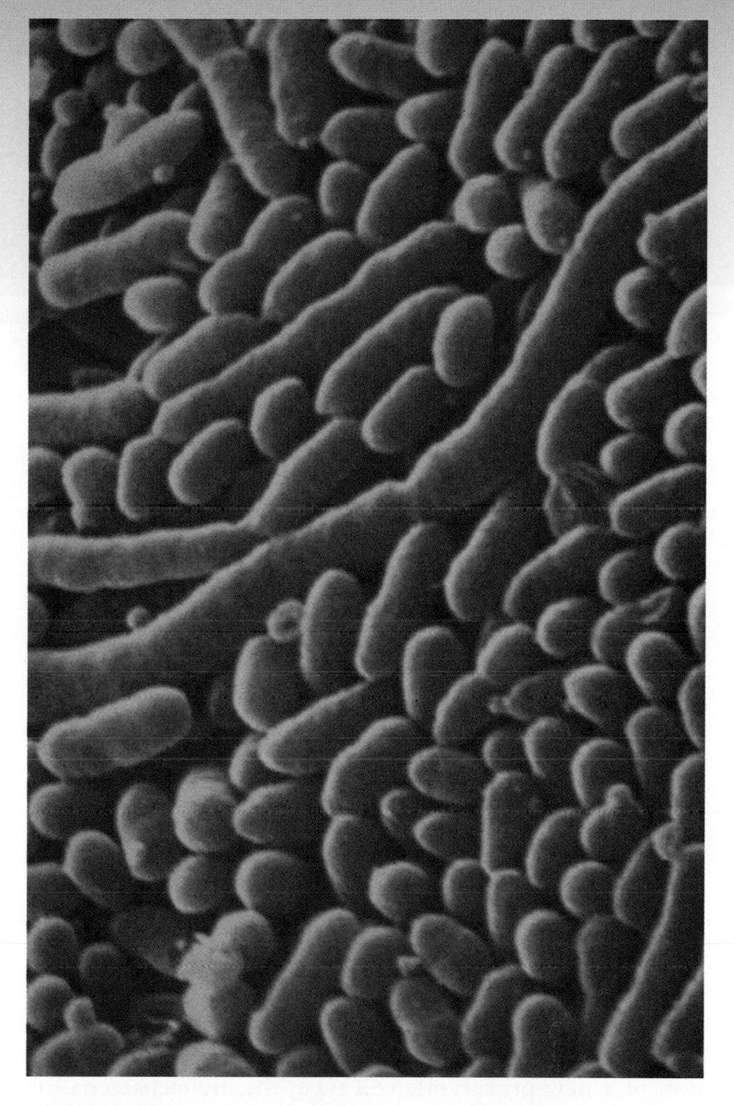

FIGURE 34.1
A colony of bacteria. With their enormous adaptability and metabolic versatility, bacteria are found in every habitat on earth, carrying out many of the vital processes of ecosystems, including photosynthesis, nitrogen fixation, and decomposition.

Concept Outline

The simplest organisms living on earth today are bacteria, and biologists think they closely resemble the first organisms to evolve on earth. Too small to see with the unaided eye, bacteria are the most abundant of all organisms (figure 34.1) and are the only ones characterized by prokaryotic cellular organization. Life on earth could not exist without bacteria because bacteria make possible many of the essential functions of ecosystems, including the capture of nitrogen from the atmosphere, decomposition of organic matter, and, in many aquatic communities, photosynthesis. Indeed, bacterial photosynthesis is thought to have been the source for much of the oxygen in the earth's atmosphere. Bacterial research continues to provide extraordinary insights into genetics, ecology, and disease. An understanding of bacteria is thus essential.

34.1 Bacteria are the smallest and most numerous organisms.

The Prevalence of Bacteria

Bacteria are the oldest, structurally simplest, and the most abundant forms of life on earth. They are also the only organisms with prokaryotic cellular organization. Represented in the oldest rocks from which fossils have been obtained, 3.5 to 3.8 billion years old, bacteria were abundant for over 2 billion years before eukaryotes appeared in the world (see figure 4.11). Early photosynthetic bacteria (cyanobacteria) altered the earth's atmosphere with the production of oxygen which lead to extreme bacterial and eukaryotic diversity. Bacteria play a vital role both in productivity and in cycling the substances essential to all other life-forms. Bacteria are the only organisms capable of fixing atmospheric nitrogen.

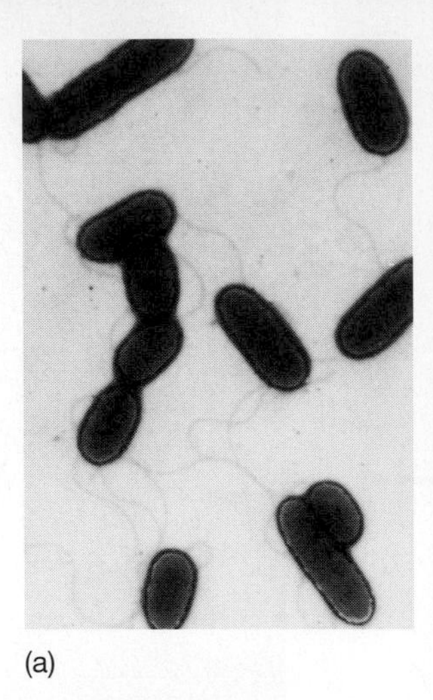

(a)

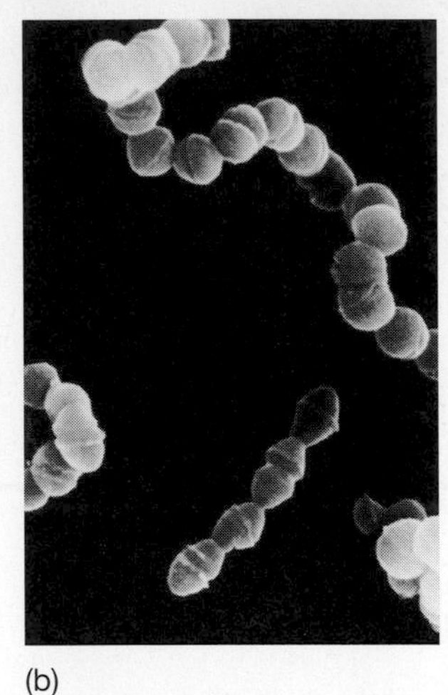

(b)

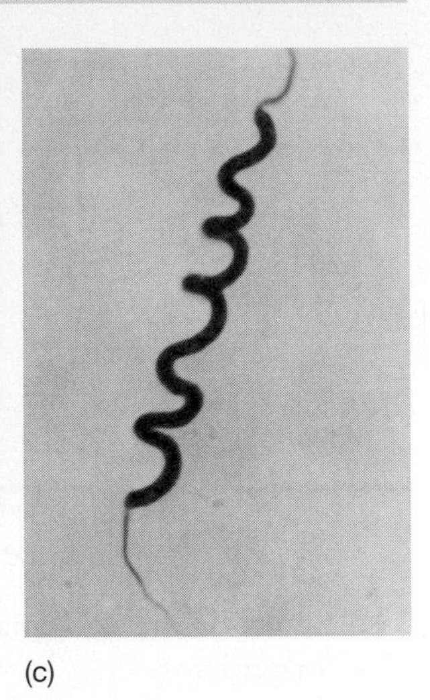

(c)

FIGURE 34.2
The diversity of bacteria. (*a*) *Pseudomonas aeruginosa*, a rod-shaped, flagellated bacterium (bacillus). *Pseudomonas* includes the bacteria that cause many of the most serious plant diseases. (*b*) *Streptococcus*. The spherical individual bacteria (cocci) adhere in chains in the members of this genus (34,000×). (*c*) *Spirillum volutans*, one of the spirilla. This large bacterium, which occurs in stagnant fresh water, has a tuft of flagella at each end (500×).

About 5000 different kinds of bacteria are currently recognized, but there are doubtless many thousands more awaiting proper identification (figure 34.2). Every place microbiologists look, new species are being discovered, in some cases altering the way we think about bacteria. In the 1970s and 80s a new type of bacterium was analyzed that eventually lead to the classification of a new prokaryotic cell type, the archaebacteria (or Archaea). Even when viewed with an electron microscope, the structural differences between different bacteria are minor compared to other groups of organisms. Because the structural differences are so slight, bacteria are classified based primarily upon their metabolic and genetic characteristics. Bacteria can be characterized properly only when they are grown on a defined medium because the characteristics of these organisms often change, depending on their growth conditions.

Bacteria are ubiquitous on Earth, and live everywhere eukaryotes do. Many of the other more extreme environments in which bacteria are found would be lethal to any other form of life. Bacteria live in hot springs that would cook other organisms, hypersaline environments that would dehydrate other cells, and in atmospheres rich in toxic gases like methane or hydrogen sulfide that would kill most other organisms. These harsh environments may be similar to the conditions present on the early Earth, when life first began. It is likely that bacteria evolved to dwell in these harsh conditions early on and have retained the ability to exploit these areas as the rest of the atmosphere has changed.

Bacterial Form

Bacteria are mostly simple in form and exhibit one of three basic structures**:** **bacillus** (plural, **bacilli**) straight and rod-shaped, **coccus** (plural, **cocci**) spherical-shaped, and **spirillus** (plural, **spirilla**) long and helical-shaped, also called spirochetes**.** Spirilla bacteria generally do not form associations with other cells and swim singly through their environments. They have a complex structure within their cell membranes that allow them to spin their corkscrew-shaped bodies which propels them along. Some rod-shaped and spherical bacteria form colonies, adhering end-to-end after they have divided, forming chains (see figure 34.2). Some bacterial colonies change into stalked structures, grow long, branched filaments, or form erect structures that release **spores,** single-celled bodies that grow into new bacterial individuals. Some filamentous bacteria are capable of gliding motion, often combined with rotation around a longitudinal axis. Biologists have not yet determined the mechanism by which they move.

Prokaryotes versus Eukaryotes

Prokaryotes—eubacteria and archaea—differ from eukaryotes in numerous important features. These differences represent some of the most fundamental distinctions that separate any groups of organisms.

1. **Multicellularity.** All prokaryotes are fundamentally single-celled. In some types, individual cells adhere to

FIGURE 34.3
Approaches to multicellularity in bacteria. *Chondromyces crocatus*, one of the gliding bacteria. The rod-shaped individuals move together, forming the composite spore-bearing structures shown here. Millions of spores, which are basically individual bacteria, are released from these structures.

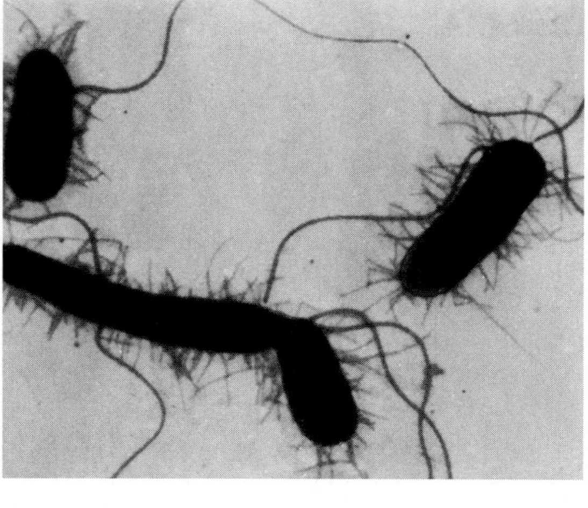

FIGURE 34.4
Flagella in the common intestinal bacterium, *Escherichia coli*. The long strands are flagella, while the shorter hairlike outgrowths are called pili.

each other within a matrix and form filaments, however the cells retain their individuality. Cyanobacteria, in particular, are likely to form such associations but their cytoplasm is not directly interconnected, as often is the case in multicellular eukaryotes. The activities of a bacterial colony are less integrated and coordinated than those in multicellular eukaryotes. A primitive form of colonial organization occurs in gliding bacteria, which move together and form spore-bearing structures (figure 34.3). Such coordinated multicellular forms are rare among bacteria.

2. **Cell size.** As new species of bacteria are discovered, we are finding that the size of prokaryotic cells varies tremendously, by as much as five orders of magnitude. Most prokaryotic cells are only 1 micrometer or less in diameter. Most eukaryotic cells are well over 10 times that size.
3. **Chromosomes.** Eukaryotic cells have a membrane-bound nucleus containing chromosomes made up of both nucleic acids and proteins. Bacteria do not have membrane-bound nuclei, nor do they have chromosomes of the kind present in eukaryotes, in which DNA forms a structural complex with proteins. Instead, their naked circular DNA is localized in a zone of the cytoplasm called the **nucleoid.**
4. **Cell division and genetic recombination.** Cell division in eukaryotes takes place by mitosis and involves spindles made up of microtubules. Cell division in bacteria takes place mainly by binary fission (see chapter 11). True sexual reproduction occurs only in eukaryotes and involves syngamy and meiosis, with an alternation of diploid and haploid forms. Despite their lack of sexual reproduction, bacteria do have mechanisms that lead to the transfer of genetic material. These mechanisms are far less regular than those of eukaryotes and do not involve the equal participation of the individuals between which the genetic material is transferred.
5. **Internal compartmentalization.** In eukaryotes, the enzymes for cellular respiration are packaged in mitochondria. In bacteria, the corresponding enzymes are not packaged separately but are bound to the cell membranes (see chapters 5 and 9). The cytoplasm of bacteria, unlike that of eukaryotes, contains no internal compartments or cytoskeleton and no organelles except ribosomes.
6. **Flagella.** Bacterial flagella are simple in structure, composed of a single fiber of the protein flagellin (figure 34.4; see also chapter 5). Eukaryotic flagella and cilia are complex and have a 9 + 2 structure of microtubules (see figure 5.27). Bacterial flagella also function differently, spinning like propellers, while eukaryotic flagella have a whiplike motion.
7. **Metabolic diversity.** Only one kind of photosynthesis occurs in eukaryotes, and it involves the release of oxygen. Photosynthetic bacteria have several different patterns of anaerobic and aerobic photosynthesis, involving the formation of end products such as sulfur, sulfate, and oxygen (see chapter 10). Prokaryotic cells can also be chemoautotrophic, using the energy stored in chemical bonds of inorganic molecules to synthesize carbohydrates; eukaryotes are not capable of this metabolic process.

Bacteria are the oldest and most abundant organisms on earth. Bacteria, or prokaryotes, differ from eukaryotes in a wide variety of characteristics, a degree of difference as great as any that separates any groups of organisms.

34.2 Bacterial cell structure is more complex than commonly supposed.

The Bacterial Surface

The bacterial **cell wall** is an important structure because it maintains the shape of the cell and protects the cell from swelling and rupturing. The cell wall usually consists of **peptidoglycan,** a network of polysaccharide molecules connected by polypeptide cross-links. In some bacteria, the peptidoglycan forms a thick, complex network around the outer surface of the cell. This network is interlaced with peptide chains. In other bacteria a thin layer of peptidoglycan is found sandwiched between two plasma membranes. The outer membrane contains large molecules of lipopolysaccharide, lipids with polysaccharide chains attached. These two major types of bacteria can be identified using a staining process called a **Gram stain**. **Gram-positive** bacteria have the thicker peptidoglycan wall and stain a purple color (figure 34.5). The more common **gram-negative** bacteria contain less peptidoglycan and do not retain the purple-colored dye. Gram-negative bacteria stain red. The outer membrane layer makes gram-negative bacteria resistant to many antibiotics that interfere with cell wall synthesis in gram-positive bacteria. In some kinds of bacteria, an additional gelatinous layer, the capsule, surrounds the cell wall.

Many kinds of bacteria have slender, rigid, helical **flagella** (singular, **flagellum**) composed of the protein flagellin (figure 34.6). These flagella range from 3 to 12 micrometers in length and are very thin—only 10 to 20 nanometers thick. They are anchored in the cell wall and spin, pulling the bacteria through the water like a propeller.

Pili (singular, **pilus**) are other hairlike structures that occur on the cells of some bacteria (see figure 34.4). They are shorter than bacterial flagella, up to several micrometers long, and about 7.5 to 10 nanometers thick. Pili help the bacterial cells attach to appropriate substrates and exchange genetic information.

Some bacteria form thick-walled **endospores** around their chromosome and a small portion of the surrounding cytoplasm when they are exposed to nutrient-poor conditions. These endospores are highly resistant to environmental stress, especially heat, and can germinate to form new individuals after decades or even centuries.

Bacteria are encased within a cell wall composed of one or more polysaccharide layers. They also may contain external structures such as flagella and pili.

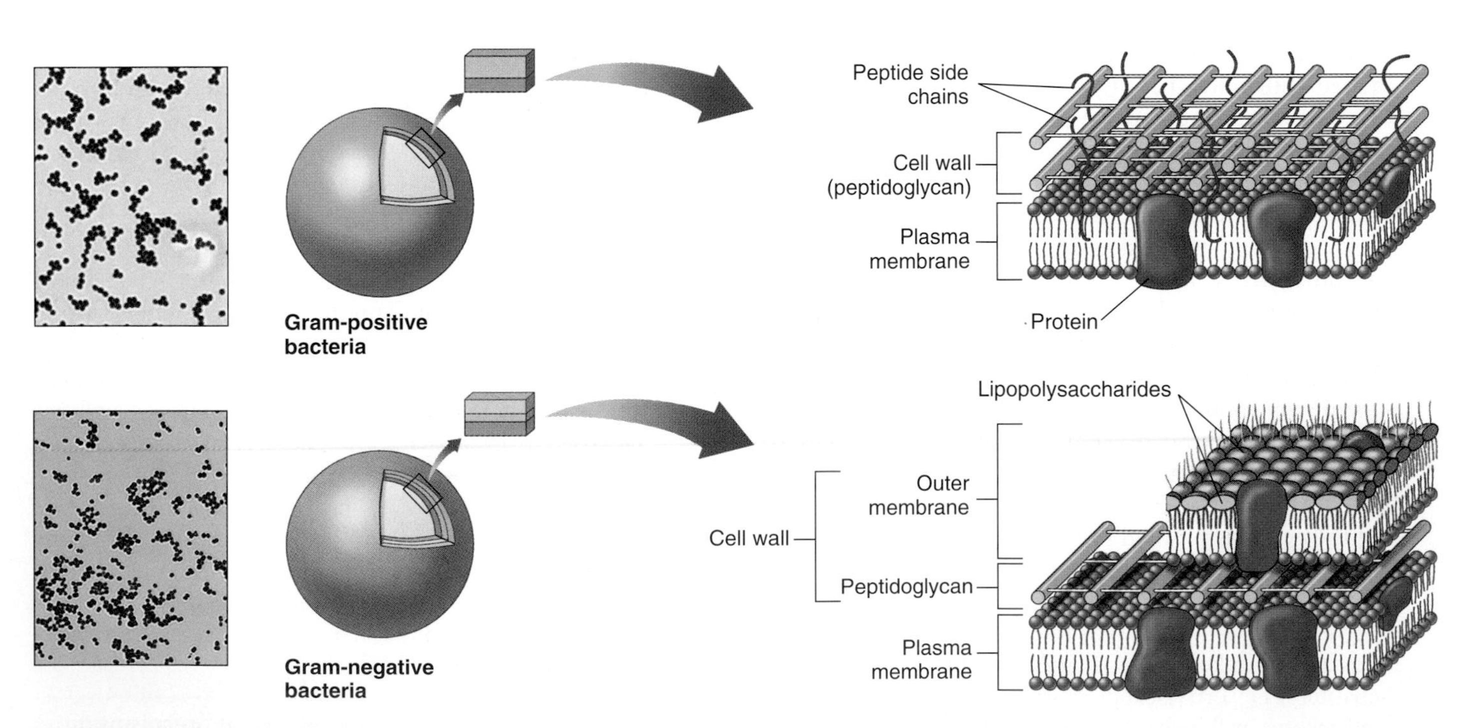

FIGURE 34.5
The Gram stain. The peptidoglycan layer encasing gram-positive bacteria traps crystal violet dye, so the bacteria appear purple in a Gram-stained smear (named after Hans Christian Gram, who developed the technique). Because gram-negative bacteria have much less peptidoglycan (located between the plasma membrane and an outer membrane), they do not retain the crystal violet dye and so exhibit the red background stain (usually a safranin dye).

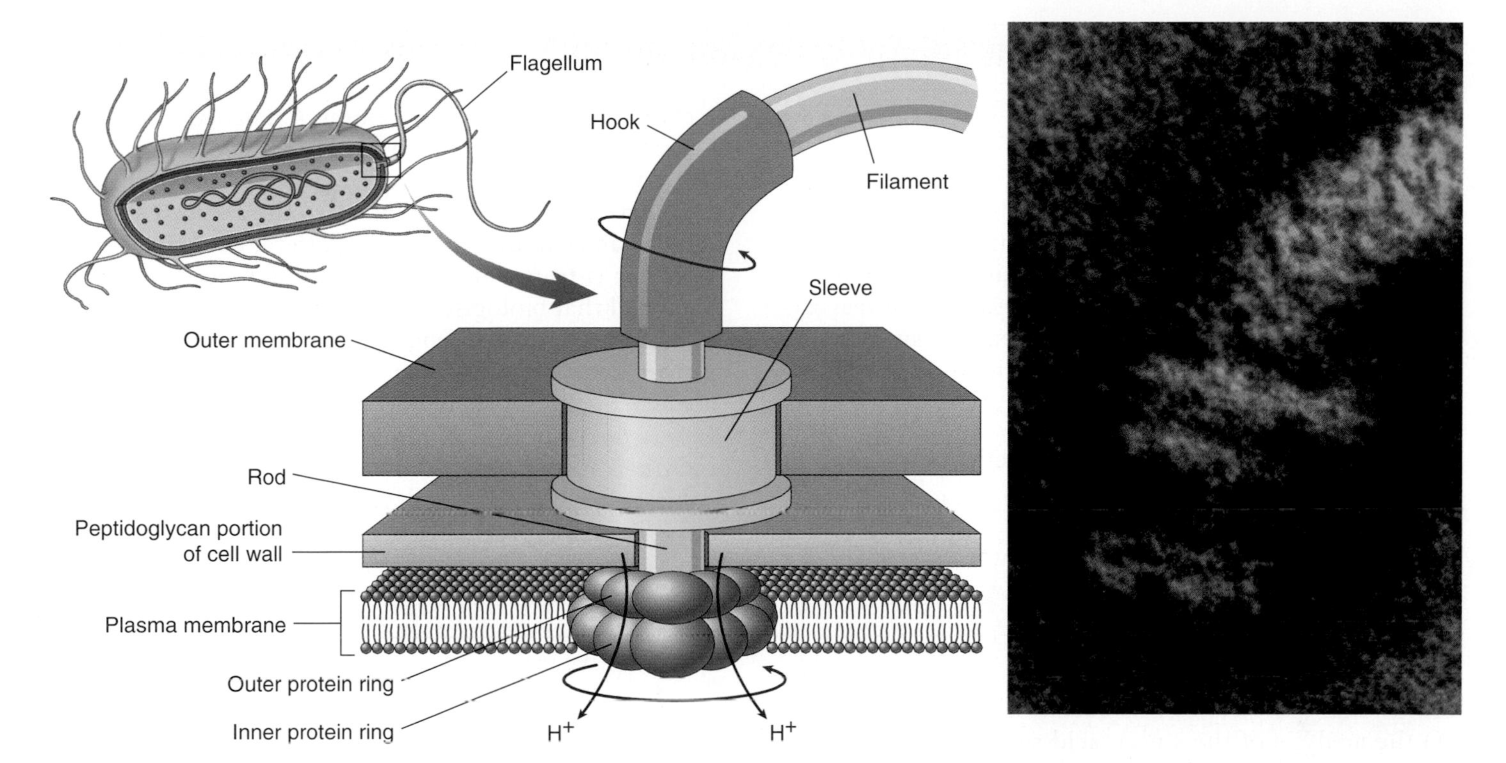

FIGURE 34.6
The flagellar motor of a gram-negative bacterium. A protein filament, composed of the protein flagellin, is attached to a protein shaft that passes through a sleeve in the outer membrane and through a hole in the peptidoglycan layer to rings of protein anchored in the cell wall and plasma membrane, like rings of ballbearings. The shaft rotates when the inner protein ring attached to the shaft turns with respect to the outer ring fixed to the cell wall. The inner ring is an H^+ ion channel, a proton pump that uses the passage of protons into the cell to power the movement of the inner ring past the outer one.

The Cell Interior

The most fundamental characteristic of bacterial cells is their prokaryotic organization. Bacterial cells lack the extensive functional compartmentalization seen within eukaryotic cells.

Internal membranes. Many bacteria possess invaginated regions of the plasma membrane that function in respiration or photosynthesis (figure 34.7).

Nucleoid region. Bacteria lack nuclei and do not possess the complex chromosomes characteristic of eukaryotes. Instead, their genes are encoded within a single double-stranded ring of DNA that is crammed into one region of the cell known as the **nucleoid region.** Many bacterial cells also possess small, independently replicating circles of DNA called plasmids. Plasmids contain only a few genes, usually not essential for the cell's survival. They are best thought of as an excised portion of the bacterial chromosome.

Ribosomes. Bacterial ribosomes are smaller than those of eukaryotes and differ in protein and RNA content. Antibiotics such as tetracycline and chloramphenicol can tell the difference—they bind to bacterial ribosomes and block protein synthesis, but do not bind to eukaryotic ribosomes.

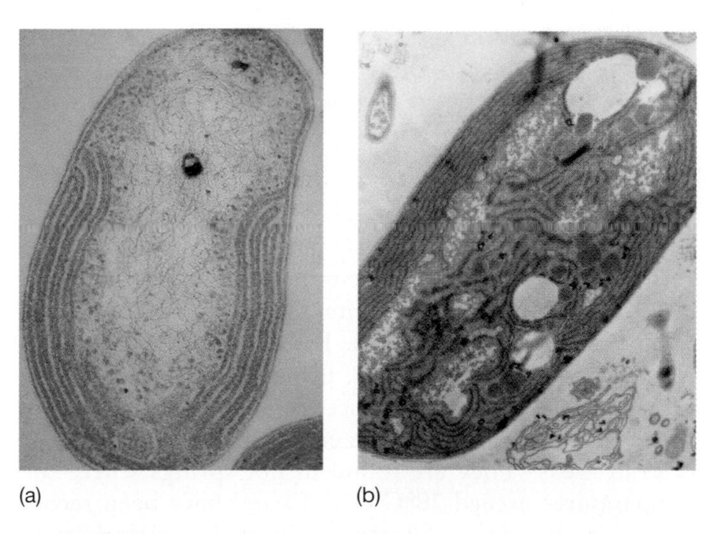

FIGURE 34.7
Bacterial cells often have complex internal membranes. This aerobic bacterium (*a*) exhibits extensive respiratory membranes within its cytoplasm not unlike those seen in mitochondria. This cyanobacterium (*b*) has thylakoid-like membranes that provide a site for photosynthesis.

The interior of a bacterial cell may possess internal membranes and a nucleoid region.

34.3 Bacteria exhibit considerable diversity in both structure and metabolism.

Bacterial Diversity

Bacteria are not easily classified according to their forms, and only recently has enough been learned about their biochemical and metabolic characteristics to develop a satisfactory overall classification comparable to that used for other organisms. Early systems for classifying bacteria relied on differential stains such as the Gram stain. Key bacterial characteristics used in classifying bacteria were:

1. Photosynthetic or nonphotosynthetic
2. Motile or nonmotile
3. Unicellular or multicellular
4. Formation of spores or dividing by transverse binary fission

With the development of genetic and molecular approaches, bacterial classifications can at last reflect true evolutionary relatedness. Molecular approaches include: (1) the analysis of the amino acid sequences of key proteins; (2) the analysis of nucleic acid base sequences by establishing the percent of guanine (G) and cytosine (C); (3) nucleic acid hybridization, which is essentially the mixing of single-stranded DNA from two species and determining the amount of base-pairing (closely related species will have more bases pairing); and (4) nucleic acid sequencing especially looking at ribosomal RNA. Lynn Margulis and Karlene Schwartz proposed a useful classification system that divides bacteria into 16 phyla, according to their most significant features. Table 34.1 outlines the major features of some of the phyla.

Kinds of Bacteria

Although they lack the structural complexity of eukaryotes, bacteria have diverse internal chemistries, metabolisms and unique functions. Bacteria have adapted to many kinds of environments, including some you might consider harsh. They have successfully invaded very salty waters, very acidic or alkaline environments, and very hot or cold areas. They are found in hot springs where the temperatures exceed 78°C (172°F) and have been recovered living beneath 435 meters of ice in Antarctica!

Much of what we know of bacteria we have learned from studies in the laboratory. It is important to understand the limits this has placed on our knowledge: we have only been able to study those bacteria that can be cultured in laboratories. Field studies suggest that these represent but a small fraction of the kinds of bacteria that occur in soil, most of which cannot be cultured with existing techniques. We clearly have only scraped the surface of bacterial diversity.

As we learned in chapter 32, bacteria split into two lines early in the history of life, so different in structure and metabolism that they are as different from each other as either is from eukaryotes. The differences are so fundamental that biologists assign the two groups of bacteria to separate domains. One domain, the Archaea, consists of the archaebacteria ("ancient bacteria"—although they are actually not as ancient as the other bacterial domain). It was once thought that survivors of this group were confined to extreme environments that may resemble habitats on the early earth. However, the use of genetic screening has revealed that these "ancient" bacteria live in nonextreme environments as well. The other more ancient domain, the Bacteria, consists of the eubacteria ("true bacteria"). It includes nearly all of the named species of bacteria.

Comparing Archaebacteria and Eubacteria

Archaebacteria and eubacteria are similar in that they both have a prokaryotic cellular structure but they vary considerably at the biochemical and molecular level. There are four key areas in which they differ:

1. **Cell wall.** Both kinds of bacteria typically have cell walls covering the plasma membrane that strengthen the cell. The cell walls of eubacteria are constructed of carbohydrate-protein complexes called peptidoglycan, which link together to create a strong mesh that gives the eubacterial cell wall great strength. The cell walls of archaebacteria lack peptidoglycan.
2. **Plasma membranes.** All bacteria have plasma membranes with a lipid-bilayer architecture (as described in chapter 6). The plasma membranes of eubacteria and archaebacteria, however, are made of very different kinds of lipids.
3. **Gene translation machinery.** Eubacteria possess ribosomal proteins and an RNA polymerase that are distinctly different from those of eukaryotes. However, the ribosomal proteins and RNA of archaebacteria are very similar to those of eukaryotes.
4. **Gene architecture.** The genes of eubacteria are not interrupted by introns, while at least some of the genes of archaebacteria do possess introns.

While superficially similar, bacteria differ from one another in a wide variety of characteristics.

Major Group	Typical Examples	Key Characteristics
		ARCHAEBACTERIA
Archaebacteria	Methanogens, thermophiles, halophiles	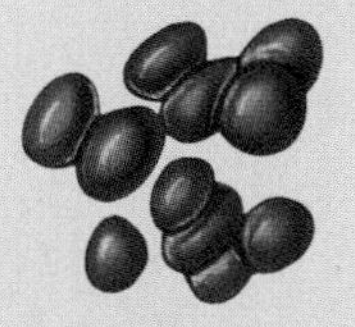 Bacteria that are not members of the kingdom Eubacteria. Mostly anaerobic with unusual cell walls. Some produce methane. Others reduce sulfur.
		EUBACTERIA
Actinomycetes	*Streptomyces*, *Actinomyces*	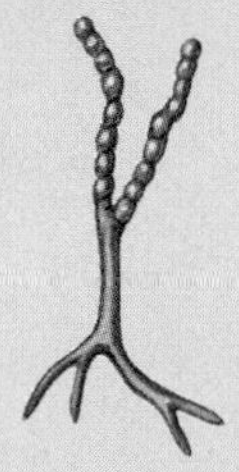 Gram-positive bacteria. Form branching filaments and produce spores; often mistaken for fungi. Produce many commonly used antibiotics, including streptomycin and tetracycline. One of the most common types of soil bacteria; also common in dental plaque.
Chemoautotrophs	Sulfur bacteria, *Nitrobacter*, *Nitrosomonas*	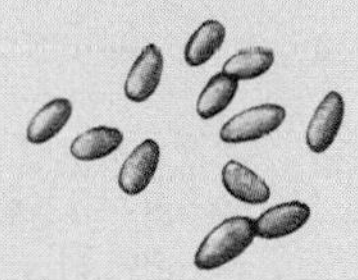 Bacteria able to obtain their energy from inorganic chemicals. Most extract chemical energy from reduced gases such as H_2S (hydrogen sulfide), NH_3 (ammonia), and CH_4 (methane). Play a key role in the nitrogen cycle.
Cyanobacteria	*Anabaena*, *Nostoc*	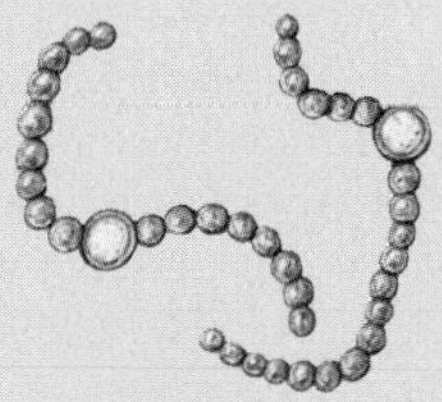 A form of photosynthetic bacteria common in both marine and freshwater environments. Deeply pigmented; often responsible for "blooms" in polluted waters.
Enterobacteria	*Escherichia coli*, *Salmonella*, *Vibrio*	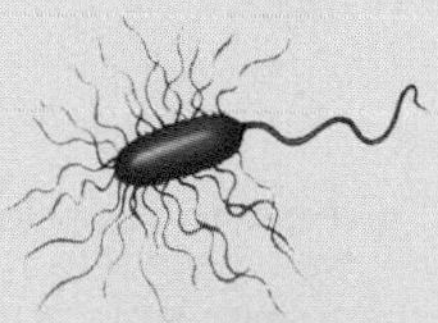 Gram-negative, rod-shaped bacteria. Do not form spores; usually aerobic heterotrophs; cause many important diseases, including bubonic plague and cholera.
Gliding and budding bacteria	Myxobacteria, *Chondromyces*	 Gram-negative bacteria. Exhibit gliding motility by secreting slimy polysaccharides over which masses of cells glide; some groups form upright multicellular structures carrying spores called fruiting bodies.
Pseudomonads	*Pseudomonas*	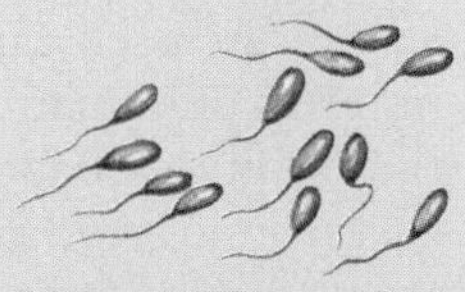 Gram-negative heterotrophic rods with polar flagella. Very common form of soil bacteria; also contain many important plant pathogens.
Rickettsias and chlamydias	*Rickettsia*, *Chlamydia*	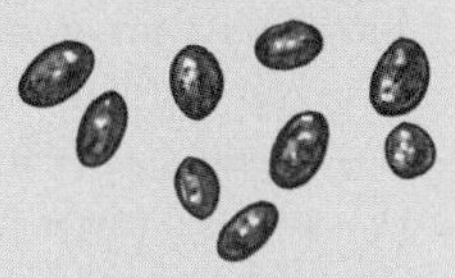 Small, gram-negative intracellular parasites. *Rickettsia* life cycle involves both mammals and arthropods such as fleas and ticks; *Rickettsia* are responsible for many fatal human diseases, including typhus (*Rickettsia prowazekii*) and Rocky Mountain spotted fever. Chlamydial infections are one of the most common sexually transmitted diseases.
Spirochaetes	*Treponema*	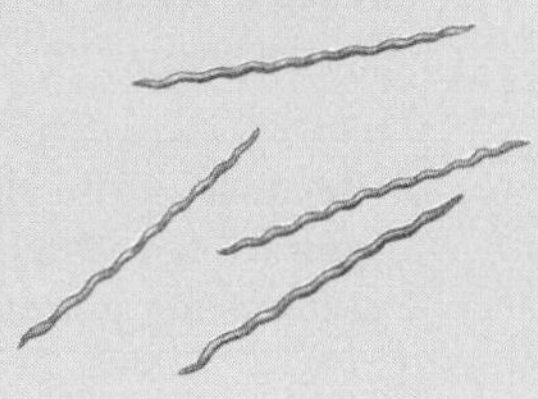 Long, coil-shaped cells. Common in aquatic environments; a parasitic form is responsible for the disease syphilis.

Table 34.1 Bacteria

Bacterial Variation

Bacteria reproduce rapidly, allowing genetic variations to spread quickly through a population. Two processes create variation among bacteria: mutation and genetic recombination.

Mutation

Mutations can arise spontaneously in bacteria as errors in DNA replication occur. Certain factors tend to increase the likelihood of errors occurring such as radiation, ultraviolet light, and various chemicals. In a typical bacterium such as *Escherichia coli* there are about 5000 genes. It is highly probably that one mutation will occur by chance in one out of every million copies of a gene. With 5000 genes in a bacterium, the laws of probability predict that 1 out of every 200 bacteria will have a mutation (figure 34.8). A spoonful of soil typically contains over a billion bacteria and therefore should contain something on the order of 5 million mutant individuals!

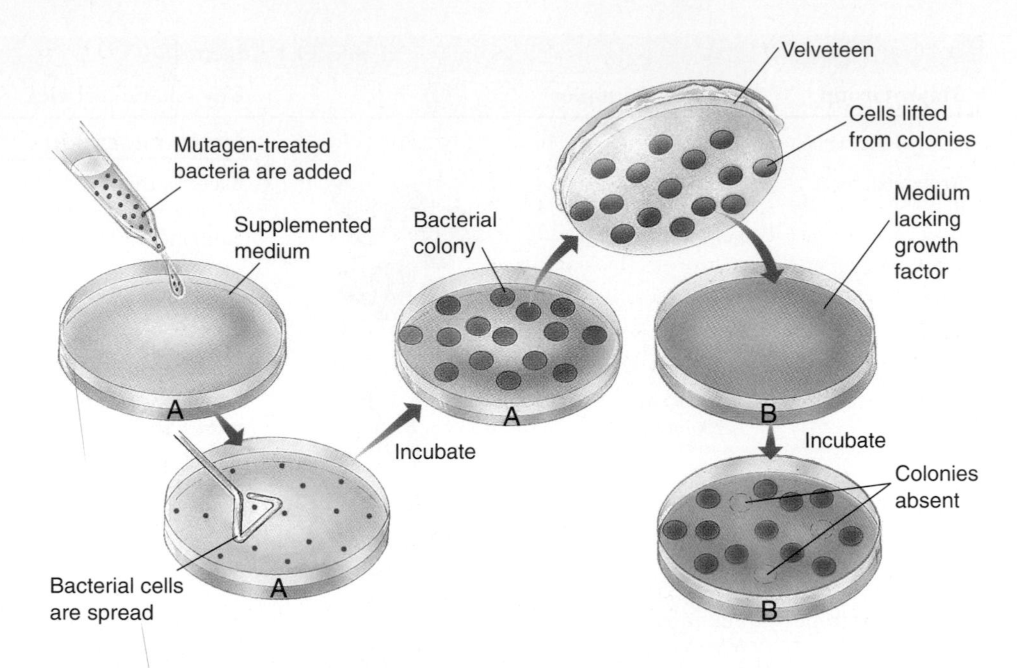

FIGURE 34.8
A mutant hunt in bacteria. Mutations in bacteria can be detected by a technique called replica plating, which allows the genetic characteristics of the colonies to be investigated without destroying them. The bacterial colonies, growing on a semisolid agar medium, are transferred from A to B using a sterile velveteen disc pressed on the plate. Plate A has a medium that includes special growth factors, while B has a medium that lacks some of these growth factors. Bacteria that are not mutated can produce their own growth factors and do not require them to be added to the medium. The colonies absent in B were unable to grow on the deficient medium and were thus mutant colonies; they were already present but undetected in A.

With adequate food and nutrients, a population of *E. coli* can double in under 20 minutes. Because bacteria multiply so rapidly, mutations can spread rapidly in a population and can change the characteristics of that population.

The ability of bacteria to change rapidly in response to new challenges often has adverse effects on humans. Recently a number of strains of *Staphylococcus aureus* associated with serious infections in hospitalized patients have appeared, some of them with alarming frequency. Unfortunately, these strains have acquired resistance to penicillin and a wide variety of other antibiotics, so that infections caused by them are very difficult to treat. *Staphylococcus* infections provide an excellent example of the way in which mutation and intensive selection can bring about rapid change in bacterial populations. Such changes have serious medical implications when, as in the case of *Staphylococcus*, strains of bacteria emerge that are resistant to a variety of antibiotics.

Recently, concern has arisen over the prevalence of antibacterial soaps in the marketplace. They are marketed as a means of protecting your family from harmful bacteria; however, it is likely that their routine use will favor bacteria that have mutations making them immune to the antibiotics contained in them. Ultimately, extensive use of antibacterial soaps could have an adverse effect on our ability to treat common bacterial infections.

Genetic Recombination

Another source of genetic variation in populations of bacteria is recombination, discussed in detail in chapter 18. Bacterial recombination occurs by the transfer of genes from one cell to another by viruses, or through conjugation. The rapid transfer of newly produced, antibiotic-resistant genes by plasmids has been an important factor in the appearance of the resistant strains of *Staphylococcus aureus* discussed earlier. An even more important example in terms of human health involves the Enterobacteriaceae, the family of bacteria to which the common intestinal bacterium, *Escherichia coli*, belongs. In this family, there are many important pathogenic bacteria, including the organisms that cause dysentery, typhoid, and other major diseases. At times, some of the genetic material from these pathogenic species is exchanged with or transferred to *E. coli* by plasmids. Because of its abundance in the human digestive tract, *E. coli* poses a special threat if it acquires harmful traits.

Because of the short generation time of bacteria, mutation and recombination play an important role in generating genetic diversity.

Bacterial Metabolism

Bacteria have evolved many mechanisms to acquire the energy and nutrients they need for growth and reproduction. Many are **autotrophs,** organisms that obtain their carbon from inorganic CO_2. Autotrophs that obtain their energy from sunlight are called *photoautotrophs,* while those that harvest energy from inorganic chemicals are called *chemoautotrophs.* Other bacteria are **heterotrophs,** organisms that obtain at least some of their carbon from organic molecules like glucose. Heterotrophs that obtain their energy from sunlight are called *photoheterotrophs,* while those that harvest energy from organic molecules are called *chemoheterotrophs.*

Photoautotrophs. Many bacteria carry out photosynthesis, using the energy of sunlight to build organic molecules from carbon dioxide. The cyanobacteria use chlorophyll *a* as the key light-capturing pigment and use H_2O as an electron donor, releasing oxygen gas as a by-product. Other bacteria use bacteriochlorophyll as their pigment and H_2S as an electron donor, leaving elemental sulfur as the by-product.

Chemoautotrophs. Some bacteria obtain their energy by oxidizing inorganic substances. Nitrifiers, for example, oxidize ammonia or nitrite to obtain energy, producing the nitrate that is taken up by plants. This process is called nitrogen fixation and is essential in terrestrial ecosystems as plants can only absorb nitrogen in the form of nitrate. Other bacteria oxidize sulfur, hydrogen gas, and other inorganic molecules. On the dark ocean floor at depths of 2500 meters, entire ecosystems subsist on bacteria that oxidize hydrogen sulfide as it escapes from thermal vents.

Photoheterotrophs. The so-called purple nonsulfur bacteria use light as their source of energy but obtain carbon from organic molecules such as carbohydrates or alcohols that have been produced by other organisms.

Chemoheterotrophs. Most bacteria obtain both carbon atoms and energy from organic molecules. These include decomposers and most pathogens.

How Heterotrophs Infect Host Organisms

In the 1980s, researchers studying the disease-causing species of *Yersinia,* a group of gram-negative bacteria, found that they produced and secreted large amounts of proteins. Most proteins secreted by gram-negative bacteria have special signal sequences that allow them to pass through the bacterium's double membrane. This key signal sequence was missing from the proteins being secreted by *Yersinia.* These proteins lacked a signal-sequence that two known secretion mechanisms require for transport across the double membrane of gram-negative bacteria. The proteins must therefore have been secreted by a third type of system, which researchers called the *type III system.*

As more bacteria species are studied, the genes coding for the type III system are turning up in other gram-negative animal pathogens, and even in more distantly related plant pathogens. The genes seem to be more closely related to one another than do the bacteria. Furthermore, the genes are similar to those that code for bacterial flagella.

The role of these proteins is still under investigation, but it seems that some of the proteins are used to transfer other virulence proteins into nearby eukaryotic cells. Given the similarity of the type III genes to the genes that code for flagella, some scientists hypothesize that the transfer proteins may form a flagellum-like structure that shoots virulence proteins into the host cells. Once in the eukaryotic cells, the virulence proteins may determine the host's response to the pathogens. In *Yersinia,* proteins secreted by the type III system are injected into macrophages; they disrupt signals that tell the macrophages to engulf bacteria. *Salmonella* and *Shigella* use their type III proteins to enter the cytoplasm of eukaryotic cells and thus are protected from the immune system of their host. The proteins secreted by *E. coli* alter the cytoskeleton of nearby intestinal eukaryotic cells, resulting in a bulge onto which the bacterial cells can tightly bind.

Currently, researchers are looking for a way to disarm the bacteria using knowledge of their internal machinery, possibly by causing the bacteria to release the virulence proteins before they are near eukaryotic cells. Others are studying the eukaryotic target proteins and the process by which they are affected.

Bacteria as Plant Pathogens

Many costly diseases of plants are associated with particular heterotrophic bacteria. Almost every kind of plant is susceptible to one or more kinds of bacterial disease. The symptoms of these plant diseases vary, but they are commonly manifested as spots of various sizes on the stems, leaves, flowers, or fruits. Other common and destructive diseases of plants, including blights, soft rots, and wilts, also are associated with bacteria. Fire blight, which destroys pears, apple trees, and related plants, is a well-known example of bacterial disease. Most bacteria that cause plant diseases are members of the group of rod-shaped bacteria known as pseudomonads (see figure 34.2*a*).

While bacteria obtain carbon and energy in many ways, most are chemoheterotrophs. Some heterotrophs have evolved sophisticated ways to infect their hosts.

34.4 Bacteria are responsible for many diseases but also make important contributions to ecosystems.

Human Bacterial Diseases

Bacteria cause many diseases in humans, including cholera, leprosy, tetanus, bacterial pneumonia, whooping cough, diphtheria and lyme disease (table 34.2). Members of the genus *Streptococcus* (see figure 34.2*b*) are associated with scarlet fever, rheumatic fever, pneumonia, and other infections. Tuberculosis (TB), another bacterial disease, is still a leading cause of death in humans. Some of these diseases like TB are mostly spread through the air in water vapor. Other bacterial diseases are dispersed in food or water, including typhoid fever, paratyphoid fever, and bacillary dysentery. Typhus is spread among rodents and humans by insect vectors.

Tuberculosis

Tuberculosis has been one of the great killer diseases for thousands of years. Currently, about one-third of all people worldwide are infected with *Mycobacterium tuberculosis*, the tuberculosis bacterium (figure 34.9). Eight million new cases crop up each year, with about 3 million people dying from the disease annually (the World Health Organization predicts 4 million deaths a year by 2005). In fact, in 1997, TB was the leading cause of death from a *single infectious agent* worldwide. Since the mid-1980s, the United States has been experiencing a dramatic resurgence of tuberculosis. TB afflicts the respiratory system and is easily transmitted from person to person through the air. The causes of this current resurgence of TB include social factors such as poverty, crowding, homelessness, and incarceration (these factors have always promoted the spread of TB). The increasing prevalence of HIV infections is also a significant contributing factor. People with AIDS are much more likely to develop TB than people with healthy immune systems.

In addition to the increased numbers of cases—more than 25,000 nationally as of March 1995—there have been alarming outbreaks of multidrug-resistant strains of tuberculosis—strains resistant to the best available anti-TB medications. Multidrug-resistant TB is particularly concerning because it requires much more time to treat, is more expensive to treat, and may prove to be fatal.

The basic principles of TB treatment and control are to make sure all patients complete a full course of medication so that all of the bacteria causing the infection are killed and drug-resistant strains do not develop. Great efforts are being made to ensure that high-risk individuals who are infected but not yet sick receive preventative therapy, which is 90% effective in reducing the likelihood of developing active TB.

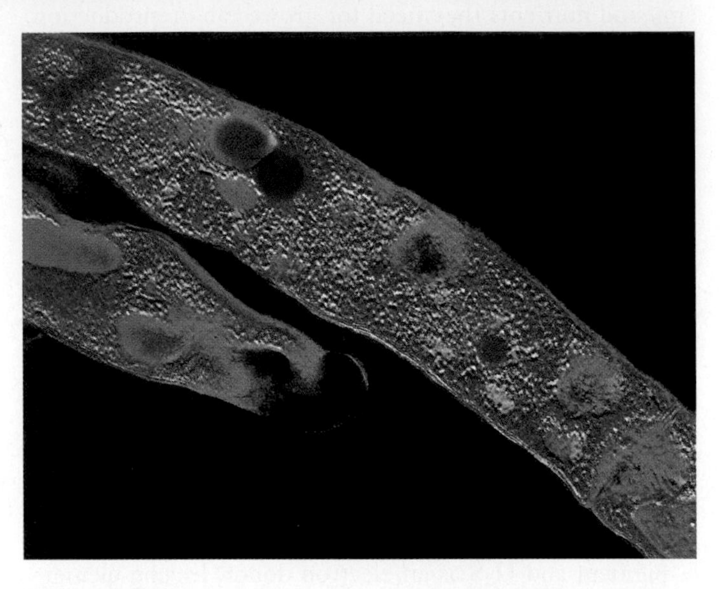

FIGURE 34.9
Mycobacterium tuberculosis. This color-enhanced image shows the rod-shaped bacterium responsible for tuberculosis in humans.

Dental Caries

One human disease we do not usually consider bacterial in origin arises in the film on our teeth. This film, or plaque, consists largely of bacterial cells surrounded by a polysaccharide matrix. Most of the bacteria in plaque are filaments of rod-shaped cells classified as various species of *Actinomyces*, which extend out perpendicular to the surface of the tooth. Many other bacterial species are also present in plaque. Tooth decay, or **dental caries,** is caused by the bacteria present in the plaque, which persists especially in places that are difficult to reach with a toothbrush. Diets that are high in sugars are especially harmful to teeth because lactic acid bacteria (especially *Streptococcus sanguis* and *S. mutans)* ferment the sugars to lactic acid, a substance that reduces the pH of the mouth, causing the local loss of calcium from the teeth. Frequent eating of sugary snacks or sucking on candy over a period of time keeps the pH level of the mouth low resulting in the steady degeneration of the tooth enamel. As the calcium is removed from the tooth, the remaining soft matrix of the tooth becomes vulnerable to attack by bacteria, which begin to break down its proteins, and tooth decay progresses rapidly. Fluoride makes the teeth more resistant to decay because it retards the loss of calcium. It was first realized that bacteria cause tooth decay when germ-free animals were raised. Their teeth do not decay even if they are fed sugary diets.

Table 34.2 Important Human Bacterial Diseases			
Disease	**Pathogen**	**Vector/Reservoir**	**Epidemiology**
Anthrax	*Bacillus anthracis*	Animals, including processed skins	Bacterial infection that can be transmitted through contact or ingested. Rare except in sporadic outbreaks. May be fatal.
Botulism	*Clostridium botulinum*	Improperly prepared food	Contracted through ingestion or contact with wound. Produces acute toxic poison; can be fatal.
Chlamydia	*Chlamydia trachomatis*	Humans, STD	Urogenital infections with possible spread to eyes and respiratory tract. Occurs worldwide; increasingly common over past 20 years.
Cholera	*Vibrio cholerae*	Human feces, plankton	Causes severe diarrhea that can lead to death by dehydration; 50% peak mortality if the disease goes untreated. A major killer in times of crowding and poor sanitation; over 100,000 died in Rwanda in 1994 during a cholera outbreak.
Dental caries	*Streptococcus*	Humans	A dense collection of this bacteria on the surface of teeth leads to secretion of acids that destroy minerals in tooth enamel—sugar alone will not cause caries.
Diphtheria	*Corynebacterium diphtheriae*	Humans	Acute inflammation and lesions of mucous membranes. Spread through contact with infected individual. Vaccine available.
Gonorrhea	*Neisseria gonorrhoeae*	Humans only	STD, on the increase worldwide. Usually not fatal.
Hansen's disease (leprosy)	*Mycobacterium leprae*	Humans, feral armadillos	Chronic infection of the skin; worldwide incidence about 10–12 million, especially in Southeast Asia. Spread through contact with infected individuals.
Lyme disease	*Borrelia bergdorferi*	Ticks, deer, small rodents	Spread through bite of infected tick. Lesion followed by malaise, fever, fatigue, pain, stiff neck, and headache.
Peptic ulcers	*Helicobacter pylori*	Humans	Originally thought to be caused by stress or diet, most peptic ulcers now appear to be caused by this bacterium; good news for ulcer sufferers as it can be treated with antibiotics.
Plague	*Yersinia pestis*	Fleas of wild rodents: rats and squirrels	Killed ¼ of the population of Europe in the 14th century; endemic in wild rodent populations of the western U.S. today.
Pneumonia	*Streptococcus, Mycoplasma, Chlamydia*	Humans	Acute infection of the lungs, often fatal without treatment
Tuberculosis	*Mycobacterium tuberculosis*	Humans	An acute bacterial infection of the lungs, lymph, and meninges. Its incidence is on the rise, complicated by the development of new strains of the bacteria that are resistant to antibiotics.
Typhoid fever	*Salmonella typhi*	Humans	A systemic bacterial disease of worldwide incidence. Less than 500 cases a year are reported in the U.S. The disease is spread through contaminated water or foods (such as improperly washed fruits and vegetables). Vaccines are available for travelers.
Typhus	*Rickettsia typhi*	Lice, rat fleas, humans	Historically a major killer in times of crowding and poor sanitation; transmitted from human to human through the bite of infected lice and fleas. Typhus has a peak untreated mortality rate of 70%.

Sexually Transmitted Diseases

A number of bacteria cause sexually transmitted diseases (STDs). Three are particularly important (figure 34.10).

Gonorrhea. **Gonorrhea** is one of the most prevalent communicable diseases in North America. Caused by the bacterium *Neisseria gonorrhoeae,* gonorrhea can be transmitted through sexual intercourse or any other sexual contacts in which body fluids are exchanged, such as oral or anal intercourse. Gonorrhea can infect the throat, urethra, cervix, or rectum and can spread to the eyes and internal organs, causing conjunctivitis (a severe infection of the eyes) and arthritic meningitis (an infection of the joints). Left untreated in women, gonorrhea can cause pelvic inflammatory disease (PID), a condition in which the fallopian tubes become scarred and blocked. PID can eventually lead to sterility. The incidence of gonorrhea has been on the decline, but it remains a serious threat.

Syphilis. **Syphilis,** a very destructive STD, was once prevalent but is now less common due to the advent of blood-screening procedures and antibiotics. Syphilis is caused by a spirochete bacterium, *Treponema pallidum,* that is transmitted during sexual intercourse or through direct contact with an open syphilis sore. The bacterium can also be transmitted from a mother to her fetus, often causing damage to the heart, eyes, and nervous system of the baby.

Once inside the body, the disease progresses in four distinct stages. The first, or primary stage, is characterized by the appearance of a small, painless, often unnoticed sore called a chancre. The chancre resembles a blister and occurs at the location where the bacterium entered the body about three weeks following exposure. This stage of the disease is highly infectious, and an infected person may unwittingly transmit the disease to others.

The second stage of syphilis is marked by a rash, a sore throat, and sores in the mouth. The bacteria can be transmitted at this stage through kissing or contact with an open sore.

The third stage of syphilis is symptomless. This stage may last for several years, and at this point, the person is no longer infectious but the bacteria are still present in the body, attacking the internal organs. The final stage of syphilis is the most debilitating, however, as the damage done by the bacteria in the third stage becomes evident. Sufferers at this stage of syphilis experience heart disease, mental deficiency, and nerve damage, which may include a loss of motor functions or blindness.

Chlamydia. Sometimes called the "silent STD," **chlamydia** is caused by an unusual bacterium, *Chlamydia trachomatis,* that has both bacterial and viral characteristics. Like a bacterium, it is susceptible to antibiotics, and, like a virus, it depends on its host to replicate its genetic material; it is an obligate internal parasite. The bacterium is transmitted through vaginal, anal, or oral intercourse with an infected person.

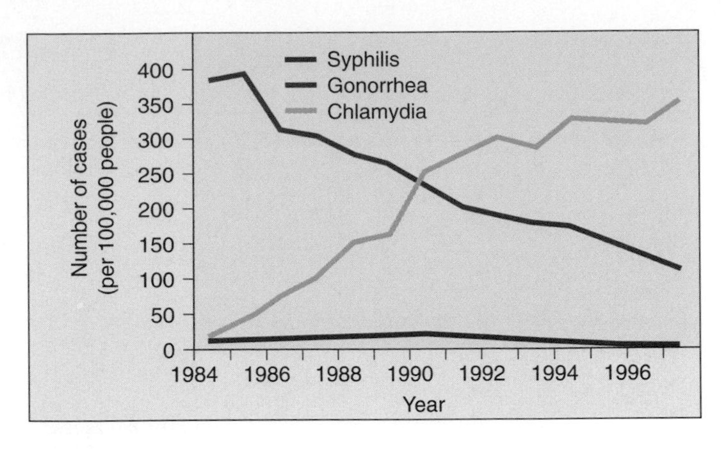

FIGURE 34.10
Trends in sexually transmitted diseases in the U.S.
Source: Data from Centers for Disease Control in Atlanta.

Chlamydia is called the "silent STD" because women usually experience no symptoms until after the infection has become established. In part because of this symptomless nature, the incidence of chlamydia has skyrocketed, increasing by more than sevenfold nationally since 1984. The effects of an established chlamydia infection on the female body are extremely serious. Chlamydia can cause pelvic inflammatory disease (PID), which can lead to sterility.

It has recently been established that infection of the reproductive tract by *Chlamydia* can cause heart disease. *Chlamydia* produce a peptide similar to one produced by cardiac muscle. As the body's immune system tries to fight off the infection, it recognizes this peptide. The similarity between the bacterial and cardiac peptides confuses the immune system and T cells attack cardiac muscle fibers, inadvertently causing inflammation of the heart and other problems.

Within the last few years, two types of tests for chlamydia have been developed that look for the presence of the bacteria in the discharge from men and women. The treatment for chlamydia is antibiotics, usually tetracycline (penicillin is not effective against chlamydia). Any woman who experiences the symptoms associated with this STD should be tested for the presence of the chlamydia bacterium; otherwise, her fertility may be at risk.

This discussion of STDs may give the impression that sexual activity is fraught with danger, and in a way, it is. It is folly not to take precautions to avoid STDs. The best way to do this is to know one's sexual partners well enough to discuss the possible presence of an STD. Condom use can also prevent transmission of most of the diseases. Responsibility for protection lies with each individual.

Bacterial diseases have a major impact worldwide. Sexually transmitted diseases (STDs) are becoming increasingly widespread among Americans as sexual activity increases.

Importance of Bacteria

Bacteria were largely responsible for creating the properties of the atmosphere and the soil over billions of years. They are metabolically much more diverse than eukaryotes, which is why they are able to exist in such a wide range of habitats. The many autotrophic bacteria—either photosynthetic or chemoautotrophic—make major contributions to the carbon balance in terrestrial, freshwater, and marine habitats. Other heterotrophic bacteria play a key role in world ecology by breaking down organic compounds. One of the most important roles of bacteria in the global ecosystem relates to the fact that only a few genera of bacteria—and no other organisms—have the ability to fix atmospheric nitrogen and thus make it available for use by other organisms (see chapter 28).

Bacteria are very important in many industrial processes. Bacteria are used in the production of acetic acid and vinegar, various amino acids and enzymes, and especially in the fermentation of lactose into lactic acid, which coagulates milk proteins and is used in the production of almost all cheeses, yogurt, and similar products. In the production of bread and other foods, the addition of certain strains of bacteria can lead to the enrichment of the final product with respect to its mix of amino acids, a key factor in its nutritive value. Many products traditionally manufactured using yeasts, such as ethanol, can also be made using bacteria. The comparative economics of these processes will determine which group of organisms is used in the future. Many of the most widely used antibiotics, including streptomycin, aureomycin, erythromycin, and chloromycetin, are derived from bacteria. Most antibiotics seem to be substances used by bacteria to compete with one another and fungi in nature, allowing one species to exclude others from a favored habitat. Bacteria can also play a part in removing environmental pollutants (figure 34.11)

Bacteria and Genetic Engineering

Applying genetic engineering methods to produce improved strains of bacteria for commercial use, as discussed in chapter 19, holds enormous promise for the future. Bacteria are under intense investigation, for example, as nonpolluting insect control agents. *Bacillus thuringiensis* attacks insects in nature, and improved, highly specific strains of *B. thuringiensis* have greatly increased its usefulness as a biological control agent. Bacteria have also been extraordinarily useful in our attempts to understand genetics and molecular biology.

Bacteria play a major role in modern medicine and agriculture, and have profound ecological impact.

FIGURE 34.11
Using bacteria to clean up oil spills. Bacteria can often be used to remove environmental pollutants, such as petroleum hydrocarbons and chlorinated compounds. In areas contaminated by the *Exxon Valdez* oil spill (rocks on the *left*), oil-degrading bacteria produced dramatic results (rocks on the *right*).

Chapter 34

www.mhhe.com/raven6/resources34.mhtml

Summary | **Questions** | **Media Resources**

34.1 Bacteria are the smallest and most numerous organisms.

- Bacteria are the oldest and simplest organisms, but they are metabolically much more diverse than all other life-forms combined.
- Bacteria differ from eukaryotes in many ways, the most important of which concern the degree of internal organization within the cell.

1. Structural differences among bacteria are not great. How are different species of bacteria recognized?

2. In what seven ways do prokaryotes differ substantially from eukaryotes?

- Enhancement Chapter: Extremophilic Bacteria

34.2 Bacterial cell structure is more complex than commonly supposed.

- Most bacteria have cell walls that consist of a network of polysaccharide molecules connected by polypeptide cross-links.
- A bacterial cell does not possess specialized compartments or a membrane-bounded nucleus, but it may exhibit a nucleoid region where the bacterial DNA is located.

3. What is the structure of the bacterial cell wall? How does the cell wall differ between gram-positive and gram-negative bacteria?

- Characteristics of Bacteria

34.3 Bacteria exhibit considerable diversity in both structure and metabolism.

- The two bacterial kingdoms, Archaebacteria and Eubacteria, are made up of prokaryotes, with about 5000 species named so far.
- The Archaebacteria differ markedly from Eubacteria and from eukaryotes in their ribosomal sequences and in other respects.
- Mutation and genetic recombination are important sources of variability in bacteria.
- Many bacteria are autotrophic and make major contributions to the world carbon balance. Others are heterotrophic and play a key role in world ecology by breaking down organic compounds.
- Some heterotrophic bacteria cause major diseases in plants and animals.

4. How do the Archaebacteria differ from the Eubacteria? What unique metabolism do they exhibit?

5. Why does mutation play such an important role in creating genetic diversity in bacteria?

6. How do heterotrophic bacteria that are successful pathogens overcome the many defenses the human body uses to ward off disease?

- Bacteria Diversity

- Scientists on Science: Marine Biotechnology

34.4 Bacteria are responsible for many diseases but also make important contributions to ecosystems.

- Human diseases caused by heterotrophic bacteria include many fatal diseases that have had major impacts on human history, including tuberculosis, cholera, plague, and typhus.
- Bacteria play vital roles in cycling nutrients within ecosystems. Certain bacteria are the only organisms able to fix atmospheric nitrogen into organic molecules, a process on which all life depends.

7. What are STDs? How are they transmitted? Which STDs are caused by viruses and which are caused by bacteria? Why is the cause of chlamydia unusual?

- Student Research: Improving Antibiotics
- *On Science* Articles:
  -Antibiotic Resistance
  -Making *E. Coli* Deadly

35

Protists

Concept Outline

35.1 Eukaryotes probably arose by endosymbiosis.

Endosymbiosis. Mitochondria and chloroplasts are thought to have arisen by endosymbiosis from aerobic bacteria.

35.2 The kingdom Protista is by far the most diverse of any kingdom.

The Challenge of Classifying the Protists. There is no general agreement among taxonomists about how to classify the protists.

General Biology of the Protists. Protista contains members exhibiting a wide range of methods of locomotion, nutrition, and reproduction.

35.3 Protists can be categorized into five groups.

Five Groups of Protists. The 15 major phyla of protists can be conveniently discussed in seven general groups that share certain characteristics.

Heterotrophs with No Permanent Locomotor Apparatus. Amoebas and other sarcodines have no permanent locomotor apparatus.

Photosynthetic Protists. The flagellates are photosynthesizers that propel themselves through the water with flagella. Diatoms are photosynthesizers with hard shells of silica. Algae are photosynthetic protists, some are multicellular.

Heterotrophs with Flagella. Flagellates propel themselves through the water. Single cells with many cilia, the ciliates possess highly complex and specialized organelles.

Nonmotile Spore-Formers. The sporozoans are nonmotile parasites that spread by forming spores.

Heterotrophs with Restricted Mobility. Heterotrophs with restricted mobility, molds have cell walls made of carbohydrate.

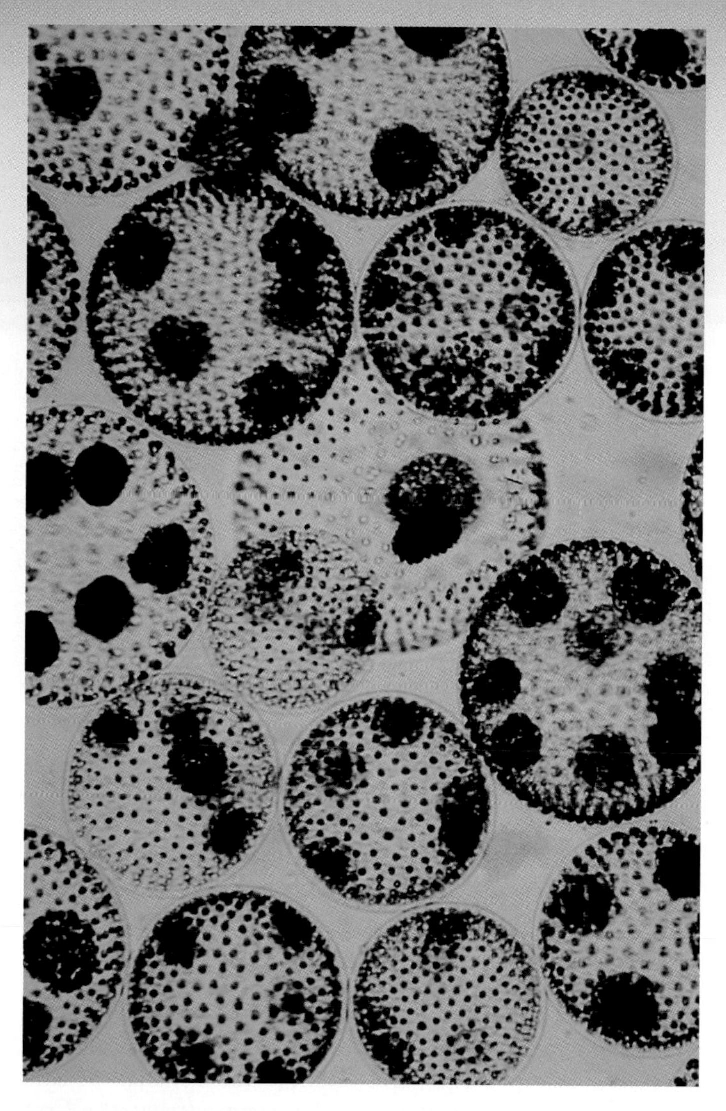

FIGURE 35.1
***Volvox*, a colonial protist.** The protists are a large, diverse group of primarily single-celled organisms, a group from which the other three eukaryotic kingdoms each evolved.

For more than half of the long history of life on earth, all life was microscopic in size. The biggest organisms that existed for over 2 billion years were single-celled bacteria fewer than 6 micrometers thick. The first evidence of a different kind of organism is found in tiny fossils in rock 1.5 billion years old. These fossil cells are much larger than bacteria (some as big as 60 micrometers in diameter) and have internal membranes and what appear to be small, membrane-bounded structures. Many have elaborate shapes, and some exhibit spines or filaments. These new, larger fossil organisms mark one of the most important events in the evolution of life, the appearance of a new kind of organism, the eukaryote (figure 35.1). Flexible and adaptable, the eukaryotes rapidly evolved to produce all of the diverse large organisms that populate the earth today, including ourselves—indeed, all organisms other than bacteria are eukaryotes.

35.1 Eukaryotes probably arose by endosymbiosis.

Endosymbiosis

What was the first eukaryote like? We cannot be sure, but a good model is *Pelomyxa palustris*, a single-celled, nonphotosynthetic organism that appears to represent an early stage in the evolution of eukaryotic cells (figure 35.2). The cells of *Pelomyxa* are much larger than bacterial cells and contain a complex system of internal membranes. Although they resemble some of the largest early fossil eukaryotes, these cells are unlike those of any other eukaryote: *Pelomyxa* lacks mitochondria and does not undergo mitosis. Its nuclei divide somewhat as do those of bacteria, by pinching apart into two daughter nuclei, around which new membranes form. Although *Pelomyxa* cells lack mitochondria, two kinds of bacteria living within them may play the same role that mitochondria do in all other eukaryotes. This primitive eukaryote is so distinctive that it is assigned a phylum all its own, Caryoblastea.

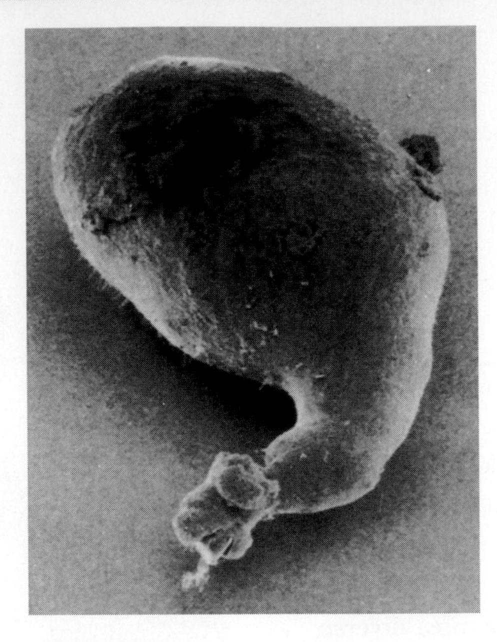

FIGURE 35.2
Pelomyxa palustris. This unique, amoeba-like protist lacks mitochondria and does not undergo mitosis. *Pelomyxa* may represent a very early stage in the evolution of eukaryotic cells.

Biologists know very little of the origin of *Pelomyxa*, except that in many of its fundamental characteristics it resembles the archaebacteria far more than the eubacteria. Because of this general resemblance, it is widely assumed that the first eukaryotic cells were nonphotosynthetic descendants of archaebacteria.

What about the wide gap between *Pelomyxa* and all other eukaryotes? Where did mitochondria come from? Most biologists agree with the theory of **endosymbiosis,** which proposes that mitochondria originated as symbiotic, aerobic (oxygen-requiring) eubacteria (figure 35.3). Symbiosis (Greek, *syn*, "together with" + *bios*, "life") means living together in close association. Recall from chapter 5 that mitochondria are sausage-shaped organelles 1 to 3 micrometers long, about the same size as most eubacteria. Mitochondria are bounded by two membranes. Aerobic eubacteria are thought to have become mitochondria when they were engulfed by ancestral eukaryotic cells, much like *Pelomyxa*, early in the history of eukaryotes.

The most similar eubacteria to mitochondria today are the nonsulfur purple bacteria, which are able to carry out oxidative metabolism (described in chapter 9). In mitochondria, the outer membrane is smooth and is thought to be derived from the endoplasmic reticulum of the host cell, which, like *Pelomyxa*, may have already contained a complex system of internal membranes. The inner membrane of mitochondria is folded into numerous layers, resembling the folded membranes of nonsulfur purple bacteria; embedded within this membrane are the proteins that carry out oxidative metabolism. The engulfed bacteria became the interior portion of the mitochondria we see today. Host cells were unable to carry out the Krebs cycle or other metabolic reactions necessary for living in an atmosphere that contained increasing amounts of oxygen before they had acquired these bacteria.

During the billion and a half years in which mitochondria have existed as endosymbionts within eukaryotic cells, most of their genes have been transferred to the chromosomes of the host cells—but not all. Each mitochondrion still has its own genome, a circular, closed molecule of DNA similar to that found in eubacteria, on which is located genes encoding the essential proteins of oxidative metabolism. These genes are transcribed within the mitochondrion, using mitochondrial ribosomes that are smaller than those of eukaryotic cells, very much like bacterial ribosomes in size and structure. Mitochondria divide by simple fission, just as bacteria do, replicating and sorting their DNA much as bacteria do. However, nuclear genes direct the process, and mitochondria cannot be grown outside of the eukaryotic cell, in cell-free culture.

The theory of endosymbiosis has had a controversial history but has now been accepted by all but a few biologists. The evidence supporting the theory is so extensive that in this text we will treat it as established.

What of mitosis, the other typical eukaryotic process that *Pelomyxa* lacks? The mechanism of mitosis, now so common among eukaryotes, did not evolve all at once. Traces of very different, and possibly intermediate, mechanisms survive today in some of the eukaryotes. In fungi and some groups of protists, for example, the nuclear membrane does not dissolve and mitosis is confined to the nucleus. When mitosis is complete in these organisms, the nucleus divides into two daughter nuclei, and only then does the rest of the cell divide. This separate nuclear division phase of mitosis does not occur in most protists, or in plants or animals. We do not know if it represents an intermediate step on the evolutionary journey to the form of mitosis that is characteristic of most

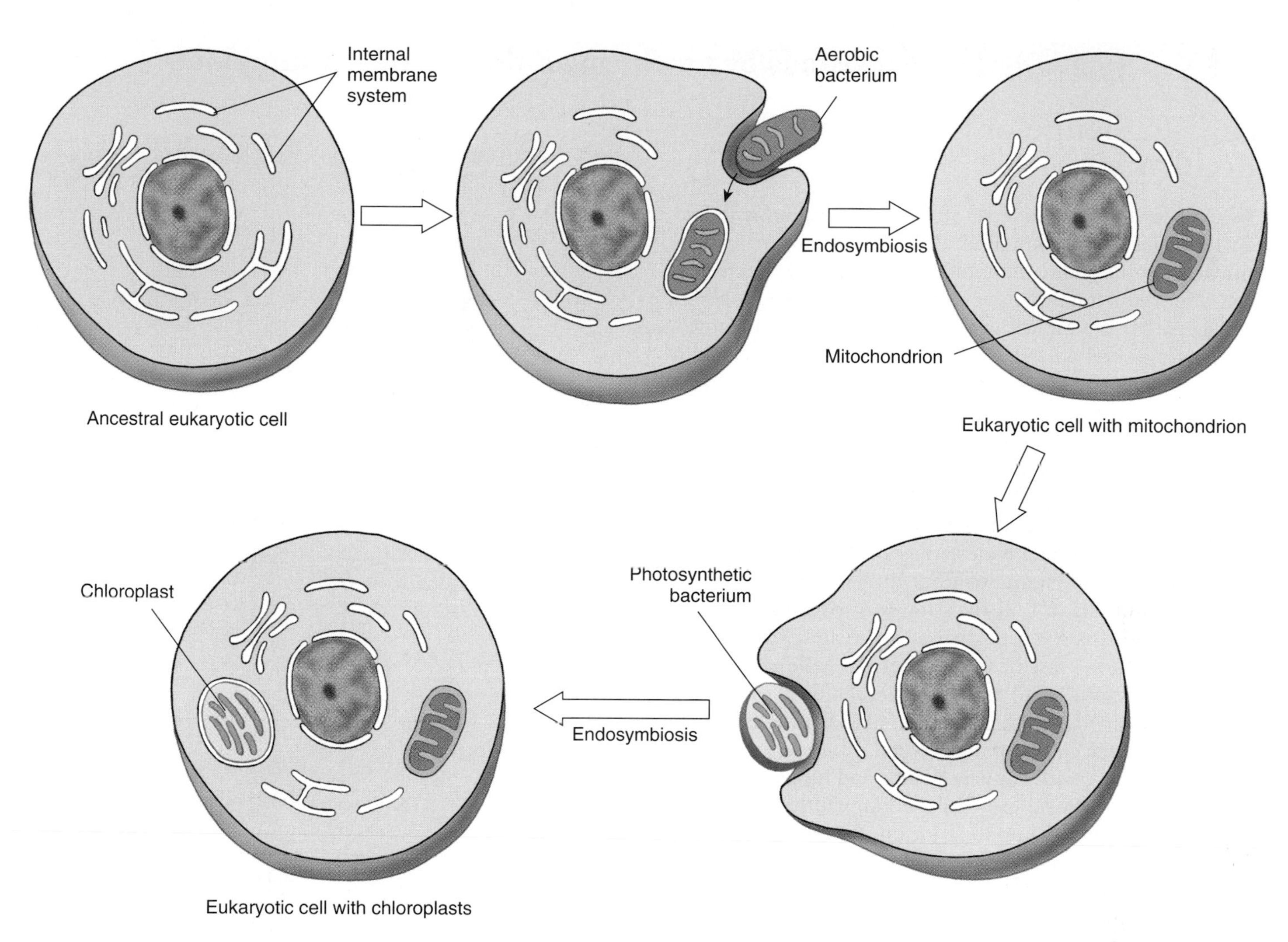

FIGURE 35.3
The theory of endosymbiosis. Scientists propose that ancestral eukaryotic cells, which already had an internal system of membranes, engulfed aerobic eubacteria, which then became mitochondria in the eukaryotic cell. Chloroplasts may also have originated this way, with eukaryotic cells engulfing photosynthetic eubacteria.

eukaryotes today or if it is simply a different way of solving the same problem. There are no fossils in which we can see the interiors of dividing cells well enough to be able to trace the history of mitosis.

Endosymbiosis Is Not Rare

Many eukaryotic cells contain other endosymbiotic bacteria in addition to mitochondria. Plants and algae contain chloroplasts, bacteria-like organelles that were apparently derived from symbiotic photosynthetic bacteria. Chloroplasts have a complex system of inner membranes and a circle of DNA. Centrioles, organelles associated with the assembly of microtubules, resemble in many respects spirochete bacteria, and they contain bacteria-like DNA involved in the production of their structural proteins.

While all mitochondria are thought to have arisen from a single symbiotic event, it is difficult to be sure with chloroplasts. Three biochemically distinct classes of chloroplasts exist, each resembling a different bacterial ancestor. Red algae possess pigments similar to those of cyanobacteria; plants and green algae more closely resemble the photosynthetic bacteria *Prochloron;* while brown algae and other photosynthetic protists resemble a third group of bacteria. This diversity of chloroplasts has led to the widely held belief that eukaryotic cells acquired chloroplasts by symbiosis at least three different times. Recent comparisons of chloroplast DNA sequences, however, suggest a single origin of chloroplasts, followed by very different postendosymbiotic histories. For example, in each of the three main lines, different genes became relocated to the nucleus, lost, or modified.

The theory of endosymbiosis proposes that mitochondria originated as symbiotic aerobic eubacteria.

35.2 The kingdom Protista is by far the most diverse of any kingdom.

The Challenge of Classifying the Protists

Protists are the most diverse of the four kingdoms in the domain Eukarya. The kingdom Protista contains many unicellular, colonial, and multicellular groups. Probably the most important statement we can make about the kingdom Protista is that it is an artificial group; as a matter of convenience, single-celled eukaryotic organisms have typically been grouped together into this kingdom. This lumps many very different and only distantly related forms together. The "single-kingdom" classification of the Protista is not representative of any evolutionary relationships; Protista is a polyphyletic group. The phyla of protists are, with very few exceptions, only distantly related to one another.

New applications of a wide variety of molecular methods are providing important insights into the relationships among the protists. Of all the groups of organisms biologists study, protists are probably in the greatest state of flux when it comes to classification. There is little consensus, even among experts, as to how the different kinds of protists should be classified. Are they a single, very diverse kingdom, or are they better considered as several different kingdoms, each of equal rank with animals, plants, and fungi?

Because the Protista are still predominantly considered part of one diverse, nonunified group, that is how we will treat them in this chapter, bearing in mind that biologists are rapidly gaining a better understanding of the evolutionary relationships among members of the kingdom Protista (figure 35.4). It seems likely that within a few years, the traditional kingdom Protista will be replaced by another more illuminating arrangement.

The taxonomy of the protists is in a state of flux as new information shapes our understanding of this kingdom.

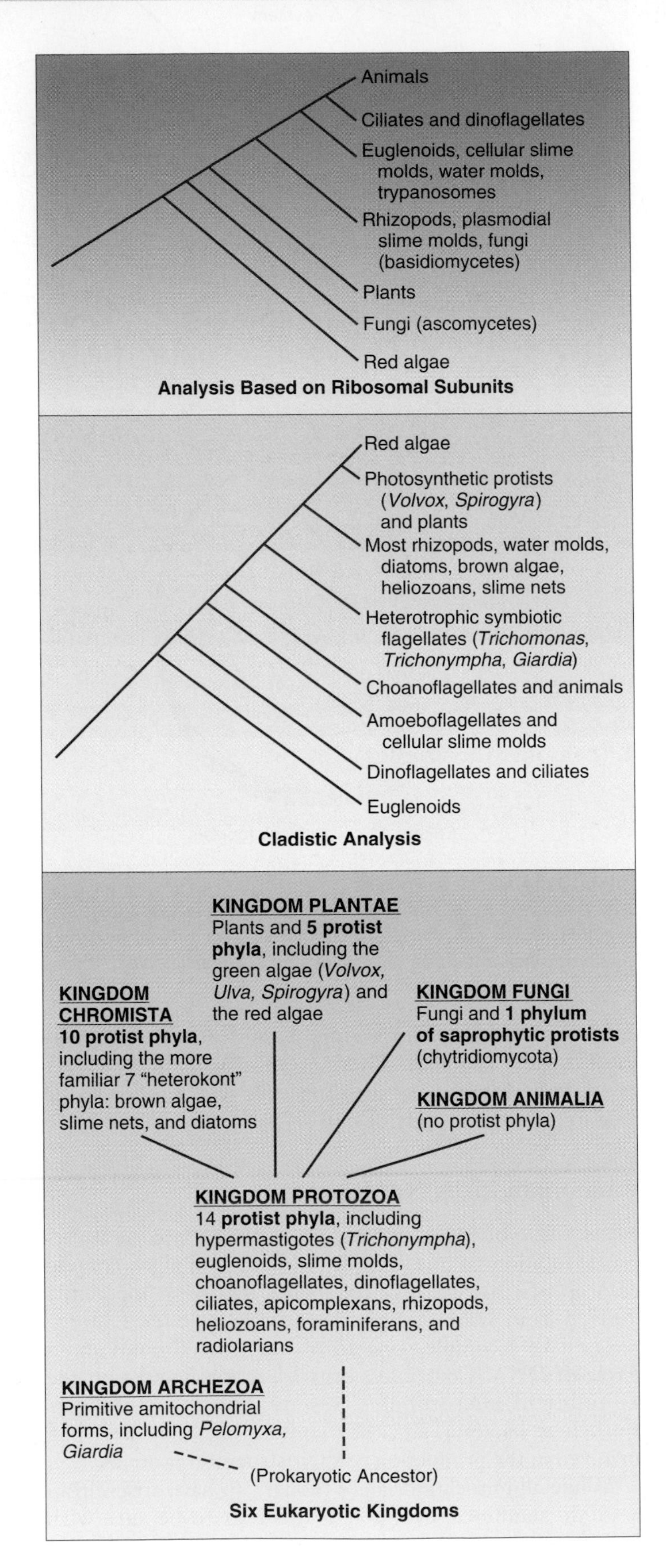

FIGURE 35.4
The challenge of protistan classification. Three different suggestions for protistan classification are presented, each adapted from the work of an authority in the field. Their great differences attest to the wide divergence of opinion within the field itself. The classification on the top is based on molecular variation in ribosomal subunits. The classification in the middle presents a cladistic analysis of a broad range of characters (including ribosomal subunits). The classification on the bottom outlines a more revolutionary reevaluation of the protists. Comparison of the three schemes reveals that some groups are commonly recognized as related (like ciliates and dinoflagellates), while the classification of others (like *Giardia*) is clearly in a state of flux.

General Biology of the Protists

Protists are united on the basis of a single negative characteristic: they are not fungi, plants, or animals. In all other respects they are highly variable with no uniting features. Many are unicellular (figure 35.5), but there are numerous colonial and multicellular groups. Most are microscopic, but some are as large as trees. They represent all symmetries, and exhibit all types of nutrition.

The Cell Surface

Protists possess a varied array of cell surfaces. Some protists, like amoebas, are surrounded only by their plasma membranes. All other protists have a plasma membrane but some, like algae and molds, are encased within strong cell walls. Still others, like diatoms and forams, secrete glassy shells of silica.

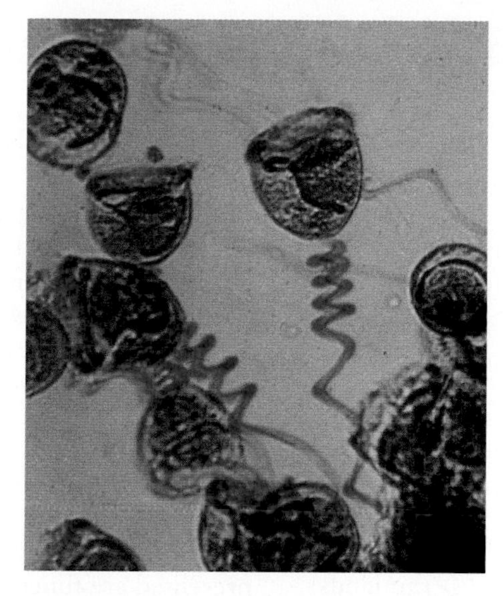

FIGURE 35.5
A unicellular protist. The protist kingdom is a catch-all kingdom for many different groups of unicellular organisms, such as this *Vorticella* (phylum Ciliophora), which is heterotrophic, feeds on bacteria, and has a retractable stalk.

Nutrition

Protists employ every form of nutritional acquisition except chemoautotrophic, which has so far been observed only in bacteria. Some protists are photosynthetic autotrophs and are called **phototrophs.** Others are heterotrophs that obtain energy from organic molecules synthesized by other organisms. Among heterotrophic protists, those that ingest visible particles of food are called **phagotrophs,** or **holozoic feeders.** Those ingesting food in soluble form are called **osmotrophs,** or **saprozoic feeders.**

Phagotrophs ingest food particles into intracellular vesicles called **food vacuoles** or **phagosomes.** Lysosomes fuse with the food vacuoles, introducing enzymes that digest the food particles within. As the digested molecules are absorbed across the vacuolar membrane, the food vacuole becomes progressively smaller.

Locomotor Organelles

Movement in protists is also accomplished by diverse mechanisms. Protists move chiefly by either flagellar rotation or pseudopodial movement. Many protists wave one or more flagella to propel themselves through the water, while others use banks of short, flagella-like structures called cilia to create water currents for their feeding or propulsion. Pseudopodia are the chief means of locomotion among amoeba, whose pseudopods are large, blunt extensions of the cell body called lobopodia. Other related protists extend thin, branching protrusions called filopodia. Still other protists extend long, thin pseudopodia called axopodia supported by axial rods of microtubules. Axopodia can be extended or retracted. Because the tips can adhere to adjacent surfaces, the cell can move by a rolling motion, shortening the axopodia in front and extending those in the rear.

Cyst Formation

Many protists with delicate surfaces are successful in quite harsh habitats. How do they manage to survive so well? They survive inhospitable conditions by forming **cysts.** A cyst is a dormant form of a cell with a resistant outer covering in which cell metabolism is more or less completely shut down. Not all cysts are so sturdy. Vertebrate parasitic amoebae, for example, form cysts that are quite resistant to gastric acidity, but will not tolerate desiccation or high temperature.

Reproduction

Protists typically reproduce asexually, reproducing sexually only in times of stress. Asexual reproduction involves mitosis, but the process is often somewhat different from the mitosis that occurs in multicellular animals. The nuclear membrane, for example, often persists throughout mitosis, with the microtubular spindle forming within it. In some groups, asexual reproduction involves spore formation, in others fission. The most common type of fission is **binary,** in which a cell simply splits into nearly equal halves. When the progeny cell is considerably smaller than its parent, and then grows to adult size, the fission is called **budding.** In multiple fission, or **schizogony,** common among some protists, fission is preceded by several nuclear divisions, so that fission produces several individuals almost simultaneously.

Sexual reproduction also takes place in many forms among the protists. In ciliates and some flagellates, **gametic meiosis** occurs just before gamete formation, as it does in metazoans. In the sporozoans, **zygotic meiosis** occurs directly *after* fertilization, and all the individuals that are produced are haploid until the next zygote is formed. In algae, there is **intermediary meiosis,** producing an alternation of generations similar to that seen in plants, with significant portions of the life cycle spent as haploid as well as diploid.

Protists exhibit a wide range of forms, locomotion, nutrition, and reproduction.

35.3 Protists can be categorized into five groups.

Five Groups of Protists

There are some 15 major phyla of protists. It is difficult to encompass their great diversity with any simple scheme. Traditionally, texts have grouped them artificially (as was done in the nineteenth century) into photosynthesizers (algae), heterotrophs (protozoa), and absorbers (funguslike protists).

In this text, we will group the protists into five general groups according to some of the major shared characteristics (figure 35.6). These are characteristics that taxonomists are using today in broad attempts to classify the kingdom Protista. These include (1) the presence or absence and type of cilia or flagella, (2) the presence and kinds of pigments, (3) the type of mitosis, (4) the kinds of cristae present in the mitochondria, (5) the molecular genetics of the ribosomal "S" subunit, (6) the kind of inclusions the protist may have, (7) overall body form (amoeboid, coccoid, and so forth), (8) whether the protist has any kind of shell or other body "armor," and (9) modes of nutrition and movement. These represent only some of the characters used to define phylogenetic relationships.

The five criteria we have chosen to define groups are not the only ones that might be chosen, and there is no broad agreement among biologists as to which set of criteria is preferable. As molecular analysis gives us a clearer picture of the phylogenetic relationships among the protists, more evolutionarily suitable groupings will without a doubt replace the one represented here. Table 35.1 summarizes some of the general characteristics and groupings of the 15 major phyla of protists. It is important to remember that while the phyla of protists discussed here are generally accepted taxa, the larger groupings of phyla presented are functional groupings.

The 15 major protist phyla can be conveniently categorized into five groups according to major shared characteristics.

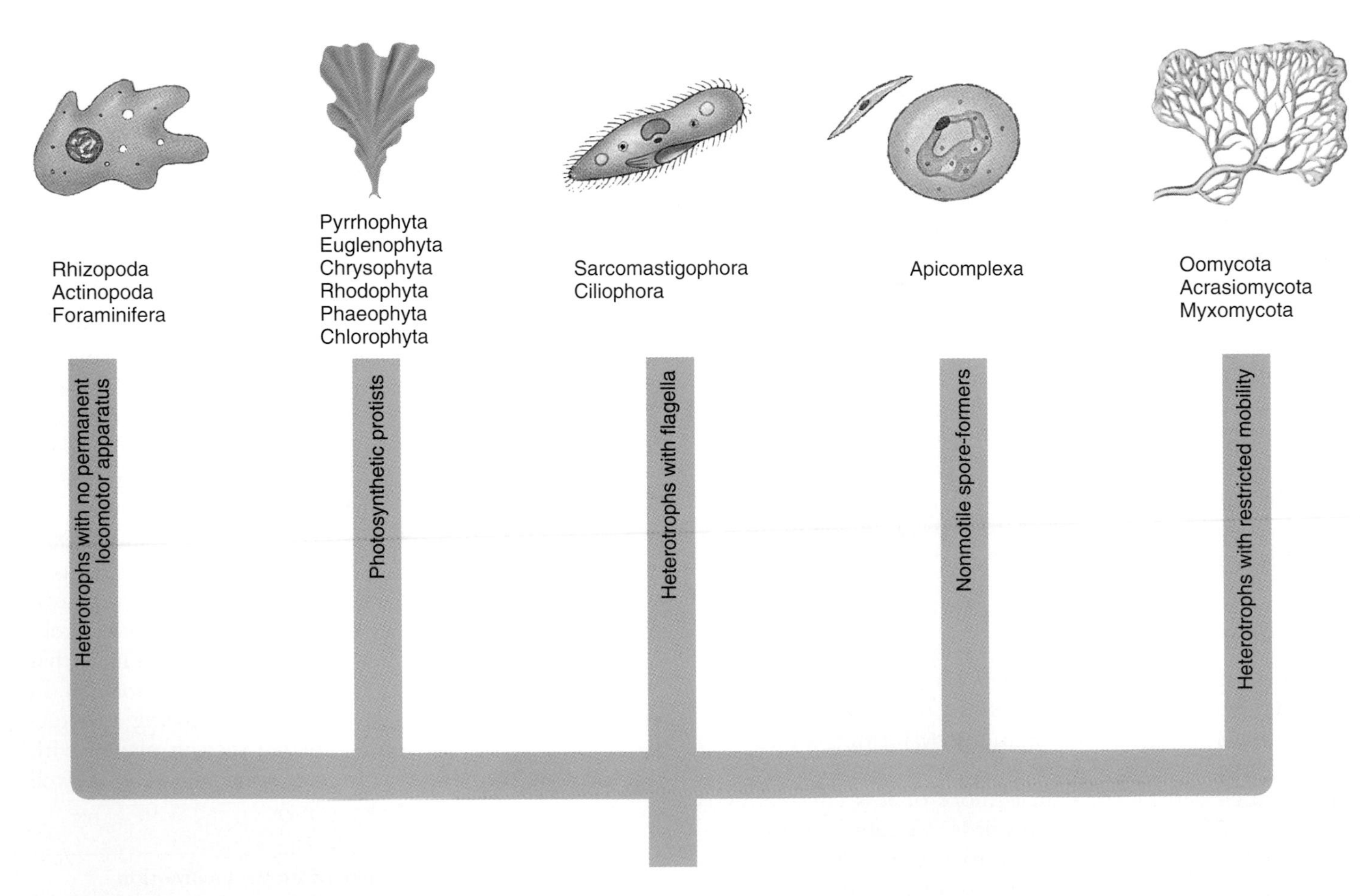

FIGURE 35.6
Five general groups of protists. This text presents the 15 major phyla of protists in five groups that share major characteristics.

Group	Phylum	Typical Examples	Key Characteristics
HETEROTROPHS WITH NO PERMANENT LOCOMOTOR APPARATUS			
Amoebas	Rhizopoda	*Amoeba*	Move by pseudopodia
Radiolarians	Actinopoda	Radiolarians	Glassy skeletons; needlelike pseudopods
Forams	Foraminifera	Forams	Rigid shells; move by protoplasmic streaming
PHOTOSYNTHETIC PROTISTS			
Dinoflagellates	Pyrrhophyta	Red tides	Photosynthetic; unicellular; two flagella; contain chlorophylls *a* and *b*
Euglenoids	Euglenophyta	*Euglena*	Some photosynthetic; others heterotrophic; unicellular; contain chlorophylls *a* and *b* or none
Diatoms	Chrysophyta	*Diatoma*	Unicellular; manufacture the carbohydrate chrysolaminarin; unique double shells of silica; contain chlorophylls *a* and *c*
Golden algae	Chrysophyta	Golden algae	Unicellular, but often colonial; manufacture the carbohydrate chrysolaminarin; contain chlorophylls *a* and *c*
Red algae	Rhodophyta	Coralline algae	Most multicellular; contain chlorophyll *a* and a red pigment
Brown algae	Phaeophyta	Kelp	Multicellular; contain chlorophylls *a* and *c*
Green algae	Chlorophyta	*Chlamydomonas*	Unicellular or multicellular; contain chlorophylls *a* and *b*
HETEROTROPHS WITH FLAGELLA			
Zoomastigotes	Sarcomastigophora	Trypanosomes	Heterotrophic; unicellular
Ciliates	Ciliophora	*Paramecium*	Heterotrophic unicellular protists with cells of fixed shape possessing two nuclei and many cilia; many cells also contain highly complex and specialized organelles
NONMOTILE SPORE-FORMERS			
Sporozoans	Apicomplexa	*Plasmodium*	Nonmotile; unicellular; the apical end of the spores contains a complex mass of organelles
HETEROTROPHS WITH RESTRICTED MOBILITY			
Water molds	Oomycota	Water molds, rusts, and mildew	Terrestrial and freshwater
Cellular slime molds	Acrasiomycota	*Dictyostelium*	Colonial aggregations of individual cells; most closely related to amoebas
Plasmodial slime molds	Myxomycota	*Fuligo*	Stream along as a multinucleate mass of cytoplasm

Table 35.1 Kinds of Protists

Heterotrophs with No Permanent Locomotor Apparatus

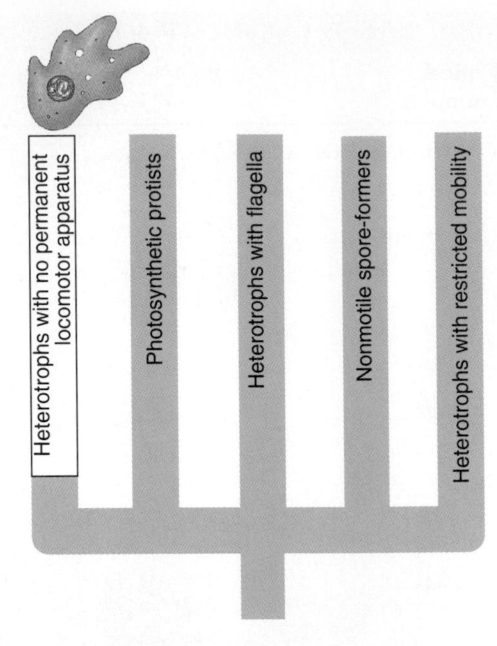

The largest of the five general groups of protists are primarily unicellular organisms with amoeboid forms. There are three principle phyla: the forams and the radiolarians have carbonate shells and the rhizopods lack shells.

Rhizopoda: The Amoebas

Hundreds of species of amoebas are found throughout the world in both fresh and salt waters. They are also abundant in soil. Many kinds of amoebas are parasites of animals. Reproduction in amoebas occurs by fission, or the direct division into two cells of equal volume. Amoebas of the phylum Rhizopoda lack cell walls, flagella, meiosis, and any form of sexuality. They do undergo mitosis, with a spindle apparatus that resembles that of other eukaryotes.

Amoebas move from place to place by means of their **pseudopods,** from the Greek words for "false" and "foot" (figure 35.7). Pseudopods are flowing projections of cytoplasm that extend and pull the amoeba forward or engulf food particles, a process called cytoplasmic streaming. An amoeba puts a pseudopod forward and then flows into it. Microfilaments of actin and myosin similar to those found in muscles are associated with these movements. The pseudopodia can form at any point on the cell body so that it can move in any direction.

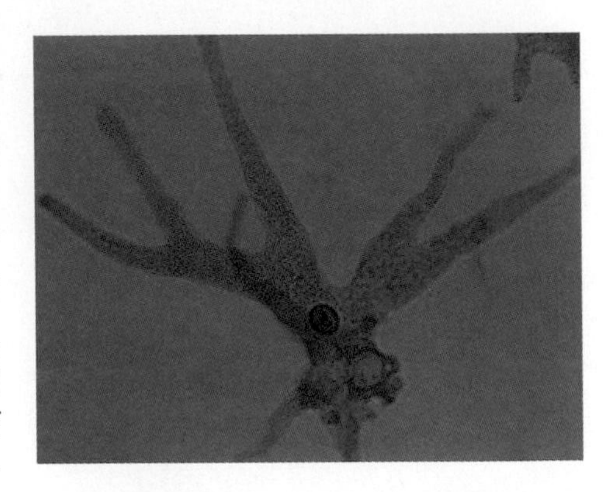

FIGURE 35.7
Amoeba proteus. This relatively large amoeba is commonly used in teaching and for research in cell biology. The projections are pseudopods; an amoeba moves by flowing into them. The nucleus of the amoeba is plainly visible.

Some kinds of amoebas form resistant cysts. In parasitic species such as *Entamoeba histolytica*, which causes amoebic dysentery, cysts enable the amoebas to resist digestion by their animal hosts. Mitotic division takes place within the cysts, which ultimately rupture and release four, eight, or even more amoebas within the digestive tracts of their host animals. The primary infection takes place in the intestine, but it often moves into the liver and other parts of the body. The cysts are dispersed in the feces and may be transmitted from person to person in infected food or water, or by flies. It is estimated that up to 10 million people in the United States have infections of parasitic amoebas, and some 2 million show symptoms of the disease, ranging from abdominal discomfort with slight diarrhea to much more serious conditions. In some tropical areas, more than half of the population may be infected. The spread of amoebic dysentery can be limited by proper sanitation and hygiene.

Actinopoda: The Radiolarians

The pseudopodia of amoeboid cells give them truly amorphous bodies. One group, however, have more distinct structures. Members of the phylum Actinopoda, often called radiolarians, secrete glassy exoskeletons made of silica. These skeletons give the unicellular organisms a distinct shape, exhibiting either bilateral or radial symmetry. The shells of different species form many elaborate and beautiful shapes and its pseudopodia extrude outward along spiky projections of the skeleton (figure 35.8). Microtubules support these cytoplasmic projections.

Foraminifera: Forams

Members of the phylum Foraminifera are heterotrophic marine protists. They range in diameter from about 20 micrometers to several centimeters. Characteristic of the group are pore-studded shells (called **tests**) composed of organic materials usually reinforced with grains of inorganic matter. These grains may be calcium carbonate, sand, or even plates from the shells of echinoderms or spicules (minute needles of calcium carbonate) from sponge skeletons. Depending on the building materials they use, foraminifera—often informally called "forams"—may have shells of very different appearance. Some of them are brilliantly colored red, salmon, or yellow-brown.

Most foraminifera live in sand or are attached to other organisms, but two families consist of free-floating planktonic organisms. Their tests may be single-chambered but more often are multichambered, and they sometimes have a spiral shape resembling that of a tiny snail. Thin cytoplasmic projections called **podia** emerge through openings in the tests (figure 35.9). Podia are used for swimming, gathering materials for the tests, and feeding. Forams eat a wide variety of small organisms.

The life cycles of foraminifera are extremely complex, involving an alternation between haploid and diploid generations (sporic meiosis). Forams have contributed massive

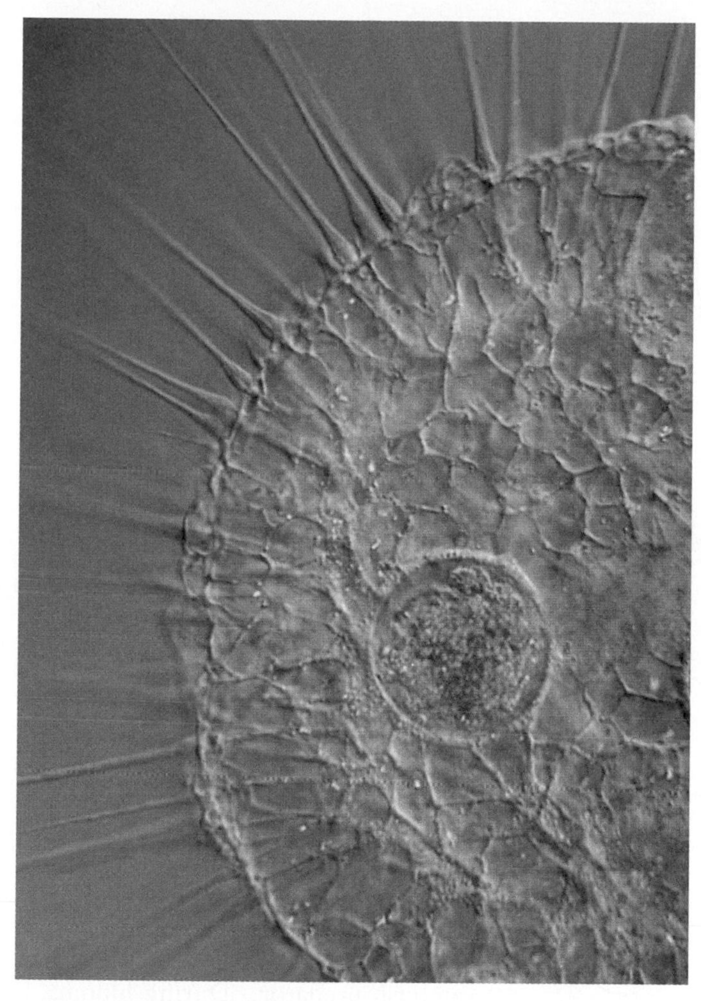

FIGURE 35.8
***Actinosphaerium*, a protist of the phylum Actinopoda (300×).** This amoeba-like radiolarian has striking needlelike pseudopods.

FIGURE 35.9
A representative of the Foraminifera (90×). A living foram with podia, thin cytoplasmic projections that extend through pores in the calcareous test, or shell, of the organism.

accumulations of their tests to the fossil record for more than 200 million years. Because of the excellent preservation of their tests and the often striking differences among them, forams are very important as geological markers. The pattern of occurrence of different forams is often used as a guide in searching for oil-bearing strata. Limestones all over the world, including the famous white cliffs of Dover in southern England, are often rich in forams (figure 35.10).

Amoebas, radiolarians, and forams are unicellular, heterotrophic protists that lack cell walls, flagella, meiosis, and sexuality. Amoebas move from place to place by means of extensions called pseudopodia. The pore-studded tests, or shells, of the forams have openings through which podia extend that are used for locomotion.

FIGURE 35.10
White cliffs of Dover. The limestone that forms these cliffs is composed almost entirely of fossil shells of protists, including coccolithophores (a type of algae) and foraminifera.

Photosynthetic Protists

Pyrrhophyta: The Dinoflagellates

The dinoflagellates consist of about 2100 known species of primarily unicellular, photosynthetic organisms, most of which have two flagella. A majority of the dinoflagellates are marine, and they are often abundant in the plankton, but some occur in fresh water. Some planktonic dinoflagellates are luminous and contribute to the twinkling or flashing effects that we sometimes see in the sea at night, especially in the tropics.

The flagella, protective coats, and biochemistry of dinoflagellates are distinctive, and they do not appear to be directly related to any other phylum. Plates made of a cellulose-like material encase the cells. Grooves form at the junctures of these plates and the flagella are usually located within these grooves, one encircling the body like a belt, and the other perpendicular to it. By beating in their respective grooves, these flagella cause the dinoflagellate to rotate like a top as it moves. The dinoflagellates that are clad in stiff cellulose plates, often encrusted with silica, may have a very unusual appearance (figure 35.11). Most have chlorophylls *a* and *c*, in addition to carotenoids, so that in the biochemistry of their chloroplasts, they resemble the diatoms and the brown algae, possibly acquiring such chloroplasts by forming endosymbiotic relationships with members of those groups.

Some dinoflagellates occur as symbionts in many other groups of organisms, including jellyfish, sea anemones, mollusks, and corals. When dinoflagellates grow as symbionts within other cells, they lack their characteristic cellulose plates and flagella, appearing as spherical, golden-brown globules in their host cells. In such a state they are called **zooxanthellae**. Photosynthetic zooxanthellae provide their hosts with nutrients. It is the photosynthesis conducted by zooxanthellae that makes coral reefs one of the most productive ecosystems on earth. Corals primarily live in warm tropical seas that are typically extremely low in nutrients; without the aid of their photosynthetic endosymbionts, they would not be able to form large reefs in the nutrient-poor environment. Most of the carbon that the zooxanthellae fix is translocated to the host corals.

The poisonous and destructive "red tides" that occur frequently in coastal areas are often associated with great population explosions, or "blooms," of dinoflagellates. The pigments in the individual, microscopic cells of the dinoflagellates are responsible for the color of the water. Red tides have a profound, detrimental effect on the fishing industry in the United States. Some 20 species of dinoflagellates are known to produce powerful toxins that inhibit the diaphragm and cause respiratory failure in many vertebrates. When the toxic dinoflagellates are abundant, fishes, birds, and marine mammals may die in large numbers.

More recently, a particularly dangerous toxic dinoflagellate called *Pfiesteria piscicida* is reported to be a carnivorous, ambush predator. During blooms, it stuns fish with its toxin and then feeds on the prey's body fluids.

Dinoflagellates reproduce primarily by asexual cell division. But sexual reproduction has been reported to occur under starvation conditions. They have a unique form of mitosis in which the permanently condensed chromosomes divide longitudinally within the confines of a permanent nuclear envelope. After the numerous chromosomes duplicate, the nucleus divides into two daughter nuclei. Also the dinoflagellate chromosome is unique among eukaryotes in that the DNA is not complexed with histone proteins. In all other eukaryotes, the chromosomal DNA is complexed with histones to form nucleosomes, which represents the first order of DNA packaging in the nucleus. How dinoflagellates are able to maintain distinct chromosomes without histones and nucleosomes remains a mystery.

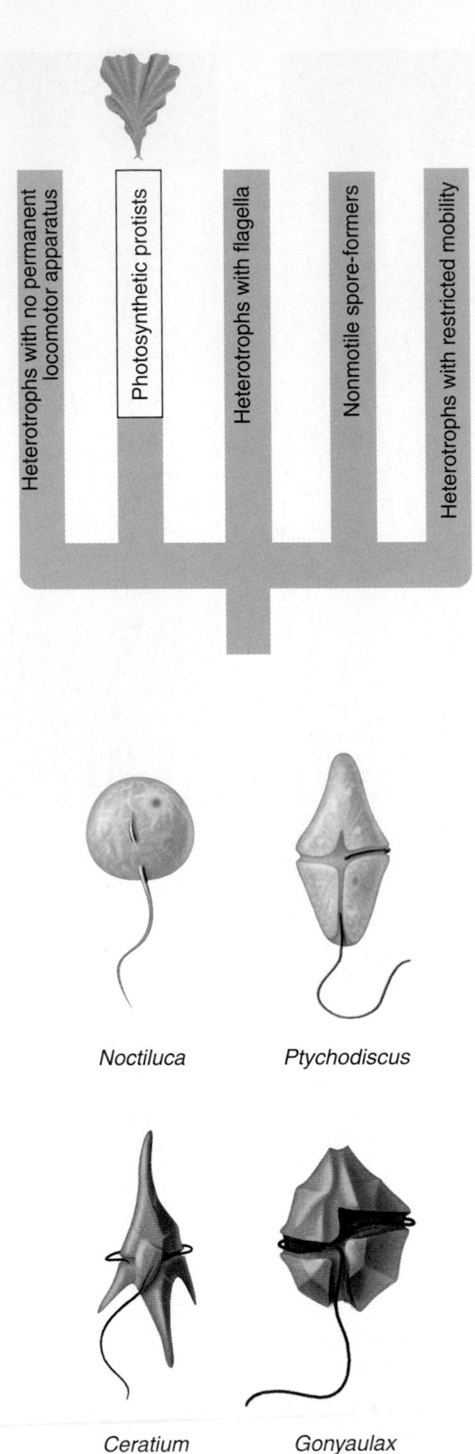

FIGURE 35.11
Some dinoflagellates: *Noctiluca*, *Ptychodiscus*, *Ceratium*, and *Gonyaulax*. *Noctiluca*, which lacks the heavy cellulose armor characteristic of most dinoflagellates, is one of the bioluminescent organisms that causes the waves to sparkle in warm seas. In the other three genera, the shorter, encircling flagellum is seen in its groove, with the longer one projecting away from the body of the dinoflagellate. (Not drawn to scale.)

Euglenophyta: The Euglenoids

Most of the approximately 1000 known species of euglenoids live in fresh water. The members of this phylum clearly illustrate the impossibility of distinguishing "plants" from "animals" among the protists. About a third of the approximately 40 genera of euglenoids have chloroplasts and are fully autotrophic; the others lack chloroplasts, ingest their food, and are heterotrophic. These organisms are not significantly different from some groups of zoomastigotes (see next section), and many biologists believe that the two phyla should be merged into one.

Some euglenoids with chloroplasts may become heterotrophic if the organisms are kept in the dark; the chloroplasts become small and nonfunctional. If they are put back in the light, they may become green within a few hours. Normally photosynthetic euglenoids may sometimes feed on dissolved or particulate food.

Individual euglenoids range from 10 to 500 micrometers long and are highly variable in form. Interlocking proteinaceous strips arranged in a helical pattern form a flexible structure called the **pellicle,** which lies within the cell membrane of the euglenoids. Because its pellicle is flexible, a euglenoid is able to change its shape. Reproduction in this phylum occurs by mitotic cell division. The nuclear envelope remains intact throughout the process of mitosis. No sexual reproduction is known to occur in this group.

In *Euglena* (figure 35.12), the genus for which the phylum is named, two flagella are attached at the base of a flask-shaped opening called the **reservoir,** which is located at the anterior end of the cell. One of the flagella is long and has a row of very fine, short, hairlike projections along one side. A second, shorter flagellum is located within the reservoir but does not emerge from it. Contractile vacuoles collect excess water from all parts of the organism and empty it into the reservoir, which apparently helps regulate the osmotic pressure within the organism. The **stigma,** an organ that also occurs in the green algae (phylum Chlorophyta), is light-sensitive and aids these photosynthetic organisms to move toward light.

Cells of *Euglena* contain numerous small chloroplasts. These chloroplasts, like those of the green algae and plants, contain chlorophylls *a* and *b*, together with carotenoids. Although the chloroplasts of euglenoids differ somewhat in structure from those of green algae, they probably had a common origin. It seems likely that euglenoid chloroplasts ultimately evolved from a symbiotic relationship through ingestion of green algae.

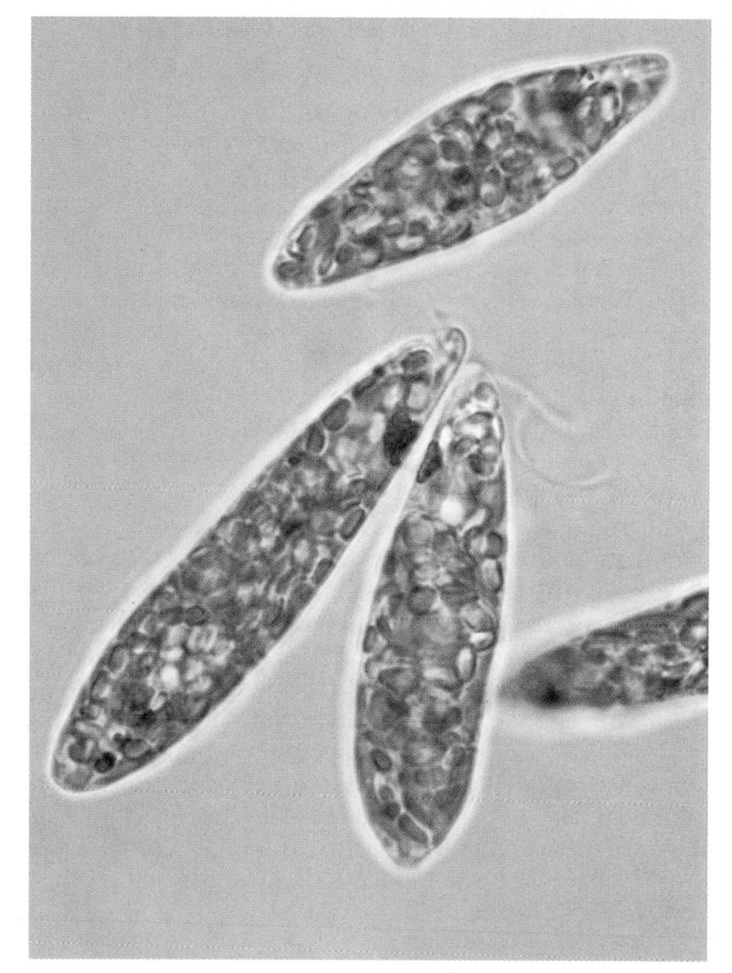

(a)

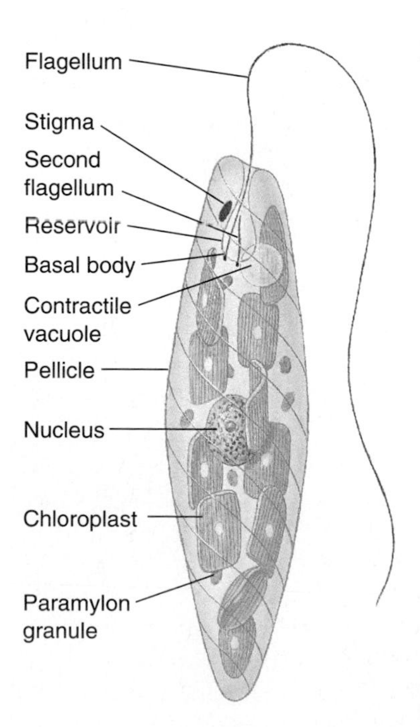

(b)

FIGURE 35.12
Euglenoids. (*a*) Micrograph of individuals of the genus *Euglena* (Euglenophyta). (*b*) Diagram of *Euglena.* Paramylon granules are areas where food reserves are stored.

Chrysophyta: The Diatoms and Golden Algae

The Diatoms. Diatoms, members of the phylum Chrysophyta, are photosynthetic, unicellular organisms with unique double shells made of opaline silica, which are often strikingly and characteristically marked. The shells of diatoms are like small boxes with lids, one half of the shell fitting inside the other. Their chloroplasts, with chlorophylls *a* and *c*, as well as carotenoids, resemble those of the brown algae and dinoflagellates. In other respects, however, there are few similarities between these groups, and they probably do not share an immediate common ancestor. Another member of the phylum Chrysophyta is the golden algae. Diatoms and golden algae are grouped together because they both produce a unique carbohydrate called chrysolaminarin.

There are more than 11,500 living species of diatoms, with many more known in the fossil record. The shells of fossil diatoms often form very thick deposits, which are sometimes mined commercially. The resulting "diatomaceous earth" is used as an abrasive or to add the sparkling quality to the paint used on roads, among other purposes. Living diatoms are often abundant both in the sea and in fresh water, where they are important food producers. Diatoms occur in the plankton and are attached to submerged objects in relatively shallow water. Many species are able to move by means of a secretion that is produced from a fine groove along each shell. The diatoms exude and perhaps also retract this secretion as they move.

There are two major groups of diatoms, one with radial symmetry (like a wheel) and the other with bilateral (two-sided) symmetry (figure 35.13). Diatom shells are rigid, and the organisms reproduce asexually by separating the two halves of the shell, each half then regenerating another half shell within it. Because of this mode of reproduction, there is a tendency for the shells, and consequently the individual diatoms, to get smaller and smaller with each asexual reproduction. When the resulting individuals have diminished to about 30% of their original size, one may slip out of its shell, grow to full size, and regenerate a full-sized pair of new shells.

Individual diatoms are diploid. Meiosis occurs more frequently under conditions of starvation. Some marine diatoms produce numerous sperm and others a single egg. If fusion occurs, the resulting zygote regenerates a full-sized individual. In some freshwater diatoms, the gametes are amoeboid and similar in appearance.

The Golden Algae. Also included within the Chrysophyta are the golden algae, named for the yellow and brown carotenoid and xanthophyll accessory pigments in their chloroplasts, which give them a golden color. Unicellular but often colonial, these freshwater protists typically have two flagella, both attached near the same end of the cell. When ponds and lakes dry out in summer, golden algae form resistant cysts. Viable cells emerge from these cysts when wetter conditions recur in the fall.

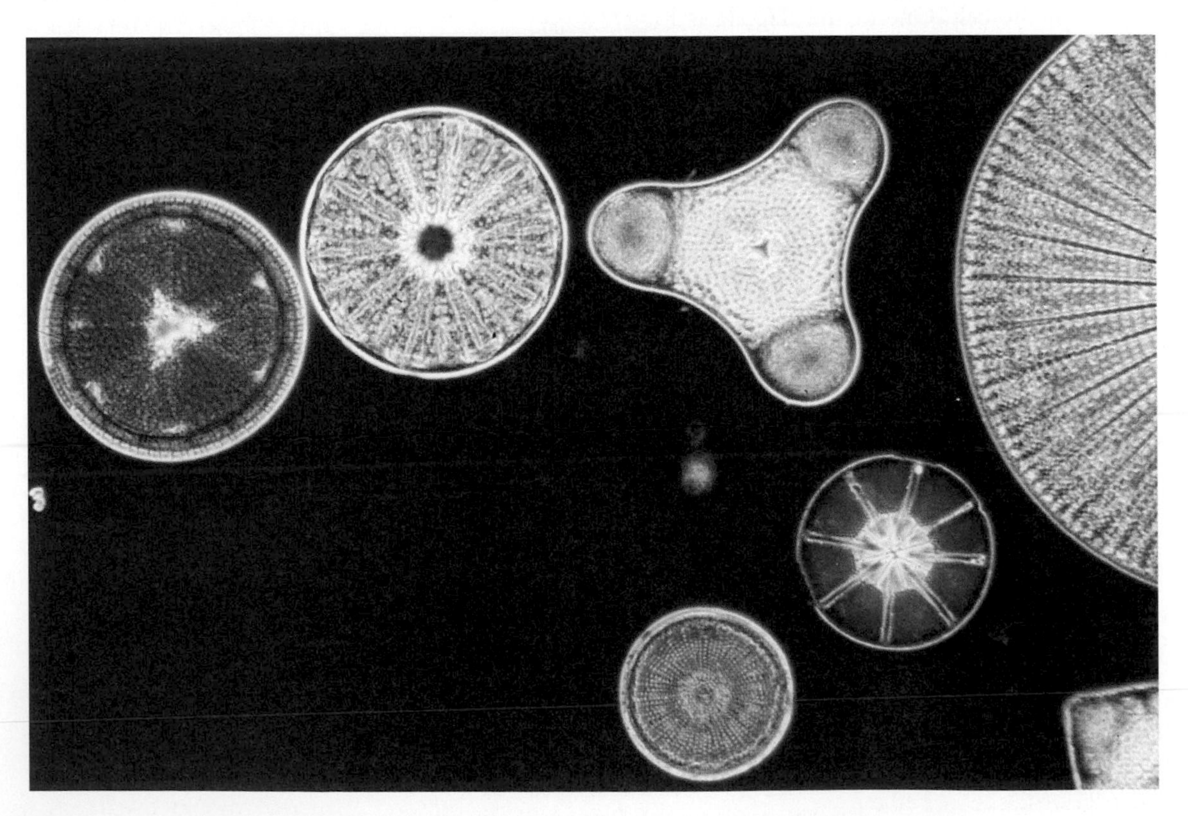

FIGURE 35.13
Diatoms (Chrysophyta). Several different centric (radially symmetrical) diatoms.

Rhodophyta: The Red Algae

Along with green algae and brown algae, red algae are the seaweeds we see cast up along shores and on beaches. Their characteristic colors result from phycoerythrin, a type of phycobilin pigment. Phycobilins are responsible for the colors of the cyanobacteria. Chlorophyll *a* also occurs with the phycobilins in red algae, just as it does in cyanobacteria. These similarities with cyanobacteria make it likely that the rhodophyta evolved when their heterotrophic eukaryotic ancestor developed an endosymbiotic relationship with a cyanobacteria which eventually gave rise to their chloroplasts.

The great majority of the estimated 4000 species of red algae occur in the sea, and almost all are multicellular.

Red algae have complex bodies made up of interwoven filaments of cells. In the cell walls of many red algae are sulfated polysaccharides such as agar and carrageenan, which make these algae important economically. Agar is used to make gel capsules, as material for dental impressions, and as a base for cosmetics. It is also the basis of the laboratory media on which bacteria, fungi, and other organisms are often grown. In addition, agar is used to prevent baked goods from drying out, for rapid-setting jellies, and as a temporary preservative for meat and fish in warm regions. Carrageenan is used mainly to stabilize emulsions such as paints, cosmetics, and dairy products such as ice cream. In addition to these uses, red algae such as *Porphyra*, called "nori," are eaten and, in Japan, are even cultivated as a human food crop.

The life cycles of red algae are complex but usually involve an alternation of generations (sporic meiosis). None of the red algae have flagella or cilia at any stage in their life cycle, and they may have descended directly from ancestors that never had them, especially as the red algae also lack centrioles. Together with the fungi, which also lack flagella and centrioles, the red algae may be one of the most ancient groups of eukaryotes.

FIGURE 35.14
Brown algae (Phaeophyta). The massive "groves" of giant kelp that occur in relatively shallow water along the coasts of the world provide food and shelter for many different kinds of organisms.

Phaeophyta: The Brown Algae

The phaeophyta, or brown algae, consist of about 1500 species of multicellular protists, almost exclusively marine. They are the most conspicuous seaweeds in many northern regions, dominating rocky shores almost everywhere in temperate North America. In habitats where large brown algae known as **kelps** (order Laminariales) occur abundantly in so-called kelp forests (figure 35.14), they are responsible for most of the food production through photosynthesis. Many kelps are conspicuously differentiated into flattened blades, stalks, and grasping basal portions that anchor them to the rocks.

Among the larger brown algae are genera such as *Macrocystis*, in which some individuals may reach 100 meters in length. The flattened blades of this kelp float out on the surface of the water, while the base is anchored tens of meters below the surface. Another ecologically important member of this phylum is sargasso weed, *Sargassum*, which forms huge floating masses that dominate the vast Sargasso Sea, an area of the Atlantic Ocean northeast of the Caribbean. The stalks of the larger brown algae often exhibit a complex internal differentiation of conducting tissues analogous to that of plants.

The life cycle of the brown algae is marked by an alternation of generations between a sporophyte and a gametophyte. The large individuals we recognize, such as the kelps, are sporophytes. The gametophytes are often much smaller, filamentous individuals, perhaps a few centimeters across. Sporangia, which produce haploid, swimming spores after meiosis, are formed on the sporophytes. These spores divide by mitosis, giving rise to individual gametophytes. There are two kinds of gametophytes in the kelps; one produces sperm, and the other produces eggs. If sperm and eggs fuse, the resulting zygotes grow into the mature kelp sporophytes, provided that they reach a favorable site.

Chlorophyta: The Green Algae

Green algae are an extremely varied group of more than 7000 species. The chlorophytes have an extensive fossil record dating back 900 million years. They are mostly aquatic, but some are semiterrestrial in moist places, such as on tree trunks or in soil. Many are microscopic and unicellular, but some, such as sea lettuce, *Ulva* (see figure 35.16), are tens of centimeters across and easily visible on rocks and pilings around the coasts.

Green algae are of special interest, both because of their unusual diversity and because the ancestors of the plant kingdom were clearly multicellular green algae. Many features of modern green algae closely resemble plants, especially their chloroplasts which are biochemically similar to those of the plants. They contain chlorophylls *a* and *b*, as well as carotenoids. Green algae include a very wide array of both unicellular and multicellular organisms.

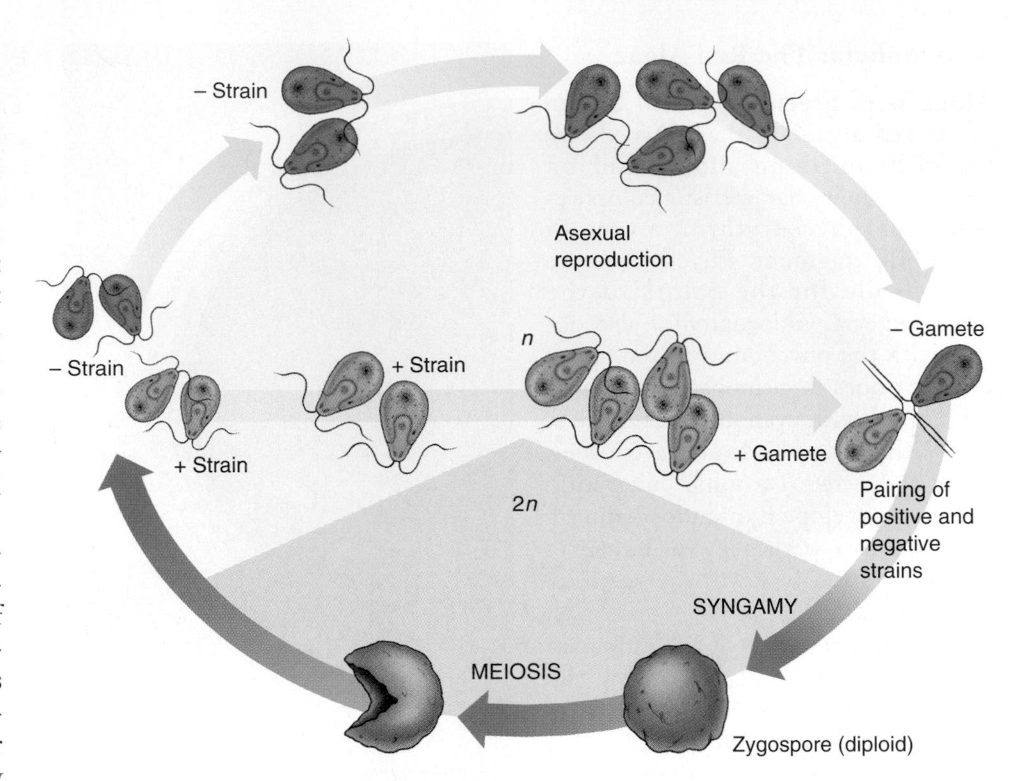

FIGURE 35.15
Life cycle of *Chlamydomonas* (Chlorophyta). Individual cells of this microscopic, biflagellated alga, which are haploid, divide asexually, producing identical copies of themselves. At times, such haploid cells act as gametes—fusing, as shown in the lower right-hand side of the diagram, to produce a zygote. The zygote develops a thick, resistant wall, becoming a zygospore; this is the only diploid cell in the entire life cycle. Within this diploid zygospore, meiosis takes place, ultimately resulting in the release of four haploid individuals. Because of the segregation during meiosis, two of these individuals are called the (+) strain, the other two the (–) strain. Only + and – individuals are capable of mating with each other when syngamy does take place, although both may divide asexually to reproduce themselves.

Among the unicellular green algae, *Chlamydomonas* (figure 35.15) is a well-known genus. Individuals are microscopic (usually less than 25 micrometers long), green, rounded, and have two flagella at the anterior end. They move rapidly in water by beating their flagella in opposite directions. Each individual has an eyespot, which contains about 100,000 molecules of rhodopsin, the same pigment employed in vertebrate eyes. Light received by this eyespot is used by the alga to help direct its swimming. Most individuals of *Chlamydomonas* are haploid. *Chlamydomonas* reproduces asexually (by cell division) as well as sexually. In sexual reproduction, two haploid individuals fuse to form a four-flagellated zygote. The zygote ultimately enters a resting phase, called the **zygospore,** in which the flagella disappear and a tough protective coat is formed. Meiosis occurs at the end of this resting period and results in the production of four haploid cells.

Chlamydomonas probably represents a primitive state for green algae and several lines of evolutionary specialization have been derived from organisms like it. The first is the evolution of nonmotile, unicellular green algae. *Chlamydomonas* is capable of retracting its flagella and settling down as an immobile unicellular organism if the ponds in which it lives dry out. Some common algae of soil and bark, such as *Chlorella*, are essentially like *Chlamydomonas* in this trait, but do not have the ability to form flagella. *Chlorella* is widespread in both fresh and salt water as well as soil and is only known to reproduce asexually. Recently, *Chlorella* has been widely investigated as a possible food source for humans and other animals, and pilot farms have been established in Israel, the United States, Germany, and Japan.

Another major line of specialization from cells like *Chlamydomonas* concerns the formation of motile, colonial organisms. In these genera of green algae, the *Chlamydomonas*-like cells retain some of their individuality. The most elaborate of these organisms is *Volvox* (see figure 35.1), a hollow sphere made up of a single layer of 500 to 60,000 individual cells, each cell with two flagella. Only a small number of the cells are reproductive. The colony has definite anterior and posterior ends, and the flagella of all of the cells beat in such a way as to rotate the colony in a clockwise direction as it moves forward through the water. The reproductive cells of *Volvox* are located mainly at the posterior end of the colony. Some may divide asexually, bulge inward, and give rise to new colonies that initially remain within the parent colony. Others produce gametes. In some species of

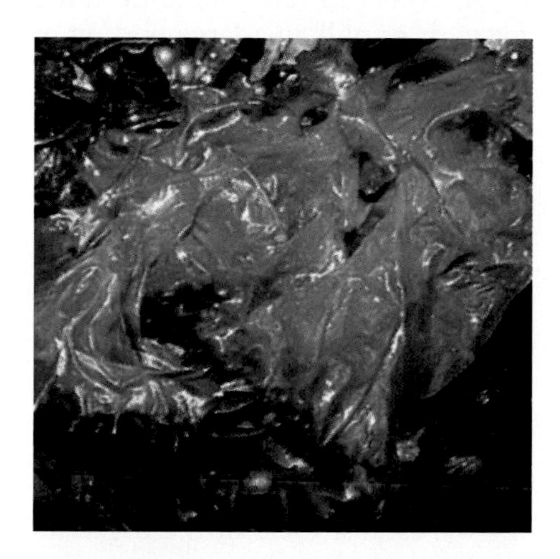

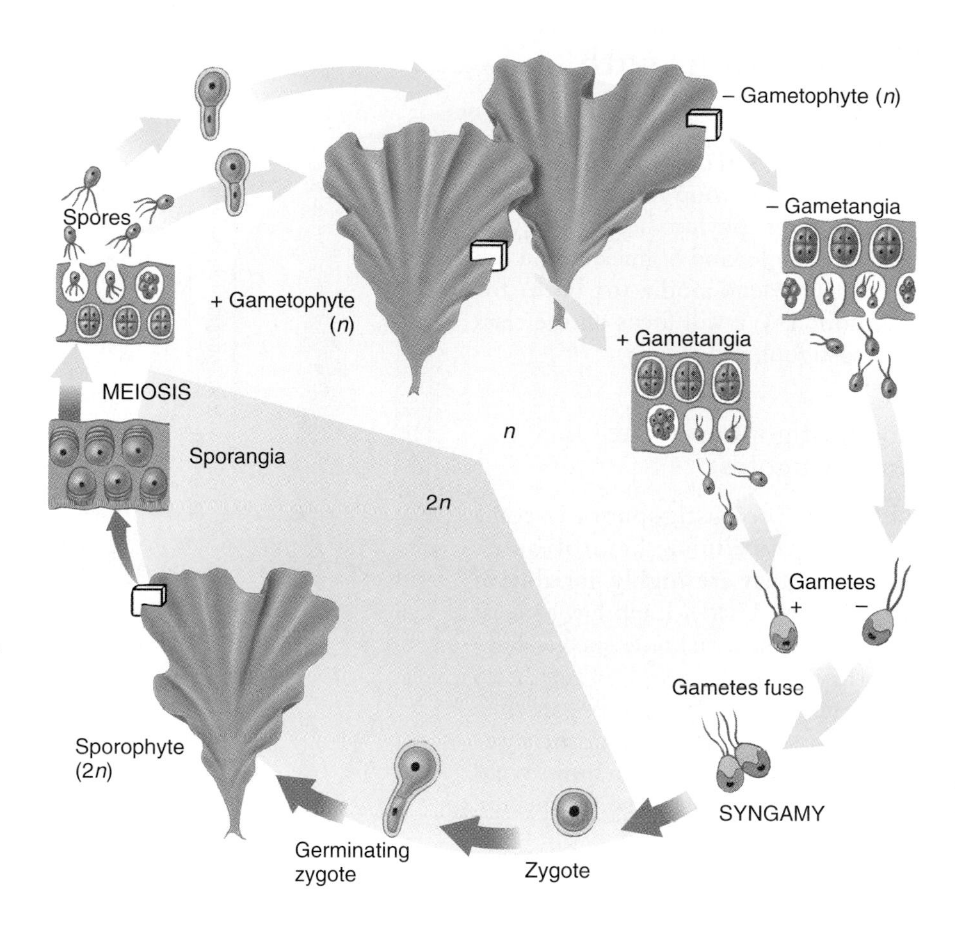

FIGURE 35.16
Life cycle of *Ulva*. In this green alga, the gametophyte and the sporophyte are identical in appearance and consist of flattened sheets two cells thick. In the haploid (n) gametophyte, gametangia give rise to haploid gametes, which fuse to form a diploid ($2n$) zygote. The zygote germinates to form the diploid sporophyte. Sporangia within the sporophyte give rise to haploid spores by meiosis. The haploid spores develop into haploid gametophytes.

Volvox, there is a true division of labor among the different types of cells, which are specialized in relation to their ultimate function throughout the development of the organism.

In addition to these two lines of specialization from *Chlamydomonas*-like cells, there are many other kinds of green algae of less certain derivation. Many filamentous genera, such as *Spirogyra*, with its ribbon-like chloroplasts, differ substantially from the remainder of the green algae in their modes of cell division and reproduction. Some of these genera have even been placed in separate phyla. The study of the green algae, involving modern methods of electron microscopy and biochemistry, is beginning to reveal unexpected new relationships within this phylum.

Ulva, or sea lettuce (figure 35.16), is a genus of marine green algae that is extremely widespread. The glistening individuals of this genus, often more than 10 centimeters across, consist of undulating sheets only two cells thick. Sea lettuce attaches by protuberances of the basal cells to rocks or other substrates. The reproductive cycle of *Ulva* involves an alternation of generations (sporic meiosis; figure 35.16) as is typical among green algae. Unlike most organisms that undergo sporic meiosis, however, the **gametophytes** (haploid phase) and **sporophytes** (diploid phase) resemble one another closely.

The stoneworts, a group of about 250 living species of green algae, many of them in the genera *Chara* and *Nitella*, have complex structures. Whorls of short branches arise regularly at their nodes, and the gametangia (structures that give rise to gametes) are complex and multicellular. Stoneworts are often abundant in fresh to brackish water and are common as fossils.

Dinoflagellates are primarily unicellular, photosynthetic, and flagellated. Euglenoids (phylum Euglenophyta) consist of about 40 genera, about a third of which have chloroplasts similar biochemically to those of green algae and plants. Diatoms and golden algae are unicellular, photosynthetic organisms that produce a unique carbohydrate. Diatoms have double shells made of opaline silica. Nonmotile, unicellular algae and multicellular, flagellated colonies have been derived from green algae like *Chlamydomonas*—a biflagellated, unicellular organism. The life cycle of brown algae is marked by an alternation of generations between the diploid phase, or sporophyte, and the haploid phase, or gametophyte.

Heterotrophs with Flagella

The phylum Sarcomastigophora contains a diverse group of protists combined into one phylum because they all possess a single kind of nucleus and use flagella or pseudopodia (or both) for locomotion. We will focus on the class Zoomastigophora.

Zoomastigophora: The Zoomastigotes

The class Zoomastigophora is composed of unicellular, heterotrophic organisms that are highly variable in form (figure 35.17). Each has at least one flagellum, with some species having thousands. They include both free-living and parasitic organisms. Many zoomastigotes apparently reproduce only asexually, but sexual reproduction occurs in some species. The members of one order, the kinetoplastids, include the genera *Trypanosoma* (figure 35.17*c*) and *Crithidia*, pathogens of humans and domestic animals. The euglenoids could be viewed as a specialized group of zoomastigotes, some of which acquired chloroplasts during the course of evolution.

Trypanosomes cause many serious human diseases, the most familiar of which is trypanosomiasis also known as African sleeping sickness (figure 35.18). Trypanosomes cause many other diseases including East Coast fever, leishmaniasis, and Chagas' disease, all of great importance in tropical areas where they afflict millions of people each year. Leishmaniasis, which is transmitted by sand flies, afflicts about 4 million people a year. The effects of these diseases range from extreme fatigue and lethargy in sleeping sickness to skin sores and deep eroding lesions that can almost obliterate the face in leishmaniasis. The trypanosomes that cause these diseases are spread by biting insects, including tsetse flies and assassin bugs.

A serious effort is now under way to produce a vaccine for trypanosome-caused diseases. These diseases make it impossible to raise domestic cattle for meat or milk in a large portion of Africa. Control is especially difficult because of the unique attributes of these organisms. For example, tsetse fly-transmitted trypanosomes have evolved an elaborate genetic mechanism for repeatedly changing the antigenic nature of their protective glycoprotein coat, thus dodging the antibodies their hosts produce against them (see chapter 57). Only a single one out of some 1000 to 2000 variable antigen genes is expressed at a time. Rearrangements of these genes during the asexual cycle of the organism allow for the expression of a seemingly endless variety of different antigen genes that maintain infectivity by the trypanosomes.

When the trypanosomes are ingested by a tsetse fly, they embark on a complicated cycle of development and multiplication, first in the fly's gut and later in its salivary glands. It is their position in the salivary glands that allows them to move into their vertebrate host. Recombination has been observed between different strains of trypanosomes introduced into a single fly, thus suggesting that mating, syngamy, and meiosis occur, even though they have not been observed directly. Although most trypanosome reproduction is asexual, this sexual cycle, reported for the first time in 1986, affords still further possibilities for recombination in these organisms.

In the guts of the flies that spread them, trypanosomes are noninfective. When they are ready to transfer to the skin or bloodstream of their host, trypanosomes migrate to the salivary glands and acquire the thick coat of glycoprotein antigens that protect them from the host's antibodies. When they are taken up by a fly, the trypanosomes again shed their coats. The production of vaccines against such a system is complex, but tests are underway. Releasing sterilized flies to impede the reproduction of populations is another technique used to try to control the fly population. Traps made of dark cloth and scented like cows, but poisoned with insecticides, have like-

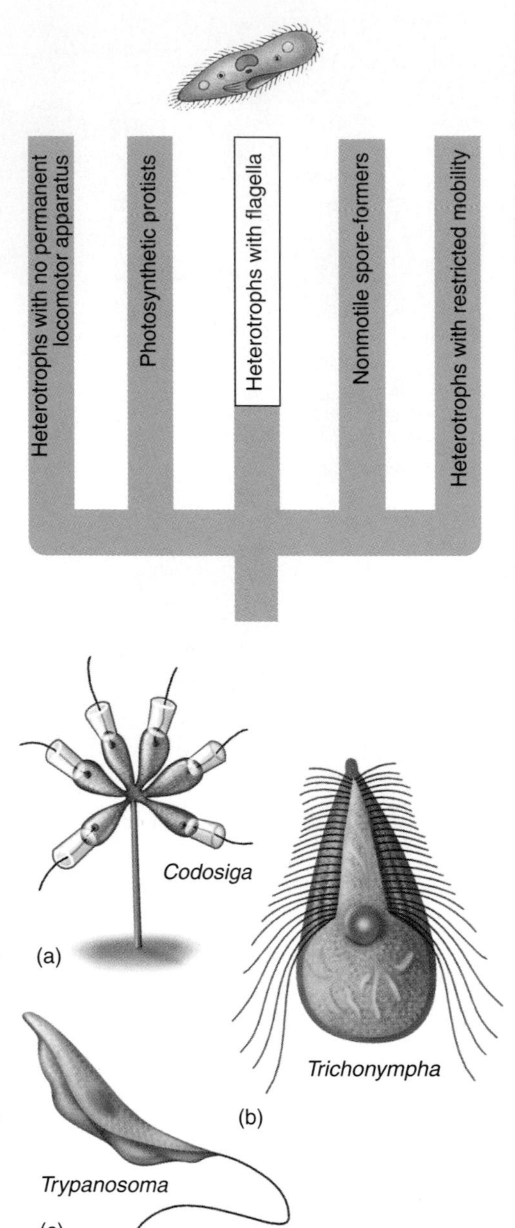

FIGURE 35.17
Three genera of zoomastigotes (Zoomastigophora), a highly diverse group. (*a*) *Codosiga*, a colonial choanoflagellate that remains attached to its substrate; other colonial choanoflagellates swim around as a colony, resembling the green alga *Volvox* in this respect. (*b*) *Trichonympha*, one of the zoomastigotes that inhabits the guts of termites and wood-feeding cockroaches and digests cellulose there. *Trichonympha* has rows of flagella in its anterior regions. (*c*) *Trypanosoma*, which causes sleeping sickness, an important tropical disease. It has a single, anterior flagellum.

wise proved effective. Research is proceeding rapidly because the presence of tsetse flies with their associated trypanosomes blocks the use of some 11 million square kilometers of potential grazing land in Africa.

Some zoomastigotes occur in the guts of termites and other wood-eating insects. They possess enzymes that allow them to digest the wood and thus make the components of the wood available to their hosts. The relationship is similar to that between certain bacteria and protozoa that function in the rumens of cattle and related mammals (see chapter 51).

Another order of zoomastigotes, the choanoflagellates, is most likely the group from which the sponges (phylum Porifera) and probably all animals arose. Choanoflagellates have a single emergent flagellum surrounded by a funnel-shaped, contractile collar composed of closely placed filaments, a unique structure that is exactly matched in the sponges. These protists feed on bacteria strained out of the water by the collar.

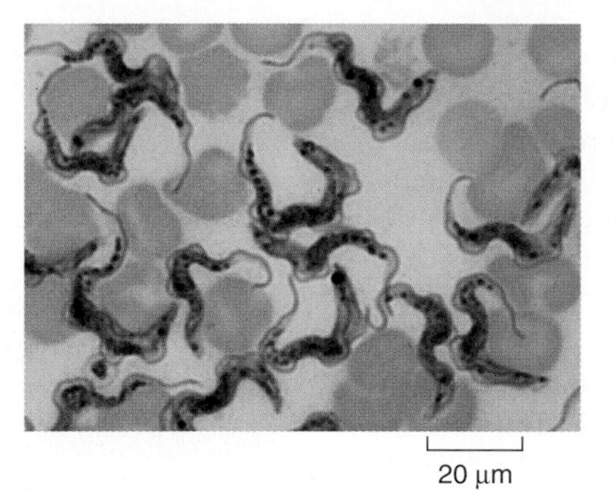

FIGURE 35.18
***Trypanosoma* is the zoomastigote that causes sleeping sickness.** (*a*) *Trypanosoma* among red blood cells. The nuclei (dark-staining bodies), anterior flagella, and undulating, changeable shape of the trypanosomes are visible in this photograph (500×). (*b*) The tsetse fly, shown here sucking blood from a human arm, can carry trypanosomes.

Hiker's Diarrhea. *Giardia lamblia* is a flagellate protist (belonging to a small order called diplomonads) found throughout the world, including all parts of the United States and Canada (figure 35.19). It occurs in water, including the clear water of mountain streams and the water supplies of some cities. It infects at least 40 species of wild and domesticated animals in addition to humans. In 1984 in Pittsburgh, 175,000 people had to boil their drinking water for several days following the appearance of *Giardia* in the city's water system. Although most individuals exhibit no symptoms if they drink water infested with *Giardia*, many suffer nausea, cramps, bloating, vomiting, and diarrhea. Only 35 years ago, *Giardia* was thought to be harmless; today, it is estimated that at least 16 million residents of the United States are infected by it.

Giardia lives in the upper small intestine of its host. It occurs there in a motile form that cannot survive outside the host's body. It is spread in the feces of infected individuals in the form of dormant, football-shaped cysts—sometimes at levels as high as 300 million individuals per gram of feces. These cysts can survive at least two months in cool water, such as that of mountain streams. They are relatively resistant to the usual water-treatment agents such as chlorine and iodine but are killed at temperatures greater than about 65°C. Apparently, pollution by humans seems to be the main way *Giardia* is released into stream water. There are at least three species of *Giardia* and many distinct strains; how many of them attack humans and under what circumstances are not known with certainty.

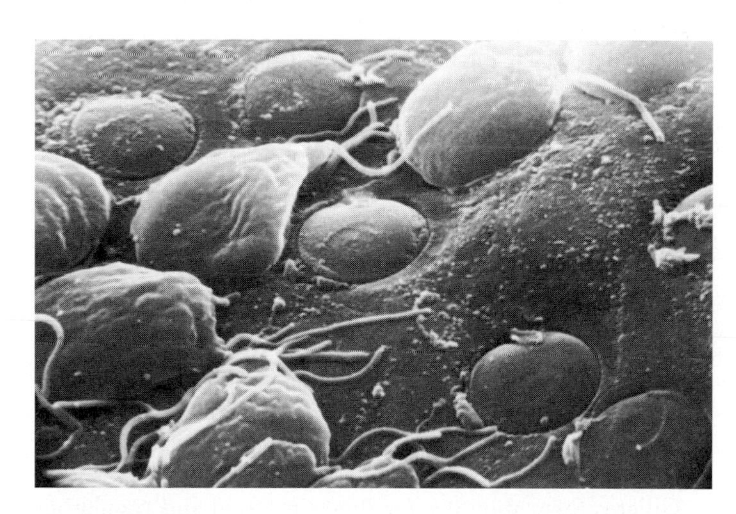

FIGURE 35.19
Giardia lamblia. *Giardia* are flagellated unicellular parasites that infect the human intestine. *Giardia* are very primitive, having only a rudimentary cytoskeleton and lacking mitochondria and chloroplasts. Sequencing of ribosomal RNA suggests that *Giardia* and *Pelomyxa*, the eukaryotes most closely related to prokaryotes, should be grouped together. The name Archezoa (Greek *arkhaios*, "ancient") has been suggested for the group, stressing its early divergence from bacteria as long as 2 billion years ago.

In the wilderness, good sanitation is important in preventing the spread of *Giardia*. Dogs, which readily contract and spread the disease, should not be taken into pristine wilderness areas. Drinking water should be filtered—the filter must be capable of eliminating particles as small as 1 micrometer in diameter—or boiled for at least one minute. Water from natural streams or lakes should never be consumed directly, regardless of how clean it looks. In other regions, good sanitation methods are important to prevent not only *Giardia* infection but also other diseases.

Ciliophora: The Ciliates

As the name indicates, most members of the Ciliophora feature large numbers of cilia. These heterotrophic, unicellular protists range in size from 10 to 3000 micrometers long. About 8000 species have been named. Despite their unicellularity, ciliates are extremely complex organisms, inspiring some biologists to consider them organisms without cell boundaries rather than single cells.

Their most characteristic feature, cilia, are usually arranged either in longitudinal rows or in spirals around the body of the organism (figure 35.20). Cilia are anchored to microtubules beneath the cell membrane, and they beat in a coordinated fashion. In some groups, the cilia have specialized locomotory and feeding functions, becoming fused into sheets, spikes, and rods which may then function as mouths, paddles, teeth, or feet. The ciliates have a tough but flexible outer covering called the pellicle that enables the organism to squeeze through or move around many kinds of obstacles.

All ciliates that have been studied have two very different types of nuclei within their cells, small **micronuclei** and larger **macronuclei** (figure 35.21). The micronuclei, which contain apparently normal diploid chromosomes, divide by meiosis and are able to undergo genetic recombination. Macronuclei are derived from certain micronuclei in a complex series of steps. Within the macronuclei are multiple copies of the genome, and the DNA is divided into small pieces—smaller than individual chromosomes. In one group of ciliates, these are equivalent to single genes. Macronuclei divide by elongating and constricting and play an essential role in routine cellular functions, such as the production of mRNA to direct protein synthesis for growth and regeneration.

Ciliates form vacuoles for ingesting food and regulating their water balance. Food first enters the gullet, which in the well-known ciliate *Paramecium* is lined with cilia fused into a membrane (figure 35.21). From the gullet, the food passes into food vacuoles, where enzymes and hydrochloric acid aid in its digestion. After the digested material has been completely absorbed, the vacuole empties its waste contents through a special pore in the pellicle known as the **cytoproct.** The cytoproct is essentially an exocytotic vesicle that appears periodically when solid particles are ready to be expelled. The contractile vacuoles, which function in the regulation of water balance, periodically expand and contract as they empty their contents to the outside of the organism.

Ciliates usually reproduce by transverse fission of the parent cell across its short axis, thus forming two identical individuals (figure 35.22*a*). In this process of cell division, the mitosis of the micronuclei proceeds normally, and the macronuclei divide as just described.

In *Paramecium*, the cells divide asexually for about 700 generations and then die if sexual reproduction has not occurred. Like most ciliates, *Paramecium* has a sexual process called **conjugation,** in which two individual cells remain attached to each other for up to several hours (figure 35.22*b,c*). Only cells of two different genetically determined mating types, *odd* and *even*, are able to conjugate. Meiosis in the micronuclei of each individual produces several haploid micronuclei, and the two partners exchange a pair of these micronuclei through a cytoplasmic bridge that appears between the two partners.

In each conjugating individual, the new micronucleus fuses with one of the micronuclei already present in that individual, resulting in the production of a new diploid micronucleus in each individual. After conjugation, the macronucleus in each cell disintegrates, while the new diploid micronucleus undergoes mitosis, thus giving rise to two new identical diploid micronuclei within each individual. One of these micronuclei becomes the precursor of the future micronuclei of that cell, while the other micronucleus undergoes multiple rounds of DNA replication, becoming the new macronucleus. This kind of complete segregation of the genetic material is a unique feature of the

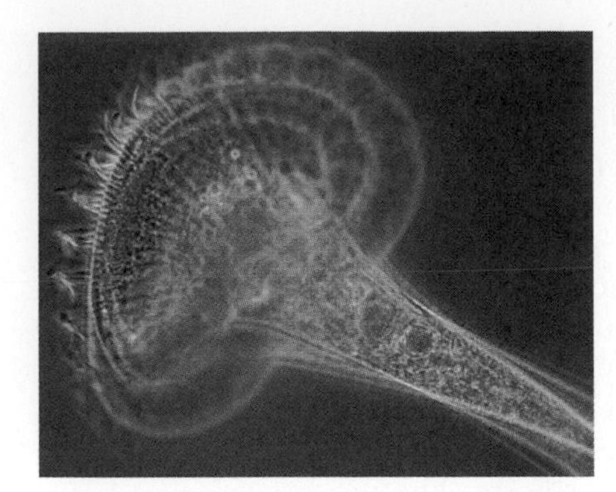

FIGURE 35.20
A ciliate (Ciliophora). *Stentor*, a funnel-shaped ciliate, showing spirally arranged cilia (120×).

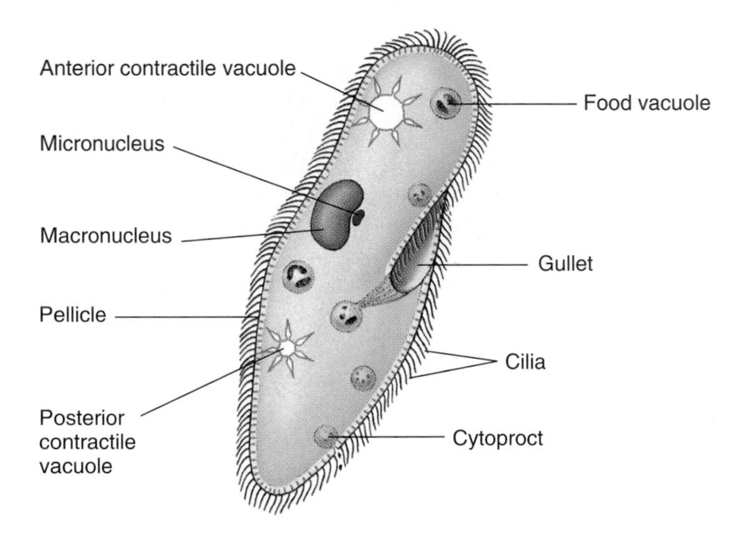

FIGURE 35.21
Paramecium. The main features of this familiar ciliate are shown.

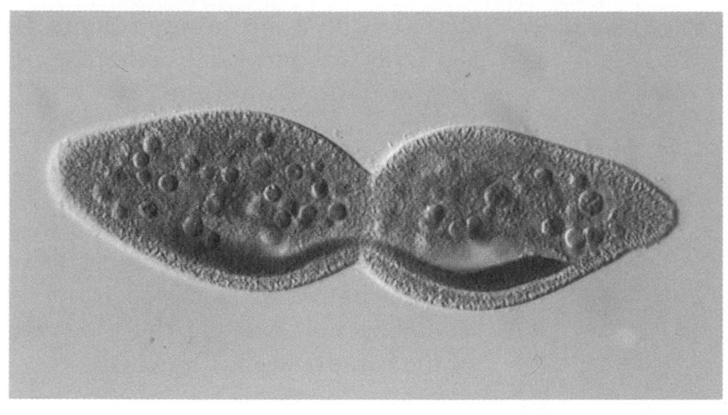

(a)

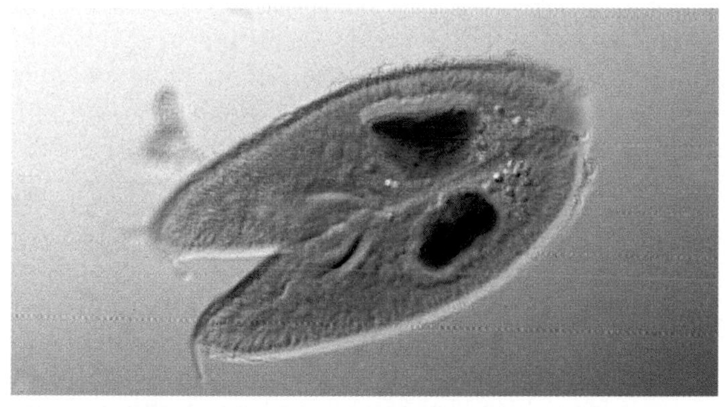

(b)

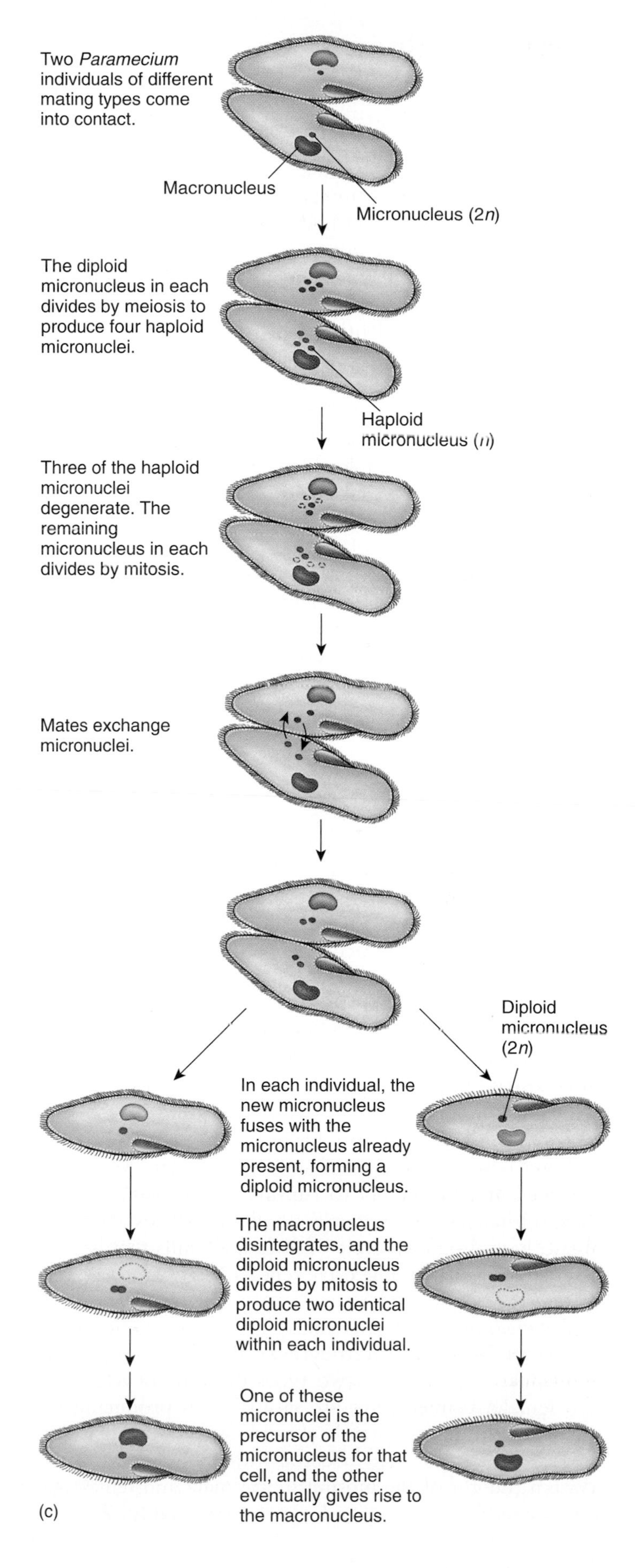

Two *Paramecium* individuals of different mating types come into contact.
Macronucleus
Micronucleus (2*n*)
The diploid micronucleus in each divides by meiosis to produce four haploid micronuclei.
Haploid micronucleus (*n*)
Three of the haploid micronuclei degenerate. The remaining micronucleus in each divides by mitosis.
Mates exchange micronuclei.
Diploid micronucleus (2*n*)
In each individual, the new micronucleus fuses with the micronucleus already present, forming a diploid micronucleus.
The macronucleus disintegrates, and the diploid micronucleus divides by mitosis to produce two identical diploid micronuclei within each individual.
One of these micronuclei is the precursor of the micronucleus for that cell, and the other eventually gives rise to the macronucleus.
(c)

FIGURE 35.22
Life cycle of *Paramecium*. (*a*) When *Paramecium* reproduces asexually, a mature individual divides, and two complete individuals result. (*b*, *c*) In sexual reproduction, two mature cells fuse in a process called conjugation (100×).

ciliates and makes them ideal organisms for the study of certain aspects of genetics.

Progeny from a sexual division in *Paramecium* must go through about 50 asexual divisions before they are able to conjugate. When they do so, their biological clocks are restarted, and they can conjugate again. After about 600 asexual divisions, however, *Paramecium* loses the protein molecules around the gullet that enable it to recognize an appropriate mating partner. As a result, the individuals are unable to mate, and death follows about 100 generations later. The exact mechanisms producing these unusual events are unknown, but they involve the accumulation of a protein, which is now being studied.

The zoomastigotes are a highly diverse group of flagellated unicellular heterotrophs, containing among their members the ancestors of animals as well as the very primitive *Giardia*. Ciliates possess characteristic cilia, and have two types of nuclei. The macronuclei contain multiple copies of certain genes, while the micronuclei contain multigene chromosomes.

Nonmotile Spore-Formers

Apicomplexa: The Sporozoans

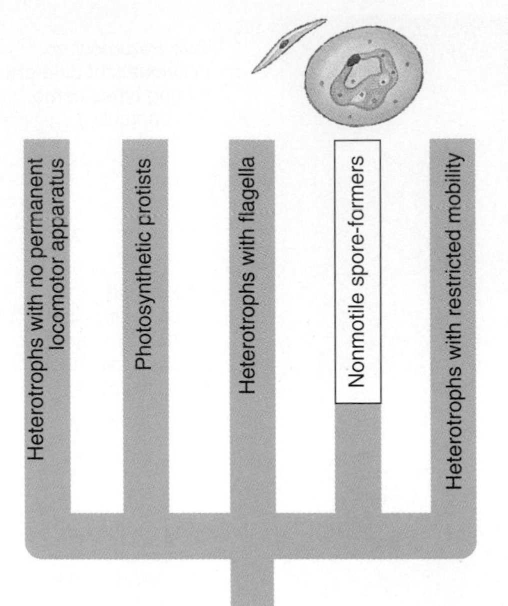

All sporozoans are nonmotile, spore-forming parasites of animals. Their spores are small, infective bodies that are transmitted from host to host. These organisms are distinguished by a unique arrangement of fibrils, microtubules, vacuoles, and other cell organelles at one end of the cell. There are 3900 described species of this phylum; best known among them is the malarial parasite, *Plasmodium.*

Sporozoans have complex life cycles that involve both asexual and sexual phases. Sexual reproduction involves an alternation of haploid and diploid generations. Both haploid and diploid individuals can also divide rapidly by mitosis, thus producing a large number of small infective individuals. Sexual reproduction involves the fertilization of a large female gamete by a small, flagellated male gamete. The zygote that results soon becomes an **oocyst.** Within the oocyst, meiotic divisions produce infective haploid spores called **sporozoites.**

An alternation between different hosts often occurs in the life cycles of sporozoans. Sporozoans of the genus *Plasmodium* are spread from person to person by mosquitoes of the genus *Anopheles* (figure 35.23); at least 65 different species of this genus are involved. When an *Anopheles* mosquito penetrates human skin to obtain blood, it injects saliva mixed with an anticoagulant. If the mosquito is infected with *Plasmodium,* it will also inject the elongated sporozoites into the bloodstream of its victim. The parasite makes its way through the bloodstream to the liver, where it rapidly divides asexually. After this division phase, **merozoites,** the next stage of the life cycle, form, either reinvading other liver cells or entering the host's bloodstream. In the bloodstream, they invade the red blood cells, dividing rapidly within them and causing them to become enlarged and ultimately to rupture. This event releases toxic substances throughout the body of the host, bringing about the well-known cycle of fever and chills that is characteristic of malaria. The cycle repeats itself regularly every 48 hours, 72 hours, or longer.

Plasmodium enters a sexual phase when some merozoites develop into **gametocytes,** cells capable of producing gametes. There are two types of gametocytes: male and female. Gametocytes are incapable of producing gametes within their human hosts and do so only when they are extracted from an infected human by a mosquito. Within the gut of the mosquito, the male and female gametocytes form sperm and eggs, respectively. Zygotes develop within the mosquito's intestinal walls and ultimately differentiate into oocysts. Within the oocysts, repeated mitotic divisions take place, producing large numbers of sporozoites. These sporozoites migrate to the salivary glands of the mosquito, and from there they are injected by the mosquito into the bloodstream of a human, thus starting the life cycle of the parasite again.

Malaria. Malaria, caused by infections by the sporozoan *Plasmodium,* is one of the most serious diseases in the world. According to the World Health Organization, about 500 million people are affected by it at any one time, and approximately 2 million of them, mostly children, die each year. Malaria kills most children under five years old who contract it. In areas where malaria is prevalent, most survivors more than five or six years old do not become seriously ill again from malaria infections. The symptoms, familiar throughout the tropics, include severe chills, fever, and sweating, an enlarged and tender spleen, confusion, and great thirst. Ultimately, a victim of malaria may die of anemia, kidney failure, or brain damage. The disease may be brought under control by the person's immune system or by drugs. As discussed in chapter 20, some individuals are genetically resistant to malaria. Other persons develop immunity to it.

Efforts to eradicate malaria have focused on (1) the elimination of the mosquito vectors; (2) the development of drugs to poison the parasites once they have entered the human body; and (3) the development of vaccines. The widescale applications of DDT from the 1940s to the 1960s led to the elimination of the mosquito vectors in the United States, Italy, Greece, and certain areas of Latin America. For a time, the worldwide elimination of malaria appeared possible, but this hope was soon crushed by the development of DDT-resistant strains of malaria-carrying mosquitoes in many regions; no fewer than 64 resistant strains were identified in a 1980 survey. Even though the worldwide use of DDT, long banned in the United States, nearly doubled from its 1974 level to more than 30,000 metric tons in 1984, its effectiveness in controlling mosquitoes is dropping. Further, there are serious environmental concerns about the use of this long-lasting chemical anywhere in the world. In addition to the problems with resistant strains of mosquitoes, strains of *Plasmodium* have appeared that are resistant to the drugs that have historically been used to kill them.

As a result of these problems, the number of new cases of malaria per year roughly doubled from the mid-1970s to

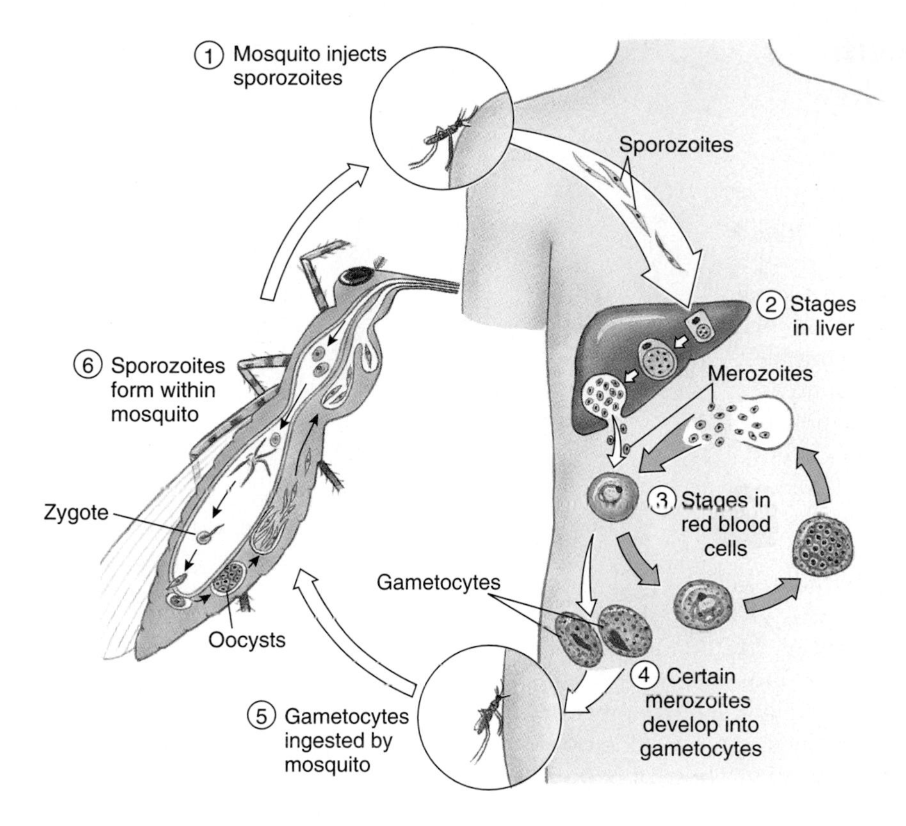

FIGURE 35.23
The life cycle of *Plasmodium*, the sporozoan that causes malaria. *Plasmodium* has a complex life cycle that alternates between mosquitoes and mammals.

the mid-1980s, largely because of the spread of resistant strains of the mosquito and the parasite. In many tropical regions, malaria is blocking permanent settlement. Scientists have therefore redoubled their efforts to produce an effective vaccine. Antibodies to the parasites have been isolated and produced by genetic engineering techniques, and they are starting to produce promising results.

Vaccines against Malaria. The three different stages of the *Plasmodium* life cycle each produce different antigens, and they are sensitive to different antibodies. The gene encoding the sporozoite antigen was cloned in 1984, but it is not certain how effective a vaccine against sporozoites might be. When a mosquito inserts its proboscis into a human blood vessel, it injects about a thousand sporozoites. They travel to the liver within a few minutes, where they are no longer exposed to antibodies circulating in the blood. If even one sporozoite reaches the liver, it will multiply rapidly there and cause malaria. The number of malaria parasites increases roughly eightfold every 24 hours after they enter the host's body. A compound vaccination against sporozoites, merozoites, and gametocytes would probably be the most effective preventive measure, but such a compound vaccine has proven difficult to develop.

However, research completed in 1997 brings a glimmer of hope. An experimental vaccine containing one of the surface proteins of the disease-causing parasite, *P. falciparum*, seems to induce the immune system to produce defenses that are able to destroy the parasite in future infections. In tests, six out of seven vaccinated people did not get malaria after being bitten by mosquitoes that carried *P. falciparum*. Although research is still underway, many are hopeful that this new vaccine may be able to fight malaria, especially in Africa, where it takes a devastating toll.

The best known of the sporozoans is the malarial parasite *Plasmodium*. Like other sporozoans, *Plasmodium* has a complex life cycle involving sexual and asexual phases and alternation between different hosts, in this case mosquitoes and humans. Malaria kills about 2 million people each year.

Heterotrophs with Restricted Mobility

Oomycota

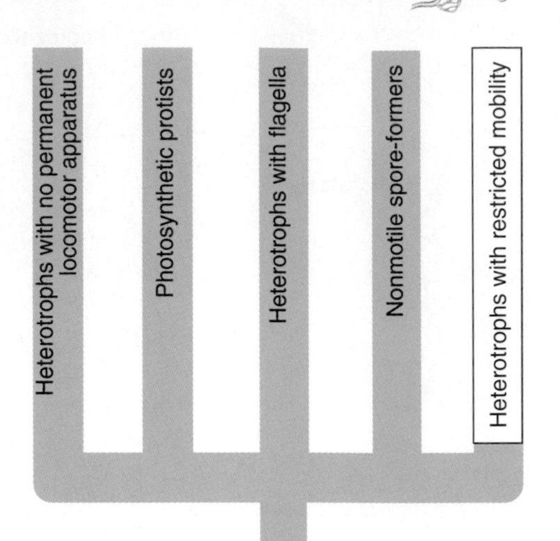

The oomycetes comprise about 580 species, among them the water molds, white rusts, and downy mildews. All of the members of this group are either parasites or **saprobes** (organisms that live by feeding on dead organic matter). The cell walls of the oomycetes are composed of cellulose or polymers that resemble cellulose. They differ remarkably from the chitin cell walls of fungi, with which the oomycetes have at times been grouped. Oomycete life cycles are characterized by gametic meiosis and a diploid phase; this also differs from fungi. Mitosis in the oomycetes resembles that in most other organisms, while mitosis in fungi has a number of unusual features, as you will see in chapter 36. Filamentous structures of fungi and, by convention, those of oomycetes, are called **hyphae.** Most oomycetes live in fresh or salt water or in soil, but some are plant parasites that depend on the wind to spread their spores. A few aquatic oomycetes are animal parasites.

Oomycetes are distinguished from other protists by the structure of their motile spores, or **zoospores,** which bear two unequal flagella, one of which is directed forward, the other backward. Such zoospores are produced asexually in a sporangium. Sexual reproduction in the group involves **gametangia** (singular, **gametangium**)— gamete-producing structures—of two different kinds. The female gametangium is called an **oogonium,** and the male gametangium is called an **antheridium.** The antheridia contain numerous male nuclei, which are the functional male gametes; the oogonia contain from one to eight eggs, which are the female gametes. When the contents of an antheridium flow into an oogonium, it leads to the individual fusion of male nuclei with eggs. This is followed by the thickening of the cell wall around the resulting zygote or zygotes. This produces a special kind of thick-walled cell called an **oospore,** the structure that gives the phylum its name. Details from the life cycle of one of the oomycetes, *Saprolegnia,* are shown in figure 35.24.

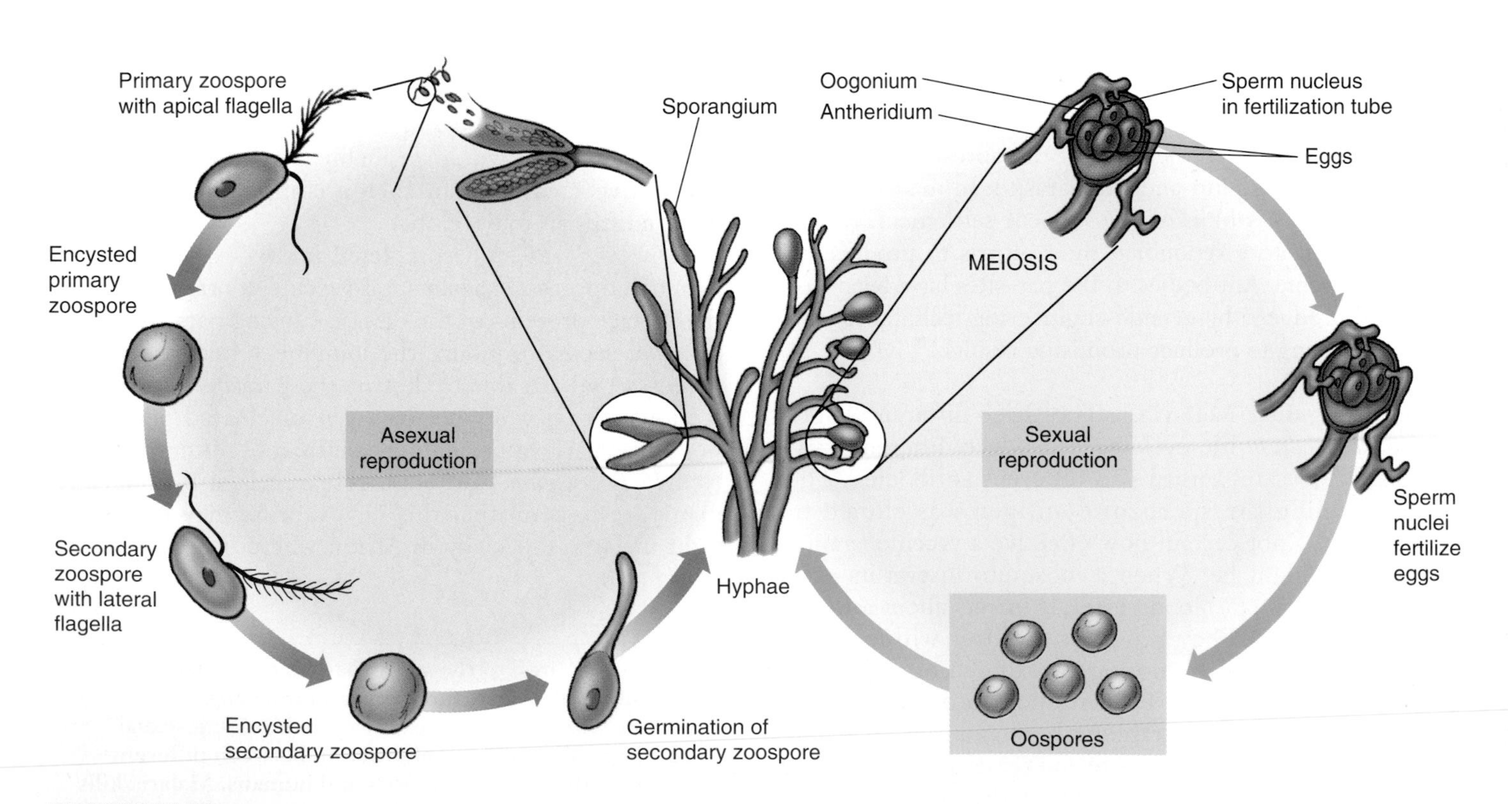

FIGURE 35.24
Life cycle of *Saprolegnia*, an oomycete. Asexual reproduction by means of flagellated zoospores is shown at left, sexual reproduction at right. Hyphae with diploid nuclei are produced by germination of both zoospores and oospores.

Aquatic oomycetes, or water molds, are common and easily cultured. Some water molds cause fish diseases, producing a kind of white fuzz on aquarium fishes. Among their terrestrial relatives are oomycetes of great importance as plant pathogens, including *Plasmopara viticola*, which causes downy mildew of grapes, and *Phytophthora infestans*, which causes the late blight of potatoes. This oomycete was responsible for the Irish potato famine of 1845 and 1847, during which about 400,000 people starved to death or died of diseases complicated by starvation. Millions of Irish people emigrated to the United States and elsewhere as a result of this disaster.

Acrasiomycota: The Cellular Slime Molds

There are about 70 species of cellular slime molds. This phylum has extraordinarily interesting features and was once thought to be related to fungi, "mold" being a general term for fungus-like organisms. In fact, the cellular slime molds are probably more closely related to amoebas (phylum Rhizopoda) than to any other group, but they have many special features that mark them as distinct. Cellular slime molds are common in fresh water, damp soil, and on rotting vegetation, especially fallen logs. They have become one of the most important groups of organisms for studies of differentiation because of their relatively simple developmental systems and the ease of analyzing them (figure 35.25).

The individual organisms of this group behave as separate amoebas, moving through the soil or other substrate and ingesting bacteria and other smaller organisms. At a certain phase of their life cycle, the individual organisms aggregate and form a moving mass, the "slug," that eventually transforms itself into a spore-containing mass, the **sorocarp.** In the sorocarp the amoebas become encysted as spores. Some of the amoebas fuse sexually to form **macrocysts,** which have diploid nuclei; meiosis occurs in them after a short period (zygotic meiosis). The sporocarp develops a stalked structure with a chamber at the top which releases the spores. Other amoebas are released directly, eventually aggregating again to form a new slug.

The development of *Dictyostelium discoideum*, a cellular slime mold, has been studied extensively because of the implication its unusual life cycle has for understanding the developmental process in general. When the individual amoebas of this species exhaust the supply of bacteria in a given area and are near starvation, they aggregate and form a compound, motile mass. The aggregation of the individual amoebas is induced by pulses of cyclic adenosine monophosphate (cAMP), which the cells begin to secrete when they are starving. The cells form an aggregate organism that moves to a new area where food is more plentiful. In the new area, the colony differentiates into a multicellular sorocarp within which spores differentiate. Each of these spores, if it falls into a suitably moist habitat, releases a new amoeba, which begins to feed, and the cycle is started again.

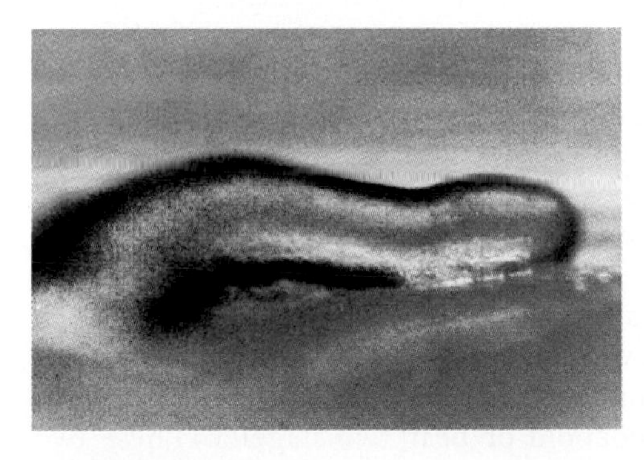

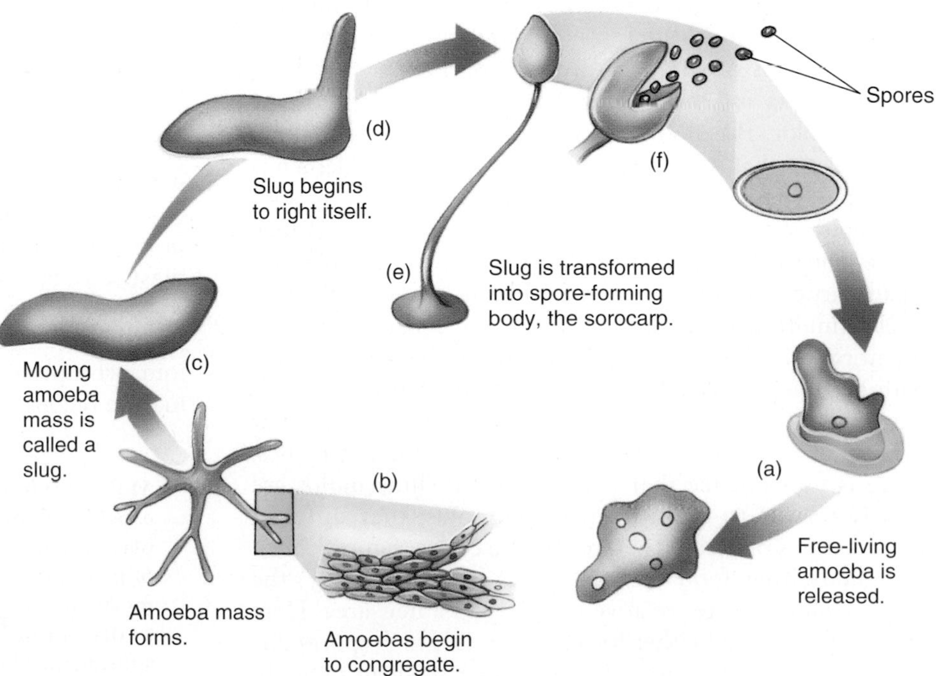
Spores
(d)
(f)
Slug begins
to right itself.
(e)
Slug is transformed
into spore-forming
body, the sorocarp.
(c)
Moving
amoeba
mass is
called a
slug.
(a)
(b)
Free-living
amoeba is
released.
Amoeba mass
forms.
Amoebas begin
to congregate.

FIGURE 35.25
Development in *Dictyostelium discoideum*, a cellular slime mold. (*a*) First, a spore germinates, forming amoebas. These amoebas feed and reproduce until the food runs out. (*b*) The amoebas aggregate and move toward a fixed center. (*c*) Next, they form a multicellular "slug" 2 to 3 mm long that migrates toward light. (*d*) The slug stops moving and begins to differentiate into a spore-forming body, called a sorocarp (*e*). (*f*) Within heads of the sorocarps, amoebas become encysted as spores.

FIGURE 35.26
A plasmodial protist. This multinucleate plasmodium moves about in search of the bacteria and other organic particles that it ingests.

Myxomycota: The Plasmodial Slime Molds

Plasmodial slime molds are a group of about 500 species. These bizarre organisms stream along as a **plasmodium,** a nonwalled, multinucleate mass of cytoplasm, that resembles a moving mass of slime (figure 35.26). This is called the feeding phase, and the plasmodia may be orange, yellow, or another color. Plasmodia show a back-and-forth streaming of cytoplasm that is very conspicuous, especially under a microscope. They are able to pass through the mesh in cloth or simply flow around or through other obstacles. As they move, they engulf and digest bacteria, yeasts, and other small particles of organic matter. Plasmodia contain many nuclei (multinucleate), but these are not separated by cell membranes. The nuclei undergo mitosis synchronously, with the nuclear envelope breaking down, but only at late anaphase or telophase. Centrioles are lacking in cellular slime molds. Although they have similar common names, there is no evidence that the plasmodial slime molds are closely related to the cellular slime molds; they differ in most features of their structure and life cycles (figure 35.27).

When either food or moisture is in short supply, the plasmodium migrates relatively rapidly to a new area. Here it stops moving and either forms a mass in which spores differentiate or divides into a large number of small mounds, each of which produces a single, mature **sporangium,** the structure in which spores are produced. These sporangia are often extremely complex in form and beautiful (figure 35.28). The spores can be either diploid or haploid. In most species of plasmodial slime molds with a diploid plasmodium, meiosis occurs in the spores within 24 hours of their formation. Three of the four nuclei in each spore disintegrate, leaving each spore with a single haploid nucleus.

The spores are highly resistant to unfavorable environmental influences and may last for years if kept dry. When conditions are favorable, they split open and release their protoplast, the contents of the individual spore. The protoplast may be amoeboid or bear two flagella. These two stages appear to be interchangeable, and conversions in either direction occur readily. Later, after the fusion of haploid protoplasts (gametes), a usually diploid plasmodium may be reconstituted by repeated mitotic divisions.

Molds are heterotrophic protists, many of which are capable of amoeba-like streaming. The feeding phase of plasmodial slime molds consists of a multinucleate mass of protoplasm; a plasmodium can flow through a cloth mesh and around obstacles. If the plasmodium begins to dry out or is starving, it forms often elaborate sporangia. Meiosis occurs in the spores once they have formed within the sporangium.

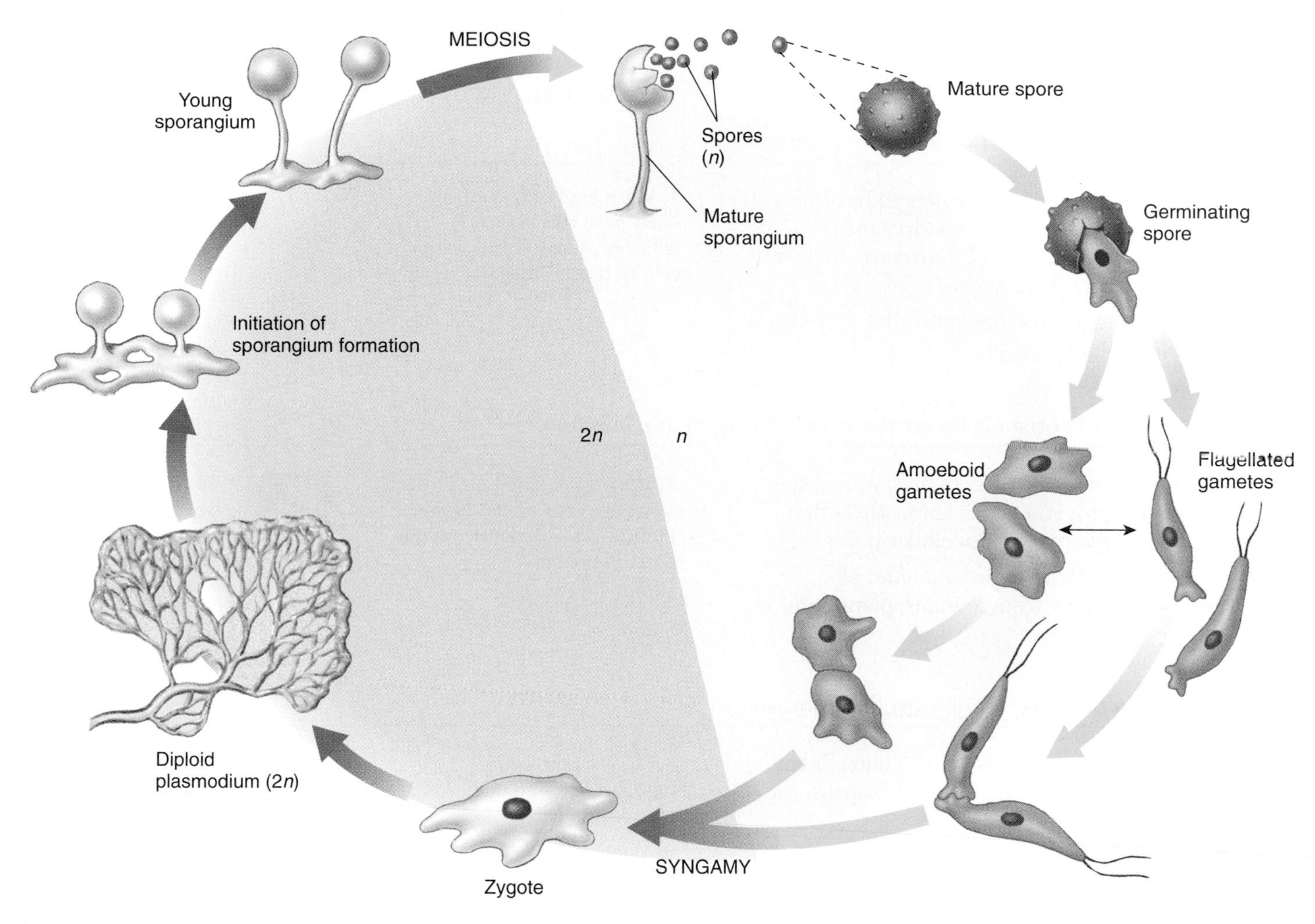

FIGURE 35.27
Life cycle of a plasmodial slime mold. When food or moisture is scarce, a diploid plasmodium stops moving and forms sporangia. Haploid spores form by meiosis. The spores wait until conditions are favorable to germinate. Spores can give rise to flagellated or amoeboid gametes; the two forms convert from one to the other readily. Fusion of the gametes forms the diploid zygote, which gives rise to the mobile, feeding plasmodium by mitosis.

FIGURE 35.28
Sporangia of three genera of plasmodial slime molds (phylum Myxomycota). (*a*) *Arcyria.* (*b*) *Fuligo.* (*c*) Developing sporangia of *Tubifera.*

Chapter 35

www.mhhe.com/raven6/resources35.mhtml

Summary | **Questions** | **Media Resources**

35.1 Eukaryotes probably arose by endosymbiosis.

- The theory of endosymbiosis, accepted by almost all biologists, proposes that mitochondria and chloroplasts were once aerobic eubacteria that were engulfed by ancestral eukaryotes.
- There is some suggestion that centrioles may also have an endosymbiotic origin.

1. What kind of bacteria most likely gave rise to the chloroplasts in the eukaryotic cells of plants and some algae?

35.2 The kingdom Protista is by far the most diverse of any kingdom.

- The kingdom Protista consists of predominantly unicellular phyla, together with three phyla that include large numbers of multicellular organisms.
- The catch-all kingdom Protista includes all eukaryotic organisms except animals, plants, and fungi.

2. Why is the kingdom Protista said to be an artificial group? How is this different from the other kingdoms?

- Characteristics of Protists

35.3 Protists can be categorized into five groups.

- The largest of the five groups are the unicellular, amoeboid protists that includes the rhizopads, the radiolarians, and the forams.
- Dinoflagellates (phylum Dinoflagellata) are a major phylum of primarily unicellular organisms that have unique chromosomes and a very unusual form of mitosis.
- Euglenoids (phylum Euglenophyta) have chloroplasts that are similar to those found in green algae and plants.
- Diatoms (phylum Chrysophyta) are unicellular, photosynthetic protists with opaline silica shells.
- Brown algae (phylum Phaeophyta) are multicellular, marine protists, some reaching 100 meters in length. The kelps contribute greatly to the productivity of the coastal sea areas.
- The zoomastigotes (phylum Sarcomastigophora) are a group of heterotrophic, mostly unicellular protists that includes Trypanosoma.
- There are about 8000 named species of ciliates (phylum Ciliophora); these protists have a very complex morphology with numerous cilia.
- The malarial parasite, *Plasmodium*, is a member of the phylum Apicomplexa. Carried by mosquitoes, it multiplies rapidly in the liver of humans and other primates and brings about the cyclical fevers characteristic of malaria by releasing toxins into the bloodstream of its host.
- The groups Oomycota and Acrasiomycota are considered to be fungus-like protists.

3. Why is mitosis in dinoflagellates unique? What are zooxanthellae?

4. What determines whether a collection of individuals is truly multicellular?

5. What unique characteristic differentiates the members of Ciliophora from other protists? What is the function of two vacuoles exhibited by most members of Ciliophora?

6. Why has it been so difficult to produce a vaccine for trypanosome-caused diseases?

7. What differentiates the oomycetes from the kingdom Fungi, in which they were previously placed? What is the feeding strategy of this phylum? Why are these organisms generally considered harmful?

- Photosynthetic Protists
- Protozoa
- Fungus-like Protists

- Art Quiz: Life Cycle of Chlamydomonas

36

Fungi

FIGURE 36.1
Spores exploding from a pore in the surface of a puffball fungus. The fungi constitute a unique kingdom of heterotrophic organisms. Along with bacteria, they are important decomposers and disease-causing organisms.

Concept Outline

36.1 Fungi are unlike any other kind of organism.

A Fungus Is Not a Plant. Unlike any plant, all fungi are filamentous heterotrophs with cell walls made of chitin.
The Body of a Fungus. Cytoplasm flows from one cell to another within the filamentous body of a fungus.
How Fungi Reproduce. Fungi reproduce sexually when filaments of different fungi encounter one another and fuse.
How Fungi Obtain Nutrients. Fungi secrete digestive enzymes and then absorb the products of the digestion.
Ecology of Fungi. Fungi are among the most important decomposers in terrestrial ecosystems.

36.2 Fungi are classified by their reproductive structures.

The Three Phyla of Fungi. There are three phyla of fungi, distinguished by their reproductive structures.
Phylum Zygomycota. In zygomycetes, the fusion of hyphae leads directly to the formation of a zygote.
Phylum Ascomycota. In ascomycetes, hyphal fusion leads to stable dikaryons that grow into massive webs of hyphae that form zygotes within a characteristic saclike structure, the ascus. Yeasts are unicellular fungi, mostly ascomycetes, that play many important commercial and medical roles.
Phylum Basidiomycota. In basidiomycetes, dikaryons also form, but zygotes are produced within reproductive structures called basidia.
The Imperfect Fungi. Fungi that have not been observed to reproduce sexually cannot be classified into one of the three phyla.

36.3 Fungi form two key mutualistic symbiotic associations.

Lichens. A lichen is a mutualistic symbiotic association between a fungus and a photosynthetic alga or cyanobacterium.
Mycorrhizae. Mycorrhizae are mutualistic symbiotic associations between fungi and the roots of plants.

Of all the bewildering variety of organisms that live on earth, perhaps the most unusual, the most peculiarly different from ourselves, are the fungi (figure 36.1). Mushrooms and toadstools are fungi, multicellular creatures that grow so rapidly in size that they seem to appear overnight on our lawns. At first glance, a mushroom looks like a funny kind of plant growing up out of the soil. However, when you look more closely, fungi turn out to have nothing in common with plants except that they are multicellular and grow in the ground. As you will see, the more you examine fungi, the more unusual they are.

36.1 Fungi are unlike any other kind of organism.

A Fungus Is Not a Plant

The fungi are a distinct kingdom of organisms, comprising about 77,000 named species (figure 36.2). **Mycologists,** scientists who study fungi, believe there may be many more species in existence, as many as 1.2 million. Although fungi have traditionally been included in the plant kingdom, they lack chlorophyll and resemble plants only in their general appearance and lack of mobility. Significant differences between fungi and plants include the following:

1. **Fungi are heterotrophs.** Perhaps most obviously, a mushroom is not green. Virtually all plants are photosynthesizers, while no fungi have chlorophyll or carry out photosynthesis. Instead, fungi obtain their food by secreting digestive enzymes onto the substrate, and then absorbing the organic molecules that are released by the enzymes.
2. **Fungi have filamentous bodies.** Fungi are basically filamentous in their growth form (that is, their bodies consist of long slender filaments called hyphae), even though these hyphae may be packed together to form complex structures like the mushroom. Plants, in contrast, are made of several types of cells organized into tissues and organs.
3. **Fungi have unusual reproductive modes.** Some plants have motile sperm with flagella. No fungi do. Most fungi reproduce sexually with nuclear exchange rather than gametes.
4. **Fungi have cell walls made of chitin.** The cell walls of fungi are built of polysaccharides (chains of sugars) and chitin, the same tough material a crab shell is made of. The cell walls of plants are made of cellulose, also a strong building material.
5. **Fungi have nuclear mitosis.** Mitosis in fungi is different from that in plants or most other eukaryotes in one key respect: the nuclear envelope does not break down and re-form. Instead, mitosis takes place *within* the nucleus. A spindle apparatus forms there, dragging chromosomes to opposite poles of the *nucleus* (not the cell, as in most other eukaryotes).

You could build a much longer list, but already the take-home lesson is clear: fungi are not like plants at all! Their many unique features are strong evidence that fungi are not closely related to any other group of organisms. DNA studies confirm significant differences from other eukaryotes.

Fungi absorb their food after digesting it with secreted enzymes. This mode of nutrition, combined with a filamentous growth form, nuclear mitosis, and other traits, makes the members of this kingdom highly distinctive.

(a)

(c)

(b)

FIGURE 36.2
Representatives of the three phyla of fungi. (*a*) A cup fungus, *Cookeina tricholoma*, an ascomycete, from the rain forest of Costa Rica. (*b*) *Amanita muscaria*, the fly agaric, a toxic basidiomycete. In the cup fungi, the spore-producing structures line the cup; in basidiomycetes that form mushrooms, like *Amanita*, they line the gills beneath the cap of the mushroom. All visible structures of fleshy fungi, such as the ones shown here, arise from an extensive network of filamentous hyphae that penetrates and is interwoven with the substrate on which they grow. (*c*) *Pilobolus*, a zygomycete that grows on animal feces. Stalks about 10 millimeters long contain dark spore-bearing sacs.

The Body of a Fungus

Fungi exist mainly in the form of slender filaments, barely visible to the naked eye, which are called **hyphae** (singular, **hypha**). These hyphae are typically made up of long chains of cells joined end-to-end divided by cross-walls called **septa** (singular, **septum**). The septa rarely form a complete barrier, except when they separate the reproductive cells. Cytoplasm characteristically flows or streams freely throughout the hyphae, passing right through major pores in the septa (figure 36.3). Because of this streaming, proteins synthesized throughout the hyphae may be carried to their actively growing tips. As a result, fungal hyphae may grow very rapidly when food and water are abundant and the temperature is optimum.

A mass of connected hyphae is called a **mycelium** (plural, **mycelia**). This word and the term *mycologist* are both derived from the Greek word for fungus, *myketos.* The mycelium of a fungus (figure 36.4) constitutes a system that may, in the aggregate, be many meters long. This mycelium grows through and penetrates its substrate, resulting in a unique relationship between the fungus and its environment. All parts of such a fungus are metabolically active, continually interacting with the soil, wood, or other material in which the mycelium is growing.

In two of the three phyla of fungi, reproductive structures formed of interwoven hyphae, such as mushrooms, puffballs, and morels, are produced at certain stages of the life cycle. These structures expand rapidly because of rapid elongation of the hyphae. For this reason, mushrooms can appear suddenly on your lawn.

The cell walls of fungi are formed of polysaccharides and chitin, not cellulose like those of plants and many groups of protists. Chitin is the same material that makes up the major portion of the hard shells, or exoskeletons, of arthropods, a group of animals that includes insects and crustaceans (see chapter 46). The commonality of chitin is one of the traits that has led scientists to believe that fungi and animals share a common ancestor.

Mitosis in fungi differs from that in most other organisms. Because of the linked nature of the cells, the cell itself is not the relevant unit of reproduction; instead, the nucleus is. The nuclear envelope does not break down and re-form; instead, the spindle apparatus is formed *within* it. Centrioles are lacking in all fungi; instead, fungi regulate the formation of microtubules during mitosis with small, relatively amorphous structures called **spindle plaques.** This unique combination of features strongly suggests that fungi originated from some unknown group of single-celled eukaryotes with these characteristics.

Fungi exist primarily in the form of filamentous hyphae, typically with incomplete division into individual cells by septa. These and other unique features indicate that fungi are not closely related to any other group of organisms.

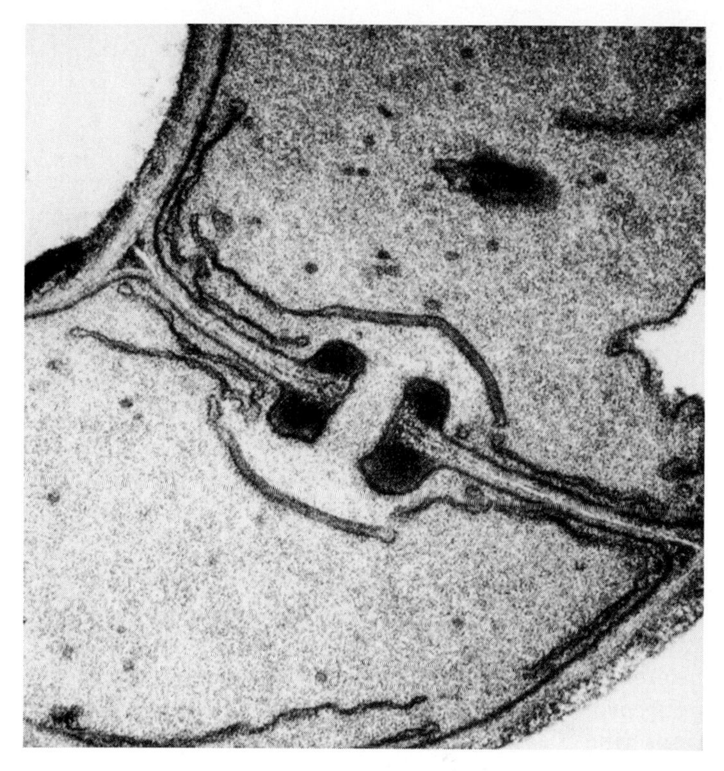

FIGURE 36.3
A septum (45,000×). This transmission electron micrograph of a section through a hypha of the basidiomycete *Inonotus tomentosus* shows a pore through which the cytoplasm streams.

FIGURE 36.4
Fungal mycelium. This mycelium, composed of hyphae, is growing through leaves on the forest floor in Maryland.

How Fungi Reproduce

Fungi are capable of both sexual and asexual reproduction. When a fungus reproduces sexually it forms a diploid zygote, as do animals and plants. Unlike animals and plants, all fungal nuclei except for the zygote are haploid, and there are many haploid nuclei in the common cytoplasm of a fungal mycelium. When fungi reproduce sexually, hyphae of two genetically different mating types come together and fuse. In two of the three phyla of fungi, the genetically different nuclei that are associated in a common cytoplasm after fusion do not combine immediately. Instead, the two types of nuclei coexist for most of the life of the fungus. A fungal hypha containing nuclei derived from two genetically distinct individuals is called a **heterokaryotic** hypha. If all of the nuclei are genetically similar to one another, the hypha is said to be **homokaryotic.** If there are two distinct nuclei within each compartment of the hyphae, they are **dikaryotic.** If each compartment has only a single nucleus, it is **monokaryotic.** Dikaryotic hyphae have some of the genetic properties of diploids, because both genomes are transcribed. These distinctions are important in understanding the life cycles of the individual groups.

Cytoplasm in fungal hyphae normally flows through perforated septa or moves freely in their absence. Reproductive structures are an important exception to this general pattern. When reproductive structures form, they are cut off by complete septa that lack perforations or have perforations that soon become blocked. Three kinds of reproductive structures occur in fungi: (1) **sporangia,** which are involved in the formation of spores; (2) **gametangia,** structures within which gametes form; and (3) **conidiophores,** structures that produce **conidia,** multinucleate asexual spores.

Spores are a common means of reproduction among fungi. They may form as a result of either asexual or sexual processes and are always nonmotile, being dispersed by wind. When spores land in a suitable place, they germinate, giving rise to a new fungal hypha. Because the spores are very small, they can remain suspended in the air for long periods of time. Because of this, fungal spores may be blown great distances from their place of origin, a factor in the extremely wide distributions of many kinds of fungi. Unfortunately, many of the fungi that cause diseases in plants and animals are spread rapidly and widely by such means. The spores of other fungi are routinely dispersed by insects and other small animals.

Fungi reproduce sexually after two hyphae of opposite mating type fuse. Asexual reproduction by spores is a second common means of reproduction.

FIGURE 36.5
A carnivorous fungus. The oyster mushroom, *Pleurotus ostreatus*, not only decomposes wood but also immobilizes nematodes, which the fungus uses as a source of nitrogen.

How Fungi Obtain Nutrients

All fungi obtain their food by secreting digestive enzymes into their surroundings and then absorbing back into the fungus the organic molecules produced by this **external digestion.** The significance of the fungal body plan reflects this approach, the extensive network of hyphae providing an enormous surface area for absorption. Many fungi are able to break down the cellulose in wood, cleaving the linkages between glucose subunits and then absorbing the glucose molecules as food. That is why fungi so often grow on dead trees.

It might surprise you to know that some fungi are predatory (figure 36.5). For example, the mycelium of the edible oyster fungus, *Pleurotus ostreatus*, excretes a substance that anesthetizes tiny roundworms known as nematodes (see chapter 44) that feed on the fungus. When the worms become sluggish and inactive, the fungal hyphae envelop and penetrate their bodies and absorb their nutritious contents. The fungus usually grows within living trees or on old stumps, obtaining the bulk of its glucose through the enzymatic digestion of cellulose from the wood, so that the nematodes it consumes apparently serve mainly as a source of nitrogen—a substance almost always in short supply in biological systems. Other fungi are even more active predators than *Pleurotus*, snaring, trapping, or firing projectiles into nematodes, rotifers, and other small animals on which they prey.

Fungi secrete digestive enzymes onto organic matter and then absorb the products of the digestion.

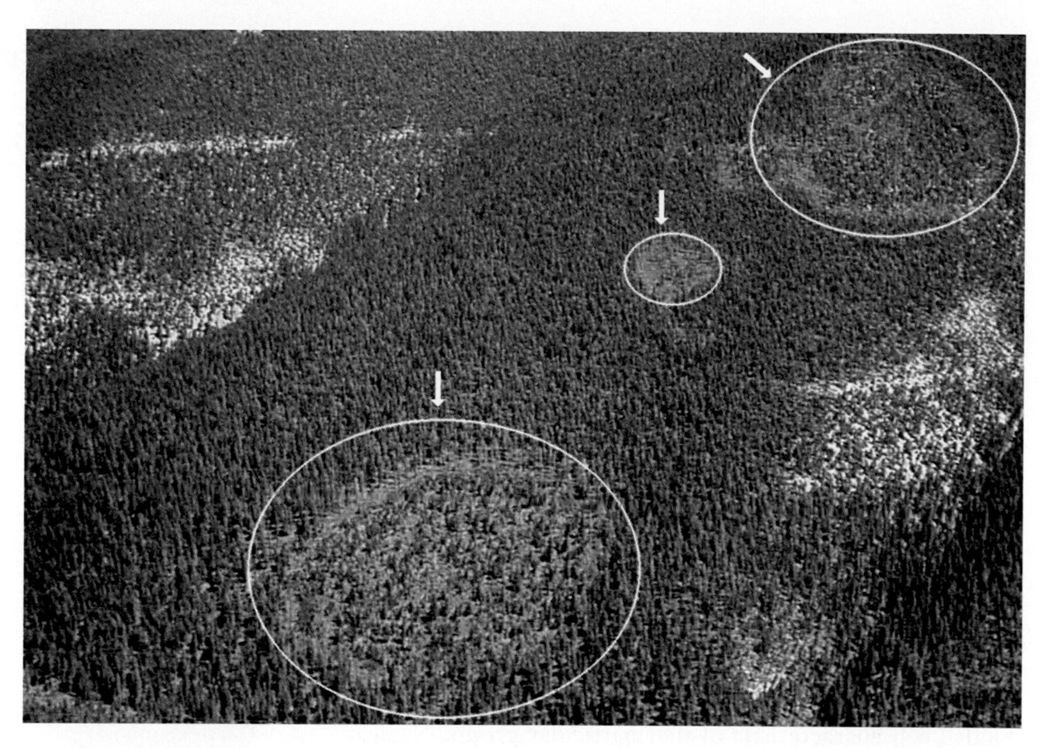

FIGURE 36.6
World's largest organism?
Armillaria, a pathogenic fungus shown here afflicting three discrete regions of coniferous forest in Montana, grows out from a central focus as a single circular clone. The large patch at the bottom of the picture is almost 8 hectares in diameter. The largest clone measured so far has been 15 hectares in diameter—pretty impressive for a single individual!

Ecology of Fungi

Fungi, together with bacteria, are the principal decomposers in the biosphere. They break down organic materials and return the substances locked in those molecules to circulation in the ecosystem. Fungi are virtually the only organisms capable of breaking down lignin, one of the major constituents of wood. By breaking down such substances, fungi release critical building blocks, such as carbon, nitrogen, and phosphorus, from the bodies of dead organisms and make them available to other organisms.

In breaking down organic matter, some fungi attack living plants and animals as a source of organic molecules, while others attack dead ones. Fungi often act as disease-causing organisms for both plants (figure 36.6) and animals, and they are responsible for billions of dollars in agricultural losses every year. Not only are fungi the most harmful pests of living plants, but they also attack food products once they have been harvested and stored. In addition, fungi often secrete substances into the foods that they are attacking that make these foods unpalatable, carcinogenic, or poisonous.

The same aggressive metabolism that makes fungi ecologically important has been put to commercial use in many ways. The manufacture of both bread and beer depends on the biochemical activities of **yeasts,** single-celled fungi that produce abundant quantities of ethanol and carbon dioxide. Cheese and wine achieve their delicate flavors because of the metabolic processes of certain fungi, and others make possible the manufacture of soy sauce and other fermented foods. Vast industries depend on the biochemical manufacture of organic substances such as citric acid by fungi in culture, and yeasts are now used on a large scale to produce protein for the enrichment of animal food. Many antibiotics, including the first one that was used on a wide scale, penicillin, are derived from fungi.

Some fungi are used to convert one complex organic molecule into another, cleaning up toxic substances in the environment. For example, at least three species of fungi have been isolated that combine selenium, accumulated at the San Luis National Wildlife Refuge in California's San Joaquin Valley, with harmless volatile chemicals—thus removing excess selenium from the soil.

Two kinds of mutualistic associations between fungi and autotrophic organisms are ecologically important. **Lichens** are mutualistic symbiotic associations between fungi and either green algae or cyanobacteria. They are prominent nearly everywhere in the world, especially in unusually harsh habitats such as bare rock. **Mycorrhizae,** specialized mutualistic symbiotic associations between the roots of plants and fungi, are characteristic of about 90% of all plants. In each of them, the photosynthetic organisms fix atmospheric carbon dioxide and thus make organic material available to the fungi. The metabolic activities of the fungi, in turn, enhance the overall ability of the symbiotic association to exist in a particular habitat. In the case of mycorrhizae, the fungal partner expedites the plant's absorption of essential nutrients such as phosphorus. Both of these associations will be discussed further in this chapter.

Fungi are key decomposers and symbionts within almost all terrestrial ecosystems and play many other important ecological and commercial roles.

36.2 Fungi are classified by their reproductive structures.

The Three Phyla of Fungi

There are three phyla but actually four groups of fungi: phylum Zygomycota, the **zygomycetes;** phylum Ascomycota, the **ascomycetes;** phylum Basidiomycota, the **basidiomycetes,** and the imperfect fungi (figure 36.7 and table 36.1). Several other groups that historically have been associated with fungi, such as the slime molds and water molds (phylum Oomycota; see chapter 35), now are considered to be protists, not fungi. Oomycetes are sharply distinct from fungi in their (1) motile spores; (2) cellulose-rich cell walls; (3) pattern of mitosis; and (4) diploid hyphae.

The three phyla of fungi are distinguished primarily by their sexual reproductive structures. In the zygomycetes, the fusion of hyphae leads directly to the formation of a zygote, which divides by meiosis when it germinates. In the other two phyla, an extensive growth of dikaryotic hyphae may lead to the formation of structures of interwoven hyphae within which are formed the distinctive kind of reproductive cell characteristic of that particular group. Nuclear fusion, followed by meiosis, occurs within these cells. The imperfect fungi are either asexual or the sexual reproductive structures have not been identified.

Sexual reproductive structures distinguish the three phyla of fungi.

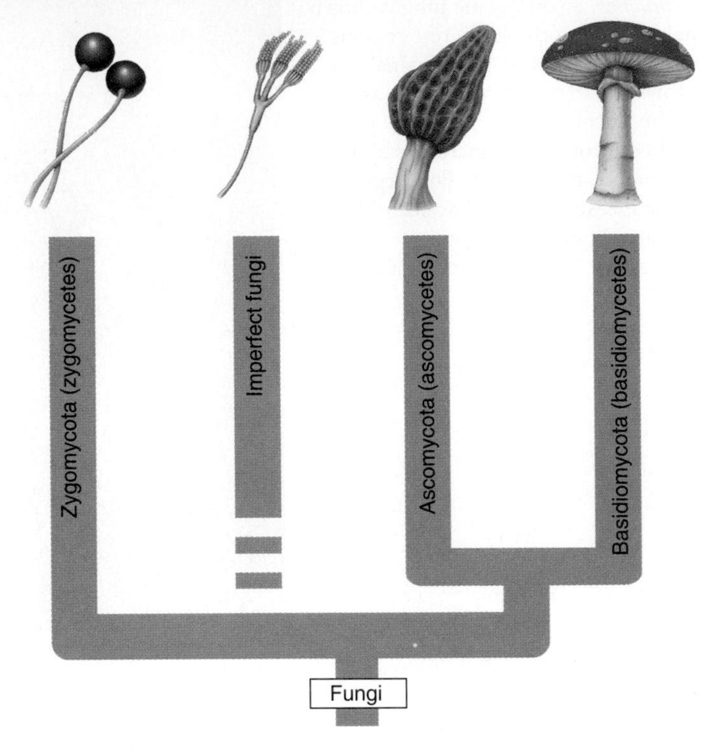

FIGURE 36.7
The four major groups of fungi. The imperfect fungi are not a true phylum, but rather a collection of fungi in which sexual structures have not been identified.

Table 36.1 Fungi

Phylum	Typical Examples	Key Characteristics	Approximate Number of Living Species
Ascomycota	Yeasts, truffles, morels	Develop by sexual means; ascospores are formed inside a sac called an ascus; asexual reproduction is also common	32,000
Imperfect fungi	*Aspergillus*, *Penicillium*	Sexual reproduction has not been observed; most are thought to be ascomycetes that have lost the ability to reproduce sexually	17,000
Basidiomycota	Mushrooms, toadstools, rusts	Develop by sexual means; basidiospores are borne on club-shaped structures called basidia; the terminal hyphal cell that produces spores is called a basidium; asexual reproduction occurs occasionally	22,000
Zygomycota	*Rhizopus* (black bread mold)	Develop sexually and asexually; multinucleate hyphae lack septa, except for reproductive structures; fusion of hyphae leads directly to formation of a zygote, in which meiosis occurs just before it germinates	1050

Phylum Zygomycota

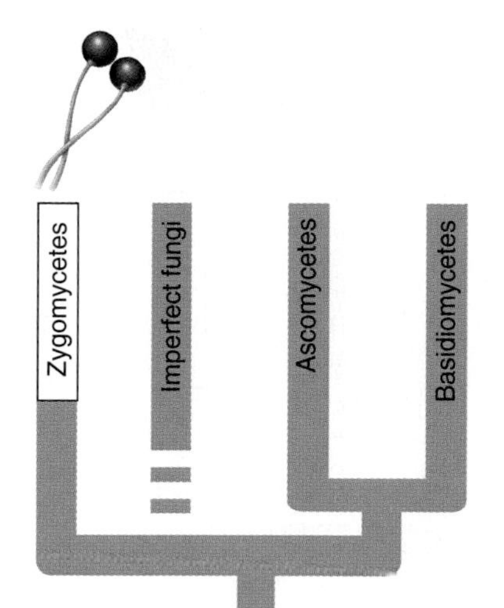

The zygomycetes (phylum Zygomycota) lack septa in their hyphae except when they form sporangia or gametangia. Zygomycetes are by far the smallest of the three phyla of fungi, with only about 1050 named species. Included among them are some of the more common bread molds (figure 36.8), as well as a variety of other microscopic fungi found on decaying organic material. The group is named after a characteristic feature of the life cycle of its members, the production of temporarily dormant structures called **zygosporangia.**

In the life cycle of the zygomycetes (figure 36.8*b*), sexual reproduction occurs by the fusion of gametangia, which contain numerous nuclei. The gametangia are cut off from the hyphae by complete septa. These gametangia may be formed on hyphae of different mating types or on a single hypha. If both + and – mating strains are present in a colony, they may grow together and their nuclei may fuse. Once the haploid nuclei have fused, forming diploid zygote nuclei, the area where the fusion has taken place develops into an often massive and elaborate zygosporangium. A zygosporangium may contain one or more diploid nuclei and acquires a thick coat. The zygosporangium helps the species survive conditions not favorable for growth. Meiosis occurs during the germination of the zygosporangium. Normal, haploid hyphae grow from the haploid spores that result from this process. Except for the zygote nuclei, all nuclei of the zygomycetes are haploid.

Asexual reproduction occurs much more frequently than sexual reproduction in the zygomycetes. During asexual reproduction, hyphae grow over the surface of the bread or other material on which the fungus feeds and produce clumps of erect stalks, called **sporangiophores.** The tips of the sporangiophores form **sporangia,** which are separated by septa. Thin-walled haploid spores are produced within the sporangia. Their spores are thus shed above the substrate, in a position where they may be picked up by the wind and dispersed to a new food source.

Zygomycetes form characteristic resting structures, called zygosporangia, which contain one or more zygotic nuclei. The hyphae of zygomycetes are multinucleate, with septa only where gametangia or sporangia are separated.

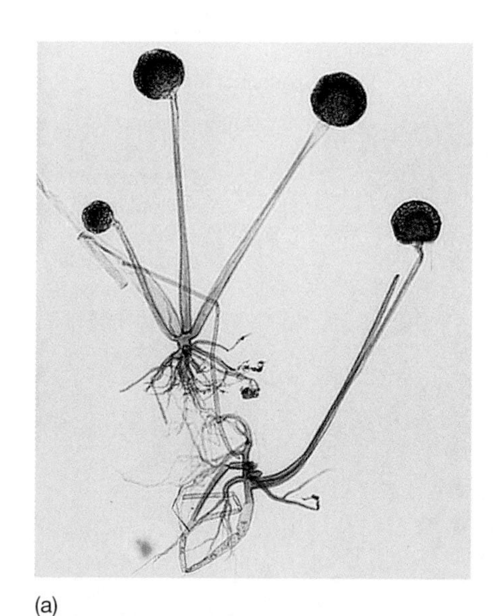

(a)

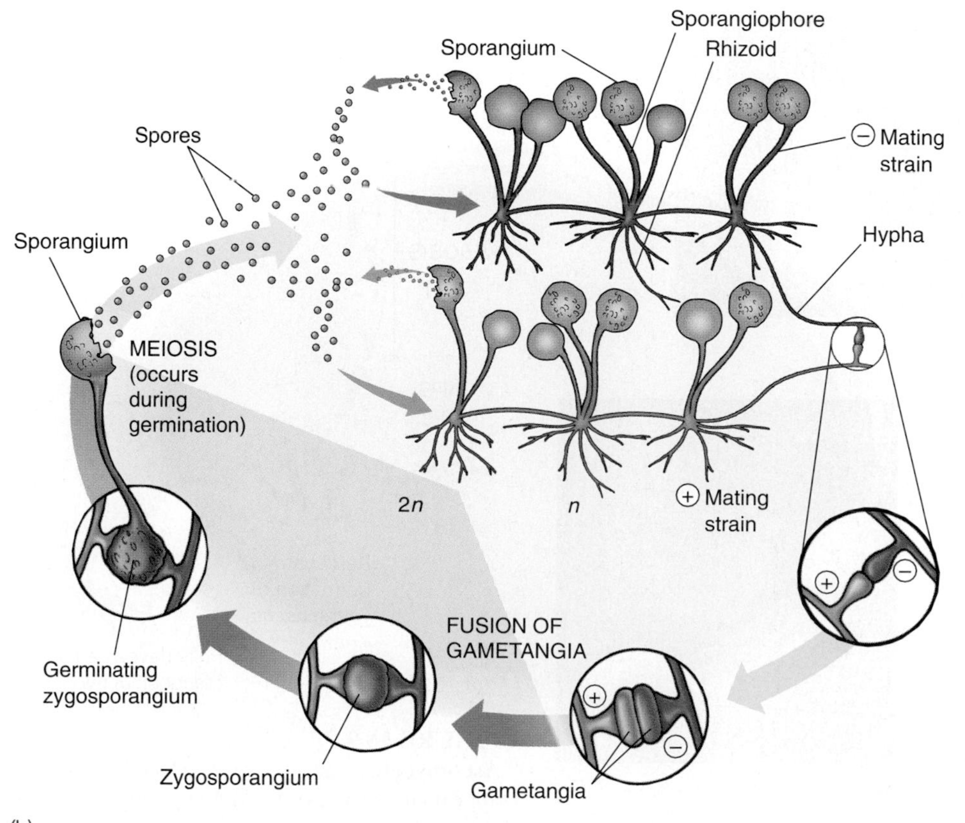

(b)

FIGURE 36.8
***Rhizopus*, a zygomycete that grows on moist bread and other similar substrates.** (*a*) The dark, spherical, spore-producing sporangia are on hyphae about a centimeter tall. The rootlike hyphae anchor the sporangia. (*b*) Life cycle of *Rhizopus*. This phylum is named for its characteristic zygosporangia.

Phylum Ascomycota

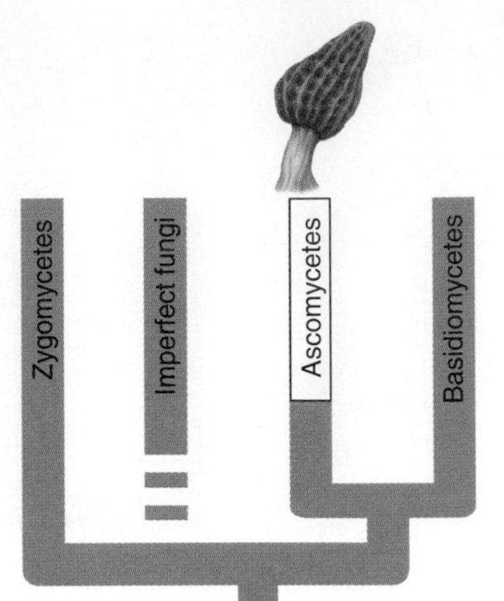

The second phylum of fungi, the ascomycetes (phylum Ascomycota), is a very large group of about 32,000 named species, with more being discovered each year. Among the ascomycetes are such familiar and economically important fungi as yeasts, common molds, morels (figure 36.9*a*,*b*), and truffles. Also included in this phylum are many serious plant pathogens, including the chestnut blight, *Cryphonectria parasitica*, and Dutch elm disease, *Ophiostoma ulmi*.

The ascomycetes are named for their characteristic reproductive structure, the microscopic, saclike **ascus** (plural, **asci**). The zygotic nucleus, which is the only diploid nucleus of the ascomycete life cycle (figure 36.9*c*), is formed within the ascus. The asci are differentiated within a structure made up of densely interwoven hyphae, corresponding to the visible portions of a morel or cup fungus, called the **ascocarp.**

Asexual reproduction is very common in the ascomycetes. It takes place by means of **conidia** (singular, **conidium**), spores cut off by septa at the ends of modified hyphae called **conidiophores.** Conidia allow for the rapid colonization of a new food source. Many conidia are multinucleate. The hyphae of ascomycetes are divided by septa, but the septa are perforated and the cytoplasm flows along the length of each hypha. The septa that cut off the asci and conidia are initially perforated, but later become blocked.

The cells of ascomycete hyphae may contain from several to many nuclei. The hyphae may be either homokaryotic or heterokaryotic. Female gametangia, called **ascogonia,** each have a beaklike outgrowth called a **trichogyne.** When the **antheridium,** or male gametangium, forms, it fuses with the trichogyne of an adjacent ascogonium. Initially, both kinds of gametangia contain a number of nuclei. Nuclei from the antheridium then migrate through the trichogyne into the ascogonium and pair with nuclei of the opposite mating type. Dikaryotic hyphae then arise from the area of the fusion. Throughout such hyphae,

(a)

(b)

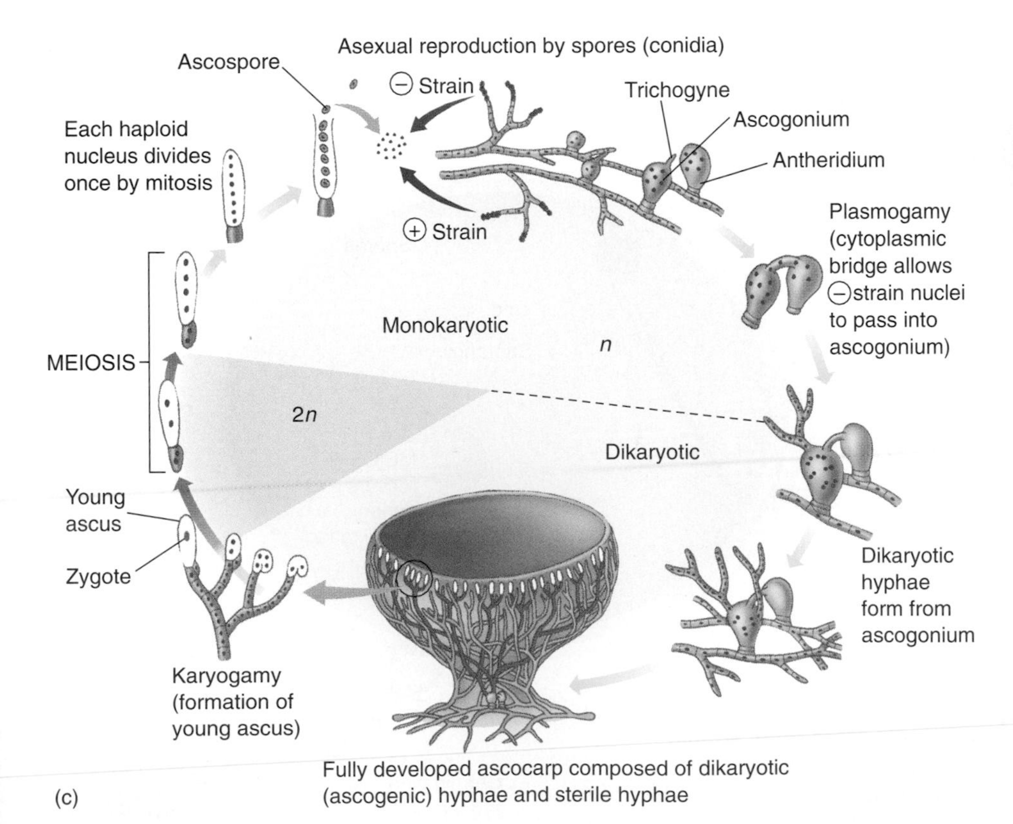

(c)

FIGURE 36.9
Ascomycetes. (*a*) This morel, *Morchella esculenta*, is a delicious edible ascomycete that appears in early spring. (*b*) A cup fungus. (*c*) Life cycle of an ascomycete. The zygote forms within the ascus.

nuclei that represent the two different original mating types occur. These hyphae are thus both dikaryotic and heterokaryotic.

Asci are formed at the tips of the dikaryotic hyphae and are separated by the formation of septa. There are two haploid nuclei within each ascus, one of each mating type represented in the dikaryotic hypha. Fusion of these two nuclei occurs within each ascus, forming a zygote. Each zygote divides immediately by meiosis, forming four haploid daughter nuclei. These usually divide again by mitosis, producing eight haploid nuclei that become walled **ascospores.** In many ascomycetes, the ascus becomes highly turgid at maturity and ultimately bursts, often at a preformed area. When this occurs, the ascospores may be thrown as far as 30 centimeters, an amazing distance considering that most ascospores are only about 10 micrometers long. This would be equivalent to throwing a baseball (diameter 7.5 centimeters) 1.25 kilometers—about 10 times the length of a home run!

Yeasts

Yeasts, which are unicellular, are one of the most interesting and economically important groups of microscopic fungi, usually ascomycetes. Most of their reproduction is asexual and takes place by cell fission or budding, when a smaller cell forms from a larger one (figure 36.10).

Sometimes two yeast cells will fuse, forming one cell containing two nuclei. This cell may then function as an ascus, with syngamy followed immediately by meiosis. The resulting ascospores function directly as new yeast cells.

Because they are single-celled, yeasts were once considered primitive fungi. However, it appears that they are actually reduced in structure and were originally derived from multicellular ancestors. The word *yeast* actually signifies only that these fungi are single-celled. Some yeasts have been derived from each of the three phyla of fungi, although ascomycetes are best represented. Even yeasts that were derived from ascomycetes are not necessarily directly related to one another, but instead seem to have been derived from different groups of ascomycetes.

Putting Yeasts to Work. The ability of yeasts to ferment carbohydrates, breaking down glucose to produce ethanol and carbon dioxide, is fundamental in the production of bread, beer, and wine. Many different strains of yeast have been domesticated and selected for these processes. Wild yeasts—ones that occur naturally in the areas where wine is made—were important in wine making historically, but domesticated yeasts are normally used now. The most important yeast in all these processes is *Saccharomyces cerevisiae.* This yeast has been used by humans throughout recorded history. Other yeasts are important pathogens and cause diseases such as thrush and

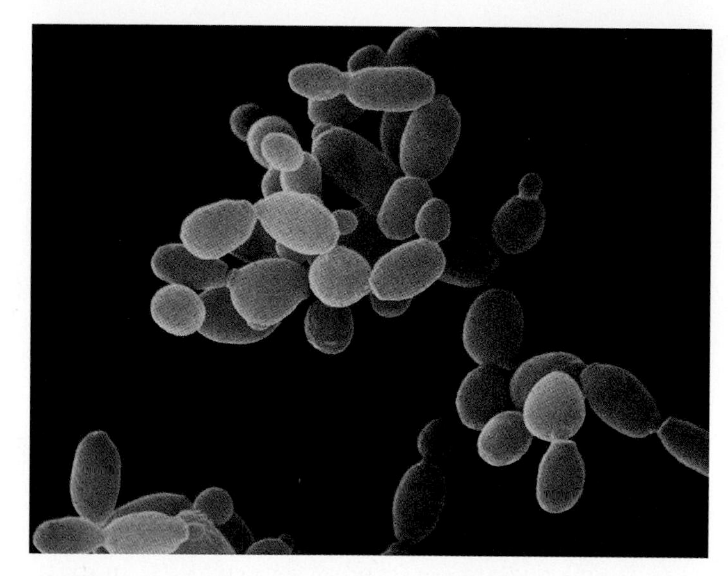

FIGURE 36.10
Scanning electron micrograph of a yeast, showing the characteristic cell division method of budding (19,000×). The cells tend to hang together in chains, a feature that calls to mind the derivation of single-celled yeasts from multicellular ancestors.

cryptococcosis; one of them, *Candida,* causes common oral or vaginal infections.

Over the past few decades, yeasts have become increasingly important in genetic research. They were the first eukaryotes to be manipulated extensively by the techniques of genetic engineering, and they still play the leading role as models for research in eukaryotic cells. In 1983, investigators synthesized a functional artificial chromosome in *Saccharomyces cerevisiae* by assembling the appropriate DNA molecule chemically; this has not yet been possible in any other eukaryote. In 1996, the genome sequence of *S. cerevisiae,* the first eukaryote to be sequenced entirely, was completed. With their rapid generation time and a rapidly increasing pool of genetic and biochemical information, the yeasts in general and *S. cerevisiae* in particular are becoming the eukaryotic cells of choice for many types of experiments in molecular and cellular biology. Yeasts have become, in this respect, comparable to *Escherichia coli* among the bacteria, and they are continuing to provide significant insights into the functioning of eukaryotic systems.

Ascomycetes form their zygotes within a characteristic saclike structure, the ascus. Meiosis follows, resulting in the production of ascospores. Yeasts are unicellular fungi, mainly ascomycetes, that have evolved from hypha-forming ancestors; not all yeasts are directly related to one another. Long useful for baking, brewing, and wine making, yeasts are now becoming very important in genetic research.

Phylum Basidiomycota

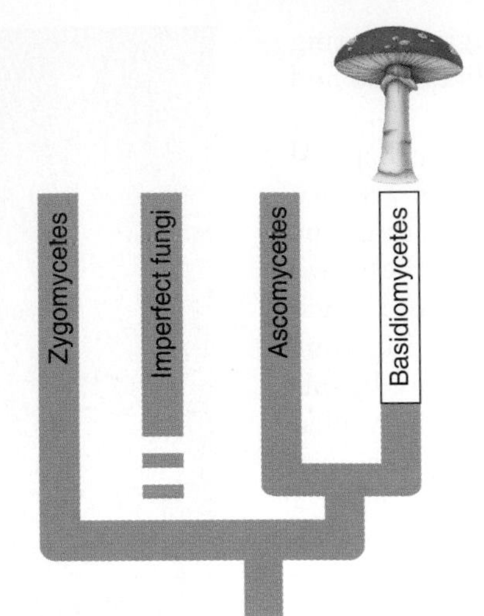

The third phylum of fungi, the basidiomycetes (phylum Basidiomycota), has about 22,000 named species. These are among the most familiar fungi. Among the basidiomycetes are not only the mushrooms, toadstools, puffballs, jelly fungi, and shelf fungi, but also many important plant pathogens including rusts and smuts (figure 36.11). Many mushrooms are used as food, but others are deadly poisonous.

Basidiomycetes are named for their characteristic sexual reproductive structure, the **basidium** (plural, **basidia**). A basidium is club-shaped. Karyogamy occurs within the basidium, giving rise to the zygote, the only diploid cell of the life cycle (figure 36.11*b*). As in all fungi, meiosis occurs immediately after the formation of the zygote. In the basidiomycetes, the four haploid products of meiosis are incorporated into **basidiospores.** In most members of this phylum, the basidiospores are borne at the end of the basidia on slender projections called **sterigmata** (singular, **sterigma**). Thus the structure of a basidium differs from that of an ascus, although functionally the two are identical. Recall that the ascospores of the ascomycetes are borne internally in asci.

The life cycle of a basidiomycete continues with the production of homokaryotic hyphae after spore germination. These hyphae lack septa at first. Eventually, septa form between the nuclei of the monokaryotic hyphae. A basidiomycete mycelium made up of monokaryotic hyphae is called a **primary mycelium.** Different mating types of monokaryotic hyphae may fuse, forming a dikaryotic or **secondary mycelium.** Such a mycelium is heterokaryotic, with two nuclei, representing the two different mating types, between each pair of septa. The maintenance of two genomes in the heterokaryon allows for more genetic plasticity than in a diploid cell with one nucleus. One genome may compensate for mutations in the other. The **basidiocarps,** or mushrooms, are formed entirely of secondary (dikaryotic) mycelium. Gills on the undersurface of the cap of a mushroom form vast numbers of minute spores. It has been estimated that a mushroom with a cap that is 7.5 centimeters across produces as many as 40 million spores per hour!

Most basidiomycete hyphae are dikaryotic. Ultimately, the hyphae fuse to form basidiocarps, with basidia lining the gills on the underside. Meiosis immediately follows syngamy in these basidia.

(a)

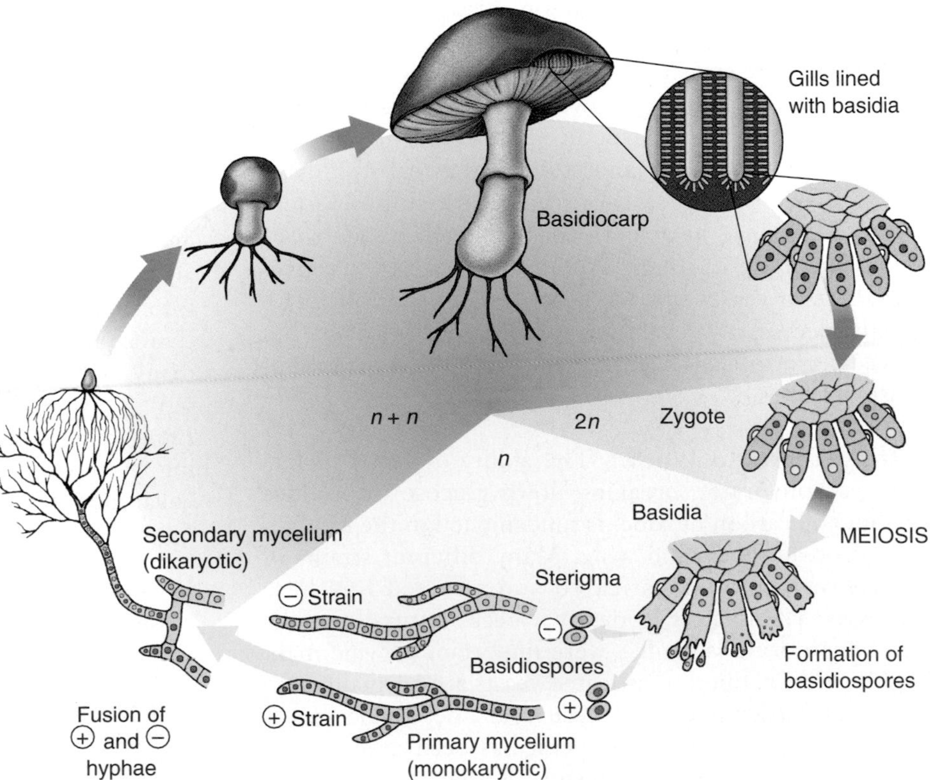

(b)

FIGURE 36.11
Basidiomycetes. (*a*) Death cap mushroom, *Amanita phalloides*. When eaten, these mushrooms are usually fatal. (*b*) Life cycle of a basidiomycete. The basidium is the reproductive structure where syngamy occurs.

The Imperfect Fungi

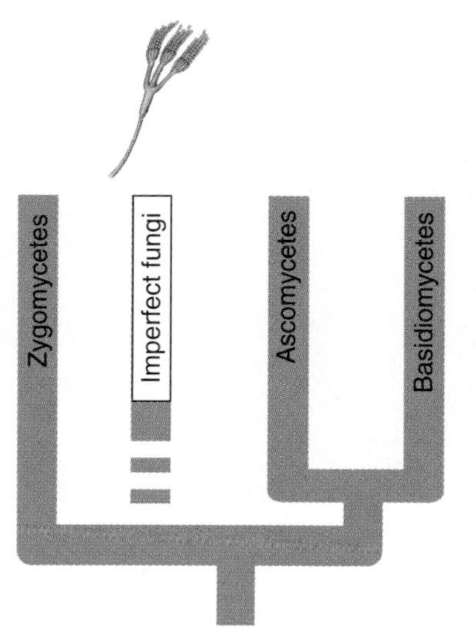

Most of the so-called imperfect fungi, a group also called deuteromycetes, are those in which the sexual reproductive stages have not been observed. Most of these appear to be related to ascomycetes although some have clear affinities to the other phyla. The group of fungi from which a particular nonsexual strain has been derived usually can be determined by the features of its hyphae and asexual reproduction. It cannot, however, be classified by the standards of that group because the classification systems are based on the features related to sexual reproduction. One consequence of this system is that as sexual reproduction is discovered in an imperfect fungus, it may have two names assigned to different stages of its life cycle.

There are some 17,000 described species of imperfect fungi (figure 36.12). Even though sexual reproduction is absent among imperfect fungi, a certain amount of genetic recombination occurs. This becomes possible when hyphae of different genetic types fuse, as sometimes happens spontaneously. Within the heterokaryotic hyphae that arise from such fusion, a special kind of genetic recombination called **parasexuality** may occur. In parasexuality, genetically distinct nuclei within a common hypha exchange portions of chromosomes. Recombination of this sort also occurs in other groups of fungi and seems to be responsible for some of the new pathogenic strains of wheat rust.

Economic Importance

Among the economically important genera of the imperfect fungi are *Penicillium* and *Aspergillus.* Some species of *Penicillium* are sources of the well-known antibiotic penicillin, and other species of the genus give the characteristic flavors and aromas to cheeses such as Roquefort and Camembert. Species of *Aspergillus* are used to ferment soy sauce and soy paste, processes in which certain bacteria and yeasts also play important roles. Citric acid is produced commercially with members of this genus under highly acidic conditions. Some species of both *Penicillium* and *Aspergillus* form ascocarps, but the genera are still classified primarily as imperfect fungi because the ascocarps are found rarely in only a few species. Most of the fungi that cause skin diseases in humans, including athlete's foot and ringworm, are also imperfect fungi.

A number of imperfect fungi occur widely on food. *Fusarium* species growing on spoiled food produce highly toxic substances such as trichothecenes. Aflatoxins, among the most carcinogenic compounds known, are produced by some *Aspergillus flavus* strains growing on corn, peanuts, etc. Most developed countries have legal limits on the concentration of aflatoxin permitted in different foods.

Imperfect fungi are fungi in which no sexual reproduction has been observed. For this reason, they cannot be classified by the standards applied to the three phyla of fungi. The great majority of the imperfect fungi are clearly ascomycetes.

(a)

(b)

(c)

FIGURE 36.12
The imperfect fungi. (*a*) *Verticillium alboatrum* (1350×), an important pathogen of alfalfa, has whorled conidiophores. The single-celled conidia of this member of the imperfect fungi are borne at the ends of the conidiophores. (*b*) In *Tolypocladium inflatum*, the conidia arise along the branches. This fungus is one of the sources of cyclosporin, a drug that suppresses immune reactions and thus assists in making human organ grafts possible; the drug was put on the market in 1979. (*c*) This scanning electron micrograph of *Aspergillus* shows conidia, the spheres at the end of the hyphae.

36.3 Fungi form two key mutualistic symbiotic associations.

Lichens

Lichens (figure 36.13) are symbiotic associations between a fungus and a photosynthetic partner. They provide an outstanding example of mutualism, the kind of symbiotic association that benefits both partners. Ascomycetes (including some imperfect fungi) are the fungal partners in all but about 20 of the approximately 15,000 species of lichens estimated to exist; the exceptions, mostly tropical, are basidiomycetes. Most of the visible body of a lichen consists of its fungus, but within the tissues of that fungus are found cyanobacteria, green algae, or sometimes both (figure 36.14). Specialized fungal hyphae penetrate or envelop the photosynthetic cells within them and transfer nutrients directly to the fungal partner. Biochemical "signals" sent out by the fungus apparently direct its cyanobacterial or green algal component to produce metabolic substances that it does not produce when growing independently of the fungus. The photosynthetic member of the association is normally held between thick layers of interwoven fungal hyphae and is not directly exposed to the light, but enough light penetrates the translucent layers of fungal hyphae to make photosynthesis possible. The fungi in lichens are unable to grow normally without their photosynthetic partners and the fungi protect their partners from strong light and desiccation.

FIGURE 36.13
Lichens are found in a variety of habitats. (*a*) A fruticose lichen, *Cladina evansii*, growing on the ground in Florida. (*b*) A foliose ("leafy") lichen, *Parmotrema gardneri*, growing on the bark of a tree in a mountain forest in Panama.

The durable construction of the fungus, combined with the photosynthetic properties of its partner, has enabled lichens to invade the harshest habitats at the tops of mountains, in the farthest northern and southern latitudes, and on dry, bare rock faces in the desert. In harsh, exposed areas, lichens are often the first colonists, breaking down the rocks and setting the stage for the invasion of other organisms.

Lichens are often strikingly colored because of the presence of pigments that probably play a role in protecting the photosynthetic partner from the destructive action of the sun's rays. These same pigments may be extracted from the lichens and used as natural dyes. The traditional method of manufacturing Scotland's famous Harris tweed used fungal dyes.

Lichens are extremely sensitive to pollutants in the atmosphere, and thus they can be used as bioindicators of air quality. Their sensitivity results from their ability to absorb substances dissolved in rain and dew. Lichens are generally absent in and around cities because of automobile traffic and industrial activity, even though suitable substrates exist.

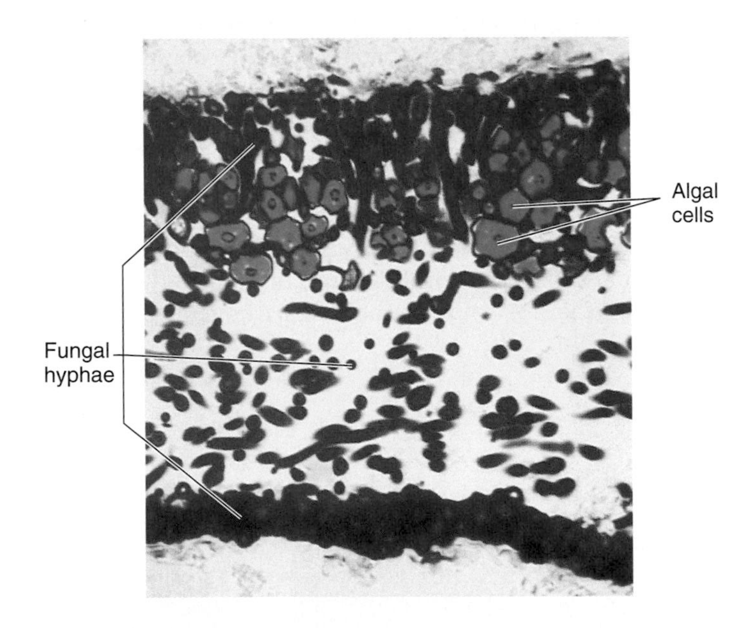

FIGURE 36.14
Stained section of a lichen (250×). This section shows fungal hyphae (*purple*) more densely packed into a protective layer on the top and, especially, the bottom layer of the lichen. The blue cells near the upper surface of the lichen are those of a green alga. These cells supply carbohydrate to the fungus.

Lichens are symbiotic associations between a fungus—an ascomycete in all but a very few instances—and a photosynthetic partner, which may be a green alga or a cyanobacterium or both.

Mycorrhizae

The roots of about 90% of all kinds of vascular plants normally are involved in mutualistic symbiotic relationships with certain kinds of fungi. It has been estimated that these fungi probably amount to 15% of the total weight of the world's plant roots. Associations of this kind are termed **mycorrhizae** (from the Greek words for "fungus" and "roots"). The fungi in mycorrhizae associations function as extensions of the root system. The fungal hyphae dramatically increase the amount of soil contact and total surface area for absorption. When mycorrhizae are present, they aid in the direct transfer of phosphorus, zinc, copper, and other nutrients from the soil into the roots. The plant, on the other hand, supplies organic carbon to the fungus, so the system is an example of mutualism.

There are two principal types of mycorrhizae (figure 36.15): **endomycorrhizae,** in which the fungal hyphae penetrate the outer cells of the plant root, forming coils, swellings, and minute branches, and also extend out into the surrounding soil; and **ectomycorrhizae,** in which the hyphae surround but do not penetrate the cell walls of the roots. In both kinds of mycorrhizae, the mycelium extends far out into the soil.

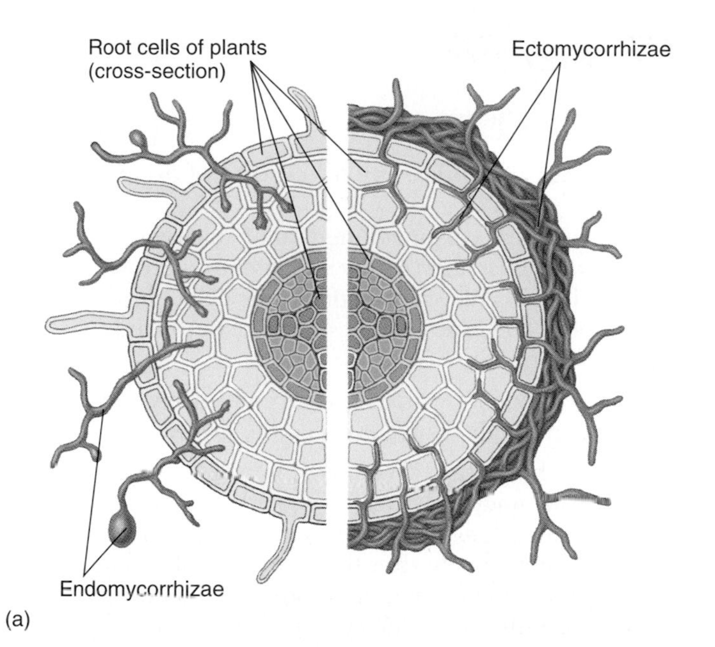

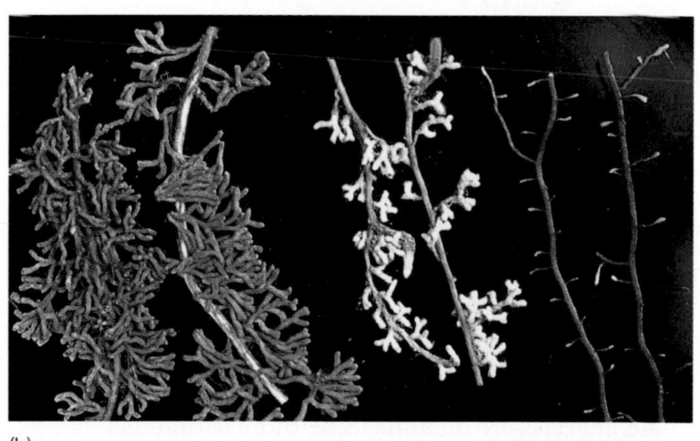

FIGURE 36.15
Endomycorrhizae and ectomycorrhizae. (*a*) In endomycorrhizae, fungal hyphae penetrate and branch out in the root cells of plants. In ectomycorrhizae, fungal hyphae do not penetrate root cells but grow around and extend between the cells. (*b*) Ectomycorrhizae on roots of pines. From left to right are yellow-brown mycorrhizae formed by *Pisolithus*, white mycorrhizae formed by *Rhizopagon*, and pine roots not associated with a fungus.

Endomycorrhizae

Endomycorrhizae are by far the more common of these two types. The fungal component in them is a zygomycete. Only about 100 species of zygomycetes are known to be involved in such relationships throughout the world. These few species of zygomycetes are associated with more than 200,000 species of plants. Endomycorrhizal fungi are being studied intensively because they are potentially capable of increasing crop yields with lower phosphate and energy inputs.

The earliest fossil plants often show endomycorrhizal roots. Such associations may have played an important role in allowing plants to colonize land. The soils available at such times would have been sterile and completely lacking in organic matter. Plants that form mycorrhizal associations are particularly successful in infertile soils; considering the fossil evidence, the suggestion that mycorrhizal associations found in the earliest plants helped them succeed on such soils seems reasonable. In addition, the most primitive vascular plants surviving today continue to depend strongly on mycorrhizae.

Ectomycorrhizae

Ectomycorrhizae (figure 36.15*b*) involve far fewer kinds of plants than do endomycorrhizae, perhaps a few thousand. They are characteristic of certain groups of trees and shrubs, particularly those of temperate regions, including pines, firs, oaks, beeches, and willows. The fungal components in most ectomycorrhizae are basidiomycetes, but some are ascomycetes. Several different kinds of ectomycorrhizal fungi may form mycorrhizal associations with one plant. Different combinations have different effects on the physiological characteristics of the plant and its ability to survive under different environmental conditions. At least 5000 species of fungi are involved in ectomycorrhizal relationships, and many of them are restricted to a single species of plant.

Mycorrhizae are symbiotic associations between plants and fungi.

Chapter 36

www.mhhe.com/raven6/resources36.mhtml

Summary | **Questions** | **Media Resources**

36.1 Fungi are unlike any other kind of organism.

- The fungi are a distinct kingdom of eukaryotic organisms characterized by a filamentous growth form, lack of chlorophyll and motile cells, chitin-rich cell walls, and external digestion of food by the secretion of enzymes.
- Fungal filaments, called hyphae, collectively make up the fungus body, which is called the mycelium.
- In many fungi, the two kinds of nuclei that will eventually undergo syngamy occur together in hyphae for a long period before they fuse. Meiosis occurs immediately after the formation of the zygote in all fungi; the zygote, therefore, is the only diploid nucleus of the entire life cycle in these organisms.

1. What is a hypha? What is the advantage to having incomplete septa?

2. What is the composition of the fungal cell wall? Why is this composition an advantage to the fungi?

3. Which fungal nuclei are diploid? Which are haploid? To what do the following terms refer: *heterokaryotic*, *homokaryotic*, *dikaryotic*, and *monokaryotic?*

- Characteristics of Fungi

36.2 Fungi are classified by their reproductive structures.

- There are three phyla of fungi: Zygomycota, the zygomycetes; Ascomycota, the ascomycetes; and Basidiomycota, the basidiomycetes.
- Zygomycetes form septa only when gametangia or sporangia are cut off at the ends of their hyphae; otherwise, their hyphae are multinucleate. Most hyphae of ascomycetes and basidiomycetes have perforated septa through which the cytoplasm, but not necessarily the nuclei, flows freely.
- Cells within the heterokaryotic hyphae of ascomycetes are multinucleate; those within the heterokaryotic hyphae of the basidiomycetes are dikaryotic. Zygotes in ascomycetes form within sac-like structures known as asci, and those in basidiomycetes form within structures known as basidia.
- Asexual reproduction in zygomycetes takes place by means of spores from multinucleate sporangia; in ascomycetes, it takes place by means of conidia. Asexual reproduction in basidiomycetes is rare.
- The imperfect fungi are not a phylum but rather, a group of fungi in which sexual reproduction has not been observed.

4. What are the three reproductive structures that occur in fungi? How do they differ?

5. Fungi are nonmotile. How are they dispersed to new areas?

6. What are the ascomycete asexual spores called? Do the nonreproductive hyphae of this division have septa?

7. To what phyla do the yeasts belong? How do they differ from other fungi? Is it more likely that this characteristic is primitive or degenerate?

8. What are the imperfect fungi? Which phylum seems to be best represented in this group? By what means can individuals in this group be classified?

- Diversity of Fungi

- Student Research: Mushroom Spore Germination

- Art Quiz: Life Cycle of *Rhizopus*

36.3 Fungi form two key mutualistic symbiotic associations.

- Lichens are mutualistic symbiotic systems involving fungi (almost always ascomycetes), which derive their nutrients from green algae, cyanobacteria, or both.
- Mycorrhizae are mutualistic symbiotic associations between fungi and plants. Endomycorrhizae, more common, involve zygomycetes, while ectomycorrhizal fungi are mainly basidiomycetes.

9. What are lichens? Which fungal phylum is best represented in the lichens?

10. What are mycorrhizae? How do endomycorrhizae and ectomycorrhizae differ?

Part

X

Plant Form and Function

A hyperaccumulaor of toxic metal. This plant of the mustard family, *Streptanthus polygaloides*, is unusual in that it accumulates large amounts of toxic nickel in its tissues. Does this serve to protect the plant?

Real People Doing Real Science

Why Do Some Plants Accumulate Toxic Levels of Metals?

All plants take up minerals from the soil. These minerals are used as cofactors in many important metabolic processes, helping enzymes carry out their catalytic duties. A few of the plant species that grow on soils that contain high levels of metals (so-called serpentine soils) are particularly interesting, as they have very high levels of metals in their tissues, even more than metal-tolerant plants do. These rare high-metal plants, termed hyperaccumulators, contain in excess of 1000 micrograms of metal per gram dry weight of tissue. Nickel (Ni) is the most accumulated metal—of 450 hyperaccumulating species of plants that have been described, 340 of them accumulate nickel.

What is going on here? Why have these hyperaccumulating plant species evolved ways to tolerate high metal levels that would kill an average plant? We do not know. Some researchers have suggested that hyperaccumulation may simply be a way to survive in a high-metal environment, with plants sequestering the metals by packaging them in vacuoles where their high concentration poses no danger to the rest of the plant cell. Other researchers have suggested that the metal taken up by hyperaccumulators makes the plants drought resistant. Still others suggest that the metal in fallen leaves and other litter may deter other plants from growing in the vicinity of a hyperaccumulating individual, lessening competition from too-near neighbors.

Few experiments have been carried out to test these hypotheses. There is one possibility that seems particularly attractive. Perhaps the metal in hyperaccumulator plants makes them toxic to herbivores. If that were so, it would be easy to see why natural selection would favor hyperaccumulator plants in heavy-metal soils: organisms which feed on these plants would die. Protection from herbivory thus offers a very tempting explanation for the evolution of hyperaccumulation in plants.

The hypothesis that hyperaccumulation of metals by plants deters herbivore consumption has been examined by Robert Boyd of Auburn University and Scott Martens of the University of California, Davis. In order to investigate this hypothesis clearly, the researchers needed a way to compare plants with and without high levels of nickel. First they selected a Ni-hyperaccumulating species of plant, a member of the mustard family that grows naturally only on Ni-rich serpentine soil. Then, in the laboratory, they attempted to grow the plant on artificial potting soil bearing no nickel. In parallel experiments they grew the plants on artificial potting soil to which they added nickel ("amended soil"). The plant grew just fine in either of these artificial soils, demonstrating that it had not evolved a *requirement* for nickel. Importantly, when they examined the tissue of the Ni-hyperaccumulator plant when it had been grown on Ni-free soil, the tissue still had the mustard oils and other chemicals characteristic of the species, but little nickel. Now they had what they needed for a successful experiment, a way to manipulate the level of nickel in plant tissue with nothing else being different.

One could imagine two ways that a plant might use high levels of nickel to protect against herbivores:

1. Mortality. Such a defense might result from insect mortality, the herbivorous insect being killed by ingesting toxic-metal-containing plant tissue. This is how the toxic elements selenium and fluorine work in defending plants that accumulate them. Selenium becomes incorporated into insect amino acids, making the proteins bearing them dysfunctional. Fluorine, incorporated into fluoroacetate, disrupts the insect's Krebs cycle. Unlike selenium and fluorine, nickel is not incorporated into an herbivore's metabolic chemicals. But its release from plant vacuoles after ingestion might directly poison many metabolic processes of a herbivore.

2. Taste. A hyperaccumulation defense against herbivores might also result from deterrence, the herbivorous insect finding the toxic-metal-containing tissue distasteful, and so learning to avoid eating the hyperaccumulating plant. The mustard oils characteristic of the mustard family (*Brassicaceae*)—you know them as the substances that give the pungent aromas and tastes to mustard, radish, and horseradish—signal the presence of distasteful chemicals to many groups of insects.

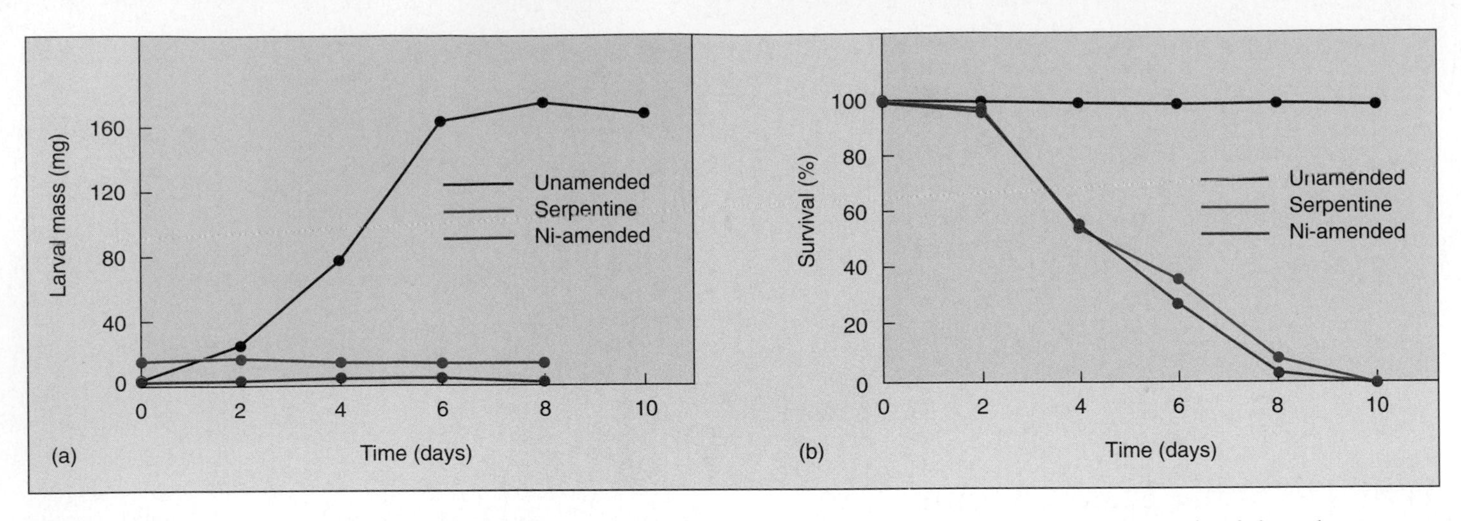

Effects on *Pieris rapae* larvae of eating Ni-hyperaccumulating plants. (*a*) Weight gain was only seen in larvae that fed on plants grown on unamended (nickel-free) potting soil and not in larvae fed on plants grown in Ni-amended soil (potting mix to which Ni had been added) or serpentine soil (which contains high levels of Ni). (*b*) Survival rates dropped dramatically in larvae that fed on plants grown in Ni-amended and serpentine soils.

The Experiment

To test the hypothesis that the hyperaccumulation of nickel (Ni) in the tissues of certain plants acts as a defense against herbivores that attempt to use the plant as a primary food source, Boyd and Martens selected for study a plant from the mustard family, *Streptanthus polygaloides*. This small plant is a Ni-hyperaccumulating annual that grows only on serpentine soils in the foothills of the Sierra Nevada in California.

Three experimental protocols were set up: (1) greenhouse-grown plants in Ni-amended soil (i.e., a potting mix to which was added 1000 mg Ni per kg mix), (2) unamended soil (potting mix with no added Ni), and (3) plants collected from natural serpentine soil rich in Ni. In each protocol, leaves were removed from the plants and fed to one of three different insects known to feed on this family of plants. The responses of the insect herbivores to the three protocols were evaluated based on food choice, on survival rates, and on weight gain over the course of the experimental treatments.

The Results

The larval stage of butterflies *Pieris rapae* and *Euchloe hyantis* were fed all three types of leaves, as were adult grasshoppers. All three insect herbivores fed in this way exhibited a disruption of growth, as determined by the measurement of body mass over the course of the experiments. The data for *Pieris rapae*, shown in graph *a* above, are representative. Weight gain was only seen in larvae and insects feeding on the unamended plants. Those herbivores fed Ni-amended or serpentine plants either maintained or lost weight.

Confirming the Mortality Hypothesis. Survival rates dropped dramatically in *Pieris rapae* larvae and grasshoppers feeding on Ni-amended or serpentine plants compared to those feeding on unamended plants (a parasitoid infection disrupted the experimental results of the *Euchloe hyantis* larvae). Again, the data for *Pieris rapae*, shown above in graph *b*, are representative.

To verify that Ni toxicity was sufficient to account for the data presented above, *P. rapae* was fed a synthetic diet containing various concentrations of Ni. Survival rates were not significantly affected with concentrations less than 500 ppm but dropped dramatically at Ni concentrations of 1000 ppm. In addition to determining the Ni toxicity in the insects, Martens and Boyd also examined the effects of Ni concentrations on plant fitness. After growing for 21 days, Ni-amended plants exhibited greater survival rates and weight gain than unamended plants which were almost completely defoliated by the larvae.

Eliminating the Taste Hypothesis. Leaves of all three plant preparations were consumed by the larvae and insects and all three species were observed eating the leaves regardless of Ni content. Based on this, Martens and Boyd could conclude that the results above were due to the Ni content in the plants and not due to feeding deterrence.

The researchers showed that the hyperaccumulation of Ni in plants can clearly serve as a defense mechanism in plants which can increase plant fitness. As with many defense mechanisms, some herbivores may have acquired counteradaptations that would allow them to continue feeding on hyperaccumulators. In order to become Ni tolerant, an organism would need to excrete or compartmentalize the Ni in its own system. An organism might also dilute the Ni concentration by consuming a mixed diet. Organisms may adapt to a diet of high Ni concentrations in other ways, one of many areas for further study.

To explore this experiment further, go to the Virtual Lab at www.mhhe.com/raven6/vlab10.mhtml

37

Evolutionary History of Plants

Concept Outline

37.1 Plants have multicellular haploid and diploid stages in their life cycles.

The Evolutionary Origins of Plants. Plants evolved from freshwater green algae and eventually developed cuticles, stomata, conducting systems, and reproductive strategies that adapt them well for life on land.

Plant Life Cycles. Plants have haplodiplontic life cycles. Diploid sporophytes produce haploid spores by meiosis. Spores develop into haploid gametophytes, by mitosis and produce haploid gametes.

37.2 Nonvascular plants are relatively unspecialized, but successful in many terrestrial environments.

Mosses, Liverworts, and Hornworts. The most conspicuous part of a nonvascular plant is the green photosynthetic gametophyte, which supports the smaller sporophyte nutritionally.

37.3 Seedless vascular plants have well-developed conducting tissues in their sporophytes.

Features of Vascular Plants. In vascular plants, specialized tissue called xylem conducts water and dissolved minerals within the plant, and tissue called phloem conducts sucrose and hormones within the plant.

Seedless Vascular Plants. Seedless vascular plants have a much more conspicuous sporophyte than nonvascular plants do, and many have well-developed conducting systems in stem, roots, and leaves.

37.4 Seeds protect and aid in the dispersal of plant embryos.

Seed Plants. In seed plants, the sporophyte is dominant. Male and female gametophytes develop within the sporophyte and depend on it for food. Seeds allow embryos to germinate when conditions are favorable.

Gymnosperms. In gymnosperms, the female gametophyte (ovule) is not completely enclosed by sporophyte tissue at the time of pollination.

Angiosperms. In angiosperms, the ovule is completely enclosed by sporophyte tissue at the time of pollination. Angiosperms, by far the most successful plant group, produce flowers.

FIGURE 37.1
An arctic tundra. This is one of the harshest environments on earth, yet a diversity of plants have made this home. These ecosystems are fragile and particularly susceptible to global change.

Plant evolution is the story of the conquest of land by green algal ancestors. For about 500 million years, algae were confined to a watery domain, limited by the need for water to reproduce, provide structural support, prevent water loss, and provide some protection from the sun's ultraviolet irradiation. Numerous evolutionary solutions to these challenges have resulted in over 300,000 species of plants dominating all terrestrial communities today, from forests to alpine tundra, from agricultural fields to deserts (figure 37.1). Most plants are photosynthetic, converting light energy into chemical-bond energy and providing oxygen for all aerobic organisms. We rely on plants for food, clothing, wood for shelter and fuel, chemicals, and many medicines. This chapter explores the evolutionary history and strategies that have allowed plants to inhabit most terrestrial environments over millions of years.

37.1 Plants have multicellular haploid and diploid stages in their life cycles.

The Evolutionary Origins of Plants

What is a plant? The term plant refers to a group of organisms that share a freshwater algal ancestor and have evolved over a 470-million-year period. The defining characteristic of plants is the protection of their embryos, essential for survival in a terrestrial environment. It is surprising that just a single species of green algae gave rise to the entire terrestrial plant lineage, from mosses through the flowering plants (angiosperms). Exactly what this ancestral alga was is still a mystery, but close relatives are believed to exist in freshwater lakes today. DNA sequence data is consistent with the claim that a single "Eve" gave rise to all plants. At each subsequent step in evolution, the evidence suggests that only a single family of plants made the transition. The shared evolutionary history with green algae has led some biologists to define three kingdoms of plants: the green plant kingdom (green algae and plants), the red plant kingdom (red algae) and the brown plant kingdom (brown algae). Fungi are not a part of this scheme. They arose later than the plants and are more closely related to metazoan animals (see chapter 44).

There are 12 living plant phyla, all of which afford some protection to their embryos. All plants also have multicellular haploid and diploid phases. The trend over time has been toward increasing embryo protection and a smaller haploid stage in the life cycle. The plants can be divided into two groups based on the presence or absence of vascular tissues which facilitate the transport of water and nutrients in plants. Three phyla (mosses, liverworts, and hornworts) lack vascular tissue and are referred to as the **nonvascular plants**. Members of 9 of the 12 plant phyla are collectively called **vascular plants,** and include, among others, the ferns, conifers, and flowering plants. Vascular plants have water-conducting xylem and food-conducting phloem strands of tissues in their stems, roots, and leaves. Vascular plants can be further grouped based on how much protection embryos have. The seedless vascular plants (ferns) provide less protection than the seeds of the gymnosperms (including conifers and cycads) and angiosperms (flowering plants) (figure 37.2). Although all seedless vascular plants are often grouped, recent evidence supports two distinct, monophyletic lineages of seedless plants: (1) club mosses and (2) ferns and horsetails. Ferns and horsetails are the closest relatives to the seed plants. About 150 million years ago the angiosperms arose with further innovations—flowers to attract pollinators and fruit surrounding the seed to protect the embryo and aid in seed dispersal. Many of these lineages have persisted. If you could travel back 65 million years to the dinosaur era, you would encounter oak, walnut, and sycamore trees!

FIGURE 37.2
Four major groups of plants. In this chapter we will discuss four major groups of plants. There are actually five distinct lineages; seedless vascular plants have two different origins, but are grouped here for simplicity. The ancestral green algae are discussed in chapter 35.

Adaptations to Land

Plants and fungi are the only major groups of organisms that are primarily terrestrial. Most plants are protected from **desiccation**—the tendency of organisms to lose water to the air—by a waxy **cuticle** that is secreted onto their exposed surfaces. The cuticle is relatively impermeable and provides an effective barrier to water loss.

This solution creates another problem by limiting gas exchange essential for respiration and photosynthesis. Water and gas diffusion into and out of a plant occurs through tiny mouth-shaped openings called **stomata** (singular, **stoma**).

The evolution of leaves resulted in increased photosynthetic surface area. The shift to a dominant diploid generation, accompanied by the structural support of vascular tissue, allowed plants to take advantage of the vertical dimension of the terrestrial environment, making the evolution of trees possible.

Plants evolved from freshwater green algae and eventually developed cuticles, stomata, conducting systems, and reproductive strategies that adapt them well for life on land.

Plant Life Cycles

All plants undergo mitosis after both gamete fusion *and* meiosis. The result is a multicellular haploid and a multicellular diploid individual, unlike us where gamete fusion directly follows meiosis. We have a **diplontic** life cycle (only the diploid stage is multicellular), but the plant life cycle is **haplodiplontic** (with multicellular haploid and diploid stages). The basic haplodiplontic cycle is summarized in figure 37.3. Brown, red, and green algae are also haplodiplontic (see chapter 35). While we produce gametes via meiosis, plants actually produce gametes by mitosis in a multicellular, haploid individual. The diploid generation, or **sporophyte,** alternates with the haploid generation, or **gametophyte.** Sporophyte means "spore plant," and gametophyte means "gamete plant." These terms indicate the kinds of reproductive cells the respective generations produce.

The diploid sporophyte produces haploid spores (*not* gametes) by meiosis. Meiosis takes place in structures called **sporangia-**, where diploid **spore mother cells (sporocytes)** undergo meiosis, each producing four haploid **spores.** Spores divide by mitosis, producing a multicellular, haploid gametophyte. Spores are the first cells of the gametophyte generation.

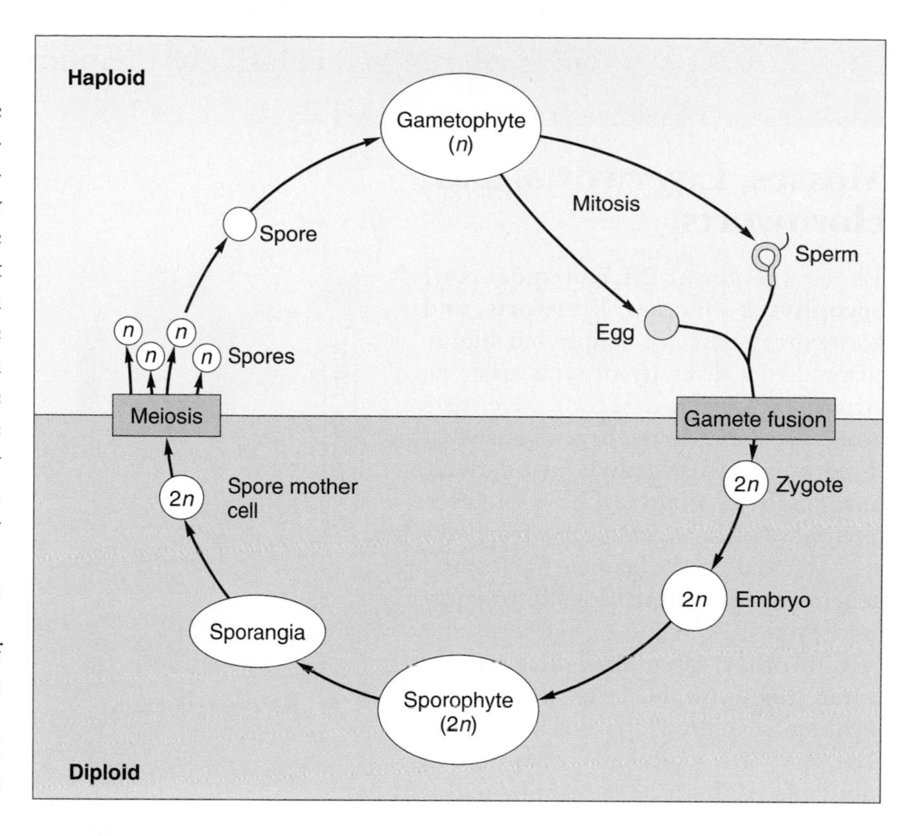

FIGURE 37.3
A generalized plant life cycle. Note that both haploid and diploid individuals can be multicellular. Also, spores are produced by meiosis while gametes are produced by mitosis.

In turn, the haploid gametophyte is produced by mitosis and is the source of gametes. When the gametes fuse, the zygote they form is diploid and is the first cell of the next sporophyte generation. The zygote grows into a diploid sporophyte that produces sporangia in which meiosis ultimately occurs.

While all plants are haplodiplontic, the haploid generation consumes a much larger portion of the life cycle in mosses than in gymnosperms and angiosperms. In mosses, liverworts, and ferns, the gametophyte is photosynthetic and free-living; in other plants it is usually nutritionally dependent on the sporophyte. When you look at moss, what you see is largely gametophyte tissue; their sporophytes are usually smaller, brownish or yellowish structures attached to tissues of the gametophyte. In all vascular plants the gametophytes are much smaller than the sporophytes. In seed plants, the gametophytes are nutritionally dependent on the sporophytes and are enclosed within their tissues. When you look at a gymnosperm or angiosperm, what you see, with rare exceptions, is a sporophyte.

The difference between dominant gametophytes and sporophytes is key to understanding why there are no moss trees. What we identify as a moss plant is a gametophyte and it produces gametes at its tips. The egg is stationery and sperm lands near the egg in a droplet of water. If the moss were the height of a sequoia, not only would it need vascular tissue for conduction and support, the sperm would have to swim up the tree! In contrast, the fern gametophyte develops on the forest floor where gametes can meet. Fern trees abound in Australia and the haploid spores fall to the ground and develop into gametophytes.

Having completed our overview of plant life cycles, we will consider the major plant groups. As we do, we will see a reduction of the gametophyte from group to group, a loss of multicellular **gametangia** (structures in which gametes are produced), and increasing specialization for life on the land, including the remarkable structural adaptations of the flowering plants, the dominant plants today. Similar trends must have characterized the evolution of seed plants over the hundreds of millions of years since a freshwater alga made the move onto land.

Plants have haplodiplontic life cycles. Diploid sporophytes produce haploid spores by meiosis. Spores develop into haploid gametophytes by mitosis and produce haploid gametes.

37.2 Nonvascular plants are relatively unspecialized, but successful in many terrestrial environments.

Mosses, Liverworts, and Hornworts

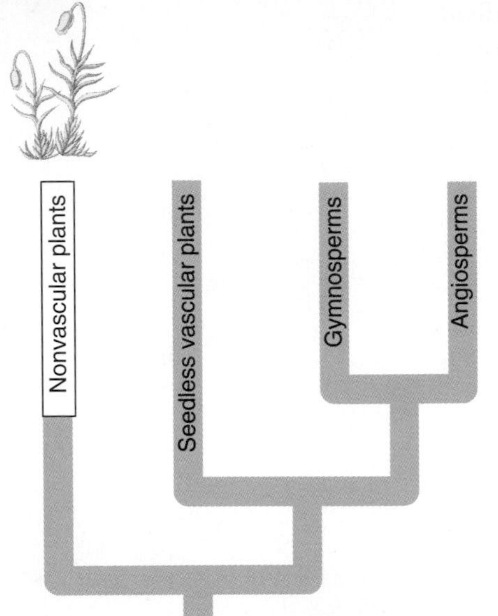

There are about 24,700 species of **bryophytes**—mosses, liverworts, and hornworts—that are simply but highly adapted to a diversity of terrestrial environments (even deserts!). Scientists now agree that bryophytes consist of three quite distinct phyla of relatively unspecialized plants. Their gametophytes are photosynthetic. Sporophytes are attached to the gametophytes and depend on them nutritionally to varying degrees. Bryophytes, like ferns and certain other vascular plants, require water (for example, rainwater) to reproduce sexually. It is not surprising that they are especially common in moist places, both in the tropics and temperate regions.

Most bryophytes are small; few exceed 7 centimeters in height. The gametophytes are more conspicuous than the sporophytes. Some of the sporophytes are completely enclosed within gametophyte tissue; others are not and usually turn brownish or straw-colored at maturity.

Mosses (Bryophyta)

The gametophytes of mosses typically consist of small leaflike structures (not true leaves which contain vascular tissue) arranged spirally or alternately around a stemlike axis (figure 37.4); the axis is anchored to its substrate by means of **rhizoids.** Each rhizoid consists of several cells that absorb water, but nothing like the volume of water absorbed by a vascular plant root. Moss "leaves" have little in common with true leaves, except for the superficial appearance of the green, flattened blade and slightly thickened midrib that runs lengthwise down the middle. They are only one cell thick (except at the midrib), lack vascular strands and stomata, and all the cells are haploid.

FIGURE 37.4
A hair-cup moss, *Polytrichum* (phylum Bryophyta). The "leaves" belong to the gametophyte. Each of the yellowish-brown stalks, with the capsule, or sporangium, at its summit, is a sporophyte.

Water may rise up a strand of specialized cells in the center of a moss gametophyte axis, but most water used by the plant travels up the outside of the plant. Some mosses also have specialized food-conducting cells surrounding those that conduct water.

Multicellular gametangia are formed at the tips of the leafy gametophytes (figure 37.5). Female gametangia **(archegonia)** may develop either on the same gametophyte as the male gametangia **(antheridia)** or on separate plants. A single egg is produced in the swollen lower part of an archegonium while numerous sperm are produced in an antheridium. When sperm are released from an antheridium, they swim with the aid of flagella through a film of dew or rainwater to the archegonia. One sperm (which is haploid) unites with an egg (also haploid), forming a diploid zygote. The zygote divides by mitosis and develops into the sporophyte, a slender, basal stalk with a swollen capsule, the *sporangium,* at its tip. As the sporophyte develops, its base is embedded in gametophyte tissue, its nutritional source. The sporangium is often cylindrical or club-shaped. Spore mother cells within the sporangium undergo meiosis, each producing four haploid spores. In many mosses at maturity, the top of the sporangium pops off, and the spores are released. A spore that lands in a suitable damp location may germinate and grow into a threadlike structure that branches to form rhizoids and "buds" that grow upright. Each bud develops into a new gametophyte plant consisting of a leafy axis.

In the Arctic and the Antarctic, mosses are the most abundant plants, boasting not only the largest number of individuals in these harsh regions, but also the largest number of species. Many mosses are able to withstand prolonged periods of drought, although they are not common in deserts. Most are remarkably sensitive to air pollution and are rarely found in abundance in or near cities or other areas with high levels of air pollution. Some mosses, such as the peat mosses (*Sphagnum*), can absorb up to 25 times their weight in water and are valuable commercially as a soil conditioner, or as a fuel when dry.

Liverworts (Hepaticophyta)

The old English word *wyrt* means "plant" or "herb." Some common liverworts have flattened gametophytes with lobes resembling those of liver—hence the combination "liverwort." Although the lobed liverworts are the best-known representatives of this phylum, they constitute only about 20% of the species (figure 37.6). The other 80% are leafy and superficially resemble mosses. Gametophytes are prostrate instead of erect, and the rhizoids are one-celled.

Some liverworts have air chambers containing upright, branching rows of photosynthetic cells, each chamber having a pore at the top to facilitate gas exchange. Unlike stomata, the pores are fixed open and cannot close.

Sexual reproduction in liverworts is similar to that in mosses. Lobed liverworts may form gametangia in umbrella-like structures. Asexual reproduction occurs when lens-shaped pieces of tissue that are released from the gametophyte grow to form new gametophytes.

Hornworts (Anthocerotophyta)

The origin of hornworts are a puzzle. They are most likely among the earliest land plants, yet the earliest hornwort fossil spores date from the Cretaceous period, 65 to 145 million years ago, when angiosperms were emerging.

The small hornwort sporophytes resemble tiny green broom handles rising from filmy gametophytes usually less than 2 centimeters in diameter (figure 37.7). The sporophyte base is embedded in gametophyte tissue, from which it derives some of its nutrition. However, the sporophyte has stomata, is photosynthetic, and provides much of the energy needed for growth and reproduction. Hornwort cells usually have a single chloroplast.

The three major phyla of nonvascular plants are all relatively unspecialized, but well suited for diverse terrestrial environments.

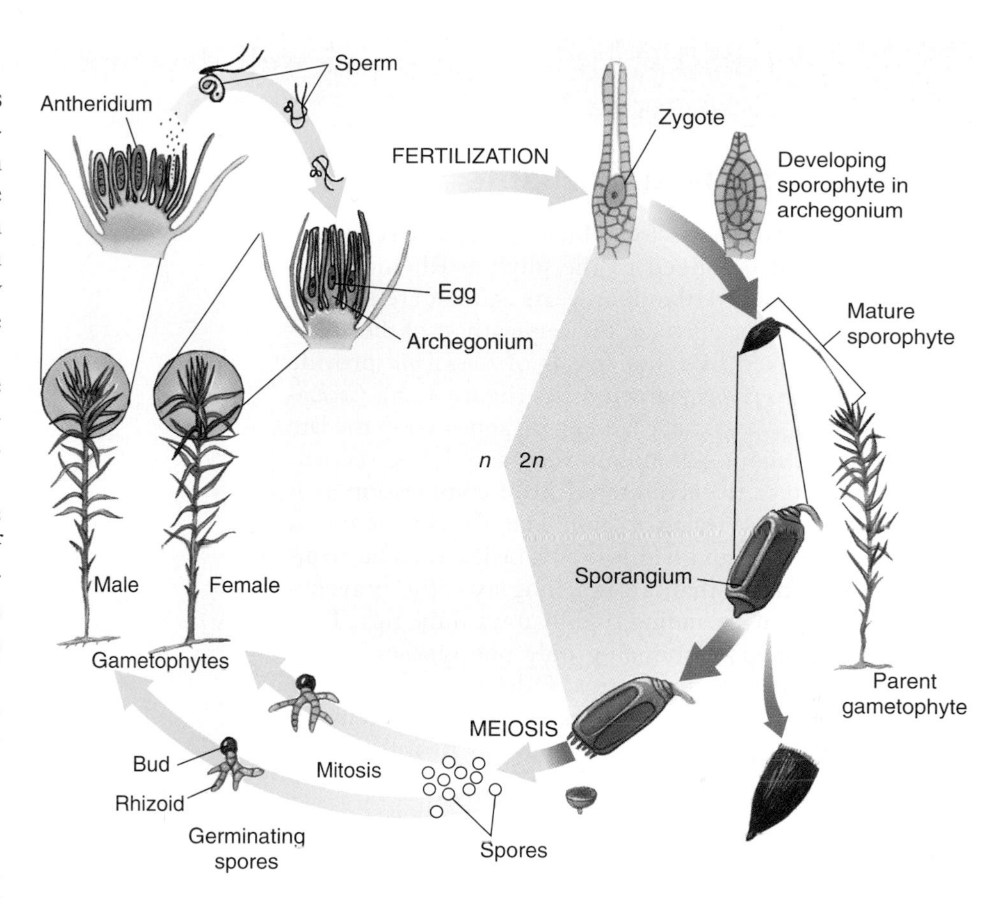

FIGURE 37.5
Life cycle of a typical moss. The majority of the life cycle of a moss is in the haploid state. The leafy gametophyte is photosynthetic, while the smaller sporophyte is not, and is nutritionally dependent on the gametophyte. Water is required to carry sperm to the egg.

FIGURE 37.6
A common liverwort, *Marchantia* (phylum Hepaticophyta). The sporophytes are borne within the tissues of the umbrella-shaped structures that arise from the surface of the flat, green, creeping gametophyte.

FIGURE 37.7
Hornworts (phylum Anthocerotophyta). Hornwort sporophytes are seen in this photo. Unlike the sporophytes of other bryophytes, most hornwort sporophytes are photosynthetic.

37.3 Seedless vascular plants have well-developed conducting tissues in their sporophytes.

Features of Vascular Plants

The first vascular plants for which we have a relatively complete record belonged to the phylum Rhyniophyta; they flourished some 410 million years ago but are now extinct. We are not certain what the very earliest of these vascular plants looked like, but fossils of *Cooksonia* provide some insight into their characteristics (figure 37.8). *Cooksonia*, the first known vascular land plant, appeared in the late Silurian period about 420 million years ago. It was successful partly because it encountered little competition as it spread out over vast tracts of land. The plants were only a few centimeters tall and had no roots or leaves. They consisted of little more than a branching axis, the branches forking evenly and expanding slightly toward the tips. They were **homosporous** (producing only one type of spore). Sporangia formed at branch tips. Other ancient vascular plants that followed evolved more complex arrangements of sporangia. Leaves began to appear as protuberances from stems.

Cooksonia and the other early plants that followed it became successful colonizers of the land through the development of efficient water- and food-conducting systems known as **vascular tissues.** The word *vascular* comes from the Latin *vasculum*, meaning a "vessel or duct." These tissues consist of strands of specialized cylindrical or elongated cells that form a network throughout a plant, extending from near the tips of the roots, through the stems, and into true leaves. One type of vascular tissue, the *xylem*, conducts water and dissolved minerals upward from the roots; another type of tissue, *phloem*, conducts sucrose and hormonal signals throughout the plant. It is important to note that vascular tissue developed in the sporophyte, but (with few exceptions) not the gametophyte. (See the discussion of vascular tissue structure in chapter 38.) The presence of a cuticle and stomata are also characteristic of vascular plants.

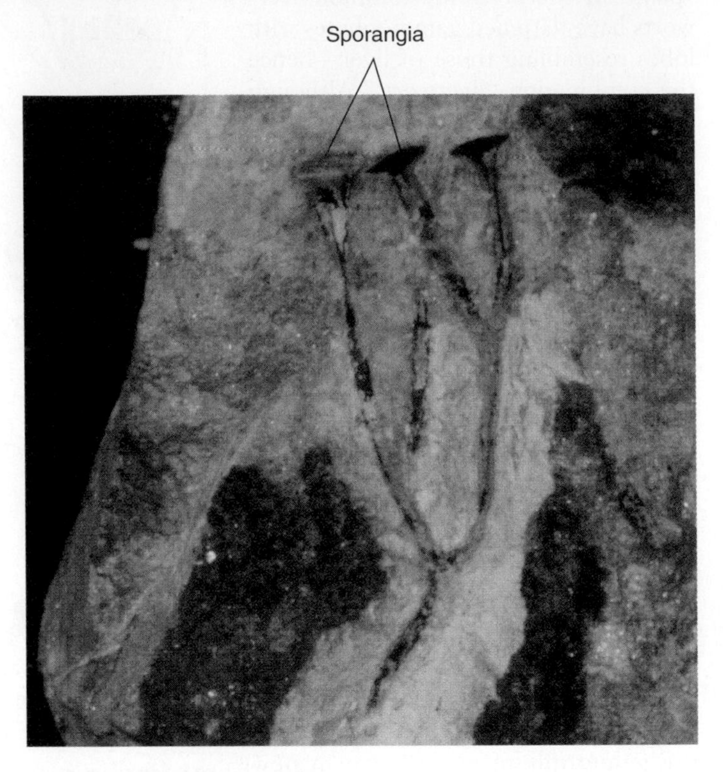

FIGURE 37.8
***Cooksonia*, the first known vascular land plant.** Its upright, branched stems, which were no more than a few centimeters tall, terminated in sporangia, as seen here. It probably lived in moist environments such as mudflats, had a resistant cuticle, and produced spores typical of vascular plants. This fossil represents a plant that lived some 410 million years ago. *Cooksonia* belongs to phylum Rhyniophyta, consisting entirely of extinct plants.

The nine living phyla of vascular plants (table 37.1) dominate terrestrial habitats everywhere, except for the highest mountains and the tundra. The haplodiplontic life cycle persists, but the gametophyte has been reduced during evolution of some phyla. A similar reduction in multicellular gametangia has occurred.

Accompanying this reduction in size and complexity of the gametophytes has been the appearance of the seed. Seeds are highly resistant structures well suited to protect a plant embryo from drought and to some extent from predators. In addition, almost all seeds contain a supply of food for the young plant. Seeds occur only in **heterosporous** plants (plants that produce two types of spores). Heterospory is believed to have arisen multiple times in the plants. Fruits in the flowering plants add a layer of protection to seeds and attract animals that assist in seed dispersal, expanding the potential range of the species. Flowers, which evolved among the angiosperms, attract pollinators. Flowers allow plants to secure the benefits of wide outcrossing in promoting genetic diversity.

Most vascular plants have well-developed conducting tissues, specialized stems, leaves, roots, cuticles, and stomata. Many have seeds which protect embryos until conditions are suitable for further development.

Table 37.1 The Nine Phyla of Extant Vascular Plants

Phylum	Examples	Key Characteristics	Approximate Number of Living Species
Anthophyta	Flowering plants (angiosperms)	Heterosporous seed plants. Sperm not motile; conducted to egg by a pollen tube. Seeds enclosed within a fruit. Leaves greatly varied in size and form. Herbs, vines, shrubs, trees. About 14,000 genera.	250,000
Pterophyta	Ferns	Primarily homosporous (a few heterosporous) vascular plants. Sperm motile. External water necessary for fertilization. Leaves are megaphylls that uncoil as they mature. Sporophytes and virtually all gametophytes photosynthetic. About 365 genera.	11,000
Lycophyta	Club mosses	Homosporous or heterosporous vascular plants. Sperm motile. External water necessary for fertilization. Leaves are microphylls. About 12–13 genera.	1,150
Coniferophyta	Conifers (including pines, spruces, firs, yews, redwoods, and others)	Heterosporous seed plants. Sperm not motile; conducted to egg by a pollen tube. Leaves mostly needlelike or scalelike. Trees, shrubs. About 50 genera.	601
Cycadophyta	Cycads	Heterosporous vascular seed plants. Sperm flagellated and motile but confined within a pollen tube that grows to the vicinity of the egg. Palmlike plants with pinnate leaves. Secondary growth slow compared with that of the conifers. Ten genera.	206
Gnetophyta	Gnetophytes	Heterosporous vascular seed plants. Sperm not motile; conducted to egg by a pollen tube. The only gymnosperms with vessels. Trees, shrubs, vines. Three very diverse genera (*Ephedra*, *Gnetum*, *Welwitschia*).	65
Arthrophyta	Horsetails	Homosporous vascular plants. Sperm motile. External water necessary for fertilization. Stems ribbed, jointed, either photosynthetic or nonphotosynthetic. Leaves scalelike, in whorls, nonphotosynthetic at maturity. One genus.	15
Psilophyta	Whisk ferns	Homosporous vascular plants. Sperm motile. External water necessary for fertilization. No differentiation between root and shoot. No leaves; one of the two genera has scalelike enations and the other leaflike appendages.	6
Ginkgophyta	*Ginkgo*	Heterosporous vascular seed plants. Sperm flagellated and motile but conducted to the vicinity of the egg by a pollen tube. Deciduous tree with fan-shaped leaves that have evenly forking veins. Seeds resemble a small plum with fleshy, ill-scented outer covering.	1

Seedless Vascular Plants

The earliest vascular plants lacked seeds. Members of four phyla of living vascular plants lack seeds, as do at least three other phyla known only from fossils. As we explore the adaptations of the vascular plants, we focus on both reproductive strategies and the advantages of increasingly complex transport systems. We will begin with the most familiar phylum of seedless vascular plants, the ferns.

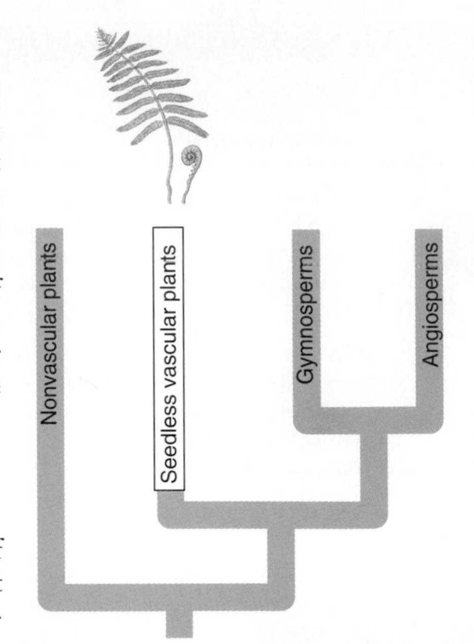

Ferns (Pterophyta)

Ferns are the most abundant group of seedless vascular plants, with about 12,000 living species. Recent research indicates that they may be the closest relatives to the seed plants. The fossil record indicates that ferns originated during the Devonian period about 350 million years ago and became abundant and varied in form during the next 50 million years. Their apparent ancestors had no broad leaves and were established on land as much as 375 million years ago.

Today, ferns flourish in a wide range of habitats throughout the world; about 75% of the species, however, occur in the tropics. The conspicuous sporophytes may be less than a centimeter in diameter—as seen in small aquatic ferns such as *Azolla*—or more than 24 meters tall and with leaves up to 5 meters or more long in the tree ferns (figure 37.9). The sporophytes and the smaller gametophytes, which rarely reach 6 millimeters in diameter, are both photosynthetic. The fern life cycle differs from that of a moss primarily in the much greater development, independence, and dominance of the fern's sporophyte. The fern's sporophyte is structurally more complex than that of the moss's; the fern sporophyte has vascular tissue and well-differentiated roots, stems, and leaves. The gametophyte, however, lacks vascular tissue.

Fern sporophytes typically have a horizontal underground stem called a *rhizome*, with roots emerging from the sides. The leaves, referred to as *fronds*, usually develop at the tip of the rhizome as tightly rolled-up coils ("fiddleheads") that unroll and expand. Many fronds are highly dissected and feathery, making the ferns that produce them prized as ornamentals. Some ferns, such as *Marsilea*, have fronds that resemble a four-leaf clover, but *Marsilea* fronds still begin as coiled fiddleheads. Other ferns produce a mixture of photosynthetic fronds and nonphotosynthetic reproductive fronds that tend to be brownish in color.

Most ferns are homosporous, producing distinctive, *sporangia*, usually in clusters called *sori*, typically on the backs of the fronds. Sori are often protected during their development by a transparent, umbrella-like covering. At first glance, one might mistake the sori for an infection on the plant. Diploid *spore mother cells* in each sporangium undergo meiosis, producing haploid spores. At maturity, the spores are catapulted from the sporangium by a snapping action, and those that land in suitable damp locations may germinate, producing gametophytes which are often heart-shaped, are only one cell thick (except in the center) and have rhizoids that anchor them to their substrate. These rhizoids are not true roots as they lack vascular tissue, but as with many of the nonvascular plants they do aid in transporting water and nutrients from the soil. Flask-shaped *archegonia* and globular *antheridia* are produced on either the same or different gametophyte.

FIGURE 37.9
A tree fern (phylum Pterophyta) in the forests of Malaysia. The ferns are by far the largest group of seedless vascular plants.

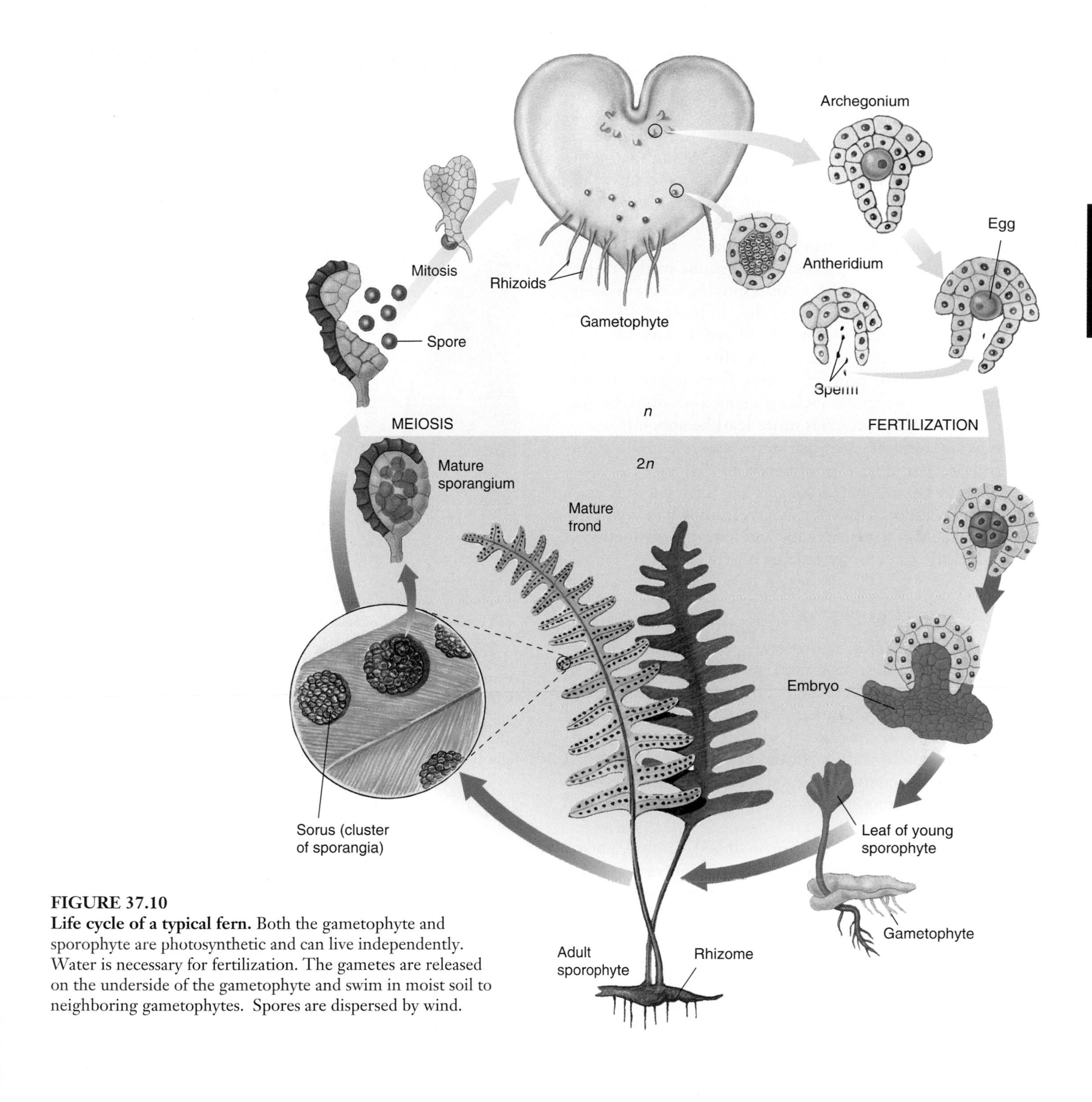

FIGURE 37.10
Life cycle of a typical fern. Both the gametophyte and sporophyte are photosynthetic and can live independently. Water is necessary for fertilization. The gametes are released on the underside of the gametophyte and swim in moist soil to neighboring gametophytes. Spores are dispersed by wind.

The sperm formed in the antheridia have flagella, with which they swim toward the archegonia when water is present, often in response to a chemical signal secreted by the archegonia. One sperm unites with the single egg toward the base of an archegonium, forming a *zygote.* The zygote then develops into a new sporophyte, completing the life cycle (figure 37.10). There are still multicellular gametangia. As discussed earlier, the shift to a dominant sporophyte generation allows ferns to achieve significant height without interfering with sperm swimming efficiently to the egg. The multicellular archegonia provide some protection for the developing embryo.

Whisk Ferns (Psilophyta)

Two other phyla of seedless vascular plants, the Psilophyta, (whisk ferns) and Arthrophyta (horsetails), have many features in common with ferns. For example, they all form antheridia and archegonia. Free water is required for the process of fertilization, during which the sperm, which have flagella, swim to and unite with the eggs. In contrast, most seed plants have nonflagellated sperm; none form antheridia, although a few form archegonia.

The ferns and whisk ferns, which occur in the tropics and subtropics, may have a monophyletic origin along with horsetails. Whisk ferns are the simplest of all extant vascular plants, consisting merely of evenly forking green stems without roots. The two or three species of the genus *Psilotum* do, however, have tiny, green, spirally arranged, flaps of tissue lacking veins and stomata. Another genus, *Tmespiteris*, has more leaflike appendages.

The gametophytes of whisk ferns are essentially colorless and are less than 2 millimeters in diameter, but they can be up to 18 millimeters long. They form parasitic associations with fungi, which furnish their nutrients. Some develop elements of vascular tissue and have the distinction of being the only gametophytes known to do so.

Horsetails (Arthrophyta)

The 15 living species of horsetails, also called scouring rushes, are all homosporous and herbaceous. They constitute a single genus, *Equisetum*. Fossil forms of *Equisetum* extend back 300 million years to an era when some of their relatives were treelike. Today, they are widely scattered around the world, mostly in damp places. Some that grow among the coastal redwoods of California may reach a height of 3 meters, but most are less than a meter tall (figure 37.11).

FIGURE 37.11
A horsetail, *Equisetum telmateia*, a representative of the only living genus of the phylum Arthrophyta. This species forms two kinds of erect stems; one is green and photosynthetic, and the other, which terminates in a spore-producing "cone," is mostly light brown.

Horsetail sporophytes consist of ribbed, jointed, photosynthetic stems that arise from branching underground rhizomes with roots at their nodes. A whorl of nonphotosynthetic, scalelike leaves emerges at each node. The stems, which are hollow in the center, have silica deposits in the epidermal cells of the ribs, and the interior parts of the stems have two sets of vertical, tubular canals. The larger outer canals, which alternate with the ribs, contain air, while the smaller inner canals opposite the ribs contain water.

Club Mosses (Lycophyta)

The club mosses are worldwide in distribution but are most abundant in the tropics and moist temperate regions. Several genera of club mosses, some of them treelike, became extinct about 270 million years ago. Members of the four genera and nearly 1000 living species of club mosses superficially resemble true mosses, but once their internal structure and reproductive processes became known it was clear that these vascular plants are quite unrelated to mosses. Modern club mosses are either homosporous or heterosporous. The sporophytes have leafy stems that are seldom more than 30 centimeters long. Lycophyta evolved independently from a second monophyletic group of seedless plants.

Ferns and other seedless vascular plants have a much larger and more conspicuous sporophyte, with vascular tissue. Many have well-differentiated roots, stem, and leaves. The shift to a dominant sporophyte lead to the evolution of trees.

37.4 Seeds protect and aid in the dispersal of plant embryos.

Seed Plants

Seed plants first appeared about 425 million years ago. Their ancestors appear to have been spore-bearing plants known as progymnosperms. Progymnosperms shared several features with modern gymnosperms, including secondary xylem and phloem (which allows for an increase in girth later in development). Some progymnosperms had leaves. Their reproduction was very simple, and it is not certain which particular group of progymnosperms gave rise to seed plants.

From an evolutionary and ecological perspective, the seed represents an important advance. The embryo is protected by an extra layer of sporophyte tissue creating the ovule. During development this tissue hardens to produce the seed coat. In addition to protection from drought, dispersal is enhanced. Perhaps even more significantly, a dormant phase is introduced into the life cycle that allows the embryo to survive until environmental conditions are favorable for further growth.

Seed plants produce two kinds of gametophytes—male and female, each of which consists of just a few cells. Pollen grains, multicellular male gametophytes, are conveyed to the egg in the female gametophyte by wind or a pollinator. In some seed plants, the sperm move toward the egg through a growing pollen tube. This eliminates the need for external water. In contrast to the seedless plants, the whole male gametophyte rather than just the sperm moves to the female gametophyte. A female gametophyte develops within an ovule. In flowering plants (angiosperms), the ovules are completely enclosed within diploid sporophyte tissue (ovaries which develop into the fruit). In gymnosperms (mostly cone-bearing seed plants), the ovules are not completely enclosed by sporophyte tissue at the time of pollination.

A common ancestor with seeds gave rise to the gymnosperms and the angiosperms. Seeds can allow for a pause in the life cycle until environmental conditions are more optimal.

A Vocabulary of Plant Terms

androecium The stamens of a flower.

anther The pollen-producing portion of a stamen. This is a sporophyte structure where microspores produced by meiosis develop into male gametophytes.

antheridium The sperm-producing structure found in the gametophytes of seedless plants and certain fungi.

archegonium The multicellular egg-producing structure in the gametophytes of seedless plants and gymnosperms.

carpel A leaflike organ in angiosperms that encloses one or more ovules; a unit of a gynoecium.

double fertilization The process, unique to angiosperms, in which one sperm fuses with the egg, forming a zygote, and the other sperm fuses with the two polar nuclei, forming the primary endosperm nucleus.

embryo sac Female gametophyte in flowering plants.

endosperm The usually triploid (although it can have a much higher ploidy level) food supply of some angiosperm seeds.

filament The stalklike structure that supports the anther of a stamen.

gametophyte The multicellular, haploid phase of a plant life cycle in which gametes are produced by mitosis.

gynoecium The carpel(s) of a flower.

heterosporous Refers to a plant that produces two types of spores: microspores and megaspores.

homosporous Refers to a plant that produces only one type of spore.

integument The outer layer(s) of an ovule; integuments become the seed coat of a seed.

micropyle The opening in the ovule integument through which the pollen tube grows.

nucellus The tissue of an ovule in which an embryo sac develops.

ovary The basal, swollen portion of a carpel (gynoecium); it contains the ovules and develops into the fruit.

ovule A seed plant structure within an ovary; it contains a female gametophyte surrounded by the nucellus and one or two integuments. At maturity, an ovule becomes a seed.

pollen grain Male gametophyte in seed plants; binucleate or trinucleate seed plant structure produced from a microspore in a microsporangium.

pollination The transfer of a pollen grain from an anther to a stigma in angiosperms, or to the vicinity of the ovule in gymnosperms.

primary endosperm nucleus The triploid nucleus resulting from the fusion of a single sperm with the two polar nuclei.

seed A reproductive structure that develops from an ovule in seed plants. It consists of an embryo and a food supply surrounded by a seed coat.

seed coat The protective layer of a seed; it develops from the integument or integuments.

spore A haploid reproductive cell, produced when a diploid spore mother cell undergoes meiosis; it gives rise by mitosis to a gametophyte.

sporophyte The multicellular, diploid phase of a plant life cycle; it is the generation that ultimately produces spores.

stamen A unit of an androecium; it consists of a pollen-bearing anther and usually a stalklike filament.

stigma The uppermost pollen-receptive portion of a gynoecium.

Gymnosperms

There are four groups of living gymnosperms (conifers, cycads, gnetophytes, and *Ginkgo*), none of which is directly related to one another, but all of which lack the flowers and fruits of angiosperms. In all of them the **ovule,** which becomes a seed, rests exposed on a scale (modified leaf) and is not completely enclosed by sporophyte tissues at the time of pollination. The name gymnosperm combines the Greek root *gymnos*, or "naked," with *sperma*, or "seed." In other words, gymnosperms are naked-seeded plants. However, although the ovules are naked at the time of pollination, the seeds of gymnosperms are sometimes enclosed by other sporophyte tissues by the time they are mature.

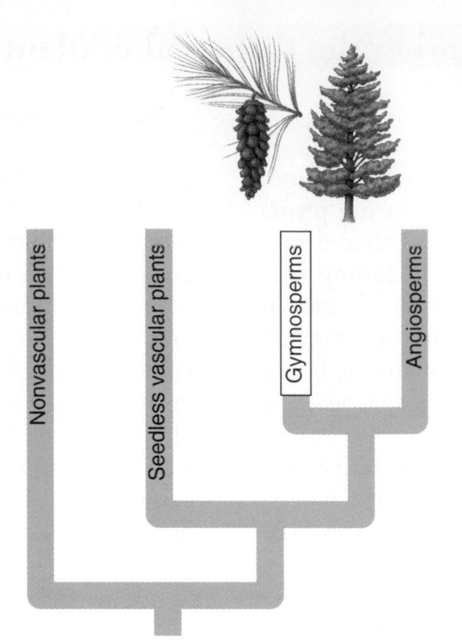

Details of reproduction vary somewhat in gymnosperms, and their forms vary greatly. For example, cycads and *Ginkgo* have motile sperm, even though the sperm are carried within a pollen tube, while many others have sperm with no flagella. The female cones range from tiny woody structures weighing less than 25 grams with a diameter of a few millimeters, to massive structures weighing more than 45 kilograms growing to lengths more than a meter.

Conifers (Coniferophyta)

The most familiar gymnosperms are conifers (phylum Coniferophyta), which include pines (figure 37.12), spruces, firs, cedars, hemlocks, yews, larches, cypresses, and others. The coastal redwood (*Sequoia sempervirens*), a conifer native to northwestern California and southwestern Oregon, is the tallest living vascular plant; it may attain nearly 100 meters (300 feet) in height. Another conifer, the bristlecone pine (*Pinus longaeva*) of the White Mountains of California is the oldest living tree; one is 4900 years of age. Conifers are found in the colder temperate and sometimes drier regions of the world. They are sources of timber, paper, resin, turpentine, taxol (used to treat cancer) and other economically important products.

Pines. More than 100 species of pines exist today, all native to the northern hemisphere, although the range of one species does extend a little south of the equator. Pines and spruces are members of the vast coniferous forests that lie between the arctic tundra and the temperate deciduous forests and prairies to their south. During the past century, pines have been extensively planted in the southern hemisphere.

Pines have tough, needlelike leaves produced mostly in clusters of two to five. The leaves, which have a thick cuticle and recessed stomata, represent an evolutionary adaptation for retarding water loss. This is important because many of the trees grow in areas where the topsoil is frozen for part of the year, making it difficult for the roots to obtain water. The leaves and other parts of the sporophyte have canals into which surrounding cells secrete resin. The resin deters insect and fungal attacks. The resin of certain pines is harvested commercially for its volatile liquid portion, called *turpentine*, and for the solid *rosin*, which is used on stringed instruments. The wood of pines consists primarily of xylem tissue that lacks some of the more rigid cell types found in other trees. Thus it is considered a "soft" rather than a "hard" wood. The thick bark of pines represents another adaptation for surviving fires and subzero temperatures. Some cones actually depend on fire to open, releasing seed to reforest burnt areas.

FIGURE 37.12
Conifers. Slash pines, *Pinus palustris*, in Florida, are representative of the Coniferophyta, the largest phylum of gymnosperms.

As mentioned earlier, all seed plants are heterosporous, so the spores give rise to two types of gametophytes (figure 37.13). The male gametophytes (pollen grains) of pines develop from microspores, which are produced in male cones that develop in clusters of 30 to 70, typically at the tips of the lower branches; there may be hundreds of such clusters on any single tree. The male cones generally are 1 to 4 centimeters long and consist of small, papery scales arranged in a spiral or in whorls. A pair of microsporangia form as sacs within each scale. Numerous microspore mother cells in the microsporangia undergo meiosis, each becoming four microspores. The microspores develop into four-celled pollen grains with a pair of air sacs that give them added buoyancy when released into the air. A single cluster of male pine cones may produce more than 1 million pollen grains.

Female cones typically are produced on the upper branches of the same tree that produces male cones. Female cones are larger than male cones, and their scales become woody. Two ovules develop toward the base of each scale. Each ovule contains a megasporangium called the **nucellus.** The nucellus itself is completely surrounded by a thick layer of cells called the integument that has a small opening (the **micropyle**) toward one end. One of the layers of the integument later becomes the seed coat. A single mega-spore mother cell within each mega-sporangium undergoes meiosis, becoming a row of four megaspores. Three of the mega-spores break down, but the remaining one, over the better part of a year, slowly develops into a female gametophyte. The female gametophyte at maturity may consist of thousands of cells, with two to six archegonia formed at the micropylar end. Each archegonium contains an egg so large it can be seen without a microscope.

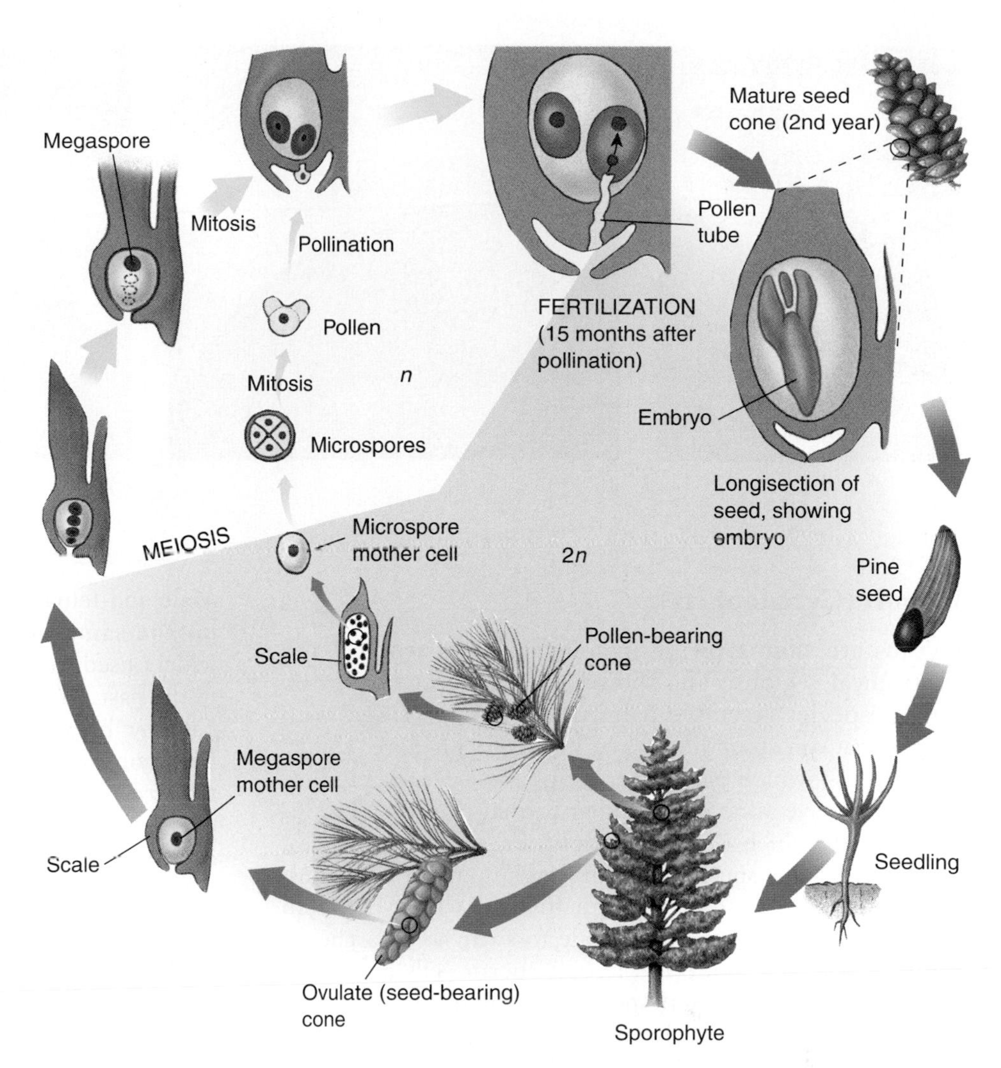

FIGURE 37.13

Life cycle of a typical pine. The male and female gametophytes are dramatically reduced in size in these plants. Wind generally disperses sperm that is within the male gametophyte (pollen). Pollen tube growth delivers the sperm to the egg on the female cone. Additional protection for the embryo is provided by the ovule which develops into the seed coat.

Female cones usually take two or more seasons to mature. At first they may be reddish or purplish in color, but they soon turn green, and during the first spring, the scales spread apart. While the scales are open, pollen grains carried by the wind drift down between them, some catching in sticky fluid oozing out of the micropyle. The pollen grains within the sticky fluid are slowly drawn down through the micropyle to the top of the nucellus, and the scales close shortly thereafter. The archegonia and the rest of the female gametophyte are not mature until about a year later. While the female gametophyte is developing, a pollen tube emerges from a pollen grain at the bottom of the micropyle and slowly digests its way through the nucellus to the archegonia. While the pollen tube is growing, one of the pollen grain's four cells, the generative cell, divides by mitosis, with one of the resulting two cells dividing once more. These last two cells function as sperm. The germinated pollen grain with its two sperm is the mature male gametophyte.

About 15 months after pollination, the pollen tube reaches an archegonium, and discharges its contents into it. One sperm unites with the egg, forming a zygote. The other sperm and cells of the pollen grain degenerate. The zygote develops into an embryo within a seed. After dispersal and germination of the seed, the young sporophyte of the next generation grows into a tree.

(a)

(b)

(c)

FIGURE 37.14 Three phyla of gymnosperms. (*a*) An African cycad, *Encephalartos transvenosus*. (*b*) *Welwitschia mirabilis*, one of the three genera of gnetophytes. (*c*) Maidenhair tree, *Ginkgo biloba*, the only living representative of the phylum Ginkgophyta.

Cycads (Cycadophyta)

Cycads are slow-growing gymnosperms of tropical and subtropical regions. The sporophytes of most of the 100 known species resemble palm trees (figure 37.14*a*) with trunks that can attain heights of 15 meters or more. Unlike palm trees—which are flowering plants—cycads produce cones and have a life cycle similar to that of pines. The female cones, which develop upright among the leaf bases, are huge in some species and can weigh up to 45 kilograms. The sperm of cycads, although formed within a pollen tube, are released within the ovule to swim to an archegonium. These sperm are the largest sperm cells among all living organisms. Several species are facing extinction in the wild and soon may exist only in botanical gardens.

Gnetophytes (Gnetophyta)

There are three genera and about 70 living species of Gnetophyta. They are the only gymnosperms with vessels (a particularly efficient conducting cell type) in their xylem—a common feature in angiosperms. The members of the three genera differ greatly from one another in form. One of the most bizarre of all plants is *Welwitschia*, which occurs in the Namib and Mossamedes deserts of southwestern Africa (figure 37.14*b*). The stem is shaped like a large, shallow cup that tapers into a taproot below the surface. It has two strap-shaped, leathery leaves that grow continuously from their base, splitting as they flap in the wind. The reproductive structures of *Welwitschia* are conelike, appear toward the bases of the leaves around the rims of the stems, and are produced on separate male and female plants.

More than half of the gnetophyte species are in the genus *Ephedra*, which is common in arid regions of the western United States and Mexico. Species are found on every continent except Australia. The plants are shrubby, with stems that superficially resemble those of horsetails as they are jointed and have tiny, scalelike leaves at each node. Male and female reproductive structures may be produced on the same or different plants. The drug ephedrine, widely used in the treatment of respiratory problems, was in the past extracted from Chinese species of *Ephedra*, but it has now been largely replaced with synthetic preparations. Mormon tea is brewed from *Ephedra* stems in the southwestern United States.

The best known species of *Gnetum* is a tropical tree, but most species are vinelike. All species have broad leaves similar to those of angiosperms. One *Gnetum* species is cultivated in Java for its tender shoots, which are cooked as a vegetable.

Ginkgo (Ginkgophyta)

The fossil record indicates that members of the Ginkgo family were once widely distributed, particularly in the northern hemisphere; today only one living species, the maidenhair tree (*Ginkgo biloba*), remains. The tree, which sheds its leaves in the fall, was first encountered by Europeans in cultivation in Japan and China; it apparently no longer exists in the wild (figure 37.14*c*). The common name comes from the resemblance of its fan-shaped leaves to the leaflets of maidenhair ferns. Like the sperm of cycads, those of *Ginkgo* have flagella. The *Ginkgo* is diecious, that is, the male and female reproductive structures of *Ginkgo* are produced on separate trees. The fleshy outer coverings of the seeds of female *Ginkgo* plants exude the foul smell of rancid butter caused by butyric and isobutyric acids. Due to this, male plants vegetatively propagated from shoots are preferred for cultivation. Because it is resistant to air pollution, *Ginkgo* is commonly planted along city streets.

Gymnosperms are mostly cone-bearing seed plants. In gymnosperms, the ovules are not completely enclosed by sporophyte tissue at pollination.

Angiosperms

The 250,000 known species of flowering plants are called angiosperms because their ovules, unlike those of gymnosperms, are enclosed within diploid tissues at the time of pollination. The name *angiosperm* derives from the Greek words *angeion*, "vessel," and *sperma*, "seed." The "vessel" in this instance refers to the carpel, which is a modified leaf that encapsulates seeds. The carpel develops into the fruit, a unique angiosperm feature. While some gymnosperms, including yew, have fleshlike tissue around their seeds, it is of a different origin and not a true fruit.

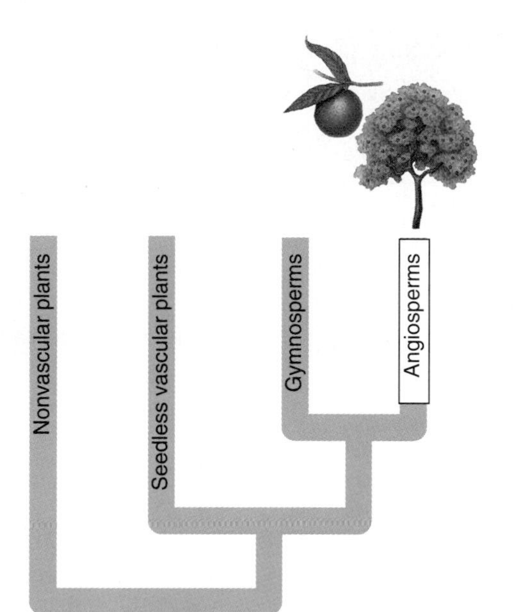

MONOCOTS

1. Seed with one cotyledon ("seed leaf").
2. Leaves with parallel veins.
3. Lateral meristems rarely occur.
4. Flower parts mostly in threes or multiples of three.

DICOTS

1. Seed with two cotyledons ("seed leaves").
2. Leaves with a network of veins.
3. Lateral meristems (cambia) present.
4. Flower parts mostly in fours or fives or multiples of four or five.

FIGURE 37.15
Comparison of monocots and dicots.

The origins of the angiosperms puzzled even Darwin (his "abominable mystery"). Recently, consensus has been growing on the most basal, living angiosperm—*Amborella trichopoda. Amborella,* with small, cream-colored flowers, is even more primitive than magnolias or water lilies, other candidates for basal ancestors. This small shrub, found only on the island of New Caledonia in the South Pacific, is the last remaining species of the earliest extant lineage of the angiosperms, arising about 135 million years ago. While *Amborella* is not the original angiosperm, it is sufficiently close that much may be learned from studying its reproductive biology that will help us understand the early radiation of the angiosperms.

Monocots and Dicots

There are two classes of angiosperms, phylum Anthophyta: the Monocotyledonae, or **monocots** (about 65,000 species) and the Dicotyledonae, or **dicots** (about 175,000 species). Dicots are the more primitive of the two classes, with monocots apparently having derived from early dicots. Included in the dicots are the great majority of familiar angiosperms—almost all kinds of trees and shrubs, snapdragons, mints, peas, sunflowers, and other plants. Monocots include the lilies, grasses, cattails, palms, agaves, yuccas, pondweeds, orchids, and irises.

Monocots and dicots differ from each other in a number of features, some are listed in figure 37.15. Monocots and dicots differ fundamentally in other ways. For example, about a sixth of all dicot species are **annuals** (plants that complete their entire growth cycle within a year); there are, however, very few annual monocots. Underground swollen storage organs, such as bulbs, occur much more frequently in monocots than they do in dicots. There are many species of woody dicots (mostly trees or shrubs), but no monocots have true wood; however, a few monocots, such as palms and bamboos, produce extra bundles of conducting tissues that give them a woody texture. Endosperm, which is usually present in mature monocot seeds, is largely absent in mature dicot seeds. Other specific differences will be presented in chapter 38.

The Structure of Flowers

Flowers are considered to be modified stems bearing modified leaves. Regardless of their size and shape, they all share certain features (see figure 37.16). Each flower originates as a **primordium** that develops into a bud at the end of a stalk called a **pedicel.** The pedicel expands slightly at the tip into a base, the **receptacle,** to which the remaining flower parts are attached. The other flower parts typically are attached in circles called *whorls.* The outermost whorl is composed of **sepals.** In most flowers there are three to five sepals, which are green and somewhat leaflike; they often function in protecting the immature flower and in some species may drop off as the flower opens. The next whorl consists of **petals** that are often colored and attract pollinators such as insects and birds. The petals, which commonly number three to five, may be separate, fused together, or missing altogether *in wind-pollinated flowers.*

The third whorl consists of **stamens,** collectively called the **androecium,** a term derived from the Greek words *andros,* "male," and *oikos,* "house." Each stamen consists of a pollen-bearing **anther** and a stalk called a **filament,** which may be missing in some flowers. The **gynoecium,** consisting of one or more **carpels,** is at the center of the flower. The term *gynoecium* derives from the Greek words *gynos,* which means "female," and *oikos,* or "house." The first carpel is believed to have been formed from a leaflike structure with ovules along its margins. The edges of the blade then rolled inward and fused together, forming a carpel.

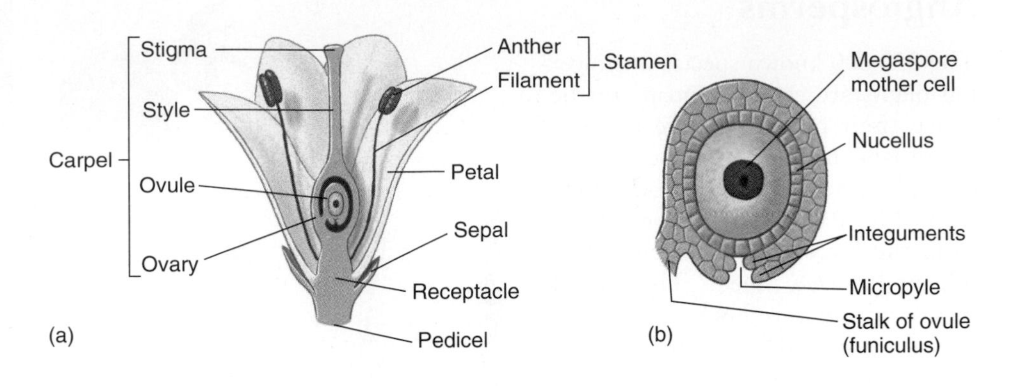

FIGURE 37.16
Diagram of an angiosperm flower. (*a*) The main structures of the flower are labeled. (*b*) Details of an ovule. The ovary as it matures will become a fruit; as the ovule's outer layers (integuments) mature, they will become a seed coat.

Primitive flowers can have several to many separate carpels, but in most flowers, two to several carpels are fused together. Such fusion can be seen in an orange sliced in half; each segment represents one carpel. A carpel has three major regions (figure 37.16). The **ovary** is the swollen base, which contains from one to hundreds of **ovules;** the ovary later develops into a **fruit.** The tip of the carpel is called a **stigma.** Most stigmas are sticky or feathery, causing pollen grains that land on them to adhere. Typically there is a neck or stalk called a **style** connecting the stigma and the ovary; in some flowers, the style may be very short or even missing. Many flowers have nectar-secreting glands called *nectaries,* often located toward the base of the ovary. Nectar is a fluid containing sugars, amino acids, and other molecules used to attract insects, birds, and other animals to flowers.

The Angiosperm Life Cycle

While a flower bud is developing, a single megaspore mother cell in the ovule undergoes meiosis, producing four megaspores (figure 37.17). In most flowering plants, three of the megaspores soon disappear while the nucleus of the remaining one divides mitotically, and the cell slowly expands until it becomes many times its original size. While this expansion is occurring, each of the daughter nuclei divide twice, resulting in eight haploid nuclei arranged in two groups of four. At the same time, two layers of the ovule, the **integuments,** differentiate and become the *seed coat* of a seed. The integuments, as they develop, leave a small gap or pore at one end—the *micropyle* (see figure 37.16). One nucleus from each group of four migrates toward the center, where they function as **polar nuclei.** Polar nuclei may fuse together, forming a single diploid nucleus, or they may form a single cell with two haploid nuclei. Cell walls also form around the remaining nuclei. In the group closest to the micropyle, one cell functions as the **egg;** the other two nuclei are called **synergids.** At the other end, the three cells are now called **antipodals;** they have no apparent function and eventually break down and disappear. The large sac with eight nuclei in seven cells is called an **embryo sac;** it constitutes the female gametophyte. Although it is completely dependent on the sporophyte for nutrition, it is a multicellular, haploid individual.

While the female gametophyte is developing, a similar but less complex process takes place in the anthers. Most anthers have patches of tissue (usually four) that eventually become chambers lined with nutritive cells. The tissue in each patch is composed of many diploid microspore mother cells that undergo meiosis more or less simultaneously, each producing four microspores. The four microspores at first remain together as a quartet or tetrad, and the nucleus of each microspore divides once; in most species the microspores of each quartet then separate. At the same time, a two-layered wall develops around each microspore. As the anther matures, the wall between adjacent pairs of chambers breaks down, leaving two larger sacs. At this point, the binucleate microspores have become **pollen grains.** The outer pollen grain wall layer often becomes beautifully sculptured, and it contains chemicals that may react with others in a stigma to signal whether or not development of the male gametophyte should proceed to completion. The pollen grain has areas called *apertures,* through which a pollen tube may later emerge.

Pollination is simply the mechanical transfer of pollen from its source (an anther) to a receptive area (the stigma of a flowering plant). Most pollination takes place between flowers of different plants and is brought about by insects, wind, water, gravity, bats, and other animals. In as many as a quarter of all angiosperms, however, a pollen grain may be deposited directly on the stigma of its own flower, and self-pollination occurs. Pollination may or may not be followed by *fertilization,* depending on the genetic compatibility of the pollen grain and the flower on whose stigma it has landed. (In some species, complex, genetically controlled mechanisms prevent self-fertilization to enhance genetic diversity in the progeny.) If the stigma is receptive, the pollen grain's dense cytoplasm absorbs substances from the stigma and bulges through an aperture. The bulge develops into a *pollen tube* that responds to chemical and mechanical stimuli that guide it to the embryo sac. It follows a diffusion gradient of the chemicals and grows down through the style and into the micropyle. The pollen tube usually takes several hours to two days to reach the micropyle, but in a few instances, it may take up to a year. One of the pollen grain's two cells, the *generative cell,* lags behind. Its nucleus divides,

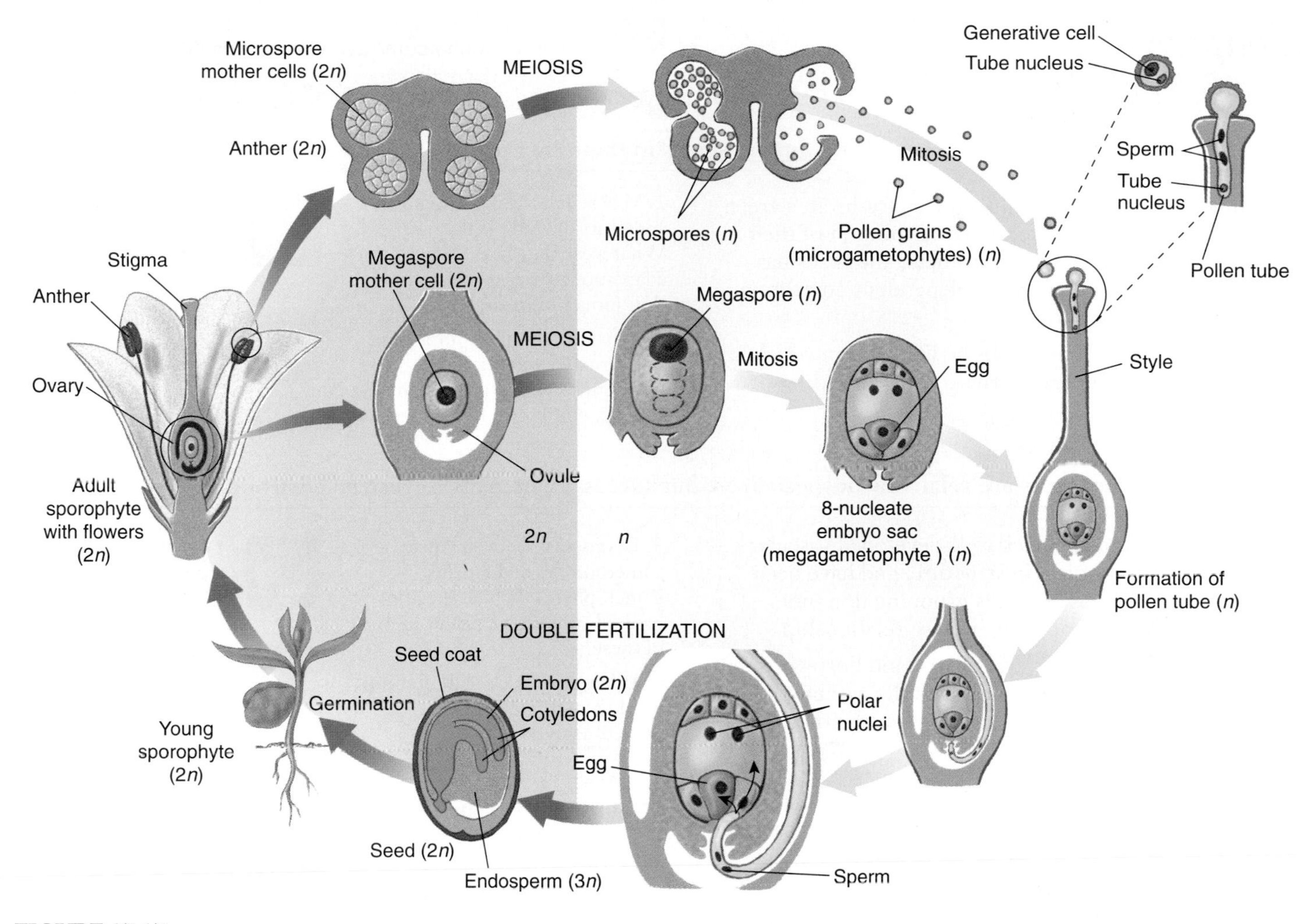

FIGURE 37.17
Life cycle of a typical angiosperm. As in pines, external water is no longer required for fertilization. In most species of angiosperms, animals carry pollen to the carpel. The outer wall of the carpel forms the fruit that entices animals to disperse the seed.

in the pollen grain or in the pollen tube, producing two sperm cells. Unlike sperm in mosses, ferns, and some gymnosperms, the sperm of flowering plants have no flagella. At this point, the pollen grain with its tube and sperm has become a mature male gametophyte.

As the pollen tube enters the embryo sac, it destroys a synergid in the process and then discharges its contents. Both sperm are functional, and an event called **double fertilization,** unique to angiosperms, follows. One sperm unites with the egg and forms a zygote, which develops into an embryo sporophyte plant. The other sperm and the two polar nuclei unite, forming a triploid primary endosperm nucleus. The primary endosperm nucleus begins dividing rapidly and repeatedly, becoming triploid **endosperm** tissue that may soon consist of thousands of cells. Endosperm tissue can become an extensive part of the seed in grasses such as corn (see figure 40.7).

In most flowering plants, it provides nutrients for the embryo that develops from the zygote; in many species, such as peas and beans, it disappears completely by the time the seed is mature. Following double fertilization, the integuments harden and become the seed coat of a seed. The haploid cells remaining in the embryo sac (antipodals, synergid, tube nucleus) degenerate. There is some evidence for a type of double fertilization in gymnosperms believed to be closely related to the angiosperms. Further studies of this and of fertilization in *Amborella,* the most basal, extant angiosperm, may provide clues to the evolution of this double fertilization event.

Angiosperms are characterized by ovules that at pollination are enclosed within an ovary at the base of a carpel—a structure unique to the phylum; a fruit develops from the ovary. Evolutionary innovations including flowers to attract pollinators, fruits to protect and aid in embryo dispersal, and double fertilization providing additional nutrients for the embryo all have contributed to the widespread success of this phylum.

Chapter 37

www.mhhe.com/raven6/resources37.mhtml

Summary	*Questions*	*Media Resources*

37.1 Plants have multicellular haploid and diploid stages in their life cycles.

- Plants evolved from a multicellular, freshwater green algae 470 million years ago. The evolution of their conducting tissues, cuticle, stomata, and seeds has made them progressively less dependent on external water for reproduction.
- All plants have a haplodiplontic life cycle in which haploid gametophytes alternate with diploid sporophytes.

1. Where did the most recent ancestors of land plants live? What were they like? What adaptations were necessary for the "move" onto land?

2. What does it mean for a plant to alternate generations? Distinguish between sporophyte and gametophyte.

- Introduction to Plants
- Plant Characteristics
- Life Cycles of Plants

- Student Research: Plant Biodiversity in New Hampshire

37.2 Nonvascular plants are relatively unspecialized, but successful in many terrestrial environments.

- Three phyla of plants lack well-developed vascular tissue, are the simplest in structure, and have been grouped as bryophytes. This grouping does not reflect a common ancestry or close relationship.
- Sporophytes of mosses, liverworts, and hornworts are usually nutritionally dependent on the gametophytes, which are more conspicuous and photosynthetic.

3. Distinguish between male gametophytes and female gametophytes. Which specific haploid spores give rise to each of these?

4. What reproductive limitations would a moss tree (if one existed) face?

- Nonvascular Plants

- Art Quiz: Moss Life Cycle

37.3 Seedless vascular plants have well-developed conducting tissues in their sporophytes.

- Nine of the 12 plant phyla contain vascular plants, which have two kinds of well-defined conducting tissues: xylem, which is specialized to conduct water and dissolved minerals; and phloem, which is specialized to conduct the sugars produced by photosynthesis and hormones.
- In ferns and other seedless vascular plants, the sporophyte generation is dominant. The fern sporophyte has vascular tissue and well-differentiated roots, stems, and leaves.

5. In what ways are the gametophytes of seedless plants different from the gametophytes of seed plants?

6. Which generation(s) of the fern are nutritionally independent?

- Seedless Vascular Plants

- Art Quiz: Fern Life Cycle

37.4 Seeds protect and aid in the dispersal of plant embryos.

- Seeds were an important evolutionary advance providing for a dormant stage in development.
- In gymnosperms, ovules are exposed directly to pollen at the time of pollination; in angiosperms, ovules are enclosed within an ovary, and a pollen tube grows from the stigma to the ovule.
- The pollen of gymnosperms is usually disseminated by wind. In most angiosperms the pollen is transported by insects and other animals. Both flowers and fruits are found only in angiosperms and may account for the extensive colonization of terrestrial environments by the flowering plants.

7. What is a seed? Why is the seed a crucial adaptation to terrestrial life?

8. What is the principal difference between gymnosperms and angiosperms?

9. If all the offspring of a plant were to develop in a small area, they would suffer from limited resources. Compare dispersal strategies in moss, pine, and angiosperms.

- Gymnosperms
- Angiosperms

- Art Quizzes:
  - -Plant Life Cycle
  - -Angiosperm Flower
  - -Angiosperm Life Cycle

38

The Plant Body

Concept Outline

FIGURE 38.1
All vascular plants share certain characteristics. Vascular plants such as this tree require an elaborate system of mechanical support and fluid transport to grow this large. Smaller plants have similar (though simpler) structures. Much of this support system is actually underground in the form of extensive branching root systems.

Although the similarities between a cactus, an orchid plant, and a tree might not be obvious at first sight, most plants have a basic unity of structure (figure 38.1). This unity is reflected in how the plants are constructed; in how they grow, manufacture, and transport their food; and in how their development is regulated. This chapter addresses the question of how a vascular plant is "built." We will focus on the diversity of cell, tissue, and organ types that compose the adult body. The roots and shoots which give the adult plant its distinct above and below ground architecture are the final product of a basic body plan first established during embryogenesis, a process we will explore in detail in chapter 40.

38.1 Meristems elaborate the plant body plan after germination.

Meristems

The plant body that develops after germination depends on the activities of meristematic tissues. Meristematic tissues are clumps of small cells with dense cytoplasm and proportionately large nuclei that act like stem cells in animals. That is, one cell divides to give rise to two cells. One remains meristematic, while the other is free to differentiate and contribute to the plant body. In this way, the population of meristem cells is continually renewed. Molecular genetic evidence supports the hypothesis that stem cells and meristem cells may also share some common pathways of gene expression.

Elongation of both root and shoot takes place as a result of repeated cell divisions and subsequent elongation of the cells produced by the **apical meristems.** In some vascular plants, including shrubs and most trees, **lateral meristems** produce an increase in girth.

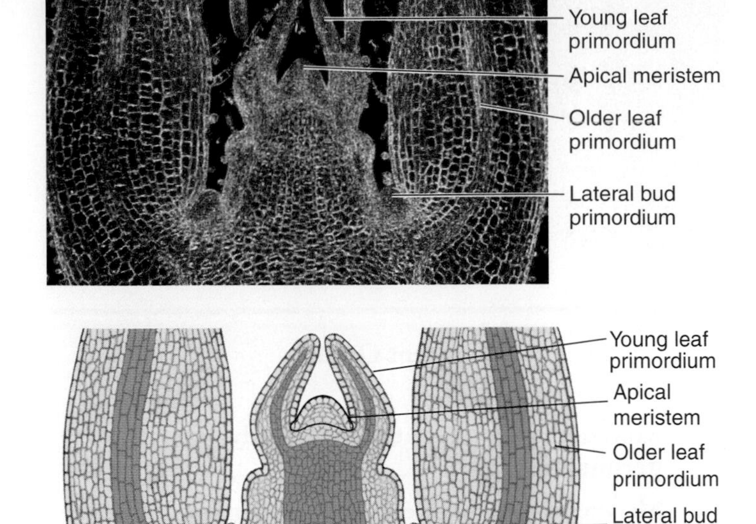

FIGURE 38.2
An apical shoot meristem. This longitudinal section through a shoot apex in *Coleus* shows the tip of a stem. Between the young leaf primordia is the apical meristem.

Apical Meristems

Apical meristems are located at the tips of stems (figure 38.2) and at the tips of roots (figure 38.3), just behind the root cap. The plant tissues that result from primary growth are called **primary tissues.** During periods of growth, the cells of apical meristems divide and continually add more cells to the tips of a seedling's body. Thus, the seedling lengthens. Primary growth in plants is brought about by the apical meristems. The elongation of the root and stem forms what is known as the **primary plant body,** which is made up of primary tissues. The primary plant body comprises the young, soft shoots and roots of a tree or shrub, or the entire plant body in some herbaceous plants.

Both root and shoot apical meristems are composed of delicate cells that need protection. The root apical meristem is protected from the time it emerges by the root cap. Root cap cells are produced by the root meristem and are sloughed off and replaced as the root moves through the soil. A variety of adaptive mechanisms protect shoot apical meristem during germination (figure 38.4). The epicotyl or hypocotyl ("stemlike" tissue above or below the cotyledons) may bend as the seedling emerges to minimize the force on the shoot tip. In the monocots (a late evolving group of angiosperms) there is often a coleoptile (sheath of tissue) that forms a protective tube around the emerging shoot. Later in development, the leaf primordia cover the shoot apical meristem which is particularly susceptible to desiccation.

The apical meristem gives rise to three types of embryonic tissue systems called **primary meristems.** Cell division continues in these partly differentiated tissues as they develop into the primary tissues of the plant body. The

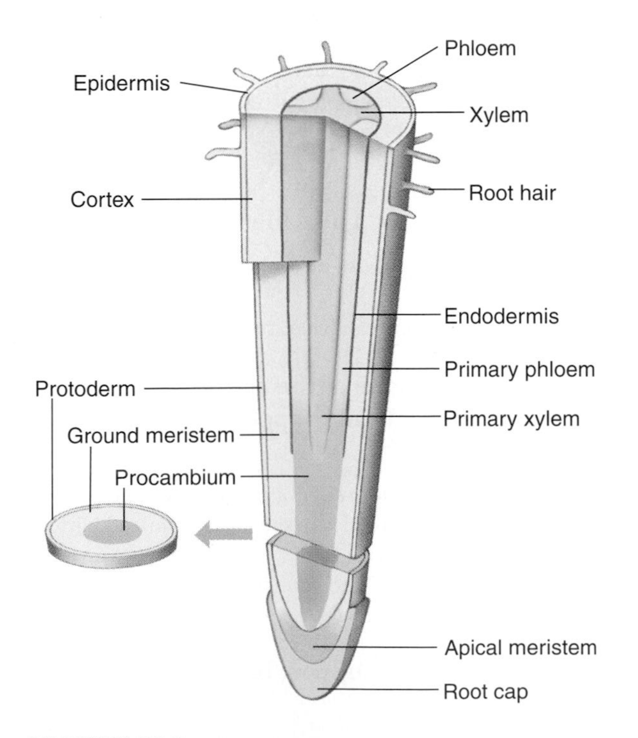

FIGURE 38.3
An apical root meristem. This diagram of meristems in the root shows their relation to the root tip.

FIGURE 38.4
Developing seedling. Apical meristems are protected early in development. (*a*) In this soybean, a dicot, a bent epicotyl (stem above the cotyledons), rather than the shoot tip, pushes through the soil before straightening. (*b*) In corn, a monocot, a sleeve of tissue called the coleoptile sheaths the shoot tip until it has made it to daylight.

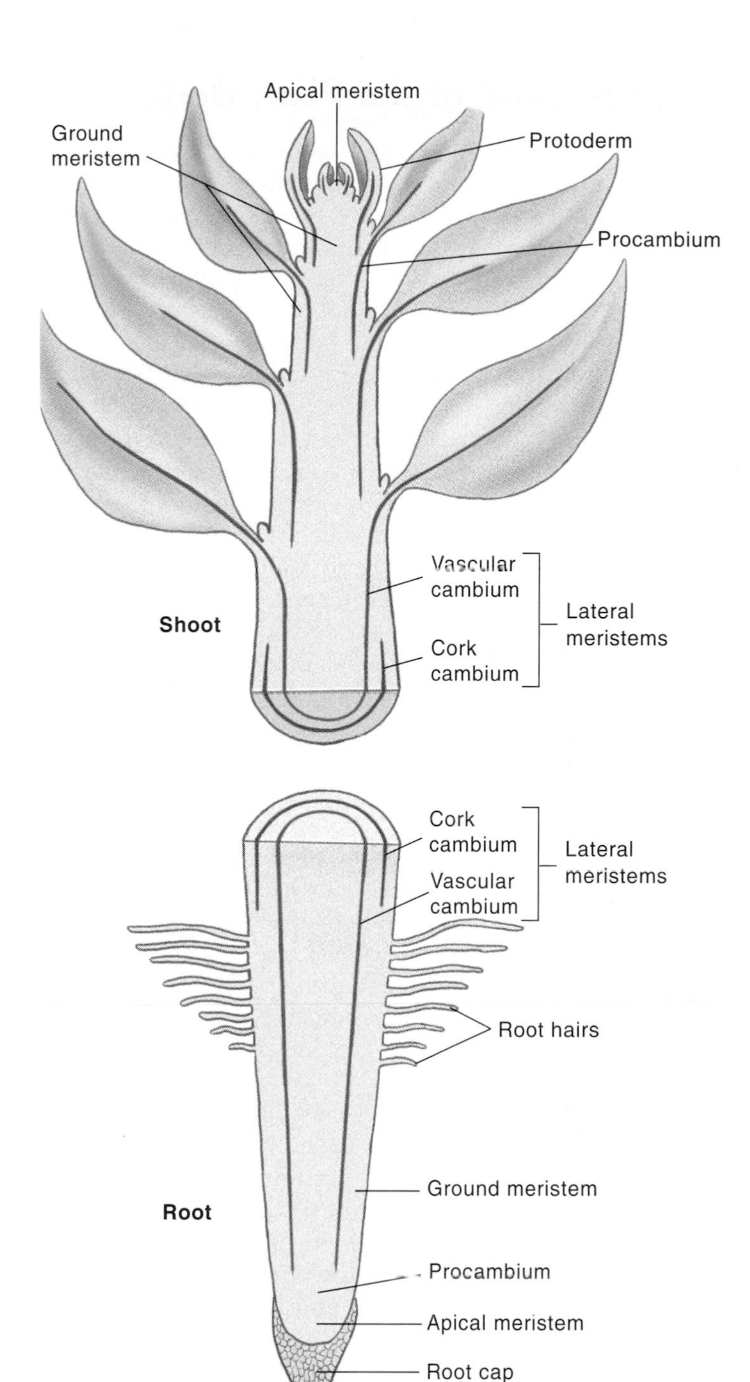

FIGURE 38.5
Apical and lateral meristems. Apical meristems produce primary growth, the elongation of the root and stem. In some plants, the lateral meristems produce an increase in the girth of a plant. This type of growth is secondary because the meristems were not directly produced by apical meristems.

three primary meristems are the **protoderm,** which forms the epidermis; the **procambium,** which produces primary vascular tissues (primary xylem and primary phloem); and the **ground meristem,** which differentiates further into ground tissue, which is composed of parenchyma cells. In some plants, such as horsetails and corn, **intercalary meristems** arise in stem internodes, adding to the internode lengths. If you walk through a corn field (when the corn is about knee high) on a quiet summer night, you may hear a soft popping sound. This is caused by the rapid growth of intercalary meristems. The amount of stem elongation that occurs in a very short time is quite surprising.

Lateral Meristems

Many herbaceous plants exhibit only primary growth, but others also exhibit **secondary growth.** Most trees, shrubs, and some herbs have active **lateral meristems,** which are peripheral cylinders of meristematic tissue within the stems and roots (figure 38.5). Although secondary growth increases girth in many nonwoody plants, its effects are most dramatic in woody plants which have two lateral meristems. Within the bark of a woody stem is the **cork cambium,** a lateral meristem that produces the cork cells of the outer bark. Just beneath the bark is the **vascular cambium,** a lateral meristem that produces secondary vascular tissue. The vascular cambium forms between the xylem and phloem in vascular bundles, adding secondary vascular tissue on opposite sides of the vascular cambium. *Secondary xylem* is the main component of wood. *Secondary phloem* is very close to the outer surface of a woody stem. Removing the bark of a tree damages the phloem and may eventually kill the tree. Tissues formed from lateral meristems, which comprise most of the trunk, branches, and older roots of trees and shrubs, are known as **secondary tissues** and are collectively called the secondary plant body.

Meristems are actively dividing, embryonic tissues responsible for both primary and secondary growth.

Organization of the Plant Body

Coordination of primary and secondary meristematic growth produces the body of the adult sporophyte. Plant bodies do not necessarily have a fixed size. Parts such as leaves, roots, branches, and flowers all vary in size and number from plant to plant—even within a species. The development of the form and structure of plant parts may be relatively rigidly controlled, but some aspects of leaf, stem, and root development are quite flexible. As a plant grows, the number, location, size, and even structure of leaves and roots are often influenced by the environment.

A vascular plant consists of a **root system** and a **shoot system** (figure 38.6). The root system anchors the plant and penetrates the soil, from which it absorbs water and ions crucial to the plant's nutrition. The shoot system consists of the **stems** and their **leaves.** The stem serves as a framework for positioning the leaves, the principal sites of photosynthesis. The arrangement, size, and other features of the leaves are of critical importance in the plant's production of food. Flowers, other reproductive organs, and, ultimately, fruits and seeds are also formed on the shoot (see chapters 40 and 42). The reiterative unit of the vegetative shoot consists of the internode, node, leaf, and axillary buds. Axillary buds are apical meristems derived from the primary apical meristem that allow the plant to branch or replace the main shoot if it is munched by an herbivore. A vegetative axillary bud has the capacity to reiterate the development of the primary shoot. When the plant has transited to the reproductive phase of development (see chapter 41), these axillaries may produce flowers or floral shoots.

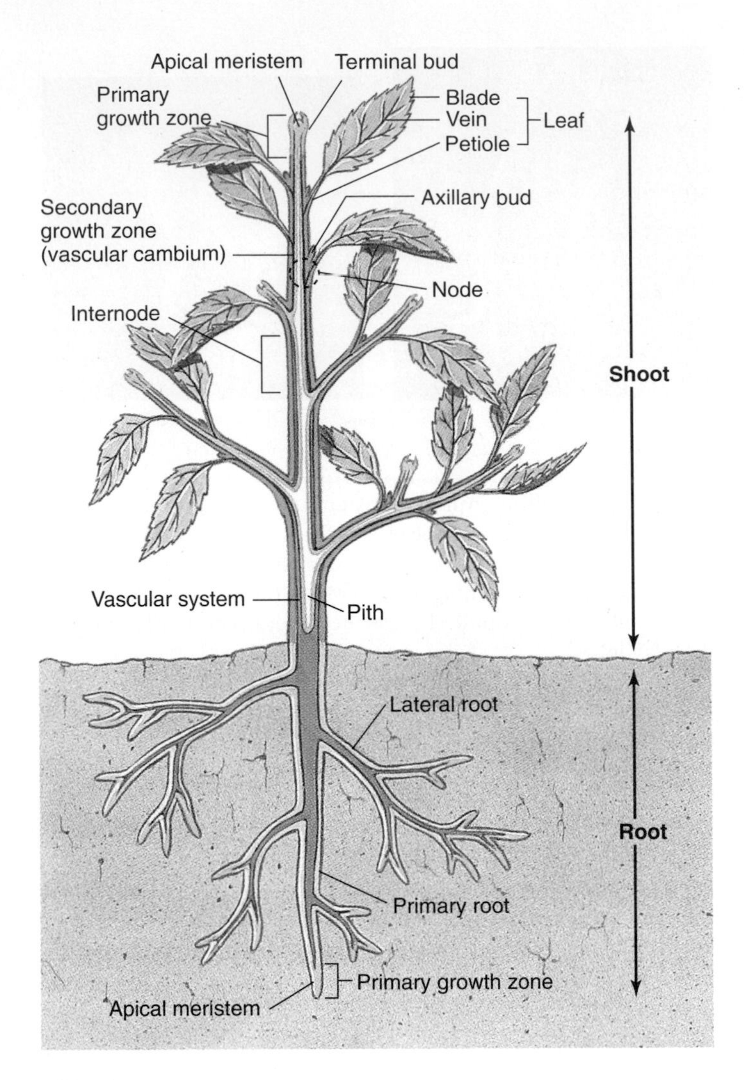

FIGURE 38.6
Diagram of a plant body. Branching in both the root and shoot systems increases the number of apical meristems. A significant increase in stem/root circumference and the formation of bark can only occur if there is secondary growth initiated by the vascular and cork cambiums (secondary meristems). The *lime green* areas are zones of active elongation; secondary growth occurs in the *lavender* areas.

Three basic types of tissues exist in plants: *ground tissue*, *dermal*, and *vascular tissue*. Each of the three basic tissues has its own distinctive, functionally related cell types. Some of these cell types will be discussed later in this chapter. In plants limited to primary growth, the dermal system is composed of the **epidermis.** This tissue is one cell thick in most plants, and forms the outer protective covering of the plant. In young exposed parts of the plant, the epidermis is covered with a fatty **cutin** layer constituting the **cuticle;** in plants such as the desert succulents, a layer of wax may be added outside the cuticle. In plants with secondary growth, the bark forms the outer protective layer and is considered a part of the dermal tissue system.

Ground tissue consists primarily of thin-walled **parenchyma** cells that are initially (but briefly) more or less spherical. However, the cells, which have living protoplasts, push up against each other shortly after they are produced and assume other shapes, often ending up with 11 to 17 sides. Parenchyma cells may live for many years; they function in storage, photosynthesis, and secretion.

Vascular tissue includes two kinds of conducting tissues: (1) **xylem,** which conducts water and dissolved minerals; and (2) **phloem,** which conducts carbohydrates—mainly sucrose—used by plants for food. The phloem also transports hormones, amino acids, and other substances that are necessary for plant growth. Xylem and phloem differ in structure as well as in function.

Root and shoot meristems give rise to a plant body with an extensive underground, branching root system and aboveground shoot system with reiterative units of node, leaf joined at the node, internode, and axillary buds.

Primary and Secondary Growth

Primary growth plays an important role in establishing the basic body plan of the organism. Some plants have secondary growth allowing for an increase in diameter. Here we will look at how the meristems give rise to highly differentiated tissues that sustain the growing plant body. In the earliest vascular plants, many of which are extinct, the vascular tissues produced by primary meristems played the same conducting roles as they do in contemporary vascular plants. However, there was no differentiation of the plant body into stems, leaves, and roots. The presence of these three kinds of organs is a property of most modern plants and reflects increasing specialization in relation to the demands of a terrestrial existence.

With the evolution of secondary growth, vascular plants could develop thick trunks and become treelike (figure 38.7). This evolutionary advance in the sporophyte generation made possible the development of forests and the domination of the land by plants. Judging from the fossil record, secondary growth evolved independently in several groups of vascular plants by the middle of the Devonian period 380 million years ago.

There were two types of conducting systems in the earliest plants—systems that have become characteristic of vascular plants as a group. *Sieve-tube members* conduct carbohydrates away from areas where they are manufactured or stored. *Vessel members* and *tracheids* are thick-walled cells that transport water and dissolved minerals up from the roots. Both kinds of cells are elongated and occur in linked strands making tubes. Sieve-tube members are characteristic of phloem tissue; vessel members and tracheids are characteristic of xylem tissue. In primary tissues, which result from primary growth, these two types of tissue are typically associated with each other in the same vascular strands. In secondary growth, the phloem is found on the periphery, while a very thick xylem core develops more centrally. You will see that roots and shoots of many vascular plants have different patterns of vascular tissue and secondary growth. Keep in mind that water and nutrients travel between the most distant tip of a redwood root and the tip of the shoot. For the system to work, these tissues connect, which they do in the transition zone between the root and the shoot. In the next section, we will consider the three tissue systems that are present in all plant organs, whether the plant has secondary growth or not.

Plants grow from the division of meristematic tissue. Primary growth results from cell division at the apical meristem at the tip of the plant, making the shoot longer. Secondary growth results from cell division at the lateral meristem in a cylinder encasing the shoot, and increases the shoot's girth.

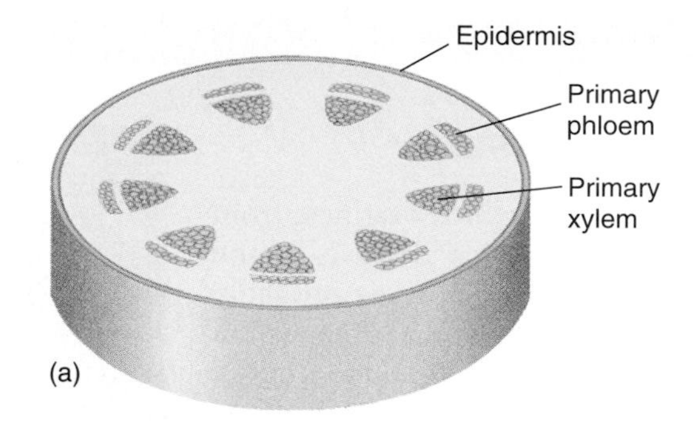

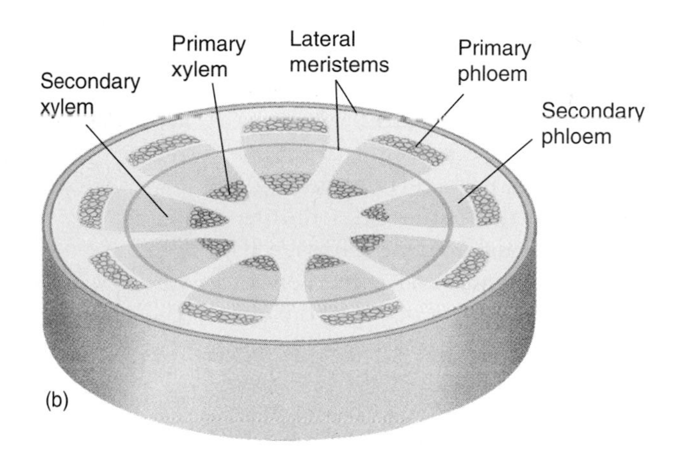

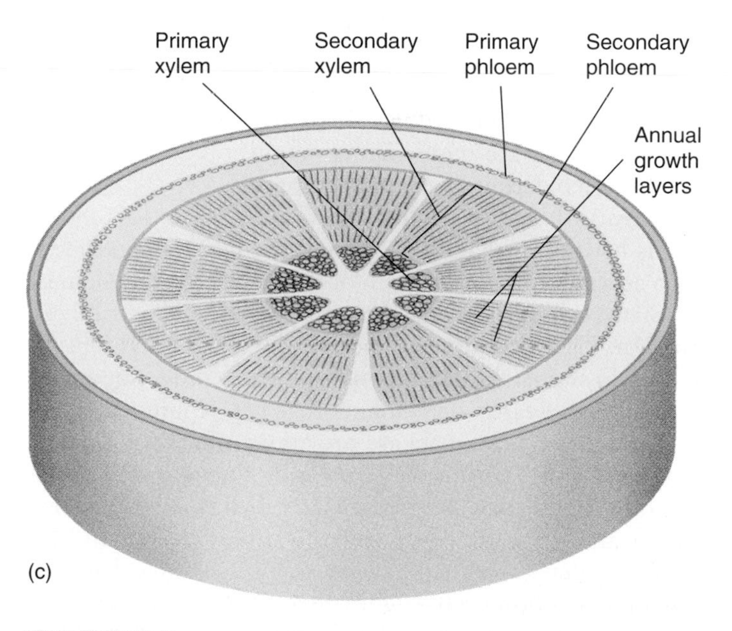

FIGURE 38.7

Secondary growth. (*a*) Before secondary growth begins in dicot stems, primary tissues continue to elongate as the apical meristems produce primary growth. (*b*) As secondary growth begins, the lateral meristems produce secondary tissues, and the stem's girth increases. (*c*) In this three-year-old stem, the secondary tissues continue to widen, and the trunk has become thick and woody. Note that the lateral meristems form cylinders that run axially (up and down) in roots and shoots that have them.

Dermal Tissue

Epidermal cells, which originate from the protoderm, cover all parts of the primary plant body. This is probably the earliest tissue system to appear in embryogenesis. The exposed outer walls have a cuticle that varies in thickness, depending on the species and environmental conditions. A number of types of specialized cells occur in the epidermis, including **guard cells, trichomes,** and **root hairs.**

Guard cells are paired sausage- or dumbbell-shaped cells flanking a **stoma** (plural, **stomata**), a mouth-shaped epidermal opening. Guard cells, unlike other epidermal cells, contain chloroplasts. Stomata occur in the epidermis of leaves (figure 38.8), and sometimes on other parts of the plant, such as stems or fruits. The passage of oxygen and carbon dioxide, as well as diffusion of water in vapor form, takes place almost exclusively through the stomata. There are between 1000 to more than 1 million stomata per square centimeter of leaf surface. In many plants, stomata are more numerous on the lower epidermis than on the upper epidermis of the leaf, which minimizes water loss. Some plants have stomata only on the lower epidermis, and a few, such as water lilies, have them only on the upper epidermis.

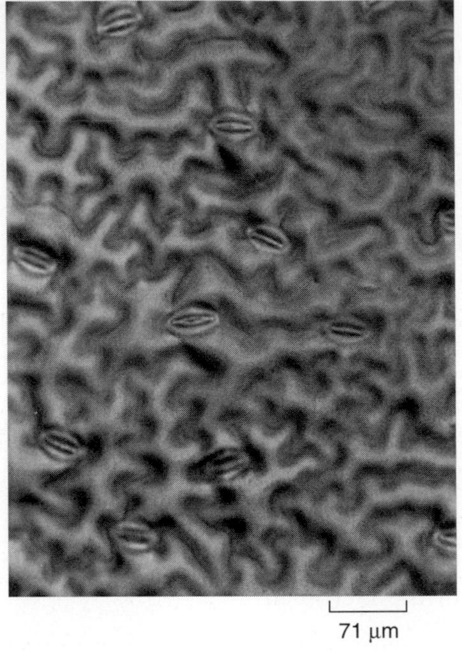

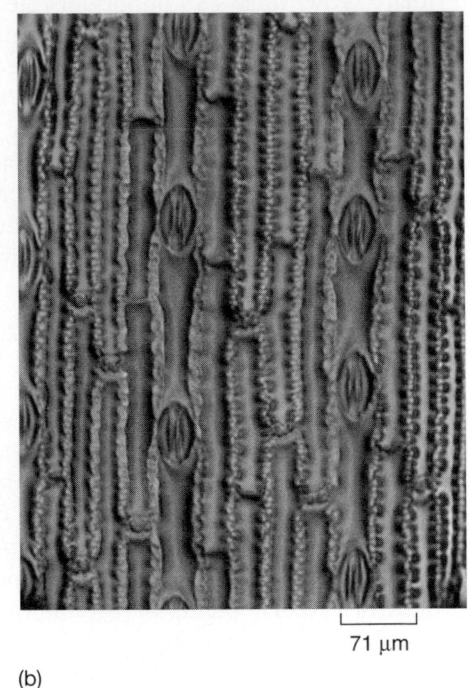

FIGURE 38.8
Epidermis of a dicot and monocot leaf (400×). Stomata are evenly distributed over the epidermis of monocots and dicots, but the patterning is quite different. (*a*) A pea (dicot) leaf with a random arrangement of stomata. (*b*) A corn (monocot) leaf with stomata evenly spaced in rows. These photos also show the variety of cell shapes in plants. Some plant cells are boxlike, as seen in corn (*b*), while others are irregularly shaped, as seen in peas (*a*).

Guard cell formation is the result of an asymmetrical cell division. The patterning of these asymmetrical divisions resulting in stomatal distribution has intrigued developmental biologists. Research on mutants that get "confused" about where to position stomata are providing information on the timing of stomatal initiation and the kind of intercellular communication that triggers guard cell formation. For example, the *too many mouths* mutation may be caused by a failure of developing stomata to suppress stomatal formation in neighboring cells (figure 38.9).

The stomata open and shut in response to external factors such as light, temperature, and availability of water. During periods of active photosynthesis, the stomata are open, allowing the free passage of carbon dioxide into and oxygen out of the leaf. We will consider the mechanism that governs such movements in chapter 39.

Trichomes are hairlike outgrowths of the epidermis (figure 38.10). They occur frequently on stems, leaves, and reproductive organs. A "fuzzy" or "woolly" leaf is covered with trichomes that can be seen clearly with a microscope under low magnification. Trichomes play an important

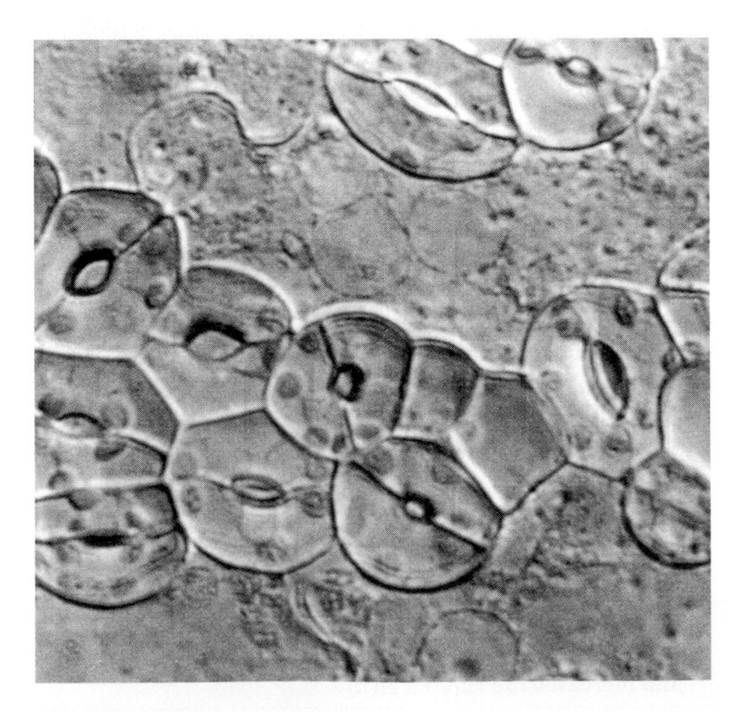

FIGURE 38.9
The *too many mouths* stomatal mutant. This *Arabidopsis* mutant plant lacks an essential signal for spacing guard cells.

role in keeping the leaf surface cool and in reducing the rate of evaporation. Trichomes vary greatly in form in different kinds of plants; some consist of a single cell, while others may consist of several cells. Some are glandular, often secreting sticky or toxic substances to deter herbivory.

Trichome development has been investigated extensively in *Arabidopsis*. Four genes are needed to specify the site of trichome formation and initiate it (figure 38.11). Next, eight genes are necessary for extension growth. Loss of function of any one of these genes results in a trichome with a distorted root hair. This is an example of taking a very simple system and trying to genetically dissect all the component parts. Understanding the formation of more complex plant parts is a major challenge.

Root hairs, which are tubular extensions of individual epidermal cells, occur in a zone just behind the tips of young, growing roots (see figure 38.3). Because a root hair is simply an extension and not a separate cell, there is no crosswall isolating it from the epidermal cell. Root hairs keep the root in intimate contact with the surrounding soil particles and greatly increase the root's surface area and the efficiency of absorption. As the root grows, the extent of the root hair zone remains roughly constant as root hairs at the older end slough off while new ones are produced at the other end. Most of the absorption of water and minerals occurs through root hairs, especially in herbaceous plants. Root hairs should not be confused with lateral roots which are multicellular and have their origins deep within the root.

In the case of secondary growth, the cork cambium (discussed in the section on stems in this chapter) produces the bark of a tree trunk or root. This replaces the epidermis which gets stretched and broken with the radial expansion of the axis. Epidermal cells generally lack the plasticity of other cells, but in some cases, they can fuse to the epidermal cells of another organ and dedifferentiate.

Some epidermal cells are specialized for protection, others for absorption. Spacing of these specialized cells within the epidermis maximizes their function and is an intriguing developmental puzzle.

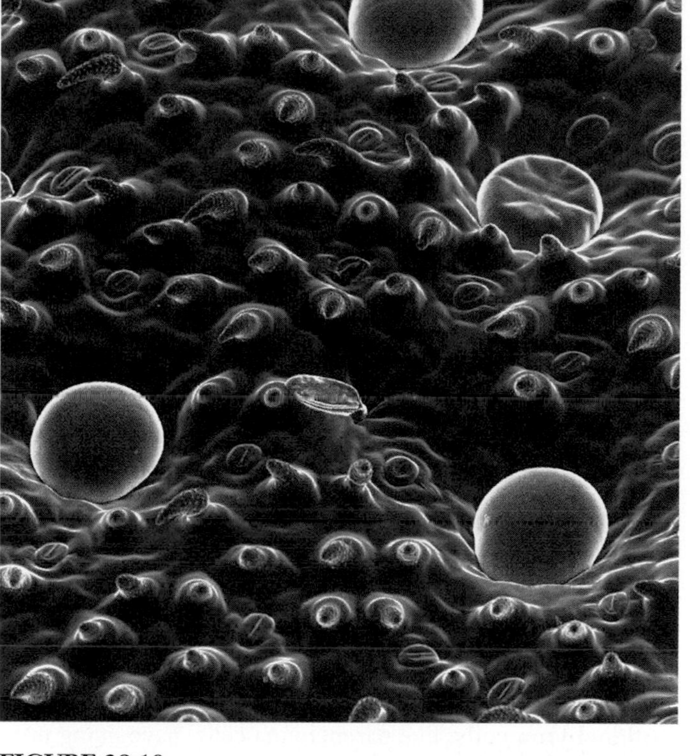

FIGURE 38.10
Trichomes. A covering of trichomes, teardrop-shaped *blue* structures above, creates a layer of more humid air near the leaf surface, enabling the plant to conserve available water supplies.

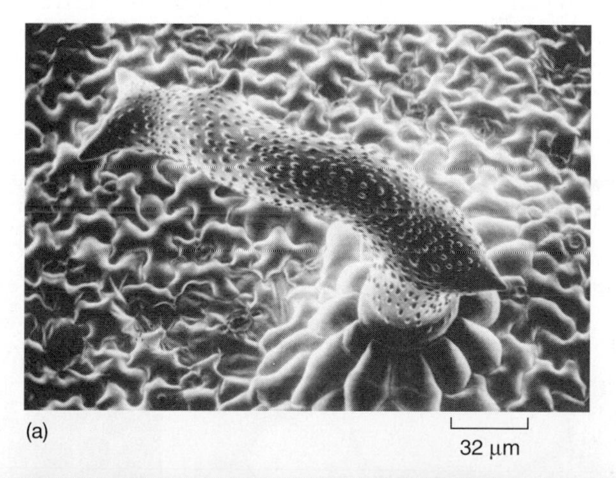

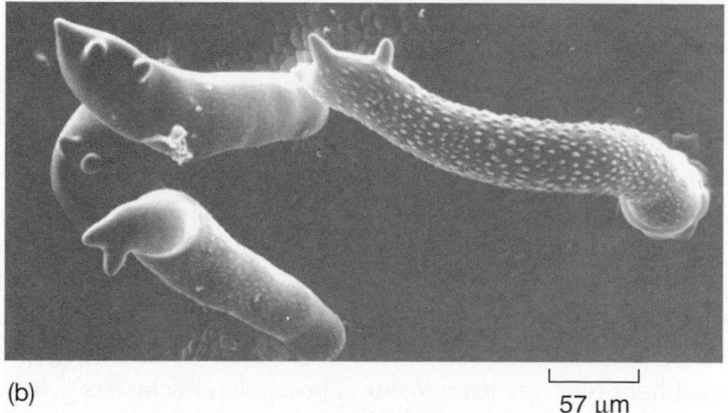

FIGURE 38.11
Trichome mutations. Mutants have revealed genes involved in a signal transduction pathway that regulates the spacing and development of trichomes. These include (*a*) *DISTORTED1* (*DIS1*) and (*b*) *DIS2* mutants in which trichomes are swollen and twisted.

Ground Tissue

Parenchyma

Parenchyma cells, which have large vacuoles, thin walls, and an average of 14 sides at maturity, are the most common type of plant cell. They are the most abundant cells of primary tissues and may also occur, to a much lesser extent, in secondary tissues (figure 38.12*a*). Most parenchyma cells have only primary walls, which are walls laid down while the cells are still maturing. Parenchyma are less specialized than other plant cells, although there are many variations that do have special functions such as nectar and resin secretion, or storage of latex, proteins, and metabolic wastes.

Parenchyma cells, which have functional nuclei and are capable of dividing, commonly also store food and water, and usually remain alive after they mature; in some plants (for example, cacti), they may live to be over 100 years old. The majority of cells in fruits such as apples are parenchyma. Some parenchyma contain chloroplasts, especially in leaves and in the outer parts of herbaceous stems. Such photosynthetic parenchyma tissue is called *chlorenchyma.*

Collenchyma

Collenchyma cells, like parenchyma cells, have living protoplasts and may live for many years. The cells, which are usually a little longer than wide, have walls that vary in thickness (figure 38.12*b*). Collenchyma cells, which are relatively flexible, provide support for plant organs, allowing them to bend without breaking. They often form strands or continuous cylinders beneath the epidermis of stems or leaf petioles (stalks) and along the veins in leaves. Strands of collenchyma provide much of the support for stems in which secondary growth has not taken place. The parts of celery that we eat (petioles, or leaf stalks), have "strings" that consist mainly of collenchyma and vascular bundles (conducting tissues).

Sclerenchyma

Sclerenchyma cells have tough, thick walls; they usually lack living protoplasts when they are mature. Their secondary cell walls are often impregnated with **lignin,** a highly branched polymer that makes cell walls more rigid. Cell walls containing lignin are said to be **lignified.** Lignin is common in the walls of plant cells that have a supporting or mechanical function. Some kinds of cells have lignin deposited in primary as well as secondary cell walls.

There are two types of sclerenchyma: fibers and sclereids. **Fibers** are long, slender cells that are usually grouped together in strands. Linen, for example, is woven from strands of sclerenchyma fibers that occur in the phloem of flax. **Sclereids** are variable in shape but often branched. They may occur singly or in groups; they are not elongated, but may have various forms, including that of a star. The gritty texture of a pear is caused by groups of sclereids that occur throughout the soft flesh of the fruit (figure 38.12*c*). Both of these tough, thick-walled cell types serve to strengthen the tissues in which they occur.

Parenchyma cells are the most common type of plant cells and have various functions. Collenchyma cells provide much of the support in young stems and leaves. Sclerenchyma cells strengthen plant tissues and may be nonliving at maturity.

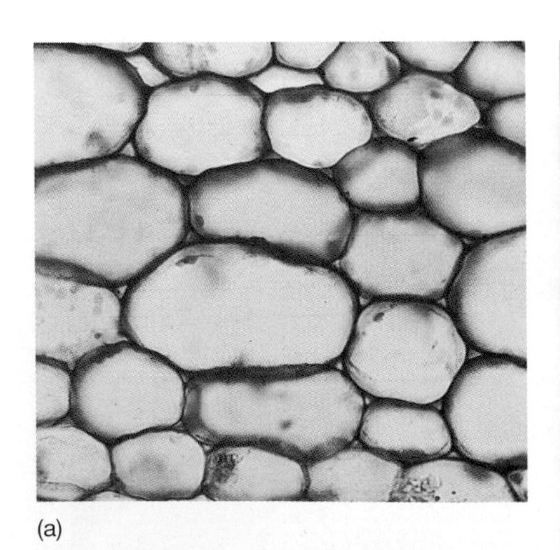

(a)

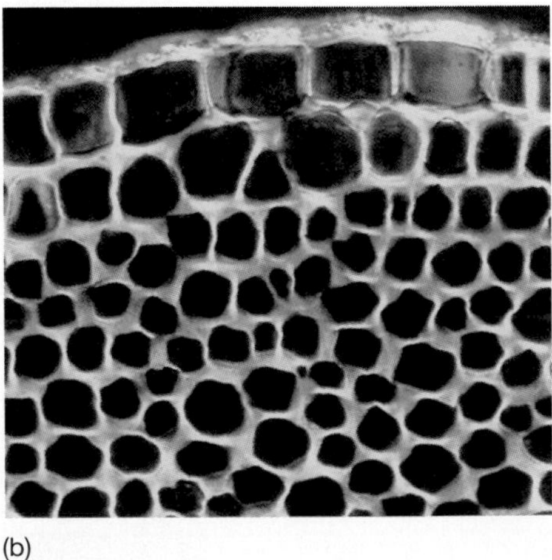

(b)

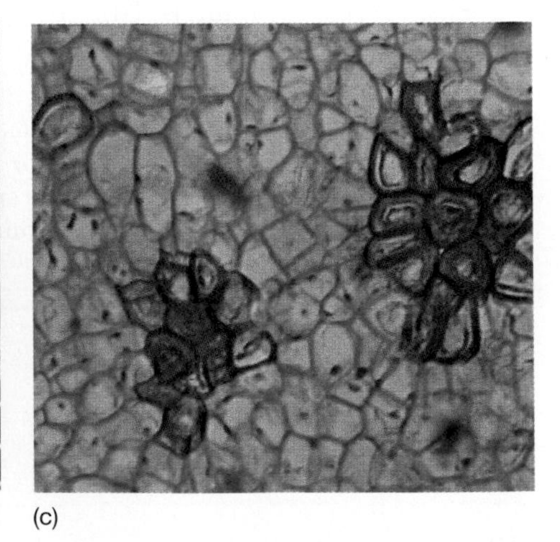

(c)

FIGURE 38.12
The three types of ground tissue. (*a*) Parenchyma cells. Only primary cell walls are seen in this cross-section of parenchyma cells from grass. (*b*) Collenchyma cells. Thickened side walls are seen in this cross-section of collenchyma cells from a young branch of elderberry (*Sambucus*). In other kinds of collenchyma cells, the thickened areas may occur at the corners of the cells or in other kinds of strips. (*c*) Sclereids. Clusters of sclereids ("stone cells"), stained red in this preparation, in the pulp of a pear. The surrounding thin-walled cells, stained light blue, are *parenchyma.* These sclereid clusters give pears their gritty texture.

Vascular Tissue

Xylem

Xylem, the principal water-conducting tissue of plants, usually contains a combination of **vessels,** which are continuous tubes formed from dead, hollow, cylindrical cells **(vessel members)** arranged end to end, and **tracheids,** which are dead cells that taper at the ends and overlap one another (figure 38.13). In some plants (but not angiosperms), tracheids are the only water-conducting cells present; water passes in an unbroken stream through the xylem from the roots up through the shoot and into the leaves. When the water reaches the leaves, much of it passes into a film of water on the outside of the parenchyma cells, and then it diffuses in the form of water vapor into the intercellular spaces and out of the leaves into the surrounding air, mainly through the stomata. This diffusion of water vapor from a plant is known as **transpiration.** In addition to conducting water, dissolved minerals, and inorganic ions such as nitrates and phosphates throughout the plant, xylem supplies mechanical support for the plant body.

Primary xylem is derived from the procambium, which comes from the apical meristem. *Secondary xylem* is formed by the vascular cambium, a lateral meristem that develops later. Wood consists of accumulated secondary xylem.

Vessel members are found almost exclusively in angiosperms. In primitive angiosperms, vessel members tend to resemble fibers and are relatively long. In more advanced angiosperms, vessel members tend to be shorter and wider, resembling microscopic, squat coffee cans with both ends removed. Both vessel members and tracheids have thick, lignified secondary walls and no living protoplasts at maturity. Lignin is produced by the cell and secreted to strengthen the cellulose cell walls before the protoplast dies, leaving only the cell wall. When the continuous stream of water in a plant flows through tracheids, it passes through **pits,** which are small, mostly rounded-to-elliptical areas where no secondary wall material has been deposited. The pits of adjacent cells occur opposite one another. In contrast, vessel members, which are joined end to end, may be almost completely open or may have bars or strips of wall material across the open ends.

Vessels appear to conduct water more efficiently than do the overlapping strands of tracheids. We know this partly because vessel members have evolved from tracheids independently in several groups of plants, suggesting that they are favored by natural selection. It is also probable that some types of fibers have evolved from tracheids, becoming specialized for strengthening rather than conducting. Some ancient flowering plants have only tracheids, but virtually all modern angiosperms have vessels. Plants, with a mutation that prevents the differentiation of xylem, but does not affect tracheids, wilt soon after germination and are unable to transport water efficiently.

In addition to conducting cells, xylem typically includes fibers and parenchyma cells (ground tissue cells). The parenchyma cells, which are usually produced in horizontal rows called **rays** by special *ray initials* of the vascular cambium, function in lateral conduction and food storage. An initial is another term for a meristematic cell. It divides to produce another initial and a cell that differentiates. In cross-sections of woody stems and roots, the rays can be seen radiating out from the center of the xylem like the spokes of a wheel. Fibers are abundant in some kinds of wood, such as oak (*Quercus*), and the wood is correspondingly dense and heavy. The arrangements of these and other kinds of cells in the xylem make it possible to identify most plant genera and many species from their wood alone. These fibers are a major component in modern paper. Earlier paper was made from fibers in phloem.

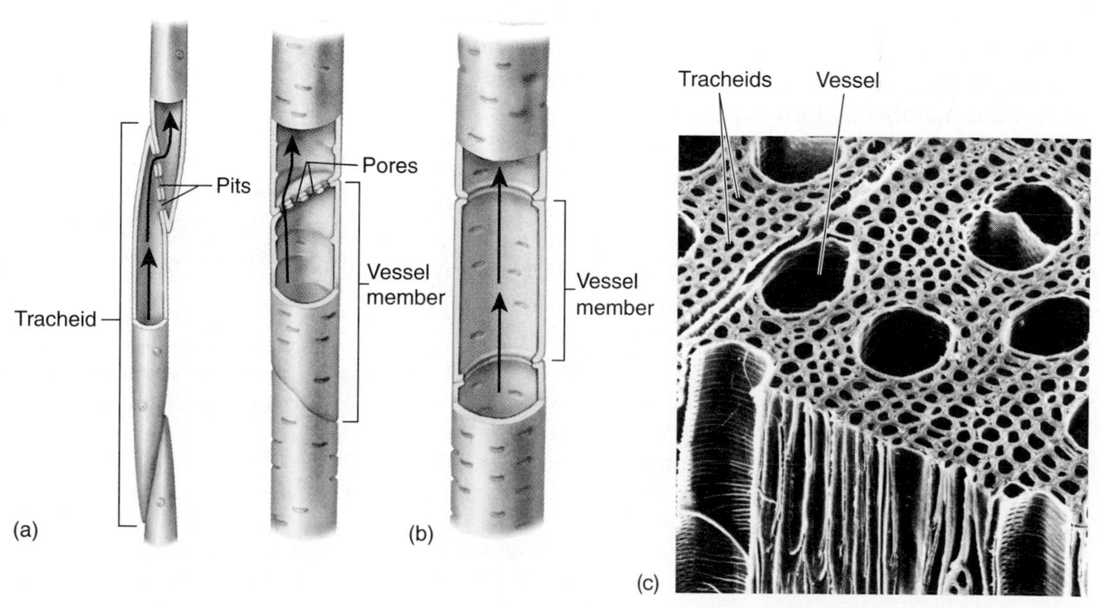

FIGURE 38.13
Comparison between vessel members and tracheids. (*a*) In tracheids, the water passes from cell to cell by means of pits, (*b*) while in vessel members, it moves by way of perforation plates or between bars of wall material. In gymnosperm wood, tracheids both conduct water and provide support; in most kinds of angiosperms, vessels are present in addition to tracheids, or present exclusively. These two types of cells conduct the water, and fibers provide additional support. (*c*) Scanning micrograph of the wood of red maple, *Acer rubrum* (350×).

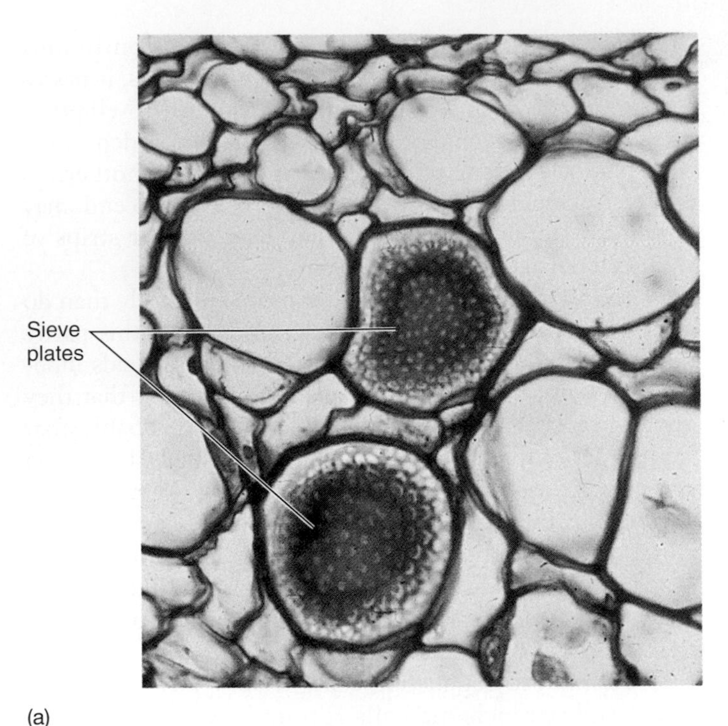

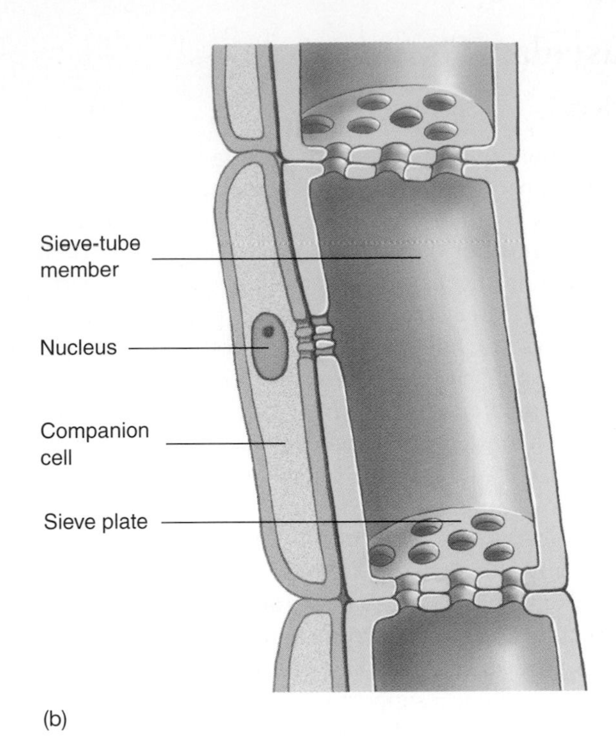

FIGURE 38.14
A sieve-tube member. (*a*) Looking down into sieve plates in squash phloem reveals the perforations sucrose and hormones move through. (*b*) Sieve-tube member cells are stacked with the sieve plates forming the connection. The narrow cell with the nucleus at the left of the sieve-tube member is a companion cell. This cell nourishes the sieve-tube members, which have plasma membranes, but no nuclei.

Phloem

Phloem, which is located toward the outer part of roots and stems, is the principal food-conducting tissue in vascular plants. If a plant is *girdled* (by removing a substantial strip of bark down to the vascular cambium), the plant eventually dies from starvation of the roots.

Food conduction in phloem is carried out through two kinds of elongated cells: **sieve cells** and **sieve-tube members** (figure 38.14). Seedless vascular plants and gymnosperms have only sieve cells; most angiosperms have sieve-tube members. Both types of cells have clusters of pores known as **sieve areas.** Sieve areas are more abundant on the overlapping ends of the cells and connect the protoplasts of adjoining sieve cells and sieve-tube members. Both of these types of cells are living, but most sieve cells and all sieve-tube members lack a nucleus at maturity.

In sieve-tube members, some sieve areas have larger pores and are called **sieve plates.** Sieve-tube members occur end to end, forming longitudinal series called **sieve tubes.** Sieve cells are less specialized than sieve-tube members, and the pores in all of their sieve areas are roughly of the same diameter. In an evolutionary sense, sieve-tube members are more advanced, more specialized, and, presumably, more efficient.

Each sieve-tube member is associated with an adjacent specialized parenchyma cell known as a **companion cell.** Companion cells apparently carry out some of the metabolic functions that are needed to maintain the associated sieve-tube member. In angiosperms, a common initial cell divides asymmetrically to produce a sieve-tube member cell and its companion cell. Companion cells have all of the components of normal parenchyma cells, including nuclei, and numerous **plasmodesmata** (cytoplasmic connections between adjacent cells) connect their cytoplasm with that of the associated sieve-tube members. Sieve cells in nonflowering plants have albuminous cells that function like companion cells. Fibers and parenchyma cells are often abundant in phloem.

Xylem conducts water and dissolved minerals from the roots to the shoots and the leaves. Phloem carries organic materials from one part of the plant to another.

38.3 Root cells differentiate as they become distanced from the dividing root apical meristem.

Root Structure

The three tissue systems are found in the three kinds of vegetative organs in plants: **roots, stems,** and **leaves.** Roots have a simpler pattern of organization and development than stems, and we will consider them first. Four zones or regions are commonly recognized in developing roots. The zones are called the **root cap,** the **zone of cell division,** the **zone of elongation,** and the **zone of maturation** (figure 38.15). In three of the zones, the boundaries are not clearly defined. When apical initials divide, daughter cells that end up on the tip end of the root become root cap cells. Cells that divide in the opposite direction pass through the three other zones before they finish differentiating. As you consider the different zones, visualize the tip of the root moving away from the soil surface by growth. This will counter the static image of a root that diagrams and photos convey.

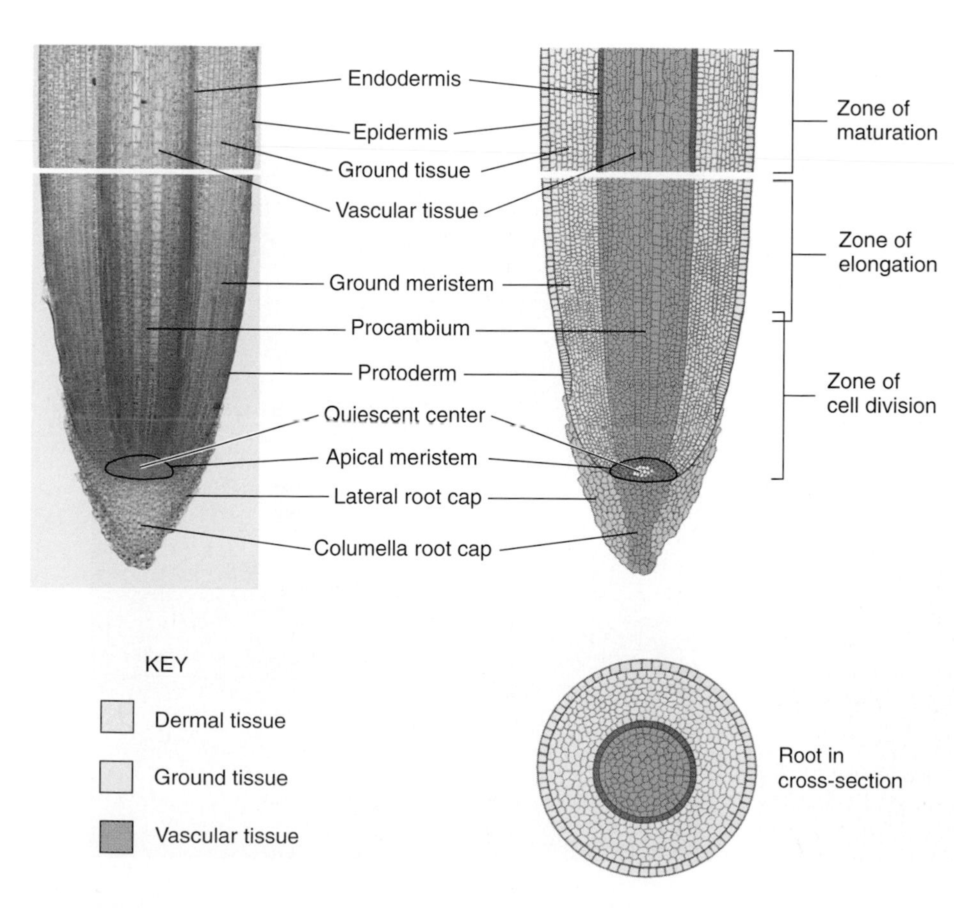

FIGURE 38.15
Root structure. A root tip in corn, *Zea mays*. This longitudinal section of a root shows the root cap, apical meristem, procambium, protoderm, epidermis, and ground meristem.

The Root Cap

The root cap has no equivalent in stems. It is composed of two types of cells, the inner columella (they look like columns) cells and the outer, lateral root cap cells that are continuously replenished by the root apical meristem. In some plants with larger roots it is quite obvious. Its most obvious function is to protect the delicate tissues behind it as growth extends the root through mostly abrasive soil particles. Golgi bodies in the outer root cap cells secrete and release a slimy substance that passes through the cell walls to the outside. The cells, which have an average life of less than a week, are constantly being replaced from the inside, forming a mucilaginous lubricant that eases the root through the soil. The slimy mass also provides a medium for the growth of beneficial nitrogen-fixing bacteria in the roots of some plants such as legumes.

A new root cap is produced when an existing one is artificially or accidentally removed from a root. The root cap also functions in the *perception of gravity*. The columella cells are highly specialized with the endoplasmic reticulum in the periphery and the nucleus located at either the middle or the top of the cell. There are no large vacuoles. Columella cells contain amyloplasts (plastids with starch grains) that collect on the sides of cells facing the pull of gravity. When a potted plant is placed on its side, the amyloplasts drift or tumble down to the side nearest the source of gravity, and the root bends in that direction. Lasers have been used to ablate (kill) individual columella cells in *Arabidopsis*. It turns out that only two columella cells are essential for gravity sensing! The precise nature of the gravitational response is not known, but some evidence indicates that calcium ions in the amyloplasts influence the distribution of growth hormones (auxin in this case) in the cells. There may be multiple signaling mechanisms because bending has been observed in the absence of auxin. A current hypothesis is that an electrical signal moves from the columella cell to cells in the elongation zone (the region closest to the zone of cell division).

The Zone of Cell Division

The apical meristem is shaped like an inverted, concave dome of cells and is located in the center of the root tip in the area protected by the root cap. Most of the activity in this *zone of cell division* takes place toward the edges of the dome, where the cells divide every 12 to 36 hours, often rhythmically, reaching a peak of division once or twice a day. Most of the cells are essentially cuboidal with small vacuoles and proportionately large, centrally located nuclei. These rapidly dividing cells are daughter cells of the apical meristem. The *quiescent center* is a group of cells in the center of the root apical meristem. They divide very infrequently. This makes sense if you think about a solid ball expanding. The outer surface would have to increase far more rapidly than the very center.

The apical meristem daughter cells soon subdivide into the three primary tissues previously discussed: *protoderm*, *procambium*, and *ground meristem*. Genes have been identified in the relatively simple root of *Arabidopsis* that regulate the patterning of these tissue systems. The patterning of these cells begins in this zone, but it is not until the cells reach the zone of maturation that the anatomical and morphological expression of this patterning is fully revealed. The *WEREWOLF* gene, for example, is required for the patterning of the two root epidermal cell types, those with and without root hairs (figure 38.16*a*). The *SCARECROW* gene is important in ground cell differentiation (figure 38.16*b*). It is necessary for an asymmetric cell division that gives rise to two cylinders of cells from one. The outer cell layer becomes ground tissue and serves a storage function. The inner cell layer forms the endodermis which regulates the intercellular flow of water and solutes into the vascular core of the root. Cells in this region develop according to their position. If that position changes because of a mistake in cell division or the ablation of another cell, the cell develops according to its new position.

The Zone of Elongation

In the *zone of elongation*, roots lengthen because the cells produced by the primary meristems become several times longer than wide, and their width also increases slightly. The small vacuoles present merge and grow until they occupy 90% or more of the volume of each cell. No further increase in cell size occurs above the zone of elongation, and the mature parts of the root, except for an increase in girth, remain stationary for the life of the plant.

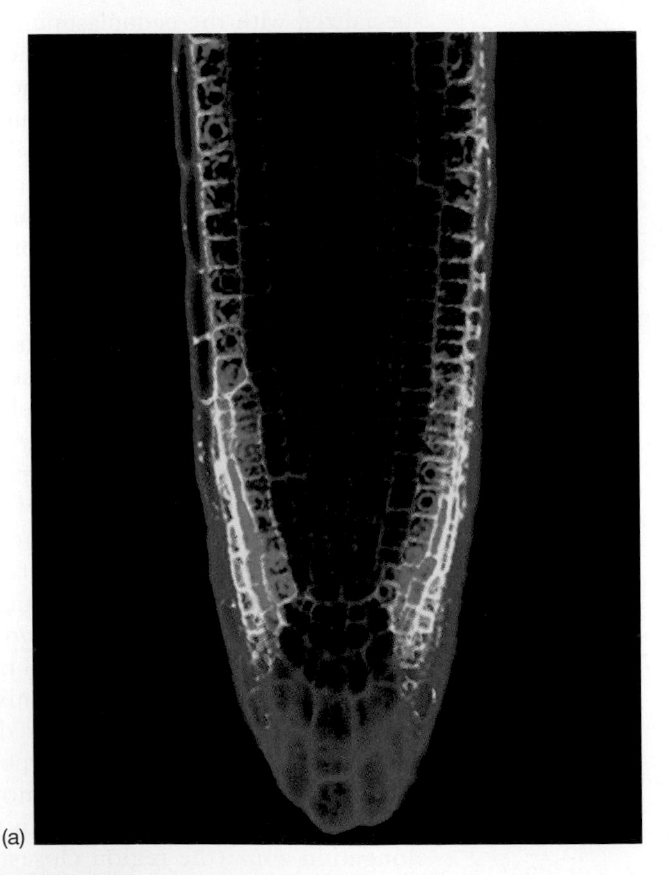

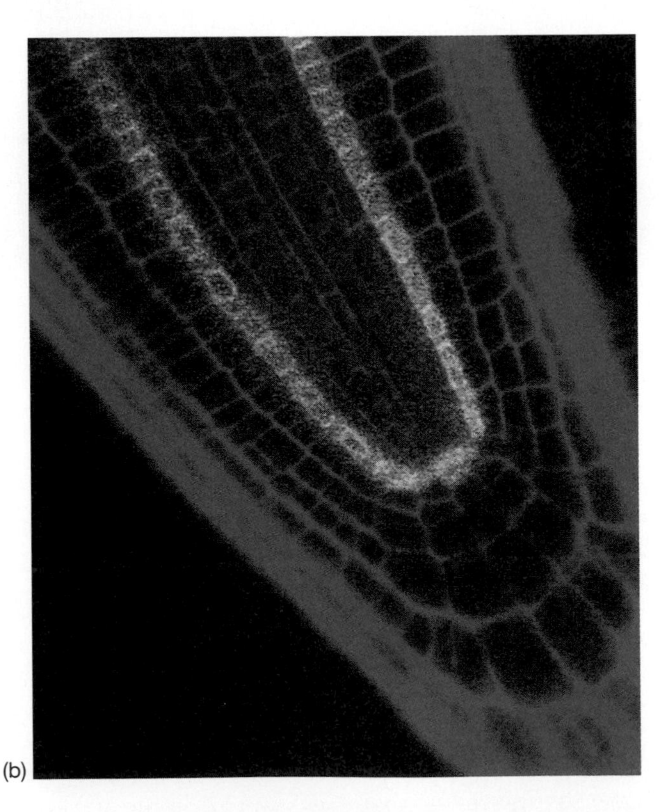

FIGURE 38.16
Tissue-specific gene expression. (*a*) Epidermal-specific gene expression. The promoter of the *WEREWOLF* gene of *Arabidopsis* was attached to a green fluorescent protein and used to make a transgenic plant. The *green* fluorescence shows the epidermal cells where the gene is expressed. The *red* was used to visually indicate cell boundaries. (*b*) Ground tissue-specific gene expression. The *SCARECROW* gene is needed for an asymmetric cell division allowing for the formation of side-by-side ground tissue and endodermal cells. These form two layers in wild-type plants, but mutants only have one cell layer (*green*) because the asymmetric cell division does not occur.

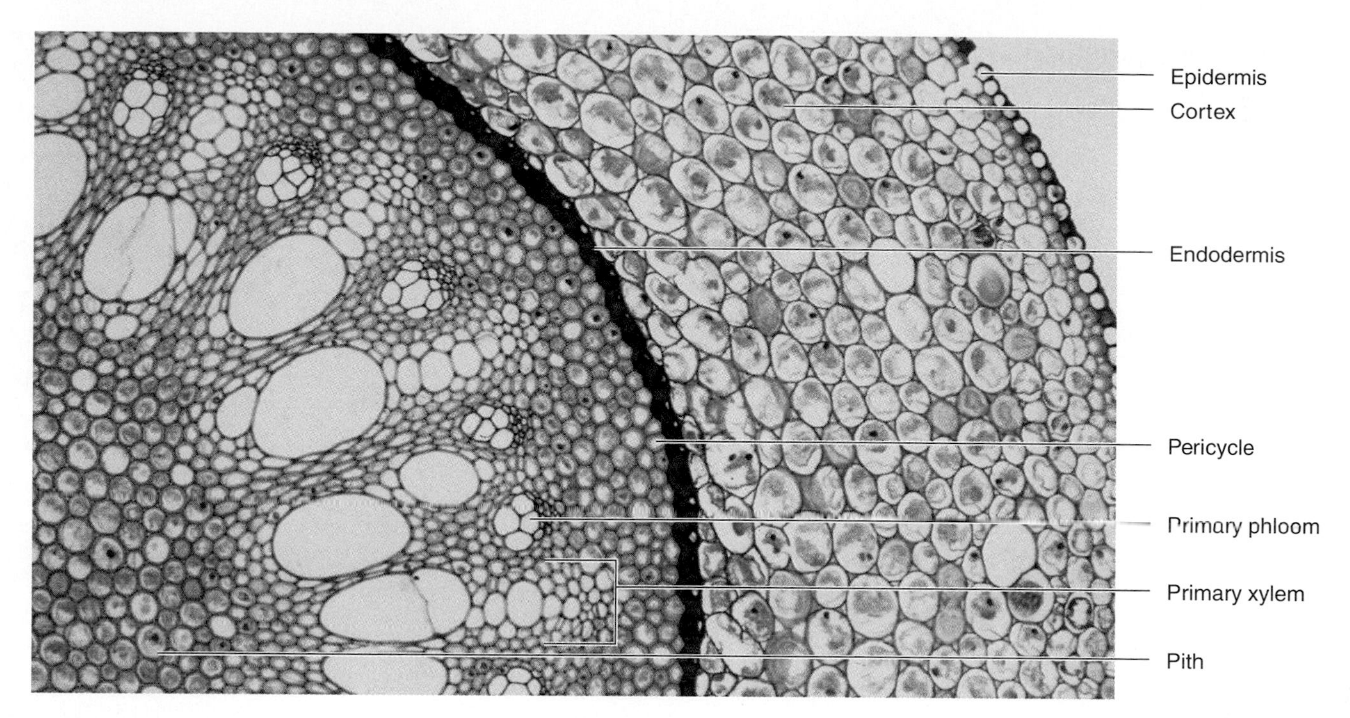

FIGURE 38.17
Cross-section of the zone of maturation of a young monocot root. Greenbrier (*Smilax*), a monocot (100×).

The Zone of Maturation

The cells that have elongated in the zone of elongation become differentiated into specific cell types in the *zone of maturation.* The cells of the root surface cylinder mature into *epidermal cells,* which have a very thin cuticle. Many of the epidermal cells each develop a **root hair;** the protuberance is not separated by a crosswall from the main part of the cell and the nucleus may move into it. Root hairs, which can number over 35,000 per square centimeter of root surface and many billions per plant, greatly increase the surface area and therefore the absorptive capacity of the root. The root hairs usually are alive and functional for only a few days before they are sloughed off at the older part of the zone of maturation, while new ones are being produced toward the zone of elongation. Symbiotic bacteria that fix atmospheric nitrogen into a form usable by legumes enter the plant via root hairs and "instruct" the plant to create a nodule around it.

Parenchyma cells are produced by the ground meristem immediately to the interior of the epidermis. This tissue, called the **cortex,** may be many cells wide and functions in food storage. The inner boundary of the cortex differentiates into a single-layered cylinder of **endodermis** (figure 38.17), whose primary walls are impregnated with **suberin,** a fatty substance that is impervious to water. The suberin is produced in bands, called **Casparian strips** that surround each adjacent endodermal cell wall perpendicular to the root's surface (figure 38.18). This blocks transport between cells. The two surfaces that are parallel to the root surface are the only way into the core of the root and the cell membranes control what passes through.

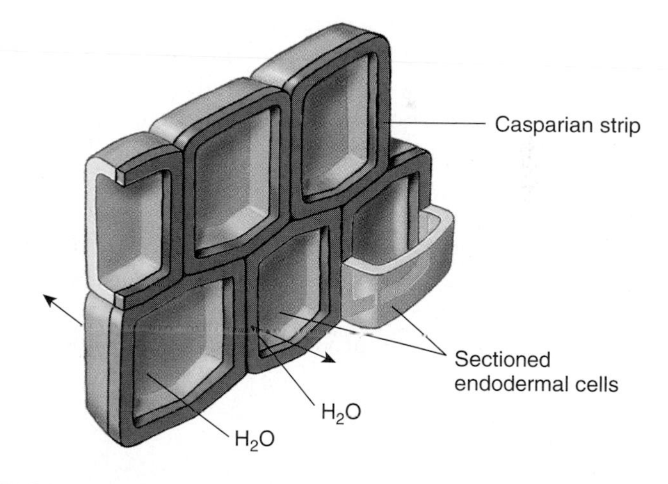

FIGURE 38.18
Casparian strip. The Casparian strip is a water-proofing band that forces water and minerals to pass through the cell membranes, rather than through the air spaces in the cell walls.

All the tissues interior to the endodermis are collectively referred to as the **stele.** Immediately adjacent and interior to the endodermis is a cylinder of parenchyma cells known as the **pericycle.** Pericycle cells can divide, even after they mature. They can give rise to *lateral* (branch) *roots* or, in dicots, to part of the *vascular cambium.*

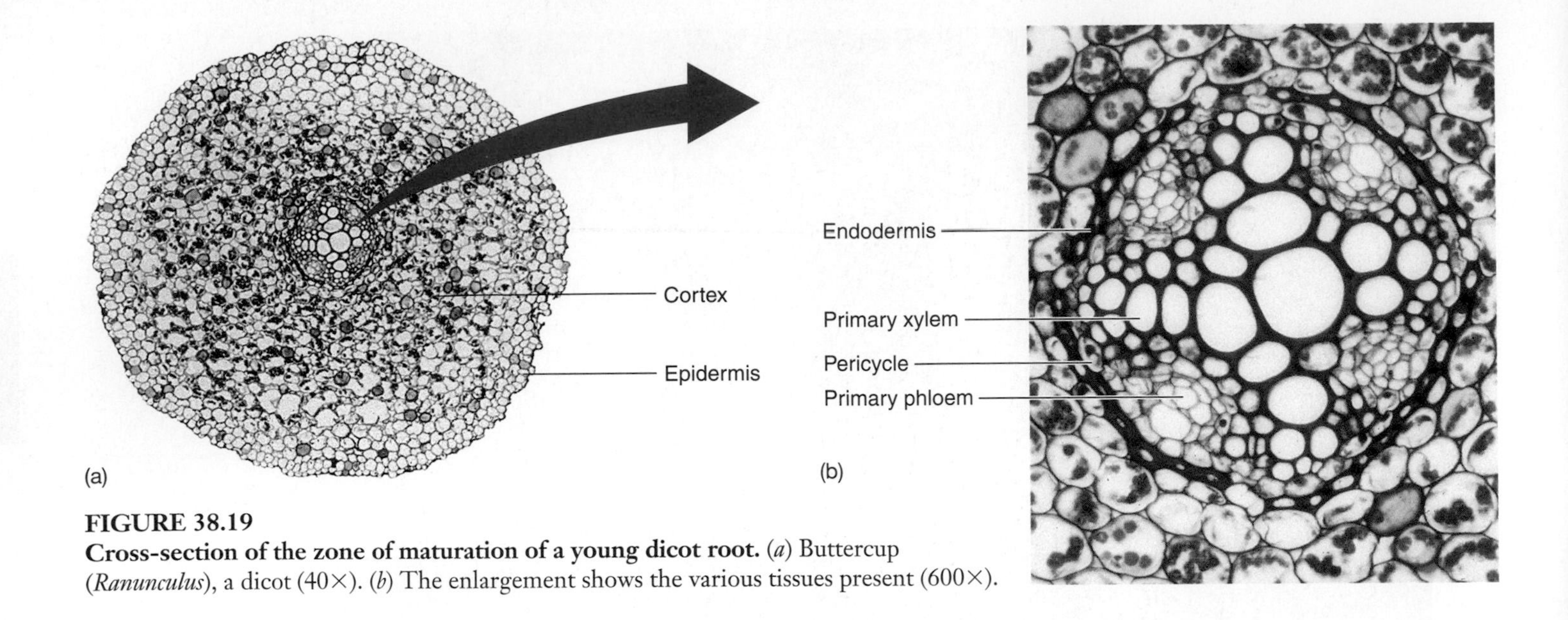

FIGURE 38.19
Cross-section of the zone of maturation of a young dicot root. (*a*) Buttercup (*Ranunculus*), a dicot (40×). (*b*) The enlargement shows the various tissues present (600×).

The water-conducting cells of the *primary xylem* are differentiated as a solid core in the center of young dicot roots. In a cross-section of a dicot root, the central core of primary xylem often is somewhat star-shaped, with one or two to several radiating arms that point toward the pericycle (figure 38.19). In monocot (and a few dicot) roots, the primary xylem is in discrete **vascular bundles** arranged in a ring, which surrounds parenchyma cells, called **pith,** at the very center of the root. **Primary phloem,** composed of cells involved in food conduction, is differentiated in discrete groups of cells between the arms of the xylem in both dicot and monocot roots.

In dicots and other plants with **secondary growth,** part of the pericycle and the parenchyma cells between the phloem patches and the xylem arms become the root vascular cambium, which starts producing **secondary xylem** to the inside and **secondary phloem** to the outside (figure 38.20). Eventually, the secondary tissues acquire the form of concentric cylinders. The primary phloem, cortex, and epidermis become crushed and are sloughed off as more secondary tissues are added. In the pericycle of woody plants, the cork cambium contributes to the bark which will be discussed in the section on stems (see figure 38.26). In the case of secondary growth in dicot roots, everything outside the stele is lost and replaced with bark.

Root apical meristems produce a root cap at the tip and root tissue on the opposite side. Cells mature as the root cap and meristem grow away from them. Transport systems, external barriers, and a branching root system develop from the primary root as it matures.

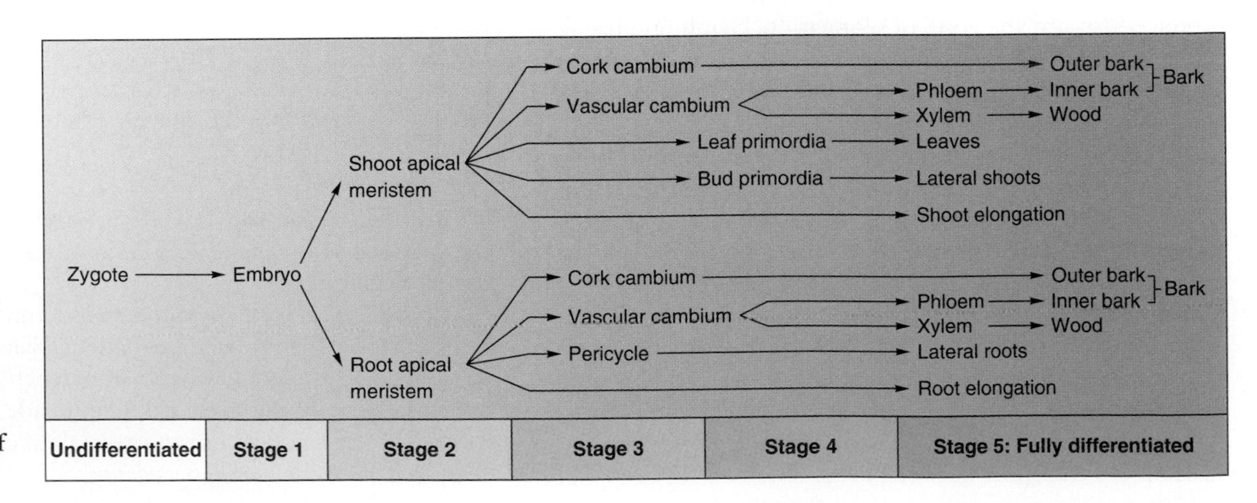

FIGURE 38.20
Stages in the differentiation of plant tissues.

Modified Roots

Most plants produce either a **taproot system** in which there is a single large root with smaller branch roots, or a **fibrous root system** in which there are many smaller roots of similar diameter. Some plants, however, have intriguing root modifications with specific functions in addition to those of anchorage and absorption.

(a)

(b)

(c)

FIGURE 38.21
Three types of modified roots. (*a*) Pneumatophores (foreground) are spongy outgrowths from the roots below. (*b*) A water storage root weighing over 25 kilograms (60 pounds). (*c*) Buttress roots of a tropical fig tree.

Aerial roots. Some plants, such as epiphytic orchids (orchids that are attached to tree branches and grow unconnected to the ground without being parasitic in any way) have roots that extend out into the air. Some aerial roots have an epidermis that is several cells thick, an adaptation to reduce water loss. These aerial roots may also be green and photosynthetic, as in the vanilla orchid. Some monocots, such as corn, produce thick roots from the lower parts of the stem; these *prop roots* grow down to the ground and brace the plants against wind. Climbing plants such as ivy also produce roots from their stems; these anchor the stems to tree trunks or a brick wall. Any root that arises along a stem or in some place other than the root of the plant is called an *adventitious root.* Adventitious root formation in ivy depends on the developmental stage of the shoot. When the shoot transitions to the adult phase of development, it is no longer capable of initiating these roots.

Pneumatophores. Some plants that grow in swamps and other wet places may produce spongy outgrowths called *pneumatophores* from their underwater roots (figure 38.21*a*). The pneumatophores commonly extend several centimeters above water, facilitating the oxygen supply to the roots beneath.

Contractile roots. The roots from the bulbs of lilies and of several other plants such as dandelions contract by spiraling to pull the plant a little deeper into the soil each year until they reach an area of relatively stable temperatures. The roots may contract to a third of their original length as they spiral like a corkscrew due to cellular thickening and constricting.

Parasitic roots. The stems of certain plants that lack chlorophyll, such as dodder (*Cuscuta*), produce peglike roots called *haustoria* that penetrate the host plants around which they are twined. The haustoria establish contact with the conducting tissues of the host and effectively parasitize their host.

Food storage roots. The xylem of branch roots of sweet potatoes and similar plants produce at intervals many extra parenchyma cells that store large quantities of carbohydrates. Carrots, beets, parsnips, radishes, and turnips have combinations of stem and root that also function in food storage. Cross sections of these roots reveal multiple rings of secondary growth.

Water storage roots. Some members of the pumpkin family (Cucurbitaceae), especially those that grow in arid regions, may produce water-storage roots weighing 50 or more kilograms (figure 38.21*b*).

Buttress roots. Certain species of fig and other tropical trees produce huge buttress roots toward the base of the trunk, which provide considerable stability (figure 38.21*c*).

Some plants have modified roots that carry out photosynthesis, gather oxygen, parasitize other plants, store food or water, or support the stem.

38.4 Stems are the backbone of the shoot, transporting nutrients and supporting the aerial plant organs.

Stem Structure

External Form

The shoot apical meristem initiates stem tissue and intermittently produces bulges (**primordia**) that will develop into leaves, other shoots, or even flowers (figure 38.22). The stem is an axis from which organs grow. Leaves may be arranged in a spiral around the stem, or they may be in pairs opposite one another; they also may occur in *whorls* (circles) of three or more. Spirals are the most common and, for reasons still not understood, sequential leaves tend to be placed 137.5° apart. This angle relates to the golden mean, a mathematical ratio, that is found in nature (the angle of coiling in some shells, for example), classical architecture (the Parthenon wall dimensions), and even modern art (Mondrian). The pattern of leaf arrangement is called **phyllotaxy** and may optimize exposure of leaves to the sun.

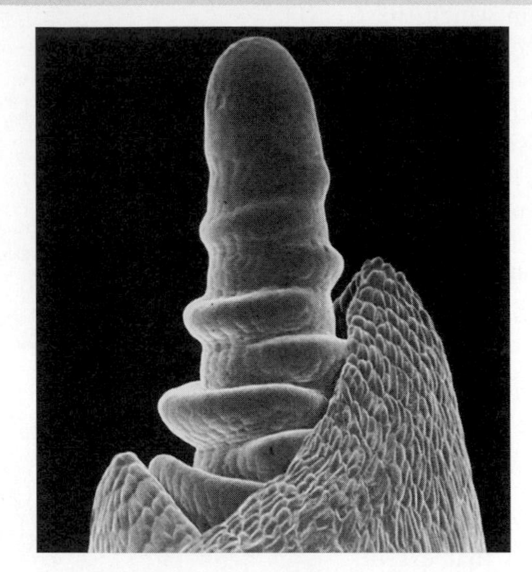

FIGURE 38.22
A shoot apex (200×). Scanning electron micrograph of the apical meristem of wheat (*Triticum*).

The *region* or *area* (no structure is involved) of leaf attachment is called a **node;** the area of stem between two nodes is called an **internode.** A leaf usually has a flattened **blade** and sometimes a **petiole** (stalk). When the petiole is missing, the leaf is then said to be **sessile.** Note that the word *sessile* as applied to plants has a different meaning than it does when applied to animals (probably obvious, as plants don't get up and move around!); in plants, it means *attached.* The space between a petiole (or blade) and the stem is called an axil. An **axillary bud** is produced in each axil. This bud is a product of the primary shoot apical meristem, which, with its associated leaf primordia, is called a **terminal bud.** Axillary buds frequently develop into branches or may form meristems that will develop into flowers. (Refer back to figure 38.6 to review these terms.)

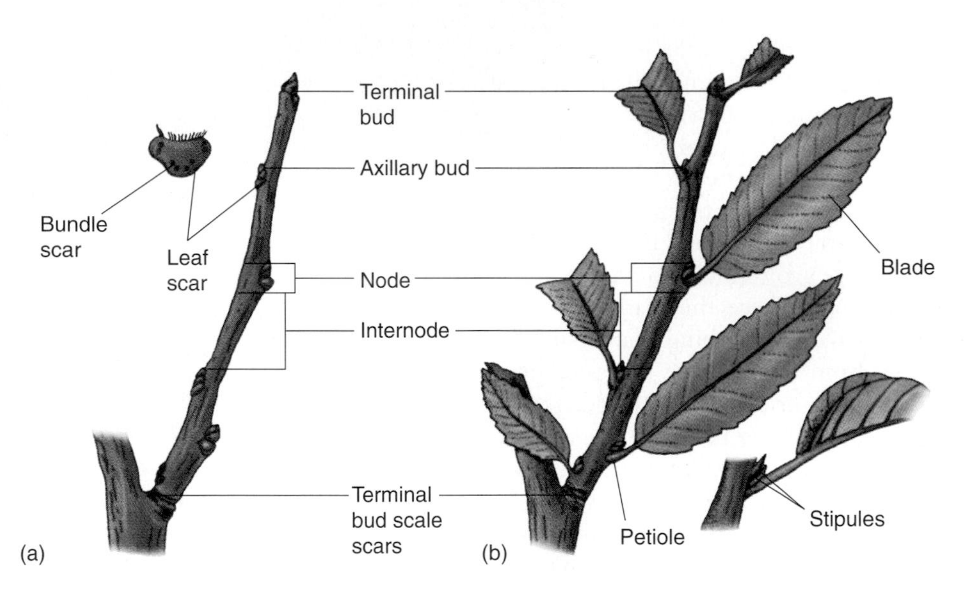

FIGURE 38.23
A woody twig. (*a*) In winter. (*b*) In summer.

Neither monocots nor herbaceous stems do not produce a cork cambium. The stems are usually green and photosynthetic, with at least the outer cells of the cortex containing chloroplasts. Herbaceous stems commonly have stomata, and may have various types of trichomes (hairs).

Woody stems can persist over a number of years and develop distinctive markings in addition to the original organs that form. Terminal buds usually extend the length of the shoot during the growing season. Some buds, such as those of geraniums, are unprotected, but most buds of woody plants have protective winter *bud scales* that drop off, leaving tiny *bud scale scars* as the buds expand. Some twigs have tiny scars of a different origin. A pair of butterfly-like appendages called **stipules** (part of the leaf) develop at the base of some leaves. The stipules can fall off and leave *stipule scars.* When leaves of deciduous trees drop in the fall, they leave *leaf scars* with tiny *bundle scars*, marking where vascular connections were. The shapes, sizes, and other features of leaf scars can be distinctive enough to identify the plants in winter (figure 38.23).

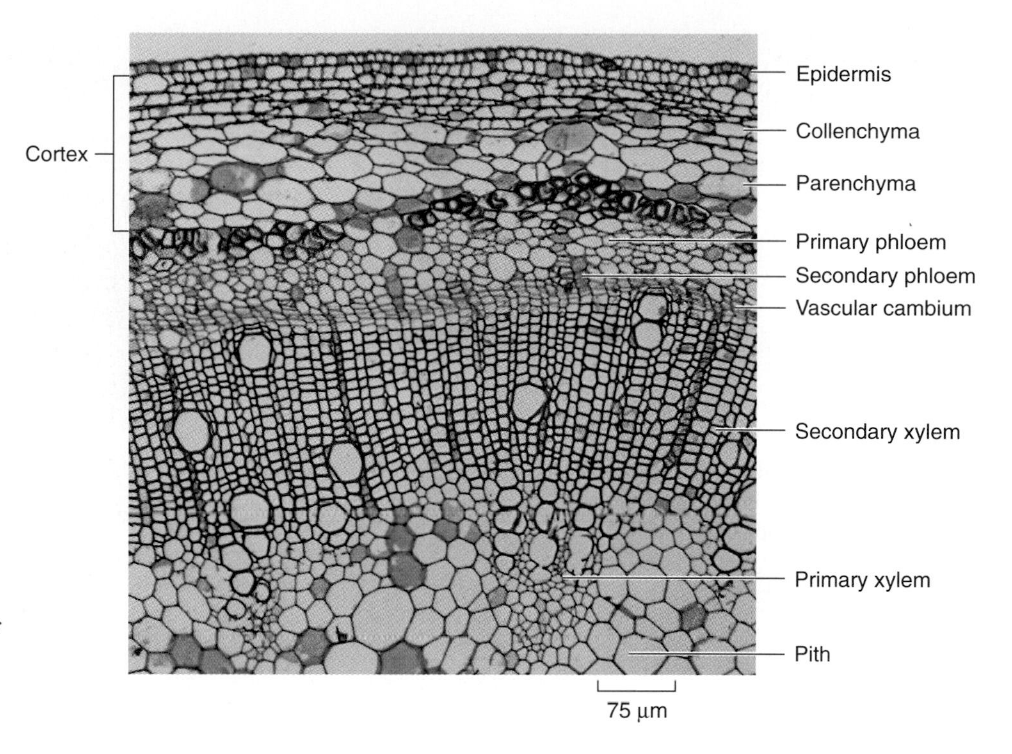

FIGURE 38.24
Early stage in vascular cambium differentiation in the castor bean, *Ricirus* (25×). The outer part of the cortex consists of collenchyma, and the inner part of parenchyma.

Internal Form

As in roots, there is an *apical meristem* at the tip of each stem, which produces *primary meristems* that contribute to the stem's increases in length. Three primary meristems develop from the apical meristem. The *protoderm* gives rise to the *epidermis.* The *ground meristem* produces parenchyma cells. Parenchyma cells in the center of the stem constitute the *pith;* parenchyma cells away from the center constitute the *cortex.* The *procambium* produces cylinders of *primary xylem* and *primary phloem,* which are surrounded by ground tissue.

A strand of xylem and phloem, called a *trace,* branches off from the main cylinder of xylem and phloem and enters the developing leaf, flower, or shoot. These spaces in the main cylinder of conducting tissues are called **gaps.** In dicots, a **vascular cambium** develops between the primary xylem and primary phloem (figure 38.24). In many ways, this is a connect-the-dots game where the vascular cambium connects the ring of primary vascular bundles. In monocots, these bundles are scattered throughout the ground tissue (figure 38.25) and there is no logical way to connect them that would allow a uniform increase in girth. Lacking a vascular cambium, monocots do not have secondary growth.

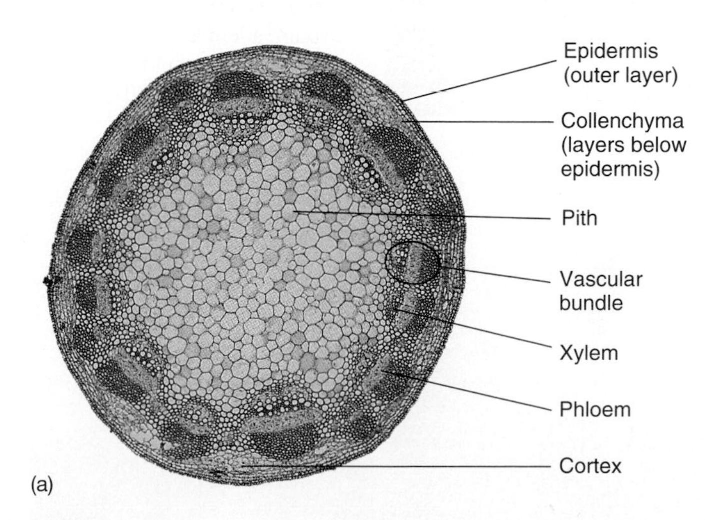

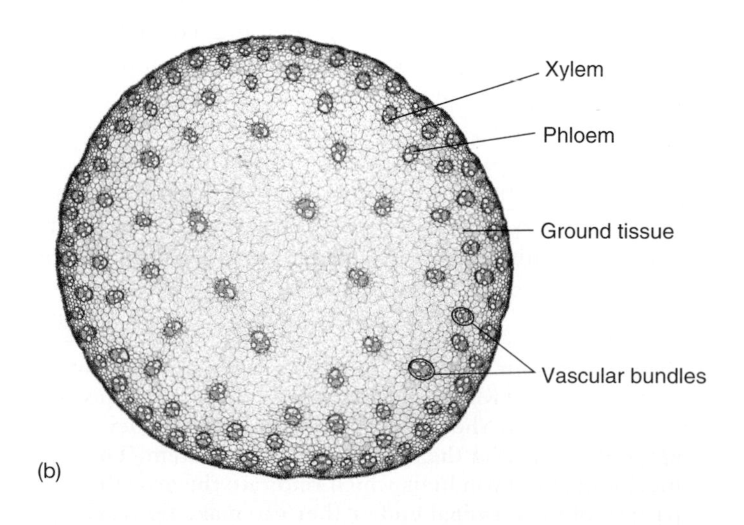

FIGURE 38.25
Stems. Transverse sections of a young stem in (*a*) a dicot, the common sunflower, *Helianthus annuus,* in which the vascular bundles are arranged around the outside of the stem (10×); and (*b*) a monocot, corn, *Zea mays,* with the scattered vascular bundles characteristic of the class (5×).

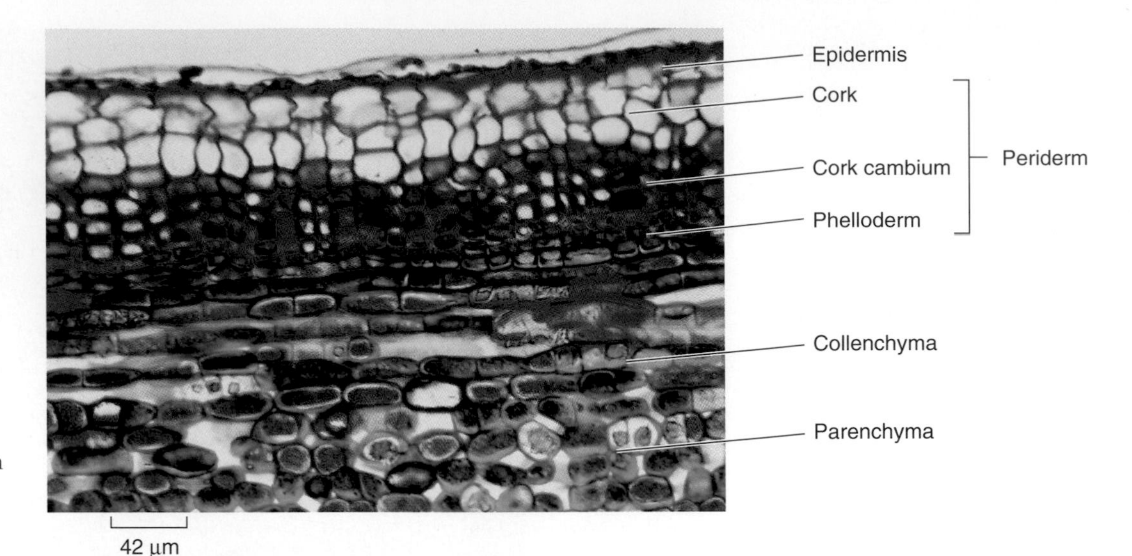

FIGURE 38.26
Section of periderm (50×). An early stage in the development of periderm in cottonwood (*Populus* sp.).

The cells of the vascular cambium divide indefinitely, producing **secondary tissues** (mainly *secondary xylem* and *secondary phloem*). The production of xylem is extensive in trees and is called wood. Rings in the stump of a tree reveal annual patterns of growth; cell size varies depending on growth conditions. In woody dicots and gymnosperms, a second cambium, the *cork cambium*, arises in the outer cortex (occasionally in the epidermis or phloem) and produces boxlike *cork cells* to the outside and also may produce parenchyma-like *phelloderm* cells to the inside; the cork cambium, cork, and phelloderm are collectively referred to as the *periderm* (figure 38.26). Cork tissues, whose cells become impregnated with *suberin* shortly after they are formed and then die, constitute the **outer bark.** The cork tissue, whose suberin is impervious to moisture, cuts off water and food to the epidermis, which dies and sloughs off. In young stems, gas exchange between stem tissues and the air takes place through stomata, but as the cork cambium produces cork, it also produces patches of unsuberized cells beneath the stomata. These unsuberized cells, which permit gas exchange to continue, are called **lenticels** (figure 38.27).

The stem results from the dynamic growth of the shoot apical meristem which initiates stem tissue and organs including leaves. Shoot apical meristems initiate new apical meristems at the junction of leaf and stem. These meristems can form buds which reiterate the growth pattern of the terminal bud or they can make flowers directly.

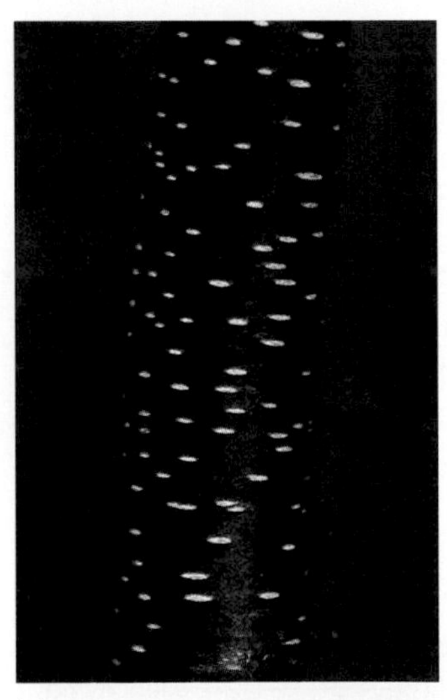

(a)

(b)

FIGURE 38.27
Lenticels. (*a*) Lenticels, the numerous, small, pale, raised areas shown here on cherry tree bark (*Prunus cerasifera*), allow gas exchange between the external atmosphere and the living tissues immediately beneath the bark of woody plants. Highly variable in form in different species, lenticels are an aid to the identification of deciduous trees and shrubs in winter. (*b*) Transverse section through a lenticel (extruding area) in a stem of elderberry, *Sambucus canadensis* (30×).

Modified Stems

Although most stems grow erect, there are some modifications that serve special purposes, including that of natural *vegetative propagation*. In fact, the widespread artificial vegetative propagation of plants, both commercial and private, frequently involves the cutting of modified stems into segments, which are then planted and produce new plants. As you become acquainted with the following modified stems, keep in mind that stems have *leaves* at *nodes*, with *internodes* between the nodes, and *buds* in the *axils* of the leaves, while roots have no leaves, nodes, or axillary buds.

Bulbs. Onions, lilies, and tulips have swollen underground stems that are really large buds with adventitious roots at the base (figure 38.28*a*). Most of a *bulb* consists of fleshy leaves attached to a small, knoblike stem. In onions, the fleshy leaves are surrounded by papery, scalelike leaf bases of the long, green aboveground leaves.

Corms. Crocuses, gladioluses, and other popular garden plants produce *corms* that superficially resemble bulbs. Cutting a corm in half, however, reveals no fleshy leaves. Instead, almost all of a corm consists of stem, with a few papery, brown nonfunctional leaves on the outside, and adventitious roots below.

Rhizomes. Perennial grasses, ferns, irises, and many other plants produce *rhizomes*, which typically are horizontal stems that grow underground, often close to the surface (figure 38.28*b*). Each node has an inconspicuous scalelike leaf with an axillary bud; much larger photosynthetic leaves may be produced at the rhizome tip. Adventitious roots are produced throughout the length of the rhizome, mainly on the lower surface.

Runners and stolons. Strawberry plants produce horizontal stems with long internodes, which, unlike rhizomes, usually grow along the surface of the ground. Several *runners* may radiate out from a single plant (figure 38.28*c*). Some botanists use the term *stolon* synonymously with runner; others reserve the term stolon for a stem with long internodes that grows underground, as seen in Irish (white) potato plants. An Irish potato itself, however, is another type of modified stem—a *tuber*.

Tubers. In Irish potato plants, carbohydrates may accumulate at the tips of stolons, which swell, becoming *tubers;* the stolons die after the tubers mature (figure 38.28*d*). The "eyes" of a potato are axillary buds formed in the axils of scalelike leaves. The scalelike leaves, which are present when the potato is starting to form, soon drop off; the tiny ridge adjacent to each "eye" of a mature potato is a leaf scar.

Tendrils. Many climbing plants, such as grapes and Boston ivy, produce modified stems knows as *tendrils*, which twine around supports and aid in climbing (figure 38.28*e*). Some tendrils, such as those of peas and pumpkins, are actually modified leaves or leaflets.

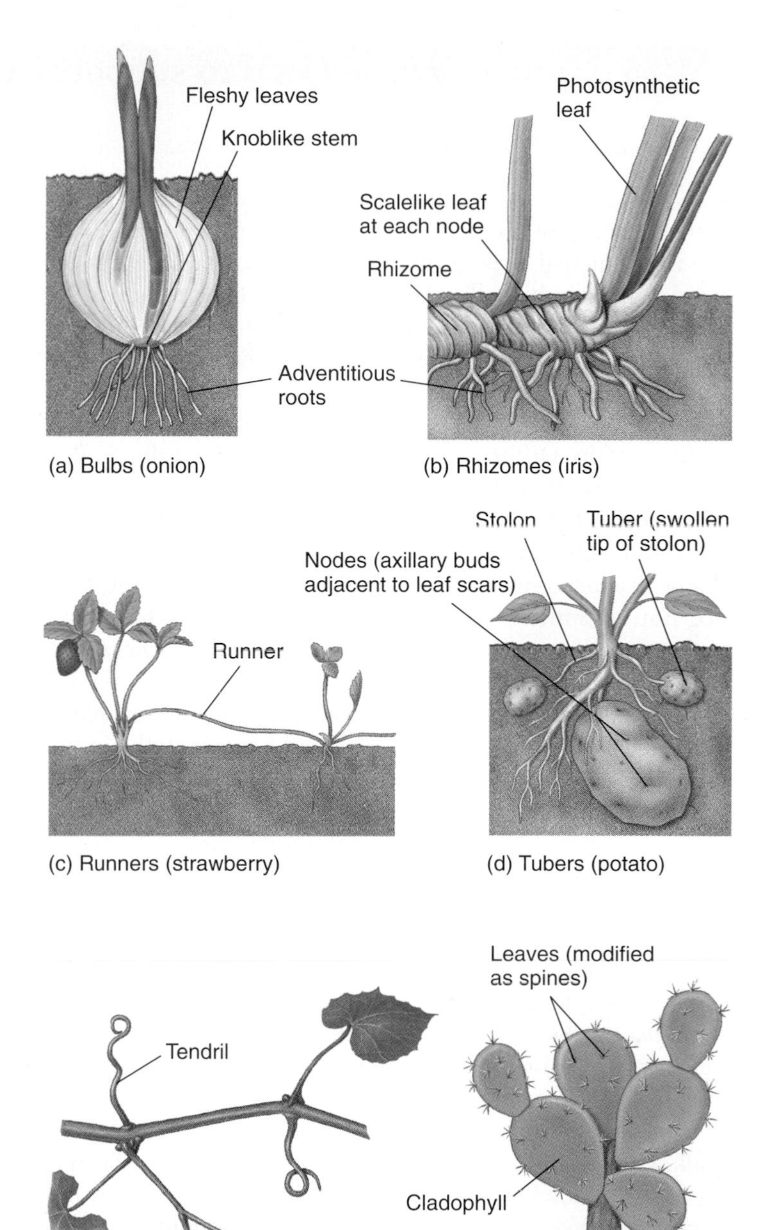

FIGURE 38.28
Types of modified stems.

Cladophylls. Cacti and several other plants produce flattened, photosynthetic stems called *cladophylls* that resemble leaves (figure 38.28*f*). In cacti, the real leaves are modified as spines.

Some plants possess modified stems that serve special purposes including food storage, support, or vegetative propagation.

38.5 Leaves are adapted to support basic plant functions.

Leaf External Structure

Leaves, which are initiated as *primordia* by the apical meristems (see figure 38.2), are vital to life as we know it. They are the principal sites of photosynthesis on land. Leaves expand primarily by cell enlargement and some cell division. Like our arms and legs, they are determinate structures which means growth stops at maturity. Because leaves are crucial to a plant, features such as their arrangement, form, size, and internal structure are highly significant and can differ greatly. Different patterns have adaptive value in different environments.

Leaves are really an extension of the shoot apical meristem and stem development. Leaves first emerge as primordia as discussed in the section on stems. At that point, they are not committed to be leaves. Experiments where very young leaf primordia in fern and in coleus are isolated and grown in culture demonstrate this. If the primordia are young enough, they will form an entire shoot rather than a leaf. So, positioning the primordia and beginning the initial cell divisions occur before those cells are committed to the leaf developmental pathway.

Leaves fall into two different morphological groups which may reflect differences in evolutionary origin. A *microphyll* is a leaf with one vein that does not leave a gap when it branches from the vascular cylinder of the stem; microphylls are mostly small and are associated primarily with the phylum Lycophyta (see chapter 37). Most plants have leaves called *megaphylls*, which have several to many veins; a megaphyll's conducting tissue leaves a gap in the stem's vascular cylinder as it branches from it.

Most dicot leaves have a flattened *blade*, and a slender stalk, the *petiole*. The flattening of the leaf blade reflects a shift from radial symmetry to dorsal-ventral (top-bottom) symmetry. We're just beginning to understand how this shift occurs by analyzing mutants like *phantastica* which prevents this transition (figure 38.29). In addition, a pair of *stipules* may be present at the base of the petiole. The stipules, which may be leaflike or modified as *spines* (as in the black locust—*Robinia pseudo-acacia*) or *glands* (as in cherry trees—*Prunus cerasifera*), vary considerably in size from microscopic to almost half the size of the leaf blade. Development of stipules appears to be independent of development of the rest of the leaf.

Grasses and other monocot leaves usually lack a petiole and tend to sheathe the stem toward the base. **Veins** (a term used for the vascular bundles in leaves), consisting of both xylem and phloem, are distributed throughout the leaf blades. The main veins are parallel in most monocot leaves; the veins of dicots, on the other hand, form an often intricate network (figure 38.30).

FIGURE 38.29
The *phantastica* mutant in snapdragon. Snapdragon leaves are usually flattened with a top and bottom side (plant on *left*). In the *phantastica* mutant (plant on *right*), the leaf never flattens but persists as a radially symmetrical bulge.

(a)

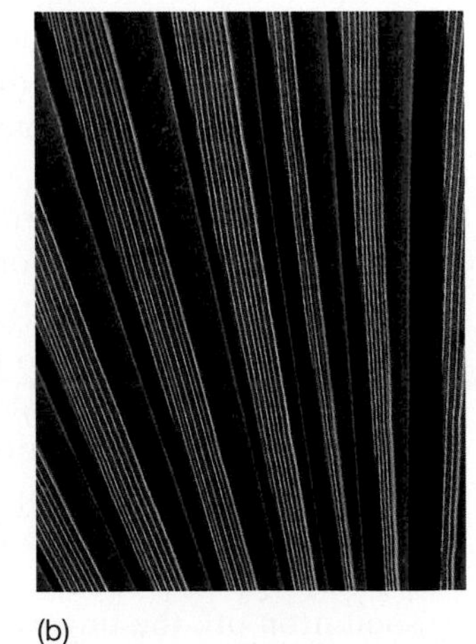

(b)

FIGURE 38.30
Dicot and monocot leaves. The leaves of dicots, such as this (*a*) African violet relative from Sri Lanka, have netted, or reticulate, veins; (*b*) those of monocots, like this cabbage palmetto, have parallel veins. The dicot leaf has been cleared with chemicals and stained with a red dye to make the veins show up more clearly.

(a)

(b)

(c)

FIGURE 38.31
Simple versus compound leaves. (*a*) A simple leaf, its margin deeply lobed, from the tulip tree (*Liriodendron tulipifera*). (*b*) A pinnately compound leaf, from a mountain ash (*Sorbus* sp.). A compound leaf is associated with a single lateral bud, located where the petiole is attached to the stem. (*c*) Palmately compound leaves of a Virginia creeper (*Parthenocissus quinquefolia*).

Leaf blades come in a variety of forms from oval to deeply lobed to having separate leaflets. In **simple leaves** (figure 38.31*a*), such as those of lilacs or birch trees, the blades are undivided, but simple leaves may have teeth, indentations, or lobes of various sizes, as in the leaves of maples and oaks. In **compound leaves**, such as those of ashes, box elders, and walnuts, the blade is divided into **leaflets.** The relationship between the development of compound and simple leaves is an open question. Two explanations are being debated: (1) a compound leaf is a highly lobed simple leaf, or (2) a compound leaf utilizes a shoot development program. There are single mutations that convert compound leaves to simple leaves which are being used to address this debate. If the leaflets are arranged in pairs along a common axis (the axis is called a *rachis*—the equivalent of the main central vein, or *midrib*, in simple leaves), the leaf is **pinnately compound** (figure 38.31*b*). If, however, the leaflets radiate out from a common point at the blade end of the petiole, the leaf is **palmately compound** (figure 38.31*c*). Palmately compound leaves occur in buckeyes (*Aesculus* spp.) and Virginia creeper (*Parthenocissus quinquefolia*). The leaf blades themselves may have similar arrangements of their veins, and are said to be **pinnately** or **palmately** veined.

Leaves, regardless of whether they are simple or compound, may be **alternately** arranged (alternate leaves usually spiral around a shoot, one leaf per node) or they may be in **opposite** pairs (two leaves per node). Less often, three or more leaves may be in a **whorl,** a circle of leaves at the same level at a node (figure 38.32).

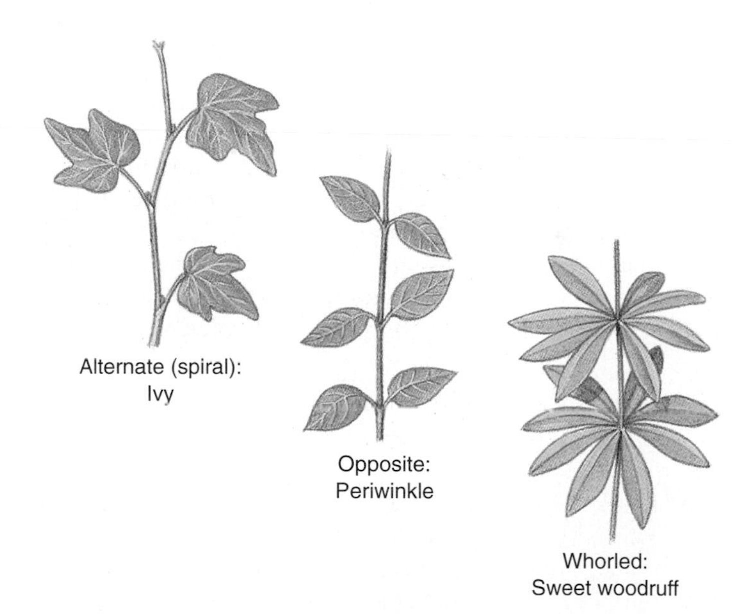

FIGURE 38.32
Types of leaf arrangements. The three common types of leaf arrangements are alternate, opposite, and whorled.

Leaves are the principal sites of photosynthesis. Their blades may be arranged in a variety of ways. In simple leaves the blades are undivided, while in compound leaves the leaf is composed of two or more leaflets.

Leaf Internal Structure

The entire surface of a leaf is covered by a transparent epidermis, most of whose cells have no chloroplasts. The epidermis itself has a waxy *cuticle* of variable thickness, and may have different types of glands and trichomes (hairs) present. The lower epidermis (and occasionally the upper epidermis) of most leaves contains numerous slit-like or mouth-shaped *stomata* (figure 38.33). Stomata, as discussed earlier, are flanked by *guard cells* and function in gas exchange and regulation of water movement through the plant.

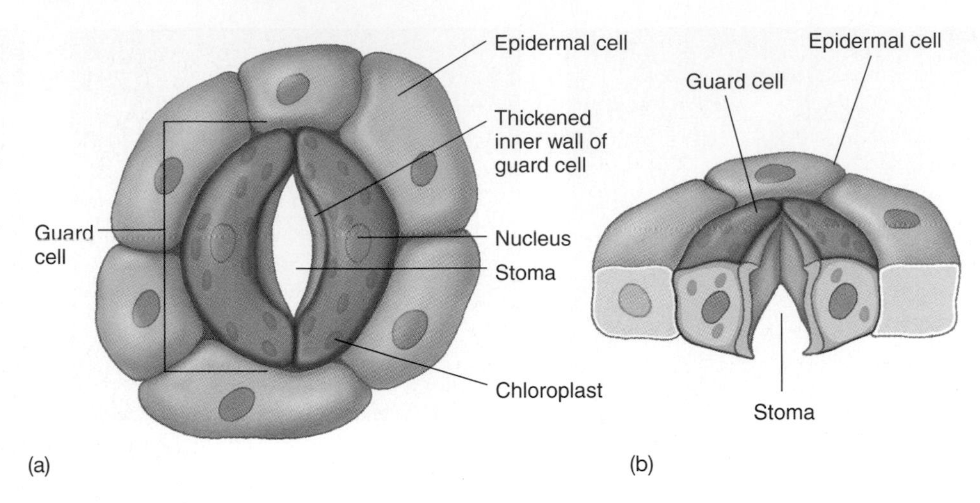

FIGURE 38.33
A stoma. (*a*) Surface view. (*b*) View in cross-section.

The tissue between the upper and lower epidermis is called **mesophyll.** Mesophyll is interspersed with veins (vascular bundles) of various sizes. In most dicot leaves, there are two distinct types of mesophyll. Closest to the upper epidermis are one to several (usually two) rows of tightly packed, barrel-shaped to cylindrical *chlorenchyma* cells (parenchyma with chloroplasts) that constitute the **palisade mesophyll** (figure 38.34). Some plants, including species of *Eucalyptus*, have leaves that hang down, rather than extending horizontally. They have palisade parenchyma on both sides of the leaf, and there is, in effect, no upper side. In nearly all leaves there are loosely arranged **spongy mesophyll** cells between the palisade mesophyll and the lower epidermis, with many air spaces throughout the tissue. The interconnected intercellular spaces, along with the stomata, function in gas exchange and the passage of water vapor from the leaves. The mesophyll of monocot leaves is not differentiated into palisade and spongy layers and there is often little distinction between the upper and lower epidermis. This anatomical difference often correlates with a modified photosynthetic pathway that maximizes the amount of CO_2 relative to O_2 to reduce energy loss through photorespiration (refer to chapter 10). Leaf anatomy directly relates to its juggling act to balance water loss, gas exchange, and transport of photosynthetic products to the rest of the plant.

Leaves are basically flattened bags of epidermis containing vascular tissue and tightly packed palisade mesophyll rich in chloroplasts and loosely packed spongy mesophyll with many interconnected air spaces that function in gas and water vapor exchange.

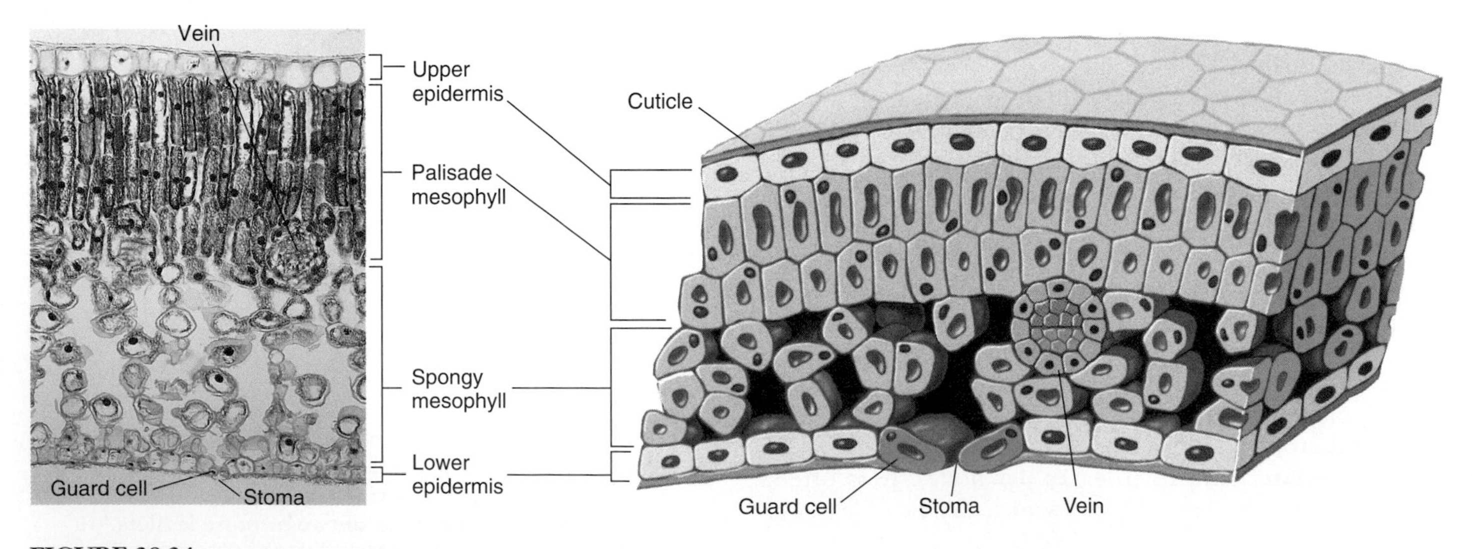

FIGURE 38.34
A leaf in cross-section. Transection of a leaf showing the arrangement of palisade and spongy mesophyll, a vascular bundle or vein, and the epidermis with paired guard cells flanking the stoma.

Modified Leaves

As plants colonized a wide variety of environments, from deserts to lakes to tropical rain forests, modifications of plant organs that would adapt the plants to their specific habitats arose. Leaves, in particular, have evolved some remarkable adaptations. A brief discussion of a few of these modifications follows.

Floral leaves (bracts). Poinsettias and dogwoods have relatively inconspicuous, small, greenish-yellow flowers. However, both plants produce large modified leaves, called **bracts** (mostly colored red in poinsettias and white or pink in dogwoods). These bracts surround the true flowers and perform the same function as showy petals (figure 38.35). It should be noted, however, that bracts can also be quite small and not as conspicuous as those of the examples mentioned.

Spines. The leaves of many cacti, barberries, and other plants are modified as **spines** (see figure 38.28*f*). In the case of cacti, the reduction of leaf surface reduces water loss and also may deter predators. Spines should not be confused with *thorns*, such as those on honey locust (*Gleditsia triacanthos*), which are modified stems, or with the *prickles* on raspberries and rose bushes, which are simply outgrowths from the epidermis or the cortex just beneath it.

Reproductive leaves. Several plants, notably *Kalanchoë*, produce tiny but complete plantlets along their margins. Each plantlet, when separated from the leaf, is capable of growing independently into a full-sized plant. The walking fern (*Asplenium rhizophyllum*) produces new plantlets at the tips of its fronds. While leaf tissue isolated from many species will regenerate a whole plant, this in vivo regeneration is unique among just a few species.

Window leaves. Several genera of plants growing in arid regions produce succulent, cone-shaped leaves with transparent tips. The leaves often become mostly buried in sand blown by the wind, but the transparent tips, which have a thick epidermis and cuticle, admit light to the hollow interiors. This allows photosynthesis to take place beneath the surface of the ground.

Shade leaves. Leaves produced where they receive significant amounts of shade tend to be larger in surface area, but thinner and with less mesophyll than leaves on the same tree receiving more direct light. This plasticity in development is remarkable, as both types of leaves on the plant have exactly the same genes. Environmental signals can have a major effect on development.

Insectivorous leaves. Almost 200 species of flowering plants are known to have leaves that trap insects, with some digesting their soft parts. Plants with insectivorous leaves often grow in acid swamps deficient in needed elements, or containing elements in forms not readily available to the plants; this inhibits the plants' capacities to maintain metabolic processes sufficient to meet their growth requirements. Their needs are, however, met by the supplementary absorption of nutrients from the animal kingdom.

Pitcher plants (for example, *Sarracenia*, *Darlingtonia*, *Nepenthes*) have cone-shaped leaves in which rainwater can accumulate. The insides of the leaves are very smooth, but there are stiff, downward-pointing hairs at the rim. An insect falling into such a leaf finds it very difficult to escape and eventually drowns. The nutrients released when bacteria, and in most species digestive enzymes, decompose the insect bodies are absorbed into the leaf. Other plants, such as sundews (*Drosera*), have glands that secrete sticky mucilage that trap insects, which are then digested by enzymes. The Venus flytrap (*Dionaea muscipula*) produces leaves that look hinged at the midrib. When tiny trigger hairs on the leaf blade are stimulated by a moving insect, the two halves of the leaf snap shut, and digestive enzymes break down the soft parts of the trapped insect into nutrients that can be absorbed through the leaf surface. Nitrogen is the most common nutrient needed. Curiously, the Venus flytrap will not survive in a nitrogen-rich environment, perhaps a trade-off made in the intricate evolutionary process that resulted in its ability to capture and digest insects.

FIGURE 38.35
Modified leaves. In this dogwood "flower," the white-colored bracts (modified leaves) surround the several true flowers without petals in the center.

The leaves of plants exhibit a variety of adaptations, including spines, vegetative reproduction, and even leaves that are carnivorous.

Chapter 38

www.mhhe.com/raven6/resources38.mhtml

Summary	Questions	Media Resources

38.1 Meristems elaborate the plant body plan after germination.

- A plant body is basically an axis that includes two parts: root and shoot—with associated leaves. There are three basic types of tissues in plants arising from meristems or embryonic cells: ground tissue, epidermis, and vascular tissue.

1. What are the three major tissue systems in plants? What are their functions?

- Art Activities:
  - -Stem Tip Structure
  - -Primary Meristem Structure

38.2 Plants have three basic tissues, each composed of several cell types.

- Epidermis forms an outer protective covering for the plant.
- Ground tissue supports the plant and stores food and water.
- Vascular tissue conducts water, carbohydrates, and dissolved minerals to different parts of the plant. Xylem conducts water and minerals from the roots to shoots and leaves, and phloem conducts food molecules from sources to all parts of the plant.

2. What is the function of xylem? How do primary and secondary xylem differ in origin? What are the two types of conducting cells within xylem?

3. What is the function of phloem?

- Vascular System of Plants

- Ground Tissue
- Dermal Tissue
- Vascular Tissue

38.3 Root cells differentiate as they become distanced from the dividing root apical meristem.

- Roots have four growth zones: the root cap, zone of cell division, zone of elongation, and zone of maturation.
- Some plants have modified roots, adapted for photosynthesis, food or water storage, structural support, or parasitism.

4. Compare monocot and dicot roots. How does the arrangement of the tissues differ?

5. How are lateral branches of roots formed?

- Art Activity: Dicot Root Structure

- Roots

38.4 Stems are the backbone of the shoot, transporting nutrients and supporting the aerial plant organs.

- Plants branch by means of buds derived from the primary apical meristem. They are found in the junction between the leaf and the stem.
- The vascular cambium is a cylinder of dividing cells found in both roots and shoots of gymnosperms and dicots. As a result of their activity, the girth of a plant increases.

6. What types of cells are produced when the vascular cambium divides outwardly, inwardly, or laterally?

7. Why don't monocots have secondary growth?

- Art Activities:
  - -Dicot Stem Structure
  - -Secondary Growth
  - -Herbaceous Dicot Stem Anatomy

38.5 Leaves are adapted to support basic plant functions.

- Leaves emerge as bulges on the meristem in a variety of patterns, but most form a spiral around the stem. The bulge lengthens and loses its radial symmetry as it flattens.
- Photosynthesis occurs in the ground tissue system which is called mesophyll in the leaf. Vascular tissue forms the venation patterns in the leaves, serving as the endpoint for water conduction and often the starting point for the transport of photosynthetically produced sugars.

8. How do simple and compound leaves differ from each other? Name and describe the three common types of leaf growth patterns.

- Art Activities:
  - -Plant Anatomy
  - -Leaf Structure

- Leaves

- Art Quizzes:
  - -Stoma
  - -Abscission Zone

39

Nutrition and Transport in Plants

Concept Outline

39.1 Plants require a variety of nutrients in addition to the direct products of photosynthesis.

Plant Nutrients. Plants require a few macronutrients in large amounts and several micronutrients in trace amounts.

Soil. Plant growth is significantly influenced by the nature of the soil.

39.2 Some plants have novel strategies for obtaining nutrients.

Nutritional Adaptations. Venus flytraps and other carnivorous plants lure and capture insects and then digest them to obtain nutrients. Some plants entice bacteria to produce organic nitrogen for them. These bacteria may be free-living or form a symbiotic relationship with a host plant. About 90% of all vascular plants rely on fungal associations to gather essential nutrients, especially phosphorus.

39.3 Water and minerals move upward through the xylem.

Overview of Water and Mineral Movement through Plants. The bulk movement of water and dissolved minerals is the result of movement between cells, across cell membranes, and through tubes of xylem.

Water and Mineral Absorption. Water and minerals enter the plant through the roots.

Water and Mineral Movement. A combination of the properties of water, structure of xylem, and transpiration of water through the leaves results in the passive movement of water to incredible heights. Water leaves the plant through openings in the leaves called stomata. Too much water is harmful to a plant, although many plants have adaptations that make them tolerant of flooding.

39.4 Dissolved sugars and hormones are transported in the phloem.

Phloem Transport Is Bidirectional. Sucrose and hormones can move from shoot to root or root to shoot in the phloem. Phloem transport requires energy to load and unload sieve tubes.

FIGURE 39.1
A carnivorous plant. Most plants absorb water and essential nutrients from the soil, but carnivorous plants are able to obtain some nutrients directly from small animals.

Vast energy inputs are required for the ongoing construction of a plant such as described in chapter 38. In this chapter, we address two major questions: (1) what inputs, besides energy from the sun, does a plant need to survive? and (2) how do all parts of the complex plant body share the essentials of life? Plants, like animals, need various nutrients to remain alive and healthy. Lack of an important nutrient may slow a plant's growth or make the plant more susceptible to disease or even death. Plants acquire these nutrients through photosynthesis and from the soil, although some take a more direct approach (figure 39.1). Carbohydrates produced in leaves must be carried throughout the plant, and minerals and water absorbed from the soil must be transported up to the leaves and other parts of the plant. As discussed in chapter 38, these two types of transport take place in specialized tissues, xylem and phloem.

39.1 Plants require a variety of nutrients in addition to the direct products of photosynthesis.

Plant Nutrients

The major source of plant nutrition is the fixation of atmospheric CO_2 into simple sugar using the energy of the sun. CO_2 enters through the stomata (openings on the surface of the leaf). O_2 is a product of photosynthesis and is an atmospheric component that also moves through the stomata. It is used in cellular respiration to release energy from the chemical bonds in the sugar to support growth and maintenance in the plant. However, CO_2 and light energy are not sufficient for the synthesis of all the molecules a plant needs. Plants require a number of inorganic nutrients (table 39.1). Some of these are macronutrients, which the plants need in relatively large amounts, and others are micronutrients, which are required in trace amounts. There are nine macronutrients: carbon, hydrogen, and oxygen—the three elements found in all organic compounds—as well as nitrogen (essential for amino acids), potassium, calcium, phosphorus, magnesium (the center of the chlorophyll molecule), and sulfur. Each of these nutrients approaches or, as in the case with carbon, may greatly exceed 1% of the dry weight of a healthy plant. The seven micronutrient elements—iron, chlorine, copper, manganese, zinc, molybdenum, and boron—constitute from less than one to several hundred parts per million in most plants (figure 39.2). The macronutrients were generally discovered in the last century, but the micronutrients have been detected much more recently as technology developed to identify and work with such small quantities.

Nutritional requirements are assessed in hydroponic cultures; the plants roots are suspended in aerated water containing nutrients. The solutions contain all the necessary nutrients in the right proportions but with certain known or suspected nutrients left out. The plants are then

Table 39.1 Essential Nutrients in Plants

Elements	Principal Form in which Element Is Absorbed	Approximate Percent of Dry Weight	Examples of Important Functions
MACRONUTRIENTS			
Carbon	(CO_2)	44	Major component of organic molecules
Oxygen	(O_2, H_2O)	44	Major component of organic molecules
Hydrogen	(H_2O)	6	Major component of organic molecules
Nitrogen	(NO_3^-, NH_4^+)	1–4	Component of amino acids, proteins, nucleotides, nucleic acids, chlorophyll, coenzymes, enzymes
Potassium	(K^+)	0.5–6	Protein synthesis, operation of stomata
Calcium	(Ca^{++})	0.2–3.5	Component of cell walls, maintenance of membrane structure and permeability, activates some enzymes
Magnesium	(Mg^{++})	0.1–0.8	Component of chlorophyll molecule, activates many enzymes
Phosphorus	($H_2PO_4^-$, $HPO_4^=$)	0.1–0.8	Component of ADP and ATP, nucleic acids, phospholipids, several coenzymes
Sulfur	($SO_4^=$)	0.05–1	Components of some amino acids and proteins, coenzyme A
MICRONUTRIENTS (CONCENTRATIONS IN PPM)			
Chlorine	(Cl^-)	100–10,000	Osmosis and ionic balance
Iron	(Fe^{++}, Fe^{+++})	25–300	Chlorophyll synthesis, cytochromes, nitrogenase
Manganese	(Mn^{++})	15–800	Activator of certain enzymes
Zinc	(Zn^{++})	15–100	Activator of many enzymes, active in formation of chlorophyll
Boron	(BO_3^- or $B_4O_7^=$)	5–75	Possibly involved in carbohydrate transport, nucleic acid synthesis
Copper	(Cu^{++})	4–30	Activator or component of certain enzymes
Molybdenum	($MoO_4^=$)	0.1–5	Nitrogen fixation, nitrate reduction

FIGURE 39.2
Mineral deficiencies in plants. (*a*) Leaves of a healthy Marglobe tomato (*Lycopersicon esculentum*) plant. (*b*) Chlorine-deficient plant with necrotic leaves (leaves with patches of dead tissue). (*c*) Copper-deficient plant with blue-green, curled leaves. (*d*) Zinc-deficient plant with small, necrotic leaves. The agricultural implications of deficiencies such as these are obvious; a trained observer can determine the nutrient deficiencies that are affecting a plant simply by inspecting it.

allowed to grow and are studied for the presence of abnormal symptoms that might indicate a need for the missing element (figure 39.3). However, the water or vessels used often contain enough micronutrients to allow the plants to grow normally, even though these substances were not added deliberately to the solutions. To give an idea of how small the quantities of micronutrients may be, the standard dose of molybdenum added to seriously deficient soils in Australia amounts to about 34 grams (about one handful) per hectare (a square 100 meters on a side—about 2.5 acres), once every 10 years! Most plants grow satisfactorily in hydroponic culture, and the method, although expensive, is occasionally practical for commercial purposes. Analytical chemistry has made it much easier to take plant material and test for levels of different molecules. One application has been the investigation of the effects of elevated levels of CO_2 (a result of global warming) on plant growth. With increasing levels of CO_2, the leaves of some plants increase in size, but the amount of nitrogen decreases relative to carbon. This decreases the nutritional value of the leaves to herbivores.

The plant macronutrients carbon, oxygen, and hydrogen constitute about 94% of a plant's dry weight; the other macronutrients—nitrogen, potassium, calcium, phosphorus, magnesium, and sulfur—each approach or exceed 1% of a plant's dry weight.

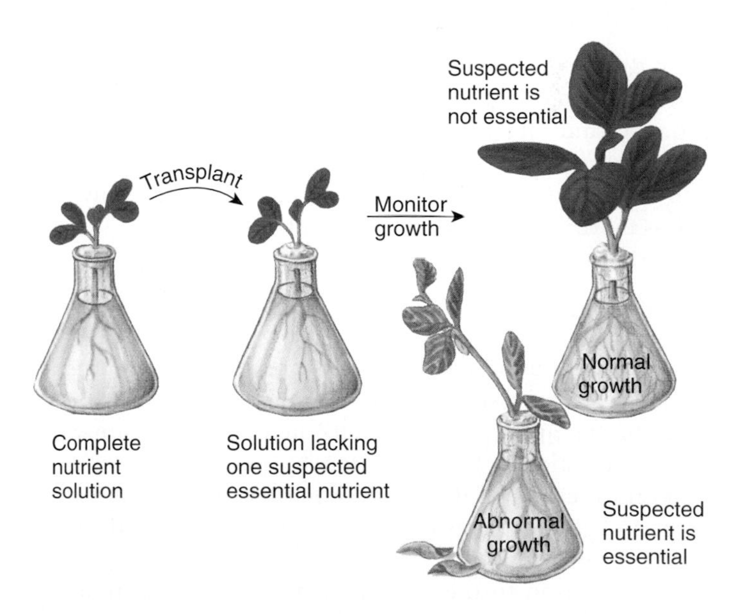

FIGURE 39.3
Identifying nutritional requirements of plants. A seedling is first grown in a complete nutrient solution. The seedling is then transplanted to a solution that lacks one suspected essential nutrient. The growth of the seedling is then studied for the presence of abnormal symptoms, such as discolored leaves and stunted growth. If the seedling's growth is normal, the nutrient that was left out may not be essential; if the seedling's growth is abnormal, the lacking nutrient is essential for growth.

Soil

Plant growth is affected by soil composition. Soil is the highly weathered outer layer of the earth's crust. It is composed of a mixture of ingredients, which may include sand, rocks of various sizes, clay, silt, humus, and various other forms of mineral and organic matter; pore spaces containing water and air occur between the particles. The mineral fraction of soils varies according to the composition of the rocks. The crust includes about 92 naturally occurring elements; table 2.1 in chapter 2 lists the most common of these elements and their percentage of the earth's crust by weight. Most elements are combined as inorganic compounds called **minerals;** most rocks consist of several different minerals. The soil is also full of microorganisms that break down and recycle organic debris. About 5 metric tons of carbon is tied up in the organisms that are present in the soil under a hectare of wheat land in England, an amount that approximately equals the weight of 100 sheep!

Most roots are found in **topsoil** (figure 39.4), which is a mixture of mineral particles of varying size (most less than 2 mm in diameter), living organisms, and **humus.** Humus consists of partly decayed organic material. When topsoil is lost because of erosion or poor landscaping, both the water-holding capacity and the nutrient relationships of the soil are adversely affected.

About half of the total soil volume is occupied by spaces or pores, which may be filled with air or water, depending on moisture conditions. Some of the soil water, because of its properties described below, is unavailable to plants. Due to gravity, some of the water that reaches a given soil will drain through it immediately. Another fraction of the water is held in small soil pores, which are generally less than about 50 micrometers in diameter. This water is readily available to plants. When it is depleted through evaporation or root uptake, the plant will wilt and eventually die unless more water is added to the soil.

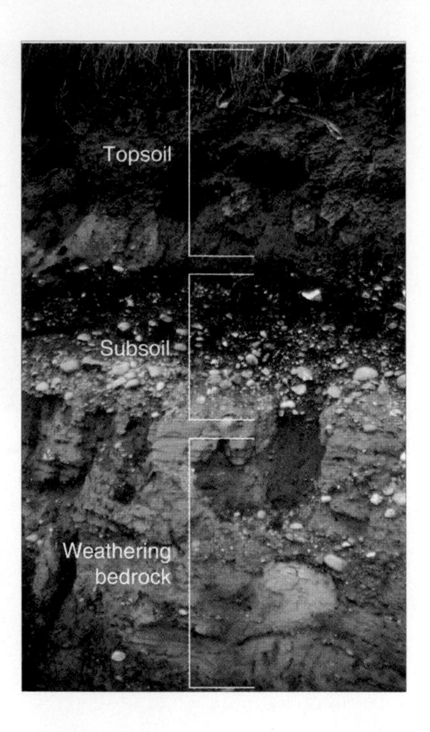

FIGURE 39.4
Most roots occur in topsoil. The uppermost layer in soil is called topsoil, and it contains organic matter, such as roots, small animals, humus, and mineral particles of various sizes. Subsoil lies underneath the topsoil and contains larger mineral particles and relatively little organic matter. Beneath the subsoil are layers of bedrock, the raw material from which soil is formed over time and weathering.

Cultivation

In natural communities, nutrients are recycled and made available to organisms on a continuous basis. When these communities are replaced by cultivated crops, the situation changes drastically: the soil is much more exposed to erosion and the loss of nutrients. For this reason, cultivated crops and garden plants usually must be supplied with additional mineral nutrients.

One solution to this is **crop rotation.** For example, a farmer might grow corn in a field one year and soybeans the next year. Both crops remove nutrients from the soil, but the plants have different nutritional requirements, and therefore the soil does not lose the same nutrients two years in a row. Soybean plants even add nitrogen compounds to the soil, released by nitrogen-fixing bacteria growing in nodules on their roots. Sometimes farmers allow a field to lie fallow—that is, they do not grow a crop in the field for a year or two. This allows natural processes to rebuild the field's store of nutrients.

Other farming practices that help maintain soil fertility involve plowing under plant material left in fields. You can do the same thing in a lawn or garden by leaving grass clippings and dead leaves. Decomposers in the soil do the rest, turning the plant material into humus.

Fertilizers are also used to replace nutrients lost in cultivated fields. The most important mineral nutrients that need to be added to soils are nitrogen (N), phosphorus (P), and potassium (K). All of these elements are needed in large quantities (see table 39.1) and are the most likely to become deficient in the soil. Both chemical and organic fertilizers are often added in large quantities and can be significant sources of pollution in certain situations (see chapter 30). Organic fertilizers were widely used long before chemical fertilizers were available. Substances such as manure or the remains of dead animals have traditionally been applied to crops, and plants are often plowed under to increase the soil's fertility. There is no basis for believing that organic fertilizers supply any element to plants that inorganic fertilizers cannot provide. However, organic fertilizers build up the humus content of the soil, which often enhances its water- and nutrient-retaining properties. For this reason, nutrient availability to plants at different times of the year may be improved, under certain circumstances, with organic fertilizers.

Soils contain organic matter and various minerals and nutrients. Farming practices like crop rotation, plowing crops under, and fertilization are often necessary to maintain soil fertility.

39.2 Some plants have novel strategies for obtaining nutrients.

Nutritional Adaptations

Carnivorous Plants

Some plants are able to obtain nitrogen directly from other organisms, just as animals do. These carnivorous plants often grow in acidic soils, such as bogs that lack organic nitrogen. By capturing and digesting small animals directly, such plants obtain adequate nitrogen supplies and thus are able to grow in these seemingly unfavorable environments. Carnivorous plants have modified leaves adapted to lure and trap insects and other small animals (figure 39.5). The plants digest their prey with enzymes secreted from various types of glands.

The Venus flytrap *(Dionaea muscipula)*, which grows in the bogs of coastal North and South Carolina, has three sensitive hairs on each side of each leaf, which, when touched, trigger the two halves of the leaf to snap together (see figure 39.1). Once the Venus flytrap enfolds a prey item within a leaf, enzymes secreted from the leaf surfaces digest the prey. These flytraps actually shut and open by a growth mechanism. They have a limited number of times they can open and close as a result.

In the sundews, glandular trichomes secrete both sticky mucilage, which traps small animals, and digestive enzymes. They do not close rapidly. It is possible that Venus flytraps and sundews share a common ancestor.

Pitcher plants attract insects by the bright, flowerlike colors within their pitcher-shaped leaves and perhaps also by sugar-rich secretions. Once inside the pitchers, insects slide down into the cavity of the leaf, which is filled with water and digestive enzymes.

Bladderworts, *Utricularia*, are aquatic. They sweep small animals into their bladderlike leaves by the rapid action of a springlike trapdoor, and then they digest these animals.

FIGURE 39.5
A carnivorous plant. A tropical Asian pitcher plant, *Nepenthes.* Insects enter the pitchers and are trapped and digested. Complex communities of invertebrate animals and protists inhabit the pitchers.

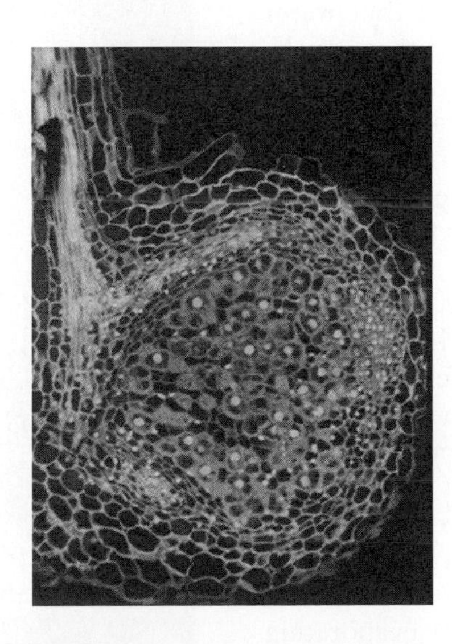

FIGURE 39.6
Nitrogen-fixing nodule. A root hair of alfalfa is invaded by *Rhizobium*, a bacterium (*yellow* structures) that fixes nitrogen. Through a series of exchanges of chemical signals, the plant cells divide to create a nodule for the bacteria which differentiate and begin producing ammonia.

Nitrogen-Fixing Bacteria

Plants need ammonia (NH_3) to build amino acids, but most of the nitrogen is in the atmosphere in the form of N_2. Plants lack the biochemical pathways (including the enzyme nitrogenase) necessary to convert gaseous nitrogen to ammonia, but some bacteria have this capacity. Some of these bacteria live in close association with the roots of plants. Others end up being housed in plant tissues created especially for them called nodules (figure 39.6). Legumes and a few other plants are capable of forming root nodules and there is a very specific recognition required by a bacteria species and its host. Hosting these bacteria costs the plant in terms of energy, but is well worth it when there is little nitrogen in the soil. An energy conservation mechanism has evolved in the legumes so that the root hairs will not respond to bacterial signals when nitrogen levels are high.

Mycorrhizae

While symbiotic relationships with nitrogen-fixing bacteria are rare, symbiotic associations with mycorrhizal fungi are found in about 90% of the vascular plants. These fungi have been described in detail in chapter 36. In terms of plant nutrition, it is important to recognize the significant role these organisms play in enhancing phosphorus transfer to the plant. The uptake of some of the micronutrients is also enhanced. Functionally, the mycorrhizae extend the surface area of nutrient uptake substantially

Carnivorous plants obtain nutrients, especially nitrogen, directly by capturing and digesting insects and other organisms. Nitrogen can also be obtained from bacteria living in close association with the roots. Fungi help plants obtain phosphorus and other nutrients from the soil.

39.3 Water and minerals move upward through the xylem.

Overview of Water and Mineral Movement through Plants

Local Changes Result in the Long-Distance, Upward Movement of Water

Most of the nutrients and water discussed previously enter the plant through the roots and move upward in the xylem. It is not unusual for a large tree to have leaves more than 10 stories off the ground (figure 39.7). Did you ever wonder how water gets from the roots to the top of a tree that high? Water moves through the spaces between the protoplasts of cells, through plasmodesmata (connections between cells), through cell membranes and through the continuous tubing system in the xylem. We know that there are interconnected, water-conducting xylem elements extending throughout a plant. We also know that water first enters the roots and then moves to the xylem. After that, however, water rises through the xylem because of a combination of factors and some exits through the stomata in the leaves.

While most of our focus will be on the mechanics of water transport through xylem, the movement of water at the cellular level plays a significant role in bulk water transport in the plant as well, although over much shorter distances. You know that the Casparian strip in the root forces water to move through cells (see figure 38.18). In the case of parenchyma cells it turns out that most water also moves across membranes rather than in the intercellular spaces. For a long time, it was believed that water moved across cell membranes only by osmosis through the lipid bilayer. We now know that osmosis is enhanced by water channels called **aquaporins.** These transport channels are found in both plants and animals. In plants they exist in vacuole and plasma membranes. There at least 30 different genes coding for aquaporin-like proteins in *Arabidopsis.* Some aquaporins only appear or open during drought stress. Aquaporins allow for faster water movement between cells than osmosis. They are important not only in maintaining water balance within a cell, but in getting water between many plant cells and the xylem. The greatest distances traveled by water molecules and dissolved minerals are in the xylem.

Once water enters the xylem, it can move upward 100 m in the redwoods. Some "pushing" from the pressure of water entering the roots is involved. However, most of the force is "pulling" caused by water evaporating **(transpiration)** through the stomata on the leaves and other plant surfaces. This works because water molecules stick to each other with hydrogen bonds (cohesion) and to the walls of the tracheid or xylem vessel (adhesion). The result is an unusually stable column of liquid reaching great heights.

FIGURE 39.7
How does water get to the top of this tree? We would expect gravity to make such a tall column of water too heavy to be maintained by capillary action. What pulls the water up?

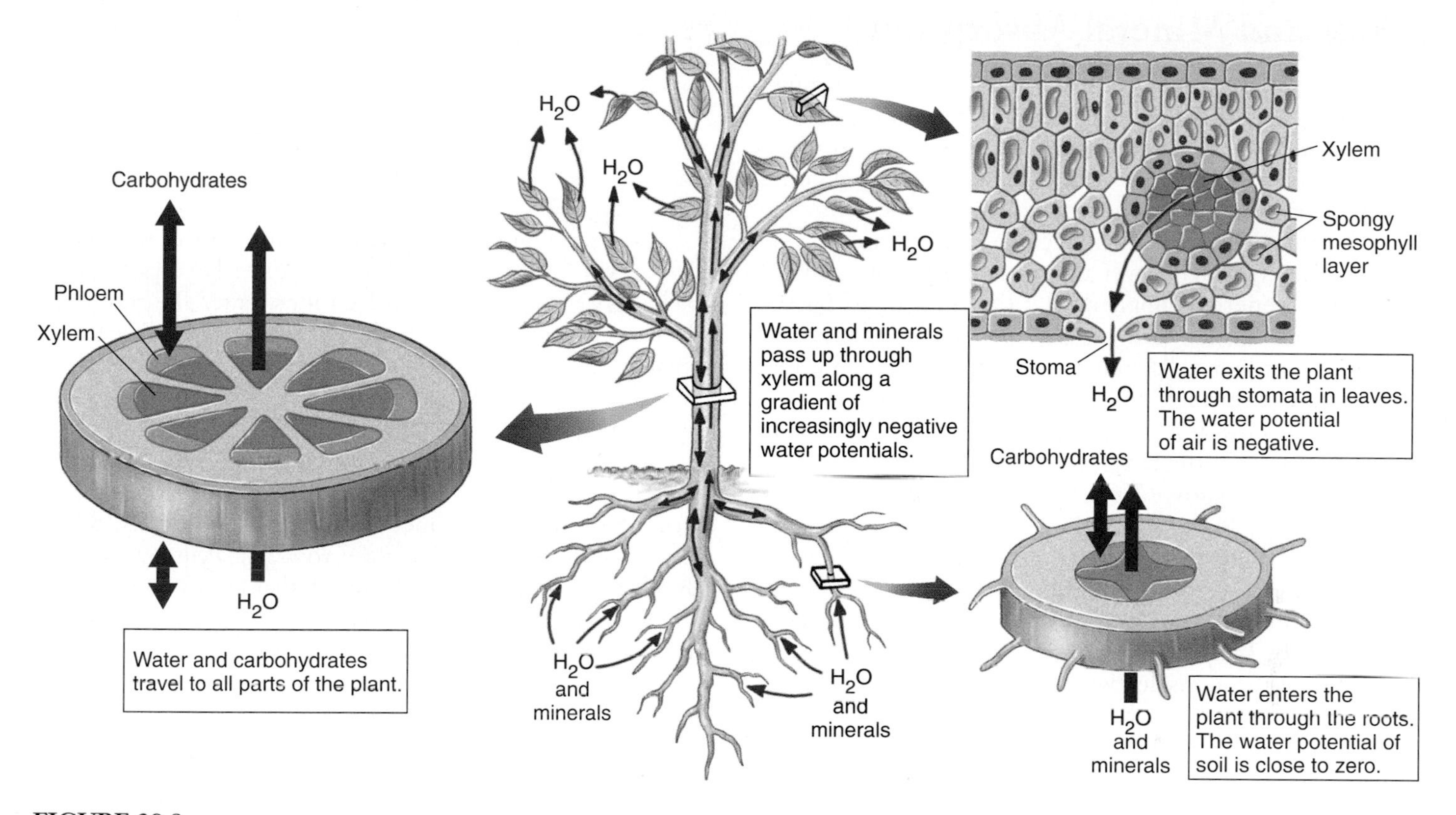

FIGURE 39.8
Water movement through a plant. This diagram illustrates the path of water and inorganic materials as they move into, through, and out of the plant body.

Water Potential

Plant biologists often discuss the forces that act on water within a plant in terms of **potentials.** The *turgor pressure*, which is a physical pressure that results as water enters the cell vacuoles, is referred to as **pressure potential** and is generally positive. Water coming through a garden hose is an example of physical pressure. There is also a potential caused by an uneven distribution of a solute on either side of a membrane, which will result in osmosis (movement of water to the side with the greater concentration of solute). By applying pressure (on the side that has the greater concentration of solute), it is possible to prevent osmosis from taking place. The smallest amount of pressure needed to stop osmosis is referred to as the **solute** (or **osmotic**) **potential** of the solution. Solute potential is generally a negative number. Water will enter a cell osmotically until it is stopped by the pressure potential caused by the cell wall. The **water potential** of a plant cell is, in essence, the combination of its pressure potential and solute potential; it represents the total potential energy of the water in a plant. Pure water without applied pressure has a water potential of zero. Water will move to the cell with the more negative water potential. Water in a plant moves along a gradient between the water potential in the soil (which can be close to zero) to successively negative water potentials in the roots, stems, leaves, and atmosphere.

Water potential in a plant regulates movement of water. At the roots the water potential can be close to zero. On the surface of leaves and other organs, water loss called **transpiration** creates a negative pressure. It depends on its osmotic absorption by the roots and the negative pressures created by water loss from the leaves and other plant surfaces (figure 39.8). The negative pressure generated by transpiration is largely responsible for the upward movement of water in xylem.

Aquaporins enhance water transport at the cellular level, which ultimately affects bulk water transport. The loss of water from the leaf surface, called transpiration, literally pulls water up the stem from the roots which have the greater water potential. This works because of the strong cohesive forces between molecules of water that allow them to stay "stuck" together in a liquid column and adhesion to walls of tracheids and vessels.

Water and Mineral Absorption

Most of the water absorbed by the plant comes in through root hairs, which collectively have an enormous surface area. Root hairs are almost always turgid because their solute potential is greater than that of the surrounding soil due to mineral ions being actively pumped into the cells. Because the mineral ion concentration in the soil water is usually much lower than it is in the plant, an expenditure of energy (supplied by ATP) is required for the accumulation of such ions in root cells. The plasma membranes of root hair cells contain a variety of protein transport channels, through which *proton pumps* (see page 120) transport specific ions against even large concentration gradients. Once in the roots, the ions, which are plant nutrients, are transported via the xylem throughout the plant.

The ions may follow the cell walls and the spaces between them or more often go directly through the plasma membranes and the protoplasm of adjacent cells (figure 39.9). When mineral ions pass between the cell walls, they do so nonselectively. Eventually, on their journey inward, they reach the endodermis and any further passage through the cell walls is blocked by the Casparian strips. Water and ions must pass through the plasma membranes and protoplasts of the endodermal cells to reach the xylem. However, transport through the cells of the endodermis is selective. The endodermis, with its unique structure, along with the cortex and epidermis, controls which ions reach the xylem.

Transpiration from the leaves (figure 39.10), which creates a pull on the water columns, indirectly plays a role in helping water, with its dissolved ions, enter the root cells. However, at night, when the relative humidity may approach 100%, there may be no transpiration. Under these circumstances, the negative pressure component of water potential becomes small or nonexistent.

Active transport of ions into the roots still continues to take place under these circumstances. This results in an increasingly high ion concentration with the cells, which causes more water to enter the root hair cells by osmosis. In terms of water potential, we say that active transport increases the solute potential of the roots. The result is movement of water into the plant and up the xylem columns despite the absence of transpiration. This phenomenon is called **root pressure.**

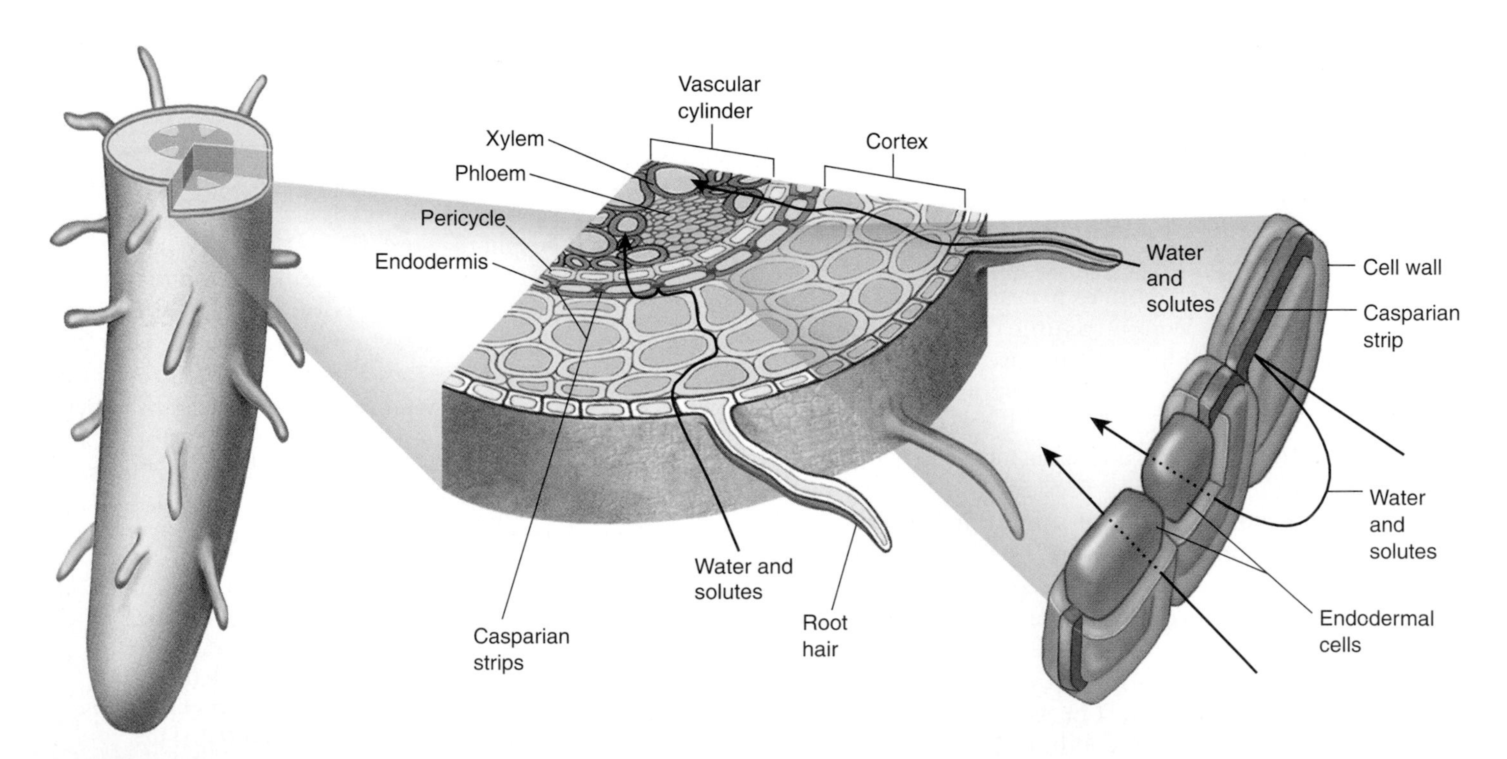

FIGURE 39.9
The pathways of mineral transport in roots. Minerals are absorbed at the surface of the root, mainly by the root hairs. In passing through the cortex, they must either follow the cell walls and the spaces between them or go directly through the plasma membranes and the protoplasts of the cells, passing from one cell to the next by way of the plasmodesmata. When they reach the endodermis, however, their further passage through the cell walls is blocked by the Casparian strips, and they must pass through the membrane and protoplast of an endodermal cell before they can reach the xylem.

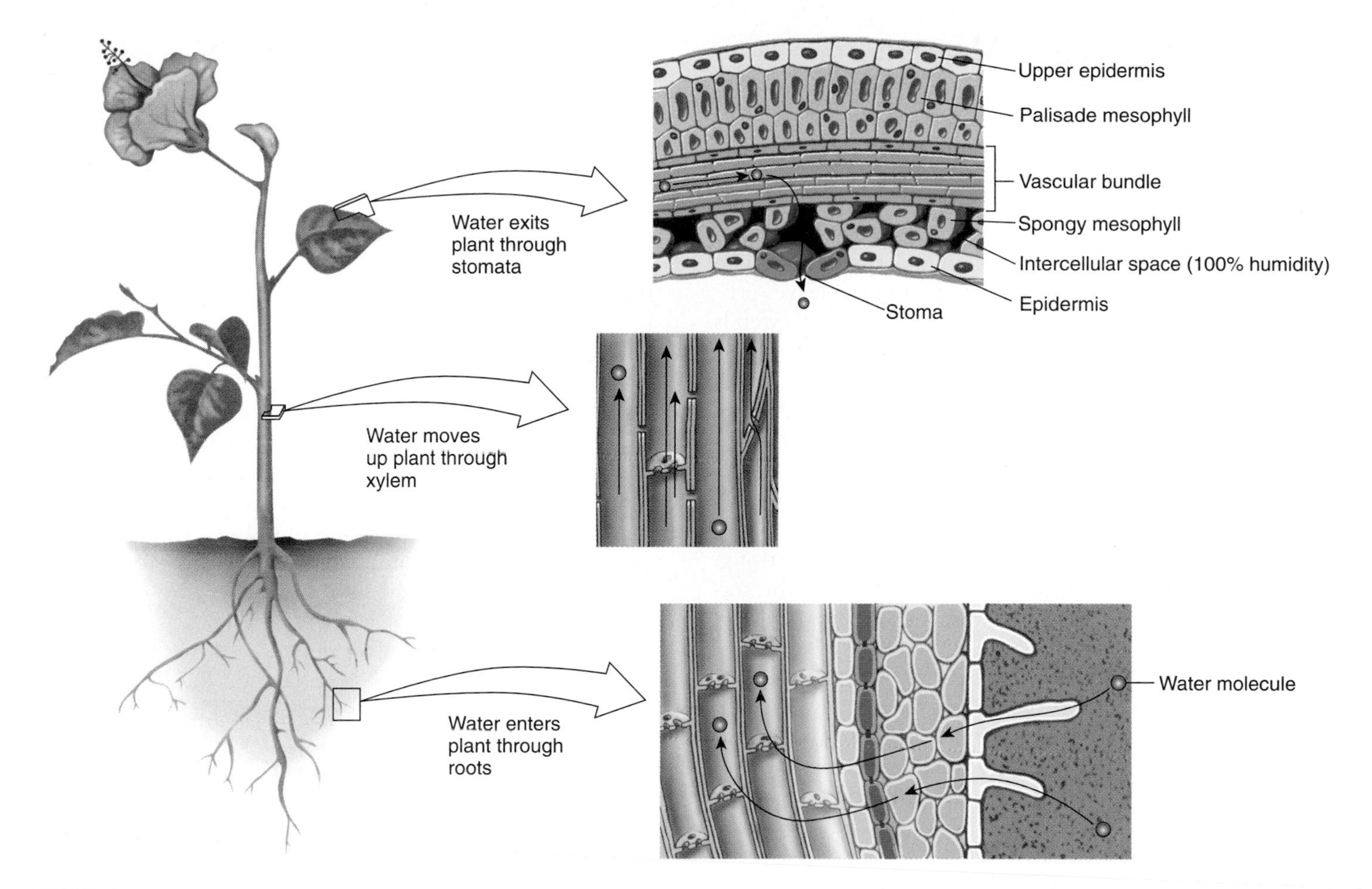

FIGURE 39.10
Transpiration. Water evaporating from the leaves through the stomata causes the movement of water upward in the xylem and the entrance of water through the roots.

Under certain circumstances, root pressure is so strong that water will ooze out of a cut plant stem for hours or even days. When root pressure is very high, it can force water up to the leaves, where it may be lost in a liquid form through a process known as **guttation** (figure 39.11). Guttation does not take place through the stomata, but instead occurs through special groups of cells located near the ends of small veins that function only in this process. Root pressure is never sufficient to push water up great distances.

Water enters the plant by osmosis. Transport of minerals (ions) across the endodermis is selective. Root pressure, which often occurs at night, is caused by the continued, active accumulation of ions in the roots at times when transpiration from the leaves is very low or absent.

FIGURE 39.11
Guttation. In herbaceous plants, water passes through specialized groups of cells at the edges of the leaves; it is visible here as small droplets around the edge of the leaf in this strawberry plant (*Fragaria ananassa*).

Water and Mineral Movement

Water and Mineral Movement through the Xylem

It is clear that root pressure is insufficient to push water to the top of a tall tree, although it can help. So, what does work? Otto Renner proposed the solution in Germany in 1911. Passage of air across leaf surfaces results in loss of water by evaporation, creating a pull at the open upper end of the "tube." Evaporation from the leaves produces a tension on the entire water column that extends all the way down to the roots. Water has an inherent tensile strength that arises from the cohesion of its molecules, their tendency to form hydrogen bonds with one another. The tensile strength of a column of water varies inversely with the diameter of the column; that is, the smaller the diameter of the column, the greater the tensile strength. Because plants have transporting vessels of very narrow diameter, the cohesive forces in them are strong. The water molecules also adhere to the sides of the tracheid or xylem vessels, further stabilizing the long column of water.

The water column would fail if air bubbles were inserted (visualize a tower of blocks and then pull one out in the middle). Anatomical adaptations decrease the probability of this. Individual tracheids and vessel members are connected by one of more *pits* (cavities) in their walls. Air bubbles are generally larger than the openings, so they cannot pass through them. Furthermore, the cohesive force of water is so great that the bubbles are forced into rigid spheres that have no plasticity and therefore cannot squeeze through the openings. Deformed cells or freezing can cause small bubbles of air to form within xylem cells. Any bubbles that do form are limited to the xylem elements where they originate, and water may continue to rise in parallel columns. This is more likely to occur with seasonal temperature changes. As a result, most of the active xylem in woody plants occurs peripherally, toward the vascular cambium.

Most minerals the plant needs enter the root through active transport. Ultimately, they are removed from the roots and relocated through the xylem to other metabolically active parts of the plant. Phosphorus, potassium, nitrogen, and sometimes iron may be abundant in the xylem during certain seasons. In many plants, such a pattern of ionic concentration helps to conserve these essential nutrients, which may move from mature deciduous parts such as leaves and twigs to areas of active growth. Keep in mind that minerals that are relocated via the xylem must move with the generally upward flow through the xylem. Not all minerals can re-enter the xylem conduit. Calcium, an essential nutrient, cannot be transported elsewhere once it has been deposited in plant parts.

Transpiration of Water from Leaves

More than 90% of the water taken in by the roots of a plant is ultimately lost to the atmosphere through transpiration from the leaves. Water moves into the pockets of air in the leaf from the moist surfaces of the walls of the mesophyll cells. As you saw in chapter 38, these intercellular spaces are in contact with the air outside of the leaf by way of the stomata. Water that evaporates from the surfaces of the mesophyll cells leaves the stomata as vapor. This water is continuously replenished from the tips of the veinlets in the leaves.

Water is essential for plant metabolism, but is continuously being lost to the atmosphere through the stomata. Photosynthesis requires a supply of CO_2 entering the stomata from the atmosphere. This results in two somewhat conflicting requirements: the need to minimize the loss of water to the atmosphere and the need to admit carbon dioxide. Structural features such as stomata and the cuticle have evolved in response to one or both of these requirements.

The rate of transpiration depends on weather conditions like humidity and the time of day. After the sun sets, transpiration from the leaves decreases. The sun is the ultimate source of potential energy for water movement. The water potential that is responsible for water movement is largely the product of negative pressure generated by transpiration, which is driven by the warming effects of sunlight.

The Regulation of Transpiration Rate. On a short-term basis, closing the stomata can control water loss. This occurs in many plants when they are subjected to water stress. However, the stomata must be open at least part of the time so that CO_2 can enter. As CO_2 enters the intercellular spaces, it dissolves in water before entering the plant's cells. The gas dissolves mainly in water on the walls of the intercellular spaces below the stomata. The continuous stream of water that reaches the leaves from the roots keeps these walls moist. A plant must respond both to the need to conserve water and to the need to admit CO_2.

Stomata open and close because of changes in the turgor pressure of their guard cells. The sausage- or dumbbell-shaped guard cells stand out from other epidermal cells not only because of their shape, but also because they are the only epidermal cells containing chloroplasts. Their distinctive wall construction, which is thicker on the inside and thinner elsewhere, results in a bulging out and bowing when they become turgid. You can make a model of this for yourself by taking two elongated balloons, tying the closed ends together, and inflating both balloons slightly. When you hold the two open ends together, there should be very little space between the two balloons. Now place

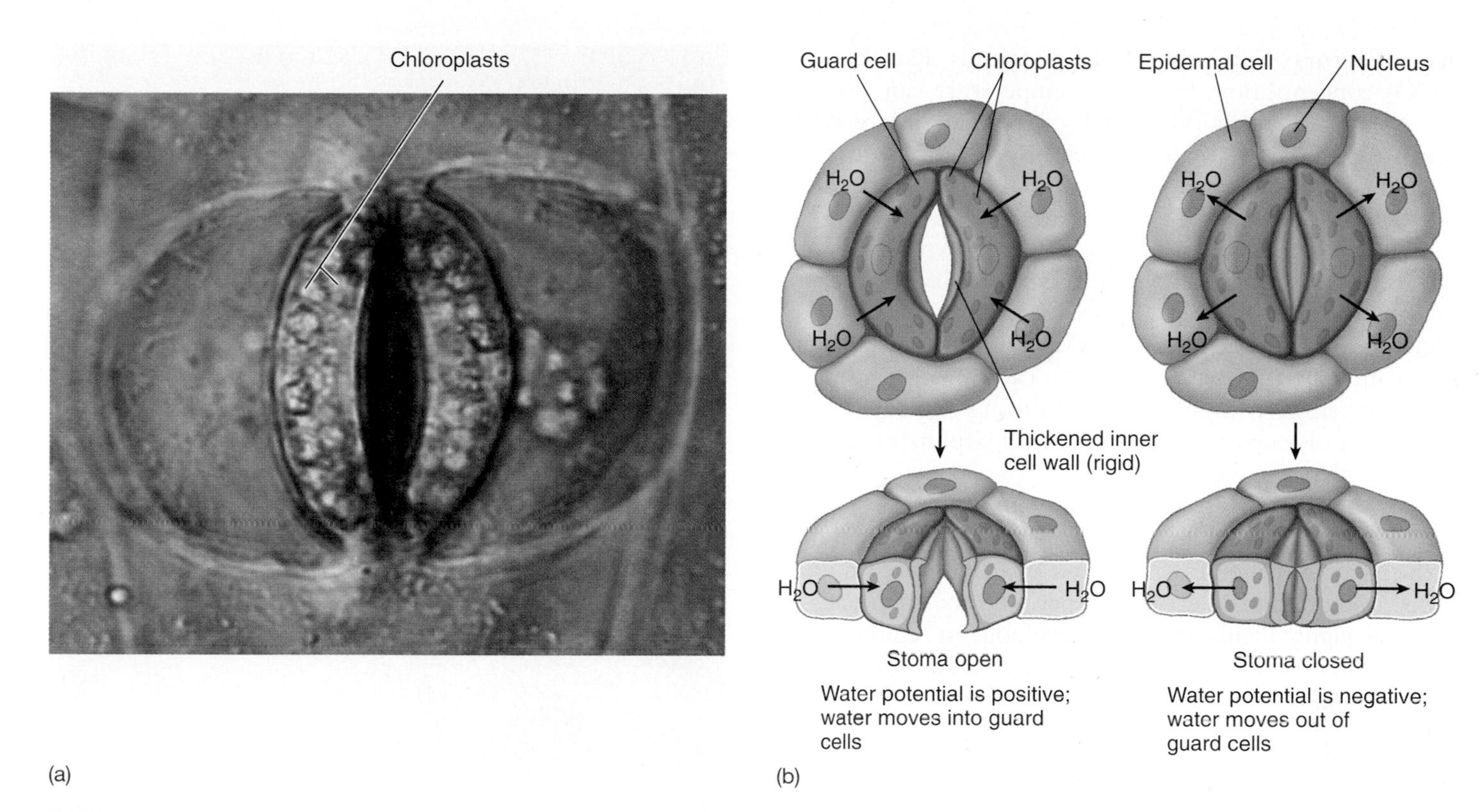

FIGURE 39.12
How a stoma opens and closes. (*a*) When potassium ions from surrounding cells are pumped into guard cells, the guard cell turgor pressure increases as water enters by osmosis. The increased turgor pressure causes the guard cells to bulge, with the thick walls on the inner side of each guard cell bowing outward, thereby opening the stoma. (*b*) When the potassium ions passively leave the guard cells, the guard cells lose water and turgor, and the stoma closes.

duct tape on the inside edge of both balloons and inflate each one a bit more. Hold the open ends together. You should now be holding a doughnut-shaped pair of "guard cells" with a "stoma" in the middle. Real guard cells rely on the influx and efflux of water, rather than air, to open and shut.

Turgor in guard cells results from the active uptake of potassium (K^+) ions. An increase in K^+ concentration creates a water potential in the guard cells that causes water to enter osmotically. As a result, these cells accumulate water and become turgid, opening the stomata (figure 39.12*a*). When K^+ passively leaves the guard cells, water also leaves. The guard cells lose turgor, and the stomata close (figure 39.12*b*). In some species, Cl^- accompanies the K^+ in and out of the guard cells, thus maintaining electrical neutrality.

When a whole plant wilts because there is insufficient water available, the guard cells may also lose turgor, and as a result, the stomata may close. The guard cells of many plant species regularly become turgid in the morning, when photosynthesis occurs, and lose turgor in the evening, regardless of the availability of water. When they are turgid, the stomata open, and CO_2 enters freely; when they are flaccid, CO_2 is largely excluded, but water loss is also retarded.

Experimental evidence is consistent with several pathways regulating stomatal opening and closing. K^+ transport through a specific channel, against a concentration gradient, is promoted by light. Blue light triggers proton (H^+) transport. The resulting proton gradient drives the opening of K^+ channels. When guard cells lose turgor, sucrose export may be another important factor.

Abscisic acid, a plant hormone discussed in chapter 41, plays a primary role in allowing K^+ to pass rapidly out of guard cells, causing the stomata to close in response to drought. This hormone is released from chloroplasts and produced in leaves. It binds to specific receptor sites in the plasma membranes of guard cells. Plants likely control the duration of stomatal opening through the integration of several stimuli. In chapter 41, we will explore the interactions between the environment and the plant in more detail.

Other Factors Regulating Transpiration. Factors such as CO_2 concentration, light, and temperature can also affect stomatal opening. When CO_2 concentrations are high, guard cells of many plant species lose turgor, and their stomata close. Additional CO_2 is not needed at such times, and water is conserved when the guard cells are closed. The stomata also close when the temperature exceeds 30° to 34°C when transpiration would increase substantially. In the dark, stomata will open at low concentrations of CO_2. In chapter 10, we mentioned CAM photosynthesis, which occurs in some succulent plants like cacti. In this process, CO_2 is taken in at night and fixed during the day. CAM photosynthesis conserves water in dry environments where succulent plants grow.

Many mechanisms to regulate the rate of water loss have evolved in plants. One involves dormancy during dry times of the year; another involves loss of leaves. Deciduous plants are common in areas that periodically experience a severe drought. Plants are often deciduous in regions with severe winters, when water is locked up in ice and snow and thus unavailable to them. In a general sense, annual plants conserve water when conditions are unfavorable, simply by going into dormancy as seeds.

Thick, hard leaves, often with relatively few stomata—and frequently with stomata only on the lower side of the leaf—lose water far more slowly than large, pliable leaves with abundant stomata. Temperatures are significantly reduced in leaves covered with masses of woolly-looking trichomes. These trichomes also increase humidity at the leaf surface. Plants in arid or semiarid habitats often have their stomata in crypts or pits in the leaf surface. Within these depressions the water vapor content of the air may be high, reducing the rate of water loss.

Plant Responses to Flooding

Plants can also receive too much water, and ultimately "drown." Flooding rapidly depletes available oxygen in the soil and interferes with the transport of minerals and carbohydrates in the roots. Abnormal growth often results. Hormone levels change in flooded plants—ethylene (the only hormone that is a gas) increases, while gibberellins and cytokinins usually decrease. Hormonal changes contribute to the abnormal growth patterns.

Oxygen-deprivation is among the most significant problems. Standing water has much less oxygen than moving water. Generally, standing water flooding is more harmful to a plant (riptides excluded). Flooding that occurs when a plant is dormant is much less harmful than flooding when it is growing actively.

Physical changes that occur in the roots as a result of oxygen deprivation may halt the flow of water through the plant. Paradoxically, even though the roots of a plant may be standing in water, its leaves may be drying out. One adaptive solution is that stomata of flooded plants often close to maintain leaf turgor.

FIGURE 39.13
Adaptation to flooded conditions. The "knees" of the bald cypress (*Taxodium*) form whenever it grows in wet conditions, increasing its ability to take in oxygen.

Adapting to Life in Fresh Water. Algal ancestors of plants adapted to a freshwater environment from a saltwater environment before the "move" onto land. This involved a major change in controlling salt balance. Since that time, many have "moved" back into fresh water and grow in places that are often or always flooded naturally; they have adapted to these conditions during the course of their evolution (figure 39.13). One of the most frequent adaptations among such plants is the formation of **aerenchyma,** loose parenchymal tissue with large air spaces in it (figure 39.14). Aerenchyma is very prominent in water lilies and many other aquatic plants. Oxygen may be transported from the parts of the plant above water to those below by way of passages in the aerenchyma. This supply of oxygen allows oxidative respiration to take place even in the submerged portions of the plant.

Some plants normally form aerenchyma, whereas others, subject to periodic flooding, can form it when necessary. In corn, ethylene, which becomes abundant under the anaerobic conditions of flooding, induces aerenchyma formation. Plants also respond to flooded conditions by forming larger lenticels (which facilitate gas exchange) and additional adventitious roots.

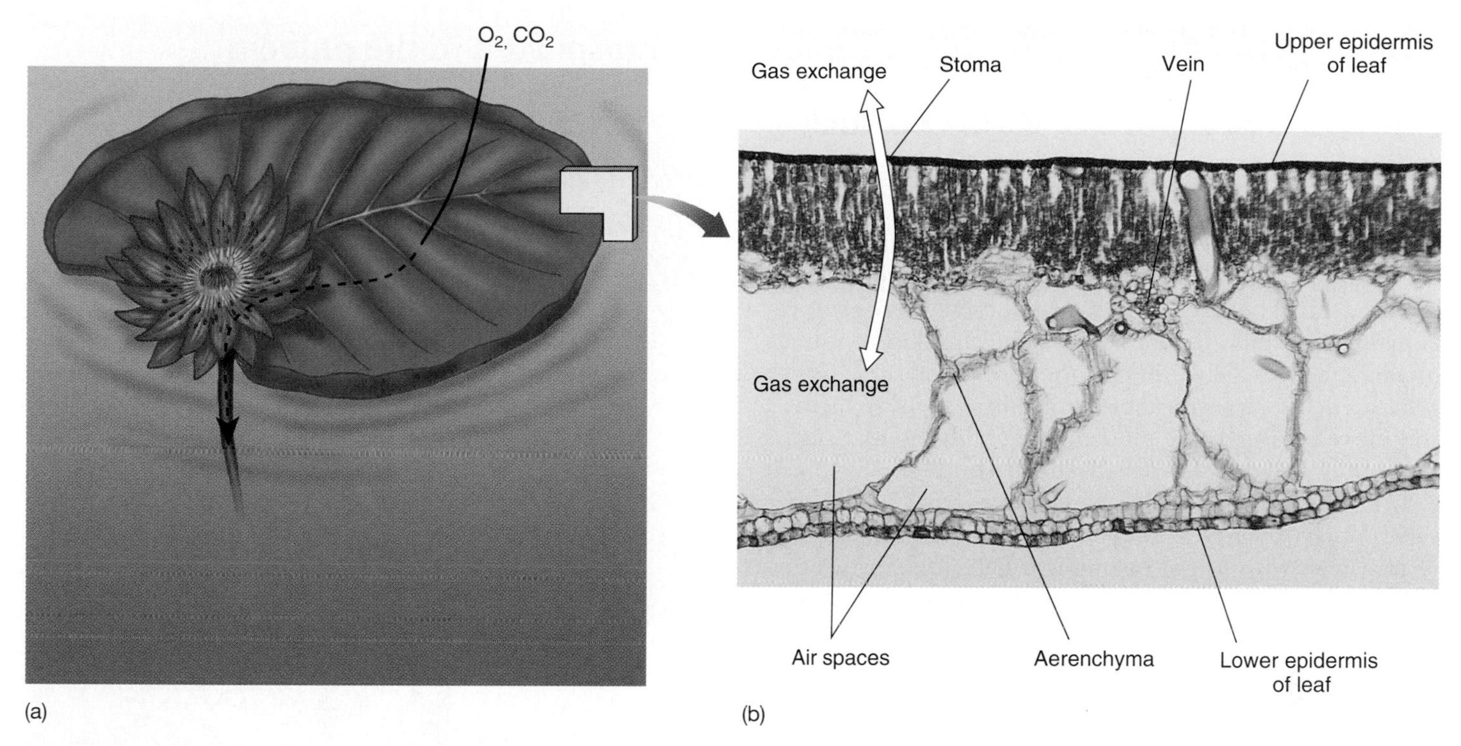

FIGURE 39.14
Aerenchyma tissue. Gas exchange in aquatic plants. (*a*) Water lilies float on the surface of ponds where oxygen is collected and transported to submerged portions of the plant. (*b*) Large air spaces in the leaves add buoyancy. The specialized parenchyma tissue that forms these open spaces is called aerenchyma. Gas exchange occurs through stomata found only on the upper surface of the leaf.

Adapting to Life in Salt Water. Plants such as mangroves that are normally flooded with salt water must not only provide a supply of oxygen for their submerged parts, but also control their salt balance. The salt must be excluded, actively secreted, or diluted as it enters. The arching silt roots of mangroves are connected to long, spongy, air-filled roots that emerge above the mud. These roots, called pneumatophores (see chapter 38), have large lenticels on their above-water portions through which oxygen enters; it is then transported to the submerged roots (figure 39.15). In addition, the succulent leaves of mangroves contain large quantities of water, which dilute the salt that reaches them. Many plants that grow in such conditions also secrete large quantities of salt or block salt uptake at the root level.

Transpiration from leaves pulls water and minerals up the xylem. This works because of the physical properties of water and the narrow diameters of the conducting tubes. Stomata open when their guard cells become turgid. Opening and closing of stomata is osmotically regulated. Biochemical, anatomical, and morphological adaptations have evolved to reduce water loss through transpiration. Plants are harmed by excess water. However, plants can survive flooded conditions, and even thrive in them, if they can deliver oxygen to their submerged parts.

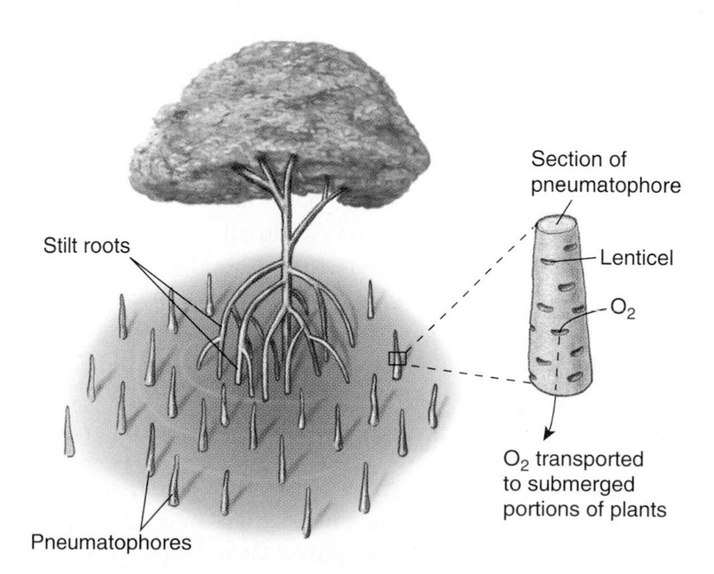

FIGURE 39.15
How mangroves get oxygen to their submerged part. Mangrove plants grow in areas that are commonly flooded, and much of each plant is usually submerged. However, modified roots called pneumatophores supply the submerged portions of the plant with oxygen because these roots emerge above the water and have large lenticels. Oxygen can enter the roots through the lenticels, pass into the abundant aerenchyma, and move to the rest of the plant.

39.4 Dissolved sugars and hormones are transported in the phloem.

Phloem Transport Is Bidirectional

Most carbohydrates manufactured in leaves and other green parts are distributed through the phloem to the rest of the plant. This process, known as **translocation,** is responsible for the availability of suitable carbohydrate building blocks in roots and other actively growing regions of the plant. Carbohydrates concentrated in storage organs such as tubers, often in the form of starch, are also converted into transportable molecules, such as sucrose, and moved through the phloem. The pathway that sugars and other substances travel within the plant has been demonstrated precisely by using radioactive tracers, despite the fact that living phloem is delicate and the process of transport within it is easily disturbed. Radioactive carbon dioxide ($^{14}CO_2$) gets incorporated into glucose as a result of photosynthesis. The glucose is used to make sucrose, which is transported in the phloem. Such studies have shown that sucrose moves both up and down in the phloem.

Aphids, a group of insects that extract plant sap for food, have been valuable tools in understanding translocation. Aphids thrust their *stylets* (piercing mouthparts) into phloem cells of leaves and stems to obtain abundant sugars there. When a feeding aphid is removed by cutting its stylet, the liquid from the phloem continues to flow through the detached mouthpart and is thus available in pure form for analysis (figure 39.16). The liquid in the phloem contains 10 to 25% dry matter, almost all of which is sucrose. Using aphids to obtain the critical samples and radioactive tracers to mark them, it has been demonstrated that movement of substances in phloem can be remarkably fast, up to 50 to 100 centimeters per hour.

While the primary focus of this chapter is on nutrient and water transport, it is important to note that phloem also transports plant hormones. As we will explore in chapter 41, environmental signals can result in the rapid translocation of hormones in the plant.

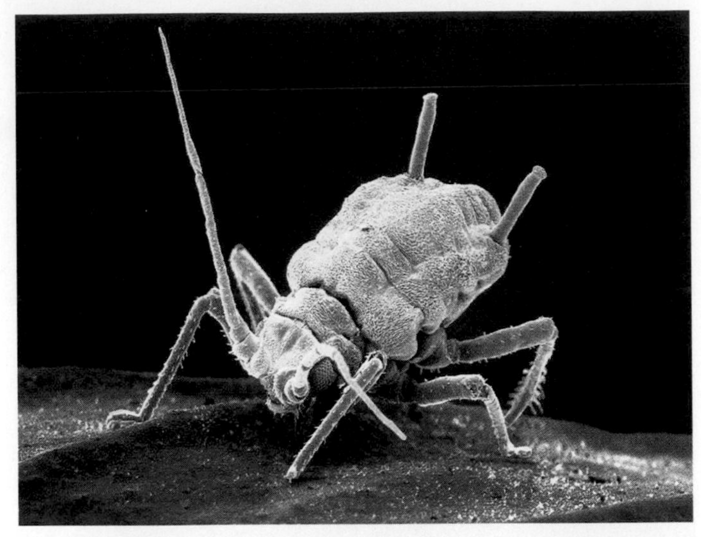
(a)

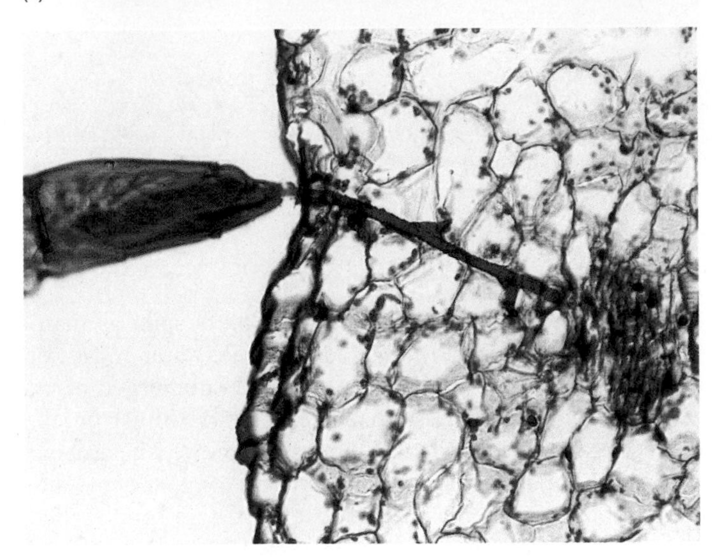
(b)

FIGURE 39.16
Feeding on phloem. (*a*) Aphids, like this individual of *Macrosiphon rosae* shown here on the edge of a rose leaf, feed on the food-rich contents of the phloem, which they extract through their piercing mouthparts (*b*), called stylets. When an aphid is separated from its stylet and the cut stylet is left in the plant, the phloem fluid oozes out of it and can then be collected and analyzed.

Energy Requirements for Phloem Transport

The most widely accepted model of how carbohydrates in solution move through the phloem has been called the **mass-flow hypothesis, pressure flow hypothesis,** or **bulk flow hypothesis.** Experimental evidence supports much of this model. Dissolved carbohydrates flow from a **source** and are released at a **sink** where they are utilized. Carbohydrate sources include photosynthetic tissues, such as the mesophyll of leaves. Food-storage tissues, such as the cortex of roots, can be either sources or sinks. Sinks also occur at the growing tips of roots and stems and in developing fruits.

In a process known as *phloem loading,* carbohydrates (mostly sucrose) enter the sieve tubes in the smallest veinlets at the source. This is an energy-requiring step, as active transport is needed. Companion cells and parenchyma cells adjacent to the sieve tubes provide the ATP energy to drive this transport. Then, because of the

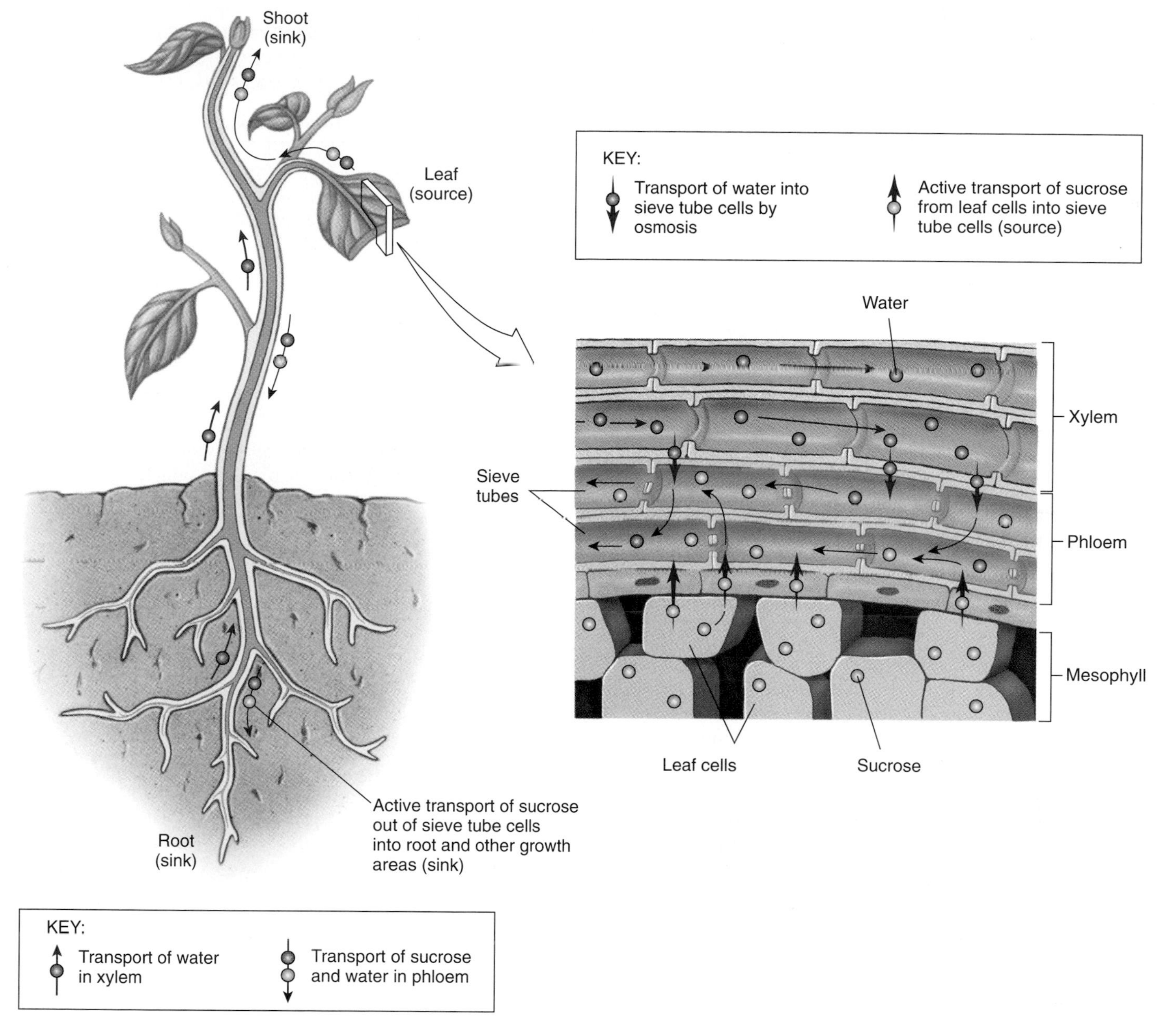

FIGURE 39.17
Diagram of mass flow. In this diagram, *red* dots represent sucrose molecules, and *blue* dots symbolize water molecules. Moving from the mesophyll cells of a leaf or another part of the plant into the conducting cells of the phloem, the sucrose molecules are then transported to other parts of the plant by mass flow and unloaded where they are required.

difference between the water potentials in the sieve tubes and in the nearby xylem cells, water flows into the sieve tubes by osmosis. Turgor pressure in the sieve tubes increases. The increased turgor pressure drives the fluid throughout the plant's system of sieve tubes. At the sink, carbohydrates are actively removed. Water moves from the sieve tubes by osmosis and the turgor pressure there drops, causing a mass flow from the more positive pressure at the source to the more negative pressure sink (figure 39.17). Most of the water at the sink diffuses then back into the xylem, where it may either be recirculated or lost through transpiration.

Transport of sucrose and other carbohydrates through sieve tubes does not require energy. The loading and unloading of these substances from the sieve tubes does.

Chapter 39

www.mhhe.com/raven6/resources39.mhtml

Summary | **Questions** | **Media Resources**

39.1 Plants require a variety of nutrients in addition to the direct products of photosynthesis.

- Plants require a few macronutrients in large amounts and several micronutrients in trace amounts. Most of these are obtained from the soil through the roots.
- Plant growth is significantly influenced by the nature of the soil. Soils vary in terms of nutrient composition and water-holding capacity.

1. What is the difference between a macronutrient and a micronutrient? Explain how a plant would utilize each of the macronutrients.

39.2 Some plants have novel strategies for obtaining nutrients.

- Some plants entice bacteria to produce organic nitrogen for them. These bacteria may be free-living or form a symbiotic relationship with a host plant.
- About 90% of all vascular plants rely on fungal associations to gather essential nutrients.

2. The atmosphere is full of nitrogen yet it is inaccessible to most plants. Why is that? What solution has evolved in legumes?

39.3 Water and minerals move upward through the xylem.

- Water and minerals enter the plant through the roots. Energy is required for active transport.
- The bulk movement of water and minerals is the result of movement between cells, across cell membranes, and through tubes of xylem. Aquaporins are water channels that enhance osmosis.
- A combination of the properties of water, structure of xylem, and transpiration of water through the leaves results in the passive movement of water to incredible heights. The ultimate energy source for pulling water through xylem vessels and tracheids is the sun.
- Water leaves the plant through openings in the leaves called stomata. Stomata open when their guard cells are turgid and bulge, causing the thickened inner walls of these cells to bow away from the opening.
- Plants can tolerate long submersion in water, if they can deliver oxygen to their submerged tissues.

3. What is pressure potential? How does it differ from solute potential? How do these pressures cause water to rise in a plant?

4. What proportion of water that enters a plant leaves it via transpiration?

5. Why are root hairs usually turgid? Does the accumulation of minerals within a plant root require the expenditure of energy? Why or why not?

6. Under what environmental condition is water transport through the xylem reduced to near zero? How much transpiration occurs under these circumstances?

7. Does stomatal control require energy? Explain.

- Uptake by Roots
- Water Movement
- Activity: Water Movement

- Student Research:
  - -Heavy Metal Uptake
  - -The Ecological and Physiological Significance of Leaf Surface Wetness

?

- Art Quiz: Mineral Transport in Roots

39.4 Dissolved sugars and hormones are transported in the phloem.

- Sucrose and hormones can move up and down in the phloem between sources and sinks.
- The movement of water containing dissolved sucrose and other substances in the phloem requires energy. Sucrose is loaded into the phloem near sites of synthesis, or sources, using energy supplied by the companion cells or other nearby parenchyma cells.

8. What is translocation? What is the driving force behind translocation?

9. Describe the movement of carbohydrates through a plant, beginning with the source and ending with the sink. Is this process active or passive?

- Nutrients

- Art Quiz: Mass-Flow Hypothesis

Part **XI**

Plant Growth and Reproduction

Arabidopsis thaliana. An important plant for studying root development because it offers a simple pattern of cellular organization in the root.

The Control of Patterning in Plant Root Development

Did you ever think of how a root grows? Down in the dark, with gravity its only cue, the very tip of the root elongates, periodically forming a node from which root branches will extend. How does the root determine the position of its branches, and the spacing between them? The serial organization of the root's branches is controlled by events that happen on a microscopic scale out at the very tip of the root, the so-called root apex. There, within a space of a millimeter or less, molecular events occur that orchestrate how the root will grow and what it will be like.

The problem of understanding how a plant's root apex controls the way a root develops is one example of a much larger issue, perhaps the most challenging research problem in modern botany: What mechanism mediates central pattern formation in the plant kingdom? Almost nothing was known of these mechanisms a decade ago, but intensive research is now rapidly painting in the blank canvas.

Much of the most exciting research on plant pattern formation is being performed on a small weedy relative of the mustard plant, the wall cress *Arabidopsis thaliana* (see photo above). With individual plants no taller than your thumb that grow quickly in laboratory test-tubes, *Arabidopsis* is an ideal model for studying plant development. Its genome, about the size of the fruit fly *Drosophila*, has been completely sequenced, greatly aiding research into the molecular events underlying pattern formation.

To gain some insight into the sort of research being done, we will focus on work being done by John Schiefelbein and colleagues at the University of Michigan. Schiefelbein has focused on one sharply defined aspect of plant root pattern formation in *Arabidopsis*, the formation of root hairs on the epidermis, the outer most layer of cells. These root hairs constitute the principal absorbing surface of the root, and their position is under tight central control.

In a nutshell, the problem of properly positioning root hairs is one of balancing cell production and cell differentiation. Cells in the growth zone beneath the surface of the root—a sheath called a meristem—are constantly dividing. The cells that are produced by the meristem go on to differentiate into two kinds of cells: trichoblasts which form hair-bearing epidermal cells, and atrichoblasts which form hairless epidermal cells. The positioning of trichoblasts among atrichoblasts determines the pattern of root hairs on the developing root.

When researchers looked very carefully at the dividing root meristem, they found that the initial cells that produce trichoblasts and atrichoblasts alternate with one another in a ring of 16 cells around the circumference of the root. As the cells divide, more and more cells are added, forming columns of cells extending out in 16 files. As the files extend farther and farther out, occasional side-ways divisions fill in the gaps that develop, forming new files.

Maintaining this simple architecture requires that the root maintains a tight control over the plane and rate of cell division. Because this rate is different for the two cell types, the root must also control the rate at which the cell types differentiate. Schiefelbein set out to learn how the root apex coordinates these two processes.

To get a handle on the process, Schiefelbein seized on a recently characterized root pattern mutant called *transparent testa glabra* (*TTG*). This mutant changes the pattern of root hairs in *Arabidopsis*, and it has been proposed that it controls whether a cell becomes a trichoblast or an atrichoblast. But does it control the rate and orientation of cell division in the root meristem epidermis?

To answer this question, Schiefelbein's team used clonal analysis to microscopically identify individual cell types in the root epidermis, and set out to see if they indeed divide at different rates, and if the *TTG* mutation affects these rates differently. If so, there must be a link between cell differentiation and the control of cell division in plants.

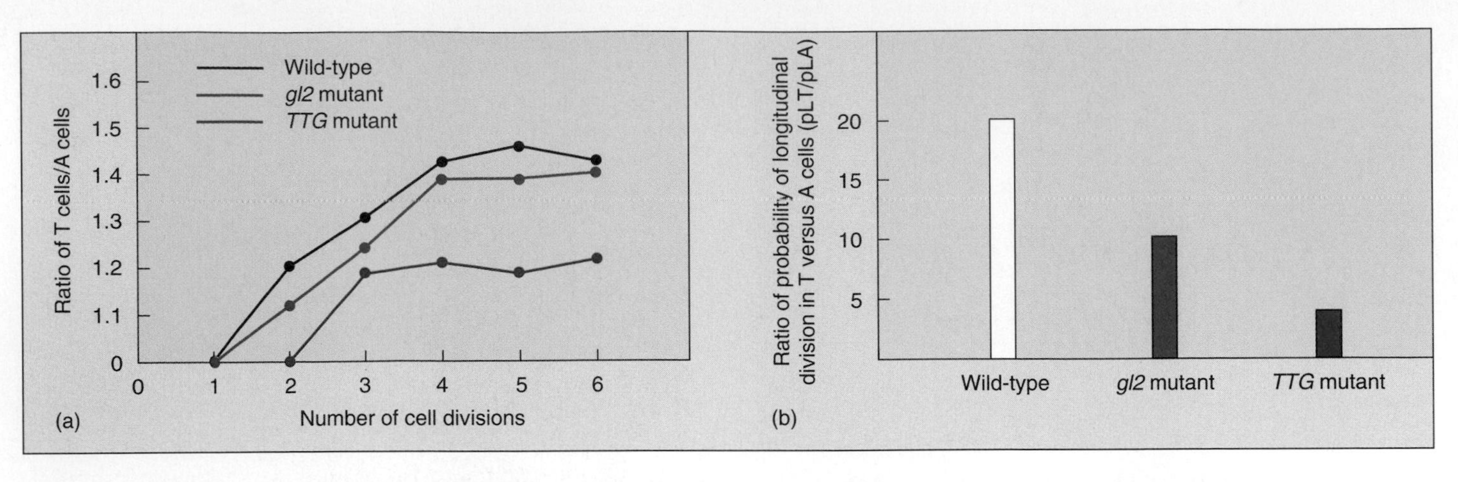

Comparing the differentiation and cell division of trichoblast (T) cells versus atrichoblast (A) cells in root epidermis. (*a*) As cell divisions proceeded, T cells and A cells were identified in the root epidermis of wild-type plants and two mutants, *gl2* and *TTG*. Comparing the ratio of T cells to A cells, there is an increase in the number of A cells compared to T cells in the *TTG* mutant. (*b*) The rate of cell division was also examined by comparing the ratio of probabilities of longitudinal anticlinal cell division in T cells and A cells among the wild-type and mutant plants. This ratio was lowest in *TTG* mutants, indicating that this mutation affects cell division.

The Experiment

Two developmental mutants of *A. thaliana* were used to investigate whether the control of cell differentiation and the rate of cell division were linked. One, *TTG*, alters early events in root epidermal cell differentiation, while the other, *glabra2* (*gl2*) acts later.

The investigators first set out to map the surface of the roots of each mutant type, as well as those of a nonmutant wild type. To avoid confusion in studying files of cells, it is necessary to clearly identify the starting point of each file of cells. To do this, roots were selected that contained clones of trichoblast and atrichoblast produced by longitudinal cell divisions perpendicular to the surface of the root. Called longitudinal anticlinal cell divisions, these clones are rare but easily recognized when stained with propidium iodide. Careful mapping of individual cells with a confocal microscope allowed investigators to determine the number and location of trichoblast and atrichoblast cells present in the epidermal tissue of each clone.

The Results

The researchers made two important observations based on their visual identification of individual trichoblast and atrichoblast cells in the various plant types examined.

1. The two cell types are produced at different rates. Among plants that had been cultured for up to six cell divisions, researchers observed a significant difference in the ratio of trichoblast (T) versus atrichoblast (A) cells following two or more cell divisions. You can readily see that the *TTG* mutant produces a significantly lower ratio of T cells to A cells compared to the wild-type plants or *gl2* mutants (see graph *a* above). This strongly suggests that *TTG* is involved in controlling the rate of cell division in the T cell file.

2. TTG controls the rate of longitudinal cell division. The research team went on to examine longitudinal cell divisions that fill in the gaps as cell division causes files of cells to extend outward from the meristem. The researchers set out to determine the probability of such longitudinal anticlinal cell division occurring in the three types of plants shown in graph *a*. The more rapidly cell files are produced, the more often longitudinal divisions would be required to fill in gaps between files. For proper root hair position to be maintained, the rate of this longitudinal division would have to be tightly coordinated with the rate of vertical division within the file.

The investigators found that longitudinal cell division, always rare, but when it did occur, was usually seen, in T cell files. Did the *TTG* mutation affect this process as well as file-extending cell divisions? This was determined by examining the ratio of the probability of longitudinal anticlinal divisions in T cells versus A cells (pLT/pLA).

Researchers compared the ratio in wild-type plants with that in the two mutants, *TTG* and *gl2*. Did the *TTG* mutation alter longitudinal division? Yes! Their results indicate at least a 60% reduction in the pLT/pLA ratio of the *TTG* mutant compared to wild type and *gl2* plants (see graph *b* above). The A file clones of the *TTG* mutants were twice as likely to exhibit this type of cell division than that seen in the wild-type or *gl2* mutants.

This observation directly supports the hypothesis that the *TTG* gene is not only required for cell division in the T cell file, but also controls longitudinal cell divisions which are characteristically more frequent in trichoblasts.

The research team concluded from these studies that *TTG* is probably the earliest point of control of root epidermis cell fate specification, and that this control most likely acts by negatively controlling trichoblast cell fate.

To explore this experiment further, go to the Virtual Lab at www.mhhe.com/raven6/vlab11.mhtml

40

Early Plant Development

Concept Outline

40.1 Plant embryo development establishes a basic body plan.

Establishing the Root-Shoot Axis. Asymmetric cell division starts patterning the embryo. Early in embryogenesis the root-shoot axis is established.

Establishing Three Tissue Systems. Three tissue systems are established without any cell movement. While the embryo is still a round ball, the root-shoot axis is established. The shape of the plant is determined by planes of cell division and direction of cell elongation. Nutrients are used during embryogenesis, but proteins, lipids, and carbohydrates are also set aside to support the plant during germination before it becomes photosynthetic.

40.2 The seed protects the dormant embryo from water loss.

How Seeds Form. Seeds allow plants to survive unfavorable conditions and invade new habitats.

40.3 Fruit formation enhances the dispersal of seeds.

How Fruits Form. Seed-containing fruits are carried far by animals, wind and water, allowing angiosperms to colonize large areas.

40.4 Germination initiates post-seed development.

Mechanisms of Germination. External signals including water, light, abrasion, and temperature can trigger germination. Rupturing the seed coat and adequate oxygen are essential. Organic reserves stored in the endosperm or cotyledon are made available to the embryo during germination.

FIGURE 40.1
This plant has recently emerged from its seed. It is extending its shoot and leaves up into the air, toward light.

In chapter 37 we emphasized evolutionary changes in reproduction and physiology that gave rise to the highly successful flowering plants (angiosperms). Chapters 38 and 39 explored the morphological and anatomical development of the angiosperm sporophyte, where most of these innovations occurred. In the next few chapters, we continue our focus on the sporophyte generation of the angiosperms. In many cases, we will use the model plant *Arabidopsis*, a weedy member of the mustard family. Its very small genome has allowed plant biologists to study how genes regulate plant growth and development. In this chapter, we will follow the development of the embryo through seed germination (figure 40.1). The next few chapters will also continue to emphasize the roles of gene expression, hormones, and environmental signals in regulating plant development and function.

40.1 Plant embryo development establishes a basic body plan.

Establishing the Root-Shoot Axis

In plants, three-dimensional shape and form arises by regulating the amount and pattern of cell division. Even the very first cell division is asymmetric, resulting in two different cell types. Early in embryo development most cells can give rise to a wide range of cell and organ types, including leaves. As development proceeds, the cells with multiple potentials are mainly restricted to regions called meristems. Many meristems are established by the time embryogenesis ends and the seed becomes dormant. Apical meristems will continue adding cells to the growing root and shoot tips after germination. Apical meristem cells of corn, for example, divide every 12 hours, producing half a million cells a day in an actively-growing corn plant. Lateral meristems can cause an increase in the girth of some plants, while intercalary meristems in the stems of grasses allow for elongation.

In addition to developing the root-shoot axis in embryogenesis, cell differentiation occurs, and three basic tissue systems are established. These are the dermal, ground, and vascular tissue, and they are radially patterned. These tissue, contain various cell types that can be highly differentiated for specific functions. These tissue systems are organized radially around the root-shoot axis.

While the embryo is developing, two other critical events are occurring. The first event is the establishment of a food supply that will support the embryo during germination while it gains photosynthetic capacity. In angiosperms, double fertilization produces endosperm for nutrition. While in gymnosperms, the megagametophyte is the food source. The second critical event is the differentiation of ovule tissue (from the parental sporophyte) to form a hard, protective covering around the embryo. The seed (ovule containing the embryo) then enters a dormant phase, signaling the end of embryogenesis. Environmental signals, such as water, temperature, and light, can break dormancy and trigger a cascade of internal events resulting in germination.

Early Cell Division and Patterning

The first division of the zygote (fertilized egg) in a flowering plant is asymmetric and generates cells with two different fates (figure 40.2). One daughter cell is small, with dense cytoplasm. That cell, which will become the embryo, begins to divide repeatedly in different planes, forming a ball of cells. The larger daughter cell divides repeatedly, forming an elongated structure called a *suspensor*, which links the embryo to the nutrient tissue of the seed. The suspensor also

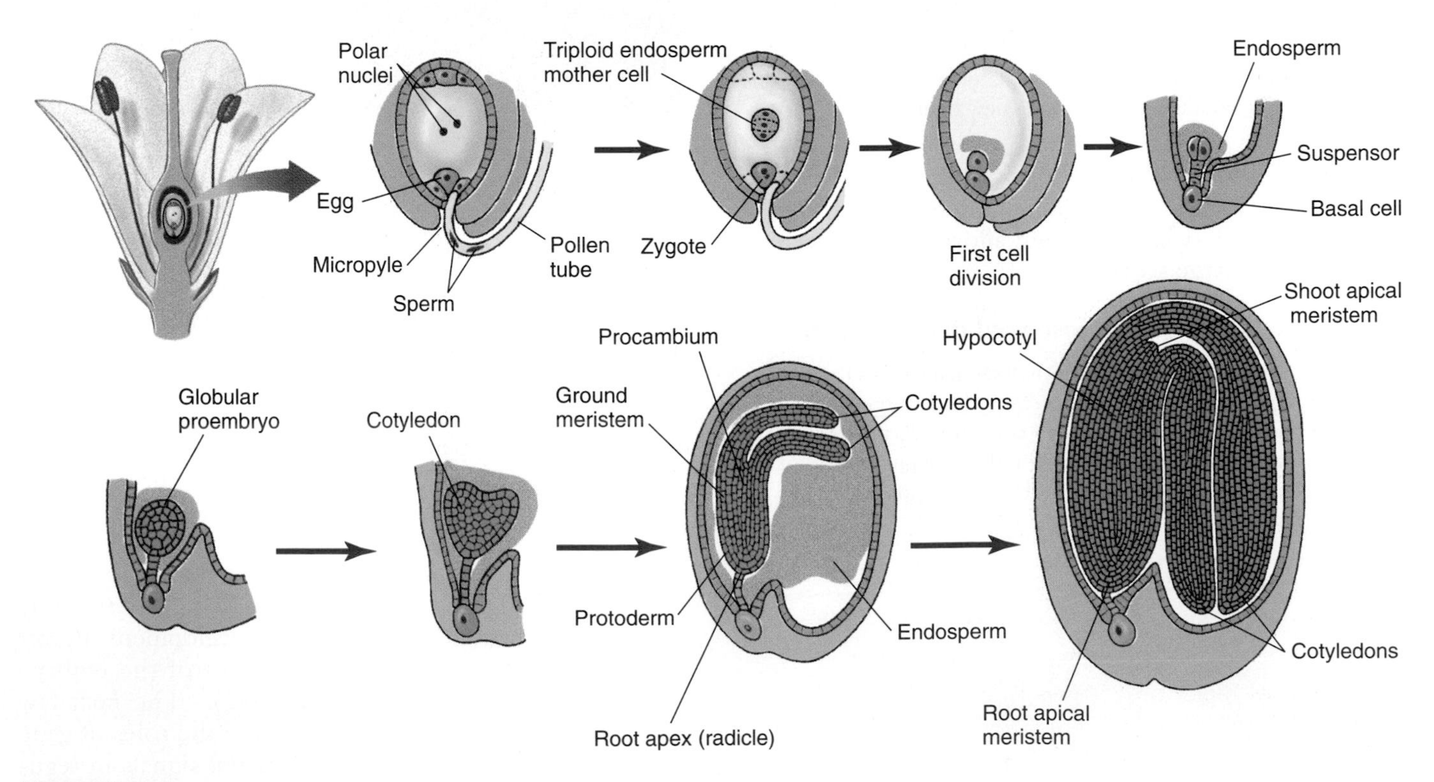

FIGURE 40.2
Stages of development in an angiosperm embryo. The very first cell division is asymmetric. Differentiation begins almost immediately after fertilization.

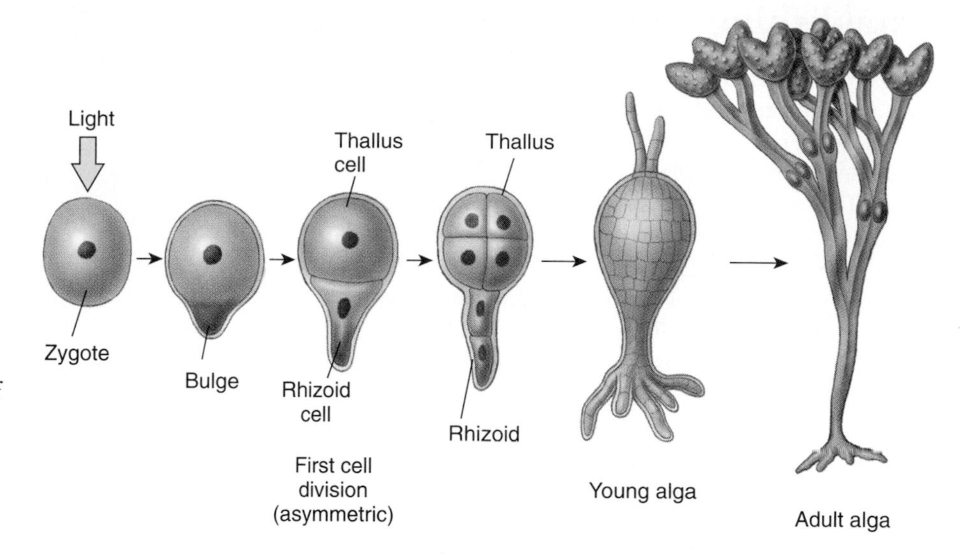

FIGURE 40.3
Asymmetric cell division in a *Fucus* zygote. An unequal distribution of material in the zygote leads to a bulge where the first cell division will occur. This division results in a smaller cell that will go on to divide and produce the rhizoid that anchors the plant; the larger cell divides to form the thallus or main plant body. The point of sperm entry determines where the smaller rhizoid cell will form, but light and gravity can modify this to ensure that the rhizoid will point downward where it can anchor this brown alga. Calcium-mediated currents set up an internal gradient of charged molecules, which leads to a weakening of the cell wall where the rhizoid will form. The fate of the two resulting cells is held "in memory" by cell wall components.

provides a route for nutrients to reach the developing embryo. The root-shoot axis also forms at this time. Cells near the suspensor are destined to form a root, while those at the other end of the axis ultimately become a shoot.

Investigating mechanisms for establishing asymmetry in plant embryo development is difficult because the zygote is embedded within the gametophyte, which is surrounded by sporophyte tissue (ovule and carpel tissue). One approach has been to use the brown alga *Fucus* as a model system. Although there is a huge evolutionary difference between brown algae and the angiosperms, early embryogenesis reveals similarities that may have ancient origins. In fucus, the egg is released prior to fertilization, so there are no extra tissues surrounding the zygote. A bulge which develops on one side of the zygote establishes the vertical axis. Cell division occurs, and the original bulge becomes the smaller of the two cells. It develops into a rhizoid that anchors the alga, and the other cell develops into the main body, or thallus, of the sporophyte. This axis is first established by the point of sperm entry, but it can be changed by environmental signals, especially light and gravity which ensure that the rhizoid is down and the thallus is up. Internal gradients are established that specify where the rhizoid will form in response to environmental signals (figure 40.3). The ability to "remember" where the rhizoid will form depends on the presence of the cell wall. Enzymatic removal of the cell wall in *Fucus* cells specified to form either rhizoids or plant body resulted in cells that could give rise to both. Cell walls contain a wide variety of carbohydrates and proteins attached to the wall's structural fibers. Attempting to pin down the identities of these suspected developmental signals is an area of active research.

Genetic approachs make it possible to explore asymmetric development in angiosperms. Studies of mutants reveal what is going wrong in development, which often makes it possible to infer normal developmental mechanisms. For example, the *suspensor* mutant in *Arabidopsis* has aberrant development in the embryo followed by embryo-like development of the suspensor (figure 40.4). Analysis of this mutant lead to the conclusion that the embryo normally prevents the suspensor from developing into a second embryo.

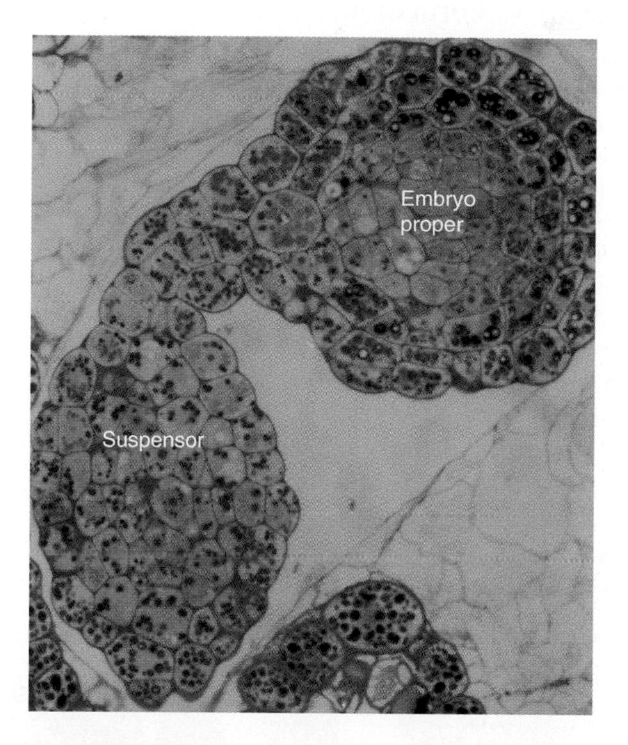

FIGURE 40.4
The embryo suppresses development of the suspensor as a second embryo. This *suspensor* mutant of *Arabidopsis* has a defect in embryo development. Aborted embryo development is followed by embryo-like development of the suspensor.

Early in embryogenesis the root-shoot axis is established.

Establishing Three Tissue Systems

Three basic tissues differentiate while the plant embryo is still a ball of cells, the globular stage (figure 40.5), but no cell movements are involved. The *protoderm* consists of the outermost cells in a plant embryo and will become *dermal tissue*. These cells almost always divide with their cell plate perpendicular to the surface. This perpetuates a single outer layer of cells. Dermal tissue produces cells that protect the plant from desiccation, including the stomata that open and close to facilitate gas exchange and minimize water loss. A ground meristem gives rise to the bulk of the embryonic interior consisting of *ground tissue* cells that eventually function in food and water storage. Lastly, *procambium* at the core of the embryo is destined to form the future *vascular tissue* responsible for water and nutrient transport.

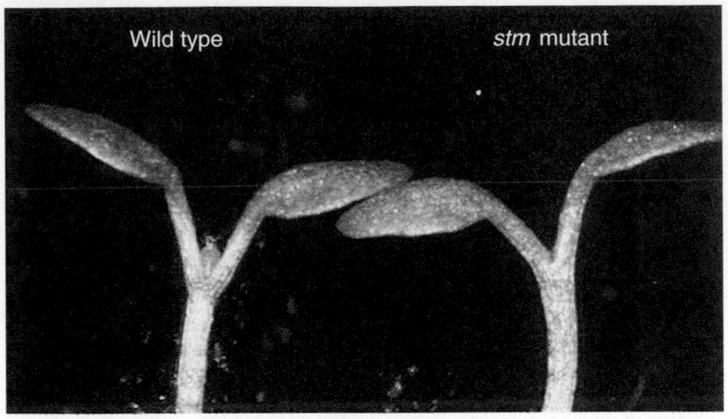

Root and Shoot Formation

The root-shoot axis is established during the globular stage of development. The shoot apical meristem will later give rise to leaves and eventually reproductive structures. While both the shoot and root meristems are apical meristems, their formation is controlled independently. This conclusion is supported by mutant analysis in *Arabidopsis*, in which the *shootmeristemless* (*stm*) mutant fails to produce a viable shoot, but does produce a root (figure 40.6). Similarly, root meristem–specific genes have been identified. For example, monopterous mutants of *Arabidopsis* lack roots. The hormone auxin may play a role in root-shoot axis formation. Auxin is one of six classes of hormones that regulate plant development and function; we will explore each of these classes of hormones in greater detail later in this unit.

As you study the development of roots and shoots after germination, you will notice that many of the same patterns of tissue differentiation seen in the embryo are reiterated in the apical meristems. Remember that there are also many events discussed earlier in this chapter that are unique to embryogenesis. For example, the *LEAFY COTYLEDON*

FIGURE 40.6
Shoot-specific genes specify formation of the shoot apical meristem. The *shootmeristemless* (*stm*) mutant of *Arabidopsis* has a normal root meristem but fails to produce a shoot meristem.

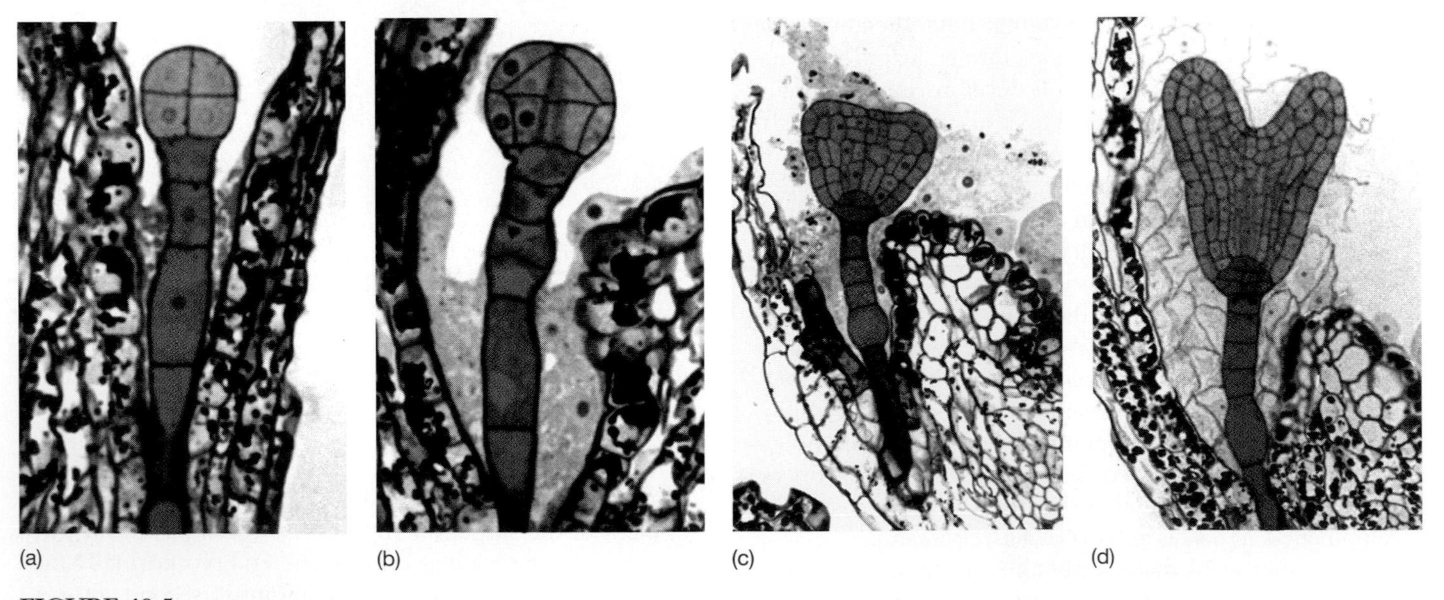

FIGURE 40.5
Early developmental stages of *Arabidopsis thaliana*. (*a*) Early cell division has produced the embryo and suspensor. (*b*) Globular stage. (*c,d*) Heart-shaped stage.

gene in *Arabidopsis* is active in early and late embryo development and may be responsible for maintaining an embryonic environment. It is possible to turn this gene on later in development using recombinant DNA techniques (see chapter 43). In that case, embryos can form on leaves!

Morphogenesis

The globular stage gives rise to a heart-shaped embryo with two bulges in one group of angiosperms (the dicots, see figure 40.5) and a ball with a bulge on a single side in another group (the monocots). The bulges are cotyledons ("first leaves") and are produced by the embryonic cells, not the shoot apical meristem that begins forming during the globular stage. This process, called *morphogenesis* (generation of form), results from changes in planes and rates of cell division. Plant cells cannot move. The form of a plant body is largely determined by the plane in which cells divide. It is also controlled by changes in cell shape as they expand osmotically after they form. Both microtubules and actin play a role in establishing the position of the cell plate which determines the direction of division. Plant hormones and other factors influence the orientation of bundles of microtubules on the interior of the plasma membrane. These microtubules also guide cellulose deposition as the cell wall forms around the outside of a new cell. This process determines the cell's final shape. For example, think of the celll as a box where four of the six sides are reinforced more heavily with cellulose; the cell will expand and grow in the direction of the two sides with less reinforcement. Much is being learned at the cell biological level about morphogenesis from mutants that divide, but cannot control their plane of cell division or the direction of cell expansion.

Food Storage

Throughout embryogenesis, starch, lipids, and proteins are produced. The seed storage proteins are so abundant that the genes coding for them were the first cloning targets for plant molecular biologists. Providing nutritional resources is part of the evolutionary trend toward enhancing embryo survival (see chapter 37). The sporophyte transfers nutrients via the suspensor in angiosperms (in gymnosperms the suspensor serves only to push the embryo closer to the gametophytic nutrient source). This happens concurrently with the development of the endosperm (present only in angiosperms, although double fertilization has been observed in the gymnosperm *Ephedra*) which may be extensive or minimal. Endosperm in coconut is the "milk" and is in liquid form. In corn the endosperm is solid, and in popcorn, expands with heat to form the edible part of popped corn. In peas and beans, the endosperm is used up during embryo development and nutrients are stored in thick, fleshy cotyledons (figure 40.7). The photosynthetic machinery is built in response to light. So, it is critical that seeds have stored nutrients to aid in germination until the growing sporophyte can photosynthesize. A seed buried too deeply will use up all its reserves in cellular respiration before reaching the surface and sunlight.

After the root-shoot axis is established, a radial, three-tissue system and a stored food supply are formed through controlled cell division and expansion.

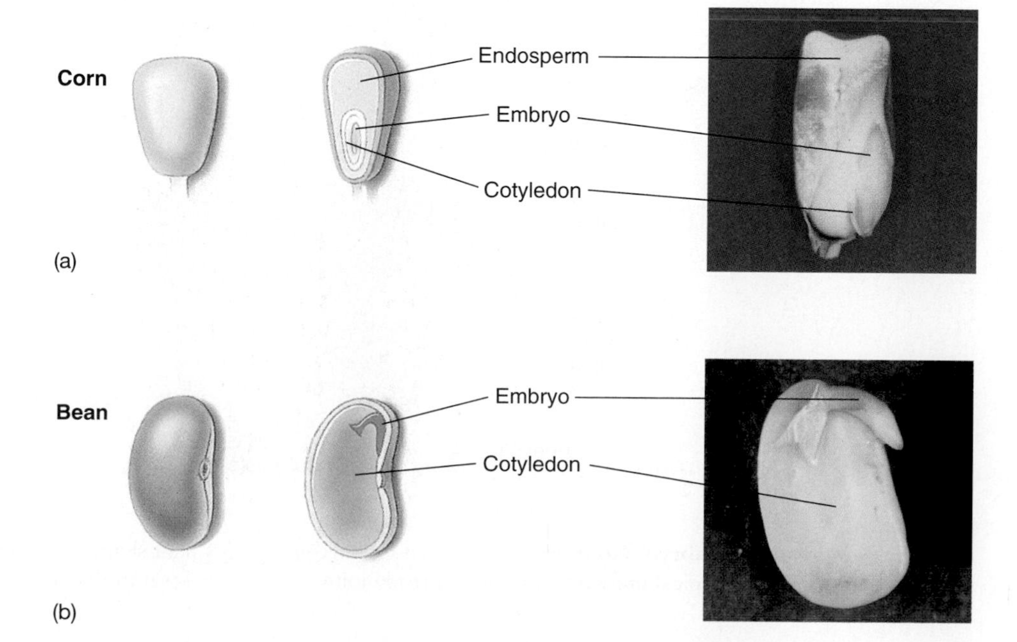

FIGURE 40.7
Endosperm in corn and bean. The corn kernel has endosperm that is still present at maturity, while the endosperm in the bean has disappeared; the bean embryo's cotyledons take over food storage functions.

How Seeds Form

A protective seed coat forms from the outer layers of ovule cells, and the embryo within is now either surrounded by nutritive tissue or has amassed stored food in its cotyledons. The resulting package, known as a *seed,* is resistant to drought and other unfavorable conditions; in its dormant state, it is a vehicle for dispersing the embryo to distant sites and allows a plant embryo to survive in environments that might kill a mature plant. In some embryos, the cotyledons are bent over to fit within the constraints of the hardening ovule wall (now the seed coat) with the inward cotyledon being slightly smaller for efficient packing (figure 40.8). Remember that the ovule wall is actually tissue from the previous sporophyte generation.

Adaptive Importance of Seeds

Early in the development of an angiosperm embryo, a profoundly significant event occurs: the embryo stops developing. In many plants, development of the embryo is arrested soon after the meristems and cotyledons differentiate. The *integuments*—the outer cell layers of the ovule—develop into a relatively impermeable seed coat, which encloses the seed with its dormant embryo and stored food. Seeds are important adaptively in at least four ways:

1. They maintain dormancy under unfavorable conditions and postpone development until better conditions arise. If conditions are marginal, a plant can "afford" to have some seeds germinate, because others will remain dormant.
2. They afford maximum protection to the young plant at its most vulnerable stage of development.
3. They contain stored food that permits development of a young plant prior to the availability of an adequate food supply from photosynthetic activity.
4. Perhaps most important, they are adapted for dispersal, facilitating the migration and dispersal of plant genotypes into new habitats.

Once a seed coat forms, most of the embryo's metabolic activities cease. A mature seed contains only about 5 to 20% water. Under these conditions, the seed and the young plant within it are very stable; it is primarily the progressive and severe desiccation of the embryo and the associated reduction in metabolic activity that are responsible for its arrested growth. Germination cannot take place until water and oxygen reach the embryo; meanwhile, the seed coat may crack by abrasion or alternate freezing and thawing. Seeds of some plants have been known to remain viable for hundreds and, in rare instances, thousands of years (figure 40.9).

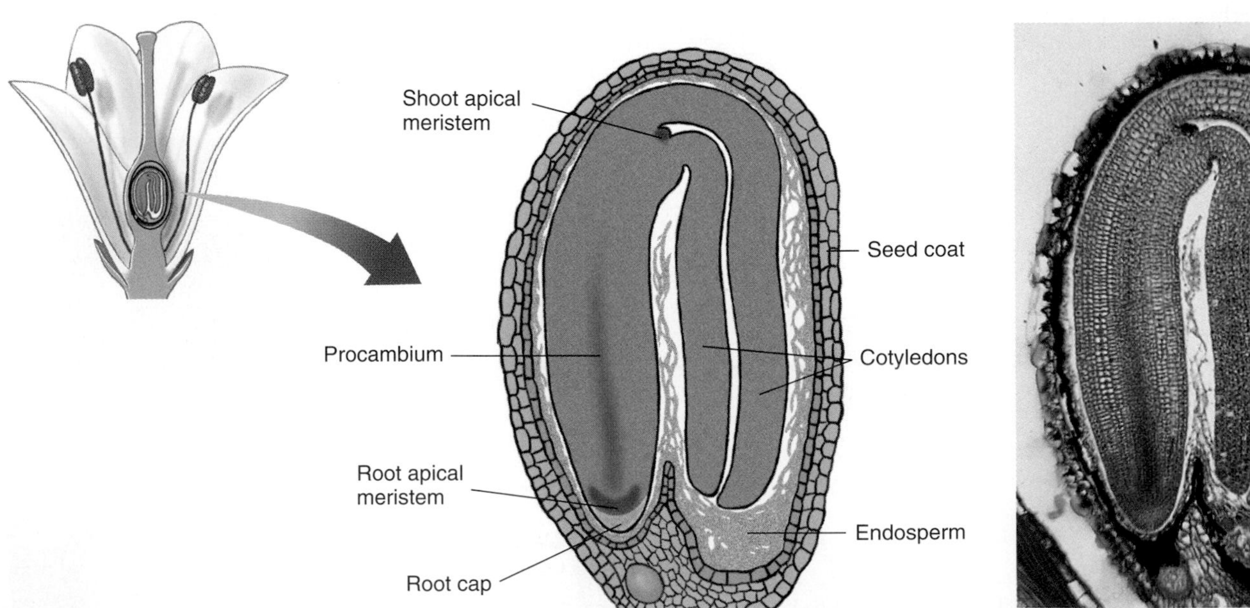

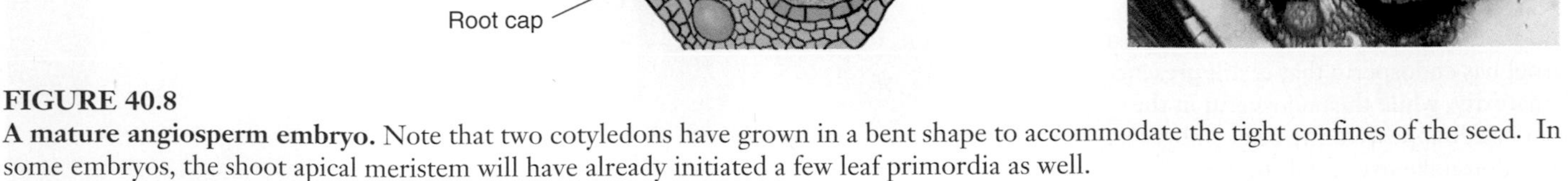

FIGURE 40.8
A mature angiosperm embryo. Note that two cotyledons have grown in a bent shape to accommodate the tight confines of the seed. In some embryos, the shoot apical meristem will have already initiated a few leaf primordia as well.

FIGURE 40.9
Seeds can remain dormant for long periods. This seedling was grown from a lotus seed recovered from the mud of a dry lake bed in Manchuria in northern China. The radiocarbon age of this seed indicates that it was formed around A.D. 1515. The coin is included in the photo to give some idea of the size.

Specific adaptations often help ensure that the plant will germinate only under appropriate conditions. Sometimes, seeds lie within tough cones that do not open until they are exposed to the heat of a fire (figure 40.10). This causes the plant to germinate in an open, fire-cleared habitat; nutrients will be relatively abundant, having been released from plants burned in the fire. Seeds of other plants will germinate only when inhibitory chemicals have been leached from their seed coats, thus guaranteeing their germination when sufficient water is available. Still other plants will germinate only after they pass through the intestines of birds or mammals or are regurgitated by them, which both weakens the seed coats and ensures the dispersal of the plants involved. Sometimes seeds of plants thought to be extinct in a particular area may germinate under unique or improved environmental circumstances, and the plants may then become reestablished.

Seed dormancy is an important evolutionary factor in plants, ensuring their survival in unfavorable conditions and allowing them to germinate when the chances of survival for the young plants are the greatest.

(a)

(b)

(c)

FIGURE 40.10
Fire induces seed germination in some pines. (*a*) Fire will destroy these adult jack pines, but stimulate growth of the next generation. (*b*) Cones of a jack pine are tightly sealed and cannot release the seeds protected by the scales. (*c*) High temperatures lead to the release of the seeds.

40.3 Fruit formation enhances the dispersal of seeds.

How Fruits Form

Paralleling the evolution of angiosperm flowers, and nearly as spectacular, has been the evolution of their **fruits,** which are defined simply as mature ovaries (carpels). During seed formation, the flower ovary begins to develop into fruit. Fruits form in many ways and exhibit a wide array of adaptations for dispersal. The differences among some of the fruit types seen today are shown in figure 40.11. Three layers of ovary wall can have distinct fates which account for the diversity of fruit types from fleshy to dry and hard. An array of mechanisms allow for the release of the seed(s) from the fruits. Developmentally, fruits are fascinating organs that contain three genotypes in one package. The seed

Follicles

Split along one carpel edge only; milkweed, larkspur.

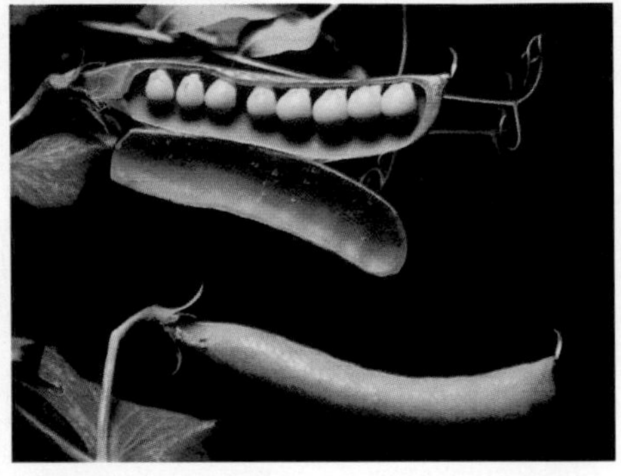

Legumes

Split along two carpel edges with seeds attached to carpel edges; peas, beans.

Samaras

Not split and with a wing formed from the outer tissues; maples, elms, ashes.

Drupes

Single seed enclosed in a hard pit; peaches, plums, cherries.

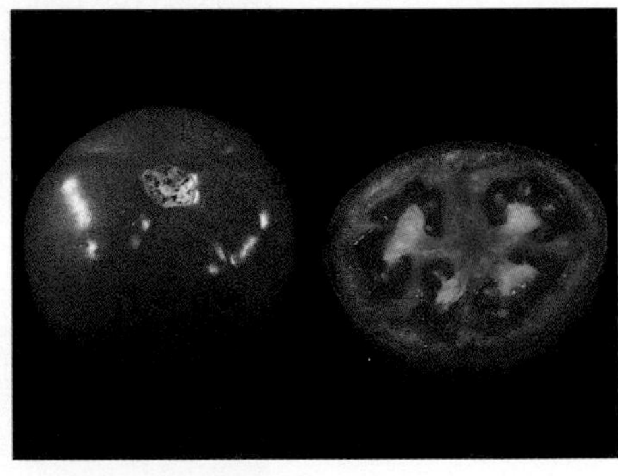

True berries

More than one seed and a thin skin; blueberries, tomatoes, grapes, peppers.

Hesperidia

More than one seed and a leathery skin; oranges, lemons, limes.

Aggregate fruits

Derived from many ovaries of a single flower; strawberries, blackberries.

Multiple fruits

Develop from a cluster of flowers; mulberries, pineapples.

FIGURE 40.11

Examples of some kinds of fruits. Distinguishing features of each of these fruit types are listed below each photo. Follicles, legumes, and samaras are examples of dry fruits. Drupes, true berries, and hesperidia are simple fleshy fruits; they develop from a flower with a single pistil. Aggregate and multiple fruits are compound fleshy fruits; they develop from flowers with more than one ovary or from more than one flower.

possesses fruit and seed coat from the prior sporophyte generation, remnants of the gametophyte generation that produced the egg, and the embryo, representative of the next sporophyte generation.

The Dispersal of Fruits

Aside from the many ways fruits can form, they also exhibit a wide array of specialized dispersal methods. Fruits with fleshy coverings, often shiny black or bright blue or red, normally are dispersed by birds or other vertebrates (figure 40.12*a*). Like red flowers, red fruits signal an abundant food supply. By feeding on these fruits, birds and other animals may carry seeds from place to place and thus transfer plants from one suitable habitat to another.

(a)

(b)

FIGURE 40.12
Animal-dispersed fruits. (*a*) The bright red berries of this honeysuckle, *Lonicera hispidula*, are highly attractive to birds. After eating the fruits, birds may carry the seeds they contain for great distances either internally or, because of their sticky pulp, stuck to their feet or other body parts. (*b*) You will know if you have ever stepped on the spiny fruits of *Cenchrus incertus*, they adhere readily to any passing animal.

Fruits with hooked spines, like those of burgrass (figure 40.12*b*), are typical of several genera of plants that occur in the northern deciduous woods. Such fruits are often disseminated by mammals, including humans. Squirrels and similar mammals disperse and bury fruits such as acorns and other nuts. Other fruits, such as those of maples, elms, and ashes, have wings which aid in their distribution by the wind. The dandelion provides another familiar example of a fruit type that is dispersed by wind (figure 40.13), and the dispersal of seeds from plants such as milkweeds, willows, and cottonwoods is similar. Orchids have minute, dustlike seeds, which are likewise blown away by the wind.

FIGURE 40.13
Wind-dispersed fruits. False dandelion, *Pyrrhopappus carolinanus.* The "parachutes" disperse the fruits of both false and true dandelions widely in the wind, much to the gardener's despair.

Coconuts and other plants that characteristically occur on or near beaches are regularly spread throughout a region by water (figure 40.14). This sort of dispersal is especially important in the colonization of distant island groups, such as the Hawaiian Islands. It has been calculated that seeds of about 175 angiosperms, nearly a third from North America, must have reached Hawaii to have evolved into the roughly 970 species found there today. Some of these seeds blew through the air, others were transported on the feathers or in the guts of birds, and still others drifted across the Pacific. Although the distances are rarely as great as the distance between Hawaii and the mainland, dispersal is just as important for mainland plant species that have discontinuous habitats, such as mountaintops, marshes, or north-facing cliffs.

FIGURE 40.14
A water-dispersed fruit. This fruit of the coconut palm, *Cocos nucifera*, is sprouting on a sandy beach. Coconuts, one of the most useful fruits for humans in the tropics, have become established on even the most distant islands by drifting on the waves.

Fruits, which are characteristic of angiosperms, are extremely diverse. The evolution of specialized structures allows fruits to be dispersed by animals, wind, and water.

40.4 Germination initiates post-seed development.

When conditions are satisfactory, the embryo emerges from its previously desiccated state, utilizes food reserves, and resumes growth. Germination is a process, but is often defined as the emergence of the radicle through the seed coat. As the sporophyte pushes through the seed coat, it orients with the environment so the root grows down and the shoot grows up. New growth comes from delicate meristems that are protected from environmental rigors. The shoot becomes photosynthetic and the post-embryonic phase of growth and development is underway.

Mechanisms of Germination

Germination is the first step in the development of the plant outside of its seed coat. Germination begins when a seed absorbs water and its metabolism resumes. The amount of water a seed can absorb is phenomenal and creates a force strong enough to break the seed coat. At this point, it is important that oxygen be available to the developing embryo because plants, like animals, require oxygen for cellular respiration. Few plants produce seeds that germinate successfully under water, although some, such as rice, have evolved a tolerance to anaerobic conditions.

A dormant seed, although it may have imbibed a full supply of water and may be respiring, synthesizing proteins and RNA, and apparently carrying on normal metabolism, may nonetheless fail to germinate without an additional signal from the environment. This signal may be light of the correct wavelengths and intensity, a series of cold days, or simply the passage of time at temperatures appropriate for germination. Seeds of many plants will not germinate unless they have been **stratified**—held for periods of time at low temperatures. This phenomenon prevents seeds of plants that grow in cold areas from germinating until they have passed the winter, thus protecting their tender seedlings from harsh, cold conditions.

Germination can occur over a wide temperature range (5° to 30°C), although certain species and specific habitats may have relatively narrow optimum ranges. Some seeds will not germinate even under the best conditions. In some species, a significant fraction of a season's seeds remain dormant, providing a gene pool of great evolutionary significance to the future plant population.

The Utilization of Reserves

Germination occurs when all internal and external requirements are met. Germination and early seedling growth require the utilization of metabolic reserves stored in the starch grains of **amyloplasts** (colorless plastids that store starch) and protein bodies. Fats and oils also are important food reserves in some kinds of seeds. These food sources can readily be digested during germination, producing glycerol and fatty acids, which yield energy through cellular respiration; they can also be converted to glucose. Depending on the kind of plant, any of these reserves may be stored in the embryo or in the endosperm.

In the kernels of cereal grains, the single cotyledon is modified into a relatively massive structure called the **scutellum** (figure 40.15), from the Latin word meaning "shield." The abundant food stored in the scutellum is used up before the endosperm during germination. Later, while the seedling is becoming established, the scutellum serves as a nutrient conduit from the endosperm to the rest of the embryo. This is one of the best examples of how hormones modulate development in plants (figure 40.16). The embryo produces gibberellic acid, a hormone, that signals the outer layer of the endosperm called the **aleurone** to produce α-amylase. This enzyme is responsible for breaking down the endosperm's starch into sugars that are passed by the scutellum to the embryo. Abscisic acid, another plant hormone, which is important in establishing dormancy, can inhibit starch breakdown. Abscisic acid levels may be reduced when a seed absorbs water.

The emergence of the embryonic root and shoot from the seed during germination varies widely from species to species. In most plants, the root emerges before the shoot appears and anchors the young seedling in the soil (see figure 40.15). In plants such as peas and corn, the cotyledons may be held below ground; in other plants, such as beans, radishes, and onions, the cotyledons are held above ground. The cotyledons may or may not become green and contribute to the nutrition of the seedling as it becomes established. The period from the germination of the seed to the establishment of the young plant is a very critical one for the plant's survival; the seedling is unusually susceptible to disease and drought during this period.

During germination and early seedling establishment, the utilization of food reserves stored in the embryo or the endosperm is mediated by gibberellic acid and abscisic acid.

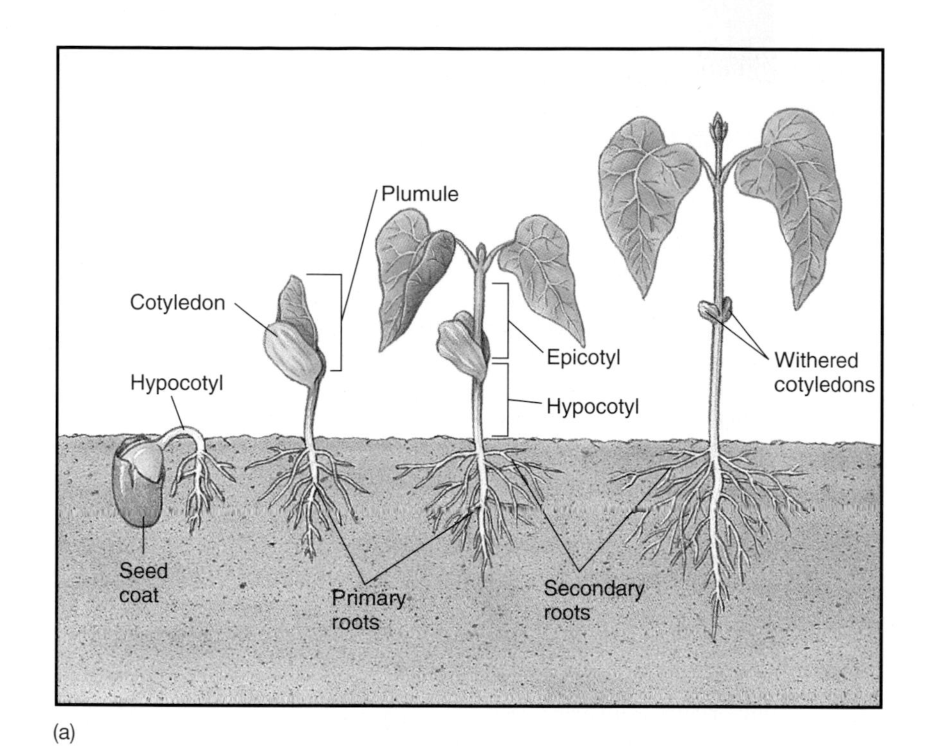

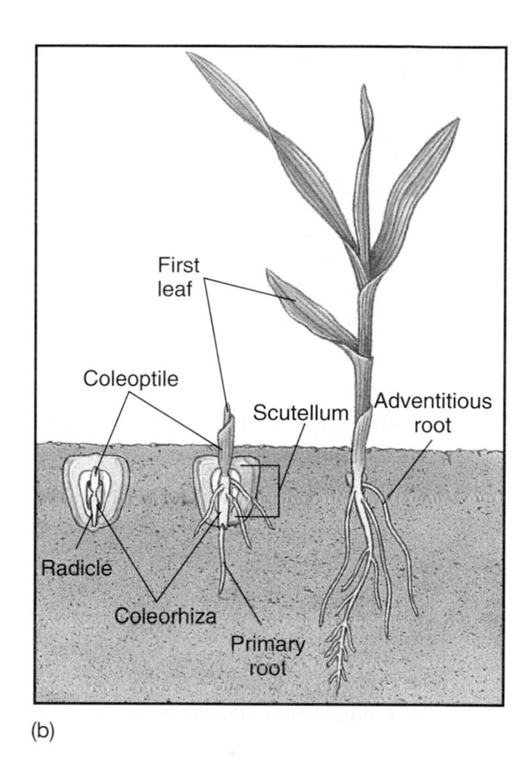

FIGURE 40.15
Shoot development. The stages shown are for a dicot, the common bean, (*a*) *Phaseolus vulgaris*, and a monocot, corn, (*b*) *Zea mays*.

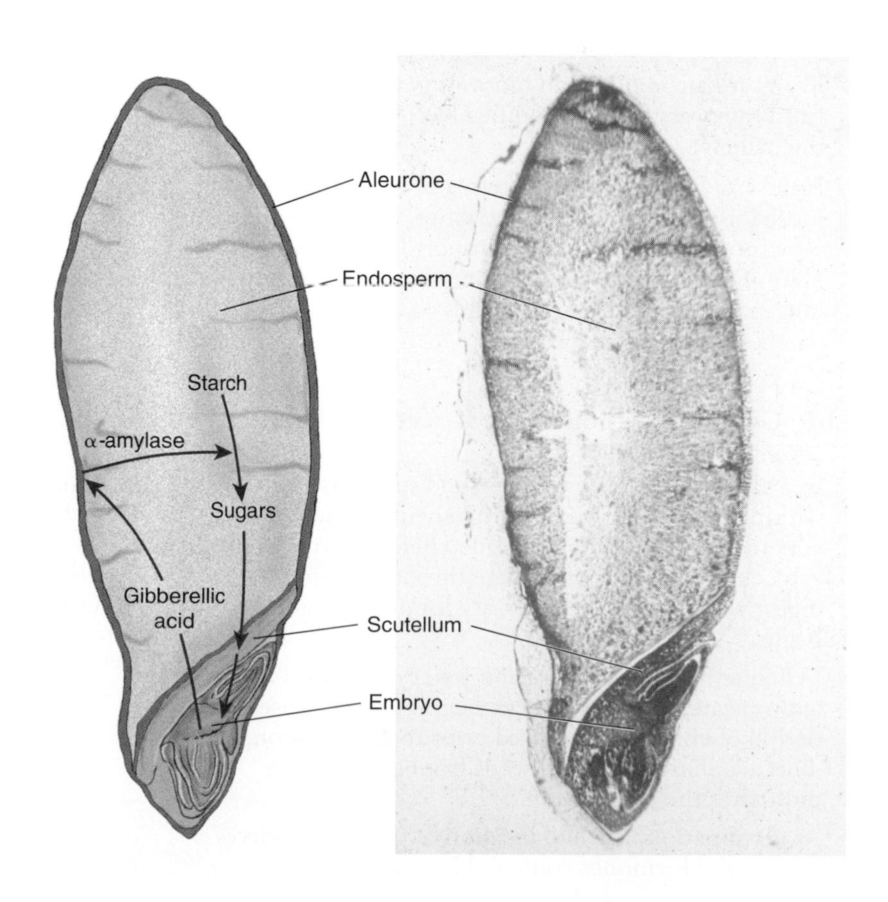

FIGURE 40.16
Hormonal regulation of seedling growth. Germinating barley embryos utilize the starch stored in the endosperm by releasing the hormone gibberellic acid (GA), which triggers the outer layers of the endosperm (aleurone layers) to produce the starch-digesting enzyme α-amylase. The α-amylase breaks down starch into sugar, which moves through the scutellum (cotyledon) into the embryo, where it provides energy for growth. A second hormone, abscisic acid (ABA), is important in establishing dormancy and becomes diluted as seeds imbibe water. When there is excess ABA, the GA-triggered production of α-amylase is inhibited.

Chapter 40

www.mhhe.com/raven6/resources40.mhtml

Summary	*Questions*	*Media Resources*

40.1 Plant embryo development establishes a basic body plan.

- Plant shape is determined by the direction of cell division and expansion.
- Three tissue systems form radially through regulated cell division and differentiation.
- Shoot and root apical meristems are established to continuously produce new tissues, which then differentiate into body parts.
- Carbohydrates, lipids, and proteins are stored for germination in the endosperm or cotyledons.

1. The pattern of cell division regulates the shape of an embryo. Describe the cell division pattern that results in the single, outer layer of protoderm in the globular stage embryo.

2. What evidence supports the claim that the shoot meristem is genetically specified separately from the root?

40.2 The seed protects the dormant embryo from water loss.

- The ovule wall (integuments) around the embryo hardens to protect the embryo as embryogenesis ends.
- Seed formation allows the embryo to enter a dormant state and continue growth under more optimal conditions.

3. Why are seeds adaptively important? Why may a seed showing proper respiration and synthesis of proteins and nucleic acids and all other normal metabolic activities still fail to germinate?

- Art Activities:
  - -Corn Grain Structure
  - -Garden Bean Seed Structure

- Embryos and Seeds

40.3 Fruit formation enhances the dispersal of seeds.

- Fruits are an angiosperm innovation that develop from the ovary wall (a modified leaf) that surrounds the ovule(s).
- Fruits are highly diverse in terms of their dispersal mechanisms, often displaying wings, barbs, or other structures that aid in their transport from place to place. Fruit dispersal methods are especially important in the colonization of islands or other distant patches of suitable habitat.

4. Why is it advantageous for a plant to produce fruit? How does the genotype of the fruit compare with the genotype of the embryo? How does the genotype of the seed wall compare with the fruit wall?

- Fruits
- Activity: Fruits

40.4 Germination initiates post-seed development.

- In a seed, the embryo with its food supply is encased within a sometimes rigid, relatively impermeable seed coat that may need to be abraded before germination can occur. Weather or passage through an animal's digestive tract may be necessary for germination to begin.
- When temperature, light, and water conditions are appropriate, germination can begin. In some cases, a period of chilling is required prior to germination. This adaptation protects seeds from germinating until after the cold season.
- At germination, the mobilization of the food reserves is critical. Hormones control this process.

5. Explain how the embryo signals the endosperm to obtain sugars for growth during germination.

6. Why does the root (actually the radicle) of the embryo emerge first?

- Germination
- Activity: Germination

41

How Plants Grow in Response to Their Environment

Concept Outline

41.1 Plant growth is often guided by environmental cues.

Growth Responses Plant growth is often influenced by light, gravity, and contact with other plants and animals.
Dormancy. The ability to cease growth allows plants to wait out the bad times.

41.2 The hormones that guide growth are keyed to the environment.

Plant Hormones. Hormones are grouped into seven classes.
Auxin. Auxin is involved in the elongation of stems.
Cytokinins. Cytokinins stimulate cell division.
Gibberellins. Gibberellins control stem elongation.
Brassinosteroids and Oligosaccharins. There are several recent additions to the list of identified plant hormones.
Ethylene. Ethylene controls leaves and flower abscision.
Abscisic Acid. Abscisic acid suppresses growth of buds and promotes leaf senescence.

41.3 The environment influences flowering.

Plants Undergo Metamorphosis. The transition of a shoot meristem from vegetative to adult is called phase change.
Pathways Leading to Flower Production. Photoperiod is regulated in complex ways.
Identity Genes and the Formation of Floral Meristems and Floral Organs. Floral meristem identity genes activate floral organ identity genes.

41.4 Many short-term responses to the environment do not require growth.

Turgor Movement. Changes in the water pressure within plant cells result in quick and reversible plant movements.
Plant Defense Responses. In addition to generalized defense mechanisms, some plants have highly evolved recognition mechanisms for specific pathogens.

FIGURE 41.1
Plant growth is affected by environmental cues. The branches of this fallen tree are growing straight up in response to gravity and light.

All organisms sense and interact with their environment. This is particularly true of plants. Plant survival and growth is critically influenced by abiotic factors including water, wind, and light. In this chapter, we will explore how a plant senses such factors, and transduces these signals to elicit an optimal physiological, growth, or developmental response. Hormones play an important role in the internal signaling that brings about environmental responses and are keyed in many ways to the environment. The effect of the local environment on plant growth also accounts for much of the variation in adult form within a species (figure 41.1). Precisely regulated responses not only allow a plant to survive pathogens and environmental shifts from day to day but also determine when a flowering plant will produce a flower. The entire process of constructing a flower in turn sets the stage for intricate reproductive strategies that will be discussed in the next chapter.

41.1 Plant growth is often guided by environmental cues.

Growth Responses

Growth patterns in plants are often guided by environmental signals. **Tropisms** (from *trope*, the Greek word for "turn") are positive or negative growth responses of plants to external stimuli that usually come from one direction. Some responses occur independently of the direction of the stimuli and are referred to as nastic movements. For example, a tendril of a pea plant will always coil in one direction when touched. Tropisms, on the other hand, are directional and offer significant compensation for the plant's inability to get up and walk away from unfavorable environmental conditions. Tropisms contribute the variety of branching patterns we see within a species. Here we will consider three major classes of plant tropisms: phototropism, gravitropism, and thigmotropism. Tropisms are particularly intriguing because they challenge us to connect environmental signals with cellular perception of the signal, transduction into biochemical pathways, and ultimately an altered growth response. Nondirectional responses such as photomorphogenesis can result in more complex changes in form.

FIGURE 41.2
Phototropism. *Impatiens* plant growing toward light.

Phototropism and Photomorphogenesis

Phototropic responses involve the bending of growing stems and other plant parts toward sources of light (figure 41.2). In general, stems are positively phototropic, growing toward a light source, while most roots do not respond to light or, in exceptional cases, exhibit only a weak negative phototropic response. The phototropic reactions of stems are clearly of adaptive value, giving plants greater exposure to available light. They are also important in determining the development of plant organs and, therefore, the appearance of the plant. Individual leaves may display phototropic responses. The position of leaves is important to the photosynthetic efficiency of the plant. A plant hormone called auxin (discussed later in this chapter) is probably involved in most, if not all, of the phototropic growth responses of plants. Photomorphogenesis is light triggered development which can include seed germination.

The first step in phototropic and photomorphogenic responses is perceiving the light. Photoreceptors perceive different wavelengths of light with blue and red being the most common. Blue light receptors are being characterized and we are beginning to understand how plants "see blue." Blue light signals many phototropic responses, while red light is more likely to signal photomorphogenesis. Much more is known about "seeing red" and translating that perception into a signal transduction pathway leading to an altered growth response. Plants possess a pigment-containing protein, **phytochrome,** which exists in two interconvertible forms, P_r and P_{fr}. In the first form, phytochrome absorbs red light; in the second, it absorbs far-red light. When a molecule of P_r absorbs a photon of red light (660 nm), it is instantly converted into a molecule of P_{fr}, and when a molecule of P_{fr} absorbs a photon of far-red light (730 nm), it is instantly converted to P_r. P_{fr} is biologically active and P_r is biologically inactive. In other words, when P_{fr} is present, a given biological reaction that is affected by phytochrome will occur. When most of the P_{fr} has been replaced by P_r, the reaction will not occur (figure 41.3). While we refer to phytochrome as a single molecule here, it is important to note that several different phytochromes have now been identified that appear to have specific biological functions.

Phytochrome is a light receptor, but it does not act directly to bring about reactions to light. The existence of phytochrome was conclusively demonstrated in 1959 by Harry A. Borthwick and his collaborators at the U.S. Department of Agriculture Research Center at Beltsville, Maryland. It has since been shown that the molecule consists of two parts: a smaller one that is sensitive to light and a larger portion that is a protein. The protein component initiates a signal transduction leading to a particular tropism. Phytochrome is present in all groups of plants and in a few genera of green algae, but not in bacteria, fungi, or protists (other than the few green algae). It is likely that phytochrome systems for measuring light evolved among the green algae and were present in the common ancestor of the plants.

Phytochrome is involved in many plant growth responses. For example, seed germination is inhibited by far-red light and stimulated by red light in many plants. Because chlorophyll absorbs red light strongly but does not absorb far-red light, light passing through green leaves of canopy trees above a seed inhibits seed germination. Consequently, seeds on the ground under deciduous plants that lose their leaves in winter are more apt to germinate in the spring after the leaves have decomposed and the seedlings are exposed to direct sunlight. This greatly improves the chances the seedlings will become established.

A second example of these relationships is the elongation of the shoot in an *etiolated* seedling (one that is pale and slender from having been kept in the dark). Such plants become normal when exposed to light, especially red light. There appears to be a link between phytochrome light perception and brassinosteroids, another group of plant hormones, in the etiolation response. Etiolation is an energy conservation strategy to help plants growing in the dark reach the light before they die. They don't green-up until there is light, and they divert energy to growing as tall as possible through internode elongation. The de-etiolated (*det2*) *Arabidopsis* mutant has a poor etiolation response. It does not have elongated internodes and greens up a bit in the dark. It turns out that *det2* mutants are defective in an enzyme necessary for brassinosteroid biosynthesis. Researchers suspect that brassinosteroids play a role in how plants respond to light through phytochrome. Thus, because *det2* mutants lack brassinosteroids, they do not respond to light, or lack of light, as normal plants do, and the *det2* mutants grow normally in the dark.

Red and far-red light also are used as signals for plant spacing. The closer plants are together, the more likely they are to grow tall and try to outcompete others for the sunshine. Plants somehow measure the amount of far-red light being bounced back to them from neighboring plants. If their perception is messed up by putting a collar around the stem with a solution that blocks light absorption, the elongation response is no longer seen.

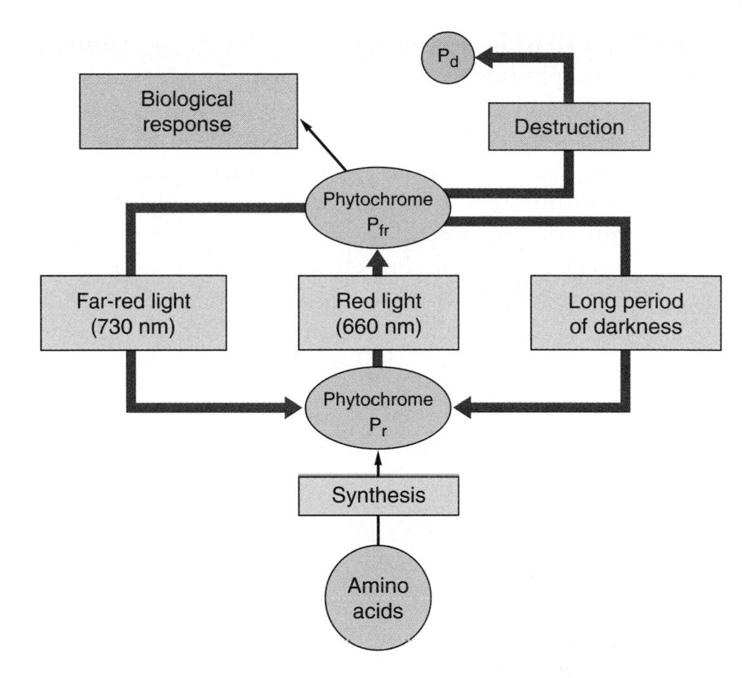

FIGURE 41.3
How phytochrome works. Phytochrome is synthesized in the P_r form from amino acids. When exposed to red light, P_r changes to P_{fr}, which is the active form that elicits a response in plants. P_{fr} is converted to P_r when exposed to far-red light, and it also converts to P_r or is destroyed in darkness. The destruction product is designated P_d.

FIGURE 41.4
Plant response to gravity. This plant (*Zebrina pendula*) was placed horizontally and allowed to grow for 7 days. Note the negative gravitational response of the shoot.

Gravitropism

When a potted plant is tipped over, the shoot bends and grows upward. The same thing happens when a storm pushes over plants in a field. These are examples of **gravitropism**, the response of a plant to the gravitational field of the earth (figures 41.1 and 41.4). Because plants also grow in response to light, separating out phototropic effects is important in the study of gravitropisms.

Gravitropic responses are present at germination when the root grows down and the shoot grows up. Why does a shoot have a negative gravitropic response (growth away from gravity), while a root has a positive gravitropic response? The opportunity to experiment on the space shuttle in a gravity-free environment has accelerated research in this area. Auxins play a primary role in gravitropic responses, but they may not be the only way gravitational information is sent through the plant. When John Glenn made his second trip into space, he was accompanied by an experiment designed to test the role of gravity and electrical signaling in root bending. Analysis of gravitropic mutants is also adding to our understanding of gravitropism. There are four steps that lead to a gravitropic response:

1. Gravity is perceived by the cell
2. Signals form in the cell that perceives gravity
3. The signal is transduced intra- and intercellularly
4. Differential cell elongation occurs between cells in the "up" and "down" sides of the root or shoot.

The first steps in perceiving gravity are being debated. Amyloplasts, plastids that contain starch, sink toward the gravitational field and may be involved in sensing gravity. Amyloplasts may interact with the cytoskeleton, but the net effect is that auxin becomes more concentrated on the lower side of the stem axis than on the upper side. The increased auxin concentration on the lower side in stems causes the cells in that area to grow more than the cells on the upper side. The result is a bending upward of the stem against the force of gravity—in other words there is a *negative gravitropic response*. Such differences in hormone concentration have not been as well documented in roots. Nevertheless, the upper sides of roots oriented horizontally grow more rapidly than the lower sides, causing the root ultimately to grow downward; this phenomenon is known as *positive gravitropism*. In shoots, the gravity-sensing cells are in the endoderm. Mutants like *scarecrow* and *short root* in *Arabidopsis* that lack normal endodermal development fail to have a normal gravitropic response. These endodermal cells are the sites of the amyloplasts in the stems.

In roots, the gravity-sensing cells are located in the root cap and the cells that actually do the asymmetric growth are in the distal elongation zone which is closest to the root cap. How the information gets transferred over this distance is an intriguing problem. Auxin may be involved, but when auxin transport is suppressed, there is still a gravitropic response in the distal elongation zone. Some type of electrical signaling involving membrane polarization has been hypothesized and this was tested aboard the space shuttle. So far the verdict is still not in on the exact mechanism.

It may surprise you to learn that in tropical rain forests, roots of some plants may grow up the stems of neighboring plants, instead of exhibiting the normal positive gravitropic responses typical of other roots. The rainwater dissolves nutrients, both while passing through the lush upper canopy of the forest, and also subsequently as it trickles down tree trunks. Such water functions as a more reliable source of nutrients for the roots than the nutrient-poor soils in which the plants are anchored. Explaining this in terms of current hypotheses is a challenge. It has been proposed that roots are more sensitive to auxin than shoots and that auxin may actually inhibit growth on the lower side of a root, resulting in a positive gravitropic response. Perhaps in these tropical plants, the sensitivity to auxin in roots is reduced.

Thigmotropism

Thigmotropism is a name derived from the Greek root *thigma*, meaning "touch." A thigmotropism is a response of a plant or plant part to contact with the touch of an object, animal, plant, or even the wind. (figure 41.5). When a tendril makes contact with an object, specialized epidermal cells, whose action is not clearly understood, perceive the contact and promote uneven growth, causing the tendril to curl around the object, sometimes within as little as 3 to 10 minutes. Both auxin and ethylene appear to be involved in tendril movements, and they can induce coiling in the absence of any contact stimulus. In other plants, such as clematis, bindweed, and dodder, leaf petioles or unmodified stems twine around other stems or solid objects.

FIGURE 41.5
Thigmotropism. The thigmotropic response of these twining stems causes them to coil around the object with which they have come in contact.

Again, *Arabidopsis* is proving valuable as a model system. A gene has been identified that is expressed in 100-fold higher levels 10 to 30 minutes after touch. Given the value of a molecular genetics approach in dissecting the pathways leading from an environmental signal to a growth response, this gene provides a promising first step in understanding how plants respond to touch.

Other Tropisms

The tropisms just discussed are among the best known, but others have been recognized. They include *electrotropism* (responses to electricity); *chemotropism* (response to chemicals); *traumotropism* (response to wounding which we discuss on page 834); *thermotropism* (response to temperature); *aerotropism* (response to oxygen); *skototropism* (response to dark); and *geomagnetotropism* (response to magnetic fields). Roots will often follow a diffusion gradient of water coming from a cracked pipe and enter the crack. Some call such growth movement *hydrotropism*, but there is disagreement whether responses to water and several other "stimuli" are true tropisms.

While plants can't move away or toward optimal conditions, they can grow. Phototropisms are growth responses of plants to a unidirectional source of light. Gravitropism, the response of a plant to gravity, generally causes shoots to grow up (negative gravitropism) and roots to grow down (positive gravitropism). Thigmotropisms are growth responses of plants to contact.

Dormancy

Sometimes modifying the direction of growth is not enough to protect a plant from harsh conditions. The ability to cease growth and go into a dormant stage provides a survival advantage. The extreme example is seed dormancy, but there are intermediate approaches to waiting out the bad times as well. Environmental signals both initiate and end dormant phases in the life of a plant. Temperature, water, and light are examples of such signals.

In temperate regions, we generally associate dormancy with winter, when freezing temperatures and the accompanying unavailability of water make it impossible for plants to grow. During this season, buds of deciduous trees and shrubs remain dormant, and apical meristems remain well protected inside enfolding scales. Perennial herbs spend the winter underground as stout stems or roots packed with stored food. Many other kinds of plants, including most annuals, pass the winter as seeds.

In some seasonally dry climates, seed dormancy occurs primarily during the dry season, often the summer. Rainfalls trigger germination when conditions for survival are more favorable. Annual plants occur frequently in areas of seasonal drought. Seeds are ideal for allowing annual plants to bypass the dry season, when there is insufficient water for growth. When it rains, they can germinate and the plants can grow rapidly, having adapted to the relatively short periods when water is available. Chapter 40 covered some of the mechanisms involved in breaking seed dormancy and allowing germination under favorable circumstances. These include the leaching from the seed coats of chemicals that inhibit germination, or mechanically cracking the seed coats, a procedure that is particularly suitable for promoting growth in seasonally dry areas. Whenever rains occur, they will leach out the chemicals from the seed coats, and the hard coats of other seeds may be cracked when they are being washed down along temporarily flooded arroyos (figure 41.6).

Seeds may remain dormant for a surprisingly long time. Many legumes (plants of the pea and bean family, Fabaceae) have tough seeds that are virtually impermeable to water and oxygen. These seeds often last decades and even longer without special care; they will eventually germinate when their seed coats have been cracked and water is available. Seeds that are thousands of years old have been successfully germinated!

Favorable temperatures, day length, and amounts of water can release buds, underground stems and roots, and seeds from a dormant state. Requirements vary among species. For example, some weed seeds germinate in cooler parts of the year and are inhibited from germinating by warmer temperatures. Day length differences near the equator and in more temperate regions have dramatic effects on dormancy. For example, tree dormancy is common in temperate climates when days are short but is unusual in tropical trees growing near the equator.

(a)

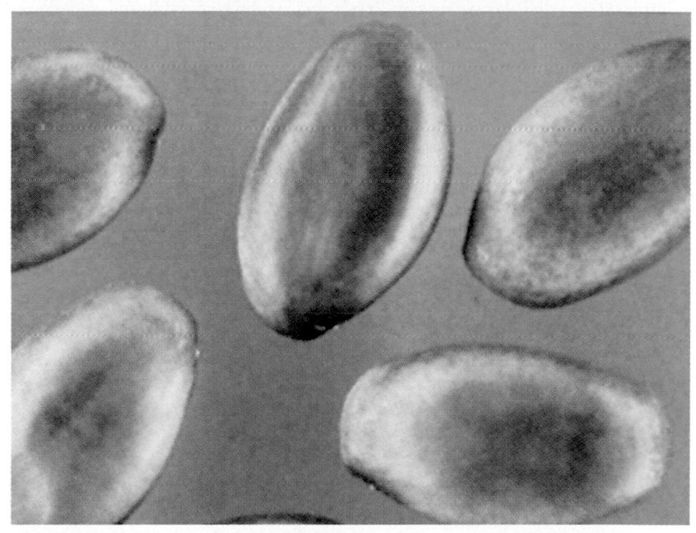

(b)

FIGURE 41.6
Dormancy is a common adaptation to periodic drought. This desert Palo Verde tree (*cercidium floridum*) (*a*) has tough seeds (*b*) that germinate only after they are cracked.

Mature plants may become dormant in dry or cold seasons that are unfavorable for growth. Dormant plants usually lose their leaves and drought-resistant winter buds are produced. Long unfavorable periods may be bypassed through the production of seeds, which themselves can remain dormant for long periods.

Plant Hormones

While initial responses of plants to environmental signals may rely primarily on electrical signaling, longer-term responses that alter morphology rely on complex physiological networks. Many internal signaling pathways involve plant hormones, which are the focus of this section. Hormones are involved in responses to the environment, as well as internally regulated development (examples of which you saw in chapter 40).

Hormones are chemical substances produced in small, often minute, quantities in one part of an organism and then transported to another part, where they bring about physiological or developmental responses. The activity of hormones results from their capacity to stimulate certain physiological processes and to inhibit others (figure 41.7). How they act in a particular instance is influenced both by the hormone and the tissue that receives the message.

In animals, hormones are usually produced at definite sites, usually organs. In plants, hormones are not produced in specialized tissues but, instead, in tissues that also carry out other, usually more obvious, functions. There are seven major kinds of plant hormones: auxin, cytokinins, gibberellins, brassinosteroids, oligosaccharins, ethylene, and abscisic acid (table 41.1). Current research is focused on the biosynthesis of hormones and on characterizing the hormone receptors that trigger signal transduction pathways. Much of the molecular basis of hormone function remains enigmatic.

Because hormones are involved in so many aspects of plant function and development, we have chosen to integrate examples of hormone activity with specific aspects of plant biology throughout the text. In this section, our goal is to give you a brief overview of these hormones. Use this section as a reference when specific hormones are discussed in the next few chapters.

There are seven major kinds of plant hormones: auxin, cytokinins, gibberellins, brassinosteroids, oligosaccharins, ethylene, and abscisic acid.

(a)

(b)

FIGURE 41.7
Plant hormones have profound effects. Plant hormones, often acting together, influence many aspects of plant growth and development, including (*a*) leaf abscission and (*b*) the formation of mature fruit.

Hormone	Major Functions	Where Produced or Found in Plant
Auxin (IAA)	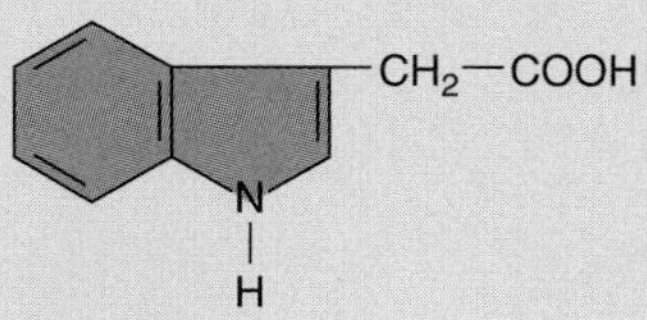 Promotion of stem elongation and growth; formation of adventitious roots; inhibition of leaf abscission; promotion of cell division (with cytokinins); inducement of ethylene production; promotion of lateral bud dormancy	Apical meristems; other immature parts of plants
Cytokinins	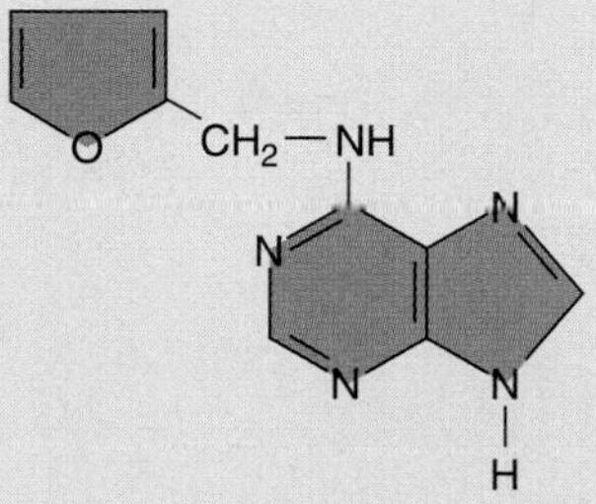 Stimulation of cell division, but only in the presence of auxin; promotion of chloroplast development; delay of leaf aging; promotion of bud formation	Root apical meristems; immature fruits
Gibberellins	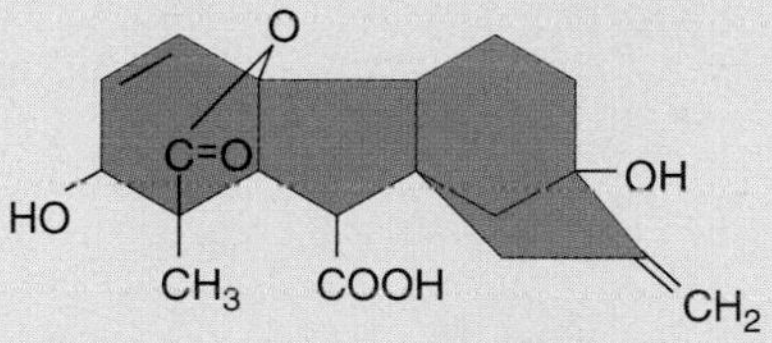 Promotion of stem elongation; stimulation of enzyme production in germinating seeds	Roots and shoot tips; young leaves; seeds
Brassinosteroids	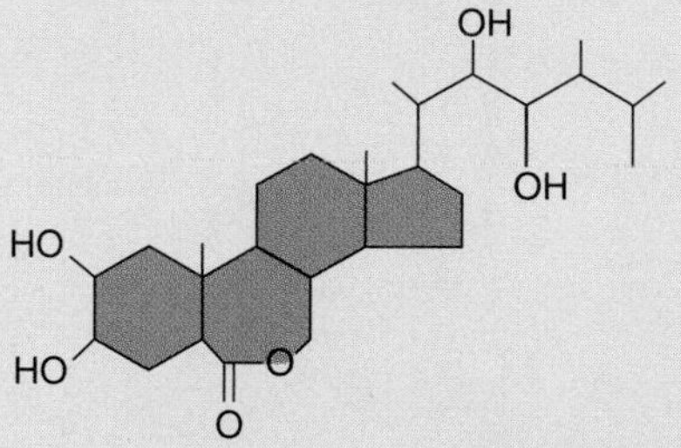 Overlapping functions with auxins and gibberellins	Pollen, immature seeds, shoots, leaves
Oligosaccharins	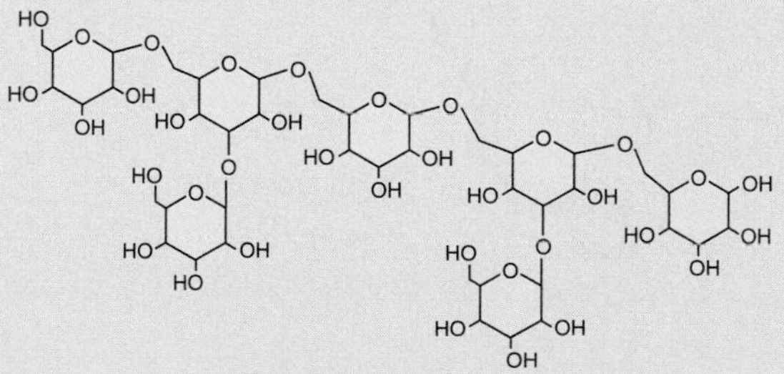 Pathogen defense, possibly reproductive development	Cell walls
Ethylene	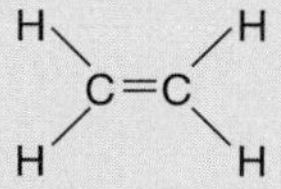 Control of leaf, flower, and fruit abscission; promotion of fruit ripening	Roots, shoot apical meristems; leaf nodes; aging flowers; ripening fruits
Abscisic acid	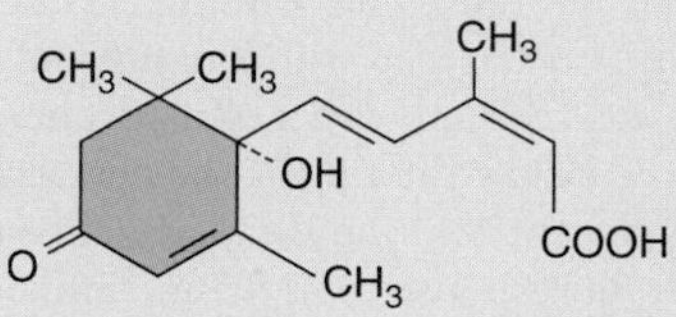 Inhibition of bud growth; control of stomatal closure; some control of seed dormancy; inhibition of effects of other hormones	Leaves, fruits, root caps, seeds

Table 41.1 Functions of the Major Plant Hormones

Auxin

More than a century ago, an organic substance known as **auxin** became the first plant hormone to be discovered. Auxin increases the plasticity of plant cell walls and is involved in elongation of stems. Cells can enlarge in response to changes in turgor pressure. Cell walls must be fairly plastic for this expansion to occur. Auxin plays a role in softening cell walls. The discovery of auxin and its role in plant growth is an elegant example of thoughtful experimental design and is recounted here for that reason. Recent efforts have uncovered an auxin receptor. Transport mechanisms are also being unraveled. As with all the classes of hormones, we are just beginning to understand, at a cellular and molecular level, how hormones regulate growth and development.

Discovery of Auxin

In his later years, the great evolutionist, Charles Darwin, became increasingly devoted to the study of plants. In 1881, he and his son Francis published a book called *The Power of Movement of Plants.* In this book, the Darwins reported their systematic experiments on the response of growing plants to light—responses that came to be known as **phototropisms.** They used germinating oat *(Avena sativa)* and canary grass *(Phalaris canariensis)* seedlings in their experiments and made many observations in this field.

Charles and Francis Darwin knew that if light came primarily from one direction, the seedlings would bend strongly toward it. If they covered the tip of the shoot with a thin glass tube, the shoot would bend as if it were not covered. However, if they used a metal foil cap to exclude light from the plant tip, the shoot would not bend (figure 41.8). They also found that using an opaque collar to exclude light from the stem below the tip did not keep the area above the collar from bending.

In explaining these unexpected findings, the Darwins hypothesized that when the shoots were illuminated from one side, they bent toward the light in response to an "influence" that was transmitted downward from its source at the tip of the shoot. For some 30 years, the Darwins' perceptive experiments remained the sole source of information about this interesting phenomenon. Then Danish plant physiologist Peter Boysen-Jensen and the Hungarian plant physiologist Arpad Paal independently demonstrated that the substance that caused the shoots to bend was a chemical. They showed that if the tip of a germinating grass seedling was cut off and then replaced with a small block of agar separating it from the rest of the seedling, the seedling would grow as if there had been no change. Something evidently was passing from the tip of the seedling through the agar into the region where the bending occurred. On the basis of these observations under conditions of uniform illumination or of darkness, Paal suggested that

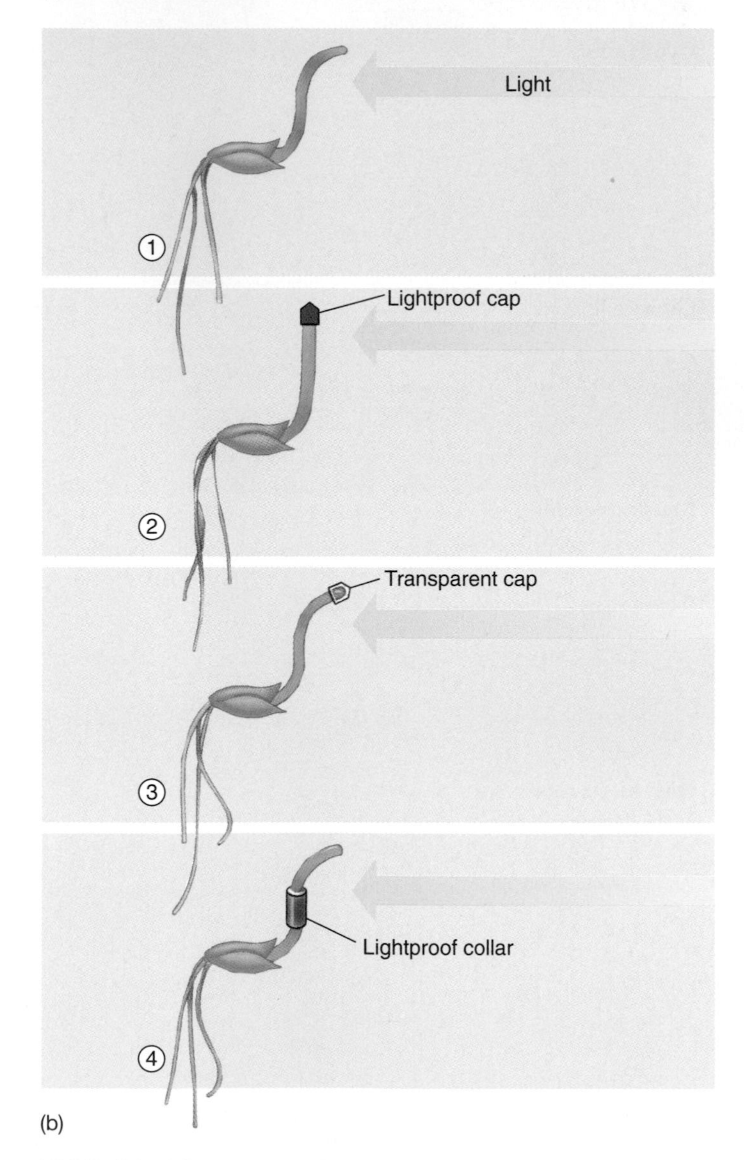

FIGURE 41.8
The Darwins' experiment. *(a)* Young grass seedlings normally bend toward the light. *(b)* The bending *(1)* did not occur when the tip of a seedling was covered with a lightproof cap *(2)*, but did occur when it was covered with a transparent one *(3)*. When a collar was placed below the tip *(4)*, the characteristic light response took place. From these experiments, the Darwins concluded that, in response to light, an "influence" that caused bending was transmitted from the tip of the seedling to the area below, where bending normally occurs.

FIGURE 41.9
Frits Went's experiment. (*1*) Went removed the tips of oat seedlings and put them in agar, an inert, gelatinous substance. (*2*) Blocks of agar were then placed off-center on the ends of other oat seedlings from which the tips had been removed. (*3*) The seedlings bent away from the side on which the agar block was placed. Went concluded that the substance that he named *auxin* promoted the elongation of the cells and that it accumulated on the side of an oat seedling away from the light.

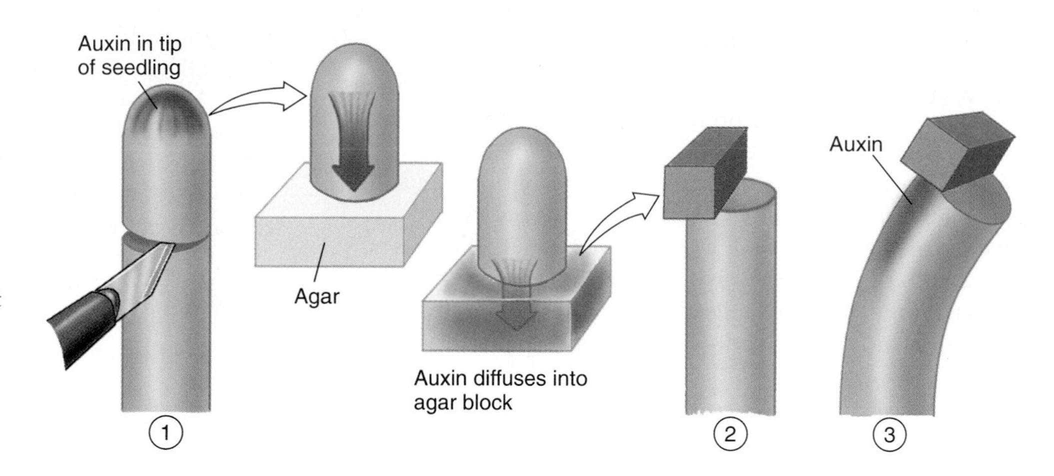

an unknown substance continually moves down from the tips of grass seedlings and promotes growth on all sides. Such a light pattern would not, of course, cause the shoot to bend.

Then, in 1926, Dutch plant physiologist Frits Went carried Paal's experiments an important step further. Went cut off the tips of oat seedlings that had been illuminated normally and set these tips on agar. He then took oat seedlings that had been grown in the dark and cut off their tips in a similar way. Finally, Went cut tiny blocks from the agar on which the tips of the light-grown seedlings had been placed and placed them off-center on the tops of the decapitated dark-grown seedlings (figure 41.9). Even though these seedlings had not been exposed to the light themselves, they bent away from the side on which the agar blocks were placed.

As an experimental control, Went put blocks of pure agar on the decapitated stem tips and noted either no effect or a slight bending toward the side where the agar blocks were placed. Finally, Went cut sections out of the lower portions of the light-grown seedlings. He placed these sections on the tips of decapitated, dark-green oat seedlings and again observed no effect.

As a result of his experiments, Went was able to show that the substance that had diffused into the agar from the tips of light-grown oat seedlings could make seedlings bend when they otherwise would have remained straight. He also showed that this chemical messenger caused the cells on the side of the seedling into which it flowed to grow more than those on the opposite side (figure 41.10). In other words, it enhanced rather than retarded cell elongation. He named the substance that he had discovered **auxin,** from the Greek word *auxein,* which means "to increase."

Went's experiments provided a basis for understanding the responses that the Darwins had obtained some 45 years earlier. The oat seedlings bent toward the light because of differences in the auxin concentrations on the two sides of the shoot. The side of the shoot that was in the shade had more auxin, and its cells therefore elongated more than those on the lighted side, bending the plant toward the light.

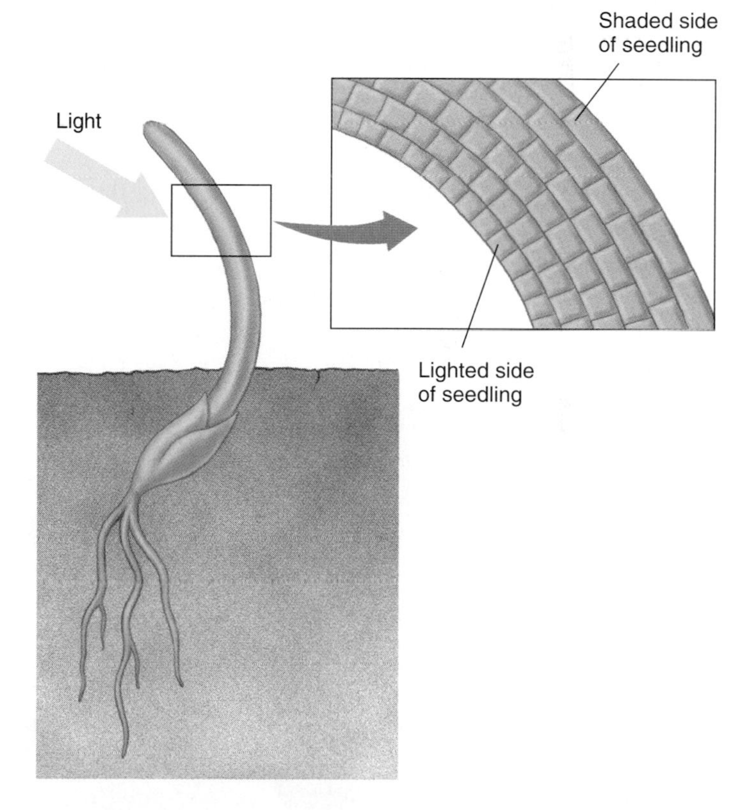

FIGURE 41.10
Auxin causes cells on the dark side to elongate. Went determined that a substance called auxin enhanced cell elongation. Plant cells that are in the shade have more auxin and grow faster than cells on the lighted side, causing the plant to bend toward light. Further experiments showed exactly why there is more auxin on the shaded side of a plant.

The Effects of Auxins

Auxin acts to adapt the plant to its environment in a highly advantageous way. It promotes growth and elongation and facilitates the plant's response to its environment. Environmental signals directly influence the distribution of auxin in the plant. How does the environment—specifically, light—exert this influence? Theoretically, it might destroy the auxin, decrease the cells' sensitivity to auxin, or cause the auxin molecules to migrate away from the light into the shaded portion of the shoot. This last possibility has proved to be the case.

In a simple but effective experiment, Winslow Briggs inserted a thin sheet of transparent mica vertically between the half of the shoot oriented toward the light and the half of the shoot oriented away from it (figure 41.11). He found that light from one side does not cause a shoot with such a barrier to bend. When Briggs examined the illuminated plant, he found equal auxin levels on both the light and dark sides of the barrier. He concluded that a normal plant's response to light from one direction involves auxin migrating from the light side to the dark side, and that the mica barrier prevented a response by blocking the migration of auxin.

The effects of auxin are numerous and varied. Auxin promotes the activity of the vascular cambium and the vascular tissues. Also, auxins are present in pollen in large quantities and play a key role in the development of fruits. Synthetic auxins are used commercially for the same purpose. Fruits will normally not develop if fertilization has not occurred and seeds are not present, but frequently they will if auxins are applied. Pollination may trigger auxin release in some species leading to fruit development occurring even before fertilization.

How Auxin Works

In spite of this long history of research on auxin, its molecular basis of action has been an enigma. The chemical structure of IAA resembles that of the amino acid tryptophan, from which it is probably synthesized by plants (figure 41.12). Unlike animal hormones, a specific signal is not sent to specific cells, eliciting a predictable response. There are most likely multiple auxin perception sites. Auxin is also unique among the plant hormones in that it is transported toward the base of the plant. Two families of genes have been identified in *Arabidopsis* that are involved in auxin transport. For example, one protein is involved in the top to bottom transport of auxin; while two other proteins function in the root tip to regulate the growth response to gravity. We are still a ways from linking the measurable and observable effects of auxin to events that transpire after it travels to a site and binds to a receptor.

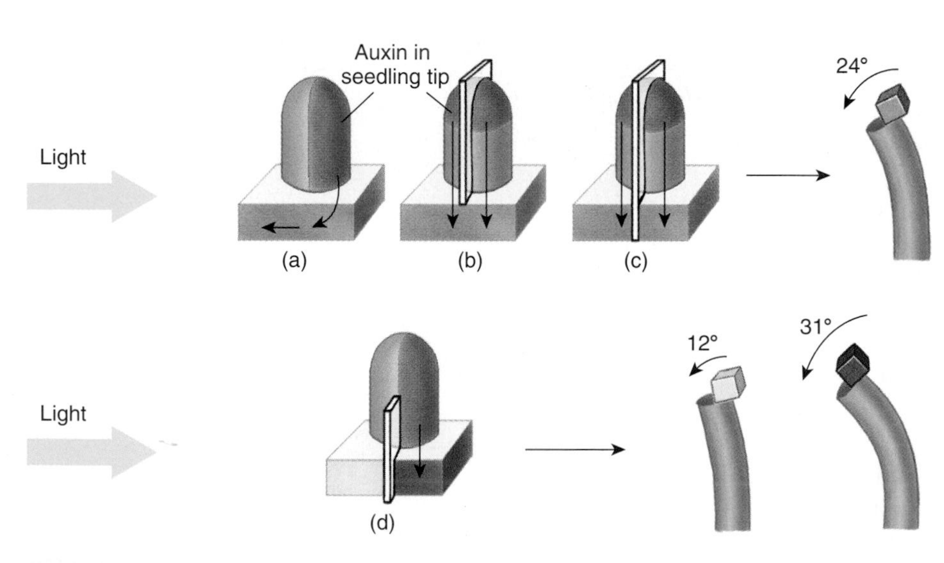

FIGURE 41.11
Phototropism and auxin: the Winslow Briggs experiments. The basic design of these experiments was to place the tip of an oat seedling on an agar block, apply light from one side, and observe the degree of curvature produced when the agar blocks were later placed on the decapitated seedlings. However, Briggs inserted a barrier in various places and noted how this affected the location of auxin. A comparison of (*a*) and (*b*) with similar experiments performed in the dark showed that auxin production does not depend on light; all produced approximately 24° of curvature. If a barrier was inserted in the agar block (*d*), light caused the displacement of the auxin away from the light. Finally, experiment (*c*) showed that it was displacement that had occurred, and not different rates of auxin production on the dark and light sides, because when displacement was prevented with a barrier, both sides of the agar block produced about 24° of curvature.

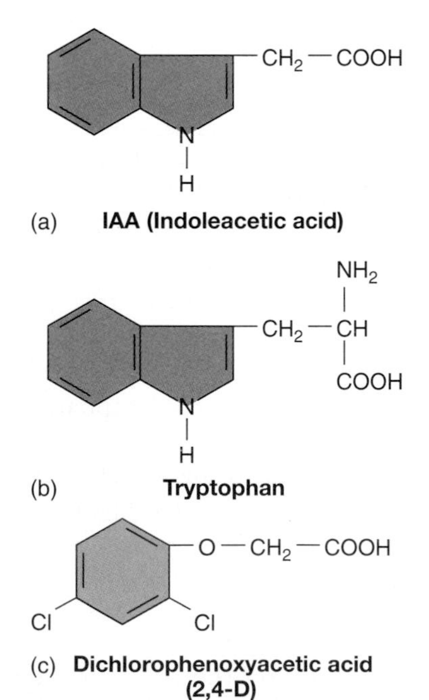

FIGURE 41.12
Auxins. (*a*) Indoleacetic acid (IAA), the principal naturally occurring auxin. (*b*) Tryptophan, the amino acid from which plants probably synthesize IAA. (*c*) Dichlorophenoxyacetic acid (2,4-D), a synthetic auxin, is a widely used herbicide.

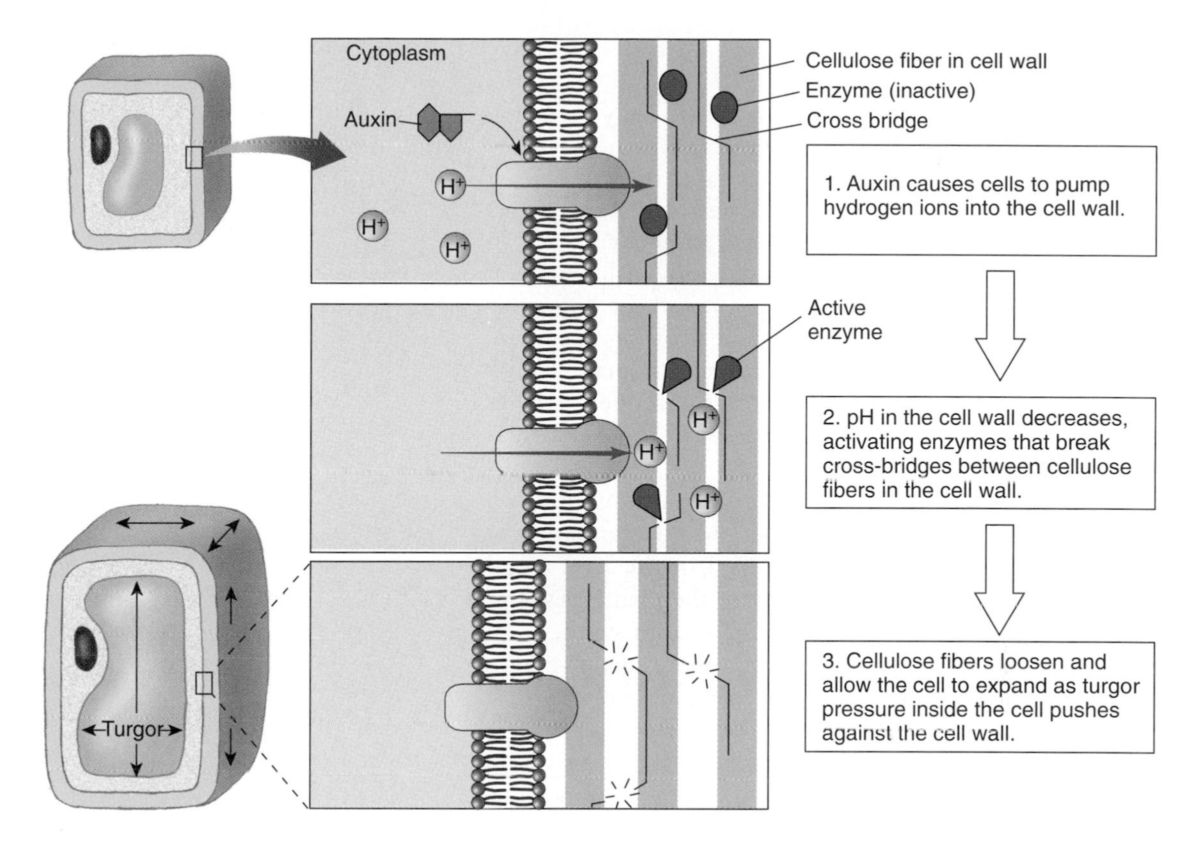

Cytoplasm
Auxin
H+
Cellulose fiber in cell wall
Enzyme (inactive)
Cross bridge
1. Auxin causes cells to pump hydrogen ions into the cell wall.
Active enzyme
2. pH in the cell wall decreases, activating enzymes that break cross-bridges between cellulose fibers in the cell wall.
Turgor
3. Cellulose fibers loosen and allow the cell to expand as turgor pressure inside the cell pushes against the cell wall.

FIGURE 41.13 Acid growth hypothesis. Auxin stimulates the release of hydrogen ions from the target cells which alters the pH of the cell wall. This optimizes the activity of enzymes which break bonds in the cell wall, allowing them to expand.

One of the downstream effects of auxin is an increase in plasticity of the plant cell wall. This will only work on young cell walls without extensive secondary cell wall formation. A more plastic wall will stretch more while its protoplast is swelling during active cell growth. The **acid growth hypothesis** provides a model linking auxin to cell wall expansion (figure 41.13). Auxin causes responsive cells to actively transport hydrogen ions from the cytoplasm into the cell wall space. This decreases the pH which activates enzymes that can break bonds between cell wall fibers. Remember that different enzymes operate optimally at different pHs. This hypothesis has been experimentally supported in several ways. Buffers that prevent cell wall acidification block cell expansion. Other compounds that release hydrogen ions from the cell can also cause cell expansion. The movement of hydrogen ions has been observed in response to auxin treatment. This mechanism results in a rapid response to an environmental signal.

Synthetic Auxins. Synthetic auxins such as NAA (naphthalene acetic acid) and IBA (indolebutyric acid) have many uses in agriculture and horticulture. One of their most important uses is based on their prevention of abscission, the process that causes a leaf or other organ to fall from a plant. Synthetic auxins are used to prevent fruit drop in apples before they are ripe and to hold berries on holly that is being prepared for shipping. Synthetic auxins are also used to promote flowering and fruiting in pineapples and to induce the formation of roots in cuttings.

Synthetic auxins are routinely used to control weeds. When used as herbicides, they are applied in higher concentrations than IAA would normally occur in plants. One of the most important synthetic auxin herbicides is 2,4-dichlorophenoxyacetic acid, usually known as 2,4-D (see figure 41.12*c*). It kills weeds in grass lawns by selectively eliminating broad-leaved dicots. The stems of the dicot weeds cease all axial growth.

The herbicide 2,4,5-trichlorophenoxyacetic acid, better known as 2,4,5-T, is closely related to 2,4-D. 2,4,5-T was widely used as a broad-spectrum herbicide to kill weeds and seedlings of woody plants. It became notorious during the Vietnam War as a component of a jungle defoliant known as Agent Orange and was banned in 1979 for most uses in the United States. When 2,4,5-T is manufactured, it is unavoidably contaminated with minute amounts of dioxin. Dioxin, in doses as low as a few parts per billion, has produced liver and lung diseases, leukemia, miscarriages, birth defects, and even death in laboratory animals.

Auxin is synthesized in apical meristems of shoots. It causes young stems to bend toward light when it migrates toward the darker side, where it makes young cell walls more plastic and thereby promotes cell elongation. Synthetic auxins have uses in agriculture and horticulture.

Cytokinins

Cytokinins comprise another group of naturally occurring growth hormones in plants. Studies by Gottlieb Haberlandt of Austria around 1913 demonstrated the existence of an unknown chemical in various tissues of vascular plants that, in cut potato tubers, would cause parenchyma cells to become meristematic, and would induce the differentiation of a cork cambium. Coconut milk promoted the differentiation of organs in masses of plant tissue growing in culture. Later it was discovered that cytokinins in coconut milk promoted the organ differentiation. Subsequent studies have focused on the role cytokinins play in the differentiation of tissues from callus.

A cytokinin is a plant hormone that, in combination with auxin, stimulates cell division and differentiation in plants. Most cytokinins are produced in the root apical meristems and transported throughout the plant. Developing fruits are also important sites of cytokinin synthesis. In mosses, cytokinins cause the formation of vegetative buds on the gametophyte. In all plants, cytokinins, working with other hormones, seem to regulate growth patterns.

Cytokinins are purines that appear to be derivatives of, or have molecule side chains similar to, those of adenine (figure 41.14). Other chemically diverse molecules, not known to occur naturally, have effects similar to those of cytokinins. Cytokinins promote growth of lateral buds into branches (figure 41.15).

Conversely, cytokinins inhibit formation of lateral roots, while auxins promote their formation. As a consequence of

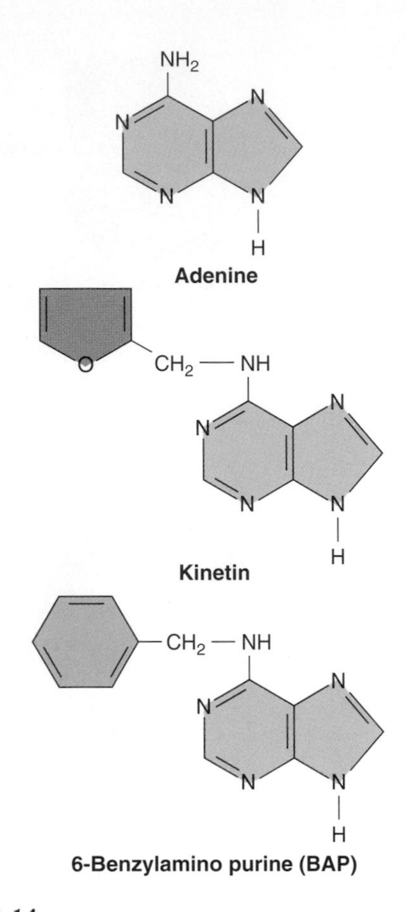

FIGURE 41.14
Some cytokinins. Two commonly used synthetic cytokinins: kinetin and 6-benzylamino purine. Note their resemblance to the purine base adenine.

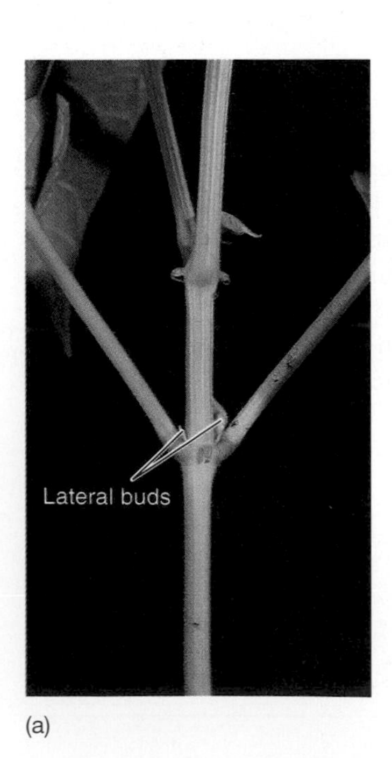

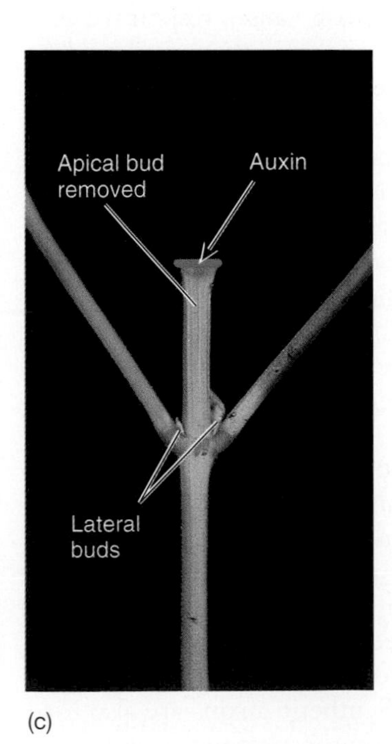

FIGURE 41.15
Cytokinins stimulate lateral bud growth. (*a*) When the apical meristem of a plant is intact, auxin from the apical bud will inhibit the growth of lateral buds. (*b*) When the apical bud is removed, cytokinins are able to produce the growth of lateral buds into branches. (*c*) When the apical bud is removed and auxin is added to the cut surface, lateral bud outgrowth is suppressed.

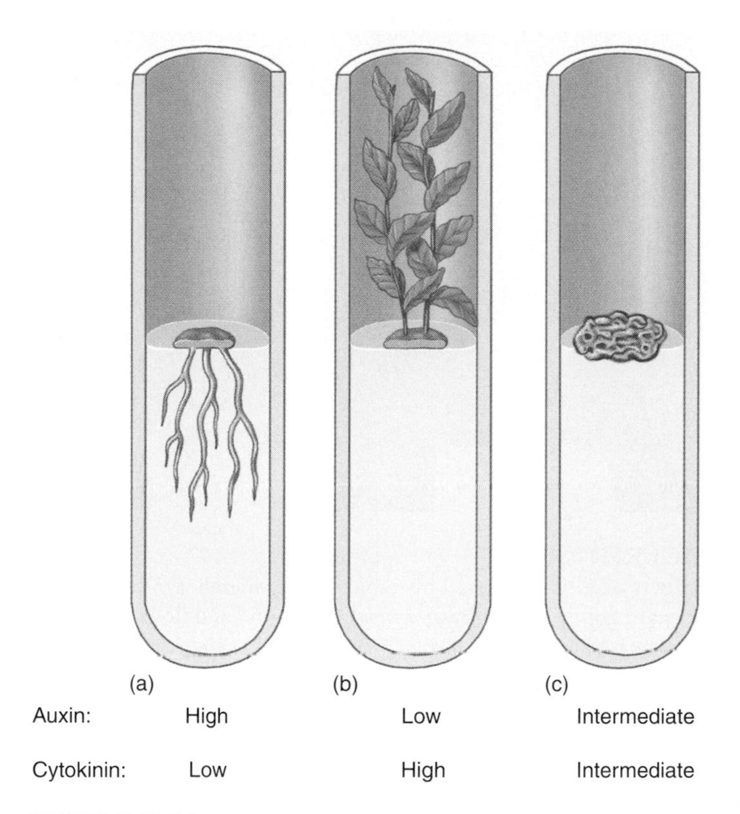

FIGURE 41.16
Relative amounts of cytokinins and auxin affect organ regeneration in culture. In the case of tobacco, (*a*) high auxin to cytokinin ratios favor root development; (*b*) high cytokinin to auxin ratios favor shoot development; and (*c*) intermediate concentrations result in the formation of undifferentiated cells. These developmental responses to cytokinin/auxin ratios in culture are species specific.

FIGURE 41.17
Crown gall tumor. Sometimes cytokinins can be used against the plant by a pathogen. In this case *Agrobacterium tumefaciens* (a bacterium) has incorporated a piece of its DNA into the plant genome. This DNA contains genes coding for enzymes necessary for cytokinin and auxin biosynthesis. The increased levels of these hormones in the plant cause massive cell division and the formation of a tumor.

these relationships, the balance between cytokinins and auxin, along with many other factors, determines the form of a plant. In addition, the application of cytokinins to leaves detached from a plant retards their yellowing. They function as anti-aging hormones.

The action of cytokinins, like that of other hormones, has been studied in terms of its effects on growth and differentiation of masses of tissue growing in defined media. Plant tissue can form shoots, roots, or an undifferentiated mass of tissues, depending on the relative amounts of auxin and cytokinin (figure 41.16). In the early cell-growth experiments coconut "milk" was an essential factor. Eventually, it was discovered that coconut "milk" is not only rich in amino acids and other reduced nitrogen compounds required for growth, but it also contains cytokinins. Cytokinins seem to be essential for mitosis and cell division. They apparently promote the synthesis or activation of proteins that are specifically required for mitosis.

Cytokinins have also been used against plants by pathogens. The bacteria *Agrobacterium*, for example, introduces genes into the plant genome that increase the rate of cytokinin, as well as auxin, production. This causes massive cell division and the formation of a tumor called crown gall (figure 41.17). How these hormone biosynthesis genes ended up in a bacterium is an intriguing evolutionary question. Coevolution does not always work to the plant's advantage.

Cytokinins are plant hormones that, in combination with auxin, stimulate cell division and, along with a number of other factors, determine the course of differentiation. In contrast to auxins, cytokinins are purines that are related to or derived from adenine.

Gibberellins

Gibberellins are named after the fungus *Gibberella fujikuroi*, which causes rice plants, on which it is parasitic, to grow abnormally tall. Japanese plant pathologist Eiichi Kurosawa investigated bakane ("foolish seedling") disease in the 1920s. He grew *Gibberella* in culture and obtained a substance that, when applied to rice plants, produced bakane. This substance was isolated and the structural formula identified by Japanese chemists in 1939. British chemists reconfirmed the formula in 1954. Although such chemicals were first thought to be only a curiosity, they have since turned out to belong to a large class of more than 100 naturally occurring plant hormones called gibberellins. All are acidic and are usually abbreviated to **GA** (for gibberellic acid), with a different subscript (GA_1, GA_2, and so forth) to distinguish each one. While gibberellins function endogenously as hormones, they also function as pheromones in ferns. In ferns gibberellin-like compounds released from one gametophyte can trigger the development of male reproductive structures on a neighboring gametophyte.

Gibberellins, which are synthesized in the apical portions of stems and roots, have important effects on stem elongation. The elongation effect is enhanced if auxin is also present. The application of gibberellins to certain dwarf mutants is known to restore the normal growth and development in many plants (figure 41.18). Some dwarf mutants produce insufficient amounts of gibberellin and respond to GA applications; others lack the ability to perceive gibberellin. The large number of gibberellins are all part of a complex biosynthetic pathway that has been unraveled using gibberellin-deficient mutants in maize (corn). While many of these gibberellins are intermediate forms in the production of GA_1, recent work shows that different forms may have specific biological roles.

In chapter 40, we noted the role gibberellins stimulate the production of α-amylase and other hydrolytic enzymes needed for utilization of food resources during germination and establishment of cereal seedlings. How is transcription of the genes encoding these enzymes regulated? GA is a signal from the embryo that turns on transcription of a gene(s) encoding hydrolytic enzymes in the aleurone layer. GA somehow enhances DNA binding proteins, which in turn allow transcription of a gene. Synthesis of DNA does not seem to occur during the early stages of seed germination but becomes important when the radicle has grown through the seed coats.

Gibberellins also affect a number of other aspects of plant growth and development. In some cases GAs hasten seed germination, apparently by substituting for the effects of cold or light requirements. Gibberellins are used commercially to space grape flowers by extending internode length so the fruits have more room to grow (figure 41.19).

FIGURE 41.18
Effects of gibberellins. This rapid cycling member of the mustard family plant *(Brassica rapa)* will "bolt" and flower because of increased gibberellin levels. Mutants such as the rosette mutant shown here (*left*) are defective in producing gibberellins. They can be rescued by applying gibberellins. Other mutants have been identified that are defective in perceiving gibberellins and they will not respond to gibberellin applications.

FIGURE 41.19
Applications of gibberellins increase the space between grapes. Larger grapes (*right*) develop because there is more room between individual grapes.

Gibberellins are an important class of plant hormones that are produced in the apical regions of shoots and roots. They play the major role in controlling stem elongation for most plants, acting in concert with auxin and other hormones.

Brassinosteroids and Oligosaccharins

Brassinosteroids

Although we've known about brassinosteroids for 30 years, it is only recently that they have claimed their place as a class of plant hormones. They were first discovered in *Brassica* pollen, hence the name. Their historical absence in discussions of hormones may be partially due to their functional overlap with other plant hormones, especially auxins and gibberellins. Additive effects among these three classes have been reported. The application of molecular genetics to the study of brassinosteroids has led to tremendous advances in our understanding of how they are made and, to some extent, how they function in signal transduction pathways. What is particularly intriguing about brassinosteroids are similarities to animal steroid hormones (figure 41.20). One of the genes coding for an enzyme in the brassinosteroid biosynthetic pathway has significant similarity to an enzyme used in the synthesis of testosterone and related steroids. Brassinosteroids have been identified in algae and appear to be quite ubiquitous among the plants. It is plausible that their evolutionary origin predated the plant-animal split.

Brassinosteroids have a broad spectrum of physiological effects—elongation, cell division, bending of stems, vascular tissue development, delayed senescence, membrane polarization, and reproductive development. Environmental signals can trigger brassinosteroid actions. Mutants have been identified that alter the response to brassinosteroid, but signal transduction pathways remain to be uncovered. From an evolutionary perspective, it will be quite interesting to see how these pathways compare with animal steroid signal transduction pathways.

Brassinolide

Cortisol

Testosterone

Estradiol

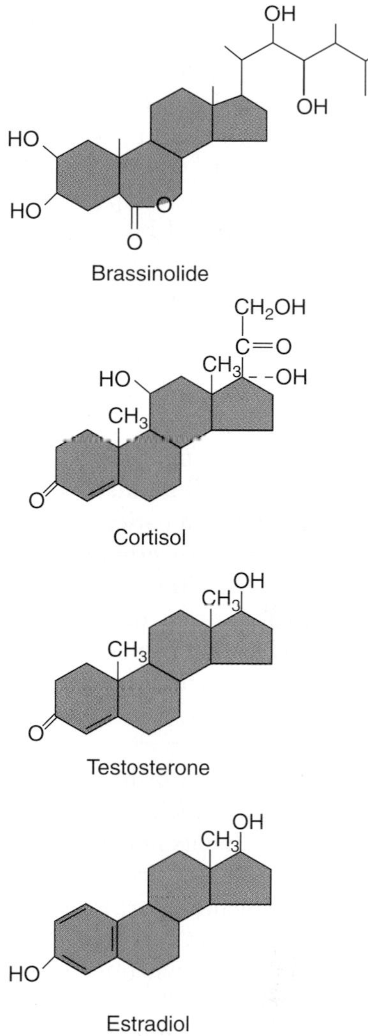

FIGURE 41.20
Brassinosteroids, such as brassinolide, have structural similarities to animal steroid hormones. Cortisol, testosterone, and estradiol are animal steroid hormones.

Oligosaccharins

In addition to cellulose, plant cell walls are composed of numerous complex carbohydrates called oligosaccharides. There is some evidence that these cell wall components (when degraded by pathogens) function as signaling molecules as well as structural wall components. Oligosaccharides that are proposed to have a hormone-like function are called oligosaccharins. Oligosaccharins can be released from the cell wall by enzymes secreted by pathogens. These carbohydrates are believed to signal defense responses, such as the hypersensitive response discussed later in this chapter. Another oligosaccharin has been shown to inhibit auxin-stimulated elongation of pea stems. These molecules are active at concentrations one to two orders of magnitude less than the traditional plant hormones. You have seen how auxin and cytokinin ratios can affect organogenesis in culture. Oligosaccharins also affect the phenotype of regenerated tobacco tissue, inhibiting root formation and stimulating flower production in tissues that are competent to regenerate flowers. How the culture results translate to in vivo systems is an open question. The structural biochemistry of oligosaccharins makes them particularly challenging molecules to study. Current research focuses on ways oligosaccharins interface with cells and initiate signal transduction pathways.

Brassinosteroids are structurally similar to animal steroid hormones. They have many effects on plant growth and development that parallel those of auxins and gibberellins. Oligosaccharins are complex carbohydrates that are released from cell walls and appear to regulate both pathogen responses and growth and development in some plants.

Ethylene

Long before its role as a plant hormone was appreciated, the simple, gaseous hydrocarbon ethylene ($H_2C{=}CH_2$) was known to defoliate plants when it leaked from gaslights in streetlamps. Ethylene is, however, a natural product of plant metabolism that, in minute amounts, interacts with other plant hormones. When auxin is transported down from the apical meristem of the stem, it stimulates the production of ethylene in the tissues around the lateral buds and thus retards their growth. Ethylene also suppresses stem and root elongation, probably in a similar way. An ethylene receptor has been identified and characterized. It appears to have evolved early in the evolution of photosynthetic organisms, sharing features with environmental-sensing proteins identified in bacteria.

Ethylene plays a major role in fruit development. At first, auxin, which is produced in significant amounts in pollinated flowers and developing fruits, stimulates ethylene production; this, in turn, hastens fruit ripening. Complex carbohydrates are broken down into simple sugars, chlorophylls are broken down, cell walls become soft, and the volatile compounds associated with flavor and scent in ripe fruits are produced.

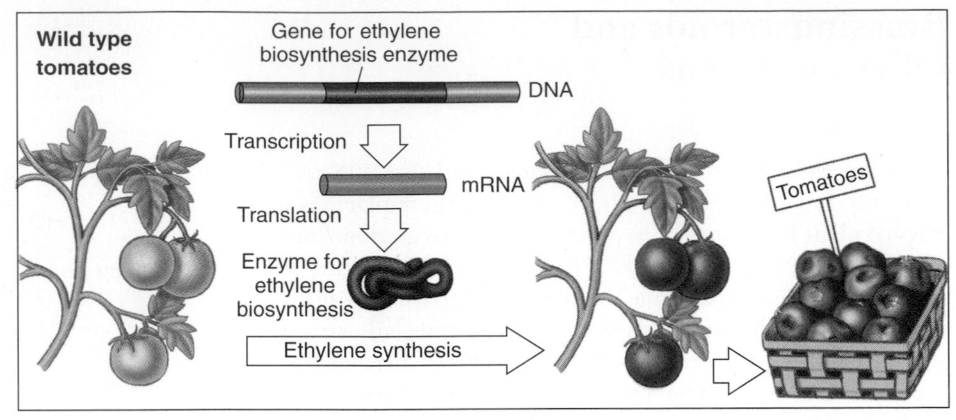

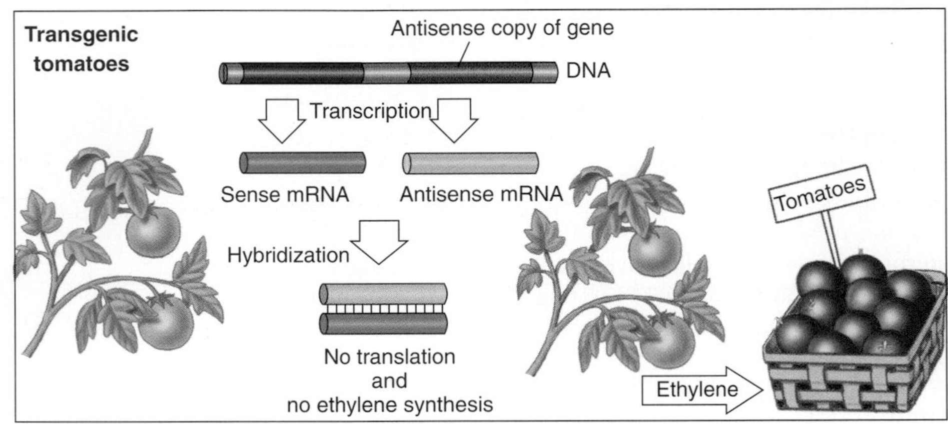

FIGURE 41.21
Genetic regulation of fruit ripening. An antisense copy of the gene for ethylene biosynthesis prevents the formation of ethylene and subsequent ripening of transgenic fruit. The antisense strand is complementary to the sequence for the ethylene biosynthesis gene. After transcription, the antisense mRNA pairs with the sense mRNA, and the double-stranded mRNA cannot be translated into a functional protein. Ethylene is not produced, and the fruit does not ripen. The fruit is sturdier for shipping in its unripened form and can be ripened later with exposure to ethylene. Thus, while wild-type tomatoes may already be rotten and damaged by the time they reach stores, transgenic tomatoes stay fresh longer.

One of the first observations that led to the recognition of ethylene as a plant hormone was the premature ripening in bananas produced by gases coming from oranges. Such relationships have led to major commercial uses of ethylene. For example, tomatoes are often picked green and artificially ripened later by the application of ethylene. Ethylene is widely used to speed the ripening of lemons and oranges as well. Carbon dioxide has the opposite effect of arresting ripening. Fruits are often shipped in an atmosphere of carbon dioxide. A biotechnology solution has also been developed (figure 41.21). One of the genes necessary for ethylene biosynthesis has been cloned, and its antisense copy has been inserted into the tomato genome. The antisense copy of the gene is a nucleotide sequence that is complementary to the sense copy of the gene. In this transgenic plant, both the sense and antisense sequences for the ethylene biosynthesis gene are transcribed. The sense and antisense mRNA sequences then pair with each other. This blocks translation, which requires single-stranded RNA; ethylene is not synthesized, and the transgenic tomatoes do not ripen. Sturdy green tomatoes can be shipped without ripening and rotting. Exposing these tomatoes to ethylene later will allow them to ripen.

Studies have shown that ethylene plays an important ecological role. Ethylene production increases rapidly when a plant is exposed to ozone and other toxic chemicals, temperature extremes, drought, attack by pathogens or herbivores, and other stresses. The increased production of ethylene that occurs can accelerate the loss of leaves or fruits that have been damaged by these stresses. Some of the damage associated with exposure to ozone is due to the ethylene produced by the plants. The production of ethylene by plants attacked by herbivores or infected with diseases may be a signal to activate the defense mechanisms of the plants. This may include the production of molecules toxic to the pests.

Ethylene, a simple gaseous hydrocarbon, is a naturally occurring plant hormone. Among its numerous effects is the stimulation of ripening in fruit. Ethylene production is also elevated in response to environmental stress.

Abscisic Acid

Abscisic acid appears to be synthesized mainly in mature green leaves, fruits, and root caps. The hormone earned its name because applications of it appear to stimulate fruit abscission in cotton, but there is little evidence that it plays an important role in this process. It is actually the ethylene that promotes senescence and abscision.

Abscisic acid probably induces the formation of winter buds—dormant buds that remain through the winter. The conversion of leaf primordia into bud scales follows (figure 41.22*a*). Like ethylene, it may also suppress growth of dormant lateral buds. It appears that abscisic acid, by suppressing growth and elongation of buds, can counteract some of the effects of gibberellins (which stimulate growth and elongation of buds); it also promotes senescence by counteracting auxin (which tends to retard senescence). Abscisic acid plays a role in seed dormancy and is antagonistic to gibberellins during germination. It is also important in controlling the opening and closing of stomata (figure 41.22*b*).

Abscisic acid occurs in all groups of plants and apparently has been functioning as a growth-regulating substance since early in the evolution of the plant kingdom. Relatively little is known about the exact nature of its physiological and biochemical effects. These effects are very rapid—often taking place within a minute or two—and therefore they must be at least partly independent of gene expression. Some longer-term effects of abscisic acid involve the regulation of gene expression. This is currently an active area of research. Having all the genes sequenced in a model plant, *Arabidopsis*, makes it easier to identify which genes are transcribed in response to abscisic acid. Abscisic acid levels become greatly elevated when the plant is subject to stress, especially drought. Like other plant hormones, abscisic acid probably will prove to have valuable commercial applications when its mode of action is better understood. It is a particularly strong candidate for understanding desiccation tolerance.

Abscisic acid, produced chiefly in mature green leaves and in fruits, suppresses growth of buds and promotes leaf senescence. It also plays an important role in controlling the opening and closing of stomata. Abscisic acid may be critical in ensuring survival under environmental stress, especially water stresses.

(a)

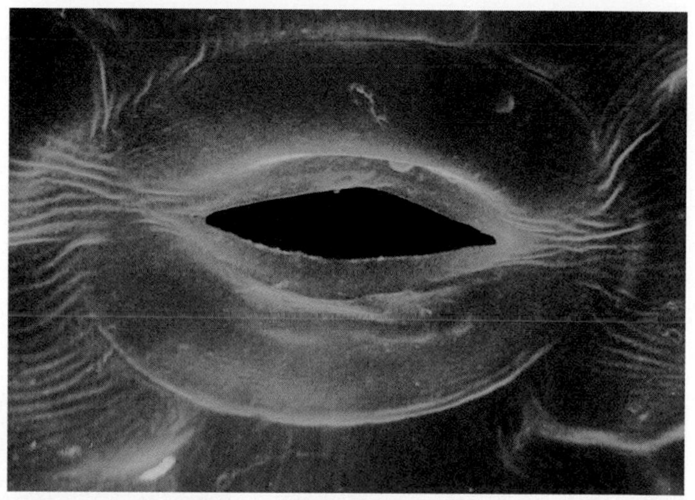

(b)

FIGURE 41.22
Effects of abscisic acid. (*a*) Abscisic acid plays a role in the formation of these winter buds of an American basswood. These buds will remain dormant for the winter, and bud scales—modified leaves—will protect the buds from desiccation.
(*b*) Abscisic acid also affects the closing of stomata by influencing the movement of potassium ions out of guard cells.

Plants Undergo Metamorphosis

Overview of Initiating Flowering

Carefully regulated processes determine when and where flowers will form. Plants must often gain competence to respond to internal or external signals regulating flowering. Once plants are competent to reproduce, a combination of factors including light, temperature, and both promotive and inhibitory internal signals determine when a flower is produced (figure 41.23). These signals turn on genes that specify where the floral organs, sepals, petals, stamens, and carpels will form. Once cells have instructions to become a specific floral organ, yet another developmental cascade leads to the three-dimensional construction of flower parts.

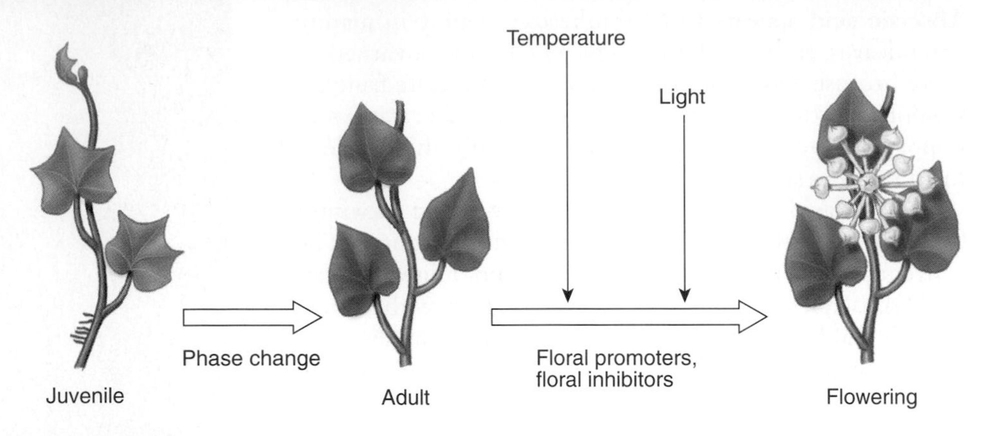

FIGURE 41.23
Factors involved in initiating flowering. This is a model of environmentally cued and internally processed events that result in a shoot meristem initiating flowers.

Phase Change

Plants go through developmental changes leading to reproductive maturity just like many animals. This shift from juvenile to adult development is seen in the metamorphosis of a tadpole to an adult frog or caterpillar to a butterfly that can then reproduce. Plants undergo a similar metamorphosis that leads to the production of a flower. Unlike the frog that loses its tail, plants just keep adding on structures to existing structures with their meristems. At germination, most plants are incapable of producing a flower, even if all the environmental cues are optimal. Internal developmental changes allow plants to obtain **competence** to respond to external and/or internal signals that trigger flower formation. This transition is referred to as **phase change**. Phase change can be morphologically obvious or very subtle. Take a look at an oak tree in the winter. The lower leaves will still be clinging to the branches, while the upper ones will be gone (figure 41.24*a*). Those lower branches were initiated by a juvenile meristem. The fact that they did not respond to environmental cues and drop their leaves indicates that they are young branches and have not made a phase change. Ivy also has distinctive juvenile and adult phases of growth (figure 41.24*b*). Stem tissue produced by a juvenile meristem initiates adventitious roots that can cling to walls. If

(a)

(b)

FIGURE 41.24
Phase change. (*a*) The lower branches of this oak tree represent the juvenile phase of development and cling to their leaves in the winter. The lower leaves are not able to form an abscission layer and break off the tree in the fall. Such visible changes are marks of phase change, but the real test is whether or not the plant is able to flower. (*b*) Juvenile ivy (*left*) makes adventitious roots and has an alternating leaf arrangement. Adult ivy (*right*) cannot make adventitious roots and has leaves with a different morphology that are arranged on an upright stem in a spiral.

you look at very old brick buildings that are covered with ivy, you will notice the uppermost branches are falling off because they have transitioned to the adult phase of growth and have lost the ability to produce adventitious roots. It is important to remember that even though a plant has reached the adult stage of development, it may or may not produce reproductive structures. Other factors may be necessary to trigger flowering.

Generally it is easier to get a plant to revert from an adult to vegetative state than to induce phase change experimentally. Applications of gibberellins and severe pruning can cause reversion. There is evidence in peas and *Arabidopsis* for genetically controlled repression of flowering. The *embryonic flower* mutant of *Arabidopsis* flowers almost immediately (figure 41.25), which is consistent with the hypothesis that the wild-type allele suppresses flowering. It is possible that flowering is the default state and that mechanisms have evolved to delay flowering. This delay allows the plant to store more energy to be allocated for reproduction.

FIGURE 41.25
In *Arabidopsis*, the *embryonic flower* gene may repress flowering. The *embryonic flower* mutant flowers upon germination. This is an aberrant flower with malformed carpels and other defective floral structures just above the root system.

The best example of inducing the juvenile to adult transition comes from overexpressing a gene necessary for flowering, that is found in many species. This gene, *LEAFY*, was cloned in *Arabidopsis* and its promoter was replaced with a viral promoter that results in constant, high levels of *LEAFY* transcription. This gene construct was then introduced into cultured aspen cells which were used to regenerate plants. When *LEAFY* is overexpressed in aspen, flowering occurs in weeks instead of years (figure 41.26). Phase change requires both sufficient signal and the ability to perceive the signal. Some plants acquire competence in the shoot to perceive a signal of a certain intensity. Others acquire competence to produce sufficient promotive signal(s) and/or decrease inhibitory signal(s).

Plants become reproductively competent through changes in signaling and perception. The transition to the adult stage of development where reproduction is possible is called phase change. Plants in the adult phase of development may or may not produce reproductive structures (flowers), depending on environmental cues.

(a)

(b)

FIGURE 41.26
Overexpression of a flowering gene can accelerate phase change. (*a*) Normally, an aspen tree grows for several years before producing flowers. (*b*) Overexpression of the *Arabidopsis* flowering gene, *LEAFY*, causes rapid flowering in a transgenic aspen.

Pathways Leading to Flower Production

Three genetically regulated pathways to flowering have been identified: (1) the light-dependent pathway, (2) the temperature-dependent pathway, and (3) the autonomous pathway. The environment can promote or repress flowering. In some cases, it can be relatively neutral. Light can be a signal that long, summer days have arrived in a temperate climate and conditions are favorable for reproduction. In other cases, plants depend on light to accumulate sufficient amounts of sucrose to fuel reproduction, but flower independently of the length of day. Temperature can also be used as a clue. Gibberellins are important and have been linked to the vernalization pathway (vernalization is the requirement for a period of chilling of seeds or shoots for flowering). Clearly, reproductive success would be unlikely in the middle of a blizzard. Assuming regulation of reproduction first arose in more constant tropical environments, many of the day length and temperature controls would have evolved as plants colonized more temperate climates. Plants can rely primarily on one pathway, but all three pathways can be present. The complexity of the flowering pathways has been dissected physiologically. Now analysis of flowering mutants is providing insight into the molecular mechanisms of the floral pathways. The redundancy of pathways to flowering ensures that there will be another generation.

Light-Dependent Pathway

Flowering requires much energy accumulated via photosynthesis. Thus, all plants require light for flowering, but this is distinct from the **photoperiodic,** or light-dependent, flowering pathway. Aspects of growth and development in most plants are keyed to changes in the proportion of light to dark in the daily 24-hour cycle (day length). This provides a mechanism for organisms to respond to seasonal changes in the relative length of day and night. Day length changes with the seasons; the farther from the equator, the greater the variation. Flowering responses of plants to day length fall into several basic categories. When the daylight becomes shorter than a critical length, flowering is initiated in **short-day plants** (figure 41.27). When the daylight becomes longer than a critical length, flowering is initiated in **long-day plants.** Other plants, such as snapdragons, roses, and many native to the

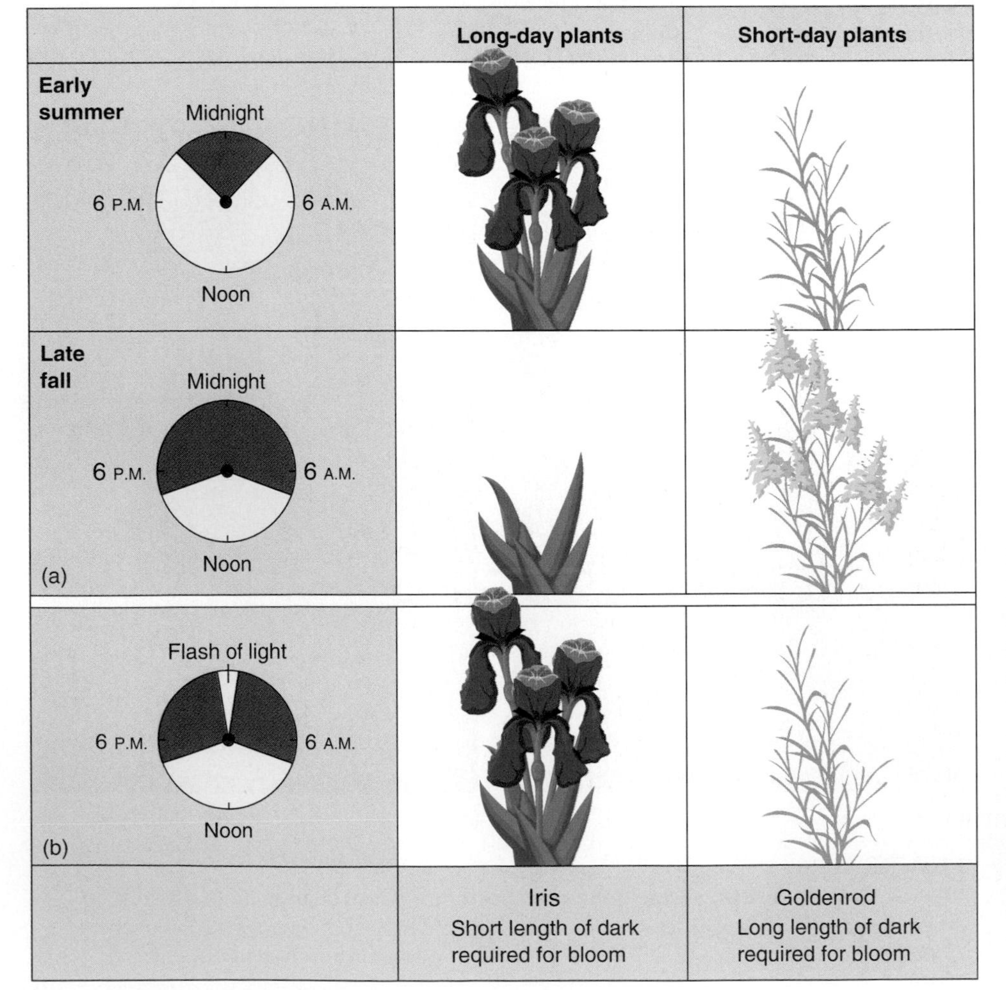

FIGURE 41.27
How flowering responds to day length. (*a*) This iris is a long-day plant that is stimulated by short nights to flower in the spring. The goldenrod is a short-day plant that, throughout its natural distribution in the Northern Hemisphere, is stimulated by long nights to flower in the fall. (*b*) If the long night of winter is artificially interrupted by a flash of light, the goldenrod will not flower, and the iris will. Although the terms refer to the length of day, in each case, it is the duration of uninterrupted darkness that determines when flowering will occur.

tropics (for example, tomatoes), will flower when mature regardless of day length, as long as they have received enough light for normal growth. These are referred to as **day-neutral plants.** Several grasses (for example, Indian grass, *Sorghastrum nutans*), as well as ivy, have two critical photoperiods; they will not flower if the days are too long, and they also will not flower if the days are too short. In some species, there is a sharp distinction between long and short days. In others, flowering occurs more rapidly or slowly depending on the length of day. These plants rely on other flowering pathways as well and are called facultative long- or short-day plants. The garden pea is an example of a facultative long-day plant. In all of these plants, it is actually the length of darkness (night), not the length of day, that is physiologically significant. Using light as a cue permits plants to flower when abiotic environmental conditions are optimal, pollinators are available, and competition for resources with other plants may be less. For example, the spring ephemerals flower in the woods before the canopy leafs out, blocking sunlight necessary for photosynthesis.

At middle latitudes, most long-day plants flower in the spring and early summer; examples of such plants include clover, irises, lettuce, spinach, and hollyhocks. Short-day plants usually flower in late summer and fall, and include chrysanthemums, goldenrods, poinsettias, soybeans, and many weeds. Commercial plant growers use these responses to day length to bring plants into flower at specific times. For example, photoperiod is manipulated in greenhouses so poinsettias flower just in time for the winter holidays (figure 41.28). The geographic distribution of certain plants may be determined by flowering responses to day length.

Photoperiod is perceived by several different forms of phytochrome and also a blue-light-sensitive molecule (cryptochrome). The conformational change in a light receptor molecule triggers a cascade of events that leads to the production of a flower. There is a link between light and the circadian rhythm regulated by an internal clock that facilitates or inhibits flowering. At a molecular level the gaps between light signaling and production of flowers are rapidly filling in and the control mechanisms are quite complex. Here is one example of how day length affects a specific flowering gene in *Arabidopsis*, a facultative long-day plant that flowers in response to both far-red and blue light. Red light inhibits flowering. The gene *CONSTANS* (*CO*) is expressed under long days but not short days. The loss of *CO* product does not alter when a plant flowers under short days, but delays flowering under long days. What happens is that the gene is positively regulated by cryptochrome that perceives blue light under long days. Cryptochrome appears to inhibit the inhibition of flowering by phytochrome B exposed to red light. Simply put, flowering is promoted by repressing a gene that represses flowering! *CO* is a transcription factor that turns on other genes which results in the expression of *LEAFY*. As discussed in the section on phase change, *LEAFY* is one of the key genes that "tells" a meristem to switch over to flowering. We will see that other pathways also converge on this important gene.

FIGURE 41.28
Manipulation of photoperiod in greenhouses ensures that short-day poinsettias flower in time for the winter holidays. Note that the colorful "petals" are actually sepals. Even after flowering is induced, there are many developmental events leading to the production of species-specific flowers.

The Flowering Hormone: Does It Exist? The Holy Grail in plant biology has been a flowering hormone, quested unsuccessfully for more than 50 years. A considerable amount of evidence demonstrates the existence of substances that promote flowering and substances that inhibit it. Grafting experiments have shown that these substances can move from leaves to shoots. The complexity of their interactions, as well as the fact that multiple chemical messengers are evidently involved, has made this scientifically and commercially interesting search very difficult, and to this day, the existence of a flowering hormone remains strictly hypothetical. We do know that *LEAFY* can be expressed in the vegetative as well as the reproductive portions of plants. Clearly, information about day length gathered by leaves is transmitted to shoot apices. Given that there are multiple pathways to flowering, several signals may be facilitating communication between leaves and shoots. We also know that roots can be a source of floral inhibitors affecting shoot development.

Temperature-Dependent Pathway

Lysenko solved the problem of winter wheat seed rotting in the fields in Russia by chilling the seeds and planting them in the spring. Planting in the spring prevented seeds from rotting in the soil over the long winter. Winter wheat would not flower without a period of chilling, called **vernalization.** Unfortunately a great many problems, including mistreatment of Russian geneticists, resulted from this scientifically significant discovery. Lysenko erroneously concluded that he had converted one species, winter wheat, to another, spring wheat, by simply altering the environment. There was a shift from science to politics. Genetics and Darwinian evolution were suspect for half a century. Social history aside, the valuable lesson here is that cold temperatures can accelerate or permit flowering in many species. As with light, this ensures that plants flower at more optimal times.

Vernalization may be necessary for seeds or plants in later stages of development. Analysis of mutants in *Arabidopsis* and pea indicate that vernalization is a separate flowering pathway that may be linked to the hormone gibberellin. In this pathway, repression may also lead to flowering. High levels of one of the gene products in the pathway may block the promotion of flowering by gibberellins. When plants are chilled, there is less of this gene product and gibberellin activity may increase. It is known that gibberellins enhance the expression of *LEAFY*. One proposal is that both the vernalization and autonomous pathways share a common intersection affecting gibberellin promotion of flowering. Weigel has shown that gibberellin actually binds the promoter of the *LEAFY* gene, so its effect on flowering is direct. The connection between gibberellin levels and temperature also needs to be understood.

Autonomous Pathway

The autonomous pathway to flowering is independent of external cues except for basic nutrition. Presumably this was the first pathway to evolve. Day-neutral plants often depend primarily on the autonomous pathway which allows plants to "count" and "remember." A field of day-neutral tobacco will produce a uniform number of nodes before flowering. If the shoots of these plants are removed at different positions, axillary buds will grow out and produce the same number of nodes as the removed portion of the shoot (figure 41.29). At a certain point in development shoots become committed or **determined** to flower (figure 41.30). The upper axillary buds of flowering tobacco will remember their position when rooted or grafted. The terminal shoot tip becomes florally determined about four nodes before it initiates a flower. In some other species, this commitment is less stable and/or occurs later.

How do shoots know where they are and at some point "remember" that information? It is clear that there are inhibitory signals from the roots. If bottomless pots are continuously placed over a growing tobacco plant and filled with soil, flowering is delayed by the formation of adventitious roots (figure 41.31). Control experiments with leaf removal show that it is the addition of roots and not the loss of leaves that delays flowering. A balance between floral promoting and inhibiting signals may regulate when

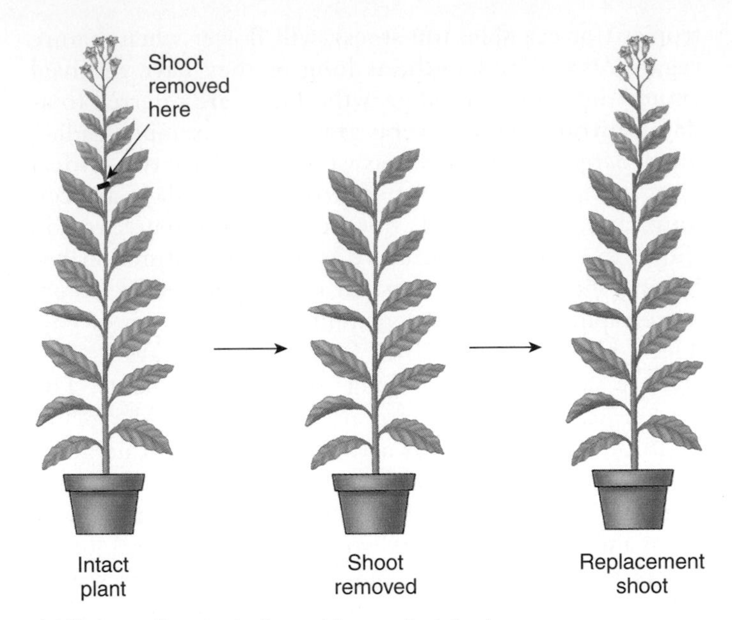

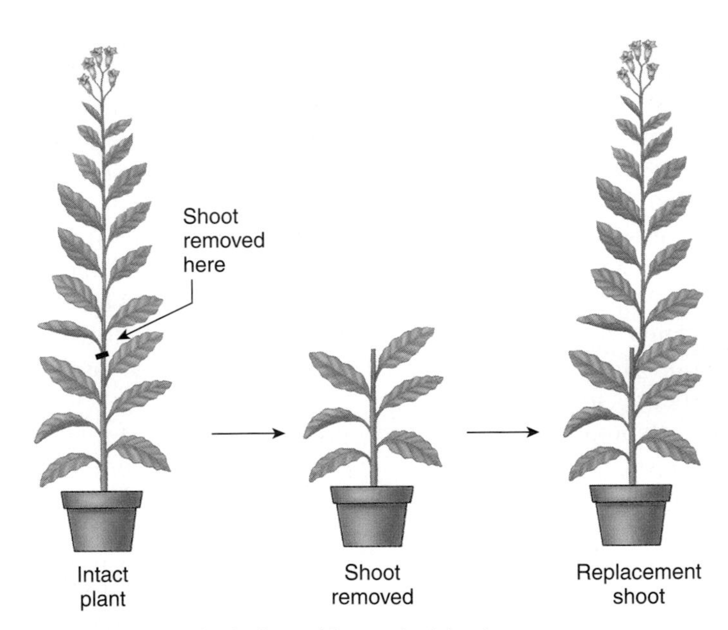

FIGURE 41.29
Plants can count. When axillary buds of flowering tobacco plants are released from apical dominance by removing the main shoot, they replace the number of nodes that were initiated by the main shoot (*a* and *b*). (After McDaniel 1996.)

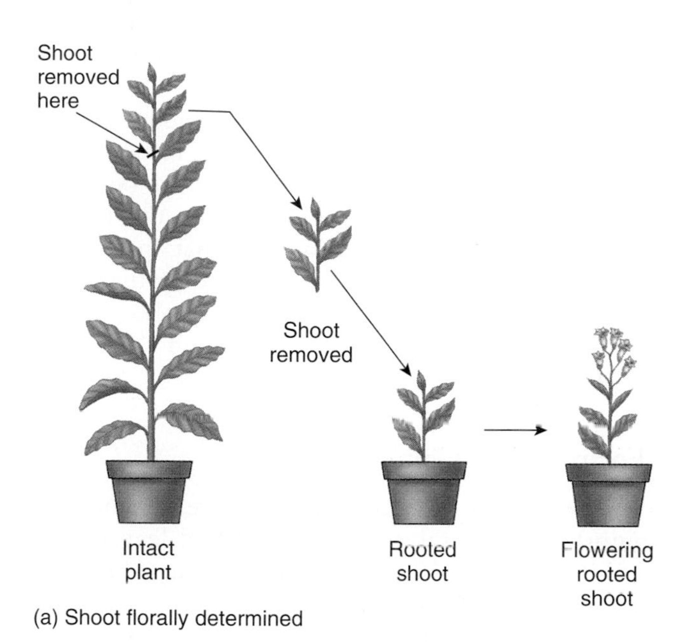

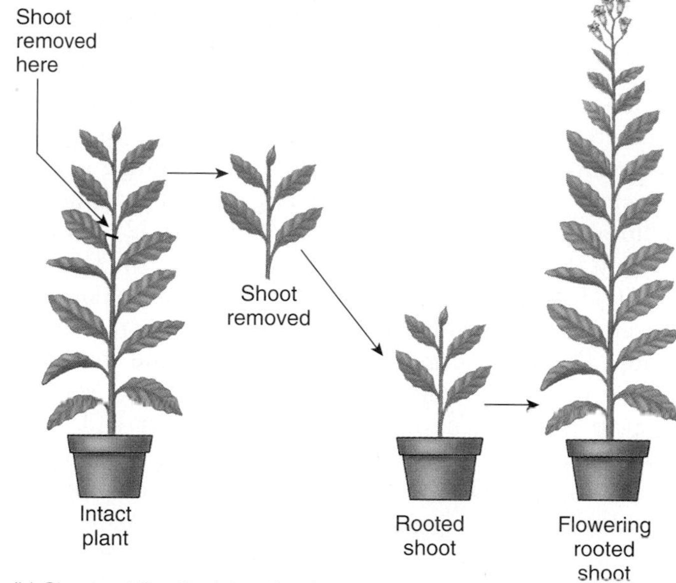

FIGURE 41.30
Plants can remember. At a certain point in the flowering process, shoots become committed to making a flower. This is called floral determination. (*a*) Florally determined shoots "remember" their position when rooted in a pot. That is, they produce the same number of nodes that they would have if they had grown out on the plant, and then they flower. (*b*) Those that are not yet florally determined cannot remember how many nodes they have left, so they start counting again. That is, they develop like a seedling and then flower. (After McDaniel 1996.)

flowering occurs in the autonomous pathway and the other pathways as well.

Determination for flowering is tested at the organ or whole plant level by changing the environment and ascertaining whether or not the fate has changed. How does floral determination correlate with molecular level changes? In *Arabidopsis*, floral determination correlates with the increase of *LEAFY* gene expression and has occurred by the time a second flowering gene, *APETALA1*, is expressed. Because all three flowering pathways appear to converge with increased levels of *LEAFY*, this determination event should occur in species with a variety of balances among the pathways.

Plants use light receptor molecules to measure the length of night. This information is then used to signal pathways that promote or inhibit flowering. Light receptors in the leaves trigger events that result in changes in the shoot meristem. Vernalization is the requirement for a period of chilling before a plant can flower. The autonomous pathway leads to flowering independent of environmental cues. Plants integrate information about position in regulating flowering and both promoters and inhibitors of flowering are important.

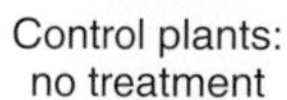

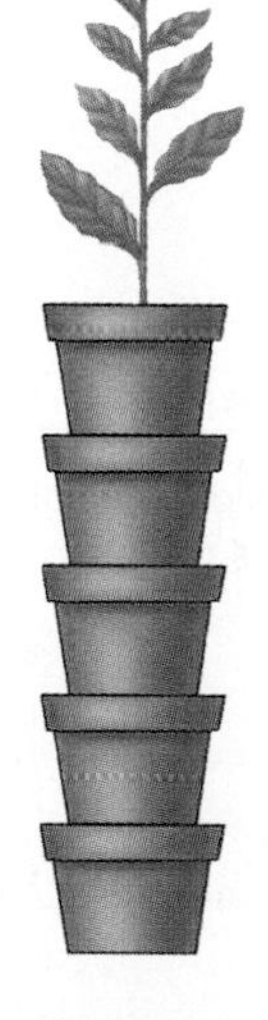

FIGURE 41.31
Roots can inhibit flowering. Adventitious roots formed as bottomless pots were continuously placed over growing tobacco plants, delaying flowering. The delay in flowering is caused by the roots, not the loss of the leaves. This was shown by removing leaves on control plants at the same time and in the same position as leaves on experimental plants that became buried as pots were added.

Identity Genes and the Formation of Floral Meristems and Floral Organs

Arabidopsis and snapdragon are valuable model systems for identifying flowering genes and understanding their interactions. The three pathways, discussed in the previous section, lead to an adult meristem becoming a floral meristem by either activating or repressing the inhibition of floral meristem identity genes (figure 41.32). Two of the key floral meristem identity genes are *LEAFY* and *APETALA1*. These genes establish the meristem as a flower meristem. They then turn on floral organ identity genes. The floral organ identity genes define four concentric whorls moving inward in the floral meristem as sepal, petal, stamen, and carpel. Meyerowitz and Coen proposed a model, called the ABC model, to explain how three classes of floral organ identity genes could specify four distinct organ types (figure 41.33). The ABC model proposes that three classes of organ identity genes (*A*, *B*, and *C*) specify the floral organs in the four floral whorls. By studying mutants the researchers have determined the following:

1. Class *A* genes alone specify the sepals.
2. Class *A* and class *B* genes together specify the petals.
3. Class *B* and class *C* genes together specify the stamens.
4. Class *C* genes alone specify the carpels.

The beauty of their ABC model is that it is entirely testable by making different combinations of floral organ identity mutants. Each class of genes is expressed in two whorls, yielding four different combinations of the gene products. When any one class is missing, there are aberrant floral organs in predictable positions.

It is important to recognize that this is actually only the beginning of the making of a flower. These organ identity genes are transcription factors that turn on many more genes that will actually give rise to the three-dimensional flower. There are also genes that "paint" the petals. Complex biochemical pathways lead to the accumulation of anthocyanin pigments in vacuoles. These pigments can be orange, red, or purple and the actual color is influenced by pH.

The Formation of Gametes

The ovule within the carpel has origins more ancient than the angiosperms. Floral parts are modified leaves, and within the ovule is the female gametophyte. This next generation develops from placental tissue in the ovary. A megaspore mother cell develops and meiotically gives rise to the embryo sac. Usually two layers of integument tissue form around this embryo sac and will become the seed coat. Genes responsible for the initiation of integuments and also those responsible for the formation of the integument have been identified. Some also affect leaf structure. This chapter has focused on the complex and elegant process that gives rise to the reproductive structure called the flower. It is indeed a metamorphosis, but the subtle shift from mitosis to meiosis in the megaspore mother cell leading to the development of a haploid, gamete-producing gametophyte is perhaps even more critical. The same can be said for pollen formation in the anther of the stamen. As we will see in the next chapter, the flower houses the haploid generations that will produce gametes. The flower also functions to increase the probability that male and female gametes from different (or sometimes the same plant) will unite.

Floral structures form as a result of floral meristem identity genes turning on floral organ identity genes which specify where sepals, petals, stamens, and carpels will form. This is followed by organ development which involves many complex pathways that account for floral diversity among species.

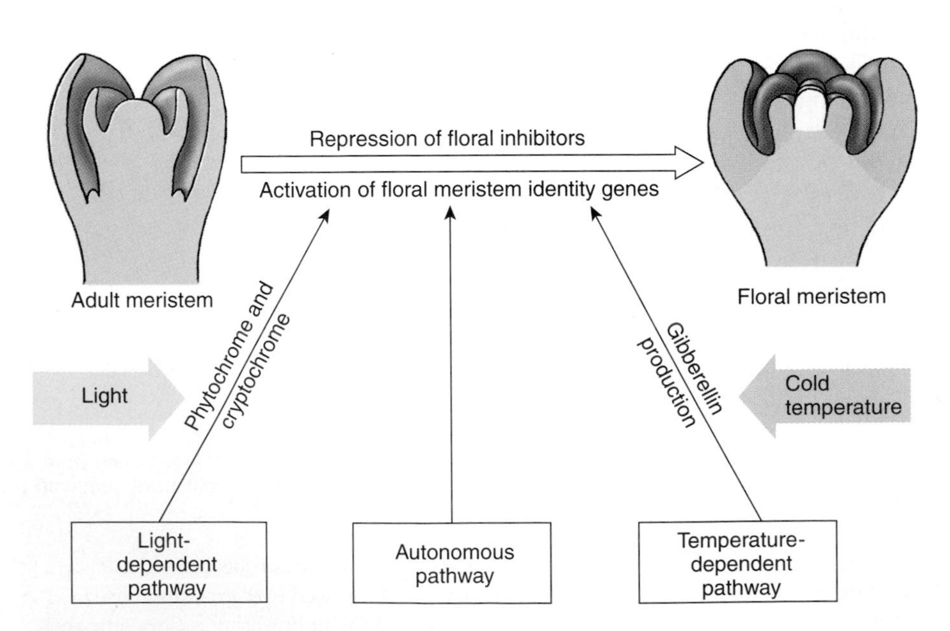

FIGURE 41.32
Model for flowering. The light-dependent, temperature-dependent, and autonomous flowering pathways promote the formation of floral meristems from adult meristems by repressing floral inhibitors and activating floral meristem identity genes.

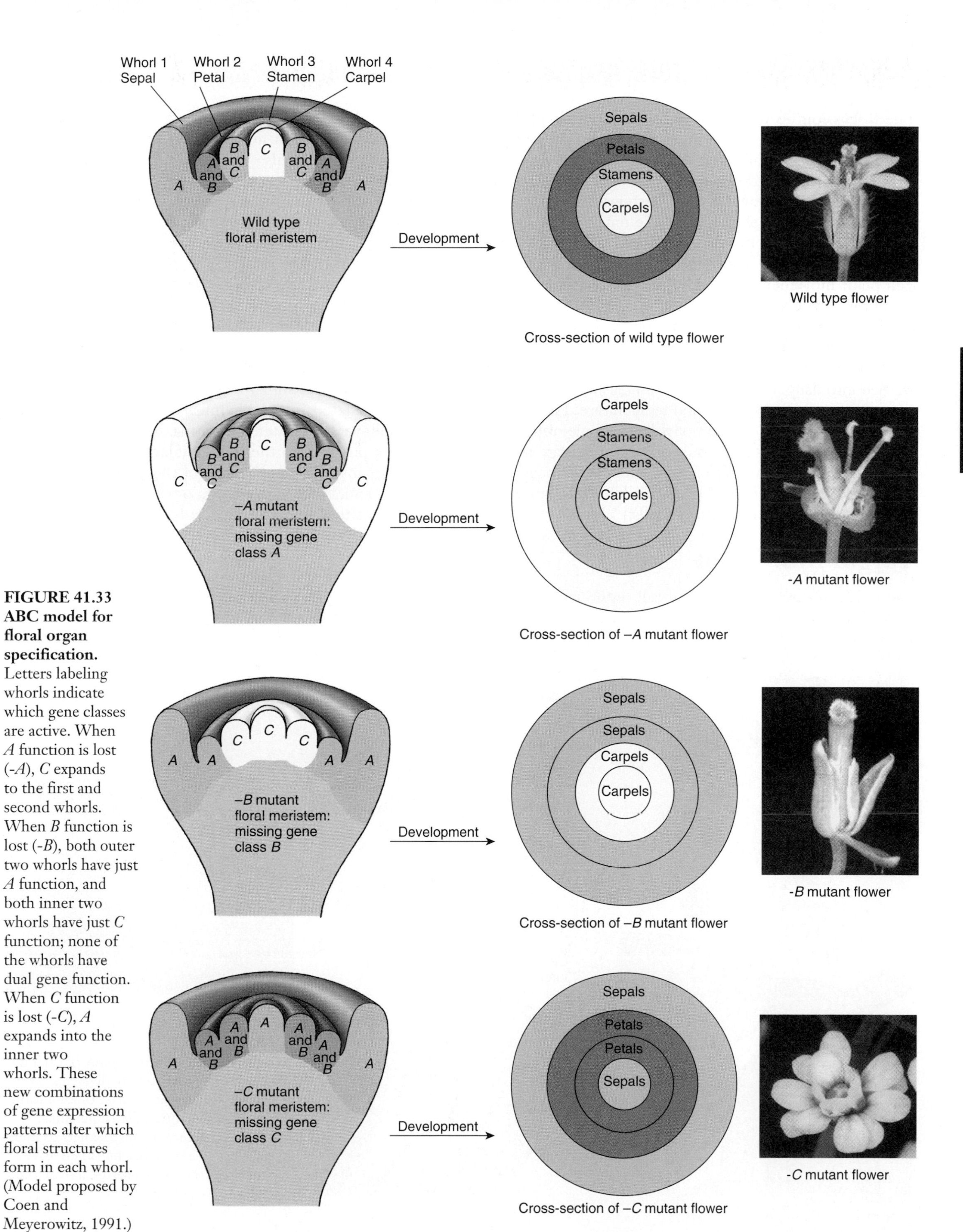

FIGURE 41.33 ABC model for floral organ specification. Letters labeling whorls indicate which gene classes are active. When *A* function is lost (-*A*), *C* expands to the first and second whorls. When *B* function is lost (-*B*), both outer two whorls have just *A* function, and both inner two whorls have just *C* function; none of the whorls have dual gene function. When *C* function is lost (-*C*), *A* expands into the inner two whorls. These new combinations of gene expression patterns alter which floral structures form in each whorl. (Model proposed by Coen and Meyerowitz, 1991.)

41.4 Many short-term responses to the environment do not require growth.

The discussion up to this point has focused on plant responses to environmental changes that involve growth and phase changes. Not all plant responses are this extensive. Larger predators, microbes, water, and wind often present a plant with rapid immediate stress. Response, to be effective, must also be immediate. There is little time for growth, and plants instead invoke a variety of other kinds of responses. Many environmental cues trigger rapid and reversible localized plant movements, for example. The rapid folding of leaves can startle a potential predator. Leaf folding can also prevent water loss by reducing the surface area available for transpiration. Some localized plant movements are triggered by unpredictable environmental signals. Other movements are tied into daily internal rhythms established by cyclic environmental signals like light and temperature. Plants lack a nervous system in the conventional sense. Some of the rapid signaling, however, is the result of electric charge moving through an organ as a wave of membrane ion exchange, not unlike that seen in animals. This is translated into movement by changing the turgor pressure of cells.

Turgor Movement

Turgor is pressure within a living cell resulting from diffusion of water into it, as discussed in chapter 39. If water leaves turgid cells (ones with turgor pressure), the cells may collapse, causing plant movement; conversely, water entering a limp cell may also cause movement as the cell once more becomes turgid.

Many plants, including those of the legume family (Fabaceae), exhibit leaf movements in response to touch or other stimuli. After exposure to a stimulus, the changes in leaf orientation are mostly associated with rapid turgor pressure changes in **pulvini** (singular: pulvinus), which are multicellular swellings located at the base of each leaf or leaflet. When leaves with pulvini, such as those of the sensitive plant *(Mimosa pudica)*, are stimulated by wind, heat, touch, or, in some instances, intense light, an electrical signal is generated. The electrical signal is translated into a chemical signal, with potassium ions, followed by water, migrating from the cells in one half of a pulvinus to the intercellular spaces in the other half. The loss of turgor in half of the pulvinus causes the leaf to "fold." The movements of the leaves and leaflets of the sensitive plant are especially rapid; the folding occurs within a second or two after the leaves are touched (figure 41.34). Over a span of about 15 to 30 minutes after the leaves and leaflets have folded, water usually diffuses back into the same cells from which it left, and the leaf returns to its original position.

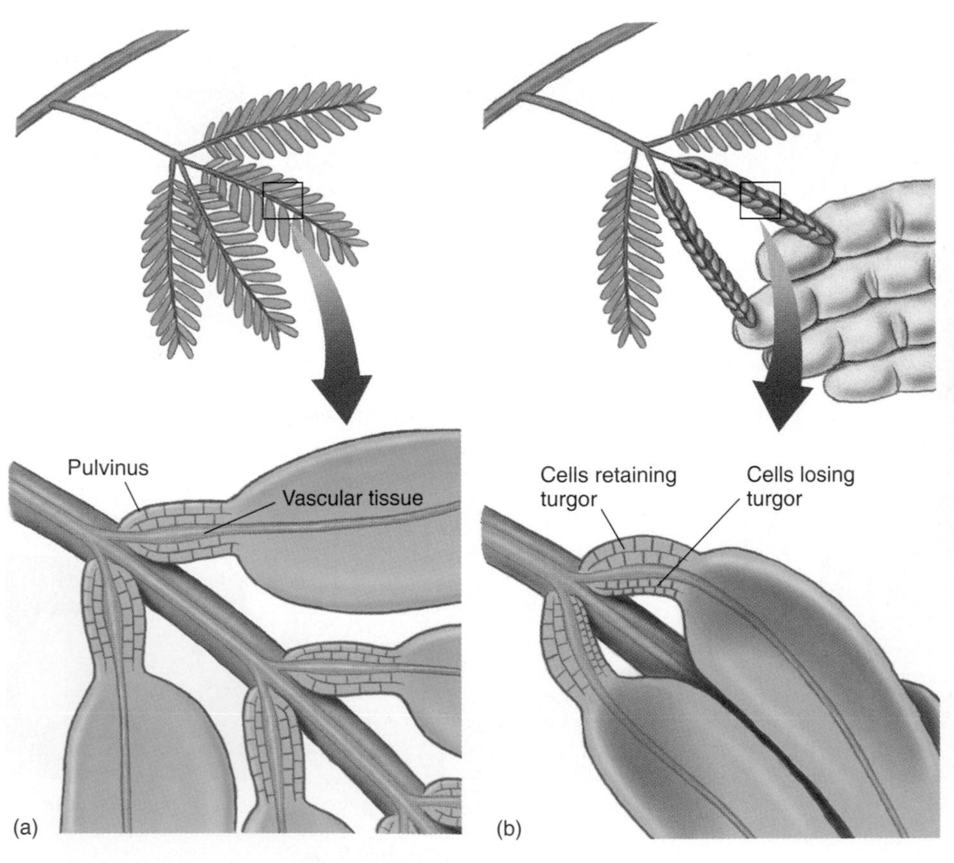

FIGURE 41.34
Sensitive plant *(Mimosa pudica)*. (*a*) The blades of *Mimosa* leaves are divided into numerous leaflets; at the base of each leaflet is a swollen structure called a pulvinus. (*b*) Changes in turgor cause leaflets to fold in response to a stimulus. (*c*) When leaves are touched (center two leaves above), they fold due to loss of turgor.

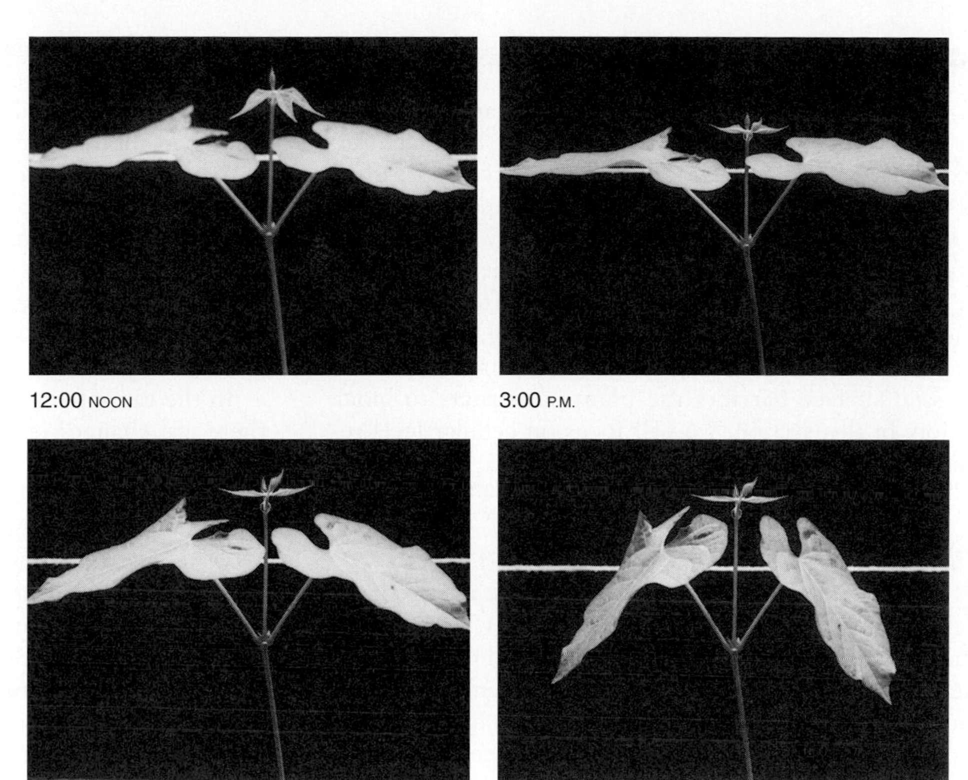

FIGURE 41.35
Sleep movements in bean leaves. In the bean plant, leaf blades are oriented horizontally during the day and vertically at night. The string behind the leaves indicates the horizontal plane.

The leaves of some plants with similar mechanisms may track the sun, with their blades oriented at right angles to it; how their orientation is directed is, however, poorly understood. Such leaves can move quite rapidly (as much as 15° an hour). This movement maximizes photosynthesis and is analogous to solar panels that are designed to track the sun.

Some of the most familiar of these reversible changes are seen in leaves and flowers that "open" during the day and "close" at night. For example, the flowers of four o'clocks open at 4 P.M. and evening primrose petals open at night. The blades of plant leaves also exhibit such a daily shift in position. The orientation of leaves and petals may be changed as a result of **turgor movements.** Bean leaves are horizontal during the day when their pulvini are turgid, but become more or less vertical at night as the pulvini lose turgor (figure 41.35). These sleep movements reduce water loss from transpiration during the night, but maximize photosynthetic surface area during the day. In these cases, the movement is closely tied to an internal rhythm.

Circadian Clocks

How do leaves know when to "sleep"? They have endogenous circadian clocks that set a rhythm with a period of about 24 hours (actually it is closer to 22 or 23 hours). While there are shorter and much longer, naturally occurring rhythms, circadian rhythms are particularly common and widespread because of the day-night cycle on earth. Jean de Mairan, a French astronomer, first identified circadian rhythms in 1729. He studied the sensitive plant which, in addition to having a touch response, closes its leaflets and leaves at night like the bean plant shown above. When de Mairan put the plants in total darkness, they continued "sleeping" and "waking" just as they did when exposed to night and day. This is one of four characteristics of a circadian rhythm: it must continue to run in the absence of external inputs. It must be about 24 hours in duration and it can be reset or entrained. (Perhaps you've experienced entrainment when traveling to a different time zone in the form of jet-lag recovery.) The fourth characteristic is that the clock can compensate for differences in temperature. This is quite unique when you consider what you know about biochemical reactions; most rates of reactions vary significantly as temperature changes. Circadian clocks exist in many organisms and appear to have evolved independently multiple times. The mechanism behind the clock is not fully understood, but is being actively investigated at the molecular level.

Turgor movements of plants are reversible and involve changes in the turgor pressure of specific cells. Circadian clocks are endogenous timekeepers that keep plant movements and other responses synchronized with the environment.

Plant Defense Responses

Interactions between plants and other organisms can be symbiotic (for example, nitrogen-fixing bacteria and mycorrhizae) or pathogenic. In evolutionary terms, these two types of interactions may simply be opposite sides of the same coin. The interactions have many common aspects and are the result of coevolution between two species that signal and respond to each other. In the case of pathogens, the microbe or pest is "winning," at least for that second in evolutionary time. In chapter 38, we discussed surface barriers the plant constructs to block invasion. In this section, we will focus on cellular level responses to attacks by microbes and animals.

Recognizing the Invader

Half a century ago, Flor proposed that there is a plant resistance gene (*R*) whose product interacts with that of a pathogen avirulence gene (*avr*). This is called the gene-for-gene model and several pairs of *avr* and *R* genes have been cloned in different species pathogenized by microbes, fungi, and insects, in one case. This has been motivated partially by the agronomic benefit of identifying genes that can be added to protect other plants from invaders. Much is now known about the signal transduction pathways that follow the recognition of the pathogen by the *R* gene. These pathways lead to the triggering of the hypersensitive response (HR) which leads to rapid cell death around the source of the invasion and also a longer-term resistance (figure 41.36). There is not always a gene-for-gene response, but plants still have defense responses to pathogens and also mechanical wounding. Some of the response pathways may be similar. Also, oligosaccharins in the cell walls may serve as recognition and signaling molecules.

While our focus is on invaders outside the plant kingdom, more is being learned about how parasitic plants invade other plants. There are specific molecules released from the root hairs of the host that the parasitic plant recognizes and responds to with invasive action. Less is known about the host response and so far the different defense genes that are activated appear to be ineffective.

Responding to the Invader

When a plant is attacked and there is gene-for-gene recognition, the HR response leads to very rapid cell death around the site of attack. This seals off the wounded tissue to prevent the pathogen or pest from moving into the rest of the plant. Hydrogen peroxide and nitric oxide are produced and may signal a cascade of biochemical events resulting in the localized death of host cells. They may also have negative effects on the pathogen, although antioxidant abilities have coevolved in pathogens. Other antimicrobial agents produced include the phytoalexins which are chemical defense agents. A variety of pathogenesis-related genes (*PR* genes) are expressed and their proteins can either function as antimicrobial agents or signals for other events that protect the plant.

In the case of virulent invaders (no *R* gene recognition), there are changes in local cell walls that at least partially block the movement of the pathogen or pest farther into the plant. In this case there is not an HR response and the local plant cells are not suicidal.

When an insect takes a mouthful of a leaf, defense responses are also triggered. Mechanical damage causes responses that have some similar components, but the reaction may be slower. Biochemically, it is distinct from some of the events triggered by signals in the insect's mouth. Such responses are collectively called wound responses. Wound responses present a challenge in designing other types of experiments with plants that involve cutting or otherwise mechanically damaging the tissue. It is important to run control experiments to be sure you are answering your question and not observing a wound response.

Preparing for Future Attacks

In addition to the HR or other local responses, plants are capable of a systemic response to a pathogen or pest attack. This is called a systemic acquired resistance (SAR). Several pathways lead to broad-ranging resistance that lasts for a period of days. The signals that induce SAR include salicylic acid and jasmonic acid. SAR allows the plant to respond more quickly if it is attacked again. However, this is not the same as the human immune response where antibodies (proteins) that recognize specific antigens (foreign proteins) persist in the body. SAR is neither as specific or long lasting.

Plants defend themselves from invasion in ways reminiscent of the animal immune system. When an invader is recognized, localized cell death seals off the infected area.

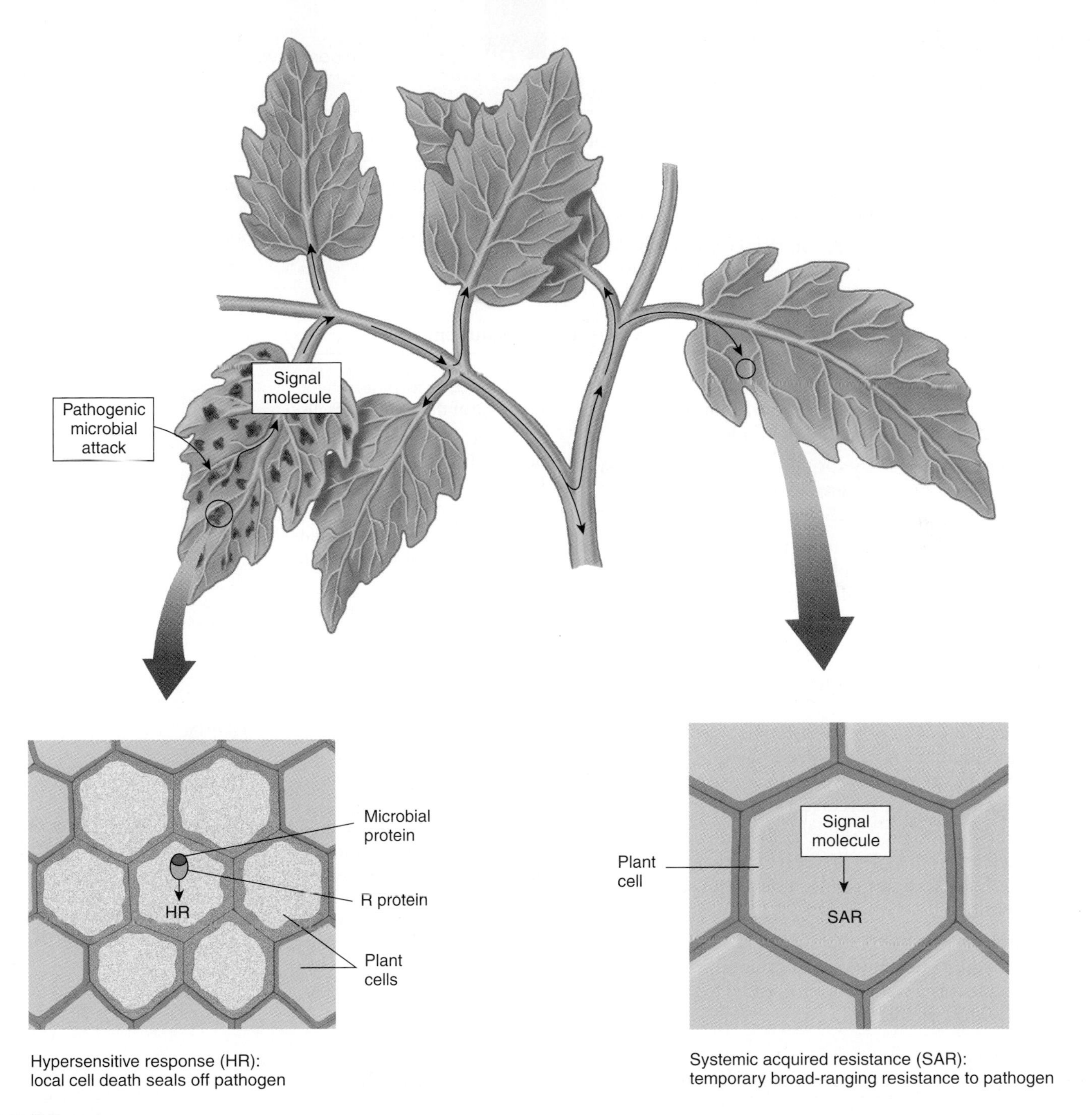

FIGURE 41.36

Plant defense responses. In the gene-for-gene response, a cascade of events is triggered leading to local cell death (HR response) and longer-term resistance in the rest of the plant (SAR).

Chapter 41

www.mhhe.com/raven6/resources41.mhtml

Summary | **Questions** | **Media Resources**

41.1 Plant growth is often guided by environmental cues.

- Tropisms in plants are directional growth responses to external stimuli, such as light, gravity, or contact.
- Dormancy is a plant adaptation that carries a plant through unfavorable seasons or periods of drought.

1. In general, which part of a plant is positively phototropic? What is the adaptive significance of this reaction?

- Plant Movement

41.2 The hormones that guide growth are keyed to the environment.

- Auxin in the shoot tip migrates away from light and promotes the elongation of plant cells on the dark side, causing stems to bend in the direction of light.
- Cytokinins are necessary for mitosis and cell division in plants. They promote growth of lateral buds and inhibit formation of lateral roots.
- Gibberellins, along with auxin, play a major role in stem elongation in most plants. They also tend to hasten the germination of seeds and to break dormancy in buds.

2. How does auxin affect the plasticity of the plant cell walls?

3. Where are most cytokinins produced? From what biomolecule do cytokinins appear to be derived?

4. What plant hormones could be lacking in genetically dwarfed plants?

- Hormones

- Student Research: Plant Growth Responses to Environmental Stress

- Art Quiz: Plant Hormones

41.3 The environment influences flowering.

- The transition of a shoot meristem from vegetative to adult development is called phase change. During phase change, plants gain competence to produce a floral signal(s) and or perceive a signal.
- The light-dependent pathway uses information from light receptor molecules integrated with a biological clock to determine if the length of night is sufficient for flowering.
- The autonomous path functions independently of environmental cues. Internal floral inhibitor(s) from roots and leaves and floral promoter(s) from leaves move through the plant.

5. A plant has undergone phase change. Although it is an adult, it does not flower. How might you get this plant to flower?

6. You have recently moved from Canada to Mexico and brought some seeds from your favorite plants. They germinate and produce beautiful leaves, but never flower. What went wrong?

- Photoperiod

- Student Research: Selection in Flowering Plants

- Art Quiz: Flowering Responses to Day Length

41.4 Many short-term responses to the environment do not require growth.

- Changes in turgor pressure reflect responses to environmental signals that can protect plants from predation, and regulate stomatal opening, among other things.
- Other reversible movements in plants are caused by changes in turgor pressure that are regulated by internal circadian rhythms.
- Plants have the ability to recognize and respond to invaders at the cellular, tissue/organ, and whole plant levels.

7. What happens in the cells of the sensitive plant (*Mimosa pudica*) when its leaves are touched?

8. In what ways can a plant protect itself from pathogenic microbes? From animals?

- Introduction to Plants

42

Plant Reproduction

Concept Outline

42.1 Angiosperms have been incredibly successful, in part, because of their reproductive strategies.

Rise of the Flowering Plants. Outcrossing resulting from animal and wind dispersal of pollen increases genetic variability in a species. Seed and fruit dispersal mechanisms allow offspring to colonize distant regions. Other features such as shortened life cycles may also have been responsible for the rapid diversification of the flowering plants.
Evolution of the Flower. A complete flower has four whorls, containing protective sepals, attractive petals, male stamens, and female carpels.

42.2 Flowering plants use animals or wind to transfer pollen between flowers.

Formation of Angiosperm Gametes. The male gametophytes are the pollen grains, and the female gametophyte is the embryo sac.
Pollination. Evolutionary modifications of flowers have enhanced effective pollination.
Self-Pollination. Self-pollination is favored in stable environments, but outcrossing enhances genetic variability.
Fertilization. Angiosperms use two sperm cells, one to fertilize the egg, the other to produce a nutrient tissue called endosperm.

42.3 Many plants can clone themselves by asexual reproduction.

Asexual Reproduction. Some plants do without sexual reproduction, instead cloning new individuals from parts of themselves.

42.4 How long do plants and plant organs live?

The Life Span of Plants. Clonal plants can live indefinitely. Parts of plants senesce and die. Some plants (annuals and biennials) reproduce sexually only once and die. Perennials continue to grow and flower year after year.

FIGURE 42.1
Reproductive success in flowering plants. Unique reproductive systems and strategies have coevolved between plants and animals, accounting for almost 250,000 species of flowering plants inhabiting all but the harshest environments on earth.

The remarkable evolutionary success of flowering plants can be linked to their reproductive strategies (figure 42.1). The evolution and development of flowers has been discussed in chapters 37 and 41. Here we explore reproductive strategies in the angiosperms and how their unique features, flowers and fruits, have contributed to their success. This is, in part, a story of coevolution between plants and animals that ensures greater genetic diversity by dispersing plant gametes widely. However, in a stable environment, there are advantages to maintaining the status quo genetically. Asexual reproduction is a strategy to clonally propagate individuals. An unusual twist to sexual reproduction in some flowering plants is that senescence and death of the parent plant follows.

42.1 Angiosperms have been incredibly successful, in part, because of their reproductive strategies.

Rise of the Flowering Plants

Most of the plants we see daily are angiosperms. The 250,000 species of flowering plants range in size from almost microscopic herbs to giant *Eucalyptus* trees, and their form varies from cacti, grasses, and daisies to aquatic pondweeds. Most shrubs and trees (other than conifers and *Ginkgo*) are also in this phylum. This chapter focuses on reproduction in angiosperms (figure 42.2) because of their tremendous success and many uses by humans. Virtually all

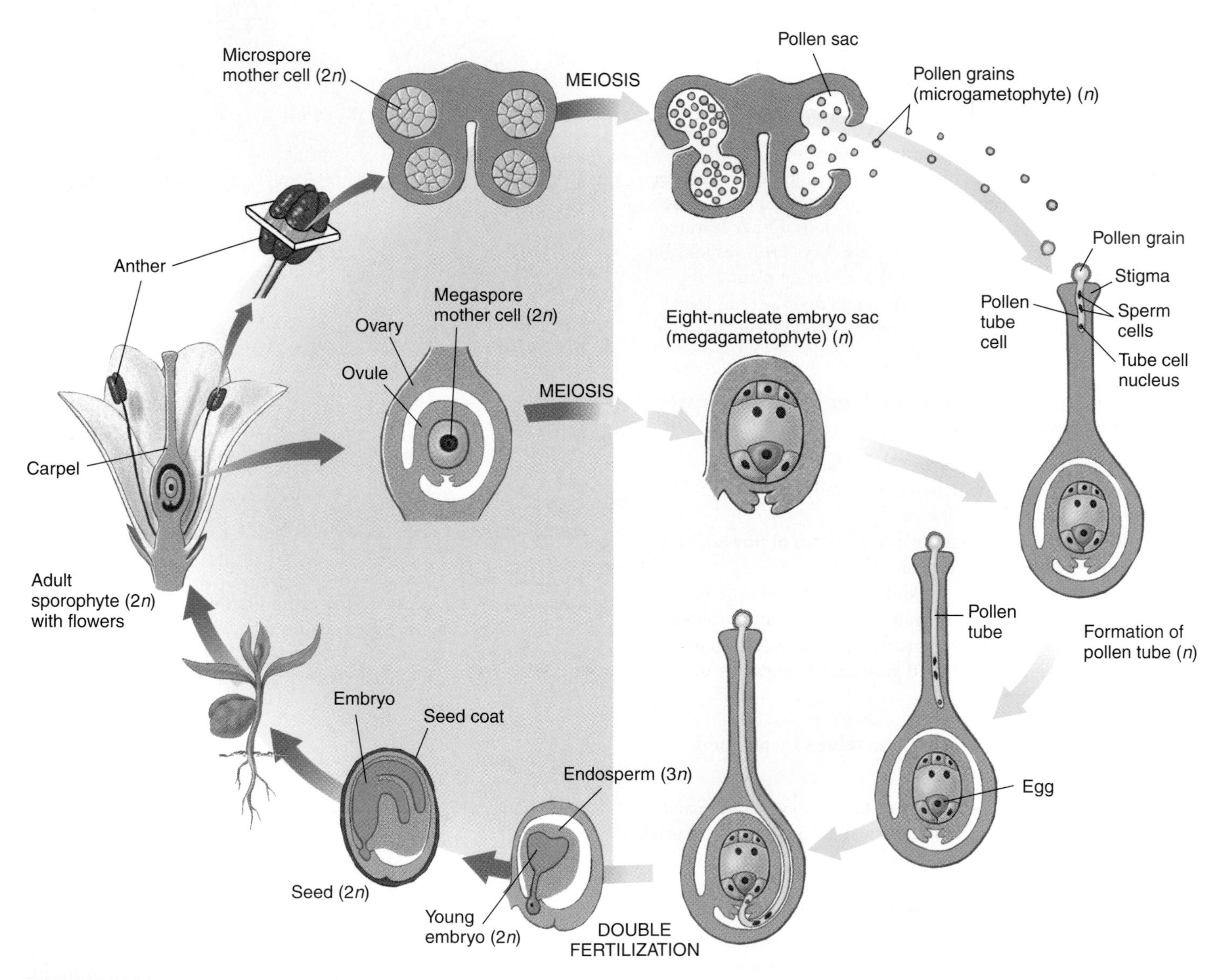

FIGURE 42.2
Angiosperm life cycle. One egg forms within the embryo sac inside the ovule, which, in turn, is enclosed in the carpel. The pollen grains, meanwhile, are formed within the sporangia of the anthers and are shed. Fertilization is a double process. A sperm and an egg come together, producing a zygote; at the same time, another sperm fuses with the polar nuclei to produce the endosperm. The endosperm is the tissue, unique to angiosperms, that nourishes the embryo and young plant.

of our food is derived, directly or indirectly, from flowering plants; in fact, more than 90% of the calories we consume come from just over 100 species. Angiosperms are also sources of medicine, clothing, and building materials. While the other plant phyla also provide resources, they are outnumbered seven to one by the angiosperms. For example, there are only about 750 extant gymnosperm species!

Why Are the Angiosperms Successful?

When flowering plants originated, Africa and South America were still connected to each other, as well as to Antarctica and India, and, via Antarctica, to Australia and New Zealand (figure 42.3). These landmasses formed the great continent known as Gondwanaland. In the north, Eurasia and North America were united, forming another supercontinent called Laurasia. The huge landmass formed by the union of South America and Africa was just south of the equator and probably had a climate characterized by extreme temperatures and aridity in its interior. Similar climates occur in the interiors of major continents at present. Much of the early evolution of angiosperms may have taken place in patches of drier and less favorable habitat found in the interior of Gondwanaland. Many features of flowering plants seem to correlate with successful growth under arid and semiarid conditions.

The transfer of pollen between flowers of separate plants, sometimes over long distances, ensures *outcrossing* (cross-pollination between individuals of the same species) and may have been important in the early success of angiosperms. The various means of effective fruit dispersal that evolved in the group were also significant in the success of angiosperms (see chapter 40). The rapid life cycle of some of the angiosperms (*Arabidopsis* can go from seed to adult flowering plant in 24 days) was another factor. Asexual reproduction has given many invasive species a competitive edge. Xylem vessels and other anatomical and morphological features of the angiosperms correlate with their biological success. As early angiosperms evolved, all of these advantageous features became further elaborated and developed, and the pace of their diversification accelerated.

The Rise to Dominance

Angiosperms began to dominate temperate and tropical terrestrial communities about 80 to 90 million years ago, during the second half of the Cretaceous Period. We can document the relative abundance of different groups of plants by studying fossils that occur at the same time and place. In rocks more than 80 million years old, the fossil remains of plant phyla other than angiosperms, including lycopods, horsetails, ferns, and gymnosperms, are most common. Angiosperms arose in temperate and tropical terrestrial communities in a relatively short time.

At about the time that angiosperms became abundant in the fossil record, pollen, leaves, flowers, and fruits of some families that still survive began to appear. For example, representatives of the magnolia, beech, and legume families, which were in existence before the end of the Cretaceous Period (65 million years ago), are alive and flourishing today.

A number of insect orders that are particularly associated with flowers, such as Lepidoptera (butterflies and moths) and Diptera (flies), appeared or became more abundant during the rise of angiosperms. Plants and insects have clearly played a major role in each other's patterns of evolution, and their interactions continue to be of fundamental importance. Additional animals, including birds and mammals, now assist in pollination and seed dispersal.

By 80 to 90 million years ago, angiosperms were dominant in terrestrial habitats throughout the world.

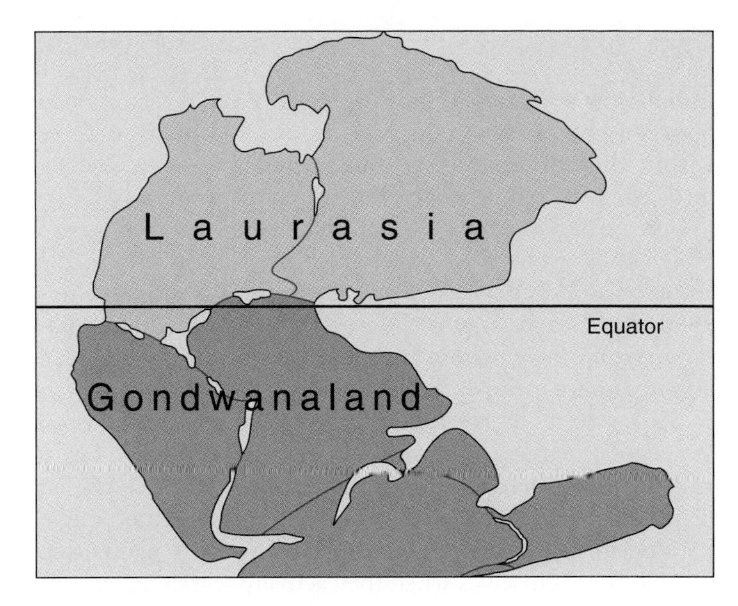

FIGURE 42.3
The alignment of the continents when the angiosperms first appeared in the fossil record about 130 million years ago. Africa, Madagascar, South America, India, Australia, and Antarctica were all connected and part of the huge continent of Gondwanaland, which eventually separated into the discrete landmasses we have today.

Evolution of the Flower

Pollination in angiosperms does not involve direct contact between the pollen grain and the ovule. Pollen matures within the anthers and is transported, often by insects, birds, or other animals, to the stigma of another flower. When pollen reaches the stigma, it germinates, and a pollen tube grows down, carrying the sperm nuclei to the embryo sac. After double fertilization takes place, development of the embryo and endosperm begins. The seed matures within the ripening fruit; the germination of the seed initiates another life cycle.

Successful pollination in many angiosperms depends on the regular attraction of **pollinators** such as insects, birds, and other animals, so that pollen is transferred between plants of the same species. When animals disperse pollen, they perform the same functions for flowering plants that they do for themselves when they actively search out mates. The relationship between plant and pollinator can be quite intricate. Mutations in either partner can block reproduction. If a plant flowers at the "wrong" time, the pollinator may not be available. If the morphology of the flower or pollinator is altered, there may be physical barriers to pollination. Clearly floral morphology has coevolved with pollinators and the result is much more complex and diverse than the initiation of four distinct whorls of organs described in chapter 41.

Characteristics of Floral Evolution

The evolution of the angiosperms is a focus of chapter 37. Here we need to keep in mind that the diversity of angiosperms is partly due to the evolution of a great variety of floral phenotypes that may enhance the effectiveness of pollination. All floral organs are thought to have evolved from leaves. In some early angiosperms, these organs maintain the spiral phyllotaxy often found in leaves. The trend has been toward four distinct whorls. A *complete flower* has four whorls of parts (calyx, corolla, androecium, and gynoecium), while an *incomplete flower* lacks one or more of the whorls (figure 42.4).

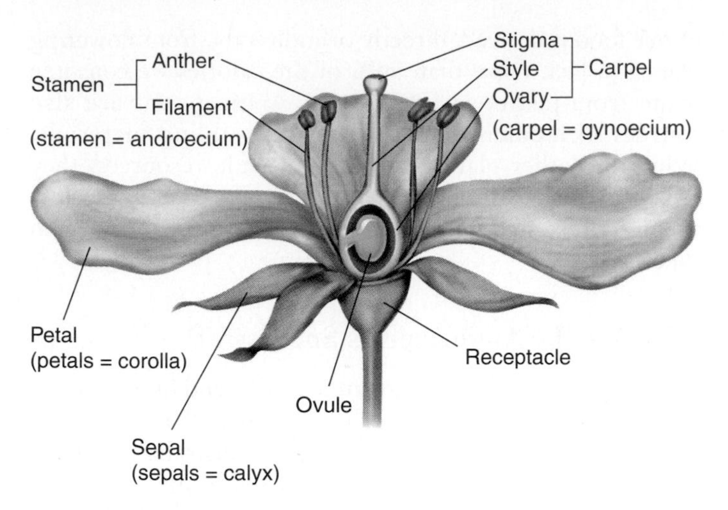

FIGURE 42.4
Structure of an angiosperm flower.

In both complete and incomplete flowers, the **calyx** usually constitutes the outermost whorl; it consists of flattened appendages, called **sepals,** which protect the flower in the bud. The petals collectively make up the **corolla** and may be fused. Petals function to attract pollinators. While these two outer whorls of floral organs are sterile, they can enhance reproductive success.

Androecium (from the Greek *andros*, "man", + *oikos*, "house") is a collective term for all the **stamens** (male structures) of a flower. Stamens are specialized structures that bear the angiosperm microsporangia. There are similar structures bearing the microsporangia in the pollen cones of gymnosperms. Most living angiosperms have stamens whose **filaments** ("stalks") are slender and often threadlike, and whose four microsporangia are evident at the apex in a swollen portion, the **anther.** Some of the more primitive angiosperms have stamens that are flattened and leaflike, with the sporangia produced from the upper or lower surface.

The **gynoecium** (from the Greek *gyne*, "woman," + *oikos*, "house") is a collective term for all the female parts of a flower. In most flowers, the gynoecium, which is unique to angiosperms, consists of a single **carpel** or two or more fused carpels. Single or fused carpels are often referred to as simple or compound pistils, respectively. Most flowers with which we are familiar—for example, those of tomatoes and oranges—have a compound pistil. In other less specialized flowers—for example, buttercups and stonecups—there may be several to many separate, simple pistils, each formed from a single carpel. **Ovules** (which develop into seeds) are produced in the pistil's swollen lower portion, the **ovary,** which usually narrows at the top into a slender, necklike **style** with a pollen-receptive **stigma** at its apex. Sometimes the stigma is divided, with the number of stigma branches indicating how many carpels are in the particular pistil. Carpels are essentially inrolled floral leaves with ovules along the margins. It is possible that the first carpels were leaf blades that folded longitudinally; the margins, which had hairs, did not actually fuse until the fruit developed, but the hairs interlocked and were receptive to pollen. In the course of evolution, there is evidence the hairs became localized into a stigma, a style was formed, and the fusing of the carpel margins ultimately resulted in a pistil. In many modern flowering plants, the carpels have become highly modified and are not visually distinguishable from one another unless the pistil is cut open.

Trends of Floral Specialization

Two major evolutionary trends led to the wide diversity of modern flowering plants: (1) separate floral parts have grouped together, or fused, and (2) floral parts have been lost or reduced (figure 42.5). In the more advanced angiosperms, the number of parts in each whorl has often been reduced from many to few. The spiral patterns of attachment of all floral parts in primitive angiosperms have, in the course of evolution, given way to a single whorl at each level. The central axis of many flowers has shortened, and the whorls are close to one another. In some evolutionary lines, the members of one or more whorls have fused with one another, sometimes joining into a tube. In other kinds of flowering plants, different whorls may be fused together. Whole whorls may even be lost from the flower, which may lack sepals, petals, stamens, carpels, or various combinations of these structures. Modifications often relate to pollination mechanisms and, in some cases like the grasses, wind has replaced animals for pollen dispersal.

While much floral diversity is the result of natural selection related to pollination, it is important to recognize the impact breeding (artificial selection) has had on flower morphology. Humans have selected for practical or aesthetic traits that may have little adaptive value to species in the wild. Maize (corn), for example, has been selected to satisfy the human palate. Human intervention ensures the reproductive success of each generation; while in a natural setting modern corn would not have the same protection from herbivores as its ancestors, and the fruit dispersal mechanism would be quite different (see figure 21.13). Floral shops sell heavily bred species with modified petals, often due to polyploidy, that enhance their economic value, but not their ability to attract pollinators. In making inferences about symbioses between flowers and pollinators, be sure to look at native plants that have not been genetically altered by human intervention.

FIGURE 42.5
Trends in floral specialization. Wild geranium, *Geranium maculatum*. The petals are reduced to five each, the stamens to ten.

FIGURE 42.6
Bilateral symmetry in an orchid. While primitive flowers are usually radially symmetrical, flowers of many advanced groups, such as the orchid family (Orchidaceae), are bilaterally symmetrical.

Trends in Floral Symmetry

Other trends in floral evolution have affected the symmetry of the flower (figure 42.6). Primitive flowers such as those of buttercups are *radially symmetrical;* that is, one could draw a line anywhere through the center and have two roughly equal halves. Flowers of many advanced groups are *bilaterally symmetrical;* that is, they are divisible into two equal parts along only a single plane. Examples of such flowers are snapdragons, mints, and orchids. Such bilaterally symmetrical flowers are also common among violets and peas. In these groups, they are often associated with advanced and highly precise pollination systems. Bilateral symmetry has arisen independently many times. In snapdragons, the *cycloidia* gene regulates floral symmetry, and in its absence flowers are more radial (figure 42.7). Here the experimental alteration of a single gene is sufficient to cause a dramatic change in morphology. Whether the same gene or functionally similar genes arose naturally in parallel in other species is an open question.

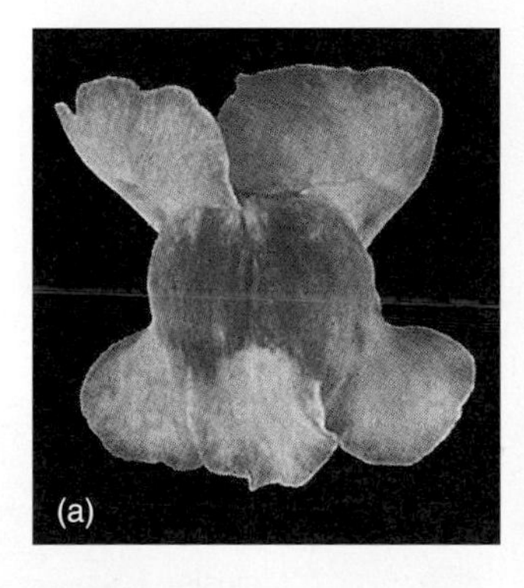

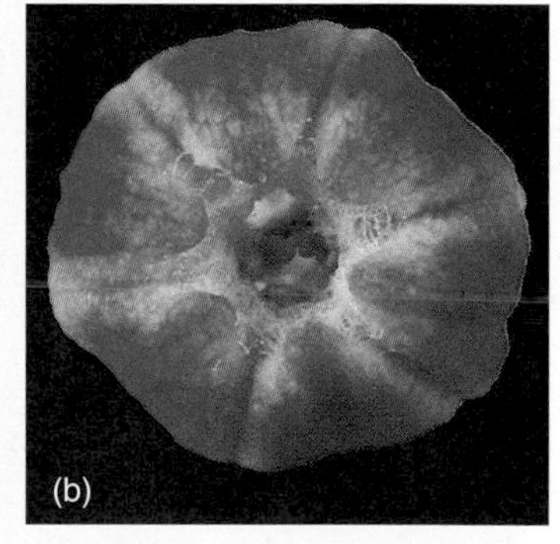

FIGURE 42.7
Genetic regulation of asymmetry in flowers. (*a*) Snapdragon flowers normally have bilateral symmetry. (*b*) The *cycloidia* gene regulates floral symmetry, and *cycloidia* mutant snapdragons have radially symmetrical flowers.

Some of the first angiosperms likely had numerous free, spirally arranged flower parts. Modification of floral parts appears to be closely tied to pollination mechanisms. More recently, horticulturists have bred plants for aesthetic reasons resulting in an even greater diversity of flowers.

Formation of Angiosperm Gametes

Reproductive success depends on uniting the gametes (egg and sperm) found in the embryo sacs and pollen grains of flowers. As mentioned previously, plant sexual life cycles are characterized by an alternation of generations, in which a diploid sporophyte generation gives rise to a haploid gametophyte generation. In angiosperms, the gametophyte generation is very small and is completely enclosed within the tissues of the parent sporophyte. The male gametophytes, or microgametophytes, are **pollen grains.** The female gametophyte, or megagametophyte, is the **embryo sac.** Pollen grains and the embryo sac both are produced in separate, specialized structures of the angiosperm flower.

Like animals, angiosperms have separate structures for producing male and female gametes (figure 42.8), but the reproductive organs of angiosperms are different from those of animals in two ways. First, in angiosperms, both male and female structures usually occur together in the same individual flower with some exceptions. Second, angiosperm reproductive structures are not permanent parts of the adult individual. Angiosperm flowers and reproductive organs develop seasonally, at times of the year most favorable for pollination. In some cases, reproductive structures are produced only once and the parent plant dies. It is significant that the germ line for angiosperms is not set aside early in development, but forms quite late, as detailed in chapter 40.

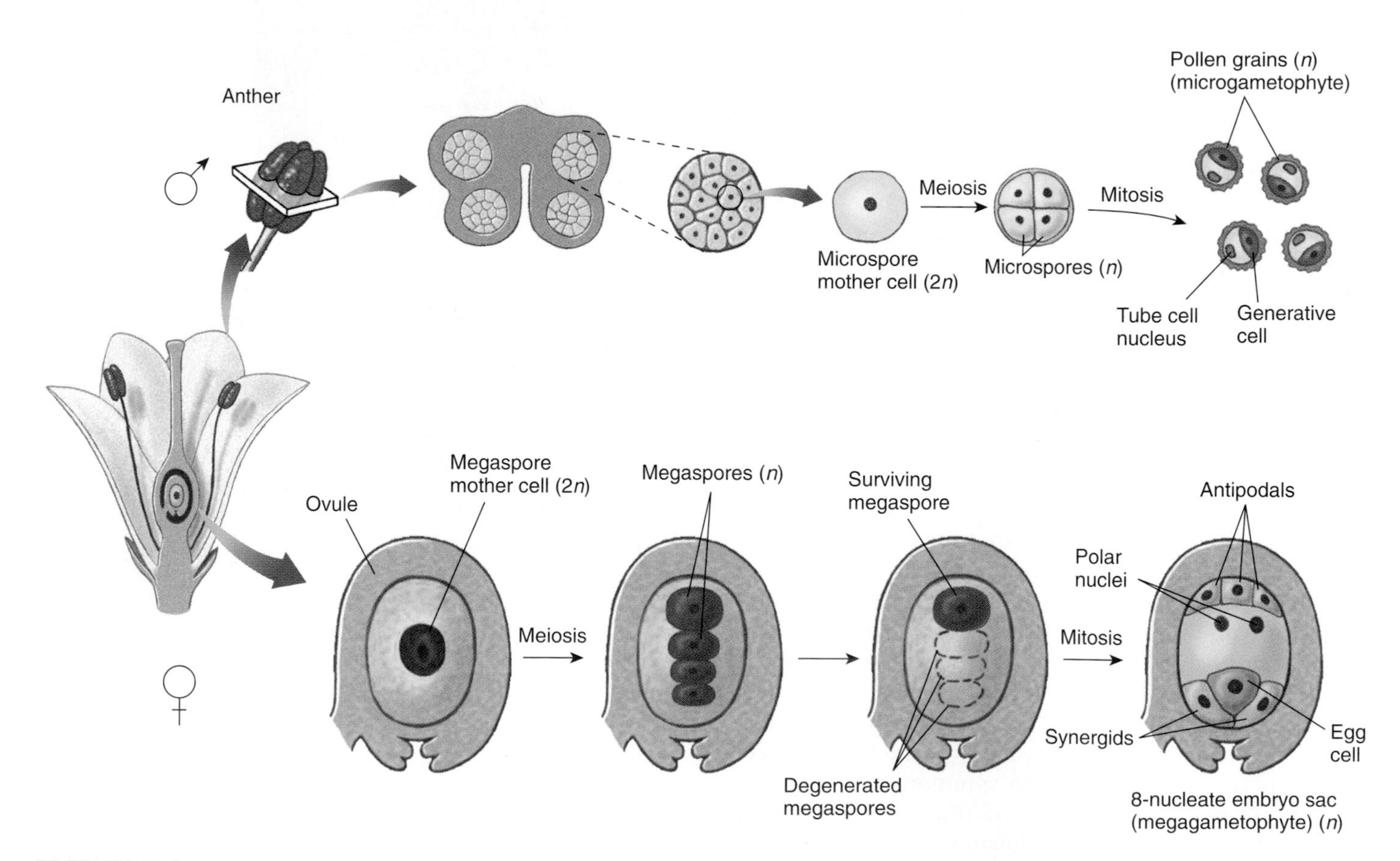

FIGURE 42.8
Formation of pollen grains and the embryo sac. Diploid (2*n*) microspore mother cells are housed in the anther and divide by meiosis to form four haploid (*n*) microspores. Each microspore develops by mitosis into a pollen grain. The generative cell within the pollen grain will later divide to form two sperm cells. Within the ovule, one diploid megaspore mother cell divides by meiosis to produce four haploid megaspores. Usually only one of the megaspores will survive, and the other three will degenerate. The surviving megaspore divides by mitosis to produce an embryo sac with eight nuclei.

FIGURE 42.9
Pollen grains. (*a*) In the Easter lily, *Lilium candidum*, the pollen tube emerges from the pollen grain through the groove or furrow that occurs on one side of the grain. (*b*) In a plant of the sunflower family, *Hyoseris longiloba*, three pores are hidden among the ornamentation of the pollen grain. The pollen tube may grow out through any one of them.

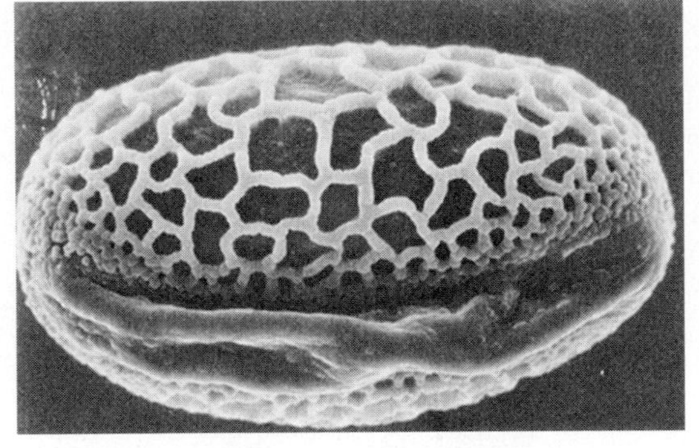
(a)

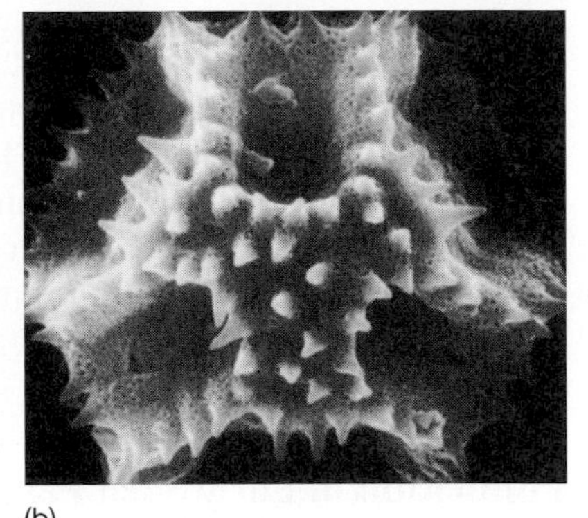
(b)

Pollen Formation

Pollen grains form in the two pollen sacs located in the anther. Each pollen sac contains specialized chambers in which the *microspore mother cells* are enclosed and protected. The microspore mother cells undergo meiosis to form four haploid microspores. Subsequently, mitotic divisions form four pollen grains. Inside each pollen grain is a generative cell; this cell will later divide to produce two sperm cells.

Pollen grain shapes are specialized for specific flower species. As discussed in more detail later in the chapter, fertilization requires that the pollen grain grow a tube that penetrates the style until it encounters the ovary. Most pollen grains have a furrow from which this pollen tube emerges; some grains have three furrows (figure 42.9).

Embryo Sac Formation

Eggs develop in the ovules of the angiosperm flower. Within each ovule is a megaspore mother cell. Each megaspore mother cell undergoes meiosis to produce four haploid megaspores. In most plants, only one of these megaspores, however, survives; the rest are absorbed by the ovule. The lone remaining megaspore undergoes repeated mitotic divisions to produce eight haploid nuclei that are enclosed within a seven-celled embryo sac. Within the embryo sac, the eight nuclei are arranged in precise positions. One nucleus is located near the opening of the embryo sac in the egg cell. Two are located in a single cell in the middle of the embryo sac and are called polar nuclei; two nuclei are contained in cells called synergids that flank the egg cell; and the other three nuclei reside in cells called the antipodals, located at the end of the sac, opposite the egg cell (figure 42.10). The first step in uniting the sperm in the pollen grain with the egg and polar nuclei is to get pollen germinating on the stigma of the carpel and growing toward the embryo sac.

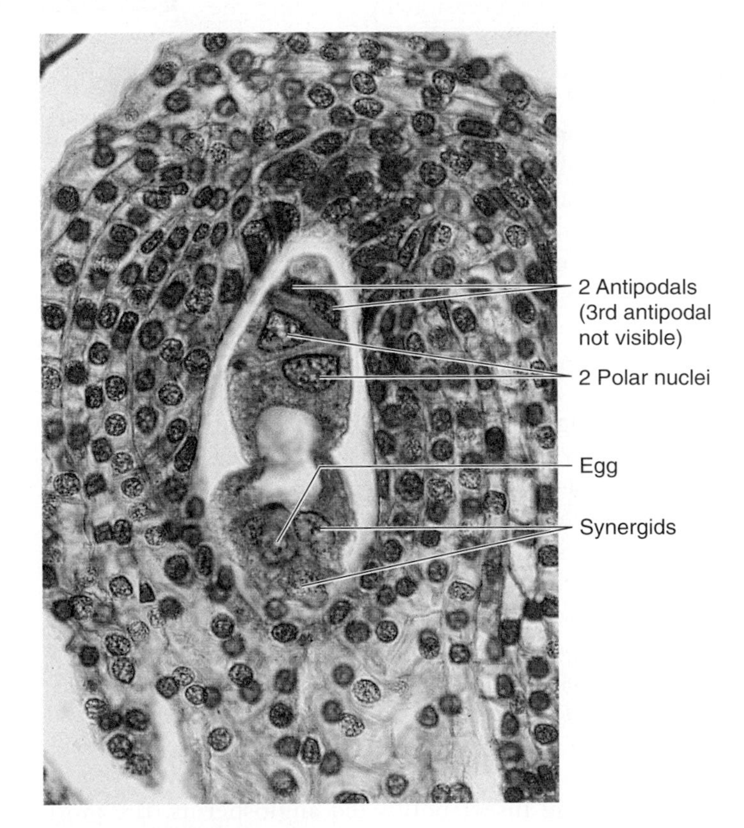

FIGURE 42.10
A mature embryo sac of a lily. The eight nuclei produced by mitotic divisions of the haploid megaspore are labeled.

In angiosperms, both male and female structures often occur together in the same individual flower. These reproductive structures are not a permanent part of the adult individual and the germ line is not set aside early in development.

Pollination

Pollination is the process by which pollen is placed on the stigma. Pollen may be carried to the flower by wind or by animals, or it may originate within the individual flower itself. When pollen from a flower's anther pollinates the same flower's stigma, the process is called *self-pollination.*

FIGURE 42.11
Pollination by a bumblebee. As this bumblebee, *Bombus* sp. squeezes into the bilaterally symmetrical, advanced flower of a member of the mint family, the stigma contacts its back and picks up any pollen that the bee may have acquired during a visit to a previous flower.

Pollination in Early Seed Plants

Early seed plants were pollinated passively, by the action of the wind. As in present-day conifers, great quantities of pollen were shed and blown about, occasionally reaching the vicinity of the ovules of the same species. Individual plants of any given species must grow relatively close to one another for such a system to operate efficiently. Otherwise, the chance that any pollen will arrive at the appropriate destination is very small. The vast majority of windblown pollen travels less than 100 meters. This short distance is significant compared with the long distances pollen is routinely carried by certain insects, birds, and other animals.

Pollination by Animals

The spreading of pollen from plant to plant by pollinators visiting flowers of specific angiosperm species has played an important role in the evolutionary success of the group. It now seems clear that the earliest angiosperms, and perhaps their ancestors also, were insect-pollinated, and the coevolution of insects and plants has been important for both groups for over 100 million years. Such interactions have also been important in bringing about increased floral specialization. As flowers become increasingly specialized, so do their relationships with particular groups of insects and other animals.

Bees. Among insect-pollinated angiosperms, the most numerous groups are those pollinated by bees (figure 42.11). Like most insects, bees initially locate sources of food by odor, then orient themselves on the flower or group of flowers by its shape, color, and texture. Flowers that bees characteristically visit are often blue or yellow. Many have stripes or lines of dots that indicate the location of the nectaries, which often occur within the throats of specialized flowers. Some bees collect nectar, which is used as a source of food for adult bees and occasionally for larvae. Most of the approximately 20,000 species of bees visit flowers to obtain pollen. Pollen is used to provide food in cells where bee larvae complete their development.

Only a few hundred species of bees are social or semi-social in their nesting habits. These bees live in colonies, as do the familiar honeybee, *Apis mellifera,* and the bumblebee, *Bombus* spp. Such bees produce several generations a year and must shift their attention to different kinds of flowers as the season progresses. To maintain large colonies, they also must use more than one kind of flower as a food source at any given time.

Except for these social and semi-social bees and about 1000 species that are parasitic in the nests of other bees, the great majority of bees—at least 18,000 species—are solitary. Solitary bees in temperate regions characteristically have only a single generation in the course of a year. Often they are active as adults for as little as a few weeks a year.

Solitary bees often use the flowers of a given group of plants almost exclusively as sources of their larval food. The highly constant relationships of such bees with those flowers may lead to modifications, over time, in both the flowers and the bees. For example, the time of day when the flowers open may correlate with the time when the bees appear; the mouthparts of the bees may become elongated in relation to tubular flowers; or the bees' pollen-collecting apparata may be adapted to the anthers of the plants that they normally visit. When such relationships are established, they provide both an efficient mechanism of pollination for the flowers and a constant source of food for the bees that "specialize" on them.

Insects Other Than Bees. Among flower-visiting insects other than bees, a few groups are especially prominent. Flowers such as phlox, which are visited regularly by butterflies, often have flat "landing platforms" on which butterflies perch. They also tend to have long, slender floral tubes filled with nectar that is accessible to the long, coiled proboscis characteristic of Lepidoptera, the order of insects that includes butterflies and moths. Flowers like jimsonweed, evening primrose, and others visited regularly by moths are often white, yellow, or some other pale color; they also tend to be heavily scented, thus serving to make the flowers easy to locate at night.

Birds. Several interesting groups of plants are regularly visited and pollinated by birds, especially the hummingbirds of North and South America and the sunbirds of Africa (figure 42.12). Such plants must produce large amounts of nectar because if the birds do not find enough food to maintain themselves, they will not continue to visit flowers of that plant. Flowers producing large amounts of nectar have no advantage in being visited by insects because an insect could obtain its energy requirements at a single flower and would not cross-pollinate the flower. How are these different selective forces balanced in flowers that are "specialized" for hummingbirds and sunbirds?

Ultraviolet light is highly visible to insects. Carotenoids, yellow or orange pigments frequently found in plants, are responsible for the colors of many flowers, such as sunflowers and mustard. Carotenoids reflect both in the yellow range and in the ultraviolet range, the mixture resulting in a distinctive color called "bee's purple." Such yellow flowers may also be marked in distinctive ways normally invisible to us, but highly visible to bees and other insects (figure 42.13). These markings can be in the form of a bull's-eye or a landing strip.

Red does not stand out as a distinct color to most insects, but it is a very conspicuous color to birds. To most insects, the red upper leaves of poinsettias look just like the other leaves of the plant. Consequently, even though the flowers produce abundant supplies of nectar and attract hummingbirds, insects tend to bypass them. Thus, the red color both signals to birds the presence of abundant nectar and makes that nectar as inconspicuous as possible to insects. Red is also seen again in fruits that are dispersed by birds.

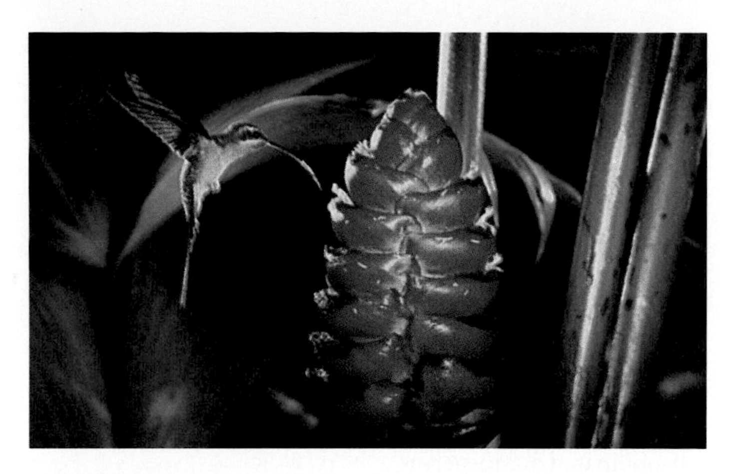

FIGURE 42.12
Hummingbirds and flowers. A long-tailed hermit hummingbird extracts nectar from the flowers of *Heliconia imbricata* in the forests of Costa Rica. Note the pollen on the bird's beak. Hummingbirds of this group obtain nectar primarily from long, curved flowers that more or less match the length and shape of their beaks.

(a) (b)

FIGURE 42.13
How a bee sees a flower. (*a*) The yellow flower of *Ludwigia peruviana* (Onagraceae) photographed in normal light and (*b*) with a filter that selectively transmits ultraviolet light. The outer sections of the petals reflect both yellow and ultraviolet, a mixture of colors called "bee's purple;" the inner portions of the petals reflect yellow only and therefore appear dark in the photograph that emphasizes ultraviolet reflection. To a bee, this flower appears as if it has a conspicuous central bull's-eye.

Other Animals. Other animals including bats and small rodents may aid in pollination. The signals here also are species specific. These animals also assist in dispersing the seeds and fruits that result from pollination. Monkeys are attracted to orange and yellow and will be effective in dispersing those fruits.

Wind-Pollinated Angiosperms

Many angiosperms, representing a number of different groups, are wind-pollinated—a characteristic of early seed plants. Among them are such familiar plants as oaks, birches, cottonwoods, grasses, sedges, and nettles. The flowers of these plants are small, greenish, and odorless; their corollas are reduced or absent (see figures 42.14 and 42.15). Such flowers often are grouped together in fairly large numbers and may hang down in tassels that wave about in the wind and shed pollen freely. Many wind-pollinated plants have stamen- and carpel-containing flowers separated among individuals or on a single individual. If the pollen-producing and ovule-bearing flowers are separated, it is certain that pollen released to the wind will reach a flower other than the one that sheds it, a strategy that greatly promotes outcrossing. Some wind-pollinated plants, especially trees and shrubs, flower in the spring, before the development of their leaves can interfere with the wind-borne pollen. Wind-pollinated species do not depend on the presence of a pollinator for species survival.

Bees are the most frequent and characteristic pollinators of flowers. Insects often are attracted by the odors of flowers. Bird-pollinated flowers are characteristically odorless and red, with the nectar not readily accessible to insects.

Self-Pollination

All of the modes of pollination that we have considered thus far tend to lead to outcrossing, which is as highly advantageous for plants as it is for eukaryotic organisms generally. Nevertheless, self-pollination also occurs among angiosperms, particularly in temperate regions. Most of the self-pollinating plants have small, relatively inconspicuous flowers that shed pollen directly onto the stigma, sometimes even before the bud opens. You might logically ask why there are many self-pollinated plant species if outcrossing is just as important genetically for plants as it is for animals. There are two basic reasons for the frequent occurrence of self-pollinated angiosperms:

1. Self-pollination obviously is ecologically advantageous under certain circumstances because self-pollinators do not need to be visited by animals to produce seed. As a result, self-pollinated plants expend less energy in the production of pollinator attractants and can grow in areas where the kinds of insects or other animals that might visit them are absent or very scarce—as in the Arctic or at high elevations.
2. In genetic terms, self-pollination produces progenies that are more uniform than those that result from outcrossing. Remember that because meiosis is involved, there is still recombination and the offspring will not be identical to the parent. However, such progenies may contain high proportions of individuals well-adapted to particular habitats. Self-pollination in normally outcrossing species tends to produce large numbers of ill-adapted individuals because it brings together deleterious recessive genes; but some of these combinations may be highly advantageous in particular habitats. In such habitats, it may be advantageous for the plant to continue self-pollinating indefinitely. This is the main reason many self-pollinating plant species are weeds—not only have humans made weed habitats uniform, but they have also spread the weeds all over the world.

FIGURE 42.14 Staminate and pistillate flowers of a birch, *Betula* sp. Birches are monoecious; their staminate flowers hang down in long, yellowish tassels, while their pistillate flowers mature into clusters of small, brownish, conelike structures.

FIGURE 42.15 Wind-pollinated flowers. The large yellow anthers, dangling on very slender filaments, are hanging out, about to shed their pollen to the wind; later, these flowers will become pistillate, with long, feathery stigmas—well suited for trapping windblown pollen—sticking far out of them. Many grasses, like this one, are therefore dichogamous.

Factors That Promote Outcrossing

Outcrossing, as we have stressed, is of critical importance for the adaptation and evolution of all eukaryotic organisms. Often flowers contain both stamens and pistils, which increases the likelihood of self-pollination. One strategy to promote outcrossing is to separate stamens and pistils.

In various species of flowering plants—for example, willows and some mulberries—staminate and pistillate flowers may occur on separate plants. Such plants, which produce only ovules or only pollen, are called **dioecious,** from the Greek words for "two houses." Obviously, they cannot self-pollinate and must rely exclusively on outcrossing. In other kinds of plants, such as oaks, birches, corn (maize), and pumpkins, separate male and female flowers may both be produced on the same plant. Such plants are called **monoecious,** meaning "one house" (figure 42.14). In monoecious plants, the separation of pistillate and staminate flowers, which may mature at different times, greatly enhances the probability of outcrossing.

Even if, as usually is the case, functional stamens and pistils are both present in each flower of a particular plant species, these organs may reach maturity at different times. Plants in which this occurs are called **dichogamous.** If the stamens mature first, shedding their pollen before the stigmas are receptive, the flower is effectively staminate at that time. Once the stamens have finished shedding pollen, the stigma or stigmas may then become receptive, and the flower may become essentially pistillate (figures 42.15 and 42.16). This has the same effect as if the flower completely lacked either functional stamens or functional pistils; its outcrossing rate is thereby significantly increased.

Many flowers are constructed such that the stamens and stigmas do not come in contact with each other. With such an arrangement, there is a natural tendency for the pollen to be transferred to the stigma of another flower rather than to the stigma of its own flower, thereby promoting outcrossing.

Even when a flower's stamens and stigma mature at the same time, genetic **self-incompatibility,** which is widespread in flowering plants, increases outcrossing. Self-incompatibility results when the pollen and stigma recognize each other as being genetically related and pollen tube growth is blocked (figure 42.17). Self-incompatibility is controlled by the *S* (self-incompatibility) locus. There are many alleles at the *S* locus that regulate recognition responses between the pollen and stigma. There are two types of self-incompatibility. Gametophytic self-incompatibility depends on the haploid *S* locus of the pollen and the diploid *S* locus of the stigma. If either of the *S* alleles in the stigma match the pollen *S* allele, pollen tube growth stops before it reaches the embryo sac. Petunias have gametophytic self-incompatibility. In the case of sporophytic self-incompatibility, such as in broccoli, both *S* alleles of the pollen parent are important; if the alleles in the stigma match with either of the pollen parent *S* alleles, the haploid pollen will not germinate.

Much is being learned about the molecular and biochemical basis of this recognition and the signal transduction pathways that block the successful growth of the pollen tube. These pollen recognition mechanisms may have had their origins in a common ancestor of the gymnosperms. Fossils with pollen tubes from the Carboniferous Period are consistent with the hypothesis that they had highly evolved pollen-recognition systems. These may have been systems that recognized foreign pollen that predated self-recognition systems.

(a)

(b)

FIGURE 42.16
Dichogamy, as illustrated by the flowers of fireweed, *Epilobium angustifolium*. More than 200 years ago (in the 1790s) fireweed, which is outcrossing, was one of the first plant species to have its process of pollination described. First, the anthers shed pollen, and then the style elongates above the stamens while the four lobes of the stigma curl back and become receptive. Consequently, the flowers are functionally staminate at first, becoming pistillate about two days later. The flowers open progressively up the stem, so that the lowest are visited first. Working up the stem, the bees encounter pollen-shedding, staminate-phase flowers and become covered with pollen, which they then carry to the lower, functionally pistillate flowers of another plant. Shown here are flowers in (*a*) the staminate phase and (*b*) the pistillate phase.

Self-pollinated angiosperms are frequent where there is a strong selective pressure to produce large numbers of genetically uniform individuals adapted to specific, relatively uniform habitats. Outcrossing in plants may be promoted through dioecism, monoecism, self-incompatibility, or the physical separation or different maturation times of the stamens and pistils. Outcrossing promotes genetic diversity.

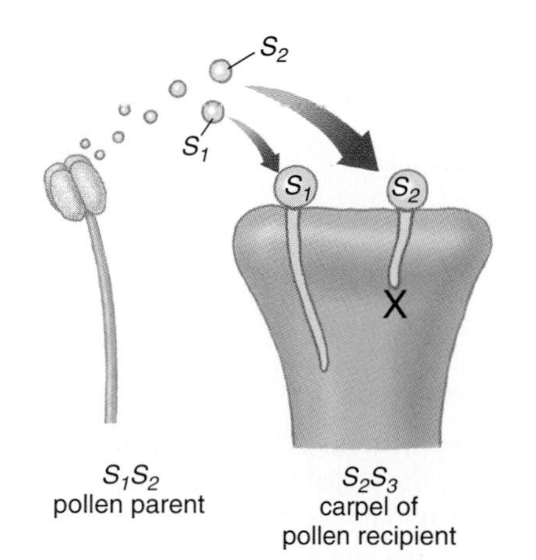

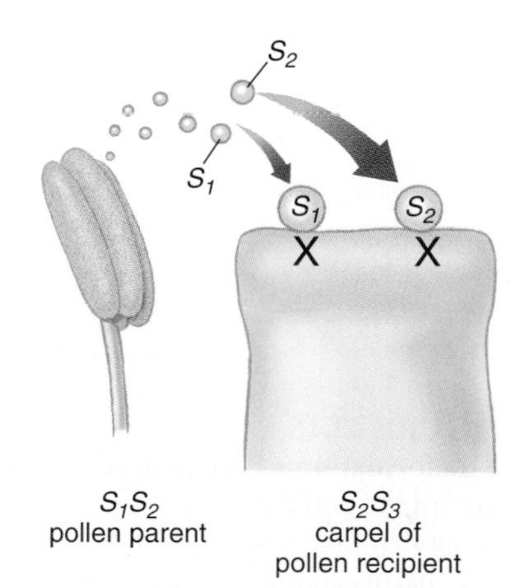

FIGURE 42.17
Self-pollination can be genetically controlled so self-pollen is blocked. (*a*) Gametophytic self-incompatibility is determined by the haploid pollen genotype. (*b*) Sporophytic self-incompatibility recognizes the genotype of the diploid pollen parent, not just the haploid pollen genotype. In both cases, the recognition is based on the *S* locus, which has many different alleles. The subscript numbers indicate the *S* allele genotype. In gametophytic self-incompatibility, the block comes after pollen tube germination. In sporophytic self-incompatibility, the pollen tube fails to germinate.

Fertilization

Fertilization in angiosperms is a complex, somewhat unusual process in which two sperm cells are utilized in a unique process called **double fertilization.** Double fertilization results in two key developments: (1) the fertilization of the egg, and (2) the formation of a nutrient substance called endosperm that nourishes the embryo. Once a pollen grain has been spread by wind, by animals, or through self-pollination, it adheres to the sticky, sugary substance that covers the stigma and begins to grow a **pollen tube** that pierces the style (figure 42.18). The pollen tube, nourished by the sugary substance, grows until it reaches the ovule in the ovary. Meanwhile, the generative cell within the pollen grain tube cell divides to form two sperm cells.

The pollen tube eventually reaches the embryo sac in the ovule. At the entry to the embryo sac, one of the synergids nuclei flanking the egg cell, degenerates, and the pollen tube enters that cell. The tip of the pollen tube bursts and releases the two sperm cells. One of the sperm cells fertilizes the egg cell, forming a zygote. The other sperm cell fuses with the two polar nuclei located at the center of the embryo sac, forming the triploid ($3n$) primary endosperm nucleus. The primary endosperm nucleus eventually develops into the endosperm.

Once fertilization is complete, the embryo develops by dividing numerous times. Meanwhile, protective tissues enclose the embryo, resulting in the formation of the seed. The seed, in turn, is enclosed in another structure called the fruit. These typical angiosperm structures evolved in response to the need for seeds to be dispersed over long distances to ensure genetic variability.

In double fertilization, angiosperms utilize two sperm cells. One fertilizes the egg, while the other helps form a substance called endosperm that nourishes the embryo.

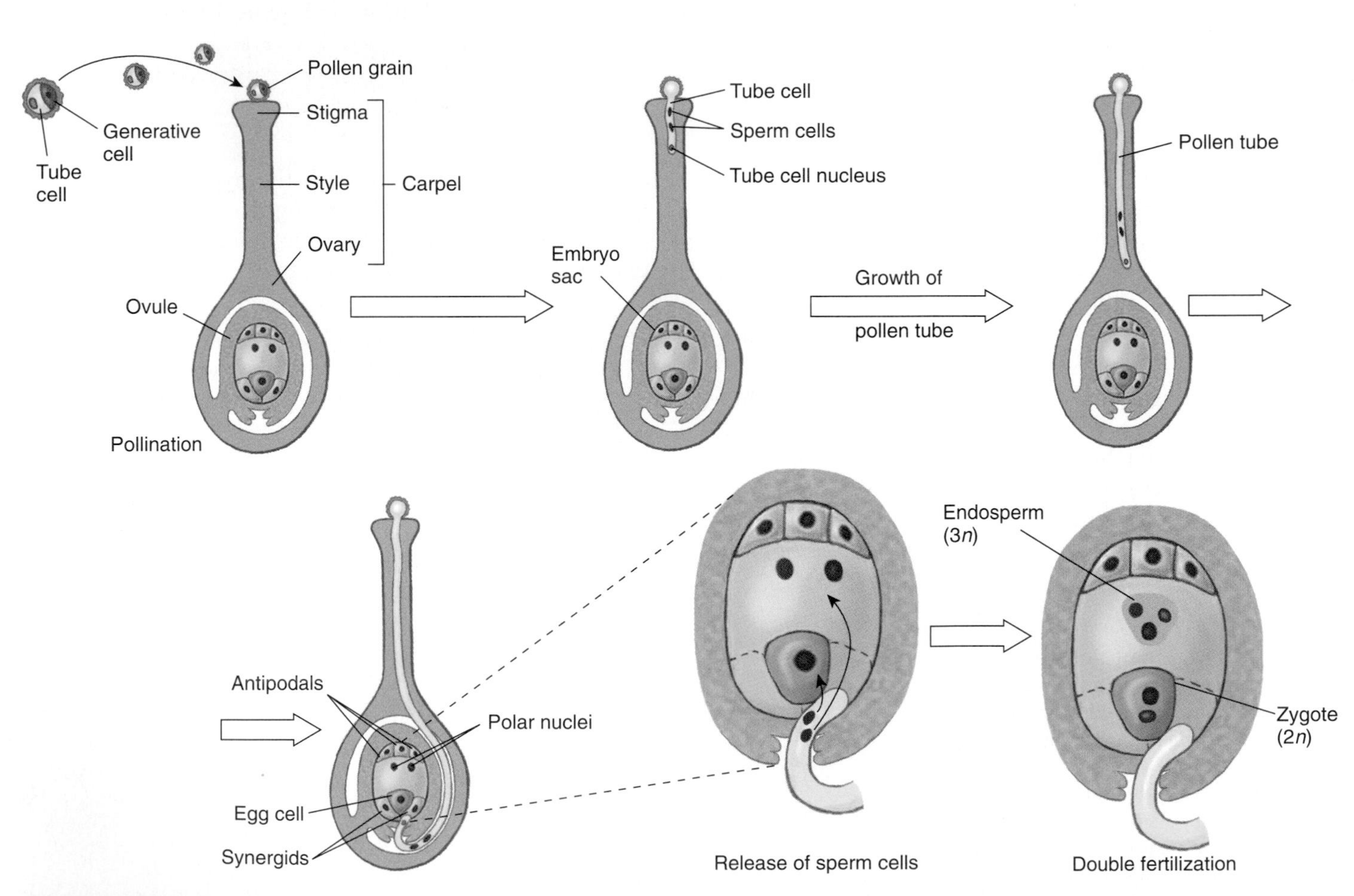

FIGURE 42.18

The formation of the pollen tube and double fertilization. When pollen lands on the stigma of a flower, the pollen tube cell grows toward the embryo sac, forming a pollen tube. While the pollen tube is growing, the generative cell divides to form two sperm cells. When the pollen tube reaches the embryo sac, it enters one of the synergids and releases the sperm cells. In a process called double fertilization, one sperm cell nucleus fuses with the egg cell to form the diploid ($2n$) zygote, and the other sperm cell nucleus fuses with the two polar nuclei to form the triploid ($3n$) endosperm nucleus.

42.3 Many plants can clone themselves by asexual reproduction.

Asexual Reproduction

While self-pollination reduces genetic variability, asexual reproduction results in genetically identical individuals because only mitotic cell divisions occur. In the absence of meiosis, individuals that are highly adapted to a relatively unchanging environment persist for the same reasons that self-pollination is favored. Should conditions change dramatically, there will be less variation in the population for natural selection to act upon and the species may be less likely to survive. Asexual reproduction is also used in agriculture and horticulture to propagate a particularly desirable plant whose traits would be altered by sexual reproduction, even self-pollination. Most roses and potatoes for example, are vegetatively propagated.

FIGURE 42.19.
Vegetative reproduction. Small plants arise from notches along the leaves of the house plant *Kalanchoë daigremontiana*.

Vegetative Reproduction

In a very common form of asexual reproduction called vegetative reproduction, new plant individuals are simply cloned from parts of adults (figure 42.19). The forms of vegetative reproduction in plants are many and varied.

Runners. Some plants reproduce by means of runners—long, slender stems that grow along the surface of the soil. In the cultivated strawberry, for example, leaves, flowers, and roots are produced at every other node on the runner. Just beyond each second node, the tip of the runner turns up and becomes thickened. This thickened portion first produces adventitious roots and then a new shoot that continues the runner.

Rhizomes. Underground horizontal stems, or rhizomes, are also important reproductive structures, particularly in grasses and sedges. Rhizomes invade areas near the parent plant, and each node can give rise to a new flowering shoot. The noxious character of many weeds results from this type of growth pattern, and many garden plants, such as irises, are propagated almost entirely from rhizomes. Corms, bulbs, and tubers are also stems specialized for storage and reproduction. White potatoes are propagated artificially from tuber segments, each with one or more "eyes." The eyes, or "seed pieces," of potato give rise to the new plant.

Suckers. The roots of some plants—for example, cherry, apple, raspberry, and blackberry—produce "suckers," or sprouts, which give rise to new plants. Commercial varieties of banana do not produce seeds and are propagated by suckers that develop from buds on underground stems. When the root of a dandelion is broken, as it may be if one attempts to pull it from the ground, each root fragment may give rise to a new plant.

Adventitious Plantlets. In a few species, even the leaves are reproductive. One example is the house plant *Kalanchoë daigremontiana*, familiar to many people as the "maternity plant," or "mother of thousands." The common names of this plant are based on the fact that numerous plantlets arise from meristematic tissue located in notches along the leaves. The maternity plant is ordinarily propagated by means of these small plants, which, when they mature, drop to the soil and take root.

Apomixis

In certain plants, including some citruses, certain grasses (such as Kentucky bluegrass), and dandelions, the embryos in the seeds may be produced asexually from the parent plant. This kind of asexual reproduction is known as *apomixis*. The seeds produced in this way give rise to individuals that are genetically identical to their parents. Thus, although these plants reproduce asexually by cloning diploid cells in the ovule, they also gain the advantage of seed dispersal, an adaptation usually associated with sexual reproduction. As you will see in chapter 43, embryos can also form via mitosis when plant tissues are cultured. In general, vegetative reproduction, apomixis, and other forms of asexual reproduction promote the exact reproduction of individuals that are particularly well suited to a certain environment or habitat. Asexual reproduction among plants is far more common in harsh or marginal environments, where there is little margin for variation. There is a greater proportion of asexual plants in the arctic, for example, than in temperate regions.

Plants that reproduce asexually clone new individuals from portions of the root, stem, leaves, or ovules of adult individuals. The asexually produced progeny are genetically identical to the parent individual.

The Life Span of Plants

Plant Life Spans Vary Greatly

Once established, plants live for highly variable periods of time, depending on the species. Life span may or may not correlate with reproductive strategy. Woody plants, which have extensive secondary growth, nearly always live longer than herbaceous plants, which have limited or no secondary growth. Bristlecone pine, for example, can live upward of 4000 years. Some herbaceous plants send new stems above the ground every year, producing them from woody underground structures. Others germinate and grow, flowering just once before they die. Shorter-lived plants rarely become very woody because there is not enough time for the accumulation of secondary tissues. Depending on the length of their life cycles, herbaceous plants may be annual, biennial, or perennial, while woody plants are generally perennial (figure 42.20). Determining life span is even more complicated for clonally reproducing organisms. Aspen trees form huge clones from asexual reproduction of their roots. Collectively, an aspen clone may form the largest "organism" on earth. Other asexually reproducing plants may cover less territory but live for thousands of years. Creosote bushes in the Mojave Desert have been identified that are up to 12,000 years old!

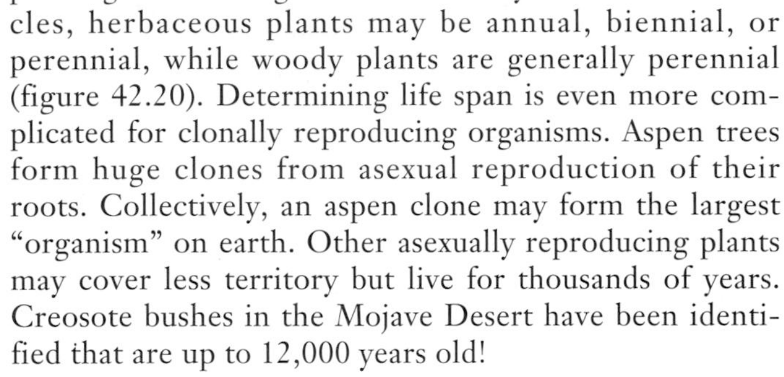

(a)

(b)

FIGURE 42.20
Annual and perennial plants. Plants live for very different lengths of time. (*a*) Desert annuals complete their entire life span in a few weeks. (*b*) Some trees, such as the giant redwood (*Sequoiadendron giganteum*), which occurs in scattered groves along the western slopes of the Sierra Nevada in California, live 2000 years or more.

Annual Plants

Annual plants grow, flower, and form fruits and seeds within one growing season; they then die when the process is complete. Many crop plants are annuals, including corn, wheat, and soybeans. Annuals generally grow rapidly under favorable conditions and in proportion to the availability of water or nutrients. The lateral meristems of some annuals, like sunflowers or giant ragweed, do produce poorly developed secondary tissues, but most are entirely herbaceous. Annuals typically die after flowering once, the developing flowers or embryos using hormonal signaling to reallocate nutrients so the parent plant literally starves to death. This can be demonstrated by comparing a population of bean plants where the beans are continually picked with a population where the beans are left on the plant. The frequently picked population will continue to grow and yield beans much longer than the untouched population. The process that leads to the death of a plant is called *senescence*.

Biennial Plants

Biennial plants, which are much less common than annuals, have life cycles that take two years to complete. During the first year, biennials store photosynthate in underground storage organs. During the second year of growth, flowering stems are produced using energy stored in the underground parts of the plant. Certain crop plants, including carrots, cabbage, and beets, are biennials, but these plants generally are harvested for food during their first season, before they flower. They are grown for their leaves or roots, not for their fruits or seeds. Wild biennials include evening primroses, Queen Anne's lace, and mullein. Many plants that are considered biennials actually do not flower until they are three or more years of age, but all biennial plants flower only once before they die.

Perennial Plants

Perennial plants continue to grow year after year and may be herbaceous, as are many woodland, wetland, and prairie wildflowers, or woody, as are trees and shrubs. The majority of vascular plant species are perennials. Herbaceous perennials rarely experience any secondary growth in their stems; the stems die each year after a period of relatively rapid growth and food accumulation. Food is often stored in the plants' roots or underground stems, which can become quite large in comparison to their less substantial aboveground counterparts.

Trees and shrubs generally flower repeatedly, but there are exceptions. Bamboo lives for many seasons as a vegetative plant, but senesces and dies after flowering. The same is true for at least one tropical tree which achieves great heights before flowering and senescing. Considering the tremendous amount of energy that goes into the growth of a tree, this particular reproductive strategy is quite curious.

Trees and shrubs are either deciduous, with all the leaves falling at one particular time of year and the plants remaining bare for a period, or evergreen, with the leaves dropping throughout the year and the plants never appearing completely bare. In northern temperate regions, conifers are the most familiar evergreens; but in tropical and subtropical regions, most angiosperms are evergreen, except where there is a severe seasonal drought. In these areas, many angiosperms are deciduous, losing their leaves during the drought and thus conserving water.

Organ Abscission

Senescence is an important developmental process that leads to the death of an organ, shoot, or the whole plant. Annual and biennial plants undergo whole plant senescence, but individual organs on any plant can also senesce and be shed. The process by which leaves or petals are shed is called **abscission**.

One advantage to organ senescence is that nutrient sinks can be dispensed with. For example, shaded leaves that are no longer photosynthetically productive can be shed. Petals, which are modified leaves, may senesce once pollination occurs. Orchid flowers remain fresh for long periods of time, even in a florist shop. However, once pollination occurs, a hormonal change is triggered that leads to petal senescence. This makes sense in terms of allocation of energy resources, as the petals are no longer necessary to attract a pollinator. On a larger scale, deciduous plants in temperate areas produce new leaves in the spring and then lose them in the fall. In the tropics, however, the production and subsequent loss of leaves in some species is correlated with wet and dry seasons. Evergreen plants, such as most conifers, usually have a complete change of leaves every two to seven years, periodically losing some but not all of their leaves.

Abscission involves changes that take place in an *abscission zone* at the base of the petiole (figure 42.21). Young leaves produce hormones (especially cytokinins) that inhibit the development of specialized layers of cells in the abscission zone. Hormonal changes take place as the leaf ages, however, and two layers of cells become differentiated. (Despite the name, abscisic acid is not involved in this process.) A *protective layer*, which may be several cells wide, develops on the stem side of the petiole base. These cells become impregnated with *suberin*, which, as you will recall, is a fatty substance that is impervious to moisture. A *separation layer* develops on the side of the leaf blade; the cells of the separation layer sometimes divide, swell, and become gelatinous. When temperatures drop, the duration and intensity of light diminishes as the days grow shorter, or other environmental changes occur, enzymes break down the pectins in the middle lamellae of the separation cells. Wind and rain can then easily separate the leaf from the stem. Left behind is a sealed leaf scar that is protected from bacteria and other disease organisms.

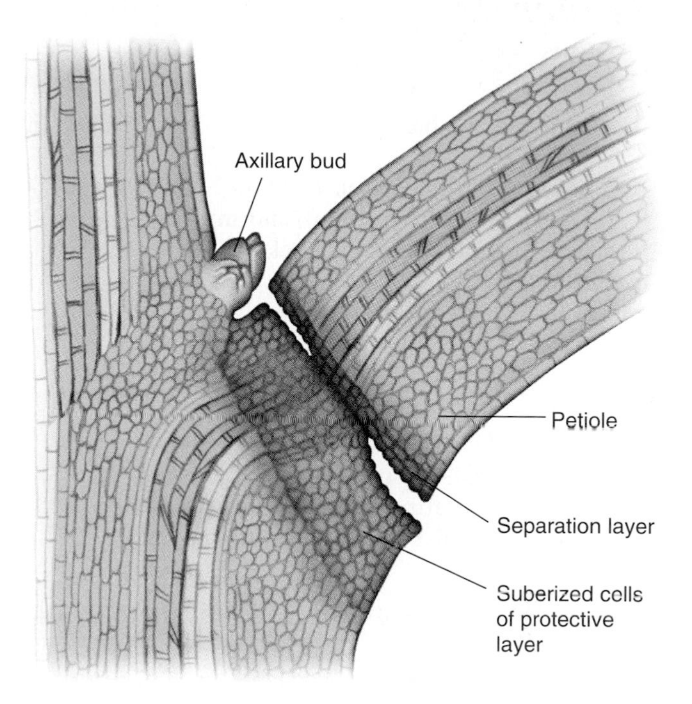

FIGURE 42.21
Leaf abscission. The abscission zone of a leaf. Hormonal changes in this zone cause abscission. Two layers of cells in the abscission zone differentiate into a protective layer and a separation layer. As pectins in the separation layer break down, wind and rain can easily separate the leaf from the stem.

As the abscission zone develops, the green chlorophyll pigments present in the leaf break down, revealing the yellows and oranges of other pigments, such as carotenoids, that previously had been masked by the intense green colors. At the same time, water-soluble red or blue pigments called *anthocyanins* and *betacyanins* may also accumulate in the vacuoles of the leaf cells—all contributing to an array of fall colors in leaves (see figure 41.7*a*).

Annual plants complete their whole growth cycle within a single year. Biennial plants flower only once, normally after two seasons of growth. Perennials flower repeatedly and live for many years. Abscission occurs when a plant sheds its organs.

Chapter 42

www.mhhe.com/raven6/resources42.mhtml

Summary | **Questions** | **Media Resources**

42.1 Angiosperms have been incredibly successful, in part, because of their reproductive strategies.

- Angiosperms have been successful because they can be relatively drought-resistant, and smaller herbaceous angiosperms have relatively short life cycles. Most important, however, are their flowers and fruits. Flowers make possible the precise transfer of pollen and, therefore, outcrossing, even when the stationary individual plants are widely separated. Fruits, with their complex adaptations, facilitate the wide dispersal of angiosperms.
- Modification of floral parts, especially petals, has been key in facilitating pollination. Bilateral symmetry has evolved independently, multiple times.

1. What characteristics of early angiosperms are thought to contribute to their success.

2. What flower whorl is collectively made up of petals? With which other plant parts are the petals of most flowers homologous?

3. What is an androecium? Of which flower parts is it composed?

- Angiosperms

42.2 Flowering plants use animals or wind to transfer pollen between flowers.

- Bees are the most frequent and constant pollinators of flowers. Insects often are attracted by the odors of flowers rather than color. Birds are attracted to red flowers, but not odors.
- Self-pollination reduces genetic variability among offspring. Outcrossing increases genetic diversity.
- Outcrossing in different angiosperms is promoted by the separation of stamens and carpels into different flowers, or even into different individuals.

4. What does it mean if a plant is dichogamous? Of what advantage is it to the plant?

5. Is it more likely that a flower visited by a social or a solitary bee will become highly specialized toward that bee? Why?

- Gamete Formation
- Fertilization

42.3 Many plants can clone themselves by asexual reproduction.

- In asexual reproduction, plants clone new individuals from portions of adult roots, stems, leaves, or ovules.
- The progeny produced by asexual reproduction are all genetically identical to the parent individual, even when they are produced in the ovules (apomixis).

6. Why would a plant capable of sexual reproduction reproduce asexually?

7. You have just cloned a gene responsible for apomixis. Several corn breeders are very interested in your gene. Why?

- Asexual Reproduction

42.4 How long do plants and plant organs live?

- Plants can live for a single season or thousands of years.
- For annual and biennial plants, reproduction triggers senescence and death.
- Asexually reproducing plants can form clones that cover huge areas and/or live for many thousands of years.
- Plant organs and shoots can senesce and die while the whole plant thrives. Organ senescence is an efficient way to maximize the use of energy resources.

8. Some plants flower once and die; others flower multiple times, reaching great heights and diameters. What are the relative advantages of the two strategies?

9. How and why does leaf abscission occur?

43

Plant Genomics

Concept Outline

FIGURE 43.1
Golden rice. Rice is the dietary staple of almost half the world's population, but it lacks vitamin A. Vitamin A deficiency leads to vision and immunity problems. Genetically engineered rice that produces vitamin A has now been developed. The rice is golden because a biosynthetic pathway has been genetically modified to produce gold-colored beta-carotene, a precursor to vitamin A. Here, white rice is mixed in with golden rice. The intensity of golden color indicates the amount of pro-vitamin A present.

By selective breeding favoring desired traits, people have been genetically modifying plants since agrarian societies began. All of our key modern crops are the result of this long effort. Today, we have even more powerful tools, recombinant DNA technologies, that are the subject of this chapter. This chapter looks ahead to the impact of these new technologies on the future of plants and our study of plant biology (figure 43.1). Both the *Arabidopsis* and rice genomes are essentially sequenced. Not only can we expect to learn much about the molecular basis of plant physiology and development from these rich databases; we will surely gain a far deeper understanding of plant evolution.

43.1 Genomic organization is much more varied in plants than in animals.

Overview of Plant Genomics

Early Approaches

While the term genetic engineering is commonly used to describe plants and animals modified using recombinant DNA technology, people have actually been genetic engineers for thousands of years. As agrarian societies formed, changes in the gene pool within crop species began. For example, seed dispersal was selected against in maize and wheat. Without the ability to disperse seed, these domesticated plants are completely dependent on humans for seed dispersal. Rice was converted from a perennial plant to an annual plant without the seed dormancy mechanisms present in wild rice. Parts of the plant that were of most dietary value to humans and domesticated animals have been selected for increased size. These include seeds, fruits, and storage organs like roots in the case of carrots. All of these changes were accomplished without knowledge of particular genes, by selecting and propagating individuals with the desired traits.

Breeding Strategies to Enhance Yield

At the beginning of the twentieth century, a growing understanding of genetics increased the rate of crop improvement. Among the most dramatic agricultural developments was the introduction of hybrid corn. As corn breeding progressed, highly inbred lines began to have decreased yield as deleterious recessive genes became homozygous. George Harrison Shull found that crossing two different inbred lines gave rise to offspring with "hybrid vigor." The yield increased fourfold! Hybrid corn now grows in almost all fields in the United States. Hybrid rice has increased yield 20%.

Breeders have now turned to specific genes to optimize food quality (see figure 43.1). Only a small percentage of the genes and their function have been identified, but we start this century with technologically powerful new ways to understand genomes.

Studying Plant Genomes

Plant genomes are complex, and analyses reveal many large evolutionary changes of the DNA sequences over time. Plants show widely different chromosome numbers and varied ploidy levels (figure 43.2). Overall, the size of plant genomes (both number of chromosomes and total nucleotide base-pairs) exhibits the greatest variation of any kingdom in the biological world. For example, tulips contain over 170 times as much DNA as the small weed *Arabidopsis thaliana* (table 43.1). The DNA of plants, like animals, can also contain regions of sequence repeats, sequence inversions, or transposable element insertions, which further modify their genetic content. Traditionally, variation in chromosome inversions and ploidy have been used to build up a picture of how plant species have evolved (figure 43.3). Increasingly, researchers are turning to studying the organization of plant DNA sequences to obtain important information about the evolutionary history of a plant species.

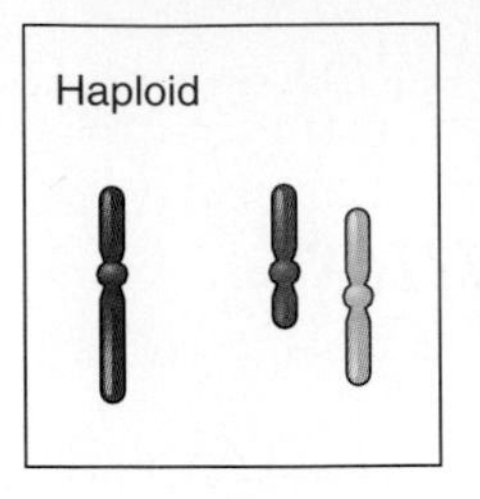

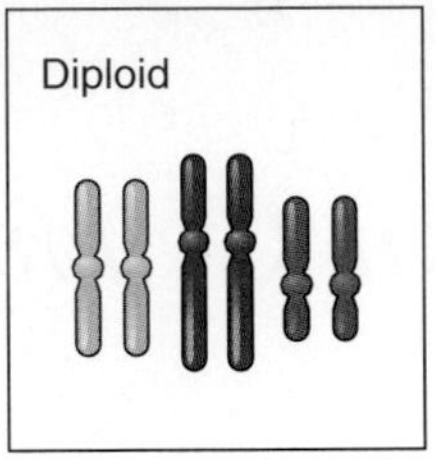

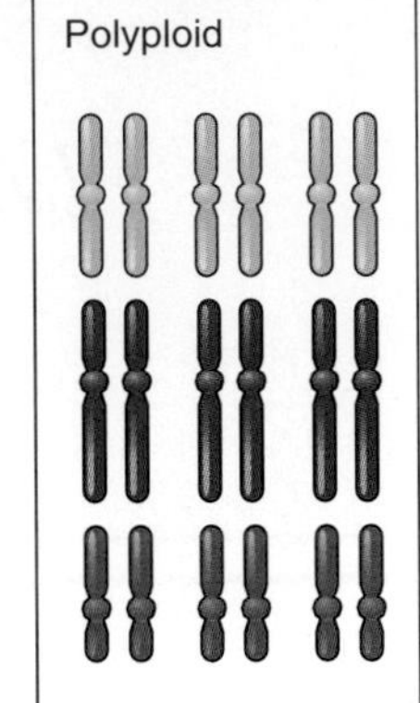

FIGURE 43.2
Chromosome numbers possible in plant genomes. Haploid: a set of chromosomes without their pairs; for example, the chromosome number present in a gamete. Diploid: a single set of chromosome pairs. Polyploid: multiple sets of chromosome pairs; for example, bananas have a triple set of chromosomes and are therefore polyploid.

Table 43.1 Genome Size of Plants

Scientific Name	Common Name	Genome Size (Millions of Base-Pairs)
Arabidopsis thaliana	Arabidopsis	145
Prunus persica	Peach	262
Ricinus communis	Castor bean	323
Citrus sinensis	Orange	367
Oryza sativa spp. *javanica*	Rice	424
Petunia parodii	Petunia	1,221
Pisum sativum	Garden pea	3,947
Avena sativa	Oats	11,315
Tulipa spp.	Garden tulip	24,704

Source: From *Plant Biochemistry and Molecular Biology*, by P. J. Lea and R. C. Leegods, eds. Copyright © 1993 John Wiley & Sons, Limited. Reproduced with permission.

People have been genetically engineering plants for centuries by selecting for desired traits. Traditionally, biologists have examined variation among plants at the chromosome level; today, researchers are focusing more of their efforts at the DNA sequence level.

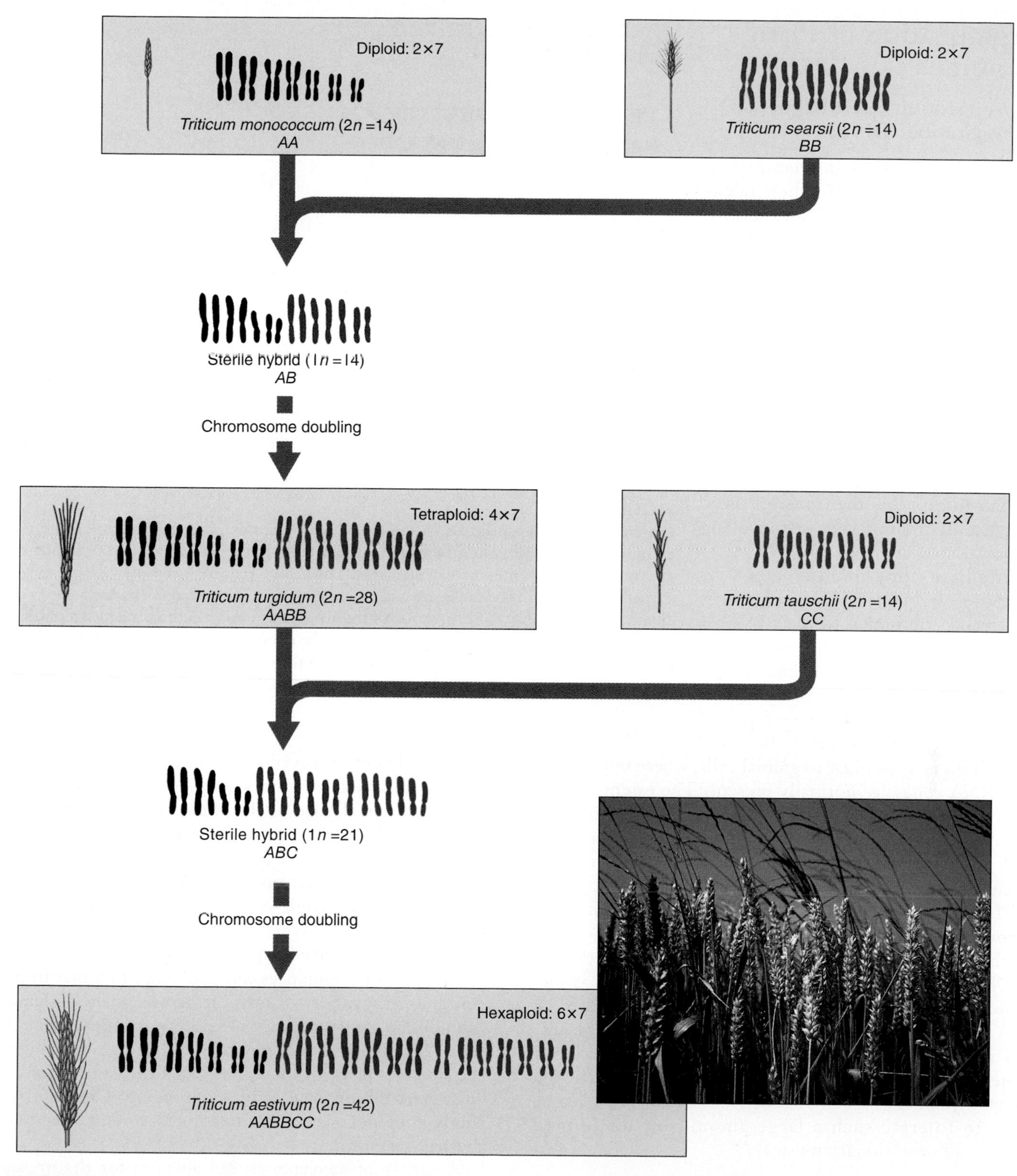

FIGURE 43.3
Evolutionary history of wheat. Domestic wheat arose in southwestern Asia in the hilly country of what is now Iraq. In this region, there is a rich assembly of grasses of the genus *Triticum*. Domestic wheat *(T. aestivum)* is a polyploid species of *Triticum* that arose through two so-called "allopolyploid" events. (1) Two different diploid species symbolized here as *AA* and *BB*, hybridized to form an *AB* polyploid; the species were so different that *A* and *B* chromosomes could not pair in meiosis, so the *AB* polyploid was sterile. However, in some plants the chromosome number spontaneously doubled due to a failure of chromosomes to separate in meiosis, producing a fertile tetraploid species *AABB*. This wheat is used in the production of pasta. (2) In a similar fashion, the tetraploid species *AABB* hybridized with another diploid species *CC* to produce, after another doubling event, the hexaploid *T. aestivum*, *AABBCC*. This bread wheat is commonly used throughout the world.

Organization of Plant Genomes

Low-, Medium-, and High-Copy-Number DNA

Most seed plants contain quantities of DNA that greatly exceed their needs for coding and regulatory function. Hence, for plants, a very small percentage of the genome may actually encode genes involved in the production of proteins. This portion of the genome which encodes most of the transcribed genes is often referred to as "low-copy-number DNA," because the DNA sequences comprising these genes are present in single or small numbers of copies. How do plants function with so much extra DNA inserted into the genome? It appears that most of these sequence alterations occur in noncoding regions.

"Medium-copy-number DNA" is composed largely of DNA sequences that encode ribosomal RNA (rRNA), a key element of the cellular machinery that translates transcribed messenger RNA (mRNA) into protein. In plant genomes, rRNA genes may be repeated several hundred to several thousand times. This is in contrast to animal cells, where only 100 to 200 rRNA genes are normally present. The extent of variability in plant genomes with respect to the number of rRNA genes and mutations in them has provided a useful tool for analyzing the evolutionary patterns of plant species.

Plant cells may also contain excess DNA in their genomes in the form of highly repetitive sequences, or "high-copy-number DNA." At present, the function of this high-copy-number DNA in plant genomes is unknown. Roughly half the maize genome is composed of such retroviral-like DNA. RNA retroviruses like the animal virus HIV use their host genomes to make DNA copies that then insert into the host genome. Clearly, the effects of some retroviruses can be lethal. How maize came to tolerate such a large amount of this foreign DNA is an evolutionary mystery.

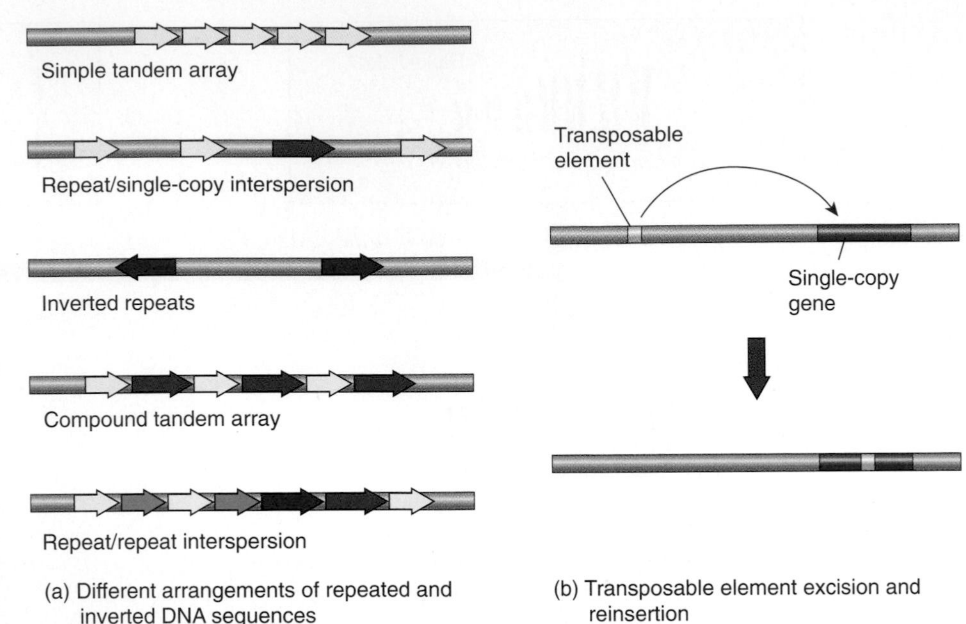

FIGURE 43.4
Organization of repeated DNA sequences and the mechanism of transposable elements in altering gene function. (*a*) Repeated DNA sequences can occur in plant genomes in several different arrangements. The arrows represent repeated DNA sequences. Arrows of the same size and color represent DNA sequences which are identical to each other. The direction of the arrowhead indicates the orientation of the DNA sequence. (*b*) Transposable elements can be a source of repetitive DNA that alters gene function. Following excision from its original location, a transposable element may reinsert in the single-copy DNA sequence comprising a gene and alter the gene's function.

Sequence Replication and Inversion

High-copy-number DNA sequences in the plant genome may be short, such as the nucleotide sequence "GAA," or much longer, involving up to several hundred nucleotides. Moreover, the number of copies of an individual high-copy repetitive DNA sequence can total from 10,000 to 100,000. There are several possibilities for how high-copy repetitive DNA sequences may be organized within a plant genome (figure 43.4*a*). Several copies of a single repetitive DNA sequence may be present together in the same orientation, in a pattern called "simple tandem array." Alternatively, repetitive DNA sequences can be dispersed among single-copy DNA in the same orientation ("repeat/single-copy interspersion") or the opposite orientation ("inverted repeats"). In addition, groups of repetitive DNA sequences can also occur together in plant genomes in a variety of possible arrangements, such as a "compound tandem array" or a "repeat/repeat interspersion." The presence of repetitive DNA can vastly increase the size of a plant genome, making it difficult to find and characterize individual single-copy genes. Characterizing single-copy genes can thus become a sort of "needle-in-the-haystack" hunt.

A variety of mechanisms can account for the presence of highly repetitive DNA sequences in plant genomes. Repetitive sequences can be generated by DNA sequence amplification in which multiple rounds of DNA replication occur for specific chromosomal regions. Unequal crossing over of the chromosomes during meiosis or the action of transposable elements (see next section) can also generate repetitive sequences.

Transposable Elements

Transposable elements, described in chapter 18, are special sequences of DNA with the ability to move from place to place in the genome. They can excise from one site at unpredictable times and reinsert in another site. For this reason, transposable elements have been called "jumping genes." Transposable elements insert into coding regions or regulatory regions of a gene. The insertion affects expression of that gene, resulting in a mutation that may or may not be detectable (figure 43.4*b*). Barbara McClintock won the Nobel Prize in 1983 for her work describing transposable elements in corn (see figure 18.23).

Due to their capacity to replicate independently and to move through the genome, transposable elements can also be involved in generating repetitive DNA sequences. This is believed to be the case with the extensive retroviral-like insertions in maize. Retention of the repetitive DNA sequence at a particular site in the genome would involve in each instance a mutation in the transposable element itself which removes its capacity to transpose.

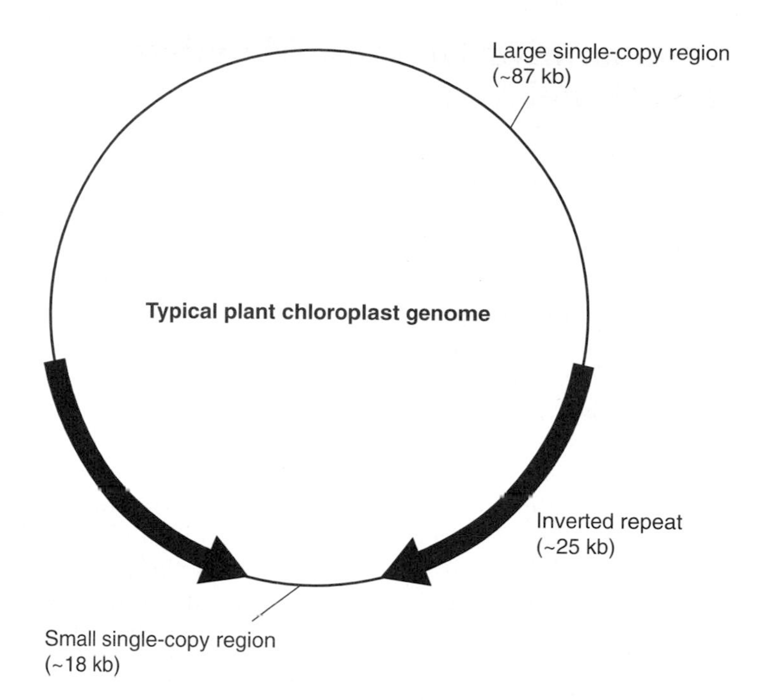

FIGURE 43.5
Chloroplast genome. A schematic drawing of a typical plant chloroplast genome indicates two regions containing single-copy genes, one containing about 87,000 nucleotides (87 kb) and another about 18 kb, and two symmetrical inverted repeats, each containing about 25 kb. Chloroplast DNA does not show recombination events that are common in the nuclear genome. It is thus a good subject for DNA phylogenetic analysis.

Chloroplast Genome and Its Evolution

The chloroplast is a plant organelle that functions in photosynthesis, and it can independently replicate in the plant cell. Plant chloroplasts have their own specific DNA, which is separate from that present in the nucleus. This DNA is maternally inherited and encodes unique chloroplast proteins. Chloroplasts are thought to have originated from a photosynthetic prokaryote that became part of a plant cell by endosymbiosis (see chapter 35). In support of this concept, research has shown that chloroplast DNA has many prokaryote-like features. Chloroplast DNA is present as a circular molecule of double-stranded DNA similar to prokaryotic chromosomal DNA. Moreover, chloroplast DNA contains genes for ribosomes that are very similar to those present in prokaryotes.

The DNA in chloroplasts of all land plants has about the same number of genes (~100), and they are present in about the same order. In contrast to the evolution of the DNA in the plant cell nucleus, chloroplast DNA has evolved at a more conservative pace, and therefore shows a more interpretable evolutionary pattern when scientists study DNA sequence similarities. Chloroplast DNA is also not subject to modification caused by transposable elements and mutations due to recombination.

Many of the proteins encoded by chloroplast DNA are involved in photosynthesis, although most of the proteins involved in photosynthesis are encoded by nuclear genes. Over time, there appears to have been some genetic exchange between the nuclear and chloroplast genomes. For example, the key enzyme in the Calvin cycle of photosynthesis (RUBISCO) consists of a large and small subunit. The large subunit is encoded in the chloroplast genome. The small subunit is encoded in the nuclear genome. The protein it encodes has a targeting sequence that allows it to enter the chloroplast and combine with large subunits. The evolutionary history of the localization of these genes is a puzzle.

A characteristic feature of the chloroplast genome is the presence of two identical inverted repeats in the DNA sequence (figure 43.5). Other DNA sequence inversions or deletions occur rarely, but when they do occur, they provide a character or a tool to analyze evolutionary relationships between plants. For instance, a large inversion in chloroplast DNA is found in the Asteraceae, or sunflower family, and not in other plant families. While previous work on the evolutionary relationships between plants has emphasized the comparative analysis of plant anatomy or morphology, there is increasing use of plant molecular data such as chloroplast DNA sequences.

Plant nuclear genomes may contain large amounts of DNA in comparison to other eukaryotic organisms, but only a small amount of this DNA represents functional genes. Excess DNA in plant genomes can result from increased chromosome copy number (polyploidy), and DNA sequence repeats. Chloroplast genomes evolve more slowly than nuclear genomes and can provide important evolutionary information.

Comparative Genome Mapping and Model Systems

Knowledge of plant genomes has been growing with the advent of new techniques to study DNA sequences. An increased understanding of plant genomes can lead to better manipulation of genetic traits such as crop yield, disease resistance, growth abilities, nutritive qualities, or drought tolerance. Multiple genes could encode each of these traits. By genome mapping model plants, plant biologists can lay a foundation for future plant breeding and for an understanding of plant evolution at the genetic level. One such model system, rice, has been chosen because it has a small genome and a high level of chromosomal conservation with other grains. In a genomic sense, "rice is wheat." The other model system that has been selected in plants is *Arabidopsis*. This small weed, a member of the mustard family, has an unusually small genome with only 20% repetitive DNA (see table 43.1) which has made it possible to sequence the entire genome. Getting down to the level of individual base pairs is a stepwise process, as described below.

RFLP and AFLP as Tools to Map Genomes and Detect Polymorphisms

The classical approach to locating genes in linear order on chromosomes involves making crosses between plants with known genes identified by mutations. The frequency of recombination is used to calculate the distance between two genetic markers (see chapter 13). The result is a genetic or linkage map. This approach is limited to genes with alleles that can be phenotypically identified. Much more of the genome can be mapped using RFLPs (restriction fragment length polymorphisms) which need not have a macroscopic phenotype. Remember, RFLPs are chunks of DNA that may contain a part of one or more genes. This approach, described in detail in chapter 19 (see figures 19.2, 19.4, 19.9, and 19.10), involves analysis of the RFLP map, or the pattern of DNA fragments, produced when DNA is treated with restriction enzymes that cleave at specific sites. RFLPs are identified by hybridizing a cloned probe to the cut DNA that has been separated by gel electrophoresis. RFLP mapping can identify important regions of the genome at a glance, while sequence data require sophisticated computer-based searching and matching systems. Currently, the most dense RFLP map is in rice where 2000 DNA sequences have been mapped onto 12 chromosomes.

Another tool that utilizes sequence variability is AFLPs, or amplified fragment length polymorphisms. Genomic DNA fragments are cut with restriction enzymes, usually *Eco*RI and *Mse*I, and then subsequently amplified using the polymerase chain reaction (PCR). The resulting PCR products, which represent each piece of DNA cut by a restriction enzyme, are separated by size via gel electrophoresis. As with RFLPs, the AFLPs are identified by hybridization with a cloned probe. The band sizes on an AFLP gel tend to show more polymorphisms than those found with RFLP mapping because the entire

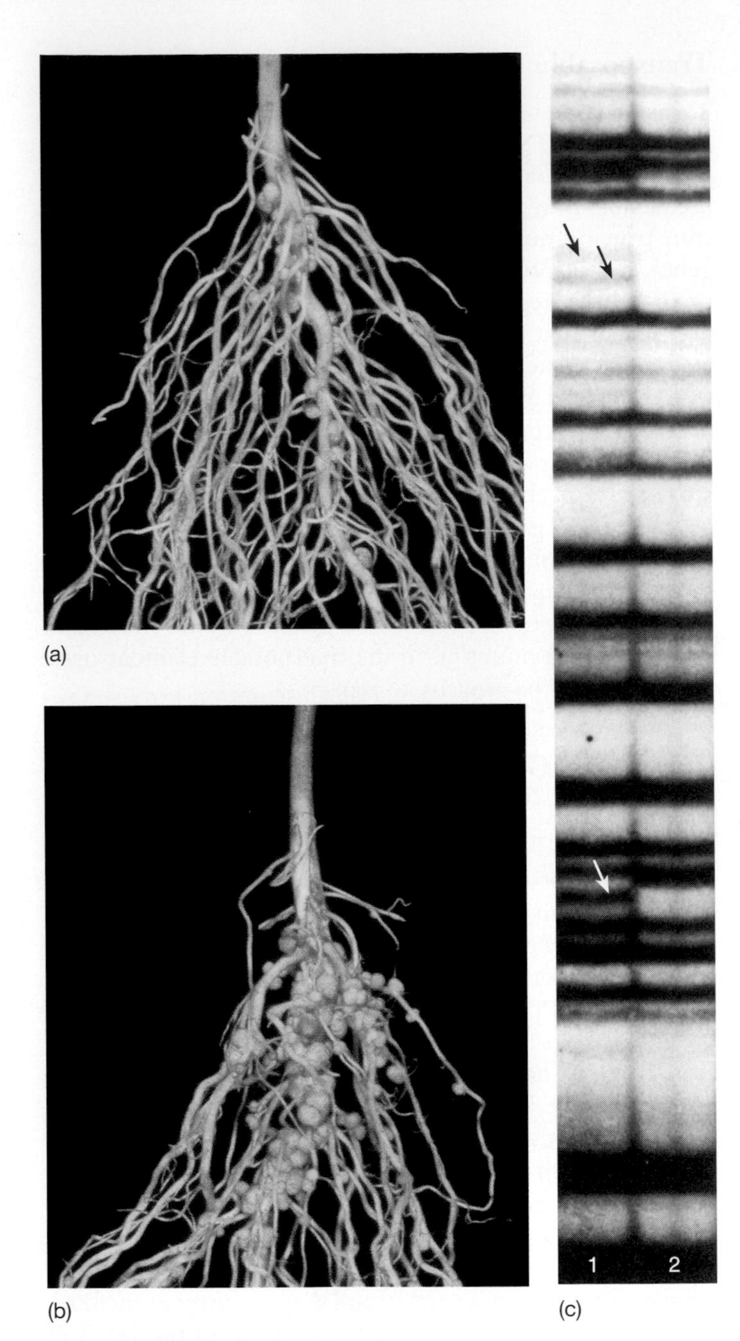

FIGURE 43.6
AFLP fingerprint pattern from normal and "hypernodulating" soybeans. It is still not known what determines the nodule number in (*a*) a normal soybean root versus (*b*) a "hypernodulating" mutant. The slight genetic differences between these plants can be evaluated by AFLP (*c*). The banding pattern changes indicate what genetic markers are linked to the "hypernodulation" mutation. Lane 1: normal soybean DNA; Lane 2: "hypernodulating" soybean DNA.

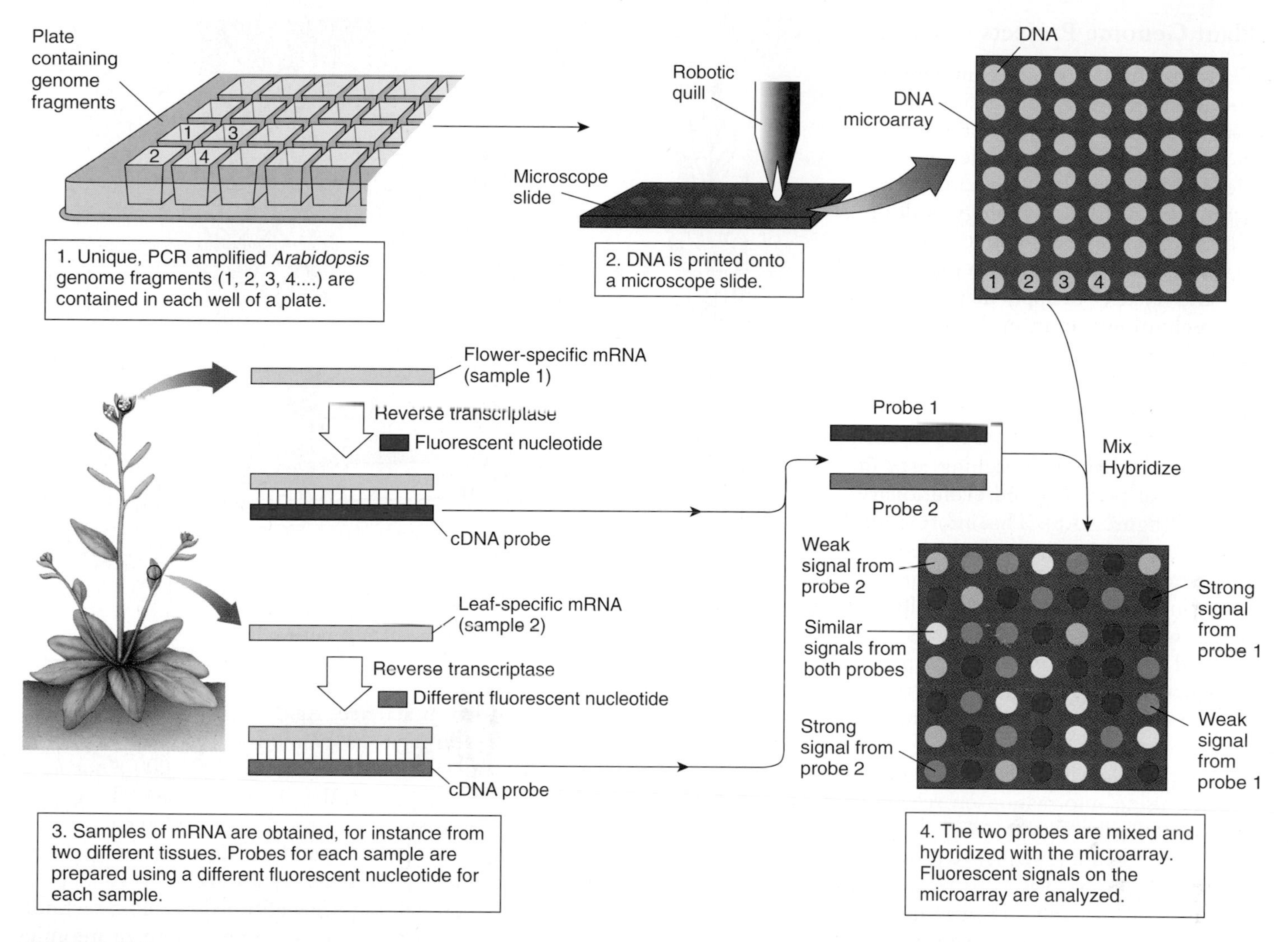

FIGURE 43.7
Microarrays. PCR is used to amplify non-redundant parts of a genome. The genome fragments will then be individually spotted on to a microscope slide by a robotic quill, creating a microarray. The microarray can then be probed with RNA from tissues of interest to identify expressed DNA. The microarray with hybridized probes is analyzed and often displayed as a false-color image. If a gene is frequently expressed in one of the samples, the fluorescent signal will be strong (*red* or *green*) where the gene is located on the microarray. If a gene is rarely expressed in one of the samples, the signal will be weak (*pink* or *light green*). A *yellow* color indicates genes that are expressed at similar levels in each sample.

genome is visible on the gel (figure 43.6). Both RFLPs and AFLPs (among many other tools for genome analysis) can provide markers of traits which are inherited from parents to progeny through crosses.

DNA Microarrays

How can DNA sequences be made available to researchers, other than as databases of electronic information? DNA microarrays are a way to link sequences with the study of gene function and make DNA sequences available to many. Also called biochips or "genes on chips," these convenient assays for the presence of a particular version of a gene were discussed in chapter 19. To prepare a particular DNA microarray, fragments of DNA are deposited on a microscope slide by a robot at indexed locations. Up to 10,000 spots can be displayed over an area of only 3.24 cm^2 (figure 43.7). The primary applications of microarrays are to determine which genes are expressed developmentally in certain tissues or in response to environmental factors. RNA from these tissues can be isolated and used as a probe for these microarrays. Only those sequences that are expressed in the tissues will be present to hybridize to the spot on the microarray.

Plant Genome Projects

The potential of having complete genomic sequences of plants is tremendous and about to be realized now that the *Arabidopsis* Genome Project is complete. This project represents a new paradigm in the way biology is done. The international effort brought together research teams with the expertise and tenacity to apply new sequencing technology to an entire genome, rather than single genes. Powerful databases are being constructed to make this information accessible to all. The completely sequenced *Arabidopsis* genome will have far-reaching uses in agricultural breeding and evolutionary analysis (figure 43.8). This information can be expected to help plant breeders in the future because the localization of genes in one plant species can help indicate where that gene might also be located in another species. In plant genomes, local gene order seems to be more conserved than the nucleotide sequences of homologous genes. Thus, the complete genomic sequence of *Arabidopsis thaliana* will facilitate gene cloning from many plant species, using information on relative genomic location as well as similarity of sequences.

FIGURE 43.8
Future directions in the genetic engineering of vegetable oils?

The rice genome sequence provides a model for a small monocot genome. Rice was selected, in part, because its genome is 6, 10, and 40 times smaller than that of maize, barley, and wheat, respectively. These grains represent a major food source for humans. By understanding the rice genome at the level of its DNA sequence, it should be much easier to identify and isolate genes from grains with larger genomes. Even though these plants diverged more than 50 million years ago, the chromosomes of rice, corn, barley, wheat, and other grass crops show extensive conserved arrangements of segments (synteny) (figure 43.9). DNA sequence analysis of cereal grains will be important for identifying genes associated with disease resistance, crop yield, nutritional quality, and growth capacity. It will also be possible to construct an approximate map of the ancestral cereal genome.

Functional Genomics and Proteomics

Sequencing the *Arabidopsis* and rice genomes represent major technological accomplishments. A new field of bioinformatics takes advantage of high-end computer technology to analyze the growing gene databases, look for relationships among genomes, and hypothesize functions of genes based on sequence. Genomics (the study of genomes) is now shifting gears and moving back to hypothesis-driven science. Again, an international community of researchers has come together with a plan to assign function to all of the 20,000 to 25,000 *Arabidopsis* genes by 2010 (Project 2010). In many ways, the goal is to ultimately answer the questions we have raised in chapters 37 through 42. One of the first steps is to determine when and where these genes are expressed. Each step beyond that will require additional enabling technology. Research will move from genomics to proteomics (the study of all proteins in an organism). Proteins are much more difficult to study because of posttranslational modification and formation of complexes of proteins. This information will be essential in understanding cell biology, physiology, development, and evolution. For example, how are similar genes used in different plants to create biochemically and morphologically distinct organisms? So, in many ways, we continue to ask the same questions that even Mendel asked, but at a much different level of organization.

Restriction fragment length polymorphisms (RFLPs) and amplified fragment length polymorphisms (AFLPs) represent important tools for mapping genetic traits in plant genomes. Due to its short life cycle, small size, and small genome, the mustard relative *Arabidopsis thaliana* is being used as a model plant for genetic studies. The genome of rice is also essentially sequenced and will be a valuable model for other monocot cereal grains such as wheat, barley, oats, and corn. Assigning function to these genes is the next challenge.

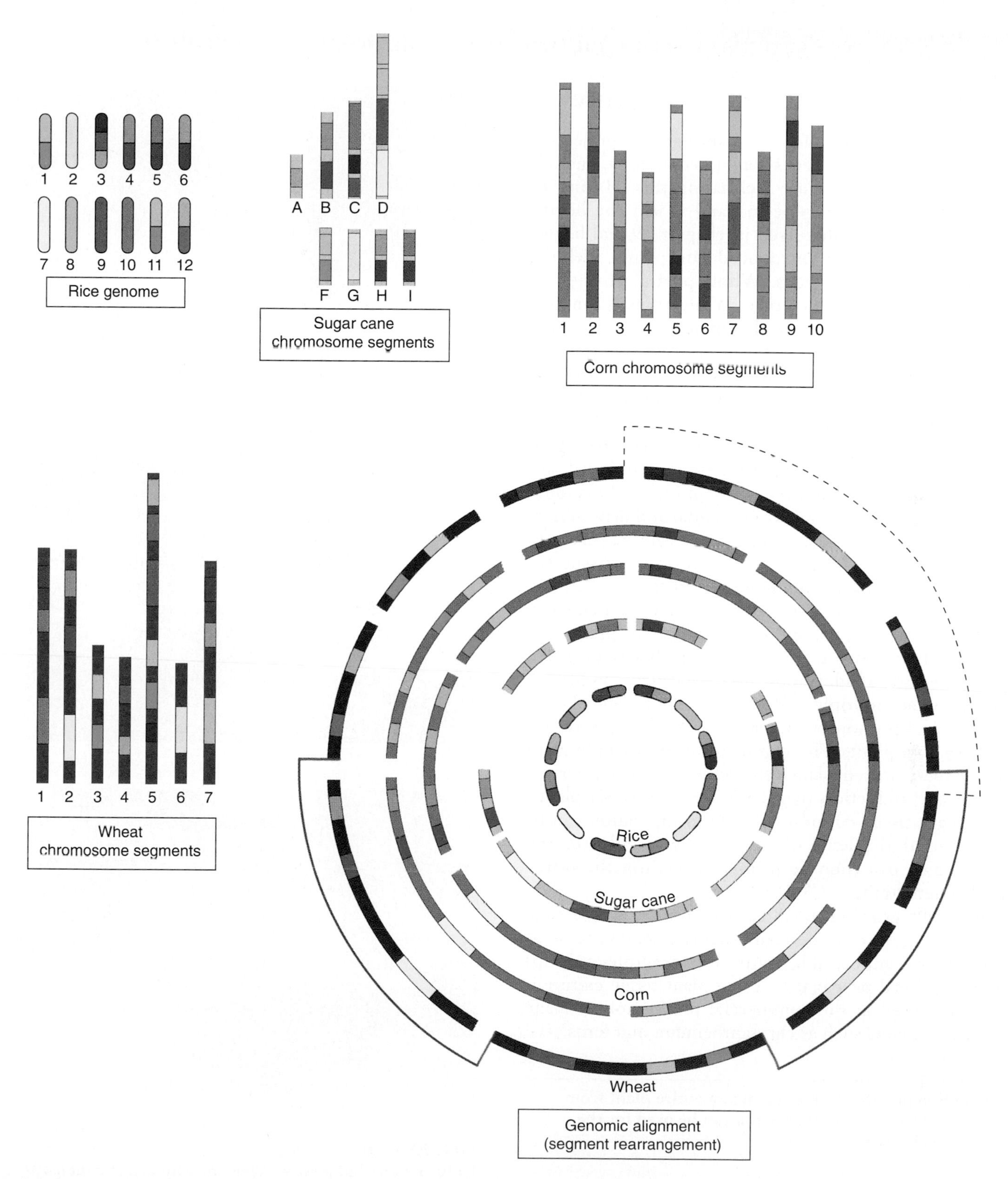

FIGURE 43.9

Grain genomes are rearrangements of similar chromosome segments. Shades of the same color represent pieces of DNA that are conserved among the different species but have been rearranged. The linear diagrams show where these conserved pieces of DNA are found on the chromosomes of individual species. The circular diagram aligns the conserved fragments, not whole chromosomes, of the grasses. By splitting the individual chromosomes of major grass species into segments, and rearranging the segments, researchers have found that the genome components of rice, sugar cane, corn, and wheat are highly conserved. This implies that the order of the segments in the ancestral grass genome has been rearranged by recombination as the grasses have evolved.

Data: G. Moore, K. M. Devos, Z. Wang, and M. D. Gale: "Grasses, line up and form a circle," *Current Biology* 1995, vol. 5, pp. 737–739.

43.2 Advances in plant tissue culture are revolutionizing agriculture.

Overview of Plant Tissue Culture

One of the major hopes for the plant genome projects is using newly identified genes for biotechnology, and advances in tissue culture are facilitating this. Having an agriculturally valuable gene in hand is just the beginning. With methods discussed in chapter 19 and below, desirable genes can be introduced into plants, yielding **transgenic** cells and tissues. Whole plants can then be regenerated using tissue culture. While animals can now be cloned, the process is much simpler in plants. Many somatic (not germ-line) plant cells are totipotent, which means they can express portions of their previously unexpressed genes and develop into whole plants under the right conditions.

The successful culture of plant cells, tissues, or organs requires utilizing the proper plant starting material, appropriate nutrient medium, and timing of hormonal treatments to maximize growth potential and drive differentiation (figure 43.10). Most plant tissue cultures are initiated from *explants*, or small sections of tissue removed from an intact plant under sterile conditions. After being placed on a sterile growth medium containing nutrients, vitamins, and combinations of plant hormones, cells present in the explant will begin to divide and proliferate. Under appropriate culture conditions, plant cells can multiply and form organs (roots, shoots, embryos, leaf primordia, and so on) and can even regenerate a whole plant. The regeneration of a whole plant from tissue-cultured plant cells represents an important step in the production of genetically engineered plants. Using plant tissue cultures, genetic manipulation can be conducted at the level of single cells in culture, and whole plants can then be produced bearing the introduced genetic trait.

Plant cells growing in culture can also be used for the mass production of genetically identical plants (clones) with valuable inheritable traits. For example, this approach of clonal propagation using plant tissue culture is commonly used in the commercial production of many ornamental plants such as chrysanthemums and ferns.

It is often possible to regenerate an entire plant from one or a few cells. Regeneration can be used for the mass production of plants.

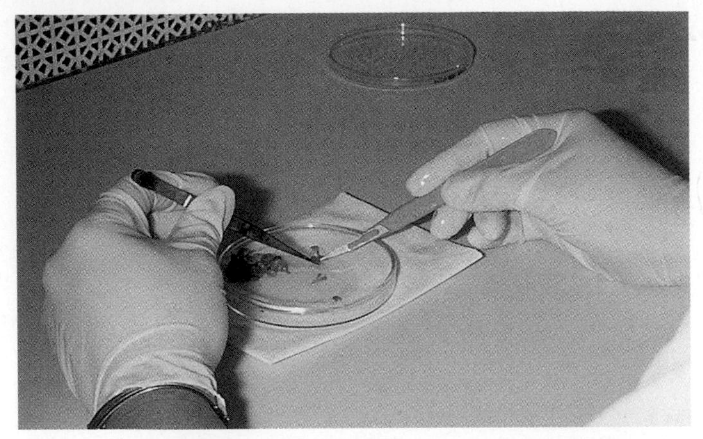

(a)

(b)

(c)

FIGURE 43.10
Culture of orchid plants. While the natural maturation of a single orchid plant may take up to seven years, commercial growers can produce thousands of cultured orchids in a relatively short time. (*a*) The apical meristem is removed from an orchid plant. (*b*) The meristematic tissue is grown in flasks containing hormone-enhanced media, and roots and shoots begin to form. (*c*) The plantlets are then separated and grown to maturity.

Types of Plant Tissue Cultures

Depending on the type of plant tissue used as the explant and the composition of the growth medium, a variety of different types of plant tissue cultures can be generated. These different types of plant tissue cultures have applications both in basic plant research and commercial plant production.

Callus Culture

Callus culture refers to the growth of unorganized masses of plant cells in culture. To generate a callus culture, an explant, usually containing a region of meristematic cells, is incubated on a growth medium containing certain plant hormones such as auxin and cytokinin (figure 43.11). The cells grow from the explant and divide to form an undifferentiated mass of cells called a callus. This unorganized mass of growing cells is analogous to a tumor. Cells can proliferate indefinitely if they are periodically transferred to fresh growth media. However, if the callus cells are transferred to a growth medium containing a different combination of plant hormones, the cells can be directed to differentiate into roots and/or shoots. This process of converting unorganized growth into the production of shoots and roots is called **organogenesis,** and it represents one means by which a whole plant can be regenerated from tissue culture cells. When a plantlet produced by organogenesis is large enough, it can be transferred to a large container with nutrients or soil and grown to maturity.

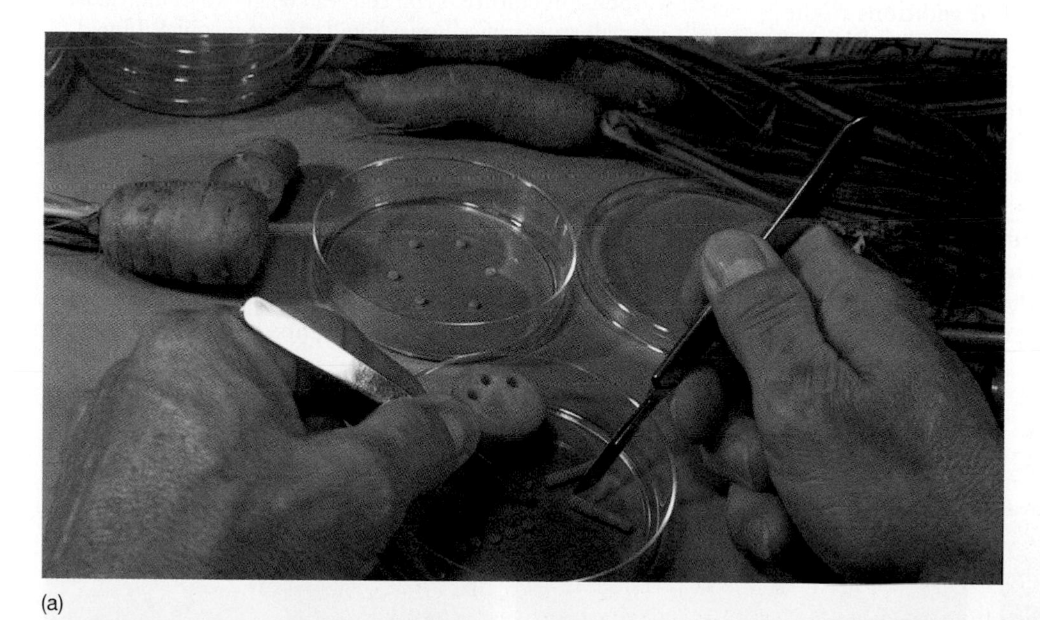

(a)

FIGURE 43.11
Callus culture. (*a*) An explant is incubated on growth media. (*b*) The cells grow and divide and form a callus. (*c*) The callus cells, grown on media containing plant hormones, differentiate into plant parts. (*d*) After the plantlet is large enough, it is grown to maturity in soil.

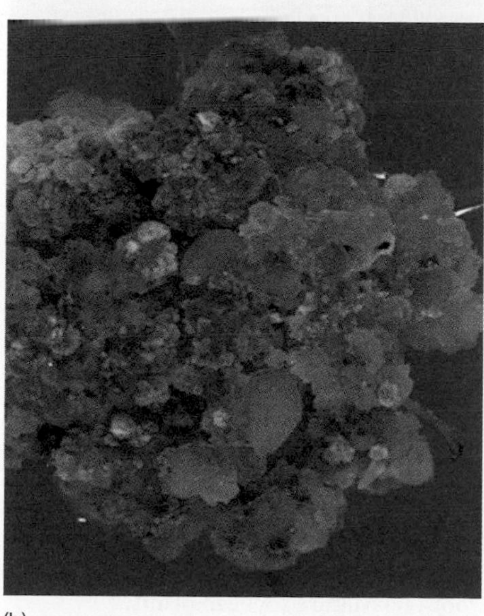

(b)

(c)

(d)

Cell Suspension Culture

Plant cell suspension culture involves the growth of single or small groups of plant cells in a liquid growth medium. Cell suspension cultures are usually initiated by the transfer of plant callus cells into a liquid medium containing a combination of plant hormones and chemicals that promote the disaggregation of the cells into single cells or small clumps of cells (figure 43.12). Continued cell growth requires that the liquid cultures be shaken at low speed to promote aeration and chemical exchange with the medium. Suspension cell cultures are often used in research applications where access to single cells is important. The suspension bath can provide an efficient means for selecting out cells with desirable traits such as herbicide tolerance or salt tolerance because the bath is in uniform contact with all the cells at once. This differs from callus culture, where only those cells in contact with the solid medium can be selected by chemical additions to the medium. Suspension cultures can also provide a convenient means for producing and collecting the chemicals plant cells secrete. These can include important plant metabolites, such as food products, oils, and medicinal chemicals. In addition, plant suspension cell cultures can often be used to produce whole plants via a process known as **somatic cell embryogenesis** (figure 43.13). For some plants, this provides a more convenient means of regenerating a whole plant after genetic engineering takes place at the single-cell level. In somatic cell embryogenesis, plant suspension culture cells are transferred to a medium containing a combination of hormones that drive differentiation and organization of the cells to form individual embryos. Under a dissection microscope, these embryos can be isolated and transferred to a new growth medium, where they grow into individual plants.

FIGURE 43.12
Cell suspension culture. Plant cells can be grown as individual cells or small groups of cells in a liquid culture medium. Liquid suspension culture of plant cells ensures that most cells are in contact with the growth medium.

FIGURE 43.13
Somatic cell embryogenesis. A large number of plants can be cloned from a single soybean seed via somatic cell embryogenesis. (*a*) Immature soybean seeds placed on culture medium. (*b*) Embryos appear on the seeds after two weeks in culture. (*c*) Four embryos at different stages of development (globular, heart, torpedo, and plantlet). (*d*) Seedlings with shoots and roots. (*e*) Mature soybean plants.

Protoplast Isolation and Culture

Protoplasts are plant cells that have had their thick cell walls removed by an enzymatic process, leaving behind a plant cell enclosed only by the plasma membrane. Plant protoplasts have been extremely useful in research on the plant plasma membrane, a structure normally inaccessible due to its close association with the cell wall. Within hours of their isolation, plant protoplasts usually begin to resynthesize cell walls, so this process has also been useful in studies on cell wall production in plants. Plant protoplasts are also more easily transformed with foreign DNA using approaches such as electroporation (see page 870). In addition, protoplasts isolated from different plants can be forced to fuse together to form a hybrid. If they are regenerated into whole plants, these hybrids formed from protoplast fusion can represent genetic combinations that would never occur in nature. Hence, protoplast fusion can provide an additional means of genetic engineering, allowing beneficial traits from one plant to be incorporated into another plant despite broad differences between the species. When either single or fused protoplasts are transferred to a culture growth medium, cell wall regeneration takes place. This is followed by cell division to form a callus (figure 43.14). Once a callus is formed, whole plants can be produced in culture.

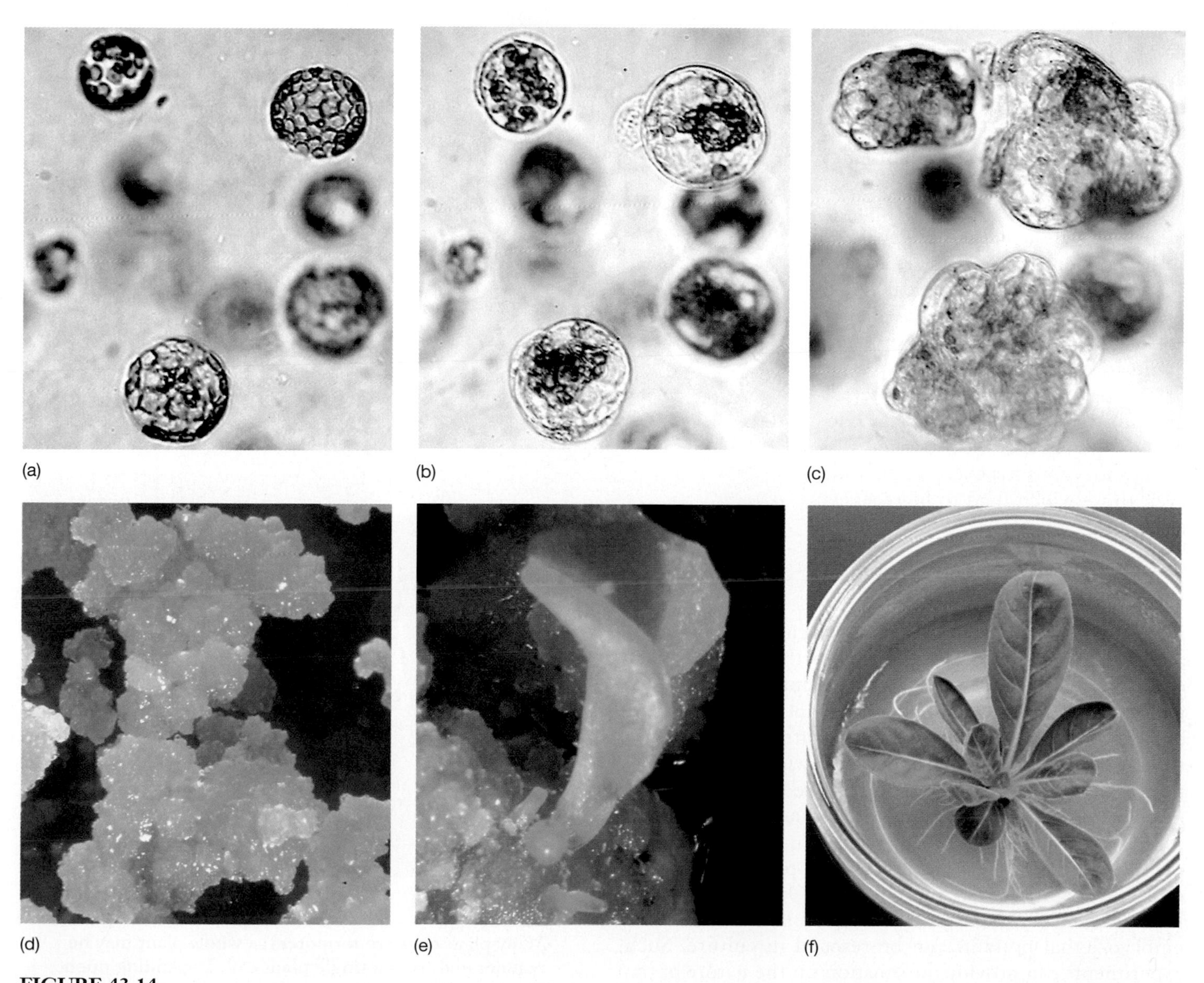

FIGURE 43.14
Protoplast regeneration. Different stages in the recovery of intact plants from single plant protoplasts of evening primrose. (*a*) Individual plant protoplasts. (*b*) Regeneration of the cell wall and the beginning of cell division. (*c, d*) Aggregates of plant cells resulting from cell division which can form a callus. (*e*) Production of somatic cell embryos from the callus. (*f*) Recovery of a plantlet from the somatic cell embryo through the process described in figure 43.13.

Anther/Pollen Culture

In flowers, the anthers are the anatomical structures that contain the pollen. In normal flower development, the anthers mature and open to allow pollen dispersal. In anther culture, anthers are excised from the flowers of a plant and then transferred to an appropriate growth medium. After a short period of time, pollen cells can be manipulated to form individual plantlets, which can be grown in culture and used to produce mature plants. The development of these plantlets usually proceeds through the formation of embryos (figure 43.15). Plants produced by anther/pollen culture can be haploid because they were originally derived from pollen cells that have undergone meiosis. However, these plants may be sterile and thus not useful for breeding or genetic manipulation. On the other hand, plants derived from anther/pollen culture can be treated at an early stage with chemical agents such as colchicine, which allows chromosome duplication. Chromosome duplication results in the conversion of sterile haploid plants into fertile diploid organisms. Under these conditions, plants can be produced that are homozygous for every single trait, even those which tend to be recessive traits.

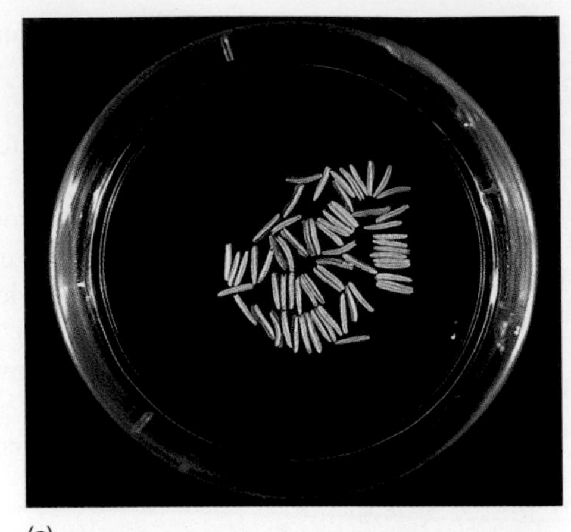

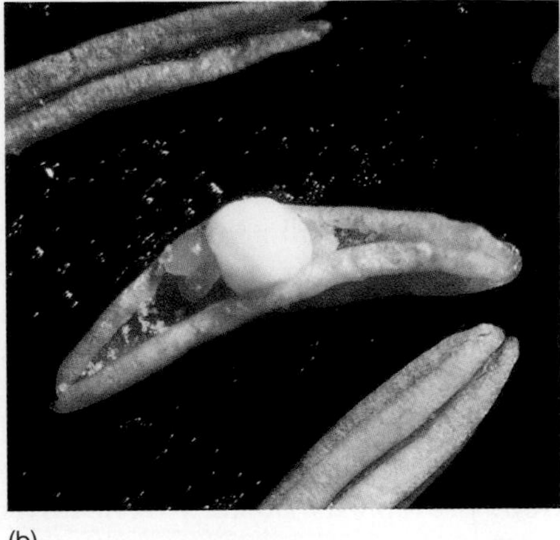

(b)

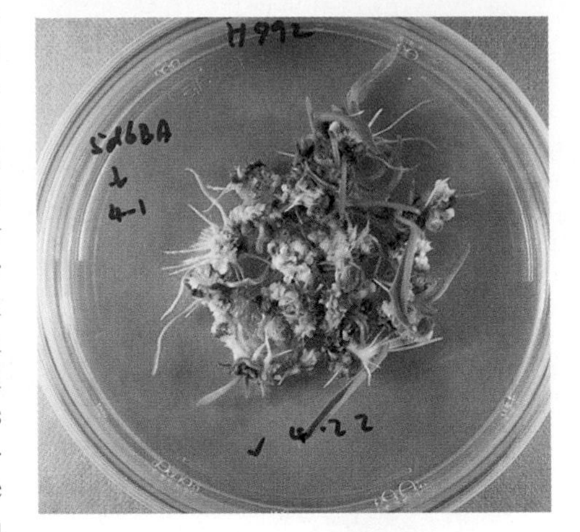

(c)

FIGURE 43.15
Anther culture. Callus formation from maize pollen. Anthers containing pollen can be regenerated on tissue culture medium. The pollen in the anthers contain a haploid set of chromosomes, which can be doubled to form a homozygous diploid cell. Regenerated homozygous diploid plants are important for plant breeding purposes. (*a*) Maize anthers in culture medium. (*b*) Callus formation from pollen. (*c*) Callus and shoot formation.

Plant Organ Culture

Plant organs can also be grown under culture conditions, and this has provided a useful tool in the study of plant organ development. For example, pollinated flowers of a plant such as a tomato can be excised and transferred to a culture flask containing an appropriate medium. Over time, the ovular portion of the plant will develop into a tomato fruit that will eventually turn red and ripen. Sections of plant roots can also be excised and transferred to a liquid growth medium. In this medium, the roots can proliferate extensively, forming both primary and secondary root branches (figure 43.16). Organ culture has been used to study embryo development mutants. Some embryo-lethal mutants can be rescued in culture. Such experiments can provide information on the nature of the defect.

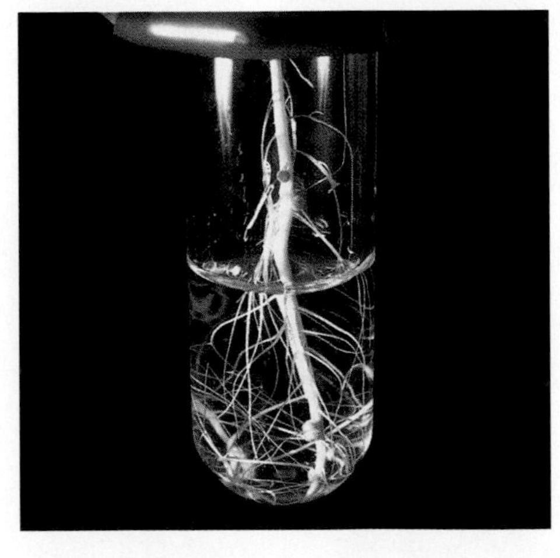

FIGURE 43.16
Plant organ culture. Plant roots growing in a liquid culture medium. From small excised sections of plant roots, the roots will grow and proliferate with extensive lateral root formation (branching).

Many plant cells are totipotent; a whole plant may be regenerated from a single plant cell. Depending upon the explant type, culture medium, and combinations of plant hormones, it is possible to grow plant cells, tissues, or organs in sterile cultures.

Applications of Plant Tissue Culture

In addition to the applications already described, plant tissue cultures have a variety of uses both in agriculture and in industry.

Suspension Cultures as Biological Factories

An important industrial application of plant tissue culture involves the use of plant cells as biological factories. Large-scale suspension cultures can be grown to produce antimicrobial compounds, antitumor alkaloids, vitamins, insecticides, and food flavors. Plant roots can also be grown in liquid culture, creating a mesh of roots that can produce a number of useful plant compounds including anesthetic and antispasmodic drugs.

Horticultural Uses

Plants with valuable traits can be mass propagated through tissue-culture cloning. In this application of plant tissue culture, hundreds or even thousands of genetically identical plants can be produced by vegetative asexual propagation from one plant source. This has been extensively used in the flower industry where genetically identical plants can be produced from a superior parent plant. Propagation of plant tissue in the sterile environment of the growth medium can also help in the production of disease-free plants, such as those cultured from the meristematic (apical dome) tissue untouched by viruses or other diseases because it is new growth. This approach has been particularly useful in the culture of disease-free orchids and raspberries.

(a)

(b)

(c)

FIGURE 43.17
Somaclonal variation. Regeneration of plants from tissue culture can produce plants that are not similar to their parents due to chromosomal alterations. This variability can be used to select plants with altered traits. These maize plants show evidence of somaclonal variation. (*a*) Yellow leaf stripe. (*b*) Dwarf maize. (*c*) Yellow leaf tip.

Somaclonal Variation

Plant tissue culture also has a problematic side effect that can be used as an asset under certain conditions. During periods of extended growth of plant cells in callus or suspension cell culture, various parts of the plant genome may become more or less "active" due to a release of control over gene expression. Transposable elements may also become more active, and chromosomal rearrangements may occur. Also, cultured tissues may end up with unusual numbers of chromosomes. Such alterations provide a new source of genetic diversity that can result in novel traits which were not even present in the original plant material used as the explant to start the cultures (figure 43.17). This increased genetic diversity following extended time in tissue culture is called **somaclonal variation.** It can be problematic if the desired goal is the propagation or production of identical plant clones. However, somaclonal variation, induced by intentionally growing plant cells in tissue culture over a longer time period, can be very useful to generate novel plants with traits not currently present in a given gene pool. Some traits can be identified at the tissue culture stage (for example, disease resistance or heat tolerance). Analysis of other traits (plant size, photosynthetic rates, and so forth) requires the regeneration of whole plants.

Plant cell, tissue, and organ cultures have important applications in agriculture and industry.

43.3 Plant biotechnology now affects every aspect of agriculture.

Plant biotechnology provides an efficient means to produce an array of novel products and tools for use by our global society. Agricultural biotechnology has the potential to increase farming revenue, lower the cost of raw materials, and improve environmental quality. Plant genetic engineering is becoming a key tool for improving crop production.

World Population in Relation to Advances Made in Crop Production

Due in a large part to scientific advances in crop breeding and farming techniques, world food production has doubled since 1960. Moreover, productivity of agricultural land and water usage has tripled over this time period. While major genetic improvements have been made in crops through crop breeding, this can be a slow process. Furthermore, most crops grown in the United States produce less than 50% of their genetic potential. These shortfalls in yield are due in large part to the inability of crops to tolerate or adapt to environmental stresses (salt, water, and temperature), herbivores, pathogens, and disease (figure 43.18).

The world now farms an area the size of South America, but without the scientific advances of the past 30 years, farmland equaling the entire western hemisphere would be required to feed the world. Nevertheless, the world population is expected to double to 12 billion by the first half of this century, and it is not clear whether current levels of food production can keep pace with this rate of population growth. Many believe the exploitation of conventional crop breeding programs may have reached their limit. The question is how best to feed billions of additional people without destroying much of the planet in the process. In this respect, the disappearance of tropical rain forests, wetlands, and other vital habitats will accelerate unless agriculture becomes more productive and less taxing to the environment. Advances in our understanding of plant reproduction are providing tools to further protect natural environments by preventing the spread of modified genes to wild populations.

Although improved farming practices and crop breeding have increased crop yields, it is uncertain whether these approaches can keep pace with the food demands of an ever-increasing world population.

FIGURE 43.18 Corn crop productivity well below its genetic potential due to drought stress. Corn production can be limited by water deficiencies due to drought during the growing season in dry climates.

Plant Biotechnology for Agricultural Improvement

It seems certain that plant genetic engineering will play a major role in resolving the problem of feeding an increasing world population. The nutritional quality of crop plants is being improved by increasing the levels of nutrients they contain, such as beta-carotene and vitamins A, C, and E, which may protect people from health problems such as cancer and heart disease. Biotechnology is now being employed to improve the quality of seed grains, increase protein levels in forage crops, and improve plant resistance to disease, insects, herbicides, and viruses. Other stresses on plants, such as heat or salt, can be improved by engineering higher tolerance levels.

Compared with approaches that rely on plant breeding, genetic engineering can compress the time frame required for the development of improved crop varieties. Moreover, in genetic engineering, genetic barriers, such as pollen compatibility with the pistil, no longer limit the introduction of advantageous traits from diverse species into crop plants. Once a useful trait has been identified at the level of individual genes, the incorporation of this trait into a crop plant requires only the introduction of the DNA bearing these genes into the crop plant genome. The process of incorporating foreign DNA into an existing plant genome is called **plant transformation.** At present, there are several approaches for plant transformation; the use of *Agrobacterium tumefaciens* in this process was described in chapter 19. This approach works best if the plant being transformed is a dicot. However, many food crops, such as the cereal grains (rice, wheat, corn, barley, oats, and so on) are monocots. Two additional plant transformation methods that can be used with both dicots and monocots are discussed in the next section.

Useful Traits That Can Be Introduced into Plants

Although plant transformation represents a relatively new technology, extensive efforts are underway to utilize this approach to develop plants and food products with beneficial characteristics. We discussed a variety of biotechnological applications for crop improvement in chapter 19. Further applications of this approach involve modifications of nutritional quality of foods, phytoremediation, production of biodegradable plastics, and using plants as "edible vaccines."

Improved Nutritional Quality of Food Crops. Approximately 75% of the world's production of oils and fats come from plant sources. For medical and dietary reasons, there is a trend away from the use of animal fats and toward the use of high-quality vegetable oils. Genetic engineering has allowed researchers to modify seed oil biochemistry to produce "designer oils" for edible and nonedible products. One approach modifies canola oil to replace cocoa butter as a source of saturated fatty acids; others modify the enzyme ACP desaturase for the creation of monounsaturated fatty acids in transgenic plants. High-lauric acid canola has been planted in several countries and used in both foods and soaps.

Attempts are also underway to modify the amino acid contents of various plant seeds to present a more complete nutritional diet to the consumer. A high-lysine corn seed is being developed; this would cut down on the need for lysine supplements that are currently added to livestock feed. Biotechnology has the potential to make plant foods healthier and more nutritious for human consumption. Fruits and vegetables, such as tomatoes, may be engineered to contain increased levels of vitamins A and C and beta-carotene, which, when included in the human diet, may help protect against chronic diseases.

Phytoremediation. Cleaning up environmental toxins to reclaim polluted land is an ongoing challenge. Genetically modified plants offer an enticing solution. Work is progressing on plants that accumulate heavy metals at high concentrations. These plants can then be harvested, followed by safe disposal of the dried plant material. Because most of their biomass is water, the dried plants allow for metals collected over a large area to be reduced to a small, contained area. Organic compounds that pose hazards to human health have the potential to be taken up by plants and broken down into harmless components. Modified biochemical pathways are being used to break down toxic substances. Modified poplars, for example, have been engineered to break down TNT.

Plants Bearing Vaccines for Human Diseases. Another very interesting application of plant genetic engineering includes the introduction of "vaccine genes" into edible plants. Here, genes encoding the antigen (for example, a viral coat protein) for a particular human pathogen are introduced into the genome of an edible plant such as a banana, tomato, or apple via plant transformation. This antigen protein would then be present in the cells of the edible plant, and a human individual that consumed the plant would develop antibodies against the pathogenic organism. Currently, researchers are trying to develop such edible vaccines for a coat protein of hepatitis B, an enterotoxin B of *E. coli*, and a viral capsid protein of the Norwalk virus. The gene from the measles virus has been introduced into tobacco as a model system and is now being introduced into lettuce and rice. This promises to be a terrific advance for tropical areas where it is difficult to keep the traditional vaccine cold (remember that proteins degrade rapidly as the temperature increases).

Genetic engineering of crop plants has allowed researchers to alter the oil content, amino acid composition, and vitamin content of food crops. Genetic engineering may also allow the production of food crops bearing "edible vaccines."

Methods of Plant Transformation

Plant Transformation Using the Particle Gun

Unlike animal cells, plant cells have a wall that presents a barrier to getting DNA into a cell. Using a "gun" to blast plant cells does not seem like a suitable method for introducing foreign DNA into a plant genome. However, it works. The *particle gun* utilizes microscopic gold particles coated with the foreign DNA, shooting these particles into plant cells at high velocity. Acceleration of the particles to a sufficient velocity to pass through the plant cell wall can be achieved by a burst of high-pressure helium gas or an electrical discharge (figure 43.19*a*). Only a few cells actually receive the foreign DNA and survive this treatment. These cells are identified with the help of a **selectable marker** also present on the foreign DNA. The selectable marker allows only those cells receiving the foreign DNA to survive on a particular growth medium (figure 43.20). The selectable markers include genes for resistance to a herbicide or antibiotic. Plant cells which survive growth in the selection medium are then tested for the presence of the foreign gene(s) of interest.

Plant Transformation Using Electroporation

Foreign DNA can also be "shocked" into cells that lack a cell wall, such as the plant protoplasts described earlier. A pulse of high-voltage electricity in a solution containing plant protoplasts and DNA briefly opens up small pores in the protoplasts' plasma membranes, allowing the foreign DNA to enter the cell (figure 43.19*b*). Ideally the DNA incorporates into one of the plant's chromosomes. Following electroporation, the protoplasts are transferred to a growth medium for cell wall regeneration, cell division, and, eventually, the regeneration of whole plants. As with the use of the particle gun, a selectable marker is typically present in the foreign DNA, and protoplasts containing foreign DNA are selected based upon their ability to survive and proliferate in a growth medium containing the selection treatment (antibiotic or herbicide). Once regenerated from electroporated protoplasts, whole plants can then be evaluated for the presence of the beneficial trait.

Plant biotechnology may play an important role in the further improvement of crop plants. The particle gun and electroporation are useful methods for introducing foreign DNA into plants.

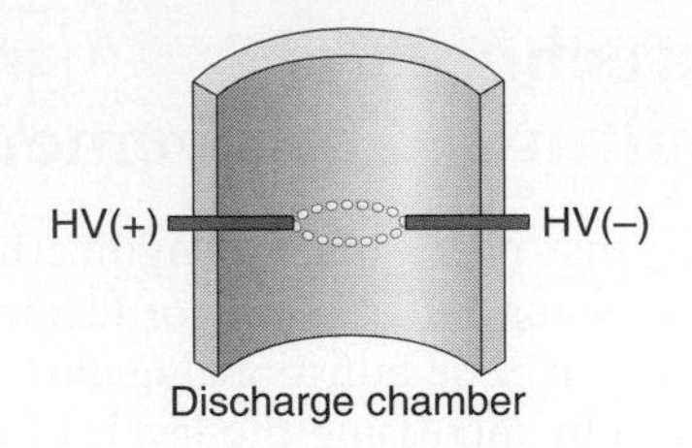

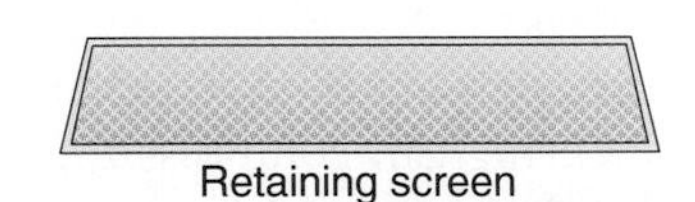

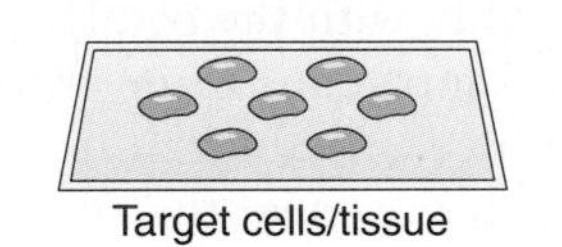

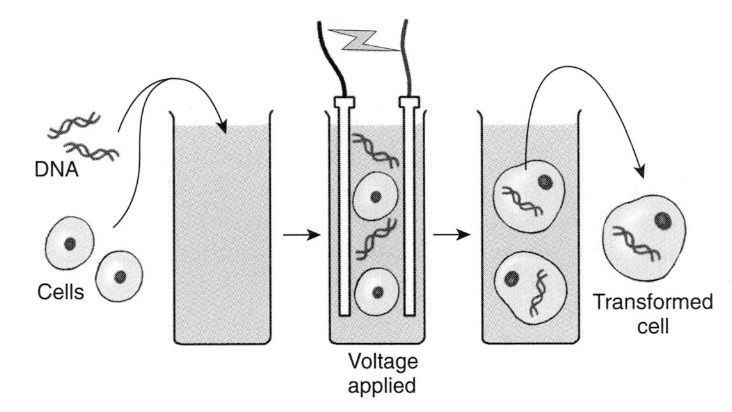

FIGURE 43.19

Methods for plant transformation. (*a*) The particle gun is one method for introducing foreign DNA into plant cells. Here an electrical discharge propels DNA-coated gold particles into plant cells or tissue. A retaining screen reduces cellular damage associated with bombardment by only allowing the DNA-coated particles to pass and retaining fragments of the mounting film. (*b*) Foreign DNA can also be introduced into plant protoplasts by electroporation. A brief pulse of electricity generates pores in the plasma membrane, allowing DNA to enter the cells.

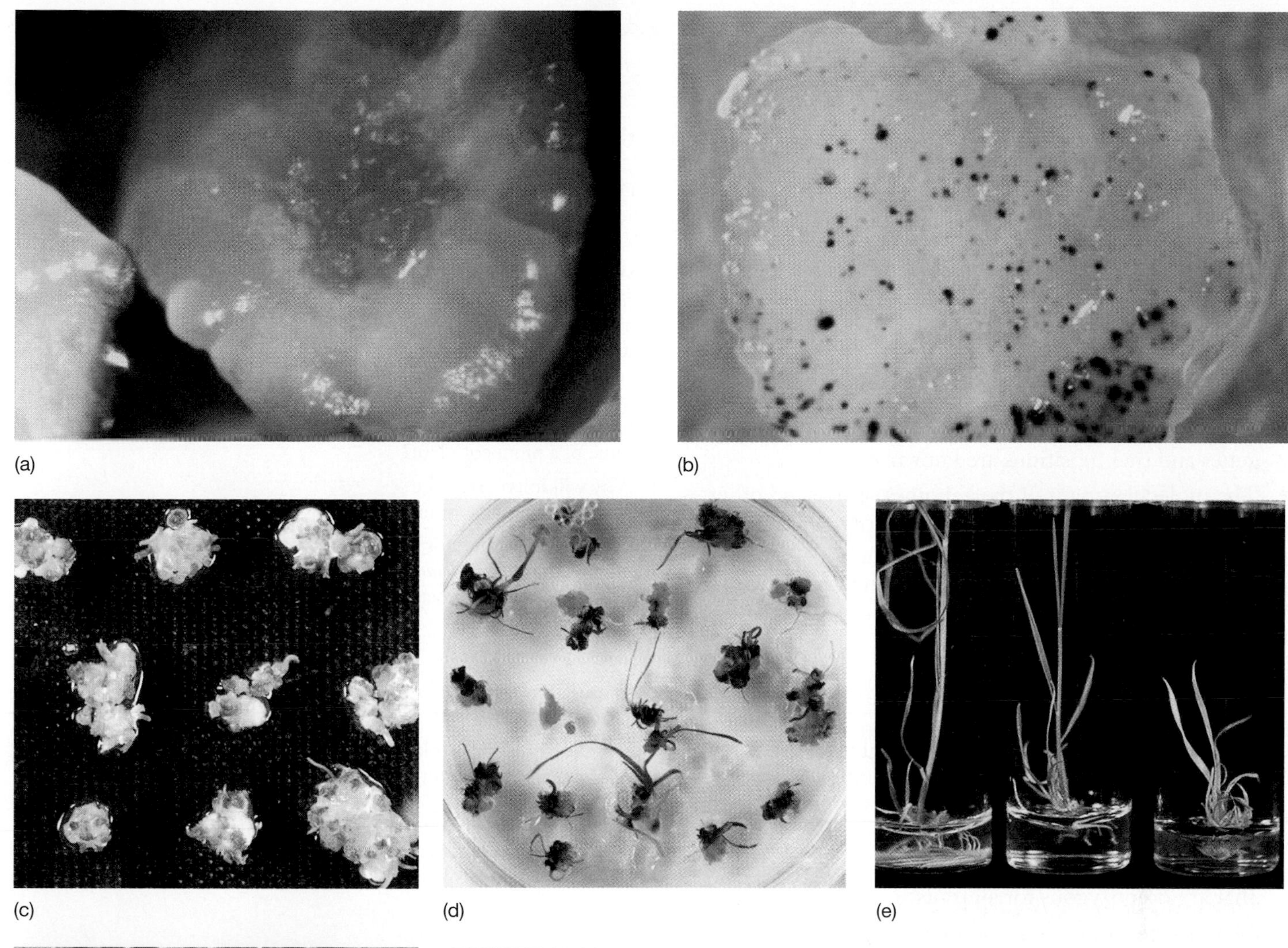

FIGURE 43.20
Regeneration after transformation with the use of a selectable marker. Stages in the recovery of a plant containing foreign DNA introduced by the "particle gun" method for plant cell transformation. A selectable marker, in this case a gene for resistance to herbicide, aids in the identification and recovery of plants containing the DNA insert. (*a*) Embryonic callus just prior to particle gun bombardment. (*b*) Following bombardment, callus cells containing the foreign DNA are indicated by color from the *gus* gene used as a tag or label on the foreign DNA. (*c*) Shoot formation in the transformed plants growing on a selective medium. Here, the gene for herbicide resistance in the transformed plants allows growth on the selective medium containing the herbicide. Nontransformed plants do not contain the herbicide resistance gene and do not grow well. (*d*) Production of plantlets from transformed plants growing on the selective medium. (*e*) Comparison of growth on the selection medium for transformed plants bearing the herbicide resistance gene (*left*) and a nontransformed plant (*right*). (*f*) Mature transgenic plants resulting from this process.

Chapter 43

www.mhhe.com/raven6/resources43.mhtml

Summary | **Questions** | **Media Resources**

43.1 Genomic organization is much more varied in plants than in animals.

- Plant genomes are very large in comparison to other eukaryotes, mainly due to a high amount of repetitive DNA.
- Plant genomes can be compared with one another by mapping the locations of certain genes or gene traits in various plants. RFLPs and AFLPs can be used to map plant DNA.
- *Arabidopsis thaliana* has a small genome, for a plant. This complete genome is essentially sequenced, so all genes and their positions are known.
- The molecular maps of the genomes of rice and other grains demonstrate remarkable similarity.
- Functional genomics and proteomics will allow us to understand and utilize the information in fully sequenced plant genomes.

1. Describe mechanisms for the generation of highly repetitive DNA in plants.

2. What characteristics of *Arabidopsis thaliana* make it useful as a model system in genetic studies and for the sequencing of its entire genome? Why is rice useful as a model system for the analysis of the genome of a monocot plant?

3. Why will microarrays be useful in functional genomics?

4. What type of questions can be asked now that the *Arabidopsis* and rice genomes are essentially sequenced?

43.2 Advances in plant tissue culture are revolutionizing agriculture.

- With the addition of appropriate combinations of plant hormones (auxin, cytokinin), plant cells in culture can be directed to form organs, embryos, or whole plants.
- Anther cultures can produce haploid plants or plants that are homozygous for all traits.
- Plant tissue culture has a number of practical applications, including the industrial production of plant chemicals, clonal propagation of horticultural plants, and the generation of disease-free plants.
- Growth of plant cells in tissue culture over extended time results in an increase in genetic variation called somaclonal variation. This variation can extend beyond the traits present in the gene pool and can generate novel genetic variations in breeding studies.

5. Describe how whole plants can be regenerated from tissue-cultured plant cells using either organogenesis or somatic cell embryogenesis. Which approach requires the use of suspension cell cultures?

6. How are plant protoplasts generated, and what is protoplast fusion? How can plant protoplasts be used to generate hybrid plants that would not occur in nature?

43.3 Plant biotechnology now affects every aspect of agriculture.

- Genetic engineering and biotechnology can be utilized to improve the quality of food crops, increase disease resistance, and improve the tolerance of crops to environmental stress.
- A key aspect of plant genetic engineering is the introduction of foreign DNA into plant cells. This can be achieved using a particle gun, electroporation, or *Agrobacterium* as a vector.

7. Describe how the particle gun and electroporation can be used to introduce foreign DNA into plant cells. Which approach requires the use of plant protoplasts? Why?

8. How can a plant be "engineered" to produce an edible vaccine ?

- Student Research: Plant Crop Protection
- Scientists on Science: Plant Biotechnology

Part **XII**

Animal Diversity

Honeybees are endotherms. By adjusting metabolic rates in their flight muscles, honeybees are able to maintain their body temperature within a narrow range.

Real People Doing Real Science

How Honeybees Keep Their Cool

One of the biggest problems terrestrial animals face is that the temperature of their environment keeps changing. Why does changing temperature present a problem? The structures of many of the body's enzymes and regulatory proteins rely on weak chemical bonds, and a change in temperature easily disrupts them, changing their shape and perturbing their function. A decrease or increase of only a few degrees can have a significant impact on metabolism, and great changes in temperature can be fatal.

Terrestrial organisms have taken two very different approaches to confronting the problem posed by changing temperatures. On the one hand, many species have evolved the ability to maintain their body temperature within a narrow range, regardless of the temperature of their surroundings. We call such organisms endotherms. Birds and mammals are the primary examples of this strategy, but certain invertebrates are also endotherms. Because the costs of maintaining a stable body temperature are considerable (90% of your food intake must be expended just to produce heat), the majority of organisms elect instead to allow their body temperature to conform closely to their surrounds. These organisms are called ectotherms.

Imagine you were standing in a high mountain meadow on a cool sunny morning, with butterflies and bees flying about among the flowers. A cloud obscures the sun, blocking its warming rays. The butterflies immediately settle to the ground and wait for the sunlight they need to keep warm enough to fly. Not the bees. They keep zipping around from flower to flower, heedless of the loss of the sun's warmth. How does the bee manage this? Bees are one of the few invertebrates that are endotherms. Aloft, they keep their body temperature relatively constant, whatever the temperature of the air through which they are flying.

How does the bee—or any endothermic animal—generate the heat needed to warm itself? About 75% of the energy content of the chemical bonds in an animal's foodstuffs is dissipated as heat during metabolism. Thus no new type of metabolic scheme is required to be an endotherm. Instead, an endotherm revs up the metabolic reactions it already has, finding ways to increase the flux through its preexisting food-burning pathways.

When a bee wants to fly, it first shivers violently, often for several minutes. This burning of glucose generates the heat needed to become airborne. Once aloft, the flight muscles generate enough heat to maintain the bee's body temperature within the range required for flight.

So what happens if the sun *stays* out, warming everything hotter and hotter? Eventually the butterflies settle back to the ground in cooler shady spots, avoiding sunlight and waiting for cooler moments. Not so the bees. They keep right on flying, again oblivious to the sun.

How do these bees avoid overheating? Researchers studying bumble bees thought they had the answer. Bumble bees have large abdomens, devoid of insulation, which radiate heat very effectively. When the bumble bee is not flying (and little heat is being produced), there is little flow of heat from the thorax. When the bumble bee starts to fly, heat is carried from the flight muscles of the thorax to the abdomen by blood (called hemolymph), which circulates in response to the body's movements. In effect, the bumble bee has an endothermic thorax and an ectothermic abdomen which acts as a heat dissipater.

However, when researchers looked at honeybees (the kind of bees flying in our mountain meadow), this explanation doesn't work. It turns out that transfer of heat between thorax and abdomen is not an important mechanism for regulating body temperature in flying honeybees. So how do honeybees keep cool while flying?

In stationary animals including birds, variation in metabolic heat production is an important mechanism of thermoregulation (witness shivering in response to cold). This mechanism, however, was thought to be unavailable to flying animals. Flight requires a huge expenditure of energy to keep the animal airborne, and the metabolic costs of flight are commonly thought to be determined by factors such as the animal's weight, wing area, flight speed, etc. How could a flying animal vary metabolism for thermoregulation and at the same time accomplish flight?

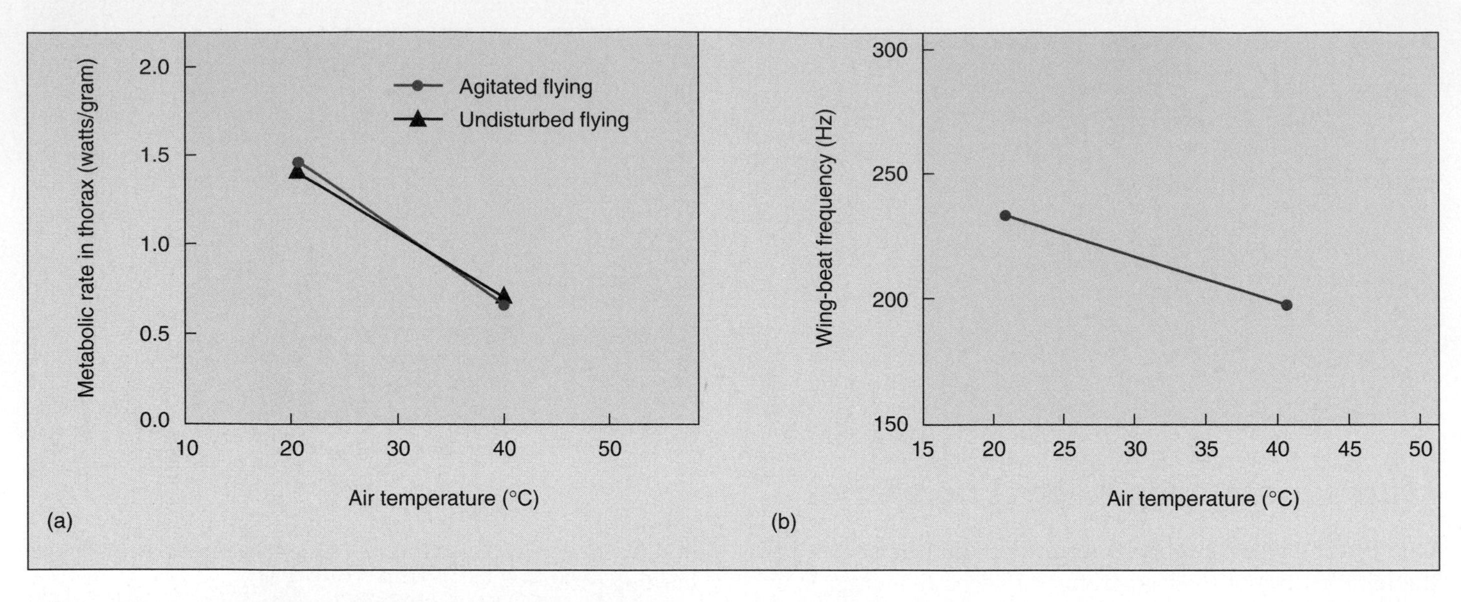

Effects of air temperature on metabolic rate and wing-beat frequency. (*a*) While the air temperature was increased from 20° to 40°C, metabolic rates of honeybees were determined by measuring carbon dioxide emissions in bees that were agitated to keep them flying (dots) and in bees that were allowed to fly undisturbed (triangles). In both cases, the metabolic rate dropped with increasing temperatures. (*b*) The wing-beat frequency was determined by measurements made with digitized tape recordings. The wing-beat frequency decreased with increasing temperatures.

The Observation

In an experiment designed to compare the flight metabolic rates of African and European honeybees, Jon Harrison (Arizona State University) and H. Glenn Hall (University of Florida) discovered that relatively small increases in air temperature were correlated with substantial decreases in flight metabolic rate and wingbeat frequency. This result suggests the hypothesis that honeybees may be able to vary their metabolic rate by changing flight muscle performance, and in this way achieve thermoregulation. However, the observed correlation might have other explanations. The air temperature variation in this study occurred as temperatures varied within and across days. Perhaps bees that fly in the afternoon, when it is warm, are simply a different age or genetic makeup than bees that fly in cooler weather. A manipulative experiment is required to rigorously test the hypothesis.

The Experiment

To test this hypothesis, Dr. Harrison and a team of researchers randomly exposed honeybees to various air temperatures between 20° to 40°C, and measured thorax and abdominal temperature, metabolic rate, and wingbeat frequency. Body temperatures were taken with a tiny, fast-responding microprobe. Metabolic rates were measured by flowing air through the bee flight chamber at a known rate, and measuring the release of carbon dioxide by the bee. Wingbeat frequency was measured using a microphone, tape recorder, and sound-editing software.

The Results

Can flying honeybees vary heat transfer between thorax and abdomen like bumblebees? If honeybees thermoregulate the thorax by varying heat transfer between thorax and abdomen, then lots of heat should be transferred from the thorax to the abdomen when the bee flies in warm air, but little heat should be transferred when the air is cold. In this experiment, thorax temperatures varied much less than air temperature. Clearly, theses honeybees did not thermoregulate like bumblebees.

Can flying honeybees vary metabolic heat production with air temperature? Flight metabolic rates decreased 40 to 50% in bees flying in 40°C air compared to 20°C air (graph *a* above). This indicates that honeybees maintain warm, relatively constant thorax temperatures by producing lots of metabolic heat at cool air temperatures, and much less metabolic heat when flying in warm air.

How can honeybees vary metabolic heat production and still hover? Harrison and his colleagues found that the wingbeat frequency of the hovering honeybees fell by 16% as air temperature increased (graph *b* above). These results suggest that honeybees may vary heat production by varying flight muscle power in a manner somewhat comparable to heat production by shivering muscles in mammals. According to this hypothesis, bees in colder air are able to utilize more metabolic fuels and produce more heat because the flight muscles work harder and use more ATP. Experiments to rigorously test this remarkable mechanism remain to be done.

To explore this experiment further, go to the Virtual Lab at www.mhhe.com/raven6/vlab12.mhtml

44

The Noncoelomate Animals

Concept Outline

44.1 Animals are multicellular heterotrophs without cell walls.

Some General Features of Animals. Animals lack cell walls and move more rapidly and in more complex ways than other organisms. The animal kingdom is divided into animals without symmetry and tissues, and animals with symmetry and tissues.

Five Key Transitions in Body Plan. Over the course of animal evolution, the animal body plan has undergone many changes, five of them significant.

44.2 The simplest animals are not bilaterally symmetrical.

Parazoa. Sponges are the most primitive animals, without either tissues or, for the most part, symmetry.

Eumetazoa: The Radiata. Cnidarians and ctenophorans have distinct tissues and radial symmetry.

44.3 Acoelomates are solid worms that lack a body cavity.

Eumetazoa: The Bilaterian Acoelomates. Flatworms are the simplest bilaterally symmetrical animals; they lack a body cavity, but possess true organs.

44.4 Pseudocoelomates have a simple body cavity.

The Pseudocoelomates. Nematodes and rotifers possess a simple body cavity.

44.5 The coming revolution in animal taxonomy will likely alter traditional phylogenies.

Reevaluating How the Animal Body Plan Evolved. The use of molecular data will most likely generate changes to traditional animal phylogenies.

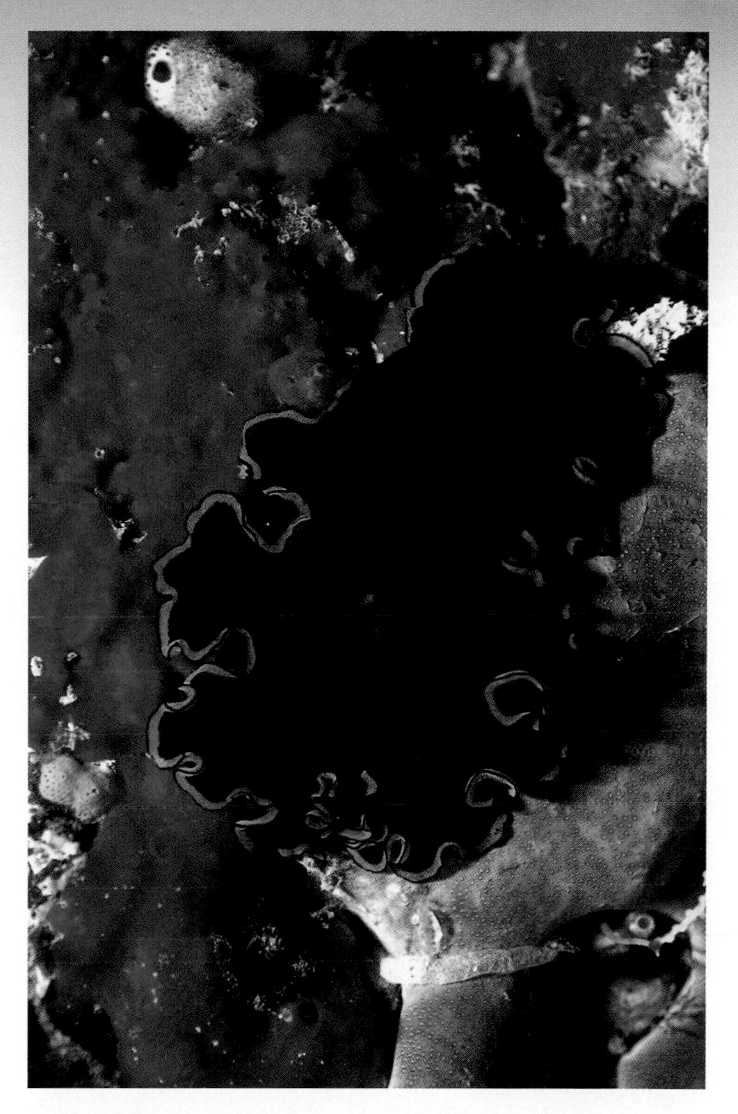

FIGURE 44.1
A noncoelomate: a marine flatworm. Some of the earliest invertebrates to evolve, marine flatworms possess internal organs but lack a true cavity called a coelom.

We will now explore the great diversity of animals, the result of a long evolutionary history. Animals, constituting millions of species, are among the most abundant living things. Found in every conceivable habitat, they bewilder us with their diversity. We will start with the simplest members of the animal kingdom—sponges, jellyfish, and simple worms. These animals lack a body cavity called a coelom, and are thus called noncoelomates (figure 44.1). The major organization of the animal body first evolved in these animals, a basic body plan upon which all the rest of animal evolution has depended. In chapters 45 through 48, we will consider the more complex animals. Despite their great diversity, you will see that all animals have much in common.

44.1 Animals are multicellular heterotrophs without cell walls.

Some General Features of Animals

Animals are the eaters or consumers of the earth. They are heterotrophs and depend directly or indirectly on plants, photosynthetic protists (algae), or autotrophic bacteria for nourishment. Animals are able to move from place to place in search of food. In most, ingestion of food is followed by digestion in an internal cavity.

Multicellular Heterotrophs. All animals are multicellular heterotrophs. The unicellular heterotrophic organisms called Protozoa, which were at one time regarded as simple animals, are now considered to be members of the kingdom Protista, the large and diverse group we discussed in chapter 35.

Diverse in Form. Almost all animals (99%) are **invertebrates,** lacking a backbone. Of the estimated 10 million living animal species, only 42,500 have a backbone and are referred to as **vertebrates.** Animals are very diverse in form, ranging in size from ones too small to see with the naked eye to enormous whales and giant squids. The animal kingdom includes about 35 phyla, most of which occur in the sea. Far fewer phyla occur in fresh water and fewer still occur on land. Members of three phyla, Arthropoda (spiders and insects), Mollusca (snails), and Chordata (vertebrates), dominate animal life on land.

No Cell Walls. Animal cells are distinct among multicellular organisms because they lack rigid cell walls and are usually quite flexible. The cells of all animals but sponges are organized into structural and functional units called **tissues,** collections of cells that have joined together and are specialized to perform a specific function; muscles and nerves are tissues types, for example.

Active Movement. The ability of animals to move more rapidly and in more complex ways than members of other kingdoms is perhaps their most striking characteristic and one that is directly related to the flexibility of their cells and the evolution of nerve and muscle tissues. A remarkable form of movement unique to animals is flying, an ability that is well developed among both insects and vertebrates. Among vertebrates, birds, bats, and pterosaurs (now-extinct flying reptiles) were or are all strong fliers. The only terrestrial vertebrate group never to have had flying representatives is amphibians.

Sexual Reproduction. Most animals reproduce sexually. Animal eggs, which are nonmotile, are much larger than the small, usually flagellated sperm. In animals, cells formed in meiosis function directly as gametes. The haploid cells do not divide by mitosis first, as they do in plants and fungi, but rather fuse directly with each other to form the zygote. Consequently, with a few exceptions, there is no counterpart among animals to the alternation of haploid (gametophyte) and diploid (sporophyte) generations characteristic of plants (see chapter 32).

Embryonic Development. Most animals have a similar pattern of embryonic development. The zygote first undergoes a series of mitotic divisions, called *cleavage,* and becomes a solid ball of cells, the **morula,** then a hollow ball of cells, the **blastula.** In most animals, the blastula folds inward at one point to form a hollow sac with an opening at one end called the **blastopore.** An embryo at this stage is called a **gastrula.** The subsequent growth and movement of the cells of the gastrula produce the digestive system, also called the gut or intestine. The details of embryonic development differ widely from one phylum of animals to another and often provide important clues to the evolutionary relationships among them.

The Classification of Animals

Two subkingdoms are generally recognized within the kingdom Animalia: **Parazoa**—animals that for the most part lack a definite symmetry and possess neither tissues nor organs, mostly comprised of the sponges, phylum Porifera; and **Eumetazoa**—animals that have a definite shape and symmetry and, in most cases, tissues organized into organs and organ systems. Although very different in structure, both types evolved from a common ancestral form (figure 44.2) and possess the most fundamental animal traits.

All eumetazoans form distinct embryonic layers during development that differentiate into the tissues of the adult animal. Eumetazoans of the subgroup Radiata (having radial symmetry) have two layers, an outer **ectoderm** and an inner **endoderm,** and thus are called **diploblastic.** All other eumetazoans, the Bilateria (having bilateral symmetry), are **triploblastic** and produce a third layer, the **mesoderm,** between the ectoderm and endoderm. No such layers are present in sponges.

The major phyla of animals are listed in table 44.1. The simplest invertebrates make up about 14 phyla. In this chapter, we will discuss 8 of these 14 phyla and focus in detail on 4 major phyla: phylum Porifera (sponges), which lacks any tissue organization; phylum Cnidaria (radially symmetrical jellyfish, hydroids, sea anemones, and corals); phylum Platyhelminthes (bilaterally symmetrical flatworms); and phylum Nematoda (nematodes), a phylum that includes both free-living and parasitic roundworms.

Animals are complex multicellular organisms typically characterized by high mobility and heterotrophy. Most animals also possess internal tissues and organs and reproduce sexually.

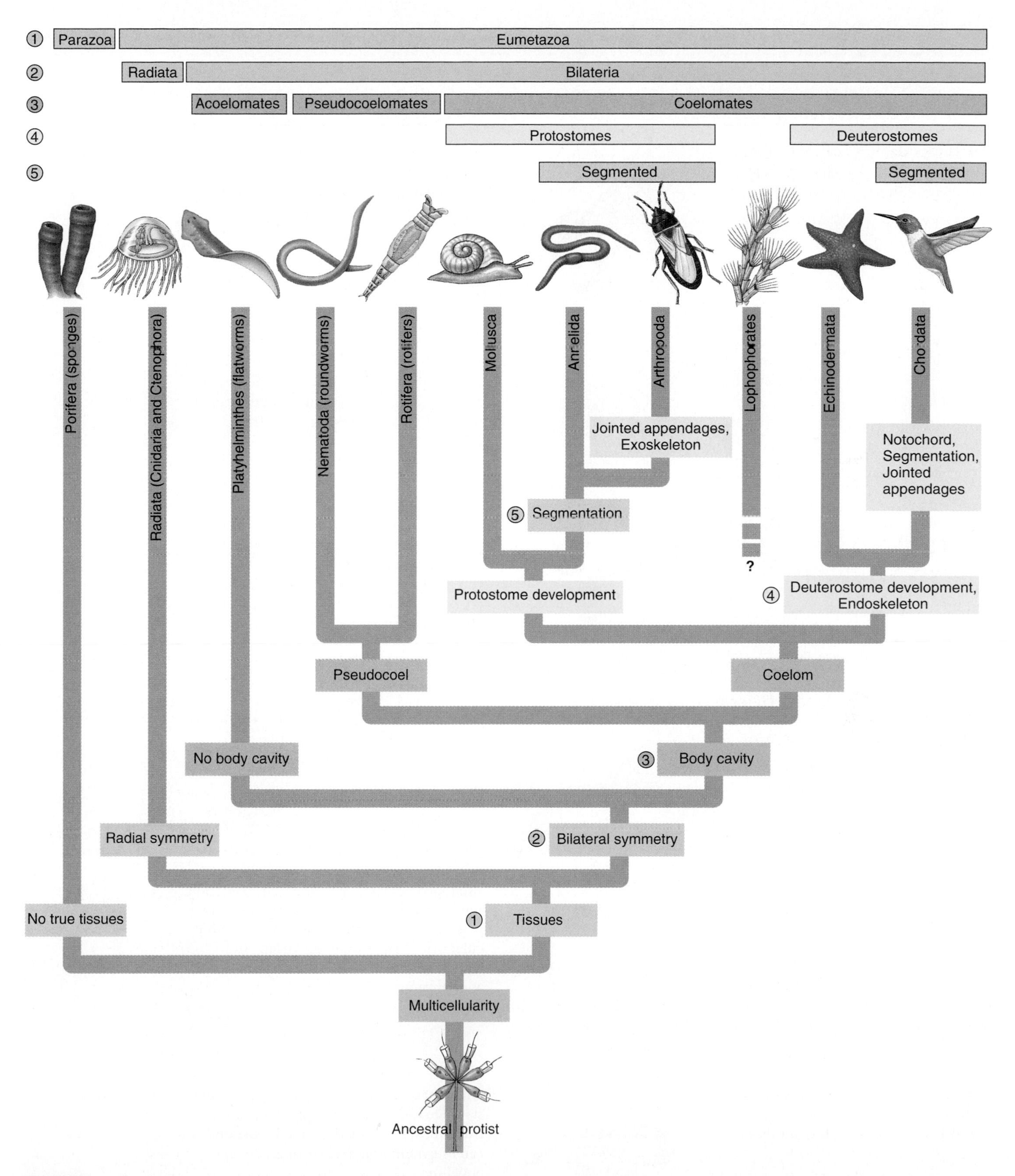

FIGURE 44.2
A possible phylogeny of the major groups of the kingdom Animalia. Transitions in the animal body plan are identified along the branches; the five key advances are the evolution of tissues, bilateral symmetry, a body cavity, protostome and deuterostome development, and segmentation.

Phylum	Typical Examples	Key Characteristics	Approximate Number of Named Species
Arthropoda (arthropods)	Beetles, other insects, crabs, spiders	Most successful of all animal phyla; chitinous exoskeleton covering segmented bodies with paired, jointed appendages; many insect groups have wings	1,000,000
Mollusca (mollusks)	Snails, oysters, octopuses, nudibranchs	Soft-bodied coelomates whose bodies are divided into three parts: head-foot, visceral mass, and mantle; many have shells; almost all possess a unique rasping tongue, called a radula; 35,000 species are terrestrial	110,000
Chordata (chordates)	Mammals, fish, reptiles, birds, amphibians	Segmented coelomates with a notochord; possess a dorsal nerve cord, pharyngeal slits, and a tail at some stage of life; in vertebrates, the notochord is replaced during development by the spinal column; 20,000 species are terrestrial	42,500
Platyhelminthes (flatworms)	*Planaria*, tapeworms, liver flukes	Solid, unsegmented, bilaterally symmetrical worms; no body cavity; digestive cavity, if present, has only one opening	20,000
Nematoda (roundworms)	*Ascaris*, pinworms, hookworms, *Filaria*	Pseudocoelomate, unsegmented, bilaterally symmetrical worms; tubular digestive tract passing from mouth to anus; tiny; without cilia; live in great numbers in soil and aquatic sediments; some are important animal parasites	$12,000^{+}$
Annelida (segmented worms)	Earthworms, polychaetes, beach tube worms, leeches	Coelomate, serially segmented, bilaterally symmetrical worms; complete digestive tract; most have bristles called setae on each segment that anchor them during crawling	12,000

Table 44.1 The Major Animal Phyla

Table 44.1 The Major Animal Phyla (continued)			
Phylum	**Typical Examples**	**Key Characteristics**	**Approximate Number of Named Species**
Cnidaria (cnidarians)	Jellyfish, hydra, corals, sea anemones	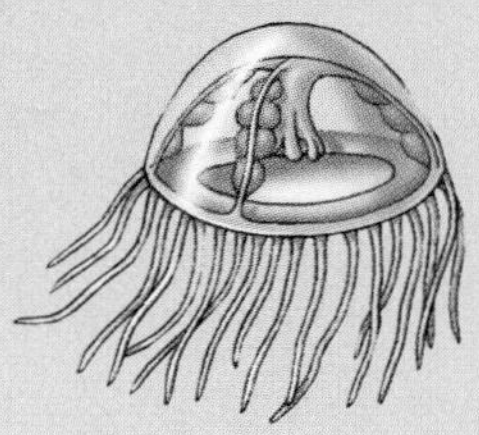 Soft, gelatinous, radially symmetrical bodies whose digestive cavity has a single opening; possess tentacles armed with stinging cells called cnidocytes that shoot sharp harpoons called nematocysts; almost entirely marine	10,000
Echinodermata (echinoderms)	Sea stars, sea urchins, sand dollars, sea cucumbers	 Deuterostomes with radially symmetrical adult bodies; endoskeleton of calcium plates; five-part body plan and unique water vascular system with tube feet; able to regenerate lost body parts; marine	6,000
Porifera (sponges)	Barrel sponges, boring sponges, basket sponges, vase sponges	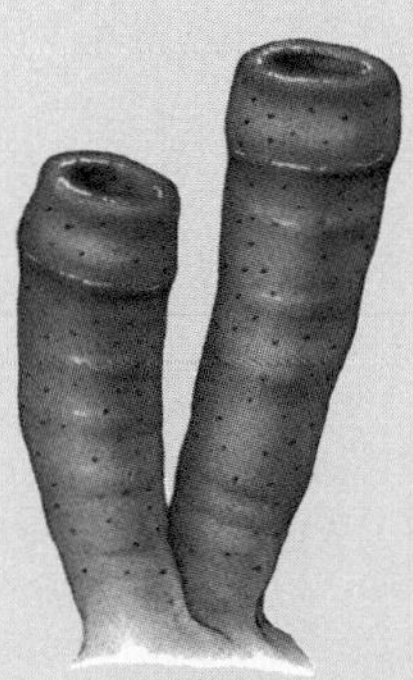 Asymmetrical bodies without distinct tissues or organs; saclike body consists of two layers breached by many pores; internal cavity lined with food-filtering cells called choanocytes; most marine (150 species live in fresh water)	5,150
Bryozoa (moss animals)	*Bowerbankia*, *Plumatella*, sea mats, sea moss	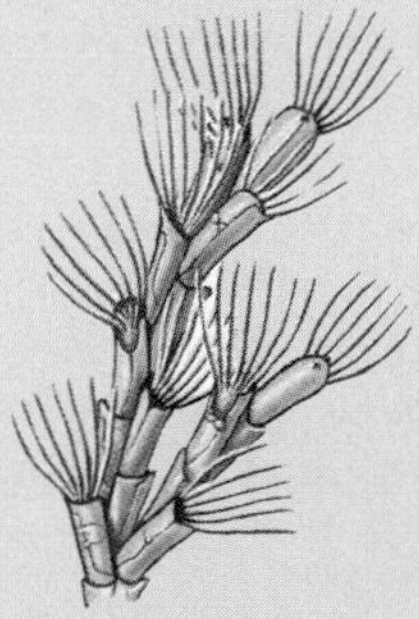 Microscopic, aquatic deuterostomes that form branching colonies, possess circular or U-shaped row of ciliated tentacles for feeding called a lophophore that usually protrudes through pores in a hard exoskeleton; also called Ectoprocta because the anus or proct is external to the lophophore; marine or freshwater	4,000
Rotifera (wheel animals)	Rotifers	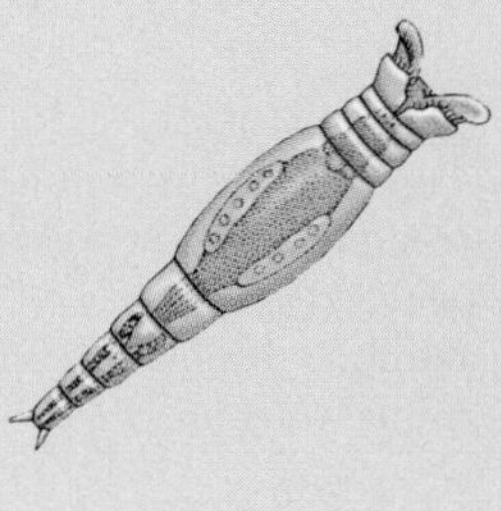 Small, aquatic pseudocoelomates with a crown of cilia around the mouth resembling a wheel; almost all live in fresh water	2,000

Table 44.1 The Major Animal Phyla (continued)			
Phylum	**Typical Examples**	**Key Characteristics**	**Approximate Number of Named Species**
Five phyla of minor worms	Velvet worms, acorn worms, arrow worms, giant tube worms 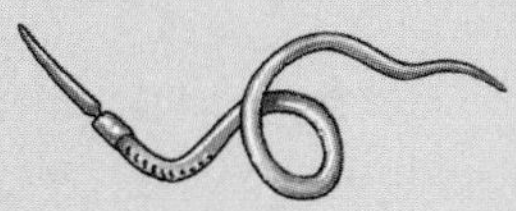 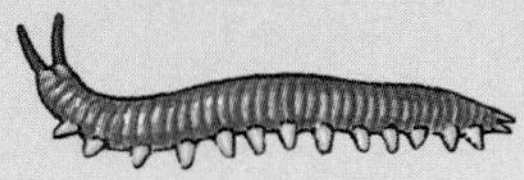	**Chaetognatha** (arrow worms): coelomate deuterostomes; bilaterally symmetrical; large eyes (some) and powerful jaws **Hemichordata** (acorn worms): marine worms with dorsal *and* ventral nerve cords **Onychophora** (velvet worms): protostomes with a chitinous exoskeleton; evolutionary relics **Pogonophora** (tube worms): sessile deep-sea worms with long tentacles; live within chitinous tubes attached to the ocean floor **Nemertea** (ribbon worms): acoelomate, bilaterally symmetrical marine worms with long extendable proboscis	980
Brachiopoda (lamp shells)	*Lingula* 	Like bryozoans, possess a lophophore, but within two clamlike shells; more than 30,000 species known as fossils	250
Ctenophora (sea walnuts)	Comb jellies, sea walnuts 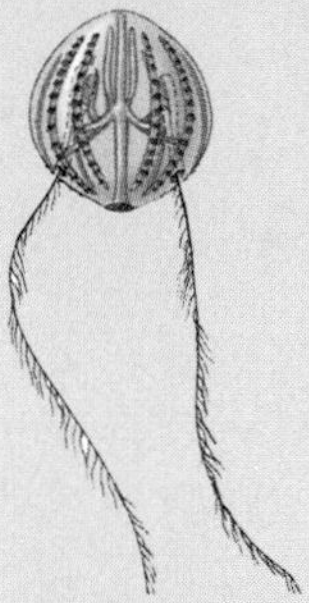	Gelatinous, almost transparent, often bioluminescent marine animals; eight bands of cilia; largest animals that use cilia for locomotion; complete digestive tract with anal pore	100
Phoronida (phoronids)	*Phoronis* 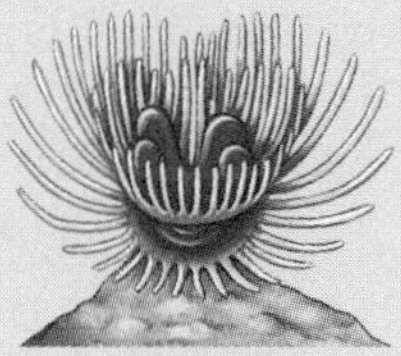	Lophophorate tube worms; often live in dense populations; unique U-shaped gut, instead of the straight digestive tube of other tube worms	12
Loricifera (loriciferans)	*Nanaloricus mysticus* 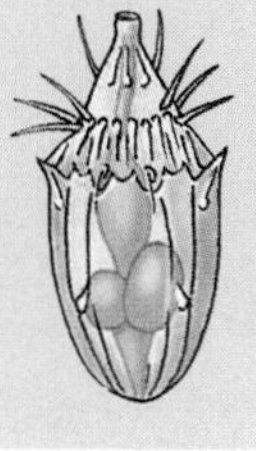	Tiny, bilaterally symmetrical, marine pseudocoelomates that live in spaces between grains of sand; mouthparts include a unique flexible tube; a recently discovered animal phylum (1983)	6

Five Key Transitions in Body Plan

1. Evolution of Tissues

The simplest animals, the Parazoa, lack both defined tissues and organs. Characterized by the sponges, these animals exist as aggregates of cells with minimal intercellular coordination. All other animals, the Eumetazoa, have distinct tissues with highly specialized cells. The evolution of tissues is the first key transition in the animal body plan.

2. Evolution of Bilateral Symmetry

Sponges also lack any definite symmetry, growing asymmetrically as irregular masses. Virtually all other animals have a definite shape and symmetry that can be defined along an imaginary axis drawn through the animal's body. Animals with symmetry belong to either the Radiata, animals with radial symmetry, or the Bilateria, animals with bilateral symmetry.

Radial Symmetry. Symmetrical bodies first evolved in marine animals belonging to two phyla: Cnidaria (jellyfish, sea anemones, and corals) and Ctenophora (comb jellies). The bodies of members of these two phyla, the Radiata, exhibit **radial symmetry,** a body design in which the parts of the body are arranged around a central axis in such a way that any plane passing through the central axis divides the organism into halves that are approximate mirror images (figure 44.3*a*).

Bilateral Symmetry. The bodies of all other animals, the Bilateria, are marked by a fundamental **bilateral symmetry,** a body design in which the body has a right and a left half that are mirror images of each other (figure 44.3*b*). A bilaterally symmetrical body plan has a top and a bottom, better known respectively as the *dorsal* and *ventral* portions of the body. It also has a front, or *anterior* end, and a back, or *posterior* end. In some higher animals like echinoderms (starfish), the adults are radially symmetrical, but even in them the larvae are bilaterally symmetrical.

Bilateral symmetry constitutes the second major evolutionary advance in the animal body plan. This unique form of organization allows parts of the body to evolve in different ways, permitting different organs to be located in different parts of the body. Also, bilaterally symmetrical animals move from place to place more efficiently than radially symmetrical ones, which, in general, lead a sessile or passively floating existence. Due to their increased mobility, bilaterally symmetrical animals are efficient in seeking food, locating mates, and avoiding predators.

During the early evolution of bilaterally symmetrical animals, structures that were important to the organism in monitoring its environment, and thereby capturing prey or avoiding enemies, came to be grouped at the anterior end. Other functions tended to be located farther back in the body. The number and complexity of sense organs are much greater in bilaterally symmetrical animals than they are in radially symmetrical ones.

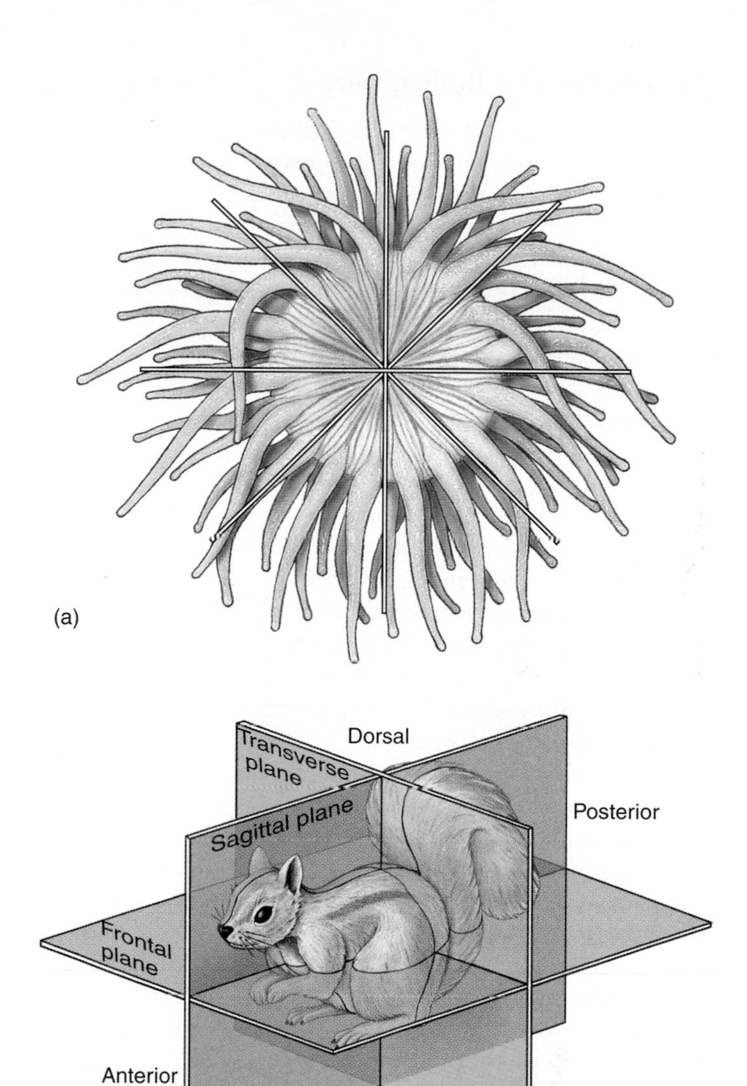

FIGURE 44.3
A comparison of radial and bilateral symmetry. (*a*) Radially symmetrical animals, such as this sea anemone, can be bisected into equal halves in any two-dimensional plane. (*b*) Bilaterally symmetrical animals, such as this squirrel, can only be bisected into equal halves in one plane (the sagittal plane).

Much of the nervous system in bilaterally symmetrical animals is in the form of major longitudinal nerve cords. In a very early evolutionary advance, nerve cells became grouped around the anterior end of the body. These nerve cells probably first functioned mainly to transmit impulses from the anterior sense organs to the rest of the nervous system. This trend ultimately led to the evolution of a definite head and brain area, a process called **cephalization,** as well as to the increasing dominance and specialization of these organs in the more advanced animal phyla.

3. Evolution of a Body Cavity

A third key transition in the evolution of the animal body plan was the evolution of the body cavity. The evolution of efficient organ systems within the animal body was not possible until a body cavity evolved for supporting organs, distributing materials, and fostering complex developmental interactions.

The presence of a body cavity allows the digestive tract to be larger and longer. This longer passage allows for storage of undigested food, longer exposure to enzymes for more complete digestion, and even storage and final processing of food remnants. Such an arrangement allows an animal to eat a great deal when it is safe to do so and then to hide during the digestive process, thus limiting the animal's exposure to predators. The tube within the body cavity architecture is also more flexible, thus allowing the animal greater freedom to move.

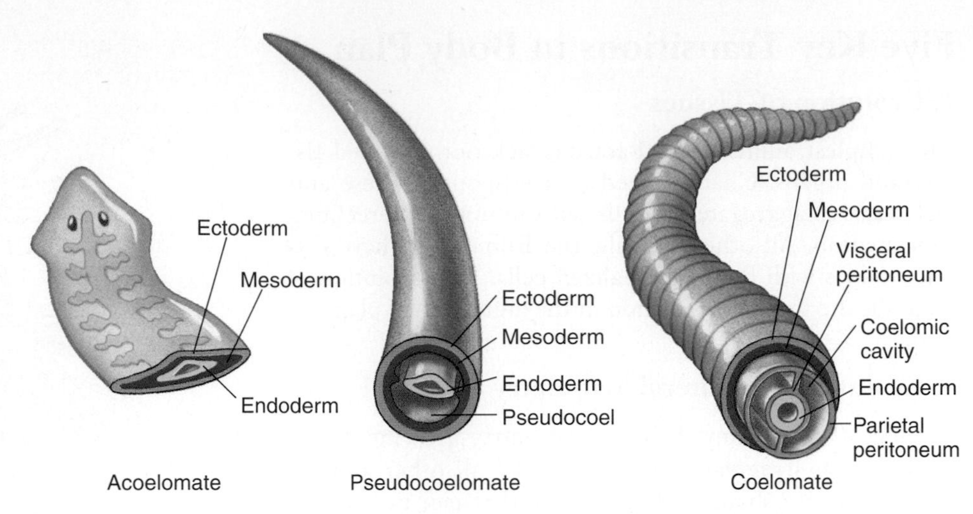

FIGURE 44.4
Three body plans for bilaterally symmetrical animals.

An internal body cavity also provides space within which the gonads (ovaries and testes) can expand, allowing the accumulation of large numbers of eggs and sperm. Such storage capacity allows the diverse modifications of breeding strategy that characterize the more advanced phyla of animals. Furthermore, large numbers of gametes can be stored and released when the conditions are as favorable as possible for the survival of the young animals.

Kinds of Body Cavities. Three basic kinds of body plans evolved in the Bilateria. **Acoelomates** have no body cavity. **Pseudocoelomates** have a body cavity called the **pseudocoel** located between the mesoderm and endoderm. A third way of organizing the body is one in which the fluid-filled body cavity develops not between endoderm and mesoderm, but rather entirely within the mesoderm. Such a body cavity is called a **coelom,** and animals that possess such a cavity are called **coelomates.** In coelomates, the gut is suspended, along with other organ systems of the animal, within the coelom; the coelom, in turn, is surrounded by a layer of epithelial cells entirely derived from the mesoderm. The portion of the epithelium that lines the outer wall of the coelom is called the **parietal peritoneum,** and the portion that covers the internal organs suspended within the cavity is called the **visceral peritoneum** (figure 44.4).

The development of the coelom poses a problem—circulation—solved in pseudocoelomates by churning the fluid within the body cavity. In coelomates, the gut is again surrounded by tissue that presents a barrier to diffusion, just as it is in solid worms. This problem is solved among coelomates by the development of a **circulatory system,** a network of vessels that carries fluids to parts of the body. The circulating fluid, or blood, carries nutrients and oxygen to the tissues and removes wastes and carbon dioxide. Blood is usually pushed through the circulatory system by contraction of one or more muscular hearts. In an **open circulatory system,** the blood passes from vessels into sinuses, mixes with body fluid, and then reenters the vessels later in another location. In a **closed circulatory system,** the blood is physically separated from other body fluids and can be separately controlled. Also, blood moves through a closed circulatory system faster and more efficiently than it does through an open system.

The evolutionary relationship among coelomates, pseudocoelomates, and acoelomates is not clear. Acoelomates, for example, could have given rise to coelomates, but scientists also cannot rule out the possibility that acoelomates were derived from coelomates. The different phyla of pseudocoelomates form two groups that do not appear to be closely related.

Advantages of a Coelom. What is the functional difference between a pseudocoel and a coelom? The answer has to do with the nature of animal embryonic development. In animals, development of specialized tissues involves a process called **primary induction** in which one of the three primary tissues (endoderm, mesoderm, and ectoderm) interacts with another. The interaction requires physical contact. A major advantage of the coelomate body plan is that it allows contact between mesoderm and endoderm, so that primary induction can occur during development. For example, contact between mesoderm and endoderm permits localized portions of the digestive tract to develop into complex, highly specialized regions like the stomach. In pseudocoelomates, mesoderm and endoderm are separated by the body cavity, limiting developmental interactions between these tissues that ultimately limits tissue specialization and development.

4. Evolution of Protostome and Deuterostome Development

Two outwardly dissimilar large phyla, Echinodermata (starfish) and Chordata (vertebrates), together with two smaller phyla, have a series of key embryological features different from those shared by the other animal phyla. Because it is extremely unlikely that these features evolved more than once, it is believed that these four phyla share a common ancestry. They are the members of a group called the **deuterostomes.** Members of the other coelomate animal phyla are called **protostomes.** Deuterostomes evolved from protostomes more than 630 million years ago.

Deuterostomes, like protostomes, are coelomates. They differ fundamentally from protostomes, however, in the way in which the embryo grows. Early in embryonic growth, when the embryo is a hollow ball of cells, a portion invaginates to form an opening called the blastopore. The blastopore of a protostome becomes the animal's mouth, and the anus develops at the other end. In a deuterostome, by contrast, the blastopore becomes the animal's anus, and the mouth develops at the other end (figure 44.5).

Deuterostomes differ in many other aspects of embryo growth, including the plane in which the cells divide. Perhaps most importantly, the cells that make up an embryonic protostome each contain a different portion of the regulatory signals present in the egg, so no one cell of the embryo (or adult) can develop into a complete organism. In marked contrast, any of the cells of a deuterostome can develop into a complete organism.

5. Evolution of Segmentation

The fifth key transition in the animal body plan involved the subdivision of the body into **segments.** Just as it is efficient for workers to construct a tunnel from a series of identical prefabricated parts, so segmented animals are "assembled" from a succession of identical segments. During the animal's early development, these segments become most obvious in the mesoderm but later are reflected in the ectoderm and endoderm as well. Two advantages result from early embryonic segmentation:

1. In annelids and other highly segmented animals, each segment may go on to develop a more or less complete set of adult organ systems. Damage to any one segment need not be fatal to the individual because the other segments duplicate that segment's functions.
2. Locomotion is far more effective when individual segments can move independently because the animal as a whole has more flexibility of movement. Because the separations isolate each segment into an individual skeletal unit, each is able to contract or expand autonomously in response to changes in hydrostatic pressure. Therefore, a long body can move in ways that are often quite complex.

Segmentation, also referred to as *metamerism*, underlies the organization of all advanced animal body plans. In some adult arthropods, the segments are fused, but segmentation is usually apparent in their embryological development. In vertebrates, the backbone and muscular areas are segmented, although segmentation is often disguised in the adult form. True segmentation is found in only three phyla: the annelids, the arthropods, and the chordates, although this trend is evident in many phyla.

Five key transitions in body design are responsible for most of the differences we see among the major animal phyla: the evolution of (1) tissues, (2) bilateral symmetry, (3) a body cavity, (4) protostome and deuterostome development, and (5) segmentation.

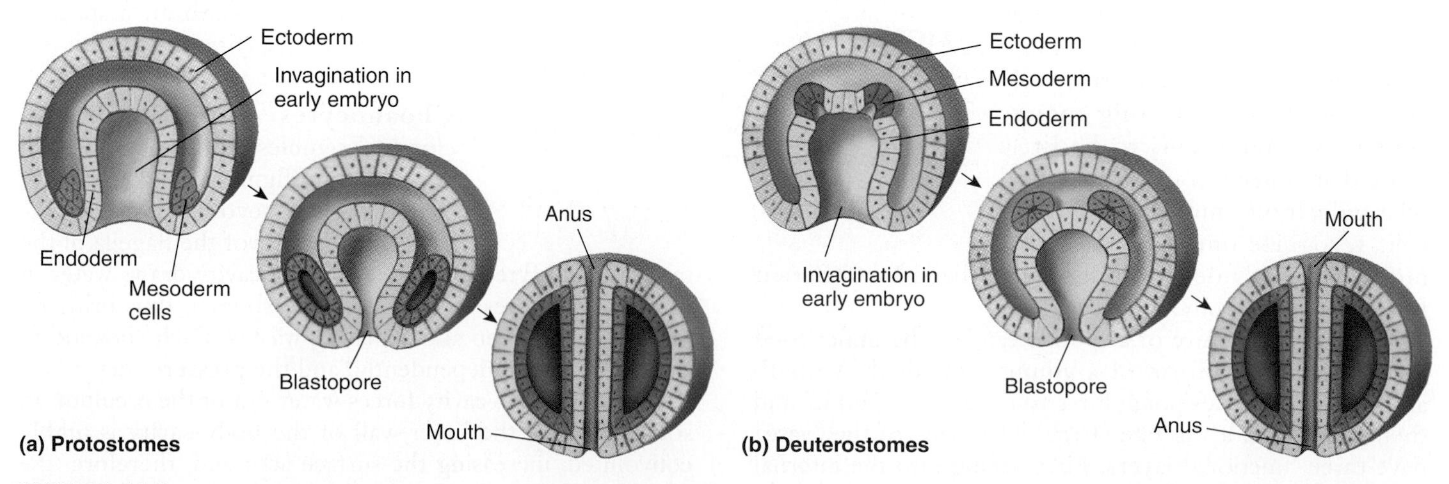

FIGURE 44.5
The fate of the blastopore. (*a*) In protostomes, the blastopore becomes the animal's mouth. (*b*) In deuterostomes, the blastopore becomes the animal's anus.

Parazoa

The sponges are Parazoans, animals that lack tissues and organs and a definite symmetry. However, sponges, like all animals, have true, complex *multicellularity*, unlike their protistan ancestors. The body of a sponge contains several distinctly different types of cells whose activities are loosely coordinated with one another. As we will see, the coordination between cell types in the eumetazoans increases and becomes quite complex.

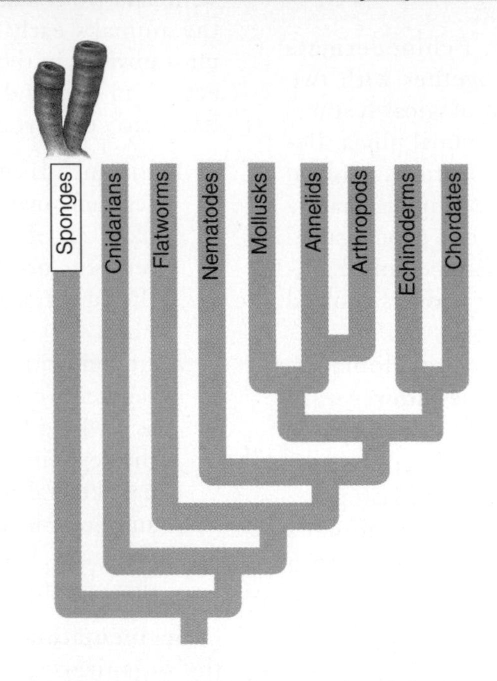

The Sponges

There are perhaps 5000 species of marine sponges, phylum Porifera, and about 150 species that live in fresh water. In the sea, sponges are abundant at all depths. Although some sponges are tiny, no more than a few millimeters across, some, like the loggerhead sponges, may reach 2 meters or more in diameter. A few small ones are radially symmetrical, but most members of this phylum completely lack symmetry. Many sponges are colonial. Some have a low and encrusting form, while others may be erect and lobed, sometimes in complex patterns. Although larval sponges are free-swimming, adults are **sessile,** or anchored in place to submerged objects.

Sponges, like all animals, are composed of multiple cell types (see figure 44.7), but there is relatively little coordination among sponge cells. A sponge seems to be little more than a mass of cells embedded in a gelatinous matrix, but these cells recognize one another with a high degree of fidelity and are specialized for different functions of the body.

FIGURE 44.6
Aplysina longissima. This beautiful, bright orange and red elongated sponge is found on deep regions of coral reefs. The osculum is an opening at the top.

The basic structure of a sponge can best be understood by examining the form of a young individual. A small, anatomically simple sponge first attaches to a substrate and then grows into a vaselike shape. The walls of the "vase" have three functional layers. First, facing into the internal cavity are specialized flagellated cells called **choanocytes,** or collar cells. These cells line either the entire body cavity or, in many large and more complex sponges, specialized chambers. Second, the bodies of sponges are bounded by an outer epithelial layer consisting of flattened cells somewhat like those that make up the epithelia, or outer layers, of other animal phyla. Some portions of this layer contract when touched or exposed to appropriate chemical stimuli, and this contraction may cause some of the pores to close. Third, between these two layers, sponges consist mainly of a gelatinous, protein-rich matrix called the **mesohyl,** within which various types of amoeboid cells occur. In addition, many kinds of sponges have minute needles of calcium carbonate or silica known as **spicules,** or fibers of a tough protein called **spongin,** or both, within this matrix. Spicules and spongin strengthen the bodies of the sponges in which they occur. A spongin skeleton is the model for the bathtub sponge, once the skeleton of a real animal, but now largely known from its cellulose and plastic mimics.

Sponges feed in a unique way. The beating of flagella that line the inside of the sponge draws water in through numerous small pores; the name of the phylum, Porifera, refers to this system of pores. Plankton and other small organisms are filtered from the water, which flows through passageways and eventually is forced out through an **osculum,** a specialized, larger pore (figure 44.6).

Choanocytes. Each choanocyte closely resembles a protist with a single flagellum (figure 44.7), a similarity that reflects its evolutionary derivation. The beating of the flagella of the many choanocytes that line the body cavity draws water in through the pores and through the sponge, thus bringing in food and oxygen and expelling wastes. Each choanocyte flagellum beats independently, and the pressure they create collectively in the cavity forces water out of the osculum. In some sponges, the inner wall of the body cavity is highly convoluted, increasing the surface area and, therefore, the number of flagella that can drive the water. In such a sponge, 1 cubic centimeter of sponge can propel more than 20 liters of water a day.

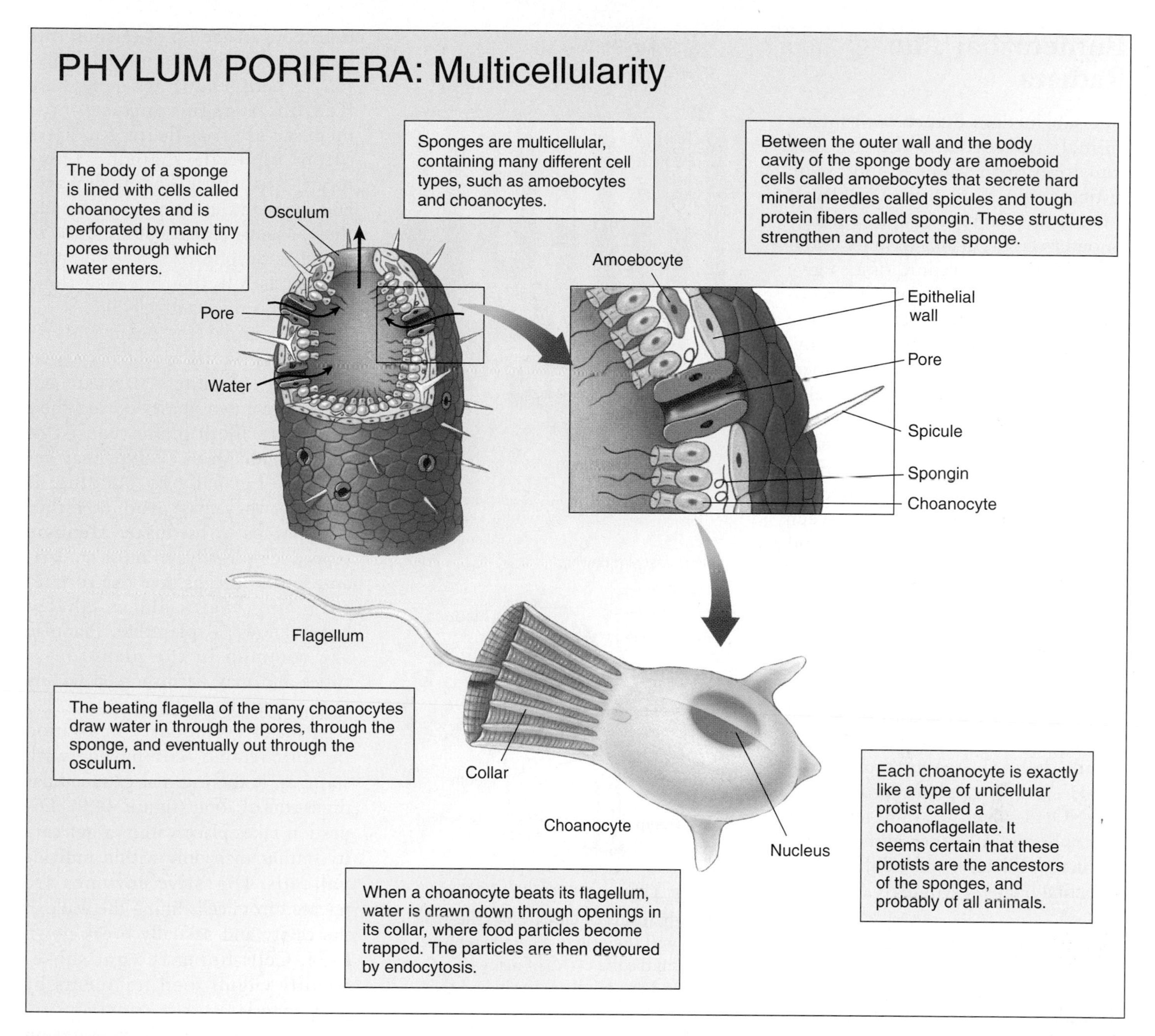

PHYLUM PORIFERA: Multicellularity
The body of a sponge is lined with cells called choanocytes and is perforated by many tiny pores through which water enters.
Sponges are multicellular, containing many different cell types, such as amoebocytes and choanocytes.
Between the outer wall and the body cavity of the sponge body are amoeboid cells called amoebocytes that secrete hard mineral needles called spicules and tough protein fibers called spongin. These structures strengthen and protect the sponge.
Osculum
Amoebocyte
Pore
Water
Epithelial wall
Pore
Spicule
Spongin
Choanocyte
Flagellum
The beating flagella of the many choanocytes draw water in through the pores, through the sponge, and eventually out through the osculum.
Collar
Choanocyte
Nucleus
Each choanocyte is exactly like a type of unicellular protist called a choanoflagellate. It seems certain that these protists are the ancestors of the sponges, and probably of all animals.
When a choanocyte beats its flagellum, water is drawn down through openings in its collar, where food particles become trapped. The particles are then devoured by endocytosis.

FIGURE 44.7
The body of a sponge is multicellular. The first evolutionary advance seen in animals is complex multicellularity, in which individuals are composed of many highly specialized kinds of cells.

Reproduction in Sponges. Some sponges will re-form themselves once they have passed through a silk mesh. Thus, as you might suspect, sponges frequently reproduce by simply breaking into fragments. If a sponge breaks up, the resulting fragments usually are able to reconstitute whole new individuals. Sexual reproduction is also exhibited by sponges, with some mature individuals producing eggs and sperm. Larval sponges may undergo their initial stages of development within the parent. They have numerous external, flagellated cells and are free-swimming. After a short planktonic stage, they settle down on a suitable substrate, where they begin their transformation into adults.

Sponges probably represent the most primitive animals, possessing multicellularity but neither tissue-level development nor body symmetry. Their cellular organization hints at the evolutionary ties between the unicellular protists and the multicellular animals. Sponges are unique in the animal kingdom in possessing choanocytes, special flagellated cells whose beating drives water through the body cavity.

Eumetazoa: The Radiata

The subkingdom Eumetazoa contains animals that evolved the first key transition in the animal body plan: distinct *tissues.* Two distinct cell layers form in the embryos of these animals: an outer ectoderm and an inner endoderm. These embryonic tissues give rise to the basic body plan, differentiating into the many tissues of the adult body. Typically, the outer covering of the body (called the epidermis) and the nervous system develop from the ectoderm, and the layer of digestive tissue (called the **gastrodermis**) develops from the endoderm. A layer of gelatinous material, called the **mesoglea,** lies between the epidermis and gastrodermis and contains the muscles in most eumetazoans.

Eumetazoans also evolved true body symmetry and are divided into two major groups. The Radiata includes two phyla of radially symmetrical organisms, Cnidaria (pronounced ni-DAH-ree-ah), the cnidarians—hydroids, jellyfish, sea anemones, and corals—and Ctenophora (pronounced tea-NO-fo-rah), the comb jellies, or ctenophores. All other eumetazoans are in the Bilateria and exhibit a fundamental bilateral symmetry.

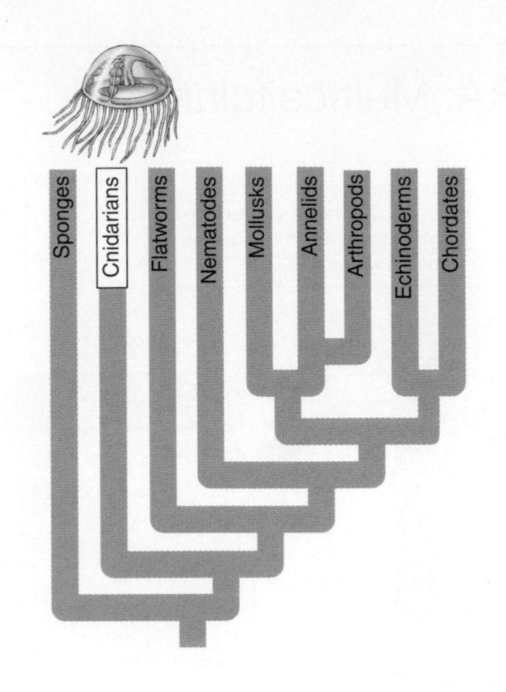

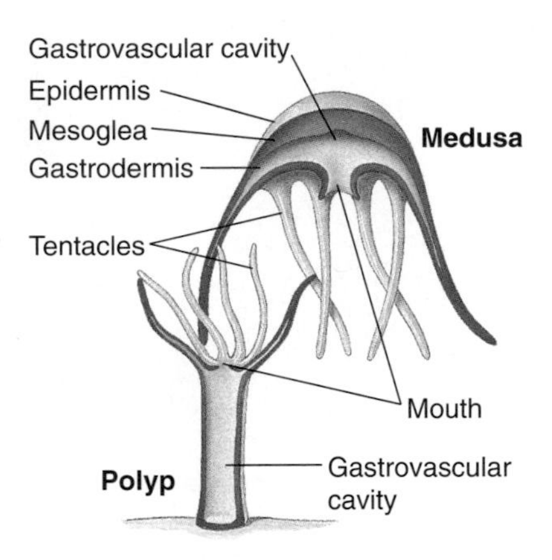

FIGURE 44.8
Two body forms of cnidarians, the medusa and the polyp. These two phases alternate in the life cycles of many cnidarians, but a number—including the corals and sea anemones, for example—exist only as polyps. Both forms have two fundamental layers of cells, separated by a jellylike layer called the mesoglea.

The Cnidarians

Cnidarians are nearly all marine, although a few live in fresh water. These fascinating and simply constructed animals are basically gelatinous in composition. They differ markedly from the sponges in organization; their bodies are made up of distinct tissues, although they have not evolved true organs. These animals are carnivores. For the most part, they do not actively move from place to place, but rather capture their prey (which includes fishes, crustaceans, and many other kinds of animals) with the tentacles that ring their mouths.

Cnidarians may have two basic body forms, polyps and medusae (figure 44.8). Polyps are cylindrical and are usually found attached to a firm substrate. They may be solitary or colonial. In a polyp, the mouth faces away from the substrate on which the animal is growing, and, therefore, often faces upward. Many polyps build up a chitinous or calcareous (made up of calcium carbonate) external or internal skeleton, or both. Only a few polyps are free-floating. In contrast, most medusae are free-floating and are often umbrella-shaped. Their mouths usually point downward, and the tentacles hang down around them. Medusae, particularly those of the class Scyphozoa, are commonly known as jellyfish because their mesoglea is thick and jellylike.

Many cnidarians occur only as polyps, while others exist only as medusae; still others alternate between these two phases during their life cycles. Both phases consist of diploid individuals. Polyps may reproduce asexually by budding; if they do, they may produce either new polyps or medusae. Medusae reproduce sexually. In most cnidarians, fertilized eggs give rise to free-swimming, multicellular, ciliated larvae known as **planulae.** Planulae are common in the plankton at times and may be dispersed widely in the currents.

A major evolutionary innovation in cnidarians, compared with sponges, is the internal extracellular digestion of food (figure 44.9). Digestion takes place within a gut cavity, rather than only within individual cells. Digestive enzymes are released from cells lining the walls of the cavity and partially break down food. Cells lining the gut subsequently engulf food fragments by phagocytosis.

The extracellular fragmentation that precedes phagocytosis and intracellular digestion allows cnidarians to digest animals larger than individual cells, an important improvement over the strictly intracellular digestion of sponges.

Nets of nerve cells coordinate contraction of cnidarian muscles, apparently with little central control. Cnidarians have no blood vessels, respiratory system, or excretory organs.

On their tentacles and sometimes on their body surface, cnidarians bear specialized cells called **cnidocytes.** The name of the phylum Cnidaria refers to these cells, which are highly distinctive and occur in no other group of organisms. Within each cnidocyte is a **nematocyst,** a small but powerful "harpoon." Each nematocyst features a coiled, threadlike tube. Lining the inner wall of the tube is a series of barbed spines.

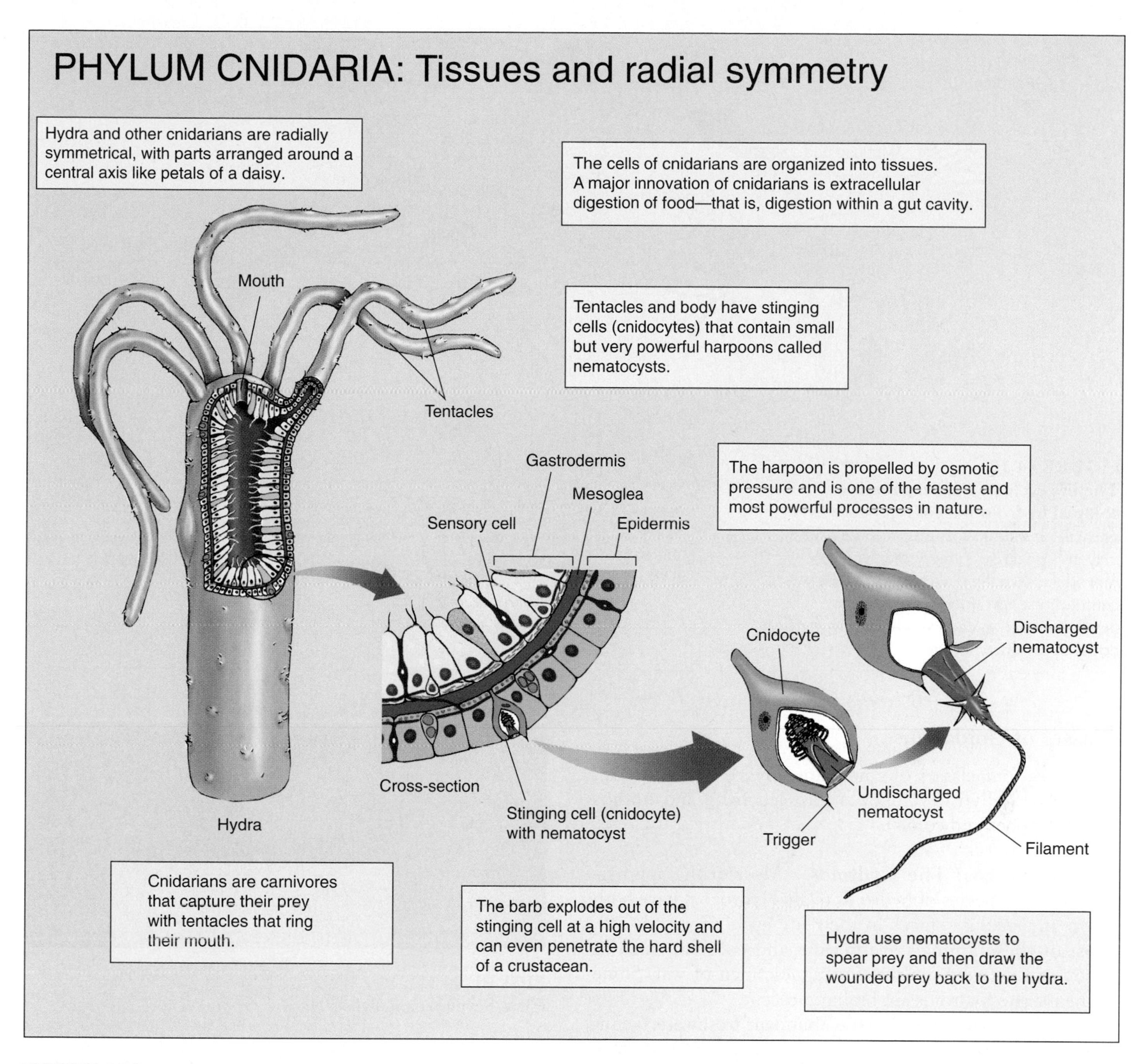
PHYLUM CNIDARIA: Tissues and radial symmetry
Hydra and other cnidarians are radially symmetrical, with parts arranged around a central axis like petals of a daisy.
The cells of cnidarians are organized into tissues. A major innovation of cnidarians is extracellular digestion of food—that is, digestion within a gut cavity.
Mouth
Tentacles and body have stinging cells (cnidocytes) that contain small but very powerful harpoons called nematocysts.
Tentacles
Gastrodermis
Mesoglea
Sensory cell
Epidermis
The harpoon is propelled by osmotic pressure and is one of the fastest and most powerful processes in nature.
Cnidocyte
Discharged nematocyst
Cross-section
Hydra
Stinging cell (cnidocyte) with nematocyst
Undischarged nematocyst
Trigger
Filament
Cnidarians are carnivores that capture their prey with tentacles that ring their mouth.
The barb explodes out of the stinging cell at a high velocity and can even penetrate the hard shell of a crustacean.
Hydra use nematocysts to spear prey and then draw the wounded prey back to the hydra.

FIGURE 44.9
Eumetazoans all have tissues and symmetry. The cells of a cnidarian like this *Hydra* are organized into specialized tissues. The interior gut cavity is specialized for extracellular digestion—that is, digestion within a gut cavity rather than within individual cells. Cnidarians are also radially symmetrical.

Cnidarians use the threadlike tube to spear their prey and then draw the harpooned prey back with the tentacle containing the cnidocyte. Nematocysts may also serve a defensive purpose. To propel the harpoon, the cnidocyte uses water pressure. Before firing, the cnidocyte builds up a very high internal osmotic pressure. This is done by using active transport to build a high concentration of ions inside, while keeping its wall impermeable to water. Within the undischarged nematocyst, osmotic pressure reaches about 140 atmospheres.

When a flagellum-like trigger on the cnidocyte is stimulated to discharge, its walls become permeable to water, which rushes inside and violently pushes out the barbed filament. Nematocyst discharge is one of the fastest cellular processes in nature. The nematocyst is pushed outward so explosively that the barb can penetrate even the hard shell of a crab. A toxic protein often produced a stinging sensation, causing some cnidarians to be called "stinging nettles."

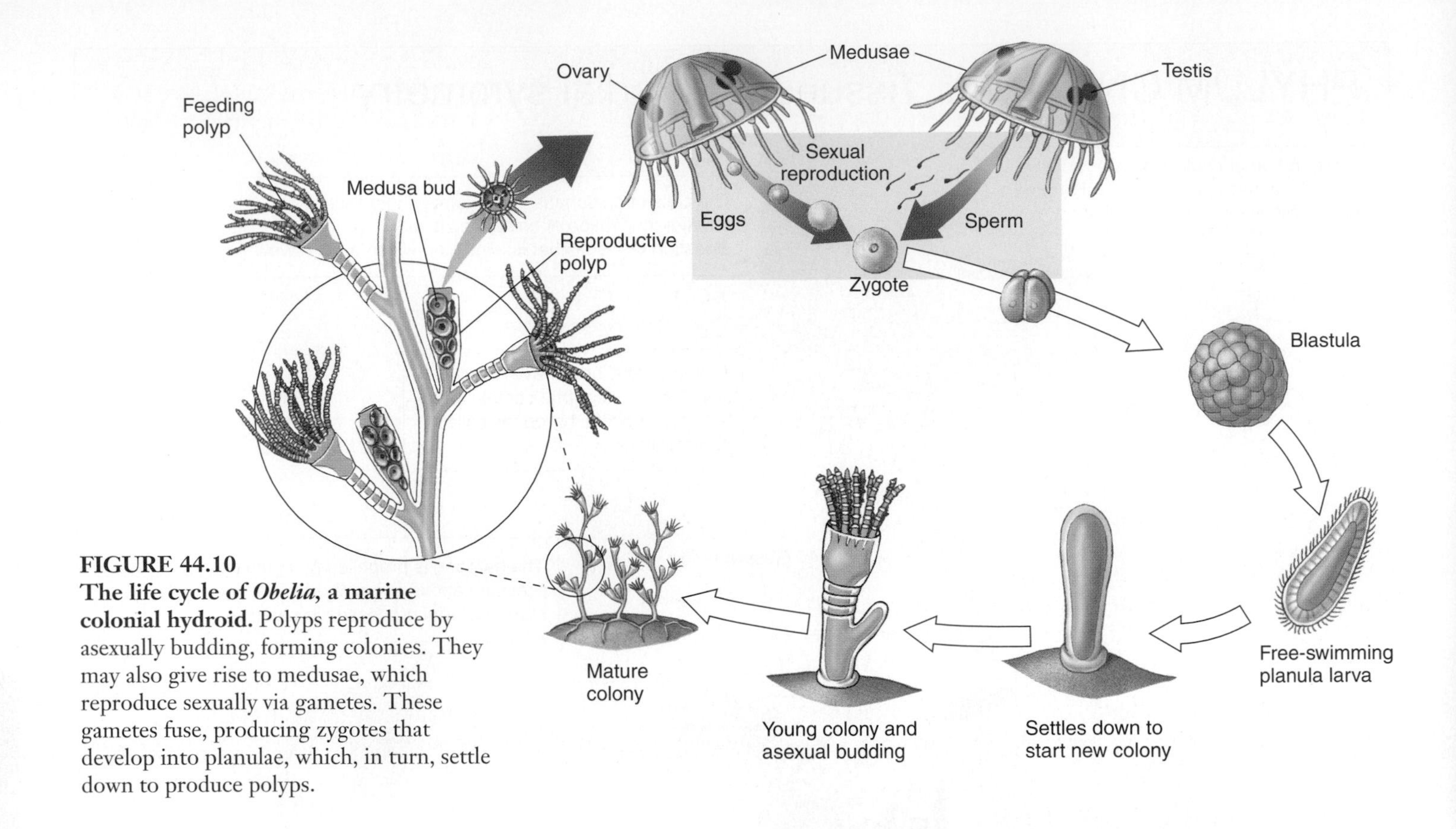

FIGURE 44.10
The life cycle of *Obelia*, a marine colonial hydroid. Polyps reproduce by asexually budding, forming colonies. They may also give rise to medusae, which reproduce sexually via gametes. These gametes fuse, producing zygotes that develop into planulae, which, in turn, settle down to produce polyps.

Classes of Cnidarians

There are four classes of cnidarians: Hydrozoa (hydroids), Scyphozoa (jellyfish), Cubozoa (box jellyfish), and Anthozoa (anemones and corals).

Class Hydrozoa: The Hydroids. Most of the approximately 2700 species of hydroids (class Hydrozoa) have both polyp and medusa stages in their life cycle (figure 44.10). Most of these animals are marine and colonial, such as *Obelia* and the very unusual Portuguese man-of-war. Some of the marine hydroids are bioluminescent.

A well-known hydroid is the abundant freshwater genus *Hydra*, which is exceptional in that it has no medusa stage and exists as a solitary polyp. Each polyp sits on a basal disk, which it can use to glide around, aided by mucous secretions. It can also move by somersaulting, bending over and attaching itself to the substrate by its tentacles, and then looping over to a new location. If the polyp detaches itself from the substrate, it can float to the surface.

Class Scyphozoa: The Jellyfish. The approximately 200 species of jellyfish (class Scyphozoa) are transparent or translucent marine organisms, some of a striking orange, blue, or pink color (figure 44.11). These animals spend most of their time floating near the surface of the sea. In all of them, the medusa stage is dominant—much larger and more complex than the polyps. The medusae are bell-shaped, with hanging tentacles around their margins. The polyp stage is small, inconspicuous, and simple in structure.

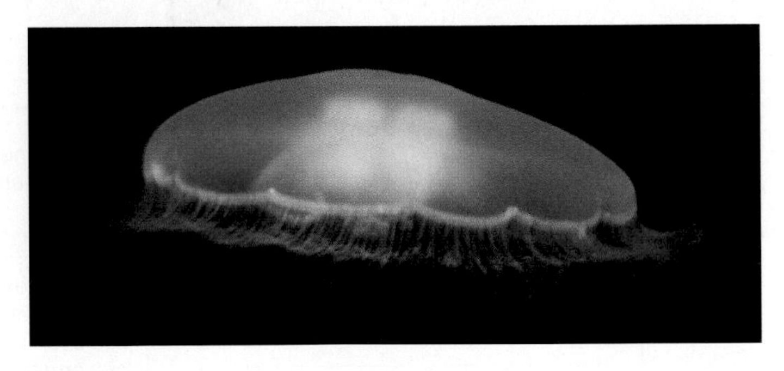

FIGURE 44.11
Class Scyphozoa. Jellyfish, *Aurelia aurita.*

The outer layer, or epithelium, of a jellyfish contains a number of specialized epitheliomuscular cells, each of which can contract individually. Together, the cells form a muscular ring around the margin of the bell that pulses rhythmically and propels the animal through the water. Jellyfish have separate male and female individuals. After fertilization, planulae form, which then attach and develop into polyps. The polyps can reproduce asexually as well as budding off medusae. In some jellyfish that live in the open ocean, the polyp stage is suppressed, and planulae develop directly into medusae.

Class Cubozoa: The Box Jellyfish. Until recently the cubozoa were considered an order of Scyphozoa. As their name implies, they are box-shaped medusa (the polyp stage is inconspicuous and in many cases not known). Most are only a few cm in height, although some are 25 cm tall. A tentacle or group of tentacles is found at each corner of the box (figure 44.12). Box jellies are strong swimmers and voracious predators of fish. Stings of some species can be fatal to humans.

Class Anthozoa: The Sea Anemones and Corals. By far the largest class of cnidarians is Anthozoa, the "flower animals" (from the Greek *anthos*, meaning "flower"). The approximately 6200 species of this group are solitary or colonial marine animals. They include stonelike corals, soft-bodied sea anemones, and other groups known by such fanciful names as sea pens, sea pansies, sea fans, and sea whips (figure 44.13). All of these names reflect a plantlike body topped by a tuft or crown of hollow tentacles. Like other cnidarians, anthozoans use these tentacles in feeding. Nearly all members of this class that live in shallow waters harbor symbiotic algae, which supplement the nutrition of their hosts through photosynthesis. Fertilized eggs of anthozoans usually develop into planulae that settle and develop into polyps; no medusae are formed.

Sea anemones are a large group of soft-bodied anthozoans that live in coastal waters all over the world and are especially abundant in the tropics. When touched, most sea anemones retract their tentacles into their bodies and fold up. Sea anemones are highly muscular and relatively complex organisms, with greatly divided internal cavities. These animals range from a few millimeters to about 10 centimeters in diameter and are perhaps twice that high.

Corals are another major group of anthozoans. Many of them secrete tough outer skeletons, or exoskeletons, of calcium carbonate and are thus stony in texture. Others, including the gorgonians, or soft corals, do not secrete exoskeletons. Some of the hard corals help form coral reefs, which are shallow-water limestone ridges that occur in warm seas. Although the waters where coral reefs develop are often nutrient-poor, the coral animals are able to grow actively because of the abundant algae found within them.

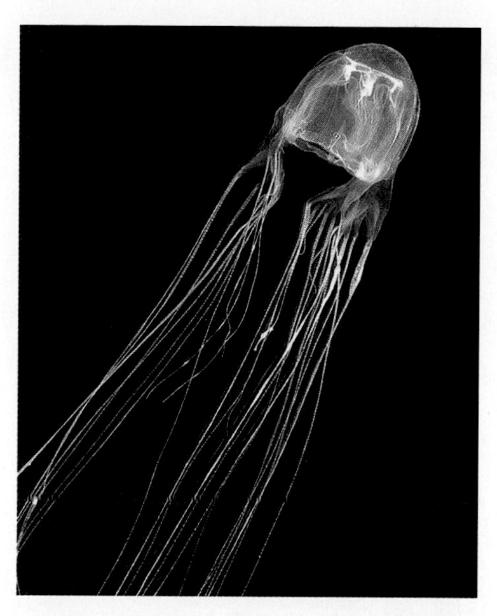

FIGURE 44.12
Class Cubozoa. Box jelly, *Chironex fleckeri*.

FIGURE 44.13
Class Anthozoa. The sessile soft-bodied sea anemone.

The Ctenophorans (Comb Jellies)

The members of this small phylum range from spherical to ribbonlike and are known as comb jellies or sea walnuts. Traditionally, the roughly 90 marine species of ctenophores (phylum Ctenophora) were considered closely related to the cnidarians. However, ctenophores are structurally more complex than cnidarians. They have anal pores, so that water and other substances pass completely through the animal. Comb jellies, abundant in the open ocean, are transparent and usually only a few centimeters long. The members of one group have two long, retractable tentacles that they use to capture their prey.

Ctenophores propel themselves through water with eight comblike plates of fused cilia that beat in a coordinated fashion (figure 44.14). They are the largest animals that use cilia for locomotion. Many ctenophores are bioluminescent, giving off bright flashes of light particularly evident in the open ocean at night.

Cnidarians and ctenophores have tissues and radial symmetry. Cnidarians have a specialized kind of cell called a cnidocyte. Ctenophores propel themselves through the water by means of eight comblike plates of fused cilia.

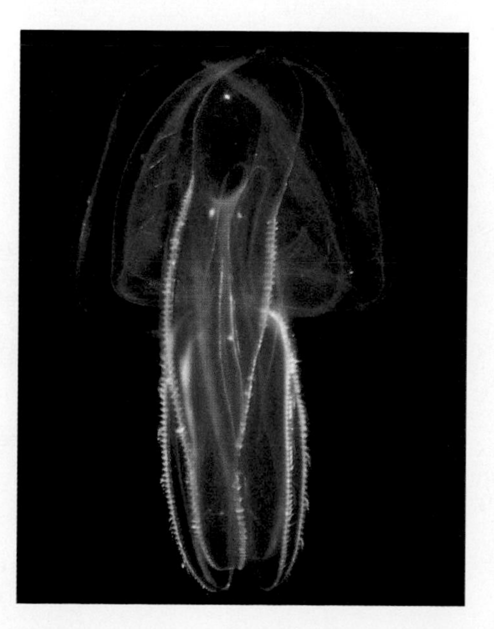

FIGURE 44.14
A comb jelly (phylum Ctenophora). Note the comblike plates along the ridges of the base.

44.3 Acoelomates are solid worms that lack a body cavity.

Eumetazoa: The Bilaterian Acoelomates

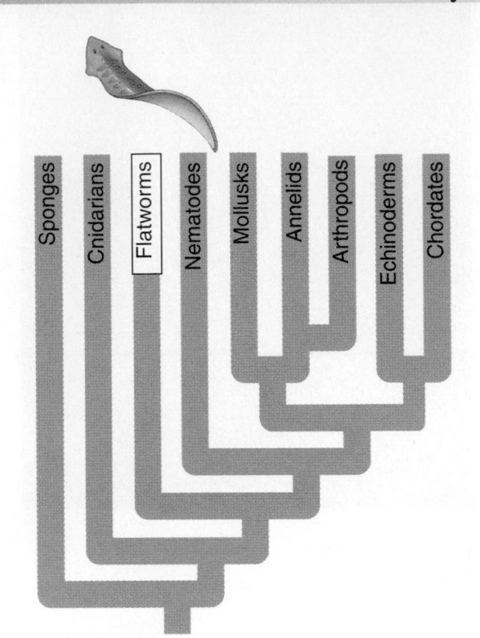

The Bilateria are characterized by the second key transition in the animal body plan, *bilateral symmetry*, which allowed animals to achieve high levels of specialization within parts of their bodies. The simplest bilaterians are the acoelomates; they lack any internal cavity other than the digestive tract. As discussed earlier, all bilaterians have three embryonic layers during development: ectoderm, endoderm, and mesoderm. We will focus our discussion of the acoelomates on the largest phylum of the group, the flatworms.

Phylum Platyhelminthes: The Flatworms

Phylum Platyhelminthes consists of some 20,000 species. These ribbon-shaped, soft-bodied animals are flattened dorsoventrally, from top to bottom. Flatworms are among the simplest of bilaterally symmetrical animals, but they do have a definite head at the anterior end and they do possess organs. Their bodies are solid: the only internal space consists of the digestive cavity (figure 44.15).

Flatworms range in size from a millimeter or less to many meters long, as in some tapeworms. Most species of flatworms are parasitic, occurring within the bodies of many other kinds of animals (figure 44.16). Other flatworms are free-living, occurring in a wide variety of marine and freshwater habitats, as well as moist places on land. Free-living flatworms are carnivores and scavengers; they eat various small animals and bits of organic debris. They move from place to place by means of ciliated epithelial cells, which are particularly concentrated on their ventral surfaces.

Those flatworms that have a digestive cavity have an incomplete gut, one with only one opening. As a result, they cannot feed, digest, and eliminate undigested particles of food simultaneously, and thus, flatworms cannot feed continuously, as more advanced animals can. Muscular contractions in the upper end of the gut cause a strong sucking force allowing flatworms to ingest their food and tear it into small bits. The gut is branched and extends throughout the body, functioning in both digestion and transport of food. Cells that line the gut engulf most of the food particles by phagocytosis and digest them; but, as in the cnidarians, some of these particles are partly digested extracellularly. Tapeworms, which are parasitic flatworms, lack digestive systems. They absorb their food directly through their body walls.

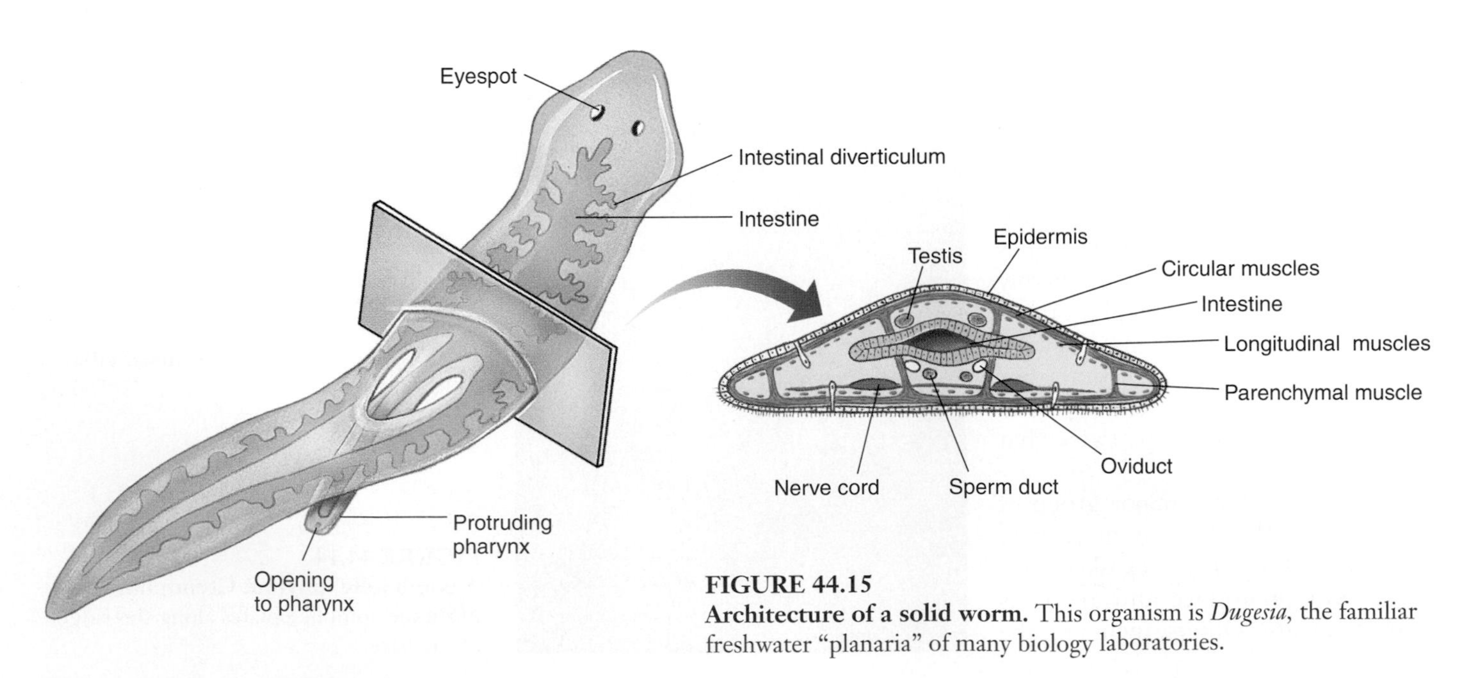

FIGURE 44.15
Architecture of a solid worm. This organism is *Dugesia*, the familiar freshwater "planaria" of many biology laboratories.

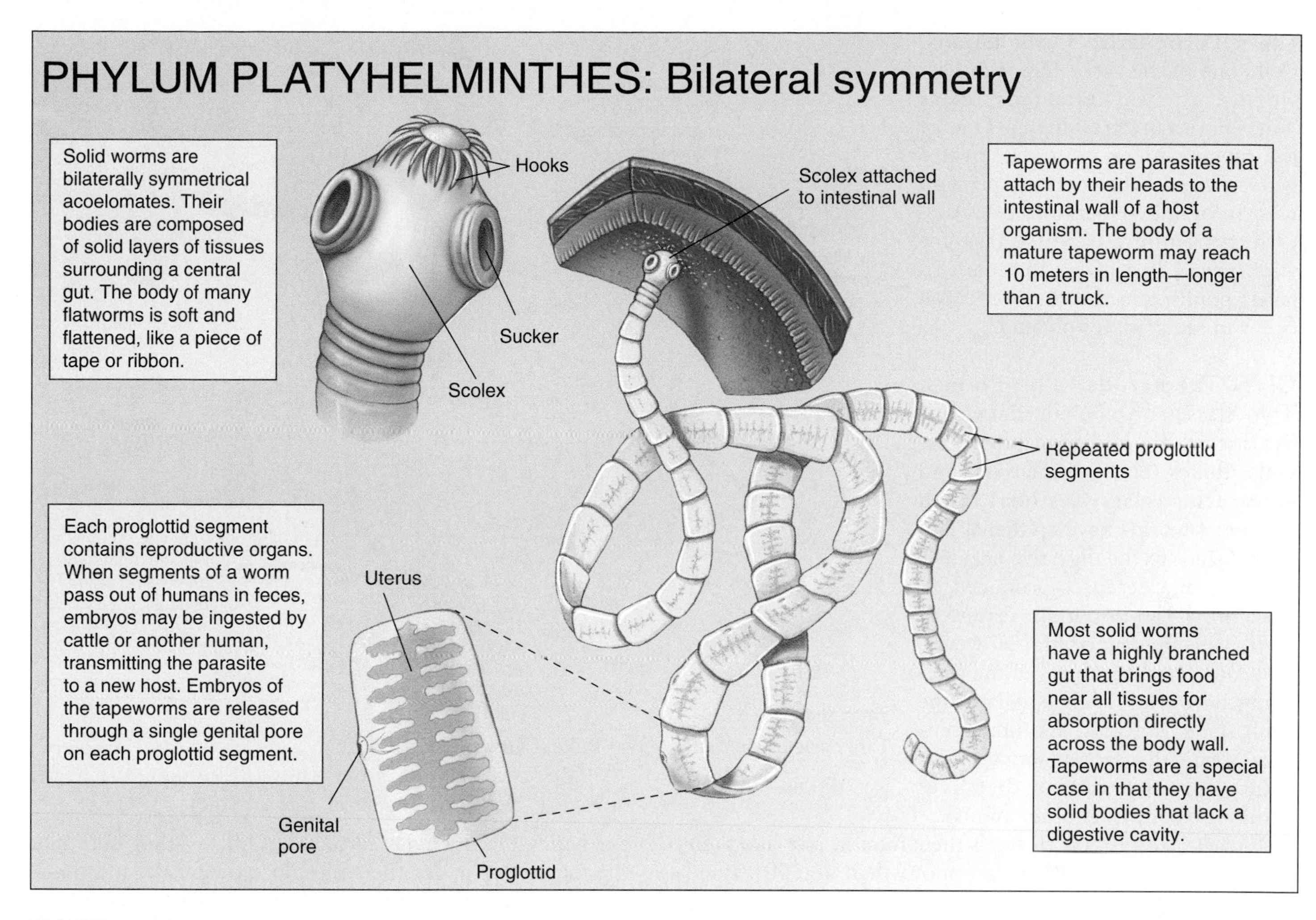

PHYLUM PLATYHELMINTHES: Bilateral symmetry
Solid worms are bilaterally symmetrical acoelomates. Their bodies are composed of solid layers of tissues surrounding a central gut. The body of many flatworms is soft and flattened, like a piece of tape or ribbon.
Hooks
Sucker
Scolex
Scolex attached to intestinal wall
Tapeworms are parasites that attach by their heads to the intestinal wall of a host organism. The body of a mature tapeworm may reach 10 meters in length—longer than a truck.
Repeated proglottid segments
Each proglottid segment contains reproductive organs. When segments of a worm pass out of humans in feces, embryos may be ingested by cattle or another human, transmitting the parasite to a new host. Embryos of the tapeworms are released through a single genital pore on each proglottid segment.
Uterus
Genital pore
Proglottid
Most solid worms have a highly branched gut that brings food near all tissues for absorption directly across the body wall. Tapeworms are a special case in that they have solid bodies that lack a digestive cavity.

FIGURE 44.16
The evolution of bilateral symmetry. Acoelomate solid worms like this beef tapeworm, *Taenia saginata*, are bilaterally symmetrical. In addition, all bilaterians have three embryonic layers and exhibit cephalization.

Unlike cnidarians, flatworms have an excretory system, which consists of a network of fine tubules (little tubes) that runs throughout the body. Cilia line the hollow centers of bulblike **flame cells** located on the side branches of the tubules (see figure 58.9). Cilia in the flame cells move water and excretory substances into the tubules and then to exit pores located between the epidermal cells. Flame cells were named because of the flickering movements of the tuft of cilia within them. They primarily regulate the water balance of the organism. The excretory function of flame cells appears to be a secondary one. A large proportion of the metabolic wastes excreted by flatworms diffuses directly into the gut and is eliminated through the mouth.

Like sponges, cnidarians, and ctenophorans, flatworms lack circulatory systems for the transport of oxygen and food molecules. Consequently, all flatworm cells must be within diffusion distance of oxygen and food. Flatworms have thin bodies and highly branched digestive cavities that make such a relationship possible.

The nervous system of flatworms is very simple. Like cnidarians, some primitive flatworms have only a nerve net. However, most members of this phylum have longitudinal nerve cords that constitute a simple central nervous system.

Free-living members of this phylum have eyespots on their heads. These are inverted, pigmented cups containing light-sensitive cells connected to the nervous system. These eyespots enable the worms to distinguish light from dark; worms move away from strong light.

The reproductive systems of flatworms are complex. Most flatworms are **hermaphroditic,** with each individual containing both male and female sexual structures. In many of them, fertilization is internal. When they mate, each partner deposits sperm in the copulatory sac of the other. The sperm travel along special tubes to reach the eggs. In most free-living flatworms, fertilized eggs are laid in cocoons strung in ribbons and hatch into miniature adults. In some parasitic flatworms, there is a complex succession of distinct larval forms. Flatworms are also capable of asexual regeneration. In some genera, when a single individual is divided into two or more parts, each part can regenerate an entirely new flatworm.

Class Turbellaria: Turbellarians. Only one of the three classes of flatworms, the turbellarians (class Turbellaria) are free-living. One of the most familiar is the freshwater genus *Dugesia*, the common planaria used in biology laboratory exercises. Other members of this class are widespread and often abundant in lakes, ponds, and the sea. Some also occur in moist places on land.

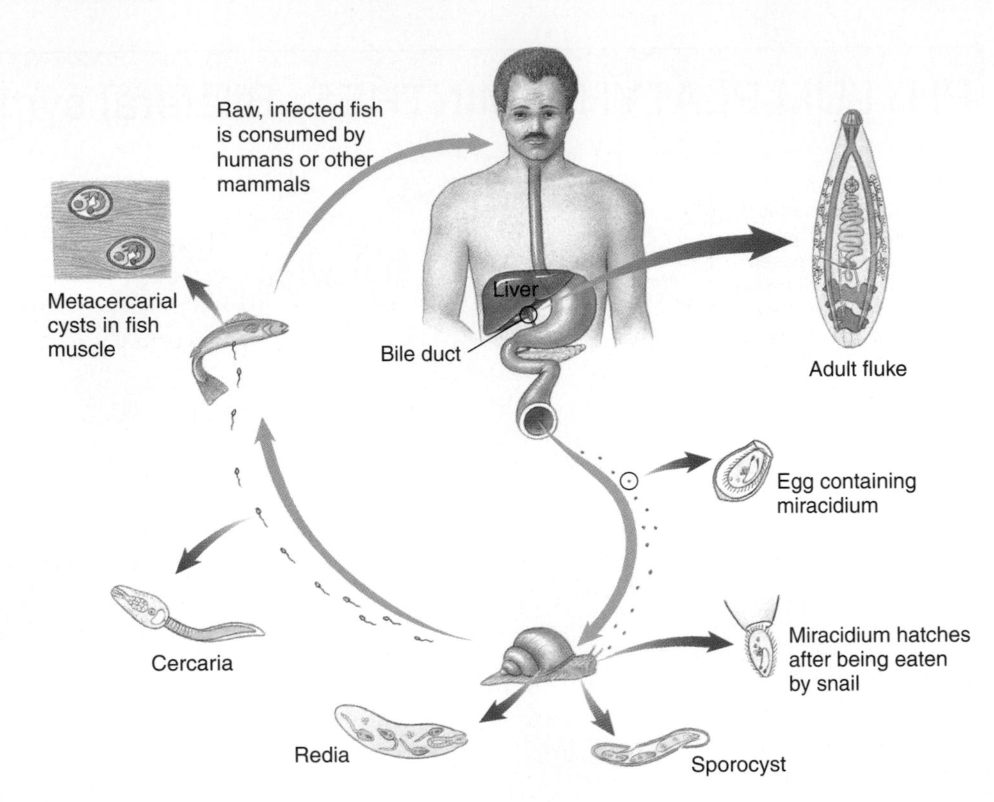

FIGURE 44.17
Life cycle of the human liver fluke, *Clonorchis sinensis*.

Class Trematoda: The Flukes. Two classes of parasitic flatworms live within the bodies of other animals: flukes (class Trematoda) and tapeworms (class Cestoda). Both groups of worms have epithelial layers resistant to the digestive enzymes and immune defenses produced by their hosts—an important feature in their parasitic way of life. However, they lack certain features of the free-living flatworms, such as cilia in the adult stage, eyespots, and other sensory organs that lack adaptive significance for an organism that lives within the body of another animal.

Flukes take in food through their mouth, just like their free-living relatives. There are more than 10,000 named species, ranging in length from less than 1 millimeter to more than 8 centimeters. Flukes attach themselves within the bodies of their hosts by means of suckers, anchors, or hooks. Some have a life cycle that involves only one host, usually a fish. Most have life cycles involving two or more hosts. Their larvae almost always occur in snails, and there may be other intermediate hosts. The final host of these flukes is almost always a vertebrate.

To human beings, one of the most important flatworms is the human liver fluke, *Clonorchis sinensis*. It lives in the bile passages of the liver of humans, cats, dogs, and pigs. It is especially common in Asia. The worms are 1 to 2 centimeters long and have a complex life cycle. Although they are hermaphroditic, cross-fertilization usually occurs between different individuals. Eggs, each containing a complete, ciliated first-stage larva, or **miracidium,** are passed in the feces (figure 44.17). If they reach water, they may be ingested by a snail. Within the snail an egg transforms into a *sporocyst*—a baglike structure with embryonic germ cells. Within the sporocysts are produced **rediae,** which are elongated, nonciliated larvae. These larvae continue growing within the snail, giving rise to several individuals of the tadpole-like next larval stage, **cercariae.**

Cercariae escape into the water, where they swim about freely. If they encounter a fish of the family Cyprinidae—the family that includes carp and goldfish—they bore into the muscles or under the scales, lose their tails, and transform into **metacercariae** within cysts in the muscle tissue. If a human being or other mammal eats raw infected fish, the cysts dissolve in the intestine, and the young flukes migrate to the bile duct, where they mature. An individual fluke may live for 15 to 30 years in the liver. In humans, a heavy infestation of liver flukes may cause cirrhosis of the liver and death.

Other very important flukes are the blood flukes of genus *Schistosoma.* They afflict about 1 in 20 of the world's population, more than 200 million people throughout tropical Asia, Africa, Latin America, and the Middle East. Three species of *Schistosoma* cause the disease called schistosomiasis, or bilharzia. Some 800,000 people die each year from this disease.

Recently, there has been a great deal of effort to control schistosomiasis. The worms protect themselves in part from the body's immune system by coating themselves with a variety of the host's own antigens that effectively render the worm immunologically invisible (see chapter 57). Despite this difficulty, the search is on for a vaccine that would cause the host to develop antibodies to one of the antigens of the young worms before they protect themselves with host antigens. This vaccine would prevent humans from infection. The disease can be cured with drugs after infection.

Class Cestoda: The Tapeworms. Class Cestoda is the third class of flatworms; like flukes, they live as parasites within the bodies of other animals. In contrast to flukes, tapeworms simply hang on to the inner walls of their hosts by means of specialized terminal attachment organs and absorb food through their skins. Tapeworms lack digestive cavities as well as digestive enzymes. They are extremely specialized in relation to their parasitic way of life. Most species of tapeworms occur in the intestines of vertebrates, about a dozen of them regularly in humans.

The long, flat bodies of tapeworms are divided into three zones: the **scolex,** or attachment organ; the unsegmented **neck;** and a series of repetitive segments, the **proglottids** (see figure 44.16). The scolex usually bears several suckers and may also have hooks. Each proglottid is a complete hermaphroditic unit, containing both male and female reproductive organs. Proglottids are formed continuously in an actively growing zone at the base of the neck, with maturing ones moving farther back as new ones are formed in front of them. Ultimately the proglottids near the end of the body form mature eggs. As these eggs are fertilized, the zygotes in the very last segments begin to differentiate, and these segments fill with embryos, break off, and leave their host with the host's feces. Embryos, each surrounded by a shell, emerge from the proglottid through a pore or the ruptured body wall. They are deposited on leaves, in water, or in other places where they may be picked up by another animal.

The beef tapeworm *Taenia saginata* occurs as a juvenile in the intermuscular tissue of cattle but as an adult in the intestines of human beings. A mature adult beef tapeworm may reach a length of 10 meters or more. These worms attach themselves to the intestinal wall of their host by a scolex with four suckers. The segments that are shed from the end of the worm pass from the human in the feces and may crawl onto vegetation. The segments ultimately rupture and scatter the embryos. Embryos may remain viable for up to five months. If they are ingested by cattle, they burrow through the wall of the intestine and ultimately reach muscle tissues through the blood or lymph vessels. About 1% of the cattle in the United States are infected, and some 20% of the beef consumed is not federally inspected. When infected beef is eaten rare, infection of humans by these tapeworms is likely. As a result, the beef tapeworm is a frequent parasite of humans.

Phylum Nemertea: The Ribbon Worms

The phylogenetic relationship of phylum Nemertea (figure 44.18) to other free-living flatworms is unclear. Nemerteans are often called ribbon worms or proboscis worms. These aquatic worms have the body plan of a flatworm, but also possess a fluid-filled sac that may be a primitive coelom. This sac serves as a hydraulic power source for their proboscis, a long muscular tube that can be thrust out quickly from a sheath to capture prey. Shaped like a thread or a ribbon, ribbon worms are mostly marine and consist of about 900 species. Ribbon worms are large, often 10 to 20 centimeters and sometimes many meters in length. They are the simplest animals that possess a **complete digestive system,** one that has two separate openings, a mouth and an anus. Ribbon worms also exhibit a circulatory system in which blood flows in vessels. Many important evolutionary trends that become fully developed in more advanced animals make their first appearance in the Nemertea.

The acoelomates, typified by flatworms, are the most primitive bilaterally symmetrical animals and the simplest animals in which organs occur.

FIGURE 44.18
A ribbon worm, *Lineus* (phylum Nemertea). This is the simplest animal with a complete digestive system.

44.4 Pseudocoelomates have a simple body cavity.

The Pseudocoelomates

All bilaterians except solid worms possess an internal *body cavity*, the third key transition in the animal body plan. Seven phyla are characterized by their possession of a pseudocoel (see figure 44.4). Their evolutionary relationships remain unclear, with the possibility that the pseudocoelomate condition arose independently many times. The pseudocoel serves as a hydrostatic skeleton—one that gains its rigidity from being filled with fluid under pressure. The animals' muscles can work against this "skeleton," thus making the movement of pseudocoelomates far more efficient than that of the acoelomates.

Pseudocoelomates lack a defined circulatory system; this role is performed by the fluids that move within the pseudocoel. Most pseudocoelomates have a complete, one-way digestive tract that acts like an assembly line. Food is first broken down, then absorbed, and then treated and stored.

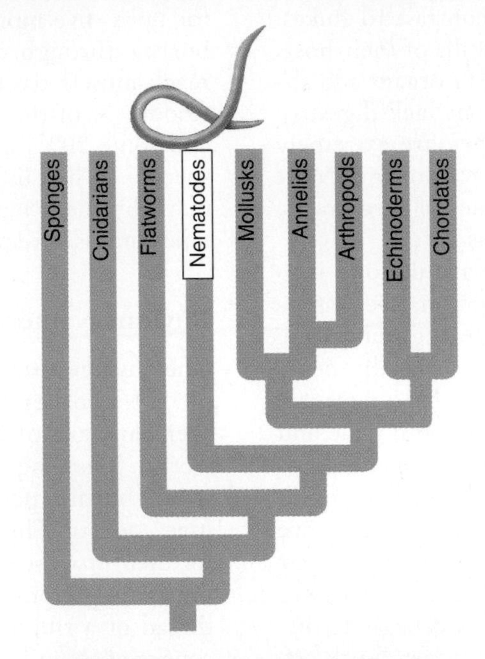

Phylum Nematoda: The Roundworms

Nematodes, eelworms, and other roundworms constitute a large phylum, Nematoda, with some 12,000 recognized species. Scientists estimate that the actual number might approach 100 times that many. Members of this phylum are found everywhere. Nematodes are abundant and diverse in marine and freshwater habitats, and many members of this phylum are parasites of animals (figure 44.19) and plants. Many nematodes are microscopic and live in soil. It has been estimated that a spadeful of fertile soil may contain, on the average, a million nematodes.

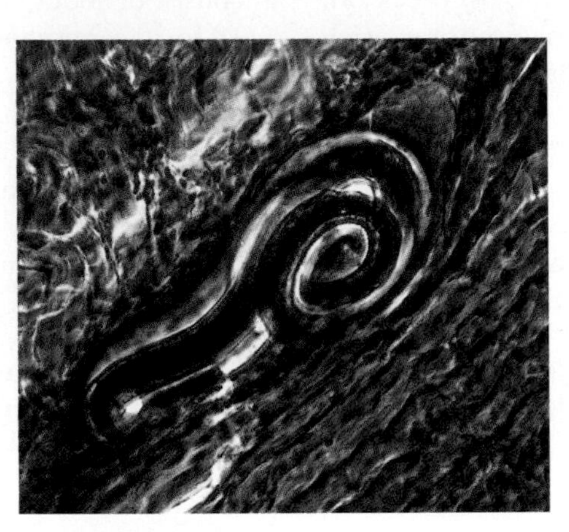

FIGURE 44.19
***Trichinella* nematode encysted in pork.** The serious disease trichinosis can result from eating undercooked pork or bear meat containing such cysts.

Nematodes are bilaterally symmetrical, unsegmented worms. They are covered by a flexible, thick cuticle, which is molted as they grow. Their muscles constitute a layer beneath the epidermis and extend along the length of the worm, rather than encircling its body. These longitudinal muscles pull both against the cuticle and the pseudocoel, which forms a hydrostatic skeleton. When nematodes move, their bodies whip about from side to side.

Near the mouth of a nematode, at its anterior end, are usually 16 raised, hairlike, sensory organs. The mouth is often equipped with piercing organs called **stylets.** Food passes through the mouth as a result of the sucking action of a muscular chamber called the **pharynx.** After passing through a short corridor into the pharynx, food continues through the other portions of the digestive tract, where it is broken down and then digested. Some of the water with which the food has been mixed is reabsorbed near the end of the digestive tract, and material that has not been digested is eliminated through the anus (figure 44.20).

Nematodes completely lack flagella or cilia, even on sperm cells. Reproduction in nematodes is sexual, with sexes usually separate. Their development is simple, and the adults consist of very few cells. For this reason, nematodes have become extremely important subjects for genetic and developmental studies (see chapter 17). The 1-millimeter-long *Caenorhabditis elegans* matures in only three days, its body is transparent, and it has only 959 cells. It is the only animal whose complete developmental cellular anatomy is known.

About 50 species of nematodes, including several that are rather common in the United States, regularly parasitize human beings. The most serious common nematode-caused disease in temperate regions is trichinosis, caused by worms of the genus *Trichinella.* These worms live in the small intestine of pigs, where fertilized female worms burrow into the intestinal wall. Once it has penetrated these tissues, each female produces about 1500 live young. The young enter the lymph channels and travel to muscle tissue throughout the body, where they mature and form highly resistant, calcified cysts. Infection in human beings or other animals arises from eating undercooked or raw pork in which the cysts of *Trichinella* are present. If the worms are abundant, a fatal infection can result, but such infections are rare; only about 20 deaths in the United States have been attributed to trichinosis during the past decade.

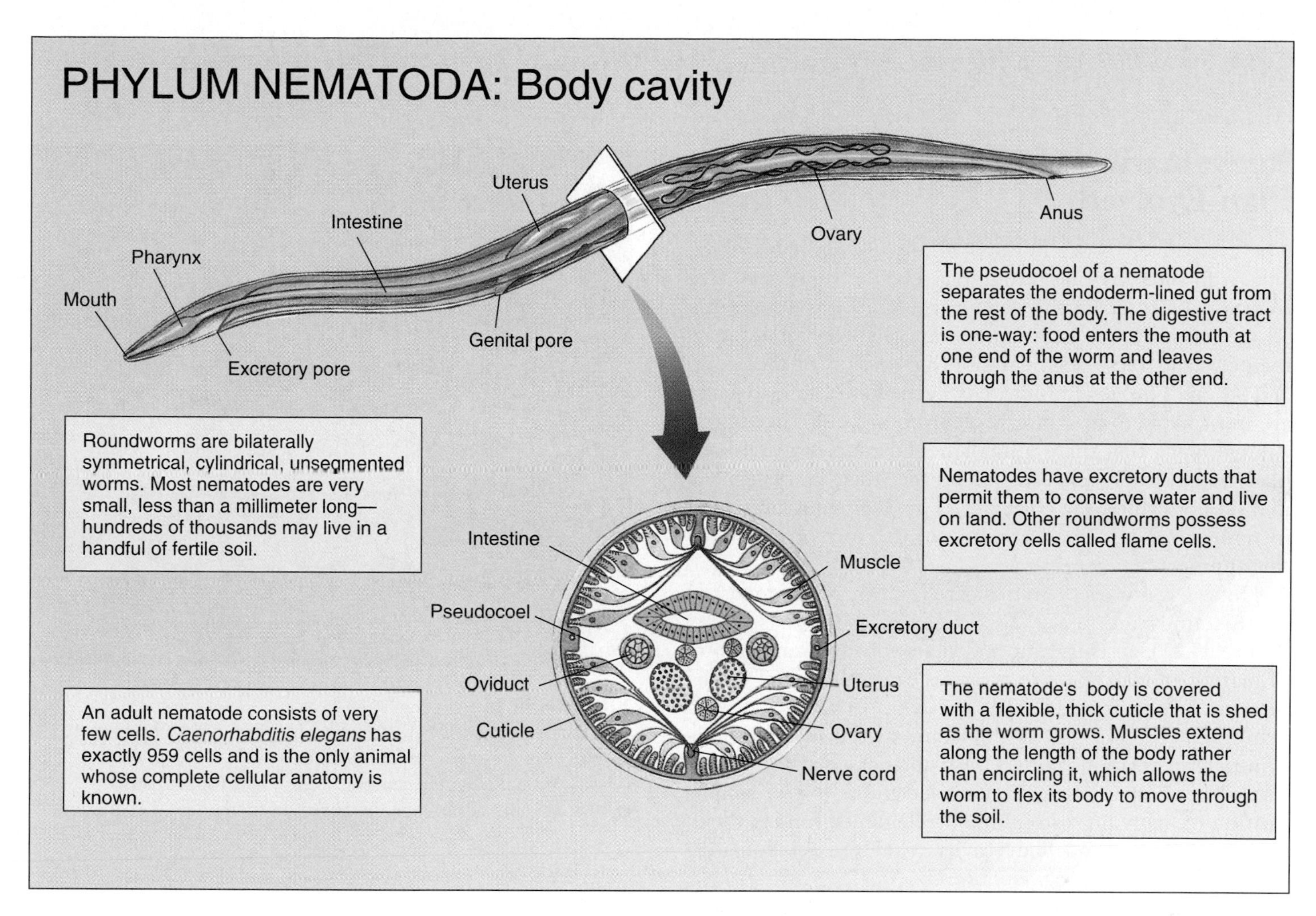
PHYLUM NEMATODA: Body cavity
Uterus
Intestine
Pharynx
Mouth
Excretory pore
Genital pore
Ovary
Anus
The pseudocoel of a nematode separates the endoderm-lined gut from the rest of the body. The digestive tract is one-way: food enters the mouth at one end of the worm and leaves through the anus at the other end.
Roundworms are bilaterally symmetrical, cylindrical, unsegmented worms. Most nematodes are very small, less than a millimeter long—hundreds of thousands may live in a handful of fertile soil.
Nematodes have excretory ducts that permit them to conserve water and live on land. Other roundworms possess excretory cells called flame cells.
Intestine
Muscle
Pseudocoel
Excretory duct
Oviduct
Uterus
Cuticle
Ovary
Nerve cord
An adult nematode consists of very few cells. Caenorhabditis elegans has exactly 959 cells and is the only animal whose complete cellular anatomy is known.
The nematode's body is covered with a flexible, thick cuticle that is shed as the worm grows. Muscles extend along the length of the body rather than encircling it, which allows the worm to flex its body to move through the soil.

FIGURE 44.20
The evolution of a simple body cavity. The major innovation in body design in roundworms (phylum Nematoda) is a body cavity between the gut and the body wall. This cavity is the pseudocoel. It allows chemicals to circulate throughout the body and prevents organs from being deformed by muscle movements.

Phylum Rotifera: Rotifers

Phylum Rotifera includes common, small, bilaterally symmetrical, basically aquatic animals that have a crown of cilia at their heads. Rotifers are pseudocoelomates but are very unlike nematodes. They have several features that suggest their ancestors may have resembled flatworms. There are about 2000 species of this phylum. While a few rotifers live in soil or in the capillary water in cushions of mosses, most occur in fresh water, and they are common everywhere. Very few rotifers are marine. Most rotifers are between 50 and 500 micrometers in length, smaller than many protists.

Rotifers have a well-developed food-processing apparatus. A conspicuous organ on the tip of the head called the corona gathers food. It is composed of a circle of cilia which sweeps their food into their mouths, as well as being used for locomotion. Rotifers are often called "wheel animals" because the cilia, when they are beating together, resemble the movement of spokes radiating from a wheel.

A Relatively New Phylum: Cycliophora

In December 1995, two Danish biologists reported the discovery of a strange new kind of creature, smaller than a period on a printed page. The tiny organism had a striking circular mouth surrounded by a ring of fine, hairlike cilia and has so unusual a life cycle that they assigned it to an entirely new phylum, Cycliophora (Greek for "carrying a small wheel"). There are only about 35 known animal phyla, so finding a new one is extremely rare! When the lobster to which it is attached starts to molt, the tiny symbiont begins a bizarre form of sexual reproduction. Dwarf males emerge, with nothing but brains and reproductive organs. Each dwarf male seeks out another female symbiont on the molting lobster and fertilizes its eggs, generating free-swimming individuals that can seek out another lobster and renew the life cycle.

The pseudocoelomates, including nematodes and rotifers, all have fluid-filled pseudocoels.

44.5 The coming revolution in animal taxonomy will likely alter traditional phylogenies.

Reevaluating How the Animal Body Plan Evolved

The great diversity see in the body plan of animals is difficult to fit into any one taxonomic scheme. Biologists have traditionally inferred the general relationships among the 35 animal phyla by examining what seemed to be fundamental characters—segmentation, possession of a coelom, and so on. The general idea has been that such characters are most likely to be conserved during a group's evolution. Animal phyla that share a fundamental character are more likely to be closely related to each other than to other phyla that do not exhibit the character. The phylogeny presented in figure 44.2 is a good example of the sort of taxonomy this approach has generated.

However, not every animal can be easily accommodated by this approach. Take, for example, the myzostomids (figure 44.21), an enigmatic and anatomically bizarre group of marine animals that are parasites or symbionts of echinoderms. Myzostomid fossils are found associated with echinoderms since the Ordovician, so the myzostomid-echinoderm relationship is a very ancient one. Their long history of obligate association has led to the loss or simplification of many myzostomid body elements, leaving them, for example, with no body cavity (they are acoelomates) and only incomplete segmentation.

This character loss has led to considerable disagreement among taxonomists. However, while taxonomists have disagreed about the details, all have generally allied myzostomids is some fashion with the annelids, sometimes within the polychaetes, sometimes as a separate phylum closely allied to the annelids.

Recently, this view has been challenged. New taxonomical comparisons using molecular data have come to very different conclusions. Researchers examined two components of the protein synthesis machinery, the small ribosomal subunit rRNA gene and an elongation factor gene (called 1 alpha). The phylogeny they obtain does not place the myzostomids in with the annelids. Indeed, they find that the myzostomids have no close links to the annelids at all. Instead, surprisingly, they are more closely allied with the flatworms!

This result hints strongly that the key morphological characters that biologists have traditionally used to construct animal phylogenies—segmentation, coeloms, jointed appendages, and the like—are not the conservative characters we had supposed. Among the myzostomids these features appear to have been gained and lost again during the course of their evolution. If this unconservative evolutionary pattern should prove general, our view of the evolution of the animal body plan, and how the various animal phyla relate to one another, will soon be in need of major revision.

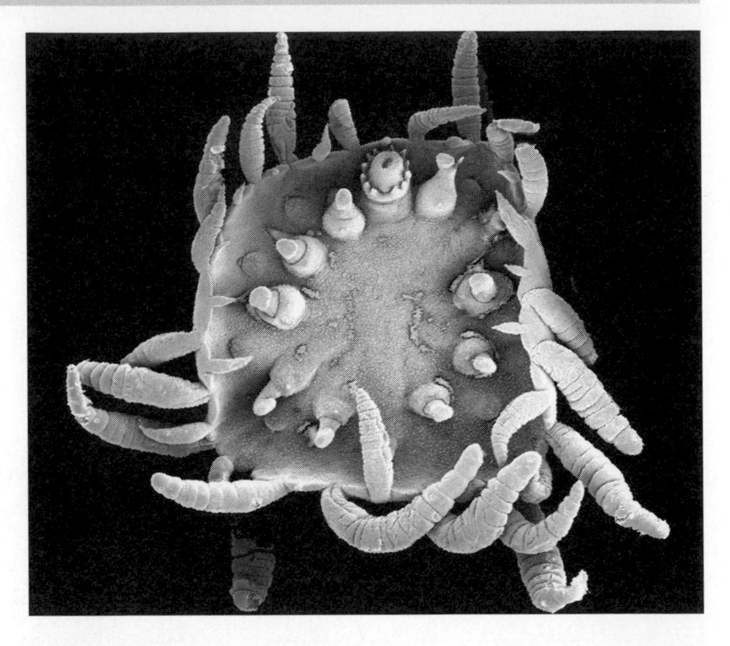

FIGURE 44.21
A taxonomic puzzle. *Myzostoma mortenseni* has no body cavity and incomplete segmentation. Animals such as this present a classification challenge, causing taxonomists to reconsider traditional animal phylogenies based on fundamental characters.

Molecular Phylogenies

The last decade has seen a wealth of new molecular sequence data on the various animal groups. The animal phylogenies that these data suggest are often significantly at odds with the traditional phylogeny used in this text and presented in figure 44.2. One such phylogeny, developed from ribosomal RNA studies, is presented in figure 44.22. It is only a rough outline; in the future more data should allow us to resolve relationships within groupings. Still, it is clear that major groups are related in very different ways in the molecular phylogeny than in the more traditional one.

At present, molecular phylogenetic analysis of the animal kingdom is in its infancy. Molecular phylogenies developed from different molecules often tend to suggest different evolutionary relationships. However, the childhood of this approach is likely to be short. Over the next few years, a mountain of additional molecular data can be anticipated. As more data are brought to bear, we can hope that the confusion will lessen, and that a consensus phylogeny will emerge. When and if this happens, it is likely to be very different from the traditional view.

The use of molecular data to construct phylogenies is likely to significantly alter our understanding of relationships among the animal phyla.

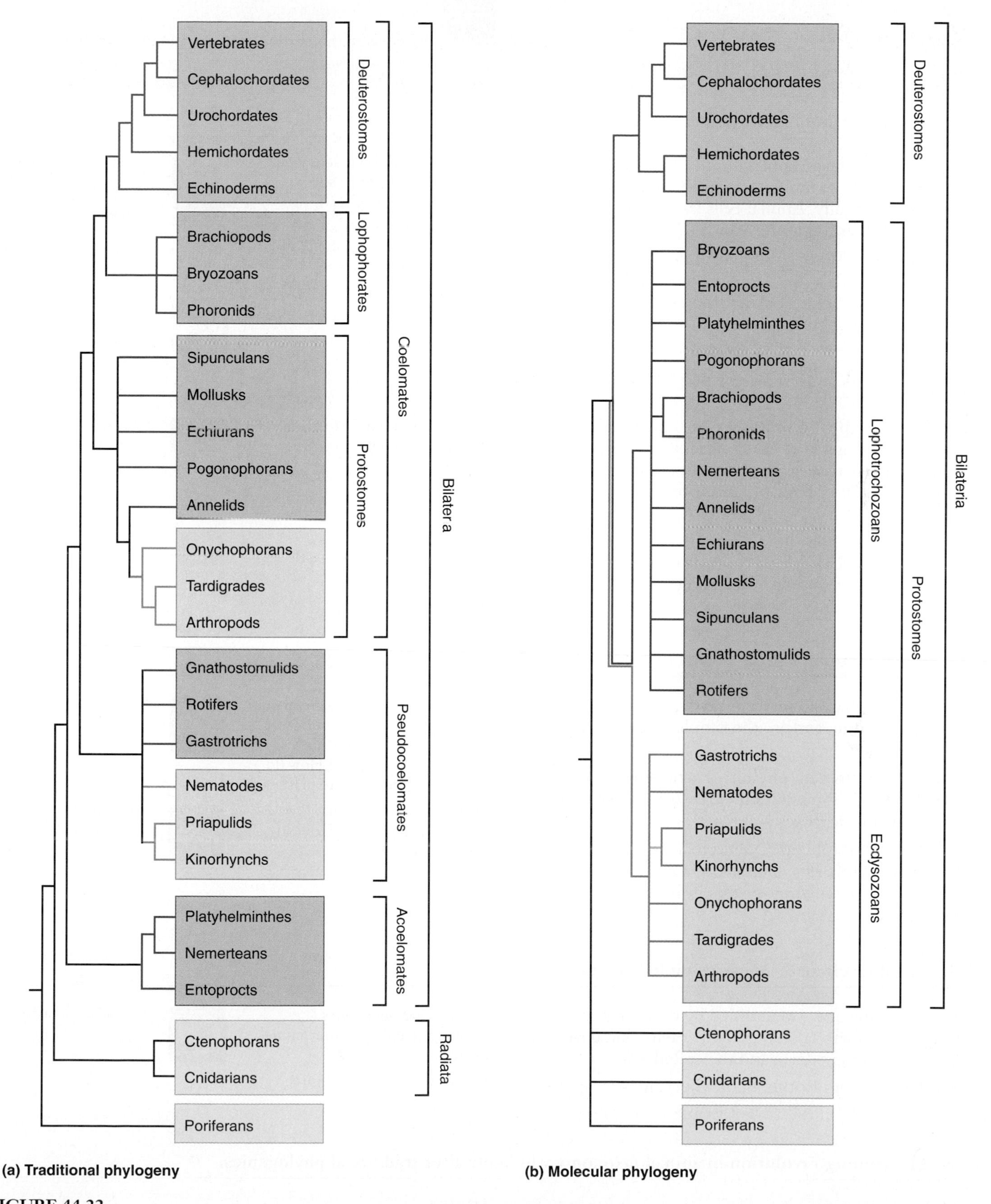

FIGURE 44.22
Traditional versus molecular animal phylogenies. (*a*) Traditional phylogenies are based on fundamental morphological characters. (After L. H. Hyman, *The Invertebrates*, 1940.) (*b*) More recent phylogenies are often based on molecular analyses, this one on comparisons of rRNA sequence differences among the animal phyla. (After Adoutte, et al., *Proc. Nat. Acad. Sci* 97: p. 4454, 2000.)

Chapter 44

www.mhhe.com/raven6/resources44.mhtml

Summary | **Questions** | **Media Resources**

44.1 Animals are multicellular heterotrophs without cell walls.

- Animals are heterotrophic, multicellular, and usually have the ability to move. Almost all animals reproduce sexually. Animal cells lack rigid cell walls and digest their food internally.
- The kingdom Animalia is divided into two subkingdoms: Parazoa, which includes only the asymmetrical phylum Porifera, and Eumetazoa, characterized by body symmetry.

1. What are the characteristics that distinguish animals from other living organisms?

2. What are the two subkingdoms of animals? How do they differ in terms of symmetry and body organization?

- Symmetry in Nature
- Posterior to Anterior
- Sagittal Plane
- Frontal to Coronal Plane
- Transvere/Cross-sectional Planes

- Characteristics of Invertebrates
- Activity: Invertebrates

44.2 The simplest animals are not bilaterally symmetrical.

- The sponges (phylum Porifera) are characterized by specialized, flagellated cells called choanocytes. They do not possess tissues or organs, and most species lack symmetry in their body organization.
- Cnidarians (phylum Cnidaria) are predominantly marine animals with unique stinging cells called cnidocytes, each of which contains a specialized harpoonlike apparatus, or nematocyst.

3. From what kind of ancestor did sponges probably evolve?

4. What are the specialized cells used by a sponge to capture food?

5. What are the two ways sponges reproduce? What do larval sponges look like?

6. What is a planula?

- Sponges
- Radial Phyla

- Art Quiz: Cnidarian Body Plan

44.3 Acoelomates are solid worms that lack a body cavity.

- Acoelomates lack an internal cavity, except for the digestive system, and are the simplest animals that have organs.
- The most prominent phylum of acoelomates, Platyhelminthes, includes the free-living flatworms and the parasitic flukes and tapeworms.
- Ribbon worms (phylum Nemertea) are similar to free-living flatworms, but have a complete digestive system and a circulatory system in which the blood flows in vessels.

7. What body plan do members of the phylum Platyhelminthes possess? Are these animals parasitic or free-living? How do they move from place to place?

8. How are tapeworms different from flukes? How do tapeworms reproduce?

- Bilateral Acoelomates

- Student Research: Parasitic Flatworms

44.4 Pseudocoelomates have a simple body cavity.

- Pseudocoelomates, exemplified by the nematodes (phylum Nematoda), have a body cavity that develops between the mesoderm and the endoderm.
- Rotifers (phylum Rotifera), or wheel animals, are very small freshwater pseudocoelomates.

9. Why are nematodes structurally unique in the animal world?

10. How do rotifers capture food?

- Pseudocoelomates

44.5 The coming revolution in animal taxonomy will likely alter traditional phylogenies.

- Molecular data are suggesting animal phylogenies that are in considerable disagreement with traditional phylogenies.

11. With what group are myzostomids most closely allied?

45

Mollusks and Annelids

Concept Outline

45.1 Mollusks were among the first coelomates.

Coelomates. Next to the arthropods, mollusks comprise the second most diverse phylum and include snails, clams, and octopuses. There are more terrestrial mollusk species than terrestrial vertebrates!

Body Plan of the Mollusks. The mollusk body plan is characterized by three distinct sections, a unique rasping tongue, and distinctive free-swimming larvae also found in annelid worms.

The Classes of Mollusks. The three major classes of mollusks are the gastropods (snails and slugs), the bivalves (oysters and clams), and the cephalopods (octopuses and squids). While they seem very different at first glance, on closer inspection, they all have the same basic mollusk body plan.

45.2 Annelids were the first segmented animals.

Segmented Animals. Annelids are segmented coelomate worms, most of which live in the sea. The annelid body is composed of numerous similar segments.

Classes of Annelids. The three major classes of annelids are the polychaetes (marine worms), the oligochaetes (earthworms and related freshwater worms), and the hirudines (leeches).

45.3 Lophophorates appear to be a transitional group.

Lophophorates. The three phyla of lophophorates share a unique ciliated feeding structure, but differ in many other ways.

FIGURE 45.1
An annelid, the Christmas-tree worm, *Spirobranchus giganteus*. Mollusks and annelids inhabit both terrestrial and aquatic habitats. They are large and successful groups, with some of their most spectacular members represented in marine environments.

Although acoelomates and pseudocoelomates have proven very successful, a third way of organizing the animal body has also evolved, one that occurs in the bulk of the animal kingdom. We will begin our discussion of the coelomate animals with mollusks, which include such animals as clams, snails, slugs, and octopuses. Annelids (figure 45.1), such as earthworms, leeches, and seaworms, are also coelomates, but in addition, were the earliest group of animals to evolve segmented bodies. The lophophorates, a group of marine animals united by a distinctive feeding structure called the lophophore, have features intermediate between those of protostomes and deuterostomes and will also be discussed in this chapter. The remaining groups of coelomate animals will be discussed in chapters 46, 47, and 48.

45.1 Mollusks were among the first coelomates.

Coelomates

The evolution of the *coelom* was a significant advance in the structure of the animal body. Coelomates have a new body design that repositions the fluid and allows the development of complex tissues and organs. This new body plan also made it possible for animals to evolve a wide variety of different body architectures and to grow to much larger sizes than acoelomate animals. Among the earliest groups of coelomates were the mollusks and the annelids.

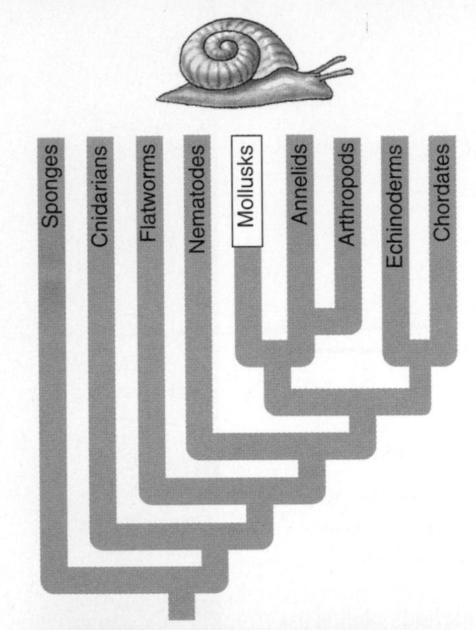

Mollusks

Mollusks (phylum Mollusca) are an extremely diverse animal phylum, second only to the arthropods, with over 110,000 described species. Mollusks include snails, slugs, clams, scallops, oysters, cuttlefish, octopuses, and many other familiar animals (figure 45.2). The durable shells of some mollusks are often beautiful and elegant; they have long been favorite objects for professional scientists and amateurs alike to collect, preserve, and study. Chitons and nudibranchs are less familiar marine mollusks. Mollusks are characterized by a coelom, and while there is extraordinary diversity in this phylum, many of the basic components of the mollusk body plan can be seen in figure 45.3.

Mollusks evolved in the oceans, and most groups have remained there. Marine mollusks are widespread and often abundant. Some groups of mollusks have invaded freshwater and terrestrial habitats, including the snails and slugs that live in your garden. Terrestrial mollusks are often abundant in places that are at least seasonally moist. Some of these places, such as the crevices of desert rocks, may appear very dry, but even these habitats have at least a temporary supply of water at certain times.

As a group, mollusks are an important source of food for humans. Oysters, clams, scallops, mussels, octopuses, and squids are among the culinary delicacies that belong to this large phylum. Mollusks are also of economic significance to us in many other ways. For example, pearls are produced in oysters, and the material called mother-of-pearl, often used in jewelry and other decorative objects, is produced in the shells of a number of different mollusks, but most notably in the snail called abalone. Mollusks are not wholly beneficial to humans, however. Bivalve mollusks called shipworms burrow through wood submerged in the sea, damaging boats, docks, and pilings. The zebra mussel has recently invaded North American ecosystems from Europe via the ballast water of cargo ships from Europe, wreaking havoc in many aquatic ecosystems. Slugs and terrestrial snails often cause extensive damage to garden flowers, vegetables, and crops. Other mollusks serve as hosts to the intermediate stages of many serious parasites, including several nematodes and flatworms, which we discussed in chapter 44.

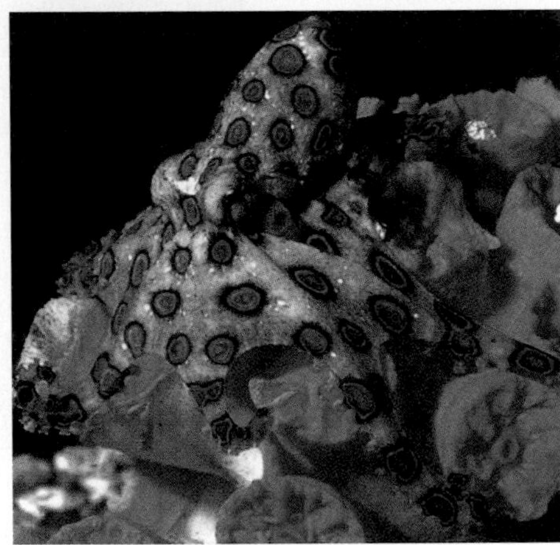

FIGURE 45.2
A mollusk. The blue-ringed octopus is one of the few mollusks dangerous to humans. Strikingly beautiful, it is equipped with a sharp beak and poison glands—divers give it a wide berth!

Mollusks range in size from almost microscopic to huge, although most measure a few centimeters in their largest dimension. Some, however, are minute, while others reach formidable sizes. The giant squid, which is occasionally cast ashore but has rarely been observed in its natural environment, may grow up to 21 meters long! Weighing up to 250 kilograms, the giant squid is the largest invertebrate and, along with the giant clam (figure 45.4), the heaviest. Millions of giant squid probably inhabit the deep regions of the ocean, even though they are seldom caught. Another large mollusk is the bivalve *Tridacna maxima*, the giant clam, which may be as long as 1.5 meters and may weigh as much as 270 kilograms.

Mollusks are the second-largest phylum of animals in terms of named species; mollusks exhibit a variety of body forms and live in many different environments.

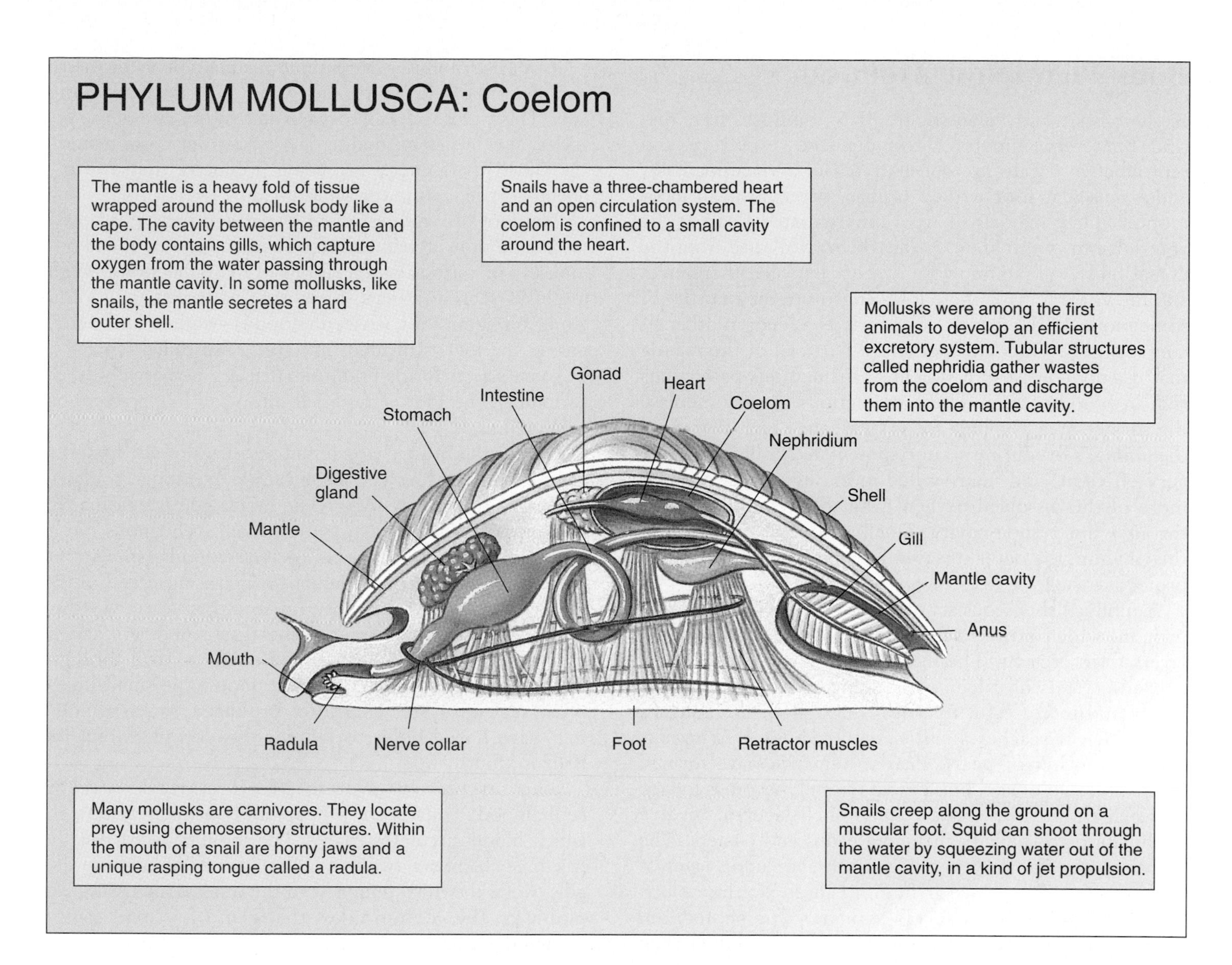

PHYLUM MOLLUSCA: Coelom
The mantle is a heavy fold of tissue wrapped around the mollusk body like a cape. The cavity between the mantle and the body contains gills, which capture oxygen from the water passing through the mantle cavity. In some mollusks, like snails, the mantle secretes a hard outer shell.
Snails have a three-chambered heart and an open circulation system. The coelom is confined to a small cavity around the heart.
Mollusks were among the first animals to develop an efficient excretory system. Tubular structures called nephridia gather wastes from the coelom and discharge them into the mantle cavity.
Gonad
Heart
Intestine
Coelom
Stomach
Nephridium
Digestive gland
Shell
Mantle
Gill
Mantle cavity
Anus
Mouth
Radula
Nerve collar
Foot
Retractor muscles
Many mollusks are carnivores. They locate prey using chemosensory structures. Within the mouth of a snail are horny jaws and a unique rasping tongue called a radula.
Snails creep along the ground on a muscular foot. Squid can shoot through the water by squeezing water out of the mantle cavity, in a kind of jet propulsion.

FIGURE 45.3
Evolution of the coelom. A generalized mollusk body plan is shown above. The body cavity of a mollusk is a coelom, which is completely enclosed within the mesoderm. This allows physical contact between the mesoderm and the endoderm, permitting interactions that lead to development of highly specialized organs such as a stomach.

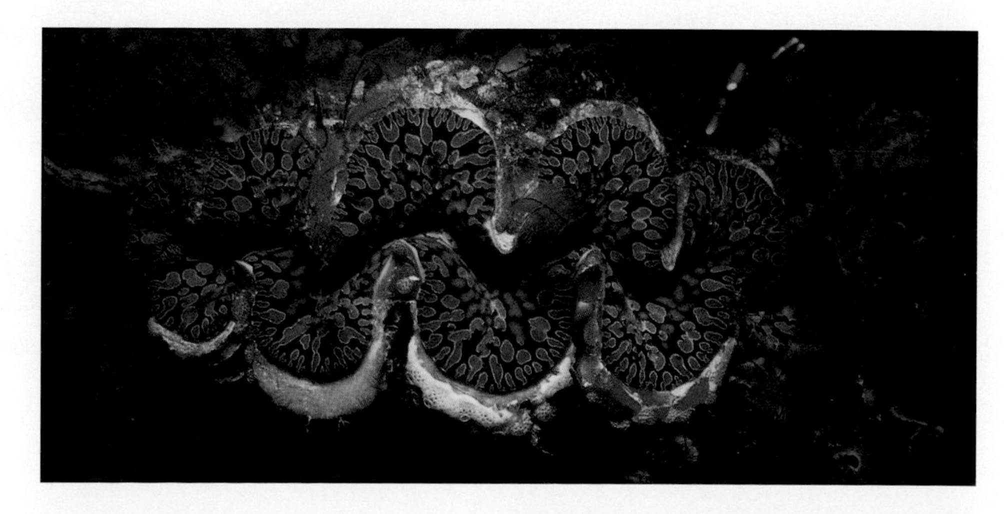

FIGURE 45.4
Giant clam. Second only to the arthropods in number of described species, members of the phylum Mollusca occupy almost every habitat on earth. This giant clam, *Tridacna maxima*, has a *green* color that is caused by the presence of symbiotic dinoflagellates (zooxanthellae). Through photosynthesis, the dinoflagellates probably contribute most of the food supply of the clam, although it remains a filter feeder like most bivalves. Some individual giant clams may be nearly 1.5 meters long and weigh up to 270 kilograms.

Body Plan of the Mollusks

In their basic body plan (figure 45.5), mollusks have distinct bilateral symmetry. Their digestive, excretory, and reproductive organs are concentrated in a **visceral mass,** and a muscular **foot** is their primary mechanism of locomotion. They may also have a differentiated **head** at the anterior end of the body. Folds (often two) arise from the dorsal body wall and enclose a cavity between themselves and the visceral mass; these folds constitute the **mantle.** In some mollusks the mantle cavity acts as a lung; in others it contains gills. **Gills** are specialized portions of the mantle that usually consist of a system of filamentous projections rich in blood vessels. These projections greatly increase the surface area available for gas exchange and, therefore, the animal's overall respiratory potential. Mollusk gills are very efficient, and many gilled mollusks extract 50% or more of the dissolved oxygen from the water that passes through the mantle cavity. Finally, in most members of this phylum, the outer surface of the mantle also secretes a protective shell.

A mollusk shell consists of a horny outer layer, rich in protein, which protects the two underlying calcium-rich layers from erosion. The middle layer consists of densely packed crystals of calcium carbonate. The inner layer is pearly and increases in thickness throughout the animal's life. When it reaches a sufficient thickness, this layer is used as mother-of-pearl. Pearls themselves are formed when a foreign object, like a grain of sand, becomes lodged between the mantle and the inner shell layer of **bivalve mollusks** (two-shelled), including clams and oysters. The mantle coats the foreign object with layer upon layer of shell material to reduce irritation caused by the object. The shell of mollusks serves primarily for protection. Many species can withdraw for protection into their shell if they have one.

In aquatic mollusks, a continuous stream of water passes into and out of the mantle cavity, drawn by the cilia on the gills. This water brings in oxygen and, in the case of the bivalves, also brings in food; it also carries out waste materials. When the gametes are being produced, they are frequently carried out in the same stream.

The foot of a mollusk is muscular and may be adapted for locomotion, attachment, food capture (in squids and octopuses), or various combinations of these functions. Some mollusks secrete mucus, forming a path that they glide along on their foot. In cephalopods—squids and octopuses—the foot is divided into arms, also called tentacles. In some *pelagic* forms, mollusks that are perpetually free-swimming, the foot is modified into wing-like projections or thin fins.

One of the most characteristic features of all the mollusks except the bivalves is the **radula,** a rasping, tongue-like organ used for feeding. The radula consists primarily of dozens to thousands of microscopic, chitinous teeth arranged in rows (figure 45.6). Gastropods (snails and their relatives) use their radula to scrape algae and other food materials off their substrates and then to convey this food to the digestive tract. Other gastropods are active predators, some using a modified radula to drill through the shells of prey and extract the food. The small holes often seen in oyster shells are produced by gastropods that have bored holes to kill the oyster and extract its body for food.

The circulatory system of all mollusks except cephalopods consists of a heart and an open system in which blood circulates freely. The mollusk heart usually has three chambers, two that collect aerated blood from the gills, while the third pumps it to the other body tissues. In mollusks, the coelom takes the form of a small cavity around the heart.

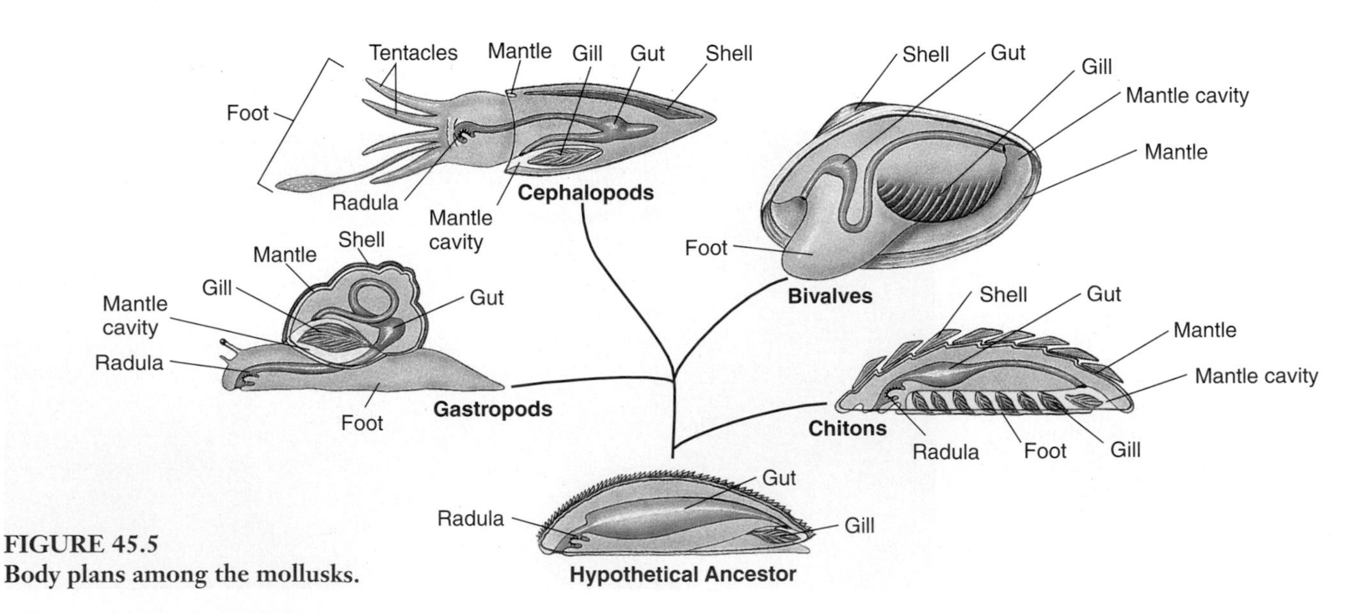

FIGURE 45.5
Body plans among the mollusks.

Nitrogenous wastes are removed from the mollusk by one or two tubular structures called **nephridia.** A typical nephridium has an open funnel, the **nephrostome,** which is lined with cilia. A coiled tubule runs from the nephrostome into a bladder, which in turn connects to an excretory pore. Wastes are gathered by the nephridia from the coelom and discharged into the mantle cavity. The wastes are then expelled from the mantle cavity by the continuous pumping of the gills. Sugars, salts, water, and other materials are reabsorbed by the walls of the nephridia and returned to the animal's body as needed to achieve an appropriate osmotic balance.

In animals with a closed circulatory system, such as annelids, cephalopod mollusks, and vertebrates, the coiled tubule of a nephridium is surrounded by a network of capillaries. Wastes are extracted from the circulatory system through these capillaries and are transferred into the nephridium, then subsequently discharged. Salts, water, and other associated materials may also be reabsorbed from the tubule of the nephridium back into the capillaries. For this reason, the excretory systems of these coelomates are much more efficient than the flame cells of the acoelomates, which pick up substances only from the body fluids. Mollusks were one of the earliest evolutionary lines to develop an efficient excretory system. Other than chordates, coelomates with closed circulation have similar excretory systems.

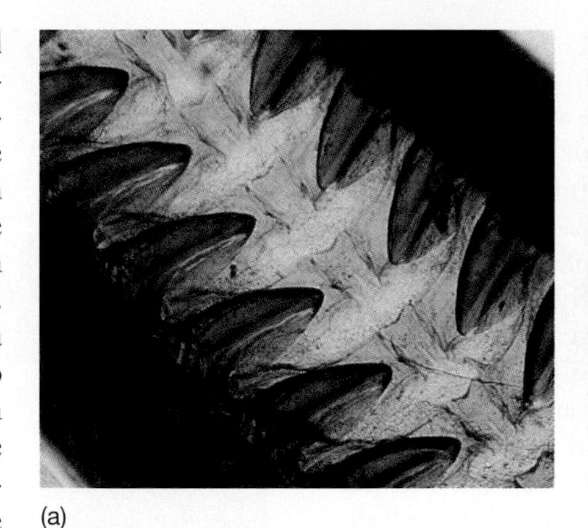

(a)

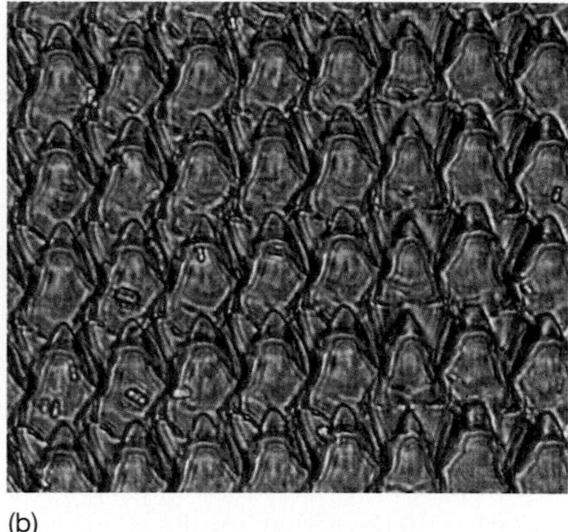

(b)

FIGURE 45.6
Structure of the radula in a snail. (*a*) The radula consists of chitin and is covered with rows of teeth. (*b*) Enlargement of the rasping teeth on a radula.

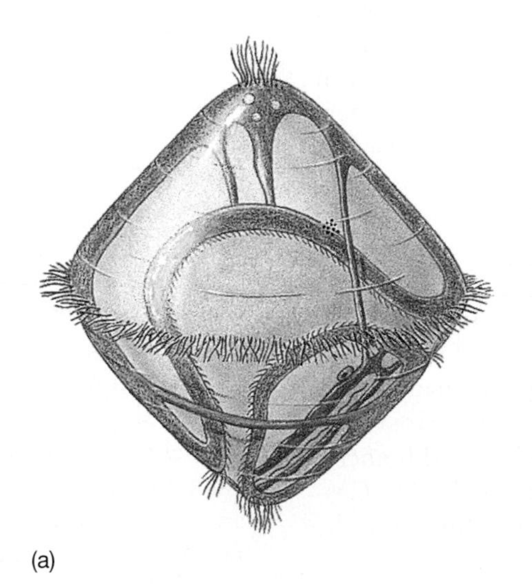

(a)

(b)

FIGURE 45.7
Stages in the molluscan life cycle. (*a*) The trochophore larva of a mollusk. Similar larvae, as you will see, are characteristic of some annelid worms as well as a few other phyla. (*b*) Veliger stage of a mollusk.

Reproduction in Mollusks

Most mollusks have distinct male and female individuals, although a few bivalves and many gastropods are hermaphroditic. Even in hermaphroditic mollusks, cross-fertilization is most common. Remarkably, some sea slugs and oysters are able to change from one sex to the other several times during a single season.

Most aquatic mollusks engage in external fertilization. The males and females release their gametes into the water, where they mix and fertilization occurs. Gastropods more often have internal fertilization, however, with the male inserting sperm directly into the female's body. Internal fertilization is one of the key adaptations that allowed gastropods to colonize the land.

Many marine mollusks have free-swimming larvae called **trochophores** (figure 45.7*a*), which closely resemble the larval stage of many marine annelids. Trochophores swim by means of a row of cilia that encircles the middle of their body. In most marine snails and in bivalves, a second free-swimming stage, the veliger, follows the trochophore stage. This **veliger** stage, has the beginnings of a foot, shell, and mantle (figure 45.7*b*). Trochophores and veligers drift widely in the ocean currents, dispersing mollusks to new areas.

Mollusks were among the earliest animals to evolve an efficient excretory system. The mantle of mollusks not only secretes their protective shell, but also forms a cavity that is essential to respiration.

The Classes of Mollusks

There are seven classes of mollusks. We will examine four classes of mollusks as representatives of the phylum: (1) Polyplacophora—chitons; (2) Gastropoda—snails, slugs, limpets, and their relatives; (3) Bivalvia—clams, oysters, scallops, and their relatives; and (4) Cephalopoda—squids, octopuses, cuttlefishes, and nautilus. By studying living mollusks and the fossil record, some scientists have deduced that the ancestral mollusk was probably a dorsoventrally flattened, unsegmented, wormlike animal that glided on its ventral surface. This animal may also have had a chitinous cuticle and overlapping calcareous scales. Other scientists believe that mollusks arose from segmented ancestors and became unsegmented secondarily.

FIGURE 45.8
A gastropod mollusk. The terrestrial snail, *Allogona townsendiana.*

Class Polyplacophora: The Chitons

Chitons are marine mollusks that have oval bodies with eight overlapping calcareous plates. Underneath the plates, the body is not segmented. Chitons creep along using a broad, flat foot surrounded by a groove or mantle cavity in which the gills are arranged. Most chitons are grazing herbivores that live in shallow marine habitats, but some live at depths of more than 7000 meters.

Class Gastropoda: The Snails and Slugs

The class Gastropoda contains about 40,000 described species of snails, slugs, and similar animals. This class is primarily a marine group, but it also contains many freshwater and terrestrial mollusks (figure 45.8). Most gastropods have a shell, but some, like slugs and nudibranchs, have lost their shells through the course of evolution. Gastropods generally creep along on a foot, which may be modified for swimming.

The heads of most gastropods have a pair of tentacles with eyes at the ends. These tentacles have been lost in some of the more advanced forms of the class. Within the mouth cavity of many members of this class are horny jaws and a radula.

During embryological development, gastropods undergo **torsion.** Torsion is the process by which the mantle cavity and anus are moved from a posterior location to the front of the body, where the mouth is located. Torsion is brought about by a disproportionate growth of the lateral muscles; that is, one side of the larva grows much more rapidly than the other. A 120-degree rotation of the visceral mass brings the mantle cavity above the head and twists many internal structures. In some groups of gastropods, varying degrees of detorsion have taken place. The **coiling,** or spiral winding, of the shell is a separate process. This process has led to the loss of the right gill and right nephridium in most gastropods. Thus, the visceral mass of gastropods has become bilaterally asymmetrical during the course of evolution.

Gastropods display extremely varied feeding habits. Some are predatory, others scrape algae off rocks (or aquarium glass), and others are scavengers. Many are herbivores, and some terrestrial ones are serious garden and agricultural pests. The radula of oyster drills is used to bore holes in the shells of other mollusks, through which the contents of the prey can be removed. In cone shells, the radula has been modified into a kind of poisonous harpoon, which is shot with great speed into the prey.

Sea slugs, or nudibranchs, are active predators; a few species of nudibranchs have the extraordinary ability to extract the nematocysts from the cnidarian polyps they eat and transfer them through their digestive tract to the surface of their gills intact and use them for their own protection. Nudibranchs are interesting in that they get their name from their gills, which instead of being enclosed within the mantle cavity are exposed along the dorsal surface (*nudi,* "naked"; *branch;* "gill").

In terrestrial gastropods, the empty mantle cavity, which was occupied by gills in their aquatic ancestors, is extremely rich in blood vessels and serves as a lung, in effect. This structure evolved in animals living in environments with plentiful oxygen; it absorbs oxygen from the air much more effectively than a gill could, but is not as effective under water.

Class Bivalvia: The Bivalves

Members of the class Bivalvia include the clams, scallops, mussels, and oysters. Bivalves have two lateral (left and right) shells (valves) hinged together dorsally (figure 45.9). A ligament hinges the shells together and causes them to gape open. Pulling against this ligament are one or two large adductor muscles that can draw the shells together.

The mantle secretes the shells and ligament and envelops the internal organs within the pair of shells. The mantle is frequently drawn out to form two siphons, one for an incoming and one for an outgoing stream of water. The siphons often function as snorkels to allow bivalves to filter water through their body while remaining almost completely buried in sediments. A complex folded gill lies on each side of the visceral mass. These gills consist of pairs of filaments that contain many blood vessels. Rhythmic beating of cilia on the gills creates a pattern of water circulation. Most bivalves are sessile filter-feeders. They extract small organisms from the water that passes through their mantle cavity.

Bivalves do not have distinct heads or radulas, differing from gastropods in this respect (see figure 45.5). However, most have a wedge-shaped foot that may be adapted, in different species, for creeping, burrowing, cleansing the animal, or anchoring it in its burrow. Some species of clams can dig into sand or mud very rapidly by means of muscular contractions of their foot.

Bivalves disperse from place to place largely as larvae. While most adults are adapted to a burrowing way of life, some genera of scallops can move swiftly through the water by using their large adductor muscles to clap their shells together. These muscles are what we usually eat as "scallops." The edge of a scallop's body is lined with tentacle-like projections tipped with complex eyes.

There are about 10,000 species of bivalves. Most species are marine, although many also live in fresh water. Over 500 species of pearly freshwater mussels, or naiads, occur in the rivers and lakes of North America.

FIGURE 45.9
A bivalve. The file shell *Lima scabra* opened, showing tentacles.

FIGURE 45.10
A cephalopod. Squids are active predators, competing effectively with fish for prey.

Class Cephalopoda: The Octopuses, Squids, and Nautilus

The more than 600 species of the class Cephalopoda—octopuses, squids, and nautilus—are the most intelligent of the invertebrates. They are active marine predators that swim, often swiftly, and compete successfully with fish. The foot has evolved into a series of tentacles equipped with suction cups, adhesive structures, or hooks that seize prey efficiently. Squids have 10 tentacles (figure 45.10); octopuses, as indicated by their name, have eight; and the nautilus, about 80 to 90. Once the tentacles have snared the prey, it is bitten with strong, beaklike paired jaws and pulled into the mouth by the tonguelike action of the radula.

Cephalopods have highly developed nervous systems, and their brains are unique among mollusks. Their eyes are very elaborate, and have a structure much like that of vertebrate eyes, although they evolved separately (see chapter 55). Many cephalopods exhibit complex patterns of behavior and a high level of intelligence; octopuses can be easily trained to distinguish among classes of objects. Most members of this class have closed circulatory systems and are the only mollusks that do.

Although they evolved from shelled ancestors, living cephalopods, except for the few species of nautilus, lack an external shell. Like other mollusks, cephalopods take water into the mantle cavity and expel it through a siphon. Cephalopods have modified this system into a means of jet propulsion. When threatened, they eject water violently and shoot themselves through the water.

Most octopus and squid are capable of changing color to suit their background or display messages to one another. They accomplish this feat through the use of their *chromatophores*, pouches of pigments embedded in the epithelium.

Gastropods typically live in a hard shell. Bivalves have hinged shells but do not have a distinct head area. Cephalopods possess well-developed brains and are the most intelligent invertebrates.

Segmented Animals

A key transition in the animal body plan was *segmentation*, the building of a body from a series of similar segments. The first segmented animals to evolve were most likely **annelid worms,** phylum Annelida (figure 45.11). One advantage of having a body built from repeated units (segments) is that the development and function of these units can be more precisely controlled, at the level of individual segments or groups of segments. For example, different segments may possess different combinations of organs or perform different functions relating to reproduction, feeding, locomotion, respiration, or excretion.

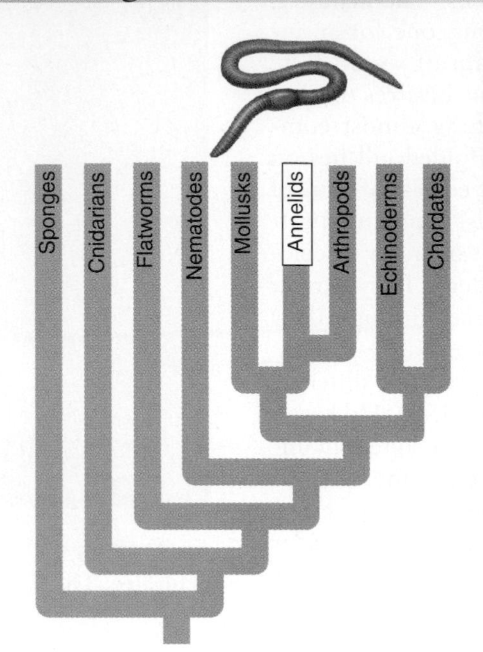

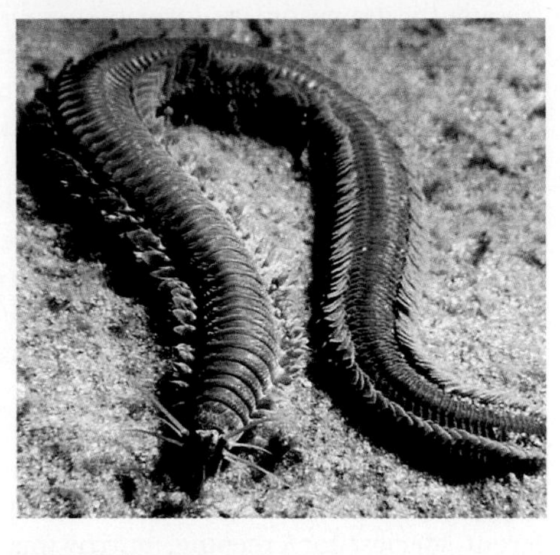

FIGURE 45.11
A polychaete annelid. *Nereis virens* is a wide-ranging, predatory, marine polychaete worm equipped with feathery parapodia for movement and respiration, as well as jaws for hunting. You may have purchased *Nereis* as fishing bait!

Annelids

Two-thirds of all annelids live in the sea (about 8000 species), and most of the rest, some 3100 species, are earthworms. Annelids are characterized by three principal features:

1. **Repeated segments.** The body of an annelid worm is composed of a series of ring-like segments running the length of the body, looking like a stack of donuts or roll of coins (figure 45.12). Internally, the segments are divided from one another by partitions called **septa,** just as bulkheads separate the segments of a submarine. In each of the cylindrical segments, the excretory and locomotor organs are repeated. The fluid within the coelom of each segment creates a hydrostatic (liquid-supported) skeleton that gives the segment rigidity, like an inflated balloon. Muscles within each segment push against the fluid in the coelom. Because each segment is separate, each can expand or contract independently. This lets the worm move in complex ways.
2. **Specialized segments.** The anterior (front) segments of annelids have become modified to contain specialized sensory organs. Some are sensitive to light, and elaborate eyes with lenses and retinas have evolved in some annelids. A well-developed cerebral ganglion, or brain, is contained in one anterior segment.
3. **Connections.** Although partitions separate the segments, materials and information do pass between segments. Annelids have a closed circulatory system that carries blood from one segment to another. A ventral nerve cord connects the nerve centers or ganglia in each segment with one another and the brain. These neural connections are critical features that allow the worm to function and behave as a unified and coordinated organism.

Body Plan of the Annelids

The basic annelid body plan is a tube within a tube, with the internal digestive tract—a tube running from mouth to anus—suspended within the coelom. The tube that makes up the digestive tract has several portions—the pharynx, esophagus, crop, gizzard, and intestine—that are specialized for different functions.

Annelids make use of their hydrostatic skeleton for locomotion. To move, annelids contract circular muscles running around each segment. Doing so squeezes the segment, causing the coelomic fluid to squirt outwards, like a tube of toothpaste. Because the fluid is trapped in the segment by the septa, instead of escaping like toothpaste, the fluid causes the segment to elongate and get much thinner. By then contracting longitudinal muscles that run along the length of the worm, the segment is returned to its original shape. In most annelid groups, each segment typically possesses **setae,** bristles of chitin that help anchor the worms during locomotion. By extending the setae in some segments so that they anchor in the substrate and retracting them in other segments, the worm can squirt its body, section by section, in either direction.

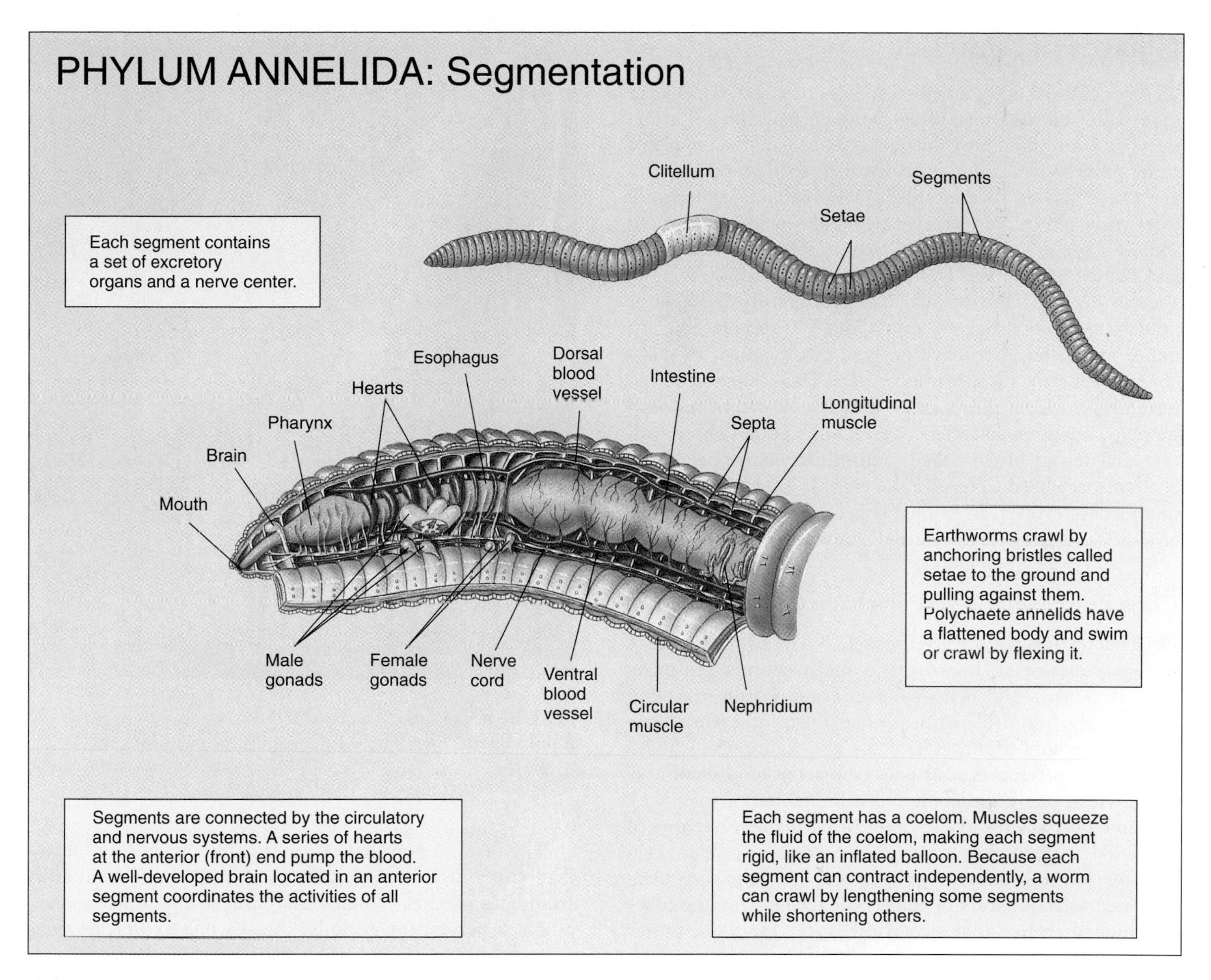
PHYLUM ANNELIDA: Segmentation
Clitellum
Segments
Setae
Each segment contains a set of excretory organs and a nerve center.
Esophagus
Dorsal blood vessel
Intestine
Hearts
Pharynx
Septa
Longitudinal muscle
Brain
Mouth
Earthworms crawl by anchoring bristles called setae to the ground and pulling against them. Polychaete annelids have a flattened body and swim or crawl by flexing it.
Male gonads
Female gonads
Nerve cord
Ventral blood vessel
Circular muscle
Nephridium
Segments are connected by the circulatory and nervous systems. A series of hearts at the anterior (front) end pump the blood. A well-developed brain located in an anterior segment coordinates the activities of all segments.
Each segment has a coelom. Muscles squeeze the fluid of the coelom, making each segment rigid, like an inflated balloon. Because each segment can contract independently, a worm can crawl by lengthening some segments while shortening others.

FIGURE 45.12
The evolution of segmentation. Marine polychaetes and earthworms (phylum Annelida) were most likely the first organisms to evolve a body plan based on partly repeated body segments. Segments are separated internally from each other by septa.

Unlike the arthropods and most mollusks, most annelids have a closed circulatory system. Annelids exchange oxygen and carbon dioxide with the environment through their body surfaces; most lack gills or lungs. However, much of their oxygen supply reaches the different parts of their bodies through their blood vessels. Some of these vessels at the anterior end of the worm body are enlarged and heavily muscular, serving as hearts that pump the blood. Earthworms have five pulsating blood vessels on each side that serve as hearts, helping to pump blood from the main dorsal vessel, which is their major pumping structure, to the main ventral vessel.

The excretory system of annelids consists of ciliated, funnel-shaped nephridia generally similar to those of mollusks. These nephridia—each segment has a pair—collect waste products and transport them out of the body through the coelom by way of specialized excretory tubes.

Annelids are a diverse group of coelomate animals characterized by serial segmentation. Each segment in the annelid body has its own circulatory, excretory, neural elements, and setae.

Classes of Annelids

The roughly 12,000 described species of annelids occur in many different habitats. They range in length from as little as 0.5 millimeter to the more than 3-meter length of some polychaetes and giant Australian earthworms. There are three classes of annelids: (1) Polychaeta, which are free-living, almost entirely marine bristleworms, comprising some 8000 species; (2) Oligochaeta, terrestrial earthworms and related marine and freshwater worms, with some 3100 species; and (3) Hirudinea, leeches, mainly freshwater predators or bloodsuckers, with about 500 species. The annelids are believed to have evolved in the sea, with polychaetes being the most primitive class. Oligochaetes seem to have evolved from polychaetes, perhaps by way of brackish water to estuaries and then to streams. Leeches share with oligochaetes an organ called a **clitellum,** which secretes a cocoon specialized to receive the eggs. It is generally agreed that leeches evolved from oligochaetes, specializing in their bloodsucking lifestyle as external parasites.

Class Polychaeta: The Polychaetes

Polychaetes (class Polychaeta) include clamworms, plume worms, scaleworms, lugworms, twin-fan worms, sea mice, peacock worms, and many others. These worms are often surprisingly beautiful, with unusual forms and sometimes iridescent colors (figure 45.13; see also figure 45.1). Polychaetes are often a crucial part of marine food chains, as they are extremely abundant in certain habitats.

Some polychaetes live in tubes or permanent burrows of hardened mud, sand, mucuslike secretions, or calcium carbonate. These sedentary polychaetes are primarily filter feeders, projecting a set of feathery tentacles from the tubes in which they live that sweep the water for food. Other polychaetes are active swimmers, crawlers, or burrowers. Many are active predators.

Polychaetes have a well-developed head with specialized sense organs; they differ from other annelids in this respect. Their bodies are often highly organized into distinct regions formed by groups of segments related in function and structure. Their sense organs include eyes, which range from simple eyespots to quite large and conspicuous stalked eyes.

Another distinctive characteristic of polychaetes is the paired, fleshy, paddlelike flaps, called **parapodia,** on most of their segments. These parapodia, which bear bristlelike setae, are used in swimming, burrowing, or crawling. They also play an important role in gas exchange because they greatly increase the surface area of the body. Some polychaetes that live in burrows or tubes may have parapodia featuring hooks to help anchor the worm. Slow crawling is carried out by means of the parapodia. Rapid crawling and swimming is by undulating the body. In addition, the polychaete epidermis often includes ciliated cells which aid in respiration and food procurement.

FIGURE 45.13
A polychaete. The shiny bristleworm, *Oenone fulgida.*

The sexes of polychaetes are usually separate, and fertilization is often external, occurring in the water and away from both parents. Unlike other annelids, polychaetes usually lack permanent **gonads,** the sex organs that produce gametes. They produce their gametes directly from germ cells in the lining of the coelom or in their septa. Fertilization results in the production of ciliated, mobile *trochophore larvae* similar to the larvae of mollusks. The trochophores develop for long periods in the plankton before beginning to add segments and thus changing to a juvenile form that more closely resembles the adult form.

Class Oligochaeta: The Earthworms

The body of an earthworm (class Oligochaeta) consists of 100 to 175 similar segments, with a mouth on the first and an anus on the last. Earthworms seem to eat their way through the soil because they suck in organic and other material by expanding their strong pharynx. Everything that they ingest passes through their long, straight digestive tracts. One region of this tract, the gizzard, grinds up the organic material with the help of soil particles.

The material that passes through an earthworm is deposited outside of its burrow in the form of castings that consist of irregular mounds at the opening of a burrow. In this way, earthworms aerate and enrich the soil. A worm can eat its own weight in soil every day.

In view of the underground lifestyle that earthworms have evolved, it is not surprising that they have no eyes. However, earthworms do have numerous light-, chemo-, and touch-sensitive cells, mostly concentrated in segments near each end of the body—those regions most likely to encounter light or other stimuli. Earthworms have fewer setae than polychaetes and no parapodia or head region.

Earthworms are hermaphroditic, another way in which they differ from most polychaetes. When they mate (figure 45.14), their anterior ends point in opposite directions, and their ventral surfaces touch. The *clitellum* is a thickened band on an earthworm's body; the mucus it secretes holds the worms together during copulation. Sperm cells are released from pores in specialized segments of one partner into the sperm receptacles of the other, the process going in both directions simultaneously.

Two or three days after the worms separate, the clitellum of each worm secretes a mucous cocoon, surrounded by a protective layer of chitin. As this sheath passes over the female pores of the body—a process that takes place as the worm moves—it receives eggs. As it subsequently passes along the body, it incorporates the sperm that were deposited during copulation. Fertilization of the eggs takes place within the cocoon. When the cocoon finally passes over the end of the worm, its ends pinch together. Within the cocoon, the fertilized eggs develop directly into young worms similar to adults.

FIGURE 45.14
Earthworms mating. The anterior ends are pointing in opposite directions.

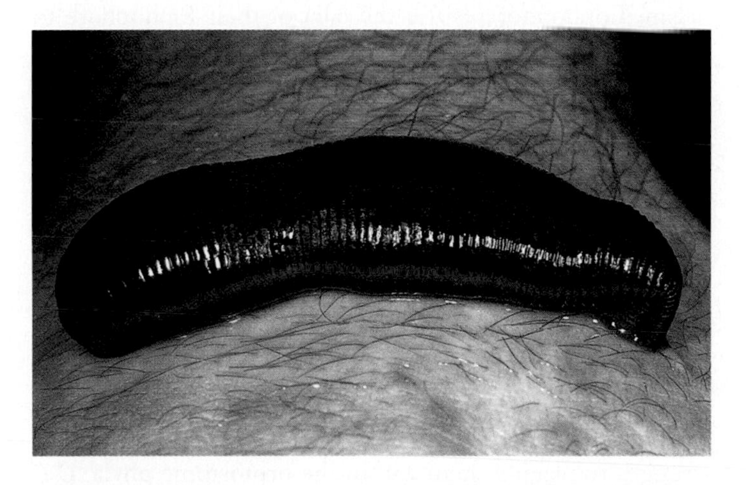

FIGURE 45.15
A leech. *Hirudo medicinalis*, the medicinal leech, is seen here feeding on a human arm. Leeches uses chitinous, bladelike jaws to make an incision to access blood and secrete an anticoagulant to keep the blood from clotting. Both the anticoagulant *and* the leech itself have made important contributions to modern medicine.

Class Hirudinea: The Leeches

Leeches (class Hirudinea) occur mostly in fresh water, although a few are marine and some tropical leeches occupy terrestrial habitats. Most leeches are 2 to 6 centimeters long, but one tropical species reaches up to 30 centimeters. Leeches are usually flattened dorsoventrally, like flatworms. They are hermaphroditic, and develop a clitellum during the breeding season; cross-fertilization is obligatory as they are unable to self-fertilize.

A leech's coelom is reduced and continuous throughout the body, not divided into individual segments as in the polychaetes and oligochaetes. Leeches have evolved suckers at one or both ends of the body. Those that have suckers at both ends move by attaching first one and then the other end to the substrate, looping along. Many species are also capable of swimming. Except for one species, leeches have no setae.

Some leeches have evolved the ability to suck blood from animals. Many freshwater leeches live as external parasites. They remain on their hosts for long periods and suck their blood from time to time.

The best-known leech is the medicinal leech, *Hirudo medicinalis* (figure 45.15). Individuals of *Hirudo* are 10 to 12 centimeters long and have bladelike, chitinous jaws that rasp through the skin of the victim. The leech secretes an anticoagulant into the wound to prevent the blood from clotting as it flows out, and its powerful sucking muscles pump the blood out quickly once the hole has been opened. Leeches were used in medicine for hundreds of years to suck blood out of patients whose diseases were mistakenly believed to be caused by an excess of blood. Today, European pharmaceutical companies still raise and sell leeches, but they are used to remove excess blood after certain surgeries. Following the surgery, blood may accumulate because veins may function improperly and fail to circulate the blood. The accumulating blood "turns off" the arterial supply of fresh blood, and the tissue often dies. When leeches remove the excess blood, new capillaries form in about a week, and the tissues remain healthy.

Segmented annelids evolved in the sea. Earthworms are their descendents, as are parasitic leeches.

45.3 Lophophorates appear to be a transitional group.

Lophophorates

Three phyla of marine animals—Phoronida, Ectoprocta, and Brachiopoda—are characterized by a **lophophore,** a circular or U-shaped ridge around the mouth bearing one or two rows of ciliated, hollow tentacles. Because of this unusual feature, they are thought to be related to one another. The lophophore presumably arose in a common ancestor. The coelomic cavity of lophophorates extends into the lophophore and its tentacles. The lophophore functions as a surface for gas exchange and as a food-collection organ. Lophophorates use the cilia of their lophophore to capture the organic detritus and plankton on which they feed. Lophophorates are attached to their substrate or move slowly.

Lophophorates share some features with mollusks, annelids and arthropods (all protostomes) and share others with deuterostomes. Cleavage in lophophorates is mostly radial, as in deuterostomes. The formation of the coelom varies; some lophophorates resemble protostomes in this respect, others deuterostomes. In the Phoronida, the mouth forms from the blastopore, while in the other two phyla, it forms from the end of the embryo opposite the blastopore. Molecular evidence shows that the ribosomes of all lophophorates are decidedly protostome-like, lending strength to placing them within the protostome phyla. Despite the differences among the three phyla, the unique structure of the lophophore seems to indicate that the members share a common ancestor. Their relationships continue to present a fascinating puzzle.

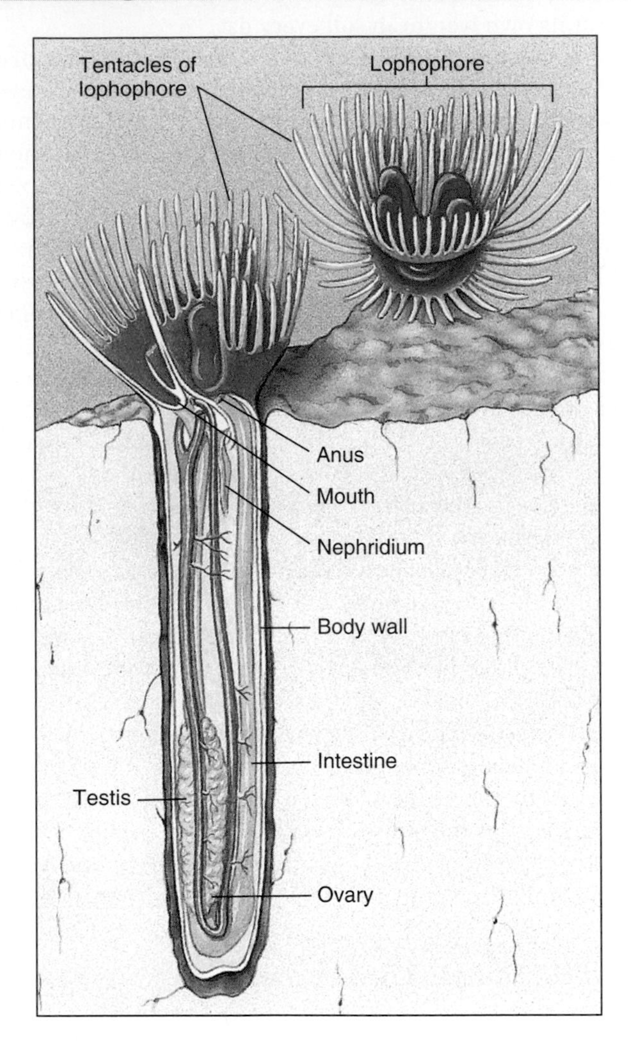

FIGURE 45.16
Phoronids (phylum Phoronida). A phoronid, such as *Phoronis*, lives in a chitinous tube that the animal secretes to form the outer wall of its body. The lophophore consists of two parallel, horseshoe-shaped ridges of tentacles and can be withdrawn into the tube when the animal is disturbed.

Phylum Phoronida: The Phoronids

Phoronids (phylum Phoronida) superficially resemble common polychaete tube worms seen on dock pilings but have many important differences. Each phoronid secretes a chitinous tube and lives out its life within it (figure 45.16). They also extend tentacles to feed and quickly withdraw them when disturbed, but the resemblance to the tube worm ends there. Instead of a straight tube-within-a-tube body plan, phoronids have a U-shaped gut. Only about 10 phoronid species are known, ranging in length from a few millimeters to 30 centimeters. Some species lie buried in sand, others are attached to rocks either singly or in groups. Phoronids develop as protostomes, with radial cleavage and the anus developing secondarily.

Phylum Ectoprocta: The Bryozoans

Ectoprocts (phylum Ectoprocta) look like tiny, short versions of phoronids (figure 45.17). They are small—usually less than 0.5 millimeter long—and live in colonies that look like patches of moss on the surfaces of rocks, seaweed, or other submerged objects (in fact, their common name bryozoans translates from Greek as "moss-animals"). The name Ectoprocta refers to the location of the anus (proct), which is external to the lophophore. The 4000 species include both marine and freshwater forms—the only nonmarine lophophorates. Individual ectoprocts secrete a tiny chitinous chamber called a **zoecium** that attaches to rocks and other members of the colony. Individuals communicate chemically through pores between chambers. Ectoprocts develop as deuterostomes, with the mouth developing secondarily; cleavage is radial.

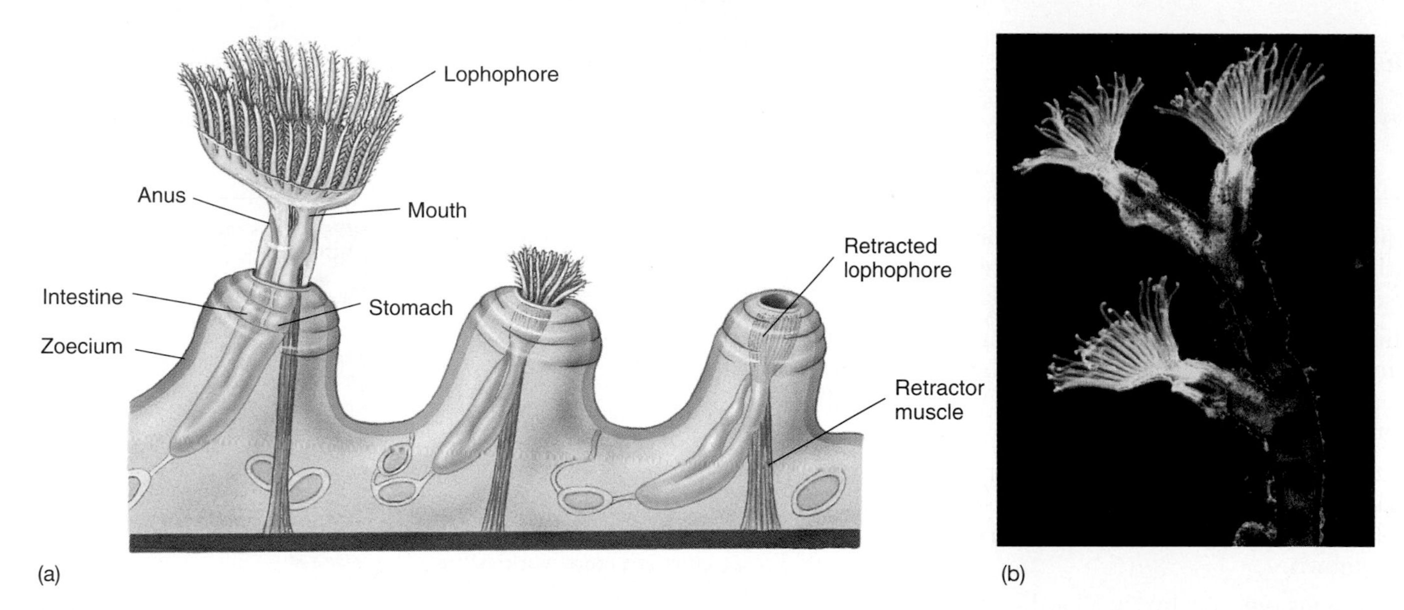

FIGURE 45.17
Ectoprocts (phylum Ectoprocta). (*a*) A small portion of a colony of the freshwater ectoproct, *Plumatella* (phylum Ectoprocta), which grows on the underside of rocks. The individual at the left has a fully extended lophophore, the structure characteristic of the three lophophorate phyla. The tiny individuals of *Plumatella* disappear into their shells when disturbed. (*b*) *Plumatella repens*, a freshwater bryozoan.

Phylum Brachiopoda: The Brachiopods

Brachiopods, or lamp shells, superficially resemble clams, with two calcified shells (figure 45.18). Many species attach to rocks or sand by a stalk that protrudes through an opening in one shell. The lophophore lies within the shell and functions when the brachiopod's shells are opened slightly. Although a little more than 300 species of brachiopods (phylum Brachiopoda) exist today, more than 30,000 species of this phylum are known as fossils. Because brachiopods were common in the earth's oceans for millions of years and because their shells fossilize readily, they are often used as index fossils to define a particular time period or sediment type. Brachiopods develop as deuterostomes and show radial cleavage.

The three phyla of lophophorates probably share a common ancestor, and they show a mixture of protostome and deuterostome characteristics.

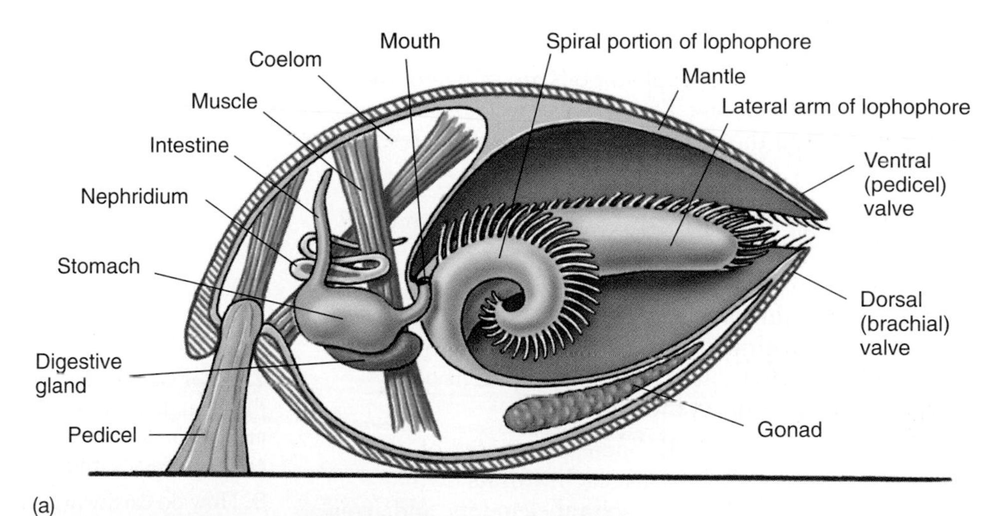

FIGURE 45.18
Brachiopods (phylum Brachiopoda). (*a*) The lophophore lies within two calcified shells, or valves. (*b*) The brachiopod *Terebratolina septentrionalis* is shown here slightly opened so that the lophophore is visible.

Chapter 45

www.mhhe.com/raven6/resources45.mhtml

Summary | **Questions** | **Media Resources**

45.1 Mollusks were among the first coelomates.

- Mollusks contain a true body cavity, or coelom, within the embryonic mesoderm and were among the first coelomate animals.
- The mollusks constitute the second largest phylum of animals in terms of named species. Their body plan consists of distinct parts: a head, a visceral mass, and a foot.
- Of the seven classes of mollusks, the gastropods (snails and slugs), bivalves (clams and scallops), and cephalopods (octopuses, squids, and nautilus), are best known.
- Gastropods typically live in a hard shell. During development, one side of the embryo grows more rapidly than the other, producing a characteristic twisting of the visceral mass.
- Members of the class Bivalvia have two shells hinged together dorsally and a wedge-shaped foot. They lack distinct heads and radulas. Most bivalves are filter-feeders.
- Octopuses and other cephalopods are efficient and often large predators. They possess well-developed brains and are the most intelligent invertebrates.

1. What is the basic body plan of a mollusk? Where is the mantle located? Why is it important in the mollusks? What occurs in the mantle cavity of aquatic mollusks?

2. What is a radula? Do all classes of mollusks possess this structure? How is it used in different types of mollusks?

3. How does the mollusk excretory structure work? Why is it better than the flame cells of acoelomates?

4. What is a trochophore? What is a veliger?

5. Do bivalves generally disperse as larvae or adults? Explain.

- Mollusks

- Student Research: Molecular Phylogenies of Gastropods

- Art Quiz: Mollusk Body Plan

45.2 Annelids were the first segmented animals.

- Segmentation is a characteristic seen only in coelomate animals at the annelid evolutionary level and above. Segmentation, or the repetition of body regions, greatly facilitates the development of specialized regions of the body.
- Annelids are worms with bodies composed of numerous similar segments, each with its own circulatory, excretory, and neural elements, and array of setae. There are three classes of annelids, the largely marine Polychaeta, the largely terrestrial Oligochaeta, and the largely freshwater Hirudinea.

6. What evolutionary advantages does segmentation confer upon an organism?

7. What are annelid setae? What function do they serve? What are parapodia? What class of annelids possess them?

8. How do earthworms obtain their nutrients? What sensory structures do earthworms possess? How do these animals reproduce?

- Annelids

- Student Research: Growth in Earthworms

45.3 Lophophorates appear to be a transitional group.

- The lophophorates consist of three phyla of marine animals—Phoronida, Ectoprocta, and Brachiopoda—characterized by a circular or U-shaped ridge, the lophophore, around the mouth.
- Some lophophorates have characteristics like protostomes, others like deuterostomes. All are characterized by a lophophore and are thought to share a common ancestor.

9. What prominent feature characterizes the lophophorate animals? What are the functions of this feature?

46

Arthropods

Concept Outline

46.1 The evolution of jointed appendages has made arthropods very successful.

Jointed Appendages and an Exoskeleton. Arthropods probably evolved from annelids, and with their jointed appendages and an exoskeleton, have successfully invaded practically every habitat on earth.

Classification of Arthropods. Arthropods have been traditionally divided into three groups based on morphological characters. However, recent research suggests a restructuring of arthropod classification is needed.

General Characteristics of Arthropods. Arthropods have segmented bodies, a chitinous exoskeleton, and often have compound eyes. They have open circulatory systems. In some groups, a series of tubes carry oxygen to the organs, and unique tubules eliminate waste.

46.2 The chelicerates all have fangs or pincers.

Class Arachnida: The Arachnids. Spiders and scorpions are predators, while most mites are herbivores.

Class Merostomata: Horseshoe Crabs. Among the most ancient of living animals, horseshoe crabs are thought to have evolved from trilobites.

Class Pycnogonida: The Sea Spiders. The spiders that are common in marine habitats differ greatly from terrestrial spiders.

46.3 Crustaceans have branched appendages.

Crustaceans. Crustaceans are unique among living arthropods because virtually all of their appendages are branched.

46.4 Insects are the most diverse of all animal groups.

Classes Chilopoda and Diplopoda: The Centipedes and Millipedes. Centipedes and millipedes are highly segmented, with legs on each segment.

Class Insecta: The Insects. Insects are the largest group of organisms on earth. They are the only invertebrate animals that have wings and can fly.

Insect Life Histories. Insects undergo simple or complete metamorphosis.

FIGURE 46.1
An arthropod. One of the major arthropod groups is represented here by *Polistes*, the common paper wasp (class Insecta).

The evolution of segmentation among annelids marked the first major innovation in body structure among coelomates. An even more profound innovation was the development of jointed appendages in arthropods, a phylum that almost certainly evolved from an annelid ancestor. Arthropod bodies are segmented like those of annelids, but the individual segments often exist only during early development and fuse into functional groups as adults. In arthropods like the wasp above (figure 46.1), jointed appendages include legs, antennae, and a complex array of mouthparts. The functional flexibility provided by such a broad array of appendages has made arthropods the most successful of animal groups.

46.1 The evolution of jointed appendages has made arthropods very successful.

Jointed Appendages and an Exoskeleton

With the evolution of the first annelids, many of the major innovations of animal structure had already appeared: the division of tissues into three primary types (endoderm, mesoderm, and ectoderm), bilateral symmetry, a coelom, and segmentation. With arthropods, two more innovations arose—the development of *jointed appendages* and an *exoskeleton.* Jointed appendages and an exoskeleton have allowed arthropods (phylum Arthropoda) to become the most diverse phylum.

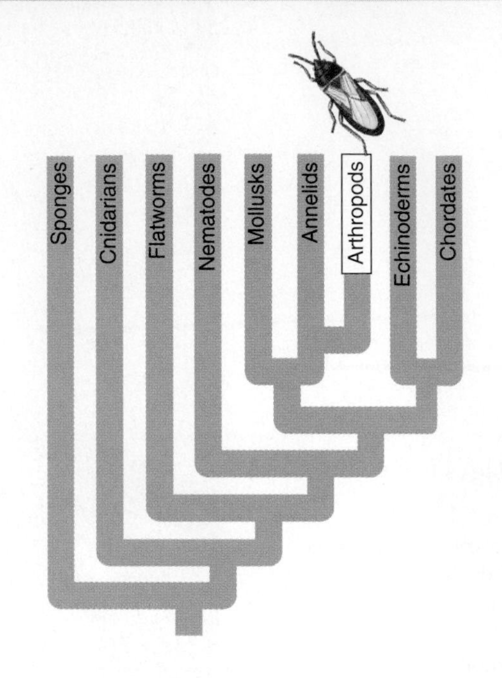

Jointed Appendages

The name "arthropod" comes from two Greek words, *arthros,* "jointed," and *podes,* "feet." All arthropods have jointed appendages. The numbers of these appendages are reduced in the more advanced members of the phylum. Individual appendages may be modified into antennae, mouthparts of various kinds, or legs. Some appendages, such as the wings of certain insects, are not homologous to the other appendages; insect wings evolved separately.

To gain some idea of the importance of jointed appendages, imagine yourself without them—no hips, knees, ankles, shoulders, elbows, wrists, or knuckles. Without jointed appendages, you could not walk or grasp any object. Arthropods use jointed appendages such as legs for walking, antennae to sense their environment, and mouthparts for feeding.

Exoskeleton

The arthropod body plan has a second major innovation: a rigid external skeleton, or **exoskeleton,** made of chitin and protein. In any animal, the skeleton functions to provide places for muscle attachment. In arthropods, the muscles attach to the interior surface of their hard exoskeleton, which also protects the animal from predators and impedes water loss. Chitin is chemically similar to cellulose, the dominant structural component of plants, and shares similar properties of toughness and flexibility. Together, the chitin and protein provide an external covering that is both very strong and capable of flexing in response to the contraction of muscles attached to it. In most crustaceans, the exoskeleton is made even tougher, although less flexible, with deposits of calcium salts. However, there is a limitation. The exoskeleton must be much thicker to bear the pull of the muscles in large insects than in small ones. That is why you don't see beetles as big as birds, or crabs the size of a cow—the exoskeleton would be so thick the animal couldn't move its great weight. Because this size limitation is inherent in the body design of arthropods, there are no large arthropods—few are larger than your thumb.

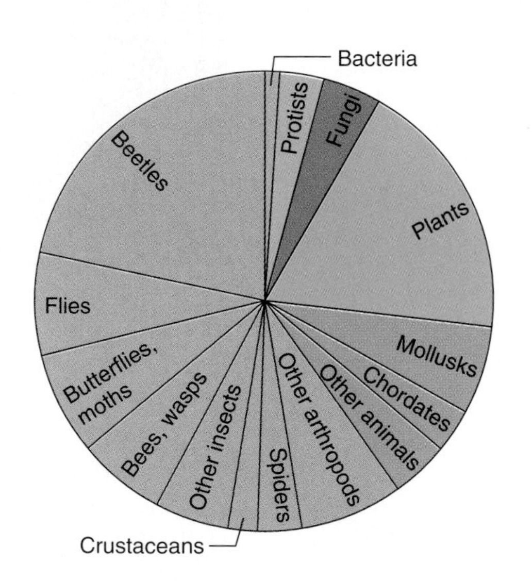

FIGURE 46.2
Arthropods are a successful group. About two-thirds of all named species are arthropods. About 80% of all arthropods are insects, and about half of the named species of insects are beetles.

The Arthropods

Arthropods, especially the largest class—insects—are by far the most successful of all animals. Well over 1,000,000 species—about two-thirds of all the named species on earth—are members of this phylum (figure 46.2). One scientist recently estimated, based on the number and diversity of insects in tropical forests, that there might be as many as 30 million species in this one class alone. About 200 million insects are alive at any one time for each human! Insects and other arthropods (figure 46.3) abound in every habitat on the planet, but they especially dominate the land, along with flowering plants and vertebrates.

The majority of arthropod species consist of small animals, mostly about a millimeter in length. Members of the phylum range in adult size from about 80 micrometers long (some parasitic mites) to 3.6 meters across (a gigantic crab found in the sea off Japan).

Arthropods, especially insects, are of enormous economic importance and affect all aspects of human life. They compete with humans for food of every kind, play a key role in the pollination of certain crops, and cause

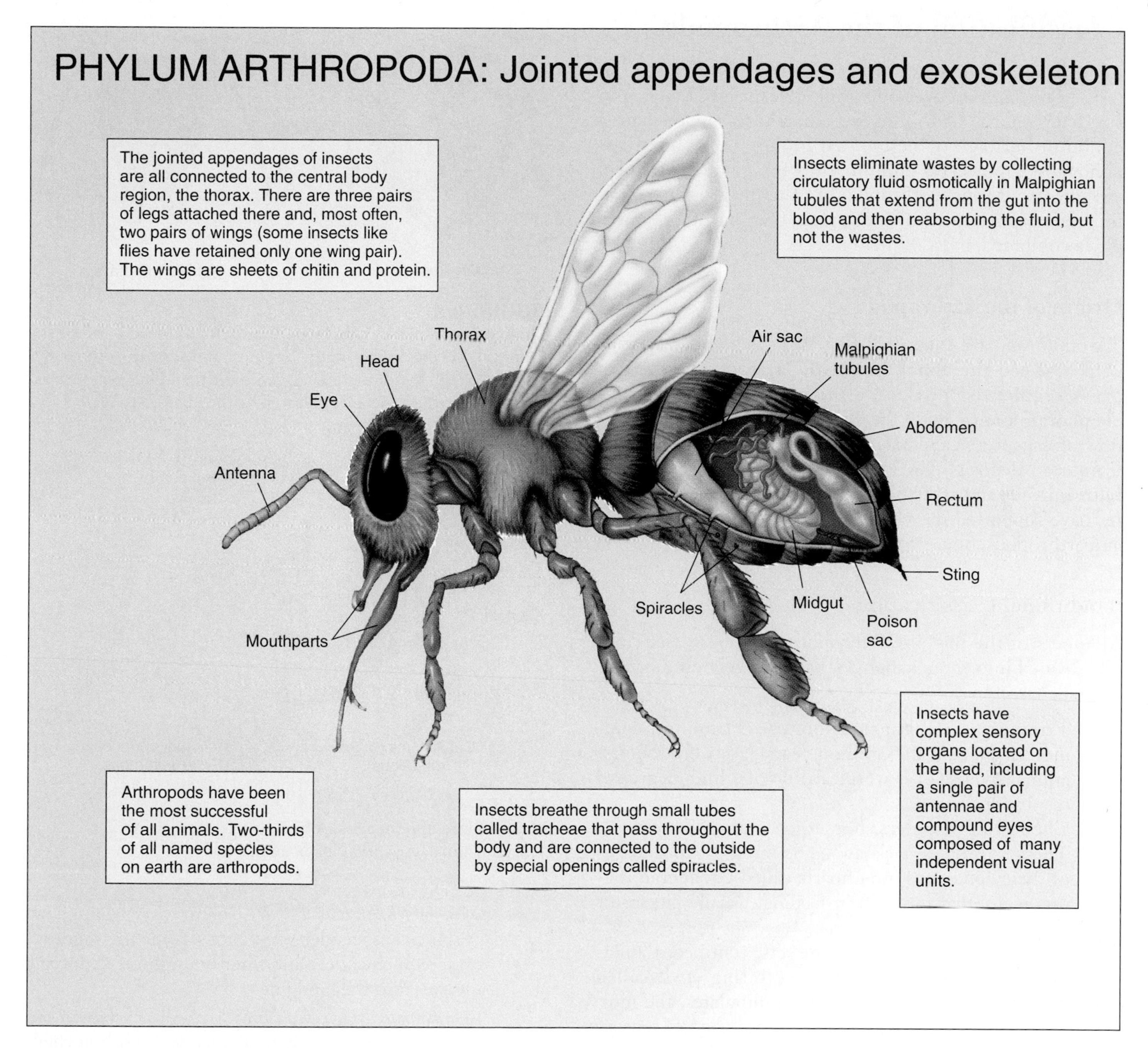

PHYLUM ARTHROPODA: Jointed appendages and exoskeleton
The jointed appendages of insects are all connected to the central body region, the thorax. There are three pairs of legs attached there and, most often, two pairs of wings (some insects like flies have retained only one wing pair). The wings are sheets of chitin and protein.
Insects eliminate wastes by collecting circulatory fluid osmotically in Malpighian tubules that extend from the gut into the blood and then reabsorbing the fluid, but not the wastes.
Thorax
Head
Eye
Antenna
Mouthparts
Air sac
Malpighian tubules
Abdomen
Rectum
Sting
Poison sac
Midgut
Spiracles
Arthropods have been the most successful of all animals. Two-thirds of all named species on earth are arthropods.
Insects breathe through small tubes called tracheae that pass throughout the body and are connected to the outside by special openings called spiracles.
Insects have complex sensory organs located on the head, including a single pair of antennae and compound eyes composed of many independent visual units.

FIGURE 46.3
The evolution of jointed appendages and an exoskeleton. Insects and other arthropods (phylum Arthropoda) have a coelom, segmented bodies, and jointed appendages. The three body regions of an insect (head, thorax, and abdomen) are each actually composed of a number of segments that fuse during development. All arthropods have a strong exoskeleton made of chitin. One class, the insects, has evolved wings that permit them to fly rapidly through the air.

billions of dollars of damage to crops, before and after harvest. They are by far the most important herbivores in all terrestrial ecosystems and are a valuable food source as well. Virtually every kind of plant is eaten by one or more species of insect. Diseases spread by insects cause enormous financial damage each year and strike every kind of domesticated animal and plant, as well as human beings.

Arthropods are segmented animals with jointed appendages. Arthropods are the most successful of all animal groups.

Classification of the Arthropods

Arthropods are among the oldest of animals, first appearing in the Precambrian over 600 million years ago. Ranging in size from enormous to microscopic, all arthropods share a common heritage of segmented bodies and jointed appendages, a powerful combination for generating novel evolutionary forms. Arthropods are the most diverse of all the animal phyla, with more species than all other animal phyla combined, most of them insects.

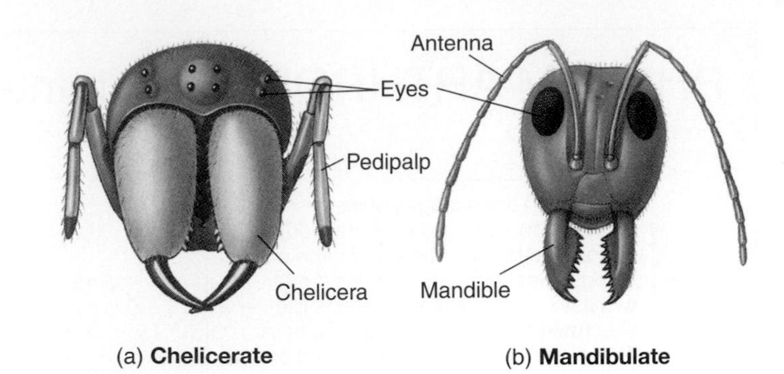

FIGURE 46.4
Chelicerates and mandibulates. In the chelicerates, such as a spider (*a*), the chelicerae are the foremost appendages of the body. In contrast, the foremost appendages in the mandibulates, such as an ant (*b*), are the antennae, followed by the mandibles.

Origin of the Arthropods

Taxonomists have long held that there is a close relationship between the annelids and the arthropods, the two great segmented phyla. Velvet worms (phylum Onychophora), known from the Burgess Shale (where upside-down fossils were called *Hallucigenia*) and many other early Cambrian deposits, have many features in common with both annelids and arthropods. Some recent molecular studies have supported the close relationship between annelids and arthropods, others have not.

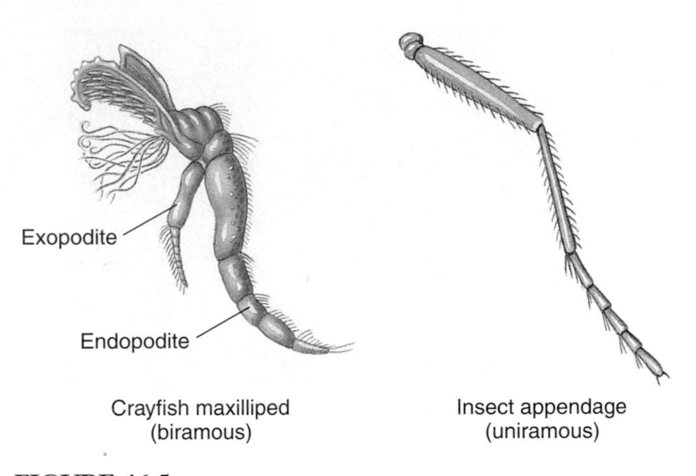

FIGURE 46.5
Mandibulate appendages. A biramous leg in a crustacean (crayfish) and a uniramous leg in an insect.

Traditional Classification

Members of the phylum Arthropoda have been traditionally divided into three subphyla, based largely on morphological characters.

1. Trilobites (the extinct trilobites). Trilobites, common in the seas 250 million years ago, were the first animals whose eyes were capable of a high degree of resolution.
2. Chelicerates (spiders, horseshoe crabs, sea spiders). These arthropods lack jaws. The foremost appendages of their bodies are mouthparts called **chelicerae** (figure 46.4*a*) that function in feeding, usually pincers or fangs.
3. Mandibulates (crustaceans, insects, centipedes, millipedes). These arthropods have biting jaws, called **mandibles** (figure 46.4*b*). In mandibulates, the most anterior appendages are one or more pairs of sensory antennae, and the next appendages are the mandibles. Among the mandibulates, insects have traditionally been set apart from the crustaceans, grouped instead with the myriapods (centipedes and millipedes) in a taxon called Tracheata. This phylogeny, still widely employed, dates back to benchmark work by the great comparative biologist Robert Snodgrass in the 1930s. He pointed out that insects, centipedes, and millipedes are united by several seemingly powerful attributes:

    A tracheal respiratory system. Trachea are small, branched air ducts that transmit oxygen from openings in the exoskeleton to every cell of the body.

    Use of Malpighian tubules for excretion. Malpighian tubules are slender projections from the digestive tract which collect and filter body fluids, emptying wastes into the hindgut.

    Uniramous (single-branched) legs. All crustacean appendages are basically biramous, or "two-branched" (figure 46.5), although some of these appendages have become single-branched by reduction in the course of their evolution. Insects, by contrast, have uniramous, or single-branched, mandibles and other appendages.

Doubts about the Traditional Approach

Recent research is casting doubt on the wisdom of this taxonomic decision. The problem is that the key morphological traits used to define the Tracheata are not as powerful taxonomically as had been assumed. Taxonomists have traditionally assumed a character like branching

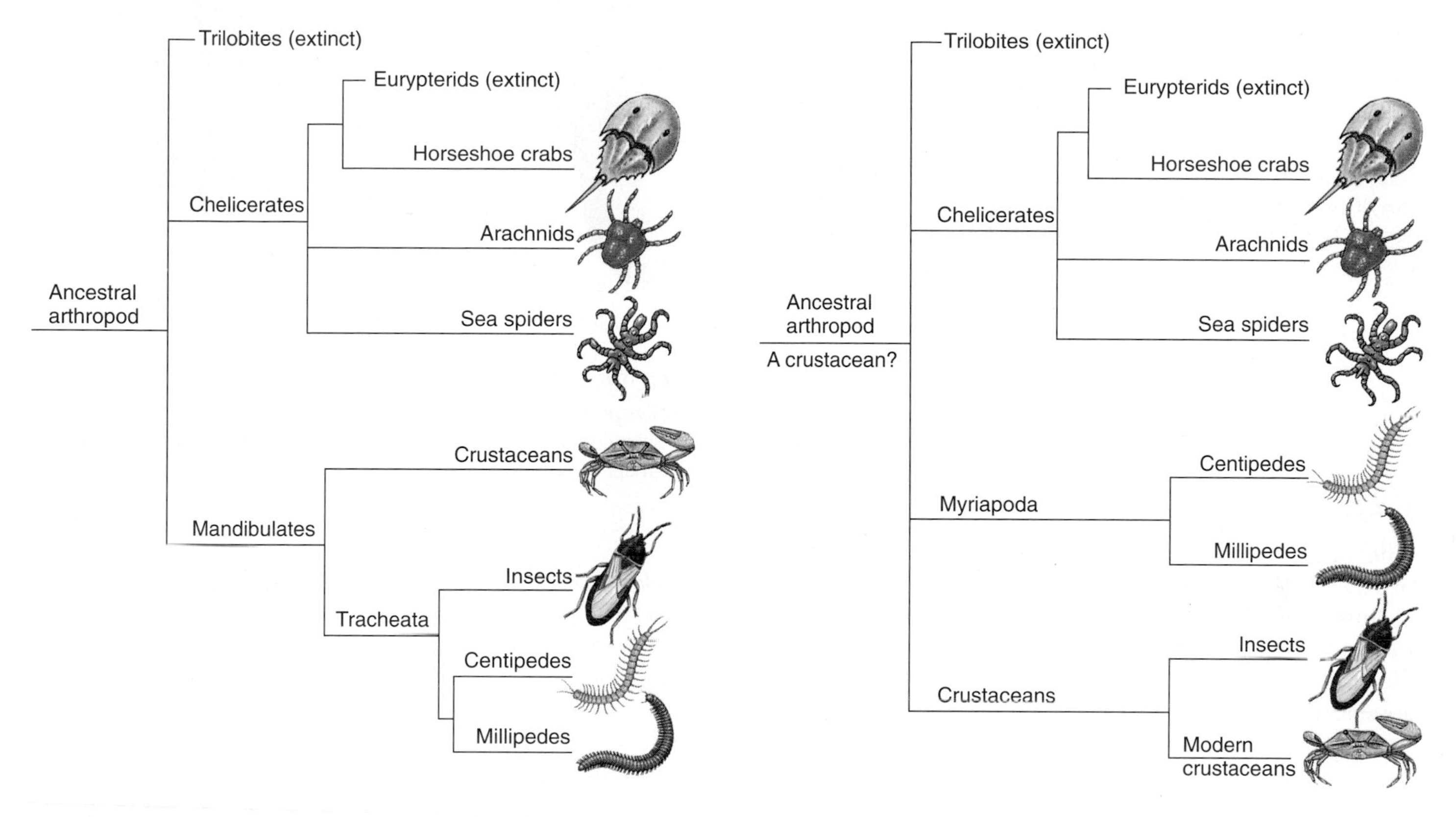

FIGURE 46.6
A proposed revision of arthropod phylogeny. Accumulating evidence supports the hypothesis that insects and modern crustaceans are sister groups, having evolved from the same ancient crustacean ancestor in the Precambrian. This implies that insects may be viewed as "flying crustaceans," and that the traditional Tracheata taxon, which places centipedes, millipedes, and insects together, is in fact a polyphyletic group.

appendages to be a fundamental one, conserved over the course of evolution, and thus suitable for making taxonomic distinctions.

However, modern molecular biology now tells us that this is not a valid assumption. The branching of arthropod legs, for example, turns out to be controlled by a single gene. The pattern of appendages among arthropods is orchestrated by a family of genes called homeotic (*Hox*) genes, described in detail in chapter 17. A single one of these *Hox* genes, called *Distal-less*, has recently been shown to initiate development of unbranched limbs in insects and branched limbs in crustaceans. The same *Distal-less* gene is found in many animal phyla, including vertebrates.

A Revolutionary New Phylogeny

In recent years a mass of accumulating morphological and molecular data has led many taxonomists to suggest new arthropod taxonomies. The most revolutionary of these, championed by Richard Brusca of Columbia University, considers crustaceans to be the basic arthropod group, and insects a close sister group (figure 46.6).

Morphological Evidence. The most recent morphological study of arthropod phylogeny, reported in 1998, was based on 100 conserved anatomical features of the central nervous system. It concluded insects were more closely related to crustaceans than to any other arthropod group. They share a unique pattern of segmental neurons, and many other features.

Molecular Evidence. Molecular phylogenies based on 18S rRNA sequences, the 18S rDNA gene, the elongation factor EF-1a, and the RNA polymerase II gene, all place insects as a sister group to crustaceans, not myriapods, and arising from within the crustaceans. In conflict with 150 years of morphology-based thinking, these conclusions are certain to engender lively discussion.

Arthropods have traditionally been classified into arachnids and other chelicerates that lack jaws and have fang mouthparts, and mandibulates (crustaceans and tracheates) with biting jaws. A revised arthropod taxonomy considers Tracheata to be the products convergent evolution, with insects and crustaceans sister groups.

General Characteristics of Arthropods

Arthropod bodies are segmented like annelids, a phylum to which at least some arthropods are clearly related. Members of some classes of arthropods have many body segments. In others, the segments have become fused together into functional groups, or **tagmata** (singular, **tagma**), such as the head and thorax of an insect (figure 46.7). This fusing process, known as **tagmatization,** is of central importance in the evolution of arthropods. In most arthropods, the original segments can be distinguished during larval development. All arthropods have a distinct head, sometimes fused with the thorax to form a tagma called the **cephalothorax.**

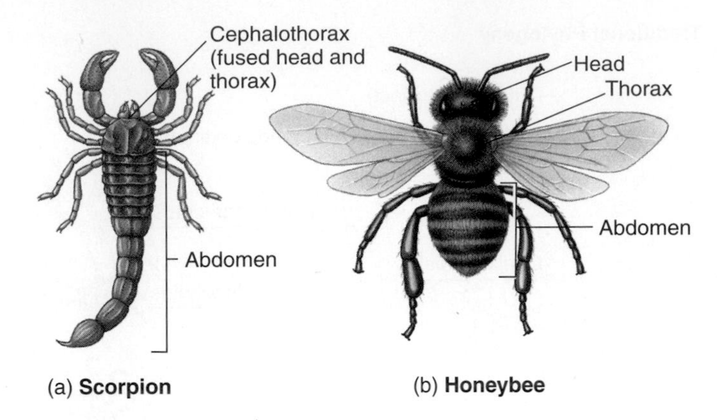

FIGURE 46.7
Arthropod evolution from many to few body segments. The (*a*) scorpion and the (*b*) honeybee are arthropods with different numbers of body segments.

Exoskeleton

The bodies of all arthropods are covered by an exoskeleton, or cuticle, that contains chitin. This tough outer covering, against which the muscles work, is secreted by the epidermis and fused with it. The exoskeleton remains fairly flexible at specific points, allowing the exoskeleton to bend and appendages to move. The exoskeleton protects arthropods from water loss and helps to protect them from predators, parasites, and injury.

Molting. Arthropods periodically undergo **ecdysis,** or molting, the shedding of the outer cuticular layer. When they outgrow their exoskeleton, they form a new one underneath. This process is controlled by hormones. When the new exoskeleton is complete, it becomes separated from the old one by fluid. This fluid dissolves the chitin and protein and, if it is present, calcium carbonate, from the old exoskeleton. The fluid increases in volume until, finally, the original exoskeleton cracks open, usually along the back, and is shed. The arthropod emerges, clothed in a new, pale, and still somewhat soft exoskeleton. The arthropod then "puffs itself up," ultimately expanding to full size. The blood circulation to all parts of the body aids them in this expansion, and many insects and spiders take in air to assist them. The expanded exoskeleton subsequently hardens. While the exoskeleton is soft, the animal is especially vulnerable. At this stage, arthropods often hide under stones, leaves, or branches.

Compound Eye

Another important structure in many arthropods is the **compound eye** (figure 46.8*a*). Compound eyes are composed of many independent visual units, often thousands of them, called **ommatidia.** Each ommatidium is covered with a lens and linked to a complex of eight retinular cells

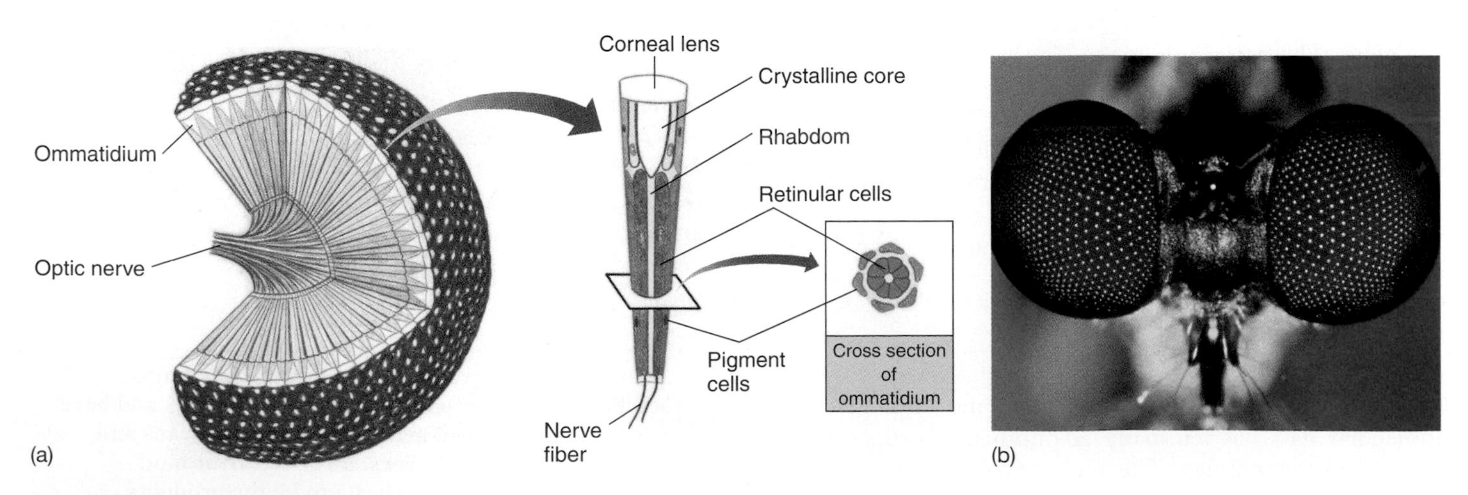

FIGURE 46.8
The compound eye. (*a*) The compound eyes found in insects are complex structures. (*b*) Three ocelli are visible between the compound eyes of the robberfly (order Diptera).

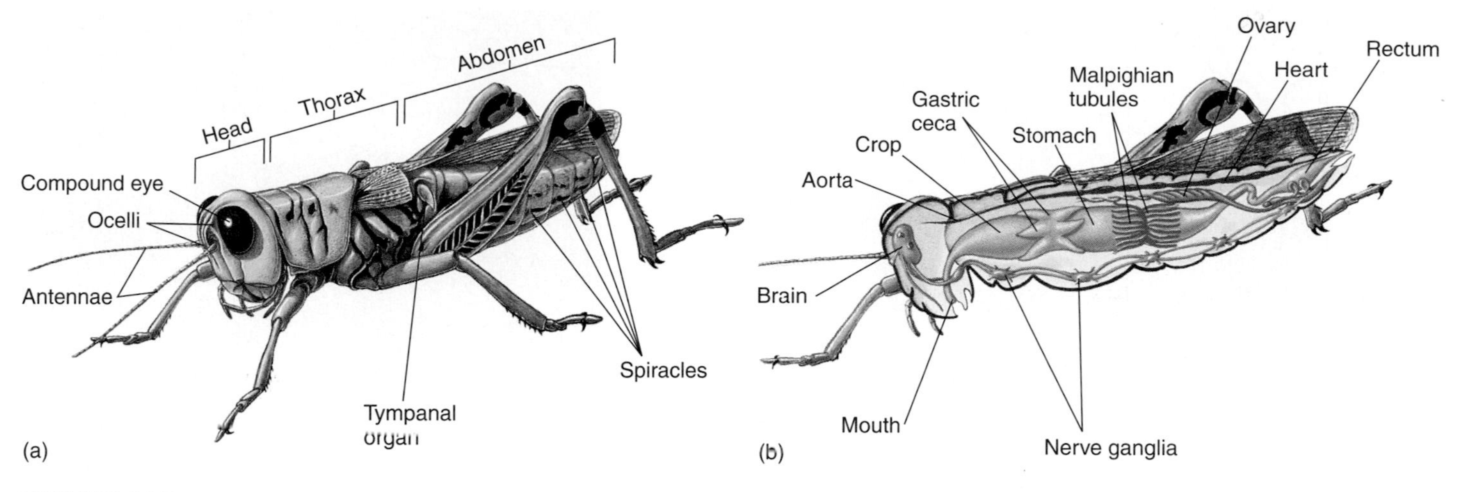

FIGURE 46.9
A grasshopper (order Orthoptera). This grasshopper illustrates the major structural features of the insects, the most numerous group of arthropods. (*a*) External anatomy. (*b*) Internal anatomy.

and a light-sensitive central core, or **rhabdom.** Compound eyes among insects are of two main types: **apposition eyes** and **superposition eyes.** Apposition eyes are found in bees and butterflies and other insects that are active during the day. Each ommatidium acts in isolation, surrounded by a curtain of pigment cells that blocks the passage of light from one to another. Superposition eyes, such as those found in moths and other insects that are active at night, are designed to maximize the amount of light that enters each ommatidium. At night, the pigment in the pigment cells is concentrated at the top of the cells so that the low levels of light can be received by many different ommatidia. During daylight, the pigment in the pigment cells is evenly dispersed throughout the cells, allowing the eye to function much like an apposition eye. The pigment in the pigment cells gives the arthropod eye its color, but it is not the critical pigment needed for vision. The visual pigment is located in an area called the rhabdom found in the center of the ommatidium. The individual images from each ommatidium are combined in the arthropod's brain to form its image of the external world.

Simple eyes, or **ocelli,** with single lenses are found in the other arthropod groups and sometimes occur together with compound eyes, as is often the case in insects (figure 46.8*b*). Ocelli function in distinguishing light from darkness. The ocelli of some flying insects, namely locusts and dragonflies, function as horizon detectors and help the insect visually stabilize its course in flight.

Circulatory System

In the course of arthropod evolution, the coelom has become greatly reduced, consisting only of cavities that house the reproductive organs and some glands. Arthropods completely lack cilia, both on the external surfaces of the body and on the internal organs. Like annelids, arthropods have a tubular gut that extends from the mouth to the anus. In the next paragraphs we will discuss the circulatory, nervous, respiratory, and excretory systems of the arthropods (figure 46.9).

The circulatory system of arthropods is open; their blood flows through cavities between the internal organs and not through closed vessels. The principal component of an insect's circulatory system is a longitudinal vessel called the heart. This vessel runs near the dorsal surface of the thorax and abdomen. When it contracts, blood flows into the head region of the insect.

When an insect's heart relaxes, blood returns to it through a series of valves. These valves are located in the posterior region of the heart and allow the blood to flow inward only. Thus, blood from the head and other anterior portions of the insect gradually flows through the spaces between the tissues toward the posterior end and then back through the one-way valves into the heart. Blood flows most rapidly when the insect is running, flying, or otherwise active. At such times, the blood efficiently delivers nutrients to the tissues and removes wastes from them.

Nervous System

The central feature of the arthropod nervous system is a double chain of segmented ganglia running along the animal's ventral surface. At the anterior end of the animal are three fused pairs of dorsal ganglia, which constitute the brain. However, much of the control of an arthropod's activities is relegated to ventral ganglia. Therefore, the animal can carry out many functions, including eating, movement, and copulation, even if the brain has been removed. The brain of arthropods seems to be a control point, or inhibitor, for various actions, rather than a stimulator, as it is in vertebrates.

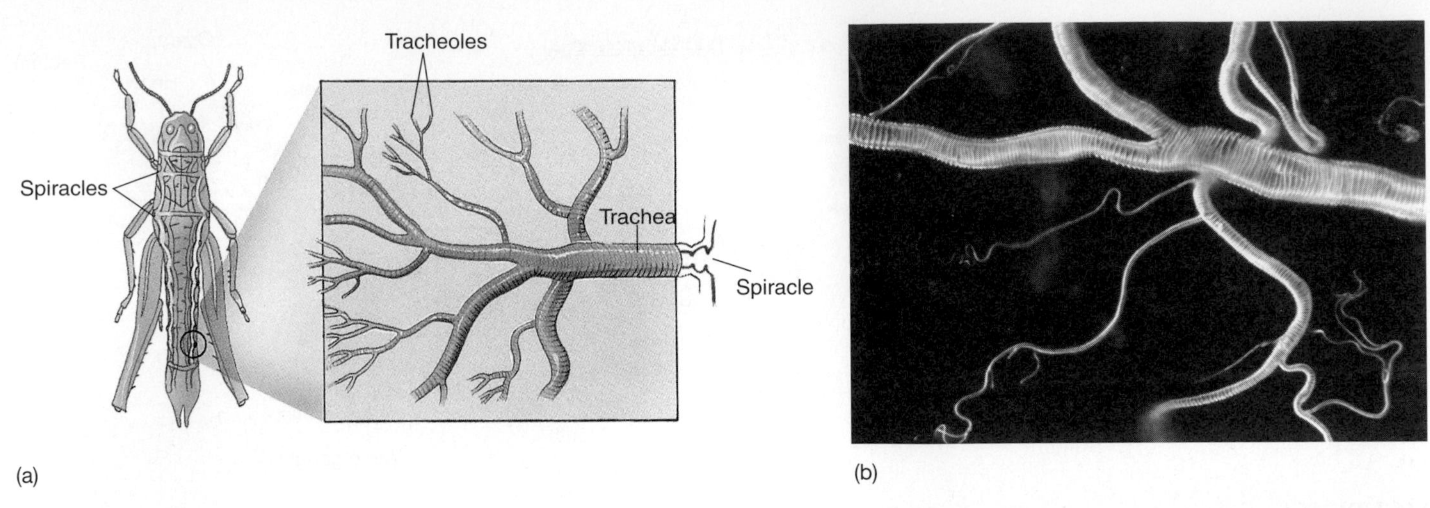

FIGURE 46.10

Tracheae and tracheoles. Tracheae and tracheoles are connected to the exterior by specialized openings called spiracles and carry oxygen to all parts of a terrestrial insect's body. (*a*) The tracheal system of a grasshopper. (*b*) A portion of the tracheal system of a cockroach.

Respiratory System

Insects and other members of subphylum Uniramia, which are fundamentally terrestrial, depend on their respiratory rather than their circulatory system to carry oxygen to their tissues. In vertebrates, blood moves within a closed circulatory system to all parts of the body, carrying the oxygen with it. This is a much more efficient arrangement than exists in arthropods, in which all parts of the body need to be near a respiratory passage to obtain oxygen. As a result, the size of the arthropod body is much more limited than that of the vertebrates. Along with the thickness of their chitin exoskeletons, this feature of arthropod design places severe limitations on size.

Unlike most animals, the arthropods have no single major respiratory organ. The respiratory system of most terrestrial arthropods consists of small, branched, cuticle-lined air ducts called **tracheae** (figure 46.10). These tracheae, which ultimately branch into very small **tracheoles,** are a series of tubes that transmit oxygen throughout the body. Tracheoles are in direct contact with individual cells, and oxygen diffuses directly across the cell membranes. Air passes into the tracheae by way of specialized openings in the exoskeleton called **spiracles,** which, in most insects, can be opened and closed by valves. The ability to prevent water loss by closing the spiracles was a key adaptation that facilitated the invasion of the land by arthropods. In many insects, especially larger ones, muscle contraction helps to increase the flow of gases in and out of the tracheae. In other terrestrial arthropods, the flow of gases is essentially a passive process.

Many spiders and some other chelicerates have a unique respiratory system that involves **book lungs,** a series of leaflike plates within a chamber. Air is drawn in and expelled out of this chamber by muscular contraction. Book lungs may exist alongside tracheae, or they may function instead of tracheae. One small class of marine chelicerates, the horseshoe crabs, have book gills, which are analogous to book lungs but function in water. Tracheae, book lungs, and book gills are all structures found only in arthropods and in the phylum Onychophora, which have tracheae. Crustaceans lack such structures and have gills.

Excretory System

Though there are various kinds of excretory systems in different groups of arthropods, we will focus here on the unique excretory system consisting of **Malpighian tubules** that evolved in terrestrial uniramians. Malpighian tubules are slender projections from the digestive tract that are attached at the junction of the midgut and hindgut (see figure 46.3). Fluid passes through the walls of the Malpighian tubules to and from the blood in which the tubules are bathed. As this fluid passes through the tubules toward the hindgut, nitrogenous wastes are precipitated as concentrated uric acid or guanine. These substances are then emptied into the hindgut and eliminated. Most of the water and salts in the fluid are reabsorbed by the hindgut and rectum and returned to the arthropod's body. Malpighian tubules are an efficient mechanism for water conservation and were another key adaptation facilitating invasion of the land by arthropods.

All arthropods have a rigid chitin and protein exoskeleton that provides places for muscle attachment, protects the animal from predators and injury, and, most important, impedes water loss. Many arthropods have compound eyes. Arthropods have an open circulatory system. Most terrestrial insects have a network of tubes called tracheae that transmit oxygen from the outside to the organs. Many arthropods eliminate metabolic wastes by a unique system of Malpighian tubules.

Class Arachnida: The Arachnids

Chelicerates (subphylum Chelicerata) are a distinct evolutionary line of arthropods in which the most anterior appendages have been modified into chelicerae, which often function as fangs or pincers. By far the largest of the three classes of chelicerates is the largely terrestrial Arachnida, with some 57,000 named species; it includes spiders, ticks, mites, scorpions, and daddy longlegs. Arachnids have a pair of chelicerae, a pair of pedipalps, and four pairs of walking legs. The chelicerae are the foremost appendages; they consist of a stout basal portion and a movable fang often connected to a poison gland.

The next pair of appendages, **pedipalps**, resemble legs but have one less segment and are not used for locomotion. In male spiders, they are specialized copulatory organs. In scorpions, the pedipalps are large pincers.

Most arachnids are carnivorous. The main exception is mites, which are largely herbivorous. Most arachnids can ingest only preliquified food, which they often digest externally by secreting enzymes into their prey. They can then suck up the digested material with their muscular, pumping pharynx. Arachnids are primarily, but not exclusively, terrestrial. Some 4000 known species of mites and one species of spider live in fresh water, and a few mites live in the sea. Arachnids breathe by means of tracheae, book lungs, or both.

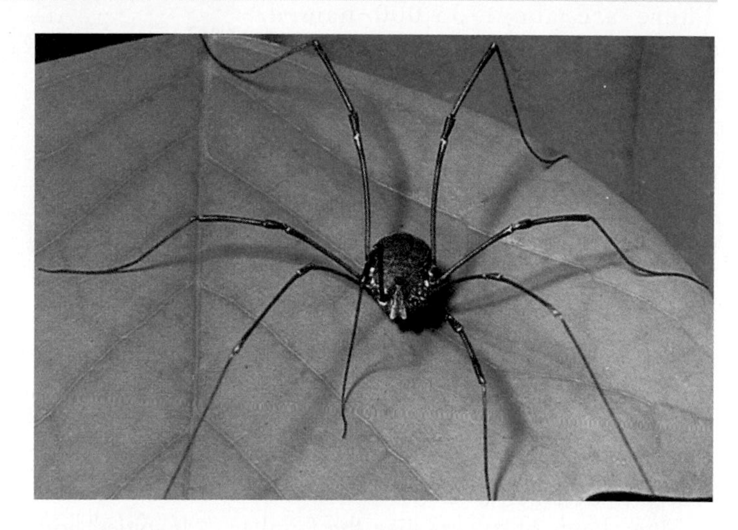

FIGURE 46.11
A harvestman, or daddy longlegs.

FIGURE 46.12
The scorpion, *Uroctonus mordax*. This photograph shows the characteristic pincers and segmented abdomen, ending in a stinging apparatus, raised over the animal's back. The *white* mass is comprised of the scorpion's young.

Order Opiliones: The Daddy Longlegs

A familiar group of arachnids consists of the daddy longlegs, or harvestmen (order Opiliones). Members of this order are easily recognized by their oval, compact bodies and extremely long, slender legs (figure 46.11). They respire by means of a primary pair of tracheae and are unusual among the arachnids in that they engage in direct copulation. The males have a penis, and the females an **ovipositor,** or egg-laying organ which deposits their eggs in cracks and crevices. Most daddy longlegs are predators of insects and other arachnids, but some live on plant juices and many scavenge dead animal matter. The order includes about 5000 species.

Order Scorpiones: The Scorpions

Scorpions (order Scorpiones) are arachnids whose pedipalps are modified into pincers. Scorpions use these pincers to handle and tear apart their food (figure 46.12). The venomous stings of scorpions are used mainly to stun their prey and less commonly in self-defense. The stinging apparatus is located in the terminal segment of the abdomen. A scorpion holds its abdomen folded forward over its body when it is moving about. The elongated, jointed abdomens of scorpions are distinctive; in most chelicerates, the abdominal segments are more or less fused together and appear as a single unit.

Scorpions are probably the most ancient group of terrestrial arthropods; they are known from the Silurian Period, some 425 million years ago. Adults of this order of arachnids range in size from 1 to 18 centimeters. There are some 1200 species of scorpions, all terrestrial, which occur throughout the world. They are most common in tropical, subtropical, and desert regions. The young are born alive, with 1 to 95 in a given brood.

Order Araneae: The Spiders

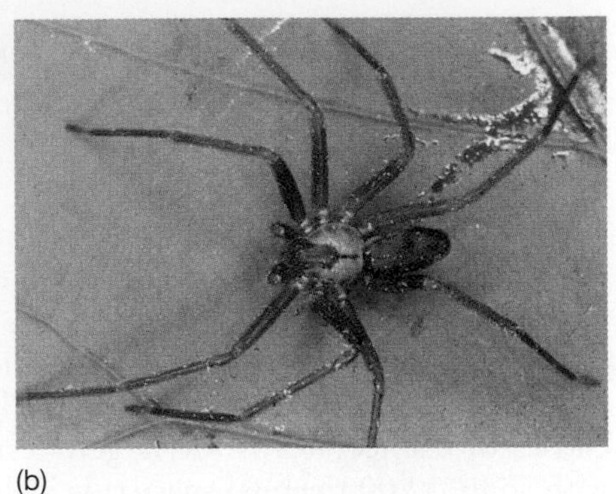

(a) (b)

FIGURE 46.13
Two common poisonous spiders. (*a*) The black widow spider, *Latrodectus mactans.* (*b*) The brown recluse spider, *Loxosceles reclusa.* Both species are common throughout temperate and subtropical North America, but bites are rare in humans.

There are about 35,000 named species of spiders (order Araneae). These animals play a major role in virtually all terrestrial ecosystems. They are particularly important as predators of insects and other small animals. Spiders hunt their prey or catch it in silk webs of remarkable diversity. The silk is formed from a fluid protein that is forced out of spinnerets on the posterior portion of the spider's abdomen. The webs and habits of spiders are often distinctive. Some spiders can spin gossamer floats that allow them to drift away in the breeze to a new site.

Many kinds of spiders, like the familiar wolf spiders and tarantulas, do not spin webs but instead hunt their prey actively. Others, called trap-door spiders, construct silk-lined burrows with lids, seizing their prey as it passes by. One species of spider, *Argyroneta aquatica,* lives in fresh water, spending most of its time below the surface. Its body is surrounded by a bubble of air, while its legs, which are used both for underwater walking and for swimming, are not. Several other kinds of spiders walk about freely on the surface of water.

Spiders have poison glands leading through their chelicerae, which are pointed and used to bite and paralyze prey. Some members of this order, such as the black widow and brown recluse (figure 46.13), have bites that are poisonous to humans and other large mammals.

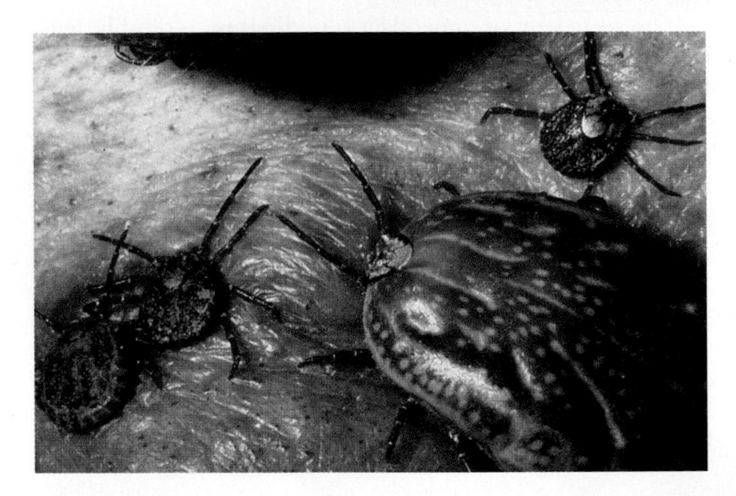

FIGURE 46.14
Ticks (order Acari). Ticks (the large one is engorged) on the hide of a tapir in Peru. Many ticks spread diseases in humans and other vertebrates.

Order Acari: Mites and Ticks

The order Acari, the mites and ticks, is the largest in terms of number of species and the most diverse of the arachnids. Although only about 30,000 species of mites and ticks have been named, scientists that study the group estimate that there may be a million or more members of this order in existence.

Most mites are small, less than 1 millimeter long, but adults of different species range from 100 nanometers to 2 centimeters. In most mites, the cephalothorax and abdomen are fused into an unsegmented ovoid body. Respiration occurs either by means of tracheae or directly through the exoskeleton. Many mites pass through several distinct stages during their life cycle. In most, an inactive eight-legged prelarva gives rise to an active six-legged larva, which in turn produces a succession of three eight-legged stages and, finally, the adult males and females.

Mites and ticks are diverse in structure and habitat. They are found in virtually every terrestrial, freshwater, and marine habitat known and feed on fungi, plants, and animals. They act as predators and as internal and external parasites of both invertebrates and vertebrates.

Many mites produce irritating bites and diseases in humans. Mites live in the hair follicles and wax glands of your forehead and nose, but usually cause no symptoms.

Ticks are blood-feeding **ectoparasites,** parasites that occur on the surface of their host (figure 46.14). They are larger than most other mites and cause discomfort by sucking the blood of humans and other animals. Ticks can carry many diseases, including some caused by viruses, bacteria, and protozoa. The spotted fevers (Rocky Mountain spotted fever is a familiar example) are caused by bacteria carried by ticks. Lyme disease is apparently caused by spirochaetes transmitted by ticks. Red-water fever, or Texas fever, is an important tick-borne protozoan disease of cattle, horses, sheep, and dogs.

Scorpions, spiders, and mites are all arachnids, the largest class of chelicerates.

FIGURE 46.15
Limulus. Horseshoe crabs, emerging from the sea to mate at the edge of Delaware Bay, New Jersey, in early May.

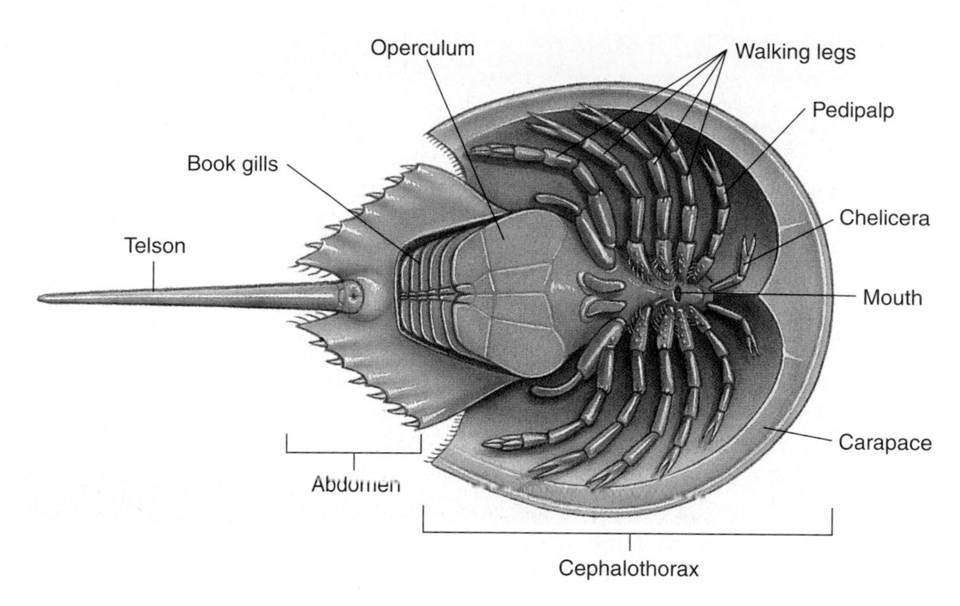

FIGURE 46.16
Diagram of a horseshoe crab, *Limulus*, from below. This diagram illustrates the principal features of this archaic animal.

Class Merostomata: Horseshoe Crabs

A second class of chelicerates is the horseshoe crabs (class Merostomata). There are three genera of horseshoe crabs. One, *Limulus* (figure 46.15), is common along the East Coast of North America. The other two genera live in the Asian tropics. Horseshoe crabs are an ancient group, with fossils virtually identical to *Limulus* dating back 220 million years to the Triassic Period. Other members of the class, the now-extinct eurypterans, are known from 400 million years ago. Horseshoe crabs may have been derived from trilobites, a relationship suggested by the appearance of their larvae. Individuals of *Limulus* grow up to 60 centimeters long. They mature in 9 to 12 years and have a life span of 14 to 19 years. *Limulus* individuals live in deep water, but they migrate to shallow coastal waters every spring, emerging from the sea to mate on moonlit nights when the tide is high.

Horseshoe crabs feed at night, primarily on mollusks and annelids. They swim on their backs by moving their abdominal plates. They can also walk on their four pairs of legs, protected along with chelicerae and pedipalps by their shell (figure 46.16).

Horseshoe crabs are a very ancient group.

FIGURE 46.17
A marine pycnogonid. The sea spider *Pycnogonum littorale* (yellow animal) crawling over a sea anemone.

Class Pycnogonida: The Sea Spiders

The third class of chelicerates is the sea spiders (class Pycnogonida), containing more than 1000 species. These animals are not often observed because many are small, only about 1 to 3 centimeters long, and rather inconspicuous. They are found in oceans throughout the world but are most abundant in the far north and far south. Adult sea spiders are mostly external parasites or predators of other animals like sea anemones (figure 46.17).

Sea spiders have a sucking proboscis in a mouth located at its end. Their abdomen is much reduced, and their body appears to consist almost entirely of the cephalothorax, with no well-defined head. Sea spiders usually have four, or less commonly five or six, pairs of legs. Male sea spiders carry the eggs on their legs until they hatch, thus providing a measure of parental care. Sea spiders completely lack excretory and respiratory systems. They appear to carry out these functions by direct diffusion, with waste products flowing outward through the cells and oxygen flowing inward through them. Sea spiders are not closely related to either of the other two classes of chelicerates.

Sea spiders are very common in marine habitats. They are not closely related to terrestrial spiders.

46.3 Crustaceans have branched appendages.

Crustaceans

The crustaceans (subphylum Crustacea) are a large group of primarily aquatic organisms, consisting of some 35,000 species of crabs, shrimps, lobsters, crayfish, barnacles, water fleas, pillbugs, and related groups (table 46.1). Most crustaceans have two pairs of antennae, three types of chewing appendages, and various numbers of pairs of legs. All crustacean appendages, with the possible exception of the first pair of antennae, are basically biramous. In some crustaceans, appendages appear to have only a single branch; in those cases, one of the branches has been lost during the course of evolutionary specialization. The **nauplius** larva stage through which all crustaceans pass (figure 46.18) provides evidence that all members of this diverse group are descended from a common ancestor. The nauplius hatches with three pairs of appendages and metamorphoses through several stages before reaching maturity. In many groups, this nauplius stage is passed in the egg, and development of the hatchling to the adult form is direct.

Crustaceans differ from insects but resemble centipedes and millipedes in that they have appendages on their abdomen as well as on their thorax. They are the only arthropods with two pairs of antennae. Their mandibles likely originated from a pair of limbs that took on a chewing function during the course of evolution, a process that apparently occurred independently in the common ancestor of the terrestrial mandibulates. Many crustaceans have compound eyes. In addition, they have delicate tactile hairs that project from the cuticle all over the body. Larger crustaceans have feathery gills near the bases of their legs. In smaller members of this class, gas exchange takes place directly through the thinner areas of the cuticle or the entire body. Most crustaceans have separate sexes. Many different kinds of specialized copulation occur among the crustaceans, and the members of some orders carry their eggs with them, either singly or in egg pouches, until they hatch.

Table 46.1 Traditional Classification of the Phylum Arthropoda

Subphylum	Characteristics	Members
Chelicerata	Mouthparts are chelicerae	The chelicerates: spiders, horseshoe crabs, mites
Crustacea	Mouthparts are mandibles; biramous appendages	The crustaceans: lobsters, crabs, shrimp, isopods, barnacles
Uniramia	Mouthparts are mandibles; uniramous appendages	Chilopods (centipedes), diplopods (millipedes), and insects

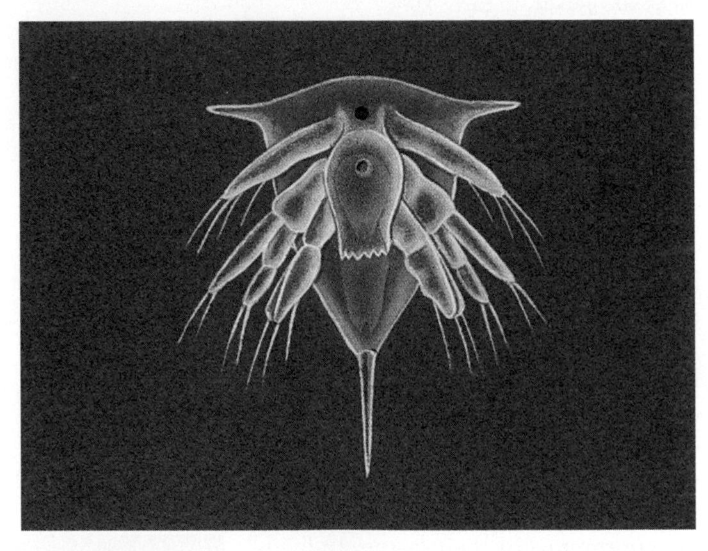

FIGURE 46.18
Although crustaceans are diverse, they have fundamentally similar larvae. The nauplius larva of a crustacean is an important unifying feature found in all members of this group.

Decapod Crustaceans

Large, primarily marine crustaceans such as shrimps, lobsters, and crabs, along with their freshwater relatives, the crayfish, are collectively called *decapod crustaceans* (figure 46.19). The term *decapod* means "ten-footed." In these animals, the exoskeleton is usually reinforced with calcium carbonate. Most of their body segments are fused into a cephalothorax covered by a dorsal shield, or **carapace,** which arises from the head. The crushing pincers common in many decapod crustaceans are used in obtaining food, for example, by crushing mollusk shells.

In lobsters and crayfish, appendages called **swimmerets** occur in lines along the ventral surface of the abdomen and are used in reproduction and swimming. In addition, flattened appendages known as **uropods** form a kind of compound "paddle" at the end of the abdomen. These animals may also have a **telson,** or tail spine. By snapping its abdomen, the animal propels itself through the water rapidly and forcefully. Crabs differ from lobsters and crayfish in proportion; their carapace is much larger and broader and the abdomen is tucked under it.

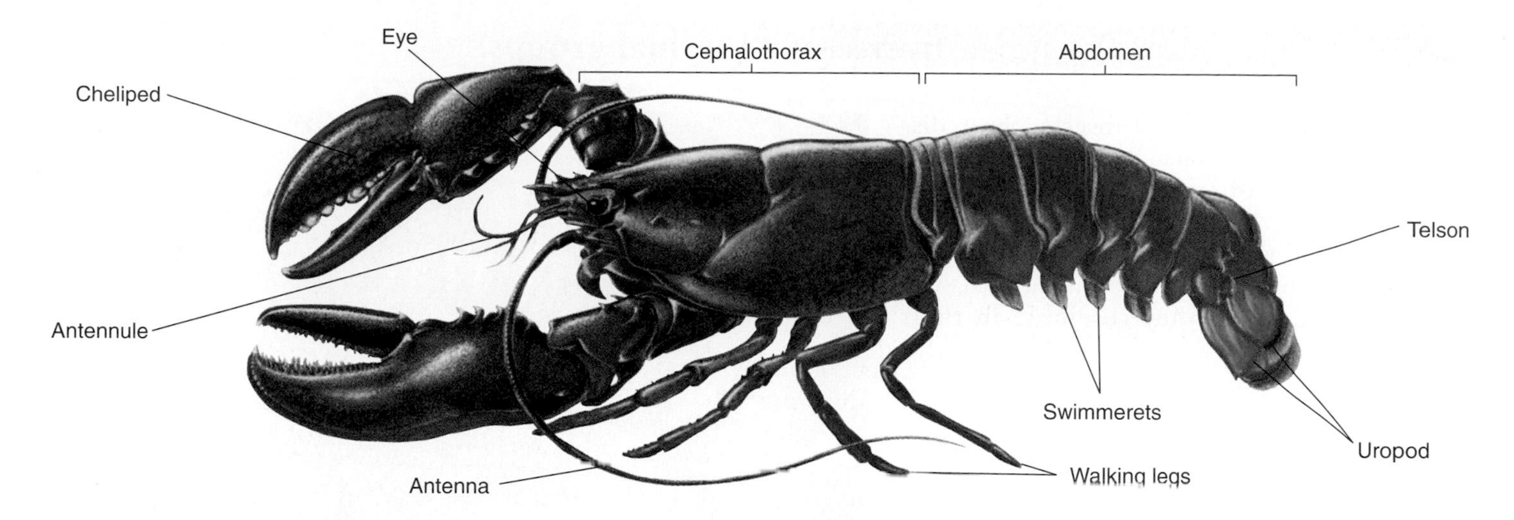

FIGURE 46.19
Decapod crustacean. A lobster, *Homarus americanus*. The principal features are labeled.

Terrestrial and Freshwater Crustaceans

Although most crustaceans are marine, many occur in fresh water and a few have become terrestrial. These include pillbugs and sowbugs, the terrestrial members of a large order of crustaceans known as the isopods (order Isopoda). About half of the estimated 4500 species of this order are terrestrial and live primarily in places that are moist, at least seasonally. Sand fleas or beach fleas (order Amphipoda) are other familiar crustaceans, many of which are semiterrestrial (intertidal) species.

Along with the larvae of larger species, minute crustaceans are abundant in the plankton. Especially significant are the tiny copepods (order Copepoda; figure 46.20), which are among the most abundant multicellular organisms on earth.

FIGURE 46.20
Freshwater crustacean. A copepod with attached eggs, a member of an abundant group of marine and freshwater crustaceans (order Copepoda), most of which are a few millimeters long. Copepods are important components of the plankton.

Sessile Crustaceans

Barnacles (order Cirripedia; figure 46.21) are a group of crustaceans that are sessile as adults. Barnacles have free-swimming larvae, which ultimately attach their heads to a piling, rock, or other submerged object and then stir food into their mouth with their feathery legs. Calcareous plates protect the barnacle's body, and these plates are usually attached directly and solidly to the substrate. Although most crustaceans have separate sexes, barnacles are hermaphroditic, but they generally cross-fertilize.

Crustaceans include marine, freshwater, and terrestrial forms. All possess a nauplius larval stage and branched appendages.

FIGURE 46.21
Gooseneck barnacles, *Lepas anatifera*, feeding. These are stalked barnacles; many others lack a stalk.

46.4 Insects are the most diverse of all animal groups.

Millipedes, centipedes, and insects, three distinct classes, are uniramian mandibulates. They respire by means of tracheae and excrete their waste products through Malpighian tubules. These groups were certainly derived from annelids, probably ones similar to the oligochaetes, which they resemble in their embryology.

FIGURE 46.22
A centipede. Centipedes, like this member of the genus *Scolopendra*, are active predators.

Classes Chilopoda and Diplopoda: The Centipedes and Millipedes

The centipedes (class Chilopoda) and millipedes (class Diplopoda) both have bodies that consist of a head region followed by numerous segments, all more or less similar and nearly all bearing paired appendages. Although the name *centipede* would imply an animal with a hundred legs and the name *millipede* one with a thousand, adult centipedes usually have 30 or more legs, adult millipedes 60 or more. Centipedes have one pair of legs on each body segment (figure 46.22), millipedes two (figure 46.23). Each segment of a millipede is a tagma that originated during the group's evolution when two ancestral segments fused. This explains why millipedes have twice as many legs per segment as centipedes.

In both centipedes and millipedes, fertilization is internal and takes place by direct transfer of sperm. The sexes are separate, and all species lay eggs. Young millipedes usually hatch with three pairs of legs; they experience a number of growth stages, adding segments and legs as they mature, but do not change in general appearance.

Centipedes, of which some 2500 species are known, are all carnivorous and feed mainly on insects. The appendages of the first trunk segment are modified into a pair of poison fangs. The poison is often quite toxic to human beings, and many centipede bites are extremely painful, sometimes even dangerous.

In contrast, most millipedes are herbivores, feeding mainly on decaying vegetation. A few millipedes are carnivorous. Many millipedes can roll their bodies into a flat coil or sphere because the dorsal area of each of their body segments is much longer than the ventral one. More than 10,000 species of millipedes have been named, but this is estimated to be no more than one-sixth of the actual number of species that exists. In each segment of their body, most millipedes have a pair of complex glands that produces a bad-smelling fluid. This fluid is exuded for defensive purposes through openings along the sides of the body. The chemistry of the secretions of different millipedes has become a subject of considerable interest because of the diversity of the compounds involved and their effectiveness in protecting millipedes from attack. Some produce cyanide gas from segments near their head end. Millipedes live primarily in damp, protected places, such as under leaf litter, in rotting logs, under bark or stones, or in the soil.

Centipedes are segmented hunters with one pair of legs on each segment. Millipedes are segmented herbivores with two pairs of legs on each segment.

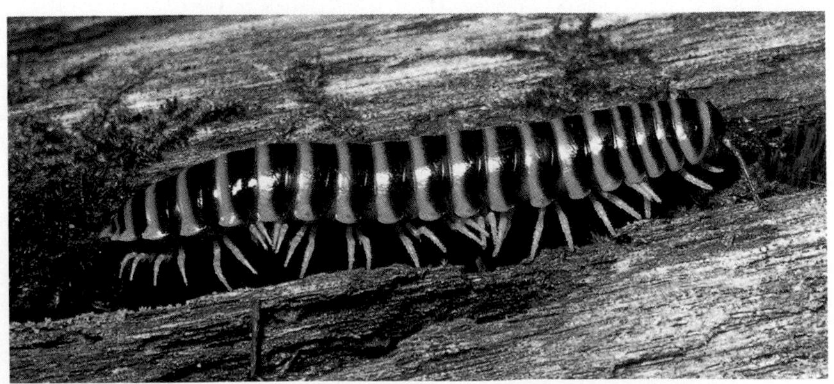

FIGURE 46.23
A millipede. Millipedes, such as this *Sigmoria* individual, are herbivores.

Class Insecta: The Insects

The insects, class Insecta, are by far the largest group of organisms on earth, whether measured in terms of numbers of species or numbers of individuals. Insects live in every conceivable habitat on land and in fresh water, and a few have even invaded the sea. More than half of all the named animal species are insects, and the actual proportion is doubtless much higher because millions of additional forms await detection, classification, and naming. Approximately 90,000 described species occur in the United States and Canada, and the actual number of species in this area probably approaches 125,000. A hectare of lowland tropical forest is estimated to be inhabited by as many as 41,000 species of insects, and many suburban gardens may have 1500 or more species. It has been estimated that approximately a billion billion (10^{18}) individual insects are alive at any one time. A glimpse at the enormous diversity of insects is presented in figure 46.24 and later in table 46.2.

(a)

(b)

(c)

(d)

(e)

(f)

FIGURE 46.24

Insect diversity. (*a*) Luna moth, *Actias luna*. Luna moths and their relatives are among the most spectacular insects (order Lepidoptera). (*b*) Soldier fly, *Ptecticus trivittatus* (order Diptera). (*c*) Boll weevil, *Anthonomus grandis*. Weevils are one of the largest groups of beetles (order Coleoptera). (*d*) A thorn-shaped leafhopper, *Umbonica crassicornis* (order Hemiptera). (*e*) Copulating grasshoppers (order Orthoptera). (*f*) Termite, *Macrotermes bellicosus* (order Isoptera). The large, sausage-shaped individual is a queen, specialized for laying eggs; most of the smaller individuals around the queen are nonreproductive workers, but the larger individual at the lower left is a reproductive male.

External Features

Insects are primarily a terrestrial group, and most, if not all, of the aquatic insects probably had terrestrial ancestors. Most insects are relatively small, ranging in size from 0.1 millimeter to about 30 centimeters in length or wingspan. Insects have three body sections, the head, thorax, and abdomen; three pairs of legs, all attached to the thorax; and one pair of antennae. In addition, they may have one or two pairs of wings. Insect mouthparts all have the same basic structure but are modified in different groups in relation to their feeding habits (figure 46.25). Most insects have compound eyes, and many have ocelli as well.

The insect thorax consists of three segments, each with a pair of legs. Legs are completely absent in the larvae of certain groups, for example, in most members of the order Diptera (flies) (figure 46.26). If two pairs of wings are present, they attach to the middle and posterior segments of the thorax. If only one pair of wings is present, it usually attaches to the middle segment. The thorax is almost entirely filled with muscles that operate the legs and wings.

The wings of insects arise as saclike outgrowths of the body wall. In adult insects, the wings are solid except for the veins. Insect wings are not homologous to the other appendages. Basically, insects have two pairs of wings, but in some groups, like flies, the second set has been reduced to a pair of balancing knobs called halteres during the course of evolution. Most insects can fold their wings over their abdomen when they are at rest; but a few, such as the dragonflies and damselflies (order Odonata), keep their wings erect or outstretched at all times.

Insect forewings may be tough and hard, as in beetles. If they are, they form a cover for the hindwings and usually open during flight. The tough forewings also serve a protective function in the order Orthoptera, which includes grasshoppers and crickets. The wings of insects are made of sheets of chitin and protein; their strengthening veins are tubules of chitin and protein. Moths and butterflies have wings that are covered with detachable scales that provide most of their bright colors (figure 46.27). In some wingless insects, such as the springtails or silverfish, wings never evolved. Other wingless groups, such as fleas and lice, are derived from ancestral groups of insects that had wings.

Internal Organization

The internal features of insects resemble those of the other arthropods in many ways. The digestive tract is a tube, usually somewhat coiled. It is often about the same length as the body. However, in the order Hemiptera, which consists of the leafhoppers, cicadas, and related groups, and in many flies (order Diptera), the digestive tube may be greatly coiled and several times longer than the body. Such long digestive tracts are generally found in

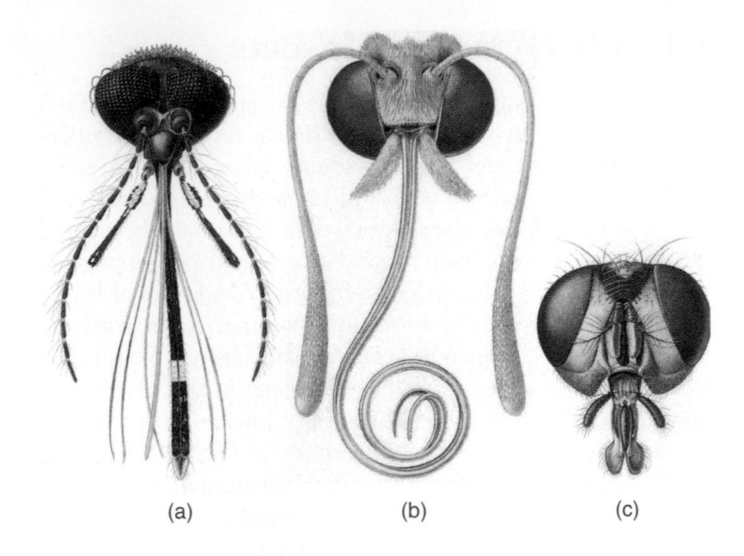

FIGURE 46.25
Modified mouthparts in three kinds of insects. Mouthparts are modified for (*a*) piercing in the mosquito, *Culex*, (*b*) sucking nectar from flowers in the alfalfa butterfly, *Colias*, and (*c*) sopping up liquids in the housefly, *Musca domestica*.

FIGURE 46.26
Larvae of a mosquito, *Culex pipiens*. The aquatic larvae of mosquitoes are quite active. They breathe through tubes from the surface of the water, as shown here. Covering the water with a thin film of oil causes them to drown.

insects that have sucking mouthparts and feed on juices rather than on protein-rich solid foods because they offer a greater opportunity to absorb fluids and their dissolved nutrients. The digestive enzymes of the insect also are more dilute and thus less effective in a highly liquid medium than in a more solid one. Longer digestive tracts give these enzymes more time to work while food is passing through.

The anterior and posterior regions of an insect's digestive tract are lined with cuticle. Digestion takes place primarily in the stomach, or midgut; and excretion takes place through Malpighian tubules. Digestive enzymes are mainly secreted from the cells that line the midgut, although some are contributed by the salivary glands near the mouth.

The tracheae of insects extend throughout the body and permeate its different tissues. In many winged insects, the tracheae are dilated in various parts of the body, forming air sacs. These air sacs are surrounded by muscles and form a kind of bellows system to force air deep into the tracheal system. The spiracles, a maximum of 10 on each side of the insect, are paired and located on or between the segments along the sides of the thorax and abdomen. In most insects, the spiracles can be opened by muscular action. Closing the spiracles at times may be important in retarding water loss. In some parasitic and aquatic groups of insects, the spiracles are permanently closed. In these groups, the tracheae run just below the surface of the insect, and gas exchange takes place by diffusion.

The **fat body** is a group of cells located in the insect body cavity. This structure may be quite large in relation to the size of the insect, and it serves as a food-storage reservoir, also having some of the functions of a vertebrate liver. It is often more prominent in immature insects than in adults, and it may be completely depleted when metamorphosis is finished. Insects that do not feed as adults retain their fat bodies and live on the food stored in them throughout their adult lives (which may be very short).

FIGURE 46.27
Scales on the wing of *Parnassius imperator*, a butterfly from China. Scales of this sort account for most of the colored patterns on the wings of butterflies and moths.

Sense Receptors

In addition to their eyes, insects have several characteristic kinds of sense receptors. These include **sensory hairs,** which are usually widely distributed over their bodies. The sensory hairs are linked to nerve cells and are sensitive to mechanical and chemical stimulation. They are particularly abundant on the antennae and legs—the parts of the insect most likely to come into contact with other objects.

Sound, which is of vital importance to insects, is detected by tympanal organs in groups such as grasshoppers and crickets, cicadas, and some moths. These organs are paired structures composed of a thin membrane, the **tympanum,** associated with the tracheal air sacs. In many other groups of insects, sound waves are detected by sensory hairs. Male mosquitoes use thousands of sensory hairs on their antennae to detect the sounds made by the vibrating wings of female mosquitoes.

Sound detection in insects is important not only for protection but also for communication. Many insects communicate by making sounds, most of which are quite soft, very high-pitched, or both, and thus inaudible to humans. Only a few groups of insects, especially grasshoppers, crickets, and cicadas, make sounds that people can hear. Male crickets and longhorned grasshoppers produce sounds by rubbing their two front wings together. Shorthorned grasshoppers do so by rubbing their hind legs over specialized areas on their wings. Male cicadas vibrate the membranes of air sacs located on the lower side of the most anterior abdominal segment.

In addition to using sound, nearly all insects communicate by means of chemicals or mixtures of chemicals known as **pheromones.** These compounds, extremely diverse in their chemical structure, are sent forth into the environment, where they are active in very small amounts and convey a variety of messages to other individuals. These messages not only convey the attraction and recognition of members of the same species for mating, but they also mark trails for members of the same species, as in the ants.

All insects possess three body segments (tagmata): the head, the thorax, and the abdomen. The three pairs of legs are attached to the thorax. Most insects have compound eyes, and many have one or two pairs of wings. Insects possess sophisticated means of sensing their environment, including sensory hairs, tympanal organs, and chemoreceptors.

Insect Life Histories

Most young insects hatch from fertilized eggs laid outside their mother's body. The zygote develops within the egg into a young insect, which escapes by chewing through or bursting the shell. Some immature insects have specialized projections on the head that assist in this process. In a few insects, eggs hatch within the mother's body.

During the course of their development, young insects undergo ecdysis a number of times before they become adults. Most insects molt four to eight times during the course of their development, but some may molt as many as 30 times. The stages between the molts are called **instars.** When an insect first emerges following ecdysis, it is pale, soft, and especially susceptible to predators. Its exoskeleton generally hardens in an hour or two. It must grow to its new size, usually by taking in air or water, during this brief period.

There are two principal kinds of metamorphosis in insects: **simple metamorphosis** and **complete metamorphosis** (figure 46.28). In insects with simple metamorphosis, immature stages are often called **nymphs.** Nymphs are usually quite similar to adults, differing mainly in their smaller size, less well-developed wings, and sometimes color. In insect orders with simple metamorphosis, such as mayflies and dragonflies, nymphs are aquatic and extract oxygen from the water through gills. The adult stages are terrestrial and look very different from the nymphs. In other groups, such as grasshoppers and their relatives, nymphs and adults live in the same habitat. Such insects usually change gradually during their life cycles with respect to wing development, body proportions, the appearance of ocelli, and other features.

In complete metamorphosis, the wings develop internally during the juvenile stages and appear externally only during the resting stage that immediately precedes the final molt (figure 46.28*b*). During this stage, the insect is called a **pupa** or **chrysalis,** depending on the group to which it belongs. A pupa or chrysalis does not normally move around much, although mosquito pupae do move around freely. A large amount of internal body reorganization takes place while the insect is a pupa or chrysalis.

More than 90% of the insects, including the members of all of the largest and most successful orders, display complete metamorphosis. The juvenile stages and adults often live in distinct habitats, have different habits, and are usually extremely different in form. In these insects, development is indirect. The immature stages, called **larvae,** are often wormlike, differing greatly in appearance from the adults. Larvae do not have compound eyes. They may be legless or have legs or leg-like appendages on the abdomen. They usually have chewing mouthparts, even in those orders in which the adults have sucking mouthparts; chewing mouthparts are the primitive condition in these groups. When larvae and adults play different ecological roles, they do not compete directly for the same resources, an advantage to the species.

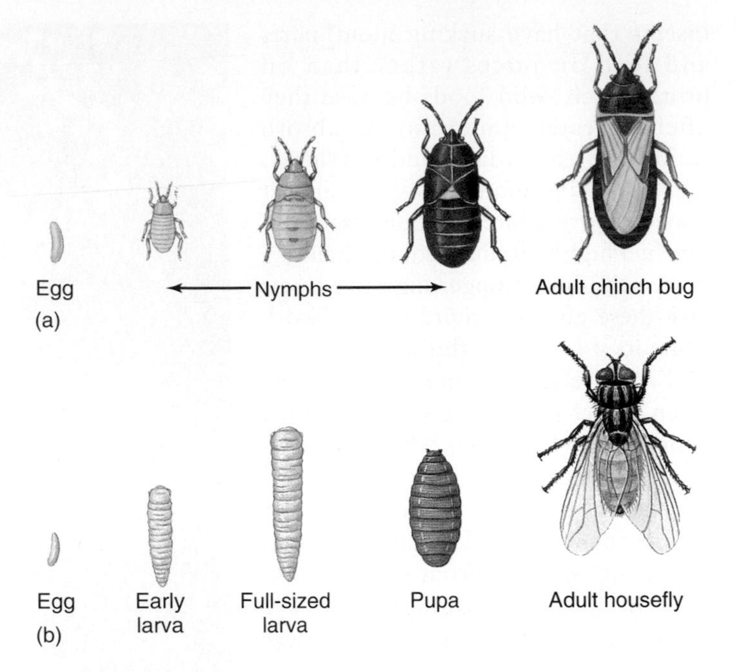

FIGURE 46.28
Metamorphosis. (*a*) Simple metamorphosis in a chinch bug (order Hemiptera), and (*b*) complete metamorphosis in a housefly, *Musca domestica* (order Diptera).

Pupae do not feed and are usually relatively inactive. As pupae, insects are extremely vulnerable to predators and parasites, but they are often covered by a cocoon or some other protective structure. Groups of insects with complete metamorphosis include moths and butterflies; beetles; bees, wasps, and ants; flies; and fleas.

Some species of insects exhibit no dramatic change in form from immature stages to adult. This type of development is called **ametabolus** (meaning without change) and is seen in the most primitive orders of insects such as the silverfish and springtails.

Hormones control both ecdysis and metamorphosis. **Molting hormone,** or **ecdysone,** is released from a gland in the thorax when that gland has been stimulated by brain hormone, which in turn is produced by neurosecretory cells and released into the blood. The effects of the molting induced by ecdysone are determined by juvenile hormone, which is present during the immature stages but declines in quantity as the insect passes through successive molts. When the level of juvenile hormone is relatively high, the molt produces another larva; when it is lower, it produces the pupa and then the final development of the adult.

Insects undergo either simple or complete metamorphosis.

Order	Typical Examples	Key Characteristics	Approximate Number of Named Species
Coleoptera	Beetles	 The most diverse animal order; two pairs of wings; front pair of wings is a hard cover that partially protects the transparent rear pair of flying wings; heavily armored exoskeleton; biting and chewing mouthparts; complete metamorphosis	350,000
Diptera	Flies	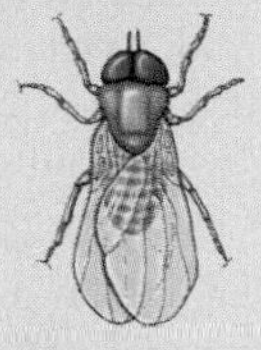 Some that bite people and other mammals are considered pests; front flying wings are transparent; hind wings are reduced to knobby balancing organs; sucking, piercing, and lapping mouthparts; complete metamorphosis	120,000
Lepidoptera	Butterflies, moths	 Often collected for their beauty; two pairs of broad, scaly, flying wings, often brightly colored; hairy body; tubelike, sucking mouthparts; complete metamorphosis	120,000
Hymenoptera	Bees, wasps, ants	 Often social, known to many by their sting; two pairs of transparent flying wings; mobile head and well-developed eyes; often possess stingers; chewing and sucking mouthparts; complete metamorphosis	100,000
Hemiptera	True bugs, bedbugs, leafhoppers	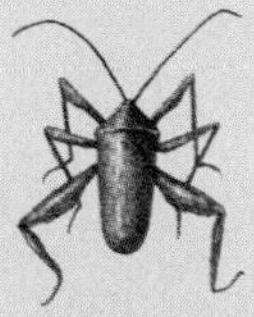 Often live on blood; two pairs of wings, or wingless; piercing, sucking mouthparts; simple metamorphosis	60,000
Orthoptera	Grasshoppers, crickets, roaches	 Known for their jumping; two pairs of wings or wingless; among the largest insects; biting and chewing mouthparts in adults; simple metamorphosis	20,000
Odonata	Dragonflies	 Among the most primitive of the insect order; two pairs of transparent flying wings; large, long, and slender body; chewing mouthparts; simple metamorphosis	5,000
Isoptera	Termites	 One of the few types of animals able to eat wood; two pairs of wings, but some stages are wingless; social insects; there are several body types with division of labor; chewing mouthparts; simple metamorphosis	2,000
Siphonaptera	Fleas	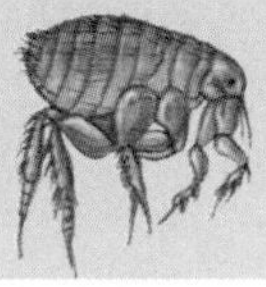 Small, known for their irritating bites; wingless; small flattened body with jumping legs; piercing and sucking mouthparts; complete metamorphosis	1,200

Table 46.2 Major Orders of Insects

Chapter 46

www.mhhe.com/raven6/resources46.mhtml

Summary	Questions	Media Resources

46.1 The evolution of jointed appendages has made arthropods very successful.

- Jointed appendages and an exoskeleton greatly expanded locomotive and manipulative capabilities for the arthropod phyla, the most successful of all animals in terms of numbers of individuals, species, and ecological diversification.
- Traditionally, arthropods have been grouped into three subphyla based on morphological characters but new research is calling this classification of the arthropods into question.
- Like annelids, arthropods have segmented bodies, but some of their segments have become fused into tagmata during the course of evolution. All possess a rigid external skeleton, or exoskeleton.

1. What are the advantages of an exoskeleton? What occurs during ecdysis? What controls this process?

2. What type of circulatory system do arthropods have? Describe the direction of blood flow. What helps to maintain this one-way flow?

3. What are Malpighian tubules? How do they work? What other system are they connected to? How does this system process wastes? How does it regulate water loss?

- Enhancement Chapter: Arthropod Taxonomy

- Art Quiz: Insect Internal Structure

46.2 The chelicerates all have fangs or pincers.

- Chelicerates consist of three classes: Arachnida (spiders, ticks, mites, and scorpions); Merostomata (horseshoe crabs); and Pycnogonida (sea spiders).
- Spiders, the best known arachnids, have a pair of chelicerae, a pair of pedipalps, and four pairs of walking legs. Spiders secrete digestive enzymes into their prey, then suck the contents out.

4. Into what two groups are arthropods traditionally divided? Describe each group in terms of its mouthparts and appendages, and give several examples of each.

46.3 Crustaceans have branched appendages.

- Crustaceans comprise some 35,000 species of crabs, shrimps, lobsters, barnacles, sowbugs, beach fleas, and many other groups. Their appendages are basically biramous, and their embryology is distinctive.

5. On which parts of the body do crustaceans possess legs?

6. How do biramous and uniramous appendages differ?

- Scientists on Science: Diving into the History of Life
- Student Research: Evolution of the Major Claw in Fiddler Crabs

- Art Quiz: Decapod Crustacean

46.4 Insects are the most diverse of all animal groups.

- Centipedes and millipedes are segmented uniramia. Centipedes are hunters with one pair of legs per segment, and millipedes are herbivores with two pairs of legs per segment.
- Insects have three body segments, three pairs of legs, and often one or two pairs of wings. Many have complex eyes and other specialized sensory structures.
- Insects exhibit either simple metamorphosis, moving through a succession of forms relatively similar to the adult, or complete metamorphosis, in which an often wormlike larva becomes a usually sedentary pupa, and then an adult.

7. How are millipedes and centipedes similar to each other? How do they differ?

8. What type of digestive system do most insects possess? What digestive adaptations occur in insects that feed on juices low in protein? Why?

9. What is an instar as it relates to insect metamorphosis? What are the two different kinds of metamorphosis in insects? How do they differ?

- Arthropods

- Student Research: Insect Behavior

- Art Quiz: Metamorphosis

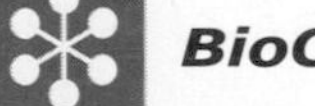

47

Echinoderms

Concept Outline

47.1 The embryos of deuterostomes develop quite differently from those of protostomes.

Protostomes and Deuterostomes. Deuterostomes—the echinoderms, chordates, and a few other groups—share a mode of development that is quite different from other animals.

47.2 Echinoderms are deuterostomes with an endoskeleton.

Deuterostomes. Echinoderms are bilaterally symmetrical as larvae but metamorphose to radially symmetrical adults.
Echinoderm Body Plan. Echinoderms have an endoskeleton and a unique water-vascular system seen in no other phylum.

47.3 The six classes of echinoderms are all radially symmetrical as adults.

Class Crinoidea: The Sea Lilies and Feather Stars. Crinoids are the only echinoderms that are attached to the sea bottom for much of their lives.
Class Asteroidea: The Sea Stars. Sea stars, also called starfish, are five-armed mobile predators.
Class Ophiuroidea: The Brittle Stars. Brittle stars are quite different from the sea stars for whom they are sometimes mistaken.
Class Echinoidea: The Sea Urchins and Sand Dollars. Sea urchins and sand dollars have five-part radial symmetry but lack arms.
Classes Holothuroidea and Concentricycloidea: Sea Cucumbers and Sea Daisies. Sea cucumbers are soft-bodied echinoderms without arms. The most recently discovered class of echinoderms, sea daisies are tiny, primitive echinoderms that live at great depths.

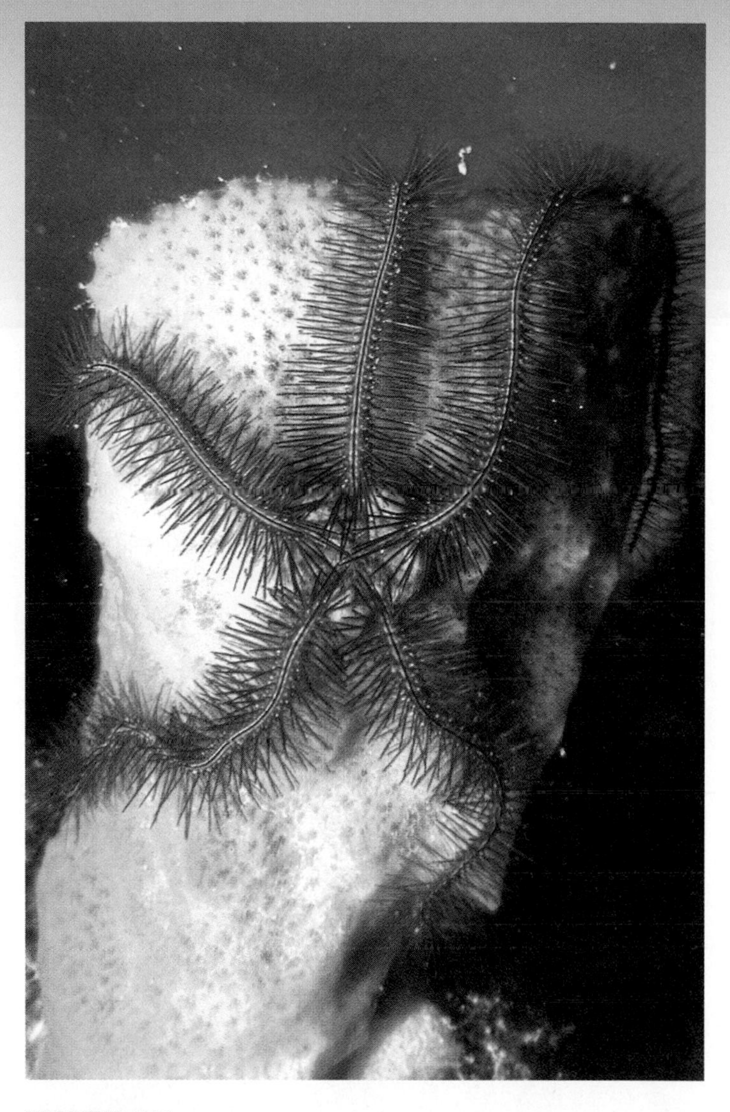

FIGURE 47.1
An echinoderm. Brittle star, *Ophiothrix*, a member of the largest group of echinoderms.

Echinoderms, which include the familiar starfish, have been described as a "noble group especially designed to puzzle the zoologist." They are bilaterally symmetrical as larvae, but undergo a bizarre metamorphosis to a radially symmetrical adult (figure 47.1). A compartment of the coelom is transformed into a unique water-vascular system that uses hydraulic power to operate a multitude of tiny tube feet that are used in locomotion and food capture. Some echinoderms have an endoskeleton of dermal plates beneath the skin, fused together like body armor. Many have miniature jawlike pincers scattered over their body surface, often on stalks and sometimes bearing poison glands. This collection of characteristics is unique in the animal kingdom.

47.1 The embryos of deuterostomes develop quite differently from those of protostomes.

Protostomes and Deuterostomes

The coelomates we have met so far—the mollusks, annelids, and arthropods—exhibit essentially the same kind of embryological development, starting as a hollow ball of cells, a blastula, which indents to form a two-layer-thick ball with a blastopore opening to the outside. Also in this group, the mouth (stoma) develops from or near the blastopore (figure 47.2). This same pattern of development, in a general sense, is seen in all noncoelomate animals. An animal whose mouth develops in this way is called a **protostome** (from the Greek words *protos*, "first," and *stoma*, "mouth"). If such an animal has a distinct anus or anal pore, it develops later in another region of the embryo. The fact that this kind of developmental pattern is so widespread in diverse phyla suggests that it is the original pattern for animals as a whole and that it was characteristic of the common ancestor of all eumetazoan animals.

A second distinct pattern of embryological development occurs in the echinoderms, the chordates, and a few other smaller related phyla. The consistency of this pattern of development, and its distinctiveness from that of the protostomes suggests that it evolved once, in a common ancestor to all of the phyla that exhibit it. In **deuterostome** (Greek, *deuteros*, "second," and *stoma*, "mouth") development, the blastopore gives rise to the organism's anus, and the mouth develops from a second pore that arises in the blastula later in development.

Deuterostomes represent a revolution in embryonic development. In addition to the pattern of blastopore formation, deuterostomes differ from protostomes in a number of other fundamental embryological features:

1. The progressive division of cells during embryonic growth is called *cleavage.* The cleavage pattern relative to the embryo's polar axis determines how the cells will array. In nearly all protostomes, each new cell buds off at an angle oblique to the polar axis. As a result, a new cell nestles into the space between the older ones in a closely packed array. This pattern is called **spiral cleavage** because a line drawn through a sequence of dividing cells spirals outward from the polar axis (figure 47.2).

   In deuterostomes, the cells divide parallel to and at right angles to the polar axis. As a result, the pairs of cells from each division are positioned directly above

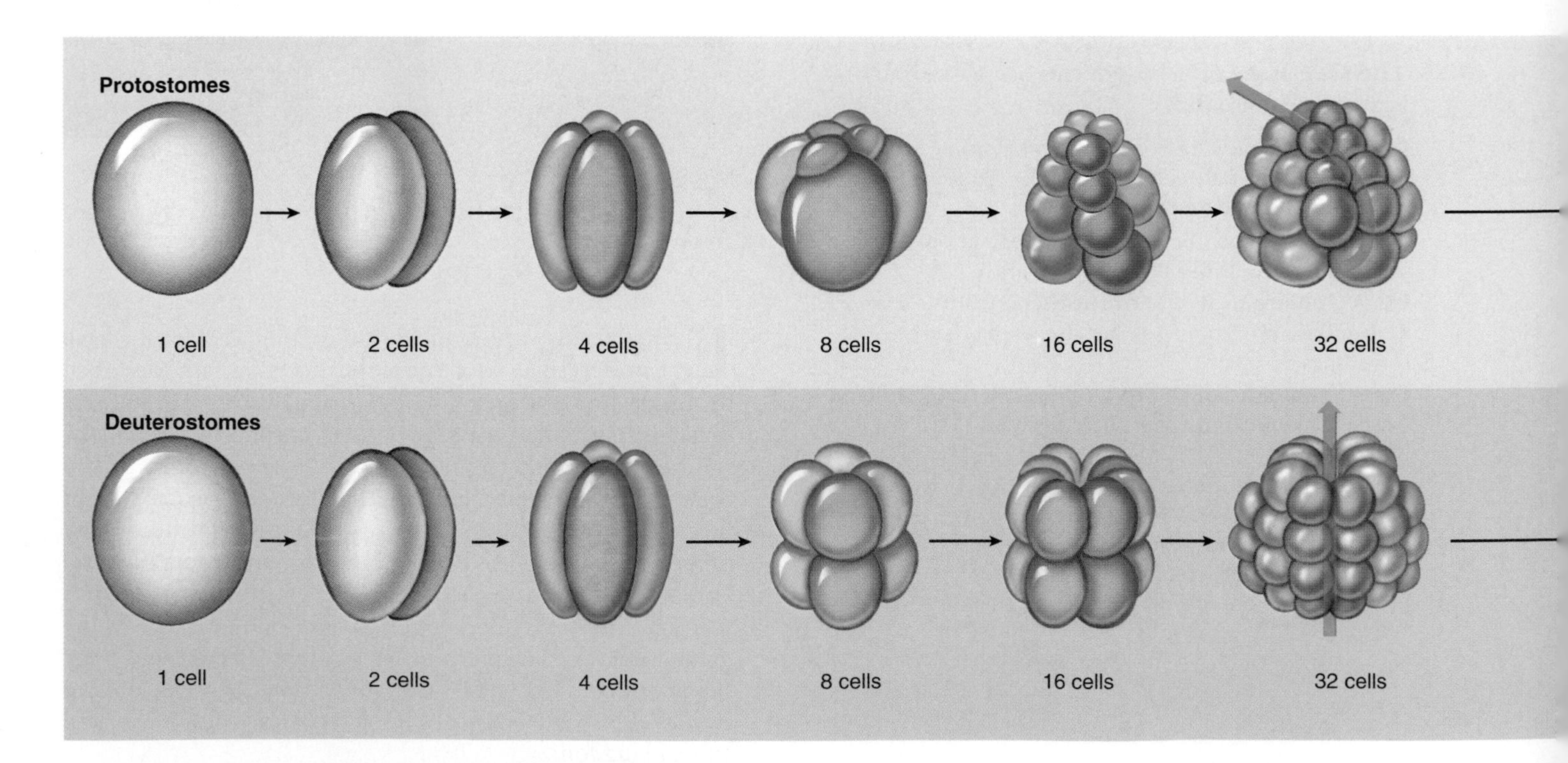

FIGURE 47.2
Embryonic development in protostomes and deuterostomes. Cleavage of the egg produces a hollow ball of cells called the blastula. Invagination of the blastula produces the blastopore and archenteron. In protostomes, embryonic cells cleave in a spiral pattern and become tightly packed. The blastopore becomes the animal's mouth, and the coelom originates from a mesodermal split.

and below one another; this process gives rise to a loosely packed array of cells. This pattern is called **radial cleavage** because a line drawn through a sequence of dividing cells describes a radius outward from the polar axis.

2. Protostomes exhibit **determinate** development. In this type of development, each embryonic cell has a predetermined fate in terms of what kind of tissue it will form in the adult. Before cleavage begins, the chemicals that act as developmental signals are localized in different parts of the egg. Consequently, the cell divisions that occur after fertilization separate different signals into different daughter cells. This process specifies the fate of even the very earliest embryonic cells. Deuterostomes, on the other hand, display **indeterminate** development. The first few cell divisions of the egg produce identical daughter cells. Any one of these cells, if separated from the others, can develop into a complete organism. This is possible because the chemicals that signal the embryonic cells to develop differently are not localized until later in the animal's development.
3. In all coelomates, the coelom originates from mesoderm. In protostomes, this occurs simply and directly: the cells simply move away from one another as the coelomic cavity expands within the mesoderm. However, in deuterostomes, whole groups of cells usually move around to form new tissue associations. The coelom is normally produced by an evagination of the **archenteron**—the central tube within the gastrula, also called the primitive gut. This tube, lined with endoderm, opens to the outside via the blastopore and eventually becomes the gut cavity.

The first abundant and well-preserved animal fossils are nearly 600 million years old; they occur in the Ediacara series of Australia and similar formations elsewhere. Among these fossils, many represent groups of animals that no longer exist. In addition, these ancient rocks bear evidence of the coelomates, the most advanced evolutionary line of animals, and it is remarkable that their two major subdivisions were differentiated so early. In the coelomates, it seems likely that all deuterostomes share a common protostome ancestor—a theory that is supported by evidence from comparison of rRNA and other molecular studies. The event, however, occurred very long ago and presumably did not involve groups of organisms that closely resemble any that are living now.

In deuterostomes, the egg cleaves radially, and the blastopore becomes the anus. In protostomes, the egg cleaves spirally, and the blastopore becomes the mouth.

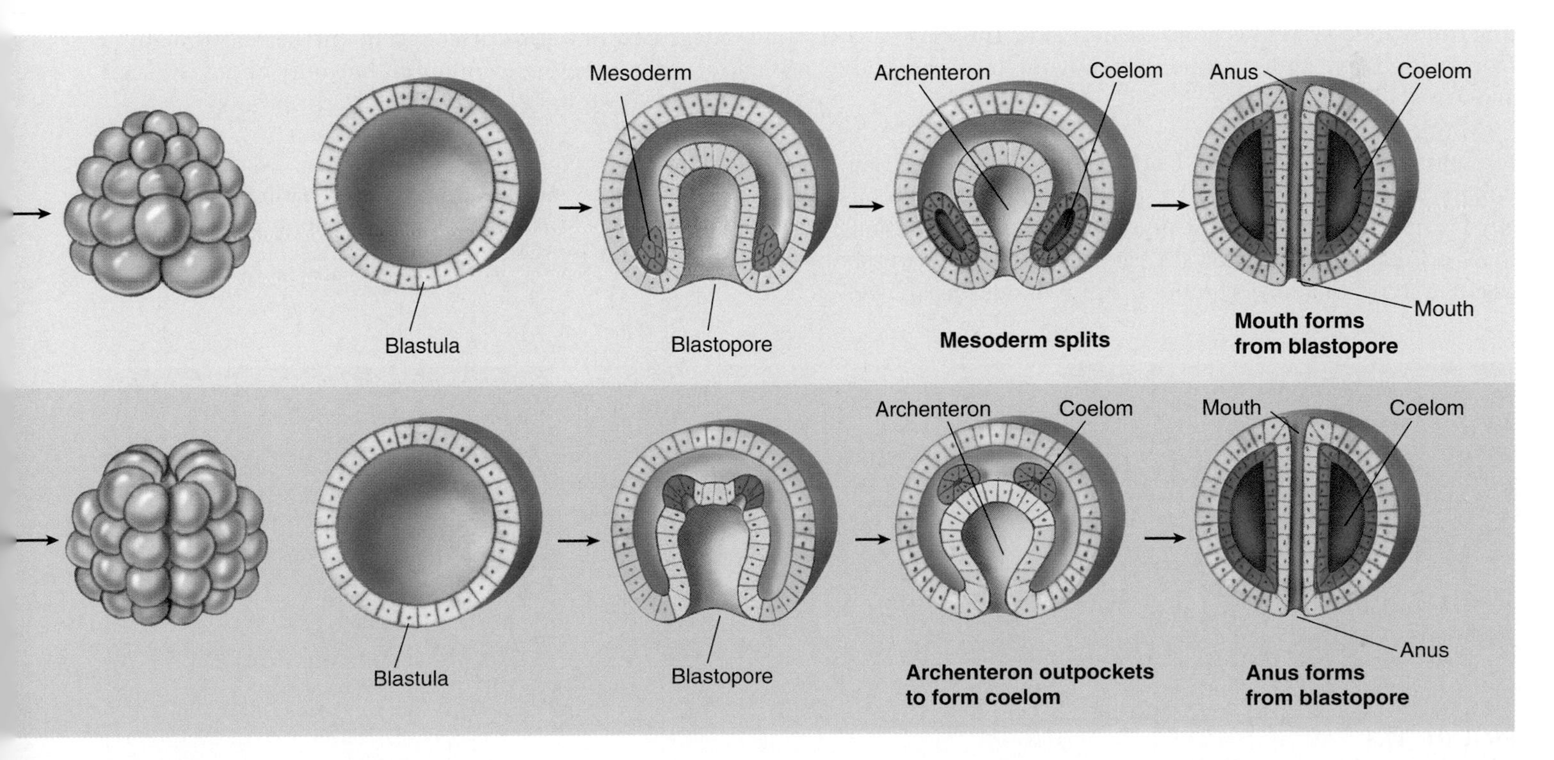

FIGURE 47.2 (continued)
In deuterostomes, embryonic cells cleave radially and form a loosely packed array. The blastopore becomes the animal's anus, and the mouth develops at the other end. The coelom originates from an evagination, or outpouching, of the archenteron in deuterostomes.

47.2 Echinoderms are deuterostomes with an endoskeleton.

Deuterostomes

Mollusks, annelids, and arthropods are protostomes. However, the echinoderms are characterized by *deuterostome development*, a key evolutionary advance. The *endoskeleton* makes its first appearance in the echinoderms also.

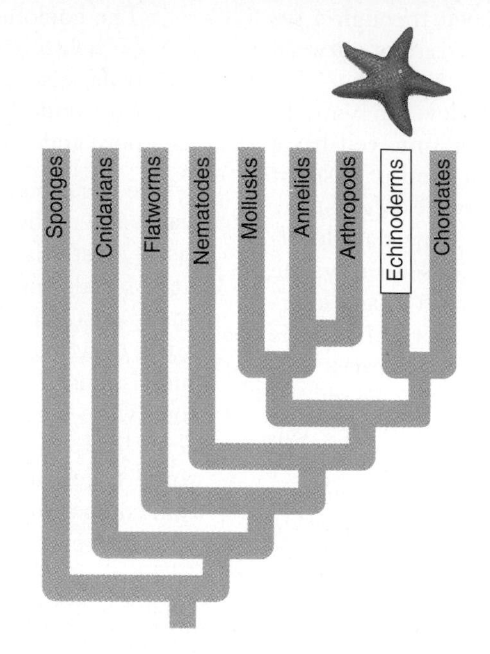

The Echinoderms

Deuterostomate marine animals called **echinoderms** appeared nearly 600 million years ago. Echinoderms (phylum Echinodermata) are an ancient group of marine animals consisting of about 6000 living species and are also well represented in the fossil record (figure 47.3). The term *echinoderm* means "spiny skin" and refers to an **endoskeleton** composed of hard calcium-rich plates just beneath the delicate skin (figure 47.4). When they first form, the plates are enclosed in living tissue and so are truly an endoskeleton, although in adults they frequently fuse, forming a hard shell. Another innovation in echinoderms is the development of a hydraulic system to aid in movement or feeding. Called a **water-vascular system,** this fluid-filled system is composed of a central ring canal from which five radial canals extend out into the body and arms.

Many of the most familiar animals seen along the seashore, sea stars (starfish), brittle stars, sea urchins, sand dollars, and sea cucumbers, are echinoderms. All are radially symmetrical as adults. While some other kinds of animals are radially symmetrical, none have the complex organ systems of adult echinoderms. Echinoderms are well represented not only in the shallow waters of the sea but also in its abyssal depths. In the oceanic trenches, which are the deepest regions of the oceans, sea cucumbers account for more than 90% of the biomass! All of them are bottom-dwellers except for a few swimming sea cucumbers. The adults range from a few millimeters to more than a meter in diameter (for one species of sea star) or in length (for a species of sea cucumber).

There is an excellent fossil record of the echinoderms, extending back into the Cambrian. However, despite this wealth of information, the origin of echinoderms remains unclear. They are thought to have evolved from bilaterally symmetrical ancestors because echinoderm larvae are bilateral. The radial symmetry that is the hallmark of echinoderms develops later, in the adult body. Many biologists believe that early echinoderms were sessile and evolved radiality as an adaptation to the sessile existence. Bilaterality is of adaptive value to an animal that travels through its environment, while radiality is of value to an animal whose environment meets it on all sides. Echinoderms attached to the sea bottom by a central stalk were once common, but only about 80 such species survive today.

Echinoderms are a unique, exclusively marine group of organisms in which deuterostome development and an endoskeleton are seen for the first time.

(a)

(b)

(c)

FIGURE 47.3
Diversity in echinoderms. (*a*) Sea star, *Oreaster occidentalis* (class Asteroidea), in the Gulf of California, Mexico. (*b*) Warty sea cucumber, *Parastichopus parvimensis* (class Holothuroidea), Philippines. (*c*) Sea urchin (class Echinoidea).

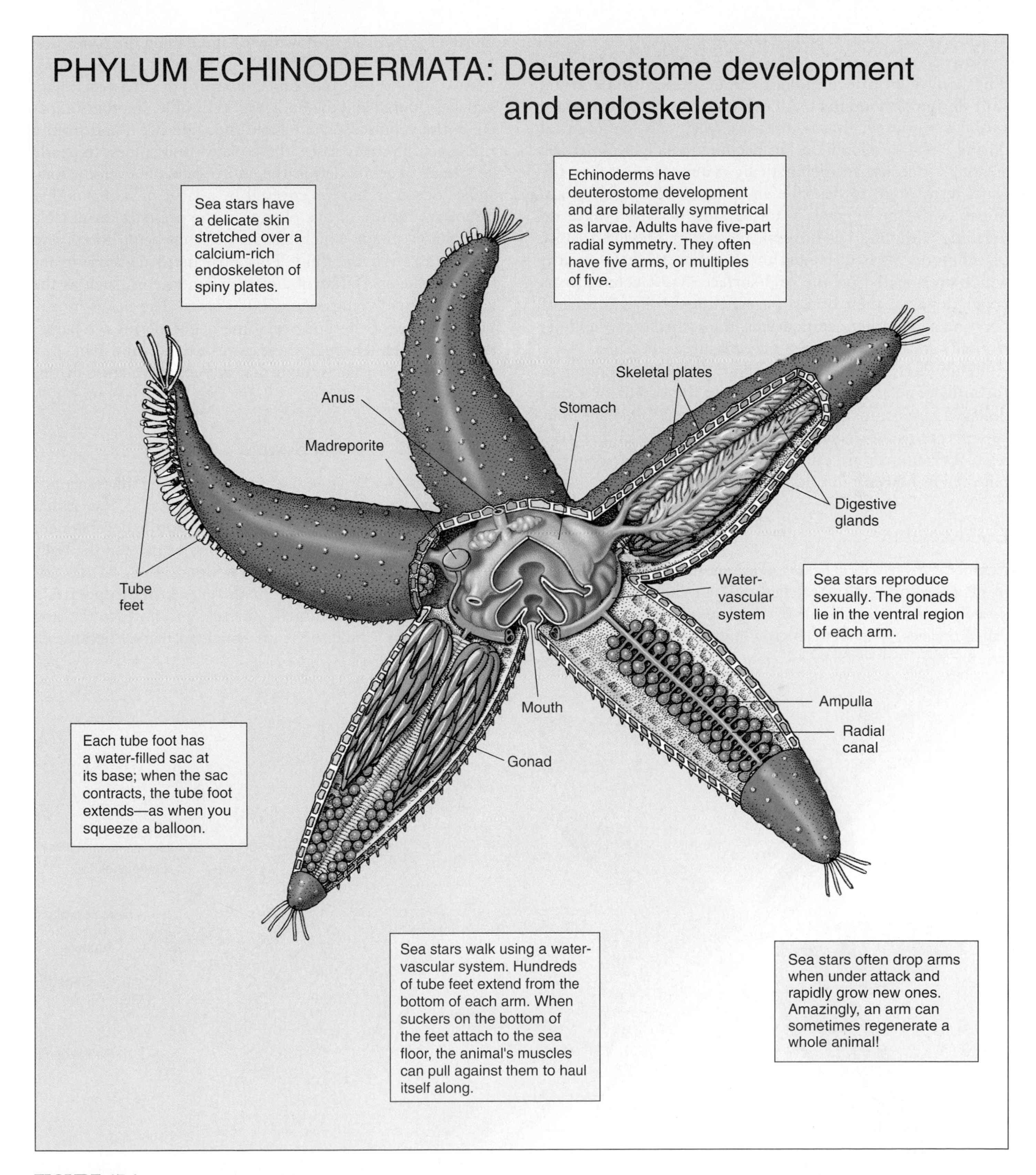

PHYLUM ECHINODERMATA: Deuterostome development and endoskeleton
Sea stars have a delicate skin stretched over a calcium-rich endoskeleton of spiny plates.
Echinoderms have deuterostome development and are bilaterally symmetrical as larvae. Adults have five-part radial symmetry. They often have five arms, or multiples of five.
Skeletal plates
Anus
Stomach
Madreporite
Digestive glands
Tube feet
Water-vascular system
Sea stars reproduce sexually. The gonads lie in the ventral region of each arm.
Ampulla
Radial canal
Mouth
Gonad
Each tube foot has a water-filled sac at its base; when the sac contracts, the tube foot extends—as when you squeeze a balloon.
Sea stars walk using a water-vascular system. Hundreds of tube feet extend from the bottom of each arm. When suckers on the bottom of the feet attach to the sea floor, the animal's muscles can pull against them to haul itself along.
Sea stars often drop arms when under attack and rapidly grow new ones. Amazingly, an arm can sometimes regenerate a whole animal!

FIGURE 47.4
The evolution of deuterostome development and an endoskeleton. Echinoderms, such as sea stars (phylum Echinodermata), are coelomates with a deuterostome pattern of development. A delicate skin stretches over an endoskeleton made of calcium-rich plates, often fused into a continuous, tough, spiny layer.

Echinoderm Body Plan

The body plan of echinoderms undergoes a fundamental shift during development. All echinoderms have **secondary radial symmetry,** that is, they are bilaterally symmetrical during larval development but become radially symmetrical as adults. Because of their radially symmetrical bodies, the usual terms used to describe an animal's body are not applicable: dorsal, ventral, anterior, and posterior have no meaning without a head or tail. Instead, the body structure of echinoderms is discussed in reference to their mouths which are located on the oral surface. Most echinoderms crawl along on their oral surfaces, although in sea cucumbers and some other echinoderms, the animal's axis lies horizontally and they crawl with the oral surface in front.

Echinoderms have a five-part body plan corresponding to the arms of a sea star or the design on the "shell" of a sand dollar. These animals have no head or brain. Their nervous systems consist of a central **nerve ring** from which branches arise. The animals are capable of complex response patterns, but there is no centralization of function.

Endoskeleton

Echinoderms have a delicate epidermis, containing thousands of neurosensory cells, stretched over an endoskeleton composed of either movable or fixed calcium-rich (calcite) plates called ossicles. The animals grow more or less continuously, but their growth slows down with age. When the plates first form, they are enclosed in living tissue. In some echinoderms, such as asteroids and holothuroids, the ossicles are widely scattered and the body wall is flexible. In others, especially the echinoids, the ossicles become fused and form a rigid shell. In many cases, these plates bear spines. In nearly all species of echinoderms, the entire skeleton, even the long spines of sea urchins, is covered by a layer of skin. Another important feature of this phylum is the presence of mutable collagenous tissue which can range from being tough and rubbery to weak and fluid. This amazing tissue accounts for many of the special attributes of echinoderms, such as the ability to rapidly autotomize body parts. The plates in certain portions of the body of some echinoderms are perforated by pores. Through these pores extend **tube feet,** part of the water-vascular system that is a unique feature of this phylum.

The Water-Vascular System

The water-vascular system of an echinoderm radiates from a **ring canal** that encircles the animal's esophagus. Five **radial canals,** their positions determined early in the development of the embryo, extend into each of the five parts of the body and determine its basic symmetry (figure 47.5). Water enters the water-vascular system through a **madreporite,** a sievelike plate on the animal's surface, and flows to the ring canal through a tube, or stone canal, so named because of

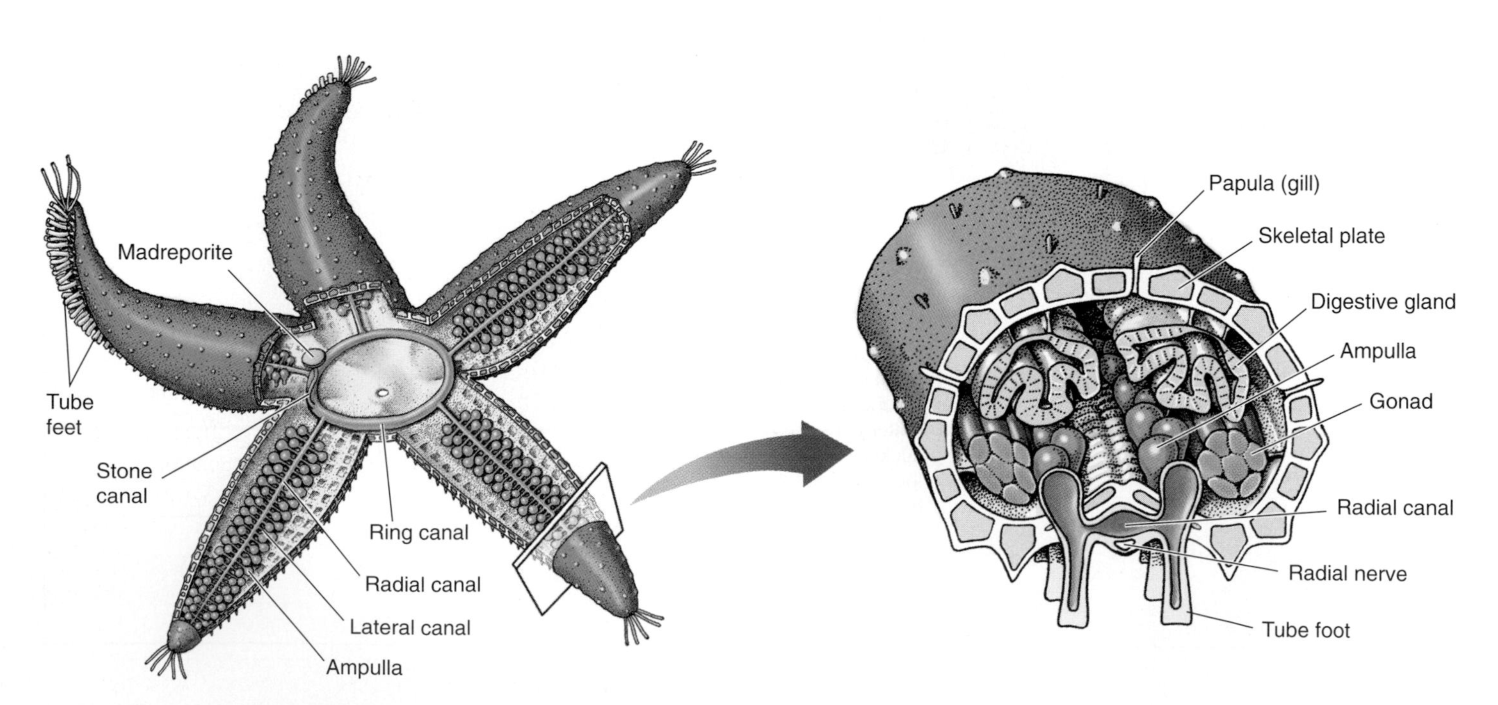

FIGURE 47.5
The water-vascular system of an echinoderm. Radial canals allow water to flow into the tube feet. As the ampulla in each tube foot contracts, the tube extends and can attach to the substrate. Subsequently, muscles in the tube feet contract bending the tube foot and pulling the animal forward.

FIGURE 47.6
Tube feet. The nonsuckered tube feet of the sea star, *Ludia magnifica*, are extended.

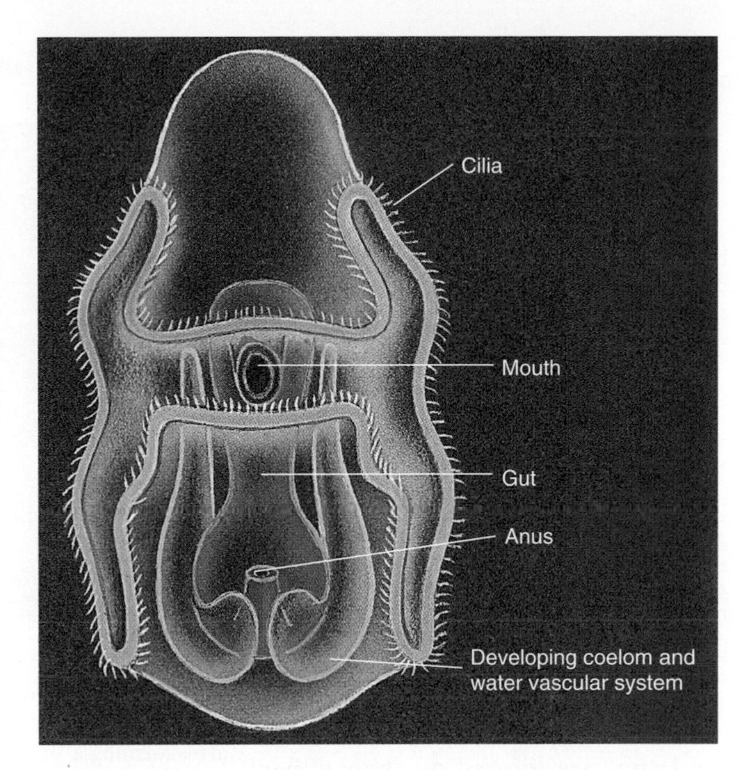

FIGURE 47.7
The free-swimming larva of an echinoderm. The bands of cilia by which the larva moves are prominent in this drawing. Such bilaterally symmetrical larvae suggest that the ancestors of the echinoderms were not radially symmetrical, like the living members of the phylum.

the surrounding rings of calcium carbonate. The five radial canals in turn extend out through short side branches into the hollow tube feet (figure 47.6). In some echinoderms, each tube foot has a sucker at its end; in others, suckers are absent. At the base of each tube foot is a muscular sac, the **ampulla,** which contains fluid. When the ampulla contracts, the fluid is prevented from entering the radial canal by a one-way valve and is forced into the tube foot, thus extending it. When extended, the foot can attach itself to the substrate. Longitudinal muscles in the tube foot wall then contract causing the tube feet to bend. Relaxation of the muscles in the ampulla allows the fluid to flow back into the ampulla which moves the foot. By the concerted action of a very large number of small individually weak tube feet, the animal can move across the sea floor.

Sea cucumbers (see figure 47.3*b*) usually have five rows of tube feet on the body surface that are used in locomotion. They also have modified tube feet around their mouth cavity that are used in feeding. In sea lilies, the tube feet arise from the branches of the arms, which extend from the margins of an upward-directed cup. With these tube feet, the animals take food from the surrounding water. In brittle stars (see figure 47.1), the tube feet are pointed and specialized for feeding.

Body Cavity

In echinoderms, the coelom, which is proportionately large, connects with a complicated system of tubes and helps provide circulation and respiration. In many echinoderms, respiration and waste removal occurs across the skin through small, fingerlike extensions of the coelom called papulae (see figure 47.5). They are covered with a thin layer of skin and protrude through the body wall to function as gills.

Reproduction

Many echinoderms are able to regenerate lost parts, and some, especially sea stars and brittle stars, drop various parts when under attack. In a few echinoderms, asexual reproduction takes place by splitting, and the broken parts of sea stars can sometimes regenerate whole animals. Some of the smaller brittle stars, especially tropical species, regularly reproduce by breaking into two equal parts; each half then regenerates a whole animal.

Despite the ability of many echinoderms to break into parts and regenerate new animals from them, most reproduction in the phylum is sexual and external. The sexes in most echinoderms are separate, although there are few external differences. Fertilized eggs of echinoderms usually develop into free-swimming, bilaterally symmetrical larvae (figure 47.7), which differ from the trochophore larvae of mollusks and annelids. These larvae form a part of the plankton until they metamorphose through a series of stages into the more sedentary adults.

Echinoderms are characterized by a secondary radial symmetry and a five-part body plan. They have characteristic calcium-rich plates called ossicles and a unique water-vascular system that includes hollow tube feet.

47.3 The six classes of echinoderms are all radially symmetrical as adults.

There are more than 20 extinct classes of echinoderms and an additional 6 with living members: (1) Crinoidea, sea lilies and feather stars; (2) Asteroidea, sea stars, or starfish; (3) Ophiuroidea, brittle stars; (4) Echinoidea, sea urchins and sand dollars; (5) Holothuroidea, sea cucumbers, and (6) Concentricycloidea, sea daisies. Sea daisies were recently discovered living on submerged wood in the deep sea.

Class Crinoidea: The Sea Lilies and Feather Stars

Sea lilies and feather stars, or crinoids (class Crinoidea) differ from all other living echinoderms in that the mouth and anus are located on their upper surface in an open disc. The two structures are connected by a simple gut. These animals have simple excretory and reproductive systems and an extensive water-vascular system. The arms, which are the food-gathering structures of crinoids, are located around the margins of the disc. Different species of crinoids may have from 5 to more than 200 arms extending upward from their bodies, with smaller structures called pinnules branching from the arms. In all crinoids, the number of arms is initially small. Species with more than 10 arms add additional arms progressively during growth. Crinoids are filter feeders, capturing the microscopic organisms on which they feed by means of the mucus that coats their tube feet, which are abundant on the animals' pinnules.

Scientists that study echinoderms believe that the common ancestors of this phylum were sessile, sedentary, radially symmetrical animals that resembled crinoids. Crinoids were abundant in ancient seas, and were present when the Burgess Shale was deposited about 515 million years ago. More than 6000 fossil species of this class are known, in comparison with the approximately 600 living species.

FIGURE 47.8
Sea lilies, *Cenocrinus asterius*. Two specimens showing a typical parabola of arms forming a "feeding net." The water current is flowing from right to left, carrying small organisms to the stalked crinoid's arms. Prey, when captured, are passed down the arms to the central mouth. This photograph was taken at a depth of about 400 meters in the Bahamas from the Johnson-Sea-Link Submersible of the Harbor Branch Foundation, Inc.

FIGURE 47.9
Feather star. This feather star is on the Great Barrier Reef in Australia.

Sea Lilies

There are two basic crinoid body plans. In sea lilies, the flower-shaped body is attached to its substrate by a stalk that is from 15 to 30 cm long, although in some species the stalk may be as much as a meter long (figure 47.8). Some fossil species had stalks up to 20 meters long. If they are detached from the substrate, some sea lilies can move slowly by means of their feather-like arms. All of the approximately 80 living species of sea lilies are found below a depth of 100 meters in the ocean. Sea lilies are the only living echinoderms that are fully sessile.

Feather Stars

In the second group of crinoids, the 520 or so species of feather stars, the disc detaches from the stalk at an early stage of development (figure 47.9). Adult feather stars have long, many-branched arms and usually anchor themselves to their substrate by claw-like structures. However, some feather stars are able to swim for short distances, and many of them can move along the substrate. Feather stars range into shallower water than do sea lilies, and only a few species of either group are found at depths greater than 500 meters. Along with sea cucumbers, crinoids are the most abundant and conspicuous large invertebrates in the warm waters and among the coral reefs of the western Pacific Ocean. They have separate sexes, with the sex organs simple masses of cells in special cavities of the arms and pinnules. Fertilization is usually external, with the male and female gametes shed into the water, but brooding—in which the female shelters the young—occurs occasionally.

Crinoids, the sea lilies and feather stars, were once far more numerous. Crinoids are the only echinoderms attached for much of their lives to the sea bottom.

Class Asteroidea: The Sea Stars

Sea stars, or starfish (class Asteroidea; figure 47.10), are perhaps the most familiar echinoderms. Among the most important predators in many marine ecosystems, they range in size from a centimeter to a meter across. They are abundant in the intertidal zone, but they also occur at depths as great as 10,000 meters. Around 1500 species of sea stars occur throughout the world.

The body of a sea star is composed of a central disc that merges gradually with the arms. Although most sea stars have five arms, the basic symmetry of the phylum, members of some families have many more, typically in multiples of five. The body is somewhat flattened, flexible, and covered with a pigmented epidermis.

Endoskeleton

Beneath the epidermis is an endoskeleton of small calcium-rich plates called ossicles, bound together with connective tissue. From these ossicles project spines that make up the spiny upper surface. Around the base of the spines are minute, pincerlike *pedicellariae*, bearing tiny jaws manipulated by muscles. These keep the body surface free of debris and may aid in food capture.

The Water-Vascular System

A deep groove runs along the oral (bottom) surface of each arm from the central mouth out to the tip of the arm. This groove is bordered by rows of tube feet, which the animal uses to move about. Within each arm, there is a radial canal that connects the tube feet to a ring canal in the central body. This system of piping is used by sea stars to power a unique hydraulic system. Contraction of small chambers called ampullae attached to the tube feet forces water into the podium of the feet, extending them. Conversely, contraction of muscles in the tube foot retracts the podium, forcing fluid back into the ampulla. Small muscles at the end of each tube foot can raise the center of the disclike end, creating suction when the foot is pressed against a substrate. Hundreds of tube feet, moving in unison, pull the arm along the surface.

Feeding

The mouth of a sea star is located in the center of its oral surface. Some sea stars have an extraordinary way of feeding on bivalve mollusks. They can open a small gape between the shells of bivalves by exerting a muscular pull on the shells (figure 47.11). Eventually, muscular fatigue in the bivalve results in a very narrow gape, sufficient enough for the sea star to insert its stomach out through its mouth into the bivalve. Within the mollusk, the sea star secretes its digestive enzymes and digests the soft tissues of its prey, retracting its stomach when the process is complete.

FIGURE 47.10
Class Asteroidea. This class includes the familiar starfish, or sea stars.

FIGURE 47.11
A sea star attacking a clam. The tube feet, each of which ends in a suction cup, are located along grooves on the underside of the arms.

Reproduction

Most sea stars have separate sexes, with a pair of gonads lying in the ventral region inside each arm. Eggs and sperm are shed into the water so that fertilization is external. In some species, fertilized eggs are brooded in special cavities or simply under the animal. They mature into larvae that swim by means of conspicuous bands of cilia.

Sea stars, also called starfish, are five-armed, mobile predators.

Class Ophiuroidea: The Brittle Stars

Brittle stars (class Ophiuroidea; figure 47.12) are the largest class of echinoderms in numbers of species (about 2000) and they are probably the most abundant also. Secretive, they avoid light and are more active at night.

Brittle stars have slender, branched arms. The most mobile of echinoderms, brittle stars move by pulling themselves along, "rowing" over the substrate by moving their arms, often in pairs or groups, from side to side. Some brittle stars use their arms to swim, a very unusual habit among echinoderms.

Brittle stars feed by capturing suspended microplankton and organic detritus with their tube feet, climbing over objects on the ocean floor. In addition, the tube feet are important sensory organs and assist in directing food into the mouth once the animal has captured it. As implied by their common name, the arms of brittle stars detach easily, a characteristic that helps to protect the brittle stars from their predators.

Like sea stars, brittle stars have five arms. More closely related to the sea stars than to the other classes of the phylum, on closer inspection they are surprisingly different. They have no pedicellariae, as sea stars have, and the groove running down the length of each arm is closed over and covered with ossicles. Their tube feet lack ampullae, have no suckers, and are used for feeding, not locomotion.

Brittle stars usually have separate sexes, with the male and female gametes in most species being released into the water and fusing there. Development takes place in the plankton and the larvae swim and feed using elaborate bands of cilia. Some species brood their young in special cavities and fully developed juvenile brittle stars emerge at the end of development.

Brittle stars, very secretive, pull themselves along with their arms.

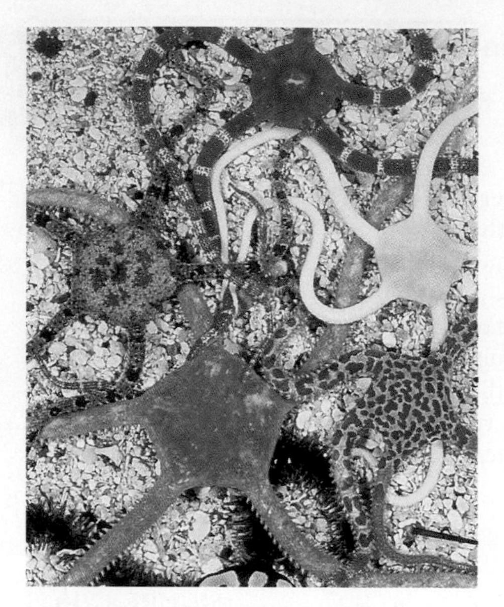

FIGURE 47.12
Class Ophiuroidea. Brittle stars crawl actively across their marine substrates.

(a)

(b)

FIGURE 47.13
Class Echinoidea. (*a*) Sand dollar, *Echinarachnius parma.* (*b*) Giant red sea urchin, *Strongylocentrotus franciscanus.*

Class Echinoidea: The Sea Urchins and Sand Dollars

The members of the class Echinoidea, sand dollars and sea urchins, lack distinct arms but have the same five-part body plan as all other echinoderms (figure 47.13). Five rows of tube feet protrude through the plates of the calcareous skeleton, and there are also openings for the mouth and anus. These different openings can be seen in the globular skeletons of sea urchins and in the flat skeletons of sand dollars. Both types of endoskeleton, often common along the seashore, consist of fused calcareous plates. About 950 living species constitute the class Echinoidea.

Echinoids walk by means of their tube feet or their movable spines, which are hinged to the skeleton by a joint that makes free rotation possible. Sea urchins and sand dollars move along the sea bottom, feeding on algae and small fragments of organic material. They scrape these off the substrate with the large, triangular teeth that ring their mouths. The gonads of sea urchins are considered a great delicacy by people in different parts of the world. Because of their calcareous plates, sea urchins and sand dollars are well preserved in the fossil record, with more than 5000 additional species described.

As with most other echinoderms, the sexes of sea urchins and sand dollars are separate. The eggs and sperm are shed separately into the water, where they fuse. Some brood their young, and others have free-swimming larvae, with bands of cilia extending onto their long, graceful arms.

Sand dollars and sea urchins lack arms but have a five-part symmetry.

Classes Holothuroidea and Concentricycloidea: Sea Cucumbers and Sea Daisies

Sea Cucumbers

Sea cucumbers (class Holothuroidea) are shaped somewhat like their plant namesakes. They differ from the preceding classes in that they are soft, sluglike organisms, often with a tough, leathery outside skin (figure 47.14). The class consists of about 1500 species found worldwide. Except for a few forms that swim, sea cucumbers lie on their sides at the bottom of the ocean. Their mouth is located at one end and is surrounded by eight to 30 modified tube feet called tentacles; the anus is at the other end. The tentacles around the mouth may secrete mucus, used to capture the small planktonic organisms on which the animals feed. Each tentacle is periodically wiped off within the esophagus and then brought out again, covered with a new supply of mucus.

Sea cucumbers are soft because their calcareous skeletons are reduced to widely separated microscopic plates. These animals have extensive internal branching systems, called respiratory trees, which arise from the **cloaca,** or anal cavity. Water is pulled into and expelled from the respiratory tree by contractions of the cloaca; gas exchange takes place as this process occurs. The sexes of most cucumbers are separate, but some of them are hermaphroditic.

Most kinds of sea cucumbers have tube feet on the body in addition to tentacles. These additional tube feet, which might be restricted to five radial grooves or scattered over the surface of the body, may enable the animals to move about slowly. On the other hand, sea cucumbers may simply wriggle along whether or not they have additional tube feet. Most sea cucumbers are quite sluggish, but some, especially among the deep-sea forms, swim actively. Sea cucumbers, when irritated, sometimes eject a portion of their intestines by a strong muscular contraction that may send the intestinal fragments through the anus or even rupture the body wall.

FIGURE 47.14
Class Holothuroidea. Sea cucumber.

Sea Daisies

The most recently described class of echinoderms (1986), sea daisies are strange little disc-shaped animals, less than 1 cm in diameter, discovered in waters over 1000 m deep off New Zealand (figure 47.15). Only two species are known so far. They have five-part radial symmetry, but no arms. Their tube feet are located around the periphery of the disc, rather than along radial lines, as in other echinoderms. One species has a shallow, saclike stomach but no intestine or anus; the other species has no digestive tract at all—the surface of its mouth is covered by a membrane through which it apparently absorbs nutrients.

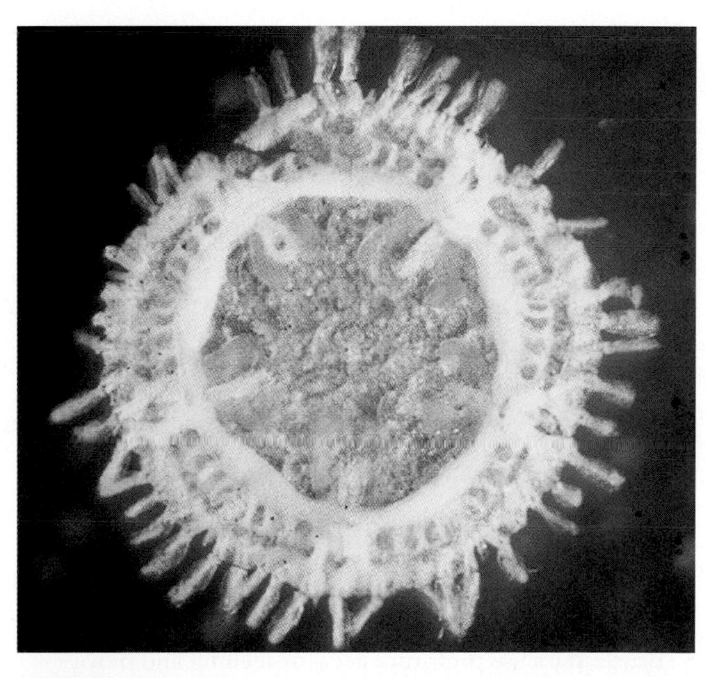

FIGURE 47.15
Class Concentricycloidea. Sea daisy.

Sea cucumbers are soft-bodied, sluglike animals without arms. The newly discovered sea daisies are the most mysterious echinoderms. Tiny and simple in form, they live at great depths, absorbing food from their surroundings.

Chapter 47

www.mhhe.com/raven6/resources47.mhtml

Summary | **Questions** | **Media Resources**

47.1 The embryos of deuterostomes develop quite differently from those of protostomes.

- The two major evolutionary lines of coelomate animals—the protostomes and the deuterostomes—are both represented among the oldest known fossils of multicellular animals, dating back some 650 million years.
- In the protostomes, the mouth develops from or near the blastopore, and the early divisions of the embryo are spiral. At early stages of development, the fate of the individual cells is already determined, and they cannot develop individually into a whole animal.
- In the deuterostomes, the anus develops from or near the blastopore, and the mouth forms subsequently on another part of the gastrula. The early divisions of the embryo are radial. At early stages of development, each cell of the embryo can differentiate into a whole animal.

1. What patterns of embryonic development related to cleavage and the blastopore occur in protostome coelomates? What patterns occur in deuterostome coelomates?

2. Which major coelomate phyla are protostomes and which are deuterostomes? How does the early developmental fate of cells differ between the two groups? How is the development of the coelom from mesodermal tissue different between them?

47.2 Echinoderms are deuterostomes with an endoskeleton.

- Echinoderms are exclusively marine deuterostomes that are radially symmetrical as adults.
- The epidermis of an echinoderm stretches over an endoskeleton made of separate or fused calcium-rich plates.
- Echinoderms use a unique water-vascular system that includes tube feet for locomotion and feeding.

3. What type of symmetry and body plan do adult echinoderms exhibit?

4. What is the composition and location of the echinoderm skeleton?

5. How do echinoderms respire? How developed is their digestive system?

47.3 The six classes of echinoderms are all radially symmetrical as adults.

- Crinoids are sessile for some or all of their lives and have a mouth and anus located on the upper surface of the animal.
- Sea stars are active predators that move about on their tube feet.
- Brittle stars use their tube feet for feeding and move about using two arms at a time.
- The endoskeletons of sea urchins and sand dollars consist of fused calcareous plates that have been well preserved in the fossil record.
- The endoskeletons of sea cucumbers are drastically reduced and separated, making them soft-bodied.
- Sea daisies are a newly described class of echinoderms with disc-shaped bodies.

6. In what two ways do members of the phylum Echinodermata reproduce? What type of larva do they possess?

7. How do sea cucumbers superficially differ from other echinoderms? How are some of their tube feet specially modified? What is the extent of their skeleton? What is the function of their unique respiratory tree? How is their reproduction different from that of other echinoderms?

- Echinoderms

48

Vertebrates

Concept Outline

48.1 Attaching muscles to an internal framework greatly improves movement.

The Chordates. Chordates have an internal flexible rod, the first stage in the evolution of a truly internal skeleton.

48.2 Nonvertebrate chordates have a notochord but no backbone.

The Nonvertebrate Chordates. Lancelets are thought to resemble the ancestors of vertebrates.

48.3 The vertebrates have an interior framework of bone.

Characteristics of Vertebrates. Vertebrates have a true, usually bony endoskeleton, with a backbone encasing the spinal column, and a skull-encased brain.

48.4 The evolution of vertebrates involves invasions of sea, land, and air.

Fishes. Over half of all vertebrate species are fishes, which include the group from which all other vertebrates evolved.
History of the Fishes. Swim bladders have made bony fishes a particularly successful group.
Amphibians. The key innovation that made life on land possible for vertebrates was the pulmonary vein.
History of the Amphibians. In the past, amphibians were far more diverse, and included many large, armored terrestrial forms.
Reptiles. Reptiles were the first vertebrates to completely master the challenge of living on dry land.
The Rise and Fall of Dominant Reptile Groups. Now-extinct forms of reptiles dominated life on land for 250 million years. Four orders survive today.
Birds. Birds are much like reptiles, but with feathers.
History of the Birds. Birds are thought to have evolved from dinosaurs with adaptations of feathers and flight.
Mammals. Mammals are the only vertebrates that possess hair and mammary glands.
History of the Mammals. Mammals evolved at the same time as dinosaurs, but only became common when dinosaurs disappeared.

FIGURE 48.1
A typical vertebrate. Today, mammals, like this snow leopard, *Panthera uncia,* dominate vertebrate life on land, but for over 200 million years in the past they were a minor group in a world dominated by reptiles.

Members of the phylum Chordata (figure 48.1) exhibit great improvements in the endoskeleton over what is seen in echinoderms. As we saw in the previous chapter, the endoskeleton of echinoderms is functionally similar to the exoskeleton of arthropods; it is a hard shell that encases the body, with muscles attached to its inner surface. Chordates employ a very different kind of endoskeleton, one that is truly internal. Members of the phylum Chordata are characterized by a flexible rod that develops along the back of the embryo. Muscles attached to this rod allowed early chordates to swing their backs from side to side, swimming through the water. This key evolutionary advance, attaching muscles to an internal element, started chordates along an evolutionary path that led to the vertebrates—and, for the first time, to truly large animals.

48.1 Attaching muscles to an internal framework greatly improves movement.

The Chordates

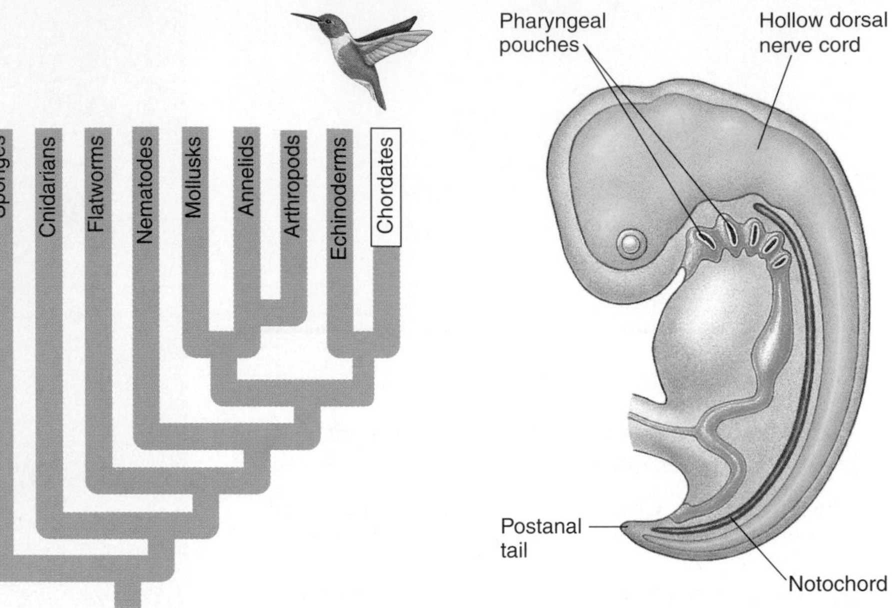

FIGURE 48.2
Some of the principal features of the chordates, as shown in a generalized embryo.

Chordates (phylum Chordata) are deuterostome coelomates whose nearest relations in the animal kingdom are the echinoderms, the only other deuterostomes. However, unlike echinoderms, chordates are characterized by a *notochord*, *jointed appendages*, and *segmentation*. There are some 43,000 species of chordates, a phylum that includes birds, reptiles, amphibians, fishes, and mammals.

Four features characterize the chordates and have played an important role in the evolution of the phylum (figure 48.2):

1. A single, hollow **nerve cord** runs just beneath the dorsal surface of the animal. In vertebrates, the dorsal nerve cord differentiates into the brain and spinal cord.
2. A flexible rod, the **notochord,** forms on the dorsal side of the primitive gut in the early embryo and is present at some developmental stage in all chordates. The notochord is located just below the nerve cord. The notochord may persist throughout the life cycle of some chordates or be displaced during embryological development as in most vertebrates by the vertebral column that forms around the nerve cord.
3. **Pharyngeal slits** connect the **pharynx,** a muscular tube that links the mouth cavity and the esophagus, with the outside. In terrestrial vertebrates, the slits do not actually connect to the outside and are better termed pharyngeal pouches. Pharyngeal pouches are present in the embryos of all vertebrates. They become slits, open to the outside in animals with gills, but disappear in those lacking gills. The presence of these structures in all vertebrate embryos provides evidence to their aquatic ancestry.
4. Chordates have a **postanal tail** that extends beyond the anus, at least during their embryonic development. Nearly all other animals have a terminal anus.

All chordates have all four of these characteristics at some time in their lives. For example, humans have pharyngeal pouches, a dorsal nerve cord, and a notochord as embryos. As adults, the nerve cord remains while the notochord is replaced by the vertebral column and all but one pair of pharyngeal pouches are lost. This remaining pair forms the Eustachian tubes that connect the throat to the middle ear.

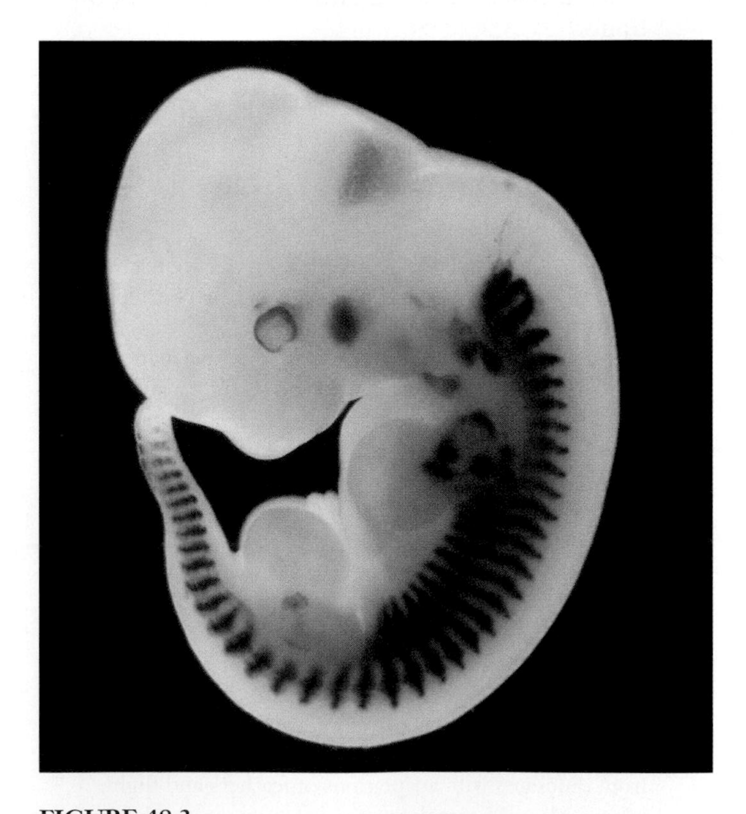

FIGURE 48.3
A mouse embryo. At 11.5 days of development, the mesoderm is already divided into segments called somites (stained dark in this photo), reflecting the fundamentally segmented nature of all chordates.

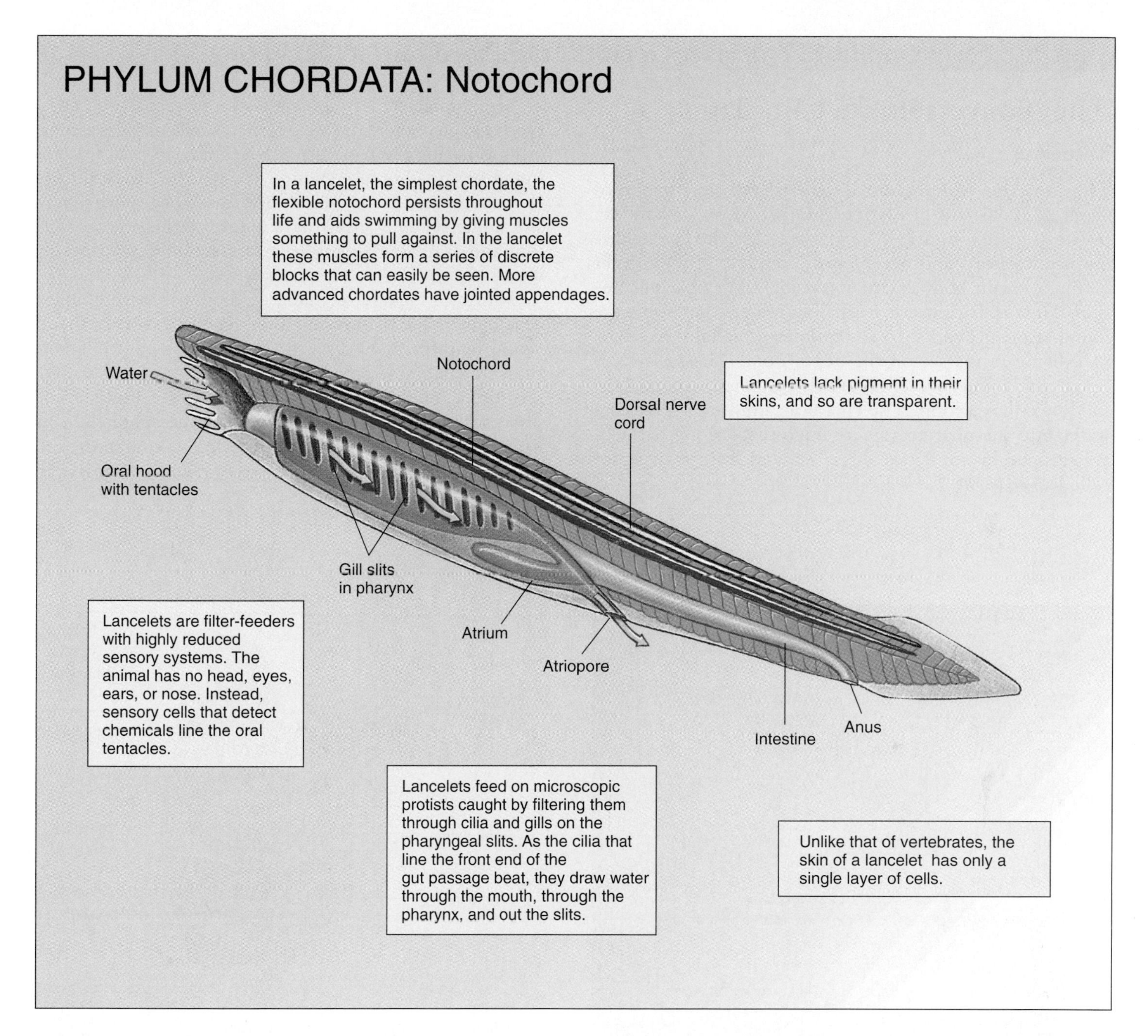

PHYLUM CHORDATA: Notochord
In a lancelet, the simplest chordate, the flexible notochord persists throughout life and aids swimming by giving muscles something to pull against. In the lancelet these muscles form a series of discrete blocks that can easily be seen. More advanced chordates have jointed appendages.
Water
Notochord
Dorsal nerve cord
Lancelets lack pigment in their skins, and so are transparent.
Oral hood with tentacles
Gill slits in pharynx
Lancelets are filter-feeders with highly reduced sensory systems. The animal has no head, eyes, ears, or nose. Instead, sensory cells that detect chemicals line the oral tentacles.
Atrium
Atriopore
Intestine
Anus
Lancelets feed on microscopic protists caught by filtering them through cilia and gills on the pharyngeal slits. As the cilia that line the front end of the gut passage beat, they draw water through the mouth, through the pharynx, and out the slits.
Unlike that of vertebrates, the skin of a lancelet has only a single layer of cells.

FIGURE 48.4

Evolution of a notochord. Vertebrates, tunicates, and lancelets are chordates (phylum Chordata), coelomate animals with a flexible rod, the notochord, that provides resistance to muscle contraction and permits rapid lateral body movements. Chordates also possess pharyngeal slits (reflecting their aquatic ancestry and present habitat in some) and a hollow dorsal nerve cord. In vertebrates, the notochord is replaced during embryonic development by the vertebral column.

A number of other characteristics also distinguish the chordates fundamentally from other animals. Chordates' muscles are arranged in segmented blocks that affect the basic organization of the chordate body and can often be clearly seen in embryos of this phylum (figure 48.3). Most chordates have an internal skeleton against which the muscles work. Either this internal skeleton or the notochord (figure 48.4) makes possible the extraordinary powers of locomotion that characterize the members of this group.

Chordates are characterized by a hollow dorsal nerve cord, a notochord, pharyngeal gill slits, and a postanal tail at some point in their development. The flexible notochord anchors internal muscles and allows rapid, versatile movement.

48.2 Nonvertebrate chordates have a notochord but no backbone.

The Nonvertebrate Chordates

Tunicates

The tunicates (subphylum Urochordata) are a group of about 1250 species of marine animals. Most of them are sessile as adults (figure 48.5*a,b*), with only the larvae having a notochord and nerve cord. As adults, they exhibit neither a major body cavity nor visible signs of segmentation. Most species occur in shallow waters, but some are found at great depths. In some tunicates, adults are colonial, living in masses on the ocean floor. The pharynx is lined with numerous cilia, and the animals obtain their food by ciliary action. The cilia beat, drawing a stream of water into the pharynx, where microscopic food particles are trapped in a mucous sheet secreted from a structure called an *endostyle.*

The tadpolelike larvae of tunicates plainly exhibit all of the basic characteristics of chordates and mark the tunicates as having the most primitive combination of features found in any chordate (figure 48.5*c*). The larvae do not feed and have a poorly developed gut. They remain free-swimming for only a few days before settling to the bottom and attaching themselves to a suitable substrate by means of a sucker.

Tunicates change so much as they mature and adjust developmentally to a sessile, filter-feeding existence that it would be difficult to discern their evolutionary relationships by examining an adult. Many adult tunicates secrete a **tunic,** a tough sac composed mainly of cellulose. The tunic surrounds the animal and gives the subphylum its name. Cellulose is a substance frequently found in the cell walls of plants and algae but is rarely found in animals. In

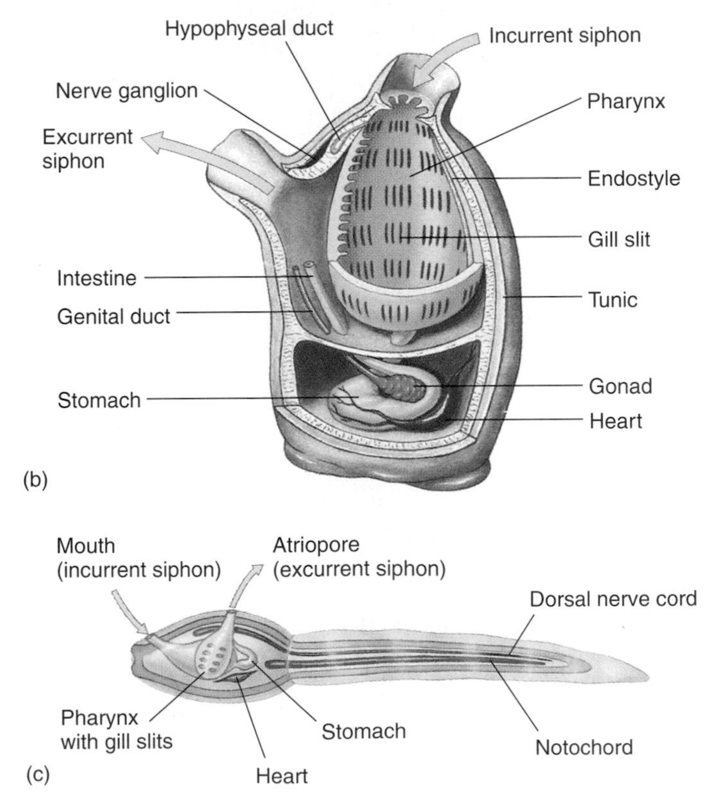

FIGURE 48.5
Tunicates (phylum Chordata, subphylum Urochordata). (*a*) The sea peach, *Halocynthia auranthium.* (*b*) Diagram of the structure of an adult tunicate. (*c*) Diagram of the structure of a larval tunicate, showing the characteristic tadpolelike form. Larval tunicates resemble the postulated common ancestor of the chordates.

colonial tunicates, there may be a common sac and a common opening to the outside. There is a group of Urochordates, the Larvacea, which retains the tail and notochord into adulthood. One theory of vertebrate origins involves a larval form, perhaps that of a tunicate, which acquires the ability to reproduce.

Lancelets

Lancelets are scaleless, fishlike marine chordates a few centimeters long that occur widely in shallow water throughout the oceans of the world. Lancelets (subphylum Cephalochordata) were given their English name because they resemble a lancet—a small, two-edged surgical knife. There are about 23 species of this subphylum. Most of them belong to the genus *Branchiostoma*, formerly called *Amphioxus*, a name still used widely. In lancelets, the notochord runs the entire length of the dorsal nerve cord and persists throughout the animal's life.

FIGURE 48.6
Lancelets. Two lancelets, *Branchiostoma lanceolatum* (phylum Chordata, subphylum Cephalochordata), partly buried in shell gravel, with their anterior ends protruding. The muscle segments are clearly visible; the square objects along the side of the body are gonads.

Lancelets spend most of their time partly buried in sandy or muddy substrates, with only their anterior ends protruding (figure 48.6). They can swim, although they rarely do so. Their muscles can easily be seen as a series of discrete blocks. Lancelets have many more pharyngeal gill slits than fishes, which they resemble in overall shape. They lack pigment in their skin, which has only a single layer of cells, unlike the multilayered skin of vertebrates. The lancelet body is pointed at both ends. There is no distinguishable head or sensory structures other than pigmented light receptors.

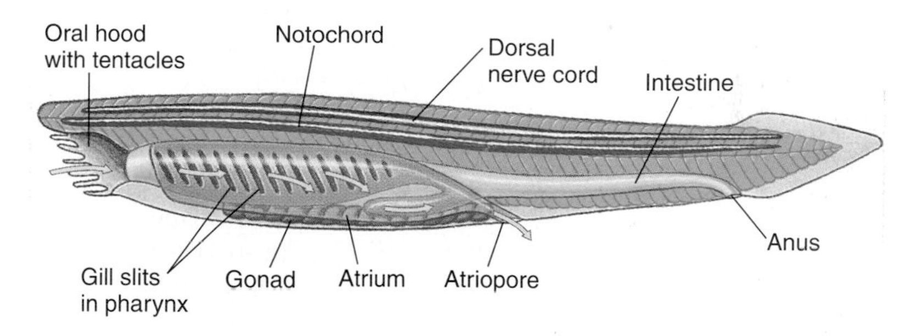

FIGURE 48.7
The structure of a lancelet. This diagram shows the path through which the lancelet's cilia pull water.

Lancelets feed on microscopic plankton, using a current created by beating cilia that lines the oral hood, pharynx, and gill slits (figure 48.7). The gill slits provide an exit for the water and are an adaptation for filter feeding. The oral hood projects beyond the mouth and bears sensory tentacles, which also ring the mouth. Males and females are separate, but no obvious external differences exist between them.

Biologists are not sure whether lancelets are primitive or are actually degenerate fishes whose structural features have been reduced and simplified during the course of evolution. The fact that lancelets feed by means of cilia and have a single-layered skin, coupled with distinctive features of their excretory systems, suggest that this is an ancient group of chordates. The recent discovery of fossil forms similar to living lancelets in rocks 550 million years old—well before the appearance of any fishes—also argues for the antiquity of this group. Recent studies by molecular systematists further support the hypothesis that lancelets are vertebrates' closest ancestors.

Nonvertebrate chordates, including tunicates and lancelets, have notochords but not vertebrae. They are the closest relatives of vertebrates.

48.3 The vertebrates have an interior framework of bone.

Characteristics of Vertebrates

Vertebrates (subphylum Vertebrata) are chordates with a spinal column. The name *vertebrate* comes from the individual bony segments called vertebrae that make up the spine. Vertebrates differ from the tunicates and lancelets in two important respects:

1. **Vertebral column.** In vertebrates, the notochord is replaced during the course of embryonic development by a bony vertebral column. The column is a series of bones that encloses and protects the dorsal nerve cord like a sleeve (figure 48.8).
2. **Head.** In all vertebrates but the earliest fishes, there is a distinct and well-differentiated head, with a skull and brain. For this reason, the vertebrates are sometimes called the **craniate chordates** (Greek *kranion*, "skull").

In addition to these two key characteristics, vertebrates differ from other chordates in other important respects:

1. **Neural crest.** A unique group of embryonic cells called the neural crest contributes to the development of many vertebrate structures. These cells develop on the crest of the neural tube as it forms by an invagination and pinching together of the neural plate (see chapter 60 for a detailed account). Neural crest cells then migrate to various locations in the developing embryo, where they participate in the development of a variety of structures.
2. **Internal organs.** Among the internal organs of vertebrates, livers, kidneys, and endocrine glands are characteristic of the group. The ductless endocrine glands secrete hormones that help regulate many of the body's functions. All vertebrates have a heart and a closed circulatory system. In both their circulatory and their excretory functions, vertebrates differ markedly from other animals.
3. **Endoskeleton.** The endoskeleton of most vertebrates is made of cartilage or bone. Cartilage and bone are specialized tissue containing fibers of the protein collagen compacted together. Bone also contains crystals of a calcium phosphate salt. Bone forms in two stages. First, collagen is laid down in a matrix of fibers along stress lines to provide flexibility, and then calcium minerals infiltrate the fibers, providing rigidity. The great advantage of bone over chitin as a structural material is that bone is strong without being brittle. The vertebrate endoskeleton makes possible the great size and extraordinary powers of movement that characterize this group.

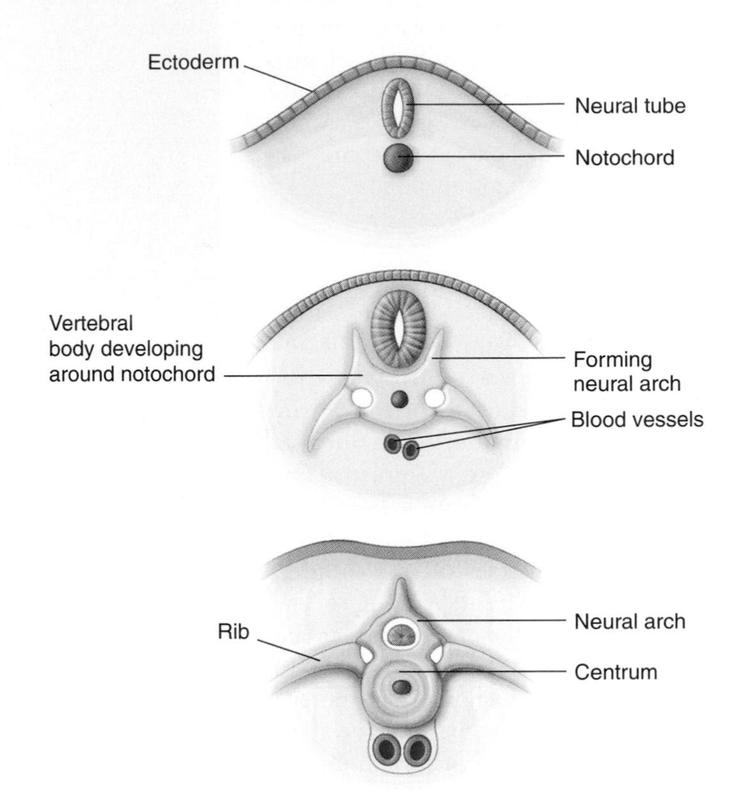

FIGURE 48.8
Embryonic development of a vertebra. During the course of evolution of animal development, the flexible notochord is surrounded and eventually replaced by a cartilaginous or bony covering, the centrum. The neural tube is protected by an arch above the centrum, and the vertebra may also have a hemal arch, which protects major blood vessels below the centrum. The vertebral column functions as a strong, flexible rod that the muscles pull against when the animal swims or moves.

Overview of the Evolution of Vertebrates

The first vertebrates evolved in the oceans about 470 million years ago. They were jawless fishes with a single caudal fin. Many of them looked like a flat hot dog, with a hole at one end and a fin at the other. The appearance of a hinged jaw was a major advancement, opening up new food options, and jawed fishes became the dominant creatures in the sea. Their descendants, the amphibians, invaded the land. Salamander-like amphibians and other, much larger now-extinct amphibians were the first vertebrates to live successfully on land. Amphibians, in turn, gave rise to the first reptiles about 300 million years ago. Within 50 million years, reptiles, better suited than amphibians to living out of water, replaced them as the dominant land vertebrates.

With the success of reptiles, vertebrates truly came to dominate the surface of the earth. Many kinds of reptiles evolved, ranging in size from smaller than a chicken to bigger

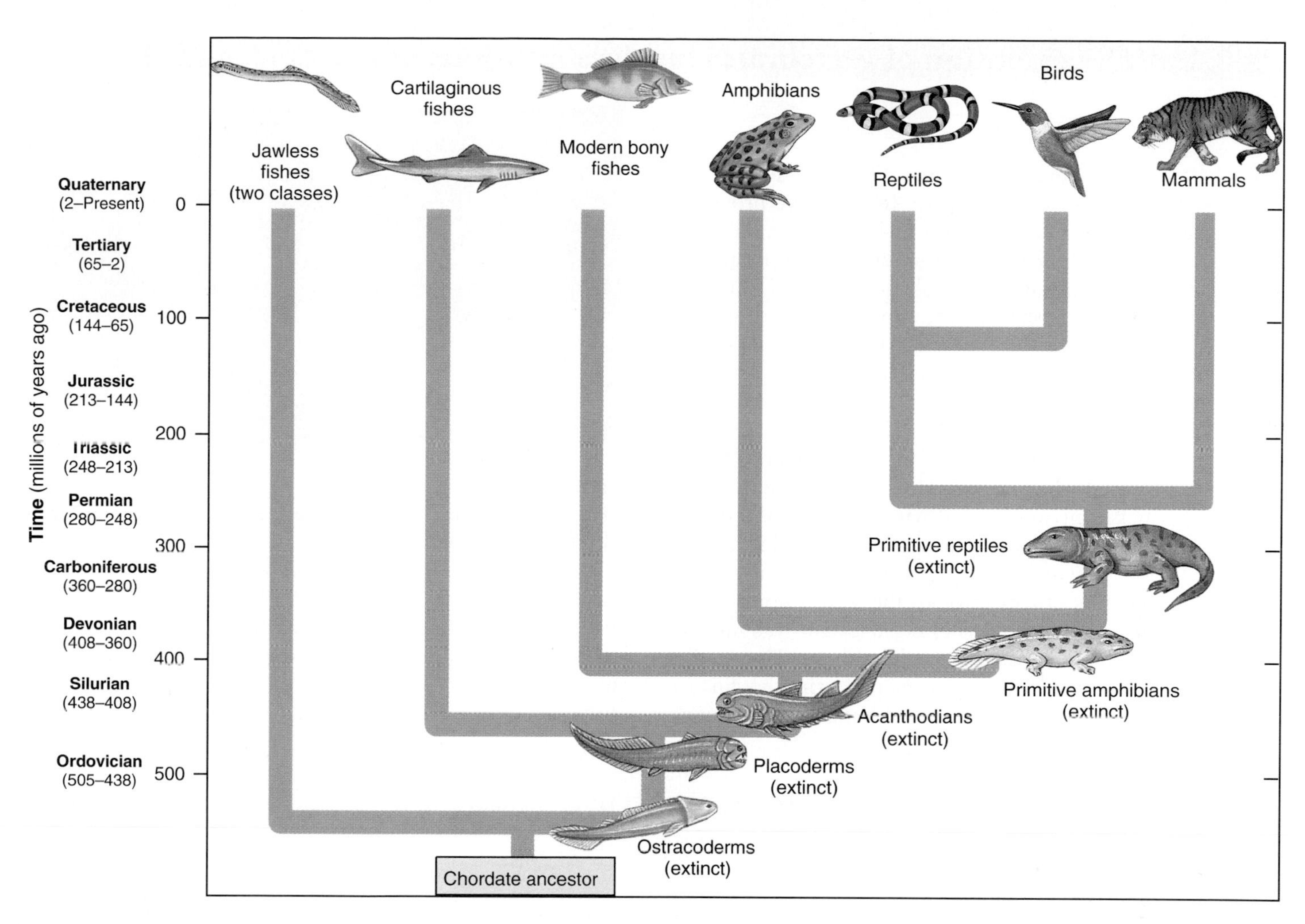

FIGURE 48.9
Vertebrate family tree. Two classes of vertebrates comprise the Agnatha, or jawless fishes. Primitive amphibians arose from fish. Primitive reptiles arose from amphibians and gave rise to mammals and to dinosaurs, which survive today as birds.

than a truck. Some flew, and others swam. Among them evolved reptiles that gave rise to the two remaining great lines of terrestrial vertebrates, birds (descendants of the dinosaurs) and mammals. Dinosaurs and mammals appear at about the same time in the fossil record, 220 million years ago. For over 150 million years, dinosaurs dominated the face of the earth. Over all these centuries (think of it—over *a million centuries!*) the largest mammal was no bigger than a cat. Then, about 65 million years ago, the dinosaurs abruptly disappeared, for reasons that are still hotly debated. In their absence, mammals and birds quickly took their place, becoming in turn abundant and diverse.

The history of vertebrates has been a series of evolutionary advances that have allowed vertebrates to first invade the sea and then the land. In this chapter, we will examine the key evolutionary advances that permitted vertebrates to invade the land successfully. As you will see, this invasion was a staggering evolutionary achievement, involving fundamental changes in many body systems.

Vertebrates are a diverse group, containing members adapted to life in aquatic habitats, on land, and in the air. There are eight principal classes of living vertebrates (figure 48.9). Four of the classes are fishes that live in the water, and four are land-dwelling **tetrapods,** animals with four limbs. (The name *tetrapod* comes from two Greek words meaning "four-footed.") The extant classes of fishes are the superclass Agnatha (the jawless fishes), which includes the class Myxini, the hagfish, and the class Cephalaspidomorphi, the lampreys; Chondrichthyes, the cartilaginous fishes, sharks, skates, and rays; and Osteichthyes, the bony fishes that are dominant today. The four classes of tetrapods are Amphibia, the amphibians; Reptilia, the reptiles; Aves, the birds; and Mammalia, the mammals.

Vertebrates, the principal chordate group, are characterized by a vertebral column and a distinct head.

48.4 The evolution of vertebrates involves invasions of sea, land, and air.

Fishes

Over half of all vertebrates are fishes. The most diverse and successful vertebrate group (figure 48.10), they provided the evolutionary base for invasion of land by amphibians. In many ways, amphibians, the first terrestrial vertebrates, can be viewed as transitional—fish out of water. In fact, fishes and amphibians share many similar features, among the host of obvious differences. First, let us look at the fishes (table 48.1).

The story of vertebrate evolution started in the ancient seas of the Cambrian Period (570 to 505 million years ago), when the first backboned animals appeared (figure 48.11). Wriggling through the water, jawless and toothless, these first fishes sucked up small food particles from the ocean floor like miniature vacuum cleaners. Most were less than a foot long, respired with gills, and had no paired fins—just a primitive tail to push them through the water. For 50 million years, during the Ordovician Period (505 to 438 million years ago), these simple fishes were the only vertebrates. By the end of this period, fish had developed primitive fins to help them swim and massive shields of bone for protection. Jawed fishes first appeared during the Silurian Period (438 to 408 million years ago) and along with them came a new mode of feeding. Later, both the cartilaginous and bony fishes appeared.

FIGURE 48.10
Fish are diverse and include more species than all other kinds of vertebrates combined.

Table 48.1 Major Classes of Fishes

Class	Typical Examples	Key Characteristics	Approximate Number of Living Species
Placodermi	Armored fishes	Jawed fishes with heavily armored heads; often quite large	Extinct
Acanthodii	Spiny fishes	Fishes with jaws; all now extinct; paired fins supported by sharp spines	Extinct
Osteichthyes	Ray-finned fishes	Most diverse group of vertebrates; swim bladders and bony skeletons; paired fins supported by bony rays	20,000
	Lobe-finned fishes	Largely extinct group of bony fishes; ancestral to amphibians; paired lobed fins	7
Chondrichthyes	Sharks, skates, rays	Streamlined hunters; cartilaginous skeletons; no swim bladders; internal fertilization	850
Myxini	Hagfishes	Jawless fishes with no paired appendages; scavengers; mostly blind, but a well-developed sense of smell	43
Cephalaspidomorphi	Lampreys	Largely extinct group of jawless fishes with no paired appendages; parasitic and nonparasitic types; all breed in fresh water	17

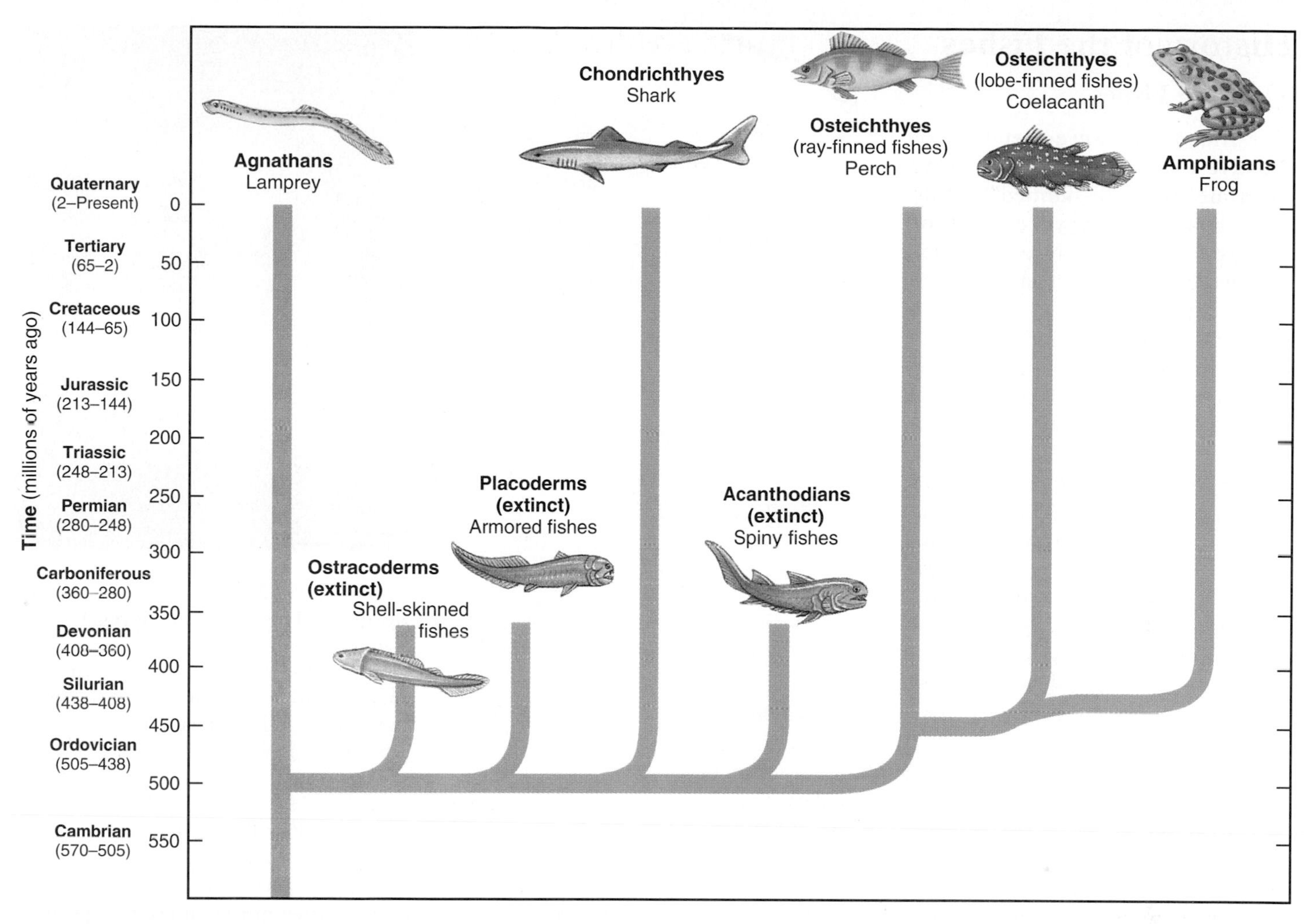

FIGURE 48.11
Evolution of the fishes. The evolutionary relationships among the different groups of fishes as well as between fishes and amphibians is shown. The spiny and armored fishes that dominated the early seas are now extinct.

Characteristics of Fishes

From whale sharks that are 18 meters long to tiny cichlids no larger than your fingernail, fishes vary considerably in size, shape, color, and appearance. Some live in freezing Arctic seas, others in warm freshwater lakes, and still others spend a lot of time out of water entirely. However varied, all fishes have important characteristics in common:

1. **Gills.** Fishes are water-dwelling creatures and must extract oxygen dissolved in the water around them. They do this by directing a flow of water through their mouths and across their gills. The gills are composed of fine filaments of tissue that are rich in blood vessels. They are located at the back of the pharynx and are supported by arches of cartilage. Blood moves through the gills in the opposite direction to the flow of water in order to maximize the efficiency of oxygen absorption.
2. **Vertebral column.** All fishes have an internal skeleton with a spine surrounding the dorsal nerve cord, although it may not necessarily be made of bone. The brain is fully encased within a protective box, the skull or cranium, made of bone or cartilage.
3. **Single-loop blood circulation.** Blood is pumped from the heart to the gills. From the gills, the oxygenated blood passes to the rest of the body, then returns to the heart. The heart is a muscular tube-pump made of four chambers that contract in sequence.
4. **Nutritional deficiencies.** Fishes are unable to synthesize the aromatic amino acids and must consume them in their diet. This inability has been inherited by all their vertebrate descendants.

Fishes were the first vertebrates to make their appearance, and today they vary considerably in appearance and habitat. They are the vertebrate group from which all other vertebrates evolved.

History of the Fishes

The First Fishes

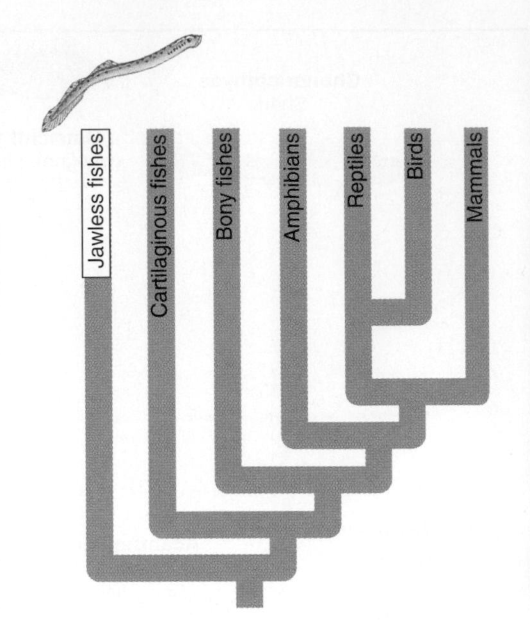

The first fishes were members of the five Ostracoderm orders (the word means "shell-skinned"). Only their head-shields were made of bone; their elaborate internal skeletons were constructed of cartilage. Many ostracoderms were bottom dwellers, with a jawless mouth underneath a flat head, and eyes on the upper surface. Ostracoderms thrived in the Ordovician Period and in the period which followed, the Silurian Period (438 to 408 million years ago), only to become almost completely extinct at the close of the following Devonian Period (408 to 360 million years ago). One group, the jawless Agnatha, survive today as hagfish and parasitic lampreys (figure 48.12).

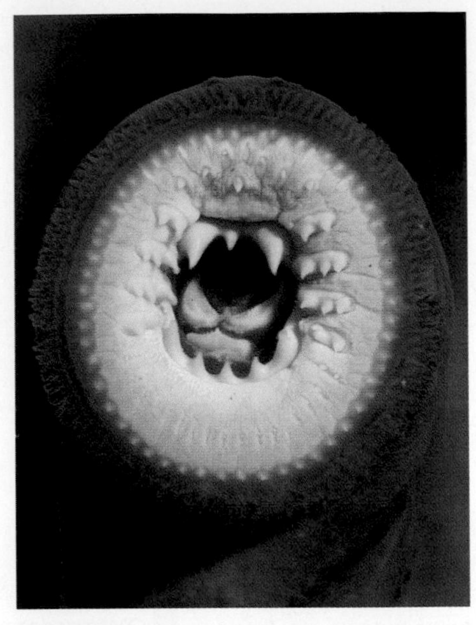

FIGURE 48.12
Specialized mouth of a lamprey. Lampreys use their suckerlike mouths to attach themselves to the fishes on which they prey. When they have done so, they bore a hole in the fish with their teeth and feed on its blood.

A fundamentally important evolutionary advance occurred in the late Silurian Period, 410 million years ago—the development of jaws. Jaws evolved from the most anterior of a series of arch-supports made of cartilage that were used to reinforce the tissue between gill slits, holding the slits open (figure 48.13). This transformation was not as radical as it might at first appear. Each gill arch was formed by a series of several cartilages (later to become bones) arranged somewhat in the shape of a V turned on its side, with the point directed outward. Imagine the fusion of the front pair of arches at top and bottom, with hinges at the points, and you have the primitive vertebrate jaw. The top half of the jaw is not attached to the skull directly except at the rear. Teeth developed on the jaws from modified scales on the skin that lined the mouth.

Armored fishes called placoderms and spiny fishes called acanthodians both had jaws. Spiny fishes were very common during the early Devonian, largely replacing ostracoderms, but became extinct themselves at the close of the Permian. Like ostracoderms, they had internal skeletons made of cartilage, but their scales contained small plates of bone, foreshadowing the much larger role bone would play in the future of vertebrates. Spiny fishes were predators and far better swimmers than ostracoderms, with as many as seven fins to aid them swimming. All of these fins were reinforced with strong spines, giving these fishes their name. No spiny fishes survive today.

By the mid-Devonian, the heavily armored placoderms became common. A very diverse and successful group, seven orders of placoderms dominated the seas of the late Devonian, only to become extinct at the end of that period. The front of the placoderm body was more heavily armored than the rear. The placoderm jaw was much improved from the primitive jaw of spiny fishes, with the upper jaw fused to the skull and the skull hinged on the shoulder. Many of the placoderms grew to enormous sizes, some over 30 feet long, with two-foot skulls that had an enormous bite.

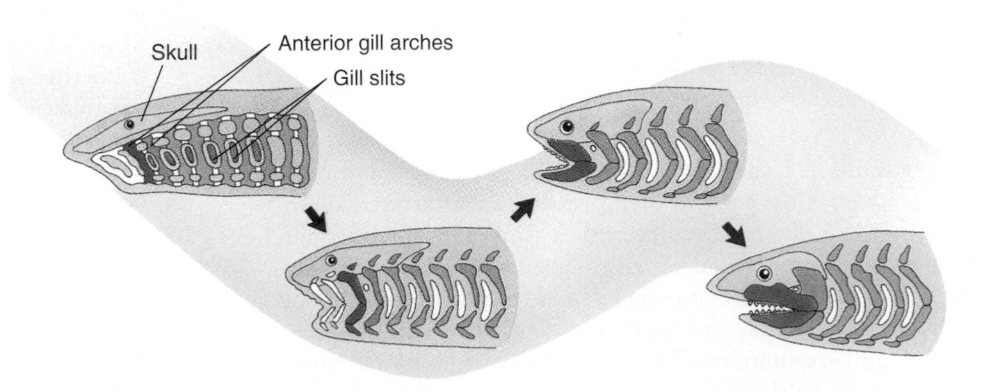

FIGURE 48.13
Evolution of the jaw. Jaws evolved from the anterior gill arches of ancient, jawless fishes.

The Rise of Active Swimmers

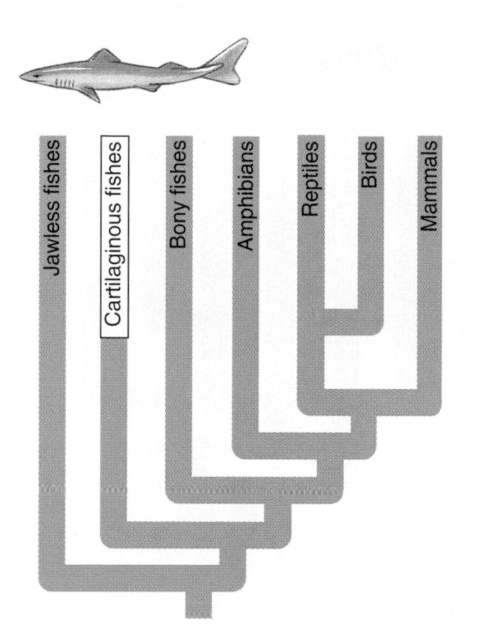

At the end of the Devonian, essentially all of these pioneer vertebrates disappeared, replaced by sharks and bony fishes. Sharks and bony fishes first evolved in the early Devonian, 400 million years ago. In these fishes, the jaw was improved even further, with the first gill arch behind the jaws being transformed into a supporting strut or prop, joining the rear of the lower jaw to the rear of the skull. This allowed the mouth to open very wide, into almost a full circle. In a great white shark, this wide-open mouth can be a very efficient weapon.

The major factor responsible for the replacement of primitive fishes by sharks and bony fishes was that they had a superior design for swimming. The typical shark and bony fish is streamlined. The head of the fish acts as a wedge to cleave through the water, and the body tapers back to the tail, allowing the fish to slip through the water with a minimum amount of turbulence.

In addition, sharks and bony fishes have an array of mobile fins that greatly aid swimming. First, there is a propulsion fin: a large and efficient tail (caudal) fin that helps drive the fish through the water when it is swept side-to-side, pushing against the water and thrusting the fish forward. Second, there are stabilizing fins: one (or sometimes two) dorsal fins on the back that act as a stabilizer to prevent rolling as the fish swims through the water, while another ventral fin acts as a keel to prevent side-slip. Third, there are the paired fins at shoulder and hip ("A fin at each corner"), consisting of a front (pectoral) pair and a rear (pelvic) pair. These fins act like the elevator flaps of an airplane to assist the fish in going up or down through the water, as rudders to help it turn sharply left or right, and as brakes to help it stop quickly.

Sharks Become Top Predators

In the period following the Devonian, the Carboniferous Period (360 to 280 million years ago), sharks became the dominant predator in the sea. Sharks (class Chondrichythes) have a skeleton made of cartilage, like primitive fishes, but it is "calcified," strengthened by granules of calcium carbonate deposited in the outer layers of cartilage. The result is a very light and strong skeleton. Streamlined, with paired fins and a light, flexible skeleton, sharks are superior swimmers (figure 48.14). Their pectoral fins are particularly large, jutting out stiffly like airplane wings—and that is how they function, adding lift to compensate for the downward thrust of the tail fin. Very aggressive predators, some sharks reached enormous size.

FIGURE 48.14
Chondrichthyes. Members of the class Chondrichthyes, such as this bull shark, are mainly predators or scavengers and spend most of their time in graceful motion. As they move, they create a flow of water past their gills, extracting oxygen from the water.

Sharks were among the first vertebrates to develop teeth. These teeth evolved from rough scales on the skin and are not set into the jaw, as yours are, but rather sit atop it. The teeth are not firmly anchored and are easily lost. In a shark's mouth, the teeth are arrayed in up to 20 rows, the teeth in front doing the biting and cutting, while behind them other teeth grow and await their turn. When a tooth breaks or is worn down, a replacement from the next row moves forward. One shark may eventually use more than 20,000 teeth. This programmed loss of teeth offers a great advantage: the teeth in use are always new and sharp. The skin is covered with tiny teeth-like scales, giving it a rough "sandpaper" texture. Like the teeth, these scales are constantly replaced throughout the shark's life.

Reproduction among the Chondrichythes is the most advanced of any fishes. Shark eggs are fertilized internally. During mating, the male grasps the female with modified fins called claspers. Sperm run from the male into the female through grooves in the claspers. Although a few species lay fertilized eggs, the eggs of most species develop within the female's body, and the pups are born alive.

Many of the early evolutionary lines of sharks died out during the great extinction at the end of the Permian Period (280 to 248 million years ago). The survivors thrived and underwent a burst of diversification during the Mesozoic era, when most of the modern groups of sharks appeared. Skates and rays (flattened sharks that are bottom-dwellers) evolved at this time, some 200 million years after the sharks first appeared. Sharks competed successfully with the marine reptiles of that time and are still the dominant predators of the sea. Today there are 275 species of sharks, more kinds than existed in the Carboniferous.

Bony Fishes Dominate the Water

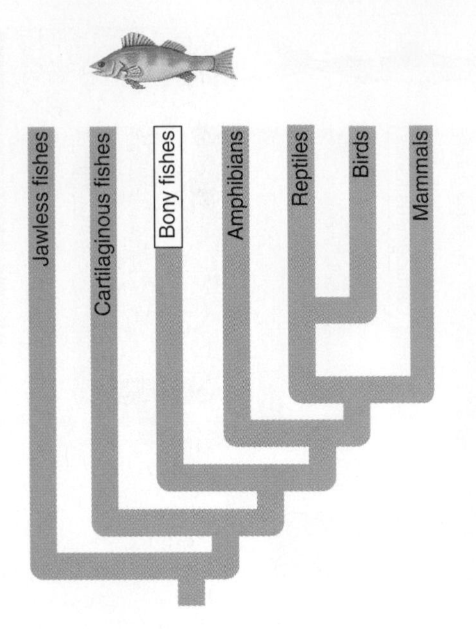

Bony fishes (members of the class Osteichthyes, figure 48.15) evolved at the same time as sharks, some 400 million years ago, but took quite a different evolutionary road. Instead of gaining speed through lightness, as sharks did, bony fishes adopted a heavy internal skeleton made completely of bone. Such an internal skeleton is very strong, providing a base against which very strong muscles could pull. The process of *ossification* (the evolutionary replacement of cartilage by bone) happened suddenly in evolutionary terms, completing a process started by sharks, who lay down a thin film of bone over their cartilage. Not only is the internal skeleton ossified, but also the external skeleton, the outer covering of plates and scales. Many scientists believe bony fishes evolved from spiny sharks, which also had bony plates set in their skin. Bony fishes are the most successful of all fishes, indeed of all vertebrates. There are several dozen orders containing more than 20,000 living species.

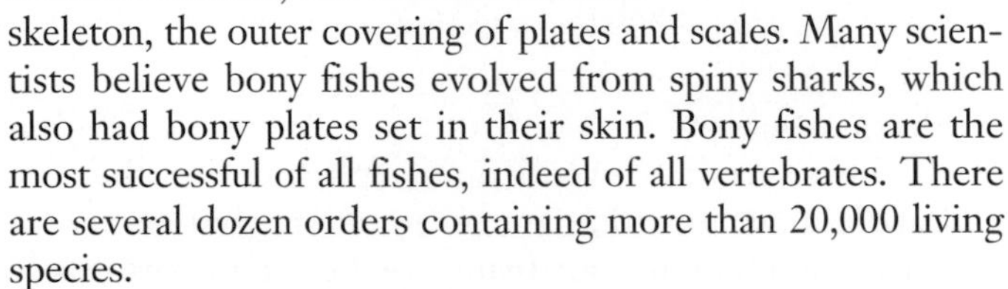

Unlike sharks, bony fishes evolved in fresh water. The most ancient fossils of bony fishes are found in freshwater lake beds from the middle Devonian. These first bony fishes were small and possessed paired air sacs connected to the back of the throat. These sacs could be inflated with air to buoy the fish up or deflated to sink it down in the water.

Most bony fishes have highly mobile fins, very thin scales, and completely symmetrical tails (which keep the fish on a straight course as it swims through the water). This is a very successful design for a fish. Two great groups arose from these pioneers: the lobe-finned fishes, ancestors of the first tetrapods, and the ray-finned fishes, which include the vast majority of today's fishes.

The characteristic feature of all ray-finned fishes is an internal skeleton of parallel bony rays that support and stiffen each fin. There are no muscles within the fins; they are moved by muscles within the body. In ray-finned fishes, the primitive air sacs are transformed into an air pouch, which provides a remarkable degree of control over buoyancy.

Important Adaptations of Bony Fishes

The remarkable success of the bony fishes has resulted from a series of significant adaptations that have enabled them to dominate life in the water. These include the swim bladder, lateral line system, and gill cover.

Swim Bladder. Although bones are heavier than cartilaginous skeletons, bony fishes are still buoyant because they possess a swim bladder, a gas-filled sac that allows them to regulate their buoyant density and so remain suspended at any depth in the water effortlessly (figure 48.16). Sharks, by contrast, must move through the water or sink, as their bodies are denser than water. In primitive bony fishes, the swim bladder is a ventral outpocketing of the pharynx behind the throat, and these species fill the swim bladder by simply gulping air at the surface of the water. In most of today's bony fishes, the swim bladder is an independent organ that is filled and drained of gases, mostly nitrogen and oxygen, internally. How do bony fishes manage this remarkable trick? It turns out that the gases are released from their blood. Gas exchange occurs across the wall of the swim bladder and the blood vessels located near the swim bladder. A variety of physiological factors controls the exchange of gases between the blood stream and the swim bladder.

Lateral Line System. Although precursors are found in sharks, bony fishes possess a fully developed lateral line system. The lateral line system consists of a series of sensory organs that project into a canal beneath the surface of the skin. The canal runs the length of the fish's body and is open to the exterior through a series of sunken pits. Movement of

FIGURE 48.15
Bony fishes. The bony fishes (class Osteichthyes) are extremely diverse. This Korean angelfish in Fiji is one of the many striking fishes that live around coral reefs in tropical seas.

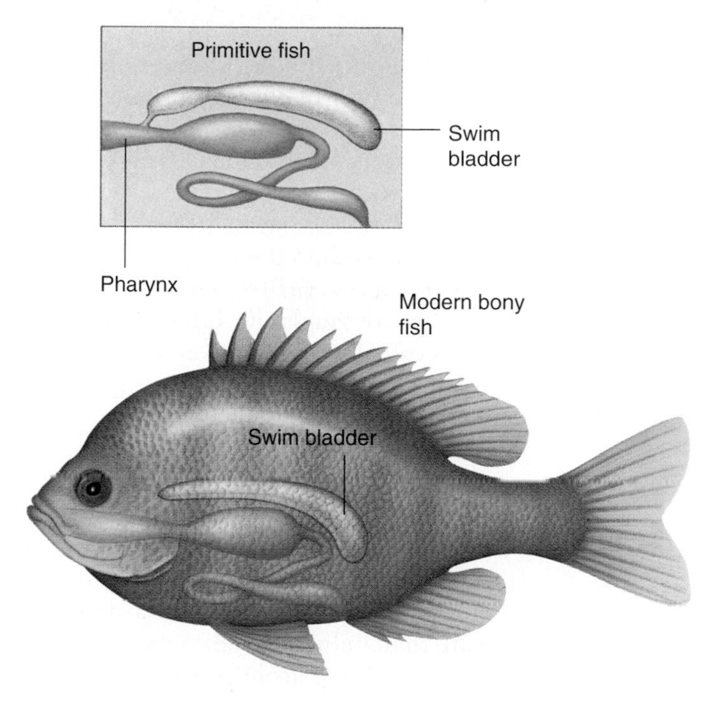

FIGURE 48.16
Diagram of a swim bladder. The bony fishes use this structure, which evolved as a ventral outpocketing of the pharynx, to control their buoyancy in water.

FIGURE 48.17
The living coelacanth, *Latimeria chalumnae*. Discovered in the western Indian Ocean in 1938, this coelacanth represents a group of fishes that had been thought to be extinct for about 70 million years. Scientists who studied living individuals in their natural habitat at depths of 100 to 200 meters observed them drifting in the current and hunting other fishes at night. Some individuals are nearly 3 meters long; they have a slender, fat-filled swim bladder. *Latimeria* is a strange animal, and its discovery was a complete surprise.

water past the fish forces water through the canal. The sensory organs consist of clusters of cells with hairlike projections called cilia, embedded in a gelatinous cap. The hairs are deflected by the slightest movement of water over them. The pits are oriented so that some are stimulated no matter what direction the water moves (see chapter 55). Nerve impulses from these sensory organs permit the fish to assess its rate of movement through water, sensing the movement as pressure waves against its lateral line. This is how a trout orients itself with its head upstream.

The lateral line system also enables a fish to detect motionless objects at a distance by the movement of water reflected off the object. In a very real sense, this is the fish equivalent of hearing. The basic mechanism of cilia deflection by pressure waves is very similar to what happens in human ears (see chapter 55).

Gill Cover. Most bony fishes have a hard plate called the operculum that covers the gills on each side of the head. Flexing the operculum permits bony fishes to pump water over their gills. The gills are suspended in the pharyngeal slits that form a passageway between the pharynx and the outside of the fish's body. When the operculum is closed, it seals off the exit. When the mouth is open, closing the operculum increases the volume of the mouth cavity, so that water is drawn into the mouth. When the mouth is closed, opening the operculum decreases the volume of the mouth cavity, forcing water past the gills to the outside. Using this very efficient bellows, bony fishes can pass water over the gills while stationary in the water. That is what a goldfish is doing when it seems to be gulping in a fish tank.

The Path to Land

Lobe-finned fishes (figure 48.17) evolved 390 million years ago, shortly after the first bony fishes appeared. Only seven species survive today, a single species of coelacanth and six species of lungfish. Lobe-finned fishes have paired fins that consist of a long fleshy muscular lobe (hence their name), supported by a central core of bones that form fully articulated joints with one another. There are bony rays only at the tips of each lobed fin. Muscles within each lobe can move the fin rays independently of one another, a feat no ray-finned fish could match. Although rare today, lobe-finned fishes played an important part in the evolutionary story of vertebrates. Amphibians almost certainly evolved from the lobe-finned fishes.

Fishes are characterized by gills and a simple, single-loop circulatory system. Cartilaginous fishes, such as sharks, are fast swimmers, while the very successful bony fishes have unique characteristics such as swim bladders and lateral line systems.

Amphibians

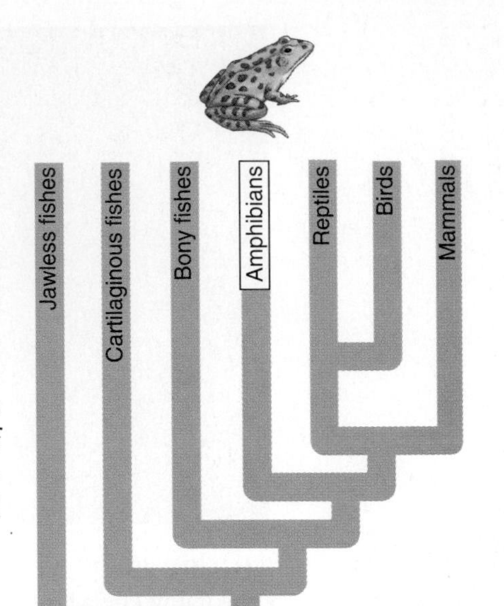

Frogs, salamanders, and caecilians, the damp-skinned vertebrates, are direct descendants of fishes. They are the sole survivors of a very successful group, the amphibians, the first vertebrates to walk on land. Most present-day amphibians are small and live largely unnoticed by humans. Amphibians are among the most numerous of terrestrial animals; there are more species of amphibians than of mammals. Throughout the world amphibians play key roles in terrestrial food chains.

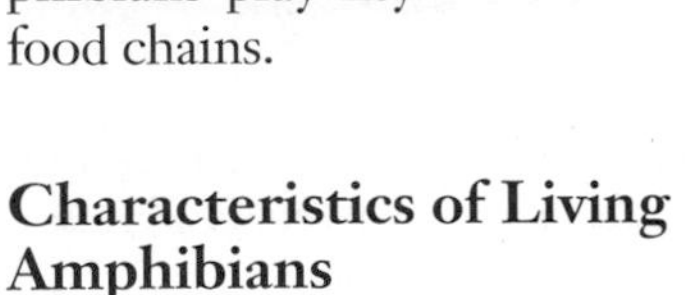

Characteristics of Living Amphibians

Biologists have classified living species of amphibians into three orders (table 48.2): 3680 species of frogs and toads in 22 families make up the order Anura ("without a tail"); 369 species of salamanders and newts in 9 families make up the order Urodela or Caudata ("visible tail"); and 168 species (6 families) of wormlike, nearly blind organisms called caecilians that live in the tropics make up the order Apoda or Gymnophiona ("without legs"). They have key characteristics in common:

1. **Legs.** Frogs and salamanders have four legs and can move about on land quite well. Legs were one of the key adaptations to life on land. Caecilians have lost their legs during the course of adapting to a burrowing existence.
2. **Cutaneous respiration.** Frogs, salamanders, and caecilians all supplement the use of lungs by respiring directly across their skin, which is kept moist and provides an extensive surface area. This mode of respiration is only efficient for a high surface-to-volume ratio in an animal.
3. **Lungs.** Most amphibians possess a pair of lungs, although the internal surfaces are poorly developed, with much less surface area than reptilian or mammalian lungs. Amphibians still breathe by lowering the floor of the mouth to suck air in, then raising it back to force the air down into the lungs.
4. **Pulmonary veins.** After blood is pumped through the lungs, two large veins called pulmonary veins return the aerated blood to the heart for repumping. This allows the aerated blood to be pumped to the tissues at a much higher pressure than when it leaves the lungs.
5. **Partially divided heart.** The initial chamber of the fish heart is absent in amphibians, and the second and last chambers are separated by a dividing wall that helps prevent aerated blood from the lungs from mixing with nonaerated blood being returned to the heart from the rest of the body. This separates the blood circulation into two separate paths, pulmonary and systemic. The separation is imperfect; the third chamber has no dividing wall.

Several other specialized characteristics are shared by all present-day amphibians. In all three orders, there is a zone of weakness between the base and the crown of the teeth. They also have a peculiar type of sensory rod cell in the retina of the eye called a "green rod." The exact function of this rod is unknown.

Amphibians, with legs and more efficient blood circulation than fishes, were the first vertebrates to walk on land.

Table 48.2 Orders of Amphibians

Order	Typical Examples	Key Characteristics	Approximate Number of Living Species
Anura	Frogs, toads	Compact tailless body; large head fused to the trunk; rear limbs specialized for jumping	3680
Caudata	Salamanders, newts	Slender body; long tail and limbs set out at right angles to the body	369
Apoda (Gymnophiona)	Caecilians	Tropical group with a snakelike body; no limbs; little or no tail	168

History of the Amphibians

The word *amphibia* (a Greek word meaning "both lives") nicely describes the essential quality of modern day amphibians, referring to their ability to live in two worlds: the aquatic world of their fish ancestors and in the terrestrial world that they first invaded. In this section, we will review the checkered history of this group, almost all of whose members have been extinct for the last 200 million years. Then, in the following section, we will examine in more detail what the few kinds of surviving amphibians are like.

Origin of Amphibians

Paleontologists (scientists who study fossils) agree that amphibians must have evolved from the lobe-finned fishes, although for some years there has been considerable disagreement about whether the direct ancestors were coelacanths, lungfish, or the extinct rhipidistian fishes. Good arguments can be made for each. Many details of amphibian internal anatomy resemble those of the coelacanth. Lungfish and rhipidistians have openings in the tops of their mouths similar to the internal nostrils of amphibians. In addition, lungfish have paired lungs, like those of amphibians. Recent DNA analysis indicates lungfish are in fact far more closely related to amphibians than are coelacanths. Most paleontologists consider that amphibians evolved from rhipidistian fishes, rather than lungfish, because the pattern of bones in the early amphibian skull and limbs bears a remarkable resemblance to the rhipidistians. They also share a particular tooth structure.

The successful invasion of land by vertebrates involved a number of major adaptations:

1. Legs were necessary to support the body's weight as well as to allow movement from place to place (figure 48.18).
2. Lungs were necessary to extract oxygen from air. Even though there is far more oxygen available to gills in air than in water, the delicate structure of fish gills requires the buoyancy of water to support them and they will not function in air.
3. The heart had to be redesigned to make full use of new respiratory systems and to deliver the greater amounts of oxygen required by walking muscles.
4. Reproduction had to be carried out in water until methods evolved to prevent eggs from drying out.
5. Most importantly, a system had to be developed to prevent the body itself from drying out.

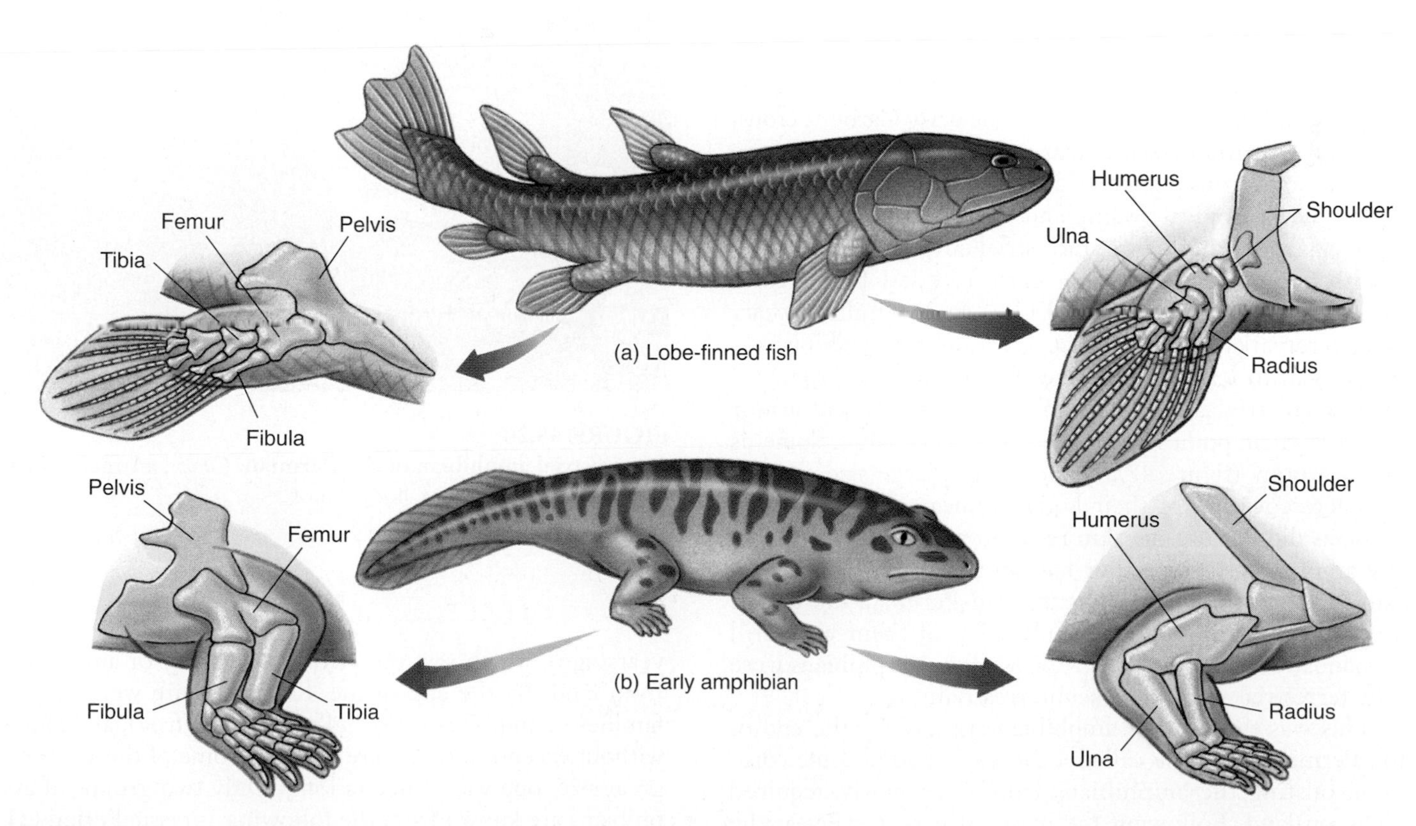

FIGURE 48.18
A comparison between the limbs of a lobe-finned fish and those of a primitive amphibian. (*a*) A lobe-finned fish. Some of these animals could probably move onto land. (*b*) A primitive amphibian. As illustrated by their skeletal structure, the legs of such an animal could clearly function on land much better than the fins of the lobe-finned fish.

The First Amphibian

Amphibians solved these problems only partially, but their solutions worked well enough that amphibians have survived for 350 million years. Evolution does not insist on perfect solutions, only workable ones.

Ichthyostega, the earliest amphibian fossil (figure 48.19) was found in a 370-million-year-old rock in Greenland. At that time, Greenland was part of the North American continent and lay near the equator. For the next 100 million years, all amphibian fossils are found in North America. Only when Asia and the southern continents all merged with North America to form the supercontinent Pangaea did amphibians spread throughout the world.

Ichthyostega was a strongly built animal, with four sturdy legs well supported by hip and shoulder bones. The shoulder bones no longer attached to the skull as in fish. The hipbones were braced against the backbone unlike in fish, so the limbs could support the animal's weight. To strengthen the backbone further, long, broad ribs that overlap each other formed a solid cage for the lungs and heart. The rib cage was so solid that it probably couldn't expand and contract for breathing. Instead, *Ichthyostega* obtained oxygen somewhat as a frog does, by lowering the floor of the mouth to draw air in, then raising it to push air down the windpipe into the lungs.

FIGURE 48.19
Amphibians were the first vertebrates to walk on land. Reconstruction of *Ichthyostega*, one of the first amphibians with efficient limbs for crawling on land, an improved olfactory sense associated with a lengthened snout, and a relatively advanced ear structure for picking up airborne sounds. Despite these features, *Ichthyostega*, which lived about 350 million years ago, was still quite fishlike in overall appearance and represents a very early amphibian.

The Rise and Fall of Amphibians

Amphibians first became common during the Carboniferous Period (360 to 280 million years ago). Fourteen families of amphibians are known from the early Carboniferous, nearly all aquatic or semiaquatic, like *Ichthyostega*. By the late Carboniferous, much of North America was covered by low-lying tropical swamplands, and 34 families of amphibians thrived in this wet terrestrial environment, sharing it with pelycosaurs and other early reptiles. In the early Permian Period that followed (280 to 248 million years ago), a remarkable change occurred among amphibians—they began to leave the marshes for dry uplands. Many of these terrestrial amphibians had bony plates and armor covering their bodies and grew to be very large, some as big as a pony (figure 48.20). Both their large size and the complete covering of their bodies indicate that these amphibians did not use the skin respiratory system of present-day amphibians, but rather had an impermeable leathery skin to prevent water loss. By the mid-Permian, there were 40 families of amphibians. Only 25% of them were still semiaquatic like Ichthyostega; 60% of the amphibians were fully terrestrial, 15% were semiterrestrial.

FIGURE 48.20
A terrestrial amphibian of the Permian. *Cacops*, a large, extinct amphibian, had extensive body armor.

This was the peak of amphibian success. By the end of the Permian, a reptile called a therapsid had become common, ousting the amphibians from their newly acquired niche on land. Following the mass extinction event at the end of the Permian, therapsids were the dominant land vertebrate and most amphibians were aquatic. This trend continued in the following Triassic Period (248 to 213 million years ago), which saw the virtual extinction of amphibians from land. By the end of the Triassic, there were only 15 families of amphibians (including the first frog), and almost without exception they were aquatic. Some of these grew to great size; one was 3 meters long. Only two groups of amphibians are known from the following Jurassic Period (213 to 144 million years ago), the anurans (frogs and toads) and the urodeles (salamanders and newts). The Age of Amphibians was over.

Amphibians Today

All of today's amphibians descended from the two families of amphibians that survived the Age of the Dinosaurs. During the Tertiary Period (65 to 2 million years ago), these moist-skinned amphibians underwent a highly successful invasion of wet habitats all over the world, and today there are over 4200 species of amphibians in 37 different families.

Anura. Frogs and toads, amphibians without tails, live in a variety of environments from deserts and mountains to ponds and puddles (figure 48.21*a*). Frogs have smooth, moist skin, a broad body, and long hind legs that make them excellent jumpers. Most frogs live in or near water, although some tropical species live in trees. Unlike frogs, toads have a dry, bumpy skin, short legs, and are well adapted to dry environments. All adult anurans are carnivores, eating a wide variety of invertebrates.

Most frogs and toads return to water to reproduce, laying their eggs directly in water. Their eggs lack water-tight external membranes and would dry out quickly out of the water. Eggs are fertilized externally and hatch into swimming larval forms called tadpoles. Tadpoles live in the water, where they generally feed on minute algae. After considerable growth, the body of the tadpole gradually changes into that of an adult frog. This process of abrupt change in body form is called **metamorphosis.**

Urodela (Caudata). Salamanders have elongated bodies, long tails, and smooth moist skin (figure 48.21*b*). They typically range in length from a few inches to a foot, although giant Asiatic salamanders of the genus *Andrias* are as much as 1.5 meters long and weigh up to 33 kilograms. Most salamanders live in moist places, such as under stones or logs, or among the leaves of tropical plants. Some salamanders live entirely in water.

Salamanders lay their eggs in water or in moist places. Fertilization is usually external, although a few species practice a type of internal fertilization in which the female picks up sperm packets deposited by the male. Unlike anurans, the young that hatch from salamander eggs do not undergo profound metamorphosis, but are born looking like small adults and are carnivorous.

Apoda (Gymnophiona). Caecilians, members of the order Apoda (Gymnophiona), are a highly specialized group of tropical burrowing amphibians (figure 48.21*c*). These legless, wormlike creatures average about 30 centimeters long, but can be up to 1.3 meters long. They have very small eyes and are often blind. They resemble worms but have jaws with teeth. They eat worms and other soil invertebrates. The caecilian male deposits sperm directly into the female, and the female usually bears live young. Mud eels, small amphibians with tiny forelimbs and no hind limbs that live in the eastern United States, are not apodans, but highly specialized urodelians.

(a)

(b)

(c)

FIGURE 48.21
Class Amphibia. (*a*) Red-eyed tree frog, *Agalychnis callidryas* (order Anura). (*b*) An adult barred tiger salamander, *Ambystoma tigrinum* (order Caudata). (*c*) A caecilian, *Caecilia tentaculata* (order Gymnophiona).

Amphibians ventured onto land some 370 million years ago. They are characterized by moist skin, legs (secondarily lost in some species), lungs (usually), and a more complex and divided circulatory system. They are still tied to water for reproduction.

Reptiles

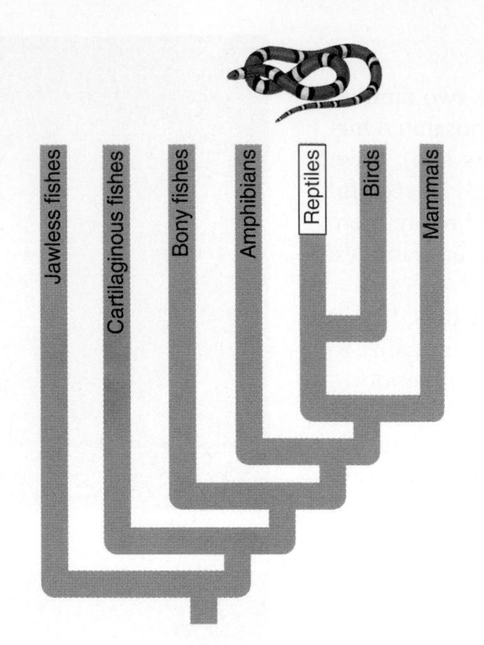

If one thinks of amphibians as a first draft of a manuscript about survival on land, then reptiles are the finished book. For each of the five key challenges of living on land, reptiles improved on the innovations first seen in amphibians. Legs were arranged to support the body's weight more effectively, allowing reptile bodies to be bigger and to *run*. Lungs and heart were altered to make them more efficient. The skin was covered with dry plates or scales to minimize water loss, and eggs were encased in watertight covers (figure 48.22). Reptiles were the first truly *terrestrial* vertebrates.

Over 7000 species of reptiles (class Reptilia) now live on earth (table 48.3). They are a highly successful group in today's world, more common than mammals. There are three reptile species for every two mammal species. While it is traditional to think of reptiles as more primitive than mammals, the great majority of reptiles that live today evolved from lines that appeared after therapsids did (the line that leads directly to mammals).

Key Characteristics of Reptiles

All living reptiles share certain fundamental characteristics, features they retain from the time when they replaced amphibians as the dominant terrestrial vertebrates. Among the most important are:

1. **Amniotic egg**. Amphibians never succeeded in becoming fully terrestrial because amphibian eggs must be laid in water to avoid drying out. Most reptiles lay watertight eggs that contain a food source (the yolk) and a series of four membranes—the yolk sac, the amnion, the allantois, and the chorion (figure 48.22). Each membrane plays a role in making the egg an independent life-support system. The outermost membrane of the egg is the **chorion,** which lies just beneath the porous shell. It allows respiratory gases to pass through, but retains water within the egg. Within, the **amnion** encases the developing embryo within a fluid-filled cavity. The **yolk sac** provides food from the yolk for the embryo via blood vessels connecting to the embryo's gut. The **allantois** surrounds a cavity into which waste products from the embryo are excreted. All modern reptiles (as well as birds and mammals) show exactly this same pattern of membranes within the egg. These three classes are called amniotes.
2. **Dry skin.** Living amphibians have a moist skin and must remain in moist places to avoid drying out. Reptiles have dry, watertight skin. A layer of scales or armor covers their bodies, preventing water loss. These scales develop as surface cells fill with keratin, the same protein that forms claws, fingernails, hair, and bird feathers.
3. **Thoracic breathing.** Amphibians breathe by squeezing their throat to pump air into their lungs; this limits their breathing capacity to the volume of their mouth. Reptiles developed pulmonary breathing, expanding and contracting the rib cage to suck air into the lungs and then force it out. The capacity of this system is limited only by the volume of the lungs.

Reptiles were the first vertebrates to completely master the challenge of living on dry land.

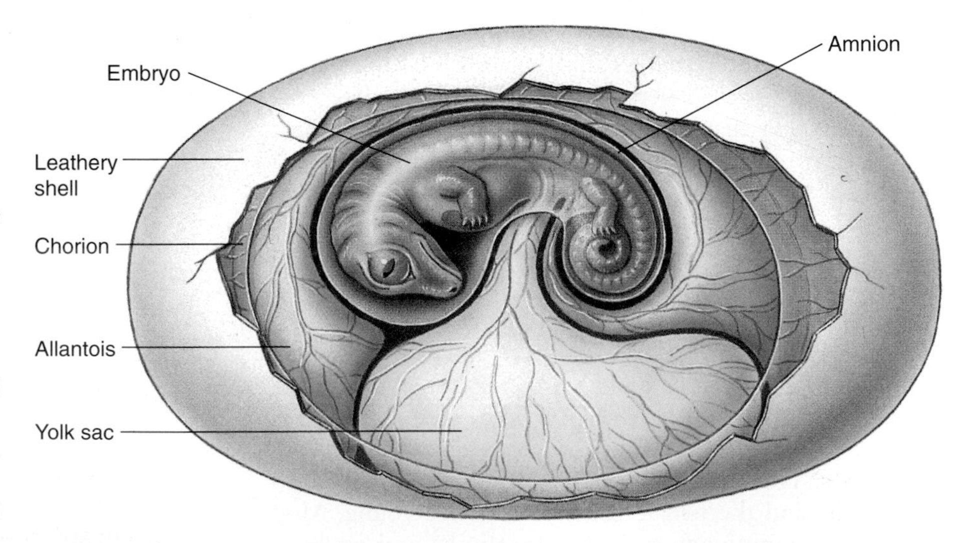

FIGURE 48.22
The watertight egg. The amniotic egg is perhaps the most important feature that allows reptiles to live in a wide variety of terrestrial habitats.

Order	Typical Examples	Key Characteristics	Approximate Number of Living Species
Ornithischia	Stegosaur	 Dinosaurs with two pelvic bones facing backward, like a bird's pelvis; herbivores, with turtlelike upper beak; legs under body	Extinct
Saurischia	Tyrannosaur	 Dinosaurs with one pelvic bone facing forward, the other back, like a lizard's pelvis; both plant- and flesh-eaters; legs under body	Extinct
Pterosauria	Pterosaur	 Flying reptiles; wings were made of skin stretched between fourth fingers and body; wingspans of early forms typically 60 centimeters, later forms nearly 8 meters	Extinct
Plesiosaura	Plesiosaur	 Barrel-shaped marine reptiles with sharp teeth and large, paddle-shaped fins; some had snakelike necks twice as long as their bodies	Extinct
Ichthyosauria	Ichthyosaur	 Streamlined marine reptiles with many body similarities to sharks and modern fishes	Extinct
Squamata, suborder Sauria	Lizards	 Lizards; limbs set at right angles to body; anus is in transverse (sideways) slit; most are terrestrial	3800
Squamata, suborder Serpentes	Snakes	 Snakes; no legs; move by slithering; scaly skin is shed periodically; most are terrestrial	3000
Chelonia	Turtles, tortoises, sea turtles	 Ancient armored reptiles with shell of bony plates to which vertebrae and ribs are fused; sharp, horny beak without teeth	250
Crocodylia	Crocodiles, alligators, gavials, caimans	 Advanced reptiles with four-chambered heart and socketed teeth; anus is a longitudinal (lengthwise) slit; closest living relatives to birds	25
Rhynchocephalia	Tuataras	 Sole survivors of a once successful group that largely disappeared before dinosaurs; fused, wedgelike, socketless teeth; primitive third eye under skin of forehead	2

Table 48.3 Major Orders of Reptiles

The Rise and Fall of Dominant Reptile Groups

During the 250 million years that reptiles were the dominant large terrestrial vertebrates, four major forms of reptiles took turns as the dominant type: pelycosaurs, therapsids, thecodonts, and dinosaurs.

Pelycosaurs: Becoming a Better Predator

Early reptiles like *pelycosaurs* were better adapted to life on dry land than amphibians because they evolved watertight eggs. They had powerful jaws because of an innovation in skull design and muscle arrangement. Pelycosaurs were **synapsids,** meaning that their skulls had a pair of temporal holes behind the openings for the eyes. An important feature of reptile classification is the presence and number of openings behind the eyes (see figure 48.27). Their jaw muscles were anchored to these holes, which allowed them to bite more powerfully. An individual pelycosaur weighed about 200 kilograms. With long, sharp, "steak knife" teeth, pelycosaurs were the first land vertebrates to kill beasts their own size (figure 48.23). Dominant for 50 million years, pelycosaurs once made up 70% of all land vertebrates. They died out about 250 million years ago, replaced by their direct descendants—the therapsids.

FIGURE 48.23
A pelycosaur. *Dimetrodon*, a carnivorous pelycosaur, had a dorsal sail that is thought to have been used to dissipate body heat or gain it by basking.

Therapsids: Speeding Up Metabolism

Therapsids (figure 48.24) ate ten times more frequently than their pelycosaur ancestors. There is evidence that they may have been endotherms, able to regulate their own body temperature. The extra food consumption would have been necessary to produce body heat. This would have permitted therapsids to be far more active than other vertebrates of that time, when winters were cold and long. For 20 million years, therapsids (also called "mammallike reptiles") were the dominant land vertebrate, until largely replaced 230 million years ago by a cold-blooded, or ectothermic, reptile line—the thecodonts. Therapsids became extinct 170 million years ago, but not before giving rise to their descendants—the mammals.

FIGURE 48.24
A therapsid. This small weaslelike cynodont therapsid, *Megazostrodon*, may have had fur. From the late Triassic, it is so similar to modern mammals that some paleontologists consider it the first mammal.

Thecodonts: Wasting Less Energy

Thecodonts were **diapsids,** their skulls having two pairs of temporal holes, and like amphibians and early reptiles, they were ectotherms (figure 48.25). Thecodonts largely replaced therapsids when the world's climate warmed 230 million years ago. In the warm climate, the therapsid's endothermy no longer offered a competitive advantage, and ectothermic thecodonts needed only a tenth as much food. Thecodonts were the first land vertebrates to be bipedal—to stand and walk on two feet. They were dominant through the Triassic and survived for 15 million years, until replaced by their direct descendants—the dinosaurs.

FIGURE 48.25
A thecodont. *Euparkeria*, a thecodont, had rows of bony plates along the sides of the backbone, as seen in modern crocodiles and alligators.

Dinosaurs: Learning to Run Upright

Dinosaurs evolved from thecodonts about 220 million years ago. Unlike the thecodonts, their legs were positioned directly underneath their bodies, a significant improvement in body design (figure 48.26). This design placed the weight of the body directly over the legs, which allowed dinosaurs to run with great speed and agility. A dinosaur fossil can be distinguished from a thecodont fossil by the presence of a hole in the side of the hip socket. Because the dinosaur leg is positioned underneath the socket, the force is directed upward, not inward, so there was no need for bone on the side of the socket. Dinosaurs went on to become the most successful of all land vertebrates, dominating for 150 million years. All dinosaurs became extinct rather abruptly 65 million years ago, apparently as a result of an asteroid's impact.

Figures 48.27 and 48.28 summarize the evolutionary relationships among the extinct and living reptiles.

FIGURE 48.26
The largest mounted dinosaur in the world. This 145-million-year-old *Brachiosaurus*, a plant-eating sauropod over 80 feet long, lived in East Africa.

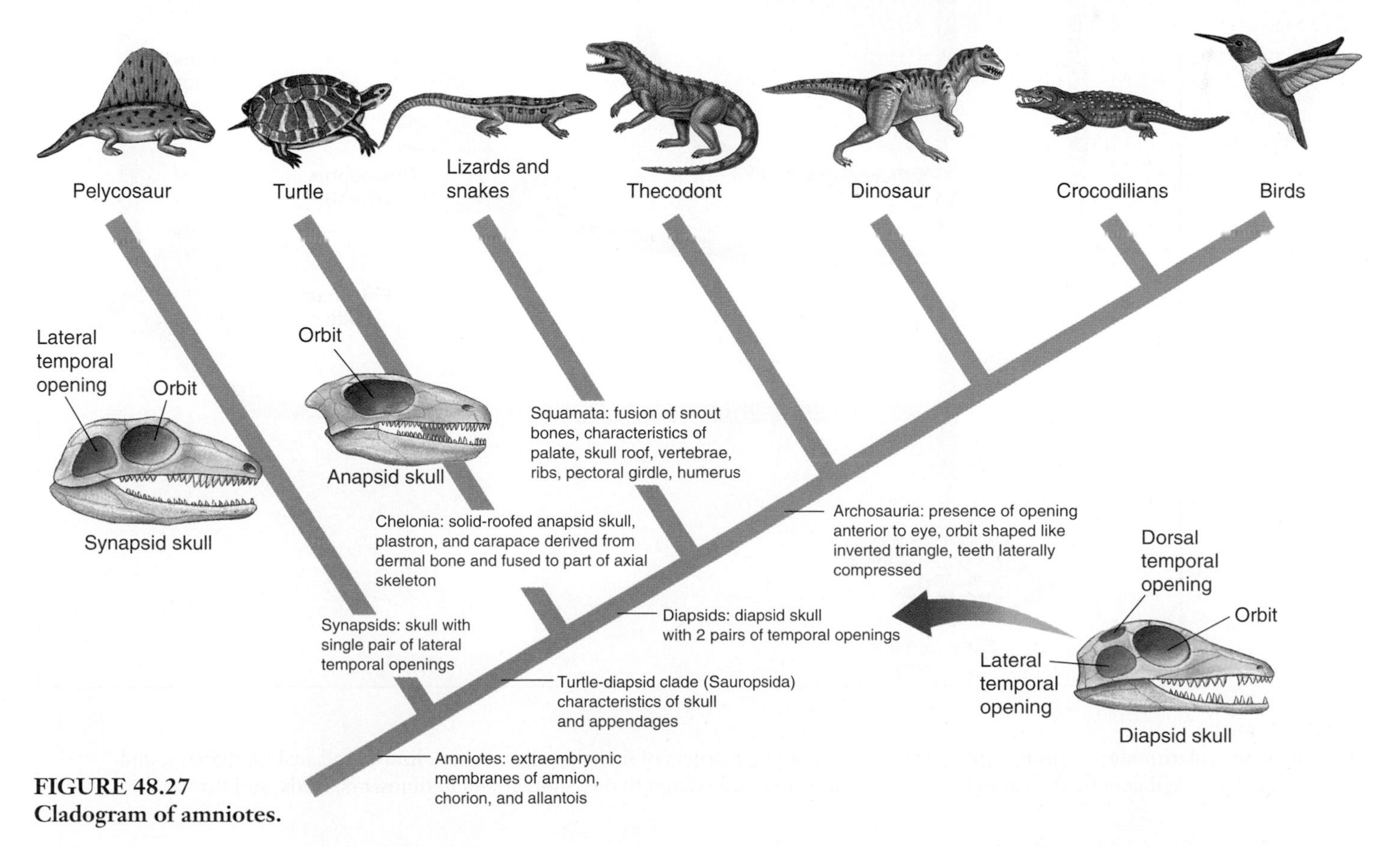

FIGURE 48.27
Cladogram of amniotes.

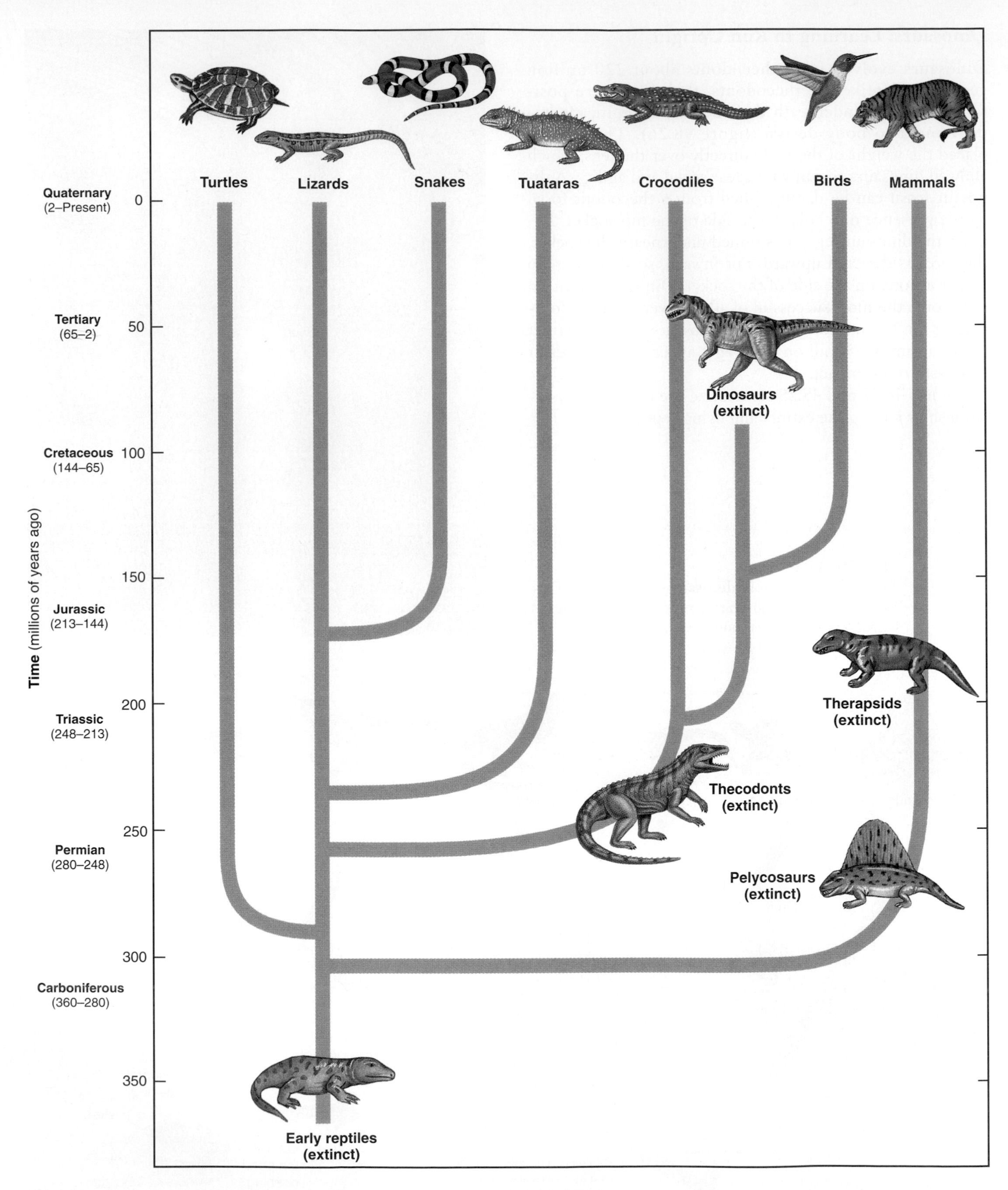

FIGURE 48.28

Evolutionary relationships among the reptiles. There are four orders of living reptiles: turtles, lizards and snakes, tuataras, and crocodiles. This phylogenetic tree shows how these four orders are related to one another and to dinosaurs, birds, and mammals.

FIGURE 48.29
A comparison of reptile and fish circulation. (*a*) In reptiles such as this turtle, blood is repumped after leaving the lungs, and circulation to the rest of the body remains vigorous. (*b*) The blood in fishes flows from the gills directly to the rest of the body, resulting in slower circulation.

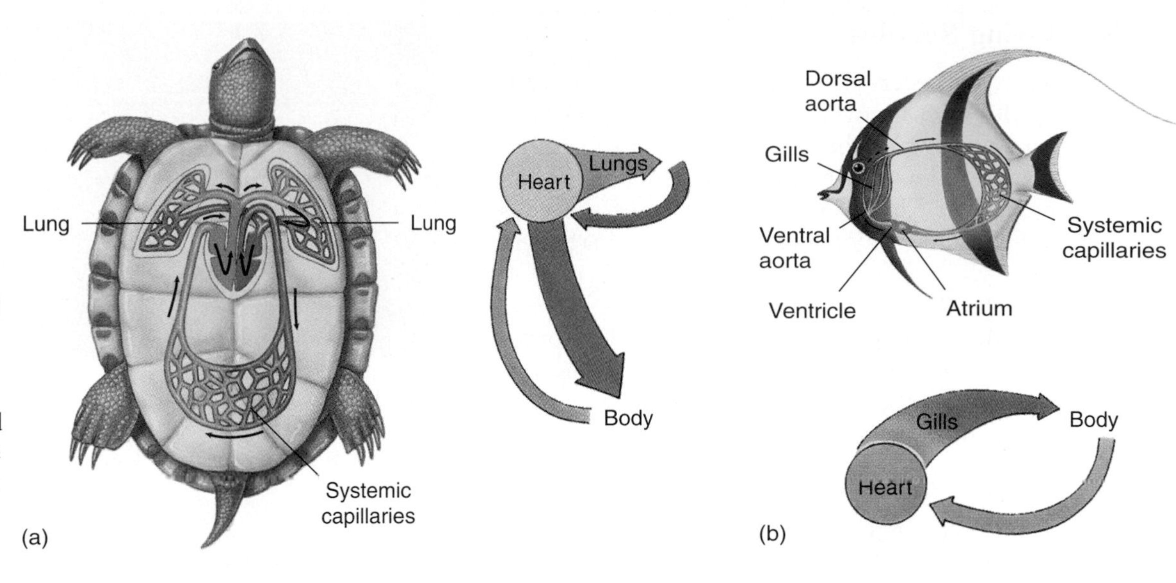

Today's Reptiles

Most of the major reptile orders are now extinct. Of the 16 orders of reptiles that have existed, only 4 survive.

Turtles. The most ancient surviving lineage of reptiles is that of turtles. Turtles have anapsid skulls much like those of the first reptiles. Turtles have changed little in the past 200 million years.

Lizards and snakes. Most reptiles living today belong to the second lineage to evolve, the lizards and snakes. Lizards and snakes are descended from an ancient lineage of lizardlike reptiles that branched off the main line of reptile evolution in the late Permian, 250 million years ago, before the thecodonts appeared (figure 48.28). Throughout the Mesozoic era, during the dominance of the dinosaurs, these reptiles survived as minor elements of the landscape, much as mammals did. Like mammals, lizards and snakes became diverse and common only after the dinosaurs disappeared.

Tuataras. The third lineage of surviving reptiles to evolve were the Rhynchocephalonts, small diapsid reptiles that appeared shortly before the dinosaurs. They lived throughout the time of the dinosaurs and were common in the Jurassic. They began to decline in the Cretaceous, apparently unable to compete with lizards, and were already rare by the time dinosaurs disappeared. Today only two species of the order Rhynchocephalia survive, both tuataras living on small islands near New Zealand.

Crocodiles. The fourth lineage of living reptile, crocodiles, appeared on the evolutionary scene much later than other living reptiles. Crocodiles are descended from the same line of thecodonts that gave rise to the dinosaurs and resemble dinosaurs in many ways. They have changed very little in over 200 million years. Crocodiles, pterosaurs, thecodonts, and dinosaurs together make up a group called archosaurs ("ruling reptiles").

Other Important Characteristics

As you might imagine from the structure of the amniotic egg, reptiles and other amniotes do not practice external fertilization as most amphibians do. There would be no way for a sperm to penetrate the membrane barriers protecting the egg. Instead, the male places sperm inside the female, where they fertilize the egg before the membranes are formed. This is called internal fertilization.

The circulatory system of reptiles is improved over that of fish and amphibians, providing oxygen to the body more efficiently (figure 48.29). The improvement is achieved by extending the septum within the heart from the atrium partway across the ventricle. This septum creates a partial wall that tends to lessen mixing of oxygen-poor blood with oxygen-rich blood within the ventricle. In crocodiles, the septum completely divides the ventricle, creating a four-chambered heart, just as it does in birds and mammals (and probably in dinosaurs).

All living reptiles are **ectothermic,** obtaining their heat from external sources. In contrast, **endothermic** animals are able to generate their heat internally. In addition, **homeothermic** animals have a constant body temperature, and **poikilothermic** animals have a body temperature that fluctuates with ambient temperature. Thus, a deep-sea fish may be an ectothermic homeotherm because its heat comes from an external source, but its body temperature is constant. Reptiles are largely ectothermic poikilotherms; their body temperature is largely determined by their surroundings. Reptiles also regulate their temperature through behavior. They may bask in the sun to warm up or seek shade to prevent overheating. The thecodont ancestors of crocodiles were ectothermic, as crocodiles are today. The later dinosaurs from which birds evolved were endothermic. Crocodiles and birds differ in this one important respect. Ectothermy is a principal reason why crocodiles have been grouped among the reptiles.

Kinds of Living Reptiles

The four surviving orders of reptiles contain about 7000 species. Reptiles occur worldwide except in the coldest regions, where it is impossible for ectotherms to survive. Reptiles are among the most numerous and diverse of terrestrial vertebrates. The four living orders of the class Reptilia are Chelonia, Rhynchocephalia, Squamata, and Crocodilia.

Order Chelonia: Turtles and Tortoises. The order Chelonia consists of about 250 species of turtles (most of which are aquatic; figure 48.30) and tortoises (which are terrestrial). They differ from all other reptiles because their bodies are encased within a protective shell. Many of them can pull their head and legs into the shell as well, for total protection from predators. Turtles and tortoises lack teeth but have sharp beaks.

Today's turtles and tortoises have changed very little since the first turtles appeared 200 million years ago. Turtles are **anapsid**—they lack the temporal openings in the skull characteristic of other living reptiles, which are diapsid. This evolutionary stability of turtles may reflect the continuous benefit of their basic design—a body covered with a shell. In some species, the shell is made of hard plates; in other species, it is a covering of tough, leathery skin. In either case, the shell consists of two basic parts. The carapace is the dorsal covering, while the plastron is the ventral portion. In a fundamental commitment to this shell architecture, the vertebrae and ribs of most turtle and tortoise species are fused to the inside of the carapace. All of the support for muscle attachment comes from the shell.

While most tortoises have a domed-shaped shell into which they can retract their head and limbs, water-dwelling turtles have a streamlined, disc-shaped shell that permits rapid turning in water. Freshwater turtles have webbed toes, and in marine turtles, the forelimbs have evolved into flippers. Although marine turtles spend their lives at sea, they must return to land to lay their eggs. Many species migrate long distances to do this. Atlantic green turtles migrate from their feeding grounds off the coast of Brazil to Ascension Island in the middle of the South Atlantic—a distance of more than 2000 kilometers—to lay their eggs on the same beaches where they hatched.

Order Rhynchocephalia: Tuatara. The order Rhynchocephalia contains only two species today, the tuataras, large, lizardlike animals about half a meter long. The only place in the world where these endangered species are found is on a cluster of small islands off the coast of New Zealand. The native Maoris of New Zealand named the tuatara for the conspicuous spiny crest running down its back.

An unusual feature of the tuatara (and some lizards) is the inconspicuous "third eye" on the top of its head, called a parietal eye. Concealed under a thin layer of scales, the eye has a lens and retina and is connected by nerves to the brain. Why have an eye, if it is covered up? The parietal eye may function to alert the tuatara when it has been exposed to too much sun, protecting it against overheating. Unlike most reptiles, tuataras are most active at low temperatures. They burrow during the day and feed at night on insects, worms, and other small animals.

FIGURE 48.30
Red-bellied turtles, *Pseudemys rubriventris*. This turtle is common in the northeastern United States.

Order Squamata: Lizards and Snakes. The order Squamata (figure 48.31) consists of three suborders: Sauria, some 3800 species of lizards, Amphisbaenia, about 135 species of worm lizards, and Serpentes, about 3000 species of snakes. The distinguishing characteristics of this order are the presence of paired copulatory organs in the male and a lower jaw that is not joined directly to the skull. A movable hinge with five joints (your jaw has only one) allows great flexibility in the movements of the jaw. In addition, the loss of the lower arch of bone below the lower opening in the skull of lizards makes room for large muscles to operate their jaws. Most lizards and snakes are carnivores, preying on insects and small animals, and these improvements in jaw design have made a major contribution to their evolutionary success.

The chief difference between lizards and snakes is that most lizards have limbs and snakes do not. Snakes also lack movable eyelids and external ears. Lizards are a more ancient group than modern snakes, which evolved only 20 million years ago. Common lizards include iguanas, chameleons, geckos, and anoles. Most are small, measuring less than a foot in length. The largest lizards belong to the monitor family. The largest of all monitors is the Komodo dragon of Indonesia, which reaches 3 meters in length and weighs up to 100 kilograms. Snakes also vary in size from only a few inches long to those that reach nearly 10 meters in length.

Lizards and snakes rely on agility and speed to catch prey and elude predators. Only two species of lizard are venomous, the Gila monster of the southwestern United States and the beaded lizard of western Mexico. Similarly, most species of snakes are nonvenomous. Of the 13 families of snakes, only 4 are venomous: the elapids (cobras, kraits,

(a)

(b)

FIGURE 48.31
Representatives from the order Squamata. (*a*) An Australian skink, *Sphenomorophus.* Some burrowing lizards lack legs, and the snakes evolved from one line of legless lizards. (*b*) A smooth green snake, *Liochlorophis vernalis.*

FIGURE 48.32
River crocodile, *Crocodilus acutus.* Most crocodiles resemble birds and mammals in having four-chambered hearts; all other living reptiles have three-chambered hearts. Crocodiles, like birds, are more closely related to dinosaurs than to any of the other living reptiles.

and coral snakes); the sea snakes; the vipers (adders, bushmasters, rattlesnakes, water moccasins, and copperheads); and some colubrids (African boomslang and twig snake).

Many lizards, including skinks and geckos, have the ability to lose their tails and then regenerate a new one. This apparently allows these lizards to escape from predators.

Order Crocodilia: Crocodiles and Alligators. The order Crocodilia is composed of 25 species of large, primarily aquatic, primitive-looking reptiles (figure 48.32). In addition to crocodiles and alligators, the order includes two less familiar animals: the caimans and gavials. Crocodilians have remained relatively unchanged since they first evolved.

Crocodiles are largely nocturnal animals that live in or near water in tropical or subtropical regions of Africa, Asia, and South America. The American crocodile is found in southern Florida and Cuba to Columbia and Ecuador. Nile crocodiles and estuarine crocodiles can grow to enormous size and are responsible for many human fatalities each year. There are only two species of alligators: one living in the southern United States and the other a rare endangered species living in China. Caimans, which resemble alligators, are native to Central America. Gavials are a group of fish-eating crocodilians with long, slender snouts that live only in India and Burma.

All crocodilians are carnivores. They generally hunt by stealth, waiting in ambush for prey, then attacking ferociously. Their bodies are well adapted for this form of hunting: their eyes are on top of their heads and their nostrils on top of their snouts, so they can see and breathe while lying quietly submerged in water. They have enormous mouths, studded with sharp teeth, and very strong necks. A valve in the back of the mouth prevents water from entering the air passage when a crocodilian feeds underwater.

Crocodiles resemble birds far more than they do other living reptiles. Alone among living reptiles, crocodiles care for their young (a trait they share with at least some dinosaurs) and have a four-chambered heart, as birds do. There are also many other points of anatomy in which crocodiles differ from all living reptiles and resemble birds. Why are crocodiles more similar to birds than to other living reptiles? Most biologists now believe that birds are in fact the direct descendants of dinosaurs. Both crocodiles and birds are more closely related to dinosaurs, and each other, than they are related to lizards and snakes.

Many major reptile groups that dominated life on land for 250 million years are now extinct. The four living orders of reptiles include the turtles, lizards and snakes, tuataras, and crocodiles.

Birds

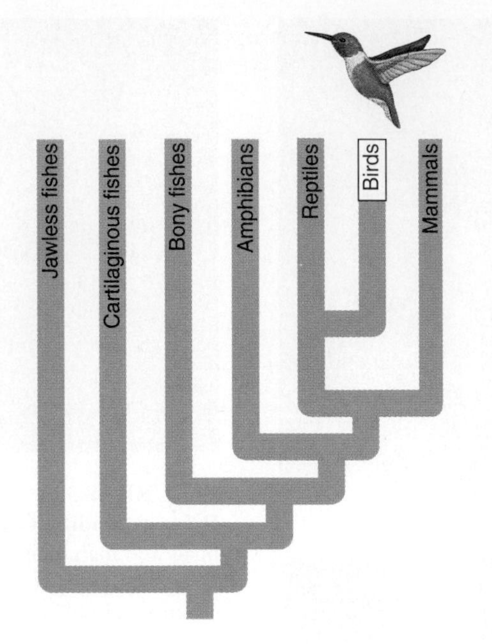

Only four groups of animals have evolved the ability to fly—insects, pterosaurs, birds, and bats. Pterosaurs, flying reptiles, evolved from gliding reptiles and flew for 130 million years before becoming extinct with the dinosaurs. There are startling similarities in how these very different animals meet the challenges of flight. Like water running downhill through similar gullies, evolution tends to seek out similar adaptations. There are major differences as well. The success of birds lies in the development of a structure unique in the animal world—the feather. Developed from reptilian scales, feathers are the ideal adaptation for flight—lightweight airfoils that are easily replaced if damaged (unlike the vulnerable skin wings of pterosaurs and bats). Today, birds (class Aves) are the most successful and diverse of all terrestrial vertebrates, with 28 orders containing a total of 166 families and about 8800 species (table 48.4).

Key Characteristics of Birds

Modern birds lack teeth and have only vestigial tails, but they still retain many reptilian characteristics. For instance, birds lay amniotic eggs, although the shells of bird eggs are hard rather than leathery. Also, reptilian scales are present on the feet and lower legs of birds. What makes birds unique? What distinguishes them from living reptiles?

1. **Feathers.** Feathers are modified reptilian scales that serve two functions: providing lift for flight and conserving heat. The structure of feathers combines maximum flexibility and strength with minimum weight (figure 48.33). Feathers develop from tiny pits in the skin called follicles. In a typical flight feather, a shaft emerges from the follicle, and pairs of vanes develop from its opposite sides. At maturity, each vane has many branches called barbs. The barbs, in turn, have many projections called barbules that are equipped with microscopic hooks. These hooks link the barbs to one another, giving the feather a continuous surface and a sturdy but flexible shape. Like scales, feathers can be replaced. Feathers are unique to birds among living animals. Recent fossil finds suggest that some dinosaurs may have had feathers.
2. **Flight skeleton.** The bones of birds are thin and hollow. Many of the bones are fused, making the bird skeleton more rigid than a reptilian skeleton. The fused sections of backbone and of the shoulder and hip girdles form a sturdy frame that anchors muscles during flight. The power for active flight comes from large breast muscles that can make up 30% of a bird's total body weight. They stretch down from the wing and attach to the breastbone, which is greatly enlarged and bears a prominent keel for muscle attachment. They also attach to the fused collarbones that form the so-called "wishbone." No other living vertebrates have a fused collarbone or a keeled breastbone.

Birds are the most diverse of all terrestrial vertebrates. They are closely related to reptiles, but unlike reptiles or any other animals, birds have feathers.

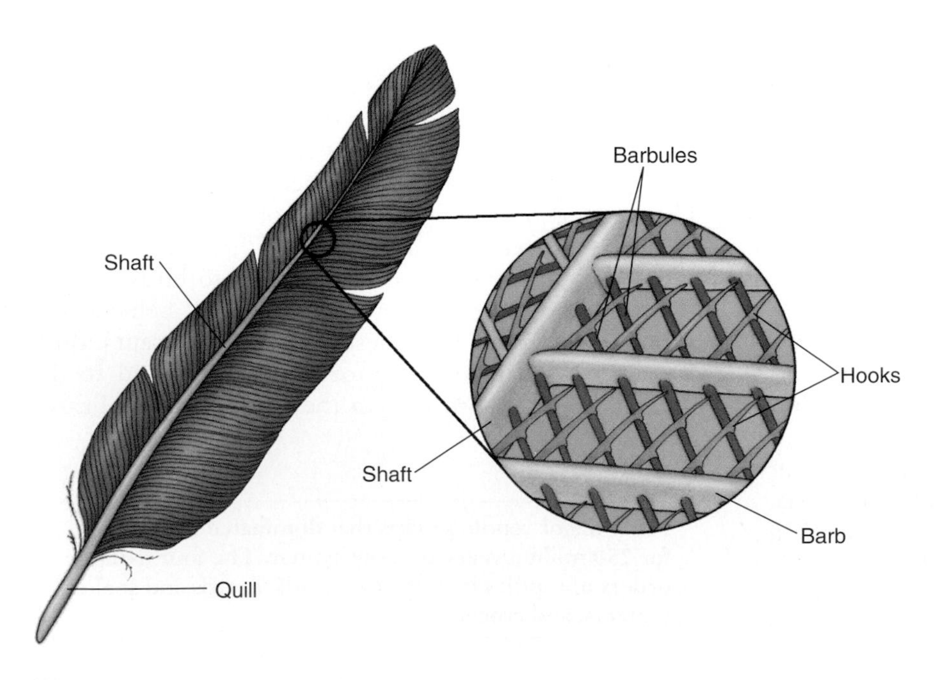

FIGURE 48.33
A feather. This enlargement shows how the vanes, secondary branches and barbs, are linked together by microscopic barbules.

Table 48.4 Major Orders of Birds			
Order	**Typical Examples**	**Key Characteristics**	**Approximate Number of Living Species**
Passeriformes	Crows, mockingbirds, robins, sparrows, starlings, warblers	Songbirds Well-developed vocal organs; perching feet; dependent young	5276 (largest of all bird orders; contains over 60% of all species)
Apodiformes	Hummingbirds, swifts	*Fast fliers* Short legs; small bodies; rapid wing beat	428
Piciformes	Honeyguides, toucans, woodpeckers	*Woodpeckers or toucans* Grasping feet; chisel-like, sharp bills can break down wood	383
Psittaciformes	Cockatoos, parrots	*Parrots* Large, powerful bills for crushing seeds; well-developed vocal organs	340
Charadriiformes	Auks, gulls, plovers, sandpipers, terns	*Shorebirds* Long, stiltlike legs; slender probing bills	331
Columbiformes	Doves, pigeons	*Pigeons* Perching feet; rounded, stout bodies	303
Falconiformes	Eagles, falcons, hawks, vultures	*Birds of prey* Carnivorous; keen vision; sharp, pointed beaks for tearing flesh; active during the day	288
Galliformes	Chickens, grouse, pheasants, quail	*Gamebirds* Often limited flying ability; rounded bodies	268
Gruiformes	Bitterns, coots, cranes, rails	*Marsh birds* Long, stiltlike legs; diverse body shapes; marsh-dwellers	209
Anseriformes	Ducks, geese, swans	*Waterfowl* Webbed toes; broad bill with filtering ridges	150
Strigiformes	Barn owls, screech owls	*Owls* Nocturnal birds of prey; strong beaks; powerful feet	146
Ciconiiformes	Herons, ibises, storks	*Waders* Long-legged; large bodies	114
Procellariformes	Albatrosses, petrels	*Seabirds* Tube-shaped bills; capable of flying for long periods of time	104
Sphenisciformes	Emperor penguins, crested penguins	*Penguins* Marine; modified wings for swimming; flightless; found only in southern hemisphere; thick coats of insulating feathers	18
Dinornithiformes	Kiwis	*Kiwis* Flightless; small; primitive; confined to New Zealand	2
Struthioniformes	Ostriches	*Ostriches* Powerful running legs; flightless; only two toes; very large	1

History of the Birds

A 150-million-year-old fossil of the first known bird, *Archaeopteryx* (figure 48.34)—pronounced "archie-op-ter-ichs"—was found in 1862 in a limestone quarry in Bavaria, the impression of its feathers stamped clearly into the rocks.

FIGURE 48.34
Archaeopteryx. An artist's reconstruction of *Archaeopteryx*, an early bird about the size of a crow. Closely related to its ancestors among the bipedal dinosaurs, *Archaeopteryx* lived in the forests of central Europe 150 million years ago. The true feather colors of *Archaeopteryx* are not known.

Birds Are Descended from Dinosaurs

The skeleton of *Archaeopteryx* shares many features with small theropod dinosaurs. About the size of a crow, its skull has teeth, and very few of its bones are fused to one another—dinosaurian features, not avian. Its bones are solid, not hollow like a bird's. Also, it has a long reptilian tail, and no enlarged breastbone such as modern birds use to anchor flight muscles. Finally, it has the forelimbs of a dinosaur. Because of its many dinosaur features, several *Archaeopteryx* fossils were originally classified as the coelurosaur *Compsognathus*, a small theropod dinosaur of similar size—until feathers were discovered on the fossils. What makes *Archaeopteryx* distinctly avian is the presence of feathers on its wings and tail. It also has other birdlike features, notably the presence of a wishbone. Dinosaurs lack a wishbone, although thecodonts had them.

The remarkable similarity of *Archaeopteryx* to *Compsognathus* has led almost all paleontologists to conclude that *Archaeopteryx* is the direct descendant of dinosaurs—indeed, that today's birds are "feathered dinosaurs." Some even speak flippantly of "carving the dinosaur" at Thanksgiving dinner. The recent discovery of feathered dinosaurs in China lends strong support to this inference. The dinosaur *Caudipteryx*, for example, is clearly intermediate between *Archaeopteryx* and dinosaurs, having large feathers on its tail and arms but also many features of velociraptor dinosaurs (figure 48.35). Because the arms of *Caudipteryx* were too short to use as wings, feathers probably didn't evolve for flight. Instead, they probably served as insulation, much as fur does for animals. Flight is something that certain kinds of dinosaurs achieved as they evolved longer arms. We call these dinosaurs birds.

Despite their close affinity to dinosaurs, biologists continue to classify birds as Aves, a separate class, because of the key evolutionary novelties of birds: feathers, hollow bones, and physiological mechanisms such as superefficient lungs that permit sustained, powered flight. It is because of their unique adaptations and great diversity that

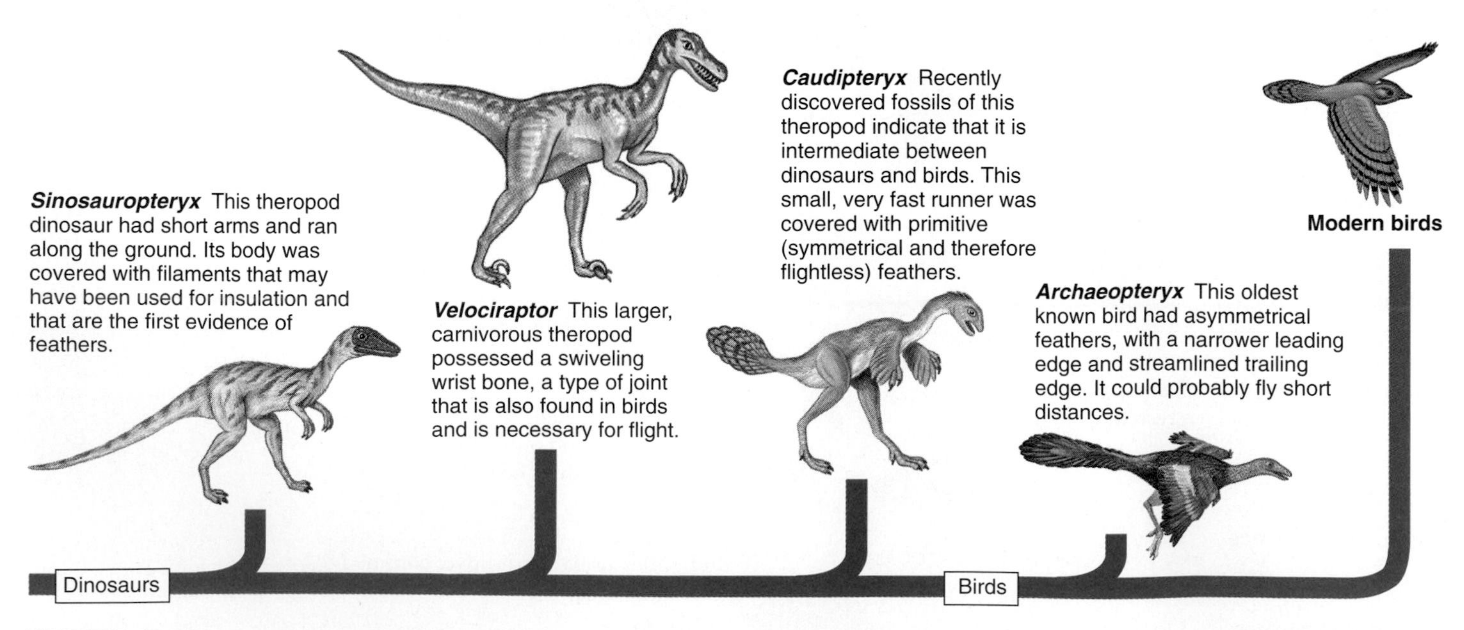

Sinosauropteryx This theropod dinosaur had short arms and ran along the ground. Its body was covered with filaments that may have been used for insulation and that are the first evidence of feathers.
Velociraptor This larger, carnivorous theropod possessed a swiveling wrist bone, a type of joint that is also found in birds and is necessary for flight.
Caudipteryx Recently discovered fossils of this theropod indicate that it is intermediate between dinosaurs and birds. This small, very fast runner was covered with primitive (symmetrical and therefore flightless) feathers.
Archaeopteryx This oldest known bird had asymmetrical feathers, with a narrower leading edge and streamlined trailing edge. It could probably fly short distances.
Modern birds
Dinosaurs
Birds

FIGURE 48.35
The evolutionary path to the birds. Almost all paleontologists now accept the theory that birds are the direct descendents of theropod dinosaurs.

birds are assigned to a separate class. This practical judgment should not conceal the basic agreement among almost all biologists that birds are the direct descendants of theropod dinosaurs, as closely related to coelurosaurs as are other theropods (see figure 48.35).

By the early Cretaceous, only a few million years after *Archaeopteryx*, a diverse array of birds had evolved, with many of the features of modern birds. Fossils in Mongolia, Spain, and China discovered within the last few years reveal a diverse collection of toothed birds with the hollow bones and breastbones necessary for sustained flight. Other fossils reveal highly specialized, flightless diving birds. The diverse birds of the Cretaceous shared the skies with pterosaurs for 70 million years.

Because the impression of feathers is rarely fossilized and modern birds have hollow, delicate bones, the fossil record of birds is incomplete. Relationships among the 166 families of modern birds are mostly inferred from studies of the degree of DNA similarity among living birds. These studies suggest that the most ancient living birds are the flightless birds, like the ostrich. Ducks, geese, and other waterfowl evolved next, in the early Cretaceous, followed by a diverse group of woodpeckers, parrots, swifts, and owls. The largest of the bird orders, Passeriformes, or songbirds (60% of all species of birds today), evolved in the mid-Cretaceous. The more specialized orders of birds, such as shorebirds, birds of prey, flamingos, and penguins, did not appear until the late Cretaceous. All but a few of the modern orders of toothless birds are thought to have arisen before the disappearance of the pterosaurs and dinosaurs at the end of the Cretaceous 65 million years ago.

FIGURE 48.36
Class Aves. This Western tanager, *Piranga ludoviciana*, is a member of the largest order of birds, the Passeriformes, with over 5000 species.

Birds Today

You can tell a great deal about the habits and food of a bird by examining its beak and feet. For instance, carnivorous birds such as owls have curved talons for seizing prey and sharp beaks for tearing apart their meal. The beaks of ducks are flat for shoveling through mud, while the beaks of finches are short, thick seed-crushers. There are 28 orders of birds, the largest consisting of over 5000 species (figure 48.36).

Many adaptations enabled birds to cope with the heavy energy demands of flight:

1. **Efficient respiration.** Flight muscles consume an enormous amount of oxygen during active flight. The reptilian lung has a limited internal surface area, not nearly enough to absorb all the oxygen needed. Mammalian lungs have a greater surface area, but as we will see in chapter 53, bird lungs satisfy this challenge with a radical redesign. When a bird inhales, the air goes past the lungs to a series of air sacs located near and within the hollow bones of the back; from there the air travels to the lungs and then to a set of anterior air sacs before being exhaled. Because air always passes through the lungs in the same direction, and blood flows past the lung at right angles to the airflow, gas exchange is highly efficient.
2. **Efficient circulation.** The revved-up metabolism needed to power active flight also requires very efficient blood circulation, so that the oxygen captured by the lungs can be delivered to the flight muscles quickly. In the heart of most living reptiles, oxygen-rich blood coming from the lungs mixes with oxygen-poor blood returning from the body because the wall dividing the ventricle into two chambers is not complete. In birds, the wall dividing the ventricle is complete, and the two blood circulations do not mix, so flight muscles receive fully oxygenated blood.

   In comparison with reptiles and most other vertebrates, birds have a rapid heartbeat. A hummingbird's heart beats about 600 times a minute. An active chickadee's heart beats 1000 times a minute. In contrast, the heart of the large, flightless ostrich averages 70 beats per minute—the same rate as the human heart.
3. **Endothermy.** Birds, like mammals, are endothermic. Many paleontologists believe the dinosaurs that birds evolved from were endothermic as well. Birds maintain body temperatures significantly higher than most mammals, ranging from 40° to 42°C (your body temperature is 37°C). Feathers provide excellent insulation, helping to conserve body heat. The high temperatures maintained by endothermy permit metabolism in the bird's flight muscles to proceed at a rapid pace, to provide the ATP necessary to drive rapid muscle contraction.

The class Aves probably debuted 150 million years ago with *Archaeopteryx*. Modern birds are characterized by feathers, scales, a thin, hollow skeleton, auxiliary air sacs, and a four-chambered heart. Birds lay amniotic eggs and are endothermic.

Mammals

There are about 4100 living species of mammals (class Mammalia), the smallest number of species in any of the five classes of vertebrates. Most large, land-dwelling vertebrates are mammals (figure 48.37), and they tend to dominate terrestrial communities, as did the dinosaurs that they replaced. When you look out over an African plain, you see the big mammals, the lions, zebras, gazelles, and antelope. Your eye does not as readily pick out the many birds, lizards, and frogs that live in the grassland community with them. But the typical mammal is not all that large. Of the 4100 species of mammals, 3200 are rodents, bats, shrews, or moles (table 48.5).

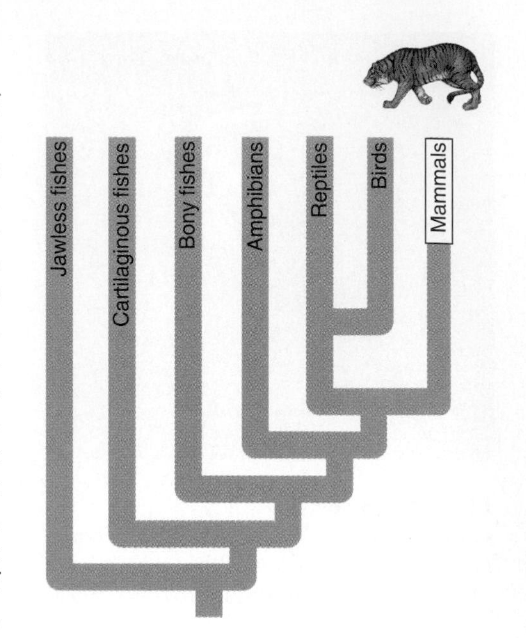

Key Mammalian Characteristics

Mammals are distinguished from all other classes of vertebrates by two fundamental characteristics that are unique to mammals:

1. **Hair.** All mammals have hair. Even apparently naked whales and dolphins grow sensitive bristles on their snouts. Evolution of fur and the ability to regulate body temperature enabled mammals to invade colder climates that ectothermic reptiles could not inhabit, and the insulation fur provided may have ensured the survival of mammals when the dinosaurs perished.

   Unlike feathers, which evolved from modified reptilian scales, mammalian hair is a completely different form of skin structure. An individual mammalian hair is a long, protein-rich filament that extends like a stiff thread from a bulblike foundation beneath the skin known as a hair follicle. The filament is composed mainly of dead cells filled with the fibrous protein keratin.

   One of the most important functions of hair is insulation against heat loss. Mammals are endothermic animals, and typically maintain body temperatures higher than the temperature of their surroundings. The dense undercoat of many mammals reduces the amount of body heat that escapes.

   Another function of hair is camouflage. The coloration and pattern of a mammal's coat usually matches its background. A little brown mouse is practically invisible against the brown leaf litter of a forest floor, while the orange and black stripes of a Bengal tiger disappear against the orange-brown color of the tall grass in which it hunts. Hairs also function as sensory structures. The whiskers of cats and dogs are stiff hairs that are very sensitive to touch. Mammals that are active at night or live underground often rely on their whiskers to locate prey or to avoid colliding with objects. Hair can also serve as a defense weapon. Porcupines and hedgehogs protect themselves with long, sharp, stiff hairs called quills.
2. **Mammary glands.** All female mammals possess mammary glands that secrete milk. Newborn mammals, born without teeth, suckle this milk. Even baby whales are nursed by their mother's milk. Milk is a fluid rich in fat, sugar, and protein. A liter of human milk contains 11 grams of protein, 49 grams of fat, 70 grams of carbohydrate (chiefly the sugar lactose), and 2 grams of minerals critical to early growth, such as calcium. About 95% of the volume is water, critical to avoid dehydration. Milk is a very high calorie food (human milk has 750 kcal per liter), important because of the high energy needs of a rapidly growing newborn mammal. About 50% of the energy in the milk comes from fat.

Mammals first appeared 220 million years ago, evolving to their present position of dominance in modern terrestrial ecosystems. Mammals are the only vertebrates that possess hair and milk glands.

FIGURE 48.37
Mammals. African elephants, *Loxodonta africana*, at a water hole (order Proboscidea).

Table 48.5 Major Orders of Mammals			
Order	**Typical Examples**	**Key Characteristics**	**Approximate Number of Living Species**
Rodentia	Beavers, mice, porcupines, rats	*Small plant-eaters* Chisel-like incisor teeth	1814
Chiroptera	Bats	*Flying mammals* Primarily fruit- or insect-eaters; elongated fingers; thin wing membrane; nocturnal; navigate by sonar	986
Insectivora	Moles, shrews	*Small, burrowing mammals* Insect-eaters; most primitive placental mammals; spend most of their time underground	390
Marsupialia	Kangaroos, koalas	*Pouched mammals* Young develop in abdominal pouch	280
Carnivora	Bears, cats, raccoons, weasels, dogs	*Carnivorous predators* Teeth adapted for shearing flesh; no native families in Australia	240
Primates	Apes, humans, lemurs, monkeys	*Tree-dwellers* Large brain size; binocular vision; opposable thumb; end product of a line that branched off early from other mammals	233
Artiodactyla	Cattle, deer, giraffes, pigs	*Hoofed mammals* With two or four toes; mostly herbivores	211
Cetacea	Dolphins, porpoises, whales	*Fully marine mammals* Streamlined bodies; front limbs modified into flippers; no hind limbs; blowholes on top of head; no hair except on muzzle	79
Lagomorpha	Rabbits, hares, pikas	*Rodentlike jumpers* Four upper incisors (rather than the two seen in rodents); hind legs often longer than forelegs; an adaptation for jumping	69
Pinnipedia	Sea lions, seals, walruses	*Marine carnivores* Feed mainly on fish; limbs modified for swimming	34
Edentata	Anteaters, armadillos, sloths	*Toothless insect-eaters* Many are toothless, but some have degenerate, peglike teeth	30
Perissodactyla	Horses, rhinoceroses, zebras	*Hoofed mammals with one or three toes* Herbivorous; teeth adapted for chewing	17
Proboscidea	Elephants	*Long-trunked herbivores* Two upper incisors elongated as tusks; largest living land animal	2

History of the Mammals

Mammals have been around since the time of the dinosaurs, although they were never common until the dinosaurs disappeared. We have learned a lot about the evolutionary history of mammals from their fossils.

Origin of Mammals

The first mammals arose from therapsids in the mid-Triassic about 220 million years ago, just as the first dinosaurs evolved from thecodonts. Tiny, shrewlike creatures that lived in trees eating insects, mammals were only a minor element in a land that quickly came to be dominated by dinosaurs. Fossils reveal that these early mammals had large eye sockets, evidence that they may have been active at night. Early mammals had a single lower jawbone. Therapsid fossils show a change from the reptile lower jaw with several bones to a jaw closer to the mammalian-type jaw. Two of the bones forming the therapsid jaw joint retreated into the middle ear of mammals, linking with a bone already there producing a three-bone structure that amplifies sound better than the reptilian ear.

Table 48.6 Some Groups of Extinct Mammals

Group	Description
Cave bears	Numerous in the ice ages; this enormous vegetarian bear slept through the winter in large groups.
Irish elk	Neither Irish nor an elk (it is a kind of deer), *Megaloceros* was the largest deer that ever lived, with horns spanning 12 feet. Seen in French cave paintings, they became extinct about 2500 years ago.
Mammoths	Although only two species of elephants survive today, the elephant family was far more diverse during the late Tertiary. Many were cold-adapted mammoths with fur.
Giant ground sloths	*Megatherium* was a giant 20-foot ground sloth that weighed three tons and was as large as a modern elephant.
Sabertooth cats	The jaws of these large, lionlike cats opened an incredible 120 degrees to allow the animal to drive its huge upper pair of saber teeth into prey.

Early Divergence in Mammals

For 155 million years, while the dinosaurs flourished, mammals were a minor group of small insectivores and herbivores. Only five orders of mammals arose in that time, and their fossils are scarce, indicating that mammals were not abundant. However, the two groups to which present-day mammals belong did appear. The most primitive mammals, direct descendents of therapsids, were members of the subclass Prototheria. Most prototherians were small and resembled modern shrews. All prototherians laid eggs, as did their therapsid ancestors. The only prototherians surviving today are the monotremes—the duckbill platypus and the echidnas, or spiny anteaters. The other major mammalian group is the subclass Theria. All of the mammals you are familiar with, including humans, are therians. Therians are viviparous (that is, their young are born alive). The two major living therian groups are marsupials, or pouched mammals, and placental mammals. Kangaroos, opossums, and koalas are marsupials. Dogs, cats, humans, horses, and most other mammals are placentals.

The Age of Mammals

At the end of the Cretaceous Period 65 million years ago, the dinosaurs and numerous other land and marine animals became extinct, but mammals survived, possibly because of the insulation their fur provided. In the Tertiary Period (lasting from 65 million years to 2 million years ago), mammals rapidly diversified, taking over many of the ecological roles once dominated by dinosaurs (table 48.6). Mammals reached their maximum diversity late in the Tertiary Period, about 15 million years ago. At that time, tropical conditions existed over much of the world. During the last 15 million years, world climates have deteriorated, and the area covered by tropical habitats has decreased, causing a decline in the total number of mammalian species. There are now 19 orders of mammals.

Characteristics of Modern Mammals

Endothermy. Mammals are endothermic, a crucial adaptation that has allowed mammals to be active at any time of the day or night and to colonize severe environments, from deserts to ice fields. Many characteristics, such as hair that provides insulation, played important roles in making endothermy possible. Also, the more efficient blood circulation provided by the four-chambered heart and the more efficient respiration provided by the *diaphragm* (a special sheet of muscles below the rib cage that aids breathing) make possible the higher metabolic rate upon which endothermy depends.

Placenta. In most mammal species, females carry their young in a uterus during development, nourishing them through a placenta, and give birth to live young. The placenta is a specialized organ within the uterus of the pregnant mother that brings the bloodstream of the fetus into close contact with the bloodstream of the mother (figure 48.38). Food, water, and oxygen can pass across from mother to child, and wastes can pass over to the mother's blood and be carried away.

Teeth. Reptiles have homodont dentition: their teeth are all the same. However, mammals have heterodont dentition, with different types of teeth that are highly specialized to match particular eating habits (figure 48.39). It is usually possible to determine a mammal's diet simply by examining its teeth. Compare the skull of a dog (a carnivore) and a deer (an herbivore). The dog's long canine teeth are well suited for biting and holding prey, and some of its premolar and molar teeth are triangular and sharp for ripping off chunks of flesh. In contrast, canine teeth are absent in deer; instead the deer clips off mouthfuls of plants with flat, chisel-like incisors on its lower jaw. The deer's molars are large and covered with ridges to effectively grind and break up tough plant tissues. Rodents, such as beavers, are gnawers and have long incisors for chewing through branches or stems. These incisors are ever-growing; that is, the ends wear down, but new incisor growth maintains the length.

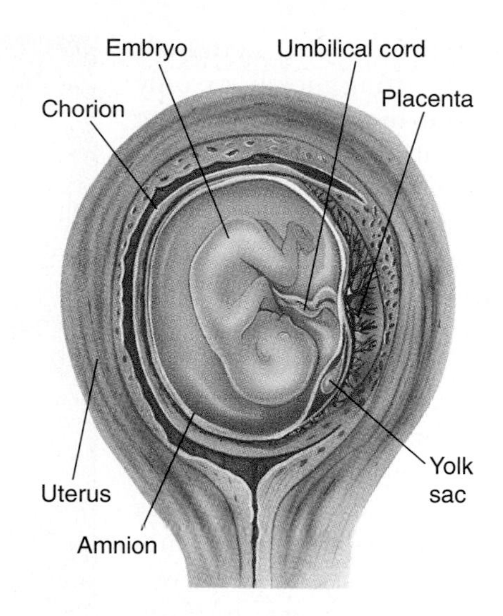

FIGURE 48.38
The placenta. The placenta is characteristic of the largest group of mammals, the placental mammals. It evolved from membranes in the amniotic egg. The umbilical cord evolved from the allantois. The chorion, or outermost part of the amniotic egg, forms most of the placenta itself. The placenta serves as the provisional lungs, intestine, and kidneys of the embryo, without ever mixing maternal and fetal blood.

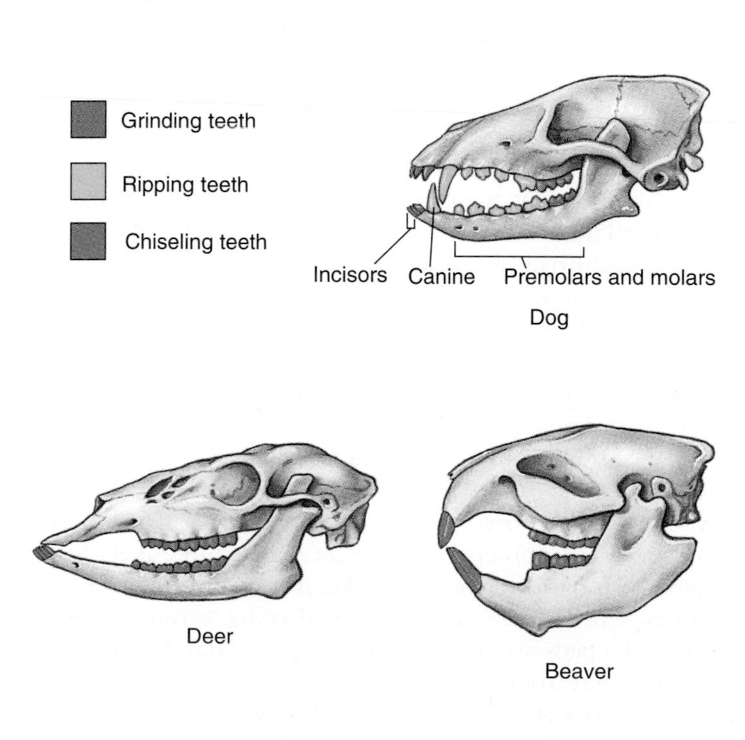

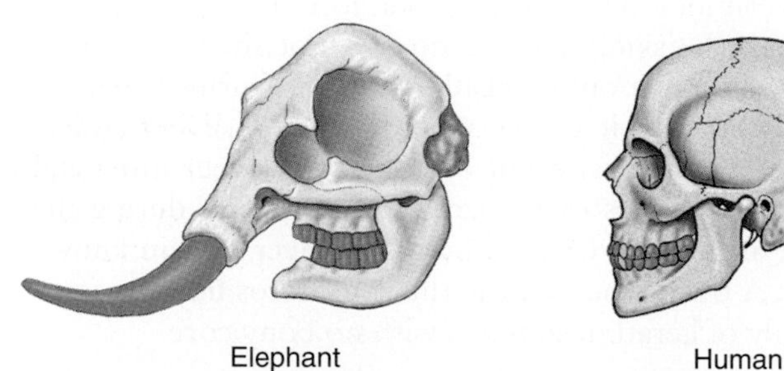

FIGURE 48.39
Mammals have different types of specialized teeth. While reptiles have all the same kind of teeth, mammals have different types of teeth specialized for different feeding habits. Carnivores, such as dogs, have *canine* teeth that are able to rip food; some of the *premolars* and *molars* in dogs are also ripping teeth. Herbivores, such as deer, have *incisors* to chisel off vegetation and molars designed to grind up the plant material. In the beaver, the chiseling incisors dominate. In the elephant, the incisors have become specialized weapons, and molars grind up vegetation. Humans are omnivores; we have ripping, chiseling, and grinding teeth.

Digesting Plants. Most mammals are herbivores, eating mostly or only plants. Cellulose, the major component of plant cell walls, forms the bulk of a plant's body and is a major source of food for mammalian herbivores. The cellulose molecule has the structure of a pearl necklace, with each pearl a glucose sugar molecule. Mammals do not have enzymes that can break the links between the pearls to release the glucose elements for use as food. Herbivorous mammals rely on a mutualistic partnership with bacteria that have the necessary cellulose-splitting enzymes to digest cellulose into sugar for them.

Mammals such as cows, buffalo, antelopes, goats, deer, and giraffes have huge, four-chambered stomachs that function as storage and fermentation vats. The first chamber is the largest and holds a dense population of cellulose-digesting bacteria. Chewed plant material passes into this chamber, where the bacteria attack the cellulose. The material is then digested further in the rest of the stomach.

Rodents, horses, rabbits, and elephants are herbivores that employ mutualistic bacteria to digest cellulose in a different way. They have relatively small stomachs, and instead digest plant material in their large intestine, like a termite. The bacteria that actually carry out the digestion of the cellulose live in a pouch called the cecum that branches from the end of the small intestine.

Even with these complex adaptations for digesting cellulose, a mouthful of plant is less nutritious than a mouthful of flesh. Herbivores must consume large amounts of plant material to gain sufficient nutrition. An elephant eats 135 to 150 kg (300 to 400 pounds) each day.

Hooves and Horns. Keratin, the protein of hair, is also the structural building material in claws, fingernails, and hooves. Hooves are specialized keratin pads on the toes of horses, cows, sheep, antelopes, and other running mammals. The pads are hard and horny, protecting the toe and cushioning it from impact.

The horns of cattle and sheep are composed of a core of bone surrounded by a sheath of keratin. The bony core is attached to the skull, and the horn is not shed. The horn that you see is the outer sheath, made of hairlike fibers of keratin compacted into a very hard structure. Deer antlers are made not of keratin but of bone. Male deer grow and shed a set of antlers each year. While growing during the summer, antlers are covered by a thin layer of skin known as velvet. A third type of horn, the rhinoceros horn, is composed only of keratinized fibers with no bony core.

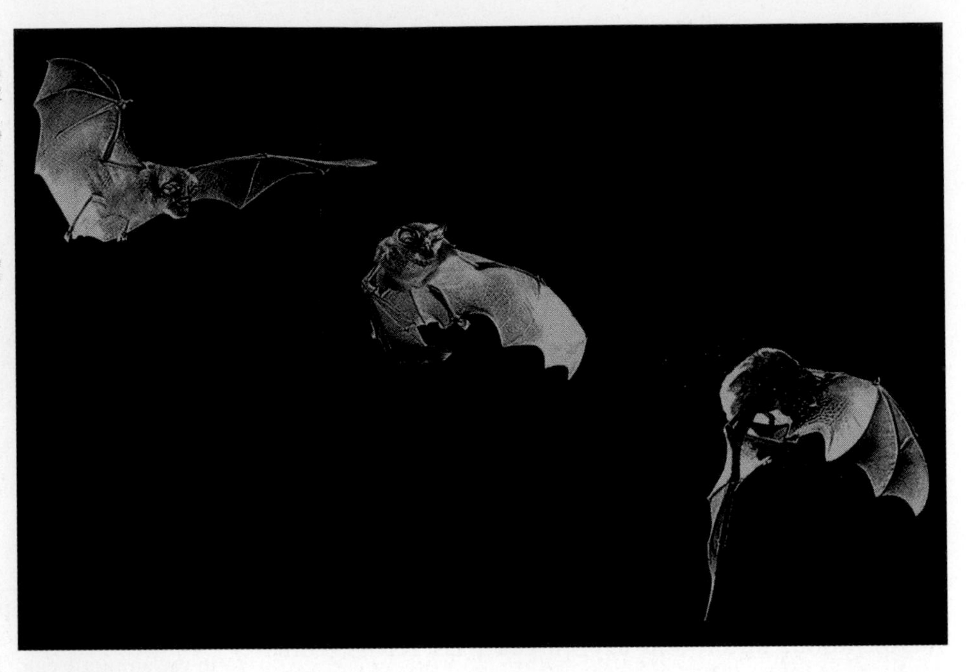

FIGURE 48.40
Greater horseshoe bat, *Rhinolophus ferrumequinum*. The bat is the only mammal capable of true flight.

Flying Mammals. Bats are the only mammals capable of powered flight (figure 48.40). Like the wings of birds, bat wings are modified forelimbs. The bat wing is a leathery membrane of skin and muscle stretched over the bones of four fingers. The edges of the membrane attach to the side of the body and to the hind leg. When resting, most bats prefer to hang upside down by their toe claws. Bats are the second largest order of mammals, after rodents. They have been a particularly successful group because many species have been able to utilize a food resource that most birds do not have access to—night-flying insects.

How do bats navigate in the dark? Late in the eighteenth century, the Italian biologist Lazzaro Spallanzani showed that a blinded bat could fly without crashing into things and still capture insects. Clearly another sense other than vision was being used by bats to navigate in the dark. When Spallanzani plugged the ears of a bat, it was unable to navigate and collided with objects. Spallanzani concluded that bats "hear" their way through the night world.

We now know that bats have evolved a sonar system that functions much like the sonar devices used by ships and submarines to locate underwater objects. As a bat flies, it emits a very rapid series of extremely high-pitched "clicking" sounds well above our range of human hearing. The high-frequency pulses are emitted either through the mouth or, in some cases, through the nose. The soundwaves bounce off obstacles or flying insects, and the bat hears the echo. Through sophisticated processing of this echo within its brain, a bat can determine not only the direction of an object but also the distance to the object.

The Orders of Mammals

There are 19 orders of mammals. Seventeen of them (containing 94% of the species) are placental. The other two are the primitive monotremes and the marsupials.

Monotremes: Egg-laying Mammals. The duck-billed platypus and two species of echidna, or spiny anteater, are the only living monotremes (figure 48.41*a*). Among living mammals, only monotremes lay shelled eggs. The structure of their shoulder and pelvis is more similar to that of the early reptiles than to any other living mammal. Also like reptiles, monotremes have a cloaca, a single opening through which feces, urine, and reproductive products leave the body. Monotremes are more closely related to early mammals than are any other living mammal.

In addition to many reptilian features, monotremes have both defining mammalian features: fur and functioning mammary glands. Young monotremes drink their mother's milk after they hatch from eggs. Females lack well-developed nipples so the babies cannot suckle. Instead, the milk oozes onto the mother's fur, and the babies lap it off with their tongues.

The platypus, found only in Australia, lives much of its life in the water and is a good swimmer. It uses its bill much as a duck does, rooting in the mud for worms and other soft-bodied animals. Echidnas of Australia and New Guinea have very strong, sharp claws, which they use for burrowing and digging. The echidna probes with its long, beaklike snout for insects, especially ants and termites.

(a)

(b)

(c)

FIGURE 48.41
Three types of mammals. (*a*) This echidna, *Tachyglossus aculeatus*, is a monotreme. (*b*) Marsupials include kangaroos, like this adult with young in its pouch. (*c*) This female African lion, *Panthera leo* (order Carnivora), is a placental mammal.

Marsupials: Pouched Mammals. The major difference between marsupials (figure 48.41*b*) and other mammals is their pattern of embryonic development. In marsupials, a fertilized egg is surrounded by chorion and amniotic membranes, but no shell forms around the egg as it does in monotremes. During most of its early development, the marsupial embryo is nourished by an abundant yolk within the egg. Shortly before birth, a short-lived placenta forms from the chorion membrane. Soon after, sometimes within eight days of fertilization, the embryonic marsupial is born. It emerges tiny and hairless, and crawls into the marsupial pouch, where it latches onto a nipple and continues its development.

Marsupials evolved shortly before placental mammals, about 100 million years ago. Today, most species of marsupials live in Australia and South America, areas that have been historically isolated. Marsupials in Australia and New Guinea have diversified to fill ecological positions occupied by placental mammals elsewhere in the world. For example, kangaroos are the Australian grazers, playing the role antelope, horses, and buffalo perform elsewhere. The placental mammals in Australia and New Guinea today arrived relatively recently and include some introduced by humans. The only marsupial found in North America is the Virginia opossum.

Placental Mammals. Mammals that produce a true placenta that nourishes the embryo throughout its entire development are called placental mammals (figure 48.41*c*). Most species of mammals living today, including humans, are in this group. Of the 19 orders of living mammals, 17 are placental mammals. They are a very diverse group, ranging in size from 1.5 g pygmy shrews to 100,000 kg whales.

Early in the course of embryonic development, the placenta forms. Both fetal and maternal blood vessels are abundant in the placenta, and substances can be exchanged efficiently between the bloodstreams of mother and offspring. The fetal placenta is formed from the membranes of the chorion and allantois. The maternal side of the placenta is part of the wall of the uterus, the organ in which the young develop. In placental mammals, unlike marsupials, the young undergo a considerable period of development before they are born.

Mammals were not a major group until the dinosaurs disappeared. Mammal specializations include the placenta, a tooth design suited to diet, and specialized sensory systems.

Chapter 48

www.mhhe.com/raven6/resources48.mhtml

Summary | **Questions** | **Media Resources**

48.1 Attaching muscles to an internal framework greatly improves movement.

- The chordates are characterized by a dorsal nerve cord and by the presence, at least early in development, of a notochord, pharyngeal slits, and a postanal tail. In vertebrates, a bony endoskeleton provides attachment sites for skeletal muscle.

1. What are the four primary characteristics of the chordates?

- Chordates

- Art Quiz: Chordate Features

48.2 Nonvertebrate chordates have a notochord but no backbone.

- Tunicates and the lancelets seem to represent ancient evolutionary Chordate offshoots.

2. What are the three subphyla of the chordates? Give an example of each.

48.3 The vertebrates have an interior framework of bone.

- Vertebrates differ from other chordates in that they possess a vertebral column, a distinct and well-differentiated head, and a bony skeleton.

3. What is the relationship between the notochord and the vertebral column in vertebrates?

- Introduction to Vertebrates

48.4 The evolution of vertebrates involves invasions of sea, land, and air.

- Members of the group Agnatha differ from other vertebrates because they lack jaws.
- Jawed fishes are active swimmers and dominant in fresh and salt water everywhere.
- The first land vertebrates were the amphibians. Amphibians are dependent on water and lay their eggs in moist places.
- Reptiles were the first vertebrates fully adapted to terrestrial habitats. Scales and amniotic eggs represented significant adaptations to the dry conditions on land.
- Birds and mammals were derived from reptiles and are now among the dominant groups of animals on land. The members of these two classes have independently become endothermic, capable of regulating their own body temperatures; all other living animals are ectothermic, their temperatures set by external conditions.
- The living mammals are divided into three major groups: (1) the monotremes, or egg-laying mammals, consisting only of the echidnas and the duck-billed platypus; (2) the marsupials, in which the young are born at a very early stage of development and complete their development in a pouch; and (3) the placental mammals, which lack pouches and suckle their young.

4. What is one advantage of possessing jaws? From what existing structures did jaws evolve?

5. What is the primary disadvantage of a bony skeleton compared to one made of cartilage?

6. What is the lateral line system in fishes? How does it function?

7. The successful invasion of land by amphibians involved five major innovations. What were they, and why was each important?

8. How does the embryo obtain nutrients and excrete wastes while contained within the egg?

9. From what reptilian structure are feathers derived?

10. How do amphibian, reptile, and mammal legs differ?

11. Exactly how would you distinguish a cat from a dog? (be specific)

- Evolution of Fish
- Fish
- Amphibians
- Reptiles
- Birds
- Mammals

- Enhancement Chapter: Dinosaurs
- *On Science* Articles:
  - -Dinosaur Hearts
  - -Dinosaur for Thanksgiving Dinner
  - -Feathered Dinosaurs
- Student Research:
  - -Phylogeny of Hylid Frogs
  - -Metamorphosis in Flatfish

- Art Quizzes:
  - -Evolution of Fishes
  - -Evolution of the Jaw
  - -Amniotic Reptile Egg
  - -Comparison of Reptile and Fish Circulation
  - -Placenta

Part XIII

Animal Form and Function

Some species of lizard breathe better than others. The savannah monitor lizard *Varanus exanthematicus* breathes more efficiently than some of its relatives by pumping air into its lungs from the gular folds over its throat.

Why Some Lizards Take a Deep Breath

Sometimes, what is intended as a straightforward observational study about an animal turns out instead to uncover an odd fact, something that doesn't at first seem to make sense. Teasing your understanding with the unexpected, this kind of tantalizing finding can be fun and illuminating to investigate. Just such an unexpected puzzle comes to light when you look very carefully at how lizards run.

A lizard runs a bit like a football fullback, swinging his shoulder forward to take a step as the opposite foot pushes off the ground. This produces a lateral undulating gait, the body flexing from side to side with each step. This sort of gait uses the body to aid the legs in power running. By contracting the chest (intercostal) muscles on the side of the body opposite the swinging shoulder, the lizard literally thrusts itself forward with each flex of its body.

The odd fact, the thing that at first doesn't seem to make sense, is that running lizards should be using these same intercostal chest muscles for something else.

At rest, lizards breathe by expanding their chest, much as you do. The greater volume of the expanded thorax lowers the interior air pressure, causing fresh air to be pushed into the lungs from outside. You expand your chest by contracting a diaphragm at the bottom of the chest. Lizards do not have a diaphragm. Instead, they expand their chest by contracting the intercostal chest muscles on both sides of the chest simultaneously. This contraction rotates the ribs, causing the chest to expand.

Do you see the problem? A running lizard cannot contract its chest muscles on both sides simultaneously for effective breathing at the same time that it is contracting the same chest muscles alternatively for running. This apparent conflict has led to a controversial hypothesis about how running lizards breathe. Called the axial constraint hypothesis, it states that lizards are subject to a speed-dependent axial constraint that prevents effective lung ventilation while they are running.

This constraint, if true, would be rather puzzling from an evolutionary perspective, because it suggests that when a lizard needs more oxygen because it is running, it breathes less effectively.

Dr. Elizabeth Brainerd of the University of Massachusetts, Amherst, is one of a growing cadre of young researchers around the country that study the biology of lizards. She set out to investigate this puzzle several years ago, first by examining oxygen consumption.

Looking at oxygen consumption seemed a very straightforward approach. If the axial constraint hypothesis is correct, then running lizards should exhibit a lower oxygen consumption because of lowered breathing efficiency. This is just what some of her colleagues found with green iguanas (*Iguana iguana*). Studying fast-running iguanas on treadmills, oxygen consumption went down as running proceeded, as the axial constraint hypothesis predicted.

Unexpectedly, however, another large lizard gave a completely different result. The savannah monitor lizard (*Varanus exanthematicus*) exhibited *elevated* oxygen consumption with increasing speeds of locomotion! This result suggests that something else is going on in monitor lizards. Somehow, they have found a way to beat the axial constraint.

How do they do it? Taking a more detailed look at running monitor lizards, Dr. Brainerd's research team ran a series of experiments to sort this out. First, they used videoradiography to directly observe lung ventilation in monitor lizards while the lizards were running on a treadmill. The X-ray video images revealed the monitor's trick: the breathing cycle began with an inhalation that did not completely fill the lungs, just as the axial constraint hypothesis predicts. But then something else kicks in. The gular cavity located in the throat area also fills with air, and as inhalation proceeds the gular cavity compresses, forcing this air into the lungs. Like an afterburner on a jet, this added air increases the efficiency of breathing, making up for the lost contribution of the intercostal chest muscles.

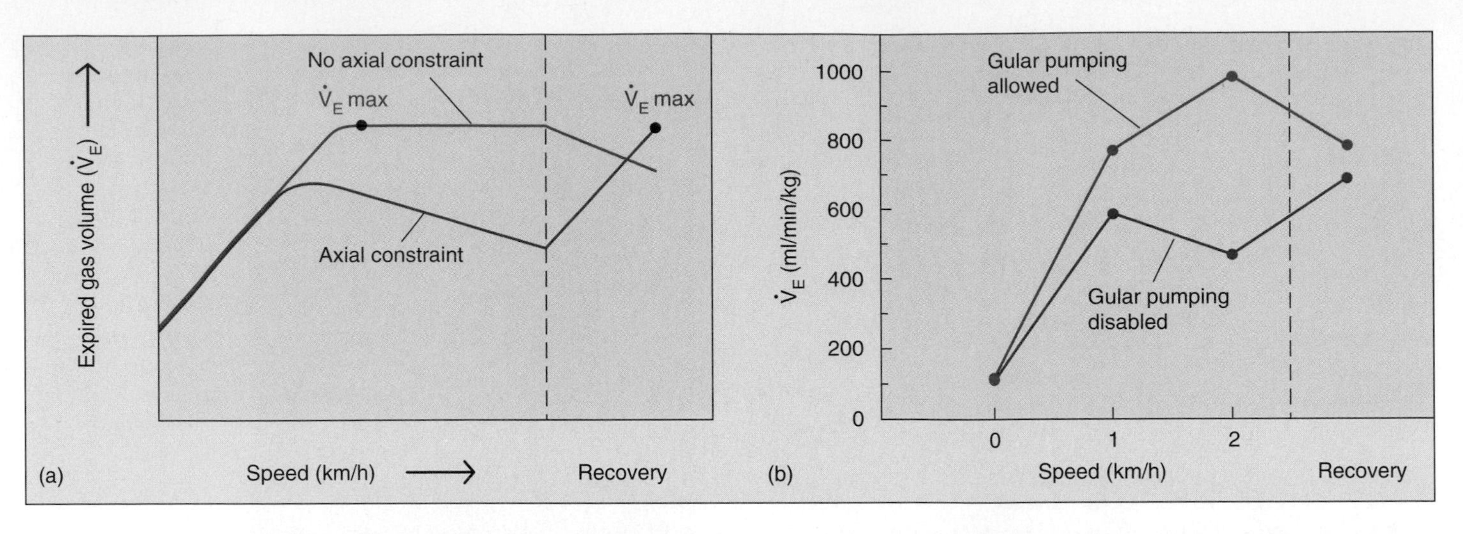

Effects of gular pumping in lizards. (*a*) THEORY: The axial constraint hypothesis predicts that, above a threshold speed, ventilation, measured by expired gas volume ($\dot{V}_E$–ml/min), will decrease with increasing speed, and only reach a maximum during the recovery period after locomotion ceases. Without axial constraint, ventilation should reach its maximum during locomotion. (*b*) EXPERIMENT: Monitor lizards typically show no axial constraint while running. Axial constraint is evident, however, if gular pumping of air is disabled. So, it seems that some species of monitor lizards are able to use gular pumping to overcome the axial constraint on ventilation.

The Experiment

Brainerd set out to test this gular pumping hypothesis. Gular pumping occurs after the initial inhalation because the lizard closes its mouth, sealing shut internal nares (nostril-like structures). Air is thus trapped in the gular cavity. By contracting muscles that compress the gular cavity, this air is forced into the lungs. This process can be disrupted by propping the mouth open so that, when the gular cavity is compressed, its air escapes back out of the mouth. The lizards were trained to run on a treadmill. A plastic mask was placed over the animal's mouth and nostrils and air was drawn through the mask. The mask permitted the measurement of oxygen and CO_2 levels as a means of monitoring gas consumption. The expired gas volume ($\dot{V}_E$) was measured in the last minutes of locomotion and the first minute of recovery at each speed. The speeds ranged from 0 km/hr to 2 km/hr. The maximum running speed of these lizards on a treadmill is 7 km/hr.

To disable gular pumping, the animal's mouth was propped open with a retainer made of plastic tubing. In parallel experiments that allow gular pumping, the same animals wore the masks, but no retainer was used to disrupt the oral seal necessary for gular pumping.

The Results

Parallel experiments were conducted on monitor lizards with and without gular pumping:

1. Gular pumping allowed. When the gular pumping mechanism was not obstructed, the $\dot{V}_E$ increased to a maximum at a speed of 2 km/hr and decreased during the recovery period (see blue line in graph *b* above). This result is predicted under conditions where there is no axial constraint on the animal (see graph *a* above).

2. Gular pumping disabled. When the gular pumping mechanism is obstructed, $\dot{V}_E$ increased above the resting value up to a speed of 1 km/hr. The value began to decrease between 1 and 2 km/hr indicating that there was constraint on ventilation. During the recovery period, $\dot{V}_E$ increased as predicted by the axial constraint hypothesis, because there was no longer constraint on the intercostal muscles. $\dot{V}_E$ increased to pay back an oxygen debt that occurred during the period of time when anaerobic metabolism took over.

Comparing the $\dot{V}_E$ measurements under control and experimental conditions, the researchers concluded that monitor lizards are indeed subject to speed-dependent axial constraint, just as theory had predicted, but can circumvent this constraint when running by using an accessory gular pump to enhance ventilation. When the gular pump was experimentally disrupted, the speed-dependent axial constraint condition became apparent.

Although the researchers have not conducted a more complete comparative analysis using the methods shown here, they have found correlations between gular pumping and increased locomotor activity. During exercise, six highly active species exhibited gular pumping for lung ventilation, while three less active species did not. It is interesting to speculate that gular pumping evolved in lizards as a means of enhancing breathing to allow greater locomotor endurance. The gular pumping seen in lizards is similar to the breathing mechanism found in amphibians and air-breathing fish. In these animals, the air first enters a cavity in the mouth called the buccal cavity. The mouth and nares close and the buccal cavity compresses, forcing air into the lungs. The similarities in these two mechanisms suggest that one might have arisen from the other.

To explore this experiment further, go to the Virtual Lab at www.mhhe.com/raven6/vlab13.mhtml

49

Organization of the Animal Body

Concept Outline

49.1 The bodies of vertebrates are organized into functional systems.

Organization of the Body. Cells are organized into tissues, and tissues are organized into organs. Several organs can cooperate to form organ systems.

49.2 Epithelial tissue forms membranes and glands.

Characteristics of Epithelial Tissue. Epithelial membranes cover all body surfaces, and thus can serve for protection or for transport of materials. Glands are also epithelial tissue. Epithelial membranes may be composed of one layer of cells or many layers.

49.3 Connective tissues contain abundant extracellular material.

Connective Tissue Proper. Connective tissues have abundant extracellular material. In connective tissue proper, this material consists of protein fibers within an amorphous ground substance.

Special Connective Tissues. These tissues include cartilage, bone, and blood, each with their own unique form of extracellular material.

49.4 Muscle tissue provides for movement, and nerve tissue provides for control.

Muscle Tissue. Muscle tissue contains the filaments actin and myosin, which enable the muscles to contract. There are three types of muscle: smooth, cardiac, and skeletal.

Nerve Tissue. Nerve cells, or neurons, have specialized regions that produce and conduct electrical impulses. Neuroglia cells support neurons but do not conduct electrical impulses.

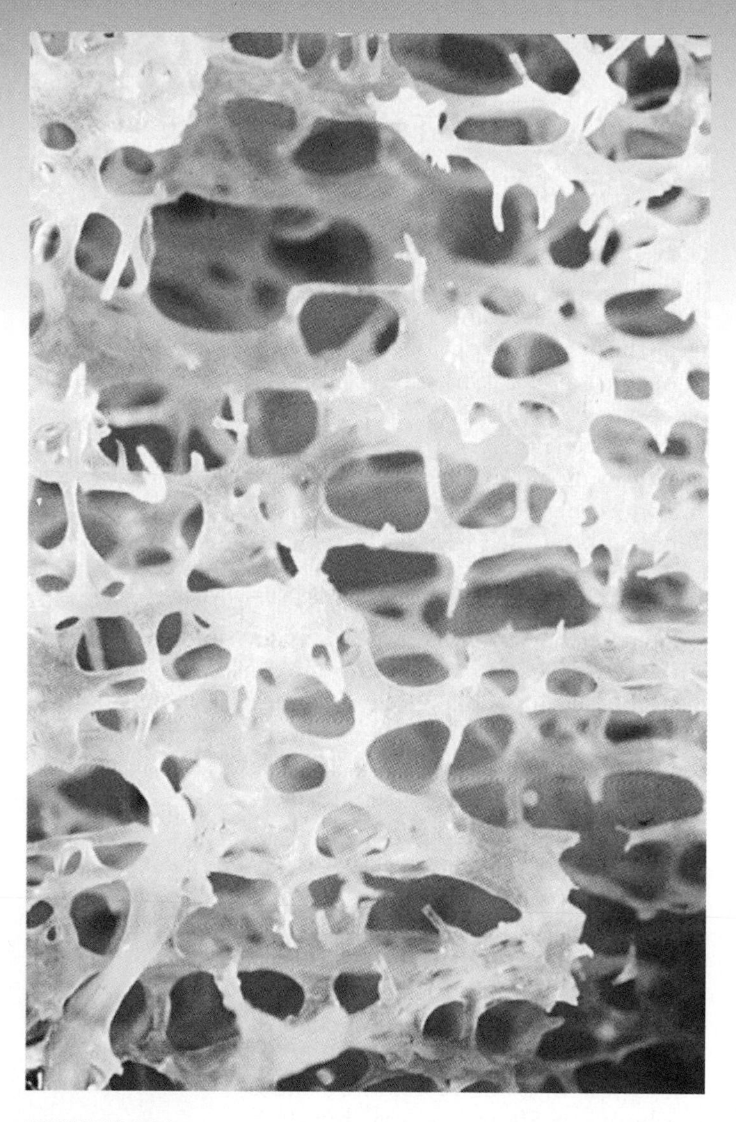

FIGURE 49.1
Bone. Like most of the tissues in the vertebrate body, bone is a dynamic structure, constantly renewing itself.

When most people think of animals, they think of their pet dogs and cats, and of the animals that they've seen in a zoo, on a farm, in an aquarium, or out in the wild. When they think about the diversity of animals, they may think of the differences between the predatory lions and tigers and the herbivorous deer and antelope, between a ferocious-looking shark and a playful dolphin. Despite the differences among these animals, they are all vertebrates. All vertebrates share the same basic body plan, with the same sorts of organs operating in much the same way. In this chapter, we will begin a detailed consideration of the biology of vertebrates and of the fascinating structure and function of their bodies (figure 49.1).

49.1 The bodies of vertebrates are organized into functional systems.

Organization of the Body

The bodies of all mammals have the same general architecture (figure 49.2), and are very similar to the general body plan of other vertebrate groups. This body plan is basically a tube suspended within a tube. Starting from the inside, it is composed of the digestive tract, a long tube that travels from one end of the body to the other (mouth to anus). This tube is suspended within an internal body cavity, the *coelom*. In fishes, amphibians, and most reptiles, the coelom is subdivided into two cavities, one housing the heart and the other the liver, stomach, and intestines. In mammals and some reptiles, a sheet of muscle, the *diaphragm*, separates the **peritoneal cavity,** which contains the stomach, intestines, and liver, from the **thoracic cavity;** the thoracic cavity is further subdivided into the *pericardial cavity*, which contains the heart, and *pleural cavities*, which contain the lungs. All vertebrate bodies are supported by an internal **skeleton** made of jointed bones or cartilage blocks that grow as the body grows. A bony *skull* surrounds the brain, and a column of bones, the *vertebrae*, surrounds the dorsal nerve cord, or *spinal cord*.

There are four levels of organization in the vertebrate body: (1) cells, (2) tissues, (3) organs, and (4) organ systems. Like those of all animals, the bodies of vertebrates are composed of different cell types. In adult vertebrates, there are between 50 and several hundred different kinds of cells.

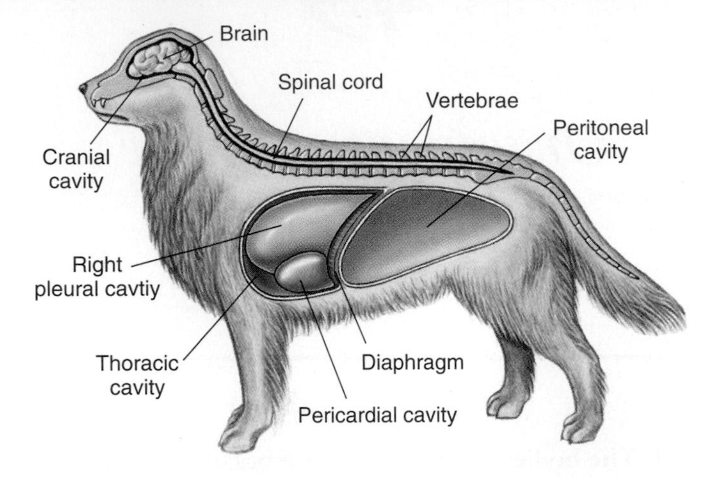

FIGURE 49.2
Architecture of the vertebrate body. All vertebrates have a dorsal central nervous system. In mammals and some reptiles, a muscular diaphragm divides the coelom into the thoracic cavity and the peritoneal cavity.

Tissues

Groups of cells similar in structure and function are organized into **tissues.** Early in development, the cells of the growing embryo differentiate (specialize) into three fundamental embryonic tissues, called *germ layers*. From innermost to outermost layers, these are the **endoderm, mesoderm,** and **ectoderm.** These germ layers, in turn, differentiate into the scores of different cell types and tissues that are characteristic of the vertebrate body. In adult vertebrates, there are four principal kinds of tissues, or *primary tissues:* epithelial, connective, muscle, and nerve (figure 49.3), each discussed in separate sections of this chapter.

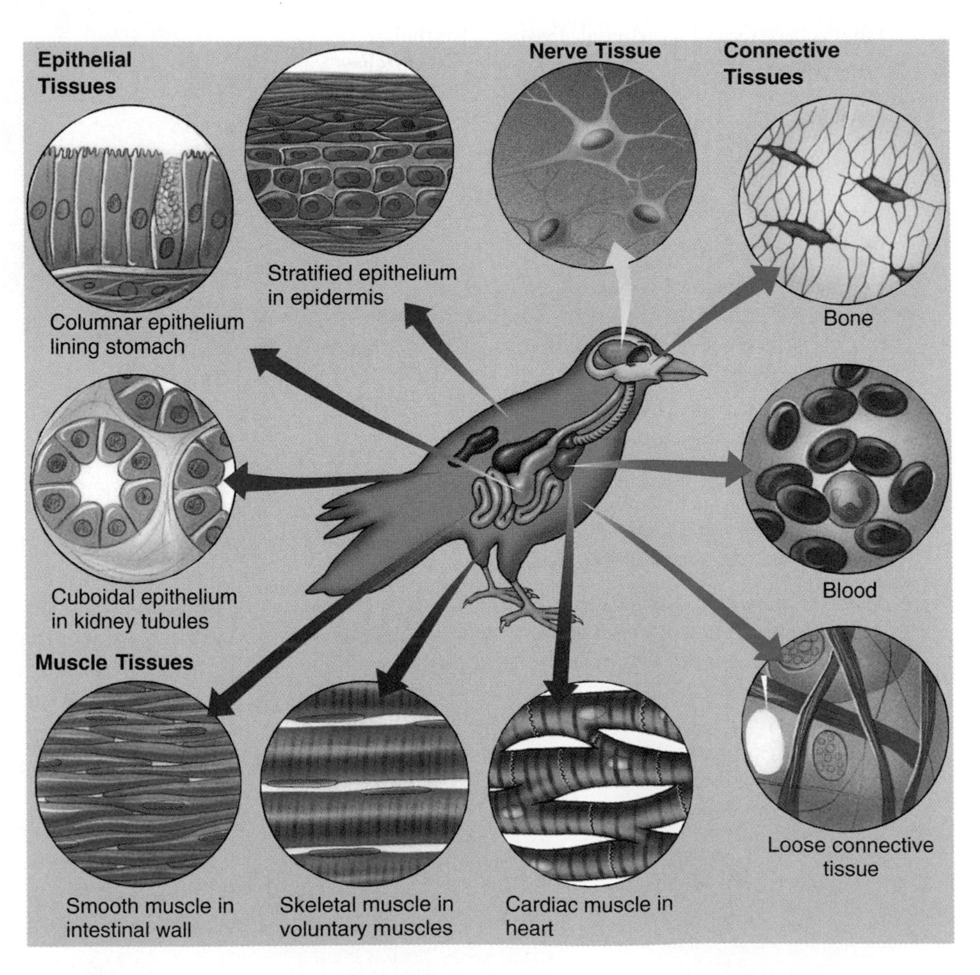

FIGURE 49.3
Vertebrate tissue types. Epithelial tissues are indicated by *blue* arrows, connective tissues by *green* arrows, muscle tissues by *red* arrows, and nerve tissue by a *yellow* arrow.

Organs and Organ Systems

Organs are body structures composed of several different tissues that form a structural and functional unit (figure 49.4). One example is the heart, which contains cardiac muscle, connective tissue, and epithelial tissue and is laced with nerve tissue that helps regulate the heartbeat. An **organ system** is a group of organs that function together to carry out the major activities of the body. For example, the digestive system is composed of the digestive tract, liver, gallbladder, and pancreas. These organs cooperate in the digestion of food and the absorption of digestion products into the body. The vertebrate body contains 11 principal organ systems (table 49.1 and figure 49.5).

The bodies of humans and other mammals contain a cavity divided by the diaphragm into thoracic and peritoneal cavities. The body's cells are organized into tissues, which are, in turn, organized into organs and organ systems.

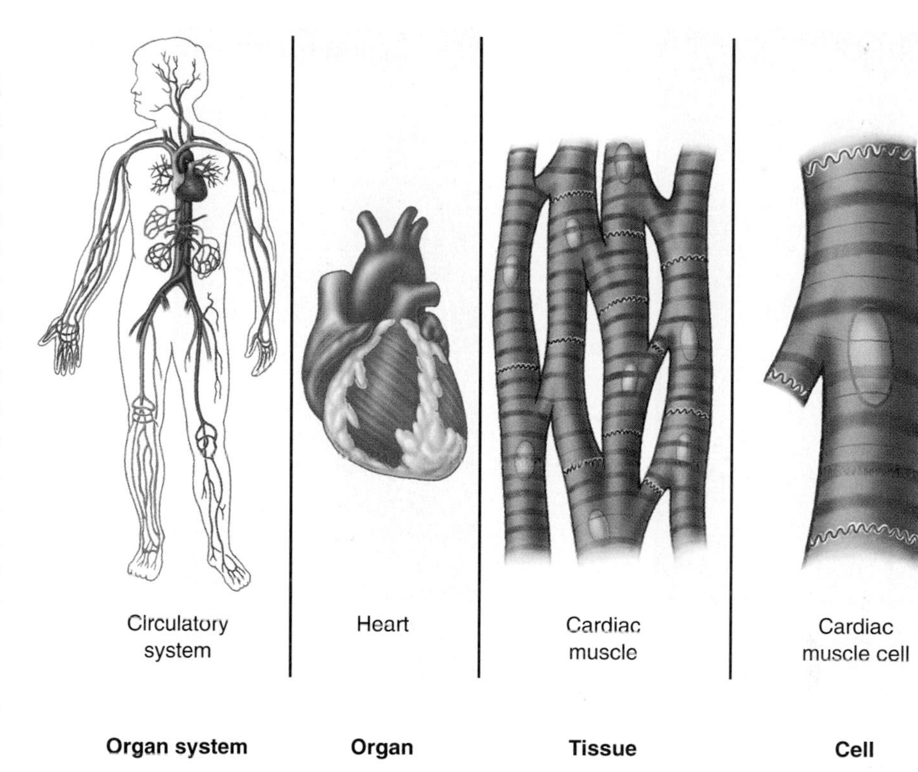

FIGURE 49.4
Levels of organization within the body. Similar cell types operate together and form tissues. Tissues functioning together form organs. Several organs working together to carry out a function for the body are called an organ system. The circulatory system is an example of an organ system.

Table 49.1 The Major Vertebrate Organ Systems

System	Functions	Components	Detailed Treatment
Circulatory	Transports cells, respiratory gases, and chemical compounds throughout the body	Heart, blood vessels, lymph, and lymph structures	Chapter 52
Digestive	Captures soluble nutrients from ingested food	Mouth, esophagus, stomach, intestines, liver, and pancreas	Chapter 51
Endocrine	Coordinates and integrates the activities of the body	Pituitary, adrenal, thyroid, and other ductless glands	Chapter 56
Integumentary	Covers and protects the body	Skin, hair, nails, scales, feathers, and sweat glands	Chapter 57
Lymphatic/ Immune	Vessels transport extracellular fluid and fat to circulatory system; lymph nodes and lymphatic organs provide defenses to microbial infection and cancer	Lymphatic vessels, lymph nodes, thymus, tonsils, spleen	Chapter 57
Muscular	Produces body movement	Skeletal muscle, cardiac muscle, and smooth muscle	Chapter 50
Nervous	Receives stimuli, integrates information, and directs the body	Nerves, sense organs, brain, and spinal cord	Chapters 54, 55
Reproductive	Carries out reproduction	Testes, ovaries, and associated reproductive structures	Chapter 59
Respiratory	Captures oxygen and exchanges gases	Lungs, trachea, gills, and other air passageways	Chapter 53
Skeletal	Protects the body and provides support for locomotion and movement	Bones, cartilage, and ligaments	Chapter 50
Urinary	Removes metabolic wastes from the bloodstream	Kidney, bladder, and associated ducts	Chapter 58

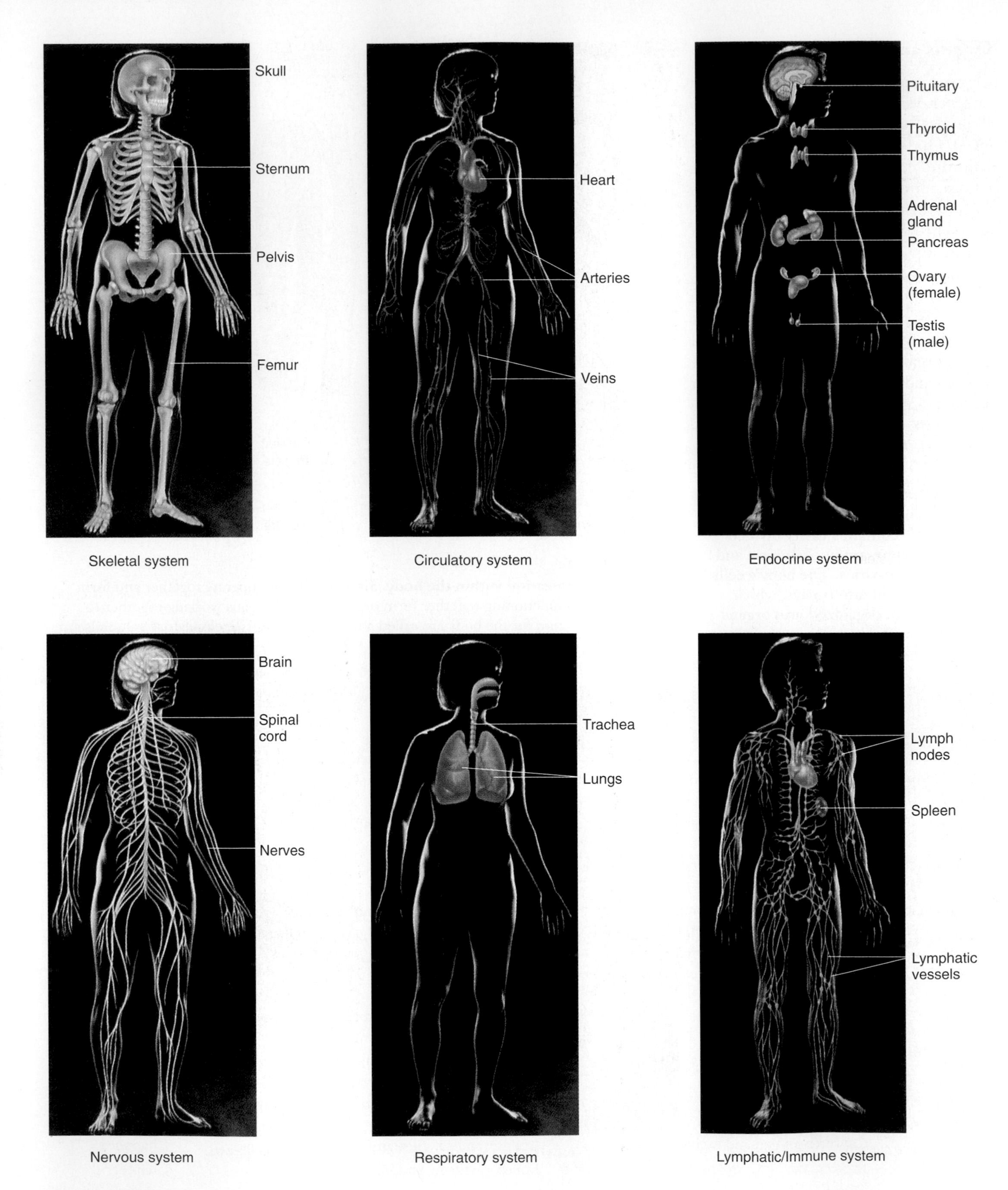

FIGURE 49.5
Vertebrate organ systems. The 11 principal organ systems of the human body are shown, including both male and female reproductive systems.

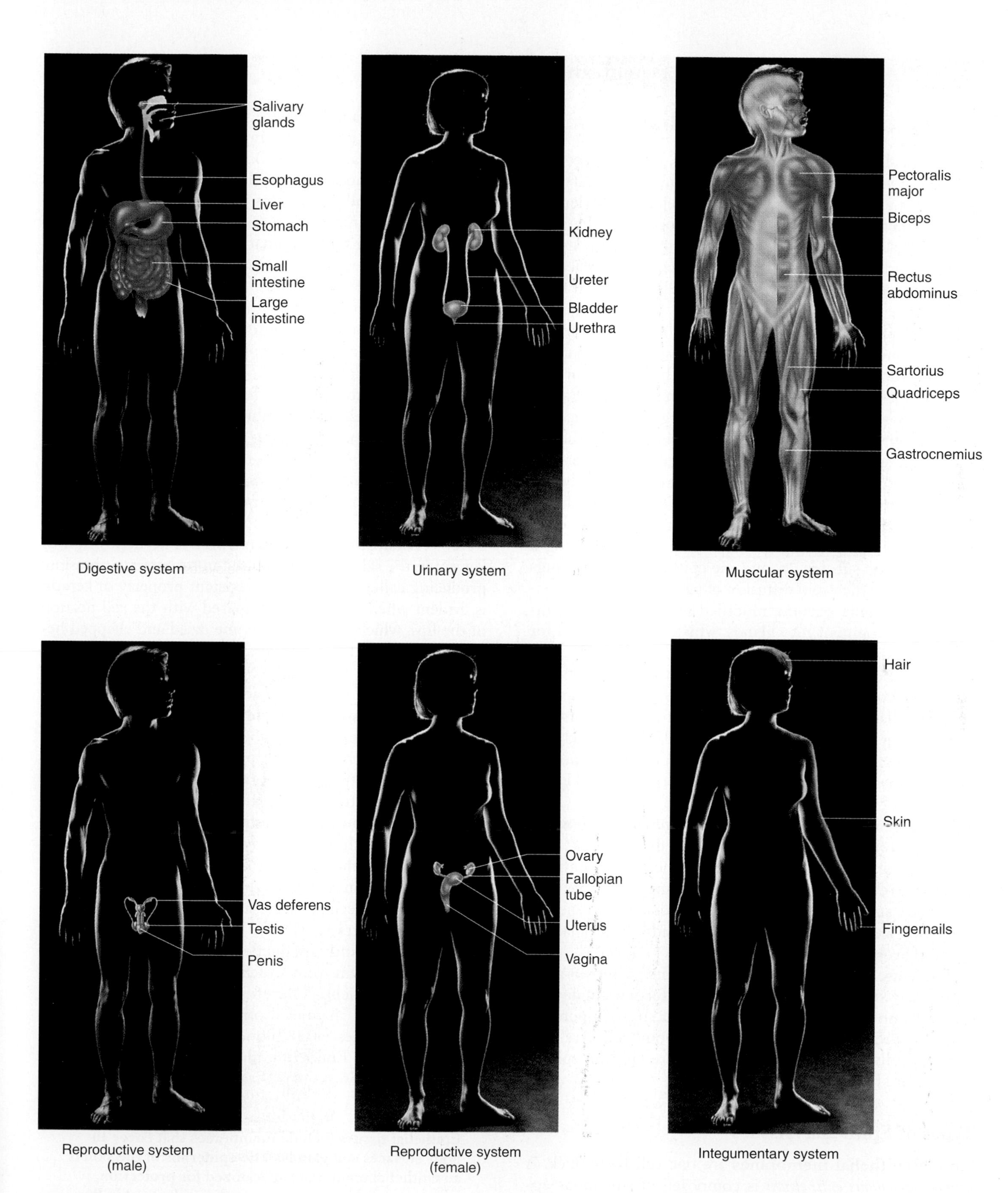
Salivary glands
Esophagus
Liver
Stomach
Small intestine
Large intestine
Digestive system
Kidney
Ureter
Bladder
Urethra
Urinary system
Pectoralis major
Biceps
Rectus abdominus
Sartorius
Quadriceps
Gastrocnemius
Muscular system
Vas deferens
Testis
Penis
Reproductive system (male)
Ovary
Fallopian tube
Uterus
Vagina
Reproductive system (female)
Hair
Skin
Fingernails
Integumentary system

FIGURE 49.5 (continued)

49.2 Epithelial tissue forms membranes and glands.

Characteristics of Epithelial Tissue

An epithelial membrane, or **epithelium,** covers every surface of the vertebrate body. Epithelial membranes are derived from all three germ layers. The epidermis, derived from ectoderm, constitutes the outer portion of the skin. The inner surface of the digestive tract is lined by an epithelium derived from endoderm, and the inner surfaces of the body cavities are lined with an epithelium derived from mesoderm.

Because all body surfaces are covered by epithelial membranes, a substance must pass through an epithelium in order to enter or leave the body. Epithelial membranes thus provide a barrier that can impede the passage of some substances while facilitating the passage of others. For land-dwelling vertebrates, the relative impermeability of the surface epithelium (the epidermis) to water offers essential protection from dehydration and from airborne pathogens (disease-causing organisms). On the other hand, the epithelial lining of the digestive tract must allow selective entry of the products of digestion while providing a barrier to toxic substances, and the epithelium of the lungs must allow for the rapid diffusion of gases.

Some epithelia become modified in the course of embryonic development into glands, which are specialized for secretion. A characteristic of all epithelia is that the cells are tightly bound together, with very little space between them. As a consequence, blood vessels cannot be interposed between adjacent epithelial cells. Therefore, nutrients and oxygen must diffuse to the epithelial cells from blood vessels in nearby tissues. This places a limit on the thickness of epithelial membranes; most are only one or a few cell layers thick.

Epithelium possesses remarkable regenerative powers, constantly replacing its cells throughout the life of the animal. For example, the liver, a gland formed from epithelial tissue, can readily regenerate after substantial portions of it have been surgically removed. The epidermis is renewed every two weeks, and the epithelium inside the stomach is replaced every two to three days.

There are two general classes of epithelial membranes: simple and stratified. These classes are further subdivided into squamous, cuboidal, and columnar, based upon the shape of the cells (table 49.2). Squamous cells are flat, cuboidal cells are about as wide as they are tall, and columnar cells are taller than they are wide.

Types of Epithelial Tissues

Simple epithelial membranes are one cell layer thick. A *simple, squamous epithelium* is composed of squamous epithelial cells that have an irregular, flattened shape with tapered edges. Such membranes line the lungs and blood capillaries, for example, where the thin, delicate nature of these membranes permits the rapid movement of molecules (such as the diffusion of gases). A *simple cuboidal epithelium* lines the small ducts of some glands, and a *simple columnar epithelium* is found in the airways of the respiratory tract and in the gastrointestinal tract, among other locations. Interspersed among the columnar epithelial cells are numerous *goblet cells,* specialized to secrete mucus. The columnar epithelial cells of the respiratory airways contain cilia on their apical surface (the surface facing the lumen, or cavity), which move mucus toward the throat. In the small intestine, the apical surface of the columnar epithelial cells form fingerlike projections called *microvilli,* that increase the surface area for the absorption of food.

Stratified epithelial membranes are several cell layers thick and are named according to the features of their uppermost layers. For example, the epidermis is a *stratified squamous epithelium.* In terrestrial vertebrates it is further characterized as a *keratinized epithelium,* because its upper layer consists of dead squamous cells and filled with a water-resistant protein called *keratin.* The deposition of keratin in the skin can be increased in response to abrasion, producing calluses. The water-resistant property of keratin is evident when the skin is compared with the red portion of the lips, which can easily become dried and chapped because it is covered by a nonkeratinized, stratified squamous epithelium.

The glands of vertebrates are derived from invaginated epithelium. In **exocrine glands,** the connection between the gland and the epithelial membrane is maintained as a duct. The duct channels the product of the gland to the surface of the epithelial membrane and thus to the external environment (or to an interior compartment that opens to the exterior, such as the digestive tract). Examples of exocrine glands include sweat and sebaceous (oil) glands, which secrete to the external surface of the skin, and accessory digestive glands such as the salivary glands, liver, and pancreas, which secrete to the surface of the epithelium lining the digestive tract.

Endocrine glands are ductless glands; their connections with the epithelium from which they were derived are lost during development. Therefore, their secretions, called *hormones,* are not channeled onto an epithelial membrane. Instead, hormones enter blood capillaries and thus stay within the body. Endocrine glands are discussed in more detail in chapter 56.

Epithelial tissues include membranes that cover all body surfaces and glands. The epidermis of the skin is an epithelial membrane specialized for protection, whereas membranes that cover the surfaces of hollow organs are often specialized for transport.

Table 49.2 Epithelial Tissue

Simple Epithelium

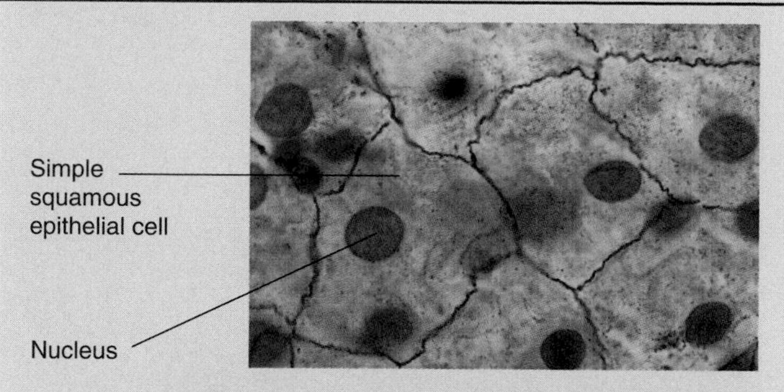

SQUAMOUS

Typical Location
Lining of lungs, capillary walls, and blood vessels

Function
Cells very thin; provides thin layer across which diffusion can readily occur

Characteristic Cell Types
Epithelial cells

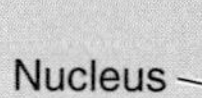
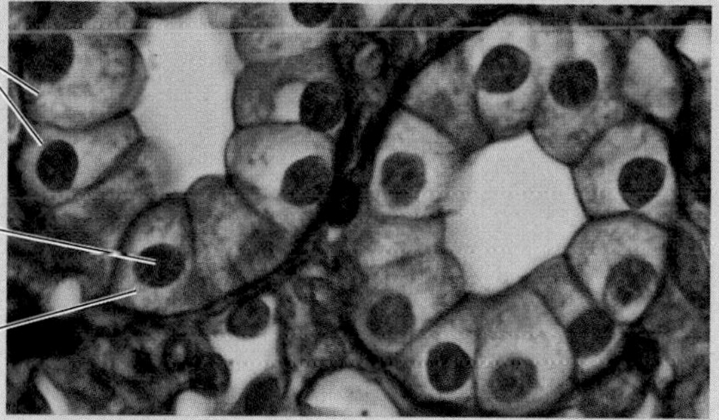
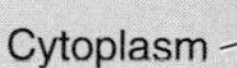

CUBOIDAL

Typical Location
Lining of some glands and kidney tubules; covering of ovaries

Function
Cells rich in specific transport channels; functions in secretion and absorption

Characteristic Cell Types
Gland cells

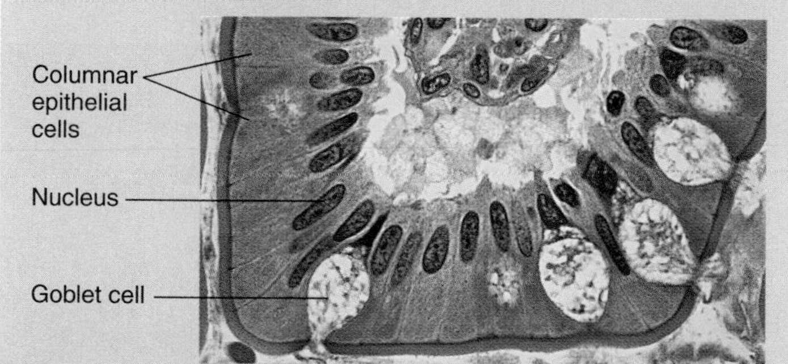

COLUMNAR

Typical Location
Surface lining of stomach, intestines, and parts of respiratory tract

Function
Thicker cell layer; provides protection and functions in secretion and absorption

Characteristic Cell Types
Epithelial cells

Stratified Epithelium

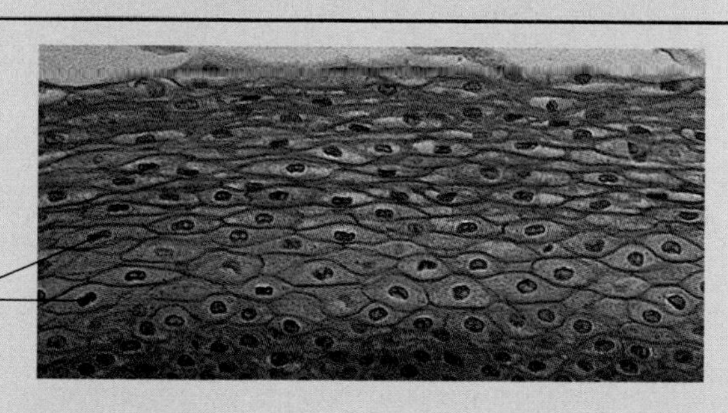

SQUAMOUS

Typical Location
Outer layer of skin; lining of mouth

Function
Tough layer of cells; provides protection

Characteristic Cell Types
Epithelial cells

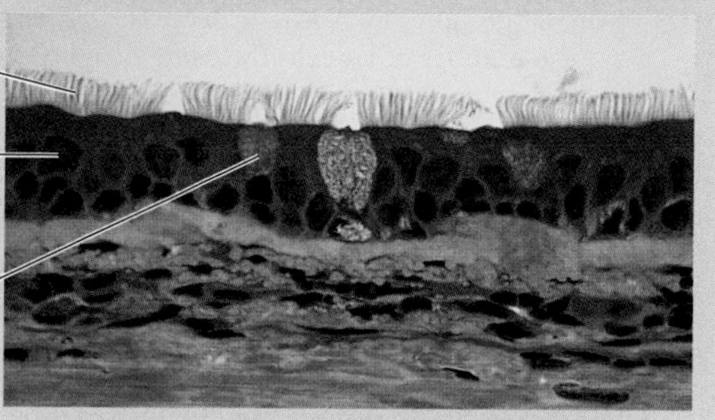

PSEUDOSTRATIFIED COLUMNAR

Typical Location
Lining parts of the respiratory tract

Function
Secretes mucus; dense with cilia that aid in movement of mucus; provides protection

Characteristic Cell Types
Gland cells; ciliated epithelial cells

Connective Tissue Proper

Connective tissues are derived from embryonic mesoderm and occur in many different forms (table 49.3). These various forms are divided into two major classes: **connective tissue proper,** which is further divided into loose and dense connective tissues; and **special connective tissues** that include cartilage, bone, and blood. At first glance, it may seem odd that such diverse tissues are placed in the same category. Yet all connective tissues do share a common structural feature: they all have abundant extracellular material because their cells are spaced widely apart. This extracellular material is generically known as the **matrix** of the tissue. In bone, the extracellular matrix contains crystals that make the bones hard; in blood, the extracellular matrix is plasma, the fluid portion of the blood.

Loose connective tissue consists of cells scattered within an amorphous mass of proteins that form a **ground substance.** This gelatinous material is strengthened by a loose scattering of protein fibers such as *collagen* (figure 49.6), *elastin,* which makes the tissue elastic, and *reticulin,* which supports the tissue by forming a collagenous meshwork. The flavored gelatin we eat for dessert consists of the extracellular material from loose connective tissues. The cells that secrete collagen and other fibrous proteins are known as *fibroblasts.*

Loose connective tissue contains other cells as well, including *mast cells* that produce histamine (a blood vessel dilator) and heparin (an anticoagulant) and *macrophages,* the immune system's first defense against invading organisms, as will be described in detail in chapter 57.

Adipose cells are found in loose connective tissue, usually in large groups that form what is referred to as *adipose tissue* (figure 49.7). Each adipose cell contains a droplet of fat (triglycerides) within a storage vesicle. When that fat is needed for energy, the adipose cell hydrolyzes its stored triglyceride and secretes fatty acids into the blood for oxidation by the cells of the muscles, liver, and other organs. The number of adipose cells in an adult is generally fixed. When a person gains weight, the cells become larger, and when weight is lost, the cells shrink.

Dense connective tissue contains tightly packed collagen fibers, making it stronger than loose connective tissue. It consists of two types: regular and irregular. The collagen fibers of **dense regular connective tissue** are lined up in parallel, like the strands of a rope. This is the structure of *tendons,* which bind muscle to bone, and *ligaments,* which bind bone to bone. In contrast, the collagen fibers of **dense irregular connective tissue** have many different orientations. This type of connective tissue produces the tough coverings that package organs, such as the *capsules* of the kidneys and adrenal glands. It also covers muscle as *epimysium,* nerves as *perineurium,* and bones as *periosteum.*

FIGURE 49.6
Collagen fibers. Each fiber is composed of many individual collagen strands and can be very strong under tension.

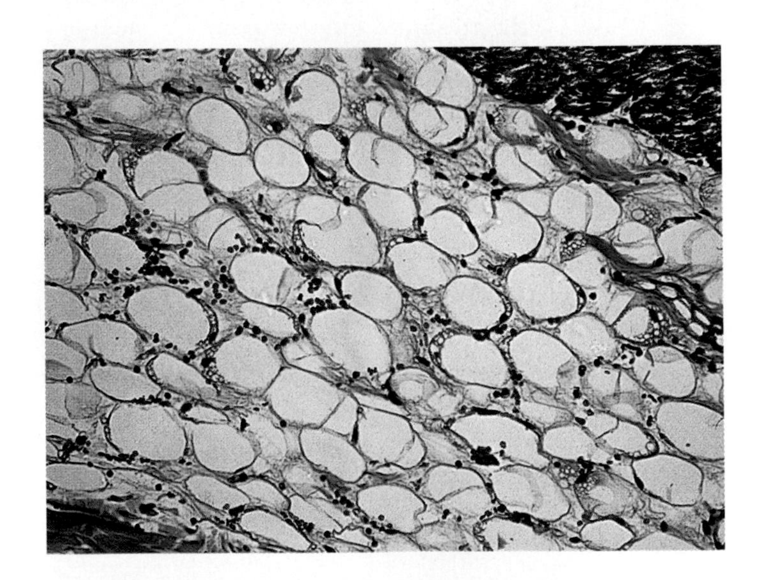

FIGURE 49.7
Adipose tissue. Fat is stored in globules of adipose tissue, a type of loose connective tissue. As a person gains or loses weight, the size of the fat globules increases or decreases. A person cannot decrease the number of fat cells by losing weight.

Connective tissues are characterized by abundant extracellular materials in the matrix between cells. Connective tissue proper may be either loose or dense.

Table 49.3 Connective Tissue

LOOSE CONNECTIVE TISSUE

Typical Location
Beneath skin; between organs

Function
Provides support, insulation, food storage, and nourishment for epithelium

Characteristic Cell Types
Fibroblasts, macrophages, mast cells, fat cells

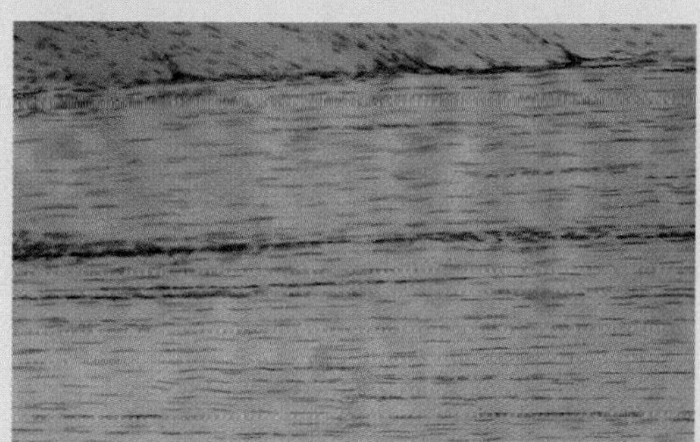

DENSE CONNECTIVE TISSUE

Typical Location
Tendons; sheath around muscles; kidney; liver; dermis of skin

Function
Provides flexible, strong connections

Characteristic Cell Types
Fibroblasts

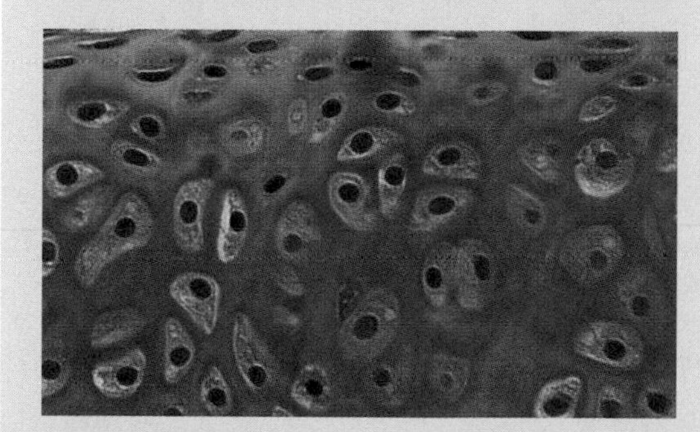

CARTILAGE

Typical Location
Spinal discs; knees and other joints; ear; nose; tracheal rings

Function
Provides flexible support, shock absorption, and reduction of friction on load-bearing surfaces

Characteristic Cell Types
Chondrocytes

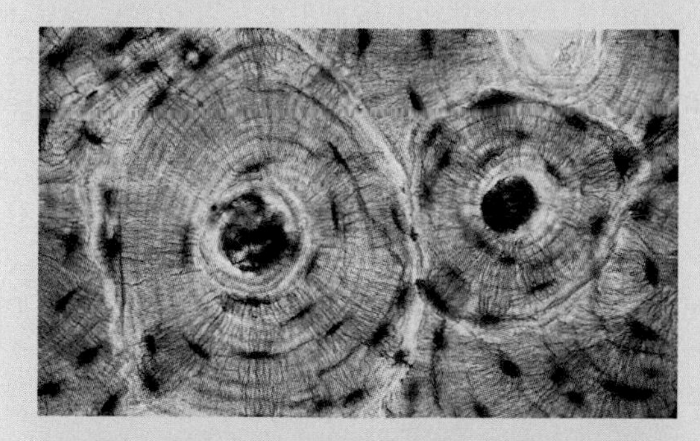

BONE

Typical Location
Most of skeleton

Function
Protects internal organs; provides rigid support for muscle attachment

Characteristic Cell Types
Osteocytes

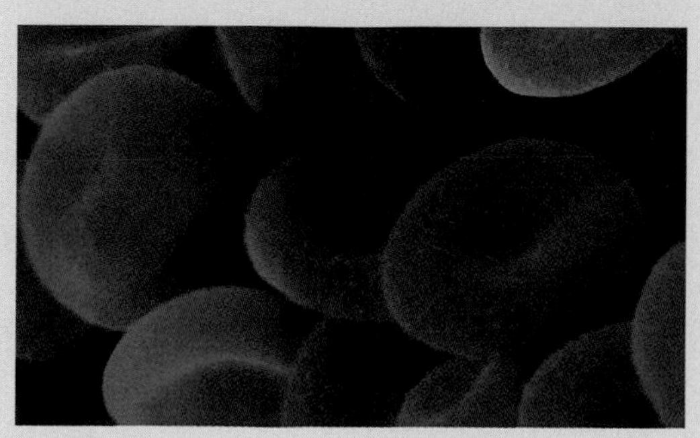

BLOOD

Typical Location
Circulatory system

Function
Functions as highway of immune system and primary means of communication between organs

Characteristic Cell Types
Erythrocytes, leukocytes

Special Connective Tissues

The special connective tissues—cartilage, bone, and blood—each have unique cells and extracellular matrices that allow them to perform their specialized functions.

Cartilage

Cartilage (figure 49.8) is a specialized connective tissue in which the ground substance is formed from a characteristic type of glycoprotein, and the collagen fibers are laid down along the lines of stress in long, parallel arrays. The result is a firm and flexible tissue that does not stretch, is far tougher than loose or dense connective tissue, and has great tensile strength. Cartilage makes up the entire skeletal system of the modern agnathans and cartilaginous fishes (see chapter 48), replacing the bony skeletons that were characteristic of the ancestors of these vertebrate groups. In most adult vertebrates, however, cartilage is restricted to the articular (joint) surfaces of bones that form freely movable joints and to other specific locations. In humans, for example, the tip of the nose, the pinna (outer ear flap), the intervertebral discs of the backbone, the larynx (voice box) and a few other structures are composed of cartilage.

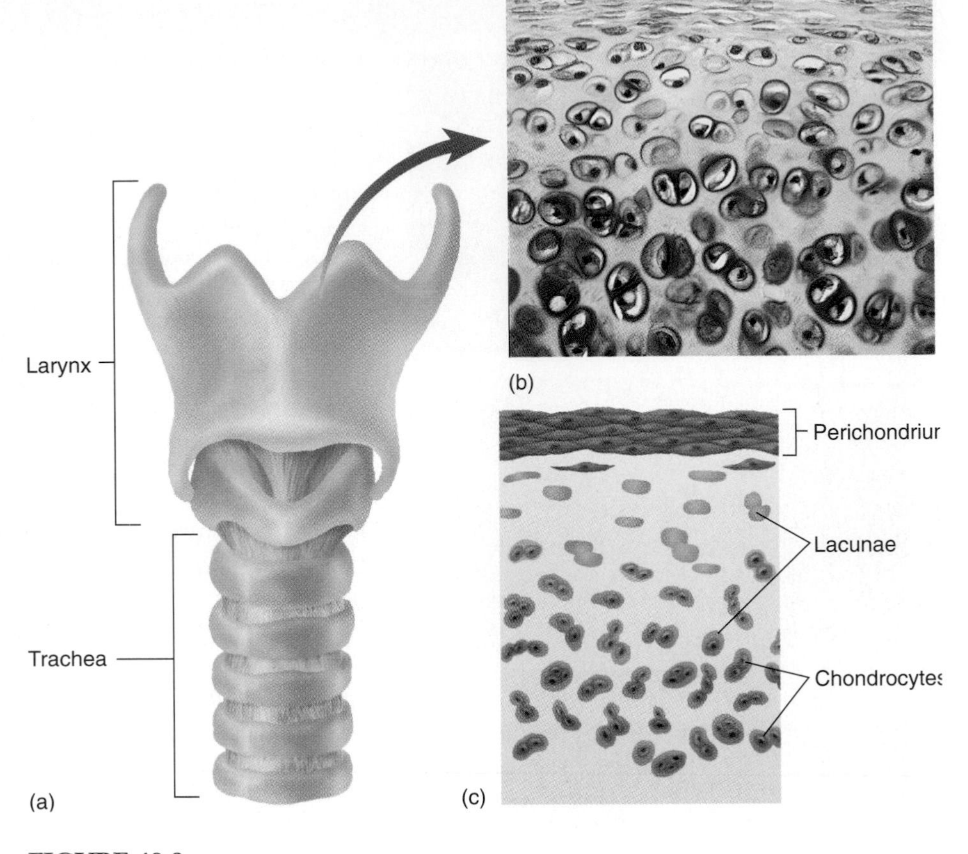

FIGURE 49.8
Cartilage is a strong, flexible tissue that makes up the larynx (voice box) and several other structures in the human body. The larynx (*a*) is seen under the light microscope in (*b*), where the cartilage cells, or chondrocytes, are visible within cavities, or lacunae, in the matrix (extracellular material) of the cartilage. This is diagrammed in (*c*).

Chondrocytes, the cells of the cartilage, live within spaces called *lacunae* within the cartilage ground substance. These cells remain alive, even though there are no blood vessels within the cartilage matrix, because they receive oxygen and nutrients by diffusion through the cartilage ground substance from surrounding blood vessels. This diffusion can only occur because the cartilage matrix is not calcified, as is bone.

Bone

In the course of fetal development, the bones of vertebrate fins, arms, and legs, among others, are first "modeled" in cartilage. The cartilage matrix then calcifies at particular locations, so that the chondrocytes are no longer able to obtain oxygen and nutrients by diffusion through the matrix. The dying and degenerating cartilage is then replaced by living bone. Bone cells, or osteocytes, can remain alive even though the extracellular matrix becomes hardened with crystals of calcium phosphate. This is because blood vessels travel through central canals into the bone. Osteocytes extend cytoplasmic processes toward neighboring osteocytes through tiny canals, or *canaliculi* (figure 49.9). Osteocytes communicate with the blood vessels in the central canal through this cytoplasmic network.

It should be noted here that some bones, such as those of the cranium, are not formed first as cartilage models. These bones instead develop within a membrane of dense, irregular connective tissue. The structure and formation of bone are discussed in chapter 50.

Blood

Blood is classified as a connective tissue because it contains abundant extracellular material, the fluid plasma. The cells of blood are erythrocytes, or red blood cells, and leukocytes, or white blood cells (figure 49.10). Blood also contains platelets, or *thrombocytes*, which are fragments of a type of bone marrow cell.

Erythrocytes are the most common blood cells; there are about 5 billion in every milliliter of blood. During their maturation in mammals, they lose their nucleus, mitochondria, and endoplasmic reticulum. As a result, mammalian erythrocytes are relatively inactive metabolically. Each erythrocyte contains about 300 million molecules of the iron-containing protein *hemoglobin*, the principal carrier of oxygen in vertebrates and in many other groups of animals.

There are several types of leukocytes, but together they are only one-thousandth as numerous as erythrocytes. Unlike mammalian erythrocytes, leukocytes have nuclei and mitochondria but lack the red pigment hemoglobin. These cells are therefore hard to see under a microscope without special staining. The names *neutrophils*, *eosinophils*, and *basophils* distinguish three types of leukocytes on the basis of their staining properties; other leukocytes include *lymphocytes* and *monocytes*. These different types of leukocytes play critical roles in immunity, as will be described in chapter 57.

The blood plasma is the "commons" of the body; it (or a derivative of it) travels to and from every cell in the body. As the plasma circulates, it carries nourishment, waste products, heat, and regulatory molecules. Practically every substance used by cells, including sugars, lipids, and amino acids, is delivered by the plasma to the body cells. Waste products from the cells are carried by the plasma to the kidneys, liver, and lungs or gills for disposal, and regulatory molecules (hormones) that endocrine gland cells secrete are carried by the plasma to regulate the activities of most organs of the body. The plasma also contains sodium, calcium, and other inorganic ions that all cells need, as well as numerous proteins. Plasma proteins include *fibrinogen*, produced by the liver, which helps blood to clot; *albumin*, also produced by the liver, which exerts an osmotic force needed for fluid balance; and *antibodies* produced by lymphocytes and needed for immunity.

Special connective tissues each have a unique extracellular matrix between cells. The matrix of cartilage is composed of organic material, whereas that of bone is impregnated with calcium phosphate crystals. The matrix of blood is fluid, the plasma.

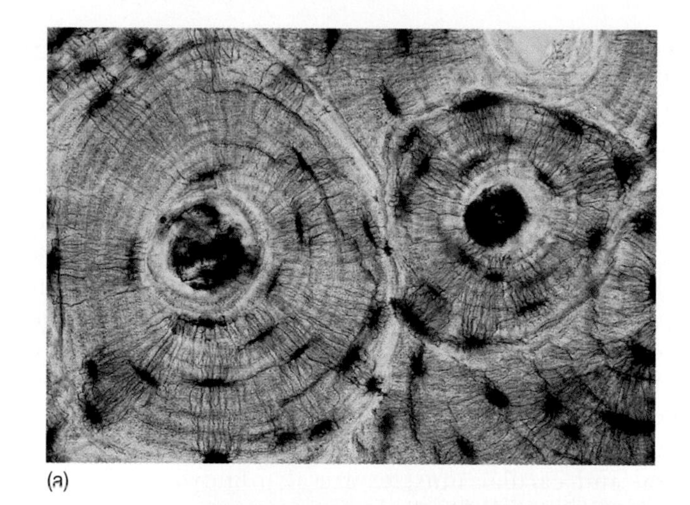

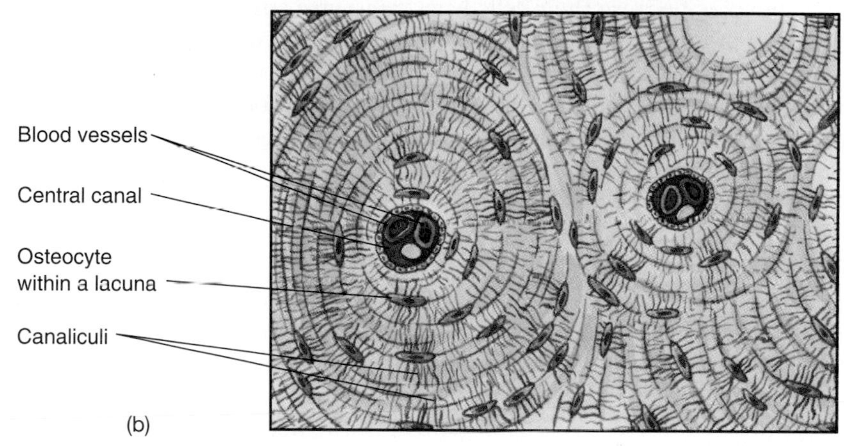

FIGURE 49.9
The structure of bone. A photomicrograph (*a*) and diagram (*b*) of the structure of bone, showing the bone cells, or osteocytes, within their lacunae (cavities) in the bone matrix. Though the bone matrix is calcified, the osteocytes remain alive because they can be nourished by blood vessels in the central cavity. Nourishment is carried between the osteocytes through a network of cytoplasmic processes extending through tiny canals, or canaliculi.

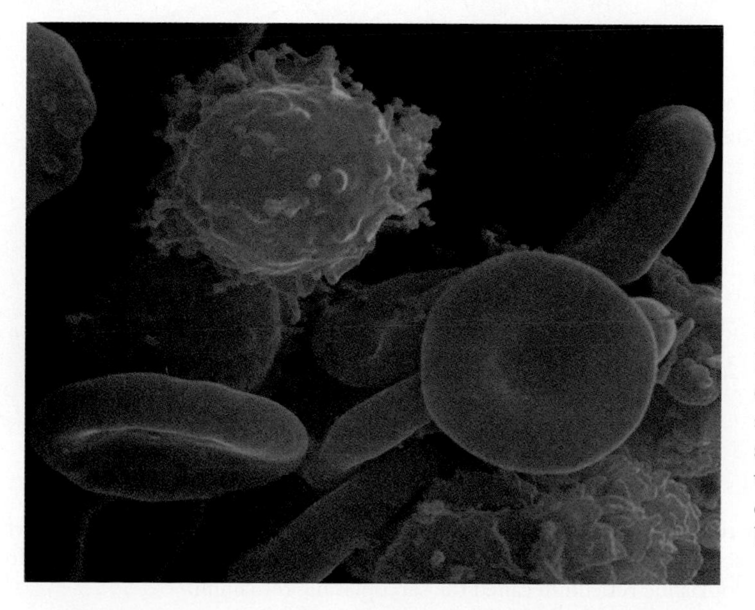

FIGURE 49.10
White and red blood cells (500×). White blood cells, or leukocytes, are roughly spherical and have irregular surfaces with numerous extending pili. Red blood cells, or erythrocytes, are flattened spheres, typically with a depressed center, forming biconcave discs.

49.4 Muscle tissue provides for movement, and nerve tissue provides for control.

Muscle Tissue

Muscle cells are the motors of the vertebrate body. The characteristic that makes them unique is the relative abundance and organization of actin and myosin filaments within them. Although these filaments form a fine network in all eukaryotic cells, where they contribute to cellular movements, they are far more common in muscle cells, which are specialized for contraction. Vertebrates possess three kinds of muscle: smooth, skeletal, and cardiac (table 49.4). Skeletal and cardiac muscles are also known as **striated muscles** because their cells have transverse stripes when viewed in longitudinal section under the microscope. The contraction of each skeletal muscle is under voluntary control, whereas the contraction of cardiac and smooth muscles is generally involuntary. Muscles are described in more detail in chapter 50.

Smooth Muscle

Smooth muscle was the earliest form of muscle to evolve, and it is found throughout the animal kingdom. In vertebrates, smooth muscle is found in the organs of the internal environment, or *viscera*, and is sometimes known as visceral muscle. Smooth muscle tissue is organized into sheets of long, spindle-shaped cells, each cell containing a single nucleus. In some tissues, the cells contract only when they are stimulated by a nerve, and then all of the cells in the sheet contract as a unit. In vertebrates, muscles of this type line the walls of many blood vessels and make up the iris of the eye. In other smooth muscle tissues, such as those in the wall of the gut, the muscle cells themselves may spontaneously initiate electric impulses and contract, leading to a slow, steady contraction of the tissue. Nerves regulate, rather than cause, this activity.

Skeletal Muscle

Skeletal muscles are usually attached by tendons to bones, so that, when the muscles contract, they cause the bones to move at their joints. A skeletal muscle is made up of numerous, very long muscle cells, called *muscle fibers*, which lie parallel to each other within the muscle and insert into the tendons on the ends of the muscle. Each skeletal muscle fiber is stimulated to contract by a nerve fiber; therefore, a stronger muscle contraction will result when more of the muscle fibers are stimulated by nerve fibers to contract. In this way, the nervous system can vary the strength of skeletal muscle contraction. Each muscle fiber contracts by means of substructures called *myofibrils* (figure 49.11) that contain highly ordered arrays of actin and myosin myofilaments, that, when aligned, give the muscle fiber its striated appearance. Skeletal muscle fibers are produced during development by the fusion of several cells, end to end. This

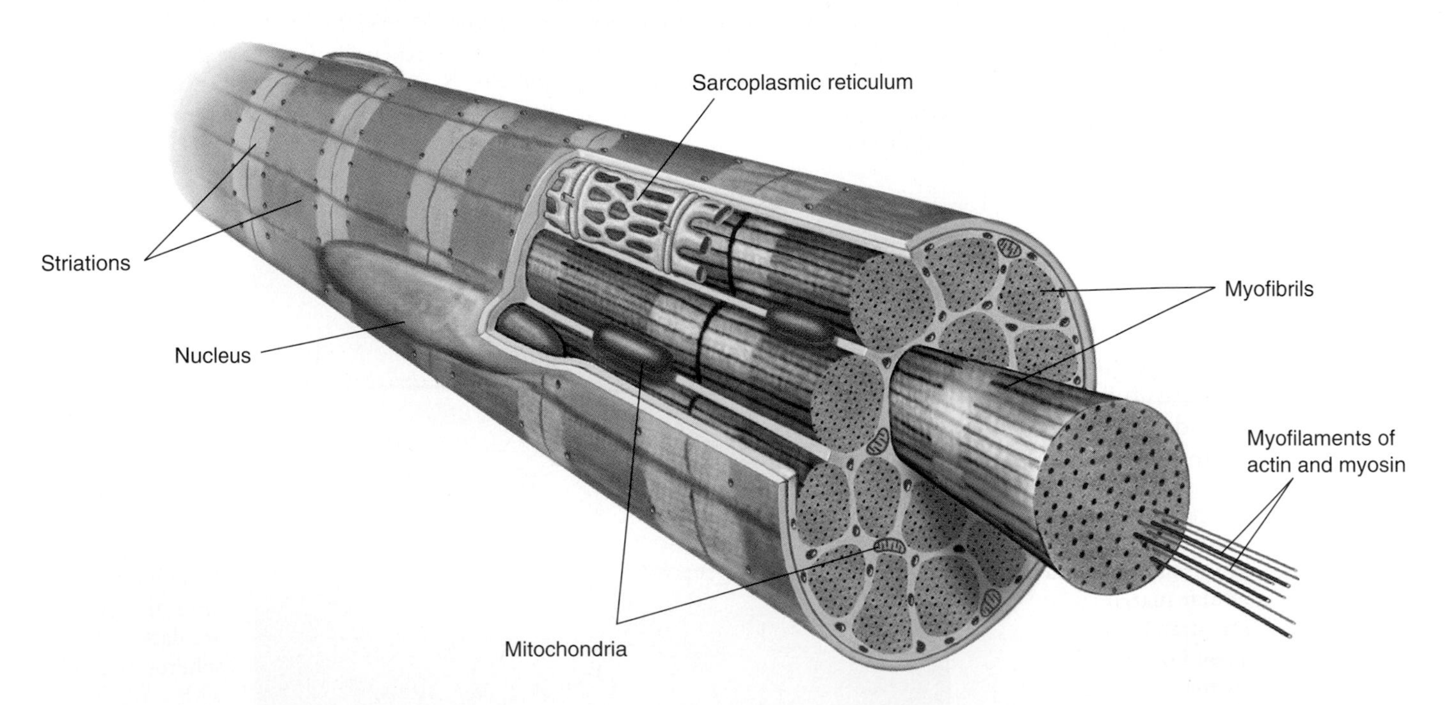

FIGURE 49.11
A muscle fiber, or muscle cell. Each muscle fiber is composed of numerous myofibrils, which, in turn, are composed of actin and myosin filaments. Each muscle fiber is multinucleate as a result of its embryological development from the fusion of smaller cells. Muscle cells have a modified endoplasmic reticulum called the sarcoplasmic reticulum.

Table 49.4 Muscle Tissue

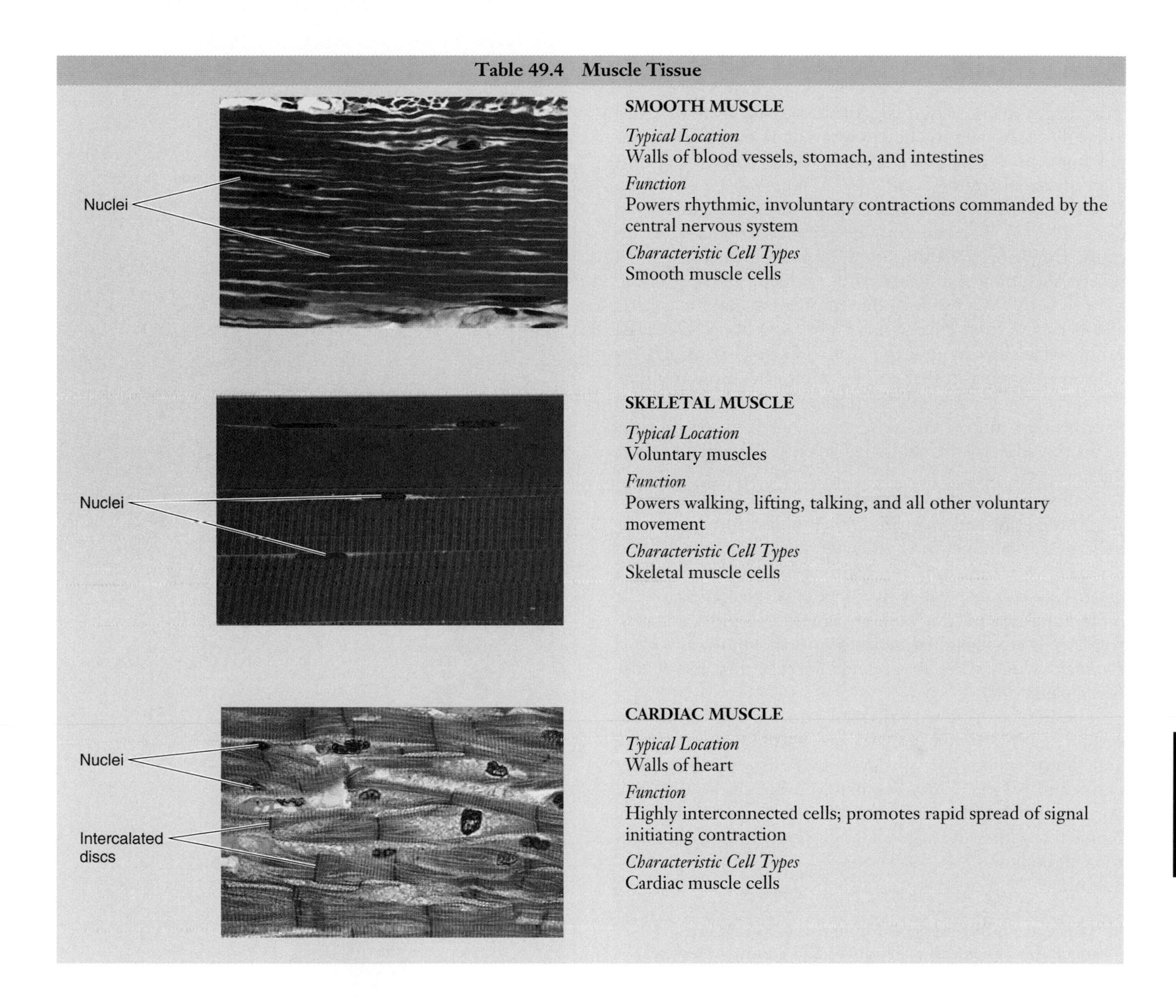

SMOOTH MUSCLE

Typical Location
Walls of blood vessels, stomach, and intestines

Function
Powers rhythmic, involuntary contractions commanded by the central nervous system

Characteristic Cell Types
Smooth muscle cells

SKELETAL MUSCLE

Typical Location
Voluntary muscles

Function
Powers walking, lifting, talking, and all other voluntary movement

Characteristic Cell Types
Skeletal muscle cells

CARDIAC MUSCLE

Typical Location
Walls of heart

Function
Highly interconnected cells; promotes rapid spread of signal initiating contraction

Characteristic Cell Types
Cardiac muscle cells

embryological development explains why a mature muscle fiber contains many nuclei. The structure and function of skeletal muscle is explained in more detail in chapter 50.

Cardiac Muscle

The hearts of vertebrates are composed of striated muscle cells arranged very differently from the fibers of skeletal muscle. Instead of having very long, multinucleate cells running the length of the muscle, cardiac muscle is composed of smaller, interconnected cells, each with a single nucleus. The interconnections between adjacent cells appear under the microscope as dark lines called *intercalated discs.* In reality, these lines are regions where adjacent cells are linked by *gap junctions.* As we noted in chapter 7, gap junctions have openings that permit the movement of small substances and electric charges from one cell to another. These interconnections enable the cardiac muscle cells to form a single, functioning unit known as a myocardium. Certain cardiac muscle cells generate electric impulses spontaneously, and these impulses spread across the gap junctions from cell to cell, causing all of the cells in the myocardium to contract. We will describe this process more fully in chapter 52.

Smooth muscles provide a variety of visceral functions. Skeletal muscles enable the vertebrate body to move. Cardiac muscle powers the heartbeat.

Nerve Tissue

The fourth major class of vertebrate tissue is nerve tissue (table 49.5). Its cells include neurons and neuroglia, or supporting cells. Neurons are specialized to produce and conduct electrochemical events, or "impulses." Each neuron consists of three parts: cell body, dendrites, and axon (figure 49.12). The cell body of a neuron contains the nucleus. Dendrites are thin, highly branched extensions that receive incoming stimulation and conduct electric events to the cell body. As a result of this stimulation and the electric events produced in the cell body, outgoing impulses may be produced at the origin of the axon. The axon is a single extension of cytoplasm that conducts impulses away from the cell body. Some axons can be quite long. The cell bodies of neurons that control the muscles in your feet, for example, lie in the spinal cord, and their axons may extend over a meter to your feet.

Neuroglia do not conduct electrical impulses but instead support and insulate neurons and eliminate foreign materials in and around neurons. In many neurons, neuroglia cells associate with the axons and form an insulating covering, a *myelin sheath*, produced by successive wrapping of the membrane around the axon (figure 49.13). Adjacent neuroglia cells are separated by interruptions known as *nodes of Ranvier*, which serve as sites for accelerating an impulse (see chapter 54).

The nervous system is divided into the central nervous system (CNS), which includes the brain and spinal cord, and the peripheral nervous system (PNS), which includes *nerves* and *ganglia*. Nerves consist of axons in the PNS that are bundled together in much the same way as wires are bundled together in a cable. Ganglia are collections of neuron cell bodies.

There are different types of neurons, but all are specialized to receive, produce, and conduct electrical signals. Neuroglia do not conduct electrical impulses but have various functions, including insulating axons to accelerate an electrical impulse. Both neurons and neuroglia are present in the CNS and the PNS.

FIGURE 49.12
A neuron has a very long projection called an axon. (*a*) A nerve impulse is received by the dendrites and then passed to the cell body and out through the axon. (*b*) Axons can be very long; single axons extend from the skull down several meters through a giraffe's neck to its pelvis.

Table 49.5 Nerve Tissue

SENSORY NEURONS

Typical Location
Eyes; ears; surface of skin

Function
Receive information about body's condition and external environment; send impulses from sensory receptors to CNS

Characteristic Cell Types
Rods and cones; muscle stretch receptors

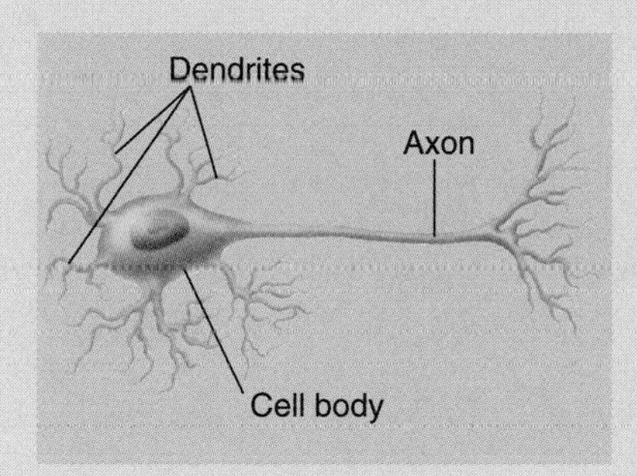

MOTOR NEURONS

Typical Location
Brain and spinal cord

Function
Stimulate muscles and glands; conduct impulses out of CNS toward muscles and glands

Characteristic Cell Types
Motor neurons

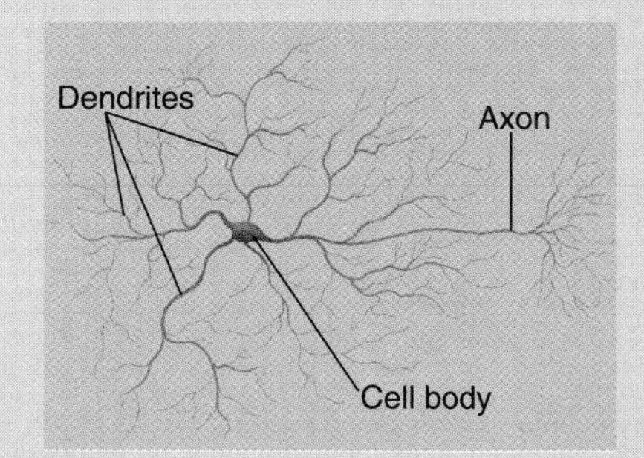

ASSOCIATION NEURONS

Typical Location
Brain and spinal cord

Function
Integrate information; conduct impulses between neurons within CNS

Characteristic Cell Types
Association neurons

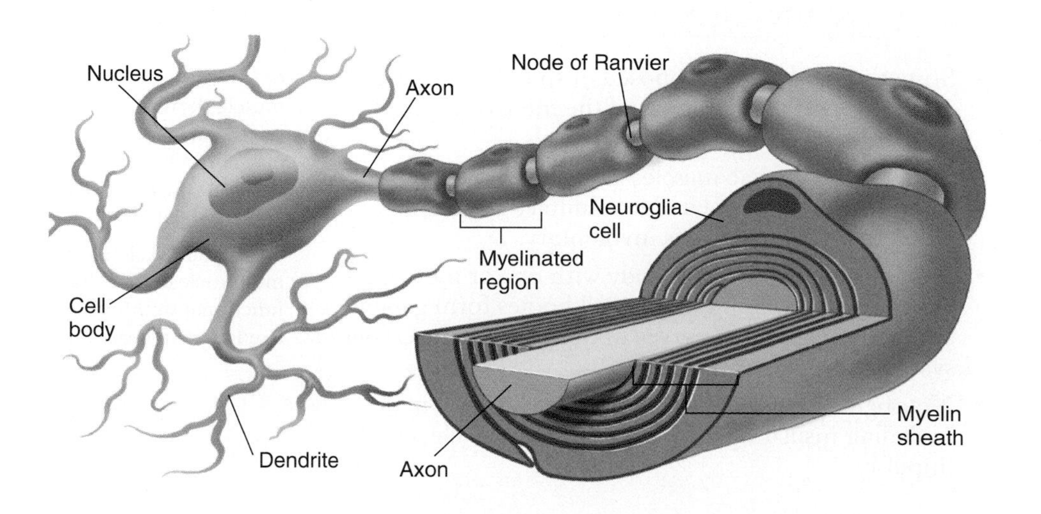

FIGURE 49.13
A myelinated neuron. Many dendrites arise from the cell body, as does a single long axon. In some neurons specialized for rapid signal conduction, the axon is encased in a myelin sheath that is interrupted at intervals. At its far end, the axon may branch to terminate on more than one cell.

Chapter 49

www.mhhe.com/raven6/resources49.mhtml

Summary	Questions	Media Resources

49.1 The bodies of vertebrates are organized into functional systems.

- The vertebrate body is organized into cells, tissues, organs, and organ systems, which are specialized for different functions.
- The four primary tissues of the vertebrate adult body—epithelial, connective, muscle, and nerve—are derived from three embryonic germ layers.

1. What is a tissue? What is an organ? What is an organ system?

- Art Activity: Mammalian Body Cavities

49.2 Epithelial tissue forms membranes and glands.

- Epithelial membranes cover all body surfaces.
- Stratified membranes, particularly the keratinized epithelium of the epidermis, provides protection, whereas simple membranes are more adapted for secretion and transport.
- Exocrine glands secrete into ducts that conduct the secretion to the surface of an epithelial membrane; endocrine glands secrete hormones into the blood.

2. What are the different types of epithelial membranes, and how do they differ in structure and function?

3. What are the two types of glands, and how do they differ in structure and function?

- Epithelial Tissue
- Epithelial Glands
- Tissues

49.3 Connective tissues contain abundant extracellular material.

- Connective tissues are characterized by abundant extracellular matrix, which is composed of fibrous proteins and a gel-like ground substance in connective tissue proper.
- Loose connective tissues contain many cell types such as adipose cells and mast cells; dense regular connective tissues form tendons and ligaments.
- Special connective tissues include cartilage, bone, and blood. Nutrients can diffuse through the cartilage matrix but not through the calcified matrix of bone, which contains canaliculi for that purpose.

4. What feature do all connective tissues share? What are the different categories of connective tissue? Give an example of each.

5. What is the structure of a ligament? How do cartilage and bone differ? Why is blood considered to be a connective tissue?

- Connective Tissue

49.4 Muscle tissue provides for movement, and nerve tissue provides for control.

- Smooth muscles are composed of spindle-shaped cells and are found in the organs of the internal environment and in the walls of blood vessels.
- Skeletal and cardiac muscles are striated; skeletal muscles, however, are under voluntary control whereas cardiac muscle is involuntary.
- Neurons consist of a cell body with one or more dendrites and one axon. Neuron cell bodies form ganglia, and their axons form nerves in the peripheral nervous system.
- Neuroglia are supporting cells with various functions including insulating axons to accelerate an electrical impulse.

6. From what embryonic tissue is muscle derived? What two contractile proteins are abundant in muscle? What are the three categories of muscle tissue? Which two are striated?

7. Why are skeletal muscle fibers multinucleated? What is the functional significance of intercalated discs in heart muscle?

- Muscle Tissue
- Nervous Tissue

- Art Quizzes:
  - -Muscle Fiber
  - -Nerve Tissue

50

Locomotion

Concept Outline

50.1 A skeletal system supports movement in animals.

Types of Skeletons. There are three types of skeletal systems found in animals: hydrostatic skeletons, exoskeletons, and endoskeletons. Hydrostatic skeletons function by the movement of fluid in a body cavity. Exoskeletons are made of tough exterior coverings on which muscles attach to move the body. Endoskeletons are rigid internal bones or cartilage which move the body by the contraction of muscles attached to the skeleton.

The Structure of Bone. The human skeleton, an example of an endoskeleton, is made of bone that contains cells called osteocytes within a calcified matrix.

50.2 Skeletal muscles contract to produce movements at joints.

Types of Joints. The joints where bones meet may be immovable, slightly movable, or freely movable.

Actions of Skeletal Muscles. Synergistic and antagonistic muscles act on the skeleton to move the body.

50.3 Muscle contraction powers animal locomotion.

The Sliding Filament Mechanism of Contraction. Thick and thin myofilaments slide past one another to cause muscle shortening.

The Control of Muscle Contraction. During contraction Ca^{++} moves aside a regulatory protein which had been preventing cross-bridges from attaching to the thin filaments. Nerves stimulate the release of Ca^{++} from its storage depot so that contraction can occur.

Types of Muscle Fibers. Muscle fibers can be categorized as slow-twitch (slow to fatigue) or fast-twitch (fatigue quickly but can provide a fast source of power).

Comparing Cardiac and Smooth Muscles. Cardiac muscle cells are interconnected to form a single functioning unit. Smooth muscles lack the myofilament organization found in striated muscle but they still contract via the sliding filament mechanism.

Modes of Animal Locomotion. Animals rarely move in straight lines. Their movements are adjusted both by mechanical feedback and by neural control. Muscles generate power for movement, and also act as springs, brakes, struts, and shock absorbers.

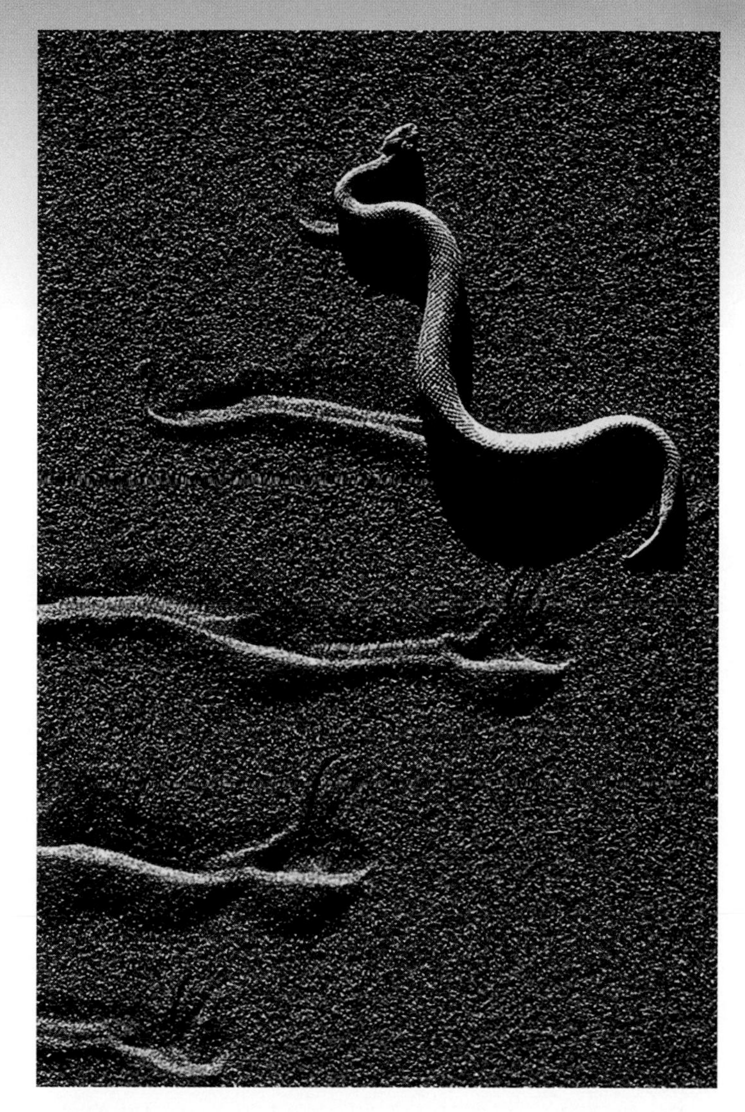

FIGURE 50.1
On the move. The movements made by this sidewinder rattlesnake are the result of strong muscle contractions acting on the bones of the skeleton. Without muscles and some type of skeletal system, complex locomotion as shown here would not be possible.

Plants and fungi move only by growing, or as the passive passengers of wind and water. Of the three multicellular kingdoms, only animals explore their environment in an active way, through locomotion. In this chapter we examine how vertebrates use muscles connected to bones to achieve movement. The rattlesnake in figure 50.1 slithers across the sand by a rhythmic contraction of the muscles sheathing its body. Humans walk by contracting muscles in their legs. Although our focus in this chapter will be on vertebrates, it is important to realize that essentially all animals employ muscles. When a mosquito flies, its wings are moved rapidly through the air by quickly contracting flight muscles. When an earthworm burrows through the soil, its movement is driven by strong muscles pushing its body past the surrounding dirt.

50.1 A skeletal system supports movement in animals.

Types of Skeletons

Animal locomotion is accomplished through the force of muscles acting on a rigid skeletal system. There are three types of skeletal systems in the animal kingdom: hydrostatic skeletons, exoskeletons, and endoskeletons.

Hydrostatic skeletons are primarily found in soft-bodied invertebrates such as earthworms and jellyfish. In this case, a fluid-filled cavity is encircled by muscle fibers. As the muscles contract, the fluid in the cavity moves and changes the shape of the cavity. In an earthworm, for example, a wave of contractions of circular muscles begins anteriorly and compresses each segment of the body, so that the fluid pressure pushes it forward. Contractions of longitudinal muscles then pull the rear of the body forward (figure 50.2).

Exoskeletons surround the body as a rigid hard case. Arthropods, such as crustaceans and insects, have exoskeletons made of the polysaccharide *chitin* (figure 50.3*a*). An exoskeleton offers great protection to internal organs and resists bending. However, in order to grow, the animal must periodically molt. During molting, the animal is particularly vulnerable to predation because its old exoskeleton has been shed. Having an exoskeleton also limits the size of the animal. An animal with an exoskeleton cannot get too large because its exoskeleton would have to become thicker and heavier, in order to prevent collapse, as the animal grew larger. If an insect were the size of a human being, its exoskeleton would have to be so thick and heavy it would be unable to move.

Endoskeletons, found in vertebrates and echinoderms, are rigid internal skeletons to which muscles are attached. Vertebrates have a flexible exterior that accommodates the movements of their skeleton. The endoskeleton of vertebrates is composed of cartilage or bone. Unlike chitin, bone is a cellular, living tissue capable of growth, self-repair, and remodeling in response to physical stresses.

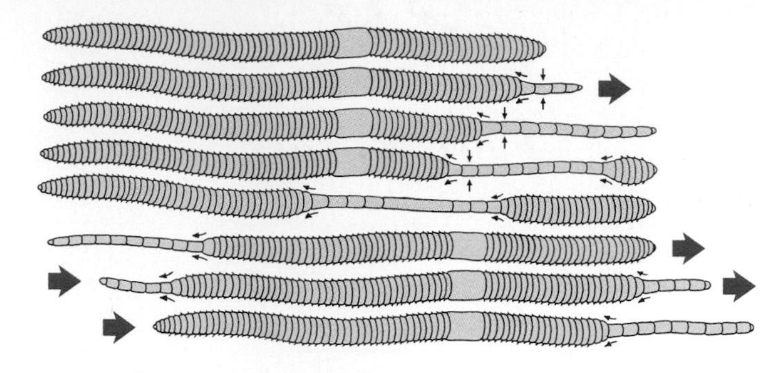

FIGURE 50.2
Locomotion in earthworms. The hydrostatic skeleton of the earthworm uses muscles to move fluid within the segmented body cavity changing the shape of the animal. When an earthworm's circular muscles contract, the internal fluid presses on the longitudinal muscles, which then stretch to elongate segments of the earthworm. A wave of contractions down the body of the earthworm produces forward movement.

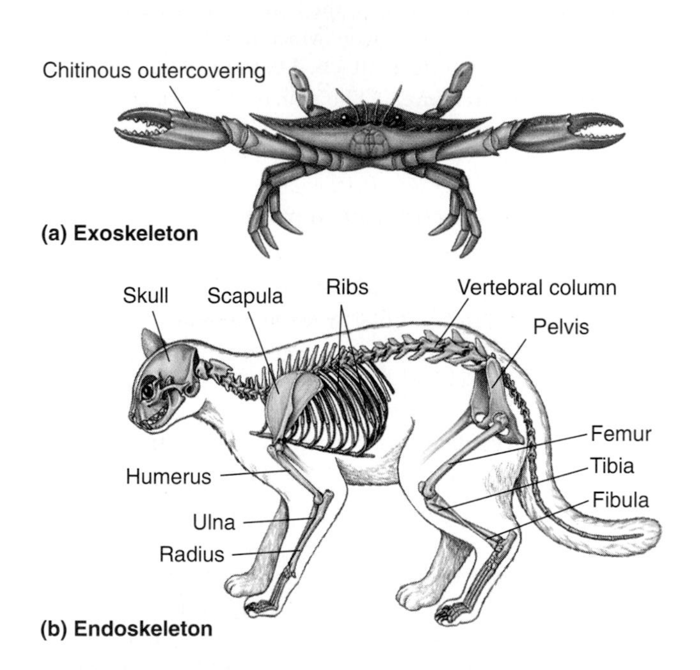

FIGURE 50.3
Exoskeleton and endoskeleton. (*a*) The hard, tough outcovering of an arthropod, such as this crab, is its exoskeleton.
(*b*) Vertebrates, such as this cat, have endoskeletons. The axial skeleton is shown in the *peach* shade, the appendicular skeleton in the *yellow* shade. Some of the major bones are labeled.

The Vertebrate Skeleton

A vertebrate endoskeleton (figure 50.3*b*) is divided into an axial and an appendicular skeleton. The axial skeleton's bones form the axis of the body and support and protect the organs of the head, neck, and chest. The appendicular skeleton's bones include the bones of the limbs, and the pectoral and pelvic girdles that attach them to the axial skeleton.

The bones of the skeletal system support and protect the body, and serve as levers for the forces produced by contraction of skeletal muscles. Blood cells form within the bone marrow, and the calcified matrix of bones acts as a reservoir for calcium and phosphate ions.

There are three types of animal skeletons: hydrostatic skeleton, exoskeleton, and endoskeleton. The endoskeletons found in vertebrates are composed of bone or cartilage and are organized into axial and appendicular portions.

The Structure of Bone

Bone, the building material of the vertebrate skeleton, is a special form of connective tissue (see chapter 49). In bone, an organic extracellular matrix containing collagen fibers is impregnated with small, needle-shaped crystals of calcium phosphate in the form of hydroxyapatite crystals. Hydroxyapatite is brittle but rigid, giving bone great strength. Collagen, on the other hand, is flexible but weak. As a result, bone is both strong and flexible. The collagen acts to spread the stress over many crystals, making bone more resistant to fracture than hydroxyapatite is by itself.

Bone is a dynamic, living tissue that is constantly reconstructed throughout the life of an individual. New bone is formed by *osteoblasts,* which secrete the collagen-containing organic matrix in which calcium phosphate is later deposited. After the calcium phosphate is deposited, the cells, now known as osteocytes, are encased within spaces called *lacunae* in the calcified matrix. Yet another type of bone cells, called osteoclasts, act to dissolve bone and thereby aid in the remodeling of bone in response to physical stress.

Bone is constructed in thin, concentric layers, or *lamellae,* which are laid down around narrow channels called *Haversian canals* that run parallel to the length of the bone. Haversian canals contain nerve fibers and blood vessels, which keep the osteocytes alive even though they are entombed in a calcified matrix. The concentric lamellae of bone, with their entrapped osteocytes, that surround a Haversian canal form the basic unit of bone structure, called a Haversian system.

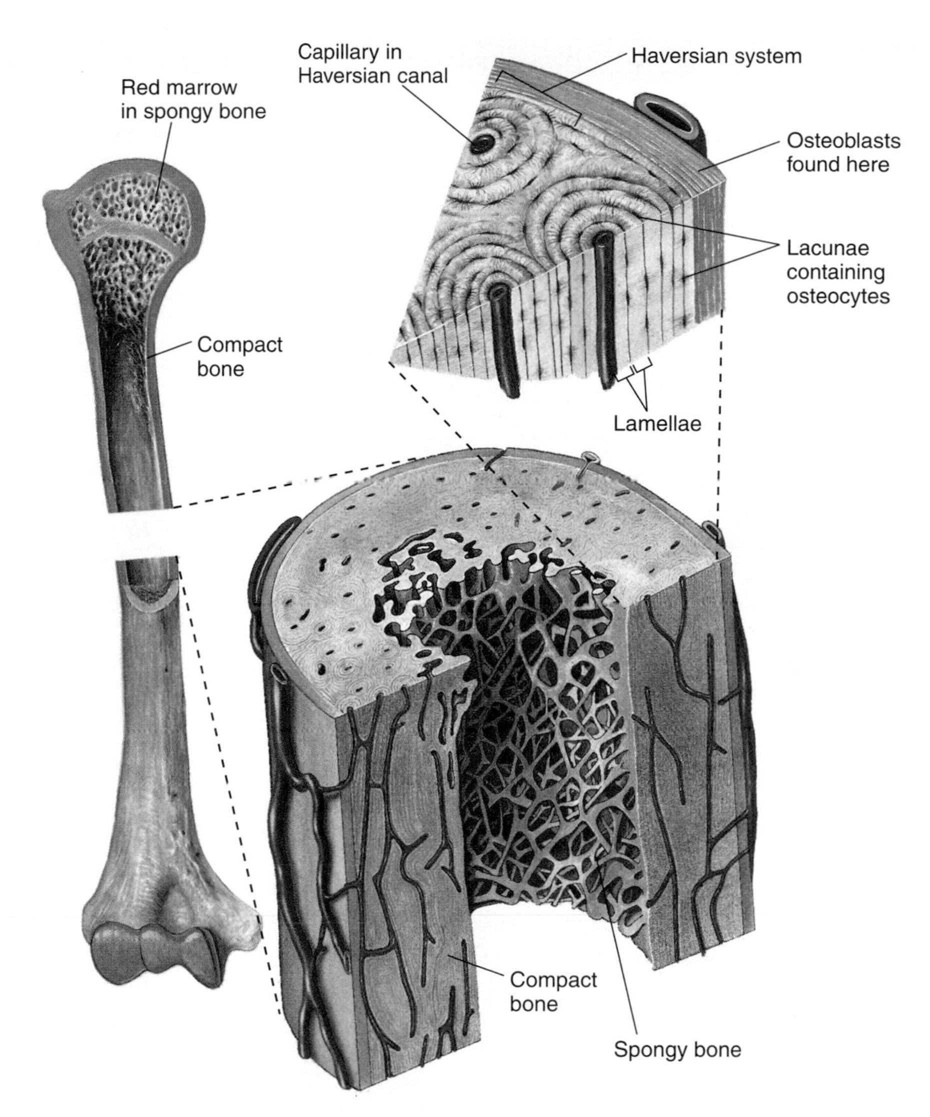

FIGURE 50.4
The organization of bone, shown at three levels of detail. Some parts of bone are dense and compact, giving the bone strength. Other parts are spongy, with a more open lattice; it is there that most blood cells are formed.

Bone formation occurs in two ways. In flat bones, such as those of the skull, osteoblasts located in a web of dense connective tissue produce bone within that tissue. In long bones, the bone is first "modeled" in cartilage. Calcification then occurs, and bone is formed as the cartilage degenerates. At the end of this process, cartilage remains only at the articular (joint) surfaces of the bones and at the growth plates located in the necks of the long bones. A child grows taller as the cartilage thickens in the growth plates and then is partly replaced with bone. A person stops growing (usually by the late teenage years) when the entire cartilage growth plate becomes replaced by bone. At this point, only the articular cartilage at the ends of the bone remains.

The ends and interiors of long bones are composed of an open lattice of bone called *spongy bone.* The spaces within contain marrow, where most blood cells are formed (figure 50.4). Surrounding the spongy bone tissue are concentric layers of *compact bone,* where the bone is much denser. Compact bone tissue gives bone the strength to withstand mechanical stress.

Bone consists of cells and an extracellular matrix that contains collagen fibers, which provide flexibility, and calcium phosphate, which provides strength. Bone contains blood vessels and nerves and is capable of growth and remodeling.

50.2 Skeletal muscles contract to produce movements at joints.

Types of Joints

The skeletal movements of the body are produced by contraction and shortening of muscles. Skeletal muscles are generally attached by tendons to bones, so when the muscles shorten, the attached bones move. These movements of the skeleton occur at joints, or articulations, where one bone meets another. There are three main classes of joints:

1. **Immovable joints** include the *sutures* that join the bones of the skull (figure 50.5*a*). In a fetus, the skull bones are not fully formed, and there are open areas of dense connective tissue ("soft spots," or *fontanels*) between the bones. These areas allow the bones to shift slightly as the fetus moves through the birth canal during childbirth. Later, bone replaces most of this connective tissue.
2. **Slightly movable joints** include those in which the bones are bridged by cartilage. The vertebral bones of the spine are separated by pads of cartilage called *intervertebral discs* (figure 50.5*b*). These *cartilaginous joints* allow some movement while acting as efficient shock absorbers.
3. **Freely movable joints** include many types of joints and are also called synovial joints, because the articulating ends of the bones are located within a *synovial capsule* filled with a lubricating fluid. The ends of the bones are capped with cartilage, and the synovial capsule is strengthened by ligaments that hold the articulating bones in place.

   Synovial joints allow the bones to move in directions dictated by the structure of the joint. For example, a joint in the finger allows only a hingelike movement, while the joint between the thigh bone (femur) and pelvis has a ball-and-socket structure that permits a variety of different movements (figure 50.5*c*).

Joints confer flexibility to a rigid skeleton, allowing a range of motions determined by the type of joint.

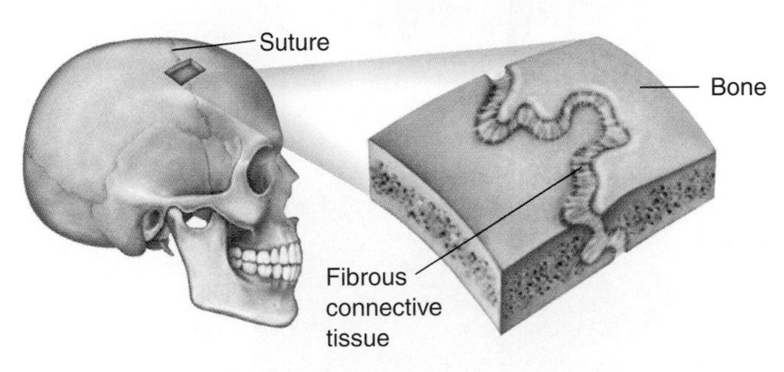

(a) Immovable joint

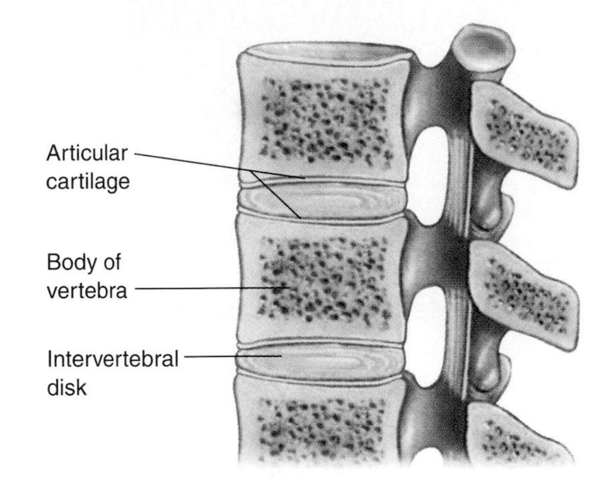

(b) Slightly movable joints

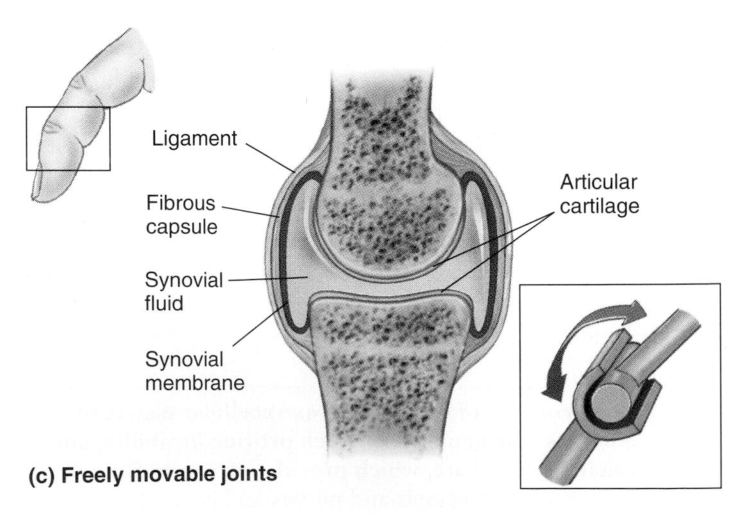

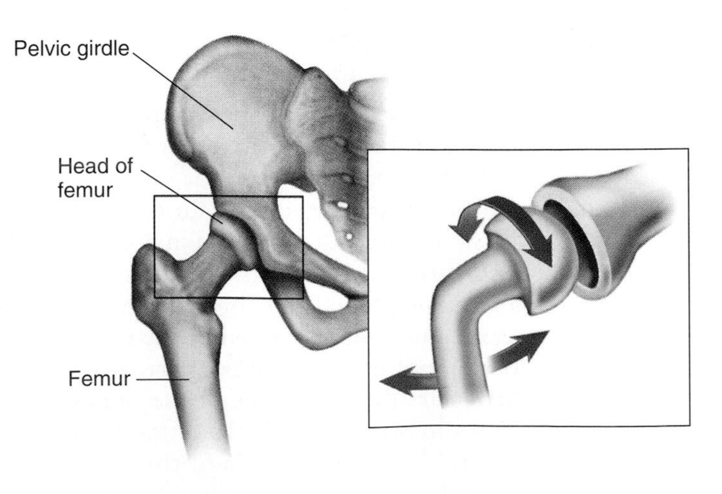

(c) Freely movable joints

FIGURE 50.5
Three types of joints. (*a*) Immovable joints include the sutures of the skull; (*b*) slightly movable joints include the cartilaginous joints between the vertebrae; and (*c*) freely movable joints are the synovial joints, such as a finger joint and or a hip joint.

Actions of Skeletal Muscles

Skeletal muscles produce movement of the skeleton when they contract. Usually, the two ends of a skeletal muscle are attached to different bones (although in some cases, one or both ends may be connected to some other kind of structure, such as skin). The attachments to bone are made by means of dense connective tissue straps called *tendons.* Tendons have elastic properties that allow "give-and-take" during muscle contraction. One attachment of the muscle, the **origin,** remains relatively stationary during a contraction. The other end of the muscle, the **insertion,** is attached to the bone that moves when the muscle contracts. For example, contraction of the biceps muscle in the upper arm causes the forearm (the insertion of the muscle) to move toward the shoulder (the origin of the muscle).

Muscles that cause the same action at a joint are **synergists.** For example, the various muscles of the quadriceps group in humans are synergists: they all act to extend the knee joint. Muscles that produce opposing actions are **antagonists.** For example, muscles that flex a joint are antagonist to muscles that extend that joint (figure 50.6*a*). In humans, when the hamstring muscles contract, they cause flexion of the knee joint (figure 50.6*b*). Therefore, the quadriceps and hamstrings are antagonists to each other. In general, the muscles that antagonize a given movement are relaxed when that movement is performed. Thus, when the hamstrings flex the knee joint, the quadriceps muscles relax.

Isotonic and Isometric Contractions

In order for muscle fibers to shorten when they contract, they must generate a force that is greater than the opposing forces that act to prevent movement of the muscle's insertion. When you lift a weight by contracting muscles in your biceps, for example, the force produced by the muscle is greater than the force of gravity on the object you are lifting. In this case, the muscle and all of its fibers shorten in length. This type of contraction is referred to as **isotonic contraction,** because the force of contraction remains relatively constant throughout the shortening process (*iso* = same; *tonic* = strength).

Preceding an isotonic contraction, the muscle begins to contract but the tension is absorbed by the tendons and other elastic tissue associated with the muscle. The muscle does not change in length and so this is called **isometric** (literally, "same length") **contraction.** Isometric contractions occur as a phase of normal muscle contraction but also exist to provide tautness and stability to the body.

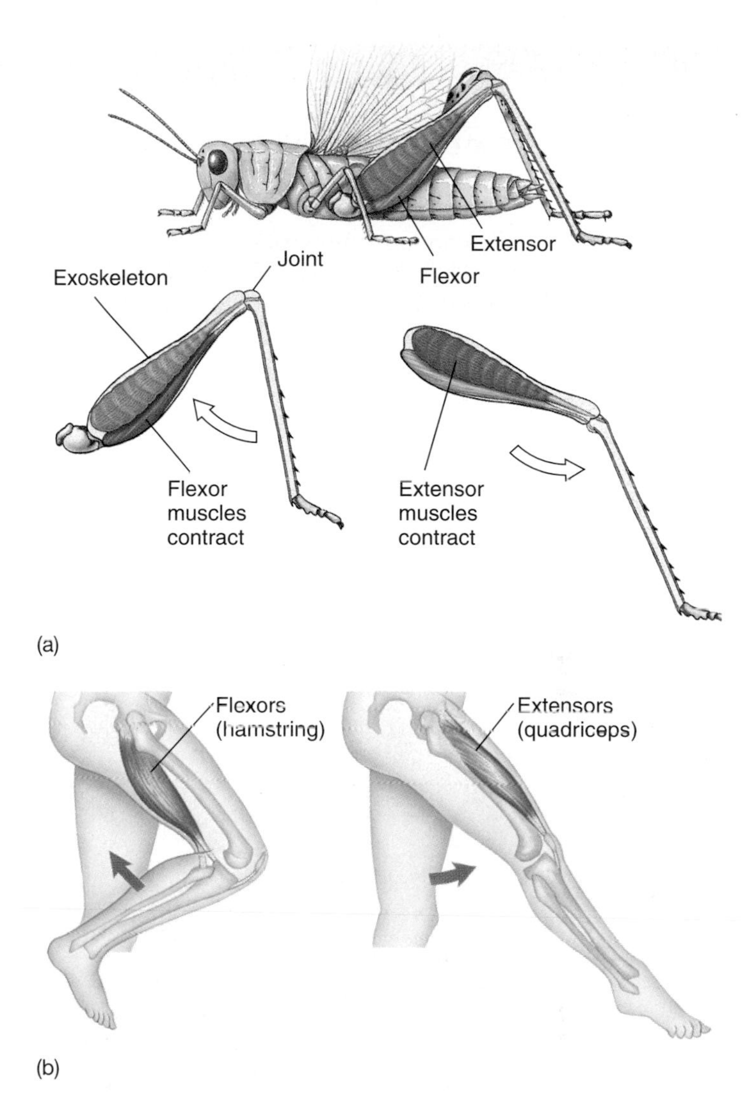

FIGURE 50.6
Flexor and extensor muscles of the leg. (*a*) Antagonistic muscles control the movement of an animal with an exoskeleton, such as the jumping of a grasshopper. When the smaller flexor tibia muscle contracts it pulls the lower leg in toward the upper leg. Contraction of the extensor tibia muscles straightens out the leg and sends the insect into the air. (*b*) Similarly, antagonistic muscles can act on an endoskeleton. In humans, the hamstrings, a group of three muscles, produce flexion of the knee joint, whereas the quadriceps, a group of four muscles, produce extension.

Synergistic muscles have the same action, whereas antagonistic muscles have opposite actions. Both muscle groups are involved in locomotion. Isotonic contractions involve the shortening of muscle, while isometric contractions do not alter the length of the muscle.

The Sliding Filament Mechanism of Contraction

Each skeletal muscle contains numerous **muscle fibers,** as described in chapter 49. Each muscle fiber encloses a bundle of 4 to 20 elongated structures called **myofibrils.** Each myofibril, in turn, is composed of **thick** and **thin myofilaments** (figure 50.7). The muscle fiber is striated (has cross-striping) because its myofibrils are striated, with dark and light bands. The banding pattern results from the organization of the myofilaments within the myofibril. The thick myofilaments are stacked together to produce the dark bands, called *A bands;* the thin filaments alone are found in the light bands, or *I bands.*

Each I band in a myofibril is divided in half by a disc of protein, called a *Z line* because of its appearance in electron micrographs. The thin filaments are anchored to these discs of proteins that form the Z lines. If you look at an electron micrograph of a myofibril (figure 50.8), you will see that the structure of the myofibril repeats from Z line to Z line. This repeating structure, called a **sarcomere,** is the smallest subunit of muscle contraction.

The thin filaments stick partway into the stack of thick filaments on each side of an A band, but, in a resting

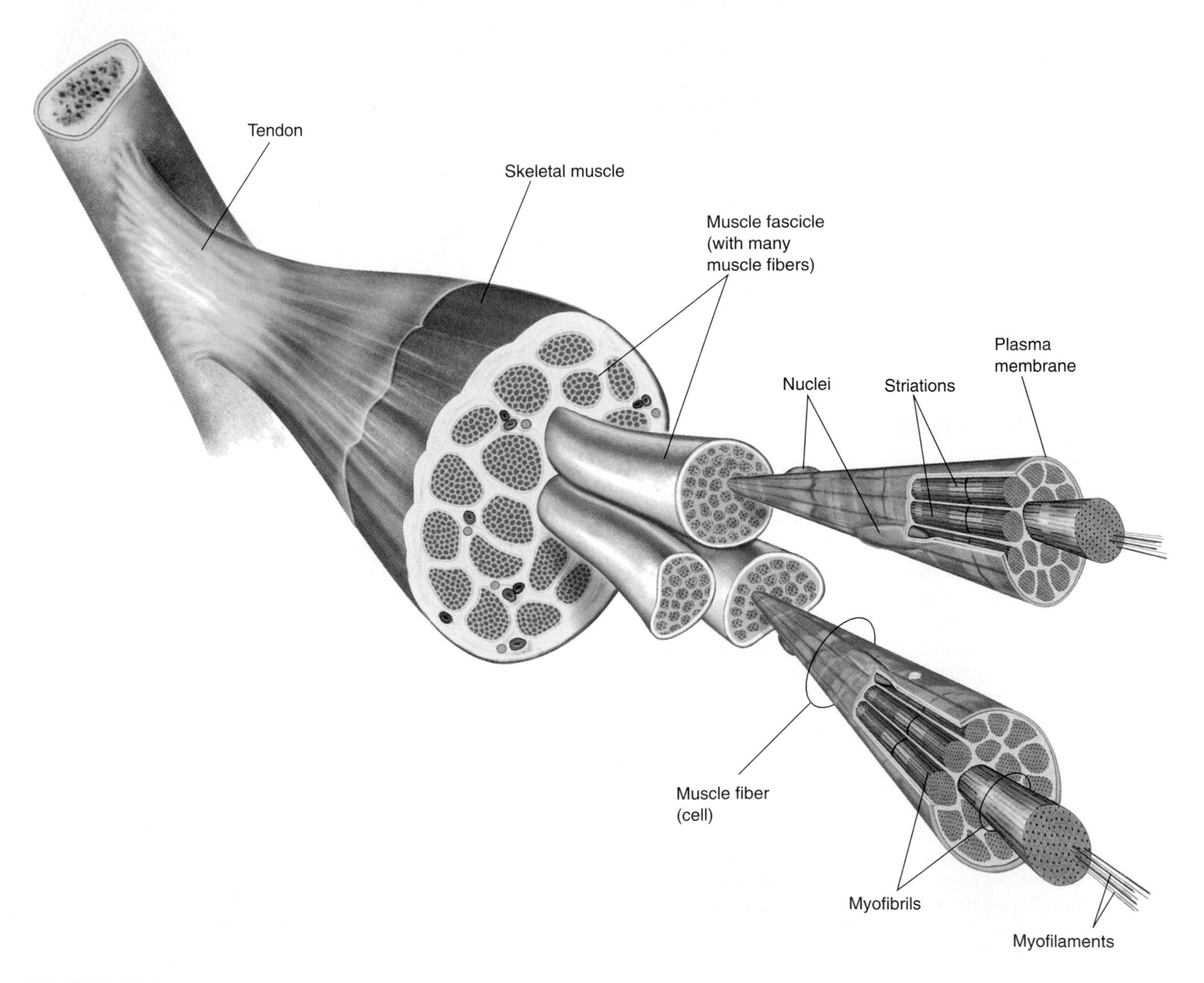

FIGURE 50.7
The organization of skeletal muscle. Each muscle is composed of many fascicles, which are bundles of muscle cells, or fibers. Each fiber is composed of many myofibrils, which are each, in turn, composed of myofilaments.

muscle, do not project all the way to the center of the A band. As a result, the center of an A band (called an *H band*) is lighter than each side, with its interdigitating thick and thin filaments. This appearance of the sarcomeres changes when the muscle contracts.

A muscle contracts and shortens because its myofibrils contract and shorten. When this occurs, the myofilaments do *not* shorten; instead, the thin filaments slide deeper into the A bands (figure 50.9). This makes the H bands narrower until, at maximal shortening, they disappear entirely. It also makes the I bands narrower, because the dark A bands are brought closer together. This is the **sliding filament mechanism** of contraction.

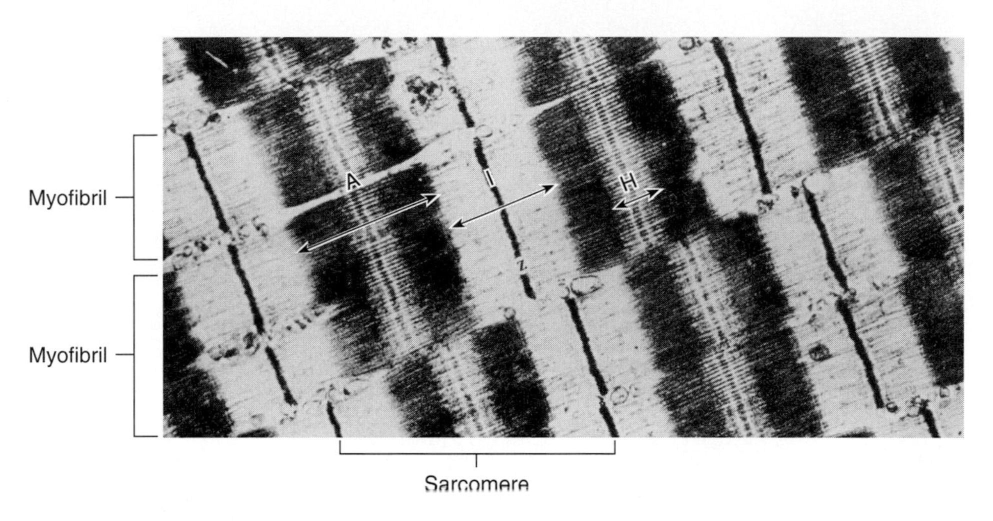

FIGURE 50.8
An electron micrograph of a skeletal muscle fiber. The Z lines that serve as the borders of the sarcomeres are clearly seen within each myofibril. The thick filaments comprise the A bands; the thin filaments are within the I bands and stick partway into the A bands, overlapping with the thick filaments. There is no overlap of thick and thin filaments at the central region of an A band, which is therefore lighter in appearance. This is the H band.

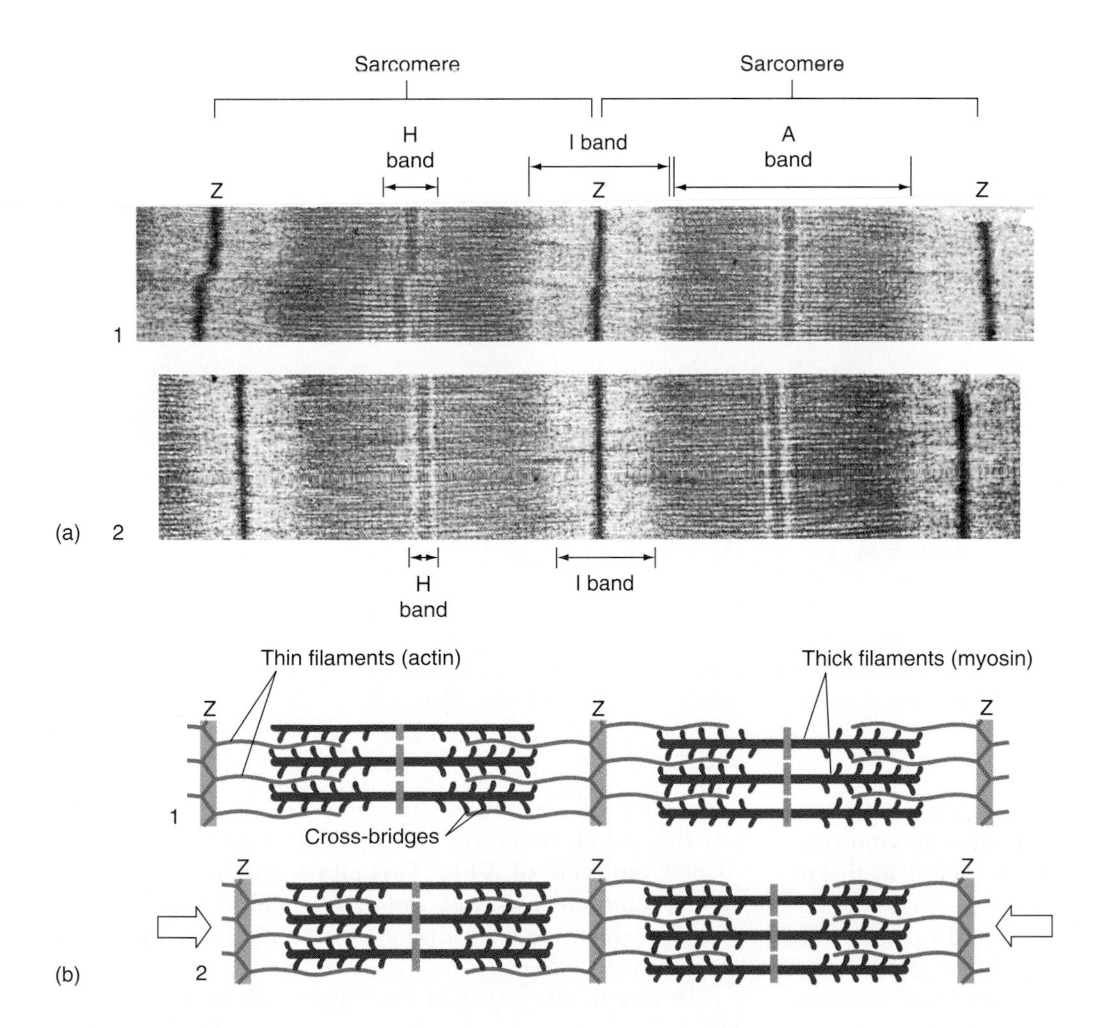

FIGURE 50.9
Electron micrograph (a) and diagram (b) of the sliding filament mechanism of contraction. As the thin filaments slide deeper into the centers of the sarcomeres, the Z lines are brought closer together. (*1*) Relaxed muscle; (*2*) partially contracted muscle.

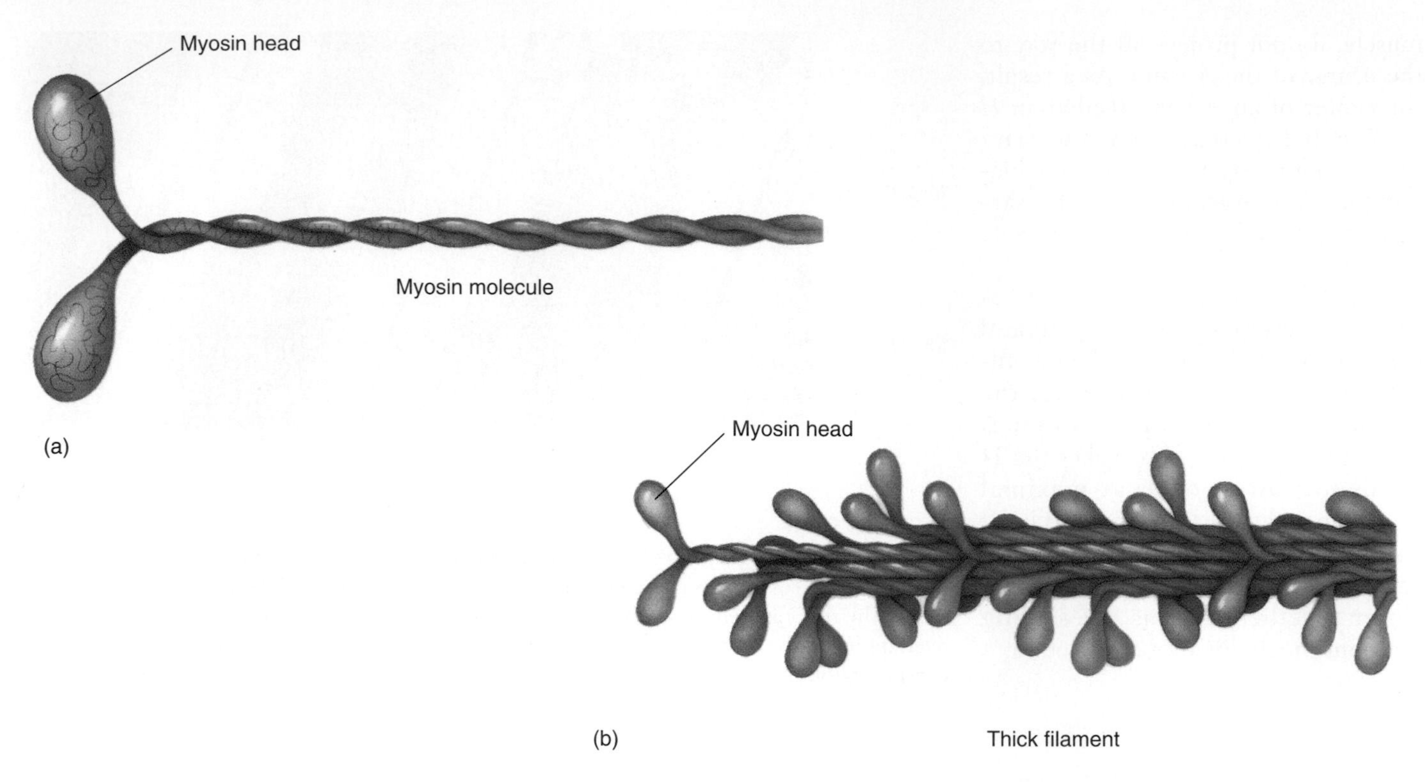

FIGURE 50.10
Thick filaments are composed of myosin. (*a*) Each myosin molecule consists of two polypeptide chains wrapped around each other; at the end of each chain is a globular region referred to as the "head." (*b*) Thick filaments consist of myosin molecules combined into bundles from which the heads protrude at regular intervals.

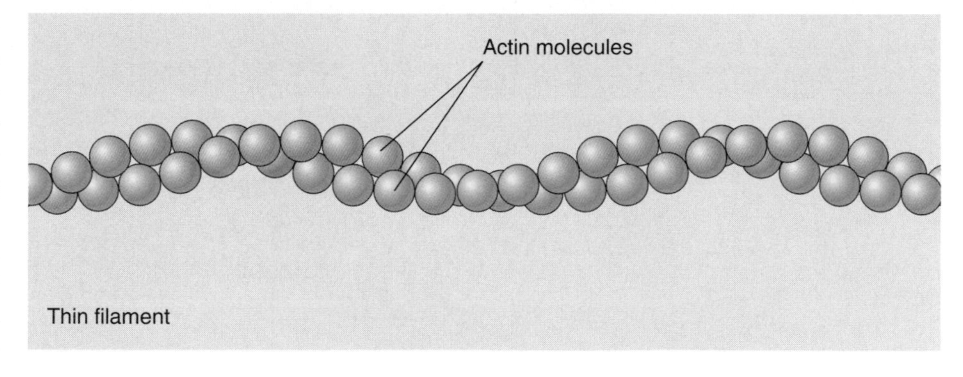

FIGURE 50.11
Thin filaments are composed of globular actin proteins. Two rows of actin proteins are twisted together in a helix to produce the thin filaments.

Electron micrographs reveal **cross-bridges** that extend from the thick to the thin filaments, suggesting a mechanism that might cause the filaments to slide. To understand how this is accomplished, we have to examine the thick and thin filaments at a molecular level. Biochemical studies show that each thick filament is composed of many **myosin** proteins packed together, and every myosin molecule has a "head" region that protrudes from the thick filaments (figure 50.10). These myosin heads form the cross-bridges seen in electron micrographs. Biochemical studies also show that each thin filament consists primarily of many globular **actin** proteins twisted into a double helix (figure 50.11). Therefore, if we were able to see a sarcomere at a molecular level, it would have the structure depicted in figure 50.12*a*.

Before the myosin heads bind to the actin of the thin filaments, they act as ATPase enzymes, splitting ATP into ADP and P_i. This activates the heads, "cocking" them so that they can bind to actin and form cross-bridges. Once a myosin head binds to actin, it undergoes a conformational (shape) change, pulling the thin filament toward the center of the sarcomere (figure 50.12*b*) in a *power stroke.* At the end of the power stroke, the myosin head binds to a new molecule of ATP. This allows the head to detach from actin and continue the **cross-bridge cycle** (figure 50.13), which repeats as long as the muscle is stimulated to contract.

In death, the cell can no longer produce ATP and therefore the cross-bridges cannot be broken—this causes the muscle stiffness of death, or *rigor mortis.* A living cell, however, always has enough ATP to allow the

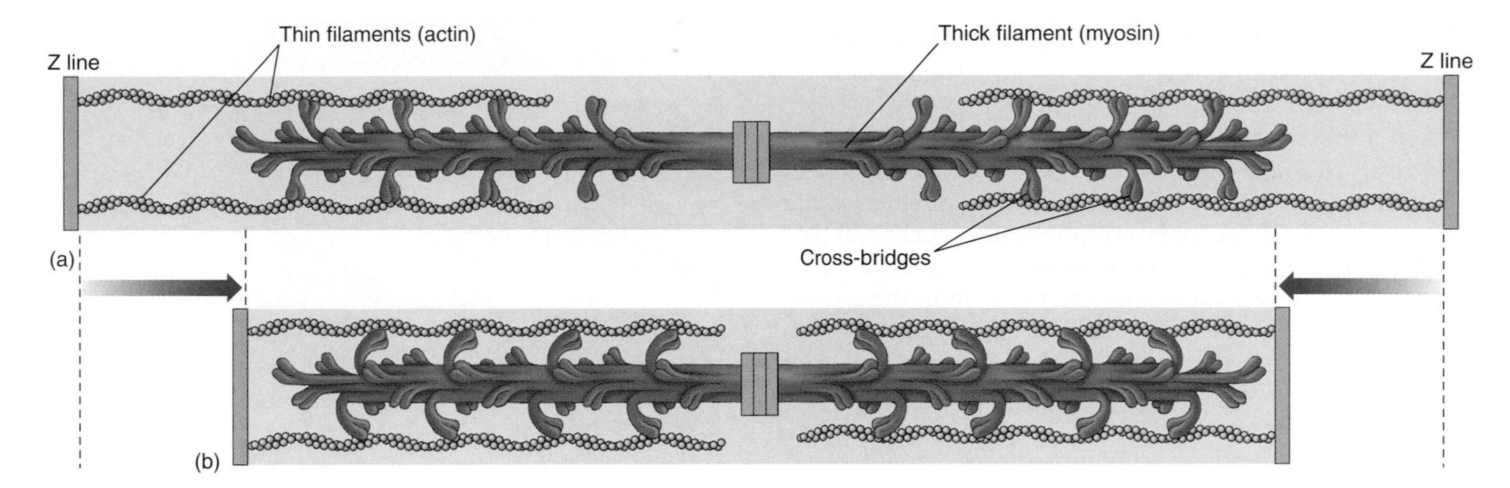

FIGURE 50.12
The interaction of thick and thin filaments in a striated muscle sarcomere. The heads on the two ends of the thick filaments are oriented in opposite directions (*a*), so that the cross-bridges pull the thin filaments and the Z lines on each side of the sarcomere toward the center. (*b*) This sliding of the filaments produces muscle contraction.

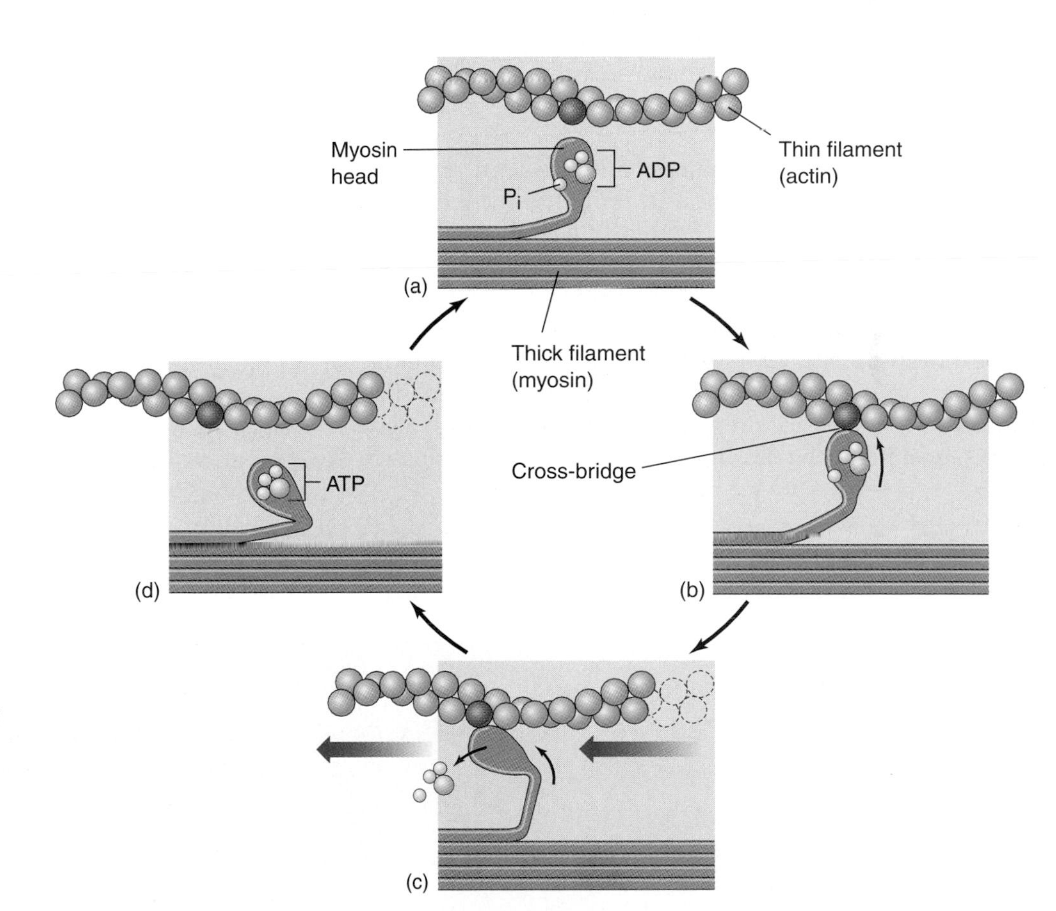

FIGURE 50.13
The cross-bridge cycle in muscle contraction. (*a*) With ADP and P_i attached to the myosin head, (*b*) the head is in a conformation that can bind to actin and form a cross-bridge. (*c*) Binding causes the myosin head to assume a more bent conformation, moving the thin filament along the thick filament (to the left in this diagram) and releasing ADP and P_i. (*d*) Binding of ATP to the head detaches the cross-bridge; cleavage of ATP into ADP and P_i puts the head into its original conformation, allowing the cycle to begin again.

myosin heads to detach from actin. How, then, is the cross-bridge cycle arrested so that the muscle can relax? The regulation of muscle contraction and relaxation requires additional factors that we will discuss in the next section.

Thick and thin filaments are arranged to form sarcomeres within the myofibrils. Myosin proteins comprise the thick filaments, and the heads of the myosin form cross-bridges with the actin proteins of the thin filaments. ATP provides the energy for the cross-bridge cycle and muscle contraction.

The Control of Muscle Contraction

The Role of Ca^{++} in Contraction

When a muscle is relaxed, its myosin heads are "cocked" and ready, through the splitting of ATP, but are unable to bind to actin. This is because the attachment sites for the myosin heads on the actin are physically blocked by another protein, known as **tropomyosin,** in the thin filaments. Cross-bridges therefore cannot form in the relaxed muscle, and the filaments cannot slide.

In order to contract a muscle, the tropomyosin must be moved out of the way so that the myosin heads can bind to actin. This requires the action of **troponin,** a regulatory protein that binds to the tropomyosin. The troponin and tropomyosin form a complex that is regulated by the calcium ion (Ca^{++}) concentration of the muscle cell cytoplasm.

When the Ca^{++} concentration of the muscle cell cytoplasm is low, tropomyosin inhibits cross-bridge formation and the muscle is relaxed (figure 50.14). When the Ca^{++} concentration is raised, Ca^{++} binds to troponin. This causes the troponin-tropomyosin complex to be shifted away from the attachment sites for the myosin heads on the actin. Cross-bridges can thus form, undergo power strokes, and produce muscle contraction.

Where does the Ca^{++} come from? Muscle fibers store Ca^{++} in a modified endoplasmic reticulum called a sarcoplasmic reticulum, or SR (figure 50.15). When a muscle fiber is stimulated to contract, an electrical impulse travels into the muscle fiber down invaginations called the **transverse tubules (T tubules**). This triggers the release of Ca^{++} from the SR. Ca^{++} then diffuses into the myofibrils, where it binds to troponin and causes contraction. The contraction of muscles is regulated by nerve activity, and so nerves must influence the distribution of Ca^{++} in the muscle fiber.

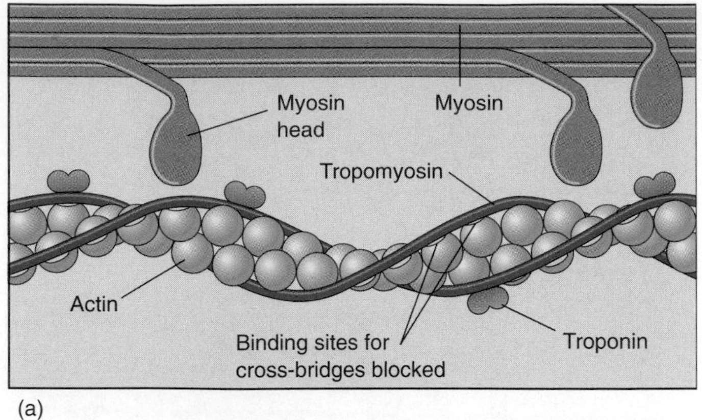

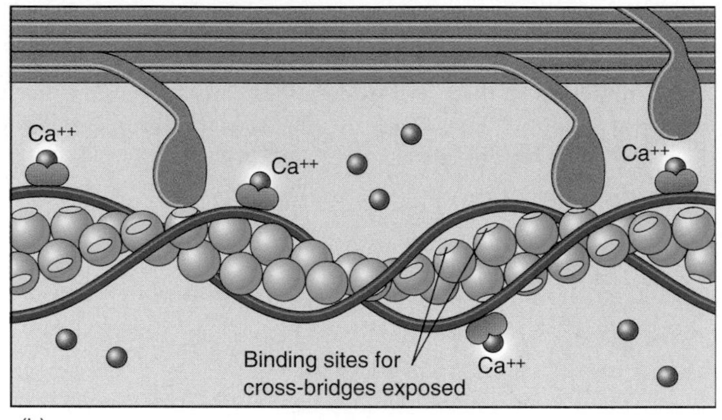

FIGURE 50.14
How calcium controls striated muscle contraction. (*a*) When the muscle is at rest, a long filament of the protein tropomyosin blocks the myosin-binding sites on the actin molecule. Because myosin is unable to form cross-bridges with actin at these sites, muscle contraction cannot occur. (*b*) When Ca^{++} binds to another protein, troponin, the Ca^{++}-troponin complex displaces tropomyosin and exposes the myosin-binding sites on actin, permitting cross-bridges to form and contraction to occur.

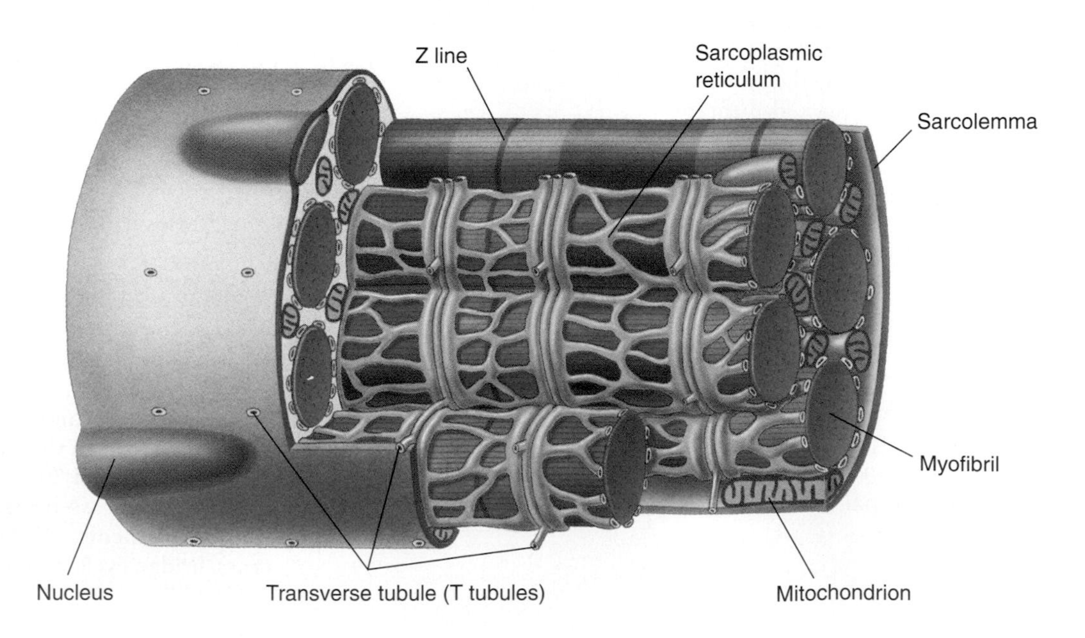

FIGURE 50.15
The relationship between the myofibrils, transverse tubules, and sarcoplasmic reticulum. Impulses travel down the axon of a motor neuron that synapses with a muscle fiber. The impulses are conducted along the transverse tubules and stimulate the release of Ca^{++} from the sarcoplasmic reticulum into the cytoplasm. Ca^{++} diffuses toward the myofibrils and causes contraction.

Nerves Stimulate Contraction

Muscles are stimulated to contract by motor neurons. The particular motor neurons that stimulate skeletal muscles, as opposed to cardiac and smooth muscles, are called *somatic motor neurons.* The axon (see figure 49.12) of a somatic motor neuron extends from the neuron cell body and branches to make functional connections, or *synapses,* with a number of muscle fibers. (Synapses are discussed in more detail in chapter 54.) One axon can stimulate many muscle fibers, and in some animals a muscle fiber may be innervated by more than one motor neuron. However, in humans each muscle fiber only has a single synapse with a branch of one axon.

When a somatic motor neuron produces electrochemical impulses, it stimulates contraction of the muscle fibers it innervates (makes synapses with) through the following events:

1. The motor neuron, at its synapse with the muscle fibers, releases a chemical known as a *neurotransmitter.* The specific neurotransmitter released by somatic motor neurons is **acetylcholine (ACh).** ACh acts on the muscle fiber membrane to stimulate the muscle fiber to produce its own electrochemical impulses.
2. The impulses spread along the membrane of the muscle fiber and are carried into the muscle fibers through the T tubules.
3. The T tubules conduct the impulses toward the sarcoplasmic reticulum, which then release Ca^{++}. As described earlier, the Ca^{++} binds to troponin, which exposes the cross-bridge binding sites on the actin myofilaments, stimulating muscle contraction.

When impulses from the nerve stop, the nerve stops releasing ACh. This stops the production of impulses in the muscle fiber. When the T tubules no longer produce impulses, Ca^{++} is brought back into the SR by active transport. Troponin is no longer bound to Ca^{++}, so tropomyosin returns to its inhibitory position, allowing the muscle to relax.

The involvement of Ca^{++} in muscle contraction is called, **excitation-contraction coupling** because it is the release of Ca^{++} that links the excitation of the muscle fiber by the motor neuron to the contraction of the muscle.

Motor Units and Recruitment

A single muscle fiber responds in an all-or-none fashion to stimulation. The response of an entire muscle depends upon the number of individual fibers involved. The set of muscle fibers innervated by all axonal branches of a given motor neuron is defined as a **motor unit** (figure 50.16). Every time the motor neuron produces impulses, all muscle fibers in that motor unit contract together. The division of the muscle into motor units allows the muscle's strength of contraction to be finely graded, a requirement for coordinated movements of the skeleton. Muscles that require a finer degree of control have smaller motor units (fewer muscle fibers per neuron) than muscles that require less precise control but must exert more force. For example, there are only a few muscle fibers per motor neuron in the muscles that move the eyes, while there are several hundred per motor neuron in the large muscles of the legs.

Most muscles contain motor units in a variety of sizes, which can be selectively activated by the nervous system. The weakest contractions of a muscle involve the activation of a few small motor units. If a slightly stronger contraction is necessary, additional small motor units are also activated. The initial increments to the total force generated by the muscle are therefore relatively small. As ever greater forces are required, more and larger motor units are brought into action, and the force increments become larger. The nervous system's use of increased numbers and sizes of motor units to produce a stronger contraction is termed **recruitment.**

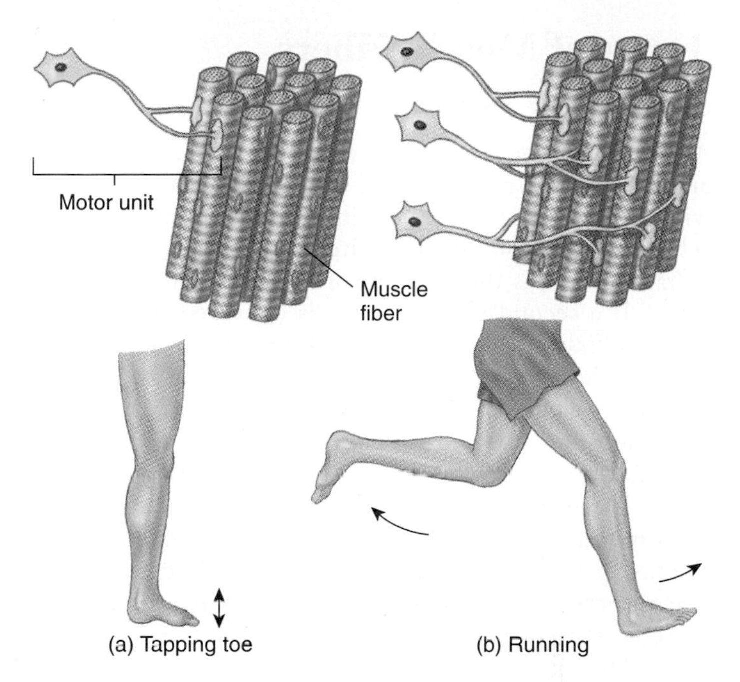

FIGURE 50.16
The number and size of motor units. (*a*) Weak, precise muscle contractions use smaller and fewer motor units. (*b*) Larger and stronger movements require additional motor units that are larger.

The cross-bridges are prevented from binding to actin by tropomyosin in a relaxed muscle. In order for a muscle to contract, Ca^{++} must be released from the sarcoplasmic reticulum, where it is stored, so that it can bind to troponin and cause the tropomyosin to shift its position in the thin filaments. Muscle contraction is stimulated by neurons. Varying sizes and numbers of motor units are used to produce different types of muscle contractions.

Types of Muscle Fibers

Muscle Fiber Twitches

An isolated skeletal muscle can be studied by stimulating it artificially with electric shocks. If a muscle is stimulated with a single electric shock, it will quickly contract and relax in a response called a **twitch.** Increasing the stimulus voltage increases the strength of the twitch up to a maximum. If a second electric shock is delivered immediately after the first, it will produce a second twitch that may partially "ride piggyback" on the first. This cumulative response is called **summation** (figure 50.17).

If the stimulator is set to deliver an increasing frequency of electric shocks automatically, the relaxation time between successive twitches will get shorter and shorter, as the strength of contraction increases. Finally, at a particular frequency of stimulation, there is no visible relaxation between successive twitches. Contraction is smooth and sustained, as it is during normal muscle contraction in the body. This smooth, sustained contraction is called **tetanus.** (The term *tetanus* should not be confused with the disease of the same name, which is accompanied by a painful state of muscle contracture, or *tetany.*)

Skeletal muscle fibers can be divided on the basis of their contraction speed into **slow-twitch,** or **type I, fibers,** and **fast-twitch,** or **type II, fibers.** The muscles that move the eyes, for example, have a high proportion of fast-twitch fibers and reach maximum tension in about 7.3 milliseconds; the soleus muscle in the leg, by contrast, has a high proportion of slow-twitch fibers and requires about 100 milliseconds to reach maximum tension (figure 50.18).

Muscles like the soleus must be able to sustain a contraction for a long period of time without fatigue. The resistance to fatigue demonstrated by these muscles is aided by other characteristics of slow-twitch (type I) fibers that endow them with a high capacity for aerobic respiration. Slow-twitch fibers have a rich capillary supply, numerous mitochondria and aerobic respiratory enzymes, and a high concentration of **myoglobin** pigment. Myoglobin is a red pigment, similar to the hemoglobin in red blood cells, but its higher affinity for oxygen improves the delivery of oxygen to the slow-twitch fibers. Because of their high myoglobin content, slow-twitch fibers are also called *red fibers.*

The thicker, fast-twitch (type II) fibers have fewer capillaries and mitochondria than slow-twitch fibers and not as much myoglobin; hence, these fibers are also called *white fibers.* Fast-twitch fibers are adapted to respire anaerobically by using a large store of glycogen and high concentrations of glycolytic enzymes. Fast-twitch fibers are adapted for the rapid generation of power and can grow thicker and stronger in response to weight training. The "dark meat" and "white meat" found in meat such as chicken and turkey consists of muscles with primarily red and white fibers, respectively.

In addition to the type I (slow-twitch) and type II (fast-twitch) fibers, human muscles also have an intermediate form of fibers that are fast-twitch but also have a high oxidative capacity, and so are more resistant to fatigue. Endurance training increases the proportion of these fibers in muscles.

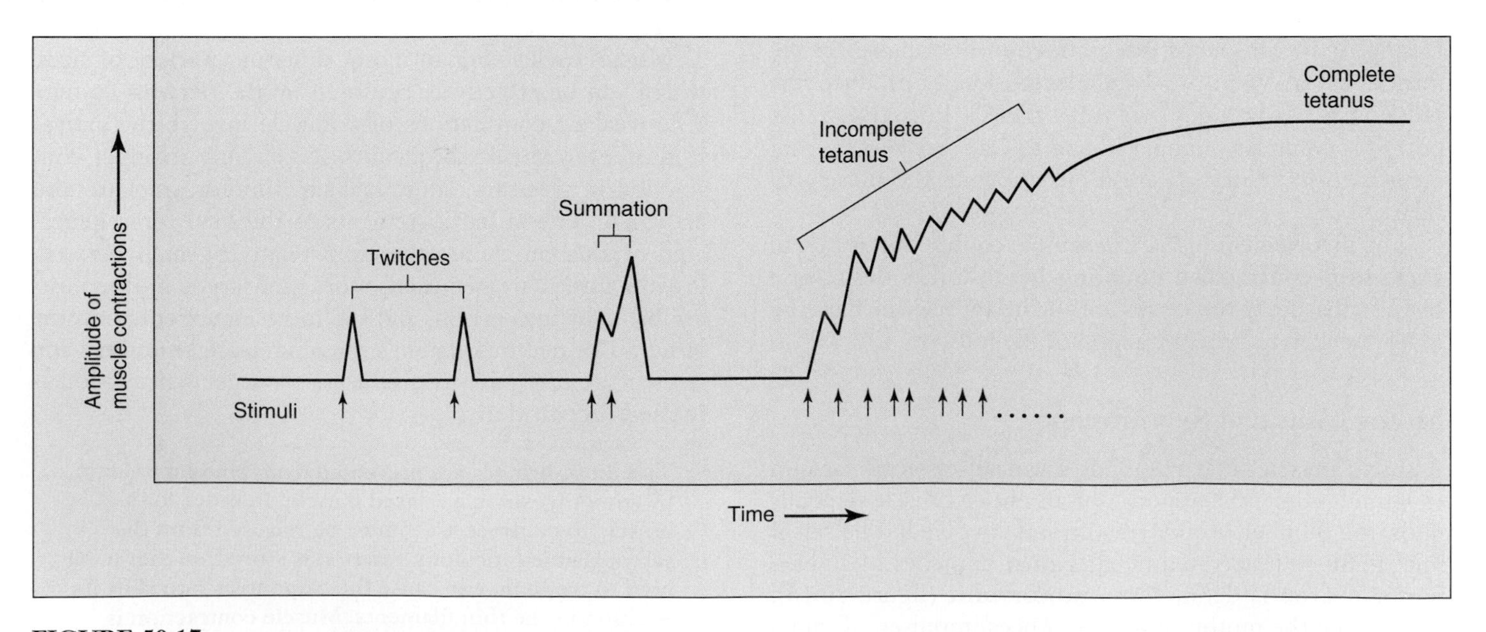

FIGURE 50.17
Muscle twitches summate to produce a sustained, tetanized contraction. This pattern is produced when the muscle is stimulated electrically or naturally by neurons. Tetanus, a smooth, sustained contraction, is the normal type of muscle contraction in the body.

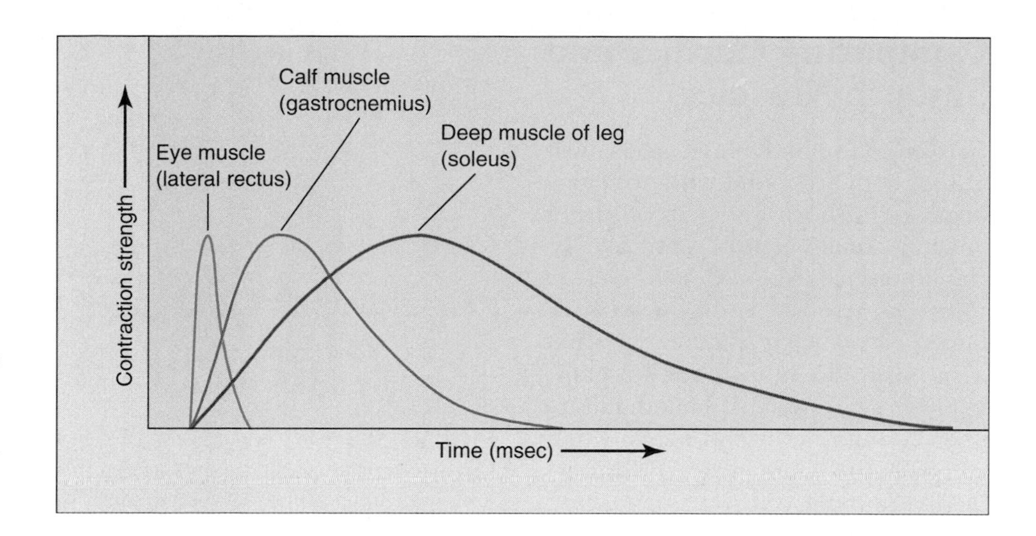

FIGURE 50.18
Skeletal muscles have different proportions of fast-twitch and slow-twitch fibers. The muscles that move the eye contain mostly fast-twitch fibers, whereas the deep muscle of the leg (the soleus) contains mostly slow-twitch fibers. The calf muscle (gastrocnemius) is intermediate in its composition.

Muscle Metabolism during Rest and Exercise

Skeletal muscles at rest obtain most of their energy from the aerobic respiration of fatty acids. During exercise, muscle glycogen and blood glucose are also used as energy sources. The energy obtained by cell respiration is used to make ATP, which is needed for (1) the movement of the cross-bridges during muscle contraction and (2) the pumping of Ca^{++} into the sarcoplasmic reticulum for muscle relaxation. ATP can be obtained by skeletal muscles quickly by combining ADP with phosphate derived from creatine phosphate. This compound was produced previously in the resting muscle by combining creatine with phosphate derived from the ATP generated in cell respiration.

Skeletal muscles respire anaerobically for the first 45 to 90 seconds of moderate-to-heavy exercise, because the cardiopulmonary system requires this amount of time to sufficiently increase the oxygen supply to the exercising muscles. If exercise is moderate, aerobic respiration contributes the major portion of the skeletal muscle energy requirements following the first 2 minutes of exercise.

Whether exercise is light, moderate, or intense for a given person depends upon that person's maximal capacity for aerobic exercise. The maximum rate of oxygen consumption in the body (by aerobic respiration) is called the maximal oxygen uptake, or the aerobic capacity. The intensity of exercise can also be defined by the lactate threshold. This is the percentage of the maximal oxygen uptake at which a significant rise in blood lactate levels occurs as a result of anaerobic respiration. For average, healthy people, for example, a significant amount of blood lactate appears when exercise is performed at about 50 to 70% of the maximal oxygen uptake.

Muscle Fatigue and Physical Training

Muscle fatigue refers to the use-dependant decrease in the ability of a muscle to generate force. The reasons for fatigue are not entirely understood. In most cases, however, muscle fatigue is correlated with the production of lactic acid by the exercising muscles. Lactic acid is produced by the anaerobic respiration of glucose, and glucose is obtained from muscle glycogen and from the blood. Lactate production and muscle fatigue are therefore also related to the depletion of muscle glycogen.

Because the depletion of muscle glycogen places a limit on exercise, any adaptation that spares muscle glycogen will improve physical endurance. Trained athletes have an increased proportion of energy derived from the aerobic respiration of fatty acids, resulting in a slower depletion of their muscle glycogen reserve. The greater the level of physical training, the higher the proportion of energy derived from the aerobic respiration of fatty acids. Because the aerobic capacity of endurance-trained athletes is higher than that of untrained people, athletes can perform more exercise before lactic acid production and glycogen depletion cause muscle fatigue.

Endurance training does not increase muscle size. Muscle enlargement is produced only by frequent periods of high-intensity exercise in which muscles work against high resistance, as in weight lifting. As a result of resistance training, type II (fast-twitch) muscle fibers become thicker as a result of the increased size and number of their myofibrils. Weight training, therefore, causes skeletal muscles to grow by **hypertrophy** (increased cell size) rather than by cell division and an increased number of cells.

Muscles contract through summation of the contractions of their fibers, producing tension that may result in shortening of the muscle. Slow-twitch skeletal muscle fibers are adapted for aerobic respiration and are slower to fatigue than fast-twitch fibers, which are more adapted for the rapid generation of power.

Comparing Cardiac and Smooth Muscles

Cardiac and smooth muscle are similar in that both are found within internal organs and both are generally not under conscious control. Cardiac muscle, however, is like skeletal muscle in that it is striated and contracts by means of a sliding filament mechanism. Smooth muscle (as its name implies) is not striated. Smooth muscle does contain actin and myosin filaments, but they are arranged less regularly within the cell.

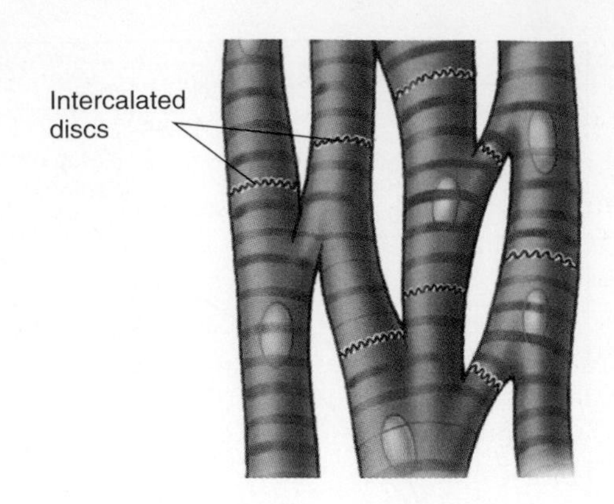

FIGURE 50.19
Cardiac muscle. Cells are organized into long branching chains that interconnect, forming a lattice; neighboring cells are linked by structures called intercalated discs.

Cardiac Muscle

Cardiac muscle in the vertebrate heart is composed of striated muscle cells that are arranged differently from the fibers in a skeletal muscle. Instead of the long, multinucleate cells that form skeletal muscle, cardiac muscle is composed of shorter, branched cells, each with its own nucleus, that interconnect with one another at intercalated discs (figure 50.19). Intercalated discs are regions where the membranes of two cells fuse together, and the fused membranes are pierced by **gap junctions** (chapter 7). The gap junctions permit the diffusion of ions, and thus the spread of electric excitation, from one cell to the next. The mass of interconnected cardiac muscle cells forms a single, functioning unit called a *myocardium.* Electric impulses begin spontaneously in a specific region of the myocardium known as the *pacemaker.* These impulses are *not* initiated by impulses in motor neurons, as they are in skeletal muscle, but rather are produced by the cardiac muscle cells themselves. From the pacemaker, the impulses spread throughout the myocardium via gap junctions, causing contraction.

The heart has two myocardia, one that receives blood from the body and one that ejects blood into the body. Because all of the cells in a myocardium are stimulated as a unit, cardiac muscle cannot produce summated contractions or tetanus. This would interfere with the alternation between contraction and relaxation that is necessary for pumping.

Smooth Muscle

Smooth muscle surrounds hollow internal organs, including the stomach, intestines, bladder, and uterus, as well as all blood vessels except capillaries. Smooth muscle cells are long and spindle-shaped, and each contains a single nucleus. They also contain actin and myosin, but these contractile proteins are not organized into sarcomeres. Parallel arrangements of thick and thin filaments cross diagonally from one side of the cell to the other. The thick filaments are attached either to structures called dense bodies, the functional equivalents of Z lines, or to the plasma membrane. Most smooth muscle cells have 10 to 15 thin filaments per thick filament, compared to 3 per thick filament in striated muscle fibers.

Smooth muscle cells do not have a sarcoplasmic reticulum; during a contraction, Ca^{++} enters from the extracellular fluid. In the cytoplasm, Ca^{++} binds to calmodulin, a protein that is structurally similar to troponin. The Ca^{++}-calmodulin complex activates an enzyme that phosphorylates (adds a phosphate group to) the myosin heads. Unlike the case with striated muscles, this phosphorylation is required for the myosin heads to form cross-bridges with actin.

This mechanism allows gradations in the strength of contraction in a smooth muscle cell, increasing contraction strength as more Ca^{++} enters the cytoplasm. Heart patients sometimes take drugs that block Ca^{++} entry into smooth muscle cells, reducing the cells' ability to contract. This treatment causes vascular smooth muscle to relax, dilating the blood vessels and reducing the amount of work the heart must do to pump blood through them.

In some smooth muscle tissues, the cells contract only when they are stimulated by the nervous system. These muscles line the walls of many blood vessels and make up the iris of the eye. Other smooth muscle tissues, like those in the wall of the gut, contains cells that produce electric impulses spontaneously. These impulses spread to adjoining cells through gap junctions, leading to a slow, steady contraction of the tissue.

Neither skeletal nor cardiac muscle can be greatly stretched because if the thick and thin filaments no longer overlay in the sarcomere, cross-bridges cannot form. Unlike these striated muscles, smooth muscle can contract even when it is greatly stretched. If one considers the degree to which some internal organs may be stretched—a uterus during pregnancy, for example—it is no wonder that these organs contain smooth muscle instead of striated muscle.

Cardiac muscle cells interconnect physically and electrically to form a single, functioning unit called a myocardium, which produces its own impulses at a pacemaker region. Smooth muscles lack the organization of myofilaments into sarcomeres and lack sarcoplasmic reticulum but contraction still occurs as myofilaments slide past one another by use of cross-bridges.

Modes of Animal Locomotion

Animals are unique among multicellular organisms in their ability to actively move from one place to another. Locomotion requires both a propulsive mechanism and a control mechanism. Animals employ a wide variety of propulsive mechanisms, most involving contracting muscles to generate the necessary force. The quantity, quality, and position of contractions are initiated and coordinated by the nervous system. In large animals, active locomotion is almost always produced by appendages that oscillate—*appendicular locomotion*—or by bodies that undulate, pulse, or undergo peristaltic waves—*axial locomotion.*

While animal locomotion occurs in many different forms, the general principles remain much the same in all groups. The physical restraints to movement—gravity and frictional drag—are the same in every environment, differing only in degree. You can conveniently divide the environments through which animals move into three types, each involving its own forms of locomotion: water, land, and air.

Locomotion in Water

Many aquatic and marine invertebrates move along the bottom using the same form of locomotion employed by terrestrial animals moving over the land surface. Flatworms employ ciliary activity to brush themselves along, roundworms a peristaltic slither, leeches a contract-anchor-extend creeping. Crabs walk using limbs to pull themselves along; mollusks use a muscular foot, while starfish use unique tube feet to do the same thing.

Moving directly through the water, or swimming, presents quite a different challenge. Water's buoyancy reduces the influence of gravity. The primary force retarding forward movement is frictional drag, so body shape is important in reducing the friction and turbulence produced by swimming through the water.

Some marine invertebrates swim using hydraulic propulsion. Scallops clap their shells together forcefully, while squids and octopuses squirt water like a marine jet. All aquatic and marine vertebrates, however, swim.

Swimming involves using the body or its appendages to push against the water. An eel swims by sinuous undulations of its whole body (figure 50.20*a*). The undulating body waves of eel-like swimming are created by waves of muscle contraction alternating between the left and right axial musculature. As each body segment in turn pushes against the water, the moving wave forces the eel forward.

Fish, reptiles, and aquatic amphibians swim in a way similar to eels, but only undulate the posterior (back) portion of the body (figure 50.20*b*) and sometimes only the caudal (rear) fin. This allows considerable specialization of the front end of the body, while sacrificing little propulsive force.

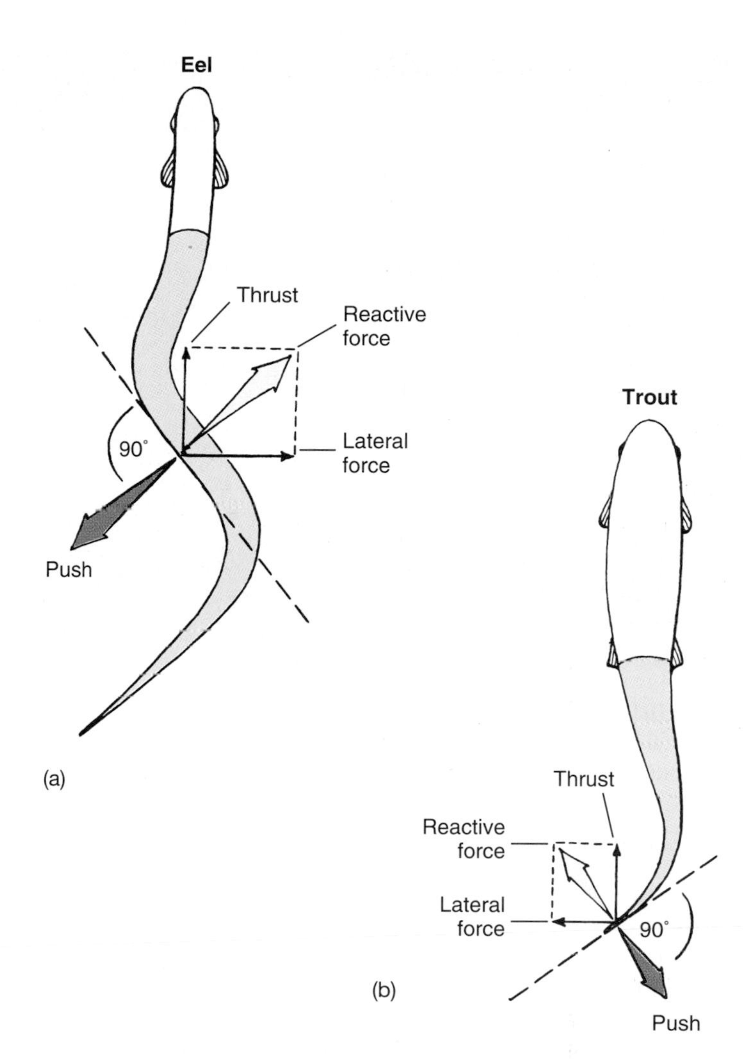

FIGURE 50.20
Movements of swimming fishes. (*a*) An eel pushes against the water with its whole body, (*b*) a trout only with its posterior half.

Whales also swim using undulating body waves, but unlike any of the fishes, the waves pass from top to bottom and not from side to side. The body musculature of eels and fish is highly segmental; that is, a muscle segment alternates with each vertebra. This arrangement permits the smooth passage of undulatory waves along the body. Whales are unable to produce lateral undulations because mammals do not have this arrangement.

Many tetrapod vertebrates swim, usually with appendicular locomotion. Most birds that swim, like ducks and geese, propel themselves through the water by pushing against it with their hind legs, which typically have webbed feet. Frogs, turtles, and most marine mammals also swim with their hind legs and have webbed feet. Tetrapod vertebrates that swim with their forelegs usually have these limbs modified as flippers, and pull themselves through the water. These include sea turtles, penguins, and fur seals. A few principally terrestrial tetrapod vertebrates, like polar bears and platypuses, swim with walking forelimbs not modified for swimming.

FIGURE 50.21
Animals that hop or leap use their rear legs to propel themselves through the air. The powerful leg muscles of this frog allow it to explode from a crouched position to a takeoff in about 100 milliseconds.

Locomotion on Land

The three great groups of terrestrial animals—mollusks, arthropods, and vertebrates—each move over land in different ways.

Mollusk locomotion is far less efficient than that of the other groups. Snails, slugs, and other terrestrial mollusks secrete a path of mucus that they glide along, pushing with a muscular foot.

Only vertebrates and arthropods (insects, spiders, and crustaceans) have developed a means of rapid surface locomotion. In both groups, the body is raised above the ground and moved forward by pushing against the ground with a series of jointed appendages, the legs.

Because legs must provide support as well as propulsion, it is important that the sequence of their movements not shove the body's center of gravity outside of the legs' zone of support. If they do, the animal loses its balance and falls. It is the necessity to maintain stability that determines the sequence of leg movements, which are similar in vertebrates and arthropods.

The apparent differences in the walking gaits of these two groups reflects the differences in leg number. Vertebrates are tetrapods (four limbs), while all arthropods have six or more limbs. Although having many legs increases stability during locomotion, they also appear to reduce the maximum speed that can be attained.

The basic walking pattern of all tetrapod vertebrates is left hind leg (LH), left foreleg (LF), right hindleg (RH), right foreleg (RF), and then the same sequence again and again. Unlike insects, vertebrates can begin to walk with any of the four legs, and not just the posterior pair. Both arthropods and vertebrates achieve faster gaits by overlapping the leg movements of the left and right sides. For example, a horse can convert a walk to a trot, by moving diagonally opposite legs simultaneously.

The highest running speeds of tetrapod vertebrates, such as the gallop of a horse, are obtained with asymmetric gaits. When galloping, a horse is never supported by more than two legs, and occasionally is supported by none. This reduces friction against the ground to an absolute minimum, increasing speed. With their larger number of legs, arthropods cannot have these speedy asymmetric gaits, because the movements of the legs would interfere with each other.

Not all animals walk or run on land. Many insects, like grasshoppers, leap using strong rear legs to propel themselves through the air. Vertebrates such as kangaroos, rabbits, and frogs are also effective leapers (figure 50.21).

Many invertebrates use peristaltic motion to slide over the surface. Among vertebrates, this form of locomotion is exhibited by snakes and caecilians (legless amphibians). Most snakes employ serpentine locomotion, in which the body is thrown into a series of sinuous curves. The movements superficially resemble those of eel-like swimming, but the similarity is more apparent than real. Propulsion is not by a wave of contraction undulating the body, but by a simultaneous lateral thrust in all segments of the body in contact with the ground. To go forward, it is necessary that the strongest muscular thrust push against the ground opposite the direction of movement. Because of this, thrust tends to occur at the anterior (outside) end of the inward-curving side of the loop of the snake's body.

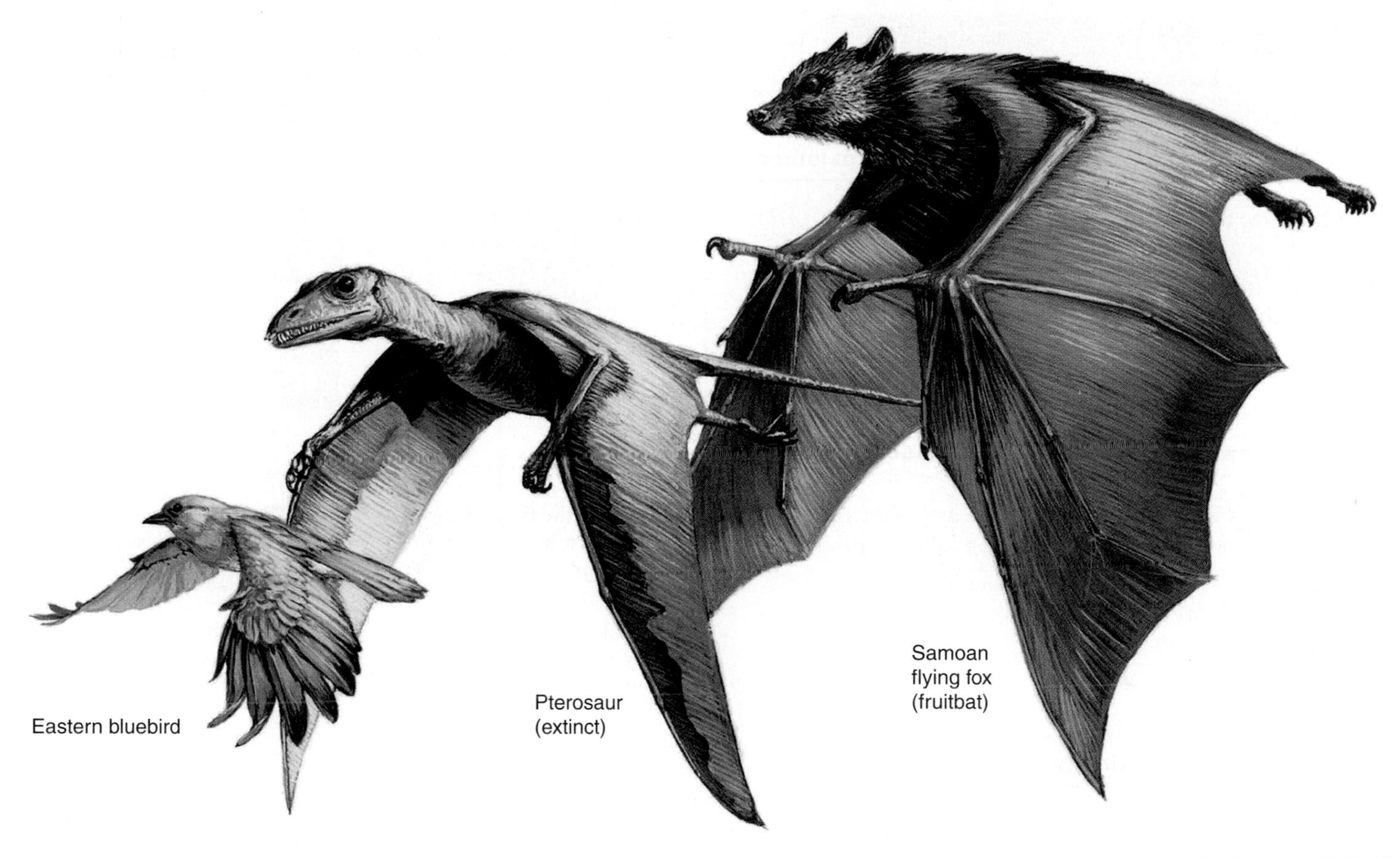

FIGURE 50.22
Flight has evolved three times among the vertebrates. These three very different vertebrates all have lightened bones and forelimbs transformed into wings.

Locomotion in Air

Flight has evolved among the animals four times: insects, pterosaurs (extinct flying reptiles), birds, and bats. In all four groups, active flying takes place in much the same way. Propulsion is achieved by pushing down against the air with wings. This provides enough lift to keep insects in the air. Vertebrates, being larger, need greater lift, obtaining it with wings that are convex in cross section. Because air must travel farther over the top surface, it moves faster, creating lift over the wing.

In birds and most insects, the raising and lowering of the wings is achieved by the alternate contraction of extensor muscles (elevators) and flexor muscles (depressors). Four insect orders (including those containing flies, mosquitoes, wasps, bees, and beetles), however, beat their wings at frequencies from 100 to more than 1000 times per second, faster than nerves can carry successive impulses! In these insects, the flight muscles are not attached to the wings at all but rather to the stiff wall of the thorax, which is distorted in and out by their contraction. The reason that these muscles can beat so fast is that the contraction of one set stretches the other, triggering its contraction in turn without waiting for the arrival of a nerve impulse.

Among vertebrates (figure 50.22), flight first evolved some 200 million years ago among flying reptiles called pterosaurs. A very successful and diverse group, pterosaurs ranged in size from individuals no bigger than sparrows to pterodons the size of a fighter plane. For much of this time, they shared the skies with birds, which most paleontologists believe evolved from feathered dinosaurs about 150 million years ago. How did they share their ecological world for 100 million years without competition driving one or the other from the skies? No one knows for sure. Perhaps these early birds were night fliers, while pterosaurs flew by day.

Such an arrangement for sharing resources is not as unlikely as it might at first appear. Bats, flying mammals which evolved after the pterosaurs disappeared with the dinosaurs, are night fliers. By flying at night bats are able to shop in a store with few other customers and a wealth of food: night-flying insects. It has proven to be a very successful approach. One-quarter of all mammal species are bats.

Locomotion in larger animals is almost always produced by appendages that push against the surroundings in some fashion, or by shoving the entire body forward by an undulation.

Chapter 50

www.mhhe.com/raven6/resources50.mhtml

Summary	Questions	Media Resources

50.1 A skeletal system supports movement in animals.

- There are three types of skeleton: hydrostatic skeletons, exoskeletons, and endoskeletons.
- Bone is formed by the secretion of an organic matrix by osteoblasts; this organic matrix becomes calcified.

1. What are the two major components of the extracellular matrix in bone? What structural properties does each component have? How do the two components combine to make bone resistant to fracture?

- Bone Structure
- Bone Function

- *On Science* Article: Running Improperly
- Bioethics Case Study: Osteoporosis

50.2 Skeletal muscles contract to produce movements at joints.

- Freely movable joints surround the articulating bones with a synovial capsule filled with a lubricating fluid.
- Skeletal muscles can work together as synergists, or oppose each other as antagonists.

2. What are the three types of joints in a vertebrate skeleton? Give an example of where each type is found in the body.

3. What is the difference between a skeletal muscle's origin and its insertion?

- Art Activity: Skeletal Muscles

- Skeleton
- Muscles

50.3 Muscle contraction powers animal locomotion.

- A muscle fiber contains numerous myofibrils that are composed of myofilaments of myosin and actin.
- There are small cross-bridges of myosin that extend out toward the actin; the cross-bridges are activated by the hydrolysis of ATP so that it can bind to actin and undergo a power stroke that causes the sliding of the myofilaments.
- When Ca^{++} binds to troponin, the tropomyosin shifts position in the thin filament, allowing the cross-bridges to bind to actin and undergo a power stroke.
- The release of Ca^{++} from the sarcoplasmic reticulum is stimulated by impulses in the muscle fiber produced by neural stimulation.
- Slow-twitch fibers are adapted for aerobic respiration and are resistant to fatigue; fast-twitch fibers can provide power quickly but produce lactic acid and fatigue quickly.
- Cardiac muscle cells have gap junctions that permit the spread of electric impulses from one cell to the next.
- Cardiac and smooth muscles are involuntary; the contractions are automatically produced in cardiac muscle and some smooth muscles.
- Animals have adapted modes of locomotion to three different environments: water, land, and air.

4. Of what proteins are thick and thin filaments composed?

5. Describe the steps involved in the cross-bridge cycle. What functions does ATP perform in the cycle?

6. Describe the steps involved in excitation-contraction coupling. What functions do acetylcholine and Ca^{++} perform in this process?

7. How does a somatic motor neuron stimulate a muscle fiber to contract?

8. What is the difference between a muscle twitch and tetanus?

9. Why can't a myocardium produce a sustained contraction?

10. How does smooth muscle differ from skeletal muscle in terms of thick and thin filament organization, the role of Ca^{++} in contraction, and the effect of stretching on the muscle's ability to contract?

11. What do all modes of locomotion have in common?

- Straited Muscle Contraction
- Actin-Myosin Crossbridges
- Muscle Contraction Action Potential
- Walking

- Muscle Cell Function
- Activity: Muscle Contraction
- Muscle Twitch Physiology
- Smooth and Cardiac Muscle

- *On Science* Article: Climbing the Walls

- Art Quizzes:
  - -Muscle Stimulation Pattern
  - -Thick and Thin Filaments
  - -Muscle Contraction
  - -Fast- and Slow-Twitch Fibers

51

Fueling Body Activities: Digestion

FIGURE 51.1
Animals are heterotrophs. All animals must consume plant material or other animals in order to live. The nuts in this chipmunk's cheeks will be consumed and converted to body tissue, energy, and refuse.

Concept Outline

Plants and other photosynthetic organisms can produce the organic molecules they need from inorganic components. Therefore, they are autotrophs, or self-sustaining. Animals are heterotrophs: they must consume organic molecules present in other organisms (figure 51.1). The molecules heterotrophs eat must be digested into smaller molecules in order to be absorbed into the animal's body. Once these products of digestion enter the body, the animal can use them for energy in cell respiration or for the construction of the larger molecules that make up its tissues. The process of animal digestion is the focus of this chapter.

Types of Digestive Systems

Heterotrophs are divided into three groups on the basis of their food sources. Animals that eat plants exclusively are classified as **herbivores;** common examples include cows, horses, rabbits and sparrows. Animals that are meat-eaters, such as cats, eagles, trout, and frogs, are **carnivores. Omnivores** are animals that eat both plants and other animals. Humans are omnivores, as are pigs, bears, and crows.

Single-celled organisms (as well as sponges) digest their food intracellularly. Other animals digest their food extracellularly, within a digestive cavity. In this case, the digestive enzymes are released into a cavity that is continuous with the animal's external environment. In coelenterates and flatworms (such as Planaria), the digestive cavity has only one opening that serves as both mouth and anus. There can be no specialization within this type of digestive system, called a *gastrovascular cavity,* because every cell is exposed to all stages of food digestion (figure 51.2).

Specialization occurs when the digestive tract, or alimentary canal, has a separate mouth and anus, so that transport of food is one-way. The most primitive digestive tract is seen in nematodes (phylum Nematoda), where it is simply a tubular *gut* lined by an epithelial membrane. Earthworms (phylum Annelida) have a digestive tract specialized in different regions for the ingestion, storage, fragmentation, digestion, and absorption of food. All higher animal groups, including all vertebrates, show similar specializations (figure 51.3).

The ingested food may be stored in a specialized region of the digestive tract or may first be subjected to physical fragmentation. This fragmentation may occur through the chewing action of teeth (in the mouth of many vertebrates), or the grinding action of pebbles (in the gizzard of earthworms and birds). Chemical digestion then occurs, breaking down the larger food molecules of polysaccharides and disaccharides, fats, and proteins into their smallest subunits. Chemical digestion involves hydrolysis reactions that liberate the subunit molecules—primarily monosaccharides, amino acids, and fatty acids—from the food. These products of chemical digestion pass through the epithelial lining of the gut into the blood, in a process known as absorption. Any molecules in the food that are not absorbed cannot be used by the animal. These waste products are excreted, or defecated, from the anus.

Most animals digest their food extracellularly. The digestive tract, with a one-way transport of food and specialization of regions for different functions, allows food to be ingested, physically fragmented, chemically digested, and absorbed.

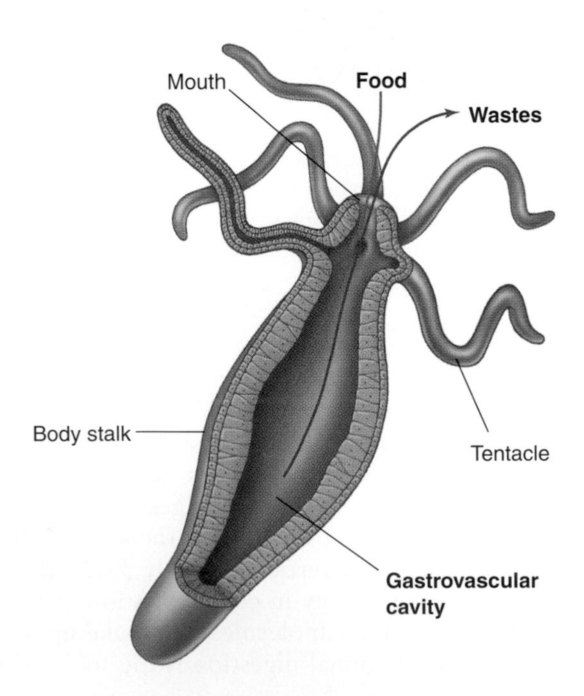

FIGURE 51.2
The gastrovascular cavity of *Hydra*, a coelenterate. Because there is only one opening, the mouth is also the anus, and no specialization is possible in the different regions that participate in extracellular digestion.

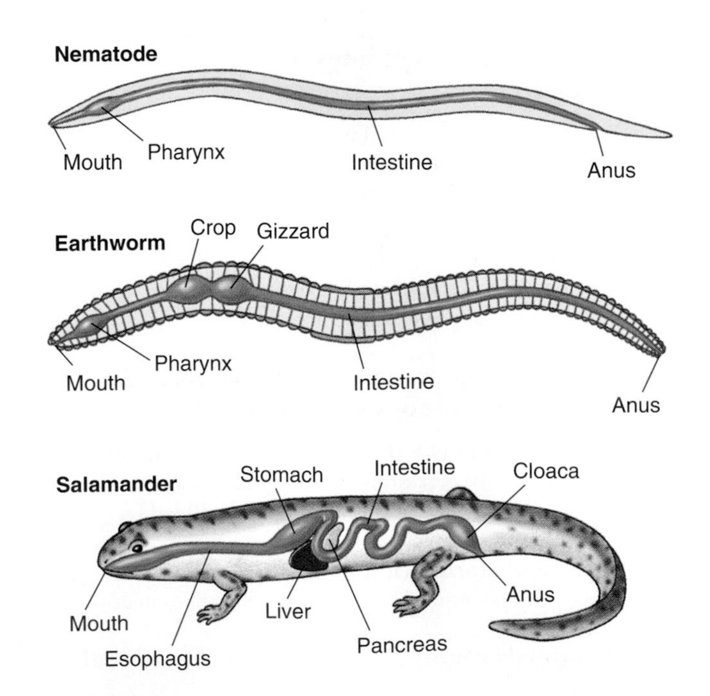

FIGURE 51.3
The one-way digestive tract of nematodes, earthworms, and vertebrates. One-way movement through the digestive tract allows different regions of the digestive system to become specialized for different functions.

Vertebrate Digestive Systems

In humans and other vertebrates, the digestive system consists of a tubular gastrointestinal tract and accessory digestive organs (figure 51.4). The initial components of the gastrointestinal tract are the mouth and the pharynx, which is the common passage of the oral and nasal cavities. The pharynx leads to the esophagus, a muscular tube that delivers food to the stomach, where some preliminary digestion occurs. From the stomach, food passes to the first part of the small intestine, where a battery of digestive enzymes continues the digestive process. The products of digestion then pass across the wall of the small intestine into the bloodstream. The small intestine empties what remains into the large intestine, where water and minerals are absorbed. In most vertebrates other than mammals, the waste products emerge from the large intestine into a cavity called the cloaca (see figure 51.3), which also receives the products of the urinary and reproductive systems. In mammals, the urogenital products are separated from the fecal material in the large intestine; the fecal material enters the rectum and is expelled through the anus.

In general, carnivores have shorter intestines for their size than do herbivores. A short intestine is adequate for a carnivore, but herbivores ingest a large amount of plant cellulose, which resists digestion. These animals have a long, convoluted small intestine. In addition, mammals called *ruminants* (such as cows) that consume grass and other vegetation have stomachs with multiple chambers, where bacteria aid in the digestion of cellulose. Other herbivores, including rabbits and horses, digest cellulose (with the aid of bacteria) in a blind pouch called the **cecum** located at the beginning of the large intestine.

The accessory digestive organs (described in detail later in the chapter) include the liver, which produces *bile* (a green solution that emulsifies fat), the gallbladder, which stores and concentrates the bile, and the pancreas. The pancreas produces *pancreatic juice*, which contains digestive enzymes and bicarbonate. Both bile and pancreatic juice are secreted into the first region of the small intestine and aid digestion.

The tubular gastrointestinal tract of a vertebrate has a characteristic layered structure (figure 51.5). The innermost layer is the mucosa, an epithelium that lines the interior of the tract (the lumen). The next major tissue layer, made of connective tissue, is called the submucosa. Just outside the submucosa is the muscularis, which consists of a double layer of smooth muscles. The muscles in the inner layer have a circular orientation, and those in the outer layer are arranged longitudinally. Another connective tissue layer, the serosa, covers the external surface of the tract. Nerves, intertwined in regions called *plexuses*, are located in the submucosa and help regulate the gastrointestinal activities.

The vertebrate digestive system consists of a tubular gastrointestinal tract, which is modified in different animals, composed of a series of tissue layers.

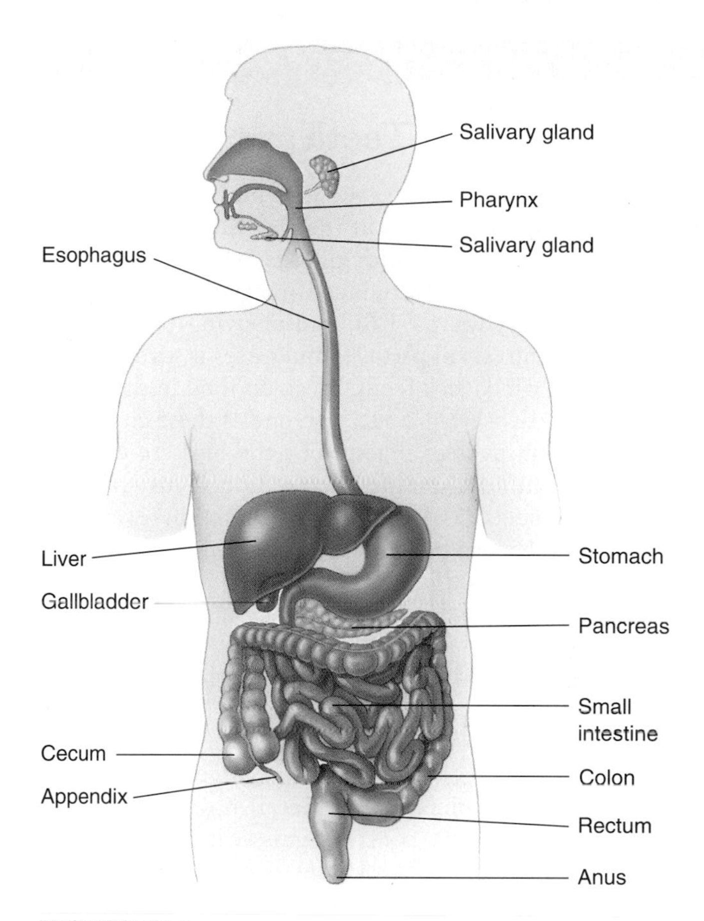

FIGURE 51.4
The human digestive system. Humans, like all placental mammals, lack a cloaca and have a separate exit from the digestive tract through the rectum and anus.

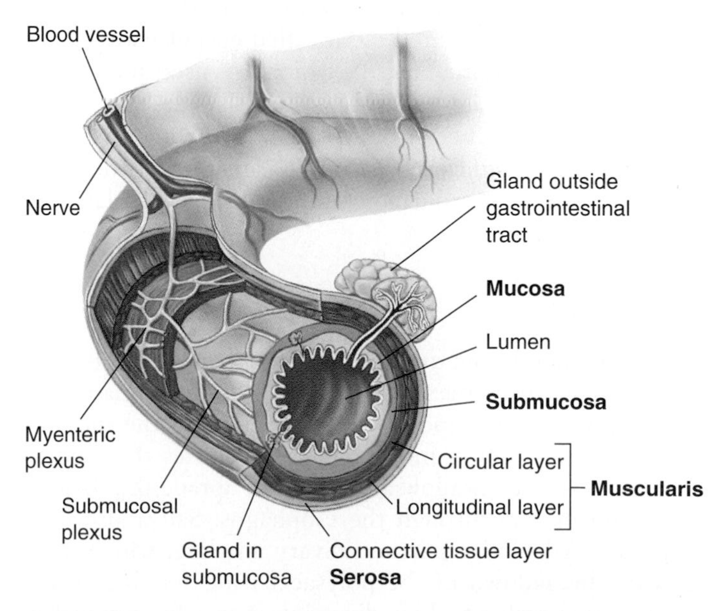

FIGURE 51.5
The layers of the gastrointestinal tract. The mucosa contains a lining epithelium; the submucosa is composed of connective tissue (as is the serosa), and the muscularis consists of smooth muscles.

51.2 Food is ingested, swallowed, and transported to the stomach.

The Mouth and Teeth

Specializations of the digestive systems in different kinds of vertebrates reflect differences in the way these animals live. Fishes have a large pharynx with gill slits, while air-breathing vertebrates have a greatly reduced pharynx. Many vertebrates have teeth (figure 51.6), and chewing (*mastication*) breaks up food into small particles and mixes it with fluid secretions. Birds, which lack teeth, break up food in their two-chambered stomachs (figure 51.7). In one of these chambers, the gizzard, small pebbles ingested by the bird are churned together with the food by muscular action. This churning grinds up the seeds and other hard plant material into smaller chunks that can be digested more easily.

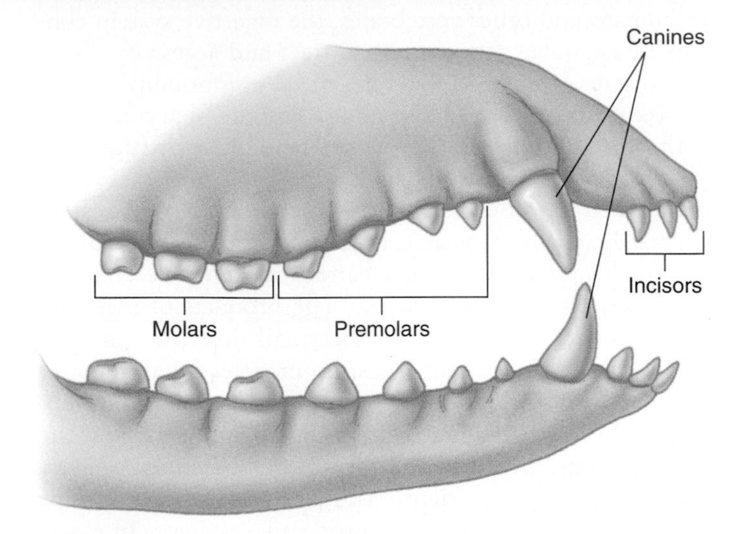

FIGURE 51.6
Diagram of generalized vertebrate dentition. Different vertebrates will have specific variations from this generalized pattern, depending on whether the vertebrate is an herbivore, carnivore, or omnivore.

Vertebrate Teeth

Carnivorous mammals have pointed teeth that lack flat grinding surfaces. Such teeth are adapted for cutting and shearing. Carnivores often tear off pieces of their prey but have little need to chew them, because digestive enzymes can act directly on animal cells. (Recall how a cat or dog gulps down its food.) By contrast, grass-eating herbivores, such as cows and horses, must pulverize the cellulose cell walls of plant tissue before digesting it. These animals have large, flat teeth with complex ridges well-suited to grinding.

Human teeth are specialized for eating both plant and animal food. Viewed simply, humans are carnivores in the front of the mouth and herbivores in the back (figure 51.8). The four front teeth in the upper and lower jaws are sharp, chisel-shaped incisors used for biting. On each side of the incisors are sharp, pointed teeth called cuspids (sometimes referred to as "canine" teeth), which are used for tearing food. Behind the canines are two premolars and three molars, all with flattened, ridged surfaces for grinding and crushing food. Children have only 20 teeth, but these deciduous teeth are lost during childhood and are replaced by 32 adult teeth.

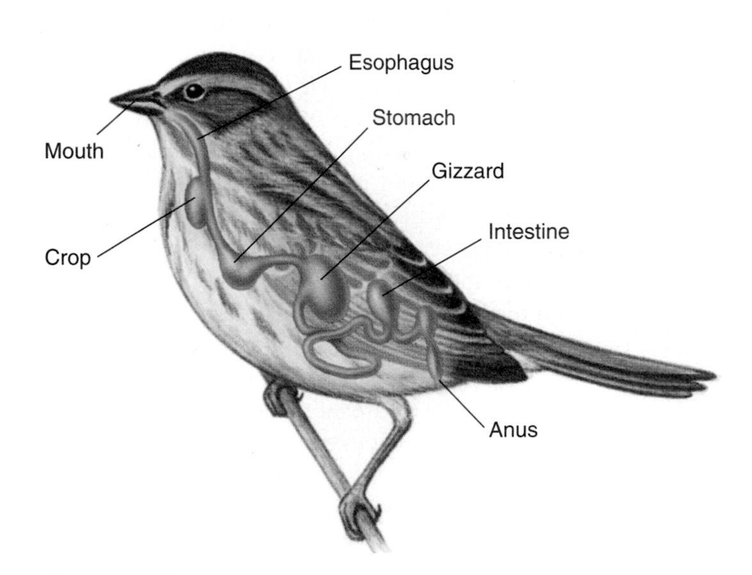

FIGURE 51.7
Birds store food in the crop and grind it up in the gizzard. Birds lack teeth but have a muscular chamber called the gizzard that works to break down food. Birds swallow gritty objects or pebbles that lodge in the gizzard and pulverize food before it passes into the small intestine.

The Mouth

Inside the mouth, the tongue mixes food with a mucous solution, saliva. In humans, three pairs of salivary glands secrete saliva into the mouth through ducts in the mouth's mucosal lining. Saliva moistens and lubricates the food so that it is easier to swallow and does not abrade the tissue it passes on its way through the esophagus. Saliva also contains the hydrolytic enzyme salivary amylase, which initiates the breakdown of the polysaccharide starch into the disaccharide maltose. This digestion is usually minimal in humans, however, because most people don't chew their food very long.

The secretions of the salivary glands are controlled by the nervous system, which in humans maintains a constant flow of about half a milliliter per minute when the mouth is empty of food. This continuous secretion keeps the mouth moist. The presence of food in the mouth triggers an increased rate of secretion, as taste-sensitive neurons in the mouth send impulses to the brain, which responds by stimulating the salivary glands. The most potent stimuli are acidic solutions; lemon juice, for example, can increase the rate of salivation eightfold. The sight, sound, or smell of food can stimulate salivation markedly in dogs, but in humans, these stimuli are much less effective than thinking or talking about food.

When food is ready to be swallowed, the tongue moves it to the back of the mouth. In mammals, the process of swallowing begins when the soft palate elevates, pushing against the back wall of the pharynx (figure 51.9). Elevation of the soft palate seals off the nasal cavity and prevents food from entering it. Pressure against the pharynx triggers an automatic, involuntary response called a reflex. In this reflex, pressure on the pharynx stimulates neurons within its walls, which send impulses to the swallowing center in the brain. In response, electrical impulses in motor neurons stimulate muscles to contract and raise the *larynx* (voice box). This pushes the *glottis*, the opening from the larynx into the trachea (windpipe), against a flap of tissue called the *epiglottis*. These actions keep food out of the respiratory tract, directing it instead into the esophagus.

In many vertebrates ingested food is fragmented through the tearing or grinding action of specialized teeth. In birds, this is accomplished through the grinding action of pebbles in the gizzard. Food mixed with saliva is swallowed and enters the esophagus.

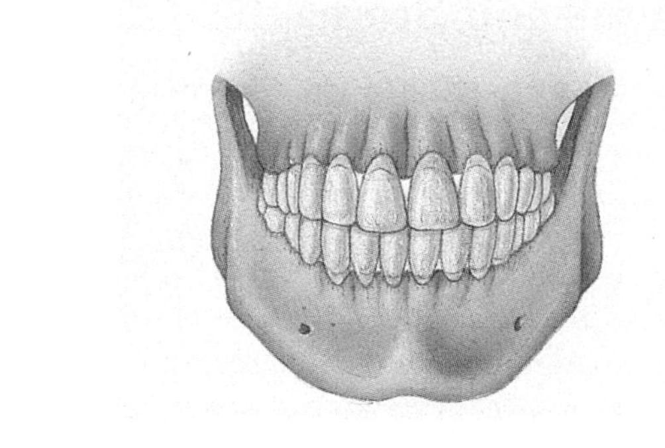
(a)

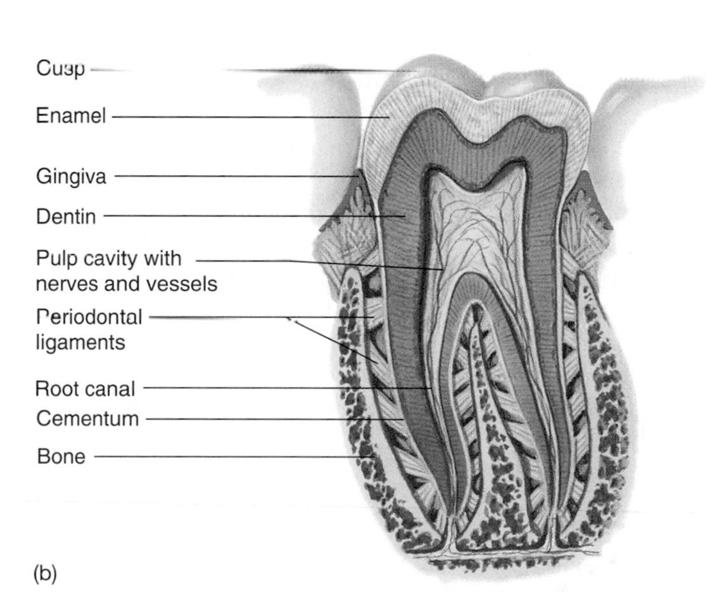

(b)

FIGURE 51.8
Human teeth. (*a*) The front six teeth on the upper and lower jaws are cuspids and incisors. The remaining teeth, running along the sides of the mouth, are grinders called premolars and molars. Hence, humans have carnivore-like teeth in the front of their mouth and herbivore-like teeth in the back. (*b*) Each tooth is alive, with a central pulp containing nerves and blood vessels. The actual chewing surface is a hard enamel layered over the softer dentin, which forms the body of the tooth.

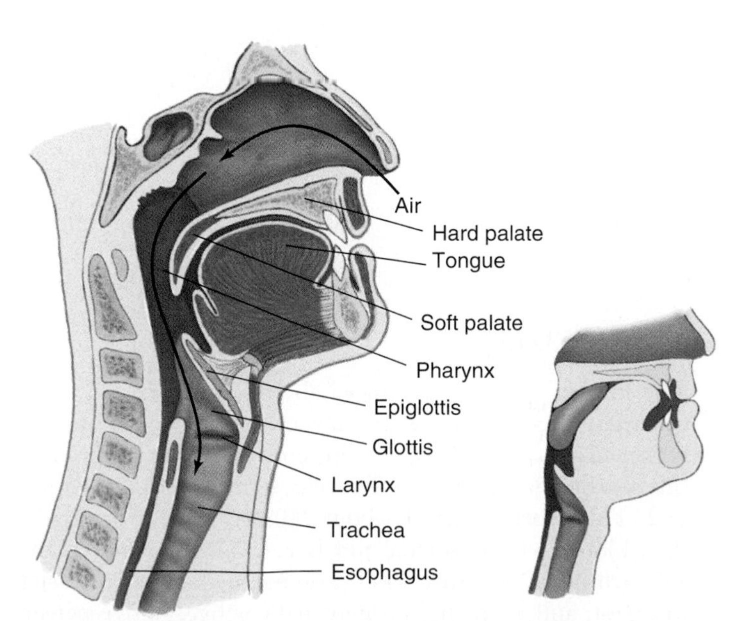

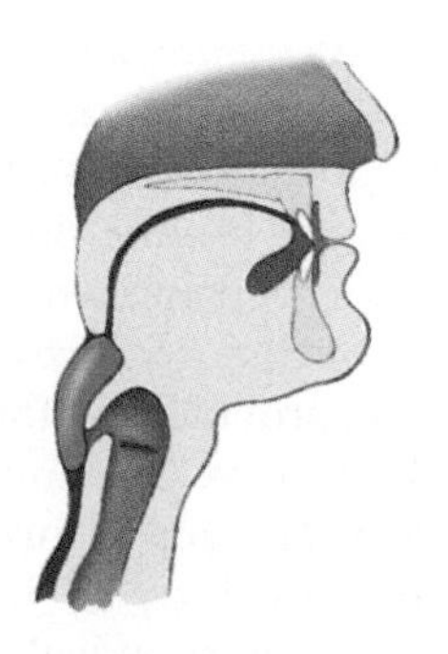

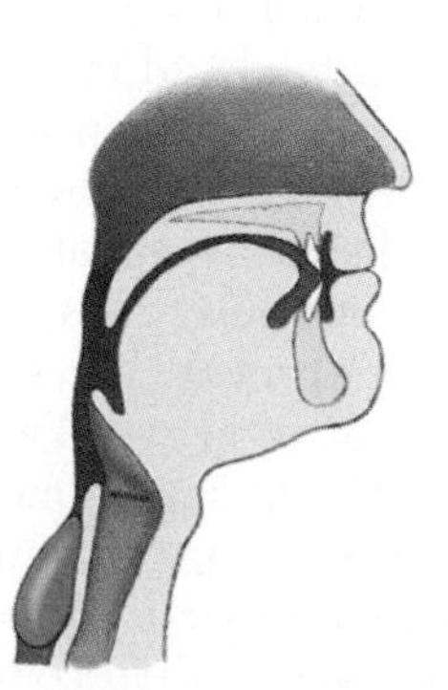

FIGURE 51.9
The human pharynx, palate, and larynx. Food that enters the pharynx is prevented from entering the nasal cavity by elevation of the soft palate, and is prevented from entering the larynx and trachea (the airways of the respiratory system) by elevation of the larynx against the epiglottis.

Esophagus and Stomach

Structure and Function of the Esophagus

Swallowed food enters a muscular tube called the esophagus, which connects the pharynx to the stomach. In adult humans, the esophagus is about 25 centimeters long; the upper third is enveloped in skeletal muscle, for voluntary control of swallowing, while the lower two-thirds is surrounded by involuntary smooth muscle. The swallowing center stimulates successive waves of contraction in these muscles that move food along the esophagus to the stomach. These rhythmic waves of muscular contraction are called peristalsis (figure 51.10); they enable humans and other vertebrates to swallow even if they are upside down.

In many vertebrates, the movement of food from the esophagus into the stomach is controlled by a ring of circular smooth muscle, or a *sphincter*, that opens in response to the pressure exerted by the food. Contraction of this sphincter prevents food in the stomach from moving back into the esophagus. Rodents and horses have a true sphincter at this site and thus cannot regurgitate, while humans lack a true sphincter and so are able to regurgitate. Normally, however, the human esophagus is closed off except during swallowing.

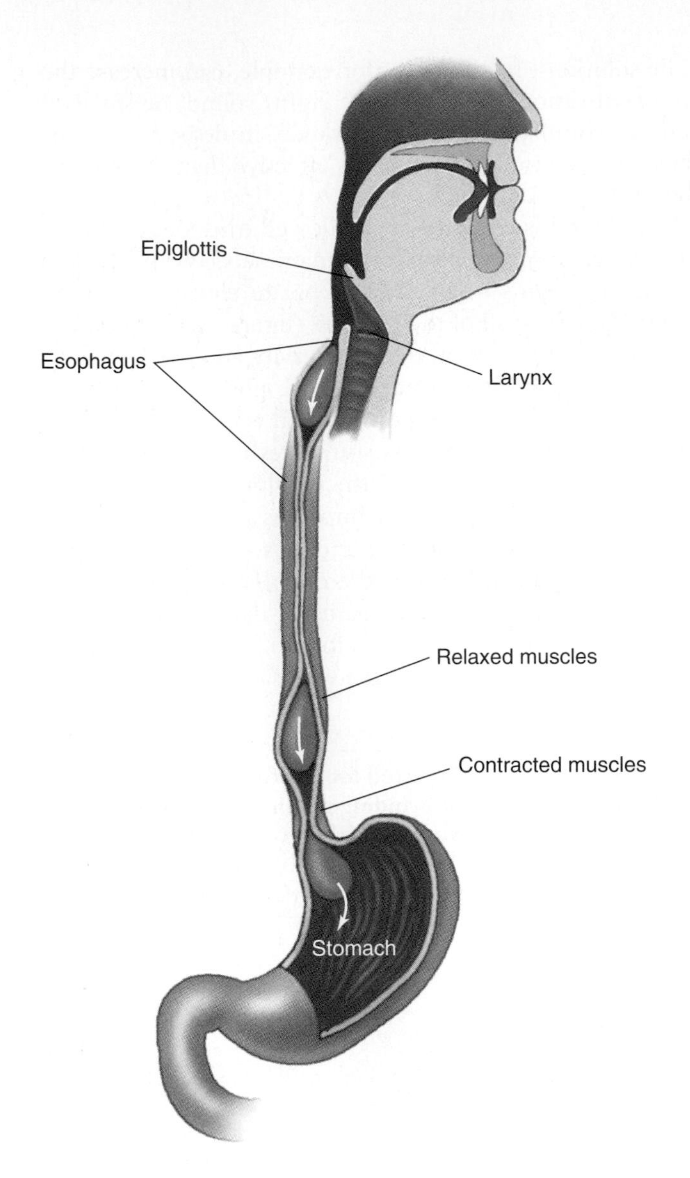

FIGURE 51.10
The esophagus and peristalsis. After food has entered the esophagus, rhythmic waves of muscular contraction, called peristalsis, move the food down to the stomach.

Structure and Function of the Stomach

The stomach (figure 51.11) is a saclike portion of the digestive tract. Its inner surface is highly convoluted, enabling it to fold up when empty and open out like an expanding balloon as it fills with food. Thus, while the human stomach has a volume of only about 50 milliliters when empty, it may expand to contain 2 to 4 liters of food when full. Carnivores that engage in sporadic gorging as an important survival strategy possess stomachs that are able to distend much more than that.

Secretory Systems

The stomach contains an extra layer of smooth muscle for churning food and mixing it with *gastric juice*, an acidic secretion of the tubular gastric glands of the mucosa (figure 51.11). These exocrine glands contain two kinds of secretory cells: parietal cells, which secrete hydrochloric acid (HCl); and chief cells, which secrete pepsinogen, a weak protease (protein-digesting enzyme) that requires a very low pH to be active. This low pH is provided by the HCl. Activated pepsinogen molecules then cleave one another at specific sites, producing a much more active protease, pepsin. This process of secreting a relatively inactive enzyme that is then converted into a more active enzyme outside the cell prevents the chief cells from digesting themselves. It should be noted that only proteins are partially digested in the stomach—there is no significant digestion of carbohydrates or fats.

Action of Acid

The human stomach produces about 2 liters of HCl and other gastric secretions every day, creating a very acidic solution inside the stomach. The concentration of HCl in this solution is about 10 millimolar, corresponding to a pH of 2. Thus, gastric juice is about 250,000 times more acidic than blood, whose normal pH is 7.4. The low pH in the stomach helps denature food proteins, making them easier to digest, and keeps pepsin maximally active. Active pepsin hydrolyzes food proteins into shorter chains of polypeptides that are not fully digested until the mixture enters the small intestine. The mixture of partially digested food and gastric juice is called chyme.

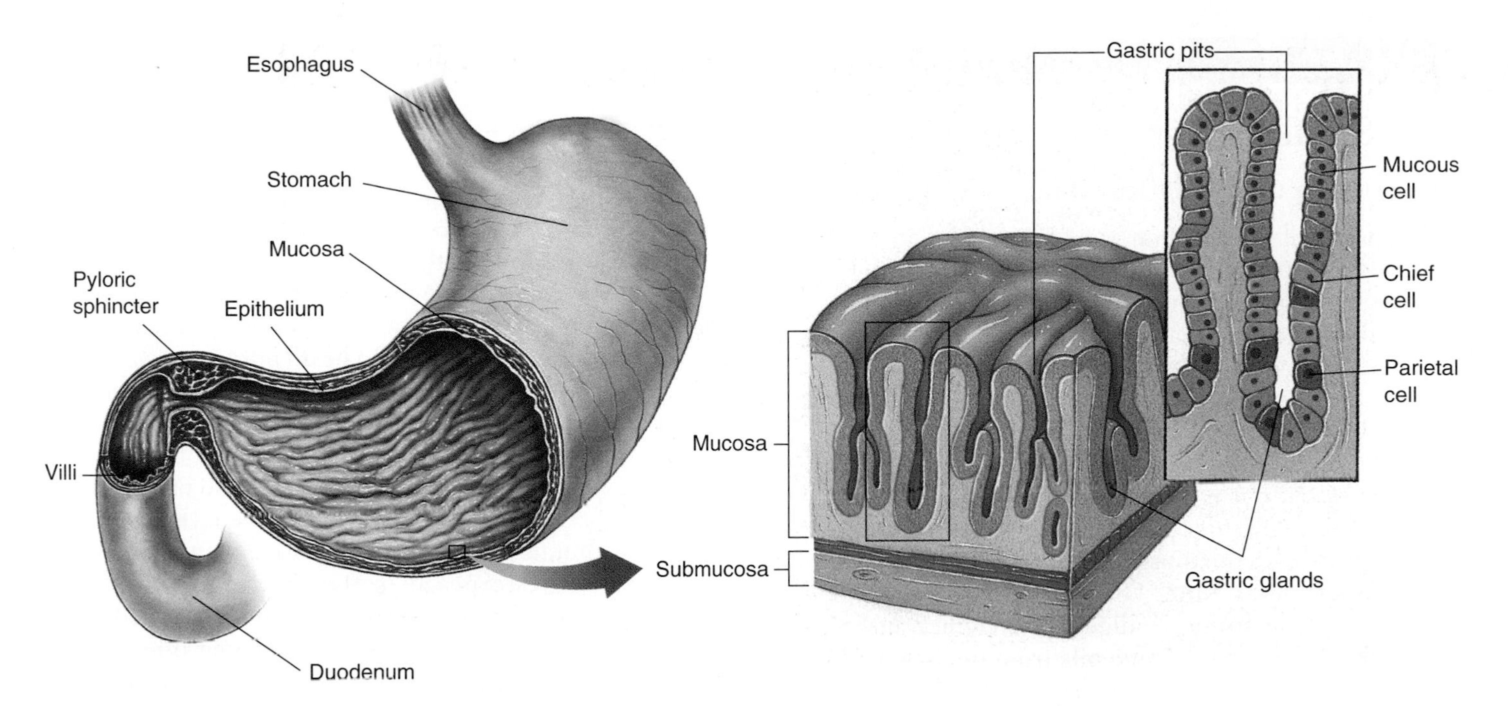

FIGURE 51.11
The stomach and duodenum. Food enters the stomach from the esophagus. A band of smooth muscle called the pyloric sphincter controls the entrance to the duodenum, the upper part of the small intestine. The epithelial walls of the stomach are dotted with gastric pits, which contain gastric glands that secrete hydrochloric acid and the enzyme pepsinogen. The gastric glands consist of mucous cells, chief cells that secrete pepsinogen, and parietal cells that secrete HCl. Gastric pits are the openings of the gastric glands.

The acidic solution within the stomach also kills most of the bacteria that are ingested with the food. The few bacteria that survive the stomach and enter the intestine intact are able to grow and multiply there, particularly in the large intestine. In fact, most vertebrates harbor thriving colonies of bacteria within their intestines, and bacteria are a major component of feces. As we will discuss later, bacteria that live within the digestive tract of cows and other ruminants play a key role in the ability of these mammals to digest cellulose.

Ulcers

Overproduction of gastric acid can occasionally eat a hole through the wall of the stomach. Such *gastric ulcers* are rare, however, because epithelial cells in the mucosa of the stomach are protected somewhat by a layer of alkaline mucus, and because those cells are rapidly replaced by cell division if they become damaged (gastric epithelial cells are replaced every 2 to 3 days). Over 90% of gastrointestinal ulcers are *duodenal ulcers.* These may be produced when excessive amounts of acidic chyme are delivered into the duodenum, so that the acid cannot be properly neutralized through the action of alkaline pancreatic juice (described later). Susceptibility to ulcers is increased when the mucosal barriers to self-digestion are weakened by an infection of the bacterium *Helicobacter pylori.* Indeed, modern antibiotic treatments of this infection can reduce symptoms and often even cure the ulcer.

In addition to producing HCl, the parietal cells of the stomach also secrete intrinsic factor, a polypeptide needed for the intestinal absorption of vitamin B_{12}. Because this vitamin is required for the production of red blood cells, persons who lack sufficient intrinsic factor develop a type of anemia (low red blood cell count) called *pernicious anemia.*

Leaving the Stomach

Chyme leaves the stomach through the *pyloric sphincter* (see figure 51.11) to enter the small intestine. This is where all terminal digestion of carbohydrates, lipids, and proteins occurs, and where the products of digestions—amino acids, glucose, and so on—are absorbed into the blood. Only some of the water in chyme and a few substances such as aspirin and alcohol are absorbed through the wall of the stomach.

Peristaltic waves of contraction propel food along the esophagus to the stomach. Gastric juice contains strong hydrochloric acid and the protein-digesting enzyme pepsin, which begins the digestion of proteins into shorter polypeptides. The acidic chyme is then transferred through the pyloric sphincter to the small intestine.

51.3 The small and large intestines have very different functions.

The Small Intestine

Digestion in the Small Intestine

The capacity of the small intestine is limited, and its digestive processes take time. Consequently, efficient digestion requires that only relatively small amounts of chyme be introduced from the stomach into the small intestine at any one time. Coordination between gastric and intestinal activities is regulated by neural and hormonal signals, which we will describe in a later section.

The small intestine is approximately 4.5 meters long in a living person, but is 6 meters long at autopsy when the muscles relax. The first 25 centimeters is the **duodenum;** the remainder of the small intestine is divided into the **jejunum** and the **ileum.** The duodenum receives acidic chyme from the stomach, digestive enzymes and bicarbonate from the pancreas, and bile from the liver and gallbladder. The pancreatic juice enzymes digest larger food molecules into smaller fragments. This occurs primarily in the duodenum and jejunum.

The epithelial wall of the small intestine is covered with tiny, fingerlike projections called villi (singular, villus; figure 51.12). In turn, each of the epithelial cells lining the villi is covered on its apical surface (the side facing the lumen) by many foldings of the plasma membrane that form cytoplasmic extensions called *microvilli.* These are quite tiny and can be seen clearly only with an electron microscope (figure 51.13). In a light micrograph, the microvilli resemble the bristles of a brush, and for that reason the epithelial wall of the small intestine is also called a brush border.

The villi and microvilli greatly increase the surface area of the small intestine; in humans, this surface area is 300 square meters! It is over this vast surface that the products of digestion are absorbed. The microvilli also participate in digestion because a number of digestive enzymes are embedded within the epithelial cells' plasma membranes, with their active sites exposed to the chyme (figure 51.14). These brush border enzymes include those that hydrolyze the disaccharides lactose and sucrose, among others (table 51.1). Many adult humans lose the ability to produce the brush border enzyme *lactase* and therefore cannot digest lactose (milk sugar), a rather common condition called *lactose intolerance.* The brush border enzymes complete the digestive process that started with the action of the pancreatic enzymes released into the duodenum.

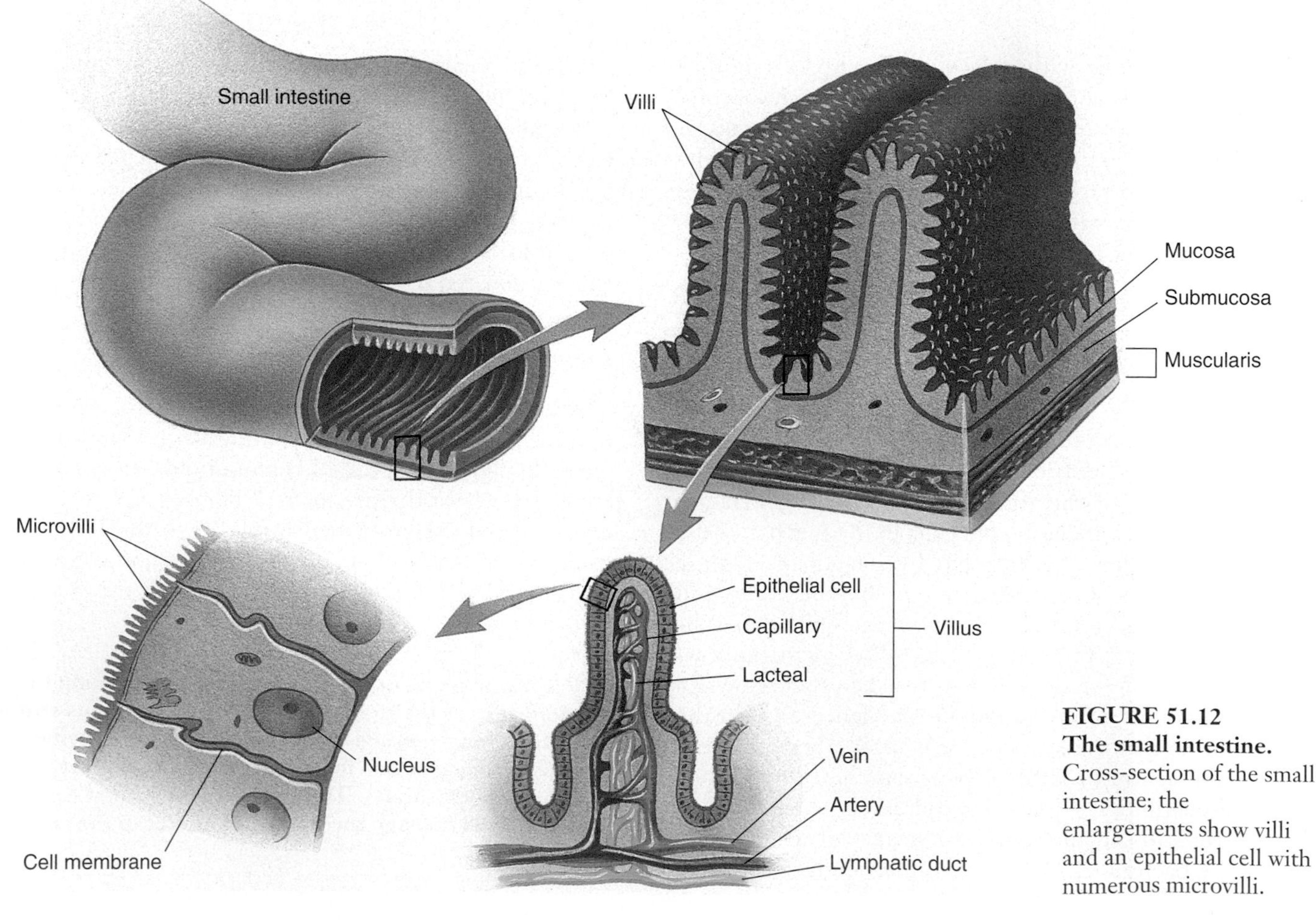

FIGURE 51.12
The small intestine. Cross-section of the small intestine; the enlargements show villi and an epithelial cell with numerous microvilli.

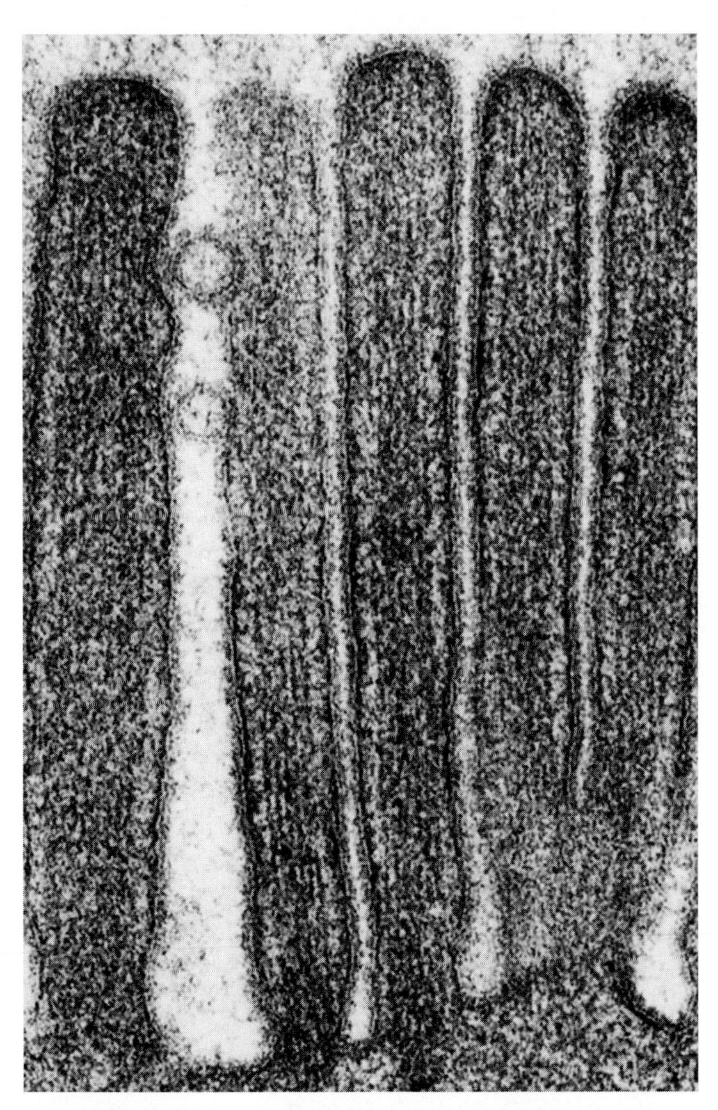

FIGURE 51.13
Intestinal microvilli. Microvilli, shown in an electron micrograph, are very densely clustered, giving the small intestine an enormous surface area important in efficient absorption of the digestion products.

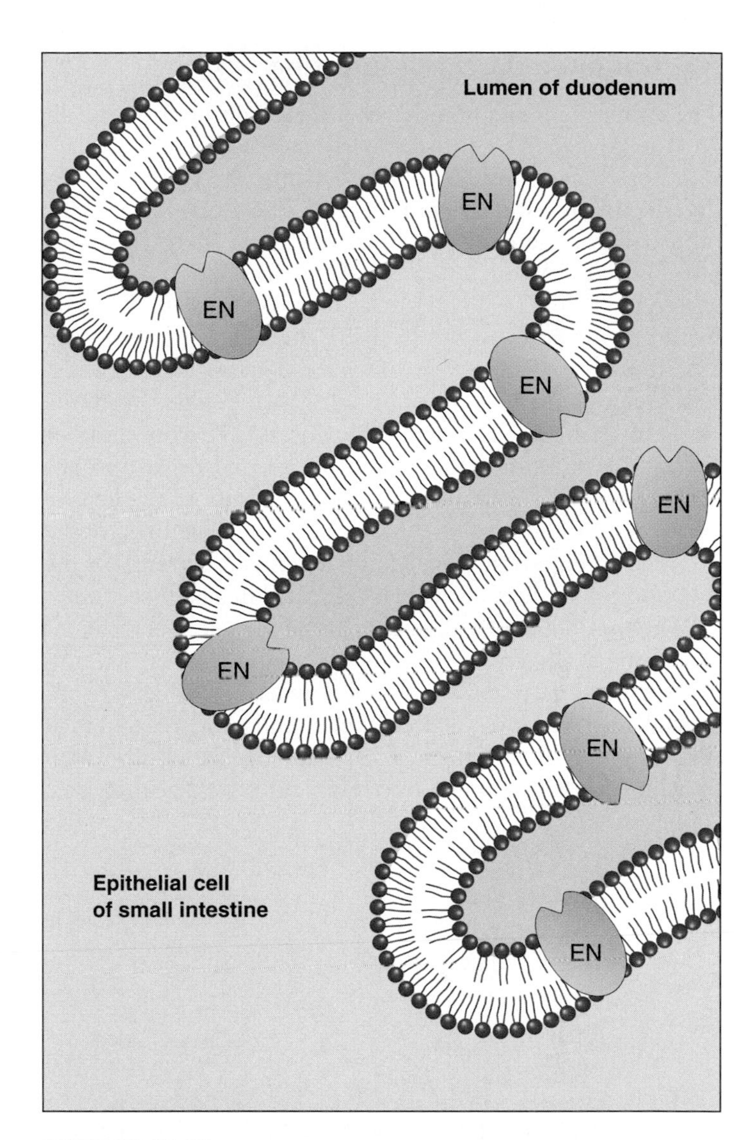

FIGURE 51.14
Brush border enzymes. These enzymes, which are labeled "EN" in this diagram, are part of the plasma membrane of the microvilli in the small intestine. They catalyze many of the terminal steps in digestion.

Table 51.1 Digestive Enzymes

Location	Enzymes	Substrates	Digestion Products
Salivary glands	Amylase	Starch, glycogen	Disaccharides
Stomach	Pepsin	Proteins	Short peptides
Small intestine	Peptidases	Short peptides	Amino acids
(brush border)	Nucleases	DNA, RNA	Sugars, nucleic acid bases
	Lactase, maltase, sucrase	Disaccharides	Monosaccharides
Pancreas	Lipase	Triglycerides	Fatty acids, glycerol
	Trypsin, chymotrypsin	Proteins	Peptides
	DNase	DNA	Nucleotides
	RNase	RNA	Nucleotides

Absorption in the Small Intestine

The amino acids and monosaccharides resulting from the digestion of proteins and carbohydrates, respectively, are transported across the brush border into the epithelial cells that line the intestine (figure 51.15*a*). They then move to the other side of the epithelial cells, and from there are transported across the membrane and into the blood capillaries within the villi. The blood carries these products of digestion from the intestine to the liver via the hepatic portal vein. The term *portal* here refers to a special arrangement of vessels, seen only in a couple of instances, where one organ (the liver, in this case) is located "downstream" from another organ (the intestine). As a result, the second organ receives blood-borne molecules from the first. Because of the hepatic portal vein, the liver is the first organ to receive most of the products of digestion. This arrangement is important for the functions of the liver, as will be described in a later section.

The products of fat digestion are absorbed by a different mechanism (figure 51.15*b*). Fats (triglycerides) are hydrolyzed into fatty acids and monoglycerides, which are absorbed into the intestinal epithelial cells and reassembled into triglycerides. The triglycerides then combine with proteins to form small particles called chylomicrons. Instead of entering the hepatic portal circulation, the chylomicrons are absorbed into lymphatic capillaries (see chapter 52), which empty their contents into the blood in veins near the neck. Chylomicrons can make the blood plasma appear cloudy if a sample of blood is drawn after a fatty meal.

The amount of fluid passing through the small intestine in a day is startlingly large: approximately 9 liters. However, almost all of this fluid is absorbed into the body rather than eliminated in the feces. About 8.5 liters are absorbed in the small intestine and an additional 350 milliliters in the large intestine. Only about 50 grams of solid and 100 milliliters of liquid leave the body as feces. The normal fluid absorption efficiency of the human digestive tract thus approaches 99%, which is very high indeed.

Digestion occurs primarily in the duodenum, which receives the pancreatic juice enzymes. The small intestine provides a large surface area for absorption. Glucose and amino acids from food are absorbed through the small intestine and enter the blood via the hepatic portal vein, going to the liver. Fat from food enters the lymphatic system.

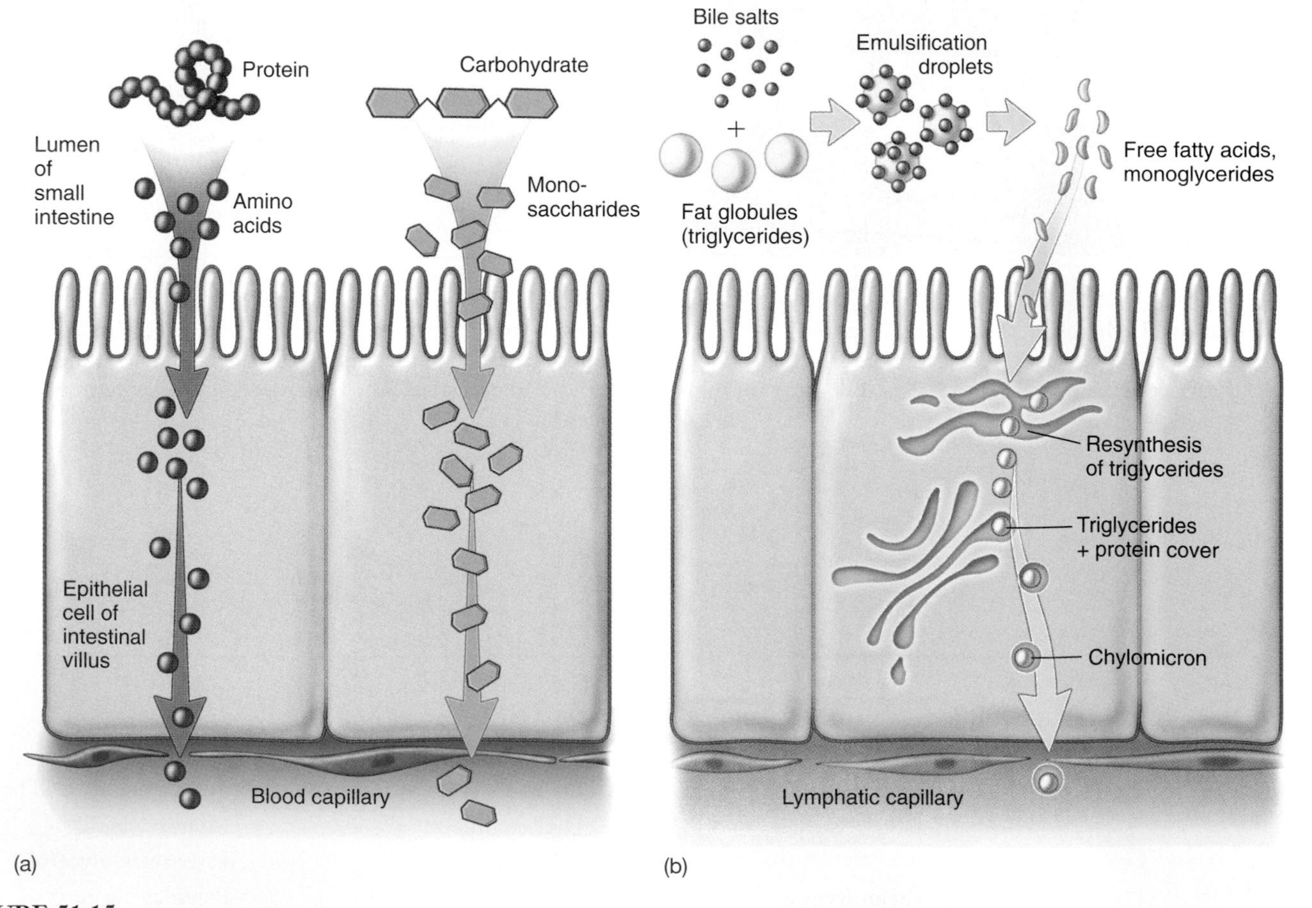

FIGURE 51.15
Absorption of the products of digestion. (*a*) Monosaccharides and amino acids are transported into blood capillaries. (*b*) Fatty acids and monoglycerides within the intestinal lumen are absorbed and converted within the intestinal epithelial cells into triglycerides. These are then coated with proteins to form tiny structures called chylomicrons, which enter lymphatic capillaries.

The Large Intestine

The large intestine, or colon, is much shorter than the small intestine, occupying approximately the last meter of the digestive tract; it is called "large" only because of its larger diameter. The small intestine empties directly into the large intestine at a junction where two vestigial structures, the cecum and the appendix, remain (figure 51.16). No digestion takes place within the large intestine, and only about 4% of the absorption of fluids by the intestine occurs there. The large intestine is not as convoluted as the small intestine, and its inner surface has no villi. Consequently, the large intestine has less than one-thirtieth the absorptive surface area of the small intestine. Although sodium, vitamin K, and some products of bacterial metabolism are absorbed across its wall, the primary function of the large intestine is to concentrate waste material. Within it, undigested material, primarily bacterial fragments and cellulose, is compacted and stored. Many bacteria live and reproduce within the large intestine, and the excess bacteria are incorporated into the refuse material, called *feces*. Bacterial fermentation produces gas within the colon at a rate of about 500 milliliters per day. This rate increases greatly after the consumption of beans or other vegetable matter because the passage of undigested plant material (fiber) into the large intestine provides substrates for fermentation.

The human colon has evolved to process food with a relatively high fiber content. Diets that are low in fiber, which are common in the United States, result in a slower passage of food through the colon. Low dietary fiber content is thought to be associated with the level of colon cancer in the United States, which is among the highest in the world.

Compacted feces, driven by peristaltic contractions of the large intestine, pass from the large intestine into a short tube called the rectum. From the rectum, the feces exit the body through the anus. Two sphincters control passage through the anus. The first is composed of smooth muscle and opens involuntarily in response to pressure inside the rectum. The second, composed of striated muscle, can be controlled voluntarily by the brain, thus permitting a conscious decision to delay defecation.

In all vertebrates except most mammals, the reproductive and urinary tracts empty together with the digestive tract into a common cavity, the cloaca. In some reptiles and birds, additional water from either the feces or urine may be absorbed in the cloaca before the products are expelled from the body.

The large intestine concentrates wastes for excretion by absorbing water. Some ions and vitamin K are also absorbed by the large intestine.

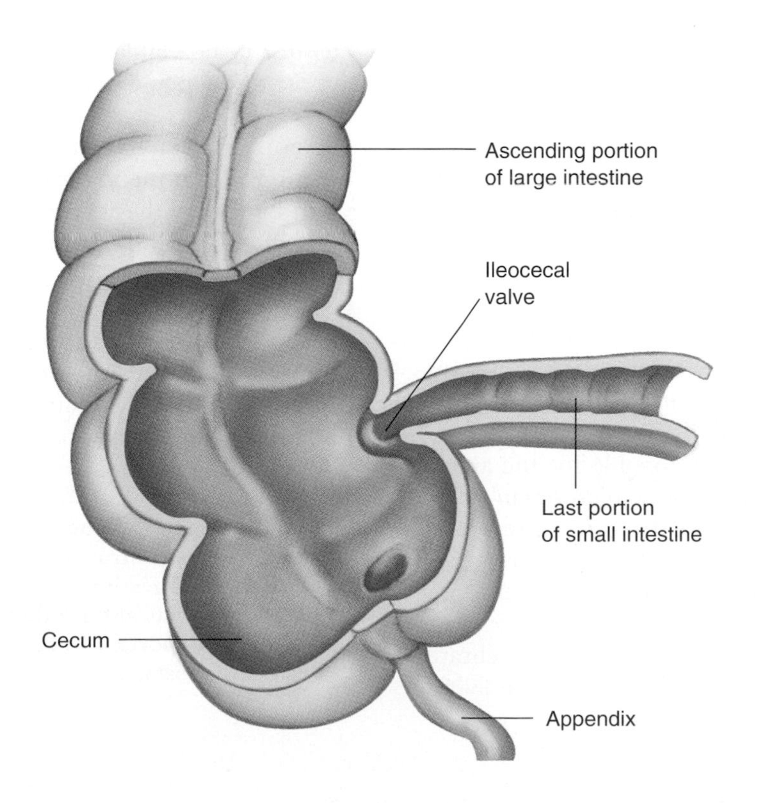

FIGURE 51.16
The junction of the small and large intestines in humans. The large intestine, or colon, starts with the cecum, which is relatively small in humans compared with that in other mammals. A vestigial structure called the appendix extends from the cecum.

Variations in Vertebrate Digestive Systems

Most animals lack the enzymes necessary to digest cellulose, the carbohydrate that functions as the chief structural component of plants. The digestive tracts of some animals, however, contain bacteria and protists that convert cellulose into substances the host can digest. Although digestion by gastrointestinal microorganisms plays a relatively small role in human nutrition, it is an essential element in the nutrition of many other kinds of animals, including insects like termites and cockroaches, and a few groups of herbivorous mammals. The relationships between these microorganisms and their animal hosts are mutually beneficial and provide an excellent example of symbiosis.

Cows, deer, and other ruminants have large, divided stomachs (figure 51.17). The first portion consists of the rumen and a smaller chamber, the reticulum; the second portion consists of two additional chambers: the omasum and abomasum. The rumen which may hold up to 50 gallons, serves as a fermentation vat in which bacteria and protozoa convert cellulose and other molecules into a variety of simpler compounds. The location of the rumen at the front of the four chambers is important because it allows the animal to regurgitate and rechew the contents of the rumen, an activity called *rumination,* or "chewing the cud." The cud is then swallowed and enters the reticulum, from which it passes to the omasum and then the abomasum, where it is finally mixed with gastric juice. Hence, only the abomasum is equivalent to the human stomach in its function. This process leads to a far more efficient digestion of cellulose in ruminants than in mammals that lack a rumen, such as horses.

In horses, rodents, and lagomorphs (rabbits and hares), the digestion of cellulose by microorganisms takes place in the cecum, which is greatly enlarged (figure 51.18). Because the cecum is located beyond the stomach, regurgitation of its contents is impossible. However, rodents and lagomorphs have evolved another way to digest cellulose that achieves a degree of efficiency similar to that of ruminant digestion. They do this by eating their feces, thus passing the food through their digestive tract a second time. The second passage makes it possible for the animal to absorb the nutrients produced by the microorganisms in its cecum. Animals that engage in this practice of **coprophagy** (from the Greek words *copros,* "excrement," and *phagein,* "eat") cannot remain healthy if they are prevented from eating their feces.

Cellulose is not the only plant product that vertebrates can use as a food source because of the digestive activities of intestinal microorganisms. Wax, a substance indigestible by most terrestrial animals, is digested by symbiotic bacteria living in the gut of honey guides, African birds that eat the wax in bee nests. In the marine food chain, wax is a major constituent of copepods (crustaceans in the plankton), and many marine fish and birds appear to be able to digest wax with the aid of symbiotic microorganisms.

Another example of the way intestinal microorganisms function in the metabolism of their animal hosts is provided by the synthesis of vitamin K. All mammals rely on intestinal bacteria to synthesize this vitamin, which is necessary for the clotting of blood. Birds, which lack these bacteria, must consume the required quantities of vitamin K in their food. In humans, prolonged treatment with antibiotics greatly reduces the populations of bacteria in the intestine; under such circumstances, it may be necessary to provide supplementary vitamin K.

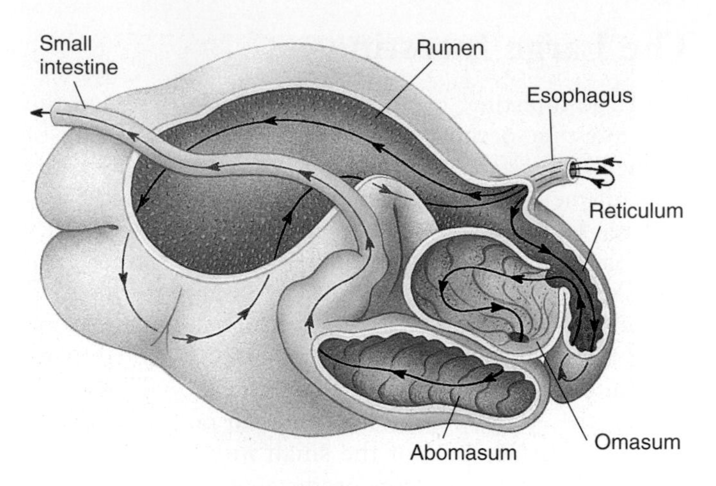

FIGURE 51.17
Four-chambered stomach of a ruminant. The grass and other plants that a ruminant, such as a cow, eats enter the rumen, where they are partially digested. Before moving into a second chamber, the reticulum, the food may be regurgitated and rechewed. The food is then transferred to the rear two chambers: the omasum and abomasum. Only the abomasum is equivalent to the human stomach in its function of secreting gastric juice.

Much of the food value of plants is tied up in cellulose, and the digestive tract of many animals harbors colonies of cellulose-digesting microorganisms. Intestinal microorganisms also produce molecules such as vitamin K that are important to the well-being of their vertebrate hosts.

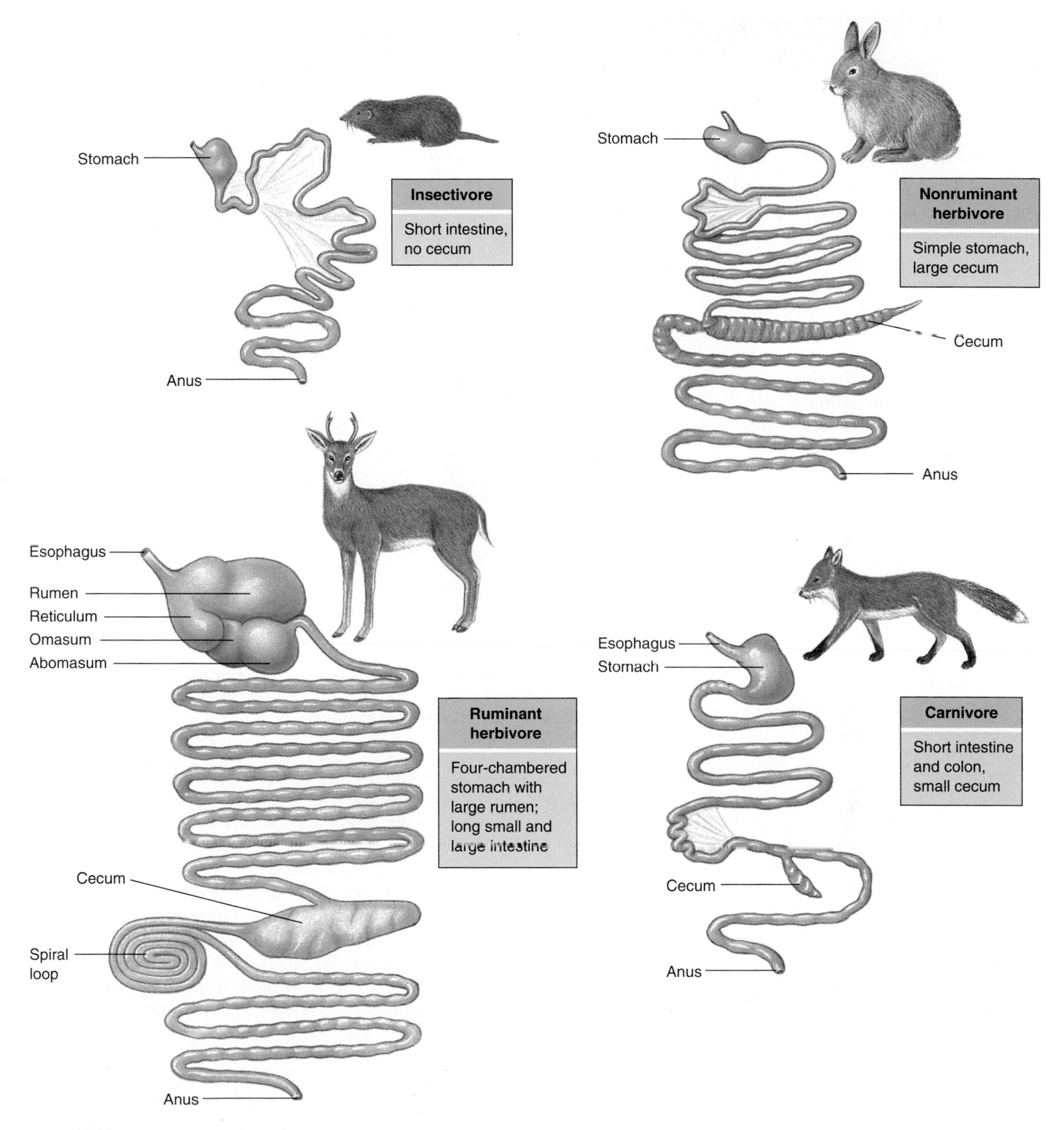

FIGURE 51.18
The digestive systems of different mammals reflect their diets. Herbivores require long digestive tracts with specialized compartments for the breakdown of plant matter. Protein diets are more easily digested; thus, insectivorous and carnivorous mammals have short digestive tracts with few specialized pouches.

51.4 Accessory organs, neural stimulation, and endocrine secretions assist in digestion.

Accessory Organs

Secretions of the Pancreas

The pancreas (figure 51.19), a large gland situated near the junction of the stomach and the small intestine, is one of the accessory organs that contribute secretions to the digestive tract. Pancreatic fluid is secreted into the duodenum through the *pancreatic duct*; thus, the pancreas functions as an exocrine organ. This fluid contains a host of enzymes, including trypsin and chymotrypsin, which digest proteins; pancreatic amylase, which digests starch; and lipase, which digests fat. These enzymes are released into the duodenum primarily as inactive zymogens and are then activated by the brush border enzymes of the intestine. Pancreatic enzymes digest proteins into smaller polypeptides, polysaccharides into shorter chains of sugars, and fat into free fatty acids and other products. The digestion of these molecules is then completed by the brush border enzymes.

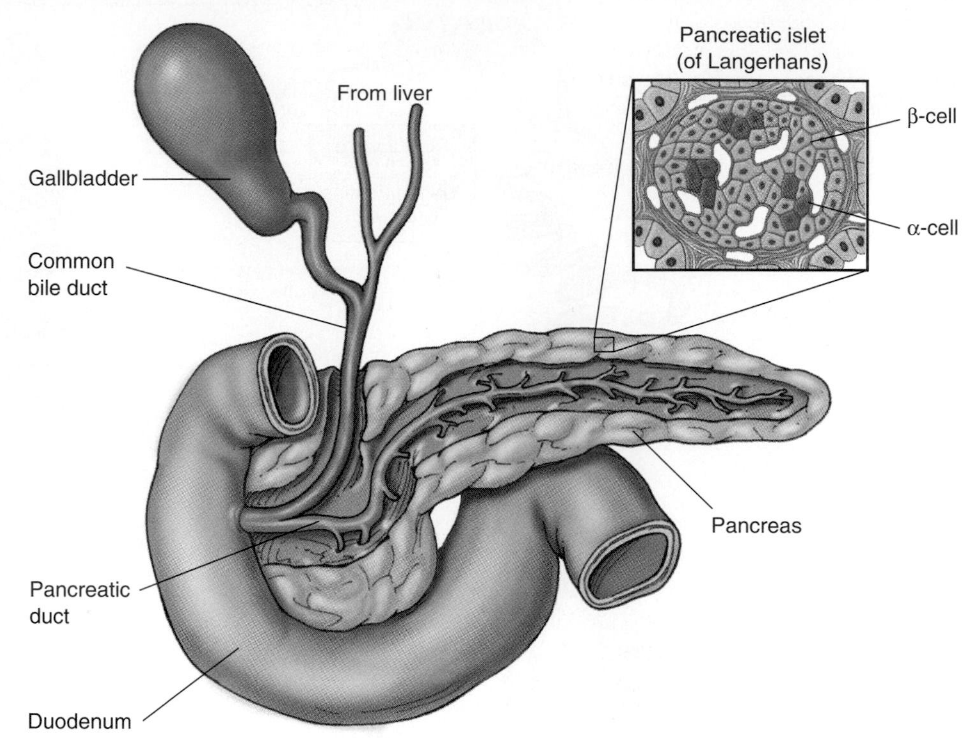

FIGURE 51.19
The pancreas and bile duct empty into the duodenum. The pancreas secretes pancreatic juice into the pancreatic duct. The pancreatic islets of Langerhans secrete hormones into the blood; a-cells secrete glucagon and β-cells secrete insulin.

Pancreatic fluid also contains bicarbonate, which neutralizes the HCl from the stomach and gives the chyme in the duodenum a slightly alkaline pH. The digestive enzymes and bicarbonate are produced by clusters of secretory cells known as *acini.*

In addition to its exocrine role in digestion, the pancreas also functions as an endocrine gland, secreting several hormones into the blood that control the blood levels of glucose and other nutrients. These hormones are produced in the **islets of Langerhans,** clusters of endocrine cells scattered throughout the pancreas. The two most important pancreatic hormones, insulin and glucagon, are discussed later in this chapter.

The Liver and Gallbladder

The liver is the largest internal organ of the body (see figure 51.4). In an adult human, the liver weighs about 1.5 kilograms and is the size of a football. The main exocrine secretion of the liver is bile, a fluid mixture consisting of *bile pigments* and *bile salts* that is delivered into the duodenum during the digestion of a meal. The bile pigments do not participate in digestion; they are waste products resulting from the liver's destruction of old red blood cells and ultimately are eliminated with the feces. If the excretion of bile pigments by the liver is blocked, the pigments can accumulate in the blood and cause a yellow staining of the tissues known as *jaundice.*

In contrast, the bile salts play a very important role in the digestion of fats. Because fats are insoluble in water, they enter the intestine as drops within the watery chyme. The bile salts, which are partly lipid-soluble and partly water-soluble, work like detergents, dispersing the large drops of fat into a fine suspension of smaller droplets. This emulsification process produces a greater surface area of fat upon which the lipase enzymes can act, and thus allows the digestion of fat to proceed more rapidly.

After it is produced in the liver, bile is stored and concentrated in the gallbladder. The arrival of fatty food in the duodenum triggers a neural and endocrine reflex (discussed later) that stimulates the gallbladder to contract, causing bile to be transported through the common bile duct and injected into the duodenum. If the bile duct is blocked by a *gallstone* (formed from a hardened precipitate of cholesterol), contraction of the gallbladder will cause pain generally felt under the right scapula (shoulder blade).

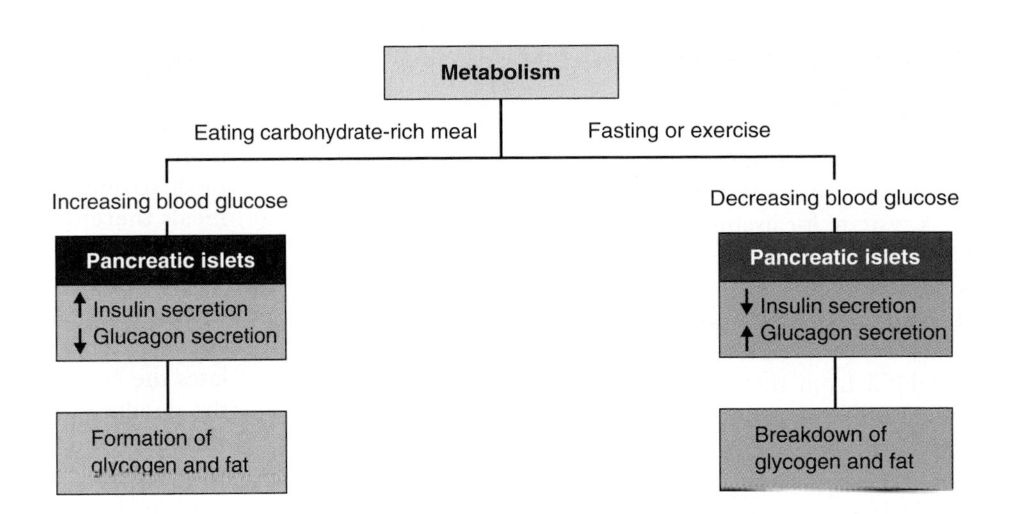

FIGURE 51.20
The actions of insulin and glucagon. After a meal, an increased secretion of insulin by the β cells of the pancreatic islets promotes the deposition of glycogen and fat. During fasting or exercising, an increased glucagon secretion by the α cells of the pancreatic islets and a decreased insulin secretion promote the breakdown (through hydrolysis reactions) of glycogen and fat.

Regulatory Functions of the Liver

Because the hepatic portal vein carries blood from the stomach and intestine directly to the liver, the liver is in a position to chemically modify the substances absorbed in the gastrointestinal tract before they reach the rest of the body. For example, ingested alcohol and other drugs are taken into liver cells and metabolized; this is why the liver is often damaged as a result of alcohol and drug abuse. The liver also removes toxins, pesticides, carcinogens, and other poisons, converting them into less toxic forms. An important example of this is the liver's conversion of the toxic ammonia produced by intestinal bacteria into urea, a compound that can be contained safely and carried by the blood at higher concentrations.

Similarly, the liver regulates the levels of many compounds produced within the body. Steroid hormones, for instance, are converted into less active and more water-soluble forms by the liver. These molecules are then included in the bile and eliminated from the body in the feces, or carried by the blood to the kidneys and excreted in the urine.

The liver also produces most of the proteins found in blood plasma. The total concentration of plasma proteins is significant because it must be kept within normal limits in order to maintain osmotic balance between blood and interstitial (tissue) fluid. If the concentration of plasma proteins drops too low, as can happen as a result of liver disease such as cirrhosis, fluid accumulates in the tissues, a condition called *edema.*

Regulation of Blood Glucose Concentration

The neurons in the brain obtain their energy primarily from the aerobic respiration of glucose obtained from the blood plasma. It is therefore extremely important that the blood glucose concentration not fall too low, as might happen during fasting or prolonged exercise. It is also important that the blood glucose concentration not stay at too high a level, as it does in people with uncorrected *diabetes mellitus,* because this can lead to tissue damage.

After a carbohydrate-rich meal, the liver and skeletal muscles remove excess glucose from the blood and store it as the polysaccharide glycogen. This process is stimulated by the hormone insulin, secreted by the β *(beta) cells* in the islets of Langerhans of the pancreas (figure 51.19). When blood glucose levels decrease, as they do between meals, during periods of fasting, and during exercise, the liver secretes glucose into the blood. This glucose is obtained in part from the breakdown of liver glycogen to glucose-6-phosphate, a process called glycogenolysis. The phosphate group is then removed, and free glucose is secreted into the blood. Skeletal muscles lack the enzyme needed to remove the phosphate group, and so, even though they have glycogen stores, they cannot secrete glucose into the blood. The breakdown of liver glycogen is stimulated by another hormone, glucagon, which is secreted by the α *(alpha) cells* of the islets of Langerhans in the pancreas (figure 51.20).

If fasting or exercise continues, the liver begins to convert other molecules, such as amino acids and lactic acid, into glucose. This process is called **gluconeogenesis** ("new formation of glucose"). The amino acids used for gluconeogenesis are obtained from muscle protein, which explains the severe muscle wasting that occurs during prolonged fasting.

The pancreas secretes digestive enzymes and bicarbonate into the pancreatic duct. The liver produces bile, which is stored and concentrated in the gallbladder. The liver and the pancreatic hormones regulate blood glucose concentration.

Neural and Hormonal Regulation of Digestion

The activities of the gastrointestinal tract are coordinated by the nervous system and the endocrine system. The nervous system, for example, stimulates salivary and gastric secretions in response to the sight and smell of food. When food arrives in the stomach, proteins in the food stimulate the secretion of a stomach hormone called gastrin (table 51.2), which in turn stimulates the secretion of pepsinogen and HCl from the gastric glands (figure 51.21). The secreted HCl then lowers the pH of the gastric juice, which acts to inhibit further secretion of gastrin. Because inhibition of gastrin secretion will reduce the amount of HCl released into the gastric juice, a negative feedback loop is completed. In this way, the secretion of gastric acid is kept under tight control.

The passage of chyme from the stomach into the duodenum inhibits the contractions of the stomach, so that no additional chyme can enter the duodenum until the previous amount has been processed. This inhibition is mediated by a neural reflex and by a hormone secreted by the small intestine that inhibits gastric emptying. The hormone is known generically as an enterogastrone (*entero* refers to the intestine; *gastro* to the stomach). The chemical identity of the enterogastrone is currently controversial. A hormone known as gastric inhibitory peptide (GIP), released by the duodenum, was named for this function but may not be the only, or even the major, enterogastrone. The secretion of enterogastrone is stimulated most strongly by the presence of fat in the chyme. Fatty meals therefore remain in the stomach longer than meals low in fat.

The duodenum secretes two additional hormones. Cholecystokinin (CCK), like enterogastrone, is secreted in response to the presence of fat in the chyme. CCK stimulates the contractions of the gallbladder, injecting bile into the duodenum so that fat can be emulsified and more efficiently digested. The other duodenal hormone is secretin. Released in response to the acidity of the chyme that arrives in the duodenum, secretin stimulates the pancreas to release bicarbonate, which then neutralizes some of the acidity. Secretin has the distinction of being the first hormone ever discovered.

Neural and hormonal reflexes regulate the activity of the digestive system. The stomach's secretions are regulated by food and by the hormone gastrin. Other hormones, secreted by the duodenum, inhibit stomach emptying and promote the release of bile from the gallbladder and the secretion of bicarbonate in pancreatic juice.

Table 51.2 Hormones of Digestion

Hormone	Class	Source	Stimulus	Action	Note
Gastrin	Polypeptide	Pyloric portion of stomach	Entry of food into stomach	Stimulates secretion of HCl and pepsinogen by stomach	Unusual in that it acts on same organ that secretes it
Cholecystokinin	Polypeptide	Duodenum	Fatty chyme in duodenum	Stimulates gallbladder contraction and secretion of digestive enzymes by pancreas	Structurally similar to gastrin
Gastric inhibitory peptide	Polypeptide	Duodenum	Fatty chyme in duodenum	Inhibits stomach emptying	Also stimulates insulin secretion
Secretin	Polypeptide	Duodenum	Acidic chyme in duodenum	Stimulates secretion of bicarbonate by pancreas	The first hormone to be discovered (1902)

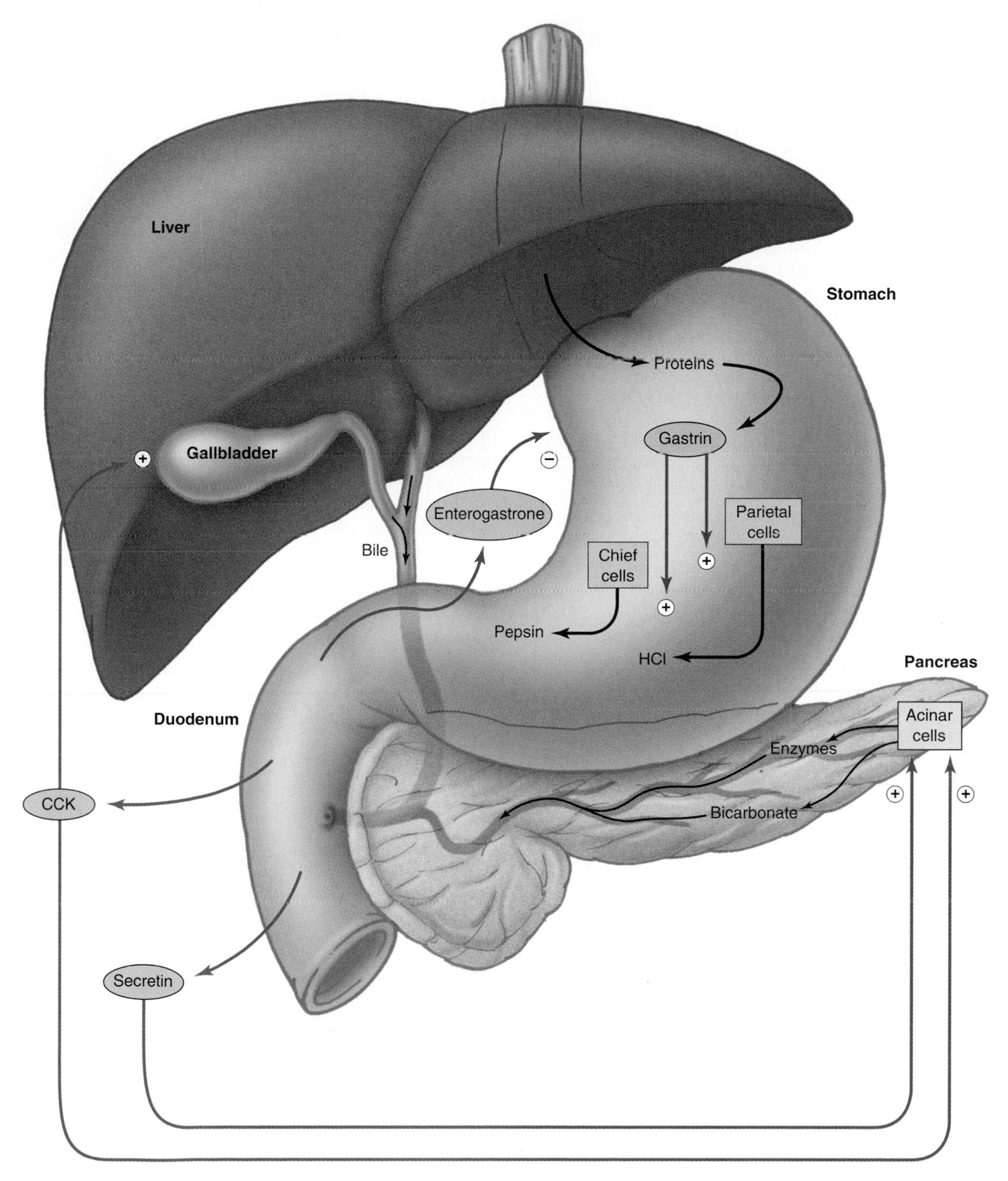

FIGURE 51.21

Hormonal control of the gastrointestinal tract. Gastrin is secreted by the mucosa of the stomach and stimulates the secretion of pepsinogen (which is converted into pepsin) and HCl. The duodenum secretes three hormones: cholecystokinin (CCK), which stimulates contraction of the gallbladder and secretion of pancreatic enzymes; secretin, which stimulates secretion of pancreatic bicarbonate; and an enterogastrone, which inhibits stomach emptying.

51.5 All animals require food energy and essential nutrients.

Food Energy and Energy Expenditure

The ingestion of food serves two primary functions: it provides a source of energy, and it provides raw materials the animal is unable to manufacture for itself. Even an animal that is completely at rest requires energy to support its metabolism; the minimum rate of energy consumption under defined resting conditions is called the basal metabolic rate (BMR). The BMR is relatively constant for a given individual, depending primarily on the person's age, sex, and body size.

Exercise raises the metabolic rate above the basal levels, so the amount of energy that the body consumes per day is determined not only by the BMR but also by the level of physical activity. If food energy taken in is greater than the energy consumed per day, the excess energy will be stored in glycogen and fat. Because glycogen reserves are limited, however, continued ingestion of excess food energy results primarily in the accumulation of fat. The intake of food energy is measured in kilocalories (1 kilocalorie = 1000 calories; nutritionists use Calorie with a capital C instead of kilocalorie). The measurement of kilocalories in food is determined by the amount of heat generated when the food is "burned," either literally, when the caloric content of food is measured using a calorimeter, or in the body when the food is digested. Caloric intake can be altered by the choice of diet, and the amount of energy expended in exercise can be changed by the choice of lifestyle. The daily energy expenditures (metabolic rates) of people vary between 1300 and 5000 kilocalories per day, depending on the person's BMR and level of physical activity. If the food kilocalories ingested exceed the metabolic rate for a sustained period, the person will accumulate an amount of fat that is deleterious to health, a condition called **obesity.** In the United States, about 30% of middle-aged women and 15% of middle-aged men are classified as obese, which means they weigh at least 20% more than the average weight for their height.

Regulation of Food Intake

Scientists have for years suspected that adipose tissue secretes a hormonal *satiety factor* (a circulating chemical that decreases appetite), because genetically obese mice lose weight when their circulatory systems are surgically joined with those of normal mice. Apparently, some weight-loss hormone was passing into the obese mice! The satiety factor secreted by adipose tissue has recently been identified. It is the product of a gene first observed in a strain of mice known as *ob/ob* (*ob* stands for "obese"; the double symbols indicate that the mice are homozygous for this gene—they inherit it from both parents). The *ob* gene has been cloned in mice, and more recently in humans, and has been found to be expressed (that is, to produce mRNA) only in adipocytes. The protein product of this gene, the presumed satiety factor, is called leptin. The *ob* mice produce a mutated and ineffective form of leptin, and it is this defect that causes their obesity. When injected with normal leptin, they stop eating and lose weight (figure 51.22).

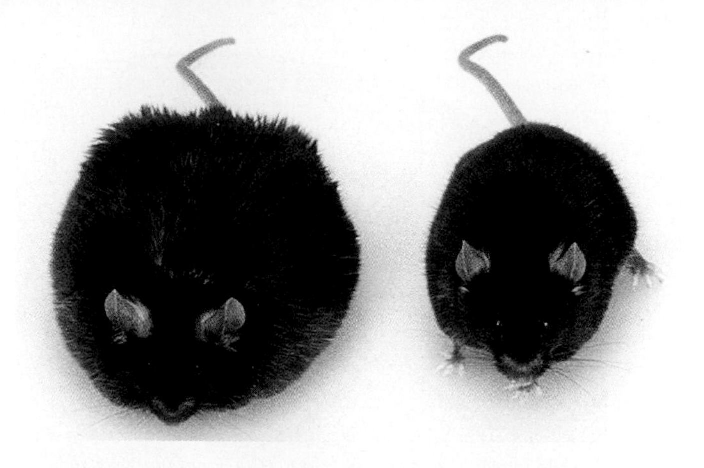

FIGURE 51.22
Injection of the hormone leptin causes genetically obese mice to lose weight. These two mice are identical twins, both members of a mutant strain of obese mice. The mouse on the right has been injected with the hormone leptin. It lost 30% of its body weight in just two weeks, with no apparent side effects.

More recent studies in humans show that the activity of the *ob* gene and the blood concentrations of leptin are actually higher in obese than in lean people, and that the leptin produced by obese people appears to be normal. It has therefore been suggested that most cases of human obesity may result from a reduced sensitivity to the actions of leptin in the brain rather than from reduced leptin production by adipose cells. Aggressive research is ongoing, as might be expected from the possible medical and commercial applications of these findings.

In the United States, serious eating disorders have become much more common since the mid-1970s. The most common of these disorders are *anorexia nervosa*, a condition in which the afflicted individuals literally starve themselves, and *bulimia*, in which individuals gorge themselves and then vomit, so that their weight stays constant. Ninety to 95% of those suffering from these disorders are female; researchers estimate that 2 to 5% of the adolescent girls and young women in the United States have eating disorders.

The amount of caloric energy expended by the body depends on the basal metabolic rate and the additional calories consumed by exercise. Obesity results if the ingested food energy exceeds the energy expenditure by the body over a prolonged period.

Essential Nutrients

Over the course of their evolution, many animals have lost the ability to synthesize specific substances that nevertheless continue to play critical roles in their metabolism. Substances that an animal cannot manufacture for itself but which are necessary for its health must be obtained in the diet and are referred to as **essential nutrients.**

Included among the essential nutrients are *vitamins*, certain organic substances required in trace amounts. Humans, apes, monkeys, and guinea pigs, for example, have lost the ability to synthesize ascorbic acid (vitamin C). If vitamin C is not supplied in sufficient quantities in their diets, they will develop scurvy, a potentially fatal disease. Humans require at least 13 different vitamins (table 51.3).

Some essential nutrients are required in more than trace amounts. Many vertebrates, for example, are unable to synthesize 1 or more of the 20 amino acids used in making proteins. These *essential amino acids* must be obtained from proteins in the food they eat. There are nine essential amino acids for humans. People who are vegetarians must choose their foods so that the essential amino acids in one food complement those in another.

In addition, all vertebrates have lost the ability to synthesize certain unsaturated fatty acids and therefore must obtain them in food. On the other hand, some essential nutrients that vertebrates can synthesize cannot be manufactured by the members of other animal groups. For example, vertebrates can synthesize cholesterol, a key component of steroid hormones, but some carnivorous insects cannot.

Food also supplies **essential minerals** such as calcium, phosphorus, and other inorganic substances, including a wide variety of *trace elements* such as zinc and molybdenum which are required in very small amounts (see table 2.1). Animals obtain trace elements either directly from plants or from animals that have eaten plants.

The body requires vitamins and minerals obtained in food. Also, food must provide particular essential amino acids and fatty acids that the body cannot manufacture by itself.

Table 51.3 Major Vitamins

Vitamin	Function	Source	Deficiency Symptoms
Vitamin A (retinol)	Used in making visual pigments, maintenance of epithelial tissues	Green vegetables, milk products, liver	Night blindness, flaky skin
B-complex vitamins			
B_1	Coenzyme in CO_2 removal during cellular respiration	Meat, grains, legumes	Beriberi, weakening of heart, edema
B_2 (riboflavin)	Part of coenzymes FAD and FMN, which play metabolic roles	Many different kinds of foods	Inflammation and breakdown of skin, eye irritation
B_3 (niacin)	Part of coenzymes NAD^+ and $NADP^+$	Liver, lean meats, grains	Pellagra, inflammation of nerves, mental disorders
B_5 (pantothenic acid)	Part of coenzyme-A, a key connection between carbohydrate and fat metabolism	Many different kinds of foods	Rare: fatigue, loss of coordination
B_6 (pyridoxine)	Coenzyme in many phases of amino acid metabolism	Cereals, vegetables, meats	Anemia, convulsions, irritability
B_{12} (cyanocobalamin)	Coenzyme in the production of nucleic acids	Red meats, dairy products	Pernicious anemia
Biotin	Coenzyme in fat synthesis and amino acid metabolism	Meat, vegetables	Rare: depression, nausea
Folic acid	Coenzyme in amino acid and nucleic acid metabolism	Green vegetables	Anemia, diarrhea
Vitamin C	Important in forming collagen, cement of bone, teeth, connective tissue of blood vessels; may help maintain resistance to infection	Fruit, green leafy vegetables	Scurvy, breakdown of skin, blood vessels
Vitamin D (calciferol)	Increases absorption of calcium and promotes bone formation	Dairy products, cod liver oil	Rickets, bone deformities
Vitamin E (tocopherol)	Protects fatty acids and cell membranes from oxidation	Margarine, seeds, green leafy vegetables	Rare
Vitamin K	Essential to blood clotting	Green leafy vegetables	Severe bleeding

Chapter 51

www.mhhe.com/raven6/resources51.mhtml

Summary | **Questions** | **Media Resources**

51.1 Animals employ a digestive system to prepare food for assimilation by cells.

- The digestive system of vertebrates consists of a gastrointestinal tract and accessory digestive organs.
- Different regions of the digestive tract display specializations of structure and function.

1. What are the layers that make up the wall of the vertebrate gastrointestinal tract? What type of tissue is found in each layer?

- Art Activity: Digestive Tract Wall

- Introduction to Digestion
- Human Digestion

51.2 Food is ingested, swallowed, and transported to the stomach.

- The teeth of carnivores are different from those of herbivores
- The esophagus contracts in peristaltic waves to drive the swallowed food to the stomach.
- Cells of the gastric mucosa secrete hydrochloric acid, which activates pepsin, an enzyme that promotes the partial hydrolysis of ingested proteins.

2. How does tooth structure vary among carnivores, herbivores, and omnivores?

3. What normally prevents regurgitation in humans, and why can't horses regurgitate?

4. What inorganic substance is secreted by parietal cells?

- Digestion
- Mouth to Stomach
- Stomach Digestion
- Ulcers

- Art Quizzes:
  - -Digestion in Birds
  - -Stomach and Duodenum

51.3 The small and large intestines have very different functions.

- The duodenum receives pancreatic juice and bile, which help digest the chyme that arrives from the stomach through the pyloric sphincter.
- Digestive enzymes in the small intestine finish the breakdown of food into molecules that can be absorbed by the small intestine.
- The large intestine absorbs water and ions, as well as certain organic molecules such as vitamin K; the remaining material passes out of the anus.

5. How are the products of protein and carbohydrate digestion absorbed across the intestinal wall, and where do they go after they are absorbed?

6. What anatomical and behavioral specializations do ruminants have for making use of microorganisms?

- Art Activities:
  - -Small Intestine Anatomy
  - -Hepatic Lobules

- Small Intestines
- Large Intestines

51.4 Accessory organs, neural stimulation, and endocrine secretions assist in digestion.

- Pancreatic juice contains bicarbonate to neutralize the acid chyme from the stomach. Bile contains bile pigment and bile salts, which emulsify fat. The liver metabolizes toxins and hormones that are delivered to it in the hepatic portal vein; the liver also helps to regulate the blood glucose concentration.
- The stomach secretes the hormone gastrin, and the small intestine secretes various hormones that help to regulate the digestive system.

7. What are the main exocrine secretions of the pancreas, and what are their functions?

8. What is the function of bile salts in digestion?

9. Describe the role of gastrin and secretin in digestion.

- Art Activity: Digestive System

- Formation of Gallstones

- Art Quiz: Actions of Insulin and Glucagon

51.5 All animals require food energy and essential nutrients.

- The basal metabolic rate (BMR) is the lowest level of energy consumption of the body.
- Vitamins, minerals, and the essential amino acids and fatty acids must be supplied in the diet.

10. What is a vitamin? What is the difference between an essential amino acid and any other amino acid?

- Bioethics Case Study: Bulimia
- *On Science* Article: Diabetes-Obesity Link

52

Circulation

Concept Outline

52.1 The circulatory systems of animals may be open or closed.

Open and Closed Circulatory Systems. All vertebrates have a closed circulation, while many invertebrate animals have open circulatory systems.

52.2 A network of vessels transports blood through thc body.

The Blood Plasma. The blood plasma transports a variety of solutes, including ions, metabolites, proteins, and hormones.

The Blood Cells. The blood cells include erythrocytes, which transport oxygen, leukocytes, which provide defenses for the body, and platelets, which function in blood clotting.

Characteristics of Blood Vessels. Blood leaves the heart in arteries and returns in veins; in between, the blood passes through capillaries, where all exchanges with tissues occur.

The Lymphatic System. The lymphatic system returns interstitial fluid to the bloodstream.

52.3 The vertebrate heart has undergone progressive evolutionary change.

The Fish Heart. The fish heart consists of a row of four chambers that receives blood in the posterior end from the body and pumps blood from the anterior end to the gills.

Amphibian and Reptile Circulation. Land vertebrates have a double circulation, where blood from the lungs returns to the heart to be pumped to the rest of the body.

Mammalian and Bird Hearts. Mammals and birds have a complete separation between the two sides of the heart.

52.4 The cardiac cycle drives the cardiovascular system.

The Cardiac Cycle. The right and left sides of the heart rest and receive blood at the same time, then pump the blood into arteries at the same time.

Electrical Excitation and Contraction of the Heart. The impulse begins in one area of the heart and is conducted to the rest of the heart.

Blood Flow and Blood Pressure. Blood flow and blood pressure depend on the diameter of the arterial vessels and on the amount of blood pumped by the heart.

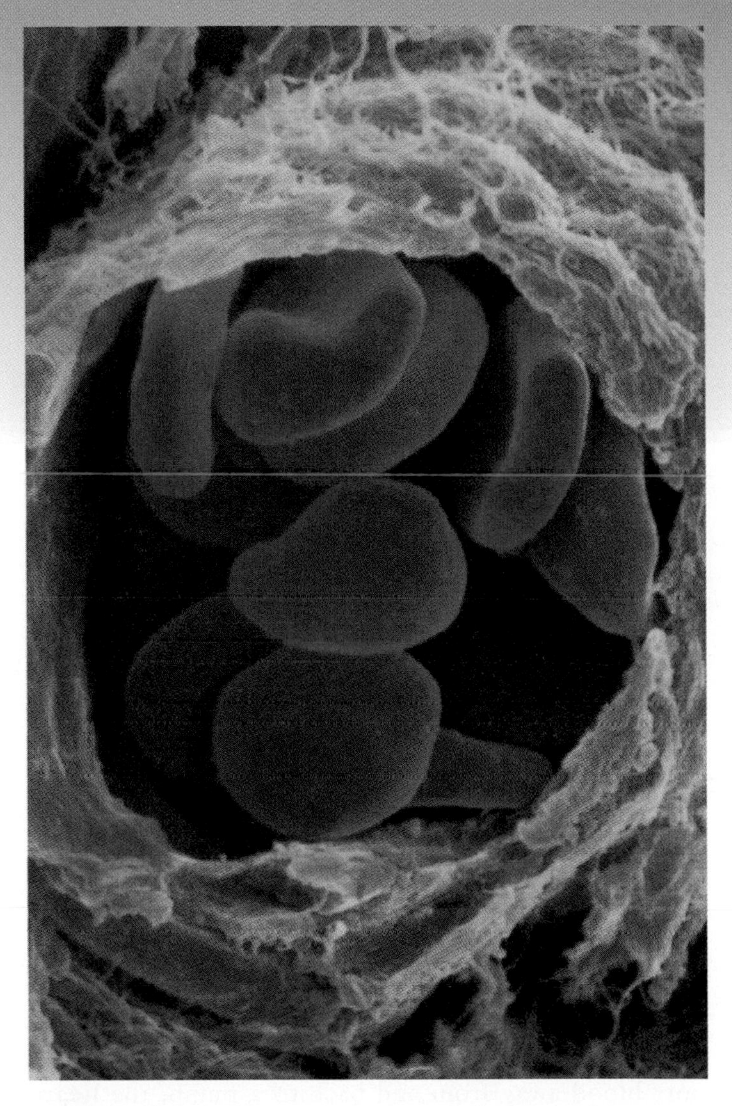

FIGURE 52.1
Red blood cells. This ruptured blood vessel, seen in a scanning electron micrograph, is full of red blood cells, which move through vessels transporting oxygen from one place to another in the body.

Every cell in the animal body must acquire the energy it needs for living from other molecules in food. Like residents of a city whose food is imported from farms in the countryside, cells in the body need trucks to carry the food, highways for the trucks to travel on, and a way to cook the food when it arrives. In animals, the circulatory system provides blood and blood vessels (the trucks and highways), and is discussed in this chapter (figure 52.1). The digestive system, discussed in chapter 51, provides the glucose (food) and the respiratory system provides oxygen necessary for metabolism (fuel to cook the food), and will be discussed in the following chapter.

52.1 The circulatory systems of animals may be open or closed.

Open and Closed Circulatory Systems

Among the unicellular protists, oxygen and nutrients are obtained directly from the aqueous external environment by simple diffusion. The body wall is only two cell layers thick in cnidarians, such as *Hydra*, and flatworms, such as Planaria. Each cell layer is in direct contact with either the external environment or the gastrovascular cavity (figure 52.2*a*). The gastrovascular cavity of *Hydra* (see chapter 51) extends from the body cavity into the tentacles, and that of Planaria branches extensively to supply every cell with oxygen and nutrients. Larger animals, however, have tissues that are several cell layers thick, so that many cells are too far away from the body surface or digestive cavity to exchange materials directly with the environment. Instead, oxygen and nutrients are transported from the environment and digestive cavity to the body cells by an internal fluid within a *circulatory system.*

There are two main types of circulatory systems: *open* or *closed.* In an **open circulatory system,** such as that found in mollusks and arthropods (figure 52.2*b*), there is no distinction between the circulating fluid (blood) and the extracellular fluid of the body tissues (interstitial fluid or lymph). This fluid is thus called **hemolymph.** In insects, the heart is a muscular tube that pumps hemolymph through a network of channels and cavities in the body. The fluid then drains back into the central cavity.

In a **closed circulatory system,** the circulating fluid, or blood, is always enclosed within blood vessels that transport blood away from and back to a pump, the **heart.** Annelids (see chapter 45) and all vertebrates have a closed circulatory system. In annelids such as an earthworm, a dorsal vessel contracts rhythmically to function as a pump. Blood is pumped through five small connecting arteries which also function as pumps, to a ventral vessel, which transports the blood posteriorly until it eventually reenters the dorsal vessel. Smaller vessels branch from each artery to supply the tissues of the earthworm with oxygen and nutrients and to transport waste products (figure 52.2*c*).

The Functions of Vertebrate Circulatory Systems

The vertebrate circulatory system is more elaborate than the invertebrate circulatory system. It functions in transporting oxygen and nutrients to tissues by the cardiovascular system. Blood vessels form a tubular network that permits blood to flow from the heart to all the cells of the body and then back to the heart. *Arteries* carry blood away from the heart, whereas *veins* return blood to the heart. Blood passes from the arterial to the venous system in *capillaries,* which are the thinnest and most numerous of the blood vessels.

As blood plasma passes through capillaries, the pressure of the blood forces some of this fluid out of the capillary walls. Fluid derived from plasma that passes out of capillary walls into the surrounding tissues is called **interstitial fluid.** Some of this fluid returns directly to capillaries, and some enters into **lymph vessels,** located in the connective tissues around the blood vessels. This fluid, now called *lymph,* is returned to the venous blood at specific sites. The lymphatic system is considered a part of the circulatory system and is discussed later in this chapter.

The vertebrate circulatory system has three principal functions: transportation, regulation, and protection.

1. **Transportation.** All of the substances essential for cellular metabolism are transported by the circulatory system. These substances can be categorized as follows:
   a. *Respiratory.* Red blood cells, or *erythrocytes,* transport oxygen to the tissue cells. In the capillaries of

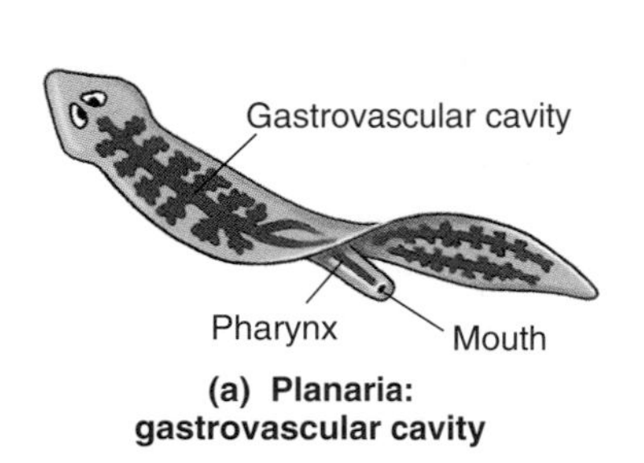

(a) Planaria: gastrovascular cavity

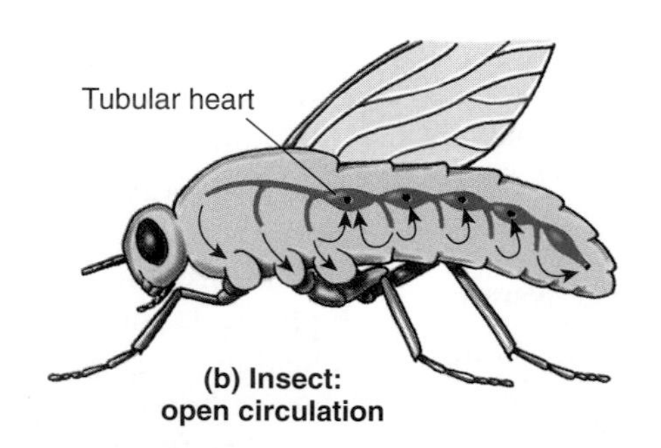

(b) Insect: open circulation

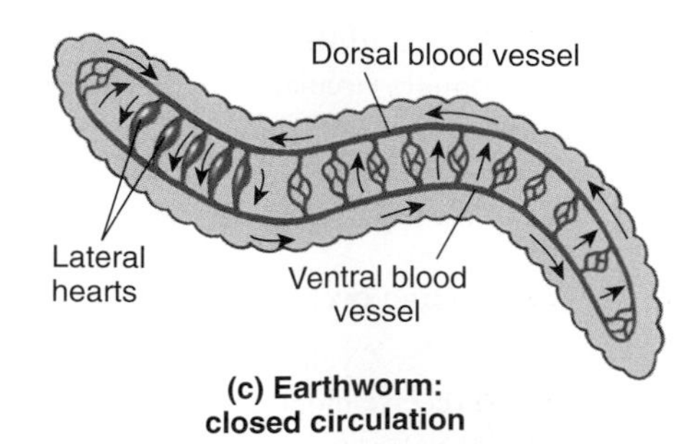

(c) Earthworm: closed circulation

FIGURE 52.2
Circulatory systems of the animal kingdom. (*a*) The gastrovascular cavity of Planaria serves as both a digestive and circulatory system, delivering nutrients directly to the tissue cells by diffusion from the digestive cavity. (*b*) In the open circulation of an insect, hemolymph is pumped from a tubular heart into cavities in the insect's body; the hemolymph then returns to the blood vessels so that it can be recirculated. (*c*) In the closed circulation of the earthworm, blood pumped from the hearts remains within a system of vessels that returns it to the hearts. All vertebrates also have closed circulatory systems.

the lungs or gills, oxygen attaches to hemoglobin molecules within the erythrocytes and is transported to the cells for aerobic respiration. Carbon dioxide, a by-product of cell respiration, is carried by the blood to the lungs or gills for elimination.

b. Nutritive. The digestive system is responsible for the breakdown of food into molecules so that nutrients can be absorbed through the intestinal wall and into the blood vessels of the circulatory system. The blood then carries these absorbed products of digestion through the liver and to the cells of the body.

c. Excretory. Metabolic wastes, excessive water and ions, and other molecules in the fluid portion of blood are filtered through the capillaries of the kidneys and excreted in urine.

2. **Regulation.** The cardiovascular system transports regulatory hormones and participates in temperature regulation.

a. Hormone transport. The blood carries hormones from the endocrine glands, where they are secreted, to the distant target organs they regulate.

b. Temperature regulation. In warm-blooded vertebrates, or **endotherms,** a constant body temperature is maintained regardless of the ambient temperature. This is accomplished in part by blood vessels located just under the epidermis. When the ambient temperature is cold, the superficial vessels constrict to divert the warm blood to deeper vessels. When the ambient temperature is warm, the superficial vessels dilate so that the warmth of the blood can be lost by radiation (figure 52.3).

Some vertebrates also retain heat in a cold environment by using a **countercurrent heat exchange** (also see chapter 53). In this process, a vessel carrying warm blood from deep within the body passes next to a vessel carrying cold blood from the surface of the body (figure 52.4). The warm blood going out heats the cold blood returning from the body surface, so that this blood is no longer cold when it reaches the interior of the body.

3. **Protection.** The circulatory system protects against injury and foreign microbes or toxins introduced into the body.

a. Blood clotting. The clotting mechanism protects against blood loss when vessels are damaged. This clotting mechanism involves both proteins from the blood plasma and cell fragments called platelets (discussed in the next section).

b. Immune defense. The blood contains white blood cells, or leukocytes, that provide immunity against many disease-causing agents. Some white blood cells are phagocytic, some produce antibodies, and some act by other mechanisms to protect the body.

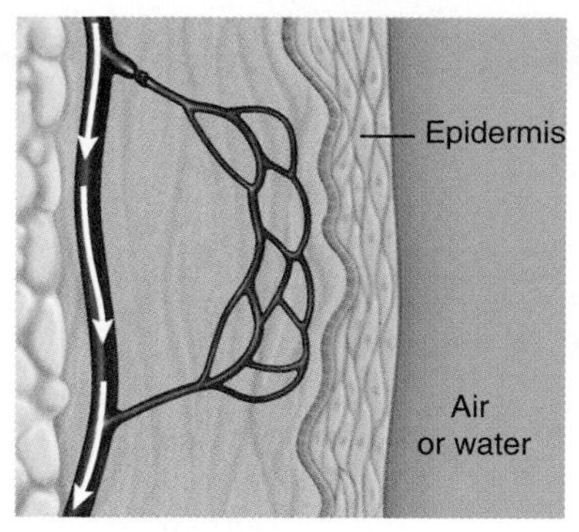

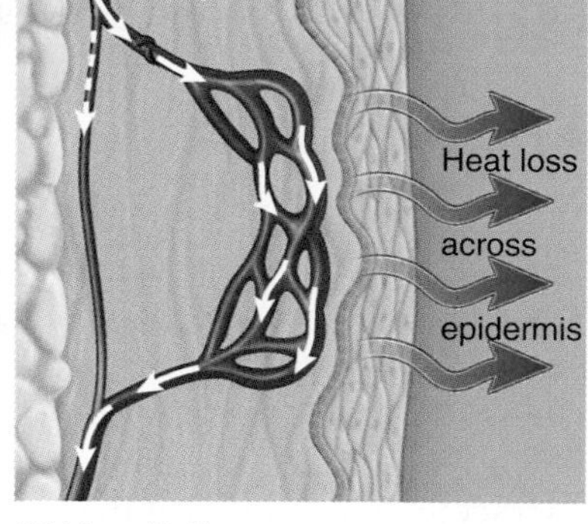

(a) Vasoconstriction (b) Vasodilation

FIGURE 52.3
Regulation of heat loss. The amount of heat lost at the body's surface can be regulated by controlling the flow of blood to the surface. (*a*) Constriction of surface blood vessels limits flow and heat loss; (*b*) dilation of these vessels increases flow and heat loss.

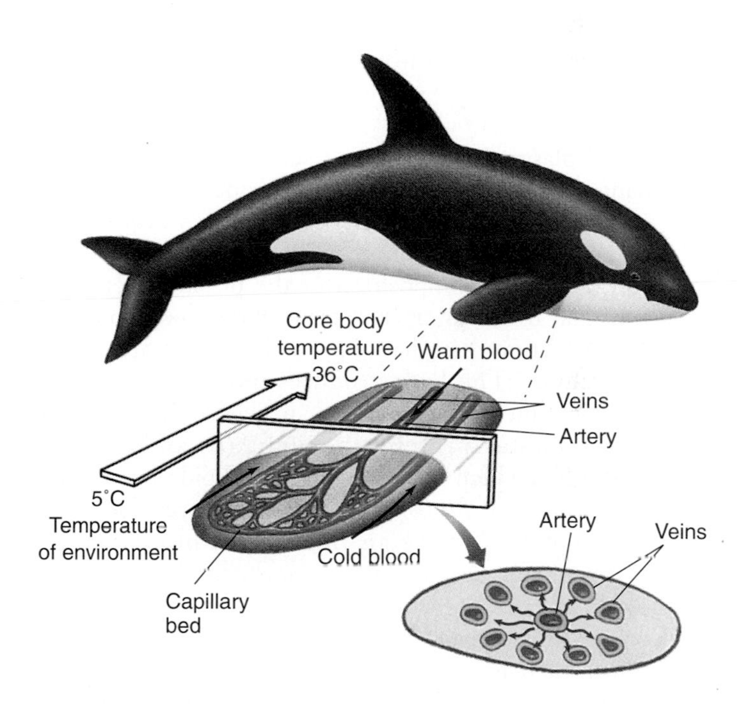

FIGURE 52.4
Countercurrent heat exchange. Many marine animals, such as this killer whale, limit heat loss in cold water using countercurrent heat exchange. The warm blood pumped from within the body in arteries loses heat to the cold blood returning from the skin in veins. This warms the venous blood so that the core body temperature can remain constant in cold water and cools the arterial blood so that less heat is lost when the arterial blood reaches the tip of the extremity.

Circulatory systems may be open or closed. All vertebrates have a closed circulatory system, in which blood circulates away from the heart in arteries and back to the heart in veins. The circulatory system serves a variety of functions, including transportation, regulation, and protection.

52.2 A network of vessels transports blood through the body.

The Blood Plasma

Blood is composed of a fluid **plasma** and several different kinds of cells that circulate within that fluid (figure 52.5). Blood platelets, although included in figure 52.5, are not complete cells; rather, they are fragments of cells that reside in the bone marrow. Blood plasma is the matrix in which blood cells and platelets are suspended. Interstitial (extracellular) fluids originate from the fluid present in plasma.

Plasma contains the following solutes:

1. **Metabolites, wastes, and hormones.** Dissolved within the plasma are all of the metabolites used by cells, including glucose, amino acids, and vitamins. Also dissolved in the plasma are hormones that regulate cellular activities, wastes such as nitrogen compounds, and CO_2 produced by metabolizing cells. CO_2 is carried in the blood as bicarbonate because free carbon dioxide would decrease blood pH.
2. **Ions.** Like the water of the seas in which life arose, blood plasma is a dilute salt solution. The predominant plasma ions are sodium, chloride, and bicarbonate ions. In addition, there are trace amounts of other ions such as calcium, magnesium, copper, potassium, and zinc. The composition of the plasma, therefore, is similar to seawater, but plasma has a lower total ion concentration than that of present-day seawater.
3. **Proteins.** The liver produces most of the plasma proteins, including **albumin,** which comprises most of the plasma protein; the alpha (α) and beta (β) **globulins,** which serve as carriers of lipids and steroid hormones; and *fibrinogen*, which is required for blood clotting. Following an injury of a blood vessel, platelets release clotting factors (proteins) into the blood. In the presence of these clotting factors, fibrinogen is converted into insoluble threads of *fibrin*. Fibrin then aggregates to form the clot. Blood plasma which has had fibrinogen removed is called **serum**.

Plasma, the liquid portion of the blood, contains different types of proteins, ions, metabolites, wastes, and hormones. This liquid, and fluids derived from it, provide the extracellular environment of most the cells of the body.

Blood cell	Life span in blood	Function
 Erythrocyte	120 days	O_2 and CO_2 transport
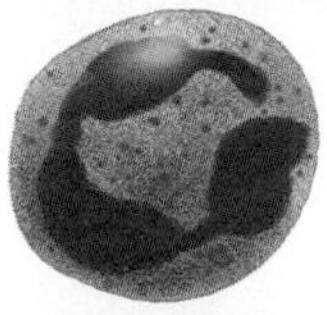 Neutrophil	7 hours	Immune defenses
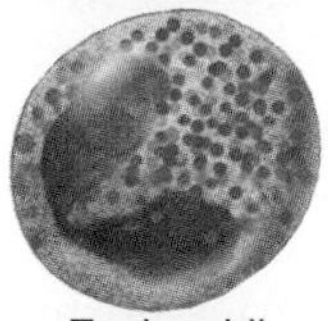 Eosinophil	Unknown	Defense against parasites
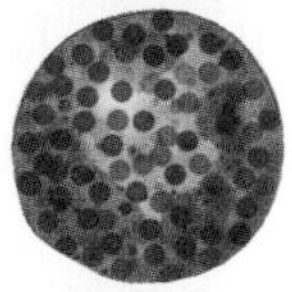 Basophil	Unknown	Inflammatory response
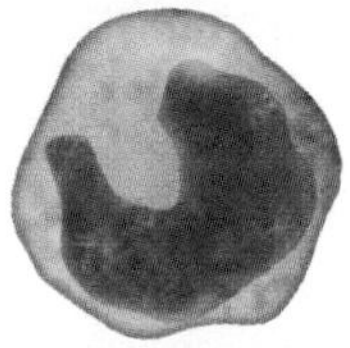 Monocyte	3 days	Immune surveillance (precursor of tissue macrophage)
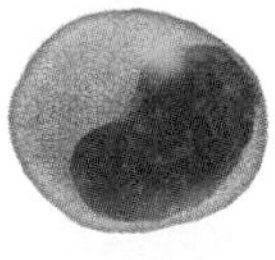 B lymphocyte	Unknown	Antibody production (precursor of plasma cells)
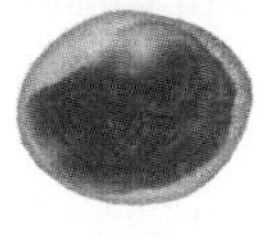 T lymphocyte	Unknown	Cellular immune response
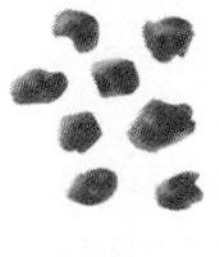 Platelets	7-8 days	Blood clotting

FIGURE 52.5
Types of blood cells. Erythrocytes are red blood cells, platelets are fragments of a bone marrow cell, and all the other cells are different types of leukocytes, or white blood cells.

The Blood Cells

Red blood cells function in oxygen transport, white blood cells in immunological defenses, and platelets in blood clotting (see figure 52.5).

Erythrocytes and Oxygen Transport

Each cubic millimeter of blood contains about 5 million **red blood cells,** or **erythrocytes.** The fraction of the total blood volume that is occupied by erythrocytes is called the blood's *hematocrit;* in humans, it is typically around 45%. A disc with a central depression, each erythrocyte resembles a doughnut with a hole that does not go all the way through. As we've already seen, the erythrocytes of vertebrates contain hemoglobin, a pigment which binds and transports oxygen. In vertebrates, hemoglobin is found only in erythrocytes. In invertebrates, the oxygen binding pigment (not always hemoglobin) is also present in plasma.

Erythrocytes develop from unspecialized cells, called *stem cells.* When plasma oxygen levels decrease, the kidney converts a plasma protein into the hormone, *erythropoietin.* Erythropoietin then stimulates the production of erythrocytes in bone marrow through a process called *erythropoiesis.* In mammals, maturing erythrocytes lose their nuclei. This is different from the mature erythrocytes of all other vertebrates, which remain nucleated. As mammalian erythrocytes age, they are removed from the blood by phagocytic cells of the spleen, bone marrow, and liver. Balancing this loss, new erythrocytes are constantly formed in the bone marrow.

Leukocytes Defend the Body

Less than 1% of the cells in human blood are **leukocytes,** or **white blood cells;** there are only 1 or 2 leukocytes for every 1000 erythrocytes. Leukocytes are larger than erythrocytes and have nuclei. Furthermore, leukocytes are not confined to the blood as erythrocytes are, but can migrate out of capillaries into the interstitial (tissue) fluid.

There are several kinds of leukocytes, each of which plays a specific role in defending the body against invading microorganisms and other foreign substances, as described in Chapter 57. **Granular leukocytes** include **neutrophils, eosinophils,** and **basophils,** which are named according to the staining properties of granules in their cytoplasm. **Nongranular leukocytes** include **monocytes** and **lymphocytes.** Neutrophils are the most numerous of the leukocytes, followed in order by lymphocytes, monocytes, eosinophils, and basophils.

Platelets Help Blood to Clot

Megakaryocytes are large cells present in bone marrow. Pieces of cytoplasm are pinched off of the megakaryocytes and become **platelets.** Platelets play an important role in blood clotting. When a blood vessel is broken, smooth muscle in the vessel walls contracts, causing the vessel to constrict. Platelets then accumulate at the injured site and form a plug by sticking to each other and to the surrounding tissues. This plug is reinforced by threads of the protein **fibrin** (figure 52.6), which contract to form a tighter mass. The tightening plug of platelets, fibrin, and often trapped erythrocytes constitutes a blood clot.

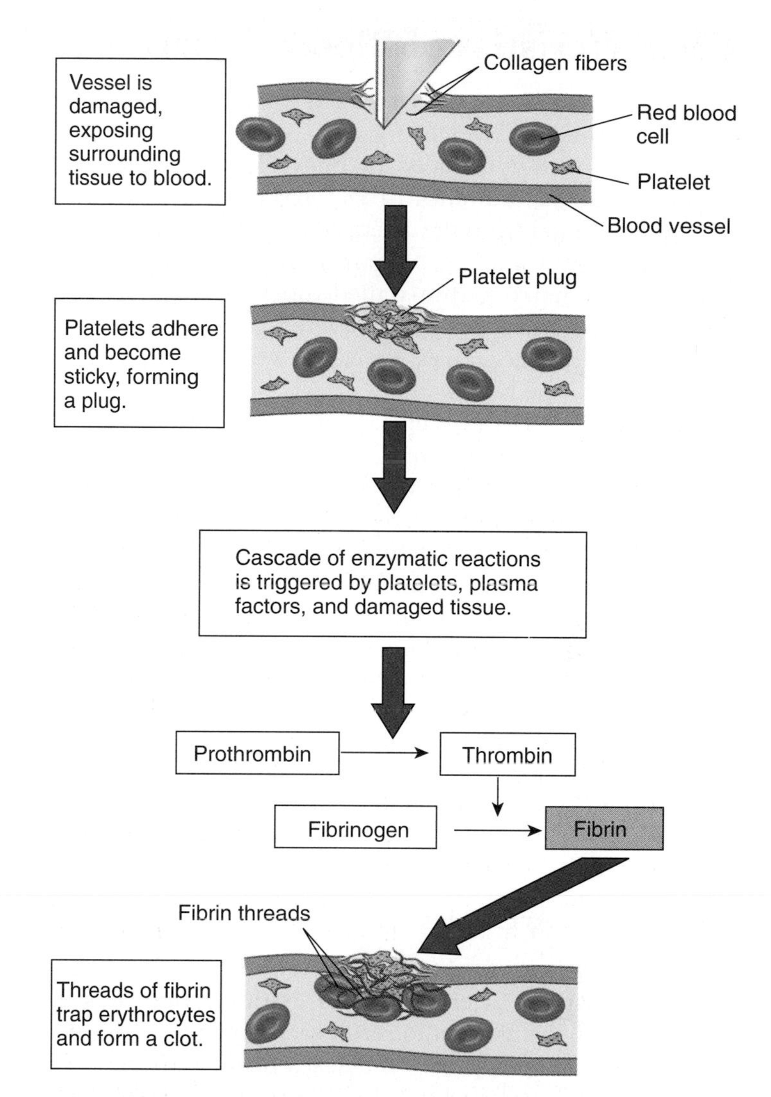

FIGURE 52.6
Blood clotting. Fibrin is formed from a soluble protein, fibrinogen, in the plasma. This reaction is catalyzed by the enzyme thrombin, which is formed from an inactive enzyme called prothrombin. The activation of thrombin is the last step in a cascade of enzymatic reactions that produces a blood clot when a blood vessel is damaged.

Erythrocytes contain hemoglobin and serve in oxygen transport. The different types of leukocytes have specialized functions that serve to protect the body from invading pathogens, and the platelets participate in blood clotting.

Characteristics of Blood Vessels

Blood leaves the heart through vessels known as **arteries.** These continually branch, forming a hollow "tree" that enters each of the organs of the body. The finest, microscopically-sized branches of the arterial trees are the **arterioles.** Blood from the arterioles enters the **capillaries** (from the Latin *capillus*, "a hair"), an elaborate latticework of very narrow, thin-walled tubes. After traversing the capillaries, the blood is collected into **venules;** the venules lead to larger vessels called **veins,** which carry blood back to the heart.

Arteries, arterioles, veins, and venules all have the same basic structure (figure 52.7). The innermost layer is an epithelial sheet called the *endothelium*. Covering the endothelium is a thin layer of elastic fibers, a smooth muscle layer, and a connective tissue layer. The walls of these vessels are thus too thick to permit any exchange of materials between the blood and the tissues outside the vessels. The walls of capillaries, however, are made up of only the endothelium, so molecules and ions can leave the blood plasma by diffusion, by filtration through pores in the capillary walls, and by transport through the endothelial cells. Therefore, it is while blood is in the capillaries that gases and metabolites are exchanged with the cells of the body.

Arteries and Arterioles

Arteries function in transporting blood away from the heart. The larger arteries contain extra elastic fibers in their walls, allowing them to recoil each time they receive a volume of blood pumped by the heart. Smaller arteries and arterioles are less elastic, but their disproportionately thick smooth muscle layer enables them to resist bursting.

The vast tree of arteries presents a frictional resistance to blood flow. The narrower the vessel, the greater the frictional resistance to flow. In fact, a vessel that is half the diameter of another has 16 times the frictional resistance! This is because the resistance to blood flow is inversely proportional to the radius of the vessel. Therefore, within the arterial tree, it is the small arteries and arterioles that provide the greatest resistance to blood flow. Contraction of the smooth muscle layer of the arterioles results in **vasoconstriction,** which greatly increases resistance and decreases flow. Relaxation of the smooth muscle layer results in **vasodilation,** decreasing resistance and increasing blood flow to an organ (see figure 52.3).

In addition, blood flow through some organs is regulated by rings of smooth muscle around arterioles near the region where they empty into capillaries. These **precapillary sphincters** (figure 52.8) can close off specific capillary beds completely. For example, the closure of precapillary sphincters in the skin contributes to the vasoconstriction that limits heat loss in cold environments.

Exchange in the Capillaries

Each time the heart contracts, it must produce sufficient pressure to pump blood against the resistance of the arterial tree and into the capillaries. The vast number and extensive branching of the capillaries ensure that *every cell in the body is within 100 μm of a capillary.* On the average, capillaries are about 1 mm long and 8 μm in diameter, only slightly larger than a red blood cell (5 to 7 μm in diameter). Despite the close fit, red blood cells can squeeze through capillaries without difficulty.

Although each capillary is very narrow, there are so many of them that the capillaries have the greatest *total* cross-sectional area of any other type of vessel. Consequently, the

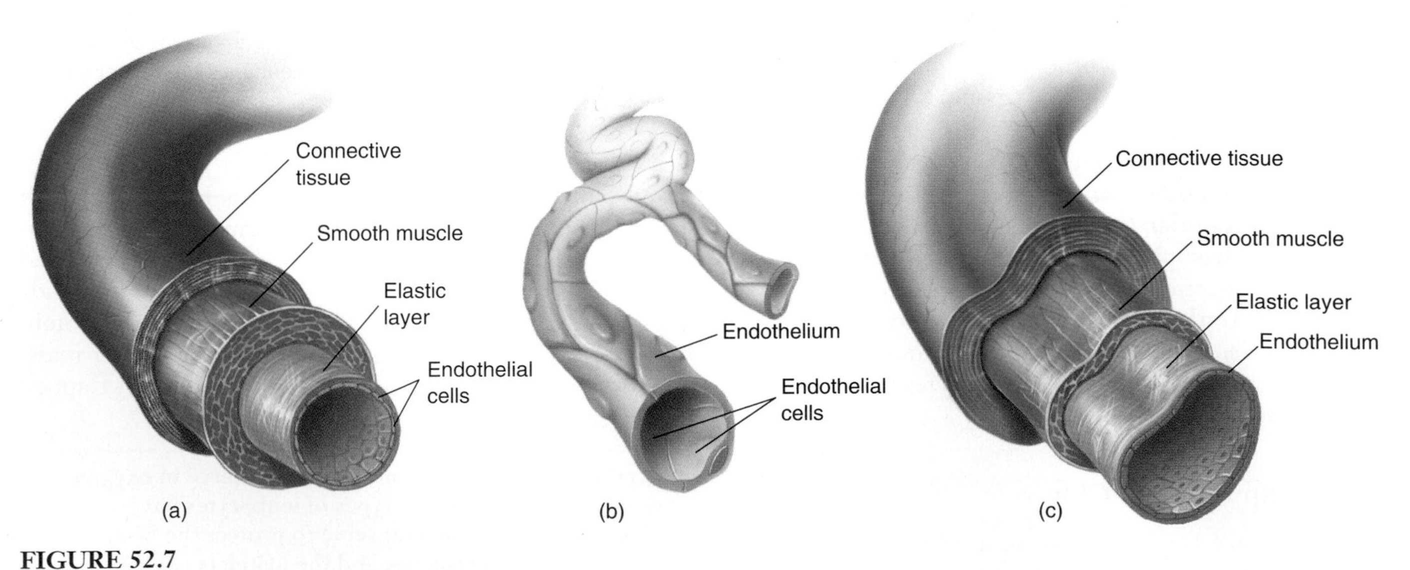

FIGURE 52.7
The structure of blood vessels. (*a*) Arteries and (*c*) veins have the same tissue layers. (*b*) Capillaries are composed of only a single layer of endothelial cells. (not to scale)

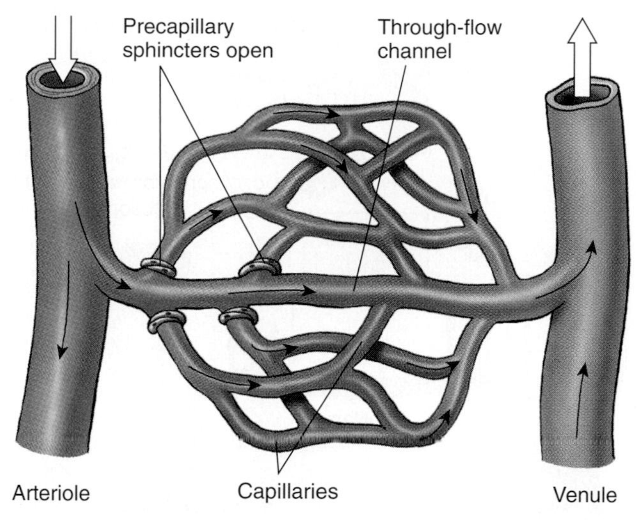

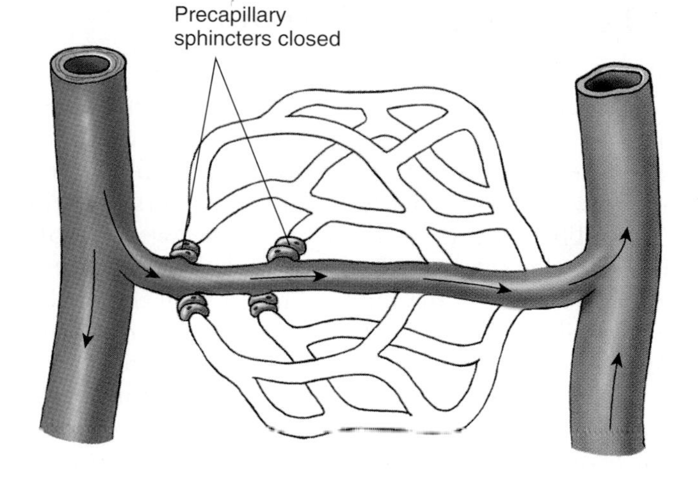

FIGURE 52.8
The capillary network connects arteries with veins. (*a*) Most of the exchange between the blood and the extracellular fluid occurs while the blood is in the capillaries. Entrance to the capillaries is controlled by bands of muscle called precapillary sphincters at the entrance to each capillary. (*b*) When a sphincter contracts, it closes off the capillary. By contracting these sphincters, the body can limit the amount of blood in the capillary network of a particular tissue, and thus control the rate of exchange in that tissue.

blood decreases in velocity as it passes through the capillary beds, allowing more time for it to exchange materials with the surrounding extracellular fluid. By the time the blood reaches the end of a capillary, it has released some of its oxygen and nutrients and picked up carbon dioxide and other waste products. Blood also loses most of its pressure in passing through the vast capillary networks, and so is under very low pressure when it enters the veins.

Venules and Veins

Blood flows from the venules to ever larger veins, and ultimately back to the heart. Venules and veins have the same tissue layers as arteries, but they have a thinner layer of smooth muscle. Less muscle is needed because the pressure in the veins is only about one-tenth that in the arteries. Most of the blood in the cardiovascular system is contained within veins, which can expand when needed to hold additional amounts of blood. You can see the expanded veins in your feet when you stand for a long time.

When the blood pressure in the veins is so low, how does the blood return to the heart from the feet and legs? The venous pressure alone is not sufficient, but several sources provide help. Most significantly, skeletal muscles surrounding the veins can contract to move blood by squeezing the veins. Blood moves in one direction through the veins back to the heart with the help of **venous valves** (figure 52.9). When a person's veins expand too much with blood, the venous valves may no longer work and the blood may pool in the veins. Veins in this condition are known as varicose veins.

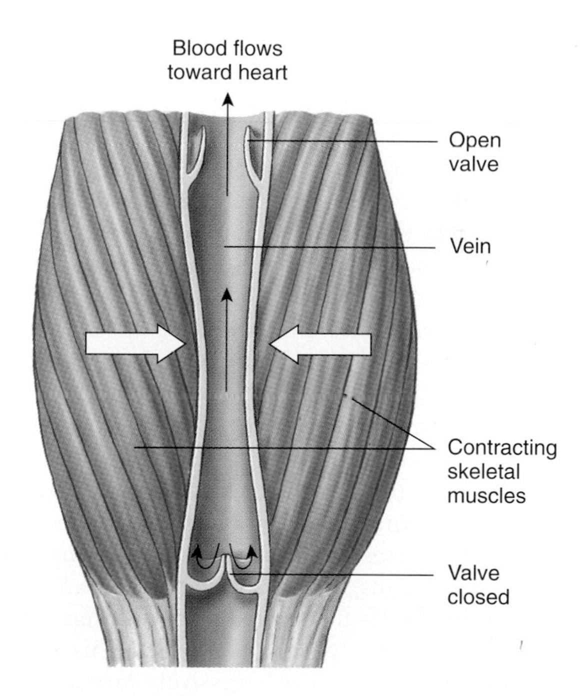

FIGURE 52.9
One-way flow of blood through veins. Venous valves ensure that blood moves through the veins in only one direction, back to the heart.

Blood is pumped from the heart into the arterial system, which branches into fine arterioles. This blood is delivered into the thinnest and most numerous of vessels, the capillaries, where exchanges with the tissues occur. Blood returns to the heart through veins.

The Lymphatic System

The cardiovascular system is considered to be a closed system because all of its vessels are connected with one another—none are simply open-ended. However, some water and solutes in the blood plasma do filter through the walls of the capillaries to form the interstitial (tissue) fluid. This filtration is driven by the pressure of the blood, and it helps supply the tissue cells with oxygen and nutrients. Most of the fluid is filtered from the capillaries near their arteriolar ends, where the blood pressure is higher, and returned to the capillaries near their venular ends. This return of fluid occurs by osmosis, which is driven by a higher solute concentration within the capillaries. Most of the plasma proteins cannot escape through the capillary pores because of their large size and so the concentration of proteins in the plasma is greater than the protein concentration in the interstitial fluid. The difference in protein concentration produces an osmotic pressure, called the *oncotic pressure*, that causes osmosis of water into the capillaries (figure 52.10).

Because interstitial fluid is produced as a result of the blood pressure, high capillary blood pressure could cause too much interstitial fluid to be produced. A common example of this occurs in pregnant women, when the fetus compresses veins and thereby increases the capillary blood pressure in the mother's lower limbs. The increased interstitial fluid can cause swelling of the tissues, or *edema*, of the feet. Edema may also result if the plasma protein concentration (and thus the oncotic pressure) is too low. Fluids will not return to the capillaries but will remain as interstitial fluid. This may be caused either by liver disease, because the liver produces most of the plasma proteins, or by protein malnutrition (*kwashiorkor*).

Even under normal conditions, the amount of fluid filtered out of the capillaries is greater than the amount that returns to the capillaries by osmosis. The remainder does eventually return to the cardiovascular system, however, by way of an *open* circulatory system called the **lymphatic system.** The lymphatic system consists of lymphatic capillaries, lymphatic vessels, lymph nodes, and lymphatic organs, including the spleen and thymus. Excess fluid in the tissues drains into blind-ended lymph capillaries with highly permeable walls. This fluid, now called **lymph,** passes into progressively larger lymphatic vessels, which resemble veins and have one-way valves (figure 52.11). The lymph eventually enters two major lymphatic vessels, which drain into veins on each side of the neck.

Movement of lymph in mammals is accomplished by skeletal muscles squeezing against the lymphatic vessels, a mechanism similar to the one that moves blood through veins. In some cases, the lymphatic vessels also contract rhythmically. In many fishes, all amphibians and reptiles, bird embryos, and some adult birds, movement of lymph is propelled by **lymph hearts.**

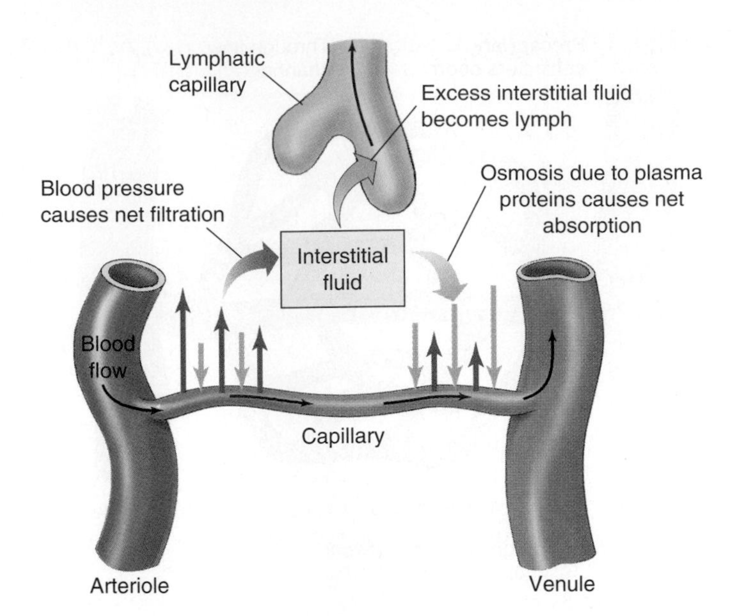

FIGURE 52.10
Plasma fluid, minus proteins, is filtered out of capillaries. This forms interstitial fluid, which bathes the tissues. Much of the interstitial fluid is returned to the capillaries by the osmotic pressure generated by the higher protein concentration in plasma. The excess interstitial fluid is drained into open-ended lymphatic capillaries, which ultimately return the fluid to the cardiovascular system.

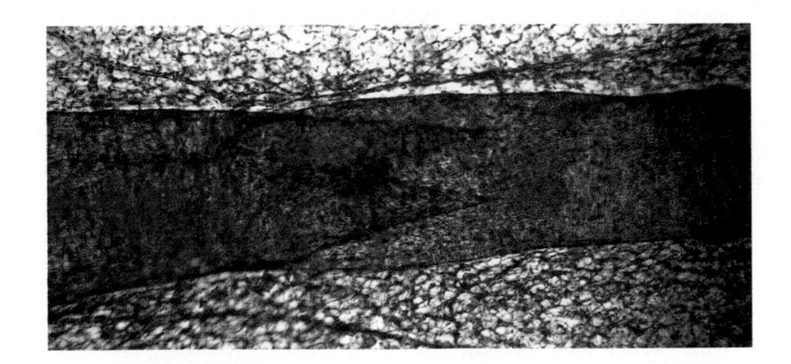

FIGURE 52.11
A lymphatic vessel valve (25×). Valves allow lymph to flow in one direction (from left to right in this figure) but not in the reverse direction.

As the lymph moves through lymph nodes and lymphatic organs, it is modified by phagocytic cells that line the channels of those organs. In addition, the lymph nodes and lymphatic organs contain *germinal centers* for the production of lymphocytes, a type of white blood cell critically important in immunity.

Lymphatic vessels carry excess interstitial fluid back to the vascular system. This fluid, called lymph, travels through lymph nodes and lymphatic organs where it encounters the immune cells called lymphocytes that are produced in these organs.

52.3 The vertebrate heart has undergone progressive evolutionary change.

The Fish Heart

The chordates that were ancestral to the vertebrates are thought to have had simple tubular hearts, similar to those now seen in lancelets (see chapter 48). The heart was little more than a specialized zone of the ventral artery, more heavily muscled than the rest of the arteries, which contracted in simple peristaltic waves. A pumping action results because the uncontracted portions of the vessel have a larger diameter than the contracted portion, and thus present less resistance to blood flow.

The development of gills by fishes required a more efficient pump, and in fishes we see the evolution of a true chamber-pump heart. The fish heart is, in essence, a tube with four chambers arrayed one after the other (figure 52.12*a*). The first two chambers—the **sinus venosus** and **atrium**—are collection chambers, while the second two, the **ventricle** and **conus arteriosus,** are pumping chambers.

As might be expected, the sequence of the heartbeat in fishes is a peristaltic sequence, starting at the rear and moving to the front, similar to the early chordate heart. The first of the four chambers to contract is the sinus venosus, followed by the atrium, the ventricle, and finally the conus arteriosus. Despite shifts in the relative positions of the chambers in the vertebrates that evolved later, this heartbeat sequence is maintained in all vertebrates. In fish, the electrical impulse that produces the contraction is initiated in the sinus venosus; in other vertebrates, the electrical impulse is initiated by their equivalent of the sinus venosus.

The fish heart is remarkably well suited to the gill respiratory apparatus and represents one of the major evolutionary innovations in the vertebrates. Perhaps its greatest advantage is that the blood that passes through the gills is fully oxygenated when it moves into the tissues. After blood leaves the conus arteriosus, it moves through the gills, where it becomes oxygenated; from the gills, it flows through a network of arteries to the rest of the body; then it returns to the heart through the veins (figure 52.12*b*). This arrangement has one great limitation, however. In passing through the capillaries in the gills, the blood loses much of the pressure developed by the contraction of the heart, so the circulation from the gills through the rest of the body is sluggish. This feature limits the rate of oxygen delivery to the rest of the body.

The fish heart is a modified tube, consisting of a series of four chambers. Blood first enters the heart at the sinus venosus, where the wavelike contraction of the heart begins.

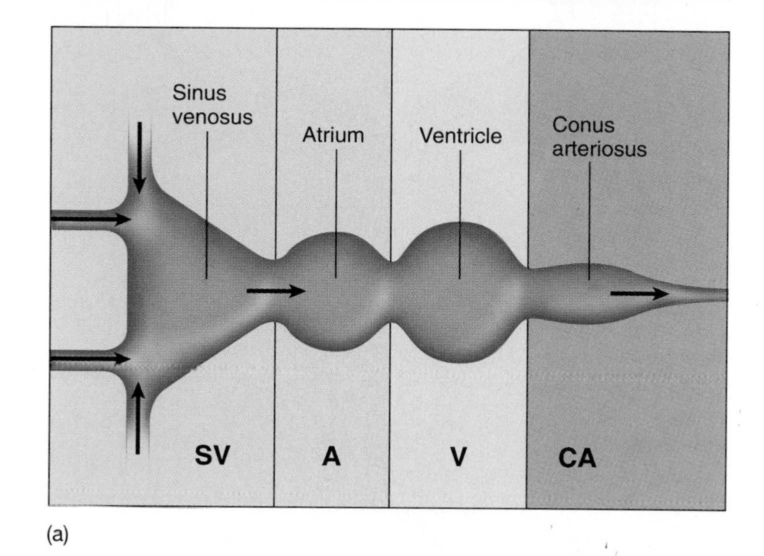

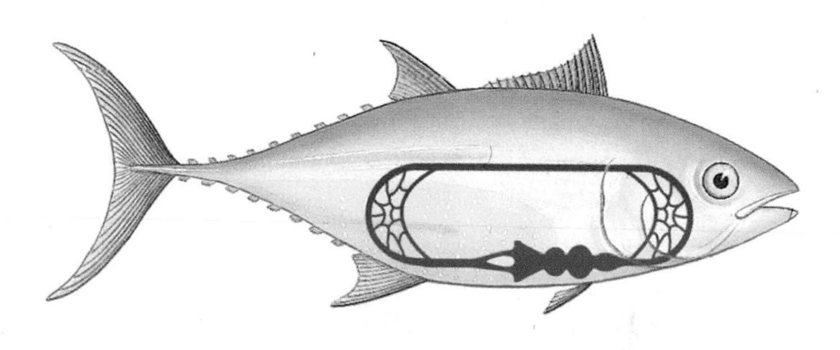

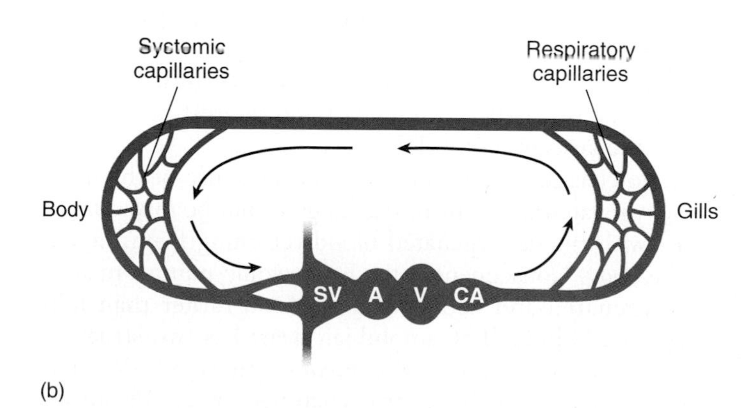

FIGURE 52.12
The heart and circulation of a fish. (*a*) Diagram of a fish heart, showing the chambers in series with each other. (*b*) Diagram of fish circulation, showing that blood is pumped by the ventricle through the gills and then to the body. Blood rich in oxygen (oxygenated) is shown in *red;* blood low in oxygen (deoxygenated) is shown in *blue.*

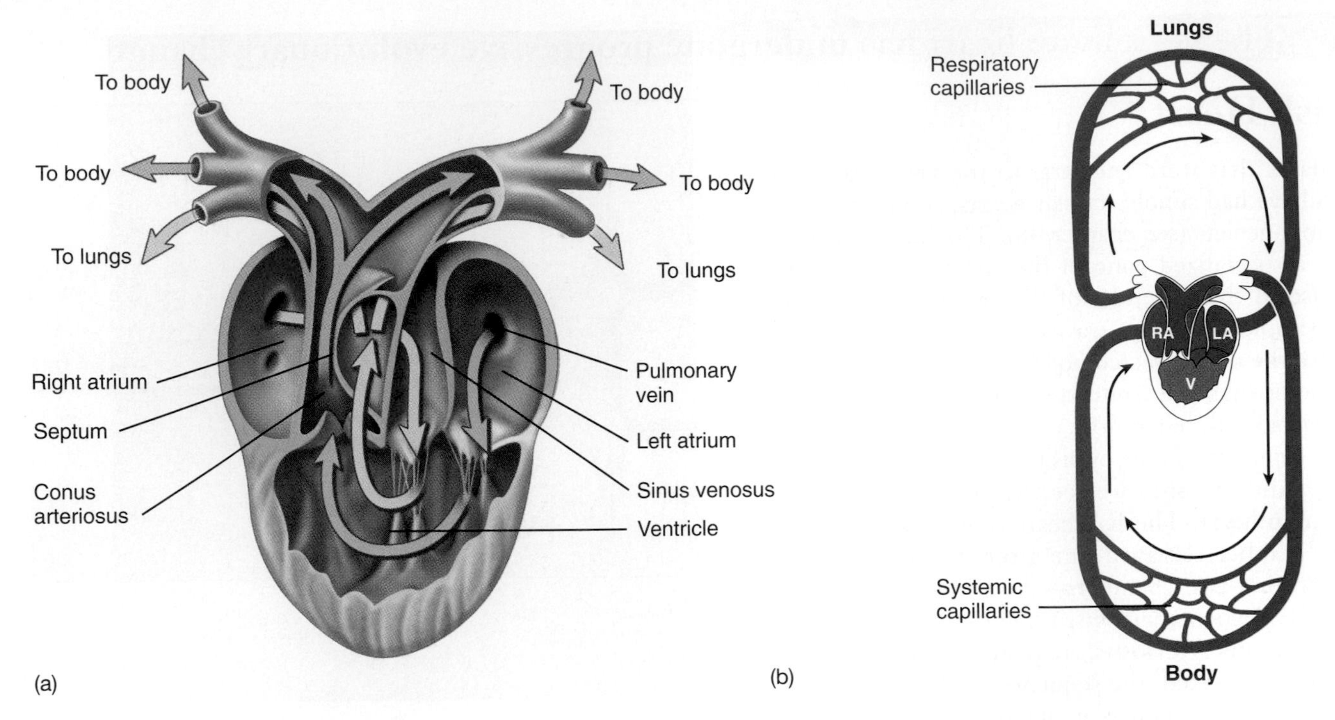

FIGURE 52.13
The heart and circulation of an amphibian. (*a*) The frog heart has two atria but only one ventricle, which pumps blood both to the lungs and to the body. (*b*) Despite the potential for mixing, the oxygenated and deoxygenated bloods (*red* and *blue*, respectively) mix very little as they are pumped to the body and lungs. The slight mixing is shown in *purple.* RA = right atrium; LA = left atrium; V = ventricle.

Amphibian and Reptile Circulation

The advent of lungs involved a major change in the pattern of circulation. After blood is pumped by the heart through the *pulmonary arteries* to the lungs, it does not go directly to the tissues of the body but is instead returned via the *pulmonary veins* to the heart. This results in two circulations: one between heart and lungs, called the **pulmonary circulation,** and one between the heart and the rest of the body, called the **systemic circulation.**

If no changes had occurred in the structure of the heart, the oxygenated blood from the lungs would be mixed in the heart with the deoxygenated blood returning from the rest of the body. Consequently, the heart would pump a mixture of oxygenated and deoxygenated blood rather than fully oxygenated blood. The amphibian heart has two structural features that help reduce this mixing (figure 52.13). First, the atrium is divided into two chambers: the right atrium receives deoxygenated blood from the systemic circulation, and the left atrium receives oxygenated blood from the lungs. These two stores of blood therefore do not mix in the atria, and little mixing occurs when the contents of each atrium enter the single, common ventricle, due to internal channels created by recesses in the ventricular wall. The conus arteriosus is partially separated by a dividing wall which directs deoxygenated blood into the pulmonary arteries to the lungs and oxygenated blood into the *aorta*, the major artery of the systemic circulation to the body.

Because there is only one ventricle in an amphibian heart, the separation of the pulmonary and systemic circulations is incomplete. Amphibians in water, however, can obtain additional oxygen by diffusion through their skin. This process, called **cutaneous respiration,** helps to supplement the oxygenation of the blood in these vertebrates.

Among reptiles, additional modifications have reduced the mixing of blood in the heart still further. In addition to having two separate atria, reptiles have a septum that partially subdivides the ventricle. This results in an even greater separation of oxygenated and deoxygenated blood within the heart. The separation is complete in one order of reptiles, the crocodiles, which have two separate ventricles divided by a complete septum. Crocodiles therefore have a completely divided pulmonary and systemic circulation. Another change in the circulation of reptiles is that the conus arteriosus has become incorporated into the trunks of the large arteries leaving the heart.

Amphibians and reptiles have two circulations, pulmonary and systemic, that deliver blood to the lungs and rest of the body, respectively. The oxygenated blood from the lungs is kept relatively separate from the deoxygenated blood from the rest of the body by incomplete divisions within the heart.

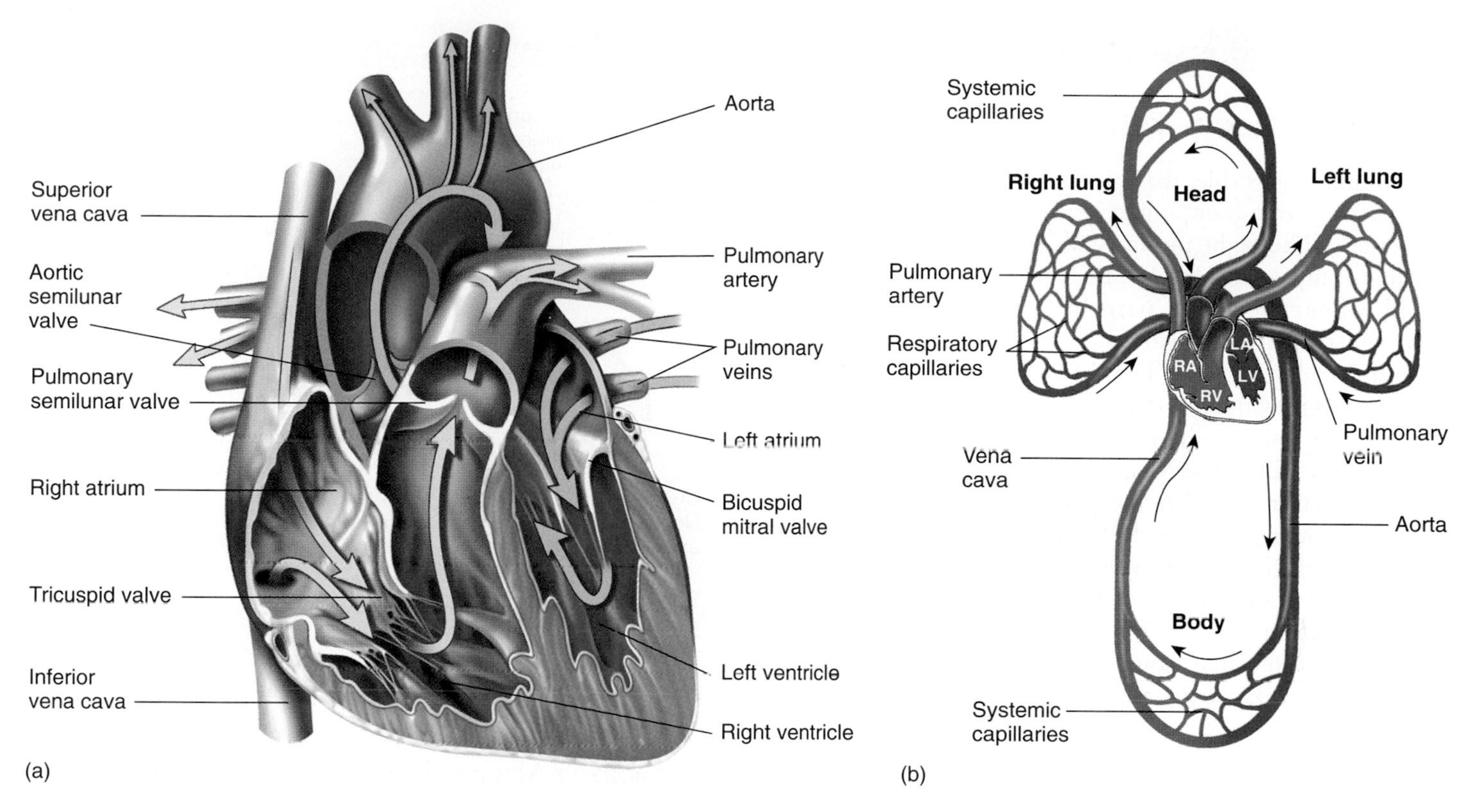

FIGURE 52.14
The heart and circulation of mammals and birds. (*a*) The path of blood through the four-chambered heart. (*b*) The right side of the heart receives deoxygenated blood and pumps it to the lungs; the left side of the heart receives oxygenated blood and pumps it to the body. In this way, the pulmonary and systemic circulations are kept completely separate. RA = right atrium; LA = left atrium; RV = right ventricle; LV = left ventricle.

Mammalian and Bird Hearts

Mammals, birds, and crocodiles have a four-chambered heart with two separate atria and two separate ventricles (figure 52.14). The right atrium receives deoxygenated blood from the body and delivers it to the right ventricle, which pumps the blood to the lungs. The left atrium receives oxygenated blood from the lungs and delivers it to the left ventricle, which pumps the oxygenated blood to the rest of the body. This completely double circulation is powered by a two-cycle pump. Both atria fill with blood and simultaneously contract, emptying their blood into the ventricles. Both ventricles contract at the same time, pushing blood simultaneously into the pulmonary and systemic circulations. The increased efficiency of the double circulatory system in mammals and birds is thought to have been important in the evolution of endothermy (warm-bloodedness), because a more efficient circulation is necessary to support the high metabolic rate required.

Because the overall circulatory system is closed, the same volume of blood must move through the pulmonary circulation as through the much larger systemic circulation with each heartbeat. Therefore, the right and left ventricles must pump the same amount of blood each time they contract. If the output of one ventricle did not match that of the other, fluid would accumulate and pressure would increase in one of the circuits. The result would be increased filtration out of the capillaries and edema (as occurs in congestive heart failure, for example). Although the volume of blood pumped by the two ventricles is the same, the pressure they generate is not. The left ventricle, which pumps blood through the higher-resistance systemic pathway, is more muscular and generates more pressure than does the right ventricle.

Throughout the evolutionary history of the vertebrate heart, the sinus venosus has served as a pacemaker, the site where the impulses that initiate the heartbeat originate. Although it constitutes a major chamber in the fish heart, it is reduced in size in amphibians and further reduced in reptiles. In mammals and birds, the sinus venosus is no longer evident as a separate chamber, but its disappearance is not really complete. Some of its tissue remains in the wall of the right atrium, near the point where the systemic veins empty into the atrium. This tissue, which is called the *sinoatrial (SA) node*, is still the site where each heartbeat originates.

The oxygenated blood from the lungs returns to the left atrium and is pumped out the left ventricle. The deoxygenated blood from the body returns to the right atrium and out the right ventricle to the lungs.

The Cardiac Cycle

The human heart, like that of all mammals and birds, is really two separate pumping systems operating within a single organ. The right pump sends blood to the lungs, and the left pump sends blood to the rest of the body.

The heart has two pairs of valves. One pair, the **atrioventricular (AV) valves,** guards the opening between the atria and ventricles. The AV valve on the right side is the **tricuspid valve,** and the AV valve on the left is the **bicuspid,** or **mitral, valve.** Another pair of valves, together called the **semilunar valves,** guard the exits from the ventricles to the arterial system; the **pulmonary valve** is located at the exit of the right ventricle, and the **aortic valve** is located at the exit of the left ventricle. These valves open and close as the heart goes through its **cardiac cycle** of rest *(diastole)* and contraction *(systole)*. The sound of these valves closing produces the "lub-dub" sounds heard with a stethoscope.

Blood returns to the resting heart through veins that empty into the right and left atria. As the atria fill and the pressure in them rises, the AV valves open to admit the blood into the ventricles. The ventricles become about 80% filled during this time. Contraction of the atria wrings out the final 20% of the 80 milliliters of blood the ventricles will receive, on average, in a resting person. These events occur while the ventricles are relaxing, a period called ventricular **diastole.**

After a slight delay, the ventricles contract; this period of contraction is known as ventricular **systole.** Contraction of each ventricle increases the pressure within each chamber, causing the AV valves to forcefully close (the "lub" sound), thereby preventing blood from backing up into the atria. Immediately after the AV valves close, the pressure in the ventricles forces the semilunar valves open so that blood can be pushed out into the arterial system. As the ventricles relax, closing of the semilunar valves prevents back flow (the "dub" sound).

The right and left **pulmonary arteries** deliver oxygen-depleted blood to the right and left lungs. As previously mentioned, these return blood to the left atrium of the heart via the **pulmonary veins.** The **aorta** and all its branches are systemic arteries, carrying oxygen-rich blood from the left ventricle to all parts of the body. The **coronary arteries** are the first branches off the aorta; these supply the heart muscle itself. Other systemic arteries branch from the aorta as it makes an arch above the heart, and as it descends and traverses the thoracic and abdominal cavities. These branches provide all body organs with oxygenated blood. The blood from the body organs, now lower in oxygen, returns to the heart in the systemic veins. These eventually empty into two major veins: the **superior vena cava,** which drains the upper body, and the **inferior vena cava,** which drains the lower body. These veins empty into the right atrium and thereby complete the systemic circulation.

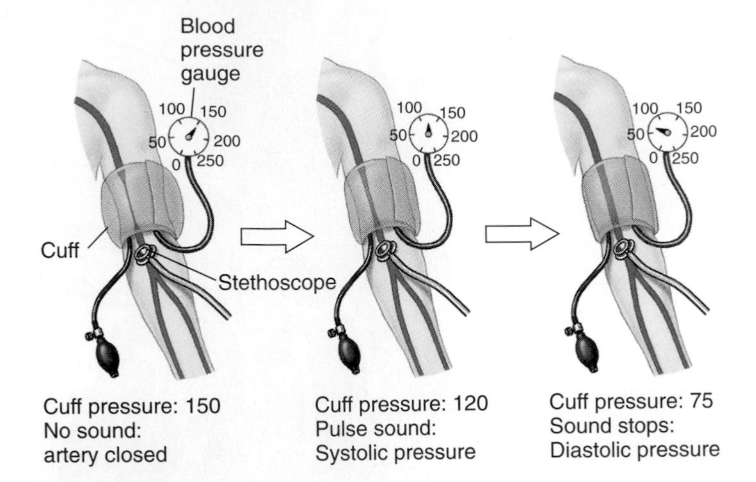

FIGURE 52.15
Measurement of blood pressure.

Measuring Arterial Blood Pressure

As the ventricles contract, great pressure is generated in the arteries throughout the body. You can tell this by feeling your pulse, either on the inside of your wrist, below the thumb or on the sides of your neck below your ear and jawbone. The contraction of the ventricles has to be strong enough to force blood through capillary beds but not too strong as to cause damage to smaller arteries and arterioles. Doctors measure your blood pressure to determine how hard your heart is working.

The measuring device used is called a sphygmomanometer and measures the blood pressure of the brachial artery found on the inside part of the arm, at the elbow (figure 52.15). A cuff wrapped around the upper part of the arm is tightened enough to stop the flow of blood to the lower part of the arm. As the cuff is loosened, blood will begin pulsating through the artery and can be detected using a stethoscope. Two measurements are recorded: the systolic and the diastolic pressure. The systolic pressure is the peak pressure during ventricular systole (contraction of the ventricle). The diastolic pressure is the minimum pressure between heartbeats (repolarization of the ventricles). The blood pressure is written as a ratio of systolic over diastolic pressure, and for a healthy person in his or her twenties, a typical blood pressure is 120/75 (measurement in mm of mercury). A condition called *hypertension* (high blood pressure) occurs when the ventricles experience very strong contractions, and the blood pressure is elevated, either systolic pressure greater than 150 or diastolic pressures greater than 90.

The cardiac cycle consists of systole and diastole; the ventricles contract at systole and relax at diastole.

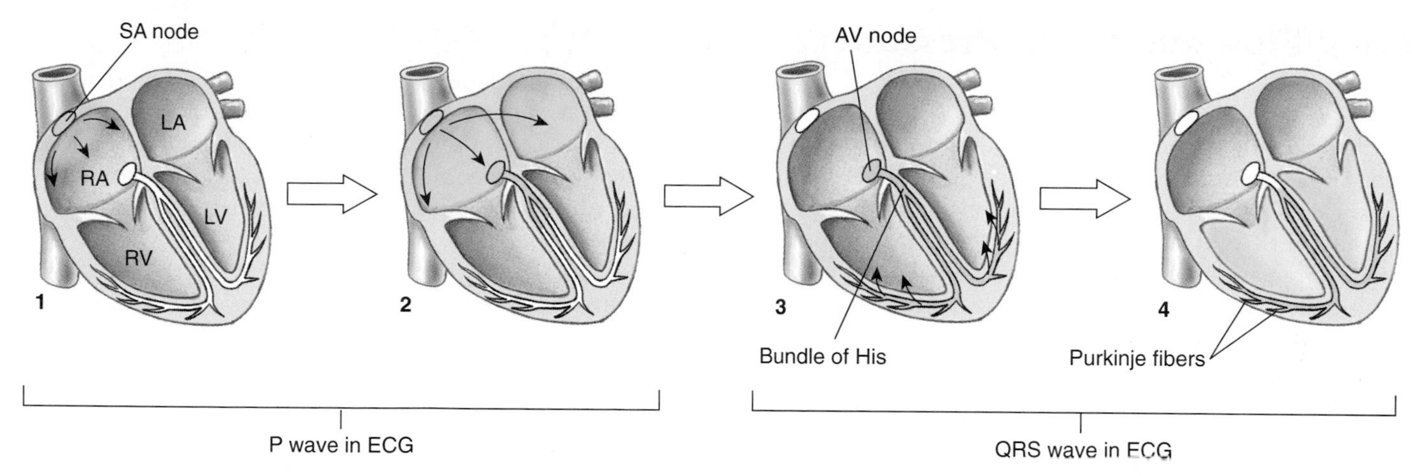

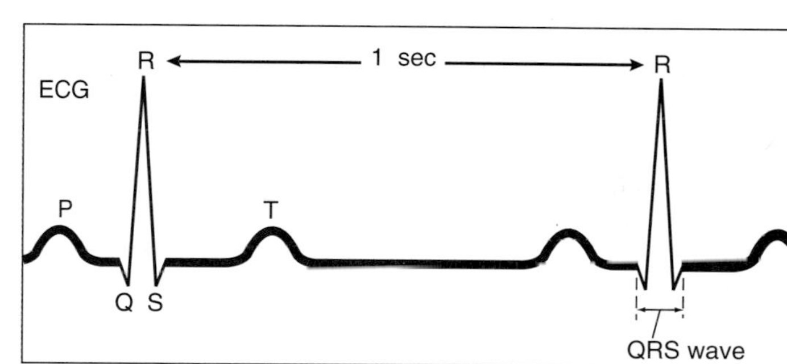

FIGURE 52.16
The path of electrical excitation in the heart. A wave of depolarization begins at the sinoatrial (SA) node. After passing over the atria and causing them to contract (forming the P wave on the ECG), the depolarization reaches the atrioventricular (AV) node, from which it passes to the ventricles along the septum by the bundle of His. Finer Purkinje fibers carry the depolarization into the right and left ventricular muscles (forming the QRS wave on the ECG). The T wave on the ECG corresponds to the repolarization of the ventricles.

Electrical Excitation and Contraction of the Heart

As in other types of muscle, contraction of heart muscle is stimulated by membrane **depolarization,** a reversal of the electrical polarity that normally exists across the plasma membrane (see chapter 54). In skeletal muscles, the nervous system initiates depolarization. However, in the heart, the depolarization is triggered by the **sinoatrial (SA) node** (figure 52.16), the small cluster of cardiac muscle cells derived from the sinus venosus. The SA node acts as a **pacemaker** for the rest of the heart by producing depolarization impulses spontaneously at a particular rate. Each depolarization initiated within this pacemaker region passes quickly from one cardiac muscle cell to another in a wave that envelops the right and left atria nearly simultaneously. The spread of depolarization is possible because the cardiac muscle cells are electrically coupled by gap junctions.

After a delay of almost 0.1 second, the wave of depolarization spreads to the ventricles. The reason for this delay is that connective tissue separates the atria from the ventricles, and connective tissue cannot transmit depolarization. The depolarization would not pass to the ventricles at all, were it not for a group of specialized cardiac muscle cells known as the **atrioventricular (AV) node.** The cells in the AV node transmit the depolarization slowly, causing the delay. This delay permits the atria to finish contracting and emptying their blood into the ventricles before the ventricles contract.

From the AV node, the wave of depolarization is conducted rapidly over both ventricles by a network of fibers called the **atrioventricular bundle** or **bundle of His.** It is then transmitted by **Purkinje fibers,** which directly stimulate the myocardial cells of the ventricles. The rapid conduction of the depolarization along the bundle of His and the Purkinje fibers causes the almost simultaneous contraction of the left and right ventricles. The rate can be increased or decreased by neural regulation or increased by the hormone epinephrine.

The spread of electrical activity through the heart creates currents that can be recorded from the surface of the body with electrodes placed on the limbs and chest. The recording, called an **electrocardiogram (ECG** or **EKG),** shows how the cells of the heart depolarize and repolarize during the cardiac cycle (see figure 52.16). A depolarization causes contraction of a muscle (including the heart), while repolarization causes relaxation. The first peak in the recording, P, is produced by the depolarization of the atria, and thus is associated with atrial systole. The second, larger peak, QRS, is produced by ventricular depolarization; during this time, the ventricles contract (ventricular systole) and eject blood into the arteries. The last peak, T, is produced by ventricular repolarization; at this time the ventricles begin diastole.

The SA node in the right atrium initiates waves of depolarization that stimulate first the atria and then the ventricles to contract.

Blood Flow and Blood Pressure

Cardiac Output

Cardiac output is the volume of blood pumped by each ventricle per minute. Because humans (like all vertebrates) have a closed circulation, the cardiac output is the same as the volume of blood that traverses the systemic or pulmonary circulations per minute. It is calculated by multiplying the heart rate by the *stroke volume*, which is the volume of blood ejected by each ventricle per beat. For example, if the heart rate is 72 beats per minute and the stroke volume is 70 milliliters, the cardiac output is 5 liters per minute, which is about average in a resting person.

Cardiac output increases during exercise because of an increase in heart rate and stroke volume. When exercise begins, the heart rate increases up to about 100 beats per minute. As exercise becomes more intense, skeletal muscles squeeze on veins more vigorously, returning blood to the heart more rapidly. In addition, the ventricles contract more strongly, so they empty more completely with each beat.

During exercise, the cardiac output increases to a maximum of about 25 liters per minute in an average young adult. Although the cardiac output has increased five times, not all organs receive five times the blood flow; some receive more, others less. This is because the arterioles in some organs, such as in the digestive system, constrict, while the arterioles in the exercising muscles and heart dilate. As previously mentioned, the resistance to flow decreases as the radius of the vessel increases. As a consequence, vasodilation greatly increases and vasoconstriction greatly decreases blood flow.

Blood Pressure and the Baroreceptor Reflex

The arterial blood pressure depends on two factors: how much blood the ventricles pump (the cardiac output) and how great a resistance to flow the blood encounters in the entire arterial system. An increased blood pressure, therefore, could be produced by an increased heart rate or an increased blood volume (because both increase the cardiac output) or by vasoconstriction, which increases the resistance to blood flow. Conversely, blood pressure will fall if the heart rate slows or if the blood volume is reduced, for example by dehydration or excessive bleeding (hemorrhage).

Changes in the arterial blood pressure are detected by **baroreceptors** located in the arch of the aorta and in the carotid arteries. These receptors activate sensory neurons that relay information to *cardiovascular control centers* in the medulla oblongata, a region of the brain stem. When the baroreceptors detect a fall in blood pressure, they stimulate neurons that go to blood vessels in the skin and viscera, causing arterioles in these organs to constrict and raise the blood pressure. This baroreceptor reflex therefore completes a negative feedback loop that acts to correct the fall in blood pressure and restore homeostasis.

Blood Volume Reflexes

Blood pressure depends in part on the total blood volume. A decrease in blood volume, therefore, will decrease blood pressure, if all else remains equal. Blood volume regulation involves the effects of four hormones: (1) antidiuretic hormone; (2) aldosterone; (3) atrial natriuretic hormone; and (4) nitric oxide.

Antidiuretic Hormone. *Antidiuretic hormone* (ADH), also called *vasopressin*, is secreted by the posterior pituitary gland in response to an increase in the osmotic concentration of the blood plasma. Dehydration, for example, causes the blood volume to decrease while the remaining plasma becomes more concentrated. This stimulates *osmoreceptors* in the hypothalamus of the brain, a region located immediately above the pituitary. The osmoreceptors promote thirst and stimulate ADH secretion from the posterior pituitary gland. ADH, in turn, stimulates the kidneys to retain more water in the blood, excreting less in the urine (urine is derived from blood plasma—see chapter 58). A dehydrated person thus drinks more and urinates less, helping to raise the blood volume and restore homeostasis.

Aldosterone. If a person's blood volume is lowered (by dehydration, for example), the flow of blood through the organs will be reduced if no compensation occurs. Whenever the kidneys experience a decreased blood flow, a group of kidney cells initiate the release of a short polypeptide known as *angiotensin II*. This is a very powerful molecule: it stimulates vasoconstriction throughout the body while it also stimulates the adrenal cortex (the outer region of the adrenal glands) to secrete the hormone *aldosterone*. This important steroid hormone is necessary for life; it acts on the kidneys to promote the retention of Na^+ and water in the blood. An animal that lacks aldosterone will die if untreated, because so much of the blood volume is lost in urine that the blood pressure falls too low to sustain life.

Atrial Natriuretic Hormone. When the body needs to eliminate excessive Na^+, less aldosterone is secreted by the adrenals, so that less Na^+ is retained by the kidneys. In recent years, scientists have learned that Na^+ excretion in the urine is promoted by another hormone. Surprisingly, this hormone is secreted by the right atrium of the heart—the heart is an endocrine gland! The right atrium secretes *atrial natriuretic hormone* in response to stretching of the atrium by an increased blood volume. The action of atrial natriuretic hormone completes a negative feedback loop, because it promotes the elimination of Na^+ and water, which will lower the blood volume and pressure.

Nitric Oxide. *Nitric oxide* (NO) is a gas that acts as a hormone in vertebrates, regulating blood pressure and blood flow. As described in chapter 7, nitric oxide gas is a paracrine hormone, produced by one cell, penetrates through membranes, and alters the activities of other

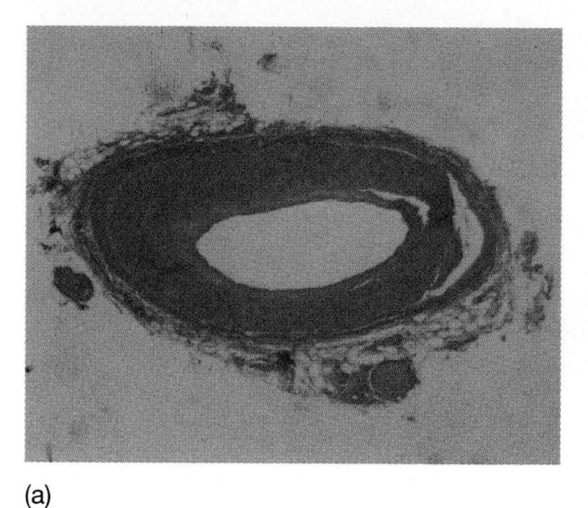
(a)

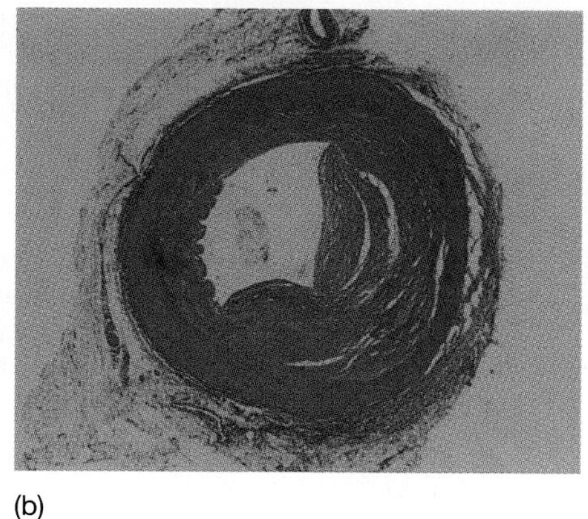
(b)

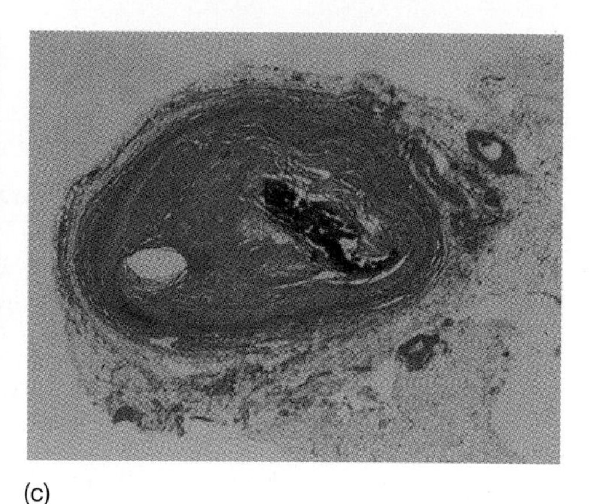
(c)

FIGURE 52.17
Atherosclerosis. (*a*) The coronary artery shows only minor blockage. (*b*) The artery exhibits severe atherosclerosis—much of the passage is blocked by build-up on the interior walls of the artery. (*c*) The coronary artery is essentially completely blocked.

neighboring cells. In 1998 the Nobel Prize for Medicine was awarded for the discovery of this signal transmission activity. How does NO regulate blood pressure? Nitric oxide gas produced by the surface endothelial cells of blood vessels passes inward through the cell layers of the vessel, causing the smooth muscles that encase it to relax and the blood vessel to dilate (become wider). For over a century, heart patients have been prescribed nitroglycerin to relieve chest pain, but only now has it become clear that nitroglycerin acts by releasing nitric oxide gas.

Cardiovascular Diseases

Cardiovascular diseases are the leading cause of death in the United States; more than 42 million people have some form of cardiovascular disease. Heart attacks are the main cause of cardiovascular deaths in the United States, accounting for about a fifth of all deaths. They result from an insufficient supply of blood reaching one or more parts of the heart muscle, which causes myocardial cells in those parts to die. Heart attacks may be caused by a blood clot forming somewhere in the coronary arteries (the arteries that supply the heart muscle with blood) and blocking the passage of blood through those vessels. They may also result if an artery is blocked by atherosclerosis. Recovery from a heart attack is possible if the portion of the heart that was damaged is small enough that the other blood vessels in the heart can enlarge their capacity and resupply the damaged tissues. **Angina pectoris,** which literally means "chest pain," occurs for reasons similar to those that cause heart attacks, but it is not as severe. The pain may occur in the heart and often also in the left arm and shoulder. Angina pectoris is a warning sign that the blood supply to the heart is inadequate but still sufficient to avoid myocardial cell death.

Strokes are caused by an interference with the blood supply to the brain. They may occur when a blood vessel bursts in the brain, or when blood flow in a cerebral artery is blocked by a thrombus (blood clot) or by atherosclerosis. The effects of a stroke depend on how severe the damage is and where in the brain the stroke occurs.

Atherosclerosis is an accumulation within the arteries of fatty materials, abnormal amounts of smooth muscle, deposits of cholesterol or fibrin, or various kinds of cellular debris. These accumulations cause blood flow to be reduced (figure 52.17). The lumen (interior) of the artery may be further reduced in size by a clot that forms as a result of the atherosclerosis. In the severest cases, the artery may be blocked completely. Atherosclerosis is promoted by genetic factors, smoking, hypertension (high blood pressure), and high blood cholesterol levels. Diets low in cholesterol and saturated fats (from which cholesterol can be made) can help lower the level of blood cholesterol, and therapy for hypertension can reduce that risk factor. Stopping smoking, however, is the single most effective action a smoker can take to reduce the risk of atherosclerosis.

Arteriosclerosis, or hardening of the arteries, occurs when calcium is deposited in arterial walls. It tends to occur when atherosclerosis is severe. Not only do such arteries have restricted blood flow, but they also lack the ability to expand as normal arteries do to accommodate the volume of blood pumped out by the heart. This inflexibility forces the heart to work harder.

Cardiac output depends on the rate of the heart and how much blood is ejected per beat. Blood flow is regulated by the degree of constriction of the arteries, which affects the resistance to flow. Blood pressure is influenced by blood volume. The volume of water retained in the vascular system is regulated by hormones that act on the kidneys and blood vessels. Many cardiovascular diseases are associated with the accumulation of fatty materials on the inner surfaces of arteries.

Chapter 52

www.mhhe.com/raven6/resources52.mhtml

Summary	Questions	Media Resources

52.1 The circulatory systems of animals may be open or closed.

- Vertebrates have a closed circulation, where the blood stays within vessels as it travels away from and back to the heart.
- The circulatory system serves a variety of functions, including transport, regulation, and protection.

1. What is the difference between a closed circulatory system and an open circulatory system? In what types of animals would you find each?

- Bioethics Case Study: Heart Transplant

52.2 A network of vessels transports blood through the body.

- Plasma is the liquid portion of the blood. A variety of plasma proteins, ions, metabolites, wastes, and hormones are dissolved in the plasma.
- Erythrocytes, or red blood cells, contain hemoglobin and function to transport oxygen; the leukocytes, or white blood cells, function in immunological defenses.
- The heart pumps blood into arteries, which branch into smaller arterioles.
- Blood from the arterial system empties into capillaries with thin walls; all exchanges between the blood and tissues pass across the walls of capillaries.
- Blood returns to the heart in veins, which have one-way valves to ensure that blood travels toward the heart only.
- Lymphatic vessels return interstitial fluid to the venous system.

2. What are the major components of blood plasma?

3. Describe the structure of arteries and veins, explaining their similarities and differences. Why do arteries differ in structure from veins?

4. What is the relationship between vessel diameter and the resistance to blood flow? How do the arterial trees adjust their resistance to flow?

5. What drives the flow of fluid within the lymphatic system, and in what direction does the fluid flow?

- Art Activities:
  - -Blood Vessels
  - -Capillary Bed Anatomy
  - -Lymphatic System
  - -Lymphoid Organs

- Lymphatic System

- Erythrocytes
- Blood
- Lymph System

- Bioethics Case Study: Artifical Blood

52.3 The vertebrate heart has undergone progressive evolutionary change.

- The fish heart consists of four chambers in a row; the beat originates in the sinus venosus and spreads through the atrium, ventricle, and conus arteriosus.
- In the circulation of fishes, blood from the heart goes to the gills and then to the rest of the body before returning to the heart; in terrestrial vertebrates, blood returns from the lungs to the heart before it is pumped to the body.

6. Describe the pattern of circulation through a fish and an amphibian, and compare the structure of their hearts. What new circulatory pattern accompanies the evolution of lungs?

- Exploration: Evolution of the Heart

- Art Quiz: Mammal and Bird Circulation

52.4 The cardiac cycle drives the cardiovascular system.

- Electrical excitation of the heart is initiated by the SA (sinoatrial) node, spreads through gap junctions between myocardial cells in the atria, and then is conducted into the ventricles by specialized conducting tissue.
- The cardiac output is regulated by nerves that influence the cardiac rate and by factors that influence the stroke volume.

7. How does the baroreceptor reflex help to maintain blood pressure? How do ADH and aldosterone maintain blood volume and pressure? What causes their secretion?

- Art Activities:
  - -Human Circulatory System
  - -Plaque

- Cardiac Cycle Blood Flow

53

Respiration

Concept Outline

53.1 Respiration involves the diffusion of gases.

Fick's Law of Diffusion. The rate of diffusion across a membrane depends on the surface area of the membrane, the concentration gradients, and the distance across the membrane.

How Animals Maximize the Rate of Diffusion. The diffusion rate increases when surface area or concentration gradient increases.

53.2 Gills are used for respiration by aquatic vertebrates.

The Gill as a Respiratory Structure. Water is forced past the gill surface, and blood flows through the gills.

53.3 Lungs are used for respiration by terrestrial vertebrates.

Respiration in Air-Breathing Animals. In insects, oxygen diffuses directly from the air into body cells; in vertebrates, oxygen diffuses into blood and then into body cells.

Respiration in Amphibians and Reptiles. Amphibians force air into their lungs, whereas reptiles, birds, and mammals draw air in by expanding their rib cage.

Respiration in Mammals. In mammals, gas exchange occurs across millions of tiny air sacs called alveoli.

Respiration in Birds. In birds, air flows through the lung unidirectionally.

53.4 Mammalian breathing is a dynamic process.

Structures and Mechanisms of Breathing. The rib cage and lung volumes are expanded during inspiration by the contraction of the diaphragm and other muscles.

Mechanisms That Regulate Breathing. The respiratory control center in the brain is influenced by reflexes triggered by the blood levels of carbon dioxide and blood pH.

53.5 Blood transports oxygen and carbon dioxide.

Hemoglobin and Oxygen Transport. Hemoglobin, a molecule within the red blood cells, loads with oxygen in the lungs and unloads its oxygen in the tissue capillaries.

Carbon Dioxide and Nitric Oxide Transport. Carbon dioxide is converted into carbonic acid in erythrocytes and is transported as bicarbonate.

FIGURE 53.1
Elephant seals are respiratory champions. Diving to depths greater than those of all other marine animals, including sperm whales and sea turtles, elephant seals can hold their breath for over two hours, descend and ascend rapidly in the water, and endure repeated dives without suffering any apparent respiratory distress.

Animals pry energy out of food molecules using the biochemical process called cellular respiration. While the term cellular respiration pertains to the use of oxygen and production of carbon dioxide at the cellular level, the general term respiration describes the uptake of oxygen from the environment and the disposal of carbon dioxide into the environment at the body system level. Respiration at the body system level involves a host of processes not found at the cellular level, like the mechanics of breathing and the exchange of oxygen and carbon dioxide in the capillaries. These processes, one of the principal physiological challenges facing all animals (figure 53.1), are the subject of this chapter.

53.1 Respiration involves the diffusion of gases.

Fick's Law of Diffusion

Respiration involves the diffusion of gases across plasma membranes. Because plasma membranes must be surrounded by water to be stable, the external environment in gas exchange is always aqueous. This is true even in terrestrial animals; in these cases, oxygen from air dissolves in a thin layer of fluid that covers the respiratory surfaces, such as the alveoli in lungs.

In vertebrates, the gases diffuse into the aqueous layer covering the epithelial cells that line the respiratory organs. The diffusion process is passive, driven only by the difference in O_2 and CO_2 concentrations on the two sides of the membranes. In general, the rate of diffusion between two regions is governed by a relationship known as **Fick's Law of Diffusion:**

$$R = D \times A \frac{\Delta p}{d}$$

In this equation,

R = the rate of diffusion; the amount of oxygen or carbon dioxide diffusing per unit of time;

D = the diffusion constant;

A = the area over which diffusion takes place;

Δp = the difference in concentration (for gases, the difference in their partial pressures) between the interior of the organism and the external environment; and

d = the distance across which diffusion takes place.

Major changes in the mechanism of respiration have occurred during the evolution of animals (figure 53.2) that have tended to optimize the rate of diffusion R. By inspecting Fick's Law, you can see that natural selection can optimize R by favoring changes that (1) increase the surface area A; (2) decrease the distance d; or (3) increase the concentration difference, as indicated by Δp. The evolution of respiratory systems has involved changes in all of these factors.

Fick's Law of Diffusion states that the rate of diffusion across a membrane depends on surface area, concentration (partial pressure) difference, and distance.

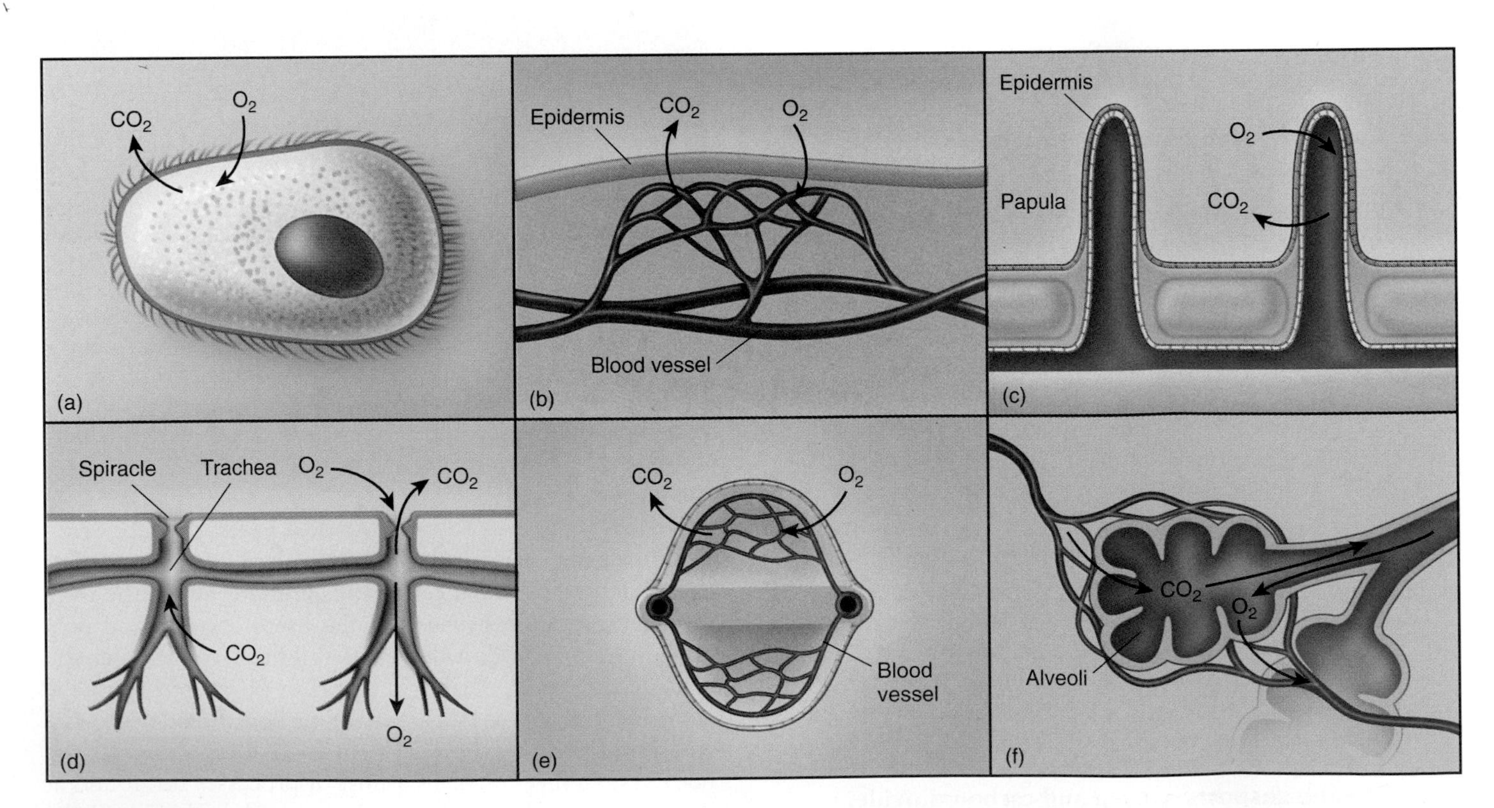

FIGURE 53.2
Gas exchange may take place in a variety of ways. (*a*) Gases diffuse directly into single-celled organisms. (*b*) Amphibians and many other animals respire across their skin. (*c*) Echinoderms have protruding papulae, which provide an increased respiratory surface area. (*d*) Inspects respire through an extensive tracheal system. (*e*) The gills of fishes provide a very large respiratory surface area and countercurrent exchange. (*f*) The alveoli in mammalian lungs provide a large respiratory surface area but do not permit countercurrent exchange.

How Animals Maximize the Rate of Diffusion

The levels of oxygen required by oxidative metabolism cannot be obtained by diffusion alone over distances greater than about 0.5 millimeter. This restriction severely limits the size of organisms that obtain their oxygen entirely by diffusion directly from the environment. Protists are small enough that such diffusion can be adequate (see figure 53.2*a*), but most multicellular animals are much too large.

Most of the more primitive phyla of invertebrates lack special respiratory organs, but they have developed means of improving the movement of water over respiratory structures. In a number of different ways, many of which involve beating cilia, these organisms create a *water current* that continuously replaces the water over the respiratory surfaces. Because of this continuous replenishment with water containing fresh oxygen, *the external oxygen concentration does not decrease along the diffusion pathway.* Although each oxygen molecule that passes into the organism has been removed from the surrounding water, new water continuously replaces the oxygen-depleted water. This increases the rate of diffusion by maximizing the concentration difference—the Δp of the Fick equation.

All of the more advanced invertebrates (mollusks, arthropods, echinoderms), as well as vertebrates, possess respiratory organs that increase the surface area available for diffusion and bring the external environment (either water or air) close to the internal fluid, which is usually circulated throughout the body. The respiratory organs thus increase the rate of diffusion by maximizing surface area and decreasing the distance the diffusing gases must travel (the A and d factors, respectively, in the Fick equation).

Atmospheric Pressure and Partial Pressures

Dry air contains 78.09% nitrogen (N_2), 20.95% oxygen, 0.93% argon and other inert gases, and 0.03% carbon dioxide. Convection currents cause air to maintain a constant composition to altitudes of at least 100 kilometers, although the *amount* (number of molecules) of air that is present decreases with altitude (figure 53.3).

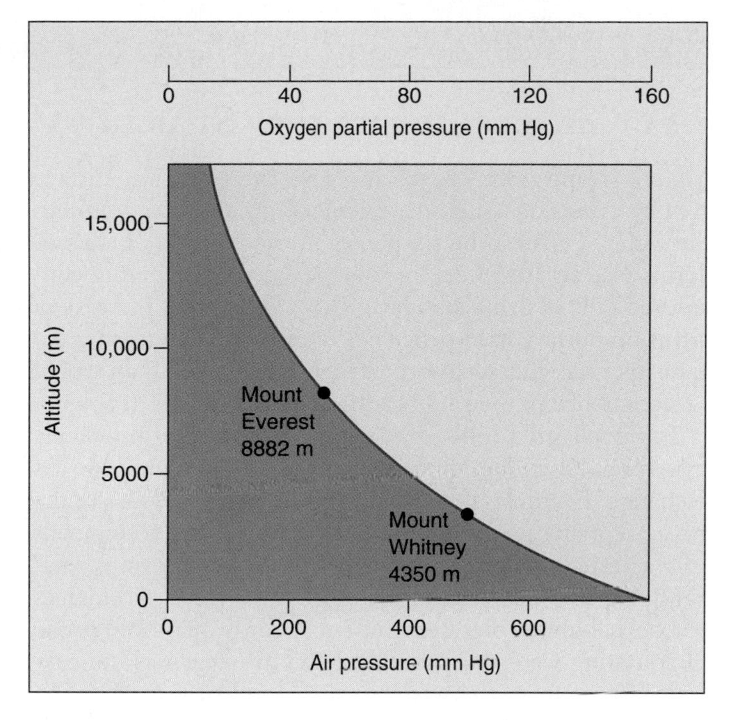

FIGURE 53.3
The relationship between air pressure and altitude above sea level. At the high altitudes characteristic of mountaintops, air pressure is much less than at sea level. At the top of Mount Everest, the world's highest mountain, the air pressure is only one-third that at sea level.

Imagine a column of air extending from the ground to the limits of the atmosphere. All of the gas molecules in this column experience the force of gravity, so they have weight and can exert pressure. If this column were on top of one end of a U-shaped tube of mercury at sea level, it would exert enough pressure to raise the other end of the tube 760 millimeters under a set of specified, standard conditions (see figure 53.3). An apparatus that measures air pressure is called a barometer, and 760 mm Hg (millimeters of mercury) is the barometric pressure of the air at sea level. A pressure of 760 mm Hg is also defined as **one atmosphere** of pressure.

Each type of gas contributes to the total atmospheric pressure according to its fraction of the total molecules present. That fraction contributed by a gas is called its **partial pressure** and is indicated by P_{N_2}, P_{O_2}, P_{CO_2}, and so on. The total pressure is the sum of the partial pressures of all gases present. For dry air, the partial pressures are calculated simply by multiplying the fractional composition of each gas in the air by the atmospheric pressure. Thus, at sea level, the partial pressures of N_2+ inert gases, O_2, and CO_2 are:

$$P_{N_2} = 760 \times 79.02\% = 600.6\text{mm Hg},$$
$$P_{O_2} = 760 \times 20.95\% = 159.2\text{mm Hg, and}$$
$$P_{CO_2} = 760 \times 0.03\% = 0.2\text{mm Hg}.$$

Humans do not survive long at altitudes above 6000 meters. Although the air at these altitudes still contains 20.95% oxygen, the atmospheric pressure is only about 380 mm Hg, so its P_{O_2} is only 80 mm Hg (380 × 20.95%), only half the amount of oxygen available at sea level.

The exchange of oxygen and carbon dioxide between an organism and its environment occurs by diffusion of dissolved gases across plasma membranes and is maximized by increasing the concentration gradient and the surface area and by decreasing the distance that the diffusing gases must travel.

53.2 Gills are used for respiration by aquatic vertebrates.

The Gill as a Respiratory Structure

Aquatic respiratory organs increase the diffusion surface area by extensions of tissue, called *gills*, that project out into the water. Gills can be simple, as in the papulae of echinoderms (see figure 53.2*c*), or complex, as in the highly convoluted gills of fish (see figure 53.2*e*). The great increase in diffusion surface area provided by gills enables aquatic organisms to extract far more oxygen from water than would be possible from their body surface alone.

External gills (gills that are not enclosed within body structures) provide a greatly increased surface area for gas exchange. Examples of vertebrates with external gills are the larvae of many fish and amphibians, as well as developmentally arrested *(neotenic)* amphibian larvae that remain permanently aquatic, such as the axolotl. One of the disadvantages of external gills is that they must constantly be moved or the surrounding water becomes depleted in oxygen as the oxygen diffuses from the water to the blood of the gills. The highly branched gills, however, offer significant resistance to movement, making this form of respiration ineffective except in smaller animals. Another disadvantage is that external gills are easily damaged. The thin epithelium required for gas exchange is not consistent with a protective external layer of skin.

Other types of aquatic animals evolved specialized *branchial chambers*, which provide a means of pumping water past stationary gills. Mollusks, for example, have an internal *mantle cavity* that opens to the outside and contains the gills. Contraction of the muscular walls of the mantle cavity draws water in and then expels it. In crustaceans, the branchial chamber lies between the bulk of the body and the hard exoskeleton of the animal. This chamber contains gills and opens to the surface beneath a limb. Movement of the limb draws water through the branchial chamber, thus creating currents over the gills.

The Gills of Bony Fishes

The gills of bony fishes are located between the *buccal* (mouth) *cavity* and the *opercular cavities* (figure 53.4). The buccal cavity can be opened and closed by opening and closing the mouth, and the opercular cavity can be opened and closed by movements of the **operculum,** or gill cover. The two sets of cavities function as pumps that expand alternately to move water into the mouth, through the gills, and out of the fish through the open operculum. Water is brought into the buccal cavity by lowering the jaw and floor of the mouth, and then is moved through the gills into the opercular cavity by the opening of the operculum. The lower pressure in the opercular cavity causes water to move in the correct direction across the gills, and tissue that acts as valves ensures that the movement is one-way.

Some fishes that swim continuously, such as tuna, have practically immobile opercula. These fishes swim with their mouths partly open, constantly forcing water over the gills in a form of **ram ventilation.** Most bony fishes, however, have flexible gill covers that permit a pumping action. For example, the remora, a fish that rides "piggyback" on sharks, uses ram ventilation while the shark swims and the pumping action of its opercula when the shark stops swimming.

There are four **gill arches** on each side of the fish head. Each gill arch is composed of two rows of *gill filaments*, and each gill filament contains thin membranous plates, or *lamellae*, that project out into the flow of water (figure 53.5). Water flows past the lamellae in one direction only. Within each lamella, blood flows in a direction that is *opposite* the direction of water movement. This arrangement is called **countercurrent flow,** and it acts to maximize the oxygenation of the blood by increasing the concentration gradient of oxygen along the pathway for diffusion, increasing Δp in Fick's Law of Diffusion.

The advantages of a countercurrent flow system were discussed in chapter 52 in relation to temperature regulation and are again shown here in figure 53.6*a*. Blood low in oxygen

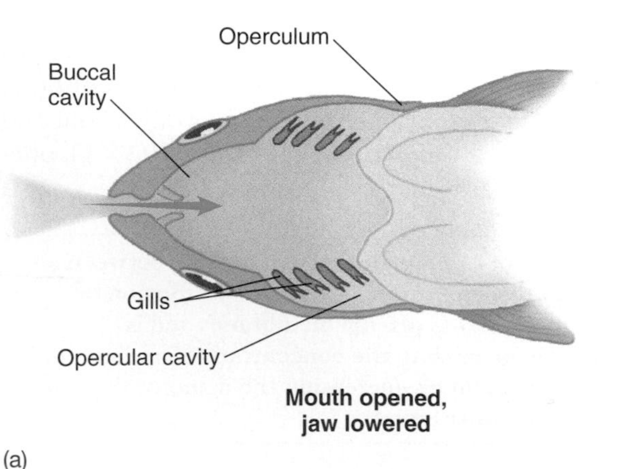

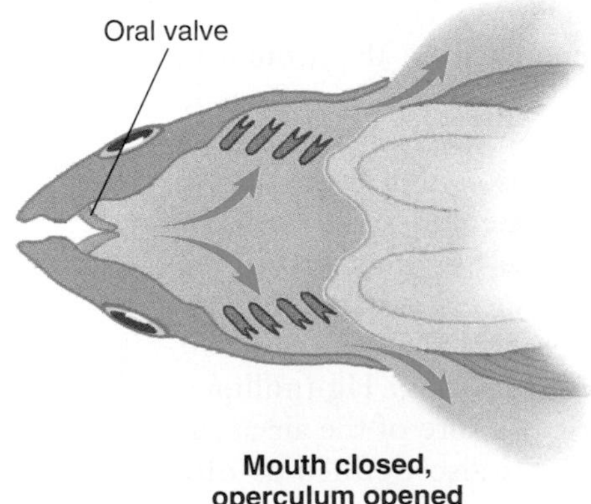

FIGURE 53.4
How most bony fishes respire. The gills are suspended between the buccal (mouth) cavity and the opercular cavity. Respiration occurs in two stages. (*a*) The oral valve in the mouth is opened and the jaw is depressed, drawing water into the buccal cavity while the opercular cavity is closed. (*b*) The oral valve is closed and the operculum is opened, drawing water through the gills to the outside.

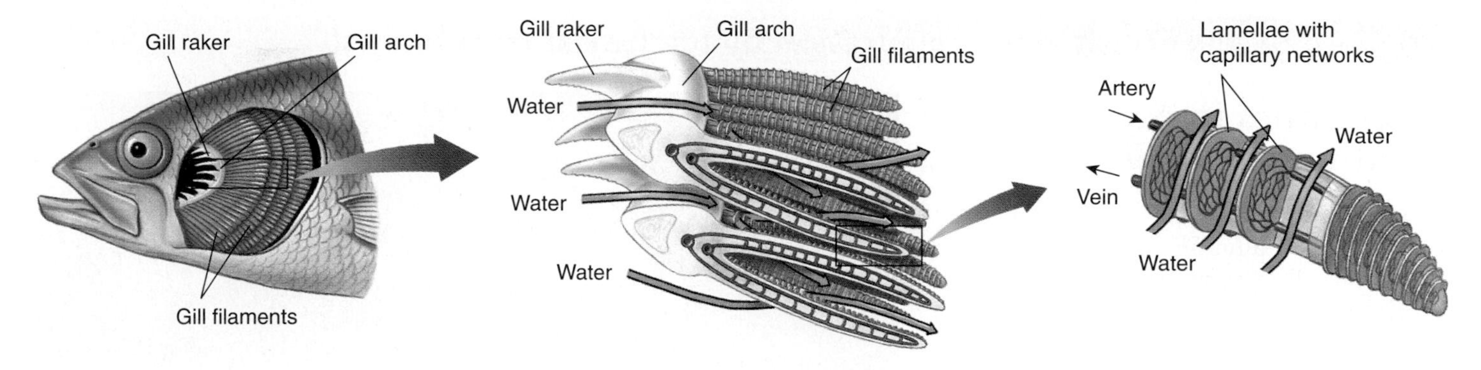

FIGURE 53.5
Structure of a fish gill. Water passes from the gill arch over the filaments (from left to right in the diagram). Water always passes the lamellae in a direction that is opposite to the direction of blood flow through the lamellae. The success of the gill's operation critically depends on this countercurrent flow of water and blood.

enters the back of the lamella, where it comes in close proximity to water that has already had most of its oxygen removed as it flowed through the lamella in the opposite direction. The water still has a higher oxygen concentration than the blood at this point, however, so oxygen diffuses from the water to the blood. As the blood flows toward the front of the lamella, it runs next to water that has a still higher oxygen content, so oxygen continuously diffuses from the water to the blood. Thus, countercurrent flow ensures that a concentration gradient remains between blood and water throughout the flow. This permits oxygen to continue to diffuse all along the lamellae, so that the blood leaving the gills has nearly as high an oxygen concentration as the water entering the gills.

This concept is easier to understand if we look at what would happen if blood and water flowed in the same direction, that is, had a *concurrent flow*. The difference in oxygen concentration would be very high at the front of each lamella, where oxygen-depleted blood would meet oxygen-rich water entering the gill (figure 53.6*b*). The concentration difference would fall rapidly, however, as the water lost oxygen to the blood. Net diffusion of oxygen would cease when the oxygen concentration of blood matched that of the water. At this point, much less oxygen would have been transferred to the blood than is the case with countercurrent flow. The flow of blood and water in a fish gill is in fact countercurrent, and because of the countercurrent exchange of gases, fish gills are the most efficient of all respiratory organs.

In bony fishes, water is forced past gills by the pumping action of the buccal and opercular cavities, or by active swimming in ram ventilation. In the gills, blood flows in an opposite direction to the flow of water. This countercurrent flow maximizes gas exchange, making the fish's gill an efficient respiratory organ.

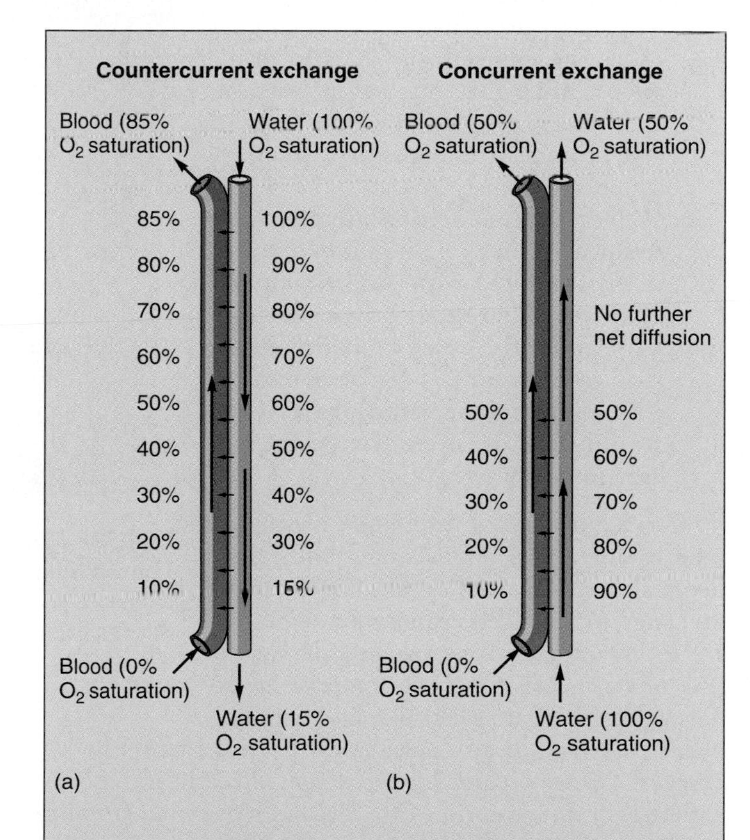

When blood and water flow in *opposite* directions (*a*), the initial oxygen concentration difference between water and blood is not large, but is sufficient for oxygen to diffuse from water to blood. As more oxygen diffuses into the blood, raising the blood's oxygen concentration, the blood encounters water with ever higher oxygen concentrations. At every point, the oxygen concentration is higher in the water, so that diffusion continues. In this example, blood attains an oxygen concentration of 85%. When blood and water flow in the *same* direction (*b*), oxygen can diffuse from the water into the blood rapidly at first, but the diffusion rate slows as more oxygen diffuses from the water into the blood, until finally the concentrations of oxygen in water and blood are equal. In this example, blood's oxygen concentration cannot exceed 50%.

FIGURE 53.6
Countercurrent exchange. This process allows for the most efficient blood oxygenation known in nature.

53.3 Lungs are used for respiration by terrestrial vertebrates.

Respiration in Air-Breathing Animals

Despite the high efficiency of gills as respiratory organs in aquatic environments, gills were replaced in terrestrial animals for two principal reasons:

1. **Air is less buoyant than water.** The fine membranous lamellae of gills lack structural strength and rely on water for their support. A fish out of water, although awash in oxygen (water contains only 5 to 10 mL O_2/L, compared with air with 210 mL O_2/L), soon suffocates because its gills collapse into a mass of tissue. This collapse greatly reduces the diffusion surface area of the gills. Unlike gills, internal air passages can remain open, because the body itself provides the necessary structural support.
2. **Water diffuses into air through evaporation.** Atmospheric air is rarely saturated with water vapor, except immediately after a rainstorm. Consequently, terrestrial organisms that are surrounded by air constantly lose water to the atmosphere. Gills would provide an enormous surface area for water loss.

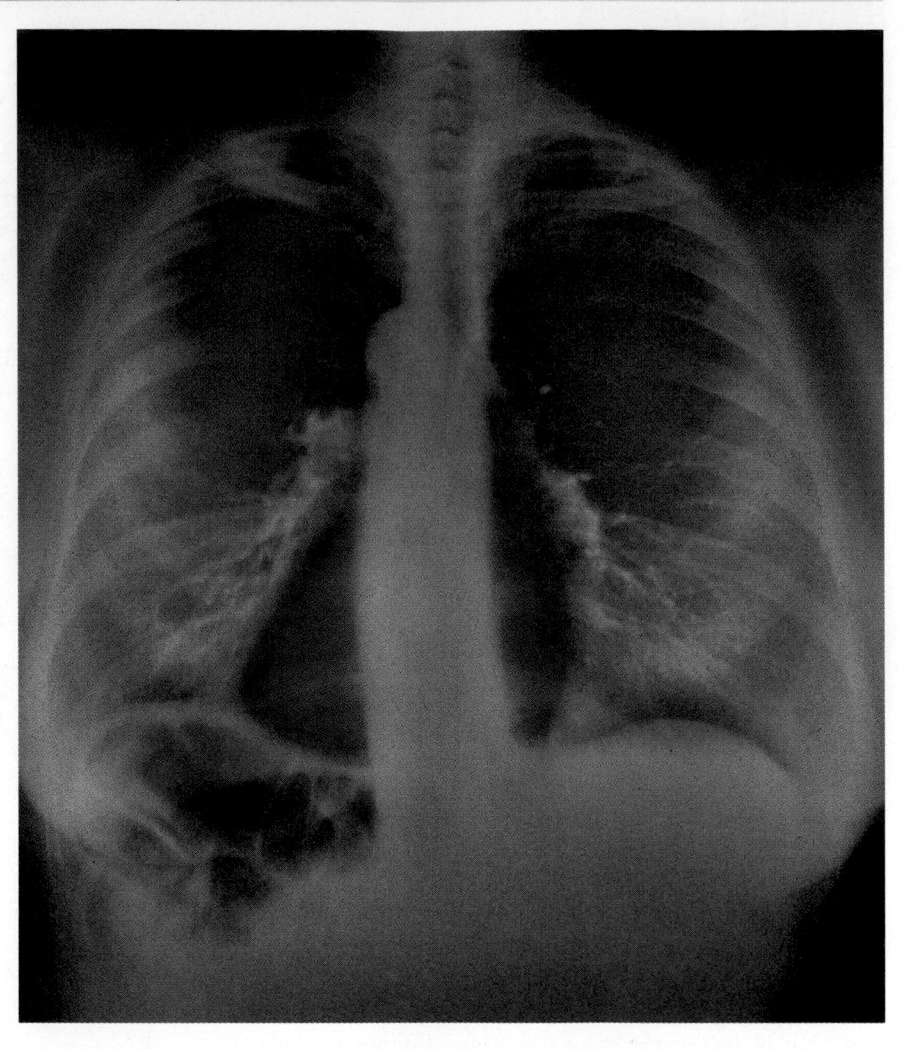

FIGURE 53.7
Human lungs. This chest X ray (dorsal view) was color-enhanced to show the lungs clearly. The heart is the pear-shaped object behind the vertical white column that is the esophagus.

Two main types of respiratory organs are used by terrestrial animals, and both sacrifice respiratory efficiency to some extent in exchange for reduced evaporation. The first are the **tracheae** of insects (see chapter 46 and figure 53.2*d*). Tracheae comprise an extensive series of air-filled passages connecting the surface of an insect to all portions of its body. Oxygen diffuses from these passages directly into cells, without the intervention of a circulatory system. Piping air directly from the external environment to the cells works very well in insects because their small bodies give them a high surface area-to-volume ratio. Insects prevent excessive water loss by closing the external openings of the tracheae whenever their internal CO_2 levels fall below a certain point.

The other main type of terrestrial respiratory organ is the **lung** (figure 53.7). A lung minimizes evaporation by moving air through a branched tubular passage; the air becomes saturated with water vapor before reaching the portion of the lung where a thin, wet membrane permits gas exchange. The lungs of all terrestrial vertebrates except birds use a **uniform pool** of air that is in contact with the gas exchange surface. Unlike the one-way flow of water that is so effective in the respiratory function of gills, air moves in and out by way of the same airway passages, a two-way flow system. Let us now examine the structure and function of lungs in the four classes of terrestrial vertebrates.

Air is piped directly to the body cells of insects, but the cells of terrestrial vertebrates obtain oxygen from the blood. The blood obtains its oxygen from a uniform pool of air by diffusion across the wet membranes of the lungs, which are filled with air.

Respiration in Amphibians and Reptiles

The lungs of amphibians are formed as saclike outpouchings of the gut (figure 53.8). Although the internal surface area of these sacs is increased by folds, much less surface area is available for gas exchange in amphibian lungs than in the lungs of other terrestrial vertebrates. Each amphibian lung is connected to the rear of the oral cavity, or pharynx, and the opening to each lung is controlled by a valve, the glottis.

Amphibians do not breathe the same way other terrestrial vertebrates do. Amphibians force air into their lungs by creating a greater-than-atmospheric pressure (positive pressure) in the air outside their lungs. They do this by filling their *buccal cavity* with air, closing their mouth and nostrils, and then elevating the floor of their oral cavity. This pushes air into their lungs in the same way that a pressurized tank of air is used to fill balloons. This is called positive pressure breathing; in humans, it would be analogous to forcing air into a victim's lungs by performing mouth-to-mouth resuscitation.

All other terrestrial vertebrates breathe by expanding their lungs and thereby creating a lower-than-atmospheric pressure (a negative pressure) within the lungs. This is called negative pressure breathing and is analogous to taking air into an accordion by pulling the accordion out to a greater volume. In reptiles, birds, and mammals, this is accomplished by expanding the thoracic (chest) cavity through muscular contractions, as will be described in a later section.

The oxygenation of amphibian blood by the lungs is supplemented by cutaneous respiration—the exchange of gases across the skin, which is wet and well vascularized in amphibians. Cutaneous respiration is actually more significant than pulmonary (lung) ventilation in frogs during winter, when their metabolisms are slow. Lung function becomes more important during the summer as the frog's metabolism increases. Although not common, some terrestrial amphibians, such as plethodontid salamanders, rely on cutaneous respiration exclusively.

Terrestrial reptiles have dry, tough, scaly skins that prevent desiccation, and so cannot have cutaneous respiration. Reptiles expand their rib cages by muscular contraction, and thereby take air into their lungs through negative pressure breathing. Their lungs have somewhat more surface area than the lungs of amphibians and so are more efficient at gas exchange. Cutaneous respiration, however, has been demonstrated in marine sea snakes.

Amphibians force air into their lungs by positive pressure breathing, whereas reptiles and all other terrestrial vertebrates take air into their lungs by expanding their lungs when they increase rib cage volume through muscular contractions. This creates a subatmospheric pressure in the lungs.

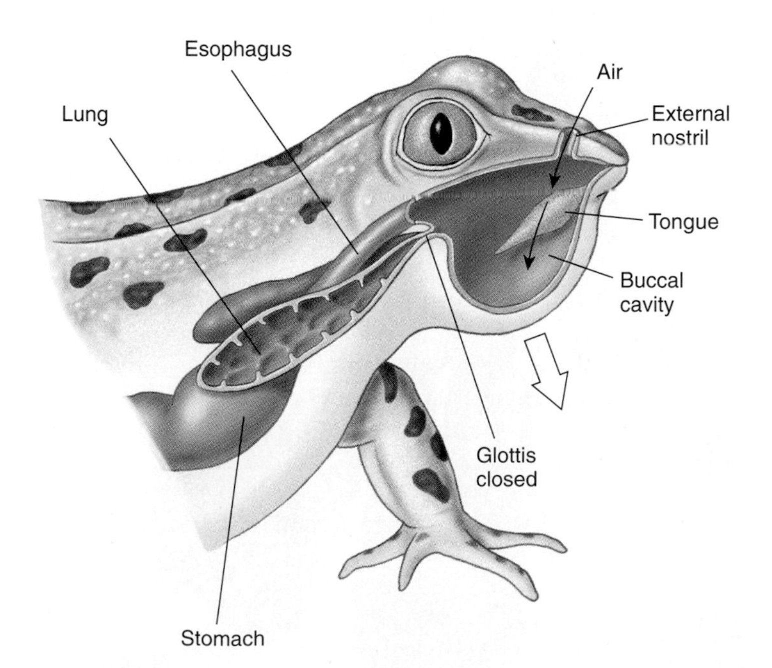

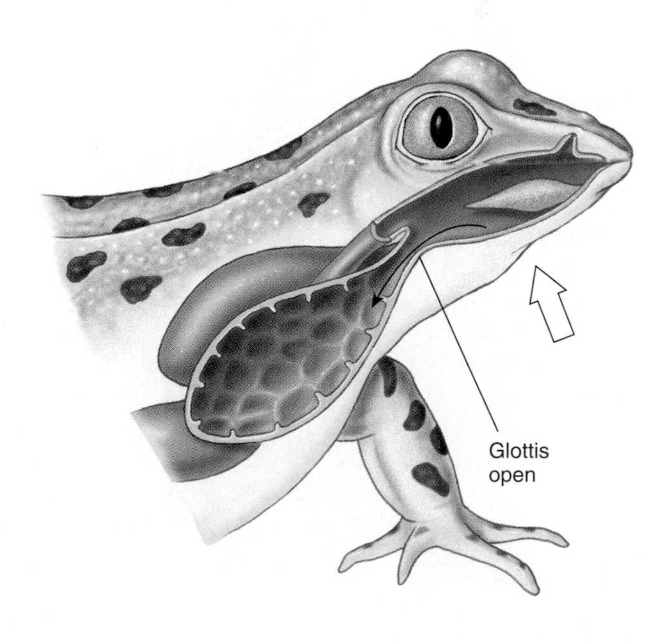

FIGURE 53.8
Amphibian lungs. Each lung of this frog is an outpouching of the gut and is filled with air by the creation of a positive pressure in the buccal cavity. The amphibian lung lacks the structures present in the lungs of other terrestrial vertebrates that provide an enormous surface area for gas exchange, and so are not as efficient as the lungs of other vertebrates.

Respiration in Mammals

The metabolic rate, and therefore the demand for oxygen, is much greater in birds and mammals, which are endothermic and thus require a more efficient respiratory system.

The lungs of mammals are packed with millions of *alveoli,* tiny sacs clustered like grapes (figure 53.9). This provides each lung with an enormous surface area for gas exchange. Air is brought to the alveoli through a system of air passages. Inhaled air is taken in through the mouth and nose past the pharynx to the **larynx** (voice box), where it passes through an opening in the vocal cords, the *glottis,* into a tube supported by C-shaped rings of cartilage, the **trachea** (windpipe). The trachea bifurcates into right and left **bronchi** (singular, *bronchus*), which enter each lung and further subdivide into **bronchioles** that deliver the air into blind-ended sacs called **alveoli.** The alveoli are surrounded by an extremely extensive capillary network. All gas exchange between the air and blood takes place across the walls of the alveoli.

The branching of bronchioles and the vast number of alveoli combine to increase the respiratory surface area (*A* in Fick's Law) far above that of amphibians or reptiles. In humans, there are about 300 million alveoli in each of the two lungs, and the total surface area available for diffusion can be as much as 80 square meters, or about 42 times the surface area of the body. Respiration in mammals will be considered in more detail in a separate section later.

Mammalian lungs are composed of millions of alveoli that provide a huge surface area for gas exchange. Air enters and leaves these alveoli through the same system of airways.

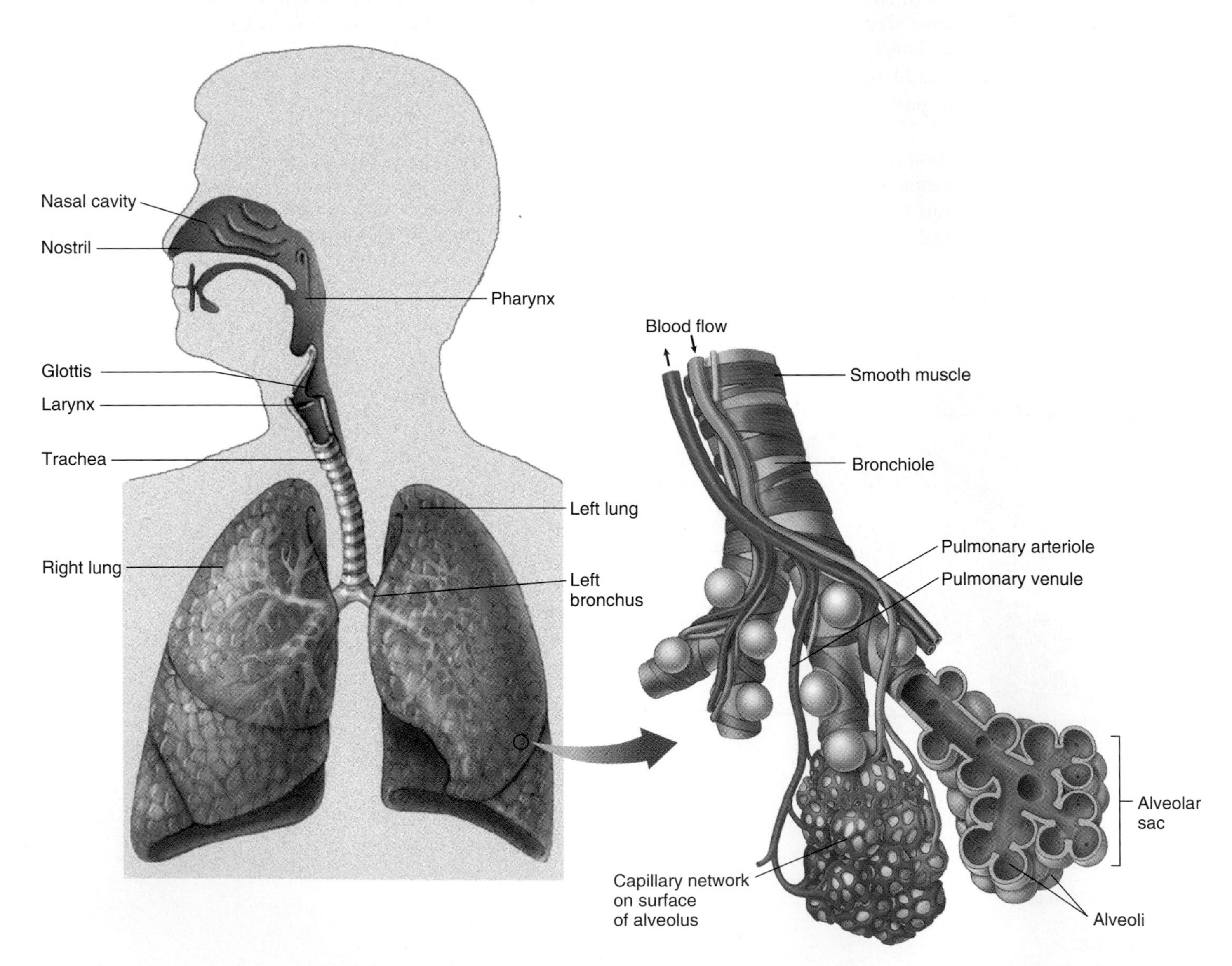

FIGURE 53.9
The human respiratory system and the structure of the mammalian lung. The lungs of mammals have an enormous surface area because of the millions of alveoli that cluster at the ends of the bronchioles. This provides for efficient gas exchange with the blood.

Respiration in Birds

The avian respiratory system has a unique structure that affords birds the most efficient respiration of all terrestrial vertebrates. Unlike the blind-ended alveoli in the lungs of mammals, the bird lung channels air through tiny air vessels called parabronchi, where gas exchange occurs (figure 53.10*a*). Air flows through the parabronchi in one direction only; this is similar to the unidirectional flow of water through a fish gill, but markedly different from the two-way flow of air through the airways of other terrestrial vertebrates. In other terrestrial vertebrates, the inhaled fresh air is mixed with "old" oxygen-depleted air that was not exhaled from the previous breathing cycle. In birds, only fresh air enters the parabronchi of the lung, and the old air exits the lung by a different route.

The unidirectional flow of air through the parabronchi of an avian lung is achieved through the action of air sacs, which are unique to birds (figure 53.10*b*). There are two groups of air sacs, anterior and posterior. When they are expanded during *inspiration* they take in air, and when they are compressed during *expiration* they push air into and through the lungs.

If we follow the path of air through the avian respiratory system, we will see that respiration occurs in *two cycles.* Each cycle has an inspiration and expiration phase—but the air inhaled in one cycle is not exhaled until the second cycle. Upon inspiration, both anterior and posterior air sacs expand and take in air. The inhaled air, however, only enters the posterior air sacs; the anterior air sacs fill with air from the lungs (figure 53.10*c*). Upon expiration, the air forced out of the anterior air sacs is exhaled, but the air forced out of the posterior air sacs enters the lungs. This process is repeated in the second cycle, so that air flows through the lungs in one direction and is exhaled at the end of the second cycle.

The unidirectional flow of air also permits a second respiratory efficiency: the flow of blood through the avian lung runs at a 90° angle to the air flow. This **cross-current flow** is not as efficient as the 180° countercurrent flow in fish gills, but it has the capacity to extract more oxygen from the air than a mammalian lung can. Because of the unidirectional air flow in the parabronchi and cross-current blood flow, a sparrow can be active at an altitude of 6000 meters while a mouse, which has a similar body mass and metabolic rate, cannot respire successfully at that elevation.

The avian respiration system is the most efficient among terrestrial vertebrates because it has unidirectional air flow and cross-current blood flow through the lungs.

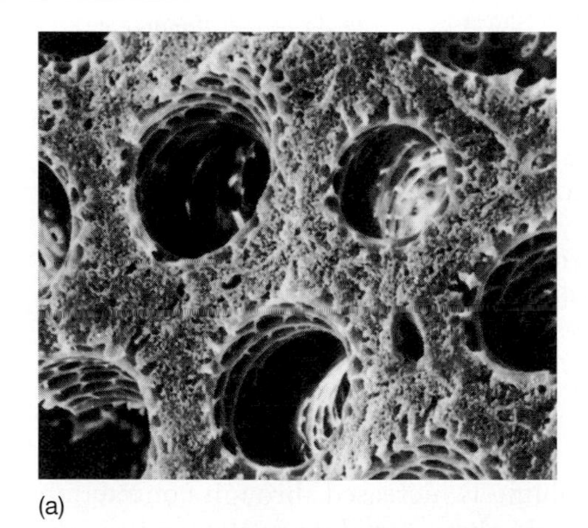
(a)

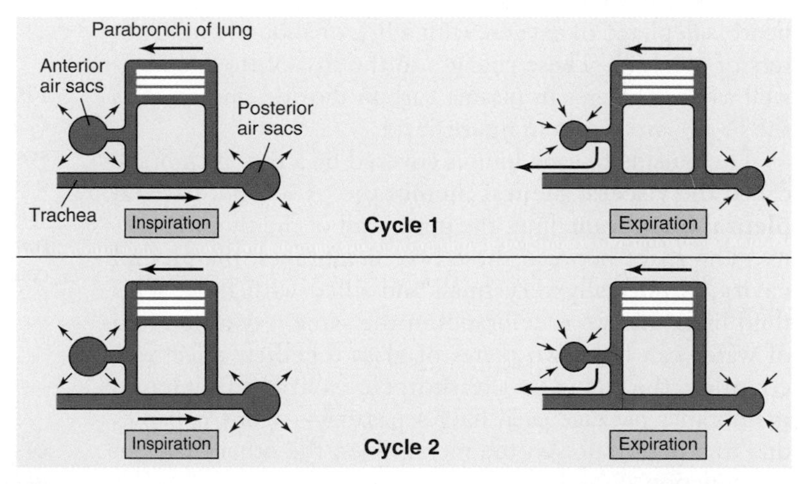

(c)

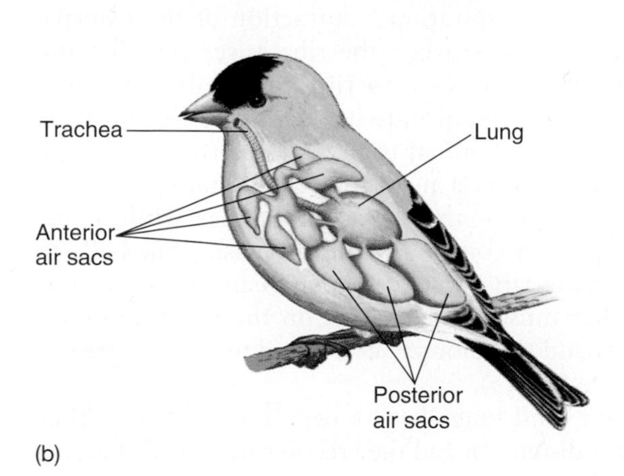

(b)

FIGURE 53.10
How a bird breathes. (*a*) Cross section of lung of a domestic chicken (75×). Air travels through tiny tunnels in the lungs, called parabronchi, while blood circulates within the fine lattice at right angles to the air flow. This cross-current flow makes the bird lung very efficient at extracting oxygen. (*b*) Birds have a system of air sacs, divided into an anterior group and posterior group, that extend between the internal organs and into the bones. (*c*) Breathing occurs in two cycles. *Cycle 1:* Inhaled air (shown in *red*) is drawn from the trachea into the posterior air sacs and then is exhaled into the lungs. *Cycle 2:* Air is drawn from the lungs into the anterior air sacs and then is exhaled through the trachea. Passage of air through the lungs is always in the same direction, from posterior to anterior (right to left in this diagram).

Structures and Mechanisms of Breathing

In mammals, inspired air travels through the trachea, bronchi, and bronchioles to reach the alveoli, where gas exchange occurs. Each alveolus is composed of an epithelium only one cell thick, and is surrounded by blood capillaries with walls that are also only one cell layer thick. There are about 30 billion capillaries in both lungs, or about 100 capillaries per alveolus. Thus, an alveolus can be visualized as a microscopic air bubble whose entire surface is bathed by blood. Because the alveolar air and the capillary blood are separated by only two cell layers, the distance between the air and blood is only 0.5 to 1.5 micrometers, allowing for the rapid exchange of gases by diffusion by decreasing *d* in Fick's Law.

The blood leaving the lungs, as a result of this gas exchange, normally contains a partial oxygen pressure (P_{O_2}) of about 100 millimeters of mercury. As previously discussed, the P_{O_2} is a measure of the concentration of dissolved oxygen—you can think of it as indicating the plasma oxygen. Because the P_{O_2} of the blood leaving the lungs is close to the P_{O_2} of the air in the alveoli (about 105 mm Hg), the lungs do a very effective, but not perfect, job of oxygenating the blood. After gas exchange in the systemic capillaries, the blood that returns to the right side of the heart is depleted in oxygen, with a P_{O_2} of about 40 millimeters of mercury. These changes in the P_{O_2} of the blood, as well as the changes in plasma carbon dioxide (indicated as the P_{CO_2}), are shown in figure 53.11.

The outside of each lung is covered by a thin membrane called the **visceral pleural membrane.** A second, **parietal pleural membrane** lines the inner wall of the thoracic cavity. The space between these two membranes, the **pleural cavity,** is normally very small and filled with fluid. This fluid links the two membranes in the same way a thin film of water can hold two plates of glass together, effectively coupling the lungs to the thoracic cavity. The pleural membranes package each lung separately—if one collapses due to a perforation of the membranes, the other lung can still function.

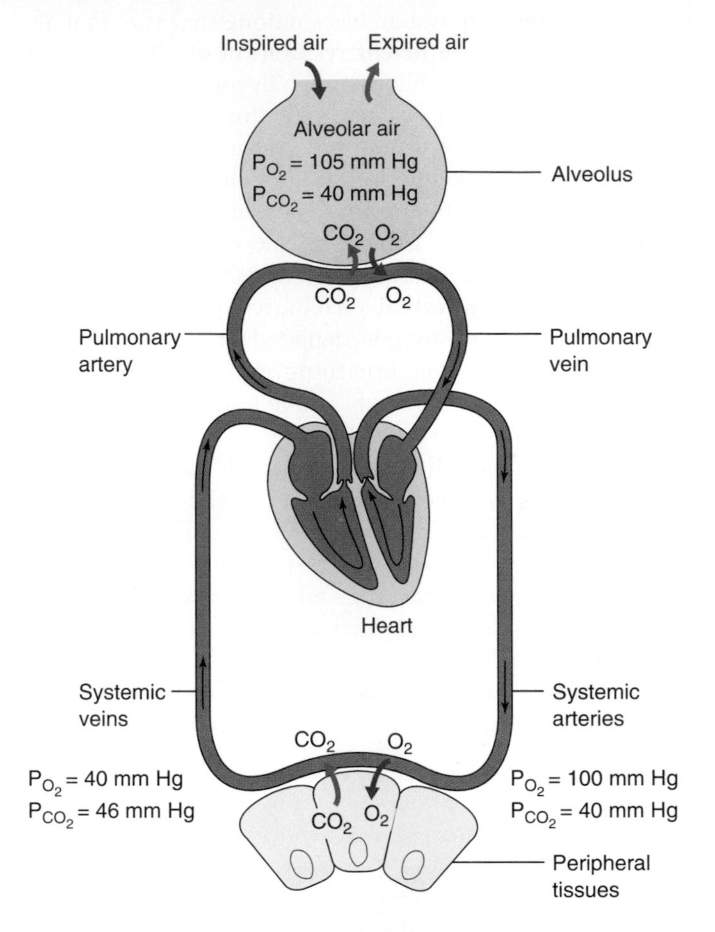

FIGURE 53.11
Gas exchange in the blood capillaries of the lungs and systemic circulation. As a result of gas exchange in the lungs, the systemic arteries carry oxygenated blood with a relatively low carbon dioxide concentration. After the oxygen is unloaded to the tissues, the blood in the systemic veins has a lowered oxygen content and an increased carbon dioxide concentration.

Mechanics of Breathing

As in all other terrestrial vertebrates except amphibians, air is drawn into the lungs by the creation of a negative, or subatmospheric, pressure. In accordance with *Boyle's Law,* when the volume of a given quantity of gas increases its pressure decreases. This occurs because the volume of the thorax is increased during inspiration (inhalation), and the lungs likewise expand because of the adherence of the visceral and parietal pleural membranes. When the pressure within the lungs is lower than the atmospheric pressure, air enters the lungs.

The thoracic volume is increased through contraction of two sets of muscles: the *external intercostals* and the *diaphragm.* During inspiration, contraction of the external intercostal muscles between the ribs raises the ribs and expands the rib cage. Contraction of the diaphragm, a convex sheet of striated muscle separating the thoracic cavity from the abdominal cavity, causes the diaphragm to lower and assume a more flattened shape. This expands the volume of the thorax and lungs while it increases the pressure on the abdomen (causing the belly to protrude). You can force a deeper inspiration by contracting other muscles that insert on the sternum or rib cage and expand the thoracic cavity and lungs to a greater extent (figure 53.12*a*).

The thorax and lungs have a degree of *elasticity*—they tend to resist distension and they recoil when the distending

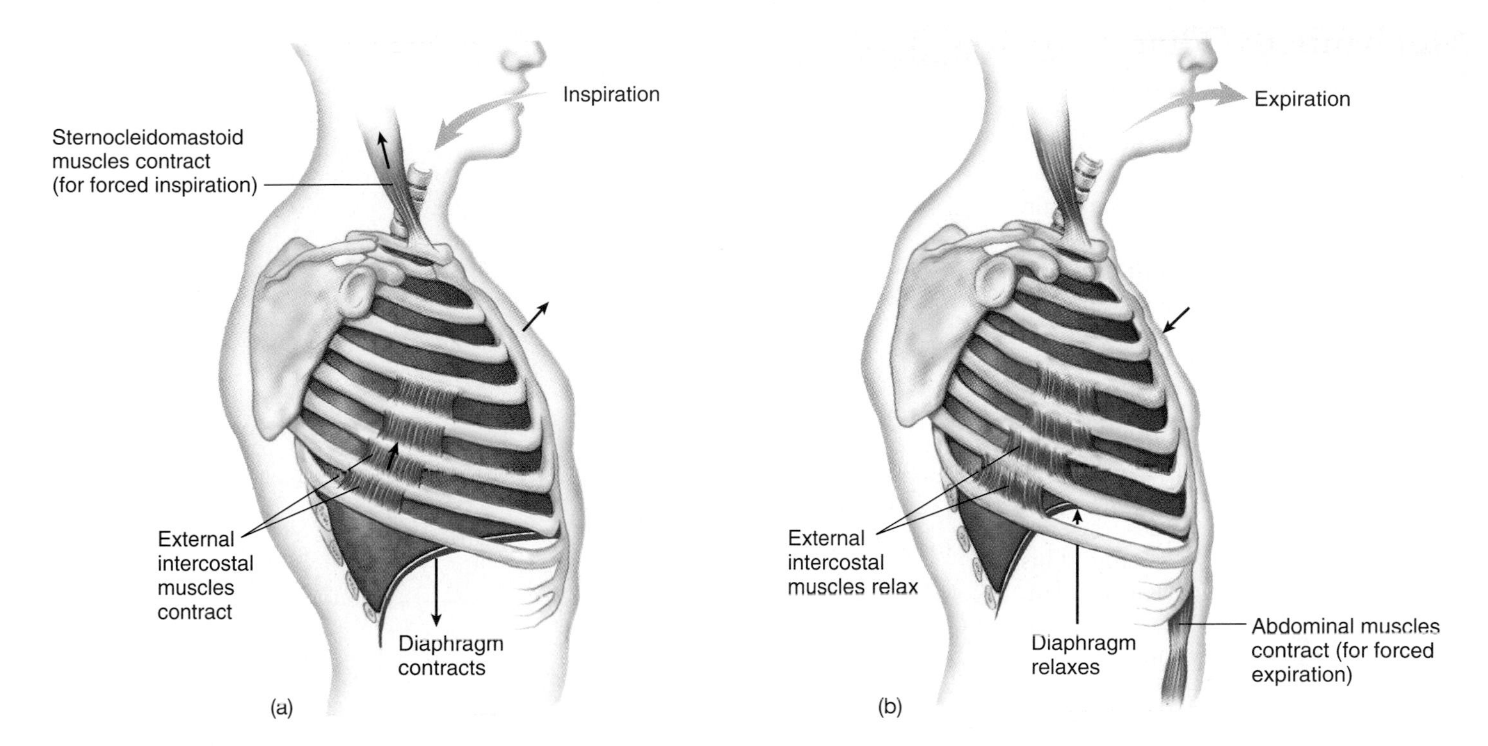

FIGURE 53.12
How a human breathes. (*a*) Inspiration. The diaphragm contracts and the walls of the chest cavity expand, increasing the volume of the chest cavity and lungs. As a result of the larger volume, air is drawn into the lungs. (*b*) Expiration. The diaphragm and chest walls return to their normal positions as a result of elastic recoil, reducing the volume of the chest cavity and forcing air out of the lungs through the trachea. Note that inspiration can be forced by contracting accessory respiratory muscles (such as the sternocleidomastoid), and expiration can be forced by contracting abdominal muscles.

force subsides. Expansion of the thorax and lungs during inspiration places these structures under elastic tension. It is the relaxation of the external intercostal muscles and diaphragm that produces unforced expiration, because it relieves that elastic tension and allows the thorax and lungs to recoil. You can force a greater expiration by contracting your abdominal muscles and thereby pressing the abdominal organs up against the diaphragm (figure 53.12*b*).

Breathing Measurements

A variety of terms are used to describe the volume changes of the lung during breathing. At rest, each breath moves a **tidal volume** of about 500 milliliters of air into and out of the lungs. About 150 milliliters of the tidal volume is contained in the tubular passages (trachea, bronchi, and bronchioles), where no gas exchange occurs. The air in this *anatomical dead space* mixes with fresh air during inspiration. This is one of the reasons why respiration in mammals is not as efficient as in birds, where air flow through the lungs is one-way.

The maximum amount of air that can be expired after a forceful, maximum inspiration is called the **vital capacity.** This measurement, which averages 4.6 liters in young men and 3.1 liters in young women, can be clinically important, because an abnormally low vital capacity may indicate damage to the alveoli in various pulmonary disorders. For example, in **emphysema,** a potentially fatal condition usually caused by cigarette smoking, vital capacity is reduced as the alveoli are progressively destroyed.

A person normally breathes at a rate and depth that properly oxygenate the blood and remove carbon dioxide, keeping the blood P_{O_2} and P_{CO_2} within a normal range. If breathing is insufficient to maintain normal blood gas measurements (a rise in the blood P_{CO_2} is the best indicator), the person is **hypoventilating.** If breathing is excessive for a particular metabolic rate, so that the blood P_{CO_2} is abnormally lowered, the person is said to be **hyperventilating.** Perhaps surprisingly, the increased breathing that occurs during moderate exercise is not necessarily hyperventilation, because the faster breathing is matched to the faster metabolic rate, and blood gas measurements remain normal. The next section describes how breathing is regulated to keep pace with metabolism.

Humans inspire by contracting muscles that insert on the rib cage and by contracting the diaphragm. Expiration is produced primarily by muscle relaxation and elastic recoil. As a result, the blood oxygen and carbon dioxide levels are maintained in a normal range through adjustments in the depth and rate of breathing.

Mechanisms That Regulate Breathing

Each breath is initiated by neurons in a *respiratory control center* located in the medulla oblongata, a part of the brain stem (see chapter 54). These neurons send impulses to the diaphragm and external intercostal muscles, stimulating them to contract, and contractions of these muscles expand the chest cavity, causing inspiration. When these neurons stop producing impulses, the inspiratory muscles relax and expiration occurs. Although the muscles of breathing are skeletal muscles, they are usually controlled automatically. This control can be voluntarily overridden, however, as in hypoventilation (breath holding) or hyperventilation.

A proper rate and depth of breathing is required to maintain the blood oxygen and carbon dioxide levels in the normal range. Thus, although the automatic breathing cycle is driven by neurons in the brain stem, these neurons must be responsive to changes in blood P_{O_2} and P_{CO_2} in order to maintain homeostasis. You can demonstrate this mechanism by simply holding your breath. Your blood carbon dioxide immediately rises and your blood oxygen falls. After a short time, the urge to breathe induced by the changes in blood gases becomes overpowering. This is due primarily to the rise in blood carbon dioxide, as indicated by a rise in P_{CO_2}, rather than to the fall in oxygen levels.

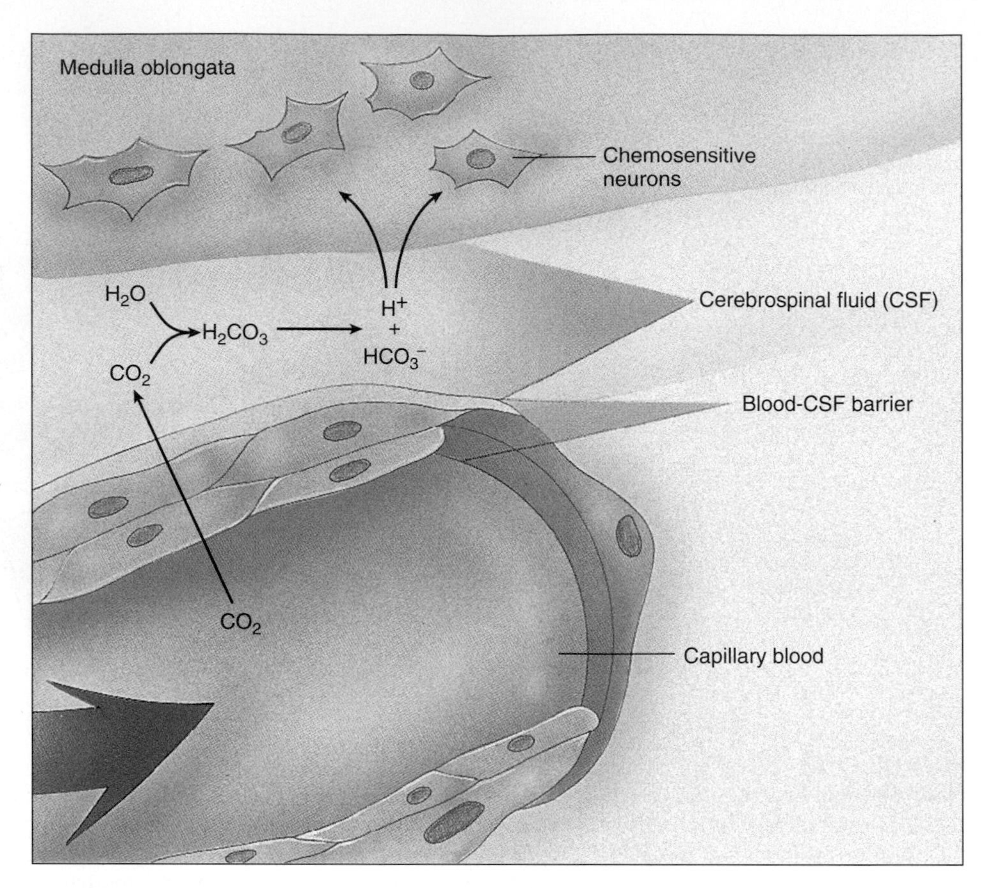

FIGURE 53.13
The effect of blood CO_2 on cerebrospinal fluid (CSF). Changes in the pH of the CSF are detected by chemosensitive neurons in the brain that help regulate breathing.

A rise in P_{CO_2} causes an increased production of carbonic acid (H_2CO_3), which is formed from carbon dioxide and water and acts to lower the blood pH (carbonic acid dissociates into HCO_3^- and H^+, thereby increasing blood H^+ concentration). A fall in blood pH stimulates neurons in the **aortic** and **carotid bodies,** which are sensory structures known as *peripheral chemoreceptors* in the aorta and the carotid artery. These receptors send impulses to the respiratory control center in the medulla oblongata, which then stimulates increased breathing. The brain also contains chemoreceptors, but they cannot be stimulated by blood H^+ because the blood is unable to enter the brain. After a brief delay, however, the increased blood P_{CO2} also causes a decrease in the pH of the cerebrospinal fluid (CSF) bathing the brain. This stimulates the **central chemoreceptors** in the brain (figure 53.13).

The peripheral chemoreceptors are responsible for the immediate stimulation of breathing when the blood P_{CO_2} rises, but this immediate stimulation only accounts for about 30% of increased ventilation. The central chemoreceptors are responsible for the sustained increase in ventilation if P_{CO_2} remains elevated. The increased respiratory rate then acts to eliminate the extra CO_2, bringing the blood pH back to normal (figure 53.14).

A person cannot voluntarily hyperventilate for too long. The decrease in plasma P_{CO_2} and increase in pH of plasma and CSF caused by hyperventilation extinguish the reflex drive to breathe. They also lead to constriction of cerebral blood vessels, causing dizziness. People can hold their breath longer if they hyperventilate first, because it takes longer for the CO_2 levels to build back up, not because hyperventilation increases the P_{O_2} of the blood. Actually, in people with normal lungs, P_{O_2} becomes a significant stimulus for breathing only at high altitudes, where the P_{O_2} is low. Low P_{O_2} can also stimulate breathing in patients with emphysema, where the lungs are so damaged that blood CO_2 can never be adequately eliminated.

Breathing serves to keep the blood gases and pH in the normal range and is under the reflex control of peripheral and central chemoreceptors. These chemoreceptors sense the pH of the blood and cerebrospinal fluid, and they regulate the respiratory control center in the medulla oblongata of the brain.

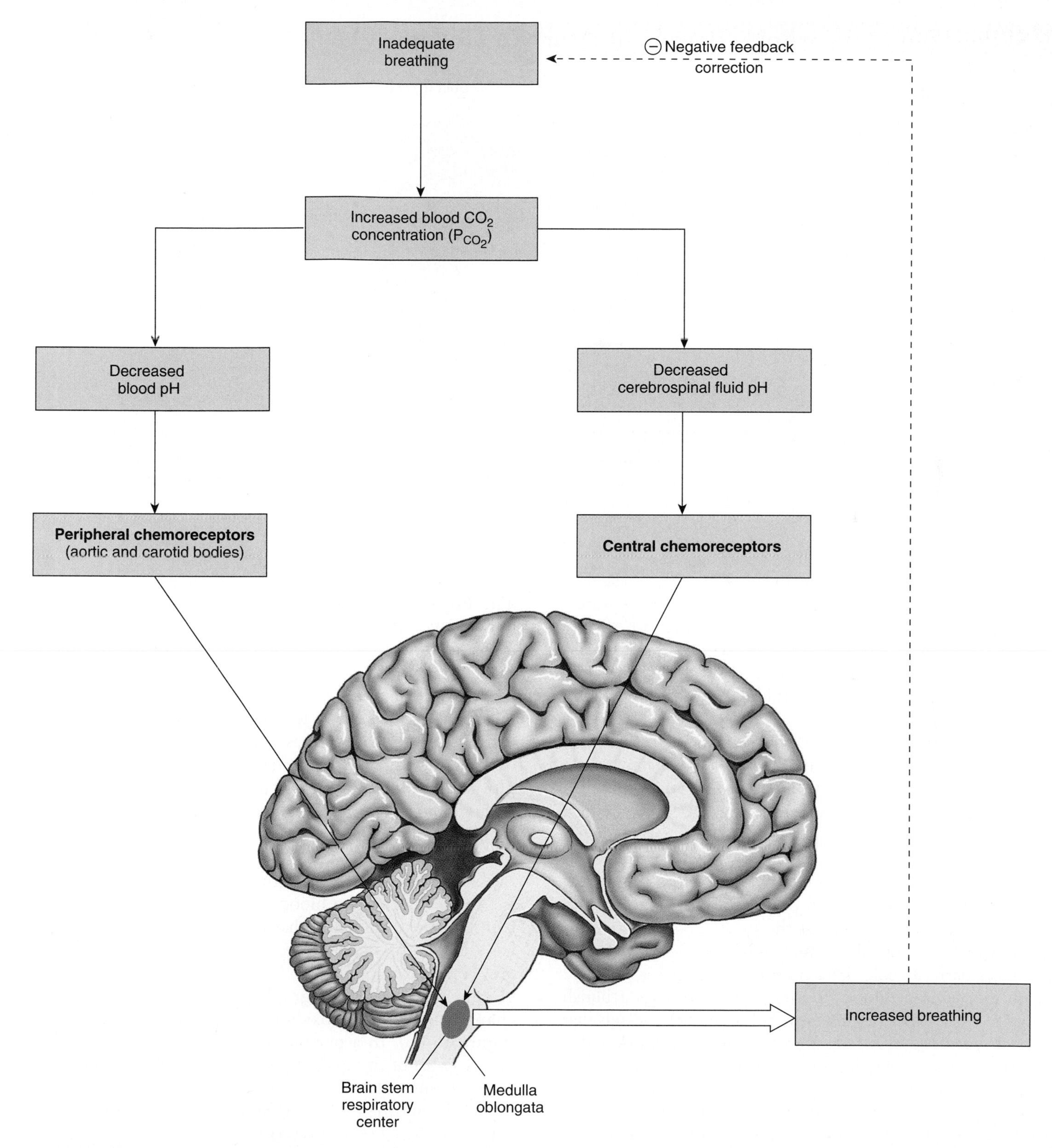

FIGURE 53.14
The regulation of breathing by chemoreceptors. Peripheral and central chemoreceptors sense a fall in the pH of blood and cerebrospinal fluid, respectively, when the blood carbon dioxide levels rise as a result of inadequate breathing. In response, they stimulate the respiratory control center in the medulla oblongata, which directs an increase in breathing. As a result, the blood carbon dioxide concentration is returned to normal, completing the negative feedback loop.

53.5 Blood transports oxygen and carbon dioxide.

Hemoglobin and Oxygen Transport

When oxygen diffuses from the alveoli into the blood, its journey is just beginning. The circulatory system delivers oxygen to tissues for respiration and carries away carbon dioxide. The transport of oxygen and carbon dioxide by the blood is itself an interesting and physiologically important process.

The amount of oxygen that can be dissolved in the blood plasma depends directly on the P_{O_2} of the air in the alveoli, as we explained earlier. When the lungs are functioning normally, the blood plasma leaving the lungs has almost as much dissolved oxygen as is theoretically possible, given the P_{O_2} of the air. Because of oxygen's low solubility in water, however, blood plasma can contain a maximum of only about 3 milliliters O_2 per liter. Yet whole blood carries almost 200 milliliters O_2 per liter! Most of the oxygen is bound to molecules of hemoglobin inside the red blood cells.

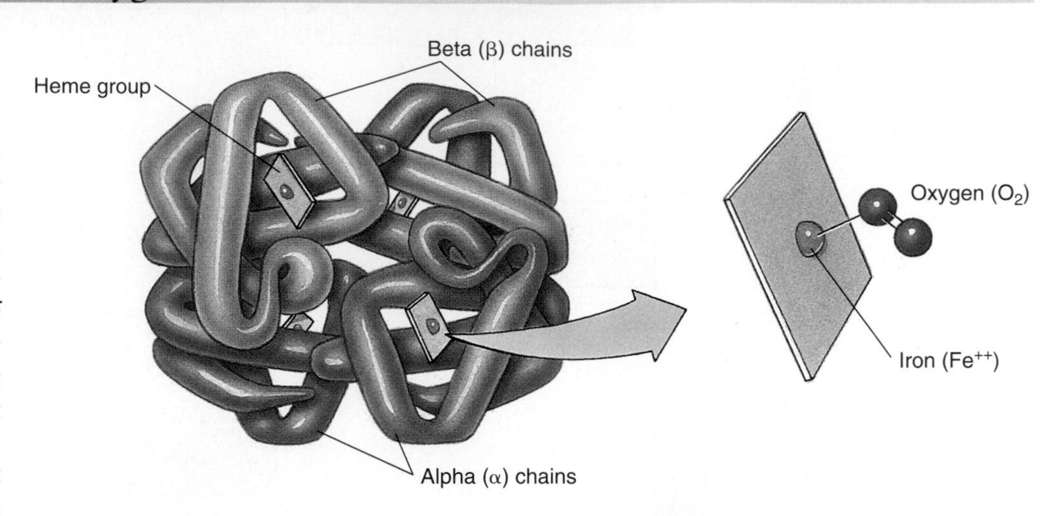

FIGURE 53.15
Hemoglobin consists of four polypeptide chains—two alpha (α) chains and two beta (β) chains. Each chain is associated with a heme group, and each heme group has a central iron atom, which can bind to a molecule of O_2.

Hemoglobin is a protein composed of four polypeptide chains and four organic compounds called *heme groups.* At the center of each heme group is an atom of iron, which can bind to a molecule of oxygen (figure 53.15). Thus, each hemoglobin molecule can carry up to four molecules of oxygen. Hemoglobin loads up with oxygen in the lungs, forming **oxyhemoglobin.** This molecule has a bright red, tomato juice color. As blood passes through capillaries in the rest of the body, some of the oxyhemoglobin releases oxygen and becomes **deoxyhemoglobin.** Deoxyhemoglobin has a dark red color (the color of blood that is collected from the veins of blood donors), but it imparts a bluish tinge to tissues. Because of these color changes, vessels that carry oxygenated blood are always shown in artwork with a red color, and vessels that carry oxygen-depleted blood are indicated with a blue color.

Hemoglobin is an ancient protein that is not only the oxygen-carrying molecule in all vertebrates, but is also used as an oxygen carrier by many invertebrates, including annelids, mollusks, echinoderms, flatworms, and even some protists. Many other invertebrates, however, employ different oxygen carriers, such as *hemocyanin.* In hemocyanin, the oxygen-binding atom is copper instead of iron. Hemocyanin is not found in blood cells, but is instead dissolved in the circulating fluid (hemolymph) of invertebrates.

Oxygen Transport

The P_{O_2} of the air within alveoli at sea level is approximately 105 millimeters of mercury (mm Hg), which is less than the P_{O_2} of the atmosphere because of the mixing of freshly inspired air with "old" air in the anatomical dead space of the respiratory system. The P_{O_2} of the blood leaving the alveoli is slightly less than this, about 100 mm Hg, because the blood plasma is not completely saturated with oxygen due to slight inefficiencies in lung function. At a blood P_{O_2} of 100 mm Hg, approximately 97% of the hemoglobin within red blood cells is in the form of oxyhemoglobin—indicated as a percent oxyhemoglobin saturation of 97%.

As the blood travels through the systemic blood capillaries, oxygen leaves the blood and diffuses into the tissues. Consequently, the blood that leaves the tissue in the veins has a P_{O_2} that is decreased (in a resting person) to about 40 mm Hg. At this lower P_{O_2}, the percent saturation of hemoglobin is only 75%. A graphic representation of these changes is called an oxyhemoglobin dissociation curve (figure 53.16). In a person at rest, therefore, 22% (97% minus 75%) of the oxyhemoglobin has released its oxygen to the tissues. Put another way, roughly one-fifth of the oxygen is unloaded in the tissues, leaving four-fifths of the oxygen in the blood as a reserve.

This large reserve of oxygen serves an important function. It enables the blood to supply the body's oxygen needs during exercise as well as at rest. During exercise, the muscles' accelerated metabolism uses more oxygen from the capillary blood and thus decreases the venous blood P_{O_2}. For example, the P_{O_2} of the venous blood could drop to 20 mm Hg; in this case, the percent saturation of hemoglobin will be only 35% (see figure 53.16).

Because arterial blood still contains 97% oxyhemoglobin (ventilation increases proportionately with exercise), the amount of oxygen unloaded is now 62% (97% minus 35%), instead of the 22% at rest. In addition to this function, the oxygen reserve also ensures that the blood contains enough oxygen to maintain life for four to five minutes if breathing is interrupted or if the heart stops pumping.

Oxygen transport in the blood is affected by other conditions. The CO_2 produced by metabolizing tissues as a product of aerobic respiration combines with H_2O to ultimately form bicarbonate and H^+, lowering the pH of the blood. This reaction occurs primarily inside red blood cells, where the lowered pH reduces hemoglobin's affinity for oxygen and thus causes it to release oxygen more readily. The effect of pH on hemoglobin's affinity for oxygen is known as the Bohr effect and is shown graphically by a shift of the oxyhemoglobin dissociation curve to the right (figure 53.17*a*). Increasing temperature has a similar affect on hemoglobin's affinity for oxygen (figure 53.17*b*) Because skeletal muscles produce carbon dioxide more rapidly during exercise and active muscles produce heat, the blood unloads a higher percentage of the oxygen it carries during exercise.

Deoxyhemoglobin combines with oxygen in the lungs to form oxyhemoglobin, which dissociates in the tissue capillaries to release its oxygen. The degree to which the loading reaction occurs depends on ventilation; the degree of unloading is influenced by such factors as pH and temperature.

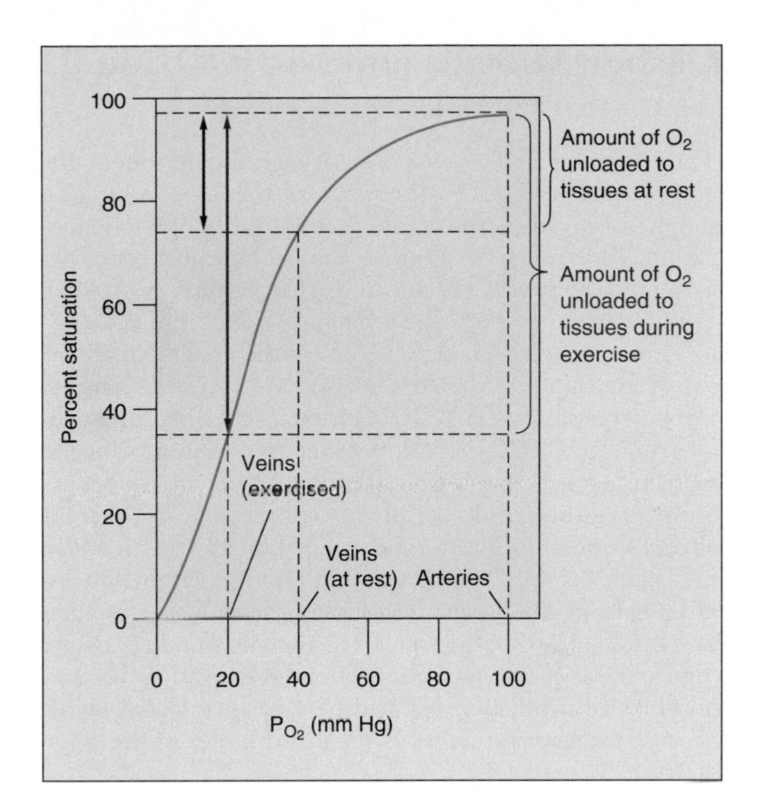

FIGURE 53.16
The oxyhemoglobin dissociation curve. Hemoglobin combines with O_2 in the lungs, and this oxygenated blood is carried by arteries to the body cells. After oxygen is removed from the blood to support cell respiration, the blood entering the veins contains less oxygen. The difference in O_2 content between arteries and veins during rest and exercise shows how much O_2 was unloaded to the tissues.

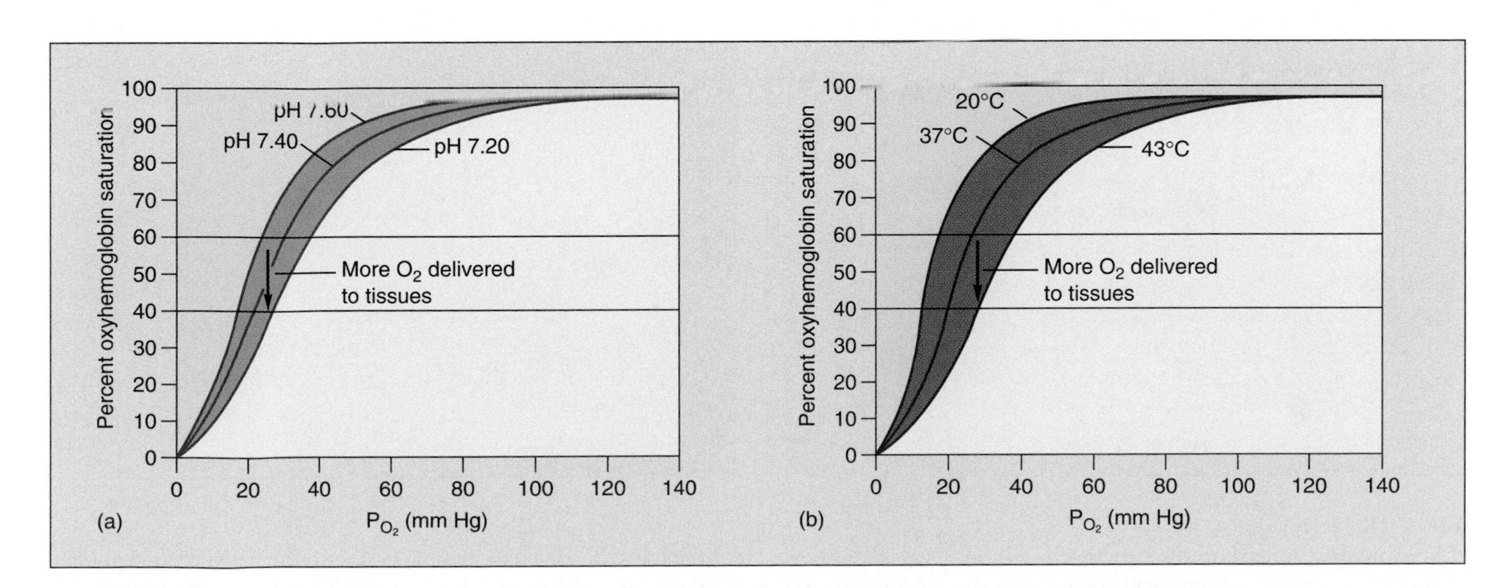

FIGURE 53.17
The effect of pH and temperature on the oxyhemoglobin dissociation curve. Lower blood pH (*a*) and higher blood temperatures (*b*) shift the oxyhemoglobin dissociation curve to the right, facilitating oxygen unloading. This can be seen as a lowering of the oxyhemoglobin percent saturation from 60 to 40% in the example shown, indicating that the difference of 20% more oxygen is unloaded to the tissues.

Carbon Dioxide and Nitric Oxide Transport

The systemic capillaries deliver oxygen to the tissues and remove carbon dioxide. About 8% of the CO_2 in blood is simply dissolved in plasma; another 20% is bound to hemoglobin. (Because CO_2 binds to the protein portion of hemoglobin, however, and not to the heme irons, it does not compete with oxygen.) The remaining 72% of the CO_2 diffuses into the red blood cells, where the enzyme carbonic anhydrase catalyzes the combination of CO_2 with water to form carbonic acid (H_2CO_3). Carbonic acid dissociates into bicarbonate (HCO_3^-) and hydrogen (H^+) ions. The H^+ binds to deoxyhemoglobin, and the bicarbonate moves out of the erythrocyte into the plasma via a transporter that exchanges one chloride ion for a bicarbonate (this is called the "chloride shift"). This reaction removes large amounts of CO_2 from the plasma, facilitating the diffusion of additional CO_2 into the plasma from the surrounding tissues (figure 53.18). The formation of carbonic acid is also important in maintaining the acid-base balance of the blood, because bicarbonate serves as the major buffer of the blood plasma.

The blood carries CO_2 in these forms to the lungs. The lower P_{CO_2} of the air inside the alveoli causes the carbonic anhydrase reaction to proceed in the reverse direction, converting H_2CO_3 into H_2O and CO_2 (see figure 53.18). The CO_2 diffuses out of the red blood cells and into the alveoli, so that it can leave the body in the next exhalation (figure 53.19).

Nitric Oxide Transport

Hemoglobin also has the ability to hold and release nitric oxide gas (NO). Although a noxious gas in the atmosphere, nitric oxide has an important physiological role in the body and acts on many kinds of cells to change their shapes and functions. For example, in blood vessels the presence of NO causes the blood vessels to expand because it relaxes the surrounding muscle cells (see chapters 7 and 52). Thus, blood flow and blood pressure are regulated by the amount of NO released into the bloodstream.

A current hypothesis proposes that hemoglobin carries NO in a special form called super nitric oxide. In this form, NO has acquired an extra electron and is able to bind to the amino acid cysteine in hemoglobin. In the lungs, hemoglobin that is dumping CO_2 and picking up O_2 also picks up NO as super nitric oxide. In blood vessels at the tissues, hemoglobin that is releasing its O_2 and picking up CO_2 can do one of two things with nitric oxide. To increase blood flow, hemoglobin can release the

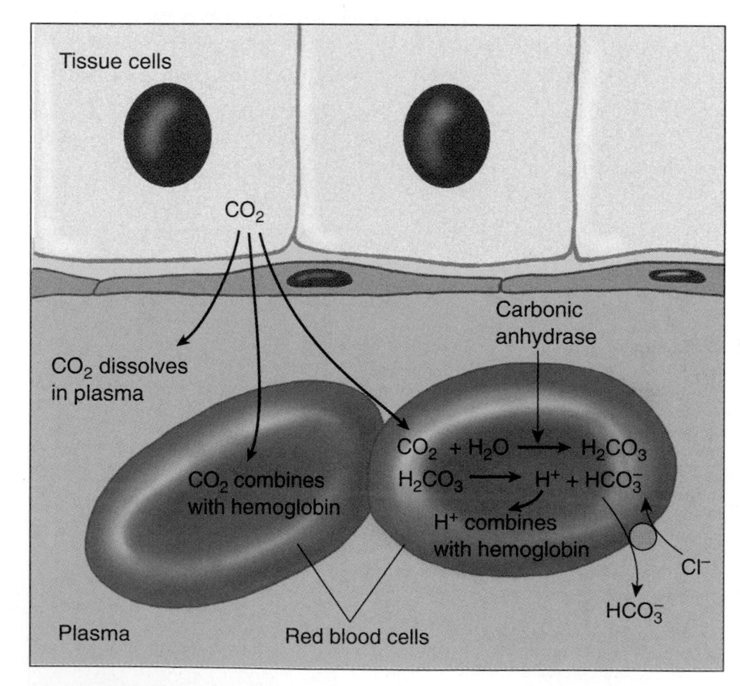

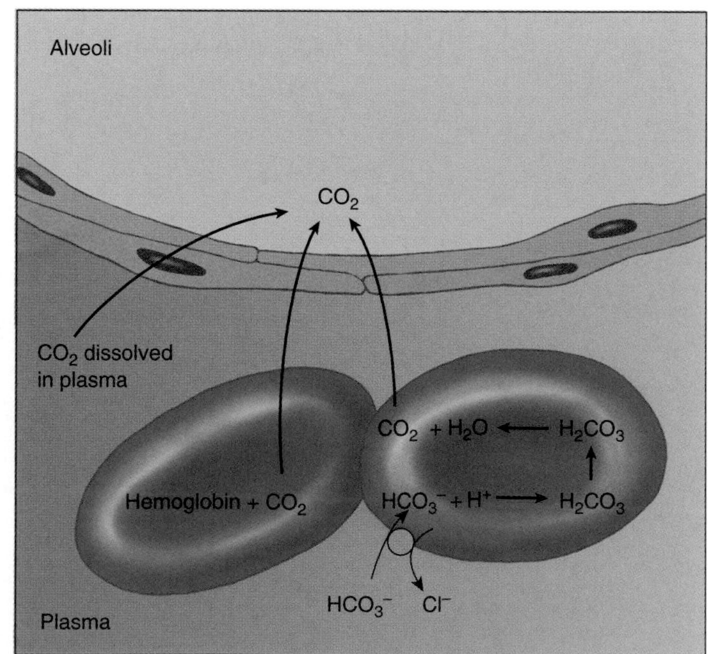

FIGURE 53.18
The transport of carbon dioxide by the blood. CO_2 is transported in three ways: dissolved in plasma, bound to the protein portion of hemoglobin, and as carbonic acid and bicarbonate, which form in the red blood cells. When the blood passes through the pulmonary capillaries, these reactions are reversed so that CO_2 gas is formed, which is exhaled.

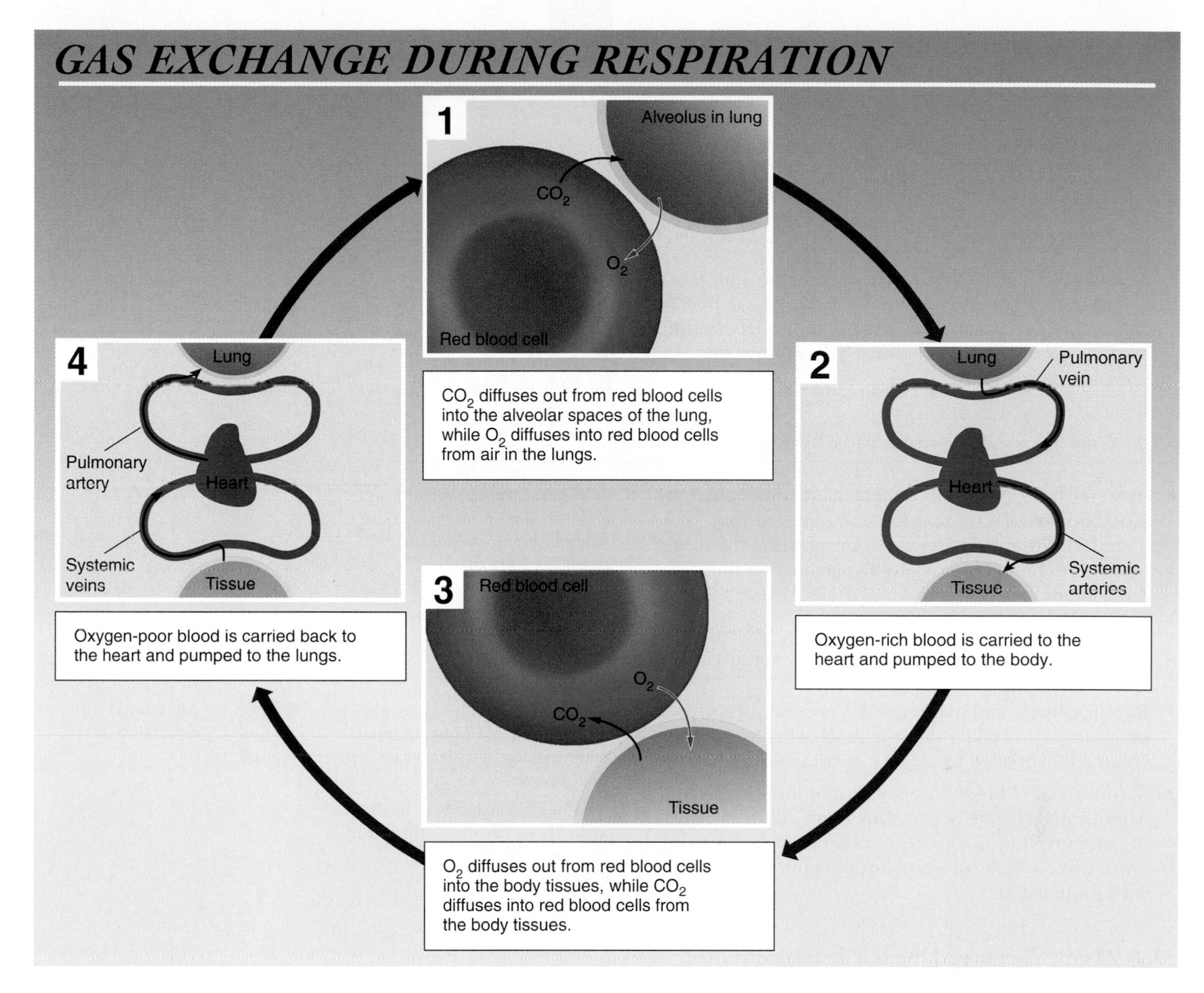
GAS EXCHANGE DURING RESPIRATION
1
Alveolus in lung
CO2
O2
Red blood cell
CO2 diffuses out from red blood cells into the alveolar spaces of the lung, while O2 diffuses into red blood cells from air in the lungs.
2
Lung
Pulmonary vein
Heart
Tissue
Systemic arteries
Oxygen-rich blood is carried to the heart and pumped to the body.
3
Red blood cell
O2
CO2
Tissue
O2 diffuses out from red blood cells into the body tissues, while CO2 diffuses into red blood cells from the body tissues.
4
Lung
Pulmonary artery
Heart
Systemic veins
Tissue
Oxygen-poor blood is carried back to the heart and pumped to the lungs.

FIGURE 53.19
Summary of respiratory gas exchange.

super nitric oxide as NO into the blood, making blood vessels expand because NO acts as a relaxing agent. Or, hemoglobin can trap any excess of NO on its iron atoms left vacant by the release of oxygen, causing blood vessels to constrict. When the red blood cells return to the lungs, hemoglobin dumps its CO_2 and the regular form of NO bound to the iron atoms. It is then ready to pick up O_2 and super nitric oxide and continue the cycle.

Carbon dioxide is transported in the blood in three ways: dissolved in the plasma, bound to hemoglobin, and the majority as bicarbonate in the plasma following an enzyme-catalyzed reaction in the red blood cells. Nitric oxide is also transported in the blood providing yet another explanation of NO actions on blood vessels.

Chapter 53

www.mhhe.com/raven6/resources53.mhtml

Summary | **Questions** | **Media Resources**

53.1 Respiration involves the diffusion of gases.

- The factors that influence the rate of diffusion, surface area, concentration gradient, and diffusion distance, are described by Fick's Law.
- Animals have evolved to maximize the diffusion rate across respiratory membranes by increasing the respiratory surface area, increasing the concentration gradient across the membrane, or decreasing the diffusion distance.

1. Approximately what percentage of dry air is oxygen, and what percentage is carbon dioxide?

2. Why is it that only very small organisms can satisfy their respiratory requirements by direct diffusion to all cells from the body surface?

- Respiration

- Art Quiz: Gas Exchange Variations

53.2 Gills are used for respiration by aquatic vertebrates.

- As water flows past a gill's lamellae, it comes close to blood flowing in an opposite, or countercurrent, direction; this maximizes the concentration difference between the two fluids, thereby maximizing the diffusion of gases.

3. What is countercurrent flow, and how does it help make the fish gill the most efficient respiratory organ?

53.3 Lungs are used for respiration by terrestrial vertebrates.

- Reptiles, birds, and mammals use negative pressure breathing; air is taken into the lungs when the lung volume is expanded to create a partial vacuum.
- Mammals have lungs composed of millions of alveoli, where gas exchange occurs; this is very efficient, but because inspiration and expiration occur through the same airways, new air going into the lungs is mixed with some old air.

4. How do amphibians get air into their lungs? How do other terrestrial vertebrates get air into their lungs?

5. What two features in birds make theirs the most efficient of all terrestrial respiratory systems?

- Art Activity: Respiratory Tract

53.4 Mammalian breathing is a dynamic process.

- The lungs are covered with a wet membrane that sticks to the wet membrane lining the thoracic cavity, so the lungs expand as the chest expands through muscular contractions.
- Breathing is controlled by centers in the medulla oblongata of the brain; breathing is stimulated by a rise in blood CO_2, and consequent fall in blood pH, as sensed by chemoreceptors located in the aorta and carotid artery.

6. How are the lungs connected to and supported within the thoracic cavity?

7. How does the brain control inspiration and expiration? How do peripheral and central chemoreceptors influence the brain's control of breathing?

- Boyle's Law
- Breathing
- Smoking Risks

- Human Breathing
- Disorders

53.5 Blood transports oxygen and carbon dioxide.

- Hemoglobin loads with oxygen in the lungs; this oxyhemoglobin then unloads oxygen as the blood goes through the systemic capillaries.
- Carbon dioxide combines with water as the carbon dioxide is transported to the lungs for exhalation.

8. In what form does most of the carbon dioxide travel in the blood? How and where is this molecule produced?

- Art Activity: Hemoglobin Module

- Hemoglobin
- Gas Exchange

Part XIV

Regulating the Animal Body

Are Pollutants Affecting the Sexual Development of Florida's Alligators?

Alligators are among the most interesting of animals for a biologist to study. Their ecology is closely tied to the environment, and their reptilian biology offers an interesting contrast to that of mammals like ourselves. Studies of alligator development offer powerful general lessons well worthy of our attention.

In no area of biology is this more true than in investigation of alligator sexual development. The importance is not because sexual development in alligators is unusual. It is not. As with all vertebrates, sexual development in alligator males—particularly development of their external sexual organs—is largely dependent on the androgenic sex hormone testosterone and its derivatives. In the alligator embryo, these steroid hormones are responsible for the differentiation of the male internal duct system, as well as the formation of the external genitalia. After the alligator's birth, androgenic hormones are essential for normal maturation and growth of the juvenile male's reproductive system, particularly during puberty.

The strong dependence of a male alligator's sexual development on androgens is not unusual—mammals show the same strong dependence. So why should researchers be interested in alligators? In a nutshell, we humans don't spend our lives sloshing around in an aquatic environment, and alligators do. Florida alligators live in the many lakes that pepper the state, and, living in these lakes, they are exposed all their lives to whatever chemicals happen to enter the lakewater by chemical spills, industrial wastes, and agricultural runoff.

The androgen-dependent sexual development of alligators provides a sensitive barometer to environmental pollution, because the androgen response can be blocked by a class of pollutant chemicals called endocrine disrupters. When endocrine-disrupting pollutants contaminate Florida lakes, their presence can be detected by its impact on the sexual development of the lakes' resident alligators. Just as the death of coal miners' canaries warn of the buildup of dangerous gas within the shaft of coal mines, so disruption of the sexual development of alligators can warn us of dangerous chemicals in the environment around us.

Catching alligators is a job best done at night. Alligators in Florida lakes, like the one shown here in the hands of Professor Guillette, seem to be experiencing developmental abnormalities, perhaps due to pollution of many of Florida's lakes by endocrine-disrupting chemicals.

One of the great joys of biological research is being able to choose research that is fun to do. Few research projects offer the particular joys of studying alligators. With State Game Commission permits, researchers go to lakes in central Florida, wait till after dark, then spend the night on the lake in small boats hand-capturing the animals. The captured animals are confined in cloth bags until sex can be determined, body measurements made, and blood samples collected, and then released (see photo above).

For over ten years, Louis Guillette of the University of Florida, Gainesville, and his students have been carrying out just this sort of research. Their goal has been to assess the degree to which agricultural and other chemicals have polluted the lakes of central Florida, using as their gauge the disruption of normal sexual development in alligators.

To assess hormonal changes that might be expected to inhibit male sexual development, Guillette's team looked at the relative ratio of androgens (which promote male development) to estrogens (which promote female development) in each captured alligator. Some male endocrine disrupters act like estrogens, while others decrease native androgen levels. In either case, the ratio of estrogen to androgen (the E/A ratio) increases, producing a more estrogenic environment and so retarding male sexual development. Particularly after puberty, the growth of male alligators' external sexual organs is very dependent upon a high-androgenic environment. Any pollutant that raises the E/A ratio would be expected to markedly inhibit this development.

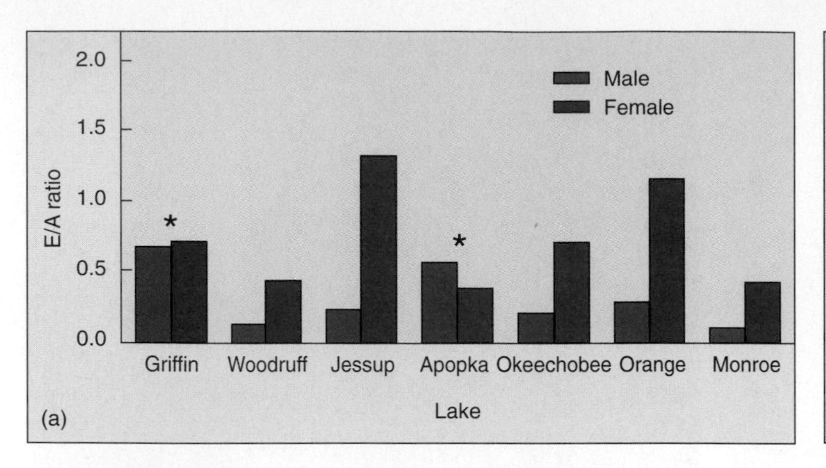

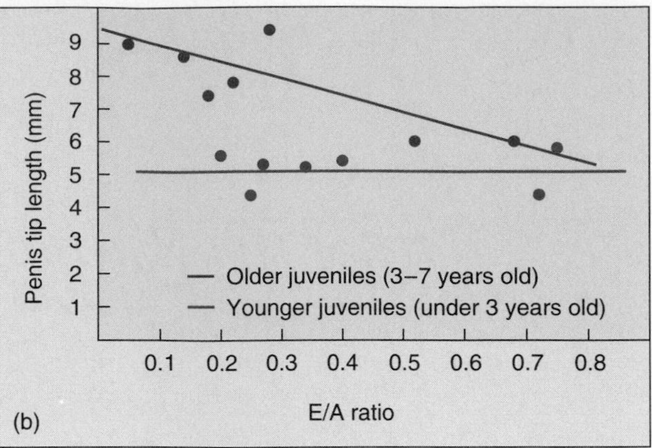

Alligator sexual development inhibited by contamination. (*a*) Ratio of estrogen/androgen (E/A) plasma concentrations in large juvenile alligators 3–7 years old. A relatively larger ratio in males (*) is atypical and indicates an estrogenic hormonal environment, as opposed to the expected androgenic hormonal environment. (*b*) Sexual development in male alligators, measured by penis length as a function of E/A ratio. In small juveniles under 3 years old, there is no apparent influence. In older juveniles 3–7 years old, there is a pronounced effect, higher E/A ratios retarding sexual development.

The Experiment

Guillette's team first looked at animals in two lakes and then expanded the research to look at animals from several other lakes. Alligators were initially collected from Lake Woodruff National Wildlife Refuge and from Lake Apopka. Lake Woodruff is a relatively pristine lake with no agricultural or industrial runoff. Lake Apopka, on the other hand, is a large eutrophic lake exposed to various agricultural and municipal contaminants. In 1980, the lake experienced a pesticide and acid spill from a chemical company, and has a history of pesticide contamination by DDT.

Clear comparisons of alligators collected from different lakes required that animals be captured as nearly as possible at the same time, to minimize possible variation due to photoperiod, temperature, and nutrition. This experimental requirement led to truly prodigious feats of alligator catching by the research team. On a single night in 1994, 40 male alligators were hand-captured from Lake Woodruff. The following night, 54 males were captured from Lake Apopka. In a broader study of seven lakes carried out the following year, 528 animals were captured during a 17-day period.

The external genitalia and total body length were measured on captured animals. Body-length is a good indicator of the age of the alligator. Alligators begin puberty at about 3 years of age, and this must be taken into account when making comparisons.

Blood samples were taken from each animal in order to determine the plasma levels of estrogen and an androgen. Investigators measured plasma concentrations of estradiol-17β (E) and testosterone (A). The researchers estimated the E/A ratio, to determine if the internal environment was androgenic or estrogenic.

The Results

In most of the seven lakes studied, female alligators showed a much higher E/A ratio than males (graph *a* above), a normal result. The exceptions are Lake Griffin and Lake Apopka, the most polluted of the lakes. The larger E/A ratio observed in male alligators caught from these two lakes indicates an estrogenic hormonal environment in these animals rather than the normal androgenic one.

Does this estrogenic environment have an impact on juvenile sexual development? Yes. Researchers observed that postpuberty juvenile males from Lake Apopka and Lake Griffith (where E/A ratios were elevated) exhibited stunted reproductive organs compared to those found in Lake Woodruff and other lakes (graph *b* above).

Prepuberty males did not show this effect, exhibiting the same size external reproductive organs whatever the E/A ratio. This is as you would expect, as organ growth occurs primarily after puberty, in response to androgenic hormones released from the testes.

A primary contaminant found in alligators' eggs in Lake Apopka is *p,p*'-DDE, a major metabolite of DDT. *p,p*'-DDE has been shown to bind to androgenic receptors, and functions as an antiandrogen in mammals. The presence of *p,p*'-DDE reduces the androgenic effect in cells, creating a more estrogenic environment.

The researchers also measured levels of plasma testosterone. The plasma levels of testosterone were significantly reduced in alligators from Lake Apopka compared to the control animals removed from Lake Woodruff. These reduced levels of plasma testosterone from the Lake Apopka alligators also act to reduce the E/A ratio, producing the observed abnormalities in reproductive structures.

To explore this experiment further, go to the Virtual Lab at www.mhhe.com/raven6/vlab14.mhtml

54

The Nervous System

Concept Outline

54.1 The nervous system consists of neurons and supporting cells.

Neuron Organization. Neurons and neuroglia are organized into the central nervous system (the brain and spinal cord) and the peripheral nervous system (sensory and motor neurons).

54.2 Nerve impulses are produced on the axon membrane.

The Resting Membrane Potential. The inside of the membrane is electrically negative in comparison with the outside.

Action Potentials. In response to a stimulus that depolarizes the membrane, voltage-gated channels open, producing a nerve impulse. One action potential stimulates the production of the next along the axon.

54.3 Neurons form junctions called synapses with other cells.

Structure of Synapses. Neurotransmitters diffuse across to the postsynaptic cell and combine with receptor proteins.

Neurotransmitters and Their Functions. Some neurotransmitters cause a depolarization in the postsynaptic membrane; others produce inhibition by hyperpolarization.

54.4 The central nervous system consists of the brain and spinal cord.

The Evolution of the Vertebrate Brain. Vertebrate brains include a forebrain, midbrain, and hindbrain.

The Human Forebrain. The cerebral cortex contains areas specialized for different functions.

The Spinal Cord. Reflex responses and messages to and from the brain are coordinated by the spinal cord.

54.5 The peripheral nervous system consists of sensory and motor neurons.

Components of the Peripheral Nervous System. A spinal nerve contains sensory and motor neurons.

The Autonomic Nervous System. Sympathetic motor neurons arouse the body for fight or flight; parasympathetic motor neurons have antagonistic actions.

FIGURE 54.1
A neuron in the retina of the eye (500×). This neuron has been injected with a fluorescent dye, making its cell body and long dendrites readily apparent.

All animals except sponges use a network of nerve cells to gather information about the body's condition and the external environment, to process and integrate that information, and to issue commands to the body's muscles and glands. Just as telephone cables run from every compartment of a submarine to the conning tower, where the captain controls the ship, so bundles of nerve cells called neurons connect every part of an animal's body to its command and control center, the brain and spinal cord (figure 54.1). The animal body is run just like a submarine, with status information about what is happening in organs and outside the body flowing into the command center, which analyzes the data and issues commands to glands and muscles.

54.1 The nervous system consists of neurons and supporting cells.

Neuron Organization

An animal must be able to respond to environmental stimuli. A fly escapes a swat; the antennae of a crayfish detect food and the crayfish moves toward it. To do this, it must have sensory receptors that can detect the stimulus and motor *effectors* that can respond to it. In most invertebrate phyla and in all vertebrate classes, sensory receptors and motor effectors are linked by way of the nervous system. As described in chapter 49, the nervous system consists of neurons and supporting cells. **Sensory** (or **afferent**) **neurons** carry impulses from sensory receptors to the central nervous system (CNS); **motor** (or **efferent**) **neurons** carry impulses from the CNS to effectors—muscles and glands (figure 54.2).

In addition to sensory and motor neurons, a third type of neuron is present in the nervous systems of most invertebrates and all vertebrates: **association neurons** (or **interneurons**). These neurons are located in the brain and spinal cord of vertebrates, together called the **central nervous system (CNS),** where they help provide more complex reflexes and higher associative functions, including learning and memory. Sensory neurons carry impulses into the CNS, and motor neurons carry impulses away from the CNS. Together, sensory and motor neurons constitute the **peripheral nervous system (PNS)** of vertebrates. Motor neurons that stimulate skeletal muscles to contract are **somatic motor neurons,** and those that regulate the activity of the smooth muscles, cardiac muscle, and glands are **autonomic motor neurons.** The autonomic motor neurons are further subdivided into the **sympathetic** and **parasympathetic** systems, which act to counterbalance each other (figure 54.3).

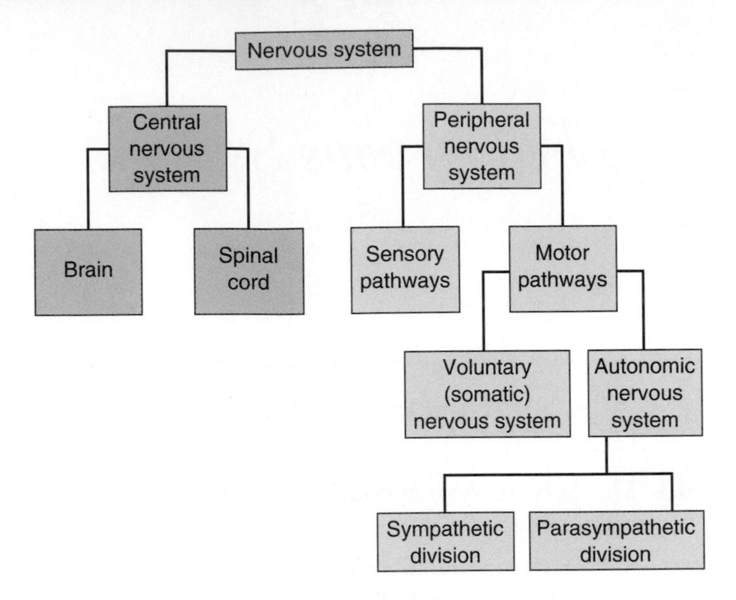

FIGURE 54.3
The divisions of the vertebrate nervous system. The major divisions are the central and peripheral nervous systems.

Despite their varied appearances, most neurons have the same functional architecture (figure 54.4). The cell body is an enlarged region containing the nucleus. Extending from the cell body are one or more cytoplasmic extensions called **dendrites.** Motor and association neurons possess a profusion of highly branched dendrites, enabling those cells to

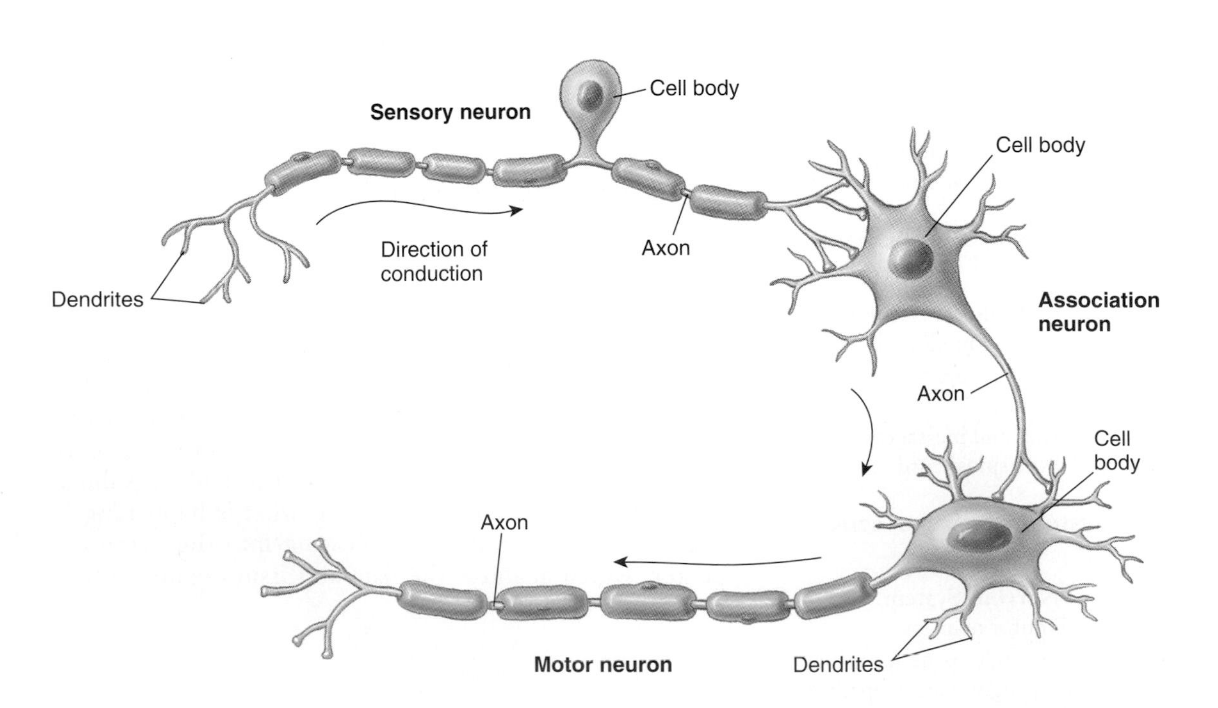

FIGURE 54.2
Three types of neurons. *Sensory neurons* carry information about the environment to the brain and spinal cord. *Association neurons* are found in the brain and spinal cord and often provide links between sensory and motor neurons. *Motor neurons* carry impulses or "commands" to muscles and glands (effectors).

receive information from many different sources simultaneously. Some neurons have extensions from the dendrites called *dendritic spines* that increase the surface area available to receive stimuli. The surface of the cell body integrates the information arriving at its dendrites. If the resulting membrane excitation is sufficient, it triggers impulses that are conducted away from the cell body along an **axon.** Each neuron has a single axon leaving its cell body, although an axon may produce small terminal branches to stimulate a number of cells. An axon can be quite long: the axons controlling the muscles in your feet are more than a meter long, and the axons that extend from the skull to the pelvis in a giraffe are about three meters long!

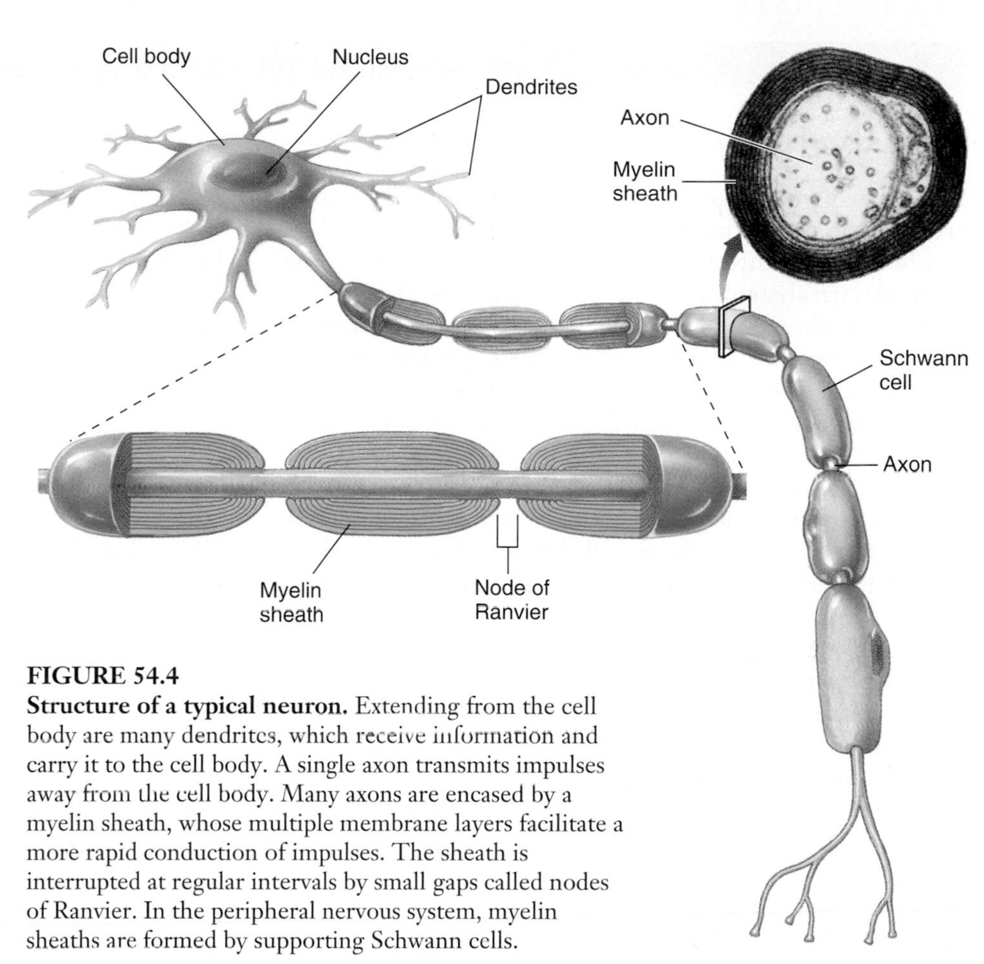

FIGURE 54.4
Structure of a typical neuron. Extending from the cell body are many dendrites, which receive information and carry it to the cell body. A single axon transmits impulses away from the cell body. Many axons are encased by a myelin sheath, whose multiple membrane layers facilitate a more rapid conduction of impulses. The sheath is interrupted at regular intervals by small gaps called nodes of Ranvier. In the peripheral nervous system, myelin sheaths are formed by supporting Schwann cells.

Neurons are supported both structurally and functionally by **supporting cells,** which are called **neuroglia.** These cells are ten times more numerous than neurons and serve a variety of functions, including supplying the neurons with nutrients, removing wastes from neurons, guiding axon migration, and providing immune functions. Two of the most important kinds of neuroglia in vertebrates are **Schwann cells** and **oligodendrocytes,** which produce **myelin sheaths** that surround the axons of many neurons. Schwann cells produce myelin in the PNS, while oligodendrocytes produce myelin in the CNS. During development, these cells wrap themselves around each axon several times to form the myelin sheath, an insulating covering consisting of multiple layers of membrane (figure 54.5). Axons that have myelin sheaths are said to be myelinated, and those that don't are unmyelinated. In the CNS, myelinated axons form the **white matter,** and the unmyelinated dendrites and cell bodies form the **gray matter.** In the PNS, both myelinated and unmyelinated axons are bundled together, much like wires in a cable, to form **nerves.**

The myelin sheath is interrupted at intervals of 1 to 2 μm by small gaps known as **nodes of Ranvier** (see figure 54.4). The role of the myelin sheath in impulse conduction will be discussed later in this chapter.

Neurons and neuroglia make up the central and peripheral nervous systems in vertebrates. Sensory, motor, and association neurons play different roles in the nervous system, and the neuroglia aid their function, in part by producing myelin sheaths.

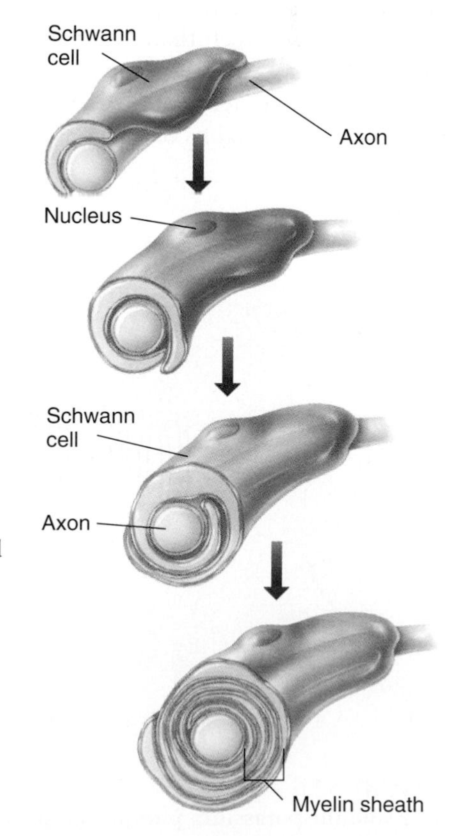

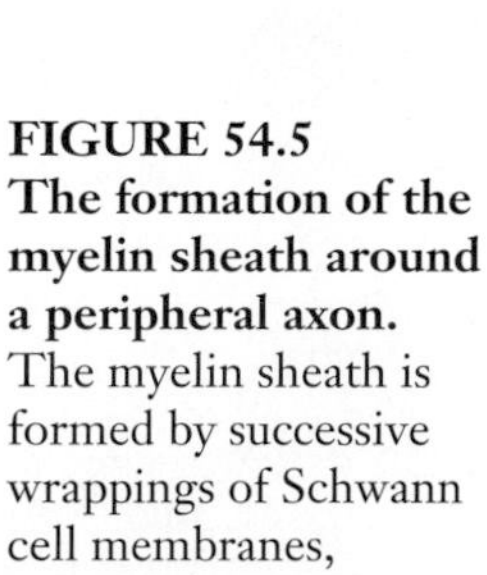
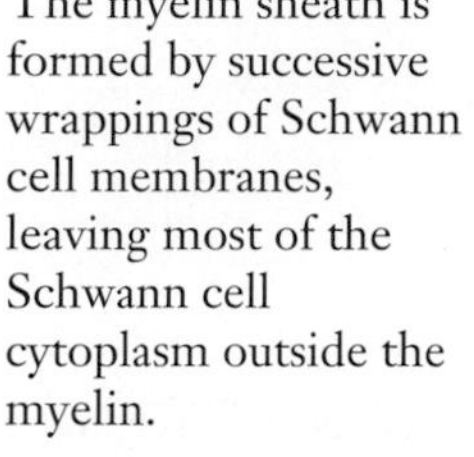

FIGURE 54.5
The formation of the myelin sheath around a peripheral axon. The myelin sheath is formed by successive wrappings of Schwann cell membranes, leaving most of the Schwann cell cytoplasm outside the myelin.

54.2 Nerve impulses are produced on the axon membrane.

The Resting Membrane Potential

Neurons communicate through changes in electrical properties of the plasma membrane that travel from one cell to another. The architecture of the neuron aids the spread of these electrical signals called nerve impulses. To understand how these signals are generated and transmitted within the nervous system, we must first examine some of the electrical properties of plasma membranes.

The battery in a car or a flashlight separates electrical charges between its two poles. There is said to be a *potential difference*, or voltage, between the poles, with one pole being positive and the other negative. Similarly, a potential difference exists across every cell's plasma membrane. The side of the membrane exposed to the cytoplasm is the negative pole, and the side exposed to the extracellular fluid is the positive pole. This potential difference is called the membrane potential.

The inside of the cell is more negatively charged in relation to the outside because of three factors: (1) Large molecules like proteins and nucleic acids that are negatively charged are more abundant inside the cell and cannot diffuse out. These molecules are called *fixed anions*. (2) The sodium-potassium pump brings only two potassium ions (K^+) into the cell for every three sodium ions (Na^+) it pumps out (figure 54.6). In addition to contributing to the electrical potential, this also establishes concentration gradients for Na^+ and K^+. (3) Ion channels allow more K^+ to diffuse out of the cell than Na^+ to diffuse into the cell. Na^+ and K^+ channels in the plasma membrane have *gates*, portions of the channel protein that open or close the channel's pore. In the axons of neurons and in muscle fibers, the gates are closed or open depending on the membrane potential. Such channels are therefore known as *voltage-gated ion channels* (figure 54.7).

In most cells, the permeability of ions through the membrane is constant, and the net negativity on the inside of the cell remains constant. The plasma membranes of muscle cells and neurons, however, are excitable because the permeability of their ion channels can be altered by various stimuli. When a neuron is not being stimulated, it maintains a *resting membrane potential.* A cell is very small, and so its membrane potential is very small. The electrical potential of a car battery is typically 12 volts, but a cell's resting membrane potential is about –70 millivolts (–70 mV or 0.07 volts). The negative sign indicates that the inside of the cell is negative with respect to the outside.

We know that the resting membrane potential is –70 mV because of an unequal distribution of electrical charges across the membrane. But why –70 mV rather than –50 mV or –10 mV? To understand this, we need to remember that there are two forces acting on the ions involved in establishing the resting membrane potential: (1) ions are attracted to ions or molecules of opposite charge; and (2) ions respond to concentration gradients by moving from an area of high concentration to an area of lower concentration.

The positively charged ions, called *cations*, outside the cell are attracted to the negatively charged fixed anions inside the cell. However, the resting plasma membrane is more permeable to K^+ than to other cations, so K^+ enters the cell. Other cations enter the cell, but the leakage of K^+ into the cell has the dominant effect on the resting membrane potential. In addition to the electrical gradient

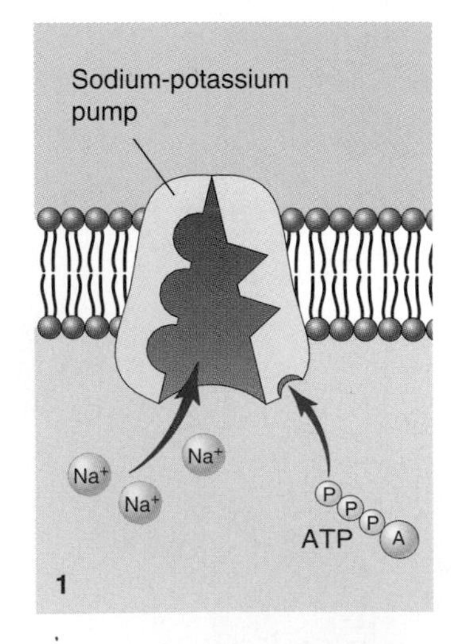

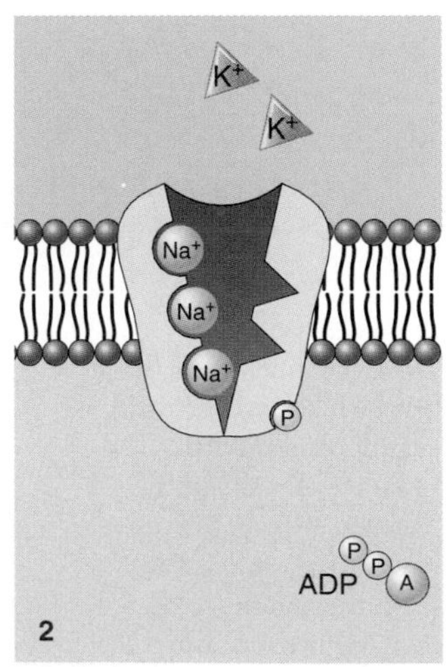

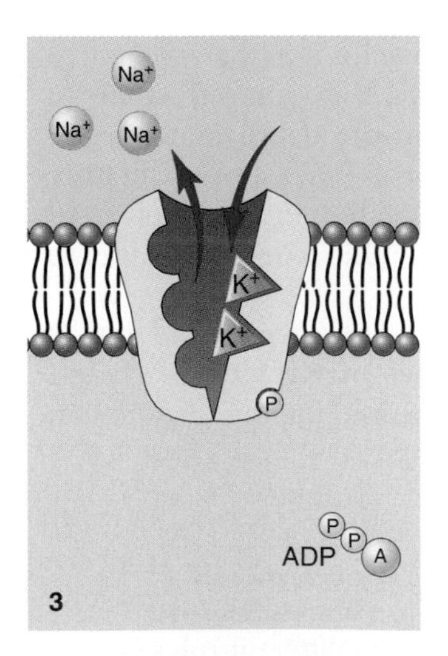

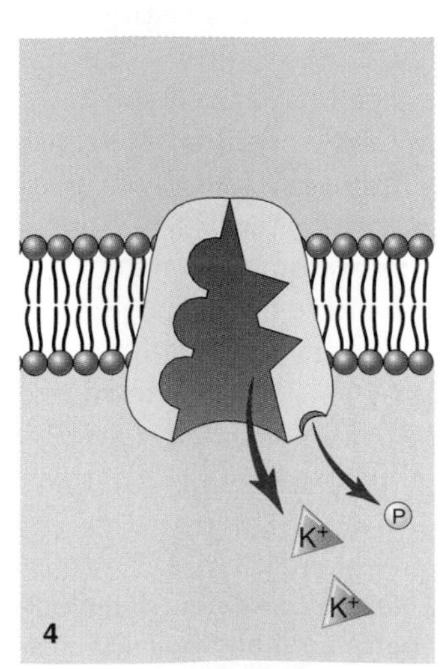

FIGURE 54.6
The sodium-potassium pump. This pump transports three Na^+ to the outside of the cell and simultaneously transports two K^+ to the inside of the cell. This is an active transport carrier requiring the breakdown of ATP.

Table 54.1 The Ionic Composition of Cytoplasm and Extracellular Fluid (ECF)

Ion	Concentration in Cytoplasm (mM)	Concentration in ECF (mM)	Ratio	Equilibrium Potential (mV)
Na^+	15	150	10:1	+60
K^+	150	5	1:30	–90
Cl^-	7	110	15:1	–70

driving K^+ into the cell, there is also a concentration gradient established by the sodium-potassium pump that is driving K^+ out of the cell. At a point, these two forces balance each other, and the voltage at which the influx of K^+ equals the efflux of K^+ is called the **equilibrium potential** (table 54.1). For potassium, the equilibrium potential is –90 mV. At –80 mV, K^+ will diffuse out of the cell and at –100 mV K^+ will diffuse into the cell.

If K^+ were the only cation involved, the resting membrane potential of the cell would be –90 mV. However, the membrane is also slightly permeable to Na^+, and its equilibrium potential is +60 mV. The effects of Na^+ leaking into the cell make the resting membrane potential less negative. With a membrane potential less negative than –90 mV, K^+ diffuses into the cell, and the combined effect bring the equilibrium potential for the resting cell to –70 mV. The resting membrane potential of a neuron can be seen using a voltmeter and a pair of electrodes, one outside and one inside the cell (figure 54.8)

When a nerve or muscle cell is stimulated, sodium channels become more permeable, and Na^+ rushes into the cell, down its concentration gradient. This sudden influx of positive charges reduces the negativity on the inside of the cell and causes the cell to *depolarize* (move toward a polarity above that of the resting potential). After a slight delay, potassium channels also become more permeable, and K^+ flows out of the cell down its concentration gradient. Similarly, the membrane becomes more permeable to Cl^-, and Cl^- flows into the cell. But the effects of Cl^- on the membrane potential are far less than those of K^+. The inside of the cell again becomes more negative and causes the cell to *hyperpolarize* (move its polarity below that of the resting potential).

The resting plasma membrane maintains a potential difference as a result of the uneven distribution of charges, where the inside of the membrane is negatively charged in comparison with the outside (–70 mV). The magnitude, measured in millivolts, of this potential difference primarily reflects the difference in K^+ concentration.

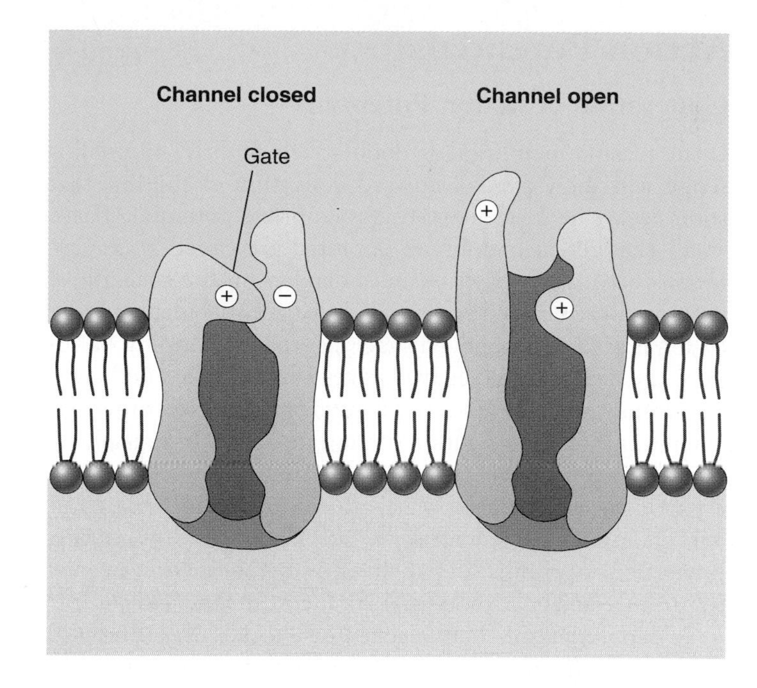

FIGURE 54.7
Voltage-gated ion channels. In neurons and muscle cells, the channels for Na^+ and K^+ have gates that are closed at the resting membrane potential but open when a threshold level of depolarization is attained.

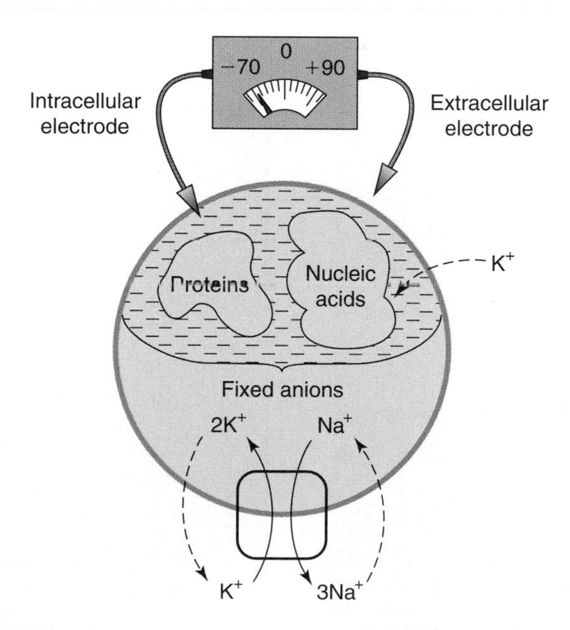

FIGURE 54.8
The establishment of the resting membrane potential. The fixed anions (primarily proteins and nucleic acids) attract cations from the extracellular fluid. If the membrane were only permeable to K^+, an equilibrium would be established and the membrane potential would be –90 mV. A true resting membrane potential is about –70 mV, because the membrane does allow a low rate of Na^+ diffusion into the cell. This is not quite sufficiently negative to prevent the outward diffusion of K^+, so the cell is not at equilibrium and the action of the sodium-potassium pumps is required to maintain stability.

Action Potentials

Generation of Action Potentials

If the plasma membrane is depolarized slightly, an oscilloscope will show a small upward deflection of the line that soon decays back to the resting membrane potential. These small changes in membrane potential are called *graded potentials* because their amplitudes depend on the strength of the stimulus. Graded potentials can be either depolarizing or hyperpolarizing and can add together to amplify or reduce their effects, just as two waves add to make one bigger one when they meet in synchrony or cancel each other out when a trough meets with a crest. The ability of graded potentials to combine is called summation (figure 54.9). Once a particular level of depolarization is reached (about –55 mV in mammalian axons), a nerve impulse, or **action potential,** is produced. The level of depolarization needed to produce an action potential is called the *threshold.*

A depolarization that reaches or exceeds the threshold opens both the Na^+ and the K^+ channels, but the Na^+ channels open first. The rapid diffusion of Na^+ into the cell shifts the membrane potential toward the equilibrium potential for Na^+ (+60 mV—recall that the positive sign indicates that the membrane reverses polarity as Na^+ rushes in).

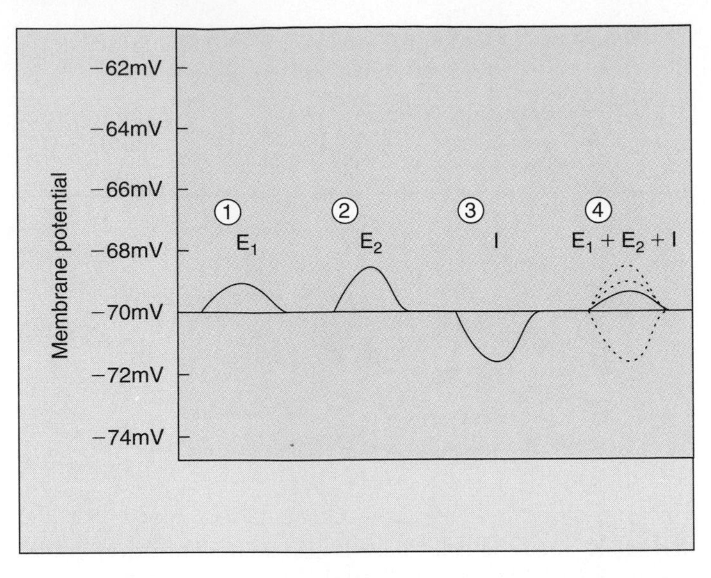

FIGURE 54.9
Graded potentials. (*1*) A weak excitatory stimulus, E_1, elicits a smaller depolarization than (*2*) a stronger stimulus, E_2. (*3*) An inhibitory stimulus, I, produces a hyperpolarization. (*4*) Because graded potentials can summate, if all three stimuli occur very close together, the resulting polarity change will be the algebraic sum of the three changes individually.

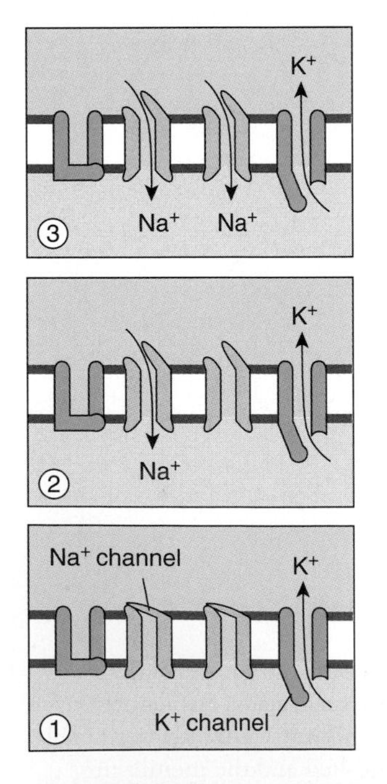

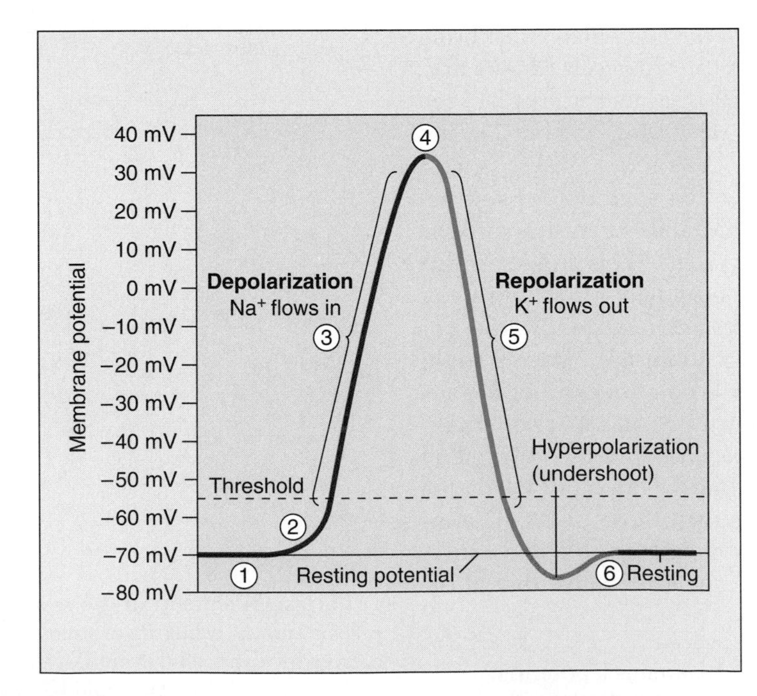

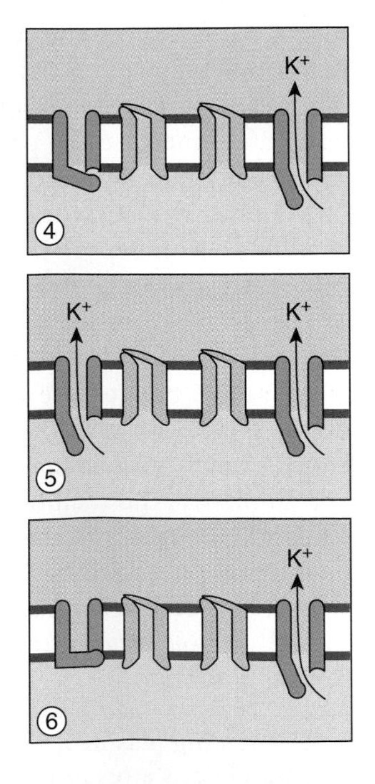

FIGURE 54.10
The action potential. (*1*) At resting membrane potential, some K^+ channels are open. (*2*) In response to a stimulus, the cell begins to depolarize, and once the threshold level is reached, an action potential is produced. (*3*) Rapid depolarization occurs (the rising portion of the spike) because sodium channels open, allowing Na^+ to diffuse into the axon. (*4*) At the top of the spike, Na^+ channels close, and K^+ channels that were previously closed begin to open. (*5*) With the K^+ channels open, repolarization occurs because of the diffusion of K^+ out of the axon. (*6*) An undershoot occurs before the membrane returns to its original resting potential.

When the action potential is recorded on an oscilloscope, this part of the action potential appears as the *rising phase* of a spike (figure 54.10). The membrane potential never quite reaches +60 mV because the Na^+ channels close and, at about the same time, the K^+ channels that were previously closed begin to open. The action potential thus peaks at about +30 mV. Opening the K^+ channels allows K^+ to diffuse out of the cell, repolarizing the plasma membrane. On an oscilloscope, this repolarization of the membrane appears as the *falling phase* of the action potential. In many cases, the repolarization carries the membrane potential to a value slightly more negative than the resting potential for a brief period because K^+ channels remain open, resulting in an *undershoot*. The entire sequence of events in an action potential is over in a few milliseconds.

Action potentials have two distinguishing characteristics. First, they follow an all-or-none law: each depolarization produces either a full action potential, because the voltage-gated Na^+ channels open completely at threshold, or none at all. Secondly, action potentials are always separate events; they cannot add together or interfere with one another as graded potentials can because the membrane enters a brief refractory period after it generates an action potential during which time voltage-gated Na^+ channels cannot reopen.

The production of an action potential results entirely from the passive diffusion of ions. However, at the end of each action potential, the cytoplasm has a little more Na^+ and a little less K^+ than it did at rest. The constant activity of the sodium-potassium pumps compensates for these changes. Thus, although active transport is not required to produce action potentials, it is needed to maintain the ion gradients.

Propagation of Action Potentials

Although we often speak of axons as conducting action potentials (impulses), action potentials do not really travel along an axon—they are events that are reproduced at different points along the axon membrane. This can occur for two reasons: action potentials are stimulated by depolarization, and an action potential can serve as a depolarization stimulus. Each action potential, during its rising phase, reflects a reversal in membrane polarity (from –70 mV to +30 mV) as Na^+ diffuses rapidly into the axon. The positive charges can depolarize the next region of membrane to threshold, so that the next region produces its own action potential (figure 54.11). Meanwhile, the previous region of membrane repolarizes back to the resting membrane potential. This is analogous to people in a stadium performing the "wave": individuals stay in place as they stand up (depolarize), raise their hands (peak of the action potential), and sit down again (repolarize).

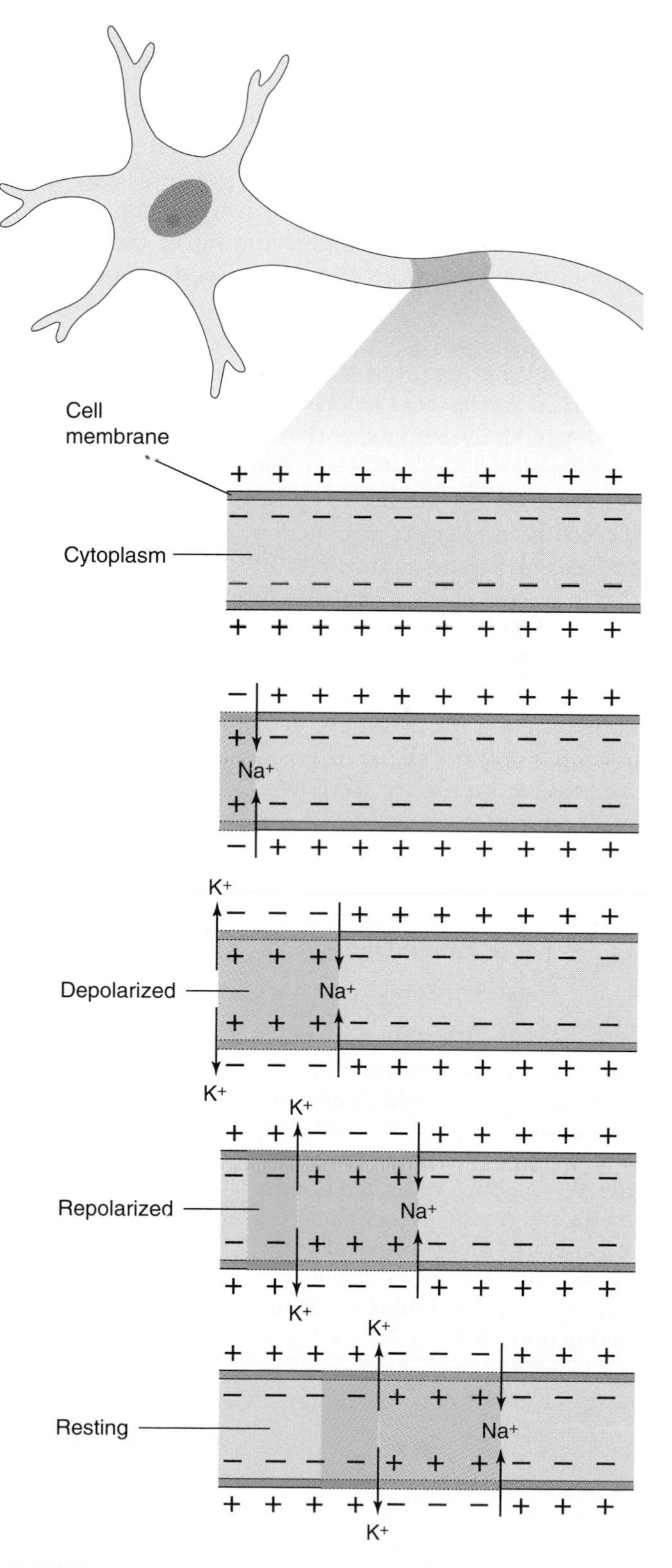

FIGURE 54.11
Propagation of an action potential in an unmyelinated axon. When one region produces an action potential and undergoes a reversal of polarity, it serves as a depolarization stimulus for the next region of the axon. In this way, action potentials are regenerated along each small region of the unmyelinated axon membrane.

Saltatory Conduction

Action potentials are conducted without decrement (without decreasing in amplitude); thus, the last action potential at the end of the axon is just as large as the first action potential. The velocity of conduction is greater if the diameter of the axon is large or if the axon is myelinated (table 54.2). Myelinated axons conduct impulses more rapidly than nonmyelinated axons because the action potentials in myelinated axons are only produced at the nodes of Ranvier. One action potential still serves as the depolarization stimulus for the next, but the depolarization at one node must spread to the next before the voltage-gated channels can be opened. The impulses therefore seem to jump from node to node (figure 54.12) in a process called **saltatory conduction** (Latin *saltare*, "to jump").

To see how saltatory conduction speeds nervous transmission, return for a moment to the "wave" analogy used on the previous page to describe propagation of an action potential. The "wave" moves across the seats of a crowded stadium as fans standing up in one section trigger the next section to stand up in turn. Because the "wave" will skip sections of empty bleachers, it actually progresses around the stadium even faster with more empty sections. The wave doesn't have to wait for the missing people to stand, simply "jumping" the gaps—just as saltatory conduction jumps the nonconduction "gaps" of myelin between exposed nodes.

The rapid inward diffusion of Na^+ followed by the outward diffusion of K^+ produces a rapid change in the membrane potential called an action potential. Action potentials are all-or-none events and cannot summate. Action potentials are regenerated along an axon as one action potential serves as the depolarization stimulus for the next action potential.

Table 54.2 Conduction Velocities of Some Axons

	Axon Diameter (μm)	Myelin	Conduction Velocity (m/s)
Squid giant axon	500	No	25
Large motor axon to human leg muscle	20	Yes	120
Axon from human skin pressure receptor	10	Yes	50
Axon from human skin temperature receptor	5	Yes	20
Motor axon to human internal organ	1	No	2

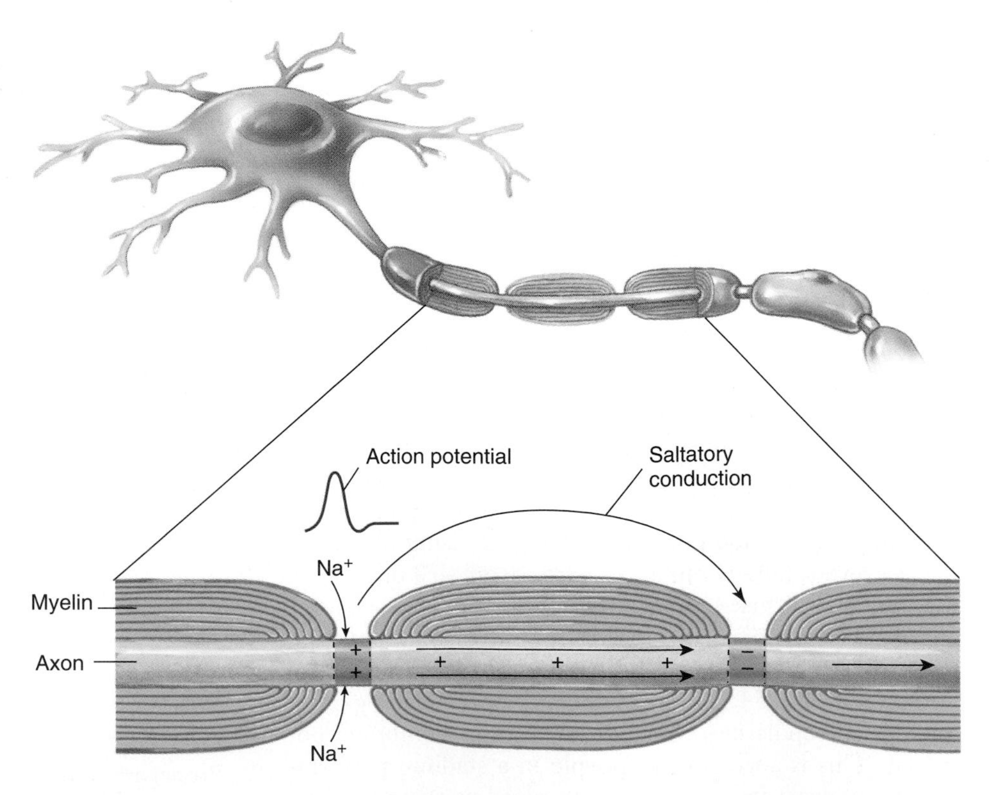

FIGURE 54.12
Saltatory conduction in a myelinated axon. Action potentials are only produced at the nodes of Ranvier in a myelinated axon. One node depolarizes the next node so that the action potentials can skip between nodes. As a result, saltatory ("jumping") conduction in a myelinated axon is more rapid than conduction in an unmyelinated axon.

54.3 Neurons form junctions called synapses with other cells.

Structure of Synapses

An action potential passing down an axon eventually reaches the end of the axon and all of its branches. These branches may form junctions with the dendrites of other neurons, with muscle cells, or with gland cells. Such intercellular junctions are called **synapses.** The neuron whose axon transmits action potentials to the synapse is the *presynaptic cell,* while the cell on the other side of the synapse is the *postsynaptic cell.* Although the presynaptic and postsynaptic cells may appear to touch when the synapse is seen under a light microscope, examination with an electron microscope reveals that most synapses have a **synaptic cleft,** a narrow space that separates these two cells (figure 54.13).

The end of the presynaptic axon is swollen and contains numerous **synaptic vesicles,** which are each packed with chemicals called **neurotransmitters.** When action potentials arrive at the end of the axon, they stimulate the opening of voltage-gated Ca^{++} channels, causing a rapid inward diffusion of Ca^{++}. This serves as the stimulus for the fusion of the synaptic vesicles membrane with the plasma membrane of the axon, so that the contents of the vesicles can be released by exocytosis (figure 54.14). The higher the frequency of action potentials in the presynaptic axon, the more vesicles will release their contents of neurotransmitters. The neurotransmitters diffuse rapidly to the other side of the cleft and bind to **receptor proteins** in the membrane of the postsynaptic cell. There are different types of neurotransmitters, and different ones act in different ways. We will next consider the action of a few of the important neurotransmitter chemicals.

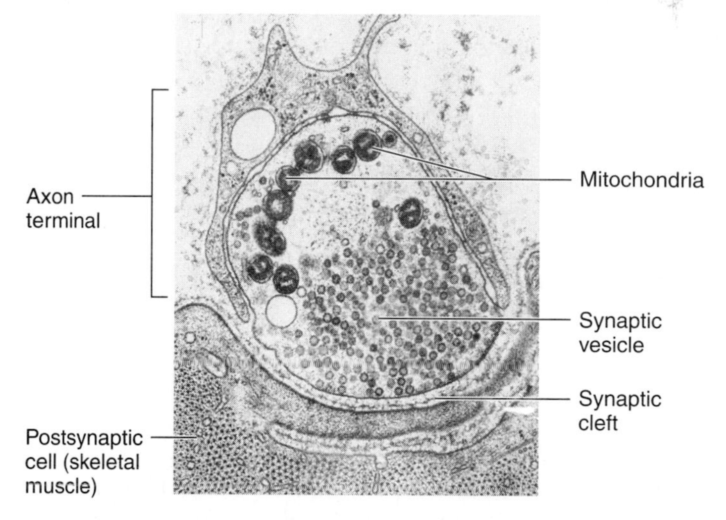

FIGURE 54.13
A synaptic cleft. An electron micrograph showing a neuromuscular synapse.

The presynaptic axon is separated from the postsynaptic cell by a narrow synaptic cleft. Neurotransmitters diffuse across it to transmit a nerve impulse.

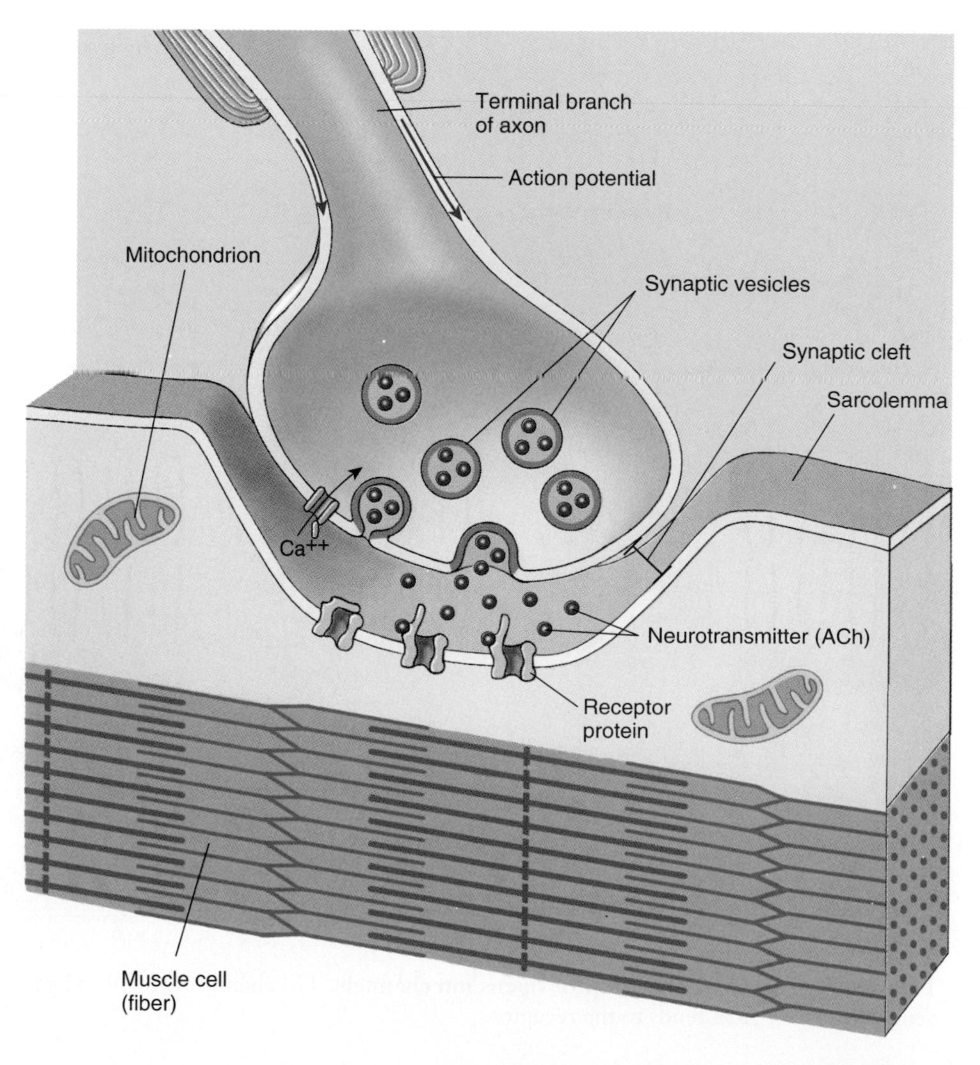

FIGURE 54.14
The release of neurotransmitter. Action potentials arriving at the end of an axon trigger the uptake of Ca^{++}, which causes synaptic vesicles to fuse with the plasma membrane and release their neurotransmitters (acetylcholine [ACh] in this case), which diffuse across the synaptic gap and bind to receptors in the postsynaptic membrane.

Neurotransmitters and Their Functions

Acetylcholine was the first neurotransmitter chemical to be discovered and is widely used in the nervous system. Many other neurotransmitter chemicals have been shown to play important roles, however, and ongoing research continues to produce new information about neurotransmitter function.

Acetylcholine

Acetylcholine (ACh) is the neurotransmitter that crosses the synapse between a motor neuron and a muscle fiber. This synapse is called a *neuromuscular junction* (figure 54.15). Acetylcholine binds to its receptor proteins in the postsynaptic membrane and thereby causes ion channels within these proteins to open (figure 54.16). The gates to these ion channels are said to be *chemically gated* because they open in response to ACh, rather than in response to depolarization. The opening of the chemically gated channels permits Na^+ to diffuse into the postsynaptic cell and K^+ to diffuse out. Although both ions move at the same time, the inward diffusion of Na^+ occurs at a faster rate and has the predominant effect. As a result, that site on the

FIGURE 54.15
Neuromuscular junctions. A light micrograph shows axons branching to make contact with several individual muscle fibers.

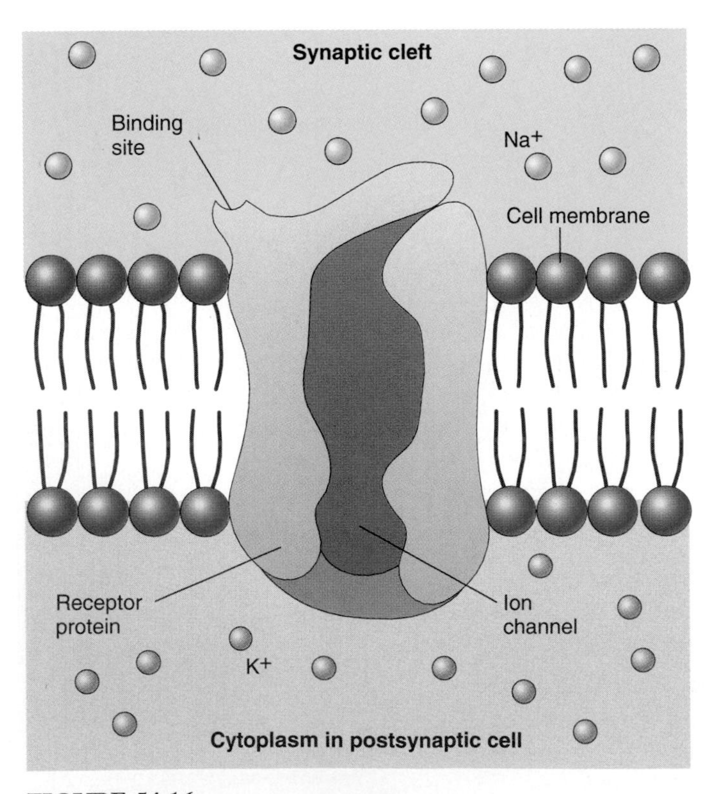

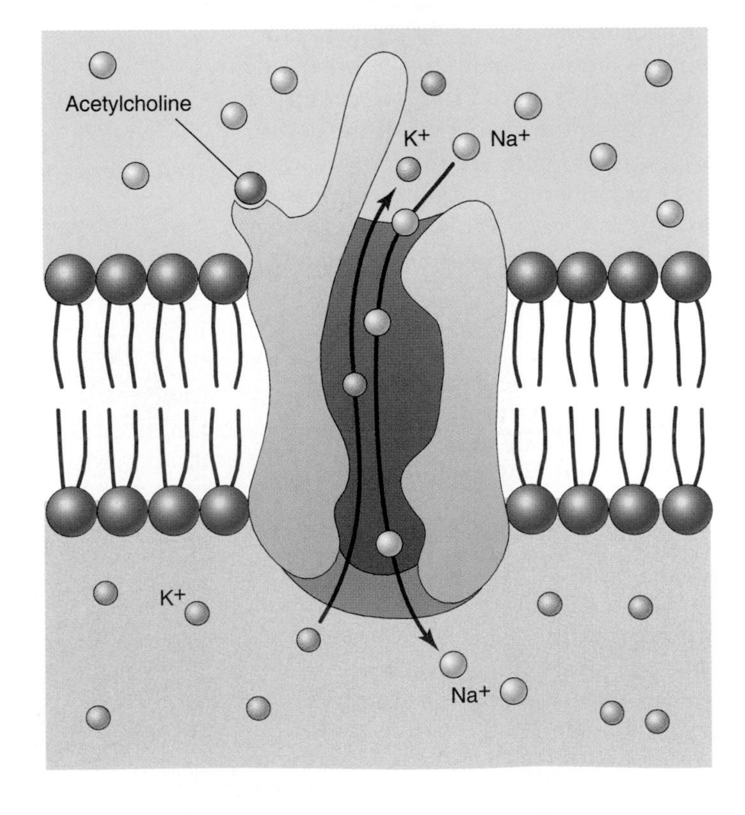

FIGURE 54.16
The binding of ACh to its receptor opens ion channels. The chemically regulated gates to these channels open when the neurotransmitter ACh binds to the receptor.

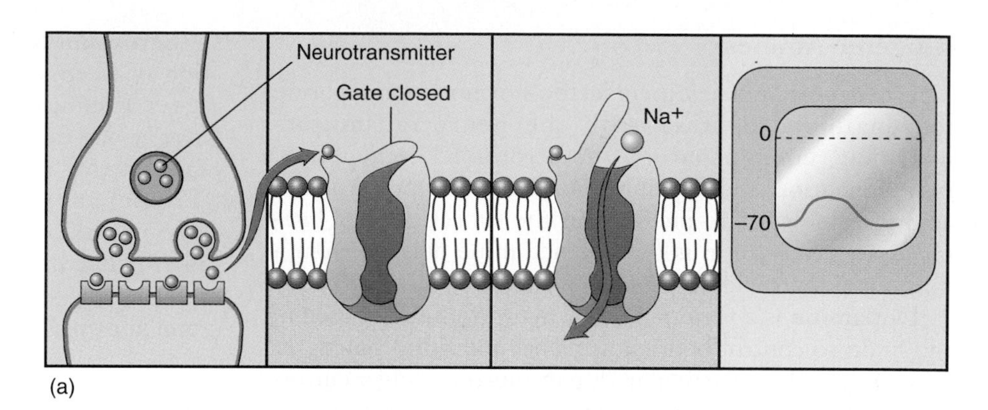

(a)

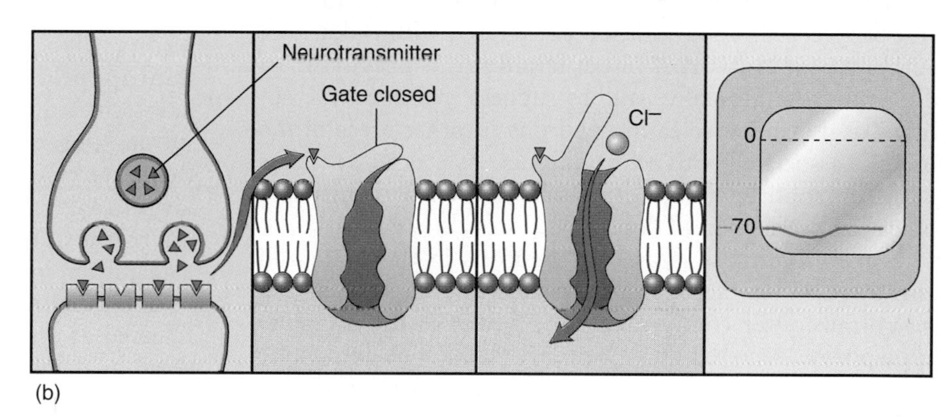

(b)

FIGURE 54.17
Different neurotransmitters can have different effects. (*a*) An excitatory neurotransmitter promotes a depolarization, or excitatory postsynaptic potential (EPSP). (*b*) An inhibitory neurotransmitter promotes a hyperpolarization, or inhibitory postsynaptic potential (IPSP).

postsynaptic membrane produces a depolarization (figure 54.17*a*) called an **excitatory postsynaptic potential (EPSP).** The EPSP, if large enough, can now open the voltage-gated channels for Na^+ and K^+ that are responsible for action potentials. Because the postsynaptic cell we are discussing is a skeletal muscle cell, the action potentials it produces stimulate muscle contraction through the mechanisms discussed in chapter 50.

If ACh stimulates muscle contraction, we must be able to eliminate ACh from the synaptic cleft in order to relax our muscles. This illustrates a general principle: molecules such as neurotransmitters and certain hormones must be quickly eliminated after secretion if they are to be effective regulators. In the case of ACh, the elimination is achieved by an enzyme in the postsynaptic membrane called **acetylcholinesterase (AChE).** This enzyme is one of the fastest known, cleaving ACh into inactive fragments. Nerve gas and the agricultural insecticide parathion are potent inhibitors of AChE and in humans can produce severe spastic paralysis and even death if the respiratory muscles become paralyzed.

Although ACh acts as a neurotransmitter between motor neurons and skeletal muscle cells, many neurons also use ACh as a neurotransmitter at their synapses with other neurons; in these cases, the postsynaptic membrane is generally on the dendrites or cell body of the second neuron. The EPSPs produced must then travel through the dendrites and cell body to the initial segment of the axon, where the first voltage-regulated channels needed for action potentials are located. This is where the first action potentials will be produced, providing that the EPSP depolarization is above the threshold needed to trigger action potentials.

Glutamate, Glycine, and GABA

Glutamate is the major excitatory neurotransmitter in the vertebrate CNS, producing EPSPs and stimulating action potentials in the postsynaptic neurons. Although normal amounts produce physiological stimulation, excessive stimulation by glutamate has been shown to cause neurodegeneration, as in Huntington's chorea.

Glycine and **GABA** (an acronym for gamma-aminobutyric acid) are inhibitory neurotransmitters. If you remember that action potentials are triggered by a threshold level of depolarization, you will understand why hyperpolarization of the membrane would cause inhibition. These neurotransmitters cause the opening of chemically gated channels for Cl^-, which has a concentration gradient favoring its diffusion into the neuron. Because Cl^- is negatively charged, it makes the inside of the membrane even more negative than it is at rest—from –70 mV to –85 mV, for example (figure 54.17*b*). This hyperpolarization is called an **inhibitory postsynaptic potential (IPSP),** and is very important for neural control of body movements and other brain functions. Interestingly, the drug diazepam (Valium) causes its sedative and other effects by enhancing the binding of GABA to its receptors and thereby increasing the effectiveness of GABA at the synapse.

Biogenic Amines

The **biogenic amines** include the hormone epinephrine (adrenaline), together with the neurotransmitters dopamine, norepinephrine, and serotonin. Epinephrine, norepinephrine, and dopamine are derived from the amino acid tyrosine and are included in the subcategory of *catecholamines.* Serotonin is a biogenic amine derived from a different amino acid, tryptophan.

Dopamine is a very important neurotransmitter used in the brain to control body movements and other functions. Degeneration of particular dopamine-releasing neurons produces the resting muscle tremors of Parkinson's disease, and people with this condition are treated with L-dopa (an acronym for dihydroxyphenylalanine), a precursor of dopamine. Additionally, studies suggest that excessive activity of dopamine-releasing neurons in other areas of the brain is associated with schizophrenia. As a result, patients with schizophrenia are sometimes helped by drugs that block the production of dopamine.

Norepinephrine is used by neurons in the brain and also by particular autonomic neurons, where its action as a neurotransmitter complements the action of the hormone epinephrine, secreted by the adrenal gland. The autonomic nervous system will be discussed in a later section of this chapter.

Serotonin is a neurotransmitter involved in the regulation of sleep and is also implicated in various emotional states. Insufficient activity of neurons that release serotonin may be one cause of clinical depression; this is suggested by the fact that antidepressant drugs, particularly fluoxetine (Prozac) and related compounds, specifically block the elimination of serotonin from the synaptic cleft (figure 54.18). The drug lysergic acid diethylamide (LSD) specifically blocks serotonin receptors in a region of the brain stem known as the raphe nuclei.

Other Neurotransmitters

Axons also release various polypeptides, called **neuropeptides,** at synapses. These neuropeptides may have a neurotransmitter function or they may have more subtle, long-term action on the postsynaptic neurons. In the latter case, they are often referred to as **neuromodulators.** A given axon generally releases only one kind of neurotransmitter, but many can release both a neurotransmitter and a neuromodulator.

One important neuropeptide is **substance P,** which is released at synapses in the CNS by sensory neurons activated by painful stimuli. The perception of pain, however, can vary depending on circumstances; an injured football

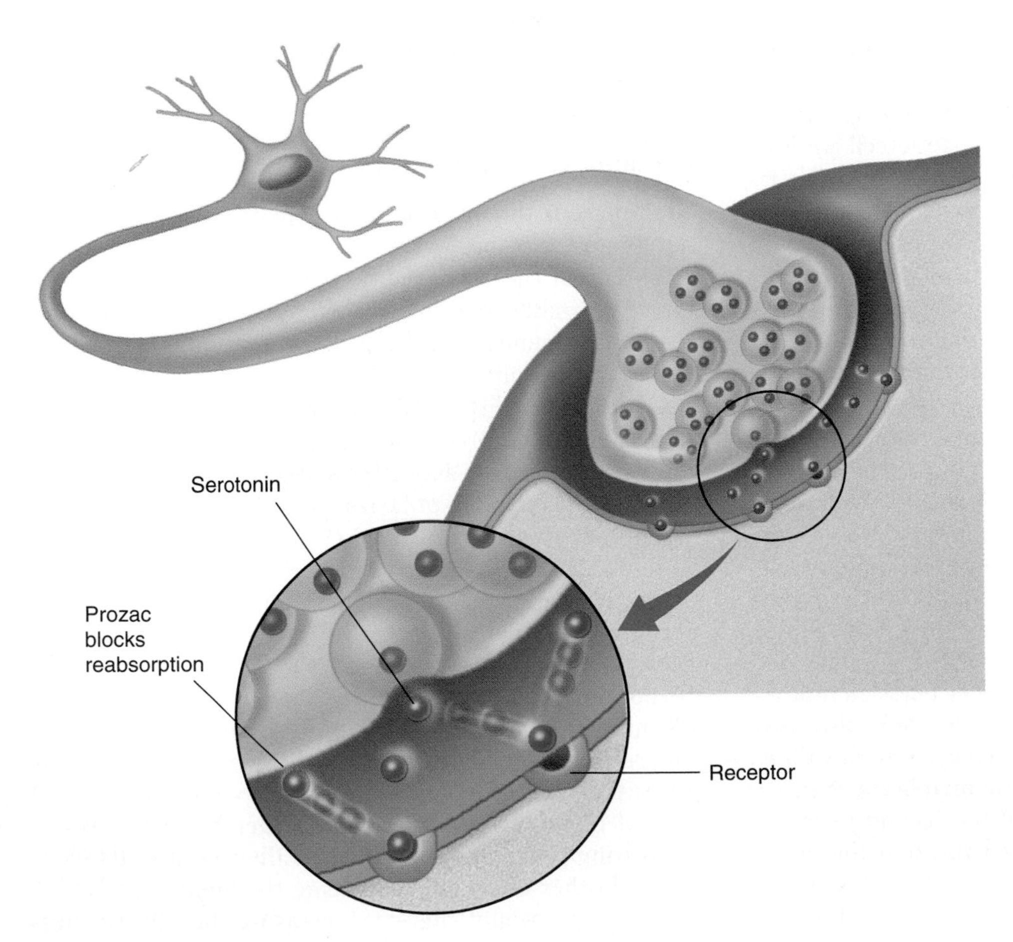

FIGURE 54.18
Serotonin and depression. Depression can result from a shortage of the neurotransmitter serotonin. The antidepressant drug Prozac works by blocking reabsorption of serotonin in the synapse, making up for the shortage.

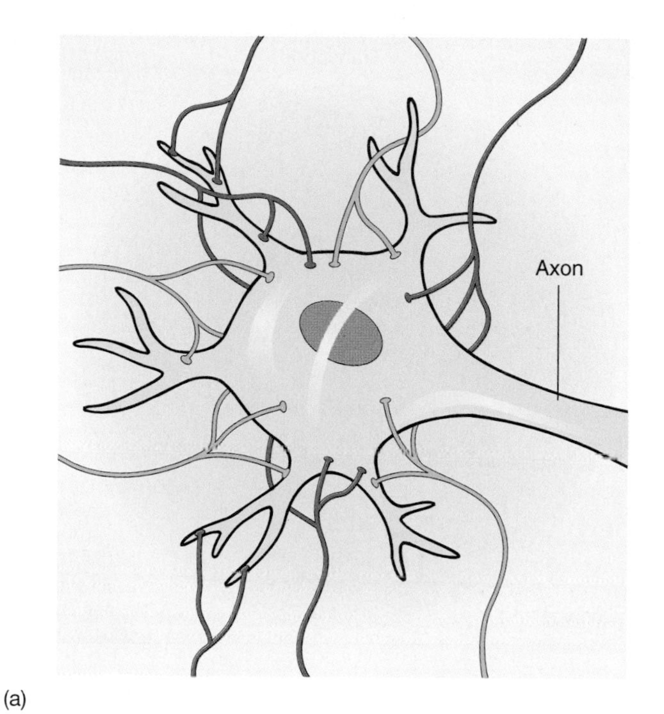

(a)

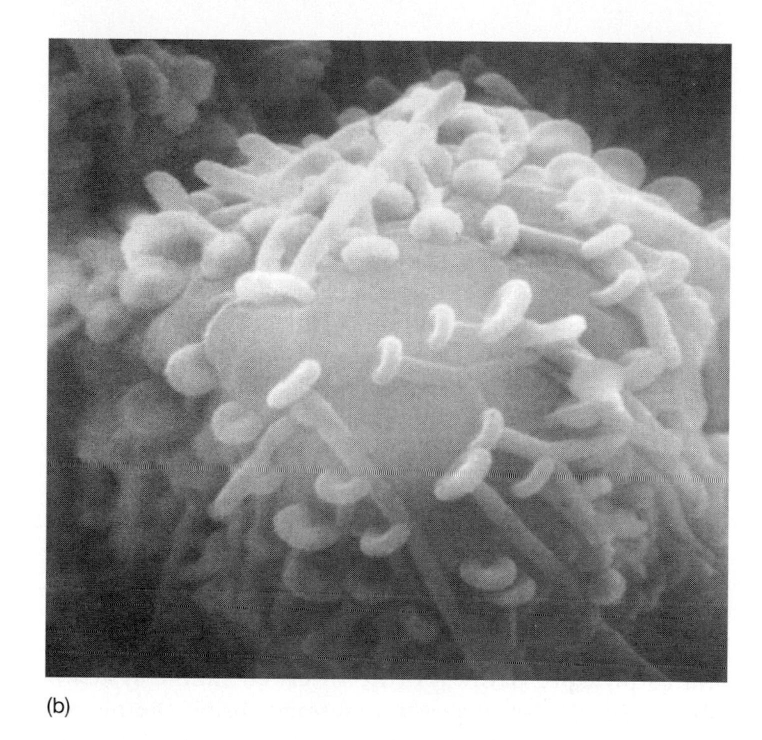

(b)

FIGURE 54.19
Integration of EPSPs and IPSPs takes place on the neuronal cell body. (*a*) The synapses made by some axons are excitatory *(blue)*; the synapses made by other axons are inhibitory *(red)*. The summed influence of all of these inputs determines whether the axonal membrane of the postsynaptic cell will be sufficiently depolarized to produce an action potential. (*b*) Micrograph of a neuronal cell body with numerous synapses (15,000×).

player may not feel the full extent of his trauma, for example, until he's taken out of the game. The intensity with which pain is perceived partly depends on the effects of neuropeptides called **enkephalins** and **endorphins.** Enkephalins are released by axons descending from the brain and inhibit the passage of pain information to the brain. Endorphins are released by neurons in the brain stem and also block the perception of pain. Opium and its derivatives, morphine and heroin, have an analgesic (pain-reducing) effect because they are similar enough in chemical structure to bind to the receptors normally utilized by enkephalins and endorphins. For this reason, the enkephalins and the endorphins are referred to as *endogenous opiates.*

Nitric oxide (NO) is the first gas known to act as a regulatory molecule in the body. Because NO is a gas, it diffuses through membranes so it cannot be stored in vesicles. It is produced as needed from the amino acid arginine. Nitric oxide's actions are very different from those of the more familiar nitrous oxide (N_2O), or laughing gas, sometimes used by dentists. Nitric oxide diffuses out of the presynaptic axon and into neighboring cells by simply passing through the lipid portions of the cell membranes. In the PNS, nitric oxide is released by some neurons that innervate the gastrointestinal tract, penis, respiratory passages, and cerebral blood vessels. These are autonomic neurons that cause smooth muscle relaxation in their target organs. This can produce, for example, the engorgement of the spongy tissue of the penis with blood, causing erection. The drug sildenafil (Viagra) increases the release of nitric oxide in the penis, prolonging erection. Nitric oxide is also released as a neurotransmitter in the brain, and has been implicated in the processes of learning and memory.

Synaptic Integration

The activity of a postsynaptic neuron in the brain and spinal cord of vertebrates is influenced by different types of input from a number of presynaptic neurons. For example, a single motor neuron in the spinal cord can receive as many as 50,000 synapses from presynaptic axons! Each postsynaptic neuron may receive both excitatory and inhibitory synapses. The EPSPs (depolarizations) and IPSPs (hyperpolarizations) from these synapses interact with each other when they reach the cell body of the neuron. Small EPSPs add together to bring the membrane potential closer to the threshold, while IPSPs subtract from the depolarizing effect of the EPSPs, keeping the membrane potential below the threshold (figure 54.19). This process is called **synaptic integration.**

Neurotransmitters and Drug Addiction

When a cell of your body is exposed to a stimulus that produces a chemically mediated signal for a prolonged period, it tends to lose its ability to respond to the stimulus with its original intensity. (You are familiar with this loss of sensitivity—when you sit in a chair, how long are you aware of the chair?) Nerve cells are particularly prone to this loss of sensitivity. If receptor proteins within synapses are exposed to high levels of neurotransmitter molecules for prolonged periods, that nerve cell often responds by inserting fewer receptor proteins into the membrane. This feedback is a normal function in all neurons, one of several mechanisms that have evolved to make the cell more efficient, in this case, adjusting the number of "tools" (receptor proteins) in the membrane "workshop" to suit the workload.

Cocaine. The drug cocaine causes abnormally large amounts of neurotransmitter to remain in the synapses for long periods of time. Cocaine affects nerve cells in the brain's pleasure pathways (the so-called limbic system). These cells transmit pleasure messages using the neurotransmitter dopamine. Using radioactively labeled cocaine molecules, investigators found that cocaine binds tightly to the transporter proteins in synaptic clefts. These proteins normally remove the neurotransmitter dopamine after it has acted. Like a game of musical chairs in which all the chairs become occupied, there are no unoccupied carrier proteins available to the dopamine molecules, so the dopamine stays in the cleft, firing the receptors again and again. As new signals arrive, more and more dopamine is added, firing the pleasure pathway more and more often (figure 54.20).

When receptor proteins on limbic system nerve cells are exposed to high levels of dopamine neurotransmitter molecules for prolonged periods of time, the nerve cells "turn down the volume" of the signal by lowering the number of receptor proteins on their surfaces. They respond to the greater number of neurotransmitter molecules by simply reducing the number of targets available for these molecules to hit. The cocaine user is now addicted (figure 54.21). With so few receptors, the user needs the drug to maintain even normal levels of limbic activity.

Is Nicotine an Addictive Drug? Investigators attempting to explore the habit-forming nature of nicotine used what had been learned about cocaine to carry out what seemed a reasonable experiment—they introduced radioactively labeled nicotine into the brain and looked to see what sort of carrier protein it attached itself to. To their great surprise, the nicotine ignored proteins in the synaptic clefts and instead bound directly to a specific receptor on the postsynaptic cell! This was totally unexpected, as nicotine does not normally occur in the brain—why should it have a receptor there?

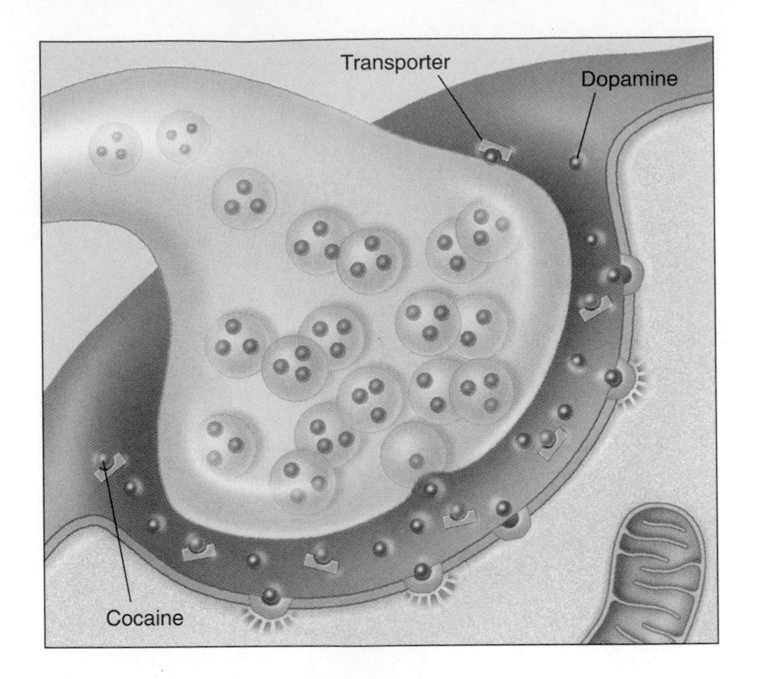

FIGURE 54.20
How cocaine alters events at the synapse. When cocaine binds to the dopamine transporters, the neurotransmitter survives longer in the synapse and continues to stimulate the postsynaptic cell. Cocaine thus acts to intensify pleasurable sensations.

Intensive research followed, and researchers soon learned that the "nicotine receptors" were a class of receptors that normally served to bind the neurotransmitter acetylcholine. There are other types of ACh receptors that don't respond to nicotine. It was just an accident of nature that nicotine, an obscure chemical from a tobacco plant, was also able to bind to them. What, then, is the normal function of these receptors? The target of considerable research, these receptors turned out to be one of the brain's most important tools. The brain uses them to coordinate the activities of many other kinds of receptors, acting to "fine tune" the sensitivity of a wide variety of behaviors.

When neurobiologists compare the nerve cells in the brains of smokers to those of nonsmokers, they find changes in both the number of nicotine receptors and in the levels of RNA used to make the receptors. They have found that the brain adjusts to prolonged exposure to nicotine by "turning down the volume" in two ways: (1) by making fewer receptor proteins to which nicotine can bind; and (2) by altering the pattern of *activation* of the nicotine receptors (that is, their sensitivity to neurotransmitter).

It is this second adjustment that is responsible for the profound effect smoking has on the brain's activities. By overriding the normal system used by the brain to coordinate its many activities, nicotine alters the pattern of release into synaptic clefts of many neurotransmitters, including acetylcholine, dopamine, serotonin, and many

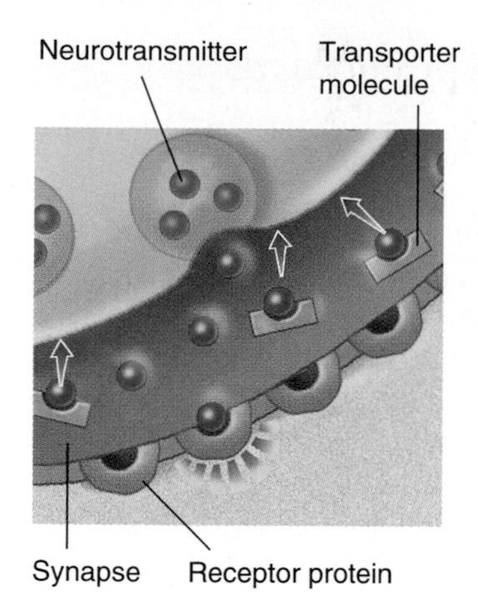

1. Neurotransmitter is reabsorbed at a normal synapse.

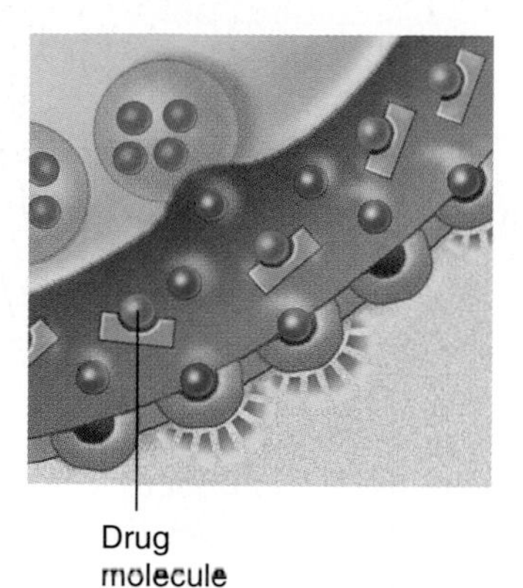

2. Drug molecules prevent reabsorption and cause overstimulation of the postsynaptic membrane.

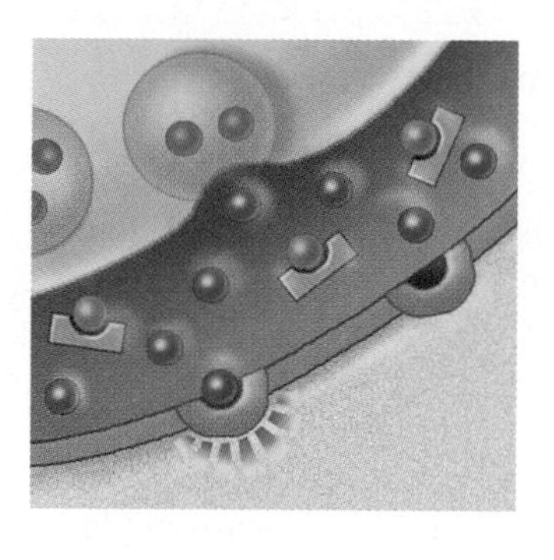

3. The number of receptors decreases.

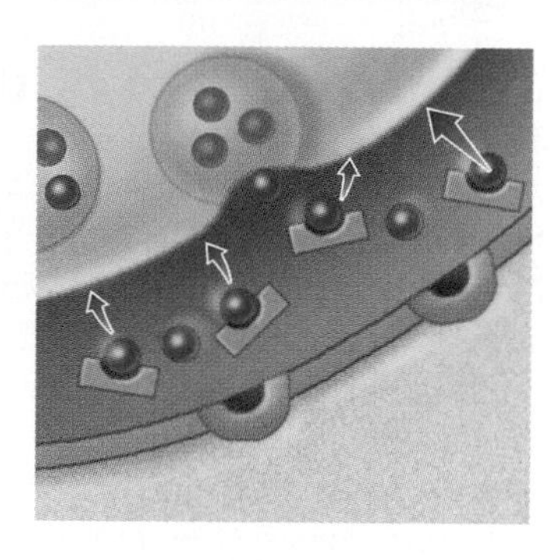

4. The synapse is less sensitive when the drug is removed.

FIGURE 54.21
Drug addiction. (*1*) In a normal synapse, the neurotransmitter binds to a transporter molecule and is rapidly reabsorbed after it has acted. (*2*) When a drug molecule binds to the transporters, reabsorption of the neurotransmitter is blocked, and the postsynaptic cell is overstimulated by the increased amount of neurotransmitter left in the synapse. (*3*) The central nervous system adjusts to the increased firing by producing fewer receptors in the postsynaptic membrane. The result is addiction. (*4*) When the drug is removed, normal absorption of the neurotransmitter resumes, and the decreased number of receptors creates a less-sensitive nerve pathway. Physiologically, the only way a person can then maintain normal functioning is to continue to take the drug. Only if the drug is removed permanently will the nervous system eventually adjust again and restore the original amount of receptors.

others. As a result, changes in level of activity occur in a wide variety of nerve pathways within the brain.

Addiction occurs when chronic exposure to nicotine induces the nervous system to adapt physiologically. The brain compensates for the many changes nicotine induces by making other changes. Adjustments are made to the numbers and sensitivities of many kinds of receptors within the brain, restoring an appropriate balance of activity.

Now what happens if you stop smoking? Everything is out of whack! The newly coordinated system *requires* nicotine to achieve an appropriate balance of nerve pathway activities. This is addiction in any sensible use of the term. The body's physiological response is profound and unavoidable. There is no way to prevent addiction to nicotine with willpower, any more than willpower can stop a bullet when playing Russian roulette with a loaded gun. If you smoke cigarettes for a prolonged period, you will become addicted.

What do you do if you are addicted to smoking cigarettes and you want to stop? When use of an addictive drug like nicotine is stopped, the level of signaling will change to levels far from normal. If the drug is not reintroduced, the altered level of signaling will eventually induce the nerve cells to once again make compensatory changes that restore an appropriate balance of activities within the brain. Over time, receptor numbers, their sensitivity, and patterns of release of neurotransmitters all revert to normal, once again producing normal levels of signaling along the pathways. There is no way to avoid the down side of addiction. The pleasure pathways will not function at normal levels until the number of receptors on the affected nerve cells have time to readjust.

Many people attempt to quit smoking by using patches containing nicotine; the idea is that providing gradually lesser doses of nicotine allows the smoker to be weaned of his or her craving for cigarettes. The patches do reduce the craving for cigarettes—so long as you keep using the patches! Actually, using such patches simply substitutes one (admittedly less dangerous) nicotine source for another. If you are going to quit smoking, there is no way to avoid the necessity of eliminating the drug to which you are addicted. Hard as it is to hear the bad news, there is no easy way out. The only way to quit is to quit.

Acetylcholine stimulates the opening of chemically gated ion channels, causing a depolarization called an excitatory postsynaptic potential (EPSP). Glycine and GABA are inhibitory neurotransmitters that produce hyperpolarization of the postsynaptic membrane. There are also many other neurotransmitters, including the biogenic amines: dopamine, norepinephrine, and serotonin. The effects of different neurotransmitters are integrated through summation of depolarizations and hyperpolarizations.

54.4 The central nervous system consists of the brain and spinal cord.

The Evolution of the Vertebrate Brain

Sponges are the only major phylum of multicellular animals that lack nerves. The simplest nervous systems occur among cnidarians (figure 54.22): all neurons are similar and are linked to one another in a web, or *nerve net*. There is no associative activity, no control of complex actions, and little coordination The simplest animals with associative activity in the nervous system are the free-living flatworms, phylum Platyhelminthes. Running down the bodies of these flatworms are two nerve cords; peripheral nerves extend outward to the muscles of the body. The two nerve cords converge at the front end of the body, forming an enlarged mass of nervous tissue that also contains associative neurons with synapses connecting neurons to one another. This primitive "brain" is a rudimentary central nervous system and permits a far more complex control of muscular responses than is possible in cnidarians.

All of the subsequent evolutionary changes in nervous systems can be viewed as a series of elaborations on the characteristics already present in flatworms. For example, earthworms exhibit a central nervous system that is connected to all other parts of the body by peripheral nerves. And, in arthropods, the central coordination of complex response is increasingly localized in the front end of the nerve cord. As this region evolved, it came to contain a progressively larger number of associative interneurons, and to develop tracts, which are highways within the brain that connect associative elements.

Casts of the interior braincases of fossil agnathans, fishes that swam 500 million years ago, have revealed much about the early evolutionary stages of the vertebrate brain. Although small, these brains already had the three divisions that characterize the brains of all contemporary vertebrates: (1) the hindbrain, or rhombencephalon; (2) the midbrain, or mesencephalon; and (3) the forebrain, or prosencephalon (figure 54.23).

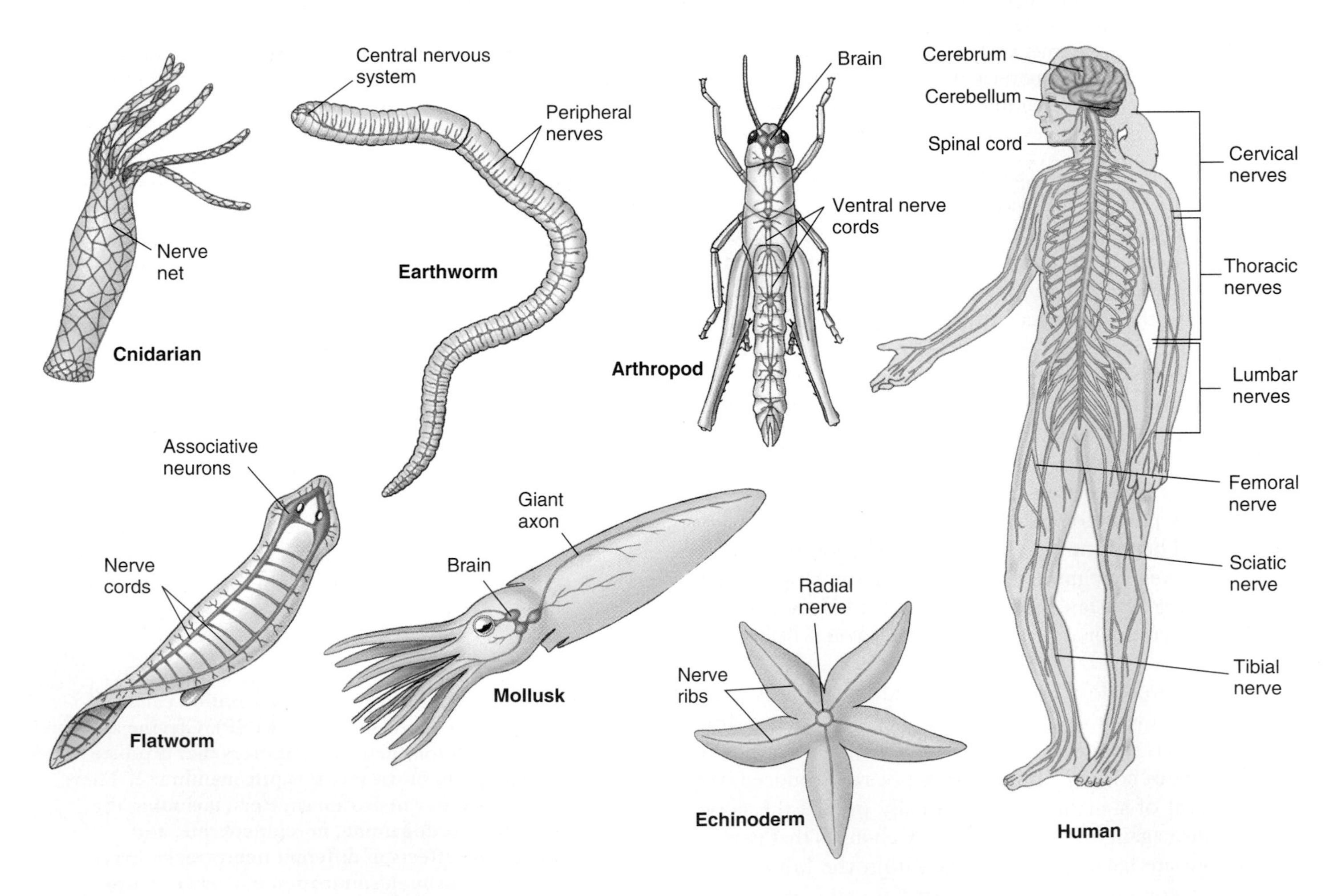

FIGURE 54.22
Evolution of the nervous system. Animals exhibit a progressive elaboration of organized nerve cords and the centralization of complex responses in the front end of the nerve cord.

The hindbrain was the major component of these early brains, as it still is in fishes today. Composed of the *cerebellum*, *pons*, and *medulla oblongata*, the hindbrain may be considered an extension of the spinal cord devoted primarily to coordinating motor reflexes. Tracts containing large numbers of axons run like cables up and down the spinal cord to the hindbrain. The hindbrain, in turn, integrates the many sensory signals coming from the muscles and coordinates the pattern of motor responses.

Much of this coordination is carried on within a small extension of the hindbrain called the cerebellum ("little cerebrum"). In more advanced vertebrates, the cerebellum plays an increasingly important role as a coordinating center for movement and is correspondingly larger than it is in the fishes. In all vertebrates, the cerebellum processes data on the current position and movement of each limb, the state of relaxation or contraction of the muscles involved, and the general position of the body and its relation to the outside world. These data are gathered in the cerebellum, synthesized, and the resulting commands issued to efferent pathways.

In fishes, the remainder of the brain is devoted to the reception and processing of sensory information. The midbrain is composed primarily of the **optic lobes,** which receive and process visual information, while the forebrain is devoted to the processing of *olfactory* (smell) information. The brains of fishes continue growing throughout their lives. This continued growth is in marked contrast to the brains of other classes of vertebrates, which generally complete their development by infancy (figure 54.24). The human brain continues to develop through early childhood, but no new neurons are produced once development has ceased, except in the tiny hippocampus, which controls which experiences are filed away into long-term memory and which are forgotten.

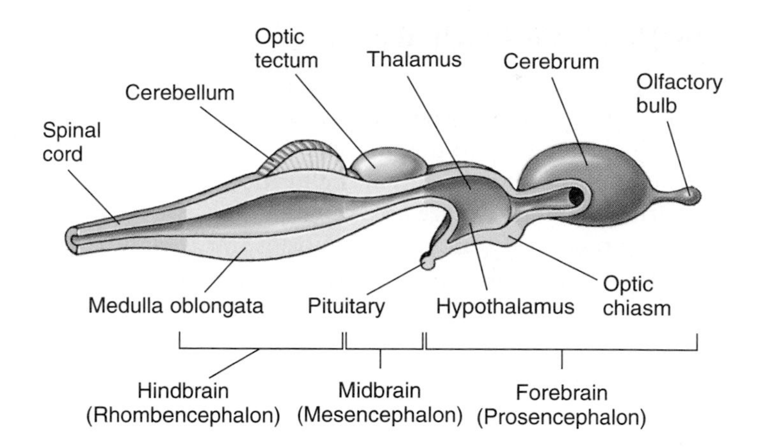

FIGURE 54.23
The basic organization of the vertebrate brain can be seen in the brains of primitive fishes. The brain is divided into three regions that are found in differing proportions in all vertebrates: the hindbrain, which is the largest portion of the brain in fishes; the midbrain, which in fishes is devoted primarily to processing visual information; and the forebrain, which is concerned mainly with olfaction (the sense of smell) in fishes. In terrestrial vertebrates, the forebrain plays a far more dominant role in neural processing than it does in fishes.

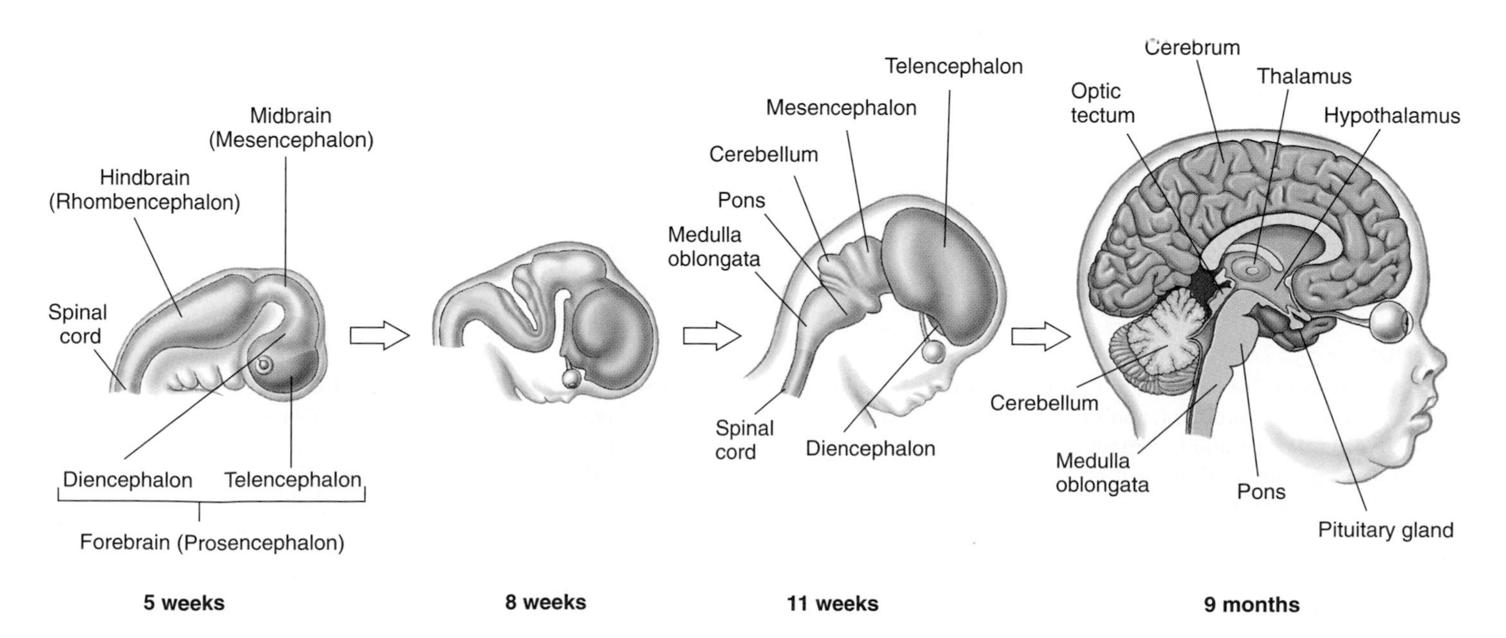

FIGURE 54.24
Development of the brain in humans. The main regions of the brain form during fetal development.

The Dominant Forebrain

Starting with the amphibians and continuing more prominently in the reptiles, processing of sensory information is increasingly centered in the forebrain. This pattern was the dominant evolutionary trend in the further development of the vertebrate brain (figure 54.25).

The forebrain in reptiles, amphibians, birds, and mammals is composed of two elements that have distinct functions. The *diencephalon* (Greek *dia*, "between") consists of the thalamus and hypothalamus. The **thalamus** is an integrating and relay center between incoming sensory information and the cerebrum. The **hypothalamus** participates in basic drives and emotions and controls the secretions of the pituitary gland. The *telencephalon*, or "end brain" (Greek *telos*, "end"), is located at the front of the forebrain and is devoted largely to associative activity. In mammals, the telencephalon is called the **cerebrum.**

The Expansion of the Cerebrum

In examining the relationship between brain mass and body mass among the vertebrates (figure 54.26), you can

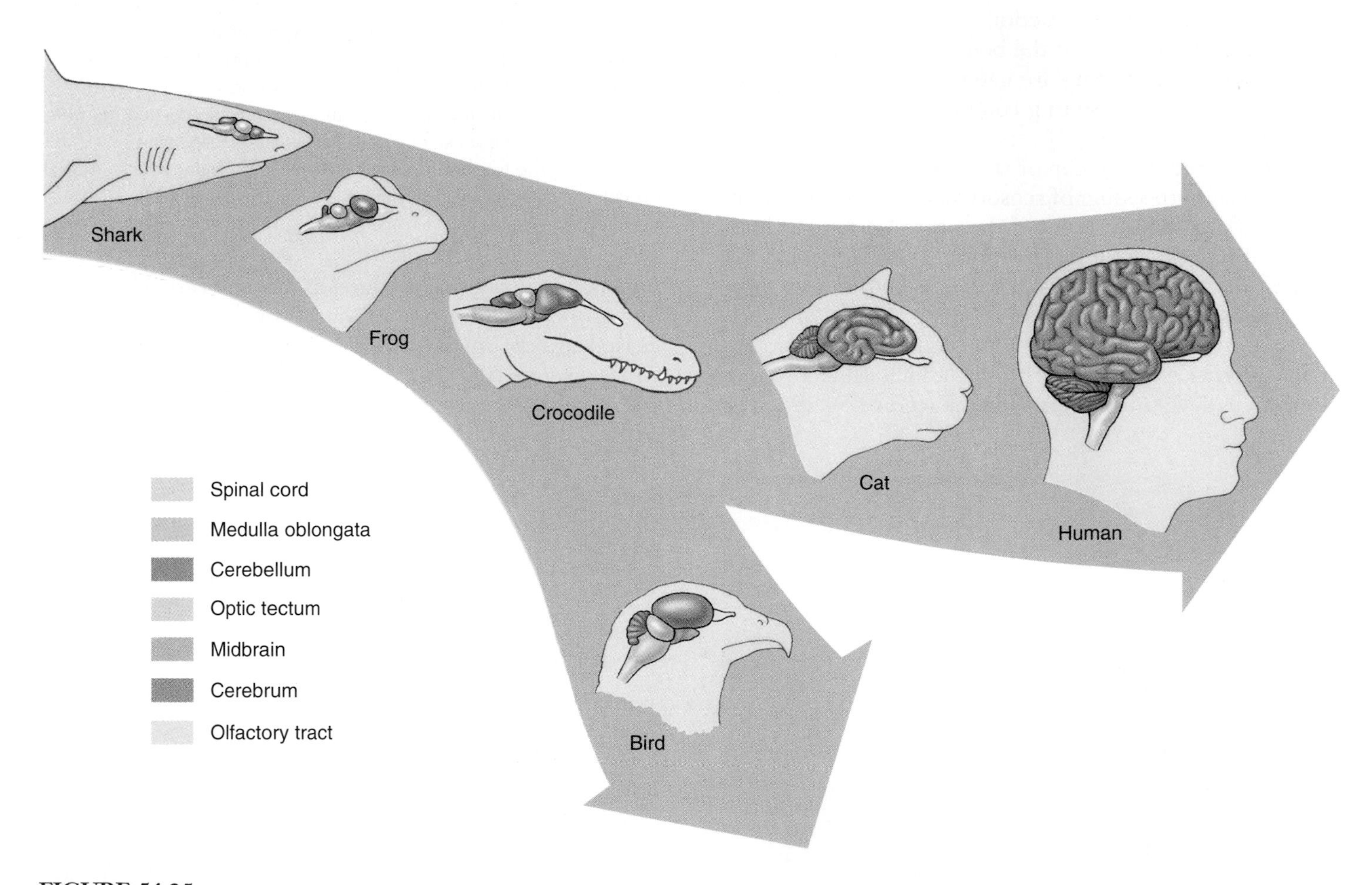

FIGURE 54.25
The evolution of the vertebrate brain involved changes in the relative sizes of different brain regions. In sharks and other fishes, the hindbrain is predominant, and the rest of the brain serves primarily to process sensory information. In amphibians and reptiles, the forebrain is far larger, and it contains a larger cerebrum devoted to associative activity. In birds, which evolved from reptiles, the cerebrum is even more pronounced. In mammals, the cerebrum covers the optic tectum and is the largest portion of the brain. The dominance of the cerebrum is greatest in humans, in whom it envelops much of the rest of the brain.

see a remarkable difference between fishes and reptiles, on the one hand, and birds and mammals, on the other. Mammals have brains that are particularly large relative to their body mass. This is especially true of porpoises and humans; the human brain weighs about 1.4 kilograms. The increase in brain size in the mammals largely reflects the great enlargement of the cerebrum, the dominant part of the mammalian brain. The cerebrum is the center for correlation, association, and learning in the mammalian brain. It receives sensory data from the thalamus and issues motor commands to the spinal cord via descending tracts of axons.

In vertebrates, the central nervous system is composed of the brain and the spinal cord (table 54.3). These two structures are responsible for most of the information processing within the nervous system and consist primarily of interneurons and neuroglia. Ascending tracts carry sensory information to the brain. Descending tracts carry impulses from the brain to the motor neurons and interneurons in the spinal cord that control the muscles of the body.

The vertebrate brain consists of three primary regions: the forebrain, midbrain, and hindbrain. The hindbrain was the principal component of the brain of early vertebrates; it was devoted to the control of motor activity. In vertebrates more advanced than fishes, the processing of information is increasingly centered in the forebrain.

Table 54.3 Subdivisions of the Central Nervous System

Major Subdivision	Function
SPINAL CORD	Spinal reflexes; relays sensory information
HINDBRAIN (Rhombencephalon)	
Medulla oblongata	Sensory nuclei; reticular activating system; visceral control
Pons	Reticular activating system; visceral control
Cerebellum	Coordination of movements; balance
MIDBRAIN (Mesencephalon)	Reflexes involving eyes and ears
FOREBRAIN (Prosencephalon)	
Diencephalon	
Thalamus	Relay station for ascending sensory and descending tracts; visceral control
Hypothalamus	Visceral control; neuroendocrine control
Telencephalon (cerebrum)	
Basal ganglia	Motor control
Corpus callosum	Connects the two hemispheres
Hippocampus (limbic system)	Memory; emotion
Cerebral cortex	Higher functions

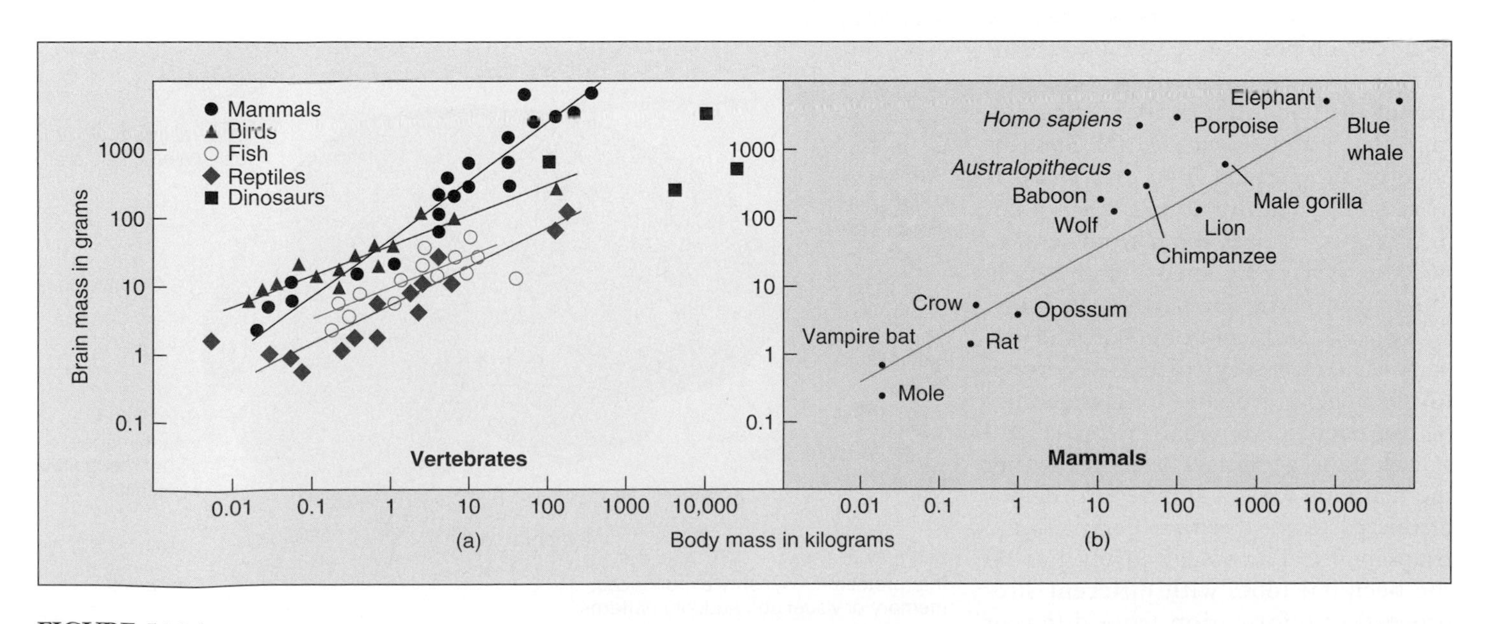

FIGURE 54.26
Brain mass versus body mass. Among most vertebrates, brain mass is a relatively constant proportion of body mass, so that a plot of brain mass versus body mass gives a straight line. (*a*) However, the proportion of brain mass to body mass is much greater in birds than in reptiles, and it is greater still in mammals. (*b*) Among mammals, humans have the greatest brain mass per unit of body mass (that is, the farthest perpendicular distance from the plotted line). In second place are the porpoises.

The Human Forebrain

The human cerebrum is so large that it appears to envelop the rest of the brain (figure 54.27). It is split into right and left cerebral hemispheres, which are connected by a tract called the *corpus callosum.* The hemispheres are further divided into the *frontal, parietal, temporal,* and *occipital lobes.*

Each hemisphere receives sensory input from the opposite, or contralateral, side of the body and exerts motor control primarily over that side. Therefore, a touch on the right hand, for example, is relayed primarily to the left hemisphere, which may then initiate movement of that hand in response to the touch. Damage to one hemisphere due to a stroke often results in a loss of sensation and paralysis on the contralateral side of the body.

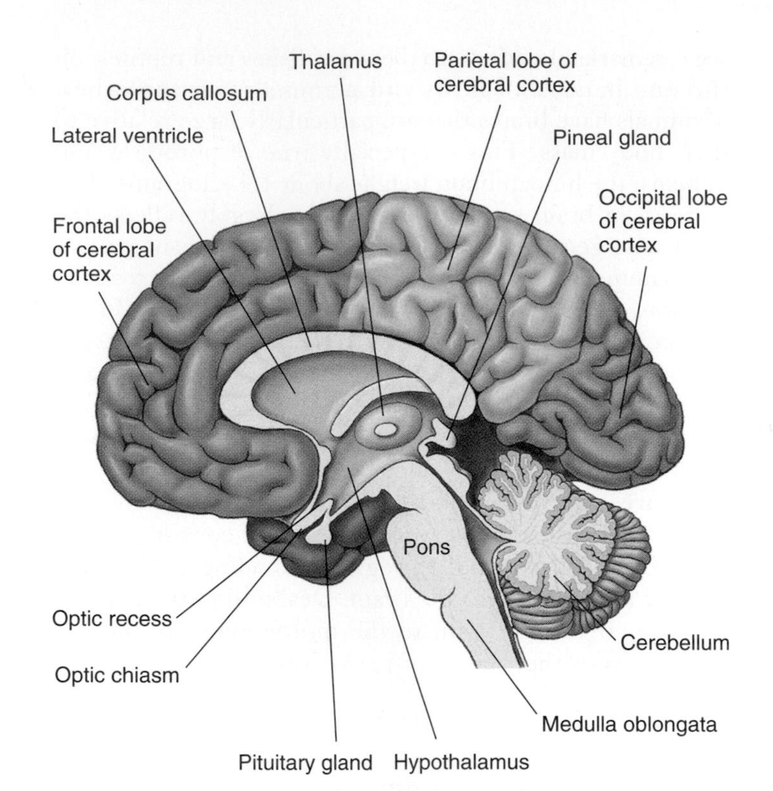

FIGURE 54.27
A section through the human brain. In this sagittal section showing one cerebral hemisphere, the corpus callosum, a fiber tract connecting the two cerebral hemispheres, can be clearly seen.

Cerebral Cortex

Much of the neural activity of the cerebrum occurs within a layer of gray matter only a few millimeters thick on its outer surface. This layer, called the **cerebral cortex,** is densely packed with nerve cells. In humans, it contains over 10 billion nerve cells, amounting to roughly 10% of all the neurons in the brain. The surface of the cerebral cortex is highly convoluted; this is particularly true in the human brain, where the convolutions increase the surface area of the cortex threefold.

The activities of the cerebral cortex fall into one of three general categories: motor, sensory, and associative. The primary motor cortex lies along the *gyrus* (convolution) on the posterior border of the frontal lobe, just in front of the central *sulcus* (crease) (figure 54.28). Each point on its surface is associated with the movement of a different part of the body (figure 54.29). Just behind the central sulcus, on the anterior edge of the parietal lobe, lies the primary somatosensory cortex. Each point in this area receives input from sensory neurons serving cutaneous and muscle senses in a particular part of the body. Large areas of the motor cortex and primary somatosensory cortex are devoted to the fingers, lips, and tongue because of the need for manual dexterity and speech. The auditory cortex lies within the temporal lobe, and different regions of this cortex deal with different sound frequencies. The visual cortex lies on the occipital lobe, with different sites processing information from different positions on the retina, equivalent to particular points in the visual fields of the eyes.

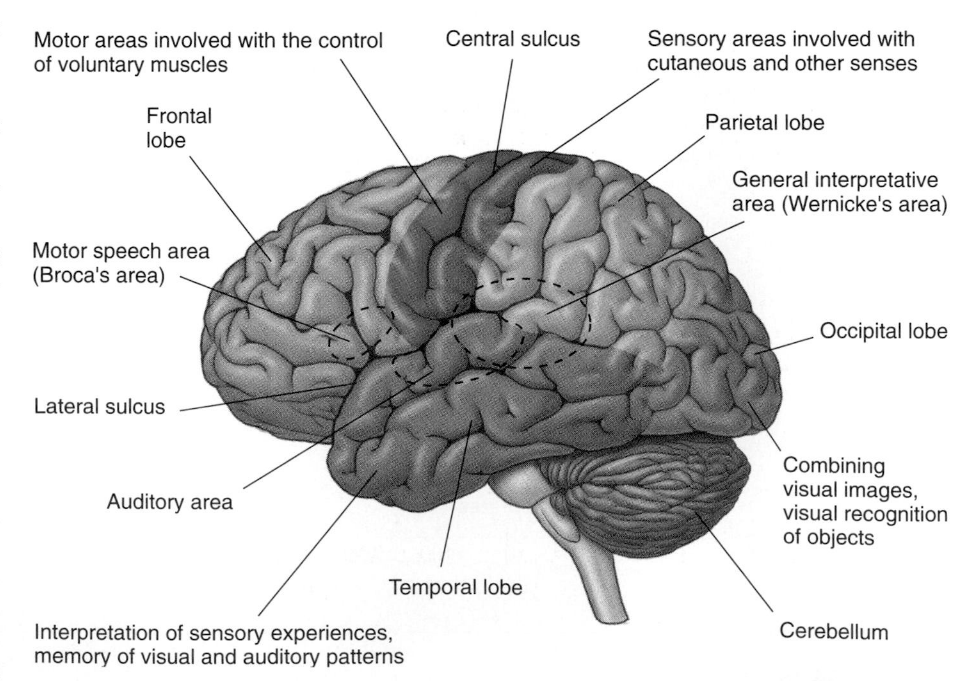

FIGURE 54.28
The lobes of the cerebrum. Some of the known regions of specialization are indicated in this diagram.

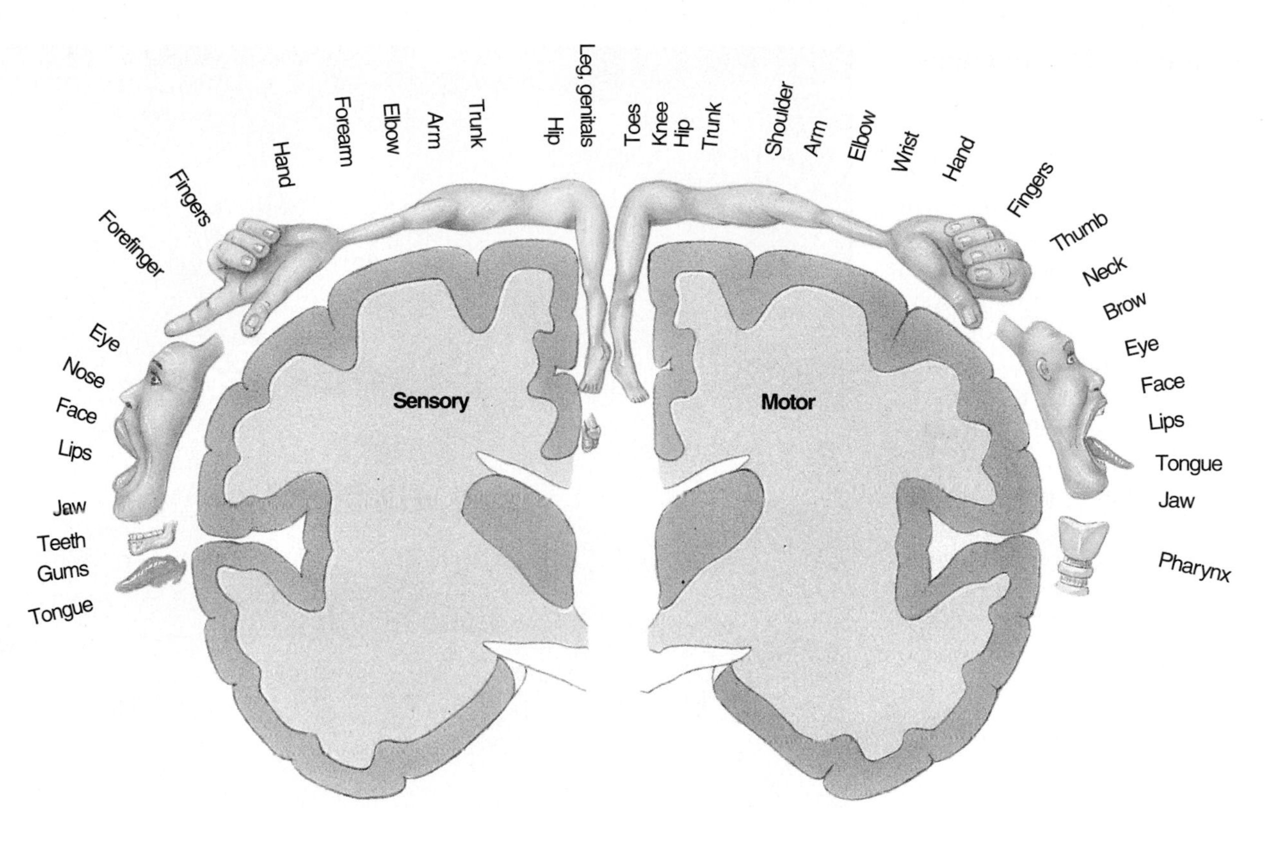

FIGURE 54.29
The primary somatosensory cortex *(left)* and the primary motor cortex *(right)*. Each of these regions of the cerebral cortex is associated with a different region of the body, as indicated in this stylized map. The areas of the body are drawn in proportion to the amount of cortex dedicated to their sensation or control. For example, the hands have large areas of sensory and motor control, while the pharynx has a considerable area of motor control but little area devoted to the sensations of the pharynx.

The portion of the cerebral cortex that is not occupied by these motor and sensory cortices is referred to as *association cortex.* The site of higher mental activities, the association cortex reaches its greatest extent in primates, especially humans, where it makes up 95% of the surface of the cerebral cortex.

Basal Ganglia

Buried deep within the white matter of the cerebrum are several collections of cell bodies and dendrites that produce islands of gray matter. These aggregates of neuron cell bodies, which are collectively termed the basal ganglia, receive sensory information from ascending tracts and motor commands from the cerebral cortex and cerebellum. Outputs from the basal ganglia are sent down the spinal cord, where they participate in the control of body movements. Damage to specific regions of the basal ganglia can produce the resting tremor of muscles that is characteristic of people with Parkinson's disease.

Thalamus and Hypothalamus

The thalamus is a primary site of sensory integration in the brain. Visual, auditory, and somatosensory information is sent to the thalamus, where the sensory tracts synapse with association neurons. The sensory information is then relayed via the thalamus to the occipital, temporal, and parietal lobes of the cerebral cortex, respectively. The transfer of each of these types of sensory information is handled by specific aggregations of neuron cell bodies within the thalamus.

The hypothalamus integrates the visceral activities. It helps regulate body temperature, hunger and satiety, thirst, and—along with the limbic system—various emotional states. The hypothalamus also controls the pituitary gland, which in turn regulates many of the other endocrine glands of the body. By means of its interconnections with the cerebral cortex and with control centers in the brain stem (a term used to refer collectively to the midbrain, pons, and medulla oblongata), the hypothalamus helps coordinate the neural and hormonal responses to many internal stimuli and emotions.

The *hippocampus* and *amygdala* are, together with the hypothalamus, the major components of the **limbic system.** This is an evolutionarily ancient group of linked structures deep within the cerebrum that are responsible for emotional responses. The hippocampus is also believed to be important in the formation and recall of memories, a topic we will discuss later.

Language and Other Functions

Arousal and Sleep. The brain stem contains a diffuse collection of neurons referred to as the *reticular formation.* One part of this formation, the reticular activating system, controls consciousness and alertness. All of the sensory pathways feed into this system, which monitors the information coming into the brain and identifies important stimuli. When the reticular activating system has been stimulated to arousal, it increases the level of activity in many parts of the brain. Neural pathways from the reticular formation to the cortex and other brain regions are depressed by anesthetics and barbiturates.

The reticular activating system controls both sleep and the waking state. It is easier to sleep in a dark room than in a lighted one because there are fewer visual stimuli to stimulate the reticular activating system. In addition, activity in this system is reduced by serotonin, a neurotransmitter we previously discussed. Serotonin causes the level of brain activity to fall, bringing on sleep.

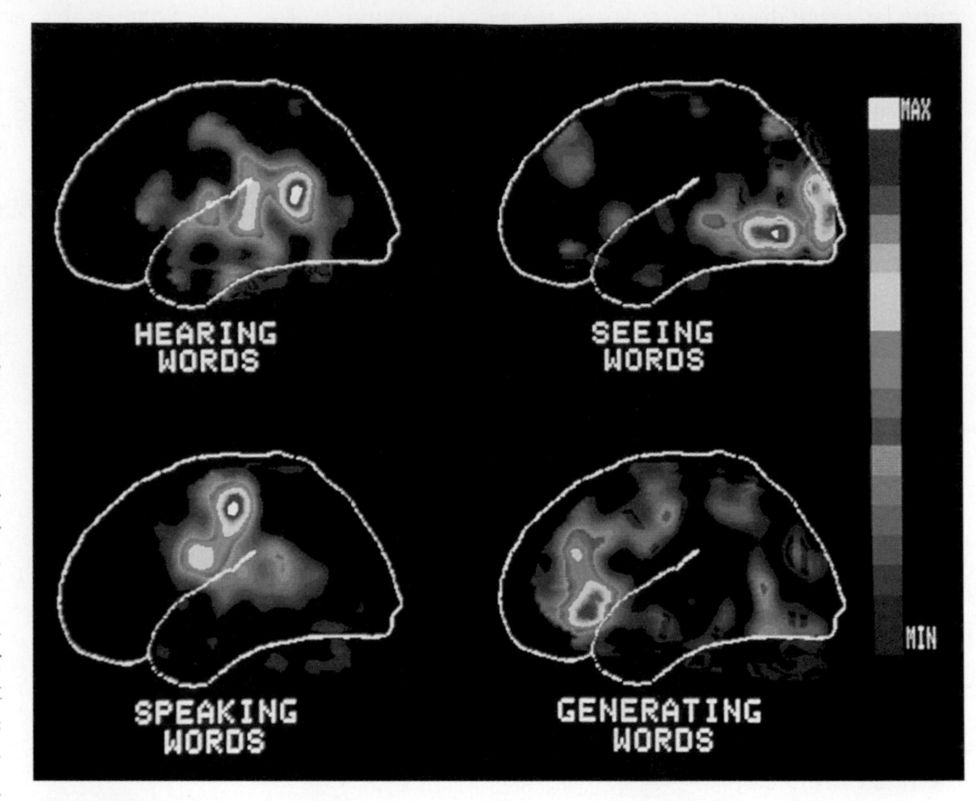

FIGURE 54.30
Different brain regions control various activities. This illustration shows how the brain reacts in human subjects asked to listen to a spoken word, to read that same word silently, to repeat the word out loud, and then to speak a word related to the first. Regions of white, red, and yellow show the greatest activity. Compare this with figure 54.28 to see how regions of the brain are mapped.

Sleep is not the loss of consciousness. Rather, it is an active process whose multiple states can be revealed by recording the electrical activity of the brain in an electroencephalogram (EEG). In a relaxed but awake individual whose eyes are shut, the EEG consists primarily of large, slow waves that occur at a frequency of 8 to 13 hertz (cycles per second). These waves are referred to as *alpha waves.* In an alert subject whose eyes are open, the EEG waves are more rapid (*beta waves* are seen at frequencies of 13 to 30 hertz) and is more desynchronized because multiple sensory inputs are being received, processed, and translated into motor activities.

Theta waves (4 to 7 hertz) and *delta waves* (0.5 to 4 hertz) are seen in various stages of sleep. The first change seen in the EEG with the onset of drowsiness is a slowing and reduction in the overall amplitude of the waves. This slow-wave sleep has several stages but is generally characterized by decreases in arousability, skeletal muscle tone, heart rate, blood pressure, and respiratory rate. During REM sleep (named for the rapid eye movements that occur during this stage), the EEG resembles that of a relaxed, awake individual, and the heart rate, blood pressure, and respiratory rate are all increased. Paradoxically, individuals in REM sleep are difficult to arouse and are more likely to awaken spontaneously. Dreaming occurs during REM sleep, and the rapid eye movements resemble the tracking movements made by the eyes when awake, suggesting that dreamers "watch" their dreams.

Language and Spatial Recognition. Although the two cerebral hemispheres seem structurally similar, they are responsible for different activities. The most thoroughly investigated example of this lateralization of function is language. The left hemisphere is the "dominant" hemisphere for language—the hemisphere in which most neural processing related to language is performed—in 90% of right-handed people and nearly two-thirds of left-handed people. There are two language areas in the dominant hemisphere. Wernicke's area (see figure 54.28), located in the parietal lobe between the primary auditory and visual areas, is important for language comprehension and the formulation of thoughts into speech (figure 54.30). Broca's area, found near the part of the motor cortex controlling the face, is responsible for the generation of motor output needed for language communication. Damage to these brain areas can cause language disorders known as *aphasias*. For example, if Wernicke's area is damaged, the person's speech is rapid and fluid but lacks meaning; words are tossed together as in a "word salad."

While the dominant hemisphere for language is adept at sequential reasoning, like that needed to formulate a sentence, the nondominant hemisphere (the right hemisphere in most people) is adept at spatial reasoning, the type of reasoning needed to assemble a puzzle or draw a picture. It is also the hemisphere primarily involved in musical ability—a person with damage to Broca's speech area in the left hemisphere may not be able to speak but may retain the ability to sing! Damage to the nondominant hemisphere may lead to an inability to appreciate spatial relationships and may impair musical activities such as singing. Even more specifically, damage to the inferior temporal cortex in that hemisphere eliminates the capacity to recall faces. Reading, writing, and oral comprehension remain normal, and patients with this disability can still recognize acquaintances by their voices. The nondominant hemisphere is also important for the consolidation of memories of nonverbal experiences.

Memory and Learning. One of the great mysteries of the brain is the basis of memory and learning. There is no one part of the brain in which all aspects of a memory appear to reside. Specific cortical sites cannot be identified for particular memories because relatively extensive cortical damage does not selectively remove memories. Although memory is impaired if portions of the brain, particularly the temporal lobes, are removed, it is not lost entirely. Many memories persist in spite of the damage, and the ability to access them gradually recovers with time. Therefore, investigators who have tried to probe the physical mechanisms underlying memory often have felt that they were grasping at a shadow. Although we still do not have a complete understanding of these mechanisms, we have learned a good deal about the basic processes in which memories are formed.

There appear to be fundamental differences between short-term and long-term memory. Short-term memory is transient, lasting only a few moments. Such memories can readily be erased by the application of an electrical shock, leaving previously stored long-term memories intact. This result suggests that short-term memories are stored electrically in the form of a transient neural excitation. Long-term memory, in contrast, appears to involve structural changes in certain neural connections within the brain. Two parts of the temporal lobes, the hippocampus and the amygdala, are involved in both short-term memory and its consolidation into long-term memory. Damage to these structures impairs the ability to process recent events into long-term memories.

Synapses that are used intensively for a short period of time display more effective synaptic transmission upon subsequent use. This phenomenon is called long-term potentiation (LTP). During LTP, the presynaptic neuron may release increased amounts of neurotransmitter with each action potential, and the postsynaptic neuron may become increasingly sensitive to the neurotransmitter. It is believed that these changes in synaptic transmission may be responsible for some aspects of memory storage.

Mechanism of Alzheimer's Disease Still a Mystery

In the past, little was known about *Alzheimer's disease*, a condition in which the memory and thought processes of the brain become dysfunctional. Drug companies are eager to develop new products for the treatment of Alzheimer's, but they have little concrete evidence to go on. Scientists disagree about the biological nature of the disease and its cause. Two hypotheses have been proposed: one that nerve cells in the brain are killed from the outside in, and the other that the cells are killed from the inside out.

In the first hypothesis, external proteins called β-amyloid peptides kill nerve cells. A mistake in protein processing produces an abnormal form of the peptide, which then forms aggregates, or plaques. The plaques begin to fill in the brain and then damage and kill nerve cells. However, these amyloid plaques have been found in autopsies of people that did not have Alzheimer's disease.

The second hypothesis maintains that the nerve cells are killed by an abnormal form of an internal protein. This protein, called tau (τ), normally functions to maintain protein transport microtubules. Abnormal forms of τ assemble into helical segments that form tangles, which interfere with the normal functioning of the nerve cells. Researchers continue to study whether tangles and plaques are causes or effects of Alzheimer's disease.

Progress has been made in identifying genes that increase the likelihood of developing Alzheimer's and genes that, when mutated, can cause Alzheimer's disease. However, the genes may not reveal much about Alzheimer's as they do not show up in most Alzheimer's patients, and they cause symptoms that start much earlier than when most Alzheimer's patients show symptoms.

The cerebrum is composed of two cerebral hemispheres. Each hemisphere consists of the gray matter of the cerebral cortex overlying white matter and islands of gray matter (nuclei) called the basal ganglia. These areas are involved in the integration of sensory information, control of body movements, and such associative functions as learning and memory.

The Spinal Cord

The spinal cord is a cable of neurons extending from the brain down through the backbone (figure 54.31). It is enclosed and protected by the vertebral column and layers of membranes called *meninges*, which also cover the brain. Inside the spinal cord there are two zones. The inner zone, called gray matter, consists of interneurons and the cell bodies of motor neurons. The outer zone, called white matter, contains the axons and dendrites of nerve cells. Messages from the body and the brain run up and down the spinal cord, an "information highway."

In addition to relaying messages, the spinal cord also functions in reflexes, the sudden, involuntary movement of muscles. A reflex produces a rapid motor response to a stimulus because the sensory neuron passes its information to a motor neuron in the spinal cord, without higher level processing. One of the most frequently used reflexes in your body is blinking, a reflex that protects your eyes. If anything, such as an insect or a cloud of dust, approaches your eye, the eyelid blinks before you realize what has happened. The reflex occurs before the cerebrum is aware the eye is in danger.

FIGURE 54.31
A view down the human spinal cord. Pairs of spinal nerves can be seen extending from the spinal cord. It is along these nerves, as well as the cranial nerves that arise from the brain, that the central nervous system communicates with the rest of the body.

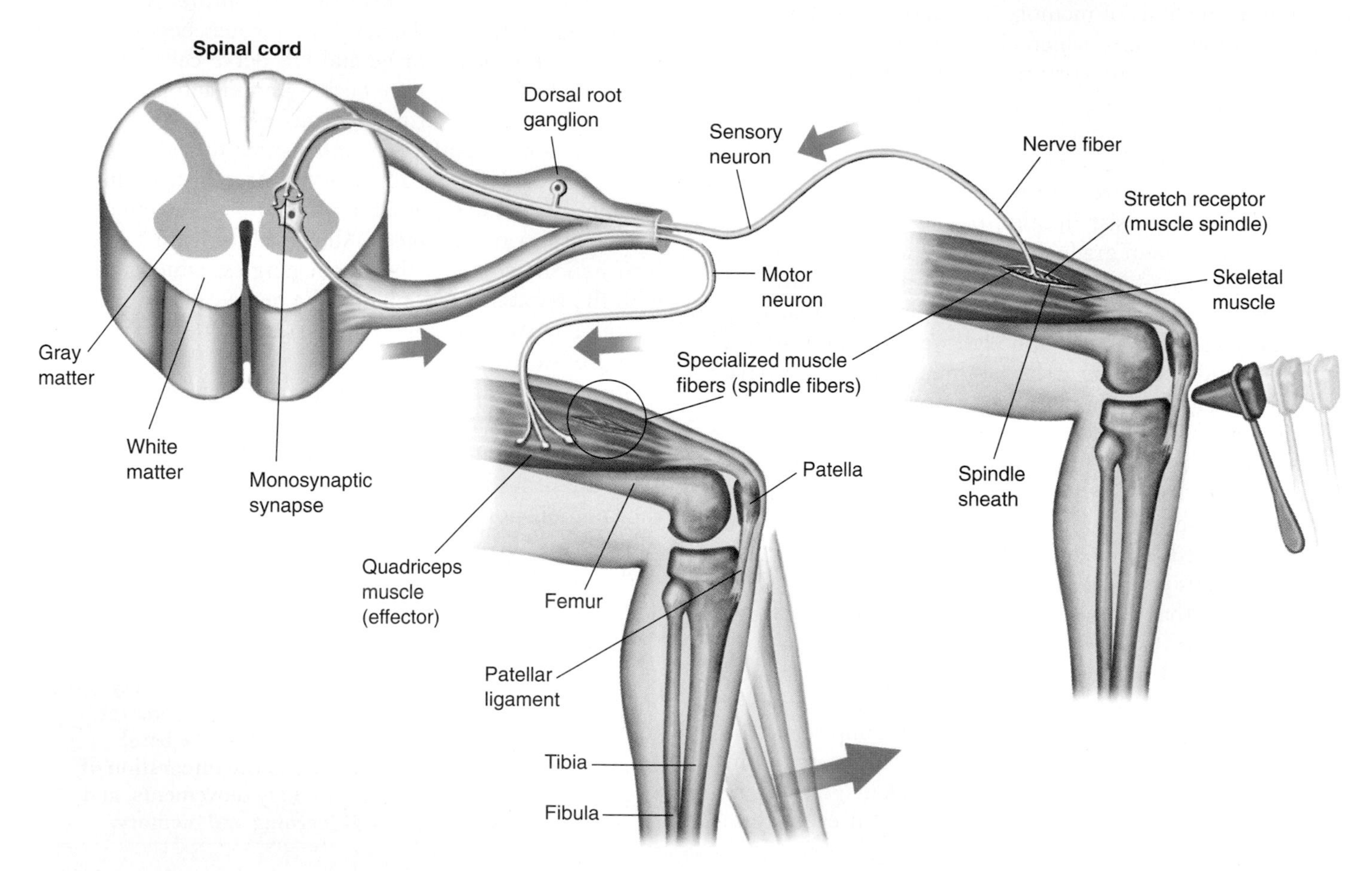

FIGURE 54.32
The knee-jerk reflex. This is the simplest reflex, involving only sensory and motor neurons.

Because they pass information along only a few neurons, reflexes are very fast. Many reflexes never reach the brain. The nerve impulse travels only as far as the spinal cord and then comes right back as a motor response. A few reflexes, like the knee-jerk reflex (figure 54.32), are monosynaptic reflex arcs. In these, the sensory nerve cell makes synaptic contact directly with a motor neuron in the spinal cord whose axon travels directly back to the muscle. The knee-jerk reflex is also an example of a *muscle stretch reflex.* When the muscle is briefly stretched by tapping the patellar ligament with a rubber mallet, the *muscle spindle apparatus* is also stretched. The spindle apparatus is embedded within the muscle, and, like the muscle fibers outside the spindle, is stretched along with the muscle. Stretching of the spindle activates sensory neurons that synapse directly with somatic motor neurons within the spinal cord. As a result, the somatic motor neurons conduct action potentials to the skeletal muscle fibers and stimulate the muscle to contract. This reflex is the simplest in the vertebrate body because only one synapse is crossed in the reflex arc.

Most reflexes in vertebrates, however, involve a single connecting interneuron between the sensory and the motor neuron. The withdrawal of a hand from a hot stove or the blinking of an eye in response to a puff of air involve a relay of information from a sensory neuron through one or more interneurons to a motor neuron. The motor neuron then stimulates the appropriate muscle to contract (figure 54.33).

Spinal Cord Regeneration

In the past, scientists have tried to repair severed spinal cords by installing nerves from another part of the body to bridge the gap and act as guides for the spinal cord to regenerate. But most of these experiments have failed because although axons may regenerate through the implanted nerves, they cannot penetrate the spinal cord tissue once they leave the implant. Also, there is a factor that inhibits nerve growth in the spinal cord. After discovering that fibroblast growth factor stimulates nerve growth, neurobiologists tried gluing on the nerves, from the implant to the spinal cord, with fibrin that had been mixed with the fibroblast growth factor. Three months later, rats with the nerve bridges began to show movement in their lower bodies. In further analyses of the experimental animals, dye tests indicated that the spinal cord nerves had regrown from both sides of the gap. Many scientists are encouraged by the potential to use a similar treatment in human medicine. However, most spinal cord injuries in humans do not involve a completely severed spinal cord; often, nerves are crushed, which results in different tissue damage. Also, while the rats with nerve bridges did regain some locomotory ability, tests indicated that they were barely able to walk or stand.

The spinal cord relays messages to and from the brain and processes some sensory information directly.

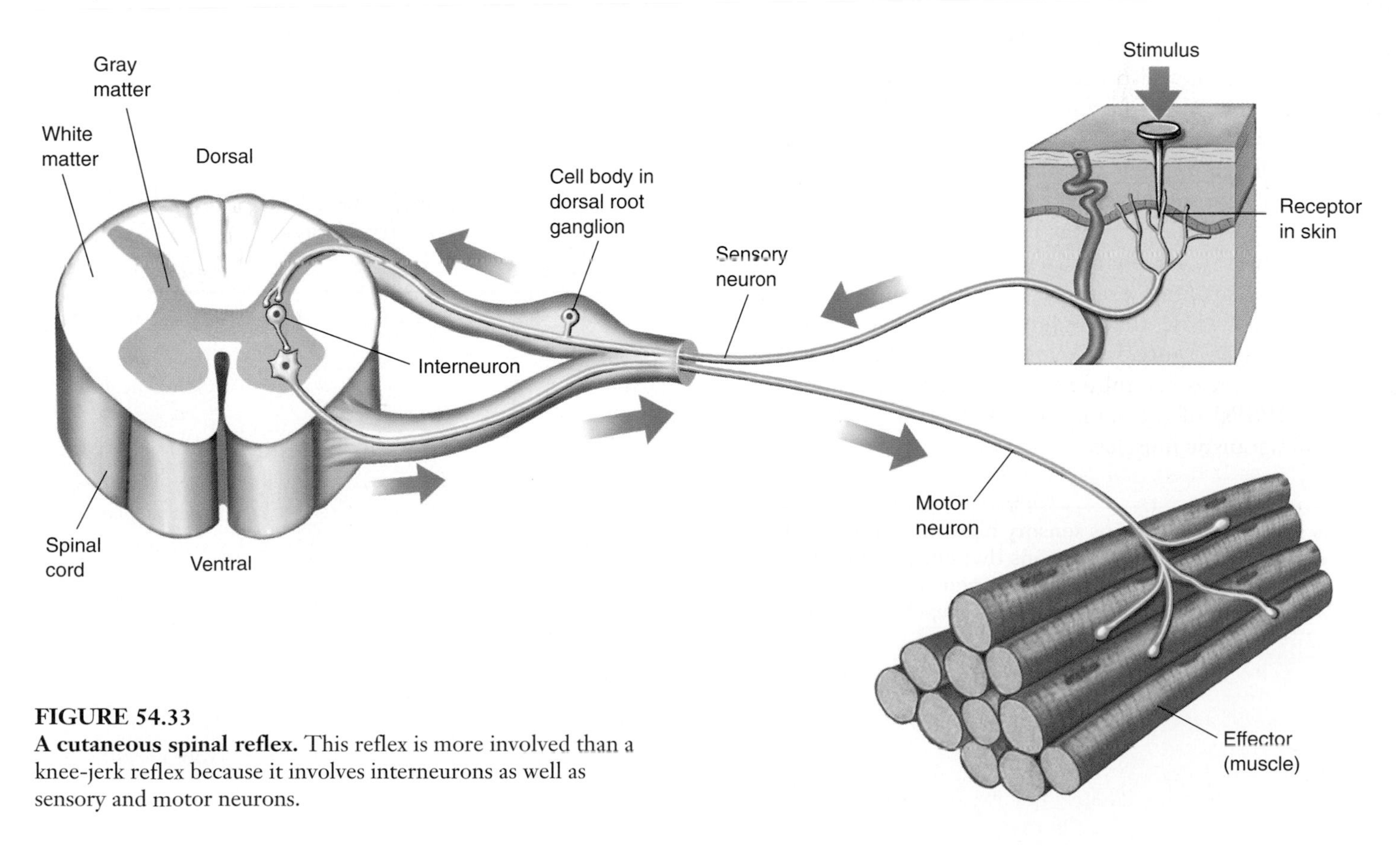

FIGURE 54.33
A cutaneous spinal reflex. This reflex is more involved than a knee-jerk reflex because it involves interneurons as well as sensory and motor neurons.

54.5 The peripheral nervous system consists of sensory and motor neurons.

Components of the Peripheral Nervous System

The peripheral nervous system consists of nerves and ganglia. Nerves are cablelike collections of axons (figure 54.34), usually containing both sensory and motor neurons. Ganglia are aggregations of neuron cell bodies located outside the central nervous system.

At its origin, a spinal nerve separates into sensory and motor components. The axons of sensory neurons enter the dorsal surface of the spinal cord and form the **dorsal root** of the spinal nerve, whereas motor axons leave from the ventral surface of the spinal nerve and form the **ventral root** of the spinal nerve. The cell bodies of sensory neurons are grouped together outside each level of the spinal cord in the **dorsal root ganglia.** The cell bodies of somatic motor neurons, on the other hand, are located within the spinal cord and so are not located in ganglia.

Somatic motor neurons stimulate skeletal muscles to contract, and autonomic motor neurons innervate involuntary effectors—smooth muscles, cardiac muscle, and glands. A comparison of the somatic and autonomic nervous systems is provided in table 54.4 and each will be discussed in turn. Somatic motor neurons stimulate the skeletal muscles of the body to contract in response to conscious commands and as part of reflexes that do not require conscious control. Conscious control of skeletal muscles is achieved by activation of tracts of axons that descend from the cerebrum to the appropriate level of the spinal cord. Some of these descending axons will stimulate spinal cord motor neurons directly, while others will activate interneurons that in turn stimulate the spinal motor neurons. When a particular muscle is stimulated to contract, however, its antagonist must be inhibited. In order to flex the arm, for example, the flexor muscles must be stimulated while the antagonistic extensor muscle is inhibited (see figure 50.6). Descending motor axons produce this necessary inhibition by causing hyperpolarizations (IPSPs) of the spinal motor neurons that innervate the antagonistic muscles.

A spinal nerve contains sensory neurons that enter the dorsal root and motor neurons that enter the ventral root of the nerve. Somatic motor neurons innervate skeletal muscles and stimulate the muscles to contract.

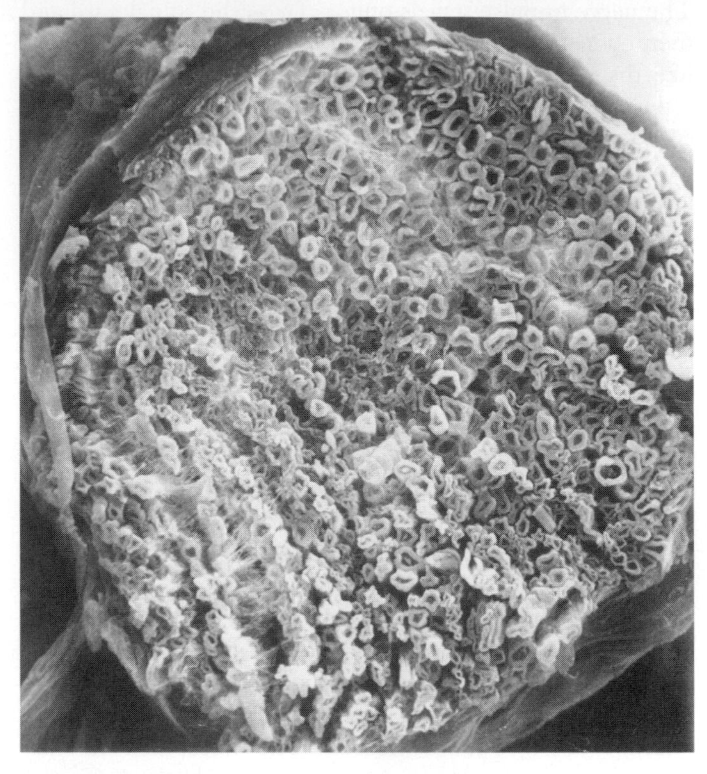

FIGURE 54.34
Nerves in the peripheral nervous system. Photomicrograph (1600×) showing a cross section of a bullfrog nerve. The nerve is a bundle of axons bound together by connective tissue. Many myelinated axons are visible, each looking somewhat like a doughnut.

Table 54.4 Comparison of the Somatic and Autonomic Nervous Systems

Characteristic	Somatic	Autonomic
Effectors	Skeletal muscle	Cardiac muscle Smooth muscle Gastrointestinal tract Blood vessels Airways Exocrine glands
Effect on motor nerves	Excitation	Excitation or inhibition
Innervation of effector cells	Always single	Typically dual
Number of neurons in path to effector	One	Two
Neurotransmitter	Acetylcholine	Acetylcholine Norepinephrine

The Autonomic Nervous System

The autonomic nervous system is composed of the sympathetic and parasympathetic divisions and the medulla oblongata of the hindbrain, which coordinates this system. Though they differ, the sympathetic and parasympathetic divisions share several features. In both, the efferent motor pathway involves two neurons: the first has its cell body in the CNS and sends an axon to an autonomic ganglion, while the second has its cell body in the autonomic ganglion and sends its axon to synapse with a smooth muscle, cardiac muscle, or gland cell (figure 54.35). The first neuron is called a *preganglionic neuron*, and it always releases ACh at its synapse. The second neuron is a *postganglionic neuron*; those in the parasympathetic division release ACh, while those in the sympathetic division release norepinephrine.

In the sympathetic division, the preganglionic neurons originate in the thoracic and lumbar regions of the spinal cord (figure 54.36). Most of the axons from these neurons synapse in two parallel chains of ganglia immediately outside the spinal cord. These structures are usually called the *sympathetic chain* of ganglia. The sympathetic chain contains the cell bodies of postganglionic neurons, and it is the axons from these neurons that innervate the different visceral organs. There are some exceptions to this general pattern, however. Most importantly, the axons of some preganglionic sympathetic neurons pass through the

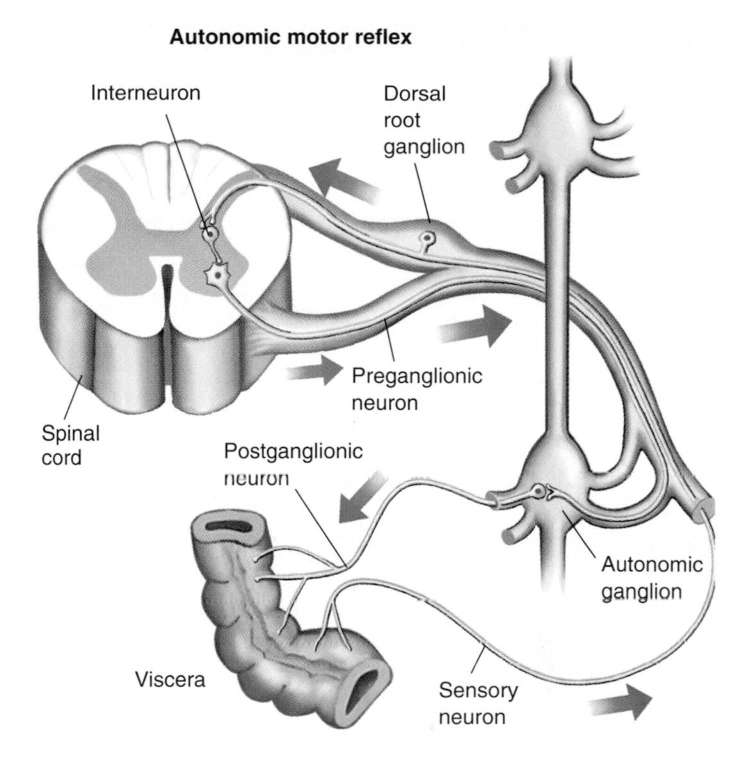

FIGURE 54.35
An autonomic reflex. There are two motor neurons in the efferent pathway. The first, or preganglionic neuron, exits the CNS and synapses at an autonomic ganglion. The second, or postganglionic neuron, exits the ganglion and regulates the visceral effectors (smooth muscle, cardiac muscle, or glands).

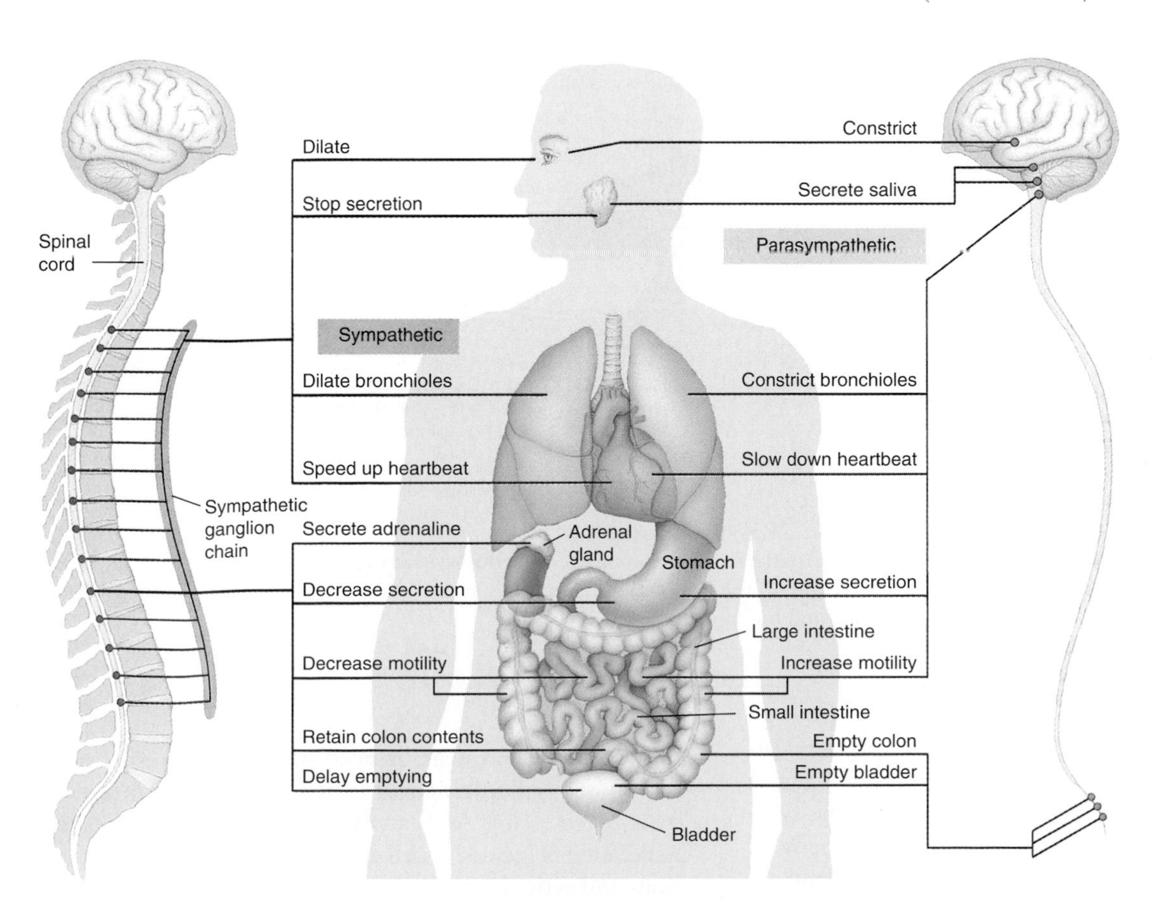

FIGURE 54.36
The sympathetic and parasympathetic divisions of the autonomic nervous system. The preganglionic neurons of the sympathetic division exit the thoracic and lumbar regions of the spinal cord, while those of the parasympathetic division exit the brain and sacral region of the spinal cord. The ganglia of the sympathetic division are located near the spinal cord, while those of the parasympathetic division are located near the organs they innervate. Most of the internal organs are innervated by both divisions.

Table 54.5 Autonomic Innervation of Target Tissues

Target Tissue	Sympathetic Stimulation	Parasympathetic Stimulation
Pupil of eye	Dilation	Constriction
Glands		
Salivary	Vasoconstriction; slight secretion	Vasodilation; copious secretion
Gastric	Inhibition of secretion	Stimulation of gastric activity
Liver	Stimulation of glucose secretion	Inhibition of glucose secretion
Sweat	Sweating	None
Gastrointestinal tract		
Sphincters	Increased tone	Decreased tone
Wall	Decreased tone	Increased motility
Gallbladder	Relaxation	Contraction
Urinary bladder		
Muscle	Relaxation	Contraction
Sphincter	Contraction	Relaxation
Heart muscle	Increased rate and strength	Decreased rate
Lungs	Dilation of bronchioles	Constriction of bronchioles
Blood vessels		
In muscles	Dilation	None
In skin	Constriction	None
In viscera	Constriction	Dilation

sympathetic chain without synapsing and, instead, terminate within the adrenal gland. The adrenal gland consists of an outer part, or cortex, and an inner part, or medulla. The adrenal medulla receives sympathetic nerve innervation and secretes the hormone epinephrine (adrenaline) in response.

When the sympathetic division becomes activated, epinephrine is released into the blood as a hormonal secretion, and norepinephrine is released at the synapses of the postganglionic neurons. Epinephrine and norepinephrine act to prepare the body for fight or flight (figure 54.37). The heart beats faster and stronger, blood glucose concentration increases, blood flow is diverted to the muscles and heart, and the bronchioles dilate (table 54.5).

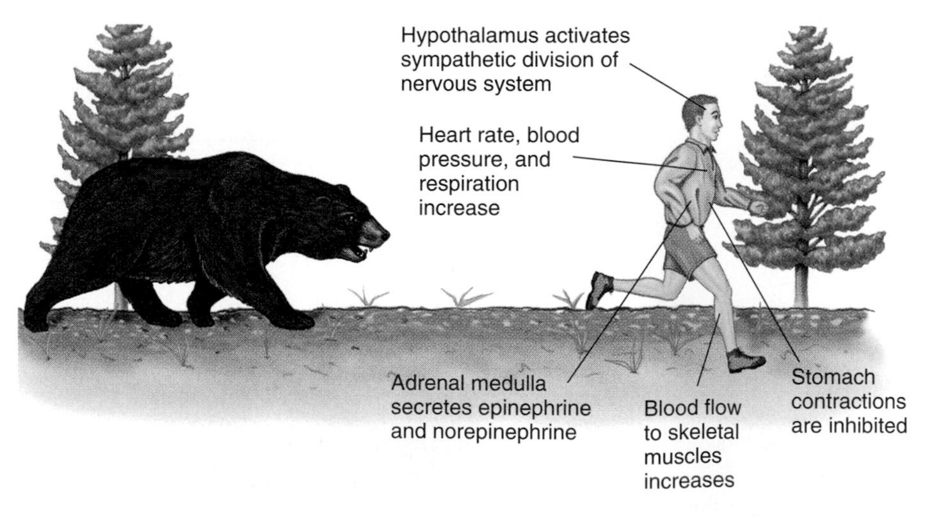

FIGURE 54.37
The sympathetic division of the nervous system in action. To prepare the body for fight or flight, the sympathetic division is activated and causes changes in many organs, glands, and body processes.

These responses are antagonized by the parasympathetic division. Preganglionic parasympathetic neurons originate in the brain and sacral regions of the spinal cord. Because of this origin, there cannot be a chain of parasympathetic ganglia analogous to the sympathetic chain. Instead, the preganglionic axons, many of which travel in the vagus (the tenth cranial) nerve, terminate in ganglia located near or even within the internal organs. The postganglionic neurons then regulate the internal organs by releasing ACh at their synapses. Parasympathetic nerve effects include a slowing of the heart, increased secretions and activities of digestive organs, and so on.

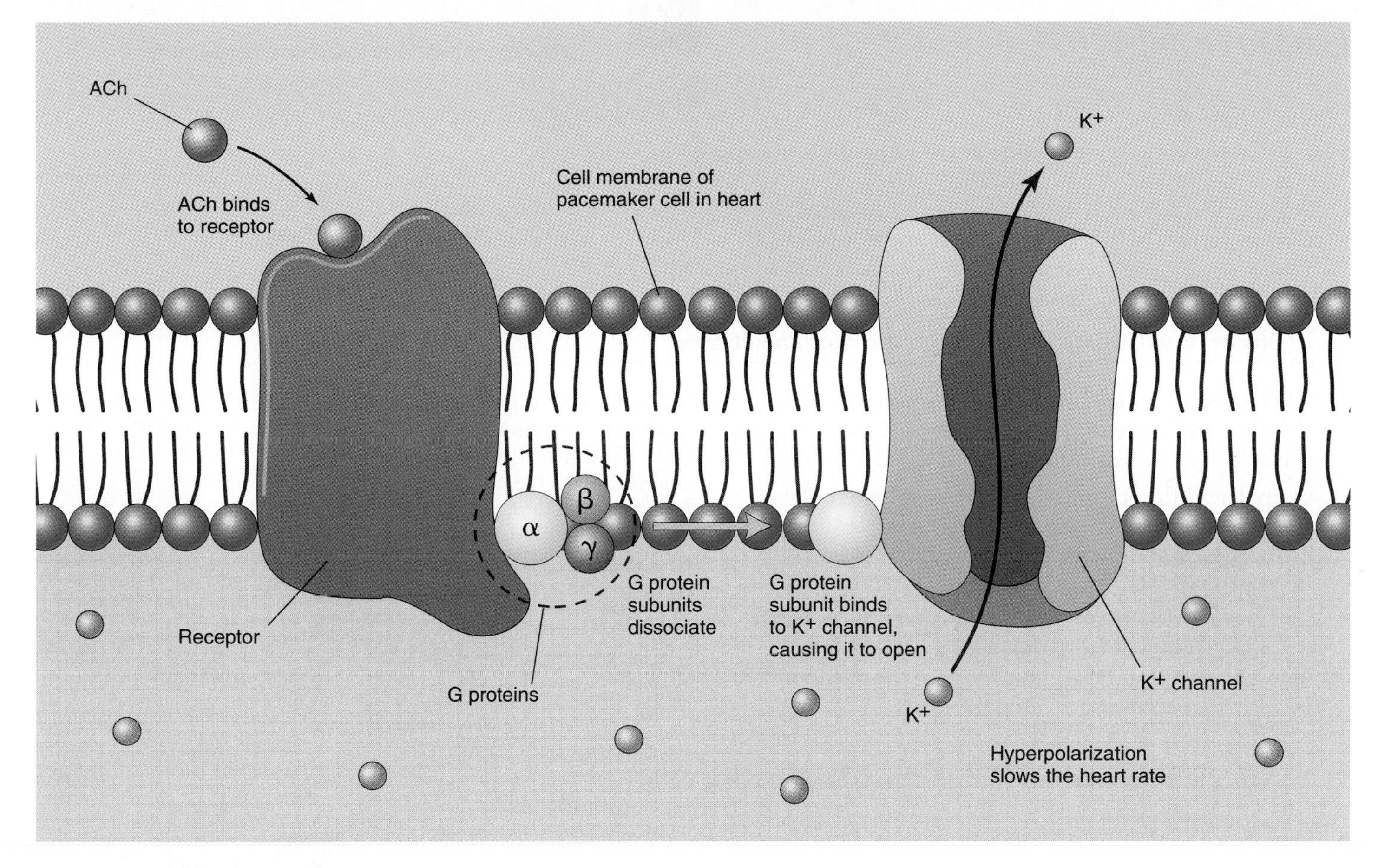

FIGURE 54.38
The parasympathetic effects of ACh require the action of G proteins. The binding of ACh to its receptor causes dissociation of a G protein complex, releasing some components of this complex to move within the membrane and bind to other proteins that form ion channels. The example shown here is the effects of ACh on the heart, where the G protein components cause the opening of potassium channels. This leads to outward diffusion of potassium and hyperpolarization, slowing the heart rate.

G Proteins Mediate Cell Responses to Autonomic Nerves

You might wonder how ACh can slow the heart rate—an inhibitory effect—when it has excitatory effects elsewhere. This inhibitory effect in the pacemaker cells of the heart is produced because ACh causes the opening of potassium channels, leading to the outward diffusion of potassium and thus to hyperpolarization. This and other parasympathetic effects of ACh are produced indirectly, using a group of membrane proteins called **G proteins** (so-called because they are regulated by guanosine diphosphate and guanosine triphosphate [GDP and GTP]). Because the ion channels are located some distance away from the receptor proteins for ACh, the G proteins are needed to serve as connecting links between them.

There are three G protein subunits, designated α, β, and γ, bound together and attached to the receptor protein for ACh. When ACh, released by parasympathetic endings, binds to its receptor, the G protein subunits dissociate (figure 54.38). Specific G protein components move within the membrane to the potassium channel and cause it to open, producing hyperpolarization and a slowing of the heart. In other organs, the G proteins have different effects that lead to excitation. In this way, for example, the parasympathetic nerves that innervate the stomach can cause increased gastric secretions and contractions.

The sympathetic nerve effects also involve the action of G proteins. Stimulation by norepinephrine from sympathetic nerve endings and epinephrine from the adrenal medulla requires G proteins to activate the target cells. We will describe this in more detail, together with hormone action, in chapter 56.

The sympathetic division of the autonomic system, together with the adrenal medulla, activates the body for fight-or-flight responses, whereas the parasympathetic division generally has antagonistic effects. The actions of parasympathetic nerves are produced by ACh, whereas the actions of sympathetic nerves are produced by norepinephrine.

Chapter 54

www.mhhe.com/raven6/resources54.mhtml

Summary | **Questions** | **Media Resources**

54.1 The nervous system consists of neurons and supporting cells.

- The nervous system is subdivided into the central nervous system (CNS) and peripheral nervous system (PNS).

1. What are the differences and similarities among the three types of neurons?

- Art Quiz: Myelin Sheath Formation

54.2 Nerve impulses are produced on the axon membrane.

- The resting axon has a membrane potential of –70 mV; the magnitude of this voltage is produced primarily by the distribution of K^+.
- A depolarization stimulus opens voltage-gated Na^+ channels and then K^+ channels, producing first the upward phase and then the repolarization phase of the action potential.
- Action potentials are all or none and are conducted without decrease in amplitude because each action potential serves as the stimulus for the production of the next action potential along the axon.

2. Which cation is most concentrated in the cytoplasm of a cell, and which is most concentrated in the extracellular fluid? How are these concentration differences maintained?

3. What is a voltage-gated ion channel?

4. What happens to the size of an action potential as it is propagated?

- Membrane Potential
- Action Potential

- Action Potential I

- Art Quizzes:
  - -Sodium-Potassium Pump
  - -Resting Membrane Potential
  - -Saltatory Conduction

54.3 Neurons form junctions called synapses with other cells.

- The presynaptic axon releases neurotransmitter chemicals that diffuse across the synapse and stimulate the production of either a depolarization or a hyperpolarization in the postsynaptic membrane.
- Depolarizations and hyperpolarization can summate in the dendrites and cell bodies of the postsynaptic neuron, allowing integration of information.

5. If a nerve impulse can jump from node to node along a myelinated axon, why can't it jump from the presynaptic cell to the postsynaptic cell across a synaptic cleft?

- Nervous Tissue
- Synapse

- Bioethics Case Study: Smoking Ban
- *On Science* Articles:
  - -Is Smoking Addictive?
  - -Nobel Prize 2000

54.4 The central nervous system consists of the brain and spinal cord.

- The vertebrate brain is divided into a forebrain, midbrain, and hindbrain, and these are further subdivided into other brain regions. The cerebral cortex has a primary motor area and a primary somatosensory area, as well as areas devoted to the analysis of vision and hearing and the integration and association of information.
- The spinal cord carries information to and from the brain and coordinates many reflex movements.

6. Where are the basal ganglia located, and what is their function?

7. How are short-term and long-term memory thought to differ in terms of their basic underlying mechanisms?

- Art Activities:
  - -Central Nervous System
  - -Spinal Cord Anatomy
  - -Human Brain

- Reflex Arc

54.5 The peripheral nervous system consists of sensory and motor neurons.

- The sympathetic division is activated during fight-or-flight responses; the parasympathetic division opposes the action of the sympathetic division in most activities.

8. How do the sympathetic and parasympathetic divisions differ in the locations of the ganglionic neurons?

- Art Activity: Peripheral Nervous System

55

Sensory Systems

Concept Outline

55.1 Animals employ a wide variety of sensory receptors.

Categories of Sensory Receptors and Their Actions. Sensory receptors can be classified according to the type of stimuli to which they can respond.

55.2 Mechanical and chemical receptors sense the body's condition.

Detecting Temperature and Pressure. Receptors within the skin respond to touch, pressure, pain, heat and cold.

Sensing Muscle Contraction and Blood Pressure. A muscle spindle responds to stretching of the muscle; receptors in arteries monitor changes in blood pressure.

Sensing Taste, Smell, and Body Position. Receptors that respond to chemicals produce sensations of taste and smell. Hair cells send nerve impulses when they are bent.

55.3 Auditory receptors detect pressure waves in the air.

The Ears and Hearing. Sound causes vibrations in the ear that bend hair cell processes, initiating a nerve impulse.

Sonar. Bats orient themselves in space by emitting sounds and detecting the time required for the sounds to bounce off objects and return to their ears.

55.4 Optic receptors detect light over a broad range of wavelengths.

Evolution of the Eye. True image-forming eyes evolved independently in several phyla.

Vertebrate Photoreceptors. Light causes a pigment molecule in a rod or cone cell to dissociate; this "bleaching" reaction activates the photoreceptor.

Visual Processing in the Vertebrate Retina. Action potentials travel from the retina of the eyes to the brain for visual perception.

55.5 Some vertebrates use heat, electricity, or magnetism for orientation.

Diversity of Sensory Experiences. Special receptors can detect heat, electrical currents, and magnetic fields.

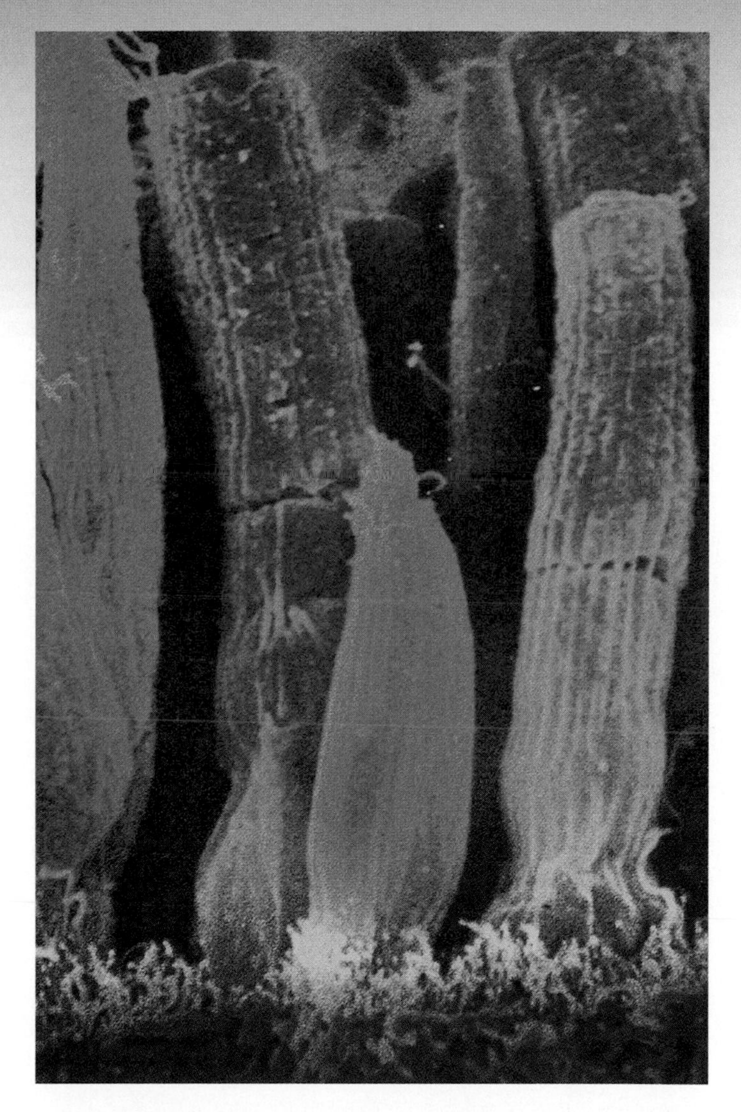

FIGURE 55.1
Photoreceptors in the vertebrate eye. Rods, the broad, tubular cells, allow black-and-white vision, while cones, the short, tapered cells, are responsible for color vision. Not all vertebrates have both types of receptors.

All input from sensory neurons to the central nervous system arrives in the same form, as action potentials propagated by afferent (inward-conducting) sensory neurons. Different sensory neurons project to different brain regions, and so are associated with different sensory modalities (figure 55.1). The intensity of the sensation depends on the frequency of action potentials conducted by the sensory neuron. A sunset, a symphony, and searing pain are distinguished by the brain only in terms of the identity of the sensory neuron carrying the action potentials and the frequency of these impulses. Thus, if the auditory nerve is artificially stimulated, the brain perceives the stimulation as sound. But if the optic nerve is artificially stimulated in exactly the same manner and degree, the brain perceives a flash of light.

55.1 Animals employ a wide variety of sensory receptors.

Categories of Sensory Receptors and Their Actions

Sensory information is conveyed to the CNS and perceived in a four-step process (figure 55.2): (1) *stimulation*—a physical stimulus impinges on a sensory neuron or an accessory structure; (2) *transduction*—the stimulus energy is used to produce electrochemical nerve impulses in the dendrites of the sensory neuron; (3) *transmission*—the axon of the sensory neuron conducts action potentials along an afferent pathway to the CNS; and (4) *interpretation*—the brain creates a sensory perception from the electrochemical events produced by afferent stimulation. We actually see (as well as hear, touch, taste, and smell) with our brains, not with our sense organs.

Sensory receptors differ with respect to the nature of the environmental stimulus that best activates their sensory dendrites. Broadly speaking, we can recognize three classes of environmental stimuli: (1) mechanical forces, which stimulate **mechanoreceptors;** (2) chemicals, which stimulate **chemoreceptors;** and (3) electromagnetic and thermal energy, which stimulate a variety of receptors, including the **photoreceptors** of the eyes (table 55.1).

The simplest sensory receptors are *free nerve endings* that respond to bending or stretching of the sensory neuron membrane, to changes in temperature, or to chemicals like oxygen in the extracellular fluid. Other sensory receptors are more complex, involving the association of the sensory neurons with specialized epithelial cells.

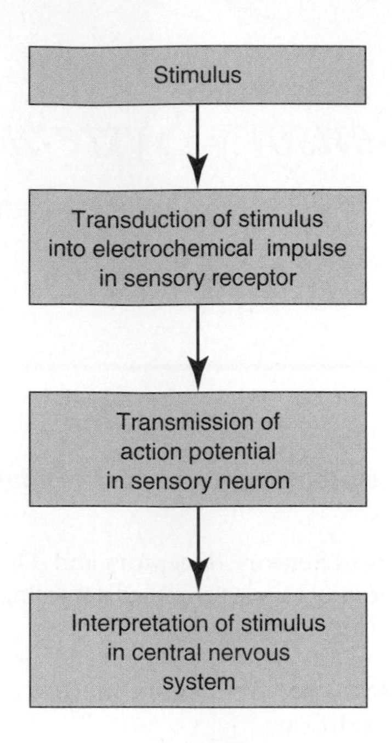

FIGURE 55.2
The path of sensory information. Sensory stimuli must be transduced into electrochemical nerve impulses that are conducted to the brain for interpretation.

Table 55.1 Classes of Environmental Stimuli

Mechanical Forces	Chemicals	Electromagnetic Energy
Pressure	Taste	Light
Gravity	Smell	Heat
Inertia	Humidity	Electricity
Sound		Magnetism
Touch		
Vibration		

Sensing the External and Internal Environments

Exteroceptors are receptors that sense stimuli that arise in the external environment. Almost all of a vertebrate's exterior senses evolved in water before vertebrates invaded the land. Consequently, many senses of terrestrial vertebrates emphasize stimuli that travel well in water, using receptors that have been retained in the transition from the sea to the land. Mammalian hearing, for example, converts an airborne stimulus into a waterborne one, using receptors similar to those that originally evolved in the water. A few vertebrate sensory systems that function well in the water, such as the electrical organs of fish, cannot function in the air and are not found among terrestrial vertebrates. On the other hand, some land-dwellers have sensory systems, such as infrared receptors, that could not function in the sea.

Sensory systems can provide several levels of information about the external environment. Some sensory systems provide only enough information to determine that an object is present; they call the animal's attention to the object but give little or no indication of where it is located. Other sensory systems provide information about the location of an object, permitting the animal to move toward it. Still other sensory systems enable the brain to construct a three-dimensional image of an object and its surroundings.

Interoceptors sense stimuli that arise from within the body. These internal receptors detect stimuli related to

Table 55.2 Sensory Transduction Among the Vertebrates				
Stimulus	**Receptor**	**Location**	**Structure**	**Transduction Process**
INTEROCEPTION				
Temperature	Heat receptors and cold receptors	Skin, hypothalamus	Free nerve ending	Temperature change opens/ closes ion channels in membrane
Touch	Meissner's corpuscles, Merkel cells	Surface of skin	Nerve ending within elastic capsule	Rapid or extended change in pressure deforms membrane
Vibration	Pacinian corpuscles	Deep within skin	Nerve ending within elastic capsule	Severe change in pressure deforms membrane
Pain	Nociceptors	Throughout body	Free nerve ending	Chemicals or changes in pressure or temperature open/close ion channels in membrane
Muscle stretch	Stretch receptors	Within muscles	Spiral nerve endings wrapped around muscle spindle	Stretch of spindle deforms membrane
Blood pressure	Baroreceptors	Arterial branches	Nerve endings over thin part of arterial wall	Stretch of arterial wall deforms membrane
EXTEROCEPTION				
Gravity	Statocysts	Outer chambers of inner ear	Otoliths and hair cells	Otoliths deform hair cells
Motion	Cupula	Semicircular canals of inner ear	Collection of hair cells	Fluid movement deforms hair cells
	Lateral line organ	Within grooves on body surface of fish	Collection of hair cells	Fluid movement deforms hair cells
Taste	Taste bud cells	Mouth; skin of fish	Chemoreceptors: epithelial cells with microvilli	Chemicals bind to membrane receptors
Smell	Olfactory neurons	Nasal passages	Chemoreceptors: ciliated neurons	Chemicals bind to membrane receptors
Hearing	Organ of Corti	Cochlea of inner ear	Hair cells between basilar and tectorial membranes	Sound waves in fluid deform membranes
Vision	Rod and cone cells	Retina of eye	Array of photosensitive pigments	Light initiates process that closes ion channels
Heat	Pit organ	Face of snake	Temperature receptors in two chambers	Receptors compare temperatures of surface and interior chambers
Electricity	Ampullae of Lorenzini	Within skin of fishes	Closed vesicles with asymmetrical ion channel distribution	Electrical field alters ion distribution on membranes
Magnetism	Unknown	Unknown	Unknown	Deflection of magnetic field initiates nerve impulses?

muscle length and tension, limb position, pain, blood chemistry, blood volume and pressure, and body temperature. Many of these receptors are simpler than those that monitor the external environment and are believed to bear a closer resemblance to primitive sensory receptors. In the rest of this chapter, we will consider the different types of interoceptors and exteroceptors according to the kind of stimulus each is specialized to detect (table 55.2).

Sensory Transduction

Sensory cells respond to stimuli because they possess *stimulus-gated ion channels* in their membranes. The sensory stimulus causes these ion channels to open or close, depending on the sensory system involved. In doing so, a sensory stimulus produces a change in the membrane potential of the receptor cell. In most cases, the sensory stimulus produces a depolarization of the receptor cell, analogous to the excitatory postsynaptic potential (EPSP, described in chapter 54) produced in a postsynaptic cell in response to neurotransmitter. A depolarization that occurs in a sensory receptor upon stimulation is referred to as a **receptor potential** (figure 55.3*a*).

Like an EPSP, a receptor potential is graded: the larger the sensory stimulus, the greater the degree of depolarization. Receptor potentials also decrease in size (*decrement*) with distance from their source. This prevents small, irrelevant stimuli from reaching the cell body of the sensory neuron. Once a threshold level of depolarization is reached, the receptor potential stimulates the production of action potentials that are conducted by a sensory axon into the CNS (figure 55.3*b*). The greater the sensory stimulus, the greater the depolarization of the receptor potential and the higher the frequency of action potentials. There is generally a logarithmic relationship between stimulus intensity and action potential frequency—a sensory stimulus that is ten times greater than another stimulus will produce action potentials at twice the frequency of the other stimulus. This allows the brain to interpret the incoming signals as indicating a sensory stimulus of a particular strength.

Sensory receptors transduce stimuli in the internal or external environment into graded depolarizations, which stimulates the production of action potentials. Sensory receptors may be classified on the basis of the type of stimulus energy to which they respond.

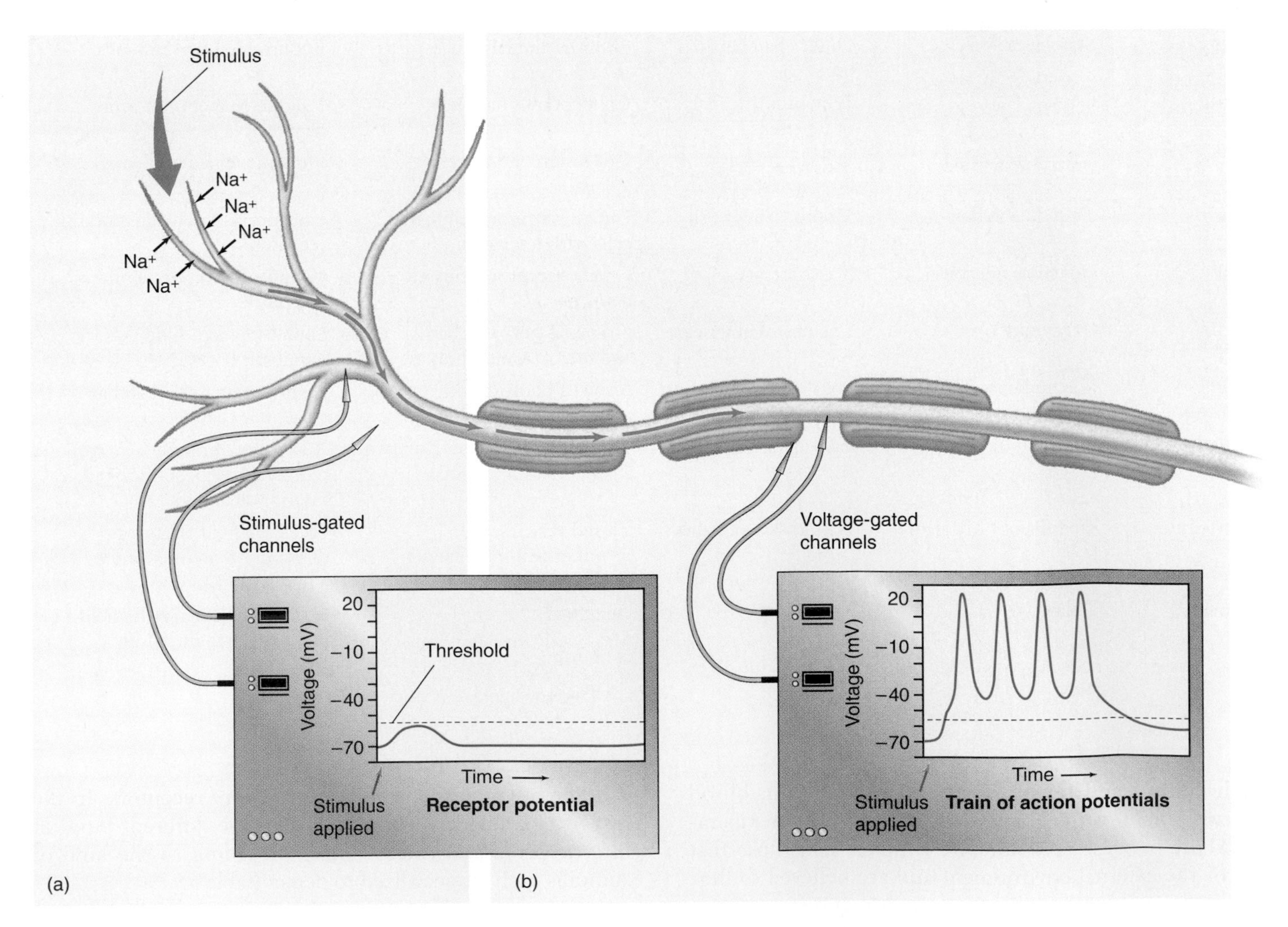

FIGURE 55.3
Events in sensory transduction. (*a*) Depolarization of a free nerve ending leads to a receptor potential that spreads by local current flow to the axon. (*b*) Action potentials are produced in the axon in response to a sufficiently large receptor potential.

Detecting Temperature and Pressure

While the receptors of the skin, called the **cutaneous receptors,** are classified as interoceptors, they in fact respond to stimuli at the border between the external and internal environments. These receptors serve as good examples of the specialization of receptor structure and function, responding to heat, cold, pain, touch, and pressure.

The skin contains two populations of **thermoreceptors,** which are naked dendritic endings of sensory neurons that are sensitive to changes in temperature. *Cold receptors* are stimulated by a fall in temperature and inhibited by warming, while *warm receptors* are stimulated by a rise in temperature and inhibited by cooling. Cold receptors are located immediately below the epidermis, while warm receptors are located slightly deeper, in the dermis. Thermoreceptors are also found within the hypothalamus of the brain, where they monitor the temperature of the circulating blood and thus provide the CNS with information on the body's internal (core) temperature.

A stimulus that causes or is about to cause tissue damage is perceived as pain. The receptors that transmit impulses that are perceived by the brain as pain are called **nociceptors.** They consist of free nerve endings located throughout the body, especially near surfaces where damage is most likely to occur. Different nociceptors may respond to extremes in temperature, very intense mechanical stimulation, or specific chemicals in the extracellular fluid, including some that are released by injured cells. The thresholds of these sensory cells vary; some nociceptors are sensitive only to actual tissue damage, while others respond before damage has occurred.

Several types of mechanoreceptors are present in the skin, some in the dermis and others in the underlying subcutaneous tissue (figure 55.4). Morphologically specialized receptors that respond to fine touch are most concentrated on areas such as the fingertips and face. They are used to localize cutaneous stimuli very precisely and can be either phasic (intermittently activated) or tonic (continuously activated). The phasic receptors include *hair follicle receptors* and *Meissner's corpuscles,* which are present on body surfaces that do not contain hair, such as the fingers, palms, and nipples. The tonic receptors consist of *Ruffini endings* in the dermis and *touch dome endings (Merkel cells)* located near the surface of the skin. These receptors monitor the duration of a touch and the extent to which it is applied.

Deep below the skin in the subcutaneous tissue lie phasic, pressure-sensitive receptors called **Pacinian corpuscles.** Each of these receptors consists of the end of an afferent axon, surrounded by a capsule of alternating layers of connective tissue cells and extracellular fluid. When sustained pressure is applied to the corpuscle, the elastic capsule absorbs much of the pressure and the axon ceases to produce impulses. Pacinian corpuscles thus monitor only the onset and removal of pressure, as may occur repeatedly when something that vibrates is placed against the skin.

Different cutaneous receptors respond to touch, pressure, heat, cold, and pain. Some of these receptors are naked dendrites of sensory neurons, while others have supporting cells that modify the activities of their sensory dendrites.

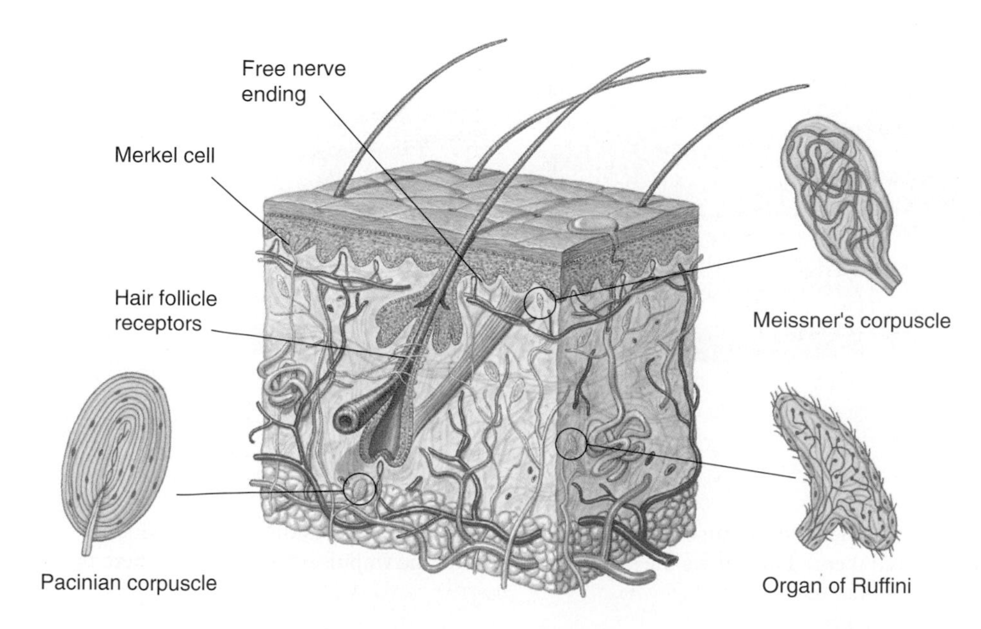

FIGURE 55.4
Sensory receptors in human skin. Cutaneous receptors may be free nerve endings or sensory dendrites in association with other supporting structures.

Sensing Muscle Contraction and Blood Pressure

Mechanoreceptors contain sensory cells with ion channels that are sensitive to a mechanical force applied to the membrane. These channels open in response to mechanical distortion of the membrane, initiating a depolarization (receptor potential) that causes the sensory neuron to generate action potentials.

Muscle Length and Tension

Buried within the skeletal muscles of all vertebrates except the bony fishes are muscle spindles, sensory stretch receptors that lie in parallel with the rest of the fibers in the muscle (figure 55.5). Each spindle consists of several thin muscle fibers wrapped together and innervated by a sensory neuron, which becomes activated when the muscle, and therefore the spindle, is stretched. Muscle spindles, together with other receptors in tendons and joints, are known as **proprioceptors,** which are sensory receptors that provide information about the relative position or movement of the animal's body parts. The sensory neurons conduct action potentials into the spinal cord, where they synapse with somatic motor neurons that innervate the muscle. This pathway constitutes the muscle stretch reflex, including the knee-jerk reflex, previously discussed in chapter 54.

When a muscle contracts, it exerts tension on the tendons attached to it. The Golgi tendon organs, another type of proprioceptor, monitor this tension; if it becomes too high, they elicit a reflex that inhibits the motor neurons innervating the muscle. This reflex helps to ensure that muscles do not contract so strongly that they damage the tendons to which they are attached.

Blood Pressure

Blood pressure is monitored at two main sites in the body. One is the *carotid sinus,* an enlargement of the left and right internal carotid arteries, which supply blood to the brain. The other is the *aortic arch,* the portion of the aorta very close to its emergence from the heart. The walls of the blood vessels at both sites contain a highly branched network of afferent neurons called **baroreceptors,** which detect tension in the walls. When the blood pressure decreases, the frequency of impulses produced by the baroreceptors decreases. The CNS responds to this reduced input by stimulating the sympathetic division of the autonomic nervous system, causing an increase in heart rate and vasoconstriction. Both effects help to raise the blood pressure, thus maintaining homeostasis. A rise in blood pressure, conversely, reduces sympathetic activity and stimulates the parasympathetic division, slowing the heart and lowering the blood pressure.

Mechanical distortion of the plasma membrane of mechanoreceptors produces nerve impulses that serve to monitor muscle length from skeletal muscle spindles and to monitor blood pressure from baroreceptors within arteries.

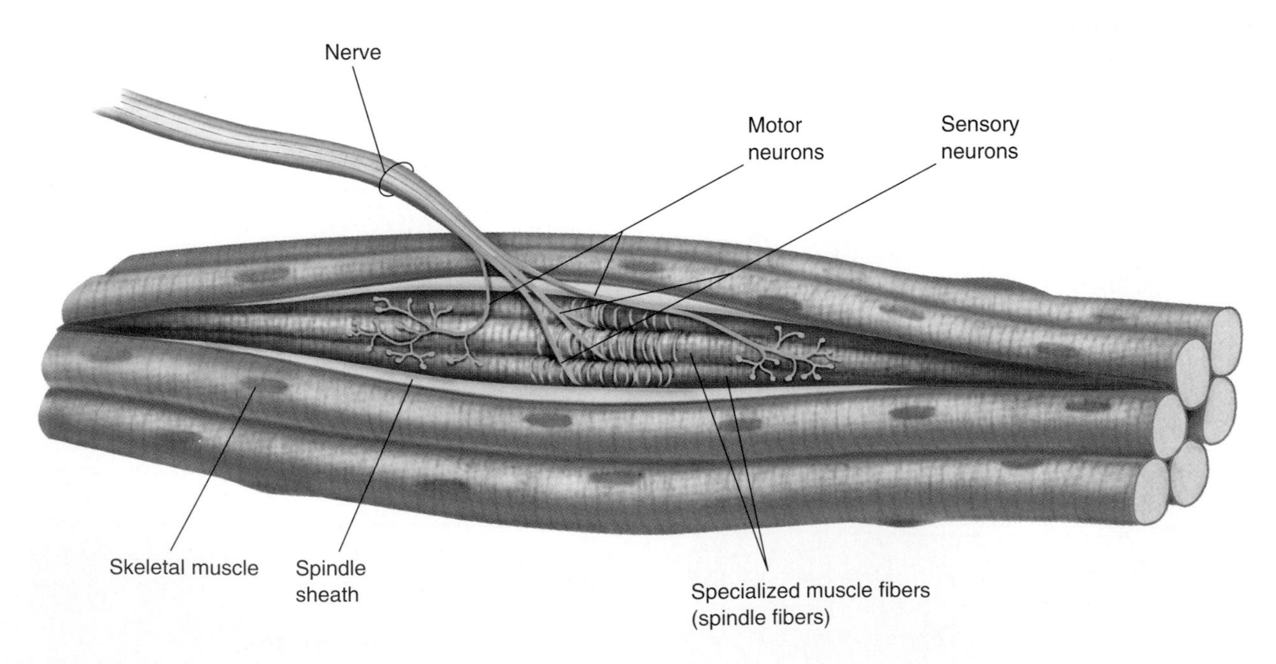

FIGURE 55.5
A muscle spindle is a stretch receptor embedded within skeletal muscle. Stretching of the muscle elongates the spindle fibers and stimulates the sensory dendritic endings wrapped around them. This causes the sensory neurons to send impulses to the CNS, where they synapse with motor neurons.

Sensing Taste, Smell, and Body Position

Some sensory cells, called **chemoreceptors,** contain membrane proteins that can bind to particular chemicals in the extracellular fluid. In response to this chemical interaction, the membrane of the sensory neuron becomes depolarized, leading to the production of action potentials. Chemoreceptors are used in the senses of taste and smell and are also important in monitoring the chemical composition of the blood and cerebrospinal fluid.

Taste

Taste buds—collections of chemosensitive epithelial cells associated with afferent neurons—mediate the sense of taste in vertebrates. In a fish, the taste buds are scattered over the surface of the body. These are the most sensitive vertebrate chemoreceptors known. They are particularly sensitive to amino acids; a catfish, for example, can distinguish between two different amino acids at a concentration of less than 100 parts per billion (1 g in 10,000 L of water)! The ability to taste the surrounding water is very important to bottom-feeding fish, enabling them to sense the presence of food in an often murky environment.

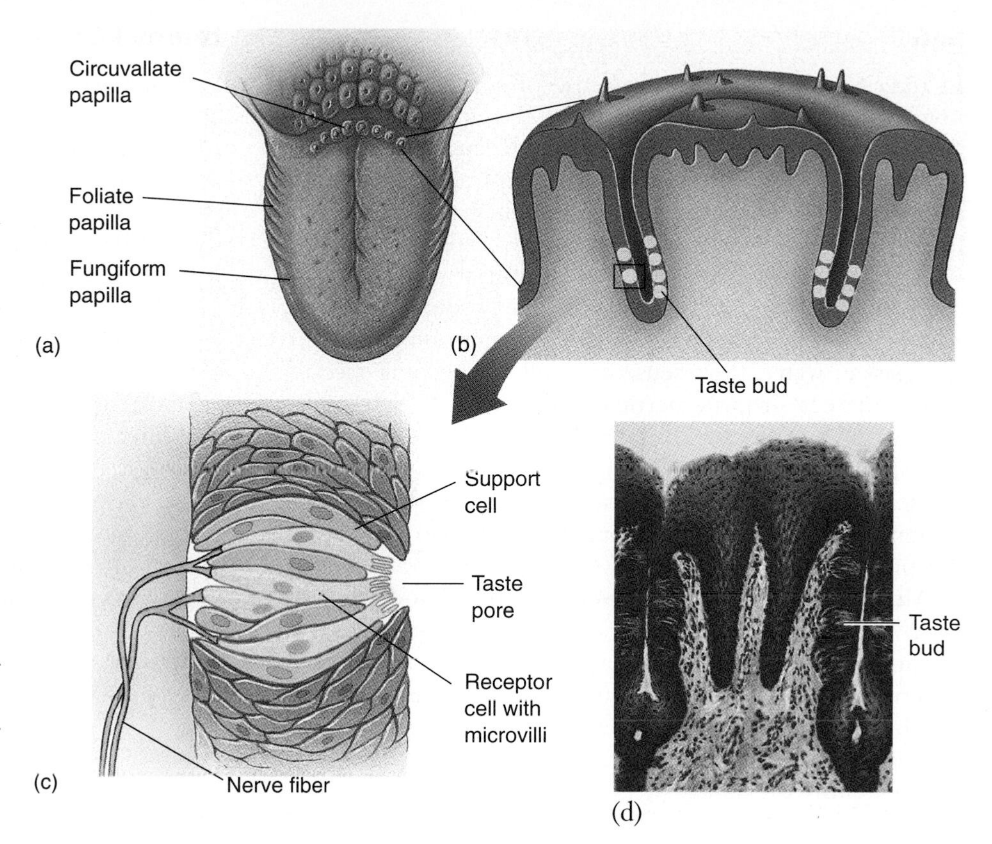

FIGURE 55.6
Taste. (*a*) Human tongues have three kinds of projections, called papillae, that bear taste buds located on different regions of the tongue. (*b*) Groups of taste buds are imbedded within the papilla. (*c*) Individual taste buds are bulb-shaped collections of chemosensitive receptors that open out into the mouth through a pore. (*d*) Photomicrograph of taste buds in papillae.

The taste buds of all terrestrial vertebrates are located in the epithelium of the tongue and oral cavity, within raised areas called *papillae* (figure 55.6). Taste buds are onion-shaped structures of between 50 to 100 taste cells, each of which have fingerlike projections called microvilli that poke up through an opening at the top of the taste bud, called the taste pore (figure 55.6*c*). Chemicals from food dissolve in saliva and contact the taste cells through the taste pore.

Within a taste bud, the chemicals that produce salty and sour tastes act directly through ion channels, and those responsible for sweet and bitter tastes bind to surface receptor proteins that trigger G-proteins to initiate changes in the cell interior that ultimately open and close ion channels. These changes in ion channels initiate signals in the sensory neurons extending from the tastebuds to the brain. There they interact with other sensory neurons carrying information related to smell.

Like vertebrates, many arthropods also have taste chemoreceptors. For example, flies, because of their mode of searching for food, have taste receptors in sensory hairs located on their feet. The sensory hairs contain different chemoreceptors that are able to detect sugars, salts, and other molecules (figure 55.7). They can detect a wide variety of tastes by the integration of stimuli from these chemoreceptors. If they step on potential food, the proboscis (the tubular feeding apparatus) extends to feed.

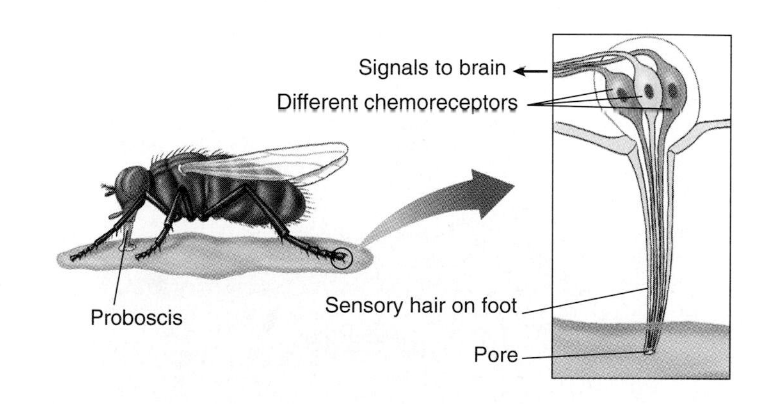

FIGURE 55.7
Many insects taste with their feet. In the blowfly shown here, chemoreceptors extend into the sensory hairs on the foot. Each different chemoreceptor detects a different type of food molecule. When the fly steps in a food substance, it can taste the different food molecules and extend its proboscis for feeding.

Smell

In terrestrial vertebrates, the sense of smell, or olfaction, involves chemoreceptors located in the upper portion of the nasal passages (figure 55.8). These receptors are bipolar neurons whose dendrites end in tassels of cilia that project into the nasal mucosa, and whose axon projects directly into the cerebral cortex. A terrestrial vertebrate uses its sense of smell in much the same way that a fish uses its sense of taste—to sample the chemical environment around it. Because terrestrial vertebrates are surrounded by air rather than water, their sense of smell has become specialized to detect airborne particles (but these particles must first dissolve in extracellular fluid before they can activate the olfactory receptors). The sense of smell can be extremely acute in many mammals, so much so that a single odorant molecule may be all that is needed to excite a given receptor.

Although humans can detect only four modalities of taste, they can discern thousands of different smells. New research suggests that there may be as many as a thousand different genes coding for different receptor proteins for smell. The particular set of olfactory neurons that respond to a given odor might serve as a "fingerprint" the brain can use to identify the odor.

Internal Chemoreceptors

Sensory receptors within the body detect a variety of chemical characteristics of the blood or fluids derived from the blood, including cerebrospinal fluid. Included among these receptors are the *peripheral chemoreceptors* of the aortic and carotid bodies, which are sensitive primarily to plasma pH, and the *central chemoreceptors* in the medulla oblongata of the brain, which are sensitive to the pH of cerebrospinal fluid. These receptors were discussed together with the regulation of breathing in chapter 53. When the breathing rate is too low, the concentration of plasma CO_2 increases, producing more carbonic acid and causing a fall in the blood pH. The carbon dioxide can also enter the cerebrospinal fluid and cause a lowering of the pH, thereby stimulating the central chemoreceptors. This chemoreceptor stimulation indirectly affects the respiratory control center of the brain stem, which increases the breathing rate. The aortic bodies can also respond to a lowering of blood oxygen concentrations, but this effect is normally not significant unless a person goes to a high altitude.

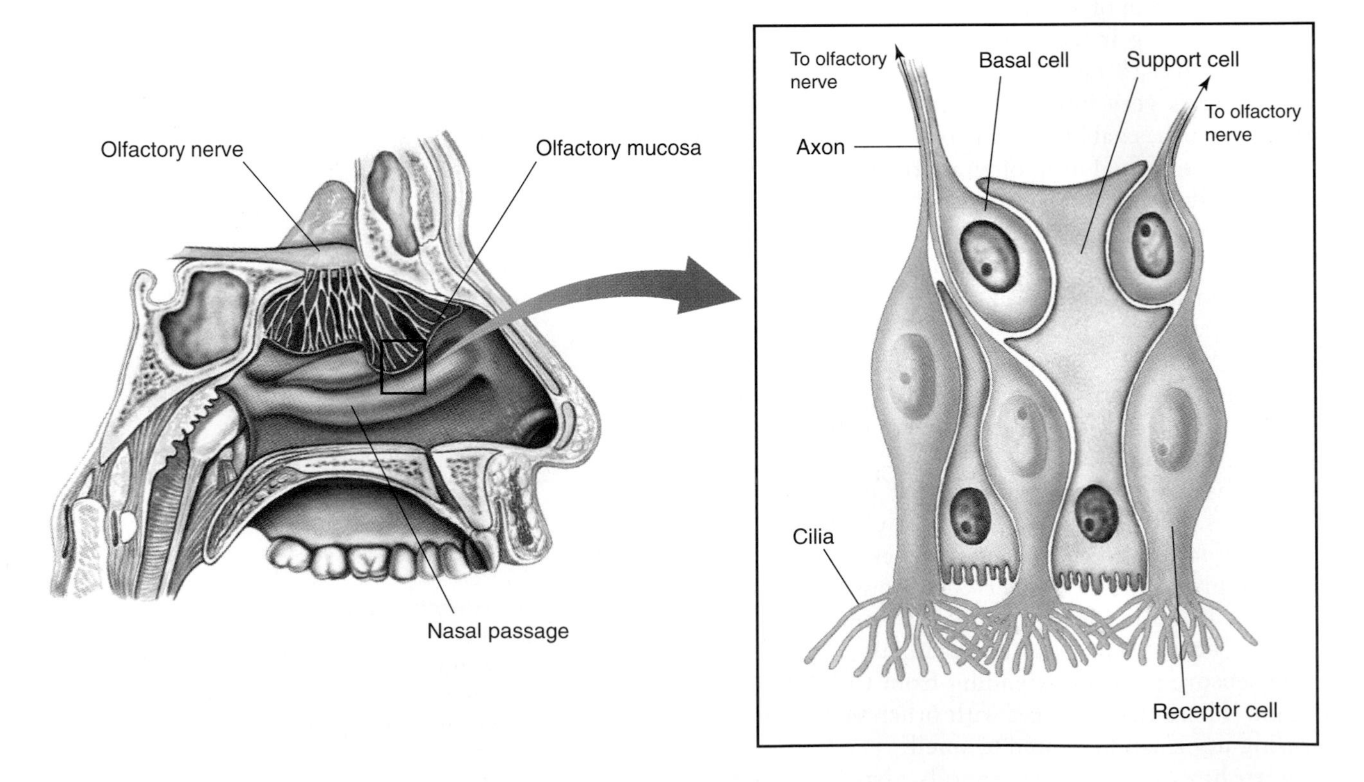

FIGURE 55.8
Smell. Humans detect smells by means of olfactory neurons located in the lining of the nasal passages. The axons of these neurons transmit impulses directly to the brain via the olfactory nerve. Basal cells regenerate new olfactory neurons to replace dead or damaged cells. Olfactory neurons typically live about one month.

The Lateral Line System

The lateral line system provides fish with a sense of "distant touch," enabling them to sense objects that reflect pressure waves and low-frequency vibrations. This enables a fish to detect prey, for example, and to swim in synchrony with the rest of its school. It also enables a blind cave fish to sense its environment by monitoring changes in the patterns of water flow past the lateral line receptors. The lateral line system is found in amphibian larvae, but is lost at metamorphosis and is not present in any terrestrial vertebrate. The sense provided by the lateral line system supplements the fish's sense of hearing, which is performed by a different sensory structure. The structures and mechanisms involved in hearing will be described in a later section.

The lateral line system consists of sensory structures within a longitudinal canal in the fish's skin that extends along each side of the body and within several canals in the head (figure 55.9*a*). The sensory structures are known as hair cells because they have hairlike processes at their surface that project into a gelatinous membrane called a *cupula* (Latin, "little cup"). The hair cells are innervated by sensory neurons that transmit impulses to the brain.

Hair cells have several hairlike processes of approximately the same length, called *stereocilia*, and one longer process called a *kinocilium* (figure 55.9*b*). Vibrations carried through the fish's environment produce movements of the cupula, which cause the hairs to bend. When the stereocilia bend in the direction of the kinocilium, the associated sensory neurons are stimulated and generate a receptor potential. As a result, the frequency of action potentials produced by the sensory neuron is increased. If the stereocilia are bent in the opposite direction, on the other hand, the activity of the sensory neuron is inhibited.

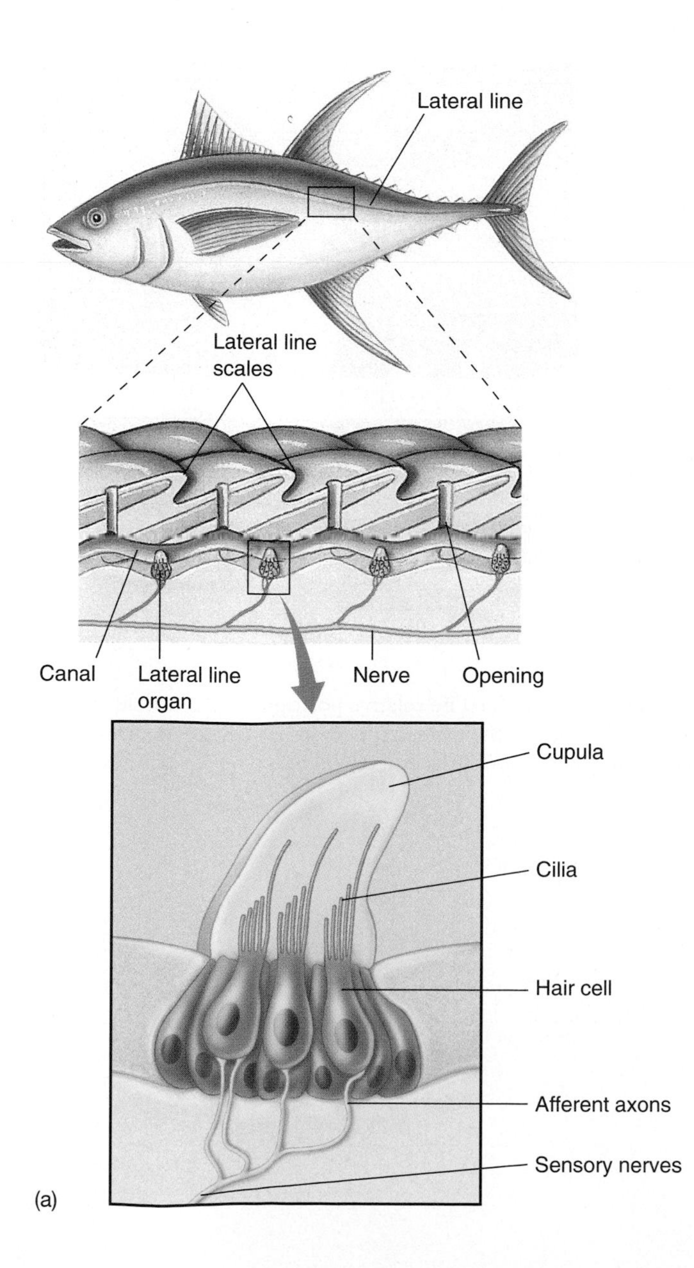

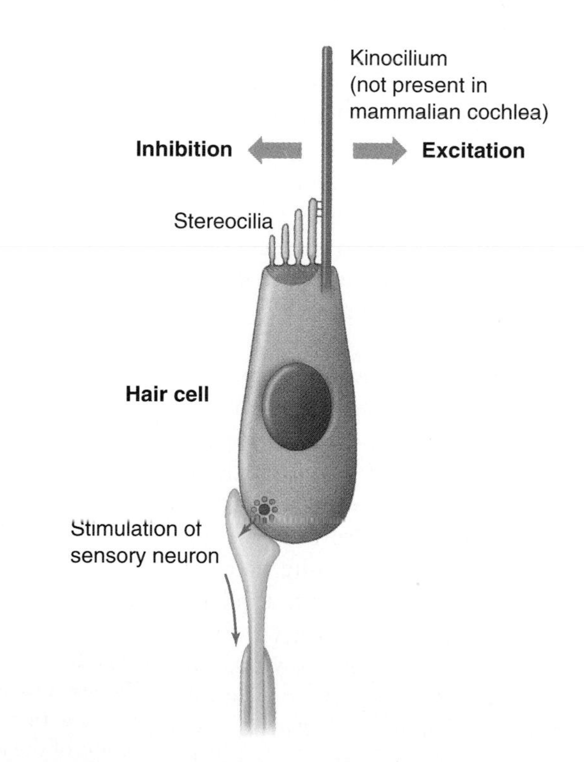

FIGURE 55.9
The lateral line system. (*a*) This system consists of canals running the length of the fish's body beneath the surface of the skin. Within these canals are sensory structures containing hair cells with cilia that project into a gelatinous cupula. Pressure waves traveling through the water in the canals deflect the cilia and depolarize the sensory neurons associated with the hair cells. (*b*) Hair cells are mechanoreceptors with hairlike cilia that project into a gelatinous membrane. The hair cells of the lateral line system (and the membranous labyrinth of the vertebrate inner ear) have a number of smaller cilia called stereocilia and one larger kinocilium. When the cilia bend in the direction of the kinocilium, the hair cell releases a chemical transmitter that depolarizes the associated sensory neuron. Bending of the cilia in the opposite direction has an inhibitory effect.

Gravity and Angular Acceleration

Most invertebrates can orient themselves with respect to gravity due to a sensory structure called a *statocyst.* Statocysts generally consist of ciliated hair cells with the cilia embedded in a gelatinous membrane containing crystals of calcium carbonate. These "stones," or *statoliths*, increase the mass of the gelatinous membrane so that it can bend the cilia when the animal's position changes. If the animal tilts to the right, for example, the statolith membrane will bend the cilia on the right side and activate associated sensory neurons.

A similar structure is found in the membranous labyrinth of the inner ear of vertebrates. The labyrinth is a system of fluid-filled membranous chambers and tubes that constitute the organs of equilibrium and hearing in vertebrates. This membranous labyrinth is surrounded by bone and perilymph, which is similar in ionic content to interstitial fluid. Inside, the chambers and tubes are filled with endolymph fluid, which is similar in ionic content to intracellular fluid. Though intricate, the entire structure is very small; in a human, it is about the size of a pea.

The receptors for gravity in vertebrates consist of two chambers of the membranous labyrinth called the utricle and saccule (figure 55.10). Within these structures are hair cells with stereocilia and a kinocilium, similar to those in the lateral line system of fish. The hairlike processes are embedded within a gelatinous membrane containing calcium carbonate crystals; this is known as an *otolith membrane*, because of its location in the inner ear (*oto* is derived from the Greek word for ear). Because the otolith organ is oriented differently in the utricle and saccule, the utricle is more

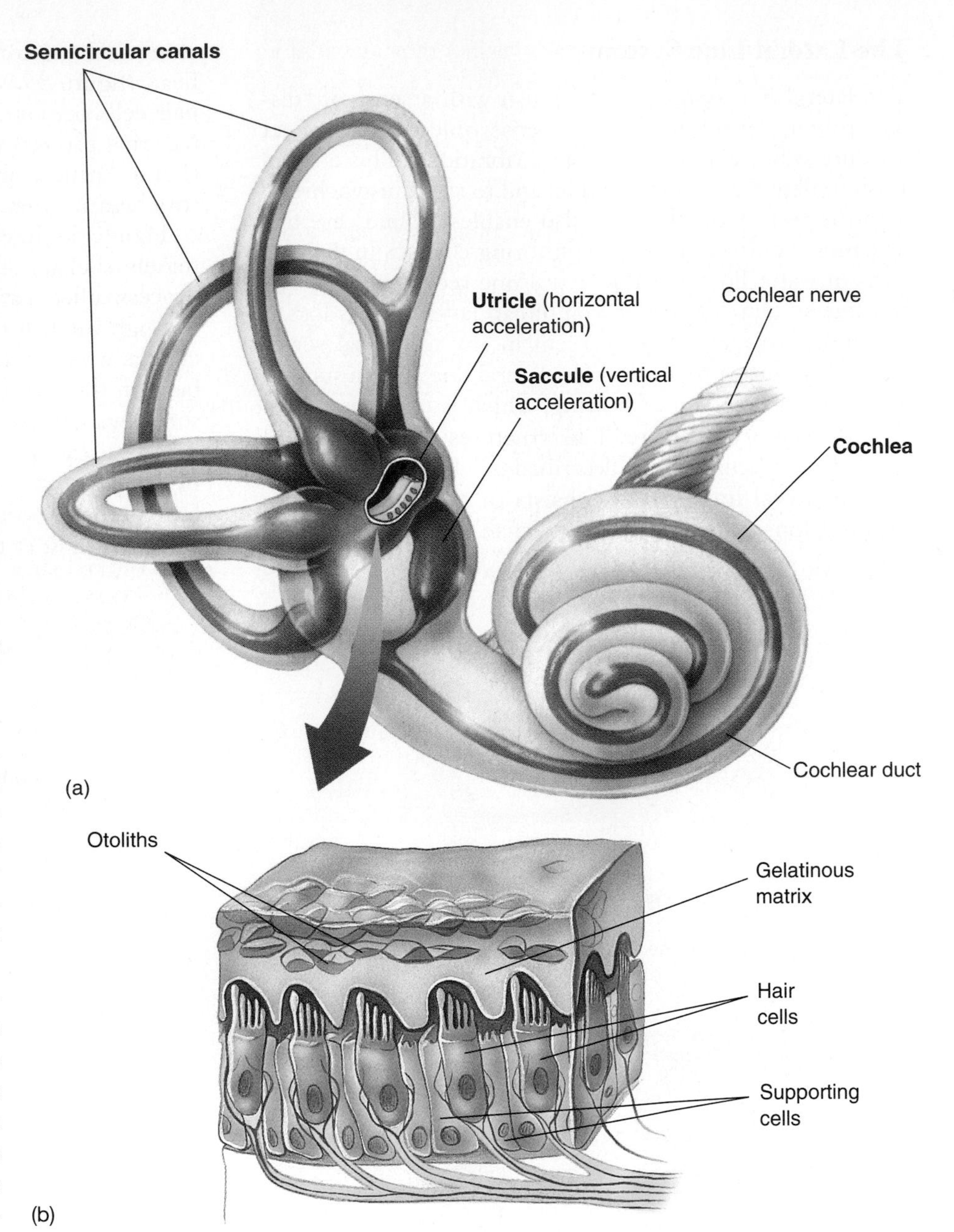

FIGURE 55.10
The structure of the utricle and saccule. (*a*) The relative positions of the utricle and saccule within the membranous labyrinth of the human inner ear. (*b*) Enlargement of a section of the utricle or saccule showing the otoliths embedded in the gelatinous matrix that covers the hair cells.

sensitive to horizontal acceleration (as in a moving car) and the saccule to vertical acceleration (as in an elevator). In both cases, the acceleration causes the stereocilia to bend and consequently produces action potentials in an associated sensory neuron.

The membranous labyrinth of the utricle and saccule is continuous with three semicircular canals, oriented in different planes so that angular acceleration in any direction can be detected (figure 55.11). At the ends of the canals are swollen chambers called *ampullae*, into which protrude the cilia of another group of hair cells. The tips of the cilia are embedded within a sail-like wedge of gelatinous material called a *cupula* (similar to the cupula of the fish lateral line system) that protrudes into the endolymph fluid of each semicircular canal.

When the head rotates, the fluid inside the semicircular canals pushes against the cupula and causes the cilia to bend. This bending either depolarizes or hyperpolarizes the hair cells, depending on the direction in which the cilia are bent. This is similar to the way the lateral line system works in a fish: if the stereocilia are bent in the direction of the kinocilium, a depolarization (receptor potential) is produced, which stimulates the production of action potentials in associated sensory neurons.

The saccule, utricle, and semicircular canals are collectively referred to as the *vestibular apparatus.* While the saccule and utricle provide a sense of linear acceleration, the semicircular canals provide a sense of angular acceleration. The brain uses information that comes from the vestibular apparatus about the body's position in space to maintain balance and equilibrium.

Receptors that sense chemicals originating outside the body are responsible for the senses of odor, smell, and taste. Internal chemoreceptors help to monitor chemicals produced within the body and are needed for the regulation of breathing. Hair cells in the lateral line organ of fishes detect water movements, and hair cells in the vestibular apparatus of terrestrial vertebrates provide a sense of acceleration.

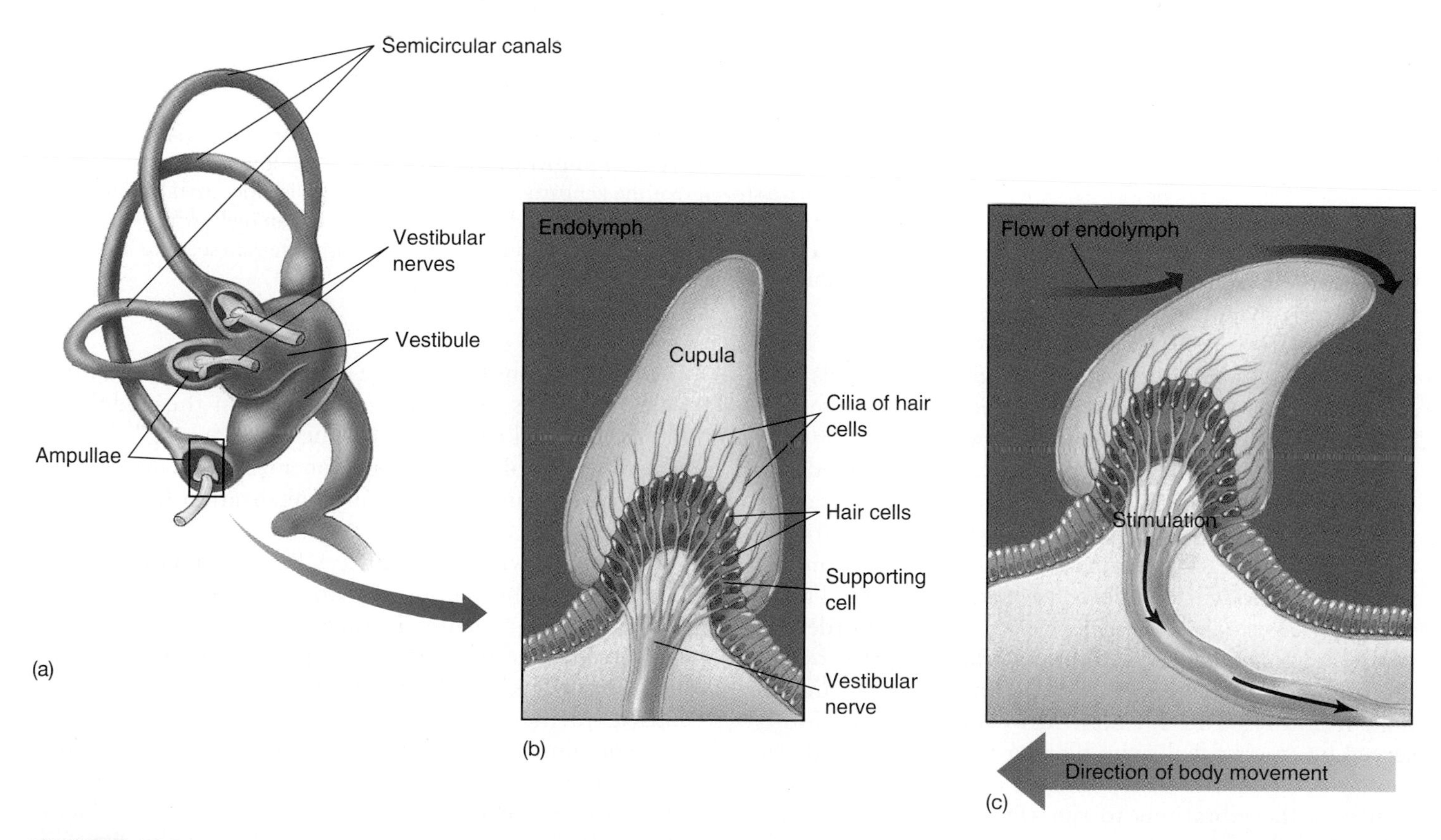

FIGURE 55.11
The structure of the semicircular canals. (*a*) The position of the semicircular canals in relation to the rest of the inner ear. (*b*) Enlargement of a section of one ampulla, showing how hair cell cilia insert into the cupula. (*c*) Angular acceleration in the plane of the semicircular canal causes bending of the cupula, thereby stimulating the hair cells.

55.3 Auditory receptors detect pressure waves in the air.

The Ears and Hearing

Fish detect vibrational pressure waves in water by means of their lateral line system. Terrestrial vertebrates detect similar vibrational pressure waves in air by means of similar hair cell mechanoreceptors in the inner ear. Hearing actually works better in water than in air because water transmits pressure waves more efficiently. Despite this limitation, hearing is widely used by terrestrial vertebrates to monitor their environments, communicate with other members of the same species, and to detect possible sources of danger (figure 55.12). Auditory stimuli travel farther and more quickly than chemical ones, and auditory receptors provide better directional information than do chemoreceptors. Auditory stimuli alone, however, provide little information about distance.

FIGURE 55.12
Kangaroo rats have specialized ears. Kangaroo rats (*Dipodomys*) are unique in having an enlarged tympanic membrane (eardrum), a lengthened and freely rotating malleus (ear bone), and an increased volume of air-filled chambers in the middle ear. These and other specializations result in increased sensitivity to sound, especially to low-frequency sounds. Experiments have shown that the kangaroo rat's ears are adapted to nocturnal life and allow them to hear the low-frequency sounds of their predators, such as an owl's wingbeats or a sidewinder rattlesnake's scales rubbing against the ground. Also, the ears seem to be adapted to the poor sound-carrying quality of dry, desert air.

Structure of the Ear

Fish use their lateral line system to detect water movements and vibrations emanating from relatively nearby objects, and their hearing system to detect vibrations that originate from a greater distance. The hearing system of fish consists of the otolith organs in the membranous labyrinth (utricle and saccule) previously described, together with a very small outpouching of the membranous labyrinth called the lagena. Sound waves travel through the body of the fish as easily as through the surrounding water, as the body is composed primarily of water. Therefore, an object of different density is needed in order for the sound to be detected. This function is served by the otolith (calcium carbonate crystals) in many fish. In catfish, minnows, and suckers, however, this function is served by an air-filled swim bladder that vibrates with the sound. A chain of small bones, Weberian ossicles, then transmits the vibrations to the saccule in some of these fish.

In the ears of terrestrial vertebrates, vibrations in air may be channeled through an ear canal to the eardrum, or tympanic membrane. These structures are part of the *outer ear*. Vibrations of the tympanic membrane cause movement of three small bones (ossicles)—the *malleus* (hammer), *incus* (anvil), and *stapes* (stirrup)—that are located in a bony cavity known as the *middle ear* (figure 55.13). These middle ear ossicles are analogous to the Weberian ossicles in fish. The middle ear is connected to the throat by the *Eustachian tube*, which equalizes the air pressure between the middle ear and the external environment. The "ear popping" you may have experienced when flying in an airplane or driving on a mountain is caused by pressure equalization between the two sides of the eardrum.

The stapes vibrates against a flexible membrane, the *oval window*, which leads into the *inner ear*. Because the oval window is smaller in diameter than the tympanic membrane, vibrations against it produce more force per unit area, transmitted into the inner ear. The inner ear consists of the cochlea (Latin for "snail"), a bony structure containing part of the membranous labyrinth called the cochlear duct. The cochlear duct is located in the center of the cochlea; the area above the cochlear duct is the *vestibular canal*, and the area below is the *tympanic canal*. All three chambers are filled with fluid, as previously described. The oval window opens to the upper vestibular canal, so that when the stapes causes it to vibrate, it produces pressure waves of fluid. These pressure waves travel down to the tympanic canal, pushing another flexible membrane, the *round window*, that transmits the pressure back into the middle ear cavity (see figure 55.13).

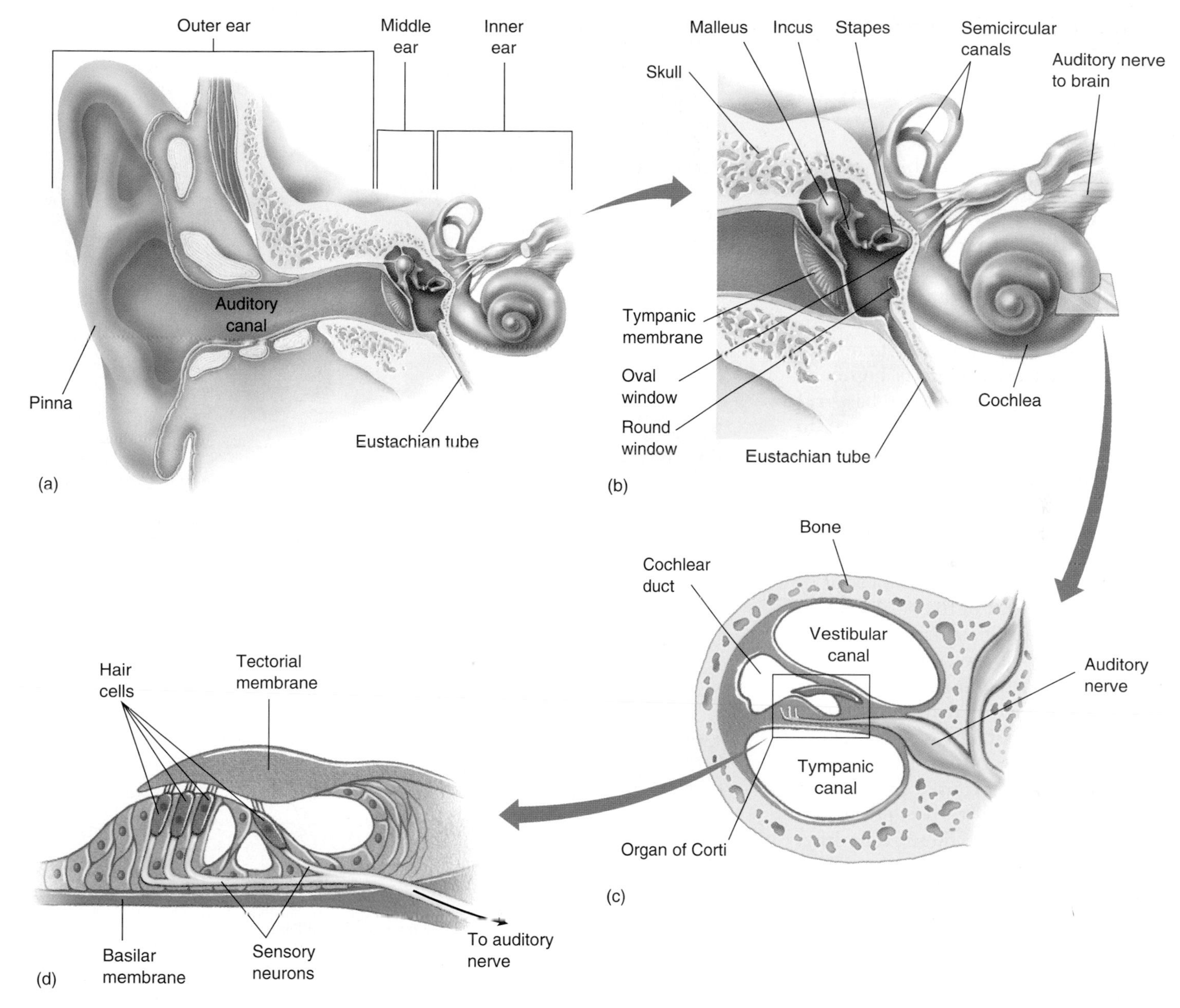

FIGURE 55.13
Structure of the human ear. (*a*) Sound waves passing through the ear canal produce vibrations of the tympanic membrane, which cause movement of the (*b*) middle ear ossicles (the malleus, incus, and stapes) against an inner membrane called the oval window. Vibration of the oval window sets up pressure waves that (*c* and *d*) travel through the fluid in the vestibular and tympanic canals of the cochlea.

Transduction in the Cochlea

As the pressure waves produced by vibrations of the oval window are transmitted through the cochlea to the round window, they cause the cochlear duct to vibrate. The bottom of the cochlear duct, called the *basilar membrane,* is quite flexible and vibrates in response to these pressure waves. The surface of the basilar membrane contains sensory hair cells, similar to those of the vestibular apparatus and lateral line system but lacking a kinocilium. The cilia from the hair cells project into an overhanging gelatinous membrane, the *tectorial membrane.* This sensory apparatus, consisting of the basilar membrane, hair cells with associated sensory neurons, and tectorial membrane, is known as the **organ of Corti.**

As the basilar membrane vibrates, the cilia of the hair cells bend in response to the movement of the basilar membrane relative to the tectorial membrane. As in the lateral line organs and the vestibular apparatus, the bending of these cilia depolarizes the hair cells. The hair cells, in turn, stimulate the production of action potentials in sensory neurons that project to the brain, where they are interpreted as sound.

Frequency Localization in the Cochlea

The basilar membrane of the cochlea consists of elastic fibers of varying length and stiffness, like the strings of a musical instrument, embedded in a gelatinous material. At the base of the cochlea (near the oval window), the fibers of the basilar membrane are short and stiff. At the far end of the cochlea (the apex), the fibers are 5 times longer and 100 times more flexible. Therefore, the resonant frequency of the basilar membrane is higher at the base than the apex; the base responds to higher pitches, the apex to lower.

When a wave of sound energy enters the cochlea from the oval window, it initiates a traveling up-and-down motion of the basilar membrane. However, this wave imparts most of its energy to that part of the basilar membrane with a resonant frequency near the frequency of the sound wave, resulting in a maximum deflection of the basilar membrane at that point (figure 55.14). As a result, the hair cell depolarization is greatest in that region, and the afferent axons from that region are stimulated to produce action potentials more than those from other regions. When these action potentials arrive in the brain, they are interpreted as representing a sound of that particular frequency, or pitch.

The flexibility of the basilar membrane limits the frequency range of human hearing to between approximately 20 and 20,000 cycles per second (hertz) in children. Our ability to hear high-pitched sounds decays progressively throughout middle age. Other vertebrates can detect sounds at frequencies lower than 20 hertz and much higher than 20,000 hertz. Dogs, for example, can detect sounds at 40,000 hertz, enabling them to hear high-pitched dog whistles that seem silent to a human listener.

Hair cells are also innervated by efferent axons from the brain, and impulses in those axons can make hair cells less sensitive. This central control of receptor sensitivity can increase an individual's ability to concentrate on a particular auditory signal (for example, a single voice) in the midst of background noise, which is effectively "tuned out" by the efferent axons.

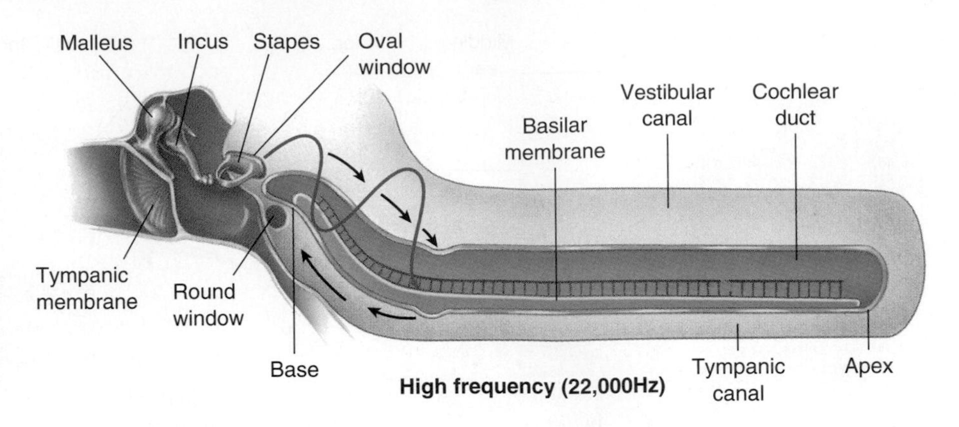

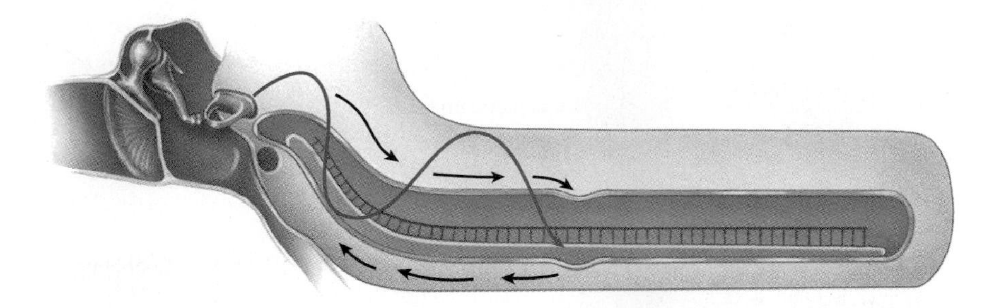

Medium frequency (2000Hz)

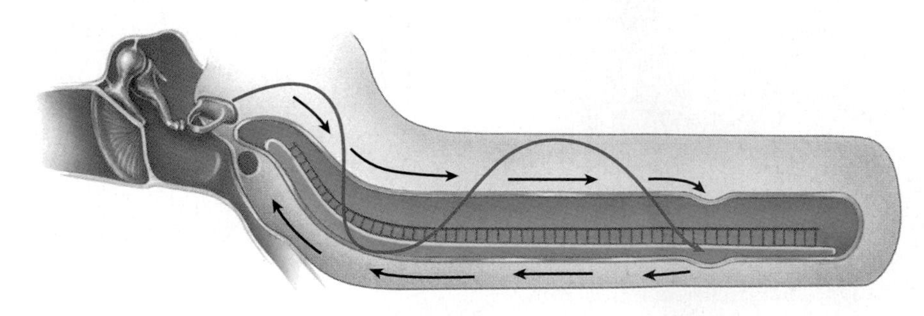

Low frequency (500Hz)

FIGURE 55.14
Frequency localization in the cochlea. The cochlea is shown unwound, so that the length of the basilar membrane can be seen. The fibers within the basilar membrane vibrate in response to different frequencies of sound, related to the pitch of the sound. Thus, regions of the basilar membrane show maximum vibrations in response to different sound frequencies. Notice that low-frequency (pitch) sounds vibrate the basilar membrane more toward the apex, while high frequencies cause vibrations more toward the base.

The middle ear ossicles vibrate in response to sound waves, creating fluid vibrations within the inner ear. This causes the hair cells to bend, transducing the sound into action potentials. The pitch of a sound is determined by which hair cells (and thus which sensory neurons) are activated by the vibration of the basilar membrane.

Sonar

Because terrestrial vertebrates have two ears located on opposite sides of the head, the information provided by hearing can be used by the CNS to determine *direction* of a sound source with some precision. Sound sources vary in strength, however, and sounds are attenuated (weakened) to varying degrees by the presence of objects in the environment. For these reasons, auditory sensors do not provide a reliable measure of *distance.*

A few groups of mammals that live and obtain their food in dark environments have circumvented the limitations of darkness. A bat flying in a completely dark room easily avoids objects that are placed in its path—even a wire less than a millimeter in diameter (figure 55.15). Shrews use a similar form of "lightless vision" beneath the ground, as do whales and dolphins beneath the sea. All of these mammals perceive distance by means of sonar. They emit sounds and then determine the time it takes these sounds to reach an object and return to the animal. This process is called echolocation. A bat, for example, produces clicks that last 2 to 3 milliseconds and are repeated several hundred times per second. The three-dimensional imaging achieved with such an auditory sonar system is quite sophisticated.

Being able to "see in the dark" has opened a new ecological niche to bats, one largely closed to birds because birds must rely on vision. There are no truly nocturnal birds; even owls rely on vision to hunt, and do not fly on dark nights. Because bats are able to be active and efficient in total darkness, they are one of the most numerous and widespread of all orders of mammals.

Some mammals emit sounds and then determine the time it takes for the sound to return, using the method of sonar to locate themselves and other objects in a totally dark environment by the characteristics of the echo. Bats are the most adept at this echolocation.

FIGURE 55.15
Sonar. As it flies, a bat emits high-frequency "clicks" and listens for the return of the clicks after they are reflected by objects such as moths. By timing how long it takes for a click to return, the bat can locate its prey and catch it even in total darkness.

Evolution of the Eye

Vision begins with the capture of light energy by photoreceptors. Because light travels in a straight line and arrives virtually instantaneously, visual information can be used to determine both the direction and the distance of an object. No other stimulus provides as much detailed information.

Many invertebrates have simple visual systems with photoreceptors clustered in an eyespot. Simple eyespots can be made sensitive to the direction of a light source by the addition of a pigment layer which shades one side of the eye. Flatworms have a screening pigmented layer on the inner and back sides of both eyespots allowing stimulation of the photoreceptor cells only by light from the front of the animal (figure 55.16). The flatworm will turn and swim in the direction in which the photoreceptor cells are the least stimulated. Although an eyespot can perceive the direction of light, it cannot be used to construct a visual image. The members of four phyla—annelids, mollusks, arthropods, and chordates—have evolved well-developed, image-forming eyes. True image-forming eyes in these phyla, though strikingly similar in structure, are believed to have evolved independently (figure 55.17). Interestingly, the photoreceptors in all of them use the same light-capturing molecule, suggesting that not many alternative molecules are able to play this role.

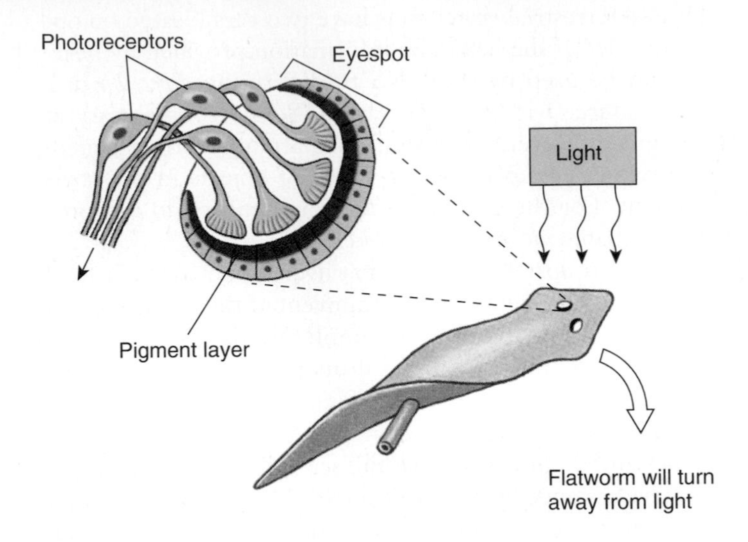

FIGURE 55.16
Simple eyespots in the flatworm. Eyespots will detect the direction of light because a pigmented layer on one side of the eyespot screens out light coming from the back of the animal. Light is thus the strongest coming from the front of the animal; flatworms will respond by turning away from the light.

Structure of the Vertebrate Eye

The eye of a human is typical of the vertebrate eye (figure 55.18). The "white of the eye" is the sclera, formed of tough connective tissue. Light enters the eye through a transparent cornea, which begins to focus the light. This occurs because light is refracted (bent) when it travels into a medium of different density. The colored portion of the eye is the iris; contraction of the iris muscles in bright light decreases the size of its opening, the pupil. Light passes through the pupil to the lens, a transparent structure that completes the focusing of the light onto the retina at the back of the eye. The lens is attached by the *suspensory ligament* to the ciliary muscles.

The shape of the lens is influenced by the amount of tension in the suspensory ligament, which surrounds the

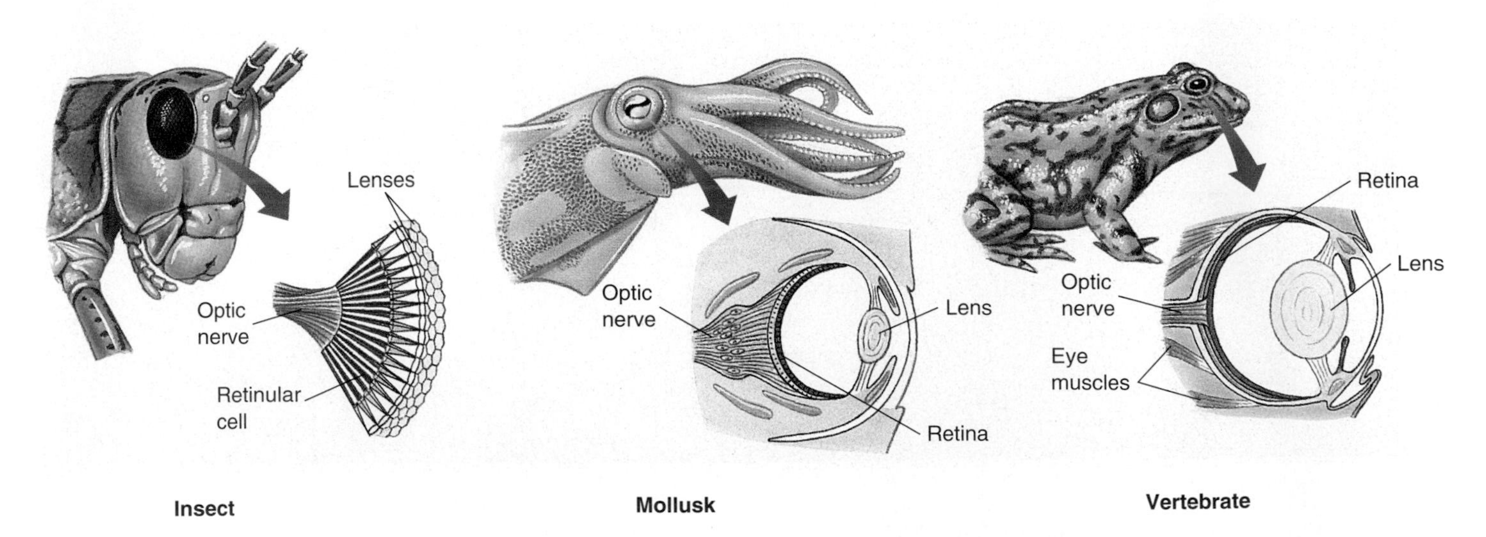

FIGURE 55.17
Eyes in three phyla of animals. Although they are superficially similar, these eyes differ greatly in structure and are not homologous. Each has evolved separately and, despite the apparent structural complexity, has done so from simpler structures.

lens and attaches it to the circular ciliary muscle. When the ciliary muscle contracts, it puts slack in the suspensory ligament and the lens becomes more rounded and powerful. This is required for close vision; in far vision, the ciliary muscles relax, moving away from the lens and tightening the suspensory ligament. The lens thus becomes more flattened and less powerful, keeping the image focused on the retina. People who are nearsighted or farsighted do not properly focus the image on the retina (figure 55.19). Interestingly, the lens of an amphibian or a fish does not change shape; these animals instead focus images by moving their lens in and out, just as you would do to focus a camera.

Annelids, mollusks, arthropods, and vertebrates have independently evolved image-forming eyes. The vertebrate eye admits light through a pupil and then focuses this light by means of an adjustable lens onto the retina at the back of the eye.

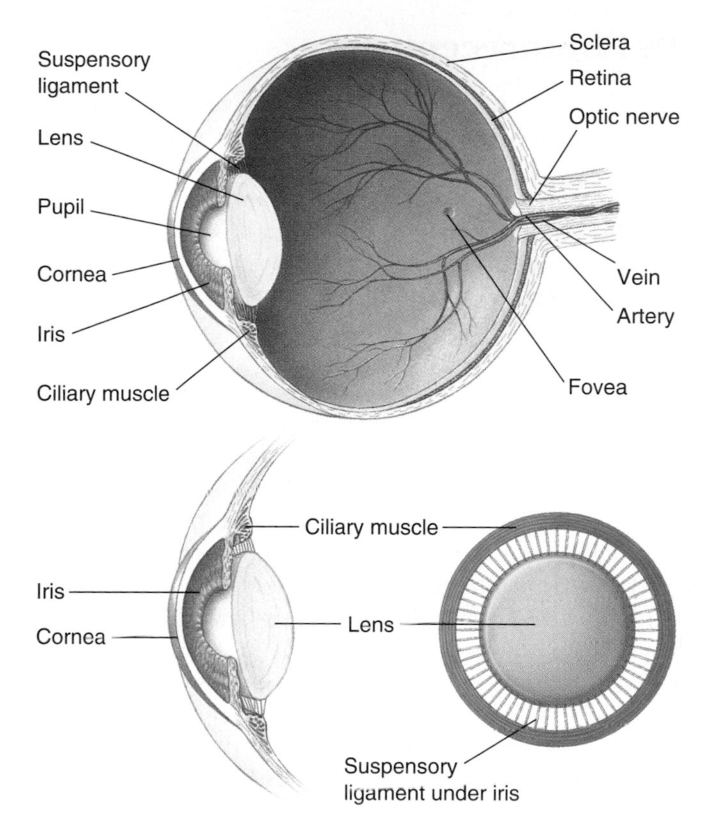

FIGURE 55.18
Structure of the human eye. The transparent cornea and lens focus light onto the retina at the back of the eye, which contains the rods and cones. The center of each eye's visual field is focused on the fovea. Focusing is accomplished by contraction and relaxation of the ciliary muscle, which adjusts the curvature of the lens.

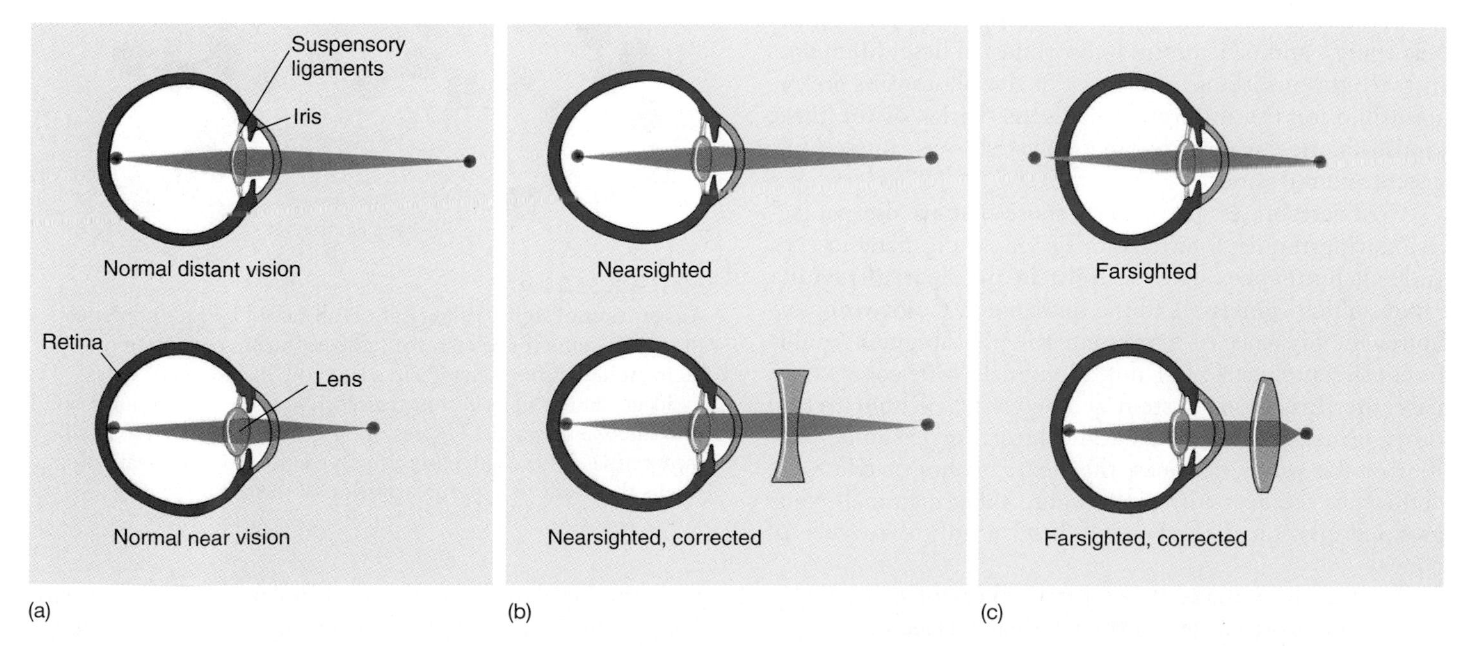

FIGURE 55.19
Focusing the human eye. (*a*) In people with normal vision, the image remains focused on the retina in both near and far vision because of changes produced in the curvature of the lens. When a person with normal vision stands 20 feet or more from an object, the lens is in its least convex form and the image is focused on the retina. (*b*) In nearsighted people, the image comes to a focus in front of the retina and the image thus appears blurred. (*c*) In farsighted people, the focus of the image would be behind the retina because the distance from the lens to the retina is too short.

Vertebrate Photoreceptors

The vertebrate retina contains two kinds of photoreceptors, called rods and cones (figure 55.20). Rods are responsible for black-and-white vision when the illumination is dim, while cones are responsible for high visual acuity (sharpness) and color vision. Humans have about 100 million rods and 3 million cones in each retina. Most of the cones are located in the central region of the retina known as the fovea, where the eye forms its sharpest image. Rods are almost completely absent from the fovea.

Rods and cones have the same basic cellular structure. An inner segment rich in mitochondria contains numerous vesicles filled with neurotransmitter molecules. It is connected by a narrow stalk to the outer segment, which is packed with hundreds of flattened discs stacked on top of one another. The light-capturing molecules, or photopigments, are located on the membranes of these discs.

In rods, the photopigment is called rhodopsin. It consists of the protein opsin bound to a molecule of *cis*-retinal (figure 55.21), which is derived from carotene, a photosynthetic pigment in plants. The photopigments of cones, called photopsins, are structurally very similar to rhodopsin. Humans have three kinds of cones, each of which possesses a photopsin consisting of *cis*-retinal bound to a protein with a slightly different amino acid sequence. These differences shift the *absorption maximum*—the region of the electromagnetic spectrum that is best absorbed by the pigment—(figure 55.22). The absorption maximum of the *cis*-retinal in rhodopsin is 500 nanometers (nm); the absorption maxima of the three kinds of cone photopsins, in contrast, are 455 nm (blue-absorbing), 530 nm (green-absorbing), and 625 nm (red-absorbing). These differences in the light-absorbing properties of the photopsins are responsible for the different color sensitivities of the three kinds of cones, which are often referred to as simply blue, green, and red cones.

Most vertebrates, particularly those that are diurnal (active during the day), have color vision, as do many insects. Indeed, honeybees can see light in the near-ultraviolet range, which is invisible to the human eye. Color vision requires the presence of more than one photopigment in different receptor cells, but not all animals with color vision have the three-cone system characteristic of humans and other primates. Fish, turtles, and birds, for example, have four or five kinds of cones; the "extra" cones enable these animals to see near-ultraviolet light. Many mammals (such as squirrels), on the other hand, have only two types of cones.

The retina is made up of three layers of cells (figure 55.23): the layer closest to the external surface of the eyeball consists of the rods and cones, the next layer contains bipolar cells, and the layer closest to the cavity of the eye is composed of ganglion cells. Thus, light must first pass through the ganglion cells and bipolar cells in order to reach the photoreceptors! The rods and cones synapse with the bipolar cells, and the bipolar cells synapse with the ganglion cells, which transmit impulses to the brain via the optic nerve. The flow of sensory information in the retina is therefore opposite to the path of light through the retina. It should also be noted that the retina contains two additional types of neurons, horizontal cells and amacrine cells. Stimulation of horizontal cells by photoreceptors at

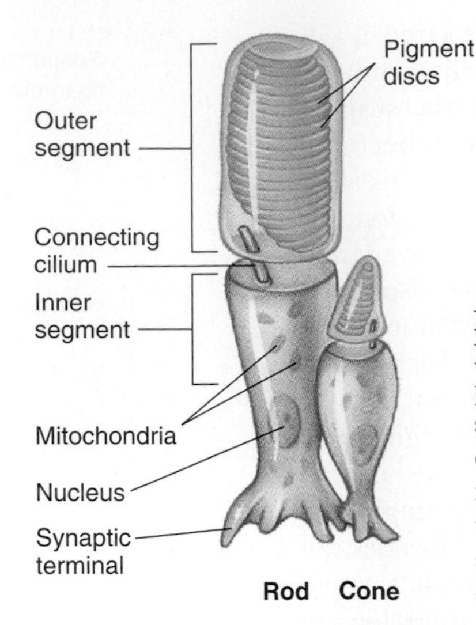

FIGURE 55.20
Rods and cones. The pigment-containing outer segment in each of these cells is separated from the rest of the cell by a partition through which there is only a narrow passage, the connecting cilium.

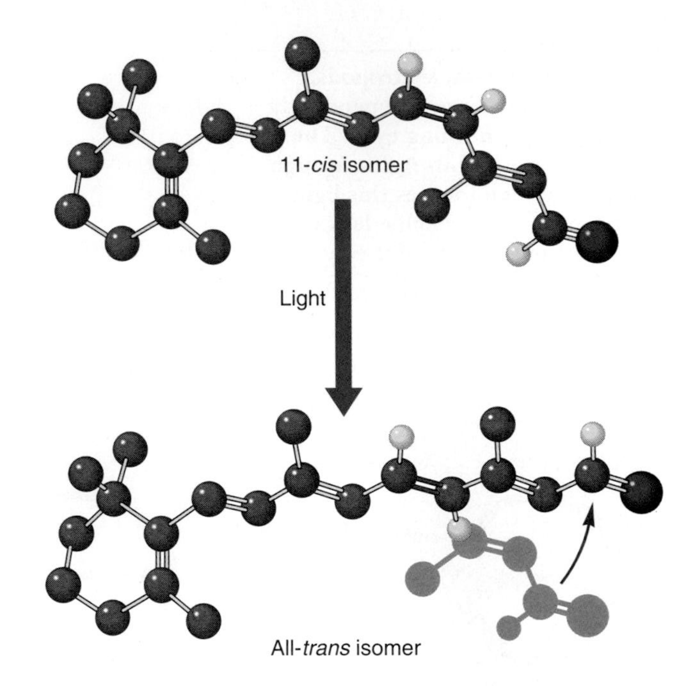

FIGURE 55.21
Absorption of light. When light is absorbed by a photopigment, the 11-*cis* isomer of retinal, the light-capturing portion of the pigment undergoes a change in shape: the linear end of the molecule (at the right in this diagram) rotates about a double bond (indicated here in *red*). The resulting isomer is referred to as all-*trans* retinal. This change in retinal's shape initiates a chain of events that leads to hyperpolarization of the photoreceptor.

the center of a spot of light on the retina can inhibit the response of photoreceptors peripheral to the center. This lateral inhibition enhances contrast and sharpens the image.

Sensory Transduction in Photoreceptors

The transduction of light energy into nerve impulses follows a sequence that is the inverse of the usual way that sensory stimuli are detected. This is because, in the dark, the photoreceptors release an inhibitory neurotransmitter that hyperpolarizes the bipolar neurons. Thus inhibited, the bipolar neurons do not release excitatory neurotransmitter to the ganglion cells. Light *inhibits* the photoreceptors from releasing their inhibitory neurotransmitter, and by this means, *stimulates* the bipolar cells and thus the ganglion cells, which transmit action potentials to the brain.

A rod or cone contains many Na^+ channels in the plasma membrane of its outer segment, and in the dark, many of these channels are open. As a consequence, Na^+ continuously diffuses into the outer segment and across the narrow stalk to the inner segment. This flow of Na^+ that occurs in the absence of light is called the dark current, and it causes the membrane of a photoreceptor to be somewhat depolarized in the dark. In the light, the Na^+ channels in the outer segment rapidly close, reducing the dark current and causing the photoreceptor to hyperpolarize.

Researchers have discovered that cyclic guanosine monophosphate (cGMP) is required to keep the Na^+ channels open, and that the channels will close if the cGMP is converted into GMP. How does light cause this conversion and consequent closing of the Na^+ channels? When a photopigment absorbs light, *cis*-retinal isomerizes and dissociates from opsin in what is known as the bleaching reaction. As a result of this dissociation, the opsin protein changes shape. Each opsin is associated with over a hundred regulatory *G proteins* (see chapters 7 and 54). When the opsin changes shape, the G proteins dissociate, releasing subunits that activate hundreds of molecules of the enzyme *phosphodiesterase*. This enzyme converts cGMP to GMP, thus closing the Na^+ channels at a rate of about 1000 per second and inhibiting the dark current. The absorption of a single photon of light can block the entry of more than a million sodium ions, thereby causing the photoreceptor to hyperpolarize and release less inhibitory neurotransmitters. Freed from inhibition, the bipolar cells activate ganglion cells, which transmit action potentials to the brain.

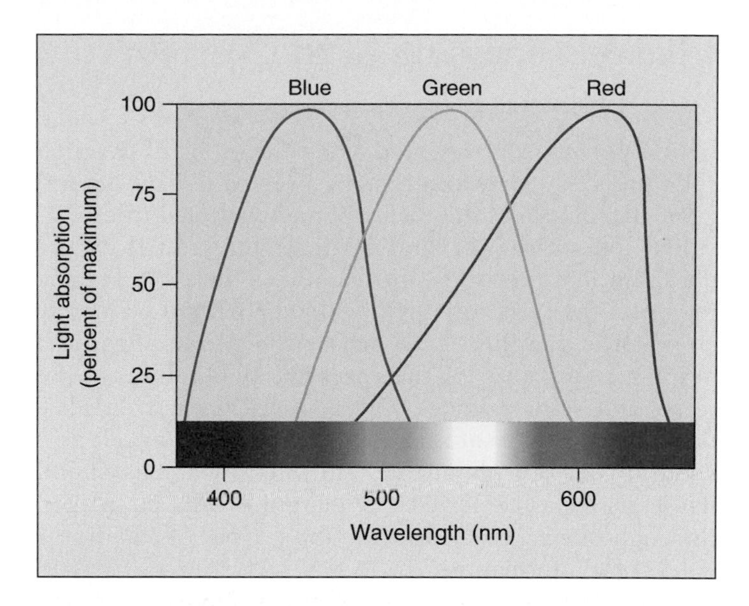

FIGURE 55.22
Color vision. The absorption maximum of *cis*-retinal in the rhodopsin of rods is 500 nanometers (nm). However, the "blue cones" have their maximum light absorption at 455 nm; the "green cones" at 530 nm, and the red cones at 625 nm. The brain perceives all other colors from the combined activities of these three cones' systems.

Photoreceptor rods and cones contain the photopigment *cis*-retinal, which dissociates in response to light and indirectly activates bipolar neurons and then ganglion cells.

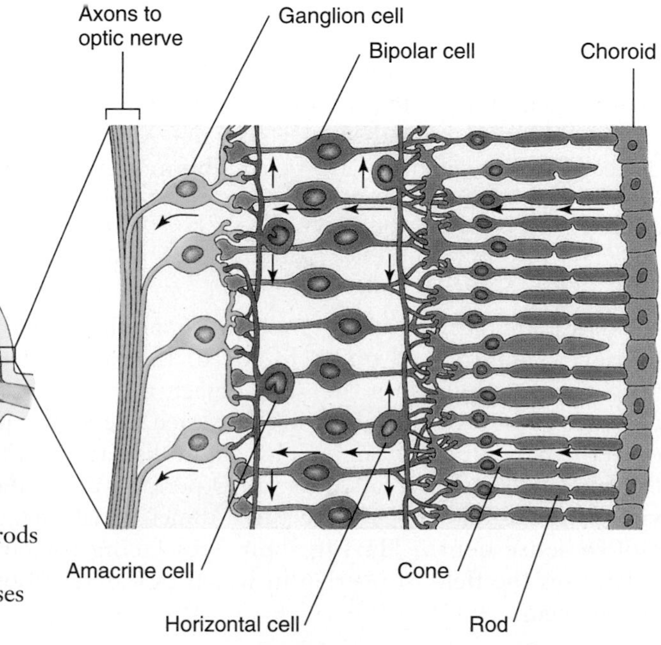

FIGURE 55.23
Structure of the retina. Note that the rods and cones are at the rear of the retina, not the front. Light passes through four other types of cells in the retina before it reaches the rods and cones. Once the photoreceptors are activated, they stimulate bipolar cells, which in turn stimulate ganglion cells. The direction of nerve impulses in the retina is thus opposite to the direction of light.

Visual Processing in the Vertebrate Retina

Action potentials propagated along the axons of ganglion cells are relayed through structures called the lateral geniculate nuclei of the thalamus and projected to the occipital lobe of the cerebral cortex (figure 55.24). There the brain interprets this information as light in a specific region of the eye's receptive field. The pattern of activity among the ganglion cells across the retina encodes a point-to-point map of the receptive field, allowing the retina and brain to image objects in visual space. In addition, the frequency of impulses in each ganglion cell provides information about the light intensity at each point, while the relative activity of ganglion cells connected (through bipolar cells) with the three types of cones provides color information.

The relationship between receptors, bipolar cells, and ganglion cells varies in different parts of the retina. In the fovea, each cone makes a one-to-one connection with a bipolar cell, and each bipolar cell synapses, in turn, with one ganglion cell. This point-to-point relationship is responsible for the high acuity of foveal vision. Outside the fovea, many rods can converge on a single bipolar cell, and many bipolar cells can converge on a single ganglion cell. This convergence permits the summation of neural activity, making the area of the retina outside of the fovea more sensitive to dim light than the fovea, but at the expense of acuity and color vision. This is why dim objects, such as faint stars at night, are best seen when you don't look directly at them. It has been said that we use the periphery of the eye as a detector and the fovea as an inspector.

Color blindness is due to an inherited lack of one or more types of cones. People with normal color vision are *trichromats*; those with only two types of cones are *dichromats.* People with this condition may lack red cones (have *protanopia*), for example, and have difficulty distinguishing red from green. Men are far more likely to be color blind than women, because the trait for color blindness is carried on the X chromosome; men have only one X chromosome per cell, whereas women have two X chromosomes and so can carry the trait in a recessive state.

Binocular Vision

Primates (including humans) and most predators have two eyes, one located on each side of the face. When both eyes are trained on the same object, the image that each sees is slightly different because each eye views the object from a different angle. This slight displacement of the images (an effect called parallax) permits **binocular vision,** the ability to perceive three-dimensional images and to sense depth. Having their eyes facing forward maximizes the field of overlap in which this stereoscopic vision occurs.

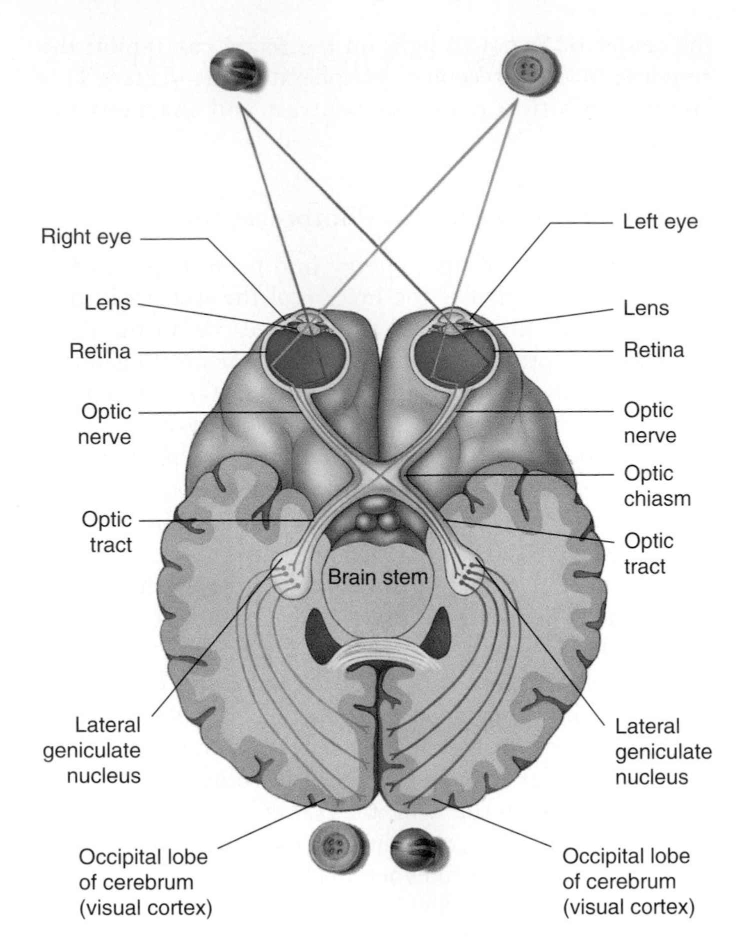

FIGURE 55.24
The pathway of visual information. Action potentials in the optic nerves are relayed from the retina to the lateral geniculate nuclei, and from there to the visual cortex of the occipital lobes. Notice that the medial fibers of the optic nerves cross to the other side at the optic chiasm, so that each hemisphere of the cerebrum receives input from both eyes.

In contrast, prey animals generally have eyes located to the sides of the head, preventing binocular vision but enlarging the overall receptive field. Depth perception is less important to prey than detection of potential enemies from any quarter. The eyes of the American Woodcock, for example, are located at exactly opposite sides of its skull so that it has a 360-degree field of view without turning its head! Most birds have laterally placed eyes and, as an adaptation, have two foveas in each retina. One fovea provides sharp frontal vision, like the single fovea in the retina of mammals, and the other fovea provides sharper lateral vision.

The axons of ganglion cells transmit action potentials to the thalamus, which in turn relays visual information to the occipital lobe of the brain. The fovea provides high visual acuity, whereas the retina outside the fovea provides high sensitivity to dim light. Binocular vision with overlapping visual fields provides depth perception.

55.5 Some vertebrates use heat, electricity, or magnetism for orientation.

Diversity of Sensory Experiences

Vision is the primary sense used by all vertebrates that live in a light-filled environment, but visible light is by no means the only part of the electromagnetic spectrum that vertebrates use to sense their environment.

Heat

Electromagnetic radiation with wavelengths longer than those of visible light is too low in energy to be detected by photoreceptors. Radiation from this *infrared* ("below red") portion of the spectrum is what we normally think of as radiant heat. Heat is an extremely poor environmental stimulus in water because water has a high thermal capacity and readily absorbs heat. Air, in contrast, has a low thermal capacity, so heat in air is a potentially useful stimulus. However, the only vertebrates known to have the ability to sense infrared radiation are the snakes known as pit vipers.

The pit vipers possess a pair of heat-detecting pit organs located on either side of the head between the eye and the nostril (figure 55.25). The pit organs permit a blindfolded rattlesnake to accurately strike at a warm, dead rat. Each pit organ is composed of two chambers separated by a membrane. The infrared radiation falls on the membrane and warms it. Thermal receptors on the membrane are stimulated. The nature of the pit organ's thermal receptor is not known; it probably consists of temperature-sensitive neurons innervating the two chambers. The two pit organs appear to provide stereoscopic information, in much the same way that two eyes do. Indeed, the information transmitted from the pit organs is processed by the visual center of the snake brain.

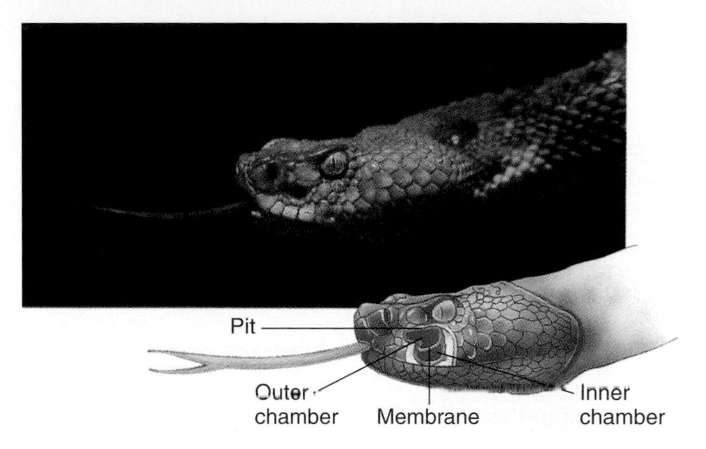

FIGURE 55.25
"Seeing" heat. The depression between the nostril and the eye of this rattlesnake opens into the pit organ. In the cutaway portion of the diagram, you can see that the organ is composed of two chambers separated by a membrane. Snakes known as pit vipers have this unique ability to sense infrared radiation (heat).

Electricity

While air does not readily conduct an electrical current, water is a good conductor. All aquatic animals generate electrical currents from contractions of their muscles. A number of different groups of fishes can detect these electrical currents. The *electrical fish* even have the ability to produce electrical discharges from specialized electrical organs. Electrical fish use these weak discharges to locate their prey and mates and to construct a three-dimensional image of their environment even in murky water.

The elasmobranchs (sharks, rays, and skates) have electroreceptors called the ampullae of Lorenzini. The receptor cells are located in sacs that open through jelly-filled canals to pores on the body surface. The jelly is a very good conductor, so a negative charge in the opening of the canal can depolarize the receptor at the base, causing the release of neurotransmitter and increased activity of sensory neurons. This allows sharks, for example, to detect the electrical fields generated by the muscle contractions of their prey. Although the ampullae of Lorenzini were lost in the evolution of teleost fish (most of the bony fish), electroreception reappeared in some groups of teleost fish that use sensory structures analogous to the ampullae of Lorenzini. Electroreceptors evolved yet another time, independently, in the duck-billed platypus, an egg-laying mammal. The receptors in its bill can detect the electrical currents created by the contracting muscles of shrimp and fish, enabling the mammal to detect its prey at night and in muddy water.

Magnetism

Eels, sharks, bees, and many birds appear to navigate along the magnetic field lines of the earth. Even some bacteria use such forces to orient themselves. Birds kept in blind cages, with no visual cues to guide them, will peck and attempt to move in the direction in which they would normally migrate at the appropriate time of the year. They will not do so, however, if the cage is shielded from magnetic fields by steel. Indeed, if the magnetic field of a blind cage is deflected 120° clockwise by an artificial magnet, a bird that normally orients to the north will orient toward the east-southeast. There has been much speculation about the nature of the magnetic receptors in these vertebrates, but the mechanism is still very poorly understood.

Pit vipers can locate warm prey by infrared radiation (heat), and many aquatic vertebrates can locate prey and ascertain the contours of their environment by means of electroreceptors.

Chapter 55

www.mhhe.com/raven6/resources55.mhtml

Summary	Questions	Media Resources

55.1 Animals employ a wide variety of sensory receptors.

- Mechanoreceptors, chemoreceptors, and photoreceptors are responsive to different categories of sensory stimuli; interoceptors and exteroceptors respond to stimuli that originate in the internal and external environments, respectively.

1. Can you name a sensory receptor that does not produce a membrane depolarization?

- Introduction to Sense Organs

55.2 Mechanical and chemical receptors sense the body's condition.

- Muscle spindles respond to stretching of the skeletal muscle.
- The sensory organs of taste are taste buds, scattered over the surface of a fish's body but located on the papillae of the tongue in terrestrial vertebrates.
- Chemoreceptors in the aortic and carotid bodies sense the blood pH and oxygen levels, helping to regulate breathing.
- Hair cells in the membranous labyrinth of the inner ear provide a sense of acceleration.

2. What mechanoreceptors detect muscle stretch and the tension on a tendon?

3. What structures in the vertebrate ear detect changes in the body's position with respect to gravity? What structures detect angular motion?

- Taste
- Smell
- Sense of Balance
- Sense of Rotational Acceleration

- Art Quizzes:
  - -Sensory Receptors in Skin
  - -Smell
  - -Semicircular Canals and Ampulla

55.3 Auditory receptors detect pressure waves in the air.

- In terrestrial vertebrates, sound waves cause vibrations of ear membranes.
- Different pitches of sounds vibrate different regions of the basilar membrane, and therefore stimulate different hair cells.
- Bats and some other vertebrates use sonar to provide a sense of "lightless vision."

4. How are sound waves transmitted and amplified through the middle ear? How is the pitch of the sound determined?

- Art Activity Human Ear Anatomy

- Bioethics Case Study: Hearing Loss and Jobs
- *On Science* Article: Going Batty

55.4 Optic receptors detect light over a broad range of wavelengths.

- A flexible lens focuses light onto the retina, which contains the photoreceptors.
- Light causes the photodissociation of the visual pigment, thereby blocking the dark current and hyperpolarizing the photoreceptor; this inverse effect stops the inhibitory effect of the photoreceptor and thereby activates the bipolar cells.

5. How does focusing in fishes and amphibians differ from that in other vertebrates?

6. When a photoreceptor absorbs light, what happens to the Na^+ channels in its outer segment?

- Art Activity Human Eye Anatomy

- Vision

55.5 Some vertebrates use heat, electricity, or magnetism for orientation.

- The pit organs of snakes allow them to detect the position and movements of prey. Many aquatic vertebrates can detect electrical currents produced by muscular contraction. Some vertebrates can orient themselves using the earth's magnetic field.

7. Why do rattlesnakes strike a moving lightbulb?

8. How do sharks detect their prey? Why don't terrestrial vertebrates have this sense?

56

The Endocrine System

FIGURE 56.1
The endocrine system controls when animals breed. These Japanese macaques live in a close-knit community whose members cooperate to ensure successful breeding and raising of offspring. Not everybody breeds at the same time because hormone levels vary among individuals.

Concept Outline

56.1 Regulation is often accomplished by chemical messengers.

Types of Regulatory Molecules. Regulatory molecules may function as neurotransmitters, hormones, or as organ-specific regulators.
Endocrine Glands and Hormones. Endocrine glands secrete molecules called hormones into the blood.
Paracrine Regulation. Paracrine regulators act within organs that produce them.

56.2 Lipophilic and polar hormones regulate their target cells by different means.

Hormones That Enter Cells. Steroid and thyroid hormones act by entering target cells and stimulating specific genes.
Hormones That Do Not Enter Cells. All other hormones bind to receptors on the cell surface and activate second-messenger molecules within the target cells.

56.3 The hypothalamus controls the secretions of the pituitary gland.

The Posterior Pituitary Gland. The posterior pituitary receives and releases hormones from the hypothalamus.
The Anterior Pituitary Gland. The anterior pituitary produces a variety of hormones under stimulation from hypothalamic releasing hormones.

56.4 Endocrine glands secrete hormones that regulate many body functions.

The Thyroid and Parathyroid Glands. The thyroid hormones regulate metabolism; the parathyroid glands regulate calcium balance.
The Adrenal Glands. The adrenal medulla secretes epinephrine during the fight-or-flight reaction, while the adrenal cortex secretes steroid hormones that regulate glucose and mineral balance.
The Pancreas. The islets of Langerhans in the pancreas secrete insulin, which acts to lower blood glucose, and glucagon, which acts to raise blood glucose.
Other Endocrine Glands. The gonads, pineal gland, thymus, kidneys, and other organs secrete important hormones that have a variety of functions.

The tissues and organs of the vertebrate body cooperate to maintain homeostasis of the body's internal environment and control other body functions such as reproduction. Homeostasis is achieved through the actions of many regulatory mechanisms that involve all the organs of the body. Two systems, however, are devoted exclusively to the regulation of the body organs: the nervous system and the endocrine system (figure 56.1). Both release regulatory molecules that control the body organs by first binding to receptor proteins in the cells of those organs. In this chapter we will examine these regulatory molecules, the cells and glands that produce them, and how they function to regulate the body's activities.

56.1 Regulation is often accomplished by chemical messengers.

Types of Regulatory Molecules

As we discussed in chapter 54, the axons of neurons secrete chemical messengers called *neurotransmitters* into the synaptic cleft. These chemicals diffuse only a short distance to the postsynaptic membrane, where they bind to their receptor proteins and stimulate the postsynaptic cell (another neuron, or a muscle or gland cell). Synaptic transmission generally affects only the one postsynaptic cell that receives the neurotransmitter.

A **hormone** is a regulatory chemical that is secreted into the blood by an **endocrine gland** or an organ of the body exhibiting an endocrine function. The blood carries the hormone to every cell in the body, but only the **target cells** for a given hormone can respond to it. Thus, the difference between a neurotransmitter and a hormone is not in the chemical nature of the regulatory molecule, but rather in the way it is transported to its target cells, and its distance from these target cells. A chemical regulator called norepinephrine, for example, is released as a neurotransmitter by sympathetic nerve endings and is also secreted by the adrenal gland as a hormone.

Some specialized neurons secrete chemical messengers into the blood rather than into a narrow synaptic cleft. In these cases, the chemical that the neurons secrete is sometimes called a **neurohormone.** The distinction between the nervous system and endocrine system blurs when it comes to such molecules. Indeed, because some neurons in the brain secrete hormones, the brain can be considered an endocrine gland!

In addition to the chemical messengers released as neurotransmitters and as hormones, other chemical regulatory molecules are released and act *within* an organ. In this way, the cells of an organ regulate one another. This type of regulation is not endocrine, because the regulatory molecules work without being transported by the blood, but is otherwise similar to the way that hormones regulate their target cells. Such regulation is called **paracrine**. Another type of chemical messenger that is released into the environment is called a *pheromone*. These messengers aid in the communication between animals, not in the regulation within an animal. A comparison of the different types of chemical messengers used for regulation is given in figure 56.2.

Regulatory molecules released by axons at a synapse are neurotransmitters, those released by endocrine glands into the blood are hormones, and those that act within the organ in which they are produced are paracrine regulators.

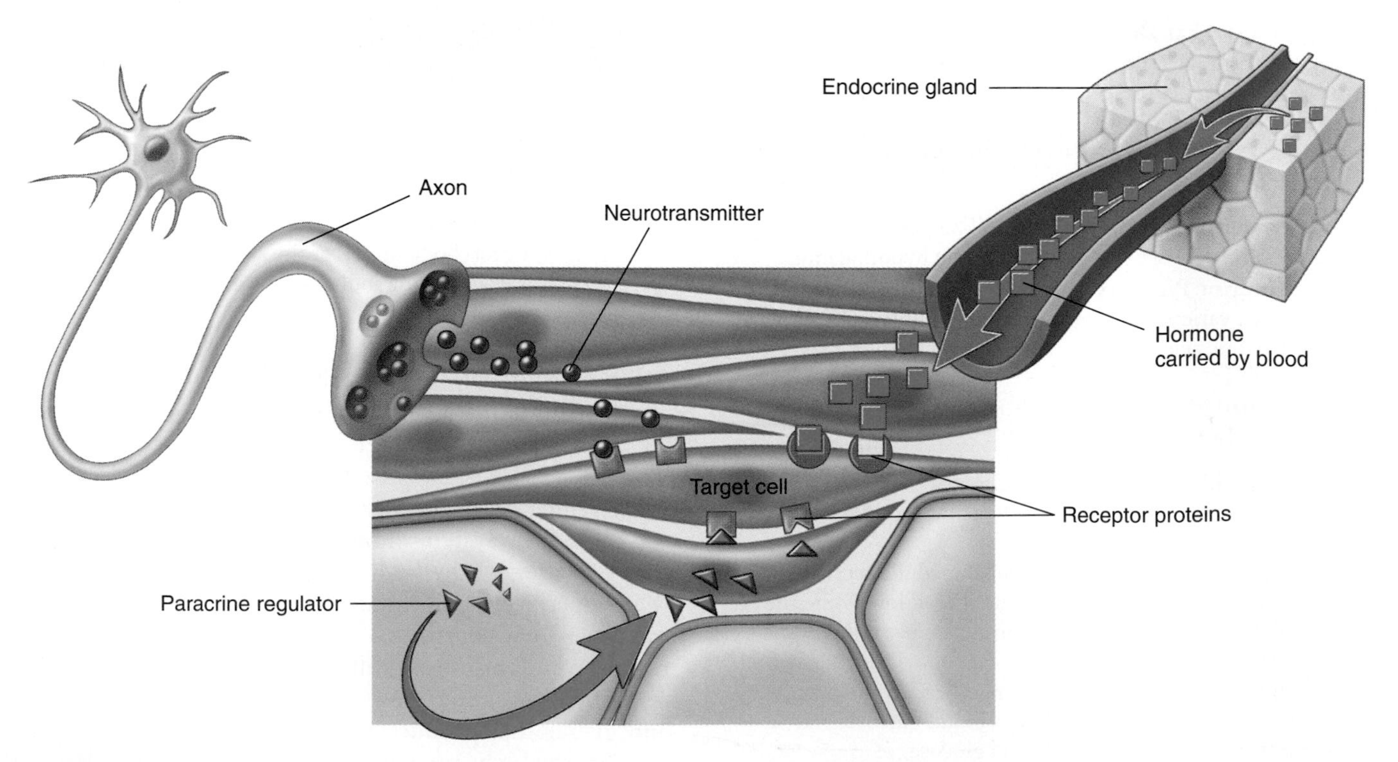

FIGURE 56.2
The functions of organs are influenced by neural, paracrine, and endocrine regulators. Each type of chemical regulator binds in a specific fashion to receptor proteins on the surface of or within the cells of target organs.

Endocrine Glands and Hormones

The endocrine system (figure 56.3) includes all of the organs that function exclusively as endocrine glands—such as the thyroid gland, pituitary gland, adrenal glands, and so on (table 56.1)—as well as organs that secrete hormones in addition to other functions. Endocrine glands lack ducts and thus must secrete into surrounding blood capillaries, unlike exocrine glands, which secrete their products into a duct.

Hormones secreted by endocrine glands belong to four different chemical categories:

1. **Polypeptides.** These hormones are composed of chains of amino acids that are shorter than about 100 amino acids. Some important examples include insulin and antidiuretic hormone (ADH).
2. **Glycoproteins.** These are composed of a polypeptide significantly longer than 100 amino acids to which is attached a carbohydrate. Examples include follicle-stimulating hormone (FSH) and luteinizing hormone (LH).
3. **Amines.** Derived from the amino acids tyrosine and tryptophan, they include hormones secreted by the adrenal medulla, thyroid, and pineal glands.
4. **Steroids.** These hormones are lipids derived from cholesterol, and include the hormones testosterone, estradiol, progesterone, and cortisol.

Steroid hormones can be subdivided into **sex steroids,** secreted by the testes, ovaries, placenta, and adrenal cortex, and **corticosteroids,** secreted only by the adrenal cortex (the outer portion of the adrenal gland). The corticosteroids include cortisol, which regulates glucose balance, and aldosterone, which regulates salt balance.

The amine hormones secreted by the adrenal medulla (the inner portion of the adrenal gland), known as **catecholamines,** include epinephrine (adrenaline) and norepinephrine (noradrenaline). These are derived from the amino acid tyrosine. Another hormone derived from tyrosine is thyroxine, secreted by the thyroid gland. The pineal gland secretes a different amine hormone, melatonin, derived from tryptophan.

All hormones may be categorized as lipophilic (fat-soluble) or hydrophilic (water-soluble). The **lipophilic hormones** include the steroid hormones and thyroxine; all other hormones are water-soluble. This distinction is important in understanding how these hormones regulate their target cells.

Neural and Endocrine Interactions

The endocrine system is an extremely important regulatory system in its own right, but it also interacts and cooperates with the nervous system to regulate the activities of the other organ systems of the body. The secretory activity of many endocrine glands is controlled by the nervous system. Among such glands are the adrenal medulla, posterior pituitary, and pineal gland. These three glands are derived from the neural ectoderm (to be discussed in chapter 60), the same embryonic tissue layer that forms the nervous system. The major site for neural regulation of the endocrine system, however, is the brain's regulation of the anterior pituitary gland. As we'll see, the hypothalamus controls the hormonal secretions of the anterior pituitary, which in turn regulates other endocrine glands. On the other hand, the secretion of a number of hormones is largely independent of neural control. The release of insulin by the pancreas and aldosterone by the adrenal cortex, for example, are stimulated primarily by increases in the blood concentrations of glucose and potassium (K^+), respectively.

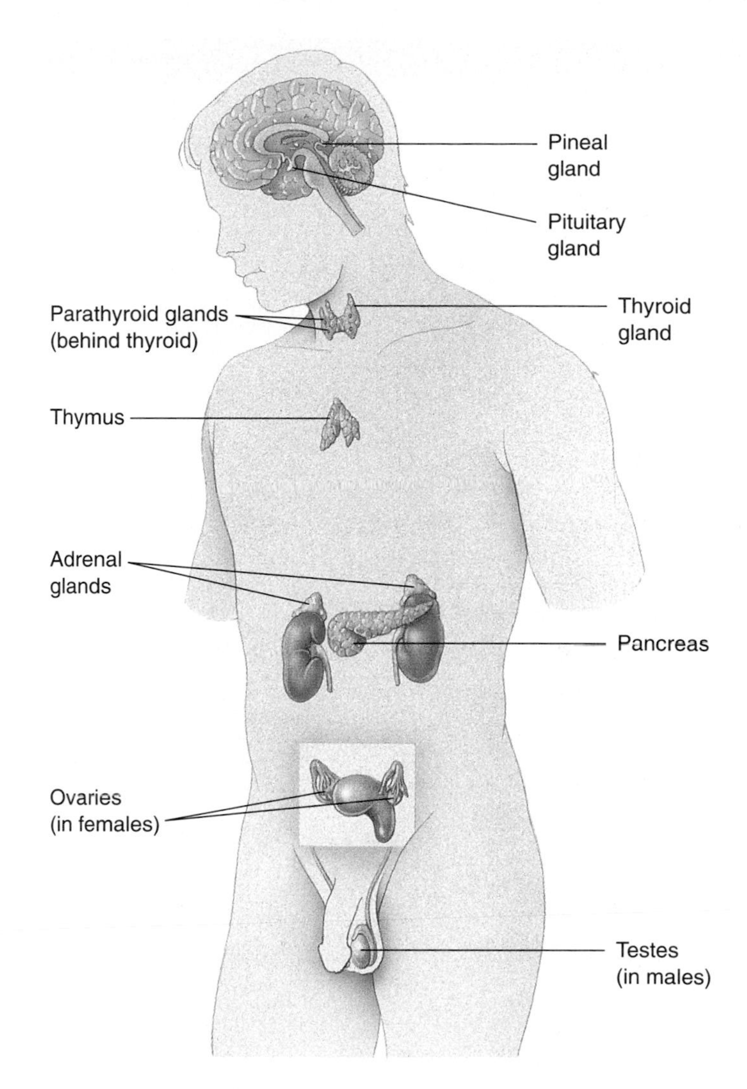

FIGURE 56.3
The human endocrine system. The major endocrine glands are shown, but many other organs secrete hormones in addition to their primary functions.

Any organ that secretes a hormone from a ductless gland is part of the endocrine system. Hormones may be any of a variety of different chemicals.

Endocrine Gland and Hormone	Target Tissue	Principal Actions	Chemical Nature
POSTERIOR LOBE OF PITUITARY			
Antidiuretic hormone (ADH)	Kidneys	Stimulates reabsorption of water; conserves water	Peptide (9 amino acids)
Oxytocin	Uterus	Stimulates contraction	Peptide
	Mammary glands	Stimulates milk ejection	(9 amino acids)
ANTERIOR LOBE OF PITUITARY			
Growth hormone (GH)	Many organs	Stimulates growth by promoting protein synthesis and fat breakdown	Protein
Adrenocorticotropic hormone (ACTH)	Adrenal cortex	Stimulates secretion of adrenal cortical hormones such as cortisol	Peptide (39 amino acids)
Thyroid-stimulating hormone (TSH)	Thyroid gland	Stimulates thyroxine secretion	Glycoprotein
Luteinizing hormone (LH)	Gonads	Stimulates ovulation and corpus luteum formation in females; stimulates secretion of testosterone in males	Glycoprotein
Follicle-stimulating hormone (FSH)	Gonads	Stimulates spermatogenesis in males; stimulates development of ovarian follicles in females	Glycoprotein
Prolactin (PRL)	Mammary glands	Stimulates milk production	Protein
Melanocyte-stimulating hormone (MSH)	Skin	Stimulates color change in reptiles and amphibians; unknown function in mammals	Peptide (two forms; 13 and 22 amino acids)
THYROID GLAND			
Thyroxine (thyroid hormone)	Most cells	Stimulates metabolic rate; essential to normal growth and development	Iodinated amino acid
Calcitonin	Bone	Lowers blood calcium level by inhibiting loss of calcium from bone	Peptide (32 amino acids)
PARATHYROID GLANDS			
Parathyroid hormone	Bone, kidneys, digestive tract	Raises blood calcium level by stimulating bone breakdown; stimulates calcium reabsorption in kidneys; activates vitamin D	Peptide (34 amino acids)

Table 56.1 Principal Endocrine Glands and Their Hormones*

*These are hormones released from endocrine glands. As discussed previously, many hormones are released from other body organs.

Paracrine Regulation

Paracrine regulation occurs in many organs and among the cells of the immune system. Some of these regulatory molecules are known as **cytokines,** particularly if they regulate different cells of the immune system. Other paracrine regulators are called **growth factors,** because they promote growth and cell division in specific organs. Examples include *platelet-derived growth factor*, *epidermal growth factor*, and the *insulin-like growth factors* that stimulate cell division and proliferation of their target cells. *Nerve growth factor* is a regulatory molecule that belongs to a family of paracrine regulators of the nervous system called **neurotrophins.**

Nitric oxide, which can function as a neurotransmitter (see chapter 54), is also produced by the endothelium of blood vessels. In this context, it is a paracrine regulator because it diffuses to the smooth muscle layer of the blood vessel and promotes vasodilation. The endothelium of blood vessels also produces other paracrine regulators, including *endothelin*, which stimulates vasoconstriction, and *bradykinin*, which promotes vasodilation. This paracrine regulation supplements the regulation of blood vessels by autonomic nerves.

The most diverse group of paracrine regulators are the **prostaglandins.** A prostaglandin is a 20-carbon-long fatty acid that contains a five-member carbon ring. This molecule is derived from the precursor molecule *arachidonic acid*, released from phospholipids in the cell membrane under hormonal or other stimulation. Prostaglandins are produced in almost every organ and participate in a variety of regulatory functions, including:

1. **Immune system.** Prostaglandins promote many aspects of inflammation, including pain and fever. Drugs that inhibit prostaglandin synthesis help to alleviate these symptoms.
2. **Reproductive system.** Prostaglandins may play a role in ovulation. Excessive prostaglandin production

Endocrine Gland and Hormone	Target Tissue	Principal Actions	Chemical Nature
ADRENAL MEDULLA			
Epinephrine (adrenaline) and norepinephrine (noradrenaline)	Smooth muscle, cardiac muscle, blood vessels	Initiate stress responses; raise heart rate, blood pressure, metabolic rate; dilate blood vessels; mobilize fat; raise blood glucose level	Amino acid derivatives
ADRENAL CORTEX			
Aldosterone	Kidney tubules	Maintains proper balance of Na^+ and K^+ ions	Steroid
Cortisol	Many organs	Adaptation to long-term stress; raises blood glucose level; mobilizes fat	Steroid
PANCREAS			
Insulin	Liver, skeletal muscles, adipose tissue	Lowers blood glucose level; stimulates storage of glycogen in liver	Peptide (51 amino acids)
Glucagon	Liver, adipose tissue	Raises blood glucose level; stimulates breakdown of glycogen in liver	Peptide (29 amino acids)
OVARY			
Estradiol	General	Stimulates development of secondary sex characteristics in females	Steroid
	Female reproductive structures	Stimulates growth of sex organs at puberty and monthly preparation of uterus for pregnancy	
Progesterone	Uterus	Completes preparation for pregnancy	Steroid
	Mammary glands	Stimulates development	
TESTIS			
Testosterone	Many organs	Stimulates development of secondary sex characteristics in males and growth spurt at puberty	Steroid
	Male reproductive structures	Stimulates development of sex organs; stimulates spermatogenesis	
PINEAL GLAND			
Melatonin	Gonads, pigment cells	Function not well understood; influences pigmentation in some vertebrates; may control biorhythms in some animals; may influence onset of puberty in humans	Amino acid derivative

Table 56.1 Principal Endocrine Glands and Their Hormones

may be involved in premature labor, endometriosis, or dysmenorrhea (painful menstrual cramps).

3. **Digestive system.** Prostaglandins produced by the stomach and intestines may inhibit gastric secretions and influence intestinal motility and fluid absorption.
4. **Respiratory system.** Some prostaglandins cause constriction, whereas others cause dilation of blood vessels in the lungs and of bronchiolar smooth muscle.
5. **Circulatory system.** Prostaglandins are needed for proper function of blood platelets in the process of blood clotting.
6. **Urinary system.** Prostaglandins produced in the renal medulla cause vasodilation, resulting in increased renal blood flow and increased excretion of urine.

The synthesis of prostaglandins are inhibited by aspirin. Aspirin is the most widely used of the **nonsteroidal anti-inflammatory drugs (NSAIDs),** a class of drugs that also includes indomethacin and ibuprofen. These drugs produce their effects because they specifically inhibit the enzyme cyclooxygenase-2 (cox-2), needed to produce prostaglandins from arachidonic acid. Through this action, the NSAIDs inhibit inflammation and associated pain. Unfortunately, NSAIDs also inhibit another similar enzyme, cox-1, which helps maintain the wall of the digestive tract, and in so doing can produce severe unwanted side effects, including gastric bleeding and prolonged clotting time. A new kind of pain reliever, celecoxib (Celebrex), inhibits cox-2 but not cox-1, a potentially great benefit to arthritis sufferers and others who must use pain relievers regularly.

The neural and endocrine control systems are supplemented by paracrine regulators, including the prostaglandins, which perform many diverse functions.

Hormones That Enter Cells

As we mentioned previously, hormones can be divided into those that are lipophilic (lipid-soluble) and those that are hydrophilic (water-soluble). The lipophilic hormones—all of the steroid hormones (figure 56.4) and thyroxine—as well as other lipophilic regulatory molecules (including the retinoids, or vitamin A) can easily enter cells. This is because the lipid portion of the cell membrane does not present a barrier to the entry of lipophilic regulators. Therefore, all lipophilic regulatory molecules have a similar mechanism of action. Water-soluble hormones, in contrast, cannot pass through cell membranes. They must regulate their target cells through different mechanisms.

Steroid hormones are lipids themselves and thus lipophilic; thyroxine is lipophilic because it is derived from a nonpolar amino acid. Because these hormones are not water-soluble, they don't dissolve in plasma but rather travel in the blood attached to protein carriers. When the hormones arrive at their target cells, they dissociate from their carriers and pass through the plasma membrane of

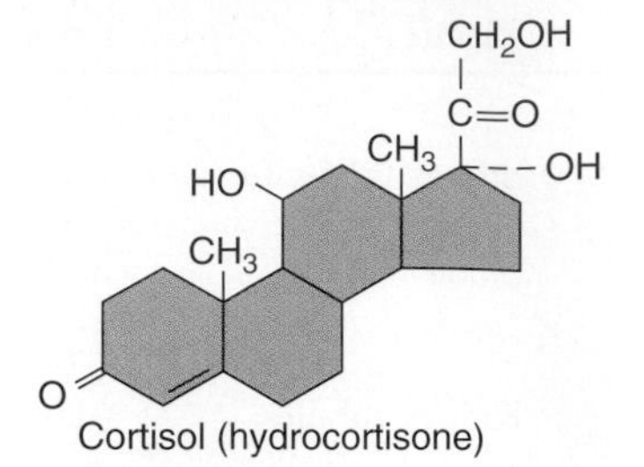

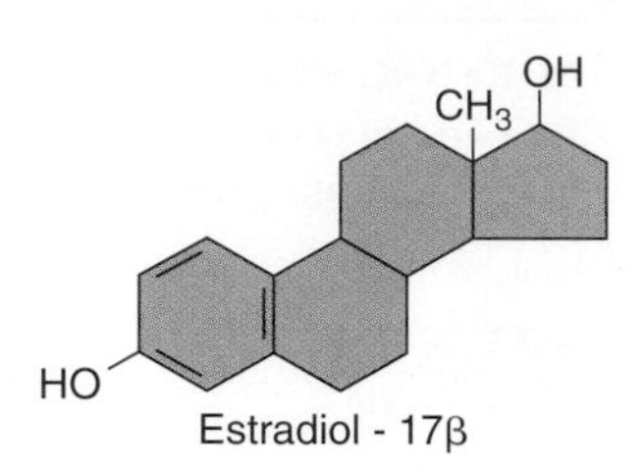

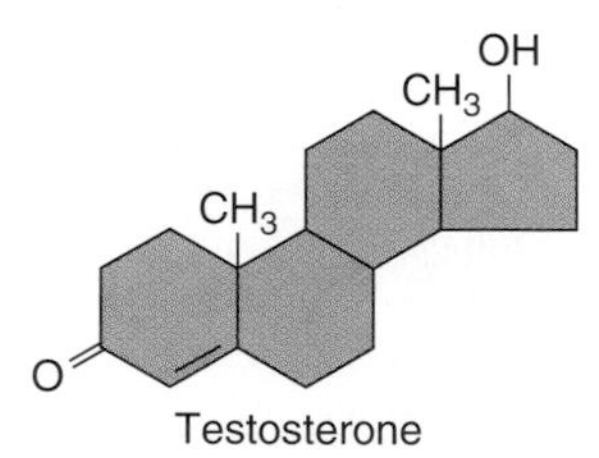

FIGURE 56.4
Chemical structures of some steroid hormones. Steroid hormones are derived from the blood lipid cholesterol. The hormones shown, cortisol, estradiol, and testosterone, differ only slightly in chemical structure yet have widely different effects on the body. Steroid hormones are secreted by the adrenal cortex, testes, ovaries, and placenta.

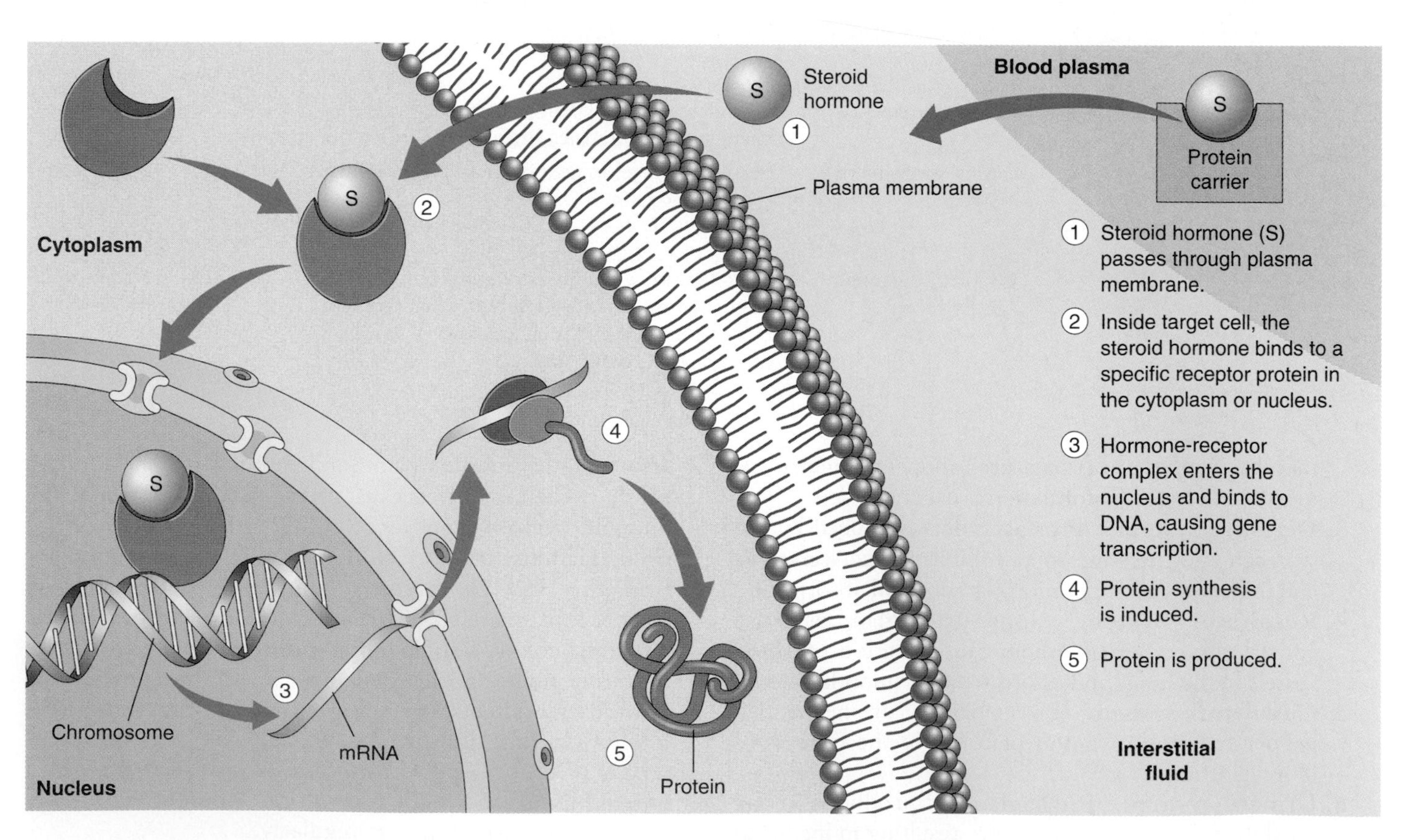

FIGURE 56.5
The mechanism of steroid hormone action. Steroid hormones are lipid-soluble and thus readily diffuse through the plasma membrane of cells. They bind to receptor proteins in either the cytoplasm or nucleus (not shown). If the steroid binds to a receptor in the cytoplasm, the hormone-receptor complex moves into the nucleus. The hormone-receptor complex then binds to specific regions of the DNA, stimulating the production of messenger RNA (mRNA).

the cell (figure 56.5). Some steroid hormones then bind to very specific receptor proteins in the cytoplasm, and then move as a hormone-receptor complex into the nucleus. Other steroids travel directly into the nucleus before encountering their receptor proteins. Whether the steroid finds its receptor in the nucleus or translocates with its receptor to the nucleus from the cytoplasm, the rest of the story is the same.

The hormone receptor, activated by binding to the lipophilic hormone, is now also able to bind to specific regions of the DNA. These DNA regions are known as the hormone response elements. The binding of the hormone-receptor complex has a direct effect on the level of transcription at that site by activating genetic transcription. This produces messenger RNA (mRNA), which then codes for the production of specific proteins. These proteins often have enzymatic activity that changes the metabolism of the target cell in a specific fashion.

The thyroid hormone's mechanism of action resembles that of the steroid hormones. Thyroxine contains four iodines and so is often abbreviated T_4 (for *tetraiodothyronine*). The thyroid gland also secretes smaller amounts of a similar molecule that has only three iodines, called *triiodothyronine* (and abbreviated T_3). Both hormones enter target cells, but all of the T_4 that enters is changed into T_3 (figure 56.6). Thus, only the T_3 form of the hormone enters the nucleus and binds to nuclear receptor proteins. The hormone-receptor complex, in turn, binds to the appropriate hormone response elements on DNA.

The lipophilic hormones pass through the target cell's plasma membrane and bind to intracellular receptor proteins. The hormone-receptor complex then binds to specific regions of DNA, thereby activating genes and regulating the target cells.

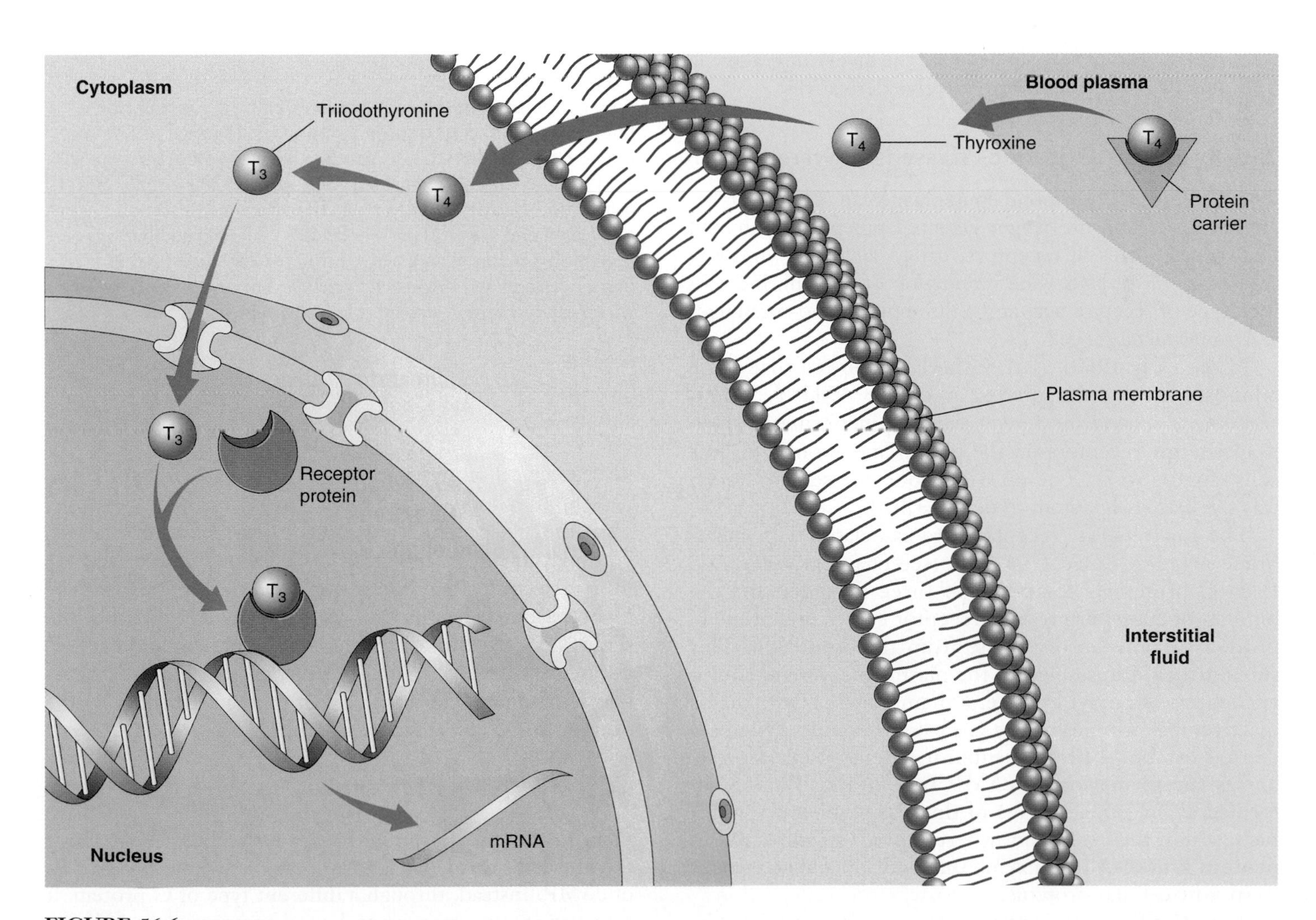

FIGURE 56.6
The mechanism of thyroxine action. Thyroxine contains four iodines. When it enters the target cell, thyroxine is changed into triiodothyronine, with three iodines. This hormone moves into the nucleus and binds to nuclear receptors. The hormone-receptor complex then binds to regions of the DNA and stimulates gene transcription.

Hormones That Do Not Enter Cells

Hormones that are too large or too polar to cross the plasma membranes of their target cells include all of the peptide and glycoprotein hormones, as well as the catecholamine hormones epinephrine and norepinephrine. These hormones bind to receptor proteins located on the outer surface of the plasma membrane—the hormones do *not* enter the cell. If you think of the hormone as a messenger sent from an endocrine gland to the target cell, it is evident that a second messenger is needed within the target cell to produce the effects of the hormone. A number of different molecules in the cell can serve as second messengers, as we saw in chapter 7. The interaction between the hormone and its receptor activates mechanisms in the plasma membrane that increase the concentration of the second messengers within the target cell cytoplasm.

The binding of a water-soluble hormone to its receptor is reversible and usually very brief. After the hormone binds to its receptor and activates a second-messenger system, it dissociates from the receptor and may travel in the blood to another target cell somewhere else in the body. Eventually, enzymes (primarily in the liver) degrade the hormone by converting it into inactive derivatives.

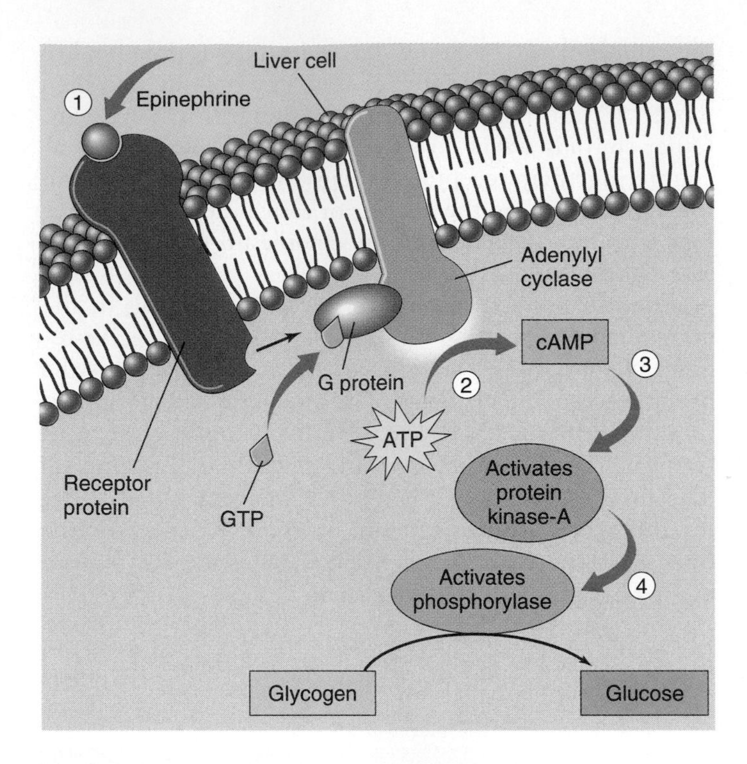

FIGURE 56.7
The action of epinephrine on a liver cell. *(1)* Epinephrine binds to specific receptor proteins on the cell surface. *(2)* Acting through intermediary G proteins, the hormone-bound receptor activates the enzyme adenylyl cyclase, which converts ATP into cyclic AMP (cAMP). *(3)* Cyclic AMP performs as a second messenger and activates protein kinase-A, an enzyme that was previously present in an inactive form. *(4)* Protein kinase-A phosphorylates and thereby activates the enzyme phosphorylase, which catalyzes the hydrolysis of glycogen into glucose.

The Cyclic AMP Second-Messenger System

The action of the hormone epinephrine can serve as an example of a second-messenger system. Epinephrine can bind to two categories of receptors, called *alpha* (α)- and *beta* (β)-*adrenergic receptors.* The interaction of epinephrine with each type of receptor activates a different second-messenger system in the target cell.

In the early 1960s, Earl Sutherland showed that cyclic adenosine monophosphate, or **cyclic AMP (cAMP),** serves as a second messenger when epinephrine binds to β-adrenergic receptors on the plasma membranes of liver cells (figure 56.7). The cAMP second-messenger system was the first such system to be described.

The β-adrenergic receptors are associated with membrane proteins called *G proteins* (see chapters 7 and 54). Each G protein is composed of three subunits, and the binding of epinephrine to its receptor causes one of the G protein subunits to dissociate from the other two. This subunit then diffuses within the plasma membrane until it encounters **adenylyl cyclase,** a membrane enzyme that is inactive until it binds to the G protein subunit. When activated by the G protein subunit, adenylyl cyclase catalyzes the formation of cAMP from ATP. The cAMP formed at the inner surface of the plasma membrane diffuses within the cytoplasm, where it binds to and activates **protein kinase-A,** an enzyme that adds phosphate groups to specific cellular proteins.

The identities of the proteins that are phosphorylated by protein kinase-A varies from one cell type to the next, and this variation is one of the reasons epinephrine has such diverse effects on different tissues. In liver cells, protein kinase-A phosphorylates and thereby activates another enzyme, phosphorylase, which converts glycogen into glucose. Through this multistep mechanism, epinephrine causes the liver to secrete glucose into the blood during the fight-or-flight reaction, when the adrenal medulla is stimulated by the sympathetic division of the autonomic nervous system (see chapter 54). In cardiac muscle cells, protein kinase-A phosphorylates a different set of cellular proteins, which cause the heart to beat faster and more forcefully.

The IP_3/Ca^{++} Second-Messenger System

When epinephrine binds to α-adrenergic receptors, it doesn't activate adenylyl cyclase and cause the production of cAMP. Instead, through a different type of G protein, it activates another membrane-bound enzyme, phospholipase C (figure 56.8). This enzyme cleaves certain membrane phospholipids to produce the second messenger, inositol

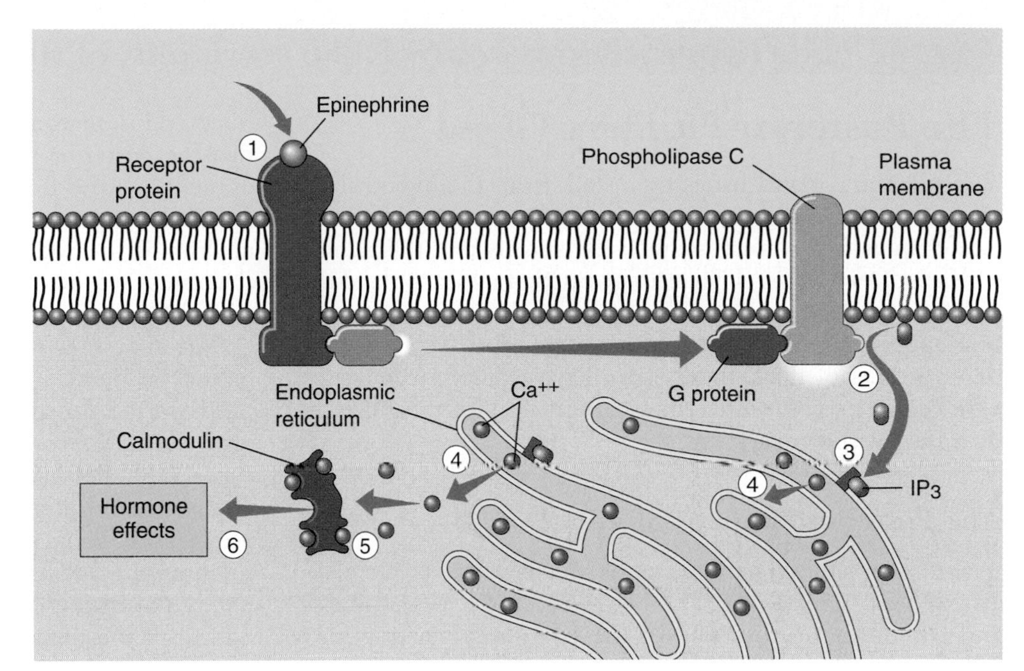

FIGURE 56.8
The IP_3/Ca^{++} second-messenger system. *(1)* The hormone epinephrine binds to specific receptor proteins on the cell surface. *(2)* Acting through G proteins, the hormone-bound receptor activates the enzyme phospholipase C, which converts membrane phospholipids into inositol trisphosphate (IP_3). *(3)* IP_3 diffuses through the cytoplasm and binds to receptors on the endoplasmic reticulum. *(4)* The binding of IP_3 to its receptors stimulates the endoplasmic reticulum to release Ca^{++} into the cytoplasm. *(5)* Some of the released Ca^{++} binds to a regulatory protein called calmodulin. *(6)* The Ca^{++}/calmodulin complex activates other intracellular proteins, ultimately producing the effects of the hormone.

trisphosphate (IP_3). IP_3 diffuses into the cytoplasm from the plasma membrane and binds to receptors located on the surface of the endoplasmic reticulum.

Recall from chapter 5 that the endoplasmic reticulum is a system of membranous sacs and tubes that serves a variety of functions in different cells. One of its functions is to accumulate Ca^{++} by actively transporting Ca^{++} out of the cytoplasm. Other pumps transport Ca^{++} from the cytoplasm through the plasma membrane to the extracellular fluid. These two mechanisms keep the concentration of Ca^{++} in the cytoplasm very low. Consequently, there is an extremely steep concentration gradient for Ca^{++} between the cytoplasm and the inside of the endoplasmic reticulum, and between the cytoplasm and the extracellular fluid.

When IP_3 binds to its receptors on the endoplasmic reticulum, it stimulates the endoplasmic reticulum to release its stored Ca^{++}. Calcium channels in the plasma membrane may also open, allowing Ca^{++} to diffuse into the cell from the extracellular fluid. Some of the Ca^{++} that has suddenly entered the cytoplasm then binds to a protein called calmodulin, which has regulatory functions analogous to those of cyclic AMP. One of the actions of calmodulin is to activate another type of protein kinase, resulting in the phosphorylation of a different set of cellular proteins.

What is the advantage of having multiple second-messenger systems? Consider the antagonistic actions of epinephrine and insulin on liver cells. Epinephrine uses cAMP as a second messenger to promote the hydrolysis of glycogen to glucose, while insulin stimulates the synthesis of glycogen from glucose. Clearly, insulin cannot use cAMP as a second messenger. Although the exact mechanism of insulin's action is still not well understood, insulin may act in part through the IP_3/Ca^{++} second-messenger system.

Not all large polar hormones act by increasing the concentration of a second messenger in the cytoplasm of the target cell. Others cause a change in the shape of a membrane protein called an ion channel (see chapters 6 and 54). If these channels are normally "closed," then a change in shape will open them allowing a particular ion to enter or leave the cell depending on its concentration gradient. If an ion channel is normally open, a chemical messenger can cause it to close. For example, some hormones open Ca^{++} channels on smooth muscle cell membranes; other hormones close them. This will increase or decrease, respectively, the amount of muscle contraction.

The molecular mechanism for changing the shape of an ion channel is similar to that for activating a second messenger. The hormone first binds to a receptor protein on the outer surface of the target cell. This receptor protein may then use a G protein to signal the ion channel to change shape.

Although G proteins play a major role in many hormone functions they don't seem to be necessary for all identified actions of hormones on target cells. In the cases where G proteins are not involved, the receptor protein is connected directly to the enzyme or ion channel.

The water-soluble hormones cannot pass through the plasma membrane; they must rely on second messengers within the target cells to mediate their actions. Such second messengers include cyclic AMP (cAMP), inositol trisphosphate (IP_3), and Ca^{++}. In many cases, the second messengers activate previously inactive enzymes.

56.3 The hypothalamus controls the secretions of the pituitary gland.

The Posterior Pituitary Gland

The **pituitary gland** hangs by a stalk from the hypothalamus of the brain (figure 56.9) posterior to the optic chiasm (see chapter 54). A microscopic view reveals that the gland consists of two parts. The anterior portion appears glandular and is called the **anterior pituitary;** the posterior portion appears fibrous and is the **posterior pituitary.** These two portions of the pituitary gland have different embryonic origins, secrete different hormones, and are regulated by different control systems.

The Posterior Pituitary Gland

The posterior pituitary appears fibrous because it contains axons that originate in cell bodies within the hypothalamus and extend along the stalk of the pituitary as a tract of fibers. This anatomical relationship results from the way that the posterior pituitary is formed in embryonic development. As the floor of the third ventricle of the brain forms the hypothalamus, part of this neural tissue grows downward to produce the posterior pituitary. The hypothalamus and posterior pituitary thus remain interconnected by a tract of axons.

The endocrine role of the posterior pituitary gland first became evident in 1912, when a remarkable medical case was reported: a man who had been shot in the head developed the need to urinate every 30 minutes or so, 24 hours a day. The bullet had lodged in his pituitary gland. Subsequent research demonstrated that removal of this gland produces the same symptoms. Pituitary extracts were found to contain a substance that makes the kidneys conserve water, and in the early 1950s investigators isolated a peptide from the posterior pituitary, **antidiuretic hormone**

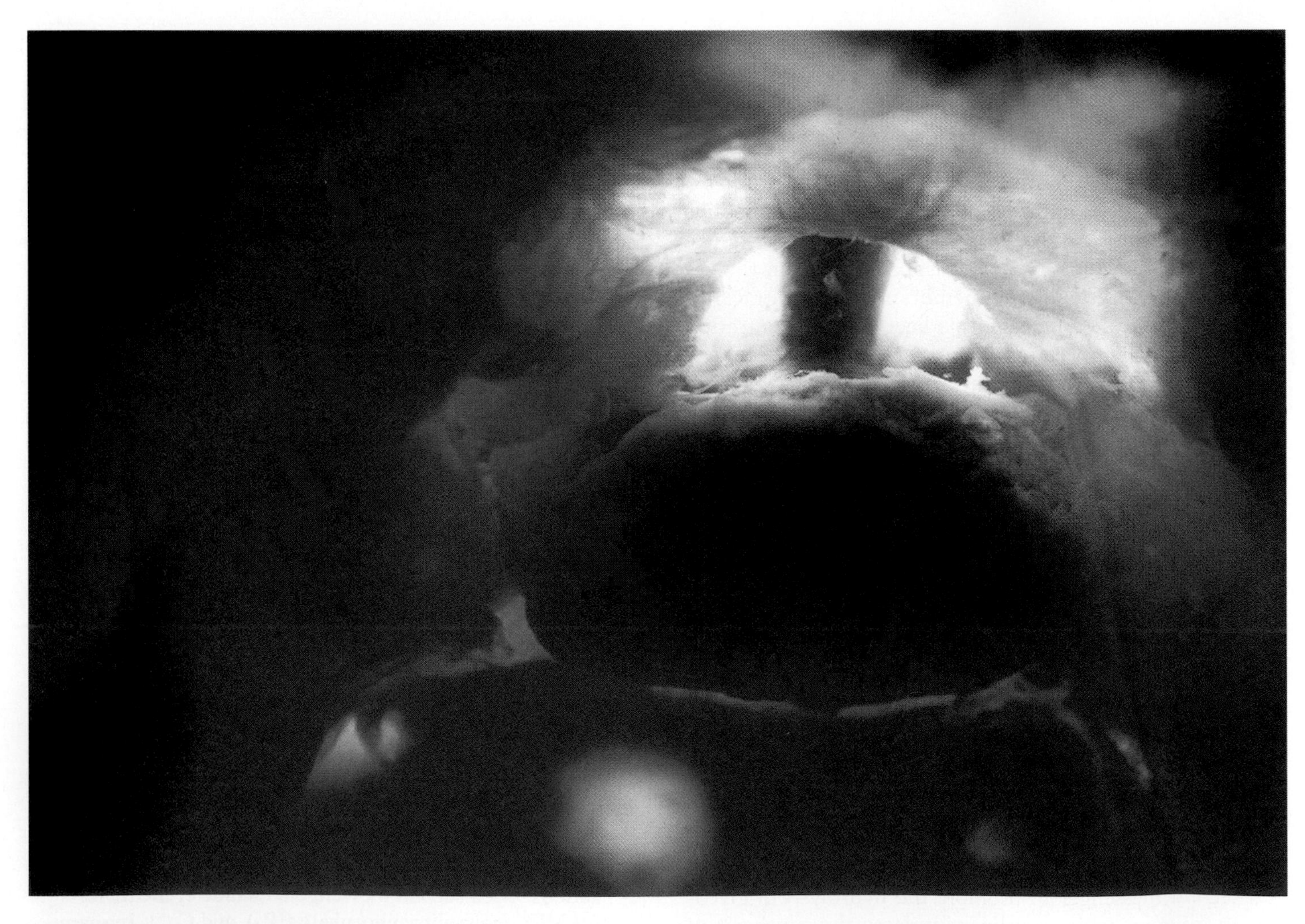

FIGURE 56.9
The pituitary gland hangs by a short stalk from the hypothalamus. The pituitary gland (the oval structure hanging from the stalk), shown here enlarged 15 times, regulates hormone production in many of the body's endocrine glands.

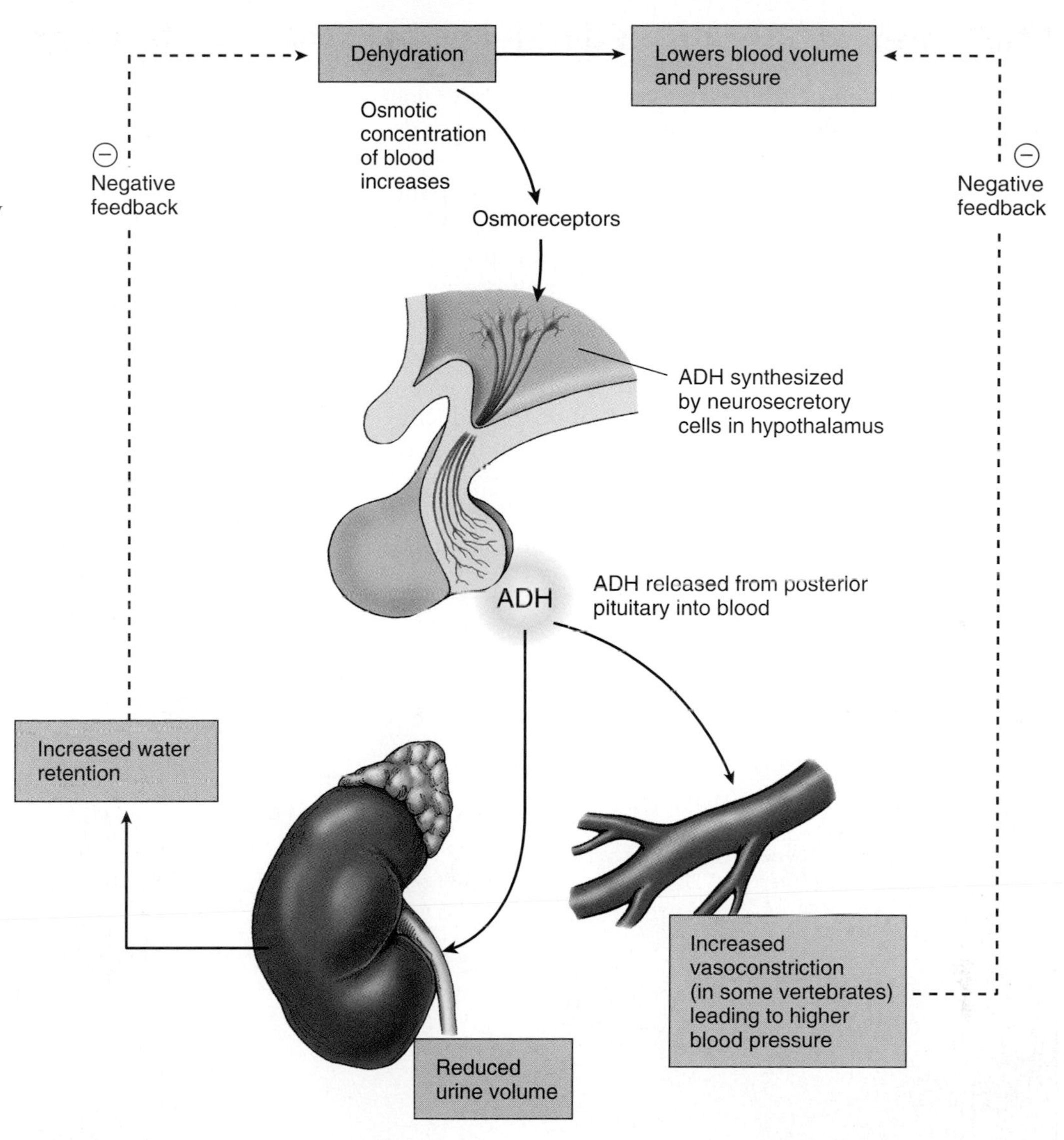

FIGURE 56.10
The effects of antidiuretic hormone (ADH). An increase in the osmotic concentration of the blood stimulates the posterior pituitary gland to secrete ADH, which promotes water retention by the kidneys. This works as a negative feedback loop to correct the initial disturbance of homeostasis.

(ADH, also known as vasopressin), that stimulates water retention by the kidneys (figure 56.10). When ADH is missing, the kidneys do not retain water and excessive quantities of urine are produced. This is why the consumption of alcohol, which inhibits ADH secretion, leads to frequent urination.

The posterior pituitary also secretes **oxytocin,** a second peptide hormone which, like ADH, is composed of nine amino acids. Oxytocin stimulates the milk-ejection reflex, so that contraction of the smooth muscles around the mammary glands and ducts causes milk to be ejected from the ducts through the nipple. During suckling, sensory receptors in the nipples send impulses to the hypothalamus, which triggers the release of oxytocin. Oxytocin is also needed to stimulate uterine contractions in women during childbirth. Oxytocin secretion continues after childbirth in a woman who is breast-feeding, which is why the uterus of a nursing mother returns to its normal size after pregnancy more quickly than does the uterus of a mother who does not breast-feed.

ADH and oxytocin are actually *produced* by neuron cell bodies located in the hypothalamus. These two hormones are transported along the axon tract that runs from the hypothalamus to the posterior pituitary and are stored in the posterior pituitary. In response to the appropriate stimulation—increased blood plasma osmolarity in the case of ADH, the suckling of a baby in the case of oxytocin—the hormones are released by the posterior pituitary into the blood. Because this reflex control involves both the nervous and endocrine systems, the secretion of ADH and oxytocin are **neuroendocrine reflexes.**

The posterior pituitary gland contains axons originating from neurons in the hypothalamus. These neurons produce ADH and oxytocin, which are stored in and released from the posterior pituitary gland in response to neural stimulation from the hypothalamus.

The Anterior Pituitary Gland

The anterior pituitary, unlike the posterior pituitary, does not develop from a downgrowth of the brain; instead, it develops from a pouch of epithelial tissue that pinches off from the roof of the embryo's mouth. Because it is epithelial tissue, the anterior pituitary is a complete gland—it produces the hormones it secretes. Many, but not all, of these hormones stimulate growth in their target organs, including other endocrine glands. Therefore, the hormones of the anterior pituitary gland are collectively termed *tropic hormones* (Greek *trophe*, "nourishment"), or *tropins*. When the target organ of a tropic hormone is another endocrine gland, that gland is stimulated by the tropic hormone to secrete its own hormones.

The hormones produced and secreted by different cell types in the anterior pituitary gland (figure 56.11) include the following:

1. **Growth hormone (GH,** or *somatotropin*) stimulates the growth of muscle, bone (indirectly), and other tissues and is also essential for proper metabolic regulation.
2. **Adrenocorticotropic hormone (ACTH,** or *corticotropin*) stimulates the adrenal cortex to produce corticosteroid hormones, including cortisol (in humans) and corticosterone (in many other vertebrates), which regulate glucose homeostasis.
3. **Thyroid-stimulating hormone (TSH,** or *thyrotropin*) stimulates the thyroid gland to produce thyroxine, which in turn stimulates oxidative respiration.
4. **Luteinizing hormone (LH)** is needed for ovulation and the formation of a corpus luteum in the female menstrual cycle (see chapter 59). It also stimulates the testes to produce testosterone, which is needed for sperm production and for the development of male secondary sexual characteristics.

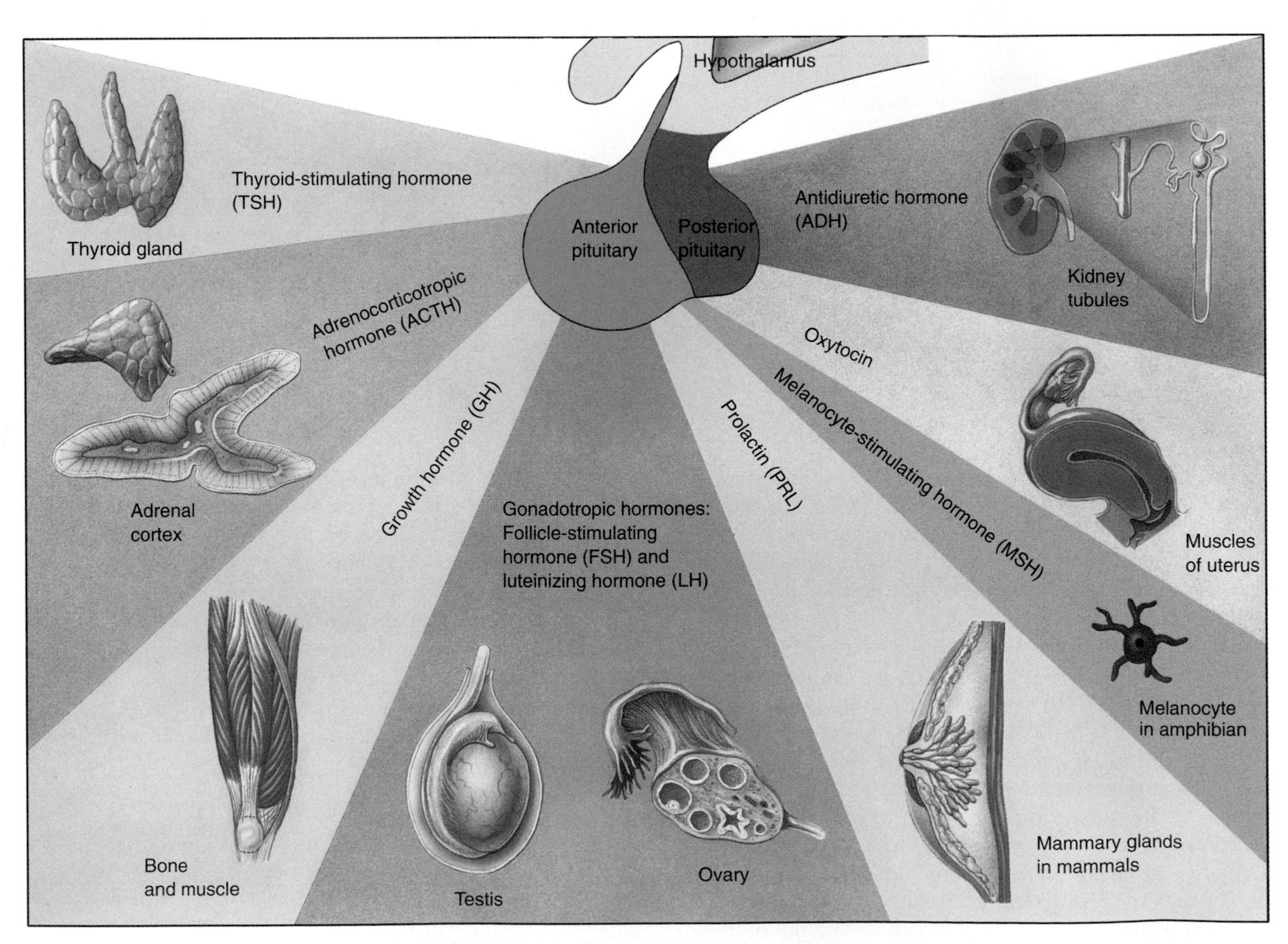

FIGURE 56.11
The major hormones of the anterior and posterior pituitary glands. Only a few of the actions of these hormones are shown.

5. **Follicle-stimulating hormone (FSH)** is required for the development of ovarian follicles in females. In males, it is required for the development of sperm. FSH and LH are both referred to as *gonadotropins.*
6. **Prolactin (PRL)** stimulates the mammary glands to produce milk in mammals. It also helps regulate kidney function in vertebrates, the production of "crop milk" (nutritional fluid fed to chicks by regurgitation) in some birds, and acts on the gills of fish that travel from salt to fresh water to promote sodium retention.
7. **Melanocyte-stimulating hormone (MSH)** stimulates the synthesis and dispersion of melanin pigment, which darken the epidermis of some fish, amphibians, and reptiles. MSH has no known specific function in mammals, but abnormally high amounts of ACTH can cause skin darkening because it contains the amino acid sequence of MSH within its structure.

Growth Hormone

The importance of the anterior pituitary gland first became understood in 1909, when a 38-year-old South Dakota farmer was cured of the growth disorder *acromegaly* by the surgical removal of a pituitary tumor. Acromegaly is a form of gigantism in which the jaw begins to protrude and other facial features thicken. It was discovered that gigantism is almost always associated with pituitary tumors. Robert Wadlow, born in 1928 in Alton, Illinois, stood 8 feet, 11 inches tall and weighed 485 pounds before he died from infection at age 22 (figure 56.12). He was the tallest human being ever recorded, and he was still growing the year he died.

FIGURE 56.12
The Alton giant. This photograph of Robert Wadlow of Alton, Illinois, taken on his 21st birthday, shows him at home with his father and mother and four siblings. Born normal size, he developed a growth hormone–secreting pituitary tumor as a young child and never stopped growing during his 22 years of life.

We now know that gigantism is caused by the excessive secretion of growth hormone (GH) by the anterior pituitary gland in a growing child. GH stimulates protein synthesis and growth of muscles and connective tissues; it also indirectly promotes the elongation of bones by stimulating cell division in the cartilaginous epiphyseal growth plates of bones. Researchers found that this stimulation does not occur in the absence of blood plasma, suggesting that bone cells lack receptors for GH and that the stimulation by GH was indirect. We now know that GH stimulates the production of **insulin-like growth factors,** which are produced by the liver and secreted into the blood in response to stimulation by GH. The insulin-like growth factors then stimulate growth of the epiphyseal growth plates and thus elongation of the bones.

When a person's skeletal growth plates have converted from cartilage into bone, however, GH can no longer cause an increase in height. Therefore, excessive GH secretion in an adult produces bone and soft tissue deformities in the condition called acromegaly. A deficiency in GH secretion during childhood results in pituitary dwarfism, a failure to achieve normal growth.

Other Anterior Pituitary Hormones

Prolactin is like growth hormone in that it acts on organs that are not endocrine glands. In addition to its stimulation of milk production in mammals and "crop milk" production in birds, prolactin has varied effects on electrolyte balance by acting on the kidneys, the gills of fish, and the salt glands of marine birds (discussed in chapter 58). Unlike growth hormone and prolactin, the other anterior pituitary hormones act on specific glands.

Some of the anterior pituitary hormones that act on specific glands have common names, such as thyroid-stimulating hormone (TSH), and alternative names that emphasize the tropic nature of the hormone, such as thyrotropin. TSH stimulates only the thyroid gland, and adrenocorticotropic hormone (ACTH) stimulates only the adrenal cortex (outer portion of the adrenal glands). Follicle-stimulating hormone (FSH) and luteinizing hormone (LH) act only on the gonads (testes and ovaries); hence, they are collectively called *gonadotropic hormones.* Although both FSH and LH act on the gonads, they each act on different target cells in the gonads of both females and males.

Hypothalamic Control of Anterior Pituitary Gland Secretion

The anterior pituitary gland, unlike the posterior pituitary, is not derived from the brain and does not receive an axon tract from the hypothalamus. Nevertheless, the hypothalamus controls production and secretion of the anterior pituitary hormones. This control is exerted hormonally rather than by means of nerve axons. Neurons in the hypothalamus secrete releasing hormones and inhibiting hormones into blood capillaries at the base of the hypothalamus (figure 56.13). These capillaries drain into small veins that run within the stalk of the pituitary to a second bed of capillaries in the anterior pituitary. This unusual system of vessels is known as the hypothalamohypophyseal portal system (another name for the pituitary is the hypophysis). It is called a portal system because it has a second capillary bed downstream from the first; the only other body location with a similar system is the liver, where capillaries receive blood drained from the gastrointestinal tract (via the hepatic portal vein—see chapter 51). Because the second bed of capillaries receives little oxygen from such vessels, the vessels must be delivering something else of importance.

Each releasing hormone delivered by the hypothalamohypophyseal system regulates the secretion of a specific anterior pituitary hormone. For example, *thyrotropin-releasing hormone (TRH)* stimulates the release of TSH, *corticotropin-releasing hormone (CRH)* stimulates the release of ACTH, and *gonadotropin-releasing hormone (GnRH)* stimulates the release of FSH and LH. A releasing hormone for growth hormone, called *growth hormone–releasing hormone (GHRH)* has also been discovered, and a releasing hormone for prolactin has been postulated but has thus far not been identified.

The hypothalamus also secretes hormones that *inhibit* the release of certain anterior pituitary hormones. To date, three such hormones have been discovered: *somatostatin* inhibits the secretion of GH; *prolactin-inhibiting factor* (*PIF*), found to be the neurotransmitter dopamine, inhibits the secretion of prolactin; and *melanotropin-inhibiting hormone* (*MIH*) inhibits the secretion of MSH.

Negative Feedback Control of Anterior Pituitary Gland Secretion

Because hypothalamic hormones control the secretions of the anterior pituitary gland, and the anterior pituitary hormones control the secretions of some other endocrine glands, it may seem that the hypothalamus is in charge of hormonal secretion for the whole body. This idea is not valid, however, for two reasons. First, a number of endocrine organs, such as the adrenal medulla and the pancreas, are not directly regulated by this control system. Second, the hypothalamus and the anterior pituitary gland are themselves partially controlled by the very hormones whose secretion they stimulate! In most cases this is an

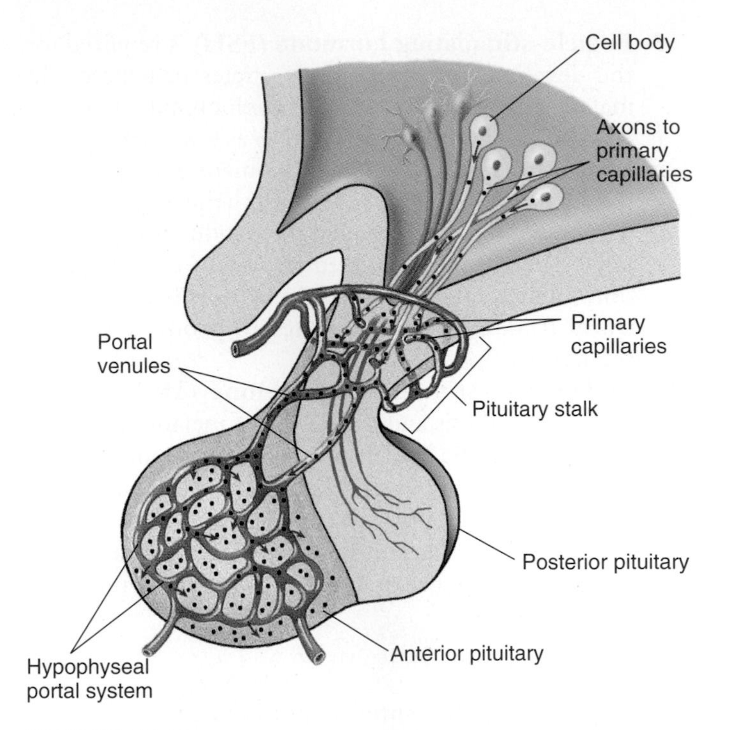

FIGURE 56.13
Hormonal control of the anterior pituitary gland by the hypothalamus. Neurons in the hypothalamus secrete hormones that are carried by short blood vessels directly to the anterior pituitary gland, where they either stimulate or inhibit the secretion of anterior pituitary hormones.

inhibitory control, where the target gland hormones inhibit the secretions of the hypothalamus and anterior pituitary (figure 56.14). This type of control system is called negative feedback inhibition and acts to maintain relatively constant levels of the target cell hormone.

Let's consider the hormonal control of the thyroid gland. The hypothalamus secretes TRH into the hypothalamohypophyseal portal system, which stimulates the anterior pituitary gland to secrete TSH, which in turn stimulates the thyroid gland to release thyroxine. Among thyroxine's many target organs are the hypothalamus and the anterior pituitary gland themselves. Thyroxine acts upon these organs to inhibit their secretion of TRH and TSH, respectively (figure 56.15). This negative feedback inhibition is essential for homeostasis because it keeps the thyroxine levels fairly constant.

To illustrate the importance of negative feedback inhibition, we will examine a person who lacks sufficient iodine in the diet. Without iodine, the thyroid gland cannot produce thyroxine (which contains four iodines per molecule). As a result, thyroxine levels in the blood fall drastically, and the hypothalamus and anterior pituitary receive far less negative feedback inhibition than is normal. This reduced inhibition causes an elevated secretion of TRH and TSH. The high levels of TSH

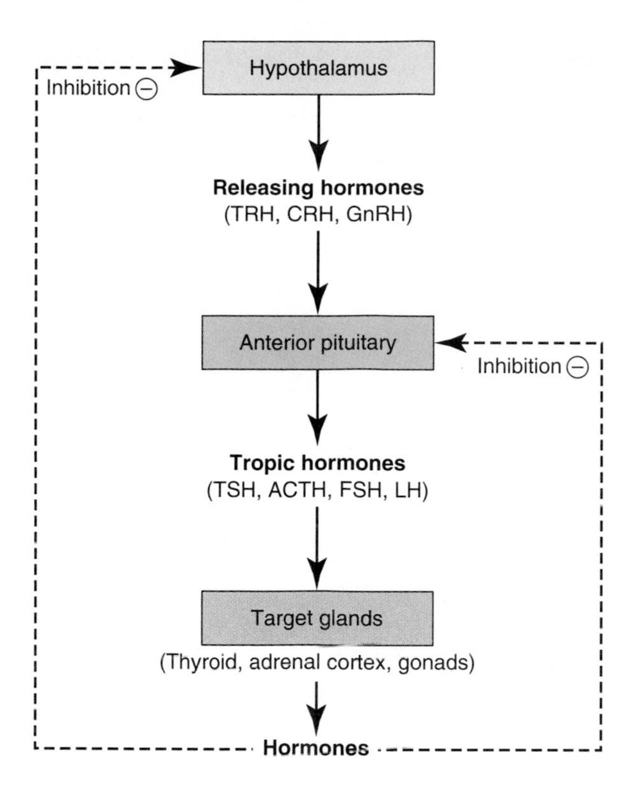

FIGURE 56.14
Negative feedback inhibition. The hormones secreted by some endocrine glands feed back to inhibit the secretion of hypothalamic releasing hormones and anterior pituitary tropic hormones.

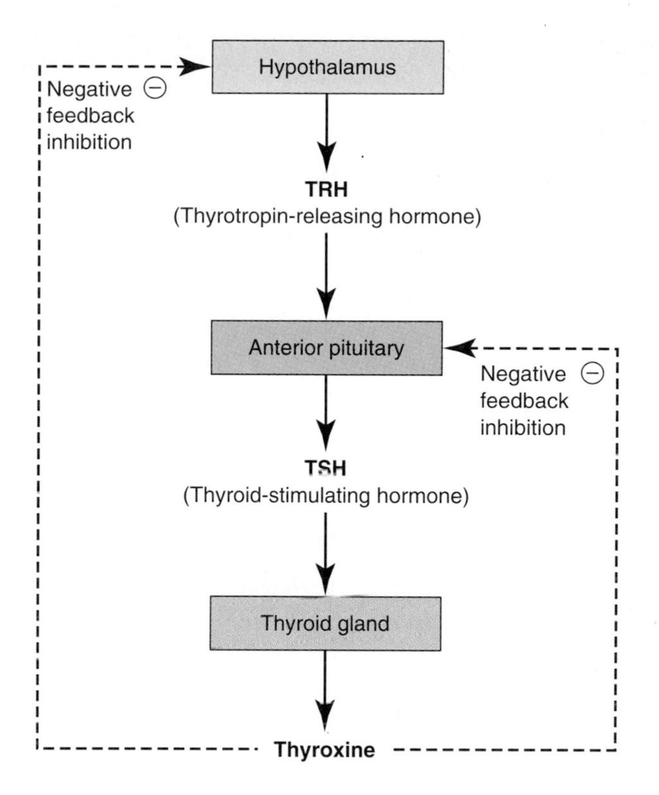

FIGURE 56.15
Regulation of thyroxine secretion. The hypothalamus secretes TRH, which stimulates the anterior pituitary to secrete TSH. The TSH then stimulates the thyroid to secrete thyroxine, which exerts negative feedback control of the hypothalamus and anterior pituitary.

stimulate the thyroid gland to grow, but it still cannot produce thyroxine without iodine. The consequence of this interruption of the normal inhibition by thyroxine is an enlarged thyroid gland, a condition known as a goiter (figure 56.16).

Positive feedback in the control of the hypothalamus and anterior pituitary by the target glands is not common because positive feedback cannot maintain constancy of the internal environment (homeostasis). Positive feedback accentuates change, driving the change in the same direction. One example of positive control involves the control of ovulation, an explosive event that culminates in the expulsion of the egg cell from the ovary. In that case, an ovarian hormone, estradiol, actually stimulates the secretion of an anterior pituitary hormone, LH. This will be discussed in detail in chapter 59.

The hypothalamus controls the anterior pituitary gland by means of hormones, and the anterior pituitary gland controls some other glands through the hormones it secretes. However, both the hypothalamus and the anterior pituitary gland are controlled by other glands through negative feedback inhibition.

FIGURE 56.16
A person with a goiter. This condition is caused by a lack of iodine in the diet. As a result, thyroxine secretion is low, so there is less negative feedback inhibition of TSH. The elevated TSH secretion, in turn, stimulates the thyroid to grow and produce the goiter.

56.4 Endocrine glands secrete hormones that regulate many body functions.

The Thyroid and Parathyroid Glands

The endocrine glands that are regulated by the anterior pituitary, and those endocrine glands that are regulated by other means, help to control metabolism, electrolyte balance, and reproductive functions. Some of the major endocrine glands will be considered in this section.

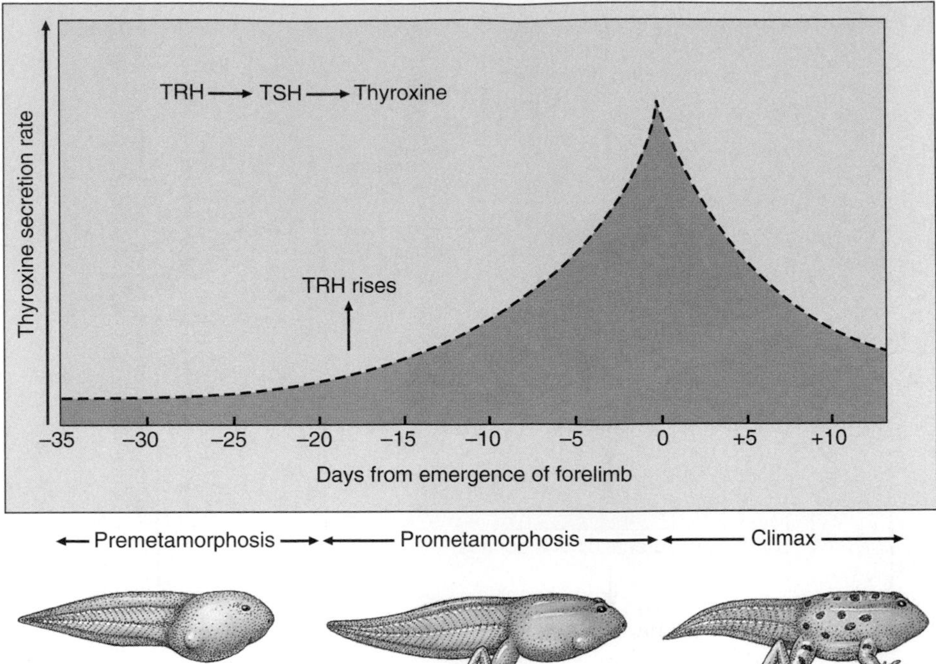

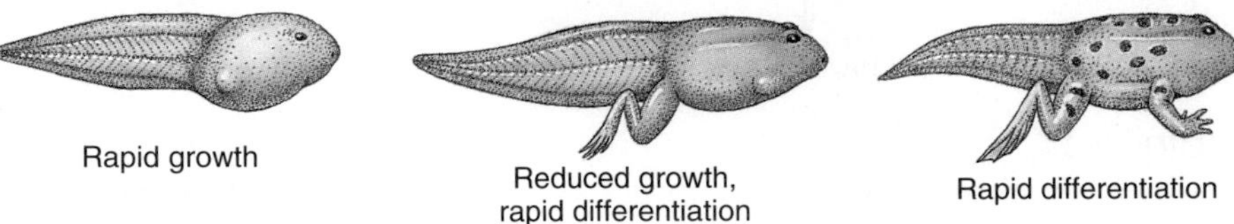

FIGURE 56.17
Thyroxine triggers metamorphosis in amphibians. In tadpoles at the premetamorphic stage, the hypothalamus releases TRH (thyrotropin-releasing hormone), which causes the anterior pituitary to secrete TSH (thyroid-stimulating hormone). TSH then acts on the thyroid gland, which secretes thyroxine. The hindlimbs then begin to form. As metamorphosis proceeds, thyroxine reaches its maximal level, after which the forelimbs begin to form.

The Thyroid Gland

The thyroid gland (Greek *thyros*, "shield") is shaped like a shield and lies just below the Adam's apple in the front of the neck. We have already mentioned that the thyroid gland secretes thyroxine and smaller amounts of triiodothyronine (T_3), which stimulate oxidative respiration in most cells in the body and, in so doing, help set the body's basal metabolic rate (see chapter 51). In children, these thyroid hormones also promote growth and stimulate maturation of the central nervous system. Children with underactive thyroid glands are therefore stunted in their growth and suffer severe mental retardation, a condition called cretinism. This differs from pituitary dwarfism, which results from inadequate GH and is not associated with abnormal intellectual development.

People who are hypothyroid (whose secretion of thyroxine is too low) can take thyroxine orally, as pills. Only thyroxine and the steroid hormones (as in contraceptive pills), can be taken orally because they are nonpolar and can pass through the plasma membranes of intestinal epithelial cells without being digested.

There is an additional function of the thyroid gland that is unique to amphibians—thyroid hormones are needed for the metamorphosis of the larvae into adults (figure 56.17). If the thyroid gland is removed from a tadpole, it will not change into a frog. Conversely, if an immature tadpole is fed pieces of a thyroid gland, it will undergo premature metamorphosis and become a miniature frog!

The thyroid gland also secretes calcitonin, a peptide hormone that plays a role in maintaining proper levels of calcium (Ca^{++}) in the blood. When the blood Ca^{++} concentration rises too high, calcitonin stimulates the uptake of Ca^{++} into bones, thus lowering its level in the blood. Although calcitonin may be important in the physiology of some vertebrates, its significance in normal human physiology is controversial, and it appears less important in the day-to-day regulation of Ca^{++} levels. A hormone that plays a more important role in Ca^{++} homeostasis is secreted by the parathyroid glands, described in the next section.

The Parathyroid Glands and Calcium Homeostasis

The parathyroid glands are four small glands attached to the thyroid. Because of their size, researchers ignored them until well into the twentieth century. The first suggestion that these organs have an endocrine function came from experiments on dogs: if their parathyroid glands were removed, the Ca^{++} concentration in the dogs' blood plummeted to less than half the normal value. The Ca^{++} concentration returned to normal when an extract of parathyroid gland was administered. However, if too much of the extract was administered, the dogs' Ca^{++} levels rose far above normal as the calcium phosphate crystals in their bones was dissolved. It was clear that the parathyroid glands produce a hormone that stimulates the release of Ca^{++} from bone.

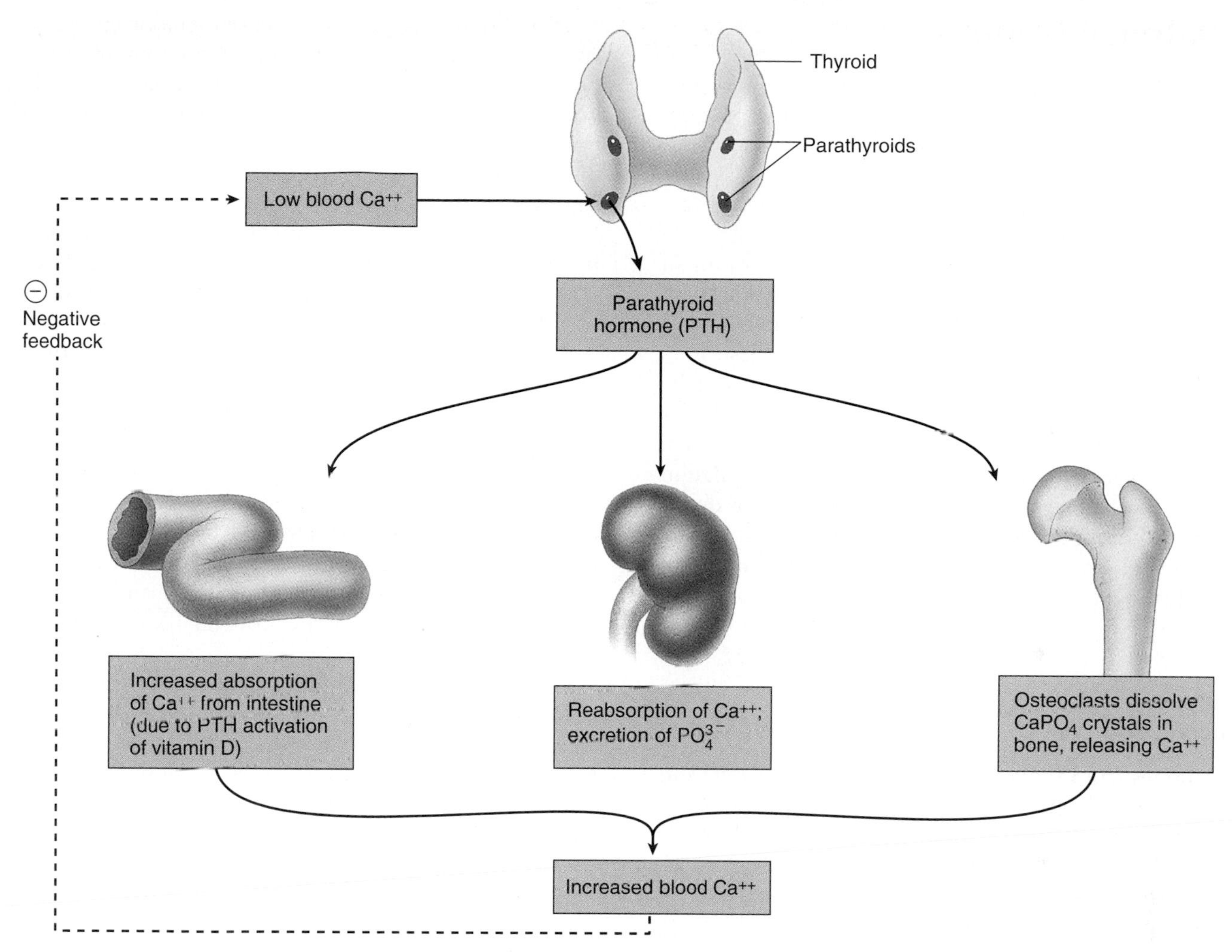

FIGURE 56.18
Regulation of blood Ca^{++} levels by parathyroid hormone (PTH). When blood Ca^{++} levels are low, parathyroid hormone (PTH) is released by the parathyroid glands. PTH directly stimulates the dissolution of bone and the reabsorption of Ca^{++} by the kidneys. PTH indirectly promotes the intestinal absorption of Ca^{++} by stimulating the production of the active form of vitamin D.

The hormone produced by the parathyroid glands is a peptide called parathyroid hormone (PTH). It is one of only two hormones in humans that are absolutely essential for survival (the other is aldosterone, which will be discussed in the next section). PTH is synthesized and released in response to falling levels of Ca^{++} in the blood. This cannot be allowed to continue uncorrected, because a significant fall in the blood Ca^{++} level can cause severe muscle spasms. A normal blood Ca^{++} is important for the functioning of muscles, including the heart, and for proper functioning of the nervous and endocrine systems.

PTH stimulates the osteoclasts (bone cells) in bone to dissolve the calcium phosphate crystals of the bone matrix and release Ca^{++} into the blood (figure 56.18). PTH also stimulates the kidneys to reabsorb Ca^{++} from the urine and leads to the activation of vitamin D, needed for the absorption of Ca^{++} from food in the intestine.

Vitamin D is produced in the skin from a cholesterol derivative in response to ultraviolet light. It is called a vitamin because a dietary source is needed to supplement the amount that the skin produces. Secreted into the blood from the skin, vitamin D is actually an inactive form of a hormone. In order to become activated, the molecule must gain two hydroxyl groups (—OH); one of these is added by an enzyme in the liver, the other by an enzyme in the kidneys. The enzyme needed for this final step is stimulated by PTH, thereby producing the active form of vitamin D known as 1, 25-dihydroxyvitamin D. This hormone stimulates the intestinal absorption of Ca^{++} and thereby helps to raise blood Ca^{++} levels so that bone can become properly mineralized. A diet deficient in vitamin D thus leads to poor bone formation, a condition called rickets.

Thyroxine helps to set the basal metabolic rate by stimulating the rate of cell respiration throughout the body; this hormone is also needed for amphibian metamorphosis. Parathyroid hormone acts to raise the blood Ca^{++} levels, in part by stimulating osteoclasts to dissolve bone.

The Adrenal Glands

The adrenal glands are located just above each kidney (figure 56.19). Each gland is composed of an inner portion, the adrenal medulla, and an outer layer, the adrenal cortex.

The Adrenal Medulla

The adrenal medulla receives neural input from axons of the sympathetic division of the autonomic nervous system, and it secretes epinephrine and norepinephrine in response to stimulation by these axons. The actions of these hormones trigger "alarm" responses similar to those elicited by the sympathetic division, helping to prepare the body for "fight or flight." Among the effects of these hormones are an increased heart rate, increased blood pressure, dilation of the bronchioles, elevation in blood glucose, and reduced blood flow to the skin and digestive organs. The actions of epinephrine released as a hormone supplement those of norepinephrine released as a sympathetic nerve neurotransmitter.

The Adrenal Cortex: Homeostasis of Glucose and Na^+

The hormones from the adrenal cortex are all steroids and are referred to collectively as *corticosteroids*. Cortisol (also called hydrocortisone) and related steroids secreted by the adrenal cortex act on various cells in the body to maintain glucose homeostasis. In mammals, these hormones are referred to as glucocorticoids. The glucocorticoids stimulate the breakdown of muscle protein into amino acids, which are carried by the blood to the liver. They also stimulate the liver to produce the enzymes needed for gluconeogenesis, the conversion of amino acids into glucose. This creation of glucose from protein is particularly important during very long periods of fasting or exercise, when blood glucose levels might otherwise become dangerously low.

In addition to regulating glucose metabolism, the glucocorticoids modulate some aspects of the immune response. Glucocorticoids are given medically to suppress the immune system in persons with immune disorders, such as rheumatoid arthritis. Derivatives of cortisol, such as prednisone, have widespread medical use as antiinflammatory agents.

Aldosterone, the other major corticosteroid, is classified as a mineralocorticoid because it helps regulate mineral balance through two functions. One of the functions of aldosterone is to stimulate the kidneys to reabsorb Na^+ from the urine. (Urine is formed by filtration of blood plasma, so the blood levels of Na^+ will decrease if Na^+ is not reabsorbed from the urine; see chapter 58.) Sodium is the major extracellular solute and is needed for the maintenance of normal blood volume and pressure. Without aldosterone, the kidneys would lose excessive amounts of blood Na^+ in the urine, followed by Cl^- and water; this would cause the blood volume and pressure to fall. By stimulating the kidneys to reabsorb salt and water, aldosterone thus maintains the normal blood volume and pressure essential to life.

The other function of aldosterone is to stimulate the kidneys to secrete K^+ into the urine. Thus, when aldosterone levels are too low, the concentration of K^+ in the blood may rise to dangerous levels. Because of these essential functions performed by aldosterone, removal of the adrenal glands, or diseases that prevent aldosterone secretion, are invariably fatal without hormone therapy.

The adrenal medulla is stimulated by sympathetic neurons to secrete epinephrine and norepinephrine during the fight-or-flight reaction. The adrenal cortex is stimulated to secrete its steroid hormones by ACTH from the anterior pituitary. Cortisol helps to regulate blood glucose and aldosterone acts to regulate blood Na^+ and K^+ levels.

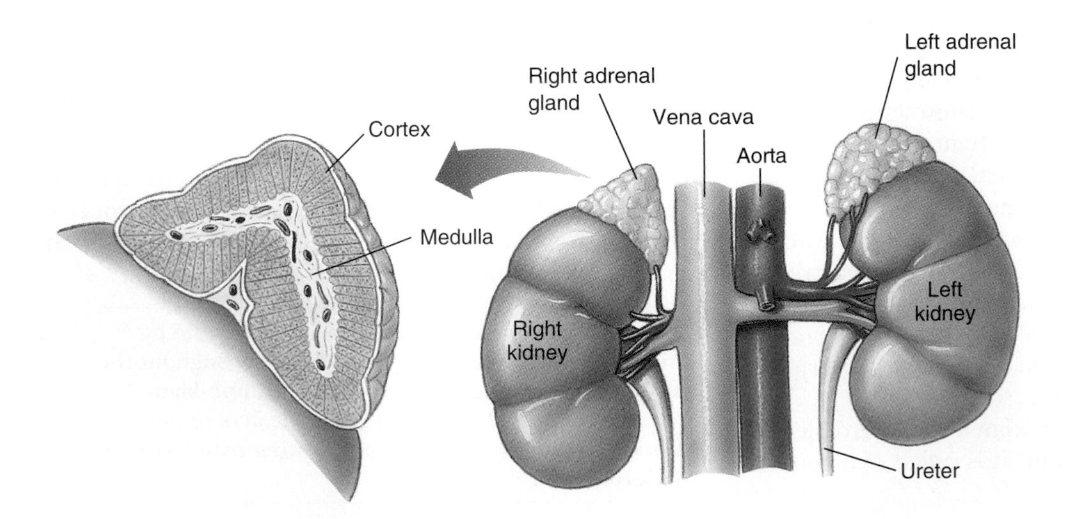

FIGURE 56.19
The adrenal glands. The inner portion of the gland, the adrenal medulla, produces epinephrine and norepinephrine, which initiate a response to stress. The outer portion of the gland, the adrenal cortex, produces steroid hormones that influence blood glucose levels.

The Pancreas

The pancreas is located adjacent to the stomach and is connected to the duodenum of the small intestine by the pancreatic duct. It secretes bicarbonate ions and a variety of digestive enzymes into the small intestine through this duct (see chapter 51), and for a long time the pancreas was thought to be solely an exocrine gland. In 1869, however, a German medical student named Paul Langerhans described some unusual clusters of cells scattered throughout the pancreas; these clusters came to be called islets of Langerhans. Laboratory workers later observed that the surgical removal of the pancreas caused glucose to appear in the urine, the hallmark of the disease diabetes mellitus. This suggested that the pancreas, specifically the islets of Langerhans, might be producing a hormone that prevents this disease.

That hormone is insulin, secreted by the beta (β) cells of the islets. Insulin was not isolated until 1922, when two young doctors working in a Toronto hospital succeeded where many others had not. On January 11, 1922, they injected an extract purified from beef pancreas into a 13-year-old diabetic boy, whose weight had fallen to 65 pounds and who was not expected to survive. With that single injection, the glucose level in the boy's blood fell 25%. A more potent extract soon brought the level down to near normal. The doctors had achieved the first instance of successful insulin therapy.

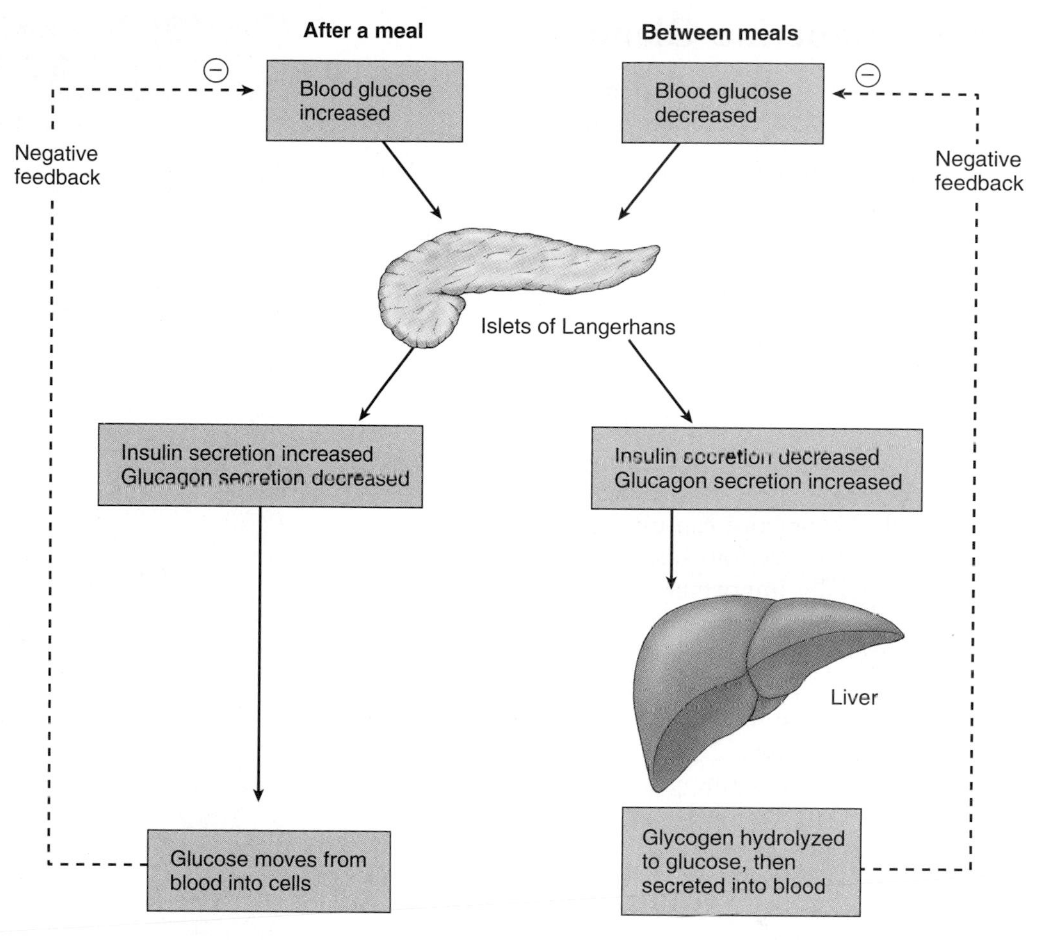

FIGURE 56.20
The antagonistic actions of insulin and glucagon on blood glucose. Insulin stimulates the cellular uptake of blood glucose into skeletal muscles and the liver after a meal. Glucagon stimulates the hydrolysis of liver glycogen between meals, so that the liver can secrete glucose into the blood. These antagonistic effects help to maintain homeostasis of the blood glucose concentration.

Two forms of diabetes mellitus are now recognized. People with *type I*, or insulin-dependent diabetes mellitus, lack the insulin-secreting β cells. Treatment for these patients therefore consists of insulin injections. (Because insulin is a peptide hormone, it would be digested if taken orally and must instead be injected into the blood.) In the past, only insulin extracted from the pancreas of pigs or cattle was available, but today people with insulin-dependent diabetes can inject themselves with human insulin produced by genetically engineered bacteria. Active research on the possibility of transplanting islets of Langerhans holds much promise of a lasting treatment for these patients. Most diabetic patients, however, have *type II*, or non-insulin-dependent diabetes mellitus. They generally have normal or even above-normal levels of insulin in their blood, but their cells have a reduced sensitivity to insulin. These people do not require insulin injections and can usually control their diabetes through diet and exercise.

The islets of Langerhans produce another hormone; the alpha (α) cells of the islets secrete glucagon, which acts antagonistically to insulin (figure 56.20). When a person eats carbohydrates, the blood glucose concentration rises. This stimulates the secretion of insulin by β cells and inhibits the secretion of glucagon by the α cells. Insulin promotes the cellular uptake of glucose into liver and muscle cells, where it is stored as glycogen, and into adipose cells, where it is stored as fat. Between meals, when the concentration of blood glucose falls, insulin secretion decreases and glucagon secretion increases. Glucagon promotes the hydrolysis of stored glycogen in the liver and fat in adipose tissue. As a result, glucose and fatty acids are released into the blood and can be taken up by cells and used for energy.

The β cells of the islets of Langerhans secrete insulin, and the α cells secrete glucagon. These two hormones have antagonistic actions on the blood glucose concentration; insulin lowers and glucagon raises blood glucose.

Other Endocrine Glands

Sexual Development, Biological Clocks, and Immune Regulation in Vertebrates

The ovaries and testes are important endocrine glands, producing the steroid sex hormones called androgens (including estrogens, progesterone, and testosterone), to be described in detail in chapter 59. During embryonic development, testosterone production in the embryo is critical for the development of male sex organs. In mammals, androgens are responsible for the development of secondary sexual characteristics at puberty. These characteristics include breasts in females, body hair, and increased muscle mass in males. Because of this, some bodybuilders illegally take androgens to increase muscle mass. In addition to being illegal, this practice can cause liver disorders as well as a number of other serious side effects. In females, androgens are especially important in maintaining the sexual cycle. Estrogen and progesterone produced in the ovaries are critical regulators of the menstrual and ovarian cycles. During pregnancy, estrogen production in the placenta maintains the uterine lining, which protects and nourishes the developing embryo.

Another major endocrine gland is the pineal gland, located in the roof of the third ventricle of the brain in most vertebrates (see figure 54.27). It is about the size of a pea and is shaped like a pine-cone (hence its name). The pineal gland evolved from a median light-sensitive eye (sometimes called a "third eye," although it could not form images) at the top of the skull in primitive vertebrates. This pineal eye is still present in primitive fish (cyclostomes) and some reptiles. In other vertebrates, however, the pineal gland is buried deep in the brain and functions as an endocrine gland by secreting the hormone melatonin. One of the actions of melatonin is to cause blanching of the skin of lower vertebrates by reducing the dispersal of melanin granules.

The secretion of melatonin is stimulated by activity of the *suprachiasmatic nucleus (SCN)* of the hypothalamus. The SCN is known to function as the major biological clock in vertebrates, entraining (synchronizing) various body processes to a circadian rhythm (one that repeats every 24 hours). Through regulation by the SCN, the secretion of melatonin by the pineal gland is entrained to cycles of light and dark, decreasing during the day and increasing at night. This daily cycling of melatonin release regulates body cycles such as sleep/wake cycles and temperature cycles. In some vertebrates, melatonin helps to regulate reproductive physiology in species with distinct breeding seasons, but the role of melatonin in human reproduction is controversial.

There are a variety of hormones secreted by nonendocrine organs. The thymus is the site of production of particular lymphocytes called T cells, and it secretes a number of hormones that function in the regulation of the immune system. The right atrium of the heart secretes atrial natriuretic hormone, which stimulates the kidneys to excrete salt and water in the urine. This hormone, therefore, acts antagonistically to aldosterone, which promotes salt and water retention. The kidneys secrete erythropoietin, a hormone that stimulates the bone marrow to produce red blood cells. Other organs such as the liver, stomach, and small intestines secrete hormones; even the skin has an endocrine function: it secretes vitamin D. The gas nitric oxide, made by many different cells, controls blood pressure by dilating arteries. The drug sildenafil (Viagra) counters impotence by causing nitric oxide to dilate the blood vessels of the penis.

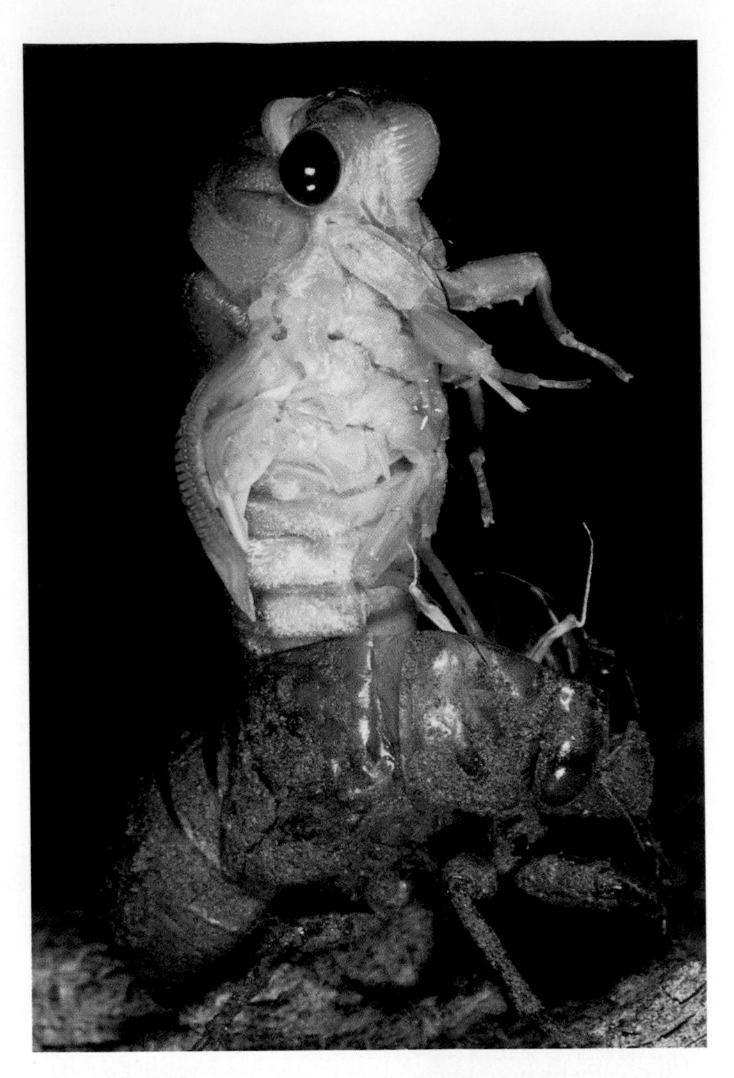

FIGURE 56.21
A molting cicada. This adult insect is emerging from its old cuticle.

Molting and Metamorphosis in Insects

As insects grow during postembryonic development, their hardened exoskeletons do not expand. To overcome this problem, insects undergo a series of molts wherein they shed their old exoskeleton (figure 56.21) and secrete a new,

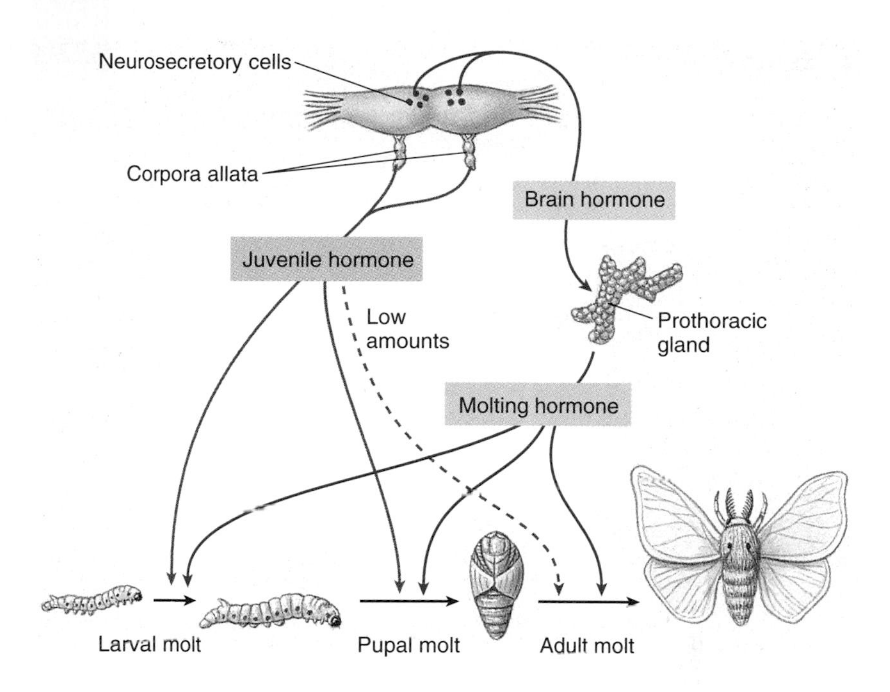

FIGURE 56.22
The hormonal control of metamorphosis in the silkworm moth, *Bombyx mori*. While molting hormone (ecdysone), produced by the prothoracic gland, triggers when molting will occur, juvenile hormone, produced by bodies near the brain called the corpora allata, determines the result of a particular molt. High levels of juvenile hormone inhibit the formation of the pupa and adult forms. At the late stages of metamorphosis, therefore, it is important that the corpora allata not produce large amounts of juvenile hormone.

larger one. In one molt, a juvenile insect, or larvae, often undergoes a radical transformation to the adult. This process is called metamorphosis. Hormonal secretions influence both molting and metamorphosis in insects. Prior to molting, neurosecretory cells on the surface of the brain secrete a small peptide, **brain hormone,** which in turn stimulates a gland in the thorax called the prothoracic gland to produce *molting hormone*, or *ecdysone* (figure 56.22). High levels of ecdysone bring about the biochemical and behavioral changes that cause molting to occur. Another pair of endocrine glands near the brain called the corpora allata produce a hormone called **juvenile hormone.** High levels of juvenile hormone prevent the transformation to the adult and result in a larval to larval molt. If the level of juvenile hormone is low, however, the molt will result in metamorphosis.

Endocrine Disrupting Chemicals

Because target cells are very sensitive to hormones, these hormones are in very low concentrations in the blood. Therefore, small changes in concentrations can make a big difference in how the target organs function. Unfortunately, scientists are now finding that the endocrine system is not sufficiently protected from the outside world. Some man-made chemicals (and even a few plant-produced chemicals) can enter the body and interrupt normal endocrine function. These chemicals may be ones we manufacture for some other purpose and accidentally leak into the environment, ones that are industrial waste products which we "throw away" into the environment, or ones we purposefully release into the environment such as pesticides. These environmental contaminants get into the food we eat and the air we breathe. They are everywhere on earth and cannot be avoided. Some can last for years in the environment, and just as long in an animal's body. Those chemicals that interfere with hormone function are called *endocrine disrupting chemicals*.

Any chemical that can bind to receptor proteins and mimic the effects of the hormone is called a hormone *agonist*. Any chemical that binds to receptor proteins and has no effect but blocks the hormone from binding is called a hormone *antagonist*. Endocrine disrupting chemicals can also interfere by binding to the hormone's protein carriers in the blood. So far, endocrine disrupting chemicals have been shown to interfere with reproductive hormones, thyroid hormones, and the immune system chemical messengers. These effects are not lethal, but may make individuals vulnerable in their environment. If they are having problems reproducing, maintaining the proper metabolic rate, or fighting off infections, then their numbers will decrease (perhaps even leading to extinction). These environmental contaminants may be harming humans in addition to other species. Laws have been passed requiring the testing of thousands of chemicals to see if they have endocrine disrupting potential, and the Environmental Protection Agency (EPA) must include this testing in their standard protocols before approving any new compounds.

Sex steroid hormones from the gonads regulate reproduction, melatonin secreted by the pineal gland helps regulate circadian rhythms, and thymus hormones help regulate the vertebrate immune system. Molting hormone, or ecdysone, and juvenile hormone regulate metamorphosis and molting in insects.

Chapter 56

www.mhhe.com/raven6/resources56.mhtml

Summary | **Questions** | **Media Resources**

56.1 Regulation is often accomplished by chemical messengers.

- Endocrine glands secrete hormones into the blood, which are then transported to target cells.
- Hormones may be lipophilic, such as the steroid hormones and thyroxine, or hydrophilic, such as amine, polypeptide, and glycoprotein hormones.
- Prostaglandins and other paracrine regulatory molecules are produced by one cell type and regulate different cells within the same organ.

1. What is the definition of a hormone? How do hormones reach their target cells? Why are only certain cells capable of being target cells for a particular hormone?

2. How do hormones and paracrine regulators differ from one another?

- Art Activity: Endocrine System

- Art Quiz: Human Endocrine System

56.2 Lipophilic and polar hormones regulate their target cells by different means.

- Lipid-soluble hormones enter their target cells, bind to intracellular receptor proteins, and the complex then binds to hormone response elements on the DNA, activating specific genes.
- Polar hormones do not enter their target cells, but instead bind to receptor proteins on the cell membrane and activate second-messenger systems or control ion channels.

3. How does epinephrine result in the production of cAMP in its target cells? How does cAMP bring about specific changes inside target cells?

- Peptide Hormone Action

- Activities:
  –Steroid Hormones
  –Peptide Hormones

56.3 The hypothalamus controls the secretions of the pituitary gland.

- Axons from neurons in the hypothalamus enter the posterior pituitary, carrying ADH and oxytocin; the posterior pituitary stores these hormones and secretes them in response to neural activity.
- The anterior pituitary produces and secretes a variety of hormones, many of which control other endocrine glands; the anterior pituitary, however, is itself controlled by the hypothalamus via releasing and inhibiting hormones secreted by the hypothalamus.

4. Where are hormones secreted by the posterior pituitary gland actually *produced*?

5. Why are the hormones of the anterior pituitary gland called tropic hormones?

6. How does the hypothalamus regulate the secretion of the anterior pituitary?

- Endocrine System Regulation

- Hypothalamus
- Human System Overview

- Art Quizzes:
  -Pituitary Control by Hypothalamus
  -Negative Feedback Inhibition

56.4 Endocrine glands secrete hormones that regulate many body functions.

- The thyroid secretes thyroxine and triiodothyronine, which set the basal metabolic rate by stimulating the rate of cell respiration in most cells of the body.
- The adrenal cortex secretes cortisol, which regulates glucose balance, and aldosterone, which regulates Na^+ and K^+ balance.
- The β cells of the islets of Langerhans in the pancreas secrete insulin, which lowers the blood glucose; glucagon, secreted by the α cells, raises the blood glucose level.

7. What hormones are produced by the adrenal cortex? What functions do these hormones serve? What stimulates the secretion of these hormones?

8. What pancreatic hormone is produced when the body's blood glucose level becomes elevated?

- Parathyroid Hormone
- Glucose Regulation

- Bioethics Case Study: Endocrine System
- *On Science* Article: Type Two Diabetes

57

The Immune System

Concept Outline

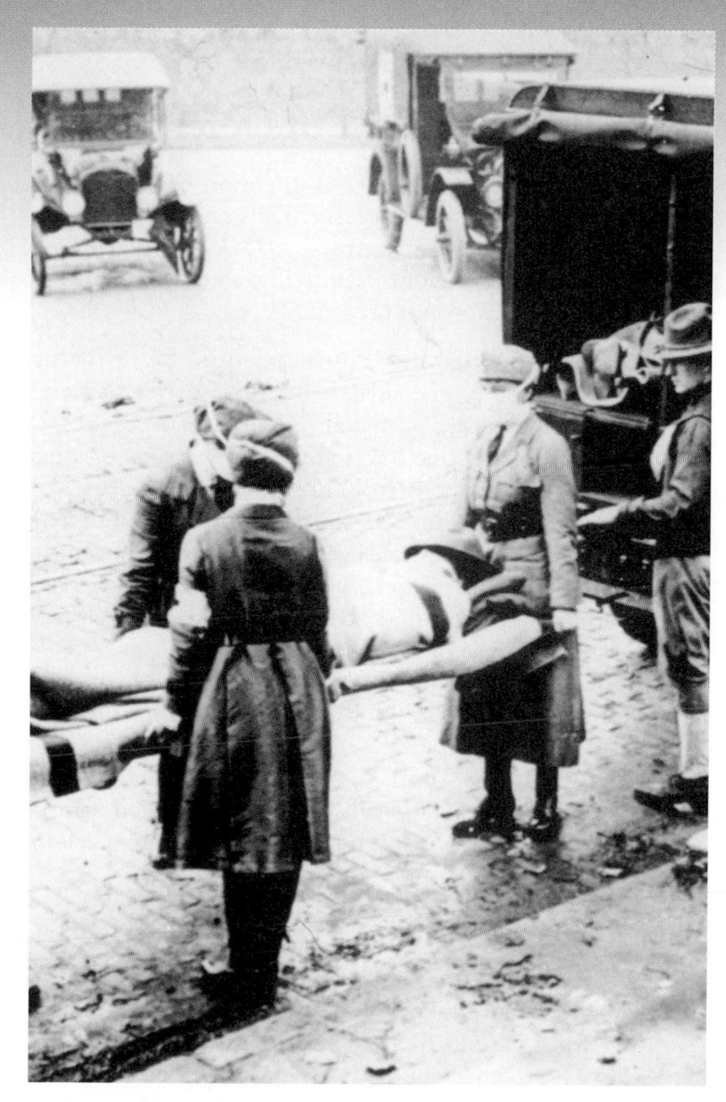

FIGURE 57.1
The influenza epidemic of 1918–1919 killed 22 million people in 18 months. With 25 million Americans infected, the Red Cross often worked around the clock.

When you consider how animals defend themselves, it is natural to think of turtles, armadillos, and other animals covered like tanks with heavy plates of armor. However, armor offers no protection against the greatest dangers vertebrates face—microorganisms and viruses. We live in a world awash with attackers too tiny to see with the naked eye, and no vertebrate could long withstand their onslaught unprotected. We survive because we have evolved a variety of very effective defenses against this constant attack. As we review these defenses, it is important to keep in mind that they are far from perfect. Some 22 million Americans and Europeans died from influenza over an 18-month period in 1918–1919 (figure 57.1), and more than 3 million people will die of malaria this year. Attempts to improve our defenses against infection are among the most active areas of scientific research today.

57.1 Many of the body's most effective defenses are nonspecific.

Skin: The First Line of Defense

The vertebrate is defended from infection the same way knights defended medieval cities. "Walls and moats" make entry difficult; "roaming patrols" attack strangers; and "sentries" challenge anyone wandering about and call patrols if a proper "ID" is not presented.

1. **Walls and moats.** The outermost layer of the vertebrate body, the **skin,** is the first barrier to penetration by microbes. Mucous membranes in the respiratory and digestive tracts are also important barriers that protect the body from invasion.
2. **Roaming patrols.** If the first line of defense is penetrated, the response of the body is to mount a **cellular counterattack,** using a battery of cells and chemicals that kill microbes. These defenses act very rapidly after the onset of infection.
3. **Sentries.** Lastly, the body is also guarded by mobile cells that patrol the bloodstream, scanning the surfaces of every cell they encounter. They are part of the **immune system.** One kind of immune cell aggressively attacks and kills any cell identified as foreign, whereas the other type marks the foreign cell or virus for elimination by the roaming patrols.

The Skin as a Barrier to Infection

The skin is the largest organ of the vertebrate body, accounting for 15% of an adult human's total weight. The skin not only defends the body by providing a nearly impenetrable barrier, but also reinforces this defense with chemical weapons on the surface. Oil and sweat glands give the skin's surface a pH of 3 to 5, acidic enough to inhibit the growth of many microorganisms. Sweat also contains the enzyme **lysozyme,** which digests bacterial cell walls. In addition to defending the body against invasion by viruses and microorganisms, the skin prevents excessive loss of water to the air through evaporation.

The epidermis of skin is approximately 10 to 30 cells thick, about as thick as this page. The outer layer, called the stratum corneum, contains cells that are continuously abraded, injured, and worn by friction and stress during the body's many activities. The body deals with this damage not by repairing the cells, but by replacing them. Cells are shed continuously from the stratum corneum and are replaced by new cells produced in the innermost layer of the epidermis, the stratum basale, which contains some of the most actively dividing cells in the vertebrate body. The cells formed in this layer migrate upward and enter a broad intermediate stratum spinosum layer. As they move upward they form the protein keratin, which makes skin tough and water-resistant. These new cells eventually arrive at the stratum corneum, where they normally remain for about a month before they are shed and replaced by newer cells from below. Psoriasis, which afflicts some 4 million Americans, is a chronic skin disorder in which epidermal cells are replaced every 3 to 4 days, about eight times faster than normal.

The dermis of skin is 15 to 40 times thicker than the epidermis. It provides structural support for the epidermis and a matrix for the many blood vessels, nerve endings, muscles, and other structures situated within skin. The wrinkling that occurs as we grow older takes place in the dermis, and the leather used to manufacture belts and shoes is derived from very thick animal dermis.

The layer of subcutaneous tissue below the dermis contains primarily adipose cells. These cells act as shock absorbers and provide insulation, conserving body heat. Subcutaneous tissue varies greatly in thickness in different parts of the body. It is nonexistent in the eyelids, is a half-centimeter thick or more on the soles of the feet, and may be much thicker in other areas of the body, such as the buttocks and thighs.

Other External Surfaces

In addition to the skin, two other potential routes of entry by viruses and microorganisms must be guarded: the *digestive tract* and the *respiratory tract.* Recall that both the digestive and respiratory tracts open to the outside and their surfaces must also protect the body from foreign invaders. Microbes are present in food, but many are killed by saliva (which also contains lysozyme), by the very acidic environment of the stomach, and by digestive enzymes in the intestine. Microorganisms are also present in inhaled air. The cells lining the smaller bronchi and bronchioles secrete a layer of sticky mucus that traps most microorganisms before they can reach the warm, moist lungs, which would provide ideal breeding grounds for them. Other cells lining these passages have cilia that continually sweep the mucus toward the glottis. There it can be swallowed, carrying potential invaders out of the lungs and into the digestive tract. Occasionally, an infectious agent, called a pathogen, will enter the digestive and respiratory systems and the body will use defense mechanisms such as vomiting, diarrhea, coughing, and sneezing to expel the pathogens.

The surface defenses of the body consist of the skin and the mucous membranes lining the digestive and respiratory tracts, which eliminate many microorganisms before they can invade the body tissues.

Cellular Counterattack: The Second Line of Defense

The surface defenses of the vertebrate body are very effective but are occasionally breached, allowing invaders to enter the body. At this point, the body uses a host of nonspecific cellular and chemical devices to defend itself. We refer to this as the second line of defense. These devices all have one property in common: they respond to *any* microbial infection without pausing to determine the invader's identity.

Although these cells and chemicals of the nonspecific immune response roam through the body, there is a central location for the collection and distribution of the cells of the immune system; it is called the lymphatic system (see chapter 52). The lymphatic system consists of a network of lymphatic capillaries, ducts, nodes and lymphatic organs (figure 57.2), and although it has other functions involved with circulation, it also stores cells and other agents used in the immune response. These cells are distributed throughout the body to fight infections, and also stored in the lymph nodes where foreign invaders can be eliminated as body fluids pass through.

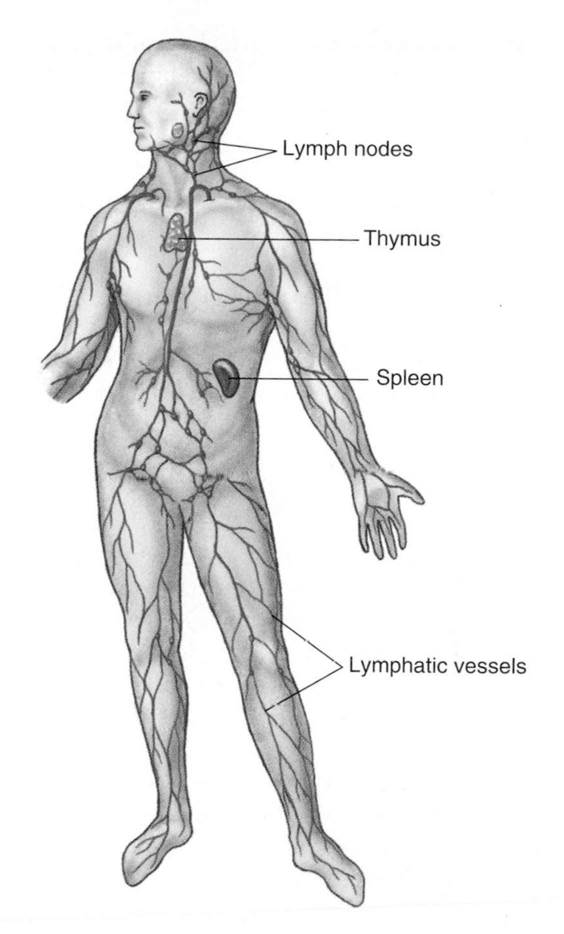

FIGURE 57.2
The lymphatic system. The lymphatic system consists of lymphatic vessels, lymph nodes, and lymphatic organs, including the spleen and thymus gland.

Cells That Kill Invading Microbes

Perhaps the most important of the vertebrate body's nonspecific defenses are white blood cells called leukocytes that circulate through the body and attack invading microbes within tissues. There are three basic kinds of these cells, and each kills invading microorganisms differently.

Macrophages ("big eaters") are large, irregularly shaped cells that kill microbes by ingesting them through *phagocytosis*, much as an amoeba ingests a food particle (figure 57.3). Within the macrophage, the membrane-bound vacuole containing the bacterium fuses with a lysosome. Fusion activates lysosomal enzymes that kill the microbe by liberating large quantities of oxygen free-radicals. Macrophages also engulf viruses, cellular debris, and dust particles in the lungs. Macrophages circulate continuously in the extracellular fluid, and their phagocytic actions supplement those of the specialized phagocytic cells that are part of the structure of the liver, spleen, and bone marrow. In response to an infection, monocytes (an undifferentiated leukocyte) found in the blood squeeze through capillaries to enter the connective tissues. There, at the site of the infection, the monocytes are transformed into additional macrophages.

Neutrophils are leukocytes that, like macrophages, ingest and kill bacteria by phagocytosis. In addition, neutrophils release chemicals (some of which are identical to household bleach) that kill other bacteria in the neighborhood as well as neutrophils themselves.

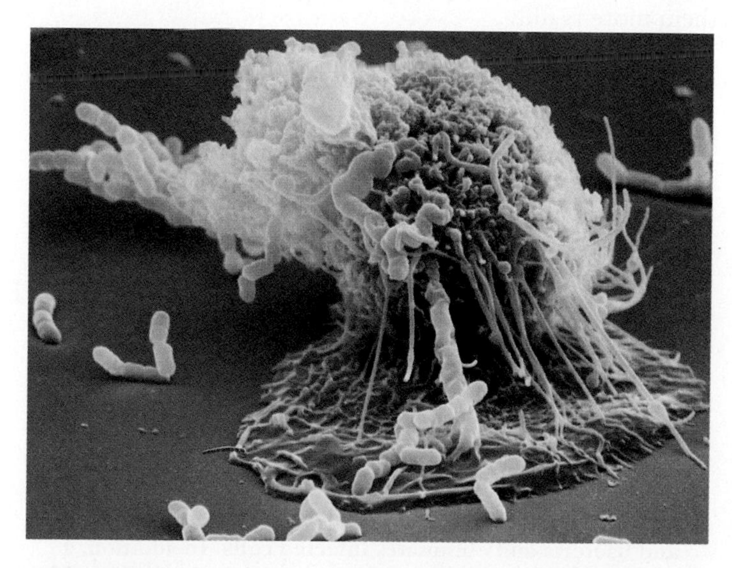

FIGURE 57.3
A macrophage in action (1800×). In this scanning electron micrograph, a macrophage is "fishing" with long, sticky cytoplasmic extensions. Bacterial cells that come in contact with the extensions are drawn toward the macrophage and engulfed.

Natural killer cells do not attack invading microbes directly. Instead, they kill cells of the body that have been infected with viruses. They kill not by phagocytosis, but rather by creating a hole in the plasma membrane of the target cell (figure 57.4). Proteins, called *perforins*, are released from the natural killer cells and insert into the membrane of the target cell, forming a pore. This pore allows water to rush into the target cell, which then swells and bursts. Natural killer cells also attack cancer cells, often before the cancer cells have had a chance to develop into a detectable tumor. The vigilant surveillance by natural killer cells is one of the body's most potent defenses against cancer.

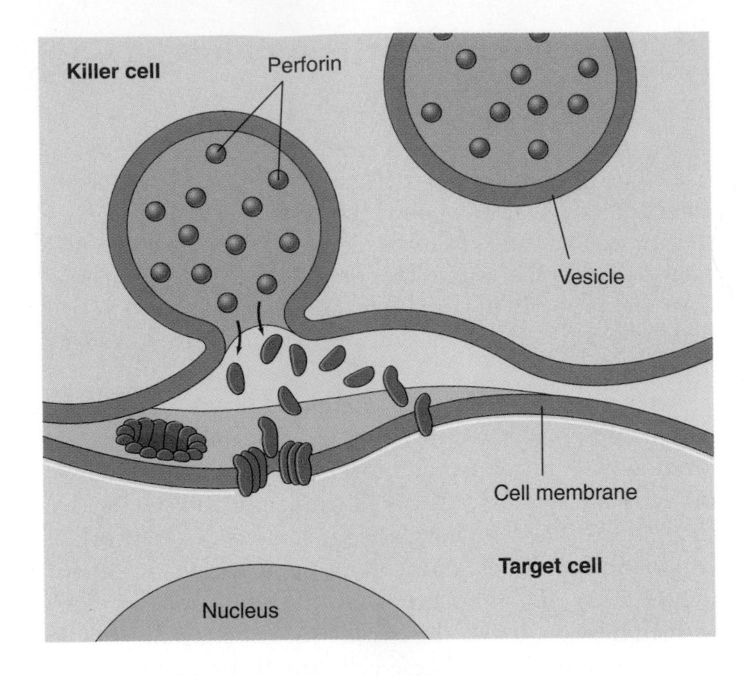

FIGURE 57.4
How natural killer cells kill target cells. The initial event, the tight binding of the killer cell to the target cell, causes vesicles loaded with *perforin* molecules within the killer cell to move to the plasma membrane and disgorge their contents into the intercellular space over the target cell. The perforin molecules insert into the plasma membrane of the target cell like staves of a barrel, forming a pore that admits water and ruptures the cell.

Proteins That Kill Invading Microbes

The cellular defenses of vertebrates are enhanced by a very effective chemical defense called the *complement system*. This system consists of approximately 20 different proteins that circulate freely in the blood plasma. When they encounter a bacterial or fungal cell wall, these proteins aggregate to form a *membrane attack complex* that inserts itself into the foreign cell's plasma membrane, forming a pore like that produced by natural killer cells (figure 57.5). Water enters the foreign cell through this pore, causing the cell to swell and burst. Aggregation of the complement proteins is also triggered by the binding of antibodies to invading microbes, as we will see in a later section.

The proteins of the complement system can augment the effects of other body defenses. Some amplify the inflammatory response (discussed next) by stimulating histamine release; others attract phagocytes to the area of infection; and still others coat invading microbes, roughening the microbes' surfaces so that phagocytes may attach to them more readily.

Another class of proteins that play a key role in body defense are interferons. There are three major categories of interferons: *alpha*, *beta*, and *gamma*. Almost all cells in the body make alpha and beta interferons. These polypeptides act as messengers that protect normal cells in the vicinity of infected cells from becoming infected. Though viruses are still able to penetrate the neighboring cells, the alpha and beta interferons prevent viral replication and protein assembly in these cells. Gamma interferon is produced only by particular lymphocytes and natural killer cells. The secretion of gamma interferon by these cells is part of the immunological defense against infection and cancer, as we will describe later.

A patrolling army of macrophages, neutrophils, and natural killer cells attacks and destroys invading viruses and bacteria and eliminates infected cells. In addition, a system of proteins called complement may be activated to destroy foreign cells, and body cells infected with a virus secrete proteins called interferons that protect neighboring cells.

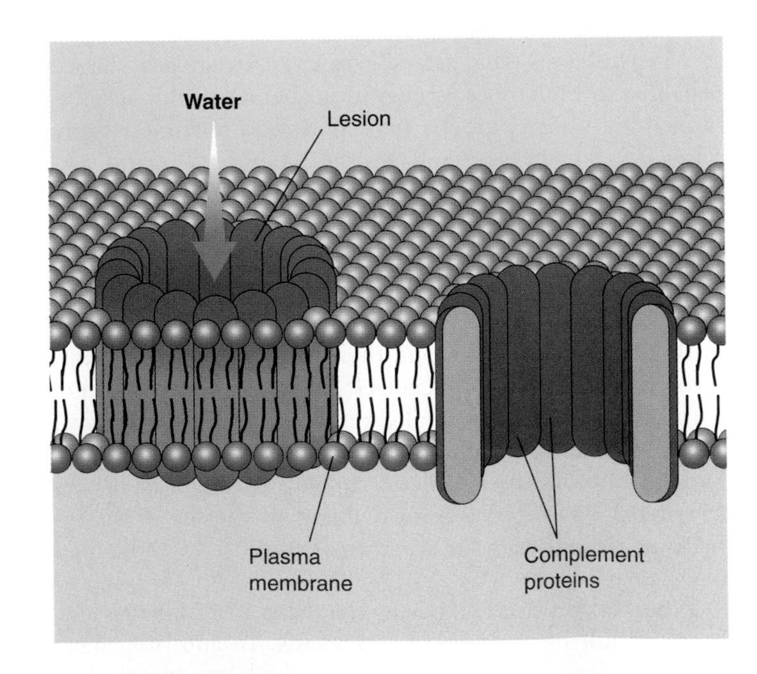

FIGURE 57.5
How complement creates a hole in a cell membrane. As the diagram shows, the complement proteins form a complex transmembrane pore resembling the perforin-lined pores formed by natural killer cells.

The Inflammatory Response

The inflammatory response is a localized, nonspecific response to infection. Infected or injured cells release chemical alarm signals, most notably histamine and prostaglandins. These chemicals promote the dilation of local blood vessels, which increases the flow of blood to the site of infection or injury and causes the area to become red and warm. They also increase the permeability of capillaries in the area, producing the edema (tissue swelling) so often associated with infection. The more permeable capillaries allow phagocytes (monocytes and neutrophils) to migrate from the blood to the extracellular fluid, where they can attack bacteria. Neutrophils arrive first, spilling out chemicals that kill the bacteria in the vicinity (as well as tissue cells and themselves); the *pus* associated with some infections is a mixture of dead or dying pathogens, tissue cells, and neutrophils. Monocytes follow, become macrophages and engulf pathogens and the remains of the dead cells (figure 57.6).

The Temperature Response

Macrophages that encounter invading microbes release a regulatory molecule called interleukin-1, which is carried by the blood to the brain. Interleukin-1 and other pyrogens (Greek *pyr*, "fire") such as bacterial endotoxins cause neurons in the hypothalamus to raise the body's temperature several degrees above the normal value of 37°C (98.6°F). The elevated temperature that results is called a fever.

Experiments with lizards, which regulate their body temperature by moving to warmer or colder locations, demonstrate that infected lizards choose a warmer environment—they give themselves a fever! Further, if lizards are prevented from elevating their body temperature, they have a slower recovery from their infection. Fever contributes to the body's defense by stimulating phagocytosis and causing the liver and spleen to store iron, reducing blood levels of iron, which bacteria need in large amounts to grow. However, very high fevers are hazardous because excessive heat may inactivate critical enzymes. In general, temperatures greater than 39.4°C (103°F) are considered dangerous for humans, and those greater than 40.6°C (105°F) are often fatal.

Inflammation aids the fight against infection by increasing blood flow to the site and raising temperature to retard bacterial growth.

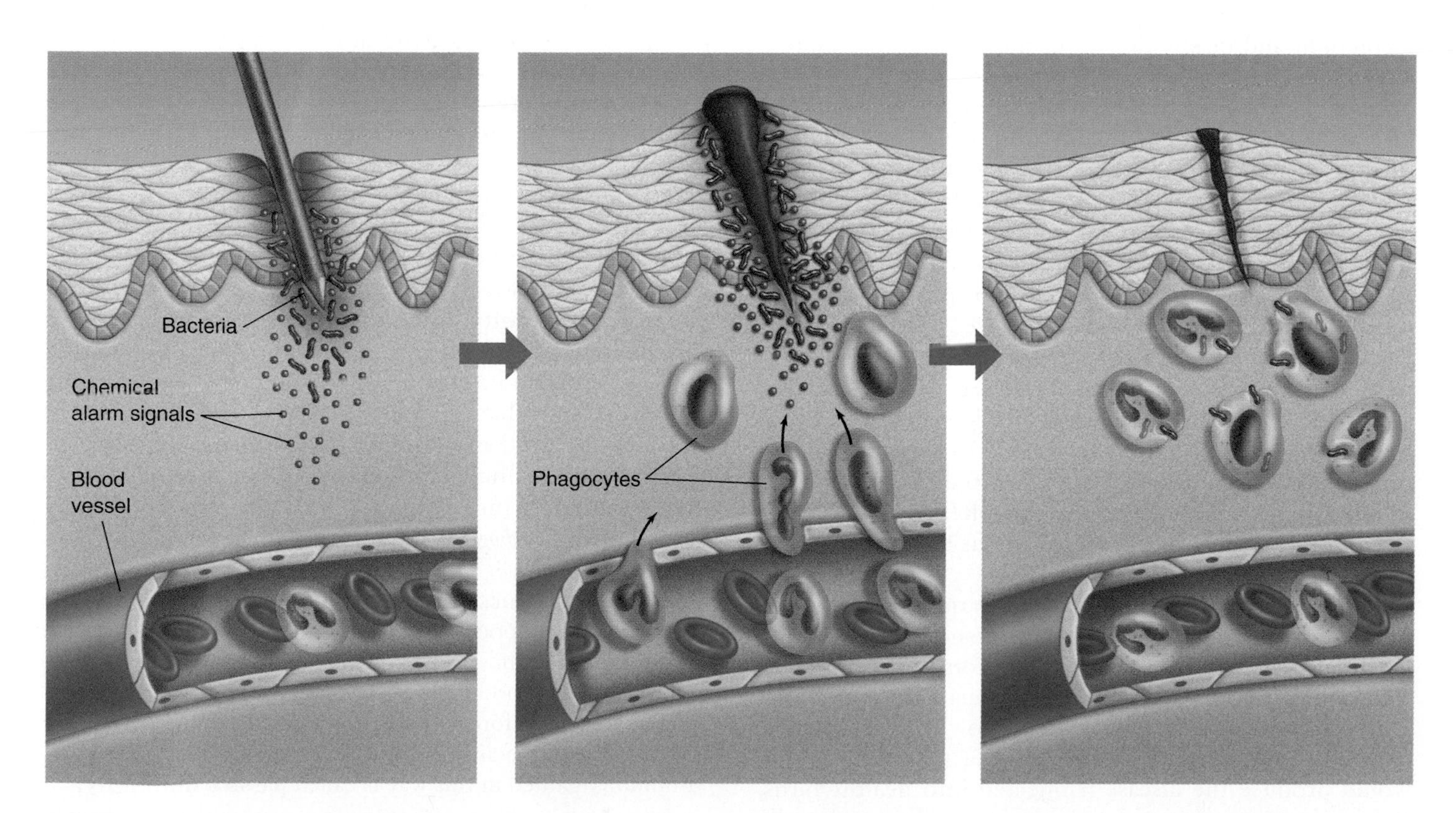

FIGURE 57.6
The events in a local inflammation. When an invading microbe has penetrated the skin, chemicals, such as histamine and prostaglandins, cause nearby blood vessels to dilate. Increased blood flow brings a wave of phagocytic cells, which attack and engulf invading bacteria.

57.2 Specific immune defenses require the recognition of antigens.

The Immune Response: The Third Line of Defense

Few of us pass through childhood without contracting some sort of infection. Chicken pox, for example, is an illness that many of us experience before we reach our teens. It is a disease of childhood, because most of us contract it as children and *never catch it again.* Once you have had the disease, you are usually immune to it. Specific immune defense mechanisms provide this immunity.

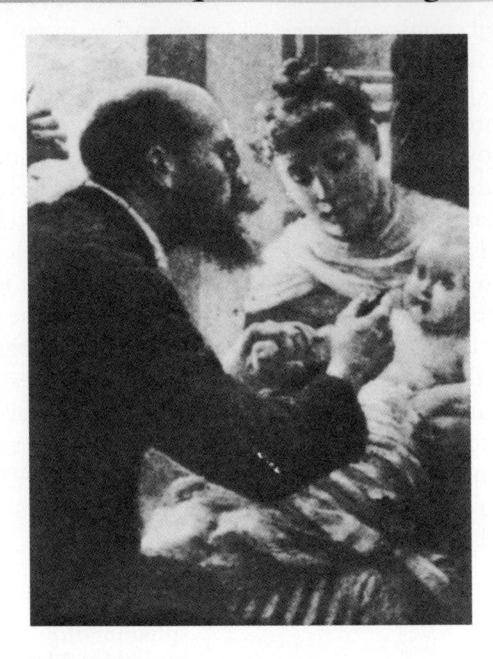

FIGURE 57.7
The birth of immunology. This famous painting shows Edward Jenner inoculating patients with cowpox in the 1790s and thus protecting them from smallpox.

Discovery of the Immune Response

In 1796, an English country doctor named Edward Jenner carried out an experiment that marks the beginning of the study of immunology. Smallpox was a common and deadly disease in those days. Jenner observed, however, that milkmaids who had caught a much milder form of "the pox" called cowpox (presumably from cows) rarely caught smallpox. Jenner set out to test the idea that cowpox conferred protection against smallpox. He infected people with cowpox (figure 57.7), and as he had predicted, many of them became immune to smallpox.

We now know that smallpox and cowpox are caused by two different viruses with similar surfaces. Jenner's patients who were injected with the cowpox virus mounted a defense that was also effective against a later infection of the smallpox virus. Jenner's procedure of injecting a harmless microbe in order to confer resistance to a dangerous one is called **vaccination.** Modern attempts to develop resistance to malaria, herpes, and other diseases often involve delivering antigens via a harmless vaccinia virus related to cowpox virus.

Many years passed before anyone learned how exposure to an infectious agent can confer resistance to a disease. A key step toward answering this question was taken more than a half-century later by the famous French scientist Louis Pasteur. Pasteur was studying fowl cholera, and he isolated a culture of bacteria from diseased chickens that would produce the disease if injected into healthy birds. Before departing on a two-week vacation, he accidentally left his bacterial culture out on a shelf. When he returned, he injected this old culture into healthy birds and found that it had been weakened; the injected birds became only slightly ill and then recovered. Surprisingly, however, those birds did not get sick when subsequently infected with fresh fowl cholera. They remained healthy even if given massive doses of active fowl cholera bacteria that did produce the disease in control chickens. Clearly, something about the bacteria could elicit immunity as long as the bacteria did not kill the animals first. We now know that molecules protruding from the surfaces of the bacterial cells evoked active immunity in the chickens.

Key Concepts of Specific Immunity

An **antigen** is a molecule that provokes a specific immune response. Antigens are large, complex molecules such as proteins; they are generally foreign to the body, usually present on the surface of pathogens. A large antigen may have several parts, and each stimulate a different specific immune response. In this case, the different parts are known as **antigenic determinant sites,** and each serves as a different antigen. Particular lymphocytes have receptor proteins on their surfaces that recognize an antigen and direct a specific immune response against either the antigen or the cell that carries the antigen.

Lymphocytes called B cells respond to antigens by producing proteins called **antibodies.** Antibody proteins are secreted into the blood and other body fluids and thus provide **humoral immunity.** (The term *humor* here is used in its ancient sense, referring to a body fluid.) Other lymphocytes called T cells do not secrete antibodies but instead directly attack the cells that carry the specific antigens. These cells are thus described as producing **cell-mediated immunity.**

The specific immune responses protect the body in two ways. First, an individual can gain immunity by being exposed to a *pathogen* (disease-causing agent) and perhaps getting the disease. This is *acquired immunity,* such as the resistance to the chicken pox that you acquire after having the disease in childhood. Another term for this process is **active immunity.** Second, an individual can gain immunity by obtaining the antibodies from another individual. This happened to you before you were born, with antibodies made by your mother being transferred to you across the placenta. Immunity gained in this way is called **passive immunity.**

Antigens are molecules, usually foreign, that provoke a specific immune attack. This immune attack may involve secreted proteins called antibodies, or it may invoke a cell-mediated attack.

Cells of the Specific Immune System

The immune defense mechanisms of the body involve the actions of white blood cells, or leukocytes. Leukocytes include neutrophils, eosinophils, basophils, and monocytes, all of which are phagocytic and are involved in the second line of defense, as well as two types of lymphocytes (*T cells* and *B cells*), which are not phagocytic but are critical to the specific immune response (table 57.1), the third line of defense. T cells direct the cell-mediated response, B cells the humoral response.

After their origin in the bone marrow, T cells migrate to the thymus (hence the designation "T"), a gland just above the heart. There they develop the ability to identify microorganisms and viruses by the antigens exposed on their surfaces. Tens of millions of different T cells are made, each specializing in the recognition of one particular antigen. No invader can escape being recognized by at least a few T cells. There are four principal kinds of T cells: inducer T cells oversee the development of T cells in the thymus; helper T cells (often symbolized T_H) initiate the immune response; cytotoxic ("cell-poisoning") T cells (often symbolized T_C) lyse cells that have been infected by viruses; and suppressor T cells terminate the immune response.

Unlike T cells, B cells do not travel to the thymus; they complete their maturation in the bone marrow. (B cells are so named because they were originally characterized in a region of chickens called the bursa.) From the bone marrow, B cells are released to circulate in the blood and lymph. Individual B cells, like T cells, are specialized to recognize particular foreign antigens. When a B cell encounters the antigen to which it is targeted, it begins to divide rapidly, and its progeny differentiate into plasma cells and memory cells. Each plasma cell is a miniature factory producing antibodies that stick like flags to that antigen wherever it occurs in the body, marking any cell bearing the antigen for destruction. The immunity that Pasteur observed resulted from such antibodies and from the continued presence of the B cells that produced them.

The lymphocytes, T cells and B cells, are involved in the specific immune response. T cells develop in the thymus while B cells develop in the bone marrow.

Table 57.1 Cells of the Immune System

Cell Type	Function
Helper T cell	Commander of the immune response; detects infection and sounds the alarm, initiating both T cell and B cell responses
Inducer T cell	Not involved in the immediate response to infection; mediates the maturation of other T cells in the thymus
Cytotoxic T cell	Detects and kills infected body cells; recruited by helper T cells
Suppressor T cell	Dampens the activity of T and B cells, scaling back the defense after the infection has been checked
B cell	Precursor of plasma cell; specialized to recognize specific foreign antigens
Plasma cell	Biochemical factory devoted to the production of antibodies directed against specific foreign antigens
Mast cell	Initiator of the inflammatory response, which aids the arrival of leukocytes at a site of infection; secretes histamine and is important in allergic responses
Monocyte	Precursor of macrophage
Macrophage	The body's first cellular line of defense; also serves as antigen-presenting cell to B and T cells and engulfs antibody-covered cells
Natural killer cell	Recognizes and kills infected body cells; natural killer (NK) cell detects and kills cells infected by a broad range of invaders; killer (K) cell attacks only antibody-coated cells

Initiating the Immune Response

To understand how the third line of defense works, imagine you have just come down with the flu. Influenza viruses enter your body in small water droplets inhaled into your respiratory system. If they avoid becoming ensnared in the mucus lining the respiratory membranes (first line of defense), and avoid consumption by macrophages (second line of defense), the viruses infect and kill mucous membrane cells.

At this point macrophages initiate the immune defense. Macrophages inspect the surfaces of all cells they encounter. The surfaces of most vertebrate cells possess glycoproteins produced by a group of genes called the **major histocompatibility complex (MHC).** These glycoproteins are called **MHC proteins** or, specifically in humans, **human leukocyte antigens (HLA).** The genes encoding the MHC proteins are highly polymorphic (have many forms); for example, the human MHC proteins are specified by genes that are the most polymorphic known, with nearly 170 alleles each. Only rarely will two individuals have the same combination of alleles, and the MHC proteins are thus different for each individual, much as fingerprints are. As a result, the MHC proteins on the tissue cells serve as self markers that enable the individual's immune system to distinguish its cells from foreign cells, an ability called **self-versus-nonself recognition.** T cells of the immune system will recognize a cell as self or nonself by the MHC proteins present on the cell surface.

When a foreign particle, such as a virus, infects the body, it is taken in by cells and partially digested. Within the cells, the viral antigens are processed and moved to the surface of the plasma membrane. The cells that perform this function are known as **antigen-presenting cells** (figure 57.8). At the membrane, the processed antigens are complexed with the MHC proteins. This enables T cells to recognize antigens presented to them associated with the MHC proteins.

There are two classes of MHC proteins. MHC-I is present on every nucleated cell of the body. MHC-II, however, is found only on macrophages, B cells, and a subtype of T cells called $CD4^+$ T cells (table 57.2). These three cell types work together in one form of the immune response, and their MHC-II markers permit them to recognize one another. Cytotoxic T lymphocytes, which act to destroy infected cells as previously described, can only interact with antigens presented to them with MHC-I proteins. Helper T lymphocytes, whose functions will soon be described, can interact only with antigens presented with MHC-II proteins. These restrictions result from the presence of coreceptors, which are proteins associated with the T cell receptors. The coreceptor known as CD8 is associated with the cytotoxic T cell receptor (these cells can therefore be indicated as $CD8^+$). The CD8 coreceptor can interact only with the MHC-I proteins of an infected cell. The coreceptor known as CD4 is associated with the helper T cell receptor (these cells can thus be indicated as $CD4^+$) and interacts only with the MHC-II proteins of another lymphocyte (figure 57.9).

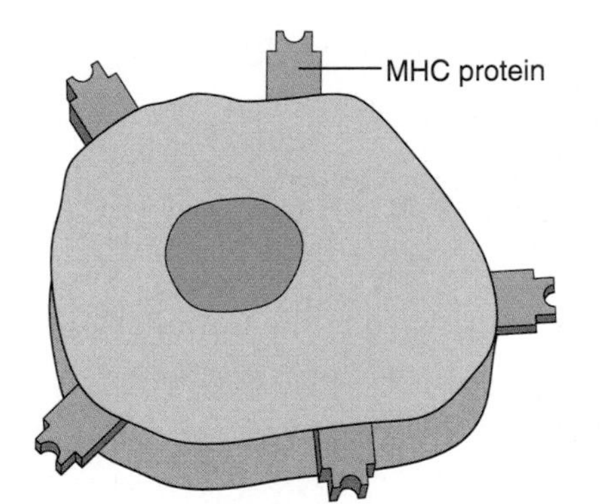

(a) **Body cell**

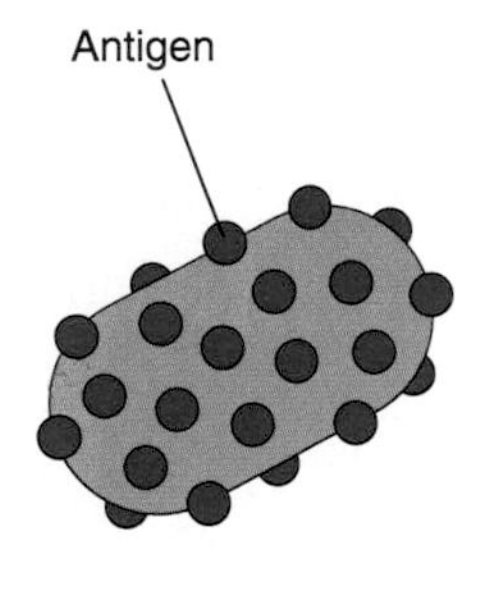

(b) **Foreign microbe**

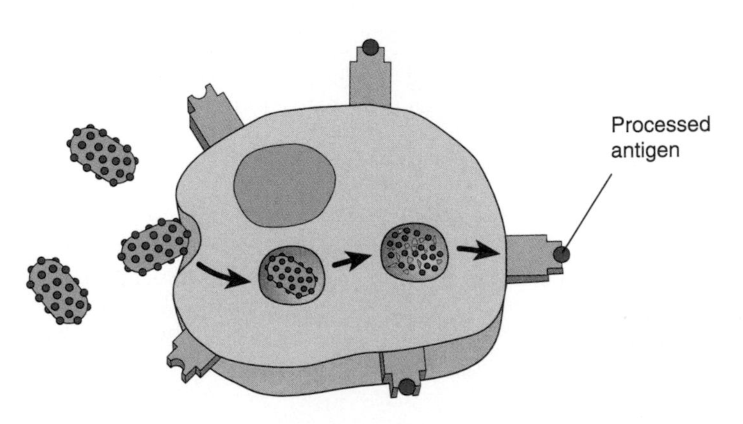

(c) **Antigen-presenting cell**

FIGURE 57.8
Antigens are presented on MHC proteins. (*a*) Cells of the body have MHC proteins on their surfaces that identify them as "self" cells. Immune system cells do not attack these cells. (*b*) Foreign cells or microbes have antigens on their surfaces. B cells are able to bind directly to free antigens in the body and initiate an attack on a foreign invader. (*c*) T cells can bind to antigens only after the antigens are processed and complexed with MHC proteins on the surface of an antigen-presenting cell.

Table 57.2 Key Cell Surface Proteins of the Immune System

	Immune Receptors		MHC Proteins	
Cell Type	**T Receptor**	**B Receptor**	**MHC-I**	**MHC-II**
B cell	–	+	+	+
CD4+ T cell	+	–	+	+
CD8+ T cell	+	–	+	–
Macrophage	–	–	+	+

Note: CD4+ T cells include inducer T cells and helper T cells; CD8+ T cells include cytotoxic T cells and suppressor T cells. + means present; – means absent.

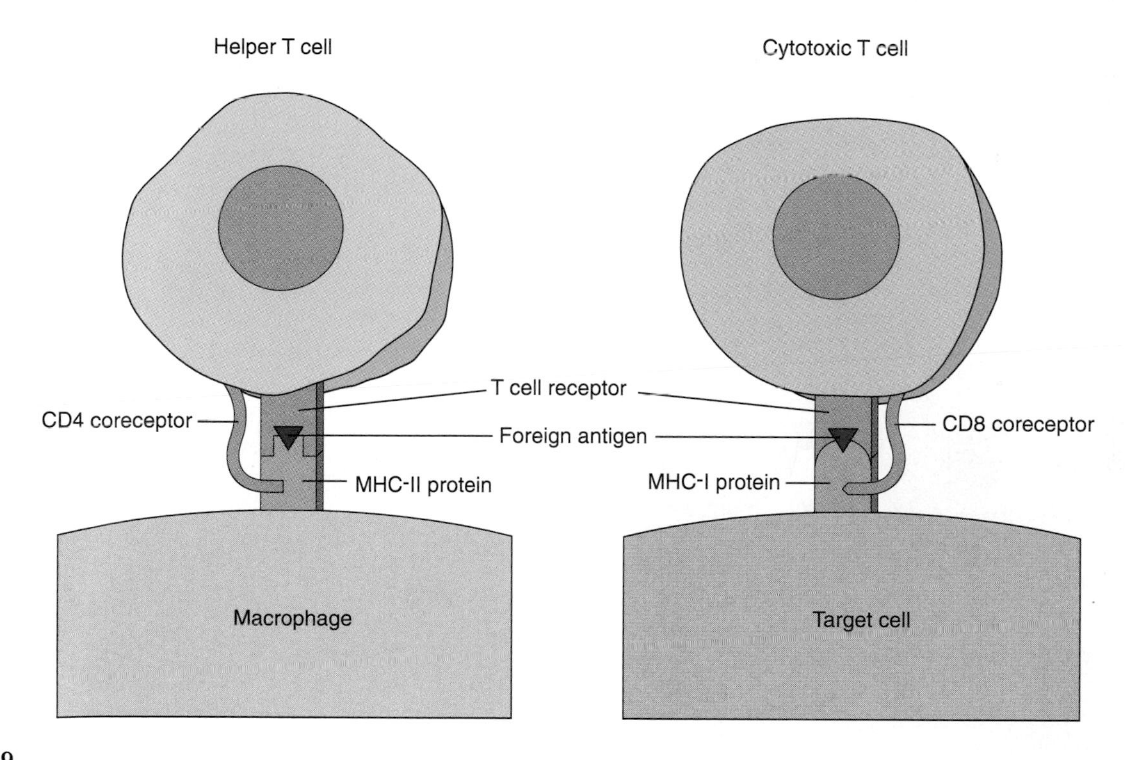

FIGURE 57.9
T cells bind to foreign antigens in conjunction with MHC proteins. The CD4 coreceptor on helper T cells requires that these cells interact with class-2 MHC (or MHC-II) proteins. The CD8 coreceptor on cytotoxic T cells requires that these cells interact only with cells bearing class-1 MHC (or MHC-I) proteins.

Macrophages encounter foreign particles in the body, partially digest the virus particles, and present the foreign antigens in a complex with the MHC-II proteins on its membrane. This combination of MHC-II proteins and foreign antigens is required for interaction with the receptors on the surface of helper T cells. At the same time, macrophages that encounter antigens or antigen-presenting cells release a protein called **interleukin-1** that acts as a chemical alarm signal (discussed in the next section). Helper T cells respond to interleukin-1 by simultaneously initiating two parallel lines of immune system defense: the cell-mediated response carried out by T cells and the humoral response carried out by B cells.

Antigen-presenting cells must present foreign antigens together with MHC-II proteins in order to activate helper T cells, which have the CD4 coreceptor. Cytotoxic T cells use the CD8 coreceptor and must interact with foreign antigens presented on MHC-I proteins.

57.3 T cells organize attacks against invading microbes.

T Cells: The Cell-Mediated Immune Response

The cell-mediated immune response, carried out by T cells, protects the body from virus infection and cancer, killing abnormal or virus-infected body cells.

Once a helper T cell that initiates this response is presented with foreign antigen together with MHC proteins by a macrophage or other antigen-presenting cell, a complex series of steps is initiated. An integral part of this process is the secretion of regulatory molecules known generally as **cytokines,** or more specifically as **lymphokines** if they are secreted by lymphocytes.

When a cytokine is first discovered, it is named according to its biological activity (such as *B cell–stimulating factor*). However, because each cytokine has many different actions, such names can be misleading. Scientists have thus agreed to use the name **interleukin**, followed by a number, to indicate a cytokine whose amino acid sequence has been determined. *Interleukin-1*, for example, is secreted by macrophages and can activate the T cell system. B cell–stimulating factor, now called interleukin-4, is secreted by T cells and is required for the proliferation and clone development of B cells. *Interleukin-2* is released by helper T cells and, among its effects, is required for the activation of cytotoxic T lymphocytes. We will consider the actions of the cytokines as we describe the development of the T cell immune response.

Cell Interactions in the T Cell Response

When macrophages process the foreign antigens, they secrete **interleukin-1,** which stimulates cell division and proliferation of T cells (figure 57.10). Once the helper T cells have been activated by the antigens presented to them by

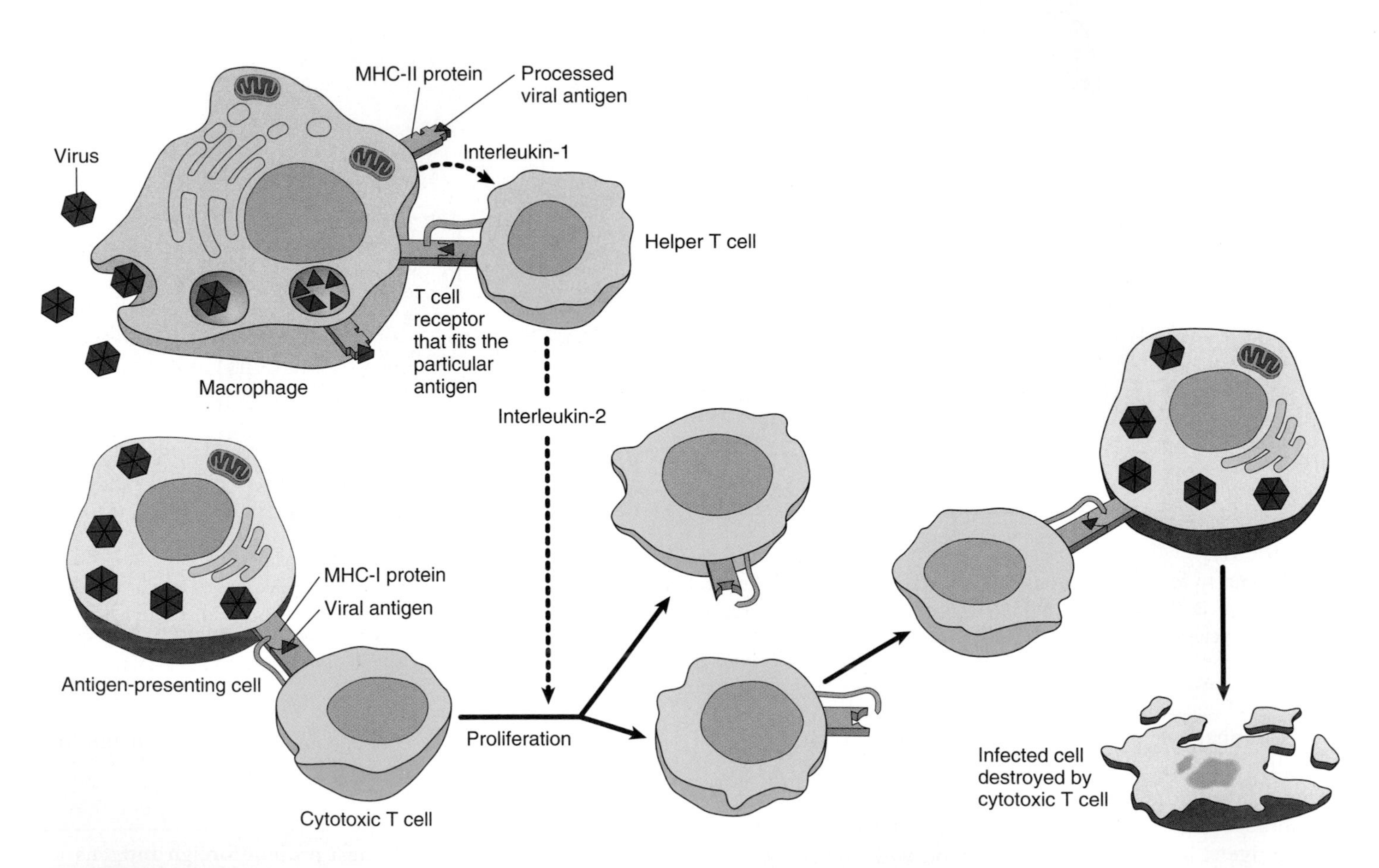

FIGURE 57.10
The T cell immune defense. After a macrophage has processed an antigen, it releases interleukin-1, signaling helper T cells to bind to the antigen-MHC protein complex. This triggers the helper T cell to release interleukin-2, which stimulates the multiplication of cytotoxic T cells. In addition, proliferation of cytotoxic T cells is stimulated when a T cell with a receptor that fits the antigen displayed by an antigen-presenting cell binds to the antigen-MHC protein complex. Body cells that have been infected by the antigen are destroyed by the cytotoxic T cells. As the infection subsides, suppressor T cells "turn off" the immune response.

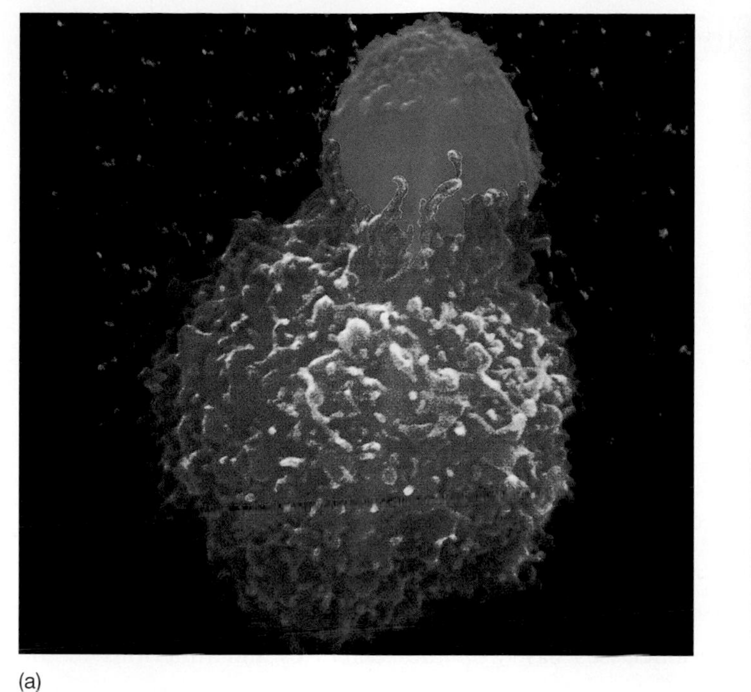

(a)

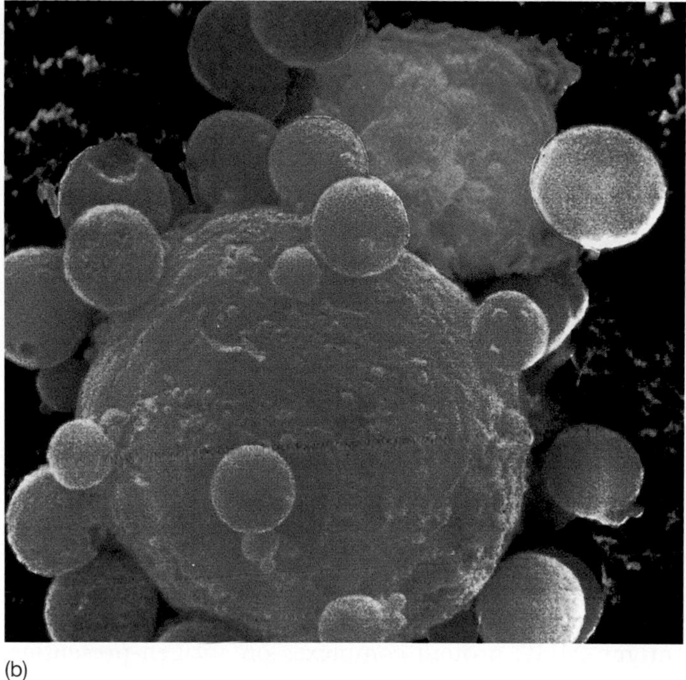

(b)

FIGURE 57.11
Cytotoxic T cells destroy cancer cells. (*a*) The cytotoxic T cell *(orange)* comes into contact with a cancer cell *(pink)*. (*b*) The T cell recognizes that the cancer cell is "nonself" and causes the destruction of the cancer.

the macrophages, they secrete the cytokines known as macrophage colony-stimulating factor and gamma interferon, which promote the activity of macrophages. In addition, the helper T cells secrete **interleukin-2,** which stimulates the proliferation of cytotoxic T cells that are specific for the antigen. (Interleukin-2 also stimulates B cells, as we will see in the next section.) Cytotoxic T cells can destroy infected cells only if those cells display the foreign antigen together with their MHC-I proteins (see figure 57.10).

T Cells in Transplant Rejection and Surveillance against Cancer

Cytotoxic T cells will also attack any foreign version of MHC-I as if it signaled a virus-infected cell. Therefore, even though vertebrates did not evolve the immune system as a defense against tissue transplants, their immune systems will attack transplanted tissue and cause graft rejection. Recall that the MHC proteins are polymorphic, but because of their genetic basis, the closer that two individuals are related, the less variance in their MHC proteins and the more likely they will tolerate each other's tissues—this is why relatives are often sought for kidney transplants. The drug cyclosporin inhibits graft rejection by inactivating cytotoxic T cells.

As tumors develop, they reveal surface antigens that can stimulate the immune destruction of the tumor cells. Tumor antigens activate the immune system, initiating an attack primarily by cytotoxic T cells (figure 57.11) and natural killer cells. The concept of **immunological surveillance** against cancer was introduced in the early 1970s to describe the proposed role of the immune system in fighting cancer.

The production of human interferons by genetically engineered bacteria has made large amounts of these substances available for the experimental treatment of cancer. Thus far, interferons have proven to be a useful addition to the treatment of particular forms of cancer, including some types of lymphomas, renal carcinoma, melanoma, Kaposi's sarcoma, and breast cancer.

Interleukin-2 (IL-2), which activates both cytotoxic T cells and B cells, is now also available for therapeutic use through genetic-engineering techniques. Particular lymphocytes from cancer patients have been removed, treated with IL-2, and given back to the patients together with IL-2 and gamma interferon. Scientists are also attempting to identify specific antigens and their genes that may become uniquely expressed in cancer cells, in an effort to help the immune system to better target cancer cells for destruction.

Helper T cells are only activated when a foreign antigen is presented together with MHC antigens by a macrophage or other antigen-presenting cells. The helper T cells are also stimulated by interleukin-1 secreted by the macrophages, and, when activated, secrete a number of lymphokines. Interleukin-2, secreted by helper T cells, activates both cytotoxic T cells and B cells. Cytotoxic T cells destroy infected cells, transplanted cells, and cancer cells by cell-mediated attack.

57.4 B cells label specific cells for destruction.

B Cells: The Humoral Immune Response

B cells also respond to helper T cells activated by interleukin-1. Like cytotoxic T cells, B cells have receptor proteins on their surface, one type of receptor for each type of B cell. B cells recognize invading microbes much as cytotoxic T cells recognize infected cells, but unlike cytotoxic T cells, they do not go on the attack themselves. Rather, they mark the pathogen for destruction by mechanisms that have no "ID check" system of their own. Early in the immune response, the markers placed by B cells alert complement proteins to attack the cells carrying them. Later in the immune response, the markers placed by B cells activate macrophages and natural killer cells.

The way B cells do their marking is simple and foolproof. Unlike the receptors on T cells, which bind only to antigen-MHC protein complexes on antigen-presenting cells, B cell receptors can bind to free, unprocessed antigens. When a B cell encounters an antigen, antigen particles will enter the B cell by endocytosis and get processed. Helper T cells that are able to recognize the specific antigen will bind to the antigen-MHC protein complex on the B cell and release interleukin-2, which stimulates the B cell to divide. In addition, free, unprocessed antigens stick to antibodies on the B cell surface. This antigen exposure triggers even more B cell proliferation. B cells divide to produce long-lived memory B cells and plasma cells that serve as short-lived antibody factories (figure 57.12). The antibodies are released into the blood plasma, lymph, and other extracellular fluids. Figure 57.13 summarizes the roles of helper T cells, which are essential in both the cell-mediated and humoral immune responses.

Antibodies are proteins in a class called **immunoglobulins** (abbreviated Ig), which is divided into subclasses based on the structures and functions of the

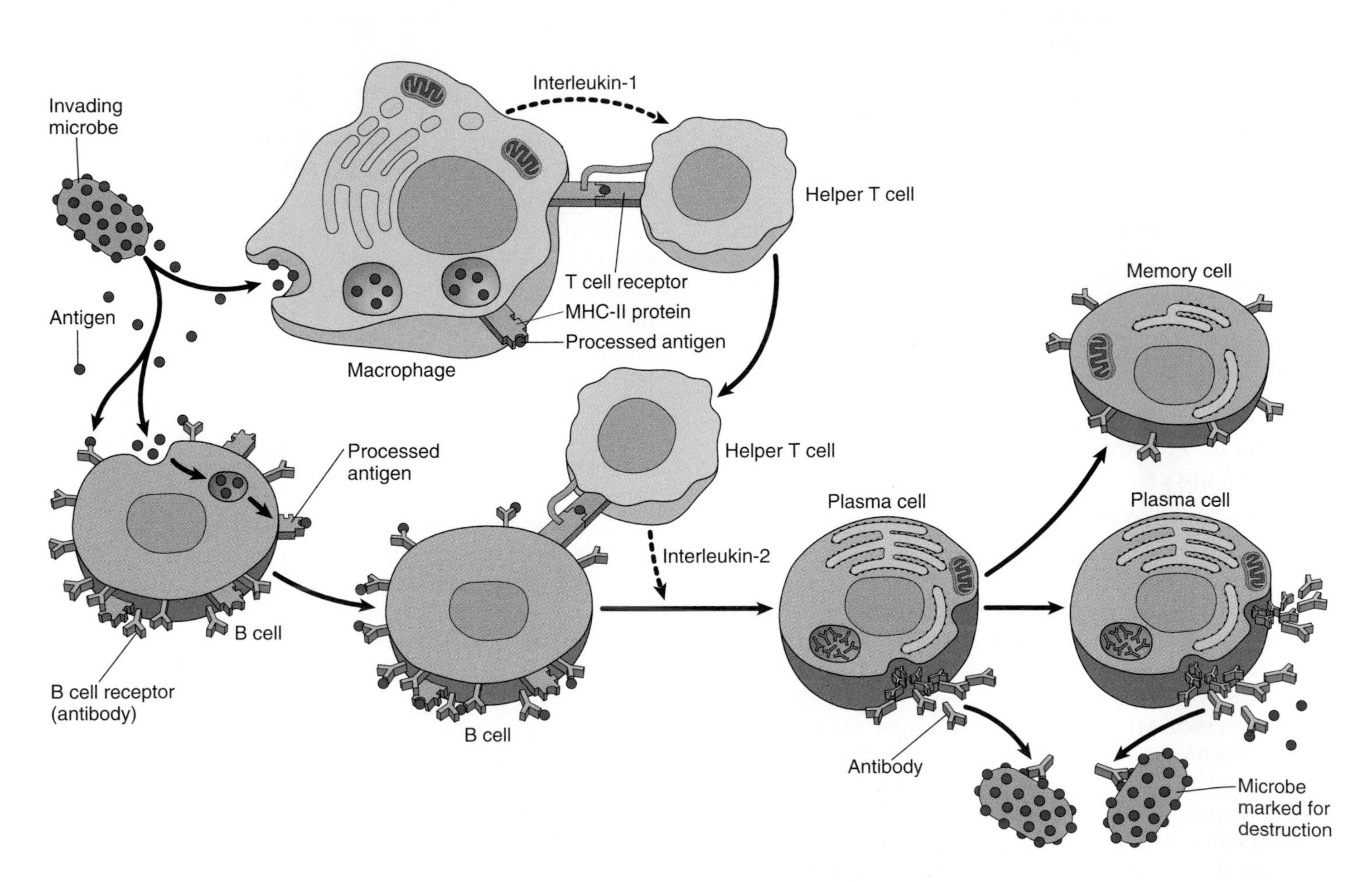

FIGURE 57.12
The B cell immune defense. Invading particles are bound by B cells, which interact with helper T cells and are activated to divide. The multiplying B cells produce either memory B cells or plasma cells that secrete antibodies which bind to invading microbes and tag them for destruction by macrophages.

antibodies. The different immunoglobulin subclasses are as follows:

1. **IgM.** This is the first type of antibody to be secreted during the primary response and they serve as receptors on the lymphocyte surface. These antibodies also promote agglutination reactions (causing antigen-containing particles to stick together, or agglutinate).
2. **IgG.** This is the major form of antibody in the blood plasma and is secreted in a secondary response.
3. **IgD.** These antibodies serve as receptors for antigens on the B cell surface. Their other functions are unknown.
4. **IgA.** This is the major form of antibody in external secretions, such as saliva and mother's milk.
5. **IgE.** This form of antibodies promotes the release of histamine and other agents that aid in attacking a pathogen. Unfortunately, they sometimes trigger a full-blown response when a harmless antigen enters the body producing allergic symptoms, such as those of hay fever.

Each B cell has on its surface about 100,000 IgM or IgD receptors. Unlike the receptors on T cells, which bind only to antigens presented by certain cells, B receptors can bind to *free* antigens. This provokes a primary response in which antibodies of the IgM class are secreted, and also stimulates cell division and clonal expansion. Upon subsequent exposure, the plasma cells secrete large amounts of antibodies that are generally of the IgG class. Although plasma cells live only a few days, they produce a vast number of antibodies. In fact, antibodies constitute about 20% by weight of the total protein in blood plasma. Production of IgG antibodies peaks after about three weeks (figure 57.14).

When IgM (and to a lesser extent IgG) antibodies bind to antigens on a cell, they cause the aggregation of complement proteins. As we mentioned earlier, these proteins form a pore that pierces the plasma membrane of the infected cell (see figure 57.5), allowing water to enter and causing the cell to burst. In contrast, when IgG antibodies bind to antigens on a cell, they serve as markers that stimulate phagocytosis by macrophages. Because certain complement proteins attract phagocytic cells, activation of complement is generally accompanied by increased phagocytosis. Notice that antibodies don't kill invading pathogens directly; rather, they cause destruction of the pathogens by activating the complement system and by targeting the pathogen for attack by phagocytic cells.

In the humoral immune response, B cells recognize antigens and divide to produce plasma cells, producing large numbers of circulating antibodies directed against those antigens. IgM antibodies are produced first, and they activate the complement system. Thereafter, IgG antibodies are produced and promote phagocytosis.

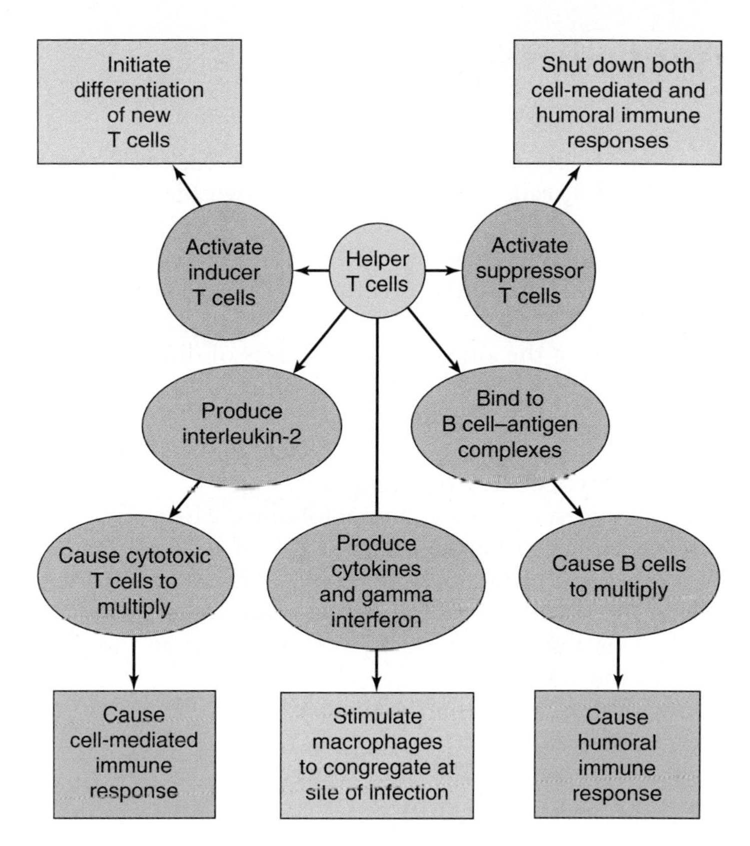

FIGURE 57.13
The many roles of helper T cells. Helper T cells, through their secretion of lymphokines and interaction with other cells of the immune system, participate in every aspect of the immune response.

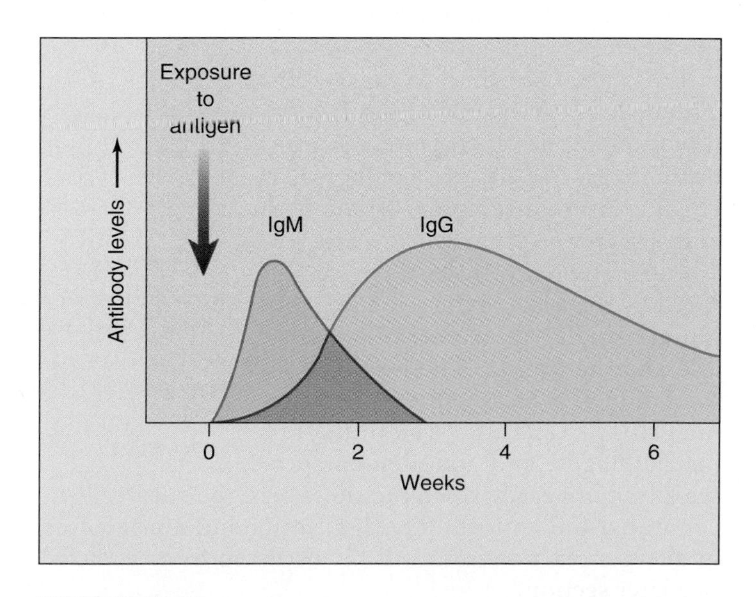

FIGURE 57.14
IgM and IgG antibodies. The first antibodies produced in the humoral immune response are IgM antibodies, which are very effective at activating the complement system. This initial wave of antibody production peaks after about one week and is followed by a far more extended production of IgG antibodies.

Antibodies

Structure of Antibodies

Each antibody molecule consists of two identical short polypeptides, called **light chains,** and two identical long polypeptides, called **heavy chains** (figure 57.15). The four chains in an antibody molecule are held together by disulfide (—S—S—) bonds, forming a Y-shaped molecule (figure 57.16).

Comparing the amino acid sequences of different antibody molecules shows that the specificity of antibodies for antigens resides in the two arms of the Y, which have a variable amino acid sequence. The amino acid sequence of the polypeptides in the stem of the Y is constant within a given class of immunoglobulins. Most of the sequence variation between antibodies of different specificity is found in the variable region of each arm. Here, a cleft forms that acts as the binding site for the antigen. Both arms always have exactly the same cleft and so bind to the same antigen.

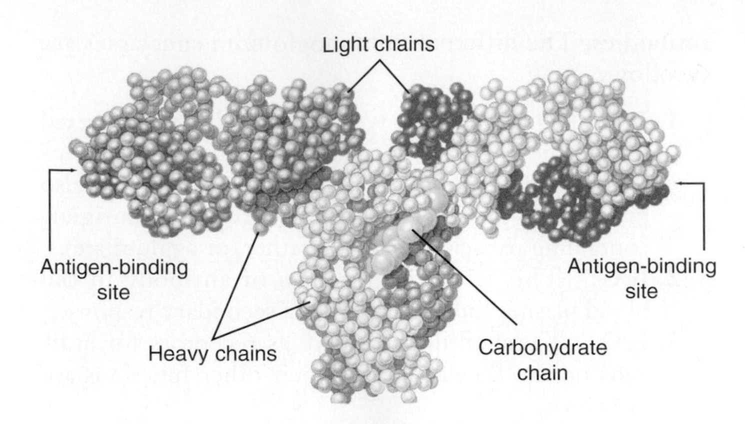

FIGURE 57.15
The structure of an antibody molecule. In this molecular model of an antibody molecule, each amino acid is represented by a small sphere. The heavy chains are colored *blue*; the light chains are *red*. The four chains wind about one another to form a Y shape, with two identical antigen-binding sites at the arms of the Y and a stem region that directs the antibody to a particular portion of the immune response.

Antibodies with the same variable segments have identical clefts and therefore recognize the same antigen, but they may differ in the stem portions of the antibody molecule. The stem is formed by the so-called "constant" regions of the heavy chains. In mammals there are five different classes of heavy chain that form five classes of immunoglobulins: IgM, IgG, IgA, IgD, and IgE. We have already discussed the roles of IgM and IgG antibodies in the humoral immune response.

IgE antibodies bind to **mast cells.** The heavy-chain stems of the IgE antibody molecules insert into receptors on the mast cell plasma membrane, in effect creating B receptors on the mast cell surface. When these cells encounter the specific antigen recognized by the arms of the antibody, they initiate the inflammatory response by releasing histamine. The resulting vasodilation and increased capillary permeability enable lymphocytes, macrophages, and complement proteins to more easily reach the site where the mast cell encountered the antigen. The IgE antibodies are involved in allergic reactions and will be discussed in more detail in a later section.

IgA antibodies are present in secretions such as milk, mucus, and saliva. In milk, these antibodies are thought to provide immune protection to nursing infants, whose own immune systems are not yet fully developed.

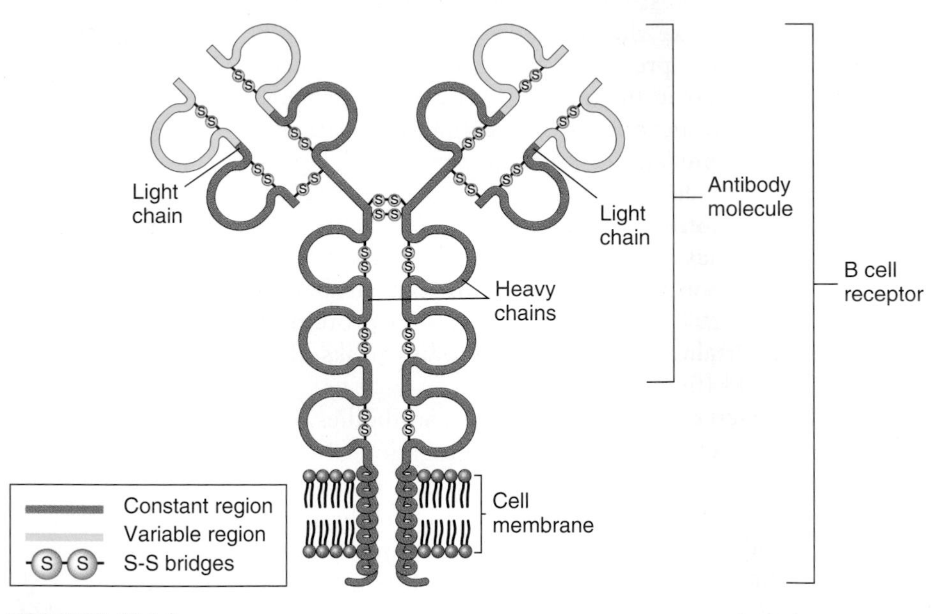

FIGURE 57.16
Structure of an antibody as a B cell receptor. The receptor molecules are characterized by domains of about 100 amino acids (represented as loops) joined by —S—S— covalent bonds. Each receptor has a constant region *(purple)* and a variable region *(yellow)*. The receptor binds to antigens at the ends of its two variable regions.

Antibody Diversity

The vertebrate immune system is capable of recognizing as foreign millions of nonself molecules presented to it. Although vertebrate chromosomes contain only a few hundred receptor-encoding genes, it is estimated that human B cells can make between 10^6 and 10^9 different antibody molecules. How do vertebrates generate millions of different antigen receptors when their chromosomes contain only a few hundred copies of the genes encoding those receptors?

The answer to this question is that in the B cell the millions of immune receptor genes do not have to be inherited at conception because they do not exist as single sequences of nucleotides. Rather, they are assembled by stitching together three or four DNA segments that code for different parts of the receptor molecule. When an antibody is assembled, the different sequences of DNA are brought together to form a composite gene (figure 57.17). This process is called **somatic rearrangement.** For example, combining DNA in different ways can produce 16,000 different heavy chains and about 1200 different light chains (in mouse antibodies).

Two other processes generate even more sequences. First, the DNA segments are often joined together with one or two nucleotides off-register, shifting the reading frame during gene transcription and so generating a totally different sequence of amino acids in the protein. Second, random mistakes occur during successive DNA replications as the lymphocytes divide during clonal expansion. Both mutational processes produce changes in amino acid sequences, a phenomenon known as **somatic mutation** because it takes place in a somatic cell, a B cell rather than in a gamete.

Because a B cell may end up with any heavy-chain gene and any light-chain gene during its maturation, the total number of different antibodies possible is staggering: 16,000 heavy-chain combinations × 1200 light-chain combinations = 19 million different possible antibodies. If one also takes into account the changes induced by somatic mutation, the total can exceed 200 million! It should be understood that, although this discussion has centered on B cells and their receptors, the receptors on T cells are as diverse as those on B cells because they also are subject to similar somatic rearrangements and mutations.

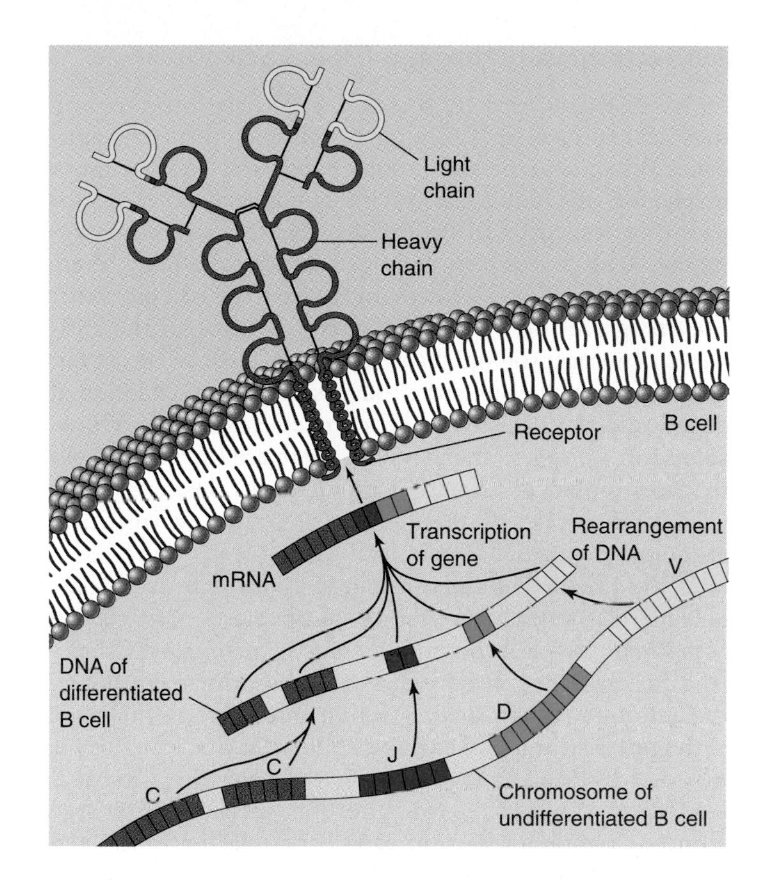

FIGURE 57.17
The lymphocyte receptor molecule is produced by a composite gene. Different regions of the DNA code for different regions of the receptor structure: *C*, constant regions; *J*, joining regions; *D*, diversity regions; and *V*, variable regions.These are brought together to make a composite gene that codes for the receptor. Through different somatic rearrangements of these DNA segments, an enormous number of different receptor molecules can be produced.

Immunological Tolerance

A mature animal's immune system normally does not respond to that animal's own tissue. This acceptance of self cells is known as **immunological tolerance.** The immune system of an embryo, on the other hand, is able to respond to both foreign and self molecules, but it loses the ability to respond to self molecules as its development proceeds. Indeed, if foreign tissue is introduced into an embryo before its immune system has developed, the mature animal that results will not recognize that tissue as foreign and will accept grafts of similar tissue without rejection.

There are two general mechanisms for immunological tolerance: clonal deletion and clonal suppression. During the normal maturation of hemopoietic stem cells in an embryo, fetus, or newborn, most lymphocyte clones that have receptors for self antigens are either eliminated (clonal deletion) or suppressed (clonal suppression). The cells "learn" to identify self antigens because the antigens are encountered very frequently. If a receptor is activated frequently, it is assumed that the cell is recognizing a self antigen and the lymphocytes are eliminated or suppressed. Thus, the only clones that survive this phase of development are those that are directed against foreign rather than self molecules.

Immunological tolerance sometimes breaks down, causing either B cells or T cells (or both) to recognize their own tissue antigens. This loss of immune tolerance results in autoimmune disease. Myasthenia gravis, for example, is an autoimmune disease in which individuals produce antibodies directed against acetylcholine receptors on their own skeletal muscle cells, causing paralysis. Autoimmunity will be discussed in more detail later in this chapter.

An antibody molecule is composed of constant and variable regions. The variable regions recognize a specific antigen because they possess clefts into which the antigen can fit. Lymphocyte receptors are encoded by genes that are assembled by somatic rearrangement and mutation of the DNA.

Active Immunity through Clonal Selection

As we discussed earlier, B and T cells have receptors on their cell surfaces that recognize and bind to specific antigens. When a particular antigen enters the body, it must, by chance, encounter the specific lymphocyte with the appropriate receptor in order to provoke an immune response. The first time a pathogen invades the body, there are only a few B or T cells that may have the receptors that can recognize the invader's antigens. Binding of the antigen to its receptor on the lymphocyte surface, however, stimulates cell division and produces a *clone* (a population of genetically identical cells). This process is known as **clonal selection.** In this first encounter, there are only a few cells that can mount an immune response and the response is relatively weak. This is called a **primary immune response** (figure 57.18).

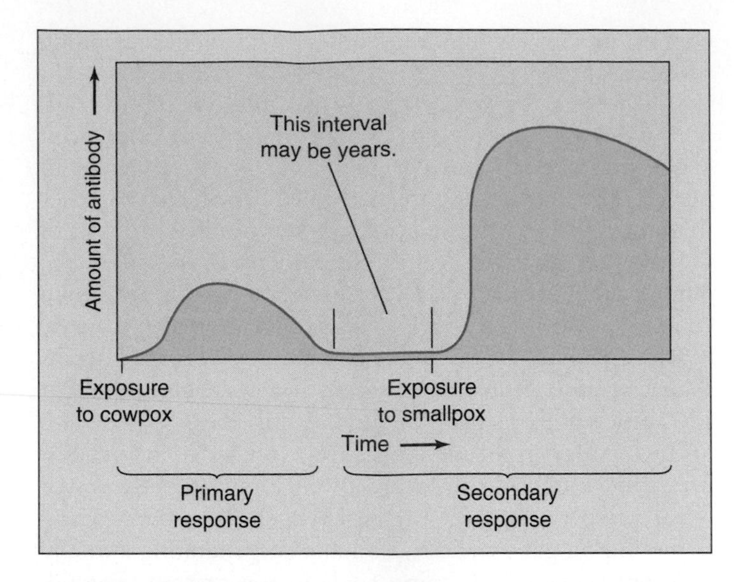

FIGURE 57.18
The development of active immunity. Immunity to smallpox in Jenner's patients occurred because their inoculation with cowpox stimulated the development of lymphocyte clones with receptors that could bind not only to cowpox but also to smallpox antigens. As a result of clonal selection, a second exposure, this time to smallpox, stimulates the immune system to produce large amounts of the antibody more rapidly than before.

If the primary immune response involves B cells, some become plasma cells that secrete antibodies, and some become memory cells. Because a clone of memory cells specific for that antigen develops after the primary response, the immune response to a second infection by the same pathogen is swifter and stronger. The next time the body is invaded by the same pathogen, the immune system is ready. As a result of the first infection, there is now a large clone of lymphocytes that can recognize that pathogen (figure 57.19). This more effective response, elicited by subsequent exposures to an antigen, is called a **secondary immune response.**

Memory cells can survive for several decades, which is why people rarely contract chicken pox a second time after they have had it once. Memory cells are also the reason that vaccinations are effective. The vaccine triggers the primary response so that if the actual pathogen is encountered later, the large and rapid secondary response occurs and stops the infection before it can start. The viruses causing childhood diseases have surface antigens that change little from year to year, so the same antibody is effective for decades.

Figure 57.20 summarizes how the cellular and humoral lines of defense work together to produce the body's specific immune response.

Active immunity is produced by clonal selection and expansion. This occurs because interaction of an antigen with its receptor on the lymphocyte surface stimulates cell division, so that more lymphocytes are available to combat subsequent exposures to the same antigen.

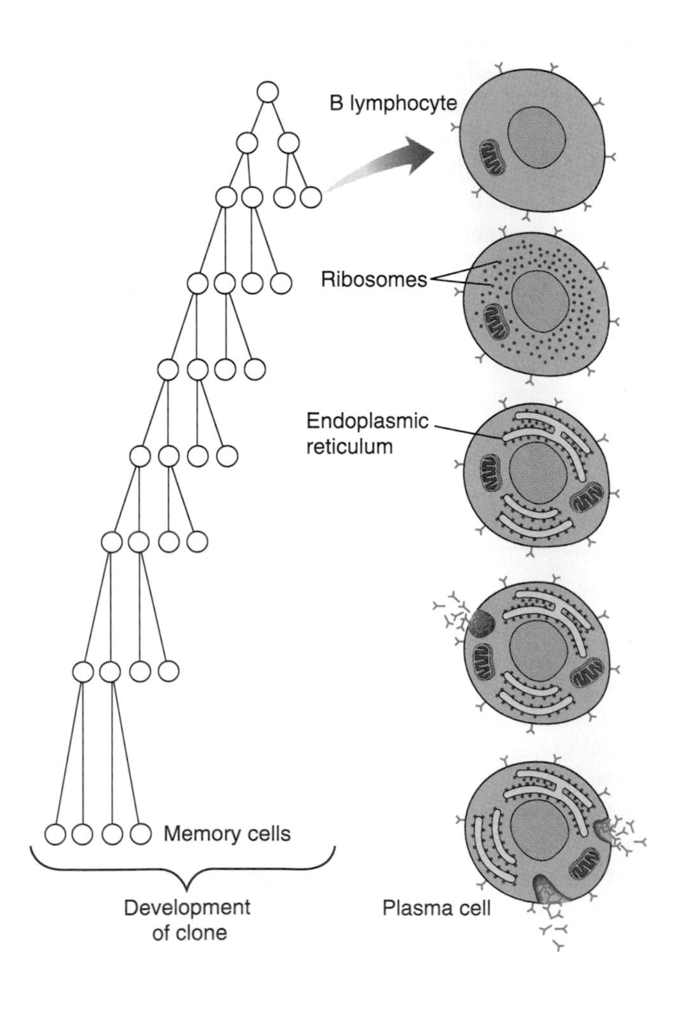

FIGURE 57.19
The clonal selection theory of active immunity. In response to interaction with an antigen that binds specifically to its surface receptors, a B cell divides many times to produce a clone of B cells. Some of these become plasma cells that secrete antibodies for the primary response, while others become memory cells that await subsequent exposures to the antigen for the mounting of a secondary immune response.

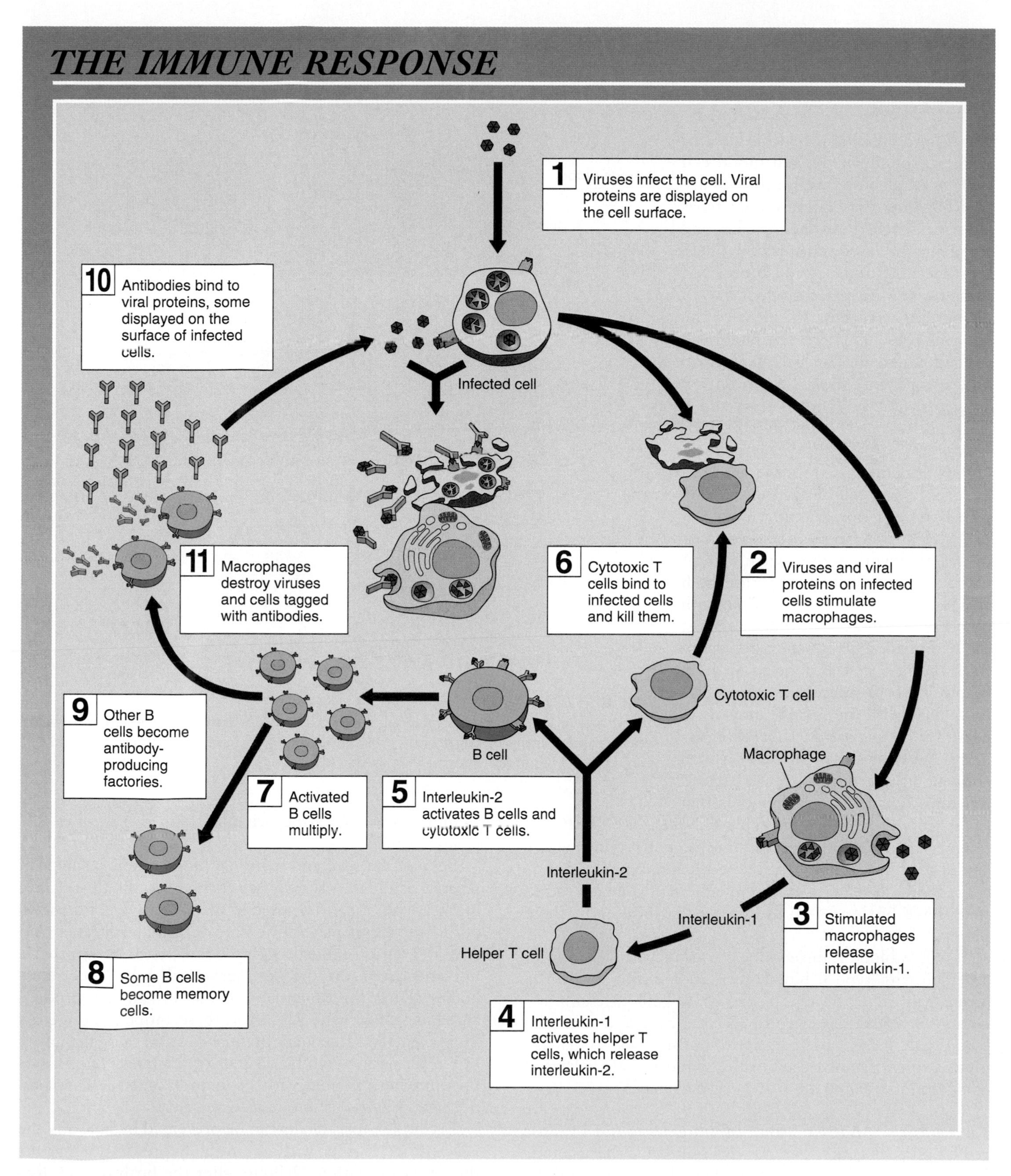

THE IMMUNE RESPONSE
1 Viruses infect the cell. Viral proteins are displayed on the cell surface.
10 Antibodies bind to viral proteins, some displayed on the surface of infected cells.
Infected cell
11 Macrophages destroy viruses and cells tagged with antibodies.
6 Cytotoxic T cells bind to infected cells and kill them.
2 Viruses and viral proteins on infected cells stimulate macrophages.
Cytotoxic T cell
9 Other B cells become antibody-producing factories.
B cell
Macrophage
7 Activated B cells multiply.
5 Interleukin-2 activates B cells and cytotoxic T cells.
Interleukin-2
Interleukin-1
3 Stimulated macrophages release interleukin-1.
Helper T cell
8 Some B cells become memory cells.
4 Interleukin-1 activates helper T cells, which release interleukin-2.

FIGURE 57.20
Overview of the specific immune response.

Antibodies in Medical Diagnosis

Blood Typing

The blood type denotes the class of antigens found on the red blood cell surface. Red blood cell antigens are clinically important because their types must be matched between donors and recipients for blood transfusions. There are several groups of red blood cell antigens, but the major group is known as the **ABO system.** In terms of the antigens present on the red blood cell surface, a person may be *type A* (with only A antigens), *type B* (with only B antigens), *type AB* (with both A and B antigens), or *type O* (with neither A nor B antigens).

The immune system is tolerant to its own red blood cell antigens. A person who is type A, for example, does not produce anti-A antibodies. Surprisingly, however, people with type A blood do make antibodies against the B antigen, and conversely, people with blood type B make antibodies against the A antigen. This is believed to result from the fact that antibodies made in response to some common bacteria cross-react with the A or B antigens. A person who is type A, therefore, acquires antibodies that can react with B antigens by exposure to these bacteria but does not develop antibodies that can react with A antigens. People who are type AB develop tolerance to both antigens and thus do not produce either anti-A or anti-B antibodies. Those who are type O, in contrast, do not develop tolerance to either antigen and, therefore, have both anti-A and anti-B antibodies in their plasma.

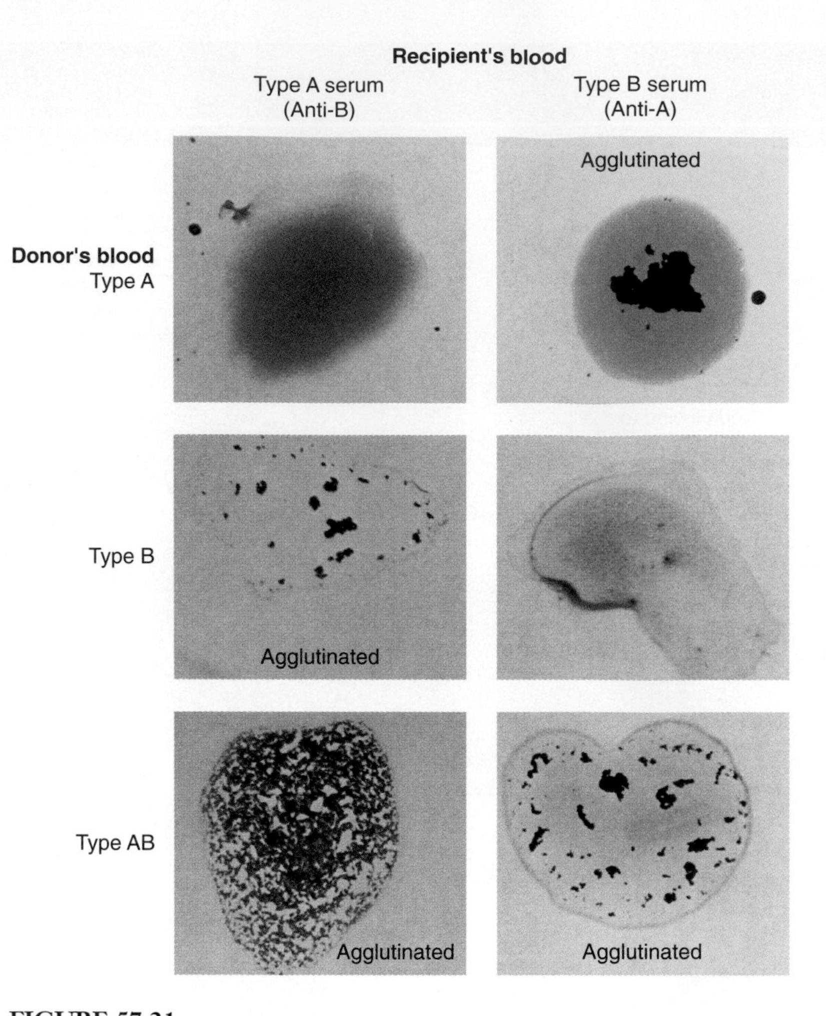

FIGURE 57.21
Blood typing. Agglutination of the red blood cells is seen when blood types are mixed with sera containing antibodies against the ABO antigens. Note that no agglutination would be seen if type O blood (not shown) were used.

If type A blood is mixed on a glass slide with serum from a person with type B blood, the anti-A antibodies in the serum will cause the type A red blood cells to clump together, or **agglutinate** (figure 57.21). These tests allow the blood types to be matched prior to transfusions, so that agglutination will not occur in the blood vessels, where it could lead to inflammation and organ damage.

Rh Factor. Another group of antigens found in most red blood cells is the *Rh factor* (Rh stands for rhesus monkey, in which these antigens were first discovered). People who have these antigens are said to be **Rh-positive,** whereas those who do not are **Rh-negative.** There are fewer Rh-negative people because this condition is recessive to Rh-positive. The Rh factor is of particular significance when Rh-negative mothers give birth to Rh-positive babies.

Because the fetal and maternal blood are normally kept separate across the placenta (see chapter 60), the Rh-negative mother is not usually exposed to the Rh antigen of the fetus during the pregnancy. At the time of birth, however, a variable degree of exposure may occur, and the mother's immune system may become sensitized and produce antibodies against the Rh antigen. If the woman does produce antibodies against the Rh factor, these antibodies can cross the placenta in subsequent pregnancies and cause hemolysis of the Rh-positive red blood cells of the fetus. The baby is therefore born anemic, with a condition called *erythroblastosis fetalis,* or *hemolytic disease of the newborn.*

Erythroblastosis fetalis can be prevented by injecting the Rh-negative mother with an antibody preparation against the Rh factor within 72 hours after the birth of each Rh-positive baby. This is a type of passive immunization in which the injected antibodies inactivate the Rh antigens and thus prevent the mother from becoming actively immunized to them.

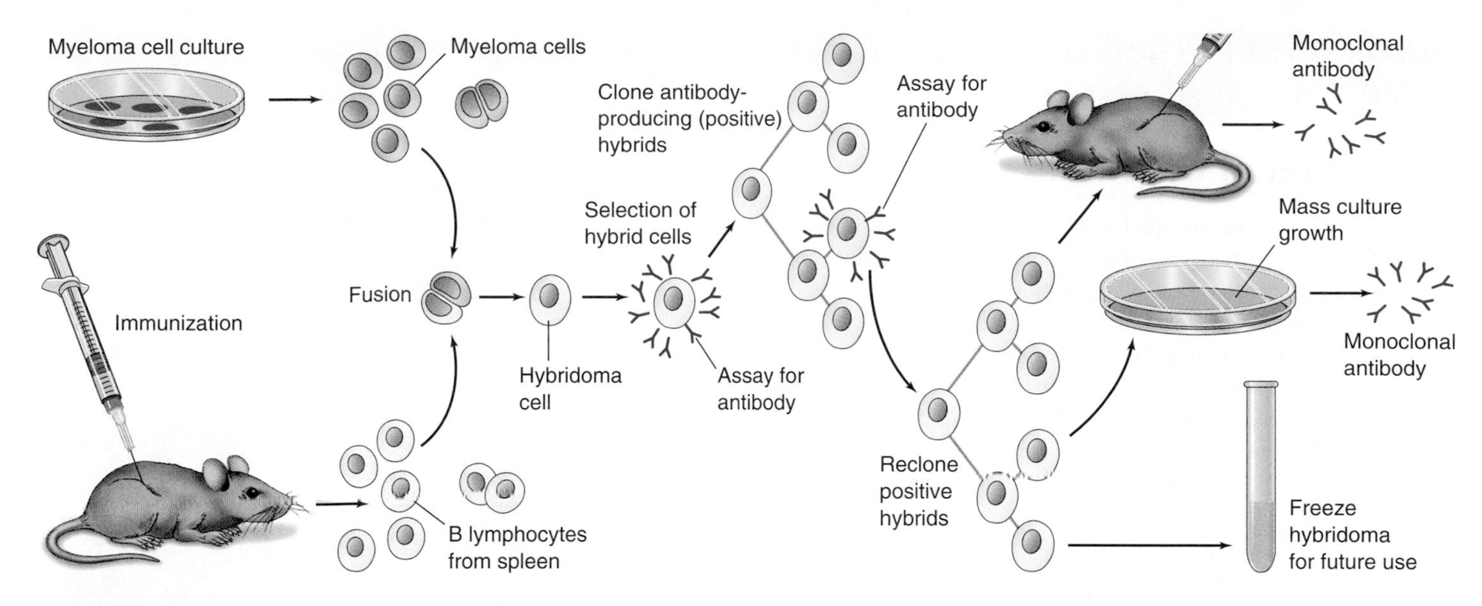

FIGURE 57.22
The production of monoclonal antibodies. These antibodies are produced by cells that arise from successive divisions of a single B cell, and hence all of the antibodies target a single antigenic determinant site. Such antibodies are used for a variety of medical applications, including pregnancy testing.

Monoclonal Antibodies

Antibodies are commercially prepared for use in medical diagnosis and research. In the past, antibodies were obtained by chemically purifying a specific antigen and then injecting this antigen into animals. However, because an antigen typically has many different antigenic determinant sites, the antibodies obtained by this method were *polyclonal;* they stimulated the development of different B-cell clones with different specificities. This decreased their sensitivity to a particular antigenic site and resulted in some degree of cross-reaction with closely related antigen molecules.

Monoclonal antibodies, by contrast, exhibit specificity for one antigenic determinant only. In the preparation of monoclonal antibodies, an animal (frequently, a mouse) is injected with an antigen and subsequently killed. B lymphocytes are then obtained from the animal's spleen and placed in thousands of different in vitro incubation vessels. These cells soon die, however, unless they are hybridized with cancerous multiple myeloma cells. The fusion of a B lymphocyte with a cancerous cell produces a hybrid that undergoes cell division and produces a clone called a *hybridoma.* Each hybridoma secretes large amounts of identical, monoclonal antibodies. From among the thousands of hybridomas produced in this way, the one that produces the desired antibody is cultured for large-scale production, and the rest are discarded (figure 57.22).

The availability of large quantities of pure monoclonal antibodies has resulted in the development of much more sensitive clinical laboratory tests. Modern pregnancy tests, for example, use particles (latex rubber or red blood cells) that are covered with monoclonal antibodies produced against a pregnancy hormone (abbreviated hCG—see chapter 59) as the antigen. When these particles are mixed with a sample that contains this hormone antigen from a pregnant woman, the antigen-antibody reaction causes a visible agglutination of the particles (figure 57.23).

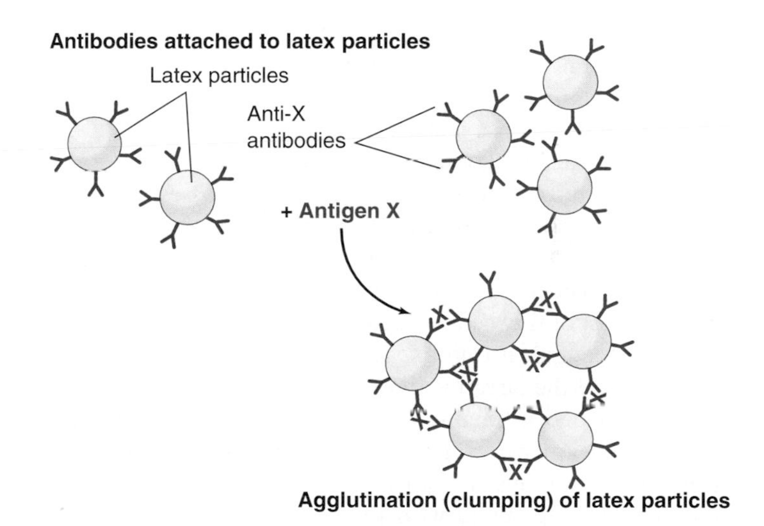

FIGURE 57.23
Using monoclonal antibodies to detect an antigen. In many clinical tests (such as pregnancy testing), the monoclonal antibodies are bound to particles of latex, which agglutinate in the presence of the antigen.

Agglutination occurs because different antibodies exist for the ABO and Rh factor antigens on the surface of red blood cells. Monoclonal antibodies are commercially produced antibodies that react against one specific antigen.

57.5 All animals exhibit nonspecific immune response but specific ones evolved in vertebrates.

Evolution of the Immune System

All organisms possess mechanisms to protect themselves from the onslaught of smaller organisms and viruses. Bacteria defend against viral invasion by means of *restriction endonucleases*, enzymes that degrade any foreign DNA lacking the specific pattern of DNA methylation characteristic of that bacterium. Multicellular organisms face a more difficult problem in defense because their bodies often take up whole viruses, bacteria, or fungi instead of naked DNA.

Invertebrates

Invertebrate animals solve this problem by marking the surfaces of their cells with proteins that serve as "self" labels. Special amoeboid cells in the invertebrate attack and engulf any invading cells that lack such labels. By looking for the absence of specific markers, invertebrates employ a *negative* test to recognize foreign cells and viruses. This method provides invertebrates with a very effective surveillance system, although it has one great weakness: any microorganism or virus with a surface protein resembling the invertebrate self marker will not be recognized as foreign. An invertebrate has no defense against such a "copycat" invader.

In 1882, Russian zoologist Elie Metchnikoff became the first to recognize that invertebrate animals possess immune defenses. On a beach in Sicily, he collected the tiny transparent larva of a common starfish. Carefully he pierced it with a rose thorn. When he looked at the larva the next morning, he saw a host of tiny cells covering the surface of the thorn as if trying to engulf it (figure 57.24). The cells were attempting to defend the larva by ingesting the invader by phagocytosis (described in chapter 6). For this discovery of what came to be known as the **cellular immune response,** Metchnikoff was awarded the 1908 Nobel Prize in Physiology or Medicine, along with Paul Ehrlich for his work on the other major part of the immune defense, the antibody or **humoral immune response.** The invertebrate immune response shares several elements with the vertebrate one.

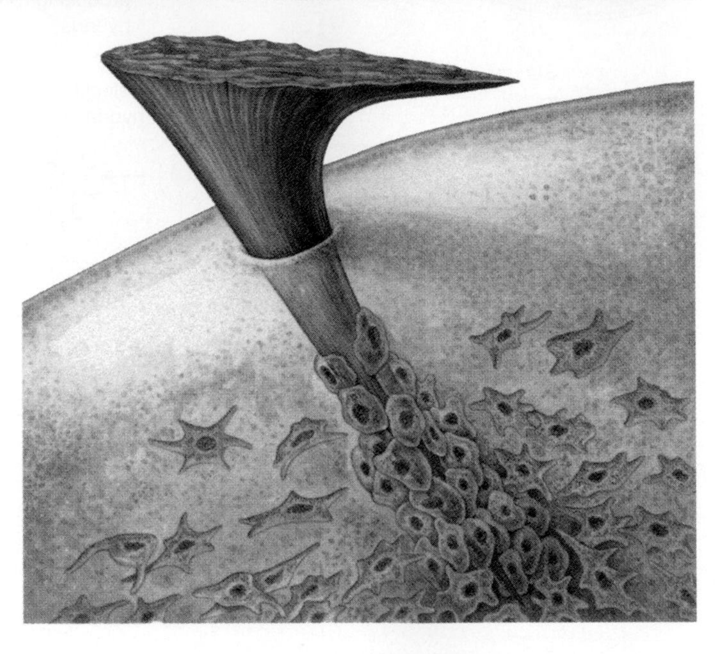

FIGURE 57.24
Discovering the cellular immune response in invertebrates. In a Nobel-Prize-winning experiment, the Russian zoologist Metchnikoff pierced the larva of a starfish with a rose thorn and the next day found tiny phagocytic cells covering the thorn.

Phagocytes. All animals possess phagocytic cells that attack invading microbes. These phagocytic cells travel through the animal's circulatory system or circulate within the fluid-filled body cavity. In simple animals like sponges that lack either a circulatory system or a body cavity, the phagocytic cells circulate among the spaces between cells.

Distinguishing Self from Nonself. The ability to recognize the difference between cells of one's own body and those of another individual appears to have evolved early in the history of life. Sponges, thought to be the oldest animals, attack grafts from other sponges, as do insects and starfish. None of these invertebrates, however, exhibit any evidence of immunological memory; apparently, the antibody-based humoral immune defense did not evolve until the vertebrates.

Complement. While invertebrates lack complement, many arthropods (including crabs and a variety of insects) possess an analogous nonspecific defense called the prophenyloxidase (proPO) system. Like the vertebrate complement defense, the proPO defense is activated as a cascade of enzyme reactions, the last of which converts the inactive protein prophenyloxidase into the active enzyme phenyloxidase. Phenyloxidase both kills microbes and aids in encapsulating foreign objects.

Lymphocytes. Invertebrates also lack lymphocytes, but annelid earthworms and other invertebrates do possess lymphocyte-like cells that may be evolutionary precursors of lymphocytes.

Antibodies. All invertebrates possess proteins called lectins that may be the evolutionary forerunners of antibodies. Lectins bind to sugar molecules on cells, making the cells stick to one another. Lectins isolated from sea urchins, mollusks, annelids, and insects appear to tag invading microorganisms, enhancing phagocytosis. The genes encoding vertebrate antibodies are part of a very ancient gene family, the immunoglobulin superfamily. Proteins in

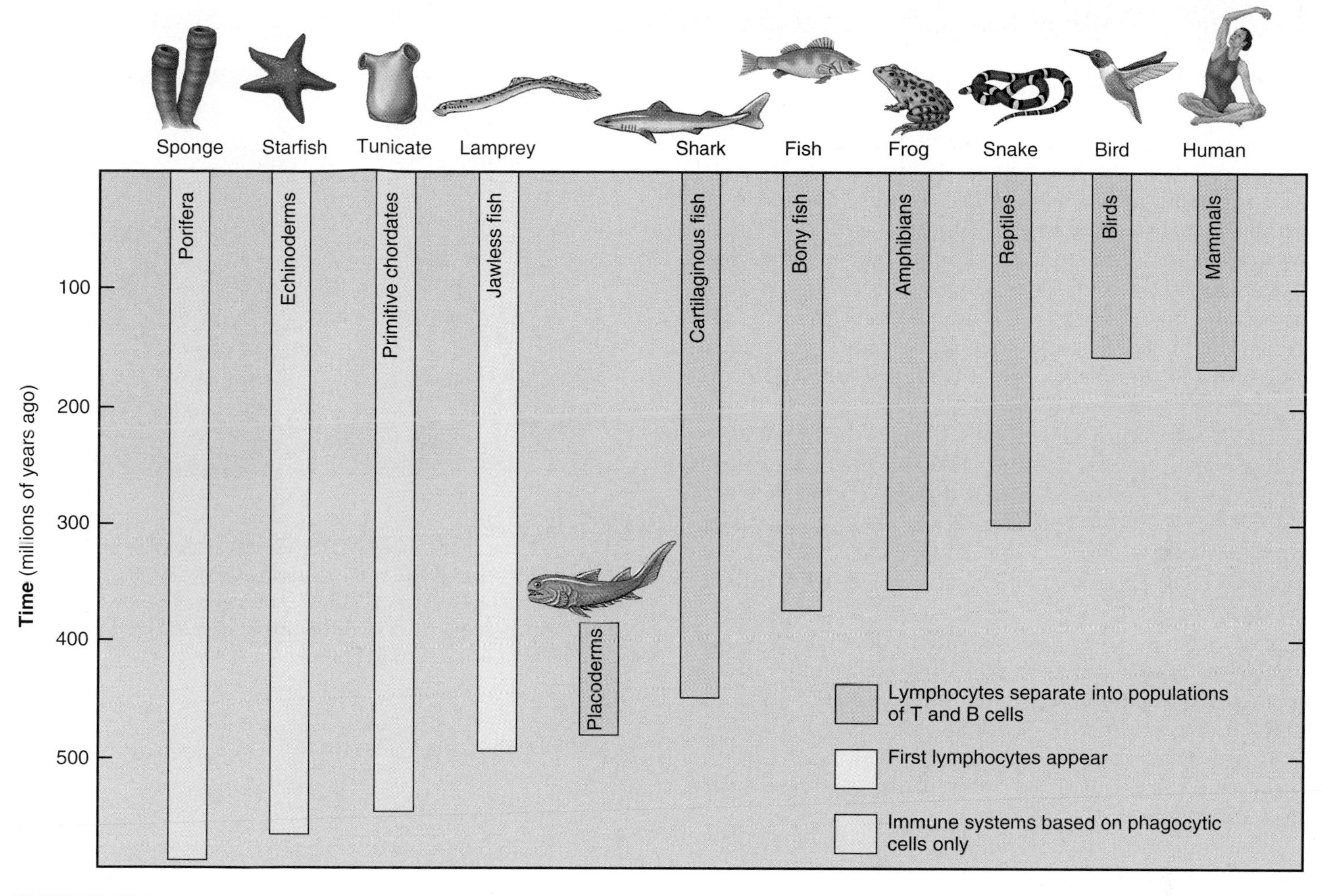

FIGURE 57.25
How immune systems evolved. Lampreys were the first vertebrates to possess an immune system based on lymphocytes, although distinct B and T cells did not appear until the jawed fishes evolved. By the time sharks and other cartilaginous fish appeared, the vertebrate immune response was fully formed.

this group all have a characteristic recognition structure called the Ig fold. The fold probably evolved as a self-recognition molecule in early metazoans. Insect immunoglobulins have been described in moths, grasshoppers, and flies that bind to microbial surfaces and promote their destruction by phagocytes. The antibody immune response appears to have evolved from these earlier, less complex systems.

Vertebrates

The earliest vertebrates of which we have any clear information, the jawless lampreys that first evolved some 500 million years ago, possess an immune system based on lymphocytes. At this early stage of vertebrate evolution, however, lampreys lack distinct populations of B and T cells such as found in all higher vertebrates (figure 57.25).

With the evolution of fish with jaws, the modern vertebrate immune system first arose. The oldest surviving group of jawed fishes are the sharks, which evolved some 450 million years ago. By then, the vertebrate immune defense had fully evolved. Sharks have an immune response much like that seen in mammals, with a cellular response carried out by T-cell lymphocytes and an antibody-mediated humoral response carried out by B cells. The similarities of the cellular and humoral immune defenses are far more striking than the differences. Both sharks and mammals possess a thymus that produces T cells and a spleen that is a rich source of B cells. Four hundred fifty million years of evolution did little to change the antibody molecule—the amino acid sequences of shark and human antibody molecules are very similar. The most notable difference between sharks and mammals is that their antibody-encoding genes are arrayed somewhat differently.

The sophisticated two-part immune defense of mammals evolved about the time jawed fishes appeared. Before then, animals utilized a simpler immune defense based on mobile phagocytic cells.

57.6 The immune system can be defeated.

T Cell Destruction: AIDS

One mechanism for defeating the vertebrate immune system is to attack the immune mechanism itself. Helper T cells and inducer T cells are $CD4^+$ T cells. Therefore, any pathogen that inactivates $CD4^+$ T cells leaves the immune system unable to mount a response to *any* foreign antigen. Acquired immune deficiency syndrome (AIDS) is a deadly disease for just this reason. The AIDS retrovirus, called human immunodeficiency virus (HIV), mounts a direct attack on $CD4^+$ T cells because it recognizes the CD4 coreceptors associated with these cells.

HIV's attack on $CD4^+$ T cells cripples the immune system in at least three ways. First, HIV-infected cells die only after releasing replicated viruses that infect other $CD4^+$ T cells, until the entire population of $CD4^+$ T cells is destroyed (figure 57.26). In a normal individual, $CD4^+$ T cells make up 60 to 80% of circulating T cells; in AIDS patients, $CD4^+$ T cells often become too rare to detect (figure 57.27). Second, HIV causes infected $CD4^+$ T cells to secrete a soluble suppressing factor that blocks other T cells from responding to the HIV antigen. Finally, HIV may block transcription of MHC genes, hindering the recognition and destruction of infected $CD4^+$ T cells and thus protecting those cells from any remaining vestiges of the immune system.

The combined effect of these responses to HIV infection is to wipe out the human immune defense. With no defense against infection, any of a variety of otherwise commonplace infections proves fatal. With no ability to recognize and destroy cancer cells when they arise, death by cancer becomes far more likely. Indeed, AIDS was first recognized as a disease because of a cluster of cases of an unusually rare form of cancer. More AIDS victims die of cancer than from any other cause.

Although HIV became a human disease vector only recently, possibly through transmission to humans from chimpanzees in Central Africa, it is already clear that AIDS is one of the most serious diseases in human history (figure 57.28). The fatality rate of AIDS is 100%; no patient exhibiting the symptoms of AIDS has ever been known to survive more than a few years without treatment. Aggressive treatments can prolong life but how much longer has not been determined. However, the disease is *not* highly contagious, as it is transmitted from one individual to another through the transfer of internal body fluids, typically in semen and in blood during transfusions. Not all individuals exposed to HIV (as judged by anti-HIV antibodies in their blood) have yet acquired the disease.

Until recently, the only effective treatment for slowing the progression of the disease involved treatment with drugs such as AZT that inhibit the activity of reverse transcriptase, the enzyme needed by the virus to produce DNA from RNA. Recently, however, a new type of drug has

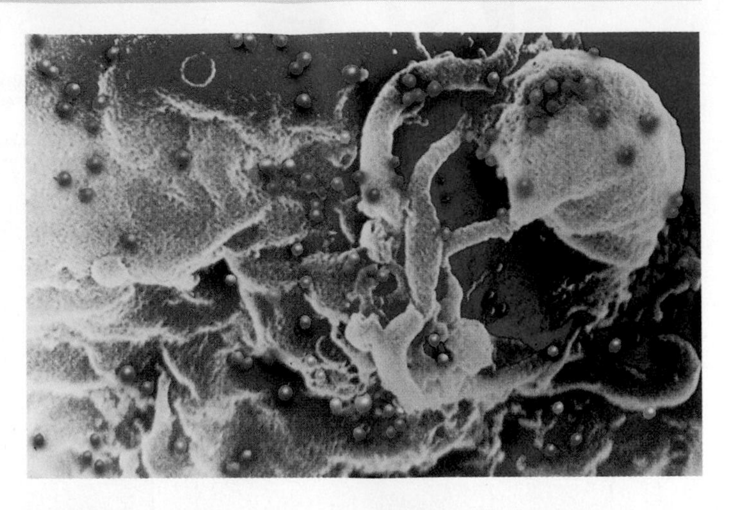

FIGURE 57.26
HIV, the virus that causes AIDS. Viruses released from infected $CD4^+$ T cells soon spread to neighboring $CD4^+$ T cells, infecting them in turn. The individual viruses, colored *red* in this scanning electron micrograph, are extremely small; over 200 million would fit on the period at the end of this sentence.

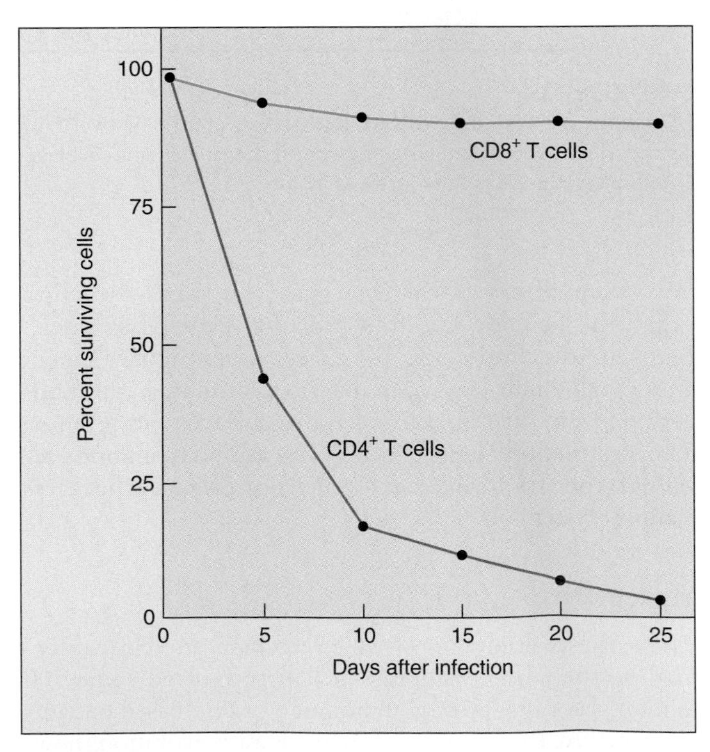

FIGURE 57.27
Survival of T cells in culture after exposure to HIV. The virus has little effect on the number of $CD8^+$ T cells, but it causes the number of $CD4^+$ T cells (this group includes helper T cells) to decline dramatically.

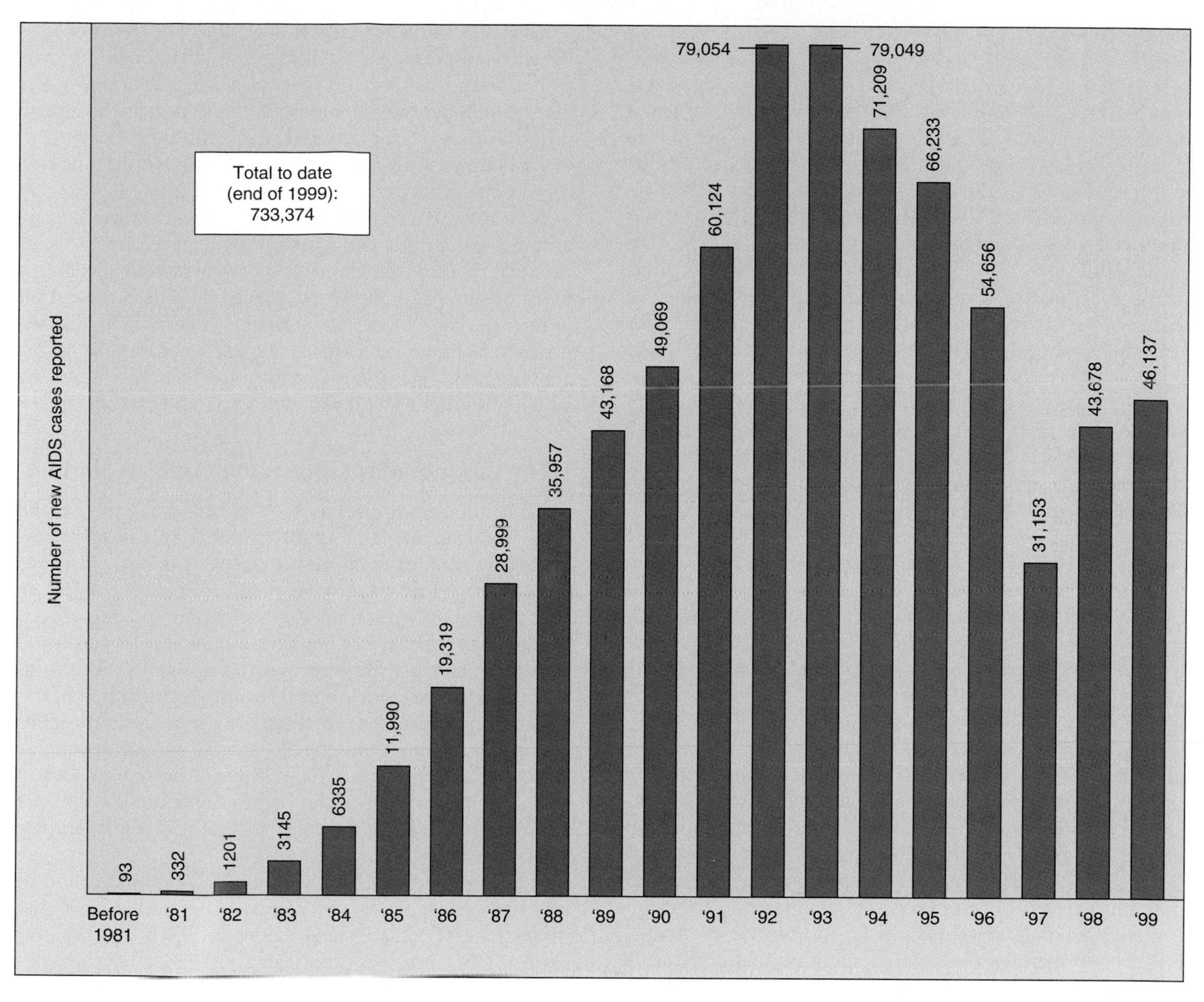

FIGURE 57.28
The AIDS epidemic in the United States: new cases. The U.S. Centers for Disease Control and Prevention (CDC) reports that 43,678 new AIDS cases were reported in 1998 and 46,137 new cases in 1999, with a total of 733,374 cases and 390,692 deaths in the United States. Over 1.5 million other individuals are thought to be infected with the HIV virus in the United States, and 14 million worldwide. The 100,000th AIDS case was reported in August 1989, eight years into the epidemic; the next 100,000 cases took just 26 months; the third 100,000 cases took barely 19 months (May 1993), and the fourth 100,000 took only 13 months (June 1994). The extraordinarily high numbers seen in 1992 reflect an expansion of the definition of what constitutes an AIDS case.
Source: Data from U.S. Centers for Disease Control and Prevention, Atlanta, GA.

become available that acts to inhibit protease, an enzyme needed for viral assembly. Treatments that include a combination of reverse transcriptase inhibitors and protease inhibitors (p. 672) appear to lower levels of HIV, though they are very costly. Efforts to develop a vaccine against AIDS continue, both by splicing portions of the HIV surface protein gene into vaccinia virus and by attempting to develop a harmless strain of HIV. These approaches, while promising, have not yet proved successful and are limited by the fact that different strains of HIV seem to possess different surface antigens. Like the influenza virus, HIV engages in some form of antigen shifting, making it difficult to develop an effective vaccine.

AIDS destroys the ability of the immune system to mount a defense against any infection. HIV, the virus that causes AIDS, induces a state of immune deficiency by attacking and destroying $CD4^+$ T cells.

Antigen Shifting

A second way that a pathogen may defeat the immune system is to mutate frequently so that it varies the nature of its surface antigens. The virus which causes influenza uses this mechanism, and so we have to be immunized against a different strain of this virus periodically. This way of escaping immune attack is known as antigen shifting, and is practiced very effectively by trypanosomes, the protists responsible for sleeping sickness (see chapter 35). Trypanosomes possess several thousand different versions of the genes encoding their surface protein, but the cluster containing these genes has no promoter and so is not transcribed as a unit. The necessary promoter is located within a transposable element that jumps at random from one position to another within the cluster, transcribing a different surface protein gene with every move. Because such moves occur in at least one cell of an infective trypanosome population every few weeks, the vertebrate immune system is unable to mount an effective defense against trypanosome infection. By the time a significant number of antibodies have been generated against one form of trypanosome surface protein, another form is already present in the trypanosome population that survives immunological attack, and the infection cycle is renewed. People with sleeping sickness rarely rid themselves of the infection.

Although this mechanism of mutation to alter surface proteins seems very "directed" or intentional on the part of the pathogen, it is actually the process of evolution by natural selection at work. We usually think of evolution as requiring thousands of years to occur, and not in the time frame of weeks. However, evolution can occur whenever mutations are passed on to offspring that provide an organism with a competitive advantage. In the case of viruses, bacteria, and other pathogenic agents, their generation times are on the order of hours. Thus, in the time frame of a week, the population has gone through millions of cell divisions. Looking at it from this perspective, it is easy to see how random mutations in the genes for the surface antigens could occur and change the surface of the pathogen in as little as a week's time.

How Malaria Hides from the Immune System

Every year, about a half-million people become infected with the protozoan parasite *Plasmodium falciparum*, which multiplies in their bodies to cause the disease malaria. The plasmodium parasites enter the red blood cells and consume the hemoglobin of their hosts. Normally this sort of damage to a red blood cell would cause the damaged cell to be transported to the spleen for disassembly, destroying the plasmodium as well. The plasmodium avoids this fate, however, by secreting knoblike proteins that extend through the surface of the red blood cell and anchor the cell to the inner surface of the blood vessel.

Over the course of several days, the immune system of the infected person slowly brings the infection under control. During this time, however, a small proportion of the plasmodium parasites change their knob proteins to a form different from those that sensitized the immune system. Cells infected with these individuals survive the immune response, only to start a new wave of infection.

Scientists have recently discovered how the malarial parasite carries out this antigen-shifting defense. About 6% of the total DNA of the plasmodium is devoted to encoding a block of some 150 *var* genes, which are shifted on and off in multiple combinations. Each time a plasmodium divides, it alters the pattern of *var* gene expression about 2%, an incredibly rapid rate of antigen shifting. The exact means by which this is done is not yet completely understood.

DNA Vaccines May Get Around Antigen Shifting

Vaccination against diseases such as smallpox, measles, and polio involves introducing into your body a dead or disabled pathogen, or a harmless microbe with pathogen proteins displayed on its surface. The vaccination triggers an immune response against the pathogen, and the bloodstream of the vaccinated person contains B cells which will remember and quickly destroy the pathogen in future infections. However, for some diseases, vaccination is nearly impossible because of antigen shifting; the pathogens change over time, and the B cells no longer recognize them. Influenza, as we have discussed, presents different surface proteins yearly. The trypanosomes responsible for sleeping sickness change their surface proteins every few weeks.

A new type of vaccine, based on DNA, may prove to be effective against almost any disease. The vaccine makes use of the killer T cells instead of the B cells of the immune system. DNA vaccines consist of a plasmid, a harmless circle of bacterial DNA, that contains a gene from the pathogen that encodes an internal protein, one which is critical to the function of the pathogen and does not change. When this plasmid is injected into cells, the gene they carry is transcribed into protein but is not incorporated into the DNA of the cell's nucleus. Fragments of the pathogen protein are then stuck on the cell's membrane, marking it for destruction by T cells. In actual infections later, the immune system will be able to respond immediately. Studies are now underway to isolate the critical, unchanging proteins of pathogens and to investigate fully the use of the vaccines in humans.

Antigen shifting refers to the way a pathogen may defeat the immune system by changing its surface antigens and thereby escaping immune recognition. Pathogens that employ this mechanism include flu viruses, trypanosomes, and the protozoans that cause malaria.

Autoimmunity and Allergy

The previous section described ways that pathogens can elude the immune system to cause diseases. There is another way the immune system can fail; it can itself be the agent of disease. Such is the case with autoimmune diseases and allergies—the immune system is the cause of the problem, not the cure.

Autoimmune Diseases

Autoimmune diseases are produced by failure of the immune system to recognize and tolerate self antigens. This failure results in the activation of autoreactive T cells and the production of autoantibodies by B cells, causing inflammation and organ damage. There are over 40 known or suspected autoimmune diseases that affect 5 to 7% of the population. For reasons that are not understood, two-thirds of the people with autoimmune diseases are women.

Autoimmune diseases can result from a variety of mechanisms. The self antigen may normally be hidden from the immune system, for example, so that the immune system treats it as foreign if exposure later occurs. This occurs when a protein normally trapped in the thyroid follicles triggers autoimmune destruction of the thyroid (Hashimoto's thyroiditis). It also occurs in systemic lupus erythematosus, in which antibodies are made to nucleoproteins. Because the immune attack triggers inflammation, and inflammation causes organ damage, the immune system must be suppressed to alleviate the symptoms of autoimmune diseases. Immune suppression is generally accomplished with corticosteroids (including hydrocortisone) and by nonsteroidal antiinflammatory drugs, including aspirin.

Allergy

The term *allergy*, often used interchangeably with *hypersensitivity*, refers to particular types of abnormal immune responses to antigens, which are called *allergens* in these cases. There are two major forms of allergy: (1) **immediate hypersensitivity,** which is due to an abnormal B cell response to an allergen that produces symptoms within seconds or minutes, and (2) **delayed hypersensitivity,** which is an abnormal T cell response that produces symptoms within about 48 hours after exposure to an allergen.

Immediate hypersensitivity results from the production of antibodies of the IgE subclass instead of the normal IgG antibodies. Unlike IgG antibodies, IgE antibodies do not circulate in the blood. Instead, they attach to tissue mast cells and basophils, which have membrane receptors for these antibodies. When the person is again exposed to the same allergen, the allergen binds to the antibodies attached to the mast cells and basophils. This stimulates these cells to secrete various chemicals, including histamine, which produce the symptom of the allergy (figure 57.29).

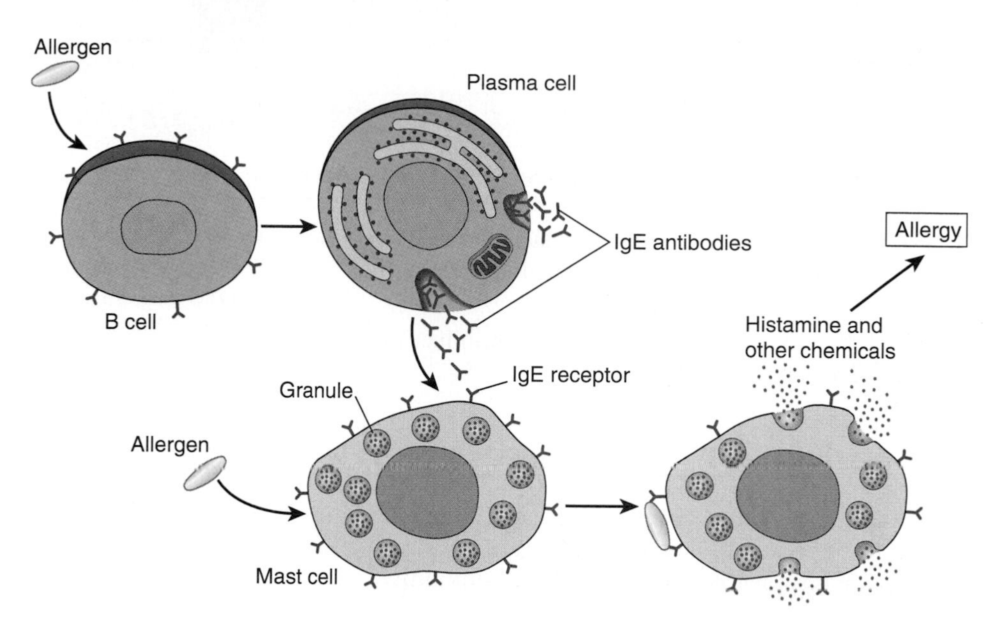

FIGURE 57.29
An allergic reaction. This is an immediate hypersensitivity response, in which B cells secrete antibodies of the IgE class. These antibodies attach to the plasma membranes of mast cells, which secrete histamine in response to antigen-antibody binding.

Allergens that provoke immediate hypersensitivity include various foods, bee stings, and pollen grains. The most common allergy of this type is seasonal hay fever, which may be provoked by ragweed *(Ambrosia)* pollen grains. These allergic reactions are generally mild, but in some allergies (as to penicillin or peanuts in susceptible people) the widespread and excessive release of histamine may cause **anaphylactic shock,** an uncontrolled fall in blood pressure.

In delayed hypersensitivity, symptoms take a longer time (hours to days) to develop than in immediate hypersensitivity. This may be due to the fact that immediate hypersensitivity is mediated by antibodies, whereas delayed hypersensitivity is a T cell response. One of the best-known examples of delayed hypersensitivity is **contact dermatitis,** caused by poison ivy, poison oak, and poison sumac. Because the symptoms are caused by the secretion of lymphokines rather than by the secretion of histamine, treatment with antihistamines provides little benefit. At present, corticosteroids are the only drugs that can effectively treat delayed hypersensitivity.

Autoimmune diseases are produced when the immune system fails to tolerate self antigens.

Chapter 57

www.mhhe.com/raven6/resources57.mhtml

Summary	Questions	Media Resources

57.1 Many of the body's most effective defenses are nonspecific.

- Nonspecific defenses include physical barriers such as the skin as well as phagocytic cells, killer cells, and complement proteins.
- The inflammatory response aids the mobilization of defensive cells at infected sites.

1. How do macrophages destroy foreign cells?

2. How does the complement system participate in defense against infection?

- Lymph System
- Nonspecific Immunity

57.2 Specific immune defenses require the recognition of antigens.

- Lymphocytes called B cells secrete antibodies and produce the humoral response; lymphocytes called T cells are responsible for cell-mediated immunity.

3. On what types of cells are the two classes of MHC proteins found?

- Scientists on Science: Integration of Science and People
- Student Research: Improving Antibodies

57.3 T cells organize attacks against invading microbes.

- T cells only respond to antigens presented to them by macrophages or other antigen-presenting cells together with MHC proteins.
- Cytotoxic T cells kill cells that have foreign antigens presented together with MHC-I proteins.

4. In what two ways do macrophages activate helper T cells? How do helper T cells stimulate the proliferation of cytotoxic T cells?

- T Cell Function

57.4 B cells label specific cells for destruction.

- The antibody molecules consist of two heavy and two light polypeptide regions arranged like a "Y"; the ends of the two arms bind to antigens.
- An individual can produce a tremendous variety of different antibodies because the genes which produce those antibodies recombine extensively.
- Active immunity occurs when an individual gains immunity by prior exposure to a pathogen; passive immunity is produced by the transfer of antibodies from one individual to another.

5. How do IgM and IgG antibodies differ in triggering destruction of infected cells?

6. How does the clonal selection model help to explain active immunity?

7. How are lymphocytes able to produce millions of different types of immune receptors?

- Activity: Plasma Cell Production

- Art Quiz: Monoclonal Antibody Production

57.5 All animals exhibit nonspecific immune response but specific ones evolved in vertebrates.

- The immune system evolved in animals from a strictly nonspecific immune response in invertebrates to the two-part immune defense found in mammals.

8. Compare insect and mammalian immune defenses.

- Phagocytic Cells

57.6 The immune system can be defeated.

- Flu viruses, trypanosomes, and the protozoan that causes malaria are able to evade the immune system by mutating the genes that produce their surface antigens. In autoimmune diseases, the immune system targets the body's own antigens.

9. How does HIV defeat human immune defenses?

10. What might cause an immune attack of self antigens?

- Abnormalities

58

Maintaining the Internal Environment

Concept Outline

58.1 The regulatory systems of the body maintain homeostasis.

The Need to Maintain Homeostasis. Regulatory mechanisms maintain homeostasis through negative feedback loops.

Antagonistic Effectors and Positive Feedback. Antagonistic effectors cause opposite changes, while positive feedback pushes changes further in the same way.

58.2 The extracellular fluid concentration is constant in most vertebrates.

Osmolality and Osmotic Balance. Vertebrates have to cope with the osmotic gain or loss of body water.

Osmoregulatory Organs. Invertebrates have a variety of organs to regulate water balance; kidneys are the osmoregulatory organs of most vertebrates.

Evolution of the Vertebrate Kidney. Freshwater bony fish produce a dilute urine and marine bony fish produce an isotonic urine. Only birds and mammals can retain so much water that they produce a concentrated urine.

58.3 The functions of the vertebrate kidney are performed by nephrons.

The Mammalian Kidney. Each kidney contains nephrons that produce a filtrate which is modified by reabsorption and secretion to produce urine.

Transport Processes in the Mammalian Nephron. The nephron tubules of birds and mammals have loops of Henle, which function to draw water out of the tubule and back into the blood.

Ammonia, Urea, and Uric Acid. The breakdown of protein and nucleic acids yields nitrogen, which is excreted as ammonia in bony fish, as urea in mammals, and as uric acid in reptiles and birds.

58.4 The kidney is regulated by hormones.

Hormones Control Homeostatic Functions. Antidiuretic hormone promotes water retention and the excretion of a highly concentrated urine. Aldosterone stimulates the retention of salt and water, whereas atrial natriuretic hormone promotes the excretion of salt and water.

FIGURE 58.1
Regulating body temperature with water. One of the ways an elephant can regulate its temperature is to spray water on its body. Water also cycles through the elephant's body in enormous quantities each day and helps to regulate its internal environment.

The first vertebrates evolved in seawater, and the physiology of all vertebrates reflects this origin. Approximately two-thirds of every vertebrate's body is water. If the amount of water in the body of a vertebrate falls much lower than this, the animal will die. In this chapter, we discuss the various mechanisms by which animals avoid gaining or losing too much water. As we shall see, these mechanisms are closely tied to the way animals exploit the varied environments in which they live and to the regulatory systems of the body (figure 58.1).

58.1 The regulatory systems of the body maintain homeostasis.

The Need to Maintain Homeostasis

As the animal body has evolved, specialization has increased. Each cell is a sophisticated machine, finely tuned to carry out a precise role within the body. Such specialization of cell function is possible only when extracellular conditions are kept within narrow limits. Temperature, pH, the concentrations of glucose and oxygen, and many other factors must be held fairly constant for cells to function efficiently and interact properly with one another.

Homeostasis may be defined as the dynamic constancy of the internal environment. The term *dynamic* is used because conditions are never absolutely constant, but fluctuate continuously within narrow limits. Homeostasis is essential for life, and most of the regulatory mechanisms of the vertebrate body that are not devoted to reproduction are concerned with maintaining homeostasis.

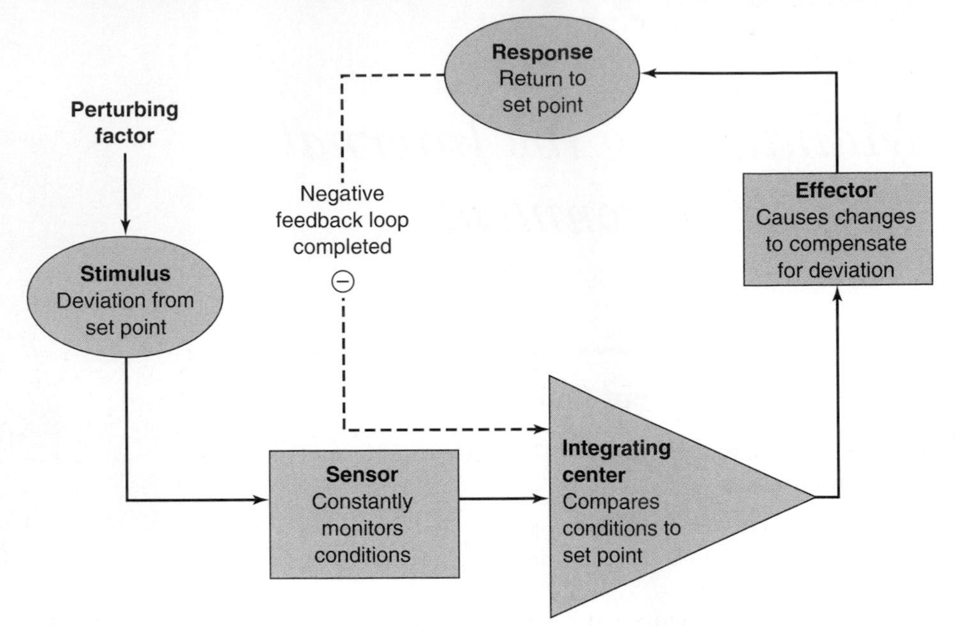

FIGURE 58.2
A generalized diagram of a negative feedback loop. Negative feedback loops maintain a state of homeostasis, or dynamic constancy of the internal environment, by correcting deviations from a set point.

Negative Feedback Loops

To maintain internal constancy, the vertebrate body must have **sensors** that are able to measure each condition of the internal environment (figure 58.2). These constantly monitor the extracellular conditions and relay this information (usually via nerve signals) to an **integrating center,** which contains the "*set point*" (the proper value for that condition). This set point is analogous to the temperature setting on a house thermostat. In a similar manner, there are set points for body temperature, blood glucose concentration, the tension on a tendon, and so on. The integrating center is often a particular region of the brain or spinal cord, but in some cases it can also be cells of endocrine glands. It receives messages from several sensors, weighing the relative strengths of each sensor input, and then determines whether the value of the condition is deviating from the set point. When a deviation in a condition occurs (the "stimulus"), the integrating center sends a message to increase or decrease the activity of particular **effectors**. Effectors are generally muscles or glands, and can change the value of the condition in question back toward the set point value (the "response").

To return to the idea of a home thermostat, suppose you set the thermostat at a set point of 70°F. If the temperature of the house rises sufficiently above the set point, the thermostat (equivalent to an integrating center) receives this input from a temperature sensor, like a thermometer (a sensor) within the wall unit. It compares the actual temperature to its set point. When these are different, it sends a signal to an effector. The effector in this case may be an air conditioner, which acts to reverse the deviation from the set point.

In a human, if the body temperature exceeds the set point of 37°C, sensors in a part of the brain detect this deviation. Acting via an integrating center (also in the brain), these sensors stimulate effectors (including sweat glands) that lower the temperature (figure 58.3). One can think of the effectors as "defending" the set points of the body against deviations. Because the activity of the effectors is influenced by the effects they produce, and because this regulation is in a negative, or reverse, direction, this type of control system is known as a *negative feedback loop.*

The nature of the negative feedback loop becomes clear when we again refer to the analogy of the thermostat and air conditioner. After the air conditioner has been on for some time, the room temperature may fall significantly below the set point of the thermostat. When this occurs, the air conditioner will be turned off. The effector (air conditioner) is turned on by a high temperature; and, when activated, it produces a negative change (lowering of the temperature) that ultimately causes the effector to be turned off. In this way, constancy is maintained.

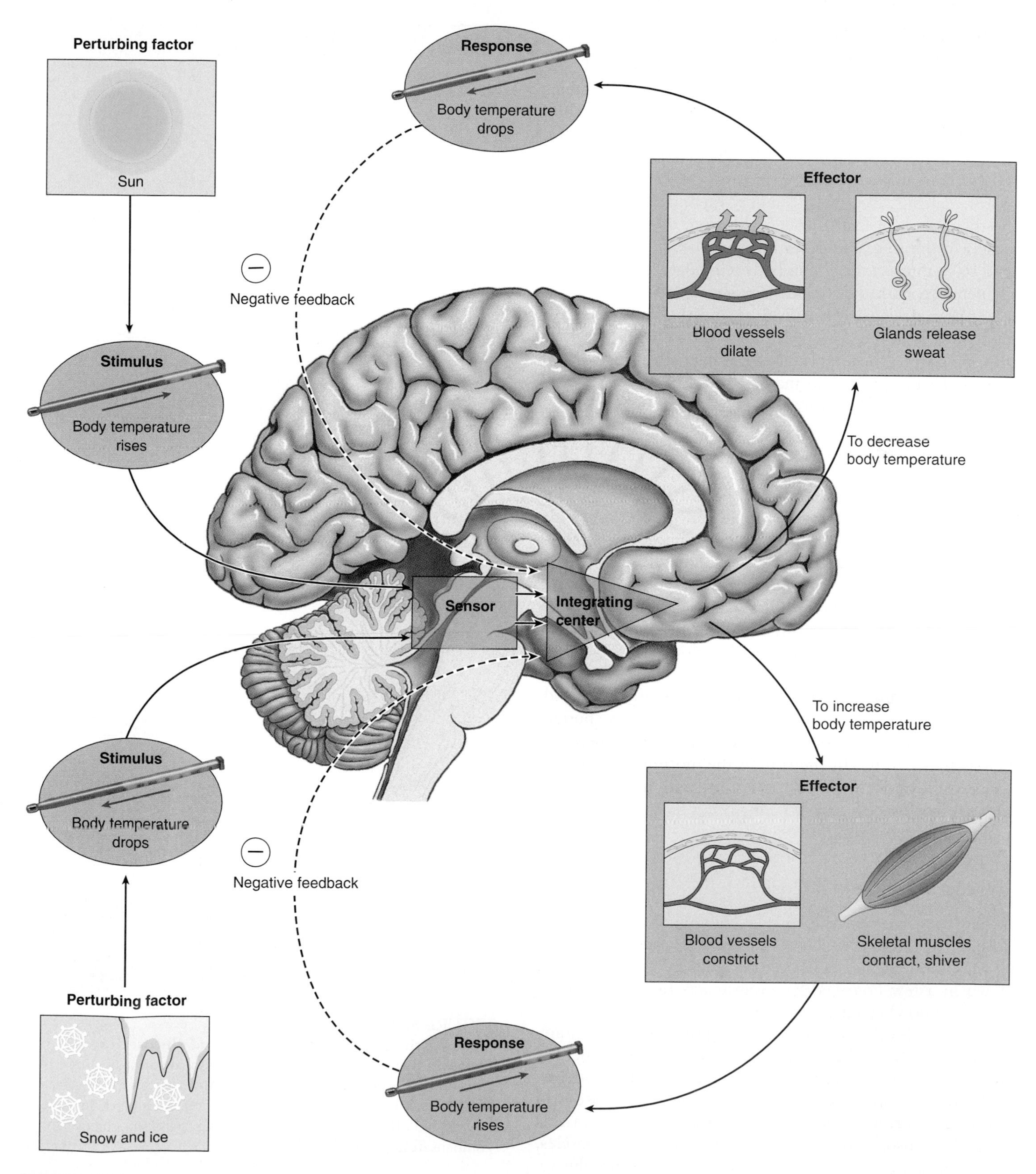

FIGURE 58.3
Negative feedback loops keep the body temperature within a normal range. An increase (*top*) or decrease (*bottom*) in body temperature is sensed by the brain. The integrating center in the brain then processes the information and activates effectors, such as surface blood vessels, sweat glands, and skeletal muscles. When the body temperature returns to normal, negative feedback prevents further stimulation of the effectors by the integrating center.

Regulating Body Temperature

Humans, together with other mammals and with birds, are *endothermic*; they can maintain relatively constant body temperatures independent of the environmental temperature. When the temperature of your blood exceeds 37°C (98.6°F), neurons in a part of the brain called the hypothalamus detect the temperature change. Acting through the control of motor neurons, the hypothalamus responds by promoting the dissipation of heat through sweating, dilation of blood vessels in the skin, and other mechanisms. These responses tend to counteract the rise in body temperature. When body temperature falls, the hypothalamus coordinates a different set of responses, such as shivering and the constriction of blood vessels in the skin, which help to raise body temperature and correct the initial challenge to homeostasis.

Vertebrates other than mammals and birds are *ectothermic*; their body temperatures are more or less dependent on the environmental temperature. However, to the extent that it is possible, many ectothermic vertebrates attempt to maintain some degree of temperature homeostasis. Certain large fish, including tuna, swordfish, and some sharks, for example, can maintain parts of their body at a significantly higher temperature than that of the water. Reptiles attempt to maintain a constant body temperature through behavioral means—by placing themselves in varying locations of sun and shade (see chapter 29). That's why you frequently see lizards basking in the sun. Sick lizards even give themselves a "fever" by seeking warmer locations!

Most invertebrates do not employ feedback regulation to physiologically control their body temperature. Instead, they use behavior to adjust their temperature. Many butterflies, for example, must reach a certain body temperature before they can fly. In the cool of the morning they orient so as to maximize their absorption of sunlight. Moths and many other insects use a shivering reflex to warm their thoracic flight muscles (figure 58.4).

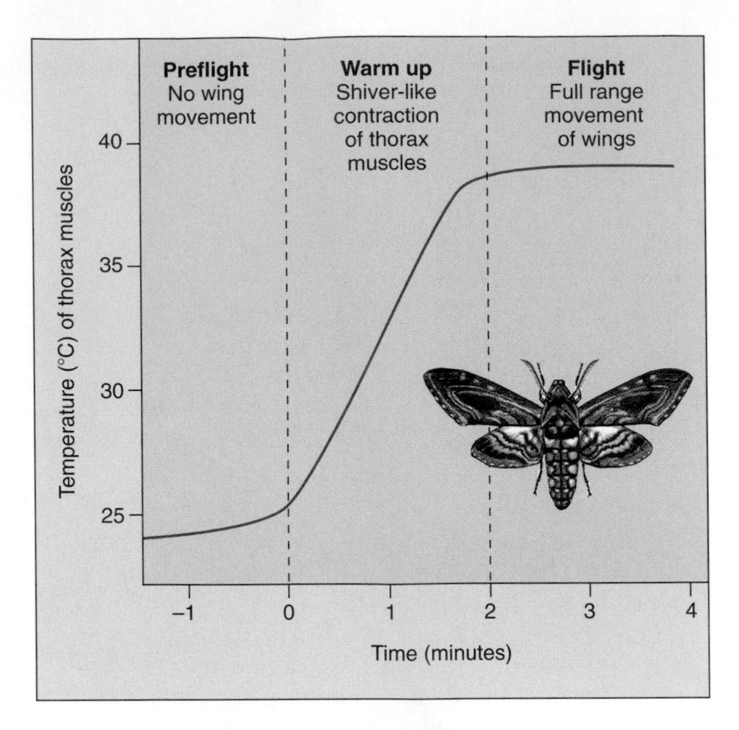

FIGURE 58.4
Thermoregulation in insects. Some insects, such as the sphinx moth, contract their thoracic muscles to warm up for flight.
Source: Data from B. Heinrich, *Science*, American Association for the Advancement of Science.

Regulating Blood Glucose

When you digest a carbohydrate-containing meal, you absorb glucose into your blood. This causes a temporary rise in the blood glucose concentration, which is brought back down in a few hours. What counteracts the rise in blood glucose following a meal?

Glucose levels within the blood are constantly monitored by a sensor, the islets of Langerhans in the pancreas. When levels increase, the islets secrete the hormone *insulin*, which stimulates the uptake of blood glucose into muscles, liver, and adipose tissue. The islets are, in this case, the sensor and the integrating center. The muscles, liver, and adipose cells are the effectors, taking up glucose to control the levels. The muscles and liver can convert the glucose into the polysaccharide glycogen; adipose cells can convert glucose into fat. These actions lower the blood glucose (figure 58.5) and help to store energy in forms that the body can use later.

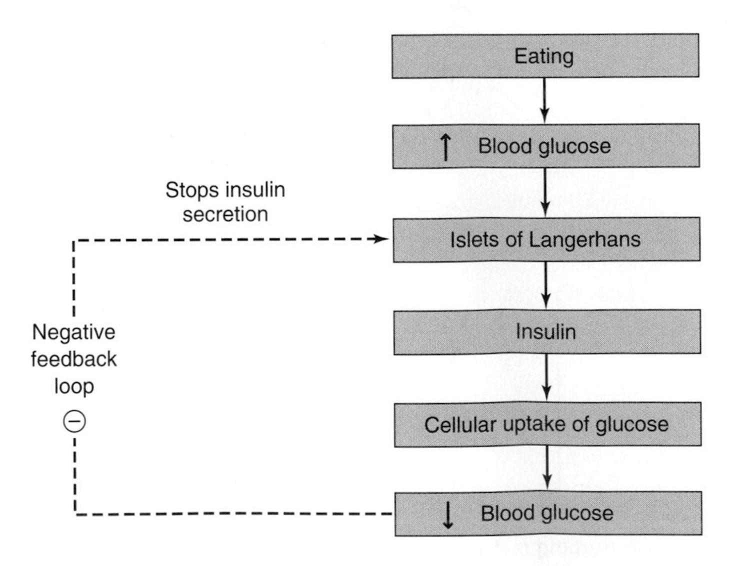

FIGURE 58.5
The negative feedback control of blood glucose. The rise in blood glucose concentration following a meal stimulates the secretion of insulin from the islets of Langerhans in the pancreas. Insulin is a hormone that promotes the entry of glucose in skeletal muscle and other tissue, thereby lowering the blood glucose and compensating for the initial rise.

Negative feedback mechanisms correct deviations from a set point for different internal variables. In this way, body temperature and blood glucose, for example, are kept within a normal range.

Antagonistic Effectors and Positive Feedback

The negative feedback mechanisms that maintain homeostasis often oppose each other to produce a finer degree of control. In a few cases positive feedback mechanisms, which push a change further in the same direction, are used by the body.

Antagonistic Effectors

Most factors in the internal environment are controlled by several effectors, which often have antagonistic actions. Control by antagonistic effectors is sometimes described as "push-pull," in which the increasing activity of one effector is accompanied by decreasing activity of an antagonistic effector. This affords a finer degree of control than could be achieved by simply switching one effector on and off.

Room temperature can be maintained, for example, by simply turning an air conditioner on and off, or by just turning a heater on and off. A much more stable temperature, however, can be achieved if the air conditioner and heater are both controlled by a thermostat (figure 58.6). Then the heater is turned on when the air conditioner shuts off, and vice versa. Antagonistic effectors are similarly involved in the control of body temperature and blood glucose. Whereas insulin, for example, lowers blood glucose following a meal, other hormones act to raise the blood glucose concentration between meals, especially when a person is exercising. The heart rate is similarly controlled by antagonistic effectors. Stimulation of one group of nerve fibers increases the heart rate, while stimulation of another group slows the heart rate.

Positive Feedback Loops

Feedback loops that accentuate a disturbance are called positive feedback loops. In a positive feedback loop, perturbations cause the effector to drive the value of the controlled variable even *farther* from the set point. Hence, systems in which there is positive feedback are highly unstable, analogous to a spark that ignites an explosion. They do not help to maintain homeostasis. Nevertheless, such systems are important components of some physiological mechanisms. For example, positive feedback occurs in blood clotting, where one clotting factor activates another in a cascade that leads quickly to the formation of a clot. Positive feedback also plays a role in the contractions of the uterus during childbirth (figure 58.7). In this case, stretching of the uterus by the fetus stimulates contraction, and contraction causes further stretching; the cycle continues until the fetus is expelled from the uterus. In the body, most positive feedback systems act as part of some larger mechanism that maintains homeostasis. In the examples we've described, formation of a blood clot stops bleeding and hence tends to keep blood volume constant, and expulsion of the fetus reduces the contractions of the uterus.

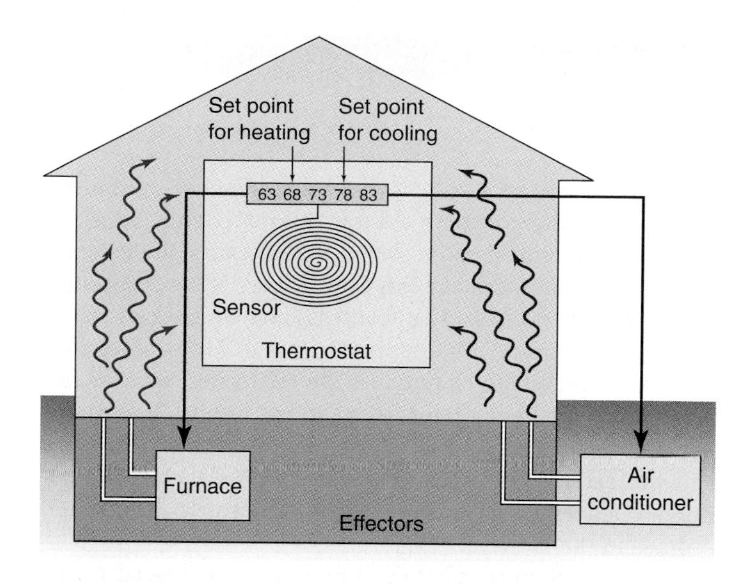

FIGURE 58.6
Room temperature is maintained by antagonistic effectors. If a thermostat senses a low temperature, the heater is turned on and the air conditioner is turned off. If the temperature is too high, the air conditioner is activated, and the heater is turned off.

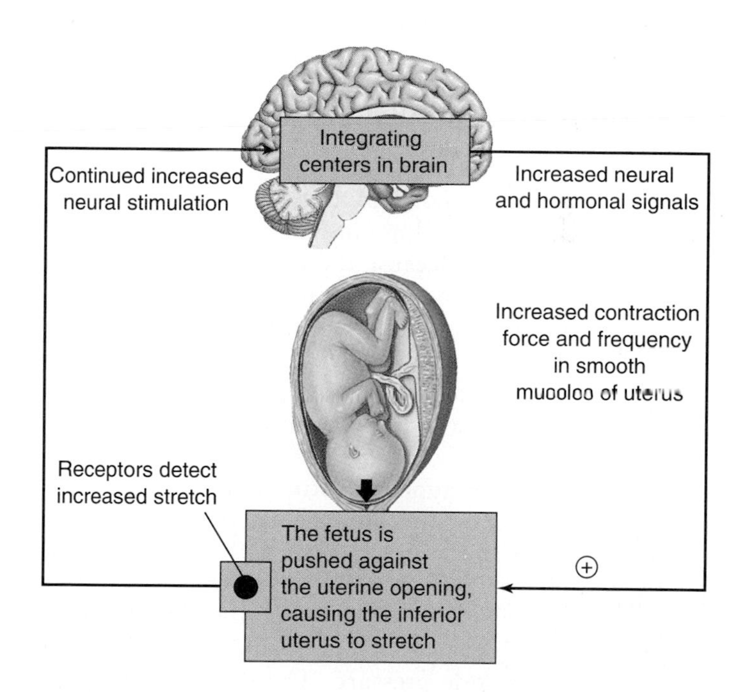

FIGURE 58.7
An example of positive feedback during childbirth. This is one of the few examples of positive feedback that operate in the vertebrate body.

Antagonistic effectors that act antagonistically to each other are more effective than effectors that act alone. Positive feedback mechanisms accentuate changes and have limited functions in the body.

58.2 The extracellular fluid concentration is constant in most vertebrates.

Osmolality and Osmotic Balance

Water in an animal's body is distributed between the intracellular and extracellular compartments (figure 58.8). In order to maintain osmotic balance, the extracellular compartment of an animal's body (including its blood plasma) must be able to take water from its environment or to excrete excess water into its environment. Inorganic ions must also be exchanged between the extracellular body fluids and the external environment to maintain homeostasis. Such exchanges of water and electrolytes between the body and the external environment occur across specialized epithelial cells and, in most vertebrates, through a filtration process in the kidneys.

Most vertebrates maintain homeostasis in regard to the total solute concentration of their extracellular fluids and in regard to the concentration of specific inorganic ions. Sodium (Na^+) is the major cation in extracellular fluids, and chloride (Cl^-) is the major anion. The divalent cations, calcium (Ca^{++}) and magnesium (Mg^{++}), as well as other ions, also have important functions and must be maintained at their proper concentrations.

Osmolality and Osmotic Pressure

Osmosis is the diffusion of water across a membrane, and it always occurs from a more dilute solution (with a lower solute concentration) to a less dilute solution (with a higher solute concentration). Because the total solute concentration of a solution determines its osmotic behavior, the total moles of solute per kilogram of water is expressed as the **osmolality** of the solution. Solutions that have the same osmolality are *isosmotic*. A solution with a lower or higher osmolality than another is called *hypoosmotic* or *hyperosmotic*, respectively.

If one solution is hyperosmotic compared with another, and if the two solutions are separated by a semipermeable membrane, water may move by osmosis from the more dilute solution to the hyperosmotic one. In this case, the hyperosmotic solution is also **hypertonic** ("higher strength") compared with the other solution, and it has a higher osmotic pressure. The **osmotic pressure** of a solution is a measure of its tendency to take in water by osmosis. A cell placed in a hypertonic solution, which has a higher osmotic pressure than the cell cytoplasm, will lose water to the surrounding solution and shrink. A cell placed in a **hypotonic** solution, in contrast, will gain water and expand.

If a cell is placed in an isosmotic solution, there may be no net water movement. In this case, the isosmotic solution can also be said to be **isotonic.** Isotonic solutions such as normal saline and 5% dextrose are used in medical care to bathe exposed tissues and to be given as intravenous fluids.

Osmoconformers and Osmoregulators

Most marine invertebrates are **osmoconformers;** the osmolality of their body fluids is the same as that of seawater (although the concentrations of particular solutes, such as magnesium ion, are not equal). Because the extracellular fluids are isotonic to seawater, there is no osmotic gradient and no tendency for water to leave or enter the body. Therefore, osmoconformers are in osmotic equilibrium with their environment. Among the vertebrates, only the primitive hagfish are strict osmoconformers. The sharks and their relatives in the class Chondrichthyes (cartilaginous fish) are also isotonic to seawater, even though their blood level of NaCl is lower than that of seawater; the difference in total osmolality is made up by retaining urea at a high concentration in their blood plasma.

All other vertebrates are osmoregulators—that is, animals that maintain a relatively constant blood osmolality despite the different concentration in the surrounding environment. The maintenance of a relatively constant body fluid osmolality has permitted vertebrates to exploit a wide variety of ecological niches. Achieving this constancy, however, requires continuous regulation.

Freshwater vertebrates have a much higher solute concentration in their body fluids than that of the surrounding water. In other words, they are hypertonic to their environment. Because of their higher osmotic pressure, water tends to enter their bodies. Consequently, they must prevent water from entering their bodies as much as possible and eliminate the excess water that does enter. In addition, they tend to lose inorganic ions to their environment and so must actively transport these ions back into their bodies.

In contrast, most marine vertebrates are hypotonic to their environment; their body fluids have only about one-third the osmolality of the surrounding seawater. These animals are therefore in danger of losing water by osmosis and must retain water to prevent dehydration. They do this by drinking seawater and eliminating the excess ions through their kidneys and gills.

The body fluids of terrestrial vertebrates have a higher concentration of water than does the air surrounding them. Therefore, they tend to lose water to the air by evaporation from the skin and lungs. All reptiles, birds, and mammals, as well as amphibians during the time when they live on land, face this problem. These vertebrates have evolved excretory systems that help them retain water.

Marine invertebrates are isotonic with their environment, but most vertebrates are either hypertonic or hypotonic to their environment and thus tend to gain or lose water. Physiological mechanisms help most vertebrates to maintain a constant blood osmolality and constant concentrations of individual ions.

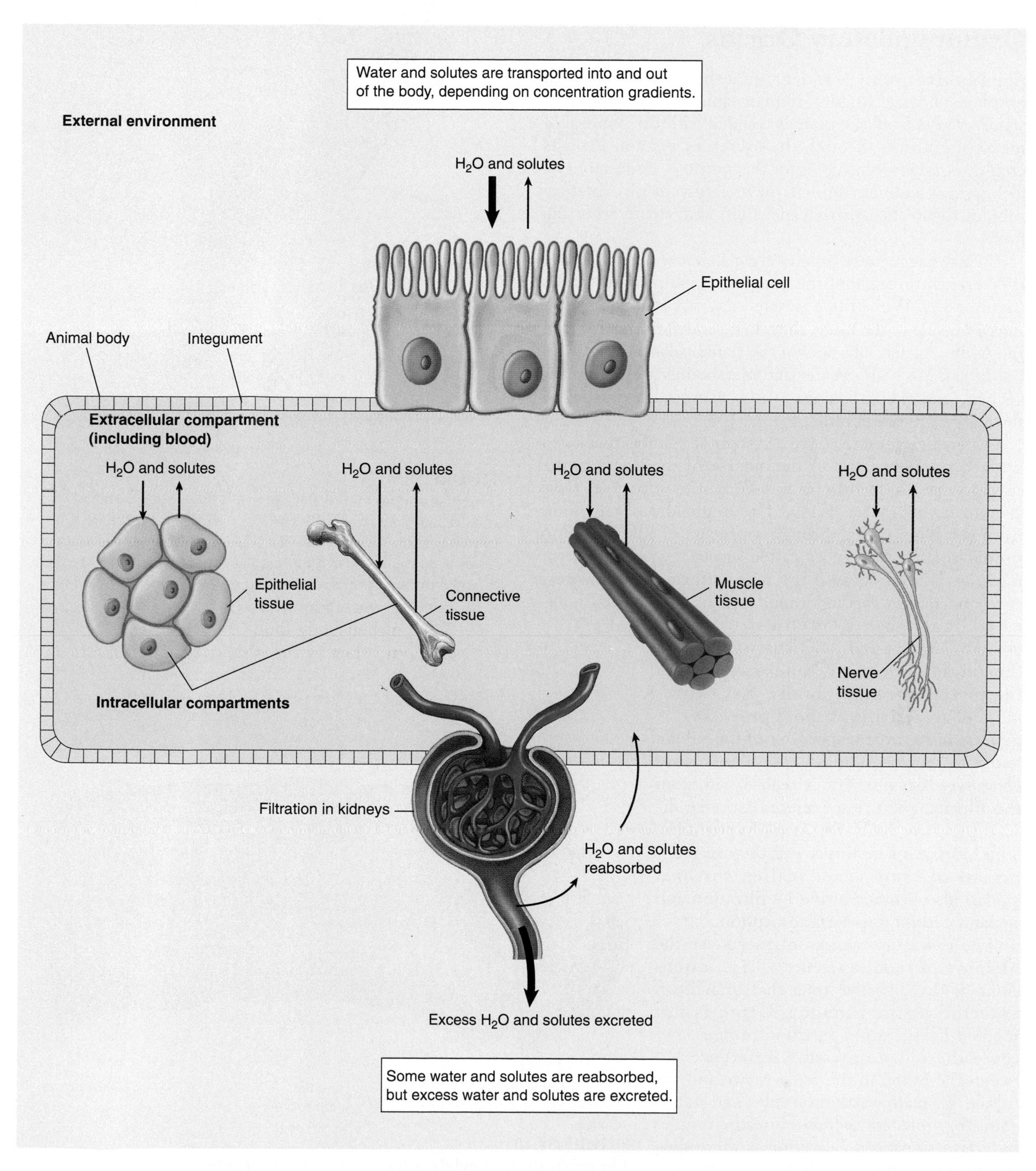

FIGURE 58.8
The interaction between intracellular and extracellular compartments of the body and the external environment. Water can be taken in from the environment or lost to the environment. Exchanges of water and solutes between the extracellular fluids of the body and the environment occur across transport epithelia, and water and solutes can be filtered out of the blood by the kidneys. Overall, the amount of water and solutes that enters and leaves the body must be balanced in order to maintain homeostasis.

Osmoregulatory Organs

Animals have evolved a variety of mechanisms to cope with problems of water balance. In many animals, the removal of water or salts from the body is coupled with the removal of metabolic wastes through the excretory system. Protists employ contractile vacuoles for this purpose, as do sponges. Other multicellular animals have a system of excretory tubules (little tubes) that expel fluid and wastes from the body.

In flatworms, these tubules are called *protonephridia*, and they branch throughout the body into bulblike **flame cells** (figure 58.9). While these simple excretory structures open to the outside of the body, they do not open to the inside of the body. Rather, cilia within the flame cells must draw in fluid from the body. Water and metabolites are then reabsorbed, and the substances to be excreted are expelled through excretory pores.

Other invertebrates have a system of tubules that open both to the inside and to the outside of the body. In the earthworm, these tubules are known as *nephridia* (*blue* structures in figure 58.10). The nephridia obtain fluid from the body cavity through a process of filtration into funnel-shaped structures called *nephrostomes.* The term *filtration* is used because the fluid is formed under pressure and passes through small openings, so that molecules larger than a certain size are excluded. This filtered fluid is isotonic to the fluid in the coelom, but as it passes through the tubules of the nephridia, NaCl is removed by active transport processes. A general term for transport out of the tubule and into the surrounding body fluids is *reabsorption.* Because salt is reabsorbed from the filtrate, the urine excreted is more dilute than the body fluids (is hypotonic). The kidneys of mollusks and the excretory organs of crustaceans (called *antennal glands*) also produce urine by filtration and reclaim certain ions by reabsorption.

The excretory organs in insects are the Malpighian tubules (figure 58.11), extensions of the digestive tract that branch off anterior to the hindgut. Urine is not formed by filtration in these tubules, because there is no pressure difference between the blood in the body cavity and the tubule. Instead, waste molecules and potassium (K^+) ions are secreted into the tubules by active transport. Secretion is the opposite of reabsorption—ions or molecules are transported from the body fluid into the tubule. The secretion of K^+ creates an osmotic gradient that causes water to enter the tubules by osmosis from the body's

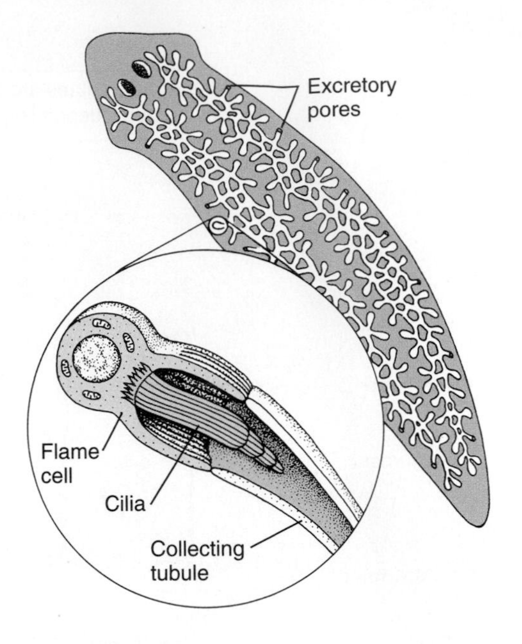

FIGURE 58.9
The protonephridia of flatworms. A branching system of tubules, bulblike flame cells, and excretory pores make up the protonephridia of flatworms. Cilia inside the flame cells draw in fluids from the body by their beating action. Substances are then expelled through pores which open to the outside of the body.

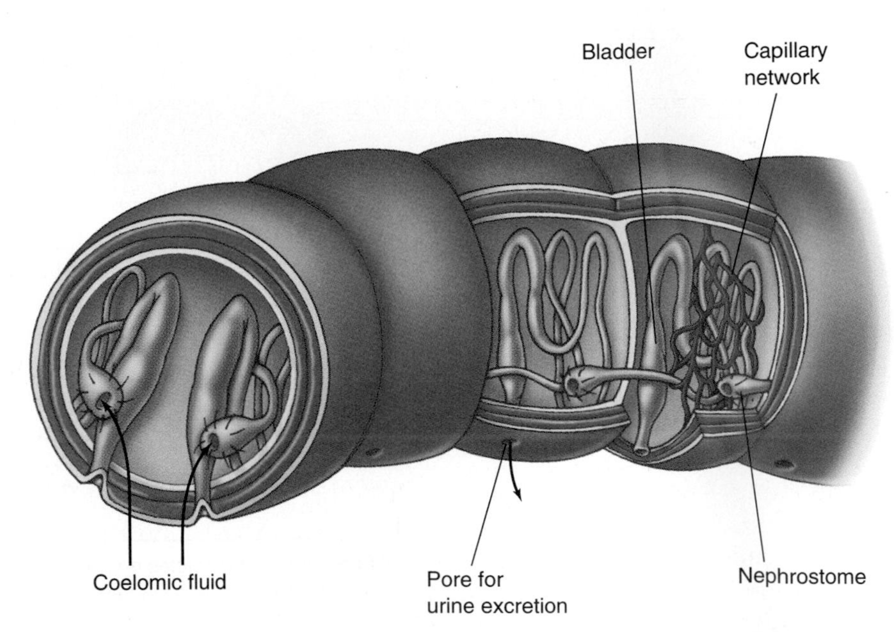

FIGURE 58.10
The nephridia of annelids. Most invertebrates, such as the annelid shown here, have nephridia. These consist of tubules that receive a filtrate of coelomic fluid, which enters the funnel-like nephrostomes. Salt can be reabsorbed from these tubules, and the fluid that remains, urine, is released from pores into the external environment.

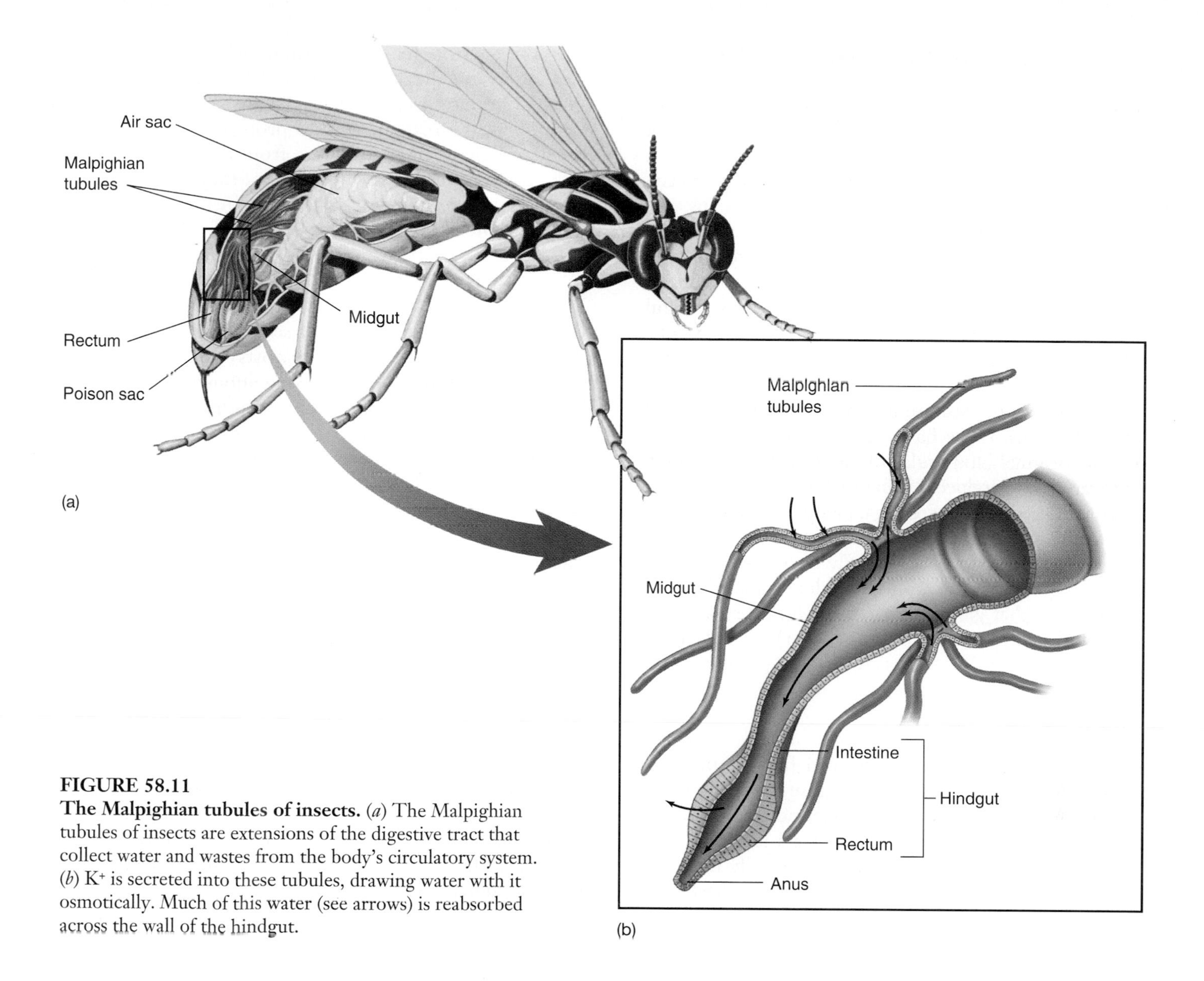

FIGURE 58.11
The Malpighian tubules of insects. (*a*) The Malpighian tubules of insects are extensions of the digestive tract that collect water and wastes from the body's circulatory system. (*b*) K^+ is secreted into these tubules, drawing water with it osmotically. Much of this water (see arrows) is reabsorbed across the wall of the hindgut.

open circulatory system. Most of the water and K^+ is then reabsorbed into the circulatory system through the epithelium of the hindgut, leaving only small molecules and waste products to be excreted from the rectum along with feces. Malpighian tubules thus provide a very efficient means of water conservation.

The kidneys of vertebrates, unlike the Malpighian tubules of insects, create a tubular fluid by filtration of the blood under pressure. In addition to containing waste products and water, the filtrate contains many small molecules that are of value to the animal, including glucose, amino acids, and vitamins. These molecules and most of the water are reabsorbed from the tubules into the blood, while wastes remain in the filtrate. Additional wastes may be secreted by the tubules and added to the filtrate, and the final waste product, urine, is eliminated from the body.

It may seem odd that the vertebrate kidney should filter out almost everything from blood plasma (except proteins, which are too large to be filtered) and then spend energy to take back or reabsorb what the body needs. But selective reabsorption provides great flexibility, because various vertebrate groups have evolved the ability to reabsorb different molecules that are especially valuable in particular habitats. This flexibility is a key factor underlying the successful colonization of many diverse environments by the vertebrates.

Many invertebrates filter fluid into a system of tubules and then reabsorb ions and water, leaving waste products for excretion. Insects create an excretory fluid by secreting K^+ into tubules, which draws water osmotically. The vertebrate kidney produces a filtrate that enters tubules and is modified to become urine.

Evolution of the Vertebrate Kidney

The kidney is a complex organ made up of thousands of repeating units called **nephrons,** each with the structure of a bent tube (figure 58.12). Blood pressure forces the fluid in blood past a filter, called the *glomerulus,* at the top of each nephron. The glomerulus retains blood cells, proteins, and other useful large molecules in the blood but allows the water, and the small molecules and wastes dissolved in it, to pass through and into the bent tube part of the nephron. As the filtered fluid passes through the nephron tube, useful sugars and ions are recovered from it by active transport, leaving the water and metabolic wastes behind in a fluid urine.

Although the same basic design has been retained in all vertebrate kidneys, there have been a few modifications. Because the original glomerular filtrate is isotonic to blood, all vertebrates can produce a urine that is isotonic to blood by reabsorbing ions and water in equal proportions or hypotonic to blood—that is, more dilute than the blood, by reabsorbing relatively less water. Only birds and mammals can reabsorb enough water from their glomerular filtrate to produce a urine that is hypertonic to blood—that is, more concentrated than the blood, by reabsorbing relatively more water.

Freshwater Fish

Kidneys are thought to have evolved first among the freshwater teleosts, or bony fish. Because the body fluids of a freshwater fish have a greater osmotic concentration than the surrounding water, these animals face two serious problems: (1) water tends to enter the body from the environment; and (2) solutes tend to leave the body and enter the environment. Freshwater fish address the first problem by *not* drinking water and by excreting a large volume of dilute urine, which is hypotonic to their body fluids. They address the second problem by reabsorbing ions across the nephron tubules, from the glomerular filtrate back into the blood. In addition, they actively transport ions across their gill surfaces from the surrounding water into the blood.

Marine Bony Fish

Although most groups of animals seem to have evolved first in the sea, marine bony fish (teleosts) probably evolved from freshwater ancestors, as was mentioned in chapter 48. They faced significant new problems in making the transition to the sea because their body fluids are hypotonic to the surrounding seawater. Consequently, water tends to leave their bodies by osmosis across their gills, and they also lose water in their urine. To compensate for this continuous water loss, marine fish drink large amounts of seawater (figure 58.13).

Many of the divalent cations (principally Ca^{++} and Mg^{++}) in the seawater that a marine fish drinks remain in the digestive tract and are eliminated through the anus. Some, however, are absorbed into the blood, as are the monovalent ions K^+, Na^+, and Cl^-. Most of the monovalent ions are actively transported out of the blood across the gill surfaces, while the divalent ions that enter the blood are secreted into the nephron tubules and excreted in the urine. In these two ways, marine bony fish eliminate the ions they get from the seawater they drink. The urine they excrete is isotonic to their body fluids. It is more concentrated than the urine of freshwater fish, but not as concentrated as that of birds and mammals.

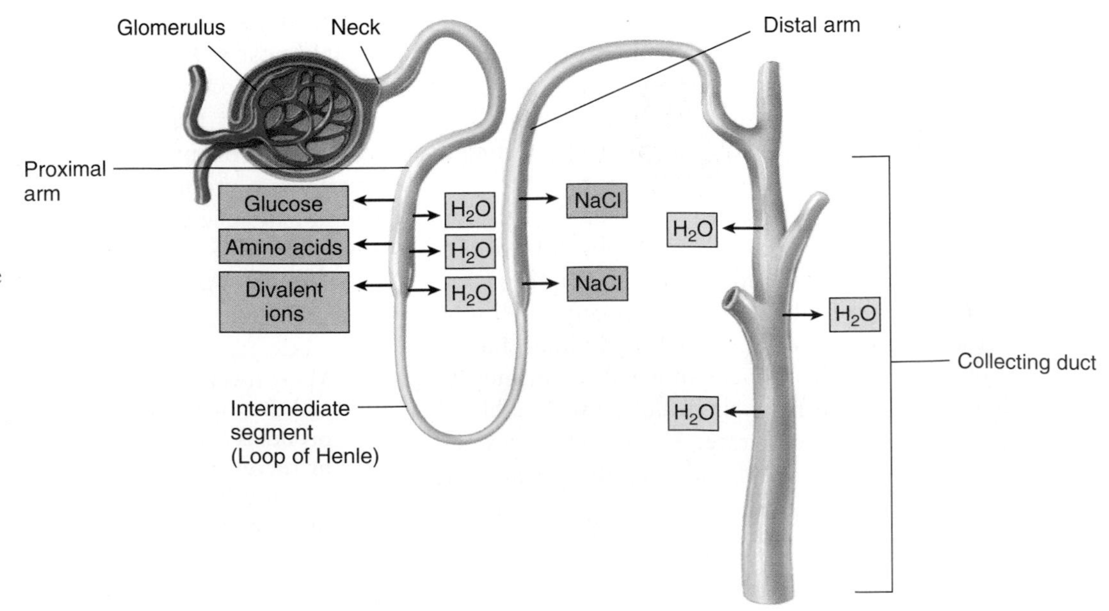

FIGURE 58.12
The basic organization of the vertebrate nephron. The nephron tubule is a basic design that has been retained in the kidneys of vertebrates. Sugars, amino acids, and divalent ions such as Ca^{++} are recovered in the proximal arm; monovalent ions such as Na^+ and Cl^- are recovered in the distal arm; and water is recovered in the collecting duct.

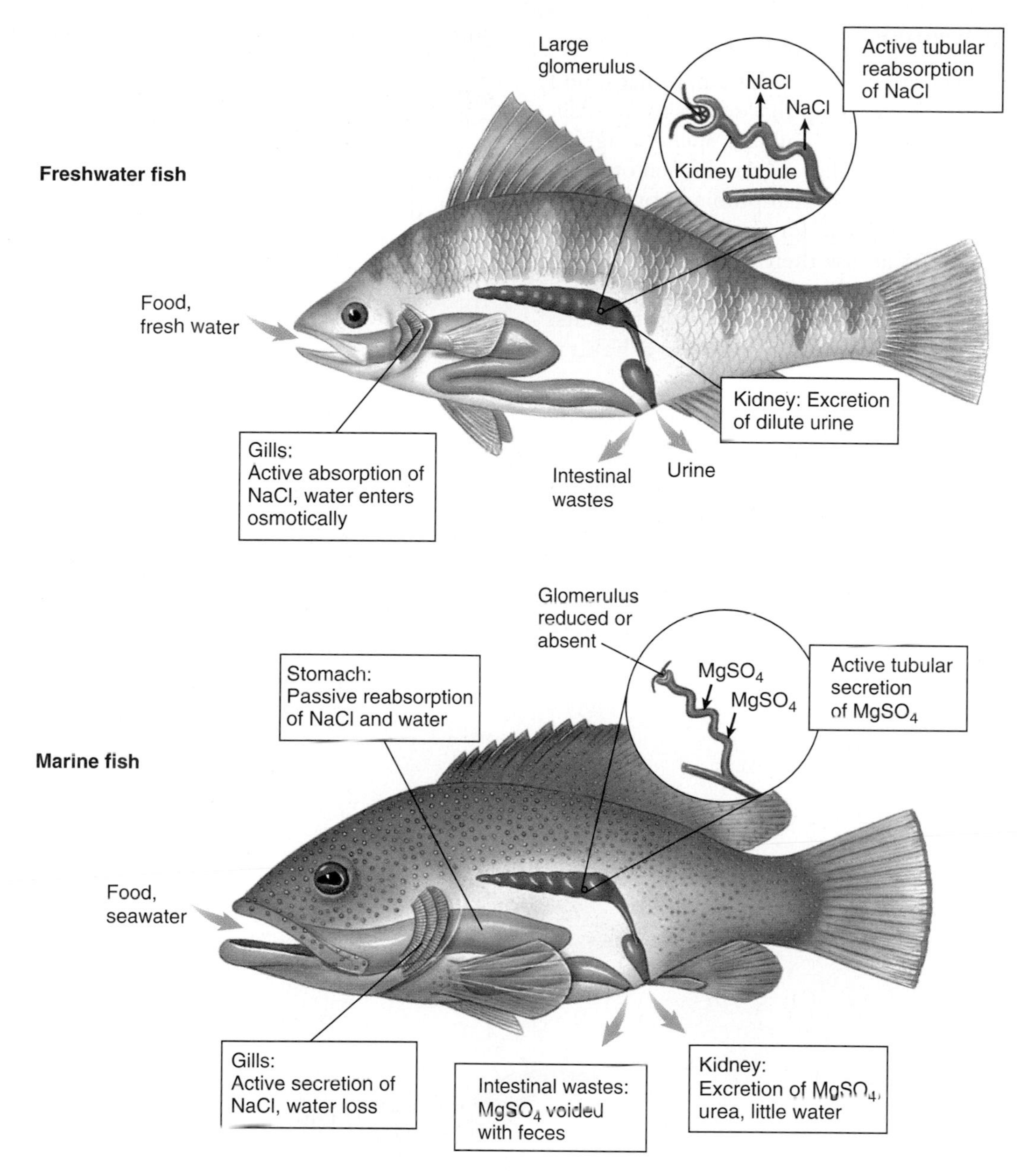

FIGURE 58.13

Freshwater and marine teleosts (bony fish) face different osmotic problems. Whereas the freshwater teleost is hypertonic to its environment, the marine teleost is hypotonic to seawater. To compensate for its tendency to take in water and lose ions, a freshwater fish excretes dilute urine, avoids drinking water, and reabsorbs ions across the nephron tubules. To compensate for its osmotic loss of water, the marine teleost drinks seawater and eliminates the excess ions through active transport across epithelia in the gills and kidneys.

Cartilaginous Fish

The elasmobranchs, including sharks and rays, are by far the most common subclass in the class Chondrichthyes (cartilaginous fish). Elasmobranchs have solved the osmotic problem posed by their seawater environment in a different way than have the bony fish. Instead of having body fluids that are hypotonic to seawater, so that they have to continuously drink seawater and actively pump out ions, the elasmobranchs reabsorb urea from the nephron tubules and maintain a blood urea concentration that is 100 times higher than that of mammals. This added urea makes their blood approximately isotonic to the surrounding sea. Because there is no net water movement between isotonic solutions, water loss is prevented. Hence, these fishes do not need to drink seawater for osmotic balance, and their kidneys and gills do not have to remove large amounts of ions from their bodies. The enzymes and tissues of the cartilaginous fish have evolved to tolerate the high urea concentrations.

Amphibians and Reptiles

The first terrestrial vertebrates were the amphibians, and the amphibian kidney is identical to that of freshwater fish. This is not surprising, because amphibians spend a significant portion of their time in fresh water, and when on land, they generally stay in wet places. Amphibians produce a very dilute urine and compensate for their loss of Na^+ by actively transporting Na^+ across their skin from the surrounding water.

Reptiles, on the other hand, live in diverse habitats. Those living mainly in fresh water occupy a habitat similar to that of the freshwater fish and amphibians and thus have similar kidneys. Marine reptiles, including some crocodilians, sea turtles, sea snakes, and one lizard, possess kidneys similar to those of their freshwater relatives but face opposite problems; they tend to lose water and take in salts. Like marine teleosts (bony fish), they drink the seawater and excrete an isotonic urine. Marine teleosts eliminate the excess salt by transport across their gills, while marine reptiles eliminate excess salt through salt glands located near the nose or the eye (figure 58.14).

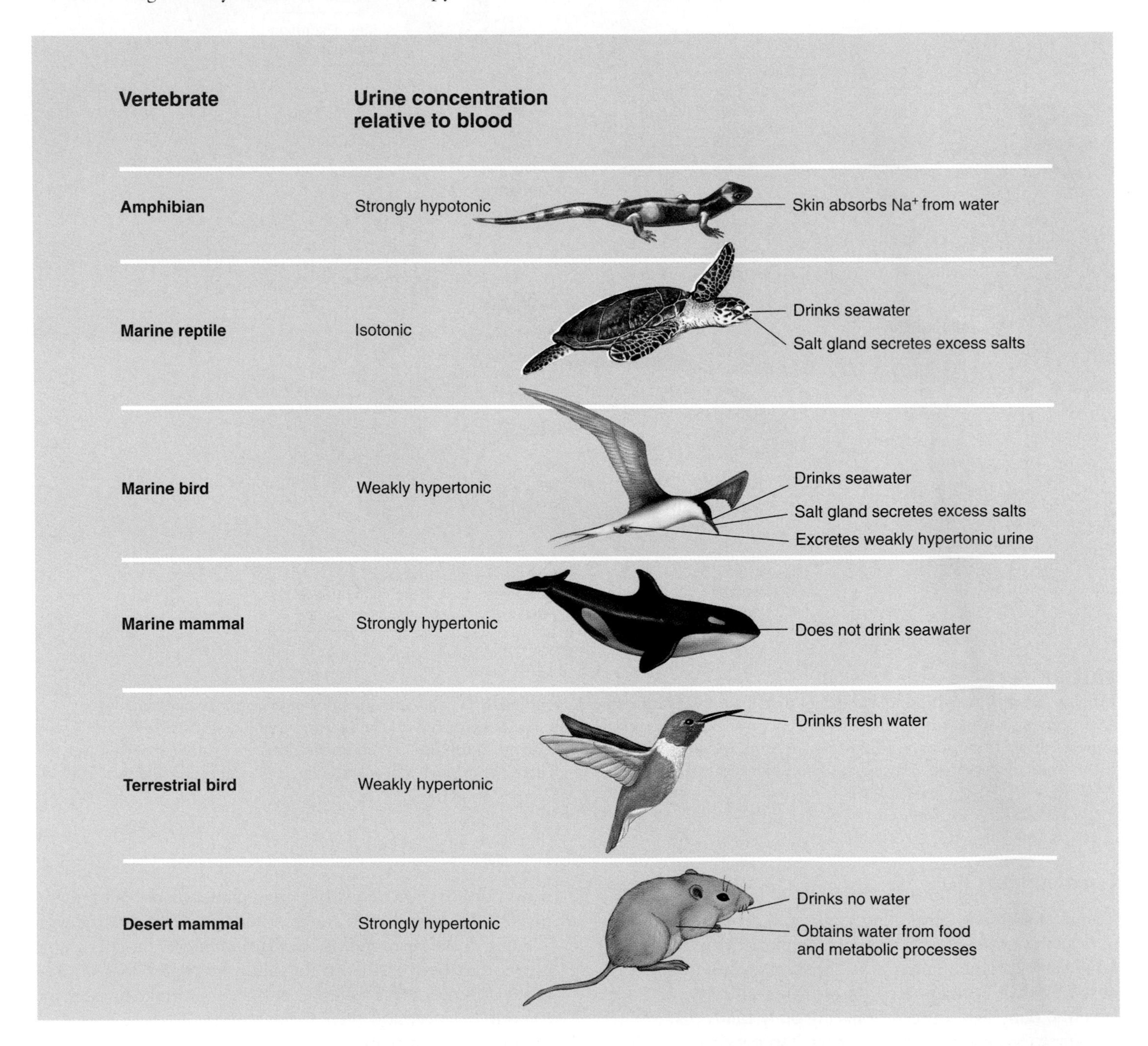

Vertebrate
Urine concentration relative to blood
Amphibian
Strongly hypotonic
Skin absorbs Na⁺ from water
Marine reptile
Isotonic
Drinks seawater
Salt gland secretes excess salts
Marine bird
Weakly hypertonic
Drinks seawater
Salt gland secretes excess salts
Excretes weakly hypertonic urine
Marine mammal
Strongly hypertonic
Does not drink seawater
Terrestrial bird
Weakly hypertonic
Drinks fresh water
Desert mammal
Strongly hypertonic
Drinks no water
Obtains water from food and metabolic processes

FIGURE 58.14
Osmoregulation by some vertebrates. Only birds and mammals can produce a hypertonic urine and thereby retain water efficiently, but marine reptiles and birds can drink seawater and excrete the excess salt through salt glands.

The kidneys of terrestrial reptiles also reabsorb much of the salt and water in their nephron tubules, helping somewhat to conserve blood volume in dry environments. Like fish and amphibians, they cannot produce urine that is more concentrated than the blood plasma. However, when their urine enters their cloaca (the common exit of the digestive and urinary tracts), additional water can be reabsorbed.

Mammals and Birds

Mammals and birds are the only vertebrates able to produce urine with a higher osmotic concentration than their body fluids. This allows these vertebrates to excrete their waste products in a small volume of water, so that more water can be retained in the body. Human kidneys can produce urine that is as much as 4.2 times as concentrated as blood plasma, but the kidneys of some other mammals are even more efficient at conserving water. For example, camels, gerbils, and pocket mice of the genus *Perognathus* can excrete urine 8, 14, and 22 times as concentrated as their blood plasma, respectively. The kidneys of the kangaroo rat (figure 58.15) are so efficient it never has to drink water; it can obtain all the water it needs from its food and from water produced in aerobic cell respiration!

The production of hypertonic urine is accomplished by the *loop of Henle* portion of the nephron (see figure 58.18), found only in mammals and birds. A nephron with a long loop of Henle extends deeper into the renal medulla, where the hypertonic osmotic environment draws out more water, and so can produce more concentrated urine. Most mammals have some nephrons with short loops and other nephrons with loops that are much longer (see figure 58.17). Birds, however, have relatively few or no nephrons with long loops, so they cannot produce urine that is as concentrated as that of mammals. At most, they can only reabsorb enough water to produce a urine that is about twice the concentration of their blood. Marine birds solve the problem of water loss by drinking salt water and then excreting the excess salt from salt glands near the eyes (figure 58.16).

The moderately hypertonic urine of a bird is delivered to its cloaca, along with the fecal material from its digestive tract. If needed, additional water can be absorbed across the wall of the cloaca to produce a semisolid white paste or pellet, which is excreted.

The kidneys of freshwater fish must excrete copious amounts of very dilute urine, while marine teleosts drink seawater and excrete an isotonic urine. The basic design and function of the nephron of freshwater fishes have been retained in the terrestrial vertebrates. Modifications, particularly the presence of a loop of Henle, allow mammals and birds to reabsorb more water and produce a hypertonic urine.

FIGURE 58.15
The kangaroo rat, *Dipodomys panamintensis*. This mammal has very efficient kidneys that can concentrate urine to a high degree by reabsorbing water, thereby minimizing water loss from the body. This feature is extremely important to the kangaroo rat's survival in dry or desert habitats.

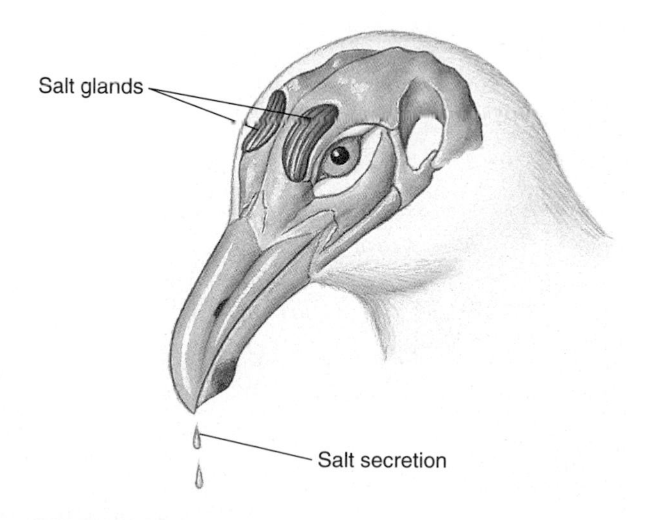

FIGURE 58.16
Marine birds drink seawater and then excrete the salt through salt glands. The extremely salty fluid excreted by these glands can then dribble down the beak.

The Mammalian Kidney

In humans, the kidneys are fist-sized organs located in the region of the lower back. Each kidney receives blood from a renal artery, and it is from this blood that urine is produced. Urine drains from each kidney through a **ureter,** which carries the urine to a **urinary bladder.** Within the kidney, the mouth of the ureter flares open to form a funnel-like structure, the *renal pelvis.* The renal pelvis, in turn, has cup-shaped extensions that receive urine from the renal tissue. This tissue is divided into an outer **renal cortex** and an inner **renal medulla** (figure 58.17). Together, these structures perform filtration, reabsorption, secretion, and excretion.

Nephron Structure and Filtration

On a microscopic level, each kidney contains about one million functioning *nephrons.* Mammalian kidneys contain a mixture of juxtamedullary nephrons, which have long loops which dip deeply into the medulla, and cortical nephrons with shorter loops (see figure 58.17). The significance of the length of the loops will be explained a little later.

Each nephron consists of a long tubule and associated small blood vessels. First, blood is carried by an *afferent arteriole* to a tuft of capillaries in the renal cortex, the **glomerulus** (figure 58.18). Here the blood is filtered as the blood pressure forces fluid through the porous capillary walls. Blood cells and plasma proteins are too large to enter

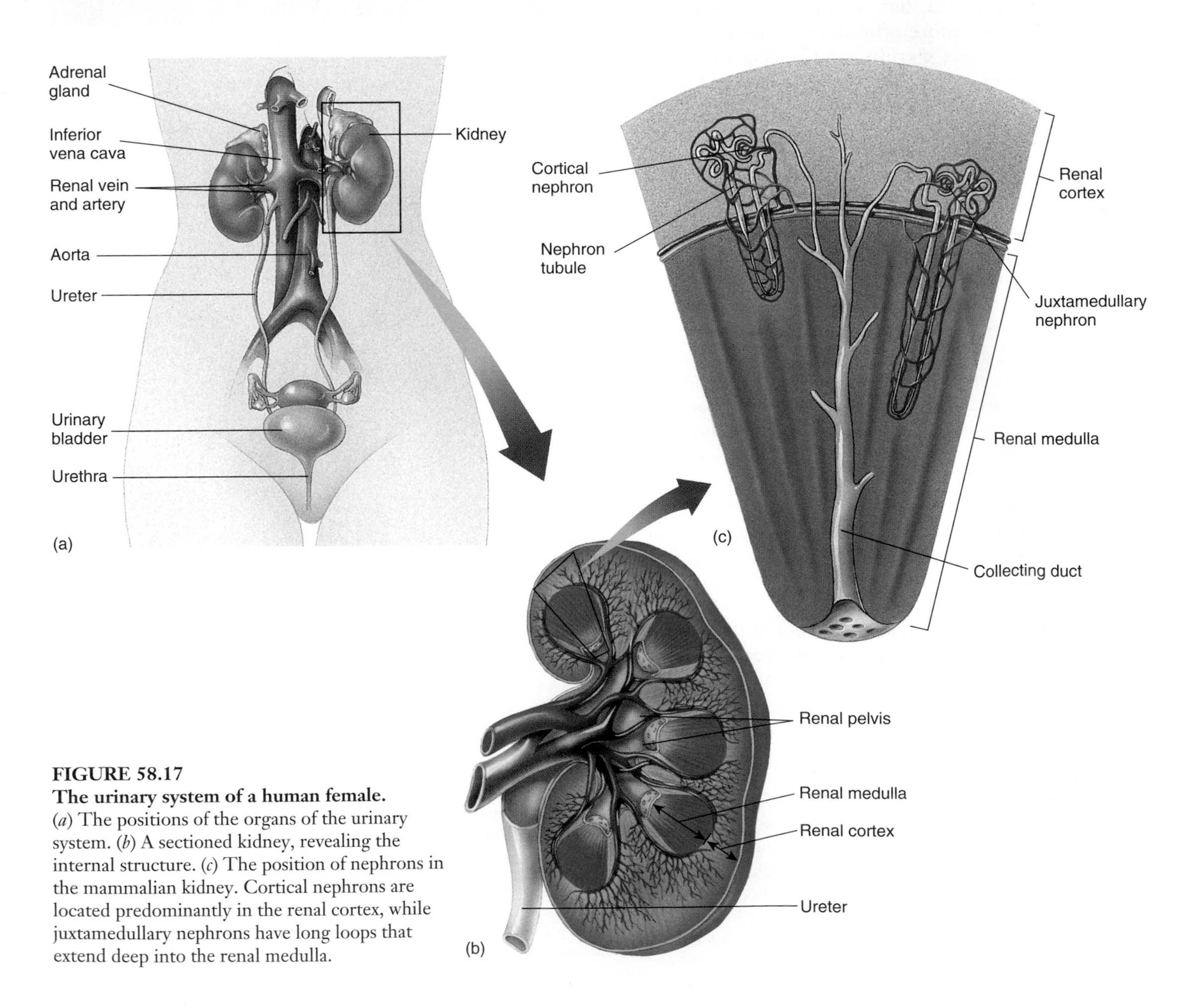

FIGURE 58.17
The urinary system of a human female.
(*a*) The positions of the organs of the urinary system. (*b*) A sectioned kidney, revealing the internal structure. (*c*) The position of nephrons in the mammalian kidney. Cortical nephrons are located predominantly in the renal cortex, while juxtamedullary nephrons have long loops that extend deep into the renal medulla.

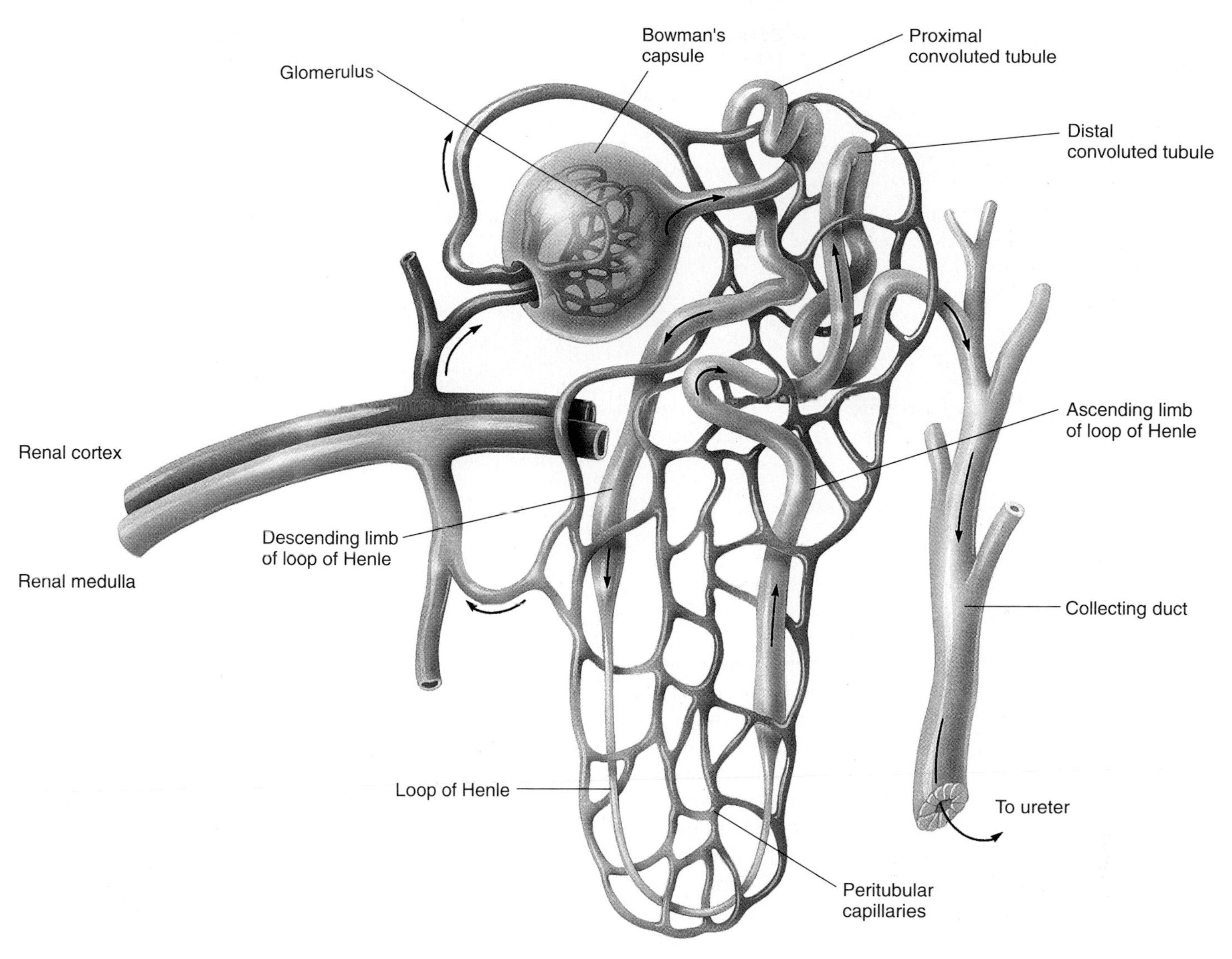

FIGURE 58.18
A nephron in a mammalian kidney. The nephron tubule is surrounded by peritubular capillaries, which carry away molecules and ions that are reabsorbed from the filtrate.

this **glomerular filtrate,** but large amounts of water and dissolved molecules leave the vascular system at this step. The filtrate immediately enters the first region of the nephron tubules. This region, *Bowman's capsule,* envelops the glomerulus much as a large, soft balloon surrounds your fist if you press your fist into it. The capsule has slit openings so that the glomerular filtrate can enter the system of nephron tubules.

After the filtrate enters Bowman's capsule it goes into a portion of the nephron called the **proximal convoluted tubule,** located in the cortex. The fluid then moves down into the medulla and back up again into the cortex in a **loop of Henle.** Only the kidneys of mammals and birds have loops of Henle, and this is why only birds and mammals have the ability to concentrate their urine. After leaving the loop, the fluid is delivered to a **distal convoluted tubule** in the cortex that next drains into a **collecting duct.** The collecting duct again descends into the medulla, where it merges with other collecting ducts to empty its contents, now called urine, into the renal pelvis.

Blood components that were not filtered out of the glomerulus drain into an *efferent arteriole,* which then empties into a second bed of capillaries called *peritubular capillaries* that surround the tubules. This is the only location in the body where two capillary beds occur in series. As described later, the peritubular capillaries are needed for the processes of reabsorption and secretion.

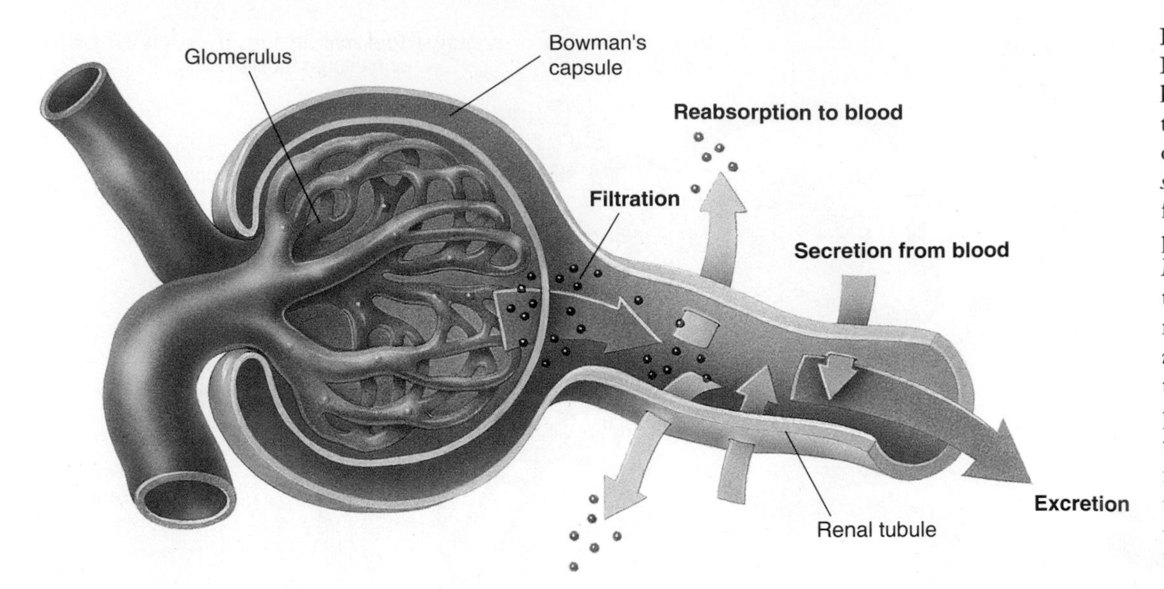

FIGURE 58.19
Four functions of the kidney. Molecules enter the urine by *filtration* out of the glomerulus and by *secretion* into the tubules from surrounding peritubular capillaries. Molecules that entered the filtrate can be returned to the blood by *reabsorption* from the tubules into surrounding peritubular capillaries, or they may be eliminated from the body by *excretion* through the tubule to a ureter, then to the bladder.

Reabsorption and Secretion

Most of the water and dissolved solutes that enter the glomerular filtrate must be returned to the blood (figure 58.19), or the animal would literally urinate to death. In a human, for example, approximately 2000 liters of blood passes through the kidneys each day, and 180 liters of water leaves the blood and enters the glomerular filtrate. Because we only have a total blood volume of about 5 liters and only produce 1 to 2 liters of urine per day, it is obvious that each liter of blood is filtered many times per day and most of the filtered water is reabsorbed. The reabsorption of water occurs as a consequence of salt (NaCl) reabsorption through mechanisms that will be described shortly.

The reabsorption of glucose, amino acids, and many other molecules needed by the body is driven by active transport carriers. As in all carrier-mediated transport, a maximum rate of transport is reached whenever the carriers are saturated (see chapter 6). For the renal glucose carriers, saturation occurs when the concentration of glucose in the blood (and thus in the glomerular filtrate) is about 180 milligrams per 100 milliliters of blood. If a person has a blood glucose concentration in excess of this amount, as happens in untreated diabetes mellitus, the glucose left untransported in the filtrate is expelled in the urine. Indeed, the presence of glucose in the urine is diagnostic of diabetes mellitus.

The secretion of foreign molecules and particular waste products of the body involves the transport of these molecules across the membranes of the blood capillaries and kidney tubules into the filtrate. This process is similar to reabsorption, but it proceeds in the opposite direction. Some secreted molecules are eliminated in the urine so rapidly that they may be cleared from the blood in a single pass through the kidneys. This rapid elimination explains why penicillin, which is secreted by the nephrons, must be administered in very high doses and several times per day.

Excretion

A major function of the kidney is the elimination of a variety of potentially harmful substances that animals eat and drink. In addition, urine contains nitrogenous wastes, such as urea and uric acid, that are products of the catabolism of amino acids and nucleic acids. Urine may also contain excess K^+, H^+, and other ions that are removed from the blood. Urine's generally high H^+ concentration (pH 5 to 7) helps maintain the acid-base balance of the blood within a narrow range (pH 7.35 to 7.45). Moreover, the excretion of water in urine contributes to the maintenance of blood volume and pressure; the larger the volume of urine excreted, the lower the blood volume.

The purpose of kidney function is therefore homeostasis—the kidneys are critically involved in maintaining the constancy of the internal environment. When disease interferes with kidney function, it causes a rise in the blood concentration of nitrogenous waste products, disturbances in electrolyte and acid-base balance, and a failure in blood pressure regulation. Such potentially fatal changes highlight the central importance of the kidneys in normal body physiology.

The mammalian kidney is divided into a cortex and medulla and contains microscopic functioning units called nephrons. The nephron tubules receive a blood filtrate from the glomeruli and modify this filtrate to produce urine, which empties into the renal pelvis and is expelled from the kidney through the ureter.

Transport Processes in the Mammalian Nephron

As previously described, approximately 180 liters (in a human) of isotonic glomerular filtrate enters the Bowman's capsules each day. After passing through the remainder of the nephron tubules, this volume of fluid would be lost as urine if it were not reabsorbed back into the blood. It is clearly impossible to produce this much urine, yet water is only able to pass through a cell membrane by osmosis, and osmosis is not possible between two isotonic solutions. Therefore, some mechanism is needed to create an osmotic gradient between the glomerular filtrate and the blood, allowing reabsorption.

Proximal Tubule

Approximately two-thirds of the NaCl and water filtered into Bowman's capsule is immediately reabsorbed across the walls of the proximal convoluted tubule. This reabsorption is driven by the active transport of Na^+ out of the filtrate and into surrounding peritubular capillaries. Cl^- follows Na^+ passively because of electrical attraction, and water follows them both because of osmosis. Because NaCl and water are removed from the filtrate in proportionate amounts, the filtrate that remains in the tubule is still isotonic to the blood plasma.

Although only one-third of the initial volume of filtrate remains in the nephron tubule after the initial reabsorption of NaCl and water, it still represents a large volume (60 L out of the original 180 L of filtrate produced per day by both human kidneys). Obviously, no animal can excrete that much urine, so most of this water must also be reabsorbed. It is reabsorbed primarily across the wall of the collecting duct because the interstitial fluid of the renal medulla surrounding the collecting ducts is hypertonic. The hypertonic renal medulla draws water out of the collecting duct by osmosis, leaving behind a hypertonic urine for excretion.

Loop of Henle

The reabsorption of much of the water in the tubular filtrate thus depends on the creation of a hypertonic renal medulla; the more hypertonic the medulla is, the steeper the osmotic gradient will be and the more water will leave the collecting ducts. It is the loops of Henle that create the hypertonic renal medulla in the following manner (figure 58.20):

1. The ascending limb of the loop actively extrudes Na^+, and Cl^- follows. The mechanism that extrudes NaCl from the ascending limb of the loop differs from that which extrudes NaCl from the proximal tubule, but the most important difference is that the ascending limb is *not permeable to water*. As Na^+ exits, the fluid within the ascending limb becomes increasingly dilute (hypotonic) as it enters the cortex, while the surrounding tissue becomes increasingly concentrated (hypertonic).

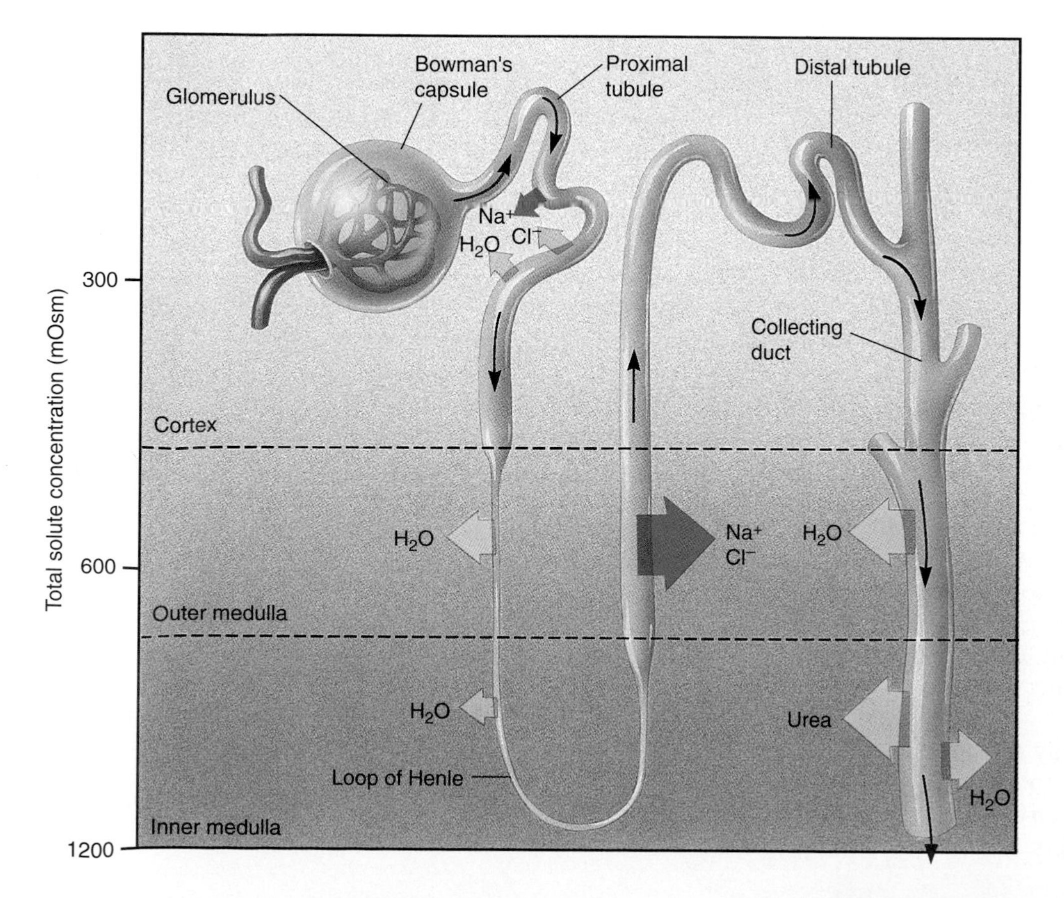

FIGURE 58.20
The reabsorption of salt and water in the mammalian kidney. Active transport of Na^+ out of the proximal tubules is followed by the passive movement of Cl^- and water. Active extrusion of NaCl from the ascending limb of the loop of Henle creates the osmotic gradient required for the reabsorption of water from the collecting duct. The changes in osmolality from the cortex to the medulla is indicated to the left of the figure.

2. The NaCl pumped out of the ascending limb of the loop is trapped within the surrounding interstitial fluid. This is because the peritubular capillaries in the medulla also have loops, called *vasa recta*, so that NaCl can diffuse from the blood leaving the medulla to the blood entering the medulla. Thus, the vasa recta functions in a countercurrent exchange, similar to that described for oxygen in the countercurrent flow of water and blood in the gills of fish (see chapter 53). In the case of the renal medulla, the diffusion of NaCl between the blood vessels keeps much of the NaCl within the interstitial fluid, making it hypertonic.
3. The descending limb is permeable to water, so water leaves by osmosis as the fluid descends into the hypertonic renal medulla. This water enters the blood vessels of the vasa recta and is carried away in the general circulation.
4. The loss of water from the descending limb multiplies the concentration that can be achieved at each level of the loop through the active extrusion of NaCl by the ascending limb. The longer the loop of Henle, the longer the region of interaction between the descending and ascending limbs, and the greater the total concentration that can be achieved. In a human kidney, the concentration of filtrate entering the loop is 300 milliosmolal, and this concentration is multiplied to more than 1200 milliosmolal at the bottom of the longest loops of Henle in the renal medulla.

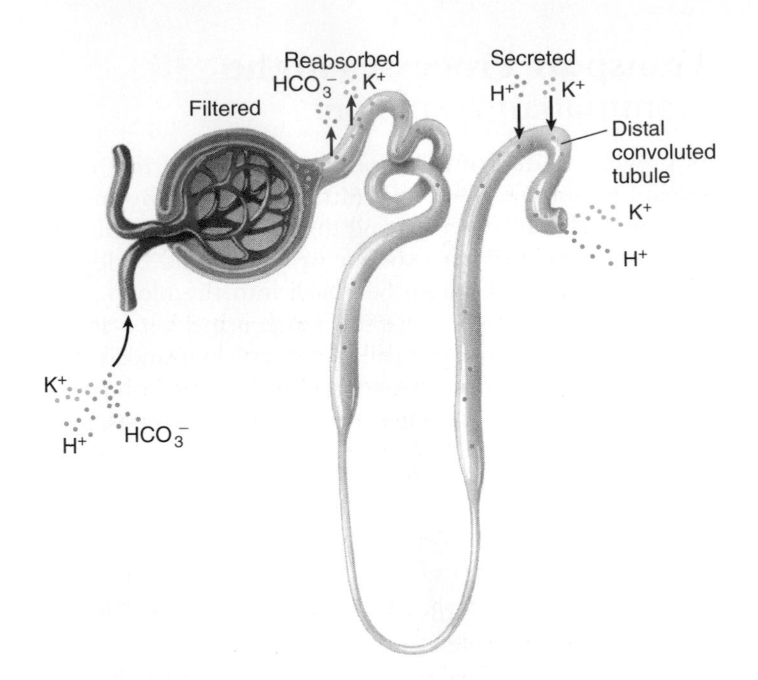

FIGURE 58.21
The nephron controls the amounts of K^+, H^+, and HCO_3^- excreted in the urine. K^+ is completely reabsorbed in the proximal tubule and then secreted in varying amounts into the distal tubule. HCO_3^- is filtered but normally completely reabsorbed. H^+ is filtered and also secreted into the distal tubule, so that the final urine has an acidic pH.

Because fluid flows in opposite directions in the two limbs of the loop, the action of the loop of Henle in creating a hypertonic renal medulla is known as the *countercurrent multiplier system.* The high solute concentration of the renal medulla is primarily the result of NaCl accumulation by the countercurrent multiplier system, but urea also contributes to the total osmolality of the medulla. This is because the descending limb of the loop of Henle and the collecting duct are permeable to urea, which leaves these regions of the nephron by diffusion.

Distal Tubule and Collecting Duct

Because NaCl was pumped out of the ascending limb, the filtrate that arrives at the distal convoluted tubule and enters the collecting duct in the renal cortex is hypotonic (with a concentration of only 100 mOsm). The collecting duct carrying this dilute fluid now plunges into the medulla. As a result of the hypertonic interstitial fluid of the renal medulla, there is a strong osmotic gradient that pulls water out of the collecting duct and into surrounding blood vessels.

The osmotic gradient is normally constant, but the permeability of the collecting duct to water is adjusted by a hormone, **antidiuretic hormone** (**ADH,** also called **vasopressin**), discussed in chapters 52 and 56. When an animal needs to conserve water, the posterior pituitary gland secretes more ADH, and this hormone increases the number of water channels in the plasma membranes of the collecting duct cells. This increases the permeability of the collecting ducts to water so that more water is reabsorbed and less is excreted in the urine. The animal thus excretes a hypertonic urine.

In addition to the regulation of water balance, the kidneys regulate the balance of electrolytes in the blood by reabsorption and secretion. For example, the kidneys reabsorb K^+ in the proximal tubule and then secrete an amount of K^+ needed to maintain homeostasis into the distal convoluted tubule (figure 58.21). The kidneys also maintain acid-base balance by excreting H^+ into the urine and reabsorbing bicarbonate (HCO_3^-), as previously described.

The loop of Henle creates a hypertonic renal medulla as a result of the active extrusion of NaCl from the ascending limb and the interaction with the descending limb. The hypertonic medulla then draws water osmotically from the collecting duct, which is permeable to water under the influence of antidiuretic hormone.

Ammonia, Urea, and Uric Acid

Amino acids and nucleic acids are nitrogen-containing molecules. When animals catabolize these molecules for energy or convert them into carbohydrates or lipids, they produce nitrogen-containing by-products called **nitrogenous wastes** (figure 58.22) that must be eliminated from the body.

The first step in the metabolism of amino acids and nucleic acids is the removal of the amino (—NH_2) group and its combination with H^+ to form **ammonia** (NH_3) in the liver. Ammonia is quite toxic to cells and therefore is safe only in very dilute concentrations. The excretion of ammonia is not a problem for the bony fish and tadpoles, which eliminate most of it by diffusion through the gills and less by excretion in very dilute urine. In elasmobranchs, adult amphibians, and mammals, the nitrogenous wastes are eliminated in the far less toxic form of **urea.** Urea is water-soluble and so can be excreted in large amounts in the urine. It is carried in the bloodstream from its place of synthesis in the liver to the kidneys where it is excreted in the urine.

Reptiles, birds, and insects excrete nitrogenous wastes in the form of **uric acid**, which is only slightly soluble in water. As a result of its low solubility, uric acid precipitates and thus can be excreted using very little water. Uric acid forms the pasty white material in bird droppings called *guano*. The ability to synthesize uric acid in these groups of animals is also important because their eggs are encased within shells, and nitrogenous wastes build up as the embryo grows within the egg. The formation of uric acid, while a lengthy process that requires considerable energy, produces a compound that crystallizes and precipitates. As a precipitate, it is unable to affect the embryo's development even though it is still inside the egg.

Mammals also produce some uric acid, but it is a waste product of the degradation of purine nucleotides (see chapter 3), not of amino acids. Most mammals have an enzyme called *uricase*, which converts uric acid into a more soluble derivative, **allantoin.** Only humans, apes, and the dalmatian dog lack this enzyme and so must excrete the uric acid. In humans, excessive accumulation of uric acid in the joints produces a condition known as *gout*.

The metabolic breakdown of amino acids and nucleic acids produces ammonia as a by-product. Ammonia is excreted by bony fish, but other vertebrates convert nitrogenous wastes into urea and uric acid, which are less toxic nitrogenous wastes.

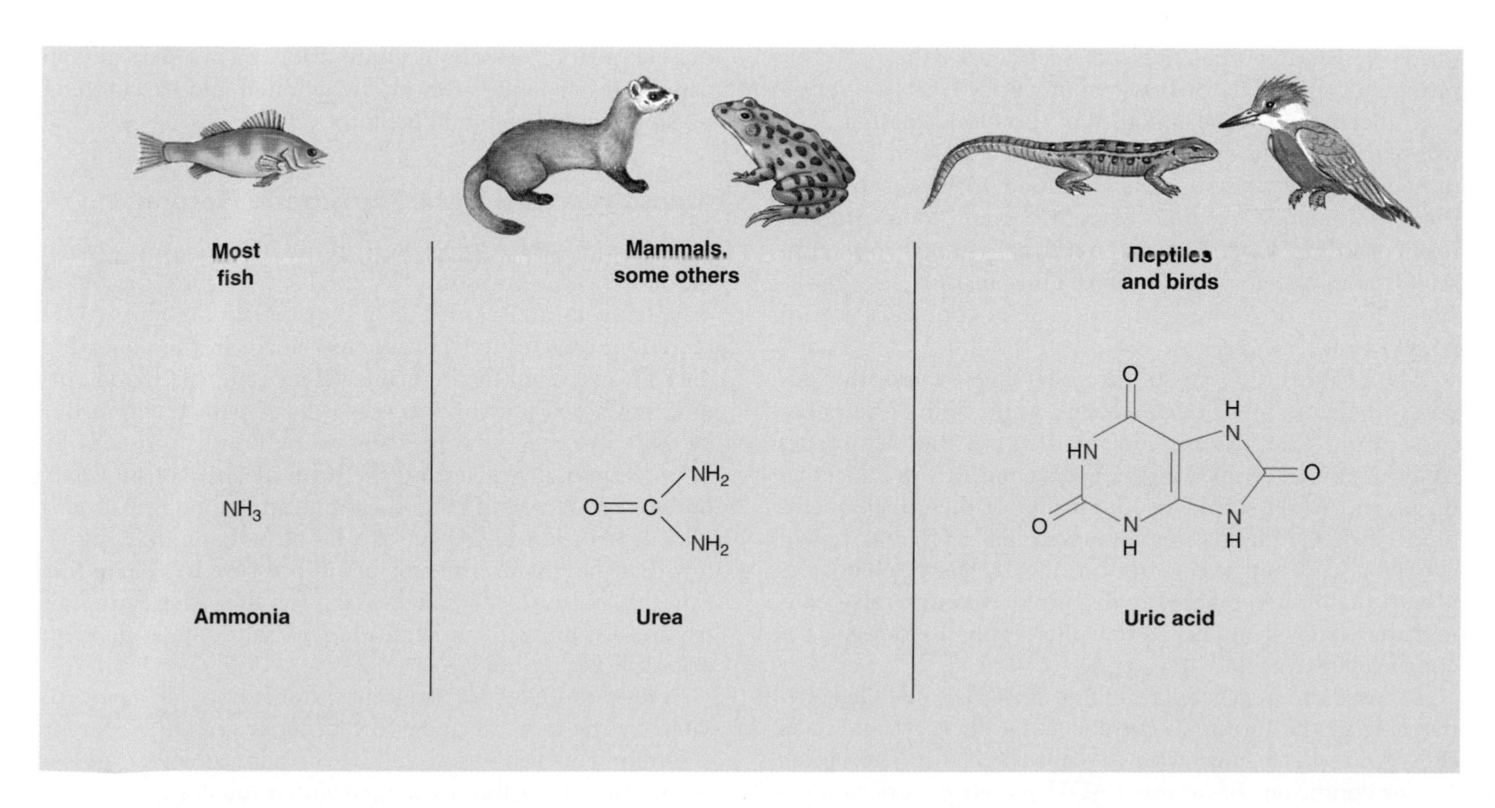

FIGURE 58.22
Nitrogenous wastes. When amino acids and nucleic acids are metabolized, the immediate by-product is ammonia, which is quite toxic but which can be eliminated through the gills of teleost fish. Mammals convert ammonia into urea, which is less toxic. Birds and terrestrial reptiles convert it instead into uric acid, which is insoluble in water.

58.4 The kidney is regulated by hormones.

Hormones Control Homeostatic Functions

In mammals and birds, the amount of water excreted in the urine, and thus the concentration of the urine, varies according to the changing needs of the body. Acting through the mechanisms described next, the kidneys will excrete a hypertonic urine when the body needs to conserve water. If an animal drinks too much water, the kidneys will excrete a hypotonic urine. As a result, the volume of blood, the blood pressure, and the osmolality of blood plasma are maintained relatively constant by the kidneys, no matter how much water you drink. The kidneys also regulate the plasma K^+ and Na^+ concentrations and blood pH within very narrow limits. These homeostatic functions of the kidneys are coordinated primarily by hormones (see chapter 56).

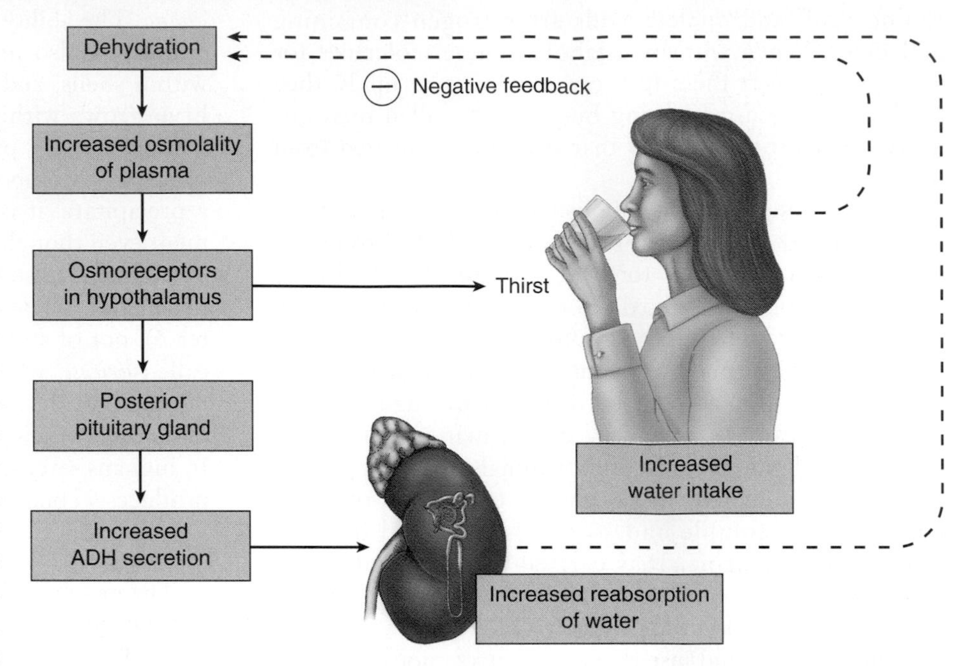

FIGURE 58.23
Antidiuretic hormone stimulates the reabsorption of water by the kidneys. This action completes a negative feedback loop and helps to maintain homeostasis of blood volume and osmolality.

Antidiuretic Hormone

Antidiuretic hormone (ADH) is produced by the hypothalamus and secreted by the posterior pituitary gland. The primary stimulus for ADH secretion is an increase in the osmolality of the blood plasma. The osmolality of plasma increases when a person is dehydrated or when a person eats salty food. Osmoreceptors in the hypothalamus respond to the elevated blood osmolality by sending more nerve signals to the integration center (also in the hypothalamus). This, in turn, triggers a sensation of thirst and an increase in the secretion of ADH (figure 58.23).

ADH causes the walls of the collecting ducts in the kidney to become more permeable to water. This occurs because water channels are contained within the membranes of intracellular vesicles in the epithelium of the collecting ducts, and ADH stimulates the fusion of the vesicle membrane with the plasma membrane, similar to the process of exocytosis. When the secretion of ADH is reduced, the plasma membrane pinches in to form new vesicles that contain the water channels, so that the plasma membrane becomes less permeable to water.

Because the extracellular fluid in the renal medulla is hypertonic to the filtrate in the collecting ducts, water leaves the filtrate by osmosis and is reabsorbed into the blood. Under conditions of maximal ADH secretion, a person excretes only 600 milliliters of highly concentrated urine per day. A person who lacks ADH due to pituitary damage has the disorder known as *diabetes insipidus* and constantly excretes a large volume of dilute urine. Such a person is in danger of becoming severely dehydrated and succumbing to dangerously low blood pressure.

Aldosterone and Atrial Natriuretic Hormone

Sodium ion is the major solute in the blood plasma. When the blood concentration of Na^+ falls, therefore, the blood osmolality also falls. This drop in osmolality inhibits ADH secretion, causing more water to remain in the collecting duct for excretion in the urine. As a result, the blood volume and blood pressure decrease. A decrease in extracellular Na^+ also causes more water to be drawn into cells by osmosis, partially offsetting the drop in plasma osmolarity but further decreasing blood volume and blood pressure. If Na^+ deprivation is severe, the blood volume may fall so low that there is insufficient blood pressure to sustain life. For this reason, salt is necessary for life. Many animals have a "salt hunger" and actively seek salt, such as the deer at "salt licks."

A drop in blood Na^+ concentration is normally compensated by the kidneys under the influence of the hormone aldosterone, which is secreted by the adrenal cortex. Aldosterone stimulates the distal convoluted tubules to reabsorb Na^+, decreasing the excretion of Na^+ in the urine. Indeed, under conditions of maximal aldosterone secretion, Na^+ may be completely absent from the urine. The reabsorption of

FIGURE 58.24
A lowering of blood volume activates the renin-angiotensin-aldosterone system. (*1*) Low blood volume accompanies a decrease in blood Na^+ levels. (*2*) Reduced blood flow past the juxtaglomerular apparatus triggers (*3*) the release of renin into the blood, which catalyzes the production of angiotensin I from angiotensinogen. (*4*) Angiotensin I converts into an active form, angiotensin II. (*5*) Angiotensin II stimulates blood vessel constriction and (*6*) the release of aldosterone from the adrenal cortex. (7) Aldosterone stimulates the reabsorption of Na^+ in the distal convoluted tubules. (*8*) Increased Na^+ reabsorption is followed by the reabsorption of Cl^- and water. (*9*) This increases blood volume. An increase in blood volume may also trigger the release of atrial natriurctic hormone that inhibits the release of aldosterone. These two systems work together to maintain homeostasis.

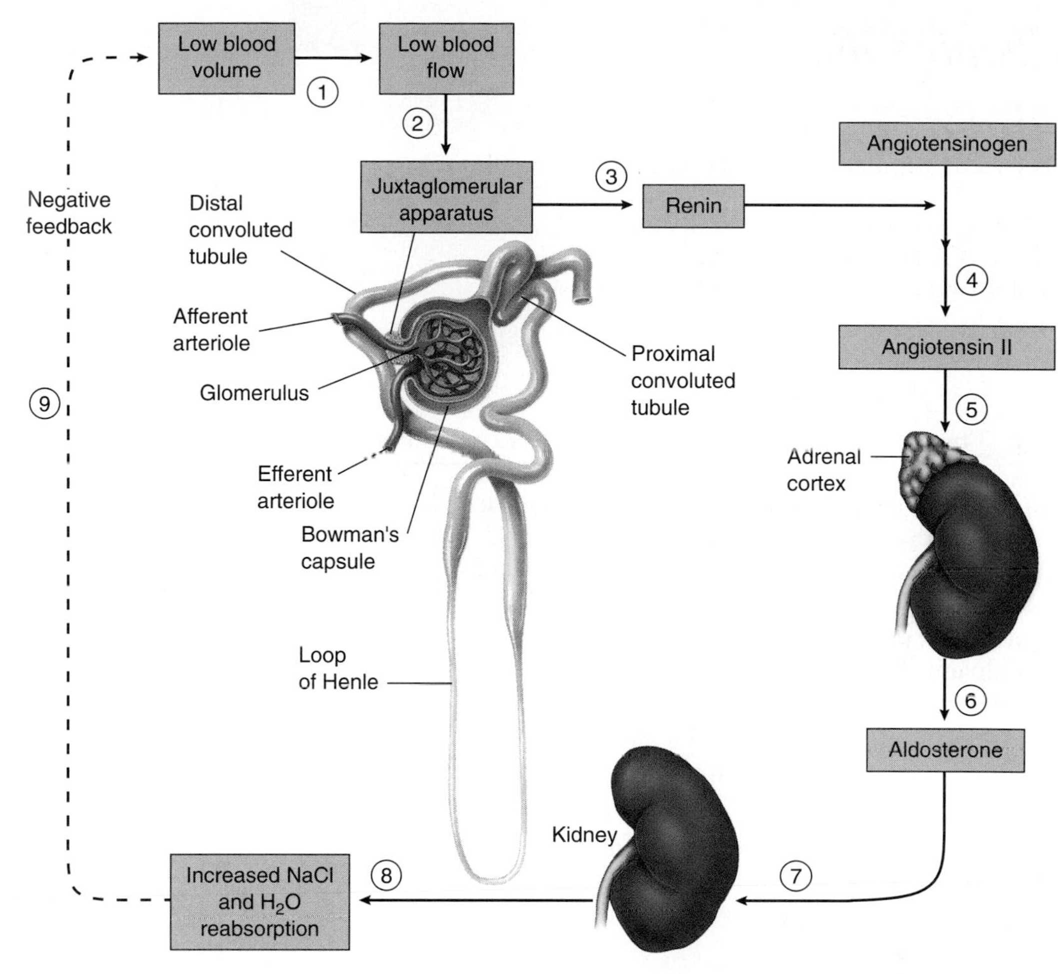

Na^+ is followed by Cl^- and by water, so aldosterone has the net effect of promoting the retention of both salt and water. It thereby helps to maintain blood volume and pressure.

The secretion of aldosterone in response to a decreased blood level of Na^+ is indirect. Because a fall in blood Na^+ is accompanied by a decreased blood volume, there is a reduced flow of blood past a group of cells called the juxtaglomerular apparatus, located in the region of the kidney between the distal convoluted tubule and the afferent arteriole (figure 58.24). The juxtaglomerular apparatus responds by secreting the enzyme *renin* into the blood, which catalyzes the production of the polypeptide angiotensin I from the protein angiotensinogen (see chapter 52). Angiotensin I is then converted by another enzyme into angiotensin II, which stimulates blood vessels to constrict and the adrenal cortex to secrete aldosterone. Thus, homeostasis of blood volume and pressure can be maintained by the activation of this renin-angiotensin-aldosterone system.

In addition to stimulating Na^+ reabsorption, aldosterone also promotes the secretion of K^+ into the distal convoluted tubules. Consequently, aldosterone lowers the blood K^+ concentration, helping to maintain constant blood K^+ levels in the face of changing amounts of K^+ in the diet. People who lack the ability to produce aldosterone will die if untreated because of the excessive loss of salt and water in the urine and the buildup of K^+ in the blood.

The action of aldosterone in promoting salt and water retention is opposed by another hormone, atrial natriuretic hormone (ANH, see chapter 52). This hormone is secreted by the right atrium of the heart in response to an increased blood volume, which stretches the atrium. Under these conditions, aldosterone secretion from the adrenal cortex will decrease and atrial natriuretic hormone secretion will increase, thus promoting the excretion of salt and water in the urine and lowering the blood volume.

ADH stimulates the insertion of water channels into the cells of the collecting duct, making the collecting duct more permeable to water. Thus, ADH stimulates the reabsorption of water and the excretion of a hypertonic urine. Aldosterone promotes the reabsorption of NaCl and water across the distal convoluted tubule, as well as the secretion of K^+ into the tubule. ANH decreases NaCl reabsorption.

Chapter 58

www.mhhe.com/raven6/resources58.mhtml

Summary | **Questions** | **Media Resources**

58.1 The regulatory systems of the body maintain homeostasis.

- Negative feedback loops maintain nearly constant extracellular conditions in the internal environment of the body, a condition called homeostasis.
- Antagonistic effectors afford an even finer degree of control.

1. What is homeostasis? What is a negative feedback loop? Give an example of how homeostasis is maintained by a negative feedback loop.

58.2 The extracellular fluid concentration is constant in most vertebrates.

- Osmoconformers maintain a tissue fluid osmolality equal to that of their environment, whereas osmoregulators maintain a constant blood osmolality that is different from that of their environment.
- Insects eliminate water by secreting K^+ into Malpighian tubules and the water follows the K^+ by osmosis.
- The kidneys of most vertebrates eliminate water by filtering blood into nephron tubules.
- Freshwater bony fish are hypertonic to their environment, and saltwater bony fish are hypotonic to their environment; these conditions place different demands upon their kidneys and other regulatory systems.
- Birds and mammals are the only vertebrates that have loops of Henle and thus are capable of producing a hypertonic urine.

2. What is the difference between an osmoconformer and an osmoregulator? What are examples of each?

3. How does the body fluid osmolality of a freshwater vertebrate compare with that of its environment? Does water tend to enter or exit its body? What must it do to maintain proper body water levels?

4. In what type of animal are Malpighian tubules found? By what mechanism is fluid caused to flow into these tubules? How is this fluid further modified before it is excreted?

- Osmoregulation

- Art Quiz: Osmotic Balance—Freshwater Fish

58.3 The functions of the vertebrate kidney are performed by nephrons.

- The primary function of the kidneys is homeostasis of blood volume, pressure, and composition, including the concentration of particular solutes in the blood and the blood pH.
- Bony fish remove the amine portions of amino acids and excrete them as ammonia across the gills.
- Elasmobranchs, adult amphibians, and mammals produce and excrete urea, which is quite soluble but much less toxic than ammonia.
- Insects, reptiles, and birds produce uric acid from the amino groups in amino acids; this precipitates, so that little water is required for its excretion.

5. What drives the movement of fluid from the blood to the inside of the nephron tubule at Bowman's capsule?

6. In what portion of the nephron is most of the NaCl and water reabsorbed from the filtrate?

7. What causes water reabsorption from the collecting duct? How is this influenced by antidiuretic hormone?

- Art Activities
  - -Urinary System
  - -Anatomy of Kidney and Lobe
  - -Nephron Anatomy in Mammalian Nephron

- Kidney Function

- Bioethics Case Study: Kidney Transplant

58.4 The kidney is regulated by hormones.

- Antidiuretic hormone is secreted by the posterior pituitary gland in response to an increase in blood osmolality, and acts to increase the number of water channels in the walls of the collecting ducts.

8. What effects does aldosterone have on kidney function? How is the secretion of aldosterone stimulated?

- Kidney Function

59

Sex and Reproduction

Concept Outline

59.1 Animals employ both sexual and asexual reproductive strategies.

Asexual and Sexual Reproduction. Some animals reproduce asexually, but most reproduce sexually; male and female are usually different individuals, but not always.

59.2 The evolution of reproduction among the vertebrates has led to internalization of fertilization and development.

Fertilization and Development. Among vertebrates that have internal fertilization, the young are nourished by egg yolk or from their mother's blood.

Fish and Amphibians. Most bony fish and amphibians have external fertilization, while most cartilaginous fish have internal fertilization.

Reptiles and Birds. Most reptiles and all birds lay eggs externally, and the young develop inside the egg.

Mammals. Monotremes lay eggs, marsupials have pouches where their young develop, and placental mammals have placentas that nourish the young within the uterus.

59.3 Male and female reproductive systems are specialized for different functions.

Structure and Function of the Male Reproductive System. The testes produce sperm and secrete the male sex hormone, testosterone.

Structure and Function of the Female Reproductive System. An egg cell within an ovarian follicle develops and is released from the ovary; the egg cell travels into the female reproductive tract, which undergoes cyclic changes due to hormone secretion.

59.4 The physiology of human sexual intercourse is becoming better known.

Physiology of Human Sexual Intercourse. The human sexual response can be divided into four phases: excitement, plateau, orgasm, and resolution.

Birth Control. Various methods of birth control are employed, including barriers to fertilization, prevention of ovulation, and prevention of the implantation.

FIGURE 59.1
The bright color of male golden toads serves to attract mates. The rare golden toads of the Monteverde Cloud Forest Reserve of Costa Rica are nearly voiceless and so use bright colors to attract mates. Always rare, they may now be extinct.

The cry of a cat in heat, insects chirping outside the window, frogs croaking in swamps, and wolves howling in a frozen northern forest are all sounds of evolution's essential act, reproduction. These distinct vocalizations, as well as the bright coloration characteristic of some animals like the tropical golden toads of figure 59.1, function to attract mates. Few subjects pervade our everyday thinking more than sex, and few urges are more insistent. This chapter deals with sex and reproduction among the vertebrates, including humans.

59.1 Animals employ both sexual and asexual reproductive strategies.

Asexual and Sexual Reproduction

Asexual reproduction is the primary means of reproduction among the protists, cnidaria, and tunicates, but it may also occur in some of the more complex animals. Indeed, the formation of identical twins (by the separation of two identical cells of a very early embryo) is a form of asexual reproduction.

Through mitosis, genetically identical cells are produced from a single parent cell. This permits asexual reproduction to occur in protists by division of the organism, or **fission.** Cnidaria commonly reproduce by **budding,** where a part of the parent's body becomes separated from the rest and differentiates into a new individual. The new individual may become an independent animal or may remain attached to the parent, forming a colony.

Sexual reproduction occurs when a new individual is formed by the union of two sex cells, or **gametes,** a term that includes **sperm** and **eggs** (or ova). The union of sperm and egg cells produces a fertilized egg, or **zygote,** that develops by mitotic division into a new multicellular organism. The zygote and the cells it forms by mitosis are diploid; they contain both members of each homologous pair of chromosomes. The gametes, formed by meiosis in the sex organs, or **gonads**—the **testes** and **ovaries**—are haploid (see chapter 12). The process of spermatogenesis (sperm formation) and oogenesis (egg formation) will be described in later sections. For a more detailed discussion of asexual and sexual reproduction, see chapter 12.

(a)

(b)

FIGURE 59.2
Hermaphroditism and protogyny. (*a*) The hamlet bass (genus *Hypoplectrus*) is a deep-sea fish that is a hermaphrodite—both male and female at the same time. In the course of a single pair-mating, one fish may switch sexual roles as many as four times, alternately offering eggs to be fertilized and fertilizing its partner's eggs. Here the fish acting as a male curves around its motionless partner, fertilizing the upward-floating eggs. (*b*) The bluehead wrasse, *Thalassoma bifasciatium*, is protogynous—females sometimes turn into males. Here a large male, or sex-changed female, is seen among females, typically much smaller.

Different Approaches to Sex

Parthenogenesis (females produce offspring from unfertilized eggs) is common in many species of arthropods; some species are exclusively parthenogenic (and all female), while others switch between sexual reproduction and parthenogenesis in different generations. In honeybees, for example, a queen bee mates only once and stores the sperm. She then can control the release of sperm. If no sperm are released, the eggs develop parthenogenetically into drones, which are males; if sperm are allowed to fertilize the eggs, the fertilized eggs develop into other queens or worker bees, which are female.

The Russian biologist Ilya Darevsky reported in 1958 one of the first cases of unusual modes of reproduction among vertebrates. He observed that some populations of small lizards of the genus *Lacerta* were exclusively female, and suggested that these lizards could lay eggs that were viable even if they were not fertilized. In other words, they were capable of asexual reproduction in the absence of sperm, a type of parthenogenesis. Further work has shown that parthenogenesis also occurs among populations of other lizard genera.

Another variation in reproductive strategies is **hermaphroditism,** when one individual has both testes and ovaries, and so can produce both sperm and eggs (figure 59.2*a*). A tapeworm is hermaphroditic and can fertilize itself, a useful strategy because it is unlikely to encounter another tapeworm. Most hermaphroditic animals, however, require another individual to reproduce. Two earthworms, for example, are required for reproduction—each functions as both male and female, and each leaves the encounter with fertilized eggs.

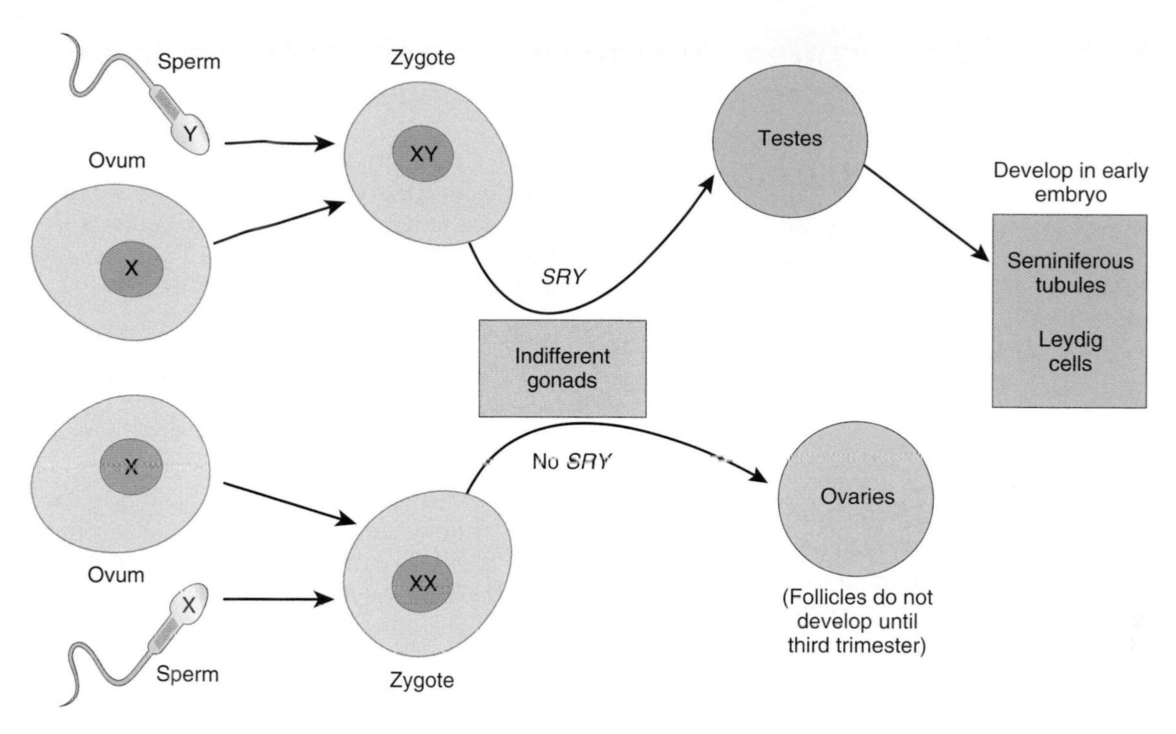

FIGURE 59.3
Sex determination in mammals is made by a region of the Y chromosome designated *SRY*. Testes are formed when the Y chromosome and *SRY* are present; ovaries are formed when they are absent.

There are some deep-sea fish that are hermaphrodites—both male and female at the same time. Numerous fish genera include species in which individuals can change their sex, a process called *sequential hermaphroditism.* Among coral reef fish, for example, both **protogyny** ("first female," a change from female to male) and **protandry** ("first male," a change from male to female) occur. In fish that practice protogyny (figure 59.2*b*), the sex change appears to be under social control. These fish commonly live in large groups, or schools, where successful reproduction is typically limited to one or a few large, dominant males. If those males are removed, the largest female rapidly changes sex and becomes a dominant male.

Sex Determination

Among the fish just described, and in some species of reptiles, environmental changes can cause changes in the sex of the animal. In mammals, the sex is determined early in embryonic development. The reproductive systems of human males and females appear similar for the first 40 days after conception. During this time, the cells that will give rise to ova or sperm migrate from the yolk sac to the embryonic gonads, which have the potential to become either ovaries in females or testes in males. For this reason, the embryonic gonads are said to be "indifferent." If the embryo is a male, it will have a Y chromosome with a gene whose product converts the indifferent gonads into testes. In females, which lack a Y chromosome, this gene and the protein it encodes are absent, and the gonads become ovaries. Recent evidence suggests that the sex-determining gene may be one known as ***SRY*** (for "sex-determining region of the Y chromosome") (figure 59.3). The *SRY* gene appears to have been highly conserved during the evolution of different vertebrate groups.

Once testes form in the embryo, the testes secrete testosterone and other hormones that promote the development of the male external genitalia and accessory reproductive organs. If the embryo lacks testes (the ovaries are nonfunctional at this stage), the embryo develops female external genitalia and sex accessory organs. In other words, all mammalian embryos will develop female sex accessory organs and external genitalia unless they are masculinized by the secretions of the testes.

Sexual reproduction is most common among animals, but many reproduce asexually by fission, budding, or parthenogenesis. Sexual reproduction generally involves the fusion of gametes derived from different individuals of a species, but some species are hermaphroditic.

59.2 The evolution of reproduction among the vertebrates has led to internalization of fertilization and development.

Fertilization and Development

Vertebrate sexual reproduction evolved in the ocean before vertebrates colonized the land. The females of most species of marine bony fish produce eggs or ova in batches and release them into the water. The males generally release their sperm into the water containing the eggs, where the union of the free gametes occurs. This process is known as **external fertilization.**

Although seawater is not a hostile environment for gametes, it does cause the gametes to disperse rapidly, so their release by females and males must be almost simultaneous. Thus, most marine fish restrict the release of their eggs and sperm to a few brief and well-defined periods. Some reproduce just once a year, while others do so more frequently. There are few seasonal cues in the ocean that organisms can use as signals for synchronizing reproduction, but one all-pervasive signal is the cycle of the moon. Once each month, the moon approaches closer to the earth than usual, and when it does, its increased gravitational attraction causes somewhat higher tides. Many marine organisms sense the tidal changes and entrain the production and release of their gametes to the lunar cycle.

The invasion of land posed the new danger of desiccation, a problem that was especially severe for the small and vulnerable gametes. On land, the gametes could not simply be released near each other, as they would soon dry up and perish. Consequently, there was intense selective pressure for terrestrial vertebrates (as well as some groups of fish) to evolve **internal fertilization,** that is, the introduction of male gametes into the female reproductive tract. By this means, fertilization still occurs in a nondesiccating environment, even when the adult animals are fully terrestrial. The vertebrates that practice internal fertilization have three strategies for embryonic and fetal development:

1. **Oviparity.** This is found in some bony fish, most reptiles, some cartilaginous fish, some amphibians, a few mammals, and all birds. The eggs, after being fertilized internally, are deposited outside the mother's body to complete their development.
2. **Ovoviviparity.** This is found in some bony fish (including mollies, guppies, and mosquito fish), some cartilaginous fish, and many reptiles. The fertilized eggs are retained within the mother to complete their development, but the embryos still obtain all of their nourishment from the egg yolk. The young are fully developed when they are hatched and released from the mother.
3. **Viviparity.** This is found in most cartilaginous fish, some amphibians, a few reptiles, and almost all mammals. The young develop within the mother and obtain nourishment directly from their mother's blood, rather than from the egg yolk (figure 59.4).

Fertilization is external in most fish but internal in most other vertebrates. Depending upon the relationship of the developing embryo to the mother and egg, those vertebrates with internal fertilization may be classified as oviparous, ovoviviparous, or viviparous.

FIGURE 59.4 Viviparous fish carry live, mobile young within their bodies. The young complete their development within the body of the mother and are then released as small but competent adults. Here a lemon shark has just given birth to a young shark, which is still attached by the umbilical cord.

Fish and Amphibians

Most fish and amphibians, unlike other vertebrates, reproduce by means of external fertilization.

Fish

Fertilization in most species of bony fish (teleosts) is external, and the eggs contain only enough yolk to sustain the developing embryo for a short time. After the initial supply of yolk has been exhausted, the young fish must seek its food from the waters around it. Development is speedy, and the young that survive mature rapidly. Although thousands of eggs are fertilized in a single mating, many of the resulting individuals succumb to microbial infection or predation, and few grow to maturity.

In marked contrast to the bony fish, fertilization in most cartilaginous fish is internal. The male introduces sperm into the female through a modified pelvic fin. Development of the young in these vertebrates is generally viviparous.

Amphibians

The amphibians invaded the land without fully adapting to the terrestrial environment, and their life cycle is still tied to the water. Fertilization is external in most amphibians, just as it is in most species of bony fish. Gametes from both males and females are released through the cloaca. Among the frogs and toads, the male grasps the female and discharges fluid containing the sperm onto the eggs as they are released into the water (figure 59.5). Although the eggs of most amphibians develop in the water, there are some interesting exceptions. In two species of frogs, for example, the eggs develop in the vocal sacs and stomach, and the young frogs leave through their father's and mother's mouths, respectively (figure 59.6)!

The time required for development of amphibians is much longer than that for fish, but amphibian eggs do not include a significantly greater amount of yolk. Instead, the process of development in most amphibians is divided into embryonic, larval, and adult stages, in a way reminiscent of the life cycles found in some insects. The embryo develops within the egg, obtaining nutrients from the yolk. After hatching from the egg, the aquatic larva then functions as a free-swimming, food-gathering machine, often for a considerable period of time. The larvae may increase in size rapidly; some tadpoles, which are the larvae of frogs and toads, grow in a matter of weeks from creatures no bigger than the tip of a pencil into individuals as big as a goldfish. When the larva has grown to a sufficient size, it undergoes a developmental transition, or metamorphosis, into the terrestrial adult form.

The eggs of most bony fish and amphibians are fertilized externally. In amphibians the eggs develop into a larval stage that undergoes metamorphosis.

FIGURE 59.5
The eggs of frogs are fertilized externally. When frogs mate, as these two are doing, the clasp of the male induces the female to release a large mass of mature eggs, over which the male discharges his sperm.

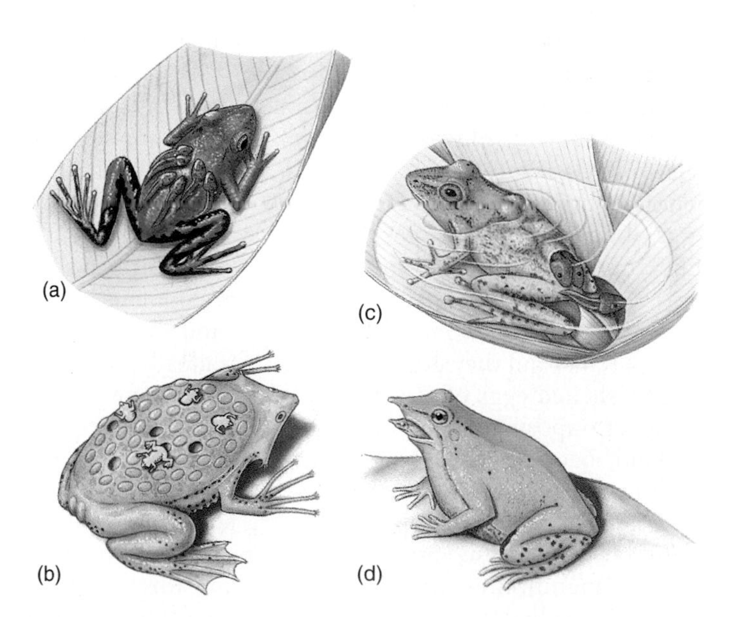

FIGURE 59.6
Different ways young develop in frogs. (*a*) In the poison arrow frog, the male carries the tadpoles on his back. (*b*) In the female Surinam frog, froglets develop from eggs in special brooding pouches on the back. (*c*) In the South American pygmy marsupial frog, the female carries the developing larvae in a pouch on her back. (*d*) Tadpoles of the Darwin's frog develop into froglets in the vocal pouch of the male and emerge from the mouth.

Reptiles and Birds

Most reptiles and all birds are oviparous—after the eggs are fertilized internally, they are deposited outside of the mother's body to complete their development. Like most vertebrates that fertilize internally, male reptiles utilize a tubular organ, the penis, to inject sperm into the female (figure 59.7). The penis, containing erectile tissue, can become quite rigid and penetrate far into the female reproductive tract. Most reptiles are oviparous, laying eggs and then abandoning them. These eggs are surrounded by a leathery shell that is deposited as the egg passes through the oviduct, the part of the female reproductive tract leading from the ovary. A few species of reptiles are ovoviviparous or viviparous, forming eggs that develop into embryos within the body of the mother.

FIGURE 59.7
The introduction of sperm by the male into the female's body is called copulation. Reptiles such as these tortoises were the first terrestrial vertebrates to develop this form of reproduction, which is particularly suited to a terrestrial environment.

All birds practice internal fertilization, though most male birds lack a penis. In some of the larger birds (including swans, geese, and ostriches), however, the male cloaca extends to form a false penis. As the egg passes along the oviduct, glands secrete albumin proteins (the egg white) and the hard, calcareous shell that distinguishes bird eggs from reptilian eggs. While modern reptiles are poikilotherms (animals whose body temperature varies with the temperature of their environment), birds are homeotherms (animals that maintain a relatively constant body temperature independent of environmental temperatures). Hence, most birds incubate their eggs after laying them to keep them warm (figure 59.8). The young that hatch from the eggs of most bird species are unable to survive unaided, as their development is still incomplete. These young birds are fed and nurtured by their parents, and they grow to maturity gradually.

FIGURE 59.8
Crested penguins incubating their egg. This nesting pair is changing the parental guard in a stylized ritual.

The shelled eggs of reptiles and birds constitute one of the most important adaptations of these vertebrates to life on land, because shelled eggs can be laid in dry places. Such eggs are known as amniotic eggs because the embryo develops within a fluid-filled cavity surrounded by a membrane called the amnion. The amnion is an extraembryonic membrane—that is, a membrane formed from embryonic cells but located outside the body of the embryo. Other extraembryonic membranes in amniotic eggs include the chorion, which lines the inside of the eggshell, the yolk sac, and the allantois. In contrast, the eggs of fish and amphibians contain only one extraembryonic membrane, the yolk sac. The viviparous mammals, including humans, also have extraembryonic membranes that will be described in chapter 60.

Most reptiles and all birds are oviparous, laying amniotic eggs that are protected by watertight membranes from desiccation. Birds, being homeotherms, must keep the eggs warm by incubation.

Mammals

Some mammals are seasonal breeders, reproducing only once a year, while others have shorter reproductive cycles. Among the latter, the females generally undergo the reproductive cycles, while the males are more constant in their reproductive activity. Cycling in females involves the periodic release of a mature ovum from the ovary in a process known as ovulation. Most female mammals are "in heat," or sexually receptive to males, only around the time of ovulation. This period of sexual receptivity is called **estrus,** and the reproductive cycle is therefore called an **estrous cycle.** The females continue to cycle until they become pregnant.

In the estrous cycle of most mammals, changes in the secretion of follicle-stimulating hormone (FSH) and luteinizing hormone (LH) by the anterior pituitary gland cause changes in egg cell development and hormone secretion in the ovaries. Humans and apes have menstrual cycles that are similar to the estrous cycles of other mammals in their cyclic pattern of hormone secretion and ovulation. Unlike mammals with estrous cycles, however, human and ape females bleed when they shed the inner lining of their uterus, a process called menstruation, and may engage in copulation at any time during the cycle.

Rabbits and cats differ from most other mammals in that they are induced ovulators. Instead of ovulating in a cyclic fashion regardless of sexual activity, the females ovulate only after copulation as a result of a reflex stimulation of LH secretion (described later). This makes these animals extremely fertile.

The most primitive mammals, the **monotremes** (consisting solely of the duck-billed platypus and the echidna), are oviparous, like the reptiles from which they evolved. They incubate their eggs in a nest (figure 59.9*a*) or specialized pouch, and the young hatchlings obtain milk from their mother's mammary glands by licking her skin, as monotremes lack nipples. All other mammals are viviparous, and are divided into two subcategories based on how they nourish their young. The **marsupials,** a group that includes opossums and kangaroos, give birth to fetuses that are incompletely developed. The fetuses complete their development in a pouch of their mother's skin, where they can obtain nourishment from nipples of the mammary glands (figure 59.9*b*). The **placental mammals** (figure 59.9*c*) retain their young for a much longer period of development within the mother's uterus. The fetuses are nourished by a structure known as the placenta, which is derived from both an extraembryonic membrane (the chorion) and the mother's uterine lining. Because the fetal and maternal blood vessels are in very close proximity in the placenta, the fetus can obtain nutrients by diffusion from the mother's blood. The functioning of the placenta is discussed in more detail in chapter 60.

Among mammals that are not seasonal breeders, the females undergo shorter cyclic variations in ovarian function. These are estrous cycles in most mammals and menstrual cycles in humans and apes. Some mammals are induced ovulators, ovulating in response to copulation.

(a)

(b)

(c)

FIGURE 59.9
Reproduction in mammals. (*a*) Monotremes, like the duck-billed platypus shown here, lay eggs in a nest. (*b*) Marsupials, such as this kangaroo, give birth to small fetuses which complete their development in a pouch. (*c*) In placental mammals, like this domestic cat, the young remain inside the mother's uterus for a longer period of time and are born relatively more developed.

59.3 Male and female reproductive systems are specialized for different functions.

Structure and Function of the Male Reproductive System

The structures of the human male reproductive system, typical of mammals, are illustrated in figure 59.10. If testes form in the human embryo, they develop *seminiferous tubules* beginning at around 43 to 50 days after conception. The seminiferous tubules are the sites of sperm production. At about 9 to 10 weeks, the Leydig cells, located in the interstitial tissue between the seminiferous tubules, begin to secrete testosterone (the major male sex hormone, or androgen). Testosterone secretion during embryonic development converts indifferent structures into the male external genitalia, the *penis* and the *scrotum*, a sac that contains the testes. In the absence of testosterone, these structures develop into the female external genitalia.

In an adult, each testis is composed primarily of the highly convoluted seminiferous tubules (figure 59.11). Although the testes are actually formed within the abdominal cavity, shortly before birth they descend through an opening called the inguinal canal into the scrotum, which suspends them outside the abdominal cavity. The scrotum maintains the testes at around 34°C, slightly lower than the core body temperature (37°C). This lower temperature is required for normal sperm development in humans.

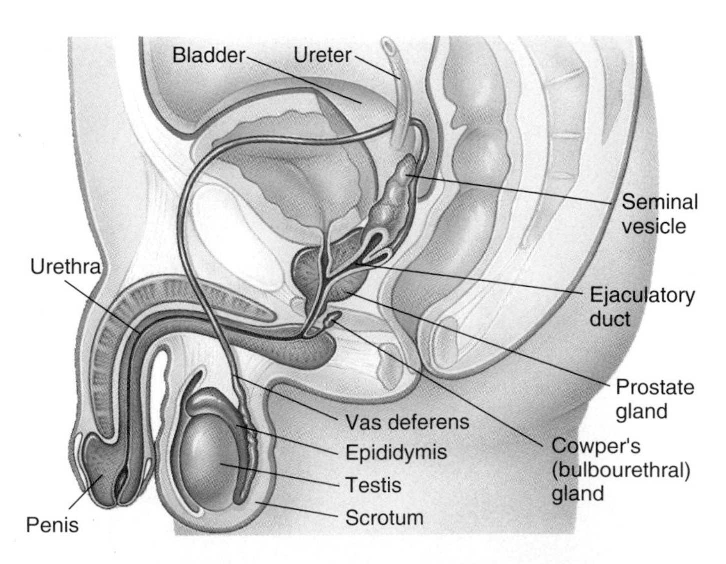

FIGURE 59.10
Organization of the human male reproductive system. The penis and scrotum are the external genitalia, the testes are the gonads, and the other organs are accessory sex organs, aiding the production and ejaculation of semen.

Production of Sperm

The wall of the seminiferous tubule consists of *germinal cells*, which become sperm by meiosis, and supporting *Sertoli cells*. The germinal cells near the outer surface of the seminiferous tubule are diploid (with 46 chromosomes in humans), while cells located closer to the lumen of the tubule are haploid (with 23 chromosomes each). Each parent cell duplicates by mitosis, and one of the two daughter cells then undergoes meiosis to form sperm; the other remains as a parent cell. In that way, the male never runs out of parent cells to produce sperm. Adult males produce an average of 100 to 200 million sperm each day and can continue to do so throughout most of the rest of their lives.

The diploid daughter cell that begins meiosis is called a *primary spermatocyte*. It has 23 pairs of homologous chromosomes (in humans) and each chromosome is duplicated, with two chromatids. The first meiotic division separates the homologous chromosomes, producing two haploid *secondary spermatocytes*. However, each chromosome still consists of two duplicate chromatids. Each of these cells then undergoes the second meiotic division to separate the chromatids and produce two haploid cells, the *spermatids*. Therefore, a total of four haploid spermatids are produced by each primary spermatocyte (figure 59.11). All of these cells constitute the germinal epithelium of the seminiferous tubules because they "germinate" the gametes.

In addition to the germinal epithelium, the walls of the seminiferous tubules contain nongerminal cells known as Sertoli cells. The Sertoli cells nurse the developing sperm and secrete products required for spermatogenesis (sperm production). They also help convert the spermatids into *spermatozoa* by engulfing their extra cytoplasm.

Spermatozoa, or sperm, are relatively simple cells, consisting of a head, body, and tail (figure 59.12). The head encloses a compact nucleus and is capped by a vesicle called an acrosome, which is derived from the Golgi complex. The acrosome contains enzymes that aid in the penetration of the protective layers surrounding the egg. The body and tail provide a propulsive mechanism: within the tail is a flagellum, while inside the body are a centriole, which acts as a basal body for the flagellum, and mitochondria, which generate the energy needed for flagellar movement.

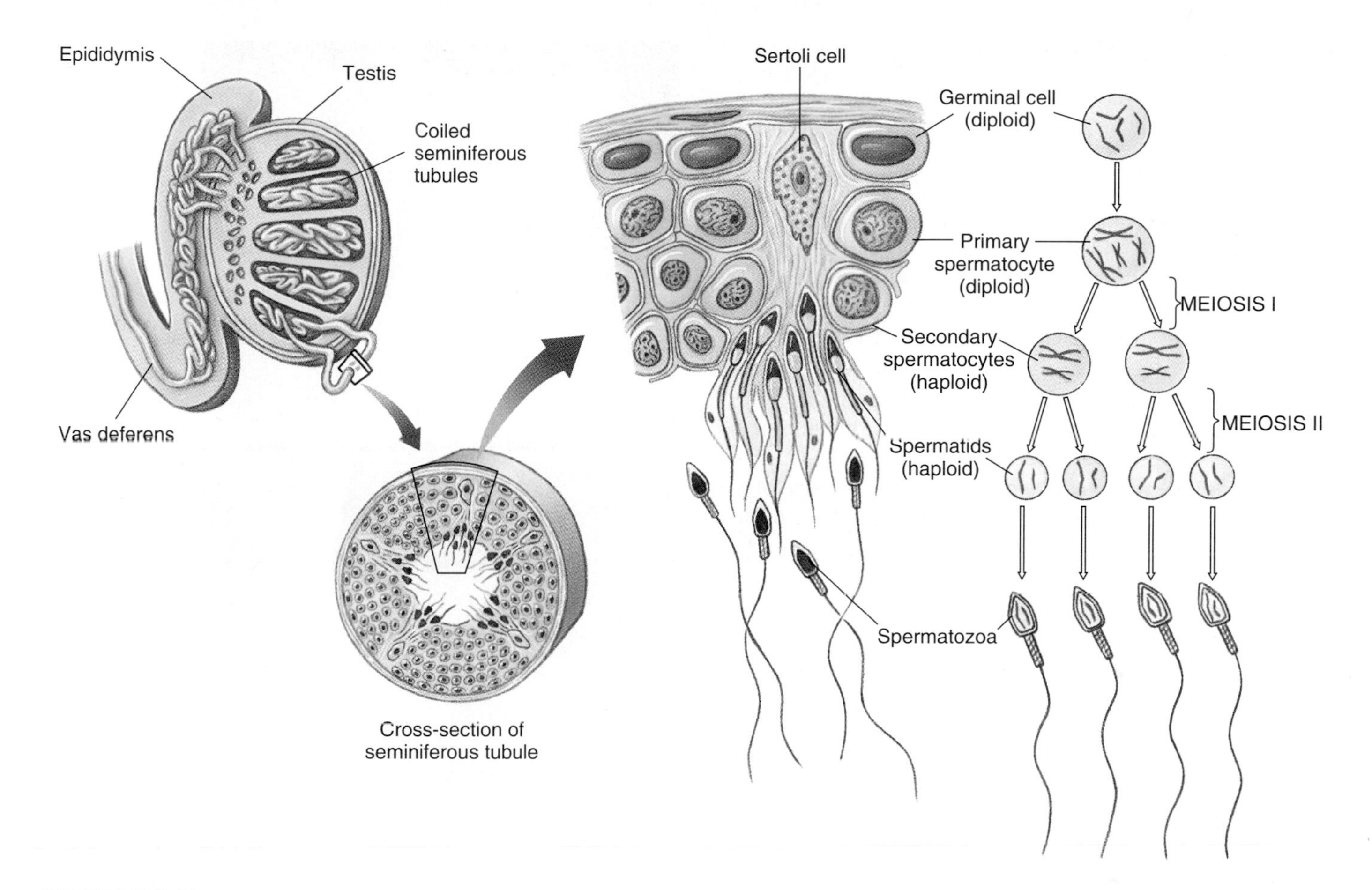

FIGURE 59.11
The testis and spermatogenesis. Inside the testis, the seminiferous tubules are the sites of spermatogenesis. Germinal cells in the seminiferous tubules give rise to spermatozoa by meiosis. Sertoli cells are nongerminal cells within the walls of the seminiferous tubules. They assist spermatogenesis in several ways, such as helping to convert spermatids into spermatozoa. A primary spermatocyte is diploid. At the end of the first meiotic division, homologous chromosomes have separated, and two haploid secondary spermatocytes form. The second meiotic division separates the sister chromatids and results in the formation of four haploid spermatids.

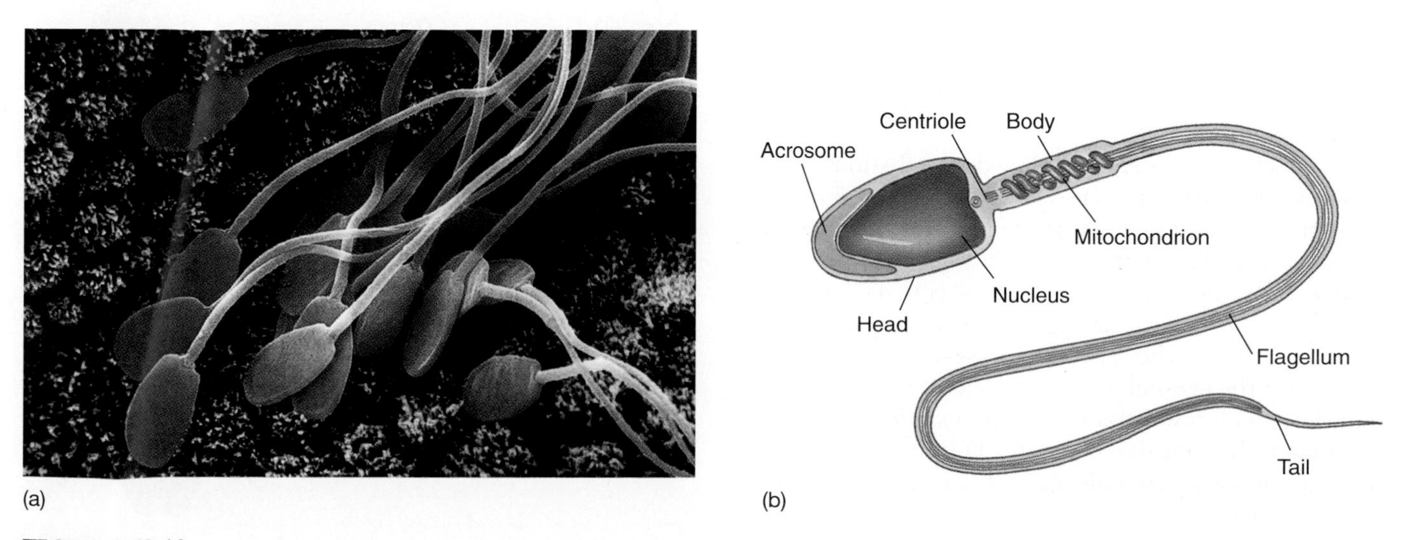

FIGURE 59.12
Human sperm. (*a*) A scanning electron micrograph. (*b*) A diagram of the main components of a sperm cell.

Male Accessory Sex Organs

After the sperm are produced within the seminiferous tubules, they are delivered into a long, coiled tube called the epididymis (figure 59.13). The sperm are not motile when they arrive in the epididymis, and they must remain there for at least 18 hours before their motility develops. From the epididymis, the sperm enter another long tube, the vas deferens, which passes into the abdominal cavity via the inguinal canal.

The vas deferens from each testis joins with one of the ducts from a pair of glands called the seminal vesicles (see figure 59.10), which produce a fructose-rich fluid. From this point, the vas deferens continues as the ejaculatory duct and enters the prostate gland at the base of the urinary bladder. In humans, the prostate gland is about the size of a golf ball and is spongy in texture. It contributes about 60% of the bulk of the semen, the fluid that contains the products of the testes, fluid from the seminal vesicles, and the products of the prostate gland. Within the prostate gland, the ejaculatory duct merges with the urethra from the urinary bladder. The urethra carries the semen out of the body through the tip of the penis. A pair of pea-sized bulbourethral glands secrete a fluid that lines the urethra and lubricates the tip of the penis prior to coitus (sexual intercourse).

In addition to the urethra, there are two columns of erectile tissue, the corpora cavernosa, along the dorsal side of the penis and one column, the corpus spongiosum, along the ventral side (figure 59.14). Penile erection is produced by neurons in the parasympathetic division of the autonomic nervous system. As a result of the release of nitric oxide by these neurons, arterioles in the penis dilate, causing the erectile tissue to become engorged with blood and turgid. This increased pressure in the erectile tissue compresses the veins, so blood flows into the penis but cannot flow out. The drug sildenafil **(Viagra)** prolongs erection by stimulating release of nitric oxide in the penis. Some mammals, such as the walrus, have a bone in the penis that contributes to its stiffness during erection, but humans do not.

The result of erection and continued sexual stimulation is ejaculation, the ejection from the penis of about 5 milliliters of semen containing an average of 300 million sperm. Successful fertilization requires such a high sperm count because the odds against any one sperm cell successfully completing the journey to the egg and fertilizing it are extraordinarily high, and the acrosomes of several sperm need to interact with the egg before a single sperm can penetrate the egg. Males with fewer than 20 million sperm per milliliter are generally considered sterile. Despite their large numbers, sperm constitute only about 1% of the volume of the semen ejaculated.

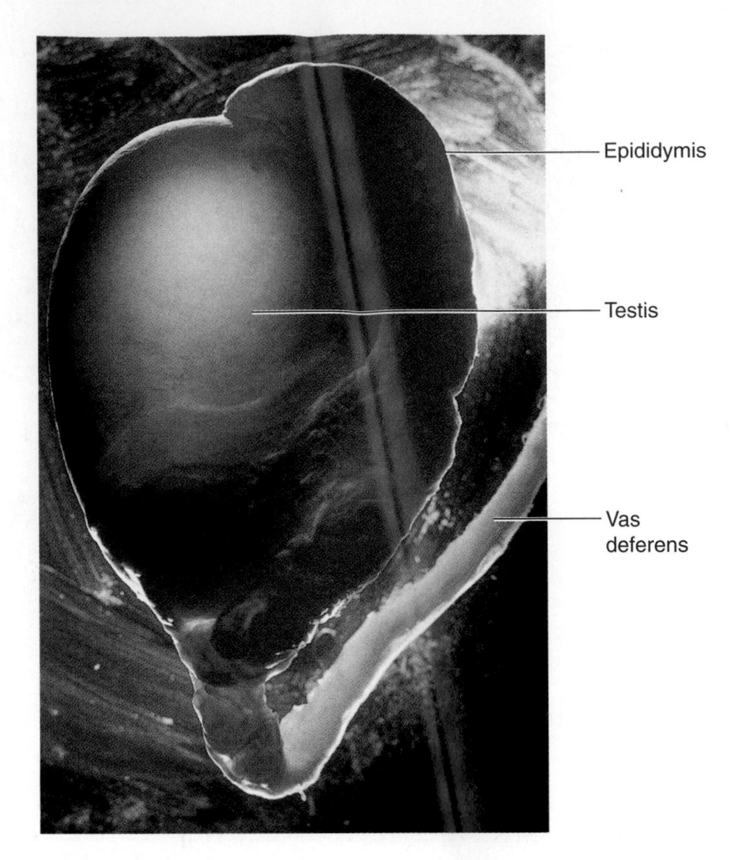

FIGURE 59.13
Photograph of the human testis. The dark, round object in the center of the photograph is a testis, within which sperm are formed. Cupped around it is the epididymis, a highly coiled passageway in which sperm complete their maturation. Mature sperm are stored in the vas deferens, a long tube that extends from the epididymis.

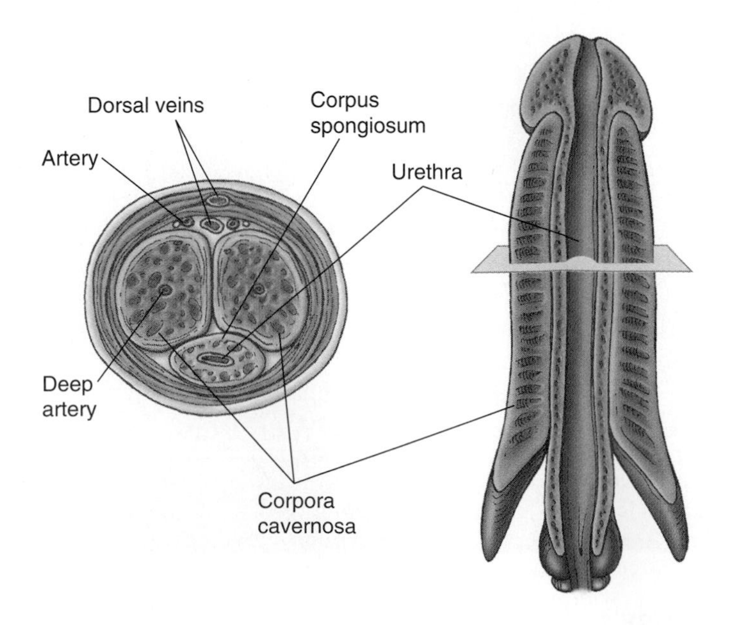

FIGURE 59.14
A penis in cross-section (*left*) and longitudinal section (*right*). Note that the urethra runs through the corpus spongiosum.

Table 59.1 Mammalian Reproductive Hormones	
MALE	
Follicle-stimulating hormone (FSH)	Stimulates spermatogenesis
Luteinizing hormone (LH)	Stimulates secretion of testosterone by Leydig cells
Testosterone	Stimulates development and maintenance of male secondary sexual characteristics and accessory sex organs
FEMALE	
Follicle-stimulating hormone (FSH)	Stimulates growth of ovarian follicles and secretion of estradiol
Luteinizing hormone (LH)	Stimulates ovulation, conversion of ovarian follicles into corpus luteum, and secretion of estradiol and progesterone by corpus luteum
Estradiol	Stimulates development and maintenance of female secondary sexual characteristics; prompts monthly preparation of uterus for pregnancy
Progesterone	Completes preparation of uterus for pregnancy; helps maintain female secondary sexual characteristics
Oxytocin	Stimulates contraction of uterus and milk-ejection reflex
Prolactin	Stimulates milk production

Hormonal Control of Male Reproduction

As we saw in chapter 56, the anterior pituitary gland secretes two gonadotropic hormones: FSH and LH. Although these hormones are named for their actions in the female, they are also involved in regulating male reproductive function (table 59.1). In males, FSH stimulates the Sertoli cells to facilitate sperm development, and LH stimulates the Leydig cells to secrete testosterone.

The principle of negative feedback inhibition discussed in chapter 56 applies to the control of FSH and LH secretion (figure 59.15). The hypothalamic hormone, gonadotropin-releasing hormone (GnRH), stimulates the anterior pituitary gland to secrete both FSH and LH. FSH causes the Sertoli cells to release a peptide hormone called inhibin that specifically inhibits FSH secretion. Similarly, LH stimulates testosterone secretion, and testosterone feeds back to inhibit the release of LH, both directly at the anterior pituitary gland and indirectly by reducing GnRH release. The importance of negative feedback inhibition can be demonstrated by removing the testes; in the absence of testosterone and inhibin, the secretion of FSH and LH from the anterior pituitary is greatly increased.

An adult male produces sperm continuously by meiotic division of the germinal cells lining the seminiferous tubules. Semen consists of sperm from the testes and fluid contributed by the seminal vesicles and prostate gland. Production of sperm and secretion of testosterone from the testes are controlled by FSH and LH from the anterior pituitary.

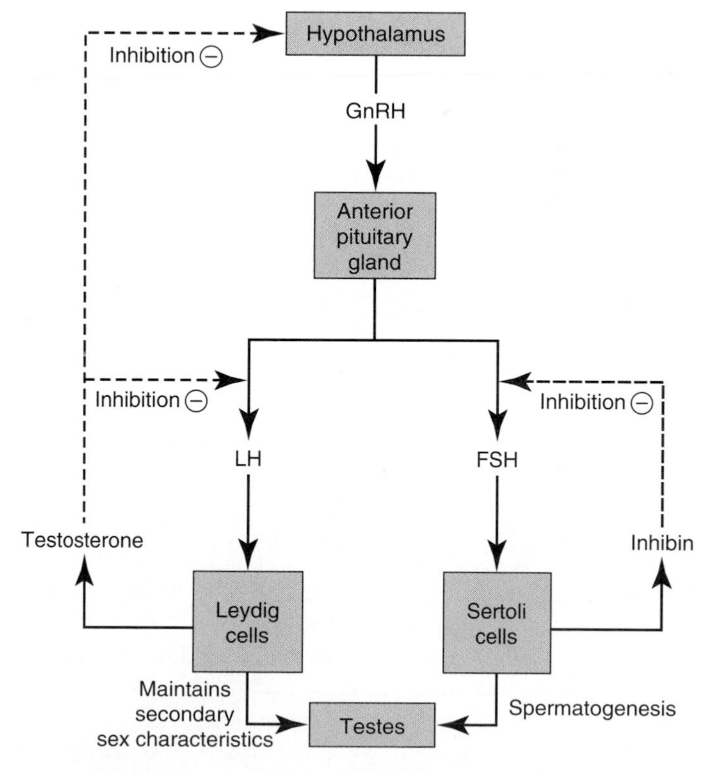

FIGURE 59.15
Hormonal interactions between the testes and anterior pituitary. LH stimulates the Leydig cells to secrete testosterone, and FSH stimulates the Sertoli cells of the seminiferous tubules to secrete inhibin. Testosterone and inhibin, in turn, exert negative feedback inhibition on the secretion of LH and FSH, respectively.

Structure and Function of the Female Reproductive System

The structures of the reproductive system in a human female are shown in figure 59.16. In contrast to the testes, the ovaries develop much more slowly. In the absence of testosterone, the female embryo develops a **clitoris** and **labia majora** from the same embryonic structures that produce a penis and scrotum in males. Thus clitoris and penis, and the labia majora and scrotum, are said to be *homologous structures*. The clitoris, like the penis, contains corpora cavernosa and is therefore erectile. The ovaries contain microscopic structures called **ovarian follicles,** which each contain an egg cell and smaller **granulosa cells.** The ovarian follicles are the functional units of the ovary.

At puberty, the granulosa cells begin to secrete the major female sex hormone estradiol (also called estrogen), triggering **menarche,** the onset of menstrual cycling. Estradiol also stimulates the formation of the **female secondary sexual characteristics,** including breast development and the production of pubic hair. In addition, estradiol and another steroid hormone, progesterone, help to maintain the female accessory sex organs: the fallopian tubes, uterus, and vagina.

Female Accessory Sex Organs

The fallopian tubes (also called uterine tubes or oviducts) transport ova from the ovaries to the uterus. In humans, the uterus is a muscular, pear-shaped organ that narrows to form a neck, the cervix, which leads to the vagina (figure 59.17*a*). The uterus is lined with a simple columnar epithelial membrane called the endometrium. The surface of the endometrium is shed during menstruation, while the underlying portion remains to generate a new surface during the next cycle.

Mammals other than primates have more complex female reproductive tracts, where part of the uterus divides to form uterine "horns," each of which leads to an oviduct (figure 59.17*b*, *c*). In cats, dogs, and cows, for example, there is one cervix but two uterine horns separated by a septum, or wall. Marsupials, such as opossums, carry the split even further, with two unconnected uterine horns, two cervices, and two vaginas. A male marsupial has a forked penis that can enter both vaginas simultaneously.

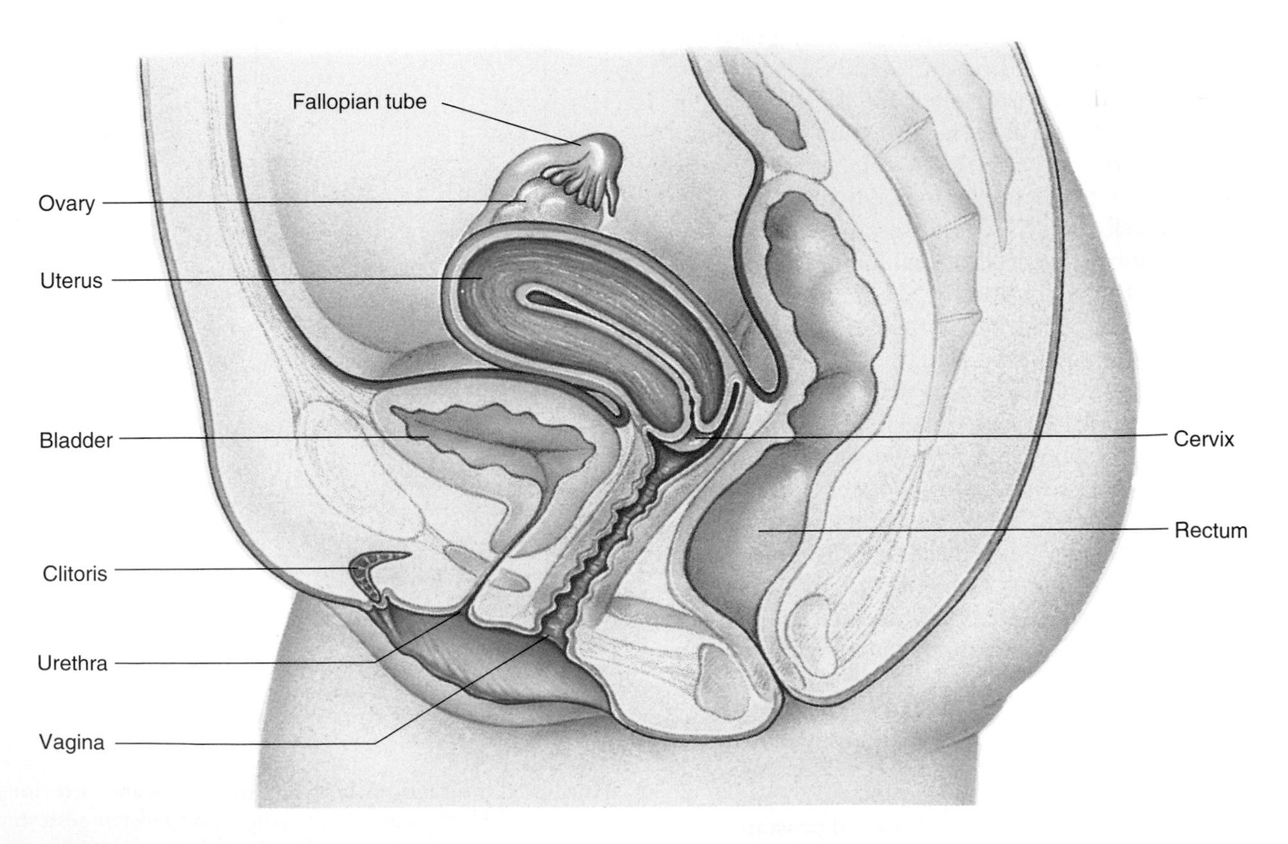

FIGURE 59.16
Organization of the human female reproductive system. The ovaries are the gonads, the fallopian tubes receive the ovulated ova, and the uterus is the womb, the site of development of an embryo if the egg cell becomes fertilized.

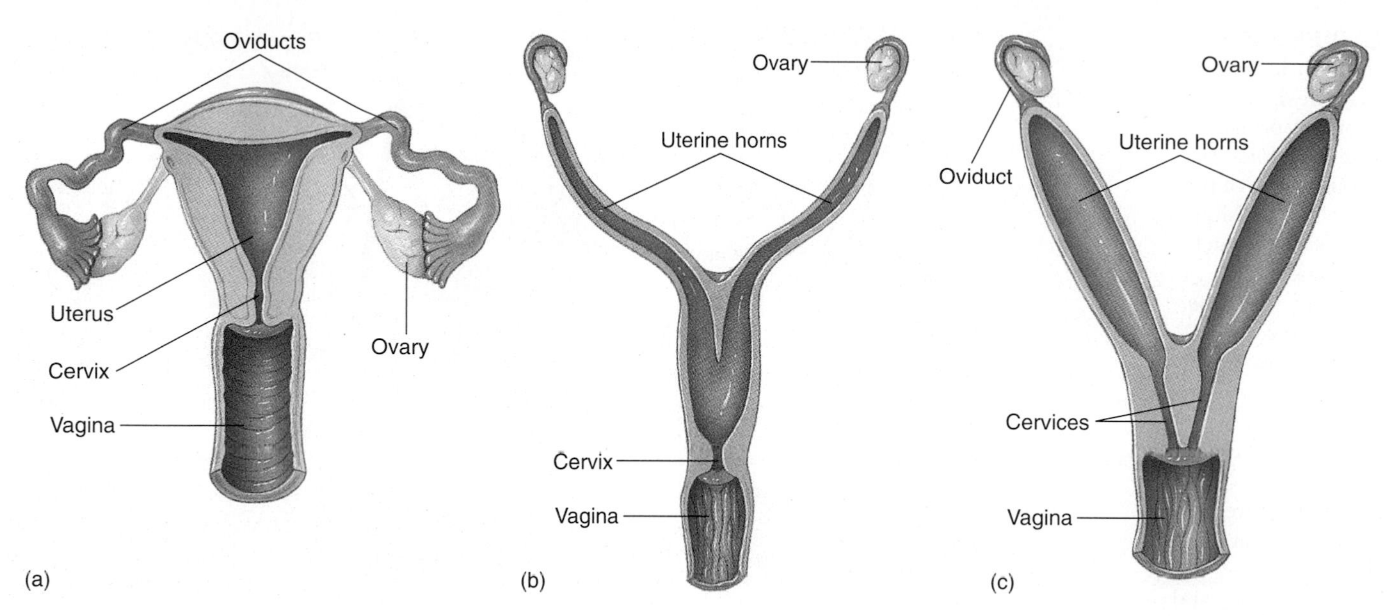

FIGURE 59.17
A comparison of mammalian uteruses. (*a*) Humans and other primates; (*b*) cats, dogs, and cows; and (*c*) rats, mice, and rabbits.

Menstrual and Estrous Cycles

At birth, a female's ovaries contain some 2 million follicles, each with an ovum that has begun meiosis but which is arrested in prophase of the first meiotic division. At this stage, the ova are called primary oocytes. Some of these primary-oocyte-containing follicles are stimulated to develop during each cycle. The human menstrual (Latin *mens,* "month") cycle lasts approximately one month (28 days on the average) and can be divided in terms of ovarian activity into a follicular phase and luteal phase, with the two phases separated by the event of ovulation.

Follicular Phase

During the follicular phase, a few follicles are stimulated to grow under FSH stimulation, but only one achieves full maturity as a tertiary, or Graafian, follicle (figure 59.18). This follicle forms a thin-walled blister on the surface of the ovary. The primary oocyte within the Graafian follicle completes the first meiotic division during the follicular phase. Instead of forming two equally large daughter cells, however, it produces one large daughter cell, the secondary oocyte, and one tiny daughter cell, called a polar body. Thus, the secondary oocyte acquires almost all of the cytoplasm from the primary oocyte, increasing its chances of sustaining the early embryo should the oocyte be fertilized. The polar body, on the other hand, often disintegrates. The secondary oocyte then begins the second meiotic division, but its progress is arrested at metaphase II. It is in this form that the egg cell is discharged from the ovary at ovulation, and it does not complete the second meiotic division unless it becomes fertilized in the fallopian tube.

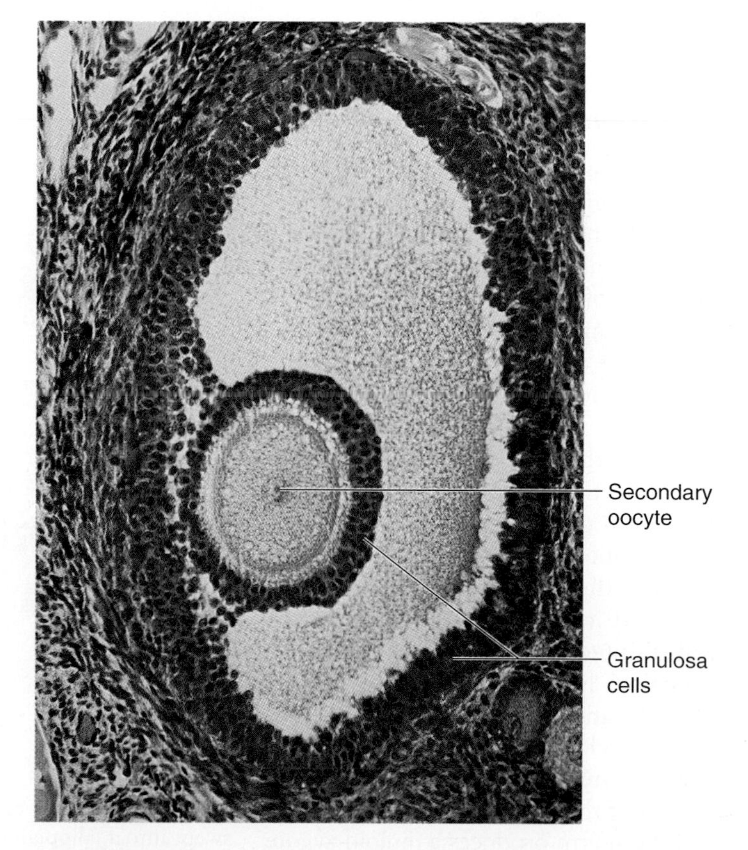

FIGURE 59.18
A mature Graafian follicle in a cat ovary (50×). Note the ring of granulosa cells that surrounds the secondary oocyte. This ring will remain around the egg cell when it is ovulated, and sperm must tunnel through the ring in order to reach the plasma membrane of the egg cell.

FIGURE 59.19
The meiotic events of oogenesis in humans. A primary oocyte is diploid. At the completion of the first meiotic division, one division product is eliminated as a polar body, while the other, the secondary oocyte, is released during ovulation. The secondary oocyte does not complete the second meiotic division until after fertilization; that division yields a second polar body and a single haploid egg, or ovum. Fusion of the haploid egg with a haploid sperm during fertilization produces a diploid zygote.

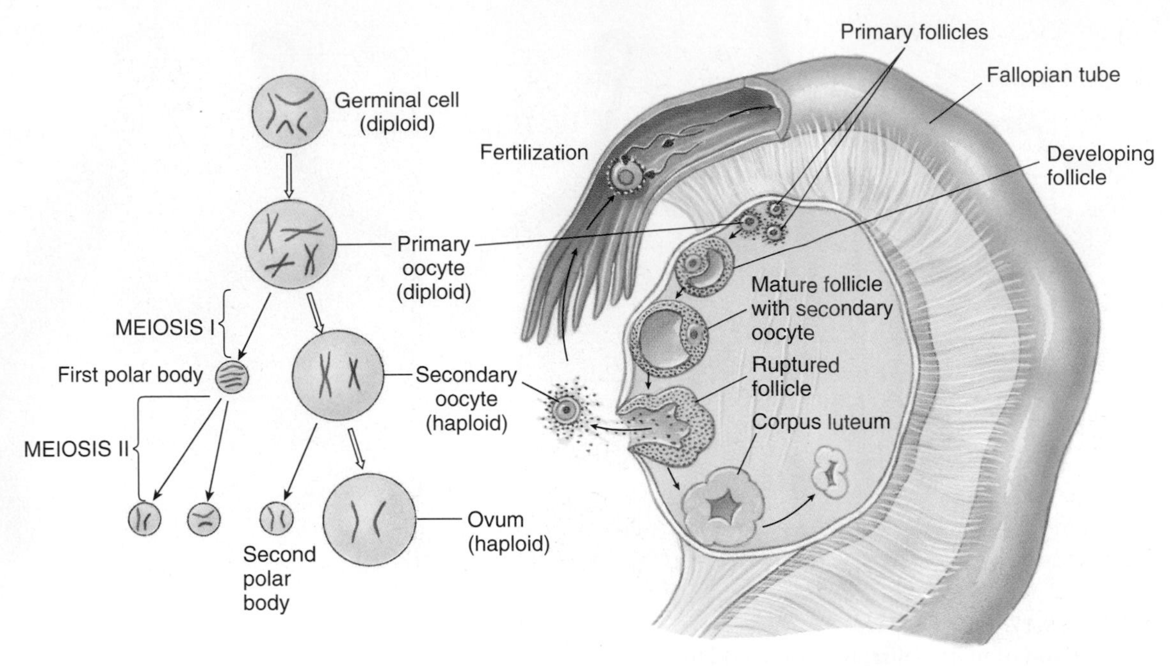

Ovulation

The increasing level of estradiol in the blood during the follicular phase stimulates the anterior pituitary gland to secrete LH about midcycle. This sudden secretion of LH causes the fully developed Graafian follicle to burst in the process of ovulation, releasing its secondary oocyte. The released oocyte enters the abdominal cavity near the fimbriae, the feathery projections surrounding the opening to the fallopian tube. The ciliated epithelial cells lining the fallopian tube propel the oocyte through the fallopian tube toward the uterus. If it is not fertilized, the oocyte will disintegrate within a day following ovulation. If it is fertilized, the stimulus of fertilization allows it to complete the second meiotic division, forming a fully mature ovum and a second polar body. Fusion of the two nuclei from the ovum and the sperm produces a diploid zygote (figure 59.19). Fertilization normally occurs in the upper one-third of the fallopian tube, and in a human the zygote takes approximately three days to reach the uterus, then another two to three days to implant in the endometrium (figure 59.20).

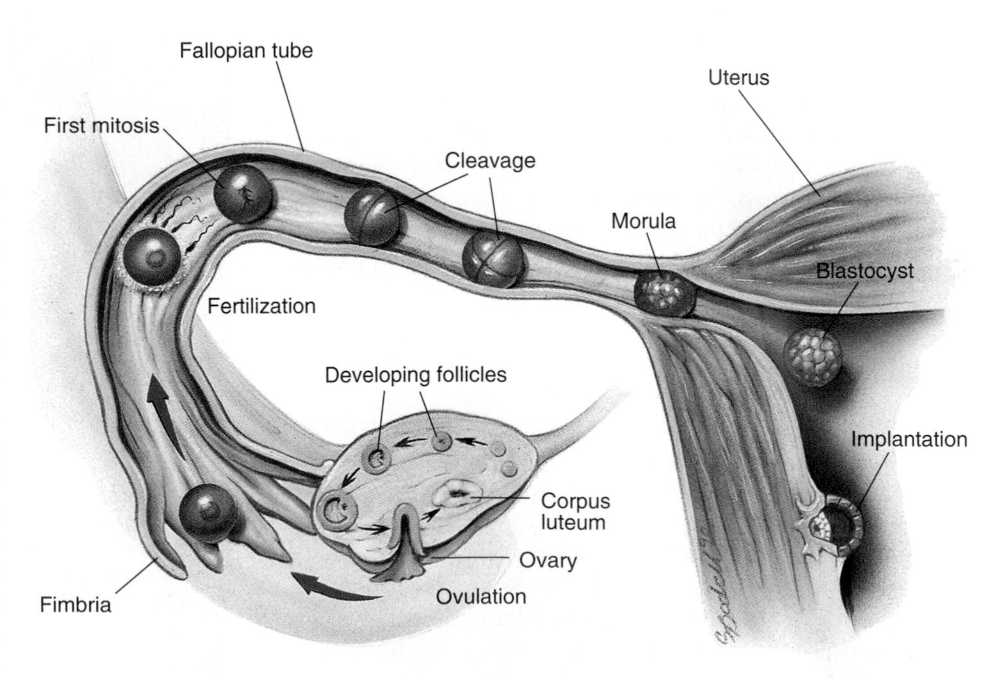

FIGURE 59.20
The journey of an egg. Produced within a follicle and released at ovulation, an egg is swept into a fallopian tube and carried along by waves of ciliary motion in the tube walls. Sperm journeying upward from the vagina fertilize the egg within the fallopian tube. The resulting zygote undergoes several mitotic divisions while still in the tube, so that by the time it enters the uterus, it is a hollow sphere of cells called a blastocyst. The blastocyst implants within the wall of the uterus, where it continues its development. (The egg and its subsequent stages have been enlarged for clarification.)

Luteal Phase

After ovulation, LH stimulates the empty Graafian follicle to develop into a structure called the corpus luteum (Latin, "yellow body"). For this reason, the second half of the menstrual cycle is referred to as the **luteal phase** of the cycle. The corpus luteum secretes both estradiol and another steroid hormone, progesterone. The high blood levels of estradiol and progesterone during the luteal phase now exert negative feedback inhibition of FSH and LH secretion by the anterior pituitary gland. This inhibition during the luteal phase is in contrast to the stimulation exerted by estradiol on LH secretion at midcycle, which caused ovulation. The inhibitory effect of estradiol and progesterone on FSH and LH secretion after ovulation acts as a natural contraceptive mechanism, preventing both the development of additional follicles and continued ovulation.

During the follicular phase the granulosa cells secrete increasing amounts of estradiol, which stimulates the growth of the endometrium. Hence, this portion of the cycle is also referred to as the **proliferative phase** of the endometrium. During the luteal phase of the cycle, the combination of estradiol and progesterone cause the endometrium to become more vascular, glandular, and enriched with glycogen deposits. Because of the endometrium's glandular appearance, this portion of the cycle is known as the **secretory phase** of the endometrium (figure 59.21).

In the absence of fertilization, the corpus luteum triggers its own atrophy, or regression, toward the end of the luteal phase. It does this by secreting hormones (estradiol and progesterone) that inhibit the secretion of LH, the hormone needed for its survival. In many mammals, atrophy of the corpus luteum is assisted by luteolysin, a paracrine regulator believed to be a prostaglandin. The disappearance of the corpus luteum results in an abrupt decline in the blood concentration of estradiol and progesterone at the end of the luteal phase, causing the built-up endometrium to be sloughed off with accompanying bleeding. This process is called menstruation, and the portion of the cycle in which it occurs is known as the **menstrual phase** of the endometrium.

If the ovulated oocyte is fertilized, however, regression of the corpus luteum and subsequent menstruation is averted by the tiny embryo! It does this by secreting human chorionic gonadotropin (hCG), an LH-like hormone produced by the chorionic membrane of the embryo. By maintaining the corpus luteum, hCG keeps the levels of estradiol and progesterone high and thereby prevents menstruation, which would terminate the pregnancy. Because hCG comes from the embryonic chorion and not the mother, it is the hormone that is tested for in all pregnancy tests.

Menstruation is absent in mammals with an estrous cycle. Although such mammals do cyclically shed cells from the endometrium, they don't bleed in the process. The estrous cycle is divided into four phases: proestrus, estrus, metestrus, and diestrus, which correspond to the proliferative, mid-cycle, secretory, and menstrual phases of the endometrium in the menstrual cycle.

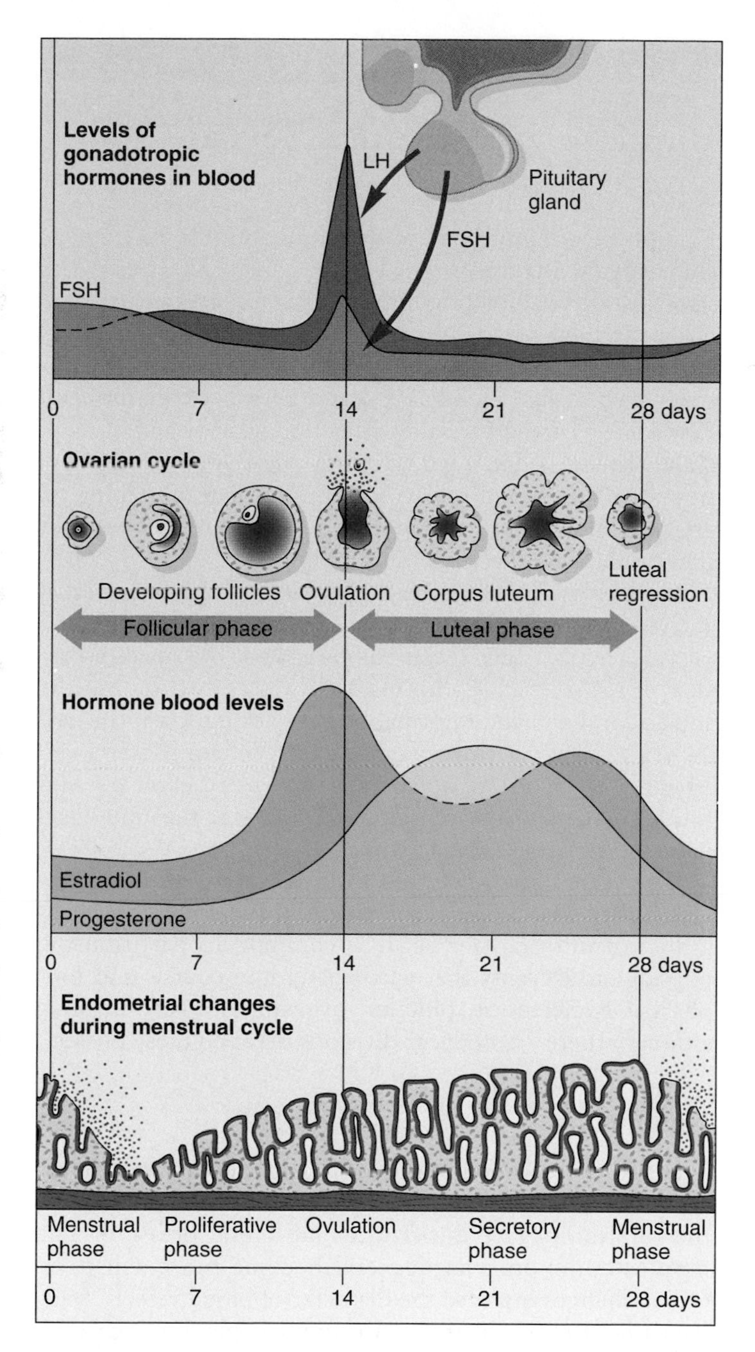

FIGURE 59.21
The human menstrual cycle. The growth and thickening of the endometrial (uterine) lining is stimulated by estradiol and progesterone. The decline in the levels of these two hormones triggers menstruation, the sloughing off of built-up endometrial tissue.

The ovarian follicles develop under FSH stimulation, and one follicle ovulates under LH stimulation. During the follicular and luteal phases, the hormones secreted by the ovaries stimulate the development of the endometrium, so an embryo can implant there if fertilization has occurred. A secondary oocyte is released from an ovary at ovulation, and it only completes meiosis if it is fertilized.

59.4 The physiology of human sexual intercourse is becoming better known.

Physiology of Human Sexual Intercourse

Few physical activities are more pleasurable to humans than sexual intercourse. The sex drive is one of the strongest drives directing human behavior, and as such, it is circumscribed by many rules and customs. Sexual intercourse acts as a channel for the strongest of human emotions such as love, tenderness, and personal commitment. Few subjects are at the same time more private and of more general interest. Here we will limit ourselves to a very narrow aspect of sexual behavior, its immediate physiological effects. The emotional consequences are no less real, but they are beyond the scope of this book.

Until relatively recently, the physiology of human sexual activity was largely unknown. Perhaps because of the prevalence of strong social taboos against the open discussion of sexual matters, no research was carried out on the subject, and detailed information was lacking. Over the past 40 years, however, investigations by William Masters and Virginia Johnson, as well as an army of researchers who followed them, have revealed much about the biological nature of human sexual activity.

The sexual act is referred to by a variety of names, including sexual intercourse, copulation, and coitus, as well as a host of informal terms. It is common to partition the physiological events that accompany intercourse into four phases—**excitement, plateau, orgasm,** and **resolution**—although there are no clear divisions between these phases.

Excitement

The sexual response is initiated by the nervous system. In both males and females, commands from the brain increase the respiratory rate, heart rate, and blood pressure. The nipples commonly harden and become more sensitive. Other changes increase the diameter of blood vessels, leading to increased circulation. In some people, these changes may produce a reddening of the skin around the face, breasts, and genitals (the sex flush). Increased circulation also leads to vasocongestion, producing erection of the male's penis and similar swelling of the female's clitoris. The female experiences changes that prepare the vagina for sexual intercourse: the labia majora and labia minora, lips of tissue that cover the opening to the vagina, swell and separate due to the increased circulation; the vaginal walls become moist; and the muscles encasing the vagina relax.

Plateau

The penetration of the vagina by the thrusting penis continuously stimulates nerve endings both in the tip of the penis and in the clitoris. The clitoris, which is now swollen, becomes very sensitive and withdraws up into a sheath or "hood." Once it has withdrawn, the clitoris is stimulated indirectly when the thrusting movements of the penis rub the clitoral hood against the clitoris. The nervous stimulation produced by the repeated movements of the penis within the vagina elicits a continuous response in the autonomic nervous system, greatly intensifying the physiological changes initiated during the excitement phase. In the female, pelvic thrusts may begin, while in the male the penis reaches its greatest length and rigidity.

Orgasm

The climax of intercourse is reached when the stimulation is sufficient to initiate a series of reflexive muscular contractions. The nerve impulses producing these contractions are associated with other activity within the central nervous system, activity that we experience as intense pleasure. In females, the contractions are initiated by impulses in the hypothalamus, which causes the posterior pituitary gland to release large amounts of oxytocin. This hormone, in turn, causes the muscles in the uterus and around the vaginal opening to contract and the cervix to be pulled upward. Contractions occur at intervals of about one per second. There may be one to several intense peaks of contractions (orgasms), or the peaks may be more numerous but less intense.

Analogous contractions take place in the male. The first contractions, which occur in the vas deferens and prostate gland, cause *emission,* the peristaltic movement of sperm and seminal fluid into a collecting zone of the urethra located at the base of the penis. Shortly thereafter, violent contractions of the muscles at the base of the penis result in *ejaculation* of the collected semen through the penis. As in the female, the contractions are spaced about one second apart, although in the male they continue for only a few seconds and are almost invariably restricted to a single intense wave.

Resolution

After ejaculation, males rapidly lose their erection and enter a refractory period lasting 20 minutes or longer, in which sexual arousal is difficult to achieve and ejaculation is almost impossible. By contrast, many women can be aroused again almost immediately. After intercourse, the bodies of both men and women return over a period of several minutes to their normal physiological state.

Sexual intercourse is a physiological series of events leading to the ultimate deposition of sperm within the female reproductive tract. The phases are similar in males and females.

Birth Control

In most vertebrates, copulation is associated solely with reproduction. Reflexive behavior that is deeply ingrained in the female limits sexual receptivity to those periods of the sexual cycle when she is fertile. In humans and a few species of apes, the female can be sexually receptive throughout her reproductive cycle, and this extended receptivity to sexual intercourse serves a second important function—it reinforces pair-bonding, the emotional relationship between two individuals living together.

Not all human couples want to initiate a pregnancy every time they have sexual intercourse, yet sexual intercourse may be a necessary and important part of their emotional lives together. The solution to this dilemma is to find a way to avoid reproduction without avoiding sexual intercourse; this approach is commonly called birth control or contraception. A variety of approaches differing in effectiveness and in their acceptability to different couples are commonly taken to achieve birth control (figure 59.22 and table 59.2).

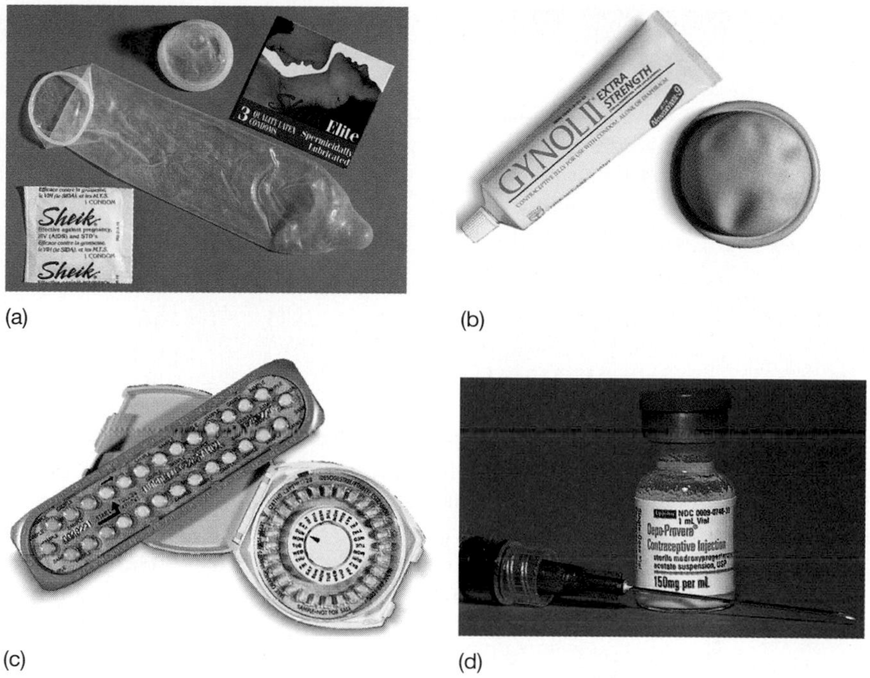

(a) (b) (c) (d)

FIGURE 59.22
Four common methods of birth control. (*a*) Condom; (*b*) diaphragm and spermicidal jelly; (*c*) oral contraceptives; (*d*) Depo-Provera.

Abstinence

The simplest and most reliable way to avoid pregnancy is not to have sexual intercourse at all. Of all methods of birth control, this is the most certain. It is also the most limiting, because it denies a couple the emotional support of a sexual relationship.

Sperm Blockage

If sperm cannot reach the uterus, fertilization cannot occur. One way to prevent the delivery of sperm is to encase the penis within a thin sheath, or condom. Many males do not favor the use of condoms, which tend to decrease their sensory pleasure during intercourse. In principle, this method is easy to apply and foolproof, but in practice it has a failure rate of 3 to 15% because of incorrect use or inconsistent use. Nevertheless, it is the most commonly employed form of birth control in the United States. Condoms are also widely used to prevent the transmission of AIDS and other sexually transmitted diseases (STDs). Over a billion condoms were sold in the United States last year.

A second way to prevent the entry of sperm into the uterus is to place a cover over the cervix. The cover may be a relatively tight-fitting cervical cap, which is worn for days at a time, or a rubber dome called a diaphragm, which is inserted immediately before intercourse. Because the dimensions of individual cervices vary, a cervical cap or diaphragm must be fitted by a physician. Failure rates average 4 to 25% for diaphragms, perhaps because of the propensity to insert them carelessly when in a hurry. Failure rates for cervical caps are somewhat lower.

Sperm Destruction

A third general approach to birth control is to eliminate the sperm after ejaculation. This can be achieved in principle by washing out the vagina immediately after intercourse, before the sperm have a chance to enter the uterus. Such a procedure is called a douche (French, "wash"). The douche method is difficult to apply well, because it involves a rapid dash to the bathroom immediately after ejaculation and a very thorough washing. Its failure rate is as high as 40%. Alternatively, sperm delivered to the vagina can be destroyed there with spermicidal jellies or foams. These treatments generally require application immediately before intercourse. Their failure rates vary from 10 to 25%. The use of a spermicide with a condom increases the effectiveness over each method used independently.

Prevention of Ovulation

Since about 1960, a widespread form of birth control in the United States has been the daily ingestion of birth control pills, or oral contraceptives, by women. These pills contain analogues of progesterone, sometimes in combination with

Table 59.2 Methods of Birth Control				
Device	**Action**	**Failure Rate***	**Advantages**	**Disadvantages**
Oral contraceptive	Hormones (progesterone analogue alone or in combination with other hormones) primarily prevent ovulation	1–5, depending on type	Convenient; highly effective; provides significant noncontraceptive health benefits, such as protection against ovarian and endometrial cancers	Must be taken regularly; possible minor side effects which new formulations have reduced; not for women with cardiovascular risks (mostly smokers over age 35)
Condom	Thin sheath for penis that collects semen; "female condoms" sheath vaginal walls	3–15	Easy to use; effective; inexpensive; protects against some sexually transmitted diseases	Requires male cooperation; may diminish spontaneity; may deteriorate on the shelf
Diaphragm	Soft rubber cup covers entrance to uterus, prevents sperm from reaching egg, holds spermicide	4–25	No dangerous side effects; reliable if used properly; provides some protection against sexually transmitted diseases and cervical cancer	Requires careful fitting; some inconvenience associated with insertion and removal; may be dislodged during intercourse
Intrauterine device (IUD)	Small plastic or metal device placed in the uterus; prevents implantation; some contain copper, others release hormones	1–5	Convenient; highly effective; infrequent replacement	Can cause excess menstrual bleeding and pain; risk of perforation, infection, expulsion, pelvic inflammatory disease, and infertility; not recommended for those who eventually intend to conceive or are not monogamous; dangerous in pregnancy
Cervical cap	Miniature diaphragm covers cervix closely, prevents sperm from reaching egg, holds spermicide	Probably similar to that of diaphragm	No dangerous side effects; fairly effective; can remain in place longer than diaphragm	Problems with fitting and insertion; comes in limited number of sizes
Foams, creams, jellies, vaginal suppositories	Chemical spermicides inserted in vagina before intercourse that prevent sperm from entering uterus	10–25	Can be used by anyone who is not allergic; protect against some sexually transmitted diseases; no known side effects	Relatively unreliable; sometimes messy; must be used 5–10 minutes before each act of intercourse
Implant (levonorgestrel; Norplant)	Capsules surgically implanted under skin slowly release hormone that blocks ovulation	.03	Very safe, convenient, and effective; very long-lasting (5 years); may have nonreproductive health benefits like those of oral contraceptives	Irregular or absent periods; minor surgical procedure needed for insertion and removal; some scarring may occur
Injectable contraceptive (medroxy-progesterone; Depo-Provera)	Injection every 3 months of a hormone that is slowly released and prevents ovulation	1	Convenient and highly effective; no serious side effects other than occasional heavy menstrual bleeding	Animal studies suggest it may cause cancer, though new studies in humans are mostly encouraging; occasional heavy menstrual bleeding

*Failure rate is expressed as pregnancies per 100 actual users per year.
Source: Data from American College of Obstetricians and Gynecologists: Contraception, Patient Education Pamphlet No. AP005.ACOG, Washington, D.C., 1990.

estrogens. As described earlier, progesterone and estradiol act by negative feedback to inhibit the secretion of FSH and LH during the luteal phase of the menstrual cycle, thereby preventing follicle development and ovulation. They also cause a buildup of the endometrium. The hormones in birth control pills have the same effects. Because the pills block ovulation, no ovum is available to be fertilized. A woman generally takes the hormone-containing pills for three weeks; during the fourth week, she takes pills without hormones (placebos), allowing the levels of those hormones in her blood to fall, which causes menstruation. Oral contraceptives provide a very effective means of birth control, with a failure rate of only 1 to 5%. In a variation of the oral contraceptive, hormone-containing capsules are implanted beneath the skin. These implanted capsules have failure rates below 1%.

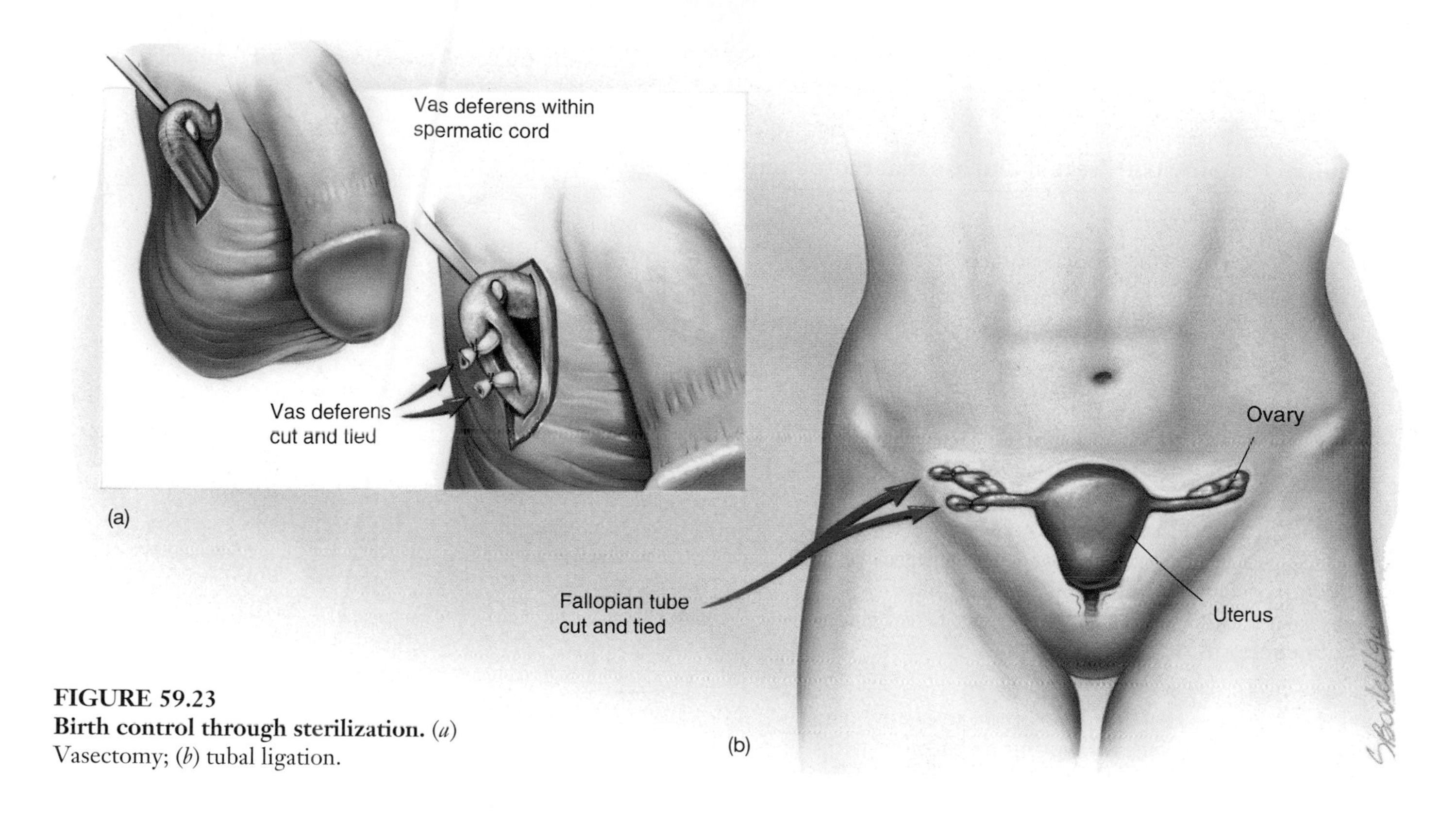

FIGURE 59.23
Birth control through sterilization. (*a*) Vasectomy; (*b*) tubal ligation.

A small number of women using birth control pills or implants experience undesirable side effects, such as blood clotting and nausea. These side effects have been reduced in newer generations of birth control pills, which contain less estrogen and different analogues of progesterone. Moreover, these new oral contraceptives provide a number of benefits, including reduced risks of endometrial and ovarian cancer, cardiovascular disease, and osteoporosis (for older women). However, they may increase the risk of contracting breast cancer and cervical cancer. The risks involved with birth control pills increase in women who smoke and increase greatly in women over 35 who smoke. The current consensus is that, for many women, the health benefits of oral contraceptives outweigh their risks, although a physician must help each woman determine the relative risks and benefits.

Prevention of Embryo Implantation

The insertion of a coil or other irregularly shaped object into the uterus is an effective means of birth control, because the irritation it produces in the uterus prevents the implantation of an embryo within the uterine wall. Such intrauterine devices (IUDs) have a failure rate of only 1 to 5%. Their high degree of effectiveness probably reflects their convenience; once they are inserted, they can be forgotten. The great disadvantage of this method is that almost a third of the women who attempt to use IUDs experience cramps, pain, and sometimes bleeding and therefore must discontinue using them.

Another method of preventing embryo implantation is the "morning after pill," which contains 50 times the dose of estrogen present in birth control pills. The pill works by temporarily stopping ovum development, by preventing fertilization, or by stopping the implantation of a fertilized ovum. Its failure rate is 1 to 10%, but many women are uneasy about taking such high hormone doses, as side effects can be severe. This is not recommended as a regular method of birth control but rather as a method of emergency contraception.

Sterilization

A completely effective means of birth control is sterilization, the surgical removal of portions of the tubes that transport the gametes from the gonads (figure 59.23). Sterilization may be performed on either males or females, preventing sperm from entering the semen in males and preventing an ovulated oocyte from reaching the uterus in females. In males, sterilization involves a vasectomy, the removal of a portion of the vas deferens from each testis. In females, the comparable operation involves the removal of a section of each fallopian tube.

Fertilization can be prevented by a variety of birth control methods, including barrier contraceptives, hormonal inhibition, surgery, and abstinence. Efficacy rates vary from method to method.

Chapter 59

www.mhhe.com/raven6/resources59.mhtml

Summary	*Questions*	*Media Resources*

59.1 Animals employ both sexual and asexual reproductive strategies.

- Parthenogenesis is a form of asexual reproduction that is practiced by many insects and some lizards.
- Among mammals, the sex is determined by the presence of a Y chromosome in males and its absence in females.

1. How are oviparity, ovoviviparity, and viviparity different?

59.2 The evolution of reproduction among the vertebrates has led to internalization of fertilization and development.

- Most bony fish practice external fertilization, releasing eggs and sperm into the water where fertilization occurs. Amphibians have external fertilization and the young go through a larval stage before metamorphosis.
- Reptiles and birds are oviparous, the young developing in eggs that are deposited externally. Most mammals are viviparous, the young developing within the mother.

2. How does fetal development differ in the monotremes, marsupials, and placental mammals?

- Student Research:
  - -Reproductive Biology of House Mice
  - -Evolution of Uterine Function

59.3 Male and female reproductive systems are specialized for different functions.

- Sperm leave the testes and pass through the epididymis and vas deferens; the ejaculatory duct merges with the urethra, which empties at the tip of the penis.
- An egg cell released from the ovary in ovulation is drawn by fimbria into the fallopian tube, which conducts the egg cell to the lining of the uterus, or endometrium, where it implants if fertilized.
- If fertilization does not occur, the corpus luteum regresses at the end of the cycle and the resulting fall in estradiol and progesterone secretion cause menstruation to occur in humans and apes.

3. Briefly describe the function of the seminal vesicles, prostate gland, and bulbourethral glands.

4. When do the ova in a female mammal begin meiosis? When do they complete the first meiotic division?

5. What hormone is secreted by the granulosa cells in a Graafian follicle? What effect does this hormone have on the endometrium?

- Art Activities:
  - -Male Reproductive System
  - -Penis Anatomy
  - -Female Reproductive System
  - -Breast Anatomy

- Spermatogenesis
- Menstruation
- Female Reproductive Cycle
- Oogenesis

59.4 The physiology of human sexual intercourse is becoming better known.

- The physiological events that occur in the human sexual response are grouped into four phases: excitement, plateau, orgasm, and resolution.
- Males and females have similar phases, but males enter a refractory period following orgasm that is absent in many women.
- There are a variety of methods of birth control available that range in ease of use, effectiveness, and permanence.

6. What are the four phases in the physiological events of sexual intercourse in humans? During the first phase, what events occur specifically in males, and what events occur specifically in females?

7. How do birth control pills prevent pregnancy?

- Penile Erection
- Vasectomy
- Tubal Ligation

- Birth Control

60

Vertebrate Development

Concept Outline

60.1 Fertilization is the initial event in development.

Stages of Development. Fertilization of an egg cell by a sperm occurs in three stages: penetration, activation of the egg cell, and fusion of the two haploid nuclei.

60.2 Cell cleavage and the formation of a blastula set the stage for later development.

Cell Cleavage Patterns. The cytoplasm of the zygote is divided into smaller cells by mitotic cell division in a process called cleavage.

60.3 Gastrulation forms the three germ layers of the embryo.

The Process of Gastrulation. Cells of the blastula invaginate and involute to produce an outer ectoderm, an inner endoderm layer, and a third layer, the mesoderm.

60.4 Body architecture is determined during the next stages of embryonic development.

Developmental Processes during Neurulation. The mesoderm of chordates forms a notochord, and the overlying ectoderm rolls to produce a neural tube.
How Cells Communicate during Development. Cell-to-cell contact plays a major role in selecting the paths along which cells develop.
Embryonic Development and Vertebrate Evolution. The embryonic development of a mammal includes stages that are characteristic of more primitive vertebrates.
Extraembryonic Membranes. Embryonic cells form several membranes outside of the embryo that provide protection, nourishment, and gas exchange for the embryo.

60.5 Human development is divided into trimesters.

First Trimester. A blastocyst implants into the mother's endometrium, and the formation of body organs begins during the third and fourth week.
Second and Third Trimesters. All of the major organs of the body form during the first trimester and so further growth and development take place during this time.
Birth and Postnatal Development. Birth occurs as a result of uterine contractions; the human brain continues to grow significantly after birth.

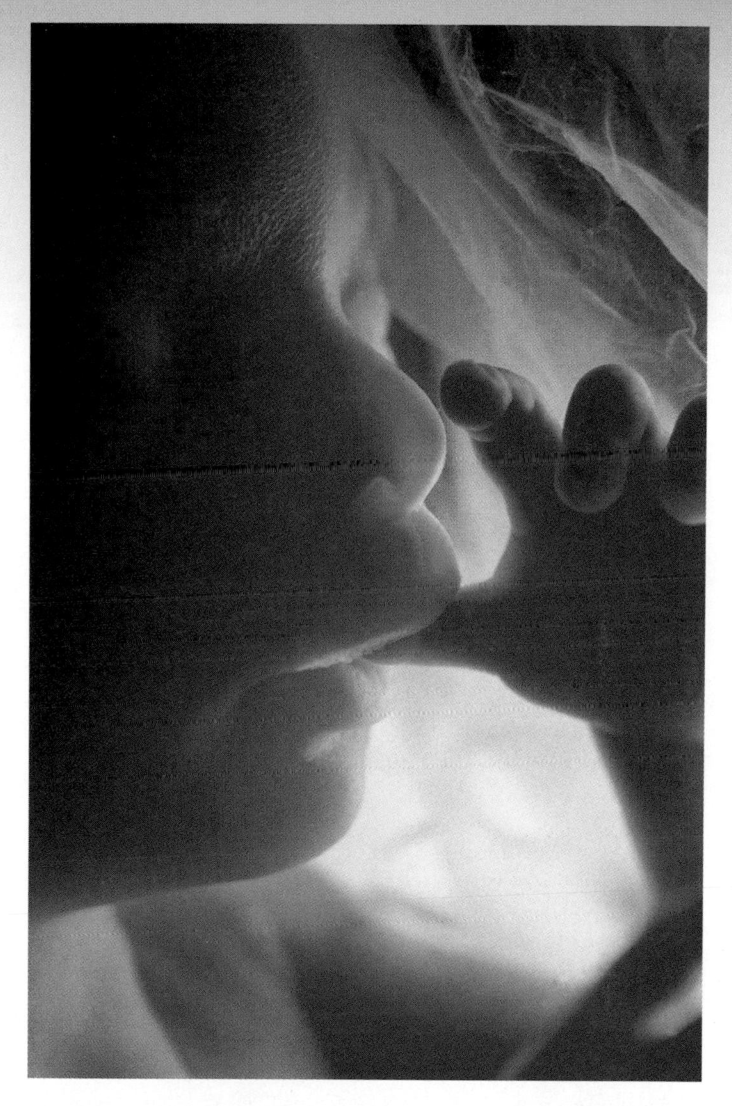

FIGURE 60.1
Development is the process that determines an organism's form and function. A human fetus at 18 weeks is not yet halfway through the 38 weeks—about 9 months—it will spend within its mother, but it has already developed many distinct behaviors, such as the sucking reflex that is so important to survival after birth.

Reproduction in all but a few vertebrates unites two haploid gametes to form a single diploid cell called a zygote. The zygote develops by a process of cell division and differentiation into a complex multicellular organism, composed of many different tissues and organs (figure 60.1). Although the process of development is a continuous series of events with some of the details varying among different vertebrate groups, we will examine vertebrate development in six stages. In this chapter, we will consider the stages of development and conclude with a description of the events that occur during human development.

60.1 Fertilization is the initial event in development.

Stages of Development

In vertebrates, as in all sexual animals, the first step in development is the union of male and female gametes, a process called **fertilization.** Fertilization is typically external in fish and amphibians, which reproduce in water, and internal in all other vertebrates. In internal fertilization, small, motile sperm are introduced into the female reproductive tract during mating. The sperm swim up the reproductive tract until they encounter a mature egg or oocyte in an oviduct, where fertilization occurs. Fertilization consists of three stages: penetration, activation, and fusion.

Penetration

As described in chapter 59, the secondary oocyte is released from a fully developed Graafian follicle at ovulation. It is surrounded by the same layer of small granulosa cells that surrounded it within the follicle (figure 60.2). Between the granulosa cells and the egg's plasma membrane is a glycoprotein layer called the zona pellucida. The head of each sperm is capped by an organelle called the acrosome, which contains glycoprotein-digesting enzymes. These enzymes become exposed as the sperm begin to work their way into the layer of granulosa cells, and the activity of the enzymes enables the sperm to tunnel their way through the zona pellucida to the egg's plasma membrane. In sea urchins, egg cytoplasm bulges out at this point, engulfing the head of the sperm and permitting the sperm nucleus to enter the cytoplasm of the egg (figure 60.3).

Activation

The series of events initiated by sperm penetration are collectively called egg activation. In some frogs, reptiles, and birds, more than one sperm may penetrate the egg, but only one is successful in fertilizing it. In mammals, by contrast, the penetration of the first sperm initiates changes in the egg membrane that prevent the entry of other sperm. As the sperm makes contact with the oocyte membrane, there is a change in the membrane potential (see chapter 54

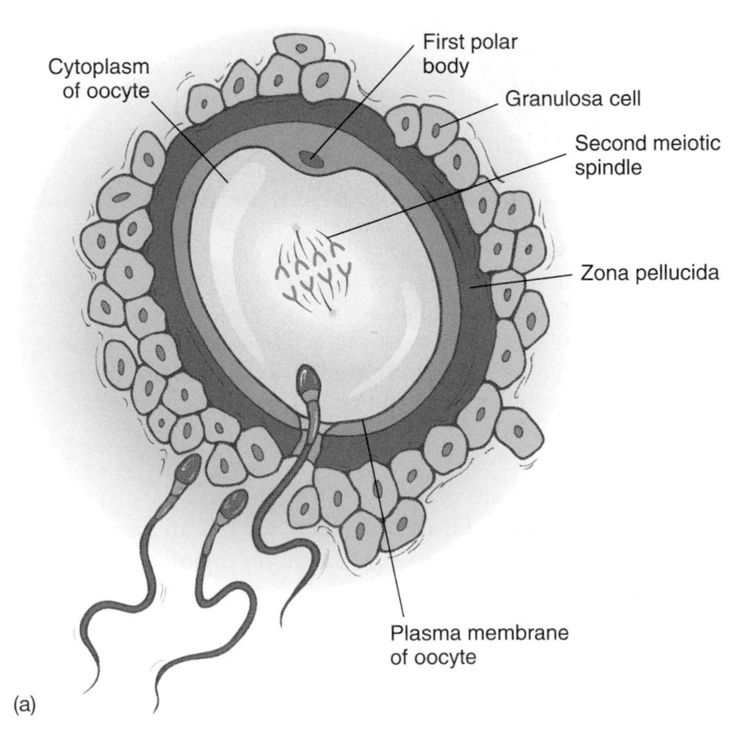

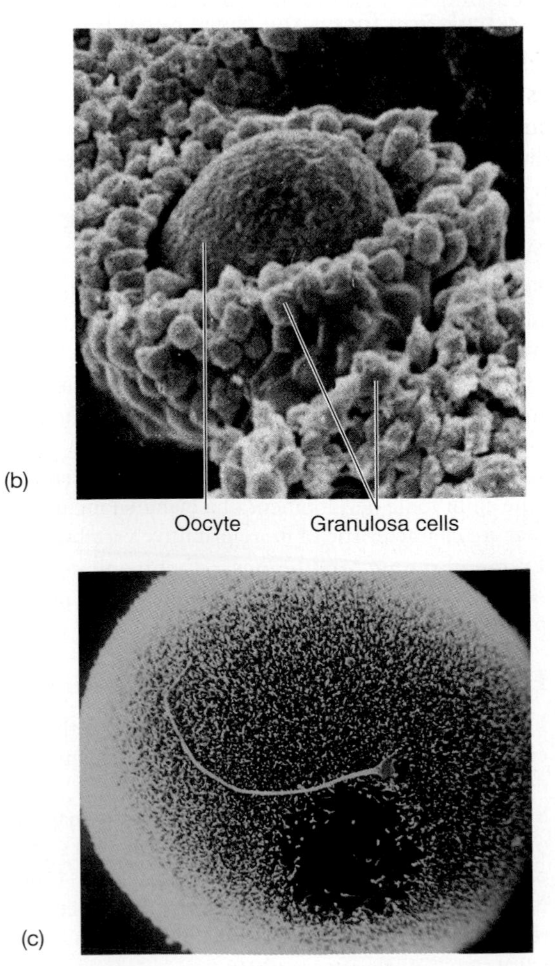

FIGURE 60.2
Mammalian reproductive cells. (*a*) A sperm must penetrate a layer of granulosa cells and then a layer of glycoprotein called the zona pellucida, before it reaches the oocyte membrane. This penetration is aided by digestive enzymes in the acrosome of the sperm. These scanning electron micrographs show (*b*) a human oocyte (90×) surrounded by numerous granulosa cells, and (*c*) a human sperm on an egg (3000×).

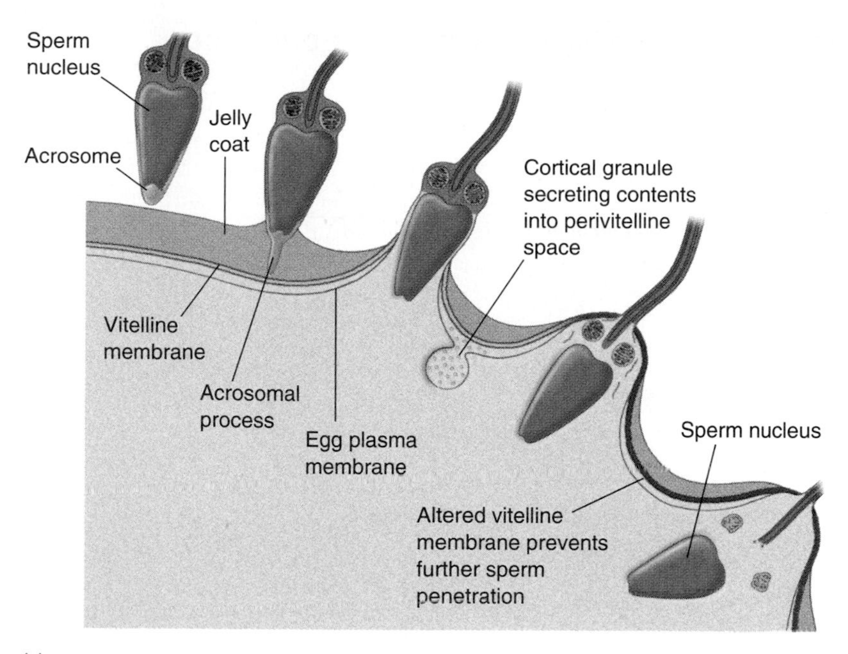

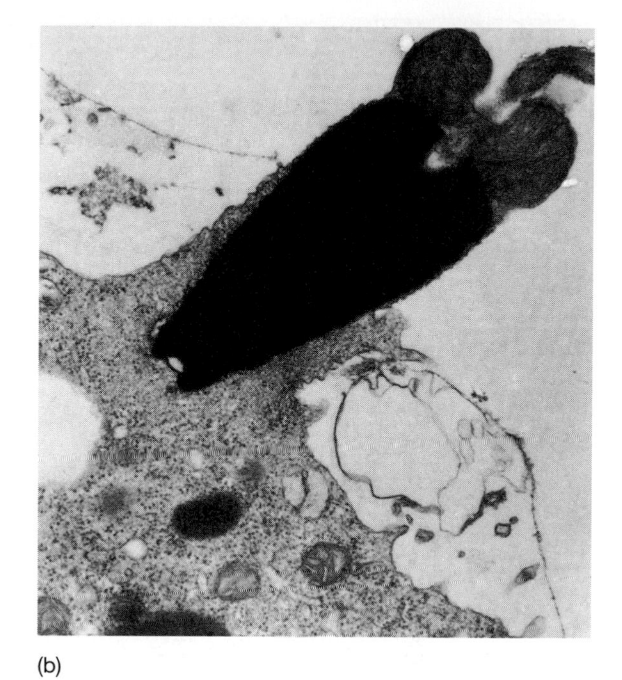

FIGURE 60.3
Sperm penetration of a sea urchin egg. (*a*) The stages of penetration. (*b*) An electron micrograph (50,000×) of penetration. Penetration in both invertebrate and vertebrate eggs is similar.

for discussion of membrane potential) that prevents other sperm from fusing with the oocyte membrane. In addition to these changes, sperm penetration can have three other effects on the egg. First, in mammals it stimulates the chromosomes in the egg nucleus to complete the second meiotic division, producing two egg nuclei. One of these nuclei is extruded from the egg as a second polar body (see chapter 59), leaving a single haploid egg nucleus within the egg.

Second, sperm penetration in some animals triggers movements of the egg cytoplasm around the point of sperm entry. These movements ultimately establish the bilateral symmetry of the developing animal. In frogs, for example, sperm penetration causes an outer pigmented cap of egg cytoplasm to rotate toward the point of entry, uncovering a gray crescent of interior cytoplasm opposite the point of penetration (figure 60.4). The position of the gray crescent determines the orientation of the first cell division. A line drawn between the point of sperm entry and the gray crescent would bisect the right and left halves of the future adult. Third, activation is characterized by a sharp increase in protein synthesis and an increase in metabolic activity in general. Experiments demonstrate that the protein synthesis in the activated oocyte is coded by mRNA that was previously produced and already present in the cytoplasm of the unfertilized egg cell.

In some vertebrates, it is possible to activate an egg without the entry of a sperm, simply by pricking the egg membrane. An egg that is activated in this way may go on to develop parthenogenetically. A few kinds of amphibians, fish, and reptiles rely entirely on parthenogenetic reproduction in nature, as we mentioned in chapter 59.

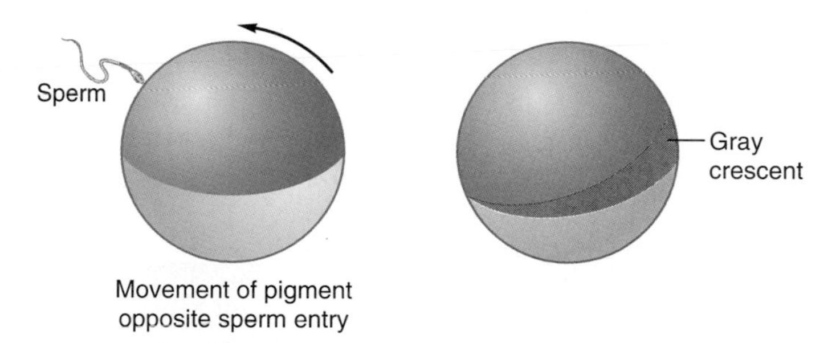

FIGURE 60.4
Gray crescent formation in frog eggs. The gray crescent appears opposite the point of penetration by the sperm.

Nuclei Fusion

The third stage of fertilization is the fusion of the entering sperm nucleus with the haploid egg nucleus to form the diploid nucleus of the zygote. This fusion is triggered by the activation of the egg. If a sperm nucleus is microinjected into an egg without activating the egg, the two nuclei will not fuse. The nature of the signals that are exchanged between the two nuclei, or sent from one to the other, is not known.

The three stages of fertilization are penetration, activation, and nuclei fusion. Penetration initiates a complex series of developmental events, including major movements of cytoplasm, which eventually lead to the fusion of the egg and sperm nuclei.

60.2 Cell cleavage and the formation of a blastula set the stage for later development.

Cell Cleavage Patterns

Following fertilization, the second major event in vertebrate reproduction is the rapid division of the zygote into a larger and larger number of smaller and smaller cells (table 60.1). This period of division, called cleavage, is not accompanied by an increase in the overall size of the embryo. The resulting tightly packed mass of about 32 cells is called a **morula,** and each individual cell in the morula is referred to as a **blastomere.** As the blastomeres continue to divide, they secrete a fluid into the center of the morula. Eventually, a hollow ball of 500 to 2000 cells, the **blastula,** is formed. The fluid-filled cavity within the blastula is known as the *blastocoel.*

The pattern of cleavage division is influenced by the presence and location of yolk, which is abundant in the eggs of many vertebrates (figure 60.5). As we discussed in the previous chapter, vertebrates have embraced a variety of reproductive strategies involving different patterns of yolk utilization.

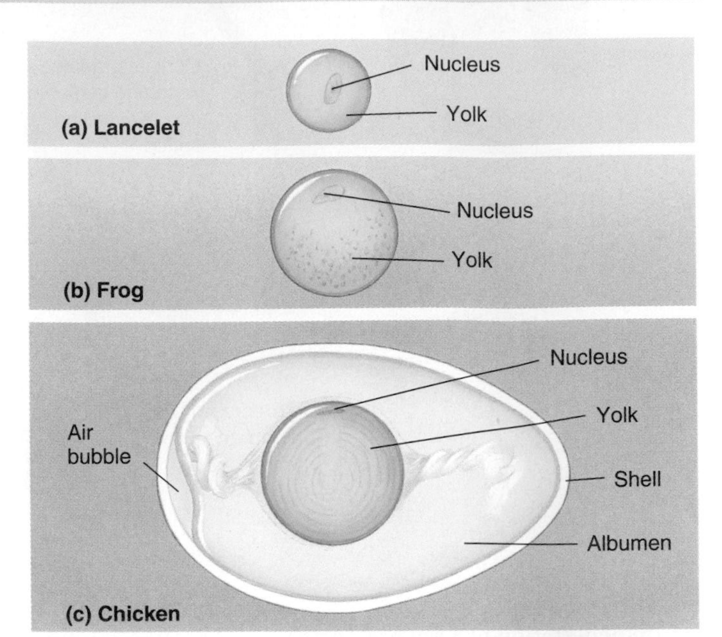

FIGURE 60.5
Yolk distribution in three kinds of eggs. (*a*) In the lancelet, a primitive chordate, the egg consists of a central nucleus surrounded by a small amount of yolk. (*b*) In a frog egg there is much more yolk, and the nucleus is displaced toward one pole. (*c*) Bird eggs are complexly organized, with the nucleus just under the surface of a large, central yolk.

Primitive Chordates

When eggs contain little or no yolk, cleavage occurs throughout the whole egg (figure 60.6). This pattern of cleavage, called holoblastic cleavage, was characteristic of the ancestors of the vertebrates and is still seen in groups such as the lancelets and agnathans. In these animals, holoblastic cleavage results in the formation of a symmetrical blastula composed of cells of approximately equal size.

Amphibians and Advanced Fish

The eggs of bony fish and frogs contain much more cytoplasmic yolk in one hemisphere than the other. Because yolk-rich cells divide much more slowly than those that have little yolk, holoblastic cleavage in these eggs results in a very asymmetrical blastula with large cells containing a lot of yolk at one pole and a concentrated mass of small cells containing very little yolk at the other. In these blastulas, the pole that is rich in yolk is called the vegetal pole, while the pole that is relatively poor in yolk is called the animal pole (figure 60.7).

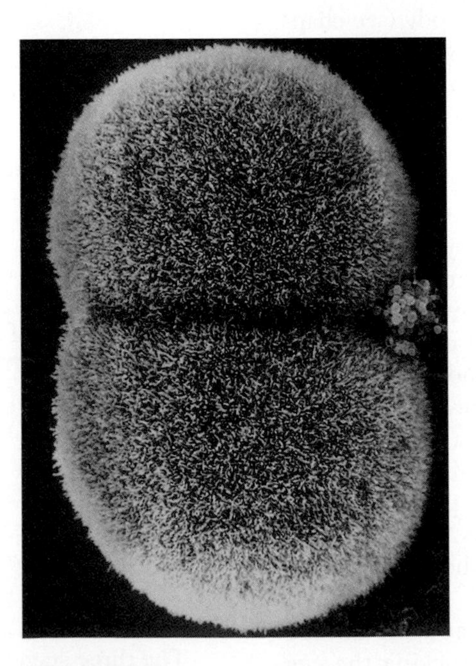

FIGURE 60.6
Holoblastic cleavage (3000×). In this type of cleavage, cell division occurs throughout the entire egg.

FIGURE 60.7
Dividing frog eggs. The closest cells in this photo (those near the animal pole) divide faster and are smaller than those near the vegetal pole.

Table 60.1 Stages of Vertebrate Development (Mammal)

	Stage	Description
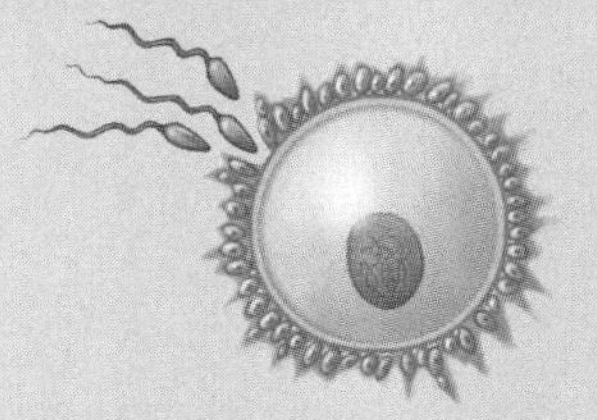	Fertilization	The haploid male and female gametes fuse to form a diploid zygote.
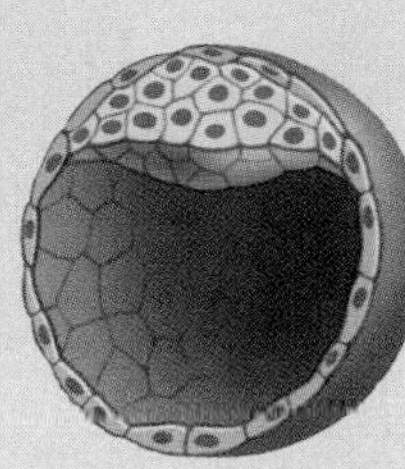	Cleavage	The zygote rapidly divides into many cells, with no overall increase in size. These divisions affect future development because different cells receive different portions of the egg cytoplasm and, hence, different regulatory signals.
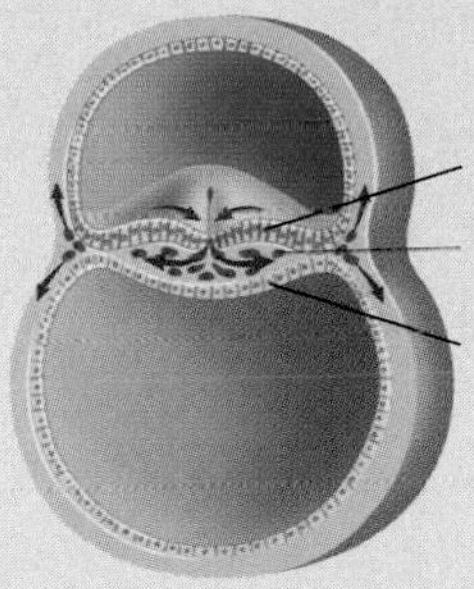 Ectoderm Mesoderm Endoderm	Gastrulation	The cells of the embryo move, forming three primary cell layers: ectoderm, mesoderm, and endoderm.
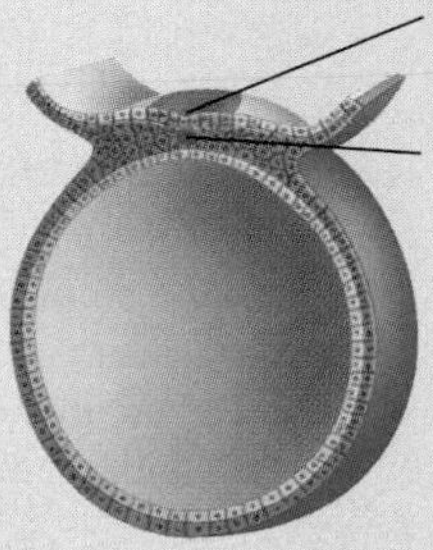 Neural groove Notochord	Neurulation	In all chordates, the first organ to form is the notochord; second is the dorsal nerve cord.
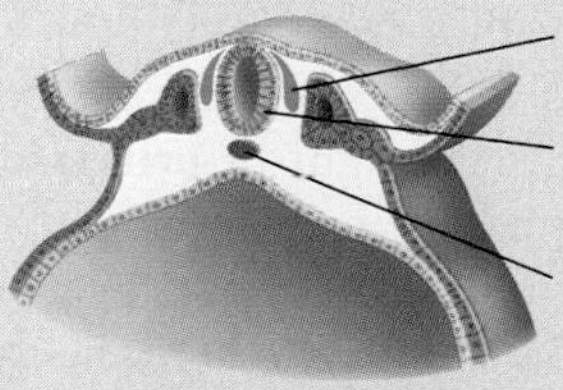 Neural crest Neural tube Notochord	Neural crest	During neurulation, the neural crest is produced as the neural cell formation tube is formed. The neural crest gives rise to several uniquely vertebrate structures.
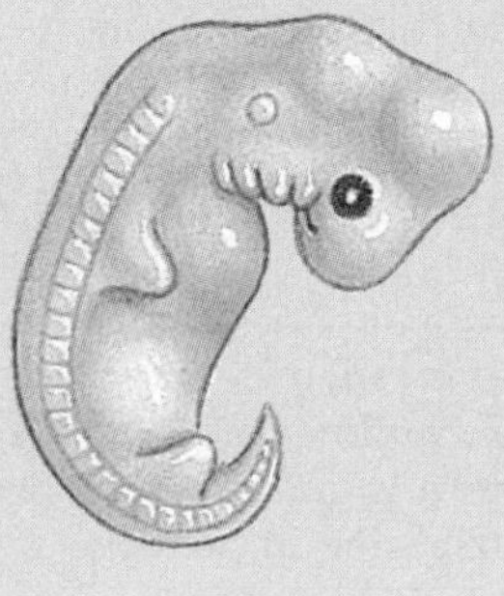	Organogenesis	Cells from the three primary layers combine in various ways to produce the organs of the body.

Reptiles and Birds

The eggs produced by reptiles, birds, and some fish are composed almost entirely of yolk, with a small amount of cytoplasm concentrated at one pole. Cleavage in these eggs occurs only in the tiny disc of polar cytoplasm, called the blastodisc, that lies astride the large ball of yolk material. This type of cleavage pattern is called meroblastic cleavage (figure 60.8). The resulting embryo is not spherical, but rather has the form of a thin cap perched on the yolk.

Mammals

Mammalian eggs are in many ways similar to the reptilian eggs from which they evolved, except that they contain very little yolk. Because cleavage is not impeded by yolk in mammalian eggs, it is holoblastic, forming a ball of cells surrounding a blastocoel. However, an inner cell mass is concentrated at one pole (figure 60.9). This inner cell mass is analogous to the blastodisc of reptiles and birds, and it goes on to form the developing embryo. The outer sphere of cells, called a trophoblast, is analogous to the cells that form the membranes underlying the tough outer shell of the reptilian egg. These cells have changed during the course of mammalian evolution to carry out a very different function: part of the trophoblast enters the endometrium (the epithelial lining of the uterus) and contributes to the placenta, the organ that permits exchanges between the fetal and maternal bloods. While part of the placenta is composed of fetal tissue (the trophoblast), part is composed of the modified endometrial tissue (called the *decidua basalis*) of the mother's uterus. The placenta will be discussed in more detail in a later section.

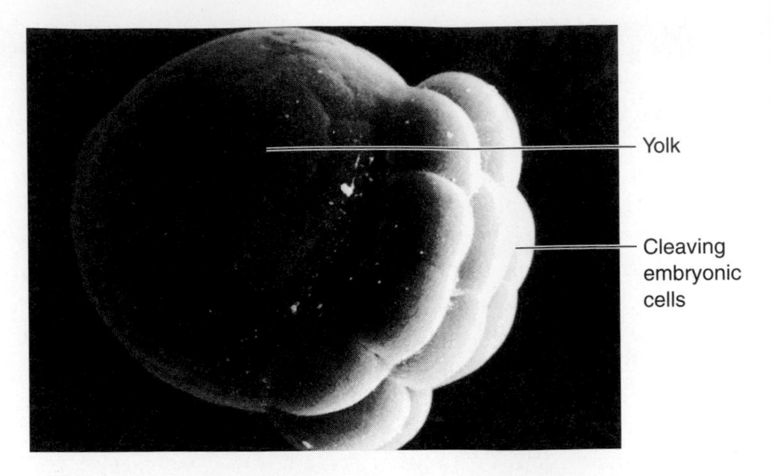

FIGURE 60.8
Meroblastic cleavage (400×). In this type of cleavage, only a portion of the egg actively divides to form a mass of cells.

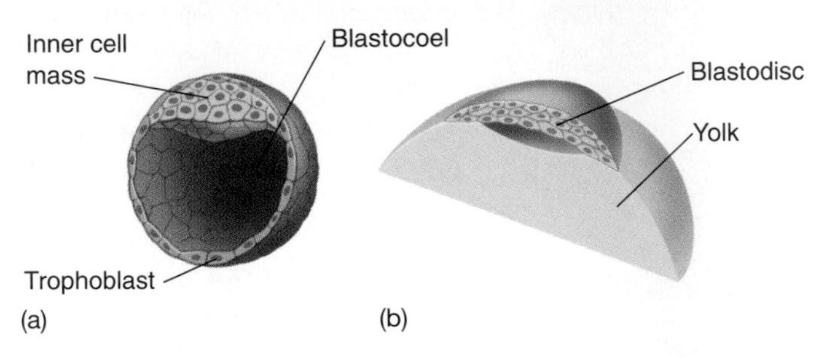

FIGURE 60.9
The embryos of mammals and birds are more similar than they seem. (*a*) A mammalian blastula, called a blastocyst is composed of a sphere of cells, the trophoblast, surrounding a cavity, the blastocoel, and an inner cell mass. (*b*) An avian (bird) blastula consists of a cap of cells, the blastodisc, resting atop a large yolk mass. The blastodisc will form an upper and a lower layer with a compressed blastocoel in between.

The Blastula

Viewed from the outside, the blastula looks like a simple ball of cells all resembling each other. In many animals, this appearance is misleading, as unequal distribution of developmental signals from the egg produces some degree of mosaic development, so that the cells are already committed to different developmental paths (see chapter 17). In mammals, however, it appears that all of the blastomeres receive equivalent sets of signals, and body form is determined by cell-cell interactions. In a mammalian blastula, called a *blastocyst*, each cell is in contact with a different set of neighboring cells, and these interactions with neighboring cells are a major factor influencing the developmental fate of each cell. This positional information is of particular importance in the orientation of mammalian embryos, setting up different patterns of development along three embryonic axes: anterior-posterior, dorsal-ventral, and proximal-distal.

For a short period of time, just before they implant in the uterus, the cells of the mammalian blastocyst have the power to develop into many of the 210 different types of cells in the body—and probably all of them. Biologists have long sought to grow these cells, called *embryonic stem cells*, in tissue culture, as such stem cells might in principle be used to produce tissues for human transplant operations. Injected into a patient, for example, they might be able to respond to local signals and produce new tissue (see chapter 19). The first success in growing stem cells in culture was reported in 1998, when researchers isolated cells from the inner cell mass of human blastocysts and successfully grew them in tissue culture. These stem cells continue to grow and divide in culture indefinitely, unlike ordinary body cells, which divide only 50 or so times and then die.

A series of rapid cell divisions called cleavage transforms the zygote into a hollow ball of cells, the blastula. The cleavage pattern is influenced by the amount of yolk and its distribution in the egg.

60.3 Gastrulation forms the three germ layers of the embryo.

The Process of Gastrulation

The first visible results of cytoplasmic distribution and cell position within the blastula can be seen immediately after the completion of cleavage. Certain groups of cells **invaginate** (dent inward) and **involute** (roll inward) from the surface of the blastula in a carefully orchestrated activity called **gastrulation.** The events of gastrulation determine the basic developmental pattern of the vertebrate embryo. By the end of gastrulation, the cells of the embryo have rearranged into three primary **germ layers: ectoderm, mesoderm,** and **endoderm.** The cells in each layer have very different developmental fates. In general, the ectoderm is destined to form the epidermis and neural tissue; the mesoderm gives rise to connective tissue, skeleton, muscle, and vascular elements; and the endoderm forms the lining of the gut and its derivatives (table 60.2).

How is cell movement during gastrulation brought about? Apparently, migrating cells creep over stationary cells by means of actin filament contractions that change the shapes of the migrating cells affecting an invagination of blastula tissue. Each cell that moves possesses particular cell surface polysaccharides, which adhere to similar polysaccharides on the surfaces of the other moving cells. This interaction between cell surface molecules enables the migrating cells to adhere to one another and move as a single mass (see chapter 7).

Just as the pattern of cleavage divisions in different groups of vertebrates depends heavily on the amount and distribution of yolk in the egg, so the pattern of gastrulation among vertebrate groups depends on the shape of the blastulas produced during cleavage.

Table 60.2 Developmental Fates of the Primary Cell Layers

Ectoderm	Epidermis, central nervous system, sense organs, neural crest
Mesoderm	Skeleton, muscles, blood vessels, heart, gonads
Endoderm	Lining of digestive and respiratory tracts; liver, pancreas

Gastrulation in Primitive Chordates

In primitive chordates such as lancelets, which develop from symmetrical blastulas, gastrulation begins as the surface of the blastula invaginates into the blastocoel. About half of the blastula's cells move into the interior of the blastula, forming a structure that looks something like an indented tennis ball. Eventually, the inward-moving wall of cells pushes up against the opposite side of the blastula and then stops moving. The resulting two-layered, cup-shaped embryo is the gastrula (figure 60.10). The hollow structure resulting from the invagination is called the archenteron, and it is the progenitor of the gut. The opening of the archenteron, the future anus, is known as the blastopore.

This process produces an embryo with two cell layers: an outer ectoderm and an inner endoderm. Soon afterward, a third cell layer, the mesoderm, forms between the ectoderm and endoderm. In lancelets, the mesoderm forms from pouches that pinch off the endoderm. The appearance of these three primary cell layers sets the stage for all subsequent tissue and organ differentiation.

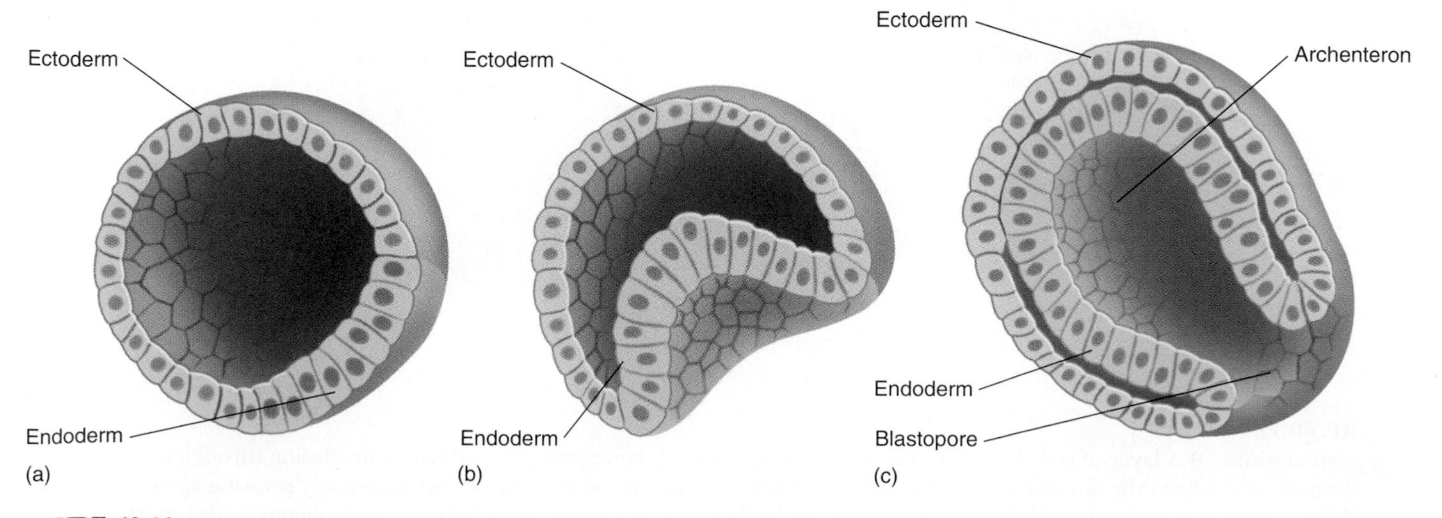

FIGURE 60.10
Gastrulation in a lancelet. In these chordates, the endoderm is formed by invagination of surface cells (*a*, *b*). This produces the primitive gut, or archenteron (*c*). Mesoderm will later be formed from pouches off the endoderm.

Gastrulation in Most Aquatic Vertebrates

In the blastulas of amphibians and those aquatic vertebrates with asymmetrical yolk distribution, the yolk-laden cells of the vegetal pole are fewer and much larger than the yolk-free cells of the animal pole. Consequently, gastrulation is more complex than it is in the lancelets. First, a layer of surface cells invaginates to form a small, crescent-shaped slit where the blastopore will soon be located. Next, cells from the animal pole involute over the dorsal lip of the blastopore (figure 60.11), at the same location as the gray crescent of the fertilized egg (see figure 60.4). As in the lancelets, the involuting cell layer eventually presses against the inner surface of the opposite side of the embryo, eliminating the blastocoel and producing an archenteron with a blastopore. In this case, however, the blastopore is filled with yolk-rich cells, forming the yolk plug. The outer layer of cells resulting from these movements is the ectoderm, and the inner layer is the endoderm. Other cells that involute over the dorsal lip and ventral lip (the two lips of the blastopore that are separated by the yolk plug) migrate between the ectoderm and endoderm to form the third germ layer, the mesoderm (figure 60.11).

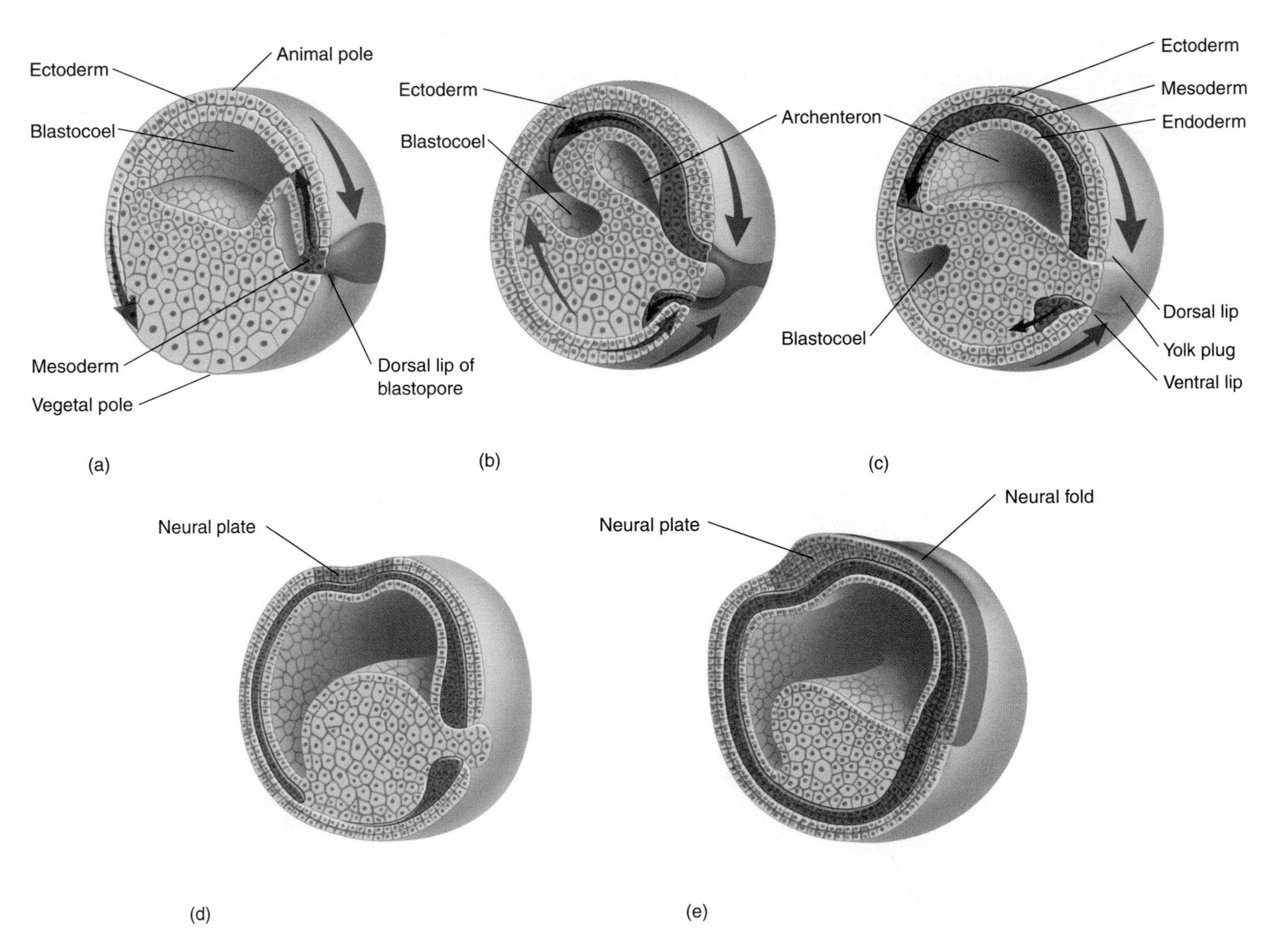

FIGURE 60.11
Frog gastrulation. (*a*) A layer of cells from the animal pole moves toward the vegetal pole ultimately involuting through the dorsal lip of the blastopore. (*b*) Cells in the dorsal lip zone then involute into the hollow interior, or blastocoel, eventually pressing against the far wall. The three primary tissues (ectoderm, mesoderm, and endoderm) become distinguished. Ectoderm is shown in *blue*, mesoderm in *red*, and endoderm in *yellow*. (*c*) The movement of cells in the dorsal lip creates a new internal cavity, the archenteron, and displaces the blastocoel. (*d*) The neural plate later forms from ectoderm. (*e*) This will next form a neural groove and then a neural tube as the embryo begins the process of neurulation. The cells of the neural ectoderm are shown in *green*.

Gastrulation in Reptiles, Birds, and Mammals

In the blastodisc of a bird or reptile and the inner cell mass of a mammal, the developing embryo is a small cap of cells rather than a sphere. No yolk separates the two sides of the embryo, and, as a result, the lower cell layer is able to differentiate into endoderm and the upper layer into ectoderm without cell movement. Just after these two primary cell layers form, the mesoderm arises by invagination and involution of cells from the upper layer. The surface cells begin moving to the midline where they involute and migrate laterally to form a mesodermal layer between the ectoderm and endoderm. A furrow along the longitudinal midline marks the site of this involution (figures 60.12 and 60.13). This furrow, analogous to an elongated blastopore, is called the **primitive streak.**

Gastrulation in lancelets involves the formation of ectoderm and endoderm by the invagination of the blastula, and the mesoderm layer forms from pouches pinched from the endoderm. In those vertebrates with extensive amounts of yolk, gastrulation requires the involution of surface cells into a blastopore or primitive streak, and the mesoderm is derived from some of these involuted cells.

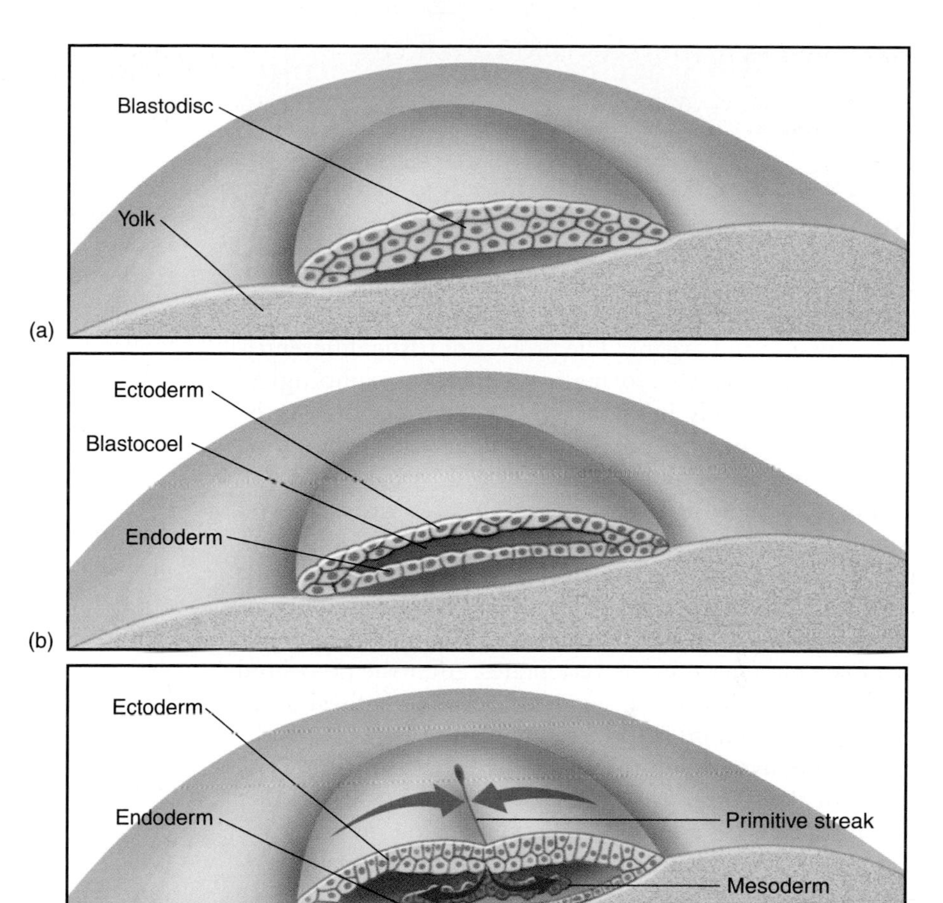

FIGURE 60.12
Gastrulation in birds. The upper layer of the blastodisc (*a*) differentiates into ectoderm, the lower layer into endoderm (*b*). Among the cells that migrate into the interior through the dorsal primitive streak are future mesodermal cells (*c*).

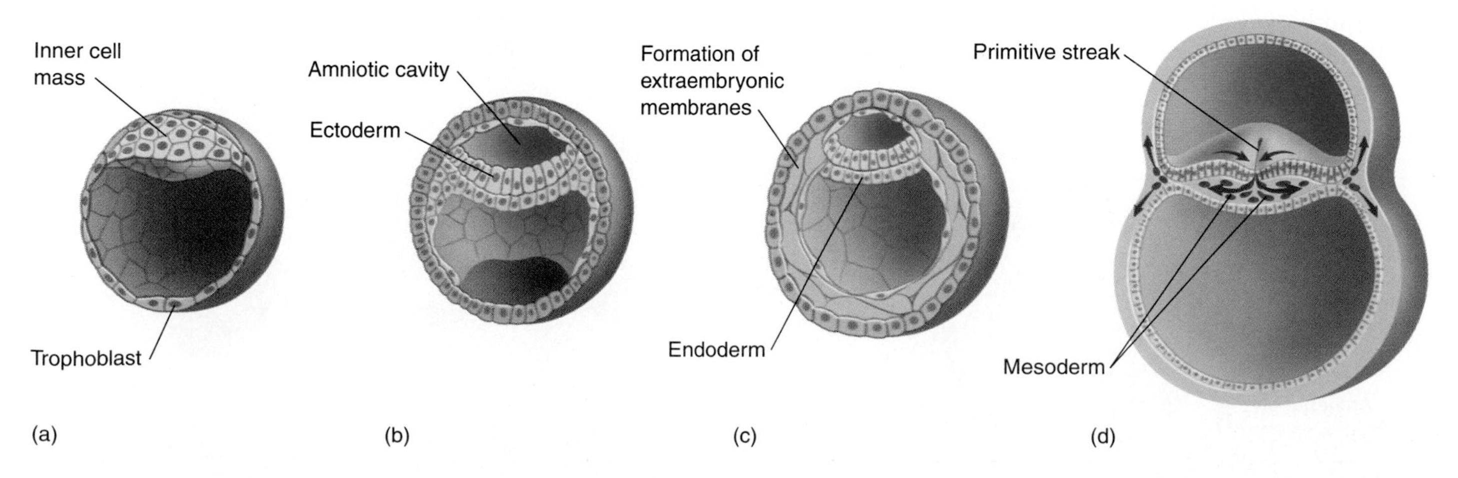

FIGURE 60.13
Mammalian gastrulation. (*a*) The amniotic cavity forms within the inner cell mass and its base. Layers of ectoderm and endoderm differentiate (*b* and *c*) as in the avian blastodisc. (*d*) A primitive streak develops, through which cells destined to become mesoderm migrate into the interior, again reminiscent of gastrulation in birds. The trophoblast has now moved further away from the embryo and begins to play a role in forming the placenta.

60.4 Body architecture is determined during the next stages of embryonic development.

Developmental Processes during Neurulation

During the next step in vertebrate development, the three primary cell layers begin their transformation into the body's tissues and organs. The process of tissue differentiation begins with the formation of two morphological features found only in chordates, the **notochord** and the hollow dorsal nerve cord. This development of the dorsal nerve cord is called **neurulation.**

The notochord is first visible soon after gastrulation is complete, forming from mesoderm along the dorsal midline of the embryo. It is a flexible rod located along the dorsal midline in the embryos of all chordates, although its function is replaced by the vertebral column when it develops from mesoderm in the vertebrates. After the notochord has been laid down, a layer of ectodermal cells situated above the notochord invaginates, forming a long crease, the **neural groove,** down the long axis of the embryo. The edges of the neural groove then move toward each other and fuse, creating a long hollow cylinder, the **neural tube** (figure 60.14), which runs beneath the surface of the embryo's back. The neural tube later differentiates into the spinal cord and brain.

The dorsal lip of the blastopore induces the formation of a notochord, and the presence of the notochord induces the overlying ectoderm to differentiate into the neural tube. The process of induction, when one embryonic region of cells influences the development of an adjacent region by changing its developmental pathway, was discussed in chapter 17 and is further examined in the next section.

While the neural tube is forming from ectoderm, the rest of the basic architecture of the body is being determined rapidly by changes in the mesoderm. On either side of the developing notochord, segmented blocks of mesoderm tissue called **somites** form; more somites are added as development continues. Ultimately, the somites give rise to the muscles, vertebrae, and connective tissues. The mesoderm in the head region does not separate into discrete somites but remains connected as **somitomeres** and form the striated muscles of the face, jaws, and throat. Some body organs, including the kidneys, adrenal glands, and gonads, develop within another strip of mesoderm that runs alongside the somites. The remainder of the mesoderm moves out and around the endoderm and eventually surrounds it completely. As a result of this movement, the mesoderm becomes separated into two layers. The outer layer is associated with the body wall and the inner layer is associated with the gut. Between these two layers of mesoderm is the coelom (see chapter 45), which becomes the body cavity of the adult.

The Neural Crest

Neurulation occurs in all chordates, and the process in a lancelet is much the same as it is in a human. However, in vertebrates, just before the neural groove closes to form the neural tube, its edges pinch off, forming a small strip of cells, the **neural crest,** which becomes incorporated into the roof of the neural tube (figure 60.14). The cells of the neural crest later move to the sides of the developing embryo. The appearance of the neural crest was a key event in the evolution of the vertebrates because neural crest cells, after migrating to different parts of the embryo, ultimately develop into the structures characteristic of (though not necessarily unique to) the vertebrate body.

The differentiation of neural crest cells depends on their location. At the anterior end of the embryo, they merge

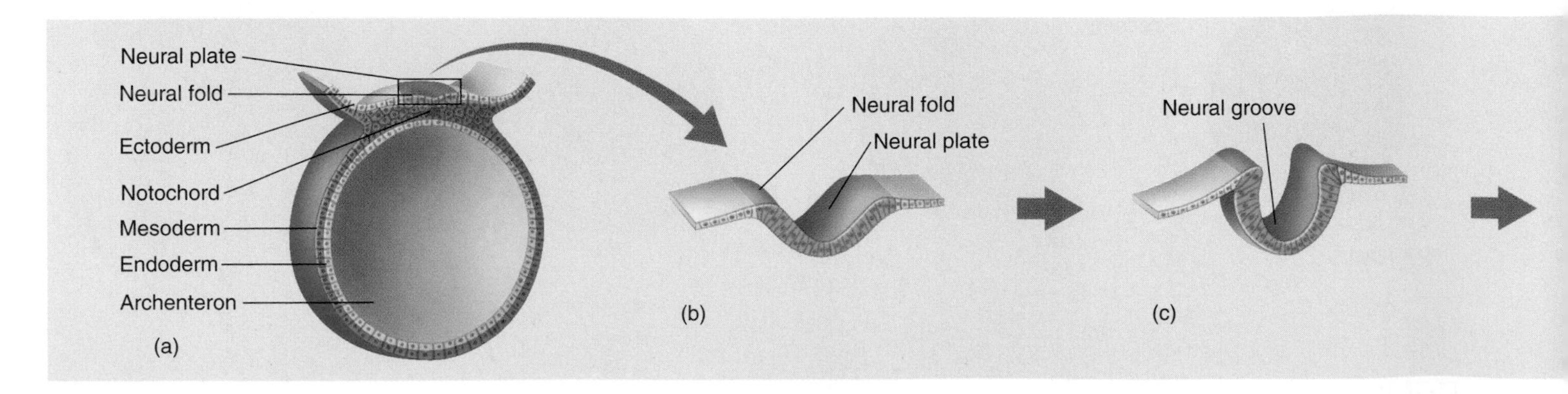

FIGURE 60.14
Mammalian neural tube formation. (*a*) The neural tube forms above the notochord when (*b*) cells of the neural plate fold together to form the (*c*) neural groove.

with the anterior portion of the brain, the forebrain. Nearby clusters of ectodermal cells associated with the neural crest cells thicken into **placodes,** which are distinct from neural crest cells although they arise from similar cellular interactions. Placodes subsequently develop into parts of the sense organs in the head. The neural crest and associated placodes exist in two lateral strips, which is why the vertebrate sense organs that develop from them are paired.

Neural crest cells located in more posterior positions have very different developmental fates. These cells migrate away from the neural tube to other locations in the head and trunk, where they form connections between the neural tube and the surrounding tissues. At these new locations, they contribute to the development of a variety of structures that are particularly characteristic of the vertebrates, several of which are discussed below. The migration of neural crest cells is unique in that it is not simply a change in the relative positions of cells, such as that seen in gastrulation. Instead, neural crest cells actually pass through other tissues.

The Gill Chamber

Primitive chordates such as lancelets are filter-feeders, using the rapid beating of cilia to draw water into their bodies through slits in their pharynx. These pharyngeal slits evolved into the vertebrate gill chamber, a structure that provides a greatly improved means of respiration. The evolution of the gill chamber was certainly a key event in the transition from filter-feeding to active predation.

In the development of the gill chamber, some of the neural crest cells form cartilaginous bars between the embryonic pharyngeal slits. Other neural crest cells induce portions of the mesoderm to form muscles along the cartilage, while still others form neurons that carry impulses between the nerve cord and these muscles. A major blood vessel called the aortic arch passes through each of the embryonic bars. Lined by still more neural crest cells, these bars, with their internal blood supply, become highly branched and form the gills of the adult.

Because the stiff bars of the gill chamber can be bent inward by powerful muscles controlled by nerves, the whole structure is a very efficient pump that drives water past the gills. The gills themselves act as highly efficient oxygen exchangers, greatly increasing the respiratory capacity of the animals that possess them.

Elaboration of the Nervous System

Some neural crest cells migrate ventrally toward the notochord and form sensory neurons in the dorsal root ganglia (see chapter 54). Others become specialized as Schwann cells, which insulate nerve fibers and permit the rapid conduction of nerve impulses. Still others form the autonomic ganglia and the adrenal medulla. Cells in the adrenal medulla secrete epinephrine when stimulated by the sympathetic division of the autonomic nervous system during the fight-or-flight reaction. The similarity in the chemical nature of the hormone epinephrine and the neurotransmitter norepinephrine, released by sympathetic neurons, is understandable—both adrenal medullary cells and sympathetic neurons derive from the neural crest.

Sensory Organs and Skull

A variety of sense organs develop from the placodes. Included among them are the olfactory (smell) and lateral line (primitive hearing) organs discussed in chapter 55. Neural crest cells contribute to tooth development and to some of the facial and cranial bones of the skull.

The appearance of the neural crest in the developing embryo marks the beginning of the first truly vertebrate phase of development, as many of the structures characteristic of vertebrates derive directly or indirectly from neural crest cells.

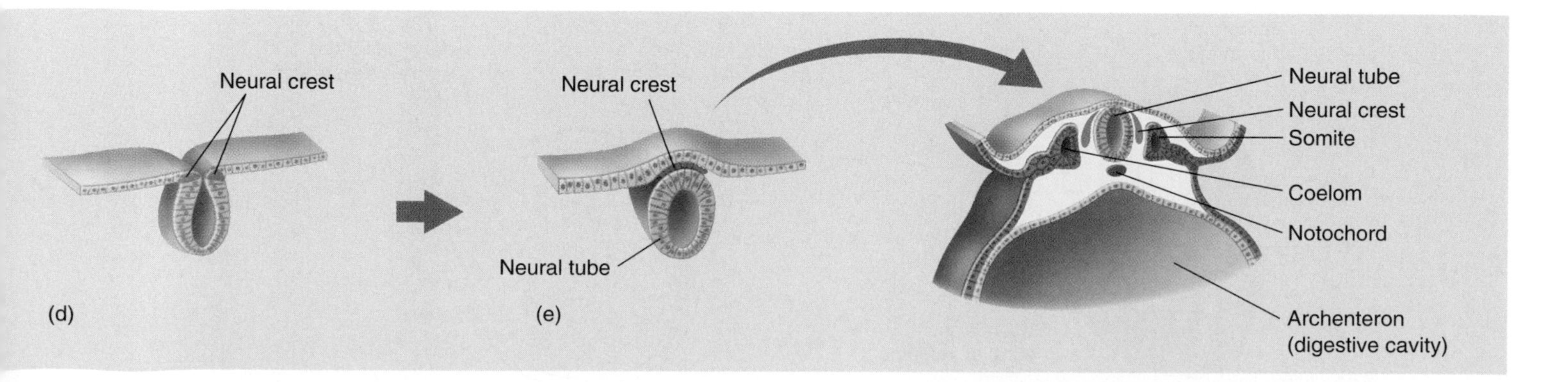

FIGURE 60.14 (continued)
(*d*) The neural groove eventually closes to form a hollow tube. (*e*) As the tube closes, some of the cells from the dorsal margin of the neural tube differentiate into the neural crest, which is characteristic of vertebrates.

How Cells Communicate during Development

In the process of vertebrate development, the relative position of particular cell layers determines, to a large extent, the organs that develop from them. By now, you may have wondered how these cell layers know where they are. For example, when cells of the ectoderm situated above the developing notochord give rise to the neural groove, how do these cells know they are above the notochord?

The solution to this puzzle is one of the outstanding accomplishments of experimental embryology, the study of how embryos form. The great German biologist Hans Spemann and his student Hilde Mangold solved it early in the twentieth century. In their investigation they removed cells from the dorsal lip of an amphibian blastula and transplanted them to a different location on another blastula (figure 60.15). (The dorsal lip region of amphibian blastulas develops from the grey crescent zone and is the site of origin of those mesoderm cells that later produce the notochord.) The new location corresponded to that of the animal's belly. What happened? The embryo developed *two* notochords, a normal dorsal one and a second one along its belly!

By using genetically different donor and host blastulas, Spemann and Mangold were able to show that the notochord produced by transplanting dorsal lip cells contained host cells as well as transplanted ones. The transplanted dorsal lip cells had acted as **organizers** (see also chapter 17) of notochord development. As such, these cells stimulated a developmental program in the belly cells of the embryos in which they were transplanted: the development of the notochord. The belly cells clearly contained this developmental program but would not have expressed it in the normal course of their development. The transplantation of the dorsal lip cells caused them to do so.

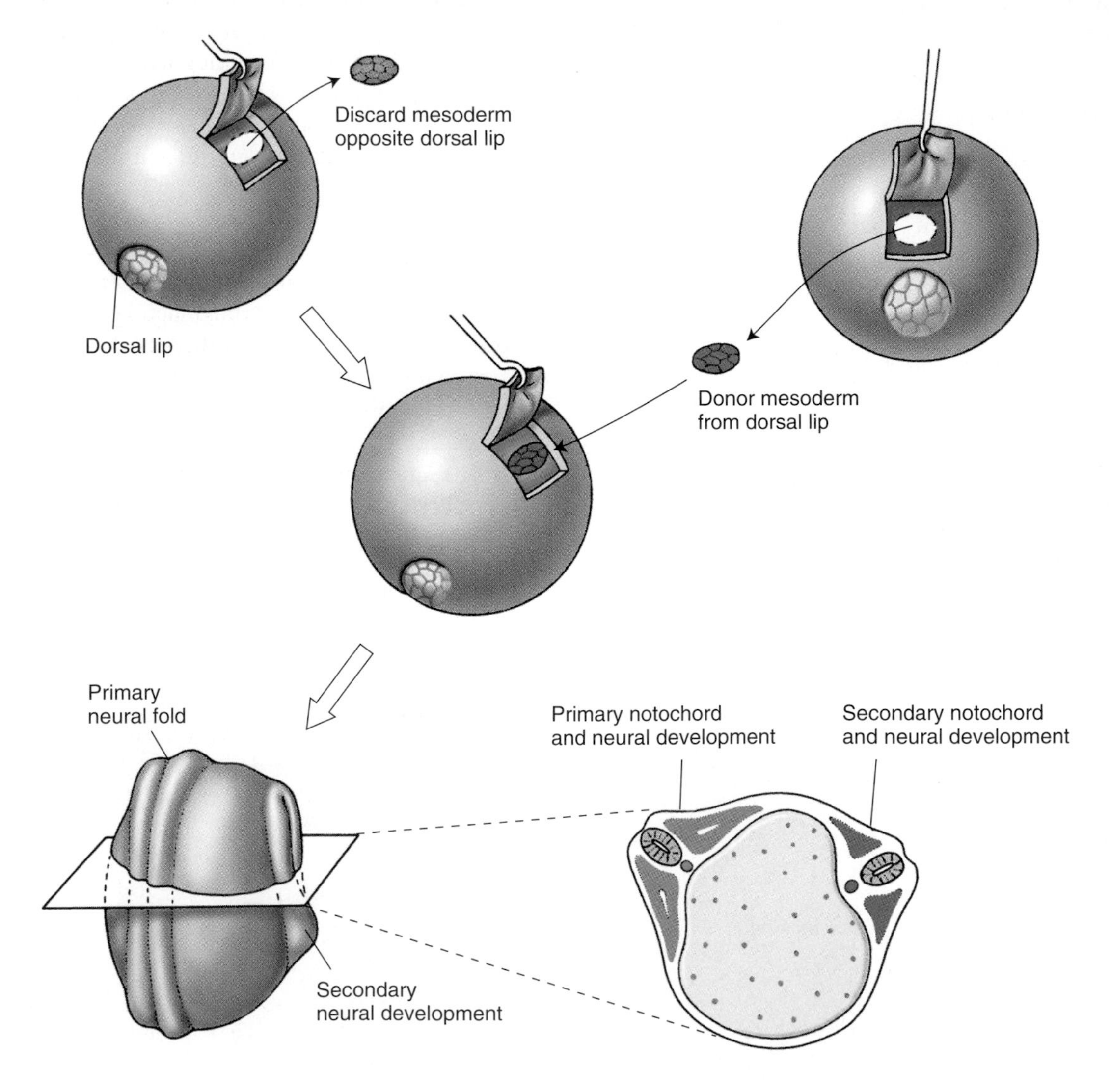

FIGURE 60.15
Spemann and Mangold's dorsal lip transplant experiment.

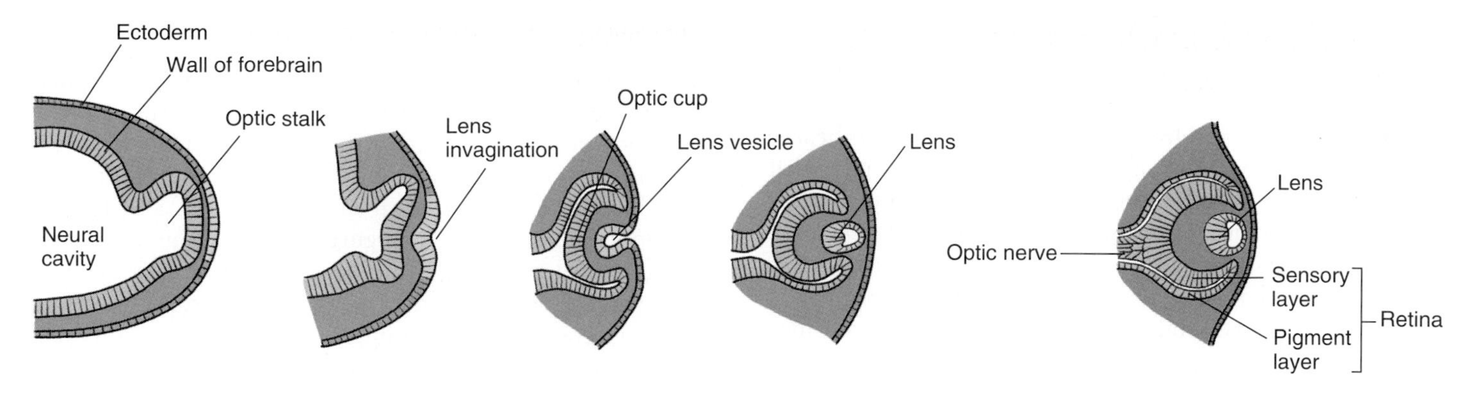

FIGURE 60.16
Development of the eye by induction. An extension of the optic stalk grows until it contacts ectoderm, which induces a section of the ectoderm to pinch off and form the lens. Other structures of the eye develop from the optic stalk.

These cells had indeed induced the ectoderm cells of the belly to form a notochord. This phenomenon as a whole is known as **induction.**

The process of induction that Spemann discovered appears to be the basic mode of development in vertebrates. Inductions between the three primary tissue types—ectoderm, mesoderm, and endoderm—are referred to as **primary inductions.** Inductions between tissues that have already been differentiated are called **secondary inductions.** The differentiation of the central nervous system during neurulation by the interaction of dorsal ectoderm and dorsal mesoderm to form the neural tube is an example of primary induction. In contrast, the differentiation of the lens of the vertebrate eye from ectoderm by interaction with tissue from the central nervous system is an example of secondary induction.

The eye develops as an extension of the forebrain, a stalk that grows outward until it comes into contact with the epidermis (figure 60.16). At a point directly above the growing stalk, a layer of the epidermis pinches off, forming a transparent lens. When the optic stalks of the two eyes have just started to project from the brain and the lenses have not yet formed, one of the budding stalks can be removed and transplanted to a region underneath a different epidermis, such as that of the belly. When Spemann performed this critical experiment, a lens still formed, this time from belly epidermis cells in the region above where the budding stalk had been transplanted.

What is the nature of the inducing signal that passes from one tissue to the other? If one imposes a nonporous barrier, such as a layer of cellophane, between the inducer and the target tissue, no induction takes place. In contrast, a porous filter, through which proteins can pass, does permit induction to occur. The induction process was discussed in detail in chapter 17. In brief, the inducer cells produce a protein factor that binds to the cells of the target tissue, initiating changes in gene expression.

The Nature of Developmental Decisions

All of the cells of the body, with the exception of a few specialized ones that have lost their nuclei, have an entire complement of genetic information. Despite the fact that all of its cells are genetically identical, an adult vertebrate contains hundreds of cell types, each expressing various aspects of the total genetic information for that individual. What factors determine which genes are to be expressed in a particular cell and which are not to be? In a liver cell, what mechanism keeps the genetic information that specifies nerve cell characteristics turned off? Does the differentiation of that particular cell into a liver cell entail the physical loss of the information specifying other cell types? No, it does not—but cells progressively lose the capacity to *express* ever-larger portions of their genomes. *Development is a process of progressive restriction of gene expression.*

Some cells become **determined** quite early in development. For example, all of the egg cells of the human female are set aside very early in the life of the embryo, yet some of these cells will not achieve differentiation into functional oocytes for more than 40 years. To a large degree, a cell's location in the developing embryo determines its fate. By changing a cell's location, an experimenter can alter its developmental destiny. However, this is only true up to a certain point in the cell's development. At some stage, every cell's ultimate fate becomes fixed, a process referred to as **commitment.** Commitment is not irreversible (entire individuals can be cloned from an individual specialized cell, as recounted in chapter 17), but rarely if ever reverses under ordinary circumstances.

When a cell is "determined," it is possible to predict its developmental fate; when a cell is "committed," that developmental fate cannot be altered. Determination often occurs very early in development, commitment somewhat later.

Embryonic Development and Vertebrate Evolution

The primitive chordates that gave rise to vertebrates were initially slow-moving, filter-feeding animals with relatively low metabolic rates. Many of the unique vertebrate adaptations that contribute to their varied ecological roles involve structures that arise from neural crest cells. The vertebrates became fast-swimming predators with much higher metabolic rates. This accelerated metabolism permitted a greater level of activity than was possible among the more primitive chordates. Other evolutionary changes associated with the derivatives of the neural crest provided better detection of prey, a greatly improved ability to orient spatially during prey capture, and the means to respond quickly to sensory information. The evolution of the neural crest and of the structures derived from it were thus crucial steps in the evolution of the vertebrates (figure 60.17).

Ontogeny Recapitulates Phylogeny

The patterns of development in the vertebrate groups that evolved most recently reflect in many ways the simpler patterns occurring among earlier forms. Thus, mammalian development and bird development are elaborations of reptile development, which is an elaboration of amphibian development, and so forth (figure 60.18). During the development of a mammalian embryo, traces can be seen of appendages and organs that are apparently relicts of more primitive chordates. For example, at certain stages a human embryo possesses pharyngeal slits, which occur in all chordates and are homologous to the gill slits of fish. At later stages, a human embryo also has a tail.

In a sense, the patterns of development in chordate groups has built up in incremental steps over the evolutionary history of those groups. The developmental instructions for each new form seem to have been layered on top of the previous instructions, contributing

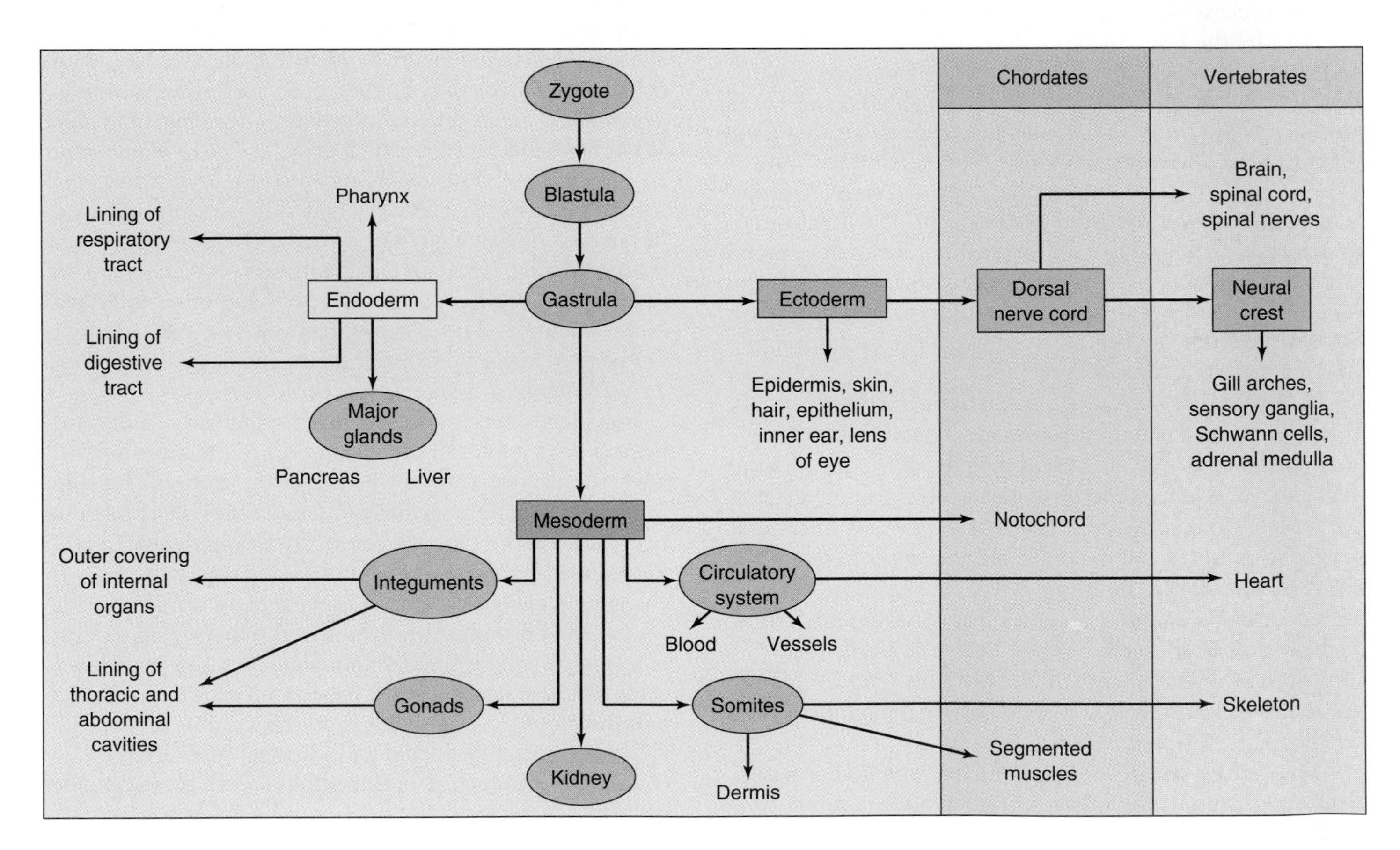

FIGURE 60.17
Derivation of the major tissue types. The three germ layers that form during gastrulation give rise to all organs and tissues in the body, but the neural crest cells that form from ectodermal tissue give rise to structures that are prevalent in the vertebrate animal such as gill arches and Schwann cells.

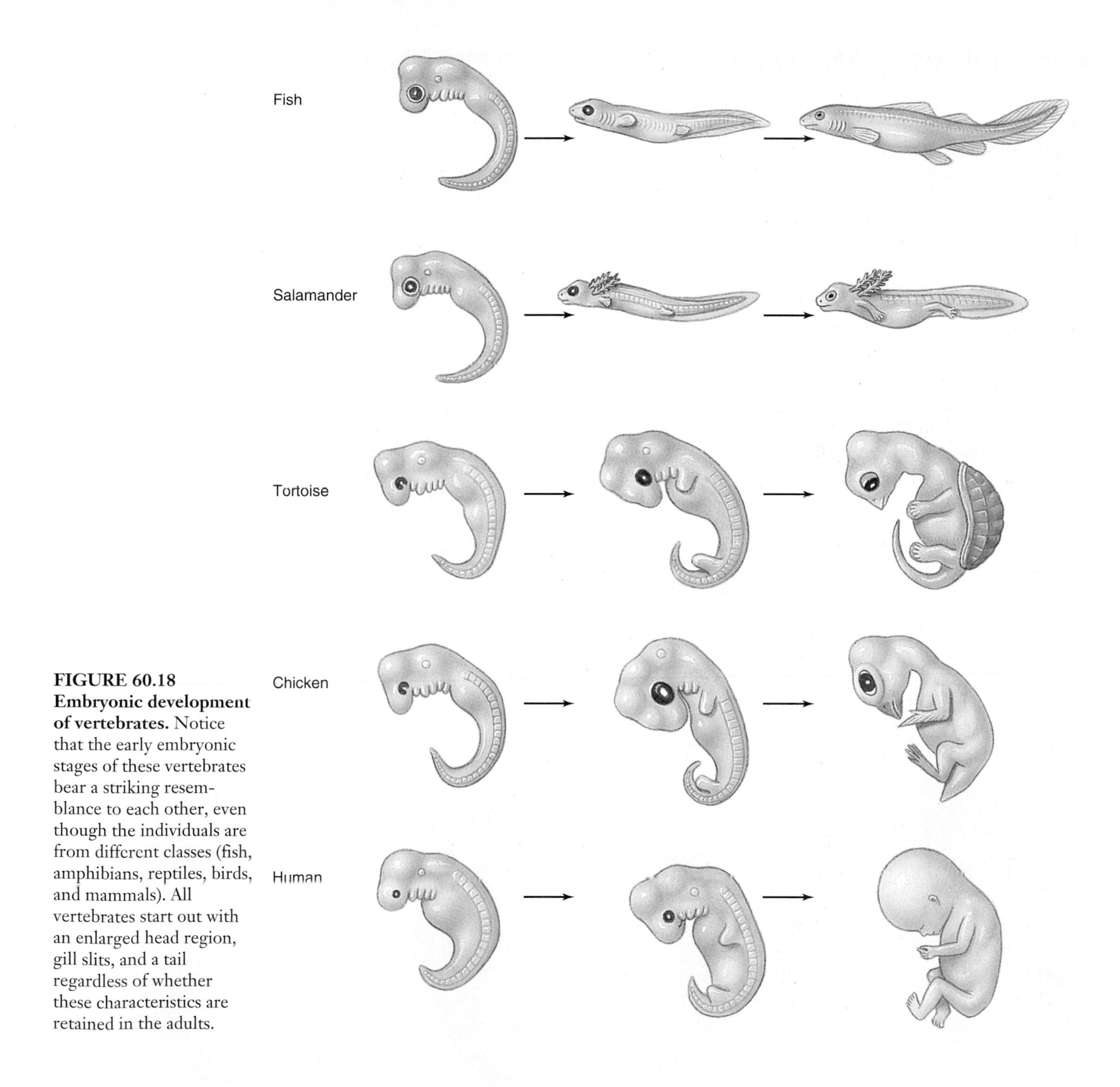

FIGURE 60.18
Embryonic development of vertebrates. Notice that the early embryonic stages of these vertebrates bear a striking resemblance to each other, even though the individuals are from different classes (fish, amphibians, reptiles, birds, and mammals). All vertebrates start out with an enlarged head region, gill slits, and a tail regardless of whether these characteristics are retained in the adults.

additional steps in the developmental journey. This hypothesis, proposed in the nineteenth century by Ernst Haeckel, is referred to as the "biogenic law." It is usually stated as an aphorism: *ontogeny recapitulates phylogeny;* that is, embryological development (ontogeny) involves the same progression of changes that have occurred during evolution (phylogeny). However, the biogenic law is not literally true when stated in this way because embryonic stages are not reflections of *adult* ancestors. Instead, the embryonic stages of a particular vertebrate often reflect the *embryonic* stages of that vertebrate's ancestors. Thus, the pharyngeal slits of a mammalian embryo are *not* like the gill slits its ancestors had *when they were adults.* Rather, they are like the pharyngeal slits its ancestors had *when they were embryos.*

Vertebrates seem to have evolved largely by the addition of new instructions to the developmental program. Development of a mammal thus proceeds through a series of stages, and the earlier stages are unchanged from those that occur in the development of more primitive vertebrates.

Extraembryonic Membranes

As an adaptation to terrestrial life, the embryos of reptiles, birds, and mammals develop within a fluid-filled **amniotic membrane** (see chapter 48). The amniotic membrane and several other membranes form from embryonic cells but are located outside of the body of the embryo. For this reason, they are known as **extraembryonic membranes.** The extraembryonic membranes, later to become the **fetal membranes,** include the amnion, chorion, yolk sac, and allantois.

In birds, the **amnion** and **chorion** arise from two folds that grow to completely surround the embryo (figure 60.19). The amnion is the inner membrane that surrounds the embryo and suspends it in *amniotic fluid*, thereby mimicking the aquatic environments of fish and amphibian embryos. The chorion is located next to the eggshell and is separated from the other membranes by a cavity, the *extraembryonic coelom.* The **yolk sac** plays a critical role in the nutrition of bird and reptile embryos; it is also present in mammals, though it does not nourish the embryo. The **allantois** is derived as an outpouching of the gut and serves to store the uric acid excreted in the urine of birds. During development, the allantois of a bird embryo expands to form a sac that eventually fuses with the overlying chorion, just under the eggshell. The fusion of the allantois and chorion form a functioning unit in which embryonic blood vessels, carried in the allantois, are brought close to the porous eggshell for gas exchange. The allantois is thus the functioning "lung" of a bird embryo.

In mammals, the embryonic cells form an inner cell mass that will become the body of the embryo and a layer of surrounding cells called the trophoblast (see figure 60.9). The trophoblast implants into the endometrial lining of its mother's uterus and becomes the chorionic membrane (figure 60.20). The part of the chorion in contact with endometrial tissue contributes to the placenta, as is described in more detail in the next section. The allantois in mammals contributes blood vessels to the structure that will become the umbilical cord, so that fetal blood can be delivered to the placenta for gas exchange.

The extraembryonic membranes include the yolk sac, amnion, chorion, and allantois. These are derived from embryonic cells and function in a variety of ways to support embryonic development.

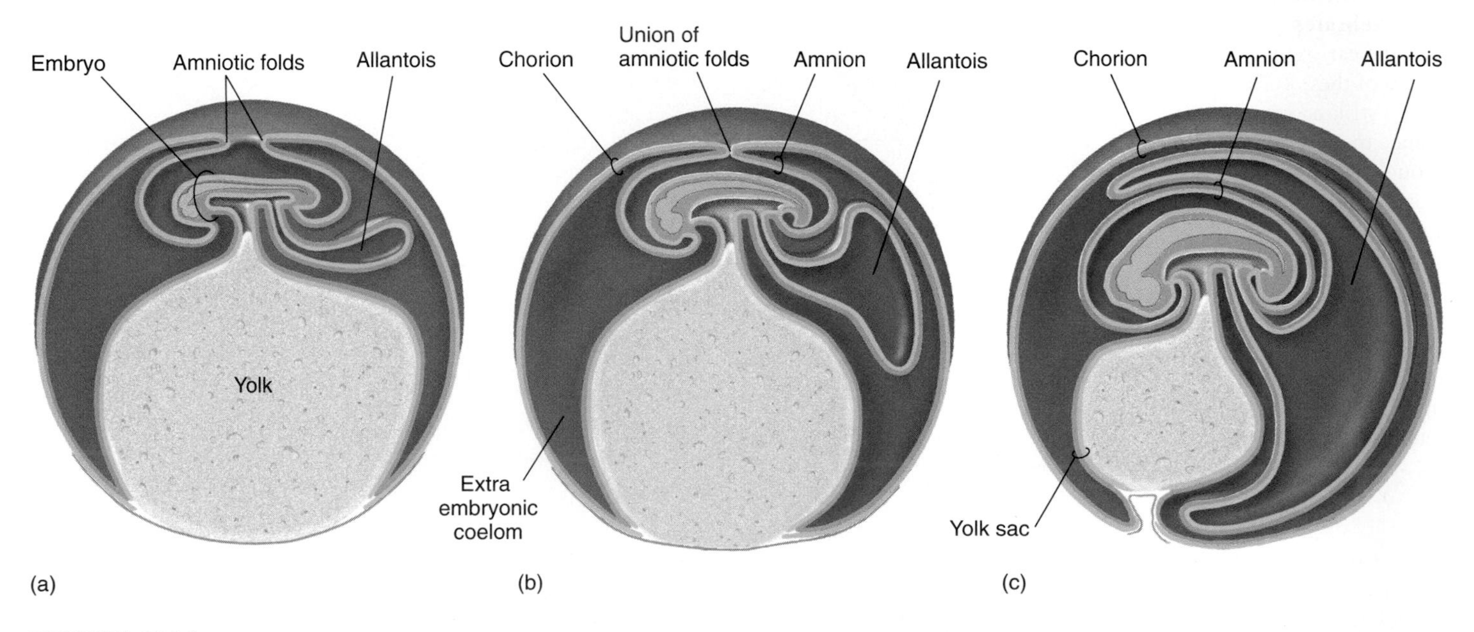

FIGURE 60.19
The extraembryonic membranes of a chick embryo. The membranes begin as amniotic folds from the embryo (*a*) that unite (*b*) to form a separate amnion and chorion (*c*). The allantois continues to grow until it will eventually unite with the chorion just under the eggshell.

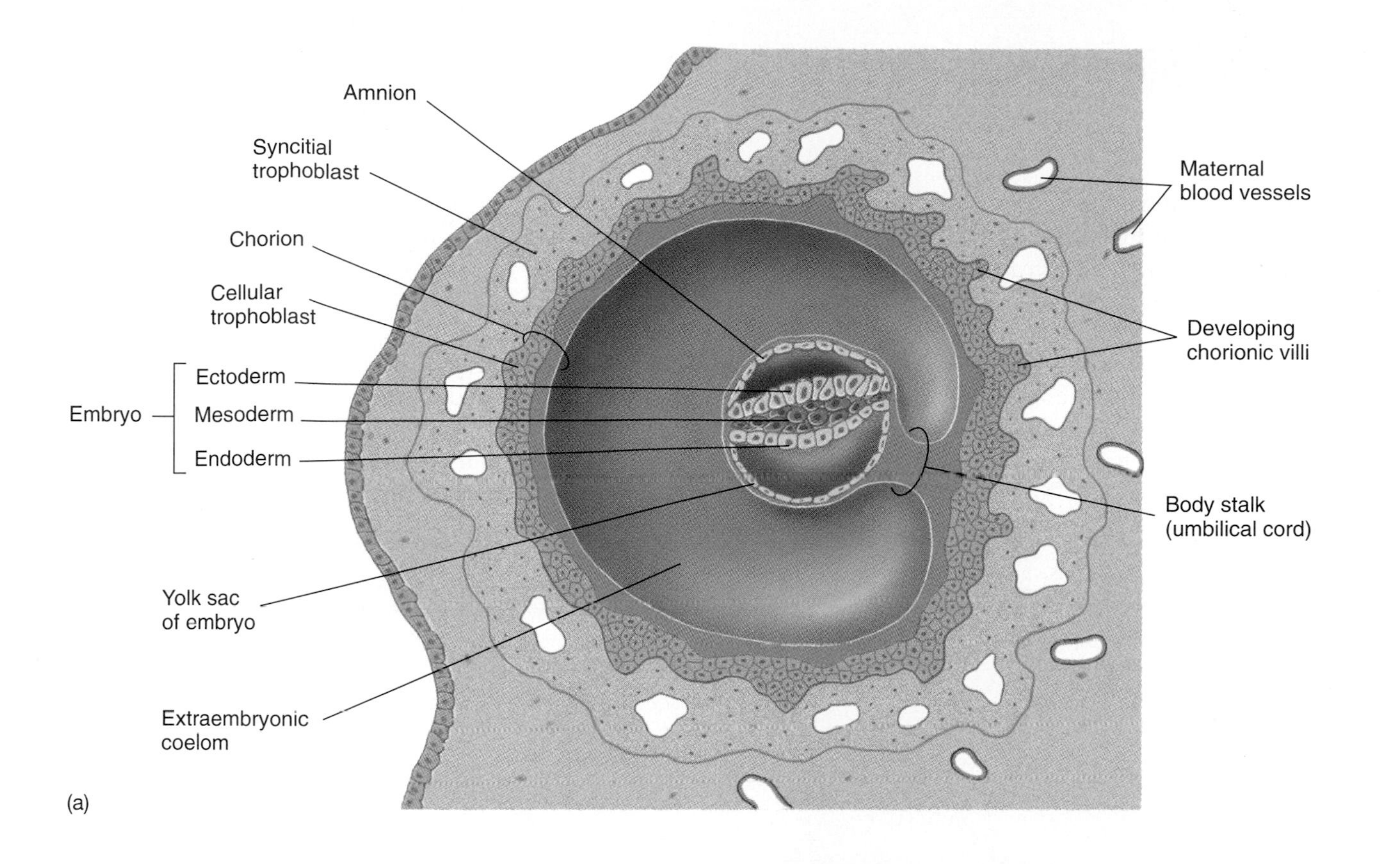

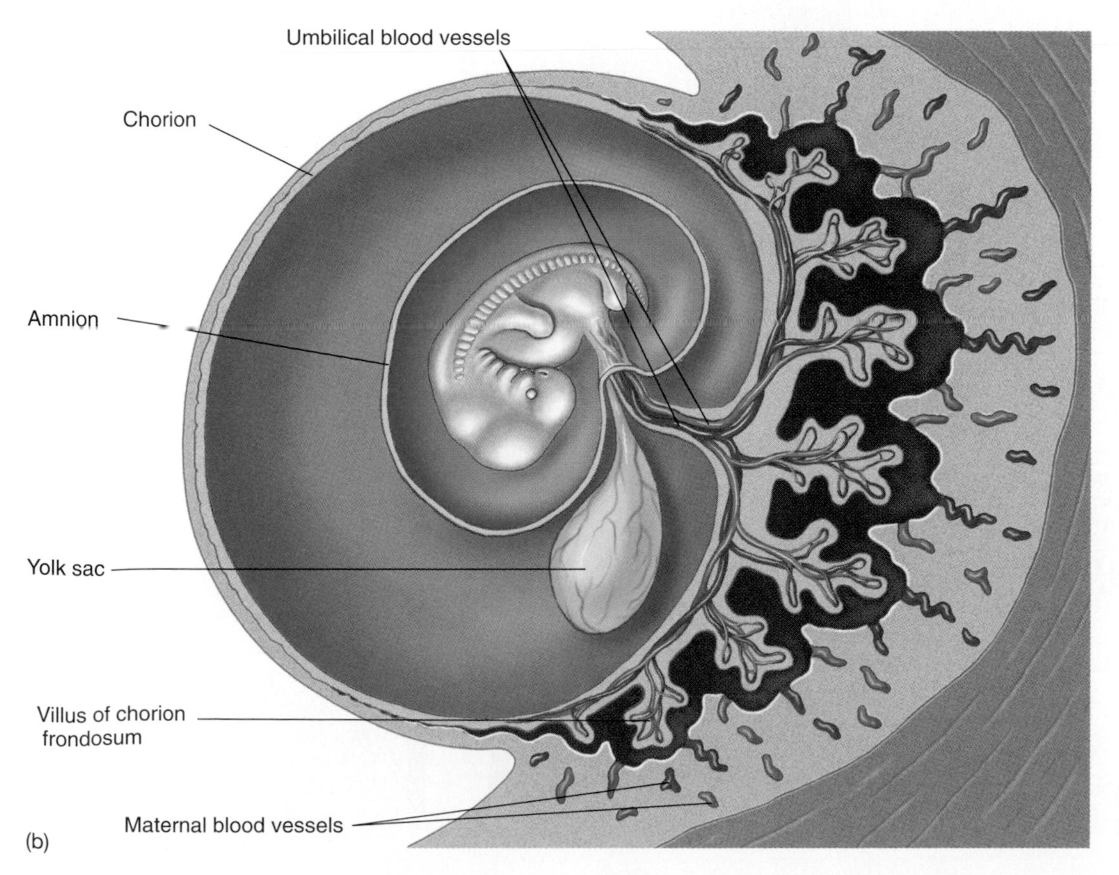

FIGURE 60.20

The extraembryonic membranes of a mammalian embryo. (*a*) After the embryo implants into the mother's endometrium (6–7 days after fertilization), the trophoblast becomes the chorion, and the yolk sac and amnion are produced. (*b*) The chorion develops extensions, called villi, that interdigitate with surrounding endometrial tissue. The embryo is encased within an amniotic sac.

60.5 Human development is divided into trimesters.

First Trimester

The development of the human embryo shows its evolutionary origins. Without an evolutionary perspective, we would be unable to account for the fact that human development proceeds in much the same way as development in a bird. In both animals, the embryo develops from a flattened collection of cells—the blastodisc in birds or the inner cell mass in humans. While the blastodisc of a bird is flattened because it is pressed against a mass of yolk, the inner cell mass of a human is flat despite the absence of a yolk mass. In humans as well as in birds, a primitive streak forms and gives rise to the three primary germ layers. Human development, from fertilization to birth, takes an average of 266 days. This time is commonly divided into three periods, called trimesters.

The First Month

About 30 hours after fertilization, the zygote undergoes its first cleavage; the second cleavage occurs about 30 hours after that. By the time the embryo reaches the uterus (6–7 days after fertilization), it is a blastula, which in mammals is referred to as a blastocyst. As we mentioned earlier, the

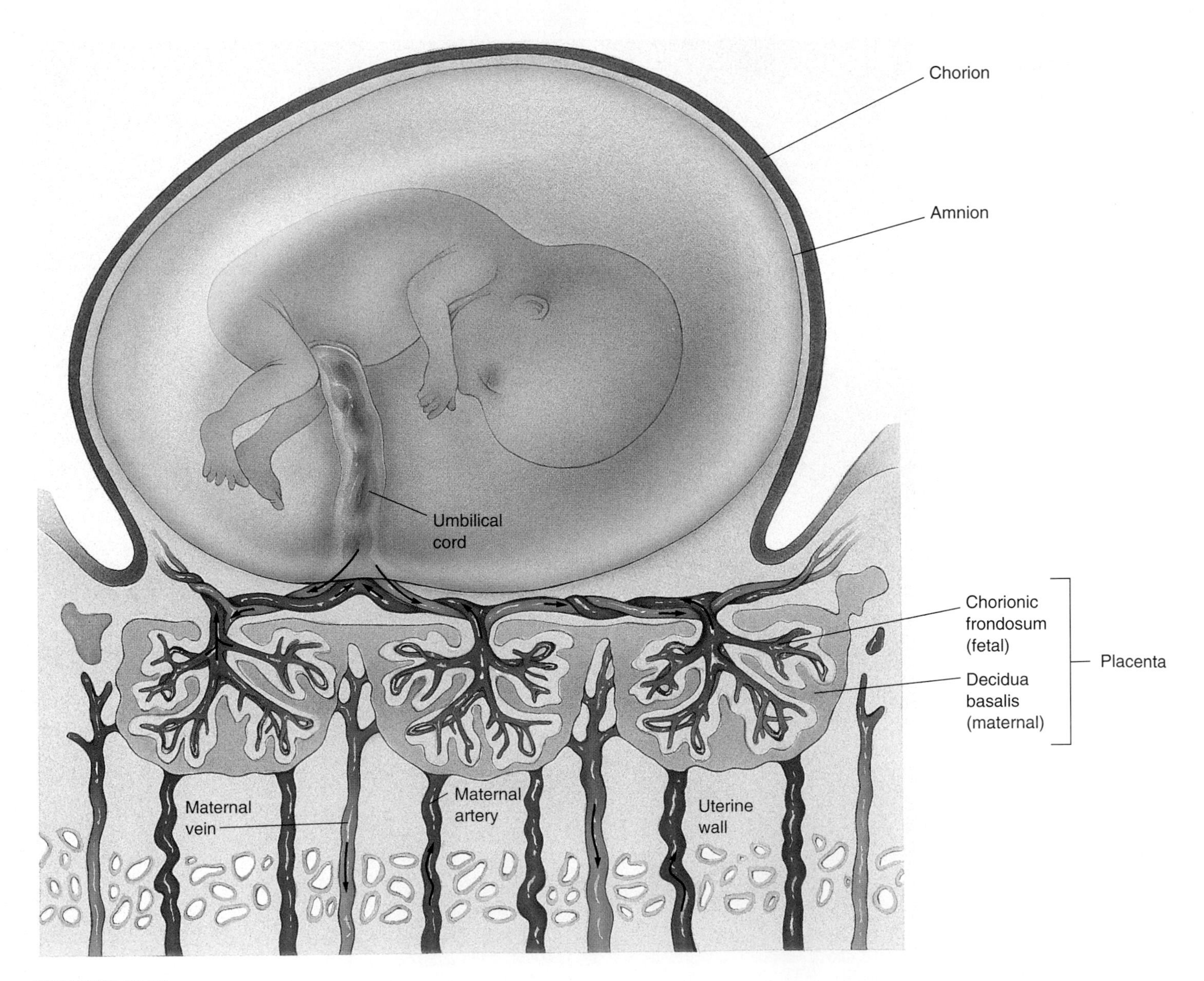

FIGURE 60.21
Structure of the placenta. The placenta contains a fetal component, the chorionic frondosum, and a maternal component, the decidua basalis. Deoxygenated fetal blood from the umbilical arteries (shown in *blue*) enters the placenta, where it picks up oxygen and nutrients from the mother's blood. Oxygenated fetal blood returns in the umbilical vein (shown in *red*) to the fetus.

blastocyst consists of an inner cell mass, which will become the body of the embryo, and a surrounding layer of trophoblast cells (see figure 60.9). The blastocyst begins to grow rapidly and initiates the formation of the amnion and the chorion. The blastocyst digests its way into the endometrial lining of the uterus in the process known as implantation.

During the second week after fertilization, the developing chorion forms branched extensions, the *chorionic frondosum* (fetal placenta) that protrude into the endometrium (figure 60.21). These extensions induce the surrounding endometrial tissue to undergo changes and become the *decidua basalis* (maternal placenta). Together, the chorionic frondosum and decidua basalis form a single functioning unit, the placenta (figure 60.22). Within the placenta, the mother's blood and the blood of the embryo come into close proximity but do not mix (see figure 60.21). Oxygen can thus diffuse from the mother to the embryo, and carbon dioxide can diffuse in the opposite direction. In addition to exchanging gases, the placenta provides nourishment for the embryo, detoxifies certain molecules that may pass into the embryonic circulation, and secretes hormones. Certain substances such as alcohol, drugs, and antibiotics are not stopped by the placenta and pass from the mother's bloodstream to the fetus.

One of the hormones released by the placenta is human chorionic gonadotropin (hCG), which was discussed in chapter 59. This hormone is secreted by the trophoblast cells even before they become the chorion, and is the hormone assayed in pregnancy tests. Because its action is almost identical to that of luteinizing hormone (LH), hCG maintains the mother's corpus luteum. The corpus luteum, in turn, continues to secrete estradiol and progesterone, thereby preventing menstruation and further ovulations.

Gastrulation also takes place in the second week after fertilization. The primitive streak can be seen on the surface of the embryo, and the three germ layers (ectoderm, mesoderm, and endoderm) are differentiated.

In the third week, neurulation occurs. This stage is marked by the formation of the neural tube along the dorsal axis of the embryo, as well as by the appearance of the first somites, which give rise to the muscles, vertebrae, and connective tissues. By the end of the week, over a dozen somites are evident, and the blood vessels and gut have begun to develop. At this point, the embryo is about 2 millimeters long.

Organogenesis (the formation of body organs) begins during the fourth week (figure 60.23*a*). The eyes form. The tubular heart develops its four chambers and starts to pulsate rhythmically, as it will for the rest of the individual's life. At 70 beats per minute, the heart is destined to beat more than 2.5 billion times during a lifetime of 70 years. Over 30 pairs of somites are visible by the end of the fourth week, and the arm and leg buds have begun to form. The embryo has increased in length to about 5 millimeters.

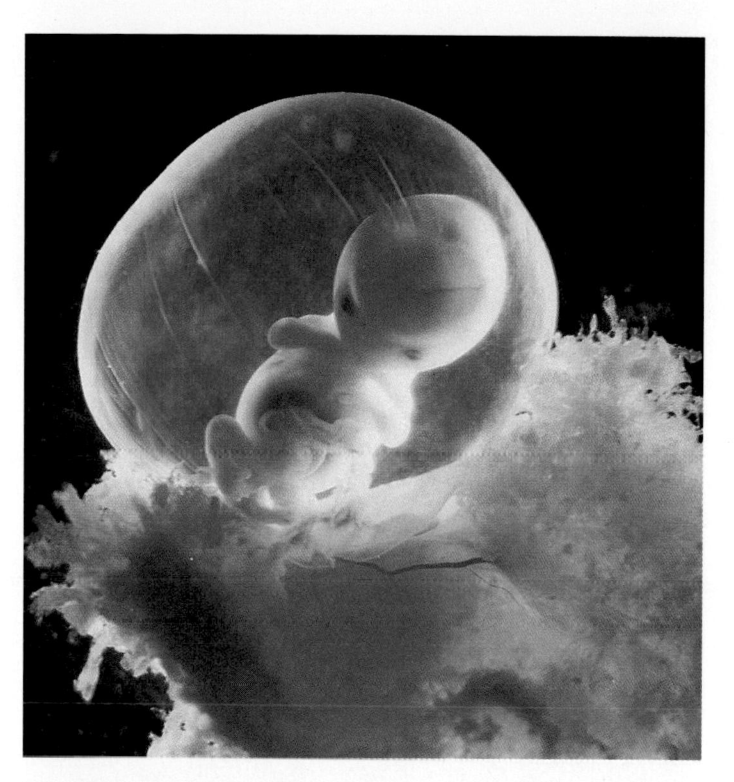

FIGURE 60.22
Placenta and fetus at seven weeks.

Although the developmental scenario is now far advanced, many women are unaware they are pregnant at this stage.

Early pregnancy is a very critical time in development because the proper course of events can be interrupted easily. In the 1960s, for example, many pregnant women took the tranquilizer thalidomide to minimize the discomforts associated with early pregnancy. Unfortunately, this drug had not been adequately tested. It interferes with limb bud development, and its widespread use resulted in many deformed babies. Organogenesis may also be disrupted during the first and second months of pregnancy if the mother contracts rubella (German measles). Most spontaneous abortions occur during this period.

The Second Month

Morphogenesis (the formation of shape) takes place during the second month (figure 60.23*b*). The miniature limbs of the embryo assume their adult shapes. The arms, legs, knees, elbows, fingers, and toes can all be seen—as well as a short bony tail! The bones of the embryonic tail, an evolutionary reminder of our past, later fuse to form the coccyx. Within the abdominal cavity, the major organs, including the liver, pancreas, and gallbladder, become evident. By the end of the second month, the embryo has grown to about 25 millimeters in length, weighs about one gram, and begins to look distinctly human.

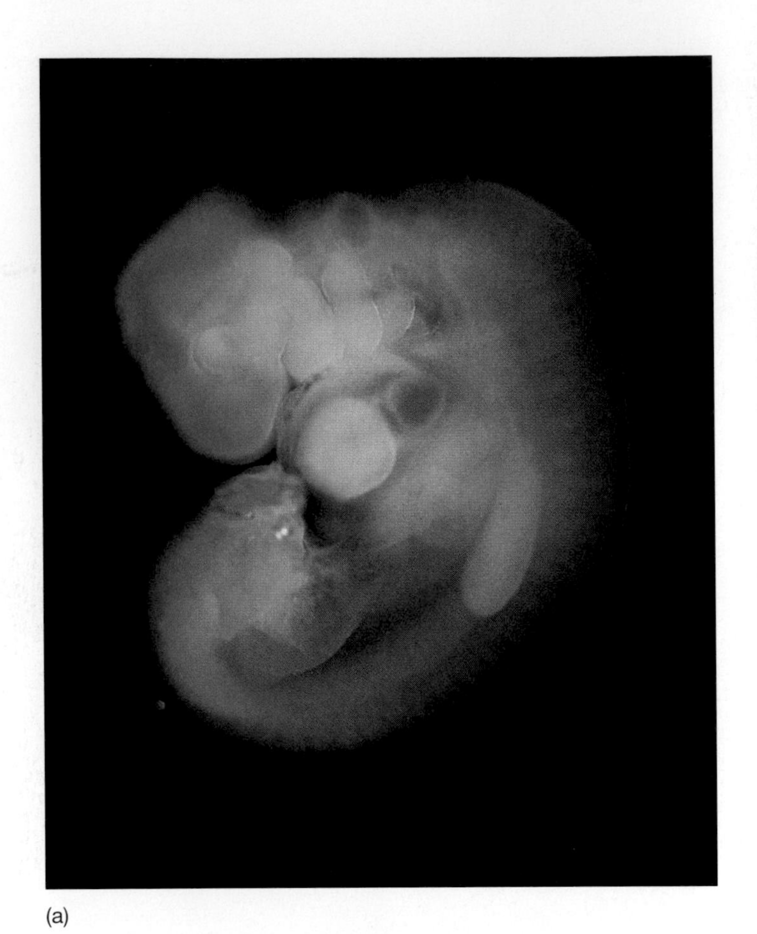

(a)

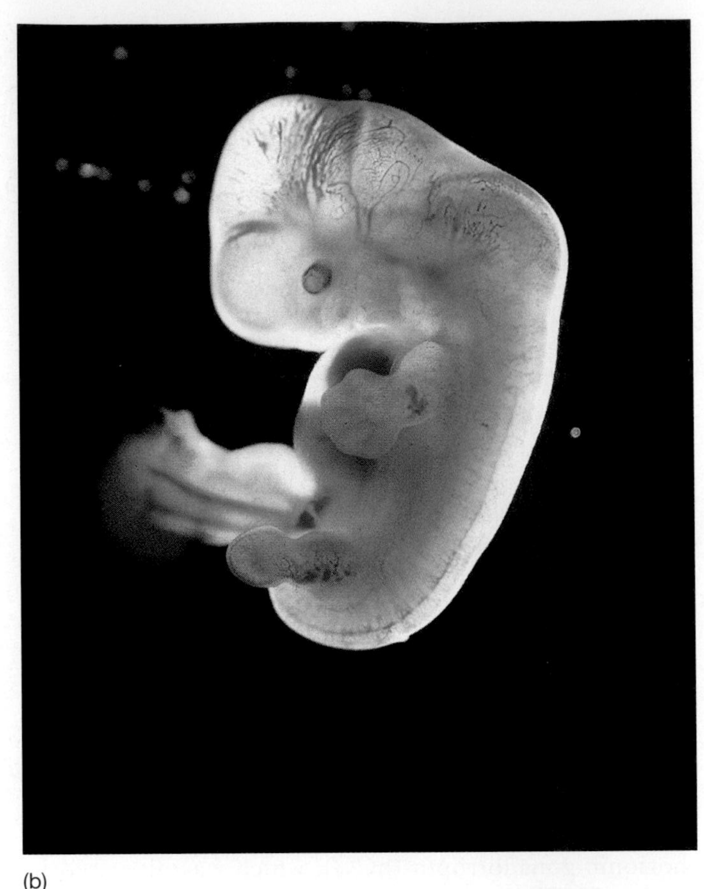

(b)

FIGURE 60.23
The developing human. (*a*) Four weeks, (*b*) seven weeks, (*c*) three months, and (*d*) four months.

The Third Month

The nervous system and sense organs develop during the third month, and the arms and legs start to move (figure 60.23*c*). The embryo begins to show facial expressions and carries out primitive reflexes such as the startle reflex and sucking. The eighth week marks the transition from embryo to fetus. At this time, all of the major organs of the body have been established. What remains of development is essentially growth.

At around 10 weeks, the secretion of human chorionic gonadotropin (hCG) by the placenta declines, and the corpus luteum regresses as a result. However, menstruation does not occur because the placenta itself secretes estradiol and progesterone (figure 60.24). In fact, the amounts of these two hormones secreted by the placenta far exceed the amounts that are ever secreted by the ovaries. The high levels of estradiol and progesterone in the blood during pregnancy continue to inhibit the release of FSH and LH, thereby preventing ovulation. They also help maintain the uterus and eventually prepare it for labor and delivery, and they stimulate the development of the mammary glands in preparation for lactation after delivery.

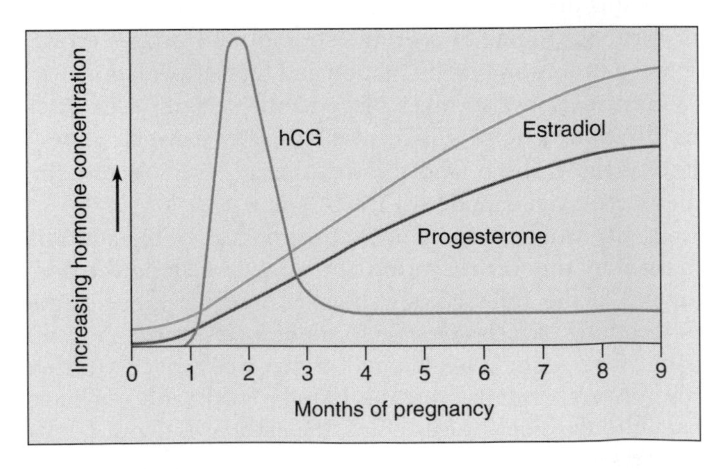

FIGURE 60.24
Hormonal secretion by the placenta. The placenta secretes chorionic gonadotropin (hCG) for 10 weeks. Thereafter, it secretes increasing amounts of estradiol and progesterone.

The embryo implants into the endometrium, differentiates the germ layers, forms the extraembryonic membranes, and undergoes organogenesis during the first month and morphogenesis during the second month.

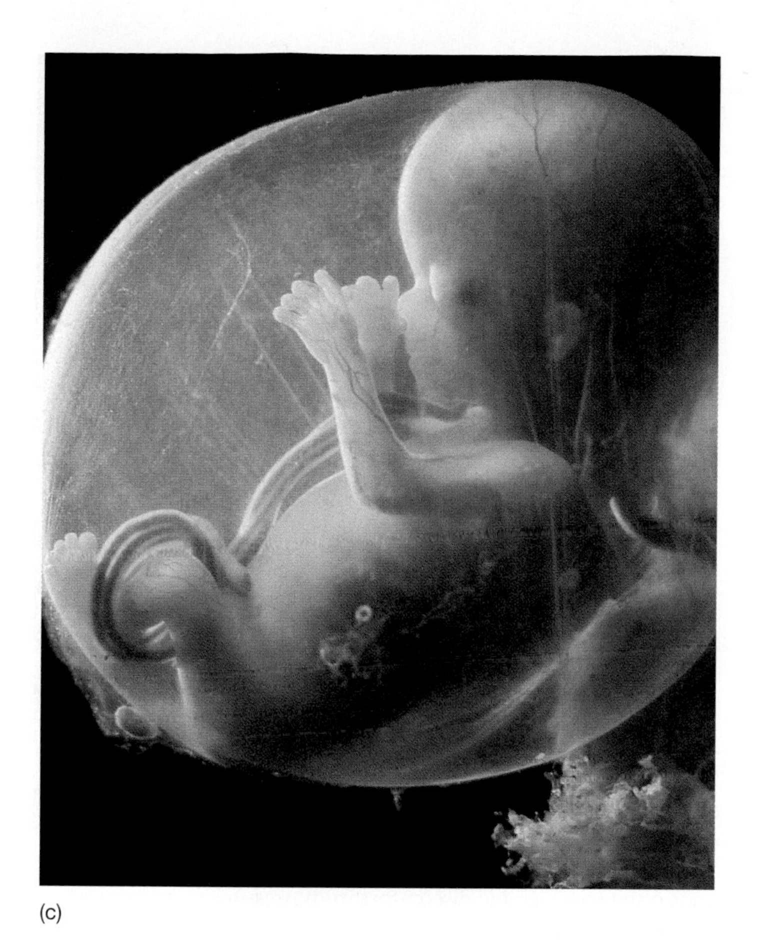
(c)

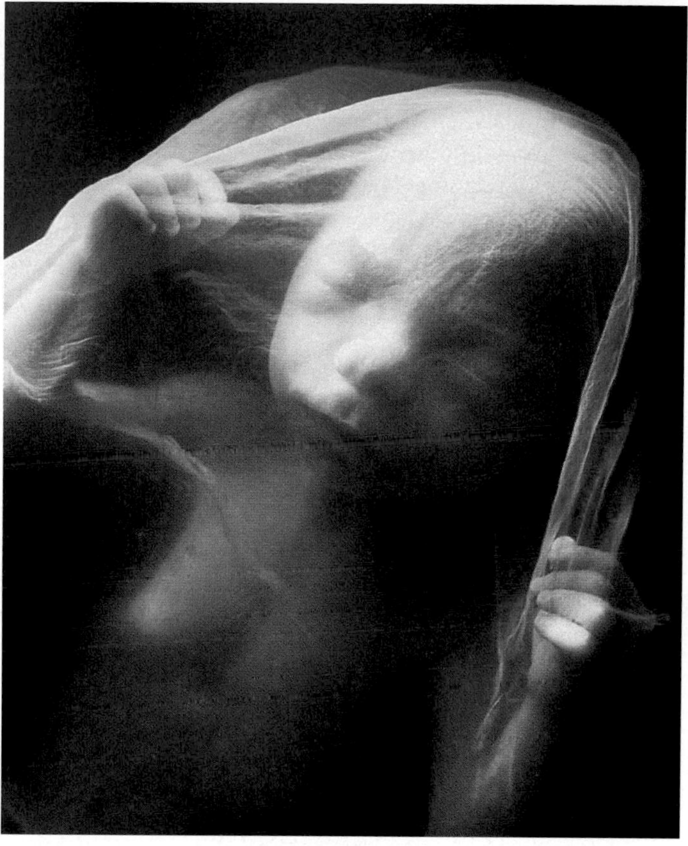
(d)

Second and Third Trimesters

The second and third trimesters are characterized by the tremendous growth and development required for the viability of the baby after its birth.

Second Trimester

Bones actively enlarge during the fourth month (figure 60.23*d*), and by the end of the month, the mother can feel the baby kicking. During the fifth month, the head and body grow a covering of fine hair. This hair, called **lanugo,** is another evolutionary relict but is lost later in development. By the end of the fifth month, the rapid heartbeat of the fetus can be heard with a stethoscope, although it can also be detected as early as 10 weeks with a fetal monitor. The fetus has grown to about 175 millimeters in length and attained a weight of about 225 grams. Growth begins in earnest in the sixth month; by the end of that month, the baby weighs 600 grams (1.3 lbs) and is over 300 millimeters (1 ft) long. However, most of its prebirth growth is still to come. The baby cannot yet survive outside the uterus without special medical intervention.

Third Trimester

The third trimester is predominantly a period of growth rather than development. The weight of the fetus doubles several times, but this increase in bulk is not the only kind of growth that occurs. Most of the major nerve tracts in the brain, as well as many new neurons (nerve cells), are formed during this period. The developing brain produces neurons at an average rate estimated at more than 250,000 per minute! Neurological growth is far from complete at the end of the third trimester, when birth takes place. If the fetus remained in the uterus until its neurological development was complete, it would grow too large for safe delivery through the pelvis. Instead, the infant is born as soon as the probability of its survival is high, and its brain continues to develop and produce new neurons for months after birth.

The critical stages of human development take place quite early, and the following six months are essentially a period of growth. The growth of the brain is not yet complete, however, by the end of the third trimester, and must be completed postnatally.

Birth and Postnatal Development

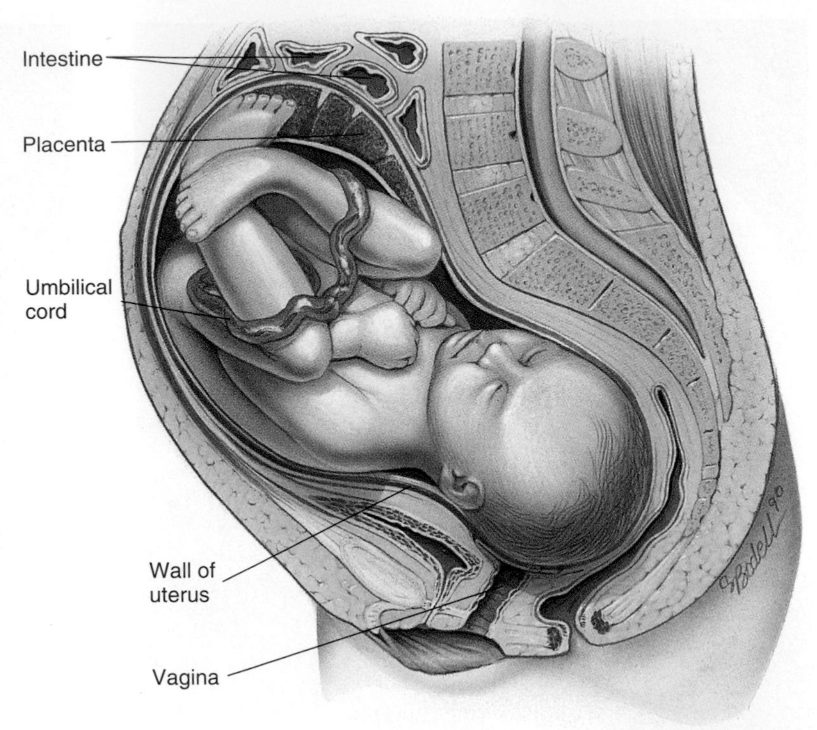

FIGURE 60.25
Position of the fetus just before birth. A developing fetus is a major addition to a woman's anatomy. The stomach and intestines are pushed far up, and there is often considerable discomfort from pressure on the lower back. In a natural delivery, the fetus exits through the vagina, which must dilate (expand) considerably to permit passage.

In some mammals, changing hormone levels in the developing fetus initiate the process of birth. The fetuses of these mammals have an extra layer of cells in their adrenal cortex, called a fetal zone. Before birth, the fetal pituitary gland secretes corticotropin, which stimulates the fetal zone to secrete steroid hormones. These corticosteroids then induce the uterus of the mother to manufacture prostaglandins, which trigger powerful contractions of the uterine smooth muscles.

The adrenal glands of human fetuses lack a fetal zone, and human birth does not seem to be initiated by this mechanism. In a human, the uterus releases prostaglandins, possibly as a result of the high levels of estradiol secreted by the placenta. Estradiol also stimulates the uterus to produce more oxytocin receptors, and as a result, the uterus becomes increasingly sensitive to oxytocin. Prostaglandins begin the uterine contractions, but then sensory feedback from the uterus stimulates the release of oxytocin from the mother's posterior pituitary gland. Working together, oxytocin and prostaglandins further stimulate uterine contractions, forcing the fetus downward (figure 60.25). Initially, only a few contractions occur each hour, but the rate eventually increases to one contraction every two to three minutes. Finally, strong contractions, aided by the mother's pushing, expel the fetus, which is now a newborn baby.

After birth, continuing uterine contractions expel the placenta and associated membranes, collectively called the afterbirth. The umbilical cord is still attached to the baby, and to free the newborn, a doctor or midwife clamps and cuts the cord. Blood clotting and contraction of muscles in the cord prevent excessive bleeding.

Nursing

Milk production, or lactation, occurs in the alveoli of mammary glands when they are stimulated by the anterior pituitary hormone, prolactin. Milk from the alveoli is secreted into a series of alveolar ducts, which are surrounded by smooth muscle and lead to the nipple (figure 60.26). During pregnancy, high levels of progesterone stimulate the development of the mammary alveoli, and high levels of estradiol stimulate the development of the alveolar ducts. However, estradiol blocks the actions of prolactin on the

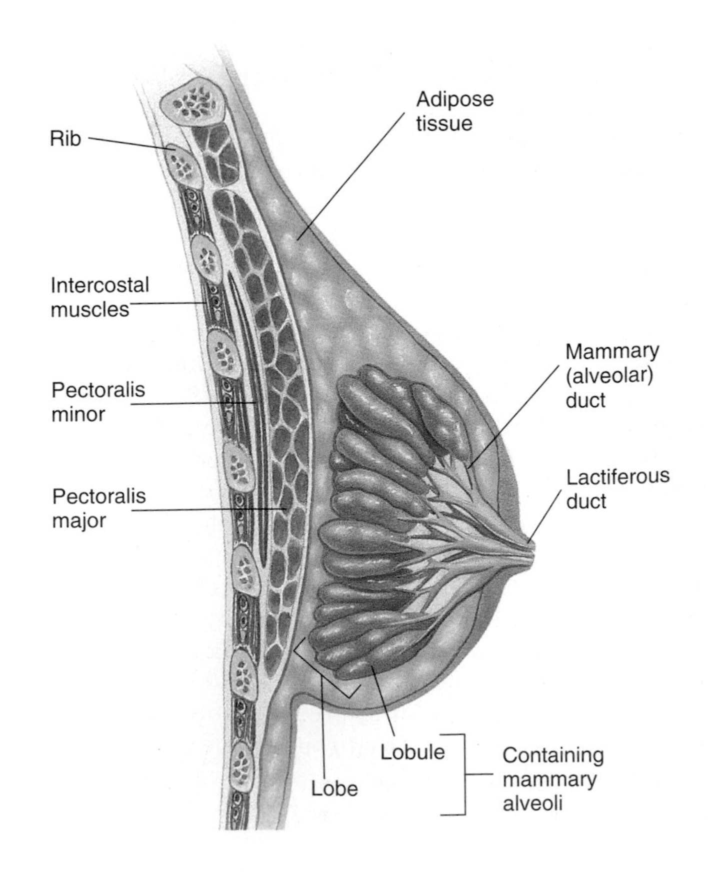

FIGURE 60.26
A sagittal section of a mammary gland. The mammary alveoli produce milk in response to stimulation by prolactin, and milk is ejected through the lactiferous duct in response to stimulation by oxytocin.

mammary glands, and it inhibits prolactin secretion by promoting the release of prolactin-inhibiting hormone from the hypothalamus. During pregnancy, therefore, the mammary glands are prepared for lactation but prevented from lactating.

When the placenta is discharged after birth, the concentrations of estradiol and progesterone in the mother's blood decline rapidly. This decline allows the anterior pituitary gland to secrete prolactin, which stimulates the mammary alveoli to produce milk. Sensory impulses associated with the baby's suckling trigger the posterior pituitary gland to release oxytocin. Oxytocin stimulates contraction of the smooth muscle surrounding the alveolar ducts, thus causing milk to be ejected by the breast. This pathway is known as the milk-ejection reflex. The secretion of oxytocin during lactation also causes some uterine contractions, as it did during labor. These contractions help to restore the tone of uterine muscles in mothers who are breastfeeding.

The first milk produced after birth is a yellowish fluid called colostrum, which is both nutritious and rich in maternal antibodies. Milk synthesis begins about three days following the birth and is referred to as when milk "comes in." Many mothers nurse for a year or longer. During this period, important pair-bonding occurs between the mother and child. When nursing stops, the accumulation of milk in the breasts signals the brain to stop prolactin secretion, and milk production ceases.

Postnatal Development

Growth of the infant continues rapidly after birth. Babies typically double their birth weight within two months. Because different organs grow at different rates and cease growing at different times, the body proportions of infants are different from those of adults. The head, for example, is disproportionately large in newborns, but after birth it grows more slowly than the rest of the body. Such a pattern of growth, in which different components grow at different rates, is referred to as allometric growth.

In most mammals, brain growth is mainly a fetal phenomenon. In chimpanzees, for instance, the brain and the cerebral portion of the skull grow very little after birth, while the bones of the jaw continue to grow. As a result, the head of an adult chimpanzee looks very different from that of a fetal chimpanzee (figure 60.27). In human infants, on the other hand, the brain and cerebral skull grow at the same rate as the jaw. Therefore, the jaw-skull proportions do not change after birth, and the head of a human adult looks very similar to that of a human fetus. It is primarily for this reason that an early human fetus seems so remarkably adultlike. The fact that the human brain continues to grow significantly for the first few years of postnatal life means that adequate nutrition and a rich, safe environment are particularly crucial during this period for the full development of the person's intellectual potential.

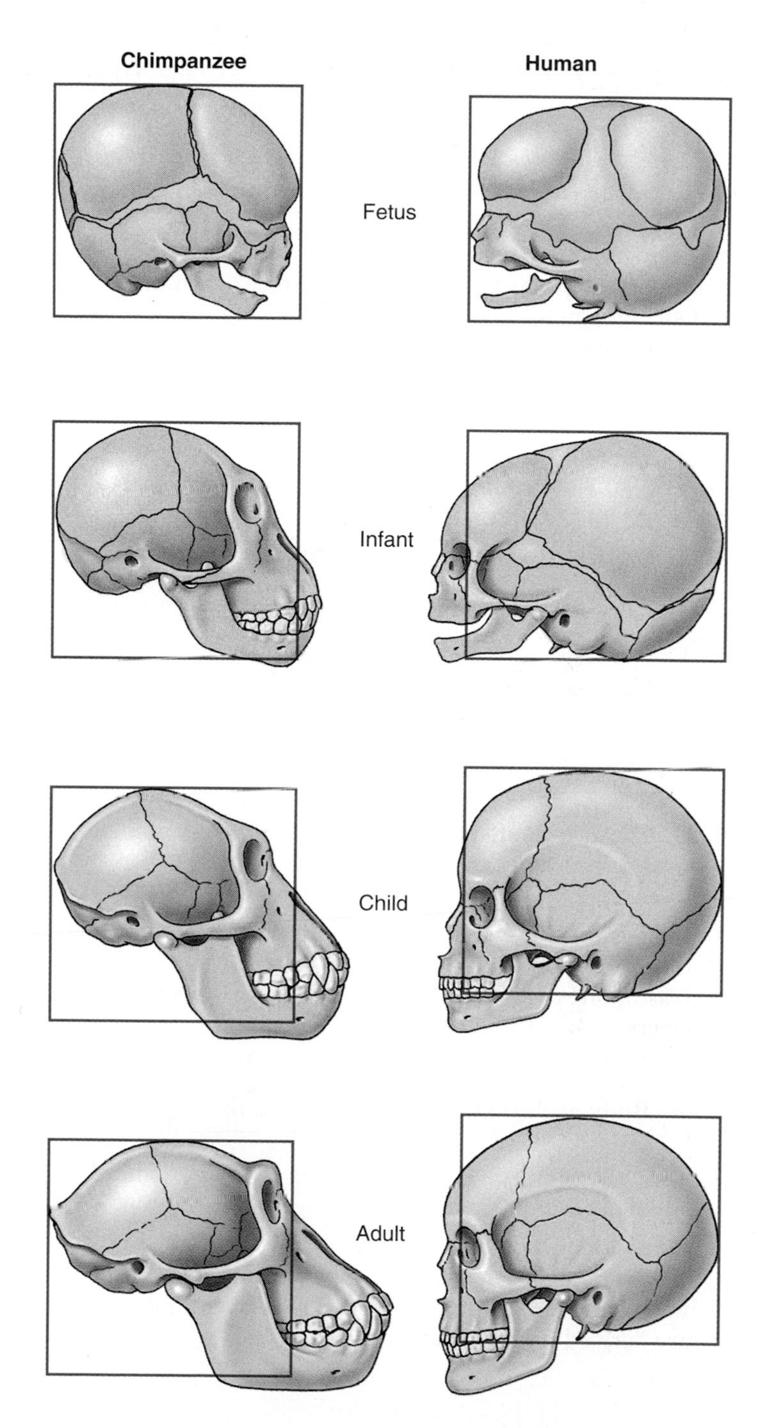

FIGURE 60.27
Allometric growth. In young chimpanzees, the jaw grows at a faster rate than the rest of the head. As a result, the adult chimpanzee head shape differs greatly from its head shape as a newborn. In humans, the difference in growth between the jaw and the rest of the head is much smaller, and the adult head shape is similar to that of the newborn.

Birth occurs in response to uterine contractions stimulated by oxytocin and prostaglandins. Lactation is stimulated by prostaglandin, but the milk-ejection reflex requires the action of oxytocin.

Chapter 60

www.mhhe.com/raven6/resources60.mhtml

Summary | **Questions** | **Media Resources**

60.1 Fertilization is the initial event in development.

- Fertilization is the union of an egg and a sperm to form a zygote. It is accomplished externally in most fish and amphibians, and internally in all other vertebrates.
- The three stages of fertilization are (1) penetration, (2) activation, and (3) nuclei fusion.

1. What happens during each stage of fertilization?

- Art Activity: Sperm and Egg Anatomy

- Fertilization

60.2 Cell cleavage and the formation of a blastula set the stage for later development.

- Cleavage is the rapid division of the newly formed zygote into a mass of cells, without any increase in overall size.
- The cleavages produce a hollow ball of cells, called the blastula.

2. What is the difference between holoblastic cleavage and meroblastic cleavage? What is responsible for an embryo undergoing one or the other type of cleavage?

60.3 Gastrulation forms the three germ layers of the embryo.

- During gastrulation, cells in the blastula change their relative positions, forming the three primary cell layers: ectoderm, mesoderm, and endoderm.
- In eggs with moderate or large amounts of yolk, cells involute down and around the yolk, through a blastopore or primitive streak to form the three germ layers.

3. What is an archenteron, and during what developmental stage does it form? What is the future fate of this opening in vertebrates?

4. How is gastrulation in amphibians different from that in lancelets?

- Art Quiz: Gastrulation—Mammal

60.4 Body architecture is determined during the next stages of embryonic development.

- Neurulation involves the formation of the notochord and dorsal hollow nerve tube.
- The formation of the neural crest is the first developmental event unique to vertebrates. Most of the distinctive structures associated with vertebrates are derived from the neural crest.

5. What structure unique to chordates forms during neurulation?

6. What are the functions of the amnion, chorion, and allantois in birds and mammals?

- Art Activity: Human Extra-Embryonic Membranes

- Cell Differentiation
- Early Development

60.5 Human development is divided into trimesters.

- Most of the critical events in human development occur in the first month of pregnancy. Cleavage occurs during the first week, gastrulation during the second week, neurulation during the third week, and organogenesis during the fourth week.
- The second and third months of human development are devoted to morphogenesis and to the elaboration of the nervous system and sensory organs.
- During the last six months before birth, the human fetus grows considerably, and the brain produces large numbers of neurons and establishes major nerve tracts.

7. How does the placenta prevent menstruation during the first two months of pregnancy?

8. At what time during human pregnancy does organogenesis occur?

9. Is neurological growth complete at birth?

10. What hormone stimulates lactation (milk production)? What hormone stimulates milk ejection from the breast?

- Human development
- Hormones and Pregnancy

- Bioethics Case Study: Critically Ill Newborns

- Art Quiz: Allometric Growth

Appendix

Classification of Organisms

The classification used in this book is presented here and explained in chapter 32. Not all phyla are included. Responding to a wealth of recent molecular data, it recognizes two separate kingdoms for the prokaryotes (bacteria), Archaebacteria and Eubacteria, and divides the eukaryotes into four kingdoms, Protista, Fungi, Plantae, and Animalia. There is no attempt to reorganize the kingdoms into domains. Viruses, which are considered nonliving, are not included in this appendix, but are treated in chapter 33. Plant and fungal divisions are the taxonomic equivalents of phyla, and are considered to be phyla in this text. For each entry in this classification, the page in the text where the organism is treated is indicated in parentheses.

Kingdom Archaebacteria

Prokaryotic bacteria; single-celled, cell walls lack muramic acid. Like all bacteria, they lack a membrane-bound nucleus, sexual recombination, and internal cell compartments. Archaebacterial cells have distinctive membranes, and unique rRNA and metabolic cofactors. Many are capable of living in extreme or anaerobic environments.

Methanogens (page 658)
Extremophiles (page 658)
Nonextreme archaebacteria (page 658)

Kingdom Eubacteria

Single-celled prokaryotes with cell walls containing muramic acid, sometimes forming filaments or other forms of colonies. Eubacteria lack a membrane-bound nucleus, sexual recombination, and internal cell compartments. Their flagella are simple, composed of a single fiber of protein. They are much more diverse metabolically than the eukaryotes. About 5000 species are currently recognized, but that is almost certainly only a small fraction of the actual number.

Actinomycetes (page 685)
Chemoautotrophs (page 685)
Cyanobacteria (page 685)
Enterobacteria (page 685)
Gliding and budding bacteria (page 685)
Pseudomonads (page 685)
Rickettsias and chlamydias (page 685)
Spirochaetes (page 685)

Kingdom Protista

Eukaryotic, primarily single-celled organisms. Eukaryotes have a membrane-bound nucleus containing chromosomes, sexual recombination, and extensive internal compartmentalization of cells. Their flagella are complex, with a 9 + 2 internal microtubular organization. Some protists have flagella, some have pseudopods, and some are sessile (immobile). They are diverse metabolically, but much less so than bacteria. Protists may capture prey, absorb their food, or photosynthesize. Reproduction among protists can be sexual, involving meiosis, but is usually asexual. The kingdom Protista is a catch-all classification, including all eukaryotes which are not plants, animals, or fungi.

Heterotrophs with no permanent locomotor apparatus

Phylum Rhizopoda (page 700)
Phylum Actinopoda (page 700)
Phylum Foraminifera (page 700)

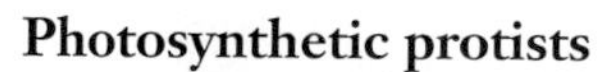

Photosynthetic protists

Phylum Pyrrhophyta (page 702)
Phylum Euglenophyta (page 703)

Phylum Chrysophyta (page 704)

Phylum Rhodophyta (page 705)
Phylum Phaeophyta (page 705)
Phylum Chlorophyta (page 706)

Heterotrophs with flagella

Phylum Sarcomastigophora (page 708)

Phylum Ciliophora (page 710)

Nonmotile spore-formers

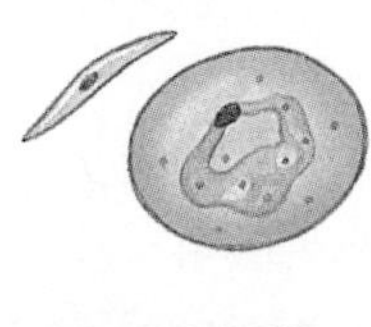

Phylum Apicomplexa (page 712)

Heterotrophs with restricted mobility

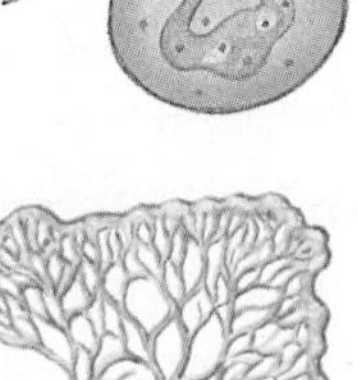

Phylum Oomycota (page 714)
Phylum Acrasiomycota (page 715)
Phylum Myxomycota (page 716)

Kingdom Fungi

Filamentous, multinucleate, heterotrophic eukaryotes with cell walls rich in chitin; no flagellated cells present. Mitosis in fungi takes place within the nuclei, the nuclear envelope never breaking down. Filaments of fungal cells called hyphae grow through the substrate, secreting enzymes and absorbing the products of their digestion. Septa between the nuclei in the hyphae normally complete only at the borders of reproductive structures. Asexual reproduction frequent in some groups. The nuclei of fungi are haploid, with the zygote the only diploid stage of the life cycle. About 77,000 named species.

Fungi with a known sexual cycle

Phylum Zygomycota (page 725)

Phylum Ascomycota (page 726)

Phylum Basidiomycota (page 728)

Fungi with an unknown sexual cycle

Fungi Imperfecti (page 729)

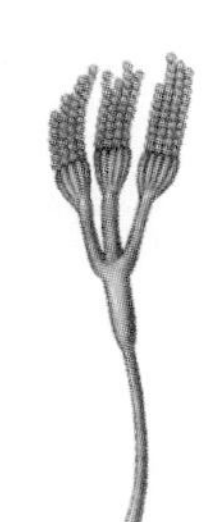

Symbiotic Associations—*fungi and a photosynthetic partner*

Lichens (page 730)

Mycorrhizae (page 731)

Kingdom Plantae

Multicellular, photosynthetic, terrestrial eukaryotes derived from the green algae, and like them containing chlorophylls *a* and *b*. Pigments are present in chloroplasts which may also store starch. The cell walls of plants have cellulose matrix and sometimes become lignified; cell division is by means of a cell plate that forms across the mitotic spindle. The vascular plants have an elaborate system of xylem and phloem conducting cells that move water through the plant body. Plants have a waxy cuticle that helps them retain water; most have stomata. All plants have an alternation of generations, many with reduced gametophyte and multicellular gametangia. About 270,000 species.

Nonvascular Plants—*mosses, liverworts, and hornworts*

Phylum Bryophyta (page 738)

Phylum Hepaticophyta (page 739)

Phylum Antherocerotophyta (page 739)

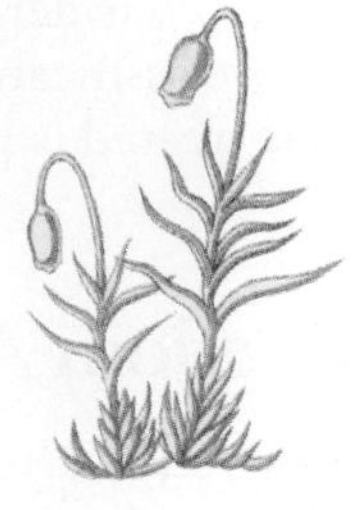

Seedless Vascular Plants—*ferns*

Phylum Pterophyta (page 742)

Phylum Psilophyta (page 744)

Phylum Lycophyta (page 744)

Phylum Arthrophyta (page 744)

Gymnosperms—*conifers and cycads*

Phylum Coniferophyta (page 746)

Phylum Cycadophyta (page 748)

Phylum Gnetophyta (page 748)

Phylum Ginkgophyta (page 748)

Angiosperms—*flowering plants*

Phylum Anthophyta (page 749)

Kingdom Animalia

Animals are multicellular eukaryotes that characteristically ingest their food. Their cells are usually flexible, without a cell wall. In all of the approximately 35 phyla except sponges, body cells are organized into structural and functional units called tissues. Cells move extensively during the development of the animal embryo; the blastula, a hollow ball of cells, forms early in this process and is characteristic of the group. Most animals reproduce sexually; their nonmotile eggs are much larger than their small flagellated sperm. The gametes fuse directly to produce a zygote and do not divide by mitosis as in plants. More than a million species of animals have been described.

Subkingdom Parazoa—*lacking definite symmetry and tissues*

- Phylum Porifera (page 884)

Subkingdom Eumetazoa—*definite symmetry and tissues*

- Phylum Cnidaria (page 886)
  - Class Hydrozoa (page 888)
  - Class Scyphozoa (page 888)
  - Class Cubozoa (page 889)
  - Class Anthozoa (page 889)

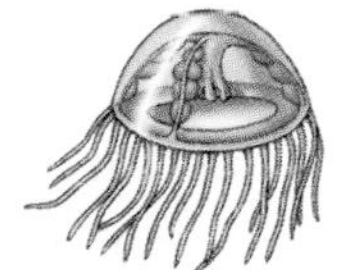

- Phylum Ctenophora (page 889)
- Phylum Platyhelminthes (page 890)
  - Class Turbellaria (page 892)
  - Class Trematoda (page 892)
  - Class Cestoda (page 893)
- Phylum Nemertea (page 893)

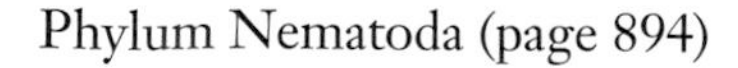

- Phylum Nematoda (page 894)
- Phylum Rotifera (page 895)
- Phylum Cycliophora (page 895)
- Phylum Mollusca (page 900)
  - Class Polyplacophora (page 904)
  - Class Gastropoda (page 904)
  - Class Bivalvia (page 904)
  - Class Cephalopoda (page 905)

- Phylum Annelida (page 906)
  - Class Polychaeta (page 908)
  - Class Oligochaeta (page 908)
  - Class Hirudinea (page 909)

- Phylum Phoronida (page 910)
- Phylum Ectoprocta (page 910)
- Phylum Brachiopoda (page 911)

- Phylum Arthropoda (page 918)
  - Class Arachnida (page 921)
  - Class Merostomata (page 923)
  - Class Pycnogonida (page 923)
  - Subphylum Crustacea (page 924)
  - Class Chilopoda (page 926)
  - Class Diplopoda (page 926)
  - Class Insecta (page 927)

- Phylum Echinodermata (page 936)
  - Class Crinoidea (page 940)
  - Class Asteroidea (page 941)
  - Class Ophiuroidea (page 942)
  - Class Echinoidea (page 942)
  - Class Holothuroidea (page 943)
  - Class Concentricycloidea (page 943)
- Phylum Chaetognatha (page 880)
- Phylum Hemichordata (page 880)
- Phylum Chordata (page 946)
  - Subphylum Urochordata (page 948)
  - Subphylum Cephalochordata (page 949)
  - Subphylum Vertebrata (page 950)
    - Class Agnatha (page 954)
    - Class Chondrichthyes (page 955)
    - Class Osteichthyes (page 956)
    - Class Amphibia (page 958)

    - Class Reptilia (page 962)

    - Class Aves (page 970)

    - Class Mammalia (page 974)

Glossary

A

ABO blood group A set of four phenotypes produced by different combinations of three alleles at a single locus; blood types are A, B, AB, and O, depending on which alleles are expressed as antigens on the red blood cell surface.

abscission (L. *ab*, away, off + *scissio*, dividing) In vascular plants, the dropping of leaves, flowers, fruits, or stems at the end of the growing season, as the result of the formation of a layer of specialized cells (the abscission zone) and the action of a hormone (ethylene).

acquired immune deficiency syndrome (AIDS) An infectious and usually fatal human disease caused by a retrovirus, HIV (human immunodeficiency virus), which attacks T cells. The affected individual is helpless in the face of microbial infections.

actin (Gr. *actis*, a ray) One of the two major proteins that make up vertebrate muscle (the other is myosin).

action potential A transient, all-or-none reversal of the electric potential across a membrane; in neurons, an action potential initiates transmission of a nerve impulse.

activation energy The energy that must be processed by a molecule in order for it to undergo a specific chemical reaction.

active site The region of an enzyme surface to which a specific set of substrates binds, lowering the activation energy required for a particular chemical reaction and so facilitating it.

active transport The pumping of individual ions or other molecules across a cellular membrane from a region of lower concentration to one of higher concentration (that is, against a concentration gradient); this transport process requires energy, which is typically supplied by the expenditure of ATP.

adaptation (L. *adaptare*, to fit) A peculiarity of structure, physiology, or behavior that promotes the likelihood of an organism's survival and reproduction in a particular environment.

adaptive radiation The evolution of several divergent forms from a primitive and unspecialized ancestor.

adenosine diphosphate (ADP) A nucleotide consisting of adenine, ribose sugar, and two phosphate groups; formed by the removal of one phosphate from an ATP molecule.

adenosine triphosphate (ATP) A nucleotide consisting of adenine, ribose sugar, and three phosphate groups; ATP is the energy currency of cellular metabolism in all organisms.

adenylyl cyclase An enzyme that produces large amounts of cAMP from ATP; the cAMP acts as a second messenger in a target cell.

adipose cells Fat cells, found in loose connective tissue, usually in large groups that form adipose tissue. Each adipose cell can store a droplet of fat (triacylglyceride).

adventitious (L. *adventicius*, not properly belonging to) Referring to a structure arising from an unusual place, such as stems from roots or roots from stems.

aerobic (Gr. *aer*, air + *bios*, life) Requiring free oxygen; any biological process that can occur in the presence of gaseous oxygen (O_2).

AFLP (amplified fragment length polymorphisms) Genomic DNA fragments are cut with restriction enzymes, and the polymerase chain reaction is used to amplify all these fragments. For diploid genomes, there will be a large number of homologous fragments that have different lengths. These fragments can be separated on gels.

alga, pl. algae A unicellular or simple multicellular photosynthetic organism lacking multicellular sex organs.

allantois (Gr. *allas*, sausage + *eidos*, form) A membrane of the amniotic egg that functions in respiration and excretion in birds and reptiles and plays an important role in the development of the placenta in most mammals.

allele (Gr. *allelon*, of one another) One of two or more alternative states of a gene.

allometric growth (Gr. *allos*, other + *ergon*, action) A pattern of growth in which different components grow at different rates.

allopatric speciation The differentiation of geographically isolated populations into distinct species.

allosteric site A part of an enzyme, away from its active site, that serves as an on/off switch for the function of the enzyme.

alternation of generations A reproductive cycle in which a haploid ($1n$) phase, the gametophyte, gives rise to gametes, which, after fusion to form a zygote, germinate to produce a diploid ($2n$) phase, the sporophyte. Spores produced by meiotic division from the sporophyte give rise to new gametophytes, completing the cycle.

altruism Self-sacrifice for the benefit of others; in formal terms, the behavior that increases the fitness of the recipient while reducing the fitness of the altruistic individual.

alveolus, pl. alveoli (L., a small cavity) One of many small, thin-walled air sacs within the lungs in which the bronchioles terminate.

amino acid The subunit structure from which proteins are produced, consisting of a central carbon atom with a carboxyl group (—COOH), an amino group (—NH_2), a hydrogen, and a side group (R group); only the side group differs from one amino acid to another.

amniocentesis (Gr. *amnion*, membrane around the fetus + *centes*, puncture) Examination of a fetus indirectly, by tests on cell cultures grown from fetal cells obtained from a sample of the amniotic fluid surrounding the developing embryo or tests on the fluid itself.

amnion (Gr., membrane around the fetus) The innermost of the extraembryonic membranes; the amnion forms a fluid-filled sac around the embryo in amniotic eggs.

amniotic egg An egg that is isolated and protected from the environment by a more or less impervious shell during the period of its development and that is completely self-sufficient, requiring only oxygen.

amyloplast (Gr. *amylon*, starch + *plastos*, formed) A plant organelle called a plastid that specializes in storing starch.

anabolism (Gr. *ana*, up + *bolein*, to throw) The biosynthetic or constructive part of metabolism; those chemical reactions involved in biosynthesis.

anaerobic (Gr. *an*, without + *aer*, air + *bios*, life) Any process that can occur without oxygen, such as anaerobic fermentation or H_2S photosynthesis.

analogous (Gr. *analogos*, proportionate) Structures that are similar in function but different in evolutionary origin, such as the wing of a bat and the wing of a butterfly.

anaphase In mitosis and meiosis II, the stage initiated by the separation of sister chromatids, during which the daughter chromosomes move to opposite poles of the cell; in meiosis I, marked by separation of replicated homologous chromosomes.

androecium (Gr. *andros*, man + *oilos*, house) The floral whorl that comprises the stamens.

aneuploidy (Gr. *an*, without + *eu*, good + *ploid*, multiple of) The condition in an organism whose cells have lost or gained a chromosome; Down syndrome, which results from an extra copy of human chromosome 21, is an example of aneuploidy in humans.

angiosperms The flowering plants, one of five phyla of seed plants. In angiosperms, the ovules at the time of pollination are completely enclosed by tissues.

animal pole In fish and other aquatic vertebrates with asymmetrical yolk distribution in their eggs, the hemisphere of the blastula, comprising cells relatively poor in yolk.

anion (Gr. *anienae*, to go up) A negatively charged ion.

anther (Gr. *anthos*, flower) In angiosperm flowers, the pollen-bearing portion of a stamen.

antheridium, pl. antheridia A sperm-producing organ.

antibody (Gr. *anti*, against) A protein called immunoglobulin that is produced by lymphocytes in response to a foreign substance (antigen) and released into the bloodstream.

anticodon The three-nucleotide sequence at the end of a transfer RNA molecule that is complementary to, and base-pairs with, an amino-acid specifying codon in messenger RNA.

antigen (Gr. *anti*, against + *genos*, origin) A foreign substance, usually a protein or polysaccharide, that stimulates an immune response.

anus The terminal opening of the gut; the solid residues of digestion are eliminated through the anus.
aorta (Gr. *aeirein*, to lift) The major artery of vertebrate systemic blood circulation; in mammals, carries oxygenated blood away from the heart to all regions of the body except the lungs.
apical meristem (L. *apex*, top + Gr. *meristos*, divided) In vascular plants, the growing point at the tip of the root or stem.
apoptosis A process of programmed cell death, in which dying cells shrivel and shrink; used in all animal cell development to produce planned and orderly death of cells not destined to be present in the final tissue.
aposematic coloration An ecological strategy of some organisms that "advertise" their poisonous nature by the use of bright colors.
aquifers Permeable, saturated, underground layers of rock, sand, and gravel, which serve as reservoirs for groundwater.
archaebacteria A group of bacteria that are among the most primitive still in existence, characterized by the absence of peptidoglycan in their cell walls, a feature that distinguishes them from all other bacteria.
archegonium, pl. **archegonia** (Gr. *archegonos*, first of a race) The multicellular egg-producing organ in bryophytes and some vascular plants.
archenteron (Gr. *arche*, beginning + *enteron*, gut) The principal cavity of a vertebrate embryo in the gastrula stage; lined with endoderm, it opens up to the outside and represents the future digestive cavity.
arteriole A smaller artery, leading from the arteries to the capillaries.
arteriosclerosis Hardening and thickening of the wall of an artery.
ascomycetes A large group comprising part of the "true fungi." They are characterized by separate hyphae, asexually produced conidiospores and sexually produced ascospores within asci.
ascospore A fungal spore produced within an ascus.
ascus, pl. **asci** (Gr. *askos*, wineskin, bladder) A specialized cell, characteristic of the ascomycetes, in which two haploid nuclei fuse to produce a zygote that divides immediately by meiosis; at maturity, an ascus contains ascospores.
asexual reproduction The process by which an individual inherits all of its chromosomes from a single parent, thus being genetically identical to that parent; cell division is by mitosis only.
aster In animal cell mitosis, a radial array of microtubules extending from the centrioles toward the plasma membrane, possibly serving to brace the centrioles for retraction of the spindle.
atrioventricular (AV) node A slender connection of cardiac muscle cells that receives the heartbeat impulses from the sinoatrial node and conducts them by way of the bundle of His.
atrium (L., vestibule or courtyard) An antechamber; in the heart, a thin-walled chamber that receives venous blood and passes it on to the thick-walled ventricle; in the ear, the tympanic cavity.
autonomic nervous system (Gr. *autos*, self + *nomos*, law) The involuntary neurons and ganglia of the peripheral nervous system of vertebrates; regulates the heart, glands, visceral organs, and smooth muscle.
autosome (Gr. *autos*, self + *soma*, body) Any eukaryotic chromosome that is not a sex chromosome; autosomes are present in the same number and kind in both males and females of the species.
autotroph (Gr. *autos*, self + *trophos*, feeder) An organism able to build all the complex organic molecules that it requires as its own food source, using only simple inorganic compounds.
auxin (Gr. *auxein*, to increase) A plant hormone that controls cell elongation, among other effects.
axon (Gr., axle) A process extending out from a neuron that conducts impulses away from the cell body.

B

bacteriophage (Gr. *bakterion*, little rod + *phagein*, to eat) A virus that infects bacterial cells; also called a *phage*.
Barr body A deeply staining structure, seen in the interphase nucleus of a cell of an individual with more than one X chromosome, that is a condensed and inactivated X. Only one X remains active in each cell after early embryogenesis.
basal body A self-reproducing, cylinder-shaped cytoplasmic organelle composed of nine triplets of microtubules from which the flagella or cilia arise.
base-pair A complementary pair of nucleotide bases, consisting of a purine and a pyrimidine.
basidiospore A spore of the basidiomycetes, produced within and borne on a basidium after nuclear fusion and meiosis.
basidium, pl. **basidia** (L., a little pedestal) A specialized reproductive cell of the basidiomycetes, often club shaped, in which nuclear fusion and meiosis occur.
basophil A leukocyte containing granules that rupture and release chemicals that enhance the inflammatory response. Important in causing allergic responses.
Batesian mimicry A situation in which a palatable or nontoxic organism resembles another kind of organism that is distasteful or toxic. Both species exhibit warning coloration.
B cell A type of lymphocyte that, when confronted with a suitable antigen, is capable of secreting a specific antibody protein.
behavioral ecology The study of how natural selection shapes behavior.
biennial A plant that normally requires two growing seasons to complete its life cycle. Biennials flower in the second year of their lives.
bile salts A solution of organic salts that is secreted by the vertebrate liver and temporarily stored in the gallbladder; emulsifies fats in the small intestine.
binary fission (L. *binarius*, consisting of two things or parts + *fissus*, split) Asexual reproduction by division of one cell or body into two equal or nearly equal parts.
binomial distribution The distribution of phenotypes seen among the progeny of a cross in which there are only two alternative alleles.
biodiversity The number of species and their range of behavioral, ecological, physiological, and other adaptations, in an area.
biomass (Gr. *bios*, life + *maza*, lump or mass) The weight of all the living organisms in a given population, area, or other unit being measured.
biome One of the major terrestrial ecosystems, characterized by climatic and soil conditions; the largest ecological unit.
bipedal Able to walk upright on two feet.
bipolar cell A specialized type of neuron connecting cone cells to ganglion cells in the visual system. Bipolar cells receive a hyperpolarized stimulus from the cone cell and then transmit a depolarization stimulus to the ganglion cell.
blade The broad, expanded part of a leaf; also called the lamina.
blastocoel (Gr. *blastos*, a sprout + *koilos*, hollow) The central cavity of the blastula stage of vertebrate embryos.
blastodisc (Gr. *blastos*, sprout + *discos*, a round plate) In the development of birds, a disclike area on the surface of a large, yolky egg that undergoes cleavage and gives rise to the embryo.
blastomere (Gr. *blastos*, *sprout* + *meros*, part) One of the cells of a blastula.
blastopore (Gr. *blastos*, sprout + *poros*, a path or passage) In vertebrate development, the opening that connects the archenteron cavity of a gastrula stage embryo with the outside.
blastula (Gr., a little sprout) In vertebrates, an early embryonic stage consisting of a hollow, fluid-filled ball of cells one layer thick; a vertebrate embryo after cleavage and before gastrulation.
Bohr effect The release of oxygen by hemoglobin molecules in response to elevated ambient levels of CO_2.
Bowman's capsule In the vertebrate kidney, the bulbous unit of the nephron, which surrounds the glomerulus.
β-oxidation In the cellular respiration of fats, the process by which two-carbon acetyl groups are removed from a fatty acid and combined with coenzyme A to form acetyl-CoA, until the entire fatty acid has been broken down.
bronchus, pl. **bronchi** (Gr. *bronchos*, windpipe) One of a pair of respiratory tubes branching from the lower end of the trachea (windpipe) into either lung.
bryophytes Nonvascular plants including mosses, hornworts, and liverworts.
bud An asexually produced outgrowth that develops into a new individual. In plants, an embryonic shoot, often protected by young leaves; buds may give rise to branch shoots.

C

C_3 photosynthesis The main cycle of the dark reactions of photosynthesis, in which CO_2 binds to ribulose 1,5-bisphosphate (RuBP) to form two three-carbon phosphoglycerate (PGA) molecules.
C_4 photosynthesis A process of CO_2 fixation in photosynthesis by which the first product is the four-carbon oxaloacetate molecule.
callus (L. *callos*, hard skin) Undifferentiated tissue; a term used in tissue culture, grafting, and wound healing.
Calvin cycle The dark reactions of C_3 photosynthesis; also called the Calvin-Benson cycle.
calyx (Gr. *kalyx*, a husk, cup) The sepals collectively; the outermost flower whorl.
capillary (L. *capillaris*, hairlike) The smallest of the blood vessels; the very thin walls of capillaries

are permeable to many molecules, and exchanges between blood and the tissues occur across them; the vessels that connect arteries with veins.

capsid The outermost protein covering of a virus.

carapace (Fr. from Sp. *carapacho*, shell) Shieldlike plate covering the cephalothorax of decapod crustaceans; the dorsal part of the shell of a turtle.

carbohydrate (L. *carbo*, charcoal + *hydro*, water) An organic compound consisting of a chain or ring of carbon atoms to which hydrogen and oxygen atoms are attached in a ratio of approximately 2:1; having the generalized formula $(CH_2O)_n$; carbohydrates include sugars, starch, glycogen, and cellulose.

carbon fixation The conversion of CO_2 into organic compounds during photosynthesis; the first stage of the dark reactions of photosynthesis, in which carbon dioxide from the air is combined with ribulose 1,5-bisphosphate.

carpel (Gr. *karpos*, fruit) A leaflike organ in angiosperms that encloses one or more ovules.

carrying capacity The maximum population size that a habitat can support.

cartilage (L. *cartilago*, gristle) A connective tissue in skeletons of vertebrates. Cartilage forms much of the skeleton of embryos, very young vertebrates, and some adult vertebrates, such as sharks and their relatives.

catabolism (Gr. *ketabole*, throwing down) In a cell, those metabolic reactions that result in the breakdown of complex molecules into simpler compounds, often with the release of energy.

catalysis The process by which chemical subunits of larger organic molecules are held and positioned by enzymes that stress their chemical bonds, leading to the disassembly of the larger molecule into its subunits, often with the release of energy.

cation (Gr. *katienai*, to go down) A positively charged ion.

cecum, also **caecum** (L. *caecus*, blind) In vertebrates, a blind pouch at the beginning of the large intestine.

cell cycle The repeating sequence of growth and division through which cells pass each generation.

cell plate The structure that forms at the equator of the spindle during early telophase in the dividing cells of plants and a few green algae.

cell surface receptor A cell surface protein that binds a signal molecule and converts the extracellular signal into an intracellular one.

cellular respiration The metabolic harvesting of energy by oxidation, ultimately dependent on molecular oxygen; carried out by the Krebs cycle and oxidative phosphorylation.

cellulose (L. *cellula*, a little cell) The chief constituent of the cell wall in all green plants, some algae, and a few other organisms; an insoluble complex carbohydrate formed of microfibrils of glucose molecules.

cell wall The rigid, outermost layer of the cells of plants, some protists, and most bacteria; the cell wall surrounds the cell (plasma) membrane.

central nervous system (CNS) That portion of the nervous system where most association occurs; in vertebrates, it is composed of the brain and spinal cord; in invertebrates it usually consists of one or more cords of nervous tissue, together with their associated ganglia.

centriole (Gr. *kentron*, center of a circle + L. *olus*, little one) A cytoplasmic organelle located outside the nuclear membrane, identical in structure to a basal body; found in animal cells and in the flagellated cells of other groups; divides and organizes spindle fibers during mitosis and meiosis.

centromere (Gr. *kentron*, center + *meros*, a part) Condensed region on a eukaryotic chromosome where sister chromatids are attached to each other after replication. *See* kinetochore.

cerebellum (L., little brain) The hindbrain region of the vertebrate brain that lies above the medulla (brain stem) and behind the forebrain; it integrates information about body position and motion, coordinates muscular activities, and maintains equilibrium.

cerebral cortex The thin surface layer of neurons and glial cells covering the cerebrum; well developed only in mammals, and particularly prominent in humans. The cerebral cortex is the seat of conscious sensations and voluntary muscular activity.

cerebrum (L., brain) The portion of the vertebrate brain (the forebrain) that occupies the upper part of the skull, consisting of two cerebral hemispheres united by the corpus callosum. It is the primary association center of the brain. It coordinates and processes sensory input and coordinates motor responses.

chelicera, pl. **chelicerae** (Gr. *chele*, claw + *keras*, horn) The first pair of appendages in horseshoe crabs, sea spiders, and arachnids—the chelicerates, a group of arthropods. Chelicerae usually take the form of pincers or fangs.

chemiosmosis The mechanism by which ATP is generated in mitochondria and chloroplasts; energetic electrons excited by light (in chloroplasts) or extracted by oxidation in the Krebs cycle (in mitochondria) are used to drive proton pumps, creating a proton concentration gradient; when protons subsequently flow back across the membrane, they pass through channels that couple their movement to the synthesis of ATP.

chiasma An X-shaped figure that can be seen in the light microscope during meiosis; evidence of crossing over, where two chromatids have exchanged parts; chiasmata move to the ends of the chromosome arms as the homologues separate.

chitin (Gr. *chiton*, tunic) A tough, resistant, nitrogen-containing polysaccharide that forms the cell walls of certain fungi, the exoskeleton of arthropods, and the epidermal cuticle of other surface structures of certain other invertebrates.

chloroplast (Gr. *chloros*, green + *plastos*, molded) A cell-like organelle present in algae and plants that contains chlorophyll (and usually other pigments) and carries out photosynthesis.

chorion (Gr., skin) The outer member of the double membrane that surrounds the embryo of reptiles, birds, and mammals; in placental mammals, it contributes to the structure of the placenta.

chromatid (Gr. *chroma*, color + L. *-id*, daughters of) One of the two daughter strands of a duplicated chromosome that is joined by a single centromere.

chromatin (Gr. *chroma*, color) The complex of DNA and proteins of which eukaryotic chromosomes are composed; chromatin is highly uncoiled and diffuse in interphase nuclei, condensing to form the visible chromosomes in prophase.

chromosome (Gr. *chroma*, color + *soma*, body) The vehicle by which hereditary information is physically transmitted from one generation to the next; in a bacterium, the chromosome consists of a single naked circle of DNA; in eukaryotes, each chromosome consists of a single linear DNA molecule and associated proteins.

cilium A short cellular projection from the surface of a eukaryotic cell, has the same internal structure of microtubules in a 9 + 2 arrangement as seen in a flagellum.

circadian rhythm An endogenous cyclical rhythm that oscillates on a daily (24 hour) basis.

cisterna A small collecting vessel that pinches off from the end of a Golgi body to form a transport vesicle that moves materials through the cytoplasm.

cladistics A taxonomic technique used for creating hierarchies of organisms that represent true phylogenetic relationship and descent.

class A taxonomic category between phyla (divisions) and orders. A class contains one or more orders, and belongs to a particular phylum or division.

classical conditioning The repeated presentation of a stimulus in association with a response that causes the brain to form an association between the stimulus and the response, even if they have never been associated before.

clathrin A protein located just inside the plasma membrane in eukaryotic cells, in indentations called clathrin-coated pits.

cleavage In vertebrates, a rapid series of successive cell divisions of a fertilized egg, forming a hollow sphere of cells, the blastula.

climax vegetation Vegetation encountered in a self-perpetuating community of plants that has proceeded through all the stages of succession and stabilized.

cloaca (L., sewer) In some animals, the common exit chamber from the digestive, reproductive, and urinary system; in others, the cloaca may also serve as a respiratory duct.

cloning Producing a cell line or culture all of whose members contain identical copies of a particular nucleotide sequence; an essential element in genetic engineering.

coacervate (L. *coacervatus*, heaped up) A spherical aggregation of lipid molecules in water, held together by hydrophobic forces.

cochlea (Gr. *kochlios*, a snail) In terrestrial vertebrates, a tubular cavity of the inner ear containing the essential organs for hearing.

codon (L., code) The basic unit of the genetic code; a sequence of three adjacent nucleotides in DNA or mRNA that codes for one amino acid.

coenzyme (L. *co-*, together + Gr. *en*, in + *zyme*, leaven) A nonprotein organic molecule such as NAD^+ that plays an accessory role in enzyme-catalyzed processes, often by acting as a donor or acceptor of electrons.

coevolution (L. *co-*, together + *e-*, out + *volvere*, to fill) The simultaneous development of adaptations, in two or more populations, species, or other categories that interact so closely that each is a strong selective force on the other.

cofactor One or more nonprotein components required by enzymes in order to function; many cofactors are metal ions, others are organic coenzymes.

collenchyma (Gr. *kolla*, glue + *en*, in + *chymein*, to pour) In plants, a supporting tissue composed of collenchyma cells; often found in regions of primary growth in stems and in some leaves.

commensalism (L. *cum*, together with + *mensa*, table) A relationship in which one individual lives close to or on another and benefits, and the host is unaffected; a kind of symbiosis.

community (L. *communitas*, community, fellowship) All of the organisms inhabiting a common environment and interacting with one another.

competitive exclusion The hypothesis that two species with identical ecological requirements cannot exist in the same locality indefinitely, and that the more efficient of the two in utilizing the available scarce resources will exclude the other; also known as Gause's principle.

complementarity The basis for copying the genetic information, where each nucleotide base has a complementary partner with which it forms a base-pair.

complement system The chemical defense of a vertebrate body that consists of a battery of proteins that become activated by the walls of bacteria and fungi.

concentration gradient The concentration difference of a substance across a distance; in a cell, a greater concentration of its molecules in one region than in another.

cone (1) In plants, the reproductive structure of a conifer. (2) In vertebrates, a type of light-sensitive neuron in the retina, concerned with the perception of color and with the most acute discrimination of detail.

conidia An asexually produced fungal spore.

conjugation (L. *conjugare*, to yolk together) Temporary union of two unicellular organisms, during which genetic material is transferred from one cell to the other; occurs in bacteria, protists, and certain algae and fungi.

contractile vacuole In protists and some animals, a clear fluid-filled vacuole that takes up water from within the cell and then contracts, releasing it to the outside through a pore in a cyclical manner; functions primarily in osmoregulation and excretion.

conus arteriosus The anteriormost chamber of the embryonic heart in vertebrate animals.

convergent evolution The independent development of similar structures in organisms that are not directly related; often found in organisms living in similar environments.

cork cambium The lateral meristem that forms the periderm, producing cork (phellem) toward the surface (outside) of the plant and phelloderm toward the inside.

cornea (L. *corneus*, horny) The transparent outer layer of the vertebrate eye.

corolla (L. *cornea*, crown) The petals, collectively; usually the conspicuously colored flower whorl.

corpus callosum The band of nerve fibers that connect the two hemispheres of the cerebrum in humans and other primates.

corpus luteum (L., yellowish body) A structure that develops from a ruptured follicle in the ovary after ovulation.

cortex (L., bark) The outer layer of a structure; in animals, the outer, as opposed to the inner, part of an organ; in vascular plants, the primary ground tissue of a stem or root.

cotyledon (Gr. *kotyledon*, a cup-shaped hollow) A seed leaf that generally stores food in dicots or absorbs it in monocots, providing nourishment used during seed germination.

crassulacean acid metabolism (CAM) A mode of carbon dioxide fixation by which CO_2 enters open leaf stomata at night and is used in photosynthesis during the day, when stomata are closed to prevent water loss.

cross-current flow In bird lungs, the latticework of capillaries arranged across the air flow, at a 90° angle.

crossing over In meiosis, the exchange of corresponding chromatid segments between homologous chromosomes; responsible for genetic recombination between homologous chromosomes.

cumulus cells Cells that surround an egg cell and nurture it.

cuticle (L. *cuticula*, little skin) A waxy or fatty, noncellular layer [formed of a substance called cutin] on the outer wall of epidermal cells.

cyanobacteria A group of photosynthetic eubacteria, sometimes called the "blue-green algae," that contain the chlorophyll pigments most abundant in plants and algae, as well as other pigments.

cyclic AMP (cAMP) A form of adenosine monophosphate (AMP) in which the atoms of the phosphate group form a ring; found in almost all organisms, cAMP functions as an intracellular second messenger that regulates a diverse array of metabolic activities.

cystic fibrosis An autosomal disorder that produces the most common fatal genetic disease in Caucasians, characterized by secretion of thick mucus that clogs passageways in the lungs, liver, and pancreas.

cytochrome (Gr. *kytos*, hollow vessel + *chroma*, color) Any of several iron-containing protein pigments that serve as electron carriers in transport chains of photosynthesis and cellular respiration.

cytokinesis (Gr. *kytos*, hollow vessel + *kinesis*, movement) Division of the cytoplasm of a cell after nuclear division.

cytoplasm (Gr. *ketos*, hollow vessel + *plasma*, anything molded) The material within a cell, excluding the nucleus; the protoplasm.

cytoskeleton A network of protein microfilaments and microtubules within the cytoplasm of a eukaryotic cell that maintains the shape of the cell, anchors its organelles, and is involved in animal cell motility.

cytotoxic T cell (Gr. *kytos*, hollow vessel + toxin) A special T cell activated during cell-mediated immune response that recognizes and destroys infected body cells.

D

dehydration reaction (L. *co-*, together + *densare*, to make dense) A type of chemical reaction in which two molecules join to form one larger molecule, simultaneously splitting out a molecule of water; one molecule is stripped of a hydrogen atom and another is stripped of a hydroxyl group (—OH), resulting in the joining of the two molecules, while the H^+ and —OH released may combine to form a water molecule.

demography (Gr. *demos*, people + *graphein*, to draw) The properties of the rate of growth and the age structure of populations.

denaturation The loss of the native configuration of a protein or nucleic acid as a result of excessive heat, extremes of pH, chemical modification, or changes in solvent ionic strength or polarity that disrupt hydrophobic interactions; usually accompanied by loss of biological activity.

dendrite (Gr. *dendron*, tree) A process extending from the cell body of a neuron, typically branched, that conducts impulses toward the cell body.

deoxyribonucleic acid (DNA) The genetic material of all organisms; composed of two complementary chains of nucleotides wound in a double helix.

depolarization The movement of ions across a cell membrane that wipes out locally an electrical potential difference.

derived character A characteristic used in taxonomic analysis representing a departure from the primitive form.

desmosome A type of anchoring junction that links adjacent cells by connecting their cytoskeletons with cadherin proteins.

diaphragm (Gr. *diaphrassein*, to barricade) (1) In mammals, a sheet of muscle tissue that separates the abdominal and thoracic cavities and functions in breathing. (2) A contraceptive device used to block the entrance to the uterus temporarily and thus prevent sperm from entering during sexual intercourse.

dicot Short for dicotyledon; a class of flowering plants generally characterized as having two cotyledons, net-veined leaves, and flower parts usually in fours or fives.

differentiation A developmental process by which a relatively unspecialized cell undergoes a progressive change to a more specialized form or function.

diffusion (L. *diffundere*, to pour out) The net movement of dissolved molecules or other particles from a region where they are more concentrated to a region where they are less concentrated.

digestion (L. *digestio*, separating out, dividing) The breakdown of complex, usually insoluble foods into molecules that can be absorbed into cells, there to be degraded to yield energy and the raw materials for synthetic processes.

dihybrid (Gr. *dis*, twice + *hibridia*, mixed offspring) An individual heterozygous at two different loci; for example A/a B/b.

dikaryotic (Gr. *di*, two + *karyon*, kernel) In fungi, having pairs of nuclei within each cell.

dioecious (Gr. *di*, two + *oikos*, house) Having the male and female elements on different individuals.

diploid (Gr. *diploos*, double + *eidos*, form) Having two sets of chromosomes ($2n$); in animals, twice the number characteristic of gametes; in plants, the chromosome number characteristic of the sporophyte generation; in contrast to haploid ($1n$).

directional selection A form of selection in which selection acts to eliminate one extreme from an array of phenotypes.

disaccharide A carbohydrate formed of two simple sugar molecules bonded covalently.

disruptive selection A form of selection in which selection acts to eliminate rather than favor the intermediate type.

diurnal (L. *diurnalis*, day) Active during the day.

division A major taxonomic group; kingdoms are divided into divisions (or phyla, which are equivalent), and divisions are divided into classes.

DNA ligase The enzyme that links together Okazaki fragments in DNA replication of the lagging strand; it also links other broken areas of the DNA backbone.

domain A distinct modular region of a protein that serves a particular function in the action of the protein, such as a regulatory domain or DNA-binding domain.

dominant An allele that is expressed when present in either the heterozygous or the homozygous condition.

double fertilization The fusion of the egg and sperm (resulting in a $2n$ fertilized egg, the zygote) and the simultaneous fusion of the second male gamete with the polar nuclei (resulting in a primary endosperm nucleus, which is often triploid, $3n$); a unique characteristic of all angiosperms.

Down syndrome A congenital syndrome caused by the presence of an extra copy of chromosome 21.

duodenum (L. *duodeni*, 12 each—from its length, about 12 fingers' breadth) In vertebrates, the upper portion of the small intestine.

E

ecdysis (Gr. *ekdysis*, stripping off) Shedding of outer, cuticular layer; molting, as in insects or crustaceans.

ecdysone (Gr. *ekdysis*, stripping off) Molting hormone of arthropods, which triggers when ecdysis occurs.

ecology (Gr. *oikos*, house + *logos*, word) The study of interactions of organisms with one another and with their physical environment.

ecosystem (Gr. *oikos*, house + *systems*, that which is put together) A major interacting system that involves both organisms and their nonliving environment.

ecotype (Gr. *oikos*, house + L. *typus*, image) A locally adapted variant of an organism; differing genetically from other ecotypes.

ectoderm (Gr. *ecto*, outside + *derma*, skin) One of the three embryonic germ layers of early vertebrate embryos; ectoderm gives rise to the outer epithelium of the body (skin, hair, nails) and to the nerve tissue, including the sense organs, brain, and spinal cord.

ectomycorrhizae Externally developing mycorrhizae that do not penetrate the cells they surround.

ectotherms Animals, such as reptiles, fish, or amphibians, whose body temperature is regulated by their behavior or by their surroundings.

electronegativity A property of atomic nuclei that refers to the affinity of the nuclei for valence electrons; a nucleus that is more electronegative has a greater pull on electrons than one that is less electronegative.

electron transport chain The passage of energetic electrons through a series of membrane-associated electron-carrier molecules to proton pumps embedded within mitochondrial or chloroplast membranes. *See* chemiosmosis.

endergonic reaction A chemical reaction in which the products contain more energy than the reactants, so that free energy must be put into the reaction from an outside source to allow it to proceed.

endocrine gland (Gr. *endon*, within + *krinein*, to separate) Ductless gland that secretes hormones into the extracellular spaces, from which they diffuse into the circulatory system.

endocytosis (Gr. *endon*, within + *kytos*, hollow vessel) The uptake of material into cells by inclusion within an invagination of the plasma membrane; the uptake of solid material is phagocytosis, while that of dissolved material is pinocytosis.

endoderm (Gr. *endon*, within + *derma*, skin) One of the three embryonic germ layers of early vertebrate embryos, destined to give rise to the epithelium that lines internal structures and most of the digestive and respiratory tracts.

endodermis (Gr. *endon*, within + *derma*, skin) In vascular plants, a layer of cells forming the innermost layer of the cortex in roots and some stems.

endometrium (Gr. *endon*, within + *metrios*, of the womb) The lining of the uterus in mammals; thickens in response to secretion of estrogens and progesterone and is sloughed off in menstruation.

endomycorrhizae Mycorrhizae that develop within cells.

endoplasmic reticulum (ER) An internal membrane system that forms a netlike array of channels and interconnections of organelles within the cytoplasm of eukaryotic cells.

endorphin One of a group of small neuropeptides produced by the vertebrate brain; like morphine, endorphins modulate pain perception.

endosperm (Gr. *endon*, within + *sperma*, seed) A storage tissue characteristic of the seeds of angiosperms, which develops from the union of a male nucleus and the polar nuclei of the embryo sac. The endosperm is digested by the growing sporophyte either before maturation of the seed or during its germination.

endosymbiotic theory Proposes that eukaryotic cells evolved from a symbiosis between different species of prokaryotes.

endothermy The ability of animals to maintain a constant body temperature.

energy The capacity to do work.

enhancer A site of regulatory protein binding on the DNA molecule distant from the promoter and start site for a gene's transcription.

enthalpy In a chemical reaction, the energy contained in the chemical bonds of the molecule, symbolized as H; in a cellular reaction, the free energy is equal to the enthalpy of the reactant molecules in the reaction.

entropy (Gr. *en*, in + *tropos*, change in manner) A measure of the randomness or disorder of a system; a measure of how much energy in a system has become so dispersed (usually as evenly distributed heat) that it is no longer available to do work.

enzyme (Gr. *enzymes*, leavened, from *en*, in + *zyme*, leaven) A protein that is capable of speeding up specific chemical reactions by lowering the required activation energy.

epicotyl The region just above where the cotyledons are attached.

epidermis (Gr. *epi*, on or over + *derma*, skin) The outermost layers of cells; in plants, the exterior primary tissue of leaves, young stems, and roots; in vertebrates, the nonvascular external layer of skin, of ectodermal origin; in invertebrates, a single layer of ectodermal epithelium.

epididymis (Gr. *epi*, on + *didymos*, testicle) A sperm storage vessel; a coiled part of the sperm duct that lies near the testis.

epistasis (Gr. *epi*, on + *stasis*, a standing still) Interaction between two nonallelic genes in which one of them modifies the phenotypic expression of the other.

epithelium (Gr. *epi*, on + *thele*, nipple) In animals, a type of tissue that covers an exposed surface or lines a tube or cavity.

equilibrium (L. *acequus*, equal + *libra*, balance) A stable condition; the point at which a chemical reaction proceeds as rapidly in the reverse direction as it does in the forward direction, so that there is no further net change in the concentrations of products or reactants.

erythrocyte (Gr. *erythros*, red + *kytos*, hollow vessel) Red blood cell, the carrier of hemoglobin.

erythropoiesis The manufacture of blood cells in the bone marrow.

estrous cycle The periodic cycle in which periods of estrus correspond to ovulation events.

estrus (L. *oestrus*, frenzy) The period of maximum female sexual receptivity, associated with ovulation of the egg.

ethology (Gr. *ethos*, habit or custom + *logos*, discourse) The study of patterns of animal behavior in nature.

eubacteria A major group of bacteria, including most of those living today, which have very strong cell walls and a simpler gene architecture than that of the archaebacteria. *See* archaebacteria.

euchromatin (Gr. *eu*, good + *chroma*, color) That portion of a eukaryotic chromosome that is transcribed into mRNA; contains active genes that are not tightly condensed during interphase.

eukaryote (Gr. *eu*, good + *karyon*, kernel) A cell characterized by membrane-bounded organelles, most notably the nucleus, and one that possesses chromosomes whose DNA is associated with proteins; an organism composed of such cells.

eutrophic (Gr. *eutrophos*, thriving) Refers to a lake in which an abundant supply of minerals and organic matter exists.

evolution (L. *evolvere*, to unfold) Genetic change in a population of organisms; in general, evolution leads to progressive change from simple to complex.

exergonic reaction A chemical reaction in which the products contain less free energy than the reactants, so that free energy is released in the reaction.

exocrine gland (Gr. *ex*, out of + *krinein*, to separate) A type of gland that releases its secretion through a duct, such as digestive glands and sweat glands.

exocytosis (Gr. *ex*, out + *kytos*, vessel) A type of bulk transport out of cells where a vacuole fuses with the cell membrane, discharging the vacuole's contents to the outside.

exon (Gr. *exo*, outside) A segment of DNA that is both transcribed into RNA and translated into protein.

exoskeleton (Gr. *exo*, outside + *skeletos*, hand) An external skeleton, as in arthropods.

exteroception (L. *exter*, outward + *capere*, to take) The condition when a sense organ is excited by stimuli from the external world.

F

facilitated diffusion Carrier-assisted diffusion of molecules across a cellular membrane through specific channels from a region of higher concentration to one of lower concentration; the process is driven by the concentration gradient and does not require cellular energy from ATP.

fat A molecule composed of glycerol and three fatty acid molecules.

feedback inhibition Control mechanism whereby an increase in the concentration of some molecules inhibits the synthesis of that molecule.

fermentation (L. *fermentum*, ferment) The enzyme-catalyzed extraction of energy from organic compounds without the involvement of oxygen.

fertilization The fusion of two haploid gamete nuclei to form a diploid zygote nucleus.

fibroblast (L. *fibra*, fiber + Gr. *blastos*, sprout) A flat, irregularly branching cell of connective tissue that secretes structurally strong proteins into the matrix between the cells.

first law of thermodynamics Energy cannot be created or destroyed, but can only undergo conversion from one form to another; thus, the amount of energy in the universe is unchangeable.

fitness The genetic contribution of an individual to succeeding generations. Relative fitness refers to the fitness of an individual relative to other individuals in a population.

fixed action pattern A stereotyped animal behavior response, thought by ethologists to be based on programmed neural circuits.

foraging behavior A collective term for the many complex, evolved behaviors that influence what an animal eats and how the food is obtained.

founder principle The effect by which rare alleles and combinations of alleles may be enhanced in new populations.

fovea (L., a small pit) A small depression in the center of the retina with a high concentration of cones; the area of sharpest vision.

free energy Energy available to do work.

free radical An ionized atom with one or more unpaired electrons, resulting from electrons that have been energized by ionizing radiation being ejected from the atom; free radicals react violently with other molecules, such as DNA, causing damage by mutation.

fruit In angiosperms, a mature, ripened ovary (or group of ovaries), containing the seeds.

functional genomics Research into the function of specific genes in an organism.

fundamental niche Also referred to as the hypothetical niche, this is the entire niche an organism could fill if there were no other interacting factors (such as competition or predation).

G

gametangium, pl. gametangia (Gr. *gamein*, to marry + L. *tangere*, to touch) A cell or organ in which gametes are formed.

gamete (G., wife) A haploid reproductive cell.

gametocytes Cells in the malarial sporozoite life cycle capable of giving rise to gametes when in the correct host.

gametophyte In plants, the haploid ($1n$), gamete-producing generation, which alternates with the diploid ($2n$) sporophyte.

ganglion, pl. ganglia (Gr., a swelling) An aggregation of nerve cell bodies; in invertebrates, ganglia are the integrative centers; in vertebrates, the term is restricted to aggregations of nerve cell bodies located outside the central nervous system.

gap junction A junction between adjacent animal cells that allows the passage of materials between the cells.

gastrula (Gr., little stomach) In vertebrates, the embryonic stage in which the blastula with its single layer of cells turns into a three-layered embryo made up of ectoderm, mesoderm, and endoderm.

gene (Gr. *genos*, birth, race) The basic unit of heredity; a sequence of DNA nucleotides on a chromosome that encodes a protein, tRNA, or rRNA molecule, or regulates the transcription of such a sequence.

gene conversion Alteration of one homologous chromosome by the cell's error-detection and repair system to make it resemble the sequence on the other homologue.

genetic drift Random fluctuation in allele frequencies over time by chance.

genome The entire DNA sequence of an organism.

genomics The study of genomes as opposed to individual genes.

genotype (Gr. *genos*, offspring + *typos*, form) The genetic constitution underlying a single trait or set of traits.

genus, pl. genera (L., race) A taxonomic group that ranks below a family and above a species.

germination (L. *germinare*, to sprout) The resumption of growth and development by a spore or seed.

germ layers The three cell layers formed at gastrulation of the embryo that foreshadow the future organization of tissues; the layers, from the outside inward, are the ectoderm, the mesoderm, and the endoderm.

gill (1) In aquatic animals, a respiratory organ, usually a thin-walled projection from some part of the external body surface, endowed with a rich capillary bed and having a large surface area. (2) In basidiomycete fungi, the plates on the underside of the cap.

glomerular filtrate The fluid that passes out of the capillaries of each glomerulus.

glomerulus (L., a little ball) A cluster of capillaries enclosed by Bowman's capsule.

glucagon (Gr. *glukus*, sweet + *ago*, to lead forward) A vertebrate hormone produced in the pancreas that acts to initiate the breakdown of glycogen to glucose subunits.

gluconeogenesis The synthesis of glucose from noncarbohydrates (such as proteins or fats).

glucose A common six-carbon sugar ($C_6H_{12}O_6$); the most common monosaccharide in most organisms.

glycocalyx A "sugar coating" on the surface of a cell resulting from the presence of polysaccharides on glycolipids and glycoproteins embedded in the outer layer of the plasma membrane.

glycogen (Gr. *glykys*, sweet + *gen*, of a kind) Animal starch; a complex branched polysaccharide that serves as a food reserve in animals, bacteria, and fungi.

glycolipid Lipid molecule modified within the Golgi complex by having a short sugar chain (polysaccharide) attached.

glycolysis (Gr. *glykys*, sweet + *lyein*, to loosen) The anaerobic breakdown of glucose; this enzyme-catalyzed process yields two molecules of pyruvate with a net of two molecules of ATP.

glycoprotein Protein molecule modified within the Golgi complex by having a short sugar chain (polysaccharide) attached.

glyoxysome A small cellular organelle or microbody containing enzymes necessary for conversion of fats into carbohydrates.

Golgi apparatus A collection of flattened stacks of membranes (each called a Golgi body) in the cytoplasm of eukaryotic cells; functions in collection, packaging, and distribution of molecules synthesized in the cell.

G protein A protein that binds guanosine triphosphate (GTP) and assists in the function of cell surface receptors. When the receptor binds its signal molecule, the G protein binds GTP and is activated to start a chain of events within the cell.

gravitropism (L. *gravis*, heavy + *tropes*, turning) Growth response to gravity in plants; formerly called geotropism.

ground meristem (Gr. *meristos*, divisible) The primary meristem, or meristematic tissue, that gives rise to the plant body (except for the epidermis and vascular tissues).

guttation (L. *gutta*, a drop) The exudation of liquid water from leaves due to root pressure.

gymnosperm (Gr. *gymnos*, naked + *sperma*, seed) A seed plant with seed not enclosed in an ovary; conifers are gymnosperms.

gynoecium (Gr. *gyne*, woman + *oikos*, house) The aggregate of carpels in the flower of a seed plant.

H

habitat (L. *hibitare*, to inhabit) The environment of an organism; the place where it is usually found.

habituation (L. *habitus*, condition) A form of learning; a diminishing response to a repeated stimulus.

haplodiploidy A phenomenon encountered in certain organisms such as wasps, wherein both haploid (male) and diploid (female) individuals are encountered.

haploid (Gr. *haploos*, single + *ploion*, vessel) Having only one set of chromosomes (n), in contrast to diploid ($2n$).

Hardy-Weinberg equilibrium A mathematical description of the fact that allele and genotype frequencies remain constant in a random-mating population in the absence of inbreeding, selection, or other evolutionary forces; usually stated: if the frequency of allele a is p and the frequency of allele b is q, then the genotype frequencies after one generation of random mating will always be $p^2(a) = 2pq(ab) + q^2(b)$.

Haversian canal After Clopton Havers, English anatomist. Narrow channels that run parallel to the length of a bone and contain blood vessels and nerve cells.

helper T cell A class of white blood cells that initiates both the cell-mediated immune response and the humoral immune response; helper T cells are the targets of the AIDS virus (HIV).

hemoglobin (Gr. *haima*, blood + L. *globus*, a ball) A globular protein in vertebrate red blood cells and in the plasma of many invertebrates that carries oxygen and carbon dioxide.

hemophilia (Gr. *haima*, blood + *philios*, friendly) A group of hereditary diseases characterized by failure of the blood to clot.

hemopoietic stem cell (Gr. *haimatos*, blood + *poiesis*, a making) The cells in bone marrow where blood cells are formed.

hermaphrodite (Gr. *hermaphroditos*, containing both sexes) An organism with both male and female functional reproductive organs.

heterochromatin (Gr. *heteros*, different + *chroma*, color) The portion of a eukaryotic chromosome that is not transcribed into RNA; remains condensed in interphase and stains intensely in histological preparations.

heterokaryotic (Gr. *heteros*. other + *karyon*, kernel) In fungi, having two or more genetically distinct types of nuclei within the same mycelium.

heterosporous (Gr. *heteros*, other + *sporos*, seed) In vascular plants, having spores of two kinds, namely, microspores and megaspores.

heterotroph (Gr. *heteros*, other + *trophos*, feeder) An organism that cannot derive energy from photosynthesis or inorganic chemicals, and so must feed on other plants and animals, obtaining chemical energy by degrading their organic molecules.

heterozygous Having two different alleles of the same gene; the term is usually applied to one or more specific loci, as in "heterozygous with respect to the *W* locus" (that is, the genotype is *W/w*).

histone (Gr. *histos*, tissue) One of a group of relatively small, very basic polypeptides, rich in arginine and lysine, forming the core of nucleosomes around which DNA is wrapped in the first stage of chromosome condensation.

holoblastic cleavage (Gr. *holos*, whole + *blastos*, germ) Process in vertebrate embryos in which the cleavage divisions all occur at the same rate, yielding a uniform cell size in the blastula.

homeobox A sequence of 180 nucleotides located in homeotic genes that produces a 60-amino acid peptide sequence (the homeodomain) active in transcription factors.

homeostasis (Gr. *homeos*, similar + *stasis*, standing) The maintenance of a relatively stable internal physiological environment in an organism, usually involves some form of feedback self-regulation.

homeotherm (Gr. *homoios*, same or similar + *therme*, heat) An organism, such as a bird or mammal, capable of maintaining a stable body temperature independent of the environmental temperature. *See* endothermy.

homeotic gene One of a series of "master switch" genes that determine the form of segments developing in the embryo.

hominid (L. *homo*, man) Any primate in the human family, Hominidae. *Homo sapiens* is the only living representative.

hominoid (L. *homo*, man) Collectively, hominids and apes; the monkeys and hominoids constitute the anthropoid primates.

homokaryotic (Gr. *homos*, same + *karyon*, kernel) In fungi, having a nuclei with the same genetic makeup within a mycelium.

homologous chromosome (homologue) One of a pair of chromosomes of the same kind located in a diploid cell; one copy of each pair of homologues comes from each gamete that formed the zygote.

homology (Gr. *homologia*, agreement) A condition in which the similarity between two structures or functions is indicative of a common evolutionary origin.

homosporous (Gr. *homos*, same or similar + *sporos*, seed) In some plants, production of only one type of spore rather than differentiated types. Compare heterosporous.

homozygous Being a homozygote, having two identical alleles of the same gene; the term is usually applied to one or more specific loci, as in "homozygous with respect to the *W* locus" (that is, the genotype is *W/W* or *w/w*).

hormone (Gr. *hormaein*, to excite) A molecule, usually a peptide or steroid, that is produced in one part of an organism and triggers a specific cellular reaction in target tissues and organs some distance away.

human immunodeficiency virus (HIV) The virus responsible for AIDS, a deadly disease that destroys the human immune system. HIV is a retrovirus (its genetic material is RNA) that is thought to have been introduced to humans from chimpanzees.

hybridization The mating of unlike parents.

hydroskeleton (Gr. *hydro*, water + *skeletos*, hard) The skeleton of most soft-bodied invertebrates that have neither an internal nor an external skeleton. They use the relative incompressibility of the water within their bodies as a kind of skeleton.

hyperosmotic The condition in which a (hyperosmotic) solution has a higher osmotic concentration than that of a second solution.

hyperpolarization An above normal negativity of a cell membrane during its resting potential.

hypersensitive response Plants respond to pathogens by selectively killing plant cells to block the spread of the pathogen.

hypha, pl. hyphae (Gr. *hyphe*, web) A filament of a fungus or oomycete; collectively, the hyphae comprise the mycelium.

hypocotyl The region immediately below where the cotyledons are attached.

hypoosmotic The condition in which a (hypoosmotic) solution has a lower osmotic concentration than that of a second solution.

hypothalamus (Gr. *hypo*, under + *thalamos*, inner room) A region of the vertebrate brain just below the cerebral hemispheres, under the thalamus; a center of the autonomic nervous system, responsible for the integration and correlation of many neural and endocrine functions.

I

imaginal disk One of about a dozen groups of cells set aside in the abdomen of a larval insect and committed to forming key parts of the adult insect's body.

immune response In vertebrates, a defensive reaction of the body to invasion by a foreign substance or organism. *See* antibody and B cell.

immunoglobulin (L. *immunis*, free + *globus*, globe) An antibody.

inbreeding The breeding of genetically related plants or animals; inbreeding tends to increase homozygosity.

inclusive fitness Describes the sum of the number of genes directly passed on in an individual's offspring and those genes passed on indirectly by kin (other than offspring) whose existence results from the benefit of the individual's altruism.

induction Processes by which one group of embryonic cells affects an adjacent group, thereby inducing those cells to differentiate in a manner they otherwise would not have.

industrial melanism (Gr. *melas*, black) Phrase used to describe the evolutionary process in which initially light-colored organisms become dark as a result of natural selection.

inflammatory response (L. *inflammare*, to flame) A generalized nonspecific response to infection that acts to clear an infected area of infecting microbes and dead tissue cells so that tissue repair can begin.

initiation factor One of several proteins involved in the formation of an initiation complex in prokaryote polypeptide synthesis.

insertional inactivation Destruction of a gene's function by the insertion of a transposon.

instar A larval developmental stage in insects.

interferon In vertebrates, a protein produced in virus-infected cells that inhibits viral multiplication.

interneuron (association neuron) A nerve cell found only in the middle of the spinal cord that acts as a functional link between sensory neurons and motor neurons.

internode In plants, the region of a stem between two successive nodes.

interoception The sensing of information that relates to the body itself, its internal condition, and its position.

interphase The period between two mitotic or meiotic divisions in which a cell grows and its DNA replicates; includes G_1, S, and G_2 phases.

intron (L. *intra*, within) Portion of mRNA as transcribed from eukaryotic DNA that is removed by enzymes before the mature mRNA is translated into protein. *See* exon.

inversion (L. *invertere*, to turn upside down) A reversal in order of a segment of a chromosome; also, to turn inside out, as in embryogenesis of sponges or discharge of a nematocyst.

ion Any atom or molecule containing an unequal number of electrons and protons and therefore carrying a net positive or net negative charge.

ionizing radiation High-energy radiation that is highly mutagenic, producing free radicals that react with DNA; includes X rays and gamma rays.

isomer (Gr. *isos*, equal + *meros*, part) One of a group of molecules identical in atomic composition but differing in structural arrangement; for example, glucose and fructose.

isosmotic solutions The conditions in which the osmotic concentrations of two solutions are equal, so that no net water movement occurs between them by osmosis.

K

karyotype (Gr. *karyon*, kernel + *typos*, stamp or print) The morphology of the chromosomes of an organism as viewed with a light microscope.

keratin (Gr. *kera*, horn + *in*, suffix used for proteins) A tough, fibrous protein formed in epidermal tissues and modified into skin, feathers, hair and hard structures such as horns and nails.

kidney In vertebrates, the organ that filters the blood to remove nitrogenous wastes and regulates the balance of water and solutes in blood plasma.

kilocalorie Unit describing the amount of heat required to raise the temperature of a kilogram of water by one degree Celsius (°C); sometimes called a Calorie, equivalent to 1000 calories.

kinesis (Gr. *kinesis*, motion) Changes in activity level in an animal that are dependent on stimulus intensity. *See* kinetic energy.

kinetic energy The energy of motion.

kinetochore (Gr. *kinetikos*, putting in motion + *choros*, chorus) Disc-shaped protein structure within the centromere to which the spindle fibers attach during mitosis or meiosis. *See* centromere,

kingdom The highest commonly used taxonomic category.

kin selection Selection favoring relatives; an increase in the frequency of related individuals (kin) in a population, leading to an increase in the relative frequency in the population of those alleles shared by members of the kin group.

Krebs cycle Another name for the citric acid cycle; also called the tricarboxylic acid (TCA) cycle.

L

labrum (L., a lip) The upper lip of insects and crustaceans situated above or in front of the mandibles.

larynx The voice box; a cartilaginous organ that lies between the pharynx and trachea and is responsible for sound production in vertebrates.

lateral line system A sensory system encountered in fish, through which mechanoreceptors in a line down the side of the fish are sensitive to motion.

lateral meristems (L. *latus*, side + Gr. *meristos*, divided) In vascular plants, the meristems that give rise to secondary tissue; the vascular cambium and cork cambium.

Law of independent assortment Mendel's second law of heredity, stating that genes located on nonhomologous chromosomes assort independently of one another.

Law of segregation Mendel's first law of heredity, stating that alternative alleles for the same gene segregate from each other in production of gametes.

leaf primordium, pl. **primordia** (L. *primordium*, beginning) A lateral outgrowth from the apical meristem that will eventually become a leaf.

lenticels (L. *lenticella*, a small window) Spongy areas in the cork surfaces of stem, roots, and other plant parts that allow interchange of gases between internal tissues and the atmosphere through the periderm.

leucoplast (Gr. *leukos*, white + *plasein*, to form) In plant cells, a colorless plastid in which starch grains are stored; usually found in cells not exposed to light.

leukocyte (Gr. *leukos*, a white + *kytos*, hollow vessel) A white blood cell; a diverse array of nonhemoglobin-containing blood cells, including phagocytic macrophages and antibody-producing lymphocytes.

lichen Symbiotic association between a fungus and a photosynthetic organism such as a green alga or cyanobacterium.

limbic system The hypothalamus, together with the network of neurons that link the hypothalamus to some areas of the cerebral cortex. Responsible for many of the most deep-seated drives and emotions of vertebrates, including pain, anger, sex, hunger, thirst, and pleasure.

lipase (Gr. *lipos*, fat + *-ase*, enzyme suffix) An enzyme that catalyzes the hydrolysis of fats.

lipid (Gr. *lipos*, fat) A nonpolar hydrophobic organic molecule that is insoluble in water (which is polar) but that dissolves readily in nonpolar organic solvents; includes fats, oils, waxes, steroids, phospholipids, and carotenoids.

lipid bilayer The structure of a cellular membrane, in which two layers of phospholipids spontaneously align so that the hydrophilic head groups are exposed to water, while the hydrophobic fatty acid tails are pointed toward the center of the membrane.

locus The position on a chromosome where a gene is located.

loop of Henle In the kidney of birds and mammals, a hairpin-shaped portion of the renal tubule in which water and salt are reabsorbed from the glomerular filtrate by diffusion.

lophophore A ring or U-shaped arrangement of tentacles, often ciliated, into which the coelom of the animal extends. The organ functions in feeding and gas exchange and is encountered in only a few phyla of small marine invertebrates.

luteal phase The second phase of the reproductive cycle, during which the mature eggs are released into the fallopian tubes, a process called ovulation.

lymph (L. *lympha*, clear water) In animals, a colorless fluid derived from blood by filtration through capillary walls in the tissues.

lymphatic system In animals, an open vascular system that reclaims water that has entered interstitial regions from the bloodstream (lymph); includes the lymph nodes, spleen, thymus, and tonsils.

lymphocyte (L. *lympha*, water + Gr. *kytos*, hollow vessel) A type of white blood cell. Lymphocytes are responsible for the immune response; there are two principal classes—B cells and T cells.

lymphokine A regulatory molecule that is secreted by lymphocytes. In the immune response, lymphokines secreted by helper T cells unleash the cell-mediated immune response.

lysis (Gr., a loosening) Disintegration of a cell by rupture of its cell membrane.

M

macroevolution (Gr. *makros*, large + L. *evolvere*, to unfold) The creation of new species and the extinction of old ones.

macromolecule (Gr. *makros*, large + L. *moleculus*, a little mass) An extremely large biological molecule; refers specifically to proteins, nucleic acids, polysaccharides, lipids, and complexes of these.

macronutrients (Gr. *makros*, large + L. *nutrire*, to nourish) Inorganic chemical elements required in large amounts for plant growth, such as nitrogen, potassium, calcium, phosphorus, magnesium, and sulfur.

macrophage A large phagocytic cell that is able to engulf and digest cellular debris and invading bacteria.

major histocompatibility complex (MHC) A set of protein cell-surface markers anchored in plasma membrane, which the immune system uses to identify "self." All the cells of a given individual have the same "self" marker, called an MHC protein.

Malpighian tubules Blind tubules opening into the hindgut of terrestrial arthropods; they function as excretory organs.

mandibles (L. *mandibula*, jaw) In crustaceans, insects, and myriapods, the appendages immediately posterior to the antennae; used to seize, hold, bite, or chew food.

mantle The soft, outermost layer of the body wall in mollusks; the mantle secretes the shell.

marsupial (L. *marsupium*, pouch) A mammal in which the young are born early in their development, sometimes as soon as eight days after fertilization, and are retained in a pouch.

mass flow The overall process by which materials move in the phloem of plants.

matrix (L. *mater*, mother) In mitochondria, the solution in the interior space surrounded by the cristae that contains the enzymes and other molecules involved in oxidative respiration; more generally, that part of a tissue within which an organ or process is embedded.

menstrual cycle (L. *mens*, month) In humans and higher primates, the cyclic changes in the ovaries and uterine endometrium; lasts about a month in humans.

menstruation Periodic sloughing off of the blood-enriched lining of the uterus when pregnancy does not occur.

meristem (Gr. *merizein*, to divide) Undifferentiated plant tissue from which new cells arise.

meroblastic cleavage (Gr. *meros*, part + *blastos*, sprout) A type of cleavage in the eggs of reptiles, birds, and some fish. Occurs only on the blastodisc.

mesoderm (Gr. *mesos*, middle + *derma*, skin) One of the three embryonic germ layers that form in the gastrula; gives rise to muscle, bone and other connective tissue, the peritoneum, the circulatory system, and most of the excretory and reproductive systems.

mesophyll (Gr. *mesos*, middle + *phyllon*, leaf) The photosynthetic parenchyma of a leaf, located within the epidermis.

messenger RNA (mRNA) The RNA transcribed from structural genes; RNA molecules complementary to a portion of one strand of DNA, which are translated by the ribosomes to form protein.

metabolism (Gr. *metabole*, change) The sum of all chemical processes occurring within a living cell or organism.

metamorphosis (Gr. *meta*, after + *morphe*, form + *osis*, state of) Process in which there is a marked change in form during postembryonic development as, for example, in tadpole to frog.

metaphase (Gr. *meta*, middle + *phasis*, form) The stage of mitosis or meiosis during which microtubules become organized into a spindle and the chromosomes come to lie in the spindle's equatorial plane.

metastasis The process by which cancer cells move from their point of origin to other locations in the body; also a population of cancer cells in a secondary location, the result of movement from the primary tumor.

methanogens Obligate, anaerobic archaebacteria that produce methane.

microarray DNA sequences are placed on a microscope slide or chip with a robot. The microarray can then be probed with RNA from specific tiisues to identify expressed DNA.

microbody A cellular organelle bounded by a single membrane and containing a variety of enzymes; generally derived from endoplasmic reticulum; includes peroxisomes and glyoxysomes.

microevolution (Gr. *mikros*, small + L. *evolvere*, to unfold) Refers to the evolutionary process itself. Evolution within a species. Also called adaptation.

micronutrient (Gr. *mikros*, small + L. *nutrire*, to nourish) A mineral required in only minute amounts for plant growth, such as iron, chlorine, copper, manganese, zinc, molybdenum, and boron.

micropyle In the ovules of seed plants, an opening in the integuments through which the pollen tube usually enters.

microtubule (Gr. *mikros*, small + L. *tubulus*, little pipe) In eukaryotic cells, a long, hollow protein cylinder, composed of the protein tubulin; these influence cell shape, move the chromosomes in cell division, and provide the functional internal structure of cilia and flagella.

microvillus (Gr. *mikros*, small + L. *villus*, shaggy hair) Cytoplasmic projection from epithelial cells, microvilli greatly increase the surface area of the small intestine.

middle lamella The layer of intercellular material, rich in pectic compounds, that cements together the primary walls of adjacent plant cells.

mimicry (Gr. *mimos*, mime) The resemblance in form, color, or behavior of certain organisms (mimics) to other more powerful or more protected ones (models).

mitosis (Gr. *mitos*, thread) Somatic cell division; nuclear division in which the duplicated chromosomes separate to form two genetically identical daughter nuclei.

monocot Short for monocotyledon; flowering plant in which the embryos have only one cotyledon, the floral parts are generally in threes, and the leaves typically are parallel-veined.

monocyte (Gr. *monos*, single + *kytos*, hollow vessel) A type of leukocyte that becomes a phagocytic cell (macrophage) after moving into tissues.

monoecious (Gr. *monos*, single + *oecos*, house) A plant in which the staminate and pistillate flowers are separate, but borne on the same individual.

monosaccharide (Gr. *monos*, one + L. *saccharum*, sugar) A simple sugar that cannot be decomposed into smaller sugar molecules.

monotreme (Gr. *mono*, single + *treme*, hole) An egg-laying mammal.

morphogen A signal molecule produced by an embryonic organizer region that informs surrounding cells of their distance from the organizer, thus determining relative position of cells during development.

morphology The form and structure of an organism.

morula (L., a little mulberry) Solid balls of cells in the early stage of embryonic development.

mosaic development A pattern of embryonic development in which initial cells produced by cleavage divisions contain different developmental signals (determinants) from the egg, setting the individual cells on different developmental paths.

motor (efferent) neuron Neuron that transmits nerve impulses from the central nervous system to an effector, which is typically a muscle or gland.

Müellerian mimicry After Fritz Müeller, German biologist. A phenomenon in which two or more unrelated but protected species resemble one another, thus achieving a kind of group defense.

multigene family A collection of related genes on a single chromosome or on different chromosomes.

muscle fiber A long, cylindrical, multinucleated cell containing numerous myofibrils, which is capable of contraction when stimulated.

mutagen (L. *mutare*, to change + Gr. *genaio*, to produce) An agent that induces changes in DNA (mutations); includes physical agents that damage DNA and chemicals that alter DNA bases.

mutant (L. *mutare*, to change) A mutated gene; alternatively, an organism carrying a gene that has undergone a mutation.

mutation A permanent change in a cell's DNA; includes changes in nucleotide sequence, alteration of gene position, gene loss or duplication, and insertion of foreign sequences.

mutualism (L. *mutuus*, lent, borrowed) The living together of two or more organisms in a symbiotic association in which both members benefit.

mycelium, pl. **mycelia** (Gr. *mykes*, fungus) In fungi, a mass of hyphae.

mycorrhiza, pl. **mycorrhizae** (Gr. *mykes*, fungus + *rhiza*, root) A symbiotic association between fungi and the roots of a plant.

myelin sheath (Gr. *myelinos*, full of marrow) A fatty layer surrounding the long axons of motor neurons in the peripheral nervous system of vertebrates.

myofilament (Gr. *myos*, muscle + L. *filare*, to spin) A contractile microfilament, composed largely of actin and myosin, within muscle. Sometimes called myofibril.

myosin (Gr. *mys*, muscle + *in*, belonging to) One of the two protein components of microfilaments (the other is actin); a principal component of vertebrate muscle.

N

natural killer cell A cell that does not kill invading microbes, but rather, the cells infected by them.

natural selection The differential reproduction of genotypes; caused by factors in the environment; leads to evolutionary change.

negative feedback A homeostatic control mechanism whereby an increase in some substance or activity inhibits the process leading to the increase; also known as feedback inhibition.

nephridium, pl. **nephridia** (Gr. *nephros*, kidney) In invertebrates, a tubular excretory structure.

nephrid organ A filtration system of many freshwater invertebrates in which water and waste pass from the body across the membrane into a collecting organ, from which they are expelled to the outside through a pore.

nephron (Gr. *nephros*, kidney) Functional unit of the vertebrate kidney; one of numerous tubules involved in filtration and selective reabsorption of blood; each nephron consists of a Bowman's capsule, an enclosed glomerulus, and a long attached tubule; in humans, called a renal tubule.

nerve A group or bundle of nerve fibers (axons) with accompanying neurological cells, held together by connective tissue; located in the peripheral nervous system.

neural crest A special strip of cells that develops just before the neural groove closes over to form the neural tube in embryonic development.

neural groove The long groove formed along the long axis of the embryo by a layer of ectodermal cells.

neural tube The dorsal tube, formed from the neural plate, that differentiates into the brain and spinal cord.

neuroethology (Gr. *neuron*, nerve + *ethos*, habit or custom + *logos*, discourse) The study of the neural basis of behavior.

neuroglia (Gr. *neuron*, nerve + *glia*, glue) Nonconducting nerve cells that are intimately associated with neurons and appear to provide nutritional support.

neuromuscular junction The structure formed when the tips of axons contact (innervate) a muscle fiber.

neuron (Gr., nerve) A nerve cell specialized for signal transmission; includes cell body; dendrites, and axon.

neurotransmitter (Gr. *neuron*, nerve + L. *trans*, across + *mittere*, to send) A chemical released at the axon terminal of a neuron that travels across the synaptic cleft, binds a specific receptor on the far side, and, depending on the nature of the receptor, depolarizes or hyperpolarizes a second neuron or a muscle or gland cell.

neurulation A process in early embryonic development by which a dorsal band of ectoderm thickens and rolls into the neural tube.

neutrophil An abundant type of granulocyte capable of engulfing microorganisms and other foreign particles; neutrophils comprise about 50 to 70% of the total number of white blood cells.

niche The role played by a particular species in its environment.

nicotinamide adenine dinucleotide (NAD^+) A molecule that becomes reduced (to NADH) as it carries high-energy electrons from oxidized molecules and delivers them to ATP-producing pathways in the cell.

nociceptor A naked dendrite that acts as a receptor in response to a pain stimulus.

nocturnal (L. *nocturnus*, night) Active primarily at night.

node (L. *nodus*, knot) The part of a plant stem where one or more leaves are attached. *See* internode.

node of Ranvier After L. A. Ranvier, French histologist. A gap formed at the point where two Schwann cells meet and where the axon is in direct contact with the surrounding intercellular fluid.

nonassociative learning A learned behavior that does not require an animal to form an association between two stimuli, or between a stimulus and a response.

nonsense codon One of three codons (UAA, UAG, and UGA) that are not recognized by tRNAs, thus serving as "stop" signals in the mRNA message and terminating translation.

notochord (Gr. *noto*, back + L. *chorda*, cord) In chordates, a dorsal rod of cartilage that runs the length of the body and forms the primitive axial skeleton in the embryos of all chordates.

nucellus (L. *nucella* a small nut) Tissue composing the chief pair of young ovules, in which the embryo sac develops; equivalent to a megasporangium.

nucleic acid A nucleotide polymer; chief types are deoxyribonucleic acid (DNA), which is double-stranded, and ribonucleic acid (RNA), which is typically single-stranded.

nucleolus (L., a small nucleus) In eukaryotes, the site of rRNA synthesis; a spherical body composed chiefly of rRNA in the process of being transcribed from multiple copies of rRNA genes.

nucleosome (L. *nucleus*, kernel + *soma*, body) The fundamental packaging unit of eukaryotic chromosomes; a complex of DNA and histone proteins in which the double-helical DNA winds around eight molecules of histone; chromatin is composed of long sequences of nucleosomes.

nucleotide A single unit of nucleic acid, composed of a phosphate, a five-carbon sugar (either ribose or deoxyribose), and a purine or a pyrimidine.

nucleus In atoms, the central core, containing positively charged protons and (in all but hydrogen) electrically neutral neurons; in eukaryotic cells, the membranous organelle that houses the chromosomal DNA; in the central nervous system, a cluster of nerve cell bodies.

O

ocellus, pl. **ocelli** (L., little eye) A simple light receptor common among invertebrates.

Okazaki fragment A short segment of DNA produced by discontinuous replication elongating in the 5′–> 3′ direction away from the replication.

olfaction (L. *olfactum*, smell) The process or function of smelling.

ommatidium, pl. **ommatidia** (Gr., little eye) The visual unit in the compound eye of arthropods; contains light-sensitive cells and a lens able to form an image.

oncogene (Gr. *oncos*, cancer + *genos*, birth) A mutant form of a growth-regulating gene that is inappropriately "on," causing unrestrained cell growth and division.

oocyst The zygote in a sporozoan life cycle. It is surrounded by a tough cyst to prevent dehydration or other damage.

oogonium The female gametangium of the Oomycota (fungi).

oospore A thick-walled zygote resulting from the fusion of Oomycote gametes.

operant conditioning A learning mechanism in which the reward follows only after the correct behavioral response.

operator A site of gene regulation; a sequence of nucleotides overlapping the promoter site and recognized by a repressor protein; binding of the repressor prevents binding of the polymerase to the promoter site and so blocks transcription of the structural gene.

operculum A flat, external, bony protective covering over the gill chamber in fish.

operon (L. *operis* work) A cluster of adjacent structural genes transcribed as a unit into a single mRNA molecule.

order A category of classification above the level of family and below that of class.

organ (Gr. *organon*, tool) A body structure composed of several different tissues grouped in a structural and functional unit.

organelle (Gr. *organella*, little tool) Specialized part of a cell; literally a small cytoplasmic organ.

osmoconformer An animal that maintains the osmotic concentration of its body fluids at about the same level as that of the medium in which it is living.

osmosis (Gr. *osmos*, act of pushing, thrust) The diffusion of water across a selectively permeable membrane (a membrane that permits the free passage of water but prevents or retards the passage of a solute); in the absence of differences in pressure or volume, the net movement of water is from the side containing a lower concentration of solute to the side containing a higher concentration.

osmotic pressure The potential pressure developed by a solution separated from pure water by a differentially permeable membrane. The higher the solute concentration, the greater the osmotic potential of the solution; also called *osmotic potential.*

osteoblast (Gr. *osteon*, bone + *blastos*, bud) A bone-forming cell.

osteocyte (Gr. *osteon*, bone + *kytos*, hollow vessel) A mature osteoblast.

outcrossing Breeding with individuals other than oneself or one's close relatives.

ovary (L. *ovum*, egg) (1) In animals, the organ in which eggs are produced. (2) In flowering plants, the enlarged basal portion of a carpel, which contains the ovule(s); the ovary matures to become the fruit.

oviduct (L. *ovum*, egg + *ductus*, duct) In vertebrates, the passageway through which ova (eggs) travel from the ovary to the uterus.

oviparity (L. *ovum*, egg + *parere*, to bring forth) Refers to a type of reproduction in which the eggs are developed after leaving the body of the mother, as in reptiles.

ovoviviparity Refers to a type of reproduction in which young hatch from eggs that are retained in the mother's uterus.

ovulation In animals, the release of an egg or eggs from the ovary.

ovum, pl. **ova** (L., egg) The egg cell; female gamete.

oxidation (Fr. *oxider*, to oxidize) Loss of an electron by an atom or molecule; in metabolism, often associated with a gain of oxygen or loss of hydrogen.

oxidative respiration Process of cellular activity in which glucose or other molecules are broken down to water and carbon dioxide with the release of energy.

oxygen debt The amount of oxygen required to convert the lactic acid generated in the muscles during exercise back into glucose.

oxytocin (Gr. *oxys*, sharp + *tokos* birth) A hormone of the posterior pituitary gland that affects uterine contractions during childbirth and stimulates lactation.

ozone O_3, a stratospheric layer of the earth's atmosphere responsible for filtering out ultraviolet radiation supplied by the sun.

P

pacemaker A patch of excitatory tissue in the vertebrate heart that initiates the heartbeat.

palisade parenchyma (L. *palus*, stake) In plant leaves, the columnar, chloroplast-containing parenchyma cells of the mesophyll. Also called *palisade cells.*

papilla A small projection of tissue.

paracrine A type of chemical signaling between cells where the effects are local and short-lived.

parapodia (Gr. *para*, beside + *pous*, foot) One of the paired lateral processes on each side of most segments in polychaete annelids.

parasexuality In certain fungi, the fusion and segregation of heterokaryotic haploid nuclei to produce recombinant nuclei.

parasitism (Gr. *para*, beside + *sitos*, food) A living arrangement in which an organism lives on or in an organism of a different species and derives nutrients from it.

parthenogenesis (Gr. *parthenos*, virgin + *genesis*, birth) The development of an egg without fertilization, as in aphids, bees, ants, and some lizards.

partial pressure The components of each individual gas—such as nitrogen, oxygen, and carbon dioxide—that together comprise the total air pressure.

pelagic Free-swimming, usually in open water.

pellicle A tough, flexible covering in ciliates and euglenoids.

peptide bond (Gr. *peptein*, to soften, digest) The type of bond that links amino acids together in proteins through a dehydration reaction.

perianth (Gr. *peri*, around + *anthos*, flower) In flowering plants, the petals and sepals taken together.

pericycle (Gr. *peri*, around + *kykos*, circle) In vascular plants, one or more cell layers surrounding the vascular tissues of the root, bounded externally by the endodermis and internally by the phloem.

periderm (Gr. *peri*, around + *derma*, skin) Outer protective tissue in vascular plants that is produced by the cork cambium and functionally replaces epidermis when it is destroyed during secondary growth; the periderm includes the cork, cork cambium, and phelloderm.

peristalsis (Gr. *peristaltikos*, compressing around) In animals, a series of alternating contracting and relaxing muscle movements along the length of a tube such as the oviduct or alimentary canal that tend to force material such as an egg cell or food through the tube.

peroxisome A microbody that plays an important role in the breakdown of highly oxidative hydrogen peroxide by catalase.

petal A flower part, usually conspicuously colored; one of the units of the corolla.

petiole (L. *petiolus*, a little foot) The stalk of a leaf.

phagocyte (Gr. *phagein*, to eat + *kytos*, hollow vessel) Any cell that engulfs and devours microorganisms or other particles.

phagocytosis (Gr., cell-eating) Endocytosis of solid particle; the cell membrane folds inward around the particle (which may be another cell) and engulfs it to form a vacuole.

pharynx (Gr., gullet) In vertebrates, a muscular tube that connects the mouth cavity and the esophagus; it serves as the gateway to the digestive tract and to the trachea.

phenetics Also called numerical taxonomy, this taxonomic technique seeks to classify organisms based on overall morphological similarity and does not take into account historical relationships between organisms.

phenotype (Gr. *phainein,* to show + *typos,* stamp or print) The realized expression of the genotype; the physical appearance or functional expression of a trait.

pheromone (Gr. *pherein,* to carry + *hormonos,* exciting, stirring up) Chemical substance released by one organism that influences the behavior or physiological processes of another organism of the same species. Some pheromones serve as sex attractants, as trail markers, and as alarm signals.

phloem (Gr. *phloos,* bark) In vascular plants, a food-conducting tissue basically composed of sieve elements, various kinds of parenchyma cells, fibers, and sclereids.

phosphodiester bond The type of bond that links nucleotides in a nucleic acid; formed when the phosphate group of one nucleotide binds to the 3′ hydroxyl group of the sugar of another.

phospholipid Similar in structure to a fat, but having only two fatty acids attached to the glycerol backbone, with the third space linked to a phosphorylated molecule; contains a polar hydrophilic "head" end (phosphate group) and a nonpolar hydrophobic "tail" end (fatty acids).

photoperiodism (Gr. *photos,* light + *periodos,* a period) The tendency of biological reactions to respond to the duration and timing of day and night; a mechanism for measuring seasonal time.

photoreceptor (Gr. *photos,* light) A light-sensitive sensory cell.

photosystem An organized collection of chlorophyll and other pigment molecules embedded in the thylakoid of chloroplasts; traps photon energy and channels it as energetic electrons to the thylakoid membrane.

phototropism (Gr. *photos,* light + *trope,* turning) In plants, a growth response to a light stimulus.

phycologist (Gr. *phykos,* seaweed) The scientist who studies algae.

phylum, pl. **phyla** (Gr. *phylon,* race, tribe) A major category, between kingdom and class, of taxonomic classifications.

phytochrome (Gr. *phyton,* plant + *chroma,* color) A plant pigment that is associated with the absorption of light; photoreceptor for red to far-red light.

pigment (L. *pigmentum,* paint) A molecule that absorbs light.

pilus, pl. **pili** Extensions of a bacterial cell enabling it to transfer genetic materials from one individual to another or to adhere to substrates.

pinocytosis (Gr. *pinein,* to drink + *kytos,* hollow vessel + *osis,* condition) The process of fluid uptake by endocytosis in a cell.

pistil (L. *pistillum,* pestle) Central organ of flowers, typically consisting of ovary, style, and stigma; a pistil may consist of one or more fused carpels and is more technically and better known as the gynoecium.

pith The ground tissue occupying the center of the stem or root within the vascular cylinder.

placenta, pl. **placentae** (L., a flat cake) (1) In flowering plants, the part of the ovary wall to which the ovules or seeds are attached. (2) In mammals, a tissue formed in part from the inner lining of the uterus and in part from other membranes, through which the embryo (later the fetus) is nourished while in the uterus and wastes are carried away.

plankton (Gr. *planktos,* wandering) Free-floating, mostly microscopic, aquatic organisms.

plasma (Gr., form) The fluid of vertebrate blood; contains dissolved salts, metabolic wastes, hormones, and a variety of proteins, including antibodies and albumin; blood minus the blood cells.

plasma cell An antibody-producing cell resulting from the multiplication and differentiation of a B lymphocyte that has interacted with an antigen.

plasma membrane The membrane surrounding the cytoplasm of a cell; consists of a single phospholipid bilayer with embedded protein.

plasmid (Gr. *plasma,* a form or mold) A small fragment of extrachromosomal DNA, usually circular, that replicates independently of the main chromosome, although it may have been derived from it.

plasmodium (Gr. *plasma,* a form, mold + *eidos,* form) Stage in the life cycle of myxomycetes (plasmodial slime molds); a multinucleate mass of protoplasm surrounded by a membrane.

plastid (Gr. *plastos,* formed or molded) An organelle in the cells of photosynthetic eukaryotes that is the site of photosynthesis and, in plants and green algae, of starch storage.

platelet (Gr. dim. of *plattus,* flat) In mammals, a fragment of a white blood cell that circulates in the blood and functions in the formation of blood clots at sites of injury.

pleiotropy Condition in which an individual allele has more than one effect on production of the phenotype.

plumule The epicotyl of a plant with its two young leaves.

point mutation An alteration of one nucleotide in a chromosomal DNA molecule.

polar body Minute, nonfunctioning cell produced during the meiotic divisions leading to gamete formation in vertebrates.

pollen tube A tube formed after germination of the pollen grain; carries the male gametes into the ovule.

pollination The transfer of pollen from an anther to a stigma.

polyandry The condition in which a female mates with more than one male.

polyclonal antibody An antibody response in which an antigen elicits many different antibodies, each fitting a different portion of the antigen surface.

polygyny (Gr. *poly* many + *gyne,* woman, wife) A mating choice in which a male mates with more than one female.

polymer (Gr. *polus,* many + *meris,* part) A molecule composed of many similar or identical molecular subunits; starch is a polymer of glucose.

polymerase chain reaction (PCR) A process by which DNA polymerase is used to copy a sequence of interest repeatedly, making millions of copies of the same DNA.

polymorphism (Gr. *polys,* many + *morphe,* form) The presence in a population of more than one allele of a gene at a frequency greater than that of newly arising mutations.

polypeptide (Gr. *polys,* many + *peptein,* to digest) A molecule consisting of many joined amino acids; not usually as complex as a protein.

polyploidy Condition in which one or more entire sets of chromosomes is added to the diploid genome.

polysaccharide (Gr. *polys,* many + *sakcharon,* sugar, from Latin *sakara,* gravel, sugar) A carbohydrate composed of many monosaccaride sugar subunits linked together in a long chain; examples are glycogen, starch, and cellulose.

polyunsaturated fat A fat molecule in which there are at least two double bonds between adjacent carbons, in one or more of the fatty acid chains.

population (L. *populus,* the people) Any group of individuals, usually of a single species, occupying a given area at the same time.

posttranscriptional control A mechanism of control over gene expression that operates after the transcription of mRNA is complete.

potential energy Energy that is not being used, but could be; energy in a potentially usable form; often called "energy of position."

precapillary sphincter A ring of muscle that guards each capillary loop and that, when closed, blocks flow through the capillary.

primary endosperm nucleus In flowering plants, the result of the fusion of a sperm nucleus and the (usually) two polar nuclei.

primary growth In vascular plants, growth originating in the apical meristems of shoots and roots; results in an increase in length.

primary immune response The first response of an immune system to a foreign antigen. If the system is challenged again with the same antigen, the memory cells created during the primary response will respond more quickly.

primary induction Inductions between the three primary tissue types—ectoderm, mesoderm, and endoderm.

primary nondisjunction Failure of chromosomes to separate properly at meiosis I.

primary productivity The amount of energy produced by photosynthetic organisms in a community.

primary structure The specific amino acid sequence of a protein.

primary tissues Tissues that comprise the primary plant body.

primary transcript The initial mRNA molecule copied from a gene by RNA polymerase, containing a faithful copy of the entire gene, including introns as well as exons.

primary wall In plants, the wall layer deposited during the period of cell expansion.

primate Monkeys and apes (including humans).

primitive streak (L. *primus,* first) In the early embryos of birds, reptiles, and mammals, a dorsal, longitudinal strip of ectoderm and mesoderm that is equivalent to the blastopore in other forms.

prions Infectious proteinaceous particles.

procambium (L. *pro,* before + *cambiare,* to exchange) In vascular plants, a primary meristematic tissue that gives rise to primary vascular tissues.

prokaryote (Gr. *pro,* before + *karyon,* kernel) A bacterium; a cell lacking a membrane-bounded nucleus or membrane-bounded organelles.

promoter A specific nucleotide sequence to which RNA polymerase attaches to initiate transcription of mRNA from a gene.

prophase (Gr. *pro,* before + *phasis,* form) An early stage in nuclear division, characterized by the formation of a microtubule spindle along the future axis of division, the shortening and thickening of the chromosomes, and their movement toward the equator of the spindle (the "metaphase plate").

proprioceptor (L. *proprius,* one's own) In vertebrates, a sensory receptor that senses the body's position and movements.

prostaglandins (Gr. *prostas,* a porch or vestibule + *glans,* acorn) A group of modified fatty acids, that function as chemical messengers.

prostate gland (Gr. *prostas,* a porch or vestibule) In male mammals, a mass of glandular tissue at the base of the urethra that secretes an alkaline fluid that has a stimulating effect on the sperm as they are released.

protein (Gr. *proteios,* primary) A chain of amino acids joined by peptide bonds.

protein kinase An enzyme that adds phosphate groups to proteins, changing their activity.

proteomics The study of all proteins in an organism.

proton pump A protein channel in a membrane of the cell that expends energy to transport protons against a concentration gradient; involved in the chemiosmotic generation of ATP.

proto-oncogene A normal gene that promotes cell division, so called because mutations that cause these genes to become overexpressed convert them into oncogenes that produce excessive cellular proliferation—cancer.

protozoa (Gr. *protos,* first + *zoon,* animal) The traditional name given to heterotrophic protists.

pseudogene (Gr. *pseudos,* false + *genos,* birth) A copy of a gene that is not transcribed.

pseudopod, or **pseudopodium** (Gr. *pseudes,* false + *pous,* foot) A nonpermanent cytoplasmic extension of the cell body.

punctuated equilibrium A hypothesis of the mechanism of evolutionary change that proposes that long periods of little or no change are punctuated by periods of rapid evolution.

pupa (L., girl, doll) A developmental stage of some insects in which the organism is nonfeeding, immotile, and sometimes encapsulated or in a cocoon; the pupal stage occurs between the larval and adult phases.

purine (Gr. *purinos,* fiery, sparkling) The larger of the two general kinds of nucleotide base found in DNA and RNA; a nitrogenous base with a double-ring structure, such as adenine or guanine.

pyrimidine (alt. of pyridine, from G. *pyr,* fire) The smaller of two general kinds of nucleotide base found in DNA and RNA; a nitrogenous base with a single-ring structure, such as cytosine, thymine, or uracil.

Q

quaternary structure The structural level of a protein composed of more than one polypeptide chain, each of which has its own tertiary structure; the individual chains are called subunits.

R

radicle (L. *radicula,* root) The part of the plant embryo that develops into the root.

radioactivity The emission of nuclear particles and rays by unstable atoms as they decay into more stable forms.

radula (L., scraper) Rasping tongue found in most mollusks.

realized niche The actual niche occupied by an organism when all biotic and abiotic interactions are taken into account.

receptor-mediated endocytosis Process by which specific macromolecules are transported into eukaryotic cells at clathrin-coated pits, after binding to specific cell-surface receptors.

receptor protein A highly specific cell-surface receptor embedded in a cell membrane that responds only to a specific messenger molecule.

recessive An allele that is only expressed when present in the homozygous condition, while being "hidden" by the expression of a dominant allele in the heterozygous condition.

reciprocal altruism Performance of an altruistic act with the expectation that the favor will be returned. A key and very controversial assumption of many theories dealing with the evolution of social behavior. *See* altruism.

reciprocal recombination A mechanism of genetic recombination that occurs only in eukaryotic organisms, in which two chromosomes trade segments; can occur between nonhomologous chromosomes as well as the more usual exchange between homologous chromosomes in meiosis.

recombinant DNA Fragments of DNA from two different species, such as a bacterium and mammal, spliced together in the laboratory into a single molecule.

reduction (L. *reductio,* a bringing back; originally "bringing back" a metal from its oxide) The gain of an electron by an atom often with an associated proton.

reflex (L. *reflectere,* to bend back) In the nervous system, a motor response subject to little associative modification; among the simplest neural pathways, involving only a sensory neuron, sometimes (but not always) an interneuron, and one or more motor neurons.

reflex arc The nerve path in the body that leads from stimulus to reflex action.

refractory period The recovery period after membrane depolarization during which the membrane is unable to respond to additional stimulation.

replication fork A Y-shaped end of a growing replication bubble in a DNA molecule undergoing replication.

repolarization Return of the ions in a nerve to their resting potential distribution following depolarization.

repressor (L. *reprimere,* to press back, keep back) A protein that regulates DNA transcription by preventing RNA polymerase from attaching to the promoter and transcribing the structural gene. *See* operator.

residual volume The amount of air remaining in the lungs after the maximum amount of air has been exhaled.

resting membrane potential The charge difference (difference in electric potential) that exists across a neuron at rest (about 70 millivolts).

restriction endonuclease An enzyme that cleaves a DNA duplex molecule at a particular base sequence, usually within or near a palindromic sequence; also called restriction enzyme.

retina (L., a small net) The photosensitive layer of the vertebrate eye; contains several layers of neurons and light receptors (rods and cones); receives the image formed by the lens and transmits it to the brain via the optic nerve.

retrovirus (L. *retro,* turning back) An RNA virus. When a retrovirus enters a cell, a viral enzyme (reverse transcriptase) transcribes viral RNA into duplex DNA, which the cell's machinery then replicates and transcribes as if it were its own.

RFLP (restriction fragment length polymorphism) Restriction enzymes recognize very specific DNA sequences. Alleles of the same gene or surrounding sequences may have base-pair differences, so that DNA near one allele is cut into a different length fragment than DNA near the other allele. These different fragments separate based on size on gels.

Rh blood group A set of cell surface markers (antigens) on the surface of red blood cells in humans and rhesus monkeys (for which it is named); although there are several alleles, they are grouped into two main types called Rh-positive and Rh-negative.

rhizome (Gr. *rhizoma,* mass of roots) In vascular plants, a usually more or less horizontal underground stem; may be enlarged for storage or may function in vegetative reproduction.

ribonucleic acid (RNA) A class of nucleic acids characterized by the presence of the sugar ribose and the pyrimidine uracil; includes mRNA, tRNA, and rRNA.

ribosomal RNA (rRNA) A class of RNA molecules found, together with characteristic proteins, in ribosomes; transcribed from the DNA of the nucleolus.

ribosome The molecular machine that carries out protein synthesis; the most complicated aggregation of proteins in a cell, also containing three different rRNA molecules.

ribozyme An RNA molecule that can behave as an enzyme, sometimes catalyzing its own assembly; rRNA also acts as a ribozyme in the polymerization of amino acids to form protein.

RNA polymerase An enzyme that catalyzes the assembly of an mRNA molecule, the sequence of which is complementary to a DNA molecule used as a template. *See* transcription.

RNA primer In DNA replication, a sequence of about 10 RNA nucleotides complementary to unwound DNA that attaches at a replication fork; the DNA polymerase uses the RNA primer as a starting point for addition of DNA nucleotides to form the new DNA strand; the RNA primer is later removed and replaced by DNA nucleotides.

RNA splicing A nuclear process by which intron sequences of a primary mRNA transcript are cut out and the exon sequences are spliced together to give the correct linkages of genetic information that will be used in protein construction.

rod Light-sensitive nerve cell found in the vertebrate retina; sensitive to very dim light; responsible for "night vision."

root The usually descending axis of a plant, normally below ground, which anchors the plant and serves as the major point of entry for water and minerals.

rumen An "extra stomach" in cows and related mammals wherein digestion of cellulose occurs and from which partially digested material can be ejected back into the mouth.

S

saltatory conduction A very fast form of nerve impulse conduction in which the impulses leap from node to node over insulated portions.

saprobes Heterotrophic organisms that digest their food externally (such as most fungi).

sarcolemma The specialized cell membrane in a muscle cell.

sarcoma A cancerous tumor arising from cells of connective tissue, bone, or muscle.

sarcomere (Gr. *sarx,* flexh + *meris,* part of) Fundamental unit of contraction in skeletal muscle; repeating bands of actin and myosin that appear between two Z lines.

sarcoplasmic reticulum (Gr. *sarx,* flesh + *plassein,* to form, mold; L. *reticulum,* network) The endoplasmic reticulum of a muscle cell. A sleeve of membrane that wraps around each myofilament.

satellite DNA A nontranscribed region of the chromosome with a distinctive base composition; a short nucleotide sequence repeated tandemly many thousands of times.

saturated fat A fat composed of fatty acids in which all the internal carbon atoms contain the maximum possible number of hydrogen atoms.

Schwann cells The supporting cells associated with projecting axons, along with all the other nerve cells that make up the peripheral nervous system.

sclereid (Gr. *skleros,* hard) In vascular plants, a sclerenchyma cell with a thick, lignified, secondary wall having many pits; not elongate like a fiber.

sclerenchyma (Gr. *skleros,* hard + *en,* in + *chymein,* to pour) A type of tissue made up of sclerenchyma cells.

scrotum (L., bag) The pouch that contains the testes in most mammals.

scuttellum The modified cotyledon in cereal grains.

secondary cell wall In plants, the innermost layer of the cell wall. Secondary walls have a highly organized microfibrillar structure and are often impregnated with lignin.

secondary growth In vascular plants, an increase in stem and root diameter made possible by cell division of the lateral meristems.

secondary immune response The swifter response of the body the second time it is invaded by the same pathogen because of the presence of memory cells, which quickly become antibody-producing plasma cells.

secondary induction An induction between tissues that have already differentiated.

secondary structure In a protein, hydrogen-bonding interactions between CO and NH groups of the primary structure.

second law of thermodynamics A statement concerning the transformation of potential energy into heat; it says that disorder (entropy) is continually increasing in the universe as energy changes occur, so disorder is more likely than order.

second messenger A small molecule or ion that carries the message from a receptor on the target cell surface into the cytoplasm.

segregation The process by which alternative forms of traits are expressed in offspring rather than blending each trait of the parents in the offspring.

selection The process by which some organisms leave more offspring than competing ones, and their genetic traits tend to appear in greater proportions among members of succeeding generations than the traits of those individuals that leave fewer offspring.

selective permeability Condition in which a membrane is permeable to some substances but not to others.

self-fertilization The union of egg and sperm produced by a single hermaphroditic organism.

semen (L., seed) In reptiles and mammals, sperm-bearing fluid expelled from the penis during male orgasm.

semicircular canal Any of three fluid-filled canals in the inner ear that help to maintain balance.

semiconservative replication DNA replication, in which each strand of the original duplex serves as the template for construction of a totally new complementary strand, so the original duplex is partially conserved in each of the two new DNA molecules.

senescent Aged, or in the process of aging.

sensory neuron A neuron that transmits nerve impulses from a sensory receptor to the central nervous system or central ganglion; an afferent neuron.

sepal (L. *sepalum,* a covering) A member of the outermost floral whorl of a flowering plant.

septum, pl. **septa** (L., fence) A wall between two cavities.

seta, pl. **setae** (L., bristle) In an annelid, bristles of chitin that help to anchor the worm during locomotion or when it is in its burrow.

sex-linked A trait determined by a gene carried on the X chromosome and absent on the Y chromosome.

sexual reproduction The process of producing offspring through an alternation of fertilization (producing diploid cells) and meiotic reduction in chromosome number (producing haploid cells).

sexual selection A type of differential reproduction that results from variable success in obtaining mates.

shoot In vascular plants, the aboveground portions, such as the stem and leaves.

sieve cell In the phloem of vascular plants, a long, slender sieve element with relatively unspecialized sieve areas and with tapering end walls that lack sieve plates.

sinoatrial (SA) node *See* pacemaker.

sinus (L., curve) A cavity or space in tissues or in bone.

sister chromatid One of two identical copies of each chromosome, still linked at the centromere, produced as the chromosomes duplicate for mitotic division; similarly, one of two identical copies of each homologous chromosome present in a tetrad at meiosis.

sodium-potassium pump Transmembrane channels engaged in the active (ATP-driven) transport of sodium ions, exchanging them for potassium ions, where both ions are being moved against their respective concentration gradients; maintains the resting membrane potential of neurons and other cells.

solute A molecule dissolved in some solution; as a general rule, solutes dissolve only in solutions of similar polarity; for example, glucose (polar) dissolves in (forms hydrogen bonds with) water (also polar), but not in vegetable oil (nonpolar).

solvent The medium in which one or more solutes is dissolved.

somatic cell Any of the cells of a multicellular organism except those that are destined to form gametes (germline cells).

somatic mutation A change in genetic information (mutation) occurring in one of the somatic cells of a multicellular organism, not passed from one generation to the next.

somatic nervous system (Gr. *soma,* body) In vertebrates, the neurons of the peripheral nervous system that control skeletal muscle.

somite (Gr, *soma,* body) One of the blocks, or segments, of tissue into which the mesoderm is divided during differentiation of the vertebrate embryo.

Southern blot A procedure used for identifying a specific gene, in which DNA from the source being tested is cut into fragments with restriction enzymes and separated by gel electrophoresis, then blotted onto a sheet of nitorocellulose and probed with purified, labeled, single-stranded DNA corresponding to a specific gene; if the DNA matching the specific probe is present in the source DNA, it is visible as a band of radioactive label on the sheet.

species, pl. **species** (L., kind, sort) A kind of organism; species are designated by binomial names written in italics.

spectrin A scaffold of proteins that links plasma membrane proteins to actin filaments in the cytoplasm of red blood cells, producing their characteristic biconcave shape.

sperm (Gr. *sperma,* seed) A mature male gamete, usually motile and smaller than the female gamete.

spermatid (Gr. *sperma,* seed) In animals, each of four haploid ($1n$) cells that result from the meiotic divisions of a spermatocyte; each spermatid differentiates into a sperm cell.

spermatozoa The male gamete, usually smaller than the female gamete, and usually motile.

sphincter (Gr. *sphinkter,* band, from *sphingein,* to bind tight) In vertebrate animals, a ring-shaped muscle capable of closing a tubular opening by constriction (such as the one between stomach and small intestine, or between anus and exterior).

spindle apparatus The assembly that carries out the separation of chromosomes during cell division; composed of microtubules (spindle fibers) and assembled during prophase at the equator of the dividing cell.

spiracle (L. *spiraculum,* from *spirare,* to breathe) External opening of a trachea in arthropods.

spongy parenchyma A leaf tissue composed of loosely arranged, chloroplast-bearing cells. *See* palisade parenchyma.

sporangium, pl. **sporangia** (Gr. *spora,* seed + *angeion,* a vessel) A structure in which spores are produced.

spore A haploid reproductive cell, usually unicellular, capable of developing into an adult without fusion with another cell.

sporophyte (Gr. *spora*, seed + *phyton*, plant) The spore producing, diploid (2*n*) phase in the life cycle of a plant having alternation of generations.

stabilizing selection A form of selection in which selection acts to eliminate both extremes from a range of phenotypes.

stamen (L., thread) The organ of a flower that produces the pollen; usually consists of anther and filament; collectively, the stamens make up the androecium.

starch (Mid. Eng. *sterchen*, to stiffen) An insoluble polymer of glucose; the chief food storage substance of plants.

statocyst (Gr. *statos*, standing + *kystis*, sac) A sensory receptor sensitive to gravity and motion.

stele The central vascular cylinder of stems and roots.

stem cell A relatively undifferentiated cell in animal tissue that can divide to produce more differentiated tissue cells.

stereoscopic vision (Gr. *stereos*, solid + *opitkos*, pertaining to the eye) Ability to perceive a single, three-dimensional image from the simultaneous but slightly divergent two-dimensional images delivered to the brain by each eye.

stigma (Gr., mark, tattoo mark) (1) In angiosperm flowers, the region of a carpel that serves as a receptive surface for pollen grains. (2) Light-sensitive eyespot of some algae.

stipules Leaflike appendages that occur at the base of some flowering plant leaves or stems.

stolon (L. *stolo*, shoot) A stem that grows horizontally along the ground surface and may form adventitious roots, such as runners of the strawberry plant.

stoma, pl. stomata (Gr., mouth) In plants, a minute opening bordered by guard cells in the epidermis of leaves and stems; water passes out of a plant mainly through the stomata.

stratum corneum The outer layer of the epidermis of the skin of the vertebrate body.

striated muscle (L., from *striare*, to groove) Skeletal voluntary muscle and cardiac muscle.

stromatolite A fossilized mat of ancient bacteria formed as long as 2 or 3 billion years ago, in which the bacterial remains individually resemble some modern-day bacteria.

style (Gr. *stylos*, column) In flowers, the slender column of tissue that arises from the top of the ovary and through which the pollen tube grows.

substrate (L. *substratus*, strewn under) (1) The foundation to which an organism is attached. (2) A molecule upon which an enzyme acts.

succession In ecology, the slow, orderly progression of changes in community composition that takes place through time.

summation Repetitive activation of the motor neuron resulting in maximum sustained contraction of a muscle.

surface tension A tautness of the surface of a liquid, caused by the cohesion of the molecules of liquid. Water has an extremely high surface tension.

surface-to-volume ratio Relationship of the surface area of a structure, such as a cell, to the volume it contains.

swim bladder An organ encountered only in the bony fish that helps the fish regulate its buoyancy through increasing or decreasing the amount of gas in the bladder via the esophagus or a specialized network of capillaries.

symbiosis (Gr. *syn*, together with + *bios*, life) The condition in which two or more dissimilar organisms live together in close association; includes parasitism (harmful to one of the organisms), commensalism (beneficial to one, of no significance to the other), and mutualism (advantageous to both).

sympatric speciation The differentiation of populations within a common geographic area into species.

synapse (Gr. *synapsis*, a union) A junction between a neuron and another neuron or muscle cell; the two cells do not touch, the gap being bridged by neurotransmitter molecules.

synapsis (Gr., contact, union) The point-by-point alignment (pairing) of homologous chromosomes that occurs before the first meiotic division; crossing over takes place during synapsis.

synaptic cleft The space between two adjacent neurons.

synaptic vesicle A vesicle of a neurotransmitter produced by the axon terminal of a nerve. The filled vesicle migrates to the presynaptic membrane, fuses with it, and releases the neurotransmitter into the synaptic cleft.

synaptonemal complex A protein lattice that forms between two homologous chromosomes in prophase I of meiosis, holding the replicated chromosomes in precise register with each other so that base-pairs can form between nonsister chromatids for crossing over that is usually exact within a gene sequence.

syncytial blastoderm A structure composed of a single large cytoplasm containing about 4000 nuclei in embryonic development of insects such as *Drosophila*.

syngamy (Gr. *syn*, together with + *gamos*, marriage) The process by which two haploid cells (gametes) fuse to form a diploid zygote; fertilization.

synteny Extensive conserved regions of DNA among different species.

systolic pressure A measurement of how hard the heart is contracting. When measured during a blood pressure reading, ventricular systole (contraction) is what is being monitored.

T

tagma, pl. tagmata (Gr., arrangement, order, row) A compound body section of an arthropod resulting from embryonic fusion of two or more segments; for example, head, thorax, abdomen.

taxis, pl. taxes (Gr., arrangement) An orientation movement by a (usually) simple organism in response to an environmental stimulus.

T cell A type of lymphocyte involved in cell-mediated immunity and interactions with B cells; the "T" refers to the fact that T cells are produced in the thymus.

telencephalon (Gr. *telos*, end + *encephalon*, brain) The most anterior portion of the brain, including the cerebrum and associated structures.

telomere A specialized nontranscribed structure that caps each end of a chromosome.

tendon (Gr. *tendon*, stretch) A strap of cartilage that attaches muscle to bone.

tertiary structure The folded shape of a protein, produced by hydrophobic interactions with water, ionic and covalent bonding between side chains of different amino acids, and van der Waal's forces; may be changed by denaturation so that the protein becomes inactive.

testcross A mating between a phenotypically dominant individual of unknown genotype and a homozygous "tester," done to determine whether the phenotypically dominant individual is homozygous or heterozygous for the relevant gene.

testis, pl. testes (L., witness) In mammals, the sperm-producing organ.

tetanus Sustained forceful muscle contraction with no relaxation.

thalamus (Gr. *thalamos*, chamber) That part of the vertebrate forebrain just posterior to the cerebrum; governs the flow of information from all other parts of the nervous system to the cerebrum.

thermodynamics (Gr. *therme*, heat + *dynamis*, power) The study of transformations of energy, using heat as the most convenient form of measurement of energy.

thigmotropism In plants, unequal growth in some structure that comes about as a result of physical contact with an object.

threshold The minimum amount of stimulus required for a nerve to fire (depolarize).

thylakoid (Gr. *thylakos*, sac + *-oides*, like) A saclike membranous structure containing chlorophyll in cyanobacteria and the chloroplasts of eukaryotic organisms.

tight junction Region of actual fusion of cell membranes between two adjacent animal cells that prevents materials from leaking through the tissue.

tissue (L. *texere*, to weave) A group of similar cells organized into a structural and functional unit.

totipotent A cell that possesses the full genetic potential of the organism.

trachea, pl. tracheae (L., windpipe) A tube for breathing; in terrestrial vertebrates, the windpipe that carries air between the larynx and bronchi (which leads to the lungs); in insects and some other terrestrial arthropods, a system of chitin-lined air ducts.

transcription (L. *trans*, across + *scribere*, to write) The enzyme-catalyzed assembly of an RNA molecule complementary to a strand of DNA.

transcription factor One of a set of proteins required for RNA polymerase to bind to a eukaryotic promoter region, become stabilized, and begin the transcription process.

transfection The transformation of eukaryotic cells in culture.

transfer RNA (tRNA) (L. *trans*, across + *ferre*, to bear or carry) A class of small RNAs (about 80 nucleotides) with two functional sites; at one site an "activating enzyme" adds a specific amino acid, while the other site carries the nucleotide triplet (anticodon) specific for that amino acid.

translation (L. *trans*, across + *latus*, that which is carried) The assembly of a protein on the ribosomes, using mRNA to specify the order of amino acids.

translocation (L. *trans*, across + *locare*, to put or place) (1) In plants, the long-distance transport of soluble food molecules (mostly sucrose), which occurs primarily in the sieve tubes of phloem tissue. (2) In genetics, the interchange of chromosome segments between nonhomologous chromosomes.

transpiration (L. *trans*, across + *spirare*, to breathe) The loss of water vapor by plant parts; most transpiration occurs through the stomata.

transposition Type of genetic recombination in which transposable elements (transposons) move from one site in the DNA sequence to another, apparently randomly.

transposon (L. *transponere*, to change the position of) A DNA sequence capable of transposition.

triacylglyceride (triglyceride) An individual fat molecule, composed of a glycerol and three fatty acids.

triploid Possessing three sets of chromosomes.

trochophore A free-swimming larval stage unique to the mollusks and annelids.

trophic level (Gr. *trophos*, feeder) A step in the movement of energy through an ecosystem.

trophoblast (Gr. *trephein*, to nourish + *blastos*, germ) In vertebrate embryos, the outer ectodermal layer of the blastodermic vesicle; in mammals it is part of the chorion and attaches to the uterine wall.

tropism (Gr. *trope*, a turning) A response to an external stimulus.

tropomyosin (Gr. *tropos*, turn + *myos*, muscle) Low-molecular-weight protein surrounding the actin filaments of striated muscle.

troponin Complex of globular proteins positioned at intervals along the actin filament of skeletal muscle; thought to serve as a calcium-dependent "switch" in muscle contraction.

tubulin (L. *tubulus*, small tube + *in*, belonging to) Globular protein subunit forming the hollow cylinder of microtubules.

tumor-suppressor gene A gene that normally functions to inhibit cell division; mutated forms can lead to the unrestrained cell division of cancer, but only when both copies of the gene are mutant.

turgor pressure (L. *turgor*, a swelling) The pressure within a cell resulting from the movement of water into the cell; a cell with high turgor pressure is said to be turgid. *See* osmotic pressure.

U

unequal crossing over A process by which a crossover in a small region of misalignment at synapsis causes two homologous chromosomes to exchange segments of unequal length.

unsaturated fat A fat molecule in which one or more of the fatty acids contain fewer than the maximum number of hydrogens attached to their carbons.

urea (Gr. *ouron*, urine) An organic molecule formed in the vertebrate liver; the principal form of disposal of nitrogenous wastes by mammals.

urethra (Gr. from *ourein*, to urinate) The tube carrying urine from the bladder to the exterior of mammals.

uric acid Insoluble nitrogenous waste products produced largely by reptiles, birds, and insects.

urine (Gr. *ouron*, urine) The liquid waste filtered from the blood by the kidney and stored in the bladder pending elimination through the urethra.

uterus (L., womb) In mammals, a chamber in which the developing embryo is contained and nurtured during pregnancy.

V

vacuole A membrane-bounded sac in the cytoplasm of some cells, used for storage or digestion purposes in different kinds of cells; plant cells often contain a large central vacuole that stores water, proteins, and waste materials.

vascular cambium In vascular plants, a cylindrical sheath of meristematic cells, the division of which produces secondary phloem outwardly and secondary xylem inwardly; the activity of the vascular cambium increases stem or root diameter.

vascular tissue (L. *vasculum*, a small vessel) Containing or concerning vessels that conduct fluid.

vas deferens (L. *vas*, a vessel + *ferre*, to carry down) In mammals, the tube carrying sperm from the testes to the urethra.

vasopressin A posterior pituitary hormone that regulates the kidney's retention of water.

vegetal pole The hemisphere of the zygote, comprising cells rich in yolk.

vein (L. *vena*, a blood vessel) (1) In plants, a vascular bundle forming a part of the framework of the conducting and supporting tissue of a stem or leaf. (2) In animals, a blood vessel carrying blood from the tissues to the heart.

veliger The second larval stage of mollusks following the trochophore stage, during which the beginning of a foot, shell, and mantle can be seen.

ventricle A muscular chamber of the heart that receives blood from an atrium and pumps blood out to either the lungs or the body tissues.

vesicle (L. *vesicula*, a little bladder) A small intracellular, membrane-bounded sac in which various substances are transported or stored.

vessel element In vascular plants, a typically elongated cell, dead at maturity, which conducts water and solutes in the xylem.

vestibular apparatus The complicated sensory apparatus of the inner ear that provides for balance and orientation of the head in vertebrates.

villus, pl. villi (L., a tuft of hair) In vertebrates, one of the minute, fingerlike projections lining the small intestine that serve to increase the absorptive surface area of the intestine.

visceral mass (L., internal organs) Internal organs in the body cavity of an animal.

vitamin (L. *vita*, life + *amine*, of chemical origin) An organic substance that cannot be synthesized by a particular organism but is required in small amounts for normal metabolic function.

viviparity (L. *vivus*, alive + *parere*, to bring forth) Refers to reproduction in which eggs develop within the mother's body and young are born freeliving.

voltage-gated ion channel A transmembrane pathway for an ion that is opened or closed by a change in the voltage, or charge difference, across the cell membrane.

W

water potential The potential energy of water molecules. Regardless of the reason (e.g., gravity, pressure, concentration of solute particles) for the water potential, water moves from a region where water potential is greater to a region where water potential is lower.

wild type In genetics, the phenotype or genotype that is characteristic of the majority of individuals of a species in a natural environment.

X

xylem (Gr. *xylon*, wood) In vascular plants, a specialized tissue, composed primarily of elongate thick-walled conducting cells, which transports water and solutes through the plant body.

Y

yolk plug A plug occurring in the blastopore of amphibians during formation of the archenteron in embryological development.

Z

zona pellucida An outer membrane that encases a mammalian egg.

zoospore A motile spore.

zooxanthellae Symbiotic photosynthetic protists in the tissues of corals.

zygomycetes (Gr. *zygon*, yoke + *mykes*, fungus) A type of fungus whose chief characteristic is the production of sexual structures called zygosporangia, which result from the fusion of two of its simple reproductive organs.

zygote (Gr. *zygotos*, paired together) The diploid ($2n$) cell resulting from the fusion of male and female gametes (fertilization).

Credits

Line Art

Part Openers

PO 1.2: Data from Autumn, Kellar, Yiching A. Liang, S. Tonia Hsieh, Wolfgang Zesch, Wal Pang Chan, Thomas W. Kenny, Ronald Fearing, and Robert J. Full, "Adhesive Force of a Single Gecko Foot-hair," *Nature*, June 2000. **PO 2.2:** Data from Wymer, Carol L., Scott A. Wymer, Daniel J. Cosgrove, and Richard J. Cyr, "Plant Cell Growth Responds to External Forces and the Response Requires Intact Microtubules," *Plant Physiology*, 1996. **PO 3.2:** Data from Webber, Andrew N., Hui Su, Scott E. Bingham, Hanno Kass, Ludwig Krabben, Matthias Kuhn, Rafael Jordan, Eberhard Schlodder, and Wolfgang Lubitz, "Site-directed Mutations Affecting the Spectroscopic Characteristics and Midpoint Potential of the Primary Donor in Photosystem I," *Biochemistry*, 1996. American Chemical Society. **PO 4.2:** Data from Rosenzweig, R. Frank, R.R. Sharp, David S. Treves, and Julian Adams, "Microbial Evolution in a Simple Unstructured Environment: Genetic Differentiation in Escherichia coli," *Genetics*, August, 1994. **PO 5.2:**Data from Ryan, Heather, E., Jessica Lo and Randall, "HIF-1(alpha) is Required for Solid Tumor Formation and Embryonic Vascularization," *The EMBO Journal*, 1998. **PO 6.2:** Data from Endler, John A. "Natural Selection on Color Patterns in Poecilia reticulata," *Evolution*, 1980. **PO 7.2:** Data from Boyce, Mark S. and C.M. Perrins, "Optimizing Great Tit Clutch Size in a Fluctuating Environment," *Ecology*. **PO 8.2:** Data from Blaustein, Andrew R., Joseph M. Kiesecker, Douglas P. Chivers and Robert G. Anthony, "Ambient UV-B Radiation Causes Deformities in Amphibian Embryos," *Proc. Natl. Acad. Sci. USA*, December 1977. **PO 9.2:** Data from Kuo, G., Q.-L. Choo, H.J. Alter, G.L. Gitnick, A.G. Redeker, R.H. Purcell, T. Miyamura, J.L. Dienstag. M.J. Alter, C.E. Stevens, G.E. Tegtmeier, F. Bonino, M. Colombo, W.-S. Lee, C. Kuo, K. Berger, J.R. Shuster, L.R. Overby, D.W. Bradley, M. Houghton, "An Assay for Circulating Antibodies to a Major Etiologic Virus of Human Non-A., Non-B Hepatitis," *Science*, April 21, 1989. **PO 10.2:** Data from Martens, Scott N. and Robert S. Boyd, "The Ecological Significance of Nickel Hyperaccumulation: A Plant Chemical Defense," *Oecologia*, 1994. **PO 11.2:** Data from Berger, Fred, Chen-Yi Hung, Liam Dolan, and John Schiefelbein, "Control of Cell Division in the Root Epidermis of Arabidopsis thaliana," *Developmental Biology*,1998. **PO 12.2:** Data from Harrison, Jon F., Jennifer H. Fewell, Stephen P. Roberts, H. Glenn Hall, "Achievement of Thermal Stability by Varying Metabolic Heat Production in Flying Honeybees," *Science*, October 4, 1996. **PO 13.2:** Data from Owerkowicz, Tomasz, Colleen G. Farmer, James W. Hicks, and Elizabeth L. Brainerd, "Contribution of Gular Pumping to Lung Ventilation in Monitor Lizards," *Science*, June 4, 1999. **PO 14.2:** Data from Louis J. Guillette, Jr., Allan R. Woodward, D. Andrew Crain, Daniel B. Pickford, Andrew A. Rooney, and H. Franklin Percival, "Plasma Steroid Concentrations and Male Phallus Size in Juvenile Alligators from Seven Florida Lakes,"*General and Comparative Endocrinology*, 1999.

Chapter 1

Box 1.1: From Howard Neverov, "The Consent" in *The Collected Works of Howard Nemerov*, 7th edition, 1981. Reprinted by permission. **Box 1.2:** Reprinted by permission of Dr. Stephen Jay Gould, Harvard University. **Fig. 1.6:** From Cleveland P. Hickman, Jr. et al., *Integrated Principles of Zoology*, 10th Edition. Copyright © 1996 McGraw-Hill Companies, Inc., Dubuque, Iowa. All Rights Reserved.

Chaper 4

Fig. 4.15: From Pennisi, "Genome Data Shake Tree of Life," *Science*, May, 1998. Copyright © 1998 American Association for the Advancement of Science.

Chapter 5

Fig. 5.2: From Alberts, et al., *Essentials of Cell Biology*, 1997. Copyright © Garland Publishing, New York, NY. Reprinted by permission.

Chapter 6

Fig. 6.10: Modified from Alberts, et al., *Molecular Biology of the Cell*, 3rd Edition. Copyright © 1994 Garland Publishing, New York, NY.

Chapter 9

Fig. 9.23: Modified from Fox, *Human Physiology*, 5th edition. Copyright © 1995 McGraw-Hill Companies, Inc., Dubuque, Iowa. All Rights Reserved.

Chapter 10

Fig. 10.13: From *Plant Physiology* by Lincoln Taiz and Eduardo Zeiger. Copyright ©1991 by The Benjamin/Cummings Company, Inc. Reprinted by permission of Addision Wesley Longman Publishers. **Fig. 10.7:** From Raven, et al., *Biology of Plants*, 5th Edition. Reprinted by permission of Worth Publishers.

Chapter 11

Fig. 11.16: From Alberts, et al., *Essentials of Cell Biology*, 1997. Copyright © Garland Publishing, New York, NY. Reprinted by permission.

Chapter 12

Fig. 12.8: From Alberts, et al., *Molecular Biology of the Cell*, 3rd Edition. Copyright © 1994 Garland Publishing, New York, NY.

Chapter 16

Fig. 16.20: From A.J.F. Griffiths et al., *Genetic Analysis*, 5th edition. Copyright © Freeman, 1993.

Chapter 17

Fig. 17.19: Source: Modified from John Kochik for Howard Hughes Medical Institute. **Fig. 17.5:** Copyright © John Kochik for Howard Hughes Medical Institute. **Text 17.24:** From H. Robert Horvitz, *From Egg to Adult*, published by Howard Hughes Medical Institute. Copyright © 1992. Reprinted by permission. **Fig. 17.24**(top): M.E. Challinor illustration. From Howard Hughes Medical Institute as published in *From Egg to Adult*, © 1992. Reprinted by permission. **Fig. 17.24**(bottom): Illustration by: The studio of Wood Ronsaville Harlin, Inc.

Chapter 21

Fig. 21.7: From Cleveland P. Hickman, Jr., et al., *Integrated Principles of Zoology*, 10th edition. Copyright © 1997 McGraw-Hill Companies, Inc., Dubuque, Iowa. All Rights Reserved. Reprinted by permission. **Fig. 21.13:** Modified from Moore et al., *Botany*, 2nd edition. Copyright © 1997 McGraw-Hill Companies, Inc., Dubuque, Iowa. All Rights Reserved. **Table 21.2:** Based on suggestions by Dr. Leslie Dendy, Dept. of Science & Technology, University of New Mexico, Los Alamos.

Chapter 22

Fig. 22.13: Redrawn from Fryer & Iles. **Fig. 22.16:** From Cleveland P. Hickman, Jr., et al., *Integrated Principles of Zoology*, 10th edition. Copyright © 1997 McGraw-Hill Companies, Inc., Dubuque, Iowa. All Rights Reserved.

Chapter 24

Fig. publication 24.14: Reprinted with permission from P.H. Harvey and R.M. Zammuto in *Nature*, 1985, 315, pp. 319-320. Copyright © 1985 Macmillan Magazines Limited.

Chapter 26

Fig. 26.11: From John Alcock, *Animal Behavior*, 4th edition. Copyright © 1989 Sinauer Associates, Inc. Reprinted by permission. **Fig. 26.16:** From John Alcock, *Animal Behavior*, 1st edition. Copyright © 1975 Sinauer Associates, Inc. Reprinted by permission. **Fig. 26.20:** From John Alcock, *Journal of Animal Behavior*, 1988. Reprinted by permission of Academic Press, Ltd., London. **Fig. 26.22:** From Cleveland P. Hickman, Jr., et al., *Integrated Principles of Zoology*, 10th edition. Copyright © 1997 McGraw-Hill Companies, Inc., Dubuque, Iowa. All Rights Reserved.

Chapter 27

Fig. 27.10 b: Source: Data from H.L. Gibbs et al., "Realized Reproductive Excess of Polygynous Red-Winged Blackbirds Revealed by DNA Markers," *Science*, 1990.

Chapter 29

Fig. 29.16: © 1997 The Economist Newspaper Group, Inc. Reprinted by permission. Further reproduction prohibited.

Chapter 31

Fig. 31.9: After Green and Sussman, 1990. **Fig. 31.10:** Data assembled by Pima, 1991. **Fig. 31.25:** Modified from Nature and Resources, Vol. XXII, 1986. Copyright © The UNESCO Press, Paris, France. Reprinted by permission. **Fig. 31.26:** From "Quetzal & the Macaw," *The Story of Costa Rica's National Parks*" by David Rains Wallace. Copyright © 1992. Reprinted by permission of Sierra Club Books.

Chapter 40

Fig. 40.3: From Ralph Quantrano, Washington University. **Fig. 40.16:** Modified from Fincher, G.B. "Molecular and Cellular Biology Associated with Endosperm Mobilization in Germinating Cereal Grains," with permission, from the *Annual Review of Plant Physiology and Plant Molecular Biology*, Volume 40. © 1989 by Annual Reviews. www.AnnualReviews.org.

Chapter 41

Fig. 41.29: After McDaniel, 1996. **Fig. 41.30:** After McDaniel, 1996. **Fig. 41.31:** After McDaniel, 1996.

Chapter 43

Fig. 43.7: Source: Modified from Kehoe Villlard, & Sommerville, "DNA Microarrays for Studies of Photosynthetic Organisms," *Trends in Plant Science*, 1999. **Fig. 43.8:** Reprinted from *Trends in Biotechnology*, Vol. 13, Copyright © 1995, page 181, with permission from Elsevier Science. **Fig. 43.19:** From *Plant Biochemistry and Molecular Biology* by P.J. Lea and R.C. Leegoods, eds. Copyright © 1993 John Wiley & Sons, Limited. Reproduced with permission.

Chapter 44

Fig. 44.3: From Cleveland P. Hickman, Jr., et al., *Integrated Principles of Zoology*, 10th edition. Copyright © 1997 McGraw-Hill Companies, Inc., Dubuque, Iowa. All Rights Reserved. **Fig. 44.17:** From Cleveland P. Hickman, Jr., et al., *Integrated Principles of Zoology*, 10th edition. Copyright © 1997 McGraw-Hill Companies, Inc., Dubuque, Iowa. All Rights Reserved. **Fig. 44.22:** a) After L. H. Hyman, *The Invertebrates*, 1940. b) After Adoutte, et al., *Proceedings of the Academy of Sciences*, April 2000.

Chapter 45

Fig. 45.16: From Cleveland P. Hickman, Jr., et al., *Integrated Principles of Zoology*, 11th edition. Copyright © 2000 McGraw-Hill Companies, Inc., Dubuque, Iowa. All Rights Reserved. **Fig. 45.18:** From Cleveland P. Hickman, Jr., et al., *Integrated Principles of Zoology*, 11th edition. Copyright © 2000 McGraw-Hill Companies, Inc., Dubuque, Iowa. All Rights Reserved.

Chapter 46

Fig. 46.6: From Cleveland P. Hickman, Jr., et al., *Integrated Principles of Zoology*, 10th edition. Copyright © 1997 McGraw-Hill Companies, Inc., Dubuque, Iowa. All Rights Reserved. **Fig. 46.28:** From Cleveland P. Hickman, Jr., et al., *Integrated Principles of Zoology*, 9th edition. Copyright © 1993 McGraw-Hill Companies, Inc., Dubuque, Iowa. All Rights Reserved.

Chapter 47

Fig. 47.5: From Cleveland P. Hickman, Jr., et al., *Integrated Principles of Zoology*, 10th edition. Copyright © 1997 McGraw-Hill Companies, Inc., Dubuque, Iowa. All Rights Reserved.

Chapter 48

Fig. 48.22: From Cleveland P. Hickman, Jr., et al., *Integrated Principles of Zoology*, 10th edition. Copyright © 1997 McGraw-Hill Companies, Inc., Dubuque, Iowa. All Rights Reserved. **Fig. 48.27:** From Cleveland P. Hickman, Jr., et al., *Integrated Principles of Zoology*, 10th edition. Copyright © 1997 McGraw-Hill Companies, Inc., Dubuque, Iowa. All Rights Reserved.

Chapter 50

Fig. 50.20: From Cleveland P. Hickman, Jr., et al., *Integrated Principles of Zoology*, 11th edition. Copyright © 2000 McGraw-Hill Companies, Inc., Dubuque, Iowa. All Rights Reserved. **Fig. 50.22:** New York Times Graphics. Reprinted by Permission.

Chapter 51

Fig. 51.18: From Cleveland P. Hickman, Jr., et al., *Integrated Principles of Zoology*, 10th edition. Copyright © 1997 McGraw-Hill Companies, Inc., Dubuque, Iowa. All Rights Reserved.

Chapter 57

Fig. 57.24: From Beck & Habicht, "Immunity and the Invertebrates" in *Scientific American*, November 1996. Reprinted by permission of Roberto Osti Illustrations.

Chapter 58

Fig. 58.13: From Cleveland P. Hickman, Jr., et al., *Integrated Principles of Zoology*, 11th edition. Copyright © 2000 McGraw-Hill Companies, Inc., Dubuque, Iowa. All Rights Reserved.

Chapter 60

Fig. 60.15: From Cleveland P. Hickman, Jr., et al., *Integrated Principles of Zoology*, 11th edition. Copyright © 2000 McGraw-Hill Companies, Inc., Dubuque, Iowa. All Rights Reserved.

Photo Credits

Part Openers

I(page 1): © Mark Mofett/Minden Pictures; **I(page 2):** Courtesy of Kellar Autumn & Ed Florance; **I(page 2 inset):** Courtesy of Kellar Autumn & Ed Florance; **II:** Courtesy of Richard Cyr; **IV:** © David M. Phillips/Visuals Unlimited; **VI(right):** Courtesy of H. Rodd; **VII:** © Hal. H. Harrison/The National Audubon Society Collection/Photo Researchers; **VIII:** © William Leonard/DRK Photo; **IX:** Courtesy of Alfred M. Prince, MD; **PO.X:** Courtesy of Robert Boyd, Dept. of Biological Sciences, Auburn University; **XI:** Courtesy of Elliot Meyerowitz; **XII:** © Stephen Dalton/Photo Researchers; **XIII:** © Zig Leszczynski/Animals Animals; **XIV:** © Howard K. Suzuki

Chapter 1

1.1: © Christopher Ralling; **1.A:** © N.H. (Dan) Cheatham/National Audubon Society Collection/Photo Researchers; **1.2**(top left): © Lennart Nilsson; **1.2**(left upper middle): From C.P. Morgan & R.A. Jersid, Anatomical Record, 166:575-586, 1970 © John Wiley & Sons; **1.2**(top center): © Tom J. Ulrich/Visuals Unlimited; **1.2**(middle bottom): © Ed Reschke; **1.2**(top right): © Robert & Jean Pollock; **1.2**(right upper middle): © John D. Cunningham/Visuals Unlimited; **1.2**(right lower middle left): © Arthur Morris/Visuals Unlimited;**1.2** (right lower middle right): © Alan Nelson/Animals Animals/Earth Scenes; **1.2(**bottom right): © Maslowski/ Visuals Unlimited; **1.5:** Mohr/ Darwin Collection; Herbert Rose Barraud, photographer. Reproduced by permission of the Huntington Library, San Marino, California; **1.7:** Courtesy of James Moore, from *Darwin*, 1992 Warner Books; **1.11:** © Mary Evans Picture Library/Photo Researchers Inc.; **Bx1.2**(left): © E.R. Degginger/Animals Animals/Earth Scenes; **Bx1.2**(right): Courtesy of Stephen Jay Gould

Chapter 2

2.1: © Irving Geis/Photo Researchers; **2.10B:** © Archive Photos; **2.11**(left): © Jurgen Schmitt/The Image Bank; **2.11**(center): © Nicholas Foster/The Image Bank; **2.11**(right): © John Kelly/The Image Bank; **2.14:** © Hermann Eisenbeiss/National Audubon Society Collection/Photo Researchers

Chapter 3

3.1: © Science Photo Library/Photo Researchers; **3.4**(top left): © Michael Pasdzior/The Image Bank; **3.4**(top right): © Manfred Kage/Peter Arnold, Inc.; **3.4:** (bottom left): © George Bernard/Animals Animals/Earth Scenes; **3.4**(bottom center): © OSF/Animals Animals/Earth Scenes; **3.4**(bottom right): © Scott Blackman/Tom Stack & Associates; **3.11A:** Courtesy of Lawrence Berkeley National Laboratory; **3.12:** © Driscoll, Youngquist, & Baldeschwieler, Caltech/SPPL/Photo Researchers, Inc.; **3.27A:** © J.D. Litvay/Visuals Unlimited; **3.28:** © Scott Johnson/Animals Animals/Earth Scenes

Chapter 4

4.1: © Visuals Unlimited; **4.2**(left): © Edward S. Ross; **4.3**(right): © Y. Arthur Bertrand/Peter Arnold, Inc.; **4.4:** © T.E. Adams/Visuals Unlimited; **4.5:** © Bob McKeever/Tom Stack & Associates; **4.9:** Courtesy of J.William Schopf, UCLA; **4.10(both):** Reproduced with permission from Dr. J. William Schopf, 1992-93, CSEOL/UCLA; **4.11:** © Dwight R. Kuhn; **4.12:** Andrew H. Knoll, Harvard University; **4.16:** NASA

Chapter 5

5.1: © Manfred Kage/Peter Arnold Inc.; **5.3**(top left): © John Walsh/Photo Researchers Inc.;

5.3(top right): © David M. Phillips/Visuals Unlimited; **5.3**(below): © David M. Phillips/Visuals Unlimited; **5.6A:** © I. M. Pope/Tom Stack & Associates; **5.7:** Courtesy of E.H. Newcomb & T.D. Pugh, U. of Wisconsin; **5.8B:** © K.G. Murti/Visuals Unlimited; **5.9B:** © Biophoto Associates/Photo Researchers; **5.10B:** Courtesy of Dr. Thomas Tillack ; **5.10C:** Photo J. David Robertson, from Charles Flickinger, Medical Cellular Biology, W.B. Saunders, 1979; **5.12:** © Ed Reschke; **5.13:** © R. Bolender & D. Fawcett/Visuals Unlimited; **5.15B:** Courtesy of Dr. Charles Flickinger, Medical Cellular Biology, W.B. Saunders, 1979; **5.18:** Courtesy of E.H. Newcomb & S.E. Frederick, U. of Wisconsin; **5.20:** From C.P. Morgan & R.A. Jersid, Anatomical Record 166:575-586 © John Wiley & Sons; **5.21B:** © Don W. Fawcett/Visuals Unlimited; **5.22:** Courtesy of Dr. Kenneth Miller, Brown University; **5.23A:** © Biophoto Associates/Photo Researchers; **F5.26:** Courtesy of Dr. James A. Spudich; **5.27B:** Courtesy of William Dentler; **5.27C:** Courtesy of William Dentler; **5.27D:** © Stanley Flegler/Visuals Unlimited; **5.28:** © Biophoto Associates/Photo Researchers; **5.29B:** © Biophoto Associates/Photo Researchers

Chapter 6

6.1: © K.R. Porter/Photo Researchers; **6.6**(left): © Keith Porter/Photo Researchers; **6.7**(right): © W. Omerod/Visuals Unlimited; **6.15**(left): © David M. Phillips/Visuals Unlimited; **6.15**(center): © David M. Phillips/Visuals Unlimited; **6.15**(right): © David M. Phillips/Visuals Unlimited; **6.18**(all): Courtesy of M.M. Perry & A.B. Gilbert, *Cell Science*, 39-257, 1979; **6.19B:** Courtesy of Dr. Birgit H. Satir

Chapter 7

7.1: © Cabisco/Visuals Unlimited; **7.18A:** © Don Fawcett/Visuals Unlimited

Chapter 8

8.1: © Robert A. Caputo/Aurora; **8.5**(both): © Spencer Grant/Photo Edit; **8.10B:** Courtesy of Dr. Lester J. Reed, University of Texas, Austin

Chapter 9

9.1: © Jane Buron/Bruce Coleman Inc.; **9.2:** © Steve Kaufman/Peter Arnold, Inc.; **9.10:** © Grant Heilman/Grant Heilman Photography; **9.A**(all): © Edward S. Ross

Chapter 10

10.1: © Grant Heilman/Grant Heilman Photography; **10.2**(left): © Manfred Kage/Peter Arnold; **10.2C:** Courtesy of Dr. Kenneth Miller, Brown University; **10.3A:** Courtesy of Hunt Institute for Botanical Documentation, Carnegie Mellon University, Pittsburgh, PA; **10.7:** © Eric V. Grave/Photo Researchers Inc.; **10.8A:** © Eric Soder/Tom Stack & Associates; **10.8C:** © Eric Soder/Tom Stack & Associates; **10.22A:** © Joseph Nettis/ National Audubon Society Collection/Photo Researchers; **10.22B:** © Clyde H. Smith/Peter Arnold

Chapter 11

11.1: © George Musil/Visuals Unlimited; **11.2:** © Lee D. Simon/Photo Researchers Inc.; **11.4:** © Biophoto Associates/ Photo Researchers Inc.; **11.6:** © Science Photo Library/Photo Researchers Inc.; **11.10:** © Dr. Andrew S. Bajer; **11.11**(all): © Dr. Andrew S. Bajer; **11.12**(both): Dr. Jeremy Pickett-Heaps; **11.13A:** © David M. Phillips/Visuals Unlimited; **11.13B:** © Dr. Guenter Albrecht-Buehler; **11.14A:** © B.A. Palevits & E.H. Newcomb/BPS/Tom Stack & Associates

Chapter 12

12.1: © L. Maziarski/Visuals Unlimited; **12.7:** Diter von Wettstein; Reproduced with permission from *Annual Review of Genetics*, v. 6, 1972, by Annual Review, Inc.; **12.9:** © B. John Cabisco/Visuals Unlimited; **12.13**(all): © C.A. Hasenkampf/ Biological Photo Service; **12.14**(right): Courtesy of Prof. William F. Morgan; **12.14**(left): Courtesy of Sheldon Wolff & Judy Bodycote

Chapter 13

13.1: © Alan Carey/The Image Works; **13.2:** George Johnson; **13.3:** © Visuals Unlimited; **13.4:** © Richard Gross/Biological Photography; **13.5,13.6,13.10:** Courtesy of V. Orel, Mendelianum Musei Moraviae, Brno; **13.11:** Courtesy of R.W. Van Norman; **13.17A:** From Albert & Blakeslee Corn and Man *Journal of Heredity*, Vol. 5, pg. 511, 1914, Oxford University Press; **13.19**(both): © Fred Bruemmer; **13.21A:** © Richard Hutchings/Photo Researchers; **13.21B:** © Cheryl A. Ertelt/ Visuals Unlimited; **13.21C:** © William H. Mullins/Photo Researchers; **13.21D:** Gerard Lacz/Peter Arnold Inc.; **13.25:** © Corbis-Bettmann; **13.27:** © SPL/Photo Researchers Inc; **13.29**(left): © Cabisco/Phototake; **13.29**(right): © Cabisco/ Phototake; **13.35:** © Leonard Lessin/Peter Arnold; **13.37:** © Hans Reinhard/Okapia/Photo Researchers; **13.38**(left): Courtesy of Loris McGavaran, Denver Children's Hospital; **13.38B:** © Richard Hutchings/Photo Researchers Inc.

Chapter 14

14.9A: © Cold Spring Harbor Laboratory; **14.10B:** From *The Double Helix* by J.D. Watson, Atheneum Press, N.Y. 1968; **14.10B:** © Barrington Brown/Photo Researchers Inc.; **14.11:** From M. Meselson and F.W. Stahl/ *Proceedings of the Nat. Acad. of Sci.* 44(1958):671; **14.15B,C:** From: *Biochemistry 4/e* by Stryer © 1995 by Lubert Stryer. Used with permission of W.H. Freeman and Company; **14.18:** © Prof. Ulrich Laemmli/Photo Researchers Inc.; **14.19A:** Courtesy of Dr. David Wolstenholme

Chapter 15

15.1: © K.G. Murti/Visuals Unlimited; **15.3:** N. Ban, P. Nissen, J. Hansen, P.B. Moore & T.A. Steitz, "The Complete Atomic Structure of the Large Ribosomal Subunit at 2.4A Resolution," reprinted with permission from *Science*, v. 289 #5481, p917, © 2000 American Association for the Advancement of Science; **15.7:** From R.C. Williams; *Proc. Nat. Acad. of Sci.* 74(1977):2313; **15.10:** Courtesy of Dr. Oscar L. Miller; **15.15B:** Courtesy of Dr. Bert O'Malley, Baylor College of Medicine

Chapter 16

16.1: Courtesy of Dr. Claus Pelling; **16.15**(both): Courtesy of Dr. Harrison Echols; **16.17A:** Courtesy of Dr. Victoria Foe

Chapter 17

17.1: © Cabisco/Visuals Unlimited; **17.3:** Photo Lennart Nilsson/ Albert Bonniers Forlag AB, A Child is Born, Dell Publishing Company; **17.4**(all): © Cabisco/Phototake; **17.6:** © Ed Lewis; **17.14A:** Dr. Christiane Nusslein-Volhard/Max Planck Institute as published in *From Egg to Adult*, © 1992 HHMI; **17.14B,C,D:** James Langeland, Stephen Paddock & Sean Carroll as published in *From Egg to Adult*, © 1992 HHMI; **17.15**(all): Courtesy of Manfred Frasch; **17.16:** Courtesy of E.B. Lewis; **17.20**(both): William McGinnis/Yale University; **17.22A:** © A. Tsiaras/Photo Researchers

Chapter 18

18.1: © David M. Phillips/Visuals Unlimited; **18.2:** © David Scharf; **18.8:** © Custom Medical Stock Photo; **18.18:** Courtesy of American Cancer Society; **18.23:** © Georgiana Silk/Corbis Outline; **18.25:** Courtesy of Dr. Charles Brinton

Chapter 19

19.1: © Stanley Cohen/Science Photo Library/Photo Researchers Inc.; **Bx19.1**(top left): Alger Meekma; **Bx19.1**(top right): David Umberger/Purdue News Service; **Bx19.1**(bottom left): Robert H. Devlin/Fisheries & Oceans Canada; **Bx19.1**(bottom right): Richard Shade; **19.4B:** Courtesy of Bio-Rad Laboratories; **19.13:** Nancy Federspiel, Stanford DNA Sequence & Tech. Center; **19.14:** Courtesy of Lifecodes Corp., Stanford, CT; **19.15:** AP/Wide World Photos; **19.17:** R.L. Brinster, U. of Pennsylvania Sch. of Vet. Med.; **19.20:** Courtesy of Monsanto; **19.23C:** AP/Wide World Photos

Chapter 20

20.1: © Dick Poe/Visuals Unlimited; **20.3:** Biological Photo Service

Chapter 21

21.1. © John Cancalosi/Peter Arnold Inc.; **21.8:** The Smithsonian Institution; **21.10**(both): © Breck P. Kent/Animals Animals/Earth Scenes; **21.20:** © Cleveland P. Hickman

Chapter 22

22.1: © G.L. Kooyman/Animals Animals/Earth Scenes; **22.4A:** © Kjell Sandved/Visuals Unlimited; **F22.4B:** © Fateh Singh Rathore/Peter Arnold Inc.; **22.4C:** © Porterfield/Chickering/ Photo Researchers Inc.; **22.5:** © Barbara Gerlach/Visuals Unlimited; **22.6**(top left): © John Shaw/Tom Stack & Associates; **22.6**(bottom left): © Rob & Ann Simpson/Visuals Unlimited; **22.6**(top right): © Suzanne L. Collins & Joseph T. Collins/ National Audubon Society Collection/Photo Researchers; **22.6**(bottom right): © Phil A. Dotson/National Audubon Society Collection/Photo

Researchers; **22.8**(top left): © Chas. McRae/Visuals Unlimited; **22.8**(bottom right): © Tom Evans/Photo Researchers; **22.8B:** Courtesy of Jonathan Losos; **22.8C**(bottom left): Courtesy of Jonathan Losos; **22.10A:** © Alan & Linda Detrick/National Audubon Society Collection/Photo Researchers; **22.10B:** © Stephen J. Kraseman/DRK Photo; **22.12**(both): © William P. Mull; **22.14:** © G.R. Roberts

Chapter 23

23.1: © National Museum of Kenya/Visuals Unlimited; **23.2:** © Alan Nelson/Animals Animals/Earth Scenes; **23.4A:** © Denise Tackett/Tom Stack & Associates; **23.4B:** © Edward S. Ross; **23.5**(top right): © Russell A. Mittermeier; **23.5** (bottom left): © Russell A. Mittermeier; **23.5**(bottom right): © Anup Shah/DRK Photo; **23.5A**(top left): © Tom & Pat Leeson/ National Audubon Society Collection/Photo Researchers; **23.7**(all): © David L. Brill 1985/National Museum of Tanzania, Dar es Salaam; **23.8**(all): © David L. Brill; **23.9A:** © 1994 Tim D. White/Brill Atlanta, housed in National Museum of Ethiopia; **23.9B:** © Bob Campbell/Corbis sygma; **23.9C:** © David L. Brill; **23.11:** Courtesy of National Museums of Kenya, photo © 1985 David L. Brill; **23.13**(all): © 1985 David L. Brill, artifact credit Musee de L'Homme, Paris; **23.15:** AP/Wide World Photos

Chapter 24

24.1: © Dick Poe/Visuals Unlimited; **24.16:** © John Shaw/Tom Stack & Associates; **24.19**(left): © Jean Vie/Gamma-Liaison; **24.19**(right): © Gianni Tortole/National Audubon Society Collection/Photo Researchers; **24.21:** Courtesy of National Museum of Natural History, Smithsonian Institution

Chapter 25

25.1: © Dave Fleetham/Tom Stack & Associates; **25.4** (all): Courtesy of J.B. Losos; **25.8** (both): © Edward S. Ross; **25.9** (both): © Lincoln P. Brower; **25.10:** © James L. Castner; **25.11:** © James L. Castner; **25.13:** © Merlin D. Tuttle, Bat Conservation International; **25.14:** © Jim Harvey/Visuals Unlimited; **25.15:** © Michael Fogden/DRK Photo; **25.16A:** © Edward S. Ross; **25.17:** © Leach, OSF/Animals Animals; **25.18A:** © E.R. Degginger/Animals Animals/Earth Scenes; **25.18B:** © Anne Wertheim/Animals Animals/Earth Scenes; **25.21:** © David Hosking/National Audubon Society Collection/ Photo Researchers; **25.23**(all): © Tom Bean; **25.24:** Reprinted with permission from *Science*, Vol. 283 Jan 22. Copyright 1999 American Association for the Advancement of Science

Chapter 26

26.1: © K. Ammann/Bruce Coleman Inc.; **26.2:** © UPI/Corbis-Bettmann; **26.3:** © Stephen Dalton/Photo Researchers; **26.5:** © William C. Dilger, Cornell U.; **26.6**(both): From J.R. Brown et al, "A defect in nurturing mice lacking...gene for fosB" *Cell* v. 86, 1996 pp 297-308, © Cell Press; **26.7**(all): © Lee Boltin Picture Library **26.8:** © William Grenfell/Visuals Unlimited; **26.9:** Thomas McAvoy, Life Magazine/© Time, Inc.; **26.10**(both): *Grzimek's Encyclopedia of Ethology* Van Nostrand-Reinhold Co.; **26.12:** © Roger Wilmshurst/The National Audubon Society Collection/ Photo Researchers; **26.14:** © B. Miller/Biological Photo Service; **26.15:** © Dwight R. Kuhn/DRK Photo; **26.17:** © Cabisco/Visuals Unlimited; **26.18** (both): © Sol Mednick; **26.19A:** © Dr. Mark Moffett/Minden Pictures; **26.20A:** © S. Osolinski/OSF/ Animals Animals/Earth Scenes; **26.21B:** © Fred Bruenner/Peter Arnold Inc.; **26.21C:** © James L. Amos/Peter Arnold Inc.; **26.24B:** © Jeff Foott/Tom Stack & Associates; **26.24A:** © Linda Koebner/Bruce Coleman; **26.25**(all): Superstock; **26.26:** Courtesy of Bernd Heinrich

Chapter 27

27.1: © Jane Burton/Bruce Coleman Collection; **27.2:** Nina Leen, Life Magazine, © Time Inc.; **27.5:** © Bios (C. Thouvenin)/Peter Arnold, Inc.; **27.7B:** © B. Chudleigh/Vireo; **27.8A:** Courtesy of James Traniello; **27.8B:** © Merllin B. Tuttle/Bat Conservation International; **27.9:** © Merlin B. Tuttle/Bat Conservation International; **27.10A:** Courtesy of T.A. Burke, Reprinted by permission from *Nature*, "Parental care and mating behavior of polyandrous dunnocks," 338: 247-251, 1989; **27.12:** © Edward S. Ross; **27.14:** © Nigel Dennis/National Audubon Society Collection/Photo Researchers

Chapter 28

28.1: © David Cavagnaro/Peter Arnold, Inc.; **28.3:** © Walt Anderson/Visuals Unlimited; **28.7A:** U.S. Forest Service

Chapter 29

29.1: NASA; **29.2:** Courtesy of William J. Hamilton III; **29.13:** © Michael Graybill & Jan Hodder/ Biological Photo Service; **29.14:** © IFA/Peter Arnold, Inc.; **29.17:** © Kenneth L. Smith; **29.19:** © David Hall/National Audubon Society Collection/Photo Researchers; **29.20A:** © Jim Church; **29.20B:** Courtesy of J. Frederick Grassel, Woods Hole Oceanographic Institution; **29.21A:** © Edward S. Ross; **29.21B:** © Fred Rhode/ Visuals Unlimited; **29.21C:** © Dwight R. Kuhn; **29.24:** © T. Henneghan/West Stock

Chapter 30

30.1: NASA; **30.3:** © Stephanie Maze/Woodfin Camp & Associates; **30.4A:** © George H. Huey/Animals Animals/Earth Scenes; **30.4B:** Courtesy of Rodale Institute; **30.8:** © Gilbert S. Grant/National Audubon Society Collection/Photo Researchers; **30.9:** NASA; **30.10A:** © Peter May/Peter Arnold Inc.; **30.10B:** © Frans Lanting/Minden Pictures

Chapter 31

31.1: © Tom & Pat Leeson/Photo Researchers; **31.2:** Photo by Tom McHugh,© Natural History Museum of Los Angeles County/Photo Researchers; **31.5:** © Edward S. Ross; **31.6:** © Still Pictures/Peter Arnold Inc.; **31.12:** © 1990 R.O. Bierregaard; **31.13:** © Paul Chesley; **31.16:** Mark Chandler, New England Aquarium; **31.17:** © Tim W. Clark; **31.18:** © Wm. J. Weber/Visuals Unlimited; **31.20:** Reprinted with permission from *Science*, Vol 282 Dec. © 1998 American Association for the Advancement of Science; **31.21**(both): Courtesy of University of Wisconsin, Madison Arboretum; **31.23A:** © Elizabeth N. Orians; **31.24:** Merlin D. Tuttle, Bat Conservation International

Chapter 32

32.1: © Hal Beral/Visuals Unlimited; **32.3**(top left): © Dwight R. Kuhn; **32.3**(top right): © Henry Ausloos/Animals Animals/Earth Scenes; **32.3**(middle left): © Heather Angel; **32.3**(middle right): © John Cancalosi/Peter Arnold, Inc.; **32.3**(bottom left): © S. Maslowski/Visuals Unlimited; **32.3F**(bottom right): © Manfred Danegger/Peter Arnold, Inc.; **32.5**(top left): © Gerard Lacz/Peter Arnold, Inc.; **32.5**(top right): © Ralph Reinhold/Animals Animals/Earth Scenes; **32.5C**(bottom left): © Grant Heilman/Grant Heilman Photography; **32.13:** © Karen Stephens/Biological Photo Service; **32.15:** © Lee D. Simon/Photo Researchers

Chapter 33

33.1: © K.G. Murti/Visuals Unlimited; **33.4A,B** © Dept. of Microbiology, Biozentrum/ Science Photo Library/Photo Researchers; **33.6:** © Scott Camazine/Photo Researchers; **33.9A:** Cynthia Goldsmith & Jackie Katz, Centers for Disease Control & Prevention; **33.10A:** © Larry Mulvehill/Photo Researchers

Chapter 34

34.1: © David M. Phillips/Visuals Unlimited; **34.2A:** © J.J. Cardamore/BPS/Tom Stack & Associates; **34.2B:** © David M. Phillips/Visuals Unlimited; **34.2C:** © Ed Reschke; **34.3:** Courtesy of Dr. Hans Reichenbach; **34.4:** © Abraham & Beachey/BPS/Tom Stack & Associates; **34.5**(both): © G.W. Willis/Biological Photo Service; **34.6:** © Julius Adler/Visuals Unlimited; **34.7A:** © W. Watson/ Visuals Unlimited; **34.7B:** © Norma J. Lang/ Biological Photo Service; **34.9:** © CNRI/SPL/ Photo Researchers; **34.11:** © Science/Visuals Unlimited

Chapter 35

35.1: © John D. Cunningham/ Visuals Unlimited; **35.2:** Courtesy of Dr. Edward W. Daniels, Argonne National Lab, the University of Illinois College of Medicine at Chicago; **35.5:** © John D. Cunningham/Visuals Unlimited; **35.7:** © John D. Cunningham/Visuals Unlimited; **35.8:** © Phil A. Harrington/Peter Arnold, Inc.; **35.9:** © Manfred Kage/Peter Arnold; **35.10:** © Richard Rowan/Photo Researchers; **35.12A:** © Michael Abby/Visuals Unlimited; **35.13:** © D.P. Wilson/Photo Researchers; **35.14:** © Gregory Ochocki/Photo Researchers; **35.16:** © John D. Cunningham/Visuals Unlimited; **35.18A:** © Manfred Kage/Peter Arnold Inc.; **35.18B:** Edward S. Ross; **35.19:** Courtesy of Stanley Erlandsen; **35.20:** © Manfred Kage/Peter Arnold, Inc.; **35.22A:** © Brian Parker/Tom Stack & Associates; **35.22B:** © Manfred Kage/Peter Arnold; **35.25:** © Genichiro Higuchi, Higuchi Science Laboratory; **35.26:** © Edward S. Ross; **35.28**(all): ©

John Shaw/Tom Stack & Associates

Chapter 36

36.1: © Bill Keogh/Visuals Unlimited; **36.2A:** © Michael & Patricia Fogden; **36.2B:** © Robert Simpson/Tom Stack & Associates; **36.2C:** © Cabisco/Visuals Unlimited; **36.3:** Courtesy of E.C. Setliff & W. L. MacDonald; **36.4:** © Kjell Sandved/Butterfly Alphabet; **36.5:** © L. West/Photo Researchers; **36.6:** © Ralph Williams/USDA Forest Service; **36.8A:** © Cabisco/Phototake; **36.9A:** © Ed Pembleton; **36.9B:** © Kjell Sandved/Butterfly Alphabet; **36.10:** © David M. Phillips/Visuals Unlimited; **36.11A:** © Alexandra Lowry/The National Audubon Society Collection/Photo Researchers; **36.12A:** © H.C. Huang/Visuals Unlimited; **36.12B:** Courtesy of Roland Wenger/Sandoz Pharma, Ltd; **36.12C:** © David M. Phillips/Visuals Unlimited; **36.13A:** © James Castner; **36.13B:** © Edward S. Ross; **36.14:** © Ed Reschke; **36.15B:** © D.H. Marx/Visuals Unlimited

Chapter 37

37.1: © Stephen J. Krasemann/ DRK Photo; **37.4:** © Edward S. Ross; **37.6:** © Kirtley Perkins/ Visuals Unlimited; **37.7:** © Kingsley R. Stern; **37.8:** Courtesy of Hans Steur, The Netherlands; **37.9:** © Edward S. Ross; **37.11:** © Edward S. Ross; **37.12:** © Edward S. Ross; **37.14A:** © Walter H. Hodge/Peter Arnold Inc.; **37.14B:** © Kjell Sandved/Butterfly Alphabet; **37.14C:** © Runk/Schoenberger/Grant Heilman Photography; **37.15:** Courtesy of Missouri Botanical Garden

Chapter 38

38.1: © Ned Therrien/Visuals Unlimited; **38.2:** © Terry Ashley/Tom Stack & Associates; **38.4A:** © Hart-Davis/Science Photo Library/Photo Researchers; **38.4B:** © R. Calentine/Visuals Unlimited; **38.8**(both): © Heidi Mullen; **38.9:** Courtesy of Fred Sack, Ohio State University; **38.10:** © Oliver Meckes/Photo Researchers Inc.; **38.11**(both): Courtesy of M. David Marks; **38.12A:** © Biophoto Associates/ Photo Researchers Inc.; **38.12B:** © Randy Moore/Visuals Unlimited; **38.12C:** © Lawrence Mellinchamp/Visuals Unlimited; **38.13C:** Courtesy of Wilfred Cote, Suny College of Environmental Forestry; **38.14A:** © Biophot; **38.15:** © E.J. Cable/Tom Stack & Associates; **38.16A:** Courtesy John Schiefelbein, from Myeong min Lee & John Schiefelbein, "WEREWolf, MYB-related Protein in Arabidopsis," *Cell* V. 99:473-483, Nov. 24, 1999; **38.16B:** Courtesy of Dr. Philip Benfey; **38.17:** Photomicrograph by G.S. Ellmore; **38.19A:** © Kingsley R. Stern; **38.19B:** Photomicrograph by G.S. Ellmore; **38.21A:** © Walter H. Hodge/Peter Arnold, Inc.; **38.21B:** Courtesy of Robert A. Schlising; **38.21C:** © Kingsley R. Stern; **38.22:** Courtesy of J.H. Troughton and L. Donaldson and Industrial Research Ltd.; **38.24:** © John D. Cunningham/Visuals Unlimited; **38.25**(both): © Ed Reschke; **38.26:** © Ed Reschke/ Peter Arnold Inc.; **38.27A:** © Ed Reschke; **38.27B:** © Jack M. Bostrack/Visuals Unlimited; **38.29:** Richard Waites & Andrew Hudson, from Phantastica: a gene required for dorsoventrality of leaves in Antirrhinum majus, *Development* 121, 2143-2154 (1995) © The Company of Biologists Limited 1995; **38.30A:** © Kjell Sandved/Butterfly Alphabet; **38.30B:** © Pat Anderson/Visuals Unlimited; **38.31A:** © Edward S. Ross; **38.31B:** © Glenn M. Oliver/ Visuals Unlimited; **38.31C:** © Joel Arrington/Visuals Unlimited; **38.34:** © Ed Reschke; **38.35**(above): © Michael P. Godomski/Photo Researchers; **38.35**(below): © Ed Reschke/ Peter Arnold, Inc.

Chapter 39

39.1: © Runk/Schoenberger/Grant Heilman Photoraphy; **39.2**(all): Courtesy of Dr. Emmanuel Epstein; **39.4:** © Michael P. Gadomski/Photo Researchers; **39.5:** © Kjell Sandved/Butterfly Alphabet; **39.6:** Courtesy of Sharon Long; **39.7:** © Richard Rowan's Collection, Inc./Photo Researchers Inc.; **39.11:** © John D. Cunningham/Visuals Unlimited; **39.12A:** © Terry Ashley/Tom Stack & Associates; **39.13:** © Grant Heilman/Grant Heilman Photography; **39.14B:** © Bruce Iverson Photomicrography; **39.16A:** © Andrew Syred/Science Photo Library/Photo Researchers; **39.16B:** © Bruce Iverson/Science Photo Library/Photo Researchers Inc.

Chapter 40

40.1: © Norm Thomas/The National Audubon Society Collection/Photo Researchers; **40.4:** Courtesy of E.C. Yeung & D.W. Meinke; **40.5**(all): Courtesy of Dr. Chun-Ming Liu; **40.6:** Courtesy of Kathy Barton; **40.7A:** © Runk/Schoenberger/Grant Heilman Photography; **40.7B:** © Kevin & Betty Collins/Visuals Unlimited; **40.8:** © Jack M. Bostrack/Visuals Unlimited; **40.9:** © David A. Priestly; **40.10A:** © Walt Anderson/Visuals Unlimited; **40.10B:** © Ed Reschke/Peter Arnold, Inc.; **40.10C:** © David Sieren/Visuals Unlimited; **40.11**(top left): © Kingsley R. Stern; **40.11**(top center): © James Richardson/Visuals Unlimited; **40.11**(top right): © Kingsley R. Stern; **40.11**(middle center): © Kingsley R. Stern; **40.11** (center far right): © Barry L. Runk/Grant Heilman Photography; **40.11** (bottom left): Courtesy of Robert A. Schlising; **40.11** (bottom right): © Charles D. Winters/ Photo Researchers; **40.11D**(middle left): © Kingsley R. Stern; **40.12A:** © Edward S. Ross; **40.12B:** © James Castner; **40.13:** © James Castner; **40.14:** © James Castner; **40.16:** Courtesy of Prof. Tuan-hua David Ho

Chapter 41

41.1: © John D. Cunningham/Visuals Unlimited; **41.2:** © Runk/Schoenberge/ Grant Heilman Photography; **41.4:** © Runk/Schoenberger/ Grant Heilman Photography; **41.5:** © John D. Cunningham/ Visuals Unlimited; **41.6A:** © T. Walker; **41.6B:** © R.J. Delorit, Agronomy Publications; **41.7A:** © Jim Zipp/The National Audubon Society Collection/Photo Researchers; **41.7B:** © Runk/ Schoenberger/Grant Heilman Photography; **41.8A:** © Prof. Malcolm B. Wilkins, Botany Dept., Glasgow University; **41.15** (all): © Prof. Malcolm B. Wilkins, Botany Dept., Glasgow University; **41.17:** © Robert Calentine/Visuals Unlimited; **41.18:** © Runk/Schoenberger/ Grant Heilman Photography; **41.19:** © Sylvan H. Wittwer/ Visuals Unlimited; **41.22A:** © John Solden/Visuals Unlimited; **41.22B:** © David M. Phillips/ Visuals Unlimited; **41.24A:** © Stephen G. Maka/DRK Photo; **41.24B:** © Heidi Mullen; **41.25:** Courtesy of Lingjing Chen & Renee Sung; **41.26** (both): Detlef Weigel & Ove Nilsson, The Salk Institute for Biological Studies; **41.28:** © Jim Strawser/Grant Heilman Photography; **41.33** (all): Courtesy of John L. Bowman; **41.34C:** © S.J. Krasemann/Peter Arnold Inc.; **41.35** (all): Courtesy of Frank B. Salisbury

Chapter 42

42.1: © Richard La Val/Animals Animals; **42.5:** © John Bishop/ Visuals Unlimited; **42.6:** © Paul Gier/Visuals Unlimited; **42.7:** Courtesy of Enrico Coen; **42.9A:** Courtesy of William F. Chissoe, Noble Microscopy Lab, U. of Oklahoma; **42.9B:** Courtesy of Dr. Joan Nowicke, Smithsonian Institution; **42.10:** © Kingsley R. Stern; **42.11:** © Edward S. Ross; **42.12:** © Michael & Patricia Fogden; **42.13** (both): © Thomas Eisner; **42.14:** © John D. Cunningham/Visuals Unlimited; **42.15:** © Edward S. Ross; **42.16A:** © David Sieren/Visuals Unlimited; **42.16B:** © Barbara Gerlach/Visuals Unlimited; **42.19:** © Jerome Wexler/Photo Researchers; **42.20** (both): © Edward S. Ross

Chapter 43

43.1: Courtesy of Ingo Potrykus & Peter Beyer, photo by Peter Beyer; **43.6** (all): Tri Dinh Vuong, PhD thesis, U. of Illinois, Urbana, 1998, "Genetic & Molecular Analysis of the rj7 Gene in Soybean Using PCR-Based Technologies"; **43.10** (all): © Kingsley R. Stern; **43.11** (all): © Runk/Schoenberger/Grant Heilman Photography; **43.12:** Courtesy of German Spangenberg, from *Plant Cell Reports* 16, 1997; **43.13** (all): Courtesy of Dept. of Crop & Soil Sciences, U. of Georgia; **43.14** (all): Courtesy of Dr. Hans Ulrich Koop, from *Plant Cell Reports*, 17:601-604; **43.15** (all): Courtesy of Prof. Jack Widholm, U. of Illinois; **43.16:** © Alfred Wolf/Explorer/Photo Researchers; **43.17** (all): Courtesy of Prof. Jack Widholm, U. of Illinois; **43.18:** © Grant Heilman/Grant Heilman Photography; **43.20** (all): Reproduced with permission from Altpeter et al, *Plant Cell Reports* 16:12-17 1996, photos provided by Indra Vasil

Chapter 44

44.1: © Denise Tackett/Tom Stack & Associates; **44.6:** © Andrew J. Martinez/Photo Researchers; **44.11:** © Gwen Fidler/Tom Stack & Associates; **44.12:** © Kelvin Aitken/Peter Arnold Inc.; **44.13:** © Daniel Gotshall; **44.14:** © David Wrobel/Visuals Unlimited; **44.18:** © Kjell Sandved/Butterfly Alphabet; **44.19:** © Biology Media/R. Knauf/Photo Researchers; **44.21:** Courtesy of Dr. Igor Eeckhaut

Chapter 45

45.1: © Kjell Sandved/Butterfly Alphabet; **45.2:** © Alex Kerstich/ Visuals Unlimited; **45.4:** © A. Flowers & L. Newman/Photo Researchers; **45.6** (both): © Kjell Sandved/Butterfly Alphabet; **45.7B:** © Kjell Sandved/Butterfly

Alphabet; **45.8:** © Milton Rand/Tom Stack & Associates; **45.9:** © E.R. Degginger/Photo Researchers; **45.10:** © James D. Watt/Animals Animals/Earth Scenes; **45.11:** © Visuals Unlimited; **45.13:** © Kjell Sandved/Butterfly Alphabet; **45.14:** © David Dennis/Tom Stack & Associates; **45.15:** © Cleveland P. Hickman; **45.17B:** © Robert Brons/Biological Photo Service; **45.18B:** © Fred Bavendam/Peter Arnold Inc.

Chapter 46

46.1: © James H. Robinson/ Animals Animals/Earth Scenes; **46.8B:** © Kjell Sandved/Butterfly Alphabet; **46.10B:** © Thomas Eisner; **46.11:** © Edward S. Ross; **46.12:** © Edward S. Ross; **46.13A:** © Rod Planck/Tom Stack & Associates; **46.13B:** © Ann Moreton/Tom Stack & Associates; **46.14:** © Edward S. Ross; **46.15:** © Frans Lanting/ Minden Pictures; **46.17:** © Heather Angel; **46.20:** © T.E. Adams/Visuals Unlimited; **46.21:** © Kjell Sandved/Butterfly Alphabet; **46.22:** © Alex Kerstich/ Visuals Unlimited; **46.23:** © Edward S. Ross; **46.24A:** © Cleveland P. Hickman; **46.24B:** © Kjell Sandved/Butterfly Alphabet; **46.24C:** © Norm Thomas/Photo Researchers; **46.24D:** © James Castner; **46.24E:** © J.A. Alcock/Visuals Unlimited; **46.24F:** © Kjell Sandved/Butterfly Alphabet; **46.26:** © John Shaw/Tom Stack & Associates; **46.27:** © Kjell Sandved/Butterfly Alphabet

Chapter 47

47.1: © William C. Ober; **47.3A:** © Alex Kerstitch/Visuals Unlimited; **47.3B:** © Randy Morse/Tom Stack & Associates; **47.3C:** © Daniel W. Gotshall/ Visuals Unlimited; **47.6:** © Kjell Sandved/Butterfly Alphabet; **47.8:** Courtesy of David L. Pawson, Harbor Branch Oceanographic Institute, Inc./National Museum of Natural History; **47.9:** © Carl Roessler/Tom Stack & Associates; **47.10:** © Daniel W. Gotshall/ Visuals Unlimited; **47.11:** © Gary Milburn/Tom Stack & Associates; **47.12:** © Kjell Sandved/Visuals Unlimited; **47.13A:** © Jeff Rotman; **47.13B:** © Daniel W. Gotshall; **47.14:** © Alex Kerstitch/Visuals Unlimited; **47.15:** © Alan Baker/Visuals Unlimited

Chapter 48

48.1: © Brian Parker/Tom Stack & Associates; **48.3:** © Eric N. Olson, PhD/The University of Texas MD Anderson Cancer Center; **48.5A:** © Rick Harbo Marine Images; **48.6:** © Heather Angel; **48.10:** © Denise Tackett/Tom Stack & Associates; **48.12:** © Tom McHugh/Photo Researchers; **48.14:** © 1993 Norbert Wu; **48.15:** © John D. Cunningham/Visuals Unlimited; **48.17:** © 1979 Peter Scoones/ Contact Press Images/Woodfin Camp & Associates Inc.; **48.20:** © Natural History Museum/J. Sibbick; **48.21:** © John Shaw/ Tom Stack & Associates; **48.21B:** © Suzanne L. Collins & Joseph T. Collins/Photo Researchers; **48.21C:** © Jany Sauvanet/Photo Researchers; **48.23:** © Natural History Museum/J. Sibbick; **48.26:** © Louis Psihoyos/Matrix International Inc.; **48.30:** © William J. Weber/Visuals Unlimited; **48.31A:** © John Cancalosi/Tom Stack & Associates; **48.31B:** © Rod Planck/Tom Stack & Associates; **48.32:** © Brian Parker/Tom Stack & Associates; **48.36:** © Art Wolfe; **48.37:** © Joe McDonald/Visuals Unlimited; **48.40:** © Stephen Dalton/ National Audubon Society Collection/Photo Researchers; **48.41A:** © B.J. Alcock/Visuals Unlimited; **48.41B:** © Fritz Prenzel/Animals Animals/Earth Scenes; **48.41C:** © Stephen J. Krasemann/DRK Photo

Chapter 49

49.1: Photo Lennart Nilsson/Albert Bonniers Forlag AB, *Behold Man*, Little Brown & Co.; **p.989**(top): © Ed Reschke; **p.989**(2nd from top): © Ed Reschke; **p.989**(3rd from top) © Ed Reschke; **p. 989**(4th from top): © Fred Hossler/Visuals Unlimited; **p.989**(5th from top): © Ed Reschke; **49.6:** © J. Gross/Science Photo Library/Photo Researchers; **49.7:** © Biophoto Associates/ Photo Researchers; **p.991**(top): © Biophoto Associates/Photo Researchers; **p.991**(2nd from top): © Cleveland P. Hickman Jr.; **p.991**(3rd from top): © Chuck Brown/Photo Researchers; **p.991**(4th from top): © Ed Reschke; **p.991**(5th from top): © Ken Edward/Science Source/Photo Researchers; **49.9**(both): © Ed Reschke; **49.10:** © David M. Phillips/Visuals Unlimited; **p.995**(all): © Ed Reschke; **49.12B:** © Joe McDonald/Visuals Unlimited

Chapter 50

50.1: © Anthony Bannister/Animals Animals/Earth Scenes; **50.8:** © Dr. H.E. Huxley; **50.9A:** © Dr. H.E. Huxley; **50.21:** © Treat Davidson/Photo Researchers

Chapter 51

51.1: © John Gerlach/Animals Animals/Earth Scenes; **51.13:** Photo Susama Ito, from Charles Flickinger, *Medical Cellular Biology*, W.B. Saunders, 1979; **51.22:** © John Sholtis/The Rockefeller University

Chapter 52

52.1: © Professors P.M. Motta & S. Correr/Science Photo Library/Photo Researchers; **52.11:** © Ed Reschke; **52.17**(all): Courtesy of Frank P. Sloop, Jr.;

Chapter 53

53.1: © Frans Lanting/Minden Pictures; **53.7:** © Chris Bjornberg/Photo Researchers; **53.10A:** Courtesy of H.R. Dunker

Chapter 54

54.1: Courtesy of David I. Vaney, University of Queensland, Australia; **54.4:** © C.S. Raines/Visuals Unlimited; **54.13:** © John Heuser, Washington University School of Medicine, St. Louis, MO; **54.15:** © Ed Reschke; **54.19B:** © E.R. Lewis, YY Zeevi, T.E. Everhart, U. of California/Biological Photo Service; **54.30:** Dr. Marcus E. Rachle, Washington University, McDonnell Center for High Brain Function/Peter Arnold, Inc.; **54.31:** Photo Lennart Nilsson/Albert Bonniers Forlag AB, *Behold Man*, Little Brown & Co.; **54.34:** © E.R. Lewis/ Biological Photo Service

Chapter 55

55.1: © Omikron/Photo Researchers; **55.6D:** © Ed Reschke; **55.12:** © Wendy Shatil/Bob Rozinski/Tom Stack & Associates; **55.15:** © Dalton, S. OSF/Animals Animals/Earth Scenes; **55.25:** © Leonard L. Rue, III

Chapter 56

56.1: © Francois Gohier/Science Source/Photo Researchers; **56.9:** Photo Lennart Nilsson/Albert Bonniers Forlag AB, *Behold Man*, Little Brown & Co.; **56.12:** © Corbis/Bettmann; **56.16:** © John Paul Kay/Peter Arnold, Inc.; **56.21:** © Robert & Linda Mitchell

Chapter 57

57.1: Courtesy of the National Library of Medicine; **57.3:** © Manfred Kage/Peter Arnold, Inc.; **57.7:** © Visuals Unlimited; **57.11**(both): © Dr. Andrejs LiepinsScience Photo Library/Photo Researchers; **57.21:** © Stuart Fox; **57.26:** © CDC/Science Source/Photo Researchers

Chapter 58

58.1: © Belinda Wright/DRK Photo; **58.15:** © Larry Brock/ Tom Stack & Associates

Chapter 59

59.1: © Michael Fogden/DRK Photo; **59.2A:** © Chuck Wise/Animals Animals/Earth Scenes; **59.2B:** © Fred McConnaughey/The National Audubon Society Collection/ Photo Researchers; **59.4:** © David Doubilet; **59.5:** © Hans Pfletschinger/Peter Arnold Inc.; **59.7:** © Cleveland P. Hickman Jr.; **59.8:** © Frans Lanting/ Minden Pictures; **59.9A:** © Jean Phillippe Varin/Jacana/Photo Researchers; **59.9B:** © Tom McHugh/The National Audubon Society Collection/Photo Researchers; **59.9C:** © Fritz Prenzel/Animals Animals/Earth Scenes; **59.12A:** © David M. Phillips/Photo Researchers; **59.13:** Photo Lennart Nilsson/ Bonniers Forlag AB, *A Child is Born*, Dell Publishing Co.; **59.18:** © Ed Reschke; **59.22**(all): © McGraw-Hill Higher Education/ Bob Coyle, photographer

Chapter 60

60.1: Photo Lennart Nilsson/Bonniers Forlag AB, *A Child is Born*, Dell Publishing Co.; **60.2B:** © P.Bagavandoss/Science Source/Photo Researchers; **60.2C:** © David M. Phillips/ Visuals Unlimited; **60.3B:** Courtesy of Dr. Everett Anderson; **60.6:** © David M. Phillips/Visuals Unlimited; **60.7:** © Cabisco/Phototake; **60.8:** © David M. Phillips/Visuals Unlimited; **60.22:** Photo Lennart Nilsson/Albert Bonniers Forlag AB, *A Child is Born*, Dell Publishing Company; **60.23**(all): Photo Lennart Nilsson/Albert Bonniers Forlag AB, *A Child is Born*, Dell Publishing Company

Index

Boldface page numbers correspond with **boldface terms** in the text. Page numbers followed by an "f" indicate figures; page numbers followed by a "t" indicate tabular material.

A

C

D

E

F

G

H

I

J

K

L

M

N

O

P

Q

R

S

T